Encyclopaedic dictionary of textile terms

Encyclopaedic dictionary of textile terms

Volume I

Mathews Kolanjikombil
Senior Textile Processing Technologist

WOODHEAD PUBLISHING INDIA PVT LTD

New Delhi

Published by Woodhead Publishing India Pvt. Ltd.
Woodhead Publishing India Pvt. Ltd.,
303, Vardaan House, 7/28, Ansari Road,
Daryaganj, New Delhi - 110002, India
www.woodheadpublishingindia.com

First published 2018, Woodhead Publishing India Pvt. Ltd.

Woodhead Publishing India Pvt. Ltd. ISBN: 978-93-85059-11-7
Woodhead Publishing India Pvt. Ltd. e-ISBN: 978-93-85059-66-7

Typeset by Bhumi Graphics, New Delhi
Printed and bound by Replika Press Pvt. Ltd.

Contents

Volume IV

Preface

It has been told the first thing that a business man should know, is—the little things of his business. This book is meant for textile personae, be it a textile student, textile chemist, a spinner, a weaver, garment maker or a merchandiser who may be an expert in his/her field but he/she may come across many terms in his/her day to day business which he/she is not familiar with but is related to his/her fieldwhich he/she should know, but nine times out of ten does not know. He/She may not have an expert in that field of that term near him/her to clear his doubt about that term. This book comes handy in such circumstances.

There is nothing scholarly in this book, but is a compilation of short easy understandable meanings of the textile terms enough to help the referrer to understand the term. I have come across dictionaries which gives the meaning of the textile terms in one sentence. But it may not be sufficient to give a full idea about it. But this book is a little different. The meaning of the terms is explained in a concise manner even with the help of diagrams or photos, wherever necessary, which is enough to clear his/her doubts. These terms and meanings have been collected right from my college days and throughout my career. I believe it is not complete, but, such as it is, the compilation is reliable. There are further terms which are being collected by the author which will be added in the next edition, probably. The author believes that he has produced a book which may be profitably consulted by all who are either interested or practically engaged in textile trade.

It has to be specially mentioned that, the students can use the present book as a reference guide for his/her immediate needs without going to many text books. For detailed study of any terms he/she can further refer books specialized in that field. I may not suggest this book for a research student. The author also has tried to explain the construction of many fabrics new or old for general knowledge.

Hope this book will be greatly accepted by textile personnel. Any suggestions or corrections are welcome, which will be included in the next editions.

Mathews Kolanjikombil

Senior Textile Processing Technologist

Bangalore

Dedication

This book is dedicated to my Father (Thomas Abraham) and my Mother (Thankamma Abraham) who solely made me what I am today.

A

2×1 Plain stitch in warp knitting: This stitch is the same as tricot stitch except that each yarn laps over to the next but one wale [See **Warp knitting, Plain (Tricot stitch)**].

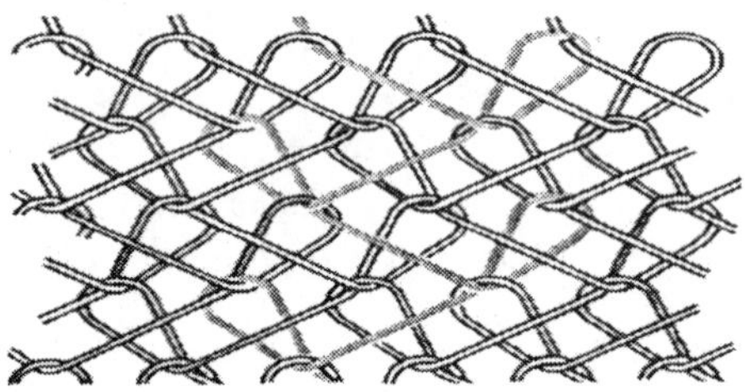

2×1 Plain stitch

2×2 Rib (A weft knitted fabric): A wide range of fabrics can be produced, depending on the set-out of the needles. The most popular is 2 × 2 rib which is made by taking every third needle out of action in each needle bed (2 : 1 needle set-out). The two sides of the rib fabric look the same. If the fabric is stretched in the width, the two rib loops will be exposed between the two face loops. It is very elastic in the width direction. Applications include cuffs and welts, pullovers and dresses.

A la grecque: French name of the meander pattern.

A la reine: French silk droguet of different coloured warp and filling.

A.B.S. (ABS): A plastic; short form for acrylonitrile butadiene styrene. ABC material is an alternative to PVC.

A.B.C. Silk: A plain weave cotton/silk fabric with cotton warp and spun silk filing. Used for under wears and lining; it comes in white and plain colours.

AAFEBR: Anaerobic Attached Film Expanded Bed Reactor, i.e. an ***expanded bed biofilm reactor*** operated anaerobically.

Aachen felting test: Test for estimating the felting character of certain wool. The wool is made into the shape of a ball and immersed in felting solution and shaken for a predetermined period and then taken and dried. The measure of the ball dia compared to the earlier dia gives the felting character of that wool.

Aal: A red dye produced from the plant of the same name, related to the **madder**

AAO process, A2O process: An adaptation of the A/O process. It is an activated sludge process that can achieve nitrification, denitrification and biological phosphorus removal as well as BOD removal. The three-stage process involves an anaerobic zone of about one hour retention followed by an anoxic zone of similar size, and finally an aerobic (oxidation) zone of 4–6 hours retention See also **Nitrification, Denitrification, A/O Process, Anoxic.**

AATCC: American Association of Textile Chemists and Colourists. Their website is www.aatcc.org. AATCC was founded to promote greater knowledge of textile dyes and chemicals, and therefore is concerned specifically with textile products. This organization works very closely with ASTM, but writes chemical-type tests. In addition to the development of test methods, AATCC sponsors scientific meetings and promotes textile education. The activities are concerned primarily with the chemical properties of textiles in contrast to ASTM's emphasis on physical properties.

AATCC fading unit (AFU): Unit for the evaluation of colour fastness to light. 20 AFU corresponds to the amount of light exposure necessary to bring about a perceptible change of shade on a specified AATCC Blue Wool LF Standard.

AATCC Blue Wool LF standard: One of a group of dyed wool fabrics distributed by AATCC for use in determining the amount of light exposure of specimens during light fastness testing.

Aba: (1) Coarse and thick, felted woollen fabric made of naturally coloured wool that is usually grey, made in Hungary, once worn by the peasants. (2) In some other countries it is a coarse fabric, made of wool and camel's hair, made in stripes.

Abaca: Manila hemp, alternative name for manila hemp called by natives in Philippines.

Abaiser: Ivory black.

Absinthe, absinth: The light-green colour of the potent liqueur of the same name which was banned in France in 1915 because of its effect on health and the performance of French troops at the beginning of World War I. It continues to be banned in France and in the United States, but is allowed in the United Kingdom where it continues to be imported since 1998. The liqueur takes on a milky colour when water is added.

Abassi: Raw cotton grown in Egypt, the staple is of good quality and white in colour, but not as strong as the other cottons.

Abat-chauvee: French word for a low grade of wool.

Abb: (1) Merino clothing warp wool, being skirting from the breech; (2) Short wool, taken from the breech and cow tail skirts of English and coarse wool, used mostly for warp; (3) Same as warp.

Abbotsford: The fabric is always slightly napped on one side, and could be made from cotton, wool, viscose, modal or acrylic, and of dress weight and usually made in muted check design.

Abdullah kani: Striped silks made in few counties.

Abee: Fabric made with cotton warp and woollen filling in Asia.

Abelmoschus: Fibre similar to jute; it is tough and does not rot in water and is used for making ropes.

Abercrombie: A Scottish tartan woven with a blue and black base and green and white over check. Traditionally all wool, but it may now contain a proportion of polyester or acrylic fibre. Used for kilts, pleated skirts, bias cut skirts, pinafore dresses and children's clothes. It is a medium–weight fabric, not usually heavy enough to be used as outer wear.

Aberdeen: Men's half hose made in Scotland of coarse grey wool.

Abeston: Incombustible flax of old Egypt, mentioned by Pliny.

Ablaque or pearl silk: A very fine raw silk from Persian counties which does not stand warm water, may not be in use now.

Abnakee rug: American hooked rug, which is made on a coarse and open jute burlap ground. Unbleached, all-wool, twilled flannel is dyed with vegetable dyes and cut lengthwise into strips of one-quarter inch width. These strips are hooked through the burlap to form the pile. The patterns are bold.

Abnormal crimp: A relative term for crimp that is either too low or too high in frequency and/or amplitude or that has been put into the fibre with improper angular characteristics.

ABNT: Associacao Brasileira de Normas Te'cnicas, Rio de Janeiro. Brazilian Technical Standards Association.

Abouchouchon: Low-grade French woollen, originally had 1,600 ends and black and white selvage.

Abraded yarn: A filament yarn in which the filament has been cut or broken to create hairiness (fibrillation) to simulate the surface character of the spun yarns. Abraded yarn is usually plied or twisted with other yarns before use.

Abrang: A fine violet striped fabric of cotton that is given a glazed finish, used to be made in India in olden days.

Abrasive wear(in carpet): Damage to carpet from foot traffic or other rubbing friction.

Abrasion: 'Wearing-off' of a textile material by rubbing against another surface.

Abrasion mark: An area where the fabric has been damaged by friction or rubbing. *See* **Chafe mark, Rub, Distortion**.

Abrasion resistance: The ability of the fibre or fabric to withstand surface wear and rubbing.

Abrasion tester: A mechanical instrument used for testing fabric resistance to destructive action of surface wear and rubbing. The best abrasion tests simulate surface wear and rubbing of the fabrics in actual use. Fabric may be tested for flat abrasion, flex abrasion and/or edge abrasion depending on the end use.

Abrawan: Next to the finest quality of the plain Dacca muslin.

Abrohany: Very sheer cotton muslin made in India. See **Mallemolles**.

Abroma hemp: Called also perennial Indian hemp, grows in the Philippine Islands, India, etc. The fibres are used for coarse cloth and for tow. See **Devil's cotton**.

Absolute: When referring to filters, it is used in reference to the micron rating of cartridge or disc filters, indicating that all particles larger than a specified size will be trapped within or on the filter and will not pass through.

Absolute alcohol: Water-free ethyl alcohol.

Absolute humidity: The mass of water vapour in a unit volume of air. See **Air humidity**.

Absolute system of measurement: Physical system of measurement.

Absolute temperature: See **Temperature**. Symbol: *T* A temperature defined by the relationship: T = c + 273.15 where c is the Celsius temperature. The absolute scale of temperature was a fundamental scale based on Charles' law applied to an ideal gas: $V = V_0$(1 + áè) where V is the volume at temperature c, V_0 the volume at 0, and á the thermal expansivity of the gas. At low pressures, when real gases show ideal behaviour, á has the value 1/273.15. Therefore, at c = –273.15 the volume of the gas theoretically becomes zero. In practice, of course, substances become solids at these temperatures. Nevertheless, the extrapolation can be used to create a scale of temperature on which –273.15 degrees Celsius (°C) corresponds to zero (0°). This scale was also known as the ideal-gas scale; on it temperature interval units were called degrees absolute (°A) or degrees Kelvin(°K), and were equal in size to the Celsius degree. It can be shown that the absolute temperature scale is identical to the thermodynamic temperature scale, on which the temperature interval unit is the kelvin.

Absolute white: In theory, a material that perfectly reflects all light energy at every visible wavelength. In practice, a solid white with known spectral reflectance data that is used as the "reference white" for all measurements of absolute reflectance. When calibrating a spectrophotometer, often a white ceramic plaque is measured and used as the absolute white reference.

Absolute zero: The zero value of thermodynamic temperature; 0 kelvin or –273.15 degrees Celsius.

Absolutely equivalent matching: A dyeing or matching whose reflectance curve is same or as close as possible over the entire visible spectrum to that of the original swatch/sample is called a an absolutely equivalent matching.

Adsorb: The process by which a liquid penetrates the solid structure of the absorbent's fibres or particles, which then swell in size to accommodate the liquid.

Absorb/absorption, in colour chemistry: Dissipation of the energy of electromagnetic waves into other forms (e.g., heat) as a result of its interaction with matter; a decrease in directional transmittance of incident radiation, resulting in a modification or conversion of the absorbed energy.

Absorbance: The ability of a substrate to transform radiant energy into a different form, usually with a rise in temperature. Mathematically, absorbance is the negative logarithm to the base of 10 of transmittance.

Absorbency: (1) See **Wettability**.

(2) The ability of one material to take up another material.

Absorbent: The capacity of a substrate to take up liquid. In the case of textile fabrics, it can be a total of the adsorption within the fibre, adsorption or adhesion on the yarn surface or intercapillary absorption in between the yarn structure in a fabric.

Absorbent compound (in textile cleaning): Sponge like particles which, when saturated with water or dry solvent and detergent and brushed in to the textile, remove soil from the textile.

Absorbent pad (cleaning of textile floor covering): A damp textile material (fabric, sponge, felt or mop) used to agitate and wipe the pile and, in the process, absorb oil.

Absorption: A process in which a gas (the absorbate) is taken up by a liquid or solid (absorbent), or in which a liquid is taken up by a solid. For example, absorption of moisture by fibres. In absorption, the substance absorbed goes into the bulk of the absorbing material. Solids that absorb gases or liquids often have a porous structure. The absorption of gases in solids is sometimes called sorption. Compare***adsorption***.

Absorption coefficient (K): See **Kubelka Munk**. The equation relates reflectance (R), absorption coefficient (K) and scattering coefficient (S).

$$2K^3 = (1 - R^2)$$

Absorption of dye (in dyeing): Absorption of dye from the dye liquor during the dyeing cycle. The absorption rate depends on the substrate and the dye; it normally takes a long time for the dye to exhaust from the dyebath.

Absorptiometer: Instrument for measuring the absorption of light similar to photometer.

Abudig: Medium fine-wool produced in Morocco.

Abutilon: Strong and glossy fibre, yielded by the abutilon species in tropical counties; used as hemp substitute for ropes.

Acacia: A greyish or greenish-yellow colour.

Acacia leucophloea: A coarse and strong bast fibre, a species of tree. Mainly used for ropes and nets before the introduction of synthetic fibres.

Academy blue: A mixture of **viridian** and **ultramarine**; a greenish-blue.

Acca: A rich mediaeval fabric made of silk, decorated with gold threads. Originally made at Acre, Syria. Used during the 14th century for church vestments and regal applications.

Acajou: A reddish-brown from the mahoganyof the same name.

Accelaerator: (1) A substance which accelerates a process. For example: Carrier which hastens the dyeing process of polyester.

(2) An instrument for making rapid laboratory determination of abrasion resistance and wear characteristics of fabrics such as pilling.

Accelarated aging (textile processing and testing): The use of controlled environmental conditions to promote rapid physical or chemical change in textile material.

Accelarated spinning performance: A method for evaluating spinning performance of a fibre, using small amount of fibre and reduced spinning time. End breakage is accentuated by spinning increasingly finer yarns or by varying yarn tension, spindle speed, yarn twist or other factors.

Accelerant: A chemical used to speed-up chemical or other processes.

Accelerated ageing: Use of controlled environmental conditions to promote rapid physical and/ or chemical change in textile material. See also **Accelarated** in textile processing and testing.

Accent colours: Highlights; sharp colours.

Acceptable quality level (AQL): Standard value, under which acceptance sampling specifications are compiled into an acceptable sampling plan.

Acceptance number: The maximum for the number of defective units in a sample that test results may not exceed without concluding, at least tentatively, that the lot fails to meet the specification.

Acceptance quality level: The maximum fraction of defective that, for the purpose of acceptance, sampling can be considered satisfactory for the process average.

Acceptance sampling: Used for acceptance for testing of a bulk finished product mainly by taking random samples for testing the required quality parameters and accepting the bulk on that basis.

Acceptance sampling lot: See **Lot**.

Acceptance testing: Testing carried out to decide whether the material meets the requirements of acceptance.

Accidental colour: The *after-image*temporarily fixed on the retina, also referred to as 'subjective colour'. When the eye concentrates on one colour and then looks away, the *complementary colour*often appears as a false or accidental colour.

Accordion fabric: Itis single jersey with the long floats held in place on the technical back by tuck stitches. It was originally developed using knit and miss pattern wheel selection. Needles required to tuck (if not selected to knit) were provided with an extra butt, in line with a tuck cam placed immediately after the pattern wheel selection.

Accordion pleats: Narrow straight pleats usually parallel to one another. See **Pleats**.

Accumulation tank (ETP): A tank used to store water or waste-water.

Accuracy of a test method: The degree of agreement between the true value of the property being tested and the average of many observations made according to the test methods, preferentially by many observers.

Acedianthrone: This is an anthraquinone dye in the group of Derivatives of anthraquinone with a fused ring structure. The example is a dye produced by treatment of 2-chloroanthrone with glyoxal sulphate, fusion with alcoholic potassium hydroxide and treatment with sulphuric acid.

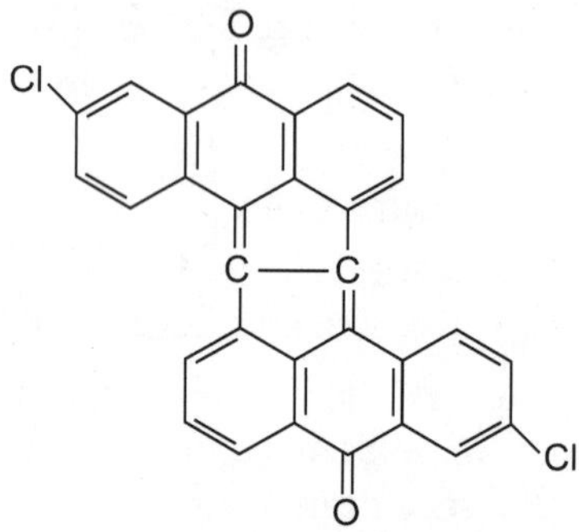

Vat red-brown RR

Acetate: See **Cellulose acetate**.

Acetate fibre: A manufactured fibre in which the fibre forming substance is cellulose acetate where not less than 92% of the hydroxyl groups are acetylated. The generic description of the fibre is triacetate. Acetate is manufactured from purified cellulose with acetic anhydride in the presence of a catalyst. It is then precipitated, purified, dried and dissolved in acetone into a highly viscous solution which is extruded through spinnerets into a column of warm air, is evaporated, leaving solid continuous filaments of cellulose acetate.

Acetate jersey: Knitted jersey fabric, plain or printed, made of acetate or triacetate fibre.

Acetate satin: Firm, medium weight satin made of acetate.

Acetic acid: An organic acid (CH_3COOH), widely used in textile applications. It is used in textile wet-processing, dyeing and printing, mainly for adjusting pH of a process including neutralisation, and in the manufacture of cellulose acetate and triacetate.

Acetic acid value: The percentage of weight of combined acetyl radical in any chemical is expressed as acetic acid.

Acetic anhydride: Anhydrous acetic acid [$(CH_3CO)_2O$] widely used in textile applications. It is used in the acetylation process in the manufacture of cellulose acetate.

Acetoacetanilide (C_6H_5-C=O-CH_2-$COCH_3$): Chemical which forms yellow pigments by reaction with diazotized m-nitro-para-toluidine, etc.

Acetone: Dimethyl ether (CH_3COCH_3). It is one of the most powerful solvents. Acetone dissolves secondary acetate and other cellulose derivatives. It is miscible with water and has a low boiling point 55–56°C.

Acetone recovery: It is the process of recovering solvent acetone from the acetate fibre or plastic manufacture. Usually recovery is done by adsorption by activated carbon and redistillation.

Acetyl: The radical CH_3CO- of acetic acid.

Acetyl cellulose: Cellulose acetate is sometimes wrongly described as *Acetyl Cellulose*.

Acetic acid value: The percentage of weight of combined acetyl radicals is expressed as acetic acid.

Acetylation: A chemical reaction whereby acetyl radical is introduced into a compound as in the conversion of cellulose to cellulose acetate.

Acetyl value: A measure of degree of esterification or combination of acetyl radicals with cellulose in acetate or triacetate products.

Aceytuni: Satin was originally known as aceytuni in ancient times.

ACFM: ACFM means actual cubic feet per minute. In air and gas streams, the ACFM is the actual flow. When corrected for pressure and temperature, ACFM can be correlated to SCFM (Standard Cubic Feet Per Minute).

Acherontic: Gloomy or dark. Derived from *Acheron* which according to Homer was one of the rivers of Hell – its water having a deathly foreboding appearance. 'Stygian' (in reference to the infernal River Styx of Greek mythology) has a similar meaning.

Achlorophyllaceous: Colourless.

Achromate: A person who is unable to detect colour.

Achromatic: Colourless object is known as achromatic.

Achromatic colour: A neutral colour that has no hue (white, grey or black).

Achromatic objectives: Optical objectives with compound lens systems having highest degree of correction for chromatic aberration.

Achromatic point: White point. Achromatic point is effective as the colour point of the lighting dependent physical ideal white.

Achromatistous: Colourless.

Achromobacterium: A genusofbacteria commonly found in activated sludge, trickling filters, denitrificationprocesses and anaerobic sludge digestion.

Achromatism: Colour blindness.

Achromatopsia, achromatopsy: The inability to see colours: a severe form of colour-blindness.

Achromatopsia: Total colour blindness.

Achromous: Colourless.

Acid: A chemical that will give a pH of less than 7 in water solution. In textile processing, many acids are used for various purposes. They include acetic acid, citric acid, hydrochloric acid, sulphuric acid, formic acid, etc. Several other compounds, such as sodium bisulphate, ammonium sulphate form acids in solution through hydrolysis.

Acid dew-point: Small amount of acids, particularly sulphuric acid, can raise the ***dew-point*** substantially. This can cause acid corrosion of components in air-pollution control devices, even though the temperature is well above 100 °C.

Acid-drop yellow: A medium yellow.

Acid green: A bright green.

Acidic Hydrogen (Acid Hydrogen): A hydrogenatom in an acid that forms a positiveion when the acid dissociates. For instance, in methanoic acid

$$HCOOH \rightarrow H^+ + HCOO^-$$

the hydrogen atom on the carboxylate group is the acidic hydrogen (the onebound directly to the carbon atom does not dissociate).

Acid-washed: Describing fabric which has been bleached to give the appearance of being worn or faded; particularly denim used for jeans – referred to as 'distressed denim'. Similar results can be obtained by other processes to produce fashion items referred to as 'stonewashed', 'bleached', 'pre-washed' or having an 'antique look'.

Acid ager: Steamer used for fixing printed/dyed material using acid steaming.

Acid binding equivalent(Acid saturation value): It is the binding equivalent of acid dyes per unit weight of protein fibres (wool, silk) as well as polyamide fibres. It depends on the respective number of free amine group in the fibre mule available to form salts.

Acid chlorination, colour fastness: See **Colour fastness to acid chlorination**.

Acid content (of felt): The number of milli-equivalents of acid present per unit weight of felt, measured under prescribed conditions.

Acid damage to cellulose: See **Hydrocellulose**.

Acid degumming of silk; Alkaline degumming of silk: Silk degumming in aqueous solutions of acids and alkalis is greatly influenced by pH and temperature. An alkaline reaction at a pH > 9 and acid reaction at pH < 2.5 ensure a rapid elimination of sericin completely after 30 min of treatment.

The temperature should not exceed 95°C to avoid weakening of the fibre. The degree of hydrolysis of fibroin also depends on pH of the bath.

Acid desizing: In this method, cotton fabric is treated with dilute sulphuric acid with a concentration of 5–10 g/1 at a temperature of about 40°C for 3–4 h. Dilute acid attacks the polymer chain of starch and due to chain cleavage of starch molecule, short water-soluble or dispersible chain segments are formed. With sulphuric acid higher than 10 g/1 and above 50~ there is always the possibility of weakening the cloth or causing holes . The treated cloth must not be allowed to dry at all, otherwise degradation of cotton will occur at the dried area.

Acid discharges (discharge resists): Discharges based on acids and their salts like citric acid or tartaric acid.

Acid donor (resin finishing): Substances added in the resin finishing, which during the polymerization process is liberated and accelerates the cross-linking.

Acid donor: A compound that hydrolyses or breaks down to yield acid. In a number of dyeing and printing processes, primarily with acid dyes, there are advantages to starting with a bath or print paste that is pH neutral or only slightly acid, but that will become strongly acid as a result of the boiling or steaming process. Acid donors, such as ammonium sulphate or ammonium tartarate are often used for this purpose.

Acid dyeable variants: Polymers chemically modified to make them receptive to acid dyes.

Acid dyeable nylon: Nylon that can be dyed with acid dyes.

Acid dyes: An anionic dye having substantivity for fibres, which contains cationic group like polyamide, silk, etc. A large class of dyes that are applied from acidic medium to protein fibres. They are usually applied from strongly acid to neutral on wool, silk and nylon, usually at a temperature near to boiling or boiling. They give bright to dull shades and fastness from poor to excellent.

Acid dyes (aggregated): One group of acid dyes when classified as per their state in solution. The actual size of the dye particle in solution varies according to the degree of aggregation. These are fast milling dyes and are characterized by poor stability, turbidity of solution (particularly when cold), low capillary rise, capacity of staining vegetable fibres, and distinct affinity towards wool when dyed in neutral bath. They may be dyed with ammonium acetate, ammonium sulphate, acetic acid and even from a neutral dye bath. They have poor levelling properties but good fastness to wet-processing. Examples are Polar Yellow R, Coomassie Navy blur RNS.

Acid dyes(equalising): Acid dyes which level well, but have poor fastness to wet treatments. Sulphuric acid is incorporated into the dyebath.

Acid dyes(milling): (*Acid-dyeing*) Acid dyes have poorer levelling properties but have better fastness to wet treatments. They are applied using a weaker acid such as methanoic (formic) or ethanoic (acetic) acid, the acid chosen depending upon the substantivity of the dye.

Acid dyes (molecularly dispersed): One group of acid dyes when classified as per their state in solution. Such dyes are characterized by high solubility, cleanliness of solution, high capillary rise, lack of affinity for vegetable fibres and their poor affinity for wool in neutral dye liquor. They are dyed with sulphuric or formic acid in the dyebath. They are easily levelling and possess poor fastness to wet-processing. Examples are Xylene Light Yellow 2G, Acid Orange GG, Azo Geranine 2GS.

Acid dyes(super-milling): (*Milling acid dyes: neutral-dyeing*); these are sometimes referred to as *super-milling* acid dyes. Their levelling properties are poor, but their fastness to wet treatments is excellent.

Acid equivalent of protein and polyamide fibres: Binding equivalent of the acid dyes per unit weight of the protein fibres, such as wool, silk, polyamide, etc. Also called acid binding equivalent, acid saturation value.

Acid fading: Fading due to the dye-fibre bond breakage under acidic atmospheric conditions. More relevant in case of reactive dyes.

Acid hydrolysis: Hydrolysis of a chemical by the addition of an acid. For example, wastes high in cellulose (e.g. waste straw or paper) can be pulped in water. The addition of a strong acid and steam (to raise the temperature) hydrolyses the cellulose in the pulp into sugars, which can then be separated out and fermented into alcohol fuels.

Acid metal complex dyes: See **Metal complex dyes**.

Acid recovery: A reclamation process in chemical processing in which the acid is extracted from a raw material, by-product or waste product. For example: In the manufacture of cellulose acetate, acetic acid is a major by-product. The recovery process consists of collecting all wash liquor containing considerable acetic acid and concentrating it by suitable method to form glacial acetic acid.

Acid resistance: The property to withstand the contact or treatment with acid (type of acid being specified-organic or inorganic) normally encountered in use.

Acid salt(acidic salt): A salt in which there is only partial replacement of the acidic hydrogen of an acid by metal or other cations. For polybasic acids, the formulae for such salts are of the type $NaHSO_4$ (sodium hydrogen sulphate) and $Na_3H(CO_3)_2.2H_2O$ (sodium sesquicarbonate). For monobasic acids such as HF the acid salts are of the form KHF_2 (potassium hydrogen difluoride). Although monobasic acid salts were at one time formulated as normal salts plus excess acid (i.e. KF.HF), it is preferable to treat them as hydrogen-bonded systems of the type K+(F–H–F)–.

Acid shock dyeing: A method of continuous dyeing of polyamide fabric using metal complex dyes. Fabric is first padded and heatset on stenter and afterwards fixed by shock treatment in a boiling bath containing 3ml/l of formic acid on jig or open width washing range.

Acid soaps: Sulphonated fatty acids (e.g., Castor oil) which is not fully neutralized.

Acid value: It is a measure of free acid content in oils, fats, waxes, etc. Estimated as the mg of pure potassium hydroxide required to neutralize 1 g of the material.

Acid yellow: A medium yellow acid dye.

Acid (colour fastness to): See **Colour fastness to acid spotting**.

Acid (strong): An acid which can virtually completely ionize in water is called strong acid. All of the acid which can be represented by HA, exists in solution as H^+ (or as H_3O^- the hydrogen ion combines with a water molecule to form a hydronium ion) and A-. Hydrochloric and sulphuric acid are common strong acids. Note that even in weak solution these are still considered as strong acids.

Acid (weak): A weak acid is a one that does not fully ionize in solution. Few of the acids remain in molecular form, rather than completely ionizing. If HA represents the acid, few would remain in solution as HA, and few would ionize to H^+ and A^-. Acetic acid and citric acid are commonly used weak acids. An acid may be technically weak, but is still capable of causing serious skin burns. See **Acid (strong)** and **Base (weak)**.

Acidic: A term describing a material having a pH of less than 7.0 in water. Or having a tendency to release a proton or to accept an electron pair from a donor. In aqueous solutions, the pH is a measure of the acidity, i.e. an acidic solution is one in which the concentration of H_3O^+ exceeds that in pure water at the same temperature; i.e. the pH is lower than 7. A pH of 7 is regarded as being neutral.

Acidic washing: Washing/soaping of dyeings or printing in acid medium to improve wash fastness.

Acidimetry: Titration.

Acidity: Property of acid and acid salts, whereby hydrogen ion is liberated due to dissociation of radicals when dissolved in water. Acidity of a water is measured by hydrogen ion concentration, usually called **pH**. Waters with pH below 7.0 are acid. In unpolluted water, acidity comes from dissolved carbon dioxide or organic acids leached from the soil. Atmospheric pollution also may cause acidity. Acid waters can corrode metal or concrete. See **Alkalinity**.

Acidity (degree): pH is the measure of hydrogen of a molecule in water; hence, will give the degree of acidity of that solution. See **pH**.

Acidolysis: Acidolysis is the degradation of the molecule in presence of an acid. In polymers, the splitting of the bond of a high polymer weight polyamide takes place resulting into a low molecular acid as follows:

$$-NH-(CH_2)_{n1}-NH-CO-(CH_2)_{n2}-CO-)\ COOH-R-COOH \rightarrow$$

Polyamide Polymer

$$-NH-(CH_2)_{n1}-NH-CO-R-COOH + HOOC-(CH_2)_{n2}-CO-$$

Acier: Steel-coloured; grey.

ACIT: Association des Chimistes de l'Industie Textle(French Association of the Chemists of the Textile Industry).

ACIMIT: Italian Association of Textile Machinery Manufacturers.

Ackermann's Green: A yellowish-green.

Acme: Variety of raw cotton whose staple is fairly long, but is of inferior quality.

Acores: A variety of unbleached linen of various fineness produced in France.

Acorn brown: The brown colour of the acorn; also simply referred to as 'acorn'.

Acoustic pressure: It is the periodic pressure variation that occurs in the sound waves. Measured in pascals (p).

Acre-foot: The volume of water required to cover one acre of land by one foot of water. It is equal to 1233 m^3.

Acri: 'Raw' cotton from Syria.

Acramin dyeing process: Process by which the pigment colour is padded along with binder and catalysts and cured in suitable machine. Pigment dyeing can be done for pale and medium shades on cotton and sometimes blends where cotton may be the major component. Even though the handle is slightly inferior in pigment dyeing, now a days, processes and binders are available which can match the handle of normal dyeing. Presently, pigment dyeing is more used for washing purposes. Here binder quantity is reduced, or partially cured, or not cured as per the washing off requirements, depth of shade or washing process adopted, etc.

Acridine yellow: A yellow coal-tar dye also used as an anti-bacterial agent.

Acritochromacy: Colourblindness.

Across shoulder (body measurements): The distance between the shoulder joint to shoulder joint across the back.

Across stitch: Embroidery stitch. Method: Bring the needle through on the lower right line of the cross and insert at the same line, taking a stitch through the fabric to lower left line. Continue to the end of the row in this way. Complete the other half of the cross. It is important that the upper half of each stitch lies in one direction. Thread should be under the needle point. Pull up the stitch to form a loop and repeat. This stitch may also be used on even-weave fabric.

Acrylic: The generic name of man-made fibres derived from polyacrylonitrile (min 85% acrylonirile units). Its chief properties are soft, woolly hand; wash and wear performance; brilliant and high shades of colours; good performance in carpeting; good wrinkle resistance. Acrylic is normally dyed with basic dyes.

Acrylic dyes: Dyes used for dyeing polyacrylonitrile fibre/yarn/fabric. They are mainly cationic dyes, disperse dyes (for pale shades) and acid dyes in certain cases.

Acrylic fibre: A manufactured fibre in which the fibre-forming substance is any long chain synthetic polymer composed of at least 85% by weight of acrylonitrile units [-CH_2-CH(CN)-] (FTC definition). Acrylic fibres are produced by two basic methods of spinning (extrusion), dry and wet. In the dry spinning method, material to be spun is dissolved is a solvent. After extrusion through the spinneret, the solvent is evaporated, producing continuous filaments which later may be cut into staple, if desired. In wet spinning, the spinning solution is extruded into a liquid coagulating bath to form filaments, which are drawn, dried, and processed. End uses of acrylic fibres include floor coverings, blankets, and apparel uses such as suiting, pile fabrics, coats, collars, linings, dresses, sweaters and shirts. Probably, the most popular synthetic fibre for yarns for hand knitting.

Acrylic jersey: A soft, warm, absorbent fabric, even warmer if the back is brushed or fleeced. Usually knitted in bright stripes as the right side is rather dull and unattractive if a plain colour is used. Used for leisure cloths, jogging suits, robes and dress gowns.

Acrylic knit: This covers a wide variety knitted fabric based on acrylic fibre but often containing small percentages of other fibres. Designs include fine and heavy ribs, lacy and crochet effects and open, often marled effect, marked effect, fabrics resembling hand knits.

Acrylic monomers: Acrylic monomers are used as finishing agents for cross linking on the fibre.

Acrylonitrile: Propene nitrile vinyl cyanide: CH_2= CH-CN. Raw material for the manufacture of Polyacrylonitrile.

Acrylic resin: Polyacrylate resins including their copolymers. See **Synthetic resins**.

Acrylic sheer: A heavy, rather stringy open-weave vision net for curtains, often in two-colour plain weave. All acrylic fibre is very soft and difficult to press, but a farmer net of this kind contains about 20% flax which improves its texture and handle. Others contain polyester which acts like scaffolding.

Acrylonitrile: A colourless, volatile, flammable liquid (CH_2= CHCN) used as a raw material in the manufacture of acrylic polymers and fibre.

Acetate: This is a manmade fibre which is not affected by moths or mildew; it is comfortable to wear and has very good draping properties. The fibre is inflammable, but many fabrics especially those made for furnishings have greatly reduced inflammability. Acetate is often mixed with other fibres including cotton, viscose and nylon producing very interesting textures and colourings. Woven and knitted fabrics are made for a wide variety of uses including dress fabrics, linings, ribbons, and furnishing materials.

Actinic degradation: See **Ultraviolet degradation**.

Actinic resistance: See **Ultraviolet resistance**.

Action stretch: A term applied for fabrics and garments that give and recover in both warp-wise and weft-wise. Action stretch is ideal for tight-fitting garments such as ski pants.

Activated alumina: Alumina (Al_2O_3) that has been treated to increase its surface area and thus to enhance its adsorptive properties. It can remove phosphate and arseniccompounds from liquid wastes and can dry gases.

Activated bio-filter: A high rate trickling filterusually followed by an aeration tank and always a sedimentation tank. Some of the sludge from the sedimentation tank is recycled to the filter. This enables a high concentration of biofilm growth in the filter and therefore the BOD loadings on the filter can be much higher than conventional design. This system may be used for a roughing filter giving about 60% removal of BOD.

Activated bio-solids: Another name for ***Activated sludge***.

Activated carbon: Amorphous form of carbon with porous internal structure having capacity to adsorb many gases, vapours and colloidal solids. Also used for decolourizing liquids and other adsorption purifications. Activated carbon is made from charcoal made from wood, peat, etc., by treatment of the charcoal with steam at 900 °C, or by with phosphoric acid at 600 °C, or in some other way to increase its porosity and its surface area. Its large surface area per unit mass (up to 1400m^2/g) enables it to adsorb trace organics and other undesirable components of a water intended for drinking, or from a malodorous air. It is expensive. It can be used in water treatment as a filter of granular activated carbon(GAC) in an activated carbon filter. Alternatively it may be added as powdered activated carbon (PAC) into the water, after which it is removed by filtration. GAC is larger than 100 um; whereas, PAC is usually less than 50um. Activated carbon can be used in air-pollution control to remove toxic organics fromgaseous effluents, such as incinerator waste gases. Activated carbon is widely used for odour control, with the odorous gas being passed through beds of GAC.

Activated carbon bed: An activated carbon filter.

Activated carbon filter, carbon adsorption bed, GAC filter: A tank used in water (or possibly wastewater) treatment, containing 1–9 m depth of granular activated carbon(GAC) through which water flows, usually as a downward flow filter, but possibly as an upflow or moving bed filter. The filter can be used to remove trace organics (e.g. pesticides) from water. The beds filter out suspended solids and so they may need backwashingpreceded with air scouring. The bed or filter may be designed for surface loading ratesof 100 to 600$m^3m^{-2}d^{-1}$. The spent activated carbon is removed from the bed or filtered and regenerated by the same process as newly manufactured activated carbon. GAC filters can be used in air-pollution control to remove toxic organics from gaseous effluents, such as incinerator waste gases. GAC filters are also used for odour control, with the odorous gas being passed through beds of GAC. GAC on which microbes are grown is known as biological activated carbon. The microbes enhance the treatment of the water or wastewater or gas.

Activated carbon in dyeing and printing: Used as an addition to wash baths when washing of dyeing or prints to prevent staining of grounds by re-depositing.

Activated carbon in waste-water treatment: Activated carbons are used as an adsorptive agent in many stages of waste-water treatment.

Activated clay: An adsorbent clay that removes colour, odour, free fatty acids, etc., from oils and tallows.

Activated oxygen bleach: A bleaching system comprising of an oxygen bleach and a bleach activator.

Activated sludge: A biomass made by continuous recirculation of the sludge from the secondary sedimentation tankto the aeration tank, thus acquiring many active aerobic bacteria, from 100 to 1000 million per millilitre of mixed liquor. The bacteria are embedded in or on a slime that forms 90% of the solids content of the sludge. The slime, mainly polysaccharides, is produced by the bacteria as they consume the BOD. Protozoa, especially ciliate protozoa, mayalso be present in the sludge, feeding on the bacteria.

Activated sludge process: A continuous, aerobic biological treatment for waste-water dating from 1913 that uses a culture of bacteria suspended in the waste-water in an aeration tank to adsorb, absorb and biodegrade the organic pollutants. Flocsare formed that may reach 0.1 mm in diameter in the aeration tank. These are kept in suspension either by air blown into the bottom of the tank (diffused air system) or by mechanical aeration. The mixture of the activated sludge and the waste-water in the aeration tank is known as the mixed liquor. The concentration of mixed liquor is known as mixed liquor suspended solids (MLSS), but is also measured as mixed liquor volatile suspended solids (MLVSS). The mixed liquor flows from the aeration tank to a sedimentation tank, where the activated sludge flocs combine together into larger particles that settle as a sludge. Most of the sludge from the sedimentation tank returns to the aeration tank. The dissolved oxygen in the aeration tank should be at least 0.5 mg/l, preferably 1–2 mg/l. The aeration tank may be designed on aeration period, preferably on F : M ratio or mean cell residence time. If the ammonia present is to be oxidised to nitrate, the plant must be designed for nitrification. In municipal waste-water, the activated sludge process is usually preceded by primary sedimentation. In the start-up of an activated sludge plant, the time needed for establishing the appropriate bacteria and protozoa can be greatly reduced by seedingthe new aeration tank with sludge from one that is working well. Activated sludge treatment demands only about one seventh of the land occupied by trickling filters. It also does not suffer from filterfliesand has little smell. The operating cost of the aeration tank can be high and skilled attention is essential. Many varieties of activated sludge treatment exist, including contact stabilisation,extended aeration, modified aeration, oxygen activated sludge.

Activation energy: Generally, the heat that is required to make a chemical reaction go. An "energy barrier" has to be overcome before a starting material can be converted into a product. The height of the barrier is directly proportional to the rate of the reaction.

Active bleach: See **Available chlorine**.

Active carbon: See **Activated carbon**.

Active content: A measure of the active matter in a solution or emulsion of textile auxiliaries. It is the percentage of the active matter of a pure substance or auxiliary, which is supplied in the diluted form in water, solvent, etc., calculated on the basis of the weight of the auxiliary.

Active earths: See **Bleaching earths**.

Active oxygen or nascent oxygen: Oxygen in the formative stage, i.e. at the moment of decomposition, once believed to be responsible to the bleaching action/decolourisation during the bleaching processes.

Active substances: A group of highly active chemical compounds which are essential to life through their control functions in organisms.

Active toxicity: A toxicity, exposure of which will result in significant response on living beings/humans shortly after exposure(typically a response is observed within 48–96 h).

Active wear: Functional clothing, for active wear and sports activities.

Activity coefficient: The activity coefficient can be estimated using the following expression, derived from the Debye-Huckel theory and known as the Gutenberg approximation

$$\text{Log } r = 0.5(Z_i)^2.I^{1/2}/(1+I^{1/2})$$

Where Z_i = charge in the ith ionic species

I = Ionic strength

Activity(of an antibacterial agent): A measure of effectiveness of the anti-bacterial agent.

Activity(of anti-dust mite agent): A measure of the effectiveness of the anti-dust mite agent.

Acrylic sheer: A heavy, rather stringy open-weave vision net for curtains, often in two-colour plain weave. All acrylic fibre is very soft and difficult to press, but a firmer net of this kind contains about 20% flax which improves its texture and handle. Others contain polyester which acts like scaffolding.

Actual blend ratio: Ratio of the different fibres in a textile material determined by analysis.

Actual calorific value: The calorific energy that is released by the combustion of a defined quantity of a material under specified test conditions. Also called *Heat release*.

Acute toxicity: A toxic effect that has clear symptoms that occur rapidly after the exposure to the toxic substance. Acute toxicity can be measured by median lethal concentration or dose.

Acylaminoanthraquinone: Dye-class derived from anthraquinone and chain-substituted and chain-linked anthraquinone derivatives. Can be manufactured by the condensation of 1,5-diaminoanthraquinone with benzoyl chloride.

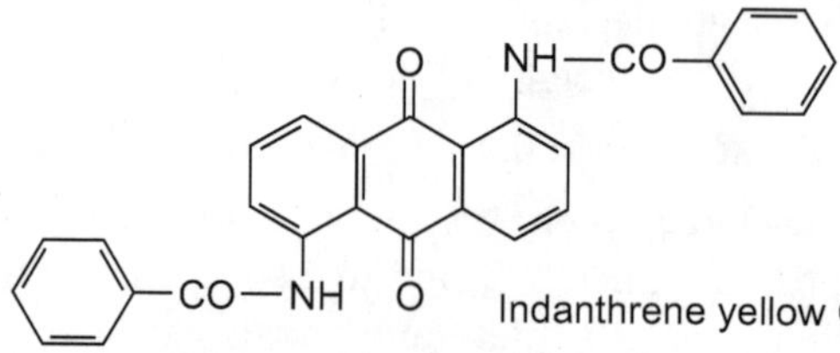

Indanthrene yellow GK

Acyanoblepsia: The inability to see the colour blue.

Acyl polyglycolethers: See **Polyglycolethers**.

Acylcellulose: Acyl substituted cellulose; e.g., Cellulose esters.

Adad: Fibres yielded by the *Ficaria ranucuioides*, a creeper in the Marshall Islands; used for clothing mats by the natives.

Adam blue: A greenish-blue.

Adapangia: Trade name for Indian raw silk.

Adarsa: A fine muslin made in India.

Adatais or addatis: Fine cotton muslin from India; the best grades are from Bengal.

Addition high-wet pick-up: A process in which damp cloth (from a preliminary washing range) is squeezed out and run through a highly specialized high wetpick-up unit without liquor.

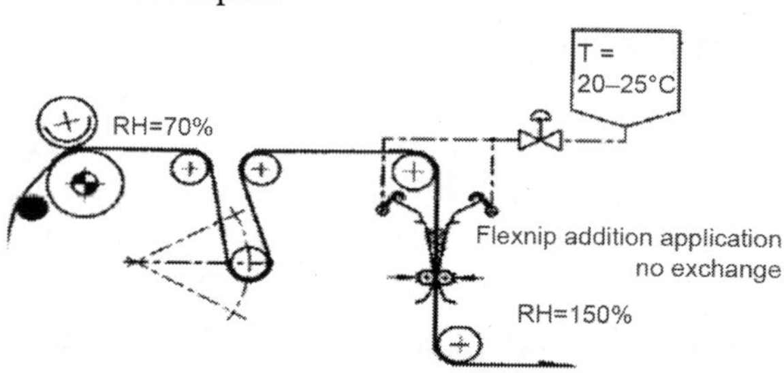

Addition pick up in the flexinip system by Kusters

This is obtained by using a very low liquor bath and rapid continuous liquor replenishing.

Addition mercerization: A cost effective mercerization process where scoured fabric with 50–70% moisture (avoiding a drying process thereby saving energy) is fed to a caustic soda bath of about 300g/l NaOH (30 °C) for mercerization process. Many squeeze points, addition of wetting agents, temperature, viscosity of the mercerizing liquor, absorbency of the fabric to be mercerized and the dwell time for the process ensures proper and even mercerisation.

Addition polymerisation: Polymerization process in which participating components are joined together due to trans-molecular rearrangements caused by shifting of hydrogen atoms to form high molecular synthetic compounds. The molecular weight of a polymer so formed is a simple sum of the molecular weights of the combined molecular units. This occurs when small molecules join together under the stimulus of a catalyst, heat or radiation to form a linear polymer usually without the elimination of a small molecule.

A	+	A	+	A	+	A	+	A	+	A	→	A	A	A	A	A	A

Monomer Monomer Monomer Monomer Homopolymer

Schematic representation of addition polymerisation

This can be of the following three types:

(a) *Free radical addition polymerization*: In this type, chains are initiated by a free-radical such as phenyl.

(b) *Cationic addition polymerization*: The active species which initiates the addition polymerization is a cation such as a proton.

(c) *Anionic addition polymerization*: The initiating species in this case is an anion such as NH~-). See **Addition polymers**.

Addition polymers: When the monomer units are repeatedly added to form long chains without the elimination of any by-product molecules, the product formed is called addition polymer, and the process involved is called addition polymerisation. The monomer units are unsaturated compounds and are usually of alkenes. The molecular formula and hence the molecular mass of the addition polymer is an integral multiple of that of the monomer units. A few examples of addition polymerisation are:

(i) $n\mathrm{CH_2 = CH_2} \rightarrow (\mathrm{–CH_2–CH_2–})n$

Ethylene Polyethylene or Polyethene

(ii) $n\mathrm{CH_2 = CHCl} \rightarrow (\mathrm{–CH_2–CHCl–})n$

Vinyl chloride Polyvinyl chloride(PVC)

(iii) $n\mathrm{CH_2 = CHCH_3} \rightarrow (\mathrm{–CH_2–CHCH_3–})n$

Propylene Polypropylene

(iv) $n\mathrm{CH_2 = CHCN} \rightarrow (\mathrm{–CH_2–CHCN})n$

Acrylonitrile Orlon

Addition test: A test for assessing the levelling behaviour of the dye, by actually giving an addition in the dyeing bath itself and then evaluating.

Addition (in dyeing): A term used to describe each corrective addition of dye(s) made during the course of dyeing which is needed to achieve the closest possible colour match to a given colour standard or pattern.

Additive: A supplementary material combined with a base material to provide special properties. For example, pigments are used as dope additives to give colour in mass dyeing.

Additive colour: A mixture of coloured lights. The three primary colours of red, green and blue, when mixed together in equal proportions, produce white light. Mixing the three additive primaries in differing amount can create any colour in the rain bow. Colour televisions use the principle of additive colour mixing.

Additive primaries: Red, green and blue light. When all three additive primaries are combined at 100% intensity, white light is produced. When these three are combined atvarying intensities, a gamut of different colours are produced. Combining two primaries at 100% produces a subtractive primary, either cyan, magenta or yellow:

100% red + 100% green = yellow

100% red + 100% blue = magenta

See **Subtractive primaries**.

Add-on: The weight of solids left on a given weight of fabric after impregnation and drying. The percentage add on is given by (w1-w) × 100/w; where w is the weight of the fabric impregnated and w1 is the weight of the fabric after impregnation and drying.

Adduct: (1) An addition product, e.g. an ethylene oxide formed from which substituted polyglycolether are formed.

(2) An unstable compound formed in between the dyes and levelling agents which, after a long time of dyeing, gradually breaks down partially or completely, there by achieving slow exhaustion of the dye and improved levelling.

Adenos: One of the best variety of cotton; it is also the name of a cotton fabric made in Syria.

Adhered water: Superficially loosely adhered water, mainly yarn and fabric interstices.

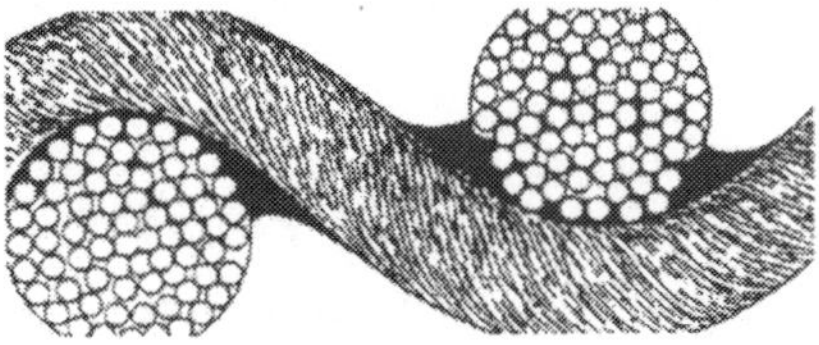

Water collects at the intersection of the warp and weft and is so tightly bound that it cannot be removed as easily as other surface water. When fabric is passed through a washing range, a layer of adhered water is formed directly on the surface of the fabric which forms an obstruction to the diffusion of dissolved soil. Only a vigorous flushing will disturb the lamellar film boundaries.

Adhesion: The force that holds different materials together at their interface and resists separation into two layers.

Adhesion, in tyre fabrics: The force required to separate a textile material from the rubber or other elastomer by a definite prescribed method.

Adhesion energy: Potential energy, at a steady temperature, released when a free liquid surface in the separation of two phases (liquid/solid) to cause an increase in free energy.

Adhesion promoters: Chemical which facilitates adhesion. These products help the smooth fibre-face of closely constructed base fabric to provide a chemical bonding site for subsequent coating. Products containing the isocyanate group are the most widely used promoters.

Adhesive activated yarns: Yarns which are having adhesive property induced during the manufacturing process itself so that to promote better adhesion to another material such as rubber and/or to allow easier processing.

Adhesive bonded non-woven fabrics: Fabrics which are produced by sticking a knop yarn or fibre on to a textile or plastic substrate.

Adhesive bonded pile coverings: Pile coverings made by laying and sticking the fibre web on to a prefabricated woven base fabric, resulting in a continuous rib structure. These ribs are sheared to give velour like appearance.

Adhesive bonded cloque: Fabric produced by sticking two woven fabrics together by applying adhesive and then laminating, one being shrunk and the other without shrinking. Subsequently, the fabric is treated with hot water or alkaline creping bath, the unshrunk layer shrinks while giving a crepe effect.

Adhesive migration: In non-woven, the movement of adhesive together with its carrier solvent in a fabric during drying, giving it a non-uniform distribution with the web usually increase to the outer layer.

Adhesive lamination: Laminating process for applying a foam material film to a textile material.

Adhesive strength or peel strength (in carpet): Measure of bond of adhesive to a surface.

Adhesives: In textiles, material which cause fibres, yarn, or fabric to stick together or to other materials.

Adhi: An Indian term for light weight fabric in plain weave. It is used as shirting in tropical countries and also as the ground fabric for ***Chikan, Kamadani***, etc.

Adia: Good quality bleached cotton cloth in Persian/gulf counties.

Adipic acid: Acid used for the manufacture of 6,6 polyamide and 1,6 diaminohexane, akyde resins and esters used as plasticisers. It is 1,4 butanedicarboxylic acid also called hexanedioic acid. $HOOC-(CH_2)_4-COOH$.

Adipic acid fibres: Polyamide fibres of the polyamide 6,6 type, produced from adipic acid and 1,6 diaminohexane (hexamethylene diamine) are also called as adipic acid fibres.

Adire: The name given by Nigerian people (Yoruba tribe) to the patterns created by reserve Africa print techniques on cotton fabrics using Indigo.

Adire "oniko": It is tie-and-die technique.

Adire "eleso": It is a technique in which plant seeds, fruit capsules, shells or stones are incorporated in the damp cloths and used for producing circular and star shaped patterns on dyed/printed fabrics.

Adire "alabere": It is a technique which employs basting and whip stitches to produce reserve patterns.

Adire "eloko": It is an African variant of the Indonesian batik technique. Instead of wax, a reserving agent "eko" is produced in the form of a paste from rice or cassava starch. The paste is applied on the surface of the fabric without any crack while drying to avoid forming indistinct lines due to the cracks as against the batik technique.

ADR: American Dyestuff Reporter.

Adjacent colours: Colours which are next to each other on the ***colour wheel***or on a painting or design, etc. The placement of two colours next to each other often results in both hues taking on a different appearance so that, to take one example, red next to yellow makes the red turn towards purple and the yellow appear green.

Adjacent fabric: Undyed RFD (ready for dyeing) fabric composed of one or several types of fibres used in the fastness tests to determine the degree of staining from a dyed or printed test specimen.

Admix: The medium that is added directly in to the batch tank of pre-coat to create a permeable filter cake. Usually used in place of body feed.

Adobe: Pinkish-red; probably so called after the sun-dried bricks of the same name used in Latin American countries; also 'adobe red'.

Adras: A glossy union fabric, half silk and half cotton, usually made with narrow stripes and a beetled finish, made in Asia.

Adrianople red: A red colour also called ***Turkey red.***

Adrianople twill: French name, synonymous with ***Turkey red*** twill.

Adsorbant substances: See **Adsorption**.

Adsorbate: See **Adsorption**. A substance that is adsorbed on a surface.

Adsorbent: See **Adsorption**. The substance on whose surface adsorption takes place.

Adsorption: A process in which a layer of atoms or molecules of one substance (adsorbate) forms on the surface of a solid or liquid (adsorbant). All solid surfaces take up layers of gas from the surrounding atmosphere. The adsorbed layer may be held by chemical bonds (chemisorption) or by weaker Van Der Waal's forces (physisorption). Taking up of a soluble substance (adsorbate) onto a solid body (adsorbant). Examples are dissolved substances on to fibres, moisture on to machine parts, gases on to solid bodies, etc.

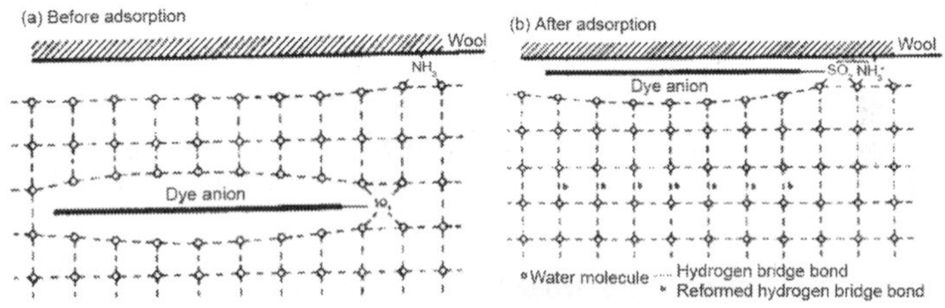

Schematic representation of dye adsorption on wool

Adularescent: Bluish. According to *Stormonth's English Dictionary* 1884 – having the whitish sheen of moonstone found on Mount Adula in Switzerland.

Adumbral: Shadowy.

Adumbration: Shading.

Adust: A scorched brown colour.

Advanced composite: Polymer, resin or other matrix material in which reinforcement is accomplished via high strength, high modulus materials in continuous filament form or in discontinuous form such as staple fibres, fibrets and in-situ dispersions. See also **Composite**.

Advancing colours: Colours in the range of yellow to red which when applied to a surface make it appear more prominent or as if it is advancing towards the viewer. See also **Receding colours**.

Aea: 'Native' Hawaiian name for cord, made of the bark of the aleurites tree.

AECC: Asociacion Espanola para el Control de la Calidad, Madrid (Spanish Quality Control Association).

AEK: Fastness Committee of AATCC.

Aene- (L): Bronze.

Aeneous: Having the colour or lustre of brass.

Aeolian: A warp rib dress material made of cotton warp and silk, man-made fibre or worsted weft. The cloth is light in weight, has a glossy finish and is piece dyed.

AEQCT: Asociacion Espanola de Quimicos y Coloristas Textiles (Spanish Association of Textile Chemists and Colourists).

Aeration: The process of adding air to a water supply for the purpose of oxidizing or mixing.

Aere: Woollen yarn fabric with a tweed-like character and a marked crocheted open-work pattern effect.

Aerobic bacteria: Microorganisms which especially thrive on atmospheric oxygen, so called aerobes are called aerobic bacteria. Opposite term ***Anaerobic bacteria***.

Aerobic degradation: Degradation of dissolved matter or solid matter into other compound by aerobic bacteria. In this process, the organic substrates are oxidized and the energy released is used for the reproduction of the microorganisms.

Aerobic denitrification: See **Anoxic denitrification**.

Aerobic digestion: The breaking down of complex organic material into simpler compounds by aerobiosisas in an aerobic digesteror in composting.

Aerobic process in effluent treatment: They are biological treatment processes that occur in the presence of oxygen. Certain bacteria that can survive only in the presence of dissolved oxygenand are known as obligate aerobes.

Aerophane: Thin, solid-coloured silk gauze, used as millinery and dress trimming.

Aerosols: Liquid droplets or solid particles dispersed in air or gases of fine enough particle size (0.01–100μm) to remain so dispersed for a period of time. Generally removed by coalescing fibres.

Aerosol dyeing: In this method of dyeing the dyes are applied to a textile material in the form of an aerosol composed of air, dye and water vapour.

Aertex: Trade name for a British cotton cellular cloth; it is cool, absorbent and comfortable. Used for making underwear, sports cloths and shirts.

Aeruginous: Having the colour of ***verdigris*** or copper-rust.

Aesthetics: In textiles, properties perceived by touch and sight, such as hand colour, lustre, drape and texture of fabrics or garments.

Aesthetics (in sewing): Refers to the appearance of the thread in the finished seam that can be effected by contrast stitching, colour matching and the sheen of the thread or the size of the thread.

Aethio, aetho, aethrio (G): Bright, fiery.

Aetzstickerei: German term for burned embroidery. The pattern is embroidered in vegetable fibreand on animal fibre foundation, or vice versa, in such a manner that it will hang together like a lace after the foundation is destroyed with chemicals which do not affect the embroidered yarn.

AFCT: Association Farancais de Controle Textile (French Association for Textile Control).

Affinity: The quantitative expression of substantivity. A measure of the force which binds dye or textile auxiliary to a substrate like fibre. It is the difference between the chemical potential of the dye in its standard state in the fibre and the corresponding potential in the dyebath. In dyeing, affinity essentially means the preferential attraction of the dye for the fibre rather than the solution of the dyebath.

Afgalane: A plain-weave all-wool dress fabric containing –(a) woollen warp with woollen weft (b) worsted warp with woollen weft. In both types, the warp has usually S and Z twist alternatively with S twist woollen weft.

Afghan: A knitted or crocheted wool blanket, made with fancy coloured stripes.

Afghan carpet: Afghanistan carpets made in Turkestan and Afghanistan of wool or goat's hair; the medium long loose pile is tied in Senna knot. The patterns are angular octagons, diamonds and sometimes stiff floral designs in red, brown, blue and white. Most of these carpets have a strong odour due to imperfect washing of the wool.

Afghaline: Originally made from Afghan wool(laine=wool). Plain weave, medium weight, all-wool woven fabric. A lightly pearled appearance due to the use of pairs of S and Z yarns, alternately in both warp and weft. Used for ladies dress material with faint, striped pattern on the back.

Afghaline

Afiume: A coarse grade of flax grown in Egypt.

AFNOR: Association Francaise de Normaisation, Paris (French Standards Association).

A-Frame: A movable batching unit in which the horizontal roller is supported by two A shaped frames. This unit is used to wind fabric in batch form for either storage or processing.

Afric, African fibre: See **Crin d'Afrique**.

African bass: Very coarse, dark, stiff and tough fibre, yielded by the leaf stalks of the bamboo palm in Africa; used for brushes; also for lines by the natives.

African fibre: Commercial name for a fibre yielded by the leaves of the palmetto in Algiers. Used for mattress stuffing.

African print: Printing technique on cotton fabrics used mainly in the countries of West and Central Africa (e.g., Congo, Nigeria, Ivory Coast, Senegal, Kenya, etc.). The prints are characterized by typical naturalistic designs coloured with usually a few frequently repeated colour combinations like golden-yellow, maroons, bottle-greens, browns, etc. Fabrics are usually used on both sides and hence, full penetration is required so that print would look same on both sides.

African silk: See **Anaphe silk**.

African violet: The pinkish-violet of the flower of the same name.

Afridi: A cotton fabric made by the natives in India and decorated with wax ornaments.

After burn time: Time consumed between the moment the ignition flame is removed and the extinction of the flame on the test specimen.

Afterchrome dyes: Dyes for wool which can be dyed by the after chrome dyeing method. See **After chroming, Chroming of dyes**.

After chroming: See ***Chroming of dyes***. In after chroming method of dyeing chrome dyes, the dyeing is done first in a similar manner to an acid dye and then treated in a chrome bath to produce the final shade. Dyeing with good light and wet fastness properties are obtained.

After coppering dyes: Some direct dyes having a hydroxyl group in an Ortho position to the azo nitrogen atom are capable of forming stable metal complexes with metal ions are often given an after treatment with copper salts to improve their light-fastness.

After coppering of direct dyes: See **After coppering dyes**.

After coppering test: A test to find out the after copper treatment of dyes. The test is as follows: Take 10ml of 30% H_2O_2 in a test tube and add 2–3 drops of sodium silicate to it. Shake well and add 1ml con. Ammonia. Shake well and wait till all the gas formed has escaped, if any, and then add the test specimen. An effervescence with the evolution of oxygen proves that the specimen is treated with copper salts.

After flame (in flammability test): Persistence of flaming of a material under specific test conditions, after the ignition source has been removed.

After flame time (in flammability test): The length of time for which a material continues to flame, under specified test conditions after the ignition source has been removed (also called duration of flame).

After glow (in flammability test): Persistence of glowing of a material, under specific test conditions, after cessation of the flame or if no flaming occurs after removal of the ignition source.

After glow time (in flammability test): The length of time for which a material continues to glow under specified test conditions after cessation of the flaming or after the ignition source has been removed (also called duration of afterglow).

After-image: The sensation of colour remaining after the stimulus has ceased where, for example, one stares at an image or object and then looks away to a white surface. The shape remains fixed for a while. Often, the colour of the original image is found to be complementary to the after-image. See also **Complementary colours***and* **Accidental colour**.

After-treatment of dyeing: After treatment of dyes are done to improve wet fastness (usually reactive and direct dyes). Most common after treatments are based on formaldehyde condensation products or polyammonium compounds. These products (cationic) can bond with anionic dyestuffs and reduce the water solubility of the dye, and dye migration is prevented, whether substantive dye or dye hydrolysate.

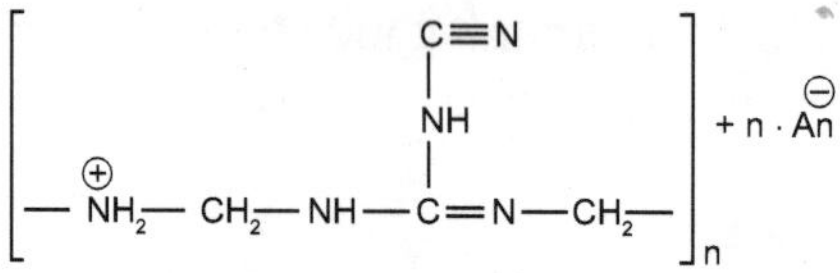

Structure in principle of a formaldehyde condensation product

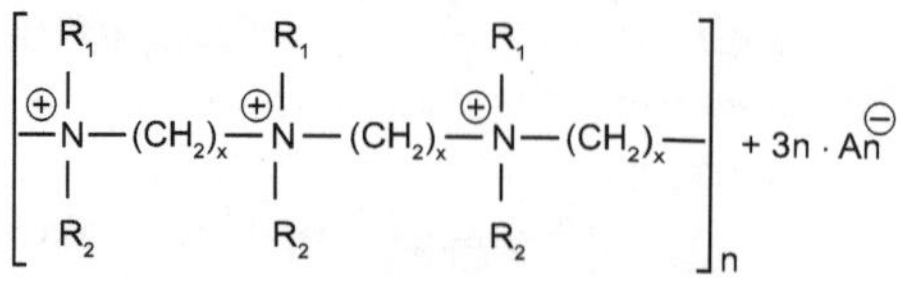

Structure in principle of a Polyammonium compound

After-treatment: Any treatment done after a particular process is completed. In dyeing, it refers to treatment given to improve properties; in non-woven, it refers to finishing process carried out after a web has been formed and bonded. Examples are embossing, creping softening, printing and dyeing.

AFU: AATCC Fading Unit.

AG: Alginate fibre.

Aglao- (G): Bright.

Agraffe: An ornamental clasp, loop or lock.

Agar agar: Agar agar is extracted from various red and brown algae which grow on the sea bed in warm temperature seas like India and China. It greatly swells in water, dissolves in hot water and sets on cooling to a jelly form at a concentration as low as 0.5%.It is used as a finishing agent, clarifying agent and thickener.

Agaric: Cotton fabric with a loop construction, like towelling.

Agave: Leaf fibres obtained from the various aloe and century plants; used for ropes and cords.

Ageing: (1) Treatment of fabric in moist steam, especially printed fabric from printing machine. Ageing is also done in the dyeing of certain colours, e.g. Aniline black.

(2) Deterioration of textile or other materials caused by gradual oxidation during storage and/or exposure to light.

(3) The oxidation stage of alkali cellulose in the manufacture of viscose rayon from bleached wood pulp.

Ageing, accelerated: See **Accelerated ageing**.

Agenois: Unbleached linen made in and around Agen, France; Obsolete.

Ager: A machine or chamber used for ageing. It can be a separate machine or a part of machine/range like pad-steam range. The chamber machine is filled with moist steam by suitable method and the dyed or printed fabric is passed through the steam filled chamber in loop form or over guide rolls to give time for the treatment. See **Ageing**.

Aggebonce: A silk embroidered cotton fabric, made in Syria.

Agglomerate: A combination or aggregation of colloidal particles, molecules, ion, dyes, particulate dirt, etc., in the form of clusters of approximately spherical shape.

Agglomeration: The tendency to form agglomerates. An indiscriminately formed cluster of particles.For example: A cluster of particle on the fibre.

Aggoned bundei: The best grade of Indian and Japanese raw silk. In European markets it is called Tani or Tanny.

Aglet: A white lacing cord.

Agra gauze: A plain weave open-set silk fabric having a gauze like appearance and given a stiff finish and is used for trimmings.

Agre' gauze: Same as **Agra gauze**.

Aggregate: Assembling several separate units into one.

Aggregate dyestuff: Occurs among other things, flat-bed screen printing due to surplus dye paste passing through the screen. The dye paste is not adsorbed by the substrate during the squeeze process, but nevertheless ends up on the printed goods when the screen is lifted and causes dyestuff aggregation there.

Aggregation: Chemical association of dissimilar molecules to form molecular associates, e.g. as occurs typically between dyes and levelling agents which possess dye-affinity.

Aggregation value or aggregation number: The number molecules in an aggregate.

Agitate: To stir or to mix, as in the case of a dyebath or solution.

Agnelin: Antiquated an obsolete name of Danish wool from Agnelin province of Denmark.

Agneline: Coarse, black, stout woollen, made with a long nap, thoroughly filled to shed water; used by the poorer classes in Europe for winter clothing.

Agra: Large and very thick knotted wool or cotton carpets, made in India. The colours are blue, green and browns.

Aguilles: Plain woven cotton cloth made in Syria.

Aguillettes: Metal-tagged laces which replace the sewn ones to attach the breeches to the doublet.

AICQ: Associazione Italiana per il Controllo della Qualita, Milano (Italian Society for Quality Control).

AICTC: Associazione Italiana de Chimica Tessile e Coloristica, Milano (Italian Association of Textile Chemists and Colourists).

Aida: Plain weave fabric with pattern appearing alternately on the face and on the reverse side; used as table cloth and awnings, etc., for table purposes, awnings, etc.

Aida canvas: Coarse yarn, open work, highly chemically finished woven fabric (cotton viscose filament) also double weave consists of 2–3 layers often with pattern on both sides.

Ailanthus: Wild silk of greyish or brownish colour, produced in ancient India.

Ailesham cloth: Fine linen cloth made in England in the Middle Ages.

Air balloon system: Another term for air injection for the untwisting and crease-free ballooning of tubular knitted fabrics, normally before passing through a squeezing.

Air blade: See **Air knife, Squeegee**.

Air blanket in dyeing system: An air pressure blanket/cushion created by compressed air and positioned above the dyebath is employed in some yarn and piece dyeing machines to reduce the liquor-to-goods ratio and prevent boiling of the liquor. This system is also useful when the goods for dyeing are less than the full capacity of the machine. Pump cavitation is also avoided by this means.

Air brushing: Blowing colour on fabric or paper with a mechanized pneumatic brush.

Air conditioning (in textile): A chemical process for sealing short fuzzy fibres into a yarn. Fabrics made from air conditioned yarns are porous because they allow more air circulation.These fabrics are cooler.

Air cushioning (in yarn dyeing): See **Air blanket in dyeing system**.

Air drying: Air drying is sometimes carried out in the open, with goods suspended by hanging over rods or wooden slates or frames. It is the simplest and most gentle and probably cheapest method of drying and involves no energy consumption.

Air entangled yarns: See **Compacted yarns**.

Air forming: A process in which air is used to separate and move fibres to fashion a web such as the Kroyer process for short fibres, usually of wood pulp.

Air-injection dyeing: A dyeing procedure employed in beam dyeing machines which involves the injection of fine bubbles of air in order to reach the dead zones in the yarn packages and disperse air contained therein.

Air-jet loom: A loom in which the weft yarn is propelled through the shed by means of a jet of air.

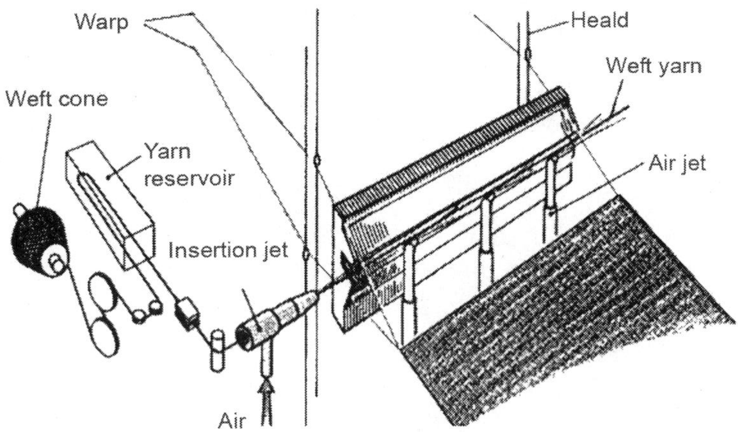

Weft insertion by jet

Air-jet yarn: Yarn made by the air-jet spinning method. The yarn more closely resembles, but is weaker than, ring-spun yarns.

Air-jet method (spinning): Draw-frame sliver (often combed) is supplied to the air-jet spinner, with twist being inserted to the fibres, largely on the yarn surface, by the vortex created in one or two air-jet nozzles, generally only one when spinning 100% cotton, the yarn structure therefore consisting of a core of largely parallel fibres and a sheath of wrapped (twisted) fibres.

Air-jet spinning: Spinning developed by m/s Murata of Japan. In this, roving is directly fed to the spinning machine through a drawing frame into the spinning zone in which two air-jets are arranged, one behind the other. In the first jet, the yarn is given a false twist effect that is untwisted by the second jet. Hence, the conditions are almost like the OE spinning. With this method, fine yarns can be spun with high velocity. The yarn is directly wound on to the bobbins. In this, pre-spinning and subsequent winding operations are omitted.

Air-jet weaving: An intermittently directed stream of compressed air, with or without additional air-jets (via relay nozzles), serves as the weft insertion medium by conveying the weft yarn through the shed.

Air-jet texturing: See **Texturising, Texturing,Airjet.**

Air-jet weaving machine: A weaving machine where air-jet weaving principle is adopted. See **Air-jet weaving**. The air-jet weaving machines are the weaving machines with the highest weft insertion performance and are considered as most productive in the manufacturing of light to medium weight fabrics, preferably made of cotton and certain man-made fibres (sheets, shirting fabrics, linings, taffetas and satins in staple yarns of man-made fibres); it has anyway to be pointed out that technically positive results are obtained at present also with heavy weight fabrics (denims) and that few manufacturers also produce machine models for terry production. These machines are the ideal solution for those who want to produce bulk quantities of customized fabric styles. The weaving widths range generally from 190 to 400 cm. As regards the multicolour weft carrier, up to 8 different wefts can be fed.

Air knife: A paste application system used in coating. The blade applies a coating paste to the free running fabric web, i.e. without support from underneath.

Air laces: These are etched-out products (burnt out products) without printing. Yarns are embroidered on the ground fabric which is then burnt out with some after treatment.

Air-laid nonwovens: Fabrics made by an air-forming process (q.v.). The fibres are distributed by air currents to give a random orientation within the web and a fabric with isotropic properties.

Air laying: A method of forming a fibre web or batt in which the fibres are dispersed into an air stream and condensed from the air stream on to a permeable cage or lattice to form the web or batt.

Air permeability: The rate of air flow through a material under differential pressure between the two fabric surfaces or the porosity, or the ease with which air passes through material. Air permeability determines such factors as the wind resistance of sailcloth, the air resistance of parachute cloth and the efficiency of various types of air filtration media. It is also a measure of warmness, or coolness of a fabric.

Air pollution: Air pollution is caused by the changes in the natural composition of atmospheric air due to the presence of smoke, soot, dust , gases, aerosols, vapours, foul smelling substances, etc.

Air retentivity: It is all the forces existing in a garment which participate in retaining the layer of air which is in close contact with the human body even under stormy weather conditions.

Air wicking (in tyre fabrics): The passage of air longitudinally along or through yarns in a fabric that has been encased and cured in rubber or any other elastomer, that is, air permeability in the plane of the fabric.

Airflow doctor blade system: Steel doctor blade system with a plastic profile making possible precise control of the colour paste application and penetration during printing by independent adjustment of the doctor blade angle and contact pressure.

Airflow dyeing machine: Soft-flow dyeing machine developed to reduce the MLR and improve dyeing efficiency. In such machines, airflow moves and controls the fabric by means of the circulating air to which finely dispersed dyes and auxiliary products are fed via jet systems. Higher fabric speeds can be used in the gas flow than is possible in a hydraulic system, since the fabric rope is carried neither in the liquor nor with the liquor, and is consequently subject to lower tension during acceleration. There are therefore different prerequisites in bath exhaustion, produced not by the liquor encompassing the textile fabric but from the dye and product preparations distributed through the textile fabric. As the liquor supply is greatly reduced in consequence, the textile fabric liquor charge is reduced, i.e. in the case of fabric movement in a gas flow, the free space between fibres is not completely filled with liquor. Depending on the physical parameters of the gas, as air with a differing water vapour content, a state of equilibrium occurs in which the injected liquor quantity is interchanged with the textile fabric moisture charge. This flows out of the treatment vessel back into the injection circuit in the same quantity as the injected liquor. The suction conditions of the injection pump are better than those of a liquor pump, and only a small liquor supply is therefore necessary.

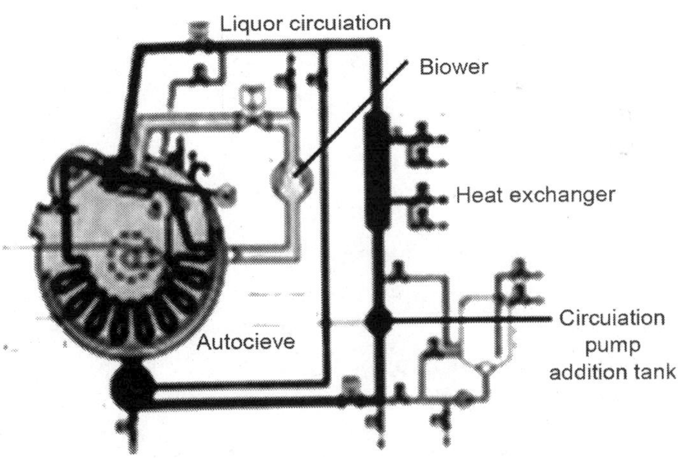

Then airflow jet dyeing machine (schematic)

Air-blue: Used to describe the light-blue of the thrush egg.

Air-laid non-woven: Fabrics made by an air-forming process (q.v.). The fibres are distributed by air currents to give a random orientation within the web and a fabric with isotropic properties.

Airplane fabric: A plain, tightly woven, water repellent fabric, traditionally made of mercerized cotton during world war II. The fabric was treated with cellulose acetate dope and used to cover the wings, tail and fuselage of airplanes. Today, similar fabrics are made from nylon and polyester or cotton blends and are used as rainwear and sportswear.

Air-supported roof: A fabric-based roofing system that is supported and held in place by air pressure.

Air-vortex spinning: Air-vortex spinning is similar to open-end spinning but, utilizes a stationary tube rather than a rotor. A high-speed air vortex is created in the tube that deposits fibres within the tube and simultaneously provides twist due to the vortex. The yarn is continuously withdrawn as in the case of open-end spinning.

Air wicking (in fabrics): The passage of air longitudinally along or through yarns in a fabric that has been encased and cured in rubber or other elastomer, that is, air permeability in the plane of the fabric.

Air-force blue: A dark-blue colour. The blue adopted by the Royal Air Force in 1919.

AIS: (1) Association Internationale de la Savonnerie (International Association of Detergent Manufacurers).

(2) Association Internationale de la Soie (International Silk Association).

Aitho (G): Burnt brown.

Aithochrous: Reddish-brown.

AITIT: Association Internationale de la Teinture et de l'Impression Textiles (European Association of Textile Dyers, Printers and Finishers).

A' Jour fabrics: Open-work woven or knitted fabrics, produced with fine yarns. The open-work effects (holes) are due to the fabric construction. In woven varieties, the ground is of plain weave and the open areas of leno weave construction. It is used for dress materialsand curtaining.

Ajamis: Calico from the Levant.

Ajiji: Cotton muslin with silk stripes, made in India.

Ajour: Ajour is a collective term for open work fabrics, woven or knitted. The open work areas are produced by the structure, e.g. leno, basket weave, missing warp or weft threads, lacy, open work embroidery. Normally, cotton with viscose fibre; a decoration used on blouses and underwear.

An example of Ajour

Akaakai: Hawaiian bulrush used for mats, baskets, etc., by the natives.

Akaroa: The ribbon tree, Plagianthus betulinus, of New Zealand, yielding a flexible, lustrous, strong lace-like bast, used for nets, lines, etc.

Akhissar: Rugs made in Asia Minor, the warp and weft are of wool, the loose pile of mohair, tied in Ghiordes knot. Red and golden brown colours are used.

Akia: A very tough bast fibre, used for ropes in Hawaii.

Akund: (Asclepias fibre, Calotropis floss) A natural cellulosic fibre belonging to vegetable hair group.Seed hairs of the Asclepias type (Africa, Asia and Central and South America). Similar to kapok fibre. It grows about 50mm length, yellowish in colour, smooth, silklike, and lustrous, but coarser and stiffer than kapok. Main uses are for stuffing upholstery and braids, selom spun or blended.

AL: (1) Alfa grass.

(2) Alginate fibres.

Alabaster: Akin to alabaster in its whiteness and the smoothness of its texture. Shakespeare's *Richard III*, Act 4 Scene 3 '*Gentle babes girdling one another Within their alabaster innocent arms*'.

Aladjas: Heavy taffeta with stripe or flower patterns, made in India.

Aladsha: Fancy cotton goods, originated in the East.

Alagai: Union silk fabric made in Asia.

Alamode: Plain-woven, glossy, light and soft silk fabric, used for scarfs, linings, millinery.

Alapeen: 18th century fabric in England, made of worsted or mohair and silk.

Alaska: A mixture yarn of long staple cotton and carded wool.

Alatcha: Yarn dyed cotton fabric, having stripes on a blue base; made in Turkestan.

Alb- (L): White.

Albert cloth: Reversible all-wool materials, each side of different colours, and so finished that no lining is required; used chiefly for overcoats; better known as "golf cloth", "plaid-back Coverts", etc.

Albanian embroidery: It is cross and goblin stitch in green, red and blue over canvas, showing conventionalized flowers or geometrical forms.

Albarazine: Spanish wool from Albaracia, Aragonia.

Albatross: (1) A soft light-weight wool or wool blend fabric in plain weave with a napped, fleecy surface that resembles in texture, the breast of albatross. It is usually light-coloured and is used in negligees, infants wear, etc.

(2) Albatross is probably a mistaken version of '*Alcatraz*', a Portuguese word for the sea-fowl (and the origin of the name of the prison island of San Francisco). Curiously, although the Alcatraz was black in colour, the word mutated to 'albatross' in order to describe the white petrel of Coleridge's *Ancient Mariner,* gainedits credibility on its journey from the 'alb' prefix which means white.

Albernus: Oriental woollen fabric, similar to camlet, imported into France through Marseilles.

Alberoni: Obsolete French and Holland camlet, made with silk and gold thread warp and silk or Angora wool filling.

Albert cloth: "Albert cloth" that is named after England's prince, is double-faced, reversible coat fabric, often with a different face and back, each side of different colours and so finished that no lining is required. Made from wool, it is a good quality cloth used to make expensive overcoats for men, sometimes with velvet collars. It is used chiefly for overcoats and better known as "golf cloth", "plaid back", etc.

Albert cord: Fancy English alpaca cloth in the 19th century.

Albert crepe: English dress fabric iijade of silk and cotton mixture.

Albescent: Becoming white.

Albesine: A grade of wool from Spain.

Albigeois: Grey, unbleached linen canvas, made in Languedoc, France.

Albissola: Italian bobbin lace having small pattern.

Albumen: Protein from animals (egg, blood, fish and milk albumen) or vegetable sources (soya). Albumen is water soluble, coagulated by water, and can be salted out from their solutions by ammonium suphate in the presence of alkali. Formerly they were used in printing as thickeners and fixing agent, but now is only of academic interest.

Albumin: Same as Albumen.

Alcanna, alcana: The reddish-orange body dye from the Oriental plant of the same name which is of the same family as the European **alkanet** plant.

Alcandra: Non-woven imitation leather made from (e.g. polyester and 40% polyurethane) microfibre batt impregnated with polyurethane. It handles like fine leather but is lightweight and easy care.

Alcatifa: Spanish trade name of fine Oriental rugs and carpets.

Alcatquen: Fine Persian knotted rugs that are inter-woven with gold thread, used over divans.

Alcyonne: Closely woven ten-leaf silk satin drapery fabric, made with single or ply warp and often with filling of a different colour from the warp.

Aldehyde green: A green dye, also referred to as **emeraldine**.

Alencon: Lightweight French cloth of silk and cotton.

Alencon lace: French needle-point lace made with net ground, the reseau resembling the Brussels pjint reseau. Early Alencon, also called point Je France, resembles Venise and Spanish point. Chief characteristic of the Alencon lace is the cordonnet, usually filled with horsehair; the design usually follows the taste of the period the lace was made in. Usually made in a floral design with a heavy corded thread outlining in it. The background is a fine mesh or net.

Alencon bar: Needle-point bar used to fill-up space, consists of a zig-zag thread covered with button-hole stitches.

Alencon ground: First made as bride and then as reseau; the bride was either plain or picotee, while the grande bride was a six-sided mesh, covered with button-hole stitches. The reseau was worked after the pattern to join it.

Alepine: (1) Twilled fabric of English origin made of silk warp and worsted filling in serge weave, and dyed in the piece, usually black. Used for mourning wear; (2) Twilled cloth of soft-spun silk warp and fine worsted filling, made in Aleppo, Syria, originally only in black, later in all colours. Imitated in England in hard spun worsted; (3) French fabric made of silk or cotton warp and soft worsted filling in single or various colours.

Aleppo: Cotton from Syria; also raw silk exported from the city of Aleppo.

Alesan: A light chestnut colour.

Alexander: A medieval striped silk that originally said to have come from Alexandria, Egypt; used for church vestments. See **Burdalisander**.

Alexandra: A plain woven cotton fabric that is found in Austria, dyed black and finished with a dress face; used for linings, underskirts, etc.

Alexandria: Fine, light-weight dress fabric made of cotton and wool with small woven design. Possibly its name derives from the very early silks and wool fabrics made in the ancient city of Alexandria, which featured small designs such as flowers and scrolls.

Alexandrine: Ancient name for fine cotton and linen cloths woven in Switzerland and France with fancy coloured patterns over white foundation. Given a silk-like finish.

Alexandrite: A valuable gemstone named after Tsar Alexander II of Russia. Alexandrite is green in colour but, appears columbine-red in artificial light. It has colour shifts from green to orange-yellow to red according to the crystal direction.

Alfa grass: Also called Esparto grass (North Africa, Spain, Sicily) is a natural cellulosic hard fibre. The fibre is extracted from their stalks by boiling an alkali and then retting. Hemp like fibre without lustre, stiff greenish-yellow, 10–14 cm in length and 90–500 um thick with low tenacity. Used for making hats, mats, as jute substitute, etc.

ALG: Alginate fibres.

Algaecide: The material that kills algae.

Algerian: A yellowish-brown.

Algerian lace: A gimp lace made of silver and gold thread.

Algerian stripes: Fabric having alternate stripes of coarse cotton and very fine silk, often with gold threads and is usually cream coloured; used for women's burnouses.

Alginate or algin: An extract of see weed used as thickener. Sodium alginate is used in textile printing pastes, and sometimes to thicken dye solutions for direct application. It is a preferred thickener of reactive dyes because it does not react with the dye. It comes in a number of variations which have somewhat different properties. Low viscosity types are more useful for reducing migration of wet dye solutions for fine line work, while high viscosity are more suitable for making printing pastes.

Alginate fibre: Fibre formed from a metallic salt (normally calcium) alginic acid, which is a natural polymer present in seaweed. Alginate fibre is soluble in water.

Alginate film: Alginate film is made from calcium alginate, which is manufactured by precipitation from a soluble alkali alginate. Used as a paper substitute in heat transfer printing.

Alginic acid: [algin; $(C_6H_8O_6)n$] A yellow-white organic solid that is found in brown algae. It is a complex polysaccharide and produces, in even very dilute solutions, a viscous liquid. Alginic acid has various uses, especially in the textile and food industry as a thickener, stabilizer and texture agent.

Alhamba quilt: A jacquard figured fabric with a plain ground weave that requires two warps. The figuring warp is usually two-ply and coloured, the ground warp singles and undyed. The weft is often made on the condenser system, soft spun and of coarde count.

Alice blue: US term for a greenish-blue after Alice the wife of Theodore Roosevelt. As Alice Blue, one of the 140 colours in the **X11 Colour Set.** It has hex code #F0F8FF.

Aliphatic compounds: Derived from aliphatic Hydrocarbons of the methane type CH_4. Typical examples of this group of compounds include, e.g. fats, oils, soaps, sugars, alcohols and fatty acids. Originally used to describe fats.

Aliphatic amino acids: H_2N-R-COOH. One of the monomers used in the production of used for nylon 6 by polycondensation.

Aliphatic di-acids: HOOC-R-COOH. One of the monomers used for the production of nylon 6.6 by polycondensation

Aliphatic di-amines: NH_2-R-NH_2. One of the monomers used for the production of nylon 6.6 by polycondensation.

Alizarin, alizarin: A synthetic red dye identified in 1820 and replacing the natural red dye from the root of the **madder** plant. It produces crimsons, greens and blues and other shades depending on the **mordant** used, in particular, **alizarin crimson**. Produced artificially since 1868 after which vineyards began to replace the redundant madder fields of Europe. This was the first natural pigment to be made synthetically. Alizarin dyes replaced aniline dyes but were themselves replaced in 1958 by **quinacridones,** which have greater **light fastness**.

Alizarin crimson: A bluish-red, serving as a substitute for **rose madder**. See **Alizarin**.

Alkali: A sub class of base, though often used to refer to any base. Alkaline is often used to refer to solution that is basic – having pH greater than 7.

Alkali binding agents: Chemicals used in naphthol dyeing for neutralization and pH adjustments of the coupling baths. Typical alkali binding agents are acetic and formic acids, aluminium, zinc and magnesium sulphates, and mono and disodium phosphates.

Alkali blue: A class of blue pigment with a very high tinting quality that is used in making printing inks.

Alkali cellulose: The primary intermediate product got by the action of strong alkali on purified cellulose, like viscose ($C_6H_9O_4$-O-Na+).

Alkali damage: Caused by alkali retention on the fabric after any treatment where caustic soda is involved. Alkali cannot be easily washed off, hence has to be neutralized with acids after the rinsing process. Alkali damages are usually dyed in darker shades because of localised mercerization due to drying of alkali at the affected place.

Alkali donors: These are chemical compounds which release OH- ions under the influence of heat during dyeing. Alkaline salts of volatile and easily decomposed organic acids are used as alkali donors.

Alkali fastness testing of dyeings and prints: See **Colour fastness to spotting, Alkali**.

Alkali insoluble impurities (in scoured wool): The oven-dry ash free alcohol extractives- free alkali insoluble substances other than vegetable matter base, such as skin cotton or other fibres, paper, string, tag, etc., pieces and paint pieces, etc.

Alkali earth metals: Collective term for calcium, barium and strontium which are classified under Group II A of the periodic table. Magnesium is also often included as alkaline earth metal.

Alkali shock process (scouring): A short period process for open width scouring of cotton and cotton/polyester, like impregnation with caustic soda, wetting agent and chelating agent followed by steaming and washing.

Alkali shock process (printing): A printing process for reactive dyes, whereby the printing is done without alkali and the fabric is treated with lye and electrolytes in an open width scouring machine (1–2 boxes) in order to fix the reactive dye. The temperature of the bath will depend on the reactivity of the reactive dye used.

Alkali shock process (dyeing): In continuous dyeing of polyester/cotton fabric the reactive dye can be fixed by caustic shock process, whereby the fabric is passed through a mangle with concentrated caustic soda.

Alkali solubility (in wool): The percent of clean wool that is soluble in a specified alkaline solution under controlled conditions of temperature and time.

Alkali stains on textile fabrics: Often described as localized mercerization. It is due to alkaline residues on cellulose, mainly due to caustic soda giving rise to darker areas in subsequent dyeings. Even if the alkaline residue is of low strength as the fabric is dried, the strength increases and localized mercerization takes place which causes that place to dye darker. See **Alkali damages**.

Alkali traces (in fabrics): It can be detected by washing the fabric with distilled water at 100 °C and adding 1–2 drops of methyl green solution (0.01g of methyl green in 100ml of distilled water). Alkaline traces in the water will cause the methyl green to lose its colour.

Alkali treatment of polyester: Alkaline pre-treatment for polyester. In this process, polyester molecules on the surface are hydrolytically split, partially dissolved, producing a silk like handle and drape together with reduced lustre and increased dye uptake. In controlled conditions of pH, temperature, and time, usually 10–20% weight loss is achieved.

Alkaline: A term used to describe a material having a pH greater than 7.0 in water.

Alkaline degumming: See **Acid degumming**.

Alkaline discharges: Discharge printing using discharge paste containing potassium carbonate or caustic soda.

Alkaline discharge-resists: Printing process for polyester. Disperse dyes which are saponified by alkali are used as ground. Alkaline lye, alkali carbonates, silicate of soda and alkali dispensers are used as alkali. By this method, wider range of disperse colours can be used.

Alkaline milling: See **Milling**.

Alkaline scour: See **Scouring**.

Alkanet: An ancient red or orange dye from the roots of the Mediterranean plant of the same name and of the same genus as the **alcanna.** Also called 'anchusa'. See **Henna**.

Alkannin: A natural bronze-coloured pigment.

Alkene suphonates: Used as anionic surfactants, these are commonly called paraffin sulfonates or secondary alkane sulphonates and are used extensively, especially in Europe. Their commercial synthesis requires a relatively high capital investment, so they do not find worldwide application in household detergents. They have high biodegradability, low toxicity and, depending on chain length and degree of sulphonation, reasonably good water solubility. Paraffin sulphonates are made by sulphoxidation or sulphochlorination of *n*-paraffins in the C_{14}-C_{17} range. About 90% of the product of sulphoxidation consists of monosulphonates, with the balance being di- and polysulphonates. Eighty to ninety percent of sulphonation occurs at the internal carbon atoms, rather than at the ends of the chain, with the sulphonate group substituted almost uniformly on secondary carbons throughout the chain.

Alkyl amines: Amines with substituted alkenes.

Alkyl suphonated: Lignin is a complex polymer found in wood which is separated from cellulose during manufacture of paper and other products by various means, including sulphonation. Lignin sulphonates are obtained as by-products of the sulphite process. These are widely disperse, water-soluble lingo sulphonic acids, with an average molecular weight greater than 100,000. They find a variety of uses as low cost emulsifiers in areas such as ore processing and oilfield chemicals. Impurities include sugars, such as glucose, mannose, xylose, and arabinose. Both dried powder and aqueous solutions are sold, usually without stringent specifications. Some purified products find their way into higher value applications, such as pesticide formulations.

Alkyl aryl suphonates (AAS): They are surface active substances with 1 or 2 aromatic rings which contain one or more alkyl chains and a sulpho groupalkyl aryl sulphonates are classified into lower molecular weights (wetting agents) and higher molecular weights (detergents). As a thumb rule, more branched the chain better the wetting power and more the molecular weight of the alkyl chain, better the detergency.

Alkyl benzene sulphonates: The alkyl benzene sulphontes are used as anionic surfactants.The term alkylbenzenesulphonate (ABS) is generally applied to the branched-chain products, which are only biodegradable with difficulty. ABS is not used in developed countries except for specialty applications where it will not reach natural waters (e.g., as an emulsifier for agricultural chemicals). As opposed to ABS, linear alkylbenzene sulphonate (LAS) is rapidly biodegradable under aerobic conditions.

Alkylphenol ethoxylates (APEs): A group of chemicals that are used in wool scouring and may be found in polluted water. Some of them are endocrine disrupters.

Alkylation: Introduction of alkyl groups into organic compounds by addition or substitution reaction.

Alkylation (degree of): Descriptive term for the quantitative proportion of alkylated groups in a molecule.

All silk crepe: The example for this type of fabric is 170 × 110 using 13/15 denier warp and 18/20 denier weft. It would be woven in plain weave. The warp and weft arrangement would be two S twist alternating with two Z twist which during processing will give the crepe effect. It is an extremely light fabric, the weight per sq. metre would be around 2 oz.

All skin fibres: Modified viscose fibres of the so called super-cord and super-fest-s type the cross section of which no longer consists, as with normal viscose fibres, of core and skin, but exclusively of a highly oriented skin as a carrier of special quality features such as increased tear strength and higher wet modulus.

Allabatis, Allibalis, Alliabat, Allibanis: Names of Indian cotton cloths, made either plain, embroidered or brocaded.

Allah haik: Moorish cloth made with equally wide stripes of nub cotton warp and fine silk warp of cream colour. Used for turbans and burnouses.

Allahabad: Large knotted wool or cotton carpets made with loose thick pile in India.

Allapine: See **Alapeen**.

Allochorous, Allochroous: Multi-coloured or changing colour.

Allochroic: Changing in colour.

Allochromatic: Referring to a change of colour; as regards minerals, having no colour.

Alleanthus: Very tough fibre yielded by the bast of the Alleanthus zeylanicus, a tree in Sri Lanka; used for ropes, nets, etc.

Allegias, Allejars: East Indian pure cotton muslins, made in plain weave, also mixed with silk or other fibres. Also called Bethilles.

Allejah: A fabric made with gold, green and white stripes that was used to make dresses in the 18th century.

Allemande: Corded French silk dresses goods and men's vesting.

Allen: Variety of raw cotton where the staple is fine and silky, measuring up to 35 mm, the lint is less than 30 percent.

Alley (Spinning): The area between the breaker carding and finisher carding machines in which alley tender works.

Allgaeu carpet(ribbon carpet, patchwork carpet): It is hand-made smooth carpet in a strong cotton yarn warp with a ribbon-like weft (fancy weft carpet) in strips of woven and knitted fabric remnants, stockings, selvedge trimmings, etc.

Alliabably: Fine cotton muslin from Dacca, Bangladesh.

Alligator: The skin of the reptile, easily recognizable by square boxy markings. Used mainly for shoes and handbags.

Allonge-perruque: French term for periwig, also called state-wig. Worn by fashionable men in the late 17th and early 18th Century. Introduced probably by Louis XIV and usually in black or dark brown shades. The periwig had very high 'horns' on top of the forehead and was extremely long, curled and flowing down the back and over the shoulders.

Allover lace: Any piece-lace with a pattern repeated regularly all over without marked divisions in the pattern, no edgings, borders, etc.

All-over designs: Designs with balanced motifs that recur regularly within the repeat unit. The motifs cover the fabric with little ground showing.

All-over print: A fabric that has a printed pattern that covers practically the whole face of the fabric.

Alloa wheeling: Heavy Scotch knitting woollen yarn.

Allowance body tolerance: The extra dimensions allowed on garment measurement over body measurements. See **Seam allowance**.

Alloy: A solid or liquid mixture of two or more metals; or of one or more metals with certain non-metallic elements formed by fusing the components.

Almagra: Deep-red ochre found in Spain.

Almanesque: A variety of cotton goods used in Argentina.

Almond: Pink or yellowish-brown as in almond blossom; sometimes the greyish-green colour of the underside of the leaves of the almond tree. Used to describe a variety of colours.

Almond black: A black pigment.

Almond green: Greyish-green.

Alneestloni: Navajo blanket made in twill weave, both sides being different.

Alni mayini: Woollen blanket made by the Indians, made with a wide black stripe in the centre and narrower stripes at each end; the corners trimmed with black tassels.

Aloe: The fibres of the agave, made into net and lace in the Philippines, Italy, Spain and Paraguay.

Aloe hemp: Trade name for the Mauritius hemp and also for the leaf fibre of certain Indian Sansevieria species.

Aloe lace: A very fragile type of lace usually made in Italy, but some comes from Philippines. Very expensive, made from the fibre of the plant –aloe.

Alost: Belgian bobbin lace that is similar to the Valenciennes, the threads of the mesh ground being twisted four or five times.

Alpaca: (1) A close relative of the Ilama and is similar to guanaco. It exclusively lives in Chile, Peru, Bolivia and the higher levels of Andes. Its wool belongs among the finer animal hair qualities.

(2) A fabric from alpaca fibre or blends, (originally a cotton cloth with alpaca filling) that is used for dresses, coats, suits and sweaters. It is also used as a pile lining for jackets and coats.

(3) The fleece and fibre produced by the alpaca, an animal of genus Ilama(lama glamapacus). The fibre is obtained from several species, namely Huacaya and Suri.

Alpaca crêpe: A soft acetate or polyester crêpe with dull surface, fairly soft to handle. Used for dresses. See also **Crêpe**.

Alpaca lustre: A costly alpaca fabric with very high finish.

Alpaca mixture: 19th century undyed English fabric, made of cotton or silk warp and alpaca filling.

Alpaca Orleans: One of the first alpaca fabrics made in England in the 19th century; woven with a cotton warp and alpaca filling.

Alpaca rayon: Usually refers to soft non-lustrous fabric made of rayon or acetate. Used for women's wear; it is composed of four ply yarns 75 denier rayon being used for warp and 100 denier acetate in the weft. It is usually woven 48 in wide in plain weave with a texture of 36 × 34. In some cases,2 ply to 8 ply yarns are used. Another example is 68 × 64, with 60s combed cotton warp and 150 denier Rayon weft. Usually woven in about 35 in width.

Alpaca stitch: A 1 × 1 purl links stitch that is knit so that so that the courses run vertically instead of horizontally as the fabric comes off the knitting machine. A garment made of alpaca stitch is not always 100% alpaca; it can be made of other natural fibres or manufactured fibres,

Alpenglow: The purple gleam on alpine snow.

Alp green: Yellow-green.

Alpha and beta amylase: The first process in wet processing of woven fabrics is the removal of starch which has been applied on the warp in the sizing process prior to weaving. This process is called desizing and is invariably done using enzymes. Enzymes are high molecular weight protein biocatalysts that are very specific in their action. Enzymes are named after the compound they break down.For example, amylase breaks down amylose and amylopectin;maltase breaks down maltose; and cellulase breaks down cellulose. For desizing starch, amylase and maltase are used. Cellulase, on the other hand, is used for finishing cotton fabrics. This will be discussed in later chapters. Amylase will degrade starch into maltose, a water soluble disaccharide and maltase will convert maltose into glucose, a simple sugar. See **Desizing**.

Alpha cellulose: One of the three forms of cellulose. Alpha cellulose has the highest degree of polymerization and is the chief constituent of paper pulp and chemical dissolving–grade pulp.

Altar cloth: A very fine sheer, crisp fabric used, as the name suggests, for church cloths.

Alternating twist: A texturizing procedure in which S and Z twists are alternately inserted in the yarn by means of special heating arrangement to give special effect for the yarn and fabric made using the yarn.

Alternation (in designing): It is a specific instance of patterning in which a sequence of repeating motifs are presented in turn (short/long, fat/thin, round/square, dark/light). See **Rhythm**.

Alum: A term for a variety of chemicals. There are several compounds that are called alum. One is aluminium potassium sulphate-picking alum. Others include aluminium sulphate, aluminium ammonium sulphate and chromium potassium sulphate. Some alums are used as mordant in dyeing.

Alum sludge: A waterworks sludge formed when alum is the coagulant. The sludge is usually conditioned by adding polyelectrolytes, thickened and then dewatered mechanically, typically in a filter plate press.

Alumina: A strongly alkaline aluminium sulphate which is insoluble in water but soluble in acids. It is used to prepare aluminium mordant for dyeing and printing. $Al_4(SO)_4(OH)_4.2H_2O$.

Alumina fibre: A refractorymade of alumina (Al_2O_3) that can withstand furnace temperatures up to 1600 °C.

Alumina gel: [$Al_4(SO_4)(OH)_{10}\cdot 2H_2O$]. A strongly alkaline aluminium sulphate which is insoluble in water but soluble in acids. It is used to prepare aluminium mordants for dyeing and printing.

Aluminium potassium sulphate: “Pickling” alum $AlK(SO_4)_2$.

Aluminium sulphate: An alum $Al_2(SO_4)_3$.

Alumnat: Cloth closely woven, black woollen fabric, made in Bohemia for the use of the clergy and alumni.

Alwan: Fine and plain woven fabric, made in Tibet of pashmina wool yarn.

Amabouk: Half bleached coarse linen, made in northern Ireland; used for sailors' shirts in earlier days; unbleached *amabouk* is used for bags.

Amadaure: Variety of Egyptian raw cotton.

Amadou brown: A dark reddish-brown.

Amadis sleeve: Tight fitting sleeve continuing on the back of the hand, invented in 1684 by Mlle le Rochois, an actress at the opera, who had unsightly arms.

Amamee: Smooth, closely woven cotton cloth from India; the coarser grades also called Tissuti and the finer Bissuti. Used for shirts, bed covers, curtains, and also for printing.

Aman: Plain woven and blue coloured cotton fabric, made in Syria.

Amana: Staple blue printed cotton fabric and knitted goods.

Amanouri: One of the best grades of Levant cottons.

Amaranth: The reddish-purple or deep crimson colour of the leaves of the Amarantus. The words amarantin, amarantine and amaranthine, signifying 'fadeless', 'immortal' or 'unwithering', refer to Pliny's imaginary and never fading amaranth flower.

Amauro- (L): Dark.

Amazon: A satin weave dress fabric with worsted warp and woollen weft, usually dyed in plain colours. It is lightly milled and raised to give a fibrous effect but, this does not conceal the twill effect of the satin weave.

Amazones: Woollen dress goods in South America.

Ambadal: A bandhani design. See ***Bandhani***.

Ambari: Hemp. Very long, light coloured, silky but harsh and strong bast fibre, yielded by the *Hibiscus cannabinus* in southern Asia. Used for cordage, coarse bagging, etc.

Amber: The clear yellow-brown or reddish-orange of the stone amber; '*amber-coloured raven*' Shakespeare's *Loves Labours Lost*,Act4**,** Scene3. Amber as a stone variesin colour and the colour term embraces a wide variety of shades.

Amber yellow: A rich ochre-coloured yellow.

Ambient conditions: See **Atmospheric conditions**.

Ambient printing process: A registered printing process of the Ambiente Co., Helsinky, of local colouration of open width woven or knitted fabric with thickened dye solution in the nip of a two bowl padder. A wide range of different lively variations in colour and designs are obtained with stripes of different widths and directions including circular, speckled and cloud like motifs.

Ambiphilic: Ambiphilic substances have double solubility, i.e. in oil as well as in water.

Ambivalent: Term used in resin finishing to describe cross linking agents with two different types of groups capable of reacting with cellulose.

Ambivalent cross-linking: The cross linking of cellulose with resin finishing agents containing different types of reacting groups.

Amblyopia: The condition of having defective vision.

Americaine: French dress goods of corded silk made with eight warps and eight picks in a repeat.

American cloth: In many places, an enamelledoilcloth for household or upholstery purposes.

American cord: See **Rat tail-cord**.

American grain count: A direct yarn numbering system for expressing linear density, equal to the mass in grains per 120 yards of sliver or roving.

Americana: Name for coarse cotton sheeting.

Americani or Amerikano: Unbleached cotton sheeting in various parts of Africa.

Americano Assilia: Name for grey cotton sheeting on the eastern coast of Africa. They came originally from the United States.

Americano Gamti: A very dark and coarse grey cotton sheeting on the eastern coast of Africa;imported from India.

Americano marduff: A stout grey cotton fabric in East Africa made of twill weave, imported from the United States or Europe; used for tents, sails and dresses by the natives.

American ulayiti: Native name of grey cotton sheeting on the eastern coast of Africa imported from Europe.

Amethyst: Violet-purple or purplish-blue (particularly in heraldry); from the Greek meaning 'preventing intoxication', a characteristic once ascribed to the stone of the same name.

Amertis: Closely woven calicos or cotton goods from India.

Amianthus: Fine, flexible asbestos; used for fireproof curtains, etc.

Amide resins: Synthetic resins of the type which includes urea resins, melamine resins and dicyanodiamide resins.

Amides: Organic compound derived from ammonia by substitution of one or more of its hydrogen atoms by organic acid groups.

Amiens or Amens: Closely woven twill cloth of hard-twist worsted, in solid colours, striped or made with patterns. Similar to the lasting but of better grade.

Amine end group: The terminating amine (-NH_2) group of a nylon polymer chain. Amine end group gives dye sites for nylon.

Amine functional cationic softeners: Long-chain amines are not water soluble at neutral and alkaline pH; however, when converted to their acid salt, they develop a cationic charge and become water soluble. They exhaust and become excellent softeners under acidic conditions. The cationic charge on a given hydrophobe is proportional to the number of amino groups; therefore, the attraction of the cationic portion to the fibre surface increases as the number of amine groups increase.

Amino acids: (Amino carboxylic acids), R-NH_2- COOH, building blocks for proteins. Comprising of a carboxylic group (–COOH), an amino group (–NH_2) and a mostly very complicated chain (e.g. fatty acid chain) which can be aliphatic or aromatic. As a result of the simultaneous presence of amino and carboxylic groups, amino acids react as both acids and bases, i.e.they are amphoteric and form Zwitterions with two opposite electrical charges which neutralise each other within the molecule (Isoelectric point).

Amino functional silicones: Amino functional silicones are made by incorporating the appropriate organo-functional chlorosilane to the reaction mix. Amino functional silicones become cationic at acid pHs and exhaust from aqueous baths.

Aminocaproic lactum: Caprolactum.

Amino carboxylic acids: See **Amino acids**.

Aminolysis: Aminolysis occurs in presence of an amine. This reaction is similar to acidolysis, where splitting of a high molecular weight polymer involves resulting into a low molecular weight polymer with an amine terminal group as follows:

$$\underset{\text{Polyamide}}{\text{-NH-}(CH_2)_n\text{ -NH-CO-}(CH_2)_{n2}}\text{ -CO- }\underset{\text{amine}}{H_2N\text{-}NH_2}\text{à}$$

$$\underset{\text{Polymer with amine terminal group}}{\text{-NH-}(CH_2)_{n1}\text{- }NH_2 + H_2N\text{ - R-NH-CO-}(CH_2)_{n2} - \text{CO-}}$$

Aminoplasts: Self cross-linking resins. Collective term still used for urea formaldehyde and melamine formaldehyde resins.

Aminopolycarboxylic acids: A group of important organic complexing agents which are important in the water softening applications. Ethylene diaminetetra acetic acid (EDTA) and Nitrilotriacetic acid (NTA) and their sodium salts are its examples.

Amiray: Native Philippine name for the ramie fibre; used for cords, threads and fabrics.

Amoer: Name of a strong silk taffeta in many parts of Italy. Same as ***gros de tours.***

Amole: Coarse, yellow fibre, yielded by the bulb of a lily-like plant in California; used for stuffing.

Amorgis: Fine linen in ancient Greece, usually dyed purple.

Amour: Linen with round or oval damask like pattern for table use, made in Caen, France. Also called La cs d'amour.

Ammonia: A gas (NH_3), often used to refer to a solution of ammonia in water, called aqua ammonia or ammonium hydroxide(NH_4OH). Ammonium hydroxide is used for pH control, mostly where the desired pH is only moderately basic. It is used in some stripping processes for acid dyes and in rinses for reactive dye on wool.

Ammonium hydroxide: See **Ammonia**.

Ammonium persulphate: White crystals soluble upto 559 g/l in water that is used in desizing, oxidizing agent in dyeing and printing.$(NH_4)_2S_2O_8$, contains 6–8% active oxygen.

Ammonium stearate: A wax-like solid soluble in alcohol, used in foam generation in coating application ($C_{17}H_{35}COOHNH_4$).

Ammoniun sulphate [$(NH_4)_2SO_4$]: Ammonium sulphate is a chemical used most commonly with acid dyes and 2: 1 pre-metallized dyes. It decomposes through hydrolysis as the bath temperature rises, slowly releasing acid. This helps to give level dyeing.

Ammonium thiocyanate: Colourless crystals hygroscopic, readily soluble in water and alcohol. It is used in dyeing silk with leuco ester dyes, dyebath additive in copper vessels as protection for copper sensitive dye and in analytical detection of trivalent iron (NH_4CNS).

Amorphous: Describing a solid that is not crystalline, i.e. one that has no long range order in its lattice. Many powders that are described as 'amorphous' in fact are composed of microscopic crystals, as can be demonstrated by X-ray diffraction. Glass is an example of true amorphous solids.

Amount (in textile testing): A measure of the thickness or pneumatic density of the test beard, proportional to the number of fibres present at various distances from the comb(s).

Ampholyte: A substance that can act as either an acid, in the presence of a strong base, or a base, when in the presence of a strong acid.

Ampholyte ion: See **Zwitterion**.

Ampholytic surface active compounds: Surface active compounds possessing two or more functional groups capable of ionization in aqueous solution which as a consequence, depending on the conditions of the medium, confer anionic or cationic properties on the product. In the broadest sense, this ionic behaviour is analogous to that of the Amphoteric compounds.

Amphoteric: Describing a compound that can act as both an acid and a base (in the traditional sense of the term). For instance, aluminium hydroxide is amphoteric: as a base [$Al(OH)_3$] it reacts with acids to form aluminium salts; as an acid [H_3AlO_3] it reacts with alkalis to give aluminates. Oxides of metals are typically basic and oxides of non-metals tend to be acidic. The existence of amphoteric oxides is sometimes regarded as evidence that an element is a metalloid. Compounds such as the amino acids that contain both acidic and basic groups in their molecules can also be described as amphoteric. Solvents, such as water, that can both donate and accept protons are usually described as amphiprotic.

Amphoteric surfactants: These form a minor group of amphoteric agents that are used mainly as lubricants, dyeing retarders and softeners. They may have both cationic and anionic sites in the same molecule, depending upon the pH of the solution. In acidic solution, some of these products will be cationic but become anionic at higher pH. At pH values close to their isoelectric point, they exist as zwitterions but are least effective as surfactants under these conditions. These chemicals have excellent stability in acidic and alkaline solutions and little sensitivity to salts.

Amritsar: Large size Indian wool rugs made in Persian designs with heavy pile.

Amritsar silk: Very thick, close and strong, and quite free from vulgar gloss of the fraudulently sized European silks made in Amritsar, India. Today Amritsar is no longer a weaving hub anymore. In earlier days, it produced many types of cloths, the two most popular types were called daryai (plain like sea or dariya) and gulbadan (striped) that featured largely scarlet stripes.

Amylase: An enzyme which catalyses the breakdown of starch. Used in desizing process, it converts starch into water soluble sugar. See **Alpha and Beta Amylase**.

Amylose: A starch made of long unbranched chains a-D-glucopyranose [which is quite similar to glucose ($C_6H_{12}O_6$), the "a" should be the Greek letter alpha]. Amyloseis a linear polymer, molecular weight range 100,000–300,000, found in the interior of the starch granule and accounts for 19 to 26% of the weight. It is soluble in hot water; however, when the solution is cooled, it will form strong hydrogen bond between adjacent chains making it difficult to re-solubilize.

Amylopectin: A starch made up of long branched chains of a-D-glucopyranose(which is quite similar to glucose ($C_6H_{12}O_6$). Amylopectinis the major component of starch and comprises the outer sheath of the granule. A highly branched, high MW polymer (1.6M), less water soluble than amylose.

Amylose: A polysaccharide consisting of linear chains of between 100 and 1000 linked glucose molecules. Amylose is a constituent of starch. In water, amylase reacts with iodine to give a characteristic blue colour.

Anabasses: Blue and white striped woollen covers.

Anacostas: Fine all-worsted English dress goods, woven in two-and-two twill with a weft face, as the number of picks is much higher than the number of ends. It is woven in grey and dyed in the piece.

Anacoste or Anacote: Worsted serge made in twill weave in France and Germany that is given a very smooth finish, used by religious orders for clothing. An inferior quality made in Netherlands is known in Japan as Saaij.

Anadendron: Very strong bast fibre of a plant in the Andaman Islands, used for nets and bowstrings by the natives.

Anaerobic contact process: Anaerobic activated sludge process: Industrial waste-water with high BOD can be treated by an anaerobic suspended growth process. The waste-water is fed into a completely mixed anaerobic digesterwhich is followed by a sedimentation or flotation tank from which the sludge is returned to the digester.

Anaerobic bacteria: The bacteria, which can live in the absence of oxygen. Sewerage sludge can be reduced to hydrocarbons by anaerobic bacteria in digestion towers in the absence of oxygen.

Anaerobic bio-degradation: Biological degradation by anaerobic bacteria in the absence of oxygen.

Anaerobic digestion: The breaking down of complex organic material into simpler compounds by anaerobiosisas in anaerobic sludge digestion.

Anaerobic process in effluent treatment: They are biological treatment processes that occur in the absence of oxygen. Bacteria that can survive only in the absence of any dissolved oxygen are known as obligate anaerobes.

Anaerobic sludge digestion: The anaerobic digestion of waste-water sludge. It is common to maintain a temperature of around 35 °C (mesophilic digestion). The process may be operated as a single stage process (e.g. slow rate anaerobic sludge digestion). But it is normally a two stage process (sometimes known as high rate anaerobic sludge digestion) with the heated primary digestion tank mixed either mechanically or by pumping the sludge gas back through the digester or by pumping around the digesting sludge. Most of the digestion takes place in the primary digester. Primary digestion is followed by an open unheated tank for secondary digestion, which allows the digestion process to finish and the digested sludge to settle and thicken. The design of the primary tank is either on retention time (e.g. 15 days) or on the loading of volatile suspended solids, VSS (e.g. 1.5 kg VSS per day per m^3 of tank). Complete digestion reduces the volatile solids in the raw sludge by up to 70%. This is equivalent to a 50% reduction in the total sludge solids, assuming that 70% of the raw sludge solids are volatile. Many of the problems of operating anaerobic digestion arise from the difficulty of maintaining an adequate population of methanogenic bacteria. They are seriously affected by small quantities of toxic materials in the sludge. If the methanogenic bacteria are killed, the sludge is left in an obnoxious half-digested state. Starting an anaerobic digester can be difficult and pH control is critical to enable the methanogenic bacteria to flourish. The ratio of the concentration of organic acids to total alkalinity(both expressed as equivalents per litre) should be under 0.5, preferably about 0.2.

Anaerobic sludge digestion tank (a.s. digester): A tank for anaerobic sludge digestion. It is normally heated to around 35 °C (mesophilic digestion) or possible 55 °C (thermophilic digestion). The sludge gas is collected and used as a fuel to heat the sludge. Primary tanks usually have scum breakers and a roof that may be floating or fixed. If the roof is floating, it forms a gasometer to hold the methane. The primary digester normally is mixed by recirculating either the sludge or the gas or both. Sludge is heated by circulating it through an external heat exchanger. Tanks are usually circular, not less than 7.5 m deep at the centre or less than 6 m diameter, but up to 35 m diameter and 14 m deep. If a floating roof is provided, it has to be gastight round the rim where it meets the tank wall.

Anaerobic suspended growth process: In waste-water treatment, a process in which anaerobes are suspended and mixed with the waste-water in an anaerobic reactor. Examples include anaerobic contact process, anaerobic sludge digestion, sequencing batch reactor, up flow anaerobic sludge blanket.

Anaerobic treatment (a. biological t.): Any treatment process that uses anaerobes to remove unwanted organic or inorganic compounds. In the treatment of waste-waters, the processes can be divided into anaerobic suspended growth process (e.g. up flow anaerobic sludge blanket) or attached growth process such as the anaerobic filter.

Analao: Philippine name for a cordage fibre.

Analogous colours: Colours, which appear close to or adjacent to each other on a **colour wheel.**

Ananong: A Philippine cordage fibre.

Anaphe silk: Also called African silk, produced by a silkworm of the Anaphe family (Africa) which feeds on fig leaves. Groups of these caterpillars make large nests and make individual cocoons inside. The nests and cocoons are made of silk. Anaphe silk is very fine and processed into bourette and schappe silk.

Analysis of variance (ANOVA): A procedure for dividing the total variation of a set of data into two or more parts, one of which estimates the error due to selecting and testing specimens and other part(s) possible sources of other variation.

Anatase: One of the three naturally occurring Titanium Dioxide. Used as white pigment in textile printing.

Anatolian: Long and medium fine wool, yielded by the caraman sheep of Anatolia. The sheep has a very heavy and flat tail. The wool is used for carpets.

Anatolian carpet: Oriental carpet originating from interior Asia. They are mainly based on geometrical designs with highly stylized plant motifs in vivid fresh colours. Figural motifs are absent.Principal regional varieties are Bergama, Brussa, Dermirchi, Ghiordes,Hereke, Kayseri,Konya, Kula, Ladik, Mekri,Milas, Mudjur, Nigde, Panderma and Yuruk carpets.

Ancelia: Union dress goods of cotton thread warp and mixed wool yarn filling, forming patterns.

Anchali: A wide ribbon-like fabric, made in India.

Anchusin: A red dye from the root of the ***alkanet*** which is also referred to as ‘anchusa’.

Andalusia: Medium fine wool from Spain.

Anderson: Very fine gingham made in Scotland.

Anemone: A pale violet.

Angiaiangika: Short and tight-fitting bodice worn by women in India from early times. Literally 'covering for the body'.

Anglaise: Plain coloured French serge, made with eight ends and four picks in a repeat.

Angle blade: Used for applying lacquer layer in screen production.

Angle of incidence: (1) The angle between a ray falling on a surface and the perpendicular (normal) to the surface at the point at which the ray strikes the surface; (2) The angle between a wave-front and a surface that it strikes.

Angle of reflection: (1) The angle between a ray leaving a reflecting surface and the perpendicular (normal) to the surface at the point at which the ray leaves the surface; (2) The angle between a wave-front and a surface that it leaves.

Angle of refraction: (1) The angle between a ray that is refracted at a surface between two different media and the perpendicular (normal) to the surface at the point of refraction; (2)The angle between a wave-front and a surface at which it has been refracted.

Angleterre: Highly finished silk taffeta made in France.

Angleterre bars: Filling in places in modern point laces, consisting of lines of threads crossed at right angles with knots or spots formed by a separate thread at each crossing.

Angleterre edge: A needle-point edging to braid or cordonnet; made with one line of point de Bruxelles loops.

Angola: (1) Yarn composed of cotton and wool; (2) Twilled, red cotton cloth; (3) A thick, soft twilled and napped woollen over-coating.

Angola brocade: A highly finished English worsted dress goods of the 19th century, woven in coloured patterns; obsolete.

Angola cloth: Diaper cotton of cream colour with rough face, used for embroidery.

Angora: (1) Angora goat hair or mohair and angora rabbit hair, although the latter is a misnomer. Angora goat hair is classified as a speciality wool fibre, which under wool labelling act is classified as wool. A 100% mohair fabric may be called wool. Angora rabbit fur, under this act is classified as hair.

(2) Woven and knitted fabric made out of angora wool. Fabrics composed solely of blends containing this animal hair are also falsely described as Angora. Used for high quality outer wear, knitwear and hand knitted articles.

(3) Hair from the angora rabbit. Often blended and mixed with wool to lower the price of the finished article or to obtain fancy or novelty effects, various weaves and knitted.

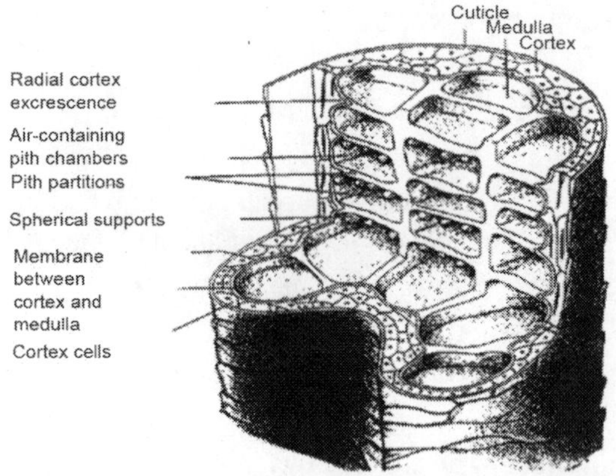

Angora fibre morphology

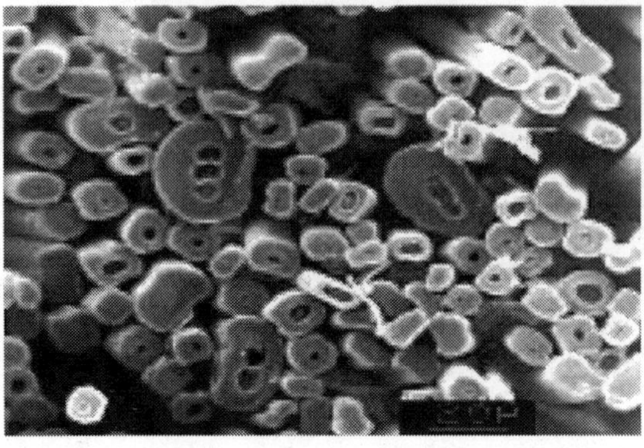

Angora fibres in cross section

Very fine, light weight, extremely warm and fluffy. Has a tendency to shed and mat with time. Must be designated as angora rabbit's hair. Uses: Used mostly in knit wear – gloves, scarves, sweaters, etc., for children and women. Also blended with wool in dress goods and suits to give a softer feel.

Angora cashmere: Soft, light, twilled mohair dress fabric.

Angora goat hair: Mohair.

Angora rabbit hair: Hair of Angora rabbits grown in Europe and East Asia. Can be considered as the finest of the hairs. The annual yield can be 250g per rabbit in two to three shearing. The hair is mostly pure white, lustrous and silk like. For technological reasons these fibres are spun with wool, a percentage of synthetic fibre or cotton. Chemically Angora fibres are keratin, like wool.

Angstrom: Symbol Å. A unit of length equal to 10–10 metre. It was formerly used to measure wavelengths and intermolecular distances but has now been replaced by the nanometre. 1 Å = 0.1 nanometre. The unit is named after Anders Ångström.

Anhydride: A compound formed by removing water from an acid or, less commonly, a base. Many non-metal oxides are anhydrides of acids; for example, CO_2 is the anhydride of H_2CO_3 and SO_3 is the anhydride of H_2SO_4.

Anhydrous: Without water. Many 'dry' chemical may contain some water as a part of the crystal structure. In addition, this can often be compensated for making up formulas, it is often more convenient to use chemicals that contain no water, that is that are anhydrous.

Anidex fibre: A manufactured fibre in which the fibre forming substance is a long chain synthetic chain composed of at least 50% by weight of one or more esters of a monohydric alcohol and acrylic acid (CH_2=CH-COOH). In 1970, anidex fibres were introduced as an elastomeric fibre by Rohm and Haas with the trade name Anim. Anidex fibres are defined as fibres containing polymers that are at least 50%of one or more polymerized acrylate esters. Anidex fibres are formed through emulsion copolymerization of acrylate esters with reactive cross-linkable co-monomers such as -methylolacrylamide. The resulting copolymer emulsion is mixed with a filler and wet spun to form a fibre which is heated to cross-link the polymer chains and provide the necessary elastomeric properties. The morphology and elastomeric action of the fibre resemble spandex and rubber, but anidex generally has a lower elongation at break than these fibres. It has a round cross section. The fibre has a specific gravity of 1.22 and a moisture regain of 0.5% under standard conditions. Anidex is reported to be more resistant to heat, light and chemicals than either spandex or rubber. Otherwise, the fibre possesses end-use properties very much 1ike those of spandex. Anidex fibres apparently did not have sufficient differences in properties to become an economic success and are no longer being produced.

Anil: The blue dye now called **indigo**. 'Anil' comes from the Sanskrit word '*nila*' meaning dark-blue which is also the root of the word 'lilac'.

Aniline: C_6H_7N; also called aniline oil, benzeneamine, aminobenzene. Aniline used to be a very commonly used chemical in the synthesis of dyes. Because of toxicity, aniline has been replaced with other compounds for most dye synthesis.

Aniline black: Insoluble oxidation dye from aniline produced on the fibre (mostly on cotton) from individual soluble components. Aniline black is the best black and unsurpassed in depth and bloom. Chlorates and chromates are used as oxidizing agents and ferrocyanides, ammonium vanadate and copper salts as oxygen carriers. During the development acid liberated which can cause damage to the fabric under inappropriate conditions. Also called single bath black, oxidation black, steam black, hanging black, Prud'homme black, Prussiate black, Koechlins black, diphenyl black.

Aniline dyes: Any of a large class of synthetic dyes made from intermediates based on or made from aniline. The term is still in use for individual brands.

Aniline salt(aniline hydrochloride): $C_6H_5NH_2$– HCl, MW 129.5. Colourless, large crystalline plates or lumps which turn dark green on prolonged storage, m.p. 192°C, readily soluble in water and alcohol. Often contains free hydrochloric acid. May be identified by the production of a green colour with methyl violet test paper.

Aniline sulphate test: For the identification of hemp resp. jute fibres: Place the fibres in a 1: 20 solution of aniline sulphate (on a boiling water bath) for 20 min, rinse and dry. The fibres are stained – a) pale yellow = hemp, or b) intense golden yellow = jute. Pure cellulosic fibres (including linen) remain unstained.

Animal fibres: Fibres of animal origin such as – (a) wool (sheep) (b) Specialty hair fibres like alpaca, camel, cashmere, guanaco, llama, mohair (angora goat), vicuna, (c) fur fibres like mink, muskrat, angora rabbit.

Animal fibres and filaments: Fibres and filaments (generally hair, fur and cocoon materials) taken from animals for the purpose of weaving, knitting or felting into a fabric. Typical animal fibres include wool, mohair, Ilama, alpaca, cashmere. Camel hair, cowhair, fur, horsehair, vicuna. The cocoon material is silk. All are nitrogenous substances. Wool and other hair fibres come in staple lengths. The original forms of silk are parallel filaments several thousand yards long.

Animal hair fibres: Angora, mohair, cashmere and other fine animal hair fibres, occupy a significant place in the textile fibres. Used mainly in high fashion garments.

Aniodic oxidation: A problem which used to occur in package dyeing. Metal tubes or package spindle used to get oxidized due to anodic action of two different metal used, i.e. metal used in machine and metal used in the tubes. This causes a higher consumption of reducing agent. Nowadays, this is not a major problem as machines are invariably made by stainless steel.

Anion: A negatively charged ion, formed by the addition of electrons to atoms or molecules. In electrolysis, anions are attracted to the positive electrode or anode. Compare ***Cation***.

Anion exchange: Replacement of acid radical ions such as $NO_3^-$by OH^- (hydroxyl ions) in water by using ion exchange.

Anionic-cationic compounds: Electrically neutral compounds.

Anionic: Those that develop a negative charge on the water solubilizing end-function in a manner similar to emulsifying agents.

Anionic dye: A dye which dissociates in water to form negatively charged colour ions.

Anion exchange capacity: The ability of a material or soil, etc., to exchange anions. In soil, it is small as compared to the cation exchange capacity.

Anionic exchanger: These are high molecular weight, polyvalent bases or their salts which have basic, negatively-charged exchange ions, i.e. interchangeable OH groups against anions or end-groups suitable for the formation of acids. The principle of the exchange reaction is as follows:

$$A - OH + HCl = A - Cl + H_2O$$

See **Ion exchanger**.

Anionic polymerisation: In anionic polymerisation, a negatively charged ion attacks on 1t electron pair of the monomer molecule and pushing it as far away as possible, i.e., to the end of the molecule. Simultaneously, it forms a sigma bond with the monomer unit. At the same time, a carbanion is also formed. This is represented as follows:

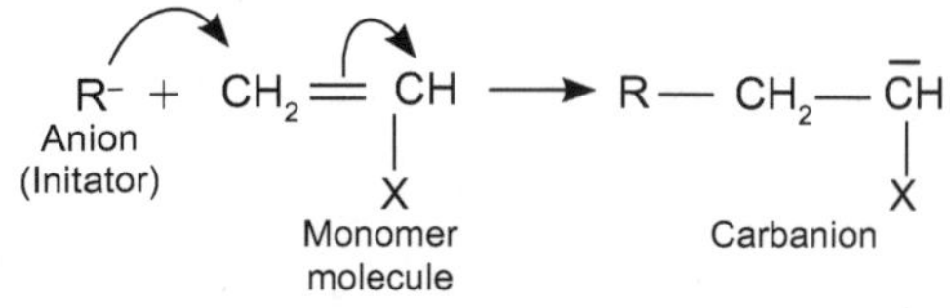

Anionic resin: An Ion-exchange material that can exchange anions, such as Cl– and OH–, for anions in the surrounding medium. Such resins are often produced by the addition of a quaternary ammonium group [$-N(CH_3)_3^+$] or a phenolic group ($-OH^-$) to a stable polyphenylethene resin.

A typical exchange reaction is:

Resin-$N(CH_3)_3$+ Cl^- + KOH → Resin–$N(CH_3)_3$+ OH^- + KCl

Anionic resins can be used to separate mixtures of halide ions. Such mixtures can be attached to the resin and recovered separately by Elution.

Anionic retarders: These are retarding agents with dye affinity and are used in the dyeing of polyacrylonitrile fibre with cationic dyes.

Anionics: Short form for anionic surface active agents.

Anionic softeners: Anionic softeners and/or surfactant molecules have a negative charge on the molecule which come from either a carboxylate group (–COO–), a sulphate group (–OSO_3–) or a phosphate group (–PO_4–). Sulphates and sulphonates make up the bulk of the anionic softeners. Some phosphates and to a lesser extent the carboxylates are used as softeners. Anionic softeners impart pliability and flexibility without making the fabric feel silky. They are used extensively on fabrics to be mechanically finished, e.g. napped, and sheared or sanforized. A good napping lubricant, for example, provides lubrication between the fabric and the napping wires yet at the same time provides a certain amount of cohesiveness between fibres. If the fibres are too slippery, the napping wires will overly damage the yarn. Sulphonated oils (e.g. Turkey red oil) impart a soft raggy hand, sulphonated tallow a full waxy hand and sulphonated fatty esters a smooth waxy hand.

Anionic surface active agents: Surface active compounds having one or several functional groups which ionize in the aqueous solution to form negatively charged organic ions that are responsible for surface-activity.

Anionic surface active compounds: See **Anionic surface active agents**.

Anionic surfactants: Surfactants that are anion active. This is the major group comprising almost 70% of commercial surfactants. Anionic surfactants are invariably the sodium salts of either aliphatic or aromatic sulphonic acids (R–SO_3Na, sulphonates), or of the sulphuric or phosphoric acid esters of alcohols (R–O–SO_3Na, sulphates; R–O–PO_3Na, phosphates). They are derived from both natural and synthetic chemicals.

Anisomophic: Non-uniformly configured. Opposite to ***Isomorphic***.

Anisotropic: Having different properties in different directions. For e.g., the capacity of a dye in solution to dye only the fibre surface or only certain internal regions of the fibre due to various reasons.

Ankle: The joint between the foot and lower leg.

Ankle girth: The circumference of the leg over the greatest prominence of the ankle.

Annatta (annotto, arnotto, roucou): A dye obtained from the pulp surrounding the seeds of the *Bixa orellana*; chiefly used in dyeing silk an orange colour, but is of a fugitive nature.

Anodendron: Very tough but fine, bast fibres of a climber (*Anodendron panic-ulatum*) in southern India and Sri Lanka; used for ropes.

Anomaloscope: An optical instrument used for the testing and classification of anomalous and defective colour vision.

Anoncillo: Fibre yielded by the bark of a species of the sour-sap in Venezuela.

Anorak: A lined and hooded hip-length jacket designed to give a degree of protection bad weather.

Anoxic denitrification: The process by which nitrate nitrogen is converted biologically to nitrogen gas in the absence of oxygen. This process is also known as aerobic denitrification.

Anoxic filter: A denitrifying filter in which a tank is filled with media (typically plastic media) and flooded, through which is fed a nitrified effluent and a source of organic food. The bacteria that grow in the filter metabolise the organic material and use the oxygen in the nitrate as the oxygen source and N_2 is produced. The effluent is thereby denitrified. Denitrification of a nitrified effluent can be achieved by locating an anoxic filter after secondary sedimentation (i.e. postanoxic). Methanol or other carbon sources may be added to the secondary effluent to act as an organic carbon source for the denitrifying bacteria. Anoxic filters can be used in the pre-anoxic location, i.e. the filter before the main biological treatment, for the removal of BOD and oxidation of ammonia. The nitrified waste-water from the main biological treatment stage is recycled back to the anoxic filter and the organic food for the denitrifying bacteria is from the BOD in the settled waste-water. Various designs of pre- and post-anoxic filters are used and the waste-water flow through the tank may be upward or downward. The depth of media is typically between 1.5 and 2 m. Some designs may require a hydraulic surge for a few minutes through the filter every few hours to stop a build-up of nitrogen gas becoming trapped in the filter and some designs may require a backwash every few days. Compare ***Anaerobic filter***.

Anoxic process: The process by which nitrate nitrogen is converted biologically to nitrogen gas in the absence of oxygen.

ANS unit: A colour difference unit in the Colour Difference Formula of Adams, Nickerson and Stulz.

ANSI: American National Standards Institute, New York. ANSI represents the United States member body in the ISO. The purpose of ANSI is to coordinate voluntary standards development and use in the United States and to serve as liaison between standards organizations in this and other countries, through the ISO.

Antelope: The colour of the antelope. With regard to the fact that there are so many different varieties of antelope and that they have a wide range of different colourings and markings, this definition is not very helpful! Perhaps referring to dusky-brown beige or a pale bronze-gold colour.

Antelope cloth: Used for embroidered waistcoats and embroidery foundation in England. There are small pinholes in the cloth placed at equal distance from each other, through which the yarn is put.

Anterine or Anterne: 18th century fabric in England made of worsted and silk or mohair and cotton.

Anthracene: A white crystalline solid, $C_{14}H_{10}$; r.d. 1.28; m.p. 215.8°C; b.p. 341.4°C. It is an aromatic hydrocarbon with three fused rings (see formula), and is obtained by the distillation of crude oils. It is a linear condensed homocyclic hydrocarbon. It forms colourless plate-like crystals which sublime readily and have a blue-violet fluorescence. Used in the manufacture of Anthraquinone, the parent substance of many synthetic dyestuffs.

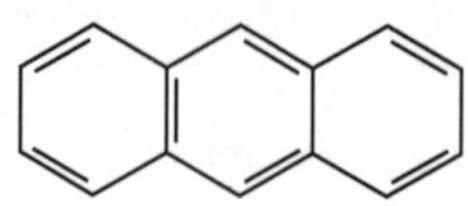

Antherea: Collective scientific name for various wild silks of Japan, China and India.

Anthistirta: Stem fibre yielded by a grass (*Anthistiria arundinacea*) in north-western India, used for cordage.

Anthanthrone: Anthraquinonoid dyes with a pyrene structure. Its example is the dye manufactured by the bromination of anthanthrone.

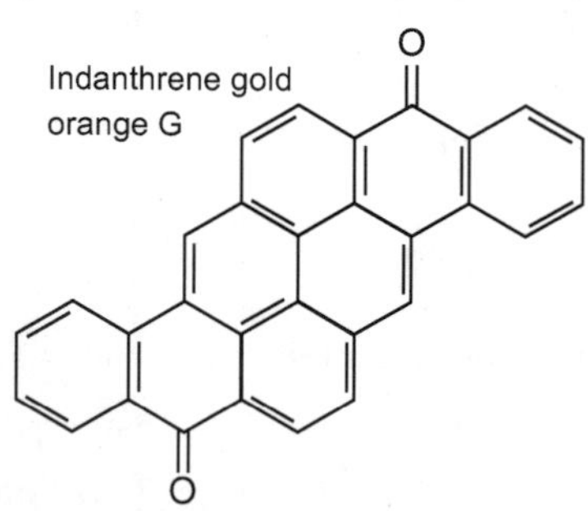

Anthrapyrimidine: An anthraquinone dye-class coming under derivatives of anthraquinone with a fused ring structure. It can be manufactured by acylation of 4-amino-1,9-anthrapyrimidine with p-chlorobenzoyl chloride in the presence of pyridine.

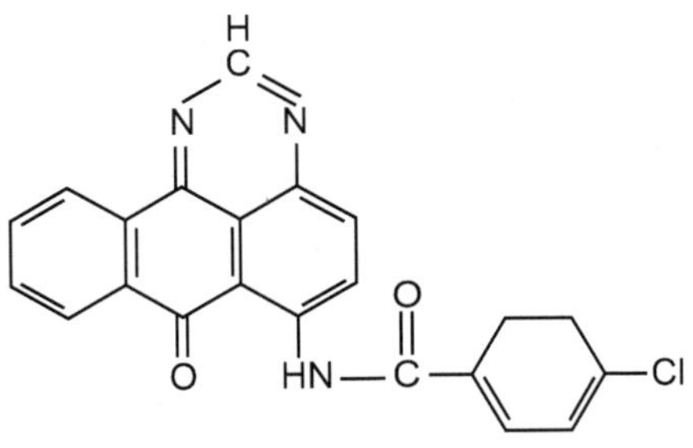

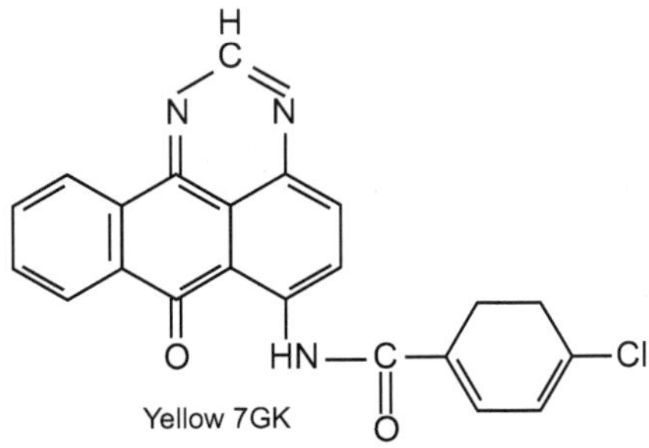

Yellow 7GK

Yellow 7GK

Anthraquinone ($C_{14}H_8O_2$): A colourless crystallinequinone; m.p. 154°C. It may be prepared by reacting benzene with phthalic anhydride. Also oxidation product of Anthracene. Outstanding reduction catalyst which increases activity of reducing agents. Used in the discharge printing of cationic dyes, mordant dyes naphthols and bromoindigo. A common arrangement of the carbonyl groups in coloured molecules gives rise to a group of compounds known as *quinones*. These may be defined as cyclohexadienediones, i.e. compounds containing two ketone carbonyl groups and two double bonds in a six-membered ring. The simplest quinones are *o*- and *p*-benzoquinones, respectively.

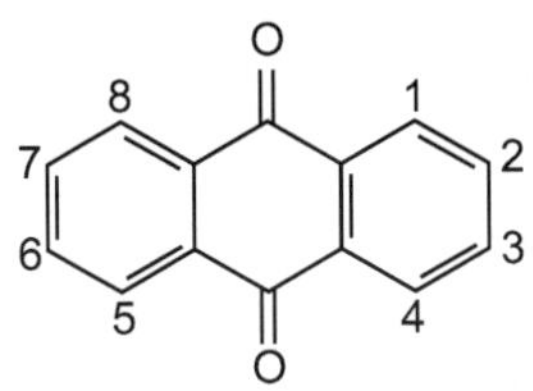

Anthraquinone dyes: Dyes which are derived from anthraquinone. They belong to *qunonoid* class of dyes. Anthrquinone derivatives are coloured when at least two further donor substituents are (-OH, -NH_2) are present in addition to the basic anthraquinone structure. Anthrquinone is one of the most important dye-classes. The detailedclassification of anthraquinone dyes are as follows:

1. *Chain-substituted and chain-linked anthraquinone derivatives* –
(a) acylaminoanthraquinones, (b) Anthraquinonylaminotriazine,
(c) Anthraquinonylamine or anthrimide.
2. *Derivatives of anthraquinone with a fused ring structure*–
(a) Anthraquinone-thiazole, (b) Anthraquinone-oxazole,
(c) Anthraquinone-carbazole, (d) Anthraquinone-acridone,
(e) Anthraquinone-N-hydroazine or indanthrone, (f) Flavanthrone,
(g) Anthrapyrimidine, (h) Acedianthrone.

3. *Anthraquinonoid dyes with a pyrene structure–*
   (a) Pyranthrone, (b) Anthanthrone, (c) Dibenzpyrenequinone.
4. *Anthraquinonoid dyes with a perylene structure –*
   (a) Dibenzanthrone or violanthrone, (b) Isodibenzanthrone or isoviolanthrone.
5. *Benzanthronoid vat dyes–*
   (a) Thiobenzanthrone, (b) Benzanthronylpyrazole-anthrone, (d) Benzanthronyl-aminoanthraquinone.
6. *Naphthalene derivatives–*
   (a) Naphthoylene-benziminazole, (b) Naphth-oquinone.

Anthraquinone-acridone: An anthraquinone dye-class coming under derivatives of anthraquinone with a fused ring structure. An example is the conversion of 2-methyl-1-nitroanthraquinone to 1-amino-2-anthraquinonecarboxylic acid, condensation with aniline, cyclization and chlorination.

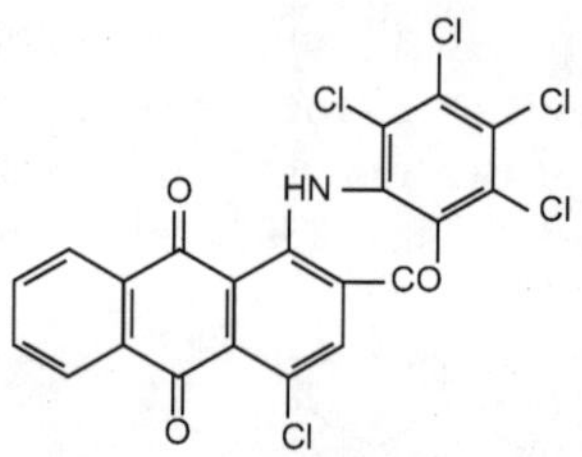

Indanthrene brilliant pink BBL

Anthraquinone-carbazole: An anthraquinone dye class coming under derivatives of anthraquinone with a fused ring structure. An example is the condensation of 4 mol. 1- amino-anthraquinone with 1,4,5,8-tetrachloroanthraquinone in the presence of copper and sodium carbonate in nitrobenzene at 205 °C, 36 h. Cyclization of the pentanthrimide so formed with aluminium chloride and sodium chloride at 160 °C, 2–3 h.

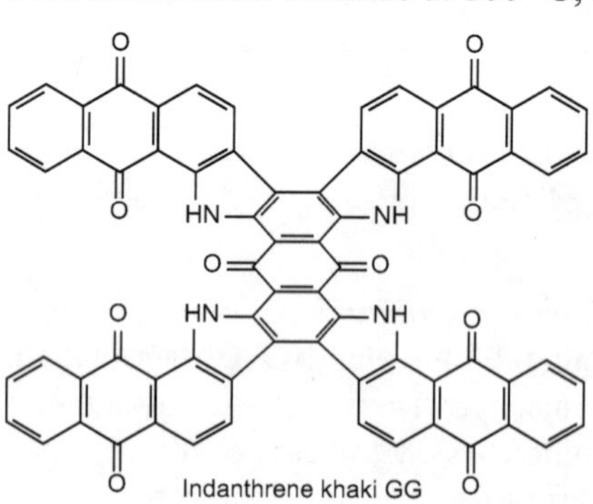

Indanthrene khaki GG

Anthraquinone-N-hydroazine or indanthrone: An anthraquinone dye-class coming under derivatives of anthraquinone with a fused ring structure. It is also called indanthrone. An example of dye is produced by the reduction of 8-nitro-1-anthraquinonesulphonic acid to 8-hydroxylamino-1-anthraquinonesulphonic acid. Conversion to indanthrone disulphonic acid followed by desulphonation.

Indanthrene blue RS

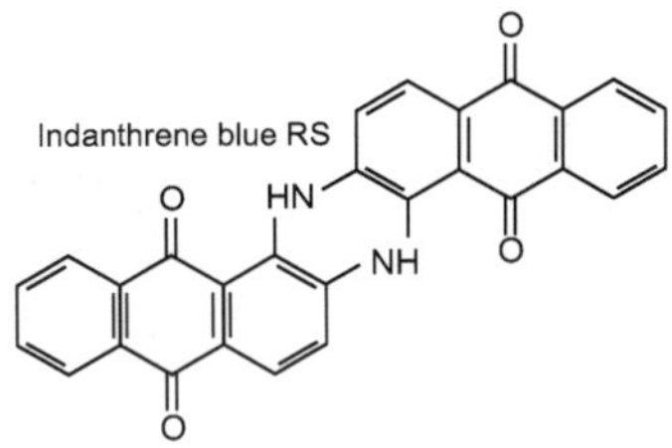

Anthraquinone-oxazole: An anthraquinone dye-class coming under derivatives of anthraquinone with a fused ring structure; condensation of 1-nitro-2-anthraquinonecarbonyl-chloride with 2-amino-3- hydroxyanthraquinone in o-dichlorobenzene (pyridine, 140 °C, 5 h) to the carbonamide; oxazole ring closure in 90% sulphuric acid, 90–105 °C, 2 h. Substitution of the nitro group in 15% ammonia at 125 °C, 12 h.

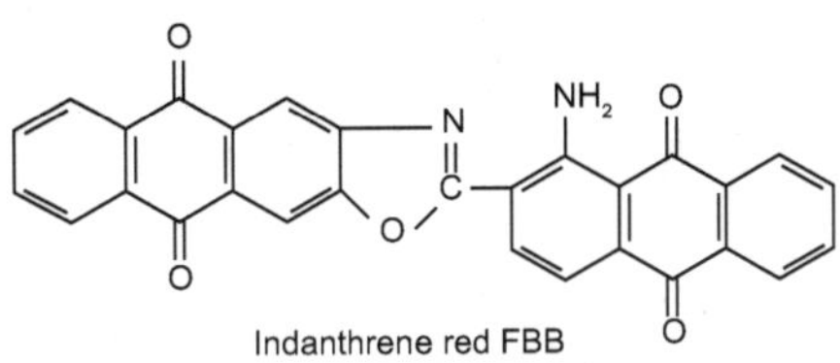

Indanthrene red FBB

Anthraquinone-thiazole: An anthraquinone dye-class coming under derivatives of anthraquinone with a fused ring structure. An example is the condensation of 1-amino-4-nitro-2-anthraquinonecarboxylic acid with 2-amino-3- chloroanthraquinone. From the mercaptan, cyclize, reduce and acylate with m-trifluoromethylbenzoyl fluoride.

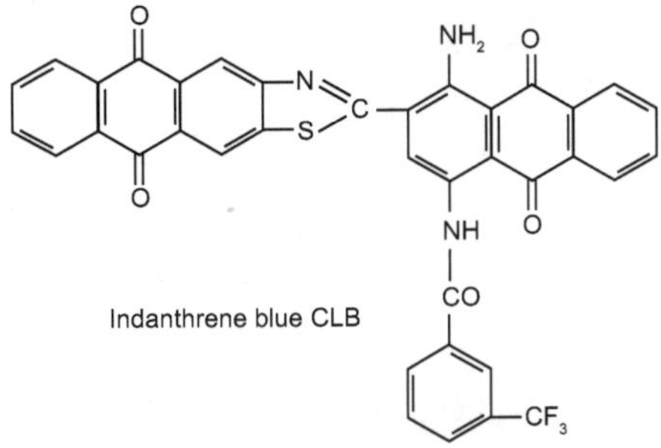

Indanthrene blue CLB

Anthraquinonylamine or anthrimide: An anthrquinone dye-class coming under chain-substituted and chain-linked anthraquinone derivatives. An example is condensation of 2,6-dichloroanthraquinone with 2 mol. 1-aminoanthraquinone in the presence of cuprous chloride and sodium acetate in nitrobenzene solution.

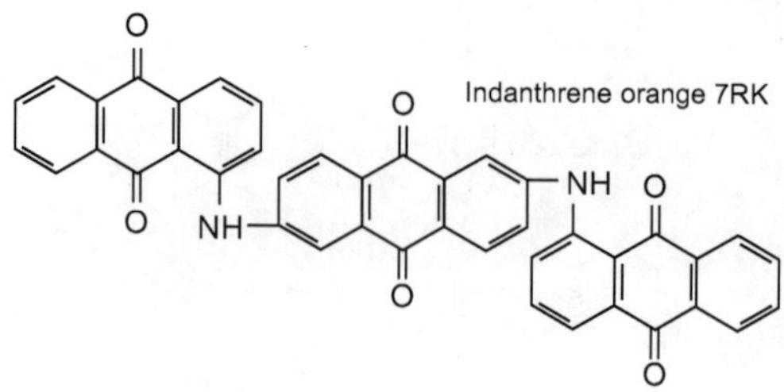

Anthraquinonylaminotriazine: A class of dye coming under anthraquinones and classified under chain-substituted and chain-linked anthraquinone derivatives. Example: from the condensation of cyanuric chloride with 2 mol. 1-amino-4-methoxyanthraquinone.

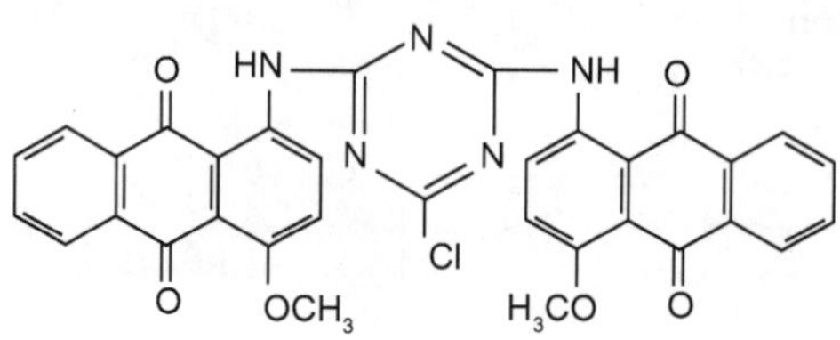

Orange 6R

Anthrimides: Anthraquinone imines produced from at least 2 Anthraquinone groups linked by an Imido group. Anthrimide structures are present e.g., in various Indanthren and Algol dyes. Among the anthrimides dianthraquinonyl-amines), only the derivatives have achieved limited importance as vat dyes.

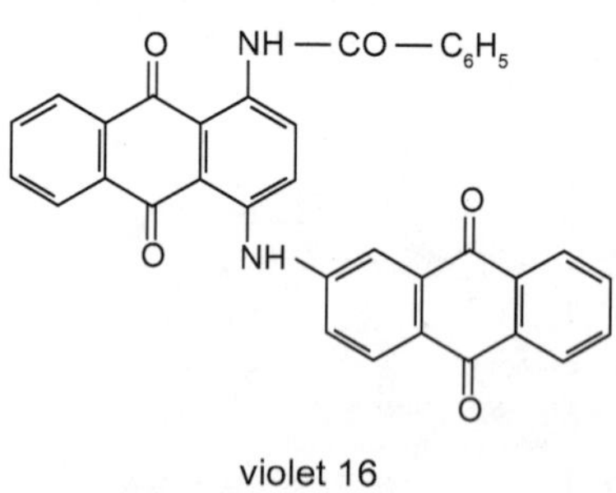

violet 16

Anti-corrosive: See **Corrosion inhibitors**.

Anti-crease finish: Resin finishing of cellulose for reducing creasing tendency to the lowest possible dimension.

Anti-diazotates: These are also called Nitrosamines. They are fairly stable *naphthol diazonium* salts, no longer capable of coupling in solid form and are derived from amines with very weak basic properties.

Anti-felting finish: A finish which prevents or delays felting. The typical scale structure of the wool fibre can be affected by means of different processes so that it is no longer detrimental in terms of felting. (1) Subtractive process: Hydrophilisation of the exocuticle A by oxidation of the 35% cystine there with hypochlorite results in the oxidised outer fibre areas containing cysteic acid being covered by a film of water when they are washed. This water film results in neutralizing the ratchet effect (as the innermost parameter of the felting process); (2) Additive process: masking the scales with polymers (forming a film-like cover). Their action is mainly based on the so-called "spot weld effect", i.e. the polymers, in condensing, glue the individual fibres together at their contact points (unsuitable for tops), providing good shrink-proof effects at low cost; (3) Combined process: The formation of ionised and hydrated groups in anti-felt finishing makes the fibre surface hydrophilic; at the same time, an electric double layer is produced. Both factors together reduce fibre contact in the aqueous medium, and prevent the longitudinal anisotrophy of the frictional characteristics responsible for felting from being effective on wool fibres

Anti-microbial: A term used to describe the action of growth inhibition or the destruction of microorganisms, e.g. as a property of man-made fibres which have been spun with anti-microbial compounds (Antimicrobial fibres) or as in Antimicrobial finishes.

Anti-microbial agent (in textiles): Any chemical material which kills or inhibits the growth of microorganisms.

Anti-microbial fibres: They are manmade fibres with anti-microbial properties which include: inhibition of microbial growth, prevention of pus formation, acceleration of wound healing and reconvalescence. Used in the production of bandages, medical gauze, cotton wool and tampons. Also called medical fibres.

Anti-microbial finishes: They are effectively produced on textiles by: (1) Addition of microbicidal substances to the spinning solution in fibre manufacture; (2) Modifications involving grafting or other chemical reactions; (3) The finishing of textiles with suitable active substances. Such substances are fixed on textile materials after a thermal treatment (drying and curing) by incorporation into polymers and resin finishing agents.

Anti-migration: Opposite of migration. Measures taken to influence, inhibit or prevent the undesired migration of dye by physical, e.g. drying, or chemical means by the addition of suitable products. Migration inhibitors.

Anti-oxidants (oxidation inhibitors): These compounds inhibit, or even prevent, oxidation processes (inhibitors). Antioxidants are of practical importance, e.g. in the form of a small addition to drying oils used in oil sizing to prevent the evolution of heat and attack on the fibre due to an excessively rapid self-oxidation or autoxidation. For this purpose, certain substances with OH and NH_2 groups capable of being oxidized such as, e.g. phenols, amines, aldehydes, ketones, benzyl alcohol, diphenylamine, hydroxydiphenylamine, hydroquinone, etc., are used.

Anti-slip finish: These finishes are used to avoid the shifting of crossing warp and weft threads in fabrics containing a low number of yarns or woven in open constructions to prevent the formation of holes, splitting of seams, etc., (slip resistance). Anti-slip finishes involve the application of anti-slip finishing agents by wet-processing.

Anti-foam: A chemical used to prevent/suppress the generation of foam.

Anti-house dust mite agent: Any chemical which kills or repels house dust mites.

Anti-redeposition: A product that prevents soil redeposition during washing off or laundering process.

Anti-slip and anti-snagging agents: These serve to reduce the slippage of open set-woven fabrics, to prevent the formation of ladders in warp knitted fabrics and the so-called snags in stockings and finished fabrics of fine synthetic yarns. The effect is based on the production of a thin, rough, adherent film on the fibre surface of the treated fabrics. Preparations of plastics (e.g., polyvinyls, polyacrylates, polymethacrylates).

Anti-adhesives: Special auxiliaries which are used to prevent the adhesion of hosiery material on metal formers during the post boarding process. Typical commercial anti-adhesives improve the elasticity and softness of the fabric.

Anti-bacterial agent: Any chemical which kills bacteria (bactericide) or restricts the multiplication, growth or activity of bacteria.Agents or chemicals applied on the fabric which prevents the action of bacteria on the fabric/chemicals, etc.

Anti-bacterial finish(Anti-microbial finishes): The application of products to textile fabrics capable of preventing the growth of various microorganisms and which, therefore, also contribute to deodorizing. Such products are subdivided into Bactericides (destruction of bacteria) and bacteriostats (inhibition of bacterial growth). Anti-bacterial finishes are effective not only against bacteria but also against Mycotoxins.

Anti-bacterial textiles: In these textiles which are grafted with antiseptic substances, the molecular chains with antiseptic agents are grafted onto the parent polymers of the raw material. The parent polymers are activated by electron discharge. In the process they open up at certain sites. The graft priming monomers are deposited at these sites. The polymer chains propagated laterally impart the bactericidal properties to the textile.

Antichlor: Any product which serves to decompose residual chlorine on the fabric following a chlorine bleach process. The residual chlorine can cause yellowness to the fabric on drying and interfere in the subsequent process like dyeing. It is very difficult to completely rinse chlorine bleach out of fabric. A rinse in a solution of antichlor, most commonly sodium bisulphite, will quickly neutralize the bleach. Other antichlors are sodium thiosulphate, sodium hydrosulphite, sulphurous acid, ammonia, etc.

Anticle finishes: Elimination of the possible loop distortion in garment pieces made from combined wool yarns during subsequent wet-processing by treatment in boiling bisulphate, sulphite.

Anti-cling finish: Finishes which will reduce the static electricity on the fabric. One way is by increasing the moisture on the fabric. These finishes are applied on fabric made from synthetic fibres in order to prevent garments from clinging to the body. See **Antistatic finish**.

Anti-cockle treatment: A mild setting treatment done on woollen knitwear to prevent distortion and cockling of the knitted structure during processing. For example: Treatment with boiling sulphite or bisulphate solutions.

Anti-cockling treatment: Elimination of possible loop distortion in garment pieces made from combed wool yarns during subsequent wet-processing (chlorine/hercosett process) by treatment in boiling sulphite or bisulphite solutions.

Anti-crease finish: Resin finishing cellulose to reduce the creasing tendency to the lowest possible dimension.

Anti-diazosulphonates: They act as components which are not capable of coupling of the type B-N=SO_3Na. After printing, this structure undergoes a rearrangement due to the action of steam or light energy into a syn-diazosulphonate which is capable of coupling with naphthols.

Anti-felting agents: Product that minimize or prevent matting and compaction of textile materials. See also **Anti-felting finishes**. In subtractive process, hydrophilisation of exocuticle (to about 35% cystine) by oxidation by Hypochlorite. The cysteric acid thus formed holds a film of water when washed which prevents felting. In additive process, material is resin finished whereby "a spot welding effect" is formed which binds the individual fibres at their contact point which prevents shrinkage.

Anti-felting finish: Process by which the wool structure; so that it is no longer having the felting quality. Three processes are available – subtractive process, additive process and combined process. See **Anti-felting agents**.

Anti-foams: (1) In many processes, formation of foam can create problem. Anti-foams are added to the liquors and printing pastes to either reduce or prevent the foam formation. For e.g., Tributyl phosphate (TBP).

(2) An additive that minimizes the formation of bubbles within or on the surface of a liquid by reducing the forces that support the bubble structure.

Anti-foamers: See **Antifoams**.

Anti-foaming agent: See **Anti-foamers**.

Anti-foams in dry-cleaning: Anti-foams are added in dry-cleaning solvent to prevent foam formation during the distillation which can happen due to various factors such as dissolved silicones.

Anti-frosting agents: Textile auxiliaries based on alkyl aryl polyglycol ethers for the prevention of frosting. See **Frosting**.

Anti-fungal: Generic name for fungicidal and fungistatic finishes.

Anti-fungal agent: An auxiliary used for Anti-fungal finish. Any chemical that kills or inhibits the growth of fungi.

Anti-greying agent: See **Anti-frosting agents**.

Anti-greying treatment: Any treatment that prevents or suppresses the greying/ageing of the fabric/garment. See **Anti-frosting agents**.

Anti-microbial: A term used to describe the action of growth inhibition, or destruction of microorganisms.

Anti-microbial agents: Bacteriostatic, bactericidal, fungistatic, fungicidal textile protective agents. Phenolic compounds, quarternary ammonium compounds and organometallic compounds containing mercury are a few of the important anti-microbial agents.

Anti-microbial fibres: Synthetic fibres made with anti-microbial properties which include: inhibition of microbial growth prevention of pus formation, acceleration of wound healing and convalescence. Also called medical fibres.

Anti-microbial textiles: An antimicrobial agent is defined as a natural or synthetic substance that kills or inhibits the growth of micro-organisms such as bacteria, fungi and algae.

Anti-migrant: An additive used in dye or pigment mixtures to prevent undesired movement or spreading of the wet dye on fabric. Anti-migrants are also used in commercial pad-batch, pad-dry, etc., dyeing to prevent uneven shading across the width of the fabric.

Anti-migration: The measures taken to influence, inhibit or prevent the undesired migration of dye by physical, e.g. drying, or by chemical means by addition of suitable agents. Opposite to migration.

Anti-migrating agent: Textile auxiliary designed for anti-migration. See **Anti-migrant**.

Anti-migration agent: See Anti-migrating agent. Textile auxiliary designed for anti-migration. For e.g., as in pigment-pad and continuous dyeing processes.Thickeners can be used as anti-migrating agent. See **Migration inhibitors**, **Anti-migrant**.

Anti-mycotic finish: Fungicidal or fungistatic finishes.

Anti-odour finish: The microbial action on the fabric/garment along with human sweat or other matters produces bad odour on the fabric/garment or on the skin. The anti-microbial finish prevents or suppresses this action and the odour, and hence the name anti-odour finish.See **Anti-microbial finish**.

Anti-oxidant: Prolonged exposure to atmosphere results in atmospheric oxygen causing the degradation of fabric, fibre, finish, etc. An anti-oxidant is a substance that retard the deterioration (of fibre, fabric, finishes, etc.) resulting from reaction with oxygen.

Anti-perspirants: An acidic salt or some other compounds (deodorants) prevent the decomposition of perspiration in the alkaline range. Such compounds (e.g. aluminium chloride 15–30% solution) along with an anti-sceptic may be used as anti-perspirants.

Anti-pill fabric finishing: The finish which prevents or reduces the pill formation during the use of the garment. Chemicals which bind the fibres so that the movement of the fibre is reduced and hence the pill formations are used as anti-pilling agents. Sometimes bio-polishing which removes all the protruding and loose fibres from the surface of the fabric which starts the pill formation are also called anti-pill finish.

Anti-pilling process: Any process intended to minimize or prevent pilling, mainly applicable to fabric made from high tenacity, man-made fibres and its blends. Pilling may be prevented by initial heat-setting, total shrinking, besides finishing with film forming products such as binder, resin finishing which restricts fibre migration.

Anti-pole: Philippine fibre, used for ropes and cords.

Antique bronze: A bronze colour imitative of that of very old objects; also 'antique gold', 'antique green' and 'antique red'.

Antique ivory: The yellowish-white colour of old ivory.

Antique white: A yellowish-white colour adopted as a colour name by Web page creators on the Internet with hex code #FAEBD7.

Antique satin: A reversible fabric, the right side of which resembles shantung slubbed in the weft direction. The wrong side resembles silk satin of an earlier century. Made of a mixture of various fibres.

Antique taffeta: A stiff-finished fabric in plain weave made to resemble fabrics of the 18th century. It may be made of doupion silk or synthetic fibres.

Antique lace: A term used to describe hand-made bobbin lace where heavy threads have been used. Designs are often irregular.

Anti-redeposition agent: Additional component of a washing agent for increasing the soil suspending property. Also called builders.

Anti-rotting agents (preservatives): Used in textile exposing to moisture to protect against rotting by microorganisms.

Anti-rheumatic: A term generally applied to textile fibres which are effective against rheumatism, e.g. pain relief achieved from the use of polyvinyl chloride fibres in underwear and quilt fillings. Wool also possesses anti-rheumatic properties.

Anti-shrink setting: Heat-setting.

Anti-slip and anti-nagging agents: Auxiliaries used to reduce the slippage of open set woven fabric, to prevent the formation of ladder in warp knitted fabrics and the so called snag in stockings and finished fabrics of fine synthetic and cotton yarns. Preparations of plastics (e.g. polyvinyls, polyacrylates, polymethacrylates and polystyrenes)and natural resins and silicic acids are used for this purpose.

Anti-slip finish: These finishes are used to prevent the shifting of crossing warp and weft threads in fabric containing low no of yarns or woven yarns or woven in open constructions to prevent the formation of holes, splitting of seams, etc. Suitable anti-slip finishes is film forming polymer dispersions, silicic acid, hydrosols. Addition of anti-static agents along with these products also may help.

Anti-snag finish for velvet: A finish applies on velvet and plush fabric to prevent the pile getting detached from the warp. It is almost like an anti-slip finish but involves only impregnating the backside of the fabrics with various synthetic resins.

Anti-soiling finish: Finishes designed to achieve active stain blocking. The term is also used to describe various technical measures applied in the finishing to keep soil away from textiles. Crease-resistant finishes, water and oil repellent finishes improve anti-soiling property.

Anti-soiling properties: The property of the textile material whereby they resist the deposition of dirt and stains.

Anti-staining properties: The property of the textile material to resist the deposition of oil and water based stains.

Anti-static agent: A reagent capable of preventing, reducing, or dissipating static electric charges that may be produced on textile material.

Anti-static fibres: To eliminate build-up of electrostatic charges on textiles (especially in textile floor covering) permanently anti-static polyamide or metallic fibres are used. Various polyamide fibres are available in the market with antistatic properties based on different principles; for e.g. embedded carbon particles, metallized fibres, copper sulphide deposition on the surface of the fibre, besides incorporation of antistatic agents in to the spinning solution, etc.

Anti-static finishes: Treatment of textiles (fabric/yarn) with special chemicals to increase the surface conductivity in order to prevent the build-up of electrostatic charges or to reduce the static property of the fabric/yarn. See **Anti-static fibres**.

Anti-static nylon: Anti-static nylon; for example, Celon can be produced for both knitted and woven lingerie fabrics. The latter have a greatly reduced tendency to cling and spark when worn. Static electricity can still be a problem because other factors, such as temperature and humidity, are involved. The use of propriety fabric softeners may help. See **Anti-static fibres**.

Anti-static properties: The ability of the textile material to disperse an electrostatic charge and to prevent the build-up of static electricity.

Anti-static in dry-cleaning: The anti-static agents used in dry-cleaning to prevent spontaneous ignition of spirit, increase its conductivity and prevent fires and explosions (resulting from the electricity generated due to friction or electrostatic charging). Anti-statics used includes anhydrous magnesium oleate products (magnesia soap), dry-cleaning soap, dry-cleaning detergents, etc.

Anti-static tricot: Fine synthetic jersey for lingerie and nightwear. May be named trilobal or filament.

Anti-swelling agent: Finishing agent to resist swelling-resin finishing agent.

Antron: A trademarked Nylon marketed by Du Pont co.

Antung (silk): Warp and weft of tussah greige; unlike honan silk, this fabric is produced on a mechanical loom. The fabric has affirmed lustrous appearance.

Antwerp: Belgian bobbin lace of bold patterns, similar to the Malines. Made either of separate sprigs connected with each other by brides or the pattern made in one and upon a ground. Usually plaited thread outlines the patterns. A characteristic design is the so called Potten Kant.

Antwerp blue: A greenish-blue variety of **Prussian blue**; also Antwerp brown and Antwerp red.

Antwerp edge: Needle-point edge to braid or cordonnet, consisting of one line of open buttonhole caught with a knot in each loop.

Antwerp pot lace: A bobbin lace which always had motifs of baskets or vasesof flowers as decoration.

A/O process: An adaptation of the activated sludge process that can achieve biological phosphorus removal as well as BOD removal. The two-stage process involves an anaerobic zone followed by an aerobic (oxidation) zone, which is used for BOD removal. The return activated sludge from the sedimentation tank is fed back to the start of the anaerobic zone. The first anaerobic zone is required in order to obtain biological phosphorus removal.

AOPs: Advanced oxidation processes.

AOX: Adsorbable organic halogen compounds. Important in waste-water treatment.

Ape skin (velveteen): Raised closely sheared or sanded cotton sateen or twill with short, velvety pile (imitation velvet). See also **Duvetine, Apricot skin**.

APEO: Alkyl Phenol Ethoxylates. Even though once it was thought that APEO is biodegradable, once the ecological failure was known (around 1988) this chemical was completely banned as wetting and washing agents used in textile industry.

Apparel textiles: The clothing or apparel market includes most garments that are worn.

Apparent wall thickness: The apparent width of a fibre wall as seen under the microscope. In the maturity test for cotton, the apparent wall thickness is assessed visually at the widest part of the fibres as a fraction of the maximum ribbon width.

Appearance: An object's or material's manifestation through visual attributes such as size, shape, colour, texture, glossiness, transparency, opacity, etc.

Appearance of creases: See **Crease retention**.

Appearance of textile end-products: The overall visual impression of a textile end product quantified by comparison of individual components with appropriate reference standards.

Appenzel: Swiss hand embroidery featuring loop-stitch.

Apple-green: A pale green; a colour used extensively in the making of Sèvres porcelain.

Applique lace: Any fabric where separate lace motifs are applied to a background.

Application prints: Prints where the design is printed on a white or ecru fabric. The ground colour of the fabric forms an inherent part of the design.

Applicator roller (metering roller, slop padding roller, tip roller): In coating by the reverse roll process, the roller running in the application direction from which the excess of molten polymer is doctored at each revolution, is described as applicator roller (in contrast to the metering roller which runs at a different speed against the direction of paste application).

Appliqué: A cut-out design or shape attached to the face of a textile material for ornamentation, usually of a different type fabric and/or shade.

Appliqué (in sewing): It is the process of sewing shapes to a background fabric using a satin stitch or by hand.

Appliqué carpet: Handmade non-pile carpet comprising of a base fabric to which the fabrics are stitched or are affixed by adhesive pieces of fabric and/ or yarn. Used for floor covering and wall covering.

Appliqué broderie: Suisse design is embroidered on white cambric or muslin laid upon satin or silk background.

Appliqué broderie perse: Coloured pieces of chintz or cretonne, representing flowers, birds, etc., are pasted on the ground and outlined with stitches.

Appliqué lace: Needle-point or bobbin sprigs applied to a machine-made ground, as for instance, modern Brussels lace.

Apolda: Fine, printed wool shawl, made in Germany.

Apposition dyeing or deposition dyeing: A method of dyeing in which the periphery of the cross section is only dyed. In printing this may be called surface print.

Apou: Transparent Chinese fabric of great lustre, made of Ramie.

AP, USP: American patent.

Appret de laine: French for the soft and permanent finish given to mercerized Italian linings, similar to the Botany worsted Italian lining.

Apricot: The orange-yellow colour of the apricot fruit.

Apricot skin: A finish (peach skin) similar to ***Ape skin***. Usually cotton warp and Schappe silk or woollen yarn weft.

Apron checks: A cotton gingham made in simple check pattern of an equal number of white and colour threads, alternating both in warp and weft. Apron checks are designated 4 × 4, 8 × 8, etc., according to the number of threads of each colour. Used for aprons.

Apron mark: See **Decatizing mark**.

Apron rods: These are sturdy wood or metal rods, one of which is attached to the warp beam, the other to the cloth beam, sometimes by cloth aprons, more often by cords. The warp is tied to the apron rod of the warp beam before beaming and to the apron rod of the cloth beam after beaming, threading and sleying.

Aprotic solvents: Inert neutral solvents (with low dielectric constant) containing no ionisable proton in the molecule, e.g. benzene, chlorobenzene and chlorohydrocarbons.

Apta: Brown fibre gained from the East Indian Bausinia, used for tows, fishing nets and coarse cloth.

AQL: Applicable quality level.

Aqua: A light greenish-blue; a colour name adopted by Web page creators on the Internet with hex code #OOFFF. See **X11 Colour Set**. The colour aqua was one of the many different colours in which lamp-posts in Notting Hill were painted for the 1999 Carnival.

Aqua-blue: Light greenish-blue.

Aqua-fortis: An old name for nitric acid, HNO_3.

Aqua-green: A light bluish-green.

Aquaculture: The farming of fish or other aquatic animals, or possibly the farming of aquatic plants.

Aquamarine: Pale bluish-green; sea-colour.

Aquarelle: A pale bluish-green. Also a technique of painting using thin transparent watercolours.

Aquatic: Concerned with water.

Aqueous: Describing a solution in water.

Aqueous emulsion: Oil in water emulsion (O/W type).

Aqueous repellency (in textiles): The characteristic of a fibre, yarn or fabric whereby it resists wetting by aqueous liquids.

Aqueous extract: See **Extract, Aqueous**.

Aqueous extract (in wool testing): The solution obtained by digesting a material with water or with sodium chloride solution to dissolve soluble materials.

AR: Aramid fibre.

Arabian: A curtain made of netting and edged with macramé lace.

Arabian crepe: Silk crepe dyed in the piece and embroidered with dots.

Arabian lace: An ecru-coloured piece-lace corded in a darker shade. Used for curtains.

Arabian stripes: Coarse cotton fabric with blue and brown weft stripes; originated in Arabia.

Arabic gum: Gum Arabic.

Arabiennes: Fancy coloured cloth made of cotton and flax in Germany for the South American trade.

Arachin fibre: A fibre belonging to the group of manmade protein fibres derived from vegetable protein arachin which is present in peanuts.

Arachne machine: A machine for producing loop bonded non-woven. The fabric is formed by knitting a series of warp yarns through a fibre web-processed on a card.

Arain: Indian silk taffeta, made with stripes or checks.

Aramide fibre: A manufactured fibre in which the fibre forming substance is a long chain synthetic aromatic polyamide out of which 85%of the amide linkages (–NHOCO–) are attached directly to two aromatic rings. The word aramid comes from "aromatic polyamide".

Aramide fibres: The word Aramide comes from "aromatic polyamide" (Polyaramide) and means that it is associated with polyamides that contain aromatic six carbon rings in the polymer chain.These make polyamides, which would otherwise soften and burn at 260 °C max, into non-melting and temperature stable compounds of a slightly yellowish colour. They can be converted into textile fibres by a special spinning process and thus form the basis of asbestos substitutes in many application areas. They are of high strength because the crystalline content is high due to the intensive interaction between the highly structured polymer molecules. The chemical properties are similar to those of Polyamide 6.6.

Aramina fibres: See **Urena fibre**.

Aran: Beige.

Arbaccio: Coarse homespun, made in Sardinia of native wool.

Arbascio: Stout brown cloth made of coarse wool or goat's hair, once imported to Italy, where it was used for raincoats.

Arch steamer: A continuous steamer as a mainly floor mounted, vertical, semi-circular type machine, through which the fabric to be steamed is passed over side rollers, particularly for two stage printing process.

Arcadian green: A pale yellow-tinged green colour.

Archangelsk flax: Fine Russian flax of long, soft, greyish fibres.

Archil: See **Natural dyes**.

Ardamu: Raw silk of Ghilan, Persia.

Ardas: Stout silk fabric from Persia.

Ardasse: Low-grade raw silk of Persia, manufactured in Europe into embroidery silk.

Ardassin: Fine Persian raw silk, same as **Ablaque**.

Area bonding: It is a method of bonding non-woven where adhesion takes place over the entire surface. See also **Point bonding**.

Area burning rate: Area of the material burned per unit time under specified test conditions.

Area method of landfilling: A common method of operating a landfill. The waste is deposited on the surface in layers of 2–3 m deep, compacted and, at the end of the day, covered with inert material or soil (see daily cover). The waste is often tipped into a cell.

Area-of-view (of colour measuring equipment): The dimensions of the surface area that a colour measuring instrument is capable of covering in a single colour measurement.

Aredas: Indian plain woven fabric made of a soft and lustrous vegetable fibre, yielded from a grass. See **Aridas**.

Areste: A rich gold cloth woven with elaborate patterns and used for church vestments in medieval England.

Argali long: Dark-grey hair yielded from the Argali sheep in Siberia, used for carpet yarns.

Argaman: An ancient purple dye made from the shellfish murex trunculus referred to in *Judges* viii: 26 and rediscovered in 1998.

Argent: Like silver; silvery-white colour; in heraldry a silver or white colour.

Argent- (L): Silver.

Argentan lace: French needle-point lace. Early specimens called point de France. The Argentan lace is similar to the Alencon and has the following characteristics: net ground with large patterns, made higher and bolder than the Alencon, closer toile than in Alencon, and a great variety of rich brides, especially the bride picotee, the ground is the grande bride, a six-sided mesh. See **Point de France, Toile, Alencon, Picotee**.

Argentella lace: Italian white needle-point similar to the Alencon, but with flat cordonnet. The designs consist of very delicate patterns spread over a net ground.

Argentine croisee: Former name of twilled dress goods and men's suiting made with silk warp and cotton filling, manufactured in France.

Argil: The reddish-brown colour of clay – argil being potter's clay.

Argol: The tartar deposited from wines completely fermented, and adhering to the sides of casks as a hard crust. When purified, it becomes *Cream of Tartar*.

Argos: Wool rug made in Argos, Greece.

Argudan: Variety of coarse raw cotton from China.

Argyl gimp: A woven figured narrow fabric having three series of weft and warp. Two series consists of three gimp cords laid flat; the ground or third series consists of two gimp cords and forms a plain weave. The two series if three gimp cords form a double weave raised pattern by passing through the warp every sixth pick alternately and returning over the top of the warp. The overall width is about 16 mm. Warp is usually made of rayon.

Argyle: A pattern consisting of diamond shapes of different colours knitted together in a fabric.

Argyll: Diamond-shaped design for knitted fabrics using two or more, usually three colours of yarn. Often used for jerseys and hosiery, but also incorporated into knit fabrics.

Aridas: Light-weight, plain-woven and solid coloured East Indian silk cloth of high lustre; made originally of fibres of grass-like plants.

Arimina: A long and strong stem fibre that is similar to jute, yielded by a species of agave in Brazil; used for ropes.

Aristo: See **Moquette carpet**.

Ariyalur: Fine, weft-faced silk satin, with weft stripes used for garments, made in India and Bangladesh.

Arm (in sewing machine): The upper visible part of the sewing machine above the material being sewn.

Arm length (in garment measurements): With the arm bent at 90 degrees and the clenched fist placed on the hip, the distance from the shoulder joint along the outside of the arm over to the elbow to the greatest prominence on outside of the wrist.

Armazine: Stout, plain-coloured, corded black silk, used in the 16th century for gowns and men's waists and later for scholastic gowns, hatbands, etc.

Armenian lace: Coarse and narrow crochet lace edgings.

Armenian edging: Narrow lace trimming with formal, often geometric, designs.

Armhole: The area of a garment through which the arm passes or into which the sleeve is fitted.

Armiak: Camel-hair cloth made in Astrakhan, also called Biaza; used for coats.

Armistice: Once fashionable English worsted fabric in orange, blue, black and other colours; now obsolete.

Armoured woven fine cloth: Woven metal filter fabric with the finest orifices (max 17uM); Used for filtration.

Armpit (in anatomy): The hollow under the junction of the arm and the shoulder.

Armscye (in garment): The opening in a garment for the attachment of a fitted sleeve.

Armscye circumference (in garment): With arm hanging down, the distance from the shoulder jointthrough the front-break point, the armpit, the back-break point and to the starting point.

Armstrong: A highland tartan fabric with green checks over a blue and black background with red lines.

Armure: (1) French term for a variety of fabrics that have an embossed or pebbled surface, imitating birds-eye effect, pique, stripes, crepe and ribs. Sometimes they also have a woven design. Made from any fibre or blends of various fibres, this is mainly a dress weight fabric.

(2) A great variety of dress goods made of Botany wool, mohair, cotton or artificial silk or combinations of these fibres, made in a small pebbled or embossed effect which is produced from warp or weft ribs.

Armure bosphore: A reversible silk armure fabric.

Armure cheviot: An all-wool, piece dyed black cheviot, used for dresses in England.

Armure royale: French silk dress goods and vesting, made with two sets of ply warps in different colours; 16 warps and 6 fillings in a repeat.

Armure victoria: A thin and sheer wool dress goods of dull-black background with delicate patterns; used for mourning.

Army cloth: (1) A number of woollen and worsted fabrics used for soldiers' uniforms; (2) Low-grade grey woollen fabric made in United States.

Army duck: Ducks made to Government specifications are called army ducks. A characteristic army duck is 30 in or wider, weighs about 12 oz per yard for 30 in width and has a construction of about 40 × 32 with 3/12s yarn in warp and weft. See **Duck**.

Aromatic di-acids: HOOC-Ar-COOH. Monomer used for the production of polyester by polycondensation (with this operation two molecules of same type or of different types are joined together to form macromolecules by removing simple secondary products as water, hydrochloric acid, alcohol).

Aromatic polyamide fibres: See **Aramide fibres**.

Arti: A very fine Indian cotton muslin.

Aroosha: A fairly strong fibre, yielded by a species of the Verbenaceae tree in India.

Arramas: A mediaeval brocade made of gold and silk.

Arras: (1) White French bobbin lace, similar to the Lille lace. It is of strong and close texture with straight edge and the mignonette as the characteristic pattern. See **Mignoette**, **Lille lace**.

(2) Worsted cloth that was produced in ancient times.

(3) Tapestry work made famous centuries ago in the city of Arras.

Arrasene: Chenille fabric made of silk or wool, specially made for embroidery.

Arrasene embroidery: Produced with Arrasene, by either drawing it through the cloth or laying it on the surface and catching down as in couching.

Array fibre length: A series of individual fibres that are arranged in order, based on the fibre length or a series of groups of fibres that represents essentially uniform lengths within a group and which are arranged in order based on the group length.

Arree: Bark fibre yielded by the Bauhinia racemostly in India and Bangladesh; used for ropes.

Arrindy: Raw silk fabric manufactured using eria silk which is strong, made and worn in India.

Arrow heads: The arrow head and crow's foot are ornamental fastenings used in fine tailoring as endings for seams, tucks, plaits, and at corners.

Arrow stitch: So called from the slanting position of threads forming it. Identical with the stem stitch.

Arscot: Fine woollen serge made in Belgium.

Arsenic traces (test for): Marsh's test for arsenic. Hydrogen gas is produced in a gas generator vessel by the action of sulphuric acid on zinc and the solution to be tested is poured in. If arsenic is present, arsenic hydride AsH_3 is formed which, after drying, is passed through a hot zone where the AsH_3 decomposes with the precipitation of arsenic as a dark lustrous film and hydrogen is burnt off.

Art delaine: A 19th century fine woollen dress goods.

Arte': Term for insect fretted silk.

Art linen: A plain weave soft finished fabric used as bleached or unbleached as a base fabric for needle work.

Art muslin: A fine cotton fabric dyed or printed and finished with a high gloss; used for upholstery.

Art serge: A fine and stout worsted serge, used in England for draping and table covers.

Art square: Large variety of small or large ingrain rugs, made square or oblong with fringe at both ends.

Artichoke green: The yellowish-green colour of the artichoke.

Artificial cellulose fibres: Manmade fibres of which the basic component is cellulose; for example, viscose fibres.

Artificial daylight: Xenon lamp is nearest to daylight. Artificial daylight term is loosely applied to light sources, frequently equipped with filters that try to reproduce the colour and spectral distribution of daylight. A more specific definition of the light source is preferred.

Artificial fibres: Any class of fibres not found in nature. It consists of both regenerated and synthetic fibres. Also called man-made fibres.

Artificial flower finish: Stiff finish predominantly done on woven viscose and other fabric imitating the petals of flowers for producing artificial flowers.

Artificial fur: Woven imitation fur from woven pile fabrics (e.g. astrakhan, karakul, sealskin) high pile plush, mainly from synthetic fibres.

Artificial leather: Fabric coated with different chemicals to imitate leather like characteristics and/or leather like surface structure, e.g. embossing. Used for most of the places where leather is used. Main types: fibre, woven fabric, film, non-woven fabric and warp knitted fabric artificial leather. Artificial leather is mainly produced from polyvinyl chloride and polyurethane.

Artificial silk: Regenerated cellulose used to be known as artificial silk

Artificial turf: A manufactured carpet having the appearance of grass.Used to replace grass in sports arenas, yard, etc.

Artist's canvas: A fabric made of cotton, linen, jute or hemp prepared with size and primed with lead specifically for artists painting grounds.

Ærugo Verdigris. Similar to **æruca**, but derived from carbonate of copper.

Arylide yellow: A light-yellow mentioned in Louis de Berniere's *Captain Corellis Mandolin* (also known as Hansa yellowand cadmium yellow). Arylide is an aromatic compound. See also **Diarylide yellow** and **Azo**.

AS: (1) Asbestos fibre.

(2) Australian standard.

As: Asbestos fibre.

Asaltus: Fleece of the wild goat and sheep in Tibet, used for shawls.

Asan: Usually a small size East Indian prayer rug with hand-knotted pile.

Asbestos: A fibrous variety of magnesium silicate. Its particular value is due to a combination of three properties, i.e. the material is stable to temperatures up to 1000 °C and Above. It is resistant to acids and also forms fibrous crystals with good tensile strength. Asbestos occurs in nature in the form of fibre bundles.

Asbestos in pollution: Many types of asbestos have been used for insulating steam pipes, etc. The most dangerous type for lungs is amphibole, particularly crocidolite, which is lavender blue. Amosite, chrysotile and tremolite are also a hazard. Thorough wetting of asbestos lagging reduces the amount of dust generated when it is stripped. A slurry of asbestos should not be allowed to dry-out but should be disposed of immediately in appropriately sealed thick bags. The USEPA MCL in drinking water is 7 MFL (million fibres/litre).

Asbestos fibre: The only mineral fibre which occurs in nature (Europe, Russia, Africa, Canada and America). It is fibrous silicate of magnesium and calcium, though often contains iron and aluminium in its composition. The colour is white to greenish. The best fibres are several inches long, rather curly, white and lustrous. The fabric is produced mainly in plain or twill weave and is used largely for industrial purposes, including protective clothing, theatre curtains, ironing board covers, and fire blankets.

ASBL: Annales Scientifiques Textiles Belges (Belgian Textile Research Centre).

Ascending design (in carpet): Pictorial carpet motifs which are mainly discernible only when seen in one direction.

Aschodur: Persian cotton fabric, dyed black with sumac and iron; used for garments by native women.

Asclepias fibre: It is a very soft and lustrous fibre, yielded by the pod of the milk-weed; used chiefly for stuffing. See **Akund**.

ASDC: Associate of the Society of Dyers and Colourists.

Aseptic: Germ-free condition; Sterile. Opposite of septic.

Ash: Whitish-grey or brownish-grey.

Asharfi buti: A popular textile design consisting of small floral discs or circles, sometimes with small patterns within the circle.

Ash content: Ash residue remaining after the complete combustion of a specific material. E.g., The ash content of the wood pulp is 0.2–0.3% and of viscose is about 0.1%.

Ashes of roses: A pinkish-grey colour.

Asiatic gums: Vegetable gums with a greater or lesser content of arabin and bassorin. Includes ghatto, karaya and shiraz gums; they are swollen in boiling water but dissolves by pressure cooking to give a thick, viscous liquid with excellent flow properties to be used in textile printing.

Ashmara: Commercial term for weak jute fibre from India.

Ashmouni: Formerly a very important variety of Egyptian cotton, has a yellowish-brown staple that is over an inch long. Little cultivated now.

Asimani: Light Indian fabric made of silk and cotton, often made with zig-zag stripes across.

Aspect ratio: (1) The ratio of length to diameter of a fibre or yarn bundle.

(2) In tyre production, the ratio of the height of the tyre to its width.

(3) In a rectangular structure, the ratio of the longer side to the shorter.

Asparagus: The green colour of asparagus; also referred to as 'asparagus-green'.

Aspero: Peruvian term for the native cotton: nailed in England, full rough Peruvian.

Asphalt: Having the brownish-black colour of asphalt; dark grey.

Asphalt overlay fabrics: See **Geo textiles**.

Asphodel: The rich yellow of the daffodil.

Asphodel green: A yellowish-green.

Assais: Strong fibre, yielded by the Assais palm of Brazil; used for cordage and coarse fabrics.

Assembled button: A decorative button consisting of combinations of similar or dissimilar materials, such as plastic and metal, and metal which have been joined together by such processes as gluing, swedging or metal stamping.

Assembled yarn: Two or more yarns wound side by side on to same package, but without twisting around each other are called assembled yarn.

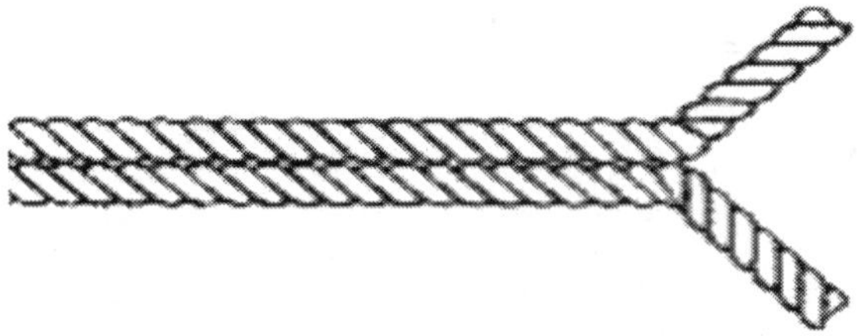

Assembly winding: The winding two turns of twist.

Assili: Egyptian raw cotton.

Assorcebunder: The lowest grade of Indian raw silk.

Association : (Lat.: *associare* = to ally with). Chemically: combination of molecules of the same type to form molecular associates such as micellar colloids. The stability of such associates varies greatly: decomposition takes place sooner or later on heating into the individual molecules. In the case of dyes, association allows a concentration of dye to form in close proximity to the fibre, e.g. an important factor for dye exhaustion on cellulose (therefore a high tendency to form associates = good substantivity). Dyes which form weak associates or none at all are acid dyes.

Association (degree of): Amount of association, which increases as the electrolyte (salt addition) content rises and falls with temperature for direct dyes. See **Association**.

Assouplissage: Softening of the degummed and bleached raw silk fibre in boiling water.

Astar: Muslin made in Asia, used in Turkey for turbans and underwear. It has three grades which are called dagbezd, thadirbezi and churumbezi; or best, medium and low.

Asteri: Calico used for lining in Persia.

Asticotine: Lightweight, filled, French woollen cloth; slightly elastic both warp and weft wise.

ASTM: American Society for Testing and Materials. The purpose of this organization is to develop standards on characteristics and performance of materials, products, systems and services. The standards developed by ASTM include test methods, specifications and definitions and usually deal with physical properties of materials. ASTM writes standard tests not only for textiles but also for virtually every other product such as steel, plastics, lumber, etc. For textiles, ASTM writes primarily physical-type tests such as methods for testing the tensile strength, abrasion resistance, twist determination, fibre maturity, denier and yarn count, among many others.

Astoli: Cotton canvas, made in Dublin according to a newly patented process. It is waterproof and good heat insulator, and does not contain rubber; it comes 56 inches wide and is used for carriage and automobile tops.

Astra Work: Consists of stars which previously have been embroidered over linen, cut out and appliquéd.

Astrakhan: (1) The skin of a stillborn or very young lamb (Originally from Astrakhan, Russia), the fine grey or black curly hair of which resemble fur.

(2) Real Astrakhan is a curly lamb's fleece (Persian). Its substitute is lambskin imitation like woven plush fabric (frieze velvet) with an uncut mohair pile warp. Also produced in jute and acetate filament yarn.

Astrakhan cloth: A thick knit or woven fabric with loops or curls on the surface. The base yarns are usually cotton or wool and the loops are made with fibres such as mohair, wool and certain synthetic fibre. The face simulated the pelt of an astrakhan lamb. Used for coats, jackets and collars. Good quality cloth contains some mohair to add lustre and curl.

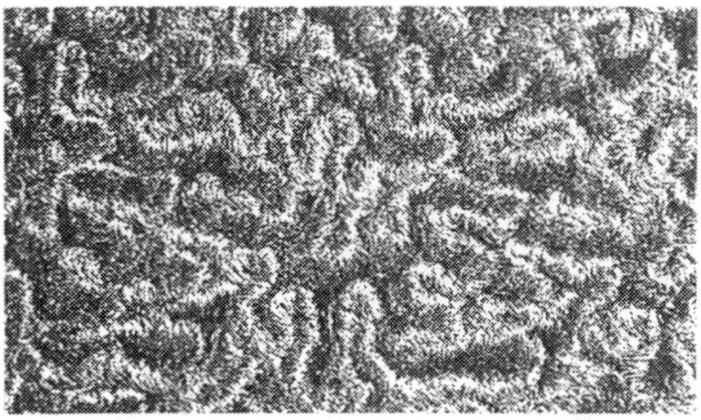

Astrakhan fur: Luxurious and expensive fur taken from young lambs in the Astrakhan area of central Asia. Used for coats, hats and trimming for men and women. See **Astrakhan**.

Astrakhan yarn: Wool crimp yarn. Forced crimping can be fixed by cross-linking to make wash fast.

Astrakin: Compound woven and warp knitted fabric; base fabric of raw crêpe georgette with a different top fabric glued on. Creping produces a cloqué-like surface.

Astro-dyed: Licensed process for dyeing non-repeat effects (space dyeing) on yarns by injection of dye liquor into the thinner layers of wound packages upto a maximum of 5 colours with special equipments. Mainly used for carpet and upholstery.

AT: Technical atmosphere, replaced in the SI system by Pa (Pascal). 1 at = 0.981 · 105 Pa = 0.981bar.

Atabi or otabi: Medieval dress goods of Arabic origin, made from the mixture of silk and cotton in various colours.

Atactic polymer: A type of polymer molecule in which substituent groups or atoms are arranged randomly above and below the backbone chain of atoms, when the latter are all in one plane.

Atansaw: A wide, commodious *chogha* like garment for wrapping around the body.

ATE process: Anthrasol-thermosol development process. A dry development process for leuco vat ester dyestuffs with the use of an activator and a levelling agent for woven cellulose and polyester cellulosic fabrics.

ATF: Arbeitsgemeinschaft Textiler Fachvereinigungen (Working group of German Textile Organisations).

ATI: Associate of Textile Institute (Professional qualification of the textile institute), Manchester, United Kingdom.

Atlas: (1) A term used in France, England, Germany and Austria for a highly finished fabric made with an eight-harness satin weave. Used for dresses and cotton-back lining. Originally made in India in plain weave, in stripe or floral in colours or gold.

(2) A light-weight satin lining made with silk face and cotton back. Used for dress fabrics, mainly evening wear. A less expensive Atlas made with a cotton weft is used as a lining fabric.

Atlas lapping fabrics (warp knitted): The loops at the atlas lapping fabric incline towards the same direction for a few courses and then incline towards the opposite direction for another few courses. Thus, horizontal stripes or coloured zig-zag pattern effects will occur on the technical face of the fabric.

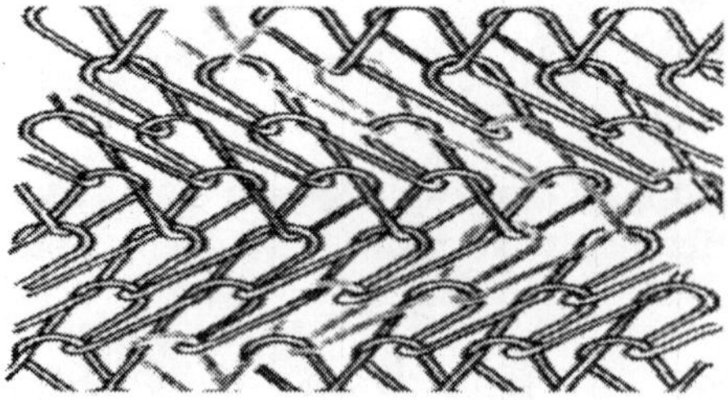

Atlas

Atlasgradle: A linen fabric in Germany and Austria, made with five-leaf satin stripes in linen and cotton back. Used for bed linen.

Atm: Abbreviation for standard atmosphere. Earlier it was standard unit of pressure, but now in SI system it is represented by Pa (Pascal) 1 atm = 1.013300000Pa = 1.0133 bar.

ATMA: Abbreviation for American Textile Machinery Association.

ATME: Abbreviation for American Textile Machinery Exhibition.

Atmosphere for preconditioning textiles, standard: The International Standards Organization has defined the following atmosphere for this purpose and there are widely accepted for textile testing and specifications.

Temperature ^{0}C			Relative Humidity %			Atmospheric Pressure
	Tolerance			Tolerance		
	Ordinary	Close		Ordinary	Close	
20	+/- 2	+/- 1	65	+/- 5	+/- 2	Between 860 and 1060
27	+/- 2	+/- 1	65	+/- 5	+/- 2	Between 860 and 1060
23	+/- 2	+/- 1	50	+/- 5	+/- 2	Between 860 and 1060

Atmosphere for testing: Air at ambient conditions of relative humidityand temperature in tests or experiments is conducted.

Atmosphere for testing textiles, standard: The atmosphere 20°C +/– 2°C and 65% +/– 2% r.h. at prevailing barometric pressure is widely used in various standards all over the world.

Atmospheric conditions: Generally refers to the relative humidity, barometric pressure and temperature existing at a given time.

Atmospheric fading: See **Gas fading**.

Atomiser: A high performance device which uses water jets to form a mist of water droplets for various purposes.

ATP: See **Adenosine triphosphate**.

ATPUL: l'Association Technique pour la Production et l'Utilisation du Lin et autres Fibres Liberiennes (French textile research organization for linen and other bast fibres).

Atrament, atramentum: Black ink. Any black colorant. A very dark-brown pigment described by Pliny and produced from calcined bones. In the ancient world, *atramentum librarium* was used as writing ink, *atramentum sutorium* for dyeing shoe leather and *atramentum pictorium* was used by artists as a varnish.

Atred: Black.

Atrorubent: Reddish-black.

Attached cushion (in pile yarn floor coverings): A material bonded to the backing fabric side of a pile yarn floor covering to provide additional dimensional stability, thickness and padding.

Attached growth process: Biological treatment process in which the microorganisms are responsible for the conversion of organic matter or other constituents in the waste-water into gases, and the cell tissues are attached to some inert medium, such as rocks, slags, or specially designed ceramic or plastic materials. Attached growth treatment process is also called fixed film process.

Attached upholstery fabric: The exterior fabric covering secured to a furniture unit by the furniture manufacturer or custom upholsterer.

Attalea: Wash fabric used in England for trimming sailors' suits.

Attalic: (1) Thread made with flat gold strip wound around a wool or linen core.

(2) Woven goods made of gold thread which were mixed with other fibres from Asian countries during the middle ages.

Attribute: A specific characteristic of a thing.

Attribute (in colour chemistry): Distinguishing characteristic of a sensation, perception or mode of appearance. Colours are often described by their attributes of hue, chroma (or saturation) and lightness.

Attribute data: Observed values or determinations which indicate the presence or absence of specific characteristic.

Attrition mills: Machines for reducing materials into smaller particles by grinding down by friction. In the manufacture of acetate and tri-acetate fibres, this equipment is used in shredding pulp before acetylation.

Atrous: Jet-black in colour.

Attushi: A fabric made from the bast fibre of the elm tree. It is a hard and rough fabric made by the Ainu tribe of Japan in ancient days.

Au fuseau: Name for reseau ground in laces when made on the pillow.

Au gratin: Brown – as regards the surface of food. Describing, in particular, a dish which has been cooked with cheese until it has become brown. Also applied to food cooked with breadcrumbs or in a white sauce cooked until brown. From the French, *gratiner*, to brown.

Au passe: Also called satin stitch or long stitch; used in all kinds of embroidery over any ground, the thread being worked across the material without any raised foundation.

Au portique: French flammability standard.

Aubergine: The purplish-blue of the vegetable, aubergine.

Auburn: A brownish-red or sometimes golden-brown colour, especially as regards to hair. Auburn derives from the Latin *albus*, meaning 'white' and originally indicated a yellowish or brownish-white colour. Its meaning changed during the 16th century when (perhaps because it was sometimes spelt 'abrun') it came to be associated with the colour brown.

Aucube: Wool rug made in Belgium.

Augsburg: Fine checked ginghams used to be made in Augsburg, Germany.

Aupoz: Native name in the Philippine Islands for the intermediate fibrous layer in the Musa textile; used for weaving sheer fabrics.

Auquili: Coarse bagging made in Syria; the better grades dyed blue or are partly white and always mixed with cotton.

Aur-, auri-, auro-, aurat- (L): Gold, golden.

Aures: Sort of caddis or stout twilled shorn woollen, known in France as fleuret.

Aurillac lace: A French bobbin lace, made with close patterns.

Auriphrygium: Ancient name for fabrics, usually silks, richly embroidered with gold.

Auripigment: A vivid yellow, also called *King's yellow* and *orpiment.*

Auroclavum: Gold tissue of the Middle Ages; used for robes worn by state dignitaries.

Australian crepe: A crinkled fabric, made with cotton warp and worsted filling.

Austria: A twilled silk umbrella cloth.

Autoclave: (1) A vessel in which textile material may be treated with steam under pressure. Facilities are often provided for creation of vacuum either before introduction of steam or subsequently and controlling temperature, pressure, etc., for precisely getting the reaction required to be carried out. Used in the manufacture of many fibres, dyes, and auxiliaries. E.g. vessel for polymerizing condensation polymers such as nylon or polyester above or below atmospheric level.

(2) An apparatus for carrying out certain finishing operation, such as pleating and heatsetting, under pressure in a superheated steam atmosphere.

Autoconer: Automatic cross-winding machine for rewinding yarn from spinning cops on to cross-wound packages. Electronic yarn clearers as well as automatic splicing of yarn breaks are state-of-the-art features. Both the spinning machine and the autoconer can be equipped with automated transport systems. Manuf.: Schlafhorst.

Autoleveller: An automatic device which is fitted to machine to improve the evenness of the output material. This is achieved by monitoring the mass per unit length and if necessary, changing the draft on the machine to compensate for any deviation from a present value.

Automated dye weighing and dispensing system: An example of such a system involves the use of coded plastic cards on the dye-powder containers to identify the dye to be weighed, weighing done automatically and transferred to the dissolving station or an integrated recipe print-out facility and transferring to a container with recipe and barcode to identify the colour. Such systems are designed to eliminate weighing errors, automatic inventory control with facility to get the stock any point of time.

Automatic batcher: A batching system connected to the stenter or any machine for that matter, batch changes are done without stopping the machine.

Automatic continuous laundering dry-cleaning: A system consists of an automatic washing machine plant, mechanical water extracting unit and an opening or drying unit, fully automatically controlled and technologically interlinked for non-operator assembly line laundering operations from loading to finishing.

Automatic feeder: A machine that feeds a steady supply of raw and dirty cotton to the carding machine.

Automatic jigger: Jiggers with automatic controls which automatically change direction of running, speed and regulate fabric tension. Fully automatic jiggers can now be programmed for the full dyeing cycle of the dyeing to be carried out including, filling, draining, chemical and dye additions, temperature control, etc.

Automatic lock slider (in zipper): A slider that provides involuntary, positive locking action on the chain when the pull is released.

Automatic pattern repeat system: Machine designed to give repeated patterns exactly the same as compared to the manual methods. In order to set the desired repeat length, a movable limit stop is positioned against a graduated scale and fixed. Fixing and repeating of the limit stop is carried out automatically.

Automatic reeling machine: It is the most sophisticated device for reeling silk with automatic cocoon feeding device, for effective labour saving. It has a denier control device to maintain constant raw-silk denier with the result, the evenness of silk reeled will be better.

Automatic weaving machines: It is a power driven loom on which the shuttles or pirns are changed automatically.

Automation (in processing): Automation in processing is to control and regulate equipment and processing, e.g. flow of goods, preparation of baths, dosing, temperature and treatment times, by automatic programming. Automation and control can be used in dyeing and printing. For example in the following processes:

- Temperature-time programmes
- Running partial or fully automatic bleaching
- Dyeing processes, including dye kitchens
- pH,
- Flow pressure/differential pressure
- Dosing of liquid chemicals.

Automotive textiles specifications: Textiles are used very extensively in cars; examples of their applications include tyres, hoses, tapes, nets, twines, ropes, V-belts, noise insulation, etc., and extending to upholstery materials and carpeting. Textile interior trim in cars is not confined to needled floorcoverings and pile non-woven covers which are chemically bonded. Laminated woven and knitted fabrics are used as seat upholstery and inside cladding. Flocking also plays a major role in car headliner products.

Automobile tyre cloth: Very strong, plain woven, stout and heavy cotton fabric, made with heavy yarn of the longest staple. Used for automobile tyres.

Autumnal colours: The colour characteristic of autumn, particularly in reference to the leaves of the trees of the English countryside with their rich variety of russets, oranges, browns, yellows and reds. In North America, the amazing colour change is referred to as 'fall foliage'. The process occurs by reason of leaves in the fall being sealed off from moisture giving rise to the **chlorophyll** in them breaking down. Their green colour thus gradually becomes masked by yellow, orange and brown pigments known as **carotenoids** also present in the leaves. There are also red and purple pigments called **anthocyanins**. The red, purples and bronzes in some trees derive from the sugar produced in the leaves being trapped by the colder temperatures. See also **Erythrophyll, Phylloxanthin** and **Xanthophyll**.

Auvergne lace: French bobbin lace of no distinct character.

Auxerre: White linen bobbin lace, made in France.

Auxochrome: Auxochromes are 'salt-forming' groups of atoms whose role, rather more loosely defined, was to provide an essential 'enhancement' of the colour.A substituent group in a dye molecule having a basic or weakly acidic character which acts as a "colour intensifier" and also confers affinity for textile fibres. For e.g. –COOH, SO_3H, -OH, $-NO_3$, $-OCH_3$, $-CH_3$ and NH_2 groups. See **Chromophore**.

Auxonne: French hemp canvas, made in various widths, stripes or checked; bleached or grey. Not very popular now.

Auxy: Fine, soft knitting wool yarn made in France and used in the finest grades of women's hosiery.

Available chlorine: Chlorine available for active bleaching in chlorine compounds. In chloride of lime (bleaching powder), for example, it is the quantity of chlorine released by the action of hydrochloric acid: $CaOCl_2 + 2HCl \rightarrow CaCl_2 + H_2O + Cl_2$. Chloride of lime has an active chlorine content of 35–40%.

Available chlorine in bleaching baths (determination of): 10 ml of the bleach liquor to be tested (diluted to approx. 50 ml with water) is titrated with 0,1n arsenic acid solution until a drop of the liquid no longer turns potassium iodide paper blue. 1 ml of 0.1N arsenic acid is equivalent to 0.003546 g of available chlorine.

Avantagee: French name for the best grade of Nankin silk.

Avasca: A coarse, durable fabric, made by the natives of Peru from alpaca; used for garments.

Avellaneous: The light-brown colour of the hazel-nut or filbert-nut.

Ave maria: Narrow French peasant bobbin lace with plaited ground and very simple patterns. The edges are made with cloth stitch. Similar to the ***Valencienne***.

Average (for a series of observations): The total divided by the number of observations.

Average degree of polymerization: Average number of monomers contained in a polymer. The average degree of polymerization is determined by osmotic methods or viscosimetry.

Average fibre diameter (in wool and other animal fibres): The average width of a group of fibres when measured on projected image.

Average molecular weight: A term used in polymer chemistry for the mean value of the entire molecular weight range.

Average stiffness: The ratio of change in stress to change in strain between two points on a stress strain diagram, particularly the points of zero stress and breaking stress.

Average toughness: See **Toughness**.

Avignon: Light French silk taffeta lining.

Avila: Spanish wool from the Province Avila.

Avocado: The green of the avocado – usually in reference to the pulp rather than the skin; also avocado-green. Perhaps from the Spanish *abocado* – a delicacy – or the Aztec, *ahuacatl*.

Avouet: Persian wool from three-year old sheep; used for rugs.

Awl: A sewing tool used for making holes in the fabric.

Awl drilling: Stacks of cut-sections of garments may be drilled or notched to guide the subsequent sewing operations. There are two drilling methods – awl needle drilling and hypodermic drilling. Awl drilling (needle drilling) uses a solid tool that rotates as it is pressed into the stack of fabric. The awl may have elements that cut a hole in the fabric stack or may simply penetrate the stack spreading or severing the yarn.

Awning cloth (tarpaulin): Coated heavy-sail cloth.

Awning stripes: (1) A design of wide even stripes;(2) A heavy canvas fabric with this design; may be yarn dyed or printed.

Axial dyeing process: Package dyeing on dyeing machines with axial liquor flow.

Axial yarn: A system of longitudinal yarns in a tri-axial braid that are inserted between bias yarns.

Axminster carpet: Machine woven carpets with cut pile in which successive weft wise rows of pile are inserted during the weaving in a prearranged colour sequence.

Aya momen: General trade term for twilled cotton fabrics in Japan, especially drill.

Ayrishke: Japanese silk brocade with flower patterns.

Ayrshire: Fine twilled woollen blanket made in England.

Azalea pink: An orange-red colour.

Azamgar: Indian fabric, made of cotton and silk in satin weave.

Azara or azera: Fine Indian muslin.

Azazul: Sheer Indian muslin, with warp stripes of wild silk.

Azlon: A generic term for regenerated protein fibres from peanut, soya bean, corn, milk, etc. They are soft and smooth and blend well with other fibres. Used for knitted garments, sportswear and coats.

Azlon fibre: A manufactured fibre in which the fibre forming substance consists of any regenerated naturally occurring proteins.

Azo-: Referring to a chemical compound which contains two nitrogen atoms with a double bond between them (–N=N–). "Azo" is used for a class of dyes based on (this) chemical structure. Azo dyes may be found among direct, acid, basic, reactive, and disperse dye classes. Dyes with one pair of nitrogen atoms azo bonded are called mono-azo. Those with two or three azo bonded pairs are called disazo (not diazo) and trisazo, respectively.

Azo dye coupling: The combination of a coupling component with a diazo compound through an azo linkage.

Azo dyes: These dyes contain the characteristic azo group –N=N– which is present to a greater or lesser degree in nearly all classes of dyes. Over half of all commercial dyestuffs belong in this general category. By varying the chemical composition, it is possible to produce acid, basic, direct, reactive, disperse and mordant dyes containing the azo chromophore. This general group of dyes is subdivided into monoazo, disazo and polyazo dyes, depending on the number of azo groups in the dye molecule. See also **Azo-**.

Azoic bases: Coupling products for development into naphthol dyes. Azoic bases are not dyes themselves but soluble intermediate product with basic character due to the presence amino groups (NH_2) which are diazotized before use when they become capable of coupling in a similar manner to azoic salts with a naphthol component to form insoluble azo dyes.

Azoic composition: A physical mixture of an azoic coupling component and a stabilized azoic diazo component which produce, in situ, an insoluble azo colourant in both cellulosic and synthetic substrates.

Azoic diazo compound: A stabilized diazonium salt of a primary acrylamine or derivative thereof which is capable of reacting with an azoic coupling component.

Azoic dye: A term generally applied to a class of dyes based on application method, sometimes called naphthol dyes. Azoic dyes are actually synthesized inside the fibre during the process of dyeing by reacting an azoic diazo component with an azoic coupling component and are not truly dyes, but insoluble pigments.

Azoic salts: Already diazotized water-soluble coupling components of the corresponding Azoic bases used for developing naphthols.

Azores: Loosely woven, thick, long haired woollens.

B

BA cotton: A flame-retardant cotton produced by fibre modification by BAP.

Babci: A type of white sisal hemp fibre from Europe.

Baby-blue, Baby blue: A pale blue shade. 'Baby blues' is a slang term for 'eyes'.

Baby flannel: A light-weight flannel fabric used for children's garments.

Baby Irish: Fine, narrow and flat Irish crochet lace.

Baby lace: The narrow lace edging of the Valenciennes type. See **Valenciennes.**

Baby pink: A pale pink.

Baby ribbon: The narrowest ribbons, usually silk in pale blue or pink colour.

Baccaba: Very strong leaf fibre, yielded by the Turu palm in Brazil; used for ropes.

Back and free stitch: A hand stitch employed for sewing lining and pockets. It involves taking a back stitch and the running stitch before the needle is removed.

Back beam: It is the horizontal crosspiece on the back of the loom around which the warp passes on its way to the warp beam.

Back bearer: Same as **Back rest**.

Back beating: An operation which involves beating or vibrating the back side of piece-dyed carpets as they leave a drying machine. Back beating improves the appearance of acrylic pile carpets, in particular.

Back coating of carpets (carpet back coating): In the case of tufted carpets, back coating is an essential element of their construction. As a result of this process, the tufted pile is firmly bonded to the base fabric on the one hand, whilst noise reduction and thermal insulation properties are achieved on the other.

Back coating (in pile yarn floor covering): A process done on tufted carpet whereby the tufted piles are firmly bonded to the base fabric. Back coating also helps in noise reduction.

Back filling: Set of weft threads, forming the back of many modern worsted cloths. It is used to provide weight to the fabric.

Back finishing: Application of finish to the backside of the fabric for material-specific reasons. Applied on ribbons, tapes, velveteen cord and plush fabrics.

Back grey: (1) The term covers a variety of special cloths used in textile finishing operation, such as a leader cloth, end cloth, back greys and intermediate greys to facilitate the processing of fabric in jig dyeing machines, decatizing machines, steamers, roller printing machine, and in screen printing for combining with fabrics (temporary lamination) which would otherwise be difficult to print satisfactory.

(2) A fabric used in printing machine in between the printed cloth and blanket to avoid colour paste staining the blanket.

Back grey (printing): In roller printing, a gummed woven fabric which runs constantly tensioned over the printing blanket and lapping covered drum and is used to support the printing blanket and the fabric to be printed.

Back grey washing: Depending on the number of passages, coverage of print design and thickness of the textile fabric being printed, used back greys contain greater or lesser amounts of print paste residues containing thickeners, dyes and chemicals. The purpose of back grey washing, therefore, is to remove these print paste residues and make the material soft, absorbent and fit for reuse again.

Back length: The dimension on a garment taken from the centre collar attaching seam to the bottom of the garment, or in the case of a coverall, to the top of the waistband.

Back notches (in sleeve): Two notches spaced ½ inches apart on the back of sleeve saline to mark points.

Back rail: See **Back rest**.

Back rest: Also called back rail, whip roll. The bar, rail or roller at the back of a loom over which the warp threads from the beam pass. See **Cloth fell**.

Back rise: The distance on a garment from the crotch to the centre back of the waistline.

Back side of fabric: The reverse side of the fabric which can also be used as the wearing side. The side of the material that faces inward in the completed object.

Back stitch: (1) A hand stitch used to seam garment parts. So called because the needle on emerging goes back to be inserted at the end of the previous stitch.

(2) Embroidery stitch.

Method – Bring the thread through on the stitch line, then take a small backward stitch through the fabric. Bring the needle through again a little in front of the first stitch, take another stitch while inserting the needle at the point where it first came through.

(3) Embroidery stitch.

Method – Bring the thread out at the right-hand side. Take a backward stitch over three threads while bringing the needle out in front of the place where the thread first emerged. Continue in this way, working form right to left in the required direction.

Back strap: A strap across the back of a garment usually fitted with a buckle or other method of adjustable fastening and used to adjust the fit of the garment, within limits, to the wearers requirements.

Back tacking: Sewing and reverse sewing at the beginning or end of a seam to secure the stitching.

Back tanning: An after treatment to improve the wet fastness of dye or printed silk or nylon by using natural tanning agents.

Back-to-back: Finishing operation of a textile fabric on a table raising/shearing/polishing machine on which both the sides of a fabric can be processed in one without turning or reversing arrangement.

Back waist length (in body measurements): It is measured from the most prominent neckbone down to the cord that marks ones natural waist.

Back wash: (1) Commercial term for wool washed on the back of a living sheep with cold water.

(2) In water softening or filtration treatment, back wash is a process by which a reverse flow of water through the unit is produced to remove the filtered sludge, etc., to make the unit ready for the next softening/filtration process.

Back warp: The warp, which along with the backfilling actually forms the second face (back) of the double, triple or quadruple fabric.

Back width (in body measurements): The distance from back-break point to back-break point.

Back wind (pirn winding): The final stage of pirn winding in which the full pirn continues to and revolves whilst the yarn traverse-guide eye carries the yarn from rose to the base of the pirn, whereupon the rotation of pirn is stopped. Consequently, backwind prevents lovely or slippery yarns from falling off the nose chase these reducing transports and handling problems and hence waste.

Back-break point: Thc location on the back of the body where the arm separates from the body.

Back coating (in textiles): An adhesive type substance applied on the back of a fabric for such purposes as locking pile yarn, etc.

Back coating (in textile floor covering): The operation of depositing bonding agent on the back of a textile floor covering to ensure its penetration into the substrate of the textile floor covering without reaching the use surface.

Back frame: The side of a fly frame on which the bobbins, from which roving is drawn into the machine, are held.

Backed cloth: A material with an extra warp or weft added for weight and warmth. Satin weave and twill-weave constructions are frequently used in the design of backed cloth because they are relatively resistant to the passage of air.

Backed cloth worsted: Backed-cloth worsteds are woven with an extra layer of warp or other filling underneath the face, usually for increased weight and bulk.

Backfilling: A solution (filler) containing various quantities of starch, china clay, talc and tallow that is applied to the backside of low cost cloth to change its hand, improve its appearance and increase in weight.

Backgrey accumulator: A system developed for large quantities of back greys in roller printing. Mainly constructed in the form of a channel accumulator on the J-box principle with capacities of 800–5000 m.

Backgrey washer: A unit for the continuous washing of back greys normally located behind or beneath a roller printing machine.

Back-grey washing machines: Normally, for washing back greys winches or heavy rope washing machines were used for back-grey washing. Open width washing machines are not that suitable due to the less agitation in these machines.

Backing: (1) A knit or woven fabric or plastic foam bonded to a face fabric.

(2) A general term for any system of yarn which interlaces on the back of a textile material.

(3) A knot or woven fabric bonded to a vinyl or other plastic sheet material.

(4) See **Carpet backing**.

(5) In woven pile floor coverings, the fabric formed by the warp and weft which supports and retains the pile yarns.

(6) A strip of material placed underside of a part of garment to act as a reinforcement.

(7) Strengthening applique or other embroidery, if the materials applied are not of the same strength as the foundation.

Backing, primary (in tufted floor covering): A coarse textile fabric into which the pile yarn is stitched and to which a back coating may be applied, or onto which a secondary backing may be laminated.

Backing fabric: (1) A support textile fabric located behind the textile artefact.

(2) A fabric to which a pile yarn is inserted, or a reinforcing layer which is adhered to the reverse side of the fabric.

Backing thickness: The distance between the upper and lower surfaces of the backing components of a finished floor covering, measured after the pile has been sheared down to the upper surface.

Backing yarn: See **Yarn**, **Backing**.

Backing (pile yarn floor covering): All materials in a pile floor-covering, other than the pile yarn.

Backing, primary, in tufted floor covering: A coarse textile fabric into which the pile yarn is stitched and to which a back coating may be applied, or onto which a secondary backing may be laminated.

Backing (secondary): (1) In tufted floor coverings a coarse textile fabric laminated to the back of a primary backing to reinforce the latter; (2) In pile-yarn floor covering – a material adhered to the backing fabric side of a pile-yarn floor covering.

Backside (in textile materials): The side of the material that faces inwards in the complimented object.

Back-sizing: See **Filler**.

Backstitch: See **Purl**. The back stitch is made by placing the needle back to the last stitch, bringing it out at the length of the last stitch, then placing the needle back into the last stitch, and so on, making the stitches follow each other without any space between. This is used in all places that are to bear great strain.

Backtacking (in sewing): Refers to the reverse feed sewing at the beginning and ending of the seam to prevent the thread from unravelling.

Backwaist length: The vertical distance along the spine from the cervical to the waist.

Backwarp: (1) The warp which, along with back filling, actually forms the second face (back) of double, triple, quadruple fabrics.

(2) Additional warp (or weft) on the back of the fabric bound or stitched to the ground structure, so that it does not interfere with the appearance of the face giving greater weight, thickness, warmth, etc.

Backwash: Reversal of a solutions flow through a filtration system. Often used as a cleansing mechanism in sand and dual media filters.

Backweft: See **Backwarp**.

Backwidth: The distance from back-break point to back-break point.

Backwinding: (1) Rewinding yarn or fibre from one type of package to other.

(2) Winding yarn as it is deknit.

Bacteria: They are the smallest unicellular microorganisms, and are only visible under the microscope with powerful magnification. They multiply rapidly by simple fission (at an average rate of one division per hour, 16 million bacteria can develop from one bacterium in a day). Types of bacteria include: (1) Spherical bacteria or coccus. These are spherical to oval in shape and 1 to2 μm in size;(2) Rod-like bacteria, including spore-forming Bacilli, with a size of 0.5 to 2, 5 : 1, 5 to 15 μm;(3) Helical bacteria: a) vibrios, curved types. Size 0.3 to 0.5 : 2 to 4 μm; b) spiral, with corkscrew-like coils. Size 1 to 3:5 to 30 μm.

Bacterial desizing: See **Bacterial diastases**.

Bacterial Amylase Unit (BAU): A measure of starch degradation as shown by the quality of an enzyme that will dextrinize one milligram of starch per minute under specified experimental conditions.

Bacterial damage of textiles: See **Microbial damage to textiles**.

Bacterial diastases: The effective enzymes (amylases and polyases) are obtained by the cultivation of subtilis and mesentericus (diastases). Bacterial diastases breakdown starch into glucose with the assistance of sodium chloride as activator. Hence, these are used in desizing of greige cloth.

Bacterial growth inhibition: See **Bactericidal finishes**; **Phenol coefficient**.

Bacterial inhibition: Bacteriostatic action.

Bacterial resistance: Resistance to the development of visible bacterial growth and accompanying odours, resulting from bacterial degradation of fibres or soil on them, as distinguished from musty fungal odours.

Bacterial staining: See **Gram's staining**.

Bactericidal fibre: Fibre used for medical applications, socks, shoe liners, etc., in which bactericides are introduced directly into the fibre matrixes as opposed to the fibre simply having a bacterial finish applied.

Bactericide: A product capable of destroying bacteria, e.g. disinfectants or preservatives.

Bactericidal finishes: These finishes are applied on fabrics to provide the fabric with active protection against the destruction of bacteria and Mycotoxins (antimicrobial finishes). Copper naphthenate, copper-8-quinolinolate, chlorinated phenoltri-alkyl tin derivatives, etc., are normally used for this purpose.

Bacteriostat: A product with bacteriostatic properties. Doesn't necessarily mean that it kills bacteria. A stat means that it may simply be slowing growth or holding the death to growth rates of bacteria (same for fungal stats) more or less in equilibrium. It inhibits bacterial growth.

Bacteriostatic: Adjective used to describe the product or properties which prevent or inhibit the growth of bacteria.

Bacteriostatic finishes: Finishes that inhibits the growth of bacteria on textiles; also called antimicrobial finishes. See also **Antifungal/Fungicidal finishes**.

Batch card (lot card): An identification card with all the details of order number, quantity, required process, recipes, width, requirements, etc., to identify and help the processor through-out the process of the goods in an order/lot etc.

Bad cast (in silk): An abruptly increased diameter in a raw silk thread. Usually caused by the improper joining of a single cocoon filament to the thread during the reeling operation.

Bad listing: See **Cut selvedge**.

Bad temple: See **Temple mark**.

Badan khas: Fine cotton muslin once produced in India.

Baden: Stout and plain woven linen fabric made in Baden, Germany.

Baden embroidery: A contrast appliqué work done in a special way. First the patterns are painted on a fine fabric, which is usually leaves and flowers, and cut out and sewn on to a contrast colour dyed foundation fabric as an appliqué work by the edges of the leaves, and petals worked around with chain stitch and the veins with herringbone or feather stitch.

Baden hemp: Superior grade of hemp, stripped by hand.

Baden lace: Good quality of bobbin lace made in Baden, Germany.

Badger brown: The grey-brown colour of the badger.

Badla: Flat metallic wire, often silver gilt, used in brocading and embroidery.

Badlan: Indian embroidery, cxccuted with flat gold or silver wire.

Baeta: A plain woven, loose, woollen fabric that is napped on one side; used in Portuguese and Spaniards.

Baffeta: (1) Indian plain and closely woven cotton fabric; formerly exported to and printed in Europe. Now it is largely displaced by calicoes, etc.; (2) Plain woven, bleached, or blue cotton fabric in the African and East Indian trade, in old times.

Baffle: A plate or deflector to provide flow distribution in a filter-housing. Primary functions are to provide uniform flow and to prevent erosion of pre-coat or bed and setting of body-feed.

Baft Ribbon: Consists of threads being glued together to form a flat tape without any weft.

Bafta or baftah: (1) Native name in East Africa for white cotton shirting or bleached cotton long cloth, made with a heavy size, imported from Europe and used for shirts, bed covers, and other home uses, etc.; (2) It is also the name given to a variety of dyed-cotton fabrics imported to African markets from various countries.

Bafta: An Indian term for fabric made with silk warp and cotton weft, used as dress material.

Baftowa: A very fine Indian cotton muslin.

Bag cloth: A woven cloth of inferior quality yarns, heavily sized to prevent any leakage/spillage of the material carried in the bags made using these fabrics. Used, as the name implies, for bags and sacks to hold dry goods such as flour, lentils, rice, wheat, etc. The size removed, the soft fabric can be used for dusters and as cleaning and moping cloth.

Bag duck: Duck fabric that is usually used for mail bags and other such uses is sometimes called bag ducks.

Bag sheeting: Originally, bag sheeting ranged from rather coarse fabrics upto fabrics having the texture of regular sheeting. Today, very fine fabrics such as print clothes are used as bag fabrics. For regular bag sheeting, coarse cotton yarns say from 12s to 18s are used. Construction ranges from 40 square to 60 square. Normal widths are 30 and 36 in or more and of 5–8 m per kg. The material is given heavy finish if required. The cloth is used to make bag containers for feed, foodstuffs, fruits, grain, salt, sugar, etc. Bleached bag sheeting is used as a substitute for towelling or for the household uses.

Bag strapping: Very broad stay tape; used by upholsterers to preserve selvedge.

Bagasse: Waste obtained from the sisal scraping, the pulp from the leaf fibres. Sugarcane waste after extracting sugarcane juice is also called Bagasse. Used as fuel for boilers.

Bagginess: A fabric defect especially of viscose and wool caused by excessive tensioning winding, distorted weft threads, etc., which gives rise to tight threads in weft direction and waviness in the fabric. This effect is further intensified by the influence of moisture.

Bagging: (1) A fabric woven in cylindrical or tubular form on an ordinary cam loom and is used for grain bags, bale covering, etc.

(2) (in garments) A term used to describe the undesirable bulging of fabric caused by extension at elbows, knees, etc., of a garment lacking in dimensional stability.

Baggy cloth: A fabric that does not lie flat on a cutting table, caused by section of tight or loose yarn in either warp or weft.

Baggy selvedge: See **Slack selvedge**.

Bagh: The word 'Bagh' literally means Garden. It is traditional embroidery work done by women of Punjab and Haryana in India to be gifted to their daughters and daughters-in-law at ceremonies connected with the birth of grand-children. The embroidery of a bagh often begins soon after the birth of a child.

Baghaitloni: Navajo blankets with a slit in the centre, made in a large variety of patterns.

Bagheera velvet: A piece-dyed velvet with a rough surface. It is used for outer wear as the rough surface makes it largely uncrushable.

Baghal Bandi: A kind of tunic or jacket, worn shorts and fastened under the armpits.

Bagir's music style: Global tailored suit brand Bagir is a revolutionary company that produces machine washable suits, waterproof suits with nano-technology, super-light suits, and more. They were the first to introduce the concept of music player functionality with ElekTex technology (Comprehensive Merchandising Support 2007). It is the first time that smart clothing technology has been applied to a suit. The development of music player clothes will expand to suits or fashion clothing from pre-existing sports or casual clothes.

Bahama: (1) A cotton from the West Indies; (2) A commercial variety of Texas cotton, the staple measuring 23–26 mm, growing in large balls, yielding one-third of lint.

Bahia: A variety of cotton having a fairly strong but harsh and wiry staple, grown in Brazil.

Bahmia: Raw cotton having a fine staple; formerly grown in Egypt, but now little cultivated.

Bail or lug (in zippers): A portion or portions of the slider to which the pull or pulls are attached.

Bailey: Variety of upland cotton, grown in the United States; the staple is considered of very good quality, measuring from 28 to 32 mm, the lint being less than 30 percent.

Baindix: Name of a Turkish cotton.

Bainin: Hand-woven woollen home-spun fabric from Ireland. It is used for coats, skirts, stoles, etc.

Baique: A coarse baize, made in Belgium.

Bairaiti: Variety of fine raw cotton from Indian sub-continent; used for Dacca muslins.

Baird: A highland tartan that is composed of blue and black stripes, narrower green stripes and bright-red lines.

Baize: (1) Originally made in Spain and used as a wall covering, this is a loosely woven, short napped fabric, made to imitate felt. It is usually dark green, plain weave and is rarely found on counter sales.

(2) A coarse woollen cloth of plain-weave construction made in United Kingdom for centuries. It is now a thick cloth with a napped surface and is rarely available on counter sales.

Baking soda: Sodium bicarbonate.

Bakhtiari carpets: Strong rustic knotted carpets from Chahar Mahal region to the south of Isfahan. They are easily identifiable by their design involving squares and diamonds made to stand out by a plane outline in the ground. These geometrical designs are decorated with animal or plant motifs and will contain ten or more designs. They are made in dark colours like deep-red, yellow-ochre, bottle-green, dark-brown or bright-blue with red or natural grounds. The finest variety is also called Bibibaff.

Bakhshis rugs: Persian rug of cotton warp and weft, the loose coarse wool pile being tied in Ghiordes knot.

Bakrabadi: Very soft and pale-coloured Indian jute.

Baku carpets: Caucasian short-pile, thin, hand-knotted carpets from the region around Baku. They have a marked geometric decoration. Generally, the field decoration is of three or four motifs in the shape of diamonds or rectangles and diamonds one above the other which alternate with hooked Greek-key motifs filled in light-red or ochre-yellow on a light-blue ground. The unusually pale colours are a characteristic of Caucasian carpets. These are also used as prayer mats. Baku carpets contain approximately 100,000 Turkish knots per metre square.

Balance (in garments): The relation of one section of garment to another in harmony with the natural attitude of the figure, especially that of back and front lengths.

Balance marks: Notches or threads in garment parts, which helps the tailor to preserve the balance of the garments by serving as guides during assembly. They are generally put at side seams, in the scye, shoulder seams, hind arm and fore arm seams and elsewhere as necessary.

Balance twist: See **Twist, balanced.**

Balance wheel: Wheel located at the end of the machine arm.

Balanced cloth: A term used to describe a woven cloth with yarns of the same linear density and same number of threads per cm in both warp and weft.

Balanced colour ways: When the colours change, but tonal relationship within the design stays the same while giving the same overall visual effect.

Balanced crepe': A fine crepe using the same yarn arrangement in warp and weft is called a balanced crepe'. A common example is to alternate 2 S twist with 2 Z twist, or 4 S twist with 4 Z twist in warp and weft. See **Crepe'**.

Balanced twists: In a plied yarn or cord, an arrangement of twist which will not cause the yarn or cord to twist on itself of kink when held in an open loop.

Balancing pond: A pond or lake, which may be artificial, that is used to attenuate the flows during storms and thereby reduce storm flows downstream of the pond and alleviate flooding. The retention time in the pond is relatively short, possibly 24 hours or less. The pond contains some water during low-flow conditions. The term 'balancing pond' may be synonymous with a detention pond, but a balancing pond may mean a combination of systems to balance stormflows, such as a detention pond followed by constructed wet-lands.

Balancing reservoir: A reservoir into which there may be a variable delivery and from which the flow is steady. It provides steady working conditions for the filters and other plant downstream of it.

Balancing tank, equalisation (equalization) basin, equalisation tank: A tank or basin designed to reduce and even-out the variations in the flow or chemical characteristics prior to the waste-water treatment plant. The tank may store the waste-water or storm-water at high flows and release at low flows. Alternatively, the tanks may be used in industry to balance-out the variations in the nature of the waste-water that emanates from differing activities. This may occur prior to onsite treatment or discharge to a sewer. Many water and waste-water treatment processes operate most efficiently at a relatively constant flow. Flow balancing can be particularly important in treating waste-waters from small communities which can generate erratic flows. See **Package waste-water treatment plants**.

Balagny cloak: A cloak or cape with wide collar, used in France, in the first half of 17th Century, named after a military hero.

Balanced stripes: A design of stripes that are even in width and spacing.

Balao: A type of raw cotton from Brazil.

Balasse stout: Old plain-woven cotton fabric from Surat.

Balassoi: Old Indian fabric, made of bast fibres of a tree.

Balastei: A gold tissue, made in Vienna and used for trimmings.

Balbriggan: A light knit fabric made on a circular knitting machine. It was originally made at Balbriggan in Ireland and used for hosiery. Now it is used for under wear, sportswear and pyjamas. Usually made from cotton, it may have a slight fleece on wrong side.

Baldacs: Rich gold-brocaded silks imported in the Middle Ages from Bagdad and Damascus, also made in Italy.

Baldachin, Baudekin, Baldaquin, or Bodkin: Rich medieval silk fabric, originally from Bagdad, made with silk warp and gold filling, often studded with precious stones; used for church vestments, state occasions, etc. Later they were made only of silk and dyed crimson.

Baldrick (French bandelier): Sword Hanger, usually decorated with exquisite embroidery (often metal thread embroidery), and worn from the right shoulder to the left hip, usually over the waistcoat or earlier bolero-style doublet, but under the coat or justacorps. Frequently worn over the coat to show off the embroidery, when the baldric had become very broad and long. The sword (rapier, later also dress-sword) hangs very low at the knees.

Bale: (1) A bag, sack, square or oblong box or package into which silk, staple fibres or tow are compressed. The common shipping and storage package for these fibres.

(2) A defined quantity of raw silk carefully wrapped in a cotton cloth and gunny cloth. A Japanese or Shanghai bale contains 60kg (Picul), a European bale contains 100kg and a Canton bale contains 48kg of raw silk. Indian bale occasionally contains 20kg raw silk.

Bale breaker: A machine used for opening cotton direct from the bale.

Bale dyeing: The process in which the bales of wool, cotton or synthetic fibre are dyed as such. The bales are placed directly in a cage designed to hold up to 4 bales and placed into a conventional kier and dyed. After dyeing, the bales are either hydro-extracted or vacuum extracted to remove the extra moisture and dyed.

Bale weights for cotton: Bale weights of cotton in different countries are as follows:

Australia = 227 kilos

Colombia = 233 kilos

Mexico = 220 kilos

Nigeria = 185 kilos

Uganda = 182 kilos

India/Pakistan = 170 kilos

South Africa = 200 kilos

Egypt = 720 lbs = 327 kilos

Sudan = 420 lbs = 191 kilos

Tanzania = 400 lbs = 181 kilos

USA = 480 lbs = 218 kilos

Balicnong: A fibre used for making cords in Philippines.

Baline: Plain woven, coarse fabric made of jute, hemp or cotton; used for bagging, stiffening for garments, or for upholstery.

Baliziei: Coarse, medium, strong leaf fibre, yielded by the Heliconia.

Balkal yarn: Yarn spun out of tussar cocoon peduncles, after they are boiled in alkaline solutions and opened up.

Ball (yarn): Yarn package.

Ball: Knitting cottons, silks, or wools are often made up in a ball or egg shape.

Ball mill: A standard equipment for reducing the particle size of water-insoluble pigments and dye-stuffs to fine particle size. It consists of a cylinder rotating on an axis partly filled with steel balls, porcelain ball or common pebbles. The controlling factors are the size of the balls, relative volume occupied by the balls, and the substance being milled, type and quality, and rate and time of rotation.

Ball warp: (1) Parallel threads in the form of a twist-less rope wound into a large ball. When wound mechanically with quick traverse, a ball warp may be made in the form of a large cylindrical package.

(2) Warps that, after being wound on a warping mill are bunched together in rope form and wound into a ball or similar form (silk).

Ball warping: Ball warping is mainly used in manufacturing of denim fabrics. The warp yarns are wound on a ball beam in the form of a tow for indigo dyeing. After the dyeing process, the tow is separated and wound on a beam. This stage is also called long-chain beaming or re-beaming.

Ballanca: Cloth made in Austria of black wool mixed with goat's hair.

Ballast: Material that is used to bring the total weight or volume of textiles to an amount specified in the procedure for processing or testing of textiles.

Balleta: A red colour woollen fabric once made formerly in Turkey.

Balling up: A defect in which loose or frayed fibres that have formed into a ball and are then woven into the fabric.

Balling up: See **Fuzz ball**.

Ballistic: A thick woven fabric that is extremely abrasion resistant and tough; has a denier of about 2000, and is used in apparel, packs and gears.

Balloon cloth: (1) A very closely plain woven cotton fabric, once used for ready-made shirts, type-writer ribbon and balloons.

(2) A balloon fabric is any fabric which forms a functional part of the lift creating envelope of floating aircraft. It is usually a simple single cloth, of lightly woven construction, and may include rib stop threads to enhance the tear resistance. It was made historically of cotton or linen doped to reduce air permeability. Modern balloon fabrics are typically polyamide or polyester based and are coated with polymers, and or laminated with a thin polymer film or films, to reduce permeability to the specific lifting gas employed and to confer protection against UV light.

Balloon: The curved path of the running yarns about the take-up package during spinning, down twisting, plying or winding, or while they are being withdrawn over-end from package under appropriate yarn winding conditions.

Balloon control rings: One more ring positioned between the ballooning eye and ring on a ring spinning machine or ring twisting machine, in order to contain the ballooning yarn. The purpose is principally to permit spinning on to the longer package at high spinning speeds without excessively high spinning tension and resultant end breaks.

Balloon fabric: A plain weave cloth having the same breaking strength in each direction. This fabric is made from fine (60s to 100s) combed yarn woven to the constructions 92 × 108 to 116 to 128. Once used for readymade shirts, for typewriter ribbons and of course, vulcanized balloon fabric is used for air cells in planes and barrage balloons.

Balloon net: Machine-made net; used for lace foundation, similar to the net used on balloons.

Ballooning: The entrapment of air either deliberately or accidentally in a fabric while processing. In knitted fabric, ballooning is done before squeezing to remove all wrinkles.

Ballymena: A sort of Irish linen.

Ballushar: An Indian silk fabric.

Balmoral: Heavy and very strong woollen fabric made in red, blue, and black stripes.

Balsa: Silky, yellowish seed hair of the Bombax tree in Central America.

Baluchari: A hand-made silk fabric for sarees or scarves made in Baluchar, near Murshidabad, India, where it was patronized by art loving *nawabs*. It was a silk weave and its solid base colours such as green, red, purple, magenta, brown and blue provided a sharp contrast to the distinctly elaborate silver/gold zari (embroidery weave) *Mughal* borders and Pallu. This technique had died out in the original place due to lack of patronage and was adopted by neighbouring Bishnapur where it prospered. The Hindu kings encouraged the weaving of *puranic* (epics) tales from the two epics *Ramayana* and *Mahabharata* which remains hallmark of today's classic *baluchari*. Woven on jacquard looms, the fabric has very fine filament as untwisted yarns are used, which exhibits opulence but is not stiff to drape. Another feature of baluchari is the white outlining of the motifs.

BAM: Bundesanstalt für Materialprüfung (German Federal Institute for Materials Testing).

Bamia: (1)Variety of Egyptian cotton, having a light-brown staple of inferior strength; (2) A strong, lustrous, white stem fibre of good quality, yielded by the wild okra in Sudan; used for making ropes by the natives.

Banana: The yellow colour of the banana skin; also 'banana-yellow'.

Banana fibre: See **Manila fibre**. Sometimes banana fibre is obtained in India from one-year-old plants by steaming the unrolled stalk and scraping off the outer skin. The fibres are pounded (wrapped up in a cloth) and cleaned. They are very durable and said to be excellent for fabrics intended for tropical wear.

Bananeiras: Strong and fairly flexible fibre, yielded by the young leaf stalks of a palm tree in Brazil; used for cordage.

Banbury plush: Woollen plush fabric, made with cotton warp and wool pile and is used for upholstery in England.

Bancal or Banker: Tapestry of green or blue colour, produced during medieval period in Britain.

Bancroft: Commercial variety of cotton from United States; the fibre measures up to 20–25 mm, grows in medium large bolls, yield 30–32% lint; it has a late maturity. Also called **Herlong**.

Band [in garments (continuous/grown-on)]: Pant panels that extend to the top of the pant and are folded over without an outside band. A separate inside band lining is sewn through the pant and has an inter-lining.

Band (pasted-on/folder-set) (in garments): A separate band sewn on the pant with stitching that shows on the outside at the top and bottom.

Band (rocap) (in garments): A separate band of body fabric sewn on and turned down so that the attaching seam is not visible. Inside, the band is a separate lining, made from picketing fabric and interlining.

Band work: (Similar to filling, jours, modes); used in needle laces to fill the centres with fancy open stitches.

Band (shade): After final inspection of the fabric, the shade of each piece is grouped together into bands which are having the same depth, hue, and tone. Shade bands are made, and each roll is identified for the particular shade band while laying for cutting and cut together to avoid shade variation in that group.

Bandage: A narrow strip of cotton or linen fabric, plain and loosely woven; used by surgeons for the purpose of bandaging.

Bandala: Native name for the harder and stronger outer fibre of the Musa textiles, in the Philippine Islands; used for cordage.

Bandana: (1) Special printing effect, producing light coloured designs over a dark foundation by destroying the dyes through bleaching agents.

(2) Cotton fabrics having white- or bright-coloured spots on a red or dark back ground; used as handkerchiefs.

(3) Cloth made of the lustrous fibres of various East Indian plants, having pale yellow patterns over a dark foundation.

Banderoles: Bolting cloth.

Bandhor rugs: Made in Asia of wool and cotton; the very thick pile is tied in Ghiordes knot.

Bandhani: The traditional tie and dye craft in northern India. Making a piece of bandhani from start to finish is intense and time consuming and in most villages; it functions as a sort of community enterprise where it is not unusual for a piece to be woven at one home, tied in another, and dyed in a third and finally embroidered in the last. The *abha* wedding costumes, a speciality of Kutch, for example, involve the women of the house – whose finger nails are maintained unusually long to speed up her work – bunching minute sections of the material, according to the pre-ordained design, and tying each bunch with waxed cotton thread. The colours are generally black and scarlet with touches of yellow; while the hems are finished with lavish gold brocade. The common colour red stands for marriage and fertility, yellow as a celebration of spring, saffron a colour preferred by yogis, and black and maroon both of which are symbols of mourning. Until recently, the dyes used were natural: madder and cochineal for red, indigo for blue, jasmine and saffron for yellow. The red is supposed to be as brilliant as possible and because of differing water qualities some towns are better suited to get a brighter shade compared to others. The different knots produce a different effect and the more miniscule the raised bandhani, the finer the artistic value of the piece. Each traditional pattern has its own name, such as the *chandokhni* and the *shikara* which are specially designed for weddings. Chokidal is a design of squares usually with elephants, while the *kambaliya* has a dotted pattern in the centre with a different design along the border. The *basant bahar* represents the flowers of spring; the *mor zad*, a peacock pattern; while *ambadal* depicts a network of branches and leaves inter-woven with a variety of birds.

Banding (heavy tow): Non-uniform distribution of filaments across the tow-band width.

Bandle: A coarse home-spun linen made on narrow hand looms in Ireland.

Bandura-wel: Very pliable cord made of the stem of the pitcher plant in Sri Lanka.

Bandy: Striped cotton fabric made in West Africa, made of waste cotton.

Bang off: See **Filling bar, coarse; filling bar, fine**.

Bangaline (Bengaline): Lightweight dress goods, woven with silk warp and heavier cotton or woollen filling, forming coloured cross ribs heavier than in poplin. Sometimes two picks are in the same shed. Often also printed. The cheaper grades are all cotton.

Bangalore: Hand woven woollen rug, made in India with a knotted pile.

Banig: Mats made of the leaves of the buri-palm, the pandang or a sedge, called tikug by the natives of the Philippines.

Bank: Another name for a yarn creel.

Bankukri: Raw cotton grown in Rajasthan, India; the staple is long and silky.

Ban-lon: Fabric made of yarns specially treated by textralised process of M/s Bancroft and sons, which adds bulk, moderate stretch, and texture to synthetic thermoplastic yarns. Fabric made from these yarns is soft, strong, easy to wash, quick drying and crease and shrink resistant.

Banner cloth: A plain weave fabric of cotton, wool or silk used for the fabrication of flags, banners and buntings. Sometimes bunting is known under this name.

Bannister harness: See **Split harness**.

Bannockburn: The main characteristic of this material is that a single colour yarn is alternated with a yarn made of two colours; the yarns are twisted together before weaving. A firmly woven cheviot tweed in 2/2 twill weave (straight twill or herringbone) having single and two-ply yarns, alternately in both warp and weft. Originally, the two-ply yarns were made by plying single yarn identical to the one used in the fabric with a white yarn for use in the warp and with a dark yarn for the use in the weft, but modern bannockburn tweeds favour the count of the coloured single yarns approximately to the resultant count of the ply yarn which are white colour and/or light-dark woollen grandelle or marl yarn, warp and weft. The name has come from a Scottish town that has been the centre of an area producing a top quality tweed for suits and coats. In some cases, it is a soft, slightly napped, twilled woollen fabric, made of cheviot wool, having double-and-twist warp and single filling.

Banos: Native name for fibre used for ropes and cords in the Philippines.

Bantine: Raw silk from Genoa.

Banuacalao: Native name for fibre used for ropes and cords in the Philippines.

Banyai system: A carpet machine knotting technique from Hungary.

Baobab: A large tree in Africa and India, yielding very strong bast fibres; once used for cord and bagging in Africa and for saddles in India.

BAP: Bromoform alkylphosphate. The BAP method involves application of an aqueous emulsion of a polymer produced from tribromomethane, $CHBr_3$ and triallyl phosphate, $(CH_2{=}CH{-}CH_2O)_3PO_4$ which forms a crosslinked polymers on cotton with a flame retardant action. See also **BA cotton**.

Bar: (1) A CGS unit of pressure equal to 10^5 (Pa). A metric unit of atmospheric pressure, equal to one million dynes per square centimetre, 100 kilopascals, 750.062 tor, 1.019 72 kilograms of force per square centimetre (kgf/cm2), or about 14.503 78 pounds per square inch (lbf/in2). The word comes from the Greek *baros*, "weighty." We see the same root in our word, barometer, for an instrument measuring atmospheric pressure. One bar is just a bit less than the average pressure of the Earth's atmosphere, which is 1.013 25 bar. In practice, meteorologists generally record atmospheric pressure in millibars (mb). In English-speaking countries, barometric pressure is also expressed as the height, in inches, of a column of mercury supported by the pressure of the atmosphere. In this unit, one bar equals 29.53 inches of mercury (in Hg). It may be used with SI units, and SI prefixes may be attached to it. Since it describes decimal multiples of the pascal, it is convenient in practical use. The millibar (symbol: mbar or mb) is a commonly used unit of pressure in meteorology.

(2) The edging of the button-hole with button-hole stitches to prevent its being torn.

(3) In needle laces, the threads connecting the various solid parts of the lace made of two or more strands and being either corded or covered with button-hole stitches. Also called bride, coxcomb, pearl, and tie. Bars are also parts of pillow and macrame laces.

(4) (in woven fabric) A band (q.v.) that runs with clearly defined edges and that differs in appearance from the adjacent normal fabric (It may be shady and may or may not run parallel with the picks). Bar is a general term that covers *pick bar*, *weaving bar*, *shade bar* and *tension bar*, etc.

Bar marking in screen printing: A problem encountered in printing on flat-bed screen printing machines, whereby the sides of the screen frames at right angles to the direction of printing exert pressure on the preceding print and thereby crushing or soiling the printed colours (screen fall-on effect). The problem can be minimized by using screen frames with a triangular profile (i.e. to reduce the area of contact with the printed fabric) and/or by covering the undersides of the screen frames with coarse grains (e.g. wood chippings of approx. 1 mm size) to break up and minimise contact with the fabric during printing.

Bar tack: See **Bar tacking**.

Bar tacking: Reinforcement by means of stitching at points of stress, e.g. button holes, pocket corners, seam ends, belt loops, etc.

Bar (pick): See **Starting place set mark, stop mark**. A prominent band in a woven fabric that has one clearly defined edge and that gradually merges into normal fabric and is caused by an abrupt change in pick spacing followed by gradual reversal to normal pick spacing. Such a bar occurs on restarting the loom without sufficient care after (1) Pick finding, (2) Uneven weaving or pulling-back, or (3) Prolonged loom stoppage. These bars may also be referred to as "standing places" or "pulling-back places" if the precise cause is unknown. In knitting, the band has several courses containing stitch lengths longer than in adjacent normal courses and has resulted from a machine stop that has caused changes in warp tension.

Bar (shade): A band that has developed a different colour from the adjacent fabric during (or subsequent to) dyeing and finishing, owing to damage to (or contamination of) otherwise normal fabric or weft yarn prior to weaving.

Bar (tension): A band composed of weft yarn that has been stretched more (or less) than the normal weft prior to or during weaving. This abnormal stretch may have been imposed during winding by faulty manipulation or by some mechanical fault in the loom; during weaving by incorrect tensioning in the shuttle; or may have arisen owing to faulty yarn having been excessively moistened at some stage and stretched more than the normal yarn under normal applied tensions. It may appear as a cockled bar in those cases where stretch has been sufficient. See also **Cockle (fabric)**.

Bar (weaving): A band that usually shades away to normal fabric at both its edges. It owes its appearance to a change in pick spacing, and may repeat at regular intervals throughout an appreciable length or even the whole length of the piece and is the result of some mechanical fault in the loom, e.g. faulty gearing in the take-up motion, bent beam gudgeons, uneven or eccentric beam ruffles, uneven bearing surfaces at some point in the let-off motion, etc. Bars of this type associated with the take-up or let-off motions are also referred to as "motion marks"

Bar (weft): A band that is solid in appearance, runs parallel with the picks and contains weft that is different in material, count, filament, twist, lustre, colour or shade from the adjacent normal weft.

Baracan, Bouracan, Berkan, or Perkan: Closely woven heavy cloth used for furniture cover or drapery, made of doubled and hard-twist worsted yarn warp and three or six-ply, hard-twist worsted filling and finished with moiré effect. It has warp ribs.

Baracan grosgrain: French term for *baracan*, with a heavier weft than warp.

Baracanee: French term for warp ribbed, plain coloured fabrics.

Baracani: Fabric made of mohair in Italy during the Middle Ages.

Baras: Coarse bagging made in Bohemia.

Barathea: (1) A silk, rayon or man-made fibre necktie fabric with a broken-rib weave or twill hopsack and a characteristic pebbly appearance. It is a rich, soft-looking fine fabric.

(2) A fine dress fabric with a silk warp and worsted filling, woven in broken filling rib which completely covers the warp.

(3) A smooth-faced worsted uniform cloth with an indistinct twilled basket weave of fine two-ply yarns

(4) Expensive fine English cloth, closely woven with a slight diagonal weave appearance and broken rib effect. Feels smooth, but has a granular appearance. Originally in worsted and silk, but now made with synthetic fibres such as viscose, acetate and polyester. Used for men's evening cloths in black and in many colours for coats and skirts for women.

Barawazi: Cotton cloth woven with dark blue, yellow and red checks, the border is in checks of red, black and yellow; used in East Africa.

Bare elastane: Elastane yarn in the so called bare state, in contrast to the wrapped form. See **Core spun lycra**.

Bare pychon ka pyjama (Hindi language): A pyjama with wide and flared legs.

Barium chloride: $BaCl_2 \cdot 2H_2O$, MW 244,34. Colourless crystals with a bitter salty revolting taste, stable to air, water-soluble (446 g/l at 20 °C; 769 g/l at 100 °C). Uses: As a de-lustring agent for viscose (two-bath method); weighting and filling agent in finishing (seldom used); reagent for the detection of sulphuric acid and sulphates (white insoluble precipitate); softening of boiler feed water (calcium sulphate hardness).

Barium hydroxide: [Baryta; $Ba(OH)_2$] A white solid, usually obtained as the octahydrate [$Ba(OH)_2.8H_2O$]. It can be made by adding water to barium oxide. Barium hydroxide is the most soluble of the group 2 hydroxides. It is used in volumetric analysis for the estimation of weak acids using phenolphthalein as an indicator. Used in the testing of degree of mercerization.

Bark: Dark brown.

Bark crêpe: Mostly atlas construction clothing crepe with a woven matt face and rough structure, in a bark-like pattern produced by embossed printing or by construction made of viscose, acetate, cotton, etc. A stiff fabric for evening wear.

Barking: Removal of bark from wood before pulping.

Bar marking in screen printing: A problem encountered in flat-bed screen printing whereby the sides of the screen frame exert pressure on the preceding print, thereby crushing or soiling the printed colours. The problem can be minimized by using screen frames of triangular profile to reduce the area of contact with the printed fabric and/or covering the underside of the screen frames with coarse grains to break-up and minimize contact with fabric while printing.

Barman lace: The name given to machine-made reproduction of French and Belgian hand-made laces.

Bar: A bar, running across the full width of a piece, which differs in appearance from the adjacent normal cloth. It may be shady or solid in appearance, is clearly defined and runs parallel with the picks. There are different types of bars like *weft bar*, *shade bar*, *pick bar*, *tension bar*, etc.

Barrackan: In Mediterranean countries, a camel's-hair cloth for men's wear; in England a sort of moleskin. Also spelled barragon.

Barras: A coarse linear fabric similar to sack cloth; originally produced in Holland.

Barb (in feathers): The primary structure of plumage emanating from the quill of a feather.

Barbadoes: West Indian raw cotton.

Barcelona: Raw cotton from Colombia.

Barcelona handkerchiefs: They are of fine twilled silk in plain colours, checks and fancy patterns.

Barcelona lace: Stitch in old needle-point, producing checkerboard effects with button-hole stitches.

Barchent: A stout, twilled cotton fabric, usually napped on the back, bleached, dyed or printed. Used in Germany, Turkey and Austria for heavy underwear, lining and also cheap dresses.

Barclay: A Highland tartan composed of alternate blue and green stripes and crossed by red lines.

Bareface: Fabrics without any nap.

Barege: (1) Earlier days a light, French homespun, dyed in the yarn or printed, made of fine, hard twist worsted warp and filling, producing a crepe-like effect. Later made with silk or cotton warp. It is a sheer fabric; used for dresses, etc.

(2) Nowadays, a light-weight dress goods woven like gauze, with fine silk warp and worsted filling; also called woollen gauze and woollen grenadine.

(3) Shawl made in France of organzine warp and worsted filling.

Barege yarn: Fine, hand-spun yarn; used for very fine gauze or veils.

Barfoul: A cotton fabric in the West African trade; used for garments.

Barhana: Lowest grade of Smyrna rug.

Barinas: Native name for fibre used for ropes and cords in the Philippines.

Barleycorn: Name for a small, reversed weave or for a weave similar to the matt weave.

Barmen lace: Braid made of odd number of yarns.

Barndenburg coat: A loose overcoat with turned back cuffs used in the fourth quarter of the 17th Century. The sleeves are made one with the rest of the garment.

Barnes: A commercial variety of late maturing American cotton of medium long staple.

Barnett: A commercial variety of cotton from Alabama; the medium late maturing staple measures 23–25 mm, forming medium size balls, yielding 30–32% lint.

Barnsley crash: Narrow crash used for towels.

Barnsley linen: Fine grey or bleached linen cloth; used for embroidery.

Barn-tine: Levant silk; used for trame and braids.

Barometer (barograph): An instrument for measuring atmospheric pressure, usually in the region of 101 330 Pa (1,0133 bar = normal atmospheric pressure).

Baroy: Native name for fibre used for ropes and cords in the Philippines.

Barpours: Dress goods used for making garments worn for mourning. It is a fine fabric made using organzine warp and very fine wool filling in serge or brocaded weave. Used for men's suiting.

Barr: Textile diffusion resistance to water vapour relative to atmospheric air (1 cm thick calm air layer at 20 °C = 1.00 barr).

Examples:

Air (atm: 0 °C) 1.05

Air (atm: 20 °C) 1.00

Air (atm: 40 °C) 0.95

Non-woven fabric (20 g/m^2) 2.20

Non-woven fabric (40 g/m^2) 4.20

Light-weight suiting fabric 8.50

Standard suiting fabric 16.40

Sweaters 20.50

Blanket material 22.50

Overcoat material 40.00

Wind-cheaters approximately 100.00

The barr value is expressed here in multiples of the barrier capacity of a 1 cm thick layer of calm air.

Barracan: Coarse, thick, strong cloth, somewhat resembling camlet; used as coating and cloaking; now made with silk and wool warp and Angora or other goat hair weft, forming warp ribs. See **Baracan.**

Barrage: Figured table linen, made in France in three qualities.

Barragon or moleskin: Sort of strong cotton fustian of coarse quality, twilled and shorn; used for men's working clothes.

Barragones: In South America, a closely woven twilled cotton trousering with narrow stripes; made in black or bleached.

Barrangan: Woollen cloth used in the Middle Ages.

Barras: Coarse, plain woven linen; used for bags.

Barratee: Silk cloth, being a variety of the Barathea.

Barre': (1) A term used for a defect seen as unintentional, repetitive visual pattern of continuous bars and width-wise stripes (so called bars) in textile fabrics, produced by weaving, due to the effect of yarns or colours or dyed appearance of the fabrics. Descriptive terms and code letters for the barre patterns. (1) **Simple**: Consisting of not more than two contrasting yarns – light, dark; lean, bulky; intermittent, uniform, etc.—spaced at regular intervals; (2) **Banded**: A simple pattern in which the contrasting yarns alternate in equal width intervals; (3) **Complex**: Consisting of two or more interspersed simple patterns.

(2) Barre effect is often occur in warp-knit fabric produced from textured polyester filament yarn dyed with disperse dyes. This is due to the structural difference of the polyester yarn due to texturizing.

(3) (in weft knitted fabrics) A clearly defined band (q.v.) or bands that run(s) full width across an open-width fabric or spirally in a tubular fabric, and differ(s) in appearance from the adjacent normal fabric as the result of variation of yarn characteristics. When the yarn is of a different colour (owing to differential dyeing) from that of the rest of the fabric, that defect is termed "barriness").

Barre': Descriptive of a fabric containing bars. See also **Bar**.

Barred witch stitch: Same as **Plaited stitch**.

Bare'ge: See **Barre'ge**.

Barre'ge: Very sheer fabric of wool and silk or cotton, used for veiling. Originates from barrage in the Pyranees. Unsuitable for clothing. Drapes well, but will not withstand strain in wear.

Barrel shaped yarn package: A precision wound barrel-shaped package for short length of yarn.

Barrier effect of wool fibre epicule: Prevention of dye from penetrating the fibre interior from dye liquor at room temperature due to the Epicuticle layer. Dye can only penetrate through the epicuticle from heated dye liquor. Chlorination treatment removes the epicuticle which is why deeper dyeing are obtained on chlorinated wool.

Barrigudo: Short, silky fibre, yielded by the pod of the Bombax ventricosa, in Brazil; used for stuffing.

Barriness: Longitudinally or transversely striped dyeing of woven or knitted polyamide, polyester or viscose fabric, due to mixed up yarn or density and tension variation in weaving or warp and weft knitting. Differences in yarn count, yarn twist, lustre or deformational differences also can cause this fault. Other reasons include chemical difference in polyamide and viscose, fibre change due to action of light and chemicals differences in drawing, crystalline structure and orientation due to varying inner-fibre fine structure during production, and these become more prominent in dyeing. The same defect can occur in woven fabric also, due to the same reasons.

Barroches: Fine, unbleached Indian cotton cloth; sort of bafta.

Barry: A flaw in the fabric, showing bars in the direction of the warp or the weft.

Barutine: An inferior silk fabric, made in Persia.

Basal layer: In fibre structure, the basal layer is a well-ordered active cell layer zone; e.g. in wool structure it is the intermediate membrane which acts as the basal layer of the internal-scale cell layer.

Bas de cotte/de jupe/de robe: This term was used in the second half of 17th Century for the lower part of the petticoat or skirt, which would be covered by the gown body.

Bas de jump: See Bas de cotte.

Bas de robe: See Bas de cotte .

Basalt fibre: A mineral fibre produced from molten basalt at about 1100–1400 °C. Cheap, very fine, highly ductile, long fibres which are thermally stable upto 1100 °C. Woven cloth used as insulating material against heat, cold and sound.

Basane: A shaggy-face French twilled woollen fabric.

Basant bahar: A bandhani design. See **Bandhani**.

Base: Chemical compound that dissociates in aqueous solution into negatively charged Hydroxyl ions and positively charged weak acid or metal anions and will produce pH more than 7 in solution. Alkali is often used synonymously, though this is not strictly true (all alkalis are bases, not all bases are alkalis).

Base cloth (needling): A textile fabric that is normally woven, which may be included within a needle felt to provide dimensional stability and strength, and in some cases to facilitate the punching operation.

Base colour: A colour which is predominant in a **colour scheme**.

Base dyeing: It is a dyeing process in which after naphtholation, the fabric is passed through the base solution when the diazotization and coupling is happened with the padded naphthols and the azoic dye is formed on the fabric.

Base exchange: See **Cation exchange**.

Base fabric: The base fabric or substrate on which the coating is done, in coated fabrics.

Base layer: The apparel in contact with your skin. The purpose of the base layer is to keep you warm/cool and dry.

Base material (carpet): Carpet primary backing.

Base printing: See **Naphthols** in direct printing. In direct printing of naphthols, the naphthols are printed first and the base is printed afterwards when the diazotization and coupling is happened at the places where the naphthols are printed where the azoic dye is formed on the fabric.

Base units: Seven base units defined in the SI system for the basic measures – metre, kilogram, second, ampere, Kelvin, mole and candela for basic measures of length, mass, time, electric current, thermodynamic temperature, amount of substance and luminous intensity. Their abbreviated forms are m, kg, s, A, K, mole and cd.

Base (strong): A strong base completely ionizes in water solution. Sodium hydroxide is the most common strong base used in textile processing. See ***Base, weak***.

Base (weak): A weak base does not fully ionize in solution. Some of the weak base remains in solution in molecular form. If BOH represent the base, some would remain in solution as BOH and some would ionize to B+ and OH-.

This makes for something of a "reservoir" action –if some of the OH-ions are used up in the reaction, more BOH would ionize. Because of this, the pH of a solution of a weak base will change by less than one for a change in concentration of the base of by a factor of 10 (for a strong base increasing the concentration by factor of 10 will increase the pH by 1). pH, unless very accurately measured, is a poor indication of how much of a weak base is in solution. Sodium carbonate and sodium bicarbonate are weak bases frequently used in textile processing. Both of the do not contain OH, but they rather hydrolyse to yield NaOH.

Bases: Chemical compounds which dissociate in aqueous solution into negatively-charged hydroxyl ions and positively-charged weak acid or metal anions. They form salts with acids = neutralization. Bases have an alkaline reaction (pH > 7) and turn red litmus paper blue. The number of ionizable OH groups determines whether a base is mono, di or polyvalent, etc. Inorganic bases are OH compounds of metals (e.g. sodium hydroxide solution NaOH = Na^{+}+ OH^{-}, strong base over 50% dissociated; ammonium hydroxide NH_4OH, weak base) which, on losing water, change into the corresponding oxides. Organic bases include e.g. amines (aniline $C_6H_5NH_2$), dye bases and numerous other water-soluble nitrogen (e.g. pyridine C_5H_5N), phosphonium and sulphonium bases.

BASF: Badiche Anilin and Sodafabrik.

BASF Combitest: Standard methods for determining the diffusion properties of disperse dye under practice-relevant conditions.

Bashofu: Probably an old fabric produced in Japan using Banana fibres. It used to be very light and white fabric, used for summer undershirts.

Basic: A term describing the substances having an alkaline nature. Bases may or may not be water soluble.

Basic colours: According to Berlin and Kay's 1969 work *Basic Colour Terms*, there are, from a linguistic standpoint, only 11 basic colours – black, blue, brown, green, grey, orange, purple, pink, red, white and yellow. Their research involving 98 languages indicated that no language has more than these 11 basic colours and that colour names evolve in languages in a particular order. In priority, comes black and white followed by red. Then comes yellow and green (in either order) and then blue and brown. The colour words of all those languages studied which had, for example, only five words for colours would always be black, white, red, yellow and green. See **Essential colours**.

Basic English colours: Basic English, devised by Charles Kay Ogden between 1926 and 1930 as an international form of English, uses only 850 words of which eight are colours, namely, black, blue, brown, green, grey, red, white and yellow.

Basic salts: Salts.

Basic dye: See **Cationic dyes**.

Basic dye: A dye that dissociates in an aqueous medium to give a positively charged coloured ion (cation) with affinity for fibres containing acidic groups.

Basic dyeable polyamide: See **Differential dyeing polyamide fibres**.

Basicity: A property of bases and basic salts whereby hydroxyl ions of a basic character are formed by dissociation in aqueous solution.

Basicity value: The basicity value plays an important role in the weighting of silk, e.g. with basic iron (III) sulphate. The basicity value is obtained by dividing the sulphuric acid content (calculated as H_2SO_4) by the iron content (calculated as metallic iron) = (H_2SO_4:Fe).

Basic fabric: Practical, down-to-earth, natural looking fabric that can be made into apparel uniquely suited to meet the needs of the 1970s lifestyles. Basic fabrics are made in plain, twill or satin weaves. The majority of which are plain and twill. Warp- and weft-knitting are the basic knits. Both natural and manmade fibres and blends are used. Important basic fabrics include broad cloth, denim, corduroy, gabardine, batiste, duck, seersucker, velvet, velveteen, oxfords, warp knits, double knits, gingham, chambray, chino, etc.

Basic weave: Basic weave of a fabric is the way the yarns are inter-locked. The weave gives the fabric its characteristics and strength. The three basic weaves are plain, twill and satin.

Basics: A term used to describe classic jeans fabric as well as standard jeans qualities which appear regularly each season, in contrast to newly developed styles ("designer jeans"). They include, e.g. flats, chambray, blue denim, canvas, toile, cord velvet, scrubbed denim, brushed denim, suedette and duvetine.

Basin: (1) A white twilled cotton cloth, similar to dimity, made with or without narrow stripes, sometimes napped on one side; used for vests; (2) A French fabric, originated in the 16th century, made of pure or cotton mixed linen, or hemp warp and cotton filling in a twill weave.

Basin refuse: It is the last parchment of cocoon which is unreelable. Normally this waste contains pupal body. It is very inferior in quality. This waste is not used in spun silk mills; however, this waste can be used for hand spinning.

Basin waste: See **Basin refuse**.

Basin royal: White stripped, fine ticking cloth made of pure linen.

Basinas: The waste silk obtained from the residue cocoon after reeling.

Basin test: Modified permeability bag test in which a fabric test specimen is formed into a suitably supported bag and filled with water.

Basine: A silk fabric in which two sets of fine warps float over every two wefts and interlace with the third.

Basine's: Cocoons with frison removed; although unpierceable by moth, are not reelable or only partly reelable and are therefore used as waste silk for spinning. Also called Pelades. (French)

Basinetto: Waste silk obtained from the inner skin of the cocoon, which remains after reeling. See also **Ricotto, Bisou**.

Basis weight: The weight of a unit area of the fabric. It can be grams per square metre or Ounces per square yard.

Basket braid: A soutache braid, made with five threads, each thread passing alternately under and over two threads.

Basket cloth: Coarse cloth woven with regular or fancy basket weave. Sometimes twill derivatives are also used. See **Abbots cloth**.

Basket stitch: A knit construction; purl and plain loops are combined with a preponderance of the purl loops in pattern courses to give a basket weave effect.

Basket weave: A variation of plain weave in which two or more warp and weft threads are woven side by side to resemble a plaited basket. Fabrics have a loose construction and a flat appearance and are used for such things as monk's cloth and drapery fabrics.

Basmas: Closely woven fine linen or cotton cloth, made in Turkey.

Basolan DC Process (BASF): Wool chlorination process under mildly acidic conditions with the sodium salt of dichloroisocyanuric acid (Basloan DC) at pH4.

Basolan SW Process (BASF): Super-wash finish for wool with prepolymer Basolan SW, which cross-links into a soft film on the fibre on drying. Can be applied by padding or exhaust.

Basques: Middle 17th Century French word for short tabs at bodices and male douglets that extended below the waist. The jackets with basques were worn in combination with skirts instead of gowns.

Bass fibre: Very strong, straight, coarse and smooth fibre, yielded by the Raphia in West Africa; used for brushes.

Basse lisse: French term for low or horizontal warp in tapestry work.

Bassinas (also pelettes and tetelettes): The interior skin of the cocoon left after the reeling; used for floret or waste silk.

Bassine: (1) Palm fibre from *Borassus piassava.*

(2) Inner parchment-like layer of the cocoon skin (also termed Pelettes, Ricati and Galettame).

Bassiness (French): French term for taffeta ribbons.

Bassora gum: This vegetable gum generally occurs in the form of transparent, light amber to brownish-red irregular lumps that are sparingly soluble in water but swells readily into a thick gelatinous slime. It is having a high content of bassorin which is used for adulteration of gum arabica and gum tragacanth.

Bast: (1) Another name for phloem.

(2) Fibrous material obtained from the phloem of jute, hemp, flax, lime, etc.

Bast and leaf fibre: Fibre derived from the inner fibrous bark and hard coarse leaves of dicotyledonous plants, such as flax, hemp, jute, and abaca.

Bast fibre: Any of certain strong woody fibres used in making rope, cordage, etc.

Bast fibre bundless: Flax stem structure.

Bast fibre tow: Flax and hemp waste, etc. Short fibres which arise in the hackling process.

Bast fibres: A sub group of natural cellulosic fibres that form bundles or strands which act as hawsers in the fibrous layer lying beneath the bark of numerous dicotyledenous plants such as (a) flax (b) ramie (c) hemp (d) jute (e) sunn (f) kenaf (g) urena.

Bast layer: Flax and hemp waste, etc. Short fibres which arise in the Hackling process.

Bast liquor: A bath liquor, used in dyeing, that is prepared by acidifying soap solutions that have been used for the degumming of silk. It greatly facilitates the production of a level shade. Artificial bast liquor is prepared from a solution of gelatine in soap solution.

Bast soap; Degumming soap: A term used for the soap used in the degumming of natural silk which contains variable proportions of silk gum removed from the silk fibres, but useful as a protein material. Used as a dyeing auxiliary with levelling properties.

Bast tape: A tape product consists of warp ends stuck together, with no weft ends produced from glazed yarn or artificial horse hair.

Bastancini: Fine, sheer and bleached linen that is finished with stiffener and bluing.

Bastard: (1) General term for substitute; (2) A woollen fabric made in England during the reign of Richard III.

Bastard aloe: Strong, leaf fibre yielded by the Aloe vivipara in North-western India; used for ropes.

Bastard: Any fabric woven in imitation of better and more expensive.

Bastard vat: Combination of the Wood vat and sodium carbonate (the first as vat preparation and the second for continuation of the dyeing process). Formerly used for the dyeing of wool with indigo.

Bastard velvet: It is placed between the velvet and plush as to fineness and length of pile.

Baste: It refers to long and temporary stitches that hold layers of fabric together. It can be done by machine or hands baste.

Bastella: A cellophane product into which narrow cotton threads have been incorporated, thereby producing a reed-like appearance. Uses: hat trimmings (braiding), etc.

Basting: Temporary stitching either by hand or machine in the preparation of work to hold garment parts (stuff and lining or any two or more parts of the work together while it is being stitched) in the correct position for subsequent permanent operation, none being left in the finished garment. It is also used as a guide for sewing, feather stitching, etc.

Basting cotton: Cotton thread, used for basting; is similar to sewing cotton except that it is weaker and is not finished as smoothly.

Basting needle: This hand-sewing needle is curved to make basting and tying quilts easy.

Basting stitches: See **Basting**.

Basto: Indian name for heavily sized bleached cotton shirting or bleached long cloth, earlier imported, used for shirts, caps, bed covers, etc.

Basuto: A commercial grade of South African mohair.

Bataloni: Made of hemp warp and cotton filling, usually dyed in light-blue colour.

Batarde: (1) Black wool used as a substitute for vicuna wool; (2) Solid coloured French serge made with 10 warps and five picks in a repeat.

Batavia: (1) French term for a twill weave forming diagonals; (2) Serge made at least with four harnesses.

Batch: (1) A group or set of usually similar materials for processing at one time as a collective unit.

(2) The process of winding cloth on a roller after it has been passed through a dye or chemical solution for a process in which the goods have to be left saturated with dye solution/chemical solution for some period of time, typically for hours, and typically at room temperature for the dye to fix onto the fibre or the chemical to react on the fibre. In commercial processes, batching mostly follows padding.

(3) The process of softening jute.

Batch process: A batch dwell process; E.g. cold pad batch process.

Batch roll: See **Batching roll**.

Batch rotation: Batching piece goods on to the roller out of the treatment liquor.

Batch steamer: Cylindrical steamers in various constructions like horizontal steamers, vertical steamers, used for discontinuous steaming process especially printed goods.

Batching centre: In yarn packages these are the tubes on which yarn is wound. In the case of piece goods, they are the metal or wooden rollers on which fabric is wound.

Batching roller: Also called batch bearer, batching roll. Cylindrical support (with square inserts) for open-width piece fabric batching.

Batching tension control: On a large batch, woven fabrics have to be batched under constant tension. The compensator roller determines the batching tension, and, via a microprocessor, regulates the batching speed.

Batch-path: Commercial term for immature jute fibre from India.

Batch-wise processing: The processing of materials as batches or lots in which the whole of each batch/lot is subjected to one stage of the process at a time.

Batenburg lace: See **Battenberg lace**.

Bates: Two commercial varieties of cotton from South Carolina, the staple measuring 24–27 mm and yielding about 33% lint.

Bates process: A cotton scouring and bleaching process in which fabric impregnated with bleaching chemicals is placed between electrodes and heated to the boil.

Bath: The liquid-filled treatment zone in dyeing, washing, finishing and dry-cleaning. The liquid medium is referred as 'Liquor'. Standing bath is one in which the bath is stationary whereas in circulating bath the liquor is circulated.

Bath brussels: Name given to the Devonshire lace, in the 18th century.

Bath coating: Light baize of wider width and long nap, comes in white or coloured; used for petticoats, bath robes, overcoats, etc.

Bath exhaustion: A term used for exhaustion at of the dye at equilibrium, generally expressed as that percentage of dye originally applied which has been removed from the bath by absorption on to a textile substrate due to the forces, if affinity.

Bath mat: An absorbent textile floor covering normally used in the bathroom as a pad on which to step when getting out of a tub or shower.

Bath robe: A long gown to be worn between bath room and dressing room.

Bath rug: A scatter rug used in the bathroom.

Bath sheet: A textile terry product with end hems or fringes and side hems or selvages that is generally much larger than bath towel.

Bath towel: A textile terry product with end hems or fringes and side hems or selvages, which is used to dry a person's body after bathing or swimming.

Bathik (Batik): Actually, it is a method of printing in which using a wax resist and dyeing a printing effect is produced. A dyeing method originated in Java using wax resists. Traditionally, wax resist is made on the cloth using special equipment (Tjanting) and the fabric is given mechanical breaking action to give the typical cracking effect while dyeing. The fabric is then dyed with one colour of the print by dip dye method. The wax is removed by dipping in hot water or soaping solution and then the dyeing is washed and dried. Again the wax resist is applied all over the other part of the fabric, except where the second colour has to be applied. This colour is then dyed and the wax is removed afterwards as explained earlier. The entire remaining colour is dyed/printed the same way till the required effect is achieved. It is very laborious and time consuming work. Hence, few colours are made by over dyeing a second colour over another colour. The resultant designs are bold and uneven and streaked where the white or cream background fabric shows through the colours. Sometimes mixtures of Batik, screen printing or block printing techniques are used to give the same effect; at the same time, the whole work is made simpler and less complicated. It is needless to say that, all the dyes used has to be cold dyeing dyes as otherwise the wax resist will not stay while dyeing.

Bathik printing: See **Batik**.

Bathochromic group: Colour intensifying group in a dye structure.

Bathrobe blanketing: A double faced fabric woven with a tightly twisted spun warp and two sets of soft spun filling yarns. The fabric is thick and warm and its filling yarns are frequently napped to produce a soft surface. Today's blankets are made of spun polyester, acrylic or polyester cotton blend.

Batiste or cambric: (1) (i) The finest grade of linen, woven plain in grey colour, then bleached; there are several grades as the batiste claire, which is very light and loosely woven, the batiste demi-claire is of stronger yarn and closer texture; the batiste hollandee is closely woven with a body; the batiste linen is still stronger; the Scotch batiste is a fine, printed cotton dress 'goods; (ii) White or coloured cotton muslin, finished with a heavy size; used for summer dresses, linings, etc.

(2) Cotton or polyester/cotton soft fine, to very fine tightly woven fabric of the lawn family in plain weave slightly heavier than voile. Often highly mercerized, bleached, dyed or printed and sometimes embroidered and wears well. Resembles nainsook but finer with a lengthwise streak. Coloured woven batiste is not batiste; it is called macozefir.

(3) Sheer fabric of manmade fibre like rayon with dobby woven stripes, jacquard patterns.

(4) A light weight woollen fabric in plain weave like cambric. It should be of all wool construction unless otherwise described.

(5) A smooth, woven fine fabric that is lighter than and very similar to nun's veilings.

Batnas: Three-coloured calico, made in India.

Baton rompu: French plain serge, made with eight warps and four picks in a repeat.

Batswing: A thick, coarse cloth, woven in grey into the shape of a seamless petticoat.

Batt: Single or multiple sheets of fibre. These are used in non-woven fabric production, felts, etc. Synonym with web.

Battenberg braid: Cotton or linen tape with picot edge; used for laces and curtains.

Battenberg lace: A coarse-type lace, often made from linen thread and used mainly in small areas, such as collars and cuffs; although, it is popular heavy curtain lace.

Battening: As the shuttle moves across the loom laying the weft yarn, it also passes through openings in another frame called reed (which resembles a comb). With each picking operation, the reed presses or battens each filling yarn against the portion of the fabric that has already been formed. The point where the fabric is formed is called the fell.

Battery: A device for holding reserve bobbins (or pirns) of weft in vertical or circular housings or shuttles in a vertical housings.

Batting: A soft, bulky assembly of fibres usually carded. Battings are sold in sheets or rolls made of cotton or polyester, and used for warm interlinings comfort stuffing, quilting, in quilted fabrics and other thermal or resiliency applications.

Batting integrity: The ability of a textile filling material to resist distortion or change when subjected multiple launderings or dry-cleaning.

Batt-on-base woven felts: Non-woven textile fabrics composed of ground fabrics and one or several fibrous webs which are consolidated and bonded with the ground fabrics.

Battlemented: In embroidery, a pattern similar to the battlement of the old fortresses.

Battleship grey: The bluish-grey colour in which battleships are often painted.

Batuz: (1) Needlework consisting of sewing upon silk as a part of the pattern to be made, very thin plates of gold or silver, these plates often being hammered into low relief. Now obsolete; (2) Silk fabrics, ornamented with hammered gold leaves, used in France during medieval times.

Baudekin: See **Baldachin**.

Bauge: A stout, thick, twilled fabric, similar to droguet, made of coarse wool in Europe.

Bauhinia: Bast fibre of various East Indian trees; dark reddish-brown, very strong; used for ropes, nets and coarse cloth.

Baume' (degrees): A frequently used but unscientific scale of measuring the relative density of liquids by hydrometry using the Hydrometers. The division in the scale of the hydrometer is based on: density of distilled water at 4 °C is 0°Be and the density of 10% sodium chloride solution is 10'Be. This separation is divided into 10 sub-divisions, and the scale is continued. The relation between specific gravity and Baume is as follows: For liquids heavier than water d = 144.38/(144.3-'Be) and for liquids lighter than water d = 144.38/(144.38+°Be).

Bave: It is the silk grey yarn as it is without degumming. It is the silk fibre complete with its natural gum, as it is withdrawn from a cocoon formed by a silk worm. It comprises two brins that are cemented together by the gum or sericin.

Bay: Reddish-brown or chestnut colour used particularly in the description of horses; hence 'bay-coloured' and 'bay-brown'.

Bay, Bayes: (1) A fabric in England made of worsted warp and woollen filling, often mixed with silk, made for clothes by religious societies, popular in 16th and 17th centuries.

(2) A loosely woven, plain woollen cloth, similar to a coarse flannel, napped on one side and made usually in white, black, red or green. It is called baigue in France.

Bayadères: (1) While pekins are formed by warp stripes, bayadère shows us stripes of different weaves running in the direction of the filling. The rules given in the pekins as to the joining of the weaves will also apply here. The warp, which was raised on the last pick of the weave must stay down wherever possible on the first pick of the following weave. The number of shafts employed must go up evenly in the repeat of each one of the weaves that go into the make-up of the bayadère.

(2) A very broad term for stripes that run cross-wise in a knit or woven fabric.

(3) Made of silk in cross-wise rib (plain or twill weave). It has brightly coloured stripes in the filling direction. Often black warp. The colour effects are usually startling or bizarre. Mostly produced in India. Name derived from the Bayadere dancing girl of India, dedicated from birth to a dancing life. The Bayadere costume includes the striped garment, a flimsy scarf or shawl, jewelled trousers, spangles, sequins, anklets. Uses: Blouses, dresses, after 5 wear.

Bayberry wax: Bayberry wax is obtained from the candleberry plant and is chemically a fat. However, its wax-like nature stems from the fact that it is mainly made up of stearic, palmitic and myristic triglycerides. It is used in textile sizing compositions because it can be easily saponified during the desizing operation.

Bayes: Plain and loosely woven wool fabric, napped on one side and obsolete. See **Bay**.

Bayeta: (1) Coarse, home-spun woollen, usually piece dyed in red, blue or green, and napped; used for ponchos, etc., by the natives of Peru and Bolivia.

(2) Full woollen fabric, usually black, navy or dark green; used for skirts in Colombia.

(3) Scarlet woollen blanket, woven by the Red Indians. It is of single strand wool which is obtained from baize by unravelling it; there is a nap on the blanket.

Bayeta de Cien Hilos: Wool flannel made in Latin Americas, having a very long nap and twill weave; the wide selvage used to have four blue stripes on white ground.

Bayeta de Faxuela: Coarse woollen baize made in Peru, dyed red, blue and green; used for ponchos, etc.

Bayeta de Pellon: Coarse wool flannel that is woven like a serge, having a very long nap; used in Latin Americas and China for bed covers, etc.

Bayette or Baguette: Loosely woven, plain, woollen fabric, similar to a coarse flannel, napped on one side, dyed in white or black shade; made in France, England, etc. In France it is also called baigue. See also **Bay**.

Bayeux lace: Closely resembles the Chantilly. In the 18th century, first silk laces were made in ecru and then in white.

Bayeux tapestry: A piece of linen that is 214 feet long and 20 inches wide, containing in 72 groups the representation in coloured wool embroidery.

Bayko: A yarn or thread having a core impregnated with a smooth metallic coating in any colour; used for weaving, knitting, embroidery, etc.

Bayonet spiral: Part of a shearing and cropping machine. The term is used to describe the upper shearing cylinder blade profile in shearing and cropping machines.

Bays: Coarse English worsted and woollen fabric that was worn by the peasants in Queen Elizabeth's time. See **Bay**.

Bayutapaux: A coarse cotton cloth with blue and white or red and white stripes, usually available in the African markets.

Bazaar dyeing: A unique ancient dyeing process still in use today with simple home produced equipments.

Bazac: A Palestinian made, evenly spun, fine, bleached cotton yarn.

BB: Short form for Batt-on-base.

BB felt cloth: Batt on base woven felts.

BC: Code designation for Boron (nitride) fibres in accordance with air travel standards.

BCF yarns: Bulked continuous filament yarns of high linear density, used as pile yarn for carpet trade, usually nylon, polypropylene or polypropylene.

Beach cloth: Very light fabric, made usually with cotton warp and mohair filling in colours and designs, having good light fastness and chlorine fastness; used for men's wear in summer.

Bead In Zippers: An enlarged section on inner edge of each tape to which scoops are affixed.

Beaded: Referring to garments decorated with beads.

Beaded seam: Beaded seams used for fine white work having a line of beading overhanded between gores, hems, or gathers. The hem along the seam should be folded on the right side, leaving a perfectly flat surface to iron on the wrong side, and finished with an ornamental stitch covering the hem.

Beaded selvage: See **Loopy selvage**.

Beaded velvet: Velvet with a cut-out pattern or a velvet pile effect, made on a Jacquard Loom. This fabric is used primarily for evening wear.

Beading: (1) On pillow lace, a simple heading; (2) Narrow, machine-made insertions, made with openwork to draw or bead a ribbon through or for trimming.

Beading lace: This describes very narrow insertion laces. These are often open-work hemstitching or faggot stitch, or they may have slits for inserting ribbons.

Beading needle: These hand-sewing needles are very long with a small eye. They range in size from 10 to 15.

Beam: (1) A cylinder of wood or metal, usually with a circular flange on each end, on which warp yarns are wound for slashing, weaving, and warp knitting.

(2) A perforated barrel on which the fabric or yarn is wound for dyeing in a beam dyeing machine. See **Beam dyeing machine**.

Beam set: One or more beams of yarn in a single shipment to be further processed together for a specific end use.

Beam setting: The process of setting of yarns or fabric made of synthetic fibre or their blends on a beam dyeing machine. The fabric or yarn is wound on the beam as in the case of dyeing and set in the machine at the dyeing temperature (say, 130–140 °C).

Beam dyeing machine: HT autoclaves for piece goods or atmospheric machines for the dyeing of warp beams. These machines are available as horizontal or vertical with various dias (say 1,000mm to 1,500mm) and widths up to 4,000mm and working pressures of 5 bar max and temperature up to approximately 135 °C. Warp yarn or piece goods are wound onto a special beam. The barrel of which is evenly perforated with holes. The dye liquor is forced through the yarn or fabric from inside to outside or vice versa by which the fixation condition of the dye is maintained in the autoclave.

Beam warping: The transferring of yarn from bobbins or cheeses onto a warp or section beam in the form of a wide sheet. Several of these beams are run through the slashing machine to make one loom beam.

Beaming: The operation of winding warp yarns onto a beam, usually in preparation for slashing, weaving, or warp knitting. Also called warping.

Beaming machine: A machine which winds the individual yarn ends from a rope-like bundle and distributes them evenly over a section beam.

Beaming the warp: It is the process of winding the warp on the warp beam. The layers are usually separated with sturdy paper or warping sticks.

Beamroll: See **Beam**.

Bear grass: Coarse and very strong fibre, yielded by the Yucca fllamentosa plant; used for twine and cordage.

Beard test: See **Test, beard**.

Bearded spring needle: See **Spring needle**.

Bearding: Fuzz on loop pile carpets, usually resulting from poor anchorage or fibre snagging.

Beater: (1) The machine which does most of the opening and cleaning work on a fibre picker and opener. Revolving at high speed, it beats against the fringe of fibre as the latter is fed into the machine.

(2) A machine used in the paper industry for opening pulp and combining additives.

(3) The beater is a frame that holds the reed. It is attached to the loom by an upright on each side that pivots to pull the reed through the warp and "beat" the weft in place at the fell of the clot.

Beater opener: Device for opening-out fabrics in rope form. The most important component is the rapidly rotating beater roller with beater blades.

Beater rollers (vibrator rollers): Employed in open-width washing machines to provide increased mechanical action on the running fabric, thereby intensifying liquor movement.

Beating machines (beaters): (1) Beating machines are used for the finishing of pile fabrics (velvet, plush, etc.) in which the back side of the fabric is subjected to a beating action which results in better alignment of the pile on the face side.

(2) A type of machines used for the cleaning of carpets and other textiles which cannot be subjected to washing treatments. Dirt is removed by an intensive beating and brushing action and suction.

Beating up: The third of the three basic motions in weaving operation of the loom in which the last pick inserted in the fabric is "beat" into position against the preceding picks or forced onto the fabric.

Bearskin: Woollen twill fabric that is heavy, thick and having a heavy nap on the face and used for coating.

Beatrice: A weft twill weave, made 4 by 1.

Beau: Ideal narrow strips of machine-made imitation of English embroidery.

Beaujeu: French hemp canvas, about 27 inches or more wide; it was used for furniture cover in olden days.

Beaujolais: French cloth made of cotton and linen.

Beaufort: A stout hemp sailcloth, made in France.

Beaupers: A woollen fabric of unknown structure, mentioned in 17th Century English writings.

Beaver cloth: It is made of high-quality wool and this material is heavy, expensive, firm-textured but soft fabric; has deep naps which are cut and laid smoothly in one direction. The lustrous nap of short fibre is produced by milling the cloth and raising the fibres. It is heavy weight woollen fabric with solid colours. Made in many combinations of double cloth weaves, it wears well. It has a soft handle with longer nap, and supposed to have the appearance of a natural beaver skin. Beaver cloth is frequently used in overcoats and winter jackets for men and women.

Beaverteen: A lighter grade of moleskin; it is piece dyed or printed to resemble worsted and napped on the back; used for men's wear.

Beby: Cotton scarfs that are usually dyed blue, made in Syria.

Beck: A dyeing vessel for dyeing fabric in rope form, consisting primarily of a tank and a reel to advance the fabric. Beck is a frequently used American term for 'open winch dyeing machine'.

Beckmann thermometer: A type of mercury thermometer which is designed to measure small differences in temperature rather than scale degrees. Beckmann thermometers have a larger bulb than common thermometers and a stem with a small internal diameter, so that a range of 5°C covers about 30 cm in the stem. The mercury bulb is connected to the stem in such a way that the bulk of the mercury can be separated from the stem, once a particular 5°C range has been attained. The thermometer can thus be set for any particular range. It is named after the German chemist Ernst Beckmann (1853–1923).

Bed (in sewing): A sewing platform which supports the needle plate and slide plate. It contains the lower sewing mechanisms.

Bed sheeting: Name originally used for plain weave cotton fabrics used for bedsheets; now used more loosely. One common construction is 64 square with 20s warp and filling. Also called muslin sheeting, it includes a wide variety of cloth. Usually starched and calendared to give appeal to cover up inferior construction, the coarsest materials border on bag sheeting. A fine muslin sheeting will have a structure 64 square with 40s count yarn [In some places, the width is given in quarters of yard (1/4 yard = 9 in). Thus, an 8 quarter sheeting is 72 in wide and 10/4 sheeting is 90 in wide].

Bed-covering: A textile product used on a bed over the sheets for warmth or decoration.

Bed lace: Sort of binding, made of white cotton, twilled or figured, employed for binding dimities. Used also for furniture, when it is printed.

Bedford cloth: See **Bedford cord**.

Bedford cord: A woven cord fabric in plain weave that is used on the face side of the cord. It is a rib weave fabric with raised length-wise cords produced by using stuffing thread in the warp. The wales are wide and stiffer yarns are usually present. There are weft floats that determine the width of the cords on the back, and wadding ends may be used to accentuate the prominence of the cords Made with carded cotton yarns, woollen or worsted yarns, viscose, acetate, polyester or combinations of these yarns. Other weaves are also used sometimes. Used for upholstery, suits, riding habits and work cloths. Light Bedford cord are used for dresses and children's clothes. Originated in Bedford, United Kingdom. Massachusetts also claims the honour, but apparently Bedford origin is more accepted.

Bedfordshire lace: English bobbin lace, introduced in the 17th Century. It is similar to the L'ille lace made with reseau ground and wavy, geometrical patterns.

Bedspread: A type of bed covering that is placed over the blankets and sheets for appearance and warmth.

Bedstout: Striped or solid coloured stout cotton, woven in four-leaved twill. Also called inlet.

Beech-green: A shade of green.

Beer: An arbitrary but customary number of threads (in Leeds 38, in Bradford 40 threads) which is sometimes taken in England as a unit to express the warp ends of a fabric.

Beer Lambert's Law: A useful measure of the strength or intensity of the colour of a dye is given by the molar extinction coefficient (e) at its L max value. This quantity may be obtained from the UV/visible absorption spectrum of the dye by using the Beer Lambert's law, i.e.

$$A = ecl$$

where A is the absorbance of the dye at a particular wave-length, e is the molar extinction coefficient at that wavelength, c is the concentration of the dye and

l is the path length of the cell (commonly 1 cm) used for measurement of the spectrum. The Beer Lambert law is obeyed by most dyes in solution at low concentrations; although when dyes show molecular aggregation effects in solution, deviations from the law may be encountered.

Bees wax: (1) Bees wax is a yellow-red-brown, pleasant odour, brittle when cold. Its melting point is 61.5–68.0 °C and is soluble in turpentine. It has white spirit and hot fats and is used in textile sizing, finishing and water repellent treatments.

(2) A dark-orange colour.

Beet: (1) Deep purple-red that is named after the vegetable of the same name.

(2) A bundle or sheaf of tied flax crop or straw.

Beetling: The process in which the round linen fibres are given a flattened effect by pounding the yarn called beetling. This makes the yarn flat and hence shining. Mask, etc., made by using beetled yarn are more lustrous and have a leather-like texture. Beetling is also used to give a thread- or linen-like appearance to cotton.

Beetling calendar: A heavy calendar, usually with 6 bowls; the lower and uppermost are heater steel bowls and the intermediate ones are elastic cotton bowls. The fabric while calendaring, passes through all the rolls producing a beetling effect.

Beetling effect: A unique effect imparted on cotton/linen fabric by calendaring in a beetling or chasing calendar.

Beetroot red: The deep reddish-purple of the vegetable, beetroot.

Beeze: Piping or cording formed at lower and inside pocket welts.

Begasse: Trade name for the short waste fibres which have their way out from the sisal hemp during the scraping process.

Beggars' velvet, also Velours de Gueux: A Lyons velvet of linen warp and cotton filling and pile.

Begonia: Yellowish-red colour found in the flower of the same name.

Beguin coarse: Stout fabric, made of natural coloured wool; used for garments by religious orders.

Behaai: Sort of Indian cotton muslin.

Beiderwand: It is originated in Europe; it was especially used in the Schleswig-Holstein area of Germany where it was used for coverlets and bed curtains.

Beiderwand coverlets were woven in North America in the nineteenth century. Historically, beiderwand coverlets most likely were woven on jacquard looms. In its construction, first and second shafts are reserved for the tie-down ends. Each block in the draft requires two shafts, so a three-block pattern requires eight shafts, six for the three blocks, plus two for the tie-down ends.

Beige: A term for cloth in the natural or undyed state; the French word for natural. The source of the terms greige, grey or gray, from which it was corrupted. They also refer to unfinished and undyed goods. The colour beige is dyed in a wide variety of textiles and is not to be confused with this definition.

Beige: A light-brown or yellowish-grey colour. Popular for fashion fabrics but 'renamed' camel, taupe, stone or sand to avoid appearing dated. Beige has the same meaning as écru and grège, each being French terms for wool or cloth in its unbleached condition and originally designating a grey colour. First recorded as a colour in 1879.

Beige damas: Natural coloured thin worsted dress goods made with Jacquard figures.

Beilik: A coarse Turkish woollen or cotton scarf.

Beistain test: Used for detection of substances containing any halogen, e.g. synthetic fibres (such as polyvinyl chloride or vinyl chloride copolymer fibres).

Bejuco: Tough, very dark coloured and coarse fibre, obtained from the bark of a creeper in South America; used for heavy cordage.

Beldia: A very heavy shrinking, coarse wool of Morocco.

Beledin: (1) A type of raw silk from the Levant; (2) An inferior grade of cotton yarn from the Levant.

Belelac or Belelais: A type of East Indian silk taffeta.

Belesmes: Coarse hemp canvas, used as ticking that is made in France.

Belgian laces: Include *Antwerp*, *Brussels*, *Malines* and *Valencienne* laces, all pillows, except the modern Brussels point-gaze. At the present, the grounds are machine made and the patterns or twigs on the pillow are man made.

Belgian: Tapestry English jute drapery with an admixture of linen.

Belgian ticking: Stout linen and cotton fabric with satin face. Used for upholstery and bedding.

Belgrade braid: Open work flat braid, made of cotton yarn, sized and glazed to imitate straw braid; used for millinery.

Belgravian embroidery: Patterns of leaves are traced on a broad braid, filled in with bugles sewn with floss silk and then the braid is cut around the edges of the leaves.

Belinge: Stout twill suiting fabric made with linen warp and woollen filling, made in France.

Bell isles: Eighteenth century woollen fabric made in England.

Bell steamer: See **Batch steamers**.

Bellacosa: Brocaded silk cloth, inter-woven with gold and silver threads which is made in imitation of the old Venetian brocades.

Bellchester: A type of English velvet.

Belle creole: A commercial variety of American cotton, having a long, strong, soft and silky staple of high percentage of yield.

Belleric, beleric: An Indian fruit producing a black dye; the dye itself.

Bellies: The coarser quality of wool from the underside of sheep.

Belt accumulator: See **Narrow fabric accumulator**.

Belt brushing machine: Finishing machine for corduroy fabric and velvet.

Belt duck: See **Number ducks**.

Belt filter press, belt press, filter belt press: A method for the mechanical dewatering of sludge. Conditioned sludge is placed onto a belt (typically 2 m wide) to allow drainage of liquid from the sludge. This belt then meets a second belt. The two belts with the sludge sandwiched between them pass round a series of rollers to introduce increasing sheer and compressive forces on the sludge. Waste-water sludge can be dewatered to 20–35% solids. It is a continuous process unlike the filter plate press.

Belt loop: A loop of material or thread which is sewn into the waist of a garment to receive a belt.

Belt screen: In water and waste-water treatment, a screen that is a sloping conveyor belt with holes in the belt to allow the water to flow through. Some solids are caught on the belt and drop off into a container at the top of the belt. The holes are often about 6 mm wide and 20 mm or more in length.

Belting: (1) A narrow, very stout warp-face fabric, woven plain or in satin weave, made of cotton or hemp, also hair, used at the waistline in skirts, waists and other garments; it comes usually black, white or grey.

(2) A narrow elastic fabric, made of cotton, wool or silk, often ornamented with woven figures, used for dress belts.

(3) A very heavy and strong narrow fabric of cotton, hemp or linen, used for power transmission. Now a days, mostly made in synthetic fibres like Nylon, Polyester, etc.

Beluchistan: Heavy rugs made of wool, goats and camel's hair, the long pile is tied in Senna knot. The design consists of geometrical figures, angular hooks, etc., in rich, dull reds, browns, blues mixed with a little white.

Belwarp: English worsted fabric with corkscrew pattern.

Bemberg: Trade mark of early cuprammonium rayon produced by the American Bemberg Company. Fabric is still labelled 'Bemberg' but must include fibre content description as well.

Benares: Light weight cotton/silk fabric from India. Usually woven with metallic threads. Often in woven design.

Bench marks: Marks placed on a specimen to define gage length, that is, the portion of the specimen that will be evaluated in a specific test.

Bending elasticity: Partly elastic partly permanent shape change of yarns and woven fabric in the bending test as a criterion of creasing, the so called angle of recovery, which occurs after folding or bending under load and subsequent relief, and is at best 180', playing an important role.

Bending length: (1) The cube root of the ratio of the flexural rigidity to the weight/unit area.

(2) A measure of the interaction between fabric weight and fabric stiffness based on how the fabric bends in one plane under its own weight (under the force of gravity).

Bending modulus: Maximum stress per unit area that a specimen can withstand without breaking when bent. For fibres, the stress per unit of linear fibre weight required to produce specified deflection of a fibre.

Bending rigidity: See **Flexural rigidity**.

Benedict-Denis reagent: A reagent used quantitative sulphur determination in wool. It is a solution of 25 g of copper nitrate, 25 g of sodium chloride and 10 g of sodium nitrate in 100 ml of water.

Bengal stripes: Bengal stripes is a name given to a type of gingham consisting of white and coloured stripes, alternatively arranged in small effects in regular order, the coloured yarn having been dyed with Indigo. They were originally made in Bengal, India, where many species of plants producing Indigo were grown.

Bengaline: (1) A cloth with more or less striking warp-rib appearance running across the fabric similar to faille, only heavier, with a fine weave and width-wise cords. Originally, bengalines were made of a silk, wool, or rayon warp with a worsted or cotton filling and used for dresses, coats, trimmings, and draperies. Modern bengalines are made with filament acetate or polyester warps. Also, some bengalines have fine spun warps with 2- and 3-ply heavier spun yarns filling cord effects. The warp rib or carded effect may be produced by employing (1) suitable thickness and setting of warp and weft threads or (2) suitable warp rib weaves or (3) a combination of (1) and (2).

(2) It originally came from Bengal, but an imitation is the familiar crosswise rib fabric-made viscose, silk, wool or cotton. Similar to poplin, or fraille in appearance, but heavier. Used for coats, suits, dresses and ribbons, can also be made as curtain fabric.

Benham's disk: A disk with black and white markings which when spun creates the illusion that the disk is coloured.

Benzanthronylamino anthraquinone: *Anthraquinone* dyes of the *Benzanthronoid vat* dyes group. Example is condensation of 3-bromobenzanthrone with 1-aminoanthraquinone in boiling naphthalene in the presence of sodium carbonate and copper oxide. Cyclization with potassium hydroxide in isobutanol.

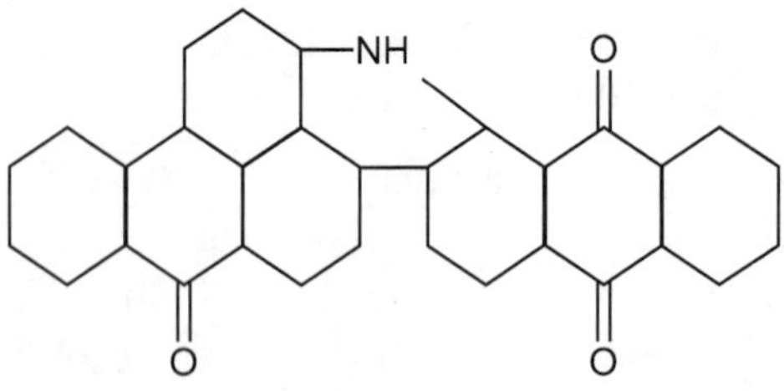

Vat olive green

Benzanthronylpyrazole-anthrone: Anthraquinone dyes of the Benzanthronoid vat dyes group. Example is the condensation of 3-bromobenzanthrone with anthrapyrazole and melting the product with alcoholic potassium hydroxide.

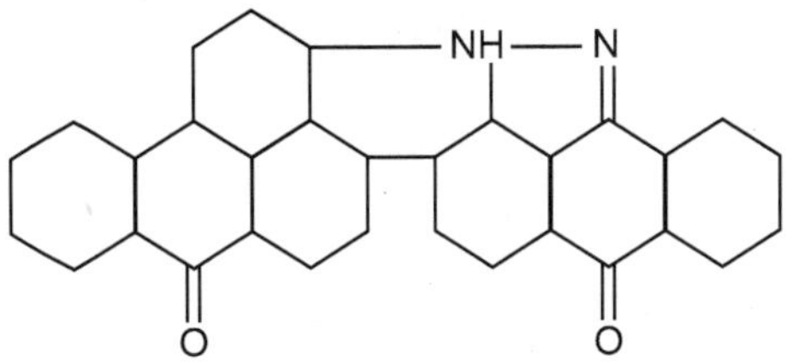

Vat navy blue R

Benzene: Colourless to light-yellow, mobile, non-polar, highly refractive liquid with a characteristic aromatic odour. C_6H_6, mol. Wt.78.06, density at 15 °C 0.885, b.p 80.15, f.p. 5.5 °C. Not to confuse with Benzine. The hydrocarbon source for many manufactured fibres.

Benzine soaps: Fat-solvent soaps of various compositions (often with high ethanol content) and consistencies (liquid, semi-solid, solid). By reducing the spontaneous combustion of benzene, they facilitate dry-cleaning in organic volatile solvents.

Benzoate fibre: Fibre with a silk-like hand-made from a condensation polymer of p-(Bhydroxyethoxy) benzoic acid.

Benzylating agents: These are used as auxiliaries in discharge printing of vat dyes to form water soluble bezlated products and leuco compounds which can be easily washed-off.

BEP: Break Even Point.

Berber carpets: Rustic knotted carpets in natural colour, hand woven by north African tribes from wool yarn, also spun by hands from local sheep wool. They contain a proportion of naturally pigmented wool and tribal motifs. Now a days, the same term is also used for carpets made of natural-coloured wool or dyed fibre with homespun appearance.

Berberine: A yellow dye obtained from the African tree of the same name.

Berdin: Wool rags in the finer machine knitted grades.

Berettino: A bluish-grey hue used in the glazing of the Italian pottery, Maiolica.

Berlin blue: A dark-blue dye, soluble in oxalic acid. It gives a bright-blue shade which is sometimes used for blueing bleached goods. Also called Prussian Blue, Paris Blue, Turnbills Blue, etc.

Berlin silk: A type of cordonnet or crocheting silk made by twisting 4–8 single yarns with a right hand twist and then twisting 3 of these with a left hand twist. The yarn is high twist and therefore round, smooth and hard.

Berthe, Bertha: Loose collar or neckline insert for women.

Beryl: Pale sea-green or greenish-blue being the colour of the stone, beryl; hence berylline – of the colour of beryl.

Besom: An edging or reinforcement around a pocket opening.

Bespoke: Clothes made to order of an individual.

Beta cellulose: One of the three forms of cellulose. It has a lower degree of polymerization that the alpha forms. With gamma cellulose it is known as hemicellulose (Also see **Alfa cellulose** and **Gamma cellulose**).

Betanin: The natural colourant made from the beetroot; see also **Beet red**.

Beverteen: A heavy cotton cloth used for men's hunting garments.

Betweens needles: These hand-sewing needles are short, fine needles with large eyes are most frequently associated with quilting, but are also used for fine hand sewing. Sizes 3 to 12 are available.

Bezold effect: The optical effect, named after its discoverer, Wilhelm von Bezold, where by changing one colour in a pattern (for example in a rug design) the whole appearance of that rug changes.

BFTC: British Fastness of Textile Committee.

Bias: (1) Direction at an angle to the warp and weft. True bias is at an angle of 45° to both warp and weft or to grainlines.

(2) See **Skewness**.

Bias (statistics): A constant or systematic error in test results.

Bias binding: A product made by cutting wide-woven fabric at an angle of 45° to the selvedge. Most bias bindings have regularly spaced joints governed by the width of the original fabric, but if converted from tubular fabric than the joints are avoided. Bias binding do not fray and will stretch and they are thus suitable for binding seams and conforming to curve contours. Most bias bindings are folded into the centre, but flat versions are also in use.

Bias fabric: A two-dimensional fabric that when oriented in the XY plane contains fibres that are aligned in a different direction, i.e. 45° to the x-axis fibres.

Bias facings: All facings around curves, such as arm holes and neck, should be a true bias which is cut by holding the warp threads diagonally across the woof threads. These strips for facings, pipings, ruffles, etc., should be cut exactly even in width. All bands, ruffles, etc., of serge, twilled, or diagonal materials should be cut across the twill and not with it, in order to have the ruffle hang well.

Bias filling: A fabric defect in which the filling yarn does not run at a right angle to the warp. The principal cause is improper processing on the stenter frame (Also see **Bow, Skew**).

Bias hem: All bias and curved edges should have the first fold basted. In cloth or silk, this first basting thread should match the material and not be taken out.

Biaxial fabric: Conventional fabric construction consisting of one warp thread system and one weft thread system.

Biaz (Linen finish suiting): It is a medium-grade cotton fabric, resembling linen in appearance. This effect is usually got on ordinary cotton yarns in the finishing process or by mercerizing the yarns or by fabric mercerizing. Usually manufactured in white. The name Biaz comes from the place it originated, called Biax in central Asia. It is pronounced as be'az, also called Linen finish suiting in a few places like United States of America.

Bibibaff: The finest quality of *Bhaktiari* carpets with a single wet thread construction.

Bice: The pale blue or green obtained from **smalt**. There is no settled view as to the origin of this word. It is thought that the word derives from the Latin '*azura debilis*' meaning a weak blue as compared with the richer pigment '*azura pura*' meaning the best blue. Ball in *The Invention of Colour* has a contrary view, namely, that in the 14th century 'bys' meant 'dark', but 'bys' eventually detached itself from the colour 'azure bys' and came to be used to describe the colour rather than its shade. 'Bice' subsequently became a generic term for a pigment based on copper.

Biceps line (in sleeves): Cross grain at bottom of cap.

Bi-chromate after treatment: Treatment of direct dyed textile materials with Bichromate to improve wet fastness properties.

Bi-chromate/Copper sulphate after treatment: Treatment of certain direct dyes with copper sulphate/bi-chromate in order to improve the wet and light fastness properties.

Bi-chrome: Having only two colours.

Bi-colour: Modest shading effect using yarns of two different shades.

Bi-component fibre: Manmade fibre combinations composed of two distinct and inseparable components having different physical or chemical structure join together as a single fibre either in layers side-by-side or in mixtures with a homogeneous distribution.

A fibre or filament consisting of a continuous matrix of one polymer in which different fibre forming polymers are dispersed as a second discontinuous phase. A fibre formed by the conjunction at a spinning jet of two fibre-forming polymers of different properties.

(a) The two components may be caused to merge approximately side by side (bilaterally), concentrically or as fibrils of one component in a matrix of the other. An example is the production of crimped fibre, e.g. a combination of polymers of different contractive properties. (b) Although formed by a natural process, wool and related animal fibres may exhibit a comparable dual structure of the cortical cells. Also called bilateral fibres, composite fibres, etc. See also **Composite fibres**.

Bi-component spinning: Bi-component spinning normally involves twisting together either a filament (sometimes water soluble) and a conventionally drafted staple (cotton) strand during the spinning operation, and is particularly attractive for the cost-effective production of superior yarns which can, for example, be woven or knitted without any further operations (i.e. eliminating plying, sizing and steaming). It also enables coarser fibres to be spun into finer yarns, reduces spinning end-breakages, allows higher winding speeds and enables yarn and fabric properties to be engineered by suitable selection of the two components and the way in which they are combined.

Bi-component yarn: Spun or filament yarn of two generic fibres or two variants of the same generic fibre.

Bi-conical package: A conical yarn package in which the traverse length is progressively reduced to produce tapered or rounded ends. Such packages are also described as tappered or pineapple cones.

Bi-constituent fibre: (1) Bi-component fibres of the matrix-fibril (M/F) type.

(2) A fibre extruded from a homogeneous mixture of two different polymers. Such fibres combine the characteristics of the two polymers into a single fibre. Bi-constituent fibres consisting fibrils of a polymer aligned parallel to the fibre length interspersed in a continuous matrix of another polymer and their structure, resembles the bilateral structure of wool. Clearly fibrillar-matrix ratio and the nature and relative dimension of the components can be altered to vary the physicochemical properties. Some of the important bi-constituent fibres are Monvelle (Nylon/ Spandex), Source (Polyester/Nylon), Chinon (Acrylonitrile/Casein), Mirafi 140 (Nylon / Olefin), Polychlal (Polyvinyl chloride/Polyvinyl alcohol), etc. New and improved types ofbi-component and bi-constituent fibres continue to find increasing application in knitting, hosiery and carpeting. See **Bi-component fibre**.

Bi-directional fabric: A fabric having reinforcing fibres in two directions, i.e. in the warp (machine) direction and filling (cross-machine) direction.

Bi-elastic: A fabric with two-dimensional elasticity in knitted/woven.

Bifilar: (1) Bi-component fibres/filaments manufactured by bifilar spinning; e.g. production of polyamide/polyester simultaneously, side-by-side without blending from one spinneret hole, accompanied by final adhesion to produce one double filament.

(2) In the sense of two thin threads (instead of one thicker one), e.g. as used in knitting/warp-knitting/warp knitting for the production of dimensionally stable, long lasting men's underwear (double-thread circular knits).

Bi-functional: A term used to represent a reactive dye which has two reactive groups in the dye molecule. See **Bifunctional reactive dye**.

Bi-functional reactive dye: A reactive dye that has more than one type of reactive group in the molecule. The reactive groups may be identical or reactive to different reactive properties, say two vinyl sulphone groups or one vinyl sulphone and one mono-chloro pyrimidine group, etc. These reactive dyes are designed to have the ability to react with the fibre in more than one way. This increases how much of the dye in the bath is actually fixed to the fibre, rather than being wasted through hydrolysis. Although the dye can react with the fibre in more than one way, the temperature used is typically around 60°C (140°F). These dyes may be preferred by industrial users because of the lower waste (less effluent treatment cost) and lower reactivity, which can mean easier process control.

Bight: The distance between the stitching line and the adjacent edge of the material. See also **Seam allowance**.

Bijar carpets: A very densely knotted broad-like stiff carpets from Persian town of the same name. Naturalistic design with floral motifs is typical of Bijar carpets. The grounds are of dark colour: Dark blue, cherry red, bottle green, but sometimes camel hair colour, while colours used for designs are very vivid.

BIL Crease resist process (built in lubrication): A special cross-linking process to preserve the natural mechanical properties of cotton by incorporating stearyl derivatives of 1, 3 dichloropropanol-2.

Bilateral: Having two sides.

Bilateral fibres: Two generic fibres or variants of the same generic fibre extruded in a side-by-side relationship. See **Bi-component fibres**.

Bilateral fibre structure: Refer to fibres where two different types of fibre structure joined together making one fibre. For e.g. in the cortical layer present in the structure of wool consisting of the para and ortho cortex, as well as the different polymer structure in the cross-section of cotton fibres. Silk which has not been degummed also reveals a bilateral structure.

Billiard cloth: Woollen cloth made from fine merino wool. Heavily milled with fibrous finish. Dyed green or red for billiard or card table tops. Plain or twill weave, the cloth is of the highest quality and must obviously be smooth and even, but must have body.

Billiard green: The deep-green of the baize on billiard/snooker tables.

Billinghames test: A test for distinguishing between abaca and sisal. Wash the sample with methylene chloride and dry. Boil the sample with 5% Nitric acid for 5–10min.Wash off the excess acid and immerse the sample in cold in 0.25n sodium hypochlorite solution for 10min. Remove the sample and dry. Abaca gives an orange colour, whereas the sisal and other leaf fibres will give only a yellow colour.

Bimli jute: See **Kenaf**.

BIN: Belgian Standard Association.

Binca or bold work cloth: A very distinctive embroidery fabric made in a range of colours. It is characterized by the fairly large square holes between the blocks of thread. Often used as beginners embroidery fabric for children.

Binche lace: A Flemish lace. A six-point star shaped ground net of snow-flake effect with a scroll pattern on it.

Binder: (1) An adhesive applied with a solvent or plastic (that can be softened) melted to bond fibres together in a web or to bind one web to another. Mainly used binder polymers are acrylic ester copolymers, and homopolymers, butadiene-acrylonitrile copolymers, styrene-butadiene copolymers, chloroprene polymers, vinyl chloride polymers, vinyl acetate polymers, vinylidene chloride polymers, natural latex, etc.

(2) A material, usually nearly colourless, that is typically used to attach a pigment to fabric. Binders are more-or-less "glue" to hold the pigment in place. Paints consist of pigments mixed with binders. Many binders used in textile paints are acrylic polymers.

(3) The binding or wrapping yarn used in the production of yarns on a hollow spindle spinning machine. The binder is usually filament, but may for certain applications be spun yarn.

Binder content: The weight of adhesive used to bond the fibres of a web together. Usually expressed as percent of fabric weight.

Binder fibres: (1) The weight of adhesive used to bond the fibres of a web together. Usually expressed as percent of fabric weight in the material.

(2) Binder fibres are used in the manufacture of composite textiles in non-woven. Synthetic fibres with thermoplastic properties or which are soluble or capable of swelling in specific solvents are used for this purpose. They are subdivided into the following categories: (1) Soluble binder fibres as polyvinyl alcohols or alginate fibres; (2) Melt-able binder fibres such as copolyamide, copolymer thermoplastic fibres with low melting point (160–190 °C);(3) Adhesive binder fibres such as cellulosic fibres, etc.

Binder film: A film of synthetic resin polymers formed around the individual fibres in a textile fabric in pigment printing and in which the colour pigment is embedded.

Binder for nonwovens: See **Binder**, **Binder fibres**.

Binder prints on polyester: Pigments are printed on polyester using binder, but they give lower rubbing fatness as compared with cotton, since here the binder is surface abraded and drawn out from the fibre interior, whereas in case of cotton residual quantities for binder remains in the fibre cavities.

Binding (edging): Usually a broad tape used as a trimming, e.g. as a decorative border for under wear aprons, national costumes, etc.

Binding site: In pile yarn floor coverings: A place at which the pile yarn is, or can be bound to the backing fabric.

Binding thread (Stitching thread): Additional warp or weft threads used in fabric to join two or more layers of fabric together or to fasten long floats of yarn to the body of the cloth.

Bioactive: Term used to describe the function of biologically active substances, e.g. Biodegradation of surfactants and Biofouling. Enzymes are also bioactive.

Bioassay: An experimental technique for measuring quantitatively the strength of a biologically active chemical by its effect on a living organism. For example, the vitamin activity of certain substances can be measured using bacterial cultures. The increase in bacterial numbers is compared against that achieved with known standards for vitamins, yolk, kidney, liver, and yeast being good.

Biocatalysts: Enzymes.

Biochemical oxidation, biological oxidation: The oxidation of a chemical by microbial processes. Numerous examples exist including bacterial oxidation of organics, sulphur oxidising bacteria or bacteria that oxidise ferrous iron salts to ferric iron salts, nitrification of ammonia to nitrate, etc. See **Aerobic biological oxidation, Anaerobic oxidation, Anoxic**.

Biochemical Oxygen Demand (B.O.D.): A standard test for estimating the degree of contamination of water supplies. The amount of oxygen taken from natural water by microorganisms that decompose organic waste matter in the water. It is therefore, a measure of the quantity of organic pollutants present. The biochemical oxygen demand is determined by measuring the amount of oxygen in a sample of water, storing the sample, and then making the measurement again five days later. It is expressed as the quantity of dissolved oxygen (in mg/l) required during stabilization of the decomposable organic matter by aerobic biochemical action.

Biodegradable: The ability of a substance to be broken down by bacteria, so that it can be returned to the environment without posing an environmental hazard.

Biodegradable organics in waste water: Composed principally of proteins, carbohydrates and fats, biodegradable organics are measured most commonly in terms of BOD (biochemical oxygen demand) and COD (chemical oxygen demand). If discharged untreated to the environment, their biological stabilization can lead to the depletion of natural oxygen resources and to the development of septic conditions.

Biodegradable textiles: Textiles as biodegraded as the final biological treatment in a chain of natural fibre polymer chemical (from raw product via textile fabric production to finishing), when enzymatic break-down process of a complex type takes place on the fibre polymers.

Biodegradation: Degradation of any material by the action of bacteria. Biodegradation is generally understood to be the sum of all processes which reduce the load or concentration of polluting substances in water courses. As soon as biodegradable pollutants enter a water course, the respiration rate increases through the reproduction of microorganisms which are universally available thereby reducing the load or concentration of pollutants. A succession of organisms is therefore built up. This heterotrophic activity is termed saprogenic (intensified decomposition of organic matter).A term commonly used in effluent treatment of textile rejects. It is important that any chemical, dye, auxiliaries, fibre, etc., should be biodegradable to conserve the environment.

Biodegradation of cellulose: The biodegradation of cellulose is caused by enzymes known as cellulases. Cellulases are produced by many microorganisms (bacteria and fungi). The most widely studied cellulases are of fungal origin, e.g., Trichoderma. The cellulose-digesting bacteria of the rumen are a complex anaerobic community. The primary reaction in the enzymatic degradation of the cellulose is hydrolysis, and degradation is a function of the available surface area and crystallinity of the cellulose. Cotton fibre is a highly crystalline form of cellulose and the ability to degrade this form of cellulose is relatively rare among microbes, and due to the solid nature and insolubility of the substrate, is not very rapid. The cotton fibre, after ginning, is about 95% cellulose with the other 5% mainly wax, pectins, and proteins. These non-cellulosic nutrients are important to the maintenance of the complex microbial communities to which extensive cellulose degradation is typically ascribed, but in cotton fibre, these non-cellulosic nutrients are in too low amounts to be of much significance.

Biodegradation of surfactants: Many surfactants used in textile industry is not biodegradable by means of bacterial oxidation involving the action of aerobic and anaerobic bacteria. Soaps, fatty alcohol sulphonates, alkylbenzene sulphonates with unbranched carbon chains, alkyl poly (oxyethylene) sulphates, fatty acid condensation products and saccharose fatty acid esters are readily biodegradable. In general, surfactants with straight C-chains (= "soft", formula I) are biodegradable, whilst surfactants with branched C-chains (= "hard", formula II) have very poor biodegradability such as, e.g. alkylaryl sulphonates of the tetrameric propylene benzene sulphonate type.

Biodegradation of textile chemicals: Biodegradability of many chemicals used in the textile industry is poor. For example certain dyes, resin finishing chemicals based on urea-formaldehyde, melamine and glyoxal resins, non-ionis polyethylene emulsions and polyvinyl alcohol. Chemicals have to be tested for biodegradability; and if not biodegradable, should be completely avoided for textile use.

Bio-filter activated sludge process: An aerobic biological treatment system that consists of an aerobic bio-filter followed by an activated sludge process (i.e. an aeration tank and a sedimentation tank). The sludge recycle from the sedimentation tank is returned either to the aeration tank or bio-filter.

Bio-finishing: See **Bio-polishing**.

Biological activated carbon: (1) Granular activated carbon on which microbes grow. The microbes enhance the treatment of the water or waste-water or gas. See **Activated carbon filter**. (2) A combination of ozonation and granular activated carbon used to remove dissolved organic material from water.

Biological oxygen demand: See **Biochemical oxygen demand**.

Biological waste-water treatment: A stage in waste-water treatment where the natural self-purification processes is used. In order to accelerate aerobic degradation, however, the water is enriched with oxygen artificially. The most frequently used treatments are the percolating filter and the activated sludge process.

Biologically active finishes: Natural fibres and regenerated natural fibres are subject to attack by various biological agents including bacteria, fungi, and insects. While synthetic fibres are not normally attacked by these biological agents, substrates of these fibres can act as a support for growth of bacteria and fungi on the fibre surface. Several metallic and organometallic salts, phenolic and anilide derivatives, and quaternary amine salts can inhibit growth of bacteria and fungi on fibres. Cellulosic most often require such treatment to prevent mildew and rot from feeding on and attacking the fibre substrate. These are called biologically active finishes.

Biomass: The mass of a biological population, usually expressed in terms of dry weight. The term is used to describe the mass of biological solids growing in a waste-water treatment process.

Biomass separator: A sedimentation tank that follows after biological treatment of waste-water and is used to settle-out the biomass that has formed in the biological reactor from the treated effluent. In a traditional treatment plant for domestic waste-water, the separator is known as secondary sedimentation tanks.

Bio-monitoring cloth: In other words, this refers to clothing with embedded devices that can monitor vital signs, electrically converting the physical values measured from the body and expressing the resulting biological phenomena into electric signals. As such, it makes health management and remote treatment possible.

Bio-polishing: Cellulase enzyme action on cotton removes protruding fibres from the surface of dyed textile products to enhance colour. This operation is referred to as bio-polishing, and cellulose preparations have been incorporated in laundry detergents to eliminate or reduce surface fuzziness on cotton goods during the wash cycle. Enzymes are used also for bio-stoning of dyed garments, as done in the garment dyeing industry. Pumice stones have been used in recent years to soften garments and to remove colour and produce unique washed-down appearances through abrasive action. However, such processes are not only environmentally unsound because of the disposal problem of ground-down stone particles after processing, but the stones are also very damaging to the processing equipment as well. The use of enzymes eliminates or substantially reduces the need for treatment with pumice stones. The effectiveness of any of these operations for the removal of the material from the substrate is dependent on the type of mechanical action during processing. This includes the abrasive action of fabric-to-fabric contact or the cascading effect of aqueous solution on the cellulosic substrate. Bio-polishing can remove the hairiness which is not even burnt off during the singeing operation.

Bio-polymer: A naturally occurring polymer such as cellulose.

Bio-scrubber: A method of biological odour control in which the malodorous air is passed up through a filter containing plastic media in a conjunction with a down flow of liquid effluent. The biofilm grows on the solid media. The biofilm adsorbs and oxidises some gaseous pollutants as the contaminated air passes through the scrubber. The effluent supplies the nutrients for microbial growth. It is also known as a bio-filter. It has been successfully used for hydrogen sulphide removal.

Biotech cotton: Biotech cottons deliver high-tech options to farmers and consumers without compromising environmental quality. Since the introduction of Bacillus thuringiensis (Bt) biotech cotton in 1996, cotton has been one of the lead crops to be genetically engineered, and biotech cotton has been one of the most rapidly adopted technologies ever. The current varieties of commercial importance address crop management or agronomic traits that assist with pest management (insect resistant) or weed control (herbicide tolerance). Nine countries representing over 59% of world cotton area allow biotech cotton to be grown: Argentina, Australia, China (Mainland), Colombia, India, Indonesia, Mexico, South Africa, and United States. Other countries (e.g., Pakistan, Brazil, Burkina Faso, Egypt, and the Philippines) are considering approving the cultivation of biotech cotton. In 2003–2004, about 21% of the world's cotton acreage and about 30% of the cotton produced in the world was biotech cotton; in 2004–2005, about 35% of world's cotton production was biotech cotton; and within five years, world biotech cotton production could be close to 50%. The initial biotech efforts have been centred on insect resistance and herbicide tolerance. Insect resistance has been conferred through the incorporation of genes from Bt that produce Bt *d*-endotoxin, a naturally occurring insect poison for bollworms and budworms. While insect resistance and herbicide tolerance are the only traits currently available in biotech cottons, a broad range of other traits are under development using modern biotechnology. These may impact the agronomic performance, stress tolerance, fibre quality, and yield potential directly. However in 2005, few of these traits are close to commercialization. As soon as new developments in bioengineered cotton for insect resistance, herbicide tolerance, stress tolerance, yield potential, improved fibre quality, etc., are available, they will be incorporated into conventional cotton varieties.

Biphenyl: A chemical used in textile dyeing e.g dyeing of Aniline Black. It is toxic to water dwelling organisms and can undergo bioaccumulation.

Bird's-egg green: A pale-green or bluish-green. See **Oxanthine**.

Birds eye: (1) A generic term describing a cloth woven on a dobby loom, with a geometric pattern having a centre dot resembling a bird's eye. Originally, bird's eye was made of cotton and used as a diaper cloth because of its absorbent qualities, but now the weave is made from a variety of fibres or fibre blends for many different end uses.

(2) A speckled effect on the back of a knit fabric resulting from the use of different colours on the face design.

(3) Also called peacock eye. Pattern is produced by regular bird's eye type weave effects. Worsted or worsted ply yarns, as far as possible of pure wool or 70–90% wool content, are used exclusively, There are several types of weaves, but an eight-shaft crepe weave is predominantly used with the yarn sequence in warp and weft of 2 dark 2 light.

Bird's eye linen: Usually made of cotton, the cloth is made with small diamond shaped weaves. Construction is about 64 × 68 with 28s warp and loosely twisted 15s filling. Made in the ranges of 18 in to 30 in width, depending on the end use. Bird's eye linen is usually bleached, but sometimes dyed in pastel shades. It is also called dobby woven diaper cloth.

Bird's eye (in knotted fabrics): An unintentional tuck stitch.

Bird's-eye weave: Floating filling yarns over warp threads in a set formation which produces this diamond shaped design. The pattern has a centre dot which resembles a bird's eye.

Bird's nest: Mat-knitted wool mat with combed-out napped surface.

Biredshend: Persian knotted carpet of very close texture, often made with motifs of palm leaves and shawl designs.

Birefingence of fibre: A term used as a measure of the orientation of the macromolecules within the fibre. The algebraic difference of the refractive index of the fibre for a plane polarized light vibrating parallel to the longitudinal axis of the fibre and the index of refraction for light vibrating perpendicular to the long axis. Dn = (Rip-RIr) Where Dn = the birefringence, Rip = the refractive index measured parallel to the fibre axis, RIr = the refractive index of the fibre measured perpendicular to the fibre axis.

Birefringence: An optical term meaning double refraction, and used in examination of manufactured fibres to measure the degree of molecular orientation effected by stretching or drawing.

Biretta: Flat cap decorated with feathers and cords.

Biretz: Double-faced woollen fabric, having ribs on one side and cashmere twilled on the reverse.

Bisage: Twice-dyed fabric.

Biscuit: The light-brown colour of a biscuit; hence 'biscuit-coloured'. From the Italian *biscotto*, 'twice-cooked'. First recorded as a colour in 1884. Also describing pottery after first being fired and prior to its decoration.

Biscuit; **Biscuit package**: A cylindrical package of yarn wound on a take-up winder as used in the extrusion of synthetic filament yarns, where the axial length of the package is small in comparison with its diameter. Biscuit may be wound on each take-up mandrel.

BISFA: Bureau International pour la Standardization de la Rayonneet de Fibres Synthetiques (International Bureau for the Standardisation of Man-made Fibres, Basle).

Bisette: (1) French term for an embroidered braid; (2) Narrow and coarse white linen pillow lace made by the peasants in Seine et Oise, France; originated in the 17th century.

BISFA Method: A method to assess crimp contraction (or crimp elasticity). This property is essentially assessed on false-twist yarns and is defined as "the contraction of a textured yarn due to crimp development, expressed as percentage rate of the stretched out (not crimped) yarn length". Method: (1) Hanks with 2500 dtex global linear mass (fil.count $\leq$ 200 dtex) with 2 cN/tex winding tensions. (2) Crimp development in oven-heated air at 120°C for 30 min (PA, PES) and cooling in standard room conditions for 12 hours. (3) 1st measurement under 2,0 cN/tex preliminary tension for 10 min (L_0). 2nd measurement under 0,01 cN/tex preliminary tension for 10 min (L_1) Calculation: $E = (L0 - L_1)/ L_0 . 100$

Bishop's lawn: A very light, fine, plain woven cotton dress goods, given a bluish starch finish. Originated in England, where it is made usually 32 inches wide, weighing five or six ounces a yard; used by the clergy.

Bismarck: A leather-brown colour.

Bismarck Brown Fr Extra: A cationic dye used for the differentiation of various textile fibres which have been subjected to different treatments, like methylene blue.

Bisonne: A French woollen fabric, made of natural grey wool used for lining garments.

Bisou: See **Bassine**.

Bisque: A pale pinkish- or yellowish-brown colour. First recorded as a colour term in 1922.

Bissonata: Coarse woollen cloth, dyed black or brown, made in France for clerical garments, in olden days

Bistre: A dark-brown colour; the brown pigment or wash used in pen and ink drawings made from boiling the soot of wood, particularly, beech wood and used by Renaissance painters. Also 'bistre brown'. Superseded by **sepia**.

Bisu; **Husks**: Silk waste remaining on cocoons at the bottom of the basin after reeling.

Bitlis: Oriental wool carpet made in Bitlis, Asiatic Turkey (Anatolia).

Bitre: Description of linen of Brabant.

Bitumen: Black **asphaltum**; an unstable brown pigment often the cause of cracking in paintings.

Bivalent: Possessing a valence of two.

Bivouac: Woollen dress goods, made with nub yarn in mixture effect.

Bivoltine: Cocoons built by silk worm race with two generations (life cycle) in a year.

Black: A very common colour for the dyer or printer, but problematic. Technically, black colour is one where all the incident light is reflected. In many dye families, such as reactive dyes, there is no such thing as a "pure" black dye: blacks are made by using mixtures of other colours, often starting with a large proportion of a navy blue. To achieve a good dark black, it typically takes much more dye than for strong shades of other colours. It is not unusual to see recommendations between 6% and 10% owg because blacks are usually mixtures, discharge of black may yield unexpected results. Coppery colours are not uncommon. Reducing agent discharge may produce different results from oxidizing agent discharge. Some dye vendors offer black dyes that are specially formulated to discharge to almost white. Sulfur dyes are extensively used commercially for black cotton fabrics, but recently there was a ban on sulphur colours in certain countries. Aniline black, not widely used now is probably the best black which can be produced on cotton.

Black board cloth: See **Broad cloth**.

Black faced: A medium long and usually kempy lustrous wool, yielded by the sheep of the same name in Scotland; used for homespun, carpets.

Black felt: The classification of felt manufactures in various shades of black.

Black fibre: Sometimes horse hair is substituted by this dark, smooth, glossy fibre obtained from the leaves of a species of caryota palm in Sri Lanka. Black fibre is the commercial name of this fibre.

Black mordant: (1) A term used for stock preparation of aniline salt used in the preparation of the impregnation of cotton fabric for the production of oxidation black. See **Black**.

(2) A mordant for logwood black pyrolignite of iron and iron (III) sulphate.

Black panel thermometer: A temperature measuring device, the sensing unit of which is coated with black colour and designed to absorb most of the radiant energy encountered in light fastness testing. It is the maximum temperature which can reach during that particular testing procedure.

Black seed: Commercial term for various American cottons with a smooth black seed.

Black standard thermometer: A temperature measuring device, the sensing unit of which is coated with a black material designed to absorb most of the radiant energy encountered in light-fastness testing and is thermally insulated by means of a plastic plate. See **Black panel Thermometer**.

Black superfine: Commercial term for the finest grade of black woollen suiting made in West of England in olden days; fulled napped, shorn and finished with a soft, lustrous face; used for dress suits.

Black thread: In flax spinning yarn spoiled by oil.

Black watch: A very dark tartan worn by the 42nd Royal Highland Regiment, composed as follows: Very dark green bar, split in the middle by a group of narrow black, navy and black lines; black stripe, half the width of the green bar; narrow blue stripe, narrow black stripe, narrow blue stripe, narrow black stripe; dark-blue stripe, half of the green in width; repeat group described between the two in reversed order.

Black (in colour chemistry): In theory, the complete absorption of incident light; the absence of any reflection. In practice, any colour that is close to this ideal in a relative viewing situation, i.e. a colour of very low saturation and very low luminance.

Blackberry: The colour of the dark-purple fruit of the *Rubus* family.

Blackcurrant: The colour of the round black berry of the *saxifrage* family.

Blackjack: Staple trade term for cotton staple containing large pieces of leaves.

Blacklight: "Longwave" ultraviolet light; wavelength typically about 365 nano-meters. Black-light lamps made with a fluorescent tube that appears very dark purple can often be found reasonably inexpensively at novelty shops and even at stationery stores where they are sold for detecting counterfeit money. Black-light lamps are useful for detecting optical brighteners in fabrics.

Black-out cloth: Different types of woven fabric, mostly highly calendared, sometimes raised on one-side dyed to a dark shade (also coated as black-out curtain 9carbon coating).

Black-out curtains: Roller blinds made from fabric composed of synthetic fibres to which one coat of carbon fabric carbon black is applied in between two polymer coatings producing a virtually total lightproof material.

Blade method of crimping: A texturizing process. The textured yarn produced with this process bore originally the mark Agilon (producer: Milliken), but at present the process is no longer in use, owing to the poor properties (low crimp stability) of the yarn.

Blade squeeze: The classical shape of squeeze in contrast to roller squeeze.

Blanc: A term used in the French dry-goods trade for everything which is bleached, irrespective of material.

Blanc de chine: Unpainted white or cream porcelain, originally made in China since the 17th century.

Blancards: A linen goods made from half-bleached flax yarn in Rouen, France. Not very common now.

Blancmange-pink: A pale pink.

Blandford lace: Fine English pillow lace of the 18th Century; made in Blandford. Now obsolete.

Blanguin: A Cuban plain woven, bleached cotton sheeting.

Blank bleach: The operation of a bleaching plant with all the chemicals but no textile material in order to check the equipment, controls, etc.

Blank dyeing: Blank dyeing is usually done to find out the influence of the actual dyeing conditions on the material in the absence of the dye which could otherwise mask any yellowing or other non-dye related problems/effects. Normally, a full dyeing cycle is done with the specific material while using all other auxiliary chemicals, etc., with omission of dye from usual recipe.

Blank print: A print produced on a textile substrate with a paste containing all the ingredients except the dye, to study the influence of the printing conditions where the presence of dye will influence the result.

Blank screen: A non-engraved screen, i.e. without coating or the design engraved on it.

Blank vat: An alkaline solution of sodium hydrosulphite as used in vat dyeing, but without the dye. This is used in stripping of dyes, stripping and levelling of vat dyes, developing of vat dyeing produced by pigment-padding method or other two-phase methods.

Blanket: (1) (i) A thick and heavily napped and twilled woollen fabric, often with an admixture of cotton; used for bed covers; horse blankets are very heavy and felted;(ii) Name for the 2–2 twill and also an eight-harness satin weave; (iii) Weaving term indicating a short length of a fabric showing one or a number of ranges of patterns made in a variety of colours.

(2) The endless strong coated fabric sheet running guided by guide rolls which is flat when it comes to the top side of the printing area, above which the flat or rotary screens are fixed and the actual printing takes place. After printing is over, the fabric goes to the drier and the blanket goes below the printing table; it is washed and dried and if necessary, gum is applied and then it comes up at the feeding end of the printing table. See **Blanket washing**.

Blanket cloth: May have been named after Thomas Banquette, a Flemish weaver living in England in the 14th century and was the first to use this material for sleeping to keep warm; but now it means any thick cloth with a heavy nap, suitable for rugs, dressing-gowns, blankets, casual coats, shawls and stoles. Warmest made in wool or worsted, but now also made from other fibres or mixtures. Not hard-wearing if made into clothing.

Blanket mark: See **Corrugation mark**, **Sanforizing mark**.

Blanket plaid: A large and vividly coloured plaid design, such as those often found on blankets

Blanket seam: A printing fault in roller printing, characterized by a light-coloured width-wise stitching mark, appearing on face side of the printed fabric, caused by the seam in the rubberized blanket beneath.

Blanket stitch and Button-hole stitch: Embroidery stitch. Method: These stitches are worked in the same way – the difference being that in button-hole stitch, the stitches are close together. Bring the thread out on the lower line, insert the needle in position in the upper line, taking a straight downward stitch with the thread under the needle point. Pull up the stitch to form a loop and repeat. This stitch may also be worked on even-weave fabric.

Blanket washer: A special compact washing machine used in roller/flat bed/ rotary printing machine for continuous washing of the printing blanket during printing operation.

Blanket (bed): See **Blanket**.

Blankets: There are many types of blankets material. It may be made of wool, rayon, cotton, dyenel or a combination of these. The principal use of blankets is for bed coverings, robes and steamer rugs. For single bed, material weighing 3 lb, with 66 × 84 in is found satisfactory. The double beds 72 × 84 with 3½ lb are found useful. The material which weighs less than 12 oz. per sq. yard is not found satisfactory as blankets. Cotton blankets are made usually with relatively fine warp and coarse filling. The filling is given a long, dense nap on both sides of the fabric. Warmness of a blanket depends on the thickness of the fabric and not on fibre content or weight. Cotton blankets lack the liveliness of wool. Blankets sheet are quite similar in construction to the heavier type flannelette. They are woven as wide as sheets and are used sometimes as top coverings. It is usually woven in plain weave, but sometimes 2/2 twill or the filling reversible weave is used. A medium weight cotton blanket will have a construction of 40 × 30 with 20s warp and 5s filling. Usually dyed in plain colours or in stripes and plaid patterns the blankets often comes with fringes.

Blanket stitch: See **Loop stitch**.

Blarney: (1) Fine Irish tweed suiting; (2) Irish woollen knitting yarn, heavier and harder than the Connaught.

Blassas: Inferior Spanish raw wool.

Blatta: Purple; also the name for silk dyed purple.

Blaze: See **Silk wadding**.

Blaze (French): The silk spun by the silkworm that serves as a scaffold to support the cocoon; also called floss silk or spelaia (Italian).

Blazer: Sports jacket. Originally a fairly light-weight jacket often striped or coloured of a sports club or school, etc., with decorative buttons and a badge bearing a coat of arms.

Blazer cloth: A type of flannel or Melton used for blazers. Usually wool, but can be mixtures. Traditionally made in stripes, now it is also made in plain colours. The right side is usually slightly napped.

Blazing: Flaming, shining. Also used as an adjective of colour especially as regards such as yellow, red and orange.

Bleach activators: Acetyl compounds which has the capacity to form peracetic acid with H_2O_2 donors in the liquor as an intermediate stage of the bleaching reaction are bleaching activators. The reaction is virtually independent on the washing and reaction temperature and proceeds preferentially at pH 9–12, at significantly lower temperature than usual. Bleaching systems are developed with sodium perborate and bleach activators such as tetra acetate glycol uril (TAGU), tetra acetyl diamine types, etc., to make high duty detergents used especially in laundering.

Bleach liquor: Sodium hypochlorite.

Bleach, Chlorine: A solution of sodium hypochlorite in water; an oxidizing bleach.

Household chlorine bleach, about 5% sodium hypochlorite, suitably diluted, can be used for whitening cellulose fabrics prior to dyeing. It is also used in some art discharge processes (this author knows of no industrial use of chlorine bleach for discharge). Chlorine bleach can damage cellulose fibres, but is safe if strength and exposure time are limited, and the bath pH is maintained at about 9.5. Goods must be thoroughly scoured prior to chlorine bleaching; otherwise impurities may form compounds that are yellow and very difficult to remove. An antichlor should usually be used to assure complete neutralization of the bleach. Do not mix chlorine bleach with acids – chlorine gas will be liberated. Do not use chlorine bleach on wool, silk or spandex - fibre damage will result. Industrially, chlorine dioxide is sometimes used for bleaching. Although it is less likely to cause cellulose fibre damage, it is not suitable for home use because the bath is very corrosive to metals, and there is significant risk of chlorine gas production. In general, chlorine bleaching is obsolescent for industrial textile preparation.

Bleach (in care of textiles): A product for brightening and aiding the removal of soils and stains from textile materials by oxidation that is inclusive of both chlorine and non-chlorine products.

Bleach (in home laundering): Products that will clean, whiten, brighten and aid the removal of soils and stains from textile materials by oxidation that is inclusive of chlorine and non-chlorine products.

Bleach (in textiles): An oxidizing or reducing agent used to partly or completely destroy natural or extraneous colouring matter in a textile, thereby leaving the textile lighter orwhiter

Bleach (oxidising): A bleach based on an oxidizing agent, such as hydrogen peroxide or sodium hypochlorite. Most general-purpose household bleaches belong to this class.

Bleach (oxygen): Bleach based on hydrogen peroxide or a chemical derivative of hydrogen peroxide. Oxygen bleaches typically are less damaging to fibres than chlorine-based bleaches. Most "colour safe" bleaches are of this type. Dry powder bleaches and detergents of this class usually use a "peroxygen" compound such as sodium perborate, sodium percarbonate or potassium monopersulphate.

Bleach, Reducing (reductive): Bleach or decolorizing compound that is based on a reducing agent. This type of bleach is typically used for discharge or stripping. Usually they are not harmful to fibres, though some processes require temperature or pH that may damage fibres. Wool is often bleached with reducing bleaches. Thiourea dioxide and sodium hydrosulphite are two compounds that find use as reductive bleaches.

Bleachability: The extent to which a particular textile can be bleached using a specific bleaching agent. Nowadays, the Whiteness Index is used for comparing the bleach obtained.

Bleached denim: Often garment processed, made up jeans washed and bleached. A precisely calculated bleaching agent is added to the wash liquor in the pre-wash which bleaches out some indigo. The garments are finally rinsed and softened as usual for the required handle, etc.

Bleaching: Any of several processes to lighten/remove the tints contained in the grey fibres and hence in the grey fabrics or yarn due to growth (in case of natural fibres) or production (in case of man-made fibres) by means of suitable bleaching agents to obtain clear whites for finished fabric or in preparation for dyeing and finishing. Commonly used bleaching agents are hydrogen peroxide or related chemicals, sodium hypochlorite or related chemicals, sodium chlorite, sodium chlorite, etc. See **Bleaching, oxidative; Bleaching, reductive**.

Bleaching agent (general): A chemical used for removing dyes or natural colouring matter on a textile material leaving white or partially white.

Bleaching agent (hydrogen peroxide): Hydrogen peroxide is used as an oxidative bleaching agent for bleaching cellulosic materials, especially cotton. This bleaching agent is more important since the other chemical, sodium hypochlorite, which is cost effective is having problem with the effluent when used. Hydrogen peroxide bleaching is carried out in the alkaline range, as per the substrate to be bleached.

Bleaching agent (sodium hypochlorite): A cost-effective bleaching agent for cellulosic material, but its AOX value as a measure of organic halogen compounds exceeds the permitted effluent pollution.

Bleaching agent (sodium chlorite): This is a very gentle bleaching agent for cotton in particular in long, liquor does not chemically damage the cellulose structure even with moderate dosing or longer treatment time. Chlorite bleach is done in acid pH of about 3.7–4.0, which enables to bleach alkali sensitive materials like polyacrylonitrile, regenerated cellulose and other fibres like cotton, viscose, acetate, etc., also.

Bleaching and catalytic damage: In peroxide bleaching of cellulosic material, the presence of metals in traces, especially iron, manganese, calcium, etc., can cause the spontaneous release of oxygen which can make local destruction of cellulosic material (pin holes) which is called catalytic damage. These meta particles can come from water, steam, alkali, stabilizer or auxiliaries. The remedy can be pre-treatment of the fabric with oxalic acid or metal chelating auxiliaries and then bleaching or bleaching along with strong chelating agents.

Bleaching auxiliaries: Products designed to improve the reliability of bleaching, accelerate the bleaching are bleaching auxiliaries. They are mainly wetting agents, stabilizers, activators, chelating agents and corrosion inhibitors are general bleaching auxiliaries.

Bleaching (degree of): Degree of whiteness obtained in a bleaching process. Whiteness index is the measure of whiteness.

Bleaching earths: They are colloidal magnesium aluminium silicate with varying iron content, used as filtering auxiliaries and decolourising dark mineral oils, fatty oils, and fat. Today there are other methods and this is not very significant.

Bleaching fastness: Colour fastness to bleaching agents, like sodium hypochlorite, peroxide, sodium chlorite, etc.

Bleaching soda: (1) Sodium dithionate or sodium hydrosulphite ; (2) A mixture of sodium carbonate and sodium silicate for water softening which is capable of binding traces of iron. Used in laundries, etc.

Bleaching stabilizers: The efficiency and effectiveness of hydrogen peroxide bleaching mainly depends on the controlled release of oxygen. If oxygen is released too fast, it can cause damage, too low a release will make the process longer and impractical. Hence stabilizers like sodium silicate (which is the best).Silicate washing off from the substrate is not that easy and hence other stabilizers like polyorganosiloxanes, sodium borate, magnesium silicate, etc., are also used.

Bleak: Pale.

Bleb: A blister or bubble on the face of a spinning jet, interrupting the extrusion of the filament from the spinneret-hole involved.

Bleb rate: The frequency of bleb formation in an extrusion operation.

Blee: Archaic term for a hue or colour.

Bleeder styles: Prints with indistinct outlines. There will be two components in the print paste of these type of prints where one component ensures the engraved area of the prints while the other component spreads beyond the print area into the unprinted area so that a contrasting colours at the edges of the printed motifs. A cotton material prepared with a naphthol if printed with a mixture of diazo-fast colour bases or salts in combination with reactive dyes will give such an effect.

Bleeding: Loss of colorant from a coloured material, say fabric or yarn in contact with liquor, (water, a solvent, or a similar liquid medium) leading to a colouration of the liquor, as a result of improper dyeing or the use of dyes of poor quality. Fabrics that bleed can cause staining of white or light shade fabrics in contact with them while wet. Usually occurs in cases where dark colours are dyed or printed with poor fastness.

Bleeding: A problem occurred during dyeing/printing where dyes with lower fastness properties are used for dark shades. When washed or undergoing any other wet treatments, the colour is bled into the liquor giving a lighter shade and in case of prints with additional problem of staining the white/dyed areas. The bleeding can also happen if proper fixation is not done.

Blench (to): To make something white or pale; to become white or pale.

Blend: (1) A yarn obtained when two or more staple fibres are combined in a textile process for producing spun yarns (e.g., at opening, carding, or drawing).

(2) A fabric that contains a blended yarn (of the same fibre content) in the warp and weft.

Blended yarn: See **Yarn (blended)**.

Blending: Process of mixing various fibres together, producing mixtures in colours or qualities. The different fibres are placed in thin layers on top of each other, taking care that no particular fibre is given prominence. The length of the fibre is a very important matter and short fibres are usually best adapted for blending. Bley Irish term for unbleached beetled linen.

Blending machine: A group of devices that are synchronized to proportion definite amounts of various grades of cotton which are to be blended together.

Blicourt: Fine French serge made in narrow widths of fine wool; used for lining garments; obsolete.

Blind chintz: Plain woven printed and calendared cottons, made in various colours and designs, mainly in stripes, and resembling Venetian blinds.

Blind fly (in garments): A material covered opening to conceal a fastening, e.g., buttons, zip, etc. A blind fly is fastened instead of left open along the finished edge.

Blind hem stitch: A complex machine stitch pattern consisting of small groups of straight or narrow zig-zag stitches separated by a side zig-zag stitch unit at consistent intervals.

Blind stitch: A stitch which does not go right through the material and hence mainly seen only one side.

Blind ticking: Made in combination of linen and cotton blind ticking is a stout twill material in any colour and stripes.

Blind twill: Trade name in England for twill showing only indistinctly.

Blinding: A marked undesirable loss of lustre of fibres after wet-processing.

Bliss tweed: Fine English woollen, similar to whipcord; used for liveries, etc.

Blister Cloth: Usually a double fabric, woven to produce blister effects.

Blister crepe': See **Cloque'**.

Blister effect: Novelty silk dress goods with Jacquard figures on crepe foundation; the crepe shrinks in the finish, producing blisters in the figures.

Blister, in bonded, fused or laminated fabric: A bulge, swelling, or similar surface condition on either face of the fabric or the backing fabric characterized by the fabric being raised from the plane of the underlying component over a limited area in bonded, fused or laminated to give a puffy appearance, mostly due to air entrapment.

Blitz: A light to medium weight worsted fabric with a filament warp and spun weft. Often has a very fine cross-wise rib. Common blends are acetate/rayon and polyester/rayon.

Block: (1) A piece of hardwood, smooth faced used for setting work during pressing.

(2) (in experimenting) A group of units that is relatively homogeneous within itself, but may differ from other group.

(3) To align warp and weft yarns at right angles, by some form of manipulation.

(4) A wooden block with designs carved on it, used in block printing. See **Block printing**.

Block (in sewing): It refers to the pieces of fabric that are sewn together to form one design element.

Block chopping: In garment, two basic approaches are used in cutting the stack of fabric into pieces of suitable size and shape to be sewn into finished products – Pattern chopping and Block chopping. *Block chopping* is a kind of rough cut that produces pieces close to the final shape, but which require some trimming cuts to bring them to the exact shape needed.

Block-co-polymer: The repeating unit consists of segments or blocks of similar monomerstied together along the macro-molecular chain.

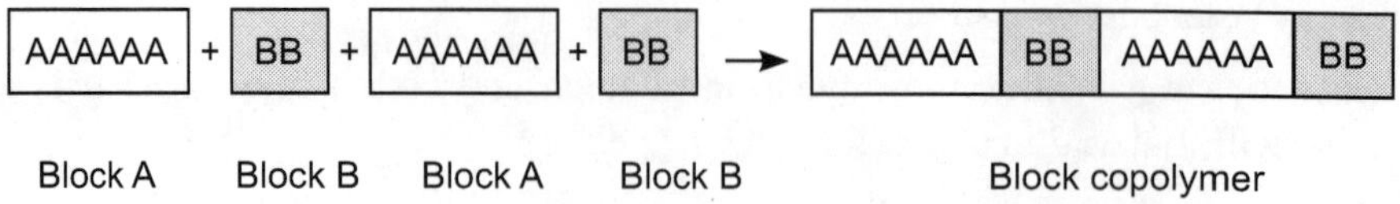

Block cutting: The manufacturing process to produce the block for hand printing. Normally, special woods are used. Pattern production by wood-working in accordance with pre-print blocks with fine contours by driving in pins and metal-sheet strips. The latter are detailed in a lime wood block by means of red hot punches and babbitting with white metal alloy which serves as a printing knife when nailed on. Pins are sometimes driven into four corners to help in the setting of different colours in a design.

Block pattern: A basic shape upon which design details can be developed.

Block polymers: These high polymers are formed from chains of different copolymers whereby, individual units are joined together in relatively long alternating sequences.

Block printing: The oldest method of printing. Blocks made of wood or metal are engraved with designs. Blocks are made where the patterns are formed like in a stamp pad, which used for printing. Each block prints only one colour. The blocks are dipped in dye solution. Only the raised portion in the blocks picks up the dye and is then pressed on the fabric, forcing the dye to be printed on the surface. Different blocks can be used for making a large motif or same blocks can be used repeatedly to give all-over design. Not used extensively, but still used in small scales. See also **Printing**.

Blocker: A frame for drying wool. A fine picture of one can be found at the cyber fibre site. This is an open frame that rests on two supports with a handle on one side. You wind the damp yarn under even tension across the frame – not trying to line anything up. Rather like winding a bobbin for weaving. After the yarn dries, you can usually slide the whole skein off of one end.

Blocking: The measurement of the development of surface tack and thermal softening point of the material.

Blocking effect in the polyamide dyeing: Mutual displacement of individual acid dyes in the dyeing of polyamide materials within the range pH 3–6. The effect is dependent on dye constitution, dye concentration, fibre saturation value and the dyeing auxiliaries used.

Blocking filters: Used in spectrophotometers for colour measurement of fluorescent white samples. The excitation of the brightening agent can be removed to a large extent with a variable U. V. filter and on the other hand with blocking filters for visible light or with a U. V. source, the values for a brightening agent alone are obtained.

Blocking out: The action of cutting many layers of material into manageable blocks which may incorporate one or more part of the garments.

Blocks for printing: For hand block-printing, blocks are made from fairly hard woods like ash, box, lime, sycamore, holly and sometimes pear wood. To facilitate the registration in the hand block-printing, brass pins are driven in at the corners of the block to indicate the position of the next repeat. See **Blocks, Block printing, Block cutting**.

Blodbende: Medieval English name for narrow silk ribbon, worn tied around the arm after bleeding.

Blond quilling: Resembles bobbin quilling, but it is made of silk, is highly sized and finished; used for frills, ruffles.

Blonde lace: Blondes bobbin or needle-point laces made originally of natural-yellow silk in the 18th Century, later also of black or coloured silk. Originally called blondes ecrus or nankings. Blondes de fantaisie had machine-made silk net ground with the design darned into. Blondes en persil with small parsley leaves strewn over the ground. Blondes des applications have bobbin or needle-point sprigs appliquéd over machine made ground. Fausse blondes are silk tulles embellished with floss silk in various stitches. A very expensive silk lace with floral designs but with boldly defined holes in the flower heads.

Blonde de Fil: Sometimes applied to mignonette lace.

Blondes en Persil: See **Blondes**.

Blood: Denotes the fineness of the wool as compared with pure merino, which is called full blood.

Blood-dark: A dark-red.

Blood-red: A dark-red colour; a rich bluish-red.

Bloom: (1) The appearance of brightness of a dyed fabric when the fabric is viewed across the top while held at eye level.

(2) In deep dyeing, the fullness or toning with a red, yellow, or blue cast, e.g., bloomy black.

(3) The colour of foam in a vat dyeing bath which in contrast to the generally quite different colour of the vatted dye appears closer to the finished dyeing (oxidised dye).

(4) The glossy finish imparted to certain woollen fabrics in the calendaring.

Blooming: The glossy finish imparted to certain woollen fabrics in the calendaring.

Blossom: A soft pink.

Blotch: (1) See **Finishing spot**.

(2) A relatively large area of uniform colour in a printed design.

Blotch (in fabric defects): An off-coloured area of any shape, caused by grease or oil.

Blotch design: A single or multi-colour printed design in which upto approximately 80–90% of the fabric is printed in one colour.

Blotch print: In direct printing, a printed design of high coverage. Also used to substitute dyed grounds for discharge print in cases where no suitable discharge dyes are available or where technical problems are encountered in coloured discharge printing. The degree of coverage of the blotch can be low, medium or high.

Blotch printing: (1) In resist printing, different colours of the design are printed and the block printing is done to cover the full design. Wherever the colour is resisted, the blotch colour won't be fixed and on the remaining place it will be fixed; (2) In normal direct printing also, the blotch printing provides a background to a design which, unlike the white grounds in normal direct printing, or the plain coloured grounds in discharge printing, consists of a distinct pattern effect. It is produced with a special printing roller either simultaneously, before or after production of the main design, e.g. for imitation weave effects, fine dots, lines, etc. It may be visible over the entire surface of the printed fabric or limited to the areas falling over other colours by resist or discharge printing.

Blotch printing machine: A roller printing machine set up to print the entire surface of a fabric with print paste or a thickened chemical solution by means of a completely engraved blotch roller (cover roller, 1000 point roller). In case of rotary machines, the same effect is done by using an unengraved rotary screen or blank screen. Nowadays, special screens are made by few manufactures for printing Blotch.

Blotch roller: See **Blotch printing machine**. Cover roller.

Blouson (French): A short jacket or top having the shape of a blouse reaching to the waist, with either a buttoned or zip front and ample sleeves. The sleeve endings are finished with elastic material.

Blow down (of boilers): A cleaning procedure for the removal of solid substances which have settled in the water reservoir of a boiler, used for steam raising purposes. The procedure is often carried out with an automatic valve operating at working pressure.

Blow ratio: See **Foam applicator**.

Blowing: The process of blowing steam through a fabric which is wound on perforated beam or roller.

Blowing room: See **Blow-room**.

Blown finish: A finish applied to wool/silk material by the process of blowing.

Blow-room; Blowing room: The room in a cotton spinning mill where opening, cleaning and blending of fibre is done.

BLRA: British Launderers Research Association.

Blue: (1) The colour of the sky or of the sea. One of the three primary colours (but not until the 16th century). The word 'blue' as the name of the colour has had an uncertain history. In some languages there is no name for the colour and it was not regarded by the ancients as a primary colour. It has been confused linguistically with the colour yellow – ***flavus*** being both the root of 'blue' and Latin for 'yellow'. In the Russian language there is no one word for blue but two words, one meaning dark-blue and the other light-blue which are regarded as different colours.

Describes any colours having wavelengths between approximately 480 and 445 nanometres. One of the three **additive primary colours**. Coloured or tinted lenses are used to correct various reading disorders including some forms of dyslexia and blue lenses apparently are the most effective. It was also a symbol of piety; associated in medieval times with the Zodiac signs Pisces and Sagittarius and with the planet Jupiter and with darkness. In English folklore, blue represents loyalty, the colour for baby boys, and supposed to bring good luck to brides who heed the superstition to wear on their wedding day '*something old, something new, something borrowed, something blue*'. Blue is the colour of the ring second from the centre in archery. Conservative – in relation to the Tory Party in the United Kingdom.

(2) English and crossbred wool, taken from the best part of the average lustre fleece; spins to 36s.

Blue bafts: Coarse muslin made in Manchester for the African trade for wearing apparel. Blue 'eating' yellow dyes: These are yellow vat dyes which accelerate the photo degradation of the blue (blue-green) vat dyes. Such dyes are only used in very low concentration along with blue dyes where it is a must.

Blue "eating" dye combinations: Catalytic photo degradation is caused by certain yellow and orange dyes on blue vat dyes. The problem occurs mainly in vat dyes.

Blue denim: Strong woven cotton fabric with a warp twill construction, dark coloured warp and light (undyed) weft. Classic fabric for jeans.

Blue flax: Trade term for flax from Bruges, Malines, Ypres, Lokeren, Ghent and other places in Belgium; it is steeped in still water and has a dark colour.

Blue jeans: See **Blue denim**.

Blue mottle: Light and sheer narrow cotton fabric made in England; it has a mottled blue effect.

Blue scale; Blue standards: A standard comparative scale from 1–8 which is used for measuring the light fastness of dyed or printed textiles. The ratings are given as per the fading of these standards along with these blue scale standards, which are dyeing on smooth wool fabric with varying degree of sensitivity to the action of sunlight; 1 = Poor, 8 = Excellent.

Blue stone: Copper Sulphate.

Blue verditer: Blue copper pigment.

Blue vitriol: Copper sulphate.

Blue Wool Scale: British standard (BS 1006-97) commonly referred to as 'the Blue Wool Scale' (BWS) adopted as an International Standard (ISO) and used by some manufacturers as the standard for measuring the **light fastness** of their pigments – BWS 7 and 8 indicating an extremely permanent pigment. See also **ASTM**.

Blue Wool LF Standard: One of the groups of dyed wool fabrics that are sensitive to the amount of light, heat and moisture to which they are exposed.

Blue-black: Black with a touch of blue.

Bluebell: The blue colour of the plant of the same name having blue bell-shaped flowers, also called 'wood hyacinth'.

Blueing agents for bleach goods: Blue colour has the masking effect on bleached yellowish grounds and hence these are padded to give the bluing effect. Blue dyes are generally known as tinting dyes (or tints) for bleached goods. See **Ultramarine blue**.

Bluette: A weft faced 2 × 2 twill weave fabric used for overalls, originally made from blue dyed yarns but now more frequently pieces are dyed. A typical construction was 20 × 16s, 44 × 108 loom state k = 10 × 27; 30 in.

Bluff edges (in garments): Edges finished without outside stitching.

Bluffing: The process of fastening down the front edges of the facings on to the forepart canvas to preserve the shape of edges finished without outside stitching.

Blueing: See **Blueing agents for bleach goods**.

Blumly linen: Linen with twill weave and printed with designs where the ground is coloured and the flowers of white, red and blue colours stand out. Made in Switzerland.

Blunk: A stout cotton or linen fabric made in England, used mainly for printing purposes.

Blanket: A greyish-blue.

Blur (to): To smear paint; to make indistinct.

Blush: Red-coloured; having the same colour as a blush.

Blush-pink: Delicate pink.

Blush rose: A delicate pink; also seen as a dark shade of rose.

Blyat: Bleaunt or Pliat rich silk dress goods of the Middle Ages.

Board (tailors): The place where the tailor works, usually in the form of a table, at which he sits when sewing and at which he stands when he is doing other works. The cutter works at the cutting board.

Boarding: A process of drying and shape setting of stockings after they have been dyed. The stocking are placed on either heated metal forms or wooden forms that are heated in a boarding machine.

Boardy: A term used to describe a fabric with a very stiff hand.

Bob: (1) A commercial variety of American cotton, the early ripening staple measuring 25–28 mm, forming medium size balls; the yield of lint is 30–32%. Also called Ozier; (2) In pillow and needle laces, the ornament connecting the bars between the patterns. Identical with the crescent crown, spine or thorn.

Bobbin: (1) A spool containing the lower thread in a sewing machine.

(2) A cylindrical or slightly tapered with or without flanges or flange, for holding slubbings, rovings or yarn. The term is usually qualified to indicate the purpose of which it is used. e.g., ring bobbin, weft bobbin, or bottle bobbin.

Bobbin cleaner: A machine that removes any remaining yarn or roving from bobbins (quills) after they have been used to the maximum possible extent in the looms, spinning achiness, or in winding machines

Bobbin fining: Trade term for machine-made laces similar to the shadow lace, usually with a heavy thread outlining the pattern.

Bobbin lace: (1) Net-like open work fabrics produced with fine threads (by hand or machine) that are fed from bobbins and worked into a pattern by twisting together and crossing of strong ply yarns.

(2) Also called Pillow Lace, it is hand-made using fancy bobbins of bone, ivory, and now wood, on a hard stuffed pillow, and using long pins with coloured glass or plastic heads to anchor the lace to the pillow. Punched paper patterns are first fixed to the pillow; it is a very old craft.

Bobbin (in sewing): A bobbin is a small spindle that is inserted into the hook of lock-stitch sewing machines. Sewing machine bobbins can be wound by the sewer or pre-wound bobbins can be purchased.

Bobbinet: (1) A fine machine-made net used as a foundation for embroidery and lace and also used for bridal veils.

(2) Net curtains or laces with at least three yarn systems. Binding or bobbinet ends which spirally wrap around warp; patterning ends are its characteristic.

(3) See **Plain net**.

Bobbin lace: See Lace, bobbin.

Bobbin net: A machine-made copy of the original bobbin lace. A machine to perform this action was first invented in the early nineteenth century and produced a hole-mesh fabric.

Bobtex yarn: Simulated spun yarns produced by embedding individual staple fibres in a thermoplastic or adhesive coating on a filament yarn. In principle, a thermoplastic polymer is applied to a filament yarn (as the strength component) and coated with staple fibres before the polymer sets in order to improve bulk and textile appearance. Yarn consolidation is achieved by means of false twist.

Bobtex: ICS yarn system.

Bocasine: A plain woven and very firm linen fabric, given a high finish; obsolete.

Boccadillos: In the Spanish and South American trade a thin, bleached and sized linen, used for shirts, etc.

Boccage: A damask table linen, made in Calvados, France.

Boccassini: Fine, muslin-like, bleached cotton cloth, made in the southern Balkan States.

BOD: Biochemical Oxygen Demand.

Bodiasse: A common Chinese silk.

Bodice: The section of women's garment covering the body from the neck to waist.

Bodkin: (1) A pointed instrument for piercing holes in the cloth.

(2) These are large needles with large eyes and are used for threading elastic or ribbon through castings. The rounded tip is used for pushing out the stitched corners of collars.

An ideal needle size for hand sewing is Pony-Crewel No. 9 because of the length and fineness. It has a large eye for easy threading.

(3) A bone or plastic instrument for removing bastings.

Body: The compact, solid or firm feel of a fabric. Usually obtained by finishing the fabric with PVA, Plasticisers, Polyvinyl acetate, etc., or a mixture along with other finishing agents to get the desired other qualities required along with body.

Body and border carpet: Wall to wall carpeting supplied in width upto a metre which is laid up to the wall skirting board.

Body blanks: Knitted panels with integral ribs, which are then cut and sewn into garments.

Body feed (in effluent treatment): The pre-coat medium that is continuously added to the filter while it is on stream. Its purpose is to create a permeable filter cake.

Body measurements (in garment construction): A body measurement that can be used to build a sizing system or to select an appropriately sized garment.

Body tolerance: See **Allowance (body)**.

Bofu: French silk dress goods of the Middle Ages.

Bogotana: White madapolam in Colombia from 28 to 35 inches wide, having a soft finish.

Bohemian flax: Fine flax of long, clean and lustrous fibres.

Bohemian lace: Bobbin or machine lace with flowing tape on a net ground as characteristic feature. Mostly made as applique lace.

Boil: In general, to heat or maintain a solution at the temperature where the vapour pressure of a liquid equals atmospheric pressure, that is, its boiling point; in dyeing sometime called “atmospheric boil” to distinguish from boiling under pressure.

In dyeing, unless otherwise specified, it is assumed that the solvent is water; therefore boiling occurs at 100°C (212°F), at sea level. At higher elevations, the boiling point will be reduced. Dissolved solids (solutes) can increase the boiling point, but it is rare for dyeing or preparation liquors to contain enough solute to raise the boiling point by more than a few degrees. In order to get aqueous (water) solutions to a temperature higher than the normal boiling point, the pressure must be increased, using a closed vessel – a “pressure cooker”. Closed pressure vessels are used extensively in dyeing of polyester with *disperse dyes*, and may be used in preparation processes such as *scouring* to 7 hasten the process. Operating temperature of around 130°C is quite common.

Boil off: See **Scouring, Boiled-off silk**.

Boiled wool: A process given to woven or knitted wool or woollen fabric to give a coarse crepe-like look by heavy felting or by putting them in a high temperature bath.

Boiled-off silk: See **Silk (boiled off)**. Silk which has all of its natural gum removed by warm solution of soap. The gum is from 17 to 25% of the weight.

Boiled-off cocoons: Undyed thrown silk/raw silk from which natural gum or sericin has been removed by boiling in soap solution. Boiling is off the process of removing sericin from silk or other gummy material from fibres like ramie.

Boiled-off-silk waste: It is degummed silk waste. However in big filatures, basin refuse is subjected to vigorous boiling in alkaline bath mainly to separate out the pupae from the silk waste and the resulting waste is also referred to as boiled-off-silk waste.

Boiler feed water: Water required for boilers, for raising steam. In normal case there is specification for boiler feed water.

Boiler horse power: A traditional unit measuring the power delivered by a boiler. The boiler horsepower is defined to be the power required to convert 30 pounds (13.61 kilograms) per hour of water at 100 °F (37.78 °C) to saturated steam at a pressure of 70 pounds per square inch gauge (482.6 kilopascals gauge). This power, about 33 471 Btu per hour or 9.8095 kilowatts, is more than 13 times the usual mechanical definition of the horsepower, but it is judged sufficient to run an engine producing one horsepower of mechanical power.

Boiler scale: A more or less hard deposits in the boiler which is caused by the use of untreated or inadequately treated boiler feed water. Boiler scales are firmly adhered to the surfaces of the plates and tubes in contact with water in boilers. If excessive, it leads to over-heating of the metal and ultimately failure. Boiler scale usually contains calcium carbonate and sulphate, magnesium salts and silicates.

Boiler (automatic): Fully automatically controlled steam boiler. This term says nothing about type of construction or type of boiler, but is applied both to flame tube, smoke tube and Benson boilers.

Boiling out: A term for yarn scouring.

Boiling water shrinkage: A test designed to measure shrinkage in a cord, yarn, or high-shrinkage fibre when it is immersed in boiling water while under a tension of 0.05 g/denier.

Bolivia: An expensive, soft, plush fabric usually of wool, and often containing a small amount of special fibre, such as Alpaca or Mohair. It is closely woven and of fine structure with a cut-in ribs running in the warp direction or in the diagonal direction. A thick and firm fabric that is generally used for coats, cloaks, but sometimes it is used as suiting material also. It comes in three separate weight groups – light, medium and heavy variety, ranging from 8 to 16 oz. per yard. Bolivia is made in a 3/3 twill weave.

Bolt of cloth: A rolled or folded length of cloth.

Bois de rose: The dark rich-brown colour of rosewood.

Bokas: A type of blue and white cotton cloth of Surat, India.

Bokhara or Tekke turkoman: (1) All-wool knotted carpet with long pile. Ground is usually of bright-red or reddish-brown with various ochre and other coloured patterns. The weave is not close, the patterns being mostly of geometrical nature; (2) A very light, transparent, natural coloured silk of plain weave.

Bokhara khilim: Made in Turkey, by embroidering square and octagonal designs on a dark red, woven foundation.

Bola: Strong and long fibre, not affected by water, yielded by the bark of some hibiscus tree in India; used for cordage.

Bolbees: (1) A coarse, bleached linen made in Normandy;(2) A light-blue coloured French linen.

Bold: As regards colour – emphasised, standing out and striking.

Bole: An orange-red or reddish-brown earth colour used as a pigment; originated from the unctuous clays of the same name.

Bolivar: A light and loosely woven all-wool flannel dyed in grey shade.

Bolivar county: A commercial variety of cotton from Louisiana, the early maturing bolls yielding about 30% of lint.

Bolivia (Elysian): Made of wool; it sometimes contains alpaca or mohair, woven in Twill – usually 3 up and 3 down. A pile weave (cut) with a diagonal pattern. Pile face which varies in depth. It is soft and has a velvety feel, usually piece dyed and has lines or ridges in the warp or in a diagonal direction on one side. It comes in light, medium and heavy weights. Uses: Cloaking and coatings and some suits.

Boll: A seed case and its contents, as of cotton or flax.

Bolduc: Tape made of linen, cotton or other yarns secured together with adhesive, without weft, often printed. Used mainly in tying of parcels etc.

Bollies: Cotton obtained from the half open or small bolls.

Bolo-bolo: A West African, very strong and long fibre that is yielded by the Honckenya ficifolia.

Bologna gauze: Finest silk gauze, craped, and dyed black; used for mourning veils; white for veils.

Bolsa: Cotton bagging made in Argentina; usually a three-harness drill.

Bolt: (1) A roll or piece of fabric of varying length.

(2) A roll of ribbon 10 yards long.

(3) It is the typical unit of fabric that is packaged and sold by manufacturers in few countries. A bolt usually consists of 12–20 yards.

Bolting cloth: An industrial cloth originally made from silk and used for screen printing and also in flour mills as a sieve. Nowadays, a variety of light-weight open fabrics characterised by their fine and uniform mesh, used for sifting flour (millers bolting cloth), voile fabric window gratings, as well as screen gauze or silk gauze for screen printing. Both warp and weft are accurately spaced and are woven in simple leno or other non-slip construction in order to maintain the mesh size. Bolting cloth contain 50–200mesh opening per sq cm in coarse qualities, 200–400 in medium qualities and up to 3000 in fine qualities.

Bolting silk: Bolting cloth of tightly twisted silk ply yarn, but also of other fibre material like cotton, flax, wool, etc.

Bolton: Counts of fine cotton yarn made of Sea Island cotton in England.

Bolton sheeting: A sheeting of 2/2 twill weave contains condenser weft.

Bombanas: Fibres yielded by the leaves of the Panama palm; used for hat braids, etc.

Bombas: Old name for Brazilian cotton.

Bombast: (1) Any soft fabric which is used for padding; hair or wool stuffing for clothing, during 16–17 century; (2) A medieval oriental cotton fabric.

Bombastic: Having a pale-yellow colour.

Bombax: (1) Cotton tree of the *Malvacea* family, the white or brownish fibre used for stuffing pillows, etc., in South America and West Indies; (2) Medieval name for Bombasine.

Bombay: Old term for a medium weight, grey cotton sheeting in Canada.

Bombay aloe: Fibre-strong fibre, yielded by the leaves of the bastard aloe in India; used for cordage.

Bombay hemp: Gambo Hemp or Bimlipatam strong fibre, yielded by a variety of hemp in India; used as substitute for jute.

Bombazet: A smooth, plain woven or twilled cloth of hard spun English worsted yarn, with single warp, finished without a glaze, found in France.

Bombazine: One of the oldest fabrics known, this was fine silk or wool fabric of plain or twill weave for formal clothes. In black colour, it was the traditional mourning cloth. It is now made from silk warp and worsted weft with imitations made from viscose or cotton.

Bombe: French term in embroidery, dressmaking, meaning puffed, or rounded.

Bombycinous: Having a pale-yellow colour.

Bombyx mori: A species of moth, cultivated silk worm of which feeds on mulberry leaves and produces the finest quality natural silk.

Bonbon: Fine French hosiery.

Bond: In general, of these bond types, hydrogen bonds are weakest, ionic bonds intermediate in strength, and covalent bonds are strongest. Van der Waals forces are something of a special case. See hydrogen bond, ionic bond, covalent bond, van der Vaals forces.

Bond strength (in bonded laminated fabrics): The tensile force expressed in g/cm of width required to separate the component layers under specified conditions.

Bonde: A loosely woven fabric with a curly, hairy surface, usually made with a jersey or stockinet body.

Bonded fabric: (1) A fabric containing two or more layers of cloth joined together with resin, rubber, foam, or adhesive, that does not significantly add to the thickness of the combined fabric, to form one ply. *NOTE 1*: In this context, a thin layer of foam is considered an adhesive when the cell structure is completely collapsed by a flame; *NOTE 2*: Normally, but not always, the backing fabric may be tricot or non-woven.

(2) See **Non-woven fabric**.

(3) See **Fabric, bonded**.

Bonded fabrics: Generic term for non-woven and non-knitted bonded textiles fabric produced from fibres (fibre composites), yarns (yarn composites) or fabrics (laminates). They are sub-divide into yarn composites, fibre composites, stitch bonded composites, swell process non-woven, fused bonded non-woven, spun-bonded non-woven, extruded composites non-woven, milled composite fabrics and knitted composites.

Bonded fibre fabric: Known as non-woven, these are made from a variety of fibres used on the same principle as in felt, but bonded together chemically as, unlike wool, they do not have inherent felting properties. Bonded fibre fabrics on non-woven have developed in many directions, which include household cloths and cleaning pads, industrial and military protective clothing, disposables and as interfacing and other haberdashery items. As non-woven do not fray and have no grain they can be used economically.

Bonded finish: Refers to a finish applied to continuous filament nylon and polyester threads which coats the fibres, giving the thread better ply security and abrasion resistance.

Bonded pile carpets: Pile carpets produced by adhesive bonding a folded material of yarns, webs or fibres to a substrate.

Bonded sewing thread: Treating continuous filament nylon or polyester with a special resin that encapsulates the filaments is called bonding. The result is a tough smooth coating that adds significantly to the thread's ability to resist abrasion and greatly enhances ply security.

Bondi process: Method of producing textile floor coverings. Yarn sheets or loosely woven fabrics are layered in zig-zag form into a low stack, which is stuck on to a backing material and cut.

Bonding: (1) A process for adhesive laminating two or more fabrics, or fabric and a layer of plastic foam. There are two methods: the flame method, used for bonding foam; and the adhesive method, used for bonding face and backing fabrics.

(2) One of the several processes of binding fibres into thin sheets, webs, or battings by means of adhesives, plastics, or cohesion (self-bonding). Also see **Needled fabrics** and **Needle loom**.

Bonding agent (in printing): In screen printing, a bonding agent is applied to the blanket on flat bed and rotary screen printing machine before coating with a thermoplast adhesive in order to ensure satisfactory bonding of the latter.

Bonding agent (in carpets): A material applied to fibre or yarn assemblies with the intention of maintaining the assembly in the required confrontation.

Bonding strength: (1) The amount of force required to delaminate a piece of woven or knitted fabric from its backing.

(2) The amount of force required to break the fusion points found in certain non-woven.

(3) The amount of force required to break the chemical bonds between atoms in molecules and crystalline salts.

(4) See **Peel adhesion**.

Bonding of textiles: Bonding of two textile materials by means of a locally applied adhesive.

Bonding with binder fibres: Specially engineered low melting point fibres are blended with other fibres in a web, so that a uniformly bonded structure can be generated at low temperature by fusion of the binder fibre with adjacent fibre.

Bond strength (of bonded fabrics): The tensile force expressed in ounces per inch (25mm) of width, required to separate the component layers under specified conditions.

Bone black: An ancient black pigment derived from carbonised bones and used to the present day.

Bone brown: Charred bone dust.

Bone white: A white pigment used since the Middle Ages consisting of the main of calcium phosphate.

Bongra: Plain woven and coarse fabric made by the natives of East India from nettle fibres.

Bonnet: A ladies head-dress covering the back of the head and having the brim in funnel form to shade the face.

Bonnet (in cleaning pile floor covering): An absorbent pad which can be mounted under a rotary shampoo machine.

Bonny blue: A blue colour, also referred to as 'Scotch blue'.

Bontane: Oblong pieces of cotton cloth, usually dyed blue and red; worn as loincloth in Africa.

Bonten: Coarse, sailor's linen with blue and white or red and white checks.

Book: (1) A bundle of skeins of raw silk. A bale of Japanese silk contains 30 books, each 2kg in weight and containing about 30 skeins; (2) A parcel of hanks of raw silk whose mass is usually 2kg.

Book cloth: The basic cloth for a book cloth may be coarse print goods or sheeting in plain weave treated with pyroxylin or starch and clay filled and calendared and used in book binding. The finished fabric may be plain or embossed. The pyroxylin treated fabrics are washable or water-proof and do not blister, chip or peel, while starch and clay filled fabrics on the other hand lose their colour when rubbed with a damp cloth.

Book fold: A method of folding finished fabric in which the fabric is first folded in half width wise, then folded back and forth in equal lengths. Finally, the fold edge on each side is folded to the inside, forming a compact bundle equal in length to one-half the width of the goods.

Book fold processing: Procedure engaged for processing some fine fabrics as viscose, silk, etc. The material is folded in the book form with laps of 90–100cm. This is then sewn with loops of twine (about 25cm long) through one selvedge only. The fabric is then hung into the scouring or soap bath horizontally using a pole which passes through this loop so that the fabric is completely immersed. Good crepe effect is produced by this method. This method is also followed for degumming and weighting of silk.

Book harness: Muslin; a light cotton muslin, upon which the figures are formed by a heavy, extra weft yarn of slack twist, the floats being cut away; used for curtains, etc.

Book muslin: Expensive white muslin-type fabric that is heavily sized. Used for stiffening bags, hats, wide belts, etc.

Bookbinders' cloth: Plain woven cotton fabric, heavily starched and glazed; used for book binding.

Booking: Coarse English woollen flannel resembling baize.

Booking bay: An 18th Century English worsted fabric.

Boomazey: Twilled Russian cotton cloth, printed on the face and napped on the back; used for shirts, dresses, etc., in the winter.

Boombi: Leaf fibre yielded by the Xerotes longifolia in Australia; used for coarse bags.

Booster: Wet steamer.

Booster boxes: American system of small boxes of cooled blank vat solutions as built-in unit to steamers, e.g. in the pad steam process.

Boot hose: A type of stocking used in the 17th Centur y, usually without feet worn in the soft fashionable boots with turned cup shaped tops (bucket tops). The lace-edge of the boot hose is turned over the boot tops. They were worn over the silk stockings to protect them in the long boots.

Borandjik: A very fine white cotton muslin, often crinkled; used for dresses by peasant women in Sertvia and Roumania.

Borassus piasaava: See **Palm fibre**.

Borato: A light-weight fabric of silk and fine wool, produced in the 17th Century in England.

Bordadillo: A floral-patterned silk taffeta fabric used for women's wear in Spain.

Bordat: Coarse cotton fabric used in lower Egypt by the poorer classes for making clothes.

Bordati: Mixture of cotton and silk made in Genoa; made plain or with flower or stripe patterns; some grades used in the Orient as loincloth in earlier days. Now not very common.

Bordeaux: A bluish-red; the colour of Bordeaux wines.

Bordeaux blue: A shade of blue.

Border: Printed or woven pattern which runs along the edge of the fabric; part of pattern in lace, which forms the outer edge.

Border (in sewing): It refers to the pieces of fabric sewn around a finished quilt top to make it larger, frame it or add design elements.

Boric acid/Borax: A mixture of boric acid/borax (sodium borate) is a commonly used non-durable flame retardant finish for cellulosic fibres. It is the safest with regard to carbon monoxide and smoke production during burning.

Boron fibre: A vapour-deposited filament made by depositing boron on a heated tungsten wire. These fibres are being developed for use in aircraft and space applications. They can be woven into fabrics.

Boron fibres: Vapour deposited filament made by depositing vapour phase boron from boron halogen compound on a heated tungsten wire. These fibres can be woven into fabrics, but found more useful in aircraft and space applications for rotors, turbines, etc.

Boron carbide fibres: Manufactured by the reaction of boron chloride and hydrogen with carbon fibres at high temperatures whereby a layer of B4C is formed on the carbon surface. Used in protective clothing against radiation, heat insulation, etc.

Boron nitride fibres: Manufactured by the reaction of boron oxide fibres with ammonia through several intermediate stages of increasing temperature (upto 1800 °C). The fibre consists of 99% BN. It is used in aircraft and space applications, heat and electrical insulations, protective clothing against atomic heatflows, etc.

Bordered fabrics: Woven or knitted fabrics with designs running parallel to the selvedge; printed, embroidered, open-work effects, etc. Used for blouses, dress fabrics, home furnishings and household textiles.

Boski: A closely woven dress material, usually from spun silk warp and weft.

Bosnia rug: Originally knitted on plain frames in widths of about 16 inches; patterns composed of straight lines.

Boss: That part of a drafting roll of largest diameter where the fibres are gripped. It may be an integral part of the roll, as in steel rolls, or it may have a covering of leather, cork, etc. In the former case, the boss is fluted.

Bostous: French fabric, woven stout of ply yarn of silk, wool or linen.

Botany twill: Cloth woven from botany quality worsted yarns in various weights. The weaves are 2 and 2 or 4 and 4 twills producing smooth fabric used for men's and women's suiting. An expensive cloth as botany is the top quality worsted yarn.

Botany wool: The term used to describe yarns, fabrics and 'tops' made from fine merino wool.

Botticelli blue: A pale greyish-blue.

Botticelli pink: A shade of magenta.

Bottle bobbin: A bobbin that has a cylindrical barrel and a conical or flanged base, and from which yarn can be withdrawn over the nose, i.e. the top of the barrel. The shape of the fully wound bobbin is that of a cylinder with a conical top.

Bottle green: The dark-green of some wine bottles; also a dark yellowish-green.

Bottom assembly: The components of the lower-most part of a slide fastener which determine whether the slide fastener will be non-separable or separable.

Bottom hair: Short, fine animal hair in the hair undergrowth, generally without any medulla, supple, elastic, high crimp and good felting properties.

Bottom stop (in zipper): A part affixed to both stringers immediately below or over the chain holding the two stringers together at the bottom and preventing the slider from leaving the chain.

Bottom thread: Refers to the under thread in a stitch formation, usually called a bobbin thread.

Boucle carpets: Carpet with uncut pile loops (loop pile carpets). Since up to approximately 1960, the pile material in loop carpets mostly consisted of hair yarn, boucle carpets were often, but incorrectly, used as a synonym for hair yarn loop pile carpets.

Bouclé: A fabric woven or knit with bouclé yarns; so called because of the uneven (looped) surface. It is usually a threefold ply yarn for worsted or woollen yarn twisted into loops or knots. A mohair loop thread is sometimes used (for lustre) in women's dress goods, sportswear and coatings. Derived from the French word 'boucle', meaning curled riblet or buckled. Boucle yarn may alternate with plain yarn, and the fabric may be all wool, rayon, cotton or a combination of fibres. Sometimes used to denote terry cloths. It is used for sportswear and coats.

Boucle' knit: Knit fabric with a boucle effect on the right side. This is achieved by the introduction of an inlaid yarn fed into the knitting machine and trapped as the knit stitch is made. The fabrics are of medium weight and are usually produced from polyester, nylon and acrylic yarns. Used for sweater dresses, sweater-type tops and jackets.

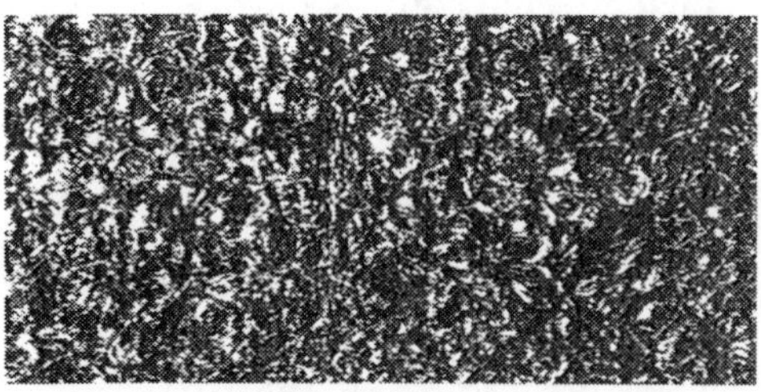

Boucle' process: See **Boucle-ondule process.**

Boucle' yarn: A novelty yarn with loops which gives fabric a rough appearance.

Compound yarn produced from coarse wool as well as cotton or man-made fibres comprising of a twisted core with an effect yarn wrapped around it so as to obtain wavy projection on its surface. Bouclé yarns may be made from wool, cotton, silk, linen, manufactured fibres, or combinations of fibres.

Boucle'-ondule' process: Carpet manufacturing process in which undulated yarn fabric or webs are laminated on to a backing fabric.

Boudary friction: Friction at low sliding speed (0.02 m/m or less) where lubrication occurs under thin film lubricant conditions.

Bougran: Also called bougrassin, bugram, tailors canvas. Gauze-like cotton or spun viscose inter-lining grey or black in colour, mostly flat in shade. Used mainly instead of excessively thick and heavy buckram.

Bound pile carpet: See **Under carpet**.

Bound seam finish: Recommended for medium to heavy weight fabrics (e.g. corduroy, fleece, etc.). It is a finish for the raw edges of a plain seam in which another fabric encloses the raw edges of one or more seam allowances. It is an excellent edge finish fir, an unlined coat or jacket. Fold under 6 mm (1/4 in.) on the long edges of the tape cut from the lining or under-lining material and press in the edges. Fold the tape length-wise a little off centre so one edge extends slightly beyond the other. Place the bias tape over the cut edge with wider side underneath. Edge stitch close to the top fold line of the tape checking to be sure the seam edge and the tape edge underneath are firmly caught in the stitching.

Bourdalou: Hat-ribbon, finer than grosgrain round the foot of the crown of hats. This trimming is sometimes finished off with a buckle and has been in use since the 17th Century.

Bourdon cord: A cord consisting essentially of a core yarn made up from a heavy central core encased in a wrapping of continuous filament yarn. It is used as a means of accentuating or outlining the motifs in lace fabric so as to give a three dimensional effect to the structure.

Bourdonette: The so called cord produced by twisting several yarns together for use as a heavy thread in lace fabric so as to simulate a bourdon cord.

Bourette: Yarn produced from carded short fibres of lower quality silk waste and combing noils in schappe spinning which is interspersed with knots and slubs. Silk noil (French).

Bourre: Scraps produced during reeling the skeins, twisting and cleaning the threads (Also termed stazza), gum waste (French).

Bourette silk: See **Bourette**.

Bourrelet: A double-knit fabric with raised loops running horizontally across the surface of the cloth, giving a rippled or corded effect.

Bournous: A long woollen cloak with hood, worn by men in the Orient.

Boutonne': Knobbly surface made by knop yarns

Bow: (1) The greatest distance, measured parallel to the selvages, between a filling yarn and a straight line drawn between the points at which this yarn meets the selvages. Bow may be expressed directly in inches or as a percentage of the width of the fabric at that point.

(2) A fabric condition resulting when weft yarns or knitted courses are displaced from a line perpendicular to the selvedge and form one or more arcs across the width of the fabric.

(3) Curvature of the warp or weft. A fabric is said to be warp-bowed or weft-bowed, according to which set of threads is curved. Weft bow may or may not extend over the full width of the fabric.

Bow straightener: A device for the correction of bow distortion in fabric webs, e.g. with curved rolls.

Bow (double): Two fabric bows arcing in the same direction, as a flattened W or M (depending on the point of view or in the opposite directions).

Bow (filling): The greatest distance measured parallel to the selvedge between a filling yarn and a straight line drawn between the points at which this yarn meets the selvedge.

Bowl: One of a pair of large rollers making a nip.

Box cloth: A thick, heavily milled woollen cloth in 2 and 2 twill, which is buff or tan-coloured cloth like melt on. Used for overcoats and suits and riding clothes.

Box crepe': A crepe made with alternating ends and picks of S twist and Z twist. Since box looms were originally used to make this crepe, it is sometimes called box crepe.

Box leather: Always known as a good quality leather it is 'boarded', i.e. the surface is broken up with parallel creases, a process only applied to top quality smooth leather. It is very expensive and used for making shoes, handbags, etc.

Box loom: A loom using two or more shuttles for weaving fabrics with filling yarns that differ in fibre type, colour, twist, level, or yarn size. The box motion is automatic, changing from one shuttle to another. Examples of fabrics made on box looms are crepes and ginghams.

Box mark: A fine line parallel to the filling caused by shuttle damage to a group of filling yarn. See **Shuttle mark**.

Box motion: A mechanism available for the weaving machines with the shuttles which allow the use of more than one supply of weft. The mechanism holds two or more shuttles and either rotates (revolving box motion) or moves vertically (drop motion to place the required shuttle in the picking position).

Box pleats: A form of pleats achieved by double folding the material alternatively in the opposite direction.

Box stain: See **Box mark**.

Boxtruck: A box mounted on wheels and used for hauling such articles as bobbins and spools from one department of a mill to another.

Bound seam finish: A finish for the raw edges of a plain seam in which another fabric encloses the raw edges of one or more seam allowances.

Boy, Boi: Coarse and loosely woven flannel lining made of cotton warp and wool and noil filling.

Boyau: Dress goods with alternated stripes of satin and rep over a figured taffeta foundation.

Boyd prolific: A commercial variety of upland cotton, the staple, maturing in medium time, measures 20–24 mm, forming small bolls and yielding 30–32% lint.

BP: British Patent, Benzoyl Peroxide.

BPC: British Productivity Council.

BPI: Bekleidungsphysiologisches Institut Hohenstein (German Institute for Clothing Physiology).

Brabant: A Belgian product using grey or half-bleached canvas, made of flax waste.

Brabant edge: Combination of Brussels and Venetian edge worked alternately; used in needle-point lace.

Brabant lace: Same as Brussels Lace.

Brabant linen or Blaams linen: Made around Ghent, Belgium, by the peasants.

Brace web (Suspended web): An elastic or rigid non-woven narrow fabric that is usually 20–30 mm in width and used for the strap of trouser supports known as braces or suspenders.

Bracken: The olive-brown or orange-brown colour of bracken.

Bracking: Trade term for the compulsory sorting or grading of flax in Russia.

Bradford stuff: Fine worsted dress goods made at Bradford, England.

Bradford system or English system: In spinning worsted yarns, the wool is oiled before it is combed, the slivers are drawn by rolls, are twisted into a rope and wound on bobbins; the yarn is composed of parallel fibres and is quite smooth. See **French system**.

Bragg long staple: A commercial variety of American cotton; the staple is long, but not of uniform length, forming large and late maturing bolls; the yield of lint is 30%.

Braid: (1) A narrow textile tubular or flat band, often used as trimming or binding, formed by plaiting several strands of yarn. The fabric is formed by interfacing the yarns diagonally to the production axis of the material.

(2) In aerospace textiles, a system of three or more yarns which are interlaced in such a way that no two yarns are twisted around each other.

Braid angle: The acute angle measured from the axis of a fabric or rope to a braiding yarn.

Braided fabric: A narrow fabric made by braids. They are frequently used in shoe laces and suspenders.

Braided fabric: A structure produced by interlacing several ends of yarn in a manner such that the paths of the yarns are no parallel to the fabric axis or crossing a number of strands diagonally so that each strand passed alternatively over or under one or more of the other strands. Usually, only narrow width is produced for cords, shoe laces, suspenders, braids for coats and uniforms.

Braided rug: The old fashioned rag rug; the strips of rags are first twisted into wide and flat braids which are wound spirally around a centre, forming a mat.

Braiding: The intertwining of three or more strands to make a cord. The strand forms a regular diagonal pattern down the length of the cord. Main fibres used in braiding are palm fibre, leaf fibre, pandanus fibre, straw fibre and bast fibres.

Braiding fibres: Fibres used for making braids. For e.g., palm fibre, bast fibre, pandanus fibre, leaf fibre and straw fibre.

Braids: Plain or figured narrow fabrics with woven in velour; also conically woven. It is used as velour or clothing (seam) protective edgings.

Bramante: White cotton cloth with stiff starch dressing and semi-glazed finish, in Colombia; it is from 26 to 35 inches wide and is coarser than the Bogotana.

Bran: The brown colour of bran.

Bran or Kleanka: A type of Russian buckram.

Branched chain polymer: A polymer having extensions of polymer chains attached to the polymer backbone. In this type of polymers, the monomeric units are linked to constitute long chains which are also called main-chains. There are side chains of different lengths which constitute branches. Branched chain polymers are irregularly packed and thus they have low density, lower tensile strength and lower melting points as compared to linear polymers. Amylopectin and glycogen are common examples of such type.

Branched chain polymer

Brand: A trade name identifying the manufacturer or product.

Brandeum: Rich silk-and-wool fabric in medieval England; used for palls, girdles, etc.

Brandom process: An English dyeing system for producing multi-coloured yarns. It is based on the principle of local dyeing of yarn in the package form.

Brandenburgs: Long, narrow, ornamental buttons, of silk or wool, on a wooden mould; used chiefly on military overcoats.

Brandon process: A process to produce adhesive bonded carpet. Pile yarn is wrapped spirally around rectangular metal rod to produce an adhesive bonded carpet. The rods are then pressed into a substrate provided with adhesive after the windings have been cut and the metal rod has been removed, a velour fabric is obtained.

Brasovian: Piece-dyed diagonal or twilled woollen fabrics, dyed usually blue; used for women's dresses in the Balkan States; imported from Austria, Hungary.

Brattice: A conveyor formerly made of inter-linked wooden slabs, but nowadays often of metals; e.g. wire mesh for the transportation of textile material through machinery in a tensionless state.

Brawels or Chiadder boraals: White and blue striped East Indian calico of loose weave; used by the African tribes for turbans, etc.

Brazen: Having the colour of brass; made of brass without shame.

Braying: Term used in England to denote the scouring of woollen fabrics after taken from the loom.

Brazilian baize: In England, a medium grade of baize-dyed orange colour.

Brazilians: Regattas woven of Brazil cotton; also madapolams made of that cotton.

Break factor: A measure of yarn strength calculated as: (1) the product of breaking strength times indirect yarn number, or (2) the product of breaking strength times the reciprocal of the direct yarn number. Example: Pound times cotton count.

Brazilwood: A vivid red colour from the dye of the tree of the genus *Caesalpinia* known as the Brazil tree which begot the name of the country rather than vice versa.

Break spinning: A direct spinning process for converting manufactured fibre tows to spun yarn that incorporates pre-stretching and tow breaking with subsequent drafting and spinning in one operation.

Break out: See **Smash**.

Breaker lap: A roll of loosely matted cotton fibres formed on the breaker picker from tufts of opened cotton.

Breaker picker: The first of the two units of older style picker machines. In this unit, the raw cotton is partially cleaned by beating and fluffing and then fed into a finisher picker.

Breaking (bast fibres): The deformation of the plant structure by flattening the stem, loosening the bond between the fibre bundles and the wood, and breaking the woody part into short pieces, to facilitate their removal from the fibre by the next process called scotching. Breaking by means of rollers is often referred to as rolling.

Breaking and opening machine: A line of machines, working as a unit, which tear apart and partially clean the matted, compressed, baled cotton.

Breaking elongation: See **Elongation at break, Breaking Extension (breaking elongation)**. The extension/elongation produced by the breaking force, i.e. the maximum force applied during a determination of breaking strength.

Breaking force: The maximum force applied to a material carried to rupture.

Breaking force (in testing sewn seams): The force that is repeatedly applied to a test specimen.

Breaking length: A measure of the breaking strength of a yarn; the calculated length of a specimen whose weight is equal to its breaking load. The breaking length expressed in kilometres is numerically equal to the breaking tenacity expressed in grams-force per tex.

Breaking load: (1) See **Tensile force**. The maximum load (or force) applied to a specimen in a tensile test carried to rupture. Use the perfect term, Breaking force.

(2) The maximum load (or force) applied to a specimen in a tensile test carried to rupture. It is commonly expressed in grams-force (kilograms-force), pounds, or newton's (Also see **Breaking strength**).

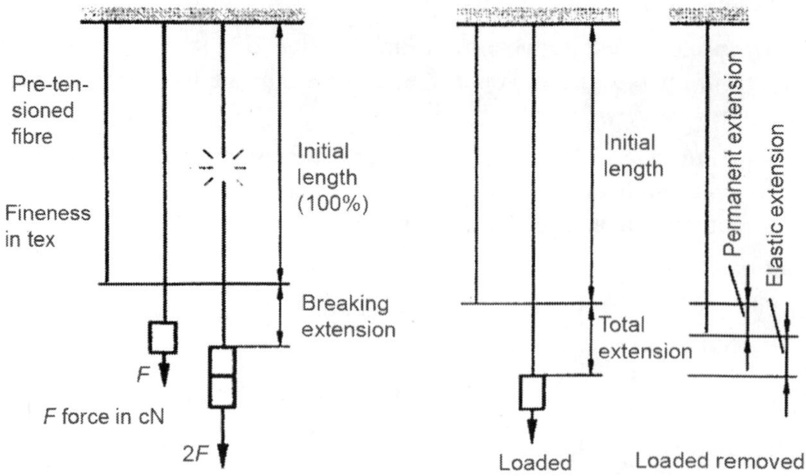

Breaking load (knot): The breaking load of yarn or a fibre specimen having a knot in the portion between the clamps.

Breaking load, loop: The breaking load of a specimen consisting of two lengths of an individual strand looped together, so that one has both its ends in one clamp of the testing machine and the other length has its both ends in the other clamp.

Breaking machine: A machine for continuously softening stiff fabrics. The fabric is drawn under tension over the edges of bars (knife edge machine) or round rollers implanted with studs.

Breaking point (on a force-elongation curve or stress-strain curve): The point corresponding with the breaking force or the breaking stress in a tensile test.

Breaking ratio: See **Break factor**.

Breaking strength: See also **Tensile strength**.

Breaking strength: (1) The maximum resultant internal force that resists rupture in a tension test. The expression "breaking strength" is not used for compression tests, bursting tests, or tear resistance tests in textiles.

(2) The load (or force) required to break or rupture a specimen in a tensile test made according to a specified standard procedure. Also see **Breaking load**.

(3) Strength expressed in terms of breaking force. The maximum tensile force observed during a test in which the specimen is stretched until it breaks.

Breaking stress: The maximum tension (expressed in Newton) developed in a specimen stretched to rupture. The force is usually related to the area of the unstrained specimen. If the actual stress, defined in terms of the area of the strained specimen, is used, then its maximum value is called "actual breaking stress".

Breaking tenacity: The tensile stress at rupture of a specimen (fibre, filament, yarn, cord, or similar structure) expressed as Newton's per tex, grams-force per tex, or gram-force per denier. The breaking tenacity is calculated from the breaking load and linear density of the unstrained specimen, or obtained directly from tensile testing machines which can be suitably adjusted to indicate tenacity instead of breaking load for specimens of known linear density. Breaking tenacity expressed in grams-force per tex is numerically equal to breaking length expressed in kilometres.

Breaking tension: Breaking tension, as defined, is independent of the acceleration due to gravity. See **Breaking stress**.

Breaking toughness: (1) Toughness up to the breaking point of a material.

(2) The actual work per unit volume or unit mass of material that is required to rupture the material.

Breaking (in flax preparation): Breaking removes the woody centre from the retted and dried flax by being passed through a series of fluted rollers. The particles of woody matter adhering to the fibres are detached by scotching.

Break-open (in testing thermal protective clothing): A material response evidenced by the formation of a hole in the fabric which allows flame to pass through the textile.

Break-out: See **Smash**.

Break-out pressure (for inflatable restraints): The pressure level during deployment which ruptures the module cover.

Breast beam: Part of loom. It is same as Front rest. See Cloth fell. It is the horizontal Cross-piece on the loom directly in front of the weaver. The woven cloth passes over the breast beam on its way to the cloth beam.

Breathable coat: A coating that repels water, but allows perspiration to pass through while allowing garments to be comfortable and waterproof. Used in garments for active wear and winter sports.

Breathability: The term is used to describe water vapour permeability, water proofness, or air permeability. The result is an overall picture heat regulatory processes which describes wear comfort of a fabric.

Breacan: Original (Gaelic) term for the tartan.

Breannes: Bleached, French linen of loose texture, similar to the BrionnB.

Breeches: Knee trousers.

Bremen blue: Blue copper pigment.

Brenne: Light-weight woollen cloth, often with silk stripes; formerly made in France.

Bresilienne: Corded French silk dress goods and vesting; also an all-wool, twilled fabric made in France in the 19th century.

Bretanas: Variety of plain woven fabrics of pure linen, or mixed with cotton; used in Latin America, imported from England.

Breton embroidery: In silver, gold and coloured silk thread over cloth, silk or mesh; to be used as borders. Flowers, geometrical designs and Breton patterns are the motifs, outlined and then filled out, usually chain and satin stitches being used.

Breton lace: A net ground with heavy embroidered designs. Usually the embroidery threads are coloured.

Bribe: (1) A locally used term for a woollen fent.

(2) English term for a piece of cloth containing flaws and cut away from the length.

Brick: The deep orange-red of brick; **lateritious**.

Brick-red: Having the colour of red brick; in the United States, a brownish- or yellowish-red colour.

Brick stitch: Used as background in antique embroideries; sort of couching, the effect being similar to a brick wall.

Bridal lace: Drawn work in the 16th and 17th centuries, worn at weddings. The designs represented coats of arms of the families.

Briddle: A narrow strip of material which is attached to the inter-lining along the roll of the lapel to hold and control it.

Bride epingle: A needle-made mesh ground for real lace, each side covered with button-hole stitches. To be found on early samples of Argentan point.

Bride Ornee: Bride covered with button-hole stitches and ornamented with picots, etc.

Bride picotee: In point laces a hexagonal buttonholed net, each side ornamented with a couple of picots or pearls.

Bridge (in button): The area between the holes partially covered by the sewing threads with dimensions varying upon design and end use.

Bridge top stop (in zipper): A part affixed immediately above the chain, holding the tops of the two stringers together and preventing the slider from leaving the chain.

Bridge water: Light English broadcloth made in the 16th century.

Brief (in designing): Describes a design project.

Briefing meeting (in designing): Where the work that is required and is stablished/when the designer finds out what is required for them.

Bright: The term applied to fibres whose lustre has not been reduced by physical or chemical means; the opposite of dull or matte.

Bright: (1) Characterized by a high degree of lustre, brilliant, opposite to dull and relatively vivid in colour.

(2) Open work effect within a wale in lace furnishings, obtained without distorting the warp thread.

(3) Descriptive of textile materials, particularly man-made fibres, the normal lustre of which has not been reduce by physical or chemical means.

Bright fibre: A fibre without or with minimal quantities of titanium dioxide; also

- semi-bright fibre: a slightly delustred fibre
- semi-dull fibre: usually terms delustred fibres with 0.25-0.5% titanium dioxide contents
- dull fibre: fibre with 0.5-1% titanium dioxide
- superdull fibre: fibre with 1-3% titanium dioxide

Bright pick: A tight pick that is usually found in a fabric containing a continuous filament weft.

Bright silk: Thrown silk, completely boiled off and dyed.

Bright yarn: A general term used in the manmade fibre industry for lustrous yarn, staple or tow. It contains no or only a trace of delusterent (TiO_2). It is compared with dull, semi-dull or extra-dull yarns or fibres.

Brightening agent: (1) See **Optical brightener**.

(2) An additive which when applied to textiles heightens their brightness.

Brightening acid: A mixture of lactic and tartaric acid. Excellent scrooping agent, especially for achieving a scroopy silk handle.

Brightness: The dimension of colour that refers to an achromatic scale, ranging from black to white. Also called lightness, luminous reflectance or transmittance (q.v.). Because of confusion with saturation, the use of this term should be discouraged.

Brighton: Not reversible diamond patterns; used on dress and other fabrics; the diamonds are alternately small and large. It is a honeycomb weave made in straight draft, the threads in a repeat being divisible by four.

Brighton embroidery: It has geometrical patterns upon coarse canvas foundation.

Brighton honey comb: Made of cotton, a fabric with typical honeycomb effect. A waffle weave constructed with depressions that are not lined up. The depressions on the face of the cloth are not matched by equal depressions on the back. This structure is often called Brighton Honeycomb.

Brilliancy: (1) Physical dimension of brightness or the impression of brightness.

(2) SI unit derived from candela per square metre. Brilliancy= Light intensity of the light source/luminous area of the light source.

Brilliant: Sparkling, shining, vivid, reflecting a large amount of light or brightness, lustrous and gleaming. See **Brilliant dyes**.

Brilliant dyes: Colorants including brilliant red, blue, orange, violet, black, yellow, green, flavine and pink.

Brilliante: Brilliante is a cotton fabric of light or medium weight, distinguished by small, detached figures, usually of geometrical or simple character, arranged on a plain ground. The figures are formed with the filling that is soft twisted. The goods are generally used for shirtwaists and dress goods.

Brilliantine: Light-weight, twilled dress goods, woven with cotton warp and lustre, worsted or mohair filling, dyed in the yarn in the same or various colours. The filling covers the face. The patterns are usually formed by floats

Brilliants: Light-weight cotton dress fabric or shirt waisting, made with fine warp and a heavier, slack twist filling which forms little, geometrical figures with floats on a plain woven ground. The fabric is often mercerized.

Brin: A single filament of silk resulting from the degumming of bave withdrawn from the cocoon. Two brins (from each gland on either side of the body of silk worm) are cemented together by gum or sericin at the spinneret of the silkworm to emerge as a bave or silk strand.

Brine: Often this means seawater or a solution of common salt, NaCl. Ordinarily, seawater has about 35,000 mg/l (3.5%) of dissolved salts. Some brines have 3,00,000 mg/l (30%) of dissolved salts. Brine is sometimes defined as water with more than 30,000 mg/l of dissolved matter.

Brins: The two single filaments of pure silk, encased in natural gum, thus forming the silk-fibre as produced by the silk worm. The two brins are produced by two glands in the worm.

Bri-nylon: A nylon fibre, used in the manufacture of carpets, lingerie, nightwear and light-weight knitted materials. Sometimes the surface is brushed to form a short pile. It is non-absorbent and hence dries quickly.

Brionne: Bleached, sheer, French linen of various grades; used for curtains.

Brique: A light brownish-red; found only in the OED and French dictionaries.

Bristle: (1) A short, stiff and coarse fibre.

(2) The hair of a hog.

Britch: The lowest grade of wool in a fleece, taken from the thighs and spins 26s to 28s.

British raised: Work short-ends of heavy wool yarn are fastened by the middle to a coarse canvas foundation and then brushed out, resembling a pile.

British Standards Institution: The British Standards Institution (BSI) is the world's leading standards organization, and facilitates the setting of standards, inspections and quality management. It is a non-profit distributing organization, independent of government, industry or trade associations, whose operating divisions (product certification, quality assurance, standards, testing and training services) are designed to further the use of standards. Manufacturers, importers and retailers rely upon BSI testing to independently assess the performance of their products for safety, reliability and quality. Its certification trade mark, the Kitemark, has been established for more than 90 years.

British gum: British gums are made by dry heating starch granules at 180 °C without acid. Some of the 1–4 acetal linkages are broken and 1–6 are formed. A more highly branched, lower molecular weight product is formed having more reducing end groups. British gums are more water soluble and produce higher solution viscosities. They are mainly used as print paste thickeners, especially for vat dyes. Other roasted starches (e.g. potato starch) are also marketed as British gum.

Brittle point: The temperature at which a polymer no longer exhibits viscoelastic properties.

Brittleness: See **Flexibility**.

Broach: Indian raw cotton, having a moderately strong and fairly clean staple of golden colour.

Broad bandpass radiometer: A relative term applied to radiometers that have a bandpass width of more than 20 nm at 50% of maximum transmittance and can be used to measure irradiance at wavelengths such as 300-400 nm or 300-800 nm.

Broad couching: Similar to the brick stitch, only the laid threads are slightly drawn together.

Broad goods: Woven fabrics that are 18 inches or more in width.

Broad rash: Heavy coating made of worsted warp and woollen filling in the 18th century in England.

Broad rib fabric: A rib fabric in which groups of three or more adjacent wales are of the same type either face or back.

Broad silk: Wide silk dress goods and linings.

Broadcloth: The term covers a variety of materials. In one way it has very little real meaning. Originally it was used to differentiate fabrics woven on broad looms say 36 in and wider. Today the term is generally used for a fine plain woven cotton cloth from 36 to 40 in wide, suitable for men's shirts and women's dress goods; when these are made from single yarns with constructions 130 × 60, it is often called single broad cloth; when ply yarn is used say 2/80s both warp and weft, it is called ply broad cloth. When in a broad cloth two-ply yarn in the warp and single yarn in the weft is used, it is called semi-broad cloth.

Broadcloth (cotton): A tightly woven, lustrous cotton or polyester/cotton blend fabric in a plain weave with a crosswise rib. It resembles poplin, but the rib is finer, and broadcloth always has more picks than poplin. The finest qualities are made with combed pima or Egyptian cotton.

Broadcloth (wool): A smooth, rich-looking, woollen fabric, usually from Merino wool, with a napped face and a twill back. Better grades have a glossy and velvety hand. Woven on wide looms so that even after processing it is fairly wide fabric, made with a high nap, which is the pressed down to give a compact look and weave is not visible. Dyed in solid shades, it can be made as an all wool fabric or worsted warp and wool filling. It may have 12–14 oz. per yard.

Broad cloth (rayon): May be composed of a filament rayon warp and spun rayon filling or it may be made entirely of spun rayon warp and weft. It comes in a variety of constructions, but one popular construction is 110 × 90 using 50s warp and filling (spun). Fabrics are usually made 40 in wide. Another construction is 90 × 70 with 100 denier rayon warp and 2/30s spun rayon filling.

Broad goods: Woven fabrics that are 18 inches or more in width.

Broad spectrum antimicrobial: An anti-microbial that effectively controls or kills at least 3 of the basic microorganism groups. This term is important to help give a specific encompassing term to the technologies that offer protection from the gamut of microorganisms, without the sometimes vague nature of the term anti-microbial, which could mean that it kills just one type or kills many types.

Broadcloth (silk): Originally, a silk shirting fabric so named because it was woven in widths exceeding the usual 29 inches. It has a high lustre and smooth finish. White, coloured or stripes, the material can be used for shirts, sportswear, pyjamas and dress material.

Broadloom: A term that refers to carpets woven in widths from 54 inches to 18 feet, as distinguished from narrow loom widths of 27–36 inches.

Broadtail: A flat, lustrous fur that is slightly wavy. The pelt of the young unborn Persian lamb.

Brocade: (1) Multi-coloured or self-coloured floral raised design, sometimes with metal threads added. Made on jacquard loom, usually in satin weave. Originally always silk, it can now be viscose, cotton, acetate, polyester and many blends. A typical construction is with 100 denier rayon warp with 250 denier bright rayon filling, 180 × 60, width 45 inches. Yarns can be dyed or even plain dyed after weaving. In true brocade, the design is produced by additional coloured threads on the weft. These pass across the back of the cloth unused and are brought to the surface when required. The wrong side is easily identified by these threads or floats. In addition, threads can be woven into the back of the fabric. Some brocades are heavy while some are crisp. Brocades are used in clothing, furnishings and upholstery.

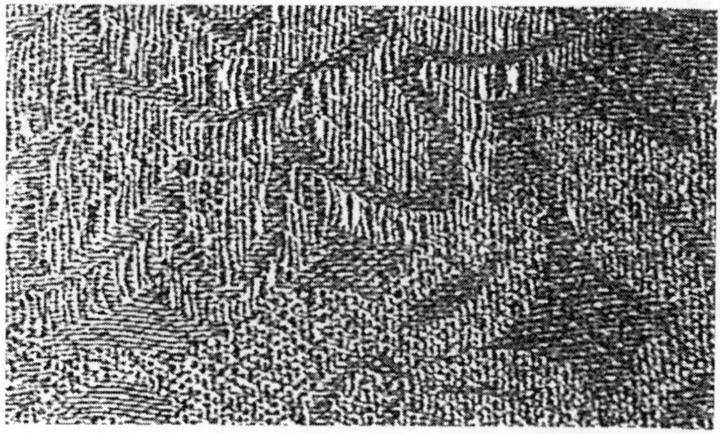

(2) (i) Originally a rich and heavy silk fabric, with flowers, foliage, figures, etc., woven in gold or silver core yarn. It was of Oriental origin; used for state and church vestments; (ii) Silk fabric having usually large flower or other pattern design woven in a colour different from the foundation, each filling thread usually reaching from selvage to selvage. The patterns are usually of an embossed character.

Brocaded satin: It has a foundation made in satin weave, the pattern being woven in another weave or made in velvet. Brocaded velvet can be either pile over pile or velvet patterns over a plain woven or other foundation. See also **Lame, Swivel and Lappet**.

Brocade embroidery: The designs traced over the foundation are outlined or also filled in with stem stitch in various coloured heavy wool or floss silk yarn.

Brocade velvet: Velvet with the pile sheared at different heights to form a design, sometimes checks, or floral or the like. Plainer designs in dark colours are used for men's jackets. It is very expensive.

Brocade: (1) A rich, Jacquard-woven fabric with an all-over inter-woven design of raised figures or flowers. The pattern is emphasized by contrasting surfaces or colours and often has gold or silver threads running through it. The background may be either a satin or a twill weave. It is often made with gold or silver threads. The design appears on the face of the fabric which is easily distinguished from the back. It is used in evening dresses, wraps, draperies, upholstery and decorative purposes. The term brocade is derived from the Spanish word that means "to figure".

(2) A term describing a cut-pile carpet having a surface texture created by mixing twisted and straight.

(3) (decoupe') An additional effect thread is inserted into the ground weft. If structures are far apart from each other due to the pattern, the floats are cut. This makes it more difficult to extract the floats on the back; the fabric weight is also reduced.

Brocading: Weaving process for introducing extra silk, gold or silver threads into the cloth to form patterns.

Brocart or Brocat: French for Brocade. Brocatelle (1) Originally a heavy, cross- ribbed furniture and drapery fabric, similar to the brocade, having Jacquard figures and usually made of silk warp and cotton filling, but also of all silk, all cotton, or all wool; (2) Modern dress goods and vesting, similar to the furniture drapery, but made in lighter weights and often all silk; (3) Silk fabric having glossy and satin figures on a less glossy taffeta ground of same colour.

Brocatelle: (1) A fabric similar to brocade with a satin or twill figure in high relief on a plain or satin background.

(2) Silk, rayon, cotton and synthetics. Woven in Jacquard – double or backed cloth. Originally supposed to be an imitation of Italian tooled leather – satin or twill pattern on plain or satin ground. It is recognized by a smooth raised figure of warp-effect, usually in a satin weave construction, on a filling effect background. True brocatelle is a double weave made of silk and linen warp and a silk and linen filling. Present-day materials may have changed from the XIIIth and XIVth Century fabrics, but they still have the embossed figure in the tight, compact woven warp-effect. While brocatelle is sometimes classed as a flat fabric, it shows patterns which stand out in "high relief" in a sort of blistered effect. A fabric made on Jacquard looms with additional weft thread of cotton, rayon or linen to bring out the pattern producing high relief. The patterns are often twill on a plain or satin background. It is used as draperies, furniture, coverings and general decorating purposes as well as all kinds of after 5 wear.

Brocatine: Brocade with raised patterns.

Broche: A large number of fabrics decorated with special threads which are introduced together with the regular warp or filling threads, but which do not form a necessary part of the structure itself. These threads are carried as floats at the back of the fabric, or are clipped away and form coloured designs. Fabric with patterns which look as they have been embroidered. Genuine broche' are made on swivel looms. Beside warp and weft, these looms have a third thread system which insert small shuttles controlled by a jacquard loom.

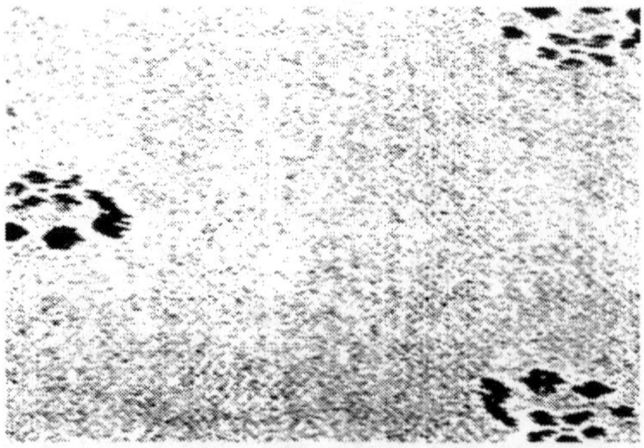

Broche/Scherli: A special type of patterned curtain, produced with additional laid-in yarns in a ground fabric. The connecting yarns between the pattern repeat would be cut-off or warped, either mechanically or by hand (Scherli). Materials used: Cotton, Polyester, Blends of Polyester/Cotton.

Broche (in carpet and floor coverings): A assemblage of yarn inserted in the same dent of the sky of the loom, the group consisting usually of two ends of the chain warp, the yarn of the stuffer warp and those of the pile.

Brochure-blue: The clear shimmering blue colour of the sea, as appearing in all travel brochure illustrations.

Broccoli green: The green shade of the vegetable broccoli.

Brode: Embroidered.

Brodequin: Embroidered hosiery.

Broderie: French name for embroidery.

Broderie anglaise: The fabric is plain weave cotton, or polyester and cotton in which shaped holes are punched and then embroidered. It can be white, or plain colours and self-coloured embroidery, sometimes contrast embroidery also. Fabric can be with less or more embroidery, can have even scalloped edge with embroidered only in the border.

Broderie en lacet: Braid stitched to satin foundation, the patterns being filled in with stitches imitating lace.

Broderie de malines: Early name for Malines lace.

Broderie de nancy: Drawn work, ornamented with coloured silk.

Brodie: A tartan, consisting of equal wide black and green stripes, both ways; these are broken up in the middle by a narrow red stripe (both ways and two still narrower stripes, one on each side of the red). Half of the green stripe is taken up in the middle by a blue stripe each way.

Broella: Coarse woollen fabric; used for clothing by the English country people in the Middle Ages.

Broken colour: A diverse term used to describe a variety of effects such as the technique of interspersing flecks of colour with another colour or colours used, in particular, by the impressionists and creating a blended effect when observed from a suitable distance; the effect produced by the random distribution of particles of colour on a surface; pure colours intermixed with black; the result of mixing one colour with another; the effect of an underlying layer of colour showing through a super-imposed layer of oil paint; a new tone of a particular colour created by juxtaposing another colour with it; the application of different shades of paint or glaze to a base coat which is broken up in a number of different ways to create a special finish for decorative purposes – see, for example, **Graining** and **Marbling**.

Broken end: A broken and untied warp thread in a fabric. There are numerous causes, such as slubs, knots, improper shuttle alignment, shuttle hitting the warp shed, excessive warp tension, faulty sizing, and rough reeds, heddles, drop wires and shuttles. See also **End out**.

Broken end: A warp-way line where a warp yarn is absent for part or all of a piece and that is caused by a warp yarn break that has not been repaired.

Broken ends: In weaving, warp ends are subjected to high loadings, the physical mechanism of which is extremely complex. Sizing is necessary to prevent warp fatigue, particularly due to repeated extension. See **End out**.

Broken filaments: Self descriptive, see **Filamentation rupture** of individual filaments (usually during winding or weaving) that results in the appearance of a fibrous or hairy surface, which may be localised or general, in a fabric made from flat continuous filament yarn. In multifilament yarn, breaks in one or more filaments.

Broken pattern: Defect in printing where the pattern is broken due one reason or other. Also used in any other design breaks such as woven or knitted.

Broken pick: A pick that is present for only part of the fabric width. A broken weft thread in a fabric. Usual causes include too much shuttle tension, weak yarn, or filling coming into contact with a sharp surface.

Broken selvage: See **Cut selvedge**.

Broken twill: A strong cotton fabric with a warp cross twill weave where the continuity of the twill is broken by design. One way to weave broken twill is to move the diagonal to the left for part of the repeat and then skip one or two sheds in the regular twill sequence and finish out the repeat in the opposite direction until all the sheds have been used. For a 2/2 twill the treadling sequence will be 1,2; 2,3; 1,4; 3,4 repeat 2,3. The fabric has a high resistance to slippage and is characterized by a particularly 'full' handle. It is used for leisure wear.

Broken weave: See **Crack**.

Broken warp: Small pieces of cut or missing warp thread.

Bromate discharges: An oxidative discharge, more of academic interest presently. Example, a coloured discharge on indigo can be obtained by printing a mixture of sodium bromate or bromite and chlorate of alumina (aluminium oxide), followed by steaming, washing of and subsequent dyeing (with alizarin).

Bromelia fibres: Leaf fibres of the *Bromeliaceae* family which are grown in tropical and subtropical regions of South America. They form a subgroup of hard fibres, which include pineapple fibres, caroa fibres, silk grass, etc. In contrast to the agave fibres they are easily extracted from the ripe leaves, and they have high fibre content. Mainly used for making ropes, braids, etc.

Bromite desizing: A process for desizing by the oxidative degradation of starch method using sodium bromate. Presently of academic importance only.

Bromothymol Blue (BTB): Used as an indicator in desizing with diastases for pH control within the optimum specified range.

Bromsia: A type of Levant raw silk.

Bronze pigments: These are manufactured from metal granules in special ball mills in the presence of lubricants to prevent cold bonding of particles. See also leafing pigments and non-leafing pigments.

Bronze: The brilliant brown colour of the metal – bronze; derived from 'Brindisi', the town in Italy. Bronze sculptures left to the elements will, by a process of oxidization, develop a patina of ***verdigris*** and turn green in colour.

Bronze powder: See **Metal powder**.

Bronze printing: This is a generic name where gold-bronze pigment prints are done with copper/zinc alloy powder, whereas the real bronze is an alloy of copper/tin. Different shades are produced by changing the copper content in the alloy. Copper colour – 100% copper; pale gold colour – 90% copper; vivid pale gold colour – 80% copper; deep gold colour – 70% copper. Also, silver bronze is produced by aluminium powder.

Bronzing of dyeing: Dyeing fault manifest as a metallic iridescent appearance associated with poor rubbing fastness. Causes can be inadequate pre-treatment, dye precipitation (e.g. Hard water), over saturation (excess dye) on the fibre and premature oxidation of vat and sulphur dyes during dyeing. A coppery lustre on the surface of a fabric which is caused by the presence of excessive dyestuff during dyeing, or by precipitation of the dyestuff during the dyeing process. See also **Gilding**.

Brookfield viscosity: Standard measurement of viscosity expressed in centipoise.

Brooks: An early maturing commercial variety of cotton from Louisiana, yielding about 31% of short staple.

Brown: (1) An early maturing commercial variety of cotton from Mississippi, yielding a short staple; (2) Wool-sorting term for about 30s quality, taken from the haunches of the fleece; and is of lower grade than "neat."

Brown: The colour of earth and of wood; in the wavelength range of approximately 620–585 nanometres. Latin had no specific word for 'brown'. The Hebrew word used for 'brown' as used in *Genesis* xxx: 32 means 'sun-scorched' and a variant form of black. Its early meaning was 'dark'.

Brown chalk: Another name for ***umber***.

Brown Egyptian: Cotton, having a long, strong, clean and easily worked staple of golden-brown colour.

Brown Holland: Originally unbleached or partly bleached plain woven linen; the latter sized and used as furniture cover or summer clothing. It comes also in dark grey and black.

Brown osnaburgs: Coarse, unbleached linen, made in Ireland and Scotland of strong hemp or linen yarn for the American trade.

Bruce: Tartan with bright-red ground, the plaid formed by groups of four (two wider in the middle and two narrower at the sides) stripes in dark green, which are together as wide as the red squares. In the middle of the red squares alternated narrow yellow and white stripes run each way.

Bruckner Solvanit Process: The Bruckner Solvanit continuous process is used for simultaneous scouring, cleaning and for shrinking. The cloth is transported by a patented figure 5-16. Continuous solvent scouring, and finishing machine (Courtesy of Sperotto Rimar, Italy). Paddle wheel which ensures thorough removal of contaminants by counter current flow of fresh perchloroethylene. Relaxation and bulking of knitted goods occur during solvent flash-off.

Bruges lace: A bar lace, originating from Gruges. Very similar in appearance to Honiton Lace, woven as a tape with a fine thread. Coarse types are used for curtains, bedcovers and tablecloths.

Bruise (in fabrics): An area that has been subjected to impact or pressure, which differs from the adjacent normal fabric. See Abrasion mark.

Brunneous: Dark brown.

Brunswick: Twilled fabric, similar to cashmere and is made of wool and cotton.

Brush: A hand held cleaning tool consisting of a base into which bristles are inserted.

Brush binding: A braid with a stiff fringe along one edge, used to bind the bottom of skirts.

Brushed acrylic: Light to medium weight woven fabric with a brushed effect on one or both sides; often printed and used for dresses, warm shirts, children's clothes, etc.

Brushed acrylic knit: Made of acrylic fabric with jersey backing with the rightside brushed into a furry surface. Used for dressing-gowns, sleeping bags, baby coats, casual jackets, toys, collars and cuffs.

Brushed cotton: Plain or printed cotton fabric which are slightly brushed on one side to add warmth. This inexpensive, creasing fabric often used for children's warm clothes, winter blouses and skirts. Brushing can be on the right or wrong side.

Brushed denim: Denim weave cloth, usually all cotton, with a brushed finish on the right side. It has a softer appearance against the harsh handle conventional denim, which is a disadvantage of denim. It is not as tough as denim but hard wearing and is used generally for shirtwaist dresses, safari-style jackets, trousers and jeans, casual shirts for men and children's clothes.

Brushed fabric: A fabric which has been brushed to raise the fibres to get a fuller handle.

Brushed nylon: Nylon Jersey fabric brushed on one side to make it more warm, but has the disadvantage of being more static than plain Nylon Jersey. Usually used only for nightwear, and sheets. Being hydrophobic, even though washes well it dries quickly.

Brushed polyester: A fine fabric with a brushed surface resembling velveteen in appearance.

Brushed yarn: A yarn has been processed to raise surface hairs in order to achieve greater bulk and softer handles. The brushing which normally takes place by abraiding the surface of the yarn with card clothing is usually performed on boucle, loop and other fancy yarn.

Brushing: (1) A finishing process in which rotating brushes raise a nap on knit or woven fabrics. Brushing is used on sweaters, scarves, knit underwear, wool broadcloths, etc.

(2) An operation in dry-finishing which serves particularly the cleaning of woven textile fabrics when fibre fly, thread or dust has to be removed. Brushing units are employed before singeing, and dyeing and printing. In the latter case it helps to remove the loose thread, dust, etc., which will hamper dyeing or printing. In the former it helps in a better singeing.

(3) Felt hats are brushed in the warm wet state in order to fix the nap in the case of long haired surfaces to produce a silken lustre and to improve the handle with the hair hats.

(4) Act of transferring newly hatched silkworms from the eggsheets on to the rearing trays, or act of removal of floss from the cooked and boiled cocoons for collections of ends for reeling.

Brushstroke: A print style in which colour looks as if it had been applied with a brush.

Brussa: Loosely woven half-silk veiling of varying weave, made in Asia in olden days.

Brussels carpet: Loop pile carpets, woven on a Wilton loom, over bladed wires; originally manufactured in Brussels. Depending on the pile materials used on these carpets, these are known as wool or hair carpets.

Brussels lace: The bobbin lace has hexagonal ground, two sides being of four threads plaited four times and four sides of two threads twisted. The outlines of the flowers and the veins of the leaves are of raised plaited cordonnet – (1) The ground is worked around the flowers, the brides and toile are very elaborate; (2) In needle point lace, the hexagonal ground is made in strips of one inch width and then joined with the raccroc stitch. The cordonnet is not buttonholed; (3) Modern Brussels lace is composed of sprigs appliqued to machine-made ground. Originally made with fine linen threads. In this lace, the motifs are made as bobbin lace and then appliquéd on to a mash ground. Now all fibres may be used and usually motifs and the mesh are machine-made.

Brussels quilling: Plain cotton net with an extra twist around the mesh; used for frills, etc.

Bruyère: A greyish-purple; the colour of heather.

BTTG: British Textile Technology Group.

Buaze: Strong, fine and durable bast fibre, yielded by the twigs of a species of securidacea in Zambesi; used for fishing net by the natives.

Bubble colouration process: A non-repeating patterned effects in polygon or ring shapes as well as grape like motifs are produced on textile fabrics from dye containing bubbles (produced by blowing air through pad liquor containing surfactant and a thickening agent. The bubbles are allowed to impinge on the backside of the fabric or they can float on the fabric surface. Discharge effects and designs produced by the application of fast colour salts on naphtholated grounds are also possible by this method.

Bubble-gum pink: A sickly gaudy pink which is a characteristic of bubble-gum.

Bubbles: Same as **Blisters**.

Buchanan tartan: Even-sided plaid with the following stripes repeated in both ways; red, narrow blue, yellow (split in the middle by a narrow blue), narrow blue, narrow green, narrow blue, narrow green and narrow blue. The wide red stripes are split in the centre by a narrow white stripe.

Buck: The lower, static, working surface of the pressing machine which may incorporate the shape or contour on to which the garment is laid.

Bucket spinning: See **Pot spinning**.

Buckinghamshire lace: A very fine hand-made lace originated in Buckinghamshire. It was characterised by a diamond mesh ground, was often narrow and was used only for trimming. Bobbin lace with plaited ground similar to that of Lille or Valence. It is made in one piece, usually in narrow widths and is used as baby lace.

Buckle stitch: In bobbin laces, the stems and fibres of the sprigs are often formed of a braid made in buckle stitch.

Buckram: A stiffened fabric of cheap cotton or hemp linen yarn mostly of natural colour. With good elastic resilience produced by impregnating a plain weave fabric with fillers and stiffeners to make the fabric stiffer. Some grades are ply fabrics with 2 face ends for each backing end and 1 face pick for each back pick. A characteristic construction is 36 × 24 with 22s yarn for the face warp and 12s for the backing warp and for face and back filling. Sometimes two layers are glued together as per the end use. It is used in lining, waistband and bookbinding, etc.

Buck skin: (1) A suiting similar in handle and appearance to, but heavier than a doeskin fabric, made from fine merino wool/cotton union closely set heavily milled, lightly raised and closely sheared on the face. A thick, stout and very durable woollen fabric woven in an eight-leaf satin weave with a warp face. It is napped, fulled thoroughly and shorn, the face being finished very smooth. Used for overcoats, riding breeches, etc. Summer buckskin is made of hard spun worsted yarn often mixed with silk. A distinction is made between summer and winter buckskin depending in the fabric weight (the latter is often produced as double fabric).

(2) A white or pinkish beige leather which originally came from the elk or deer, but is now also obtained from sheep. It is strong but supple. Used for clothes, shoes, gloves, belts, etc.

(3) A heavy satin weave fabric, often of fine merino wool with a smooth face.

Buckskin weave: An eight-leaf warp satin weave, each warp crossing over six picks and carried under two picks.

Bud green: A yellowish-green colour.

Buff: A pale yellowish-brown colour; a light yellow; of the colour of buff leather which having regard to its proximity to human skin colour which gave rise to the phrase 'in the buff' meaning 'naked'. Also 'buff-coloured'.

Buffalo checks: A bold checks pattern with bocks of two or three contrasts in colours. Often red and black in twill weave

Buffalo cloth: Heavy twilled woollen over coating with a long nap.

Buffalo wool: Very fine wool, growing among the coarse hair on the buffalo. Used for fine felt hats, shawls, etc. It is available in small quantities now.

Buffer: A general term for a solution containing an acid reacting salt and a base or an alkaline reacting salt and an acid. In chemistry, a compound that resists change in pH when moderate amounts of acid or base are added to a solution of it. pH of such solutions is not changed significantly; it changes on the dilution or on addition of H+ or OH- (acids or bases). Buffers help to keep the pH of a dye-bath from changing significantly as the process progresses. Sodium acetate in such buffer.

Buffer solution: See **Buffer**.

Buffline: A trousering of linen and cotton.

Buffinei: A coarse English woollen fabric that existed in the 16th and 17th centuries and was used for clothing.

Buggy: A lining from the neck down across the back of an otherwise unlined coat.

Bugis: Fine and plain woven cotton fabric used for scarfs in East India. It has a checked or striped border along one selvage only.

Build up: A term applied to substantivity of dye for a textile material. It refers to the ability of a dye to produce deep shades.

Builders: Substances which increase the effectiveness of soaps or detergents by acting as softeners (water softeners) or sequestering or buffering agents. Builders are generally salts such as borates, silicates, phosphates, sodium chloride, sodium sulphate, etc. Sodium metasilicate (Na_2SiO_3. $5H_20$) also acts as a detergent and buffer and assists other chemicals to penetrate into the lignin mass of cellulose materials. In general, the function of the builders is to drive the soap from water phase to fabric/water interface and consequently increase the concentration of soap on the fabric. It is inexpensive, but its addition in the scouring bath increases the ash content of cotton.

Bukhra carpets: Extremely fine-hand knotted carpets made chiefly by nomads of the Turkoman tribe who live on the Trans-Caspian steppes. The motifs used in Bukhara carpets are by far and away from those used in Oriental carpets. It is an octagonal motif with slightly rounded angles, having blue, black or sometimes green with red and orange. The ground and border colours are always red with wide variations from violet to orange.

Build up behaviour of dyes: The capacity of a particular dye to form deep shades. The comparison of different dyes can be done by studying the graph showing the depth against time in a standard dyeing procedure. See **Build up**.

Bulac: A Philippine species of cotton; used for cordage.

Bulbul: Fine cotton muslin made in India.

Bulgarian embroidery: These are made in various bright coloured silks, cottons and gold or silver thread over a coarse, undressed black or white muslin and it is alike on both sides.

Bulk density: Apparent mass per unit volume of the fabric layer.

Bulk development: Any of various relaxation treatments to produce maximum bulk in textured or latent crimp yarns or in fabrics made there from. The essential conditions are heat, lubrication, movement and the absence of tension. Bulk development may be accomplished during wet-processing or may be a separate operation such as hot-air tumbling, steam-injection tumbling or dry cleaning.

Bulk sample: In the sampling of bulk material: One or more portions which (1) are taken from material that does not consist of separately identifiable units and (2) can be identified after sampling as separate or composited units.

Bulk shrinkage: A measure of potential stretch and power of stretch yarns or a measure of textured-set yarns.

Bulk yarn: A yarn that has been prepared by mechanical, physical or chemical treatment in such a way as to have greater covering power or apparent volume than that of a conventional yarn of equal linear density and of the same basic material with normal twist.

Note 1: Staple yarn – The increased bulk may be obtained by the use of bi-component fibres and/or by blending together during yarn spinning, fibres of high and low potential shrinkage. Example, high bulk acrylic yarns. During subsequent hot/or cold wet-processing, the greater contraction of high shrinkage fibres causes the yarn to buckle; thus increasing the bulkiness of the yarn.

Note 2: In case of continuous filament yarn, the bulk may be obtained by texturing the yarn.

Bulked yarn: A yarn that has been finished to have a noticeably greater bulk. This treatment can be mechanical, physical or chemical.See also **Textured yarn**.

Bulky yarn: A yarn that has a greater apparent density than its real density, i.e. the yarn has a diameter larger than that indicated by the count, in comparison to yarn of a similar fibre type.

Bull's wool: Coarse and low grade wools are sometimes termed as Bulls wool.

Bullion: Originally gold or silver lace, also thick gold wire braid for uniforms.

Bullion knot: Round spots made by laying down the metal or silk thread in small flat coils, used for ground in embroideries.

Bullion lace: Made of gold or silver thread.

Bulobulo: A coarse Philippine fibre; used for cordage.

Bump grey: A cloth which is purchased to be used as back grey.

Bump seam: A printing fault characterized by a light coloured width-wise stitching mark, caused by stitching the ends of the bump grey.

Bump top; Bumped top: A package made by press packing the layers of horizontally coiled sliver.

Bun ochra: Fine, white and strong bast fibre of the Caesar weed in India; used for twine and bags.

Buna rubbers: Synthetic dye developed in 1910 and manufactured by polymerization of butadiene with finely divided sodium as catalyst.

Bunch (Yarn): A length of yarn initially wound, with a restricted traverse, on to the base of a pirn which serves as a supply of a weft yarn in an automatic shuttle loom from the time the need for weft.

Bunch: Contains 130,000 yards of linen yarn.

Bunch: A defect in the yarn, characterized by a segment not over 6mm in length that shows an abrupt increase in diameter caused by more fibres matted at that place.

Bundle: Linen yarn measure, equal to 60,000 yards and weighing 10 pounds.

Bundle: Handkerchief of linen and also cotton, made in England with plaid patterns in dark-blue colour.

Buntal fibre: It is obtained from the stalks of the unopened leaves of the Bun palm in the Philippines; it is a very light fibre and is used for hat braids.

Bunte salt dyes: A special group of sulphur dyes which due to transformation of the mercapto groups into thiosulphato groups are converted into a soluble form, which has no affinity for cellulose. However, the addition of barium hydrosulphide during dyeing causes a gradual reverse transformation to take place into a form having affinity for cellulose.

Bunting: A soft, flimsy, loose-textured, plain open weave cotton or woollen cloth most frequently used in ceremonial flags; it resembles scrim. Cheaper short-life flags are made from cotton bunting, fabric that resembles Cheese cloth. Usually dyed in plain bright colours as required in flags; colours may not be even fixed sometimes. Even though bunting was originally made from cotton or worsted yarns, but today's flags are made primarily from nylon or acrylic fibres. It is not used for clothing.

Burgundy: The reddish-purple colour of Burgundy wine – less blue than **Bordeaux**. Hence 'Burgundy red'.

Burette: (1) Fabric of southern Italy with warp of Bologna silk, filling tram silk; (2) Stiff, transparent linen; used in Italy for drawn work.

Burberry: (1) Light-weight mercerized and water-proofed cotton fabric, with fine warp ribs, used for raincoats; (2) Fancy and twilled woollen cheviot suiting.

Burda: Arabic name for black and white warp striped fabrics used for garments in North Africa.

Burdalisander: Medieval silk fabric in various coloured stripes and believed to originate from northern Africa. See **Alexander**.

Burden stitch: Used in embroideries to reproduce flesh. The silk threads are laid down evenly and are caught in even distances.

Bureau of Indian Standards: The Bureau of Indian Standards (BIS), the national standards body of India, is involved in the development of technical standards (popularly known as Indian Standards), product quality and management system certifications and consumer affairs. Apart from setting standards for textiles, the organization sets standards for other major industries in the country. It resolves to be the leader in all matters concerning standardization, certification and quality.

Buried end carpet: See **Under carpet**.

Buried pile design: A design formed by high-low-pile tufting with alternate needles threaded with difficult colours so that the high pile forms an area of one colour overlying the low (buried) pile of different colour.

Buried pile yarn (for coated pile yarn floor covering): After shearing and removal of the tuft legs some portion of the pile tuft elements are remained in the Pile yarn floor coverings. These are called Buried pile yarn.

Burel coarse: Brown coloured English woollen fabric used for clothing, etc., by the poorer classes in the 13th and 14th centuries. Now it is obsolete.

Burgos: A calico, dyed blue and printed in fancy colours; made once in India.

Buri raffia: A finer and weaker fibre than raffia that is obtained from the unopened leaves of the Buri palm in the Philippines and is used for coarse bagging.

Buring crane: An all-worsted English fabric of the 18th century.

Burity: Fine and lustrous leaf fibre, obtained from the Muriti palm of Brazil; used for ordage, hats and baskets.

Burl: (1) A wool trade term for an imperfection either (1) a small knot or lumps in a fabric or (2) a small cellulosic or synthetic fibre impurities in a fabric.

(2) To remove an imperfection.

Burlap: A coarse, heavy, plain weave fabric constructed from singles yarn of jute. The name originally meant a cleaning cloth, which implies hard-wear. Used for bags, upholstery lining, in curtains and draperies.

Burling: (1) The process of removing loose threads and knots from fabrics with a type of tweezers called a burling iron.

(2) The process of correcting loose tufts and replacing missing tufts following carpet construction.

Burly wood: A tan colour – one of the colours in the **X11 Colour Set**. Hex code #DEB887.

Burmese ruby: A pinkish-red colour.

Burn: To undergo combustion.

Burned area: The area of a material that has been destroyed by combustion or pyrolysis under specified conditions. See also **Damaged area**.

Burning behaviour of textiles: Pyrolysis is the destruction into breakdown products by heat, and is the first stage of every burning process. Not only does the fabric itself burn but so too do its pyrolysis products and this determines the burning behaviour.

Burning rate: The speed at which a fabric burns. It can be expressed as the amount of fabric affected per unit time, in terms of distance or area travelled by the flame, afterglow, or char.

Burning time: See **After-burn time**.

Burn-out: A brocade-like pattern effect created on the fabric through the application of a chemical, instead of colour, during the burn-out printing process (Sulphuric acid, mixed into a colourless print paste, is the most common chemical used). Many simulated eyelet effects can be created using this method. In these instances, the chemical destroys the fibre and creates a hole in the fabric in a specific design, where the chemical comes in contact with the fabric. The fabric is then over-printed with a simulated embroidery stitch to create the eyelet effect.

However, burn-out effects can also be created on velvets made of blended fibres in which the ground fabric is of one fibre like polyester, and the pile may be of a cellulosic fibre like rayon or acetate. In this case, when the chemical is printed in a certain pattern, it destroys the pile in those areas where the chemical comes in contact with the fabric, but leave the ground fabric unharmed.

Burn-out lace: See **Burn-out styles**.

Burn-out paste: In burnt-out printing, a paste containing chemicals capable of dissolving or destroying one of the fibre components in a blended fabric is used. Fabrics resembling lace have been produced in this way for shirting and other fashion articles. For e.g., sodium dihydrogen phosphate is a virtually universal burn-out agent for fabrics containing fibre mixtures with polyester. Sodium hydroxide solution is used to burn out the wool/silk components in mixtures with polyester or polyamide; protein fibres are broken down hydrolytically under conditions of high pH which results in their complete removal. For ready reference, Aluminium sulphate can dissolve cotton, viscose, linen, while polyester, polyamide, triacetate, acetate, linen, wool and silk are resistant to it. Sodium hydrogen phosphate will destroy cotton, linen, viscose, polyamide, acetate, triacetate while wool, and silk mare resistant to it. Bezoic peroxide will destroy acetate and triacetate while cotton, viscose, linen, polyester, polyamide, wool and silk are resistant to it. Caustic soda dissolves wool and silk while cotton, viscose, linen, acetate, triacetate, polyester, polyamide, etc., are resistant to it.

Burn-out printing (de'vore printing): It is the process of printing a design on flat fabrics composed of different fibre types with a paste containing a chemical capable of dissolving or destroying one of the fibre components. Fabric resembling lace have been produced in this way for shirting and other fashion articles.

Burn-out styles (devoré styles, à jour effects, chemical embroidery): These styles are produced on fabrics composed of different types of fibres by the (local) action of chemicals capable of destroying a selected fibre followed by its subsequent removal. The effects include all-over or printed open-work designs on blend fabrics or burn out lace produced by machine embroidery on a subsequently burnt-out base fabric of a different fibre type and burn-out velvet effects. For e.g., a treatment in cold 70% sulphuric acid can be used to dissolve-out the cotton in polyester/cotton fabrics (with no effect on the polyester). See also **Burt-out pastes**.

Burnt: A yellowish-brown colour; hence, 'burnt-coloured'.

Burnt-almond: A light brown colour.

Burnt gas fumes, Colour fastness to: Colour fastness to atmospheric contaminants; burnt gas fumes.

Burnt ochre: Light-brown pigment resulting from treatment of ochre by fire or having the appearance of having been scorched.

Burnt orange: A reddish-orange colour.

Burnt-out: Patterned or plain woven, blended, usually polyester/cotton fabric to which acid is applied at selected areas to remove or burn-out some fibre (cotton). The fibres are mixed so as only certain parts of the design are burnt-out and they become transparent, while the rest remains opaque creating a very attractive appearance. Used for blouses, dresses and curtains.

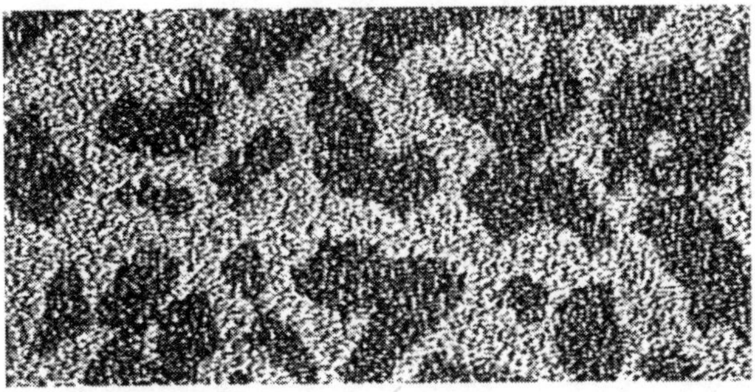

Burnt-out embroidery: A machine embroidery technique producing fabric comprising embroidered motifs. This embroidery technique creates simulated true lace. Embroidery is applied to a base material (paper or linen cambric). The embroidered motifs are linked by the required number of cross-over stitches so that the subsequent burning-out of the base material produces lace material (or motifs) in the form of piece goods.

Burnt-out velvet: Velvet fabric with designs produced by fibre selective chemical burn-out action (Veloures de'vorent).

Burnt sienna: Dark reddish-brown pigment resulting from treatment of sienna by fire or having the appearance of having been scorched. Also referred to as 'burnt terra di Sienna'.

Burnt sugar caramel: A deep yellowish-brown.

Burnt velvet: Velvet with a short, dense and tightly bound pile on which a pattern is created by selective removal of the pile, allowing the ground weave to show through. Used in dresses, blouses and for formal wear for ladies.

Burnt velvet

Burnt umber: Reddish-brown pigment resulting from the treatment of umber by fire or roasting.

Burr: A device that assists in loop formation on circular-knitting machines equipped with spring needles.

Burr-wool waste: Mass removed by the burr guard of cads or burr pickers having a very short fibre and full of burrs or seeds.

Burrah: A plain woven fabric used by African natives for garments characterized by the heavy colours. It is either dark blue or made in narrow-blue and black stripes with border stripes in other colours and a centre stripe in red or yellow. There are also headings at both ends of the piece.

Burring: Process of removing the burrs from the fleece, by means of hooks.

Burry wool: The fibres of which are entangled with burrs.

Burry blanket: A plain woven and slightly napped blanket.

Burst stitch damage: Faults occurring in making-up (significantly more serious in the case of solvent cleaned fabric) especially with wool and wool/ synthetic double jersey fabrics. Cause: fibre puncturing by needle points during sewing, stitches bursting due to the use of needle sizes unsuitable for the fabric density, or synthetic fibre yarn melting.

Bustian: Medieval worsted fabric used for dresses, made with three treadles.

Bursting strength: (1) The ability of a material to resist rupture by pressure.

(2) The force required to rupture a fabric by distending it with a force applied at right angles to the plane of the fabric under specified condition. Bursting strength is a measure widely used for knit fabrics, non-woven fabrics and felts where the constructions do not lend themselves to tensile tests. The two basic types of bursting test methods are the inflated diaphragm method and hydraulic bursting method.

(3) Defined as the multi-directional resistance to rupture of a circular fabric specimen. Testing of flat materials (textile fabrics, paper, films, etc.) is carried out under two-dimensional stress by applying a load perpendicular to the test surface.

Business plan: A plan for the proposed business which includes projected sales.

Bust: See **Bust girth**.

Bust girth: The circumference of the body over the fullest part of the breasts and parallel to the floor.

Bust point to bust point: The distance across the front from the apex of one breast to the apex of the other.

Bustle: Wire or whale-bone frame over which the rear of the skirt was gathered. Also see **'Cule de Paris'**.

Buta: Means literally, a plant. A floral motif, derived generally from Persian sources, much used in Indian textile design and traditionally rendered as a flowering plant with a curling bud at the top. The motif is also sometimes reduced to a floral pattern designed within the form of the plant.

Butadiene: It represents the simplest example of the Dienes: CH_2=CH–CH=CH_2. Butadiene is the starting material for the manufacture of butadiene rubbers (Buna rubbers).

Butadiene-acrylonitrile rubber: Synthetic rubber consisting of a copolymer of Butadiene with Acrylonitrile: -CH_2-CH=CH-CH_2-CH_2-CH-CH_2-CH=CH-CN.

Butane tetracarboxylic acid (BTCA): BTCA reacts with cellulose hydroxyls to form ester cross-links. The reactions of carboxylic acids with cellulose are rather old knowledge and have been practiced commercially for many years to make a variety of fibres and plastics from wood pulp. Using dibasic acids to cross-link cotton is also old technology. Problems with this in the past were excessive strength loss, poorer DP performance and poor durability to laundering. Certain phosphorous based salts were effective catalyst for the reaction. Because this too is classified as non-formaldehyde, there was a great deal of interest in this approach. The exact mechanism of the reaction is not known, nor is it known why the cross-linked products are so much more resistant to hydrolysis than are other cellulose esters.

Butcher rayon: This fabric used to be called Butcher linen, but it no longer contains linen yarns. See **Butcher linen**.

Butcher's linen: A plain weave, stiff and heavy fabric with thick and thin yarns in both the warp and the filling. The fabric was originally made of linen but is now duplicated in 100% polyester or a variety of blends such as polyester/rayon or polyester/cotton. Used for overalls and protective coats.

Buti: A diminutive of buta, very commonly used in Indian textile design.

Buts: In the Bible, denotes fine linen.

Butt: To level the root ends of flax straw at any stage of processing by vibrating it upright on a flat surface, either by hand or mechanically.

Butt seam sewing machine: Serves to compensate differences in elongation (= smooth seam), chiefly used in front of the entry of stenters, calendars, padders, raising and shearing machines, and also for stitching fabric into a tube for washing and dyeing. The seam is flat and can be undone by simply pulling the thread.

Butter muslin: A cheap open-weave cotton cloth, very soft. Used in the production of diary-products, for straining jelly, etc., but it is also the perfect cloth with which to press. It will shrink when first dampened for use. It is useful to have a couple of pieces, each one metre in length, so that one can be kept dry, while the other is damp.

Buttercup yellow: The pale yellow of buttercup petals.

Butterfly valve: A valve in a pipe that has a disc the size of the internal diameter of the pipe. When the disc is in the same direction as the flow the valve is fully open. The disc can be turned through 90° to shut off the flow.

Butter-nut: (1) The brownish-grey colour of the butter-nut.

(2) The coarse brown twilled homes-pun cloth woven of wool prior to the Civil War—coloured brown with dye from the butter-nut or walnut tree; used for men's wear and for decorative purposes.

Butterscotch: A yellowish-brown colour.

Button: (1) It is a porcelain material used as a slub catcher in reeling devises. It removes excess water and helps in forming a smooth circular raw silk. It also helps in improving cleanness of raw silk by preventing the defects to pass along with the yarn. Buttons with different orifice die are used for different deniers of raw silk.

(2) Lumps of fibres collected on the warp during weaving.

(3) See **Buttons**.

Button hole (eyelet): Formed by a contoured patch of zig-zag stitching, followed by a cut – a portion of which is circular. Eyelet buttonholes are usually used on heavy fabrics and/or with large buttons. A gimp or cord is usually contained within the stitches to provide a reinforcement along the edge of the hole.

Button hole (straight): Formed by two pairs of straight, parallel rows of zig-zag stitching, followed by a single, straight knife cut. Each end of the row of stitching is secured by a bartack.

Button-hole stitch: A complex machine stitch pattern made by coordinated motions of needle feed, appearing as very close stitches forming a narrow rectangle of stitching that is usually composed of four stitch segments, one on side and on each of the two ends of the rectangle.

Button-hole stitch bars: Embroidery stitches. Method: Bars are used in cut-work and Richelieu Work. Make a row of running stitches between the double lines of the design as a padding for the button-hole stitch where a single line bar occurs. Take a thread across the space and back, securing with a small stitch and button-hole stitch closely over the loose threads without picking up any of the fabric. Button side stitch round the shape, keeping the looped edge of the stitch to the inside, then cut away the fabric from behind the bar and round the inside of the shape. Where a double line or a broad bar is required between shapes or sometimes for stems of flowers, when the fabric is to be cut away on each side, make a row of running stitches along the centre, then button-hole stitch along one side, spacing the stitches slightly. Button-hole stitch along the other side into the spaces left by the first row. The fabric is then cut away close to the button-hole stitch, leaving a strong broad bar

Button-hole stitch with picot: Work as for ordinary button-hole stitch until a picot is required, then hold the thread down with the left thumb and twist the needle three times round the thread. Still holding the thread securely, pull the working thread until the twisted threads are close to the button-hole stitch, then make a button-hole stitch into the last loop.

Button-hole twist: Thick, loose cotton thread, made of 24 strands; used to strengthen the edges of buttonholes.

Button hole (built-in): These are made with a combination of zig-zag stitching and bar tacks. Most zig-zag machines have a built-in mechanism that stitches this type of buttonhole in two or four steps. The four steps are: zig-zag forward, bar tack, zig-zag in reverse, bar tack. A two-step buttonhole combines a forward or backward motion with a bar tack. Consult your machine manual for specific directions, because each machine varies. The advantage of this buttonhole is that it allows you to adjust the density of the zig-zag to suit the fabric and size of the buttonhole. Use spaced zig-zag stitches on bulky or loosely woven fabrics, closer stitches on sheer or delicate fabrics.

Button hole (over-edge): Over-edge button holes are an adaptation of the built-in or one-step buttonhole. This buttonhole is stitched with a narrow zig-zag, cut open and then stitched a second time so that the cut edge is over-edged with zig-zag stitches. The over-edge buttonhole looks like a hand-worked buttonhole. It is a good choice when the interfacing is not a close colour match to the fashion fabric.

Button hole (one-step): Buttonholes are stitched all in one step, using a special foot and a built-in stitch available on some machines. They can be stitched with a standard-width zig-zag, or a narrow zig-zag for light-weight fabrics. The button is placed in a carrier in back of the attachment and guides the stitching, so that the buttonhole fits the button perfectly. A lever near the needle is pulled down and stops the forward motion of the machine when the buttonhole reaches the correct length. All buttonholes are of uniform length, so placement is the only marking necessary.

Button holes (universal attachment): These are made with an attachment which will fit any machine, including a straight-stitch machine. The attachment has a *template* which determines the size of the buttonhole. This method also offers the advantage of uniform button-hole length and adjustable zig-zag width. The *key-hole* buttonhole, used on tailored garments or heavy fabrics, can be made using this attachment. The keyhole at one end of the buttonhole provides space for the shank. If buttonholes do not have to be re-spaced because of pattern alterations, make the buttonholes after attaching and finishing the facings but before joining to another garment section. This way there is less bulk and weight to handle at the machine.

Buttons: A knob, disc, or similar object which when forced through a narrow opening or button hole, fastens one part of a garment or other flexible substrate to other. Specified by design, size, colour, and type such as brass, melamine, or pearl, buttons are either shanked (attached by passing threads through the shank's eye) or holed (attached by passing threads through the button's holes).

Butyl rubber (isobutylene-isoprene rubber): Synthetic rubber obtained by copolymerizing isobutylene and isoprene. Properties: good fabric adhesion and abrasion resistance.

Butyrolactam (2-pyrrolidone): A monomer for the manufacture of polyamide-4, starting material for Polyvinyl pyrrolidone (dye-stripping agents, adhesives, binders).

Buyers and selectors: The people responsible for deciding which styles are going to be available as a part of the product range.

Buz: (1) A plain woven grey cotton fabric made in Central Asia; used for shirts and drawers, made 11 inches wide; (2)As given in the Bible, it means 'byssus'.

Bw: Cotton.

Byssus silk: (1) Old Greek name for fine sheer linen fabric of white or yellowish colour, woven in Egypt; (2) Long and silky hair-like beard of some sea mussels; used for gloves and stockings in southern Italy. See **Mussel silk**.

Byzance stitch: In canvas embroidery, producing a close zig-zag pattern.

Byzantine carpet: Similar to Brussels carpet with the addition of metal threads at intervals.

Byzantine embroidery: Applique embroidery, heavy worsted or floss silk yarn being laid and sewn on to leather or other thick ground, often ornamented with pieces of cloth.

C

C: Abbreviation for Celcius, Carbon.

c* : Abbreviation for chromaticity.

CA: Acetate fibres, Acetate fibres deacetylated (saponified).

Ca$^+$: Calcium.

Caaporopy: Produced from a species of plant from the Urtica family this fibre is fine, flax-like, used by the native people to make ropes, etc. The plant is usually seen in Paraguay.

Caballeros: Name in French for a spanish raw merino wool, used in large quantities in France.

Cabbage green: The green of the cabbage leaf.

Cabesa: Spanish raw wool from Estramadura, Spain.

Cabinet steamer: A special type of steamer used in particular processes as in the steaming of vigourex printing or tops printing and in pleating process. The machines have boxes or trays/drawers which can take steam pressure upto 1.25 – 1.5 bar. In case of Pleating the fabric is hung inside the boxes for steaming.

Cabinet: A basic part of the manufactured-fibre spinning machine where, in dry spinning, the filaments become solidified by solvent evaporation and, in melt spinning, the filaments are solidified by cooling.

Cable: (1) The heaviest ropes, over 10 inches in circumference, made usually of hemp, coir, the yarn, strand and hawser (usually three in number), being given alternately a right hand and left hand twist before made into cables; (2) A variety of corduroy, having broad cords.

Cable- laid rope: Formed by three ropes of three-strand yarn, twisted into one, alternately to the right and left hand.

Cable net: It is made of heavy cotton yarn with large mesh; used in England for draping and curtains in olden days.

Cable Stitch: (1) An embroidery stitch.This stitch is worked from left to right, bring the thread through on the line of the design. Insert the needle a little to the right on the line and bring the needle out to the left midway between the length of the stitch, with the thread below the needle. Work the next stitch in same way but with the thread above the needle. Continue in this way, alternating the position of the thread. This stitch may also be worked on even weave fabric. (2) A knit effect produced by crossing a group of stitches over a neighboring stitch group.

Cable thread: Sewing thread, composed of three ply threads, which are twisted together against the original twist.

Cable twist: A construction of thread, yarn, cord, or rope in which each successive twist is in the direction opposite the preceding twists; i.e., and S/Z/S or Z/S/Z construction

Cable webbing: Twilled webbing, woven with heavy round thread.

Cabled yarn: A yarn formed by twisting together two or more plied yarns.

Cablet: A trade term for cables measuring smaller circumference.

Cabling: The process of doubling cotton strands and twisting them alternately to right and left hand direction in making cotton thread.

Cabo negro: A dark, straight, smooth and glossy fibre, similar to horsehair, yielded by the leaves of a palm in the Philippines; used for ship cordage in olden days.

Cabot: General term for coarse grey or bleached, heavily sized cotton sheeting in the Balkan States and Asiatic Turkey.

Cabuja or Cabulla: Central American and West Indian name for the Mauritius hemp

Cabuya: Hemp growing in Colombia, Peru and Central America. Yields very strong fibre for ropes, bags, etc.

Cacao brown: The brown of the cacao bean used for making cocoa and chocolate.

Caceres: Spanish raw wool of medium quality.

Cacharado: A plain woven linen fabric from Spain.

Cachemire shawl: The finest French make is made with woven coloured figures on a foundation of fine cachemire wool and organzine warp and very fine and close Cachemire filling. Genuine cashemire shawl is made in one colour, and then embroidered.

Cachemirette: Originally from England; a trousering and coating, made of cotton or silk warp and woollen filling; occasionally it is fulled. It is woven in twill and given a clear face, while the back is napped.

Cactus green: The green of the cactus.

CAD in textile printing: Colour management in a print shop consists of the engraving section and the colouration section. A cad system (computer aided designing) with the help of a scanner takes the design which has to be modified in terms of shape and colour to enable on screen working.

Cadaverous: Pale.

Caddis: (1) Worsted lace and ribbon; (2) Stout and very durable twilled woollen fabric of various colours, used in France.

Cadene: Coarse and durable rug made in Asia. Has a long pile and is sewn together of narrow strips.

Cadet blue: A bluish grey or strong blue colour; sometimes referred to as 'cadet grey'.

Cadet cloth: Bluish grey, or other Military specific colours, strong, fulled and shorn woollen fabric; used for uniforms in military schools.

Cadilla: Light brown coloured, very long, lustrous, strong bast fibre, yielded by the Urena plant in South America; used for bags, ropes.

Cadillon: French twilled woollen cloth.

Cadmium green: Made by mixing cadmium yellow and viridian.

Cadmium lemon: A bright greenish yellow.

Cadmium lithopone: A pale yellow.

Cadmium orange: A bright orange made from cadmium sulphide. See **greenockite**.

Cadmium red: A bright red primary colour made from a mixture of cadmium sulphide and cadmium selenide and first marketed in 1910.

Cadmium vermilion: A bright red.

Cadmium yellow: A fine yellow in many varieties first made in 1820 from cadmium sulphide and used by artists extensively in the 19th century. Sometimes referred to 'aurora yellow'. Also 'cadmium lemon yellow' which has a more orangy hue.

Cadmopone yellow: A pale yellow also called cadmium lithopone

Cadis: (1) All wool, lightweight French serge, made about 22 inches wide, obsolete; (2) Very fine all-wool French cloth, slightly fulled; obsolete. (3) Coarse twilled worsted fabric; used by the Scotch Highlanders.

Cadiz stitch: In needle-point laces, consists of successive rows of buttonhole stitches.

Cadoxen solution: A solution for the determination of fluorescent Brightening agent on cellulosic material and determination of cellulosic material in synthetic fibre yarns especially cellulose/ polyvinyl alcohol mixtures. It contains 5% Cadmium 28-30% ethylene diammine and 0.5 moles caustic.

Caen: A grade of French serge made of wool.

Caesi- (L): Bluish-grey.

Café: Coffee colour.

café au lait: The colour produced by mixing milk with coffee.

Cafe con leche: In Paraguay white ponchos with brown stripes.

Caffa: (1) A 16th century English silk fabric, the patterns of which were either woven or printed with hand blocks; (2) An Indian fancy calico.

Caffard: (1) Satin made in various coloured stripes, the warp being of silk and the filling of silk or wool; obsolete; (2) French suiting, made of all-wool or wool and linen; used by the country people.

Caiana: Sort of raw cotton from Brazil.

Cain-coloured: Red or reddish yellow possibly in reference to the blood of Able shed by his brother Cain; having red hair; Shakespeare's *Merry Wives of Windsor* Act 1 Scene 4.

Cajun: The fibre produced from the leaves of the plant Furcraea cubesis. It is a strong leaf fibre similsr to sisal. The plant is found in central America.

Cake, in effluent treatment: The accumulation of solids in the medium, on the surface of the percolator on the septum.

Cake Dyeing: Dyeing of synthetic, viscose, acetate, triacetate filaments on Pack system dyeing machines with cylindrical material carriers or creel system dye tubes. Cake dyed material is mostly used for multicoloured products since the dye levelness through dyeing is a problem by this method.

Cake space, in effluent treatment: The volumetric space available in a filter to support the formation of cake.

Caked: Flaw in the fabric, consisting of hardened portions of size.

Calabria: Italian cotton, having a medium long, fairly strong but irregular staple of dull white colour; contains much leaf.

Caladaris: East Indian calico made with black or red stripes.

Calais lace: See **Leavers lace / Calais lace**

Calcareous: Lime coloured.

Calcined: Synonym for anhydrous. Salts are described as 'calc' (e.g. Sodium sulphate, sodium carbonate) have lost their water content through heating or roasting (calcining).during calcination.

Calcium hardness: See **Water hardness salts.** All hardness is usually expressed in total calcium hardness.

Calcium hypochlorite: (chlorinated lime, bleaching lime, bleaching powder) $Ca(OCl)_2$, salt of the hypochlorous acid (HOCl) MW. 127. White loose powder having typical chlorine smell, vividly attracts carbon dioxide from the air, (under decomposition, runs into a greasy mass; store dry, cool and air tight). Good calcium hypochlorite contains 75.8% active chlorine. Used for bleaching cellulosic materials, chlorination of wool etc.

Calcium Hydroxide: **Slaked lime, $Ca(OH)_2$.** Used in water treatment (lime-soda process) Not very stable. It looses on molecule of water forms **Calcium Oxide.** CaO.

Calcium oxide: Also called Quick lime, Caustic lime, Unslaked lime. It absorbs CO_2 readily from atmosphere to form cal. Carbonate. With water, with intense heating (up to approx. 450°C), forms highly alkaline calcium hydroxide $Ca(OH)_2$:

$$CaO + H_2O \rightarrow Ca(OH)_2 + 63.69 \text{ kJ}.$$

However, calcium oxide is only slightly water-soluble (at 15°C 1.3 grammes per litre, 100°C 0.65 grammes per litre) to form "lime water"; under the action of water, "slaked lime" results (calcium hydroxide); the watery white, milky suspension is called "milk of lime". Used for kiering agents, boiling agents (cotton), vat dyeing (zinc-lime vat), calcium mordant in dyeing (cotton); water purification (limesoda process); construction engineering, furnace lining, reference value for calcium hardness etc.

Calendering: Calendering is a process where fabric is compressed by passing it between two or more rolls under controlled conditions of time, temperature and pressure. A calendar is a machine consisting of two or more massive rolls which are compressed by means of hydraulic cylinders applying pressure at the journals. One roll is considered the pattern roll and is responsible for the finished appearance of the fabric while the other roll is called a bowl and serves as the pressure back-up for the pattern roll and also serves to transports the fabric through the machine. There are many types of calendars, each designed to impart specific effects to cloth. The composition of the rolls, number of passes, temperature controls, moisture control and pressure can vary to fit the desired effect. For example, the pattern roll can be engraved and serve to emboss a three dimensional pattern into the fabric. The engravings can be shallow or deep depending on the desired effect. The pattern roll can be smooth, made of steel or nylon to give the fabric a high lustre and sheen. The backing bowls can be made from corn husks, kraft paper, hard or soft rubber and deform to receive the pressure of the pattern roll. In calendaring, the yarns are flattened and become more oval in shape. This causes them to spread in two dimensions and closes up the fabric structure, leaving less open spaces between the yarn crossovers. In the process, the fabric becomes thinner and more lustrous. The reason fabrics are calendared is to improve aesthetics. The major fabric changes are: (i) Reduced fabric thickness. (ii) Increased fabric lustre, (iii) Increased fabric cover, (iv) Smooth silky surface feel, (v) Reduced air porosity and (vi) Reduced.

Calender: A multi roll (2-16 rollers or bowls) machine in which heavy bowls rotate in contact under mechanical pressure, for calendaring operation. The principle of operation is based on at least two adjacent calendar bowls one or more being made of metal (steel, iron) suface of which is ground, polished, chromium plated or engraved can be heated if required and the other one being a resilient bowl. Calendering effect depends on various factors like number of bowls, bowl material, bowl pressure (50-100 tonnes), metal bowl temperature (upto 300'C)rotastional speed and degree of friction, fabric stage (dry, damp, wet etc.) number of passages etc. NOTE: The bowls may be unheated or one may be a thick-walled steel shell heated internally. All bowls may rotate at the same surface speed, or one highly polished and heated bowl may rotate at a higher surface speed than the rest. In certain specialised machines, e.g. for knitted goods, two adjacent bowls may be heated, or, in the case of a laundry calender, one bowl works against a steam chest shaped to the curvature of the bowl. (See also **friction calendering**).

Calender bonding: A method of making thermally bonded nonwoven fabric in which calendar rollers are used to apply heat and pressure to a fibre web thuis causing bonding by softening or melting of the heat sensitive material.

Calender coating: Method used when coating the 4-roller calender with rubber, etc. To obtain good rubber coatings the rubber sheet is pulled into two rolls by calender rollers running at the same speed, and then on the last calender roller fabric is pressed onto the rubber sheet by means of a rubber roller.

Calender crease: Typical crease which appears on both sides of tubular knitted fabric which has been passed in the flat state through a calender. The calendar crease, like the ladder, can be used as an orienting line for slitting.

Calender, felt: A felt used in a sanforising machine which gives a sheen to the fabric after the sanforising operation. The felt usually runs over a large dia heated steel roll in between which the fabric is passed through to grt the desirable feel/look to the fabric. There are independent felt calendars also available for specific purposes.

Calendering: A mechanical finishing process for fabrics to produce special effects, such as high lustre, glazing, moiré, smoothen, flattening and embossed effects. In this operation, the fabric is passed between heated rolls under pressure.

Calendering finish: See **Calendering.** A smooth finish obtained by passing the fabric between heavy bowls of a calender, which results in the fabric being flattened so as to close the interstices between the yarns.

Calendering roll: (1) The main cylinders on a calender. (2) Smooth or fluted rolls used on various fibre-processing machines such as pickers and cards to compress the lap or sliver as it passes between them.

Calf fur: A coarse fur with a sheen not often used in its natural state, but dyed and cut to imitate others furs, such as pony, leopard, orcelot etc.

Calf girth, In body measurements: The maximum circumference around the leg between the knee and ankle parallel to the floor.

Calf Leather: A very soft pliable leather of top quality. Used as the basis of good patent leather because it takes a high polish.

Calgon T: A trade name for sodium hexametaphosphate It is sodium hexametaphosphate without additives. Water softener products sold at retail under the Calgon name usually contain additives such as fragrances and surfactants

Calibrate: To determine the characteristic of an instrument or tool and mark it at one or more places corresponding to the exact values of an established standard scale, for regular use.

Calibration: The act or process of calibrating ; the recorded relationship resulting in from calibrating

Calibration cotton standards: Cotton standards taken from a blended bulk source on which fibre properties have been determined under the International Calibration Cotton Standards program.

Calibration curve: A measuring curve which is recorded for comparison with a reference curve.

Calibration dyeings, in colorimetry: Calibration dyeings are done in selected substrate by selected class of dye to feed the calibration measured using spectrophotometer, which in turn is the basis for the recipe formulation. Hence it is very important that the dyeings are as perfect as possible. These dyeings are done in either 6, 8 or more concentration as the software can accept. E.g. for 6 concentration dyeing are 0.01, 0.05, 0.2,0.8, 1.5, 2 or 0.1, 0.5, 1.5, 3.5, 6.0,8.0 and in 8 concentration 0.005, 0.025, 0.1, 0.25, 0.5, 1.0, 1.5, 2.0 or 0.03, 0.2, 0.8, 2.0, 4.0, 6.0, 8.0, 10.0. Generally concentration should cover the usual concentrations the dyes are generally used. A blank dyeing is also made to measure and feed the data along with these calibration dyeings.

Calico: Cheap plain-weave cotton fabric often printed coarser than lawn. One of the oldest fabrics, it is named after Calicut, a place in Kerala, India where it was produced and exported. It is in the same grouping as Percale but coarser and poorer quality. Firmly woven and very strong.

Washes and wears well, used for night wear and, sheeting. Unbleached calico has the characteristic dreamish look of a natural fabric, with darker flecks of fibre in it. Used for under sheets , mattress covers, aprons, dresses, quilts etc.

California: Fancy English alpaca fabric in the 19th century.

California blanket: Originally made in San Francisco, made of fine wool, with a long soft nap.

Calimanco or Calmanco or Callimanco: A plain or striped stout worsted in England made in imitation of the camel hair cloth with single warp and finished with a high lustre made in the 19th century; used for petticoats and chair seats.

Calloose hemp: Extracted from the stem of Urtica plant grown in Sumatra, this fibre is strong and coarse, hence used for cordage.

Calmande: A wool fabric made in all wool structure or wool warp and silk filling and sometimes with wool mixed with silk or goats hair in the warp mainly for the figured Calamande, usually of satin weave but also made with rib effect with high gloss many were made of natural coloured wool, but also dyed or striped. Used for mens clothes, skirts, dresses etc. Originated in France or Denmark and is made in various widtrhs. It was made single or double faced. See also **Calimanco.**

Calmuc or Kalmuk: (1) Wool yielded by the Calmuc sheep in Central Asia; (2) Loose woven, twilled woollen made of loose twist yarn, fulled and finished with a long nap, used as winter dress goods.

Calorie (cal): The CGS unit of heat energy. This calorie (also called a gram calorie or small calorie) is the amount of heat required at a pressure of one atmosphere to raise the temperature of one gram of water by one degree Celsius. Unfortunately, this varies with the temperature of the water, so it is necessary to specify which degree Celsius is meant. A traditional choice was the degree from 14.5°C to 15.5°C; raising the temperature of water through this range requires 4.1858 joules, a quantity called the 15° calorie. Another choice produces the thermochemical calorie, equal to exactly 4.184 joules. More common today is the international steam table calorie, or IT calorie for short, defined by an international conference in 1956 to equal exactly 4.1868 joules, exactly 1.163 milliwatt hours, or about 0.003 968 32 British thermal units (Btu). The name of the unit comes from the Latin *calor*, heat.

Calorie (kcal or Cal): A common name for the MKS unit of heat energy. This unit is properly called the **kilocalorie**; it is also called the kilogram calorie or large calorie. It is often (but certainly not always!) distinguished from the small calorie by capitalizing its name and symbol. The large calorie, or rather kilocalorie, is the amount of heat required at a pressure of one atmosphere to raise the temperature of one kilogram of water by one degree Celsius. Since this is 1000 times as much water as mentioned in the definition of the small calorie, the kilocalorie equals 1000 small calories, 4.1868 kilojoules, 3.9683 Btu, or 1.163 watt hours. (These conversions assume the IT calorie is in use; see previous entry.) These are the "calories" that joggers are trying to get rid of, the ones we gain by eating. The use of the same term "calorie" for two different-size units is endlessly confusing, but we seem to be stuck with it.

Calorific value, heating value, CV: The amount of heat given off by a fuel when it burns completely. There are two values. The larger or gross CV is slightly too high because it includes heat from the steam, given up when the water formed in the burning is condensed. The lower or net CV gives a truer value of the heat that can be used. The calorific value of landfill gas is usually in the range 16 to 20 MJ/m^3. The calorific value of domestic solid waste varies from about 4 MJ/kg for the food fraction to 16 MJ/kg for paper and over 30 MJ/kg for plastics.

Calquiei: Indian silk taffeta made with double-and-twist or printed yarn.

Calorimetry: The process of measuring quantities of absorbed or evolved heat, often used to determine specific heat.

Calum couree, Kalamkari: An East Indian printed chintz. The fabric is usually printed with vegetable colours and hence limited colours. Predominantly brown and olive related colours will be more.

Cam: A rotating or sliding piece or projection used to impart timed or periodic motion to other parts of a machine. It is used chiefly as a controlling or timing element in machines rather than as part of a power transmission mechanism. Cams are particularly important in both knitting and weaving machinery.

Cam-locks, in knitting machine: The cam-locks are a cam system which gives the necessary working information to the individual needles; they include a fixed part, working as support, and movable cams, which can be divided into raising cams and lowering or knock-over cams. The different parts of the cams are chamfered; their profiles are curvilinear to make the needle move smoothly.

Cam lock slider, In Zipper: The portion or portions on the pull that extend through a window or windows to effect a locking action in cam lock sliders.

Cam, bolt in knitting: Can be caused to descend into the cam track to control the element butts or be withdrawn out of action so that the butts pass undisturbed across its face; it is mostly used on garment-length machines to produce changes of rib set-outs.

Cam, Counter in knitting: See **Cam, upthrow.**

Cam, guard: Cam guard are often placed on the opposite side of the cam-race to limit the movement of the butts and to prevent needles from falling out of track.

Cam, stitch in knitting: Controls the depth to which the needle descends, thus controlling the amount of yarn drawn into the needle loop; it also functions simultaneously as a knock-over cam.

Cam, raising in knitting: Causes the needles to be lifted to either tuck, clearing, loop transfer or needle transfer height, depending upon machine design.

Cam, swing in knitting: It is fulcrummed so that the butts will be unaffected when it is out of the track and it may also be swung into the track to raise the butts.

Cam, upthrow or counter: Takes the needles back to the rest position and allows the newly-formed loops to relax. The stitch cam is normally adjustable for different loop lengths and it may be attached to a slide together with the upthrow cam, so that the two are adjusted in unison.

Camayeux: Silk cloth of coloured warp and black filling, the latter alternately shot in single and double.

Camayeux effects: Light and dark effects in the same shade which are achieved at the same time in the same dye bath, on the same fibres with a different affinity for dyestuffs. This can be achieved on wool by using treated/ untreated wool to make yarn (e.g. chlorinated/ nonchlorinated wool; mordanted/premordanted wool) woven and knitted fabric followed by dyeing with selected chrome dyes. Chlorinated and premordanted wool has a much higher dye affinity than untreated wool thus producing light dark effect. Same way in silk using degummed yarn, mordanted yarn, etc.

Cambayes: Strong East Indian cotton fabric of coarse structure, finished to resemble linen.

Camblet: 18th and 19th century English and French, plain woven or twilled fabric, made with single or double warp of wool mixed with silk or goat's hair- It was woven in the gray and dyed in the piece; used for cloaks. Originally came from the Orient, where it was made of Angora hair. Compare with **Camlet.**

Camblettee: A 19th century pure worsted fabric in England.

Camboulas: Southern French fabric, made of ply cotton warp and woollen filling.

Cambr'e fabric: Very light, sheer, veiling-like linen used in Italy.

Cambreaine: French term for plain woven very fine cottons and sheer linens, similar to the cambric; originally imported from various parts of Asia.

Cambrian tweed: A rough woollen cloth made in the uplands of mid Wales, from the local hardy sheep. Plain weave, hope sack or herringbone weaves are the most usual and the yarns are either dyed, or more commonly left in their natural colours of white, grey and black and mixed together. Used for hard wearing cloths such as mens suits, sports jackets, Norfolk skirts, and jackets. The coarse yarn is made into rugs.

Cambric: Heavier than lawn, this is a fine, firm closely woven plain weave fabric finished with size to give a slight shine on the right side. At one time it could make for linen but is now cotton. Mainly in plain colours and bleached. Usually 30s hard twist yarns are used in the warp and filling. Used for childrens cloths, nightwear, blouses and handkerchiefs.

Cambric muslin: Fine cotton fabric, bleached and glazed, used for underwear.

Cambridge blue: The light blue Eton colour adopted, it is said, in haste in the 1836 annual Cambridge and Oxford varsity boat race by the Cambridge cox having forgotten his team's own colours. See **blue** (n).

Camel cloth: A term now used to describe any camel coloured clothing, with a soft feel and slight pile. May be wool, wool/acrylic, or other mixtures.

They are cheap imitations fof real camel hair, but some are very good quality and expensive. Sometimes made as reversible cloth with crean colour on the other side. Fabrics used for coating, scarves and rugs. Camel hair fabrics are virtually all made with woollen-spun yarns.

Camel hair: Natural protein fibres from two-humped camels from the Bactrian or dromedary (Asia, Africa, Australia).The annual yield is about 3-4kg. The hair is about 50-60 mm long and about 70 um fineness. The undercoat of the camel hair is considered to be one of the fine animal fibre and is about 60-100mm long and 14-26um fineness and sandy to light brown in colour, sometimes dark brown or black. Camel hair is mostly used in the natural colour itself as this is difficult to over dye. It is like wool and gives worth without weight. Very expensive, often mixed with sheep's wool to reduce cost. Un blended fibre used to make knitwear and underwear, whereas in blend form it is used for ladies dress material, blanket, woven cloth coating etc.

Camel hair cloth: Usually denotes a fabric wholly made using camel hair. Because the hair of the camel, especially the fine undercoat, is of higher quality and more expensive than sheep's wool it is often imitated using wool dyed to camel shades. Wool fabrics dyed to camel shades are piece-dyed with a solid-shade appearance whilst genuine camel hair cloths can be distinguished by their slightly mottled appearance which has high fastness, especially to rubbing and to light exposure, and they also withstand standard dye stripping tests. Finishes include velour (upright napped or laid pile) or as a pilot cloth with long brushed pile. Weight between 450–750 g, providing good thermal insulation. Used in high quality outerwear.

Cameline: (1) A coarse medieval fabric, made of camel hair in twill weave, similar to the cashmere; (2) Woollen dress goods in fine basket weave and finished with a soft nap.

Camelot: An imitation Camlet fabric produced in wool and cotton.

Camelot baracane: French term for baracan made with the warp heavier than the weft. See **Baracan.**

Camelotee: 18th century French fabrics of smooth finish made with pebbled or granulated effect, produced by the coloured filling.

Camera coarse: Loosely woven French linen, unbleached or dyed yellow.

Camera work: Photographs attached to a linen ground and surrounded with embroidered flowers and scrolls.

Camerick: 16th century name in England for cambric.

Cameron of Erracht: Highland tartan, made as follows: Wide dark green bar, split with a group of four narrow red stripes in the center (the outer stripes being wider than the inner ones), a black stripe (about one-third of the green), a very narrow red stripe, a dark blue bar, as wide as the green, split by a pair of the wider red stripes with a yellow stripe between (the three not touching), a very narrow red stripe, a wide black stripe.

Cameron of Lochiel: A Highland tartan with a red ground, the stripes arranged as follows: Narrow white, edged with narrower blue on both sides, the large, even sided red square is traversed both warp and weft wise by two wide stripes, narrow white stripes, edged with narrower blue on each side. The next large red square is uneven sided, traversed by the two blue stripes continued from the even sided square and these are crossed by two narrower green stripes.

Camientries: Wool fabric made in Norwich, England, in the 17th century.

Camlet: A fine lustrous yarn resembling camel hair or a corresponding plain weave fabric made from linen or cotton with worsted or silk plied yarns.

Cammaka or Camoca or Camak: Fine medieval fabric, believed to be of camel's hair and silk.

Camocato: A Damask satin made in China.

Camogi: Long and fairly strong vegetable fibre from Southeast Africa; used as substitute for hemp.

Camoyard: Twilled French fabric, made of goats' hair.

Camouflage: The disguise of animals, personnel or equipment by the use of such colours as make the object in question appear to merge with its surroundings. Derived from the French 'camoufler' – to disguise. It is an offence to wear camouflage clothing in Barbados unless the wearer is part of the Defence Force. See **cryptic colouring.**

Camouflage dyeing and prints: Dyeings and print on fabrics which is used as outer wear garments in armed forces. Print/dyeing should meet highest fastness requirements and sometimes need with special finish requirements like, UV absorbent, soil release/ repellent finish etc.

Campane or Campaigne: (1) Narrow French bobbin lace of the 18th century, made of flax or silk with scalloped edge; (2) Gold, silver or silk fringe with bellshaped small tassels; used as dress trimming.

Campanula: A violetish-blue after the flower of the same name. Used by the art theorist Charles Blanc (1813-1882).

Campbell of Argyll: A Highland tartan, made with wide green bars and twice as wide blue bars. The green bars are alternately split with a yellow and white stripe. The blue bars are outlined on both sides with a wide and two very narrow stripes (on the inside) of black.

Campbell of Breadalbane: A Highland tartan, arranged as follows: Large checks are formed by two wide black stripes and a twice as wide blue bar between (the latter split in the center by a narrow black stripe). Between these groups are two solid green bars (as wide as the blue) with a narrower black stripe between. The wide green bars are split with a narrow yellow line in the center.

Campbell of cawdor: An even sided Highland tartan, made as follows: A wide black stripe, wide dark blue stripe, narrow red stripe (edged with black lines), wide blue stripe, wide black stripe, wide green stripe, narrow pale blue stripe with black lines on each edge, wide green stripes.

Campbell of Loudon: Even sided Highland tartan composed as follows: Very wide dark blue bar with a black stripe on each side and two narrow black lines in the middle, wide green bar, the same width as the pure blue bar, the green bars are alternately split with a narrow white or yellow stripe, edged with a black line.

Campbell of Strachur: A Highland tartan, composed of wide green and narrower black stripes, and crossed by yellow lines.

Campbell twill: A weave used extensively in the fine woollen and worsted trade. Also called Mayo twill.

Camping tentage: Any portable temporary shelter or structure designed to protect persons from the elements all or a portion of the covering which is made of fabric or other pliable materials.

Canadaris: Chinese and East Indian pure silk or cotton mixed satin, made with fancy stripes or window plaids.

Canadas: French woollen blankets.

Canapina: Strong and silky fibre of Argentine, yielded by the abutilon plant; used for cordage.

Canard: A dark blue.

Canary: Bright yellow resembling the colour of the canary's plumage.

Canary green: A dark yellow colour.

Canary yellow: The bright yellow of the canary.

Candela (cd): The SI base unit for measuring the intensity of light. *Candela* is the Latin word for "candle." The unit has a long and complicated history. Originally, it represented the intensity of an actual candle, assumed to be burning whale tallow at a specified rate in grains per hour. Later this definition was replaced with a definition in terms of the light produced by the filament of an incandescent light bulb. Still later a standard was adopted that defined the candela as the intensity of 1/600 000 square meter of a "black body" (a perfect radiator of energy) at the temperature of freezing platinum (2042 K) and a pressure of 1 atmosphere. This definition has also been discarded, and the candela is now defined to be the luminous intensity of a light source producing single-frequency light at a frequency of 540 terahertz (THz) with a power of 1/683 watt per steradian, or 18.3988 milliwatts over a complete sphere centered at the light source. The frequency of 540 THz corresponds to a wave length of approximately 555.17 nanometers (nm); normal human eyes are more sensitive to the green light of this wavelength than to any other. In order to produce 1 candela of single-frequency light of wavelength l, a lamp would have to radiate $1/(683V(l))$ watts per steradian, where $V(l)$ is the relative sensitivity of the eye at wavelength l. Values of $V(l)$, defined by the International Commission on Illumination (CIE), are available online from the Colour and Vision Research Laboratories of the University of California at San Diego and the University of Tübingen, Germany.

Candid- (L): White. The root of the word 'candid' (developing its meaning from 'white' or 'pure') and of 'candidate' – candidates having worn white togas in Roman times. See **toga praetexia**.

Candle filter: (1) A small filter interposed between the spinning pump and spinning jet to effect final filtration of the spinning solution prior to extrusion. (2) Sewarage filter system.

Candle Turbidimeter: A device principally used to measure high turbidity water with the results expressed in Jackson Turbidity Units (JTU) or Farmazine Turbidity Units (FTU). The JTU is measured with light scattering.

Candle water temperature: The temperature of the water surrounding the candle filter or within the heating jacket during fibre extrusion.

Candle wick yarn: A coarse folded yarn usually of 100%, ring or open end yarn.

Candlewick fabric: A thick tufted pile fabric imitating an early handicraft when the wicks if the candles were used for decoration. The extra yarns are added to a loosely woven fabric of muslin construction and the thicker yarn is threaded through the backing in straight line or patterns, leaving spaces in between. The loops are then cut leaving thick tufts firmly wedged in the backing. An unbleached muslin bed sheeting (also called Kraft muslin) used as a base fabric on which a chenille effect is formed by application of candle wick (heavy plied yarn) loops, which are then cut to give the fuzzy effect and cut yarn appearance of true chenille yarn. May be uncut also. (True chenille is a cotton, wool, silk, or rayon yarn which has a pile protruding all around at slight angles and stimulates a caterpillar. Chenille is the French word for caterpillar). Uses: Bedspreads, drapes, housecoats, beach wear.

Candy-coloured: Having a shade of pink. Also 'candy pink'.

Canebrake: Cotton grown on the southern central prairie of Alabama, the staple is strong, measuring about 11-16 inches.

Canezou: Short small jacket, spencer type, for women; later with a wide shoulder collar.

Cannage: A fault in silk cloths due to, for example, a warp tension effect. It is prevalent in taffetas and takes the form of ripples of alternate high and low lustre, distributed irregularly over the cloth.

Cannamazos: Spanish term for various grades of unbleached linens.

Cannelas: A cinnamon-brown colour.

Cannele: (1) Silk fabric, made with two sets of warps, one single and the other ply yarn. The single warp forms the foundation, the ply yarn floating over eight fillings, forming ribs. It is also made double faced by dividing the heavy warps and letting them float alternately both over the face and the back; (2) Fabrics woven with several picks in the same shed, forming short ribs which appear alternately on the face and the back, imitating the joints of the bamboo.

Cannele weave and repp weaves: Are in their construction related to the taffeta, and are used mostly in the form of stripes as an additional ornament to a fabric. The threads going into the composition of these effects exchange continually from taffeta interlacing to floating over a certain number of threads, and must be introduced either in warp or filling close enough to make the floats cover up the taffeta work entirely, and thus enable the material used to show up with the full brilliancy it possesses. Cannele effects can be produced in two distinct ways. One is to let every individual thread work alternately taffeta and float, while in the other method one thread weaves always taffeta, and a second thread is used for the cannele exclusively. These latter threads must come from a separate warp, which is introduced to embellish the ground or taffeta part of the fabric. The floating threads can either stitch all on one pick and so form a continuous cut line, or be divided in groups, of which one will bind in the middle of the floats of the other group.

Cannette: (1) French make single chappe yarn; (2) A fine wide cotton fabric, made with warp cords; used in the Philippines for dresses.

Cannetilles: (1) Lace or military braid made of gold or silver thread. See **Bullion**; (2) French dress goods and furniture drapery, having warp ribs, made with two sets of warps and eight leaves and eight fillings in a repeat.

Canotier: French term for dress goods; used for yachting and boating costumes.

Canourge: French woollen serge.

Canques: Cotton shirting in China.

Cantala fibre: Agave fibres (Indonesia, Philippines, Mexico) similar to sisal but softer and more elastic obtained from the species of plants Agave cantala.

Cantaloupe: A light shade of yellow after the melon of the same name.

Canterbury: English fancy dress goods of silk warp and cotton filling.

Cantiloon or Cantaloon: An 18th century pure worsted, made in England; used for women's wear.

Canton crepe: (1) This is crepe with filling yarns to produce a pebbly surface, originally a silk from Canton, now often made with viscose or polyester. It is durable, due to the high twist yarn, and washes easily. Often in white or may be piece dyed. Used for blouses and dresses.

(2) A sturdy silk dress material with a pebbly crepe surface with six S threads and six Z threads in arrangement and two S picks and two Z picks, originally made from canton silk, it is now made with 150-200 denier flat rayon in the warp and crepe rayon in the weft. Six picks in the warp and crepe rayon in the weft. A typical construction of light weight canton crepe is 96 x 46 with 150 denier rayon warp and 200 denier rayon filling. Another for a heavier fabric is 124 x 52 with same yarns as in the light weight fabric. If 150 denier rayon is used in the filling, four to six picks are added to the construction. Used for coat linings, dress goods, gowns, negligees, pajamas, scarves etc.

Canton flannel : A medium to heavy cotton fabric flannel is sometimes called Canton flannel, after the Chinese city where it originated. It is usually made with twill weave with 2/1 twill suitable for fabric which has to be napped on the back and 2/2 twill for the fabric to be napped on both sides with a soft filling yarn. A typical construction is 74 x 44 with 20s warp and 10s filling usually made 30 in. wide, it is used for a warm interlinings, pocket lining in top coats, gardening gloves, some baby clothes, sometimes used as warm lining for other clothes etc. It is soft and absorbent; may be used unbleached or dyed. Washes well; very inflammable.

Canton satin: See **Satin.**

Canton silk: A raw silk reeled from very small domesticated silk cocoons in south china. The fibre is regular, soft and lustrous but light, hairy and weak.

Cantoon: A very stout, compact cotton fabric having fine diagonal wales on the face and napped in the back. It is woven in a 3-and-3 diagonal twill, having great many picks in an inch; used for riding breeches.

Canvas: Of the colour of canvas.

Canvas: A strong firm fabric (flax, cotton, viscose) woven with square interstices using hard-twist warp and weft threads. Sometimes called Duck. Used as carpet knotting, tapestry, shoes, sails awnings and suchlike and as stiffening material. It can also be open-weave, in various weights and in this form used as embroidery base fabric.

Canvas fabric

Canvas, in garment: A made up interlining for a garment.

Canvas Interfacing: This is made in various fibres and weights, but is generally heavier than other types of interfacings. May be 'sewn-in' or 'iron-on'. It is durable and very malleable under a hot iron with moisture. Most types can be easily shaped to the body and give excellent structure to a tailored garment. Used for coats and jackets.

Canvas duck: See **Number ducks**.

Canvas making: The sewing together of all the pieces which form the interlining of a jacket or coat either by hand or machine.

Cap ease, in sleeves: Fullness in cap between front and back notches that is eased to garment.

Cap height, in sleeves: Distance midway between biceps line and shoulder point.

Cap seam line, in sleeves: Seam line that outlines the top of the cap.

Cap spinning: A system of spinning employing a stationary, highly polished metal cap just large enough to fit over the take-up bobbin, which revolves at a high rate of speed. The cap controls the build and imparts sufficient tension to the yarn for winding. The yarn is twisted and wound onto packages simultaneously.

Capacitance: The measure of the ability of a nonconductor to store electrical energy by means of the potential difference across the surfaces of the nonconductor.

Capacity: Maximum value for which a machine, instrument, or a measuring instrument is designed.

Capah Damask: A rich, purple coloured silk damask mentioned in the times of Henry VIII.

Cape: Circular covering for the shoulders; the inside yoke of a coat.

Cape Coat: An overcoat with a cape and with or without sleeves.

Cape leather: Originally this name came from Cape of Good Hope sheep, but it mainly come from USSR. It is firm and hard-wearing, and is usually give a glaze or shiny finish. Used mainly for gloves.

Cape Wools: From South Africa; the staple is fine, silky but short and dirty; used for woollens.

Capella: A 2/1 twill weave using mulberry silk fabric and generally used as dress material. Capet, face to face.

Capillarity: Generic term for all physical phenomena which occur in consequence of the Surface tension of liquids in capillaries and cracks and in porosity

Capillary action: A process by which liquids are drawn through the fabric and into pores found between fibres and yarns.

Capsicin: The red pigment in cayenne pepper.

Capiton: A coarse grade of waste silk in France.

Capmeal: Coarse woollen fabric of the 18th century, made in England.

Capot: French men's wear for sailors clothes, waterproof coats, etc.; made very strong with napped and fulled face.

Cappadlne: Silk waste yielded by the inner side of the cocoon.

Caprolactam: A white, crystalline, cyclic amide ($C_6H_{11}NO$) which yields ε-amino-caproic acid on hydrolysis and is used as a raw material in the manufacture of nylon 6.

Capucine: A dark yellowy orange.

Caraco: Short tailed jacket for women, similar to a tailcoat.

Caracul: Asiatic variety the broad tail sheep which owes its name to the town of Kara Kulin eastern Bokhara, but it is also reared in Persia for example and now in south west Africa also. The caracul sheep gives a very coarse wool.

Carragheen: A sea weed usually called Irish moss, pearl moss, kelp. Contains approx. 80% mucilaginous substance content. Dried seaweed (North Atlantic), can be bleached to light colours. The substance once used as finishing agent after soaking in cold water and boiling to dissolve, to get "natural finish" for cotton = supple-voluminous stiffness, with no real hardness; also for wool, also used as a clarifying, stiffening or gelling agent.

Caraguata: Very long, soft, silky leaf fibre of good strength, yielded by the Bromelia plant in South America; used for ropes.

Caramel: A pale brown colour; the colour of toasted sugar. The dark brown colouring used as a colouring agent in food (E150).

Caraua: Fine glossy fibre, yielded by a species of the Bromelia in Brazil; used for ropes.

Caravonica: Tree-cotton, produced by the crossing of long staple Mexican cotton with a coarse long staple Peruvian growing in India, Mexico, etc. Will grow in hot climate with not too much rain. There are two varieties: (a) wool caravonica, adapted for mixing with wool and (b) silk caravonica, which has long, lustrous strong fibre.

Carcassonnes: Light woollen dress goods, made in Carcassonnes, France.

Carbamates: Carbamates are a family of related compounds that also react with Formaldehyde to form N-methylol derivatives. A general structure of the starting compound is shown in the box below. This structure is also called a urethane so carbamates are simple urethanes. The alkyl group (R-) can be methyl, ethyl, propyl or hydroxyethyl. The methyl and ethyl carbamates are carcinogenic and no longer used. The propyl and hydroxyethyl are safe and are used today. Carbamates react with formaldehyde to form N-methylol derivatives. They can react with up to two moles as shown below. The reaction is difficult to drive to completion and the equilibrium is such that the best that can be done is 1.7 to 1.8/1. This leads to products that have high free formaldehyde.

Carbamide: Urea.

Carbohydrates: This large and important group of organic vegetable products was given its name at a time when, of the chemical construction of these substances, virtually only the total formula $(C_x(H_2O)y$ was known, which is invalid for all carbohydrates in the light of current knowledge, and, incidentally, also covers other compounds which do not represent carbohydrates, e.g. acetic acid $(C_2H_4O_2)$, lactic acid $(C_3H_6O_3)$ etc.

Carbon chain fibres: Synthetic fibres having only carbon atom in the main chain of the macromolecule such as polyacrilonitrile, polyvinyl alcohol, ployvinyl chloride, polyolefin, polyflourinated fibres etc.

Carbon-arc lamp: A type of fading lamp which utilizes an arc between two carbon electrodes as the source of radiation.

Carbon black: Can be made by thermal decomposition of acetylene, residue from burning coal tar, tar oils, naphthalene, natural gases etc. It is deep black very fine carbon n the form of graphite crystals, (2-6nm India.) It is completely insoluble in water and used for making jet black pigments for mass colouration, Indian ink, typewriter ribbon, leather etc.

Carbon emissions trading: A system in which a company that emits pollutants at a lower level than its target is able to trade and sell this to companies that cannot economically reach their target emission level. This is part of the ***Kyoto Protocol 1997***.

Carbon fibre: A high-tensile fibre or whisker made starting from regenerated fibre (Viscose filaments with or without drawing) and synthetic fibres (polycarbonate, polamide, polyacrylonitrile, polyvinyl alcohol) which are first of all oxidized by pyrolysis and the carbonized (300-1600°C) and 'graphitsed' at 1500-3000°C. Carbon fibres so formed contain 80-90% carbon. Fibres may be 7-8 microns India.

Carbonate Hardness: The hardness caused by carbonates and bicarbonates of calcium and magnesium in water. The amount of hardness equivalent to the alkalinity formed and deposited when water is boiled. In boilers carbonate hardness is readily removed by the blow down.

Carbonised and neutralised wool: A term descriptive of scoured wool processed to destroy cellulosic impurities by treating with a mineral acid or an acid salt, drying and baking crushing and dusting out the embrittled cellulosic matter followed by neutralization of the acidified wool.

Carbonising, in wool: Employed for removing vegetable fibres and impurities (burrs, straw etc.) from wool. Generally sulphuric acid of 7% concentration (occasionally hydrochloric acid 4.7%, aluminium or magnesium chloride solution (7-11%) are used for this purpose. After acid treatment centrifuge to 40% moisture, dry and bake at 140^{0}C, shake to remove the carbonized material and neutralize. Aluminium and magnesium chloride is less harmful to wool.

Carbonising: A chemical process of elimination of cellulosic material (yarn or fibre from a polyester admixture. Carbonising is also essential when the raw stock is mainly composed of rags or waste (dry carbonising with HCl gas at 80°C). In fact, with this type of material the carbonising process eliminates any vegetal residue in the staple after scouring, due to the good resistance of wool to the action of acids which, on the contrary, destroy cellulose, and to the strong dehydrating action of the acid which provokes a weight loss that cannot be exactly evaluated in advance. Carbonising can be carried out also on staple fibres, yarns and fabrics. Washed and sometimes piece-dyed fabrics as well as gray fabrics can also be carbonised. The operating conditions necessary for the carbonising process are the following: the fibres are soaked with H_2SO_4 (25.4 ° Bé or 4.6 %), squeezed by means of two cylinders and then dried in a stenter at 85 - 90°C, for 30 - 60 minutes. Hot air concentrates the acid by evaporation, as a result dehydrating and hydrolysing the cellulosic matter. Finally fibres are carefully washed to remove completely any residual acidity, which could affect the fibre and subsequent operations. A series of washing processes also includes a neutralisation treatment with sodium acetate. During the fabric process, a dry beating process to remove the carbonised vegetal residues from the fabric texture precedes the washing phase.

Carbonising agent: In accordance with TEGEWA nomenclature, those textile auxiliary agents which facilitate and accelerate penetration of the carbonizing acid into the vegetable impurities in the wool, promoting destruction of the vegetable impurities. Carbonizing agents also enable the quantity of acid and the duration of carbonizing to be reduced, thus protecting the wool fibre.

Carbonising, fastness to: Assessment of the effect on the shade of a dye by the carbonizing process, by which foreign vegetable fibres are removed from wool by means of diluted sulphuric acid and aluminium chloride solutions. When acetate or polyamide fibres are to be removed from wool, carbonising must be effected with aluminium chloride. Test implementation:

15 mins at room temperature with

(a) 60% sulphuric acid solution,

(b) 60% aluminium chloride solution,

with a 20 : 1 liquor ratio, squeezing/centrifuging to 80%. Drying: approx. 30 min. at 60°C. Carbonizing: 15 min at 105 °C for the sulphuric acid process, and 15 min at 115°C for the aluminium chloride process, rinsing test specimens in cold flowing water for 5 min. Then dry half at 60°C in hot air, and neutralise the other half – with sodium carbonate in the case of the sulphuric acid process, and with ammonia after using the aluminium chloride process. Rinsing, drying. Grey scale assessment of the two halves.

Carbonizing chamber: Part of the drying device for temperatures above 100°C used in carbonisation.

Carbonyl Dyes/Pigments: The chemical class of colorants which is second in importance to azo dyes And pigments is characterized by the presence of a carbonyl (C=O) group, which may be regarded as the essential chromophoric unit. The vast majority of carbonyl dyes and pigments contain two or more carbonyl groups which, as illustrated in, are linked to one another through a conjugated system, frequently an aromatic ring system.

Carborundum: (Silicon carbide), SiC. Manufactured at above 2000°C from carbon and quartz. Characteristics: Multicoloured (reddish, greenish blue, black) crystals, extremely chemical and heat resistant, similar hardness to diamonds, density 3.2. Application: As grinding and polishing material, for the manufacture of fire resistant bricks and electrical resistors.

Carboxyl extra group: It produces amino acids e.g. in keratin molecules. The carboxyl extra group determines the behaviour of wool and silk towards alkalis, by reacting as acids.

Carboxylic acids: Carbonaceous organic acids, which are made up of Alkyls and the characteristic carboxyl group – COOH, e.g. CH_3–COOH (acetic acid), derived from the hydrocarbon CH_4 (methane), also as Polycarboxylic acids. Mono, di-, tri- etc., aliphatic, aromatic, saturated and unsaturated carboxylic acids exist. Opposite _ Mineral acids.

Carboxymethyl cellulose (CMC): Carboxymethyl cellulose is made by the reaction of sodium chloroacetate with cellulose. Lumbering by-products, namely stumps, limbs etc. are ground-up, soaked with alkali and made to react with sodium chloroacetate. The degree of substitution can be controlled up to a maximum of 3 carboxymethyl groups per anhydroglucose unit. For textile sizes, the DS is usually 1.5. It is colourless water soluble cellulose ether which swells and forms a gel when stirred into water and forms a viscous colloidal solution.

Carboxymethyl starch: Thickener used in printing of Polyester with disperse dyes and for vat dyes. Reaction product of starch with chloroacetic acid (in the presence of alkali), also mixed ether with other etherification substances (dimethylsulphate, ethylene oxide or propylene oxide). Substitution or etherification degree 0.3–0.5.Carboxymethyl starch has had which Cold water soluble carboxymethyl starches are generally highly substituted, in which some of the hydroxyl groups replaced with carboxymethyl groups. It also improves solubility and the stability of pastes and has some resistance to chlorine bleach, and may be used for thickening bleach for use in discharge techniques. Monagum is one trade name.

Carcinogen: An agent that causes or promotes the development of cancer in animals, such as tobacco smoke or ionizing radiation.

Card: (1) A machine used in the manufacture of staple yards. Its functions are to separate, align, and deliver the fibres in a sliver form and to remove impurities. The machine consists of a series of rolls, the surfaces of which are covered with many projecting wired or metal teeth. Short staple systems employ flat strips covered with card clothing rather than small rolls. Distinction has to be made between the flat card (short staple system) and the roller card (worsted and woollen systems). In the former the prepared cotton lap is opened up to single-fibre state, the fibres extended and assembled into the Card sliver. (2) The seed head of the teazle plant or more commonly an assembly of wires protruding from a base structure, both used for raising or napping the surface of a cloth, generally a woollen cloth.

Card clothing: Used to cloth the flats of cotton cards (i.e., cylinder and rolls or flats.) and the rollers of woollen and worsted cards and also as card fillet for the rollers of napping and raising machines. The clothing consists of either wire teeth set in a foundation fabric or rubber, or narrow serrated metal flutes which are spirally arranged around the roll. The metallic wire has the appearance of band-saw blade. This foundation is a composite multi-ply structure consisting of woven cotton back and a maximum stability and a pliable rubberized blanket face. Flexible ply ensures that the tip of the heavily stressed pins always return to their initial angle.

Card conversion efficiency: The efficiency of the carding process, expressed as a percentage obtained from ratio of sliver output to staple input.

Card fettlings: Wool Waste removed from carding engine cylinders. It is often matted and dirty.

Card sliver: Semi-manufacture stage in the spinning od cotton/viscose. A web of fibres in the ribbon form comprising of fibres in the parallel form.

Carda: Inferior silk fabric of the 13th century, believed to have been used for lining.

Carded wool: Scoured wool which has been processed through a carding machine.

Carded yarn: Manufacturing term for yarn produced on woollen spinning system characterized by their bulkiness and more or less hairy surface. A cotton yarn that has been carded but not combed. Carded yarns contain a wider range of fibre lengths and, as a result, are not as uniform or as strong as combed yarns. They are considerably cheaper and are used in medium and course counts.

Carders: Also known as hand carders (as opposed to "drum carders"). Some of the carders have curved backs, some straight backs. There is some belief that the reason why modern hand cards have the curved backs is because they were modeled after museum pieces. Unfortunately, the museum pieces were warped (curved). Early plans for carders show the straight backs. A more critical requirement is that the cards have offset handles. This will save wear-and-tear on your hands.

Cardigan: In knitted fabric a stitch which is used to shape underwear or is employed in knitting jackets. In a cardigan stitch, the first course knit tuck loops on the front bed and plain loops on the back. The next course is the opposite. The tuck stitch emphasise the plain loops. Cardigan is heavier and

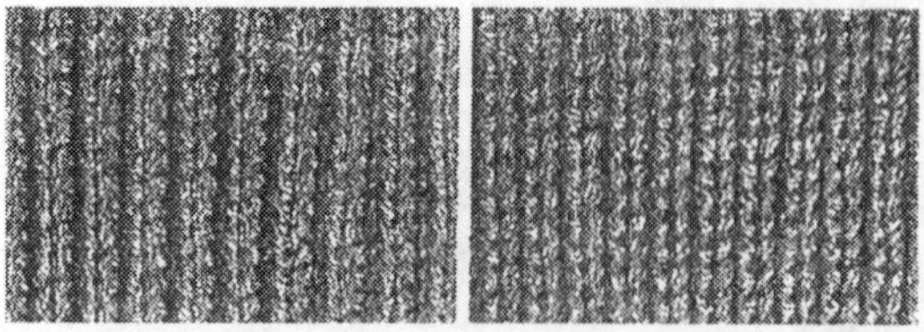

Cardigan Half cardigan

heavier and wider than 1x1 rib. In half cardigan (Royal rib) a course of 1x1 rib is followed by one with plain loops on the front bed and tuck loops on the back. The face side shows prominent wale ribs whilst the back looks like cardigan. The voluminous cardigan and half cardigan structures are used mainly for thick winter pullovers, scarves and berets.

Cardigan Jacket: A close fitting ribbed woollen or worsted body jacket, with or without sleeves.

Cardinal (red): The scarlet colour of a cardinal's vestments.

Carding: The process of untangling and partially straightening fibres by passing them between two closely spaced surfaces which are moving at different speeds, and at least one of which is covered with sharp points. Thus converting a tangled mass of fibres to a filmy web.

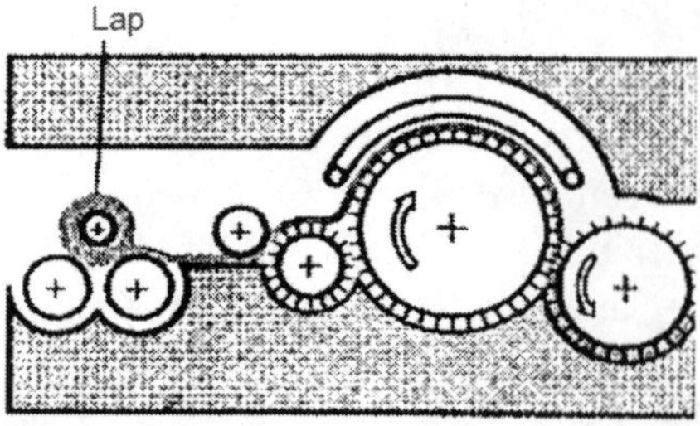

Card ,Schematic

Carding Wools: Wools that are too short to be treated by wool combing and must be processed into woollen yarns. Synonymous with "clothing wool".

Care instructions, in textiles: A series of directions describing which care practices should refurbish a product without adverse effects and warning for those care practices expected to have the a harmful effect.

Care label: A label that gives the directions for refurbishing a product.

Care labelling of textiles: Care labeling is to provide the customer/end user/ purchaser of a textile product with the information on the fibre composition and care of the textile. In the care of the textiles normally wash method, ironing temperature, whether dry cleaning allowed etc. mainly in symbol form so that customer of any language can understand them.

Care procedure, for consumer care textile products: A process by which products or specimens may be treated for soil removal and aesthetic improvement.

Career apparel: Garments, the style and performance of which are designed for various end uses as to be suitable for on-the-job wear in a variety of business and professions.

Carisol: A French thin open canvas used for embroidery foundation. Also called **Creseau**.

Carlett: Wool fabric made in Norwich, England, in the 18th century.

Carlowne: A worsted suiting made in Britain in the 17th century. Not common now.

Carmeline: Medium grade of the vicuna wool; used in France.

Carmelite: French all-wool, light, plain woven dress goods, made of natural wool and fulled; similar to bunting; used for clothing by the Carmelite monks.

Carmine: (1) A crimson colour.

(2) An ancient red, orange or crimson pigment originally derived from the **kermes** insect and subsequently from the **cochineal** insect. Used, amongst other things to manufacture eye **make-up**. See **cremosin**.

Carnadine: **Carnation** colour.

Carnauba wax: (palm wax). Most important vegetable wax, which is obtained from the surface of the Carnauba palm leaf (South America). Brittle, fragile, greyish yellow, slight lustre. Types: fatty grey, refined (paraffin additive), also bleached and brightened. Meltpoint 80–86°C, light qualities lower, refined 70–72°C.

Carnage, tear drop, teariness: Local difference in light reflectance caused by variations caused by variations in curvature of warp crimp. The fault occurs in plain weave fabric made with a continuous filament warp and may rise if the warp is too stiffy or if the warp tension is too low.

Carnation: Rose-pink, flesh-coloured. '*A' could never abide carnation- 'twas a colour he never liked*' Shakespeare's *Love's Labour's Lost Act 2 Scene 1.*

From *carneus* the Latin for flesh-coloured. It was not until the 20th century that 'carnation' also came to be used as the colour of the flower although having regard to the many varieties of the flower in modern times the colour term used in this context has little precision.

Carnauba: Leaf fibre yielded by the carnauba palm in Brazil; used for ropes, cordage, hats, baskets.

Carnelian: A pale reddish-brown or ruby colour after the semi-precious stone of the same name; see **sard.** Also called **cornelian**.

Carnestolendas: Silky seed hair yielded by a tree in Venezuela.

Carnival Lace: A 16th century reticella, similar to the bridal lace in make and worn at festive occasions.

Caroa, Craua, Croa, Coroa fibre: A fibre from the leaf of a plant Neoglazovia variegate grown in central and south America. The fibre is creamy white in colour strong, flexible and soft to touch. Compared to Jute Caroa is 3 times stronger, and the woven material is half as light as similar material made from jute. Can be used for making suitings, coatings as well as for twine, rope, rugs, paper etc.

Carob: Active content of Locust bean flour (the fruit of the locust or carob tree). Contains high molecular weight carbohydrates and hemicelluloses, mainly galactomannan.

Carocolillo: Cotton yarn dyed red with the carocol shell in the West Indies.

Carolina Pride: A commercial variety of early maturing cotton from US, the staple measuring 23-26 millimeters and forming clustered bolls; the yield is about 31 percent. Also called **Early Carolina,** since the same is grown in North Carolina.

Caroline: (1) Plain French serge dress goods, made with eight leaves and eight picks in a repeat; (2) Bleached linen of medium grade, made in Silesia.

Caroline Plaid: An early 19th century English dress goods, three-quarters wide, made with cotton warp and worsted filling in plaid design.

Caroset: General term for various French flannels twilled on both sides and thoroughly fulled. Also sort of melton, made in the Vendee.

Carousel printer: Screen printing machine for the automated printing of the finished textile using carousel system.

Carousel screen printing Machine: A semi-automatic screen printing machine on which screens are accommodated in a rotating frame, which runs between two short tables, so that two fabrics can be printed next to each other by two printers by swinging in the screens.

Carpet: A floor covering having a textile use surface formed from yarns or fibres projecting from a substrate and not designed as rugs. Thick and strong floor covering, reversible or otherwise woven, knitted or felted, made of wool, cotton, hemp, etc. It is made in widths which are sewed together to cover the entire floor. They are classified into different groups.

(a) *Type of manufacture:* knotted, woven, knitted, stitch-bonded, tufted, needle-felt, flocked, bonded carpet.

(b) *Surface:* flat, pile, bouclé (loop pile), velour (cut pile) carpet.

(c) *Colour:* plain, multi-coloured carpet.

(d) *Supply form:* rugs, roll goods, carpet tiles.

(e) *Origin:* Oriental, Balkan, Bavarian, Dutch carpets. Certain marks of origin have become generic names.

(f) *Application:* bedroom, living room, workplace; indoor, outdoor, contract, automotive carpets.

Carpet ageing: Artificial ageing of carpets, achieved by chlorination, which brings about an intensification of the appearance of colour, pile and lustre, usually associated with simultaneous damage to the fibre.

Carpet back coating: See **Back coating of Carpets.**

Carpet beetles: Beetles whose larvae feed on and cause damage to wool carpets. Textile parasites

Carpet binding: (1) A tape to bind the edges of floor coverings; comes in wool or cotton.

(2) The edge-binding of cut, tufted floor coverings especially which is laid wall to wall.

Carpet module: The textile floor covering sections usually having dimensions of less than 1 m2.

Carpet pile: See **Pile layer**.

Carpet pile brush, in Carpet cleaning: A hand operated brush having long, semi-rigid bristles intended specifically for erecting the pile of small areas of carpet.

Carpet square: A term for a carpet in rectangular form, with or without a border, at least 1.83m (traditionally 2 yards)at the shortest dimensions and and normally loosely laid.

Carpet sweeper: A manually controlled machine which has rotary brushes and and which is used for light surface of cleaning of carpets and rugs.

Carpet thread: Heavy three-ply sewing thread; used for joining carpets.

Carpet tile: A textile floor covering material supplied in small sizes, popularly 18" × 18" or 50 cm × 50 cm. Carpet tiles are supplied for being directly 'glued' to the floor or for so called 'free laying' Tiles intended for installation by being directly 'glued' to the subfloor are required to be accurately cut to shape and size and to be relatively fray-free. For for the 'free laying' tile carpet should have a dimensional stability of 0.2-0.4% under changes of humidity and temperature. To achieve this, the tile incorporates a special backcoating comprised of layers of, for example, bitumen, glass, PVC or atactic polypropylene.

Carpet Wool: Coarse, harsh, strong wool that is more suitable for carpets than for fabrics. Very little of this type is produced in the U.S. Some of the choicer carpet wools are used to make tweeds or other rough sport clothing. Some breeds, like Karakul, are mainly used for rugs.

Carpet, Axminster: Cut pile carpet produced by inserting during weaving, successive rows of tuft having colours arranged according to a predetermined order.

Carpet, body: Plain or unboardered patterned carpet in piece form, traditionally ¾ yard or 0.69m or 1 yard (0.91m) wide mainly used for making up into layer areas by seaming or otherwise joining edge to edge.

Carpet, bonded pile: Pile carpet in which the pile consists, for e.g. of a pleated sheet of textile yarns or fibres, secured to a support by coating or adhesive, or directly to an adhesive which forms the support.

Carpet, bonded pile: A textile floor covering with a pile use surface secured to a substrate by adhesion.

Carpet, braided without pile: Textile floor covering without pile made of machine braided cords sewn together.

Carpet, broad loom: Descriptive of carpet made 1.83m or more in width (Traditionally 2 yards).

Carpet, brussels: A loop pile carpet, woven on wilton loom over unbladed wires.

Carpet, chenille axminster: Carpet in which the pile is composed of so called 'chenille' yarns (chenille fur) manufactured on a special machine and inserted during the final weaving.

Carpet, chenile axminster: A carpet in which the pile is composed of so called 'chenille' yarns (chenille fur) manufactured on a special machine and inserted during the final weaving.

Carpet, cord: A low level loop pile carpet with pronounced rows of loops in weft direction.

Carpet, double plush: See **Carpet, face-to-face**.

Carpet, face to face: Carpet made on a loom which produces simultaneously, face to face two substrate joined by a pile yarn. Two cut pile carpets are made by cutting the pile yarns between the two substrates.

Carpet, face to face bonded pile: Pile carpet manufactured on a machine which produces face textile floor coverings, with pile forming material passing alternately from one substrate to another, when it is fixed by adhesive. Two cut pile carpets are made by cutting the pile forming material between the two substrates.

Carpet, flocked: Carpet obtained by projecting the fibres on to a support coated with adhesive.

Carpet, gripper axminster: Carpet manufactured on a jacquard loom in which the tufts of yarn cut after selection from appropriate creels are inserted at the point of weaving by grippers.

Carpet, gripper spool axminster: Carpet manufactured on a loom in which the yarn for each weft ways row is wound on a separate spool for spool axminster weaving. Tufts served from the yarn are inserted at the point of weaving by 'grippers' as in gripper axminster.

Carpet, hand knotted pile: Carpet obtained by knotting the pile forming yarns by hand.

Carpet, 'Jacquard Weave': Pile carpet produced on a jacquard loom and having a design obtained by the use of the creel (frames), each supporting the bobbins of pile yarn of different colours. The successive selection of the appropriate frames during weaving forms the required design.

Carpet, knitted pile: Carpet made on either (a) warp knitting machine (Rachet type) or (b) weft knitting machine (Wildman type).

Carpet, knotted pile: Carpet produced by knotting the pile forming yarns around warp yarns of the substrate.

Carpet, machine knotted pile: Carpet manufactured by automatic knotting.

Carpet, needle pile: Pile carpet produced by needling of the materials and bonding by physical or chemical process or by a combination of both. The pile may be cut or looped.

Carpet, pile: **woven pile carpet.** Pile carpet produced on a loom in one or more operations.

Carpet, 'plain weave': Woven pile carpet with uniform pile length containing generally a single colour.

Carpet, spool axminster: Carpet manufactured on a loom on which the yarn of each weft row are wrapped on a separate spool, according to the required design. The tufts are separated from the yarns presented at the pint of weaving after inserting the substrate.

Carpet, stitched on pile: Pile carpet in which the pile is stitched on to the substrate by stitching.

Carpet, tufted pile: Carpet in which the pile yarns are fixed in a previously manufactured base and then secured by coating as an adhesive.

Carpets, weft-patterned: Machine-woven carpets, the pattern of which is produced by the relevant insertion of coloured weft yarns.

Carpet without pile: The textile floor covering without pile made on a loom.

Carpet, without pile, formed of bonded textile: Textile floor covering without pile, in which the use surface is composed of entangled textile material bonded together by a mechanical, physical or chemical process or by a combination of two or more of these processes.

Carpet yarns: Filament yarns for the carpet sector are produced mainly by the BCF process, which produces in one operation from the granulate via extruder spinning, drawing, texturing, interlacing and winding. The crimp structure is obtained by braking the filaments in a heated chamber (stuffer box), producing yarns in the 500 to 4000 dtex count range. Maximum production speed is 4000 m/min (polyamide) and 3000 m/min (polypropylene).

Carpettes: Coarse French twilled bagging for raw wool, made of tow, unbleached and striped.

Carpmeals: A stout and coarse English woollen fabric; used for clothing in the northern parts of England.

Carriage, in knitting machine: The carriage carries out a double function and can be used:

(a) To select the needles and make them raise or lower to form the stitch; (b) To select and drive the thread guides which feed the needles. The carriage is made up of two metal plates linked by a stiff bridge; the plates work individually and simultaneously on the front and the rear needle-beds. The plates include cam locks bearing the drive and control systems of the needles, i.e. the cams. When the carriage moves right- or leftwards, the thread guides are locked individually by the corresponding pistons on the upper part of the bridge that are manually driven. Brushes are also mounted on the carriage to ensure a smooth needle latch opening and grant a correct feeding of the yarn.

Carriage lace: A woven narrow fabric on the face of which is an uncut pile design, generally incorporating one weft, but two or three wefts are used sometimes for further ornamental effects. It is used to give finish to the upholstery of railway carriages and other forms of transport.

Carriage weight, effective: See **Effective carriage weight**.

Carradars: Indian gingham with multi-coloured narrow stripes.

Carranclanes: Narrow ginghams in the Philippines, coming mostly in checks and plaids.

Carre: Chequered.

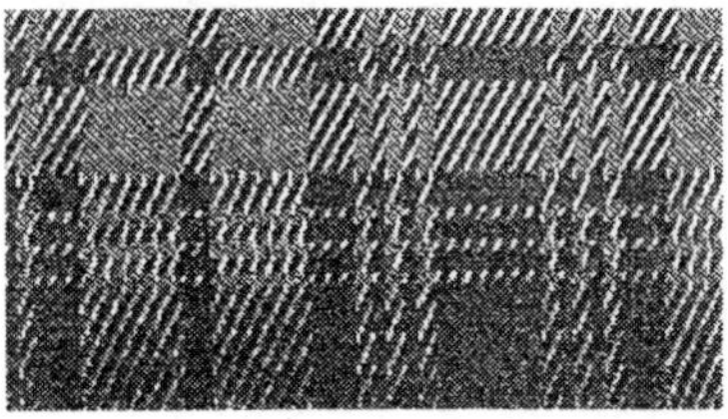

Carreau: French term for checks.

Carrick: Coat with several stepped shoulder layers.

Carrickmacross Lace: (1) An old Irish traditional type of lace made with heavy yarns in an almost crochet type of work. It is characterized by many loops. The sprigs consist of fine lawn with buttonholed edges, connected with each other by bride's picotees making guipure. (2) The sprigs are bobbin made and appliqued over a machine made ground.

Carrier with respect to disperse dyes: A chemical that aids dyeing disperse dyes on polyester at moderate temperature, like at or around the boiling pint of water. In order to dye polyester with disperse dye in a reasonable time at the boil, it is necessary to use a carrier. Exactly how the carrier works seems to be a matter of some controversy, but it may work by swelling the fibres so that the dye can penetrate. The carrier will eventually evaporate from the fibre after dyeing is complete. Carriers are obsolescent in industrial process, partly because they are quite noxious and environmentally undesirable.

Carrier residues: Since residues of carriers can impair the light-fastness of dyeings, particularly on polyester, they are best removed through reductive final scouring with soda lye and sodium dithionite.

Carrier rollers: Provided in order to prevent slippage of the rollers.

Carrier stains: Stains formed to the precipitation of carriers in the dye bath. Usually carrier is used in the emulsified form in the dye bath, which can break due to change in the pH or any other condition which keeps them in the emulsified form. Carrier stains are difficult to correct.

Carrier system: Relates to yarns (dyeing machine suspension system) for ladies' hose, which are treated in specific hose dyeing machines suspended over sticks in bundles. Sometimes also usable for hank yarn.

Carrier, in braiding machinery: Those parts of the braiding machine that holds the package of yarn thread or cord and carries the yarn when the machine is operated.

Carrot orange: The yellowish-orange colour of the carrot previously referred to as **carrot red.** Carrots have not always been orange in colour. The Egyptians from 2000 BC grew purple carrots and the Romans ate purple and white carrots. Black, red, green and yellow carrots were also grown but Dutch carrot-growers favoured their chosen orange colour which became the standard colour for carrots from around the 16th century. The orange colouring derives from **beta-carotene.** Purple carrots are now being grown in England.

Carroting: The steeping of hare and rabbit hair in mordant. A term from the hat-making industry. The best mordant for this is the mercury nitrate mordant, where mercury is not released (incorporated into the hair keratin).

Carthagena: West Indian cotton, grown from American seed; has a fairly strong staple.

Carthamin (e): The red dye of the safflower.

Carthamus: Safflower, an annual plant cultivated in South Europe, Egypt and Asia, for the red dye from its flowers.

Cartisane: Parchment used for padding the cordonnet or the foundation of the point laces and also to pad raised embroidery patterns.

Cartouche: French gros de Tours, the warp is made in repeated ombre, divided by stripes at equal distance. It was often given a moiré finish; obsolete.

Cascara: Native Peruvian and Bolivian name for a cloth-like bast of the couratari tree; used for blankets, clothing, ropes, etc.

Case: A shipping unit, usually a carton, box, bale or other container holding a number of yarn packages.

Case cord: Continuous filament case cord yarns and steel cord yarns are woven to form the load-bearing plies forming the carcass or body of a vehicle or aircraft tyre or heavyweight conveyor belting, drive belts etc.

Casement cloth: As the name implies, this is for casements or curtains. Usually cotton, it is plain weave fabric in white or cream, with the weft often predominating on the surface. It is soft and drapes well. The term is now generally used for a wide variety of simple weave plain curtaining, made from practically any fibre and in any colour, and in various weights. It is not particularly hard wearing, so not used for other furnishings. May shrink, can be lined.

Casha: A fabric woven of wool with some Cashmere goats wool; similar to flannel in appearance, but the cashmere adds softness. Used for over coats for adults and children.

Cashemire de soie: A fabric made of very fine Silk to imitate a Cashmere fabric, by giving a high grade effects, which is given a finish to look like cashmere fabric.

Cashgora: Cashgora means the fine bottom hair of the cashgora goat, which is the result of cross-breeding female cashmere goats and male angora goats in New Zealand. It was only in the 19th century that the finest bottom hair was discovered in some of these wild goats. Cross-breeding trials of these animals were first undertaken in order to reduce the proportion of coloured bottom hair, since it is with the genes of the male angora goats that the feature "white colour" is passed on to descendants.

Cashmere, in roving, yarn or fabric: Cashmere hair or products made therewith having a cashmere coarse-hair content or not exceeding a specified maximum percentage by length.

Cashmere: (1) Very fine and soft wool yielded by the cashmere goat. The undergrowth is one of the very finest animal fibres known, called pashmina, while the longer hair is somewhat coarser. It is used for the best grades of Indian shawls, rugs and also dress goods. See **Cashmere hair. Cashmere wool.** (2) Woollen cashmere is a fabric similar to cheviot. To make softer handle earlier the wool used to be mixes with the silky hair of cashmere goat and hence the name. A characteristic construction of 24 square with 21/2 run woollen yarn in both wrp and weft, with a grey width of 70 in. and finished width of 60 in. and 12–14 oz. per yard. To get the soft handle now a days so many different mixing are used. (3) A closely woven, soft, fine and light dress fabric, made with single cotton or wool warp and fine Botany filling in a 2-1 weft face twill. The colour eit'ect is usually mixture; (4) Soft, loosely woven twilled cotton dress fabric, printed to imitate the mixed woollen. (5) knit goods, made of slack twist Saxony or other fine worsted yarn.

Cashmere atlas: All-wool, highly finished weft satin dress goods, the filling is finer and of slacker twist than the warp.

Cashmere cloth: Cashmere is made from the wool of Cashmere goat. It is warm and soft, silky and lustrous, with a smooth, slippery, luxurious handle.

Used for dress fabrics and coatings and suitings. Sheep's wool is often mixed with cashmere to make it less expensive but makes it harder. Soft botany wool is often made up in imitation of cashmere, but cannot be labeled as such.

Cashmere coarse hair: Those fibre in cashmere hair having a width greater than 30 um.

Cashmere coarse hair content: The total length of the cashmere coarse hair that are present , expressed as a percentage of the total length of the of all the cashmere hair fibres; that is , the percentage by length of cashmere coarse hair in cashmere hair.

Cashmere down: Those fibre in Cashmere hair having a width of 30 μm or less.

Cashmere hair: Fibres produced by a form of goat (Capra hircus) indigenous to Asia known as Cashmere goat.

Cashmere shawl: One of the finest textiles known, made by the natives of Kashmir and Tibet on hand looms of pashmina wool (see). Some of them are embroidered, showing the pattern alike on both sides, others are only dyed in one or more colours. The French cashmere shawl is woven on jacquard loom in multi-coloured designs, which show only on one side. The shawls made at the present in Kashmir are inferior in colour and design. The warp is two-ply while the filling is single yarn. The latter is wound around small sticks used as a shuttle.

Cashmere wool: See **Cashmere**.

Cashmerette: (1) A twilled, soft and lustrous cotton fabric, often lightly napped, made to imitate cashmere; (2) Fancy woollen for waistcoats; obsolete. (3) Twilled dress goods of spun silk warp and woollen filling.

Casimii: An all-wool, twilled summer dress goods, made in narrow, diagonal twill without any fulling, usually of worsted warp and woollen filling.

Casinetes: Cheap cotton trouserings in Bolivia, Peru, etc.

Casing with Heading, in garment making: A piece of fabric is drawn up with a drawstring or elastic held in a tunnel made by applying a strip of fabric to the surface leaving a border at the top or bottom for decoration.

Casket cloth: Light, black fabric, made of wool and cotton; used for covering caskets

Cassel brown: A brown pigment made from lignite and named after the town of the same name now called Kassel; also Cassel green and ***Cassel yellow***.

Cassel yellow: A yellow pigment also called Montpellier yellow.

Cassimere: A general term for all-wool fabrics woven plain or twilled, coarse or fine, of woollen yarn. The pattern is always woven plain and distinct and the cloth is never napped. Cassimere (a) Variety of plain woven or twilled woollen or worsted fabrics, in checks, stripes, etc., soft finished but not napped; used for men's wear; (b) Twilled woollen or cotton dress goods, made with two sets of warps in various colours and given a soft finish.

Cassimere twill: An even sided, four harness twill weave, each thread passing alternately over and under two threads.

Cassimerette: Cheaper grade of cassimere.

Cassinett: Fancy English alpaca fabric in the 19th century.

Cassinet: English twilled stout trousering and waistcoating in various colours, made of fine cotton warp and woollen yarn dyed in the wool. It is slightly fulled and calendered. The double cassinet is made with alternate woollen and cotton yarn filling. Three quarters of the wool filling is thrown on the face and three-quarters of the cotton filling is on the back of the cloth.

Cassius: A purple pigment.

Cast: A rough, coarse, bad-bred fleece of wool.

Cast: A pale shade or tinge of colour; a shade or colour; a dash of colour imposed on another.

Cast off position, in knitting: The needle is moved further downwards, the needle latch is closed by the old loop.

Castalogne: Fine woollen blanket made in France and Spain.

Castane- (L): Brown, chestnut.

Castel Branco: A good Portuguese carpet wool.

Castellamare: A type of raw cotton from Italy.

Castilla: Coarse homespun or imported woollen, having a long nap; used for shawls and cloaks by the natives of Peru.

Castillian: A deep red.

Casting out: A term used in weaving to indicate that some hooks of the jacquard machine or some healds on certain shafts of a dobby loom are not used.

Castle Wheel: The flyer is usually mounted above the wheel, which means less floor space is used.

Castor: (1) Heavy, stout, fulled and calendered broadcloth overcoating.

(2) Referred to in the *Daily Mail* of 5 June 1923. A dark brownish grey or a light brown beaver colour named, it is hoped, after the star in the constellation, Gemini, and not after the unctuous malodorous secretion of beavers.

Castor beaver: A heavy, milled, face-finished, all-wool cloth lighter in weight than ordinary beaver.

Castor oil: A very important raw material for textile auxiliary before the advent of new synthetic surfactants. The major fatty acid (90%) in this oil is ricinoleic acid. Hydrogen bonding through the hydroxyl group on C is responsible for the oil's high viscosity and solubility in alcohol. The oil is sulfated to become an important anionic surfactant (turkey red oil). Ricinoleic acid is also the starting material for making sebacic acid and capryl alcohol. It is cleaved with sodium hydroxide. Sebacic acid is a C, dibasic acid used to make Nylon 6, 10.

Castor oil soaps: The soaps made using castor oil.

Castravane: Raw silk from Asia Minor; used for braids.

Casual wear: Synonymous with Leisure wear, progressing in the mid-1970s to cover unconventionally more flamboyant garments combining greater comfort with more youthful style features, being lighter in weight and more yielding to the body, with highlighted stitching (of collars, pockets, belts etc.) and above all easy-care properties.

Cat eyes: See **Pin holes.**

Cat stitch: The cat stitch or herringbone stitch is an alternate slanting back stitch, the needle being placed first to the right and then to the left. This stitch must be worked evenly to be effective. It is used to finish flannel seams and hems, fasten down linings, opened seams, and canvas facings and featherbone, in millinery—in fact, this stitch is one of the most useful in sewing. The catch stitch is a variation of the cat stitch. Instead of pointing the needle towards the chest, the stitch is taken parallel with the chest. It is used for about the same purposes as the cat stitch. As with the outline stitch, the cat stitch is worked from the worker.

Catablattion: Purple coloured silk fabric of the Middle Ages.

Catacaos: A variety of the tree cotton from Peru.

Catalowne: Same as Buffyn made with double and twist yarn and a different coloured filling.

Catalyst: A substance that alters the rate of a chemical reaction without itself being changed chemically in the reaction. The catalyst can, however, undergo physical change; for example, large lumps of catalyst can, without loss in mass, be converted into a powder. Small amounts of catalyst are often sufficient to alter the rate of reaction considerably. A positive catalyst increases the rate of a reaction and a negative catalyst reduces it. Homogeneous catalysts are those that act in the same phase as the reactants (i.e. in gaseous and liquid systems). For example, nitrogen (II) oxide gas will catalyze the reaction between sulfur(IV) oxide and oxygen in the gaseous phase. Heterogeneous catalysts act in a different phase from the reactants. For example, finely divided nickel (a solid) will catalyze the hydrogenation of oil (liquid).

The function of a catalyst is to provide a new pathway for the reaction, along which the rate-determining step has lower activation energy than in the uncatalyzed reaction. A catalyst does not change the products in an equilibrium reaction and their concentration remains identical to the concentration of the products in an uncatalyzed reaction; i.e. the position of the equilibrium remains unchanged. The catalyst simply changes the rate at which equilibrium is attained. to the negatively charged electrode of decreasing ionic radius. Promethium. Catalysts are not common in dyeing, but are used in fabric preparation and finishing.

Catalysts in resin finishing: The cross-linkers commonly used in resin finishing are used only with the addition of catalysts, which act as cross-linking reaction triggers and as reaction accelerators. The action of these catalysts is based on the freeing of hydrogen ions. Catalysis (acid), breaks the carbon/oxygen linkage of the N-methylol groups, and a reaction takes place with the hydroxyl groups of the cellulose with the discharge of water. The different catalysts used in different type of cross linking process:

Type of Finishing	Type of Catalyst	pH
Dry Cross	Linking Metal salts, Ammonium Salts	3-6
Moist Cross	Linking Mixed Catalysts, Free Acids	1-2.5
Wet cross	Linking Free Acid	0-1.5
	Free Alkali	10-12

Catalytic fading: Damage of cotton fabric due to the localised catalytic action of peroxide on cotton thereby the cellulose is destroyed (Pin holes) or oxidative changes on cellulose, which causes further problems in dyeing etc. This is mainly due to the iron contamination in cotton or in water, steam or any other material used in processing/bleaching.

Catasfittulum: A medieval fabric of unknown structure.

CATCC: Canadian Association of Textile Chemists and Colorists.

Catch: An under lap of material providing a backing to an opening to take buttons or zip tape for fastening e.g. flycatch, button catch.

Catechu: (cachou). From the Indian "Kati" plant and "chuana" = distil. Extract of the aqueous extract of the Indian tanner's acacia. No longer used brown dye, formerly used for tent and covering fabrics, as the fabric was simultaneously endowed with natural waterproofing, increased weight and a fuller, harder handle.

Catenary length: The difference between the length of the shortest and longest component of a plied yarn or cabled chord after twisting.

Caterpillar: Point Rich Italian needlepoint lace of the 17th century, the caterpillar-like patterns are outlined with raised cordonnet and connected with brides picotees.

Catawba: A late maturing commercial variety of cotton from US, the staple measuring 22-25 millimeters, the yield being 35-36 per cent.

Catechu brown: A dye (also called **cutch**) containing **tannin** made from the bark of certain trees found in Asia and used in the tanning industry and in dyeing. The shade of brown derived from **catechu**.

Catena: White bast fibre, yielded by a species of the Heliocarpus tree in Mexico.

Catgut: An open, plain woven linen fabric, made of hardspun yarn and sized; used for embroidery.

Cathay: A bluish purple.

Cathay: An Indian striped satin.

Catherine wheel: An ornament in embroidery to fill up round holes.

Cation: Positively charged ion. Many chemicals used in textile processing are described as cationic, meaning that when the compound ionizes in solution, it is the positively charged ion that is "functional". Many surfactants are cationic, as are many chemicals used as fabric finishes. (Also see **Anion**.)

Cation exchange: See **Cationic exchanger.** Replacement of one ***cation*** by another—e.g. replacing calcium and magnesium by sodium or hydrogen ions in ***water softening*** by ***ion exchange***.

Cation exchange capacity, CEC: The ability of soil to exchange cations. It is measured in milliequivalents per 100 g of soil and ranges from 2 to 3 meq/100 g for sand up to 200 or more meq/100 g for clay and organic matter.

Cationic: Those that develop a positive charge on the water solubilizing end.

Cationic aftertreatment: Many dyes like direct, reactive which are having poor wet fastness, especially washing, wet ironing, cross-dyeing, water, perspiration fastness, are given caionic treatment to improve the fastness.

Principle: conversion of the anions of the dyestuff on the fibre into sparingly soluble or insoluble salts of the dyestuff acid.

Cationic dye: A dye which dissociates in aqueous solution to give a positively charged coloured ion.

Cationic dyeable polyester: Polyester fibres can be modified into cationically dyeable polyester fibres by the copolycondensation of terephthalic acid and 5-sodium sulphoisophthalic acid with ethylene glycol.

Cationic exchanger: Cationic exchangers (base exchangers). These are the most important ion-exchangers for water softening and are mainly high molecular weight polyvalent solid acids or their salts. They have acidic, positively charged exchange ions, i.e. acidic groups (SO_3H, COOH, OH) with mobile H+ or Na^+, e.g. of the type $-SO_3Na^+$ or $-SO_3H^+$.

Cationic pretreatment: This treatment – also called mordanting – is used with cellulose fibres in order to make them absorptive to cationic dyestuffs. Such mordants are tannin, sulphuretted phenols etc.

Cationic resin: An Ion excange material that can exchange cations, such as H+ and Na+, for ions in the surrounding medium. Such resins are used for a wide range of purification and analytical purposes. They are often produced by adding a sulfonic acid group ($-SO_3-H^+$) or a carboxylate group (–COO–H+) to a stable polyphenylethene resin. A typical exchange reaction is: Resin–$(SO_3-H)^+$ + NaCl → Resin–$(SO_3-Na)^+$ + HCl They have been used to great effect to separate mixtures of cations of similar size having the same charge. Such mixtures can be attached to cationic resins and progressive elution will recover them in order of decreasing ionic radius.

Cationic softeners: Cationic softeners are ionic molecules that have a positive charge on the large part of the molecule. The important ones are based on nitrogen, either in the form of an amine or in the form of a quaternary ammonium salt. The amine becomes positively charged at acidic pHs and therefore functions as a cationic material at pH below 7. Quaternary ammonium salts, retain their cationic nature at all pHs. The important types will be described in this section. An important quality of cationic softeners is that they exhaust from water onto all fibres. When in water, fibres develop a negative surface charge, setting up an electronic field for attracting positively charged species. These forces causes the cationic softener to deposit in an oriented fashion, the positive end of the softener molecule is attracted to the fibre surface forcing the hydrocarbon tail to orient outward. The fibre now takes on low energy, nonpolar characteristics; therefore, the fibre has the lowest possible coefficient of friction. Cationics are highly efficient softeners. The ionic attraction causes complete exhaustion from baths and the orientation on the fibre surfaces allows a monolayer to-be as effective as having more lubricant piled on-top.

Cationic surfactants: These comprise a small group of surfactants that are mainly quaternary ammonium or pyridinium salts. As for anionic surfactants, these may be derived from natural fatty acids or from synthetic chemicals. They are used mainly as softening or dyeing after treatment and retarding agents. The long hydrophobic 'tails' of cationic agents, deposited as a film on the fibre surface, lubricate it and provide a much smoother and softer handle.

Cationically dyeable polyamide: See **Differential dyeing polyamide fibres.**

Catridge filter: A filter device used in effluent treatment etc. , usually disposable, filtering in the range of 0.1 micron, and usually 2 inches to 4 inches (51 to 102 mm) in diameter and 6 inches to 60 inches (152 to 1524 mm) in length.

Cattivella: Italian silk dress goods, made of fine floret silk .

Cattle hair: (cows' hairs). The most valuable from the technical spinning standpoint is the tail hair, similarly also curry comb hairs. In contrast is tannery waste hair, which is mainly used for felt and cheap coverlets etc., highly enriched with lime salts which have to be removed prior to finishing (dyeing etc.).

Cattle hair cloth: Commercial name for a stout twilled fabric, containing wool waste and some calf hair, heavy nap is raised by gigging; used for carriage robes.

Caungeantries: English changeable fabric, made of worsted and silk in the 16th century.

Caustic: Referring to a chemical that will "burn" skin; may be acid or alkali; in dyeing, caustic is also often used as an abbreviation for caustic soda (sodium hydroxide)

Caustic alkalies: Mainly Caustic soda; Potassium hydroxide.

Caustic discharges: See **Alkaline discharges**.

Caustic recovery plants: In the mercerising machine the fabric is treated with strong caustic soda for the process of mercerization and the balance caustic soda is washed off, by different methods as per the machine design. This caustic soda is usually evaporated in Caustic recovery plants to a required strength so that it can be reused. Usually caustic recovery plants work on less than atmospheric pressure to save energy (steam).

Caustic soda boil: A method of scouring cotton. In this process cotton fabric is boiled with a solution of 10 to 20 g/1 (3 to 6% o.w.f.) caustic soda in a kier with a liquor ratio of 3:1. In continuous scouring about 30 g/1 of caustic soda is added in the pad-bath with a liquor pick-up of about 100%. In batch process the piece goods are treated in 4-6% sodium hydroxide for 8 h at 130⁰C (30 lb/in^2). Loose cotton for bandage and sanitary cotton, where absorbency is the prime importance, are boiled at 80 lb/in 2. In continuous processes, it is possible to decrease the time of post impregnation steaming to about 2 min. at a temperature of 130-135°C with sodium hydroxide solution of 40-60 g/1.

Caustic shock process: Reactive prints fixation process without using steam. Not very frequently used, the process is as follows: print fabric with printing paste, dry, pass for 20–60 s in open width through a strong, boiling alkali bath containing electrolytes (roller vat), and then wash off. Variants as steam-wet fixation processes for viscose: brief pre-steaming for 3–5 min before the caustic shock process produces a deeper shade on viscose.

Caustic soda: See **Sodium Hydroxide**.

Causticaire method: An old method for assessing the ripeness of cotton by treatment with alkalis of different concentrations.

Causticaire linear density, in cotton testing: The linear density (micrograms/ inch or millimeter) of cotton fibres estimated from air flow measurements made before after treating with NaOH solution.

Causticising: Brief treatment of woven cellulosic fabrics with approx. 15–30°Baumé caustic soda without tension. The lye temperature is kept low to advantage approx. 10–20°C. Causticizing shrinks and compacts the woven fabric, and increases dye uptake in dyeing and printing particularly with reactive dyes.

Causticizing number: Test process which is based on the differing affinity of a mixture of green and red substantive dyes for untreated and causticized or mercerised cotton. Cannot be used in the presence of unripe fibres. The difference can be determined in figures by colorimetry.

CAV: Critical Add-on Value.

Cavalleys: Fine dyed cotton fabric, imported to England from Central America in the 18th century.

Cavalry twill: A rugged, very strong, smooth surfaced twill cloth with a clean double twill line. It is made with a pronounced 63 or 70 or 75 degree steep twill weave with a construction of about 100 x 60 using 2/36s worsted yarns in the warp and 6 run woollen yarn in the filling. Usually 60 in. finished width; it will have around 14 oz. per yard weight. In USA it is called Elastique. Traditionally this is a medium weight worsted or woollen cloth for hard wearing clothes such as riding breeches and men's trousers. Now it is also made from cotton or viscose for sportswear.

Cavitation: In the context of textiles, cavitation is a problem which occurs in HT dyeing especially beam dyeing. In HT dyeings vapour pressure at 85–100°C increases so quickly that normal centrifugal pump action is no longer guaranteed. On the suction side, i.e. in the low pressure zone, liquid vapour tension can then result in the formation of vapour bubbles, reducing pump efficiency, which can fall to zero.

Cayenne: (1) A type of raw silk from Guyana: (2) Lightweight, loosely woven, unbleached French linen.

CC: (1) Coconut Fibre, (2) Cupro Fibres.

CCM: Abbreviation for: Computer Colour Matching.

CCS: Abbreviation for: Comisiade Standardisonea; Rumanian Standards Institute

Ceara: Mexican and Brazilian raw cotton, having a fairly clean, medium strong, good, cohesive staple of dull white colour.

Ceba: Fine and silky cotton, grown in Mexico on the Ceba tree.

Cedar: A reddish-brown colour.

Cefiros: (1) Light cotton cambric in Colombia v often made with white or blue warp and white filling; (2) Cotton zephyr in Venezuela.

Ceiba fibre: A fibre used for stuffing pillows mattresses etc. Yellowish, silky seed hair, grown on the Bomibax tree in South America.

Celadon: A pastel green also referred to as Celadon green – Celadon, according to Partridge being a rustic lover in French pastoral poetry. Used in glazing Chinese porcelain. Celadonite is another name for **terre verte**.

Celafibre: Cellulose acetate fibre made in cut staple form. It is useful for blending with other fibres to make medium-weight fabric, blankets, etc. When mixed with other fibres in woven fabrics it may have a reduced tendency to creasing.

Celestial blue: A greenish blue.

Celestine: Sky blue.

Celestine: A woollen fabric, made in England under Edward VI., said to have been of light blue colour, and made with wide selvage.

Cellobiase: An enzyme occurring in malt, germinating spinach, oats etc., which degrades Cellobiose to glucose. It is encountered inter alia as a secretion of certain bacteria particularly in fish nets, etc. See **Microbial damage to textiles.**

Cellobiose: Disaccharide produced on partial hydrolysis of cellulose, made up of two glucose units joined by a β,1,4 linkage. See **Cellulose.**

Cellophane: Transparent cellulose film. Practically impermeable for air, bacteria, organic solvents, grease, etc. Application: as packaging, bottle tops, adhesive tape, as safety glass,

textile fashion material and basketwork, etc. (Cellulose skin, Cellulose film)

Cellosolve: $C_2H_5O(CH_2)_2OH$, (ethylene glycol monoethyl ether), colourless liquid, practically odourless, miscible with hydrocarbons and water. Combustible. MW 90, density 0,93, b.p. 130–138°C, flashpoint 40°C, evaporation number 43. Does not dissolve acetate fibres. Uses: solvent in dyeing and printing for cationic, acid and indigo dyes, solvent for nitrocellulose, natural and synthetic resins.

Cellular: (1) Resembling or consisting of cells; multicellular = having many cells. (2) A term describing a fabric constructed so as to have close and orderly distribution of hollows or holes. In woven fabric this can be achieved by (a) honey comb (b) leno or (c) mock leno weaves.

Cellulase: An enzyme that catalyzes the breakdown of cellulose Cellulase enzymes are used for de-pilling and defibrillation of cotton fabrics. They can also be used to permanently soften cotton fabrics, as an aid in or replacement for stone washing of denim, and in methods for"peaching" cotton fabrics.

Cellulose: A polymer of a very large numbers of units, each of the general formula $C_6H_{10}O_5$ Cellulose is a structural polysaccharide made by plants. Essentially, units very similar to glucose are assembled into huge molecules that form strong fibres. Among cellulose textiles are cotton, linen, ramie, jute and hemp. Rayon is a man-made cellulose fibre (actually regenerated cellulose - natural cellulose is the starting material). Each unit of the cellulose molecule has a number of hydroxyl (–OH) groups. These are the binding sites for reactive dyes.

Cellulose acetate: (Acetate cellulose, acetyl cellulose, cellulose esters of acetic acid).Cellulose in which most of the hydroxyl (-OH) groups of cellulose have been replaced with acetyl ($-OOCCH_3$) groups Cellulose acetate is usually dyed with disperse dyes at a temperature of around 80°C to 90°C. It is quite easily damaged in alkaline conditions and loses its desirable lustre if boiled.

Cellulose acetate in textile printing: Used as thickening agent and substance carrier (Metal powders) which is fixed by heat (steaming), e.g. in Pigment printing.

Cellulose derivatives: Cellulose derivatives are mainly Cellulose ethers and esters. Examples of cellulose esters are cellulose acetates, Cellulose diacetates, Cellulose triacetates etc. and that of ethers are Cellulose methyl ether (Methyl cellulose), dimethyl cellulose, trimethyl cellulose. Ethers are used depending on the degree of alkylation, as excellent stiffening agent for sizes, finishes, print thickeners, also emulsifiers. Organic cellulose esters are very important for the manufacture of chemical fibres, monofilaments and foils, such as e.g. acetate cellulose and triacetate fibres;

Cellulose, degree of polymerization: The Average degree of polymerization of various cellulosic fibres is as follows:

Cotton approx.	3000–6500
Ramie	(>) 2700
Flax	2500–9200
Hemp	2200
Cotton, spun approx.	1200
Wood pulp	700–3000
Cupro filament	400–600
Viscose filament	(>) 200–460
Acetate filament	175–360

Cellulose esters: See **Cellulose derivatives**.

Cellulose ethers: See **Cellulose derivatives**. produced as a result of ethylene oxide and a small amount of alkali acting on cellulose. See **Oxyethylated cellulose.**

Cellulose glycol ether: (glycol cellulose, polyethylated cellulose),

$$\text{Cell-O-}C_2H_4OH \text{ see Cell-O}(C_2H_4O)_x\text{–}C_2H_4OH.$$

Cellulose glycolate: See **Carboxymethylcellulose.**

Cellulose methyl ether: (cellulose methyl ether), When the 3 OH groups existing in the cellulose molecule per glucose unit, the etherification using methoxy groups CH_3O can lead to mono-, di- and trimethyl celluloses.

Cellulose nitrate: Cellulose nitrates are sometimes erroneously called as Nitrocellulose. This is not a Nitro compounds but a cellulose nitrate. A typical ester of cellulose, with characteristic sensitivity when boiling with acids and alkalis. Not a homogeneous chemical compound, but a combination of various highly nitrified cellulose molecules. See **Nitrocellulose.**

Cellulose, oxyethylated: See **Oxyethylated cellulose, Cellulose glycol ether.**

Cellulose synthetic fibres: Old term for Regenerated fibres like Viscose. Now obsolete term.

Cellulose triacetate: See **Cellulose acetate, Triacetate Fibres.**

Cellulose trinitrate: (guncotton; nitrocellulose) A highly flammable substance made by treating cellulose with a nitric-sulfuric acid mixture. Cellulose trinitrate is used in explosives and in lacquers. It is an ester of nitric acid (i.e. not a true nitro compound).

Cellulose xanthate: It is an intermediate product in the manufacture of viscose from natural cellulose like wood pulp. First the cellulose is converted to alkali cellulose (Soda cellulose). This on reaction with carbon disulphide gives Cellulose xanthate. When Xanthate is dissolved in caustic soda Viscose solution is formed which then spun into Viscose fibres.

Cellulosic fibres: Cellulose fibres which contains mainly of cellulose units. This can be divided into three types. (a) Natural cellulosic fibres; (b) Cellulosic fibres, regenerated; (c) Modifications, such as e.g. Modal fibres.

Celsius scale: A temperature scale which places the freezing point of water at zero degrees, and the boiling point of water 100 degrees; formerly called centigrade.

To convert to Fahrenheit: $°F = (°C \times 9/5) + 32$ To convert from Fahrenheit: $°C = (°F - 32) \times 5/9$.

Celtic: A term used in Scotland to denote the hopsack weave, with a satin base.

CEMATEX: Abbreviation for Comité Européen des Constructeurs de Matériel Textile (European Committee of Textile Machinery Manufacturers), Zürich. Founded in 1953. Amongst other functions, CEMATEX is responsible for organizing the ITMA international textile machinery exhibition

Cemetery-kula carpet: Prayer mats with typical designs showing small houses, cypresses and pines in yellow ochre, pale red and light green on a dark blue background.

CEN: Abbreviation for: Comité Européen de Coordination des Normes (European Standardization Committee) with headquarters in Paris.

CENATRA: Abbreviation for: Centre National d'Assistance Technique et de Recherche Appliquée (Belgian Research Centre for Fabric Care and Drycleaning), Anvers/Antwerpen. Liquidated in January 1978.

Cendati: A medieval silk fabric made in Italy.

Centre Fibre: In bobbin laces the main vein of the leaves.

Centre loop: Kink

Centering mark: See **Clip mark.**

Centiliter (cl or cL): **A** common metric unit of volume. One centiliter equals 10 cubic centimeters; this is about 0.61024 cubic inch, 0.3318 U.S.fluid ounce or 0.3519 British fluid ounce. In the kitchen, a centiliter is roughly equal to 2 U.S. teaspoons (or 0.704 British tablespoonful).

Centimeter (cm): (1) The basic unit of distance in the former CGS version of the metric system, equal to 0.01 meter. One centimeter is about 0.393 700 787 inch. (2) An obsolete name for the statfarad (approximately 1.11 picofarad), the CGS electrostatic unit of capacitance.

Centimeter (cm) (c): An obsolete name for the abhenry, the CGS electromagnetic unit of inductance. The abhenry is the same as the nanohenry.

Centimeter of mercury (cmHg): A traditional unit of pressure equal to 10 mmHg, 1.333 22 kilopascal, or about 0.193 pounds per square inch.

Centre back waist length, in body measurements: The vertical distance along the spine from the cervicale to the waist.

Centre front waist length, in body measurements: The vertical distance from the neck baseline at the centre front to the waist level.

Centre stitching: See **Double cloth.**

Centring, in designing: When a fabric design is organised in such a way that is balanced about the middle line of a fabric in a vertical direction.

Centrifugal damping machine: Centrifuge in which the material is pre-wetted or wetted during running, and where the wetting fluid is collected once more.

Centrifugal dust collector: A **centrifugal scrubber** or **cyclone** or **fan collector**, etc.

Centrifugal scrubber: A **wet scrubber** in which the gas swirls because it has been injected tangentially or has been directed around by stationary or moving swirl vanes.

Centrifuge: Also called hydro-extractors. Centrifuges are used for Water extraction (dewatering, pre-drying) of textile materials. Values of approx. 15% for residual moisture content can be achieved depending on the type of textile fibre. Most centrifuges have electric drives for speeds of approx. 750–1200 rpm and are generally provided with automatic control over various ranges. Centrifuges are also for removing water from crosswound yarn packages on individual heads.

Centrifuge dyeing: A process for dyeing, bleaching, washing, softening and centrifuging in one machine. The centrifuge, is suitable for the treatment of loose fibres (loose stock) as well as hanks of yarn in processing liquors. Most types of fibres in common use can be dyed by this process.

Centring device: (1) A designed roll or a guiding device to guide the fabric from trollies or batches to the processing machinery especially stenters straight into it without causing creases or skew.

(2) An electronic device installed at the fabric entry of a flat-bed screen printing machine to balance out any irregularities in the dimensions of woven patterns for printed hand towels, neckwear and headscarves by the formation of transverse folds in the border separation areas before the fabric is glued to the printing blanket.

Ceramic fibres: Industrially produced organically based fibres; artificial Mineral fibres. These include silicium oxide and silicium carbide fibres, and aluminium oxide and aluminium silicate fibres for example. They are all resistant to high temperature, and find use in special technical fields.

Cere Cloth: A cloth treated with wax; once used for wrapping dead bodies.

Cerise: After the French for cherry but not necessarily the same colour as the fruit. Often a bluish-red or a purplish red.

Cerule-, ceruleo- (L): Blue, sky blue, but in Roman times probably designated a yellow colour.

Cerulean blue: A vivid light greenish blue pigment made from oxides of chromium cobalt and tin discovered in 1860 by the firm George Rowney and Son.

Cerviacale To bust point, in body measurements: The distance from the cervicale around the base of the neck and down to the bust point.

Cerviacale to wrist, in body measurement: With the arm bent, the distance from the cervical to the shoulder joints, along the outside of the arm, over the elbow to the greater prominence on outside of the wrist.

CETIH: Abbrev. for: Centre d'Etude Technique de l'Industrie de l'Habilement, Paris and Lyon (French Centre for Technical Studies of the Clothing Industry);

Ceylon: A colour woven fabric for blouses and shirtings made from a cotton warp and cotton wool weft, containing little cotton.

CF: Abbreviation for carbon fibres.

CF Value: See **Comfort factor.**

Cha: Very light and thin Chinese silk cloth, made in plain weave but usually printed with very closely placed floral patterns; used for summer garments by the natives.

Chacart: Indian calico, printed with bright; multi-coloured checks.

Chadar or Chadder: (1) Plain woven cotton fabric; used for garments in Arabia. It has a very wide blue or black warp stripe on one side with two narrow white weft stripes appearing four times at intervals at the ends. The "width is about 40 inches and the length 100 inches; (2) East Indian shawl, made of wool or goat's hair, having woven figures on a solid ground; (3) Trade name in East Africa for half bleached cotton fabrics, imported from India; used for loin cloth.

Chadder ulaya: Native name in East Africa for half bleached, bordered cotton fabrics imported from England; used for loin cloth.

Chafe' mark: See **Abrasion mark**. In silk goods displacement of the fibres caused when the cloth is wet.

Chafe mark: (bruised place) A localized area where a fabric has been damaged by friction.

Chafed: A flaw in the cloth, caused by chafing the yarns.

Chafer babrics in tyre fabrics: A woven fabrics usually coated with unvulcanised rubber, which is laid around the bed of a tyre before curing.

Chafer fabric: External strengthening of the shoulder of car tyres, where the tyre and the wheel rim come into contact. In tyres with inner tubes flexibility, (shear) stability and adhesion are required; with tubeless tyres air has also to be prevented from escaping between the tyre and the wheel rim. Fabrics with coatings of adhesive applied to them are used.

Chaferconne: Fine Indian linen fabric, printed in colours (often with hand blocks) ; used for scarfs, mufflers and handkerchiefs.

Chaff: A component of trash in cotton in the form of a heterogeneous assortment of vegetable fragments most of them being small pieces of leaf, leaf bract and stalk. A component in cotton of trash (q.v.) in the form of a heterogeneous assortment of vegetable fragments, most of them being small pieces of leaf, leaf bract (a small form of leaf growing beneath the boll) and stalk. NOTE: Broken fragments of twig and small branches, particularly when brittle, may be broken up further in ginning and are then also regarded as "chaff". Another component of chaff is the silvery lining of the boll interior, sometimes termed "shale".

Chaffy Wool: Wool containing a considerable amount of chaff - finely chopped straw.

Chagrin fabric: (1) English book cloth, made of cotton; (2) Silk dress goods with pebbled surface resembling leather. (3) Narrow rich braid, made of silk or gold thread.

Chagual gum: Vegetable gum from Chile and Peru with a high content of _ Bassorin: It is found in various forms: voluminous pieces, crystalline, cylindrical and hollow, up to 15 mm in thickness. Chagual gum is only partly soluble in water. The insoluble component swells to form a crystal-clear gel with high refrangibility and low adhesive properties.

Chaguar: Strong leaf fibre, yielded by a species of the wild pineapple in Argentine; used by the natives for cordage, hammocks, .bags and also for weaving a sort of cuirass.

Chain, in stenter: On stenters: pin links are linked together into an endless chain for fabric transport purposes. Two chains (vertical or horizontal return) take up the fabric by means of pins or clips, passing it through the stenter processing zone

Chain Boulee: Part of the macrame, made by knitting two threads into a cord.

Chains, chemical: In long molecule chemicals **combinations** of atoms (chain atoms or chain units) covalently bound together in a row, and which are the basic structural elements of a series of molecules See **Chain molecules**.

Chain Cotton: Trade name for Brazilian cotton.

Chain fork: See **Lucet.**

Chain growth polymers: Chain growth polymerisation is a process of successive addition of monomer units to the growing chain by a chain mechanism. The monomer unit gets converted to some active intermediate species by a small amount of initiator such as organic peroxide or an acid or a base. Depending upon the conditions, the intermediate species may be free radical or an ion, and it reacts with other monomer unit to form still bigger intermediate species. The monomer units are, thus, successively added to intermediate species by a chain process. The chain growth polymerisation of ethane involving free radical initiation is given below:

$$\text{(i) Initiator} \xrightarrow{\text{Split up}} \underset{\text{Free radical}}{\dot{A}}$$

$$\text{(ii) } \dot{A} + \underset{\text{Monomer}}{CH_2{=}CH_2} \longrightarrow \underset{\text{Intermediate species}}{A - CH_2 - \dot{C}H_2}$$

$$\text{(iii) } A - CH_2 - \dot{C}H_2 \quad \underset{\text{Monomer}}{CH_2{=}CH_2} \longrightarrow \underset{\text{Bigger intermediate species}}{A - CH_2 - CH_2 - CH_2} \quad \dot{C}H_2$$

Chain, in Zippers: The linkage formed by interlocking the scoops of two stringers.

Chain length: The greatest molecule length produced with the greatest possible extension of the macromolecule (zig-zag chain) whilst maintaining bond angle and the atom spacing. Polymer bulk properties may be strongly dependent on the size of the polymer chain. Like any molecule, a polymer molecule's size may be described in terms of molecular weight or mass. In polymers, however, the molecular mass may be expressed in terms of degree of polymerization, essentially the number of monomer units which comprise the polymer. For synthetic polymers, the molecular weight is expressed statistically to describe the distribution of molecular weights in the sample. This is because of the fact that almost all industrial processes produce a distribution of polymer chain sizes. Examples of such statistics include the number average molecular weight and weight average molecular weight. The ratio of these two values is the polydispersity index, commonly used to express the "width" of the molecular weight distribution.

The maximum length of a polymer chain is its contour length.

Chain linearity: The simplest form of polymer molecule is a straight chain or linear polymer, composed of',a single main chain. The flexibility .of an unbranched chain polymer is characterized by its persistence length.

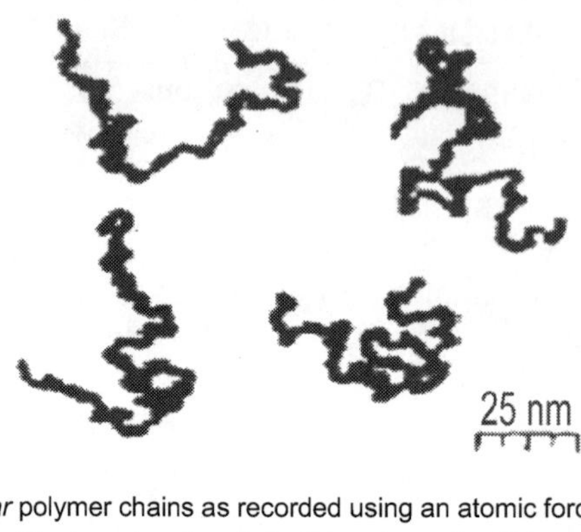

Appearance of real *linear* polymer chains as recorded using an atomic force microscope on surface under liquid medium. *Chain* contour length for this polymer Is -204 nm, thickness Is -0.4 nm

A branched polymer molecule is composed of a main chain with one or more substituent side chains or branches.

Chain plan: Weaving technology term, same as Lifting plan, Peg plan.

Chain stitch: A stitch formed with one or more needle threads and characterized by interlooping. One or more loops of thread are passed through the material and secured by intralooping with a succeeding loop or loops after they are passed through the material. The chain stitch when perfectly done should look like the stitch made by a single-thread machine. This stitch is made by taking the thread toward the worker, and before the needle is drawn out of the cloth the thread is held by the thumb under the point of the needle, as in a buttonhole, making a loop. The needle is inserted in the last loop for the next stitch. The chain stitch is used in modern embroidery as an outline and for darning, but in old embroidery, the outline and chain stitches were used for filling as well. They are found in Persian, Indian, and Italian Renaissance work. Like the feather stitch, the chain stitch is worked towards the worker.

Chain thickness, In Zippers: The measurement from the front to the back of the chain.

Chain twill: An ordinary 3 × 3 twill.

Chain Twist: Yarn first made two-ply to which a third ply is added with a reverse twist.

Chain warp, in floor coverings: Warp threads usually woven in pairs alternating over and under the shotts which are then bound and which enclose the stuffer yarns and the tufts or loops forming the pile.

Chain weave: Piece dyed worsted fabric, finished without any nap, woven in double corkscrew.

Chain width, In Zipper: The measurement from the shoulder to shoulder of the interlocked scoops.

Chain, in Zipper: An assemblage formed by interlocking several elements of two stringers.

Chain-polymerization: See **Addition Polymerization.**

Chain-transfer: A reaction in which a free-radical abstracts an atom or group of atoms from a solvent, initiator, monomer or polymer.

Chain, on stenters: The unit in stenter on which the clip is fitted which in turn holds fabric on both the selvedge and taken through the Chambers and to the delivery side of the stenter. Normally there will be two chains in on stenter which is driven by one motor or two separate motors.

Chainette: A tubular cord produced on a small circular knitting machine using more than 20 needles.

Chair web: A woven narrow fabric. Traditionally 50-85 mm. (2-31/2 in. wide) for supporting the upholstery springs of the chairs , couches etc.

Chaising calenders: the arrangement of rollers in chaising calenders allow the fabric to be wound several times inside the calender in a number of layers ranging from 5 to 13. In this way the pressure is not exerted directly by the flat surfaces of the cylinders but by the fabric itself. The warp and weft yarns carry out a progressive rounding action; the result is a precious lustrous effect, a more bulky, full and compact hand, similar to the one obtained with mangling calenders.

Chakhi: A warp faced fabric, made of silk warp and cotton filling; used in Egypt.

Chakmak: A Turkish fabric, made of silk, often with cotton filling, and interwoven with gold thread.

Chalk: A grey white.

Chalk: Calcium Carbonates. Used for marking. A piece of soft rock used for drawing or writing; a filler or **extender** used in manufacturing pigment or priming canvases.

Chalk Stripes: General term for narrow white stripes broader than pin stripes over a dark ground. Usually woven wool fabrics for suits and costumes. Stripes may be partially obscured by raising. Can also be printed

Chalk, tailors: A piece of flat pipe clay, either square or triangular in shape used marking cloth, colloquially known as clay.

chalk-white: The white of chalk.

Challis: Originally made with silk warp and worsted weft, Challis is made in plain weave. A soft light weight fabric in slightly open plain weave with a soft handle like voile. This light weight fabric, is either dyed or printed. Nowadays Challis name is used for other fibre fabrics also. Worsted yarn makes a beautiful dress weight challis. It may also be wool with polyester or polyamide and sometimes with spun rayon also. Cotton challis has a slight nap to achieve the soft finish. Used for full veils, dresses, blouses, children's clothes, baby dresses. It gathers well and takes unpressed pleats. May be plain or printed; may need lining in paler colours. It crushes in wear but sheds creases overnight. A typical construction may be 50 square to 70 × 60 with 30s spun rayon warp and 14s to 30s spun rayon weft. Another construction of 66 × 50 with 30s warp and 20s filling is also challis. The width varies from 40-56 in.

Chalon: All-wool, twilled light English suiting and dress fabric, calendered on the face, and usually made in black, about one yard wide.

Chalybd- (L): Steel.

Chalybeous: Having a dark blue metallic colour.

Chalys: See **Challis.**

Chamare: Decorative loose coat with large collar and broad sleeves.

Chambers: An old commercial variety of cotton from US, the late maturing staple measuring 22-25 millimeters, the yield of lint being 32 per cent.

Chambery: Very light French dress material made of silk or manmade fibre warp and worsted weft in twill weave.

Chamblette: Seventeenth century English pure or silk mixed worsteds, watered or plain.

Chambord: French mourning dress goods, made of all-wool. The warp sometimes contains silk or is made of cotton. It is woven with ribbed effect, often with alternating heavy and light ribs.

Chambray: (1) A shade of blue. (2) Plain weave, durable cotton fabric, with the mottled look, similar to denim, of white weft and coloured warps. Striped and checked as well as plain and even sometimes in dobby designs. A characteristic construction is 72 × 56 with a width of 30-40 in. depending on the quality 30s carded yarns or 40s combed yarns may be used. Medium and heavy chambrays are used for men's work shirts while lighter grades are used for women's and children's clothes, shirts, pyjamas.

Chambray Gingham: Fine cotton gingham, given a glossy finish; comes mostly in solid colours.

Chambre' fabric: Chambre' are made of 50 : 50 polyester/cellulose in which the warps are made of cellulose and the wefts of polyester. They are dyed with reactive dyes first, the polyester remaining undyed, and then printed with disperse dyes.

Chamfering: (1) Removing the edges of cross-wound bobbins before dyeing in order to avoid dark patches. Done by hand or using chamfering machines.

(2) A slight tapering off done on the rubber rolls of a mangle while grinding to take care of the bending of the roll when pressure loading is done and to get even pressing on the fabric.

Chamois: The yellowish-tan colour of a chamois leather.

Chamois: Originally the skin of the chamois, the small mountain goat, but now also from deer, sheep and goats. The skins are dressed with oil and give a suede finish on both sides. It is very soft and easy to sew. Fabric is napped, sheared, and dyed to simulate chamois leather. It is stiffer than kasha and thicker, softer and more durable than flannelette. Must be designated as "cotton chamoise-colour cloth". Uses: Dusters, interlining, storage bags for articles to prevent scratching.

Chamoisette: Made in cotton, rayon and nylon etc., knitted, double knit construction. Characteristics: A fine, firmly knit fabric. Has a very short soft nap. Wears well. Nylon chamoisette is more often called "glove silk". Uses: Gloves.

Champ: Same as **Fond.**

Champagne: Yellowy-white or yellowy-pink; a 1920's mink colour.

Champion Cluster: An old commercial variety of late maturing American cotton, the staple measuring 25-28 millimeters, the yield of lint being 30-31 per cent.

Chance causes: Factors generally numerous and individually of relatively small importance, which contribute to variation, but which are not feasible to detect or identify.

Chandar: Unbleached, bleached, dyed or printed cotton cloth, sold in India earlier.

Chandokhni: A bandhini design. See **Bandhani.**

Chandul: East Indian fibre, yielded by the Lepuranda succidora; used for bagging.

Change in filling: See **Mixed filling.**

Change in length on twisting: The difference in length occurring due to the twisting of a specimen.

Change in length on untwisting: The difference in length occurring due to the untwisting of a specimen.

Change of shade, in colourimetry: For colourimetric change of shade evaluation, the fastness rating is calculated from the colour difference between the dyeing subjected to the fastness test and the original dyeing.

Change, process: A generic term used for the alteration of weight, etc. of fabric, yarn or fibre subjected to a certain process.

Changeable: Effect A colour effect extensively applied in silk fabrics, produced usually by weaving the cloth of different coloured, yarn dyed warp, and weft.

Changeant: Fabrics made of natural silk or synthetic fibres, in which the warps and wefts are different colours. Filament yarn fabrics are usually piece dyed in a single bath process, viscose or cupro being the first fibre and acetate

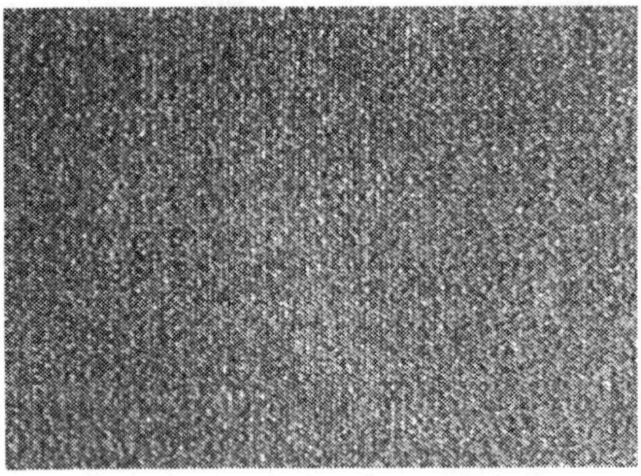

filament the other. The changeant effect is most effective in plain weave fabrics when the moving light-dark contrast of the individual components effectively enhances the changing colour effect to give an attractive interchange when the angle of incidence of the light changes. Used as clothing, overcoat or lining material.

Channel: A narrow passage formed between plies of material in a garment or by attatchment of an additional ply of material, usually to house a draw cord.

Chantilly lace: Originally a silk bobbin lace with a very fine spidery ground. Well spaced isolated decorative motifs are applied in a heavier thread and outlined. Iy is now made by machine in nylon, viscose and mercerized cotton. Often used for bridal veils.

Chapands (French): Waste silk from native reeling akin to charka silk waste.

Chaperon: Close fitting hood with shoulder collar and tail.

Chaple silk yarn: Silk yarn spun from inferior cocoons and silk spinners waste.

Chappe: See **Schappe Silk.**

Char length: Length of the torn, carbonised area in Flammability testing (vertical test). It is determined after the recorded afterglow period on the same test sample. 89 mm is the required average limit value for textiles with flame retardant finishes (mean value of 10 measurements); the max. individual value is 114 mm.

Character: The evenness, distinctiveness, and uniformity of crimp characteristic of their respective wool classes. A well-bred wool of 'good character' will usually show a pronounced crimp and distinct staple formation.

Charactertistic: A property of items in a sample or population which, when measured, counted or otherwise observed, helps to distinguish between the items.

Charara: Long staple cotton grown in Egypt.

Charcoal: A dark grey colour.

Charcoal brown: A dark brown.

charcoal grey: A dark grey.

Chardonnet silk: Thick soft silk yarn consisting of two or more strands twisted together in one direction, each strand being made of a number or singles loosely twisted in the opposite direction.

Chardonnet, H.: Father of rayon, whose research in producing what was then called artificial silk from nitrocellulose, paved the way of industrial manufacture of rayon. (French - 1839–1924).

Charkha: A simple hand operated Indian crude contrivance for reeling raw silk. It conists of a large cooking cum reeling pan where boiling water is kept. The cocoons are cooked in it and filaments collected in a bunch after brushings are passed through a hole on an ordinary thread guide device. Later the thread is crossed with another co-thread for forming a chamon type croissure. Then it is passed through a distributor before it is wound on to standard wooden reel. Four ends are maintained in this device. One person sitting near the cooking pan manipulates the cocoon cooking amd reeling.

Charkha silk: Raw silk on reeled on charkha.

Charkha silk waste: It is the silk waste produced during reeling on charkha

Charkhana: Checked muslin of cotton and silk or pure cotton in India, four-ply coloured threads, forming the checks.

Charleston: Womens dress for evening wear with straight unwaisted line, often with fringes; named after the fashionable dance.

Charly: It is a Copper man of size 50 at the Hohenstein clothing physiology institute and can imitate the temperature regulation of a person (Clothing physiology): the legs can be moved at the hip and knee joints and the arms can be moved at the shoulders and elbows, it can sit, lie, stand and also imitate walking and running movements. It is heated by warm water (pipe system). External evaporation can be measured with water (body cooling, sweat evaporation). A complex control system is used for comparative clothing studies, physiological work measurements etc.

Charmelaine: A soft worsted fabric in twill or satin weave with a lustrous face and matt reverse. The lustre is developed by shearing and pressing. Used for women's outerwear.

Charmante satin: A double wefted fabric, the face being a ½ twill and the back a weft sateen developed from the thick low twist weft.

Charmelaine: Womens dress goods of very fine worsted yarn in satin or diagonal repp weave, back of fabric with long warp floats gives elegant draping properties Uses: for party and evening dresses dyed in piece. A 1/2 twill fabric made from botany wool warp and weft that has a rtwill nep effect by virtue of a high sett two fold fine warp and a low set thick hard twisted weft is a Charmelaine.

Charmeuse: A silk, cotton or viscose satin weave fabric with a dull back and a semi lustrous right side, may sometimes be a mixture of fibres. Creasing fabric, it is not hard wearing. It is soft in feel, drapes very well and used in ladies wear.

Charmoy: Indian ribbed silk fabric.

Charring: The formation of carbonaceous residue as the result of pyrolysis or incomplete combustion.

Chartreuse: (1) A pale apple-green colour with a yellow tinge; a colour name adopted by Web page creators on the Internet; see **X11 Colour Set**. It has hex code #7FFF00.

(2) A very light silk satin crepe with a pebbled effect on the back; the face is finished with a high lustre.

Charvet: Soft fabric of silk or acetate in a diagonal rib weave, often with crosswise stripes of alternate dull and satin finish. Used for making ties.

Chase: The conical part of the body of yarn in cop, bobbin or pirn form on which the thread is coiled during the traversing cycle.

Chasing calendar: Chasing calendars are similar to swizzing calendars. The major difference is that the thread-up is such that the cloth makes several passes through the nips before it exits to a take-up roll. This is done by having cloth pass over chasing rolls which feed it back through the nips. The cloth is compressed against itself with as many as 5 to 6 layers being in a nip. This gives the cloth a thready-linen appearance and a soft special feel.

Chasselas: Cotton cloth in the West African coast

Chasseur-blue: The dark blue colour of the jacket of the French Chasseur soldier.

Chassum: A term popular in west Bengal, India, to denote the silk waste obtained during reeling of cocoons from polyvoltinne races, for example, Bengal Chassum.

Chatoyant: French for shot colours.

Chats: Fulled French woollen, made with white warp and blue or black filling; Obsolete.

Chattan: An old Highland tartan, it is composed as follows: narrow black and white line; wide dark green stripe; white line; yellow, red (split with black line) and yellow stripes of equal width, these three together being as wide as the green; white line; gray stripe, somewhat narrower than green; black stripe, being half the width of the gray; red stripe, same as black ; yellow stripe, same as gray, split with a white line; repeat, in reversed order, group described between two; red bar, one third in width of entire group.

Chaussettes: An old white thread hosiery made in Vitre, France, by the peasants.

Chavonnis: A good quality sheer Indian cotton muslin.

chay, chaya, choya, shaya, shaii: A red dye made from the root of the Indian plant *oldenlandia umbellata* which is of the same family as the **madder** plant.

Chayong: A corded silk velvet from China.

Chebka: Hand-made lace from Tunis, made with Moorish designs; used for dresses and furniture.

CHC process: Caustic soda treatment for polyester knitgoods to improve the surface characteristics.

Check canvas: Open embroidery canvas containing double and single threads, both in the warp and filling, which produce a plaid-like effect.

Check pattern fabric: (Shepard's checks) Colour pattern obtained in a fabric by using 4 pale and 4 dark threads in the warp and weft, woven in twill. Pied de poule fabric is made by using the same colour sequence, but panama weaves, for men's shirts.

Check spikes: Free standing steel spikes 15-20 cm high which are placed at important marking points on a lay, usually the corners of check patterns in order to ensure that the checks match each other when cut.

Check spring: A spring with a loop at one end through which the thread passes through. It takes up the slack in the needle thread during stitch forming and assists the take up in setting the stitch.

Checked muslin: White muslins with coloured cord stripes or checks and plaids; used for dresses, curtains, etc.

Cheda: Indian term for any fabric used as leader cloth etc.

Cheek: The end flange of a filled mangle bowl.

Cheese: (1) A cylindrical package of yarn cross wound on to a flangeless support. (2) Cheeses refer to the spirals of pencil roving produced on the large mechanized carders. The fibres can be knit as is or can be spun up.

Cheese cloth: A gauze made from soft and loosely woven cotton of low count and therefore sheer. Its natural colour is cream but it can be bleached or dyed. It is of a plain weave, sometimes heavily sized, and has rough finish and wrinkled look. A typical construction may vary from 20 × 12 to 48 × 44, 30s carded cotton warp and 40s carded cotton weft. An all-cotton fabric, cheese cloth is one of the cheapest woven fabric. It is not hardwearing but washes well. Originally used for wrapping cheese and for making polishing cloths,

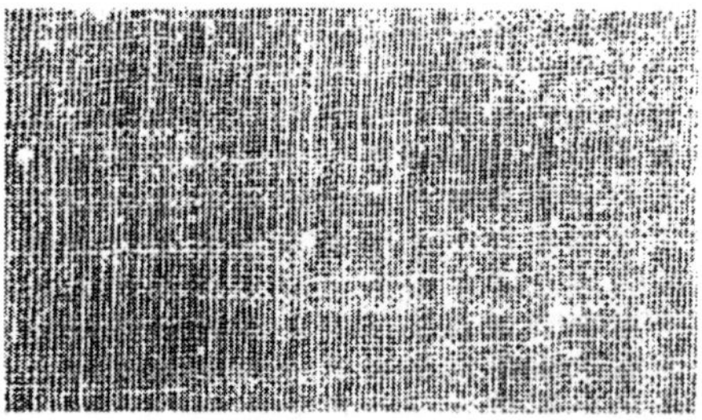

and sometimes as a mounting or underlining fabric. Low grade cheese cloth nowadays is used as dust cloths other uses are in bandages, in tobacco fields to diffuse strong sunlight and other uses. Hence this cloth is also called Tobacco cloth. Higher grade cheese cloth, often using two ply yarns are called scrim and are used as an inexpensive curtain and apparel fabric. When it became a fashion fabric, the quality was improved and the texture made slightly closer. Fashion cheese cloth contains some polyester.

Cheese warp: Parallel threads in the form of a leased twistless rope wound mechanically with a quick traverse on to a spool in the form of a large cheese.

Chelais: Plain woven cotton fabric; used for loin cloths and scarfs in East India; it is made with a checked or striped border and a wide heading.

Chelate: An inorganic complex in which a ligand is coordinated to a metal ion at two (or more) points, so that there is a ring of atoms including the metal, i.e. a metal-complex. The process is known as Chelation.

Chelate resist printing: A process for the production of resist prints on polyester whereby functional groups such as –OH, –NH, –COOH, >C=O or >NH in disperse dyes are able to form chelates with heavy metal cations such as Cu, Cr, Co, Fe or Al from the corresponding heavy metal salts. Penetration of disperse dyes into the polyester fibre is prevented by chelate formation. The best results are obtained with copper compounds in the following order: copper acetate, formate, sulphate and phosphate.

Chelating agent: See **Sequestering agent.** A molecule usually organic, which is soluble in water and undergoes reactions with metal ions to hold them in solution. A number of naturally occurring organic materials in water have chelating ability, such as humic acid, and lignin. Due to their chelating abilities, some organic materials interfere with water softening process.

Chelation: See **Chelate.**

Chemical affinity: It is the driving force behind chemical reactions that results in mutual absorption and reactivity, such as e.g. between a fibre and a dyestuff or textile auxiliary (Affinity, dye affinity). The affinity between reactants generally increases with differences in the electrochemical characteristics (Electrokinetic interfacial potential).

Chemical crimping: Crimping of synthetic fibres by chemical methods

Chemical decatizing: Finishes of a chemical nature to impart form retention, in contrast to normal decatizing which involves mechanical and/or physical treatments

Chemical embroidery: See **Burnt out styles.**

Chelem: A variety of Sisal from Yucatan.

Chelos: East Indian calico shirting, printed with fancy coloured checks or plaids.

Chemical fibre: One which is not made by nature, as the vegetable and animal fibres are, but which are produced artificially through chemical process, like the artificial silk.

Chemical finish: Chemical material other than colorants and residual processing chemicals added to textiles to impart desired functional or aesthetic properties to the textile product.

Chemical finishing: Finishing of fabric using various chemicals to impart particular handle, softness, sheen, weight.

Chemical fixation: Stabilization of a polymer state by crosslinking with bifunctional reagents.

Chemical oxygen demand: The COD value is the formal oxygen requirement in the reaction with the strong oxidizing agent, potassium dichromate. To determine the COD of all oxidizable organic components, the sample is boiled in a strong sulphuric acid medium with an excess of potassium dichromate ($K_2Cr_2O_7$). After 120 min. under reflux, the solution is cooled and back titrated with ferrous ammonium sulphate ($(NH_4)_2Fe(SO_4)_2$) solution. The amount of potassium dichromate consumed is proportional to the amount of oxidizable organic material in the sample.

Chemical pad: Padder used exclusively for the application of chemicals Chemical stability

Chemical wash in rug cleaning: A specialised process for oriental rugs.

Chemically enhanced backwash, CEB: The use of chemicals in the backwashing of **microfiltration** or **ultrafiltration** membranes. The chemicals that are used depend on the type of material that has to be removed from the filter and the type of filter material. Examples include hydrochloric acid, hypochlorous acid, hydrogen peroxide, sodium hypochlorite or sodium hydroxide. This process is also known as clean in place (CIP) or chemical clean.

Chemically enhanced primary sedimentation: The addition of ***alum*** or ***lime*** or ferrous sulphate or polymers to a wastewater can enhance the removal of solids to give up to 80% removal from domestic wastewater during ***primary sedimentation.*** The ***surface loading rate*** of the primary sedimentation tanks can be increased to 50m3m_2d_1 at maximum flow.

Chemically precipitated sludge: Sludge that is produced in a water or wastewater treatment works when ***coagulation*** has been used to help sedimentation or clarification of suspended matter.

Chemichromism: Chemichromism is a term which has been introduced to describe a change in colour, caused by a chemical conversion, induced by some external stimulus, for example exposure to light, heat or electric current. Probably the most familiar and longest-established use of reversible colour change chemistry is found in titration indicators. As an example, the azo dye, methyl orange changes from orange to red in the pH range 3—4, due to conversion into the protonated species.

Chemise: Light women's dress with high waist, puffed sleeves, décolletage, train.

Chenille: From the French word for a caterpillar, this is exactly what it resembles. A yarn is made, with pile protruding all round, by cutting a ladder like fabric into strips, so that the ends of the soft twisted yarns loosen and form pile-like fringe. This fringed yarn is then woven in a gauze weave in a gauze

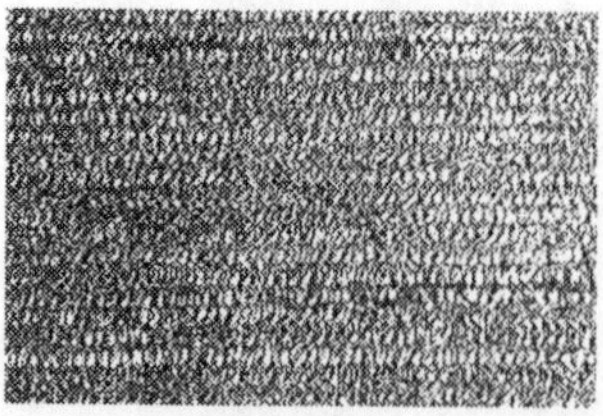

weave to make a fabric with a pile on one or both sides. Fibres used may be silk, viscose, cotton, wool or mixtures. Used for furnishings, curtains, chair covers and specially finished chenille is also made for dresses and coats.

Chenille: These hand sewing needles are with large eyes and sharp points are often used for silk ribbon embroidery Sizes range from 18 to 24.

Chenille axminster carpets: A cut pile carpet, made by binding printed chenille thread from side to side, to the foundation.

Chenille cloth: A fabric made with cotton wool or silk yarn warp and chenille filling; used for millinery.

Chenille lace: French needlepoint lace of the 18th century; it has a hexagonal silk net ground with patterns outlined with white chenille.

Chenille shawl: Made originally in England with chenille weft.

Chequer stitch: Used in bobbin lace springs as filling for berries and flowers.

Cherolee: Indian striped cottons.

Cherconnee: Indian silk and cotton taffeta; made with stripes and checks.

Cherquemolle: An old Indian fabric made of bast fibres mixed with silk.

Cherry: Commercial variety of early maturing cotton from South Carolina, the small bolls yielding 30-32 per cent lint, the staple measuring 18-22 millimeters.

Cherry: Bright red; the colour of cherries; hence cherry-coloured. See **Cerise**.

Cherry-red: Of the colour of the red cherry.

Cheshire: Plain woven bleached cotton fabrics, made in England, 36 inches wide, with 76 ends and 88 picks per square inch; used for calicoes.

Chessboard Canvas: Stout white embroidery canvas with alternate Checks, woven plain, and in honeycomb.

Chest: See **Chest girth.**

Chest girth, in body measurements: The circumference of the body over the shoulder blades, under the arms and across the upper chest.

Chest piece: An additional layer or layers of interlining in the front shoulder area.

Chestnut: Reddish-brown after the name of the shell of the chestnut.

Chestnut-brown: A reddish-brown

Cheverett: A worsted suiting made in Britain in the Seventeenth Century. Obsolete.

Cheviot: Rough, fuzzy surfaced woollen suiting similar to tweed. The name was originally given to the rough wool from the sheep on the cheviot Hills. It is plain or twill weave, which is very hard wearing because of the rough,

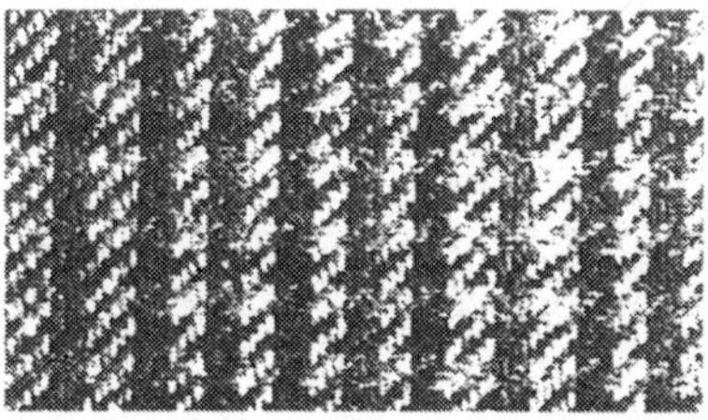

uneven yarn, but looses shapes in wear. Excellent for men's hard wearing suits, and sporting clothes, golfing skirts etc. Lighter cheviots are made from finer and softer woo, made in plain weave and in right hand or broken twill weave. A good use for Cheviot is for sports jackets.

Cheviot Britch: Britch wool from the cheviot fleece.

Cheviot fabrics: Twill, zig-zag twill, plain and fancy woven fabric using Cheviot wool but today mostly made from crossbred wools, in other words wools of a relatively coarse nature. Cheviot fabrics are made from either worsted or woollen yarns. Cheviot fabrics are primarily used in menswear.

Cheviot Shirting: Stout, twilled cotton shirting made of coarse yarn; the pattern consists of small dobby designs or fancy warp stripes, formed by single warps and ribs by double warps. It comes usually in blue or brown, while the filling is white.

Chevrette: Hair obtained from the skin of the Angora kid by a liming process.

Chevron: French serge dress goods, made with eight leaves and four picks in a repeat.

Chevron stitch: An embroidery stitch. Method: Bring the thread through on the lower line at the left side, insert the needle a little to the right on the same line and take a small stitch to the left emerging half-way between the stitch being made. Next, insert the needle on the upper line a little to the right and take a small stitch to the left as at insert the needle again on the same line a little to the right and take a small stitch to the left, emerging at centres. Work in this way alternately on the upper and lower lines. The stitch may also be worked on even weave fabric.

Chianti: A dark cerise colour after the wine from Tuscany.

Chica: Orange-red dye from South American plant.

Chichi Rugs: All-wool rugs made in Caucasia; the pile is of medium length, tied in Ghiordes knot. The colours are dark, blue is often used. The pattern consists of irregular and varied figures in the field, of crosses, conventionalized flowers and geometrical designs. There is a wide border of several stripes. Both ends are finished with a narrow knotted fringe.

Chicopee nonwovens: Stabilization of nonwoven fabrics by water jets (making holes); a binder is also necessary.

Chiffon: A plain weave, soft, filmy material. It may be made from silk of the finest lightly twisted yarn, woven in gum condition, degummed after weaving. It is also made in nylon, viscose, cotton, and polyester. Although thin and sheer it is fairly strong, but it may give at the seams.

Used for loose fitting garments, such as blouses, nightwear and as an evening overdress on top of a dress of satin, taffeta, lamp shades, millinery and for decoration purposes etc. Not expensive, silk chiffon is softest, polyester chiffon is slightly firmer, and less transparent. To manufacture chiffon S and Z twist crepe yarns are used in the warp and weft. Fine 50 denier yarns can be used in square construction ranging from 60 square up to 80 square.

Chiffon, silk: A very light, sheer, open mesh fabric made from hard twisted yarns. Silk chiffon is made from raw silk in both warp and weft directions The yarns are highly twisted and creped , ranging from 50-100 twists per inch (TPI).Warp of down silk or greige, weft of crepe with an S twist, slightly transparent article with a bark effect along the length.

Chiffon net: In England a very fine grade of black silk net; used for laces.

Chiffon twist: Hard twisted, single, raw silk thread having twist in the range of 2000 to 4000 twists per metre (TPM) used for making chiffons, mousseline de soie (French) etc.

Chiffon velvet: A very light, soft velvet with little substance to it. It is floaty and effective but very difficult to sew. It has a short cut pile.

Chijimi: A narrow Japanese silk fabric of solid colour; used for drapery.

Chikan: Hand embroidered fine cotton muslin in India.

Chikankari: A subtle, sophisticated style of embroidery traditionally worked with white threads on fine white cotton. Since the colour palette was earlier limited to white on white, it was through the incorporation of different stitches that varied effects created on fabrics. Some of these stitches are worked on the surface of the fabrics, some from reverse, while others pull at the threads to create a net effect. Now, chikankari is worked with different colours on coloured fabric is well suited for summer wear.

Chikti: Soft, strong, glossy 'fibre, similar to jute, yielded by the Triumfetta rhomboidea in India.

Chikun: Fibrous bast of the Indian nettle tree, used for clothing by the native tribes.

Chilima: Very strong bast fibre, yielded by the Bombax tree in Peru.

Chilkaht: Blanket woven 'by the Chilkaht Indians in Alaska of the hair of the mountain goat. The colours are yellow, black, white, blue and red, usually with a black border around. The blanket has a deep fringe at the bottom and narrow at the sides. The designs usually consist of eyes and faces.

Chilli-red: A strong red colour

Chimayo: Woollen blankets woven in New Mexico and Mexico by the Indians. The design consists of straight cross stripes in blue, black, red and white; formerly woven in two pieces which were sewn together, now made in one piece of two-ply yarn.

Chin: Brocaded silk fabric made in China.

China: Jute Bast fibre of the Indian mallow.

China blue: A lavender blue.

China crepe: See **Crepe de Chine.**

China grass: Sometimes called Grass Linen. An old Chinese fabric woven from vegetable or bast fibres. An uneven weave, but very fine transparent fabric once used a great deal in china as a basis for embroidered table linen. Very similar in appearance to pineapple cloth.

China Mull: Very soft, light, plain woven fabric, made of cotton and silk.

China silk: (1) A white raw silk yarn of superior quality from north China.

(2) See **Japanese silk.** Originally hand woven in China of silk from the Bonabyx mori. Very soft and extremely lightweight but fairly strong. Irregularities of threads caused by the extreme lightness and softness are characteristic of the fabric. Uses: Mostly for linings and underlinings, and could be used for blouses.

(3) A plain weave, lustrous, light weight silk fabric produced in china and Japan. The threads are irregular and soft. It is dyed or printed and used for blouse, lingerie, linings etc. The fabric was originally hand- woven in china of mulberry silk, as early as 1200BC. The term raw has been applied to machine –made imitations in other countries.

China Steam Filature: Very brilliant, strong and white and regular raw silk, reeled in 13/15 deniers in northern China on modern machinery. Graded in 3 classes; Nos. 1 and 2 for organzine; No. 3 for tram.

Chinchilla: A woollen knitted fabric with a napped surface.

Chinchilla cloth: This is made to simulate the fur of Chinchilla rodent. It is a wool coating, in thick twill weave with a napped surface rolled into little curly balls by a special rubbing machine. Its chief characteristic is its spongy feel; it is not heavy or dense. Fairly expensive. Used only for coat and capes.

Chine: General term for various designs or colours printed in faint and indefinite outlines on the warp before the weaving of certain fabrics, as taffeta, gros de Tours, serges and satins. In warp pile velvet the design printed on the warp is about six times the length of the pattern seen in the finished fabric, owing to the "take-in" during the weaving.

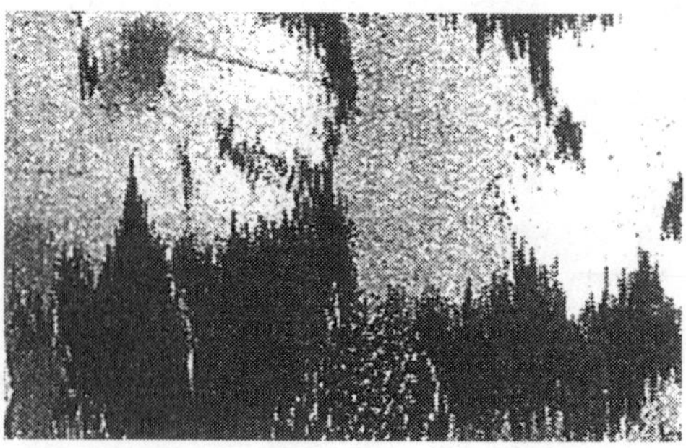

Chine' printing: Warp printed (shadow printed) See **Warp printing.**

Chinese blue: A blue pigment made from mixing **cobalt blue** with **flake white**. Synonymous with **Prussian blue.** Also Chinese white, orange and yellow; all originating in China according to Partridge although **Prussian blue** was discovered in Berlin.

Chinese Burr: Triumfetta pilosa of Queensland, Australia; yields a clean, etrong, lustrous, soft and fairly resilient fibre, similar to jute.

Chinese Jute: White, lustrous, fairly strong 'fibre, yielded by the Indian mallow.

Chinese mink: See **Mink.**

Chinese red: A chrome red pigment. Obsolete name for **cinnabar**.

Chinese rugs: Antique Chinese rugs were made with the Persian or Senna knot, making possible a large number of knots per cm and a close shearing of the pile. Later, rugs were made with large, coarse designs and a deep pile. In more recent years, Chinese rugs have been made in plain colours as well as in the floral and dragon patterns characteristic of Chinese design. The pile of Chinese rugs is unusually deep, making these rugs remarkably durable.

Chinese vermilion: A vivid yellow-red pigment.

Chinese white: A white pigment prepared from zinc oxide and possibly so named after the fine porcelain from China. This was the first permanent opaque white water colour. See **zinc oxide**.

Chinese yellow: A vivid yellow pigment made from arsenic.

Ching: A vivid greenish-blue.

Ching Ma: Silky and strong fibre in China, yielded by the abutilon plant; used for cordage.

Chino: An all-cotton twill or plain weave fabric using combed mercerized yarn. It is bleached or dyed, a hard wearing fabric and hence used as uniforms for armed forces in many countries when dyed appropriately to pass the specification for fastness etc.

Chinoline-blue: A blue colour made from quinoline.

Chintz: Good quality cotton fabric usually with a large printed design of flowers, birds etc. It is a fine closely woven cloth, finished and calendared in order to resist dirt. An average construction of chintz may be 60 × 60 with highly twisted 25s warp and loosely twisted 30s weft. Used for curtains, slip covers, etc. When the fabric is given permanent finishes it can be used for dress materials.

Chintz braid: Cotton braid, printed and finished like chintz, used for dress trimming.

Chintz calendar: A Calendar for the production of high gloss and closing of fabric yarns. The finish effect is produced by bringing the cloth into contact with a heated, polished, chilled-iron bowl which is travelling at a faster speed than the cloth itself. Three-bowl heavy friction calenders are suitable for highly glazed finishes. The bottom bowl is usually made of close grained cast-iron, the middle bowl of cotton (which is of greater diameter than the others to allow for wear; the top, or glazing bowl, is made of highly polished chilled-iron and is heated by steam or gas.

Chintz finish: High glaze finish produced by friction calendaring cellulosic fabrics previously impregnated with wax or paraffin emulsions, dried, and adjusted to appropriate residual moisture content. Permanent chintz effects are produced on fabrics to which resin finishes (mainly melamine-formaldehyde compounds) have been applied prior to friction calendaring, followed by subsequent curing.

Chintz prints: (partial or local chintz). Localized glazed effects produced by screen or roller printing, e.g. single, multicolour or non-coloured designs with high glaze on white, black or multicoloured grounds without glaze. Pigments are used as colorants. The chintz effects are resistant to washing up to 40^0C.

Chintzing: A process in weaving, by which the colours in the extra weft yarns are replaced with other ones in horizontal sections of the pattern.

Chioggia lace: Coarse, Italian bobbin lace, similar to the early Flanders laces.

Chip: Commercial form of plastics and fibre polymer starting materials (granules).

Chip colouration: See **Mass coloration.**

Chip, fibre: See **Fibre chip.**

Chips: Chips of synthetic fibre polymers (polyamide, polyester). Produced by shredding tapes of the polymerisate melt/polycondensate melt, to reduce to the required size. Undyed or dyed (chip dyeing), they are re-melted and extruded during spinning (known as the chip-melt-spin method).

Chique: Inferior French raw silk; used for threads

Chiquechique: Very strong, durable and light fibre, yielded by the leaves of a palm in Venezuela; used for brooms, ropes, and cordage.

Chirimen: Japanese lusterless silk crepe, the warp and filling consisting of equally thick yarns; the filling yarn is twisted to the right and to the left; the fabric is dyed in the piece.

Chisholm: Even sided Highland tartan, made as follows: Wide bright red and narrower dark green bars alternating. The red bars are split with a pair of white lines in the middle. The green bars are edged with three blue stripes (wider on the outside, narrower inside), and are split with red line in the middle.

Chiton: Women's and men's belted linen robe used by ancient Greek, gathered into many folds and secured at the shoulders.

Chitrak: Turkish fabric, made with cotton warp and silk filling.

Chitrang: Bast fibre, yielded by the Sterculia wightii in China; used for cordage.

Chits: Plain and close woven fine calico from India.

Chlaina: Simple woollen men's robe used by ancient Greek, draped over the back and shoulders, clasped at the right shoulder.

Chlamis: Short woollen men's cloak, thrown over the left shoulder and clasped at the right .

Chlidema square: Border squares entirely made from 0.69m body carpet where additional designing procedures are used to produce the border and corner design.

Chloramine: A compound consist of and ammonia gas which retains its bactericidal qualities for a longer time than does free chlorine.

Chlorinated water Fastness: Test for assessing the resistance to chlorinated water. A sample is wet out in distilled water, dewatered to 100% water content and placed in an aqueous solution containing 20 mg/g of active chlorine (from sodium hypochlorite) at a liquor ratio of 100 : 1 for 4 hours at 20°C. Assessment with the grey scale.

Chlorination of wool: Treatment of wool with chlorine compounds to alter the surface to make the wool less prone to shrinkage in washing and to improve dye uptake Chlorination, sometimes alone and sometime with subsequent application of special polymers, alters the surface of wool so that the scales are largely prevented from causing the "ratcheting" action that makes the motion of one strand of wool "one way" with respect to others whose scales point in the opposite direction. By preventing this ratcheting, wool fabric is much less likely to shrink when laundered. Chlorination also improves dye uptake in printing processes, but at the expense of wash fastness. Chlorination is typically done using organic chlorine compounds, sometimes the same as those used for chlorination of small swimming pools. From an operational point of view, the best results can be obtained with the combination of two different and complementary treatments: the first one is an oxidising treatment followed by a special treatment with cationic resins. The first treatment is the traditional chlorination process, carried out using: -NaClO in presence of strong inorganic acids (sulphuric acid)

– Cl_2 gas

– Chlorine organic salts (sodium salt of dichloroisocyanuric acid) which, in acid solution, releases chlorine.

The second treatment is carried out by applying special resins enhancing the anti-felting effect (PA . epichlorohydrin or cationic resins polysiloxanes).

An antichlor treatment with $NaHSO_3$ must be carried out subsequently to eliminate any residue of Cl_2 that might remain on fibres. New treatments to be carried out on tops or fabrics (plasma treatment) are now being studied as an alternative to chlorination.

Chlorination of water: The addition of small amounts of free chlorine, usually 0.2 to 2.0 ppm to render water bacteriostatic in a water supply.

Chlorination/resin finish: Antifelting finish of wool on the base of synthetic resins which is preceded by a chlorination.

Chlorine: Light green; of the colour of vegetation.

Chlorine bleach: A bleach that releases the hypochlorite ion in solution, for example, sodium hypochlorite.

Chlorine/chlorate discharges: These are examples of oxidative discharges:1.. Acid-chloride of lime process for discharging Turkey red and indigo. 2. Chlorate-red prussiate of potash process for discharging indigo, indigoid vat dyes, cationic dyes and mordant dyes.

Chlorine content of bleach liquors: Same as **Available chlorine**.

Chlorine dioxide bleach: Not a very common process now, it finds only occasional application. It is bleaching with an aqueous solution of chlorine dioxide which, during the process, is freshly prepared on a continuous basis in special equipment.

Chlorine fastness: See **Hypochlorite bleach fastness.**

Chlorine-Hercosett process: Antifelting finish for wool based on the application of synthetic resins to prechlorinated wool, e.g. See **Hercosett process.**

Chlorine resistance of resin finishes: The resistance of resin finished textiles to available chlorine, e.g. in bleaches (Chlorine retention) is largely dependent on the chemical type of resin finish applied. Finishes based on e.g. DMPU, are resistant to chlorine.

Chlorine retention: The capacity to hold chlorine the behaviour of finished textiles towards baths that contain active chlorine (so-called chlorine washing, usually as an addition to the actual washing process to improve the whitening effect or to disinfect

Chlorite bleach: Bleaching with Sodium chlorite. In special cases chlorite bleach is used for cotton, but are to be done cautiously. The concentration of the effective bleaching agent, chlorous acid, reaches its maximum at pH 2.5–3.5. Below pH 3.0, fibre damage becomes very noticeable. The decomposition of the weak acid, $HClO_2$, generates the strong acids HCl + $HClO_3$, which hydrolyse the cellulose. Hence the optimum conditions are in the range of pH 3.0–4.0.

Chlorite bleach fastness: Used for assessing the stability to sodium chlorite bleach solutions. (a) Light: treatment with 1 g/l sodium chlorite (80%) at pH 3.5 (addition of acetic acid) for 60 min. at 80°C, liquor ratio 50:1. Subsequently rinsed for 10 min. in running water. (b) Severe: with 2.5 g/l sodium chlorite (80%). Assessment with the grey scale.

Chloro- (G): Green

Chloro-: Prefix for compounds with complex bound chlorine.

Chloro(-fibres): Chlorinated polyvinyl fibres are sometimes wrongly called as Chlorofibres. See **Polyvinyl chloride fibres; Polyvinylidene chloride fibres.**

Chlorocruorin: A green pigment present in some marine creatures.

Chlorophanous: Having a yellowish appearance.

Chloroflourocarbons(CFCs): They are derivatives of methane or ethane, where the hydrogen is wholly or partially replaced by fluorine or fluorine and chlorine. With a low boiling point, non-flammable and of relatively low toxicity, chlorofluorocarbons are used as coolants, propellants for aerosols and blowing agents for foams, degreasing agents and formerly as solvents in special dry cleaning plant for furs, leather and delicate articles. There are restrictions for the use of of CFC as drycleaning agents as a large proportion of tetrachloroethane (hexachloroethylene) is decomposed in the troposphere, the lowest atmospheric layer. It is assumed that the chlorofluorocarbons or at least a part thereof, penetrate undecomposed through to the layer above, the stratosphere, and reach the ozone layer. The effect of the ultra-violet rays which are strong at this level is thought to cause chlorine atoms to separate off from the fully halogenated chlorofluorocarbons (i.e. those without hydrogen atoms), which then break down the ozone molecules to leave molecular oxygen. Thus the ozone layer is reduced and its effectiveness as a filter against UV radiation is weakened. Consequently, UV radiation to the earth's surface is stronger, which could cause biological changes. CFC 11 (trichlorofluoromethane) and CFC 113 (trichlorotrifluoroethane) are thought to pose particular risks as chlorine-containing and fully halogenated CFCs.

Chlorovinyl fibres: See **Polyvinyl chloride fibres.**

Chocolate: A deep brown, the colour of chocolate, although it embraces many shades. Also **chocolate brown**.

Chocolate brown: A dark brown colour; a food additive producing a brown colouring and used particularly in chocolate cake mixes (E155). Also called 'Brown HT'.

Choel: Fibre used for cordage extracted from young roots of Btea tree grown in India, but not very common now.

Choice: 1. Merino clothing wool taken from the best part of the neck of a fine fleece; 2. In the woollen trade the third quality of wool, taken from the middle of the sides.

Chokidal: A Bandhani design. See **Bandhani**

Choke, or choke tie: It is a very tight tie made around all of the threads in the warp—usually this is done on the warping board or reel.

Cholet: (1) Unbleached light French linen, finished without any dressing; (2) Fancy coloured handkerchiefs; (3) Light weight linen with narrow blue and yellow or blue and red stripes.

Choli: Blouse, or a ladies dress, covering the upper part of the body, in India

Chom: Native name in Yucatan for the strong, silky leaf fibre of the wild pineapple.

Chongkwen: Light silk taffeta, made in China, about 27 inches or wider; used for umbrellas.

Chop: A trade mark or ticket on a book of raw silk indicating its grade and name of the produce or shipper.

Choppat: Indian light weight silk taffeta.

Chopped strand, in glass textiles: A strand made from short predetermined lengths of cut continuous filament and used as a reinforcing material.

Choquettes: French term for cocoons of diseased silkworms.

Chord modulus, in a stress strain curve: The ratio of the change in stress to the change in strain in between two specified points on the curve.

Chosheb: In the Bible textiles having patterns woven with gold on a variegated ground.

Chouca: Strong leaf fibre, yielded by species of the aloe in South America; used for cordage.

Chow: A vivid blue.

Chowtar: Indian cotton muslin.

Chroma/Chromaticity: The intensity or saturation level of a particular hue, defined as the distance of departure of a chromatic colour from the neutral (gray) colour with the same value. In an additive colormixing environment, imagine mixing a neutral gray and a vivid red with the same value. Starting with the neutral gray, add small amounts of red until the vivid red colour is achieved. The resulting scale obtained would represent increasing chroma. The scale begins at zero for neutral colours, but has no arbitrary end. Munsell originally established 10 as the highest chroma for a vermilion pigment and related other pigments to it. Other pigments with higher chroma were noted, but the original scale remained. The chroma scale for normal reflecting materials may extend as high as 20, and for fluorescent materials it may be as high as 30.

Chroma C*: Chroma measure. As can be seen from CIELAB 1976 uniform colour space, three-dimensional, with colours located using cylindrical coordinates as follows: L* – the lightness coordinate (the same as in L*a*b*)

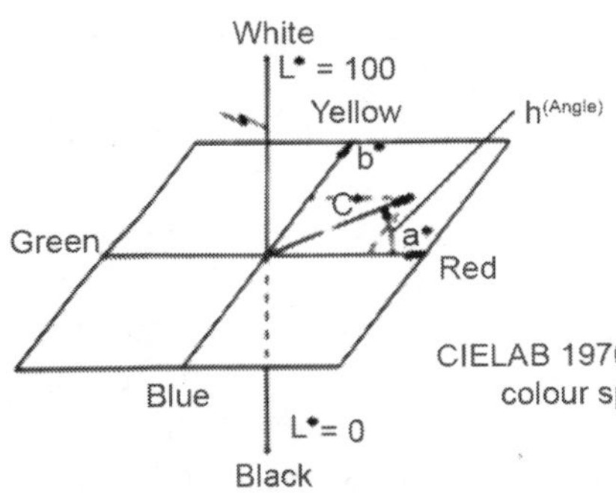

CIELAB 1976 uniform colour space

C* – the chroma coordinate, the perpendicular distance from the lightness axis h* – the hue angle, expressed in degrees, with 0° being a location on the + a* axis, continuing to 90° for the + b* axis, 180° for – a*, 270° for – b*, and back to 360° (or 0°).

From a mathematical standpoint these are the polar coordinates C*, neutral h calculated from the cartesian coordinates a*, b*. "C*" is the abbreviation of chroma and "h" of hue. Many CIE system users prefer the L*C*h* method of specifying a colour, since the concept of hue and chroma agrees well with visual experience.

Chroma component DC*: Chroma difference part of the total colour difference between two colour samples. With the aid of the colour coordinates of two colour samples, it is possible to compute exact colour differences. The first magnitude calculated is the spatial distance between the two colour points. This distance is termed the total colour difference, abbreviated DE* according to CIE. The total colour difference can be further subdivided into hue component DH*, chroma component DC* and Lightness component DL*. The metric hue difference (DH*) is the colour difference, in distance units, due to the hue angle (Dh*) difference. DH* is used in the total colour difference computation, where all terms are distances (not angles), as follows:

Chroma $C^* = [a^{*2} + b^{*2}]^{1/2}$ and Hue angle h = tan-1 (b*/a*)

Chroma measure C*: Same as **Chroma C*.**

Chromate discharges: The discharge print pastes contain sodium bichromate and alkali. Oxidative discharges for the production of coloured discharges on indigo grounds etc.

Chromate dyeing process: Single bath chroming process for chrome dyes. (1) Wool: selected chrome dyes (mordant dyes) are dyed together with a chromium salt in the same bath. This process is used especially in cases where iron or copper is present in the water or in the dyeing equipment. (2) Wool/ cotton unions: special dye mixtures (wool/ cotton chrome dyes) are dyed together with a chromium salt in the same bath.

Chromatic: Perceived as having a hue — not white, grey or black.

Chromatic circle: This is a circular representation of the Hue of so-called chromatic colours which are formed by combining additive colour mixtures. The chromatic circle is restricted to a maximum of 200 different pure shades which can be produced e.g. using perceived equal saturation and brightness values. The commonly used 24-section chromatic circle contains 8 principal colours each in 3 levels (light, saturated/full, dark) and each ordered with the so-called colour coefficient (also referred to as a colour point), starting with yellow (colour coefficient 1, 2, 3) through orange (4–6), red (7–9), violet (10–12), ultramarine blue (13–15), ice-blue (16–18), sea-green (19-21) to leaf green (22–24).

Chromatic colours: So called pure spectral colours in yellow, red, blue, green and intermediate Shades; in practice mostly mixed with white and/or black. Chromaticity diagram.

Chromaticity coordinate Curves: See **Chromaticity co-ordinates.** Chromaticity co-ordinated of tristimulus values X,Y,Z are

$$x = X / (X + Y + Z) \text{ (Yellow-blue Coordinates)}$$

$$y = Y / (X + Y + Z) \text{ (Red-green Coordinates)}$$

$$z = Z / (X + Y + Z)$$

Figure below shows the CIE chromaticity diagram with the spectrum colours forming the boundary line and the neutral colours in the centre for the Standard Illuminant D65/10°. The coordinates x, y are also called the chromaticity

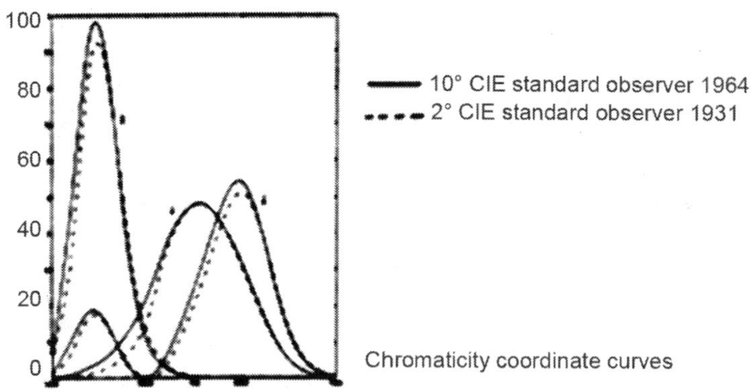

Chromaticity coordinate curves

coordinates. These coordinates can be calculated for different illuminants, i.e. the corresponding neutral points, from the tristimulus values of the illuminants. The chromaticity diagram is used in setting tolerances, for example signal light colours (green, amber, red and white) in rail, road or life jacket lamps. The diagram is also used in gamut mapping – in colour monitor colours (R, G, B), for different dye classes, etc. – while chromaticity coordinates are used in many applications, e.g. whiteness index calculations

Chromaticity co-ordinates (CIE): The ratios of each of the three tristimulus values X, Y and Z in relation to the sum of the three — designated as x, y and z respectively. They are sometimes referred to as the trichromatic coefficients. When written without subscripts, they are assumed to have been calculated for illuminant C and the 2° (1931) standard observer unless specified otherwise. If they have been obtained for other illuminants or observers, a subscript describing the observer or illuminant should be used. For example, x10 and y10 are chromaticity coordinates for the 10° observer and illuminant C.

Chromaticity diagram (CIE): A two-dimensional graph of the chromaticity coordinates (x as the abscissa and y as the ordinate), which shows the spectrum locus (chromaticity coordinates of monochromatic light, 380-770nm). It has many useful properties for comparing colours of both luminous and non-luminous materials.

Chromatography: (Chromatic adsorption analysis). Separation procedure for isolating individual components from mixtures of substances with similar constitutions (related derivatives, isomers and the like) but having different adsorption power. Can be used for the analysis of mixed dyestuffs. A distinction is made between different kinds of chromatography depending on the physio-chemical basis of the method.

Chromatometablepsy: Colour-blindness.

Chromatophore: A pigment-containing cell in the deeper layers of the skin of some animals producing changes in skin colour. In the fish called the Osbeck such cells enable it to change from red to white in 8 seconds. Many fish and animals can change their colouring. This is achieved in some species (including the African clawed frog) by the release of a hormone which changes the position of the dark pigment **melanin** contained in chromatophores and which variously makes the creature appear dark or colourless. See **chameleon**.

Chromatopsia: An inability to determine or distinguish certain colours.

Chrome: A reddish-yellow shade.

Chrome dyeing: Afterchrome dyes are still widely used in wool dyeing in all stages of wool processing. The benefits are good to excellent colour fastness in all process and wet fastness requirements, good leveling and a low price level. Traditionally, dark shades are dyed, particularly black and navy blue.

Chrome dyes: Dyes for wool that use chromium compounds as mordants These dyes are part of the broader category of mordant dyes for wool.

Chrome green: A permanent yellowy green the result of mixing **chrome yellow** with **Prussian blue**. It is called by many names including **Brunswick green**, **cinnabar green**, **Milori green**, zinc green and **chromium oxide** from which it is made. It is also referred to as, but should not be confused with, **viridian green**.

Chrome leather: An inexpensive leather that has been tanned with chromium salts. It is used in the shoe industry.

Chrome orange: A reddish orange also referred to as Derby red, Victoria red or Persian red. See **lead chromes.**

Chrome red: A deep scarlet red also called **Chinese red** and Derby red.

Chrome yellow: Brilliant yellow produced by mixing sodium chromate and lead nitrate both of which are colourless.

Chrome tanning: Animal hide is tanned into "wet blue" after cleaning by cross-linking the collagen chains with chromium. The expression describes wet, swollen leather that is subsequently dyed and re-fatted. The tanning serves to convert rot-susceptible hide into rot-resistant leather, so that the hide fibres no longer swell or shrink due to the absorption or loss of water, and become more temperature resistant and more resistant to the effect of chemicals.

Chromed wool: Stabilisation of woollen yarn against fungus and bacteria by blind chroming using the single-bath chroming process. The same applies to an even greater extent to afterchromed wool, which has shown a resistance after 75 days during a soil burial test.

Chroming: The coating of copper print rollers with a layer of chrome to protect against damage and chemical influence.

Chroming of dyes: A process used to form chromium complexes in the fibre, which develops the final shade and improves light-fastness, especially wet fastness. A prerequisite is the presence of complex-forming groupings in the dye molecule, mainly OH- and COOH-groups in the ortho position. Dyes with these groupings form complexes with trivalent chrome, with the displacement of the H atoms in the complex-forming groups

Chroming methods: See **Chroming of dyes.** (1) Pre-mordant process: In this method the wool is mordanted before the dye application. With wool (mordant dyes; chromium dyes) the application of the dye onto the premordanted wool provides immediate permanent colouring and therefore makes shading easier. However, only a relatively small number of dyes are available for this process. The process is fairly time consuming, provides little fibre protection and is hardly ever used an is of academic interst only now. (2) Single-bath chrome dyeing process: the simultaneous application of dye and chromium salt to wool, and therefore the dyeing and development of the shade is made possible in a slightly acidic bath. 3. After-chroming dyeing process: achieves the good process and end-use fastness for wool. Dyeing takes place in a slightly acidic bath. After complete exhaustion of the bath, a hexa-valent chromium salt is added (usually potassium dichromate), and the chromium pigments build up on the fibre during the rest of the dyeing process. The final colour is only seen after successful chroming, which makes shading more difficult.

Chromium: (1) The grey metallic colour of the metal of the same name.

(2) The metallic element that forms the basis of a number of compounds often used as mordants, or as part of the molecular structure of pre-metallized dyes Potassium dichromate is a common mordant chemical. Chromium compounds are toxic and some are carcinogenic. They must be handled with care and understanding of the risks involved.

Chromophore: It is proposed that dyes contain two types of group which are responsible for their colour- chromophore and auxochrome. The first of these is referred to as the chromophore, which is defined as a group of atoms principally responsible for the colour of the dye. It is understood that the chromophore is commonly an electron-withdrawing group, that auxochromes are usually electron-releasing groups and that they are linked to one another through a conjugated system. A particular chromophore structure may be found in a variety of dye classes and in pigments See also **Auxochrome.**

Chronic toxicity: Exposure that will result in sublethal response over a long term, often 1/10 of the lifespan or more.

Chronic value (ChV): Geometric mean of the **No Observed Effect Concentration** (NOEC) and **Lowest Observed Effect concentration** (LOEC) from partial and full cycle tests and early life-stages-tests.

Chrysal: Golden.

Chrysalis: The form taken by the silk worms in the passive stage of development between worm and moth. It is dark brown and fragments of it can often be in the silk waste especially in noils.

Chrysaniline: A golden-yellow dye also called 'aniline yellow'.

Chryso-, Chrys- (G): Golden or golden yellow.

Chrysoclavus: Rich Byzantine fabric, made of white or purple silk, embroidered with large gold nail heads; worn by state dignitaries.

Chrysolite green: Of the colour of chrysolite, namely, yellowish green (although varying to dark green).

Chrysoprase: A bright golden green or an apple green after the beryl gemstone of the same name which derives from the amalgam of **chryso-** (gold) and **praseo-** (leak green).

Chuddah: Solid coloured wool shawl, made in India of pashmina in very fine twill on hand looms.

Chuddei: See **Chadar.**

Chuna: South American sheep, descendant of the Spanish merino, yields long wool.

Chundri: Indian name for Bandhana. These tie-dyed cotton or silk dress used in native costumes. Usually red fabric with extremely coarse mainly white yellow and green patterns. See **Bandhana**.

Chung-shan chou silk: A chinese silk fabric woven with ply yarn yarn in the warp. It was developed in Nanking as a substitute for imported light weight woollen fabrics. It is in plain or patterned weave and was sold under the trade name " Yi Shing Kung"

Chuncu: Bast fibre yielded by the Caladium tree in Peru.

Chunri: Piece dyed cotton fabric in India, dyed like the bandanna handkerchiefs.

Chuquelas: Striped Indian taffetas, made of silk and cotton.

Church Laces: Needle point laces of the 17th century; used for ecclesiastical purposes, originally with designs of Biblical character.

Churi (Chungri & Chungdi): An Indian term originally for plain weave Indian Cotton and silk sarees dye by the knot tying method.

Chute Fabric: A broad term referring to parachute fabric, fragmentation bomb chute fabric, flare chute fabric and minelaying chute fabric.

Chuzen: Japanese process for manufacturing reserve dyeings with indigoid vat dyes or naphthols on folded wollen fabrics that are printed with a reserved paste, compressed in to a block and has dye liquor poured over them. Used as kimono matrials.

Cibacron F, FN dyes: Cibas exhaust dyeing flourotriazine based reactive dyes. Presently Huntsman and the reactive brands are known under the name Novocrons.

Cibafast W/ Tinofast CEL: Brand name for UV absorber for photochemical wool protection developed by CSIRO IWS and Ciba. Presently liscence lies with different company.

CI: Colour Index.

Ciciclia: Silk or half-silk damask dress goods figured with flowers over satin foundation; made in Asia Minor.

Ciclatoun, Siglaton, or Cyclas: Medieval golden fabrics in Europe; originated from Persia.

CIE: (Commission Internationale de l'Eclairage with headquarters in Paris) – The International Commission on llumination, the primary international organization concerned with colour and colour measurement.

CIE 1931 Standard observer: The research results of Guild and Wright led the CIE in 1931 to define the colorimetric Standard Observer. The primary reference stimuli used were a red spectral colour of wavelength 700 nm, a green of wavelength 546.1 nm and a blue of wavelength 435.8 nm. The test persons had to match all spectral colours in stepwise fashion by varying the intensities of these three primaries. The spectral sensitivity of the eye can be deduced from the intensities required for the different spectral colours. The result of these experiments led to the three sensitivity curves with the aid of which the behaviour of the human eye in the presence of

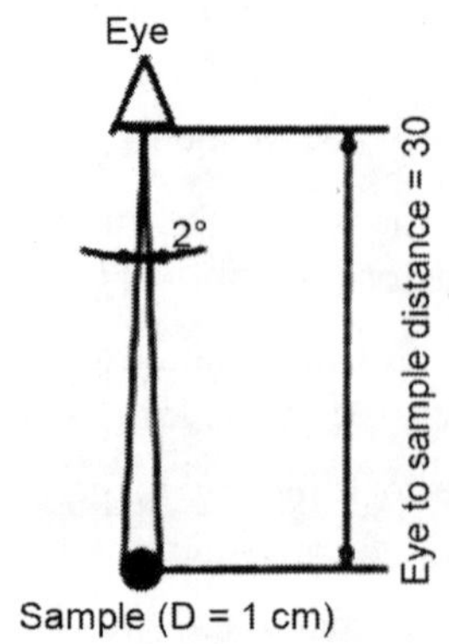

Small field of vision CIE 1931 observer

light of different wavelengths can be described. In Figure below the three sensitivity curves are shown as broken lines; they define the so-called 2° Standard Observer or the CIE 1931 Observer. These three curves are called the chromaticity coordinate curves. The red-sensitive curve is symbolized by x^- the green-sensitive curve by y^- and the blue-sensitive one with z^-.

CIE 1964 Standard observer: Based on new findings of W.S. Stiles carried out for a larger field of view of 10°, in 1964 the CIE defined the 10° Standard Observer incorporated in most colorimetric programs today. The differences between the two definitions of Standard Observer can be seen from Figure above (see **CIE 1931 observer**). The larger field of view of 10° corresponds better to colour matching conditions in actual practice. Figure below shows the size of the areas observed under a field of vision of 2° and 10°. For an eye-to-sample distance of 30 cm, the two fields of vision correspond to a circular area with a diameter of 1 and 5 cm, respectively.

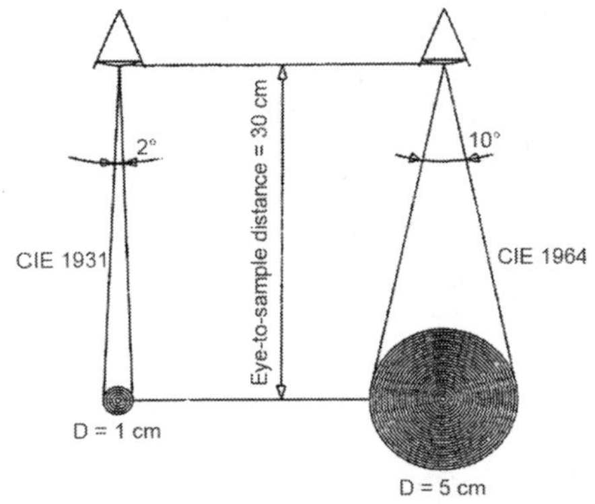

Small and large field field of vision

CIE 1976 L*a*b* colour space: A uniform colour space utilizing an Adams-Nickerson cube root formula, adopted by the CIE in 1976 for use in the measurement of small colour differences

CIE 1976 L*u*v* colour space: A uniform colour space adopted in 1976. Appropriate for use in additive mixing of light (e.g., colour TV).

CIE 1976 L*a*b* equation: A commonly used equation which transforms CIE tristimulus values into a threedimensional opposed colour space. Generally abbreviated as CIELAB

CIE Chromaticity Diagram: Although any colour under a given illuminant can be unambiguously described by chromaticity co-ordinatess X,Y,Z one has to learn how to interpret these values meaningfully and work out their relation to the perceptual magnitudes of hue, chroma and lightness. In order to obtain a clear presentation, more or less complicated mathematical transformations

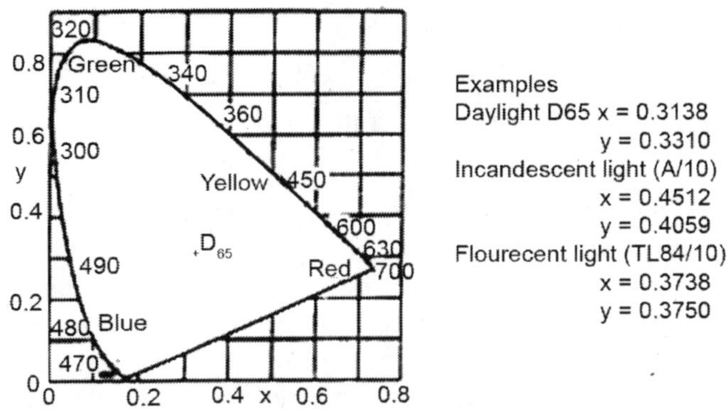

CIE chromaticity diagram

must be carried out. The oldest type of such representation, which is occasionally still used today, is the CIE chromaticity Diagram, often termed the Colour triangle. The colour coordinates are calculated as a measure of the yellow-blue component of the colour sensation and a measure of the green-red component in the following manner:

x (Yellow – Blue coordinate) = X/ (X+Y+Z) and *y* = Y/ (X+Y+Z) (Red – Green coordinates)

CIE chromaticity coordinates: See **Chromaticity co-ordinates.**

CIE chromaticity diagram: See **Chrmaticity diagram**.

CIE daylight illuminants: See **Daylight illuminants (CIE).**

CIE luminosity function (y): See **Luminosity function (CIE).**

CIE standard illuminant: See **Standard illuminants(CIE).**

CIE standard observer: See **Standard observer (CIE).**

CIE tristimulus values: See **Trstimulus values (CIE).**

CIE Triangle: The CIE chromaticity Diagram, often termed the Colour Triangle.See **Chromaticity coordinates**.

CIE Whiteness Formula: CIE recommendation for standard identification of the degree of whiteness. The formula should only be used for a relative comparison of similar samples measured at the same time on the same device because of the technical difficulty in measuring fluorescent dyes (absolute precision, precision of adapting the sample lighting to the spectral distribution of daylight). In relation to D65, 10 degrees, the CIE whiteness formula is:

$$W = Y + 800.(0.3138 - x) + 1700.(0.3310 - y)$$

As well as the whiteness formula, a formula for identifying the tint of white samples is recommended:

$$T = 900 \cdot (0.3138 - x) - 650 \cdot (0.3310 - y)$$

T is positive for samples with a green tint, and negative for samples with a red tint.

CIELab: Colour difference units (evaluation) obtained from colour measurementprocedures and calculated with the aid of colour difference formulaare generally expressed in CIELab units today. According to CIELab system , the CIELab co-ordinates L, a, b are represented in colour space as mutually perpendicular with L representing the lightness from (white to black) with a representing red/green and b representing yellow/blue colour defferences respectively.

CIELAB (1976) (or CIE L*a*b*, CIE Lab): Colour space in which values L*, a* and b* are plotted using Cartesian coordinate system. Equal distances in the space approximately represent equal colour differences. Value L* represents lightness; value a* represents the red/green axis; and value b* represents the yellow/blue axis. CIELAB is a popular colour space for use in measuring reflective and transmitive objects.

CIETEX: Conference Internationale de l'Enseignment Textile (International onference on Textile Education.), Villeneuve d'asq-Flers, France.

Cigar: The brown colour of the cigar.

Cigarette test: Flammability test.

CIM: Computer Integrated Manufacturing.

Cimmerian: Grey, ashen.

Cincinatti Red: An intense orange-red.

Cinereous: Grey, ashen; ash grey; especially of birds.

Cinnabar green: A non-permanent green made from **cinnabar**; also said to be produced by mixing.

Cadmium yellow (or sometimes **chrome yellow**) with **Prussian blue.** Referred to also as **chrome green**, zinnober green and green cinnabar.

Cinnabar red: A permanent vermilion.

Cinnabar, cinnebar, zinnober: Mercuric sulphide used in crystalline form as a red pigment since ancient times; the colour itself, a vermilion; see also **cinnabar green** and **cinnabar red.** Cinnabar was widely used by the Chinese since the third millennium BC and was formerly referred to as **Chinese red**.

Cinnamon: The yellowish-brown of cinnamon; sometimes reddish-brown or greyish-brown tinged with red. Also the colour name given to deep brown diamonds having little value. See **diamond colour**.

Cinq Trous: French lace made with a five sided mesh ground.

Ciporovica: Bulgarian tufted rug, made of native wool on hand looms.

Circassian: English worsted fabric of the 19th century.

Circassian Rugs: All-wool rugs, made in Asia Minor; the short and loose pile is tied in Ghiordes knot, both ends are 'finished with a narrow, knotted fringe; usually elementary colours ara being used, white extensively.

Circlet: Ring of metalor flowers etc, worn around the head or brow, in ramanesque, gothic.

Circular bend: Simultaneous, multidirectional deformation of a fabric in which one face of a flat specimen becomes concave and the other becomes convex.

Circular knit: Thin plain knit fabric made on circular knitting machine.

Circular knit goods: Fabric knitted in tubular form which is mostly also finished in this form and only afterwards slit open if necessary. Circular knit plie fabric: (nicky) Plush fabric of smooth or textured yarns with closed loops (plush loops) similar to terry fabric.If loops are cut, Cut-pile plush is obtained. Ground fabric is of cotton but sometimes reinforced with polyamide or polyester yarns, and pile is of cotton. Special shearing machine is used to the piles which are of about 3-4 mm size.

Circular knife cutting machine: Unit with which printed fabrics etc, particularly printed nonwovens are cut or parted to a specified width.

Circular-knit pile fabric: (Nicky) Plush cloth, of smooth or textured yarns, with closed loops (plush loops), similar to Terry fabric. If the loops are cut, Cut-pile plush is obtained. The ground of the fabric is usually cotton, possibly reinforced with polyamide or polyester yarns, the pile of cotton. Loop size 3–4 mm. Special shearing machines are used to cut the pile (loops). Before shearing, the goods are tumbled.

Circular knitting: The process of knitting the fabric is produced in circular or tubular form on circular knitting machines where the needles are fixed on needle bed which are circular.

Circular knitting machine: A knitting machine incorporates needles which are moved individually by cams acting on the needle butt; they are used for producing weft knits and are subdivided into circular knitting machines and flat-bed knitting machines. The needles used can be latch needles or compound needles. The needles are fixed on needle-bed of a knitting machine and it can be can be flat or circular. The latter ones are called circular knitting machines. Normally circular loop-wheel machines are used for the production of weft knit fabrics, which only use spring-beard needles.

Circular-knit Fabric: Fabric manufactured on a Circular knitting machine.

Circulating liquor machine: A machine for processing the textile where the liquor is in circulation. Eg. Jet dyeing machine, Beam dyeing machine. Here the fabric can be stationary or moving.

Ciré: (Fr. wax, waxed; also oilskin rainwear). A brilliant patent leather effect produced by the application of wax, heat and pressure. Fabrics are mostly impregnated with paraffin or wax emulsions followed by a passage through the Friction calender. The result is a supergloss almost metallic in appearance. (A shiny surface given by a hot calendering alone to a fabric made from thermoplastic fibres is also called ciré).

Cire' finishes: See **Embossed cire' finishes.**

Cire' calender: Calender with slightly engraved embossing bowl for getting calendered effect on the calendered material.

Cire' effects: Embossing.

Ciré: The French word for 'waxed', it was originally a waxed fabric used for shrouds. This treatment of wax in under heat and pressure the fabric surface makes it very lustrous and smooth. Fabric after impregnation in wax emulsion and immediately passed through a frictin calendar. It may be applied to silk. Not readily available nowadays.

CIRFS: Comite' International de la Rayonne et des Fibres Synthe'tiques (International Rayon and Synthetic Fibres Committee). Paris, France.

Cirrho- (G): Tawny, orange.

Cisele': A grainy plain colour silk crepe' fabric with a scroopy handle and good draping property, unlike a Moss crepe in appearance. Shrink resistant, crease resistant and water repellent properties are imparted by finishing.

Cisele' velvet: It is a velvet with pattern formed by a contrast in cut and unscut loops. See **Velvet.**

Cisterns: Open pits constructed by brickwork or concrete, lined with ceramic tiles, used in processing as temporary storage units for piling down fabric in rope form between the individual process stage of alkaline scouring or bleaching.

CITEN: Comite' International de la Teinture et du Nettoyage. (European professional Organisation of Dyers and dry Cleaners.

Citric acid: A solid organic acid; $HOCCOOH(CH_2\ COOH)_2$ (or C_6H8O_7); 2-hydroxy-1,2,3-propanetricarboxilic acid. Citric acid is sometimes used in dyeing as an alternative to other chemicals such as acetic acid. It is convenient to store and handle, but may be more expensive. It is a weak acid but can produce pH in the range of 2 to 3. It can act as a sequestering agent for some metals, so it may be inappropriate for some metal-containang dyes such as premetallized organic dyes. It is used as a resist in some printing processes with reactive dyes on cellulosic fibres, acting by maintaining a pH that prevents the dye from fixing to the fibre. Used as neutralizing agent in special cases and also in pigment printing.

Citreous: A greeny-yellow; lemon-coloured.

Citrin-, citro- (L): Lemon, yellow.

Citron: A browny yellow from mixing green and orange. Also 'citron yellow' a colour often used by Vincent Van Gogh.

Civil: Coarse but strong bast fibre, yielded 'by a species of the Malvaceae in Mexico.

CL: Polyvinyl chloride fibres

Cladding: Any item of covering material which is applied to the head or buck of a pressing or under pressing machine.

clair de lune: A pale blue glaze for ceramics.

Claires: Fine, open batiste, made in France.

Clamp: The part of the testing machine used to grip the specimen firmly during the testing.

Clan tartan: The plaids of the various highland clans of Scotland.

Clandian: Fine, lightweight woollen summer dress goods, consisting of alternating stripes of a dark colour interchanged with a number of lighter coloured threads and stripes of mixture or mottled colours.

Claret: Reddish-violet; of the colour of claret wine; a red having the same hue as bordeaux wine. From the Latin *clarus* meaning 'clear' from which the words 'clarity', 'clarinet' and 'clarion' all derive.

Clasp cloak: Circular cut shoulder cloak,held at the front with a clasp in Gothic period

Class-one Wool: Merino sheep produce the best wool which is relatively short, but the fibre is strong, fine, and elastic and has good working properties. Merino fibre has the greatest amount of crimp of all wool fibres and has a maximum number of "scales": two factors which contribute to its superior warmth and spinning properties. These sheep produce class one wool.

Class-two Wool: Class-two wools are not quite as good as the Merino wool, but this variety is nevertheless a very good quality wool. It is 50-200 mm in length, has a large number of "scales", and has good working properties. This class of sheep originated in England, Scotland, Ireland and Wales.

Class-three Wool: These fibres are about 100-455 mm long, are coarser, and have fewer "scales" and less crimp than Merino and Class-two wools. As a result, they are smoother, and therefore, they have more lustre. These wools are less elastic and resilient. They are nevertheless of good quality to be used for clothing. This class of sheep originated in the United Kingdom

Class-four Wool: These fibres are from 25-400 mm long, are coarse and hair like, have relatively few "scales" and little crimp, and therefore, more smoother and more lustrous. This wool is less desirable, with the least elasticity and strength

Classical: The second best grade of raw silk.

Classification of dyes. There are two types of classification of dyes: One based on chemical composition and other based on uses.

Classification according to chemical constitution to textile usage	
Azo dyes	Acid dyes
Anthraquinone dyes	Azoic dyes
Heterocyclic dyes*	Basic dyes
Indigoid dyes	Direct dyes
Nitro dyes	Disperse dyes
Phthalocyanine dyes	Mordant dyes
Polyme thine dyes	Pigments
Stilbene dyes	Reactive dyes
Sulphur dyes	Sulphur dyes
Triphenylmethane dyes	Vat dyes
*Includes a number of different subgroups containing heterocyclic system (only the most important subgroups in each classification are given).	

Classification of Wool by Fleece: Wool shorn from young lambs differs in quality from that of older sheep. Also, fleeces differ according to whether they come from live or dead sheep, which necessitates standards for the classification of fleeces.

Classiques: French, stout, all-cotton sheetings.

Clauder – Weldon machine: One of the earliest loose cotton dyeing machine. It consists of a trough with a slope at one side and contains the dye solution. A cylindrical perforated drum of of copper is provided in the trough. The flanges at the drum carry a shaft The drum is divided into four compartments with per forated partitions and each of the compartment is provided with a manhole for filling and removing the material. The liquor is heated in the trough while the drum is rotated. The liquor enters the drum and the cotton is dyed. Fingers provided in side the drum on its shaft prevents the slippage of the cotton and keep it in open state.

Clavi: Decorative strips used as edge trimmings, mostly purple dyed in ancient Rome during Byzantine period.

Claws: Clips.

Clay: (1) A light brown. (2) A name given to serges, worsteds, and diagonals woven after a process of J. & P. Clay of Haddersfield, England. Old term.

Clays: Originally English worsted serge for men's wear, made of slack twist warp and filling, woven with six-harness twill, forming very flat diagonals and finished with a soft 'but clear handle.

Clear Finish: Fabrics having the nap removed from the face, showing the weave.

Clean-finish seam-finish: A finish for the raw edges of the seam allowances of a plain seam, in which the raw seam allowance is folded under the edge stitched.

Clean Content (Wool): The amount of clean, scoured wool remaining after removal of all vegetable and other foreign material.

Clean wool fibre present, in raw wool: the weight of wool base present adjusted to a moisture content of 12% , an alcohol extractives content of 15% and a mineral matter content of 0.5%.

Cleaning agent: A chemical compound or formulation of several compound which loosens, disperses, dissolves, or emulsify soil to facilitate removal by mechanical action.

Cleaning efficiency (CE), in spinning: The cleaning efficiency (CE) is the percentage of the impurities removed from the fibre mass. It is calculated as per the following formula,

$$CE = 100.(W_{in} - W_o)/W_{in}$$

where W_{in} and W_o are the mass values of the impurities in the feedstock and the processed material, respectively at the input and output to a machine or a sequence of machines, and CE is the cleaning efficiency.

Cleaning head, in carpet cleaning: A vacuum head modified with spray nozzles for cleaning solution application. Some types include a powered brush unit to facilitate wetting and soil release.

Cleanness, in silk: The degree of cleanness in raw silk panels, determined on the basis of incidence of cleanness defects classified as super major, major and minor defects by using official standard photographs for cleanness. Cleanness is expected as percentage.

Cleanness defects, in silk: defects found in raw silk such as waste, slugs, bad casts, knots, cork screws, loops and loose ends.

Clear Finishing: Usually, worsteds are not brushed, but closely sheared to give the fabric a clean face and crisp feel. This is called clear finishing. See also **unfinished worsteds**.

Clearing position (knitting position), in knitting: The needle butt is still pushed upwards by the cam and the old loop is on the needle stem and behind below the needle latch.

Clearing agent: A chemical compound formulation of several compounds which loosen, disperse, dissolve or emulsify soil to facilitate removal by mechanical action.

Clearing (Yarn): The process of removing imperfections from a yarn. The fault normally replaced by a yarn joint knot knot or splice.

Clem: A very small dart which is sewn and not cut.

Clematis: A bright bluish violet similar to the colour of the flower of the same name.

Cleopatra: A vivid greenish-blue.

Clergy tartan: A tartan formerly used by the Scotch Highland clergy for every day wear. It consisted of wide dark blue and black bars of even width. The -black bars were split in the middle with a fine green stripe. The blue bars were alternately split by a pair of fine green stripes or five fine green stripes. The latter arranged in pairs at the edge and one in the middle. All green stripes were edged with a fine black line.

Clevyl: The trade name for a range of flame resistant fabrics manufactured by Societe Rhovyl and which are based on Rhovyl's PVC chlorofibre. Used for furnishings in public buildings, offices etc. On account of their exceptional safety and also because of the wide range of eaves and effects that is possible.

Cling test: Measuring instrument which measures the adhesion of a test fabric charged with static electricityby friction on ametal plate Time taken for the test specimen to fall is measured.

Clingy: A term used in USA to describe the undesireable adhesion of the fabric to the body during wear.

Clinquant: Flat gold braid for military uniforms.

Clip: One season's yield of wool from a single angora goat.

Clip, in sewing: Clip is to make small cuts in fabric to allow ease on curves or corners.

Clip mark: A defective lighter or darker mark on the on the fabric when heat setting of synthetic goods, due to the subsequent variable dye uptake. It can be enlarged pinholes, tears or shift marks along the edge of the fabric.

Clippings: Irregular small pieces of cloth, the waste of the cutting room; used for patch work, quilts, or are reworked into shoddy.

Clips: Continuous holding devices for anchoring woven fabric selvedge (stenter) during drying, heat setting shrinking, mecerising etc. enabling fabric to be tensioned or relaxed as necessary.

Clip spots: Small spots of extra warp and/or weft between which the floating material is clipped or sheared of after weaving.

Clip mark: A visible deformation near the edge of a fabric parallel with the lengthwise direction caused by pressure excerted by a clasping device on a clip stenter frame.

Clo: Unit of thermal resistance defined as the insulation required to keep a resting man (producing heat at the rate of $58w/m^2$) comfortable in an environment at 21°C, air movement of 0.1m/s or roughly the insulation value of the typical indoor clothing. One unit of clo is equal to 0.155 Km^2/W.

Cloaking: Large variety of woollen, worsted or silk fabrics used for cloaks or coats; usually of heavier weight, as chinchillas, sibeline, etc.

Clochepied: Three-thread organzine for gauze; made by twisting first two threads together, then adding the third thread.

Clock: Embroidery at the ankles of the hosiery, consisting of two lines meeting at an angle, one line going up the leg, the other toward the toe. In French is called coin.

Clock Reel: A device for winding hanks of yarn. Some come with various kinds of counters. An image of a clock reel can be found at the Illinois State Museum site.

Cloque: The French word for blistered. The fabric with raised figures regularly or irregularly is aptly called by this name. Originally this fabric was made out of silk but now it is made in acetate or polyester yarns nand are imitations of the original stuff. In the new form the blistering is more permanent thanthe originals. There are three types of cloques: I. Woven cloque, II Chemical cloque, III. Bonded cloque. Used for evening cocktail wear, some millinery, negligees, etc.

Cloqué fabric: See **Cloque.**

Cloque woven: A compound or double fabric with a figured blister effect brought about by the use of yarns of different character or twist which respond to different ways to finishing treatments.

Close Leaf: In bobbin-made sprigs leaves entirely filled out with cloth stitches.

Close Stitch: In needle-point laces all buttonhole stitches made without any loops.

Closed-face fabric: A face or shell fabric of closed construction so that no open face areas appear.

Closed feather stitch: An embroidery stitch in the shape of a zig-zag stitch in between two embroidered lines. This stitch is worked along two parallel lines. Bring the thread through at starting point and, with the thread under the needle. stitch from to the opposite side to the next point corresponding to the starting point. Swing the thread over to the left and, with the thread under the needle, take a stitch from point opposite to the starting point and to the next point opposite to the second point corresponding to the starting point. Repeat these two stitches.

Closed lap: In knitting, if an overlap and the next underlap are made in opposite directions to each other, a closed lap is then produced. A closed lap is heavier, less extensible, and more compact than an open lap.

Closing: A sewing operation to to seal a bagged out garment section.

Closures: Items used to close openings in apparel and other consumer textile products, i.e. buttons, buckles, hook and eye, snaps and zippers.

Cloth: A piece of textile fabric but especially one designed for apparel, domestic, or industrial use. This word is used to describe any type of fabric, but perhaps most used in connection with woollens and worsteds- the suitings and coatings.

Cloth (in garment): In garment industry the principal outside fabric is called as the cloth; a piece of fabric suitable for specified use.

Cloth of Areste: Medieval rich fabric woven with gold figures; used for church vestments.

Cloth beam: Serves as a take up device on the loomfor the finished (woven) textile. Also called Cloth roller. See **Cloth fell.**

Cloth blanket: Made in plain weave and slightly napped.

Cloth of bruges: Gold brocaded silk fabric; used for church vestments in England during the Middle Ages.

Cloth of Cologne: Medieval fabric made at Cologne, Germany, with gold patterns over blue or other coloured silk foundation.

Cloth control: Another term for Take-up.

Cloth Embroidery: In Asiatic countries work made of joining together various shaped and coloured small pieces of cloth with the aid of fancy stitches.

Cloth fell: A term in weaving technology. During the weaving operation, the warp unwound from the weavers beam passes round the backrest (back bearer)and comes to the heald frames (harnesses), which are responsible for separating the warp sheet for the purpose of shed formation, it then passes through the reed (swinging frame in front of the heddles) which holds the thread at uniform spacing and is also responsible beating up the last inserted pick. All the thread unite again in the fabric at the point of cloth fell. The cloth then passes over the front rest (breast beam) round the take-up roller and wound over the cloth roller (cloth beam or merchandise beam).

Cloth of Gold: Fabrics mentioned in old manuscripts, made entirely of gold wire, narrow flat strips of gold, or mixed with other fibres, as silk.

Cloth guide: Fixed guiding apparatus, which operates on a length of fabric passing over it.There are retarder guiders, expander guiders, or deflector guides depending in the use.

Cloth Ingrain: Medieval English worsted, worn by the rich.

Cloth of Pall: Rich medieval silk fabric dyed in crimson; used for church vestments.

Cloth of Tars: A costly medieval fabric, believed to 'be of silk and cashmere wool; usually dyed purple.

Cloth of Raynes: Fine medieval linen, used for shirts and bed linen.

Cloth Serge: An 18th century English serge, made of worsted warp and woollen filling.

Cloth Stitch: In bobbin laces interlacing the threads like the weaving of a cloth.

Clothes Moth: The larvae of this moth eats wool and other protein fibres. Various articles on fibre pests can be found here.

Cloud point: Cloud point is defined as the temperature at which a 1% solution of a nonionic surfactant forms a cloud of insolubility. Cloud point is influenced by the structure of the hydrophobe and the degree of ethoxylation. Temperature too has an influence on hydrogen bonds, as the temperature increases, H-bonds rupture. At some elevated temperature, enough of them will have broken to cause a cloud of insolubility to form. Eventually the surfactant will separate into an insoluble layer floating on water. Cloud point of a given hydrophobe is influenced by the number of ether units available to hydrogen-bond with water. Cloud pint occurs in the case of nonionic surfactant only.

Cloud point index: Cloud point index corresponds to the water in ml that will turn cloudy a solution of 1g of the surfactant in 10ml of n propyl alcohol at 30 C. The method has good reproducibility but is very sensitive.

Cloudiness: (1) In a weft-knitted fabric, a defect that consists of ill-defined areas of varying density attributable to the use of yarn of irregular thickness. (2) In webs and slivers, a defect that consists of ill-defined areas of varying density. (3) In a dyed fabric, a defect that consists of random, faintly defined areas of varying density. (4) In a bleached fabric, a defect that consists of opaque patches, usually visible only in transmitted light. (5) Defect in yarn, showing thin and thick places.

Cloudy Yarn or cloud yarn: (1) Fancy ply yarn, having flakes held by the strands; (2) Ply yarn with irregular twist, made by alternately twisting the different coloured strands around each other.

Cloudy dyeing: Unlevel, mottled appearance of a dyeing.

Cloudy Wool: Wool that is off-colour. It may be due to wool becoming wet while poorly stored in a pile.

Clouties: English linen cloth.

Clove: English measure of wool, equal to 7.7 pounds.

Clove brown: A greyish brown.

Clover: A pinkish purple.

Club check: Small checks in quiet two coloured effects.

Cluett process: Compressive shrinking process. Most of the shrinking machine for woven runs on this method.

Cluny guipure: Hand-made lace, the patterns being worked over square netting ground, similar to the darned net laces.

Cluny lace: A close, heavy bobbin lace made in France and Belgium, often with a wheel design. Its heaviness limited its use to collar decoration in dress but it was much used for mats, doilies and table-ware.

Cluny tapestry: Stout, thick fabric made in England with woollen warp and silk filling, forming warp cords; used for hangings.

Clydella: Trade name for a traditional, soft woven mixture of 81 % and 19% wool in plain or print. It wears well, is soft and warm and washes easily. Used for shirts, blouses, nightear, babies and childrens clothes.

CMC: (1) Carboxy Methyl Cellulose: Critical Micelle Concentration.

(2) *(Colour Measurement Committee of the Society of Dyes and Colourists of Great Britain)* - Organization that developed and published in 1988 a more logical, ellipse-based equation based on L*C*h° colour space for computing DE (see **delta E***) values as an alternative to the rectangular coordinates of the CIELAB colour space.

(3) Critical Lamellar Concentration. See **Micelle.**

CMC unit, in colour difference Evaluation: A measure of acceptability expressed in terms of the boundary for the CMC acceptability ellipsoid of $\Delta Ecmc = 1.0$.

CM Cotton: Cotton treated with monochloro acetic acid and then sodium hydroxide is conveted in to CM cotton which is having a 'starched' appearenceand handle / or CM cotton which can easily disintegrate in water.

CMD: Modal fibres

CMT: Abbrev. for Cut, Make and Trim. A section of industry in which a contractor is supplied with materials and designs in order to produce a garment for the principal.

CMYK (or CYMK): An acronym for **c**yan, **m**agenta, **y**ellow and blac**k** – the four colours which are used in colour photography and printing.The letter 'K' is used to avoid confusion with the 'B' in **RGB**and also refers to black as the '**k**ey' or '**k**eyline' colour. Although black can be obtained from mixing the other three colours it is produced in printing by using a black ink. CMYK uses the **subtractive process**. Compare RGB.

CNS: Spanish Standards Organisation.

CO: Cotton.

Coagulant: Chemical added in the water and waste water applications to cause formation of flocs that absorb, entrap, or otherwise bring together suspended matter defined as colloids. Compounds of iron and aluminium are generally used to form flocs to allow removal of turbidity, bacteria, colour, and other finely divided matter from water and waste water.

Coagulation: Coagulation or flocculation of colloidally dissolved particle by accumulation, flocculation and small lump formation and sttling to the bottom as an insoluble gel. This Process is extensively used in Treatment of Textile effluent.

Coagulation bath: Bath used for manufacturing protein fibres. In the production of the Man made Protein fibre the protein of vegetable of animal origin is used. The fibre is spun by wet spinning from coagulated bath of protein by adding suitable coagulant e.g., Sulpuric acid, sodium sulphate.

Coalescence: Confluence of the droplets of an Emulsion. In most cases, aggregation precedes coalescence.Coalescence leads to breaking of the emulsion.

Coal Tar Colours: Colours obtained by distillation and chemical treatment from coal tar, a product of coal during the making of gas. There are over 2,000 colours in use.

Coarse end: A warp yarn with a larger diameter than normally used in the fabric.

Coarse filling: See **Coarse pick.**

Coarse filling bar: See **Filling bar, coarse.**

Coarse pick: One or more weft/filling yarn with larger diameter than normally used in the fabric.

Coarse thread: see **Coarse end.**

Coarse tow: The entire fibre of flax, after the shives have been removed.

Coarse Wool: Wool that has a Blood grade of 1/4 or Common OR a numerical count grade of 44's, 45's, or 48's, OR a micron count above 31. Coarse wool may have as few as 1 to 5 crimps per inch.

Coat: An outer garment which covers at least the upper half of the body, have sleeves and a front opening , and is usually worn over another garment , such as a dress or a shirt.and often as in work wear, described further by length. E.g. ¾ th or full length.

Coated cotton: A thin covering of polyurethane on cotton or viscose backing producing a soft showerproof fabric. May be plain or printed. Used for shorts, anoraks, raincoats, jackets etc.

Coated fabric: A flexible material composed of textile fabric produced with one or both surfaces, a layer or layers of firmly adherent coating material, with coating that are homogeneous or cellular in structure, E.g. Foam laminated Textiles, Artificial leather.

Coated Fleeces: Some wool producers coat their fleeces that cut down on the amount of vegetable matter and weathering.

Coating: One or two sided application of coating compounds or foam films to woven fabric, knitted goods or nonwoven fabrics or yarns by means of coating machine, roatary screen prining or by means of sparay, hot melt and transfer coating methods.

Coating, Calender: A type of roller coating that is actually a laminating operation. The coating is formed into a sheet, then joined with the fabric.

Coating, cast: A method by which resinous materials such as vinyl are coated onto a fabric and cured by heated casting drums.

Coating, dip: The process of passing a fabric through a solution of resin or elastomer, then through squeeze rolls to remove excess and leave a thin surface layer on the base fabric. In this process, both sides can be coated in one pass.

Coating, direct: The simplest method of coating, this procedure involves spreading the coating with a knife. The moving fabric substrate is usually supported by a roller or a sleeve. The gap between the knife and the fabric determines coating thickness.

Coating Compounds: Are used for coating purpose on textile base fabrics like, starch paste, cellulose derivatives, natural and synthetic rubber, plastics like polymethacrylate, polyamide, polybutadiene, polyisobutylene, polyurethane, polyvinyl acetate, polyvinylchloride, foamable plastics and elastomers for special coatings.

Coating knife: Also known as doctor blade is a steel blade sharpened at the application end or a rubber blade which is less precise in it application.

Coating machine: Used for coating of coating compounds on to the fabric. Consists of compound trough, coating head with adjustable height and guide roll systemand dotor blades.

Coating machine for rotary screens: The machine used for coating the photographic chemicals (for exposure by light) or other coating chemicals (for laser or jet engraving) on the screen.

Cobalt: Indicating pigments made from salts of cobalt, for example, **cobalt blue** and **cobalt green**. The word cobalt comes from the German word *kobolds* meaning a ghostand was used in Bohemia to describe the pigment because it was considered to be plagued by spirits.The blue colouring agent in **smalt**.

Cobalt blue: A deep greenish-blue or purplish-blue similar to **Prussian blue**. Used in the staining of glass. A synthetic pigment also known as **Thénard's blue, Dresden blue** and **King's blue** first made in 1802 by the chemist J-L Thénard (1777-1857).

Cobalt green: A bright yellowish-green discovered in 1780 by Rinmann and also referred to as 'Rinmann's green'.

Cobalt turquoise: A blue made from cobalt titanium oxide.

Cobalt violet: A very light-toned toxic violet pigment first used in 1860 and superseded by manganese violet.

Cobalt yellow: A yellow pigment otherwise known as **aureolin**.

Cobbler: In England piece goods which have been returned because of unsatisfactory make or dye.

Cobbling: Redyeing of defective pieces of woollen fabrics that are off shade or are uneven in shade.

Cobweb: (1) Name applied to very sheer hosiery; (2) A commercial variety of late maturing cotton from 'Mississippi, the very fine and silky staple 'measuring 35-40 millimeters; the yield is '28-29 per cent.

Cob-web finish: A synthetic pilling effect achieved on stitch bonded fabric for coatlining through raising, steaming, and subsequent tumble drying.

Cobourg or Coburg: An English fabric introduced after the marriage of Queen Victoria; made of closely placed silk or cotton warp and worsted filling, woven in a 2/1 cashmere twill weave; used for coat lining and dress goods. They come dyed in the piece or printed.

Coccin- (L): Scarlet.

Cochineal: A fiery red Natural dyes (mordant dye) consisting of the dried bodies of the female insects of Coccus cacti, another species of the shield-louse family, largely cultivated in Mexico. 1 kg of dye = approx. 140 000 insects.

Cochineal red: The red colour of **cochineal.**

Cochinillin: The colouring matter of **cochineal** also called **carmine**.

Cochran: A commercial variety of cotton from Georgia, the staple measuring 35-40 millimeters; the yield is 32-33 percent.

Cockle, in fabric: The crimped, rippled, wavy or pebbled appearance of a fabric where distortion of the structure has occurred as the result of non-uniform relaxation or shrinkage.

Note: This defect may result from variations in the tension of the ends or picks at the time of weaving, from variations in the degree of stretch imposed on the yarn during earlier processes or from the differences in contraction of two or more yarns used accidentally or intentionally in the fabric. The defect may be distributed over a large area of fabric or may be confined to isolated stripes, bars or streaks.

Cockled bar: See **Tension bar.**

Cockled yarn: Irregular, thick, uneven, lumpy spots in the yarn. Spun yarn in which some fibres do not lie parallel to the other fibres but instead are curled and kinked, forming a rough and uneven surface on the yarn. The general cause is fibre overcut to the extent that the drafting rolls catch and hold both ends of the fibre at the same time while attempting to draft, resulting in slippage or breakage.

Cockles, in yarn: Irregular, thick, uneven lumps.

Cockling: A crimpiness/curliness or pucker in yarn or fabric usually caused by lack of uniform quality in the raw material used improper tension on yarn in weaving, or weaving together yarns of different numbers.

Cockling, in knitting: An irregular surface effect caused by a loop distortion.

Cockspurr willey: See **Willey.**

Cocoa, cocoa brown: The colour of cocoa; a colour particularly popular in the 1930's.

Coconut oil: The major fatty acids found in this oil are lauric (48%) and myristic (18%). Both acids have shorter length chains (C12 and C14) and are saturated. The laurel chain length is ideal for high foaming soaps. Its potassium salt has higher water solubility acids and is used to make liquid soap products.

Cocoon: An oval shell, consisting of raw silk, spun by the silkworm around itself. The outer and inner layers can be used only for floss or spun silk, while about 10 per cent of the weight can be reeled off in a single filament, averaging about 300 yards in length. About I2 pounds of cocoons will yield one pound of raw silk. If the chrysalis is permitted to pierce the cocoon and escape, the eilk can be used only for spun silk.

Cocoon perce's: Pierced or damaged cocoons whch cannot be normally reeled (also termed cocoon piques).

Cocoon strippings; Blaze; Keba: The first thread secreted by the silk worm when it finds a place to form its cocoon.

Cocoonage: Appliance used for mounting mature silkworms to enable them to spin cocoons.

Cocos Fibei: Same as **Coir.**

Cocuiza: A strong, smooth fibre, yielded by the leaves of the Furcraea gigantea in Venezuela; used for ropes, bags, etc.

COD: Chemical Oxygen Demand.

COD In G/Kg of fabric: Pollution Index.

Codilla: The scutching tow, a coarse byproduct or waste of the flax and hemp.

Codpiece: Pouch in the crotch of mens breeches used in late gothic and renaissance period.

Coefficient of friction: The ratio of the tangential force that is required to start or to maintain uniform relative motion between two contacting surfaces to the perpendicular force holding them in contact.

Coefficient of length variation: A measure of fibre length distribution. The standard deviation expressed as a percentage of mean length.

Coefficient of variation, (CV): A measure of the dispersion of observed values equal to the standard deviation for the values divided by the average of the values; may be expressed as a percentage of the average (CV%).

Coefficient of variation in unevenness: (Vu) The square root of the average of the squares of the deviation of linear density from the average linear densityexpressed as a percentage of the average linear density within the tested length of the strand.

Coeur fleuri: French ticking and linen, woven with small geometrical figures and bleached.

COFREET: Comite Francaise de l'Etiquetage pour l'Entretien des Textiles. French committee for care labeling.

Coffee: Having the colour of coffee.

Coffin cloth: Made with cotton warp and woollen filling in plain weave, dyed black; used for lining coffins and for shrouds.

Cognac: A pale brown; the colour of Cognac brandy.

Cogware: A coarse English woollen fabric of the 15-th century, made like a frieze.

Cohesion: The force that holds fibres together during yarn manufacturing or processing. It is usually a function of lubricant (type and amount) and fibre crimp.

Cohesion, in silk: The degree of agglutination of cocoon filaments forming raw silk thread, determined by subjecting the raw silk threads to friction in a cohesion tester and counting the number of strokes required to spread and open out constituent filaments.

Cohesive force: In sliver and top testing. The load or force required to overcome the cohesion of a lot specimen held in a fixed position between two slowly separating clamps.

Coiler: A delivery device that deposits a sliver in to a cylindrical can in the form of helical coils so as to permit easy withdrawal with the minimum of fibre disturbance.

Coiling: The depositing of sliver into cylindrical cans in helical loops. This arrangement permits easy removal for further processing.

Coir: Reddish brown, strong, stiff and elastic fibre, yielded by the outer shell of the coconut; used for mats, cordage, etc.

Colbertan lace: Coarse French lace named after Colbert, with ground of square meshes.

Colcothar: A reddish-brown iron oxide pigment; used as theatrical rouge and to polish gold and silver. See **Jeweller's rouge**.

Cold colours: As regards painting, those colours such as blue, green and grey suggesting coldness compared with **warm colours** such as orange, red and yellow. Cold colours appear to retreat towards the background. Also referred to as 'cool colours'.

Cold dyeing dyes: A term denoting any dye which can dye at room temperature. Reactive dyes of high reactivity like dichloro triazine type. Due to the high reactivity of these dyes it can be dyed in cold (Room Temperature) from an alkaline bath.

Cold pad batch bleach: Peroxide bleaching process characterised by a high concentration of hydrogen peroxide (> 40 ml/l) and a long reaction time (24 h). Underthese conditions, bath stabilisation is particularly important, and silicates, phosphates or organic stabilizers can be used for this purpose. Hot washing off with the addition of alkaline surfactants must be implemented in order to obtain absorbent fabric. Dwell time can be reduced (to 4–6 h) if a 1 minute steaming process follows storage. Peroxide-padder-dwell process.

Cold pad batch dyeing of reacive dyes: Principle: dye and alkali solution are prepared separately, and reach the padder trough only after passing through a mixer. Impregnated fabric dwells a few hours batched in open width, during which the dye is fixated. The so-called CPB pump is used as a metering, conveyance and mixing unit.

Cold peroxide bleaching: See **Cold pad batch bleach.**

Colibri: A bluish-green also known as 'humming-bird'.

Collapse Yarn: Collapse yarn is (usually) an over spun single, dried under tension (see **"blocking"**) that is then knit or woven. When the item is moistened, the yarn returns to its original elastic state. See **"balanced yarn"**.

Collar: (1) A neck band, upright or turned over a coat, dress, shirt etc.

(2) Band of lining, lace or other material which completes the upper part of a garment around the neck.

Collar (Banded): The visible or panel portion of the collar is cut separately and attached to the neckband portion. This is normal dress shirt construction.

Collar (convertible): The panel or visible portion of the collar and the neckband portion are cut as one piece, but folded once along the length to produce the appearance of a banded collar.

Collar (Lined): A collar made by placing a piece of interlining between the two pieces of body fabric.

Collar (one piece): A collar constructed from a single piece of fabric with the center fold forming the outer edge.

Collar (padding): Attaching the under-collar to canvas with several rows of blindstitching.

Collar (sandwich): A collar which has the top-collar inserted between the canvas and the under-collar.

Collar (topstitched): A collar with an added row of stitching along the folded edges.

Collar (two-piece): A collar formed by joining two identical pieces, inverting and sometimes topstitching along the folded edges.

Collar stand: That part of a collar whch joint the neck edge of the garment and terminates at the crease of the collar where the fall commences.

Collar velvet: Fine silk velvet of very short pile; used for collars.

Collar, Asymmetrical Straight Band: A straight band of fabric with the fastening to one side. There is a placket supporting the button fasten.

Collar, Bateau: A straight band placed on a wide neckline.

Collar, Bishop: A shirt-type collar with points extending into tab shapes and no fastening down the front.

Collar, Cossack: A narrow band around the neck that is fastened on one side. It is similar to the asymmetrical straight band.

Collar, Mandarin/Nehru: An Oriental collar that is a straight band, but that opens at the front and has curved edges. Also popular in India but generally the edges are straight rather than curved.

Collar, Peter Pan: A flat, curved shaped collar with no stand.

Collar, Polo/Turtle/Roll: A wide piece of fabric, usually knitted or cut on the cross, that folds back to form a roll neck collar.

Collar, Ring: A band of fabric that is shaped in towards the neck.

Collar, Straight Band: A collar made of a strip of fabric encircling the neck.

Collecting surface, in the rotor of an open end spinning: That portion of the internal surface of the rotor, often in the form of a grove, in which the fibres are condensed for assembly intom yarn.

Collette: Unbleached canvas of medium grade.

Colligative properties: Properties of a solution which are dependent on the number of solute molecules present.

Colloid: A substance of very fine particle size, typically between 0.1 and 0.001 microns in diameter suspended in liquid or dispersed in gas. Typically removable by reverse osmosis, distillation or ultrafiltration.

Colombe: The grey of the dove.

Colombiana: In the Philippine Islands worsted piece dyed lastings.

Colombo: Yarn Second grade of coir yarn; made in Sri Lanka; used for ropes and Mats.

Colloresine process: The two stage printing process for vat dyes is based on printing vat dyes from a thickener (originally Colloresine thickener DK) which undergoes coagulation on subsequent application of alkali and heat (steam). After printing and drying, the goods are padded with an alkaline solution (sodium carbonate or caustic soda) of a reducing agent such as Rongalite C (sodium formaldehyde sulphoxylate) after which they are steamed, with or without intermediate drying, in a conventional steamer for 3–8 minutes. Due to the high degree of coagulation of the thickener, no marking-off or smearing of the printed colours occurs in the steamer even though the fabric is in contact with the guide rollers on both sides. This process has now been almost completely replaced by the more economical Flash ageing.

Colour: One aspect of appearance; a stimulus based on visual response to light, consisting of the three dimensions of hue, saturation and lightness. The particular appearance or attribute of anything visible to the eye varying according to the wavelength of light reflected from its surface to the viewer. Colours are seen by humans as combinations of the three **primary colours**. Colours are distinguished from each other by reference to **brightness, hue** and **saturation**; a particular hue, such as red or blue, also referred to as **chromatic colour** or **spectral colour;** the variety of tone or quality of the music of voices or instruments; a dye or paint or other substance whereby colour can be imparted; an appearance of truth or semblance of something, see **colourable; vividness, interestingness**.

(a) The characteristic of the visual sensation that enables the eye to distinguish differences in its quality, such as may be caused by differences in the spectral distribution of light rather than by differences in spatial distribution or fluctuations with time.

(b) As (a) above, but applied directly to the stimulus or the source (primary or secondary) giving rise to the sensation. (For brevity, the stimulus is often referred to as the colour).

(c) The property of an object or stimulus or quality of visual sensation, distinguished by its appearance of redness, greenness, etc., in contradistinction to whiteness, greyness, blackness (i.e. chromatic colour is contradistinctinve to achromatic colour).

Colour: The US spelling of the word ‘colour’. In British English spelling most compounds of the word ‘colour’ include the letter ‘u’ as in ‘colouring’, ‘colourless’ and ‘colourant’. However, some derivatives omit the ‘u’ as in ‘coloration’, ‘colorimeter’ and ‘decolorize’.

Colour (of an object): The particular visual sensation caused by light emitted by, transmitted through or reflected from the object.

Colour abrasion: Colour changes in localized areas of a garment due to differential wear, such as the knees of blue jeans. Often evident in cross-dye shades of blends where durable press treatments are applied. Colour abrasion is often called “frosting”.

Colour affinity: See **Dye affinity.**

Colour anomaly: Anomalous colour vision.

Colour attribute: A three-dimensional characteristic of the appearance of an object. One dimension usually defines the lightness, the other two together define the chromaticity.

Colour balance: The relationship between different colours in juxtaposition.

Colour bleeding: The loss of colour from the fabric to water, dry cleaning solvent or any other liquid medium when immersed in that liquid with subsequent colouring of that medium..

Colour blindness: The inability to recognise or distinguish any or all of the colours red, green and blue. Colour blindness affects males more than females in the proportion of 20 to 1. Whilst the human eye has red, blue and green **cones** almost all other mammals (the exceptions are apes and Old World monkeys) have only blue and green (or blue and yellow) **cones** although dolphins and whales cannot see the colour blue. Giraffes are unable to differentiate between yellow, orange and green and, apparently, horses, bulls, hedgehogs and mice, amongst others, cannot distinguish colour, having no cones in the retinas of their eyes. Parrots can group objects by reference to their colour. Total colour-blindness consists of **rod** vision with the ability to see only shades of dark and light with no perception of the colours of the **spectrum.** See **dichromacy, trichromacy, achromatopsia** and **monochromacy**.Severe form of Defective colour vision. There are two type of colour blindedness. (a)Partial colour blindedness, (b) Total blindedness.

Colour box: A rectangular trough-shaped container for print paste in a roller printing machine.Print paste is transferred from the colour box via a furnisher roller to the engraved printing roller and to the fabric as a print.

Colour change, as used in colourfastness testing: a change in colour of any kind whether a change in lightness, hue or chroma or any combination of these, discernible by comparing the test specimen with a corresponding untested specimen.

Colour change grey scales: These scales consist of five pairs of grey coloured material numbered from 1 to 5. Number 5 has two identical greys, number 1 grey scale shows the greatest contrast, and numbers 2, 3 and 4 have intermediate contrasts. After appropriate treatment the specimen is compared with the original untreated material and any loss in colour is graded with reference to the grey scale. When there is no change in the colour of a test specimen it would be classified as '5'; if there is a change it is then classified with the number of the scale that shows the same contrast as that between the treated and untreated specimens.

Colour code or coding: A method of indicating differences by the use of different colours in a large number of situations. Examples include, flags used for signalling in shipping; traffic signals; railway signalling; pavement and street markings; safety warnings on machinery and equipment; warnings on vehicles carrying dangerous substance; directional indications; sizes of clothes; the wiring of electrical plugs and teaching children how to read. An interesting application has been adopted in Gwynnedd Hospital in Wales where bracelets of different colours are worn according to whether patients wish to speak Welsh, English or both. Using colour and colour coding is a powerful method of enhancing memory. See also **colour wires, colour symbolism, compass colours** and **signal red**.

Colour combination: The conception of a scheme for the effective combining of different colours in design, decoration or clothing etc.

Colour computer: Together with a scanner, corrects hue and colour value in connection with an optoelectronic scanning process.

Colour constancy: The ability of a coloured object to give the same general colour impression which viewed under different illuminations, the observer having been chromatically adapted in each case.

Colour contrast, in textiles: A general term for a visible colour difference between two adjascent areas.

Colour coordinate system: Chromaticity diagram with the components of red (x), green (y), blue (z) as co-ordinates.

Colour co-ordinated: Particularly as regards clothing and home furnishings, all aspects effectively combining or harmonising in accordance with a **colour scheme** or plan.

Colour correction: An electronically controlled process which takes place on a scanner after colour seperstion with the aid of the colour computer.

Colour craft process: See **space dyeing**. Principle is based on an ombre dyeing of wool hanks (5-6 shades per hank circumferance) after dyeing the yarns are twisted to produce multicoloured effects with 40-50 shades.

Colour difference formula: See **colour difference**. A mathematical value is used to correct the variations in colour in colourimetry.Many formulas have been derived such as Judd formula, Nickerson-Stultz formula, McAdam formula, Friele MacAdam Chickering formila, Simon-Goodwin charts, Adam Nickerson Stultz (ANS) formula, CIELab formula, The CMC formula, the last two are the most used today.

Colour differences: Two identical colours will have the identical CIE tristimulus value X,Y and Z in the CIE colour coordinate system. Different

CIE tristimulus values correspond to two different clours. But the difference in colour coordinate may not match with the colour difference. Hence using a formula the defference in colurco-ordinates can be recalculated into one number that is a measure of the visual colour difference. This measure also is known as the colour difference DE.

Colour dynamics: The concept of the movement of colours. Charles Henry in his *Cercle Chromatique in 1889* regarded red as moving upwards, blue as moving to the left and yellow as moving to the right. Colours have often been employed as an indication of direction (North, South, East and West) particularly in Mexico although without any degree of uniformity. See also **colour symbolism** and **compass colours**.

Colour embossed finishes: (Emboss printing) also represent a valuable option for creating sophisticated designs.

Colour fastness: It refers to the resistance of a dyed or printed textile to various types of influenzes to which they are normally exposed during the manufacturing process and practical use like light, perspiration, atmospheric gases, or washing that can remove or destroy the colour. A dye may be reasonably fast to one agent and only moderately fast to another. Textile materials often must meet certain fastness specifications for a particular use. Colour fastness tests are done in the prescribed methods and assessed for changes in shade and also staing on undyed fabrics as described in th test method with the help of grey scale.

Colour fastness to acids: This fasness is for assessing the resistance to the effects of dilute organic or mineral acids. Three types are used with increasing severity of effect. The Standards apply to all fibre types. Treatment types: a) 100g/l crystallized tartaric acid; b) 300 g/l acetic acid 100%; c) 50 g/l sulphuric acid 60^{0}Bé. Acetate is only tested with a) and c). Acid dilutions are prepared with distilled water. Test samples are treated with two drops of the dilutions. Assess any change in colour after 10 min reaction time at room temperature. Assessment: grey scale/change of shade.

Colour fastness to burnt gas fumes G02: This method determines the resistance to fading of dyed fabrics to the nitrogen oxides produced during the combustion of butane. In the test the sample and control fabrics are exposed to the fumes of a butane burner in a test chamber. The test is continued until the control fabric has faded to a given extent. The sample is then assessed for fading.

Colour fastness to cross-dyeing: See **Fastness to cross-dyeing.**

Colour fastness to Chlorinated water (swimming-pool water): The specimen is agitated in a wash-wheel for 1h at 27°C in a weak solution of sodium hypochlorite (either 20mg/l active chlorine for towels or 100 mg/1 for swimwear), dried and assessed for colour change

Colour fastness to cross-dyeing: See **Fastness to cross dyeing.**

Colour fastness to degumming: See **Fastness to degumming.**

Colour fastness to dry-cleaning: The sample to be tested is enclosed in an undyed cotton bag with 12 steel discs. The bag is then agitated in a wash-wheel for 30 min in perchloroethylene at 30^{0}C. The specimen is then removed, dried and then assessed for colour change. The colour of the remaining solvent is also compared with that of unused solvent to detect any staining.

Colour fastness to dry heat: In order to test the colour fastness of dyes to heat setting, ironing and contact heat, The Fixotest apparatus (Heraeus) has been developed. The Fixotest involves examining the effect of heated metal surfaces on the textile dyes under investigation.

The test fabric is sandwiched between undyed adjacent fabric and then placed between the heated plates of the apparatus. After testing, the colour change and any transfer of the dye onto the adjacent fabric are assessed.

Colour fastness to hot pressing: This fastness test involves the assessment of the resistance of the colour of textiles to the effects Pleating and Setting in Dry Heat of Dyeing and Prints. Both the Draft ISO Recommendation and DIN 54060 describe the testing of dyes at three different experimental temperatures (150, 180 and 210°C) over 30 s, in which the contact pressure exerted on the test piece is 4 ±1 kPa. These standards apply to synthetic fibres and blended fabrics containing synthetic fibres, but not to determining the colour change that occurs with synthetic resin finishing processes (permanent press process or thermosol process).

Colour fastness to milling: See **Fastness to milling.**

Colour fastness to nitrogen oxides GOl: This method determines the resistance to fading of dyed fabrics by gases produced during the combustion of gas, coal, oil, etc. In the test the sample is exposed to nitric oxide inside a special test chamber. At the same time a control sample is exposed and the test is allowed to proceed until this fades to a given extent. The sample is then assessed for fading.

Colour fastness to perspiration This method is intended for the determination of the resistance of the colour of textiles of all kinds and in all forms to perspiration. Composite specimens are treated in solutions containing histidine, drained and placed in a Perspirometer or equivalent apparatus. The specimen and the undyed cloths are dried separately. The change in colour of the specimens and the staining of the undyed cloths are assessed with standard grey scales. Solutions 1 Alkaline solution, *freshly prepared,* containing 0.5 g histidine monohydrochloride monohydrate, 5 g sodium chloride and 2.5 g disodium hydrogen orthophosphate ($Na_2HPO_47H_2O$) *per litre,* brought to pH 8 with 0.1 N sodium hydroxide.

2 Acid solution, *freshly prepared,* containing 0.5 g histidine monohydrochloride monohydrate, 5g sodium chloride and 2.2 g sodium dihydrogen orthophosphate (NaH_2PO_4-$2H_2O$) *per litre,* brought to pH 5.5 with 0.1N sodium hydroxide.

Colour fastness to rubbing/ Crocking: This test is done basically to asses the ransfer of dye from one fabric o another while washing, wearing or any other use. "Crocking" is understood to be colour transfer from the surface of a dyed fabric to an adjoining fabric of the same type or to another surface by rubbing; "rubbing fastness" describes the dyeing shade change. A crockmetre is used for testing where the dyed fabric is rubbed against an undyed fabric either wet or dry (which rubs to and fro 10 times in 10 sec. over a distance of 10 cm with a load of 900 g). Hence there are two types of rubbing fastness, viz., Rubbing Fastness wet, Rubbing fastness dry. See **Rubbing fastness testing.**

Colour fastness to rubbing, organic solvents: This test provides an indication of the resistance of the shade to organic solvents on rubbing. Testing is carried out with the Crockmeter (which rubs to and fro 10 times in 10 s over a distance of 10 cm with a load of 900 g). A cotton rub fabric is wetted in the organic solvent and squeezed to 100% solvent content before carrying out the test. Dry in warm air (< 60°C). Both the change in colour and the degree of staining are assessed with the grey scales.

Colour fastness to salt water: See **Fastness to sea water.**

Colour fastness to Seawater: The composite specimen is wetted out in sodium chloride solution (30g/l) and placed in a perspirometer in an oven for 4h at 37°C. It is then separated, dried and assessed for colour change of the test fabric and staining of the adjacent fabric.

Colour fastness to soda boiling: See **Fastness to soda boiling.**

Colour fastness to water: In this test composite specimens are wetted out in distilled water at room temperature. They are then placed in a perspirometer in an oven at 37°C for 4h, removed and dried. They are then assessed for colour change of the test fabric and staining of the adjacent fabric.

Colour fastness to storage conditions (Colour fastness during storage): (1) Mainly used for Formaldehyde fastness, as formaldehyde from packaging causes colour change under certain circumstances. Less often for sulphur dyeings which split off sulphuric acid, and result in fibre damage. (2) Sublimation fastness during storage.

Colour filter: A thin layer of coloured glass or other transparent material which limits the transmission of certain wavelengths while allowing others to penetrate.

Colour forecasting: The selection of ranges of colours that are deemed to be those that will be wanted for a particular product/market at a particular time in the future.

Colour gate: This is a term for a retractable component in roller printing machines. The colour application unit (trough, doctor, counter doctor) is stored in the colour gate, thus reducing the machine set-up times and enabling the colour way to be changed more rapidly.

Colour grading: The act of identifying a specimen by a colour grade or colour score that is specific to the colour and the material graded.

Colour Index (CI): Well-known English reference book dealing with dyes (trade names, constitution), optical brighteners, reduction agents and developers. Publisher: Society of Dyers and Colourists. Each structure is assigned a name according to chemical composition. Each dye is assigned a number according to its class and shade. A correlating structure number is given when available. The Colour Index contains information on dye structures, classifications, manufacturers and processes. Many "pure" dyes, that is, dyes which are a single colour, rather than a mixture, have Colour Index names (for example, the reactive dye Turquoise MX-G is designated in the Colour Index as Reactive Blue 140). This name is normally independent of the manufacturer. The Colour Index is a multi-volume publication.

Colour intensity: Intensity is the strength of a colour. This may also be referred to as a colour's saturation or brightness. A pure hue can be fully saturated or less saturated without altering the hue itself, only its strength. Another important factor is the reflection of light from different fibres, yarns and fabrics. Fibres with high lustre such as silk – and the many man-made fibres that imitate the characteristics of silk – give the impression of full saturation; whereas opaque fibres, such as wool, generally give colours with less saturation.

Colour kitchen: Work room for the preparation of print paste thickening and the formulations of colour pastes. The way in which the colour kitchen is organised in many cases determines the productivity of the entire printing plant. Normally coloukitchen will consists of chemical store, dyestore, weighing area, dissolving plant, manual or auto dispensing unit.

Colour lakes: Metal-dyestuff compounds which arise e.g. during cationic dyeing and mordant dyeing processes or when developing dyes using metallic salts (Chroming of dyes). The low solubility of this type of compound and their strong adhesion in or on the fibre determines the relative high fastness to processing and wear resistance of such dyes. Pre-prepared colour lakes are Colour lake substances.

Colour lamp, in colour determination of cotton with a colour Meter: A lamp with a specific energy out put function used in conjunction with special tristimulus filters to obtain a esired response function.

Colour location: (Colour point). The **Chromaticity** diagram includes colour locations of different Type of colour. Colour locations of the same shade are located on straight connecting lines between the achromatic point (= colour location of the physical ideal white) and the demarcations of the chromaticity diagram. As the hue moves away from the achromatic point, the colour becomes more saturated.

Colour measurement: Physical measurement of light radiated, transmitted or reflected by a specimen under specified condition and mathematically transformed into standardized colourimetric terms. These terms can be correlated with visual evaluations of colours relative to one another.

Colour meter: An instrument which measures the fibre sample colour as presented in the viewing window, in terms of the tristimulus values Y and Z and transmits these values to the IC/TC for further processing.

Colour model: A colour-measurement scale or system that numerically specifies the perceived attributes ofcolor. Used in computer graphics applications and by colour measurement instruments.

Colour order systems: Systems used to describe an orderly three dimensional arrangement of colours. Three bases can be used for ordering colours: 1) an appearance basis (i.e., a psychological basis) in terms of hue, saturation and lightness; an example is the Munsell System; 2) an orderly additive colour mixture basis (i.e., a psychophysical basis); examples are the CIE System and the Ostwald System; and 3) an orderly subtractive colour mixture basis; an example is the Plochere Colour System based on an orderly mixture of inks.

Colour quality: This is a term used in colorimetry for three-dimensional changes in direction of a colour relating to : Hue; Saturation; Lightness

Colour sample: (Sample pattern). Colour sample as per order for: Sample dyeing; Colour matching of dyed/printed samples.

Colour scheme: A combination of colours carefully selected, in particular, for furnishings, interior decoration, clothes etc to co-ordinate or, which when juxtaposed, appear to go well together. Research indicates that colour in the working environment exerts a significant influence on our working life. See, for example, **green room**.

Colour sensation: The characteristic of the visual sensation which enables the eyes to distinguish differences in its quality, such as may be caused by differences in spectral distribution of light rather than by differences in the spatial distribution of fluctuations with time.

Colour seperation: (1) This is a process in the automatic production of colour separation from a multi-coloured sample with the aid of a scanner. After passing through a sampling drum, the coloured beam of light is captured by an optical head which disperses it via screens, prisms, interference filters and colour separation filters into 3 reference beams in yellow, red and blue.

(2) Transfer of individual colours from a colour submission to the printing rollers and/or screens from a pattern submission in textile printing. Manually prepared, it is a time-consuming operation. Resolution of the original pattern drawing as a pure contour film for each colour with the rest of the surface blacked out by hand painting.

(3) In screen printing, a seperate screen is required for each colour within the design. Colour seperation is the process whereby individual films are made for each colour. Colours may be seperated by hand or by computer to produce films with opaque areas where the colour is to be printed, and clear or translucent areas where the colour is not to be printed.

Colour simulator Represents hues for digital transmission that can be subsequently visualised or used as numerical values. Reproduction by numerical data on digital display.

Colour space: When colour is assessed on the basis of reflectance measurements, it is common to consider the three relevant attributes of perception of colour as hue, chroma (or saturation) which is the 'colourfulness' or richness of the colour, and lightness, which refers to the amount of reflected light. These three attributes may be described using the concept of colour

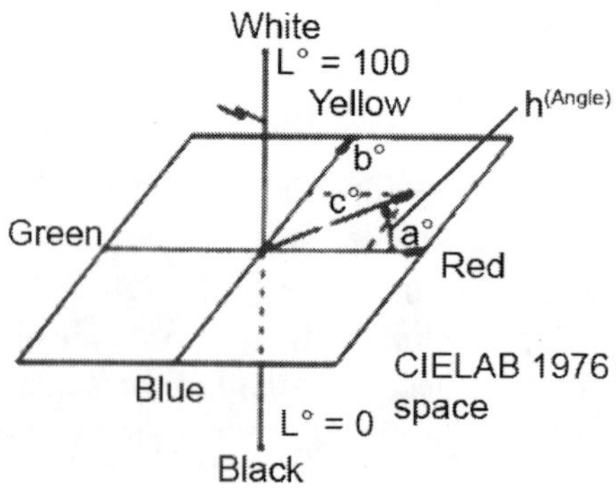

CIELAB 1976 uniform colour space

space, which shows the relationships of colours to one another and which illustrates the three-dimensional nature of colour. The day light colour of the opaque specimens are represented by points in space in terms of three colour scales: Reflectance Rd, and the chromaticity co-ordinates for redness or greenness +/-a, and yellowness or blueness +/-b.

Colour specification: Tristimulus values, chromaticity coordinates and luminance value, or other colour-scale values, used to designate a colour numerically in a specified colour System.

Colour splashes: Colour splashes due to the rapid lifting of a printing screen doctored with excessive printing paste. The cause is screen or fabric sticking, and the local splashing of printing paste on pulling loose.

Colour stability, in coated glass textiles: The ability of the applied coating to resist fading from exposure to sunlight and water.

Colour staining: The undesireds pick up of colour by a fabric: (a) when immersed in water, drycleaning solvewnt or similar liquid medium, that contains dyestuffs or colouring materials not intended for colouring the fabric, or (b) by direct contact with other dyed material from which the colour is transferred by bleeding or sublimation.

Colour stimulus: This is the luminous radiation recorded by the eye that, according to its energy distribution, stimulates the colour-sensitive cones on the retina. The eye functions as a grid-like arrangement of three types of receptive organs each having different spectral sensitivity (Eye sensitivity to hue, chroma and lightness). Expressed mathematically, the eye now operates in such a way that it multiplies the colour stimulus function for each wavelength by the corresponding CIE tristimulus values x, y, z and obtains the three effects caused by integration as colorimetric measure coordinates X, Y, Z (Colorimetric measures), which then combine in the perception of a uniform colour effect, the colour stimulus.

Colour strength: A measure of the ability of a dye to impart colour to other materials. Colour strength is evaluated by light absorption in the visible region of the spectrum.

Colour stripper: A chemical used to remove some or all of the dyestuffs from a fibre, yarn, or fabric so that a dyeing defect can be corrected, a shade lightened, or another colour applied.

Colour technologist: Anyone whose job it is to involve himself with the technicalities of colour in many different disciplines including paint mixing, dyeing, printing, film, television broadcasting and Internet web-site building.

Colour temperature: A measurement of the colour of light radiated by a black body while it is being heated i.e.,Thermal radiation, expressed in Kelvin.

This measurement is expressed in terms of absolute scale, or degrees Kelvin. Lower Kelvin temperatures such as 2400K are red; higher temperatures such as 9300K are blue. Neutral temperature is white, at 6504K. As the temperature increases, the waves emitted become shorter. The colour temperature scale covers the entire wavelength region from infrared to ultra-violet. Short wavelengths (bluish light) have a high colour temperature, long wavelengths (reddish light) a low colour temperature.

Colour transfer belt: Used in Vigoureux printing to transfer the print paste from the frame to the engraved roller.

Colour tolerance: This is an allowed colour deviation from a standard.

Colour trough: Contains the printing paste on printing machines or dye solution on dyeing machines. Usually made of copper or stainless steel.

Colour value: The value of a colour is its clarity or purity. Pure colours have no added white, black, or grey and are often termed as brights; whereas colours with an added amount of white are known as tints. Colours with an added amount of black are known as shades and those with an added amount of grey are known as tones and are often referred to as muted. The value of each primary, secondary and tertiary colour can be altered, according to the percentage of tonal value added. Values alone can be used as colours with no added hue and these are known as achromatics. Their generic names are white, grey and black. Grey may be of any combination of white and black and fashion names are often given, such as slate and silver.

Colour vision: Light entering the eye falls on the net-like arrangement rods and cones of the optic nerves (retina) causing Colour stimulus.

Colour way: (1) Colouration, colour effect. (2) Textile printing: colour composition with the number of colours determined by the design. Any required number of colourways can be produced per design.

Colour wedge: In flat-screen or rotary screen printing, a wedge of print paste is built by the action of the squeegee (Squeegee contact angle). The squeegee pressure is important in controlling the pushthrough of the colour wedge.

Colour wheel: The visible spectrum's continuum of colours arranged in a circle, where complementary colours such as red and green are located directly across from each other.

A circle of touching colours in a diagrammatic form. One particular circle used to help mix paints comprises 12 colours including the three **primary colours**, the three **secondary colours** and the six **tertiary colours** in the following order: **YELLOW**>yellow-green>**GREEN**>green-blue>**BLUE**>blue-purple >**PURPLE**>red-purple>**RED**>orange-red>**ORANGE**>yellow-orange.

Colour wheels are used to correct colours on scanners and monitors. Many wheels or colour circles have been devised including those of Sir Isaac Newton (1642-1727), Wilhem Ostwald (1853-1932), Johannes Itten (1888-1967), Johann Wolfgang von Goethe (1749-1832), Moses Harris (1731-1785), Ignaz Schiffermüller, Philipp Otto Runge (1777-1810), Ogden Rood (1831-1902) Albert Munsell (1858-1918) and Michel- Eugène Chevreul (1786-1889). See **Ostwald circle**.

Columbine: The colour of a dove. '*A kind of violet colour*' Samuel Johnson.

Colour yield; Tinctorial yield: The depthof colour obtained a standard weightof dye is applied to a substrate under specified conditions. Or quantity of dye fixed on the textile substrate in % by weight of the quantity of dye used.

Colour, wool: The actual colour of the wool. In industry a bright white to cream is most desirable; canary stains, brown or black stains are undesirable.

Colourant: Standard term for chromophoric materials. This includes dyestuffs, colour lakes, pigments, toners, waxes, phosphors.etc.

Colorant staining: The unintended pickup of colorant by a substrate due to (a) exposure to a coloured or contaminated liquid medium, or (b) direct contact with dyed or pigmented material, from which colorant transfers by sublimation or mechanical action (as in crocking).

Colouration: An arrangement of colours, in particular, the markings on animals; the appearance of an object as determined by the light emitted or reflected from its surfaces; to be distinguished from 'colour'.

Coloured: Having or possessing colour. The term 'coloured' used to describe people of African or Asian descent is offensive and politically incorrect. Although there remains uncertainty as to the correct terminology, the word 'Black' is used extensively in the UK. The preferred term in the USA for people of Negroid origin is 'Afro-American', the words 'Negro' and 'black' being no longer acceptable. See **dark-skinned**.

Coloured discharges: Printing done on dyed ground, where the dye used is dischargeable, with print paste containing reducing agents (e.g. Rongolite C) along with other dyes which is not affected by the discharging agent and can be fixed in the printing conditions will give coloured discharge on that dyed ground. Wherever motifs are printed with discharging agents only (no dyestuff) you will get white discharge.

Coloured fibre, in wool top: Any fibre the colour or shade of which differs from the normal colour or shade of the fibre mass of the sample.

Coloured resists: In contrast to white resists, where the ground dyeing or the ground printing isprevented from fixing locally, and the white ground thus emerges, in the case of coloured resists another dyestuff is simultaneously fixed at this point.

Colourimeter: (1) An optical measurement instrument that responds to colour in a manner similar to the human eye — by filtering reflected light into its dominant regions of red, green and blue.

(2) An instrument for measuring the concentration of a known substance in solution by comparing the liquid's colour with standard colours.

(3) A device that specifies colour by measuring the intensities of the three primary colours that compose the colour under study.

Colourimetry: Any technique for evaluating a given colour in terms of standard colours.

Colourist, in designing: Designers who predict colour trends and put together palettes of colours for specific seasons and product groups. Other colourists will work further down the design processline, colouring work produced by other designers to create different and alternative colourways.

Colourway: Alternative colouring of a design on paper or in/on fabric.

Colquhoun: A Highland tartan, consisting of the following: Black bar, dark blue bar, twice as wide, split by a pair of black lines in the center, black bar, narrow white stripe, two dark green bars, each as wide as the black, separated by a red line; a narrow white stripe.

Colthorp pride: A commercial variety of late maturing cotton from US; the staple measures 28-32 millimeters; the yield is 28-30 per cent.

Comaca: Silky, yellowish seed hair of the Bombax tree in British Guiana.

Comashes: A fabric mentioned in 17th century English manuscripts, as imported from Turkey; structure and composition unknown.

Comb doctor: Version of a Doctor roller, in which a flat doctor blade has comb-like indentations.

Combed top: Semi-finished product of Worsted spinning. Sliver-like web of parallel wool fibres obtained after passage through the so-called comber.

Combed sliver: A continuous band of untwisted fibre, relatively free of short fibre and trash, produced by combing card sliver.

Combed yarn: A yarn produced from fibres that have been carded or (Or prepared) and combed sliver.

Comber/brusher, in textile testing: An instrument which prepares the test beared of fibres for length, length infirmity, strength, and elongation measurements by combing the test specimen to remove loose or unclamped fibres , and by brushing the clamped fibres to remove the fibre crimp and smooth the test beared of the fibre.

Combination fabric: A fabric containing: (a) different fibres in the warp and filling (e.g., a cotton warp and a rayon filling), (b) ends of two or more fibre in the warp and/or filling, (c) combination yarns, (d) both filament yarn and spun yarn of the same or different fibres, or (5 filament yarns of two or more generic fibre types. Combination fabrics may be either knit or woven. They should not be confused with blend fabrics. Although blend fabrics also contain more that one fibre, the same intimately blended spun yarn is present in both warp and filling.

Combination milling: Alkaline and then acid milling. Provides good felting and excellent handle. Alkali and acid resistant milling agents are necessary.

Combination process: Process in which two operating stages are combined (cost-saving). Examples: sizing + dyeing; desizing + bleaching; bleaching + optical whitening; bleaching + dyeing; shrinking + dyeing; non-crease finishing + dyeing.

Combination yarn: A piled yarn containing two or more yarns that vary in fibre composition, content, and/or twist level; or plied yarn composed of both filament yarn and spun yarn.

Combine fabric: Laminated Fabrics.

Combined heald and Hairiness: An arrangement in which the ends in a jacquard fabric that are producing a basic weave for ground can be controlled by healdshafts are operated by tappets or a dobby, while the ends that are figuring will be controlled by the jacquard. The system allows the jacquard to have a greater figuring scope.

Combing Process: Process by which the short fibre (noil) are separated from the long fibres and the latter are straightened out and laid parallel to each other. Combing wool is at least 1% inches long, of good strength and used for worsteds.

Combing wool: Wool that is strong and strictly of combing length, that is, 2in. (50cm) or more.

Combined fabric: (Laminated fabric) A material composed of two or more layers, at least one of which is a textile fabric, bonded closely together by means of an added adhesive, or by adhesive properties of one or more of the component layers.

Combing: A step subsequent to carding in cotton and worsted system processing which Straightens and parellelising of the fibres and extracts neps, foreign matter, and short fibres, impurities by using a comb or combs assisted by brushes or rollers.. Combing produces a stronger, more even, more compact, finer, smoother yarn.

Combing Machine: Machine used for combing operation. See **Combing.**

Combing, Dry: The preparing and combing of wool to which no oil has been added.

Combing in Oil: The preparing and combing of wool to which oil has been added to facilitate the manipulation of the fibres.

Combing Wool: Wools having sufficient length and strength to comb. According to industry standards, the length of fibres for strictly fine combing must be over 2.75 in., with an increase in length as the wool becomes coarser.

Combi-squeegee: Combination of a magnetic pressure-application doctor roller and a print paste blade similar to a doctor knife in contact with it, the blade being attached to a print paste feed pipe. The combi-squeegee combines the advantages of both doctor types: effective nip and deep penetration. – Manufacturer: Joh. Zimmer.

Combourg ordinaire: Coarse French linen.

Combustible: Capable of burning.

Combustible textile: A textile that will ignite and burn or that will give off vapours that will ignite and burn when subjected to external sources of ignition.

Combustion: Combustion is the reaction of a substance with oxygen (generally in the presence of an oxygen concentration equal to that of air), whereby heat is generated with a consequent burning or Smoulder.

Combustion characteristic of fibres: See **The Burning Behaviour Of Textile**

Comeback wool: Yielded by the sheep which is the crossing of a half-bred ewe with a merino ram; this is an English trade term.

Comfort: Performance parameter of apparel referring to wearability. Encompasses such properties as wicking, stretch, hand, etc.

Comfort stretch: The term given to the freedom of movement experienced in the wearing of a garment that contains spandex, or has stretch engineered into a yarn through mechanical stretch construction.

Comforter: A bedcovering assembly, consisting of an insulating filler secured between two layers of fabric, used primarily to reduce heat loss.

Comfort finishes: Finishes applied to textiles composed of synthetic fibres to reduce the build-up of electrostatic charges and soiling, as well as to improve hydrophilic properties and soil release.

COMITEXTIL: Comité de Coordination des Industries Textiles de la CEE (Coordination Committee for the Textile Industries of the European Community), Brussels, Belgium

COMMENTEX: International Committee for Fabric Care Meetings, with headquarters in Paris.

Commercial allowance: An arbitrary allowance eaqual to the commercial moisture regain plus a specific allowance for finish formally adopted for use with the oven dry weight used in calculating the commercial or legal weight of a fibre shipment or delivery or making certain calculations.

Commercial composition- in wool: The percentage by weight of wool basc moisture and other non wool base components in wool to which specific commercial designation is applied.

Commercial designation- In wool: A term applied to a lot wool in a stated form and having a specified commercial composition.

Commercial laundering: A process by which the textile products or specimens may be washed, bleached or dried and tpressed typically at higher temperatures, high pH and longer times than used for home laundering.

Commercial mass: Billed mass as determined by a generally accepted method or as agreed upon between the purchaser and seller.

Commercial moisture content: (in wool) The moisture calculated as a percentage of the weight of the wool , top, noils, yarn, fsabric etc. in the 'as is' condition; that is, containing whatever moisture , oil, grease, or other extraneous matter that may be present.

Commercial moisture regain: An arbitrary value formally adopted as the regain to be used with the oven dry weight when calculating (a) the linear density (b) commercial or legal weight of a shipment or delivery of any specific textile material (c) the weight of a specific component in the analysis of blends.

Commercial moisure regain: An arbitrary value adopted as the moisture regain to be used in calculating the commercial or legal weight of a fibre shipment.

Commercial Standards: "Recorded voluntary standards of the trade." Each country through their Bureau of Standards or other equivalent organisation issues Commercial Standards which are not laws, but are important as accepted voluntary benchmarks of performance and quality by the industry. These standards are usually referred to by number, and spell out test procedures and minimum performance guidelines.

Commercial weight: (1) In natural fibres, the dry weight of fibres or yarns plus the commercial moisture regain. (2) In manufactured fibres, the dry weight of staple spun yarns or filament yarns after scouring by prescribed methods, plush the commercial moisture regain

Commingled yarn: In aerospace textiles, two or more continuous multifilament yarns, the filaments of which have been intermixed with each other without adding twist or otherwise disturbing parallel relationship of the combined filaments. Usually consists of a reinforcing yarn, such as graphite or glass, and a thermoplastic matrix yarn.

Commission dyeing: Textile dyehouse which dyes textile fabric to the order of another firm without owning the fabric itself. Also called Commission finishing.

Commodore: Stout English navy drill.

Common: One of the U.S. Grades of wool. It is next to the coarsest grade, and derives its name because it presumably comes from sheep of common ancestry.

Common twill: An even sided four-harness twill weave, each thread passing over and under two threads alternately.

Compact spinning Process: A term generally referring to a spinning process carried out using any one of the several small spinning machines of compact design offered by equipment vendors as "packaged" units in which spinning and subsequent processing (drawing, crimping, cutting, etc.) are linked.

Compacted yarns: Air-jet interlaced yarns. Since the entanglement serves only as a substitute for twist, the degree of interlace or tangle is not as great as in air-jet bulked yarns.

Compacting (Shrinkproofing): Controlled residual shrinkage is an important quality parameter for many fabrics. For example, excessive shrinkage is undesirable for fabrics to be made into garments. Here, the residual shrinkage should be less than 2% otherwise the garment will not fit after it is laundered. Drapes cut to floor length will draw up from the floor and detract from their appearance unless the residual shrinkage is controlled. Before launching into the mechanical methods of reducing shrinkage, it will be instructive to discuss the causes of fabric shrinkage.

Compaction calendar: A compaction calendar has an adjustable gap between the pattern roll and bowl. This type of calendar is used to make filter media of certain thickness.

Compactor: A machine developed by Fabric Research Laboratories which is used to compact fabrics or to produce warp-stretch fabrics by means of forced crimp and/or shrinkage of the warp yarn.

Compatible shrinkage: A term used for bonded fabrics to indicate that the face fabric and lining have similar shrinkage. This is necessary to avoid puckering.

Compatibility, in textile dyeing: Propensity of individual dye components in a combination shade to exhaust at similar rates resulting in a buildup of shade that is constant, or nearly constant, in hue throughout the dyeing process.

Compensator roller: (dancing roller). Roller which moves up and down on movable bearings for equalising the tension on open-width piece goods between continuous operation finishing machines.

Complex machine stitch pattern, in sewing: A machine stitch pattern formed when two or more simple machine stitch patterns are combined in one repeating unit.

Complex seam: A seam made in two or more steps.

Complexing agent: (metal complexing agent, inactivating additive, sequestering agent, chelating agent). Products which are qualified by characteristic functional groups to change metal ions by Complex formation, and to improve their solubility in a specific medium. This occurs by the coordinative bonding of unwanted metal ions in the soluble complex in order to avoid metal precipitation in the liquor and on the textile product, and to dissolve metal compounds already deposited on the textile product.

Compliance: The ability of a fibre to yield under stress; the ratio of the change in strain to the change in stress that produces it; the reciprocal of the textile modulus.

Complimentary colour: If rays of light visible to the eye meet coloured compounds, the absorption of specific ray zones occurs out of the numerous light pulsations, and the coloured compound appears in its socalled complementary colour . Pairs of colours which are opposite each other on the **colour wheel** and which enhance each other, for example, red and green, yellow and purple and blue and orange. See **secondary colours**. Two complementary colours, such as blue and yellow, when juxtaposed produce a perfect contrast. A colour will appear most vibrant when juxtaposed with its complementary. Complementary colours always consist of one of the **warm colours** and one of the **cold colours** and either a **primary colour** and a **secondary colour** or two of the **tertiary colours.** See **split complementary colours**.

Complements: Two colours that create neutral gray when combined. On a colour wheel, complements are directly opposite from each other: blue/yellow, red/green and so on.

Component, in pile yarn floor coverings: The individual yarn or fabric elements into which a pile yarn floor covering can be separated.

Component: **in textile fibre polymers**: A polymer with distinguishable properties.

Component fibres: Consist of two or more different high-polymer types, e.g. Bicomponent fibres.

Component of variance: A part of total variance identified with a specified source of variability.

Components, for pile yarn floor coverings: The individual yarn or fabric elements into which a pile yarn floor covering can be dissected.

Composite: (1) An article or substance of two or more constituents, generally, with reinforcing elements dispersed in a matrix or continuous phase.

(2) Hard or soft constructions in which the fibres themselves are consolidated to form structures rather than being formed into yarns. Rigidity of these con structions is controlled by the density, the modulus of the load-bearing fibres, and the fraction of fusible fibres. Strength is cont adhesion and shear-yield strength of the matrix unless fibres are bonded in a load-transferring matrix.

(3) A structure made by laminating a nonwoven fabric with another nonwoven, with other materials, or by impregnating a nonwoven fabric with resins.

Composite fibres: Fibres composed of two or more polymer types in a sheath-core or side-by-side (bilateral) relation. See **Bicomponent fibres**.

Composite yarn: A yarn composed of both staple and continuous filament components e.g. core spun and wrap spun.

Composition cloth: Waterproofed cotton or linen duck; used for bags, covers, etc.

Compound fabric: (woven) a generic term for layered fabric in which separate layers or plies, each with its own warp and weft are produced simultaneously and stitched together in one weave process.

Compound feed: A feed mechanism comprising synchronized drop feed and needle feed.

Compressibility: Refers to the ease of reducing the bulk of fabric, carpet, batting, or other material. May be high or low, soft or hard.

Compression fabric: A high tenacity stretch fabric which, when in a close fitting garment, provides muscles with a firm compression fit that lessons vibrations, reduces fatigue, and keeps muscles energized. The fabric is usually made in a knit construction, using a series of gradient fibres with an open knit inner surface to create a moisture transfer environment.

Compression moulded button: A button or button blank which is produced by compression moulding.

Compression moulding: The method of moulding of a material already in a confined cavity by applying pressure and usually heat.

Compression stretch: The name given to the expansive stretch that is created by the spandex fibres used in the development of a compression fabric.

Compressive shinkage: Mechanical process for applying a Pre-shrunk finish and reducing residual shrinkage in length in textiles. One of the factors affecting the suitability for purpose of a textile is its ability not to shrink when it gets wet or is washed. Dimensional stability is required. The shrinkage of woven and knitted fabrics is partly attributable to tensions imposed during the manufacturing cycle. These tensions are released in finishing.

Compund needle, Compound feed knitting: The needle is the basic element of loop formation. There are three most commonly used types of needle (picture 6): the latch needle (a) the spring-beard needle (b) and the compound needle (c) The compound needle includes (inside the internal groove of the stem) a sliding closing element, which opens and closes the hook through its upward and downward motion.

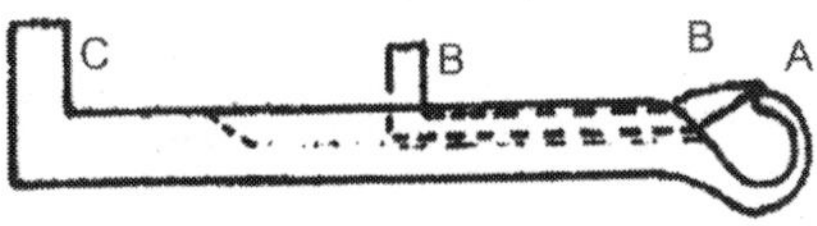

Computer colour matching: Instrumental colour formulation as colouring process based on recipe calculations using the spectrophotometric properties of dyestuffs and fibres. Necessary details concerning reflectance curves are obtained by measuring so-called calibration dyeings.

Computer aided design (CAD) systems: Computer softwarew that has been developed to assist in the designing process.

Concept selling: This is where products are sold on a lifestyle basis.

Con tallies: Inferior French silk, made of floret silk.

Condensate: Water obtained through evaporation and subsequent condensation. It can be applied to other liquids also. Normally the water resulting from condensing plant steam originally generated in the boiler. Water condensed in a water still is called distillate.

Condensation polymer: A polymer obtained when the compounds used in its formation react together, with the elimination of a further compound such as water, formaldehyde or hydrochloric acid. The reaction is called cond. ensation and the product formed is called condensation polymer. As the process involves the elimination of by-product molecules, the molecular mass of the polymer is not the integral multiple of the monomer units. For example, nylon-66 is a condensation polymer of hexamethylene diamine and adipic acid.

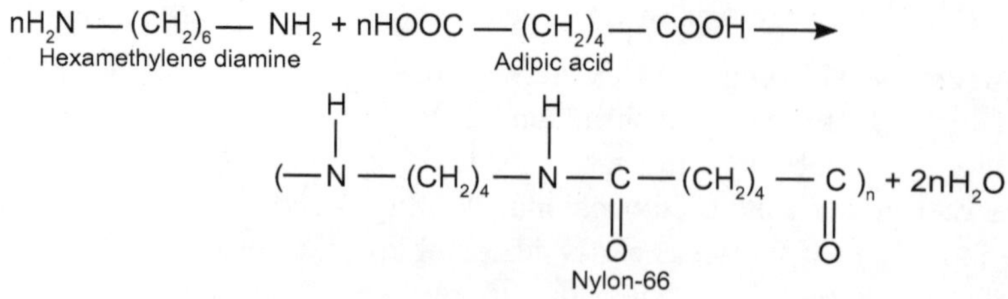

Some other examples of condensation polymers are:

Dacron or Terylene (Polyester): It is a polyester fibre, made by the esterification of terephthalic acid with ethylene glycol. In places, it is known as terylene, whereas in some other places it is called decron.

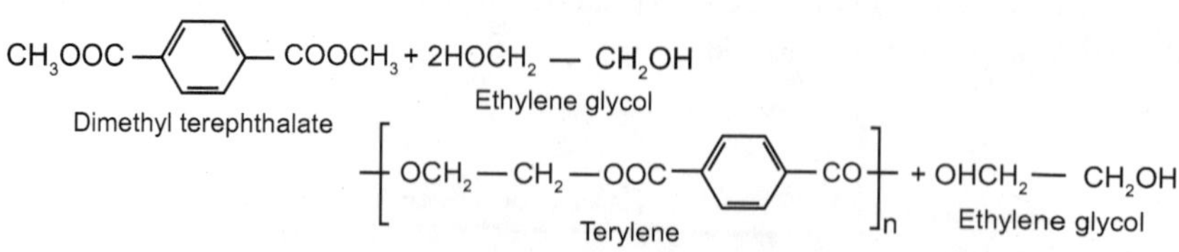

Bakelite: It is a polymer of phenol and formaldehyde.

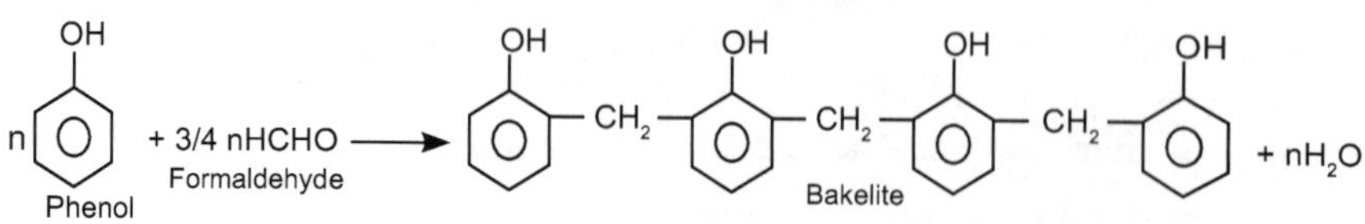

Condensation polymerization: A polymerization process yielding a product in which the repeating unit has fewer atoms that the monomer or monomers. Generally, the separation of Condensation Polymerisation (schematic) water

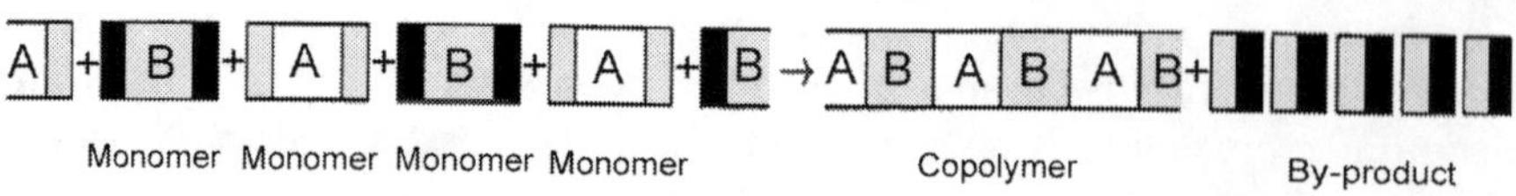

or some other simple substance occurs as a result of the reaction, e.g., ethylene glycol in polyester production.

Condense dye: A dye which during or after application, reacts covalently by itself or other compounds other than substrate, to form a molecule of greatly increased in size.

Condensed card: A roller and clearer type card as distinct from a flat card which converts fibrous raw material to slubbing by means of a condenser.

Condensed notation in waeving: See **Patter draft.** Pattern Drafts also can be represented numerically. A new system of notation codifies information of the weave type, warp lifting pattern, no. of ends lifting together, and the move number. Suitable for EDP.

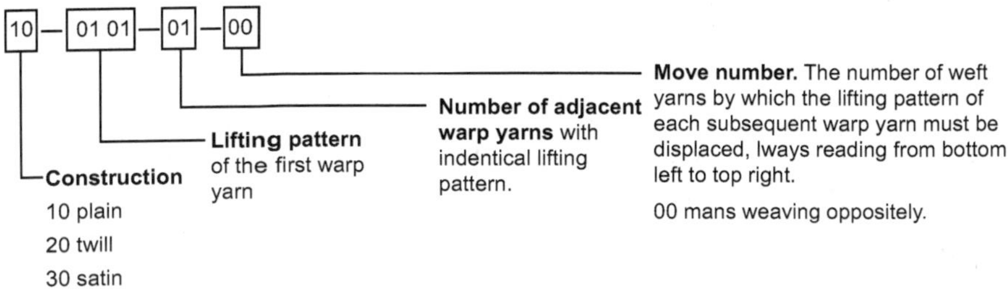

Condensor roving waste: Waste produced during the process of condenser carding and also in the roving process.

Condensor spun: Description of yarn spun from slubbings.

Condenser yarn: (1) Low grade cotton yarn (waste yarn) in high linear densities usually made from East Indian (Bengal) cotton (Imitation yarn).

(2) Generally, woollen yarn produced by the socalled double condenser spinning process, spun from roving collected from strips of card web by rubbing in a tape condenser.

Condition: See **Condition, conditioning.**

Condition, Conditioning: Bringing of a specimen or testing material to moisture equilibrium with a specified atmosphere.

Condition, standard: See **Standard condition.**

Conditional bundle system, garment: A production system in which bundle pass from storage to a operator and then back to storage for allocation to the next operation.

Conditioned weight (Silk): The weight of raw silk obtained by adding to itd oven dry weight, 11 per cent of the oven-dry weight.

Conditioner tube: A tube supplied with steam or hot air surrounding a melt spun threadline and located between extrusion and wind up whose purpose is to control the fine structure of the yarn.

Conditioning: A process of allowing textile materials (staple, tow, yarns, and fabrics) to reach hygroscopic equilibrium with the surrounding atmosphere.

Materials may be conditioned in a standard atmosphere (65%RH, 70°F) for testing purposes or in arbitrary conditions existing in manufacturing or processing areas.

Conductive textile materials: Conductive textile materials are textiles with conductivity. These textile materials are used for intercepting electron waves or for a fuel cell's electrolyte film. Conductive textile materials let electrons and ions go through, just like an electric wire. Conductive textile material is basic and essential to the growth of smart-wear technology.

Conductivity: The property of a substance's ability to transmit electricity, say water. Measured by a conductivity meter, and described in microSiemens/cm.

Cone: (1) Are used for changing the cones of conical cross-wound bobbins, e.g. to replace stainless steel cones with cardboard cones for further processing.

(2) The yarn package obtained when the yarn is wound on a disposable cone core.

(3) Light sensitive cells in the retina of the eyes of many vertebrates providing the means of detecting colour. They contain pigments and respond to bright light whereas rods respond only to dim light. There are over 5 million cones in each human eye, but relatively few blue cones so that, for example, blue text placed on a black background is particularly difficult to read. See also **colour-blindness**.

Cone holder: It allows ends from multiple cones to be wound at the same time. A screw eye, hook, or other device directs each thread so it unwinds above its cone without upsetting the cone or becoming tangled with threads from other cones.

Cone mill: Dye kitchen device for mixing and milling printing thickener and prepared printing pastes.

Coney (Rabbit): A soft, long-haired cheap fur, oftenused to imiytate more expensive ones. Not hard wearing.

Confidence interval: An interval estimation of a population parameter computed so that the statement " the population parameter lies in this interval "will be tue, on the average, in a stated proportion of the times such statements are made.

Confidence Level: The stated proportion of times the confidence interval is expected to include the population parameter.

Confidence limit: The two statistics that define the ends of a confidence interval.

Conformance, in testing methods: The agreement of test results of a specimen with requirements of a specification.

Conglomerate: (Lat.). Mixture of numerous individual particles, mainly of different or similar types.

Congo red: Red direct dye (Böttger 1884). Benzidine-bis-1-naphthylamine-4-sulphonate

Congo red paper: Colour indicator on acid traces, colour change interval pH 3 (bluish violet) to 5.2 (reddish orange, brownish).

Congress canvas: Strong, open face cotton or woollen canvas made of hard spun thread; used for embroidery.

Coning: The transfer of yarn from skeins or bobbins or other types of packages to cones.

Conjugate fibre: A two-component fibre with specific ability to crimp on hot or hot/wet treatment because of differential shrinkage.

Conjugate yarn: A yarn made from conjugate filaments.

Conker-coloured: Reddish-brown (used in *The Evening Standard* 12.8.99 to describe leather boots).

Connecting ring, In Zipper: A device shaped like the letter'D' used to secure special pull having mor than one component in its design, to bail or lugof the slider.

Connaught: Open cotton canvas for embroidpry; See **Basket Cloth.**

Connaught yarn: Soft, fine, loose woollen yarn for knitting: made originally in Ireland.

Conservation: The examination, preservation and restoration of cultural objects of cultural objects with minmal sacrifice of their and historic integrity.

Conservator: A person whose activity involves the science, technology, and documentation associated with the conservation of cultural objects.

Consignment: All the material of a specific type and quality received by a customer against a particular order.

Consistency in wool: The uniform distribution of all the fibre characteristics within each lock and throughout the entire fleece.

Consolidation: Application of heat and pressure to form composite structures.

Consolidation dimensional change: See **Dimensional change, consolidation.** The dimensional change that occurs when a fabric is gently agitated in water to overcome all the frictional constraints in it after it has been allowed to relax in water without agitation.

Constant-Rate-of-extension tensile testing machine (CRE): A tensile testing machine designed in such a way as to keep the rate of increaseof length of the specimen is uniform against time during the testing process.

Constant-rate-of-load tensile testing machine (CRL): A tensile testing machine designed in such a way as to keep the rate of increaseof the load being applied to the specimen is uniform with time during the testing process.

Constant-rate-of-traverse textile testing machine (CRT): A tensile testing machine in which the pulling clamp moves at a uniform rate and the load is applied through the other clamp which moves appreciably to actuate a weighing mechanism, so that the rate of increase of load or elongation is dependent upon the extension characteristic of specimen.

Constitution: A variety of corduroy, having broad races.

Constructed textiles: Includes woven textiles, knitted textiles, laces and carpets.

Construction, in weaving: The pattern of interlacing of warp and weft yarns is called a cloth construction.

Constructional units, in pile floor coverings: The needles, pitch, rows, shot etc. into which the warp and weft yarns are commonly grouped.

Constructive avoidance: Time spent on work that is neither important nor urgent, in preference to urgent and important work.

Consultant designer: Designer employed by a company usually on a part-time basis to advice on design matters.

Consumer care: of consumer textile products, cleaning and maintenance procedures as customarily undertaken by the ultimate user.

Consumer's risk: in acceptance sampling, the probability of accepting a lot when the process average is at the limiting quality level.

Consumer textile product: A textile product intended to satisfy human wants and needs. Textiles not falling into the categories of apparel, furnishing, household and industrial; tents back packs may be referrd to as consumer textiles, awnings, umbrellas, and luggage are also often classed in this category.

Contact angle: The angle between the surface of a liquid and the surface of a partially submerged object or the container at the line of contact. The smaller the contact angle, the greater the wettability of the solid.

Contact drier: The most common example of contact drying is Drum dryer (also called can driers, cylinder driers etc). With this system, the fabric moves forward arranged on several heated drums. The drying temperature ranges between 120-130° C and the cylinders are heated Cylinder drier with fabric

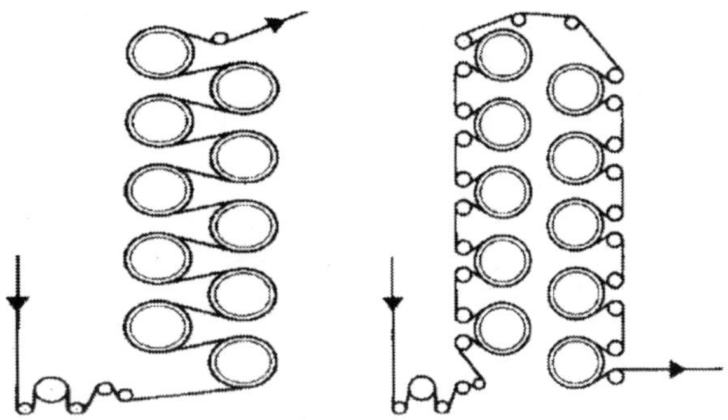

contact on both or one side by means of steam at a pressure of 1-3 atm. This very efficient and low-cost drying system is particularly suitable for flat fabrics, with slightly evidenced structure, which cannot be easily affected by tension during feeding. Used for intermediate drying and for light finishes; this system is not suitable for durable thermosetting resin finishes. See **Contact drying**.

Contact drying: Drying is implemented here by contact with heated cylinders or plates. High temperature contact process; Drying systems.

Contact printer: In the half-tone process, the screen, which is necessary for better adhesion of the printing paste in the engraving, and for preventing the doctor blade from "dipping into" the engraving, is produced either in photography by inserting a glass/engraving contact screen, or by direct use of contact screen films, or also by screen films on a contact printer. See **Half-tone screens.**

Container: A receptacle designed to hold a material, or to give integrity to the material.

Continuous ager: So-called continuous ager for printed fabrics and large-scale roduction, mainly for short or long steaming times. There are various steamer chamber designs, e.g.:

(i) Fabric transport by means of running rails or a few larger diameter idling rollers in the case of festooned fabrics (festoon steamer), especially for screen printing. (ii) With guide rollers top and bottom as a rapid steamer (flash ager). (iii) Similar to I, festoon design but with a continuous pin chain, as a screen print festoon steamer.

Continuous cure, in garment making: A method of curing durable press garments which uses a moving conveyor system to carry garments into and out of the curing oven. Also known as continuous oven.

Continuous decatizing: Dry decatizing The continuous decater has one steaming cylinder and one cooling cylinder. An endless decating apron carries the fabric around the steaming cylinder and around the cooling cylinder. The fabric is continuously moving so the time element of the process is affected by the speed of the machine - being somewhat less than the batchwise semi-decating process. Nonetheless, excellent results are obtained on many fabrics. The fabric is exposed to steam in either expanded or cooled form (suction principle) or controlled humidity is applied to the cloth in the treatment zone and this is converted into steam by applying pressure and heat.

Continuous desizing: Suitable for woven cotton and blended cotton/synthetic fabrics. Although enzymatic starch decomposition is time-consuming, continuous desizing can be effected with bacteria-diastases (other amylases are unsuitable): fabric is impregnated with hot enzyme solution and deposited in a heated J box for 3–5 min. Liquid absorption must be high (100% and over).

Continuous dyeing: A process by which a fabric is continuously dyed by a suitable method. Examples are pad-dry, pad steam method of dyeing of vats, reactives, sulphur on cotton, E-control dyeing of reactives on cotton, thermosol dyeing of disperse dyes on polyester or polyester/cotton.

Continuous dyeing Plants: Depending on class of dyestuff, they comprise padder, dryer (cylinder, hot flue dryers), continuous steamer and open-width washing machines. Continuous dyeing range: Machinery range for padding, setting/steaming/drying, wash-off of suitable dyestuffs (e.g. disperse, vat and reactive dyestuffs) in correspondingly large dye batches for increased productivity and economy in dyestuffs, chemicals and energyusage. Process generally uses a liquor ratio of 1:1, i.e. the pad-liquor pick-up = 100% = 100 kg dye liquor/100 kg fabric.

Continuous element: A configured element formed continuously along a length of monofilament.

Continuous element Zipper: A zipper consisting of two continuously formed elements, each attatched to one of the opposing edges of two tapes, which are engaged and disengaged by the movement of a slider.

Continuous filament: See **Filament.**

Continuous filament yarn: See **Filament yarn.** A yarn composed of one or more filaments that run essentially the whole length of the yarn.

Continuous filament yarns, in glass textiles: A strand made from shortpredetermined lengths of cut continuous filament and used as a reinforced material.

Continuous filament yarn: A yarn made of filaments that extend substantially through out the length of the yarn.

Continuous open width range: An open width processing machine where many processes can be done continuously. Basically it is many machines joined together which can run continuously.

Continuous polymerization In polymer manufacture, linkage of the various stages of polymerization so that materials flow without interruption from the addition of raw materials to delivery of the finished polymer from the system. Extrusion as film, chip or fibre may be linked to a continuous polymerization line. Because there is no break in the process while the transition from low molecular weight to high occurs, multiple stage reaction vessels may be required and accurate process control is critical.

Continuous rope treatment plant: A range for the continuous pretreatment, bleaching, dyeing and aftertreatment in rope form, preferably of cotton knitted fabrics.

Continuous steamer: See **Continuous ager.**

Continuous systems: The operations are carried out by means of a series of machines; every machine carries out always and solely the same process. Every machine is assembled according to specific production requirements. A system like this entails high start-up costs and a complex setup but once the system has started, it requires a smaller staff and grants excellent repeatability and high output rates; continuous systems are therefore suitable for manufacturing large lots of products with the highest cost-efficiency.

Continuous variate: A variate that is a measurement based on a scale that is assumed to be continuous.

Continuous yarn felting: A process where by slivers, rovings, slubbings are felted on a continuous basis. This is achieved by passing wool-rich material through a unit where it is agitated in an aquous medium where the felting takes place. The process is used to produce a yarn or consolidate a spun yarn.

Contouche: Women's outer garment for house, street and travel wear; fully tailored with deep folds at the back (Watteau pleats)

Contour stitching: Automatic sewing along or near the edge of a garment part. The shape may be predetermined or edge following.

Contour surface: See **Hand of fabric.**

Contract carpet: (Fitted carpet, wall-to-wall carpet, textile floor covering). Textile floor covering in the form of fitted carpet (usually wall-to-wall carpet) in widths or carpet tiles (glued or self-adhesive) or as fitted carpet.

Contract cloths: A name for "uniform cloths".

Contract furnishings: Funishing used in offices and public buildings such as schools, hotels and hospitals etc.

Contract textiles: Sometimes also termed uniform textiles, or named according to their intended end-use. Collective term for all trouser, blouse, jacket and coat cloths for authorities such as postal, railway and customs services, armed forces, police etc.

Contraction: See **Take up, yarn; Take up, twist.**

Contrast: The level of variation between light and dark areas in an image.

Contrast colour: See **Complimentary colour.**

Contrasting colours: Colours opposite to each other on the **colour wheel**, for example, red and green; blue and orange. See **complementary colours**.

Control under cover garment: A garment having a known history, the performance of which in a specific end use has been established previously and which is used as standard of comparison.

Control limits: Predetermined ranges based on the variability of past observations between which the instrument data for a test must fall to be considered valid.

Control textile: A textile having a known history, the performance of which in a specific endues has been established previously, and which is used as a standard of comparison.

Controlled colouration: In order to meet the demand for shorter lead times, a radical change in philosophy has developed in which an expectation of Right-First-Time production through the control of the coloration process is now the norm for progressive dyeing operations. Efficiency of production, elimination of the cost of non-conformance (whilst minimising the impact on the effluent loading) has led to the development of "Controlled Coloration." Controlled Coloration is the fusion of three technologies viz. dyestuff technology, application technology and management technology which must complement each other.

Convent: Plain woven woollen dress goods with two or three-ply warp and single filling, made in solid colours and in stripes, mixtures, etc.

Convent cloth: A very light dress goods, made with wool warp and silk filling, having a pebbled face.

Conventional allowance: The percentage, that in the calculation of commercial weight and yarn count or linear density, is added to the oven dry weight of the textile material, which has been previously washed off all the finishing. For such material conventional allowance is arbitrarily chosen according to commercial practice and includes the moisture regain and the normal finish that is added to impart satisfactory textile qualities.

Conventional blanket: A blanket woven in either a plain or twill weave that is napped on both sides.

Converted fabric: A finished fabric as distinguished from greige fabric.

Converter: An individual or organization which buys greige fabrics and sells them as a finished product to cutters, wholesalers, retailers, and others. The converter arranges for the finishing of the fabric, namely bleaching, mercerizing, dyeing, printing, etc., to the buyers' specifications.

Converting (conversion): The production, from a filament tow or tows of a staple sliver in such a way that thje essentially parallel arrangement of the filament is maintained.

Conversion print: Multi-colour overprint, e.g. produced by means of overprinting rollers, where colour change is achieved by the overprint print paste taking over the development and fixation of the pre-print print paste, places not overprinted therefore remaining colourless, and, without pre-printing, patterns of the second print paste developing in the intrinsic shade of the second colour print. These kinds of patterning effect presuppose dyes which can be mutually resisted and discharged, and also possess different development and fixation characteristics.

Conveyor belt shrinking machine: To relax woven or knitted fabric, the fabric is placed on a perforated conveyor belt, which runs through a tunnel. Vibrating elements or bursts of steam keep the fabric appropriately moving.

Conveyor drier: Circulating air drier as (a) drier for loose material which is carried through the drying room on brattice conveyor; (b) piece goods drier, in which the goods are carried through the drier on a conveyor belt. Particularly for knits and other cloths in piece form that are susceptible to tension.

Convict stripes: Cotton fabrics made in England for the export trade; have one inch wide filling tripes in black over white ground.

Convolution: (1) An irregular spiral or twisted condition characteristic of mature cotton fibre. It is visible under a microscopic. The finer fibres are generally more twisted than the coarser fibres.

(2) Coil and curl in certain types of textured yarns which provide bulkiness to the yarn.

Cooker's waste: Waste produced during the cooking or boiling of cocoons. It forms major part of (floss) waste produced during reeling. It is used in the spun silk mills.

Cooking (Silk): The process by which the silk cocoons are treated in boiling water for a short time with a view to softening the filament for easy unwinding during the subsequent reeling process.

Cooking salt: See **Sodium chlorite.**

Cool colours: See **Cold Colours** and compare with **warm colours**; distant colours such as blue, green and violet which appears to move away from the viewer and recede into the background. See **advancing colours.** Also referred to as 'soft colours'.

Cooling cylinder: An open or closed cylinder filled with water, over which the cloth is passed through to accelerate cooling.

Co-ordinating designs: Designs developed to be used together. Sometimes these will contain similar motifs but on different scales; usually the same colour palette will be utilised through out. Groups of co-ordinates are often developed and presented as a 'story'.

Coordination Polymerisation: In coordination polymerisation reactions the monomer molecules used are generally dienes and olefines. These reactions are catalysed by organo-metallic compounds. In such polymerisation, a monomer-catalyst complex is formed between the monomer and organometallic compound. A coordination bond is used between a carbon atom of the monomer and the metal atom of the catalyst, thus formation of monomer-catalyst complex takes place. The polymerisation process proceeds as follows:

where, M = Transition metals such as Ni, Cr, Mo, V, Ti or Rh.

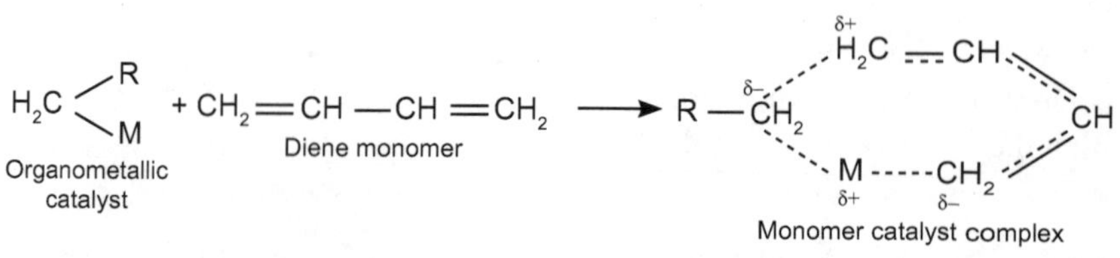

Coothay: East Indian satin made with coloured stripes.

Cop: (1)Term used for a yarn package spun on a spinning frame/ mule spindle. The package generally has tapered ends (Biconical package) and is cop-wound. (2) A headless tube upon which yarn or thread is wound. (3) Filling yarn wound upon a tapered tube(generally paper)

Cop tube: Tube onto which a Cop of yarn is spun, of cylindrical or slightly tapered form or as a former for a "hollow cop" with or without base. Cops used to be made in cardboard, alloy and steel, either varnished or unvarnished. Nowadays plastics-coated steel, some with textured surface (smooth, matt, grooved, "orangepeel").

Copenhagen blue: A bright purplish blue; a light blue. Also referred to as 'copen blue'.

Copolycondensation: In polycondensation reaction, a pair of difunctional monomers is brought about to produce a homopolymer. Whereas in copolycondensation reaction, reactants are more than a single pair and resulting in a copolymer.

If A and B are functional groups capable of undergoing condensation reactions, the simple polycondensation and co-polycondensation can be shown as follows:

In simple polycondensation reaction:

A -R-A+ B-R'-B ~ A -R-AB-R'-B + Byproduct

In co-polycondensation reaction :

A -R-A+ B-R'-B+A -R"-A ~ A -R-AB-R"-BA -R"-A+ Byproduct

Copolymer: A polymer in which the repeating units are not all the same. Polymers consisting of two or more Monomers (Copolymerization) for creating new product characteristics. Usually, but not always, copolymers are formed from two or more starting materials.

Beads of different kind (representing monomers of more than one chemicals)

Representing co-polymer molecule

In copolymers (a) Alternating copolymers possess regularly alternating monomer residues, (b) Periodic copolymers have monomer residue types arranged in a repeating sequence, (c) Random copolymers have a random sequence of monomer residue types, (d) Statistical copolymers have monomer residues arranged according to a known statistical rule, (e) Block copolymers have two or more homopolymer subunits linked by covalent bonds. Block copolymers with two or three distinct blocks are called diblock copolymers and triblock copolymers respectively.

Copolymer fibres: (multipolymer fibres). Synthetic fibres manufactured according to the principle of Copolymerization. They are produced for the purpose of achieving or enhancing specific properties e.g. dye affinity, hydrophilic/hydrophobic characteristics, thermostability, etc. Important examples include modacrylic fibres, modified polyvinyl chloride fibres as polyvinyl chloride copolymers and polyvinylidene-dinitrile fibres.

Copolymerisation: Many monomers can form polymers not only with monomers of their own type but also with other monomers. This opens up the range of possibilities enormously. For example, radical polymerization converts styrene to polystyrene, methyl methacrylate to polymethyl methacrylate and vinyl acetate to polyvinyl acetate.

Copper chelation value (CuCV): The milligrams of copper sulphate pentahydrate chelated by one gram of a chelating agent or product containing a chelating agent.

Copper-green: A generic name for green pigments made from copper including **Scheele's green**, **chrysocolla**, mineral green, **Montpellier green**, **verditer** and verdigris.

Copper number: Expression of the degree of damage especially of overbleached fabrics (Oxycellulose), which have not been boiled during or after bleaching. Principle: determination of reductive capacity by treating the test specimen with Fehling's solution, dissolution of the deposited cuprous oxide, and titrimetric determination as to how much copper has been precipitated on 100 g of test material: unchanged cellulose (copper number) 0.2–0.3, pure oxycellulose 14.2.

Copper Sulphate: $CuSO_4$; also called copper (II) sulphate (read copper two sulphate) or cupric sulphate. A copper compound sometimes used as a mordant, especially with natural dyes. Certain direct dyes are made more lightfast by after-dyeing treatment with copper sulphate.

Coppering: Copper sulphate after-treatment of direct dye dyeings mainly to improve fastness properties.

Coppering dyes: Direct dyes which have to be after-treated with copper salts in order to achieve the correct shade. See **Aftercoppering dyes.**

Copper-plate printing: A method of intaglio printingthat uses engraved copper plates.

Copper sulphate after treatment: Of direct dyeings, for the purpose of significantly improving light fastness, and mainly at the same time too the washing fastness of appropriately suitable dyeings (Aftercoppering dyes).

Coppery: Incorporating a coppery hue as in 'coppery-brown.... flowers'.

Coptic clothes: Various linen, woollen and mixed fabrics, formerly used as mummy wrappings and recovered from old Egyptian tombs. The weave is plain or twilled, the latter often showing striped and geometric designs.

Copyright: Exclusive right, granted by law, to control the use of an original piece of artwork or design.

Coquelicot: The red of the poppy; brilliant red; a colour term used by Jane Austin. Also called **ponceau**.

Coquita: Strong bark fibre, yielded by the Jubaea spectabilis, a palm tree in Chile; used for ropes.

Corabell: A light orange coral colour tinted with crimson.

Corah: East Indian light, washable silk dress goods of natural cream or white colour.

Coral: A deep pink or reddish orange.

Coral pink: A deep orange-pink; sometimes a medium pink or a yellowish pink

Coral red: Bright red or a deep pink.

Coral stitch: In embroidery a stitc.h having a stem from which short and paraaiel branches start out at an angle.

Coralline point: Italian needle-point lace, similar to the Venise point, having coral like trailing patterns.

Coram: Bleached German linen, made very stout of heavy warp and finer filling.

Corbeau: Dark green.

Corinthian pink: A purplish pink.

Cork: A beige colour.

Corkscrew twills: A special group of interleaved twill called corkscrew twills are usually woven with an odd no. of shafts so that the cloth will be either warp faced or weft faced twill and also not an even twill

Corckscrew-worsted goods: So-called from its fancied resemblance to the twists of the corkscrew.

Cord: (1) Term applied to woven rib fabric with weft yarn floats creating pronounced cord-like raised lengthway ridges; a plain fabric without pile warps or pile weft (contrast with Corduroy).

(2) String-like round braid (Braid), in contrast to tape braid. Cord tape braid: tape braid edged with cord.E.g. Paracute cords.

(3) (a) The product formed by twisting together two or more plied yarns as used in tyres. (b) A rib on the surface of a fabric (e.g., corduroy and whipcord.)

(4) Characteristic woven rib running the length of the material worsted or wollen yarn in the warp and wollen yarn in the weft. Cotton is also used as the warp threads . For sports and leisure wear.

Cord, in zippers: A strand of of multiple yarns for use in forming a bead.

Cord, of glass fibre: A strand made by combining multiple ends of filament yarns , including cabled yarns, primarily for structural applications.

Cord carpet: A low level loop pile carpet with pronounced rows of loops in the weft direction.

Cord de Chine: A lightweight fabric of botany warp and eilk filling; two warp ends are in each reed, forming cords.

Cord fabric: Cord fabric should not be confused with corduroy. It is woven in plain weave the cord effect in warp direction is given by a rib weave. A typical construction is 90 x 60. In the warp there can be 6 ends of 30s running in plain weave, then there can be 4 ends of 2/40s running parallel to form the cord. This pattern is repeated in the full width of the fabric. The filling is 35s count. There can be a lot of variations as the corded yarn can be put white and the other yarns can be put black and also in the weft white so that the cord will be standing out as white. Also in the weft sometimes textured yarns can be used. It is used as mens and women's wear. The same name is used for industrial fabric made for the production of tyres. This cord fabric consists of heavy cable yarn held together by a few picks of filling. A charactewristic construction is 20 x 3. The warp may consist of 3/5/22s (i.e. 3 ends of 5/22s twisted together. Nowadays high tenacity filament rayon or nylon is used. See **Bedford cord.**

Cord Stitch: In embroidery and needle laces, a thread twisted around by another thread.

Cord twist: The amount of twist in a cord made from two or more single or plied yarn. Cord, in Zipper.

Cord, steel: A formed structure made by twisting together two or more steel filaments or steel strands.

Cord, tyre: A twisted or formed structure composed of two or more single or plied industrial yarns. Each individual single or plied yarn element in a cordhas some nominal twist, direction of twist, length and tension. The direction of twist used to combine the single or plied yarn elements into a cord structure is in a direction opposite to that used in the yarns.

Cord twist: The amount of twistin a cord made from two or more single or plied yarns; expressed in turns per inch/metre.

Cordat: (1) Strong French canvas, made of tow; (2) Heavy, all-wool coarse French serge, thoroughly fulled.

Corded seam: A corded seam is similar to the piped seambut since the piping tube is stuffed with a cord; it stands out from the seamline. It is often used to edge necklines, short sleeves, and waistlines, and is especially popular when sewing home furnishing.

Corded cloth: Fabrics having ribs running lengthwise (produced by the warp) crosswise (produced by the weft) or diagonally (produced by a twill weave).

Corded dimity: English dimity, made of goat's hair.

Corded selvage: See **Loopy selvedge.**

Corded velvet: Corduroy

Cordelat: (1) Stout twilled woollen cloth, made around Aure, France. (2) Coarse, long napped woollen dress goods from Languedoc, France. (3) Light loosely woven woollen, finished like a flannel, from Beauvais, France.

Cordelia Lace: Lace, having patterns outlined with a heavy thread over a net foundation.

Corderette: 18th century woollen fabric in England.

Cordettes: French cloth made of hemp for headdress.

Cordetum: A coarse fabric; used in medieval England.

Cordington: A heavy boucle.

Cordoncillos: Coarse, plain woven, bleached or unbleached cotton fabric; used for garments by the poorer classes in Mexico.

Cordonnets (silk): Coarse double twists consists of several greige threads with a preliminary twist of 400-800 twists per metre and subsequent doubling od two or three threads followed by 30-600 twists per metre. Cordonnet yarns are mostly used for sewing threads but sometimes as knitting yarn as well.

Cordonnet en Laine: Cord, made of wool or camel's hair; used for upholstery, etc.

Cordovan: A heavy leather that may sometimes be available for sewing. It is from horsehide and is most common leather used for coats, jackets, boots, shoes etc.

Cordurette: Plain woven woollen fabric, made with weft ribs.

Corduroy: Courduroy, a filling pile fabric, has a ribbed velvet face with cut weft pile. It is warp-way rib structureproduced by weaves with weft floats, can also be modified to pique' form.A strong cotton yarn fabric characterized by pile ribs running along the length. The ground fabric usually plain weave

or twill, and the final finishing lays the pile at a slight sland, in one direction. Today often courduroy is referred to as Cord. A typical construction is 60 x 200 with 2.30s warp and 16s weft. These construction and weave wary and also the width. The piles or cords are technically called wales and when the wales are narrow, it is called pin wale corduroy. Approximate weight per square metre is aroung 6-8oz. All corduroy is very hard wearing, washable, and comfortable to wear. It is used as work clothes, leisure clothes and childrens clothes etc.

Core: A filament or strand that serves as an extended axis about which other elements has to be wound.

Core, in sampling fibre packages: The portion of wool or other fibre obtained by using a sampling tube.

Core sampling: The taking of cores from packages of textile fibres for sampling purposes.

Core spinning The process of making a corespun yarn. It consists of feeding the core yarn (an elastomeric filament yarn, a regular filament yarn, a textured yarn, or a previously spun yarn) into the front delivery roll of the spinning frame and of covering the core yarn with a sheath of fibres during the spinning operation.

Core-spun yarn: A compound structure in which a filament or strand serves as an axis around which a cover of either loose fibre or a yarn is wound. A yarn produced at the spinning frame by feeding a yarn through the delivery rollers only, simultaneously with the spinning of the staple fibres (q.v.). Note:

The yarn fed through at the delivery rollers only is usually known as the "core", and the other component is known as the "wrapper". The core may be of continuous-filament yarn or of spun yarn. If the core is of spun yarn, the direction of its twist is usually the same as that of the complete yarn. Core-spun yarns are made for decorative purposes or, more commonly, for strengthening the wrapper for facilitating subsequent processes. When used for strengthening, the core may, after it has served its purpose, be removed by solvent or other chemical action, e.g. the removal of calcium alginate filament yarn by an alkaline scour or of a cotton yarn by carbonising. The core is often retained for strengthening the resultant fabric as is the case if nylon or polyester continuous-filament yarns are used.

Core or Core Spun Thread, in sewing thread: A thread construction made by wrapping a cotton or polyester staple wrapper around a continuous filament core of polyester. Two or more of the core yarns are then plied to make a core spun sewing thread.

Core Yarn: The internal member of a core spun, covered or fancy yarn readily seprate from the covering fibres or yarn. See **Covered yarn.**

Core-bulked yarn: See **Textured yarn**.

Core, in sampling: The portion of wool or other fibre obtained from a package by use of sampling tube.

Core-spun yarn: A yarn made by twisting fibres around a filament or a previously spun yarn, thus concealing the core. Core yarns are used in sewing thread, blankets, and socks and also to obtain novelty effects in fabrics. The lycra corespun yarn is extensively used in woven lycra fabrics.

Cored braid, in rope: A hollow braid construction, either plain or twill, the centre of which is filled with yarns which are not braided.

Corkscrew: (1) A weave composed of the regular twills of 40 degrees, the minimum number of harnesses being five and the maximum 13; (2) A warp faced fabric, woven in fancied re-semblance to a corkscrew; the best grades have French yarn worsted warp, while the filling can be of cotton or wool; used for men's wear and shoe tops; (3) Flaw in doubled yarns, consisting of one yarn being loosely coiled around the other.

Cork screw yarn: Warp-fabric produced in a steep twill weave.

Corkscrew, in raw silk: A place in the yarn in which one or more cocoon filaments are longer than the reminder and give the appearance of a spiral form.

Corkscrew twist: A fabric defect consisting of a crimped, rippled, wavy, pebbled, or cockled area in the fabric spoiling the uniformity of the texture.

Cork screw, Wool: Very fine worsted with a close nap, woven in corkscrew. The nap prevents tiny spots developing. Dyed in piece form. For evening shirts, professional cloth for miscicians, waiters etc.

Corn-coloured: Resembling the colour of corn.

Corn oil: Corn oil has a high unsaturated acid content (46% oleic and 42% linoleic acids). It is used extensively for cooking, as salad oil and for making oleomargarine. Because of the high amount of poly-unsaturated, its use leads to lower cholesterol levels in humans.

Cornelian red: The red colour reflected by the stone, cornelian or cornaline. See **Carnelian**.

Cornflower Blue: The darkish blue of the cornflower – one of the 140 colours in **the X11 Colour Set**. It has hex code #6495ED.

Cornsilk: A beige colour – one of the colours in the **X11 Colour Set**. It has hex code #FFF8DC.

Corojo: Long and strong leaf fibre, yielded by the corojo palm in Central and South America; used for ropes.

Coronation cloth: Fulled mixture suiting in black, blue and red colours, brought out at the coronation of Edward VII. and George II.

Corporate identity: Corporate image.

Corpse grey: A macabre colour description requiring no further explanation.

Corrugation mark: A fabric defect consisting of a crimped, rippled, wavy, pebbled, or cockled area in the fabric spoiling the uniformity of the texture.

Corset Jean: Very strong, stout jean, made of pure cotton or linen, in twill or broken twill weave, in white, and for corsets, etc.

Corset lace: A narrow cotton braid; used to lace corsets with.

Corsicaine: French silk dress goods, made with printed warp. It has small squares placed on the bias over black or coloured ground.

Cortega: A white, tough bast fibre in Panama; used for cordage.

Cortex: (Lat. = cover, bark, shell) In mammalian hair fibres : the principal body of the fibre made up of elongated cells., (i) In wool structure, the cortical or spindle cell layer. (ii) In silk structure, the cortical layer in the fibroin filament and the fibroid interior of the sericin layer.

Corteza del Damajuhato: 'A fibrous bast, yielded by the Couratari tree in Brazil; used for clothing.

Cortical Cells: The spindle shaped cells forming the inside structure of a fibre.

Corvine: Having the characteristics of a crow – in particular, a glossy black colouring.

Co-spun yarn: Produced by the simultaneous spinning of two different types of polymer to give an enclosed mixed fibre.

Cosmos: Fibres recovered from flax or jute rags by tearing them up.

Co-spun yarn: Produced by the simultaneous spinning of two different types of polymer to give an enclosed mixed fibre.

Cost price: That price that essentially will cover any costs involved and ensure that no loss is made.

Cot: The covering material used on various fibre-processing rolls, especially drawing rolls. Leather, cork, rubber, and synthetic materials are frequently employed.

Cote de Cheval: Cotton, wool or silk fabric, made with warp ribs, similar to the Bedford cord.

Cote Menue: French serge dress goods, made with 10 leaves and five picks in a repeat.

Cote Piquee: Solid coloured French serge having eight leaves and eight picks in a repeat.

Cote Syrienne: Solid coloured French serge, having eight leaves and eight picks in a repeat.

Cotelaine: Bleached, corded muslin in French.

Côtelé: (Fr.: côte = rib). Worsted and cotton fabric with fine compact warp ribbed or striped textured appearance, sometimes in patterns or checks etc., commonly called bengaline (England and U.S.). French term for corduroy, although in German trade terminology generally used for finely constructed fabrics having cord-like warp-way ribs. End-uses: ladies wear, suits

Coteline: Warp ribbed dress goods, the heavy cotton ribs alternating with four or six fine warp threads. The filling is often in two colours; used for summer dresses. It is often printed in colours on white foundation; also made in silk.

Coti: Americano Bed ticking in Chile, usually stiffened.

Cotillion: Black and white striped woollen dress fabric.

Coton: Azul /Stiffened denim, made with blue warp and white filling in Chile.

Cottage basin: It is a device used for reeling silk, designed (Indian) on the principle of Japanese multi end reeing machine. It is superior to charkha reeling device. It normally consists of 6 ends.

Cottage industry: Domestic system of manufacture where textiles were spun, knitted or woven in the home.

Cotte: Long women's dress or men's long tunic used by Romans during gothic period.

Cotted: A fleece that contains fibrer those are matted (or "felted") together.

Cotton: The seed hair of plants in the Gossypium botanical genus of the mallow family. After flowering the ovary located in the calyx develops into an elongated capsule which bursts open to allow its seed hairs to billow out. A cotton capsule contains about thirty seeds and each seed is host to 2000 to 7000 seed hairs. The most important quality attribute is the staple length. This is generally between 18 and 42 mm. The separate grades are divided into the following four classes according to their maximum staple length: short staple < 26 mm, medium staple 26–29 mm, long staple 30–38 mm, extra-long staple > 39 mm.

Cotton Back: Silk fabrics, mostly satins, made with cotton back.

Cotton Bagging: Very coarse and heavy jute bagging; used for baling cotton.

Cotton Blanket Cloth: Very soft, napped, all-cotton cloth, made with two sets of warp and one filling, the face and the back of the cloth being reverse of each other; used for bathrobes, blankets, etc.

Cotton broadcloth: A lightly woven cotton, usually mercerized. Handles like medium weight cotton or linren fabrics.

Cotton Checks: In England blue and white checked or striped, all-cotton cloth.

Cotton colour diagram: A digram showing the colour ranges of standards officially established by the US Deoartment of agriculture for the various grades of cotton in relation to the scales of reflectance on the vertical axis and yellowness on the horizontal axis.

Cotton cord web appearance: The visual effect obtained by viewing a sample of web produced by a card operating at normal production speed placed on a black board of designated size.

Cotton count: The yarn numbering system based on length and weight originally used for cotton yarns and now employed for most staple yarns spun on the cotton, or short-staple, system. It is based on a unit length of 840 yards, and the count of the yarn is equal to the number of 840- yard skeins required to weigh 1 pound. Under this system, the higher the number, the finer the yarn. (Also see **Yarn number**.)

Cotton crepe': Cotton crepe is often made in the plain weave with ightly twisted single S twist yarn in the wrp and weft. The cloth is woven in much higher width than the normal finished width, and allowed to shrink while processing, therby the yarns tend to pucker which gives the characteristic effect. Warp crepe is woven using a warp in which 2 ens od S twist alternate with 2 ends of Z twist

Cotton damask: Jacquard-patterned cotton fabric with satin ground, for bed linen and table linen. The clarity of the figurative designs is achieved by the technique of warp and weft satin weave. The silky lustre is intensified by a special finish.

Cotton Ferrets: Cheap bindings or unsized tapes in gray or black.

Cotton fibre: A unicellular, natural fibre composed of almost pure cellulose. As taken from plants, the fibre is found in lengths of 3/8 to 2 inches. For marketing, the fibres are graded and classed for length, strength, and colour.

Cotton flouncing: Plain weave, fine cotton fabrics, embroidered with cotton or viscose yarns with scalloped or flounced edge down one side. Creases easily.

Cotton jersey: Knitted cotton, plain or printed. It is cool and light weight very absorbant and comfortable. Used for sports and casual wear, summer dresses and blouses , chldrens clothes.

Cotton maturity: The degree of fibre wall development.

Cotton mousseline: See **Butter muslin, Mousseline.**

Cotton ratine: A loosely woven plain weave fabric with a rough surface produced by twisting heavy and fine cotton yarns at various tewnsions. Usually dyed and polished for use as a furnishing material, though it is loosely woven for use as chair covers.

Cotton regatta: A plain weave cotton similar in appearance of deniom, but softer and with a closer weave. Produced in deniom colours, blue, grey, pink etc. but definitely a dress weight fabric. Uses for crisp topstitched dresses, unlined jackets, shorts.

Cotton satin: Cotton woven in satin weave to produce a soft fabric with a sheen on the right side. Produced in various weights and usually printed. A dress fabric, but also very popular as curtain fabric also.

Cotton shantung: A plain weave cotton fabric made from irregular cotton yarns. The nubs or slubs are soft and are weak ares. Used mainly for curtains and bedspreads.

Cotton stand: A process originally used for manufacturing cotton fibre into yarn, and now also used extensively for producing spun yarns of manufactured fibres, including blends. Processing on the cotton system includes the general operations of opening, picking, carding, drawing, roving, and ring or mule spinning in the production of carded yarns. For combed yarns, three steps, culminating in combing, are included after the carding operation. There have been many modifications of this process, especially in recent years for the so-called "long draft," or "Casablancas," system. The cotton system is also proving to be the basis of many hybrid systems for handling wool yarns and for manufacturing other long-staple yarns.

Cotton system: A spinning system adapted to fibres less than 2.5" (65mm) in length.

Cotton velvet: A short pile cotton fabric, with additional warp interlacing between two basic fabrics. The pile is afterwards cut and the fabric is separated. Not as luxurious as the other velvet but hard wearing. Suitable for trousers for example.

Cotton viscose: Soft warm light weight fabrics. May be knitted or woven. May also be slightly brushed on one side. A wide variety of plain and printed fabrics. Used for tops, tee shirts, casual dresses, cardigan style jackets etc.

Cotton waste: Material removed from seed cotton, ginned lint, or stock in process by any cleaning or processing machinery and usually consisting of undesirable fibres or a mixture of cotton fibres with foreign matter.

Cotton wax: Constituent of raw cotton (Raw cotton composition) at 0.4–0.8%. Melting point 70–75°C, saponification number 65. In particular free Fatty acids of higher molecular weight ([carnuba wax alcohol-] fatty acid ester), 50–55% free, unsaponifiable high molecular alcohol and hydrocarbons.

Cotton wool: Usually built up in layers (of nonwoven fleece) and condensed, made of fibrous material open to a single fibre, whereby the fibres are held together only by their natural adhesion. Application: upholstery padding; industrial padding, glued wadding (glued on one side) for wardrobe inserts; medical padding for bandage and surgical dressings.

Cotton worsted: Smooth finished twilled cloth, made of hard twist 'cotton yarn woven and finished to imitate worsted cloth; used for cheap clothing, overcoats.

Cotton, acetylated: Acetylated cotton.

Cotton, aminoethylated: Aminoethylated cotton.

Cotton, ran: Ginned lint that has not been subjected to any textile manufacturing process.

Cottonade: Cottonade Originally plain, also serge or twill woven all-cotton fabrics made with single yarns and heavy filling, made in solid colours, checks, stripes, plaids, etc., always dyed in the yarn; used for dress goods, table cloths, etc., and the stronger grades for trousers. Stout cotton cloth in imitation of woollen or worsted; used for men's trousers.

Cottonette: (1) Knitted cotton fabric, cut and made up into bathing suits. (2) Cotton fabric, containing one-third wool.

Cottonee: Cotton back silk satin, made in Turkey.

Cottonised flax: By cottonizing of Bast fibres with a cotton-like short staple is produced with high strength and linen- to silk-like handle. Due to their smoothness, they are usually used in combination with other fibres (cotton, spun viscose, wool etc.), in the cotton spinning system. They are especially suited to fab rics that are very hardwearing and wash-resistant

Cottonizing: "To make similar to cotton", i.e. cottonizing Bast fibres. Term for chemical retting by alkali scouring for example or acidic chlorine bleaching, the result being the exposure of individual fibre cells cemented and bundled by vegetable glue, which are spun by the 2 or 3-cylinder spinning process into yarns like cotton (Cottonized flax).

Cottonseed oil: Cottonseed oil contains 27% oleic, 50% linoleic and 21% palmitic acid. It too is used for food purposes like corn oil.

Cotty Wool: Wool that has matted or felted on the sheep's back. Caused by insufficient wool grease being produced by the sheep, usually due to breeding, injury, or sickness. This type of defective wool is more common in the medium to coarse wools. The fibres cannot be separated without excessive breakage in manufacturing

Couching: (1) In embroidery heavy threads are laid on the foundation and fastened to it by another finer thread. It is used especially in church embroidery. It is flat or raised; (2) A very thick, two-strand thread made of mercerized cotton, wool or silk; used for embroidery and fancy work.

Couleur de rose: Rose coloured.

Count, cotton: See **cotton count.**

Count, metric: See **Metric count.**

Count: (1) A numerical designation of yarn size indicating the relationship of length to weight. (Also see **yarn number**). (2) The number of warp yarns (ends) and filling yarns (picks) perinch in a woven fabric, or the number of wales and courses per inch in a knit fabric. For example,a fabric count of 68 x 52 indicates 68 ends per inch in the warp and 52 picks per inch in the filling.

Count - in knitted textiles: The no. of wale loops and course loops per 25 mm (per one inch).

Count- In woven textiles: The no. of earp yarns and weft yarns per inch counted when the cloth is in Zero tension and is free from creases.

Count, Fabric: The number (counted Units) of warp yarns and weft yarns per inch.

Count of reed, reed number: (1) The number of dents per unit width of reed. number of dents per ionch; (2) Number of dents per two inches; (3) Number of groups of 20 dents per 36 inches (4) The number of dents per 10cm. The recommended unit is no. of dents/cm.

Count, worsted: See **worsted count.**

Count, yarn: See **Yarn count.**

Count-strength product: See **skein break factor.**

Counter current, in drying: The drying medium (hot air, steam, hot gases etc.) flows against the direction of the fabric movement (Uniflow principle in drying; Circulating air drier).

Counter current, Washing/scouring Process: The scouring liquor flows against the direction of the fabric movement (See **Counter current washing process**).

Counter current principle: Diffusion-controlled processes such as scouring are affected both by temperature and an existing concentration gradient. In an open width wash range, it is preferable to bring the clean, fresh detergent solution into contact with the goods being processed at end of their dwell in the range so that the cleanest process liquor meets the cleanest goods in process. By using a counterflow system, the goods first entering the range with the highest level of soiling meet the process liquor with the highest rate of soiling and this soiling is then carried out of the system by the liquor flow.

Counter flow/current Washing process: Continuous Flow washing plant with counter-current water supply (bath current) and fabric transport: water supply in the last spray zone and drain-off in the prewash section. Counter current washing processes allow savings in water usage and optimisation of detergent and heating energy.

Counterpile rollers: The actual raising rollers in the Raising elements of raising machines.

Counting glass: A small mounted magnifying glass for examining fabric. The base of the mount generally incorporates a unit of measurement or as one cm. square, one inch square or cross shaped aperture of carious dimensions, convenient for counting ends and picks or courses and wales in a fabric.

Couple: To combine two or more components, usually a phenol or aryl amine with a diazonium salt to form an azo compound.

Couple, coupling: Operation in dyeing: the bottomed naphthol colour is coupled to the dyestuff with diazotized azoic base or the dyeing salt. Coupling in power transmission: A connection between two coaxial shafts to form one length of shafting and to transmit torque, e.g. from one shaft to another, to gear wheels, pulleys etc. Often also used to transmit power to a machine tool from a motor.

Coupling of dyeing machines: Dyeing machines are coupled in order to extend the capacity of existing dyeing machines, e.g. for large batches. This can be done by parallel coupling or cross-coupling. Existing valves can of course be regulated in such a way that each machine can also operate individually (i.e. uncoupled).

Couratari: In South America a fibrous bast of the same tree; used for blankets, clothing, clothes, etc.

Courimari: In Venezuela the fibrous bast of the Couratari tree; used for blankets, clothes, etc.

Course, in knitted fabrics: The series of successive loops lying crosswise of the knitted fabric, that is, lying at right angle to a line passing through the open throat to the close end of the loop.

Course: A seies of successive loops lying cross-wise in a knitted fabric, that is, lying at right angles to a line passing through the open throat to the close end of the loops.

Course count: the number of courses in a knit fabric per unit length measure.

For example: courses per inch.

Course counter: An instrument or unit (e.g. Tube-tex stitchcounter) to count the courses in a knitted fabric. In many shrinkage machines by counting number of courses per unit length can give indication of the shrinkage dine in the compactor/ or even the GSM. It is possible to optically photocopying the 'highlights' (more illuminated dots in a unit length) and count them because the loops of the courses are aligned three dimensionally by the hooks in the production of knitted fabrics.

Course density: Number of courses per centimetre of fabric length (course density): this is the number of horizontal rows composed of stitches per unit length of fabric in the lengthwise direction (generally, one cm fabric length is taken).

Course length: (weft knitted) the length of yarn in a weft knitted course.

Course length: The amount of yam used in forming all the knitdoops in one course of a knitted fabric. Also called run-in.

Course, in knitted textiles: A row of loops across the width of the fabric.

Course, weft knitted: A row of loop across the width of the fabric.

Courte Pointe: Bed quilts, made in 'France of pique or calico and stuffed with cotton batting

Courtelle: Courtalds' trade name for their acrylic fibre. Used alone for clothes, carpets, and blankets. It blends well with other fibres and mixtures are used for all type of clothings and furnishing fabrics. Crease resistant, strong, washable, unaffected by moth and mildew.

Courtrai: French bobbin lace similar to the Val, the threads of the mesh ground 'being twisted three and a half times.

Coutance: Strong, French ticking made of plain hemp.

Coutil, Coutille: A strong cotton fabric made in herringbone weave or reverse twist twill weaves. It is a close textured cloth, sometimes patterned and used for corsetry, and sometimes used as a strong lining fabric or tropical suiting Bleached or dyed, woven in 2/1 warp faced twill usually in herringbone stripes which is used for corsets. A typical cotton construction is 28s x 20s; 117 x 68; 5 1/2 oz./yd^2

Coutils de Brin: Coarse French ticking.

Coutil Facon de Bruxelles: French coutil, made with narrow stripes.

Covalent bond: a chemical bond where a pair of electrons is shared relatively equally between two atoms in the compound Covalent bonds are formed between the fibres and reactive dyes. These are the strongest type of chemical bond, and are responsible for the excellent washfastness of reactive dyes.

Coventry blue: A blue dye providing good fastness of colour; thus used to describe a person who shows great loyalty. Takes its name from a blue thread once made in Coventry. See **true blue**.

Cover: (1) The degree of evenness of thread spacing.

(2) The degree to which underlying structure is concealed by the surface material, as in carpets, the degree to which pile covers backing.

(3) The row of loops or stitches running across a knit fabric, corresponding to the filling in woven fabrics.

Cover: The outside layer of fibre which forms the surface of the yarn.

Cover, in woven fabric: The cover defines as the area of 1 cm2 of the fabric which is actually covered by the warp and weft yarns. It is useful and practical indication of the fabric's permeability to light, gasses, liquids and solid particles.

Cover, in yarns: The outside layer of fibres that form the surface of a yarn.

Coverall: A one piece type of lagged work wear often capable of being fastened at wrists and ankles.

Cover factor, in knitting: (CF), Knitting. A term used for the comparative assessment of knitted fabric density. The cover factor CF is calculated from the number of knitted courses c, the number of knitted wales w, and the yarn count Nm. $CF = (c+w)/(Nm)^{1/2}$

Cover factor, in weaving: The fraction of the surface area that is covered by yarns assuming round yarn shape. It indicates the extent to which the area of a woven fabric is covered by either the warp or weft (filling) threads. It is the ratio of fabric surface occupied by yarns to the total fabric surface. Thus for any woven fabric two cover factors are involved, i.e. the warp cover factor and the weft (filling) cover factor.

Cover factor (Knitted fabrics): Tightness factor (cover factor) indicates the looseness of a knitted fabric, i.e., increase in tightness factor results in tighter fabrics. It is defined as the ratio of the area covered by the yarn in one loop to the area occupied by that loop. Tightness factor ranges between 10 and 20 for most weft knitted structures, the most suitable for plain knitted fabrics is around 14–15.

Cover Roller: See **Blotch printing machine**.

Cover stock: A permeable fabric used hygiene products to cover and containan absorbent medium.

Coverd yarns: A compound structure consisting of a readily sepserable core surrounded by a cover framede by one or more spun or filament yarn.

Coverspun spinning: Covers pun is both the name of a spinning system and a yarn. In this system staple rovings are drafted in a conventional manner, and then the roving is passed into a hollow vertical spindle. On the outside of the spindle a filament yarn, from a cylindrical spool rotating at 20,000 to 30,000 rpm, is fed into the top of the hollow vertical spindle with the roving. The rotating filament spindle causes the filament to wrap around the core of the staple to produce a wrapped yarn consisting of 80% to 95% staple. Polyester filaments are usually used to wrap pure yarns or blends of cotton, wool, nylon, and acrylic staple to form sewing threads or yarns for textile substrate production.

Covering stitches: A type of chain stitch produced on a twin or multi needle machines used for covering seams.

Coverstock: A lightweight nonwoven material used to contain and conceal an underlying core material. Examples are the facing materials that cover the absorbent cores of diapers, sanitary napkins, and adult incontinence products.

Covert: A twill-woven woollen cloth, sometimes with fulled face, sometimes sheared to imitate whipcord a short top coat.

Covert, cotton: A sturdy fabric, resembling denim woven in 2/1 twill. A characteristic construction is 66 x 46 with 12s mock twist warp and 12s solid coloured weft. The weft yarn is dyed, usually the same shade as the darker colour of the mock twist in the warp. Made with two shades of colour e.g. (Medium and light brown). The warp is 2 ply (1 light; 1 dark) adn filling 1 ply (dark or same as warp). Very rugged and closely woven. Has a mottled or speckled effect. First used as a hunting fabric. Has a clear finish and hard texture. Wears exceptionally well and has a smart appearance. Light in weight. Uses: For overcoating for both men and women. It is also made waterproof and used a great deal in rain water.

Covert, wool: Made with stock dyed yarns in D & T (double and twist) or mock twists for the warp, such as dark brown and light brown which produces a mottled beige effect.. The darker shade of the two is used for the solid coloured weft. A typical construction is 44 x 36 using 3 run woollen warp and weft. The weight is about 16-18 oz. per yard. The cloth is lightly napped. Covert cloth is also made in cotton, viscose and acetate with the characteristic speckled effect. Tjis is lighter in weight than the overcoating and is used for suits suits and rain coats

Covert, worsted: It is made as in the case of woollen covert, with 80 x 50, using 2/30s worsted warp and 16s worsted weft. The weave may be a reinforced satin weave or 3/2, 63 degree twill weave with a fabric weight of 12-14 oz. per yard in 60 in. width.

Cowslip: A light yellow.

Cow tail: In worsted sorting means coarse staple taken from the tail end of the low lustre fleece; spins 26s to 28s.

Cowhide: Occasionally used for ready-made clothes, but not usually available for sewing.

Cowoven fabric: A hand device used to stretch carpets in a small area.

Cox royal arch: Commercial variety of early maturing cotton from Georgia, the staple measuring 25-28 millimeters; the yield is 30-32 percent.

Coxcomb: A bluish-red.

CPB: abbrev. for: **Cold pad batch process.**

CRA: Crease recovery angles.

Crab: A hand device used to stretch carpets in a small area.

Crabbing: The process of heating wool or hair fabrics, under tension, in a hot or boiling liquid, then cooling under tension, to provide the fabric with dimensional stability. The process is done for undyed woven fabrics (e.g. ladies and menswear fabrics in worsted, Cheviot and mohair) (a) for setting the cloth to restrict excessive creasing and felting in subsequent dyeing, and (b) for removing oils and additives.

Crabbing machine: Machine used for Crabbing operation/process. Even though batch wise and continuous machines are used, present day most of the processors use continuous crabbing machines.

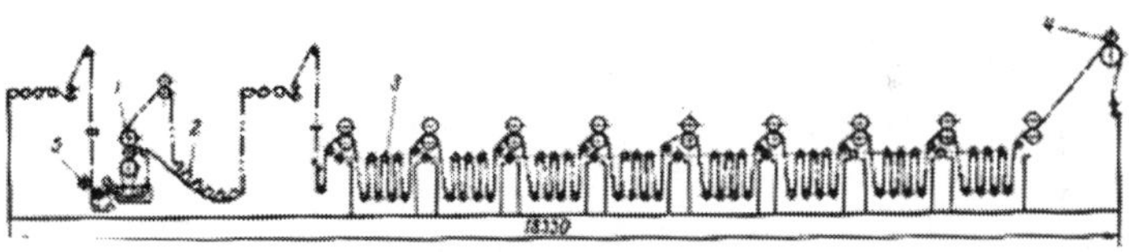

Semi-continuous crabbing

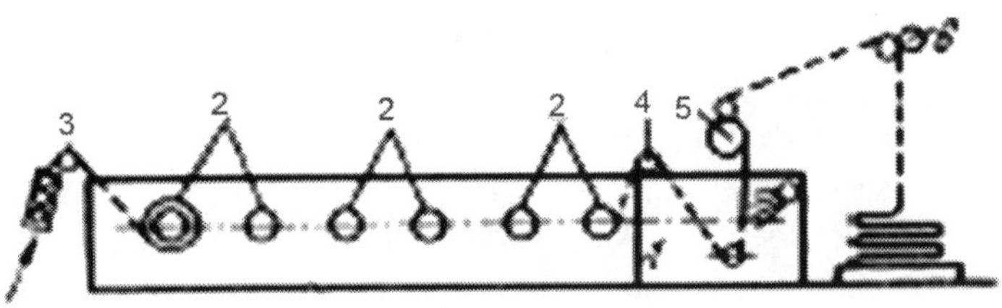

Continuous carbbing

Crack: A defect in a woven fabric consisting of an open fillingwise streak extending partly or entirely across the fabric.

Crack mark: A sharp break or crease in the surface of a coated or laminated fabric.

Crack mark, in bonded/laminated fabrics: A sharp break or crease in the surface contour of either the face fabric or the backing fabric that becomes evident when the bondedor laminated composite is rolled, bent, draped or folded.

Craft dyeing: It is the artistic use of dyes and processes of application to produce special objects of beauty and artistic value. He (or she) endeavors to produce colours and colour combinations on a fabric more with reference to the particular use of the material being dyed, and at- the same time trying to put into the work some form of artistic expression, using the colour as a component of design.

Crafts man: An artisan who is skilled in creating new cultural objects.

Cramming: Where more threads are packed into some areas of the fabric than others.

Cramoisy: Crimson.

Cranberry red: The dark red of the cranberry.

Crane-feather: Having the grey colour of a crane's feathers.

Cranston: A highland tartan, composed of dark blue and light blue stripes and red and blue lines.

Crapaud: French for mispick.

Crapaudaille: Fine silk crepon in France.

Crape: (1) In the 18th century a woollen fabric in England, made either crimped for deep mourning or smooth; (2) The extremely fine knitting on some of the 'Shetland shawls. (3) see **Crepe**.

Craping: Process of rendering a fabric crimpy or crapy. The most important processes are: (a) The filling is composed of right hand twist and left hand twist yarns, interchanged at every two or four picks; (b) Certain warp threads are wound on a separate beam and held much slacker than the rest of the ends, forming stripes of crepe; (c) Cotton is treated with caustic soda, wool or silk with concentrated sulphuric acid for a short time at certain places, producing crepe el feet; (d) Warp and filling made of different fibres with different shrinkage; (e) Combinations of these processes; (f) Produced by causing the fabric to adhere at certain parts to stretched rubber bands which, when released, will cause the crinkle. (g) Produced by using yarns of various degree of twist.

Craquele': (Cloqué). Fabric with relief texture woven with synthetic fibres. Generally produced bymixing high-shrinkage and low-shrinkage synthetic yarns woven in fancy design, the effect being obtained by appropriate finishing. At high temperature (in hot aqueous solution or steam atmosphere) the highshrinkage yarn contracts and acting in association with the low-shrinkage yarn produces the relief effect. Similar effects can also be obtained by topical application of fibre-swelling or fibre-contracting agents (alkalis, phenols, cresols etc.).

Crash: (1) Fabric (usually Linen) in plain weave or twill, or variations on twill, with a rough texture due to thick uneven yarns used in the weft. Traditionally, it is creamy beige. Browm mercerized cotton, or occasionally wool, may be used for the warp in place of linen. Mainly used for towels curtains and embroidery.

(2) Originally unbleached, rough, coarse palin weave linen fabric made with irregular yarns. Nowadays made by intentionally creasing and fixing the creases in finishing. Now this name is used for any coarse fabric with a rough, irregular surface made from thick, uneven yarns.

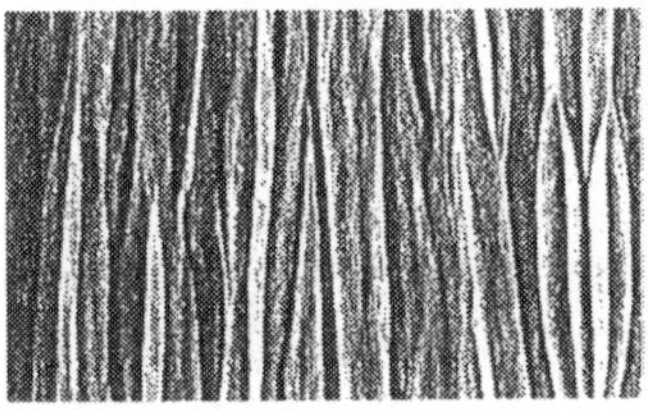

(3) (a) Plain or twilled gray, bleached or checked cotton and jute cloth, sized and calendered to resemble linen; used for towels, suits, etc.; (b) Light, plain woven, coarse linen fabric, made of uneven, slack twist yarn; used for summer suits, towels. (c) Coarse plain woven light woollen fabric made of rough hard spun yarn. Dyed in the piece or made in mixture effect.

(4) A special fabric used for curtain. Totally synthetic material, produced by crushing, stamping, thermal or chemical treatment for a creased-look. Crash-fabrics have only limited dimension stability, hanging-out is possible.

(5) Lengthways creases Introduced in by finishing.

Crash towel: A plain weave nonterry product with hems or selvedges which has rough texture caused by uneven yarns.

Cravenette: Cloths treated and finished before weaving by an improved process which renders thcm rainproof. A secret process owned by the Cravenette Company and by Priestly & Company of England and the United States.

Crawford: (1) A commercial variety of early ripening cotton from South Carolina, the staple measuring 23-27 millimeters; the yield is 32-33 per cent;

(2) A highland tartan, composed of green stripes over a crimson red ground and narrow white lines.

Crazy Quilt: Made by joining irregular pieces of cloth together with various fancy stitches, without arranging the pieces into any pattern.

CRE: abbreviation for Constant-Rate_of-Extension.

Crea: A more or less stiffened bleached cotton cloth in Chile.

Crea para sabanas: Bleached cotton sheeting in South America.

Cream Damask: Linen damask made of partly bleached yarn.

Cream of tartar: See **Potassium bitartrate**.

Creamed linen: Linen yarn partly bleached.

Creas Stout: Bleached linen canvas made of hard spun yarn; the pieces made 30 metres long in Germany and Austria.

Crease: A break or line in a fabric generally caused by a sharp fold. Creases may be either desirable or undesirable, depending upon the situation. A crease may be intentionally pressed into a fabric by application of pressure and heat and sometimes moisture.

Crease mark: A visible deformation left in the fabric after a crease has been incompletely removed during fabric processing or ironing.

Crease, pressed in: A crease intentionally inserted on a fabric usually by application of pressure, heat and moisture.

Crease recovery: See **Wrinkle recovery.**

Crease recovery angle: (uncreasing angle). Dimension for determining the effect of non-crease finishing (Crease resistance). It is a measure of the recovery power of a fabric. It is the angle which results after a specific time when a straight yarn test specimen, folded through 180°C, is released after prior loading.

Crease resist finish: See **Anti-crease finish.**

Crease resistant: A fabric which is resistant to crease formation during wearing or washing or any proess of crease formation. The level of crease resistance is assessed by crease recovery or DP ratings or any other methods.

Crease resistant finishing: See **Anti-crease finish.**

Crease resistance: (fastness to creasing). Resistance of textiles to creasing in terms of wet and dry crease resistance. Not only measurement of the wet and dry crease recovery angle, but also the actual wash and wear behaviour of a woven fabric is used for qualitative assessment. Temperature, air humidity, period of loading and recovery time has a great effect.

Crease retension: The ability of a fabric to maintain an inserted crease made during the garment making like pleats in womens skirts and the folds in the garments. Opposite of crease resistance. Usually the garments are treated with the resin precondensates and then cured during the required creases are set in. Crease retention can be measured subjectively or by the relation of a crease in a subsequent state to the crease in the initial state. Crease retention may be strongly dependent on the conditions of use, e.g., normal wear, washing or tumble-drying.

Crease, pressed in: See **Crease retention.** The creases are pressed in during the curing process of the resin precondensates to retain the same in the garments.

Crease-resistant: A term used to describe a fabric treated chemically to improve its resistance to and recovery from wrinkling.

Creasing: Sensitivity of individual fibres to continuous flexure, creasing, pressing, squeezing across the fibre direction, in woven fabrics transversely to warp and weft ends.

Creasing angle: See **Creasing; Crease recovery angle.**

Creasing Inhibitors: Textile auxiliaries for the prevention of _ Running creaser, especially when dyeing in rope form.

CRECIT: Centre de Recherches d'Essai et de Contrôles Scientifiques et Techniques pour l'Industrie Textile (Scientific Technical Research Centre of the Belgian Textile Industry), Tournai, Belgium.

Creeks: General trade name for various cottons of light body, measuring from

11-6 to 1% inches in length.

Creel: (1) A framework arranged to hold slivers, rovings, or yarns so that many ends can be withdrawn smoothly and evenly without tangling.

(2) A similar device used to aggregate sub-tows to tow in the manufactured staple processing, especially polyester.

Creel, magazine: This kind of creel is used when several warps of similar type must be prepared in sequence that is when large lots of similar yarns need to be processed. Level with each tensioner, two bobbins are positioned: one operating and the other as reserve.

Creel, mobile: this creel type is similar to the standard creel, but is formed by trolleys which can be taken individually out of the creel. The bobbins are creeled up on each trolley outside the creel. During the creeling up of a series of trolleys, the second series of trolleys is brought back to the outside of the creel to feed the warper. This reduces considerably the waiting time. The mobile creel comes in handy especially when there is insufficient room to permit the use of two standard creels.

Creel, Swivel frame: this type of creel was designed as a variation of the mobile creel to enable the creeling up of bobbins which, owing to their heavy weight (5 to 25 kg), cannot be pinned on trolleys. Each bobbin holder is double-sided: the threads are unwound from one side, while a new series of bobbins is creeled up on the other side.

Creel, V-shaped: In this creel type, the creel boards are assembled in form of endless chains. While warping is carried out from the outer sides using the already creeled up bobbins, the subsequent yarn lot can be creeled up on the empty spindles positioned inside the creel. This interior room serves at the same time as storage and bobbin exchange station. The yarn lot can be changed by simply pushing a button, which starts the electrically drive of the chains.

The empty bobbins move towards the inside of the creel, the full bobbins towards the outside.

Creeling: The mounting of supply packages in a creel to feed fibre to a process, i.e., beaming or warping.

Creep: See **Delayed deformation and Permanent deformation**.

Creepage: A relaxation shrinkage that occurs under normal conditions of storage.

Crefeld Velvet: Light German velvet made of silk and cotton; used for dresses, millinery, etc.

Crehuela: Light osnaburg in Venezuela.

Creme de violette: A deep violet.

Cremosin: Cremosin, cremesin, cremoysin, cremsin, cremysyn, cremysy, cremasie and qerinasi are all obsolete forms of **crimson** and **carmine**.

Crenele: The edge of the dress or material, made like a battlement.

Creoulo: Raw cotton, grown on trees in Brazil.

Crepe: (1) It refers to any type or weight of fabric characterized by a crinkling surface obtained by the use of (a) hard twist filling yarns (b) chemical treatment (c) crepe weaves or (d) embossing. Crepe fabric can be made with any fibre, natural or synthetic, or from mixtures or blends. (2) (Lat: crispus = puckered).

Textile fabrics with a typically wavy, grained or puckered surface (patterned). There are a variety of crepe fabrics. They are all light weight fabric with irregular surface. The crepe effect can be nade with weave, special finish or by combination of these. Blanced crepe is made with alternate 2 S twist with 2 Z twist or 4 S twist with 4 Z twist in the warp and weft. Flat crepe can be made with a typical construction of 130 x 75 with 100 denier dull rayon in the warp and 100 denier bright rayon in the filling. It is called flat crepe because in this the fabric, the yarns are not given any special twist. It is usually woven in the plain weave, with two picks of the S twist alternating with 2 picks of Z twist in the weft. The weight of the fabric will be around 3 oz. per square yard. Sand crepe is made in 90 x 70 woven in small broken crepe weave, one of the yarns is 150 denier dull acetate and the other a two ply yarn consisting of untwisted 120 denier dull acetate plied with crepe twist 100 senier bright rayon. This is one of the many possible construction and yarn combinations. All silk crepe : 170 × 110; 13/15 denier warp and 18/20 denier weft; plain weave. The warp and weft arrangement will be 2 s twist alternating with 2 Z twist; fabric weight will be around 2 oz. per square yard. Alapca Crepe is a soft, dull acetate or rayon fabric made to simulate worsted crepe.

Crepe' back satin: Satin weave with crepe yarn filling which can be used with either side as the right side. Yarns are silk or polyester. The fabric looks like crepe on one side and satin on the other. A crepe back satin may be a warp flush sartin with a crepe filling that makes the back of the fabric look like a crepe.

Crepe' charmeuse: A fine, soft (as well as stiffer) crêpe in twill or plain weave (one warp and two weftsystems, i.e. a crêpe backing weft and a crêpe face weft). The face exhibits a satin crêpe effect, and the reverse a shaded crêpe effect. Use: for elegant womenswear.

Crepe' chiffon: A printed, true crêpe with a chiffon character.

Crepe' cloque: See **Cloqué.**

Crepe crepe: The warp has a larger number of twists than in ordinary crepe, producing a very deep crepe effect.

Crepe' de chine, yarn: A silk yarn made from raw silk filaments with a more or less left-handed twist which have been very sharply twisted together to the right. Hard twisted yarn, usually with 1600-2500 TPM, generally made from 3-5 raw silk threads. It is used as weft in crepe-de-chine.

Crepe' de chine, material: Probably the best known silk fabric. A soft, supple dress fabric in plain weave with e.g. a raw silk warp and a crêpe weft (produced from two S and Z highly twisted alternating yarns). It is usually either piece-dyed or printed.

Crepe de Dante: Crepe, made of silk and wool warp and silk, lisle and wool weft.

Crepe d'Espagne: Very light fabric, made with silk warp and fine woal filling oven like gauze.

Crepe de Laine: French thin woollen dress goods, made in plain weave and slightly creped.

Crepe de Sante: Porous and close woven crepe with a coarse face, made of undyed wool and silk; used for underwear. Also made of cotton or linen.

Crepe fabric: A fabric, characterized by a crinkled, puckered or pebbly surface with highly twisted yarns in the weft and sometimes in warp or both. Similar effect is also obtained by using normal twisted yarn and crepe weave.

Crepe faconne: A crêpe fabric with small woven patterns, See **Façonné.**

Crepe, finished: Crepe effect can also be achieved in finishing. Caustic soda solution may be printed onto cotton fabrics in a stripe or other pattern. The printed portions shrink and cause the fabric to cockle. Commercial styles: Plisse, blister, bark crepe , seer sucker.

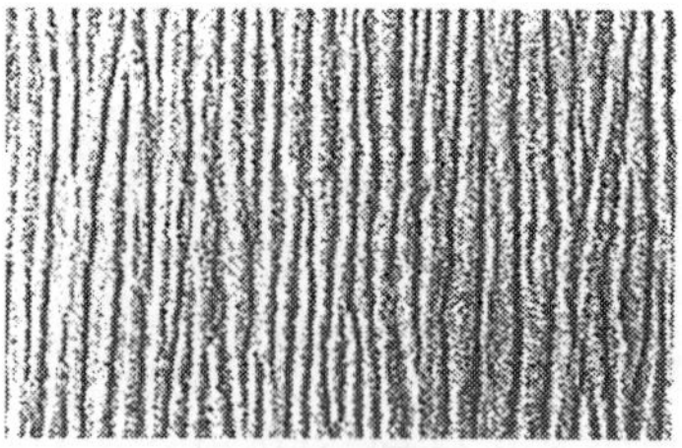

Bark Crepe

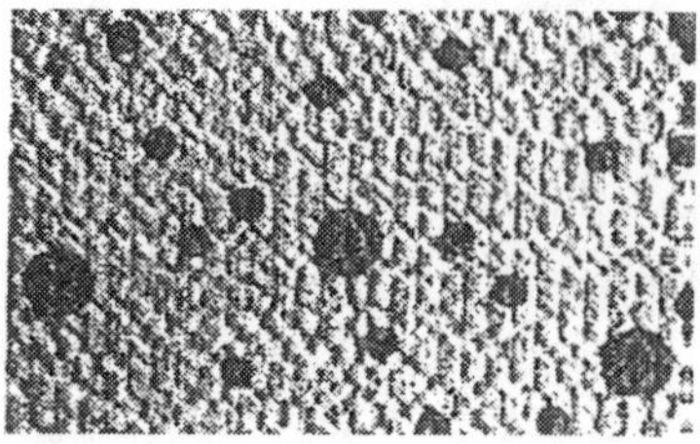

Plisse

Crêpe flamisol: A so-called matt crêpe fabric made from e.g. a warp delustred in spinning and a crêpe.

Crepe, Full: Is made with crepe yarns in both directions (Fig). The weave is either a palin or

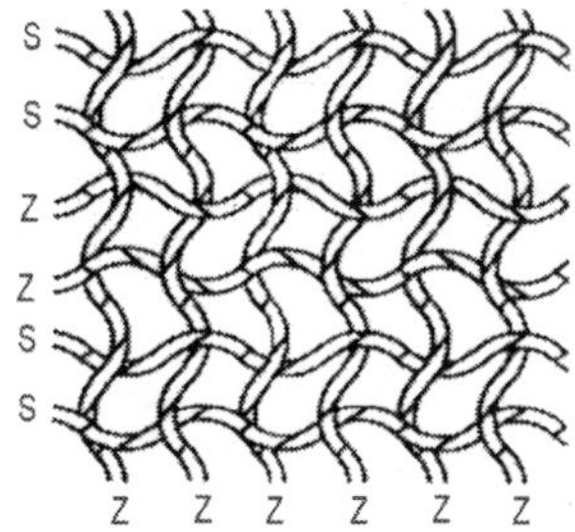

Yarn arrangement in Full Crepe

crepe construction. Commercial styles Georgette, chiffon.

Cre'pe geogette: A fine, matt crêpe fabric, similar to voile (gauze-like). It has a fine irregular surface with a grainy, dry, firm handle. The warp and weft consist of highly overtwisted (2000–3600TPM) crêpe yarns (ply yarns), with alternating pairs of S and Z twist yarns in a plain weave construction. The fabric is tightly woven but nevertheless transparent. It is dyed in the piece or printed. Uses: blouses, dresses, skirts, gowns and millinery.

Crepe, Half: has crepe yarns only in one direction usually the weft. See **Crepe fabrics.**

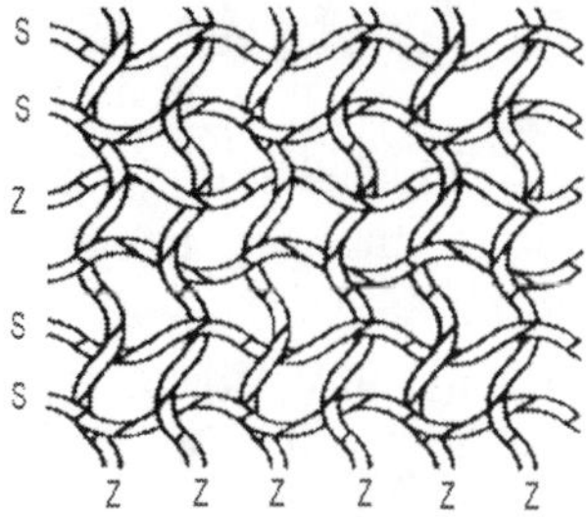

Yarn arrangement in half crepe

Crepe' georgette, silk: Warp of crepe threads with 2s/2z twist, weft also of crepe threads with 2s/2z twists. Lightweight, heavy, sheer fabric. Has quite a bit of stiffness and body. gives excellent wear. Has a dull, crinkled surface. Achieved by alternating S and Z yarns in a high twist in both warp and filling directions. Georgette has a harser, duller, more crinkled feel and appearance than crepe de chine. Uses: After 5 wear and dressy afternoon and weddings, lingerie, scarves, etc. Same uses as crepe de chineFabric is soft and slightly kinky.

Crepe gloria: A crêpe fabric having pleat-like lengthwise folds in a plain weave construction.

Crepe' jersey: A crêpe fabric resembling jerseyknit produced in a plain weave construction (with a crêpe weft consisting of 2 Z-twist and 2 S-twist yarns) used for womenswear fabrics.

Crepe knitted: A fabric characterized by a crinkled or puckered surface. The effect may be produced by a variety of ways for example by the use of S and Z high twist yarns, by the use of particular knitted construction , or by chemical or thermal treatment to provide a differential shrinkage in the finished fabric.

Crepe' lavable: A more tightly plain woven, filament crepe fabrics with crepe yarns in the warp and normal yarns in the weft, example of an easily washable crepon or crêpe de Chine with only a slight crêpe effect. Used for dresses and blouses.

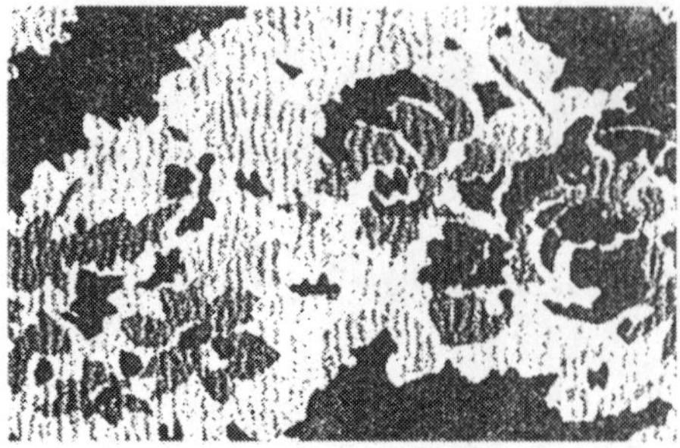

Crepe lisse: Very light highly finished crepe, made in gauze weave, slightly stiffened, comes in pale colours; the warp has fewer twists than in the crepe crepe; used for dresses and trimmings.

Crepe marocain: A light, finely ribbed, puckered crêpe (similar to crepon or crêpe de Chine) with a crimped, grainy surface, (slightly wavy weft), in plain weave, also patterned (jacquard fabrics). Generally warp of greige threads, weft of crepe multi twisted of threads with 2S/2Z twist.

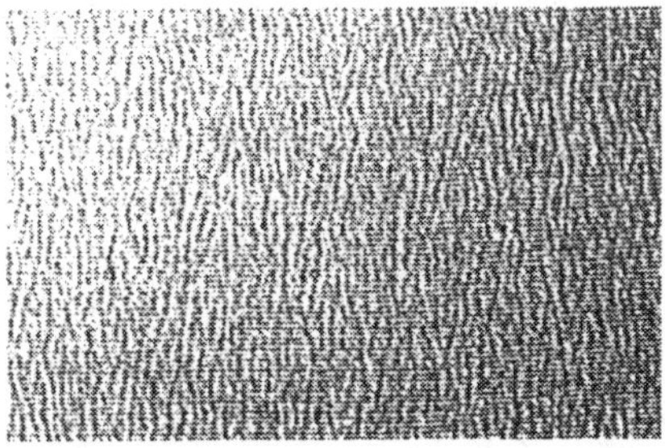

Crêpe marocaise: A lustrous crepe comprising a viscose warp and viscose filament or spun yarn or crepe wool crepe yarn weft, lightly crimped appearance in taffeta or rib weave.

Crepe meteor: A highly finished silk crepe.

Crepe morette: Light weight crinkled fabric, made of fine hard spun worsted warp, widely set and a heavier slack twist filling.

Crepe' mosseline: A transparent, light crêpe fabric with a wool-like character produced in a so-called sand crêpe weave.

Crepe onde: A crêpe with yarns consisting of folded crêpe yarns with chemically spun viscose slub filaments.

Crepe pique: A matt crêpe fabric with a piqué-like character.

Crepe' reversible: A fine, double-sided, patterned crêpe fabric (produced mainly with jacquard effects); the face is e.g. matt with a crêpe character and the reverse is smooth and lustrous (similar to satin), other variations also exist.

Crepe romain: The finest crêpe fabric, similar to crêpe Georgette, with a check-like surface, in a panama weave construction (therefore also referred to as panama crêpe), with two alternating warp and weft threads.

Crepe satin: Silk, A generally small-figured crêpe with an ultra-bright face and a low elegent lustre reverse, the latter having a ribbed effect; the ground fabric is in plain weave and the pattern effects in satin weave using warp of greige thread and weft of greige twisted thread with 2 S-twist and 2 Z-twist threads.

Crepe satin faconne: Threads as in crepe satin but with a jacquard weave. The patterning produced by weaving method lends itself to rich effects which are very mich in demand, especially when over printed.

Crepe, Flat: Also called French Crepe or Lingerie Crepe but not exactly the same. It is the flattest of all the crepes with only a very slight pebbled or crepe effect hard twist alternating 25 × 22 in filling; warp has ordinary twist. It is very soft and pliable, which makes it good for draping. It is very light weight - 2 times as many ends as picks. It may be white, coloured, or printed. Most of it launders well. Uses: Accessories, blouses, dress goods, negligees, pyjamas and other pieces of lingerie and linings.

Crepe, silk: They all have a pebbled, rough feel and appearance. Yarns have a high twist in the filling or the warp or both. Most crepes launder well with care. Crepe-back satin, satin-back crepe, crepe-satin, or satin-crepe. Satin weave on the face and a crepe effect on the back obtained with twisted crepe yarns in the filling - 2 or 3 times as many ends as picks per inch. It is a soft fabric which is reversible. It is usually piece dyed. Very interesting effects can be obtianed in a garment by using both sides, in different parts. E.g. the crepe side for the body and trim or binding with the satin part up. Uses: Dresses, blouses, linings, after 5 wear

Crepe suzette: Synonym for Crepon Georgette.

Crepe yarn: See **Crêpe yarns.** (a) A highly twisted (Generall 1200-4000 TPM) yarn which may be used in the production of crepe' fabrics in woven or knitted form. (b) A balanced hand knitting yarn with a plited appearance produced by twisting together 3-4 two folded yarns normally using 2s and 2z twists. See **Yarn, crepe.**

Crepe/Georgette yarn: These are highly twisted, tightly folded yarns, produced in S or Z twist, used to produce crêpe fabrics in various patterns thereby giving the fabric its characteristic puckered crêpe effect after the so called Crêping.

Crepe-de-chine (French Yarn): See **Crepe de chine yarn**, **Crepe de chine material**.

Crepe, weft knitted: An irregular surfaced palin or rib faced weft knitted fabric usually constructed for example, from knit float, or knit tuck loops introduced in a predetermined random or other order.

Crepe, warp knitted: A double faced warp knitted fabric , the construction of which contains more rows of cleared stitches per unit of lrength on one side than the other.

Crepe-de-chine fabric: See **Crepe de chine material.**

Crepine: (1) Black or coloured French silk dress goods with very small dotted patterns; (2) A fringe.

Crepeing: Development of a special surface effect for Crêpe products in the chemical finishing stage with completely tension-free treatment. Up to 40% fabric contraction in the crêpe yarn system direction.

Crepoline: Light sheer fabrics of silk or cotton, woven to produce a rib-like crepe effect warp wise.

Crepon fabric: It is another of the crepe group of fabrics but stronger. These are highly twisted, tightly folded yarns, where S or Z twist yarns is put according to a plan and another method is to use yarns with varying twists, used to produce crêpe fabrics in various patterns thereby giving the fabric its characteristic puckered. crêpe effect after the socalled creping. A typical construction is to use 90 × 50 to 125 x×55 structure using 100-150 denier rayon in warp and 150 – 200denier crepe twist rayon in the weft.A dress fabric in crepe but heavier and more rugged than the average crepe. Usually made in silk or Rayon. Crepe effect appears in the direction of warp and is produced either by different degrees of twist in the yarn , by employing only right or left hand twist yarns or by weaving some warp yarns slacker than the others.

Crepon Georgette: A georgette in which all the weft yarn have been twisted in the same direction.

Crepon Givre: A twilled crepon (see) with frosted effect.

Crepon yarn: A crêped material which only has crêpe yarns of one twist direction in the weft or both the warp and weft, or even 3 Z-twist threads alternating with 2 S-twist threads, which gives rise to the typically wavy folds. It is generally produced in plain weave.

Crescent: In needle-point lace the usually crescent shaped part of the flower pattern surrounded by raised cordonnet.

Crescentin: Cloth made of waste silk in France.

Creseau: (1) See **Carisol**; (2) Twilled French woollen, napped on both sides, similar to the kersey.

Crespine: Silk net in the 16th century; used as covering for the hair.

Crespinette: Silk hair net in France.

Cresson: A shade of green resembling the colour of water-cress; also 'cress-green'.

Crete: A fancy braid, composed of two groups of straight threads holding a heavy, trailing cord, forming picot edges.

Crete lace: Bobbin lace made of various coloured slack twist silk or flax in geometrical designs which are outlined with coloured thread. Similar to Torchon.

Cretonne: First produced in Cre'tonne, a village in Normandy, hence the name. A printed or plain strong cotton fabric, woven with a fine cotton warp

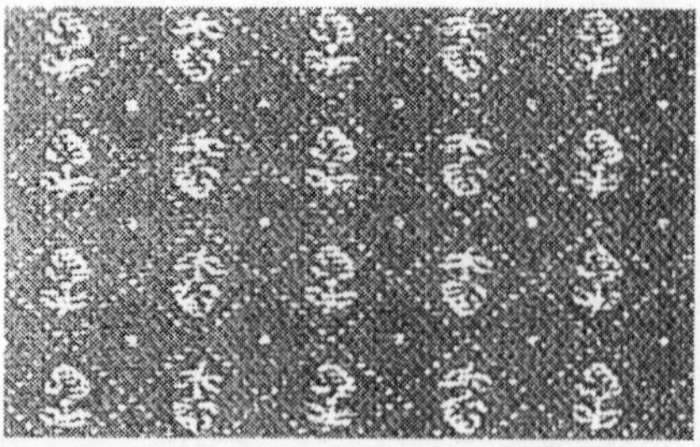

and a thick weft spun from waste cotton. The weave may be plain or twill. In appearance, cretonne is similar to Chintz but doesnort have a glazed treatment. Fairly inexpensive. Used for curtain, loose covers, bed covers, upholstery, bed linen, shirts etc. See **Chintz.**

Crevette: The deep pink of the shellfish of the same name.

Crewel: Loosely twisted, fine two-ply worsted yarn for embroidery.

Crewel lace: Narrow edging, made of crewel yarn.

Crewel needles: These hand sewing needles have a sharp point with a long narrow eye. They are used with embroidery floss, silk ribbon and heavier fabrics. Size 1 to 10

Crewel stitch: Same as stem stitch

CRF: Crease resistant finish.

Cricketing: Fine twilled flannel dress goods; used in England for sporting costumes.

Crimp: (1) Waves and curves exist in virtually all staple fibres and are especially evident under the microscope. The nature of the crimps and their frequency vary widely. Crimps hold fibres apart thus producing numerous air cavities in the yarn which influence to varying degrees its bulk, softness of handle and insulation properties. The waviness of a fibre, i.e. the condition in which the axis of a fibre under minimum external stress departs from a straight line and follows a simple or a complex or an irregular wavy path.

Note: (a) In its simplest form, crimp is uniplanar and regular, i.e. it resembles a sine wave, but it is frequently much more complicated and irregular. An example of three-dimensional crimp is helical.

(b) Crimp may be expressed numerically as the number of waves (crimps) per unit length, or as the difference between the distance between two points on the fibre when it is relaxed and when it is straightened under suitable tension, expressed as a percentage of the relaxed distance. (2) The difference in distance between two points on an unstretched fibre and the same two points when the fibre is straightened under specified tension. Crimp is expressed as a percentage of the unstretched. (3) The difference in distance between two points on a yarn as it lies in a fabric and the same two points when the yarn has been removed from the fabric and straightened under specified tension, expressed as a percentage of the distance between the two points as the yarn lies in the fabric. The waviness or distortion of a yarn owing to interlacing in the fabric.

$$\% \text{ Crimp} = [(l_y - l_{fab})/l_{fab}] \times 100$$

Note: (a) In woven fabric, the crimp is measured by the relation between the length of the fabric test specimen and the corresponding length of yarn when it is removed therefrom and straightened under suitable tension. The crimp may then be expressed numerically as a percentage or as a ratio, i.e. the ratio of yarn length to fabric length. In both methods, fabric length is the basis. (b) Although this definition could logically be applied to knitted fabrics or fabrics of pile construction, it is usual to employ special terms, e.g. stitch length, terry ratio.

Crimp, as applied to a continous-element Zipper: The predetermined formation of the monofilament cross section at the point where the continuous element is interlocked.

Crimp amplitude: The height of displacement of the fibre from from its uncrimped condition.

Crimp contraction: (or crimp retraction). Defined as the contraction in length of a previously textured yarn caused by the formation of crimp in individual filaments under specified conditions of crimp development.

Crimp crepe: The term is not quite correct. What is meant is fabrics with a crêpe-like effect a) like the crepon printed effect on cotton or wool, or b) with the use of crimp yarn.

Crimp depth, in grease wool locks: The perpendicular distance between the peak and valley of the waves.

Crimp deregistering: The process of opening a tow band by causing the peaks and valleys of the crimp to lay randomly rather that uniformly.

Crimp development medium, for testing textured yarn: An environment that allows the temporary set of fibre crimp to be overcome and that allows the filaments to assume their permanently set configuration.

Crimp effects: See **Crepon printed effects.**

Crimp elongation: Max. stretch or elongation of a crimped yarn/thread (or fibre) under tensile loading up to the neutralisation of crimp or up to the start of effective material elongation. Is expressed as a ratio to original length, and serves as a measure of volume increase due to texturing, i.e. high crimp elongation is equivalent to large volume increase.

Crimp energy: The amount of work required to uncrimp a fibre.

Crimp, fibre: (1) The waviness of a fibre expressed as waves or crimps per unit length. (2) The difference in the length of a fibre when lying in an unstretched codition and when stretched under a specific tension, expressed as a percentage of the initial length.

Crimp frequency: The crimp level, or number of crimps or full waves per unit length (inch) of a fibre related to the straightened length of the fibre/yarn or tow.

Crimp, in textile fabrics: The difference in distance between two points on a yarn as lies in a babric and same two points when yarn has been removed from the fabric and straightened under a specified tension expressed as a percentage of the distance between the points as the yarn lies in the fabric.

Crimp index: (crimping arc count). Number of crimping arcs/10 cm, independent of count. Controlled by texturing in the case of textured yarns.

Crimp, latent: A crimp that inherently present in the specially prepared fibres or filaments and which can be developed by a specific treatments such as thermal relaxation or tensioning and subsequent relaxation.

Crimp Modulus: Force required to stretch crimped (textured) yarn by 100%: e.g. at its lowest with polyamide 6 (2.1), polyamide 6.6 (2.4), at its highest with polyester (3.5). polyester (3.5).

Crimp Process: Designation for the IWS (International Wool Secretariat) process for inducing crimp in smooth wools, especially those of New Zealand origin.

Crimp recovery: A measure of the ebility of a yarn to reurn to the original crimped state after being subjected to tension.

Crimp retraction: See **Crimp contraction.**

Crimp setting: An aftertreatment to set the crimp in yarn or fibre. Usually heat and steam are used, although the treatment may be chemical in nature.

Crimp stability: (1) The ability of a textured yarn to resist the destruction of its crimp by mechanical and thermal stress, Crimp stability is normally expressed as the ratio of values of crimp retraction measure after and before specified mechanical and/or thermal treatment of the yarn.

Crimp stability: (2) The ability of a textured yarn to retain its crimp, measured before and after mechanical or thermal stress imposed under specified conditions.

Crimp, staple: See **Staple crimp.**

Crimp width-increse Wool locks: The average distance between two mid points of two adjascent valleys and the two waves.

Crimp Yarn, in fabrics: The difference in distance between two points on a yarn as it lies in a fabric and the same two points when the yarn has been removed and straightened under specific tension expressed as a percentage of the initial distance.

Crimp, types of: In man made fibres crimps are produced during the manufacturing stage using different methods, unlike in natural fibres it is ocuuring in the fibre itself. Different types of crimps are : (a) Stuffer-box crimping: Crimp-free filaments in tow form are by mechanical means or fluid-jet forcibly injected into a stuffer-box where they buckle and fold to form crimps (optionally followed by crimp stabilization by steam/hot air). (b) Bicomponent fibre crimp: Contrived crimp imitating the crimp structure of wool (paracortex/orthocortex) produced with Bicomponent fibres comprising components of differering shrinkage characteristics. Shrinking treatment then creates a highly stable threedimensional form of crimp (e.g. Superbulk-PP). (c) Hot-water crimp: It is created by immersing filaments in a relaxed state in hot water (caused by unequal shrinkage of fibre matrix and sheath) in the form of irregular three-dimensional arcuate crimp.

Crimped fibres: Wool-like crimped man-made fibres after texturing.

Crimped viscose twill: Bulked viscose yarn woven into plain colour fabric of medium weigh. The bulking adds warmth. The fabric creases slightly. It has a soft feel rather like Flannelette. Used for childrens clothes nightwear.

Crimped yarn: See **Bulk yarn (textured).**

Crimping: The process of imparting crimp to tow or filament yarn.

Crimps: Plain woven cotton cloth, made in England for the export trade. Certain warp threads, wound on a separate beam, are held slack, forming crinkled warp stripes.

Crimson: A bluish-red. See **cremosin**. Originally made from the **kermes** insect.

Crimson lake: A synonym for **carmine**.

Crin: (1) French term for horse hair; (2) Silk, heavier than cocoon silk, obtained by killing the silk worm and emptying its glands containing the silk substance.

Crin d'Afrique': (crin végétal, African fibre). A coarse variety of Palm fibres (*Chamaerops humilis*) with a similar texture to horsehair native to North Africa and Southern Spain. Used for upholstery and braids. It is usually black, dyed with cationic dyes.

Crin vegetal: French name for "vegetable horsehair," a stiff, strong and durable leaf fibre yielded by the scrub palmetto; used as substitute for animal hair, stuffing, etc. See also **Artificial horsehair. Crin d'Afrique.**

Crinkle yarn: See **Textured yarn, Crinkle process**.

Crinkle: (1) Crimp, e.g. the production of crimp yarn by the texturing of yarn hanks made from bicomponent yarns (hot air treatment in a textile tumbler).

(2) A wrinkled or puckered effect in fabric. It may be obtained either in the construction or in the finishing of the fabric.

Crinkle process: Texturing process (knit-set-deknit process) for polyamide and polyester textured yarns. The yarns are used in knitgoods, jerseys and hosiery garter tops.

Crinkling: In the sense of uneven waves, for example through excessive stretching of seams, edges, folds, pleats during ironing.

Crinoline: (1) A loosely woven stiff cotton, similar to cheese cloth, with hair filling. It may now be produced from synthetic fibre and stiffened. Made in white, grey and black. Used as an interfacing or stiffening material for hats.

(2) A stiff, heavily sized fabric, like buckram used as an interlining, or horse hair or wire or to support areas such as the edge of a hem.

Criolla: Native South American sheep yielding coarse wool.

Cristal: Dress goods, made with fine silk warp and wool filling which form alternate fine and heavy (often irregular) ribs.

Cristaline: Loosely woven silk dress goods.

Critical application value: In a low add on easycare finishing system, the amount of finishing liquor which must be applied to a given fabric to avoid a non uniform distribution of cross linking after drying and curing.

Critical chain length: The minimum chain-length required for the entanglement of the polymer chains.

Critical defect: A serious defect that judgement and experience indicate is likely to prevent the usage or the proper performance of a product from its intended purpose.

Critical difference: The observed difference between two test results, which should be considered significant at the specified probability level.

Critical length: see **Breaking length.**

Critical micelle concentration: (CMC, ck). Characteristic concentration of surfactants in solutions, above which the number of micelles increases to the extent where the concentration dependency of specific physical properties is rapidly changed. The critical micelle concentration is determined by the intersection of two extrapolated curves, which describe the property of the solution as a function of concentration.

Critical Radiant Flux: The lowest intensity of radiant heat that will cause a floor covering to propogate flame.

CRL: Abbreviation for **Constant-Rate-of-Loading.**

Croceo- (L): Saffron-coloured. See **saffron**.

Croceous: Saffron-coloured. Also 'crocreal'.

croco- (L): Orange.

Crochet Knit: Fancy openwork knit in a variety of colours and patterns, often produced in traditional designs, e.g. Aran. The fabric resembles crochet but is made on a knitting machine which works on the same principle as Rashet machines; however they are narrower and the needles are generally horizontal. Any man-made fibres may be used but they are mainly acrylic. Used for sweaters, shirts, dresses, braids, trims and scarves.

Crochet quilt: Twilled or Jacquard figured, bleached cotton quilt, woven with one set of warp and one set of filling.

Crocheting: The interlocking of loops from a single thread with a hooked needle. Crocheting can be done either by hand or by machine.

Crocking: Transfer of colour from dyed or pigmented fabric by rubbing as a result of insufficient dye penetration or fixation, the use of improper dyes or dyeing methods, or insufficient washing and treatment after dyeing operation. Wet crocking refers to transfer of colour from a piece of dyed fabric to another piece of fabric, or to an undyed area of the same fabric, while the fabric is wet. Dry crocking means the same, except that the fabric is dry.

Crockmeter: Standard test equipment for testing the Colour fastness to rubbing of dyed or printed textiles. Traditionally hand-operated, electronically-controlled crockmeters have also become available recently.

Crocodile: Crocodile skin ia very expensive and only used for top quality shoes and hand bags.It is difficult to tan. It can be distinguishred from Alligator by the fact that the marking are mor uneven.

Crofting: Scotch term for bleaching linen on the grass.

Croise': (French croisé = twill). A twill weave cotton fabric (generally 2/2), smooth on the face side, and heavily brushed on the reverse to a varying degree. It is generally extremely difficult to differentiate between finette and croisé. The former is generally more finely ribbed, the latter more heavily ribbed. There is, however, an intermediate designation as well (croisé-finette). It is available in the market bleached white, piece-dyed in delicate colours, or printed. Uses: nightwear, skirts and winter wear.

Croissure: A technique of twisting raw silk over itself in the process of reeling with the object of consolidating the different filaments constituting the raw silk, rounding off the angularities of the thread and squeezing off the extra moisture and gum.

Croissure pulleys: They are guide pulleys used in reeling machine for formation of croissure.

Crompton axminster: Rug or carpet, having a tufted pile, made on power loom.

Crooked cloth: See **Baggy cloth.**

Croop: Silk, especially after immersion in a weak acid, when compressed and rubbed, gives out a peculiar rustling sound, which is known as 'croop'.

Cropping: See **Shearing.**

Croquis: A balanced design of print. A croquis should give the impression that would be seen if a frame were placed over any section of the finished cloth. Although not in repeat, a croquis will give the feeling of being in repaeat. Croquis is a french word meaning sketch.

Cross-back width: See **Back width.**

Cross-chest width: in body measurements: The distance from the front breakpoint to front break point

Cross direction: The width dimension, within the plane of the fabric that is perpendicular to the direction in which the fabric is being produced by the machine.

Cross dyeing effect; Cross dye effect: Variation in dye pick up between yarns or fibres, resulting from their inherent dye affinities.

Cross dyeing: (1) When two components in a blends are dyed in two different shadeswe get a final cross dyed fabric.

(2) When dyeing one component in a blend the other component is tinted or dyed which has to be removed before dyeing the second component.

Cross lapping: The production of a nonwoven web or batt from a fibre web by traversing it to and fro across a lattice moving a tright angles to the direction of traverse.

Cross-Linked Polymers or Network Polymers: In this type of polymers, the monomeric units are linked together to constitute a three dimensional network. The links involved are called cross links. Cross-linked polymers are hard, rigid and brittle because of their network structure. Common examples of this type of polymers are bakelite, formaldehyde resin, melamine, etc.

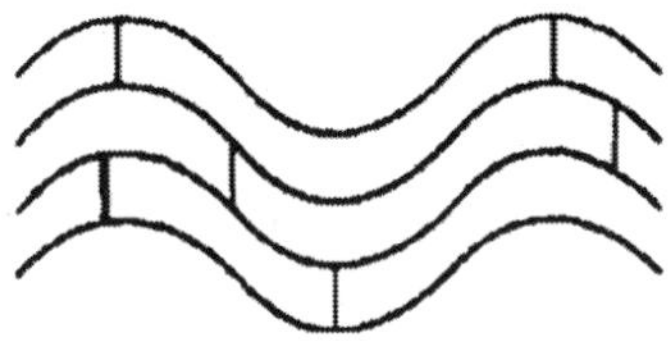

Cross-linked

Cross linkers: Chemicals which`build linear molecules of macromolecular chemicals with reactive sites into networks of three-dimensional structure by forming intermolecular bridges. See **Cross-linking.**

Cross-chest width, in body mesurements: The distance from front break point to front break point.

Cross Linking: Connecting of linear molecular chains to form three-dimensional networks, e.g. of synthetic resins in Resin finishing. A polymer-like reaction of two macromolecules with bifunctional bridge-type cross-links causes covalent bonding of polymers via their side chains.

Cross-linking resins: Amongst the reactive long chain polymers used for textile applications are found acrylamides and methacrylamides with N-methylol or N-methylol ether groups. Acrylamide and its N-methylol compound are easily soluble in water. The some what lower solubility of methacrylamide and its methylol compound is of benefit for emulsion polymerisation. Reactive long chain polymers used for textile applications containing carboxyl groups as found in acrylic acid or methacrylic acid. They are used to stabilize dispersion systems, provide adhesion to substrates, change mechanical properties and alter the reaction towards solvents after application. Carboxyl groups are also involved in the cross-linking process. The reaction of mixed polymers having only carboxyl groups with hexamethoxymethylmelamine suggests that the crosslinking here occurs via a methyl ester bridge.

Cross migration: When dyeing blend fabrics containing polyester fibres, the second fibre is often heavily stained by the disperse dyestuffs used for the polyester which is called cross migration. It will be necessary to use a two-bath dyeing process or a subsequent clearing bath for a single bath.

Cross over point: Crossing point between warp and weft yarns in woven fabrics.

Cross stich: Embroidery stitch. Method: Bring the needle out at the lower right-hand side, insert the needle four threads up and four threads to the left and bring out four threads down, thus forming a half cross stitch, continue in this way to the end of the row. Complete the other half of the cross as shown. Cross Stitch may be worked either from left to right, as shown, or from right to left. It is important that the upper half of all the stitches lie in one direction.The cross stitch is worked on linen, scrim, canvas, or any open-meshed material. If done on a flat, smooth surface, it will be necessary to work over canvas, afterwards drawing out the canvas threads. The canvas should be well basted on the material, the warp threads of the canvas lying *perfectly straight* on a line with the warp threads of the material on which the pattern is worked. The stitches should always run the same way. If the first ground stitches are made from left to right, from bottom towards the top, the cross stitches should be made from right to left from the top towards the bottom. All the ground stitches run one way and the cross stitches in the opposite way.

Cross Stripes: Stripes running weft wise.

Cross-dye effect: See **Cross dyeing effect.**

Cross flow membrane filtration in water/effluent treatment: A separation of the components of a liquid by semipermeable membrane through the application of pressure and parallel flow to the memnbrane surface. Includes the processes of reverse osmosis, ulgtrafiltration, nanofiltration and microfiltration.

Cross-linkers: Chemicals which build linear molecules of macromolecular chemicals with reactive sites into networks of three-dimensional structure by forming intermolecular bridges. See **Cross-linking.**

Cross-linking: The stabilization of cellulosic or manufactured fibres through chemical reaction with certain compounds in such a way that the cellulose or manufactured polymer chains are bridged across or "crosslinked." Cross-linking improves such mechanical factors as wrinkle resistance. Random cross-linking in manufactured polymers is undesirable and leads to brittleness and loss of tensile strength.

Cross-section: The shape of an individual filament when cut at right angles to its axis.

Normal shapes for manufactured fibres vary, e.g., round (nylon, polyester, polypropylene, and some acrylics), serrated or crenular (viscose rayon, acetate, and triacetate), bean-shaped (some acrylics and modacrylics). The shaped of manufactured fibres can be modified by changing the shape of the holes in the spinneret. Cross-sectional variants are produced intentionally in wide variety of shapes for different physical effects such as change in lustre or hand, improved resistance to soiling, etc. Examples are trilobal (T and Y) and other multilobal shapes (cruciform, K, X, pentalobal, star, etc.), I-beam, ribbon, square, triangular, elliptical, hollow, and many others.

Cross-tuck, plain (Pique) (weft knitted): A plain weft knitted fabric in which the construction is based on a repeat of either two or multiples of two courses. The original construction consisted of knitted loop and single course tuck stitches arranged to give a two couse repeat where tuck loop alternated with knitted loop within each course and between one course and another.

Cross tuck (Pique') face

Back

Other construction incorporates two or more course tuck stitches when the tuck loops alternates with knitted loops within the course and between one course and another.whole courses of knitted loops may be interspaced between the courses of the tuck and knitted loops.

Cross weaving: Where warp threads are crossed with each other, as in the gauze or leno.

Crossbred: Middle grade wool, taken from sheep crossed with merino. The fibre is usually of good length, coarser than merino.

Crossing warp: Term for those warp ends which are deflected in the gauze weave and are wound around the straight war.ps.

Crossings: Sections where the warp and weft patterns combine more by chance than by design.

Crotch, in anatomy: The body area adjascent to the vertex of the included angle between the legs.

Crotch depth: See **True rise.**

Crotch height: See **Inside-leg length.**

Crotch seam: The short seam from the back of the pants fly to the inseam.

Crottle, crotal: Having the golden-brown colour of lichen.

Crown, in anatomy: The top of the head.

Crows feet: A fabric defect consisting of breaks or wrinkles of varying degrees of intensityand size, resembling bird's footprints in shape, and occurring during wet processing of fabrics.

Crow weave: An uneven sided, four harness twill weave, the warp crossing over three filling threads and depressed under one.

Crowsfoot weave: A broken twill weave 1-up 3-down or 3-up 1-down with two ends to the right and two ends to the left , commonly reffered to as a four-harness satin or bocken crow.

Crow feet weave: A broken twill weave 1-up and 3-down or 3-up and 1-down with two ends to the right and two ends to the left, commonly referred as a4-harness satin or broken crow.

Crown Lace: English needlework of Queen Elizabeth's reign, having royal crowns in the design.

Crown Lining: Sized tarlatan or lighter weight of crinoline; used for lining women's hats.

Croydon: Plain woven stout and bleached cotton sheeting in England, given a stiff and glossy finish with the aid of size.

Crown, in anatomy: The top of the head.

CRSIT: abbrev. for: Centre de Recherches de la Soierie et des Industries Textiles (French Research Centre of the Silk and Textile Industry), Lyon, France.

CRTM: abbrev. for: Centre de Recherches Textiles de Mulhouse (French Textile Research Centre of Mulhouse), Mulhouse, France.

Crudillo gallo: Linen cloth made in Spain.

Cruel: Another name in the 17th century for Caddis

Crumb cloth: Coarse and heavy damask in gray, made to be embroidered in coloured yarns around the patterns.

Crutchings: Wool which was removed same time before the shearing, in order to improve the growth of the fleece.

Crushed feathers: Feathers and feather fibre resulting from curling , crushing, or chopping feathers without removing the quill.

Crushed feathers: Feathers and feather fibre

Crushing, In Carpet: Matting of carpet fibres.

Cushioned Backed Carpet: Carpet, which has a cushion as a part of the backing.

Crutched Wool: Wool that has been clipped from rear end and udder area of ewes in the early spring to prevent collection of manure and fly strike.

Crystal: A very fine highly finished woollen oif white colour formerly made in England for the export trade; used for nun's clothes; now obsolete.

Crystal gum: (1) Dextrines. (2) Industrial gums.

Crystalline: Made up of crystals. The term crystalline applies to sections of all chemical fibrer, which consists of alternate crystalline and amorphous (noncrystalline) regions. These regions are influenced by manufacturing conditions and to some extent can be controlled. The degree of crystallinity influences the physical properties of fibres.

Crystalline growth: (1) The expansion and development of a crystal. The process involves diffusion of the crystallizing material to special sites on the surface of the crystal, incorporation of the molecules into the surface at these sites, and diffusion of heat away from the surface of the crystal. (2) The transformation of disoriented molecules, usually of the same substance, to a higher state of order. This process generally occurs rapidly for small molecules; however, the process is slow for polymer molecules and is arrested at temperatures below the glass transition temperature.

Crystalline melting point *(T m):* This is the range of melting temperature of the crystalline domain of a polymer sample and is accompanied by change in polymer properties. It is also the first-phase transition when the solid and liquid phases are in equilibrium.

Crystalline polymers: A polymer with an ordered structure, which has been allowed to disentangle and form a crystal.

Crystallinity: (1) The state of quality of being crystalline. (2) The extent to which a polymer exists in a lattice structure.

Crystallization: The formation of highly-ordered substances (crystals) from solutions or melts. In polymers, crystalline areas are interspersed with amorphous areas in a lattice-like network. (Also see **Microlattice**).

CS: abbrev. for: Commercial Standard (USA).

CSC900T: Social Standards in textiles and clothing introduced in China by CNTAC in 2005.

CSN: Ceskoslovenskych Norem (Czechoslovakian Foreign Standard).

CSIRO: Commonwealth Scientific and Industrial Research Organization, East Melbourne, Australia.

CT: Triacetate fibres, Standard abbrev. for textile fibres.

CTA: Triacetate fibres, Standard abbrev. For textile fibres.

CTNSS: Centre for Thai National Standard Specifications, Bangkok, Thailand.

CTTN: Centre Technique de la Teinture et du Nettoyage (French Technical Centre for Dyeing and Drycleaning).

CU: Cupro fibres (Standard abbrev. for textile).

Cuban Bast: A fine, soft but strong cloth like bast of the Hibiscus elatus in Cuba; used for millinery braids, etc.

Cuban hemp: Strong smooth fibre, yielded by the leaves of the Furcraea cubensis of Central and South America.

Cubi hopi: Indian, name for the aromatic sumac; used for baskets, which can hold water.

Cubex: Wash test machine especially for determining the felting shrinkage of wool fabrics.

Cubica: Fine, thin English worsted serge, usually red; used for linings, suiting, etc. See also **Plainback**.

Cucumber green: The dark green colour of the rind of the cucumber.

Cudbear, Archil: See **Natural dyes.** A violet dyestuff, obtained from a stone moss; used on animal fibres.

Cuero de Diablo: Stout cotton denim in Colombia.

Cuir Laine: French, wool winter dress goods, similar to the ratine,, woven in a twill.

Cuirtain Fine: twilled fabric, made of white wool and used for clothing in medieval Scotland.

Cultivated: Indian and other Asiatic countries silks, made of "cultivated" silk worm as compared with the "wild" silk.

CUP: Cupro fibres.

Cuff (lined): A cuff with interlining placed between the two pieces of body fabric.

Cuff (one-piece): A two-ply cuff formed by folding over a single piece of fabric, usually with a lining in between.

Cuff (topstitched): A cuff with an added row of stitching along the folded edges.

Cuff (two-piece): A cuff in which two identical pieces of fabric, usually with a lining in between, are joined by a seam along the edge, then turned and sometimes topstitched near the folded edges.

Cuir: A yellowish-brown; the colour of leather.

Cuite: Nett silk from which the gum has been completely removed. See **bright silk.**

Cuite silk: Degummed natural silk.

Cul de Paris: Emphasised rear of womens outer dress, using a frame or padding; See also **Paris tail, French tail, Bustle.** Used in Neo rococco, Victorian period.

Culottes: Knee breeches; at first quite broad, later close fitting, baroque, rococo period.

Cumquat, kumquat: The yellowy-orange of the cumquat fruit.

Cupreous: The reddish brown colour of copper; copper-coloured; a bronze colour.

Cumulative toxicity: Effect on an organism caused by successive exposures.

Cunningham: A Highland tartan, composed of wide black stripes and narrow red, black and white lines over a red ground.

Cup seaming: A type of sewing that joins edges (usually two selvedges) by means of a chain stitch.

Cupra: The term used originally, and still in use, to describe fibres of regenerated cellulose obtained by the cuprammonium process.

Cuprammonium: Cuprammonium hydroxide.

Cuprammonium fluidity: It is well-known that purified natural cellulose dissolves in cuprammonium hydroxide solution to give a solution that is viscous even at low concentrations. Cellulose that has been partially hydrolyzed or oxidized dissolves to give less viscous (more fluid) solutions at the same concentration.

Cuprammonium hydroxide: (cuoxam, Schweizer's reagent), $[Cu(NH_3)_4](OH)_2 \cdot 3H_2O$. Dark blue solution. Production: add a saturated solution of copper(II) sulphate ($\cdot 5H_2O$) in distilled water to 25% ammonia; thoroughly rinse the precipitate with distilled water, and dry. The resultant powder is durable for a long period, and is dissolved for use with 25% ammonia. Use: for the chemical identification of cellulose fibres and silk, and also for the quantitative determination of the degree of polymerisation (DP) of cellulose fibres.

Cuprammonium rayon: See **Cupro fibres.**

Cuprammonium silk: See **Cupro fibres.**

Cupric acetate: (acetic copper, neutral copper acetate, copper acetate), u$(CH_2COO)_2 \cdot H_2O$. Molecular weight 199.67; density 1.88. Easily soluble bluish green crystals. Use: resisting agent for indigo resist products; oxygen carrier with aniline black.

Cupric chloride: (hydrochloric copper), $CuCl_2 \cdot 2H_2O$, molecular weight 170.52. Greenish crystals, easily soluble (colour green, blue diluted). Use: highly effective oxygen carrier for aniline oxidation black.

Cupro: Now the official designation for rayon produced by the Cuprammonium method. A very soft silky fibre. It is often still referred to as Bemberg, the name of the company first making it.

Cupro fibres: A class of Cellulosic fibres, regenerated, produced by the uprammonium process from raw materials containing cellulose (predominantly cotton linters, also wood pulp.

Cuprous: (1) Chemical: copper (I) compounds. (2) Sometimed used for denoting Cupro fibres.

Curing: (1) A process in which heat treatment for several minutes or heat treatment for short time a higher temperature (flash curing) or long time treatment at lower temperature and high regain is done to effect chemical reactive to fix a dye or chemical.

(2) See **Vulcanization.**

Curing conditions: Cross-linking conditions in the Cross-linking of permanent finishing agents.

Curing curve: The graphic representation of the achievable dry crease angle (Crease recovery angle) relative to the cross-linking temperature with constant cross-linking times for a specific permanent finish agent.

Curing machine: Used for the continuous heat treatment of textiles, particularly for curing (cross-linking) permanent finish agents (Dry cross-linking process). Designed on the hot flue principle, with driven rollers, generally stepless speed control, for temperature measuring and regulating systems, various infra-red radiators being located upstream for quicker fabric heating.

Curing machine, Festoon type: Condensing machine with a large fabric content and tension-free fabric guiding for (benzine-free) pigment printed knitted fabrics. Festoon system with special insulation-covered carrier rods rotating during fabric run.

Curing process: Usual term in resin finishing for the dry cross-linking process (See **Cross-linking**).

Curl: See **Snarl, Kink.**

Curl or loop: A thin basic yarn with a thicker one twisted with it, but fed at a different speed so that regular loops are joined. This type of yarn can be made into a variety of fabrics.

Curl yarn: (1) A type of yarn which presents curls or loops of various sizes all along its surface. It is usually produced as follows: Two threads, a thick and a thin are twisted together, the thin being held tightly and the thick thread slackly twisted around it. This two-fold yarn is then twisted in the reverse direction with another thin thread, this untwisting throwing up the thick thread as a loop, the two fine threads holding the loops firmly.

(2) Usually a three-strand yarn; first a thin strand (well stretched) having curls and nubs, made by winding slackly a thick thread around a thinner one, which is well stretched, after which a third strand, also thin, is wound around the whole, holding the thick yarn, which forms the curls; made in cotton, wool or silk; used for ratine, etc. See also **Loop Yarn.**

Curled pile: Long wool or silk pile having a curl, usually found in artificial fur.

Curled selvage: Self descriptive.

Curragh Lace: See **Irish Point.**

Currelles: English faibric, made of worsted and silk in the 16th century.

Curtain Serge: Stout woollen serge; used for drapery.

Curve, force-deformation: See **Force-deformation curve.**

Curve, force-elongation: See **Force-elongation curve.**

Curve, force-extension: See **Force-extension curve.**

Curve, load-deformation: See **Load-deformation curve.**

Curve, operating characteristic: See **Operating characteristic curve.**

Curve, stress-strain: See **Stress-strain curve.**

Curve, tensile hysterisis: See **Tensile-hystrisis curve.**

Curved blade applicator: (CBA). A low pick-up application system. Liquor is applied over the entire width of a stationary curved blade and forms a fine film which is stripped off by the fabric. Unused liquor at the edges is recirculated for reuse.

Curved expander rollers: The name describes the curved form of these Expander rollers. A rubber hose runs on the curved axis, and is held at regular intervals by several ball bearings. The expansion effect results from the varying distances which goods have to travel.

Cushion: A padded pillow of barrel or some other shape; used as foundation in bobtxin lace making to which the pricked pattern is pinned.

Cushion Stitch: Similar to cross stitch in Berlin work.

Cushion, attached: See **attached cusion**.

Cushion, for inflatable restraints: The inflatable fabric envelope portion of a module.

Cushion, attatched, in for yarn floor covering: A cushioning material such as foam or rubber or urethane etc. adhered to the backing fabric side of a pile yarn floor covering to provide additional dimensional stability, thickness and padding.

Cusick drape test: In the drape test the specimen deforms with multi-directional curvature and consequently the results are dependent to a certain amount upon the shear properties of the fabric. The results are mainly dependent, however, on the bending stiffness of the fabric. In the test a circular specimen is held concentrically between two smaller horizontal discs and is allowed to drape into folds under its own weight. A light is shone from underneath the specimen and the shadow that the fabric casts, is traced onto an annular piece of paper the same size as the unsupported part of the fabric specimen. The stiffer a fabric is, the larger is the area of its shadow compared with the unsupported area of the fabric. To measure the areas involved, the whole paper ring is weighed and then the shadow part of the ring is cut away and weighed. The paper is assumed to have constant mass per unit area so that the measured mass is proportional to area. The drape coefficient can then be calculated using the following equation: Drape coefficient =Mass of shaded area X 100/Total mass of paper ring % The higher the drape coefficient the stiffer is the fabric.

Cut: (1) An American term for guage. The number of needles per inch in the circumference of the cylinder or dial of a knitting machine.

(2) (Na Cut). An American yarn numbering system used for woollen yarn measurement. 1 Cut = 1 hank at 300 yards/lb. 1 cut=0.605Nm.

Cut and sewn knitwear: Garments made from pieces cut from panels or lengths of fabric, usually with integral ribs.

Cut Cashmere: Twilled woollen dress goods, having fine runs warp wise often of a different colour.

Cut fabric: Usually denoting tubular knits or Fents, made up using patterns and sewn together to form garments.

Cut-Make-and-Trim: Usually denoted as CMT.

Cut, in asbestoes and glass yarns: The no. of 100 yard lengths of yarn per pound. An indirect yarn numbering system.

Cut, in wool yarns: The number of 300 yard lengths of yarn per pound. An indirect yarn numbering system.

Cut- as appied to woven fabric: A length of approximately 60 yards in grey.

Cut fabric: Of tubular knits or Fents, made up using patterns and sewn together to form garments.

Cut film in screen making: When producing hand-painted Colour separation designs, smooth surfaced patterns may be produced through screens. Transparent film is used which has a similarly transparent laminate coating and the pattern is traced with a knife on the laminated side. The laminated layer is then removed, without damaging the base film, wherever the pattern should be absent when printing.

Cut length: Staple length.

Cut length in sewing: Cut length in sewing is the total length measurement that includes all seam allowances.

Cut loop, pile: See **Pile, cut loop.**

Cut mark: An indication on the weavers warps of a precise length of material generally a piece length or a fraction of the piece length.

Cut off: **in Zippers,** the measurement of a separate element from the head side to the pocket side of the leg.

Cut-on-cross: Fabric that is cut so that the warp runs horizontally across the garment piece.

Cut-on-fold: Fabric that is doubled, then cut.

Cut pick: See **Broken pick.**

Cut pile carpets: Tufted carpets, the pile of which consists of loops which have been cut open; in contrast to Loop pile carpets.

Cut pile: Which is cut open after the loops were formed in the process of weaving.

Cut pile floor coverings: A floor covering in which the pile is formed of tufts in the form of cut loops.

Cut pile plush: A knitted plush fabric produced by cutting the loops of Circular-knit pile fabrics. The production of this article involves high finishing costs (up to 40% of the final material cost). The following example represents a typical processing route: Slitting of tubular knitgoods – presteaming – pre-cropping – washing – hydro-extraction – drying – heat-setting – dyeing – tumbler drying – cropping – stentering. The total weight loss in pre-cropping and final cropping is 20–25%.

Cut presser, in warp knitting: A presser with an edge castellated so that only selected bearded needles are pressed. The shape of the pressing edge depends up on the pattern.

Cut selvage: Cuts or breaks that occur in the selvedges only. Same as **Broken selvedge or Damaged selvedge.**

Cut selvedge: See **Cut Selvage.**

Cut selvedge suction remover: A device used in machines along with the selvedge cutter device to suck away the cut off edge ouside the gummed area at the selvedge avoiding any more curling of the main fabric. Used on all reroll cutting, transverse cutting, laminating and coating machines, extruders, stenters, printing machines etc.

Cut strip test: in fabric testing, a strip test in which the specimen is cut to the specified testing width.

Cut work: Consists of partly filling with loops and stitches the various spaces cut into linen. It is of ancient origin, evidently coming from Greece. Called also Greek lace and reticella.

Cut-asbestoes: The number of 100 yards length/lb (7000 grains) or 91.4m/ lb (453.6 g)of asbestose yarn.

Cuticle: (Lat. cuticula = skin). In general use a biological term meaning the epidermis or superficial skin. In a textile context the term cuticle is used for the thin surface membrane of the cotton fibre and especially of hair and wool fibres or simply The surface layer of animal hair fibres consisting of flat overlapping scales.

Cutin: (1) (Suterin). A component (3.5–4%) in the primary wall of cotton. It is apparently still awaiting elucidation and cannot yet be defined more precisely.

(2) (Cutin)A working description in research for a substance enveloping the Cuticle of the wool hair which apparently consists of chemically highly resistant Keratin of high density. The terms epicuticle, exocuticle and endocuticle are used as alternatives.

Cut-Make- And Trim see **CMT.** A section of the clothing industry in which a contractor is supplied with materials and designs in order to produce garments for a principal.

Cut-off, In Zipper: The measurement of scoop from the head side to the pocket side of the legs.

Cut-pile carpets: Tufted carpets, the pile of which consists of loops which have been cut open; in contrast to Loop pile carpets.

Cut-pile plush: A knitted plush fabric produced by cutting the loops of Circular-knit pile fabrics. A usual process route is as follows: Slitting of tubular knitgoods – presteaming – pre-cropping –washing – hydro-extraction – drying – heat-setting – dyeing – tumbler drying – cropping – stentering. The total weight loss in pre-cropping and final cropping is 20–25%.

Cuts: Trade term for short length of fabrics, less than a bolt.

Cuttance: Fine heavy and stout silk satin of India, with bright coloured woven stripes and cotton back; used for upholstery.

Cutting: Stout cotton cloth with flower or trailing patterns.

Cutting Thread: Which forms the furrows dividing the rifes of corded fabrics. It is interlaced in plain weave with all the other threads and is stretched if it runs warp ways. Often it is of a different colour than the ribs.

Cuttle: (1) To place fabric in loose traverse folds, usually in open width. (2) To fold finished fabric darn in middle and place it in traverse folds of predetermined length.

Cuttle down: Generally refers to the cooling down of woollen piece goods by laying up in piles.

Cuttling: The folding of the falbric after finishing.

Cutts sizing system: A minimum application process based on the nip-padding principle. The warp is no longer immersed in the sizing bath. Instead, the sizing agent to be used is applied at higher concentration by means of an engraved roller.

CV: Viscose fibre, Standard abbrev. for textile fibres,

CY: Modal fibre, Standard abbrev. for textile fibres.

Cyan: A greenish blue colour Cyan is "officially" the subtractive primary. It is the bluish colour that is used in most printing processes, such as computer inkjet printing. The blue colour that appears in those little coloured squares that are sometimes found near the fold on advertising flyers, or on the bottom or flap of printed packages is cyan, or at least, cyan as approximated by the printing ink used. In the popular MX reactive dye family, Turquoise MX-G is a very good cyan.

Cyanine dyes: See **Polymethine dyes.**

Cyclamen: The red or pink or reddish-purple colour of the cyclamen flower.

Cycle length, in braided rope: The distance, poarellel to the rope axis, of the strand to make one revolution around the rope.

Cyclone: (1) Or depression (weather). (2) A region of low atmospheric pressure, with its minimum at the centre. North of the equator the winds blow counter-clockwise round the centre; south of the equator, clockwise. Cyclones usually accompany rain and storms, unlike anticyclones. (3) cyclone separator. The use of centrifugal force to separate particles from a gas stream, such as flue gases. The particles are thrown to the outside by the induced motion of the gases and impinge on the surface of the cyclone. They then settle to the base from where they are removed. There are two spirals within a cyclone, a downwardly directed spiral in the outer part and a more rapid inner spiral which is directed upwards to the outlet for the gases. Cyclones may be tangential in which the gases are admitted at a tangent, or axial in which the gases are admitted at the top and a swirling motion imparted by vanes. They may also be used in banks. If used in parallel, high volume passage is allowed, if in series the efficiency of separation is increased although volume passage is not. Efficiency falls off rapidly with particles below about 50um and generally 10um is the minimum particle diameter collected, although in high efficiency cyclones this may be reduced to about 5um. Cyclones are less efficient than electrostatic precipitators or fabric filters, but they are inexpensive, versatile and, being without moving parts, generally trouble free. See **Cyclonic spray scrubber dense medium cyclone, multicyclone, wet cyclone.**

Cyanoethylation of celluloses: Applies particularly to cotton and involves treatment with acrylonitrile. Main reactions:

$$\text{Cell–OH} + CH_2\text{=CH–C=N} \rightarrow \text{Cell - O}CH_2CH_2\text{– CN}$$

Cyclone: An installation for dust extraction, e.g. from raising, emerizing, shearing, singeing and other machines which generate dust. The air to be extracted is pumped into large vertical funnels with the aid of powerful fans where it circulates downwards in the form of a vortex. The dust is separated from the dust-laden air by centrifugal forces and friction on the walls of the funnel and spins slowly to the mouth of the cyclone at the bottom from where it is collected in bags for subsequent disposal.

Cylinder dryer: Cross-wound package dryer, in which the packages to be dried are exposed to different drying conditions creeled in segments of a cylinder or lying on perforated bottoms. Cylinder drying machines are also called cylinder dryer. See **Cylinder drying machines.**

Cylinder drying machine: Probably the cheapest drying machine for drying openwidth farics especially woven fabric. A simple drying machine which uses the contact drying rinciple to dry fabrics in open width. During contact transfer, heat is transferred by contact between the textile material and the heated cylinder surfaces through thermal conduction. The material may be dried by cylinder contact on one or both sides.

Cylinder fulling machine: (milling stock, bumper). Used for the dry removal of hydrocellulose after Carbonizing. Similar in design to the Rotary milling machine.

Cylinder mercerizing: This is a more compact and faster system compared to the Chain mercerizing. Cylinder mercerising does not allow the contraction of the warp because the fabric is drawn in on the cylinders. The contraction of the filling yarns is also prevented thanks to the tension produced by the simultaneous action of the cylinders and of the fabric wetting. Cylinder mercerising machines are also used for flat knits. Also called chainless mercerizing.

Cylinder roller milling machines: The most popular method of Milling involves treatment in a rotary milling machine, in which woollen piece goods are processed in rope form and drawn through the machine between two rollers, where considerable pressure may be exerted by the top roller. The material then passes into a trough or spout (stuffing channel) where it is compacted.

Cylindrical package, cheese: A Yarn package in which the surface of the wound yarn runs parallel to the axis of its supporting package centre or tube (only suitable for cross-wound packages). With single-end conical cylindrical yarn packages, one end of the package is built up conically.

Cylindrical wet scrub extraction: A carpet cleaning method in which (a) a cleaning agent is sprayed on to the carpet, and (b) soil and cleaning agenrt are removed by the macine which feeds water into the counter rotating brushes.

Cypress: Dark grey.

Cyprus: (1) Fine silk gauze, originally from Cyprus; usually dyed black and used as mourning veil. See **Cypress Cloth**.

(2) Gold thread in the Middle ages, having a flat gold strip wound around a silk core.

Encyclopaedic dictionary of textile terms

Encyclopaedic dictionary of textile terms

Volume II

Mathews Kolanjikombil
Senior Textile Processing Technologist

WOODHEAD PUBLISHING INDIA PVT LTD

New Delhi

Published by Woodhead Publishing India Pvt. Ltd.
Woodhead Publishing India Pvt. Ltd.,
303, Vardaan House, 7/28, Ansari Road,
Daryaganj, New Delhi - 110002, India
www.woodheadpublishingindia.com

First published 2018, Woodhead Publishing India Pvt. Ltd.

Reprinted, 2019, 2020

Woodhead Publishing India Pvt. Ltd. ISBN: 978-93-85059-11-7
Woodhead Publishing India Pvt. Ltd. e-ISBN: 978-93-85059-66-7

Typeset by Bhumi Graphics, New Delhi
Printed and bound by Replika Press Pvt. Ltd.

Contents

Volume IV

Preface

It has been told the first thing that a business man should know, is—the little things of his business. This book is meant for textile personae, be it a textile student, textile chemist, a spinner, a weaver, garment maker or a merchandiser who may be an expert in his/her field but he/she may come across many terms in his/her day to day business which he/she is not familiar with but is related to his/her fieldwhich he/she should know, but nine times out of ten does not know. He/She may not have an expert in that field of that term near him/her to clear his doubt about that term. This book comes handy in such circumstances.

There is nothing scholarly in this book, but is a compilation of short easy understandable meanings of the textile terms enough to help the referrer to understand the term. I have come across dictionaries which gives the meaning of the textile terms in one sentence. But it may not be sufficient to give a full idea about it. But this book is a little different. The meaning of the terms is explained in a concise manner even with the help of diagrams or photos, wherever necessary, which is enough to clear his/her doubts. These terms and meanings have been collected right from my college days and throughout my career. I believe it is not complete, but, such as it is, the compilation is reliable. There are further terms which are being collected by the author which will be added in the next edition, probably. The author believes that he has produced a book which may be profitably consulted by all who are either interested or practically engaged in textile trade.

It has to be specially mentioned that, the students can use the present book as a reference guide for his/her immediate needs without going to many text books. For detailed study of any terms he/she can further refer books specialized in that field. I may not suggest this book for a research student. The author also has tried to explain the construction of many fabrics new or old for general knowledge.

Hope this book will be greatly accepted by textile personnel. Any suggestions or corrections are welcome, which will be included in the next editions.

Mathews Kolanjikombil

Senior Textile Processing Technologist

Bangalore

Dedication

This book is dedicated to my Father (Thomas Abraham) and my Mother (Thankamma Abraham) who solely made me what I am today.

D

D and K: Short form for Damaged and Kept. Denotes lengths of fabrics spoilt in processing.

D/Y ratio: The ratio of speed of movement (D) of a friction surface or surfaces in a friction spindle used in false twist texturising to the axial speed (Y) of untwisted yarn

D65: The CIE standard illuminate that represents a colour temperature of 6504K. This is the colour temperature most widely used in graphic arts industry viewing booths. See **Kelvin (K)**.

D: (1) Symbol for: (a) Debye; (b) Dielectric constant; (c) Diffusion coefficient; (d) Deuterium.

(2) (a) Unit prefix for Deci- (one tenth), factor 1/10; (b) Abbreviation for dextrorotatory (optical activity).

DA: Dye accessible. The term is used, for e.g., in connection with the bilateral structure of the cortex cell layer in wool (paracortex/orthocortex).

Da: Symbol for deca- (prefix indicating 10, as in decametre), factor 10.

Daba: A coarse, plain woven, unbleached cotton fabric, made in Asia.

Dabbing: Faults in silk fabrics with a brushed face (white or mealy marks) are rendered invisible by dabbing with a felt pad lightly impregnated with paraffin oil. Also on Dabbing machines.

Dabbing machine: Machines used for Dabbing. Dabbing is done continuously or intermittently by oiled felt rollers of these machines which are traversed by a cam mechanism across the cloth. Once an area has been treated the cloth moves on. See also **Dabbing**.

Dabbing print: Surface printing in the form of a stamp for printing industrial textiles using pigment printing, e.g. trademarks on drive belts and carpet backing. If necessary, the substrate to be printed should be pre-treated with primers (adhesion agents) or by corona treatment in order to overcome the polymer incompatibility between the substrate and the pigment printing binder.

Dabouis: The Indian name for a narrow, bleached cotton cloth made in eastern India; used for calico.

Dacca: Silk; Embroidered silk.

Dacca twist: Fine English plain or twilled calico; used for sheets, underwear, etc.

Dacey: Coarse wild silk, produced by an East Indian silk worm from six to eight times a year.

Dacron: The trade name for an American polyester fibre. Trade name of a polyester fibre. As with any other polyester it can be blended with various fibres. As blends or pure fibre it dyes well and also bleaches well. Used in the manufacture of various dress material or as furnishing filling pillows, etc. Since it is a hydrophobic fibre, it washes well and drip dries.

Dado: A usually two-coloured, printed cotton; used for wall hangings in Italy; usually has a cream ground.

Dagges: Used in gothic period, a cut out or a sewn – on flaps of fabric at the edges of a garment.

Daghestan rugs: Made in Caucasia; the web and the short and close pile being of wool and tied in Ghiorde's knot; warp and weft in wool. The warp thread, which is often quite thick and brown in colour, can easily be seen on the back of the rug. The medium deep pile is of good quality, but the wool is lustreless. The field is completely divided into fairly narrow diagonal bands (5–10 cm wide) of different alternating colours. The most frequently used colours are blue in various shades, yellow, white and a faded green. These bands are closely decorated with small squares of different colours. At opposite corners of each square is a hooked motif. The decoration of these bands is completed by stylized roses or eight-pointed stars scattered in haphazard fashion. The ground colours are red and blue. The Turkish knot is used with a density of approximately 1,50,000 knots per m2. The design is almost always geometrical, mosaic patterns with many angular hooks being used in blue, red, yellow and ivory colours without any shading. Both ends are finished with a narrow selvage and knotted fringe; the sides are finished with a very narrow coloured selvage.

Dags: Parts of the fleece, consisting of matted fibres and dirt.

Daisy stitch or detached chain stitch: Works in the same way as chain stitch, but fashions each loop at the foot with a small stitch. This stitch may be worked singly or in groups to form flower petals.

Daka: Unbleached cotton muslin, made in Turkestan; the finer grades are used for turbans, the coarser for lining.

Dalecarlian lace: Very strong Swedish bobbin lace, made and worn by the peasant as starched ruffles. It is of buff colour.

Dalmatian lace: Coarse, narrow bobbin lace, made in Dalmatia, Austria, by the peasant women.

Dalmatic: Unbelted outer robe, for the ruling classes of the ancient Roman in Byzantine period with wide arms and coloured lengthways straps.

Dalzell: A Highland tartan, composed of green stripes over red ground, split with white narrow lines.

Damage factor (S): It is used to assess the effects of chemical treatments on cellulosic textile goods. It is calculated from the average Degree of polymerization (DP) before and after the treatment. The damaging effect of any treatment on celluloses with different degrees of polymerization can be compared through the damage factor (S) which is calculated with the following formula:

$$S = \log \{(2000/DP2) - (2000/DP1) + 1\}3.3219$$

Where, DP1 = average degree of polymerization of the material being tested before the damaging treatment; and DP2 = average degree of polymerization after the damaging treatment

Damajagua or Majagua: A fine, strong cloth-like bast that is yielded by a species of the mellow in South America; used for baskets, clothing and ropes.

Damaged feathers: The feathers that have been broken, damaged by insects, by mildew or rot, or otherwise materially injured.

Damaged selvage: See **Cut selvedge**.

Damas: French for damask.

Damascaffard: Tapestry damask, made of wool, silk waste, cotton, etc., and given a high finish on the face.

Damas chine: French silk damask dress goods, made with printed warp.

Damasen Dorure: French silk damask with gold flower designs.

Damaslisere: Silk damask, the Jacquard design being outlined in gold.

Damascene lace: Made of sprigs and lace braid joined with corded bars without any fillings. It is a modern adaptation of the Honiton pillow lace.

Damask: A very old type of fabric first made of silk in damascus. Elaborately woven on a Jacquard loom, the fabric has satin floats on a warp satin background. The surface designs run in the opposite direction from those in the background. It is now woven in different fibres and different weights and is mostly used in furnishings. Made with large weaving patterns on jacquard looms, it comes bleached. A guideline construction is 72 × 72 using 26s warp and weft. The 5 harness satin weave is usually used. In very fine damask construction, a face and back warp and 7 harness satin weave may be used. Such fabrics are called double damask. It comes in dyed shades also. Used as table linen, towels. It is rarely found now as a dress fabric. Most damask is self-toned, that is the warp and weft are in the same colour; the design creates the interest. Similar to brocade, but flatter and reversible, damask is used for napkins, tablecloths, draperies, and upholstery.

Damask satin: Double satin, the ground and the figures formed by warp and weft satin.

Damask stitch: In embroidery, a variety of ***Satin stitch,*** it takes in four horizontal threads of foundation or two stitches in a slanting direction and over two upright threads. The remaining second lines of the damask stitch are taken over the two lower threads of the upper line and two new threads instead of all the threads being new.

Damask (false): If a four-shaft broken twill structure is used in two blocks formed by areas of 1/3 twill next to areas of 3/1 twill, all woven in a broken treadling, then the cloth is called false damask or twill damask. The twill for false damask may be broken by way in which the blocks are threaded or by changing the treadling order.

Damask (twill): See **Damask (false)**.

Damasquette: An 18th Century silk brocade of Venice, showing floral designs in gold thread which was rolled out flat under heavy pressure, after taken from the loom, forming a continuous gold surface.

Damasse': (a) French shawl, made with combed wool warp and filling with large flower designs; (b) French for fabrics having both the ground and the large patterns woven in satin weave but of various coloured or lustred yarn; (c) In French, general term for fabrics woven on a Jacquard loom. See also **Ouvre**.

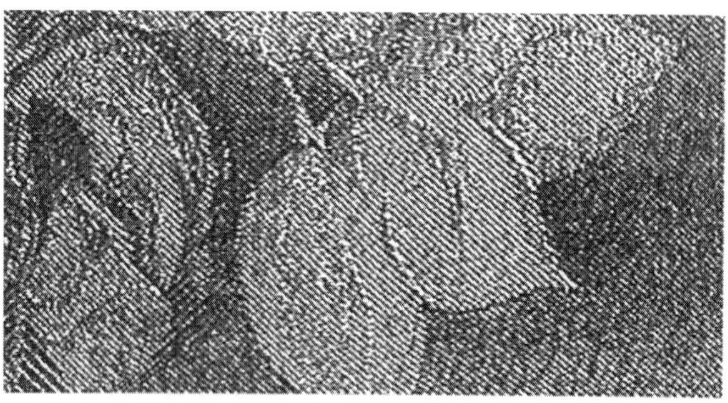

Damassin: Brocade or damask fabric with gold or silver patterns woven in to it. Originally from Venice of 17th Century, later made in France. The gold or silver threads are flattened-out (after the weaving) under heavy pressure, forming the design in a continuous surface.

Damask (woven pattern): Damask is a fine filament woven cloth, which is made by Jaquard-weaving machines (loom). There are floral and graphic patterns with glossy and dull areas possible. Damask is a high-quality decoration cloth with a wide range of use. All fibres can be used.

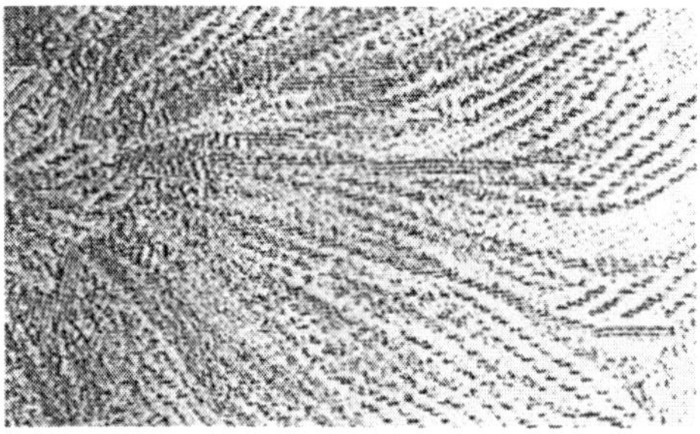

Damier: French for large square checks in solid colours.

Damp wool: Wool that has become damp or wet before or after bagging and may mildew. This weakens the fibres and seriously affects the spinning properties.

Dampening (tyre cord): The relative ability to absorb energy and deaden oscillation after citation.

Damping cloth: A piece of absorbent fabric which is wetted and laid over a garment during pressing so as to improve the press ability of the fabric/ garment.

Damping of textiles: Damping of textiles are done at various stages of processing either to help in the particular process in various ways such as lubricants, or as a participants in a reaction or sometimes as a catalyst. Example: The dampening of the fabric before sanforizing, whereby the moisture on the fabric help in slippage of the yarn which in turn help in the shrinkage of the fabric.

Dancer roll: See **Compensator roll**.

Danish cloth: Plain woven, stout cotton dress goods that are made in solid colours and slightly sized.

Danish embroidery: The lace-like white embroidery on the edges of handkerchiefs.

Doublet: Term for the men's close-fitting buttoned jacket worn over the shirt and under the coat or as an outer garment; used in gothic and renaissance period.

Darale: East Indian unbleached cotton fabric having a red stripe parallel with the selvage; used for garments by the native women.

Dardanelles canvas: Coarse Turkish canvas; used for sails and farmers' suits.

Dari: Thick, coarse and very durable and washable woven cotton carpet of small dimension; made in India.

Darida: Indian cloth, made in various colours of vegetable fibre, not common now a days.

Darin: A canvas fabric once made in France using coarse hemp yarn.

Darned lace or darned netting: Plain square or other net used for ground Roseau and is decorated with stitches, also called open lace Siena point and point Conte; in the latter, the stitches being counted.

Darnet: A French wool and silk fabric of the l0th century, brocaded with gold.

Darning: The filling up of holes, rents, etc., in textiles by means of forming a new texture with the thread and needle. Darning can be done on machine or by hand.

Darning needles: These hand-sewing needles are large needles with very large eyes. Wool darners are used when working with heavy fibres, wide ribbons and wool darning yarns. Sizes 14–19 for wool and 1–9 for cotton are available.

Dart: A wedge or diamond shape removed from the surface of a garment part by stitching or by cutting and stitching to contribute to the shape of the garment.

Dart (cut-in): An open dart cut in approximately 12" under the armhole.

Dart (front or double): An additional closed dart located toward the front edge of the garment, used to get maximum waist suppression.

Dart (panel): A panel sewn full length to the front that is used for waist suppression.

Darya: A rough faced, stout, natural coloured fabric, made of wild silk in East India.

Daryai: See **Amritsar silk**.

Dazwischen: German name for a type of curtain fabric. See **In between**.

Date palm: These leaf fibres are used for ropes, bags, hats, etc., in Asia Minor.

Datil: Leaf fibres yielded by a species of the cocoanut palm in Brazil; used for baskets, hats, etc.

Dauglin: Coarse Philippine fibre; used for cordage.

Davidson: A Highland tartan composed as follows over a dark green ground – the stripes running both warp and weftwise; a group of black stripes with two narrow stripes at the edges and a wide bar between, the latter split by a narrow strip of red in the centre; a bar of dark green, about the width of the 'black group, split in the middle with a black stripe (of the same width as the edge stripes in the black group) ; a group of navy stripes, of the same dimensions as the black, the centre bar likewise split by a red stripe; followed by a green bar as above and repeat.

Daylight colour: The colour or shades as seen in the daylight. Since the shade has to be seen at any time or anywhere, daylight tubes are used which give almost the daylight effect artificially. For example: D65 lights.

Daylight illuminants (CIE): Series of illuminant spectral power distribution curves based on measurements of natural daylight and recommended by the CIE in 1965. Values are defined for the wavelength region 300–830nm. They are described in terms of the correlated colour temperature. The most important is D65 because of the closeness of its correlated colour temperature to that of illuminant C, 6774K. D75 bluer than D65 and D55 yellower than D65 are also used.

Daylight lamp: See **D65 lamp/light.**

dB: Decibel. See **Sound level.**

DD: Differential dyeing.

D.D.: American abbreviation on textile care labels for drip drying (hang only). Applies to the garments made from self-smoothing fabrics.

DDT [dichlorodiphenyltrichloroethane; $(ClC_6H_4)_2CH(CCl_3)$]: A colourless crystalline organic compound that was once widely used as an insecticide. It is very stable and tends to accumulate in the soil, and passes up the food chain to accumulate in the fatty tissues of carnivorous animals. Its systematic name is 1,1-bis(4-chlorophenyl)- 2,2,2-trichloroethane.

De Fundato: Medieval silk, dyed purple and having a gold net pattern.

De-aereation: Removal of all undissolved gasses and part of the dissolved gases from spinning liquor before extrusion.

Dead cotton: (1) A small nep of cotton fibres which is gathered on the surface of the fabric and which is different in colour from the surrounding fabric.

(2) An extreme form of immature cotton with a very thin fibre wall. These fibres are usually weak, brittle, and lacking in twist or convolutions which easily forms 'neps' and it lacks in lustre which gives the 'dead' appearance especially after dyeing.

Dead frame yarn, in the Wilton type pile floor covering: Yarn of the same fibre composition as the pile yarn that is included in the backing and which never appears in the face of the floor covering.

Dead time: Dynamic parameter of transfer elements in automation systems; the time between turning on an input signal in a system at rest before the output signal begins to alter distinctly (for humans: reaction time).

Dead wool: Wool taken from sheep that have died of natural causes.

Dead yarn: The pile yarn in a multi-frame Wilton carpet which lies completely flat in the substrate and is not being raised by the jacquard mechanism.

Deanionised water: Water treated by ion exchange so that the anions (e.g. NO^{3-}) have been removed.

Deareation (in spinning): The removal of all undissolved gases (bubbles) and part of the dissolved gases from spinning liquor before extrusion to avoid any problem while extrusion and fibre formation.

Dearing: A late maturing commercial variety of cotton, yielding very large percentage of lint; the staple measures 20–25 millimetres.

Debage: French dress fabric, made with lustre wool warp of natural colour and dyed woollen filling.

Debuani: Native East African trade name for cotton fabrics, woven with red, yellow, black, or blue stripes, borders or checks; used for turbans.

Deburring: A process in wool yarn manufacturing for extracting burrs, seeds and vegetable matters from raw wool. Deburring is done mechanically on a burring machine.

Deca-: (Gk.: *deka*= ten). Unit prefix denoting ten times = 10; e.g. 1 daN = 10 N.

Decarbonation: The process of removing CO_2 from water typically using compact towers or air scrubbers.

Decating: The American word for Decatizing. See **Decatizing**.

Decating mark: A crease mark or impression extending weftwise across the cloth near the beginning or end of the piece. See **Apron mark, leader mark**.

Decationised (decationized) water: Water treated by ion exchange so that the cations (e.g. Ca^{2+}) have been removed.

Decatising: Same as Decatizing. See **Decatizing.**

Decatizing: Decatizing is normally the last finishing process for some fabrics. It is a method of steaming fabric between two layers of cotton press cloths. The process is used to: (a) Improve the hand and drape; (b) Brighten the colours and enhance natural lustre, assist in setting the finish or refinish fabrics after sponging or cold water shrinkage. Decating is a normal step for many wool and wool blend fabrics. It is an effective mechanical softening treatment resulting in a luxurious, soft and smooth handle. The process involves the fabric to be wound over a perforated cylinder and pass the steam through it from inside to outside or outside to inside. The process is also effective on acetate, acrylic, rayon, spun polyester and other synthetic blends.

Decatizing, fastness to: Resistance of the dye/dyeing to decatizing operation. Wind a sample sandwiched between cotton-cloth brushed on both sides around a perforated decatizing cylinder through which dry saturated steam is blown for 15 min; a) light decatizing factor: steam pressure 147 kPa, temperature 110°C; b) intense decatizing factor: steam pressure 245 kPa, temperature 127°C. Dry specimen in warm air (< 60°C). Simultaneously test dyed reference fabric. When assessing the dyed reference fabric using the grey scale, light decatizing should result in yellowing to grade 4, and intense decatizing yellowing to grade 3, which is more yellow.

Dechets: French for silk waste.

Dechlorination: (a) Removal of excess chlorine after chlorination, usually by adding sulphur dioxide

$$HOCl + SO_2 + H_2O = H_2SO_4 + HCl$$

Where, sulphur dioxide is not available sodium thiosulphate can be used; (b) The addition of a chemical to soil from contaminated land in order to remove or destroy organochlorine compounds. It has been used to treat soils contaminated with ***PCBs***.

Decibel (dB): A customary logarithmic measure most commonly used (in various ways) for measuring sound. The human ear is capable of detecting an enormous range of sound intensities. Furthermore, our perception is not linear. Experiment shows that when humans perceive one sound to be twice as loud as another, in fact the louder sound is about ten times as intense as the fainter one. For this reason, sound is measured on logarithmic scales. Informally, if one sound is 1 bel (10 decibels) “louder” than another, this means the louder sound is 10 times louder than the fainter one. A difference of 20 decibels corresponds to an increase of 10 × 10 or 100 times in intensity. The beginning of the scale (0 decibels) can be set in different ways, depending on exactly which aspect of sound is being measured. For sound intensity (the power of the sound waves per unit of area) 0 decibels is equal to 1 picowattper square meter; this corresponds approximately to the faintest sound that can be detected by a person who has good hearing. A quiet room has a normal sound intensity of around 40 decibels, ten thousand times louder than the faintest perceptible sound, and a thunderclap may have an intensity of 120 decibels, a trillion times louder than the faintest sound. For sound pressure (the pressure exerted by the sound waves) 0 decibels equals 20 micropascals (μPa) RMS, and for sound power 0 decibels sometimes equals 1 picowatt. In all cases, one decibel equals about 0.115 129 neperand *d* decibels equal *d*(ln 10)/20 nepers. See also **dB-**.

Decigram (dg): A metric unit of mass equal to 100 milligrams or about 1.5432 grains.

Deciliter (dl or dL): A common metric unit of volume equal to 0.1 litre or 100 cubic centimetres. A decilitre contains 6.10237 cubic inches, 3.38140 US fluid ounces, or 3.519 British fluid ounces. The decilitre is similar in size to the gill, an old English unit of volume.

Decimetre (dm): A fairly common metric unit of distance equal to 10 centimetres or 3.9370 inches. The decimetre is very close to the hand, a traditional English unit.

Decitex (dtex): A common metric unit of yarn density equal to 0.1 tex, 0.9 denier, or 0.1 milligram per metre. This unit was previously called the **drex**.

Deco: Designs popular on 20s and 30s with bold outlines and streamlined shapes.

Deconvolution index: According to Calvert and Clibbens, this is the criterion for assessing mercerization. It is determined by counting out approximately 800 fibres (0.1–0.2 mm long) under a microscope. The percentage number of fibres with no twist corresponds to the deconvolution index. Assessment: 9–15 = unmercerized cotton, 60–70 = mercerised cotton (for piece goods, 25–40, usually fewer in the weft direction than in the warp direction).

Decorative fabric: A broad class of fabrics used for drapery, upholstery, etc, in the field of interior decoration. Any fibres may be employed in decorative fabrics, such as cotton, silk, wool, man-made fibre, linen, plastics, and all weaves are used as well as printing, embossing etc.. Popular decorative fabrics include tapes tries, damasks, cretonnes, chintzes, brocades, tweeds etc.

Decorative seams: This is the first of the two classes of seam, which, in the old British Standard, were not regarded as seams at all and were given the name 'stitching'. The main use of the seam is for decorative sewing on garments where multiple rows of stitches are sewn through one or more layers of fabric. These several layers can be folds of the same fabric. The simplest seam in the class has decorative stitching across a garment panel. One row would have little effect but multi-needle stitching is common. The main decorative seams are:

(a) *Channel seam*: Channel seams are used as decorative seams in skirts, blouses, dresses, suits, coats and children's clothes made of heavy- or medium-weight fabrics. Pin and machine-tack on the seamline, leaving one long thread at each end. Press as stitched and then press the seam open. Clip the machine tacking on one side at four or five stitch intervals. Cut an underlay of the same or contrasting fabric 2.5 cm (1 inch) wider than the two seam edges. Working from the wrong side, centre the underlay over the seam and pin in position. Tack from the right side, stitch along each side an equal distance from the seam depression. The distance may be from 6.0 to 1.2 cm (1/4 to 1/2 inch), gauge the distance with the presser foot – if more than 6 cm (1/4 inch), mark both stitching lines with tacking Remove the machine tacking by pulling the long thread, and remove any short threads that are visible with tweezers.

(b) *Slot seam*: The slot seam is desirable when one would like to add a strip of fabric in a matching or contrasting colour to the garment. This also adds a little bit of "spice" to an otherwise plain seam. Sew a plain seam with a basting stitch. Press the seam open. Cut a strip of matching or contrasting fabric, the same width as the two edges. Lay the strip in place as illustrated. Pin if necessary. Turn the garment over with correct side of fabric up (with strip still in position). Using the presser foot as a guide, stitch 1/4 inch away from seamline with a regular machine-stitch. Repeat on opposite side seamline. Press the seam in place. Remove the basting stitch.

(c) *Corded seam*: The corded seam is a decorative seam or edge and can be used as a design feature on your garment. Place one piece of fabric on the sewing table with correct side up. Place raw edge of cording even with even edge of fabric. Place second piece of fabric with correct side down over cording with allowance edges even. Using a cording foot (or zipper foot) stitch along the edge of cording (stitch on the seamline).

(d) *Piped seam*: Piped seam is a decorative seam in which an additional component or piping is inserted between the two layers of fabric preferably of different fabric or material just to add an additional feature in the garment.

(e) *Fagotting*: Fagotting is a decorative seam which is done by hand. Here two folded edges of different pieces are joined together by Herringbone stitch or other decorative stitch with same coloured embroidery thread as the colour of the fabric.

Decorative stitching: The sewing of thread to accent a pocket, collar or some other part of the garment. Usually, this thread is a different colour than the body fabric.

Decorticating: A mechanical process for separating the woody matter from the bast fibre of such plants as ramie and hemp.

Decoupe: Extra yarn patterns from which the long floats have been cut out.

Decortication of polyester: This treatment is aimed at providing a silky-smooth hand to polyester fabrics (a few years ago this process was also used to obtain microfilaments by increasing the fibre fineness), a lustrous effect and an enhanced drapability. The best results can be obtained with fabrics produced with coarser yarns. The open-width decortication process can be carried out on jiggers or beam dyeing machines; rope decortication is performed on jet or overflow systems (batch systems). Decortication is carried out after scouring and heat setting; it is better to carry out a heat setting treatment also after the decortication process. Operating conditions applied: the process is carried out at a temperature varying from 90–95 °C to 120–130 °C for 20–35 minutes, with 30–50 g/l of NaOH 36°Bé. Once the process has been completed, the fabric is washed and neutralised.

Decrease: In bobbin lace, knitting, crocheting decrease means diminishing the patterns by using less stitches or loops.

Decrystallised cotton: Cotton treated with zinc chloride, caustic soda, etc., to reduce the degree of crystallization.

Dehbid: Persian all-wool knotted rugs, similar to Kermanshah, but of inferior quality.

Deep dyeing: The fibres which have higher dye uptake than the normal fibres when dyed together.

Deep dyeing variants: Polymers that have been chemically modified to increase their dye ability. Fibres and fabrics made therefrom can be dyed to very heavy depth.

Deep emboss printing: Coloured embossed finishes in which the embossing roller on the embossing calendar operates simultaneously as a printing roller. Immediately before embossing, the upper areas of the heated relief roller are supplied with dye paste by means of a colour embosser situated at the cloth entry end and this paste then spreads straightaway into the embossed, deep parts of the fabric.

Deep dyeing fibres: See **Deep dyeing**.

Deep embossed finishes: See **Embossed deep finishes**.

Deep pinning: Conspicuous pin-stenter marks in the body of the fabric (i.e. clear of the selvedge), and that therefore reduce the usable width of the fabric.

Defect: Specific for inflatable restraints, an imperfection in a cut piece of the fabric that judgement and experience indicate is likely to result in either hazardous or improper deployment of the inflatable restraint module in which the imperfection is incorporated.

(a) *Article* – A fault that reduces the ability of the article to perform its intended function or if it were to appear in a prominent position in the article, would readily be seen and objected to by an ordinary person who might contemplate purchasing the article in a retail shop.

(b) *Fabric* – A fault that if it were to appear in a prominent position in a garment or manufactured article made from the fabric, would readily be seen and objected to by an ordinary person who might contemplate purchasing such a garment or manufactured article in a retail shop.

Defect (in sewing): It refers to a quality imperfection found in the thread. Thread defects include – slubs, knots, neps, slack twists, corkscrew twists and single kinks.

Defective: A description of a unit or group of a unit that does not meet the unit or group tolerance. A test sample or a test specimen or a set of test specimens which fail in one or more respects to comply with the relevant requirements of the specification or standard reference sample.

Defects (in inspection and grading): The departure of non-conformance of some characteristics from its intended level or state.

Defibrillation: Individual fibres can become partially separated from the yarns that make up a fabric, poking up and degrading the appearance of the fabric. Fibrillation can result from long wet processing, especially if agitation is vigorous. With cellulose fibres, *cellulase* enzymes are now commonly used to remove fibrils from dyed garments by a process called bio-polishing.

Defoamers: Chemicals or auxiliaries used for reducing the foam developed in a process due to any reason. Silicones are good defoamers, but with disadvantages. Now a days, many defoamers are developed - both organic and inorganic.

Deformation: A change of shape of a material caused by forces of compression, shear, tension or torsion. Deformation may be immediate or delayed and the latter may be recoverable or non-recoverable.

Deformation (delayed): Deformation which is time dependent and in exhibited by material subject to a continuing force.

Deformation (immediate elastic): Recoverable deformation which is essentially independent of time.

Deformation (in testing): A change in the shape of a testing specimen.

Deformation (permanent): The net long term change in dimension of a specimen after deformation and relaxation under specified conditions.

Degasifiers: Gases, such as carbon dioxide, can readily be stripped from water by passing the water down a tower with air being forced up through the tower.

Degradation: The loss of desirable physical properties by a textile material as a result of some processes or physical/chemical phenomenon.

Dégradé: (1) An ombré effect with alternating light/dark zones, where the shade gradually changes in contrast to the uninterrupted dark/light/dark sequence (ombré). Such effects can be achieved by interlacing suitably dyed yarns or alternatively by liquor feed pipes which have holes of varying diameter. These enable dye liquor and water to flow into the trough of the padding machine and create a variable dyestuff concentration in the trough.

(2) Reductions in tonal strength of colour.

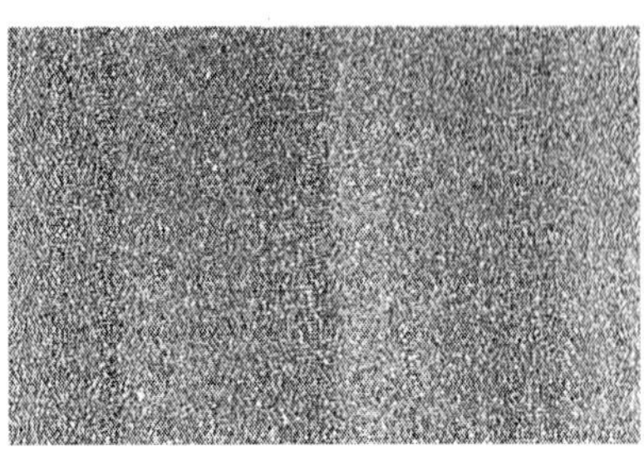

Degreasing: (a) The removal of grease and other extraneous matters by an aqueous or solvent process. E.g., Degreasing of rotary screens; (b) The removal of natural fats waxes, grease, oil and dirt from any textile material by extracting with an organic solvent.

Degree (° or deg): (1) The standard unit of angle measure, equal to 1/360 circle, 60 minutes, 3600 seconds, or about 0.017 453 293 radian. So far as we know, this unit was introduced by the Greek geometer Hipparchus of Nicaea (ca. 180-ca. 125 BC), who developed the first trigonometric tables.

(2) Any of the various units of temperature. In the United States of America, the unmodified unit "degree" generally means the degree Fahrenheit; in other countries, it means the degree Celsius.

(3) A unit measuring the hardness of water. Water is called "hard" if it contains a high concentration of mineral salts, especially calcium carbonate. This concentration can be expressed clearly in parts per million (ppm) or in milligrams per litre (mg/l). But starting in the mid-19th Century, several uses of "degree" for water hardness became well-established. In the United States of America and Britain, hardness is measured in **Clark degrees**, named for the scientist who devised the first reliable test for water hardness. The Clark degree is defined as 1 part of calcium carbonate per 70 000 parts of water; this is about 14.3 parts per million (ppm), 17.1 mg/l, or 1 grain per gallon (gpg, another popular measure of hardness). The French, properly decimal as always, used a degree equal to exactly 10 ppm, while the German degree was equivalent to 17.8 ppm.

Degree Baumé (°B or °Bé): A unit of relative density, as read on a type of hydrometer invented by the French chemist Antoine Baumé (1728–1804). Two scales are used, depending on whether the liquid is lighter than water or heavier than water. For lighter liquids, the relative density d in degrees Baumé is related to specific gravity S by the formula $d = (144.3/S) - 144.3$; for heavier liquids the formula is $d = 144.3 - (144.3/S)$.

Degree Celsius (°C): A metric unit of temperature. The Celsius temperature scale is named after the Swedish astronomer and physicist Anders Celsius (1701–1744), who used a similar scale. The freezing point of water (at one atmosphere of pressure) was originally defined to be 0 °C, while the boiling point is 100 °C. Thus the Celsius degree is 1/100 of the difference between these two temperatures (The scale actually used by Celsius was inverted, so that 0 °C was the boiling point of water and 100 °C the freezing point.) In the SI system, the Celsius scale is defined so that the temperature of the triple point of water (the temperature at which water can exist simultaneously in the gaseous, liquid, and solid states) is exactly 0.01 °C, and the size of the degree is 1/273.16 of the difference between this temperature and absolute zero (the temperature at which all molecular motion ceases). For practical purposes, this is equivalent to the original definition. See also **Degree Fahrenheit, Kelvin**.

Degree centigrade (°C): An older name for the degree Celsius. The Celsius scale is called "centigrade" because it has 100 (centi-) gradations between the freezing point and boiling point of water. About 1850 scientists began calling the unit "degree Celsius" in honour of its (presumed) inventor.

Degree Fahrenheit (°F): A traditional unit of temperature still used customarily in the United States of America. The unit was defined by the

German physicist Daniel G. Fahrenheit (1686–1736), who also invented the mercury thermometer. Fahrenheit set 0° at the coldest temperature he could conveniently achieve using an ice and salt mixture, and it is said he intended to set 96° as the temperature of the human body. (He was off a little there; normal oral temperature for humans is between 98 °F and 99 °F.) On this scale, the freezing point of water (at normal sea level atmospheric pressure) turned out to be about 32 °F and the boiling point about 212 °F. Eventually the scale was precisely defined by these two temperatures. 1°F equals 5/9 °C, but in converting between scales we have to be careful to adjust the zero points as well. To convert a temperature in °F to the Celsius scale, we must first subtract 32° and then multiply by 5/9. In the other direction, to convert a temperature in °C to the Fahrenheit scale, we must first multiply by 9/5 and then add 32°.

Degree Kelvin (°K): An obsolete name for the kelvin. In the International System, temperatures on the absolute temperature scale are stated in kelvins, not in degrees Kelvin.

Degree of crystallinity: The amount, expressed as a percentage by mass, or linear polymer that is generally present in a crystalline form, the remainder of the polymer being present in an amorphous state.

Note: There are several methods used for the determination of the amount of crystalline polymer in a man-made fibre. The results obtained differ according to the method used, so comparisons should be limited to one method of measurement.

Degree of esterification: The extent to which the acid groups of terephthalic and/or other acids have reacted with diols to form ester groups in polyester polymer production.

Degree of freedom (for a set): The number of values that can be assigned arbitrarily and still get the same value for each of one or more statistics calculated from the set of data.

Degree of orientation: The extent to which macro molecules composing a fibre or film lie in a predominant direction. In the case of fibres, the extent to which the molecules in a fibre lie in the direction of the fibre length.

Degree of polymerisation: The average number of repeating units in the individual macromolecules in a polymer.

Degree of staining grey scales: A piece of untreated, unstained, undyed cloth is compared with the treated sample that has been in contact with the test specimen during the staining test and a numerical assessment of staining is given. A different set of grey scales is used for measuring staining. A rating of 5 means that there is no difference between the treated and untreated material. If the result is in between any two of the contrasts on the scale, a rating of, for example, 3–4 is given and rating 1 shows a white and a grey sample. The other numbers show geometrical steps of contrast between white and a series of greys.

Degree of water hardness: A measure of water hardness salts available in a water sample.

Degummed natural silk: The silk yarn or fabric which has been gone through the degumming process only and now other process retaining the natural colour of the silk.

Degumming: Boiling off of silk. The removal of Sericin from silk fibre by a controlled hot alkaline soap or enzyme treatment. Degumming is done either at the yarn or fabric stage and is done normally at the first stage wet processing equivalent to desizing and scouring of cotton fabric/yarn.

Dehydration: (a) Removal of water from a substance; (b) Removal of the elements of water (i.e. hydrogen and oxygen in a 2:1 ratio) to form a new compound.

DEK: Deutsche Echtheitskommission (German Fastness Committee).

Dekagram or Decagram (dag): A common metric unit of mass, the dekagram is frequently used in European food recipes. One dekagram is equal to 10 grams, 0.01 kilogram, or 0.352 739 66 ounce. The symbol **dkg** sometimes used for this unit is incorrect.

Dekaliter or Decaliter (daL or dal): A metric unit of volume equal to 10 litresand comparable to the English peck. The dekaliter is equal to about 2.641 72 US liquid gallons, 1.135 10 US pecks, or 2.199 69 British imperial gallons (1.099 85 British pecks). The symbol **dkL** sometimes used for this unit is incorrect.

Dekameter or Decameter (dam): A common metric unit of distance equal to 10 meters (about 32.8084 feet). The symbol **dkm** sometimes used for this unit is incorrect.

Dekanewton or Decanewton (daN): A fairly common metric unit of force equal to 10 newtons. The dekanewton is equal to 1 megadyne, to 1.019 716 kilograms of force (kgf) or kilopounds (kp), to 2.248 09 pounds of force (lbf), and to 72.3301 poundals. In engineering, the dekanewton is a convenient substitute for the kilogram of force or kilopond, since it is nearly equal to those units.

Delaine: A light weight, all wool fabric, in plain weave, ornamented by printing. (a) French term, meaning "of wool"; (b) Trade term for combing Merino wool of medium fine quality; (c) Wool raised in United States is said to have the strongest staple of all wools; (d) Originally a light weight and plain woven French fabric, made of all-wool and dyed in the piece; at the present it is made in England of a mixture of cotton and wool and often printed.

Delamination: Separation of backing from carpet.

Delamination strength: The tensile force required to separate the component layers of floor covering under specified force.

Delave': Describes washed-out or faded denim. Delave' can be used instead of washed-out to describe jeans made from this fabric.

Delay cure: A process where the curing step is delayed until the garment has been made and pressed.

Delayed curing: See **Delay cure**.

Delayed deformation: Deformation that is time-dependent and is exhibited by material subjected to a continuing load. Delayed deformation may be recoverable following removal of the applied load.

Delhi embroidery: Rich Indian embroidery, made in chain stitch, in gold, silver and silk over satin or other silk ground.

Delicate or gentle cycle (in washing): A washing cycle in which agitation is slow and time is reduced.

Delinere: French bleached linen of medium quality for the household.

Deliquescent: A hygroscopic compound is one that absorbs so much water from the atmosphere that a solution eventually forms.

Delft unit: See **DU**.

Delta (D or Δ): A symbol used to indicate deviation or difference.

Delta E*, Delta e*: The total colour difference computed with a colour difference equation (ΔEab or ΔEcmc). In colour tolerance, the symbol DE is often used to express Delta Error.

Delustreing: Subduing or dulling the natural lustre of a textile material by chemical or physical means. The term often refers to the use of titanium dioxide or other white pigments as delustrants in textile materials.

Delustreing in spinning: Adding of delustreing chemicals at the spinning stage of synthetic fibres so that the resultant fibre is already delustred and requires no delustreing in the further processing.

Delustrant: A substance that can be used to dull the lustre of a manufactured fibre. Often a pigment such as Titanium dioxide.

Demeshek: In the Bible means silk.

Demi- : A traditional prefix meaning 1/2. The prefix is derived from the Latin *dimedius*, meaning a cut in the middle of something.

Demi drape: Fine French woollen cloth of light weight, fulled lightly, shorn and calendared.

Demi londres: Loosely woven and lightly fulled French woollen, usually left un-finished.

Demi lustre:(a) Wool of fairly fine staple, not as long and lustrous as lustre wool; included are Cotswold, Devon, Roscommon, Romney, Wensleydale; used for yarns of 36s to 46s count; (b) Yarns of 36s–46s count, made of such wool; (c) See **Lustre Orleans**.

Demi mousseline: Very light French cambric of cotton or linen yarn, made plain or striped.

Demi ostade: Medieval Dutch woollen fabric, finished with a napped face without fulling.

Demineralisation: The process of removing minerals (Ca^{2+}, Mg^{2+}, Fe^{3+}, Cu^{2+}) from water, and fabric is known as demineralization. Usually it is done on fabric through deionization and chelation and water by deionisation, reverse osmosis, or distillation. Deionisation mechanism is as follows: Free metal and its oxide is converted to its ionic form by protonic treatment (demineralization) and this ionic form is water soluble $Fe \rightarrow Fe^{2+} + 2e$ and the following reaction takes place $4Fe^{2+} + O \rightarrow 4Fe^{3+} + 2O^{2-}$) In fabric processing, this metal ion in solution is chelated with a suitable wetting agent so that this does not interfere in the bleaching process.

Demirdji rugs: All-wool rugs, made in Asia Minor; the loose and medium long pile is tied in Ghiordes knot. The designs and colours show great variety.

Demittons: Stout English cotton cloths for the oriental trade.

Denaturation: The process of making a product undrinkable. See **Denatured**.

Denatured: A product that has been made undrinkable, e.g. alcohol (Ethanol, denatured) or sodium chloride for exemption of tax and reduced price for industrial purposes.

Denier: A weight-per-unit-length measure of any linear material. Officially, it is the number of unit weights of 0.05 grams per 450 metre length. This is numerically equal to the weight in grams of 9,000 metres of the material. Denier is a direct numbering system in which the lower numbers represent the finer sizes and the higher numbers the coarser sizes. In the US, the denier system is used for numbering filament yarns (except glass), manufactured fibre staple (but not spun yarns), and tow. In most countries outside the US, the denier system has been replaced by the tex system. The following denier terms are in use:

(a) *Denier per filament* (*dpf*): The denier of an individual continuous filament or an individual staple fibre, if it were continuous. In filament yarns, it is the yarn denier divided by the number of filaments.

(b) *Denier* (*yard*): The denier of a filament yarn. It is the product of the denier per filament and the number of filaments in the yarn.

(c) *Denier* (*total*): The denier of a tow before it is crimped. It is the product of the denier per filament and the number of filaments in the tow. The total denier after crimping (called crimped total denier) is higher because of the resultant increase in weight per unit length.

(d) *Denier variation*: Usually variation in diameter, or other cross-sectional dimension, along the length of a filament or bundle of filaments. It is caused by malfunction or lack of process control in fibre manufacturing and degrades resulting fabric appearance or performance.

Denim: A firm 2 × 1 or 3 × 1 twill-weave fabric, often having a whitish tinge, obtained by using white filling yarns with coloured warp yarns, generally white weft and blue warp. Lighter weight denims with softer finish are made in a variety of colours and patterns and are used for sportswear. This fabric has its origin in the traditional blue (indigo-dyed) overall cloth made and worn in Nimes; hence, the name 'de Nimes'. It is stiff and unyielding, but softens in wear and is very strong and hard wearing, but fades to whitish patches at creases or points of strain, hem edges, etc., and creases easily. A typical is 2/1 left hand twill weave with 66 × 40, carded cottonyarn in counts around 10s dyed usually in indigo or navy in the warp and in count around 16s in the filling. Used for dungarees, casual style clothing such as jeans, skirts, light weight jackets, suits and shorts, children's clothes, work clothes, men's and women's sportswear and for protective clothing. Made from all-cotton or cotton with polyester, cotton with modal, or cotton with polyester and viscose, etc. A very common construction is epi 68, ppi 52, yarn number both warp and weft 20/1.

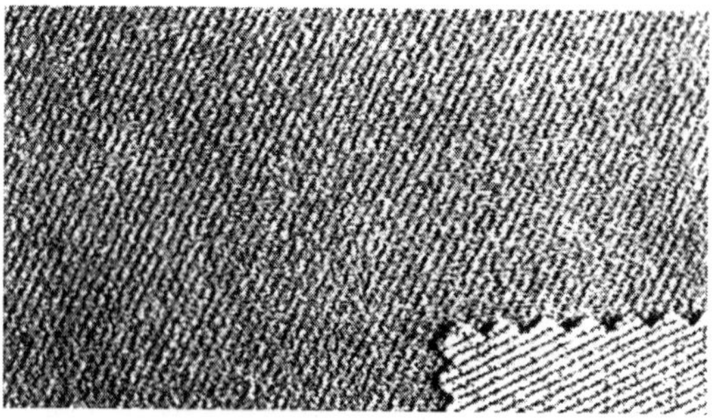

Denim cord: A denim look fabric made from white and coloured yarn, with a slight Bedford cord effect produced by raised ribs. It is a mixture of polyester and cotton fibres, and creases less than all-cotton.

Denim finishing: Various processes done on garment, fabric, yarn and even in manufacturing stages to get the appearance and finish of the ultimate product which can be fabric or garment mainly, as per the fashion. It can include chemical shrinkage control, sanforizing, liquid ammonia treatment, stone wash, enzyme wash, chemical spray, etc. These are more important as the denim fabric undergoes vey less wet-processing.

Denitrification (in effluent treatment): The biological process by which nitrate is reduced to nitrogen and other gaseous end products. The removal of oxygen from nitrates in anoxic or anaerobic conditions, resulting eventually in gaseous nitrogen, which bubbles off and is thus removed. Many heterotrophic bacteria found in aerobic biological processes can adapt to using the oxygen that is combined in nitrates, if the dissolved oxygen is very low or zero (***respiration*** under anoxic conditions). The sludge in the bottom of a secondary sedimentation tank has little oxygen and with a nitrified effluent it may rise to the surface because of denitrification, as the nitrogen bubbles lift the sludge. In activated sludge treatment, denitrification can be achieved as with the AAO process or the Bardenpho process or the modified Ludzack–Ettinger process where the nitrified wastewater is recycled back to the anoxic tank at the start of the treatment process. Alternatively, an anoxic filter can be used for denitrification. If the anoxic treatment process is located after secondary sedimentation, methanol or other carbon sources may be added to the secondary effluent to act as organic carbon sources for the denitrifying bacteria. The denitrification of a nitrified effluent can also be achieved in ***biological aerated filters*** or ***biological fluidised beds***. ***Biological fluidized beds*** can be used to remove nitrate from drinking water. The denitrification process can occur in rivers when a nitrified effluent meets a polluted river water. *See* **Nitrification–denitrification**.

Denmark satin: Stout English worsted satin, dyed black and finished with a high lustre; used for slippers.

Density: Mass (weight) per unit volume. Knowing the density of one material makes it possible to reasonably approximate a required weight of the substance by measuring its volume which is more convenient; in case of liquids, approximations can be calculated in case of powder solids also. Mass of solution/Unit of volume, kg/m^3.

Density (bulk): See **Bulk density**.

Density (fibre): See **Fibre density**.

Density (fibres): Mass per unit volume of the solid matter of which the fibre is made of.

Density (linear): Mass per unit length.

Dent: A unit of a reed comprising of a reed wire and the space between the adjusted wires.

Dental floss: Slack spun strong silk yarn; used to clean the teeth with.

Dents per unit width: For woven pile yarn floor covering, the number of binding sites per unit width; dents being the reed spaces through which the warp yarns pass in the loom or the metal strips in the reed which form these spaces.

Dent (in screen printing): A damage mark on the rotary screen usually by handling where the rounding of the screen is not perfect.

Dent mark (in printed goods): A visible damage on the print due to the mark produced due to the dent on the screen.

Dentele: French for scalloped or festoon edge.

Dentelle: French term for lace since the 17th Century.

Dentelle de Fil: French term for laces, made of one-ply thread.

Dentelle a la Vierge: French bobbin lace with double ground, similar to the **Ave Maria lace** (see).

Denting: Denting means drawing the warp thread through the dent as required by the reed plan and this determines more accurately the width and ends per inch of the fabric.

Dents: French for scallops or points, as in edges of laces.

Dents per unit width (in woven floor coverings): The number of binding sites per unit width; dents being the reed spaces through which the warp yarn pass in the loom or metal strips in the reed that form these spaces.

Deodorant: A substance used to remove or mask an unpleasant odour. Products suitable for use as deodorants include, inter alia, oxidizing agents (products containing chlorine or peroxide), adsorbents, fragrant finishes, chlorophyllin, etc. See **Fragrant finishes; Antiperspirants**.

Deodorisation: It refers to the bonding of odorous substances by sealing in the odour-emitting molecules (clathrate), a process which has not yet been fully explained. Odour molecules are bound according to their structure and thus rendered undetectable to the senses. There is a quantitative interrelationship between the odour-emitting substance and the deodorant. Zinc soaps are an example of a group of deodorants.

Deodorizing: Making a material odourless by odour suppression, masking, destruction (oxidizing agents) or adsorption (adsorbents). In solvent processes (dry-cleaning, solvent finishing, solvent sizing) deodorizing is carried out at the end of the process, after drying, by subjecting the goods to a current of fresh air in order to ventilate them and make them odourless.

Deodorising treatment: A procedure for removing substances responsible for generating undesirable odours from textiles.

Deora: Indian jute having a very dark, strong and coarse fibre; used for ropes.

Dependent variable: A variable that may respond to a change in another variable.

Depitching: The process of removing oily matters, tar, etc., from wool. It can be done by solvent extraction or by other methods.

Deployment (for inflatable restraints): The sequence of events related to the activation of a module.

Depolymerization: The opposite of Polymerization, i.e. the breakdown of polymers into their monomers, dimers, etc.

Depreciation: Amortisation; Service life of textile plant and equipment.

Depth gauges: Mechanical or optical devices which are used, e.g. to check the engraving depth when manufacturing printing rollers.

Depth (in dyeing): The deepness of colour quality in a shade, an increase in which is associated with increase in the quantity of the colourant present.

Depth: The colour quality (q.v.), an increase in which is associated with an increase in the quantity of colorant present, all other conditions (viewing, etc.) remaining the same.

Depth of shade (DOS): Ratio of weight of dye to weight of goods dyed, usually expressed as percentage; amount of dye owg Depth of shade (DOS), in these terms, is not really a very good way of comparing the darkness or intensity of colour of finished fabrics, due to inherent differences in the hues of different dyes within a family, differences between dye families, and differences due to the nature of the fabric. Dye manufacturers' shade cards are typically show one or two depths of shade for a particular dye, often between 1% and 4%, except for black, which is typically 3% to 6%.

Depth: The deepness or colour quality in a shade, an increase in which is associated with the increase in the quantity of colorant present.

Derbent rugs: Medium and large sized rugs, made in Caucasia. They are all-wool, the warp occasionally of goat hair, with long soft pile tied in Ghiordes knot and having a fine lustre. The weave is loose. The designs are geometrical, star patterns often occurring. The chief colours are red, yellow and blue. The ends are finished with a long, knotted fringe.

Derby: Hosiery knitted with six ribs on the face and three on back alternatively.

Deregistering (CRIMP): Process of disordering or disaligning the crimp in a tow band to produce bulk. See **Threaded roll process**.

Derived unit: A measurement unit defined in terms of the BASE UNITS of a system of measurement, and not directly from a standard value of the quantity it measures. For example, the newton is a unit of force defined in base SI UNITS as a kilogram. Metre second^{-2} (kgms^{-2})

Derries: Cotton dress goods, made in brown and blue colours in India.

Desalination: Any of various techniques for removing the salts (mainly sodium chloride) from seawater to make it fit for drinking, irrigation, use in water-cooled engines, and for making steam for steam turbines. There are various methods. Those based on distillation depend on a cheap source of heat energy, of which solar energy is the most promising, especially in hot climates. Flash evaporation (evaporation under reduced pressure) and freezing to make pure ice are other methods, as are electro-dialysis, ion exchange, reverse osmosis, and the use of molecular sieves.

Design: A complete plan of the character of a textile fabric, giving both its construction and the decorative pattern.

Design right: Legislation to protect original designs for goods, products and packages against copying without the need of a breach of copyright.

Designed: Made specifically to some kind of plan.

Desize/Bleach one bath: A so-called combination process whereby the oxidising agents used for bleaching will also degrade starch and polyvinyl alcohol under certain circumstances. It is dependent on the type of size used.

Desizing: A process by which size from the greige fabric is removed. Desizing is an important step prior to dyeing fabric, since size can interfere with dye uptake. Some size materials wash out easily. Starch is commonly used for size, and can be quite difficult to remove. Amylase enzymes are often used industrially for starch removal. Few sizes can be readily removed by hot water washing, especially synthetic starches.

Desizing agent: A chemical or enzyme used for desizing a woven fabric.

Desizing polyacrylic acid: By converting the -COOH to -COONa, water solubility of the polymer is increased and hydrogen bonding with the fibre is overcome. Polyacrylic can readily be desized by saturating fabric with alkali and washing out the solubilized size.

Desizing PVA: One of the advantages of PVA is that a dried film will redissolve in water without having to degrade it first. Fabrics sized with PVA are desized by first saturating with water containing a wetting agent (for rapid penetration) and then by heating in a steamer or J-box (to hydrate the film). The desizing step is completed by rinsing in hot water to complete the removal.

Desizing with acids: Mineral acids will hydrolyse starch by attacking glucosidic linkages. Acid hydrolysis lowers the molecular weight and eventually reduces starch to glucose. Hydrochloric and sulfuric acids can be used. One problem with acid desizing is that cellulose fibres are also degraded which is why the method is not used much. One advantage with using acids is that cotton fibres can be demineralized more easily. Insoluble salts are solubilized by acids making the removal of such troublesome metals such as iron more thorough.

Desizing with oxidizing agents: Sodium hypochlorite, sodium hypobromite and sodium or potassium persulfate will also degrade starch. The reaction opens hydroglucose rings at the - 2,3-hydroxyls, converting them into carboxylic acid groups. This method is not used much either because it too degrades cellulose fibres. Hypochlorite is used for bleaching however.

Desk research: Where information is gleaned from work already carried out by other individual or groups.

Desmolases: A small sub-group of enzymes which are capable of rupturing C–C- or C–H-bonds generally preceded by or accompanied by simultaneous oxidation.

Desorption: The process in which the sorbed material is released from the material concerned, as desorption of moisture from the fibres or the process of removing an adsorbed material from the solid on which it is adsorbed. The reverse of sorption, absorption, adsorption, or both.

Desulphurising: An after-treatment to remove sulphur from newly spun viscose rayon by passing the yarn through a sodium sulphide solution.

Detached chain stitch: See **Daisy stitch**.

Detergency: The cleansing power of surface-active compounds (surfactants) or detergents in developing a washing or cleansing effect often associated with the emulsification of fatty soils and the prevention of soil redeposition.

Detergency efficiency: Detergency efficiency is a measure of the extent to which chemicals and foreign substances can be removed from the goods in a specific situation. However, this parameter is not representative of the overall efficiency of a machine.

Detergent: These are surfactants which do not precipitate in hard water and have the ability to remove soils from solid surfaces. Over and above reducing water's surface tension, detergents must adsorb onto the soil's surface to aid in spontaneous release. Detergents must also keep the soil suspended to prevent redeposition.

Detergent booster: An additive to improve the performance of detergents. Usually based on fatty acid (mono) ethanolamides R–CO–HN–C_2H_4OH. These products improve foam stability (foam control) and detergency properties, and also protect the skin. It is a heavy duty detergent booster.

Detergents: A group of substances that improve the cleansing action of solvents, particularly water. The majority of detergents, including *soap*, have the same basic structure. Their molecules have a non-polar hydrocarbon chain (tail) that lacks an affinity for water molecules and is therefore said to be hydrophobic (water-fearing or repelling). Attached to this tail is a small polar group (head) that has an affinity for water molecules and is said to be hydrophilic (water-loving or attracting). Detergents reduce the surface tension of water and thus improve its wetting power by anchoring their hydrophilic heads in the water with their hydrophobic tails protruding above it. Consequently the water surface is broken up, enabling the water to spread over the material to be cleaned and penetrate between the material and dirt to which it has been exposed. With the assistance of agitation, the dirt can then be floated off. At the same time, the hydrophobic tails of the detergent molecules dissolve in grease and oils. The protruding hydrophilic heads of the molecules repel each other, causing the grease or oil to roll up and form tiny drops, which float off into the water to form an emulsion. Unlike soaps, synthetic detergents which are often derived from petrochemicals do not form insoluble scums with hard water. Synthetic detergents are of three types –

(a) *Anionic detergents* form ions consisting of a hydrocarbon chain to which is attached either a sulfonate group, –SO_2–O–, or a sulphate group, –O–SO –O–. Positive ions of the type $RNH3^+$, in which R has a long hydrocarbon chain.

(b) *Non-ionic* detergents are complex chemical compounds called *ethoxylates*. They owe their detergent properties to the presence of a number of oxygen atoms in one part of the molecule, which are capable of forming hydrogen bonds with the surface water molecules, thus reducing the surface tension of the water.

Determination process: The act of carrying-out the series of operations specified in the method whereby a single value is obtained.

Determination tolerance: As specified in a test method, the exactness with which a determination value is to be calculated and recorded.

Determination value: The numerical quantity calculated by means of the test method equation from the measurement values obtained as directed in a test method.

Developed dyes: See **Dyes**.

Developer (coupling components for diazo dyes): These are usually phenols, naphthols and amines which couple with previously diazotised dyes. This process leads to an increase in the size of the molecule and this significantly improves the wet-fastness properties of the dye.

Developing: A stage in dyeing or printing in which leuco compounds, dyes, or dye intermediates are converted to the final, stable state or shade.

Developing dyestuff: All dyes consisting of a pair of components such as Naphthols in dyeing and printing. Dyestuffs for coupling are seldom used.

Devil's cotton: A strong and silky white bast fibre, yielded by the abroma in India; used for cordage.

Devil's nettle: Fine, white and strong fibre, yielded by a species of nettle in Assam; used for cordage and coarse fabrics.

Devonia ground: A filling in Duchesse lace, consisting of irregularly spreading plaited bars and ornamented with pivots.

Devonia lace: Species of the **Honiton lace** (see), the flower petals or butterfly wings being made in raised relief effect.

Devonshire kersey: White English kersey, originated in the early part of the 15th Century. Also called washers and wash whites.

Devonshire lace: Bobbin lace of old origin, influenced in the 17th Century by Flanders. Some time ago Greek torchon laces with geometric designs, black laces, similar to the **Honiton** (see) and, also trolley laces were made at Devonshire.

Devore': A technique for "sculpting" fabric using *etchants* or burn-out paste.

Dew point: The temperature below which a gas begins to condense as a liquid at a given pressure and the atmosphere is cooled. Thus in air, it is the temperature at which the air becomes saturated when cooled with no further addition of moisture or change in pressure.

Dewatering of sludge: Removal of water from sludge. Various processes may be used, including mechanical dewatering of sludgeor sludge drying beds. The typical 65–80% water content of dewatered sludge varies with the method of dewatering and the nature of the sludge.

Dextrine: Dextrine is a modified starch, made by heating dry starch with a mineral acid. White dextrin is made by heating at moderate temperatures and yellow dextrine is made by heating at higher temperatures with less acid. The degree of hydrolysis is higher than for thin boiling starch, so dextrine solutions have lower viscosities. Dextrin is used as a thickener for some textile printing processes, and as a resist for some direct application dyeing techniques (it generally isn't usable for immersion dyeing because it washes off too easily). See **Dextrine**, **Modified starch**.

Dextrine (modified starch): It is made by partially breaking down starch to smaller molecules using acids and/or heat. There are several different dextrins with different properties. Dextrin may also called British gum, but some workers regard them as distinct.

Dextrines: Degradation products of starch, yellowish to white friable granules or powders (crystal gum, British gum, roasted starch, burnt farina). They are further broken down (liquefied) to maltose or glucose by the action of diastases.

Dextrins: Intermediate products obtained in the transformation of Õ starch with the following properties: a) External characteristics: progressively low viscosity; b) Iodine reaction: blue-violet, red-violet, red, reddish-brown (achromatic); c) Solubility: cold water soluble; d) Sizing mechanism: hardens, strongly conglutinated dextrin films (size, proofing) are transparent and decrease rapidly in strength and elasticity with progressive decomposition.

DGN: Dirección General de Normas (General Office for Standards, including textiles).

DGQ: Deutsche Gesellschaftfür Qualität (German Association for Quality, with textile division), Frankfurt, Germany.

Dha: Native hemp from Senegal; used for "artificial" wool, by treating it with nitric acid and solution of natrium superoxide, which render the fibre transparent and crinkled.

Dhaka muslins: Very fine fabric woven by Dhaka artisans. In its heyday, Dhaka muslin just weighed just 3.8 g per square yard. The fabric was aptly called as fabrics which could not be seen or felt. With Mughal court as its chief patron, it was titled '*mulmulkhas*' or King's muslins and at 2.7 pounds per yard (Rs 27,000 at today's prices). Esteemed for its sheer web, Dhaka muslin hard traits that earned it lyrical names like *bafta-hawa* (woven air) *ab-i-rawan* (running water – it became invisible in water) and *shabnam* (evening dew – if laid wet on grass, it could not be seen). In 19th Century this technique was in decline, but there is record that a Dhaka muslin woven for Prince of Wales weighed 5.4g per sq.yard – 42 per cent heavier than Shah Safi's gift.

Dharwar: Indian raw cotton having a moderately strong and fairly clean staple of golden colour.

Dhollerah: Indian raw cotton, having a fairly strong and cohesive fibre of dull whitish colour; it is usually very dirty.

Dhooties: Originally a reversible hand woven cotton fabric, having a plain body and bright coloured borders and heads, woven with coloured filling where the coloured border stripes are; used for garments, scarfs and turbans in India and Africa. Now often made with the filling for the white or grey body, interlacing with the coloured borders. It is a soft and light cloth of which the body is woven in plain weave.

Dhoti: It is a traditional Indian dress worn on the lower part of the body. It is one piece of unstitched cloth draped over the hips and length. It is worn in different ways in different parts of India and worn by both men and women.

Dhurrie: Indian thick cotton drapery, made with warp ribs and broad weft

stripes in blue or red.

Di- and Tri- arylcarbenium dyes: Di- and triarylcarbenium dyes belong to the class of polymethine dyes and can be considered as branched polymethines. The branches are created by two aryl rings, in which the polymethine chain is incorporated, and by another R group bonded to the central (meso) methine carbon atom. *Triarylmethine dyes* are usually divided into mono-, di-, and triaminotriarylmethine dyes. In some di- and tri-arylmethine dyes, the ring carbon atoms *ortho*to the central methine carbon atom are bonded via a heteroatom to form a heterocyclic six-membered ring. These include the acridine, xanthene, and thioxanthenedyes.

Diabiki: Medieval Arabic silk fabric with brocaded flowers.

Diacetate: A term used to describe fibres made from propanone-soluble ethanoate (acetate). The ISO generic term is acetate.

Diablefuerte: Bedford cords and corduroys in South American countries.
Diagonal: A worsted cloth with prominent diagonal ridges.

Diagonal flametest: See **Flammability test**.

Diagonal stitch: In embroidery, a couched gold or silk thread fastened down with diagonal stitches.

Diagonals: Textile fabrics characterized by (often fairly wide) stripes running in a diagonal direction. The term refers only to the pattern effect and not to any particular fabric quality. Mainly described as broad twill or stitched twill fabrics.

Dial: In a circular-knitting machine, a circular steel plate with radially arranged slots for needles. A knitting machine equipped with both a dial and a cylinder (q.v.) can produce double-knit fabrics.

Dial height (in knitting): The distance between the bottom edge of the dial section on a hitting machine at its perimeter from the corresponding upper edge of the cylinder at its perimeter.

Diamantee: Silk vesting made with two-ply warp, 20 leaves and eight fillings in a repeat.

Diamantine: (a) English twilled worsted dress goods with a high finish; obsolete; (b) 18th Century woollen fabric in England.

Diameter, Average fibre (in wool and other animal fibres): The average width of a group of fibres when measured on a projected image.

Diamine: A compound with two amino groups. Hexamethylene diamine, one of the intermediates in the manufacture of nylon 66 salt is an example of this chemical type.

Diamond: (a) A twill produced toy reversing the direction of the twill to form squares; (b) Matchings from the sides of down fleeces of 48/50 quality; (c) Worsted warp yarn, made by twisting a 4s with a 2/36s with eight turns per inch and then reverse twisting this with a 2/36s strand with four turns per inch; (d) Commercial variety of short staple, upland cotton in America.

Diamond braid: A soutache braid made with five threads, each thread passing alternately under and over two threads.

Diamond draught diaper: In Ireland, eight-leaf, twilled, linen diaper cloth, with diamond patterns.

Diamond stitch: In embroidery, rows of silk thread couched down by silk and gold thread running diagonally at opposite directions and forming small diamonds.

Diamond (in zippers): The wedge-shaped portion of slider between the throats.

Diammonium phosphate and Phosphoric acid: Phosphorus based flame retardants function in the condensed phase. Non-durable, semi-durable and durable treatments can be obtained with phosphorus based compounds. The presence of calcium ions negates the activity of phosphorous compounds. Whereas the ammonium salts decompose thermally into phosphoric acid by the loss of ammonia, the calcium salts do not. Presumably, the calcium salts are not volatile and buffer the acidity of phosphoric acid so the generation of char is diminished.

Diamond-wound package: See **Cross-wound package**.

Diamond (in zipper): A wedge shaped portion of a slider between the throats.

Diaper: (a) Rich silk fabric of the Middle Ages, made with single coloured patterns; used for church vestments, dresses, etc.; originated from the Orient; (b) A twill raising three warps in succession, followed by a pick after which the weave is reversed; also a pattern formed by twilled checks, originally with a five-leaf satin weave; (c) White linen or cotton fabric made with small diamond or bird's eye pattern, in the above described twill weave; used for towels, children's dresses, etc.; (d) A grade of damask linen, made in Ireland; (e) Stout and twilled cotton drill, made with coloured checks.

Diaper cloth: Diaper cloth or diaper gauze is similar to cheese cloth; but in this fabric, loosely twisted yarns are used for maximum absorbency. Usually it has got a construction of 44 × 44 with 28s carded cotton warp and weft. The fabric must be free of impurities.

Diaphragm: Porous dividing wall which (a) is water and ion permeable, (b) impedes diffusion from one side to the other (clarification), (c) renders impossible a mechanical combination of the liquids separated by the diaphragm.

Diastases: They are a more or less temperature-sensitive enzyme (Amylases) which converts insoluble carbohydrates (starch) via dextrines into soluble sugars (maltose, glucose). Malt diastase has a strong liquefying and saccharifying action. Bacterial diastase hydrolyzes the glucoside bonds in starch and has a strong liquefying and weak saccharifying action. Pancreatic diastases have a strong saccharifying action. The degradation of starch by diastases is always accompanied by various stages of degradation at the same time proceeding via dextrine to maltose resp. glucose.

Diazo-: An atomic grouping of the type –N=N– in Aryl compounds.

Diazo compound: They are characterized by the presence of the azo group (–N=N–). Isomeric diazo compounds are represented by the mostly unstable syndiazo compounds and the stable anti-diazo compounds (both in the cis-form).

Diazo fast salts: See **Fast colour salts**.

Diazo stabilizers: These are dispersing agents used in the diazo developing baths. A naphtholated fabric when it is padded with the diazotized base, some of the naphthols is invariably bled to the padding liquor. This naphthols forms pigment with the diazotized base and redeposits on the surface fabric, thus hampering the rubbing fastness. The diazo stabilizers, makes the pigments into a finely dispersed form and prevents the deposition on to the fabric.

Diazonium compound: Diazotized bases having a salt-like character (from hydrochloric, sulphuric or nitric acids). Diazonium compounds dissolve in water like ammonium compounds with a neutral or alkaline reaction. They are formed, e.g. as unstable intermediate compounds by diazotization of an amino group in a direct dye.

Diazonium compound in stabilised form: They are stabilized form of complex Diazonium compounds. They are called Fast colour salts. They can be stable tetrachlorozincates, stable tetrafluoroborates of diazonium compounds. Another approach is the introduction of naphthalene Disulphonate in diazonium compounds.

Diazotizable dyes: Includes direct dyes for cellulosic fibres containing one or several diazotizable amino groups. Limited shade gamut. Good wet-colour fastness, but only poor to moderate light fastness. Used for cheap washable articles, corduroy, hosiery and umbrella fabrics. Principle – Diazotization. Analogous to diazotizable disperse dyes for acetate, triacetate, polyester and mostly deep shades.

Diazotization: The reaction of an aromatic amine (e.g., aniline) with nitrous acid at low temperatures (below 5°C). During diazotization, the nitrogen atom resulting from the formation of nitrous acid (HNO_2). The acid is prepared *in situ* by reaction between nitric acid and sodium nitrite and combines with the nitrogen atom contained in the amino groups (NH_2) of the diazo dye to form a new compound which forms a new dye with higher fastness properties on the fibre after coupling or developing with a dye developer.

Dibah or Dibadj: Durable medieval silk brocade dress goods of Persian origin.

Dibasic: Term used to describe acids with two active acid hydrogen atoms which can be ionized and replaced by metals, e.g. sulphuric acid H_2SO_4 (sodium sulphate Na_2SO_4), phosphorous acid H_3PO_3 (sodium hydrophosphate Na_2HPO_3).

Dibenzanthrone or violanthrone: These are anthraquinonoid dyes with a perylene structure. Example is the dye manufactured by condensation of 9-anthrol with glycerol in the presence of sulphuric acid and fusion of the so formed benzanthrone with potassium hydroxide.

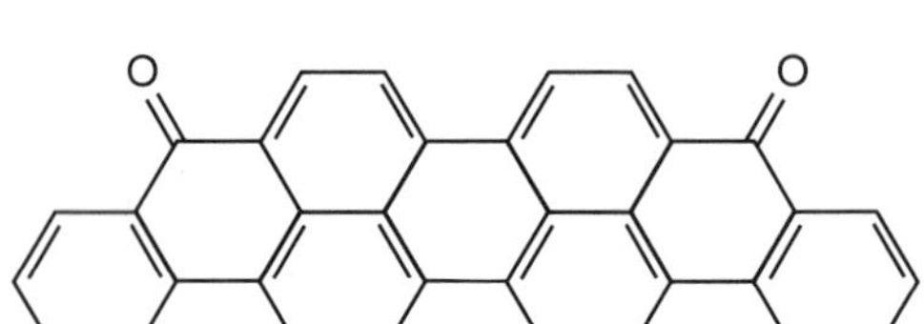

Vat dark blue BO

Dibenzpyrenequinone: These are anthraquinone dyes coming under the group-Anthraquinonoid dyes with a pyrene structure. Example is the dye manufactured by ring closure of 3-benzoylbenzanthrone with aluminium chloride and an oxidizing agent.

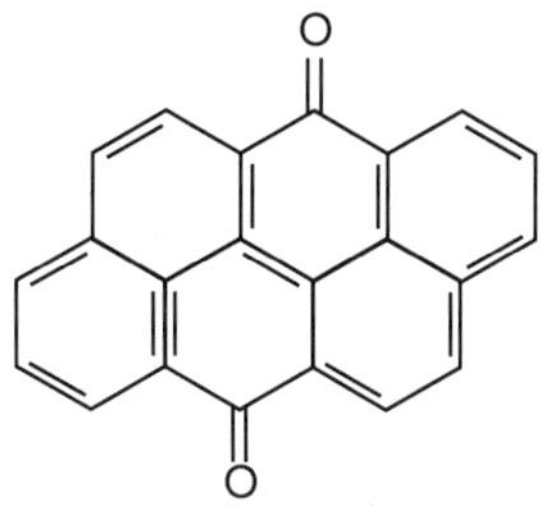

Vat golden yellow GK

Dicel: Courtald's trade name for its acetate fibre. It is derived from cellulose, it dyes well and has exceptionally good draping qualities. It wears well and has good crease recovery properties and is resistant to moths and mildew. Used for a wide variety of fabrics for clothing and furnishings including taffeta, brocade, lining fabric, velvet and in knit form for lingerie.

Dichromatic: Having or consisting of only two colours.

Dichromatism: Colour blindness.

Dickson: Commercial variety of early maturing American cotton, the staple measuring 23–26 millimetres and forming large clustered bolls; the yield is 31-32 per cent; also called Simpoon.

DID: Drawing-in draft. The drawing-in draft indicates the pattern of the arrangement of warp threads on different heald frames. Wherever possible, the ends which are to be woven similarly should be drawn through the same heald frame. There are different types of DID – straight draft, pointed draft, skip draft, fancy drafts, etc.

DID diagram: Drawing-in-draft diagram indicates which warp end is attached to which harness. Vertical column in the DID diagram represent the warp threads and the horizontal rows represent the harness frames which are sequentially numbered from bottom to top. See **DID.**

Die swell: The increase in diameter that procures as a visco-elastic melt or solution emerges from a die or spinneret hole.

Dielectric breakdown constant: In an electrical insulating material, the voltage at which electrical breakdown occurs, i.e., the voltage at which current will flow and/or the material melts.

Dielectric breakdown voltage: In an electrical insulating material, the voltage at which electrical breakdown occurs, i.e., the voltage at which current will flow and/or the material melts.

Dielectric constant: Measure of the ability of a dielectric material to store electrical potential energy under the influence of an electric field, measured by the ratio of the capacitance of a condenser with the material as the dielectric to its capacitance with a vacuum as the dielectric.

Dielectric strength: The average voltage gradient at which electrical failure or breakdown occurs. Expressed in volts per mil.

Diethylene glycolated DMDHEU (ULF): Glycolated DMDHEU is made by reacting two moles of diethylene glycol, (DEG), with one mole of DMDHEU. It is often referred to as a ULF (ultra-low formaldehyde) reactant. This product evolved from the successful experiences noted when DEG was added to the finish formulation. Precapping before it is applied insures that there are no pendent N-methylol groups so the HCHO release is even lower. The DEG still functions as a leaving group during cure so crosslinking with cellulose is not severely impaired.

Differential drop feed: A type of drop feed having two dogs arranged in tandem which move differentially. Material can be gathered or stretched by adjusting the stroke of the feed dog in front of the needle, so that it is longer or shorter respectively than the stroke of the feed dog between the needles.

Differential dyeing: Differential dyeing technology for synthetic fibre mixtures composed of normal and/or chemically modified polyacrylonitrile, polyester or polyamide fibres. Single bath piece dyeing is employed to produce solid shades, similar shades but in different depths, as well as multi-colour and contrast effects. Advantages include the avoidance of stockholding of dyed yarn, easier adaptation to changes in fashion and more rapid delivery of finished orders. *Applications* – predominantly in the carpet sector (tufted carpets) and also for woven and knitted fabrics (shirts, blouses, hosiery, beachwear, nightwear and sweaters).

Differential dyeing (Fibres): Descriptive of fibres of the same generic class, but that have potentially different dyeing properties from those of the normal fibre.

Differential dyeing behaviour of cotton: The tendency of cotton fibres to adsorb and retain selectively varying proportions of different dyes from a binary dyebath.

Differential dyeing polyamide fibres: There are many differential dyeing polyamides that are available: a) Dyeable with anionic dyes: R – Regular type normal dyeing; D – Deep type dyes more deeply than regular polyamide; L – Low type very low dye uptake; b) Dyeable with cationic dyes: B – Basic type dyeable with cationic (basic) dyes.

Differential shrinkage (in zippers): The difference in longitudinal dimensional change between the zipper tape and the fabric to which the zipper is attached.

Differential thermal analysis: A method of determining the temperature at which thermal events occur in a material undergoing continuous heating.

Diffusion: (a) A more or less gradual movement of molecules or ions through a solution or fibre as a result of the existence of a concentration gradient or repulsive or attractive forces; (b) The random movement of gas molecules.

Diffusion coefficient: A proportionality factor which gives the number of moles of a dissolved substance diffusing through a cross-sectional area of 1 sq.cm per unit time for a given temperature and solvent when the concentration gradient is –1.

Diffusion rate (in dyeing): Diffusion rate is the rate at which dye is transported to the interior of the fibre material. A high diffusion rate leads to rapid establishment of dyeing equilibrium and rapid levelling-out of irregularities in dye absorption. It also facilitates washing off the hydrolysed dye. Increased dye substantivity is frequently combined with lower diffusability. The diffusion rate doubles with a temperature increase of 10.20 °C.

Difluorochloropyrimidine (bis-FCP): Water soluble bifunctional cross-linking agent containing sulphonic acid groups based on difluoropyrimidine. It is used on wool.

Digested sludge: Wastewater sludge, which has undergone anaerobic sludge digestion or passed through an aerobic digester, should have a fibrous structure and a peaty smell that is not unpleasant, unlike raw sludge, but it still may contain some pathogenic organisms.

Digital device equipped clothing: Digital-devised clothing involves an application of digital devices, such as music players or cell phones, to clothing. It is now regularly sold in many countries.

Digital printing: See **Ink-jet-printing**.

Digitizer: Electronic pen for design processing in electronic equipment for the production of colour separations.

Dieppe lace: French bobbin lace of various fineness, originally made both in black and white, now of white flax thread with a ground of three threads. It is similar to the Valenciennes.

Diluent: A solid or liquid chemical used to dilute another. Sodium sulphate is a common diluent for dry dyes. Since the manufacturing of dyes does not always result in exactly the same strength of dye from batch to batch, manufactures routinely add a diluent to adjust the batch to a standardized strength. Diluents may make up a large portion of dyes designed to yield pale shades. For example, a dye that yields dark red at 2% o.w.g. may be diluted to yield a pale pink using 2% o.w.g. of the diluted dye. Diluting dry dyes in the way is done primarily as a convenience to the dyer.

Dimakso: Arabian term for raw silk.

Dimensional change: (1) The change in length and width of fabric which is subjected to specified condition/tests.

(2) The change in dimension due any processing or garment treatment. Mainly the changes in warp and weft way or width and length ways.

Dimensional change (consolidation): The dimensional change that takes place when a knitted fabric is gently agitated in water to overcome all frictional constraints in the fabric after it has been immersed in water without agitation to cause the relaxation dimensional change.

Dimensional change (due to felting): The irreversible dimensional change that occurs in a relaxed fabric when it is subjected to agitation in laundering.

Dimensional change (due to pressing and finishing of garments): The change in dimensions of a fabric caused by pressing and finishing during the garment manufacture.

Dimensional change (due to pressing and curing): It is self-explanatory.

Dimensional change in boiling water (felt): The change in length and width with any associated change in thickness produced by immersion in boiling water under specified conditions.

Dimensional restorability: The ability of a fabric to be returned to its original dimensions after laundering or dry cleaning, expressed in percent. For example, 2% dimensional restorability means that although a fabric may shrink more than this in washing, it can be restored to within 2% of its original dimensions by ordinary home pressing methods.

Dimensional stability to dry cleaning: The British Standard method [13] requires the use of a commercial dry cleaning machine. In the test, the sample is prepared and marked out according to BS 4931 [6]. The total load used is 50kg for each cubic metre of the machine cage made up of specimen plus makeweights. The solvent to be used is tetrachloroethylene containing 1 g/l of surfactant in a water emulsion, 6.5 l of solvent being used for each kilogram of load. The machine is run for 15 min at 30 °C, the sample is rinsed in solvent and then dried by tumbling in warm air. The sample is then given an appropriate finishing treatment, which in most cases will be steam pressing, and it is then reconditioned and measured again.

Dimensional stability to dry heat: This test is intended to predict the behaviour of fabrics when heated in a hot press such as those used in various garment manufacturing processes including fusing and transfer printing. The ISO method [14] recommends that the samples are preconditioned at a low relative humidity before conditioning in the standard atmosphere. The samples are then marked and the dimensions in between two points determined. They are then placed in a press heated to 150 °C under a pressure of 0.3 kPa for 20s. The samples are conditioned and measured again so that the dimensional change in each direction can be calculated.

Dimensional change (relaxation): The dimensional change that occurs when a knitted fabric is immersed in water without agitation and the strains and tensions put in to the fibre, yarn or fabrics during previous processing stages as spinning, knitting or finishing are relieved.

Dimensional change in thickness (in felt): The increase in thickness of the felt due to immersion in boiling water measured under specified conditions. See **Consolidation dimensional change**.

Dimensional stability: This is the stability of a textile, especially garments, in terms of its dimensions and shape due to any factor. Dimensional stability is tested for various factors as steam pressing, hot pressing, washing, etc.

Dimensional stability (carpet test): A supplementary test for the dimensional change of carpet tiles and needle felts to determine their suitability for use in humid environments. A customary accelerated test is used which involves 2 h dry heat at 60 °C followed by a 2 h wetting treatment, drying at 60 °C for 24 h and final conditioning for 48 h in a standard test atmosphere at 20°C/65% relative humidity. For special cases, more elaborate testing in climatic chambers is employed. The results are expressed as the percentage change in relation to the initial measurement made before the test.

Dimethoxymethyl-4,5-dimethoxyimidazolidone: $O=C-N_2-(CH_2-OCH_3)_4$. A resin finishing agent that is used for low-formaldehyde finishes.

Dimethoxymethylurea: $O=C(NH_2)_2(CH_2OCH_3)_2$ Resin finishing agent with better storage life than the free methylol compound. Finishes are less stiff, but otherwise similar to Dimethylolurea.

Dimethoxymethyluron: N,N′-bis (methoxymethyl) uron. It is produced from urea, formaldehyde and methanol. Because of their good stability, methoxymethyl compounds are used as reactant resins. Chlorine retention is dependent the purity. Used in co-application with buffered Methoxymethylmelamine.

Dimethyl-4,5-dihydroxyethyleneurea: (N,N′- dimethyl-4, 5-dihydroxyethyleneurea). Formaldehyde-free resin finishing agent.

Dimethylfluidsilicones:Dimethylfluidsaremadefromdimethyldichlorosilane. The reaction conditions can be controlled to vary the number of repeat dimethyl siloxane units within the polymer. As the number increase, the viscosity increases so there is a range of commercially available fluids with varying viscosity fluids can be emulsified to make stable water dispersions for use as finishes. Fluids are water clear and do not discolour with heat or age. They impart soft silky hands to fabrics. In addition to softening, dimethyl fluids render fabrics somewhat water repellent; however being fluids, they are not durable.

Dimethyl formamide (DMF): $HCON(CH_3)_2$, colourless, mobile, liquid. Density 0.952, b.p. 153 °C. Solvent for high polymers, natural and synthetic resins. Also used as a selective solvent.

Dimethyl terephthalate: [p-$C_6H_4(COOCH_3)_2$] An intermediate used in the production of polyethylene terephthalate, the polymer from which polyester fibres and resins are made.

Dimethyl-4,5, Dihydroxyethylene urea (DMeDHEU): DMeEU is the reaction product of dimethyl urea and glyoxal. The -4,5- ring hydroxyls are reactive with cellulose and form the same ether links as do the N-methylol groups. Even though the electronic configuration around the ring hydroxyls are the same as the pendant N-methylol groups, the reaction is much more sluggish because of steric hindrance and therefor require harder cures and more and stronger catalyst. Even with all this, the DP results are not near as good as the more conventional reactants. The product is more expensive; however, the fabric can truthfully be labelled as formaldehyde-free.

Dimethyloldihydroxy ethylene urea: See **Dimethyl-4,5, Dihydroxyethylene Urea (DMeDHEU)**.

Dimethylol Ethylene Urea (DMEU): The starting material for making dimethylolethylene urea (DMEU) is ethylene urea. A 5-membered heterocyclic, nitrogen compound (imidazolidone-2). It is made by reacting ethylene diamine with urea. Ethylene urea contains 2 NH- groups capable of reacting with formaldehyde and forms a bifunctional product. Since there are no other active hydrogen sites, the N-methylolated product cannot self-condense.

Dimethylol 5-propylene urea: (N,N′-dimethylol-5-hydroxy-hexahydro-2-pyrimidone). It is produced mainly by reaction of urea, 1,3-diamino-2-propanol and formaldehyde. Important reactant resin. The finishing effects are roughly similar to those obtained with the unsubstituted Dimethylolpropyleneurea (DMPU), but more reactive and somewhat more stable to hydrolysis.

Dimethylol glyoxal monourea: See **Dimethyloldihydroxyethyleneurea (DMDHEU)**.

Dimethylolhexahydro-2-pyrimidone: See **Dimethylolpropyleneurea (DMPU)**.

Dimethylolpropyleneurea (DMPU): N, N′-dimethylolhexahydro-2-pyrimidone. DMPU is produced mainly by reaction of urea, propylene diamine and formaldehyde. It is an important reactant resin for finishes with very good resistance to chlorine and fastness to washing at the boil, especially on white goods. Disadvantages – impairs the light fastness of textiles dyed with reactive or substantive dyes, DMPU finishes are sensitive to hydrolysis.

Dimethyloltriazone: Dimethylolhexahydrotriazone. It is also known as a triazone resin. It is produced inter alia by reaction of urea, formaldehyde and a primary amine. Important reactant resins (dimethylolethyltriazone, dimethylolhydroxyethyltriazone). Finishes produced with dimethyloltriazones have low chlorine retention and tend to have a fishy odour which makes an after-wash usually necessary. Reactant resin with good resistance to chlorine and hydrolysis. It does not impair the light fastness of dyed or printed fabrics. It is particularly suitable for resin finishes on cotton. On polyester/cotton blends, the finishes have a firm handle.

Dimethylol Urea (DMU): Condensation product of urea and formaldehyde. A preferred resin finishing agent, available commercially in powder or paste form and is very reactive. It is used to produce finishes with good crease resistance, non-shrink properties and high elastic resilience, especially on viscose. The finishes have only moderate fastness to washing at the boil and are not chlorine resistant. DMU has limited storage life.

Dimethylol-4,5-Dihydroxyethylene urea (DMDHEU): DMDHEU is the workhorse durable press finish. This and some of its modified versions account for over 85% of all crease resistant chemicals consumed today. DMDHEU achieved this prominent role in 1961 when delay cure processing came into being. In the trade, DMDHEU is often referred to as the Glyoxal resin. This jargon came as a way to distinguish it from DMEU, in that glyoxal was used to make the starting monomer. The starting heterocycle is made by reacting stoichiometric amounts of urea and glyoxal. The reaction is straightforward and can be carried out in regular laboratory glassware. The methylolation step is also straightforward. While the synthesis is shown in two steps, commercially DMDHEU is made directly in one step. Urea, formaldehyde and glyoxal are all combined together and heated. The extent of reaction is followed by monitoring the free formaldehyde content. The product is sold as 45% solution.

Dimethyl polysiloxanes with reactive groups: Linear dimethyl polysiloxanes containing various end groups, possibly with additional isolated reactive side groups. As silicone cross-linking agents, polymers are obtained which are capable of producing finishes on textiles with high resistance to solvents, temperature, washing, good extensibility, high tensile strength and resistance to rubbing as well as a soft and elastic handle. Uses – for coatings, as silicone elastomers for finishes on woven and knits produced from fully synthetic fibres and synthetic fibre/cotton blends. Also used for low-shrink finishes on wool.

Dimity: The name comes from the Greek word meaning 'double thread' and the fabric is made from double or treble yarns, which creates ridges in plain areas. It is a fine or medium weight fabric, usually made from cotton and wears well. The basic weave of dimity is plain weave. The characteristic cards are made by letting several ends run parallel, that is, by putting two or three ends on the same harness or through same heddle. A typical construction is 90 × 60 with 60s warp and 80s weft. Combed yarns are used, the cloth is usually mercerized, bleached or dyed. It is often made in checks. Used for nightwear, children's cloth, curtains, etc.

Dimity binding: It has smooth edges and raised patterns.

DIN: German Institute for Standardisation The predecessor of the German Institute for Standardisation (DIN) was the Hindenburg programme set up in 1917. It is based in Berlin, Germany, and has about 4,700 committees which draw on roughly 25,000 specialists. It has adopted about 25,000 standards.

DIN colour chart: A system of colour filters (hues) which may be used for the determination of colours under standard illuminant C in terms of their appearance and characteristics (hue, chroma and lightness). Used for the systematic arrangement of colour collections (colour atlases), for the determination of colour tolerances, the evaluation of re-dyeing, and price calculations.

Dingy wool: Wool that is dark or greyish in colour and generally heavy in shrinkage. May be caused by excessive yolk, poor farming conditions or parasites.

Dinner jacket: An informal dress coat without tails, somewhat like a sack, for dinner wear, men's evening parties, but not where ladies are present, as at balls, operas, etc. Also called Cowes coat and tuxedo.

Dioxan: (1,4-diethylene dioxide), $C_4H_8O_2$ A colourless, oily, neutral liquid with a faint and not unpleasant odour. Very toxic (suspected carcinogen). Miscible with water and virtually all organic solvents for printing thickeners. Solvent for cellulose acetate, nitrocellulose, cellulose ethers, chlorinated rubber, resins (mostly acrylate and methacrylate polymers), oils, fats, waxes.

Dip: (1) A chemical (rubber) compound applied to cords or other fabrics designed to improve adhesion of the textile material to an elastomer.

(2) Immersion of a relatively short duration of a textile in some processing liquid. The term is usually used in connection with a padding or slashing process.

(3) A material applied to cords or other fabrics designed to improve the adhesion of the textile material to an elastomer.

(4)(a)The depth of liquor in the inner cylinder of a rotating-cage washing machine. (b) A term sometimes used to describe the treatment of cellulosic material with a chemic (q.v.).

DIP process: The abbreviation of "DIP" stands for diazotization in the paste (Naphthols in direct printing). It represents a variation of base printing: for preference liquid bases are stirred into a thickener together with sodium nitrite. A second thickener contains a medium strength acid, e.g. phosphoric acid. By gradually mixing the two stock thickeners together, the fast colour base is diazotized without the otherwise usual addition of ice (for cooling). Due to metered addition, no nitrous gases are released by this method. The print paste with the diazotized fast colour base must not be neutralized and is immediately ready for printing.

Dip dyeing: See **Dyeing, dip**.

Dip/extract method: To dye hank yarns and cheeses with reactive dyes by the cold pad-batch method. Cheeses are impregnated with the dye bath on a centrifuge. Interest is restricted to space dyeing.

Dip penetration: The degree of saturation through a tire cord after impregnation with an adhesive.

Dip pick-up: The dip or dip components present on the textile cord or other fabric after processing.

Dip pick up (in glass cords): The amount of dip solids present as supplied.

Dip treating: The process of passing fibre, cord, or fabric through an adhesive bath, followed by drying and heat-treating of the adhesive-coated fibre to obtain better adhesion.

Diphenyl: (biphenyl, phenyl benzene), $C_{12}H_{10}$, or $(C_6H_5)_2$, cyclic hydrocarbon containing two benzene rings. Starting material for the manufacture of benzidine.

Diphenylmethane dyes: Diphenylmethane and triphenylmethane dyes are monomethine dyes with two or three terminal aryl groups, of which at least one, but preferably two or three, are substituted by a donor group para to the methane carbon atom. The most important donor is the amino group.

Direct application: Usually used to mean a method where a dye solution is locally applied to areas of fabric, such as by painting, squirting, spraying, stamping, etc.

Direct beaming: The yarn cones are mounted in a large creel. Each yarn can be tensioned individually as they are led together to the beaming machine and wound onto a beam with parallel sides. Usually the beams are not large enough to accommodate all the yarns needed in the loom, so they have to be combined to form the final weavers beam. This is usually done at the sizing machine. Direct beaming is economical for large runs of standard fabrics and when all the yarns are of the same colour.

Direct beaming machine

Direct cabling: A system of producing a twist-balanced folded yarn from balanced (or untwisted) single yarns. The process is usually carried out on a modified two-for-one twisting machine.

Direct coating: The application directly on to the web of moderate and high viscosity polymers, both in the form of solutions, concentrated dispersions and in fused, thermoplastic state. This mainly applies to coating processes, as used in the manufacture of textile plastics composites. Following this process, material can be added indirectly using another web (textile, foam or film).

Direct costs: Those costs that can be directly attributed to a product or service. They can include raw materials, labour costs for manufacturing, as weaving, knitting, finishing, etc. (per metre, for example, for treating a fabric for flame retardancy), and distribution costs.

Direct dye: A dye-class based on application method, which is essentially by immersion of the fibre in a solution of dye without the need for other chemicals to bond the dye to the fibre (though other chemicals may aid exhaustion). Direct dyes have high substantivity, but bond weakly to fibres and therefore, usually have poor wash fastness. Light fastness varies from poor to very good. A post-dyeing fixative is often used to improve wash fastness. Brightness can be limited in direct dyes, because brightness often is associated with small molecules and small molecules tend to make poor direct dyes. "Household" dyes, of the sort that are sold in grocery stores, typically are union dyes which contain a direct dye and an acid dye.

Direct engraving process: Digital engraving like laser engraving, inkjet engraving, etc., where intermediate processes like exposure developing, etc., are absent.

Direct esterification: In the production of polyethylene terephthalate, the process in which ethylene glycol is reacted with terephthalic acid to form $H_3COOC - C_6H_4 - COOCH_3$ monomer with the generation of water as a by-product.

Direct printing: See **Printing**.

Direct spun: (a) A term used to explain filaments or yarn produced by direct spinning; (b) Descriptive of woollen yarns spun on a mule onto weft bobbins.

Direct style: A style of printing by which dyes are directly printed on to dyed or white ground and fixed by aging (steaming), curing or any other suitable methods.

Direct warping: The transfer of yarn from a package creel directly to the beam. See also **Sectional warping** and **Beaming**.

Direct yarn numbering system: A system that expresses yarn number in linear density.

Direction of slippage (in yarn slippage testing): The line of movement parallel to either weft or warp on a woven fabric in which minimum force is required to produce yarn slippage.

Direction of twist: The right or left direction of the helix formed in a twisted strand as indicated by superimposition of the capital letter 'S' or 'Z'. See **Twist (direction of)**.

Direction (width-wise): See **Width-wise direction**.

Directionally oriented fabric: Rigid fabric constructions containing inlaid warpor fill yarns held in place by a warp-knit structure. Used in geotextiles and coated fabrics.

Disc centrifuge (nozzle bowl c.): A common industrial centrifuge which produces a thick slurry rather than a sludge cake. The liquid moves up through a stack of conical discs, the solids moving to the outside of the centrifuge. They have been used to thicken waste activated sludge, but they cannot dewater sludge to a cake as this blocks the gaps between the discs.

Disc filter: A type of vacuum filter.

Disc test: An in-rubber test used to predict the fatigue resistance of tire cords and other industrial yarns.

Discharge: Localized removal of dye from fabric. Discharge is used to remove dye from fabric in printing and similar processes. Often another illuminating dye is applied to the area that is discharged. Sometimes the new colour is included in the discharge paste. Few dyes are quite easy to discharge, others are very difficult to discharge. This can be the case even within a single dye family (commercially dyed black fabric is notorious for producing unpredictable results in discharge. This is because most black dyes are mixtures of several colours). Industrial discharge processes use reducing agents such as thiourea dioxide, sodium formaldehyde sulphoxylate, zinc formaldehyde sulphoxylate or tin chloride. Art dyers will often use household chlorine bleach for discharge techniques (this author knows of no commercial use of chlorine bleach for discharge). Some discharge techniques decolorize the dye, but leave remnants fixed to the fibre; while others break the bond between the dye and the fibre.

Discharge print paste: The discharge print paste consists of a thickened discharging agent which may contain further additives, e.g. white pigments, discharge-resistant dyes if required, and other necessary chemicals, auxiliaries, etc. See **Printing**.

Discharge printing: (1) See **Printing**.

(2) A style of printing in which pre-dyed or pre-padded fabric is printed with a paste containing chemicals capable of destroying the ground colour to produce white effects (white discharges) in the printed areas, or coloured effects (coloured discharges) with the addition of a discharge- resistant dye.

Discharge printing assistants: These are the chemicals added along with the discharge paste which help in discharges in many ways. Mainly, these are products based on quaternary ammonium compounds. Hydrotropic auxiliaries accelerate or make possible the discharge process during dye fixation and also dissolve cleavage products thereby facilitating their removal in subsequent after-washing. Wetting agents are also employed in discharge print pastes or pad liquors to reduce the surface tension between water and air so that the aqueous medium can penetrate air-filled capillary spaces in the fibres more easily. White effects produced by white discharges can also be improved by additions of zinc oxide and titanium dioxide (often as mixtures of both) as well as fluorescent brightening agents, to the discharge print paste. Thiodiglycol is also used as a dye solvent and hydrotropic agent, whilst glycerol is used as a hygroscopic agent and solvent.

Discharge resist printing: In contrast to normal resist printing, fixation of the dye to be resisted is not only prevented but also simultaneously destroyed by the discharging agent.

Dischargeable grounds: Dyeing made with dyes suitable for the production of white or coloured discharges. Direct, reactive and naphthol dyes are mainly used on cellulosic fibres; acid, metal-complex, disperse and cationic dyes are used on the appropriate synthetic fibres. An essential prerequisite is good discharge ability of the ground dyeing. Since there is a direct relationship between discharge ability and depth of shade, dye selection must be made by reference to recommendations given in the dye pattern card relative to the 1/1 standard depth.

Discharging: A destruction of colourant or mordant present on a cloth (dyed) by chemical means to leave a coloured or white pattern.

Discharging agents: The chemical used in discharge prints to discharge/ remove the dyed colour on the fabric which is being discharge printed.

Discoloured pick: See **Mixed end or Filling**.

Discontinuous fibres: Tow and staple fibre.

Discontinuous or batch-type systems: In this system/process, all the operations are carried out on a single machine; it is therefore necessary to load the machine, carry out the treatments following a predetermined cycle, unload the machine and finally wash it thoroughly before starting a new cycle. This working process is extremely flexible and is suitable for processing small lots: for example, it is possible to a carry out a scouring treatment on a single machine, then a bleaching one followed by a dyeing process. For the production of large lots, the discontinuous process is labour-intensive, i.e. it requires many operators to load and unload the material; it also entails long processing times and results that can vary from one batch to another.

Discordant colour schemes: Most colour schemes are considered to be harmonious, but there are few that are considered to be discordant – this does not necessarily mean undesirable. There is also a recognised rule of harmony or discord when broken from a natural order of colour. When using the complementary pair yellow and violet, the yellow is naturally lighter. If white is added to the violet and black to the yellow, this changes the natural order and is considered to be discordant. Another example of discord is where two colours do not share the same value, for example pink and brown. Pink is red and white; brown is orange and black. Again the results may not necessarily be undesirable.

Discrete variate: A variate that is a measurement based on a scale that has a limited or finite number of steps, such as count, a rating scale, or a ratio of success to total observations.

Diseased wool: The wool that is taken from the skin of a sheep that died from disease.

Disinfectants: Chemicals suitable for disinfecting, e.g. in particular oxidizing agents in the form of chlorine carriers (thermochemical disinfection of textiles), phenol derivatives, bactericidal finishes. It is probable that these agents mainly work by destroying the cytoplasm membrane of the bacteria.

Disinfection: Destruction of microorganisms that cause disease (bacteria).

Disinfection in the laundry: It is carried out in hospitals and similar institutions in pressure vessels using superheated steam and hot air, or in a washing machine using a disinfectant washing process.

Disodium hydrogen phosphate: Acid sodium pyrophosphate ($Na_2H_2P_2O_7$) fine powder, non-hygroscopic. Density 1.848. Solubility in water at 20 °C up to 120/140 g/l. The solution has an acidic reaction at pH 4,2. It gives clear acidic solutions even in hard water. As a builder it improves the washing and foaming efficiency of detergents. Removes undesirable iron and manganese compounds in washing, dyeing, etc. (especially in iron dyeing vessels). Other uses include bleaching stabilizers and flame-retardant finishes.

Disodium phosphate: Sodium phosphate, dibasic (DSP, sodium orthophosphate, secondary), Na_2HPO_4. Colourless translucent crystals or white powder with a saline taste. Soluble in water, very soluble in alcohol. Hygroscopic, converted to sodium pyrophosphate at 240 °C. Textile applications include weighting of silk, dyeing and printing, flame-retardant finishes.

Dispersant: A dispersing agent, often of a surface active chemical, that promotes formation of a dispersion or maintains a state of dispersion by preventing settling or aggregation.

Disperse dye: A dye that is almost totally insoluble in water. Disperse dyes exist in the dye bath as a suspension or dispersion of microscopic particles, with only a tiny amount in true solution at any time. They are the only dyes that are effective for “normal” polyester. Few types are used for nylon and acetate. Polyester is dyed with disperse dyes by boiling with carrier chemicals, or by heating the liquor to about 130°C, which requires elevated pressure (like a pressure cooker). Thermosol dyeing, where the fabric is padded with dye liquor then dried and heated to about 200°C for about 90 seconds, is also used for polyester and for colouring the polyester component of poly-cotton blends. Disperse dyes on polyester are generally very wash-fast and resistant to bleaching. Nylon can be dyed at or below 100°C without the use of a carrier, but wash-fastness is only moderate. Disperse dyes are also used for sublimation printing of synthetic fibres and are the colorant used in crayons and inks sold for making "iron-on" transfers.

Dispersability: The degree to which particles can be broken down to some minimum size such that they will pass through the interstices of a standard filter paper.

Dispersing agents: These function in a manner similar to emulsifying agents. The difference is that solid particulate matter, rather than insoluble oils, is dispersed. The nature of the lyophilic part of the surfactant molecule must be such that it adsorbs onto the particle's surface. Surfactant molecules must be matched with their intended use.

Dispersion: (a) A system consisting of finely divided particles and the medium in which they are distributed; (b) Separation of light into colours by diffraction or refraction; (c) A qualitative estimation of the separation and uniform distribution of fibres in the liquid during the production of a wet-formed non-woven fabric.

Dispersion (degree of): See **Dispersivity**.

Dispersion medium: Continuous phase of a dispersion.

Dispersion spinning: Aprocess in which the polymers that tend to an infusible, insoluble and generally intractable character (e.g. polytetrafluoroethylene) are dispersed as fine particles in a carrier such as sodium xanthate solutions that permit extrusion into fibres after which the dispersed polymer is made to coalesce by heating process, the carrier being removed either by a heating or by a dissolving process.

Dispersivity: In Latin, the word *dispersio* means spread; degree of dispersion of a (colloidal) substance in an aqueous solution.

Dispersed phase: The dissolved or distributed substance (also known as a dispersoid). **Dispersion medium (also known as a dispersing agent or dispersant)**: The dispersant or solvent substance.

Disperse system: Combination of a dispersion medium and a disperse phase. The degree of dispersion can fluctuate considerably, depending on the particle size of the dispersed substance. Important factors affecting the degree of dispersion are – particle size, molecule size, chemical constitution, concentration ratio, thermal influence, pH influence, influence of the addition of foreign materials (electrolyte, chemical reactions, physical processes, etc.).

Displacement resist: Displacement effect. A ground dye initially applied to a textile substrate is removed (displaced) by subsequent local application of chemicals, thickener and/or dye (or their combinations).

Disposable textiles: Articles of a more or less textile-like character intended for once-only or short-life usage, such as disposable laundry (e.g. bed linen, table linen, hand towels, underwear, diapers); disposable garments (e.g. coloured or printed holiday and party garments); disposable work wear (e.g. hospital gowns, painters' overalls); disposable sanitary goods, etc. Short-term serviceability ranges from "disposal after use" to "multi-wash/dry-clean" goods.

Dissimilation: In Latin, the word *dissimilis* means 'not alike'. The breakdown and consumption of nutrients to provide energy and simple compounds for intermediary metabolism. It is the opposite of assimilation. Also a term used in waste-water technology to describe the conversion of materials whereby energy is provided by the oxidation of organic compounds for the formation of new organic matter.

Dissociation (ionization): The electrolytic breakdown of acids, bases and salts on solution in water. This process involves the formation of ions, i.e. electrically charged particles – anions (-) and cations (+) from electrically neutral molecules. In practice, therefore, acidic reaction is due to electro-positively charged cations (H+ ions) and alkaline reaction is due to electro-negatively charged anions (OH- ions).

Dissociation (degree of): It corresponds to the respective (actual) strength of reaction (e.g., of acids, bases and salts). The degree of dissociation is of great practical importance since it is a term which denotes the strength and not the content of, e.g., an acid. The degree of dissociation can be used to determine the equilibrium constant for dissociation, an important factor in ascertaining the extent of a chemical process.

Dissolved air flotation (DAF):A treatment process in which water or wastewater or sludge is saturated with air and then flows to a tank where the air comes out of solution in tiny bubbles less than 100um that floats the sludge and enables it to be skimmed off. ***Coagulants*** may be added to help form ***flocs***. Two methods of DAF have been used. The less common is vacuum flotation, in which a short period of aeration at atmospheric pressure is followed by application of vacuum in a separate tank. The common technique takes a small part of the flow to an air saturator in which water, wastewater or sludge is aerated for a few minutes at 3–4 atmospheric pressures. When the aerated and pressurized flow re-joins the main flow, the dissolved air coming out of solution attaches itself to the flocculated particles or sludge and makes them float.

Dissolved inorganics in wastewater: In organic constituents such as calcium, sodium and sulphate are added to the original domestic water supply as a result of water use and may have to be removed if waste water is to be reused.

Dissolved oxygen (DO): Oxygen dissolved in water is needed for the existence of ***aerobic*** water life, but the solubility of oxygen in water is low. The maximum at 0 °C is only 14.6 mg/l at sea level, falling to 10mg/l at 16 °C. At chlorides concentrations of 5000 mg/l (0.5%), the solubility falls to 9.5 mg/l at 16 °C; with 10 000 mg/l of chlorides, to 9 mg/l. All this applies at atmospheric pressure. But when the external gas pressure increases many fold, as occurs in the deep shaft system, the solubility of oxygen rises and much more oxygen dissolves. In steam boilers, dissolved oxygen causes corrosion and must be removed from feed water before it enters the feed pump. The damage done by oxygen can be especially serious in boilers operating at pressures above about 17 atm. DO is measured by dissolved oxygen electrodes or by the Winkler test.

Dissolving pulp: A specially purified form of cellulose made from wood tissue.

Distillate: See **Condensate**.

Distortion (in fabric): A general term for a visible defect in the texture of the fabric caused by the uneven movement of the warp or weft in the fabric.

Distortion (in textile battings): Defects such as wholes, lumps, or thin area caused by movement of fibres.

Distressed (finish): Describes a finish that disturbs the surface of the fabric, giving it a used, beaten, or uneven appearance. Often done through sand or stone washing after the fabric has been pigment dyed.

Distribution length: In fibres, a graphic or tabular presentation of the proportion or percentage (by number or by weight) of fibres having different lengths.

District check: Acategory of small check designs, sometimes with contrasting over plaids, originally of Scottish origin. Glen plaids are part of this category.

Disturbed place: An area where the weave of the fabric has been disarranged, without damage to the yarns.

Divi-Divi: The dried pods of *Caesal pinacoriaria* growing in the West Indies and South America; they contain 20–35% tannin and a brown colouring matter.

Divided thread-line extrusion: Spinning of two separate thread lines from one spinneret.

Divider: Used in liquor compartments to decrease the liquor ratio, e.g. in padding machines including full width treatment machines and cheese dyeing machines.

Divinylsulphone compounds: They are derived from divinylsulphone CH_2=CH–SO_2–CH=CH_2, mainly with masked vinylsulphone groups, e.g. R–O–CH –CH_2–SO_2–CH_2–CH_2–O–R, where R = e.g. –SO_2Na_3,–CH_3, –H; symmetrical and asymmetrical divinylsulphone compounds. Used as resin finishing agents, e.g.: Bis (b-hydroxyethyl) sulphone; Disodium salt of tris-(bsulphoethyl) sulphonium.

Diz: The small tool that is used to help form and even top in wool combing. Traditionally, a diz was made out of carved horn. Very nice ones can also be formed (or bought) out of wood. A cheap, non-classy alternative is to trim a piece of plastic and punch or drill a hole in the middle of it. It has been seen done with the bottom corners of a plastic milk jug or a crescent cut from a section of PVC pipe.

Djule: Knitted Persian wool rugs with long haired back.

DMC: Dimethylol (mono) carbamate.

DMDHEU: Dimethylol dihydroxyethyleneurea. See **Dimethylol dihydroxyethylene urea**.

DMEU: Dimethylol ethyleneurea. See **Dymethylol ethylene urea.**

DMF: Dimethyl formamide. See **Dimethyl formamide**.

DMPU: Dimethylol propyleneurea. See **Dimethylol propylene urea**.

DMT: (a) Dimethyl terephthalate; (b) Dimethylol triazinone.

DMU: Dimethylol urea. See **Dimethylol urea**.

DNS: Sodium sulphate.

DO: Dissolved oxygen.

DOB: Damenoberbekleidung (women's outerwear).

Dobbies: Dobbies are used for the production of plain or flat fabrics, that is, of fabrics characterized by maximum 28–32 threads in the weave repeat. Dobbies can be divided as follows:

According to the working principle: (a) Hattersley dobbies; (b) Rotary dobbies.

According to the raising motion of the heald frames: (a) Dobbies with positive drive; (b) Dobbies with negative drive.

According to the card reading system: (a) Dobbies with endless Pattern Card; (b) Dobbies with Magnetic Card

Dobbies (Hattersley): The Hattersley dobbies are the ones which control the movement of the heald frames through rods and rocker levers.

Dobbies (rotary): The rotary dobbies attain the raising and lowering of the heald frame through rotating members.

Dobby: (1) A mechanism of controlling the movement of the heald shaft of a loom. It is required when the number of heald shafts or the number of picks in a repeat of a pattern, or both, are beyond the capacity of the tappet shedding.

(2) A loom equipped with a dobby.

(3) An area where the weave of the fabric has been disarranged, without damage to the yarns. A fabric woven on a dobby loom.

Dobby mechanism: A shedding mechanism on looms or weaving machines (weaving). The harness motion can be either negative or positive by means of a controlled lifting gear. In negative-dobby shedding, the shafts are raised by the dobby and lowered by some form of spring under motion. In positive-dobby shedding, the dobby machine both raises and lowers the shafts. A distinction is made between single and double lift as well as between closed and open shed dobby machines.

Dobby shedding: A mechanism attached to a loom for controlling the movement of the shafts through which the warp yarns are threaded.

Dobby weave: A weave that has a pattern that requires the use of a dobby mechanism during weaving. Dobby designs have small figures such as dots, geometric designs and floral patterns woven into the fabric. The design is produced by a combination of two or more basic weaves and the loom may have up to 32 harnesses. Examples of dobby weave are shirting madras, pique, huck towelling. See also **Dobby**.

Doctor blade: Stripping device in the form of a fixed blade which extends across the entire treatment width in steel, as a wedge-shaped, tapered vulcanite strip or as a roller squeegee (squeegee systems in printing). It is used to form an angle of incidence on a moving substrate by means of an incline. When moving, this then enables a highly viscous paste (for printing or coating) to become fluid briefly as a result of its intrinsic viscosity due to the shear forces which arise, and this can therefore be applied.

Doctor blade: (1) A metal knife that cleans or scrapes the excess dye from engraved printing rollers, leaving dye paste only in the valleys of engraved areas. Also used to describe other blades that are used to apply materials evenly to rollers or fabrics.

(2) A straight-edge whose function is to remove surplus material (e.g. filling, coating material and printing paste) from the surface of the rollers used to apply the material or from a fabric surface to which the material was applied.

(3) In effluent treatment, it is a sharp and hard blade that cuts the cake off the surface of a filter. Usually found on a rotary vacuum pre-coat or metal edge type filter.

Doctor roller: As opposed to the conventional blade used in rouleaux printing (stripping knife) and flatbed and rotary screen printing (doctor knife and roll doctor), which usually have a straight edge or surface, the doctor roller takes the form of a roller with surface engraving, fluting or helical grooving of varying depth. However because of the contact pressure on the padding roller, only a minor quantity of the liquor that is originally taken out of the trough remains on this doctor roller. Other forms include the comb doctor squeegee blade system in printing.

Doctor streak: A wavy white or coloured streak in the warp direction on the printed cloth made by a defective (damage in the sharpness in roller printing) doctor blade or imperfectly set doctor blade on the printing machine.

Doctors flannel: A soft, medium weight and inexpensive flannel. The fibres are mixed or wool. In white or red only. The type of fabric that could be used for interlining for warmth, lining children's clothes, nurse's cloaks, etc.

Doeuma: Very wide coarse gingham, often with large check design, having a medium soft finish; used in Turkey for bed and table covers, curtains, skirts, etc.

Doeskin: (1) A very close woven fabric from highest quality merino wool. It is made of high count, high twist worsted yarns woven with a 5 or 8 harness satin weave and given a suede finish. The cloth is milled and raised and given a napped finish on the right side. It is very soft and luxurious, with a sheen; expensive. Made in dress weight or heavier. Used for tailored suits, topcoats, sports jackets, leisure dresses and dresses and men's overcoats.

(2) A well-known glove-leather from lambs. It is readily available and very soft and easy to sew.

Doeskin finish: A soft low nap that is brushed in one direction. Cloth with this type of finish is used on billiard tables and in men's wear.

Doeskin rayon: Rayon twill or small satin which is napped or suede on one side. Used for suits, coats and sportswear.

Doff: A set of full bobbins produced by one machine (a roving frame, a spinning frame, or a manufactured filament-yarn extrusion machine).

Doffer: (1) The last or delivery cylinder of the card from which the sheet of fibres is removed by the doffer comb.

(2) An operator who removes full bobbins, spools, containers, or other packages from a machine and replaces them with empty ones.

Doffer comb: (1) A reciprocating comb, the teeth of which oscillate close to the card clothing of the doffer to strip the web of fibres from the card.

(2) See **Doffer comb.**

Doffing cylinder: Part of a combing machine. It is a wire-tooth-covered drum that helps to strip the cotton in a light film from the carding drum and which is stripped in turn by the doffing comb.

Doffer loading: Fibres imbedded so deeply into the doffer wire clothing that the doffer comb cannot dislodge them to form a traveling web.

Doffing: The operation of removing full packages, bobbins, spools, roving cans, caps, etc., from a machine and replacing them with empty ones.

Doffing comb: See **Doffer comb**.

Doffing cylinder: Part of a combing machine. It is a wire-tooth-covered drum that helps to strip the cotton in a light film from the carding drum and which is stripped in turn by the doffing comb.

Doffing tube (in rotor spinning): An extension to the navel to guide the withdrawn yarn towards the delivery rollers.

Dog hair: Once a valuable commercial ware (hat felt). Nowadays only of limited regional significance, e.g. for arts and crafts. Productive breeds include St Bernard, Newfoundland, chow-chow, poodle and spitz. It is spun alone or with wool.

Dogs-legs: A selvedge that varies in width.

NOTE: Variations in weft tension or lack of control of the warp ends within the selvedge may result in such unevenness. Pulled-in selvedge are caused by pulling in of the edges by isolated tight picks. Dog-legged selvedge are the result of the characteristic gradual change in weft tension that occurs as some types of weft pirn are unwound, regular changes in selvedge width being present at each pirn change.

Dogskin: Heavy weft plush fabric, the pile being formed by mohair yarn.

Doggy wool: Wools that have no character and show the results of lack of breeding. These wools are usually short, coarse and lacking in feel.

Dolichos: Fine and silky stem fibre, yielded by the Dolichos trilobus in China and India; used for cloth.

Dolly: (1) A strip of fabric rolled up tightly and bound with tape or string at one end. It is used like a brush for applying moisture during pressing of a garment.

(2) A term used for various types of faller washing machines consisting of rectangular troughs in which the goods to be treated in a detergent solution are placed. Rotating beaters or fallers are lifted by cams and allowed to fall into the trough. At the same time, the trough moves backwards and forwards. The machine is also known as a tom-tom and is used by finishers of knitwear, hosiery and lace.

Dolly varden: Light cotton or silk dress goods, printed with highly coloured flowers.

Domestic furnishing: Furnishing used in the house.

Domestic washing (household laundering): The washing of domestic household laundry.

Domestico: (1) Liso grey cotton sheeting in Colombia.

(2) Crudo grey cotton sheeting in Venezuela.

Domestics: Cotton goods, shirting and sheeting made in America.

Domet: Imitation flannel made from cotton warp and cotton or cotton/wool weft, finished with a raised surface on both sides and used for linings, skirting and pyjamas having an approximate weight of 84g/m. See also **Domette**.

Domet flannel: See **Domet**.

Domette: A very soft open weave fabric of wool and cotton with a long nap on the surface. Used for interfacing coats and jackets in which hair canvas is to be used. The domette is placed against the chest area of the coat and the hair canvas on top. The soft Domette prevents the hair from penetrating to the right side of the garment and avoid showing a hard ridge. Domette is also used as a padded interlining for quilting.

Domette flannel: Made of cotton plain and twill. Also spelled domet. Generally made in white. Has a longer nap than on flannelette. Soft filling yarns of medium or light weight are used to obtain the nap. The term domett is interchangeable with "outing flannel" but it is only made in a plain weave. Both are soft and fleecy and won't irritate the skin. Any sizing or starching must be removed before using. Outing flannel is also piece-dyed and some printed and produced in a spun rayon also. Uses: Mostly used for infants wear, interlinings and polished cloths.

Donchery: Stout French woollen serge.

Donegal: (1) Originally a rough woollen, hand-made Irish tweed typified by the use of several colours and particularly the use of white nubs, in the cloth. The word Donegal is now used to describe almost any plain weave tweedy looking cloth with white nubs in it. Fibres that are used include acrylic, cotton, viscose and mixtures and blend of these.

(2) A plain-weave tweed fabric made of wool and characterized by brightly-coloured randomly distributed slubs or flecks introduced at intervals into the weft yarn before spinning. The touches of colour are scattered throughout the cloth and show up usually against a light grey or natural coloured ground. Donegal is named after the county where it was first produced as a coarse woollen suiting in northwest Ireland. Originally, it was woven entirely by hand, but today power looms are used for its manufacture.

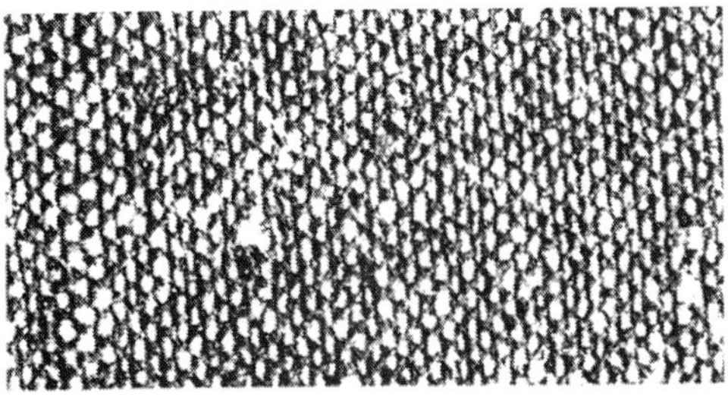

Donegal tweed: Named after an Irish county. Originally a woollen yarn fabric from knopped hand spun yarns. Now a general term to cover coarse woollen yarn fabrics in plain weave for men's and women's country wear. See **Tweed**.

Donkey: A pressing board in which an upper padded board is supported above the base. It is designed so that the parts of a garment may be pressed without creasing garment as a whole.

Donna Maria:(a) French silk fabric that is used for dresses, neckwear, etc.; there are eight leaves and twelve fillings in a repeat; (b) Very light and sheer silk fabric, used for veils by religious orders.

Donor: Latin word '*donare*' means to give. A term used in chemistry when the supply of certain functional groups is necessary for the initiation of a reaction, i.e. an atom, ion or molecule that provides a pair of electrons in forming a coordinate bond. Example: A proton donor is a compound which donates a proton for the initiation of a reaction (e.g. trans-esterification). In the case of hydrogen bonds, the water donor donates hydrogen ions and is therefore known as a proton donor.

Donskoi: Trade name for wools from Southern Russia; the staple is coarse, straight and loose, without kemps; used for carpets, etc.

Donsu: Silk Damask made in Japan.

Doorea: Fine Dacca muslin.

Dope: (1) See **Sizing**.

(2) See **Spinning solution**.

Dope dyed: See **Dyeing** and **Mass-coloured**.

Dope dyeing: Colouration of the polymer prior to manufacture of the fibre. This is really a misnomer, since the colorants are almost always pigments: "mass pigmentation" is a more accurate term. Some synthetic polymers such as polypropylene cannot be dyed after being made into fibres and coloration by adding pigments to the melted material is the only method available. Pigmenting prior to making fibres can also produce *wash fastness* and light fastness that is higher than can be obtained with any dyeing process.

Dope dyeing: In this system of dyeing, the colour is added in the dope itself before the synthetic yarn is spun there by dyed fibre is made in or in process itself. This eliminates the dyeing process in the fibre, yarn or fabric stage.

Doriah: (a) Grey, bleached or dyed blue or black, plain woven cotton goods, finished soft or hard; used for outer garments by the natives in Egypt; (b) A plain woven bleached cotton cloth in Arabia with prominent raised stripes produced by placing the warp ends closer.

Dornick twill: When you combine the idea of a broken twill with that of a reversing twill by using a threading that progresses in one direction so that the place at which the change occurs is staggered rather than pointed or symmetrical. This common twill is called a dornick twill.

Dornock or Dornek: (a) A coarse English linen diaper with check patterns; used for the table and for drapery; (b) Inferior damask of the 15th Century, made of wool, silk and gold.

Dorset: A fairly long and fine English wool.

Dorsetteen: A fabric once made in Britain using fine worsted warp and silk weft.

Doru: (1) Persian khilim (see), woven with solid cross stripes. (2) A long rope with which the thick woollen coat worn by the Gaddis is secured around the waist.

Dorure: French term for gold braids, tresses, etc.

DOS: Depth of shade.

Dosia: Chinese non-fashioned woollen hosiery.

Dosimetry: A method of studying colouring processes in molecular dye solutions, whereby a dye concentration that is constant over time is achieved using automatically-operating burettes, the replenishment from which over time provides the desired dye absorption curve. This makes it possible to perform a single dyeing test (without the need for analysis of the dyed material). Dosimetry is therefore regarded as a rational, fast and precise method (approximately 1% statistical error compared to conventional analysis). Principle: Concentration loss in a dye bath caused by dye absorption causes a potential change in a colorimeter, which activates the automatic burette via a controller, i.e. injects concentrated dye solution into the dye bath until the original bath concentration is restored.

Dosing tank: A tank in a waste-water treatment works in which wastewater collects until there is enough in it to operate the next treatment process. Where trickling filter in fluent is spread by distributor sturned by the reaction of the water flowing out of them, dosing tanks ensure that the flow is fast enough to rotate the distributors. When the dosing tank is full, it empties quickly through the dosing siphon. There is thus no flow except when there is enough influent to rotate the distributor.

Dosuti: Strong cotton cloth made in India of two-ply warp and filling in plain weave.

Dot coating: Process to produce hot melt adhesive materials using dot/pattern type paste coating of the non-woven fabric base by printing silk screen printing screens in the rotary screen printing method. In the dot process polyamide dots, for example, are "printed" onto the length of fabric in a wet-on-wet process. This process is particularly suited to non-woven bonded or siliconized fabrics. The moisture is removed from the dot in the drying chamber; it is baked at the same time.

Dot: A design dominated by circular spots which may be of any size, printed or woven on to the fabric. Small dots are often called pin dots, medium to large dots may be called aspirin dots and polka dots or coin dots.

Dot stitch: See **Point de Pois**.

Dotted Swiss: The original fabric as made in Switzerland is a sheer cotton fabric with crisp finish. The distinguishing feature is self-colour or contrast woven spots made from soft yarn and brought on the right side of the fabric. Woven dots are made by a lappet warp and a good grade Dotted Swiss may have a texture of 70 × 50 with 60s combed cotton warp and weft. The ground fabric is woven with a plain weave. The dots are made by introducing picks of very loose yarn or roving at regular intervals. The loose yarn is raised over 2–3 ends and then allowed to float on the back for about half inches more or less depending on the spacing wanted in between the spots. Floats on the back are sheered-off while finishing. A starch or permanent resin finish is given. The dull crisp finish is mainly a Swiss textile finish, but the name now applies to many fabrics that are fine and spotted. It is usually cotton, but may contain polyester and is hard wearing. Used for waisting, summer dress goods, curtains, baby clothes, etc.

Double: A process of twisting two or more single yarns together, as a rule in an opposite direction to the twist given the single yarns. See also **Doubling**.

Double back: A secondary backing glued to the back of carpet, usually to increase dimensional stability. See **Double backing**.

Double backing (double back): A secondary backing glued or laminated to the back of a carpet, usually to increase dimensional stability.

Double bonds: Atomic bonds (unsaturated); conjugated double bonds. A covalent bond between two atoms that includes two pairs of electrons, one pair being the single bond equivalent (the sigma bond) and the other forming an additional bond, the pi bond (ð bond). It is conventionally represented by two lines, for example $H_2C{=}O$. See **Multiple bond; Orbital**.

Double bow: Two fabric bows arcing in the same direction, as in a flattened M or W depending on the viewing angle.

Double braid (in rope): A braided construction consisting of two hollow braided ropes, one inside the other.

Double breasted: Coat or other garment tap over fronts having buttons and button holes on each forepart.

Double canvas: An open embroidery canvas usually made of cotton yarn. It is plain weave but with the warp and weft threads arranged in pairs, allows quite large holes to alternate with very small ones. It is a rigid canvas stiffened with size and is available in several weights.

Double carded: Yarn which has gone through the carding process twice.

Double cassinet: Satin, made of cotton warp and alternate cotton and woollen filling.

Double cloth: Cloth compound fabrics consisting of two component fabrics with two warp and two weft systems woven at the same time in a single operation. They are bound together by interchanging warp and weft threads between the face and back fabrics, or by using a special binder thread (e.g. ulster). The weave of the two fabrics can be different. They are produced in order to give the material greater weight or to achieve differently coloured face and back sides (e.g. double face or ulster fabrics). Used for blankets, coats and duffle coats.

Double cloth construction: Two fabrics are woven in the loom at the same time, one fabric on top of the other, with binder threads holding the two fabrics together. The weave on the two fabrics can be different.

Double cure process: A two-stage permanent press process. *Stage 1*: Pre-cure in the presence of swelling agents; *Stage 2*: Post-cure after garment making. This method allows the pleats, folds, etc., to be permanent.

Double damask: See **Damask**.

Double dyed fabric: Usually the blended fabric dyed in two different fabric is called double dyed. In many cases like polyester/cotton, second dyeing does not affect the first dyeing, but in blends like cotton/wool second dyeing on cotton can affect the wool (first dyeing) as almost all cotton dyes wool also.

Double end: Two ends woven as one in a fabric. A double end may be intentional for fabric styling, or accidental, in which case a fabric defect results.

Double face satin: See **Satin**.

Double faced carpet: See **Tree-ply carpet**.

Double Fold: A material which has been folded along the middle of its length normally with its face side inwards, so that the selvedges are in line.

Double hooked bow: One hooked bow at each side of the fabric that are in the opposite directions.

Double hosiery: Knit with single thread, has the toe or heel usually reinforced by the addition of another thread.

Double jersey: Although traditionally wool, which has become very expensive, this may now be made of acrylic fibre. It is a heavy fabric in plain colours and is characterized by the fact that both right and wrong sides are identical. Used for women's trousers, jackets, warm dresses and light coats.

Double knit: Made on machines with two rows of needles so that both sides look the same. Firm textured fabric suitable for skirts, trousers and jackets.

Double knit fabric: A fabric produced on a circular-knitting machine equipped with two sets of latch needles situated at right angles to each other (dial and cylinder).

Double knitting yarn: A plied hand-knitting yarn with a diameter of a botany yarn of 4/8 worsted (resultant count 440 tex).

Double looped terry cloth: See **Terry cloth**.

Double mercerisation: Double mercerization of yarn (only applicable to high quality yarns). By double mercerization, any variations in the mercerizing effect on the yarn which may occur in single mercerization are avoided. The process is nevertheless very expensive.

Double pick (in woven fabrics): Two picks wrongly placed in the same shed.

Double pique: Double pique is made on rib machines by a selection of knitted loops and floats. The floats greatly reduce the width extensibility. This allows the fabric to be processed like a woven material without losing the comfortable wearing characteristics of a knit. Unlined fabrics have a tendency to bulge. Also known as weveknit, rodier and overknit.

Double pique (face and back)

Double plush (double velvet, double plush carpets): Produced by the double velvet technique to obtain two completely identical fabrics in design and colour which are bound to each other by shared pile yarns. These pile yarns are finally separated in the weaving machine by a knife giving two identical plush, velvet or carpet materials.

Double refraction: See **Birefringence**.

Double reverse bow: Two fabric bows arcing in opposite directions. See **Double hooked bow**.

Double roving spinning: A system in which two rovings are fed to each spindle of a ring frame, so to produce a pseudo two-fold or two-strand yarn. The rovings are separated in the drafting system by means of special guides and the two drafted strands are combined after the drafting system.

Double selvege: See **Rolled selvedge, Curled selvedge**.

Double serge: See **Serge**.

Double shed: In carpet weaving machines, two sheds are formed from the pile and binder warps, i.e. an upper and a lower shed, through which the weft is inserted.

Double-sided transfer paper: It is a transfer printing paper which has been printed on both sides to enable the simultaneous transfer printing of two fabric webs by using the sandwich principle (fabric/paper/fabric).

Double split complimentaries: Apart from using additional values of one or more of the complementary hues to make a tetrad colour combination, there are two specific types of combinations using complementaries. The first is known as a double split complementary, where one of the hues (either side of the first hue) is used and one of the colours on one side of the complementary hue is used. This is usually identified by an oblong within the hub of the colour wheel.

Double spot coating: A dot-coating process in which each dot consists of a lower melting core and a higher melting outer shell. The advantage is that, in pressing, the adhesive is prevented from sinking into the fabric thereby achieving a stronger bond.

Double squeegee: A squeegee system for wide printing widths in which the print paste is applied between two squeegee blades. Used in screen printing and screen printing machines.

Double stitch (double loop): Combination of stitch and loop.

Double-stitched seam-finish: A finish for the raw edge of a plain seam in which another row of machine stitching is made about 1 cm (3/8 in) away from the cut edge. Seam allowance is trimmed close to the second stitching. A double-stitched seam finish is used where a narrow seam means ease and comfort in wearing and a better fit. Used for crotch and under arm seams, other served areas and for the knitted fabrics. For knits, a narrow zig-zag stitch is used for both rows.

Double-topstitched seam finish: It is a decorative way to hold seams of a bulky fabric flat and open. A plain seam is pressed to open. Even-hand basting is sewn along the edge of the each seam allowance to hold the seam open and flat while stitching. Topstitch is done on the right side of the garment leaving an equal distance on each side of the seamline crease.

Double velvet weaving machines: While loop pile velvets are produced with wire weaving technique, which is quite labour intensive, cut-pile velvets are produced on particular machines which permit to obtain two pieces of fabric at the same time. These machines are equipped with a 3 position shedding device (dobby or Jacquard machine), so as to form two overlapped and properly spaced out sheds and to permit to the pile warp to tie up the two fabrics together. Into each of the two shed a weft is inserted, usually by means of a pair of superimposed rods driven by the same gear. The pile warp is subsequently cut directly on the machine through a blade with horizontal traverse motion, thus forming the pile on both fabrics, which are then wound up separately. This system is largely used today owing to its high performance.

Double weave: Double weave fabrics are obtained by using five or more set of yarns. The most common types of double cloth have two set of warp and two set of weft yarns with an extra yarn interlacing both the cloths. The double cloth has additional bulk, strength and warmth. Examples are coatings, blankets, double brocade and brocatelle. Cutting the yarns that hold the two cloths together yields two separate cut-pile fabrics.

Double weft carpets: Woven carpets where each row of tufts is tied up by two wefts.

Double welt seam: A complex seam formed on the inside of the object with one trimmed raw seam edge enclosed and two rows of stitching visible on the face side.

Double worsted cloth: This name is given to the cloth because of the double-faced weave. A type of compound warp fabric consisting of two warp thread systems, but only one weft thread system. The warp and the weft are usually fine two-fold yarns in the case of worsted fabrics.

Fine, dense fabrics, sometimes worsted, usually in muted shades with discreet weave, yarn and silk effects; with a deliberately smooth finish. Dyed in loose stock, yarn and piece. Used in classic suits and formal wear.

Doublet: A sort of jacket prevalent during 15th–17th century.

Doubling: (a) The process of twisting together two or more singles or plied yarns, i.e. plying; (b) The operation of combining two or more strands without twisting; (c) A defect formed by forming a yarn from more than the required number of component strands; (d) A British term for twisting; (e) The term doubling is sometimes used in a sense opposite to singling. This is unintentional plying; (f) A yarn, considerably heavier than normal, produced by a broken end becoming attached to and twisting into another end; (g) A process for combining several strands of sliver, roving or yarn in yarn manufacturing.

Doubling machine: A machine which folds cloth to half or quarter of its original width.

Doubling machine (folding machines): During the lengthwise passage of a fabric through such a machine, it is folded in the centre to half its width and either batched on rolls (doubling and batching machine) or plaited down (doubling and plaiting machine) as desired.

Doup: A special type of heddle, used in conjunction with ordinary heddles on the harnesses of a loom to cross and uncross warp filaments (in both a horizontal and vertical plane) when weaving cross thread tissues, such as guaze weave or leno weave.

Doupion silk fabric: Silk fabric made with dupion yarn.

Douppion: (1) A fabric with an uneven surface and rough texture made from yarns that are irregular in thickness. It was originally the term used to describe the particular type of silk yarn which was spun from silk of two cocoons nested together. The two silks were not separated in spinning and were therefore uneven. Douppion may still be made from silk, but it is also made from viscose, acetate, and other synthetic yarns which can be manufactured in uneven thickness, and is a firm, medium weight fabric of plain weave. Used for women's summer suits, wedding outfits, etc.

(2) A rough or irregular yarn made of silk reeled from double or triple cocoons.

Fabrics of douppioni have an irregular appearance with long, thin slubs. Douppioni-like yarns are now being spun from polyester and/or rayon staple.

Douppioni (doupion, dupion): Silk fabric with a characteristic irregular slubby texture. The effect is obtained by the use of irregular, raw, rough silk, reeled from double cocoons in the weft. The double cocoons are the result of two cocoons having been spun by two silkworms too close together. Nowadays, schappe silk slub ply yarns or raw tussah silk is often used in the weft. A schappe silk ply yarn is used in the warp.

Dowlas: Originally, a plain-weave coarse linen fabric. Now often used for a cotton fabric produced from coarse yarns and finished to imitate linen in handle and lustre.

Down: The soft fine feathers which grow under the contour feathers of ducks, geese and other waterfowl to prevent loss of heat. The eider duck has been the principle source for textiles as a filling material. (a) Pure down (also known as "feather-free original duck or goose down") which is the softest quill-free plumage consisting of core down without any feather content; (b) Down, similar to pure down, but with a permissible content of 10% small feathers by weight; (c) Feathery down, like pure down, but with a permissible content of maximum 50% small feathers by weight; (d) Half down, which consists of feathers with at least 50% down content and occupies approximately half the volume of pure down since the latter is considerably lighter than feathers. Uses: Filling material for pillows, quilts, eiderdowns, etc. (approximately 30 geese produce 1 kg of down).

Down (pure): See **Down**.

Down proof batiste (maco down proof): A fine and tightly woven fabric made from particularly fine and uniform long-staple cotton (maco). Combed yarns are mainly used. Special finish – feather and down proof. Qualities – grey fabric and dyed. Used for tickings.

Down proof fabric: A woven cotton fabric of approximately 120 g/m2 made from very fine yarns (usually of maco cotton) and extremely tightly woven. It is used for pillows, quilts, etc., filled with down. Down proof fabric is very prone to creasing and the crease marks are highly visible after subsequent dyeing; it is also difficult to wash off because of the tightly woven construction. As a rule, it is only produced in plain weave and supplied in its natural colour for this reason. These fabrics are generally given a wax finish to make them down proof.

Down fibres: Detached barbs from down; plumules and detached barbs from the basal end of waterfowl feather quill shafts that are indistinguishable from the barbs of down.

Down twist: This is one of the two terms sometimes used when discussing plying. This refers to an "S-Twist". Alden maintains that people get so hung up trying to remember whether an S-Twist is spun clockwise, that they lose track of process. It really doesn't matter whether your singles are spun S or Z, you just need to ply them in the opposite direction.

Down wool: It is also called 'Hill wool'. Wool of medium fineness produced by such breeds as the Southdown and the Shropshire. These sheep are distinguished by their fine and curly wool of short staple, which is especially adapted for making loose, rough, moss-like, felted, carded yarns for the production of clothing. These wools are lofty and well-suited for woollen. Much of the down wool runs 1/4 to 3/8 blood in quality. This can be a great wool for felting.

Down (feathery): See **Down**.

Down (half): See **Down**.

Down (nestling): A down not fully developed with a sheath and with soft barbs emanating from the sheath.

Down draft metier: A dry-spinning machine in which the airflow within the drying cabinet is in the same direction as the yarn path (downward).

Down grade: In quality control, the lowering of the grade and/or value of a product due to the presence of defects.

Down proof cambric: Cream or white plain weave cotton of cambric weight with a wax or glazed surface to make it feather- or down-proof. Used for pillow coverings, cushion covers, duvet covers mattress covers.

Down time: Operationally dependent component of operating time.

Down twister: A cap, ring or flyer twisting frame.

Down twisting: A process for inserting twist into yarn in which the yarn passes downward from the supply package (a bobbin, cheese or cone) to the revolving spindle. The package or packages of yarn to be twisted are positioned on the creel and the ends of yarn are led downward through individual guides and stop motions to the positively driven feed roll and from there to the revolving take-up package or bobbin, which inserts twist.

Downright: Wool sorting term in the woollen trade, meaning the wool taken from the lower parts of the sides of a fleece. It is next to the finest sort.

Downs: English short and medium wools which are fuzzy; used for hosiery, woollens, etc.; included are Dorset, Oxford, Shropshire, Wiltshire, Southdown, Hampshire.

Dowtherm: Trademark of Dow Chemical Company for a series of heat transfer media. Dowtherm jackets are used around molten polymer processing lines.

Dozens: English kersey, worn under Henry VIII.

DP: Degree of Polymerization.

DP rating: The fabric must have a smooth appearance after it is laundered dried (line dried or tumble dried). Fabric smoothness is referred to as Durable press rating (DP rating) and is judged on a 1 to 5 scale. A 5 DP rating is the highest (most smooth) and a 1 is the most wrinkled. A 3.5 rating is considered commercially acceptable.

DPB: (a) Deutsches Bundespatent (German Federal Patent), now DB; (b) Dibutyl phthalate.

Drab cloth: Thick and strong woollen over-coating of grey colour in England.

Drabbet: Coarse, twilled linen duck, white or coloured, made in England.

Draft: In weaving, a pattern or plan for drawing-in.

Draft: (1) To reduce the mass per unit length of a fibrous assembly by attenuation.

(2) When drafting, the degree of reduction of linear density is usually expressed as either the ration of the input mass per unit length to the output mass per unit length or as the ratio of the surface speeds of the machine components which bring about the drafting.

(3) Fine and all-wool, warp ribbed cloth; used in churches.

Draft, in garmenting: (1) Constructed plan of the garment.

(2) The application of body or garment measurements to a flat plane.

Drafting: See **Drawing**.

Draft ratio: The ratio between the weight and length of fibre fed into various machines and that delivered from the machines in spun yarn manufacture. It represents the reduction in bulk and weight of stock, one of the most important principles in the production of yarn from staple fibres.

Drafting: The process of feeding the fibres from your drafting hand (usually your left) to your spinning hand (usually your right).

Drafting triangle: The small triangles of fibres that are formed between your drafting hand and your fibre hand. This should never be longer than the fibre length. Also called a “drafting triangle”.

Drafting zone: A section in many spinning machines which reduces the cross-section of a fibrous strand (sliver, slubbing or roving) by drawing in order to decrease the linear density and bring the fibres into parallel alignment at the same time.

Drag: See **Fibre cohesion**.

Dragged in filling: See **Pulled in filling**.

Drainage fabric: See **Geotextiles**.

Drake cluster: A commercial variety of early maturing American upland cotton, the staple measuring 22–25 millimetres; the yield is 31–32 percent.

Dralon: Trade name for an acrylic fibre. It is mothproof, has good resistance to acids and alkalis and is unaffected by sunlight. It is washable, but hardly absorbs any moisture and is therefore quick-drying. It does not shrink, felt or pill. It may be blended with other fabrics. Used for upholstery and furnishings. Often as pile fabrics and for net curtaining.

Dram: In England and the United States, the weight in drams of 1,000 yards of silk yarn is the unit of measurement. See **Denier**.

Drap: French for woollen.

Drapd'alep: French dress goods and men's suiting, made of silk warp and woollen filling and fulled; obsolete.

Drapd'alma: A fine, close, flat-ribbed, twilled fabric of wool or silk and wool, finished on but one side.

Drapd'Arest: Rich silk cloth of the Middle Ages; used for church vestments.

Draps de baye: Stout French mourning dress goods, made with long napped shaggy surface.

Drap de beaucamp: A stout and coarse French twilled suiting, made of linen warp and woollen filling in solid colours.

Drapbresilienne: French serge dress goods, made of pure silk or mixed with wool.

Drap de Chasse: A plain woven women's suiting, made with fine silk warp and heavy cotton filling, forming prominent cross ribs.

Drapd'Ete: A fine, light worsted fabric, woven in longitudinal cords.

Drap de Milord: Various kinds of French serge dress goods, made with 12 leaves and from 6 to 12 picks in a repeat.

Drap de Pauvre: Coarse serge, made of natural reddish-brown, undyed and unsecured wool. Formerly extensively used in France for clothes by the poorer classes.

Drap pique: Quilted silk fabric, made with fancy figures; used for men's vests.

Drap de prince: Silk or wool serge suiting made with eight leaves and three picks in a repeat.

Drap de Silesie: Light weight, all-wool French dress goods and fulled.

Drap de soie: (a) French term for heavy silk dress goods; (b) Various, closely woven serge dress goods and men's suiting, originally made in black colour.

Drapd'ete: A fine, light-weight twilled worsted dress goods, with warp ribs; used for summer garments, etc.

Drapedredon: Light, soft and warm winter coating, made of the finest wool, completely fulled.

Drapey: Refers to a fabric with good drape or supple and falls easily into graceful folds when hung or tailored.

Drapgeraldine: Heavy woollen dress goods, made in dark mixtures.

Draping: To hang or to adorn the body form with loose fabric and to obtain a body fitted garment by using adequate sewing techniques.

Drapnatte: Fulled woollen cloth of English origin, made in solid or several colours and having a raised nap on the back.

Drap imperial: Fine French dress goods, woven plain of wool and cotton.

Drap royal: (a) Lightweight twilled French woollen dress goods, made of organzine warp and very fine worsted filling in small cross ribs; (b) Obsolete, lightweight, fulled and printed French woollen cloth; used for vests, etc.

Drapsanglier: Rough faced French mourning dress goods, made of all wool in loose weave.

Drapsoleil: French dress fabric of high lustre, made of wool with wide weft ribs.

Drapzephii: Fine French woollen dress goods, similar to **Cachemire**.

Drape: The way the material of the garment hangs when worn. A term to describe the way a fabric falls while it hangs; the suppleness and ability of a fabric to form graceful configurations.

Draper: In French, a dealer in cloths; also a clothier, haberdasher or dry goods merchant. Also used in English.

Drapery: (a) Textiles, used for upholstery, curtains, hangings, etc., are generally known as drapery; (b) Dry goods are generally termed drapery in Britain.

Draps chats: French woollen dress goods, made of white wool warp and coloured wool-filling, dyed black in the piece.

Drapscroises: French for twilled cloths.

Drap de Dame: Very light, soft, all-wool French dress goods, similar to the flannel or Spanish stripes, slightly fulled.

Draps de Gobelin: Fine French woollen dress goods, dyed scarlet red.

Draps de Gros: Bureau Coarse French woollen suiting, dyed black, grey, or white; used by the lower class people.

Drawback: (1) A crossed end; an end broken during warping that when repaired was not free or was tied in with an adjacent end or ends overlapping the broken end. The end draws or pulls back when unwound on the slasher.

(2) A weave distortion characterized by tight and slack places in the same warp yarn.

(3) See **Sticker**.

Draw cord: A chord which passes through a channel and allow the garment to be gathered.

Draw crimping: See **Draw-texturing**.

Draw down: The amount by which manufactured filaments are stretched following extrusion. Also see **Drawing**.

Draw frame blends: Blends of fibres made at the draw frame by feeding in ends of appropriate card sliver. This method is used when blend uniformity is not a critical factor.

Draw pin: A stationary guide or pin which by inducing a localized change in yarn tension and/ or temperature may be used to stabilize the position of the draw point or neck in some processes of drawing of a manmade fibre yarns. *Note*: For the drawing of some fibres, e.g. polyester, a heated pin may be used and with other types the pin is not normally heated.

Draw ratio: (1) The ratio of the speed of the first and second pull roll stands, used to orient the flat polyolefin monofilament during the manufacture.

(2) The ratio of final to original length per unit weight of material like yarn, laps, slivers, slubbings, rovings, etc., resulting from drawing. Also see **Draft ratio** and **Draw down**.

Draw roller: Cloth-handling roller driven by cogs; See **Rollers**.

Draw sizing: A system linking draw warping and sizing in a continuous process. A typical system includes the following elements: (a) Creel, (b) eyelet board, (c) warp-draw machine, (d) inter-mingler, (e) tension compensator and break monitor, (f) sizing bath, (g) dryers, (h) waxing and winding units.

Draw sliver: The loose, untwisted strand of cotton fibres which is the product of the drawing frames.

Draw textured yarn: A yarn made by a process in which the drawing stage of man-made yarn manufacture in wholly or partially combined with the texturizing process on one machine.

Draw texturing: In the manufacture of thermoplastic fibres, the simultaneous process of drawing to increase molecular orientation and imparting crimp to increase bulk.

Draw thread (Lace): The removable threads included in the construction of lace either to act as a temporary support for certain parts of the pattern or to hold together narrow widths or units that are separated subsequently by their removal.

Draw threads: In knitting, threads that are introduced that, on removal, allow separation of pieces of fabric with welt edges.

Draw tow: A zero-twist bundle of continuous filaments that has been stretched to achieve molecular orientation (Tows for staple and spun yarn application are usually crimped).

Draw twisting: The operation of stretching continuous filament yarn to align and order the molecular and crystalline structure in which the yarn is taken up by means of a ring-and-traveller device that inserts a small amount of twist (usually 1/4 to 1/2 turn per inch) into the drawn yarn.

Draw warping: A process in which a number of thread lines, usually 800–2000 ends of POY feedstock, are oriented under essentially equal mechanical and thermal conditions by a stretching stage using variable speed rolls, then directly wound onto the beam. This process gives uniform end-to-end properties.

Draw wind: To orient a filament yarn by drawing it and then to wind it on to a package in an integrated process without imparting twist.

Draw winding: The operation of stretching continuous filament yarn to align or order molecular and crystalline structure. The drawn yarn is taken up on a parallel tub or cheese, resulting in a zero-twist yarn.

Drawers: Under-trousers.

Drawing: Operation by which the slivers are blended, levelled and by drafting reduced to the stage of sliver or roving suitable for spinning. In cotton spinning, the term is only applied to processing at the draw-frame. Various system of drawing are practiced in worsted spinning, but the machinery development and the greater use of man-made staple fibres, the differences are becoming less distinct. Most modern drawing gets in corporate three passages of pin drafting and a roving process. The system differs mainly within the means of fibre control between the major pairs of drafting rollers and the type of output package.

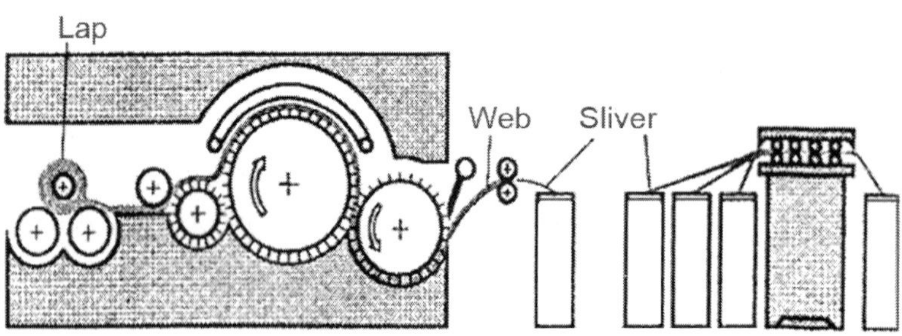

Carding Drawing (schematic)

Drawing: (1) The process of attenuating or increasing the length per unit weight of laps, slivers, slubbings, or rovings.

(2) The hot or cold stretching of continuous filament yarn or tow to align and arrange the crystalline structure of the molecules to achieve improved tensile properties.

Drawing, in processing: The process of stretching a material to increase the length per unit mass.

Drawing (cold): The drawing of synthetic filaments or films without the intentional application of external heat. Free drawing of filaments or films at a neck is also referred to as cold drawing, even though this may be carried out in a heated environment.

Drawing (hot): A term applied to the drawing process of synthetic films with the intentional application of external heat to support the drawing operation

Drawing frame: A machine in which several strands of sliver are combined into one strand and drawn out so that the combined strands approximate the weight and size of any one of the original strands.

Drawing (in hook): A tool similar to a crochet hook, used to draw the individual warp filaments through the heddles and reed of a loom.

Drawing (synthetic polymers): The stretching to near the limit of plastic flow of synthetic fibres of low molecular orientation. *NOTE*: This process orientates the polymer chains in the amorphous areas of the fibres in the direction of its length and this results in an increase in the crystalline regions. The ways of carrying out this process are by "hot drawing" or "cold drawing" (i.e. with or without the intentional application of heat).

Drawing in: (1) Forming a running stitch drawn to achieve gathering or gauging of the material hoiding foids in various forms of garments as in gowns of barristers for example.

(2) In weaving, the process of threading warp ends through the eyes of the heddles of the harness and the dents of the reed. The order in which this is done varies according to the weave and the nature of the fabric to be produced.

Drawing machine: Depending on the styles of the produced fabrics and on the company's size, this operation can be carried out manually, by drawing- in operating in pairs (a time consuming activity which requires also skill and care), or by using automatic drawing-in machines. The drawing-in begins by placing the weaver's beam, the harness and the row of healds on the proper anchor brackets, then the drawing-in program is typed in on the computer and the machine is started. A sort of long needle picks up in sequence the threads and inserts them with only one movement into the drop wires, the healds and the reed dents, which are selected each time and lined up to that purpose. The machine can be used with the usual types of healds, drop wires and reeds and can process a wide range of yarn types and counts, from silk yarns to coarse glass fibre yarns. The drawing-in speed can in optimum conditions exceed 6,000 threads/hour.

Drawing process: The process of drawing in the yarn through the healds. See **Drawing in, Drawing machine**.

Drawing roll cleaner: A pad of felt or similar material attached to the underside of the cleaner box cover which serves to wipe away the dust and lint that collects on the drawing rollers as they draw out the roving or sliver.

Drawing rollers: Two or more pairs of rollers, each pair of which rotates at a higher speed than the preceding pair, serving to draw out or attenuate the roving or sliver passing between them.

Drawing-in frame: A frame for holding a beam of warp strands, harnesses and reeds, so that the strands may be drawn easily through the harnesses and reeds in a specified order.

Drawn selvege: See **Tight selvedge**.

Drawn tow: A zero-twist bundle of continuous filaments that has been stretched to achieve molecular orientation (tows for staple and spun yarn application are usually crimped).

Drawn work: Made by pulling out certain threads of loosely woven linen and fastening the remaining threads together with various fancy stitches into patterns.

Drawn yarn: Extruded yarn that has been subjected to a stretching or drawing process which orientates the long-chain molecules of which the yarn is composed, in the direction of the filament axis. *NOTE*: On further stretching, such yarn possesses elastic extension as compared with the plastic flow of undrawn yarn.

Draw-spinning: A process for spinning partially or highly oriented filaments in which the orientation is introduced prior to the first forwarding or collecting device.

Draw-texturing: A process in which the drawing stage of man-made yarn manufacture is combined with the texturing process on one machine. The drawing and texturing stages may take place in separate, usually consecutive, zones of a machine (sequential draw texturing) or together in the same zone (simultaneous draw-texturing).

Draw-texturing: See **Draw-texturization.**

Draw-texturization: A process applied in the spinning of thermoplastic manmade fibres to achieve optimum performance characteristics. Drawing involves a post-stretching of the freshly spun filaments (by 400–1200% depending on the type of synthetic fibre). The process brings about a deformation of the polymer structure and, at the same time, an increased orientation of the chain molecules in the direction of the fibre's longitudinal axis (Texture).

The long-chain molecules are pulled into alignment so that they become more closely packed alongside one another (chain packing) thereby forming crystalline regions (crystallites) which increase the fibre's tenacity through secondary valency forces (dye affinity is reduced in highly oriented fibres).

Draw-twist: To print a filament yarn by drawing it and then to twist it in integrated sequential stages.

Draw-warping: A process for the preparation of warp beams or section beams from a creel of packages of partially oriented yarn in which the traditionally separate stages of drawing and beaming are machine, also termed draw-beaming, warp drawing.

Draw-wind: To orient a filament yarn by drawing it, and then to wind it on to a package in an integrated process without imparting twist

Draw-warping: A process in which a number of thread lines, usually 800 to 2000 ends of POY feedstock, are oriented under essentially equal mechanical and thermal conditions by a stretching stage using variable speed rolls, then directly wound onto the beam. This process gives uniform end-to-end properties.

DREF (false twisting method): In this system a sliver, preferably produced by a draw frame, is drafted on a high draft drawing system. The drawn sliver is aerodynamically split on leaving the drawing system by being separated by force lateral to the roving axis. The slivers obtained in this way are combined by conventional means and twist is imparted by a twisting unit.

Dresden: (a) 18th Century woollen fabric in England; (b) Small flower design in pastel shades, usually in warp prints.

Dresden point lace: A fine linen fabric with threads withdrawn and embroidered to form a square mesh design.

Dresden ribbon: Silk or cotton ribbon, in various widths, the warp is printed in delicate coloured flower patterns before weaving.

Dress: (1) A clothing, partially outerwear.

(2) A type of adjustment made to one side of the trousers to improve it.

Dress carrier apparel: Career apparel which is not generally subject to abusive wear and for which appearance is more important attribute than durability.

Dress faced: Woollen fabrics having a fulled and slightly napped face.

Dress glove: A covering for the hand, often extending part way up the yarn, worn formally for formal or dress occasions.

Dress shirt (for boys): A shirt made with a specific collar size or numerical size, and designed to be worn with a tie and jacket.

Dress shirt (for men): A shirt made with a specific collar size and sleeve length where appropriate, and designed to be worn with a tie and jacket.

Dresser or drum (in sectional warping): The dresser or drum is composed of a big sheet steel cylinder with a precisely turned outer surface which bears at its end a series of slope control rulers (knives), which form a cone with variable taper. There are, however, also warping machines with fixed taper. The dresser is the creel element on which the threads coming from the creel and guided by the carriage are wound, section after section. The initial taper ratio serves as a support for the various thread layers superimposing each other during warping, and acts, therefore, practically as a backing flange.

Dressing (flax): A combing process applied to pieces of line flax fibre to parallelize the strands, remove naps or bunches of entangled fibres, and square the ends of the pieces by pulling or breaking fibres that protrude from the ends.

Dressing (lace): The process of applying stiffening agent or softening agents on stretching lace, net, or lace-furnishing products, etc., and drying. The application can be done on either a running or a stationary frame.

Dressing (silk): Size made of gum, starch, china clay, etc., to stiffen cotton, linen and silk cloths.

Dressing: A finishing process, consisting in sizing the fabric with gums, rice water, oil, etc., in order to give lustre, weight or stiffness to the fabric. See **Sizing**.

Dressmaker shears: Scissor used in cutting of the fabric (meant for sewing) with a bent handle which cuts the fabric while it is flat on the cutting table.

Dreze machine: One of the earlier loose cotton dyeing machine. It consists of a cylindrical cage with a false bottom and provided with a central puffer tube. A steam injector system is provided at the bottom. The material to be dyed is packed in the cage which is then lowered into the outer vessel. The circulation and heating of the dye liquor are effected by injecting live steam which throws liquor upwards into the central puffer tube and later against a metal shield and thus percolates through the loose cotton package to the bottom and again thrown upwards by the injector. This machine can also be used for dyeing wool.

Drill: A warp-faced piece-dyed twill weave fabric that is occasionally satin weave (satin drill, mattress twill), usually cotton, linen, linen/cotton, made in different weights, usually medium to heavy weight, densely woven fabric that has a stout texture and a higher number of threads per centimetre in the warp than in the weft. The commonest fabric known as drill is in white or dark working colours and is thick and hard wearing fabric. Various terms are used as per the field of application such as bed drill, table drill, trousering twill, sacking twill, cotton drill, etc. It finds use in grey, bleached dyed or printed state for a variety of application like work wear, pocketing, shoe lining, uniforms, bookbinding, coated fabrics, industrial fabrics, ticking suiting, overalls, protective clothing, etc. *NOTE*: Some drills are made with five-end satin weave and it is recommended that these be called satin drills.

Drilled embroidery (Bohrware): It is made on hand machine or shuttle machine by cutting holes in the cambric and embroidering the cut edges.

Drillette: In England, a variety of weft faced, light weight drill, woven in a three, four or five shaft twill.

Drilling: A drill fabric. General term for various twilled cotton stuffs used for lining. Sometimes called "drill".

Drip dry: A garment washing process, where the textile materials which are resistant to disturbance of fabric structure and appearance during wear and washing process. They require minimum of ironing or pressing.

Dripping: In testing thermal protective clothing, a material response evidenced by flowing of the fibre polymer.

Drive ratio: Ratio of wheel diameter to flyer whorl diameter (or bobbin whorl on a bobbin lead wheel). Governs how much twist you get in the yarn for each treadle. To measure wheel ratios, set up wheel, tie a bright-coloured piece of yarn to the flyer arm and adjust the treadle until it is at the bottom of its movement. Slowly rotate the wheel while counting the flyer revolutions until the treadle returns to its original point. The bright yarn tied to the flyer arm just makes it easier to count.

Droguet: (a) Collective term for various cotton wool and silk cloths, made in France and England. It is made plain, woven or twilled and patterned, usually slightly fulled; (b) Obsolete term for various ribbed French worsted dress goods.

Drop: (1) That part of a bed covering that hangs perpendicular to the floor.

(2) In body measurements, the difference between the chest girth and the waist girth.

Drop lea: The strong web which handmade rugs are started with.

Drop printer: Jet printing system for manufacturing sharp-edged prints (100 lines/inch). When leaving the end of the nozzle, the dye solution droplet is given a defined electrical charge (16 possible charge levels). Depending on the charge level, the droplet is either withdrawn, in which case no dye is applied, or electrostatically deposited over a width of 4 mm in a certain position (nozzles arranged at 4 mm intervals across the width of the material).

Drop spindle: A spindle that hangs freely from the fibre source (as opposed to a supported spindle). Probably so named by people who have not added enough twist.

Drop stitch: (a) An open design made in knitting by removing some of the needles at set intervals; (b) A defect in knit fabric.

Drop wires: A stop-motion device utilizing metal wires suspended from warp or creeled yarns. When a yarn breaks, the wire drops and the switch is activated that stops the machine.

Dropped stitch: In knitted fabrics, an unknitted stitch.

Dropped stitches: A defect in knit cloth characterized by recurrent cuts in one or more wales of a length of cloth.

Drosin: Suiting made of silk waste and worsted yarn. Made in Holland.

Drugget: (a) Plain, twilled or corded English fabric of the 18th Century, made with worsted warp and woollen filling; (b) Printed and felted woollen fabric; used for floor covering.

Druid: Cotton duck in England and Australia.

Drum drying machines: See **Drying systems, Sieve drum drier, Cylinder drying machines, Tumbler drier**.

Drum dyeing machines: Consist of a perforated and reversible rotating drum, sometimes consisting of several compartments. Liquor ratio 1:8–1:10. Finished items, such as hosiery and fully-fashioned garments, and piece goods, such as uncut and cut plush fabrics, towelling and raised goods, can be dyed or finished.

Drum washing machines: In a high-speed washer (see Fig) with special drum, the goods are advanced without conveyor belt, air cushion or pressure belt and the kinetic energy transformed into impact or compression forces.

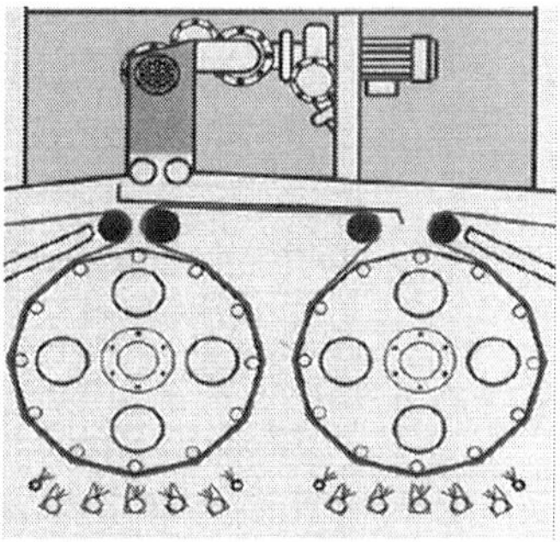

Double drum washing Unit

The washing and relaxing effects on fabrics into impact or compression forces. The washing and relaxing effects on fabrics and blends with wool content are therefore achieved with short processing times, thus eliminating the undesirable running creases which often result from wet finishing of textile substrates in rope form.

Drummond: (a) Originally a twilled English worsted in Oxford grey, woven with dou'oie and twist yarn; (b) A Highland tartan with red as predominating colour and stripes running as follows, each way – A wide field of red, split with two pairs of narrow dark-blue lines, the centre stripe between the two pairs being about one third the width of the wide side stripes. The next group of stripes is as wide as both pairs of blue stripes, the centre has red stripes and one side also has red stripe and consists of the following: Narrow pale-blue line, narrow red line, dark blue stripe, narrow red line, narrow green line, narrow red line, wide green bar (being of the same width as the wide red bars mentioned above in the red group), narrow red line and narrow dark blue line. This is followed by a wide red stripe, after which the entire group described between the two is repeated in a reversed order.

Dry cleanable button: A button that can be solvent cleaned without damage such as dissolving or loss of finish.

Dry cleaning: Cleaning of garments/fabrics using solvents (other than water). A commercial process by which soil may be removed from textile products, or parts thereof, in a machine which uses petroleum, perchloroethylene, or fluorocarbon solvents.

Dry-cleaning detergent (dry-cleaning intensifier): These are used in dry-cleaning in normal organic solvents to considerably increase the cleaning action with maximum removal including water-soluble impurities. New concept introduced in 1953, replacing Benzine soap in terms of composition, efficiency and economic viability.

Dry cleaning fastness: Fastness of a fabric against dry-cleaning process. A bag of cotton twill containing 12 steel discs and the test sample is treated for 30 min in the appropriate machine at 30 °C in 200 ml tetrachloroethylene (or other solvents) without added reagent, then spun and dried at maximum 65 °C. In commercial dry-cleaning, consideration is also given to the following fastnesses.

Dry cleaning solvents: These are generally classified into two groups– (a) Flammable solvents: Light petroleum distillates, white spirit, Stoddard solvent and heavy petroleum spirit; (b) Non-flammable solvents: Chlorinated hydrocarbons such as trichloroethylene, tetrachloroethylene (perchloroethylene), 1,1,1-trichloroethane and trichlorotrifluoroethanes (solvents R11 and R113).

Dry (fabric): Dry fabric is the one that feels lacking in surface moisture or natural lubricant. Cotton are often referred to feel dry.

Dry combed cop: A wool top containing not more than 1% of fatty matter based on the oven dry fat-free weight as tested by the IWTO method which specifies soxhelet extraction with dichloromethane.

Dry crease recovery angle: Crease recovery angle; a measure of the tendency of a fabric to crease when dry.

Dry cross-linking: Cross-linking of cellulose in a de-swollen, i.e. dry, state with resin-finishing agents in the presence of resin-finishing catalysts at elevated temperatures (120–190 °C) within a few minutes or seconds (resin finishing). Practical significance as dry cross-linking process with methylol compounds in the presence of acid-active catalysts, such as the classic curing process, the: shock-cure process and the permanent-press process.

Dry cylinders: See **Drying cylinders**.

Dry extraction cleaning: A method in which an absorbent compound is dispersed over the surface of a textile product by hand or machine, thoroughly brushed through pile, allowed to dry, and removed by suction.

Dry foam extraction cleaning: A process by which a highly aerated, low moisture content shampoo is brushed through textile floor covering pile or applied to the surface of upholstery.

Dry filling: The application of finishing chemicals to dry fabric, usually by padding.

Dry finish: (a) Final finishing;(b) Specifically relates to a dry finish applied to garments following dry-cleaning; it is used as a final finish for suits, wool garments, overcoats, skirts, which have become limp through wear and from the effects of dry cleaning, especially those items which have been finished with solvent-soluble finish (softener). Application of a dry finish has the effect of enabling garments to stay clean for longer;(c) Term used for the mechanical processing of textile fabrics (dry finishing of fabrics), e.g. opening, brushing, decatizing, embossing, calendaring, polishing, pressing, pleating, raising, shearing, etc.

Dry forming: The production of fibre webs by methods that do not use water or other liquids, i.e. air-laying or carding.

Dry goods: General term in United States of America for all textiles.

Dry laid non-woven fabric: A fabric made from a fibre web or batt by carding and/or air laying followed by any type of bonding process.

Dry laying: A process for forming a fibre web or batt by carding and/or air laying.

Dry mercerising: Mercerizing of dry cotton or cotton/viscose fabrics. Mostly carried out on grey goods (mercerizing of grey goods). Disadvantages – the finish is not as good as on scoured fabrics; heavy contamination of the caustic soda solution (size, lignin, etc.); addition of expensive wetting agent necessary (effluent problems). The advantage, however, lies in the mercerizing of blended fabrics with regenerated cellulose; as with dry mercerizing, the swelling and dissolving range of 12–13°Bé caustic soda where these fibres are particularly at risk is not undergone twice, as happens in the mercerizing of wet goods.

Dry scrubber: A method for cleaning gases in which no liquid is used. Odours in air can be scrubbed out by passing through activated carbon filters (GAC filters). An acid gas can be removed by spraying dry powder into the gases. The powder collects the unwanted gaseous pollutant and the powder is removed in a fabric filter or other device for particulate removal. For example, calcium hydroxide may be sprayed into the gases when cooled below 160 °C to collect SO_2 and HCl from the gas as calcium sulphate and calcium chloride.

Dry setting: See **Heat setting**.

Dry solvent: Any organic solvent used to dissolve another material.

Dry spinning: A spinning process for man-made fibres in which the polymer, dissolved in an organic solvent, is converted into filaments by extrusion through spinnerets and subsequent evaporation of the solvent in warm air which allows the bundle of spun filaments to harden. The dry spinning principle is used for the production of acetate, triacetate, acrylic and polyvinyl chloride (PVC) fibres.

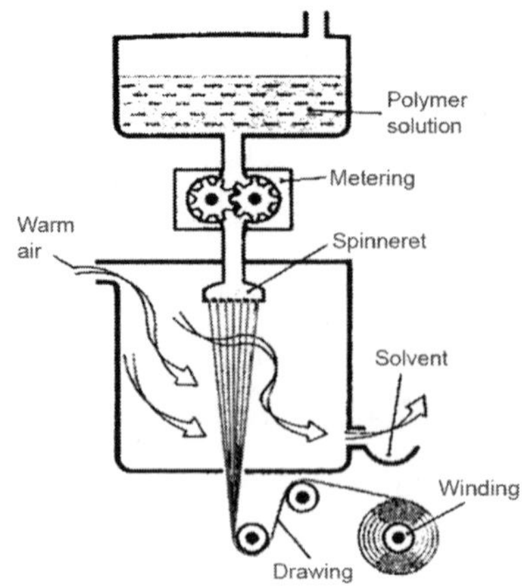

Dry spinning (schematic)

Dry spun: (a) Descriptive of a worsted yarn produced from a dry-combed top or of synthetic or blended yarn spun on similar machinery; (b) Descriptive of coarse linen yarn spun from air-dry roving (q.v.); (c) Descriptive of man-made filaments, the coagulation of which is effected by evaporation of the solvent from the spinning solution. See also **Wet-spun (b)**.

Dry spun flax: See **Dry spun.**

Dry steam: Steam without water droplets. Contrast with **Wet steam**.

Dry strength: Tensile strength.

Dry-cleaning: Dry-cleaning generally means the washing of garments or textile materials using a solvent which is usually recovered by suitable means. It is done in suitably designed machines and can involve washing in an appropriate liquor ratio using dry-cleaning solvents, spot cleaning if necessary, and pressing of the articles before delivering to the customer as per agreement.

Dryer creases: Sharp folds or lines running in any direction in a laundered or dried specimen. *NOTE*: Dryer creases are an unintended result of restricted movement of specimens in the washer or the dryer.

Drier (can): See **Contact drier**.

Drier (cylinder): See **Contact drier**.

Drier (drum): See **Contact drier**.

Dry crease recovery: Crease recovery angle; a measure of the tendency of a fabric to crease while it is dry.

Drying: A process to free a textile material of moisture, thermal hydro-extraction. Drying systems for textiles are based on, e.g. convection, contact heat, radiation, through-flow or combustion drying.

Drying cylinder: Any of a number of heated revolving cylinders for drying fabric or yarn. They are arranged either vertically or horizontally in sets, with the number varying according to the material to be dried. They are often internally heated with steam and Teflon-coated to prevent sticking of chemicals and dyes.

Dry-laid non-woven: Non-woven web made from dry fibre. Usually refers to fabrics from carded webs versus air-laid non-woven which are formed from random webs.

DS process: A direct stabilizing process for textile fabrics developed by Kannegiesser and Stockhausen. Instead of using adhesive interlinings, the outer material is coated in stripe form.

Dsedim or Jedim: Oriental rug, made of several strips of coloured knitted wool sewed together.

Dsujnabe or Jujnabe: Central Asiatic knotted wool rugs.

Dye recipe formulation: Formulation of a recipe for dyeing a particular shade. This can be done manually from personal experience of a dyer. Nowadays, recipe formulations are done with the help of a software where the formulation can be done based on dyeing parameters required like light fastness, dyes cost, etc.

DU: An arbitrary unit for the proteolytic activity of enzymes and biologically active washing agents, e.g. the enzyme preparation has a protease activity of 1000 DU/g if 1 ml of a 2% solution of the enzyme preparation gives an extinction difference of 0.400 under experimental conditions.

Dual anchor: Name for reactive dyes which have two or more reactive groups in the dye molecule. These kinds of dual anchor reactive dyes can have two "anchors" to the OH groups of cellulose molecules, i.e. for example, enter into a dual reactive bond (bifunctional).

Dubahi: A calico in Persia; used for clothing.

Ducape: A plain woven medium fine, soft ribbed silk fabric; used in the 17th Century in England.

Duchese lace: See **Duchess lace**.

Duchese satin: See **Duchess**.

Duchess: An 8–12 shaft satin. It is a dress fabric. Very fine yarns are used, particularly in the warp with more ends/inch than picks. The material is string, has a high lustre and texture, and it is firm. Usually 36 inches wide. Characterized by grainy twill on back. Satin Faconne jacquard figured fabric with an all-satin weave background. Various types of striping effects are obtained. Jacquard figure are on a satin ground. Formerly with a grey warp and a tram weft. Now produced from silk, acetate or viscose filaments, also with a fine cotton weft for dress fabrics and blouses.

Duchess lace: A bar lace with a design of bands or tapes, which is very fine and is worked with fine threads. A very old type bobbin lace.

Duchess satin: See **Duchese satin**.

Duchester: Sort of English velvet.

Duck: (1) A variety of heavy, strong, compact, firm plain weave cotton or linen fabrics of canvas appearance in plain weave. The warp yarns are in pairs and form a distinctive effect. Mass per square yard about 06–50 oz. There are two types of duck – (a) Plied yarn duck with plied yarn in both warp and weft; (b) Flat ducks: Ducks having the warp of two single yarns woven as one and either single or plied filling yarn. Used for sails, awnings, etc.

(2) A closely woven heavy plain-weave fabric of mass per sq. yd., 6–50 oz., traditionally made from cotton or linen yarns, and similar to canvas. *NOTE*: The terms "canvas" and "duck" have become almost synonymous and are often qualified by terms which indicate the use of the fabric, e.g. navy canvas, artist's canvas, duck suiting and belting duck.

Duck (plied); flat: Plied yarn duck is a compact, firm and heavy plain weave cotton fabric which has plied yarn in both warp and filling. Flat duck has a warp of two single yarns woven as one and a filling of either single or plied yarn.

Duck eye: See **Pin holes, Spinning**.

Duck (waste): These are cheaper grades of flat ducks. A characteristic construction can be 60 × 24 with 16s warp and 3s filling made from cotton waste.

Duck canvas: A strong, firm, tightly woven, durable fabric usually of cotton but sometimes of linen, hemp or other fibres. It is usually plain weave, but sometimes with a crosswire rib. It is produced in a variety of weights and used in a variety of products such as tents, awnings, sails, upholstery, footwear, jackets and trousers

Duck-chair canvas: A fabric in plain, repp or twill weave, made of vegetable or man-made fibres or a combination of any of these fibres. Traditionally widths do not exceeded 500 mm (20 in) and weights have not been less than 6 oz./sq. yard.

Ducks flat: Another class of fabric usually used for tents and water proof clothing is made from single yarns. A typical construction being 76 × 30 with 12s warp and 8s weft, tightly woven in plain weave.

Ducks (light weight): Lightweight ducks are made for cloth shoes or sneakers. A characteristic construction for this type is 50 × 36 with 2/16s warp and 2/20s weft.

Duffel (duffle, duffields): A heavy and low-quality woollen cloth, napped on both faces. Napped duffel is similar to eskimo and is mainly woven in a 4-end warp cross twill construction. Other duffel qualities are given a Melton finish. Used for short coats and certain types of military or contract cloths.

Duffle: See **Duffel**.

Dul: Very strong and fine fibre, yielded by the stem of a climber in Ceylon.

Dull: (1) Noun. Low in lustre, that is showing relatively little tendency to concentrate the reflected light in specific directions.

Dull: (2) Adj. (Matt); (a) Descriptive of textile materials the normal lustre (q.v.) of which has been reduced by physical or chemical means; (b) The colour quality, an increase in which may be compared with the effect of the addition of a small quantity of neutral grey dye to the dyestuff and such that a colour match cannot be made by adjusting the strength.

Dull colour: Consists of grey elements, i.e. white and black, in addition to other colours. The opposite dull colour is clear colour.

Dull fibre: Fibre in which suitable chemicals or pigment is added or the structure of the fibre is made in such a way as to appear less bright.

Dumb singles: The finest of Mie reeled silk yarns; it has no twist.

Dumb-waiter: Rope Trade name for untarred cables or cordage of hemp.

Dumbells: A defect frequently seen in wet-formed non-woven fabrics; an unusually long fibre will become entangled with groups of regular-length fibres at each end, thus producing a dumbbell-shaped clump.

Dumican: Embroidery work made by the old Araucanians in Chile.

Dumobin: A fine grade of Scotch plaid.

Dunbai: A Highland tartan, made with green stripes and black lines on a red ground.

Duncan: Commercial variety of late maturing long staple cotton from Georgia, growing in large bolls.

Dunchee: Strong and very elastic bast fibre, yielded by the *Sesbania Aculeata* in India; used for ropes and as a substitute for hemp.

Dundas: A Highland tartan, the stripes running both ways as follows: Black stripe; navy blue stripe, twice as wide as the 'black and split in the centre by a narrow black line; black stripe, as wide as the first one; a wide field of dark green (made a little narrower than the above mentioned black, blue and black stripes combined) and split in the centre by a group of red, green, dark blue and green lines.

Dundee: (a) Soft, smooth, twilled woollen and having very little nap; (b) Coarse burlaps, crashes, bagging, etc., made of jute, flax and hemp in D., Scotland.

Dungaree: (a) Similar to denim, but the warp and weft are usually in the same colour. Hard earing cotton fabric for dungarees, overalls and jeans children's cloth; (b) A navy blue jean, used in England for sailors clothes; made also for the export trade, chiefly in India. The filling is dyed in the yarn, while the warp is white; (c) Blue drill in South Africa.

Dunging: A process in the mordanting of cotton fabric resulting in fixing the mordant to the fabric. Various sodium, carbonate of ammonium, etc., are used. It is obsolete.

Duotuft: Special tufting technique in which some of the pile loop that has already been tufted is re-engaged. This part of the pile loop is then pulled into the next pile loop. This improves cover whilst being relatively economic in pile yarn consumption.

Dupatta: (1) A veil-like unstitched piece of cloth worn by Indian women, draped loosely around the upper part of the body.

(2) Fine East Indian cotton muslin; used for veils and shawls.

Duplex machines: (a) In general, two machines running in tandem; (b) Special sanforizing or Montfort's compressive shrinkage machines with two felt calenders; (c) Special roller printing machines for printing both sides of a textile fabric.

Duplex printing: Double-sided patterning is possible using the duplex printing process, which can be carried out in both screen and roller printing. There are two screens and/or rollers per pattern and these print the pattern identically on both sides of the fabric. This is predominantly used for blankets and towels.

Duplosolv process: A continuous two-stage process for desizing fabrics which have been sized with polymer sizes. In the first stage, the size is swollen with water (80–150% based on weight of fabric) and brought into solution. In the second stage, this solution is displaced by tetrachloro-ethylene. The size can be regenerated from the aqueous solution. It is manufactured by Brückner.

Duplicate (in experimenting or testing): To repeat a run so as to produce a duplicate.

Dupont bleaching system: Cotton and cotton/synthetic fibre blends are bleached with hydrogen peroxide using J-boxes as reaction dwell chambers. After impregnation with bleach liquor, the goods are heated to the bleaching temperature by steam in a steam chamber immediately prior to entry to each J-box.

Dupont minute bleaching system: A two-stage continuous scouring and bleaching process. (a) Alkali scour stage; (b) Peroxide bleach stage. Each stage involves a 2 minute steaming treatment in a normal Mather-Platt continuous steamer.

Dupont shrinkage test (boiling test): A test method for determining the dimensional stability of textiles made from synthetic fibres from treatment in boiling wash liquors or for determining the efficiency of heat setting. A square of specific dimensions is drawn on the material, after which it is boiled for 30 minutes in soft water in a washing machine, hydro-extracted, allowed to dry in air and the square measured again. The difference in the dimensions of the square before and after the wash test provides a measure of the dimensional stability of the material.

Durability: A relative term for the resistance of a material to loss of physical properties or appearance as a result of wear or dynamic operation.

Durable finish: Any type of finish that is reasonably resistant to normal usage and to washing or dry-cleaning (or both).

Durable press: A term describing a garment that has been treated so that it retains its smooth appearance, shape and creases or pleats in laundering. In such garments, no ironing is required, particularly if the garment is tumble-dried. Durable press finishing is accomplished by several methods; two of the most common are the following – (a) A fabric that contains a thermoplastic fibre and cotton or rayon, may be treated with a special resin that, when cured, imparts the permanent shape to the cotton or rayon component of the fabric. The resin-treated fabric may be pre-cured (cured in finishing and subsequently pressed in garment form at a higher temperature to achieve the permanent shape) or post-cured (not cured until the finished garment has been sewn and pressed into shape). In both cases, the thermoplastic fibre in the garment is set in the final heat treatment. This fibre, when heat-set, also contributes to the permanence of the garment shape, but the thermoplastic component of the blend is needed for strength since the cotton or rayon component is somewhat degraded by the durable-press treatment; (b) Garments of a fabric containing a sufficient amount of a thermoplastic fibre, such as polyester, nylon, or acrylic, may be pressed with sufficient pressure and time to achieve a permanent garment shape. See **Ease-of-care, Permanent finish,** and **Wash and wear**.

Durable press: Having the ability to retain substantially the initial shape, flat seams, pressed-in creases and unwrinkled appearance during use and after laundering or dry-cleaning.

Dura colour: The trade name for a process applied to curtain fabrics which guarantees that they will not fade.

Duralized: A commercial term for the effect produced on a fabric, preferably of pure cotton, by treatment in liquid ammonia (duralized process). Its manufacturer are Cluett, and Peabody & Co. (licensor).

Duro (durometer): A measure of the hardness of plastic, rubber, or similar material, made an instrument called a durometer. These instruments test the material with a steel pin, and the depth of the indentation made by the pin is read by a calibrated spring meter. There are various versions of the durometer, designed for materials of different kinds; to be called durometers, all versions must conform to standards set by standards organizations such as the American Society for Testing and Materials (ASTM). Durometer scales run from 0 to 100, with larger numbers indicating harder material. The different scales are identified by letters, including A, B, C, D, O, DO, and OO. However, since only one type is used for a particular material, the scale designation is often omitted. Typical notations used are "50 duro", "65 durometer A", and "D40 duro".

Duroplastics (Duromers): Plastics (synthetic resins) made from three-dimensionally cross-linked macromolecules that can be thermally hardened by means of cross-linkage. The thermal hardening process is irreversible (compared to thermoplastics). These materials cannot be melted and are insoluble in organic solvents, e.g. phenol-formaldehyde resins of the Bakelite type. Also logically includes rubber vulcanisation.

Duree quilt: Made with large patterns formed by coarse thread on a plain woven ground; made bleached or in colours.

Duria: Striped cotton muslin from India, made with a two-ply warp. **Durois**: Stout French coating of worsted yarn, made with a smooth finish. **Durry**: Cotton carpets woven by the natives in India.

Dust: Suspended particulate matter in gases or in the air that is able to remain in suspension for a relatively longer time due to its low particle size.

Dust extraction units: It is used for printing machines in order to remove lint, loose fibres and other loosely adhering matter from the surface of textile fabrics to be printed (which could impair printing quality, e.g., local resists and screen blocking problems, etc.). A dust extraction unit is installed, e.g., immediately before the printing zone.

Dust eliminator; Lint eliminator (for printing): There must be no dust particles on the print ground while it passes through the printing zone, as these would act as resist agents. The ground is therefore cleared of dust by suction or adhesive techniques.

Dustprone fabrics: Goods that are prone to dust due to the fact that quantities of non-bound finishing agents (kaolin, delustring agents etc.) are too high.

Dust ruffle: A fabric which (a) lies flat over the box spring under the mattress on bed, and (b) has a pleated, tucked, or gather drop to the floor.

Duster: (1) Thin cotton or linen fabric, made dust-proof; used for coats.

(2) A light overcoat of linen, mohair, or other thin fabric, once extensively worn while traveling or driving.

Dusting: The second step in commercial wool processing (after sorting). The purpose is to remove as much dirt and sand as is possible before scouring.

Dutch camlet: Holland dress goods, warp and filling made of combed wool and rabbits' hair; the warp is two-ply and thinner than the filling, which forms cross-ribs.

Dutch carpet: Made with worsted warp which runs over and under a single filling; otherwise similar to the Venetian carpet.

Dutch tape: Trade term in England for linen tapes, the width graded from 11 to 15d, according to the number of threads supposed to be in it.

Duvedelain: See **Velour**.

Duvetine (duvetyn, duvetyne): It is derived from the French word *duvet* which means "down". Originally of cotton warp and spun silk weft. Sometimes made of fine woollen yarns. Constructed of six-, seven-, or eight-shaft satin weave and face finished to give a smooth, plush texture. The down-like surface, produced by several napping, shearing and brushing operations is characteristic of duvetine fabrics. Today it is made with a cotton warp and spun rayon or synthetic filling, or sometimes entirely of cotton. A characteristic texture will be 70 × 150, with 2/60s warp and 25s weft. Since practically the entire surface of the cloth is formed by the sift weft, it can be given an extremely dense suede finishing, resembling that of the fine velvet. In comparison to velours, the surface pile is shorter and less pressure-sensitive. Duvetine is mainly dyed in the piece. Soft handle and appearance similar to suède, drapes well, wears well, but spots easily. Used for ladies' outerwear, costumes and coats.

Duvetyne: See **Duvetine**, **Duvetyn**.

Duvetyn: Similar to Doeskin, but more velvety and lighter in weight. The soft velvet-like surface is made by napping, then shearing and brushing the right side. The weave is twill and usually wool, but may be other fibres also.

Dwell chambers: In order to complete the process, in many finishing stages it is necessary to batch the fabric or knitwear impregnated with chemicals for a period of time. This can take place in chambers of air, steam or gas. Two transport systems are used for this: conveyor belts and roller conveyors. The conveyor belts can be arranged in one or several levels above one another. This is the most common design. In addition, there are carousel type devices and towers, in which the conveyor belt is arranged in the form of a spiral. Since the time of reaction is given in the chambers, they are called dwell chambers.

Dwell process: Bleaching and dyeing process for fabrics and knitwear, whereby goods impregnated with chemicals or dyestuffs are rolled up in open width and are stored for the necessary length of time at room temperature or in heated chambers (e.g. pad roll chamber).

Dwell time: Phase of the processing cycle during which the actual reaction takes place, e.g. at the final temperature reached.

Dye: Dyes are coloured compounds which are absorbed or adsorbed by the fibre from a solution or suspension where they are subsequently fixed and are thus used for the coloration of textiles. The term also applies to certain colourless compounds (optical brighteners) and those which only form an insoluble dye through combination on or in the fibre (e.g. naphthols and turkey red). Between these groups are the mordant dyes. The final colour obtained on the fibre with these dyes is formed by applying the dye to a textile previously treated with a metal salt (mordant) or vice versa (colour lakes). Pigments do not fit this definition since they are water-insoluble colorants having no affinity for textile fibres; and for this reason, have to be applied with a binder capable of bonding to the fibre. Consequently, pigments must be regarded as a special case in textile coloration in textile terms, a soluble colorant that attaches in molecular form to the fibres, as opposed to a pigment, which exists as much large particles that are attached to the fibre with a binder. Dyes get classified by the application technique used and by their chemical structure. A class of dye based on chemical structure may have members in several different application classes. See acid dye, azonic dye, basic dye, direct dye, disperse dye, reactive dye, sulphur dye, vat dye. All commercial dyes are organic chemicals.

Dye absorption index (fibre affinity index SF): Fibre characteristic which gives information on the maximum dyestuff affinity of an acrylic fibre for cationic dyestuffs.

Dye activator: One dye seller's name for an *alkali* intended for use with reactive dyes; believed to be pure *soda ash*. This term is somewhat misleading. In the case of most reactive dyes on cellulosic fibres, it is the fibre and not the dye that is "activated" (an exception to this is *vinyl sulphone* dyes).

Dye affinity: In simple terms, dye affinity (also neutral affinity) is the dye absorbing capacity (of textile fibres). It varies between wide limits depending on the state of equilibrium between the fibre and the dye in solution. A number of factors play an important role in this process, e.g. liquor ratio, dye characteristics, additions to the dye bath (such as acids, salts and dyeing auxiliaries), dyeing time, dyeing temperature, and the degree of purity, extent of drawing and degree of crystallinity of the fibre being dyed, as well as the number of dye bonding groups available in the fibre.

Dye ager: A horizontal continuous-dyeing machine for all conventional textile fabrics (including difficult qualities such as cotton velvet and lining fabrics) with all classes of dye, by the pad-steam process and the wet-in-wet process of vat dyeing voluminous fabrics in which dye is applied on a preceding padder and the reducing agent padded on directly before the ager entry using a special applicator unit. A high degree of reproducibility is achieved by the ager.

Dye batch: A batch of textile material dyed in a single process cycle (charge).

Dye bath preparation: It is the preparation of a dye bath incorporating additives and chemicals or dyeing assistants using dispersing agents, sometimes solvent accelerators, and in the case of disperse dyestuffs also carriers, ethoxylates, etc., usually in pourable form as dye pastes of varying viscosity. Weighing apparatus may be connected to the dissolving plant.

Dye beam: A perforated cylinder (usually of stainless steel) on which woven or knitted textile fabric to be dyed in a beam dyeing machine is wound.

Dye beam batcher: Special batching unit for batching textile fabrics on to dye beams before dyeing and debatching after dyeing. The debatching operation takes place in the wet state.

Dye carrier: See **Carrier**.

Dye cations: See **Dye ions**.

Dye classification: See **Classification of dyes, Dyes (Classification of)**.

Dye diffusion: (a) The mechanism by which dyes dissolved in the aqueous liquor diffuse through the water-filled pores in the fibres and are adsorbed on to the walls of the pores at the same time. This case applies particularly to cellulosic fibres and porous acrylic fibres; (b) The mechanism by which dissolved or dispersed dye enters the fibre through the less ordered and amorphous regions of a polymer substance. The rate of diffusion of the dye molecule is determined in this case by the mobility of the segments in the polymer chain molecules. The segment mobility starts at the glass transition temperature at which point the non-crystalline regions of the fibres or polymers are transformed from a glass-like into a viscoelastic state. Free-volume diffusion is particularly dominant in the case of polyamide.

Dye equilibrium: See **Dyeing equilibrium**.

Dye fixation efficiency in exhaust dyeing: An estimate of the average proportion of dye actually fixed on a textile substrate in exhaust dyeing. The same study can be made by investigating the proportion of unfixed dye remaining after the completion of dyeing. The proportion of dye lost in spent dye baths and rinse water are as follows:

(a)	Sulphur dyes	30–40%
(b)	Reactive dyes	20–50%
(c)	Vat dyes	05–20%
(d)	Direct dyes	05–30%
(e)	Naphthol dyes	05–10%
(f)	Disperse dyes	08–20%
(g)	Metal-complex dyes	02–05%
(h)	Cationic dyes	02–03%
(i)	Acid dyes	07–20%

Dye fixing agent: A cationic complex forming resin which is applied on the fabric dyed with direct and reactive dyes to improve wash fastness. The direct dye forms a complex with the fixing agent which is less soluble in water and hence reduce the bleeding and thus improves the fastness of the direct dyeing. In case of reactive dyeing, it works in a little different way. In reactive dye, the wash fastness is affected mainly by the hydrolysed dye which acts like a direct dye, which is complexed with the dye fixing agents. The dye fixing agents are also used in printed fabric washing to avoid tinting the white ground, discharges white area or lightly coloured printed area. The washed off colour, as soon as coming out of the dark coloured printed area, forms complex with these fixing agents thereby avoiding tinting white ground, etc.

$$\text{R-SO}_3\text{Na (direct dye)} \rightarrow \text{R-SO}_3^- + \text{Na}^+$$

$$\text{X–N}^+\text{...-OOC-CH}_3 \text{ (fixing agent)} \rightarrow \text{X-N}^+ + \text{-OOC-CH}_3$$

$$\text{RSO}_3^- + \text{X-+N} \rightarrow \text{RSO}_3^-\text{....}^+\text{N}$$

Dye fleck: (a) An imperfection in fabric caused by residual undissolved dye; (b) A defect caused by small sections of undrawn thermoplastic yarn that dye deeper that the drawn yarn.

Dye ions: Dye ions, which represent by far the greatest proportion of the dye molecule, and which are responsible for the dyeing properties at the same time, dissociate in aqueous solution. The resultant dissociated ions of the considerably smaller fraction of the molecule, which have opposite charges, are referred to as counter ions. All dyes which form dye salts or dye acids contain both types of ion together.

Dye jigger: See **Jigger**.

Dye kitchen: A part of the dye house where the dye or chemical is weighed, mixed along with necessary auxiliary, dissolved or emulsified or made into paste form, in case of printing, and made ready to be supplied to the bleaching, dyeing, printing or finishing machines. The dye kitchen can be automatic or manual or semi-automatic. See **Colour kitchen**.

Dye liquor: The liquid that contains the dye and the reagents necessary for dyeing. The dye liquor or solution consists of liquid medium (usually soft water), the dye and the additives necessary for the dyeing process in question such as levelling agents, retarding agents, padding auxiliaries and dispersants, etc. Concentrated liquor is used for padding textile fabrics. Dilute liquor is used for exhaust dyeing.

Dye migration: (a) Reduction of depth of colour in wool dyeing. Dyestuff migrates across to adjacent undyed wools. Migration test –it is a test for migration propensity of a dye; (b) In union dyeing, the migration of individual dyes to specific fibres, e.g., acrylic/wool mixes – cationic dyestuffs to acrylic, anionic dyestuffs to wool.

Dye migration in drying: The movement of dye molecule from one face to another face or on the surface or form one place to another place of any textile material during the drying of the same. This undesirable phenomenon causes previously level dyeing to become uneven due to incorrect drying. This migration phenomenon (migration during intermediate drying) mainly arises when dyeing with dyestuffs with low or no fibre affinity (but also in the case of proofing, synthetic resin finishes or similar). Typical dyeing processes include in particular the undeveloped pigment padding and leuco vat ester dyeings, non-coupled naphthols, and also disperse and similar direct dyeing.

Dye mist: A dye liquor which has been atomized by passing it through fine jets.

Dye mixtures: Dyestuff formulations pre-constituted by the dyestuff manufacturer for various fibre blends, for example:

(a) Wool/cellulose – mixtures of acid dyes and direct dyes with neutral or weakly acid affinity.

(b) Wool/polyester – mixtures of acid dyes and/or 1:2 metal-complex dyestuffs and/or reactive and disperse dyes.

(c) Cellulose/polyester – mixtures of vat and/or leuco vat ester dyes and/or reactive and/or sulphur dyes and disperse dyes.

Dye recipe formulation: Before dyeing a fabric, the most cost-effective recipe must be established, i.e. with the fewest possible or least expensive dyes and the shortest possible dyeing time. Thus, a dyeing process must be found which could minimise the recipe costs (dyes and auxiliaries) and the dyeing time for the particular dyeing machines available. The selected dye recipe must also be capable of achieving the desired requirement profile in terms of colour matching and fastness properties against the approved sample. A record is kept of the established recipes for each article number and the quantities are "scaled up" appropriately for a specific order.

Dye range: A dye range is usually a continuous dyeing machine consisting of a dyeing unit, an after-treatment unit, a soaper and a drying unit running in tandem. Its example is a pad steam unit which will have a padder, steamer, oxidizing unit, soaper, and drier.

Dye resisting agent: See **Dye retarding agent**.

Dye sites: Functional groups within a fibre that provide sites for chemical bonding with the dye molecule. Dye sites may be either in the polymer chain or in chemical additives included in the fibre.

Dye solubility: It is the extent to which the dye dissolves in the dyeing media, usually water. The dye solubility is affected by the other additions in the dyebath, like salt, auxiliaries and such chemicals. Hence, the dye solubility has to be checked along with other chemicals/auxiliaries in the dye bath for the type of dyeing to be followed. Dye solubility is important in pad dyeing methods where dye solutions of very high concentrations have to be used, especially in case of dark shades.

Dye springs: Dye centres which intermesh elastically and can be compressed slightly. Often used in the place of perforated tubes for dye packages in order to achieve better dye penetration.

Dye stain (defect): A discrete area of a colour different from that of the adjacent parts of the fabric.

Dye stick: See **Dyeing stick**.

Dye streak: See **Dye streak**.

Dye strength: The colour value of a particular dye, especially in dyeing. In the manufacture of dyes, the strength of dye from different batches is maintained constant by adding dispersants, electrolytes or by mixing different batches. Commercial dyes are always checked for dye strength in various supplies by the dyes to avoid dyeing variations.

Dye tubes: Tubes on which yarn is spooled for dyeing in the form of yarn packages (dyeing tubes for yarn dyeing), disposable dye tubes made of plastic, flexible tubes (flexible either axially or radially) for yarn with pronounced shrinking properties, non-disposable dye tubes made of plastic such as Biko dye tubes (which can also be made from stainless steel) and dyesprings.

Dye tube (split): Flexible dye tube made from perforated stainless steel, slit open in order to make it radially flexible.

Dye uptake: (a) The dye uptake is the quantity of dye absorbed by the fibre;

(b) Substantivity, exhaustion (diffusion) model.

N Dye-weave process: Polychromatic dyeing process in which the dye is supplied by moving nozzles. The jets of dye solution are directed onto the surface of an angled plate, from where they impinge upon the substrate and are then squeezed out on a vertical two-bowl padder.

Dye winch (paddle-dyeing machine, paddle vat): Dye vat of hemispherical or 3/4-spherical form made of wood, copper or stainless steel with a rotatable hank winch or paddle as a "moving device" sitting horizontally and to one side on the top. The ends of the winch or paddles extend down into the top of the liquor and when put into operation, they provide continuous movement to the liquor and material.

Dye works: Dye manufacturing plant (dye factory).

Dye (definition of): See **Dye**.

Dyeability: The capacity of fibres to accept dyes. As a prerequisite for the use of textile fibres, one of their most important properties is dyeability. In practice, the dyeability of a textile fibre is determined by the rate of dyeing and the degree of saturation which can be achieved. For example, the different dyeing properties of mercerized cotton compared to non-mercerized cotton can be characterized by results from the time of half dyeing (i.e. the reciprocal of the rate of dyeing) and the saturation concentration of the substantive dye. The range of application of some synthetic fibres has been, and still is, limited either because they have poor dyeability or the problem of dyeing these fibres has not yet been satisfactorily solved. For example, when 21/2 acetate fibres were first introduced, they were not successful for a long time because they could not be dyed with the known dye classes at that time.

Dye bath exhaustion: The adsorption of the dye from the dyebath on to the fibre. Usually some accelerators are added to the dye bath to increase the exhaustion of the dye. Example is the addition of the electrolyte in the direct and reactive dyeing of cotton. The exhaustion of the dye bath depends on the dye, fibre and exhaustion accelerators, bath temperature, etc. The dyebath exhaustion is proportional to the dye fixation.

Dyed and printed: The process of dying the ground shade and printing. Dyed print can be a direct style printing (over print) of a discharge printing. When over printing, the ground shade can have an effect on the final print shade.

Dyed in the grease: Large number of low grade worsteds are dyed as they come from the loom, without being scoured first. Dyeing in the grease is used only for black and blue colours.

Dye house: Dyestuff formulations. An industrial (textile dye house) or commercial plant (garment dyeing) for the dyeing of textiles.

Dye house laboratory: A laboratory under the control of the dyehouse mainly intended for carrying out laboratory dyeing and tests in addition to dyeing related analytical procedures in many cases.

Dyeing: The process of applying a comparatively permanent colour to fibre, yarn or fabric by immersing in a bath of dye. Dyeing is carried out in a dyehouse with dyestuffs and is used in (textile) finishing. The process depends on the type of fibre and form of the material in dyeing vessels, dyeing apparatus or machines. The dye is applied to a given substrate by a technological process, mainly consisting of deposition or adsorption and fixation (both processes being carried out simultaneously or in succession). If necessary, the process is preceded by pre-treatment and followed by post-treatment of the substrate.

Dyeing (batik): A resist-dyeing process in which portions of a fabric are coated with wax; during the dyeing process, only the uncovered areas take up dye. The process can be repeated so that several colours are used. Batik dyeing is often imitated in machine printing.

Dyeing (chain): A method of dyeing yarns and fabrics of low tensile strength of tying them end-to-end and running them through the dyebath in a continuous process.

Dyeing (cross): A method of dyeing blend or combination fabrics to two or more shades by the use of dyes with different affinities for the different fibres.

Dyeing (high temperature): A dyeing operation in which the aqueous dyebaths are maintained at temperatures greater than 100 °C by use of pressurized equipment. Used for many manufactured fibres.

Dyeing (ingrain): Term used to describe yarn or stock that is dyed in two or more shades prior to knitting or weaving to create blended colour effects in fabrics.

Dyeing (mass coloured): A term to describe a manufactured fibre (yarn, staple, or tow) that has been coloured by the introduction of pigments or insoluble dyes into the polymer melt or spinning solution prior to extrusion. Usually, the colours are fast to most destructive agents.

Dyeing (muff): A form of yarn dyeing in which the cone has been removed.

Dyeing (pad): (1) See **Dyeing, Yarn dyeing**.

A form of dyeing whereby a dye solution is applied by means of a padder or mangle.

Dyeing (piece): The dyeing of fabrics "in the piece," i.e., in fabric form after weaving or knitting as opposed to dyeing in the form of yarn or stock.

Dyeing (pressure): Dyeing by means of forced circulation of dye through packages of fibre, yarn, or fabric under super atmospheric pressure.

Dyeing (reserve): (a) A method of dyeing in which one component of a blend or combination fabric is left undyed. The objective is accomplished by the use of dyes that have affinity for the fibre to be coloured, but not for the fibre to be reserved; (b) A method of treating yarn or fabric so that in the subsequent dyeing operation the treated portion will not be dyed.

Dyeing (gel): Passing a wet-spun fibre that is in the gel state (not yet at full crystallinity or orientation) through a dyebath containing dye with affinity for the fibre. This process provides good accessibility of the dye sites.

Dyeing (short liquor): A term used to describe any yarn or piece dyeing in which the liquor ration has been significantly reduced. The technique was designed to save water and energy.

Dyeing (skein): The dyeing of yarn in the form of skeins or hanks.

Dyeing (solution): See **Dyeing, Mass-coloured**.

Dyeing (solvent): A dyeing method based on solubility of a dye in some liquid other than water, although water may be present in the dyebath.

Dyeing (space): A yarn-dyeing process in which each strand is dyed with more than one colour at irregular intervals. Space dyeing produces an effect of unorganized design in subsequent fabric form. The two primary methods are knit-de-knit and warp printing.

Dyeing (stock): The dyeing of fibres in staple form.

Dyeing (thermal fixation): A process for dyeing polyester whereby the colour is diffused into the fibre by means of dry heat.

Dyeing (union): A method of dyeing a fabric containing two or more fibres or yarns to the same shade so as to achieve the appearance of a solid coloured fabric.

Dyeing (yarn): The dyeing of yarn before the fabric is woven or knit. Yarn can be dyed in the form of skeins, muff, packages, cheeses, cakes, chain-wraps and beams.

Dyeing accelerator: See **Carriers**.

Dyeing auxiliaries: The chemicals or a mix of chemicals used along with the main dye or chemicals used in the processing to help in the process or help in the effectiveness of the chemicals used in one way or other, etc. The process can be preparation, bleaching dyeing, printing, finishing, correction of processes or washing or any processes done in the process house. Normally, the details of formulations of the auxiliaries are kept as trade secret. Some groups of auxiliaries will include, mordanting, fastness improving agents, levelling agent, dyeing oils, wetting agents for dyeing and dyeing oils, dye solubilizing and dispersing agents, fibre protections agents, wetting agent, soaping agents, retarding agent, over boiling inhibitor, stripping agent, stripping auxiliary, etc.

Dyeing curves: See **Reflectance**.

Dyeing equilibrium: A concentration equilibrium between the amount of dye on the fibre and the amount of dye in the dyebath which becomes established by the end of every dyeing process.

Dyeing machines: Devices in which the material which is to be dyed (as distinct from dyeing machine, circulating) is agitated in more-or-less stationary liquor, for example –padder, rope-winch vat, jigger, continuous dyeing machine, paddle dyeing machine, star frame and, among others, modern processors such as the pad-jigger, pad-roll, pad-steam, pad-wet and pad-winch processes.

Dyeing parameters: Dyeing parameters are all the factors exercising an influence on the dyeing process.

Dyeing solution: See **Dye liquor**.

Dyeing sticks(dye stick): Dyeing sticks are used for the following purposes –(a) In yarn-dyeing for suspending the hanks (hardwood dyeing sticks made from hazelnut, hickory, etc., with a synthetic resin/ebonite coating as well as aluminium or toughened glass, bent into a U-shape); (b) In garment dyeing for moving/turning the goods during dyeing.

Dyeing time: In the practical sense, the term "dyeing time" must include the entire treatment time required for the batch, i.e. the times required for the pre-treatment, clearing, rinsing and after-treatment stages in addition to the actual dyeing time.

Dyeing vat (beck, tub): These are four-sided vessels for dyeing, bleaching, etc., made of wood, metal, earthenware, tiled or similar for hank yarn and loose stock; may be fitted with two squeeze rollers or with winch for piece dyeing. Heated by means of open and/or closed steam pipe.

Dyeing (batik): See **Batik dyeing**.

Dyeing (cross): See **Cross dyeing**.

Dyeing (high temperature): See **High temperature dyeing**.

Dyeing (ingrain): Ingrain dyes is an incorrect description of group colours produced in *situ* on the fibre by coupling, or reaction of one or more intermediates. The process involved in this method of the production of dye on the fibre is called ingrain dyeing. An example is the naphtolation and coupling process.

Dyeing (jet): See **Jet dyeing**.

Dyeing (methods of): These have been the subject of continual development over the years, especially because of the need to rationalize or for reasons related to colour levelness and dye compatibility or because of new problems appearing with new substrates. Thus, there is a whole series of possibilities for variation such as short-liquor dyeing, foam applications, thermosol processes, jet dyeing (jet-dye machines), overflow dyeing, (overflow dyeing machines), HT dyeing processes, pad-steam dyeing processes and continuous wet-steam processes among others. In order to optimise the dyeing process, the dyeing procedure is controlled according to a program and process control systems are used to control the operations.

Dyeing (muff): Muff is dyed both in accordance with the packing system and in accordance with the creel system in the circulation machine. (a) Short muff – height approximately 15 cm, 650–800 g; (b) Long muff – height approximately 25 cm, 1440–1700 g. See **Muff**.

Dyeing (mass coloured): Mass colouration is also called spin dyeing or jet dyeing. The method involves the mixing of finely distributed dye pigments which can withstand the temperature and process of spinning, to the spinning solution or melted synthetic polymers.

(a) *Continuous process* – dyes (liquid or in paste form) are dispersed in binders (polymers of different molecular weights or polyethylene glycols of high molecular weight) and are injected into the molten polymer using additional extruder, which directly supplies the main extruder, in which polymer granules are melted.

(b) *Discontinuous process for polyester* – in a unit of production for the manufacture of polyester granules, the polyester chips are dried in a vacuum for several hours at 140–180°C in rotary driers or tumblers and at the end of this process specially selected dyes are added; they are fixed on polyester granules via thermos-diffusion. Suitable process for the manufacture of small quantities of spin-dyed fibres/threads.

(c) *Discontinuous process for polyamide*– polyamide granules are simultaneously washed and dyed in the aqueous medium. After several hours drying at 180°C, it is melted under the same conditions as colourless granules. Granule dyeing is carried out in a machine with bath circulation. Used, for example, in polyamide for upholstery fabrics, carpets, floor carpeting, automobile articles; in polyester for upholstery fabrics, menswear, work clothing.

Dyeing oils: A now largely obsolete term for wetting agents used in dyeing which confer additional softening effects on the textile material.

Dyeing parameters: These include all the factors exercising an influence on the dyeing process.

Dyeing (piece): Dyeing of textile in the piece form. See **Piece dyeing**.

Dyeing (reserve): See **Reserve dyeing, Resist dyeing**.

Dyeing (short liquor): Dyeing of textile in low liquor ratio. Short liquor dyeing is advantageous in many ways such as saving of water, energy, time of treatment, reduction in pollution, etc. See **Short liquor dyeing**.

Dyeing (skein): Dyeing of textile yarn in the skein form. See **Skein dyeing**.

Dyeing (solvent): Dyeing of textile using a solvent as the dyeing medium instead of water. In solvent dyeing, usually the solvent is recovered. See **Solvent dyeing**.

Dyeing (space): See **Space dyeing**.

Dyeing (stock): Dyeing of textile in the loose stock or fibre form is called stock dyeing. See **Stock dyeing**.

Dyeing (union): Dyeing of textile containing wool. See **Union dyeing**.

Dyeing (yarn): Dyeing of textile in the yarn form. See **Yarn dyeing**.

Dye liquor: It is also called dyebath. The dye liquor consists of liquid medium (usually soft water), the dye and the additives necessary for the dyeing process in question such as levelling agents, retarding agents, padding auxiliaries and dispersants, etc. Concentrated liquor is used for padding textile fabrics. Dilute liquor is used for exhaust dyeing.

Dyer's spirit: Aqua fortis, 10 parts; sal ammoniac, 5 parts; tin, 2 parts; all these are dissolved together.

Dyes (acid): A class of dyes used on wool, other animal fibres, and some manufactured fibres. Acid dyes are seldom used on cotton or linen since this process requires a mordant. Acid dyes are widely used on nylon when high wash-fastness is required. In a few cases, even higher wash-fastness can be obtained by after-treatment with fixatives.

Dyes (aniline): Dyes derived chemically from aniline or other coal tar derivatives.

Dyes (anthraquinone): Dyes which have anthraquinone as their base and the carbonyl group (>C=O) as the chromophore. Anthraquinone-based dyes are found in most of the synthetic dye classes.

Dyes (azo): Dyes characterized by the presence of an azo group (-N=N-) as the chromophore. Azo dyes are found in many of the synthetic dye classes.

Dyes (azoic): See **Dyes, Naphthol dyes, Dyes, Naphthol**.

Dyes (basic): A class of positive-ion-carrying dyes known for their brilliant hues. Basic dyes are composed of large-molecule, water-soluble salts which have a direct affinity for wool and silk and can be applied to cotton with a mordant. The fastness of basic dyes on these fibres is very poor. Basic dyes are also used on basic-dyeable acrylics, modacrylics, nylons, and polyesters, on which they exhibit reasonably good fastness.

Dyes (cationic): See **Dyes, Basic dyes, Dyes, Basic**.

Dyes (developed): Dyes which are formed by the use of a developer. The substrate is first dyed in a neutral solution with a dye base, usually colourless. The dye is then diazotized with sodium nitrate and an acid and afterwards treated with a solution of *B*-naphthol, or a similar substance, which is the developer. Direct dyes are developed to produce a different shade or to improve wash-fastness or light fastness.

Dyes (direct): A class of dyestuffs which are applied directly to the substrate in a neutral or alkaline bath. They produce full shades on cotton and linen without mordanting and can also be applied to rayon, silk and wool. Direct dyes give bright shades, but exhibit poor wash-fastness. Various after-treatments are used to improve the wash-fastness of direct dyes, and such dyes are referred to as "after-treated direct colours."

Dyes (disperse): A class of slightly water-soluble dyes, originally introduced for dyeing acetate and usually applied from fine aqueous suspensions. Disperse dyes are widely used for dyeing most of the manufactured fibres.

Dyes (fibre reactive): A type of water-soluble anionic dye having affinity for cellulose fibres. In the presence of alkali, they react with hydroxyl groups in the cellulose and thus are liked with the fibre. Fibre-reactive dyes are relatively new dyes and are used extensively on cellulosic when bright shades are desired.

Dyes (macromolecular): A group of inherently coloured polymers. They are useful both as polymers and as dyes with high colour yield. The chromophores fit the recognized CI classes, i.e., azo, anthraquinone, etc., although not all CI classes are represented. Used for mass dyeing, hair dyes, writing inks, etc.

Dyes (metallised):A class of dyes which have metals in their molecular structure. They are applied from an acid bath.

Dyes (naphthol): A type of azo compound formed on the fibre by first treating the fibre with a phenolic compound. The fibre is then immersed in a second solution containing a diazonium salt that reacts with the phenolic compound to produce a coloured azo compound. Since the phenolic compound is dissolved in caustic solution, these dyes are mainly used for cellulose fibre; although, other fibres can be dyed by modifying the process.

Dyes (pre-metallized): Acid dyes that are treated with coordinating metals such as chromium. This type of dye has much better wet-fastness than regular acid dye. Pre-metallized dyes are used on nylon, silk and wool.

Dyes (Sulphur): A class of water-insoluble dyes that are applied in a soluble, reduced form from a sodium sulphide solution and are then re-oxidized to the insoluble form on the fibre. Sulphur dyes are mainly used on cotton for economical dark shades of moderate to good fastness to washing and light. They generally give very poor fastness to chlorine.

Dyes (vat): A class of water-insoluble dyes which are applied to the fibre in a reduced, soluble form (leuco compound) and then re-oxidized to the original insoluble form. Vat dyes are among the most resistant dyes to both washing and sunlight. They are widely used on cotton, linen, rayon, and other cellulosic fibres.

Dye spot: See **Dye stain**.

Dyestain: An area of discolouration due to uneven absorption of a colourant.

Dyestreak: An unintended stripe in a fabric due to uneven absorption of a colourant.

Dyestuff (classification of): (a) For cotton, viscose, modal fibres and linen – direct (substantive) dyes: anionic dyes which exhaust directly on to the fibre, i.e. without pre-mordanting. They have only moderate wash fastness and some have very high light fastness.

- Reactive dyes: Anionic dyes which exhaust directly and react chemically with the fibre. Consequently, they are very fast to washing and offer a brilliant range of shades.
- Vat dyes: These are water-insoluble pigments which are dissolved (vatted) by reduction and dye the fibre in the reduced form. After subsequent oxidation, the original pigment is restored on the fibre. Vat dyes are very fast to washing and light (Indanthren) but there are no brilliant shades.
- Sulphur dyes: some sulphur dyes are available as unmodified pigments which are dissolved by reduction during dyeing, others are available in a pre-reduced form. Subsequent oxidation restores the original pigment on the fibre (analogous to vat dyes). Economical, but the shade gamut is restricted to browns, olives, dark blues and blacks.
- Azoic colorants: the dye is formed on the fibre in two stages, i.e. by pretreatment with a naphthol followed by a second treatment with a diazonium salt. High fastness properties.
- Pigments: these are deposited on the fibre and fixed mechanically with binders. They have low rub fastness but are very fast to light. They are used for dyeing pale shades and for printing all depths of shades.

(b) For wool, silk and polyamide fibres: – Acid dyes: anionic dyes. These include acid levelling dyes with moderate wash but high light fastness and milling dyes with high wash fastness.

- Chrome dyes: anionic dyes which are bound to the fibre in the form of chromium omplexes by means of an after treatment with chromium salts. Very high wash fastness.
- Metal-complex dyes: anionic 1 : 1 or 1: 2 metal dye complexes. They are characterized by high wash fastness and very high light fastness but include no brilliant shades.
- Reactive dyes: see 1, used especially for machine washable wool.

(c) Dyes for acetate, triacetate and polyester fibres:

- Disperse dyes: these are non-ionic dyes which are only sparingly soluble in water and are therefore applied as dispersions.

(d) For acrylic fibres: Cationic dyes: these are characterized by high wash and light fastness.

- Another classification system for dyestuffs is based on their chemical structure.

Dyestuff formulation: Depending on dyestuff type (reactive, disperse, vat, direct dyestuff) and application type (exhaust dyeing, printing, continuous), dyestuff formulations contain specific additives which give the commercial form the best possible properties in respect of storage stability, handling, results, etc. The usual additives are as follows:

(a) Dispersants, Surfactants – lignin sulphonates, naphthalene sulphonate, formaldehyde addition compounds, ethylene oxide-propene oxide-copolymers.

(b) Salts – sodium sulphate, sodium chloride.

(c) Dust-bonding agents – mineral oils, paraffin oils (+additives).

(d) Antifoams –ditertiary acetylene glycols.

(e) Antifreezes – glycerine, glycols.

(f) Thickeners carboxymethyl – cellulose polyacrylates.

(g) Buffer systems – phosphate, acetate.

Dynamic adhesion: The ability of a cord-to-rubber bond to resist degradation resulting from flexure.

Dynamic range: An instrument's range of measurable values, from the lowest amount it can detect to the highest amount it can handle.

Dynapoint process: A continuous computer-controlled process for manufacturing tufted carpets with intricate patterns from undyed yarn. The carpet is dyed as it is tufted, and the colours and patterns are clearly visible through the primary backing of the carpet.

Dynel: Trade name for acrylic fibre, which has the advantage of being flame resistant. It is warm handling and bulk well, and is often used in fur fabrics.

E

E Modulus: Modulus of elasticity. See **Tensile elasticity**.

EA: Elastane fibres.

Equivalent: A term applied to two or more methods, procedures, etc., expected to give the same average results.

Earth colours: The colours found in nature. Natural earth displays itself in a vast array of different colours dependent on the chemicals or minerals present. For example, carbon or magnetite will produce a black earth, cobalt, magnesium or zinc has a blue effect; iron silicate or ferrous oxide has a green effect; iron oxide or lithium a red effect; feldspar, kaolin and quartz produce a whitish earth and limonite and ferric oxide produce a yellow effect; the whole range of oranges, ochres, russets, umbers and browns.

Earth Flax: Same as **Asbestos**.

Ease-of-care: A term used to characterize fabrics that, after laundering, can be restored to their original appearance with a minimum of ironing or other treatment. An ease-of-care fabric generally wrinkles only slightly upon laundering. See **Durable press** and **Wash and wear**.

East improved: A commercial variety of late maturing cotton from Georgia, yielding 31–32% of long staple.

Eastern pulled wool: Wool is pulled from the skins after it has been loosened, usually be a depilatory. Pulled wool should not be confused with dead wool.

Easy care: See **Ease-of-care**.

Easy care finishes: In order to eliminate creasing (wrinkling) of cotton fabrics during wear and service (washing, drying, ironing) as far as possible, i.e. to make ironing or pressing superfluous if possible, cotton materials are treated with cross-linking agents. See **Wash and wear finishing**.

Eau de javelle (eau de Javel): Named after the district of Javel in Paris. Originally a clear aqueous solution of potassium hypochlorite also containing potassium chloride. Used for bleaching cellulosic fibres. Potassium hypochlorite solution has now been largely substituted by a cheaper but similar sodium hypochlorite solution under the same name.

Eau de nil: Light green; literally 'water of the Nile' and hence called Nile green.

Eaude labarraque: Sodium hypochlorite solution named after its French discoverer. Also, commercially available as Eau de Javelle.

Ebonite: Hard rubber.

Ebru' process: The precursor to Aquagraphics. This is an 800-year-old Turkish printing method which uses animal waste products including ox bile and boiled papyrus. Method as for aquagraphics.

Eburnean: Having the colour of or resembling ivory.

Ecaille: (a) French silk dress goods made with ply warps, six leaves and 32 fillings in a repeat; (b) Fish scale like work, made of flat quills over silk or velvet.

ECE: Europäische Convention für Echtheitsprüfung (European Convention on Fastness Testing).

Echantillon: Clipping; sample of a textile fabric.

Ecobalance: Comprises the description and listing of the primary effects of specific processes on the environment, i.e. on air, water and soil, taking additional account of energy and raw material requirements. An eco-balance should express by means of objectively produced figures the effects on humans, animals, plants and their environment occurring over the entire life cycle of a product.

Eco-labelling of textiles: For identifying the physiological and toxicological properties of the product in wear, disposal or recycling of "used" articles of clothing, ecology and human ecology of the production process.

Ecolabels: An ecolabel with well-founded, verifiable criteria provides the industry with the motivation for ecologically optimising textile products, security for the trade and confidence for the consumer in the textile product purchased. In this way, it fulfills the producer's obligations as to information on the one hand, and the buyer's safety requirements on the other.

Ecology: (Greek: oikos = dwelling; logos = science). Concept of the science of relationships in the living organism/environment system introduced in 1866 by Ernst Hackel. Ecological equilibrium is based on the natural capacity for self-regulation between different links of a long-term relationship (xenobiosis). This equilibrium is affected, disrupted or even broken up by massive environmental pollution. Ecological problems upset nature's equilibrium.

Economy trough: Padding trough normally supplied with high performance padders. Low liquor content to prevent tailing problems, etc. The fabric immersion path is designed to be as long as possible and the trough liquor volume is reduced by means of a divider.

Econ-Tex: This Babcock scouring machine looks like a roller vat at first glance (Fig. 1). It has actually taken over the roller vat's vertical fabric guiding system in single and double loop feed, with the possibility of scouring sections with different fabric contents, which can be adapted to different function definitions.

Ecossais: French and English dress goods made of all worsted, also with cotton warp with bright coloured plaid designs.

E-control dyeing: The E-control concept has been developed jointly by Monforts and Zeneca colours to provide a simple, rapid and economical continuous coloration process with minimum chemical usage. E-control brings together the innovation of the Thermex Hot Flue from Monforts with Procion reactive dye technology from Zeneca. The innovation utilises the physical laws of water evaporation from cellulose to provide the optimum temperature and moisture conditions within the Thermex Hot Flue drier, ideal for the efficient fixation of the specially selected Procion dyes. The control and maintenance of the humidity level in the drying process avoids the use of large quantities of chemicals such as urea and sodium chloride (salt).

Eco-tex standard 100: An ecological quality mark should embrace the following ideas – (a) a textile product, the nature of which is free of harmful substances and is skin-compatible;(b) a method of production (fibre cultivation, fabric production, finishing and making-up) which is perfectly ecologically harmless and environmentally compatible; (c) easy to dispose of in order to get rid of the used textile product by recycling, decomposition or combustion.

Ecru: (French for raw, unbleached). In general, the colour of fibres, yarns, or fabrics (especially of linen) that have not been subjected to processes affecting their natural colour, i.e. untreated, unbleached, undyed. Specifically: (a) ecru linen, unbleached linen; (b) ecru silk (bast silk) is barely degummed silk (maximum 4% loss) – only grease, wax and resinous substances removed.

Ecru lace: The geometrical designs are composed of ecru coloured plain and crinkled tape, connected with braids.

Ecru silk: Has only the most soluble part of the natural gum removed together with the colouring pigments.

ED: See **Elastodiene fibres**.

EDCO: Editing Committee. Special committee responsible for the editing of ISO recommendations.

Edelweiss lace: Another name for Aetzstickerei.

Edge abrasion resistance: See **Flexing abrasion resistance**.

Edge crimping: See **Texturing, Edge crimping method**.

Edge curling: A fabric defect which arises in the processing and finishing of woven and knitted fabrics due to faulty construction. In woven fabrics, it is due to excessive tension of warp threads in the selvedges (tight selvedges) or an unsuitable woven construction. Knit goods, in general, are particularly prone to edge curling.

Edge curling of knit goods: Problem encountered in processing knit goods in the open width state. The cause is mainly due to tightness of the fabric edges and is largely prevented by Edge gumming.

Edge gluing device: Device for gluing the edges of open-width knitted fabrics.

Edge guider (Expander): Selvedge guiding device; Edge uncurling device; Edge guiding sensors; Overfeed devices; Selvedge monitor, device to remove the fabric using clip openers or to lift the fabric from stenter pins and fabric take-off mechanism.

Edge gumming: A necessary measure to prevent the Edge curling of knit goods. The edge adhesive is applied from a special edge gluing device by spraying and drying, or by roller application from the edge gluing units on a stenter.

Edge roll: The curl that develops on the edge of a single-knit fabric preventing it from lying flat.

Edge-stitched seam-finish: A finish for the raw edges of a plain seam in which machine stitching is placed close to the raw edge of each seam allowance.

Edge stitching: (French: piquer = to quilt), lockstitch-type sewing of padding and its permanently elastic binding with the outer fabric (by means of an edge stitching machine).

Edge to centre levelness: Side-to-centre levelness of shade across the width of a textile fabric. An important quality parameter in continuous dyeing which can only be ensured by completely uniform dye uptake across the entire fabric width in pad dyeing.

Edge to edge unevenness: See **Listing**.

Edge trimmers (with suction device): Used to trim the needled edges of the run of cloth and to simultaneously aspirate the trimmed selvedges on exit from stentering and drying machines.

Edge uncurling device: Device used to remove the curl from selvedges that have a tendency to roll and make processing difficult.

Edge wire: A length of wire or monofilament yarn, drawn in through a heald and working as a warp thread at one or both edges of the warp, in a narrow fabric loom.

Edge-stitched seam-finish: A finish for the raw edges of a plain seam in which machine stitching is placed close to the raw edge of each seam allowance. See ***Top-stitching***.

Edging: (a) A narrow strip of material attached to another fabric or made up article by one edge usually for decorative purposes.

(b) Overstitching along the edge of the fabric to avoid fraying or ornamentation.

Edging (Fr.): Binding, edge binding and border on woven textiles and knitwear.

Edisto: The best grade of Sea Island cotton that has a very long, fine uniform and silky staple.

Edredon: Pine soft woollen woven in three-leaf twill, made with a filling thicker than the warp.

EDTA: Ethylene diamine tetra acetic acid. It is important in inorganic chemistry because the carboxylate ion can act as a ligand in forming complexes. EDTA is a hexadentate ligand, able to coordinate to a metal ion at the four COO– groups and the two nitrogen atoms. It is able to form octahedral complexes with certain metal ions. Complexing agent which can be used as a water softening agent by direct addition to hard water.

EEE pretreatment system (eee = efficiency, economy, ecology): Sandoz, Basle, Switzerland, offers the eee system, a series of tried and tested process combinations for all pretreatment requirements, which comply with strict quality control measures.

Effect side (weft knitted fabric): The side of a fabric intended to use outermost in a garment or other construction.

Effect thread: Yarns inserted in a fabric which are sufficiently different fibre, count or construction to form or enhance a pattern.

Effect yarn: Warp or weft yarn introduced into the cloth for the only purpose to produce certain effects, as in colour, knots, loops, etc.

Effective carriage mass: In CRL-type tensile testing machine – The force actually applied to a specimen by mass of carriage, plus added masses.

Effective carriage weight: A load actually applied to a fabric specimen by mass of the carriage plus added weight when used.

Effective fibre length (in tensile testing): The estimated length of the specimen subjected to a strain equal to that observed for the true gauge length.

Effective fibre length (in vibroscope test for linear density): That portion of the fibre free to vibrate between fixed supports or holders.

Effective gauge length (in tensile testing): The estimated length of the specimen subjected to a strain equal to that observed for the true gauge length.

Effective insulation ratio (in thermal transmittance of textile): The increase in insulation afforded by the fabric in comparison to the uncovered test plate under specified conditions of test.

Effectivity Factor(EF): The effectivity factor is extremely useful for the (relative) evaluation of the finish effect of various resin treatments:

$$EF = (A - B)/C$$

A = Dry crease recovery angle of resin finished fabric

B = Dry crease recovery angle of non-resin finished (unfinished) fabric

C = Loss of breaking strength of warp in %

Optimum values lie between 2.5 and 3.5. EF values > 4 indicate under-cured goods; values < 2 indicate over curing.

Efficiency of cleaning (of spinning machinery): During the cleaning operation of the fibres, some unavoidable fibre loss occurs during mechanical cleaning. When considering this fibre loss, normally the effective cleaning (EC) of a machine or a sequence of machines are considered

$$EC = 100.\ (W_t - W_f)/W_t$$

Wt = mass of waste; Wf = mass fibre in the waste.

Effilure: French term for unravelling of edges of the cloth.

Efflle: French for fringed.

Efflorescence: Problem of finishing agents crystallizing out as powdery grey deposits on textile surfaces.

Effluent treatment (preliminary): Removal of waste water constituents such as rags, sticks, floatables, grit and grease that may cause maintenance or operational problems with the treatment operations, process and ancillary systems.

Effluent treatment (primary): Removal of a portion of the suspended solids and organic matter from the wastewater.

Effluent treatment (advanced primary): Enhanced removal of suspended solids and organic matter from waste water. Typically accomplished by chemical addition or filtration.

Effluent treatment (secondary): Removal of biodegradable organic matter (in solution or suspension) and suspended solids. Disinfection is also typically included in the definition of conventional secondary treatment.

Effluent treatment (tertiary): Removal of residual suspended solids (after secondary treatment) usually granular medium filtration or micro-screens. Disinfection is also typically a part of tertiary treatment. Nutrient removal also is usually included in this definition.

Effluent treatment (advanced): Removal of suspended materials remaining after normal biological treatment when required for various water reuse applications.

Eggshell: A pale beige; the colour of the shell of a hen's egg.

Egyptian blue: An ancient rich purplish-blue or turquoise colour made by heating copper, lime and quartz, also known as 'Egyptian frit'. Perhaps the first synthetic pigment. Sometimes called 'Pompeian blue'.

Egyptian cloth: Cotton cloth for wool embroidery.

Egyptian cotton: Fine top-quality cotton, originally from the Nile region; Plain weave soft cotton made into fabrics that could be plain or printed. It is used for making expensive baby clothes, blouses, summer dresses, nightwear, etc.

Egyptian flax: Very long and coarse fibre of reddish hue.

Egyptian lace: An expensive ornamental lace that is hand knotted and often elaborated with beads.

Egyptian red: A dark red colour.

Egyptian tape: A closely woven tape, traditionally made from a R20/2 tex (2/60s) Egyptian cotton yarn.

Egyptienne: (a) French silk dress goods, solid coloured or in various colours; obsolete;(b) Dress goods made of wool, camel's hair and waste silk;(c) Obsolete; dress goods with narrow satin stripes over ribbed ground.

Eiderdown:(a) A soft, elastic knitted fabric, made of thick and soft spun yarn; it is heavily napped on one side; (b) A loose and thick woollen fabric with a deep nap on one side.

Eider: Very soft knitting yarn, made of fine wool.

Eight harness satin: A warp-faced or filling faced weave illustrating that the entire face of the fabric surface that is covered with warp or filling yarn, respectively.

Eight lock fabric(weft knitted): A self-colour or two-colour double faced non-jacquard interlock based fabric, intended for use in either knitwear or as a double jersey fabric.

Eight lock: It is a 2 × 2 version of interlock that may be produced by using an arrangement of two long and two short needles, provided all the tricks are fully cut through to accommodate them and knock-over bits are fitted to the verges to assist with loop formation on adjacent needles in the same bed.

Eis wool or Ice wool: Fine, two-ply worsted knitting yarn.

Ejoo: Very strong and dark coloured fibre, found at the base of the leaves of the sago palm in Malacca. It does not rot in water and is used for cordage, cables, etc.

Ell: 11/4 yards.

Enkatherm: A high temperature flame-proof fibre which is based on poly (terephthaloyl oxalic-bis-amidrazone) (PTO). This fibre is produced by wet-spinning and then drawn.

Eksuti: Cotton cloth made in India of single warp and filling.

EL: Elastane fibres.

Elance suede: A light weight suede suitable for shirts, dresses, etc.

Elastane: A term used to describe fibres that are composed of synthetic linear macromolecules and having in the chain at least 85% by (mass) of segmented polyurethane groups. They rapidly revert substantially to their original strength.

Elastane fibre: Manufactured fibres composed of at least 85% segmented polyurethane (Polyurethane elastomer fibres). Elastane fibres can be extended reversibly up to 500–700%. Compared to natural rubber fibres, they have the following advantages: almost white in colour (even transparent elastane fibres are commercially available), higher strength, good resistance to oils and cosmetics, more easily dyed. Structure: the macromolecules of elastane fibres contain hard (highly inextensible) segments alternating with soft (highly extensible) segments. Used for foundation and swimwear fabrics. They are lighter and longer lasting than rubber.

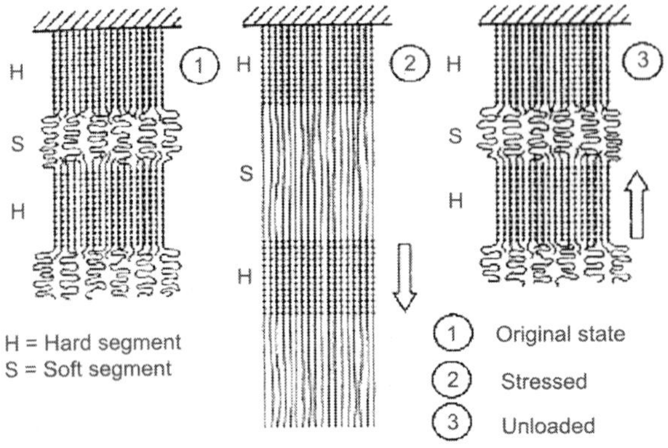

Elasticity: As in the case of strength, a distinction is made between the various types of elastic stress – Tensile elasticity, Bending elasticity and Torsional elasticity. In each case, the elasticity may be permanent or temporary. Inferences can be drawn from these parameters concerning dimensional changes in textiles which are of practical value with regard to their serviceability.

Elastic band: It consists of textile warp threads and elastic threads which are generally not covered by spinning, positioned between these at specific intervals.

Elastic binding: A light weight, elastic and narrow fabric of usually 09–25 mm wide; often with a scallop or picot edge and normally with a useful elongation of 120–160%.

Elastic cord: It has a rubber core with fine silk thread wounded around; used for men's hats.

Elastic fabric: A textile fabric made from an elastomer either alone or in combination with other textile material.

Elastic flannel: English knitted goods with raised nap on the face; used for women's garments.

Elastic gore: An elastic tape made with rubber threads; used for the sides of shoes, usually in black and tan.

Elastic in a casing: A piece of fabric is drawn up with elastic through a casing. See ***Self-casing***.

Elastic limit: The greatest stress that can be applied to a material without permanent deformation.

Elastic recovery: Recovery capacity of fibres. The degree to which the fibre, yarn, or cord return to their original size and shape after deformation from stress.

Elastic tape: A textile tape containing rubber or other elastomers to permit rubber-like stretch at least in one direction.

Elastic webbing: Textile webbing containing rubber or other elastomers to permit rubber-like stretch at least in one direction.

Elastication (in garment making): Fabric that is drawn up with elastic so that it is flexible rather than fixed.

Elasticity: The property of a material by virtue of which it tends to recover its original size and shape immediately after removal of force causing the deformation.

Elastics: Narrow fabrics with some of the warp thread made of elastomeric yarn; used for belts, suspenders, shoes, etc.

Elastik: A light weight, sheer, unbleached linen fabric, given a strong finish which is produced in Austria; used for underlining.

Elastin: Connective tissue protein, major component of elastic fibres as in blood vessels, ligaments, etc.

Elastique: A fabric with a high warp sett having a steep double twill line, used for making trousers, jackets and coats. This fabric is also referred to under a series of other names. For example, cavalry twill, whipcord, etc. In the 19th Century, the fabric was made in fine merino wool for over coating.

Elasticized fabric: A fabric that contains elastic threads. Such fabrics are used for girdles, garters and similar items.

Elastomer: Elastic polymers. As originally defined, this term refers to a group of synthetic thermosetting high polymers (such as polyisobutylene and its copolymers, polyurethane, "Thiokol", etc.) having properties similar to those of vulcanized natural rubber; namely, the ability to be stretched repeatedly to at least twice their original length at room temperature and to retract very rapidly to approximately their original length when released. The term was later extended to include uncross-linked polyolefins that are thermoplastic.

Elastodiene fibres: Generic name for man-made fibres based on polyisoprene or polymerized dienes (polybutadiene), with or without vinyl polymers. Extensibility characteristics similar to Elastane fibres.

Elastomer fibres: See **Elastane fibres**; **Elastodiene fibres**.

Elastomer filaments: See **Elastane fibres**; **Elastodiene fibres**.

Elastomeric yarn: A non-textured yarn which can be stretched repeatedly at room temperature to at least twice its original length and which after removal of the tensile force will immediately and forcibly return to the approximate of its original length.

Elatche: Indian cotton and silk cloth with check patterns.

Elbow (in anatomy): The joint which articulates between the upper arm and the lower arm.

Elbow dart (in sleeves): Dart at elbow to allow ease for movement.

Elbow girth (in body measurements): With the arm bend at 90° and the clenched fist placed on the hip, the circumference of the elbow.

Elbow line (in sleeves): Cross grain at the elbow.

Eleanor blue: A blue colour made popular by Eleanor Roosevelt.

Electric blue: A brilliant light blue colour.

Electric velvet: English cotton or silk velvet, made with light coloured dots powdered over a dark coloured foundation.

Electrical conductance (G): It is the reciprocal of electrical resistance R. The derived unit is the Siemens (unit symbol: S).

Electrical conductivity: (a) A measure of the ease of transporting electric charge from one point to another in an electric field;(b) The reciprocal of resistivity.

Electrical finish: A finish designed to increase or maintain electrical resistivity of a textile material.

Electrical resistance test (volume resistivity): A test method for measuring the static charge on carpet or other textile material. In this test we measure the electrical resistance of a carpet – a property which is indirectly connected with its antistatic behaviour. A direct current voltage is applied via two electrodes in contact with the carpet which allows a direct measurement to be made of the volume resistivity. When the electrodes are placed on the back and face sides of the carpet, the mass or volume resisting force is measured; however, if they are placed on the surface only, a measurement of the surface resistance is involved.

Electrical resistivity: The resistance of longitudinal electrical flow through a uniform rod of unit length and unit cross-sectional area.

Electrode: Any part of an electrical device or system that emits or collects electrons or other charge carriers. An electrode may also be used to deflect charged particles by the action of the electrostatic field that it produces. Positive pole = anode, negative pole = cathode. See also **Half cell**.

Electrolyte: Substances which undergo partial or complete dissociation into positive and negative ions in solution (in water or other solvent). Alternatively, they may be molten ionic compounds that also allow ions to move freely. Liquid metals in which conduction is by free electrons rather than ions are not classified as electrolytes. The solution then acts as a conductor of electricity. The most important groups of electrolytes are acids, bases and salts. Colloids with electrical properties are known as colloidal electrolytes.

Electrolytic bleaching: Vegetable or animal fibres are either immersed in a bleach bath or saturated with bleach liquor, and subjected to the action of a medium to high frequency alternating current.

Electro-m-process: An electro-chemical process for wastewater treatment. It involves breaking down water by electrolysis. Impurities are oxidised or reduced by the resultant oxygen and hydrogen at the nascent stage. It may be supplemented by the perco-filter process (a series of multi-layer adsorption and filtration processes operated under 5 bar pressure), giving water practically of drinking quality.

Electromagnetic spectrum: The massive band of electromagnetic waves which passes through the air in different sizes, as measured by wavelength. Different wavelengths have different properties, but most are invisible and some completely undetectable to human beings. Only wavelengths that are between 380 and 720 nano-meters are visible, producing light. Waves outside the visible spectrum include gamma rays, x-rays, microwaves and radio waves.

Electronic engraving: An electronic scanning and engraving system which converts the reflected light from a design sketch by means of an optical sensor into electrical signals which are transmitted to the engraving system via a computer.

Electroplated button: Plastic button that has been made conductive by chemical treatment followed by the electroplating of metallic coating. See **Spray printing**.

Electrospray process: This process is more rapid than mechanical spray printing methods and allows the production of finer details. Not yet applied in textile printing except, to some extent, carpet printing, because of high investment costs and the lack of genuine textile printing effects. On the other hand, it has found wide application in coding systems and the production of printed letters and policies, etc.

Electrostatic decay half-life (in textiles): The time in minutes for the maximum voltage induced on the textile to be reduced to one half of the maximum voltage by the various decay mechanisms – conduction and ionization of the air.

Electrostatic flocking: The process of applying a flock to an adhesive coated substrate in a high voltage electrostatic field.

Electrostatic propensity: The capacity of a non-conducting material to acquire and hold an electrical charge by induction (by means of corona discharge) or by triboelectric means (rubbing with another material).

Elmendorf tear tester: A test apparatus originally designed to determine the tearing strength of paper, but which is now specified as a standard test (BS 4253:1967 (1986); ASTM D 1424-83; M & S P-29) to measure the tearing strength of very lightweight fabrics and resin-finished apparel fabrics. A trapezoidal fabric test specimen is employed.

Element (in zippers): The device designed for interlocking, capable of being affixed along the edge of a tape.

Elementary colours: Leonardo da Vinci (1452–1519), the master of the Renaissance, considered that the six most important colours fell into a hierarchy and that each indicated one of the elements as follows:

(a) White – light

(b) Yellow – earth

(c) Green – water

(d) Blue – air

(e) Red – fire

(f) Black – darkness.

See also **essential colours** and **primary colours**.

Elephant: Light grey colour; sometimes a dark grey. Also 'elephant grey'.

Elephant towelling: Rough faced, coarse and strong cotton or linen cloth; used for towels, embroideries and drawn work in England.

Elevated temperature dyeing machine: See **High temperature dyeing**.

Elevation of boiling point: A colligative property of solutions in which the boiling point of a solution is raised relatively to that of the pure solvent. The elevation is directly proportional to the number of solute molecules introduced rather than to any specific aspect of the solute composition. The proportionality constant, *k*B, is called the *boiling-point elevation constant* or sometimes the *ebullioscopic constant*. The relationship is D*t* = *k*B*C*m

Where, *t* is the rise in boiling point and *C*m is the molar concentration; the units of *k*B are kelvins kilograms moles–1 (K kg mol–1). The property permits the measurement of the relative molecular mass of involatile solutes. Compare with ***Depression of freezing point***.

Elken: Heavy cotton duck or sail cloth produced in Romania and Bulgaria.

Ellementes: 17th Century English worsted.

Elliot: A Highland tartan with a navy ground. Far apart from each other are pairs of nut brown stripes with a wide blue stripe between the half of each pair. This blue stripe is split in the centre by a narrow red stripe.

Ellsworth: Commercial variety of late maturing cotton from North Carolina. The staple measuring 21—24 millimetres; the yield is 30–33 percent.

Elongated twill: It has an angle more than 40 degrees.

Elongation: The ratio of the change in length of a material during the application of tension to the original length of the rope when new.

Elongation: (a) An increase in length.

NOTE: The increase may be expressed – (i) in units of length, (ii) as a percentage of the initial length or (iii) as a fraction of the initial length.

(b) The deformation in the direction of load caused by a tensile force. Elongation is measured in units of length (e.g., millimetres, inches) or calculated as a percentage of the original specimen length. Elongation may be measured at any specified load or at the breaking load.

Elongation at break: (1) The increase in length corresponding to the breaking force.

(2) The quotient of the change in length (elongation) and the original length obtained in a tensile strength test. In practical tests, it is the increase in length when the last component of the specimen breaks.

(3) The elongation corresponding to the breaking force.

Elongation at breaking load: See **Elongation at break**.

Elongation at breaking load (in fibre strength testing of cotton): The elongation corresponding to the maximum load and expressed as a percentage of the 1/8 inch (3.2mm) gauge length.

Elongation at rupture: The increase in length corresponding to the rupture of the last component of the specimen. Elongation at rupture may be equal to elongation at breaking load in most cases, but can be more also.

Elongation (non-elastic): See **Non-elastic elongation**.

Elongation percent: The increase in length calculated as a percentage of the original length. See **Percent elongation**.

Elongation (recoverable): See **Recoverable elongation**.

Elongation (residual): See **Residual elongation**.

Elongation (total): See **Total elongation**.

Elongation (working): See **Working elongation**.

Eluent (eluting agent): An auxiliary consisting of pure solvent or a multi-component solvent mixture used in chromatography.

Elysee: Applique work, the floral patterns being cut out of two sateen of different colours and then embroidered.

Elysian fabric: A thick or soft woollen fabric in which extra weft is floated to the surface and subsequently burst in the finishing treatment. The floats may be planned for various patterns, such as twills and waved effects. The ground structure may be a single weave or a double weave fabric.

Embauba: Very tough bast fibre yielded by the trumpet tree in Brazil; used for bags.

Embira: Several kinds of bast fibres in Brazil; used for nets by the natives.

Emboss (relief or surface printing): It offers various application options for producing single or multi-coloured print designs alongside or combined with permanent embossed finishes.

Embossed ciré finishes: Produced not by fluted engraving but by relief engraving with a smooth surface which involves operating against smooth, relatively hard mating rolls (ciré calender with a heated engraved steel roll, a cotton roll and to retain the smooth finish, a heated plain steel roll), in other words without forming a reverse image. The design is "fused in" between the smooth faces of the embossing cylinder and the cotton roll (e.g. into acetate fabric). With moiré relief engraving: ciré moiré finish.

Embossed deep finishes: Deep embossed finishes – simulated weave designs produced by relief engraving (e.g. honeycomb piqué), geometric motifs, figured designs and the like.

Embossed finishes: Embossed finishes – simulated lace and embroidery and burnt-out patterns (finely perforated contours and burnt-out designs), also delightful combination effects (burning-out and embossing, or printing and burning-out and embossing) on acetate and synthetic textiles, produced on the same calendar with steel roll relief engraving and at high temperatures (electric heating).

Embossed lustre finishes: Lustre finishes – vary in extent and definition, ripple lines follow the design or fill out the design areas framed by contours. See **Embossing**.

Embossed moiré finish: Varied matt-lustre patterned finishes created by fluted embossing (moiring).

Embossed satin: It is a satin fabric, usually heavier ones, on which designs are permanently embossed by different methods, embossing now is invariably permanent. Use more often as wedding gown fabric.

Embossed silk finish: Not produced by the fabric design, but rather light reflectance resulting from imprinting fine ripple lines ranging from a soft, silky matt shimmer to a high lustre, associated with virtually all possible handle variations (via the schreiner calender).

Embossed transfer printing paper: Transfer printing paper with an embossed effect which is "permanently" fixed in the textile material after transfer printing.

Embossed velvet: Ordinary pile velvet, but a pattern is introduced by pressing the pile flat in places.

Embossing: Also known as embossed finish, goffering, gauffering, fluting, schreiner finish, silk finish, high-lustre finish, impressing of patterns on normally smooth textile surfaces, produced under high pressure and at high temperature on the embossing calender, schreiner calender, etc., between a heated embossing cylinder and a mating or elastic counter roll.

Embossing calendars: Embossing calendars are normally two or three roll calendars with one engraved roll and one or two bowls. The patterns range from polished rolls or cire' to very deep floral patterns. Moire is a watered appearance which resembles paper after it has been wet with water. The moire effect can be obtained by using a moiré pattern embossing roll. Thermoplastic fabrics can be permanently embossed with heated rolls and the effect can withstand repeated laundering. Natural fibres are more difficult to emboss and usually starch is needed for the embossing to take; however, this effect is not durable to laundering. Certain melamine resins can be added prior to embossing and when properly cured, the embossing effect is more durable.

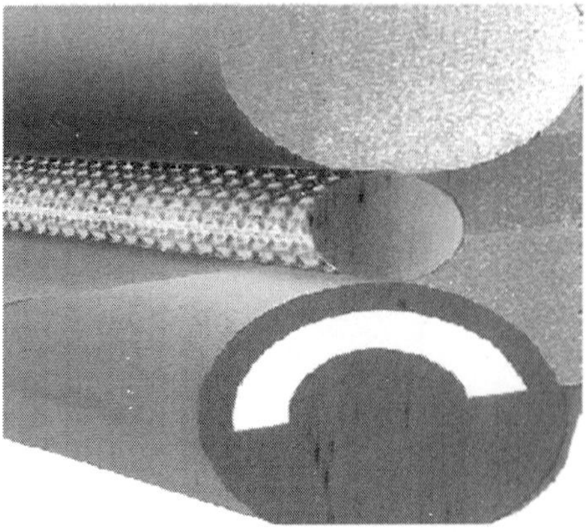

Embossing

Embossing felt: Usually tubular wool felt, 7–9 mm thick multiple fabric, for heated cylinders, ironing presses, etc.

Embossing foil: To produce shadow effects, transparent embossing foils with a granular surface are used for colour separations. They are laid on the design and the drawing is transferred with a suitable chalk. The shadow effects are created by applying heavy and light pressure when tracing.

Embroideries: See **Embroidery**.

Embrittlement: The formation of a brittle residue as the result of pyrolysis or incomplete combustion.

Embroidered chiffon: An open lace-effect chiffon often containing viscose with nylon or polyester. The added embroidery may be worked in cotton, viscose or sometimes metal thread.

Embroidered flouncing: Any plain fabric with an embroidered scalloped edge.

Embroidered knit: A lightweight synthetic, usually polyester knit fabric in plain colours with gold or silver metal embroidery; often in border designs.

Embroidery: The description came originally from an old English word meaning 'border', and was used for early decoration on ecclesiastical robes which were, presumably, worked in border designs. Now it is used for ornamental designs worked on a fabric, mostly light weight, with threads in contrasting or matching colours, often in yarn of a different fibre from the fabric, including metal. The fabric varies considerably in the fibre content and weight and also width. Uses mainly confined to saris and evening wear. Embroidery may be done either by hand or by machine. Originally a needlework of antique origin consisting in executing designs with thread, yarn or other flexible material on a textile or leather ground. It differs from lace that while embroidery always requires a ground to work on, which is essential part of the needlework, lace has no such ground or if it is built up on any ground (like the needle lace on a pricked pattern) it is not part of the fabric. The largest part of the embroidery now in use is produced by machinery.

Embroidery buttonhole: The embroidery buttonhole stitch has many possibilities and many variations. It is worked from left to right instead of from right to left as in a buttonhole. The thread from the work is carried under the point of the needle from left to right, just the reverse of the buttonhole. This stitch is used on flannel and in embroidery of all kinds; it may be padded or worked flat and the stitches may be taken a distance apart or near together.

See **Zig Zag Twill.**

Embroidery: Cotton slack twist two strand cotton thread; used for embroidery.

Embroidery lace: A lace construction obtained by working with a suitable stitching thread on a pre-existing ground of bobbinet, tulle, net or lace in order to produce an ornamental effect on that ground. See **Lace**, **Embroidery**.

Embroidery linen: Linen fabric woven of even and round thread; used for embroidery.

Embroidery quilt: It has large patterns formed by coarse thread on a fine and plain-woven foundation.

Embroidery silk: Loosely twisted silk yarn composed of a number of untwisted threads.

Emerald green: A dark yellowish-green, originally made from copper aceto-arsenate (ite) which was also used as an insecticide. Used extensively by the artist Paolo Veronese (1528–1588) to a point where the colour was called vert Paul Véronèse in France. It was not commercially produced until 1822. Also called Schweinfurt (h) green which in turn is referred to as Paris green, Vienna green, imperial green and Brunswick green. Though toxic, it was used in the manufacture of wallpaper and was considered in the 1860's to be the cause of a number of deaths resulting from arsenic fumes.

Emerizing: Surface character processing of woven or knitted fabric. Through emerizing, the fabric structure is masked by a fine pile without destroying the knitted or woven construction. The effect is dependent on yarn and fabric structure. The shorter and finer the fibres in the yarn, the easier it is to produce a better pile. The finishing stages preceding emerizing also have an effect. This mechanical finishing produced by passing the fabric over a series of emery covered rollers produces suede like finishing.

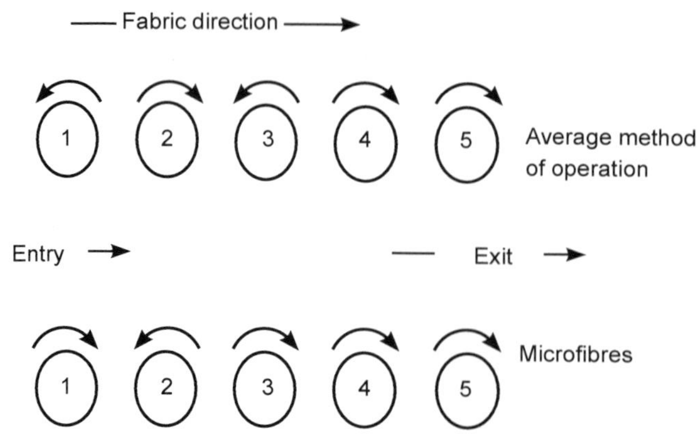

Direction of emerizing roller rotation relative to fabric direction, for optimum emerizing effects

Emery: A mineral of corundum (aluminium oxide, Al_2O_3) containing some magnetite (Fe_3O_4), hematite (Fe_2O_3), or spinel ($MgAl_2O_4$). It is extremely hard and is used as an abrasive and polishing material.

Emery cloth: Strong cotton or linen fabric, coated with powdered emery; used for polishing.

Emission: What is sent out – "emitted", not only pollutants, but also the scent from a rose, or radio waves in broadcasting.

Emissive object: An object that emits light. Emission is usually caused by a chemical reaction, such as the burning gasses of the sun or the heated filament of a light bulb.

Empire blue: A greenish-blue colour.

Empirical test for water repellency (non-standardized test): A rapid and demanding assessment for water repellency often used by garment manufacturers. Literally, a handful of water is thrown against a hanging fabric sample. There must be no water penetration or wetting out of the fabric surface. Few fabrics are considered likely to pass this test.

Empress cloth: A heavy dress goods with napped or corded surface, named for the Empress Eugenia; sometimes called Electrol cloth or Beretz.

Empress fabric: A finely ribbed, backed, double-faced, woollen fabric in 1/2 twill face weave.

Empress gauze: Fine and flower figured fabric, made of all-silk or silk and linen.

Emulsification: The preparation of a finely distributed suspension or emulsion of mutually immiscible liquids.

Emulsifier: A product which makes the formation of an Emulsion possible or easier; e.g. in the production of emulsified textile lubricants, finishes and preparations, etc., to achieve specific effects.

Emulsifying agents: These are surfactants which convert water-insoluble oils into stable, aqueous suspensions. The lyophilic part of the surfactant molecule is absorbed by the oil droplet and the lyophobic head is oriented outward, surrounding the droplet with a hydrophilic sheath. Ionic surfactants add another dimension to the stability of emulsions; they set up a charge-charge repulsion field which adds to keeping the droplets separated.

Emulsion: A suspension of finely divided liquid droplets in a second liquid, i.e., oil in water or vice versa (small droplets with a diameter range 10^{-5}–10^{-7} centimetres). Emulsions are classed as lyophobic (solvent-repelling and generally unstable) or lyophilic (solvent attracting and generally stable).

Emulsion pad process: A pad-dyeing process in which low solids water-in-oil emulsions are used as thickeners for the pad liquor. Disperse, acid and metal complex dyes are applied to synthetic fibres, and vat dyes to cellulosic fibres. The process has not found wide application in practice.

Emulsion polymerisation: A three-phase reaction system consisting of monomer, an aqueous phase containing the initiator and colloidal particles of polymer. Polymerization takes place in the colloidal phase. The process enables the production of very high molecular weights at increased polymerization rates. Only applicable to addition polymers.

Emulsion spinning: The process of spinning synthetic polymers in dispersion form, then heating to coalesce the dispersed particles. Normally, a matrix polymer provides support until coalescence is completed.

Emulsion thickener: A thickener for textile printing consisting of emulsifier, water and a water immiscible liquid (chiefly kerosene or white spirit). Depending on the type of emulsifier, either a water-in-oil or, as generally preferred in practice, an oil-in-water emulsion is formed.

Enamelled cloth: Plain woven, coarse cotton fabric coated with varnish; used for upholstery. See **American cloth**.

Encapsulation: A process in which the fibres of a fabric are coated with a filmy substance to create certain high-performance qualities, such as breathability.

Encapsulated dyes: Dyestuffs of various classes (cationic, acid, reactive, disperse and vat dyes) which have been encapsulated in fine granular form with hydrophilic film-forming substances of high molecular weight (e.g. gelatine, pectin, agar, methylcellulose or polyacrylic acid) by the microcapsule technique. The microcapsule diameter lies within the range of 10–200µm. Dyes in this form can be applied together with reactive compounds without any initial reaction; reaction eventually takes place during a subsequent steaming or other thermal treatment. Special effects, e.g. in Speck printing.

End (in fabric): An individual warp yarn (single or ply) or cord.

End: (1) An individual sliver, roving, yarn or cord.

(2) (a) An individual warp thread (weaving); (b) Each passage of a length of fabric through a machine as, e.g. in jig-dyeing.

(3) (a) Fabric – A length of finished fabric of less than a customary unit (piece) length. (b) Finishing

- (i) Each passage of a length of fabric through a machine, for example in jig-dyeing
- (ii) A joint between pieces of fabric caused by, for example, damage or short lengths in weaving, or damage in bleaching, dyeing and finishing;

(c) Spinning – An individual strand.(d) Weaving – An individual warp thread.

End and End: A plain weave fabric with a warp yarn of one colour, alternating with a warp yarn of white or a second colour. Often the weft yarns alternate with the same two colours while forming a mini check design.

End down: A break of a warp thread in the loom that, if not corrected, leads to a fault. See also **Broken end**.

End group: The chemical group at the end of the long chain of a synthetic polymer. Often the end group is the point of reaction with other chemicals dyes, etc., and hence will have an important say in the characteristic of the polymer/fibre. A liner polymer will have two end groups per molecule and the branched polymers with 'n' branch points will have n+2 end groups.

End-to-end uniformity: A term used for piece dyed goods which exhibit the same shade throughout the entire length of the batch, in contrast to defective dyed goods where ending or tailing problems have occurred during dyeing. See **Ending**.

End out: A void caused by a missing warp yarn. See **Broken end**.

End piece: A fabric threaded on the processing machine onto which the fabric which has to be processed will be stitched, and as the machine is running, this fabric is taken out and threaded on the machine when the machine is to be stopped and left on the machine itself.

End ring: Metal ring introduced into each end of a rotary printing screen to obtain a stable cylinder. End rings are bonded to the screen with a special adhesive.

End ring gluing device: A vertical or horizontal unit designed for the gluing of end ring to rotary printing screens.

End to end uniformity: A term used for piece-dyed goods which exhibits the same shade throughout the entire length of the batch, in contrast to defective dyed goods where ending or tailing problems have occurred during dyeing.

Ending: (1) A dyeing fault consisting of a change in colour from one end of a length of fabric to the other or a difference in colour between the bulk and the end of a length of fabric, commonly used with reference to batch-dyed material.

(2) A break of a warp thread in the loom that, if not corrected, leads to a fault. See also **Broken end**.

Endless conveyer: Continuous conveyor for piece goods. An enclosed track (rail) consisting of an electrically driven endless chain or endless rope.

Endless felt: See **Tubular felt**.

Endless screen: Emulsion is heated and allowed to cool back to room temperature. The electrolyte precipitated out from the saturated aqueous solution is filtered off, washed, and determined analytically.

End out: A defect caused due to a warp thread broken in between in a woven fabric.

Endring: Metal ring introduced into each end of a rotary printing screen to obtain a stable cylinder. End rings are bonded to the screen with a special adhesive.

End ring and gluing device

End-use (in wear testing): The use of which the textile product (garment, cloth, etc.) is intended for.

End-use: The final use of a product (textile) at the customer's end.

Energy absorption: The energy required to break or elongate a fibre to a certain point.

Energy-to-break: The total energy required to rupture a yarn or cord.

Engaeantes: Multiple lace flounces or ruches at the sleeve.

Engageantes: Lace cuffs with two or three rows of ruffles, finishing women's gown sleeves in Louis XIV's time. They were still worn in the 18th century with gowns a' la franchise.

Engineered design: Those designs in which the design is worked in such a way to fit the shape of the intended product. Examples of engineered designs include towels, rugs, duvets and head scarves.

Engineer's cloth: This includes all the fabrics used as uniforms, protective clothes, etc. It will include dungaree, denim, twills, moleskin and the like.

English: A collective term used in the 18th Century for processes involving the local application of indigo directly in the vetted form (orpiment).

English black: A term used to describe a black dyed cloth with logwood.

English blue: A collective term used in the 18th Century for processes involving the local application of indigo directly in the vetted form.

English chintz: The 19th Century name for printed fabrics, made of all linen or linen warp and cotton filling in England.

English combs: The multi-pitch (commonly 4-pitch) hand combs used in preparing top.

English embroidery or eyelet embroidery: The pattern is formed of a number of eyelets which are embroidered in buttonhole stitches all around the edges. It comes mostly in white, but is also available in other colours.

English gum twill: A silk cloth manufactured from low twist yarns with diagonal rib weave. If it is printed, it is called Foulard.

English net: Made with hexagonal mesh.

English number: Cotton Ne_c.

English pink: Most curiously, English pink is the name of a brilliant yellow colour. But this curiosity is not, as one might expect, a manifestation of English eccentricity for each of ***French pink***, ***Italian pink*** and ***Dutch pink***, (as well as ***Bacon's pink***) are all yellow shades – so named because the process of manufacture resembles the one that is used to make **lakes**. There are also green pinks, brown pinks and rose pinks the last of which was probably instrumental in shifting the meaning of the word 'pink' from the process to the colour itself. Also, sometimes a pink colour.

English point: A 19th Century needle-point lace similar to the Spanish point.

English print: A French method similar to Orbis printing, using polychromatic rollers which have the required design on the surface made up of a pattern of dye platelets (number of colours optional). Printing (depending on colour layer thickness, 3,000–25,000 m of cloth, 20–40 m/min) is carried out on damp cloth (prewet, with additional additives after dyeing) with simultaneous steam injection, followed by usual fixing.

English rain sook: See **Rain sook**.

English red: Red iron oxide, also referred to as **Indian red, Pompeian red, Venetian red** and **Persian red**.

English rib: See **Rib fabric, Weft knitted**.

English tulle: See **Bobbinet**.

English welt: See **Roll welt**.

English yarns: Worsted yarns spun in oil according to the Bradford system.

Engraved roll applicator: Engraved rolls similar to ones used for rotogravure printing can be utilized as low-water finish applicators. In this technique, an etched metal roll revolves in a trough of finish. A "doctor" blade wipes off the surface leaving liquid only in the engraved depressions. The liquid in the engravings wick into the fabric as the fabric passes between the engraved roll and a pressurized backing roll. Engraved roll applicators have advantages over kiss rolls because wet pick-up is not influenced by changing fabric speeds, the depth of the engravings are responsible for metering the liquid. One disadvantage is that the roll must be changed, causing machine down-time, when one wants to change wet pick-up.

Engraving: Engraving can be classified into two types; (a) *Mill engraving* – This can be described as a predominantly mechanical process. The pattern is firstly engraved manually onto soft metal rollers/plates. The design will be transferred onto hardened steel rollers, the circumference of which corresponds to the repeat of the design; (b) *Pantograph engraving* – The pattern is transferred to the printing roller using a special machine, the so-called pantograph. The engraved plates are used as a model. Several engraving levers corresponding to the number of repeats are actuated across the width of the roller using a pantograph. A diamond is inserted into the tip of each of these levers. Using these diamonds, the pattern is engraved into a coat of lacquer with which the roller is evenly coated. The outlines are then retraced. The hatchings are later added, using a hatching plate. The roller scored in this way is etched more deeply in a nitric acid bath. A further development of this is the photo-electronic pantograph, which consists of two separate units. In the first unit, there is a roller to which the design is attached. The roller to be engraved in stored in the second unit. The design is in black/white and is scanned using a photoelectric cell. The light impulses obtained are converted into electric impulses. These are then chiselled relative to their intensity. The chisel impressions are made onto a roller which has been coated in lacquer and their close location forms into lines. Up to three rollers can be engraved at one time using this method; (c) *Photo-engraving* – Enables smooth surfaces and short runs to be printed. Firstly, colour separations are prepared from the design, i.e. each colour is transferred separately onto a film with an opaque colour. A reproduction camera is then used to copy the appropriate screen or hatching into the design by means of a screen plate or hatching plate.

Engrelure: French term for footing, or the part of the lace edging which is sewn to the cloth.

Enile: Hand knotted wool rug of Asia Minor, made with a thick pile.

Ensembles: See ***Tweed***.

Ensign: The dark blue colour of naval uniforms.

Ensign cloth: Plain–weave cotton used for making flags.

Entanglement: The extent or degree to which the filaments in a yarn can be interlocked and cannot be readily separated.

Entanglement (in processing): A term used to express the tangling of fabric in the rope form which prevents smooth running of the material in winch, jet, etc.

Entangling:(a) A method of forming a fabric by wrapping and knotting fibres in a web around each other, by mechanical means, or by the use of jets of pressurized water, so as to bond the fibres. Also see **Hydro-entangling** and **Spun-laced fabrics**. (b) See **Intermingling**.

Entering: The process of threading each warp yarn on a loom beam through a separate drop.

Enthalpy: Thermodynamic variable (H), defined as the sum of the internal energy and the displacement energy (pV) of a system, being the product of the volume under constant external pressure (p): $H = E + pV$. The enthalpy is equal to the heat content under constant pressure. Chemical reactions in a system at constant temperature and constant pressure only take place spontaneously towards a reduction in the free enthalpy G ($\Delta G<0$).

Entoilage: The mesh ground in laces.

Entre large: Medium width linen in the French trade.

Entredeux lace or embroidery: It is used as insertion being edged on both sides with some other material.

Entropy: (a) The entropy of a system is a measure of a lack of information or, simply, a measure of disorder. For calculation purposes, the entropy is subdivided into separate amounts capable of addition. The two most important are the thermal and configurational amounts. Entropy is used in defining the second law of thermodynamics; in irreversible (i.e. self-contained) processes in an isolated system the entropy always increases. (b) A process in which a transfer of energy takes place can only continue spontaneously if the entropy of the system and the surroundings increases overall.

Entry zone: In a stenter or drying machine, it is the area in which the edges of a textile fabric are taken up on pins or held by clips.

Enveloppe: Packing canvas in France.

Envers: French word for the back of the cloth.

Environmental conditions in testing: The atmosphere in which specified moisture levels, temperature ranges and concentrations of gases are controlled.

Enzymatic bio washing: Denim jeans manufacturers for many years have washed their garments with pumice stones to achieve a soft-hand as well as a desirable look called "stone wash". Jeans processors typically have found that they can achieve this stone washed look by using cellulose enzyme instead of stones. Enzymatic stonewashing has thus become more widespread over the last two years. As the name indicates, celluloses hydrolyse cellulose, partly yielding soluble products such as short-chain polysaccharides and glucose.

Enzymatic bio-finish: Enzymatic bio-washing. See **Enzymes**.

Enzymatic degumming of silk: Enzyme which can hydrolyse the sericin is classified as proteolytic enzymes. The proteolytic enzymes cleave the peptide/ amide linkages and convert them into amino acid. Mainly there are three types of proteolytic enzymes such as zinc protease (e.g. carboxy peptidase A), serine protease (Chymotrypsin, Trypsin and Thrombin) and thiol protease (acts as cystine residue in the protein). The function of proteolytic enzymes in their degree of degumming depends on the pH of the bath and the optimum activity is found to be different at different pH for different enzymes. Usually enzymatic desizing of silk is a two-stage process. In pre-degumming stage, the cloth is treated with a solution containing soda-ash (1 g/l) at 95 °C for 20 min at a liquor ratio of 30. In the second stage, further degumming is carried out by treatment in a solution containing protein enzyme (0.06–0.1 g/l), non-ionic detergent (1 g/l) for 30 min at 55–60 °C. It is usually impossible to achieve full degumming of silk by enzymes. A short-time treatment in a third bath containing soda-ash or soap may be given toremove the remaining sericin.

Enzyme: The name is derived from the Greek word 'zymae', which means leaven/sour dough. Enzymes similar to the inorganic catalysts (Fig. 1) are so called biocatalysts. Technically impure enzyme products are diastases, for example. All enzyme names have the suffix 'ase' (amylase, cellulase, maltase and protease); enzyme decomposition products have the suffix 'ose'(dextrose, maltose and glucose).

Enzyme desizing: Enzyme desizing is the most widely practiced method of desizing starch. Enzymes are high molecular weight protein bio-catalyst that are very specific in their action. Enzymes are named after the compound they break down, for example, ***Amylase*** breaks down any lose and amylo pectin; ***Maltase*** breaks down maltose; and ***Cellulase*** breaks down cellulose. For desizing starch, amylase and maltase are used. Cellulase, on the other hand, is used for finishing cotton fabrics. Amylase will degrade starch into maltose, a water-soluble disaccharide and Maltase will convert maltose into glucose, a simple sugar.

Enzyme washed: A fabric or garment which has been treated with an enzyme to bring a used or worn ('vintage') look to them. Usually cellulose enzyme is used which acts on the cellulose to give this appearance. The final effect is the stone washed effect, but enzyme wash is less detrimental to the fabric when done in the recommended way.

Enzyme washing: See Enzyme bio-washing.

Eolienne: Named from the Greek word *Æolus*, the god of the winds. Originally, a soft weft-ribbed dress fabric with a silk (or more seldom, cotton) weft and a worsted warp in a plain-weave construction. The cloth is a lightweight zephyr fabric and may be classed as a very fine poplin. Now also produced from man-made fibres, e.g. acetate, rayon.

Epichlorohydrin rubber: Synthetic rubber elastomer based on Epichlorohydrin.

Properties – shore hardness 40–90°, moderate to good adhesion strength and abrasion resistance, moderate to good resistance to diluted acids (moderate to concentrated acids), resistance to aliphatic hydrocarbon solvents is excellent and good to aromatic hydrocarbon solvents, resistant to oxidation and sunlight, excellent resistance to ozone.

Epidermis (in mammalian hair fibres): The outside or surface layer of the fibre consisting of flat, irregular, horny cells or scales.

Epidermis (in wool, flax): In wool structure, the term is used in the widest sense to mean the scale cell covering of the fibre comprising the scale-cell double layer together with the epicuticle. In flax stem structure, the epidermis is regarded as the external peripheral layer consisting of an outermost thin skin which covers the cuticle.

Epigamic colours: The colours displayed by animals during and as part of the process of courtship. The male South Australian cuttlefish can take on the colouring of a female cuttlefish to be able to mate by stealth with a rival's sexual partner. Once it pounces, it reveals the black and white stripes of the male. Researchers at Naples University have found that when a telephone is positioned near to squid, the squid changes colour. The 19th Century naturalists believed that almost all animal colour patterns developed for the purpose of **camouflage**. However, it is clear that many animals (particularly fish) change their body colour for a variety of other purposes, including locating members of the same species (see **episematic**), to frighten the enemy and courtship as referred earlier. Slow colour change is often effected by means of hormones, whereas more rapid change is effected by the nervous system (as in the case of humans while blushing). See also **Camouflage**.

Epinard: The dark green colour of the vegetable spinach. The word *épinard* being the French word for spinach. There are many other French words which take on a close resemblance to the English equivalent word once the initial ‘é‘is replaced by the letter ‘s’ – for example, école, écriture, épellation, épice, éponge, état, étage and étranger.

E’pinge: Very lightweight fabric in a plain close weave, traditionally worsted, but may now contain other fibres. Used for women’s dresses and suits.

Epingle’: A smooth ladies’ dress fabric (silk and cotton) with a pattern of alternating delicate and heavy weft ribs.

Epingline: A warp ribbed dress goods, made with silk warp and worsted filling with a pebbled face.

Epipolic: Fluorescent (optics).

Epitropic fibres (epitropics, antistatic epitropic fibres): Polyamide or polyester fibres containing embedded particles of carbon to make them electrically conductive. They are used as permanent antistatic fibres in 2–5% blends with other fibres for floor coverings, home textiles, protective clothing and technical fabrics.

E’ponge: A woollen or cotton fabric, open textured, with a spongy feel, and light in comparison to its thickness. Used for dresses, children’s dresses, robes, sports shirts.

Eponge: Woven of loop yarn warp and plain filling in plain weave; used for dresses, etc. It is made of cotton, wool or silk.

Epoxides: Organic compounds (cyclic ethers) containing one or more epoxide groups:

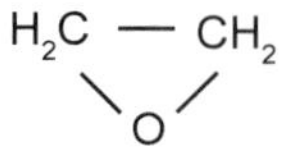

The epoxide group is a reactive group for cross-linking reactions, e.g. with cellulose in the presence of acidic (zinc fluoroborate) or alkaline (caustic soda liquor) catalysts. The application as resin finishing agents (e.g. diglycidyl ether, ethylene glycol derivatives, butanediol ether, glycerol diglycidyl ether) is problematic due to the toxicity of these compounds.

Epoxy functional silicones: Epoxy functional groups can be incorporated into silicone polymers by incorporating the appropriate group into the silicone polymer back bone. Epoxy functionality offers a non-silicone cross-linking mechanism along with the ability to react with fibrehydroxyls. These softeners are more durable to repeated laundering.

Epoxy lining: A method of rehabilitating a water pipe, tanks, etc. The pipe/tank is cleaned by boring or scraping and then 1 mm of epoxy coating is sprayed on to the inside of the pipe. The application is monitored and allowed to cure for 16 hours. Pipe diameters up to 1 m can be rehabilitated using this technique. The technique does not solve leakage problems. Epoxy linings are also used to line the insides of pre-cleaned concrete sewers that may be prone to corrosion due to the nature of the wastewater.

Epoxy resin: In textiles, a compound used in durable-press applications for white fabrics. It provides chlorine resistance, but causes loss of tensile strength. These are thermosetting resins based on the reactivity of the epoxides:

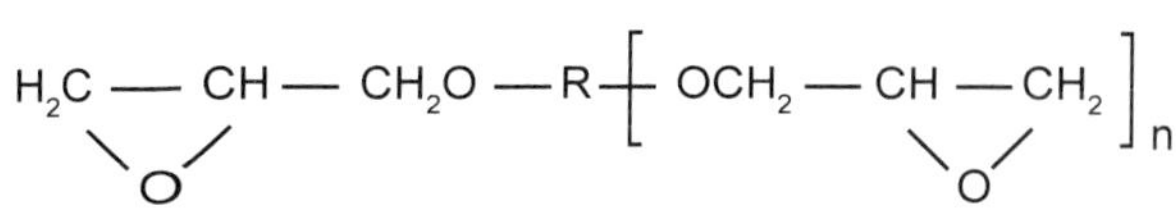

Epoxy resins are polyaddition products formed as epoxides with polyhydric alcohols, hexahydrophthalic acids, hydantoin, cyanuric acid, but mainly (up to 90%) from Epichlorohydrin and Bisphenol.

Equivalent per million (e.p.m.): The concentration of a substance in mg/l, divided by its ***equivalent weight***. This is the same as its milli-equivalent per litre, if the solution has a specific gravity of 1.

Equivalent single yarn number: See **Yarn number, Equivalent single**.

Equivalent weight (equivalent): Of an element or compound, the weight that will combine or react with or replace one atomic weight of hydrogen. For an acid, the equivalent weight is the molecular weight divided by the number of hydrogen atoms in the acid that react with an alkali.

Ercildun: A very fine, lustrous, soft and elastic wool of pure white colour, yielded by sheep in Tasmania; used for the finest grades of dress fabrics.

Ergonomics: (Gk.: *ergon* = work). The study of the relationship between workers and their environment.

Ergonomic seaming: This apparel construction technology is aimed at maximizing comfort and ease of movement. The key feature of this seaming technology is that the seams are constructed ergonomically. Therefore, the seams flow according to the body's natural movements, regardless of the type of activity engaged in by the wearer. The seams are placed away from potential pressure points, in order to maximize comfort and movement.

Eri or Eria silk: A high quality wild silk produced by the *Phylosamia ricini* larva which is found principally in East Asia. The larva feeds on the castor oil plant. This silk is highly regarded in schappe spinning.

Erision control fabrics: See **Geotextiles**.

Erskine: A Highland tartan, consisting of the following colours in a repeat – wide red bar, split in the centre with a pair of narrow and closely placed green lines; a dark green bar, of the same width as the red, split as the latter with a pair of red lines.

Ermine: A soft, fine white fur from a type of weasel with distinctive black tail tip. It will discolour after time. Used as a trimming on ceremonial gowns.

Erne-coloured: Having the colour of an eagle. Also 'eagle hued'.

Eruc: A strong fibre, yielded by the Corypha urens in the Philippines; used for cordage.

Erythro- (G): Red.

Erythrophores: Chromatophores with a red pigment.

Escalier: (a) Ladder tape; (b) Lace, made with ladder effect.

Escoba: Fine, strong bast fibre, yielded by the sida plant in Venezuela; used for ropes.

Escobadura: Fine and white bast fibre, yielded by the Pavonia spinifex in Argentina.

Escot: A comparatively coarse French dress goods, made of hard twist worsted yarn in serge weave and dyed in the piece.

Escurial: (a) Spanish needle-point lace, similar to the Venise; (b) Rich design in trailing and winding pattern; used for laces and embroideries.

ESI: Ethiopian Standards Institution, Addis Ababa.

Eskimo: An all-wool, relatively heavy weight (approximately 800–1000 g) double-faced fabric (with 2 warps and 2 weft systems) mainly in cross-twill construction. Named after the inhabitants of Greenland. The face side is constructed mainly from fine count woollen yarns and the back side from coarser woollen yarns containing a proportion of reclaimed wool. The face side has a finely napped surface. Chiefly dyed in dark shades (black, navy blue). Used for winter coats.

Espagnolette: French fabric, originally made of Spanish merino wool, later of other carded wool in plain or twilled weave, finished smooth or with nap raised on either or both sides and made similar to Melton. It was dyed in pieces and used for trousers, etc.

Esparto: Commercial name for a fine, transparent and clear leaf fibre, yielded by the esparto grass in Algiers, Spain, etc.; used for cords, sandals, baskets, also for carpets, etc.

Essential colours: See **Basic colours, Basic English colours, Bible colours, Elementary colours, Five original colours, Four colour theory, Primary colours, Rainbow, Simple colours**.

Essequibo: Cotton trade name for Brazilian cotton.

Estamenes: A loosely woven 2 × 2 twill worsted dress fabric with a rough finish. Similar to Serge, but having a rough, nappy face resembling cheviot.

Ester interchange: See **Transesterification**.

Estampados: Ordinary cotton prints in Latin American countries.

Esterases: Hydrolytic Enzymes belonging to the main group of hydrolases, which attack esters with the uptake of water, splitting off the acid (cleavage of C–O bonds). In the closest sense, esterases react specifically only with water-soluble esters.

Esterification: The chemical process of combining an acid and an alcohol to form an ester. Cellulose acetate is an ester formed by the reaction of acetic acid and the hydroxyl groups of cellulose. Polyethylene terephthalate, the most common fibre-forming polyester, is a product of esterification of teraphthalic acid with ethylene glycol.

Estopilla: In South America, batiste-like cloth, made of fine linen yarn.

Estrella: A crepe-like, plain woven fabric, made with silk warp and very hard-spun Botany filling, consisting alternately of two rigs.

ETAD: Ecological and Toxicological Association of the Dyestuffs Manufacturing Industries. As in all sectors of industry, manufacturers of dyestuffs are confronted with questions relating to work safety and environmental safety.

Etamine: (a) Yarn – a term used mainly for multiple, soft, gassed ply-yarns (produced from cotton). (b) Fabric – a lightweight, open, thin gauze plain weave fabric, usually with a leno or mock leno construction. Used for net-curtaining, dress goods, etc., produced from cotton, viscose, wool (combed yarns) and silk. The name is derived from the French word *etamine, meaning* muslin, cheesecloth).

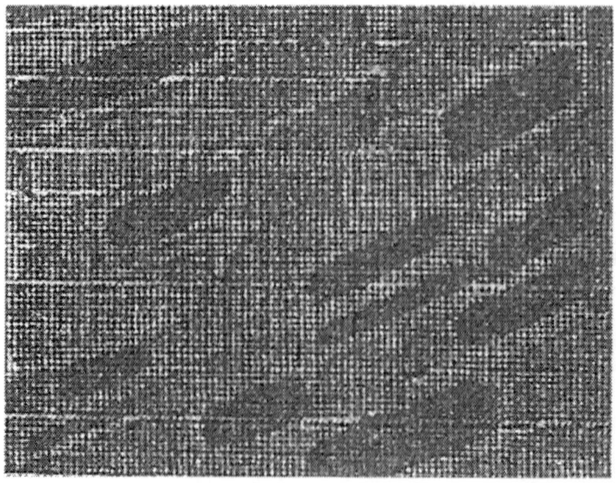

Etamine: Name usually used for porous open cloth like in Leno weave. Similar to Eponge, but thinner.

Etamine glace: Made of organzine warp and worsted filling; also known as Challis.

Etchant: Chemicals used for etching.

Etching: Diamond slivers are used to etch (scratch) the pattern into the lacquered roll on the pantograph. See **Printing, Burn-out printing**.

Etching machine: In pantograph engraving, a machine for etching copper printing rollers in an acid bath. In photo-engraving, a machine in which the unexposed lacquer is dissolved out of the printing roller surface. In case (a) a simple ether (e.g. methyl, or dimethyl ether, CH_3OCH_3) and in case (b) a mixed ether (e.g. ethyl methyl ether, $C_2H_5OCH_3$) is obtained.

Ether sulphates: These are used as anionic surfactants having typical structure –

$$CH_3(CH_2)_{10}CH_2O(CH_2CH_2O)_4SO_3\text{- }Na+$$

Alkyl ether sulphates, also called alcohol ethoxy sulphates, are prepared by addition of one to four oxyethylene groups to an alcohol which is then sulphated. Oxyethylation enhances water solubility and foaming over the analogous alcohol sulphate, giving a product useful in shampoos and in liquid and powdered detergents. The raw material for these products can be either natural fatty alcohols or primary or secondary synthetic alcohols, usually of C_{12}-C_{18} chain length. The analogous alkylphenol ether sulphates are found in industrial applications. Ether sulphates are not as sensitive to water hardness as are other anionic surfactants.

Ethers: Ethers are the anhydrides of alcohols. They are formed from (a) 2 molecules of the same and (b) 2 molecules of different alcohols, with the elimination of water, e.g. by dehydrating alcohols with sulphuric acid. In case (a) a simple ether (e.g. methyl, or dimethyl ether, CH_3OCH_3) and in case (b) a mixed ether (e.g. ethyl methyl ether, $C_2H_5OCH_3$) is obtained.

Ethical trading initiative: A United Kingdom based initiative to promote best practices in social standards in textiles and clothing.

Ethnic: Designs with elements suggesting or imitating the garments or garments itself; traditionally used by a group of people or culture.

Ethnic design: Traditional type of designs of non-western origin.

Ethoxylated fatty amines: These are chemicals used as levelling agents in textile dyeing. These are used for all types of dyes. It can be represented by the following general formula –

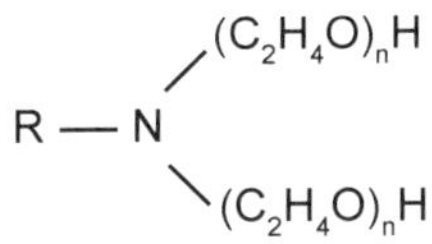

They are regarded as non-ionic, but exhibit weakly cationic behaviour in acidic liquors. Since the non-ionic character is predominant, they can also be applied together with anionic compounds.

Ethoxylated non-ionic softeners: Many polyethylene glycolated hydrophobes are oily or waxy in nature and function as non-ionic fabric softeners and fibre lubricants. They are important components of fibre spin finishes because of their dual ability to lubricate and function as anti-stats. Additionally, they are easily removed in downstream processing. There are two main routes for making this family of products, direct ethoxylation of the hydrophobe, or the reaction of fatty acids with polyethylene glycols. The former method gives mainly mono-fatty derivatives, whereas the second method gives a mixture of mono and di-fatty derivatives.

Ethoxyl ether of cellulose: Ethoxyl ether of cellulose can be obtained by the action of ethylene chlorohydrin on alkali cellulose according to the equation

$$[C_6H_7O_2(OH)_3]_n + nCH_2ClCH_2OH \xrightarrow{NaOH}$$

$$[C_6H_7O_2(OH)_2OCH_2CH_2OH]_n + nNaCl + nH_2O$$

The introduction of small amounts of ethoxyl groups into the cellulose macromolecule causes considerable changes in its solubility. Even when there is one ethoxyl group per 4 or 5 elementary units of the macromolecule, the preparations of ethoxyl cellulose are soluble in dilute alkali solutions.

Ethridge: Commercial variety of late maturing cotton from Louisiana, the fine and glossy staple measuring 28–30 millimetres.

Ethyl acetate (acetic ether, acetic ester, ethyl ethanoate): $CH_3COOC_2H_5$, MW 88, density 0,9, b.p. 77 °C. A colourless, readily flammable liquid with a pleasant fruity odour; less volatile than ethyl ether; water soluble up to 8,6%. Miscible with water (poor stability), alcohol, ether, chloroform, white spirit, etc. Uses: spotting agent for oil, grease, resin, fruit, and grass stains in particular.

Ethyl ether (ether, diethyl ether, ethyl oxide, diethyl oxide, sulphuric ether, ethoxyethane): $C_2H_5OC_2H_5$, MW 74, density 0,72, b.p. 35 °C. A colourless, extremely volatile and mobile liquid. Very flammable (vapours form highly explosive mixtures with air). Characteristic aromatic (narcotic) odour, burning and sweet taste. Slightly soluble in water up to approximately 7%. Miscible with alcohol, chloroform, benzene and other solvents, as well as in concentrated hydrochloric acid. Uses: Highly effective spotting agent (for oils, fats, resins, fruit and grass stains); does not affect the colour of dyed goods (provided it is not heated).

Ethylene: A petroleum derivative (C_2H_4) that is the raw material for polyethylene.

Ethylenediamine:1,2-diaminoethane, NH_2–CH_2–CH_2–NH_2, MW 60,1, density 0,9, b.p. 116°C. A colourless, alkaline liquid with an ammoniac odour, soluble in water and alcohol. Uses: With copper salts a good solvent for cellulose (ethylene diamine copper solution); production of wash-resistant finishes with low wet crease resistance on regenerated cellulose fibres (60–90% solution); production of crease-resistant finishes, chelating agents, cationic textile auxiliaries, etc.

Ethylene diamine tetra acetic acid (EDTA; ethylene bis imino diacetic acid; ethylenedinitrilo- tetraacetic acid): $(CH_2COOH)_2NCH_2CH_2N(CH_2COOH)_2$. An organic complexing or chelating agent of the aminopolycarboxylic acid type. As an ion exchanger, it completely eliminates water hardness salts and heavy metals in water. More expensive than polyphosphates, but more effective and more stable than the latter; EDTA is effective at all temperatures and is resistant to alkalis and boiling temperatures. It can even bind magnesium in caustic alkali solutions, and dissolves rust and corrosion residues as well as already-formed lime soaps, but does not bind metallic iron.

Ethylene glycol: A viscous, sweet, colourless liquid, (CH_2OHCH_2OH), (ethylene alcohol; glycol; 1,2-ethanediol; 1,2-dihydroxyethane). Density

1,11–1,14, b.p. 197,5 °C. A clear, colourless, syrupy liquid. Sweet taste, very hygroscopic principal uses are as intermediate in the manufacture of polyester fibres and as automobile antifreeze.

Ethylene glycol monoethyl ether: (2-ethoxyethanol; Cellosolve), $C_2H_5O(CH_2)_2OH$. MW 90, density 0,93, b.p. 130–138 °C, flashpoint 40 °C, evaporation number 43. Water soluble liquid, miscible with organic solvents. Does not dissolve acetate fibres. Uses: Solvent in dyeing and printing for cationic, acid and indigo dyes; solvent for nitrocellulose, etc.

Ethylene oxide:1,2-epoxyethane, oxirane, colourless gas at room temperature, liquid at approximately 12 °C, soluble in organic solvents, miscible with water. Extremely reactive with mobile H atoms in hydroxyl and amino groups to form numerous technically valuable products used in textile finishing, such as ethylene glycol and ethylene glycol derivatives of all types. Its principle use is for polymerization to 1,2-epoxide polymers.

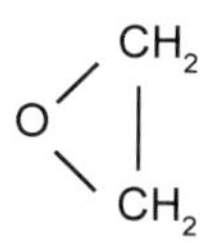

Ethylene oxide adducts: Addition product of ethylene oxide in fatty alcohols, fatty acids and amines. Used alone or in combination as a detergent, fabric conditioner, anti-static agent, emulsifying agent, etc.

Ethylene urea (cyclic ethylene urea): Trivial name for imidazolidone-2 and derivatives, ethylene urea resins. As dimethylol compounds, ethylene urea are important reactant resins, e.g. DMEU, DMDHEU. Linear ethylene ureas are reaction products derived from alkyl isocyanates and ethylene imine, e.g. octadecylethylene urea.

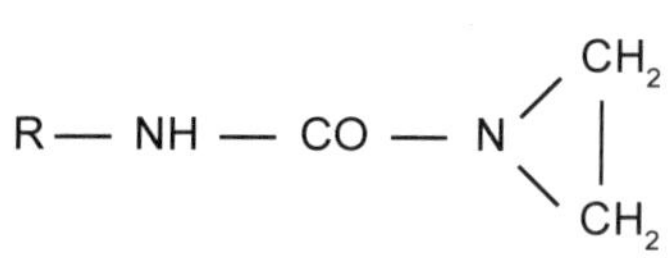

Etiolin: A yellow pigment found in plants which grow in the dark.

Etoffe: French word for cloth.

Etramee: A French hemp cloth.

Etshigo silk: Very thin Japanese silk.

Etun: In the Bible, it denotes linen.

Eulan mothproofing: The development of mothproofing and beetle-protection agents for wool is relatively old and has its origins in the coincidental observation that wool dyed with Martius yellow became resistant to moth attack, i.e. mothproof.

Eureka: A commercial variety of medium maturing American cotton, the strong fine and glossy staple measuring up to 40 millimetres, forming large bolls; the yield is 28–30 percent.

Evaluation period: The period of time an item is used before being evaluated on the specific performance properties.

Even running: Cotton containing fibres of even length.

Evenness (in wool fibre): This term refers to the uniformity of the fibre throughout the fleece.

Evenness testing: Determination of the variation in weight per unit length and thickness of yarns or fibres aggregates, such as roving, sliver or top.

Even weave linen: Fine linen cloth of fairly open construction in plain weave. It is used for embroidery.

Ever glaze: It is a permanent glazed fabric, the process of which is trade marked by Joseph Bancroft and sons. The finish is fast for washing and dry cleaning. The process also guarantees below 2 % shrinkage and soil resistance together with glazing.

Evergreen: A yellowish-green colour.

Everlasting: An all-wool or cotton-filling fabric with fine and very close and almost waterproof surface, woven in satin weave; used for shoe tops, gaiters, etc., and formerly for suits. It is very durable.

Ewes' wool: Shorn from female sheep.

Evlan: Cellulosic modified viscose fabric (courtalds). Developed for carpet industry and having property of resistance to dirt. Now it is also used for upholstery and wall coverings. Often used as blended with nylon, wool, acrylics.

Excessive clearer waste: A higher than normal amount of short and regular fibres that becomes attached to the drafting rolls and is transferred to the clearer brushes to accumulate in abnormal amounts until they are removed manually.

Excelsior: A commercial variety of American cotton, the staple measuring 26–30 millimetres; the yield is 33–35 percent.

Exfoliation: See **Lousiness**.

Exhaust dyeing (discontinuous systems): The dye is dissolved or dispersed in the dyeing liquor. The material is immersed in the dyeing liquor and is removed only when the dye has mostly transferred onto the textile to be dyed, distributed homogeneously, well penetrated into the fibre and fixed. At the end of the process, the material is washed or rinsed to remove the unfixed dye.

Exhaustion: During wet processing, the ratio at any time between the amount of dye or substance taken up by the substrate and the amount originally available. See Bath exhaustion.

Exhaustion (in water treatment): In water softening or ion exchange, the point where the resin can no longer exchange additional ions of the type the process was designed for.

Exhaustion curve: A graphical representation of the progressive exhaustion of a dyebath (bath exhaustion). It corresponds to the quantity of dye taken up by the fibre.

Exhaustion process (exhaust dyeing methods): Processes which involve desorption and absorption of dyes and textile auxiliaries from dye baths due to substantivity (affinity) for a textile substrate.

Exocuticle: Term given to the Intermediate membrane of the wool fibre situated between the epicuticle and endocuticle.

Exomis: Worn by ancient Greek, a short robe, used mainly as men's working clothing, which left the right shoulder bare.

Exophilic (Greek): Directed outwards, e.g., the tendency of a hydrophobic molecular group not to enter into the homogeneous aqueous phase.

Exothermic: Denoting a chemical reaction in which heat is evolved (i.e. heat flows from the system or the temperature rises). Combustion is an example of an exothermic process. Compare **Endothermic**.

Exotic wools: Wools from North Africa, Iran and Asiatic countries (China, India, Mongolia).

Exotics: Braided hats or hat bodies produced from various leaf fibres of the bast fibre group from East Asia, Africa, and Central and South America.

Expander: Cloth guiders as attachments on equipment and machines which feed with the grain. Serves to reduce the shrinkage in width, to form good selvedges and to prevent formation of lengthwise creases. Simple expanders are stays produced from smooth tubes.

Expander (bowed): Bowed and stationary rod which stretches the goods widthways depending on the curve of the rod's setting angle.

Expander rollers: Rollers made from any material which, from the centre, has grooves or threads which run opposite to each other, and which have progressive/constant gradient, to stretch piece goods to their full width on stentering and drying machines (expanders).

Expanding comb (in sectional warping): The expanding comb has the aim of placing all threads on a width corresponding to the beam width and the aim of maintaining them in order and without entanglements. The comb has a zig-zag shape, which permits its adjustment to the different beam widths. The comb has two traverse motions – a horizontal motion to deposit uniformly the threads on the beam and a vertical motion to avoid the local wear of the dents. A blowing system ensures that the comb remains constantly free from dust.

Expansion tank: Tank on high-temperature equipment used to equalize pressure and absorb fluids at temperatures above 100°C.

Expansion vessel: Expansion tank for HT equipment.

Exposed tape width (in zipper): The part of the tape extending beyond the shoulders of the interlocking elements to the outer tape edge.

Expression: Expression is a term used to express the efficiency of a nip, or padder or a squeezer. It is expressed as the percentage of water taken up by a nip or padder or squeezer.

$$\text{Expression} = \frac{(\text{Wt of the fabric after the padding-Wt of fabric before padding}) \times 100}{\text{Wt of fabric before padding}}$$

Extended length: The length of a face pile yarn required to produce one inch of tufted carpet.

Extender (fillers, diluents): Normally specific, inert additives used for standardizing the type or the commercial form, concentrated, also used for diluting and cheapening. Usual with dyestuffs (powder products e.g. + neutral salts, dextrine. hydrotropic substances, dispersing agents, preservatives and the like, with liquid products also + water etc.) and with tensides (e.g. + sodium sulphate, water, alcohol and the like).

Extensibility: That property by virtue of which a material can undergo extension or elongation following the application of sufficient force.

Extension: (1) The ratio of the change in length of a rope during the application of tension to the length of the rope immediately before application of that load.

(2) The change in length or width of a material due to stretching.

(3) Elongation/Initial length percentage. Breaking extension is the extension percentage at the breaking point.

Extension percentage: This measure is the strain expressed as a percentage rather than a fraction.

Extension (force): The force required to stretch a material.

Extension recovery chart (in elastic material testing): A continuously plotted graph of tension versus extension resulting from an extension-recovery cycle.

Extension recovery cycle (in tension testing): The continuous extension of a specimen with a momentary hold at a specified extension, followed by a controlled rate of return to zero extension.

Extra classical: Trade name for the very best silk yarn, made of the choicest cocoons of the year.

Extra super: A standard grade of ingrain carpet, having 1,000 warp ends in a yard and 13% pairs of filling in an inch.

Extract: A term used for a preparation containing the active principle or concentrated solution of a material, e.g. an aqueous extract.

Extract printings: See **Printing**, **Discharge printing**.

Extract (aqueous): (1) The solution obtained by digesting a material in water. See **Aqueous extract**.

(2) Extracting a substance in another substance using water; usually distilled water to avoid the interference of any other matter contained in the water.

Extracta: Open-width washing machine with vertical zig-zag cloth passage (more efficient washing effect compared to horizontal arrangement). The additional division of an Extracta compartment enables different processes to be carried out with appropriate liquors in the individual sections.

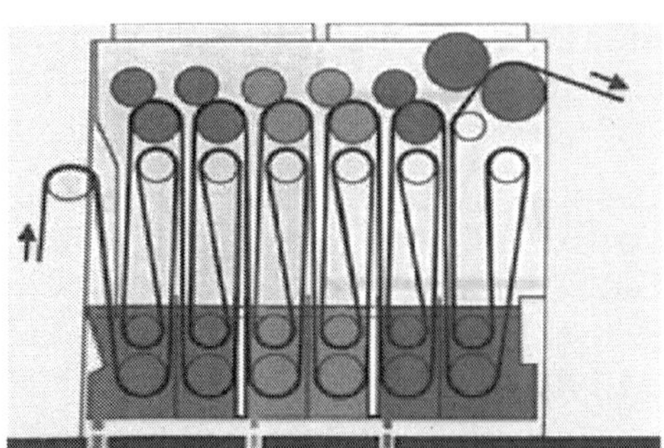

Benninger design of washing unit; it consists of a 5-m washing chamber with 2 vertical roller pairs for double fabric immersion. Available with 3, 4 and 6 chamber compartments (15, 20, 30 m fabric dwell); each compartment has an integral squeeze unit with 2 or 3 t pressure. Can be combined with e.g. a pre-treatment steamer.

Extractable matter: Non-fibrous material in or on a textile, not including water, which is removable by a specified solvent or solvents, as directed in a specified procedure.

Extractable: The material that can be removed from textiles by means of a solvent (in many cases, water).

Extraction: Removal of one substance from another, often accomplished by means of a solvent.

Extraction cleaning: A general term for a number of refurbishing methods in which the cleaning agent is delivered on to the textile product, agitated and simultaneously removed by suction.

Extraction cleaning (dry foam): See **Dry foam extraction cleaning**.

Extraction cleaning (hot water): See **Hot water extraction cleaning**.

Extraction cleaning (rotary): See **Rotary extraction cleaning**.

Extraction cleaning (steam): A deprecated term. See more preferred term Hot water extraction cleaning.

Extraneous fibres: Faults arising from the presence of foreign fibres in a textile material. Generally present in low proportion and occurring as a contamination caused either by erroneous fibre mixing, residues from other processes or fibre fly from packaging material, etc., that may be twisted, woven or knitted into a fabric. Consequences – irregular fabric appearance, colour differences, bleeding of dyes, etc. Evidence only microscopic.

Extruded composites: Flexible, net-like textile fabrics (bonded fabrics), produced by extruders from thermoplastic polymer melts for felt underlays, camouflage nets and for carpet anti-skid protection, etc.

Extruder: (a) Generally a machine in which molten or semi-soft materials are forced under pressure through a die to form continuous tubes, sheets, or fibres. It may consist of a barrel, heating elements, a screw, ram or plunger, and a die through which the material is pushed to give it shape. (b) In fibre manufacture, the machine that feeds molten polymer to an extrusion manifold or that which first melts the polymer in a uniform manner then feeds it to manifold and associated equipment for extrusion. Also see **Screw melter**.

Extrusion coating: A process used for the production of fusible interlinings. Principle – a molten thermoplast adhesive is extruded from a wide slitted jet and the resultant film is pressed on to the pre-warmed substrate (fabric or non-woven) surface with the aid of cooled rollers.

Extrusion ratio: See **Spin-stretch ratio**.

Extrusion: Speed of the melted polymer.

Eyelash: A type of fringing that is usually metallic or plastic, woven or knitted in the fabric.

Eyelet: (a) A series of small holes made to receive a string or tape. A buttonhole stitch is worked around the holes; (b) A type of yarn guide used on a creel;(c) A fabric style with areas of cut-outs surrounded by stitching.

Eyelet embroidery: The ground fabric is fine count plain weave batiste. Before the actual embroidering process begins, the required eyelet effects are stamped in (openwork effect). The holes are then embroidered with grey ply yarns according to the pattern. Eyelet or batiste embroidery receives its snow-white effect by bleaching. These embroideries are also offered coloured or additionally printed.

Eyelet stitch: Embroidery stitch.

Method – work a row of small running stitch round the circle. Pierce the centre with a stiletto and fold back the ragged edge. Closely overcast the folded edge and running stitch. Trim away any ragged edge at the back. Holes may be cut across the centre both ways and the cut ends folded back, instead of piercing with a stiletto.

Eyeletting: See **Eyelet embroidery**.

Elastodiene: A term used to describe fibres composed of natural or synthetic polyisoprene, or composed of one or more dienes polymerized with or without one or more vinyl monomers which rapidly revert substantially to their original length after extension to three times of that length.

Extrusion: (1) The process by which often the plastic, molten or dissolved state of polymer of the fibre is spun into filaments. The material is forced through the holes of a spinneret or die at a controlled rate and dried into fibres. There are five different methods of spinning or extrusion. In practice, this method or a combination of methods are used to produce the final product (fibre). (2) See **Spinning**.

Eyelet plate: A cross bar attached to the end of a creel immediately in front of each row of spindles. It is perforated with the same number of holes, as there are spindles in the row and serves to guide the individual ends from the packages on the spindles to the wrapping machine.

F

Face (of the fabric): Usually the right side or the better looking side of the fabric, or the side of the fabric which is meant to be used in making the garment for wearing. Sometimes the wrong side is also used for making garments, such as denim.

Fad: A fashion or style which is short lived. A fad will not have any lasting impact on the future fashion. They are suddenly seen everywhere for a short period and then suddenly vanish.

Faggoting: See **Fagoting**.

Falling bands: Linen or lace collars (or combined) with two distinct ends hanging down over the chest. It is also called Rabat or Hanging Collars. It is used by both men and women.

Fargul: A type of jacket.

Farji: Name of a jacket, possibly a long over coat like garment without sleeves, or with very short sleeves open in the front and worn usually over pyjama.

Farshi pyjama: A wide legged pyjama that trails on the ground, sometimes completely covering the feet. It is worn often with kurta or angarakha.

Fashion forecast: The prediction of the future fashion by fashion experts which gives the fashion trend of a specific period of time like fall, winter, spring, summer, etc. These could be trends of garment styles, colours, footwear, etc.

Fast fashion: A term to describe the way in which trends and products alter rapidly, with several changes occurring throughout the traditional spring/summer, autumn/winter seasons. Criticism of this approach is that the clothing becomes disposable and therefore uses excessive resources and ultimately contributes to landfill.

Fatuhi: A vest or a jacket without sleeves, with light padding of cotton wool and quilted.

Faux fur: A woven or knitted fabric of any fibre to look like animal fur. Acrylic and modacrylic are very common.

Faux leather (false leather): Fabric made to imitate animal leather. Generally of polyurethane laminate.

Faux linen: Fabric which looks like linen, made using slubbed yarn and later given finish to give linen handle. Often inexpensive.

Faux shearling: Fabric made to imitate shearling- the pelt of a sheep with wool in place.

Faux silk: A fabric commonly made with polyester, and woven/knitted and finished with drape, shine and softness to imitate silk. Often dyed into ecru shade to give the natural silk look.

Faz-vi: A jacket without sleeves.

Fibre ultimate: One of the unit botanical cells into which leaf and bast fibres can be disintegrated.

Fibrefill: Fibres specially used as fillers for making pillows, mattresses, sleeping bags, quilts, comforters and paddings for outer wear. These are specially engineered fibres or sometimes recycled materials.

Filament blend yarn: A filament yarn which contains separate filaments of two distinct types; the filaments being more or less randomly blended over the cross section of the yarn.

Filamentation: Breakage of filaments in filament yarn which causes fibrous or hairy appearance on the surface of yarn package or fabric.

Filler fabric: A rubber-coated cross-woven fabric which is placed around the bead section assembly of a tyre and serves to reinforce the joint between apex and casing plies. In all metallic radial-ply tyres, this filler often consists of a ply of wire cords.

Filler: A North American term for filling. See **Filler**.

Filling bobbin: A tapered core on which filling/weft yarn is wound for use in the shuttle.

Filling yarn: Filling is also called woof or weft. It is the yarn which is used for weaving a fabric in which it is interlaced through the warp yarns.

Finished fabric: A fabric which has gone through the full process inclusive of finishing and ready for garment production.

Finisher picker: The second of the two units of older style picker machines. This unit receives partially cleaned cotton in the form of a lap from a breaker picker and completes the cleaning and fluffing process.

Firm: A fabric is called firm when it is relatively solid with compact texture, having good body and reduced drape.

Flame stitch: A zig-zag design that suggests a flame.

Flash spinning: A modification of the accepted dry-spinning method in which a solution of a polymer is extruded at a temperature well-above the boiling point of the solvent such that on emerging from the spinneret evaporation occurs so rapidly that the individual filaments are disrupted into a highly fibril form.

Flat (spinning): A flat piece of material covered with a special type of card clothing, held in a horizontal position on the flat top type of carding machine. Working in conjunction with a carding drum, it aids in opening the cotton fibres.

Flat crepe: The flattest all crepes with very little crepe or pebbled effect hard twist alternating 25 × 22 in filling, warp has ordinary twist. It is very soft and pliable which makes it good for draping. It may be white coloured or printed and light weight. It is also called French crepe. It is used for accessories, negligees, pyjamas, lingerie, blouses, dress goods and linings.

Flat fabric: A two-dimensional woven or knitted fabric that has no pile loops.

Flat knitting machine: A weft knitting machine having straight needle beds carrying independently operated latch needles. Rib machines have two needle beds which are opposed to each other in inverted V formation. Purl machines have two needle beds horizontally opposed in the same plane.

Flat metal yarn: A yarn consisting of one or more continuous lengths of metal strip or incorporating one or more continuous lengths as a major component. A notable example is a singles metal yarn in banknotes, which may be 0.50 mm (0.02 in) wode and 0.08mm (0.003in) thick. For this purpose, it must be without twist, i.e., flat throughout its length in the banknote. Analysis of the metal is proof of the authenticity of a banknote, twist inserted I flat metal yarns may form irregular facets, which reflect light accordingly to give decorative effects in fabrics.

Flat screen printed: A material which is printed using a flat screen. Flat screens will have advantages over rotary printing machine in the fact that in flat screen, there is more design flexibility; where as in rotary screen printing, the objects have to be adjusted as per repeat.

Flat screen printing: The method of printing where flat screens are used to transfer the designs on to the fabric. In this type of printing, a separate flat screen is created for each colour. The open mesh part of the screen corresponds to the area to be printed in that colour. The areas where colour is not required to pass are blocked. Dye paste is forced through the open mesh area with squeegee. The fabric is then moved or the screen replaced to allow printing of the next colour. In flat screen printing, the screens are in the form of flat panels; the width of which will be same as the repeat of the pattern.

Flat screens: The screens used in flat screen printing. See **Flat screen printing**.

Flat setting: The setting of fabric at open width. The term is particularly used in the finishing of woven wool fabrics, where setting is usually affected by steaming under pressure.

Flax fibre bundle: One of the aggregates of ultimate fibre that run from the base of the stem up to the top of the branches of flax straw. They are each composed of large numbers of ultimate fibres overlapping each other.

Flax fibre strands: Flax fibre after removal from the plant, consisting in the cross section of more than one ultimate fibre.

Flax yarn bundle: The standard length by which wet-spun flax yarns are bought and sold. The 'bundle' traditionally contained 60000 yards (about 55000 m) of yarn.

Flax-spun: A term applied to staple yarn which has been prepared and spun on machinery originally designed for spinning yarns from flax.

Fleecy fabric: A weft knitted fabric composed of three separate yarns, a ground yarn of normal count, finer binding yarn which is held into the fabric at close interval by the binding yarn. The fleecy yarn appears on the back of plain knitted fabric and presents an ideal surface for brushing or raising.

Flock (wool): Waste fibres obtained from wool during the different finishing process.

Floral (design): A design where the motifs are dominated by flowers.

Fluff: Lint or short fibres that has accumulated on a knitting machine and become incorporated in the fabric.

Fluoro fibre: A term used to describe fibres composed of linear macromolecules made from fluorocarbon aliphatic monomers.

Fly frame: One of the several machines which progressively combine two strands of partially processed roving into one, draw out the combined strands until they are of prescribed weight, and twist them loosely in order to give them sufficient strength to withstand subsequent operations.

Fly shot loom: A multi-piece loom for weaving narrow fabrics in which each shuttle is knocked through the open shed by means of a peg fixed in a slide. The term is also sometimes applied to single-head narrow fabric looms.

Foam backed fabric: A combined fabric usually having two layers, one of which is of cellular plastics material.

Foam lamination: A method of making breathable laminated fabric which is mainly used for footwear. In this method, a layer of polyurethane, PVC, or latex is bonded to the fabric with adhesive or fused to the fabric by heat.

Fontange: A bow on a ladies headdress worn in the late 17th and early 18th Centuries, named after the Duchess de fontanges. The term is sometimes incorrectly applied for the whole headdress.

Forehead cloth: Name used for a band covering the forehead of women used in 16th and 17th Century. It may be triangular in shape.

Forming trumpet: A short and flared metal tube at the front of a carding machine. The film of cotton fibres from the machine is passed through the tube which forms them into a sliver. A similar device is at the front of a warping machine. It is used for gathering a number of strands of yarn into a loose and untwisted rope in ball warping.

Forte of a garment: A term used to indicate the strong point of the garment.

French terry: A knit jersey with loops on one side. Sometimes napped to make fleece.

Frey: Threads which come out from the fabric during handling.

Friction twisting: The generation of false twist by a device in which the yarn lies in contact with one or more surfaces of high friction driven in a direction at substantial angle to the yarn axis. In practice, friction disks, belts, or bushes are commonly used.

Fugitive tint: A colour used as tint to identify textile materials during handling, generally in spinning. The colour should be easily and completely removed during the normal scouring or dyeing process and should not interfere in any wet processes.

Fuji: A light-weight plain weave fabric originally of silk, but now usually of polyester bi-component yarn which gives the appearance of a subtle texture on the surface; used in blouses and dresses.

Fujiette: A medium-weight fabric with filament yarn warp and a spun weft. Usually it has a fine cross-wise rib. It is commonly found in rayon and blends of acetate and rayon; used for blouses and dresses.

Full Cardigan: A variation of 1 × 1 Rib stitch manufactured with 2 sets of needles and by alternate knit and tuck on one course and tuck and knit on the next course. The garments that are shaped wholly or in part by widening and/ or narrowing by loop transference to increase or decrease the number of wales.

Fulled: See **Milled**, **Milling**. **Fabinet**: A term used for a type of Artificial leather produced from natural fibres that have been bonded together with a binding agent.

Fabric accumulators (cloth accumulators): Storage units which are, in a few cases, identical with Dwell units in processing machines. They include, conveyor-type accumulators, scray accumulators, roller accumulators and open-width accumulators.

Fabric bagging: A fabric defect which can be caused, e.g. by milling. As a result, the longer interwoven warp or weft threads are milled to a greater extent than the shorter interwoven crossing threads.

Fabric batching machines: See **Batching machines**.

Fabric construction: The details of structure of fabric. It includes information as style, width, type of knit of weave, threads per inch in warp and fill and weight of goods.

Fabric content (cloth content): The fabric content of a textile processing machine or unit expressed either by length or by weight, e.g. the length of fabric (in metres) actually present in a continuous processing range (e.g. 15 m of fabric per roller vat), or in batch-wise processing as the charge of material being processed (e.g. 100 kg of fabric per tube in a jet-dyeing machine).

Fabric crimp: The angulation induced between a yarn and woven fabric via the weaving or braiding process.

Fabric crimp angle: The maximum acute angle of a single weaving yarn's direction measured from a plane parallel to the surface of the fabric.

Fabric dip (for tyre fabrics): A chemical composition which is applied to a textile cord or fabric to improve its adhesion to rubber compounds.

Fabric folding machine: Machine for folding piece goods in layers of defined width and stack height.

Fabric guiders (cloth guiders): Devices for the sideways control of fabric passage in a pre-programmed horizontal or vertical direction. Guiding elements with axially-movable profiled slats.

E+L fabric guider

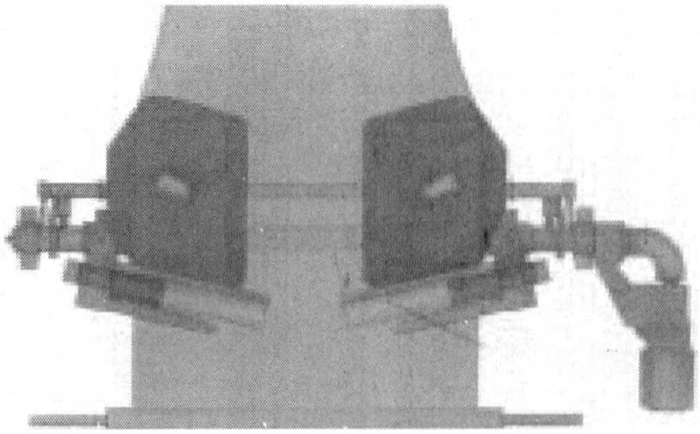

Fabric guide for knitter

Fabric handle: A large number of subjective and objective physiological and physical characteristics are involved in the determination of fabric handle. From various studies on correlations between the handle characteristics determined by subjective and objective means, it has been found that the measurable characteristics of stiffness, bulk and surface smoothness play a particularly important role in determining the handle of a textile material.

Fabric impregnation: Dipping of the fabric in the liquor and allowing proper absorption of the liquor into the body of the fabric.

Fabric inspection machine: Different types of machines for only inspection or for inspection and rolling, plaiting or packing, etc., of grey, finished or fabric at any stage.

Fabric length: The length between two fully interlaced weft threads measured parallel to the warp at both ends of the fabric in a relaxed condition. The fabric length does not include the zones at the start or end of the piece.

Fabric opening machine: See **Tubular fabric slitting machine, Rope opening machine**.

Fabric relaxation: See **Relaxation**.

Fabric roof-system: A system of coated fabric or laminated fabric along with support cables, edge ropes, clamps, neoprene, roof drains, arch wear strips and anchor bolts that constitutes the outside top covering of a building.

Fabric set: The number of warp threads per inch or other convenient unit.

Fabric sliver knitted: See **Sliver knitted fabric**.

Fabric slitting machine: See **Cutting of textile fabrics**.

Fabric slitting equipment: Used for separating the running fabric webs in the middle and into strips. Installed on production machines.

Fabric softener: (a) An auxiliary or a chemical used for improving the softness of the final fabric. It can be one which can be applied by padding (usually on a stenter or special pad application machines) or by exhaustion. Various softeners are having different effect on the handle of fabric and a few give multiple effects along with softening (a classic example is polyethylene emulsion, it gives softening, body, waxy feel and improvement in the tear strength of the fabric).

(b) Washing aftertreating product for preventing boardy handle resulting from washing in a rotary washing machine with a lot of mechanical action and drying in still air. Provides considerable improvement to the handle of washes, provides full softness and makes synthetic fabrics antistatic by using fabric softener in the form of emulsions of mainly quaternary ammonium compounds (such as distearyl dimethyl ammonium chloride), as a dilute aqueous emulsion, in the final rinse.

(c) Special softening agents (softeners) in liquid ready-for-use form, sometimes containing fluorescent brightening agents. Applied in the final rinse after washing in commercial laundries and in household washes in order to achieve a soft handle, antistatic effect, and easier ironing.

Fabric specifications/making particulars: Full instructions as to how to make the fabric.

Fabric stabiliser: Resin or latex treatment for scrims used in coated fabric manufacture to stabilize the scrim for further processing.

Fabric stability (in vinyl coated glass screening and louver cloth): The property denoting the ability to resist slippage of yarn segment in one direction over the yarn segments in the opposite direction.

Fabric straighteners: Devices for straightening distorted fabric webs. See **Weft straightening**.

Fabric stretch: The increase in length of a specimen, if a fabric resulting from a load applied under specified conditions.

Fabric take down: In knitting machine: fabric take down equipment has three main functions, fabric spreading, fabric tensioning and fabric winding. On circular knitting machines, fabric is knitted into a tube form on needles. The fabric is then stretched into a flat form by the spreader located above the take down system. The fabric is then pulled down by the fabric tensioning system (fabric take down roller) and wound into fabric roll by the fabric winding system

Fabric tension control: A system for tension-free running of woven fabrics in continuous machines. The control functions are designed so that drives (DC motors or geared servomotors) as well as brakes (pneumatic or electric) can be controlled. An additional pneumatic unit is available for use with control systems for pneumatic brakes.

Fabric (union): Blended fabric containing wool.

Fabric width: Commercially conventional widths.

130 – 150 cm wool fabrics for outerwear

80 – 90 cm silk materials

75 – 200 cm bedding

40 – 130 cm household linen

80 – 130 cm tablecloths

20 – 50 cm handkerchiefs

80 – 140 cm materials for industrial clothing

60 – 140 cm velvets

70 – 80 cm shirting, trend toward standardisation of 150 cm.

Fabric with nap: They are fabrics that have a texture or a definite "one-way" pattern to them, like suede, corduroy, velvets and velours. When using these fabrics, the pattern should always be placed on the fabric using the "with nap" layout.

Fabric (airplane): A plain weave fabric, well-woven, practically without imperfections, of mercerized two-ply yarns of width 36–56 in with maximum mass of 4.5 oz/yd and yarns per inch warp and weft 80–84.

Fabric, Backing (in tufted floor covering): The fabric on to which the pile yarn is stitched or a fabric on to which the pile are laminated in pile floor coverings.

Fabric, bonded: A lacquered fabric structure wherein a face or shell fabric is joined to a backing fabric, such as tricot, with an adhesive that doesnot significantly add to the thickness of the combined fabrics.

Fabric, braided: A structure produced by interlacing several ends of yarns in a manner such that the paths of the yarns are not parallel to the fabric axis.

Fabric, combustible: A fabric after ignition will continue to burn at a moderate to fast rate.

Fabric, highly combustible: A fabric which after ignition will flash or burn at a relatively high rate.

Fabric, impregnated: A fabric in which the interstices between the yarns are completely filled with the impregnating compound throughout the thickness of the fabric, as distinguished from sized or coated materials where the interstices are not completely filled.

Fabric, knitted: A structure produced by interloping one or more ends of yarn or comparable material.

Fabric, laid: A fabric made without the weft yarn, the parallel warp yarns being held together by means of binding material such as to rubber latex.

Fabric, laminated: A layered fabric structure wherein a face or outer fabric is joined to a continuous sheet material such as polyurethane foam in such a way that the identity of the continuous sheet material is retained, either by flame method or by an adhesive and this in turn normally, but not always, is joined on the back with a backing fabric such as tricot.

Fabric, non-combustible: A fabric which will neither ignite nor give off vapours that will ignite, when subjected to an external source of ignition.

Fabric, non-woven: A textile structure produced by bonding or interlocking of fibres, or both, accomplished by mechanical, chemical, thermal or solvent means and combination thereof.

Fabric, stable: A textile fabric which does not change in dimensions significantly with multiple passes through measuring devices. See **Stable fabric**.

Fabric, stretch: See **Stretch fabric**.

Fabric (textile): A planar structure consisting of yarns or fibres.

Fabric, treated: A fabric to which a finish has been applied in order to change the original level of a specific property or properties, for example: Water repellence, crease recovery, etc.

Fabric, union: A fabric made from yarns of two or more fibre types.

Fabric, woven: A planar structure produced by interlacing two or more sets of yarns, fibres, roving or filaments where elements pass each other essentially at right angles, and one set of elements is parallel to the fabric axis.

Fabricate (in button): The conversion of a blank in to a complete button.

Face: The side of a fabric that is intended to be used outermost.

Face goods: Usually all-wool fabrics; a full dense nap is raised on the face after a thorough fulling and is brushed down, resulting in a smooth face. Doeskin is such a fabric.

Face weight: Total amount of face yarn in the carpet, usually expressed in ounces per square yard.

Face-cloth: The term used to describe a smooth and plain-weave cloth that is luxurious looking with a nap on one side. Best quality face-cloth is wool, but offers may be acrylic or viscose. The fabrics are made in various weights. It is used mainly for coats and jackets, but inexpensive fabrics may be used for robes and dressing gowns.

Face finished fabric: A fabric in which the face side is finished selectively. Example, Raising.

Face side (in textile materials): The side of the material that is outward in the completed object.

Face-to-face carpet (double plush carpets): Carpets that are manufactured as a "sandwich" in a single operation, whereby the pile is attached to two substrates, one above the other. The pile weft forms the pile between the two warps, and two cut pile carpets are produced by cutting the pile yarns between the two substrates direct on the loom. See **Double plush carpets**.

Face-to-face tournay carpets (double tournay): A face-to-face carpet produced on jacquard looms.

Face (in button): That portion which will expose after attaching to the substrate.

Faced hem: A facing or faced hem is also used as a protection to the edge of a garment. A true bias or fitted facing should be used for a facing if the edges of the garment are curved. An extension hem is one in which the whole width of the hem is used.

Face-finished fabric: A fabric in which the finishing agent or a mechanical finished is applied on the face side of it. Chemicals can be applied on the desired side by coating machines, by nip rolls, etc. Mechanical finishes can be done only on the required side of the fabric only.

Fasciated spinning: In fasciated spinning, long staple sliver without twist is introduced into a limited space and subjected to a torque jet operating at right angles to the flow of the sliver thereby imparting a false twist to the sliver. As the sliver exits, the torque jet it rapidly untwists, and the outer fibres tend to break away from the sliver and wrap around the inner sliver to give a strong yarn consisting of mostly parallel fibres with some fibres tightly twisted around the outside.

Faciated yarn: A staple fibre yarn which by virtue of its manufacturing technique consists of a core of essentially parallel fibrous which are bound together by wrapper fibres. The current technique of manufacture is often referred to as jet spinning.

Facing: A lining or trim that protects the edges of a garment, especially at collars, cuffs and front closings.

Facing silk: A fine lustrous fabric of silk used for facing, usually corded satin, twill weave or barathea.

Faconne: (1) A fancy material with a small pattern (including velvets and plush fabrics) of varying fibre types for scarves, blouses and ladies' cocktail or evening dresses.

(2) French word for figured or 'fancy weave'. It is used in relation to textiles to describe jacquard fabric. Generally, plain colour fabrics of soft floppy crepe with satin or taffeta effect patterns are called Faconne. Designs are always small, often shiny on dull background. Popular in the thirties when it was made of silk or ' artificial silk', it returned to fashion again in the seventies, made form viscose, acetate, polyester or blends. Used for blouses, soft dresses, evening wear, etc.

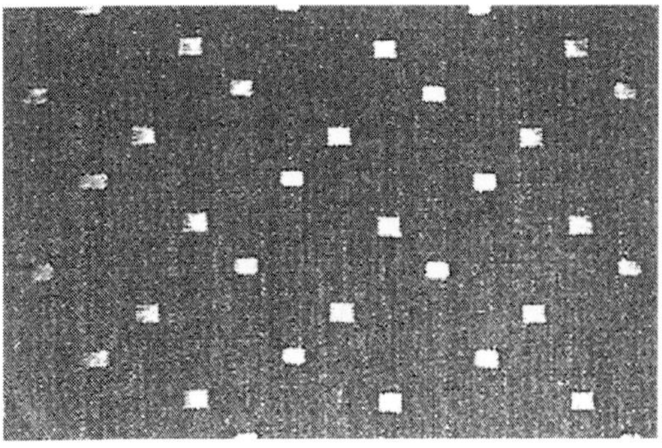

Faconne velvet: A patterned velvet of various fibres with the pattern produced by the burnt-out method of design.

Facultative process: A biological treatment (ETP) process in which the organisms are indifferent to the presence of oxygen. The organisms are called facultative micro-organisms.

Fade (in fastness testing): Any change in the colour of an object caused by light or contaminants in the atmosphere. Example, Burnt gas fumes. The change in colour may be in tone, depth or both.

Fade: Colloquially, a reduction in depth of the colour of an object, by any means.

Fade-o-meter: Laboratory testing device used to determine the fastness of a coloured textile material to light exposure. See **Light fastness**.

Fading: Loss of colour in dyeing and prints as a result of Photochemical reactions.

Fag end: (a) Untwisted end of a rope; (b) Unfinished end of the piece of cloth.

Fagara silk: Wild silk, produced by the Attacus atlas in India.

Fagoting: Fancy way of joining the seams of sheer, light fabrics, producing an openwork, the connecting thread running either in zig-zag line or in ladder effect (French fagoting); in embroidery consists in drawing out few of the threads and tying the cross threads.

Failine: French woollen serge.

Faille: A fine, cross-ribbed and poplin-like dress and coat fabric (silk, half-silk, viscose filament and staple fibres) in taffeta weave. A typical construction of this fabric is 180 × 60, with 75 denier rayon in the warp and 300 denier acetate in the filling. Another is 100 × 50 with 100 denier rayon warp and 16s spun acetate filling. It is sometimes made in cotton, simulating the look of the original silk fabric of that name. Such faille is given a glossy finish to imitate the lustre of the original silk fabric.

Faille Français: Pure silk, delicate and fine-ribbed.

Faille-luisant: Iridescent sheen.

Faille (luxury): Fine-ribbed viscose filament with lustrous and matt sides.

Faille-mirage: Moiré appearance, viscose filament in the warp and metallic yarn in the weft. **Faille-reversible (faille-satin)**: 2-sided effect, lightly ribbed and smooth on the reverse.

Failletine: Similar to ***Faille Français***, very fine ribbed and delicate.

Faille de Chine: A very rich, soft and all-silk faille.

Faille crêpe: A smooth, rich fabric resembling Crepe de Chine, but heavier. The fibre may be silk, but is usually more synthetic. Used for blouses, evening wear negligees.

Faille français: A soft, lustrous silk of wider cord than grosgrain, but narrower than ottoman.

Faille taffeta: A taffeta weave fabric made from silk or synthetic fibres, but with pronounced cross-wise rib effect that resembles faille.

Faille (heavy): Usually used for upholstery purposes, the warp of the fabric consists of fine silk or synthetic filament yarns or of fine mercerized cotton. The filling will be heavy two-ply cotton yarns.

Faille (ribbon): An adopted French term, at one time signifying a fabric in a 2/2 weave, also called double shot. It is woven typically with 20 picks per cm. In present day usage, it is nearly always in plain weave and a correspondingly thicker weft.

Faille (tissue): 200 × 50 construction with a 75 denier warp and a 2/75 denier filling.

Failure: An arbitrary point beyond which a material ceases to be functionally capable of its intended use.

Fair isle knit: Knitted fabric with coloured patterns resembling Fair Isle. The fibre is usually acrylic and the fabric has a great deal of elasticity. It is inclined to 'seat' or lose its shape under strain, hence garments made by this material should not be tight.

Fake fur: See **Fur fabric**.

Falding: Coarse English woollen, similar to the frieze; used for clothing and furniture cover in the 14th Century.

Faliegrijn: Cross ribbed, solid coloured wool camlet; used by the Dutch peasant women.

Falkland islands: Somewhat coarse cheviot wool; used for dress goods and knit goods.

Fall height: Function of action in application baths, e.g. the washing mechanism in drum washing machines (drum partitions). Fall height = drum diameter in m – liquor level in the drum in m.

Fall plate fabric (warp knitted): A fabric made on a Raschel warp knitting machine using one needle bar, involving the use of a solid metal plate (fall plate or chopper bar) to push the newly formed laps of the pattern bars down below the spoons of the open latches to be cast off with the ground laps of the previous course.

Fall weight: Trade term for medium weight coatings, dress goods and suiting.

Fall wool: Wool shorn in the fall following 5–6 months of growth.

Fall-on in printing: A partial overprint of a pre-printed print paste by another print paste, e.g. a pre-printed black outline overprinted with a coloured print paste, or the overprinting of one coloured print paste by a print paste of a different colour = combination shade.

Fallblech: Three-dimensional optical fabric, made with high-pattern yarns and used as curtains.

Fibres used: Polyester, Polyester/Acrylic blends.

Fallen wool: Wool originating from dead sheep. It is generally removed from the skin using lime before tanning. Its affinity to dyes differs from that of virgin wool.

Fallers: (1) Protrusions on a drum interior surface which turn and submerge the item being treated during drum rotation, thus increasing the mechanical action.

(2) Straight pinned bars employed in the control of fibres between drafting rollers.

(3) Curved arms that are fixed to two shafts on a mule carriage and carrying the faller wires.

Fallow: Reddish-yellow or yellowish-brown; the colour of ploughed and unsown land, namely, pale or ruddy brown.

False brocade: In contrast to a genuine brocade, this is a woven fabric produced on normal looms where the brocade effect is formed as a rule by a second warp or weft thread system with figuring threads. “False” brocade produced on looms without a swivel embroidery slay can be recognized by its low mechanical resistance – the threads are easily drawn out of the fabric.

False damask: See **Damask (false)**.

False reed: A false reed is composed of short wires with a loop at the top which are threaded on a band of wires and is used in addition to an ordinary reed to keep the threads of a fibrous or high-set warp separated. The wires of false reed are passed through the warp behind the ordinary reed and divide the warp into three or four threads.

False satin: In designing a broken twill, a three-shaft twill cannot be broken because skipping one shed reverses the direction of the diagonal and skipping two sheds brings us back to the first shed again. One four-shaft twill, 3/1 twill, when woven as broken twill, is called false satin because it is warp-faced and the twill diagonals characteristic of simple twills are almost eliminated.

False selvedge chain: Small support belt (linen warp) for men’s ready-to-wear products.

False twist: Turns inserted in opposite directions and in equal numbers in adjacent elements of yarn, silver (q.v.) or similar aggregations of fibres or filaments, and that are characterised by their temporary nature.

Note: False twist may be used as follows –

(i) To produce effects, e.g.

(a) the entanglement of fibres while false-twisted;

(b) a measure of permanence to the twisted form by heat-setting the false-twisted yarns.

(ii) To assist processing, e.g.

(a) the passage of sliver from "noble" comb to can;

(b) the attenuation of rovings (q.v.) on a condenser ring frame.

False twist direction: The direction, S or Z of twist generally by a false twisting device upstream of itself.

False twist spindle: A device for inserting false twist into a running yarn, sliver or similar assembly of fibres or filaments.

False twisting: A twisting operation applied to an intermediate position on a yarn or other similar continuous assembly of fibres, so that no net twist can be inserted, as direct from twisting at the end of a yarn where real twist is inserted.

False twist texturing: See **Texturing (false twist)**.

Falseh: Strong, harsh, wiry fibre, yielded by a species of Grewia in Persia; used for ropes, etc.

Famis: Silk cloth with gold threads interwoven, made for the Levant in France.

Fancies: Fabrics outside of the staples.

Fancy: A type of coarse flannel, napped on both sides, in linen or twill weave, of cotton or cotton-viscose blend. Double fancy is the name given to goods that are typically two-tone, e.g. with one side pink and the other a mottled blue. Used for cold-weather underwear.

Fancy cloth: Generally, any cloth where through the warp or filling either in one or more colour patterns are produced.

Fancy ply yarn: A collective term for particularly fancy yarns with knots, loops, etc. Special effect yarns are produced on twist frames with additional feeder rolls which supply the main feeder roll at varying speeds.

Fancy print: A term used for all types of printed articles which do not fit into the category of traditional Africa print styles. Fancy prints are usually produced with designs having large motifs. They are the cheap mass-produced printed fabrics in West and Central African countries.

Fancy yarn: See **Yarn (fancy)**.

Fangchow: Very light silk foulard, made in China; it is about 22 inches wide.

Fanpak: Closely woven soft wool flannel in China.

Fantaisie: (a) French for fancy (see); (b) A coarse and inferior grade of silk in France.

Fantasia: Dress goods in Italy.

Farasdanga: Fine cotton cloth, made on hand looms in India.

Fard: Indian printed cotton; used for floor covering.

Farmer satin: A lining of cotton chain or warp and wool filling, finished with a high lustre that is also called Italian cloth.

Farquharson: A Highland tartan.

Composition– wide dark green bar, one-eighth of which is taken up by a yellow stripe in the centre; black stripe made as wide as half of the above; two narrow dark blue and black lines follow; dark blue bar, with a red stripe in the middle, the latter measuring the same in width as the yellow; repeat group, described between the two in reversed order; wide dark green bar with yellow in the centre, as described above is repeated; black stripe, half the width of the green bar; dark blue, somewhat narrower than the black; group of narrow black, red and black stripes; repeat stripes, described between the last two in reversed order.

Farthingale: Earliest hoop dress; a conical frame of thin, flexible switches was worn below the underskirt. Mainly used by Spanish.

Farrell prolific: A commercial variety of very prolific American cotton, the staple measuring 30–35 millimetres; the yield is 28–30 percent.

Fash: Cloth clippings or fringe produced in England.

Fashion adoption: The take up of new styles.

Fashion colours: (1) Colours within a colour range, perceived as being fashionable or 'of that moment'.

(2) Those colours which the haute couture fashion houses have determined as the colours of the season.

Fashioned: Hosiery and underwear; knitted flat and shaped by means of dropped stitches, forming flat edges.

Fashioning: The process of shaping a fabric during knitting by increasing or decreasing the number of needles in action. Fashioning is used in manufacturing hosiery, underwear, and sweaters.

Fast back: (1) Name for welts where the wadding filling is more or less interwoven with the warp;

(2) Trade term for quilts, having the stitching warp interwoven on the back.

(3) Marseilles bleached cotton quilt, having two sets of warps and two sets of fillings, woven plain on the face and back of the fabric, and having a heavy wadding weft between the two cloths. It is woven in embossed patterns.

Fast colour bases: Coupling products for the development of Naphthol AS dyes (Hoechst). They are not dyes, but soluble intermediate products with a basic character due to the presence of

Amino groups. Before use, these amino groups must first be diazotized before they can couple, exactly like Fast colour salts, with Naphthol AS components to form insoluble dyes.

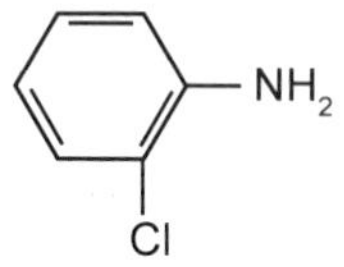

Fast yellow G base

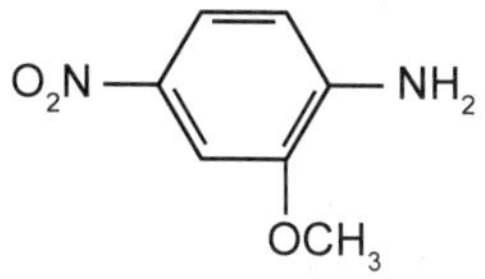

Fast bordeaux B base

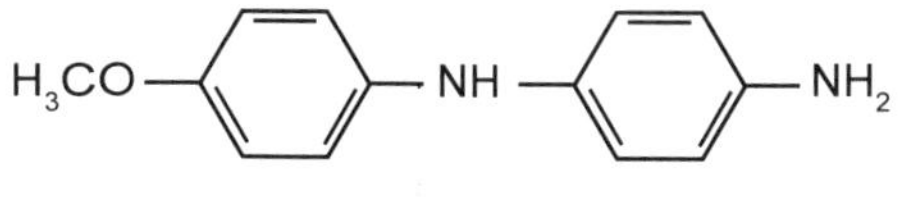

Variamine blue B base

Fast colour salts: These are already diazotized, water-soluble coupling products (so-called stabilized diazo compounds) of the corresponding fast colour bases, used for the development of Naphthol AS dyes (Hoechst). Compared to fast colour bases, the salts offer advantages of better keeping properties, easy solubility, greater reliability and simpler application.

FAST system of fabric assessment: The FAST system (Fabric Assurance by Simple Testing) was developed in Australia by the CSIRO. This system is used to determine the properties of woollen fabrics affecting both the way the fabric performs during cutting and making up and the visual appearance of clothing fabrics when worn. The system comprises three simple instruments and a testing method and was developed principally for use by fabric manufacturers, finishers and dressmakers.

(a) FAST-1 consists of a compression measurement device that measures the thickness of the fabric under two different loads.

(b) FAST-2 is a device for measuring flexural rigidity which measures the bending length of the fabric (the bending length can be used to determine the flexural rigidity).

(c) FAST-3 is a device for measuring the elasticity, measuring the elongation of the fabric under two different loads in the warp and weft direction and diagonally.

(d) FAST-4 is a measuring method for the determination of the dimensional stability of the fabric (relaxation shrinkage and hygral expansion).

Fastness: Resistance of a dyeing or a dye for a particular process. The process can be one which is done during the processing of the fabric or after the consumer usage.

Fastness commission: The German organisation (EK = Echtheits-Kommission) for colour fastness. Most member countries of the International Organisation for Standardisation(ISO) support its Technical Committee 38 (TC38) which was made responsible for all matters concerning textiles and, within TC38, Sub- Committee 1 (SC1), which became specifically responsible for all colour fastness test methods through the national committees of ECE countries.

Fastness improvement: See **Fastness improving agents**.

Fastness improving agents: Textile auxiliaries or chemicals used to improve the colour fastness of dyeing and prints by means of an after-treatment. For this purpose, products are used which form virtually insoluble compounds with dyes on the fibre (especially direct and acid dyes, as well as wool/cotton union dyeing). Other products are used to remove unfixed dye from the fibre (e.g. in naphthol dyeing).

Fastness standards (colour fastness standards): The drawing up of international standards for colour fastness determinations is an ongoing activity of ISO in cooperation with various national standards organizations. It is recommended that guidelines issued by national fastness testing commissions for determining the colour fastness of textiles should only contain slight amendments from the ISO standards.

Fastness test specimen: IT is used in fastness testing, often together with adjacent fabric. *Guidelines* – only textile materials in the flat state are suitable for testing. Yarns are knitted or otherwise converted into a flat form for this purpose. Loose fibre is converted into a fleece. Minimum size of test specimen is 10 cm × 4 cm. The adjacent fabric specimen and the test specimen, of equal size, are joined together by sewing before the fastness test.

Fastness testing: The purpose of colour fastness testing is to determine the resistance of dyed and printed textiles to various agencies at different stages of the manufacturing process (colour fastness to processing) as well as in subsequent use (colour fastness of textiles in use). Examples of the former category include – colour fastness to kier boiling; chlorination; bleaching (sodium chlorite); bleaching (hypochlorite), bleaching (hydrogen peroxide); decatizing; degumming; carbonizing; mercerizing; vulcanizing; soda boiling; cross-dyeing; stoving; metals in the dyebath; potting; milling; pleating; sublimation; resin finishes, etc., whilst examples of the latter category include colour fastness to light; weathering; washing; water; perspiration; rubbing; burnt gas fumes; chlorinated water (swimming bath water); seawater; ozone in the atmosphere.

Fastness to acid spotting E05: The specimen is spotted with either acetic, sulphuric or tartaric acid which is rubbed into the fabric with a glass rod to form a 20mm diameter spot. The specimen is then dried and compared with the grey scale for colour change.

Fastness to alkali spotting E06: This is carried out as the above test for acid spotting while using a 10% sodium carbonate solution.

Fastness to bleaching with sodium hypochlorite NO1: In this test, the sample is wetted out and placed in sodium hypochlorite solution of a standard strength (2.0 g/1 available chlorine pH 11) and liquor ratio for 1 h; it is rinsed, immersed in hydrogen peroxide solution (30%) for 10 min, rinsed again, dried and assessed for change in colour.

Fastness to bleaching with peroxide N02: A composite specimen is placed in a test-tube with standard bleaching solution (composition, temperature and time vary for different fibres, e.g. wool in hydrogen peroxide solution for 2h). It is then removed, rinsed, squeezed out, opened out, dried and then assessed for colour change of the test fabric and staining of the adjacent fabric.

Fastness to carbonising X02: The composite specimen is immersed in sulphuric acid solution (50 g/1) for 15 min at room temperature; removed, squeezed, dried and then baked for 15 min at 105 °C. The colour change is then assessed. A control specimen is treated at the same time to ensure that the test has been correctly performed. This is a fabric that has been dyed to a given formula and that should change colour to a known degree, if the procedure has been carried out correctly.

Fastness to cross dyeing (wool): Colour fastness to cross-dyeing is determined in accordance with DIN 54 049 or BS 1006 X07. In this test, composite specimens are treated in different dye baths containing all the chemicals except the dye (e.g. sodium sulphate and acetic acid for wool). It is than rinsed and dried. The composite specimens are then separated and assessed for colour change of the test fabric and staining of the adjacent fabric.

Fastness to decatizing E1O: The specimen is wrapped around a perforated steam cylinder for 15min (mild or severe treatment as required) together with a test control specimen. The control specimen is one that has been dyed to a given formula and that should change colour to a known degree, if the procedure has been carried out correctly. The specimen is then dried and assessed for colour change.

Fastness to degumming: This test determines the resistance of dyeing or prints to the effects of soap solutions used for the degumming of raw silk. Test specimens, sandwiched between undyed adjacent fabrics, are gently boiled under reflux in a soap solution (7 g/l) for 10 min (liquor ratio 100:1). After this treatment, the test specimens are rinsed in distilled water, followed by rinsing for 10 min in cold tap water, than dried in air (< 60 °C). Change of colour and staining of adjacent fabrics are assessed with the standard grey scales.

Fastness to hot air vulcanisation: Fastness of dye to the effects of a typical caoutchouc mixture and its decomposition products during hot air vulcanization. The sample is placed on the non-vulcanized bed of caoutchouc and suspended in the heating cabinet for 30 min at 125 °C with air circulation. The test piece is allowed to cool and removed from the caoutchouc. The change should be assessed immediately and after conditioning for 4 hours at 20 °C and 65% relative humidity.

Fastness to hot water (fastness to crabbing): Resistance to hot water. Test piece is wrapped around a glass rod, treated with 100 ml of acetic acid solution (7 mg/l of pure acetic acid) in a round flask for 30 min at 70 °C and dried at max 60 °C without rinsing. Assess using grey scale.

Fastness to milling (alkaline): Determines the resistance to alkaline milling conditions.

(a) Mild (light): The test specimen is treated in mechanical washing machines with 10 stainless steel balls for 30 min at 40 °C with 10 g/l soap (liquor ratio 3:1), after which water at 40 °C is added until a liquor ratio of 100:1 is reached, and the treatment iscontinued for a further 10 min followed by rinsing and drying at a temperature up to 60 °C.

(b) Strong (severe): Treatment as in (a) above, but with 50 stainless steel balls and 50 g/l soap as well as 10 g/l sodium carbonate (anhydrous) for 2 h at 40 °C followed by a further 10 min at 40 °C at a liquor ratio of 100:1.

Assessment: Change of colour and degree of staining are assessed visually by comparison with the standard grey scales. Establishing dyeing recipes to select dyes with adequate fastness properties.

Fastness to milling (acidic): It determines resistance to acidic milling conditions.

(a) Mild (light): The test specimen is treated with 1 ml/l concentrated sulphuric acid at 60 °C for 1 hat a liquor ratio of 40:1.

(b) Strong (severe): Treatment with 1 ml/l concentrated sulphuric acid and/or 5 ml/l acetic acid 30% at 90 °C for 30 min. The test specimen is constantly moved in the liquor and pressed with a glass rod every 2 min to ensure complete penetration during this period. After rinsing, the test specimen is opened from three sides and dried at a temperature up to 60 °C.

Assessment: Change of colour and degree of staining are assessed visually by comparison with the standard grey scales. Establishing dyeing recipes to select dyes with adequate fastness properties.

Fastness to organic solvents: Resistance of dyed and printed textiles to the action of organic solvents, as used in dry cleaning. The guideline for all fibres is treat test subject for 30 min at room temperature (liquor ratio 1:40) separately in each organic solvent. In manual testing, push the test subject to the bottom every 2 minutes, using a glass rod. Then squeeze out any excess solvent and dry in warm air at 80 °C.

Assessment: Colour change and bleeding, using grey scales.

Fastness to potting E09: The composite specimen is rolled around a glass rod, tied with thread, boiled for 1 h in water; the dyed specimen is then separated from the undyed cloth, dried and assessed for colour change of the test fabric and staining of the adjacent fabric.

Fastness to Pressing XII: Fastness to pressing can be carried out under three different conditions, dry, damp or wet.

(a) Dry – the specimen is placed on dry undyed cotton adjacent fabric and pressed for 15 s with a heated press (a domestic iron may be used, cotton 200 ± 20 °C; wool 150 ± 20 °C). It is then assessed immediately after a 4h interval for colour change and the adjacent fabric for staining.

(b) Damp – the dry specimen is placed on dry cotton adjacent fabric, covered with wet cotton adjacent fabric and pressed for 15s. It is then assessed as before.

(c) Wet – the specimen is wetted and placed on dry cotton adjacent fabric and then covered with wet cotton adjacent fabric and pressed for 15s. It is then assessed as before.

Fastness to sea water: Textile dyeing resistance to sea water. Wet the test piece and accompanying fabric with solution (distilled water containing 30 g/l of sodium chloride). Place test piece between two plastic plates in apparatus such as the Hydro-test and expose for 4 hours at 37 °C. Dry the sample and the accompanying fabric separately in warm air (<60 °C). *Assessment*: Colour change and bleeding using the appropriate grey scales. AATCC test 106–1972: 30 g/l of sodium chloride and 5 g/l of anhydrous magnesium chloride, 38 °C, 18 hours (DIN 54 007 or ISO 105).

Fastness to soda boiling: Resistance of dyes. The resistance of dyeing or prints to the effect of dilute boiling sodium carbonate solution. Test methods are applicable to cellulosic fibres. Two treatments with and without the addition of a reduction inhibitor are used.

(a) 10 g/l anhydrous sodium carbonate and 4 g/l sodium m-nitrobenzenesulphonate in distilled water.

(b) 10 g/l anhydrous sodium carbonate, without addition, in distilled water. The test specimen is rolled around a glass rod and boiled for 1 h in the test solution under reflux after which it is unrolled, rinsed for 10 min, squeezed, and dried at 60 °C.

Assessment: Standard grey scale for assessment of change of colour.

Fastness steaming Ell: This test is carried out to determine the effect that the steaming stage in printing has on the colours. In the test, steam is passed at atmospheric pressure through the composite specimen for 20 min, the adjacent fabric is then assessed for staining.

Fastness to super-heated steam vulcanisation: Treat test subject for 30 min at room fastness of colour to the effects of a typical caoutchouc mixture and its decomposition products during vulcanization using open steam that (a) is prevented from penetrating the test material, (b) penetrates the test material. The sample is placed on a non-vulcanized bed of caoutchouc and wrapped around a steel pipe. Accompanying fabric, cotton and material impermeable to steam and water are placed on top (in condition a) or only accompanying fabric and cotton (in condition b). This assembly is treated for 20 minutes in pre-heated autoclaves at a heating jacket temperature of 141 °C and vessel temperature of 138 °C, then air cooled.

Assessment: Colour change and staining onto accompanying fabric using grey scales.

Fastness to washing: Fastness to washing is one of the properties of a dyeing of importance to the consumer. Most washing tests are carried out at relatively low temperatures for short times and the dye–fibre system does not come to equilibrium. Most testing standards include several tests of increasing severity. These tests have successively higher washing temperatures, increasing wash times, possibly decreasing liquor ratios, and the more severe tests may have higher levels of mechanical agitation as well as small amounts of chlorine bleach in the wash. Few tests are carried out in typical domestic washing machines, others use a wash wheel.

Table 6.1 Conditions of the ISO washing fastness tests

	Temperature Test (°C)	Time (min)	Steel balls	Chemicals
ISO 1	40	30	0	Soap
ISO 2	50	45	0	Soap
ISO 3	60	30	0	Soap + Na_2CO_3
ISO 4	95	30	10	Soap + Na_2CO_3
ISO 5	95	240	10	Soap + Na_2CO_3

Fastness to wear(wear fastness properties): In colour fastness testing, a distinction is made between those colour fastness properties that are of particular relevance to the end consumer, e.g. colour fastness to light, washing, chlorinated water, sea water, water spotting, perspiration, rubbing, dry cleaning, etc., i.e. properties which relate to the performance of a dyed or printed textile in use, and those which are only important in textile manufacturing processes, e.g. colour fastness to decatizing, stoving, sublimation, vulcanizing, etc.

Fastran process (dyeing): A process for the continuous dyeing of loose wool in which dye fixation is achieved by high frequency heating.

Fastran process (printing): A wet transfer printing process for wool and polyamide in which transfer paper screen printed with water-soluble dyes is brought into contact with pre-wetted textile material in sandwich form under defined conditions of pressure in a calendar and subjected to the action of a high-frequency current. As a result, dye migrates from the paper to the textile material and is fixed thereon in a quasi-steaming process. An after-treatment is still necessary however.

Fat liquor: Fat-liquoring agent used to treat leather after tanning in order to achieve the desired softness, suppleness, fullness, improved tensile strength and resistance to water.

Fat liquoring: See **Tanning**.

Fat quarter: It is a quarter yard of fabric that is cut in a particular way. First a half yard piece is cut and then cut across the length so it ends up 18 by 22 inches. This is a convenient size for appliqué projects or for incorporating many different fabric pieces in your quilt project.

Fatigue: Refers to the resistance of a material to weakening or failure during alternate tension-compression cycles, i.e., in stretch yarns, the loss of ability to recover after having been stretched.

Fathom: Standard measure of length for ropes; one Fathom is six feet.

Fatigue load (in testing sewn seams): The force that is repeatedly applied to a test specimen.

Fats and waxes: Chemically, fats and waxes are esters of fatty acids; fats are triesters of glycerine and waxes are monoesters of fatty alcohols. Fats, also known as triglycerides, are abundantly produced by nature as vegetable oils (corn, olive, coconut, linseed, castor and soy bean oil) and, as fatty deposits in animals (beef, mutton, pork and fish). Marine animals produce both fats and waxes while land based animals produce only fats. Another source of waxes is vegetable matter, predominately the hard shiny outer coating on tropical leaves.

Fatty acids: They are having the following general formula:

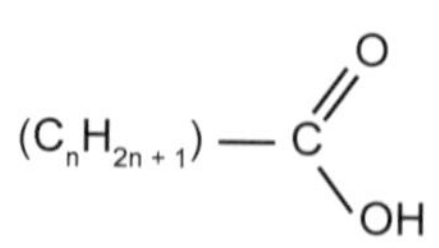

Most saturated fatty acids have wax-like properties. Stearic palmitic and myristic acids function as waxes. Fatty acids of low and frequently high molecular weight, containing more than 20 carbon atoms, are made use of for particular applications. Fatty acids with 12 and 14 carbon atoms are primarily of interest for synthetic detergents. By contrast, fatty acids with 16 and 18 carbon atoms, although still of interest on account of their detergent-building properties are, however, chiefly used for the production of synthetic softeners, sizes, finishing agents, etc. Soaps are obtained by treatment of fatty acids with alkaline liquors.

Fatty acid condensation product: (a) Condensed oils – these products are based on the principle of condensation and saponification, e.g. polyricinoleic acid with sulphonic acid groups and a reduced content of free carboxyl groups.

(b) "Masked" fatty acid (sulphonate) products – these are products in which the carboxyl group is protected by masking, i.e. by blocking the carboxyl group in its former position at the end of the long fatty acid chain.

Fatty alcohol derivatives: With sulphonic acid groups in the middle position, fatty alcohol sulphates, fatty alcohol esters, pyrophosphoric acid esters, condensates with sulphonic acids, fatty alcohol ethoxylates, etc. Properties: most have high stability with good wetting, levelling, cleaning, and softening properties.

Fatty alcohol sulphates: There are two types. (a) The type of sulphate that contains –C–O–SO_3Na as a fatty ester of sulphuric acid with the sulpho group connected to carbon through an oxygen atom. (b) Fatty alcohol sulphates (primary alkyl sulphates) are formed by the esterification of fatty alcohols with acids, e.g. sulphuric acid (after neutralization). In textile these have many uses. A range of properties are obtained from mixtures of fatty alcohol sulphates, e.g. "synthetic detergents" with combined wetting, levelling, washing, softening and dye penetrating properties. They are also used as brightening agents, delustrants, degumming and lubricating agents, besides oxidative bleach stabilisers, emulsifiers, etc.

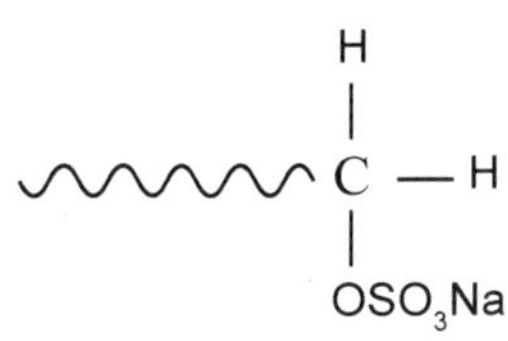

Fatty alcohols:When the complex R–COOH (Fatty acid) is reduced in the high pressure hydrogenation of fats to R–CH_2OH, as a result of which higher fatty alcohols with C8—18 are formed from the corresponding fatty acids. Fatty alcohols are insoluble in water, but soluble in many organic solvents. Together with soaps, turkey red oils, fatty alcohol sulphates and other solvents; they are used in textile auxiliaries as softeners and finishing agents. With paraffin they are also used for water repellent finishes, etc. Low molecular weight fatty alcohols are useful as antifoams. The most technically important fatty alcohols are Lauryl alcohol, Dodecyl alcohol, Myristyl alcohol, Tetradecyl alcohol, Palmityl alcohol, Cetyl alcohol, Stearyl alcohol, Octadecyl alcohol.

Fatty amines: Higher Alkylamines of the type(R–CH_2–NH_2).

Uses: Synthesis of surface-active compounds; corrosion inhibitors; as the acetates for emulsifiers and textile auxiliaries. Secondary fatty amines (distearylamine) are used as starting materials for the manufacture of laundry softeners. An interesting class of fatty amines are the hydroxy amines (cationic surfactants).

Fatty amidoamides (softeners): Aminoamides are made by the condensation of polyamines with fatty acids. Ethylene diamine, N,N-diethylethylene diamine and diethylene triamine are examples of polyamines that are condensed with fatty acids. Usually, the fatty acids are commercial grades such as would be derived from tallow or coconut oil. The products would then carry a generic name such as tallow amino-amides or coco amino-amides. The amino-amides are neutralized with a variety of acids and sold as the salt. Acetic acid, hydrochloric acid, sulphuric acid and citric acid salts of many of them are commercially available for use as softeners. The acid salts are water soluble or water dispersible, making them much easier to use.

Fatty aminoesters (softeners): Aminoesters are made by reacting alkanol amines with fatty acids. Aminoesters containing one or more amine groups are commercially available. These materials also become cationic under acidic conditions, the strength of the cationic charge is proportional to the number of amino groups. Examples of alkanol amines are ethanol amine, diethanol amine and hydroxyethyl-ethylene diamine. Disadvantage of esters is poor hydrolytic stability under alkaline conditions.

Faulty thread: A thread or pieces of thread which are coarse, fine, irregular owing to higher or lower twist or to other twist direction, of different colour, with two or three ends.

Faulty weft: A weft or pieces of weft which are coarse, fine, irregular (slubs, etc.), twisted, reversed, with different twist, of different colour, double weft.

Fautunn: In Chinese markets, a brocaded or cross-ribbed fabric having silk warp and a heavier worsted filling.

Fauve: Vivid or bright in colour after the Fauvists – a group of French artists [including Henri Matisse (1869-1948)] who were renowned for their celebration of colours in art. Their vibrant paintings elevated colour to a point where it became the single most important element in their work.

Faux uni: Giving illusion of a single colour, though made with multi-coloured yarns.

Faveur: Narrow coloured ribbon produced in France.

Fawn: Light yellowish-brown colour.

Fawn canton: A twilled fabric, made with cotton warp and wool filling; used in the manufacture of waterproof cloth.

Fayence prints: Cotton fabrics printed with indigo paste, which afterward is fixed in an alkaline bath.

Fayetta: A fine, thin and soft dress goods, made of organzine warp and fine woollen filling in a close warp plush twill weave; used for dresses, umbrella covers, etc.

FBA: Fluorescent Brightening Agent.

FCR: [German] abbreviation for Forschungsstelle Chemischreinigung (Research Unit for Drycleaning).

FCT: Fog chamber technique.

Fear nought: A stout thick woollen fabric with a heavy appearance, used chiefly as a covering for port-holes and doors of powder magazines and also for scarves and coats. See **Willey**.

Feather cloth: Coating, made by weaving soft feathers into undyed wool cloth; the filling is made of wool and feathers twisted together.

Feather edge: Same as **Picot**.

Featheredge braid: Picot braid of white cotton or linen; used for laces.

Feather fibre: Detached barbs of feathers which are not joined or attached to each other.

Feather proof: Feather proof finishing of duvets and eiderdowns.

Feather stitch: A stitch used in embroidery, imitating feather by branches from a main vein.

Feather (crushed): Feathers and feather fibre resulting from curling or chopping feathers without removing the quill. See **Crushed feather**.

Feather (damaged): Feathers that have been broken or damaged by insects, by mildew or rot or otherwise injured. See **Damaged feather**.

Feathers: The outgrowth forming the contour and external covering of fowl.

Feathers (land fowl): See **Non-waterfowl feathers**.

Feathers (nestling): Immature feathers in which the barbs are held together and covered by a sheath. See **Nestling feathers**.

Feathers (non-water): Feathers derived from chickens, turkeys or other land fowls. See **Non-water fowl feathers**.

Feathers (quill): Quill feathers. Feathers are over 100 mm (4 inches) in length or which have a quill point exceeding 9.5 mm (6/16 inches) in length.

Feathers (waterfowl): Water fowl feathers which are derived from ducks, geese or both.

Feather proof finishing of duvets (eiderdowns): Finishes for this purpose are mainly based on the application of thermoplastic synthetic resins followed by a calendar treatment. In order to produce a satisfactory finish, a distinction must be made between bed ticking and down proof fabrics and, in the case of the filling material, between down and feathers.

Feather stitch: Single, double and triple feather or coral stitches may be made very ornamental and are used in all kinds of sewing and on all materials. They are always made towards the worker; the stitches being taken alternately to the right and left of the line of the design. The thread should always be carried under the needle as in a buttonhole stitch. The design may be varied by taking the stitches diagonally or straight, by making them close or separated, etc.

Feed roller: A roller that forwards a yarn to a subsequent processing or take up stage.

Feed unit: The device which presents the feed stock to the opening roller by either a feed roller and feed plate combination or interacting feed rollers in an open-end spinning machine.

Feed unit (processing): A delivery system for feeding non-batched textile fabrics in open-width, rope or plaited form.

Feed water or feed: The input solution to a treatment/purification system including the raw water supply prior to any treatment.

Feeder density (in knitting): This is the number of feeders through which yarn is fed, per inch diameter of circular knitting machine. For example, a circular knitting machine with 90 feeders and a diameter of 30 inches has a feeder density of 3.

Feeding addition: A finishing term for the regular feeding of finishing liquors during continuous processing.

Feel: The term describing the physical character of a cloth when handled. (Sensation or perception of the textiles by touch. E.g., Soft Feel, Harsh Feel etc.)

Feeler motion: A mechanical device to detect when the weft on a pirn in a shuttle is becoming exhausted.

Fehling's solution: The Fehling's solution is freshly prepared from 69,278 g of copper (II) sulphate and Rochelle salt solution (173 g potassium sodium tartrate tetrahydrate in 400 ml water + 100 g sodium hydroxide made up to 500 ml). Both solutions must be mixed before use. Used for detecting hydrocellulose, oxycellulose and photo cellulose as well as sugars (analysis of finishes). The test is carried out by heating the material with 10% of a freshly prepared Fehling's solution to the boil on a water bath. An approximately quantitative evaluation is made with the aid of a colorimeter. The colour of the copper (I) oxide which is precipitated onto the affected area of the textile is pink to brick-red.

Feille: A glossy, soft, finely-ribbed silk-like woven fabric made from cotton, silk, or manufactured fibres. Made in crosswise rib weave, it has a definite crosswise rib effect. It is a very soft material that drapes well. It is finer than gross grain, but in that family – ribs are also flatter than in grosgrain. Some belongs to the crepe family. It is rather difficult to launder and will give good wear if handled properly. Has a lustrous finish. Uses: Dresses, blouses, soft evening purses, some dressy coats.

FEK (Greek): Französische Echtheits Kommission (French Fastness Commission) Paris. Technical and professional organizations.

Feldspar: A creamy or yellowish-white; the mineral feldspar is white or red in colour.

Fell: (a) The end of a piece of fabric that is woven last; (b) In weaving, the last filling pick laid in the fabric at any time.

Felled seam: A fell is a seam hemmed down to the goods to protect the raw edge. It is usually made in night dresses, drawers, corset covers, etc. Baste with the piece farthest from the worker extended one-eighth of an inch beyond the other and sewed *with the grain* of the goods, beginning at the widest part of any bias. Press the seam with the nail on the right side, turn the wide edge down flat to cover the raw edge and line of sewing, and hem flat either by hand or machine. Care should be taken to keep the seam flat on the right as well as on the wrong side. If the felling is done with the machine hemmer, the wide edge must be on the opposite side. The seam may be basted with both edges; even if preferred, cutting off one edge after stitching.

Felling silk: A two-ply silk thread with a left hand twist.

Fellmongered wool: Skin wool.

Fellmongering: The removal of the wool from sheep pelts by any process or the process of gathering wool from the fleeces of a dead sheep.

Felt: A textile structure characterized by densely matted condition or interlocking and consolidation of most or all of its constituent fibres it is composed of which achieved by the interaction of a suitable combination of mechanical energy, chemical action, moisture and heat but without the use of weaving, knitting, stitching, thermal bonding, or adhesives. There are two broad classes of felt – (a) fabrics relying for their construction upon the ability of the constituent fibres to matt together in order to form a composite body with neither warp or weft; (b) Materials having a woven fabric structure. Best quality felt is rigid and doesnot drape, so it is not very useful as dress fabric. It does not wear well, and it loses shape as it has no elasticity. Best quality felt is made from good quality wool, but other qualities are made by using viscose, cotton and kapok, and waste as well. The result is more brittle than wool. Used for hats, protective coverings, etc., and is useful as it has no grain.

Felt calender: A special type of calender usually, but not necessarily, (it can be a separate unit also) at the delivery end of a stenter used especially for linings.

Felt cloth: See **Felt** and **Felt fabric**.

Felt fabric: (1) This is an ordinary woven fabric, usually cotton or wool, but many other fibres also, that have been milled on the surface to produce a felt like appearance which disguises the weave beneath. The fabric may be of any weight, and the process is sometimes applied to cheap base fabric to give it more warmth and interest.

(2) A term used to describe a milled felt of a similar type to "Feltine" which is not woven, bonded or stitched. It is made, for e.g., from wool fleece (with/ without man-made fibres) which has been bonded into a web by felting in a plate felting machine followed by subsequent milling.

Felt ground: See **Needle-punched carpets**.

Felt jacket: Analogous to Tubular felt, for pulling over working rollers, or as back cloth, etc.

Felt mat: Hand-crafted flat carpet made of firmly felted natural wool, figured effects are worked in, natural-coloured felt sections; used as carpet and wall hangings.

Felt press: See **Pressing**.

Felt sleeve: See **Tubular felt**.

Felt (needled): A textile composed of natural, man-made or a combination of natural and man-made fibres physically interlocked by the action of a needle loom, with or without combination of other textile fabrics and with or without suitable combination of mechanical work, moisture, chemical action and heat, but without weaving, knitting, stitching, thermal bonding or adhesives.

Felt, needled, unsupported: A textile material composed of natural, man-made or a combination of natural and man-made fibres physically interlocked by the action of a needle loom with or without suitable combination of mechanical work, chemical action, moisture and heat, but without weaving, knitting, stitching, thermal bonding or adhesive bonding.

Felt (part wool): A felt composed of wool fibres and one or combination of natural or manmade fibres.

Felt (wool): A textile composed of wholly wool fibres interlocked by the inherent felting properties of wool and produced by a suitable combination of mechanical work, chemical action, moisture and heat, but without weaving, knitting, stitching, thermal bonding or adhesive.

Feltabilty: The degree to which a sample of fibres will consolidate in a specified time by interlocking under prescribed combinations of mechanical action, chemical action, moisture and heat.

Felted yarn: A wool-rich yarn produced by felting sliver, slubbings, rovings, or yarn.

Feltine: Proper name of a milled felt which has no woven structure and no chemical bonding, the textile fabric instead being created by the intensive felting of loose virgin wool. Feltine is piece dyed and may be impregnated. The brilliancy of the colours is typical. These materials are produced to defined thickness and handle (stiffness). Their use is traditional in certain types of clothing.

Felting: An undesired process of superficial "felting" of woollens or an intentional process. This is used to mean the processing of loose fibres to form a dense and durable fabric, whereby theoretical felt formation sequences (a) movement of the open knot (above) in direction of the fixed knot (below) by the action of the elastic, operative fibre; (b) transfer of model (a) to multiple fibres. Mechanical effects (pressure, impact, friction, wear) aided by moisture and warmth achieve a homogeneous combination of the fibre material. Milling is similar to felting.

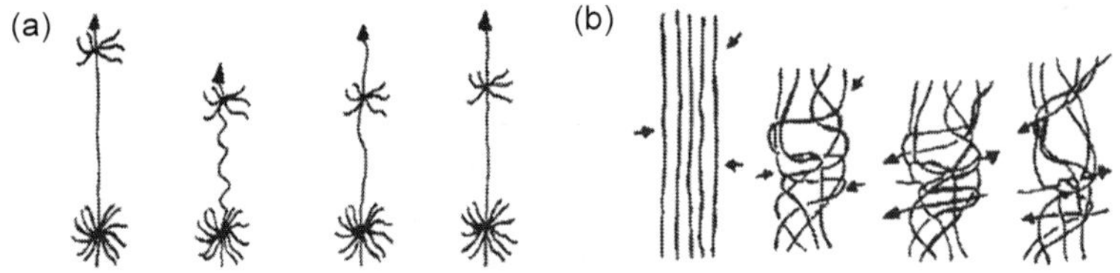

Felting density: See **Felting (degree of)**.

Felting needle: A barbed needle mounted in a needle loom to cause fibre reorientation and entanglement during needling.

Felting propensity of wool: This is a characteristic found in animal fibres, which shrink under certain conditions. The causes of felting are not precisely understood. There are many familiar felting theories; the most important theories are discussed here.

We referred earlier to individual characteristics of wool, and these are reflected in the felting theories: (a) Primary (causal) characteristics such as surface structure, frictional behaviour (DFE), lengthways or cross elastic characteristics; (b) Secondary characteristics such as fineness, length, crimp, grease content; (c) External factors, which represent the conditions during felting and milling, such as pH, temperature, milling auxiliaries.

Felting shrinkage: In contrast to **Relaxation shrinkage**, felting shrinkage results primarily from the frictional properties of the component fibres which cause them to migrate within the structure under conditions of moisture. This behaviour is normally considered to be significant only for fibres having scales on their surface such as wool and which can be counteracted by shrinkproof finishing. Adding synthetic fibres to wool counteracts the felting shrinkage. Effective improvement is achieved only with additions from 40–60%.

Felting, degree of: Measure of the tendency to felt according to the Aachen felting test, expressed in terms of felt density in g/cm^3.

Fents: Finished textile piece goods in short lengths (as metre goods) for retail sales. See **Cut fabric**.

Feraghan: Small Persian rugs of cotton warp and weft and close and short wood pile tied in Senna knot. The design consists usually of small fish in a blue field and trailing vine in the borders. Many saddlebags are also made this way.

Fergusson: A Highland tartan, made as follows – two heavy black stripes with a still wider dark blue stripe between; a green field, almost as wide as the above three stripes combined. This green field is split in the centre with a group of black, white and black lines, the two side green bars thus formed being again split in the centre by a single narrow red line.

Ferrous (ferric): Iron is generally 2, 3 or 6 valent iron(II) ferrous hydroxide is quite a strong base, iron(III)ferrous hydroxide is a weak base (has the effect of an extremely weak acid against strong bases).

Fesse: A light blue colour.

Festoon ager: A steaming chamber for short print runs. The fabric is transported in loops suspended on rotating upper rollers so that there is no roller contact with the printed side.

Festoon drier: See **Short loop drier**.

Festoon dwell chamber: A dryer in which cloth is suspended in loops over a series of supporting horizontal poles and carried through the heated chamber in this configuration. See **Festoon store**.

Festoon scouring machine: Mezzera continuous scouring machine. Machine used in silk and artificial silk finishing for creping and shrinking, also used on knitted fabrics.

Festoon steamer (steaming chamber): A chamber-type steamer of brickwork or cast-iron construction suitable for long runs (1000 m and above). The fabric is transported through the roof-heated steam zone suspended in loops from support rods attached to an endless moving chain. Disadvantages of this type of steamer are (a) extraordinarily large dimensions (up to 10 m in height), (b) high steam consumption and (c) difficulty in achieving uniform steam circulation.

Festoon stitch: Identical with the so-called "buttonhole stitch".

Festoon store: Acts as a relaxing unit. Fabric is introduced via an adjustable slit and suspended on supporting rods. Open-width submerged fabric passage, tension-free, does not cause creasing. External direct and indirect heaters, automatic temperature control.

Festoon tape: It is thin, batiste-like edging and bordering tape with a dog-legged selvedge and decoration resembling eyelet embroidery.

Festucine: Greenish-yellow straw colour.

FEU: A unit of cargo capacity, especially for container ships. These ships carry cargo in standard metal boxes, called containers that can be transferred easily to trains or trucks. FEU is an abbreviation for "forty-foot equivalent unit." One FEU represents the cargo capacity of a standard container 40 feet long, 8 feet wide, and (usually) a little over 8 feet high. One FEU equals roughly 25 register tons [see Ton (c)] or 72 cubic meters.

Feuille: A light green colour.

Feuillemorte: Having the brown colour of dead leaves. Also referred to as filemot, philimot, philomot, philamot, phyllamort, phylamort and other variations. Also 'dead-leaf brown'.

FFT applicator: Union Carbide and Gaston County developed a low-water finish applicator based on metastable foams. The finish bath and air are continuously metered into a foam generator and the foam is routed through distributing heads directly onto the fabric. Being metastable, the foam immediately breaks on contact with the fabric and the fabric becomes wet with finish solution. The wet pick-up is regulated by controlling the ratio of liquid feed rate into the foam generator with the cloth speed.

FI: Fique fibre.

FIANN (French): Fédération Internationale des Associations Nationales de Normalisation (International Federation of National Standards Associations).

Fibei: (a) Filaments used as raw material for textile fabrics are divided as follows

i. animal fibres, as wool, hair, silk,

ii. vegetable fibres, as cotton, linen, jute, hemp, ramie, etc.,

iii. mineral fibres, as asbestos,

iv artificial fibres, as glass, metal threads, various artificial silks, etc.

(b) Yarn, made of twisted and prepared paper; used for rugs and mats.

Fibre: Spelling of fibre used in The United States of America.

Fibre (in textiles): A generic term for any one of the various types matter of smallest unit of any textile material, that form the basic elements of a textile and that is characterized by having a length at least 100 times its diameter. The so-called raw fibre (fibres) is converted by spinning into threads (yarns of various types).

Fibre length: (a) Crimpled length – The extent [see (b) below] of crimped fibre substantially freed from external restraint and measured with respect to its general axis of orientation. (b) Fibre extent – The distance in a given direction between two planes (each perpendicular to the given direction) that just enclose the fibre without intersecting it. *Note*: If the fibre is in a sliver (q.v.) (or yarn, roving, etc.) and the direction of the extent is not specified, the "given direction" is to be taken as the axis of the sliver. It should be noted that the extent of a fibre is a variable property that differs from the straightened length of the fibre according to circumstances; thus in a card web, for example, where the fibres are in a state of considerable disarray, the extent of a fibre after it has been passed through one or more drawing processes. If, for any reason, a fibre is subject to a stretching force, its extent in the direction of the force may be greater than its straightened length. (c) Staple length – A measurement by which a sample of fibrous raw material is characterised according to its technically most important fibre length. *Note*: The staple length of wool is usually taken as the length of the longer fibres in a hand-prepared tuft or "staple" in its naturally crimped and wavy condition (see crimp). In cotton, on the other hand, the staple length corresponds very closely to the modal or most frequent length of the fibres when measured in a straightened condition.

Fibre affinity: Textile auxiliaries with fibre and dye affinity in wool dyeing. See **Affinity; Dye affinity**.

Fibre architecture: The spatial arrangement of fibres in the preform. Each architecture has a definite repeating unit.

Fibre attraction: Forces acting between soil particles and textile fibres – (a) mechanical retention, (b) chemical bonding (stains), (c) adsorption, (d) electrostatic attraction. The attachment of fat-free soil particles (pigment soil) is mainly due to geometrical factors, i.e. dimensional relationships between fibre surface roughness and pigment particles in aqueous dispersion which are not based on energetic bonding (e.g. adsorption, absorption).

Fibre beard: In length testing of fibres – Fibres caught randomly on a comb which are subsequently straightened and parallelised without stretching or damaging.

Fibre blend: A mixture of two or more fibres. The fibres can be mixed in the fibre stage or as yarn in the twisting of yarn or in the manufacture of fabric. The percentage of each fibre in the mixture is often called blend percentage. Examples are Polyester/cotton blend (60/40), cotton/lycra (92/8), etc.

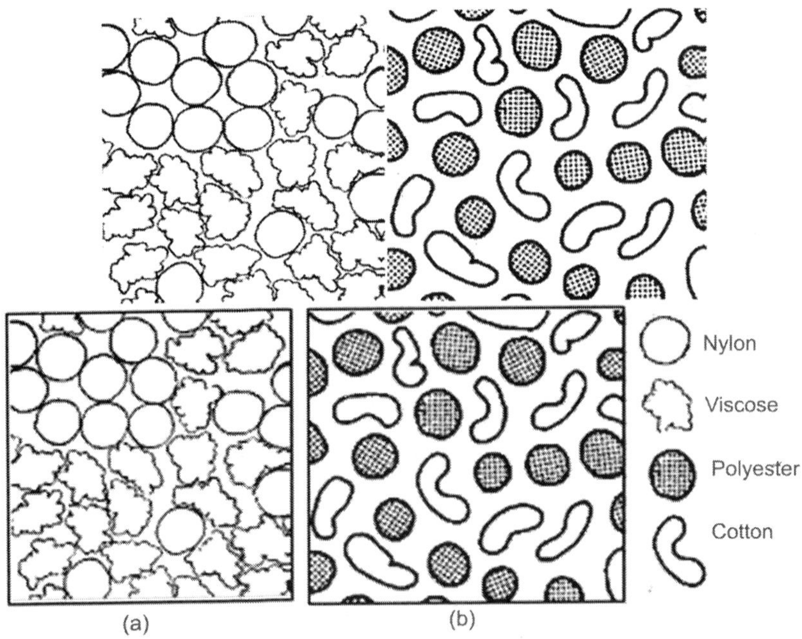

Blend of Nylon and viscose (A) and cotton polyester (B) (cross section enlarged)

Fibre bonding device: Heat bonding of thermoplastics.

Fibre bundle strength: The tensile strength of a cotton fibre bundle at zero gauge length per unit cross-section. It is determined in accordance with the **Pressley index** (see).

Fibre composites: Flexible, porous fabrics (bonded fabrics) consisting partially or entirely of textile fibres formed by the bonding of non-woven.

Fibre damage in cotton (cellulose): Fibre damage can happen due to various reasons, especially due to processing and action of various chemicals. In case of cotton (cellulose), it can be measured by the presence of hydrocellulose, oxycellulose, photo cellulose and pyrocellulose. These can be identified and determined by the following characteristics –

(a) reducing action,
(b) determining the fluidity,
(c) dye uptake of basic groups (oxycellulose), alkali solubility,
(d) Fehling solution: Brick red colour,
(e) Ammoniacal silver solution: red-brown colour,
(f) Lead-cochineal solution: Blue colour (Oxycellulose),
(g) Methylene blue reaction: Blue colour,
(h) Nessler reagent: yellow colour turning grey later,
(i) Schiffs reagent: red colour, Berlin blue reaction,
(j) Boil in caustic soda solution: blue to brown colour,
(k) and can be quantitatively determined by Copper number: Hydrocellulose,
(l) Permanganate number: Oxycellulose and hydrocellulose,
(m) Methylene blue number : Oxycellulose, etc.

Fibre density: Mass per unit volume of the solid matter of which a fibre is composed and measured under specified conditions.

Fibre distribution: In a web, the orientation (random or parallel) of fibres and the uniformity of their arrangement.

Fibre dust: Small fibre particles dispersed in the atmosphere.

Fibre engineering: The production of tailor-made fibres which are structural modifications of basic manmade fibres types to meet specific requirements, i.e. fibre construction tailored to produce special types of fibre with preprogramed properties in accordance with exact physico-chemical principles (molecular engineering).

Fibre fineness: The mean fibre diameter which is usually expresses in microns.

Fibre length: The staple length of the fibre. On combing wools, this is often 3–8 inches, on the down wools 1.5–3.0 inches. With cotton, it may be 0.25–1.00 inches long. Bast fibres, likes flax, may have a staple length of 36 inches.

Fibre (manmade): They are synthetically manufactured. There are two types of manmade fibres:

(a) Produced from natural polymers –(i) of vegetable origin; (ii) of animal origin.

(b) Produced from synthetic polymers –(i) polycondensation fibres; (ii) polyaddition fibres; (iii) polymer fibres; (iv) bicomponent fibres.

Fibre migration: Fibre migration, more correctly lateral fibre migration, popularly refers to the radial movement of a fibre within the yarn cross-section. Fibre migration (variable helix angle at different positions along the fibre length) determines the yarn structure and properties and can be characterised by (a) mean fibre radial position (b) migration amplitude (c) mean migration intensity (i.e. rate of change of radial position). See **Migration**.

Fibre modification: Some of the fibre modification used by processors are (a) weight reduction of polyester fibres, (b) acetylated cotton and (c) affinity modification.

Fibre moisture content: Moisture absorption of fibres; moisture regain.

Fibre number: The linear density of a fibre expressed in suitable units as tex, denier, millitex, etc.

Fibre placement: In general, refers to how the piles are laid into their orientation, i.e., by hand, by a textile process, by a tape layer, or by a filament winder. Tolerances and angles are specified. Microprocessor-controlled placement that gives precise control of each axis of motion permits more intricate winding patterns than are possible with conventional winding and is used to make composites that are more complex than usual filament-wound structures.

Fibre protective agents: Fibre protective agents are based on, e.g., protein degradation products, fatty acid protein condensation products, ammonium salts of alkyl sulphates and alkyl sulphonates as well as lignin sulphonic acid salts. Products of this type are used particularly for the protection of animal fibres in bleaching, dyeing and stripping In the wet processing of cellulosic fibres, fibre protective agents provide protective screening against oxidative/ reductive changes in vat dyeing and in oxidative bleaching, etc.

Fibre reactive dye: See **Reactive dyes**.

Fibre rope: A rope produced primarily from textile fibres.

Fibre saturation value: It is a measurement of the dye up-take potential of a fibre.

Fibre shape: Cross section of fibre.

Fibre sheath: The covering layer of a fibre.

Fibre silk: Commercial term for good quality of artificial silk.

Fibre softening range: See **Softening range**.

Fibre surface area: It includes both the external and, predominantly, internal surfaces of the fibre. The latter is dependent on the numerous capillary-like canals and inter-micellar spaces. The surface area of 1 g of cotton is approximately 5000 m^2, and the area of wet-table surfaces is approximately 100–400 m^2 for 1 kg of fibres. It is accordingly much greater than the surface area of the liquor at the fibre sites where surfactant molecules responsible for the wetting process are located.

Fibre texture: Fibre texture describes the state of order or orientation of the macromolecules in fibre structure. It can be understood by the presence of crystal aggregates with individual arrangements of crystallites in such a way that all the crystallites possess a definite lattice structure in approximately the same direction, i.e. a symmetry of the fibre axis exists which is detectable by X-ray crystallography. Fibre texture is characterized in other respects by the anisotropy of fibre crystallite orientation, i.e. higher orientation in the direction of the longitudinal axis due to primary valency bonds (which means greater strength, lower extensibility and swell ability as well as easier cleavability) and lower orientation in the transverse direction due to secondary valency bonds (which means lower strength and higher swell ability).

Fibre thickness: The average diameter of the fibre.

Fibre tow: See **Tow**.

Fibre transport tube: See **Fibre channel**.

Fibrefill: Virgin man-made fibres especially engineered as to linear density, cut-length, and crimp for use as a textile filling material such as pillows. This is not a good choice for your quilt interior.

Fibre length array: A series of individual fibres that are arranged in order based on fibre length or a series of groups of fibres that represent essentially uniform lengths within a group and which are arranged in order based on group length.

Fibre (birefringence): The algebraic difference of the refractive index of the fibre for plane polarized light vibrating parallel to the longitudinal axis of the fibre and the refractive index for light vibrating perpendicular to the long axis.

Fibre (channel): A component of an open-end spinning machine through which the fibres are conveyed by an air current from the opening device to the rotor.

Fibre(chip): In manmade textiles – staple fibres that are massed together as a unit and that maintain a single geometry or alignment.

Fibre cohesion: The resistance to separation of fibres in contact with one another.

Fibre (crimp): (1) The waviness of a fibre expressed as waves or crimps per unit length.

(2) The difference in distance between two points on the fibres as it lies in a non-stretched condition and the same two points when the fibres is straight.

Fibre (manmade): These are fibres manufactured from chemicals. There are many types:

(a) Produced from natural polymers –(i) of vegetable origin, (ii) of animal origin. The specific properties of viscose and modal fibres may be discussed under this category. The raw material for the manufacture of these fibres is cellulose. Whilst the chemical structure is identical to that of cotton, important differences in the physical structure of both these fibre types can be identified.

(b) Produced from synthetic polymers –(i) polycondensation fibres, (ii) polyaddition fibres; (iii) polymer fibres; (iv) bicomponent fibres. The synthetic man-made fibres include the three important polymer groups of polyester, polyamide and polyacrylonitrile. Although of somewhat lesser importance in terms of quantity, the polyolefins such as polypropylene, polyvinyl derivatives and elastomers based on polyurethane also deserve to be mentioned here.

Fibre (soft or bast): See **Soil or bast fibre**.

Fibre (hard): The non-elastic yarn wound around an elastic core in covered elastic yarns. See **Hard fibre**.

Fibre (hard or leaf): Elongated fibre strands from leaves or leaf stems of plants usually used for making rope, twine, net, etc. See **Hard or leaf fibre**.

Fibre (high tenacity): Manmade fibre having exceptional breaking strength. See **High tenacity fibre**.

Fibre (man-made): A general name for synthetic fibres produced from fibre forming substances which may be (a) polymers synthesized by man from simple chemical compound, (b) modified or trans formed natural polymers and (c) glasses.

Fibre (natural): A class name for various general of fibres of (a) animal, (b) mineral, or (c) vegetable origin.

Fibre (regenerated): Manmade fibres made from naturally occurring polymers. For example, Viscose.

Fibre(soft or bast): Flexible elongated strands from the inner bark of various plants usually used for making twine, nets and even woven fabrics.

Fibre, Synthetic: A fibre that has been manufactured by synthesis of suitable linear polymer followed by extrusion of the molten or solution of the material through a spinneret

Fibre (textile): (a) General – A generic term for the various types of matter that form the basic elements of textile fabrics and other textile structure; (b) Specific – A unit of matter that is characterized by having a length at least 100 times its diameter or width and which can be spun into a yarn or made into a fabric by interlacing in a variety of methods including weaving, knitting, braiding, felting, and twisting. See **Textile fibre**.

Fibre-bonded floor covering: A textile floor covering which is composed of entangled textile materials bonded together by a mechanical, physical or chemical process (example – needle punching, stitch bonding, heat treatment, or resin impregnation) or by combination of two or more of these processes.

Fibreset process: A resin finishing process for cotton and polyester/cotton fabrics in which a reactive prepolymer is applied together with N-methylol compounds. Polymerization and crosslinking of the applied finish is achieved by treatment in the vapour of a low boiling chlorinated hydrocarbon.

Fibrets: Very short (<1mm), fine (diameter <50μm fibrillated fibres that are highly branched and irregular resulting in very high surface area. Fibrets can be produced from a number of substances including acetate, polyester, nylon, and polyolefins. By selection of polymer type and incorporation of additives, they can be engineered to meet a range of specialized requirements.

Fibreweigh: An instrument having a capacity for weighing a 51 grain (3.30 g) specimen with a sensitivity of at least 0.2% of the mass being weighed.

Fibrids: Synthetic binding elements with a fine structure resembling that of “wood pulp”, i.e. a netted filamentary or fibrillar structure, produced from polyamide, polyester, polyacrylonitrile, etc. They are used as a binding medium after the extraction of water fromsynthetic fibres on paper making machines.

Fibril:A tiny thread-like element of a synthetic or natural fibre.

Fibrillae: Specks visible on the surface of silk yarns.

Fibrillar proteins: High molecular weight polypeptides consisting of long threadlike or chain molecules (e.g., Silk fibroin, Wool keratin). They are formed from Amino acids.

Fibrillated film tow: An assembly of textile films.

Fibrillated film yarn: Yarn produced by fibrillation film that has been converted into a longitudinally fibrillated structure.

Fibrillated yarn: A yarn produced by the process of fibrillation.

Fibrillating film: A polymer film in which molecule orientation has been induced by stretching to such a degree that it is capable of being converted into yarn or twine by manipulation.

Fibrillating roller: A pinned roller used in fibrillation.

Fibrillation: The act or process of forming fibrils. The act of breaking up a fibre, plastic sheet, or similar material into the minute fibrous elements from which the main structure is formed.

Fibrils: (a) A very fine cellulosic fibre (Ø = 0.4 μm) formed from lamellar filaments resp. individual filaments (approximately 100 fibrils); (b) Fine fibrils have also been detected in cellulosic fibres under the electron microscope; (c) In wool structure, the fibrils are spindle shaped, tightly packed, twisted, fine fibres in the so called spindle cells. These fibrils are composed of fibril bundles which themselves are composed of individual fibrils; (d) By analogy with wool, individual fibrils and fibril bundles are likewise present in the structure of silk. In the case of silk, however, fibrils are also understood to include the individual small fibres projecting from the surface of silk fabrics formed by the action of abrasion.

Fibrogram (in testing cotton fibres for length): The curve representing the second accumulation of length distribution of fibres sensed by the length measuring instrument in scanning the fibre beard.

Fibroin: Structural component (from Fibroinogen) of silk (comprising 71–82%), in the form of a double fibroin filament which is covered by sericin in the form of silk gum. Contains the amino acids alanine, glycine, serine comprising approximately 75% as essential components of high molecular order crystallites, whereas with tyrosine the remaining amino acids which represent the low-molecular portion are the presumed components of the amorphous areas, which may include the "cuticulin".

Composition ($C_{15}H_{23}N_5O_6$):49% carbon, 25% oxygen, 19% nitrogen and 6% water. Fibroin is insoluble in water, can be hydrolysed to form amino acids and detectable as a polypeptide in a similar way to wool keratin, but without the cysteine content.

Fibroin fibre: A man-made protein fibre produced from silk fibroin. A fibroin solution is obtained by the dissolution of silk wastes which can be spun into fibre again. Fibroin fibre has no commercial importance.

Fibroinogen: Soluble and water-extractable protein that is a precursor of silk fibroin. After standing for several hours, a fibrous precipitate forms spontaneously as a linear aggregation of individual fibrils (length up to 0.35 μm, diameter 0.01 μm). Can be considered to be analogous to Prekeratin of wool.

Fibronaire: An instrument that determines the micronaire reading of raw cotton fibres using the 'porous plug' air flow technic.

Fibronia: Proprietary name for a fibre made of grass; It is carded with shoddy and used for carpets or belting yarns.

Fibula: Pinned clasp for holding robes together in ancient period and middle ages.

Ficelle-coloured: Having the colour of twine.

Fick's equation: It describes the diffusion processes for the quantitative diffusion of dye in terms of rate of transfer per unit area within, e.g., a fibre

$$n = \frac{-D\,dc}{dx}$$

From this equation, the term $\frac{dc}{dx}$ is a concentration gradient, which represents an important prerequisite for the Diffusion.

Fitchu: Cloth square for covering the décolletage used by the English.

Fid: A wooden or hard plastic tapered tool used as an aid in rope splicing.

FID: (a) (French) Abbreviation for "Fédération International de Documentation", international association for documentation, with headquarters in The Hague, Holland; (b) Abbreviation for Flame ionization detector.

Fiddle string: A particular form of tight end or tight pick that becomes evident only after wet processing.

Field grey: The grey colour of military uniforms particularly those worn by the German infantry.

Field research: Information found directly- by going out and looking; by asking questions to relevant individuals and groups; by using mailed questionnaires, setting up focus groups and interviewing people.

Field of vision: In the resting position, the human field of vision with both eyes extends from an angle of approx. 65π to both sides of the direction of view and can measure approx. 50π in the upper and lower directions. The powers of vision are greatest in the centre of the field of vision and lowest at its periphery. This fact can be explained by the higher image definition in the Fovea central is where the greatest number of retinal cones are located which, in contrast to the rods (lightness/ darkness receptors), are responsible for colour vision. Towards the periphery, the capacity for colour perception decreases greatly, and the sensitivity to lightness/darkness increases due to the higher number of rods.

Fiery: Having the appearance of flames or fire as in 'fiery-red'. 'Firy' is an obsolete form.

Fiesta: A reddish shade of orange.

Fiesta pink: A bright pink colour.

FIFT: (German) Abbreviation for Forschungs Institut für Textiltechnologie (Chemnitz), Research Institute for Textile Technology, Chemnitz.

Figuartoes: Wool fabric, made in Norwich, England, in the 17th Century.

Figure: Large weave patterns (figured).

Figured casement: A casement cloth in which patterns has been introduced by weaving.

Figured fabric: A fabric in which patterns or motifs are woven using dobby or jacquard or any other mechanism.

Figured velvet: A patterned velvet with some pile cut and some uncut, but also with areas of backing fabric visible.

Fiji silk: Crisp silk, heavier than jap and therefore much more expensive. Usually found in natural colours and cream. Used for making blouses and dresses.

Fil a-fill: Dark and light yarns alternate in both warp and weft. With a 2/2 twill constructions a miniature staircase effect results.

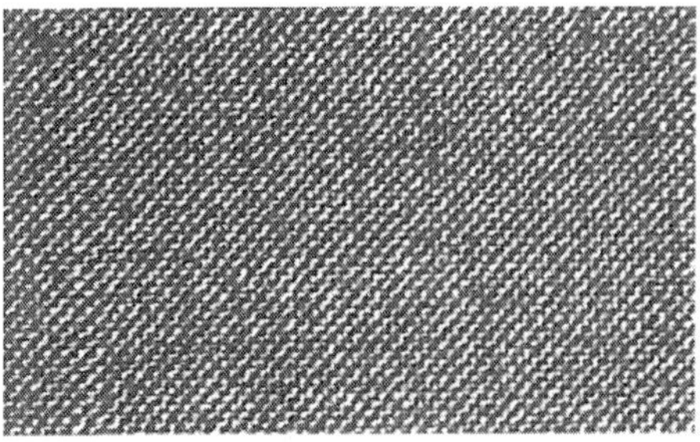

Fil-a-fil

Filamel: See **Antistatic tricot**.

Filament: A fibre of indefinite length. Also used in connection with the now obsolete term

Flat filament

Artificial silk

Filament blended yarn: A filament yarn which contains separate filaments of two distinct types, the filament being more or less randomly blended over the cross section of yarn.

Filament count: The number of individual filaments that make up a thread or yarn.

Filament fibre: A fibre of indefinite or extreme length, e.g., silk filament, which runs from 300 to 1400 yards (274 to 1280 mn) and more in length: the extreme length of filament permits their being used in yarn without twist or with very low twist, and they are generally made into yarn without the spinning operation required of the shorter staple fibres, such as wool, cotton, man-made fibre staple. Yarns made of filaments are usually smoother than yarns made of staple.

Filament number: The linear density of a filament expressed in units such as denier or tex. See **Fineness**.

Filament winding: In the fabrication of composites, the process of placing reinforcing fibres over a rotating form (mandrel) to make the product shape. Prepare fibres or dry fibres that are treated in a resin bath immediately prior to winding may be used. The wound form can be cured or consolidated after the fibre winding is complete to product specifications.

Filament yarn: A yarn composed of (continuous) filaments assembled with or without twist. Yarn which is composed of filament yarns produced from one or more filaments are preferably referred to as monofilament yarn or multifilament yarn, respectively.

Filament (in steel cord): The individual element in a steel strand or cord.

Filament (in textiles): A continuous fibre of extremely long length.

Filament (steel): See **Steel filament**.

Filamentation: Damage to multi-filament yarns that results in broken filaments.

Filemot: The colour of a dead leaf. See **Feuillemorte**.

Fillet: A long, narrow strip of wire card clothing with which the doffer and cylinder of the card are spirally wrapped.

Filet lace: A mesh fabric of square design with some squares blocked into form a design.

Filet net: Any type of square ground-mesh of lace; a woven net.

Filler: Non-fibrous material, such as insoluble clays or gypsum, together with starches, gums, etc., added to a fabric to increase its weight or to modify the appearance or handle of the fabric.

Filler (in knitting): This is the yarn length which lies between successive sinker loops in the vertical direction.

Filler yarn: See **Stuffer yarn**.

Filling: In a woven fabric, the yarn running from selvage to selvage at right angles to the warp. Each crosswise length is called a pick. In the weaving process, the filling yarn is carried by the shuttle or other type of yarn carrier.

Filling band (in woven fabrics): A visual defect across the width due to a change occurring in the yarn for a large number of picks. See **Barre', Mixed filling**.

Filling band: See **Filling bar, Weft bar**.

Filling bar (in woven fabrics): A visual defect across the width of the fabric which contains a limited number of picks of different appearance than normal.

Filling bar (coarse): A weft bar seen along the full width or a portion of the cloth due to the use of yarn of a larger diameter than normally used in the fabric or due to more number of picks than the normal picks used in the fabric.

Filling bar (fine): The opposite of filling bar, coarse. A bar caused due to yarns of diameter less than that normally used in the fabric or less picks than normally used in the fabric.

Filling barre': See **Barre'**.

Filling bow: See **Bow**.

Filling crepe: Filling crepe is made by weaving a plain warp with crepe yarn weft. Box crepe is made using alternating ends and picks of S twist and Z twist. Since these are made on box looms, these crepes are called box crepe. See **Crepe**.

Filling end: A smaller than normal diameter warp end.

Filling elongation and tension: Stretch or tension measured at right angles to the warp direction of the fabric.

Filling fibre: Used to fill cushions, duvets, sleeping bags, etc. Nowadays, it mainly consists of synthetic fibres.

Filling finish: A type of finish used to increase the weight of a textile fabric by the application of so-called filling agents (which are cheaper than the textile material).The finish also produces a "compacting" effect and a fuller handle which may or may not be resistant to washing. Weighting.

Filling material (of feathers or down): The contents of an industry product containing feathers or down of any kind or type with or without other natural or synthetic materials.

Filling skew: See **Skew**.

Filling snarl: See **Snarl**.

Filling (mixed): See **Mixed filling**.

Filling (run out): See **Broken pick**.

Filling tests (in cloth testing): Tests in which the filling yarns are torn.

Filling to filling seam: A sewn seam in which the yarns in the filling direction on both sides of the sewn seams are perpendicular to the seam.

Filling warp: Consists of coarse warp threads which provide fabrics with a coarse fullness and allow less interlacing. Filling warps are supplied for weaving on separate beams.

Fill leakage (in comforters): Either partial or total penetration of the stuffing material through the outer or face fabric.

Film: Any thin layer of a substance, e.g. size film, synthetic resin film, paint film, etc., which can also be removed from the surface of the parent substance and is likewise referred to as a film.

Film-forming agents: Polymers which form coherent film after evaporation of their solvents or dispersants. The minimum film forming temperature is important in textile applications.

Film yarn: See **Slit-film yarn**.

Final finishing: (a) All finishing processes after dyeing or certain specific mechanical finishing processes for coloured woven fabrics (dry finishing) – rubberizing and proofing, waterproofing, drying, shearing, pressing, decatizing, crease-proofing, calendaring, embossing and suchlike; (b) Post-finishing of dry-cleaned garments – Dry finishing of fabrics;(c) Final finishing means that all the chemical finishing treatments applied to a textile fabric after dyeing and/or printing which provide fitness for purpose resp. the desired wear characteristics, e.g. wash-and-wear finishes, easy-care finishes, etc.

Filter: A true filter separates solids from a gaseous or liquid flow that is passing through it. The solid media or woven fabric in the filter allows air or water to pass through, but retain the solids. Filters may remove particle sizes smaller than the pore size due to adsorption to the surface of the filter medium. In air pollution control, particles may be removed by ***fabric filters*** or ***depth filters***. In water pollution control, ***rapid deep bed filters*** and ***slowsand filters*** remove particles throughout the filter body. ***Microfiltration*** is an example of a depth filter used in water treatment. ***Surface filters*** in water and wastewater treatment include ***microstrainers and drum filters***. ***Ultrafiltration*** and ***nano-filtration*** can filter out large molecules from water. Filters used for sludge dewatering include the ***belt press filter, filter plate press*** and ***vacuum filter***. The ***trickling filter*** and ***biological aerated filter*** are not true filters, because the treatment of the wastewater is by adsorption, absorption and biodegradation in ***the biofilm***.

Filter aid: A powder added to a solution to be filtered that forms a porous bed to improve filtration.

Filter cake: (a) Dewatered sludge from a belt filter press, filter plate press, vacuum filter, etc. It has a friable consistency; (b) In air pollution control, the dust collected on a fabric filter.

Filter cloth: Any cloth used for filtering purposed. Nylon, polyester, vinyon, PBI, and glass fibres are often used in such fabrics because they are not affected by most chemicals.

Filter fabric: See **Geotextiles, Filter cloth**.

Filter medium: The permeable material that separates particles from a fluid passing through it.

Filter plate press, filter press, plate press, pressure filter:A collection of 50–100 parallel, ridged, rectangular, recessed, cast iron plates which can be separated or bolted together. The internal surfaces of the plates are covered with a textile sheet, the filter cloth. Plate presses are used in many industries, including dewatering wastewater sludge and water treatment sludge. The sludge is invariably conditioned before pressing. After the plates have been joined together, the sludge is pumped into the press at a pressure of 5–7 bar which is maintained for 6 to 18 hours. The liquor passes through the cloth as pressure is applied and drained away through holes in the plates. When the filtration cycle is complete the plates are separated to release the sludge cake which is 25–35% solids. The press effluent is highly polluting. This type of dewatering device has the disadvantage that it is a batch process, unlike the ***belt filter press***. The ***membrane press*** is a development of the filter plate press.

Filter system: The combination of a filter and associated hardware required for the filtration process.

Filtrate: Liquid that has passed through a ***filter*** and so has lost some of its suspended solids.

Filtration: Cleaning of water or air by ***filters, air filters***, etc. ***Adsorption*** on the medium is important in both types.

Filtration rate: The volume of liquid that passes through a given area in a specified time. Usually expressed as gallons per squarefoot per minute (or hour).

Final twist: The number of turns per unit length in a single yarn component of a plied yarn or the plied yarn component of a cabled yarn as the component lies in the more complex structure.

Findings: (a) Miscellaneous items attached to garments and shoes during manufacture. Included are buttons, hooks, snaps, and ornaments; (b) Miscellaneous fabrics in garments such a zipper tapes, linings, pockets, waistbands and facings.

Fine structure of fibres: The fine structure of a fibre denotes the composition of the smallest components such as the chemical constituents (polymer chains), their bonding into crystalline and non-crystalline (amorphous) regions and fibrils. The colloidal structure when spinning a synthetic fibre or during the biosynthesis of a natural fibre affects the fine structure of the fibre and its morphology, i.e. the whole structure is composed of its smallest components.

Fine end: A warp yarn smaller in diameter than normally used in the fabric.

Fine filling: See **Thin filling**.

Fine pick: See **Thin filling**.

Fine rib fabric: 1 × 1 rib fabrics are double-face knit goods. The fabric has a similar appearance on both sides and single wales of face loops alternate with single wales of back loops. Rib fabrics are very elastic width-wise and are therefore used for cuffs, collars and trimmings as well as underwear.

Fine wool: Wool, mostly from merino sheep, which is used in good quality clothing and fabrics.

Fineness: A relative measure of size, diameter, linear density, mass per unit length expressed in a variety of units.

Fineness (of textile fibres): A relative measure of size, diameter, linear density, or mass per unit length expressed in a variety of units.

Fineness of natural fibres: This describes the 'thickness' of textile fibres and is usually expressed in dtex as a rule (linear density of fibres and yarns) or, in special cases, in λm (diameter) or, the micronaire value for cotton. The fineness of fibres is an important factor which has a direct influence on yarn properties with regard to strength, spin ability limits, yarn character and uniformity. The usual fineness ranges of the main natural fibres are:

Cotton 1.5 – 5.0dtex

Flax 10 – 40 dtex

Wool 02 – 15 dtex

Silk 01 – 04 dtex

The fineness of man-made fibres is controlled by the spinning process.

Fineness of synthetic fibres: The man-made fibre industry has been successful in producing finer and finer fibres. In the filament yarn sector, fibres of 1.0–0.3 dtex have been produced by European fibre manufacturers, i.e. the so-called micro-fibres. In the meantime, Japanese developments have been aimed at fibres with a fineness of <0.3dtex. Such fibres are logically described as so-called super micro-fibres. In the staple fibre sector, fibre fineness of up to 2.4 dtex for polyester have been available for some time.

Finger mark: An irregular spot showing variation in picks per inch for a limited width.

Finger press: It is utilizing your thumb and forefinger to flatten or press small fabric sections without an iron.

Finger test: Used to evaluate dye penetration in transfer-printed textile floorcoverings. The complete back-coated and printed carpet sample is wrapped around the little finger. If no white fibres are visible, penetration is considered to be satisfactory, even for carpet laying on staircases which is one of the most critical carpet applications.

Finger-tip towel: A textile product with fringes and side hems or selvedges which is smaller than a hand towel generally used for decorations and to dry hands.

Fingering yarn: A plied hand-knitting yarn normally produced in a 2, 3, or 4 ply to a specific diameter. The count of the 4 ply yarn is half that of a double knitting yarn.

Fingram: Eighteenth century coarse English and Scotch serge.

Finishing, Special: A general term for fabric finishes imparting special properties such as water repellency, crease, stain and flame resistance.

Finish: Can be either chemicals that change the fabric's aesthetic and/or physical properties or changes in texture or surface characteristics brought about by physically manipulating the fabric with mechanical devices. It can also be a combination of the two. Final treatment, Proofing, Dressing; (more generally, also) Improving, Processing; (often also) Ironing, Pressing, etc. (cleaned or dyed clothing).Terms used broadly, as follows, to include added materials, the process employed (finishing) and the final result:

(a) A substance or a mixture of substances added to textile materials at any stage to impart desired properties.
(b) The type of process, physical or chemical, applied to produce a desired effect.
(c) Such properties, for example smoothness, drape, lustre, or crease resistance, produced by (i) or (ii) above (or both).
(d) The state of the textile material as it leaves a previous processor.

Note: The mechanical operations of spinning, weaving and knitting, though they may largely determine the result, are excluded.

Finish (verb): To apply or produce a finish.

Finish (in garmenting): It refers to any means of completing the raw edge of the garment.

Finish, velour: The fabric is moistened and an erect and loose nap is raised on the face, then dried and lightly shorn.

Finish, Saxony: The fabric is first fulled properly, then rolled and stretched, gigged, shorn short, then brushed and pressed.

Finish, serge : The goods are carefully burled and singed, crabbed and scoured and shorn, producing a clear face.

Finish, melton: The goods are fulled, steamed, dyed and shorn and finally lightly brushed.

Finish, Kersey: After thorough burling, the goods are fulled, gigged and the nap laid. The fine grade goods are then steamed and pressed, while the lower grades are given a water finish.

Finish, beaver: The goods are fulled, the lower grades also weighted with flocks, gigged and the nap well laid, then steamed and finally shorn short.

Finish, calico: The cloth is singed, bleached, boiled off and printed.

Finisher: Proofer, dresser.

Finished (for glass laminates): A descriptive term for woven fabrics that have passed through a treating procedure which is compatible with a resin matrix or facilitates manufacturing or both.

Finish (in buttons): The surface conditions or texture.

Finished: The completion of all manufacturing process/operations.

Finished fabric weight: Mass per unit area expressed in grams per square metre (ounces per square yard) grams per unit linear metre (ounces per linear yard) or inversely as metre per kilogram (linear yards per pound) or square metre per kilogram (square yards per pound).

Finished goods: Fabric which has been gone through all the required processes and finished ready for inspection and cutting.

Finished (in textile floor covering material): The completion of all manufacturing operations.

Finished pile yarn floor coverings: Pile yarn floor covering which has undergone all steps of the manufacturing process.

Finished yield (in knitted fabrics): The number of finished square yards per pound (square metres per kilogram) of finished fabric.

Finishing: It is the last manufacturing step in the production of textile fabrics. As an integral part of wet processing, finishing is the operation where the final fabric properties are developed. Finishing is not restricted to wet processing alone since any operation for improving the appearance or usefulness of a fabric after it leaves the loom or knitting machine can be considered a finishing step. Besides washing, this includes desizing, steaming, setting and the production of various effects such as Calendaring, raising, etc., as well as treatments to improve serviceability properties such as water-repellent, shrink-resistant, wrinkle-resistant and soil-release finishes, etc. Finishing is understood to involve all those processes which cannot be classified under dyeing and printing.

Finishing agents: Auxiliaries applied on the fabric to achieve different finishes.

Finishing bar: Uneven appearance across the entire width of the fabric, usually caused by the machine stoppage during processing.

Finishing mangles: The padding mangles which are used in the finishing machines. Even though the expression of the mangle is not very critical, still the mangle has to be of good quality, with average expression, since the expression will also affect the efficiency of the drying part of the finishing range.

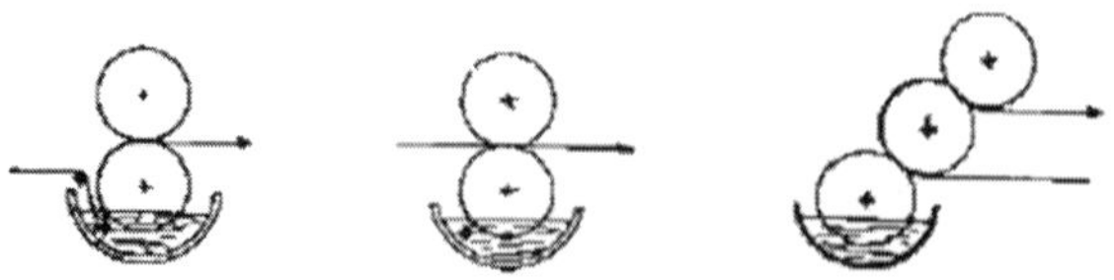

Finish padder types [immersion path in a trough (left),slop padder (centre), inclined padder (right)]

Finishing spot: A discoloured area on the fabric.

Finish stains: Such stains are usually stiffer than the surrounding material and are mainly round, oval or of matt appearance with a distinct dark ring. They may be removed by (a) a suitable detergent (with glycerol and, if necessary, ammonia); (b) 1% ammonia solution. Care is necessary in the case of poor colour fastness. Ring removal: water/acetone. If necessary, matt areas (on wool) can be removed by covering with a piece of damp cloth after drying followed by hot ironing.

Finishing stitches: Stitches which are used to finish raw edges of fabric are finishing stitches. For example – Button hole stitch.

(a) *Overcasting*: This is a method of sewing over two edges of fabric to bind then together and to prevent them from fraying. To make this stitch, take two layers of fabric together. Now insert the needle at the point where you have to start overcasting stitch at a distance of one-fourth inch after taking it from above the fabric.

(b) *Double overcasting*: This is also a stitch used for finishing of raw edges. To make this stitch, first do overcasting and then repeat it again, but now from the night opposite side.

(c) *Overhand stitches*: This stitch is used to hold unfinished edges of fabric together and to finish them. The method to form this stitch is same as overcasting. The only difference is that we do not maintain any distance between two stitches. This is a very strong stitch and it also gives a beautiful effect when a strong, secure and invisible seam is needed.

(d) *Button hole/blanket stitches*: This stitch is used to make button holes also to finish and to hold unfinished edges of fabric together. And since the edges of the blankets are used to be finished with the help of this particular stitch, it is also known as blanket stitch. To form this stitch, bring the thread out on the lower line, insert the needle in position in the upper line while taking a straight downward stitch to form a loop and repeat. This stitch may also be worked on even weave fabric.

Finishing machine stitches: Seam edges are given finishing machine stitches (finished) to prevent them from fraying in use and during washing or dry cleaning; to improve the appearance of the inside of your garment; and to strengthen the same itself. Seams in a garment should be finished as the garment is made and before being crossed by another seam. The seam finish must be without bulk, so that it is not visible when the garment is worn and does not from a ridge on the right side when the garment is pressed. Main finishing machine stitches are –

(a) *French seam*: A French seam is a "seam within a seam" and can be used on sheer fabrics or lingerie. It is not recommended for curved seams.

(b) *Mock French seam*: Use this seam in fine and lightweight fabrics, children's clothes and blouses, in place of the conventional French seam. Both straight and curved seams may be finished in this way.

(c) *Flat felled seam*: The flat-felled seam provides a neat attractive finish to both sides of the garment. This type of seam is usually used for sport clothes and reversible garments.

(d) *Welt seam*: This seam finish method allows the plain seam to be "detailed" with top stitching in the same colour thread as the garment or in a contrasting colour thread.

(e) *Open welt seam*: This seam forms a small tuck and emphasizes a construction detail. This, of course, will add interest to a plain fabric.

(f) *Bias-bound seam*: Trim notches from seam edges, wrap bias binding around it (use ready – made binding, or cut your own from bias binding fabric). Stitch does to edge of top fold, catching underneath fold in stitching. Bias binding is especially good for finishing seams in an unlined jacket or coat. Fold and press the bias strip so that the under-section is slightly wider and this will help to ensure the stitching through both edges of the binding.

(g) *Broad seam*: This seam is used to give broad turnings or folds in a garment to make tunnels for cards in salwar (Indian traditional lower wear) or pyjama, or to give bottom turnings in a garment. Fold the margin by 0.5 cm and then take a broad fold towards the wrong side of the fabric and give a plain straight seam, stitching the fold back onto the fabric. The seam is grain from the wrong side of the garment.

(h) *Pinked seam*: Use a pinked seam only on fabrics. This will not fray. After stitching, a plain seam, pink the seam edges with pinking shears. Remove only the seam edge. Cut with a medium stroke; do not fully open the shears or cut right to the point. Press the seam in the same position as it was stitched. And then press it open.

(i) *Pinked and stitched seam*: This seam finish may be used on almost any fabric which would fray. Stitch a plain seam. Place a line of stitching about .6 cm (1/4 inch) from the seam edge. Using a 1 – 1.5 mm (20) stitch length. Pink the edges. The stitching will prevent the seam from fraying and curling.

(j) *Edge–Mattering stitched seam*: Use to give finish to on seams on lightweight and medium-weight fabrics as well as on unlined jackets and coats. Stitch a plain seam. Press the seam flat, and then press it open. Then turn the edges of the seams 0.5 cm alone margins and give a straight seam. This will finish the seam edges on both sides. Pink the edges; then fold them under 0.3–0.6 cm stitches will help you keep this stitching straight and parallel to the edge.

(k) *Seams overcast by hand*: If the fabric has a tendency to fray, a machine finish would be too harsh. Finish the seam edges by hand. Stitch a plain seam and press it open. Make slanting stitches over the seam edge by hand, about 6 cm (1/4 inch) apart and about 0.3 cm (1/8 inch) in depth, depending on the weight of the fabric. Do not pull the threads too tightly.

(l) *Zig-zag seam*: Zig-zag stitching is an ideal seam finish for jersey, double knit and other fabrics with 'give' because the stitch is as flexible as the fabric. Either the plain or multi-stitch zig-zag may be used. After stitching a plain seam, press it open. Select a plain zig-zag stitch, medium stitch with and between 1.5 and 1 mm (15 and 20) stitch length. Stitch near the seam edge, but not over it. Press and trim off the seam edge close to the stitching.

(m) *Herringbone seam*: A herringbone seam is used where the garment is interfaced or mounted. This is an excellent means of preventing seam edges from rolling when the garment is dry-cleaned. Use it in shoulder seams and notched collars and where collar and facings join. Trim the interfacing or mounting seam allowances close to the stitching. Press the plain seam open. Then herringbone the seam edges by hand. Working from left to right, with the needle pointed to the left, catch only one or two threads in the seam allowance, then only a single thread outside the seam allowance and in the interfacing or mounting. Alternate the stitches along the seamline. The stitches will not show from the right side.

(n) *Seams that cross*: Seams that cross occur at the shoulder line. Waistline and underarm and where darts join a seam or tuck. Press the seams open and finish the edges as required by the fabric. To ensure accurately matched seams that cross, pin with a fine needle, with only the point nipped into the fabric at the stitching line; then pin on each side on the seamline. The needle prevents one seam from slipping beyond the other during stitching and will not mark delicate fabrics. Trim away the excess seam allowance at the point where the seams cross.

Finishing wash: A treatment carried out after the production of textile fabrics in order to remove substances present in, or applied to, the fibre during previous stages of processing and thereby achieve a further improvement in the properties of the material (e.g. scouring of wool fabrics). In the discontinuous finishing wash, the material remains in a machine and wash/rinse liquors are introduced and exchanged as required by the process.

Finishing, Basic or Regular: All processes required to convert gray (or greige) goods into merchandise saleable to the consumer or ultimate processor. Basic finishes are mechanical and/ or chemical. Finishes that appeal to the eye include shedding, singeing, brushing, beetling. mercerization, stentering, calendaring, moireeing, pressing, embossing, crimping, glazing, polishing, inspection and repair. Finishes that appeal to touch include napping, weighting, coaling, sizing or dressing and starching. Functional or Special finishes are other types of Finishes.

Finishing, Functional: Treatments that are applied to fabrics to make them better suited for specific uses. These finishes include: Absorbent Finish, Air Conditioning (fabric), Anti-bacterial Finish, Crease-resistant, Durable Crease, Durable Press, Flame-resistant, Metalizing, Mildew-resistant, Moth Repellent, Shrinkage Control, Wash-and-wear, Water Repellent, Water-proofing.

Finishing, Regular: See **Finishes, Basic.**

Fineness of fibres: A relative measure of size, diameter, linear density, or mass per unit length expressed in a variety of units.

Fique fibre (Cabuya): Very strong, straight, smooth fibre, yielded by the leaves of the Furcraea gigantea of South America; used for bagging, etc. See **Mauritius hemp**.

Fir green or fir-green: A dark muted green.

Fire engine red: A bright vivid red.

Fire hydrant: A branch pipe, usually in a street, connected to the water main, which enables a fire hose to be connected to the main. In low fire risk areas, hydrants may be placed at every 100–150 m in high risk areas, sometimes as close as 30 m. The minimum flow from one hydrant should be 1.4m^3/min, but 25m^3/min may be needed in a highly populated area; this flow is obtained by grouping several hydrants.

Fire (as related to textile flammability): An uncontrolled conflagration in which materials are destroyed by burning as evidenced by flames of varying size and shape, and a high intensity heat source of 5kw or greater, such as burning contents of a room, a burning basket, burning building, or forest fire.

Fire-orange: A reddish-orange colour.

Fire resistant fibres: Fibres based on inorganic materials (mineral fibres) take first place here, e.g. glass, ceramic or asbestos. High temperature fibres.

Fire retardance: The resistance to combustion of a material when tested under specified conditions.

Fireproof: This term relates to the property of remaining unchanged through the action of fire, e.g. as in the case of asbestos. All other textile fibres can be given aflame-retardant finish, but these are not normally fireproof. Developments in the spinning technology of carbon fibres have resulted in the production of fibres with heat resistance up to 2500 °C which therefore qualify as fireproof fibres. Fireproof properties begin at 1500 °C; the most common oven temperatures are between 1500–2000 °C.

Fire-retardant treatment: Any process such as spraying, padding, dipping, brushing, or otherwise applying a material onto the pile of a carpet or rug at any stage of manufacture, which has as its primary or secondary purpose the reduction its flammability.

Firmness factor: A term derived from cloth setting theories. It takes both of the thread spacing relative to the yarn diameter (cover factor) and the frequency of the interlacing.

Fique: A fibre from the leaf of the plant Furicraea.

Fish: A dart cut at the waist of a garment to give a closer fit.

Fishbone stitch: An embroidery stitch. This stitch is useful for filling small shapes. Bring the thread through at one point and make a small straight stitch along the centre line of the shape. Bring the thread through again at a point slightly lower and away from the initial point and make a sloping stitch across the central line at the base of first stitch. Bring the thread through at a point at the opposite side to the second point and make a similar sloping stitch to overlap the previous stitch. Continue working alternatively at each side until the shape is filled.

Fish test: Procedure for monitoring the contamination of bodies of water. The average lifespan of certain speciès of fish under such conditions is used as a measure of the contamination of the water. Specially bred fish are released at given intervals of time into a section of water (or tributary) which is isolated by wire netting. The occurrence of dead fish is counted and from the mean lifespan the degree of contamination is derived. See ***Lethal limit; Toxicology***.

'Fished' wool: Short-fibre wool, which has been "fished out" of the wool washing plant. It is usually returned for process in the woollen yarn spinning system.

Fisherman's knot: See **Weaver's knot**.

Fisherman's lace: See **Point Pecheur**.

Fisheye cloth: A fabric with a woven diamond-effect pattern, constructed in a similar way to birds eye, but larger. It is often a woollen cloth, but may contain other fibres such as polyester.

Fishnet (weft knitted): A weft knitted structure, knitted on machines having facilities for activating one set of needle individually. The fabric structure is a combination of float stitches and plated stitches. A common fishnet structure is knitted by plating a thick yarn with a thin yarn in such a way that alternate stitches in the fabric consists of plain plated loops in which the thick yarn and the thin yarn have been knitted up together. The stitches between the plated stitches are knitted from the thin yarn only, whilst the thick yarn is floated at the back of these stitches. The fabric is knitted on the basis that where, in one course, odd needles knit the plain stitches from the thin yarn only, this is done by the even needles at the next course, and so on. This produces an openwork structure.

Fitted carpets: Carpets as floor coverings, in roll form, for wall-to-wall fitting. See **Contract carpet**.

Fitted sheet (in textiles): A product usually made with boxed corners, sometimes elasticized with shape and size to conform to the contours of the mattress and used for covering the mattress on a bed.

Five O' clock: Fine damask linen table cloth; used in England.

Five original colours: In painting, the colours red, green, yellow, blue and purple are sometimes referred to as the 'original colours' (in addition to black and white). Aristotle's theories indicated that all other colours could be produced by adding black or white to these five colours. In the 17th Century, Guido Scarmiglioni and Robert Boyle regarded black, white, blue, red and yellow as the five simple colours from which all others were derived. See **Essential colours**.

Fivette: A lightweight French cotton or woollen lining, woven in diagonal twill.

Fixation: Formation of the "final" bond between the dye and fibre. The bond type formed between the fibre and the dye varies with the type of dye and the fibre. As examples, reactive dyes fix by covalent bonding while acid dyes fix by a variety of mechanisms such as ionic bonding and hydrophobic forces. Disperse and vat dyes are fixed in the fibre largely by physical entrapment of insoluble dye within the fibre. The bond that causes final fixation is not necessarily the same type of bond is first made as the dye *exhausts* onto the fibre.

Fixation accelerator: A fixation accelerator increases the diffusion of a disperse dye in a print thickener to such an extent that it is no longer a rate-determining step in the rate of reaction during the fixation of prints on polyester. A fixation accelerator thus speeds up, or lowers the temperature required for reaction. Moreover, the fibre structure is changed by a fixation accelerator in such a way that diffusion of dye into the substrate is improved without the accelerator itself remaining behind. Fixation accelerators also have wetting and dye-dissolving properties.

Fixation temperature: The temperature range in which dyes become firmly anchored (fixed) to an appropriate textile material.

Fixation time: The time span required for a dye to become firmly anchored (fixed) to a textile substrate at a specific **Fixation temperature** (See).

Fixative (in dyeing): A chemical that helps improve wash fastness of dyed fabric. A few types of dyes do not bond strongly to fibres, and will wash out over time. ***Direct dyes*** are notorious for this. Fixatives applied after dyeing can help, although some will degrade light fastness or cause shade changes. Some fixative contain a small amount of formaldehyde, which is now regarded as carcinogenic. Much care is warranted in use of these fixatives, and garments should be thoroughly washed before being worn. Many of the newer fixatives do not contain formaldehyde. The mechanisms by which fixatives work vary, but typically form large, relatively insoluble complexes with dye molecules inside the fibre. ***Soda ash*** is sometimes called a fixative for reactive dyes, but it actually creates the high-pH conditions that allow the reactive dye to bond directly to cellulose fibres, and is not a fixative in the accepted sense.

Fixed fees: Fees for a project that do not alter.

Fixed mean integrator (in yarn testing): An integrator that uses deviations from a fixed mean namely, the true linear of the mean density for the specimen passed during its operation.

Fixed retainer (in zippers): A device permanently attached to the retainer pin at the bottom of one stringer.

FL: (a) Finnish Patent; (b) See **Polyfluoroethylene fibres**.

FLA: See **WRAP/FLA**.

Fl: Flax.

Flagging (in sewn seams): Refers to a sewing problem caused by the fabric moving up with the needle as the needle rises from the bottom of its travel causing poor loop formation and leading to skipped stitches or thread breakage.

Flag: Similar to the dark blue of the flower iris; violet.

Flagging (in inspection): Marking by thread, plastic flag, etc. A fabric at the selvedge to show a minor mistake across the width of the fabric which can be seen/checked in the next process easily.

Flake yarn: Two-ply cotton, wool, or silk yarns, having flakes or nubs in a different colour at certain intervals.

Flambé: The yellow colour of the yellow iris.

Flambé: A crimson and blue glaze originating in China in the Sung Dynasty (968–1280), giving the appearance of flames.

Flame: (1) A bright yellowish-red akin to the colour of flame. Also called 'flame-red'.

(2) A hot luminous zone of gas or matter in gaseous suspension or both that is undergoing combustion and is relatively constant in size and shape and that produce relatively low heat flux. A region in which combustion occurs in the gas phase accompanied by the evolution of heat and visible light.

Flame (as related to ignition of textiles): A controlled, hot, luminous zone of gas or matter in gaseous suspension, or both, of constant size and shape that is undergoing combustion as evidenced by a low intensity heat flux of less than one kw, such as candle flame or match flame.

Flame (retardant chemicals): Most of the processes developed for cellulose are all based on achieving a permanent reduction of combustibility through the *in situ* insolubilization on the fibre (covalent reactions with the hydroxyl groups of cellulose and/or polymerization) of organic phosphorus compounds in the presence of nitrogen. It has been recognized that nitrogen contributes to the efficacy of organic compounds. The high nucleophilic character of nitrogen and the high electrophilic character of phosphorus in flame-retardant chemicals are important conditions for the achievement of an effective nitrogen/phosphorus synergy.

Flame laminating: A process used for bonding woven or knitted textiles to a layer of foam which is melted by application of a flame.

Flame lamination: See **Laminating**.

Flame pink: A dark orangey-pink colour.

Flame-proof finishes: Flame-retardant finishing.

Flame-red: A bright red colour.

Flame resistance: The property of a material whereby flaming combustion is prevented, terminated, or inhibited following application of or non-flaming source of ignition, with or without subsequent removal of the ignition source.

Flame resistant: Fabrics treated with special chemical agents or finishes to make them resistant to burning. Today many fabrics achieve this property by using fibres that have this property built directly into the polymer. A fabric is considered flame resistant, if it passes federal specifications for specific end-uses.

Flame resistant: The fabric which is resistant to fire which can be treated for fire resistance or naturally having the property.

Flame retardancy: The property of inhibiting or suppress the propagation of flame.

Flame retardant: A chemical applied to a fabric, or incorporated into the fibre at the time of production, which significantly reduces a fabric's flammability.

Flame retardant finishes: The finishes which impart the flame retardant property to the material when applied to it.

Flame retardance: The property of a textile material, either inherent or by virtue of an applied finish, to inhibit or suppress the propagation of flame.

Flame retardant: This term should not be used as an adjective, except in the terms "Flame retardant treated" and "Flame retardant treatment".

Flame retardant: A chemical used to impart flame resistance.

Flame retardant treated: Having received a flame retardant treatment.

Flame retardant treatment: A process for incorporating or adding flame retardant(s) to a material or product.

Flame-retardant fibres: Fibres which are developed with flame retardant property during the manufacturing stage itself.

Flame-retardant finishing: The oft-quoted expression "flameproof" is incorrect, since only certain fibre materials can be said to be flameproof or inflammable, mainly those of inorganic origin. Polyacrylonitrile fibres and cellulose fibres are the easiest to ignite and burn. The flame proofing procedure therefore mainly concentrates on these textiles.

Flame-retardant treated: Materials which are treated with flame retardant finish.

Flame spread: The propagation of a flame away from the source of ignition.

Flame-spread time: The time taken by a flame on a burning material to travel a specified distance under specified conditions.

Flame-yellow: A bright orangey-red colour.

Flamingo: A reddish-orange shade.

Flamingo pink: A strong pink colour; an orange-pink colour.

Flamme': (1) (a) A woollen dress fabric made from printed yarns; (b) Plain weave fabric produced from yarn dyed linen warp and cotton weft. Used for table cloth and curtains.

(2) Irregular appearance produced by using slub yarns.

Flamme

Flammability: Those characteristics of a material that pertain to its relative ease of ignition and relative ability to sustain combustion.

Flammability testing: Various (defined) ignition sources are used in this test. The textiles to be tested are subjected to these sources for a specific time. The textile finisher's problem is that

each country has its own testing methods that stipulate different test subject positions and ignition sources, e.g. glowing cigarettes, different kinds of gas flame, even alcohol flames, waste paper baskets filled with wood shavings, electrical radiators, wooden cribs, hot metal nuts, cigarette lighters and matches. In principle, the combustion time after the source has been removed or burned off is used as a criterion for measuring and evaluation, as is the after-glow, and above all the size of the destroyed area of the test subject.

Flammability tests: The following demands have been placed on test methods for flammability:

(a) Determination of flammability (from a small flame to independent ignition with flames); (b) Fire propagation speed (flame spread rate, burning rate); (c) Determination of the heat of combustion generated; (d) Tendency to form burning melt droplets; (e) Flammability of home textiles at high ambient temperatures and high ventilation; (f) Determination of smoke and/or (toxic) gas development; (g) Determination of heat shock behaviour.

Flammable textile: Any combustible textile that burns with flame.

Flanders flax: Very fine flax of long, white, soft fillers, grown in Belgium.

Flamme- (L): Flame-colour.

Flammeous: Having the colour and other qualities of flame – a reddish-orange colour.

Flamme' yarn: (a) Hank-printed yarn, e.g. for shirt flannel; (b) Dyeing – Dying of ombré or shaded effects; (c) Spinning – a yarn spun with irregular thicknesses (slubs, nepps, etc.).

Flameproof: To render a textile incapable of propagating flame beyond the edges of a charred area produced by the application of a specific test flame.

Flange dart: A dart that is only partly stitched, the rest of it being laid free rather like a pleat. This example shows an armhole with flange dart.

Flange lock slider (in zipper): A slider with notches in the flanges of the slider that block the shoulders of the elements when the sringers are pulled apart, thus preventing further separation of the chain.

Flanges (in zippers): The edges of the slider formed to contain the chain.

Flannel (as applied to bed sheeting): A napped fabric used in the fabrication of sheeting products.

Flannel: (1) A soft wool fabric with a felted surface (degree of felting can be varied) woven in twill; cheaper cloth also woven in plain. Colouration can be obtained by melanging, for men's suiting. Not to be confused with flannelette, the surface of which is raised. The term may be used to describe the old woollen underwear material which was rather coarse and scratchy, but it is more likely nowrefers to a plain, dull-surfaced, slightly fuzzy cloth. In all-wool or worsted fibre, top-class flannel is of plain or twill weave, the yarns milled during manufacture. A good wool flannel construction is 56 × 56 with a 5 run warp and filling, in 2/2 twill weave. Lower quality woollen flannels are made with coarser yarn in plain weave. The finest flannels are made of soft spun worsted yarns. It is soft, warm and elastic with all the properties of woollen cloth and makes well-tailored dresses. A small percentage of polyester is added sometimes and with viscose addition cheaper varieties are made.

(2) From the numerous test methods for the determination, collective term for single-sided or two-sided napped fabric made from wool, cotton, union fabric or viscose filament; usually with plain or twill weave. The soft, supple handle is characteristic and is created by milling, with the clear nap still allowing the weave pattern to be clearly seen. The napping process causes the colours to merge gently into each other; the fibre nap is fairly short in these

materials, and is sometimes evened out by emerizing. Wool flannels are either made from worsted or woollen yarn, usually soft-spun from fine merino, i.e. easily milled wools. Flannels are mainly light mixture shades (grey and beige shades), light weight (approximately 240–440 g/m^2) materials. Variously named according to fibre type, weave or end-use (wool, twill, crepe, dress, blouse or shirt flannel). Type of finish: light to medium Melton finish. Used for dress, costume, suit and coat materials (wool) and sports shirts or pyjamas (cotton).

Flannel (cotton): (a) *Light weight* – lightweight cotton flannels are made with plain weave are often called flannelettes. Texture 44 × 44 with 22s warp and soft twist 12s filling. The fabric may be woven in grey or piece dyed or printed before napping. Low priced flannelette is sometimes made by using short fibres, such as comber waste in the filling.

(b) *Medium weight* – It is also sometimes called canton flannel, after the Chinese city where it originated. A typical construction will be 74 × 44 with 20s warp and 10s filling. Woven in 2/1 twill for the fabric to be napped on the back and 2/2 for fabric to be napped on both sides.

Flannel (worsted): Twill weave. Made in a variety of weights. More closely woven and harder than wool flannel. Can have a very slight nap on one side. It tailors very well. Presses well and holds a hard crease. *Uses:* Men's suits, jackets and trousers, women's coats, suits, skirts and tailored dresses.

Flannel hems: The French hem is used for table linen. Fold as in an ordinary hem, then fold the hem back on the right side and overhand the edge formed, taking fine stitches. Press the hem flat from the right side.

Flannel seams: Flannel seams should be stitched, opened and pressed flat, either on the right or wrong side of the garment. If on the right side, taffeta ribbon should be basted over the seam, so that the raw edges of flannel will not show, and cat stitched or buttonhole stitched on both sides of the ribbon, or any fancy stitch (not too long) may be used. This is the Dorothy seam. For the seam on the wrong side, the edges should be cat stitched with fine thread. Any ornamental stitch may be used on the right side of the seam. Always press flannel seams and hems before finishing. Flannel should never be hem stitched.

Flannelette: A light weight imitation of wool flannel. It is made from cotton, viscose, modal or mixtures. Soft yarns are used so that the fuzzy warm surface is easily achieved. Made in plain colours or stripes or prints. Creases easily, but wears well. A heavy, soft material with a napped finish, usually only on one side. In cheaper qualities, the nap comes off. Launders well, easy to manipulate and is warm to wear. There are many types in the market. It may be bleached, dyed, printed, or woven in coloured stripes. Uses: Infants' and children's wear, men's, women's and children's sleeping wear, pocket linings, quilts, shirting.

Flannelette sheeting: Slightly brushed cotton fabric. Brushing is done for warmth of the sheeting. Highly inflammable unless treated for flammability.

Flanelle de Chine: A plain woven French fabric, made of all-wool and given a very smooth face. It is dyed with indigo in the piece, the selvage being left white.

Flannellette: Twilled, lightweight washable cotton fabric, made of slack twist single yarns; it is woven in stripe effects or printed on the face and has a short nap raised on the back; used for kimonos, house dresses. See **Flannel, (light weight).**

Flannelette sheeting: Fabric made of cotton and slightly brushed for warmth. Highly inflammable.

Flap: A shaped piece of material which provides a covering for a pocket mouth or is for ornamentation.

Flare: A sudden burst of flame.

Flare (in cord): The spreading of the filaments at the cut end of a strand or the spreading of the strand ends at the cut end of a cord.

Flaring: Descriptive of bright colours.

Flash ageing: A modern form of the two-phase printing process. The flash ageing process is characterized by the use of short steaming times at high temperatures. Principle – thickened vat dyes are printed without fixation chemicals, dried, padded with an alkaline solution of a reducing agent (caustic soda liquor and sodium dithionite), then steamed immediately without intermediate drying for 20–60 s at 120–140 °C in a special rapid steamer (rainbow or arch steamer, tower steamer).

Flash ager: The machine designed to do flash ageing process. See **Flash ageing**.

Flash point (FP): The lowest temperature at which sufficient vapour builds up over a flammable liquid to form a combustible mixture with air near the surface of the liquid. The application of a small naked flame at the flash point temperature produces a momentary flash only, since the temperature is still too low for sustained combustion.

Flash spun fabric: A non-woven formed from the fine fibrillation of a film by the rapid evaporation of solvent and subsequent bonding during extrusion.

Flat: See **Wrong draw**.

Flat bed: A type of bed which provides a flat unbroken surface with the table on which it is mounted.

Flat carpet: (a) Smooth carpet, a carpet without pile or napped surface as per DIN 61151. Examples are coir and sisal mats, Gobelin tapestry; (b) Carpets without pile or nap such as hand-woven carpets, Kilim carpets, and carpets with extraneous decorative effects (allgaeu carpets, applique carpets, skin rugs, felt mats with and without decorative materials incorporated, or mechanically woven carpets in plain weave made from hair yarn, wool, coir, sisal, cotton).

Flat crepe: See **Crepe**.

Flat duck: Duck having warp of two single yarns woven as one and either single or plied filling yarn. See **Duck**.

Flat-felled seam: A complex seam formed on the outside of an object with raw edges enclosed and two rows of machine stitching visible on the face side. It is a thick seam, so avoid using it with bulky fabrics.

Flat foam: A smooth, non-structured, foam carpet back coating.

Flat goods: Knit goods made in flat stitch without any ribs or fashioning.

Flat iron: A pressing iron used for the manual Ironing of garments.

Flat plate printing: Flat plate printing is a method that first appeared in Europe in the mid-eighteenth century. This process involves images being incised or engraved onto a metal plate (usually copper) and is known as an intaglio method. Dye is applied to the surface of the plate; the excess is then scraped away, leaving the remainder only within the incised lines. This dye is then transferred to cloth by applying pressure.

Flat point lace: Made perfectly flat without any padded or raised parts.

Flat press: A hydraulic press used for Pressing in which transverse-folded layers of woven fabric (or, more rarely, knitgoods) are interleaved in the open-width state with heated or cold press papers and the pile so formed is subjected to pressure. Heating is achieved by inserting pressboards heated by electrical resistance wires at intervals throughout the batch.

Flat screen printing: A discontinuous form of Screen printing in contrast to continuous Rotary screen printing using Flat screens. Flat screen printing on long tables (hand screen printing) is the original form of screen printing and is still used for printing short runs of high quality prints. In manual method the length of textile fabric is glued or pinned on to the printing table and then printed by hand with flat screens using squeegees. Fully-automatic screen printing carriages take care of screen transport, stopping at the repeat point, and electrically- driven squeegee operation.

Flat screens: Screens made using mesh or bolting cloth and designs engraved on them, to be used in flat screen method of printing.

Flat sheet (in textiles): A flat, hemmed product, usually rectangular, used for covering the mattress on a bed and used for sleeping on under.

Flat stitch (in embroidery): Take a small stitch alternatively on each side of the shape to be filled with the point of the needle always emerging on the outside line of the shape. Two lines may be drawn down the centre of the shape as a guide for the size of the stitch. The stitches should be close together and fold into one another.

Flat yarn: By slitting a polyolefin flat sheet, flat yarns of a rectangular cross-section are formed. These can be spool wound for weaving and knitting; they may be processed directly onto weaving or knitting machines. Mono- and multi-filament forms can also be combined with conventional yarns. Twisted/doubled flat yarns have the character of round yarns or round twist yarns. Using another method, by means of fibrillation or capillarization, split film yarns can be used to create flat yarns as pseudo-multi-filaments, or similar connected in a net-like manner, with further processing via twisting/doubling (split round yarns, split twist) or crimp texturing and twisting (split crimp yarn, e.g. for carpet pile). Similarly also from tape yarns (a) directly into flat yarns (for warp beams or cross-wound bobbins) or (b) likewise via splitting fibrillation (split or fibrillated yarns) and crimp-textured and twisted (split crimp yarn). A further process involves passing the yarns through a subsequent processing of tape yarns, is through splitting through fibrillation, crimping or fibre cutting into split fibres (split non-woven) and by twisting into split fibre yarn. Used for bast fibre substitutes, binder warp/backing fabric for tufted carpets, needle-felt carpets, coating fabric, pile for outdoor carpets, furnishing fabrics, automobile/furniture upholstery fabric, wall coverings, the packaging sector (for textiles and sacks, slip-proof/stackable through the application of wax), work wear, rainwear, high-visibility clothing, tablecloths, awnings, tarpaulins, tent materials, filter fabrics, technical purposes.

Note: The term is still used in respect of these yarns after a small amount of twist has been introduced by subsequent processing, e.g. as in over-end winding.

Flax: (1) Of the colour of flax, a yellowish-beige colour.

(2) Bast fibres belonging to the natural cellulosic fibres (up to 120 cm long) with a smooth surface structure and increased natural lustre. Preformed in

the stem of the flax plant (small seed form = fibre flax, large seed form = oil flax; also hybrids; different fibre content) as described under flax long fibre structure. Microscopic: cylindrical, smooth, narrow lumen, transverse fracture is typical, wall displacement, knotty. A slender annual plant, the bast fibre of which is called linen. The plants are pulled before ripe, retted (submerged in water until the woody parts of the stalk are decomposed), after which the straw is crushed, broken between rollers, scutched (causing the removal of the woody parts with revolving blades) and hackled (or drawn through iron combs). The fibre ranges from creamy to greenish or bluish colour before bleached. The most important grades of the flax are: In Russia the motchenetz and slanetz (see each), the first containing the following varieties: Pochocon, Uglitz, Rieff, Jaropol, Stepurin. The varieties of the slanetz are called Bejetsky (usually best), Krasmoholm, Troer, Kashin, Gosposky, Nerechta, Wologda, Jaroslaw, Gresowetz, Kostroma. The flax marketed through Riga, Dunabourg, and Kowno, is graded K, HK, PK, HPK, SPK, HSPK, ZK, GZK, HZK, RG. The Archangelsk (Russia) grades for dew retted toales are first, second, third, and fourth crown, and first and second zabrack. The marks for Dorpat and Pernau (Russia) flax are: IX, E, OD, D, HD, R, G. The marks for water retted flax from Hoffs (Russia) are: HD, PHD, FPHD, SFPHD, XHDX, XRX. The marks for Pskoff are: OD, PWW, OW, O, OO, OOO, PI, PII, PHI. The marks for the flax from Reval and Dorpat, exported in bobbins, are: GR, HD, D, OD, OOD. The Petrograd marks are classed, comprising: Fabrichng, Otbornoi, first crown, second crown; superior siretz, comprising Polochno, Fabrichng, Otbornoi, first crown, second crown; medium siretz; common siretz, comprising Otbornoi, first crown, second crown. The Koenigsberg (Germany) marks are: FWPCM, FGPOM, WPC1M, LPCM, FPCM, PCM, PI, P2. In Belgium, the Courtrai flax is graded I/III, II/III, I/IV, II/IV, I/V, II/V, VI; the Flemish or blue flax grades are II/IV, I/V, II/V, VI, VII, VIII, IX. The flax from Fumes and Bergues is marked A, B, C, D; from Zealand, IX, VIII, VII, VI; from Friesland, D, B, Ex, F, Fx, Fxx, G, Gx, Gxx, Gxxx; the Walloon flax is marked II, III, and IV. The flax from Ireland and France is known by the names of the counties and district it comes from. Dutch flax is graded II/IV, I/V, II/V, VI, VII, VIII, IX. In chemical properties the pure bleached linen is similar to cotton, the following tests serving to distinguish unbleached linen from cotton: treated with olive oil, linen becomes translucent and cotton opaque and white; when burned the ends of the cotton fibres are tufted, those of the linen round; treated with concentrated solution of caustic soda both cotton and linen shrink and curl, the former becoming grayish, the latter yellow in colour; treated with concentrated sulphuric acid and then with diluted ammonia water, linen remains unchanged, while cotton becomes soluble in water. In fabrics the tests are: if it is a white fabric a piece is boiled in 50 per cent solution of caustic soda, which renders the cotton pale clear yellow and the linen dark yellow. A sample of the fabric is washed in a solution of cyanin in alcohol, then rinsed and after treated with thin solution of sulphuric acid, after which the linen will become blue, while the cotton stays white. Under the microscope the linen fibre discloses regular cylindrical or polygonal cells, with many transverse joints and a lumen in the centre.

Flax Plant

Ripe Flax Field

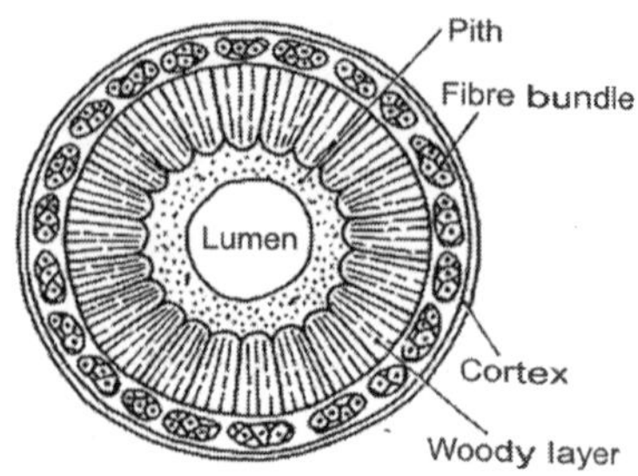

Flax canvas: A few tailoring and embroidery canvases are referred to as flax canvas.

Flax fibre: See **Linen**.

Flax tow: Flax fibre removed from carding engine cylinders. It is often matted and dirty.

Flavanthrone: Group of dyes coming in the Anthaquinone class and are derivatives of anthraquinone with a fused ring structure. Example is the dye manufactured by treatment of 2-aminoanthraquinone in nitrobenzene with titanium tetrachloride or antimony pentachloride at 100–175 °C.

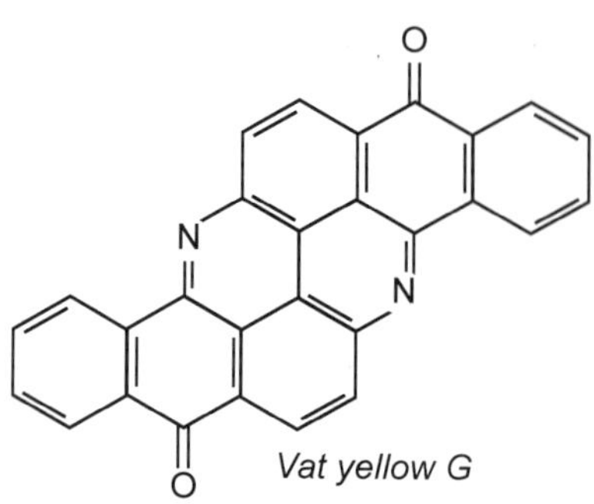

Vat yellow G

Flavid: Yellow.

Flecked: Yarn or cloth with a flaw caused by a spot of some strange colour.

Fleece: The wool of one sheep as obtained by shearing.

Fleece(in wool): The fibre coat that covers a sheep.

Fleece(a): A heavy wool or wool mixture with a long soft nap. The cloth may be woven or knitted or woven on the pile principle with a dense pile of highly resilient wool. The long nap or pile provides air spaces which make the fabric very warm and the basic weave is not visible from the right side. Different qualities are available and the fabric may be composed of stock or skein dye yarns or it may be piece dyed. The weight of the fabric will be around 15–25 oz. per yard. The nap wears out on cheaper cloth. Mainly used for overcoats.

Fleece(b): The skin of an animal, e.g. sheep, goat, to be used for making clothes.

Fleece wool: See **Fleece**.

Fleeced: A term applied to fabric that has a napped surface. The process is often applied to knitted fabrics.

Fleecy: A word used to describe the feel and implied warmth of any fuzzy surfaced fabric. The fabric may be cheap cotton or acrylic, or it may be wool. Often used as a lining fabric and there for the term 'fleecy lining' is often mentioned.

Fleecy fabric: (a) Term in the hosiery trade for fabrics having at the back a thick yarn which is brushed to raise a pile. (b) Any woven apparel cloth raised on one or both sides and made to resemble a fleece.

Flesh: A moderate pink. Also 'flesh pink'.

Flemish Holland: Stout, plain woven, unbleached linen fabric, made in Belgium.

Fleur de Soie: High grade French satin. The face is a satin de Lyon (2-1 twill) with a backing interlaced on the 12-shaft satin principle.

Fleur volante: Loops or other ornaments on the outer edge of the cordonnet.

Flexibility: That property of a material by virtue of which it may be flexed or bowed repeatedly without undergoing rupture.

Flexibility: The extent of the ability of a textile to be flexed or bowed repeatedly without being ruptured/broken.

Flexometer: American NBS testing device for determining bending resistance and evaluating crease-resistant finishes. The torque and the angle of the folded test subject are used to determine the amount of work that is required for folding and unfolding the subject. The bending resistance level during unfolding is used as a measure of crease resistance.

Flexual rigidity: (1) Resistance to bending. The flexural rigidity is the ratio of the small change in bending moment per unit width of the material to the corresponding small change in curvature:

Flexural rigidity G = $M \times$ C.C.C $\times$ 9.807 $\times$ 1/1000000uNm

Where, C = bending length (mm), M = fabric mass per unit area (g/m^2).

(2) The couple on either end of a strip of unit's width bent into unit curvature in the absence of any tension.

Flick: In England, the nap raised on flannelette and other fabrics.

Flicker: A hand tool that looks like a small hand card on a long handle. To use it, hold one end of a lock of wool in your left hand rested on your thigh and "flick" the tool up and down with your other hand catching the end of the fibre. This will open the lock and make it easier to spin. It is recommended to wear a sturdy pair of jeans or place a leather pad on your left thigh.

Flipper: Name given to tension-free open-width washing and relaxing machine for woven and knitted fabrics. Control by free-floating rollers with no mechanical drive, super-light rotation due to hydrostatic buoyancy. Almost tension-free cloth passage produces relaxing action due to positive control by floating hollow rollers, whose hydrostatic drive produces extremely

light rotation and therefore low-tension material movement. Gravity driven by the downwards path of the liquor-saturated cloth.

Flix courts: A French linen fabric.

Floak coating: A method of flocking for the manufacture of fabrics in which textile fibres are bonded at one of their ends to a textile or non-textile base material by means of an adhesive.

Float (in woven fabric): The portion of a warp or weft yarn that extends unbound over two or more filling or warp yarns.

Float: A defect in which a warp or filling yarn extends unbound over the ends with which it should be interlaced. This happens due to inaccurate drawing-in or improper shedding.

Float stitch: This is the yarn length which extends over stitches in the horizontal direction and is limited by a tuck stitch or a knit stitch. Structures having float stitches tend to be narrower and have less width-wise elasticity. Float stitch of more than four successive float stiches in the vertical direction of the fabric or more than six adjacent float stiches in the horizontal direction of the fabric should not be used, otherwise the knitting process will not be possible due to yarn breakage or stitch ravel. Weft: This is the yarn length which lies between plain and purl stitch in the horizontal direction.

Floc: Coagulated groupings of formerly suspended particles which then settle by gravity.

Flocculant: A chemical or physical reagent such as a polyelectrolyte, which promotes flocculation. It acts as a bridge to bind the suspended particles together. A flocculant may act also as a coagulant. Compare ***flocculent***.

Flocculation: The combining of the particles of a finely divided precipitate, such as a colloid, into *flocculent* clumps that sink and are easier to filter off. See **Coagulation**.

Flocculator: A stirrer used in mechanical flocculation.

Flocculator clarifier: A tank used in the treatment of water which consists of both a ***flocculation*** zone and a separation (***clarification***) zone to remove the flocs from the water. Various types exist, normally based on ***solids contact clarifier*** (e.g. ***sludge blanket clarifier,*** and ***solids recirculation clarifier***). Smaller or older treatment works may have tanks with a completely separate flocculation zone and settling zone, sometimes known as one pass (or single pass) flocculator clarifiers. These may be rectangular or circular and up to 50 m diameter. In circular clarifiers, a central flocculation zone is surrounded by the settling zone. The ***surface loading rate*** is from 24 to 60m^3 per day per m^2 of surface area, depending on raw water quality.

Flocculator DAF unit: See **Dissolved air flotation**.

Flocculent: Description of particles in water or wastewater or sludge that form ***flocs***.

Floches: Fine sewing silk produced in France.

Flock: (1) A material obtained by reducing textile fibres to fragments as by cutting, tearing, or grinding, to give various degrees of comminution.

(2) Many fabrics may be flocked and their characteristic is that the right side of the fabric has tufts of fibres added in dots or patterns. The fabric is light and floaty, the flocking may be in contrast or self-colour and it may be worked in a different type of fibre. The flocks are normally stuck on with an adhesive which is applied to the fabric in a pattern. The surplus flock fibres are afterwards blown off. Flocked nylon is often used for sarees.

Flock (to): Short pieces of monofilament fibre, i.e. short fibre material made of viscose, polyamide, polyester, polyacrylonitrile, acetate or natural fibres (wool, cotton) and frequently also so-called metal flock made from aluminium, brass or bronze powder prepared for flocking and flock printing. The most frequently-used fibre titres are in the range of 0.9–22 dtex, and the usual fibre lengths are between 0.5 and 2 mm. "Glow" or reflective flock is always prepared by static electrical means using electrolytes, and therefore conducts electricity to a certain extent.

Flock print: The velvet-like relief pattern is obtained by printing the design with adhesive and then flocking with small snippets of fibres. It is used for making dresses and blouses.

A flock print

Flock printing: Once the adhesive has been applied to the entire surface of the substrate, i.e. including the motif, and this has been produced in the flocking zone using templates, sharpness and resolution power are established by flocking. Flocking may be performed in a downward or upward direction with the parameters of the flocking process determining the sharpness of the contours of the flocked motif.

Flock silk: The outer irregularly spun cocoon layers (approximately 25% of the material) regarded as best raw material for schappe' silk (also called Frison, Stursa, Kibizzo).

Flock transfer printing: A process for the individual application of flock patterns to e.g. ready-made garments such as T-shirts. No after-cleaning is required although the preparation is laborious as far as the printing technique is concerned: a flocked release paper is screen printed with a single or multicolour design first of all. Next, a cover adhesive is then also applied by screen printing. It is most important here to ensure that the flock is only surface printed with adhesive in this operation, i.e. the adhesive layer must not penetrate through the flock. Immediately afterwards, a hot-melt granulate is strewn on to the adhesive layer whilst still wet before intermediate drying. Excess granulate is removed by brushing after the material has been dried.

Flock yarn: Velvet-like polyamide-based compound yarn, manufactured using a carrier thread with a firmly bonded adhesive sheath consisting of almost right-angled radial flocking. The flock cannot be removed without damaging the core thread, which will mean high abrasion resistance, resistance to shedding, resistant of high frequency fusion, easy to clean, tetrachloro ethane resistant. Used to add texture to woven fabric, warp-knitted fabric and knitwear with high-bulk yarns, as an appearance, to impart visual, handle and abrasion-resist properties for raised and velvet-like areas, particularly car upholstery, upholstered furniture and decorative materials.

Flocked blanket: A blanket made with a fish-net type scrim sandwiched between two thin layers of foam with flock adhered to the outside of the foam.

Flocked fabrics: See **Flocked Textiles**.

Flocked textiles: These are defined as fabrics in which textile fibres are bonded at one of their ends to a textile or non-textile base material by means of an adhesive.

Flocked carpet: Contract carpet produced by flocking on to a substrate consisting of a ground fabric, non-woven, or rubber sheeting. The applied flock is in the form of, e.g. short cut polyamide fibres.

Flocking: The application of short fibres to a base fabric by the direct printing of adhesive on to the fabric in the desired areas and then sticking the fibres to these areas.

Floconne': In French, the word *flocon* means "flock". A term used for thick, soft double fabrics (900–1000 g/m^2) made from carded yarns produced mainly with high-grade merino wools. The characteristic appearance is achieved by means of a separately picked 3rd weft yarn, the flock weft, which floats lengthways, and is subsequently raised (it is therefore also referred to as "napped cloth") which gives rise to flock-like fibre tufts on the surface of the fabric. Depending on the weave, it is produced with rib, diagonal or herringbone effects; wool or piece-dyed. Used for high quality men's coats.

Flocculation: Flocculation is the term used to describe the aggregation of smaller particles in to larger particle that can be removed by sedimentation and filtration. Flocculation is brought about by Brownian movement, differential velocity gradients and differential settling in which large particles overtake smaller particles and form larger particles.

Floor covering: An essentially planar material, having a relatively small thickness in comparison to its length or width, which is laid on a floor to enhance the beauty, comfort and utility of the floor.

Floor covering (finished pile yarn): See **Finished pile yarn floor covering**.

Floor covering (pile yarn): See **Pile yarn floor covering**.

Floorcloth: In England, a coarse and stout, plain woven hemp or flax fabric, heavily coated with varnish; used as inexpensive floor covering.

Florameda: A fabric mentioned in 17th Century English manuscripts, believed to have been woven with flower patterns.

Florence brown: A greyish shade of red.

Florentine: Heavy grey woven, 3/1 twill cotton fabric used for overalls and uniforms, having a construction 38 × 19 and 37 × 49 tex cotton yarn.

Florentine brown: A copper colour, also called Hatchett's brown and Roman brown.

Florentiner-Tulle: Flimsy and embroidered tulle for curtains.

Floret or Flor linen: An open face cotton or linen canvas, both warp and weft yarns being equally spaced from each other.

Floret: (a) Figured wool satin, made with a high finish; used in England for dresses; (b) French term for brocaded silks; (c) Yarn made of the best kind of silk waste.

Floretta: (a) See **Floss silk**; (b) Very fine, plain woven and bleached Belgian linen.

Florida: (a) A very fine grade of Sea Island cotton, having a uniform, long, fine and silky staple; (b) Trade name in Romania for a printed madras; (c) In Austria, a coarse, bleached, plain woven cotton fabric, with a heavy size finish; used for stiffener in collars, cuffs and shirt bosoms.

Flooring material: Any pliable planar structure used as a base surface in camping tentage, but excluding such things as rugs or carpets placed in the tent that are not integral parts of the item.

Florence: Very light, plain woven silk lining, the warps and wefts being placed the same distance apart, made with single warp. Sometimes mixed with wool.

Floret silk: Schappe (silk).

Floss (Duvet, Fiochetti, Pelosität): Partially fluffy silk, caused by stray drops of secretion from the silk worm or as a result of aberrant spinning process, localized fluffing during the degumming process, effects of wear, etc.

Floss: (1) Very short tangled fibre of waste silk.

(2) A silky embroidered yarn.

Flots: In laces, several rows of picots partly covering each other.

Flounce: A ruffled drop on a bed covering.

Flouncing: Fabrics of any type or fibre which have one selvedge shaped in scallops. The edge is finished off, usually with embroidery.

Flouncing laces and embroideries: These are made as yard goods intended for flounces on dresses.

Flourishing thread: Lustred linen thread; used for embroidering.

Flow dyeing: An old method of dying for achieving cheaper dyeing. The warp or weft yarn is immersed in a strong dye, dried without rinsing and woven with non dyed weft or warp yarn. The fabric is simply passed through hot water on the padder, whereby dye is transferred

to the non-dyed material. Not practiced now.

Fluff: Down.

Fluff formation: Pile formation.

Fluff threads: Refers to texturized polyester or nylon threads. These threads provide excellent seam coverage when sewn in serger or overlock machines.

Fluid bed process: In this process sand or glass pearls of 100–200 μm diameter are used as a heat transfer medium from the heat source to the material being treated. In the case of materials produced from synthetic fibres, the fluid-bed process is used for heat-setting, thermosol dyeing, the curing of pigment dyeings and prints, the curing of synthetic resins and drying operations. The material being treated passes through the mass of hot sand/glass pearls in a V-shaped chute.

Fluidity(of a cellulose solution): A measure of the ease of flow or motion of a solution, and therefore an indicator of the molecular weight of the cellulose.

Fluidization: The suspension of a finely divided solid, e.g. a catalyst, in an upward flowing liquid or gas. This suspension mimics many properties of liquids including, allowing the solids to 'float' in it and achieve a uniform temperature with the fluid medium. Fluidized beds so prepared are used industrially to carry out catalytic reactions and to minimize the production of pollutants from coal-fired furnaces.

Flume: Variety of Egyptian flax which yields coarse fibres.

Fluorocarbon polymers: Group of chemicals including, Polytetrafluoroethylene, fluoroalkyl esters of acrylic acid which are used in soil-release and stain-release finishes. They are used as synthetic resins and is important in the area of coating of rolls which are non sticking and withstand temperatures from 180 – 240 ^{0}C.

Nonstick surfaces are obtained by spraying these synthetic resins on to metal rollers which are used e.g. in sizing and finishing machines.

Fluocarbon resins: Polymers of Fluorocarbons. See Polytetrafluoroethylene; Polytrifluorochloroethylene.

Fluorocarbon chemicals: See Perfluoro compounds.

Fluorocarbon fibres, Fluorofibres: According to textile labeling regulations and the EDP code system, fluorocarbon fibres are classified as Polyfluoroethylene fibres.

Fluorocarbon finishing: Finishing of textile with fluorocarbon based polymers for water and stain repellency. Fabric finished with Fluoropolymers modify the surface properties of these textiles with regard to staining and dirt repellency Whereas paraffin or silicone-based water repelling only has a water-repelling effect on the textile fibres, fluoric chemicals repel both water and oil. Compared to conventional waterproofing, fluoric chemicals have the advantage of being resistant to washing and dry-cleaning. The "breathability" and therefore wearer comfort and handle of the fabric are not affected.

Fluorocarbon plastics: See **Fluorocarboin resins.**

Fluorocarbons: These are hydrocarbons in which the hydrogen atoms have been substituted by fluorine atoms. Mixed fluorocarbons contain other halogens in addition to fluorine, e.g. chlorine Chlorofluorocarbon. Fluorocarbon polymers are thermoplastic Fluorocarbon resins.

Fluorofibres: See **Fluorocarbon fibres.**

Fluorescence: When the molecule absorbs light, it is excitedfrom the lowest vibrational level in its ground state (S1) to a range of vibrational levels in the singlet first excitedstate (S2).

Fluorescence occurs if the molecule then emits light as it reverts from this level to various vibrational levels in the groundstate. Non-radiative processes, the most important of which is generally collisional deactivation, also gives rise to dissipation of energy from the excitedstate. As a result, there will be a reduction in the intensity of fluorescence andin many cases, it will be absent altogether. A type of Luminescence in which substances exposed to a light source emit light of another wavelength or colour.

Fluorescent brightening agent (optical brightening agents, brighteners, white dyes): Organic compounds which add fluoresce in a colour and is complementary to that of the textiles and can be used to compensate the natural colour. The brightening effect comes from fluorescence, which depends on the property of certain organic compounds to convert short wavelength light, in the 300–400 nm range, into longer, 400–500 nm, wavelengths.

Fluorescent dyes: (a) A group of organic dyestuffs which, when applied to a textile material, give rise to fluorescence under UV radiation (some in daylight; dayglow colours); (b) Range of so-called luminous dyes that are highly fluorescent in daylight and under UV light. Suitable for coloration of plastic mass (polyvinyl chloride, celluloid, urea formaldehyde resins, lacquers), used in a similar way in pigment printing.

Fluorescent lamp: A glass tube filled with mercury gas and coated on its inner surface with phosphors. When the gas is charged with an electrical current, radiation is produced. This, in turn, energizes the phosphors, causing them to glow.

Fluorescent UV lamp: A lamp in which radiation at 254 nm from a low pressure mercury arc is transformed to longer wavelength UV by a phosphor.

Fluorescent whitening agents (fluorescent brightening agents): A fluorescent brightening agent is a compound which, when applied to a textile material, absorbs the short wavelength electromagnetic radiation (UV light) which is invisible to the human eye and converts it into visible light of longer wavelength which is emitted either as violet, pure blue or greenish blue.

Fluorescent quenching: Removal of Optical brightening agent from a material which is already applied with it.

Flush: Same as **Float**.

Flushed printing: Overprinting with extremely fine colour transitions and colour mixtures, usually on silk, viscose, or triacetate; e.g. on viscose, triacetate with azo print or print with cationic dye. Dry the print, steam for 30 minutes using wet steam and then treat with hydrophilic polymers (e.g. polyacrylate) for 5–10 minutes. Later rinse and dry.

Flux (in water/effluent treatment): In crossflow filtration, the unit membrane throughput, usually expressed in volume per unit time per area, such as gallons per day per ft^2 or litres per hour per m^2.

Fly: (1) A material covered opening to conceal a fastening in a garment.

(2) Fibres which fly out into the atmosphere during carding, drawing, or spinning or any other textile processes.

(3) Unwanted loose fibres on a textile material. It is caused by the attraction of fibre dust (mainly as a result of electrostatic attraction).

Fly ash: The light ash from an ***incinerator***, coal burning electricity generating plant, etc., that is carried off in the ***flue gas***. It can be removed from the flue gas by ***fabric filters*** or ***electrostatic precipitators***. Fly ash may be considered as a hazardous waste due to toxic contaminants being carried over from the incinerator and collected in the ash. It may contain activated carbon that is sometimes injected into the gas stream to remove toxic components of incinerator gases. The activated carbon is then collected as a particulate with the fly ash.

Fly catch: An under lap of a material providing a backing to an opening to take buttons or zip tape for fastening, etc. Fly catch, Button catch, etc.

Fly catcher: Devices used to clean fibre dust, fly, loose pieces of yarn, etc., from machinery in order to prevent possible malfunctions due to their accumulation which could impair machine functional performance. Manual or electrically-operated fly catchers areavailable.

Fly stitch: Embroidery stitch. Bring the thread through at the top left, hold it down with the left thumb, insert the needle to the right on the same level, a little distance from where the thread first emerged and take a small stitch downwards to the centre with the thread below the needle. Pull through and insert the needle again below the stitch at the centre and bring it through in position for the next stitch. This stitch may be worked singly or in horizontal rows or vertically.

Flyer: A rotating device that adds twist to the slubbing or roving and winds the stock onto a spindle or bobbin in a uniform manner. See **Loom fly**.

Flyer bearings: Holds the flyer, same material as used in wheel bearings.

Flyer lead: A single band drives the flyer. The bobbin has an adjustable friction band to slow it.

Flyer spinning: A spinning system in which yarn passes through a revolving flyer leg guide on to the package. Yarn is wound on by flyer and spinning package rotating at slightly different speeds.

Flying filament: A loop of non-crimped filament protruding from the surface of a false twist draw textured yarn. The fault is caused by a filament failing to become crimped during texturing or subsequently losing its crimp.

Flyline: The line of the fly fastening from the centre of the fork to the top of the waist band on a pair of trousers.

F:M ratio (Food and Micro-organism ratio), organic loading, sludge loading rate, SLR: The number of kg of ***BOD5*** per day applied to an ***activated sludge*** plantper kg of (dry) activated sludge. F:M calculation can be based on ***MLSS*** (mixed liquor suspended solids) or ***MLVSS*** (mixed liquor volatile suspendedsolids).

For the United Kingdom:

$$\text{F:M} = \frac{\text{BOD}_5 \text{ of wastewater (mg/l)} \times \text{flow of waste water (m}^3\text{/d)}}{\text{MLVSS (mg/l)} \times \text{aeration tank volume (m}^3\text{)}}$$

For the United States of America:

$$F{:}M = \frac{BOD_5 \text{ of wastewater (mg/l)} \times \text{flow of waste water } (m^3/d)}{\text{MLSS (mg/l) aeration tank volume } (m^3)}$$

The units of F: M are days^{-1}. As about 70% of the sludge is volatile, the MLVSS is about 0.7 × MLSS. Therefore, the F:M based on MLSS is approximately 0.7 × F:M based on MLVSS for the same BOD, flow and aeration tank volume. It is the common design criteria for calculating the size of an activated sludge aeration tank.

The micro-organisms in the activated sludge process metabolise faster in warmer climates and so the F:M is higher. Nitrificationis a slow microbial process and so the F:Ms are reduced.

FNA: Fachnormenausschuß (German) (Technical Standards Committee).

FNM: Fachnormenausschuß Materialprüfung (German) (Technical Standards Committee for Material Testing).

Foam: Foams are micro-heterogeneous colloidal systems in which gases are distributed within a liquid or solid dispersing agent. See **Foam applicator**.

Foam application: Processes used for the low wet pick-up of concentrated foamed finishing liquors, especially for wash and wear finishes on textile fabrics with a possible pick-up of 40% and mainly between 10–20%. Foamed liquors are applied to the textile directly then brought to the desired thickness by means of squeeze rollers or a knife coater, e.g. two-sided foam application in a horizontal 2-bowl padder, one-sided application to carpets and foam application with an air-roll squeegee.

Foam applicators: Foam (a collection of bubbles) is created when air is whipped into a liquid. Since a bubble is a sphere of air entrapped by a thin layer of liquid, a volume of foam can be considered a quantity of liquid diluted by air. The volume of foam produced by an amount of liquid is called the blow ratio. Blow ratio is a way of describing foam density since collections of large bubbles result in low density foams, whereas small bubbles yield higher density foams. Not all liquids foam when whipped with air. Pure water alone will not foam, a surfactant must be added to lower the surface tension before it will foam.

Foam backing: Back coating of carpets.

Foam bonding: A method of making non-woven fabrics in which a fibre web or batt is treated by the application of a liquid in the form of a foam.

Foam breaking: See **Defoamer; Anti-foams; Dry-cleaning anti-foams**.

Foam cleaning: A process in which prepared foam is applied to a textile product, scrubbed in, allowed to dry and encapsulated soil is removed by suction.

Foam coating: The process can be carried out on certain foam application machines. A foam coating system consists of the following elements: mixer for mixing the compound with the required volume of air, roll-coater, drying unit, pressing cylinder, kiss-roll or padder and curing oven.

Foam decay: Occurs by self-drainage of a foamed liquid. The self-drainage rate (collapse rate) of a foam can be established by determination of the foam decay half-life period.

Foam degumming: A process for the degumming of natural silk under mild conditions. Instead of immersing hanks of yarn in the soap bath, they are suspended in the soap foam produced by the introduction of live steam into the bath. The foam action has a tendency to dissolve and eliminate sericin. This method, however, is not popularly adopted.

Foam density: The foam density is calculated according to the formula: (volume of solution consumed foam density)/(volume of foam)

Foam dyeing: Dyeing processes based on the use of finely porous foam, either as a dyeing medium itself or as a dye carrier. Foam dyeing can be carried out by exhaust and continuous methods. Foaming of the dye liquor serves to increase the liquor volume and is hindered by the presence of electrolytes.

Foam finishing technology (FFT technology): A foam application in which precisely metered quantities of foamed chemicals, etc., are applied to a textile substrate. Penetration of the porous textile fabric by foam results in a uniform distribution of the concentrated solution of chemicals at a minimum liquor pick-up of approximately 10%. Migration problems are also eliminated at the same time. The applied foam breaks immediately on contact with the fabric. Foam density is below 0.2 g/cm^3. In contrast to padder applications, the liquor pick-up can be considerably reduced to 10–25% of the fabric weight with this process. Since the greatest part of the water required in conventional padder applications is substituted by air in this process, energy consumption in subsequent drying is also reduced.

Foam formation: (a) Unwanted formation of foam during wet-processing operations. Causes problems in dyeing and printing, etc., and should be prevented. See foam as a source of problems in textile wet processing; (b) Foam formation as a desired effect for applications in dyeing and finishing. When solutions of substances with low surface activity are used for foam generation (e.g. short-chain alcohols or acids), the resultant foam is often only stable for a few seconds. Foams with considerably greater stability are produced with substances capable of causing a considerable reduction in the surface tension of water (e.g. soaps, non-ionic, cationic and amphoteric surfactants). See **Foam application**.

Foam generators: Nowadays, foam technology represents an important technique in the carpet and textile industries for various stages of finishing. The high and varied demands placed on these foams, which are not only used in the carpet industry, call for mixer concepts which can be adapted to the actual operating conditions as far as performance and control are concerned. The foam generators that are used to produce foam of high quality which supports the foam application and finished products. Foam generators are of two types, static and dynamic. In static mixers, air and liquor are brought into contact with each other in a mixing head containing glass beads or Raschig rings. A dynamic mixer consists of a stator and a rotor. This system provides control of foam supply quantity, air dosing and gelling agent quantities according to the speeds of the coating plant. The quality of the foamed compounds stands or falls on the quality of air dosing. The smallest fluctuations in air content will change the litre weight of the foam and consequently reduce the quality of the end product. It is therefore necessary to apply a very rapid air measurement with a high degree of linearity.

Foam (in cleaning of textiles): A frothy mass of fine bubbles generated by whipping or agitating a suitable surface active agent.

Foam intensifier: The auxiliary used in the foam generators or otherwise to intensify the foam formation. It is usually a chemical which increases the foaming power of the surfactant itself.

Foam lamellae: Liquid interfaces between individual bubbles of Foam. Foam is therefore said to possess a "lamellar structure". Due to capillary action in foam lamellae, strong attraction forces for soil particles are released, as a result of which the latter are loosened (by mechanical action such as liquor movement, friction, pressure and squeezing) and removed from the fibre surface. The foam lamellae contain a film of liquid between the surface pellicule layers which consist of a monomolecular surfactant film, i.e. so-called inter-lamellar liquid.

Foam padder: The foam padder is a foam dyeing unit for carpets which does not require a foam generator. A dye liquor is applied to the carpet pile with the aid of a kiss-roll and subsequently foamed *in situ* by mechanical action on passing between a rotating cylinder of rods and a pressure plate at the back of the carpet. See **Foam applicator**.

Foam printing: A textile printing variant of low wet pick-up in which a large proportion of the water in print pastes is substituted by air; thickening agents are superfluous and therefore do not need to be washed out. The advantages are reduced soiling of water during after washing and lower energy requirements in drying and steaming. The main disadvantage is the need for a separate foam generator for each print paste in the design and the resultant increase in costs. Printing of outlines is also difficult with foam print pastes.

Foam rubber (latex foam): An expanded foam produced from rubber or synthetic rubber. It consists of approximately 15% rubber and 85% entrapped air or gas cells. Foam rubber is of low weight, odourless, elastic, dimensionally stable, absorbs body transpirations and conducts them away, prevents heat accumulations, provides noise insulation, is resistant to parasites, washable and has a long service life (especially synthetic foam rubber, as well as natural foam rubber which is protected against the action of sun and air).

Foam sizing: The process of foam sizing calls for a change in the auxiliaries by which the "add on" to the yarns is applied. An emulsion which typically contains 20% solids or "additives" and 80% water is converted into a foam by the careful introduction of air, the viscosity of which is adjusted empirically in such a way that a good wet add-on is achieved.

Foam stabilizers: Special products used to increase foam stability.

Foam stability: The ability of a foam to remain in a stable condition.

Calculation: (foam volume after x min)/(foam volume immediately after preparation)

Foam suppressants: Products used to prevent the formation of foam or considerably reduce foam stability. In general, a foam suppressant should (a) be sparingly soluble in water, so that it is already present at low concentration on the lamella surface, (b) have a positive spreading coefficient relative to the surfactant solution, and (c) should form quasi-gaseous interfacial films. See **Defoamers**, **Antifoams**.

Foam tear: A condition wherein the foam portion of a laminated fabric ruptures prior to the failure of the bond.

Focus groups: Groups of customers/potential customers brought together by market researchers to find out likes/dislikes, etc.

Foil: A term in general use for materials in thin, sheet-like form, e.g. aluminium, plastic foil, etc. In this context, therefore, a plastic foil is a flat, autonomous, and more or less elastic material with a thickness of approximately 0.01–0.1 mm, whilst the still thicker forms are described as plates largely because of their stiffness. The term foil should not be confused with ***Film***.

Fold marks: These are usually seen as length-wise creases which can cause considerable problems, e.g. in the case of dyeing sensitive polyester fabrics that are thermo-plastic, or alkaline treatment in cellulose processing. For cellulose, similar conditions may be found during bleaching processes but, due to the lower quantities of alkali, heavy creasing is rare and is easier to remove.

Fold (book): The fabric is doubled selvedge to selvedge, and folded back and forth upon itself in predetermined lengths. When the piece is completed, the fold edges on each side are folded once more upon themselves so that the fold edges are inside forming a compact package as long as on half width of the goods. See **Book fold**.

Fold (shoe): The fabric is folded from both ends into 12-16 folds to the piece, the length of fold depending upon the length of the piece. See **Shoe fold**.

Folded selvedge: See **Curled selvedge**.

Folded yarn: Another term for plied yarns. See **Plied yarn, Yarn, Plied**.

Folder: (1) A sewing machine attachment which folds material as it is fed through, so that it is sewn into the desired configuration.

(2) A sample booklet to show samples of fabric print, finish, etc.

Folding hems: A hem is a fold of goods twice folded to protect a raw edge. The first turn or fold of the hem is the most important. It should be straight and even, *folded to a thread*, for upon it depends the beauty of the hem. The hem should always be turned towards the worker and creased firmly, but never pleated along the fold. First crease the narrow fold, then crease the second fold the desired width, marking by a measure and baste not too near the edge. The first fold *along* the *woof* threads should be at least one-fourth of an inch in width, as the woof threads give or stretch more than the warp threads; otherwise it will not lie flat.

Folding machine: Used (a) for make-up of piece goods on completion of finishing or (b) for individual laundry items. Can be directly connected to a packaging machine (Fabric folding machine).

Folding of yarns: The doubling of single yarns as a partial stage in the process of Twisting. The folding process represents a stage in the production of ply yarns.

Folk weave: Coarse yarn, loosely woven fabrics with a woven pattern often including several colours. Fabrics are woven on dobby or jacquard looms. Yarns used are cotton, acrylic, viscose. Patterns are usually striped with possibly diamonds or other small patterns between them. These fabrics do not crease easily and are used for bedcovers, curtains, loose covers and cushion covers.

Fond: In hand-made laces, the ground over which the patterns are spread out. It is either a mesh or is made up of brides.

Fondant pink: A pastel red colour.

Fond de Neige: A fancy ground in old laces, consisting of braids with groups of knots on them.

Fond d'or: French term for brocades with gold ground.

Fontage: Head-dress with stiff, folded lace ruches at the front and standing up like organ pipes.

Foot traffic units, for pile floor covering: The number of passes by human walkers over a specific group of carpet samples.

Footing: (a) The edge of the lace flouncing which is sewn to the fabric; (b) The repairing of holes in the foot of stockings.

Footman: The vertical connection between the treadle and the crank.

Forbes: A Highland tartan, made as follows – A wide dark green bar, split in the centre by a group of black, white and black lines; black stripe, measuring half the width of the green; a navy blue bar (as wide as the green), split in the centre with a pair of black stripes, placed close together.

Force: A physical influence exerted by one body on another which produces acceleration of bodies that are free to move and deformation of bodies that are not free to move.

Force at break: See **Breaking force**.

Force-at-rupture: The force applied to amaterial immediately preceding rupture.

Force at specified elongation (FASE): The force associated with a specific elongation on the force- extension or force-elongation curve.

Force-deformation curve: A graphical representation of thee force and deformation relationship of a material under conditions of compression, shear, tension, or torsion.

Force-elongation curve: A graphical representation of thee force and elongation relationship of a material under tension.

Force-extension curve: A graphical representation of the force and extension relationship of a material under tension.

Fore: The fronts of jackets, coats and similar garments.

Forecasting: Predicting fashion trends.

Forecasting services: Publishers and other organisations offering fashion prediction advises.

Fore-runner: A length of fabric used in processing to lead a piece of fabric through the equipment and so enable the piece to be processed from end to end, with minimum wastage. Other names like leader cloth, cheda, wrapper cloth, etc., are also used.

Foreign matter (in cotton): Non-lint material, commonly referred to as waste or thrash such as dust, sand, seed-coat fragments, leaves, and stems normally present in raw and partially processed cotton.

Forest-green: An olive-green colour, slightly darker than ***Lincoln green***; sometimes a yellowish-green or a dark muted-green.

Forest wool: Fibres extracted from pine needles; used for coarse blankets, mixed with cotton or wool.

Forestieri: Various coloured, mostly scarlet red and fulled woollens produced in Egypt.

Forget-me-not blue: The pale blue colour of the forget-me-not flower – a symbol of constancy.

Forfars: Unbleached coarse and heavy linen, made in England; used for towels.

Fork: The region of the garment corresponding to the fork of the human body. Commonly the place at which inside seam join; also called crotch, cratch.

Fork quantity: The allowance of material in bifurcated garments provided for the thickness through the trunk from front to back at the crotch.

Formal balance (in designing): Where the design elements are almost equally distributed.

Formaldehyde determination: The formaldehyde released by hydrolysis is extraordinarily dependent on the test conditions. Total formaldehyde is understood to be the total amount of formaldehyde detectable in finishing agents, finishing liquors and on textile fabrics. It represents the sum of free and combined formaldehyde. The combined formaldehyde is subdivided, in turn, in methylol and methene compounds. The determination of formaldehyde released from textiles by hydrolysis is also important. After hydrolysis in the distillate, the total formaldehyde can be determined iodometrically. The determination of free formaldehyde is mainly carried out by the sulphite method and, in a different manner, by the hydroxylamine method. Methylol formaldehyde can be determined together with the free formaldehyde. The determination of free and hydrolytically-released formaldehyde on textiles can be carried out according to various methods. However, these test methods determine not only the actually available free formaldehyde, but also the formaldehyde released by partial hydrolysis in the preparation of the sample.

Formaldehyde: H_2-C=O; (methanal), MW: 30, density 0.81. A readily polymerizable colourless gas. It is commercially available as an aqueous solution with a pungent and characteristic odour; 30–40% solution (formalin) which usually contains 10% methanol to inhibit polymerization. *Properties*: powerful reducing action; causes proteins to coagulate (principle of tanning) and the hardening of casein fibres; it is a powerful germicide either in solution or vapour form.

Uses: After-treatment of substantive dyeing (improved wash fastness); wool protection agent; hardening of casein fibres (during wet finishing); cross-linking agent in resin finishing (formaldehyde finishing); manufacture of resin finishing agents (methylolation); addition to naphthol dyeings; preservative; component in disinfectant solutions; component in numerous fillers, adhesives, etc.

Formaldehyde after-treatment of direct dyeing: Improves the colour fastness of suitable direct dyes to water, washing and perspiration. Suitable direct dyes contain free amino or oxy groups; mainly black dyes.

Formaldehyde content in resin finishing agents: The free formaldehyde content results from the state of equilibrium which is dependent on the chemical constitution, concentration and temperature.

Formaldehyde fastness: A form of 'colour fastness to storage conditions' test for finished and packed coloured textiles to possible formaldehyde-containing adhesives in packing paper, pleating paper, etc.

Test procedure: A dyed or printed textile specimen is exposed in the dry state for 24h to rising formaldehyde vapours (contained in a small open bowl under a glass bell jar or in closed boxes). The grey scale is used to evaluate change of shade.

Formaldehyde finishes: Cross-linking processes for cellulose with formaldehyde or so-called Formaldehyde-liberating compounds in the presence of acid catalysts. The formaldehyde is mainly applied from aqueous solution by the wet crosslinking process. Such finishes were originally applied to textiles composed of regenerated cellulose to reduce fibre swelling, increase wet tensile strength and improve shrink resistance; also used for the wash-and-wear finishing of cotton. Disadvantages: severe losses in tensile strength and abrasion resistance.

Formaldehyde free finishes: These finishes are based on the use of compounds which do not contain formaldehyde or do not release formaldehyde due to the action of heat or during the storage of finished fabrics (low-formaldehyde resin finishing) such as 4,5-dihydroxy-1,3-dimethylurea and 4,5-dihydroxy-1,3-bisethyleneurea. A finish with reaction products of acrylamide and glyoxal or glutaraldehyde gives poorer effects than usual. A comparison of both basic structures of a low-formaldehyde type finish and a formaldehyde-free finish clearly shows that the number of reactive hydroxyl groups for cross-linking with cellulose in low-formaldehyde glyoxalmonoureins is approximately double that in the formaldehyde-free dimethyl dihydroxy ethylene urea.

Formaldehyde-free finishing agents: These are, more precisely, formaldehyde-free resin finishing agents used to produce formaldehyde-free finishes on cellulosic materials. Cross-linking agents are available for the resin finishing of cellulose which contain no formaldehyde and do not release formaldehyde in the finishing process or during storage of the finished goods.

Formaldehyde release: Formaldehyde release, not to be confused with free formaldehyde, is the amount of formaldehyde that escapes from a fabric into the atmosphere. It is determined by the mason jar method, AATCC Test Method 112. The method calls for suspending 1 gram of fabric in a sealed quart mason jar containing distilled water in the bottom. The jar is incubated for 20 hours at 49° C. The formaldehyde that collects in the water is analysed and the results are reported in ppm or μg/g fabric. There are other fabric test methods used for determining formaldehyde in fabric. For example, the Japanese method calls for titrating the free formaldehyde in the fabric by a cold procedure. The box below shows a comparison of the results obtained on the same piece of fabric tested by the different analytical procedures. The results obtained by the mason jar method are much higher than the Japanese method, so the reader is cautioned to be sure of the details of what is being reported before jumping to conclusions.

Formaldehyde (tests for):Simple preliminary test (e.g., on fabrics containing formaldehyde-based finishes)– briefly boil a 03–04 cm^2 fabric specimen in a test tube with 5 ml approximately 2% sulphuric acid in the presence of formaldehyde, formaldehyde vapour escapes during this test and can be easily recognized by its odour. Other preliminary tests can also be carried out with Nessler's solution or Fehling's solution. More precise methods of formaldehyde detection are based on, e.g. chromotropic acid, Schiff's reagent, carbazole-sulphuric acid and xanthydrol.

Formaldhyde resins: Compounds which contain formaldehyde mainly in the form of methylol compounds which may also be etherified, and which are capable of cross-linking reactions (self-crosslinking resins; reactive resins). The term "formaldehyde resin" is mainly understood to refer only to the self-crosslinking resins such as aminoplasts and phenoplasts.

Formaldehyde scavengers: These are substances which are capable of binding formaldehyde. They are used variously in resin finishing, especially in low formaldehyde finishes, e.g. urea, ethylene urea. In the case of formaldehyde scavengers such as urea, ethylene urea, dicyandiamide and hexamethylenetetramine, adjustment of the pH at the same time is of considerable importance. Hydrazides and carbohydrazides are also recommended, among other products, to reduce the content of free formaldehyde from glyoxal cross-linking agents. Ethylene glycol and diethylene glycol also reduce the release of formaldehyde and do not cause chlorine retention.

Form D process: A formaldehyde finish for cotton in the partially-swollen state (moist cross-linking process) with formaldehyde dissolved in organic solvents in the presence of water.

Form V process (formaldehyde vapour process): A formaldehyde finish for cotton in the gas phase which takes advantage of the fact that pure formaldehyde is a gaseous compound. Formerly, gaseous hydrochloric acid was used as a catalyst.

Form W process (formaldehyde wet process): A formaldehyde finish for cotton in the swollen state (wet cross-linking process). Formaldehyde dissolved in water is allowed to react with cotton in the presence of aqueous hydrochloric acid.

Formasan dyes: Formazan dyes bear a formal resemblance to azo dyes, since they contain an azo group, but have sufficient structural dissimilarities to be considered as a separate class. The most important formazan dyes are the metal complexes, particularly copper complexes, of tetradentate formazans. They are used as reactive dyes for cotton.

Formasul: A trade name for Sodium sulphoxylate formaldehyde.

Formatted piece method: Used to determine edge/centre evenness (tailing; ending) in rotary screen printing. A specified quantity of material is mechanically cut into pieces of exactly the same format. The pieces are weighed, printed on the right or left side or centre of the backing cloth using a non-patterned rotary screen and then re-weighed. The dye paste quantity differences that are measured provide a way of gauging whether the printing mechanism and the screen produce level results.

Formed yarn: See **Plied yarn**.

Formic Acid: HCOOH; An organic *weak acid* Formic acid is commercially often used with acid dyes for wool. It produces a pH lower than that of acetic acid, but higher than that of sulphuric acid. It is a pungent, flammable liquid. Formic acid is not easy for art dyers to procure.

Fota: Thick, strong and durable cotton cloth, made in India; used for garments for native women.

Fougeres: Various linen and canvas cloths made in France as coarse packing canvas or bleached household linen.

Foulard: A soft, printed, light weight, plain or twill weave fabric, which can be made of silk, acetate, viscose, polyester, or triacetate, very fine cotton, very fine worsted. Made in Twill, two up two down. Very soft and light fabric. Noted for its soft finish and feel. It is usually printed with small figures on a dark or light background. Similar to Surah and Tie silk, but finer. It was originally imported from India. The printed pattern usually take the form of small figures on contrasting background colour. Used for soft-styled dresses and blouses, robes, scarves and men's light dressing gowns, neckwear of all kinds. First made for the handkerchief trade.

Foulard (in designing): Small patterns repeated directly under and across from one another at measured interval. The name comes from the French word for scarf or necktie.

Foulardin: A light, plain woven cotton fabric, starched and highly finished; used for sleeve lining that is made in Austria.

Foulardtherm process: A process developed by Ciba-Geigy for the application of fluorescent brightening agents to polyester; also applicable to polyamide, polyvinyl chloride, acetate and triacetate. Fixation of the fluorescent brightening agent is achieved with hot air (150 °C) in the heat-setting zone of stenter driers.

Foule': In French language, the word *fouler* means "milled". A soft, serge-like worsted fabric with a rough, fibrous Melton-like surface. Its characteristic appearance is produced by prewashing and subsequent raising and milling of the face side; the weave (usually a 2 up, 2 down twill) can still be easily recognized on the back side of the fabric. Used as a ladies' dress fabric, either plain-dyed or as coloured checks. Also used for men's outer wear, mostly piece-dyed in black.

Foule

Fouling: Irreversible adsorption of dyestuffs.

Fouling (in effluent treatment): In cross flow filtration, the reduction of flux that is attributed to a build-up of solids on the surface of a membrane.

Foundation garments: The requirements for these materials involve consideration of the following four factors –

(a) easy-care properties
(b) fashion
(c) wearing comfort
(d) functionality

The nature of the chosen finishing route and the selection of finishing products are very much dependent on the above quality standards as well as on the respective substrate (the parameters include, e.g. fabric weight, weave, type of fibre blend). In the case of knit goods for so-called "elastic articles" in the swimwear and sportswear sectors, as well as foundation garments, blends of polyamide filament yarns and polyurethane filaments have established themselves. The dyes normally used are anionic (acid) dyes, disperse dyes, 1:2 metal-complex dyes, reactive disperse dyes, after-chrome dyes.

Four-colour printing: A subtractive printing technique based on the half-tone process with the three primary colours of yellow, magenta and cyan. A fourth printing in black or grey is added to enhance the final result. See ***Trichromatic printing***.

Four-way stretch: A fabric that stretches both on the crosswise and lengthwise grains of the fabric. It is the same as two-way stretch.

Fourre: Another French term for matelassé.

Fox: (1) A reddish- or yellowish-brown; a colour in Winifred Nicholson's 1944 '*Chart of Colours*'.

(2) This is fur fabric, blue-grey, red or silver in colour, the tail is often used as trimming and an interesting effect can be achieved by adding the fur from the paws.

FP (fp): Freezing point.

FPT textiles: American term signifying Fire Proof Textiles. Textile meeting the legal fire prevention regulations in the United States of America.

FR additive: Flame-retardant additive.

FR fibres: Flame-retardant fibres.

FR finishes: Flame-retardant finishing.

Fractional solubility of dyes: A method used to investigate the homogeneity of dyes which is based on the different solubility of individual dyes in various solvent mixtures. The solvents mainly used are water, ethanol, ether, benzene, methanol, carbon tetrachloride, etc., or mixtures such as aniline and amyl alcohol (pentanol) 1:1; ethanol and chloroform 3:2; dichlorohydrin and carbon tetrachloride 1:3.

Fractionated dyeing: Dye uniformity investigation, based on the fact that the dye mixtures can be identified by the different rates at which the individual dyes are absorbed. For example, fresh pieces of material are dyed in succession in a normal dye bath (extremely inaccurate method) for instance for about 1–2 or 15 minutes. After the dye having a quick absorption rate has been fully exhausted, the dye with a slow absorption rate remains and can now also be dyed out separately. The conditions can be varied with regard to temperature, liquor pH, fibre material, etc.

Fragrant finishes: Prepared fragrance complexes. These complexes impart a fragrance for the fabric or garment which has been made out of these finishes. The basic materials are natural or synthetic odours with a characteristic smell of their own (for example, phenyl acetaldehyde and cinnamon alcohol are the odours in hyacinths and cinnamon alcohol or geraniol or nerol are the odours in roses). Additional substances for rounding off a more delicate fragrance are esters, aldehydes and alcohols. Fixing agents reinforce the uniform overall effect, reduce the vaporization pressure of the easily volatile components, slow down evaporation and increase permanence. Modern technology has made finishes which have prolonged fragrance even after many washes, and there are finishes which have been developed to be reloaded with the fragrance during the washing/laundering process.

Frailejon: White or rust coloured woolly fibre taken from the surface of the leaves of the Frailejon in Venezuela, Colombia and Ecuador.

Framboise: The red colour of the raspberry.

Frame: Sheets of pile yarn normally of one colour which are fed to the jacquard loom, i.e. Wilton or Gripper axminster.

Frame tape: A stout, half bleached linen or cotton mixed tape in England.

Frame wheel: The flyer is usually mounted above the wheel, which means less floor space is used. Also called a "castle wheel".

Frame yarn: Pile yarn in a Wilton-type carpet.

Framework: Knitted fabrics made from horizontally knitted or weft threads in plain, rib or pearl stitches.

Franella: A napped cotton cloth in Paraguay; canton flannel and coloured and printed flannelettes in Chile.

Franneru: Flannelette in Japan.

Fraser: A Highland tartan with a red ground, composed as follows – A group of two dark green (on the outside) and two dark blue stripes (on the inside), separated from each other by narrow red lines; a red field made as wide as the above group and split in the middle by a narrow white line.

Fray: Unravel.

Frayed weft: A weft yarn with broken filaments resulting from abrasion during weft insertion, winding or excessive tension.

Frazadas: Cotton blankets in Latin-American countries.

Free formaldehyde: Free formaldehyde is defined as the uncombined monomeric formaldehyde that exists in finish solutions. Usually, its sources are stoichiometric excesses that may have been included in the resin manufacturing step to drive the reaction and/or the amount arising from the equilibrium established by the particular reactants. Free formaldehyde is determined by suitable analytical titration procedures.

Free liquor: The state of a liquor which is difficult to control under practical conditions, since the "free" volume is always dependent on the type of material/quantity of material. On the other hand, terms such as "short or long liquor" represent definite relationships between liquor volumes – weight of material, i.e. the ***Liquor-to-goods-ratio***.

Free-Radical Copolymerisation: In the free-radical copolymerisation reaction, free-radical initiator, R, is used with monomers Ml and Mu and mixed well for polymerisation. The free-radical R* attacks on M_1 and M_2. This is an initiation step as

$$R^*+Ml \rightarrow RM^*_1$$
$$R^*+M2 \rightarrow RM^*_2$$

In the propagation step, the above chain can further attack on MI or M2.

$$RM^*_1 + M_1 \rightarrow RM_1M_1^*$$
$$RM^*_1 + M_2 \rightarrow RM_1M_2^*$$
$$RM^*_2 + M_1 \rightarrow RM_2M_1^*$$
$$RM^*_2 + M_2 \rightarrow RM_2M_2^*$$

Free wools: Usually means wool that is free from defects, such as vegetable matter.

Freezing point: The temperature at which a liquid is in equilibrium with its solid phase at standard pressure and below which the liquid freezes or solidifies. This temperature is always the same for a particular pure liquid and is numerically equal to the melting point of the solid.

French back: It is a cloth with corded twill backing in a weave different from that one used for the face of the cloth, which has a clear finish. The back weave is usually made with a yarn that is inferior to the worsted yarn used for the face of the cloth; often backing is cotton. Weight per yard will be around 15-20 oz. This cloth is seldom used today.

French back serge: See **Serge**.

French beige: A light brown colour.

French blue: A greenish-blue; sometimes a purplish-blue colour.

French cambric: The finest grade of cambric.

French chalk: Talc.

French combing wool: Wools that are intermediate in length between strictly combing and clothing. French combs can handle fine wools from 1.25:2.5 inches in length. Yarns that have been produced with the French combing method are combed dry(without oil added). See **Bradford spinning**. The yarn is softer and loftier than the Bradford (worsted) yarn.

French crêpe: Used for negligees and underwear. It is composed of flat warp yarn and fine voile twisted weft. The voile twist which is comparatively low, adds roundness to the weft yarn. Construction ranges from 92 × 50 to 150 × 104 with 075–150 denier rayon yarn in the warp and weft. Now usually made of polyamide. See **Crepe**.

French facing: The inside facing of a coat, extending around the armholes.

French flannel: Soft and very slightly napped twilled wool fabric, made with stripes, checks or solid colours; used for men's and women's wear.

French foot: Hosiery made with a seam in the middle of the sole.

French green: A yellowish-green colour.

French grey: A bluish-grey colour.

French hem: The French hem is used for table linen. Fold as in an ordinary hem, then fold the hem back on the right side and overhand the edge formed while taking fine stitches. Press the hem flat from the right side.

French knots: French knots are used in connection with other stitches for borders enclosed in outline and chain stitches, in initials, centres of flowers, and as a filling-in stitch. The simplest method is of taking a small back stitch, bringing the thread from the eye of the needle under the point from right to left and drawing the needle perpendicularly from the cloth. Place the needle back of the knot and bring the point out in the place where the next knot is to be made. The size of the thread will determine the size of the knot.

French navy: A shade of navy.

French ochre: An orange-yellow colour.

French pink: A pale yellow. See **English pink**.

French purple: A purple dye, very popular in the 1850's, made from lichens and instrumental in giving rise to the word 'mauve' which moved from describing the dye to designating the colour of the dye. See **Mauve**.

French seam: (1) A complex seam formed on the inside of the object with both raw edges enclosed and no stitching rows visible on the face side. A French seam is sewed twice—first on the right side as near the raw edge as possible. Cut off all frayed edges, turn the material by folding *on the seam* or line of sewing, so the seam is folded inside and the second sewing is on the wrong side below the raw edges. This is not a good seam for underwear worn next the body, as it leaves a ridge on the wrong side, but it is useful for skirts of thin material, etc. It is more easily made than a fell.

(2) A seam in which two pieces of material are superimposed, stitched together and the folded over and stitched again to cancel the edges, thus producing a flat folded seam with only one row of stitching visible.

French serge: A very superior quality serge used for women's tailored clothes.

French ultramarine: See **Ultramarine**.

French system: French system used in spinning worsted yarns. The wool is combed dry, the sliver is drawn between rolls, but the fibres are parallel and not twisted until the actual spinning. The yarn is not as smooth as that made according to the Bradford system.

French yellow: A yellowish- or orange-brown colour.

Frequency distribution (of a sample): A table giving for each value of a discrete variate, or for each group of values of a continuous variate, the corresponding number of observations.

Frequency distribution (of a population): A function that, for a specific type of distribution, given for each value of a random discrete variate, or each group of values of a random continuous variate, the corresponding probability of occurrence.

Fresco (Ital.: fresco = fresh, new): (1) (a) *Fresco twist yarn* – high-twist, 2 to 6-fold yarn, composed of relatively fine individual yarns, produced by multiple twisting stages.

(b) *Fresco fabric* – light-weight, worsted or woollen fabrics, in plain weave, woven with the twist yarn of the same name; relatively harsh, slippery handle (due to the high twist), open set; usually yarn dyed, but also piece dyed. The special attribute is the high air permeability and considerable resistance to creasing. A distinction is made between:

(i) Fine worsted frescos (Tropicals), 220–330 g/m^2; all-wool or mixed with polyester (55%); smooth finish.

(ii) Woollen frescos (Fresco imitations), not as fine, approximately 500 g/m^2, usually only made from 2-fold yarn in crossbred wool; clear finish. Used for summer suits and dresses.

Fresco

(2) An orange-red colour.

Fribs: Short second cuts of wool from merino fleece.

Fribby wool: Wool containing an excessive amount of second cuts and/or sweat locks.

Friction: The resistance to the relative motion of one body sliding, rolling, or flowing over another body with which it is in contact.

Friction calendars: In the treatment of cotton fabrics, the gloss finish imparted by calendar bowls rotating at the same speed only is insufficient to achieve a high surface glaze with a greater closing of the yarns. For this purpose, friction calendars are used in which a highly polished chromium-plated steel bowl travels at a higher circumferential speed than the fabric itself, so that the fabric receives a lower or higher degree of glazing depending on the preselected advance speed (friction) of the calendar bowl.

Friction calendar compactors: A method of compacting fabrics with calendar rolls. The fabric passes between two metal cylinders, one cylinder rotates faster than the other. The fabric is restrained by shoes that are positioned against the cylinders. The fabric delivery cylinder rotates faster than the take-off cylinder and the action is similar to stuffing a string into a straw. The friction causes filling yarns to move closer together and a loss of fabric length.

Friction calendaring: As the name implies, friction calendars apply a friction force to the face of the fabric. This is done by driving the pattern roll faster than the support bowl. Friction is created by speed differentials ranging from 5% to 100%, so it is necessary to have a strong fabric to withstand the strains. The process of friction produces a high degree of lustre on one side and the final effect is similar to ironing with a hot iron.

Friction marks: These defects are formed, e.g. in the dyebath by friction between the surface of a fabric and the walls of the dyeing vessel and/or other hard surfaces. Fabrics of acetate and silk are particularly susceptible to this kind of defect. Appearance of affected areas: darker, matt and roughened surface, fabric surface bulges on one side; often accompanied by tears and holes. Individual fibres exhibit pressure marks. Some improvement may be possible by rubbing the affected areas with a paraffin-gasoline solution.

Friction spinning: In friction spinning, twist is inserted into the yarn by a rolling action due to frictional forces generated between, for example, two perforated rollers and the yarn surface, it being relatively easy to produce core-spun yarns by introducing a filament yarn which is then wrapped by a sheath of staple fibres. Friction spinning can also be classified under the broad category of 'open-end spinning', since the sliver feed is 'broken' (i.e. discontinuous) prior to yarn formation (twist insertion). Friction spun yarn are weaker and more twist lively than other yarns, although plying can reduce these drawbacks.

Friction texturizing: Texturing method in which appropriate over twist is applied to the running thread using overlapping friction discs (torsion discs, friction discs). The required false twist is achieved by subsequent counter twisting and twist setting. Up to 1960, a rigidly mounted spindle with axial drive was used (friction elements). As entirely uniform, reproducible speeds could not be obtained, spring-mounted magnetic spindles are now used (instead of rigidly mounted friction components).

Frictional constraint: The force imposed by the multitude of fibre to fibre contacts within a fabric.

Frise: (1) Made of rayon, most popular, also mohair and silk, and synthetics. The ground or backing yarns are usually made of cotton. Sometimes jute or hemp are combined with cotton. Made usually with uncut loops in all-over pattern. It is sometimes patterned by shearing the loops at different lengths. A few are made with both cut and uncut loops in the form of a pattern. *Uses*: Upholstery, also used widely as transportation fabric by railroads, buses, and airplanes. Frise is also spelled Frieze, but frieze really refers to a rough, fuzzy, rizzy, boardy woollen over-coating fabric which originated in Friesland

Holland. Often used for over-coating material for soldiers. Much alteration is given to the cloth. Irish frieze is quite popular and more reliable and is called "cothamore".

(2) Surface with curls or fine loops, made from boucle' yarns.

(3) (a) Originally the finest grade of linen made in Friesland, Holland; (b) A French term meaning "curled"; (c) A coarse ratiné fabric made with slub yarns in a plain weave; (d) A looped pile fabric, usually of uncut loops which may be cut at various heights to provide a pattern; (e) A coarse, stout, cotton or linen fabric made in a plain weave with a flat, wiry texture and a pronounced rib; (f) This term is sometimes used for terry or bouclé fabrics; (g) Light patches occurring in viscose crepe in the warp direction consisting of flat, deformed, viscose capillaries, that are in fact dyed but reflect much more light because of their flat shape. Visible under the microscope as short, thin, and apparently white capillaries that can be proven to originate from the feedstock and can no longer be rectified. These patches are created during the viscose spinning process and appear to be due to some kind of local stretching.

Frieze: (1) A heavy woollen fabric made out of shoddy and other recycled wool, originally used for servicemen's overcoats. Cloth weight ranges from 22 to 30 oz. per yard and a common stock may be 67% of wool and 33% of shoddy and recycled wool.

(2) (a) Hand-knitting and craft yarns with brushed woolly finish; (b) Soft, thick, woollen fabric in plain or twill weave, can also be double cloth with two warp and weft systems, generally with pepper-and-salt effect on face and checked on reverse. Mossy thick hairy nap, either heavily raised on both sides or lightly raised on reverse only. Friezé cloth is similar to velour, but with rougher longer nap. Distinction may be made between randomized nap (with non-directional lay) and brushed nap (with directional lay). Used for ladies winter skirts, jackets and coats; (c) Thick and soft woollen material with raised and laid nap (milled and raised) or may be closely cropped.

Frieze Flannel: Heavy and twilled flannel that is made of cotton and wool in mixture effect.

Friezing: A process used for the production of local patterning on a pile fabric where the pile has usually been formed by raising. The fabric passes between a fixed friezing bed and a moving friezing plate mounted above it which is clothed with sandpaper, plush, cloth or rubber.

Fringe: (a) An edging or border of loose threads, tassels or loops. The edging or border may be produced by the constituent threads or by threads added to a fabric after weaving or knitting. (b) A trimming (narrow fabric) that has, on one or both edges, cut or looped weft threads that form a decorative edge, and that are sometimes bunched or knotted together to increase the decorative effect.

Note: (a) Tassels, balls or other adornments may be added. (b) The part of the fringe comprising both warp and weft is known as the heading. (c) The part of the fringe containing only weft is known as the skirt.

Friquette: A machine-made lace, having fine and clear meshes and made of fine thread. It is made as a veil and also embroidered in a light flower design along the edge.

Frise carpet: A cut pile carpet of twisted yarns in solid colour or produced with various-coloured yarns. According to few standards, a carpet with alternating closed and cut loops.

Frisette: Fine fabric of which the face is covered with small loops and is made of cotton and wool in Holland.

Frison (silk); Flock silk (Bourette): The first silk waste obtained in the course of cocoon preparation for reeling by agitation in a water bath.

Frisonette: A low grade of frison.

Frisons, Kibisu, Knubbs, Strussa: The first waste obtained in the process of reeling cocoons. It is composed of a tangled leading end of the filament that is removed before it begins to reel properly.

Frisure: (a) Core thread with a gold or silver thread wound around; used in passementeries; (b) Thin gold or silver wire braid for military uniforms.

Frizzing: Name for the chinchilla finishing process in England.

Frog: Ornamental facings, made of braid for uniforms, pyjamas, etc.

Front (stitched down): A front that has a double-turned hem that is stitched down full length of the front. The term may also refer to the shell (outside) front of self-goods.

Front falls: Trousers, not made with a single fly opening, but with a wide front flap, buttoning on both sides.

Front high-hip (in body measurements): The distance from one imaginary side seam to another imaginary side seam at the high-hip level.

Front loading washing machine (Front loading washer): Cylinder washing machine with a one-piece internal cylinder that is supported at one end and has a loading/unloading opening at the front end, usually with a transparent door made from heat-resistant safety glass. Capacity is 08–50 kg of dry material.

Front notch (in sleeves): One notch on front of the sleeve seam line to mark match point.

Front rest: Weaving technology term. See **Cloth fell**.

Front-break point (in anatomy): The location on the front of the body where the arm separates from the body.

Frosted wool process: A process for the degreasing and removal of all impurities from raw wool by cooling to –10 to –17 °C, followed by beating to break off the solidified brittle foreign matter (grease, burrs, sand, etc.). Burrs become so brittle in this treatment that they are easily pulverized by attrition.

Frosting: (1) A change in colour in a limited area of a fabric caused by abrasive wear.

(2) (a) Sandwich effects in dyeing; (b) Whitish sheen on carpets containing wool and

polyamide pile that have been continuously dyed by the pad-steam process. When the padded carpet enters the steamer, condensing steam rinses dye liquor from the tips of the pile; thus, giving it a whitish appearance. A remedial measure is to add film and foam-forming agents to the padding liquor. Foam pushes the dye liquor back towards the tips of the pile and thus, compensating for the loss of dye.

Frosting in carpet printing (tippiness): Rather than remaining uniformly distributed along the individual pile fibres, in the printing of carpets (especially those with a high pile) the print paste may flow from the tips of the pile towards the carpet backing before the dye fixation stage. This results in a depletion of dye at the tips of the pile, i.e. frosting.

Frotte': Imitation terry made with loop yarns.

Froth dyeing: A variant of foam dyeing. *Principle* – The process is carried out in an open vat as the dyeing vessel. Foam is produced by the vigorous boiling of a dye liquor containing soap or other high-foaming detergent. The dye liquor occupies approximately 20% of the dye vessel volume, and the material is dyed by suspension in the foam above the dye liquor.

Froth flotation (flotation): An industrial technique for separating the required parts of ores from unwanted gangue. The mineral is pulverized and mixed with water, to which a frothing agent is added. The mixture is aerated, after which particles of one constituent are carried to the surface by reactions and as a result of chemical decomposition at high temperatures. Free radicals are extremely reactive and can be stabilized and isolated only under special bubbles of air, which adhere to it preferentially. The froth is then skimmed off. Various additives are also used to modify the surface properties of the mineral particles (e.g., to increase the adherence of air bubbles). Some effluent treatment plants use this method to separate precipitated matters from the treated effluent.

Frotte': French word for towelling.

Frotté: Name for (a) *Yarn*: 5–7 fold-effect twist yarn in form of knop yarn, loop yarn, in cotton, viscose, wool (woollen spun); (b) Plain-weave fabric with smooth warp and knop yarn weft, irregular, contoured surface; brushed, rough finish. Sold as piece goods or by the metre; also used in outerwear.

Frotté/Bouclé: A drape fabric. Frotté and bouclé-fabrics are produced with loose yarn which have nap-, bow- or loop- effects. For this reason, a structured fabric is created with small loops or naps. All fibres can be used. Frotté should not be confused with towelling cloth, where the loops through the weave are built up by a third thread.

Frowzy wool: A waste, lifeless-appearing, dry, harsh wool, lacking in character.

Fruit fibres: Fibres of the coconut fibre type.

FSDC: Fellow of the Society of Dyers and Colourists, Bradford, United Kingdom, Professional qualification.

FSW: Forschungsstelle Wäscherei (Laundry Research Unit), Hohenstein.

FTC: (a) Federal Trade Commission. A United States government customer protection agency which, amongst other activities, has been responsible for the definitions of most of the internationally valid manmade fibre classifications; (b) Fastness Tests Commission.

FTCC: Fastness Test Coordinating Commission, the British Fastness Tests Commission.

FTI: Fellow of the Textile Institute, Manchester, United Kingdom, Professional qualification.

FTU: Formazine Turbidity Units. A measure of turbidity – by a nephelometer.

Fuchsia: Purplish-red colour, as in the fuchsia flower, named after the German biologist Leonard Fuchs (1501–1566). According to a MORI poll referred to on BBC television on 23rd September, 1998, fuchsia (with white and grey colour) are the most unpopular colours in the United Kingdom.

Fuchsia pink: A purplish-pink colour.

Fuchsin B: A cationic dye used *inter alia* as a special indicator for anionic substances: aqueous. Fuchsin solution + anionic solution = colour change from red to cornflower blue.

Fud: Waste of the wool carding machine; used after mixing with shoddy.

Fuel gas: Natural gas; Liquefied gas.

Fuel oil: Liquid fuel, mainly from mineral oil processing. Production and use almost exclusively restricted to EL fuel oil (extra light) which is mainly used for space heating and is also used for heating processes in commerce and industry, and S fuel oil (heavy), which is the preferred fuel for most industrial heating processes. L fuel oil (light) and M fuel oil (medium heavy) are the exclusive products of tar oil processing and only play a quantitatively minor role.

Fugitive colours: Dyes that fade, especially those that lose colour relatively quickly when exposed to natural light.

Fukuji: Hand-woven Japanese cotton goods, made both in plain and twill weave, usually like a fine duck, ply yarns being used; police and schoolboys' uniforms are made of this material.

Fukusa: A square piece of silk, often richly embroidered or decorated; used in Japan to cover boxes containing presents.

Fuligin- (L): Sooty.

Full bath treatment: A dyeing process carried out at a long liquor ratio by Exhaust dyeing, in contrast to the Foulardtherm process. See **Foulardtherm process.**

Full bottom wig: High, long and curly wig for men.

Full box: Trade term for a very loose effect in overcoats.

Full cover print: A printed design which covers the entire surface of a textile fabric.

Full depth shade in printing: The maximum depth of shade which can be achieved with a normally engraved screen or printing roller.

Full-dress: Black "claw-hammer" coat, low cut vest and trousers to match.

Full gauze: Weave having the warps cross other warps in both directions.

Full penetration in dyeing: The full penetration dyeing of fibres or textile materials occurs when a completely uniform distribution and penetration of the fibre occurs, i.e. when the interior of individual fibres is fully penetrated (not ring-dyed) and when the closely packed fibre material in slivers are completely dyed.

Full regular: Hosiery and underwear having hand-knitted seams.

Full set: A term that indicates that all guide eyes in a guide bar each have a yam from the warp.

Full-toned: A colour with an intense hue.

Full(to): To treat or beat cloth for the purpose of cleansing and thickening it.

Full-bust girth: See **Bust girth**.

Full-cut: Not tapered.

Fulled: Woollen fabric which is napped and then felted or made thicker and more compact by shrinking.

Fullers earth: Impure clay (refined and pulverized) with a greyish-green to yellowish-brown colour and a greasy feel, often of variable composition (e.g. 11–18% aluminium oxide, 44–42% silica, 5–4% lime, 2% magnesia, 10–6% iron oxide, 5% sodium carbonate, 3% water). *Uses*: after-washing and milling of woollen materials which have been washed in soap (especially for fine cloths, white and semi-coloured flannel, woollen blankets, etc.).

Fulling: The operation of shrinking and felting a woollen fabric to make it thicker and denser. Also called "milling". One can also full woollen yarn to produce a lovely knitting yarn (If one is going to weave with the same yarn, wait until the fabric has been woven).

Fulling agent: A chemical, usually a surfactant, that acts as a lubricant during the process of fulling.

Fulv- (L): Tawny.

Fulvid: Reddish-yellow, tawny, yellowish-grey or yellowish-brown colour.

Fully fashioned: A term used for knit goods produced on a flat-knitting machine that have been shaped by adding or reducing stitches. The material has ribbed edges (not cut) as it leaves the knitting machine. Fully fashioned flat-knitted materials are referred to as true to shape articles. Fully fashioned knit goods are used, e.g., for socks, stockings, tights, underwear, outerwear, pullovers, etc. Fully fashioned fabrics are the opposite of ***Cut fabric***.

Fully fashioned dyeing: Garment dyeing.

Fully fashioned knitwear: Knitwear from garment pieces shaped (fashioned) on the knitting machine to the correct size and shape required.

Fully-washable textiles: Finished textiles which can be washed in a washing machine without excessive shrinkage, creasing or other significant deterioration in quality. The term is particularly applicable to blend fabrics of polyester/cotton or polyester/acrylic (Machine-washable wool; Easy-care finishes).

Fulpat: Commercial term in India for weak and gummy, immature jute fibre, cut before flowering; it has a good colour.

Function: A mathematical expression describing the relation between variables; an expression having a definite value, or values, when special values are assigned to other quantities called the arguments, or independent variables of the function.

Functional clothing: Clothing that is designed to be light, comfortable, pleasant to wear and easy-care; e.g., Active wear; sportswear.

Fundamental units: The base units of a measurement system, especially those that deal with force and motion. In most systems, these are the units for length, mass, and time, plus at least one electrical unit. The fundamental units in the SI are considered to be the meter, the kilogram, the second, and the ampere.

Fungicidal finishes: These finishes are applied especially to textiles which frequently come into contact with the ground, soil, wind and weather, such as tenting, truck tarpaulins, fire hoses, etc. Suitable preservatives for such articles include, in particular, organic acid salts of zinc, copper, and mercury, phenols, chlorophenols, anilides of aromatic acids, sodium salts of chlorocresols, chloroxylenols, tri and pentachlorophenols, copper salts of cellulose glycollic acid, as well as quaternary ammonium bases, etc.

Fungicide: The material which kills fungi.

Fungistat: Material which inhibits fungal growth.

Funiculi: Plural of funiculus.

Funiculus: The slender stalk of an ovule or seed by which a growing seed is attached to the placenta of the boll.

Fur: It refers to Animal fibres.

Fur and leather cleaning: Garment cleaning terms; (a) *Fur cleaning* – non-liquid cleaning or drum with dry particle compound and fur glazing or fur ironing; (b) *Leather cleaning* – special leather care methods are required on suede, leather and plastic goods.

Fur beaver: A long napped cloth imitation fur.

Fur dyes: Metallizable types. These dyes are applied on the same principle as metallizable wool dyes and produce fast dyeing on fur skins by brushing or dipping. The classical dyeing technique is based mainly on the application of aromatic amine bases at a temperature of 30 °C from which dyes are synthesized on the fur hairs by oxidation (probably by a process analogous to Aniline black) and fixed with the aid of metal mordants. Appropriate azo dyes capable of exhausting on to mordants are also used in fur dyeing.

Fur fabric: See **Plush**.

Fur finishing: The finishing of fur materials covers all the processes and technical operations which affect the original character of the fur. These changes are mainly concerned with the fur's natural colouring.

Furnisher roller (colour furnisher roller): Used in roller printing to supply the engraved printing roller with print paste from the colour box in which it rotates. Both brush furnisher rollers and smooth rubber-coated rollers are used. See **Brush furnisher roller**.

Furnishing fabric: Collective term for printed, plain-dyed or jacquard-woven textile fabrics for room decoration purposes, e.g. upholstery fabrics, curtains, drapes, etc. Dyes with very high colour fastness to light are required for these applications.

Furniture covering: A general term for attached upholstery fabric, slipcovers and throws.

Furniture cord: Two pairs of case cord and two grimp cords, the four components being twisted together.

Furniture unit (in upholstered furniture): A complete single piece of upholstered seating, such as a sofa, love seat, lounge chair, rocker, or recliner.

Fuscescent: Having a dark brownish colour.

Fuschia: A bluish-red colour named after a flower which is named after Herr Fuchs. See **Magenta**.

Fusco-piceous: Reddish-black colour.

Fusco-testaceous: A reddish-brown colour.

Fuscous: Brown, dingy, dark, sombre; also a brownish-grey colour.

Fused bonded non-woven: Bonded fabrics which have been bonded, e.g. through the loops of the warp threads with a thermoplastic compound.

Fused collar fabric: See **Fusible fabric, Fusible interlining**.

Fused fabric: A type of bonded fabric made by adhering a single fusible fabric to another fabric, such as for use in an interlining.

Fusible fabric: A utilitarian fabric which has a thermoplastic adhesive applied to one side, sometimes in a pattern of dots, so that the surface can be bonded to another fabric surface by the use of heat and pressure.

Fusible fibres: These are used in the form of fibres, threads and monofilaments to achieve adhesive-free consolidation/permanent bonding of textile fabrics. After thermal fusion with the aid of an ironing press, hot air, etc., the cohesive strength is intense and permanent. Fusible fibres are of particular interest for non-woven, textile floorcoverings and in the making-up of garments. See **Binder fibres**.

Fusible interlining: Interlining produced with a fusible adhesive coating by the powder-dot process.

Fusible seaming: A textile bonding technique in which fabrics are joined by fusion of the thermoplastic fibres in the fabric or a thermoplastic resin that has been applied to the fabric.

Fusible web: It is a special fabric used to bind two pieces of fabric together. The fusible web is placed between the two fabrics and ironed to adhere the fabrics together.

Fusing: The process of adhering the fusible fabric to another applying heat, pressure or both.

Fusion bonded: A method of creating carpet pile construction by adhering loops or lengths of yarn to the face of primary backing.

Fussy cut (in sewing): It is when you cut a piece of fabric for a quilt so that a feature of the print is on the centre of the piece.

Fustian: (1) A general term used to describe a group of cotton fabrics, usually heavily wafted, raised on one or both sides and produced in various weaves, e.g. twill fustian, satin fustian, corduroy fustian. The description also refers to the end-use, e.g. bed-fustian, lining fustian, dress fustian.

(2) An old Egyptian cotton or linen fabric, supposed to date from Roman times. Fustian is now used to describe a range of cotton fabric including pile fabrics.

(3) (a) Medieval stout and heavy worsted fabric produced in England; (b) In the I5th Century, a printed fabric, made with linen warp and cotton tilling, produced in England; (c) A variety of closely woven and heavy cotton fabrics, being either cut pile goods or having a stout weft face. They are woven with one set of warp and two sets of filling. They are known as velveteen, swansdown, moleskin, beaverteen, canton, or diagonal corduroy and imperial. They are mostly used for clothes.

Fustic: (a) Brazil wood, Cuba wood, fustic wood – tropical dyewoods; hard, solid, light yellow or pale lemon-yellow, turning darker with age. Available commercially in block form, rasping and powder; (b) Yellow dyes of the flavone type.

Futahaba: Kanakin grey shirting in Japan, made about 30 inches wide; used also for printing.

Futako: Trade term in Japan for striped cotton fabrics (used mostly for kimonos), made with ply warp.

Fuzz: A term analogous to lint, used mainly to describe small hair, fibres, etc., projecting from, or lying on, the surface of a textile fabric.

Fuzz ball: Loose and frayed fibres which have formed into a ball and have then been woven or knitted into the fabric.

Fuzzy: Characterised by hairy appearance due to broken fibres or filaments.

Fuzzy gout: A gout is also called fuzzy gout because it is formed due to the entanglement of fuzz during weaving.

FWA: Fluorescent whitening agent (Fluorescent brightening agents).

Fyrol 76: It is an oligomeric phosphonate containing vinyl groups. The finish is applied with N-methylol acrylamide with a free radical initiator (potassium persulphate), dried and cured.

G

G: Symbol for giga, i.e. 10^9. A prefix used in the metric system to denote one thousand million times, e.g. 10^9 joules = 1 gigajoule (GJ).

g: Symbol for gram, a metric unit of mass equal to one thousandth of a kilogram.

1 g = the weight of 1 cm^3 of water at + 4°C.

1 cm^3 = 1 g = 1000 mg.

1000 cm^3 = 1 l = 1 kg (of water).

1000 l = 1 m^3 = 1000 kg = 1 t.

Gabardine: The word gabardine (alternative spelling in United Kingdom = gaberdine) is a collective term for certain woven fabrics and also describes a type of woven pattern. Gabardine is generally a fine, tightly woven, warp-faced fabric, and is one of the most popular apparel fabrics. The warp density is generally considerably greater than the weft density and the diagonal cord or rib with characteristic distinct steep twill line caused by a construction that has many more ends than picks per inch, closely set, and raised. A characteristic texture is 100 × 50 with 18s yarn in the warp and weft. Fairly hard twisted yarns are used so that the face of the fabric may be clear. Long floats make the diagonal lines, and there are short floats between the wales. This effect is often accentuated by singeing. The fabric is light weight and due to the close weave and steep slope of the rib, gabardine (especially gabardine worsted) is amazingly water repellent. Other fibres used include cotton, viscose, wool, silk, polyester and cotton. Tightly woven cotton gabardine after waterproof finish is sometimes used for raincoats. An inexpensive gabardine can be made of 88 × 44 with a 2/1 twill weave. Used for outerwear, including raincoats, anoraks and trousers and for women's skirts, and suits. It is very hard wearing.

Gaberdine: Basically same as Gabardine. A firm twill fabric (e.g., with worsted warp and cotton weft) with warp predominating in the surface; used for dress and suiting cloths and light shower proof over coatings. All cotton gaberdine is also used for similar garments. E.g., This may be made of 24s worsted warp and weft with a 100 × 56 with 3/2, 63 degree twill weave. The best worsted gabardine is a rugged fabric made of ply yarn. Made in a 2/2 twill, the cloth may have a construction of 100 × 50 with 2/40s high twist worsted warp and weft.

Gaberum: An Indian cotton fabric, made with check patterns.

GAC: Granular Activated Carbon.

GAC filter: See **Activated carbon filter**.

Gajj: Indian satin of high finish; used for dresses by the native women.

Gage: See **Gauge**.

Gaiting (in warp knitting): The setting of a guide bar one or more needle spaces to the right or left in order to increase the pattern possibilities.

Gaiting (interlock): The opposed alignment of one set of needles with the other on a knitting machine.

Gaiting (in warp knitting): The setting of a guide bar one or more needle spaces to the right or left in order to increase the pattern possibilities.

Gaiting (purl): The opposed alignment of tricks of two needle beds lying in the same pane on a machine equipped with double-headed needles.

Gaiting (rib): The alternate alignment of one set of needles with the other on a machine equipped with two sets of needles arranged to knit rib fabric.

Gala: Coarse cotton cloth; used for servants' clothes in Scotland.

Galanga gum: A medium quality grade of Senegal gum.

Galatea: A cotton fabric in twill weave, made plain, or simple stripe patterns, for nurses' uniforms, children's clothes and linings. It is given a hard lustrous finish which repels dirt.

Galb- (L): Yellow.

Galettame: See **Bassine**.

Galettame (Italian): Waste silk, obtained from the inner skin of the cocoon, which is left over after the reeling process.

Galette: Pierced, fuzzy or very thin cocoons; used for waste silk; also, the silk made of such cocoons.

Galettes: Lustrous silk taffeta produced in France, made of hard twist silk waste.

Galiclia: Pile cotton carpet, made in India.

Gallini: Egyptian cotton derived from the Sea island, having a very strong, long staple of light golden colour.

Gallipoli: Sort of raw cotton grown in Southern Italy.

Gallon (gal): (1) A traditional unit of liquid volume, derived from the Roman *galeta*, which originally meant a pailfull. Gallons of various sizes have been used in Europe ever since Roman times. In the United States, the liquid gallon is legally defined as exactly 231 cubic inches; this is equal to the old English **wine gallon**, which originated in medieval times but was not standardized until 1707, during the reign of Queen Anne. A few scholars believe the wine gallon was originally designed to hold 8 troy pounds of wine. The United States gallon holds 4 liquid quarts or exactly 3.785 411 784 litres; a United States gallon of water weighs about 8.33 pounds. American colonists were also familiar with the Elizabethan **beer and ale gallon**, which held 282 cubic inches (4.621 litres).

(2) A historic British unit of dry volume still used implicitly in the United States of America. In the United States, the term "gallon" is not used in dry measure, but if it were it would be equal to ½ peck, or 4 dry quarts, or 268.8025 cubic inches, or approximately 4.404 884 litres. This unit is the English **corn** or **grain gallon**, standardized during the reign of Elizabeth I in the 16th Century. The earliest official definition of a dry gallon in Britain is a 1303 proclamation of Edward I, where the gallon is defined as the volume of 8 pounds of wheat; the current U.S. "gallon" contains about 7.5 pounds of wheat. Grain gallons have tended to be larger than liquid gallons throughout the history of British units, apparently because they were based on heaped rather than "struck" (levelled) containers. A container in which grain has been heaped above the top will hold as much as 25% more grain, and the traditional corn gallon is in fact 16.4% larger than the wine gallon

(3) Currently the British use a gallon larger than either of the two American gallons. The Imperial Weights and Measures Act of 1824 established a new unit for all volumes, liquid or dry, replacing all the other gallons in previous use in Britain. The **imperial gallon**, designed to contain exactly 10 pounds of distilled water under precisely defined conditions, holds exactly 4.54609 litres or approximately 277.4194 cubic inches. The imperial gallon equals 1.20095 US liquid gallons (British wine gallons) or 1.03206 US dry gallons (British corn gallons).

(4) A traditional unit of volume in Scotland equals to 4 Scots quarts. This is almost exactly 3 British imperial gallons, 3.6 US liquid gallons, or 13.63 litres.

Gallon (imperial): In Britain and Canada, a gallon is approximately 4.55 litres. An imperial gallon of water weighs approximately 10 pounds.

Gallon (USA): In the USA, a gallon is approximately 3.79 litres. A US gallon of water weighs approximately 8.3 pounds.

Galloon: (1) Narrow tape or binding of cotton, wool or silk, showing usually fancy weave; used for trimming dresses, uniforms, also for lacing. The English widths are: Two penny, four penny, sixpenny, eight penny, and ten penny, the old penny pieces having been taken as gauge.

(2) A narrow band of cord, embroidery, braid, etc., used as a decorative trimming. See **Braid**.

Galloon lace: Any lace fabric with a finished scalloped edge on both sides.

Galvano screen: See **Rotary screens for screen printing**.

Galway: (a) Coarse, thick, scarlet coating, made in Ireland of Irish wools; (b) Also, a thick and coarse red flannel.

Gambiei: A fast brown dye, obtained from the leaves, flowers and wood of various acacia trees in India, marketed as a dark coloured paste.

Gambier: A Chinese textile speciality consisting of deep black clothing suitable for everyday use, made from non-degummed, heavily-weighted butterfly silk (lightweight base fabric, plain weave) glaze-calendered on one or both sides (with polished stone rollers). Highly resistant to wear (washable, suitable for dry cleaning). The deep black is produced by piece dyeing with a reaction product from a gambier-like tanning substance (juice of the so-called kaki fruit) and a river sludge containing iron.

Gamboge/camboge: A strong yellow; reddish-yellow colour. A yellow resin pigment originating from Cambodia and used in England from the beginning of the 17th Century. Replaced in 1851 by **Aureolin**.

Gambroon: (a) Strong, light and twilled union linen or union worsted; used for summer garments; a men's over coating, called G., originated in England in the 19th Century; made with ply yarn, composed of cotton and worsted strands; (b) A twilled linen lining.

Gamelotte: Fibre yielded by the 'Fim bristylis complanata' in Sri Lanka; used for ropes and cordage.

Gamma cellulose: One of the three forms of cellulose. With beta cellulose it is called hemicellulose. See **Alpha Cellulose and Beta Cellulose**.

Gamut: A term used in the dyeing and finishing industry for colour scale or palette (e.g. dye ranges illustrated in pattern cards).

Ganse: French for round or ridged soutache, or braid of silk, gold or silver.

Gantt chart: Simple horizontal bar chart that graphically displays the time relationships of the stages in a project. Each step in the project is represented by a line or block spaced on the chart in the time period in which it is to be undertaken. When completed, the Gantt chart shows the flow of activities in a sequence, as well as those that can be under way at the same time.

Garber: Commercial variety of upland American cotton from Alabama, the staple measuring 18-22 millimetres; the yield is about 34%.

Garden carpets: Oriental carpets with designs representing gardens, e.g. the garden carpet of Chosroes, a Persian carpet of the Sassanid dynasty (AD 224–641) in which the whole design depicts a garden in springtime.

Gare: Glossy and hair-like fibres grown on the legs of the sheep; they look and behave like kemp.

Garment blank: Panel of fabric knitted for cut and sewn knitwear production.

Garment dyeing (fully fashioned dyeing): For garment dyeing, textile material passes through all stages of manufacturing (with the exception of dyeing) up to the completed garment stage in order that garments can be stored and dyed in appropriate shades on demand. The dyeing of completed garments in the greige state has become increasingly important in recent years, especially in the leisurewear and sportswear sectors. Because of the reduction in lead times between garment dyeing and traditional dyeing methods (e.g. 4/5 days compared to 2/3 weeks), the dyeing of ready-made garments has proved so successful that the market place now has a tremendous demand for leisurewear and casualwear garments produced by the garment dyeing route, and these types of garments are very much colour and fashion orientated.

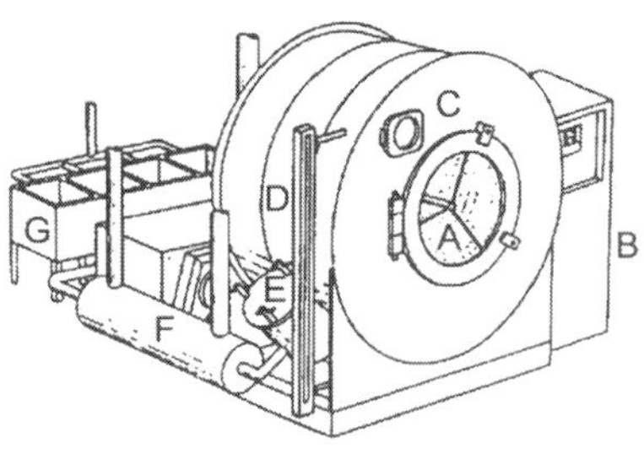

A garment dyeing machine

Garment leather: Leather manufactured from skins and hides subjected to various tanning and dressing treatments suitable for the production of leather garments of all kinds.

Garment manufacturing: A term for the standard production of garments, garment pieces and various items of clothing.

Garment-making accessories: These include all the various items necessary to produce a completed garment, e.g. underarm pads, interlinings, linings, belts, buttons, sewing thread, zip fasteners (zip fastener tapes), buckles, shoulder pads, edgings, stiffeners, waddings, etc.

Garment making technology: Even though the normal garment making involves cutting, arranging, pre-fabrication, intermediate ironing, assembling, final ironing, press finishing, this process-related sequence is not suitable for discerning technological trends in manufacturing technology. Accordingly, it may be better to subdivide it into the following production terms:

(a) *Original forming* – injection moulding, e.g. buttons, creation of endless threads (polymer shape forming).

(b) *Re-shaping* – ironing, styling, shape fixing, hot forming, knitting to shape, etc.

(c) *Separating* – cutting, stamping, pinking and weldstamping.

(d) *Joining* – sewing, fixing, gluing, welding, riveting, tacking and clipping.

(e) *Changing the material characteristics* – heating, steaming, drying and cooling.

(f) *Coating* – printing, applying adhesive.

A precise study of this system shows that a manufacturing process changes the work piece in order to achieve a certain state and a certain appearance. As a rule, tools and machines are necessary for this. Through mechanisation, automation and control, the individual elements are linked together, in conjunction with planning and organisation, to form garment technology.

Garment steamer: A steaming unit for garments; usually on steaming dummies.

Garment twist: A rotation, usually lateral, between different panels of a garment resulting from the release of latent stresses during laundering of the woven or knitted fabric forming the garment. Twist may also be referred to as torque or spirality.

Garnetted yarn: A yarn that has little bits ("garnets") of other fibres carded in. Usually the garnets are of a different colour, but they can also be from a different fibre.

Garnetting: A process for reducing various textile waste materials to fibre by passing them through a machine called a garnett that is similar to a card.

Garniture: French word for trimming.

Garter-blue: The blue colour of the ribbon of the Knights of the Garter. A dark blue shade though previously light blue colour.

Garter webbing: A form of elastic narrow fabrics, sometimes multi-coloured and characterized by selvedges that form a frill on relaxation. It may also be a medium strength elastic narrow fabric for supporting socks or stockings.

Gas fading: A change in shade of dyed or printed fabric caused by chemical reaction between certain disperses dyes and gaseous contaminants in the atmosphere, chiefly acidic gases from fuel combustion (NOx; SO_2). Colour fastness to atmospheric contaminants: nitrogen oxides; Burnt gas fumes.

Gas fastness: Colour fastness to atmospheric contaminants: nitrogen oxides; Burnt gas fumes.

Gas fume fading: Colour fastness to atmospheric contaminants: nitrogen oxides; Burnt gas fumes. An irreversible change in hue which occurs when textile, particularly cellulose acetate and triactetate dyed with certain blue anthraquinone disperse dyes are exposed to oxides of nitrogen which arise from, for example, gas or storage heaters.

Gassed yarn: Yarn or fabric, which underwent the gassing process.

Gassing: An industrial defibrillation process for yarn, where yarn travelling at high speed passes through a gas flame in order to burn off fibres which poke out to reduce its fuzz and give it a greater sheen. The term gassing is usually reserved for treatment of yarn. Singeing is the term usually used for fabric. See **Singeing**.

Gassing machines (yarn singeing machines): Machines used for the singeing of yarns.

Gathered frill: A narrow, straight piece of cloth is drawn up and fixed within a seam or at the hem of a garment to create a decorative finish.

Gathering: Gathering, gauging, casing, etc., are used for drawing up the fullness of skirts, ruffles, flounces, etc., into a given space. The running stitch is used for these. For gathering, the cloth is held in the same manner as for running. The needle, ordinarily, need not be taken out of the work, the stitches being pushed back over the eye as they are made; but for running long skirt seams in delicate material which would crinkle at the line of sewing and roughen the seam, the needle should be drawn through and the line of sewing smoothed on the thread at each needle-full of stitches.

Gathers on a band: Fabric is drawn up, arranged and fixed within a flat band.

Gathering threads: Either done by hand by hand or else the longest stitch on a sewing machine is used with the tension suitably adjusted. Two rows of stitching are urn through a piece of fabric, then the threads are drawn up to create gathers.

Gattar: Indian satin, made with silk warp and cotton filling.

Gatti: Indian cotton fabric with diamond patterns.

Gaufre: The French word *gaufre* means'waffle, emboss'. Flat fabrics in plain-weave, twill, or satin construction as well as plush materials and velvets with an embossed waffle-like pattern (embossing) or a low lying pile (= velvets). Gaufre fabrics include, e.g. Moiré fabrics; Reversible, etc.

Gaufrer: French for pressing various embossed patterns on fabrics with hot calenders during finishing. Used for ladies outerwear.

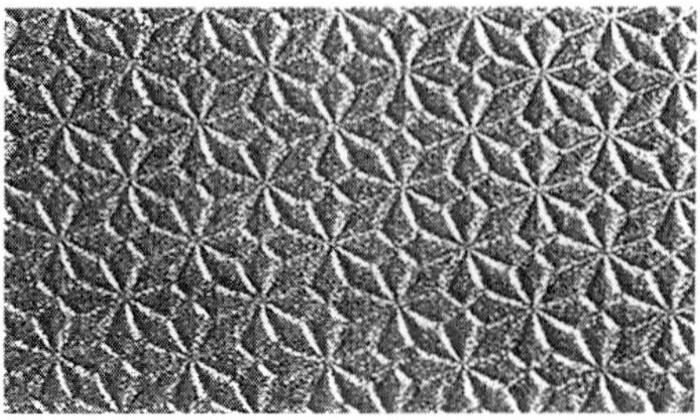

An example of gaufre fabric

Gauge (ga): (1) A traditional unit measuring the thickness of sheet metal. Larger gauge and umbers represent thinner metal: 10 gauge represents a thickness of 0.1345 inch (3.416 millimetres) and each increase of 1 in the gauge number corresponds to a reduction of about 10% in the thickness. A table is provided.

(2) A traditional unit measuring the thickness of plastic film. For this purpose, 1 gauge equals 0.01 mil or 10-5 inch (0.254 micrometre).

(3) A unit expressing the fineness of a knitted fabric, equal to the number of loops per 1.5 inches (38.1 millimetres). The same unit is also used to express the size of the knitting needles used to create a fabric of that fineness.

(4) A traditional unit measuring the diameter (or the cross-sectional area) of a wire. Various wire gauge scales have been used in the US and Britain. In traditional scales, larger gauge numbers represent thinner wires. (For very thick wires, repeated zeros are used instead of negative numbers, so gauges 00, 000, and 0000 represent -1, -2, and -3, respectively.) In the American Wire Gauge (AWG) scale, 0000 gauge represents a wire having a diameter of 0.46

inch and 36 gauge represents a diameter of 0.005 inch (5 mils). Diameters for the other gauges are obtained by geometric interpolation, meaning that the ratio between successive diameters is a constant, except for necessary round off. Thus *n* gauge wire has a diameter of .005·92[(36-*n*)/39] inches. The metric wire gauge number is equal to 10 times the diameter of the wire, in millimetres; thus, a metric 8 gauge wire has diameter 0.8 millimetres. A table of wire gauge equivalents is provided.

(5) A traditional unit measuring the interior diameter of a shotgun barrel. The gauge of a shotgun was the number of lead balls, each of a size just fitting inside the barrel, that were required to make up a pound. In other words, if a lead ball weighing 1/12 pound just fit in the barrel of a shotgun, then it was a 12-gauge shotgun. Today, the internal diameters for each gauge number are read from a table.

(6) (a) A measure of the number of needles per unit length in knitting and hosiery. The greater the number, the closer and finer the knit. (e.g., raschel machine = 2English inches; fully-fashioned knitting machine = 1.5 English inches; circular hosiery knitting machine = 1 English inch); (b) The number of wales per inch in a knitted fabric.

(7) (a) Pitch – The distance measured horizontally between spindles drive drum centres or adjacent yarns on any multi-position machine; (b) A generic term for various measurement instruments such as pressure or thickness gauges; (c) The number of needles per given distance in a knitting machine; (d) The thickness of the knitting needle in the shank and the hook; (e) The number of wales per inch in a knit fabric; (f) On spinning or twisting frames, the distance from the centre of one spindle to the centre of the next spindle in the same row.

Gauge length: The gauge length is the original length of that portion of the specimen over which the strain or change of length is determined.

Gauge (Gage) length (effective): See **Effective gauge length**.

Gauge (Gage) length (in tensile testing): The length of a specimen measured between the points of attachment to clamps while under uniform tension.

Gauge (Gage) length (nominal): See **Nominal gage length**.

Gauge (Gage) (of tufting machine): The average centre line distance between the needles.

Gauge (in full fashioned hosiery): A measure of fineness expressing the number of needles per 1.5 inch on the needle bar.

Gauge (in knitted fabrics): A measure of fineness expressing the number of needles per unit of width (across the wales).

Gauge (knitting): A term giving a notional indication of the number of needles per unit length along a needle bed or needle bar of a knitting machine. The current practice is a common unit length of 1 'English' inch (25.4 mm) that is used for all types of warp and weft knitting machines.

Gauge (in warp knitting): For simplex, tricot, Milanese, number of needles per English inch; for raschel, kayloom, twice the number of needles per English inch.

Gauge length: The length of a specimen measured between the points of attachment to the clamps while under uniform tension. See **Gauge, Knitting**.

Gauge length, effective (in tensile testing): The estimated length of the specimen subjected to a strain equal to that observed for the true gauge length.

Gauge (linking): A term specifying the spacing of points in linking and point seaming achenes and expressed as the number of elements per unit of length.

Gauge (English): Points in an arc of 38.1 mm (1½ in).

Gauge (American): Points in an arc of 25.4 mm (1 in). The term 'gauge' and 'points', respectively are used to define the two systems, e.g., an English 36 gauge linking machine has 24 points per inch and an American 24 point linking machine has 24 points per inch.

Gauge (wire): Used with an extra filling yarn during weaving, this type of standing wire controls the height of the fabric pile.

Gauging: The gauging stitch is usually longer on the face than on the back, draws the material up into distinct plaits, making it easy to dispose of the fullness neatly, regularly and securely by over handing the top edge of each plate to the bottom edge of the band. The right side of the skirt and the right side of the belt are placed against each other and each gather over sewed to the belt. The space into which the material is to be gathered determines the length of the long stitch. The succeeding rows of stitches should be *directly* under those of the first.

Gauging (gathering): The shortening of the material length by drawing together and holding succession of small folds or material by stitching machine gathering or gauging is achieved by the use of a special press fast or attachment.

Gaugoli: See **Indian yellow**.

Gauze: Light weight, open-texture, net-like woven fabrics named after the Palestinian town of Gaza where they were originally used for veils. They mostly consist of thick yarns with a full finish or fine yarns with a veil-like appearance.

Gauze flannel: Very light and striped shirting flannel, having silk warp.

Gauze weave: A weave used in very thin fabrics, e.g. gauze, or for the production of woven effects (see Fig.). With a full leno weave, the warp threads are made to cross one another between the picks. In simple leno weaving, one warp thread (generally referred to as a crossing or leno end) is caused to lift alternately on one side and then on the opposite side of the neighbouring thread (usually referred to as the standard end) thereby producing "crossed" or "open" sheds. The leno ends are also drawn through a second heald which changes its position between the individual weft insertions. Advantage of leno weaves is that the weft is anchored and hence resistant to slippage.

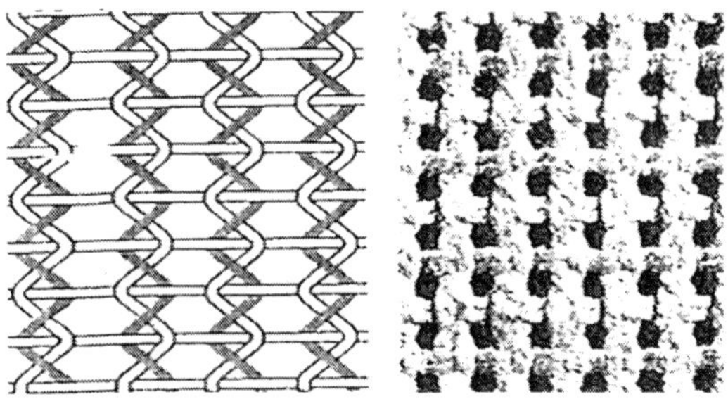

Gauze or Leno

Gaws: In a few places, a term denoting thin places in the cloth.

Gaze Barege: Very light dress goods, made of silk warp and wool filling or of all wool. It is often printed.

Gaze bluter: Very fine and light silk gauze; used for sifting flour.

Gaze brillantine: Very light, French silk dress goods, with a high lustre.

Gaze de Paris: Very light French silk dress goods, made of fine organzine warp and trame filling.

Gaze lisse: French gauze, very light and sheer, made of natural silk yarn.

Gaze ondee: Very light dress goods or trimming fabric, made of organzine warp and filling of "ondee" silk.

Gaze Tour Anglais: French for leno.

Gazzatum: Fine and sheer silk or linen gauze of the Middle Ages, said to have originated in Gaza, Asia.

Gear crimping: See **Texturing**.

Gebanga: Leaf fibre yielded by the Corypha gebanga, a palm in Java; used for making cloth, nets, ropes and bags.

Gebauer kier: The unit consists of two concentric vessels on one inside and perforated inner jacket on the other. The heated scouring liquor is injected at the top between the two concentric walls. In this system, the flow of liquor is radial instead of vertical circulation and thus, the cloth more or less floats. The pressure difference between top and bottom is eliminated.

Gedda gum: See **Gum Arabic**.

Gel: A gelatinous, glassy, easily deformable mass, comprising of a disperse system of at least two components, i.e. of a colloidally distributed solid substance (e.g., gelatine; silicic acid) and a liquid (lyogel; xerogels).

Gel dyeing (wet tow dyeing): Here, polyacrylonitrile fibres which have been produced by the wet spinning method are passed, still in the swollen state, through an aqueous dye bath (40–60%) during or immediately after drawing. A subsequent drying (heat treatment at 105–150°C) closes the pores of the fibre surface, by which the dye is fixed.

Gel spinning: With this technique, it is possible to produce fibres with high tenacity and modulus from polymers with flexible molecular chains. The main characteristics of this process are the use of polymers having ultra-high molecular weights and the processing of dilute spinning solutions. The strength of the fibre increases with increasing molecular weight. The low concentration of the spinning solution allows a high draw ratio to be applied to the fibre. A linear increase in modulus with draw ratio exists.

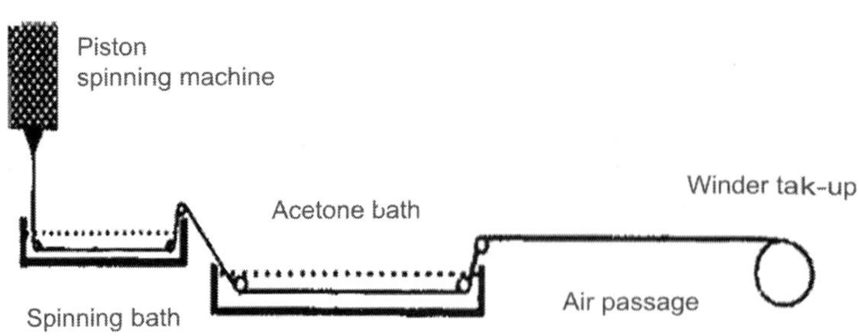

Schematic representation of gel spinning

Gelatine: The pure glue obtained by hydrolysis of collagen by boiling hides, cartilages and bones. Available as thin flakes or powder.

Properties – Transparent, glassy, lustrous material; colourless, odourless and tasteless. Water-soluble after prior swelling giving more or less viscous solutions. The addition of formaldehyde reduces water-solubility.

Quality evaluation – 1 g in 100 ml water must still set to a jelly above 0°C.

Uses – Finishes, sizes and printing thickeners.

Genappe: Ply and hard twist gassed yarn, made of worsted, mohair or alpaca, often mixed with silk; used for fringes, etc.

Genapping: Process of gassing worsted yarns.

Generic class (as used in textile fibres): A group having similar chemical compositions or specific chemical characteristics.

Generos: Bleached cotton sheeting produced in Latin-America.

Generos crudos: General trade term for grey cotton goods produced in Latin-America.

Geneva embroidery: Made by sewing velvet bands to form checks on coarse canvas foundation and filling out the square with coloured silk or wool.

Genghis: Rugs made by the nomad Turkmans west of Persia. They are small, heavy, all-wool rugs, the warp made sometimes of goat's hair, the long loose pile is tied in Ghiordes knot. The design shows crude geometrical patterns in which white colour is used extensively. The sides are finished with selvage and the ends with selvage and fringe.

Genoa: In England, the one-and-two twill used in various heavy materials.

Genoa embroidery: The patterns are worked with a corded outline over fine muslin, this being cut away from between the patterns.

Genoa lace: (a) Imitation Cyprus lace of gold, now obsolete; (b) Fine bobbin made collars, handkerchiefs and fichus made in the 17th Century; (c) See ***Macrame***.

Genoa velvet: An elaborate velvet of satin ground fabric with a multi-coloured pile. The pile is sometimes left uncut.

Gentian: The reddish-blue of the plant from the genus *gentiana*.

Gentian violet: A violet dye derived from coal-tar and used also as an antiseptic.

Geogrid: Manufactured polymer constructions characterized by large openings made by one of the following methods – (a) coating woven or knit products to form a grid; (b) welding oriented strands to form a grid; (c) punching holes in flat sheets then drawing them to align the polymer molecules. Used for soil stabilization, drainage, and erosion control applications. See **Geotextiles**.

Geomembrane protective fabric: See **Geotextiles**.

Georgia Prolific: Commercial varieties of short staple upland cotton from Georgia.

Georgienne: French silk dress goods, made with ply warp. There are 10 or 16 leaves and 10 or 16 fillings in a repeat.

Georgette: (1) A filmy fabric with a crepe appearance, woven with very high twisted yarns. It is a springy fabric, plain or printed, and made from a variety of yarns, including, wool, silk polyester, nylon. Georgette does not crease readily and polyester-georgette will hardly crease at all. Used for blouses, dresses (must be lined), evening wear, etc.

(2) A sheer, lightweight, somewhat transparent open-texture fabric with a fairly pronounced crêpe effect, usually produced in a plain weave with crêpe yarns from almost all types of fibres. It is generally woven with alternating pairs of S- and Z-twisted yarns in both warp and weft. The handle and appearance are characteristic, i.e. the fabric has a crinkly texture and is harsher and more opaque than chiffon.

Georgette crepe: It is a sheer and dull textured fabric which is known more as georgette than georgette crepe. A typical construction is 60 × 50 (to 80 × 80) using 75 denier Rayon (or silk) in both warp and weft, with two S twist alternating with two Z twist in both warp and filling.

Geotextile: Manufactured fibre products made into fabrics of various constructions for use in a wide variety of civil engineering applications, including several described below.

Asphalt overlay fabrics: Fabric systems installed between the old and new asphalt layers during pavement resurfacing. The fabric absorbs the tack coat sprayed on the old surface thus, forming a permanent moisture barrier to protect the subgrade from strength loss due to water intrusion. The fabric system also helps retard reflective cracking by serving as a flexible layer to diffuse stress.

Drainage fabrics: Fabrics used as filter media or separators in subsurface drainage systems. The fabric is installed around the drainage pipe, or coarse stone in the drain; it allows water to pass freely from the soil to the drain, but prevents soil particles from migrating into the drainage system. These fabrics are also used as outer coverings in prefabricated drainage composites and serve the same function as a filtering medium.

Erosion control fabrics: Fabrics used in the stabilization of embankments and the containment of silt run-off from erodible slopes. In embankment stabilization, the fabric functions as a filter medium behind stabilizing rip-rap revetments. In siltation control, the fabric acts as a filter to contain silt while allowing excess water to drain freely. In turf reinforcement, the mat is used to retain soil while allowing roots and stems to grow through. In fabric-forming systems for the construction of revetments, a double-layer, water-permeable fabric is positioned, then pumped full of structural grout. These systems are alternatives to rip-rap.

Geomembrane protection fabrics: Fabric systems used in the construction of landfills to protect the membrane liner from puncture and installation damage. The fabric may be installed on one or both sides of the liner. It can also be used in the leachate collection system since it is permeable to fluids and gases.

Reinforcement fabrics: Fabric system used in the construction of steep slopes and retaining walls. By stabilizing the soil mass, they reduce the stress on the retaining wall with corresponding decrease in load-bearing requirements for the wall design. In slope reinforcement, the stabilization permits steeper slope construction than would be possible, based on soil properties.

Subsurface stabilization fabrics: Fabrics used in the construction on access roads, railroads, parking and storage areas over soft and unstable soil. The fabric is placed between the subgrade and the stabilizing fill material where it provides separation of subsoil and fill, filtration of moisture at the subsoil/fill interface, and added tensile reinforcement of the compacted fill.

Geranium: The vivid red of the geranium flower, a colour popular in the 1950's.

Geranium pink: A deep pink colour.

German hardness degree (°d). : An obsolete unit of measurement for water hardness. It is not an SI unit. Now replaced as a unit of measurement by mmol/l. Water hardness units.

German Institute for Standardization: With its 81 standards committees, the German Institute for Standardization (DIN) is the responsible standardization body of the Federal Republic of Germany. Within the framework of its terms of reference, DIN is the German member in the European and international standardization organizations. DIN represents the interests of German standardization at both the national and international levels. Maintaining a collection of more than 38,000 documents, DIN offers standards covering everything from textiles to standard methods for the analysis of contaminants in water to screws and bolts. Standards offered by DIN provide companies and individuals doing business in Germany or with German businesses a solid basis for quality, safety, and minimum functionality expectations.

German prints: Medium or coarse cotton print goods in South Africa, especially in blue with white patterns and a good finish.

German stitch: Stitch in embroidery used where the foundation can be exposed; it is composed of tapestry and tent stitches alternating parallel with part of the ground between them.

German serge: A serge fabric of the 18th Century, made with worsted warp and woollen filling.

German wool: See **Berlin wool.**

Germantown yarn: Thick and slack twist woollen yarn, made four-ply and used for hand-knitting and other fancy work, made originally in Germantown, Pa., United States of America.

Germuset: Damask made in Asia with cotton and silk warp and coloured patterns.

GF: See **Glass fibres.**

GH: Gesamtwasserhärte (German) (Water hardness units).

Ghagi: Highly finished Indian/Bangladeshi silk satin; used for dresses by the native women.

Gharbasti: Indian fabric, made of cotton warp and wild silk filling.

Gharchola: The type of bandhani saris made in places of Jamnagar, Saurashtra, India. The gharchola traditionally made for weddings has checks in gold thread with tie and dye pattern within the squares and are crafted by the local Khatris of Bhuj. Motifs within each compartment range from elephants, parrots and flowers to dancing girls. Other traditional patterns include the *Baavan bagh* (fifty two gardens) and scenes from *Krishna Ras Leela* (Krishna's dances with lovers). See **Bandhani (Sari).**

Ghatto gum: Asiatic gums.

Ghent lace: Narrow bobbin lace, similar to Valenciennes, the threads of the square mesh ground being twisted two and one-half times.

Ghetee: Indian bleached cotton cloth, gassed and given a high finish.

Ghiordes knot: In oriental rugs, the ends of the hand knotted pile alternate with every two threads of the warp. This knotting produces less knots to the square inch than the Senna knotting.

Ghiordes: Ghiordes rugs are the cream of Turkish manufacture. They are rugs of ancient origin and their design has affinities with the very beautiful and rare 16th Century prayer mats. They are named after the town of the same name in western Turkey, some 100 km from the port of Izmir. Ghiordes rugs are woven with 150,000–300,000 knots/m^2. Various colours are used for Ghiordes rugs.

Ghost print: A defect in transfer printing which takes the form of unwanted (usually paler) "ghost" design images on the printed fabric due to the transfer of disperse dye from one layer of the rolled printed transfer paper to the layer of paper lying immediately on top of it. Such ghosting effects are dependent on the dye, thickener, paper quality and the storage conditions (temperature) of the printed transfer paper.

Ghosting in transfer printing: A defect in sublimation transfer printing due to inadequate storage stability of the transfer printing paper. During the storage of transfer printing papers, an undesired migration of dyes can occur, which cause disruptions in the sublimation can transfer printing process and defects in the transfer printed fabric.

Gifu dyeing process for polyester: A Japanese dyeing method which employs phenol as a swelling agent and organic or inorganic acids at high concentration.

Gigging: The process of raising a nap on fabrics by means of a teazle machine.

Gilan: A grade of Persian raw silk.

Gilham: A Chinese silk dress goods.

Gill: A non-metric English and American unit of liquid measure, equal to one quarter of a pint.

Gilling: A commercial process called "pin drafting" used to produce top fibres. See **Pin drafting**.

Gimian: Fine velvet rugs, made in Asia Minor.

Gimp: (1) A helically wrapped core yarn in the form of a stiff cord. Often consists of cotton or wire wrapped with silk or viscose yarn for use in embroidery, lace or trimmings. Gimp can be used to sew buttons on clothing.

(2) A special thread used to support and raise the buttonhole stitching.

(3) A narrow strip of fabric used for trimming furniture; often has a stiffening of wire or heavy cord running through it.

Gingham: A yarn dyed fabric, with the yarns balanced in strength and usually cotton or polyester/cotton. Regular gingham is made in stripes, checks and plaids from yarns dyed with fast colours. It is woven with a plain weave, a typical construction being, 66 × 54 with 25s warp and 35s filling. Very fine ginghams are sometimes called Madras ginghams or Zephyr ginghams and are made with a construction of 100 × 80 worth 50s combed yarn in the warp and weft. A firm and very hard wearing cloth with stripes or checks woven in; white is always used with a different colour. Medium ginghams are favourable for women's and children's dresses, table clothes, napkins, curtains, etc., and fine ginghams are used for blouses and summer dresses.

Ginghamet: Cotton muslin with striped or figured patterns.

Gingeras: An Indian silk cloth.

Ginned lint: Cotton fibres that have been separated from their seeds by ginning but not subjected to any further processing after ginning. It is also called raw cotton, i.e. at ginnery it is called ginned lint and when it is received in a textile mill it is called raw cotton.

Ginning: A process for separating cotton fibres from the seed in such an order that the fibres may be spun into yarns. Machines used for ginning may be either saw gins or roller gins.

Giselle: A sheer fabric in France, woven of worsted yarn.

Givrene: Silk, acetate or polyester fabric that is very similar in appearance to Grosgrain.

GL: Glass Fibres.

Gl: Gill.

Glace: Dress goods, made of fine and well stretched cotton warp and mohair filling in plain colours or figured.

Glace': (a) A term used for glazed yarn or polished thread; (b) Fabrics – a high-lustre fabric, a lining fabric with a lustre-imparting finish, a lustrous satin lining, etc.; (c) A woven wool fabric; (d) A lightweight British fabric with a cotton warp and mohair or lustre worsted weft produced in a plain weave. In French, the word *Glacer* means "to glaze".

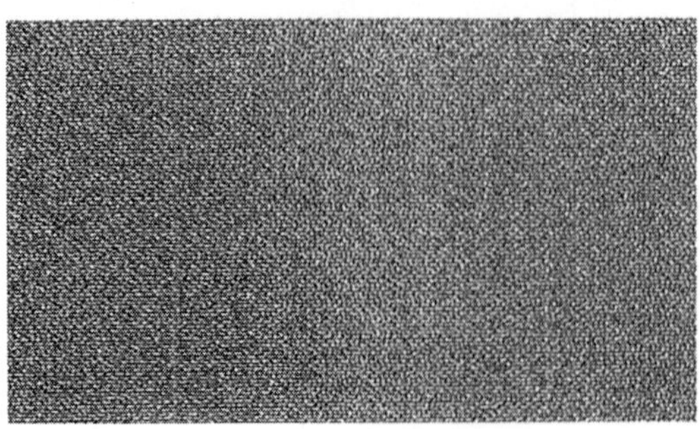

Glace, batiste: Fine, plain woven cotton made lustrous, transparent, and stiff by parchmentising. Used for blouses, trimmings and evening wear.

Glace' leather: Glossy leather.

Glanzstoff: German copyrighted name for an artificial silk.

Glass: An alkali calcium silicate. Fusion product of silica (silicon dioxide), calcium oxide (lime), and sodium carbonate (soda ash) will produce common lime-soda glass. Potash glass is obtained by substituting soda ash with potassium carbonate (potash).

Glass batiste: A glass-like, almost transparent and usually stiff batiste fabric. The character of this pronounced stiff and glassy material, produced by a special chemical treatment, is also resistant to washing. As far as cotton materials are concerned, which were formerly used exclusively for this purpose, the finish consists essentially of mercerization, leaching, and a treatment in concentrated sulphuric acid.

Glass cloth: A plain weave cloth made from twisted linen yarns, often woven in stripes or checks, but sometimes printed. The fabric is made without any surface fuzz, yet it readily absorbs moisture because it is of medium weight. It is very durable. Used to make tea towels for drying glass and china.

Glass count: An indirect yarn numbering system, equal to the number of 100 yards of lengths per pound.

Glass fibre: A manufactured fibre in which the fibre-forming substance is glass. In the continuous filament process, glass marbles are melted in an electric furnace and the liquid flows in fine streams through small orifices at the bottom of the melting chamber. The resultant filaments are caught and drawn by a high-speed draw winding mechanism. In the staple fibre process, the streams of molten glass are attenuated into fibres by jets of high-pressure steam or air. These fibres are gathered on a revolving drum and then wound on tubesto form staple fibre sliver or bands that can be drafted, twisted and plied.

Characteristics – Glass fibre is incombustible and will tolerate heat up to 1000°F without material damage. Potential strength is not realized in woven fabrics or even in yarns, because the fibre is brittle and fracture points may develop, but nevertheless, very high tensile strength is obtained in woven fabrics and is retained at elevated temperatures. The fibre originally was difficult to colour, but methods have been developed to accomplish this. Moisture absorption is low. Electrical and insulation resistance is high.

End uses – Glass fibre is used for heat and electrical insulation, filter cloth in the chemical and dye industries, reinforcing belts in tires, novelty fabrics, tablecloths, and fireproof draperies. Because of its brittleness, it is not used in wearing apparel or in household fabrics that have to withstand frequent flexing. Because of the relatively low price, glass textiles have, in the meantime, found wide application, e.g., in chemistry, chemical technology, insulation of machines, pipes and containers, in sound insulation and in electronics.

Glass plate test: A commonly used test to check the correct vatting of vat dyes in dyeing as well as the rate of re-oxidation after dyeing (greening).

Procedure – A glass plate is dipped in the prepared vat and immediately removed; if vatting is correct, the dye liquor should remain clear as it drains off the plate without revealing any solid dye particles. Indigo vats should, e.g. drain off clear yellow and only begin to green after 20–30 s, i.e. colour change to blue.

Glass roving: A form of glass fibre consisting of a multitude of parallel glass filaments without any significant twist.

Glass transition temperature: The temperature above which a material changes from a brittle, "glassy" nature to a rubbery nature; on cooling, the material changes back to glassy synthetic fibres, such as polyester and nylon show this change of physical character. The rate of dye uptake increases dramatically when the fibre is near or above the glass transition temperature. If the glass transition temperature is above the boiling point of water, as it is with polyester, the dyeing rate is extremely slow even at the boil.

Glass-rubber transition temperature: See **Glass transition temperature**.

Glass wool: See **Glass fibres**.

Glass yarn: A collective term for multifilament, spun and textured yarns produced in a range of different yarn thicknesses from glass fibres.

Glass yarn (continuous): Continuous yarn made of glass filaments that extends substantially throughout the length of the yarns. See ***Continuous glass yarn***.

Glass yarn (fabricated): See **Fabricated glass yarn**.

Glass yarn (staple): Yarn made from glass fibres that are nominally 200–375 mm in length.

Glass yarn (vinyl coated): Continuous filament glass yarn coated with a pigment and plasticized vinyl chloride resin. See **Vinyl coated glass yarn**.

Glass (textile): Glass in a form suitable for spinning or weaving. It is manufactured in two common forms, (a) filaments and (b) staple fibres of spin able length.

Glauber's salt: Sodium sulphate. In the dye industry, few use "sodium sulphate" to refer to the anhydrous form, and "Glauber's salt" only to refer to the decahydrate form ($Na_2SO_4 \cdot 10H_2O$). Used as an electrolyte in dyeing especially for turquoise and such similar shades.

Glaze finish: Refers to a finish put on 100% cotton threads made from starches, waxes or other additives. This coating is then brushed to give the thread a smooth surface. A glaze finish protects the thread during sewing giving better ply security and abrasion resistance.

Glazes (sewing thread): This is a process in which cotton threads are treated with starches and special chemicals under controlled heat and then polished to a high lustre. The glazed process results in a thread with a hard finish that protects the thread from abrasion and enhances ply security.

Glazes yarn: See **Glacé**.

Glazing: (a) A finishing process that produces a smooth, highly polished, or lustrous surface on a fabric, such as chintz. The fabric is treated with starch, glue, paraffin, or shellac, then friction calendered. Synthetic resins are used for a more permanent finish; (b) A shiny fabric appearance produced unintentionally, e.g., by pressing at excessive temperature.

Glazing Calendar: See **Friction Calendar**.

Glen Checks or Glen Urquhardt: Originally Scotch cheviots and home spun, made in combination of two checks of different sizes, having an equal number of threads in each size. The checks are usually only in two colours, often having additional coloured ornamental threads. The fabric is made now in woollens and worsteds for men's wear.

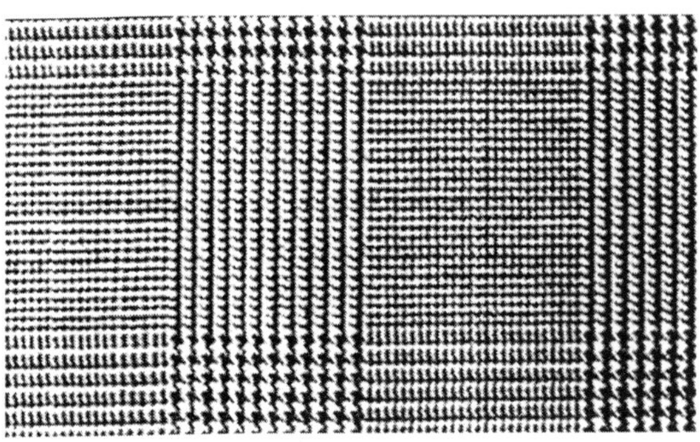

Glen Check

Glen urquhart plaid: The name comes from a Scottish clan. The cloth is also sometimes called overplaid, because the blocks of check weave are arranged at right angles, and then there is a larger outliner check effect over this, often in another colour. It may be wool or worsted and is made in various weights from light suiting to over coating. An ornate plaid, used for golfing clothes, cloaks, men's sporting hats and overcoats.

Glengarry: All-wool, mottled English tweed.

Glenshee: A plain weave mercerized cotton or linen made in a wide variety of weights and used for all types of embroidery.

Glenshee crash: A rough, uneven plain weave fabric of equal quantities of cotton and linen fibre. Used mainly for curtains and loose covers and as a base for embroidery.

Glissade: A closely woven satin-weave cotton fabric that is polished in finishing. It is made in plain colours, an uninteresting fabric used only where cotton lining is needed.

Glitter pigments: A distinction is made here between polyester and polyoxide glitter pigments. They are produced by the comminution of metallized foils followed by sieving. Because of the risk of blocking screens or engraved rollers in printing, these pigments are only suitable for screens with a maximum of 9 threads per cm or engravings with a minimum depth of 30/100. Other problems may be caused by the print pastes separating out during printing. For this reason, the use of high viscosity, short print pastes is recommended.

Glitter threads: Man-made fibres with a ribbon shaped cross-section in coarse counts, mainly produced in a variety of spun-dyed colours.

Gloria: A thin and very closely woven fabric, made with silk warp and worsted or cotton filling in a three harness warp face twill weave; it is dyed in the piece, and used for umbrella covering and also dress goods.

Gloria silk: Smooth and closely woven fabric; warp – cotton or (originally) organzine silk; weft – schappe silk or fine worsted yarn; in twill or satin weave for linings, umbrella cloth, etc.

Gloria suede: A firm and heavy suede for jackets and coats.

Glorietta: A lighter version of gloria made of cotton.

Gloss: The luminous fractional reflectance of a fabric specimen in the specular direction. An additional parameter to consider when determining a colour standard, along with hue, value, chroma, the texture of a material and whether the material has metallic or pearlescent qualities. Gloss is an additional tolerance that may be specified in the Munsell Colour Tolerance Set. The general rule for evaluating the gloss of a colour sample is the higher the gloss unit, the darker the colour sample will appear. Conversely, the lower the gloss unit, the lighter a sample will appear. Gloss is measured in gloss units, which use the angle of measurement and the gloss value (e.g. 60° gloss

29.8). A 60° geometry is recommended by the American Society for Testing and Materials (ASTM) D523 standard for the general evaluation of gloss.

Gloss (specular): See **Specular gloss**.

Gloss value: A measure of lustre for e.g., textile filaments for before delustering (e.g. 40) and with increasing delustering additives (to below 4) or for raw cotton (15–17) before mercerising and with increasing density of caustic liquor after mercerization (up to 30° Bé = 76, after which it decreases again).

Glossing: A mechanical after-treatment given to hanks of silk, linen, cotton and synthetic fibre yarns to loosen up the yarn structure, achieve increased lustre and a softer handle. Originally, wet hanks were placed over a wooden peg and wrung out.

Glossop: Plain woven bleached cotton fabric, made in England in pieces that are 50 yards long and one yard wide, with 76 ends and 88 picks in a square inch; used for calicoes.

Glossy printing: See **Chintz prints, Lacquer printing**.

Glove silk: A fine warp knit silk fabric made in basic colours and used mainly for glove lining, although it was once also used for lingerie.

Glovers' wool: The wool which was removed with lime from the skin of slaughtered sheep.

Glow: Visible and flameless combustion of the solid phase of a material. Burning without forming flames, but with visible light.

Glucose (dextrose; grape sugar): A white crystalline sugar ($C_6H_{12}O_6$) occurring widely in nature. Very hygroscopic, readily soluble in water up to 817 g/l at 15°C (anhydrous glucose) and up to 878 g/l (glucose containing water of crystallization). The oxidation of glucose results in the formation of gluconic acid. Like other monosaccharides, glucose is optically active: most naturally occurring glucose is dextrorotatory. Like all known sugars, it exists in two isomeric α and β-forms, i.e. α-glucose and β-glucose. Both forms differ only in the position of the OH group relative to the 1st carbon atom.

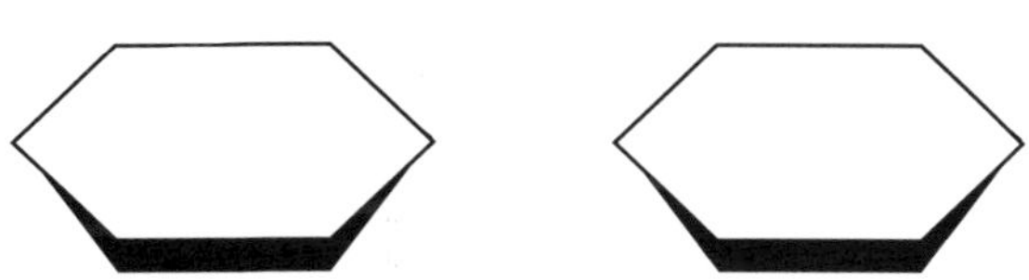

Glue: Most familiar are glues derived by boiling or steaming animal hides, tendons or bones which are high in collagen. Glues may be classified in order of quality as follows – isinglass, hide glue, leather glue, cartilage glue, bone glue and fish glue.

Glued carpets: These fall into two categories; (a) Hand bonded, e.g., appliqué carpets. (b) Machine bonded, e.g., flocked carpet as bonded bouclé carpets, bonded chenille carpets and similar products.

Glued composite fabrics: Imitations in the style of woven cloqué fabrics are produced (Adhesive bonded textiles) by, e.g. the superimposition of a crêpe fabric on top of a flat fabric (taffeta) and bonding them together locally. They can also be produced to a pattern, if desired, through application of the bonding agent by screen or roller printing. After damping, blister-like motifs appear in the upper fabric as in, e.g. adhesive-bonded cloqué. Fibre webs can also be included in a similar manner.

Glued seam: A seam formed by an adhesive.

Glueing carriage: Used to glue fabrics on to the screen printing tables for printing. The liquid glue is applied by means of a roller applicator and the fabric is brought into contact with glue film by means of pressure rollers.

Glueing unit: A device for the application of glue on the printing blankets on screen printing machines.

Glycerol:(glycerine; glycyl alcohol;1,2,3-propanetriol; 1,2,3,-triol)A trihydric alcohol, $HOCH_2CH(OH)CH_2OH$. Glycerol is a colourless sweet-tasting viscous liquid. It is widely distributed in all living organisms as a constituent of the glycerides, which yield glycerol when hydrolysed. MW 92. A clear, colourless (pure 30°Bé, lightly coloured 20°Bé), odourless, hygroscopic and very sweet-tasting syrupy liquid. It is combustible. Miscible with water, alcohol and acetic acid; insoluble in ether, white spirit, benzene and chloroform. Used as a universal solvent for salts, sodium carbonate (soda), sodium tetraborate (borax), potassium iodide, alkalis, alkaline earths, certain metal oxides, water-insoluble lime soaps, etc., dyes which are insoluble in alcohol become soluble in alcohol, if previously mixed into a paste with glycerol.

Applications – As a hygroscopic agent (sizing; softening finishes), dye solvent in textile printing in amounts up to 100–130 g/kg of print paste. Component in print pastes for discharge printing (e.g. for wool and silk). Spotting agent (solvent action improved when warm).

Glycidol: (2,3-epoxy-1-propanol), colourless liquid, b.p. 162 °C, soluble in water. As a glycerol anhydride, the simplest alcohol with an epoxy group, glycidol is used as a reactive solvent for epoxy resins, textile auxiliaries (levelling agents, dye-binding agents, etc.) and for the stabilization of halogenated hydrocarbons.

Glycine fibres: Regenerated protein fibre from the soya bean protein, glycine. Developed in the 1930s, but no longer relevant today due to the inadequate strength values (5–7cN/tex, wet strength 50%).

Glyoxal (ethane diol): HCO-HCO, a low melting point yellow crystalline solid, CHOCHO, m.p. 15°C colourless liquid, b.p. 162°C. It is a dialdehyde. It gives off a green vapour when heated and burns with a violet-coloured Ûame. Glyoxal is made by the catalytic oxidation of ethane diol and is used to harden the gelatin employed for photographic emulsions. It is soluble in water.

Properties – Mild odour, readily rearranged into polyglyoxal, soluble in water tetrahydroxydioxan), alcohol, ether and other solvents, powerful reducing action. Used in the manufacture of resin finishing agents, e.g. DMDHEU, tetramethylol acetylene diurea, glyoxal resins. Glyoxal can also be used under mild curing conditions for formaldehyde-free finishes; catalyst – aluminium sulphate and polyhydric acids, shock catalysts. The addition of ethylene glycol prevents yellowing.

Glyoxal cross-linking agents: A term in common usage for resin finishing agents which are representative of the reaction products of Glyoxal with urea in particular, e.g. Dimethylol dihydroxy ethyleneurea.

Gminder fabrics: Hard-wearing fabrics of various types (similar to linen/ cotton unions) produced from cotton and cottonized hemp (process developed by the firm Gminder, Reutlingen, Germany).

Gnafi: Very fine mat, made of the leaves of the pandanus; used for garments in Tonga Islands.

Goats' hair: Some of them, like the cashmere, are among the finest fibres and are used for the costliest fabrics; the mohair is yielded by the Angora goat; the alpaca is taken from a Peruvian goat; common goats' hair is coarse, thick and does not felt well and is used for coarser rugs and fabrics.

Gobelin: (1) French tapestry. Genuine Gobelin tapestries are hand-woven and have a similarity to Rib fabrics with coloured designs (occasionally with gold and silver threads). Imitation Gobelin tapestries are produced with jacquard designs (3–5 warp and 3–4 weft systems) or by printing. Used as tablecloths, heavy upholstery and wall coverings, etc.

(2) A tapestry fabric, made of wool or silk, the filling entirely covering the closely set warp threads (which form tine ribs), and show various designs with human animal floral, etc., figures. The genuine Gobelin is made by hand with the use of needles or small, flat shuttles, the different coloured yarns reaching

only as far as there is call for them in the design, instead of extending from selvage to selvage. The fabric is free of all nap or nubs and the pattern is shown in its completeness, but reversed on the back. The Gobelin was invented by a man of that name in Paris in the 15th Century, and the factory purchased from his descendants by the government of Louis XIV. This factory still turns out the finest hand-made Gobelin; (a) A loom woven fabric, made with heavy warp and fine filling, printed afterwards to imitate genuine Gobelin; used for curtains, hangings, and drapery; (b) An obsolete silk and wool dress fabric with pastel coloured brocaded figures.

Gobelin stitch: Also termed as ***Tapestry stitch***; it is used in gobelins and embroideries. The straight stitches are passing across and over a padding of braid, thus being raised.

Godet roll: Roll used for transporting and controlling the movement of bundles of fibres and yarns in the processing of these materials.

Goghari: Variety of raw cotton grown in Baroda, India. The staple is white and woolly.

Gold dust or Tennessee: An early maturing commercial variety of a very prolific upland cotton; the staple, measuring 25–28 millimetres, forms small bolls; yield of lint is 32–34%. It is also called King.

Gold purple reaction: A very sensitive and specific test for oxycellulose. A test specimen is soaked in a weak acetic acid solution of stannous chloride [tin (II) chloride] and then placed in gold chloride solution. Oxycellulose absorbs tin (II) oxide, and as a result assumes a red-brown colour.

Gold thread: Thread covered by core spinning only or with combination of electroplating gold alloys to get the gold appearance for the thread. Nowadays, synthetic materials with gold colour are also called gold threads; sometimes not necessarily of gold colour also.

Gold tissue: Transparent metal cloth that is very soft and luxurious, consisting of metal warp in gold colour and silk or synthetic weft. Used mainly for evening dresses and sarees, etc.

Golden moss: Trade name for a yellowish, curly and soft fibre, yielded by the leaf stalks of the tree fern in China; used for stuffing.

Goldthwait test: A differential dyeing test developed by Goldthwait et alas a practical method for estimating the maturity of cotton. The test is especially helpful in mill situations for detecting the presence of thin-walled or immature fibres which often cause problems in processing. The test employs a mixture of Chlorantine Light Green BLL (C.I. Direct Green 26) and Direct Fast Red 5B (C.I. Direct Red 81), two dyes with different dyeing properties and contrasting colours. Mature fibres absorb the red preferentially and immature fibres absorb the green.

Golf coat: A red flannel coat for golf playing, single breasted sack style.

Golf cloth: Double faced woollen over coating, the two sides being different in colour and pattern.

Golf hose: Heavyweight wool stockings, made with fancy patterns at the top.

Golfers: Reclaimed wool from pullovers and knitted garments.

Golgas: Two faced printed flannel.

Gomuti: Dark coloured and strong fibres that are yielded by the base of the leaves of the sago palm in Malacca; used for cordage.

Goods, heavy: See **Heavy goods**.

Goose: A tailor's iron.

Gordon: A Highland tartan, made as follows in a repeat – Dark green bar, split in the middle with a group of black, yellow and black lines; black stripe, half the width of the green bar; blue line, black line, blue line, black line; blue stripe, half the width of the green; repeat group described between the two, in a reversed order.

Gordon cord: In European countries, a twilled cord weave.

Gore: A wedge shaped piece of material inserted into a garment part to obtain width at a specific place.

Gorevan rugs: Fine Persian rugs of medium size, made with cotton warp and the close wool pile are tied in Ghiordes knot. The design is usually a centre medallion with curved outlines on a cream coloured field.

Gorge: The edge of the fore part to which the collar stand or neckband is joined.

Gospodsky: Trade term in Russia for a grade of carefully cultivated, retted and scotched flax.

Gossamer: (a) A very soft and fine silk gauze used for veils; (b) Light silk fabric that is water proofed and used for wraps.

Gossypium: Generic scientific name for cottons.

Gothrough: A machine-made lace.

Gouache print: A fashion print in the style of a water colour painting.

Gout: Foreign matter trapped in a fabric by accident, usually lint or waste.

Gouzlieh: A striped, heavy cotton shirting, with small dobby designs, or twilled produced in Turkey; used for long outer robes. See ***Cheviot***.

Governo: Sort of raw cotton from Brazil.

Goza variety: Variety of raw cotton from Afghanistan. The staple is greyish and harsh.

Gpd: Gallons per day.

Gpg: Grains per gallon.

Grab strength test: See **Grab test.**

Grab test: A tensile strength test in which the central part of the width of the specimen is gripped in the clamps.

Grab test (modified): See **Modified grab test**.

Gradation (in designing): Employs a series of motifs patterned to relate to one another through a regular progression of steps. This may be a gradation of shape or colour. A few shape gradations may in fact create a sequence of events, not unlike a series of images in a comic strip. See **Rhythm**.

Grade: To assign a numerical value based on number, size and severity of defects seen during visual inspection.

Grade (in warp knitting): A term used to indicate the defect index evaluation of fabric determined by the number of defects per unit, for example per pond, per linear yard or per square yard.

Grade (in wool and mohair): A numerical designation used in classifying wool and mohair in their raw, semi-processed and processed forms based on average fibre diameter and variation of fibre diameter.

Gradel (Grad'l): Dense twill canvas, made using strong thread (of linen, as half-ticking), plain or multi-coloured; also cotton and linen cloths in herringbone weave, the latter often having a striped pattern.

Grade: A measurement used in knitted garments that reflects the size of the needles used to knit the garment. The larger the gauge, the smaller the needle the finer the knit.

Grades: Classes into which cotton, wool and other fibres are classified according to length, evenness, strength, colour, etc., of the fibre. For the grades, see under the various fibres (cotton, wool, flax, silk, etc.).

Grades of chemicals: Chemicals are graded according to purity and appropriateness for end use. Common grades include "reagent" or "analytical" (very pure, for laboratory work, very expensive, often in small packages), "U.S.P." or "B.P." (United States Pharmacopeia or British Pharmacopeia, usually for use in or as drugs, quite high purity, medium price, small to very large packages), "*food grade*" and "*technical grade*". Dyes for textile use are rarely graded, and are often quite crude and impure, compared with other chemicals.

Grading: (1) The classification of cotton, wool and other fibres according to the strength, length, evenness, etc., of the staple.

(2) The process of producing a range of patterns of different sizes from a master pattern.

Grading (finished fabric): Inspecting and grading the finished fabric as per predetermined level of faults allowed in each grade.

Graphite fibre: Although the terms carbon and graphite are used interchangeably to describe these fibres. Graphite fibres are more accurately defined as fibres that are 99% carbonized, while the term carbon is used for any fibre carbonized to 93–95% or more. See **Carbon fibre**.

Graft copolymer (graft copolymers): Polymers to which side chains from other building blocks are subsequently "grafted" on to the backbone of an already formed macromolecule (which may be obtained by polymerization, polycondensation or polyaddition, or as already existing polymers from natural sources resp. modified natural products).

Graham: A Highland tartan, one repeat made as follows – Dark green bar with a narrow, pale blue stripe near the edge; black stripe, measuring one third of the green bar; navy blue bar, as wide as green bar and split in the centre with a black stripe as wide as the pale (blue; black stripe, measuring one-third of the green bar).

Graham of Menteith: A Highland tartan, composed of green bars, split by white lines, running over a black and blue ground.

Grain: (1) The direction of the warp in a fabric. Hence, cutting straight grain means cutting with the grain; cross grain means cutting across the grain.

(2) As a unit of mass (weight), approximately 64.8 milligrams.

Grain (in yarn spinning): A direct yarn numbering system for sliver, top, or roving, equal to the mass in grains of 120 yards.

Grain (in measuring mass): 1/7000 lb avoirdupois.

Grain d'Orge: (a) Fine, bleached and figured French linen; obsolete; (b) Solid coloured French serge dress fabric, made with eight leaves and six picks in a repeat; (c) A very strong, durable, twilled woollen cloth in France, originally made with designs imitating seeds, hence the name. It is dyed in the piece.

Grain de Poule: French serge dress goods, having eight leaves and six picks in a repeat.

Grainline: It is the lengthwise thread in the fabric, parallel to the selvage and the lengthwise arrow on the pattern.

Grainline (in sleeves): Lengthwise grain of the material. It generally extends from shoulder point to the back of the wrist.

Grains grossiers: Coarse French ticking.

Grains per cubic foot: A unit sometimes used for measuring the amount of solids suspended in smoke or air:

7000 grains = 1 lb, or 15 432 grains 1 kg, or 15.43 grains = 1 g, and $1 m^3 = 35.31 ft^3$.

Consequently,

$$1 \text{ g/m}^3 = \frac{15.43}{35.31} = 0.437 \text{ grain/ft}^3$$

$$1 \text{ grain/ft}^3 = 2.28 \text{ g/m}^3$$

Grain leather: The term used to describe all leather that has had a finish applied to the right side. Most leather skins available for sewing are referred to as grain leathers.

Gram calorie: Calorie.

Gram equivalent: The quantity of a substance by weight which corresponds to a gram atom of hydrogen.

Gram stain: A stain in which bacteria are classified by whether they retain or lose their primary stain.

Grams per litre: The most commonly used measure of concentration in the textile bleaching, dyeing, printing and finishing industries and indicates the number of grams of a solid substance dissolved in 1 litre of a liquid.

Granada: An old term meaning ‘grained’, the fabric is fine worsted cloth with a face finish on the right side. The weave has a broken appearance. The cloth is normally black. It is used for women’s coats and men’s lightweight overcoats.

Grand Lez: All-wool, white army coating in France.

Grand lion: Figured table linen, made in France.

Grande venise: Very fine damask table linen having large flowers for design, made in France and Holland.

Grandine: 18th Century woollen fabric in England.

Grandrelle: (a) A ply yarn spun of strands of different colours; (b) Cotton shirting, made in warp satin weave. The two-ply warp contains different coloured strands.

Granite: An irregular, mottled and pebbled effect in the weave, produced by an irregular wide twill.

Granite weave: Broken 2/2 twill, woven with a light warp and dark weft (or vice versa), is sometimes called granite weave because of its resemblance to evenly speckled granite rock.

Grant: A Highland tartan with a red ground, the repeat made as follows – Wide red field, split by two pairs of narrow dark blue lines; narrow pale blue and narrow red lines of same width as blues; dark blue stripe, three times as wide as narrow 'blue lines; dark green field of same width as red field, with red, green, red lines (of same width as narrow blue lines) at each side, and red, blue, red, blue, red lines (of same width as narrow blue stripes) in the middle; repeat group described between the two, in reversed order.

Graphic tufting machine: Tufting machine, capable of producing patterns.

Grass cloth: A fine, smooth, linen woven in checks of blue and white, red and white, etc., used for dish towels; also a thin dress material of ramie and cotton, etc.

Gratel: Twilled, coloured linen cloth, made in Germany.

Gravure print: An intaglio printing process using etched copper cylinders or plates. In order to achieve a perfectly smooth substrate surface as a prerequisite for gravure printing, it is necessary to apply a ground coat before printing (Ramisch- Kleinewefers), i.e. before print paste transfer takes place by contact between the inked roller and the substrate using an applied pressure The process is also used for the production of high quality printed papers for transfer printing.

Gravity separation: In the treatment of gaseous or liquid or solid wastes, gravity is the basis of settling. For air pollution control, gravity based processes include air classification, gravity settling chambers, etc. In water and wastewater treatment, gravity based processes include clarification, sedimentation, gravity thickening, etc.

Gravity settling chamber (gravity settler): A long ***expansion chamber*** that allows particles to settle from a gaseous effluent. It is mainly used for particles greater that 100 m and is very inefficient for particles less than 10 m. The efficiency can be increased by the incorporation of a spray within the chamber (known as a gravity spray chamber) or by adding baffles that vary the flow direction.

Gravity thickening: The thickening of sludge by allowing the sludge to settle is common in the treatment of water and wastewater sludge. The thickening tank often has a stirrer which slowly rotates, helping the sludge to settle and thicken. The supernatant from sludge thickening of wastewater sludge can be high in BOD and suspended solids. The design is based on ***solids loading rate*** and varies from 10 to 100 kg of dry solids/d per m^2 of tank, depending on the type of sludge.

Gray: Unbleached and undyed cotton or linen fabric. See **Grey** *and* **Greige**.

Gray, Gy: The SI unit for the amount of energy radiation absorbed by a material, i.e. the absorbed dose. One gray is the 1 joule of radiation energy absorbed by 1 kg of matter.

Gray: The USA spelling of grey.

Gray goods: A fabric in the condition in which it leaves the loom or knitting machine, i.e. before any bleaching, dyeing or finishing treatment has been given to it. See also **Greige goods, Loom-state fabrics.** *Note*: In few countries, particularly on the North American continent, the term "greige" (or griege) is used. For woven goods, the term "loom state" is frequently used as an alternative. In the linen and lace trades, the term "brown goods" is used.

Gray fabric: See ***Gray goods***.

Gray wash: A process in the bleaching of cotton piece goods, consisting of soaking the fabric in water and keeping it wet for a couple of hours.

Gray wool: Fleeces with a few dark fibres, a rather common occurrence in the medium wools produced by down or black-faced breeds.

Grayson: A commercial variety of prolific and early ripening upland cotton, the staple measuring 23–25 millimetres; the yield is 34–36%.

Grey felt: A blend of white fibres with naturally coloured or dyed fibres, or both, that has an overall gray appearance.

Grease dyeing: Process used in dyeing serge and cotton warp woollen cloths without scouring the first.

Grease trap: A small *skimming* tank typically with a retention time of 10–30 min, large enough to hold and cool the grease delivered in the wastewater from premises generating high quantities of grease, such as restaurants and a range of other food sector establishments. Only the wastewater from the grease producing areas should go to the grease trap. Other wastewater from the premises, particularly from toilets, is not connected to the grease trap. It should be regularly emptied. The inlet is below the surface and the outlet is at the bottom, normally separated from the inlet by a baffle that holds back the grease.

Grease wool: Wool taken from the living sheep and which has not been commercially scoured.

Grebe cloth: Long napped cotton cloth; used for underwear.

Grecian: (a) A name for a huckaback weave; used for coarse towels; (b) Woven quilt, made of bleached ply yarn of low count for warp and filling. The pattern being geometrical diapers.

Greek lace: Cut or drawn work embellished with various stitches or additions of needle-point lace; also called Reticella.

Green: The colour of growing grass. From the old English 'gréne' and the old Teutonic root 'grô' from which we derive 'grass' and 'to grow'. Green is said to have more variations than any other colour and yet, despite the richness of our language, there is a woefully inadequate supply of words to describe the variety of greens in the British landscape. It ranges from approximately 575 to 500 nanometres. It is one of the three ***additive primary colours***. A symbol of hope; associated in medieval times with the Zodiac signs Libra and Taurus and with the planet Venus. The holy colour of Islam and used on the flags of many Muslim countries. Associated once with fertility and springtime and now with environmentalism. Why the surgeons' gowns are usually green in colour and first aid kits are also often green? Is it perhaps that (as *chromotherapists* believe) the colour green can help to staunch the flow of blood? One of the 140 colours in the *X11 Colour Set* having hex code #008000. In English folklore, green is widely supposed to be unlucky especially as regards items of clothing – '*wear green and you will soonwear black*'. Why this should be so is uncertain but is sometimes explained by reference to the fact that green is the colour taken to be worn by fairies who punish others for wearing it. Research in Münster has found that green police cars generate little respect in Germany – green having been adopted in Germany for this purpose as a rejection of militarism. It was found that blue and white or blue and red patrol vehicles had a much more powerful effect.

Green linen: Trade term for linen cloth woven of unbleached or "green" yarn.

Green cotton: A brand name for T-shirts produced in accordance with specific ecological requirements. Manufacturers of these shirts which are labelled with a number and the inscription "green" are bound by a self-obligation to fulfil the following conditions:

– only cotton grown under ecologically acceptable conditions may be used;
– only the most environmentally-friendly wet processing treatments are to be used.

Green gold: A name for sisal fibres.

Green tack: A term used in fabric bonding for the preliminary bond created in the first stage of curing by the wet adhesive process. At this point, the bond is not fully cured and hence is "green".

Green unripe cotton: Cotton seed hair which has developed too late on the outer skin of the cotton seed or which originated from cotton balls which have been harvested too early. These immature cotton fibres are thin walled and generally have only few convolutions in contrast to mature fibres. The natural green colour of immature cotton fibres cannot be destroyed by the usual bleaching processes. Cotton with a high proportion of immature fibres is therefore unsuitable for white goods.

Green yarn: Trade term for undressed jute or unbleached linen yarn.

Grège raw silk: The raw silk strands or filaments (reeled from several cocoons), i.e. silk still covered by sericin (gum).

Greige (French): Grey-beige; North American equivalent to British grey. In North America, greige is often used to describe loom state fabric that is unbleached, contains size and lubricants, and may be a bit dirty. Greige goods are made ready for dyeing by singeing, desizing, scouring and usually bleaching.

Greige fabric: Loom state fabric or fabric which has not undergone any wet processing after weaving.

Greige goods: Textile fabric that have received no bleaching, dyeing or finishing treatment after being produced by any textile process.

Greige thread: Undyed or unfinished sewing thread in the state of following final plying or equivalent steps in processing sequence, such as extruding, or braiding.

Greige yield (in knitted fabrics): The number of finished square metres per kilogram (yards per pound) of greige fabric.

Greiz-geraer fabrics: Greiz-Geraer ladies' dress fabrics, obsolete German brand name for fine worsted fabrics made of merino wool, plain coloured or patterned, and in designs following Paris and Milan fashions. In the "Greiz-Geraer genre", 80% of the ladies' dress fabrics were woven in worsted spun yarns and 20% in pure silk.

Grenada: (a) A variety of Indian raw cotton; (b) Light fabric, made with black cotton warp and mohair or alpaca filling in a five-leaf weft satin weave.

Grenade: (a) French table linen, made in plain weave or with small dobby design; obsolete; (b) A fine fabric, made in France of wool and silk; used for table cover.

Grenadine: (1) Term used to describe an outer garment like cloak, but now refers to a loosely woven gauze-type fabric. It may be cotton, polyester and nylon acrylic or mixtures. Used for making filmy dress or blouse fabric.

(2) A yarn made from three very highly basic yarns. The fibre may be silk, cotton or synthetic. The yarn is woven into a variety of fabrics of different weights including, voile, organdie and satin.

(3) Silk yarn produced from doubled twisted organzine, but with a higher degree of twist organzine silk; (a) A term used for various fabrics. Grenadine that is produced with a silk warp and worsted or mercerized cotton weft is called satin grenadine which is executed in a satin weave. The term also applies to a lightweight, almost transparent, type of Étamine (silk/mercerized cotton) and fine loosely-woven women's dress materials similar to marquisette, or damask type fabrics. The term has also been used for fine stockings (viscose filament), from which crêpe stockings (yarns with a half voile twist) were developed (1933).

Grenadine crepon: All-wool black dress goods, having open check patterns and ribs.

Grenadine (silk): Silk yarn from two or more threads, similar to organzine, but more strongly twisted, 1200–1800 twists per metre for the pre-yarn, 1000–1800 for the subsequent yarn. The threads have very closed effect, show high abrasion resistance and are used as both warp and weft material or special weaves as well as ladies hosiery.

Grenadine yarn: See **Yarn (silk); Grenedine (silk).**

Grenfell cloth: Closely woven reversible twill fabric, mainly used for raincoats, because the close twill weave causes it to be water repellent. The fibres used are worsted or polyester or blends; similar to gabardine in appearance.

Grex: An obsolete direct yarn numbering system for fibre yarn or other textile strand equal to the mass in grams per 10000 m.

Grey: (1) British term equivalent to North American *greige*.

(2) The colour of lead, of ash and of the hair of the middle-aged; a mixture of black and white. A bleak colour, but perhaps resurrected by the Chinese government's decision to paint the buildings of Beijing grey as part of its bid to stage the 2008 Olympic Games – 'Grey matches our climate, cultural background and tradition' says Beijing as reported by *The Times* (7.11.00) in what must be one of the very few Leaders devoted to a colour. An achromatic hue. An example of how colours are used as nouns occurs in Marks & Spencer's explanation of why their 1999 half year pre-tax profits halved to £546.1 million – '*we bought too much grey*'. A colour used in this way, that is in a referential sense, is intended to include all shades of the colour, whereas a person using a colour term in an adjectival or descriptive sense has a particular

hue in mind. The part of the brain which perceives and processes grey-scale images is different from that part of the brain which deals with colour. The United States spelling is 'gray'.

Grey goods: The term refers to any fabric from the loom, regardless of colour or fibre content, that has been woven but has not yet undergone any finishing treatment either wet or dry. Many companies produce only grey goods, sending them to specialists to be finished.

Grey fabric: Loom state fabric which has not undergone any wet processing.

Grey scale: An achromatic scale ranging from black through a series of successively lighter grey to white. Such a series may be made up of steps which appear to be equally distant from one another (such as the Munsell Value Scale), or it may be arranged according to some other criteria such as a geometric progression based on lightness. Such scales may be used to describe the relative amount of difference between two similar colours. In textile mainly two types of grey scales are used –

(a) *Grey scale for the assessment of staining*: Consists of a scale of 5 pairs of grey and white coloured plates in which each pair of plates indicates a different degree of contrast with the 5 full steps being arranged in geometrical progression. The number of the contrast pair in the grey scale (from 1 to 5) which corresponds as closely as possible to the degree of staining that is noted and serves as the fastness rating. Half steps are provided in the grey scale to improve the accuracy of assessment –

i = heavy staining,
ii = moderate staining,
iii = medium staining,
iv = slight staining,
v = no staining.

(b) *Grey scale for the assessment of change in colour*: Consists of a scale of 5 pairs of grey-coloured plates in which each pair of plates indicates a different visible degree of contrast. In a like manner to that in (1) above, the visual difference between the grey scale and the test specimen is compared and rated as follows –

i = very poor (severe change),
ii = poor,
iii = fairly good,
iv = good,
v = very good (negligible change).

Besides the numerical rating for change in colour, the rating may be supplemented by a letter to indicate an accompanying change of hue or brightness, e.g. D = duller, Br = brighter, R = redder, Y = yellower, Bl = bluer, G = greener.

Grey thread: Undyed or unfinished sewing thread in the state following final plying or equivalent step in the processing sequence such as extruding or braiding.

Grey tufted carpets: Grey form tufted carpets, back coated only and kept in the store to reduce the lead time in supply. When the order is received, the grey tufted carpets are taken and cut into the required sizes, dyed, finished, and dispatched to the customer.

Gridelin: Violet grey; greyish-violet colour. From the French *gris-de lin* (grey of flax). Also a mixture of red and white.

Griffin: Commercial variety of cotton from America, yielding the finest and longest staple known; the bolls are large, yielding 28–29% lint.

Grin: A flaw in the cloth which results from the warp rib showing through the covering threads.

Grin (in sewn seams): To stress a seam, so that the individual seam can be seen.

Grinning in printing: A term used in textile printing to describe a printing defect in the form of small, unprinted areas in a design caused by the printing of colours bordering each other with screens or printing rollers where the engraving allowance for the bordering colours has not been sufficiently generous to ensure an unbroken border between the adjacent colours thus allowing small white (unprinted) areas to "grin" through.

Grinning in dyeing: In dyeing, lighter appearance of a fibre type in dyeing mixed fibre fabric. In printing, irregular or light-coloured, unprinted places, poorly wetted out as a result of too highly concentrated thickener (Print definition).

Grinning: Gapping of seam under stress, usually due to inadequate thread tension at sewing.

Grinny cloth: A cloth with an unsatisfactory cover. It is sometimes called Hungry cloth. See **Hungry cloth**.

Grip (in tensile testing): To hold, grasp, or secure, for example, to grip the specimen by the jaws of the clamps.

Gripper axminster: A machine-woven velour carpet in which the differently coloured dyed yarns for the warp pile are supplied as required from bobbins carried on a creel behind the weaving machine. Jacquard operated yarn carriers present the appropriate coloured yarns to the grippers which grasp the required length of pile, which is then severed. The grippers, holding the tufts, then swing down to insert the tufts of yarn at the point of weaving across the entire width of the carpet. The maximum number of colours is usually limited to 8, but may be as high as 12. See **Carpet**.

Gripper looms: Shuttleless looms. These looms employ a projectile with a jaw that grips the end of the filling yarn during the insertion of the pick.

Gripper shuttle machine: A weaving machine in which the weft thread is gripped by jaw(s) that are fitted in a projectile which is then propelled through the shed.

Gripper, Gripper web, Gripper tape: A narrow fabric, typically 30 mm wide, which may be woven or knitted, having one or more raised stripes in the warp direction formed of a high friction material. It is used in the inside of the waist band of skirts and trousers to keep it in place.

Gris: Grey.

Gris: Grey cotton goods in the Philippines.

Gris fer Bleute: French army cloth; used for coats.

Gris- brun: French military cloth, made about 48 inches wide after fulling; made of long staple wool, one-third white and two-thirds black wool, in natural colour, being mixed.

Grisaille: (a) French for grey mixture (pepper and salt) effects; (b) Plain woven French dress goods, made with a black and white printed warp and worsted filling.

Grise- (L): Grey.

Grisette: Originally light, grey coloured dress goods, made in France of mixture of silk, cotton and wool, or of all-wool, and used for garments by the lower classes. Later made in all colours and in very good qualities, similar to etamines. G is the original but obsolete name for a strong, wool etamine.

Griseous: Bluish-grey colour.

Grist: (1) In Scotland, the standard size for ropes, meaning one inch diameter and three strands, each of them twisted together of 20 yarns.

(2) The yards (or meters) per pound (YPP). A finished yarn that came up to 890 YPP, one pound of yarn would equal 890 yards. The grist (or "count") may range from 300 yds/lb to 3,000,000 yds/lb for a single filament of silk (theoretically).

Grit: In air pollution, particles larger than 75 m. In wastewater, grit includes sand, gravel, ashes, metal, glass, earth from vegetable washing, etc. It settles more easily than the ***organic*** solids in wastewater and usually increases in amount after rain. In desert areas, the grit content in waste water may be high because of wind-blown sand. In cold climates, gritting of roads in icy weather increases the grit content of ***storm water***. The volume of grit removed from domestic waste water is normally 15–30cm^3 per person per day.

Gritty wool: Wool with a dry and hard feel, owing to the presence of sand.

Grizzle: Grey.

Grog: English technical term for any fabrics woven ends together, irrespective of the construction of the fabric.

Gros or Grosse: French for stout and thick fabrics. Also a name for large variety of cross-ribbed fabrics.

Gros Forts: Strong, stout French linen; used in upholstery.

Gros d'Afrique: Plain woven, all-silk dress goods. The warp consists of single threads of ecru silk and double or triple strands of boiled silk. It is double-faced with a velvety and cross-ribbed effect.

Gros d'Afrique Corde: Double-faced, plain woven dress goods, made of two silk warps, one being a single thread of ecru silk. The other double or triple strand of boiled silk. The filling is of heavy, loosely twisted cotton yarn. The effect is heavy and velvety cross-ribs.

Gros d'Alger: French silk fabric, made with two sets of warps; it has cross-ribs.

Gros de Berlin: Cross-ribbed French silk fabric made with two sets of warp; also made of alpaca.

Gros de Chine: Cross-ribbed French silk fabric, made with two sets of warp and heavy filling.

Gros d'Ete: Silk fabric, with two sets of warps. It has cross-ribs.

Gross De Londres: A cross-ribbed fabric with alternating heavy and fine ribs, or ribs of different colours. Made in plain weave and plain weave variations. A glossy finish is applied and is fairly stiff to handle. The fibres used are viscose actetate, triacetate, polyester or silk. Used for millinery and also for evening dresses.

Gros de Lyon: Cross-ribbed French silk fabric, made with a heavy filling and two sets of warps, one containing one third, the other two-thirds of the warp ends.

Gros de Messine: Silk dress goods with fine ribs and organzine warp, which forms the face.

Gros de Naples: Plain woven silk fabric of Italian origin, made with ply warp of organzine and a heavier two-ply (filling, forming cross ribs; used for coats, hats, etc.).

Gros d'Oran: French silk brocade; used for dresses.

Gros d'Orleans: Twilled ribbed fabric, made with two sets of filling, one being of the same colour as the warp and the other a glazed yarn.

Gros de Suez: Finely ribbed silk lining for hats.

Gros de Suisse: French silk fabric, having cross-ribs on the face. It is made with two sets of warps and heavier cotton filling in plain weave.

Gros de Tours: (a) Jacquard figured cotton bed spreads in Argentine; (b) Plain woven cross-ribbed French fabric, made of silk and other materials, the ribs being formed by two picks, the warp having two or three plies.

Gros de Tours weaves: In this weave, the working of the warp is the same as in taffeta, except that instead of one pick, two or more are inserted in the same shed. It is mostly used in selvedges, where it serves to give more firmness to the edge of an otherwise loosely woven cloth, and prevents the weaving ahead of the edge in a tight weave. Gros de Tours is sometimes used, especially when cotton or wool filling is employed, with a view to lay two picks nicely side by side, whereas a thread entered two ply with the taffeta weave will always receive some twist, which may disturb the perfect evenness of the fabric.

Gros des Indes: Plain woven, all-silk dress goods, made with two warps – one single and the other double or triple, and two fillings, one very fine, the other reeled from 8 to 10 natural strands, resulting in a cross-ribbed effect.

Grosgrain: (1) A heavy fabric with prominent ribs, grosgrain has a dressy appearance and is used in ribbons, vestments, and ceremonial cloths.

(2) A plain weave cloth with a prominent rib effect. The fibres may be silk, worsted, cotton or viscose, acetate, nylon. The cords are heavier than those in poplin and rounder than those are in Faille, and it is a fairly stiff fabric, made only in plain colours. Used in formal clothes, such as ceremonial gowns and church vestments, but also for ribbons (Synthetic), hard wearing

Grosse chainette: Solid coloured plain French serge, made with eight leaves and six picks in a repeat.

Grosse cote: Solid coloured plain French serge, having eight leaves and eight picks in a repeat.

Grosse draperie: French term for all woollen fabrics which are felted or shrunk, irrespective of the fineness, weave or other characteristics.

Grosse Grenadine: French serge, made with 12 leaves and four picks.

Grotto: A strong greenish-blue colour.

Grotto blue: A vivid greenish-blue colour.

Ground: See **Fond**.

Ground colour: A term describing the plain background colour against which a design is created.

Groundwater: Water contained in the soil or rocks below the ***water table,*** i.e. in the ***zone of saturation***. It is the immediate source of well water. If too much is drawn off the water table, ***piezometric surface*** is lowered.

Group (in upholstered furniture): A number of individual upholstered furniture units that are related by one or more physical characteristics such as styling, colour, shape or covering.

Growth: (1) An increase in one or more dimensions of an object or material.

(2) The increase in length of a specimen caused by the application of a continuing load or force under specified conditions.

Growth (of a fabric): The difference between the original length of a specimen and its length after the application of the specified load for a prescribed time, and the subsequent removal of the load.

Growth (in tyre cord): The increase in length of a specimen caused by the application of a continuing load or force under specified conditions.

Gru Gru: Very fine and soft fibre yielded by the leaves of the palm of same name in the West Indies.

GSB test: Test for determining the maturity of cotton, otherwise called the Red/Green Test or simply Goldthwait test, devised by Goldthwait, Smith and Barnett.

Guana: (a) Silky and yellowish seed hair of the Bombax tree found in Cuba; (b) Fabric made of the bark of the lace tree found in Jamaica.

Guaxima: Very strong fibre yielded by the *Urena lobata* in Brazil; used for making bags and ropes.

Guayanilla: Variety of white, lustrous and strong raw cotton from the West Indies.

Guanaco: The yellowish-brown hair of the guanaco, a llama-like animal that roams wild over a large area of South America. Guanaco belongs to the group of natural protein fibres. Only the under hairs are of textile value. The fibre is extremely soft and silky, the finer hairs are approximately 20 μm and the coarse hairs approximately 75 μm in diameter.

Guar or Guar gum: A plant-derived gum – Guar gum is used as a thickener in few printing pastes for acid dyes and in *etching* paste for cellulose fibres. It is reasonably resistant to acids. Guar is unsuitable for use with reactive dyes, since it will itself react with the dye. Modified guar gums are also used.

Guar gum (Guaran): Asiatic type of gum (*Cyanopsis tetragonoloba*), fraction of the endosperm ground from the seeds (diameter approximately 5 mm) of the guar plant, a grey-white powder comprising 65% mannose and 35% galactose (*Galactomannan*), aqueous-colloidal neutral solution, has an approximately 8-fold starch thickening power and is used as a fine powdery thickener in carpet printing and dyeing processes, amongst other things.

Guard hair: The long, stiff and usually coarse fibre which projects from the woolly undercoat of a mammal's pelt.

Guembipi: (a) Fibre yielded by the stem of a species of Philodendron in Brazil; used for ropes; (b) A dark coloured, very strong and durable bark fibre, yielded by a creeper of the *Aroideae* family in Paraguay; it will not rot; used for ropes.

Guendje: See **Genghis rugs**.

Guernsey: A close-fitting knitted woollen shirt.

Guibert: Stout bleached French linen shirting.

Guibray: In France, a thick cotton yarn; used for wicks.

Guide bar: A mechanism on a warp-knitting machine that directs warp threads to the latch needles.

Guide rod: Fixed rod for guiding the cloth.

Guide roller: Device, fitted on textile machines, for guiding the cloth by means of an idler roller which rotates freely as the cloth is passed over it. The purpose of guide rollers is to ensure the correct passage of the cloth, i.e. at low tension, and free of creases (poor guide rollers are the main cause of creasing).

Guides: Fittings of various shapes for controlling the path of a thread line.

Guignet's green: A bright bluish-green colour patented in 1859 by the chemist Guignet, C. E. and sometimes referred to as permanent green. See ***Viridian***.

Guillotine: Cutting device that consists of a single blade that descends between guides for chopping fibres, plastic strands, etc.

Guimhas: Cotton or woollen ribbons produced in Latin America.

Guimp: (a) In laces, a heavy thread placed on the edge of the sprigs; (b) The design or pattern of the lace.

Guinea: Having the colour of a gold guinea coin.

Guinea cloth: A soft, napped cotton fabric of England, woven with two sets of warp, about a yard wide and dyed indigo blue. It is sold in the West African markets.

Guipure: (a) Originally gold and silver lace made with the bobbins or the needle, the patterns being formed by heavy cords padded with parchment (called cartisane) or by a thick thread. It was also called parchment lace; (b) Tape laces in the 16th and 17th Century with the outline of the patterns formed of needle-point or bobbin made tape over a coarse round meshed ground, occasionally ornamented with brides ; (c) At the present, laces with large patterns without any brides or mesh ground are called guipures.

Guipure lace: Although now entirely made by machine, this was once handmade. It is Italian type of lace. It is composed of thick, heavy patterns, often with quite large motifs, joined together with bars or brides. The ground fabric is made from nylon acetate or sometimes cotton. The thick embroidery is worked and then ground fabric is dissolved with chemicals and that will not affect the embroidery.

Guipure de Flandres: Old Flanders bobbin lace with raised patterns.

Guipure renaissance: Embroidery and applique work, composed of cheese cloth, cord and sewing silk; used for mats, etc.

Gulbadan: See **Amritsar silk**.

Gulbani: A very light and transparent Indian cloth made of silk and interwoven with gold thread.

Gull: A very old name for a shade of grey.

Guly: Red.

Gulnagai: Fine, plain cotton muslin of India.

Gum: A term covering a wide range of substances. Strictly, gums are carbohydrate high polymers, either soluble or dispersible in water, that are derived from vegetable origins. Loosely, the term gum is used to mean resins, saps, natural rubber, chicle, starch, cellulose derivatives, and many other products. In textile printing, the term refers to print-paste thickeners.

Gum Arabic: Senegal gum, acacia gum, African vegetable gums which contain up to 80% Arabic acid. Most highly prized is gum kordofan from Sudan (2 cm droplet length, slightly yellowish or of a deep amber colour, rarely colourless). Much less pure is ***Geddahgum*** from the area around Aden (grainy, deep yellow, brown to black; when dissolved leaves quite heavy insoluble sediment), similar in appearance and quality to Moroccan Mogador gum.

Gum (silk): Sericin.

Gum spots: Print mistake. The spots where the gum is not washed off properly.

Gum tragacanth: See **Tragacanth**.

Gumming: A Highland tartan, composed of the following on a red ground – A narrow dark green stripe, two wide green bars and another narrow green stripe, separated from each other by red stripes half the width of the narrow green stripes. The wide green bars are about four times as wide as the green stripes and are split in the centre by a single white line. Each group of green stripes and bars is alternated with a wide red surface, split with a narrow black stripe.

Gummy wool: Scoured wool that still has some yolk in it.

Gun patch: A reinforced patch is placed on the shoulder – here for protection when shooting.

Gunclub check: A wool fabric with distinctive checks. The checks are usually woven in the greens, browns and sometimes worth an over-check. Used as sports jackets, plus fours, golfing skirts, etc.

Gunmetal: The dull blue-grey colour of gunmetal; a dark grey colour. Also 'gunmetal blue'.

Gunn: (a) A commercial variety of short staple upland cotton from the Mississippi; (b) A Highland tartan, the repeat made as follows – wide, dark green bar, split in the middle by a single narrow red stripe; black stripe, half the width of the green; very fine green line; navy blue bar, same width as green bar, split in the middle by a very fine green line; another very fine green line; black stripe, measuring half the width of the green bar.

Gunny: Very coarse and open, plain woven bagging, made of jute, also of hemp. See **Bagging**.

Gur or Gurra: Indian coarse white muslin.

Gusset: A piece of material let into a garment to strengthen, shape, or enlarge it.

Gut (thread): See **Stuffer yarn**.

Gutta: French for "resist", which forms a barrier on the fabric to create a pattern. May be water soluble, solvent soluble, or a heat fixable resin.

Gutta percha (trans-polyisoprene): The name is derived from Malay getah = gum + percha = name of a tree from which it is extracted. A natural latex product, similar to rubber, from the latex of any of several tropical trees of the *sapotaceous* genera palaquium and payena. It may be used for painting on silk in the form of a solution with a honey-like consistency to produce resists on the fibre (it forms a film on drying which is impervious to water and dyes) and thus makes separation lines in coloured designs possible.

Guttar: Indian satin made with silk face and cotton back.

Guzieh: An inferior and plain woven Indian cotton cloth.

H

H ions: See **Hydrogen ions**.

HA: (a) Hairs; Standard abbreviation for textile fibres; (b) Hemp.

Ha: Hemp; standard abbreviation for textile fibres. See **HA**.

Haberjet: Coarse and medieval woollen broadcloth, made in England. It is said to have been worn by the monks.

Habit cloth: Medium weight suiting material made of wool in satin weave with a fine warp and a thicker weft and a dress face finish. Used for suits and winter dresses.

Habitat: The natural environment or location of a plant or animal.

Habutae: See **Pongee**.

Habutai: (1) Japanese word meaning 'soft and light'. Originally used for Japanese waste silk, but now used for silk material made on powerlooms in plain or twill weave. It is heavier than traditional Chinese silk, but similar. Very light weight and soft. Sold by weight measure known "momme" (1 momme = 3.75 g). Made from waste silk that can be twisted. It is piece dyed or printed and sized. Has many defects in the cloth which has a "shot-about" appearance, but this does not affect the cloth. It comes from Japan – originally woven in the gum on Japanese hand looms. Lighter than shantung, but heavier than silk.

Uses: Dresses, coats, shirting, lamp shades, lingerie, curtains, men's jackets, women's shirt waisters and skirts.

(2) This is a cheap silk which is excellent as a soft, lightweight lining fabric for evening dresses, wedding dresses, etc. It is used as lining for people who find synthetic lining uncomfortable or allergic.

Hackles: The comb for dressing flax or hemp.

Hackling: A mechanical process given to flax stems after scutching. Its purpose is to remove short fibre, motes (or neps) and non-fibrous material. Hackling involves drawing the fibres through inclined steel combs, each successive comb being finer than the previous one. As a result, the coarse bundles of fibre are separated into finer bundles and the fibres are thereby aligned parallel to one another. The outcome is approximately 20–70% hackled flax (long fibre), 60–20% flax tow (short fibre) and remaining motes. See **Flax processing**.

Haddat: Square printed cotton cloth; used as head covering in Asia Minor.

Hadjai: A hand-spun gold thread in India, used for embroidery.

Haifa: Native Algerian name for esparto.

Haining wool: A Chinese fine grade carpet wool.

Hair: Natural animal fibre other than sheep wool or silk.

Hair canvas: A woven interfacing material in various weights. Coarse goat hair combined with wool, cotton or rayon is used in the weft. Used in coats and jackets made from suiting and coatings.

Hair cloth: Cotton, polyester, linen or other fibre with which horse hair has been mixed to make it strong and inflexible. Used in upholstery, and also as chest canvas in tailoring.

Hair-dye: Dye which is used to colour hair. In ancient times, the Greeks used **saffron** to dye their hair. The Romans used vinegar and fermented leeches as a black dye with the caution to keep oil in the mouth during use to prevent one's teeth also going black. Blonde hair was more often the fashion until Elizabeth I created a move towards red hair. See **Dissembling colour** and **Tattoo**. A few modern hair dyes, particularly those containing para-phenylene diamine, are known to be capable of affecting people who suffer from allergies (and possibly of causing anaphylactic shock) making it essential to carry out a skin test before use. Researchers at the University of Southern California have also found that permanent hair-dyes can increase the risk of women getting bladder cancer. Even semi-permanent hair-dyes (especially those in dark colours) have been linked with an increased risk of certain cancers.

Hair fibres: This group includes all animal fibres with the exception of wool (natural protein fibres).

Hair papilla: See **Human hair**.

Hairas: Yarn made of coarse oriental wool and has little lustre.

Hairbine: An 18th Century English woollen fabric.

Haircloth: (1) A very resilient in the weft direction cloth, woven of horse hair of the horse's mane or tail which does not form a continuous yarn and is not twisted (or sometimes camelhair), from which it takes its name, for weft with cotton, wool or linen warp. It clings to the outer cloth because of its rough surface texture. It is woven in plain or satin weave and is used for interlining, for facings, furniture cover, front part of suits, jackets and coats made from medium to heavy weight fabrics, etc. Used also as chest canvas in tailoring.

(2) An imitation of the real horsehair cloth is made by using hard-spun cotton yarn entirely, which is heavily sized to give stiffness. This fabric is used as dress interlining.

Haircord: A plain woven cotton fabric characterized by fine rib lines in the warp direction created by alternate coarse and fine ends or by having two or more end weaving as one alternately with a single end. A possible construction is 28 ends per cm by 26 picks per cm; 2 ends 18 tex and one end 21 tex cotton warp and 21 tex soft spun cotton weft.

Haircord carpet: A low level loop pile carpet with pronounced rows of loops in the weft direction. The pile is 100% hair.

Hairiness: A term used to describe the projection of free floating fibre ends or loops from the surface of a fibre assembly. It is a desirable characteristic in raised, brushed, fleecy fabrics (confers good heat retention properties). However, hairiness reduces the strength of yarns.

Hairiness (in raw silk): The condition of the yarn when there are numerous loose ends of cocoon filaments projecting from the surface.

Hair line: Woollen or worsted dress goods or men's wear, made with very fine continuous stripes, produced by single warp yarns crossed by a filling of the same colour. The fabric is given a clear finish; (b) Fishing line made of horsehair.

Hair net: Made of silk or human hair with large mesh; used to prevent the hair getting disarranged.

Hairline: Very fine lines.

Hairy: See **Fuzzy**.

Haitien: Plain woven silk dress fabric, made with fine warp and heavier filling, alternately in white and coloured picks.

Hakamaji: A Japanese cotton fabric made on hand looms, mostly in plain weave. It comes usually in blue and white stripes and has cross ribs. It is used for making men's skirts.

Hakir: Strong warp striped cloth in East India, made with silk warp and cotton filling.

Hakistery: Black and white ground prints, used in Persia for clothing.

Half backs: Woollens partly made like backed goods.

Half basket: A variation of basket weave is usually woven with double warp ends but single picks. The vertical lines in this cloth, where the weft passes over the double ends, are called ribs or cords.

Half back stitch: The half-back stitch is made by taking one stitch and placing the needle half way back, then bringing it out twice the length of the stitch and placing the needle half way back each time from where the last stitch ended. The appearance on the right side will be of regular space as in the running stitch.

Half bleach: A term used to describe a bleaching effect which is adequate for goods to be subsequently dyed (but not for whites).

Half-blood: American designation of wool compared in fineness to the full blooded merino as standard.

Half damask: Made in England of silk and cotton or silk and wool.

Half-dyeing(time of): The time taken in minutes for a textile substrate to absorb 50% of the dye it will absorb at equilibrium under the same conditions. As a dye characteristic, it is less suitable for the selection of combination dyes for dyeing acrylic fibres with cationic dyes. In this case, the Compatibility value K, is a much more reliable parameter.

Half emulsion: Combination of an Emulsion thickener and a conventional thickener (e.g. alginate). Used in direct printing for the application of most dye classes with the exception of indigo and oxidation dyes.

Half Milano (ripple): A weft knitted fabric. In half Milano, a course of 1 × 1 rib is followed by a course in which only the back needle bed knits plain. The loops in the front bed are held over and are extended. The extensibility in the width is reduced by the plain course. If more than one course of plain loop is knitted, either on the front or back needle beds, the result is a ripple fabric. The plain courses are thrown in to relief (without special patterning). These fabrics are used for pullovers and waist coats.

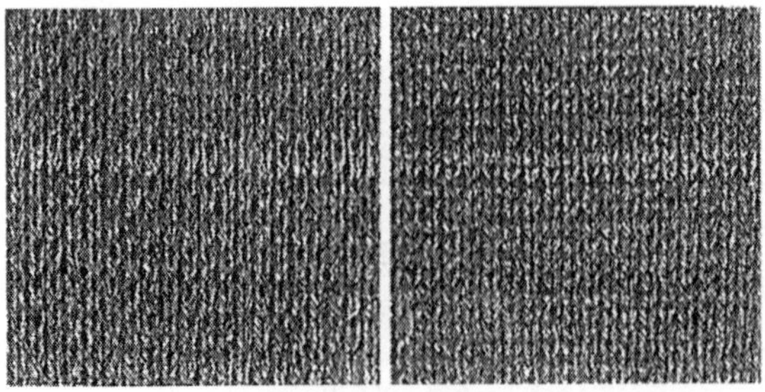

Half Milano face and back

Half pick: A fabric fault is caused when a pick is broken and is not removed in weaving.

Half resists: See **Half-tone resists; Resist printing**.

Half silks: Silk fabrics made partly (warp or filling) of cotton.

Half-tint: Intermediate tone between light and dark.

Half-tone: Tone-in-tone effects in printing, i.e. a range of tones or tonal gradations in addition to the full depth tone with a sharp or gradual transition from the latter. Half-tone process.

Half tone resists: Resist effects produced by pre-printing a colourless print paste, i.e. containing no dye (with or without the inclusion of resist agents) followed by overprinting with a normal coloured print paste. A lighter colour is produced in those areas of the printed design where the second print paste falls on the first. Pre-printing with a high solids content thickener alone will already give a certain resist effect which can be further intensified by the addition of a white pigment (e.g. titanium dioxide, zinc oxide) or a white pigment incorporated into a plastics dispersion (matt white).

Halina: Coarse and checked woollen cloth with long hair on the right side, made in Austria.

Hallencourt: Twilled French table linen.

Hallescrues: Strong unbleached linen made in France.

Halluin: Coarse serge used by the French army.

Halo: (a) A narrow pale (often blurred) outline surrounding a coloured discharge or resist print caused by the outward diffusion of the resist or discharge chemicals into the ground; (b) A ring surrounding the area of a stain on a textile material following attempts at stain removal by local application of a spotting agent (organic solvent). The ring or halo thus formed is caused by the outward diffusion of colouring matter, soil particles, grease, etc.

Halo: Off-shade colour fringe effect, often seen in direct application dyeing when used mixed-colour dyes. It is common for different colour dyes of the same family to migrate at different rates when solutions are applied to fabric. For example, a purple made by mixing Red MX-8B (usually called fuchsia) and Turquoise MX-G tends to produce strong turquoise halos. If a drop of dye solution is applied to soda ash pre-soaked fabric, the red component quickly stops moving outward, but the turquoise continues to migrate outward much longer, resulting in a too-red centre with a turquoise halo. This can be very difficult to control, other than by selection of the dyes used for mixtures. Because black dyes are almost always mixtures, haloing often occurs with black.

Halogenation: A reaction in which a halogen atom is introduced into a molecule. Halogenations are specified as *chlorinations*, *brominations*, *fluorinations*, etc., according to the element involved. There are several methods.

(a) Direct reaction with the element using high temperature or ultraviolet radiation:

$$CH_4 + Cl_2 \rightarrow CH_3Cl + HCl$$

(b) Addition to a double bond:

$$H_2C{:}CH_2 + HCl \rightarrow C_2H_5Cl$$

(c) Reaction of a hydroxyl group with a halogenating agent, such as PCl_3:

$$C_2H_5OH + PCl_3 \rightarrow C_2H_5Cl + OH-$$

(d) In aromatic compounds, direct substitution can occur using aluminum chloride as a catalyst:

$$2C_6H_6 + Cl_2 \rightarrow 2C_6H_5Cl$$

(e) Alternatively in aromatic compounds, the chlorine can be introduced by reacting the diazonium ion with copper(I) chloride:

$$C_6H_5N_2^+ + Cl^- \rightarrow C_6H_5Cl + N_2$$

Hamadan carpets: Especially hardwearing rustic mesomeric resonance (Methyl sulphinyl carbanions). Knotted carpets from the region around Hamadan which rises above the Iranian tableland at some 2000 m above sea level. Hamadan and the villages of the province are the most important centres of carpet production in Iran. In the main, they have a typical brown natural colour from the wool and camel hair used in theirchlorine, bromine, iodine and fluorine. They react with metals to form ionic halide salts, e.g. iron chloride $FeCl_3$. Their hydrogen compounds are called hydrohalic acids, e.g. hydrochloric acid, HCl. Halogens production. Hamadan carpets are, on the whole, rather mundane as far as decoration and weave are concerned. The materials, by contrast, are of a very good quality. Approximately 250,000 Turkish knots per m^2.

Hambourgeoise: (a) French silk dress goods of the 18th Century. It had a taffeta foundation with napped patterns; (b) A French double-faced silk fabric of the 19th Century, made with stripes. Hamburg Point Drawn work ornamented with coloured silk thread.

Hamamlik: Turkish name for Oriental bath rugs, usually square.

Hamme: (a) Woollen dress fabric made from printed yarns; (b) Plain weave fabric produced from yarn died linen warp and cotton weft, used for table cloth and curtains.

Hammer mill: A device used in the chemical industry for crushing and grinding solid materials at high speeds to a specified size. The impact between the particles, grinding plates, and grinding hammers pulverizes the particles. Hammer mills can be used for a greater variety of soft material than other types of grinding equipment. Compare **Ball mill**.

Hamburg: Wool; glossy embroidery wool yarn in England.

Hamburgo: Americano name for unbleached cotton sheeting in the Canary Islands.

Hamidieh: Fabric made of silk and cotton in Syria.

Hammer blow effect: A defect which occurs in various types of textile fabrics, e.g.,

(a) A generally structure-related colours which are not in level in cotton velvet, especially in the form of "lumps" caused by the presence of a large proportion of motes (these can be detected under the microscope as hard fibre tufts adhering to seed husk residues) which disrupt the pile uniformity.

(b) A "hammered" surface appearance in crêpe fabrics produced from man-made filament yarns by the change of direction of highly-twisted yarns as well as changes between flat and crêpe yarns.

(c) A moiré appearance in plain weave wool fabrics after crabbing.

Hammock cloth: Strong, soft cloth, usually woven of all-cotton with bright coloured warp in plain or fancy weave. Used for hammocks, drapery, etc.

Hamouli: Raw cotton, formerly grown in Egypt; now little cultivated.

Hanabishi: A Japanese silk fabric with gold brocaded diaper pattern.

Hand (or handle): The feel of a fabric. The tactile qualities of a fabric, e.g., softness, firmness, elasticity, fineness, resilience, and other qualities perceived by touch. Hand of a fabric is quite subjective, and often difficult to describe. Both chemical and mechanical treatments are frequently used to alter the hand of a fabric. See **Softener** and **Hand-builder**.

Hand block printing: The cloth to be printed is attached to the resilient surface of a strong and relatively short table (approximately 8 m long). The actual printing operation is carried out with hand-printing blocks on which the design is formed in relief. The printer presses the block on to the surface of a print paste contained in a vessel known as a "swimming tub" or "sieve" in two opposite directions to ensure an even coating of colour paste on the block. When the block is correctly positioned on the cloth, it is struck with the fist or a printer's mallet in order to transfer the print paste from the block to the cloth. This process is then continued as necessary by repositioning the printing block on the cloth each time in accordance with the design repeat.

Hand builders: The purpose of applying hand builders is to add bulk, weight or stiffness to a fabric. For few fabrics, this change must be permanent and withstand washing and dry-cleaning. In other applications, the change is temporary so hand builders are classified as either durable or non-durable. A treatment that makes a fabric feel firmer or stiffer. Hand builders are essentially the opposite of fabric softeners. They may be applied to make a fabric easier to manage in sewing, or to make it feel more robust (as for work clothing). Hand builders may be temporary, such as starch, or permanent, such as polymer resins. Hand builders can interfere with dyeing. Some are impossible to remove.

Hand coater: A simple and effective system for the manual production of colour strike-offs on paper (or textile material) with the aid of short stainless steel rods. The quantity of colour paste applied is determined by the circumferential grooves in the rod surface which range from fine to coarse.

Hand creasing test: By means of the so-called fist method, the fabric to be tested is crumpled, wrinkled, pressed or folded for a few seconds in the tightly clenched fist and assessed, after opening the fist, whether the fabric recovers immediately from its creased state, or only after some time by hanging or lying, or whether the creases can only be removed after smoothing out or ironing.

Hand knotted carpets: Carpets produced by the manual knotting of pile yarn tufts in a warp yarn substrate. See **Knotted carpets**.

Hand loom: It is a loom in which every operation is performed manually.

Hand modifying finishes: Finishes which modify the handle of the fabric.

Hand overcast seam-finish: A finish for the raw edges of a plain seam in which hand stitches are taken over the raw edges of each seam allowance (overcast stitches) about 3 mm (1/6 in) deep and spacing about 6 mm (1/6 in) apart. Ideal for elegant or sheer fabrics like organdy, chiffon, silk, and velvets which deserves that special touch.

Hand-painted mordanted cotton: Cotton fabrics coloured by patterning with mordants manually and then dyeing.

Hand printing blocks: Blocks which are used in hand printing. The blocks are equivalent to screens in screen printing or roller in roller printing. Probably the roller in roller printing is a development of blocks for continuous printing. The blocks are dipped carefully in inks (dyes) and printed on fabrics laid in a table. The setting of the design is done using pin points at the corner of the blocks.

Hand screen printing: Textile printing carried out with flat screens on long tables up to 100 m in length. The process is mainly limited to the production of high quality prints today as it is more labour-intensive and less productive than screen-printing by machine.

Hand sewing machine: (1) Sewing machine that is normally used at home, the drive is done by the hand. It is probably rare now.

(2) Used in wet processing departments for joining loose piece ends, e.g. after taking fabric quality control samples at various stages of processing, joining successive batches of fabric at processing machines, etc. The majority of these machines produce a seam joined by the chain stitch.

Hand wash dip dry: A washing instruction, meaning to be washed by hand only and not in machine, no wringing or machine spinning (centrifuge), and to be dip dried only. Usually this instruction is given for very fine, loose woven, and soft weak fabrics and fabric which have chances for distortion, say silk fabrics.

Hand-washed Wool: Wool washed before it is shorn from the sheep.

Hand-builders: See **Handle**.

Handewarpes: Coloured English woollen fabric of the 16th Century.

Handkerchief designs: Engineered prints to fit into a square of fabric.

Handkerchief linen: Also called handkerchief lawn, it describes the finest cotton or linen lawn, or cambric, used for handkerchief. Made in plain colours or printed or even checks. In addition to handkerchiefs, it is also used as baby clothes, christening gowns and table linen.

Handle: Hand or handle are the terms used to describe how a fabric drapes around an object or feels to the touch. When the fabric becomes stiffer or bulkier, the hand of the fabric is said to be built. Chemicals that accomplish this are called hand-builders. When the hand is made to drape more or to feel silkier, the fabric is said to have been softened. Chemicals that do this are called softeners.

Handle: The subjective assessment of the roughness, smoothness, harshness, pliability, thickness, etc., of a textile material and that is obtained by the sense of touch. The handle of a textile fabric is determined by numerous different parameters. In commission finishing, the finisher is usually presented with a sample of finished fabric by the customer, in addition to a colour sample, which he is required to match in bulk production. Comparisons between the sample and the production batch are, however, subjective. It is for this reason that efforts have been made to find precise methods for the measurement of handle, e.g., nozzle test; handle-o-meter, drape test. In Japan, attempts have been made to evaluate the handle of a textile fabric by measuring 15 mechanical and physical properties. Kawabata system of handle assessment; FAST system of fabric assessment.

Handle modifier (hand modifiers): Descriptive term for a group of textile auxiliaries, the primary function of which is to modify the handle of textile fabrics. They represent a sub-group of textile finishing agents. See **Handle-modifying finishes**.

Handle-o-meter: An instrument for evaluating the handle of textile fabrics including the determination of flexibility, compressibility, extensibility, elastic resilience, density, surface friction and thermal properties. The test procedure involves placing the textile material over a 6.35 mm wide slit where a rotating load arm brings pressure to bear on the sample in the slit. A load-measuring device determines the various resistance values of the fabric sample, the sum of which can be read off an electrical display.

Hand washing: The gentlest form of home laundering using hand manipulation without the use of a machine or device such as a scrubbing board.

Hand wheel; balance wheel: A wheel located at the end of the machine arm that controls the movement of the take-up lever and needle bar.

Handwoven: Fabric woven using a hand loom.

Hang pick: A pick caught on a warp yarn knot for a short distance producing a triangular shaped hole in a fabric.

Hang shot: See **Hang pick**.

Hanger: A looper loop fixed on a garment for hanging up.

Hanging black: See **Aniline black**.

Hanging sleeve: It is a tube of fabric sewn to the wrong side of a finished quilt so a dowel can be slipped inside the sleeve and the quilt can be hung easily onto a wall.

Hank: (1) (a) A synonym for skein – a textile material in a coiled form; (b) A definite length of sliver, slubbing, roving or yarn, e.g. in metric system it is 1000 metres; (c) A synonym for count as applied to sliver, slubbing or roving.

(2) Unit of measuring yarn; silk hank being 1,000 yards; worsted hank 560 yards long; cotton hank 840 yards; woollen skein 1,520 yards; linen and jute 3,600 yards. For both cotton and wool, these traditional hanks are equal to 7 leas or to 12 cuts. In the United States, however, a hank of woollen yarn is generally 1600 yards (1463 meters). In retail trade, a hank is often equal to 6 or 7 skeins of varying size.

Hank dyeing: Dyeing of yarn in the hank form. There are machines available for dyeing hanks, usually called a cabinet dyeing machine. Since it is labour intensive, this method is now adopted for yarns which are difficult to dye in the cone or cheese form. E.g., silk.

Hank dyeing machines: Machines used for dyeing yarn in the hank form. The dyeing of yarn in hank form is the oldest method of yarn dyeing. Traditional rectangular-type machines in which the dye liquor is heated by open or closed-coil steam pipes, and the liquor is circulated over a weir and through the yarn by means of a reversible impeller mounted at one side of the machine, are still widely used for the dyeing of carpet yarns (hanks up to 5 kg and machine capacity up to 4000 kg). More recently, vertical cylindrical dyeing machines (with or without pressure) in which the hanks of yarn are contained in special cages, represent alternatives to the rectangular-type machines.

Hank Dyeing Machine

Hank yarn printing machine: A machine used for flammé yarn printing on hanks made of wool, polyamide, staple fibre, cotton and silk. With the automatic Timmer hank yarn printing machine, printing is effected in four phases. In every phase, one quarter of the total circumference of the hanks is printed. Movement in longitudinal hank direction is automatic, so that the next quarter can be printed in the next phase. Movement in transverse direction and stopping are also automatic. This means that the hanks are passed four times through four pairs of rolls.

Hank, skein, reel: The traditional unit of length for yarn in hank form. In metric notation, a hank (consisting of 70 turns, each of 1.428 m circumference) 1/10 of a standard hank = 100 m. For ply-yarns, different numbers of turns are used. In general, hanks are loosely fastened at various positions around the bundle of yarns with tie-bands to maintain the integrity of the hank and prevent tangling during processing. If the tie bands are too tight, the yarn in these parts of the hank will resist the dye during dyeing and thereby resulting in undyed patches.

Hanolchade: Navajo blanket made with black and white stripes with small diamonds of blue and red in the black stripes. Worn usually by the tribal chiefs.

Harakake: Moori; name for the New Zealand flax.

Harami: Large size Indian rugs made for the mosques.

Harateen: A cloth used for furnishing in 18th and 19th Century, when it was considered to be interchangeable with moreen. It was made with a worsted warp and a thicker worsted weft to form horizontal ribs, then finished by watering and stamping.

Hard COD: The COD remaining after effective biological treatment. It is a measure of the organic material that is non-biodegradable or only very slowly biodegradable.

Hard detergent: A synthetic detergent that does not easily biodegrade and so may cause foaming in rivers at concentrations as low as 0.3 mg/l. Modern soft detergents are biodegradable in biological treatment, and so have greatly reduced foaming problems in rivers.

Hard colours: See **Warm colours**.

Hard crepe: Plain woven, light silk crepe, dyed black and gummed; used for mourning.

Hard end: See **Tight twist end**.

Hard fibre: Stiff and elongated fibres obtained from leaves or stems of plants. As they are coarse and stiff, they are used in matting and industrial products.

Hard shot: See **Tight twist end**.

Hard size: (1) Sections of cloth containing an excessive quantity of size.

(2) A condition found in areas of fabric where the warp contains an excessive quantity of sizing.

Hard soap (household soap): A soap with an approximately 60–70% fatty acid content used for domestic washing and cleaning, etc.

Hard spun or twist: Yarn spun with more revolutions per inch than usual.

Hard waste: See **Waste, hard**.

Hard water (in textile conservation): Water having a concentration of more than 60 ppm (3.50 gr/gal), hardness calculated as calcium carbonate.

Hardanger cloth: A plain weave cotton fabric of coarse construction. It is prepared in such a way that the weave is easily visible. Used for embroidery, especially the type where threads have to be counted. The threads are usually arranged in pairs.

Hardness: Hard water is water that will not readily form a lather with soap owing to the presence of dissolved calcium, iron, and magnesium compounds. Such compounds can react with soap to produce insoluble salts, which collect as solid scum. They can interfere with some chemical processes in preparation and dyeing. Soap, for example, forms insoluble compounds when it reacts with hardness minerals. Calcium carbonate or calcium hydroxide, formed when soda ash or sodium hydroxide is added to hard water, are quite insoluble and will precipitate from solution. The effectiveness of the cleansing solution is thus reduced. Hardness of water is of two types, temporary hardness and permanent hardness. Only temporary hardness can be removed by boiling the water. The precipitate can be deposited on fabric, adversely affecting the hand of the fabric. Sequestrate chemicals can be added to hard water to reduce the effect of the hardness ions. Sodium hexametaphosphate is generally the preferred chemical for this purpose for many dyeing processes. Hardness of water is usually expressed by assuming that all the hardness is due to dissolved calcium carbonate, which is present as ions. It can be estimated by titration with a standard soap solution or with EDTA. See also **Permanent hardness; Detergents; Temporary hardness; Water softening**.

Hardness (in water): Dissolved salts of calcium, magnesium and other cations which destroy the action of soap; expressed as parts per million (ppm) or grains per gallon (gr/gal) of calcium carbonate.

Hard rubber (ebonite): A hard form of Rubber in which the latex molecules have been cross-linked with 30–35% sulphur by curing at 100–150 °C. Lime or magnesia is used as activator. *Properties* – Black, hard and brittle mass which on heating becomes sufficiently plastic to be moulded into shapes. Hard rubber is resistant to water, chlorine, alkalis, most acids, gasoline, oils and alcohols.

Uses – Tank linings, pipes, handles, cocks and valves, etc.

Hards: The coarse fibres separated from flax in scutching.

Hardwickia: A tough, pliable bast fibre of India used for cordage.

Hare: The hair texture of hare is soft and woolly and can be spun with wool, producing a similar result to a wool and rabbit mixture.

Harlem checks: Linen from Holland with blue or red window plaid.

Harlequin: (a) Large plaid checks in more than two colours; (b) 18th Century checked woollen fabric in England.

Harmala red: A red colouring agent from the harmala or harmel plant.

Harn: In England, a coarse and low grade flax yarn.

Harrateen: All-worsted English fabric of the 18th Century.

Harness: A frame holding the heddles in position in the loom during weaving.

Harness chain: A mechanism used to control the vertical movements of the harness, or shaft, on a loom.

Harness skip: See **Float**.

Harris tweed: A homespun, all-wool tweed, of soft feel and peaty odour, made originally on the island of Lewis Harris, Uist and Barra and other islands off Scotland, using the best native blackface or cheviot wool in natural colours or dyed with vegetable dyes. The fabric is still hand-woven, even today, on the islands of Lewis, Harris tweed is made with carded yarns produced from pure Scottish virgin wool in twill or plain weave with a loose construction. Used for overcoats.

Harsh fibre: Fibre that is rough or coarse to the touch, but not fused or bonded filaments.

Harvard: A somewhat hard washable cotton shirting, woven mostly in 2 and 2 twill with coloured warp and white weft, forming stripes or zig-zag lines which cover a large part of the fabric.

Harvester duck: See **Duck**.

Haslock: Scotch term for the finest part of the fleece, taken from the throat.

Hasp: Linen or jute yarn measure, equal to 3,600 yards. See **Hank**.

Hatters' plush: Made with fine silk pile; used for men's and women's hats.

Haul-off roller: The first driven roller around which an extruded yarn passes after leaving the spinneret and whose surface speed determines the spin-stretch ratio.

Haute couture: Literally means 'high fashion'.

Haute lisse: French for high warp tapestry, viz. made with the warp placed in vertical position.

Haute nouveautee: French for novelty fabrics.

Havana: A dark shade of brown resembling Havana cigars and formerly applied to breeds of brown cats and rabbits. Also 'Havana-brown'.

Hawking: A process in dyeing. Several pieces of the fabric are sewed together end-to-end and passed between two rubber rolls, constantly kept under the surface of the dye in the vat.

Hawkins: Early maturing, short stapled commercial variety of American cotton, the fibre measuring 18–22 millimetres; the yield of lint is 32–34 percent.

Hawser twist: The construction of cabled yarn, coed, or rope in which the single and first-ply twist are in the same direction and the second-ply twist are in the opposite direction, and S/S/Z or Z/Z/S construction.

Hay (a): (1) A light olive-green colour.

(2) A Highland tartan that is composed of green stripes over a red ground, narrow white and yellow lines splitting the field.

Hays China: A late maturing commercial variety of cotton from Mississippi, the fine and lustrous staple measuring over 30 millimetres; the yield is about 28–30 percent.

Hayti: Variety of Indian raw cotton.

Hazardous waste: Waste that is dangerous because it is toxic or inflammable or explosive, etc. In the United Kingdom, hazardous waste is based on the hazardous wastes listed in the ***European Waste Catalogue*** (*see* **Hazardous Waste Regulations**). In the United States of America, hazardous waste is a solid or semi-solid waste that poses a threat to public health or the environment. The waste must be specifically listed as a hazardous waste by the ***USEPA*** or exhibit one or more of the characteristics of hazardous wastes (ignitability, corrosiveness, reactivity, and/or toxicity).

Hazel: A brown colour tinged with red.

Hazen colour unit: A colour scale for determining the colour of chlorinated hydrocarbon solvents based on a platinum-cobalt stock solution designated as 500 Hazen units.

Hazen stock solution: 1.245 g potassium chloroplatinate + 1 g cobalt chloride ($CoCl_2 \cdot 6H_2O$) dissolved in 100 ml concentrated hydrochloric acid and made up to 1000 ml with distilled water. Dilution of 1:100 with distilled water = 5 Hazen; 3:100 with distilled water = 15 Hazen. A chlorinated hydrocarbon control is contained in a small glass tube of the same diameter and filled to the same level with/without a comparative solution. The sample is viewed from above against a white background with the exclusion of side light.

Hazen-Williams formula: A formula used in evaluating ***pipe flow*** in water distribution systems:

$$V = 0.849 \text{ C R0.63S0.54}$$

Where V = velocity (m/s), R = hydraulic mean radius (m), S = hydraulic gradient and C is Hazen-Williams roughness coefficient.

HB yarn: High-bulk yarn.

HCB: Hexachlorobenzene.

HCFCs: Hydrochlorofluorocarbons.

HCH: Hexachlorocyclohexane.

HDPE: High-density polyethylene.

HE: Henequen fibre.

Head: The upper, movable and working surface of the pressing machine which may incorporate the shape which mates with that of the back. The head comes into contact with the garment with or without pressure.

Head circumference: See **Head girth**.

Head dress: Various confections made from stiffened cloth, veils, padded rolls, jewels, brooches.

Head dye: See **Illuminating dye**.

Head end: (a) The beginning of a new piece of fabric in the loom that bears appropriate identification; (b) A small sample of fabric that may be submitted to a customer for approval.

Head girth (in body measurements): The maximum circumference of the head above the ears.

Head rail: Shoulder shaped head square used in the gothic period.

Head (in zippers): The portion of the element that engages the pocket.

Head (in zipper): The portion of the scoop/element that engages the pocket.

Headband: A tape normally 12–22 mm wide, used in book binding and having a coloured piped or 'beaded edge woven integrally. Headbands are calibrated 0–6 according to the size of the head, traditionally of a diameter from 1.5 to 6 mm.

Heading: (a) That edge of the lace flouncing which is sewn to the garment; (b) Trade name for that end of the piece of bolt of cloth which is on the outside; (c) In short length fabrics, both ends of the material that are usually decorated with stripes.

Heald: Part of loom. Same as Heald.

Health crepe: See **Crepe de sante**.

Heart: The core strand of a rope around which other strands are twisted.

Heart yarn: The centre of a core yarn.

Heat: Energy transferred as a result of a temperature difference. The term is often loosely used to mean internal energy (i.e. the total kinetic and potential energy of the particles). It is common in chemistry to define such quantities as *heat of combustion*, *heat of neutralization*, etc. The unit is usually the kilojoule per mole (kJ mol–1). By convention, *.H* is negative for an exothermic reaction. Molar enthalpy changes stated for chemical reactions are changes for standard conditions, which are defined as 298 K (25 °C) and 101325 Pa (1 atmosphere).

Heat durability: The extent to which a material retains its useful properties at ambient air conditions, following its exposure to a specified temperature and environment for a specified time and its return to the ambient air conditions.

Heat durable: Having heat durability.

Heat exchangers: Devices that enable the heat from a hot fluid to be transferred to a cool fluid without allowing the two fluids to come into contact. The normal arrangement is for one of the fluids to flow in a coiled tube through a jacket containing the second fluid. Both the cooling and heating effect may be of benefit in conserving the energy used in a chemical plant and in controlling a process.

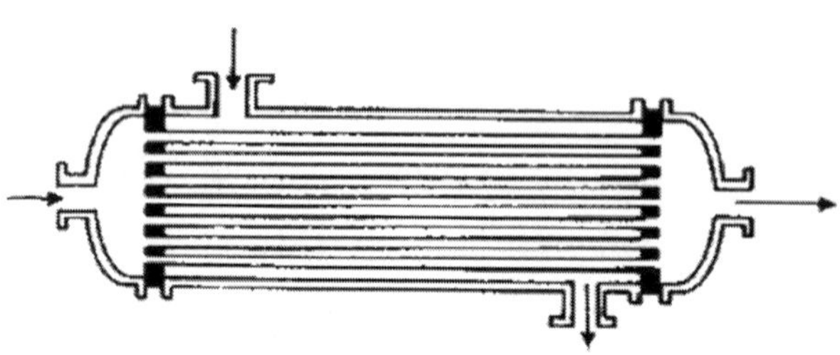

Schematic diagram of a shell and tube heat exchanger

Heat flux: The thermal intensity indicated by the amount of power per unit area.

Heat losses: Radiation heat losses from dye liquors and dyeing machines.

Heat-measuring papers: See **Thermometer papers**.

Heat of atomization: The energy required in dissociating one mole of a substance into atoms. See **Heat**.

Heat of combustion: The energy liberated when one mole of a substance burns in excess oxygen. See **Heat; Heat capacity**.

Heat of crystallization: The energy liberated when one mole of a substance crystallizes from a saturated solution of that substance.

Heat of dissociation: The energy required to dissociate one mole of a substance into its constituent elements.

Heat of formation: The energy change when one mole of a substance is formed from its elements. See **Heat**.

Heat of a liquid: The quantity of heat required to raise the temperature of 1 kg of a liquid from 0 °C to the boiling point (expressed in J resp. kJ).

Heat of neutralization: The energy liberated when one mole of an acid or base is neutralized.

Heat of reaction: The energy change that happens when molar amounts of given substances react completely. See **Heat**.

Heat of solution: The energy change that happens when one mole of a substance is dissolved in a given solvent to infinite dilution (in practice, to form a dilute solution). Bond in the hydrogen molecule could be described by quantum mechanics. Heitler extended this work to a more complicated level.

Heat of vaporization (heat of evaporation): The heat required in J/g to convert 1 g of a substance from the liquid to the gaseous state at the boiling point. The heat of vaporization for water is high, as is also that of alcohol because of the strong intermolecular forces (hydrogen bonds) in the liquid. By contrast, non-polar solvents have considerably lower heats of vaporization.

Heat recovery: Heat can be recovered from waste incinerators typically by the use of boilers to generate steam. The steam may then be used for electric power generation or for district heating systems. The equipment required for effective heat recovery can be expensive.

Heat resistance: The extent to which a material retains useful properties as measured during exposure of the material to a specified temperature and environment for a specified time.

Heat resistant: Having heat resistance.

Heat resistant finishes: Finishes which are intended to improve the resistance of textiles (strength, etc.) against the effects of high temperatures below the point of ignition. Such finishes are important for ironing cloths, paper machine felts and, possibly, tyre cords. Heat-resistant finishes for cotton include, e.g., acetylation, treatment with dicyandiamide, etc.

Heat retention (thermal retention): This is an important serviceability characteristic of a garment (Wesel triangle). It follows from the thermal conductivity coefficient of textile fibres, that all fibres, without exception, are better conductors of heat than air. Consequently, the heat retentivity of textile fabrics and similar materials is largely dependent on the respective pore volume and air content (porosity).

Heat retentivity: See **Heat retention**.

Heat sealing: A thermoplasticizing operation representing a primary stage in the bonding of non-woven, collar interlinings and the fusing of fusible interlinings, etc. The temperature and duration of thermal bonding depends on the type of resin coating and the end-use of the substrate.

Heat-set garments: Heat-set garments are made from circular-knit fabrics where the desired body shape has been permanently set on heated three-dimensional metal manikins. Collars and sleeves are welded to the main garment. This process eliminates the need for side seams and darts.

Heat setting: The effect of heat setting is greater in fibres with an increasingly hydrophobic character and is carried out on polyolefin, polyester, polyurethane, polyacrylonitrile, polyamide, triacetate, acetate and viscose fibres (the fibres are listed in order of decreasing hydrophobic character). The objectives of heat setting processes include structure homogenization and the elimination of internal tensions within the fibre resulting in reduced shrinkage, improved dimensional stability, reduced creasing propensity and reduced edge-curl in woven and knitted fabrics. To this extent, the process may be better described as thermal relaxation.

Fibre	Min T. °C	Max. T. °C	Time (in seconds)
Polyamide PA 6.6	170	210	15–40
Polyamide PA 6	160	180	15–40
Triacetate	160	180	15–40
Acrylic (PAC)	160	180–200	15–40
Elastomers	170	180–200	15–40

Heat setting of carpet yarns: Polyamide carpet yarns are set continuously in hot air at 205 °C or in equipment where heat setting is carried out in saturated steam. See also **Carpet yarns (setting)**.

Heat setting zone: The dry zone in stenter driers for the heat setting of fabrics composed of synthetic fibres or their blends with special supplementary heating.

Heat shrinkage: A change in dimensions of fabric specimen exposed to heat. A negative shrinkage is designated as growth or expansion or elongation.

Heat stability: The stability of dye dispersions/solutions to treatment temperatures in dyeing, for example especially when dyeing under HT conditions. The term must not be confused with dispersion stability since, under the action of heat, changes in the dye molecule can also occur which may result in reduced dye yields.

Heat stabilized: A term used to describe fibre or yarn heat-treated to reduce the tendency of the fibre to shrink or elongate under load at elevated temperature. See **Heat resistance.**

Heat transfer: See **Convection**.

Heat transfer coefficient: See **Thermal transmittance**. The capacity of a material (or combination) associated with heat transport to oppose the passage of heat through it with greater or lesser resistance. See **Heat transfer**. A low heat transfer coefficient (k) corresponds to good heat insulating properties. In contrast to thermal conductivity, this parameter is not related to the length but to the heat flow per unit time through unit area.

Heat transfer medium: Heat transfer liquids for circulating hot oil unit used in the textile industry and as heat transfer fluids in heat exchanger. They are used exclusively in liquid form.

Heat transfer number: A numerical expression of heat retention.

Heat transfer oil: See **Heat transfer medium**.

Heat transfer oil circulation plant: See **Circulating hot oil unit**.

Heat transfer printing: See **Transfer printing**.

Heat transfer printing with cationic dyes: Acrylic fibres, which only absorb sublimable disperse dyes at low colour yields, as well as cationic-dyeable polyester, can be transfer printed with transfer papers produced with cationic dyes in brilliant shades at acceptable colour yields. Improved colour yields are possible by means of a pretreatment with a 2% ammonium sulphate solution (3–5 min at 88–98 °C) as a result of which the sodium ions contained in the fibre are exchanged by ammonium ions. See **Transfer printing**.

Heat transmission: The heat transfer coefficient is that quantity of heat which can be transferred from a hotter to a cooler medium through a metal surface of 1 m^2 per hour per 1 °C difference in temperature. Heat transfer plays a considerable role in, e.g. the thermosol dyeing process.

Heat transmission coefficient: The heat transmission coefficient is a characteristic for the heat exchange or heat transfer between the boundary layer of a body and the adjacent medium (e.g. air). A high heat transfer coefficient corresponds to a rapid heat exchange. See **Heat transfer**.

Heat transmission resistance: See **Heat transfer coefficient**.

Heat transmission surface: See **Heating surface**.

Heat tube roller: A device to improve the heating of fabrics in thermosol dyeing ranges. Heat tube roller are installed in hot flues which transfers heat to the fabric by contact with the uniformly heated roller as well as hot air circulated through recessed nozzles in the roller surface.

Heather: A yarn that is spun using pre-dyed fibres. These fibres are blended together to give a particular look. (For example, black and white may be blended together to create a grey heathered yarn.) The term, heather, may also be used to describe the fabric made from heathered yarns.

Heather mixtures: This phrase describes a combination of greens and purples, or Scottish heather colours, found in both woollen cloth and knitting wool. The colours are often used in equal amounts so that no one colour predominates.

Heather yarn: A term describing mottled or melange-type yarns.

Heating gas: See **Natural gas**.

Heatsetting: Heat treatment of fabrics containing thermoplastic synthetic fibres. Stabilizes fabric by reducing shrinkage and distortion.

Heavy chemicals: Inorganic chemicals for industrial purposes produced in bulk quantities and often in a relatively impure state (technical grade), e.g. acids, alkalis, salts, chlorine, etc.

Heavy coating: See **Heavy duty waterproof finishes**.

Heavy duty detergent: A universal detergent suitable for all washing processes and usually for all wash temperatures.

Heavy duty detergent booster: Intensifier for exceptional circumstances (difficult washing conditions) or where the laundry is particularly soiled, e.g. more intense soiling appearing locally, such as soiled edges on collars and wristbands. Generally, pasty concentrate of suitable surfactants and compounds. See **Detergent booster**.

Heavy duty waterproof finishes: A term used for waterproof finishes applied to tarpaulins and similar articles to provide resistance to water pressure.

Heavy end: (1) See **Coarse end; Thick end.**

(2) The higher boiling fraction in distillation. **Heavy filling**: See **Coarse pick**.

Heavy filling bar: See **Filling bar (coarse)**.

Heavy goods: Name applied in Calcutta quotations to material called 'sacking' in the United States.

Heavy metals: The term "heavy metal" is understood to apply to metallic elements with a density greater than 6 g/cm3. Effluents from textile wet-processing plants can be contaminated with heavy metals such as copper, cobalt, nickel and chromium. They may be separated off by precipitation or the use of ion exchangers specifically designed for heavy metals.

Heavy pick: See **Coarse pick**.

Heavy wool: Wool that has considerable grease or dirt and will have a high shrinkage in scouring.

Hecto (Gr.): Metric prefix denoting 100 = 102, e.g. hectolitre (hl), a measure of capacity equivalent to 100 litres.

Heddle: A cord, round steel wire, or thin flat steel strip with a loop or eye near the centre through which one or more warp threads pass on the loom so that the thread movement may be controlled in weaving. The heddles are held at both ends by the harness frame. They control the weave pattern and shed as the harnesses are raised and lowered during weaving.

Heddle eyes: Each of the heddles or healds contains aperture, like the eye of a needle, through which one end may be threaded. These apertures are called heddle eyes. See **Heddle**.

Heer: Linen and jute yarn measure, equal to 600 yards.

Height (in body measurements): The vertical distance from the crown of a standing subject to the soles of the feet.

Helanca: A Swiss elastic fibre made from nylon and polyester. Fabrics made from it have two-way stretch and are very hard wearing and washable. These are used for sportswear, ski-wear, swimsuits, trousers and panels in maternity clothes.

Helianthine: Identical to Methyl orange.

Helio: A bright orange colour.

Helio-Klischograph: An electronic engraving system used in the production of gravure printing rollers for heat transfer printing. Prints from Helio-

Klischograph engravings produced by etching can be used to advantage on synthetic fibres, especially on fine woven fabrics with a smooth surface. This procedure is currently employed by print works already in possession of a Helio-klischograph and/or those who produce paper for transfer printing with such engravings, and are also engaged in the printing of textiles themselves. In addition, the process is suitable for independent printers with appropriate design selections and colorations.

Helix (wool, hair fibres): An elementary building block of the Macrofibrils in animal fibres with a screw-like structure. According to Pauling, the microfibrils are formed with a central helix in the chain molecules of wool keratin in order to combine with others by means of a right-hand twist to form a unit. The helix has a high elastic recovery and can be pulled apart (breakdown of stabilizing hydrogen bonds) and relaxed again (reformation of hydrogen bonds), a property which is referred to as reversible denaturation.

Helix angle: (a) The angle formed by the path of a ply and the major axis in a yarn or tire cord; (b) The angle between the tangent to a yarn and the minor axis of the package on which it is wound. Also called wind angle.

Helminthosporin: A maroon-coloured pigment from fungus.

Hem: An edge to a piece of textile fabric (edge reinforcement) made by folding the raw edge under and stitching it to the body of the fabric.

Hemming: The process of producing a hem.

Hemstitch: A decorative edging stitch, usually for a hem, in which the cross threads are stitched in groups. Hemstitching is used for ornament in making hems and tucks. The first step in hemstitching is the drawing of threads. Rubbing the cloth along the line of threads to be drawn will make the drawing easier if the cloth is sized. After the threads are drawn, the hem is turned and basted even with the lowest edge of the drawn space. Insert the needle into the edge of the hem and material, taking up a cluster of threads bring the thread under the needle to form a buttonhole stitch or make a simple stitch in the edge of the fold. The number of threads drawn and the number in a cluster must be determined by the coarseness or fineness of the material, the greater number being drawn and taken in fine material. There are several methods of hemstitching, but the results are about the same.

Hem (wired):The hem provides a tunnel through which wire can be threaded. The wire can be light and flexible, forming slight flutes; or it can be firmer, giving the effect of a steel hoop.

Hematine (hematein, haemateïn): See **Logwood**.

Hematoxylin (haematoxylin): See **Logwood**.

Hematron: An electronic device for the control of rope length and milling shrinkage on rotary milling machines.

Hemicellulose: Besides cellulose and lignin, hemicellulose is a component of wood and to a slight extent, also of natural cellulosic fibres. It consists of an ill-defined group of various complex polysaccharides. Hemicellulose has a lower molecular weight than cellulose (degree of polymerization of 150 or less) and lower chemical resistance. It is only sparingly soluble or insoluble in water, and is broken down by plant ferments into sugars, but does not reduce Fehling's solution.

Hemicyanine dye: In the hemicyanine dyes, one of the charged terminal atoms is part of a heterocyclic ring, and the other is a nitrogen atom directly linked to the methine chain. The lowest vinylogous hemicyanine dyes (cyclomethine dyes) do not contain a methine chain, and the donor and acceptor groups are in the same ring system.

Hemp: Flax and hemp are bast fibres. It is derived from the stem of the hemp plant (*cannabis sativa*) 50–250 cm tall, similar to jute, accompanied by a fair amount of woody matter. When spun, it is rather like flax but thicker and coarser. It is a very strong, lustrous, and very durable but harsh fibre and is used in the manufacture of carpets, rugs, ropes, etc., but has limited use because bleaching is difficult. The best grades are fine and white; used chiefly for cordage, twine and sail cloth. The following commercial varieties are cultivated: The common hemp, the Bologne hemp (known also the piemontese or great hemp), the Chinese hemp (called Japanese hemp in California), the Smyrna hemp, the small hemp and the Kentucky hemp. The Italian hemp is the best, with the following principal grades: Gorgiola (G), Gorgiola Bolognese (GB), primo cordaggio extra (PICE), primo cordaggio (PC), primo basso (PB), secondo basso (SB), Napoli extrissimo (N), secondo cordaggio (SC), terzio basso (TB), quarto basso (QB), and strappature. In Russia, the mark for the longest and best hemp is RH, for shorter SF1SPOH, for the shortest (.pass hemp) SFSPPH. A good grade of Polish hemp is marked G.FSPRH. The current qualities in Petrograd are: Clean, out shot and half clean. The hemp marketed through Konigsberg (Germany) is classed as clean, cut and schicking. The numbering for the fine hemp yarns is the same as for linen yarn; for rope yarn the number gives the number of threads required for one of the three strands found in a rope of three-inch circumference.

Hemp tow: The short fibre stock of varying lengths removed in breaking, scutching, hackling, or combing. It is often very gnarled, matted and very irregular.

Hemp line: Fine, straight, parallel hemp strands produced by hackling long hemp fibres.

Henequen: A vegetable fibre, closely resembling sisal, belonging to the Hard fibres group. Henequen is produced from the leaves of the Mexican plant (*agave fourcroydes*).

Henna: Reddish-brown colour.

Hepepetwan: Rich and double-faced Chinese silk satin.

Henrietta: Fine, diagonal, twilled dress fabric, made with silk warp and fine worsted filling which makes it resemble cashmere cloth. Once in great demand, but it has not been very much in evidence for years.

Heart: Very durable all-wool rugs made in Persia and Afghanistan, the medium long pile is tied in Ghiordes knot. The design is of floral and flash patterns or of a centre medallion. The colours are blue, red and yellow.

Hercules braid: (1) A flat braid made with seven threads, each thread passing alternately under and over three threads; (2) Very wide flat braid, made cxf mohair, wool or silk; used for trimming on dresses and uniforms.

Hereke carpets: Especially valuable knotted carpets from the time of the last Turkish sultans (beginning of the 19th to the 20th Century). The carpets are woven with a wool or silk pile on a fine cotton or silk ground fabric. 3, 00,000–4,00,000 Turkish knots or over 1,000,000 Persian knots/m2. The design consists of tightly arranged strewn flowers as well as, to some extent, motifs inspired by the French rococo period.

Herez rugs: Fine Persian rugs made with cotton web and close wool pile tied in Ghiordes knot. The design is usually a centre medallion with straight outlines and floral patterns in blue and various reds. The border is in light colour.

Heris: Persian camel's hair rug, made with old patterns in dark brown.

Herlong: A late maturing commercial variety of prolific cotton from Alabama, Georgia, etc., the staple, measuring 22–25 millimetres, forms medium size, round bolls; the lint yield is 30–32 percent.

Hernarvi: French dress goods, similar to Grenadine (only lighter), made of silk and wool.

Herringbone: (1) A combination of twill weaves in which the direction of twill is so reversed (usually by drafting) as to produce stripes that resemble herring bones.

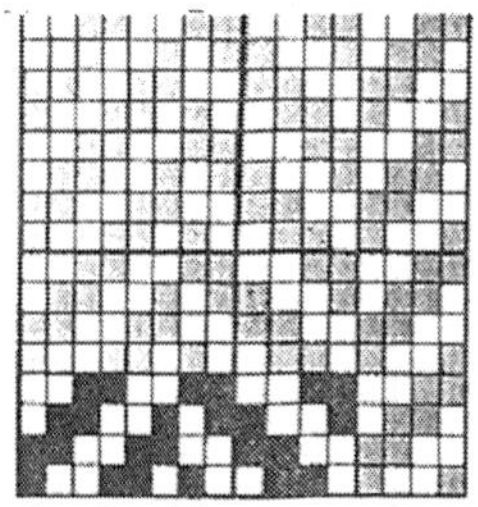

Patter Draft

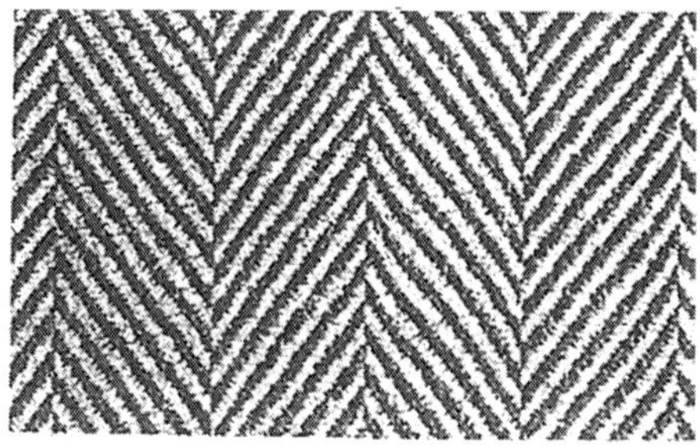

Herringbone

(2) The name is often applied to any woollen cloth made in a herringbone design, i.e. with a broken twill effect in two colours producing an up and down striped effect. Any fibres can be used for making a herringbone cloth.

Herringbone Stitch: Bring the needle out on the lower line at the left side and insert on the upper line a little to the right, taking a small stitch to the left with the thread below the needle. Next, insert the needle on the lower line a little to the right and take a small stitch to the left with the thread above the needle. These two movements are worked throughout. For the best effect, the fabric lifted by the needle and the spaces between the stitches should be of equal size. This stitch can be laced with a matching or contrasting thread. Use a round pointed needle for lacing and do not pick up any of the fabric. Herringbone stitch may also be worked on even weave fabric. See **Plaited stitch**.

Herringbone weave: Herringbone is executed in a broken twill weave with alternating twill direction in a balanced chevron pattern that resembles the skeleton of a herring.

Herringbone (pointed 2/2 straight or symmetrical): A 2/2 twill in which symmetrical reversals take place in the weft direction is called a symmetrical or pointed 2/2 herringbone. This structure can be woven on a straight draw on four shafts at any time simply by changing the direction of the treadling progression.

Hessian: A coarse, heavy and plain-weave fabric of jute or similar yarn. Its natural colour is brownish. It is used for wrappings, wall hangings, etc.

Hessians: Name applied in United Kingdom and India, and on the Continent to material called 'burlap' in the United States.

Hetero bi-reactive dyes: Depending on their reaction mechanism during reaction with the fibre, reactive dyes with a single reactive group are subdivided into two types, i.e.: (a) substitution type, (b) addition type.

Bifunctional reactive dyes such as C.I. Reactive Black 5 (20 505) can also react twice with the fibre (bi-reactive, bifunctional). Reactive dyes with two different reactive groups, which can react according to both the addition as well as the substitution mechanism, are heterobire active. They react both in the sense of a nucleophilic substitution as well as in the sense of a nucleophilic addition at a double bond.

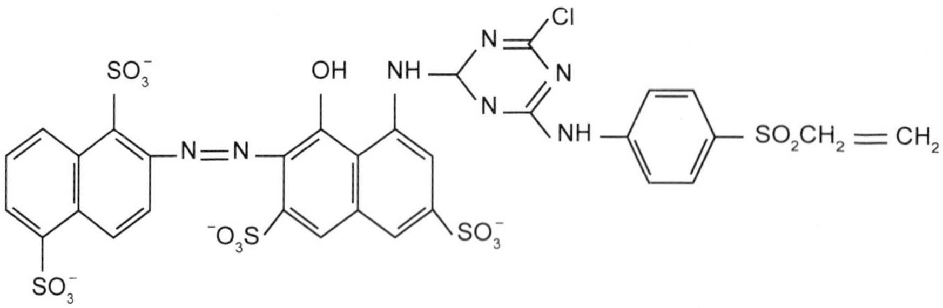

Vinyl sulphone/monochlorotriazine

Hetero yarns: Core-spun yarns, i.e. yarns produced from two filaments combined in spinning which, e.g. contract differently in thermal treatments to give a texturing effect (e.g. heteroyarns from triacetate and polyamide). Heteroyarns are not bicomponent fibres.

Heterofil fibres: Core-sheath bicomponent fibres in which a core of polyamide 6.6 is surrounded by a sheath of polyamide 6. Due to their different melting points, nonwovens and needlefelt floorcoverings can be permanently bonded by a simple thermal treatment (the fibres being bonded at cross-over points through sintering of the outer sheath) without the need for a binder.

Heterofilament: See **Heterofil fibres**.

Heterogeneous: Composed of dissimilar, unrelated or differing parts or elements; e.g. from a dyeing point of view, the wool fibre has heterogeneous properties since, in contrast to the cortical cells with high dye affinity, the outer scale cells have a very much lower affinity due to their different composition.

Heteropolymerisation: See **Polymerization**.

Heteroyarns: Core-spun yarns, i.e. yarns produced from two filaments combined in spinning which, e.g. contract differently in thermal treatments to give a texturing effect (e.g. heteroyarns from triacetate and polyamide). Heteroyarns are not bicomponent fibres.

Hexachlorophene (dihydroxyhexachlorodiphenylmethane): White, free-flowing powder; odourless; insoluble in water; soluble in alcohol, acetone, dilute alkalis and soaps. It does not cause skin irritation and is a powerful bactericide. Prevents bacterial decomposition of perspiration.

Hexachlorophene test (colorimetric method): A colour which can be measured in a colorimeter is produced from a solution of Hexachlorophene in methanol with an ammoniacal-alcoholic solution of copper chloride. The test is likewise suitable for the detection of pentachlorophenol and trichlorophenol.

Hexafluoropropene: Teflon.

Hexalin(hexahydrophenol, cyclohexanol): Used in soap making for the incorporation of grease-solubilizing solvents; spotting agent; thinner for paints, lacquers and varnishes, etc.

Hexamethylenediamine: 1,6-diaminohexane, [$H_2N(CH_2)6NH_2$]. It is used in the polymerization reaction to form nylon 66.

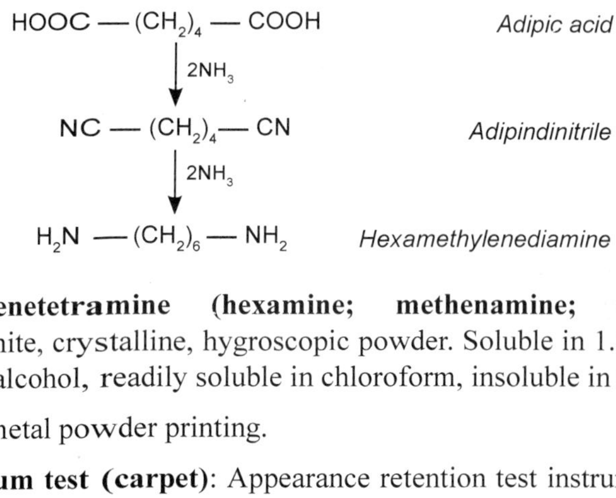

Hexamethylenetetramine (hexamine; methenamine; urotropine): $(CH_2)_6N_4$. White, crystalline, hygroscopic powder. Soluble in 1.5 parts water and 10 parts alcohol, readily soluble in chloroform, insoluble in ether.

Uses – e.g., metal powder printing.

Hexapod drum test (carpet): Appearance retention test instrument using a hexapod in a drum.

Hexoses: Carbohydrates with six carbon atoms –sugars; carbohydrates.

Hi Kanakin: In Japan, cotton cambric dyed Turkey red.

Hiapu: Light Chinese cloth, made of ramie.

Hickory: A dark grey-brown colour.

Hickory: A very durable stout but pliable cotton trousering and shirting, made with coloured stripes in the warp and white filling, woven in warp twill weave.

Hickory cloth: This cloth resembles ticking, but is lighter in weight and not so firmly woven. It is made from cotton or polyester/cotton. It is used for protective clothing, overalls, etc.

High-bulk yarn: See **Bulk yarn (textured).**

High-bust girth: See **Chest girth**.

High-hip girth (in body measurements): The circumference of the body at the point approximately 7.5 cm (3 in) below the waist and parallel to the floor.

High-low pile: A pile construction characterized by the presence of two or more pile heights. High-low pile carpets sometimes combine looped and cut-surface yarns.

High-shrink fibres: High shrink fibres are produced to increase the bulkiness of the fibre. Physically modified types of synthetic fibres which undergo shock shrinkage (by approximately 40–45%) in boiling water or hot air (heat-setting temperature 185 °C). Normal acrylic and polyester fibres are used in blends with pre-shrunk fibres for the production of high bulk yarns. See **Textured yarns**.

High-shrink staple: Staple with a higher degree of potential shrinkage than regular staple of the same generic fibre. When blended with regular staple and treated (in yarn or fabric form) to induce shrinkage, it produces a high degree of bulk in the product.

High-speed spinning: The demands for higher quality, increased productivity and lower production costs which began in the 1970's, led to modifications in spinning technologies. In the classic melt-spinning process, filament extrusion speeds of approximately 1000 m/min are employed. The desired physical and textile properties are achieved in subsequent drawing. Initially, the combined spin-draw-wind process represented an alternative to this technology. Filament yarns produced by this method gave similar properties to those produced by the classic technologies.

High-tech fibres: Due to the irregularities, partial crystallinity and the molecular weights, the common synthetic filaments and fibres with breaking strengths of 3–10 cN/dtex and emodulus of approximately 35–130 cN/dtex, which are presently used in great quantities, achieve only approximately 10– 25% of the theoretically possible strength or E-modulus. By using materials with an extremely high molecular weight and drawing the filaments to the limits of possibility, a much closer approach can be made to the theoretically possible strengths and E-modulus. In the case of polyethylene, molecular weights of 2.10^6 with filament strengths of 30–35 cN/dtex have been achieved. Compared to the classical synthetic fibres, these so-called "high-tech fibres" are of increasing importance for technical textiles (geotextiles, filter materials, composites, protective clothing, etc.). The properties of these fibres (with differences in relative importance) include high tensile strength, high thermal stability, low combustibility, high chemical resistance and good electrical resistance.

High capacity yarn packages: Yarn packages with a maximum diameter of 300 mm weighing up to 3.6 kg which are used particularly for wool and acrylic yarns. They offer advantages of shorter liquor ratios, e.g. down to 4.5:1, and a doubling of the dyeing machine capacity.

High concentration bleach: Continuous bleach process in which high quantities of chemicals are applied by impregnation resp. padding, and bleaching takes place by a subsequent dwell stage or steaming.

High density: A term to describe a material with heavier than normal weight per unit volume. See **Density**.

High density polyethylene: H. D. polythene, HDPE. A thermoplastic plastic used extensively since the 1970's. It has a density of 0.955 g/cc which is greater than medium or low density polyethylene. It is light and flexible and it can suffer some surface damage without rupturing and affecting performance. It is resistant to corrosion and easy to weld together into long lengths. It is used for water pipes or landfill liners. It is also present in municipal waste because HDPE is used for bottles for household chemicals. It is also used for sewers or in the rehabilitation of sewers. See **Polyethylene**.

High frequency current: An alternating current with over 1,00,000 oscillations (frequencies) per second. Whilst the tension (volt) can certainly be high at these frequencies, the current (ampere) is greatly reduced.

High frequency driers: High frequency driers employ high frequency electromagnetic waves at frequencies from v = 105–1012 Hz with corresponding wavelengths from λ= 3 km (radio waves) up to 300 μm (Hertzian waves). However, officially designated frequencies

have been allocated for industrial applications in order to avoid possible interference with radio communications. Frequencies approved by international agreements for industrial use are:

13.56 MHz ± 05%

27.12 MHz ± 0.60%

40.68 MHz ± 0.04%

Most high frequency driers work with voltages of 10,000–30,000 V at an operating frequency of 13.56 or 27.12 MHz.

High frequency drying (HF-drying, RF-drying, dielectric drying): Drying takes place between two capacitor plates supplied with a constant high frequency AC voltage. The textile material is heated and dried by high frequency electrical oscillations whereby the depth of penetration is independent of material thickness so that HF-drying is particularly suitable for drying yarn packages. High frequency drying can also be used for warp sheets, tops, loose stock, hanks and garments.

High frequency dye fixation: Dye fixation with the aid of high frequency energy. For high frequency applications, the form of the electrodes is of decisive importance.

High grade synthetic resins: Pure synthetic resins free of fillers. Mainly phenolic plastics, used for a variety of applications.

High loft: General term for a fibre structure containing more air than fibre. Specifically, a lofty, low-density, non-woven structure that is used for applications such as fibrefill, insulation, health care, personal protection and cleaning material.

High modulus: A term that refers to a material with a higher than normal resistance to deformation. Also see **Modulos.**

High modulus (aramid): Those aramid yarns with an initial modulus of at least 400 gf/den (35 N/tex).

High modulus fibres: See **High wet modulus fibres (HWM fibres).**

High pH finish: A finish, basic in nature rather than acid or neutral, that is applied to yarn or fibre.

High pressure atomizers: Atomisers high pressure atomizers are used in sanforizing (zero-zero) machines to moisten the fabric which acts as a lubricant during the shrinking operation. These atomisers employ centrifugal force high performance jets for lubricant emulsions subjected to liquid pressures between 50–150 bar and higher produced by high pressure pumps.

High pressure drier: These driers were very common for drying yarn packages in which the packages, supported on a carrier frame, are hydro-extracted/dried with high velocity hot air. Disadvantage of these driers are that they are high energy consuming and presently other type of driers like RF driers are more common.

High pressure dyeing: See **High temperature dyeing**.

High pressure kier: The prototype of modem high pressure kier with multi-tubular heater is often referred to as Walsh's kier and considered to be the best. A 2 ton capacity kier is most common which is about 9 inches high and 6.5 inches in diameter and mounted on R.C.C. columns. The kier is generally made for a maximum likely working press sure 40 lb/in^2 (2.812 kg/cm^2) at

141~ and is composed of mild steel plates with countersunk rivet heads inside to give a smooth surface. The goods in rope form are piled either manually or by mechanical plaiting on the perforated false bottom. The cloth can be saturated with alkaline solutions containing caustic soda (2–3%), sodium silicate (0.75–1.0%), sodium carbonate (0.5–1.0%) and wetting agent (0.1%) during piling. Sufficient scouring liquor is then admitted to the kier from below to sweep out entrapped air from the goods to avoid the formation of oxycellulose. The lid is then closed, but the air valve is kept open. Steam is turned on and liquor is circulated by centrifugal pump (C) coupled to an electric motor. High pressure kier operates at about 130°C (2.109 kg/cm^2) for about 8 to 10 h with a liquor ratio between 5:1 and 6:1. After scouring, the drain cock is opened at reduced pressure (5 to 6 lb/in^2) and equivalent amount of hot water is admitted from the top. After circulating cold water, the goods are removed and washed in a rope washing machine. The liquor is withdrawn from the bottom of the kier and forced through the heater and spread over the top of the load by the distributor.

High pressure steam: Steam at a pressure greater than 2 bar. Both saturated steam and superheated steam can be handled in this form. The difference between the two is given by the respective temperature.

High pressure steamer: See **HT pressure steamer**.

High rise: Term applied to trousers cut high in the waist.

High shrinkage yarns: High-bulk yarns containing 30–40% High-shrink fibres, also used in blends with wool. See **Textured yarns**.

High temperature amylases: These are enzymes used for desizing at higher temperatures. These starch-degrading enzymes obtained by microbiological processes can be active at working temperatures of 100 °C and above, thereby requiring extremely short reaction times.

High temperature contact process: A process used for surface bonding of fibrous webs, coated and milled or needled felts, back-coated and tufted carpets on a contact drying machine. The process is based on the principle of direct contact drying between the coated side of the substrate and the cylinder surface at high temperature.

High temperature dyeing (HT dyeing): The dyeing of textiles at temperatures above 100°C in pressurized equipment high temperature dyeing machines. The chief advantage is prevention of pump cavitation and consequent secure liquor circulation. High temperature dyeing has specific advantages for the dyeing of various fibres, especially the low-swelling or non-swelling synthetic fibres without reactive groups, i.e. above all polyester and triacetate fibres and their blends with other fibres.

High temperature dyeing machines (HT dyeing machines): These machines offer the possibility of a closed dyeing system suitable for all dyeing temperatures under pressure (besides temperatures near the boil below 100°C) to achieve the objectives of high temperature dyeing by raising the boiling point of the dye liquor and eliminating pump cavitation problems.

High temperature fibres: Organic and inorganic fibres such as, ceramic fibres, glass fibres, aluminium silicates and metal fibres. A collective term especially for fire-resistant fibres and heat-resistant fibres whose strength only falls to zero under normal atmospheric conditions at approximately 3000°C.

High temperature incineration: Incineration at temperatures in excess of 1100°C. It is used for incineration of hazardous wastes in a two stage incinerator. The waste is fed to the primary chamber and the gases produced go to a secondary combustion chamber which is in excess of 1100 °C. The gas stream is retained for a minimum of one second in the second chamber before being extensively treated in an air pollution control plant.

High temperature light fastness tests: In addition to the usual requirements (surface uniformity, shade conformity, rubbing fastness, fibre comfort), extremely high requirements have now been placed on the high temperature light fastness of textiles for automobile interiors. It has been found that the ***grey scale*** is more suitable as an evaluation standard than the co-exposed blue standards.

High tenacity: High breaking strength. See **High tenacity fibres**.

High tenacity fibre: A manmade fibre either (a) belonging to a generic class of fibres having exceptional breaking strength; or (b) having a breaking strength significantly greater than the average strength of other (regular tenacity) fibres in the same generic class and of equivalent linear densities.

High tower: A commercial variety of cotton from Alabama, having a medium long staple.

High visibility fabrics: Fabrics which contain fluorescent materials in order to make the wearer visible in dim and dark lights. These fabrics have the ability to reflect on-coming lights which cause them to glow in the dark.

High wet modulus fibres (HWM fibres): Modified regenerated cellulosic fibres of the modal fibres group. The strength and wet modulus of these fibres are similar to cotton. They have lower strength but higher extension than polynosic fibres. Used alone or in blends for easy-care textiles, etc.

High wet pickup: With conventional padding techniques, however, it is difficult to exceed a maximum pick-up of 70–90%. Higher applications of liquor are described as high wet pick-up (wet on dry) or addition impregnation (wet on wet). There are special methods machines available to make wet pick up to 130-160%.

Highly sulphonated oils: See **sulphonated surfactants**.

Hilliard: A commercial variety of upland cotton, the staple measuring about 23–24 millimetres; the yield is 34–36 percent.

Himalaya: Like shantung in appearance and weight, but made from slub cotton yarns. Easier to sew than silk shantung.

Himation: Large and rectangular woollen outer robe for men and women.

Hinrop: Indian silk fabric, made with flower patterns.

Hip (in anatomy): The laterally projecting region formed by the lateral parts of the pelvis and the upper part of the femur together with the flesh covering them.

Hip girth (in body measurements): The maximum circumference of the body at the level of maximum prominence of the buttocks.

Hirschburger process: A space dyeing method for yarn sheets. The working speed of the winding head is between 600 and 1000 m per min, and can be varied in three stages. The dwell time in the steamer is 1–5 min and depends on the kind of winding (close winding or widely spaced winding), i.e. it depends on the speed of winding and transport as well as the thickness of the material.

Histogram: A graphical representation of a frequency distribution characterized by (a) values or groups of values of a property plotted on one axis and (b) columns perpendicular to that axis with areas proportional to the count of the frequency with which each value or group of value occurs.

Hitch back: See **Draw-back**, **Sticker**.

HL process: A process for dry cleaning establishments in which a dry cleaning detergent/water emulsion may be added directly to the cleaning liquor with a dry cleaning detergent concentrate.

HM process (H = Habis Textil AG, M = Monforts): A plant for the curing of finishes and fixation of prints on woven fabrics in textile finishing. It combines a rapid heat-up phase in hot air and a short dwell time in a steam atmosphere.

Hoar: Greyish-white colour, especially as regards frost.

Hobble skirt: Very narrow and long skirt; very difficult to walk in.

Hocirunck: A plain woven cotton fabric produced in East Africa, dyed in cinnamon brawn; used for outer garments by the native men. Similar to ***Khudurangi***.

Hockle (in rope): A strand kink in a rope causing yarn displacement in the strand resulting in rope deformation and damage.

Hog: Wool yielded by one-year old sheep which has not been shorn previously.

Hogget wool: Hogget wool comes from sheep that is 12–14 months old and that have not been previously shorn. The fibre is fine, soft resilient and mature, and has tapered ends. Hogget wool is a very desirable grade of wool and because of its strength, it is used primarily for the warp yarns of fabrics.

Holbein embroidery: Outline embroidery for table cloths, towels, etc.; both sides made alike.

Holbein rugs: Oriental rugs or carpets mainly from the Bergama area in Western Turkey and named after the painter Hans Holbein the Younger, who portrayed a few examples in his pictures.

Hole etcher: A variant of the burn-out print style. The print pastes containing an acidic component are applied to 100% cotton in order to burn holes out of the fabric in specific areas corresponding to the printed design. Fraying of the burned-out holes in the fabric during washing is avoided by selecting tightly woven fabrics for this style of printing. The problem does not arise if the burned-out areas are surrounded by a border printed with pigment colours since, in this case, the polymer film of the pigment binder prevents frayed edges by bonding the exposed fibres.

Hole (in fabric): An imperfection where one or more yarns are sufficiently damaged to create an aperture.

Hole (in milling): A technical term for a fabric defect (hole) caused by sudden high pressure or impact during fulling or milling (rotary cylinder, perforated roller).

Hole spacing (on a button): The distance from the centre of one hole to another.

Holes (tow): In tow opening processes, partial or complete filament breakage within a confined spread of tow, usually circular or oval in shape. Not to be confused with splitting or partial crimp deregistration, which are linear.

Holi: In Arabian countries, a woollen fabric, similar to Baracan; used for men's clothing. See also **Baracan**.

Holland: (1) General term for a great variety of light weight and coloured cotton goods made in Greece.

(2) Finished with a stiff finish, plain woven cloth, made from linen or cotton yarn. Medium weight fabric mainly used in upholstery for an under covering, as it is from and hard wearing.

(3) A canvas interfacing.

Holland cloth: A completely filled woven fabric having a smooth gloss finish on both sides, used as separating medium for sheeted rubber compounds.

Holland finish(for cotton): A glazed or unglazed finish consisting of oil and a filling material applied to cotton material to make them opaque or semi-opaque, resembling a beating linen fabric formerly known as 'Hollands'.

Hollandas: Coarse, starched cotton fabrics in the Philippines, made with black, blue or red warp stripes on white ground and white filling.

Hollands: In Cuba, a finely striped linen; used for children's dresses and for summer trousers.

Hollie point: Old needle-point church lace with scriptural patterns. See **Holy Lace**.

Hollow braid (in rope): A braided construction of either plain or twill braid, having an empty centre.

Hollow cop (tubular cop): A Cop of yarn with no internal support, i.e. a form of yarn package consisting of wound yarn only without any bobbin or tube. The yarn is unwound from the inside of the hollow cop.

Hollow cut: Cotton corduroy woven first with even pile, the runs between the ribs are hollowed out with a shearing machine. It is also made by holding the knife at various angles while cutting the corduroy. Also known as velvet cord.

Hollow filament fibres: Manufactured and continuous filament fibres, having voids created by introduction of air or other gases in the polymer solution or by melt spinning through specially designed spinnerets.

Hollow spindle spinning: Also called wrap spinning. A system of yarn formation in which the feed stock (sliver, roving) is drafted and the drafted non-twisted strand is wrapped with a yarn as it passes through a rotating hollow spindle. The binder or wrapping yarn is mounted on the hollow spindle and is unwound and wrapped around the core because of the rotation of the spindle. This technique may be used for producing a range of warp spun yarns or fancy yarns.

Hollow squeegee: A squeegee system used on flat-bed screen printing machines consisting of a pair of parallel rubber squeegee blades mounted in a single support. The print paste held between the two blades is transported across the printing screen by the rear blade only during each pass, the leading blade being raised at the same time. Thus, the leading blade for the first pass becomes the rear blade for the subsequent pass. See **Double squeegee**.

Hollow filament yarn: See **Hollow filament fibres**.

Holly green: Olive-green colour.

Homage: A deep ultramarine blue.

Homard: The pinkish-red of the lobster.

Home: A Highland tartan composed of dark blue and black stripes, split by red and green lines.

Home furnishings: Together with household textiles, home furnishings account for approximately 30% of all textile fibre consumption. Apart from woven articles such as upholstery materials, carpets and bed covers, various special techniques are also used for the manufacture and finishing of home furnishings.

Home laundering: A process by which textile products or parts of them may be washed, bleached, dried and pressed by any customary method designed for use in a residence, or non-professional use.

Homespun: (1) Plain-weave coarse fabrics originally produced in the British Isles from hand-spun Cheviot wool, generally undyed and characterized by their uneven quality which was due to the hand weaving and spinning. Today, homespun is frequently used as a general term for loosely woven fabrics produced from very coarse, rough-haired, mottled or even slubby woollen yarns mainly in plain-weave constructions, often with a light Melton finish. Used for lightweight suits, men's jackets, costumes and coats.

(2) This fabric was hand woven in many countries; but now it is made on power looms with plain weave, made of wool and having a coarse appearance. Hard wearing serviceable cloth made with cotton or wool in plain weave. Coarse and rugged yarn is used. Originally an undyed woollen cloth that is spun into yarn and woven at home by peasants and country folk worldwide. It has substantial appearance and serviceable qualities. Made with irregular and slightly twisted uneven yarns. It has a spongy feel with a hand-loomed tweedy appearance. Genuine homespun is produced in a very limited quantity and much powerloom cloth is sold as genuine homespun. Many qualities made – the best is an ideal rough-and-ready type of cloth.

Uses – Coats, suits, separates and sportswear.

Homiak: Good grade of home reeled silk in Central Asia. It is exported to other countries.

Homienchow: Fancy silk fabric in plain weave, made of spun silk in China and is about 23 inches wide.

Homogeneous: Composed of similar or identical parts or elements. A composition in which the components are of the same kind. A homogeneous system consists of a single phase, e.g. a common solution such as a salt or sugar solution.

Homogenization: Intimate mixing, e.g. of importance in the preparation of particularly fine emulsions.

Homogenizers: High-speed mixers in which coarse and polydisperse emulsions are transformed into almost monodisperse systems by subjecting the liquid to high shear forces.

Homopolymers: Natural or synthetic polymers derived from a single type of building block (monomer) in contrast to copolymers. Poly (styrene), for example, is composed only of styrene monomers, and is therefore is classified as a homopolymer.

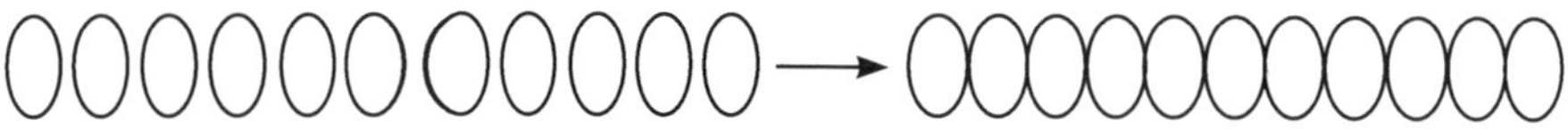

Beads of same kind (representing monomers of same chemical) — Representing homo-polymer molecule

Honan: Silk fabric made from silk yarn that is obtained from wild silkworms raised in the Honan area of China, which is the only type of wild silk that dyes evenly. The fabric is light and rough textured. Nowadays made from silk or from manmade synthetics in plain weave. It is the best grade of wild silk. Very similar to "pongee", but finer. It should not be fitted too tightly. *Uses* – Dresses, ensembles, blouses and lingerie.

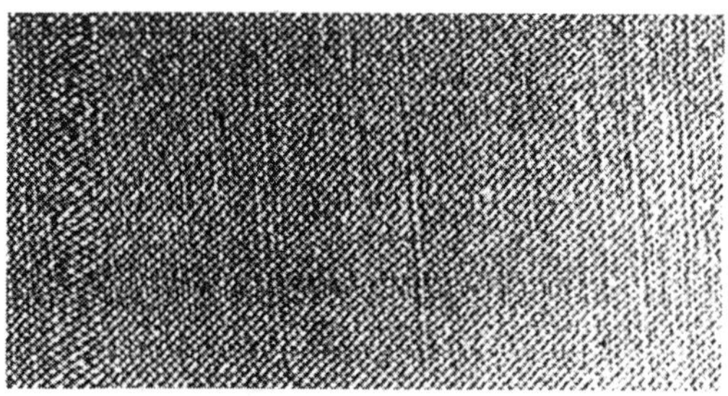

Honan/Henan fabric (silk): Fabric of tussah yarn woven on a hand loom warp of raw geige or tussah greige and weft of either raw tussah greige or tussah tram. The predominant general impression is of linen, as relatively strong warp and weft bars appear. This fabric is also often called raw silk. In accordance with new designation, it is also called Honan or Henan.

Honey: Greyish-yellow colour; also 'honey-coloured'.

Honeycomb fabric: Cotton fabric in which the warp and weft threads form ridges and hollows to give a cellular appearance which may be evident on one or other or both faces of the fabric. The honeycomb weave is achieved with groups of short running warp and weft floats. The structure makes the fabric more absorbent than other fabrics. It can be dyed, printed or with woven patterns. Once made only from worsted yarn, it is now made with cotton and is used for children's clothes, dresses, kitchen curtains, table clothes, and comes in plain or printed forms.

Honeycomb: (a) Hexagonal mesh found in laces; (b) Patterns similar to cells on cotton or other fabrics, made with fine warp and much coarser filling, producing reversible effect; used on towels; also on some dress goods.

Honeycomb canvas: Bleached cotton canvas, made in square honeycomb effect; used for embroidery.

Honeycomb backing (embossed foam, structured foam): Back coating of carpets with foam having an embossed honeycomb-like pattern.

Honeycomb reseau: In laces, a ground composed of diamond shaped squares.

Honeycomb screen: An arrangement with a honeycomb grid used in the transfer of designs on to rotary printing screens during exposure to a light source in which oblique incident rays from all directions are, for the most part, screened out by the honeycomb grid so that they have no effect on the copying process.

Honey dew: (1) A yellowish-pink; an orangey-pink colour.

(2) A descriptive term for secretions on cotton hairs caused by whitefly or aphids during cotton growing. The sticky deposit is composed of various sugars. The spread and intensity of honeydew infestation almost always has a direct correlation with the use of various insecticides.

Honeysuckle: Pale pinkish-yellow; a yellowish-brown colour.

Honghow silk: Piece dyed silk foulard, made in China. It is about 29 inches wide and used for wrapping.

Hongkong seam finish: A finish for the raw edges of a plain seam in which a binding fabric (bias- 1.0–1.5 in or 2.5–3.8 cm. strip from a lightweight fabric like lining or underlining) encloses the raw edge of each seam allowance; the binding fabric has one raw edge cut and the other raw edge exposed. Stitch one edge of the bias strip to the seam allowance, right side together and stitch 6 mm (¼ in) from the cut edge. Then fold the bias strip over the raw edge to the underside of the seam allowance. Stitch in the crease of the first row of stitching to secure the back edge of the bias strip to the seam allowance. Trim away the extra bias.

Hongroise: Plain French serge dress goods of eight leaves and four picks in a repeat.

Honiton lace: A handmade lace that originated in Devonshire. It is characterized by the design of flowers and leaves on a very fine mesh ground.

Hook and loop fastener: A touch fastener, comprised of two flexible mating strips, the surface of one mating strip being covered with tiny, stiff protrusions shaped like hooks which engage the other mating strip which is covered with pliable loops.

Hook: A device which acts as a loop taker and carries the needle thread around the bobbin containing the lower thread in order to form the lock stitch.

Hook and loop fastener: A touch fastener comprised of two flexible mating strips, the surface of the mating strip is covered with tiny, stiff protrusions shaped like hooks which engages the other mating strip which is covered with pliable loops.

Hook reed: See **Reed**.

Hooked bow: A fabric condition in which the filling or course yarns are in the proper position for most of the fabric width, but are pulled out of alignment at one side of the fabric.

Hop pocketing: Jute bagging which is used for hop bags and is similar to tarpaulin.

Hopsack: Originally hopsack was a coarse double yarn sacking, made from jute or hemp, and used as sacking by hopgrowers. Now the name has been given to a plain-weave suiting weight cloth made by using two yarns in each direction. It can be made in synthetic or natural fibres; but if made of synthetic, the fibres are usually bulked to produce the traditional cloth. It may be used for curtain and furnishing.

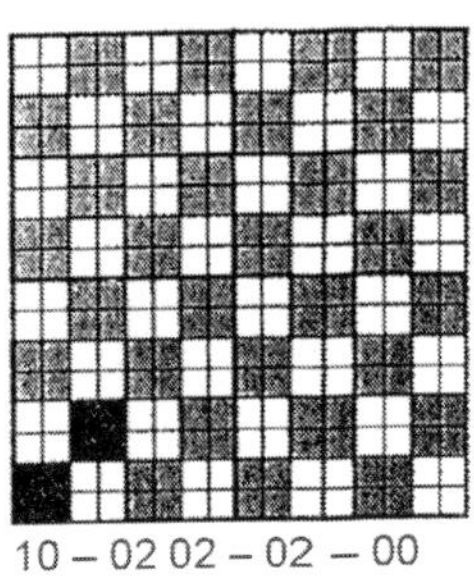

Pattern Draft

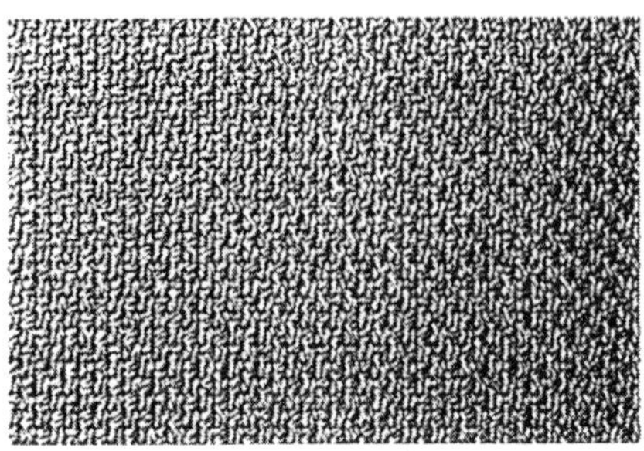

Hopesack

Hopsack weave: Also called mat weave, matt weave and panama weave. A modification of a plain weave in which two or more warp yarns weave together as one and two or more weft yarns weave together as one.

Note – The basic hopsack weaves may be modified in a number of ways such as: (a) by introducing additional interlacing to give a firmer cloth, e.g. stitched hopsack weave; (b) by arranging the interlacing in diagonal lines, e.g. twilled hopsack weave.

Hopsacking: A plain woven canvas dress fabric of wool.

Hop-sacking: The English name for American basket weave. In woollen and worsted trade, hopsacking refers to a suiting fabric made with the 2/2 basket weave. It comes in wide range of textures and weights. A characteristic hopsacking material may be made of 2/8 worsted warp and filling. Instead of coarse worsted, 3-run woollen yarn may be used; a characteristic texture is 24 × 24. Also, cotton warp and a soft spun woollen filling may be used for hopsacking.

Horauf process: A space dyeing process for yarn sheets in which yarns from cross-wound packages/cops are passed through a dye liquor applicator. The amount of dye applied to the yarn can then be reduced to the desired liquor pick-up by means of a patented procedure (which is similar to the OPI method). The yarn is then conducted to a winding machine (Buddecke system) to obtain a continuous, hank-like wound package consisting of individual layers of yarn held by four endless conveyor bands. These layers are transported along a bar over a distance of 2.7 m through a steamer in a tension-free state to fix the dyes. The winding rate can be varied between 400 and 1200 rpm. The dwell time in the steamer is 7.25 min for yarn which is generally closely packed (16 threads per cm), the winding speed being 600 rpm. The dwell time is only 3.6 min if the yarn is spaced wide apart during winding (4 threads per cm) at a winding speed of 300 rpm. After steaming, the yarn is drawn off from the carrier and wound on bobbins.

Horizon blue: A light greenish-blue colour.

Horizontal emerizing machine: For woven fabrics with emerizing rolls and a dust bag filter for dust extraction from the machine. Fabric feed and delivery via batch rolls.

Horizontal flame test: See **Flammability tests**.

Horizontal line: See **Ring**.

Horizontal migration: Selective horizontal migration is due to the different migration rates of the components in a dye combination. It causes dichroism which results in an irregular and uneven appearance of the dyed fabric.

Horizontal organisations: Companies (often small but not always) specializing in one production process such as knitting, dyeing, printing and finishing.

Horizontal open-width washing machines: This design of washing machine has wash compartments in which the guide rollers are arranged in vertical banks between which the fabric passes upwards in the open-width state in a series of almost horizontal loops with liquor flowing from top to bottom.

Horizontal pad: A horizontal pad with a controlled gap between the rollers is a variation of knife coating. Banks of foam can be placed in the crevice between the rollers and the fabric so when the fabric passes downward, the roller will act as the knife allowing a measured amount of foam to pass through. This method applies foam to both sides of the fabric at the same time.

Horizontal padder (horizontal foulard): A padder, without a liquor trough, in which fabric is padded through a wedge of liquor held in the nip between two horizontally arranged padder rolls.

Horizontal stenter: A stenter with an air-cushion nozzle system for the drying and heat-setting of textile fabrics. The machine can be incorporated into an integrated finishing line for the finishing of woven and knits and is provided with a double batching system on giant batch rolls at the delivery end for fully continuous operation.

Horizontal two-bowl padder: A padder in which piece goods are run vertically downwards through the nip of a horizontal padder containing the liquor in between the two bowls.

Horizontal washer (tangential washing machines): Industrial long drum washing machines mostly with Pullman divisions (horizontal separators) for larger units and a lengthwise (tangential) opening for loading/unloading.

Horrocks: Old English calico, named after its maker.

Horse cloth: Double-faced twilled jute; the back is often made of wool. It is usually single-coloured or striped; used as horse blanket.

Horse hair: Coarse and stiff elastic hairs which are usually white, brown and black in colour. Natural horsehair is obtained from the tail of the horse (60–80 cm in length, 80–400 m) or the mane (25–45 cm in length, 50– 200 m). Long lustrous hair; sometimes used in furniture seats, underlinings, etc. It is spun into yarn by gluing the hair to a cotton binding thread and then twisting it. The glue is insoluble. According to another system, two fine cotton threads are twisted around the hair.

Hose: A tubular woven fabric for covering liquid under pressure. Hose is manufactured both in underlined and lined forms. When unlined, the weave is plain and the material is generally flax or hemp with a weaving density so arranged that, when the fibre swells on wetting, the fabric becomes tight enough to reduce percolation under pressure, to a negligible amount. For the lined hose, fibres other than flax or hemp may be used in a plain or twill weave. Light weight hose woven from synthetic yarns mainly incorporate an independent tubular plastic lining which is introduced during a weaving process or afterwards.

Hosiery: Can mean all types of knitted fabrics, and goods made from knitted fabrics.

Hosiery goods: Knitted fabrics.

Hot dissolving method: Dissolving method for naphthols in piece dyeing and in printing for generating insoluble azo compounds on the fibre material. The naphthols are made into a paste with, for example, Turkey-red oil and caustic soda solution and then made into a solution with boiling water. Other classes of dyes in powder form, such as direct, acid, reactive and cationic dyestuffs, are applied according to the hot dissolving method.

Hot dyeing dyes: Vat dyes which exhibit their highest affinity for cellulosic fibres at 50–60 °C and are applied by the IN dyeing method. They are referred to as hot-dyeing dyes to distinguish them from so called cold-dyeing vat dyes which have the highest affinity for cellulosic fibres at 25–30 °C and are applied by the IK dyeing method.

Hot orange: An intense orange colour.

Hot tack: Cohesive strength of a coating when molten.

Hot melt sizing: Process developed by Burlington, USA, for sizing with a thermoplastic, water soluble polymer (hot melt), which is applied via a roller at 135–155°C. Each individual warp thread runs in a groove.

Advantage – Energy saving on conventional methods.

Disadvantages – High investment costs, expensive sizing material, non-recoverable, additional operation of assembling.

Hot mercerisation: Mercerisation with an alkaline solution of approximately 60 °C on raw goods causes less change in dye up-take on dyeing. Swelling of the fibres is less, and thus hot and cold mercerisation is often combined. Since swelling is an exothermic process, the hot impregnation of cotton fabrics promotes the even penetration of the less swollen mass of fibres with caustic soda solution, while a subsequent reduction in temperature initiates the swelling process.

Hot notcher: A machine with a heated blade used for making position marks on the edge of cut fabric parts.

Hot pink: An intense pink colour; a colour particularly popular in the 1950's.

Hot water extraction cleaning: A process in which a heated solution of detergent is sprayed on to the textile material and immediately removed by a wet suction nozzle behind the spray-head.

Hot-air shrinkage: Generally, the reduction in the dimensions of a fabric, yarn, or fibre induced by exposure to dry heat. Specifically, a fundamental property of fibres.

Hotflue: Hot-air drying machine for fabric, preferred for intermediate drying. Passage of fabric is usually either vertical, alternating up and down through a drying area with circulating hot air, or diagonally in an up and down zig-zag fashion with double-sided air flow (air jets) from above and below.

Hot-head press: A pressing machine capable of generating high temperatures and pressures. Used for pressing and processing permanent-press fabrics.

Hot-melt adhesive: A solid material that melts quickly upon heating, then sets to a firm bond upon cooling. Use of this type adhesive provides almost instantaneous bonding.

Hottenroth number: A measure of degree of ripening of viscose. See also **Salt figure**. Hottenroth number is expressed as the number of millilitres of 10% ammonium chloride solution that is necessary to add to a somewhat diluted viscose solution to induce the incipient coagulation under standard conditions.

Hot water extraction (in carpet cleaning): A method of cleaning carpet by injecting a heated cleaning solution into the pile and quickly removing the solution and soil by vacuum. Hot water extraction is often erroneously called steam cleaning. Hot water is used at a temperature of 60 ± 3°C (140 ± 5 °F). These temperatures are far below the temperature of steam of 100°C (212°F). See also **Steam cleaning**.

Hounds tooth check or Hounds tooth: A four pointed star check design in a broken twill weave. The checks are medium sized and one of the colours is often white. It is used for men's sports suiting. Particular form of shepherd check shaving short diagonal extensions at corners. These patterns also can be printed. It is used in suits and costumes. See **Pepita**.

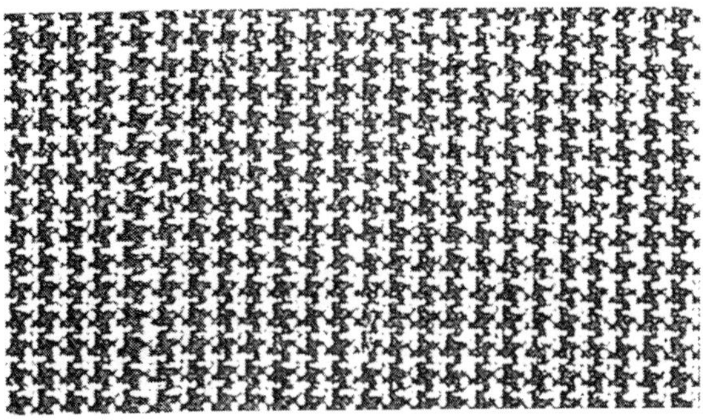

Hounscot say: A 17th Century English worsted.

Houppelande: Cloak-like outer garment for men and women; long and heavily folded, mostly open at the front and belted.

House flannel: Lightweight flannel, made with cotton warp and woollen filling; used for household purposes.

Household textiles: Textile products used within the home, except furnishings.

Included are bedsheets, pillow cases, towels, blankets and table clothes.

Howell: (a) An early ripening commercial variety of upland cotton from Louisiana, the staple measuring up to 25 millimetres; the yield of lint is 34–36 percent; (b) General trade term for T cloths in Greece.

HPD dyeing process for sulphur dyes: A hotpad-dry continuous dyeing process for lightweight cotton or regenerated cellulose fabrics. The liquor contains dye, sodium sulphydrate, caustic soda, glucose and thiourea. After padding, the material is dried or steamed and then after-treated.

Hsiu: Chinese embroidery; the designs are filled out with coloured silk or gold and silver threads.

HT beam dyeing machine: High temperature dyeing machines in which piece goods can be treated horizontally at temperatures above 100 °C (up to 140 °C). The fabric is wound on perforated beams; the liquor is pumped through in either direction.

HT bleach: Peroxide bleaching at temperatures of 110–120 °C on HT machines autoclaves or in special pressure steamer for cotton piece goods.

HT bleaching process: By using pressure reaction chambers, e.g. Vaporloc (Mather & Platt) with hydrogen peroxide, it is possible to obtain a good bleaching effect on cotton and cotton/polyester blends in two minutes.

HT dyeing machine: Any dyeing machine like jig, jet dyeing, winch, beam dyeing machine, etc., which can dye at high temperatures (able to withstand high pressure).

HT fabric beam dyeing machine: High temperature dyeing machines.

HT jigger: A jigger for the high temperature dyeing (HT-dyeing) of textile fabrics in open-width under pressure (at pressures up to 4 bar). Now it is largely replaced by HT beam dyeing machines.

HT loop steamer: Used as a universal continuous steamer. Characteristic features are low-tension cloth guide, high productivity and the facility for operating with both saturated steam (100 °C) and superheated steam (up to 180 °C). This allows synthetic fibres (such as polyester, triacetate) to be continuously steamed. In addition, HT loop steamers are ideally suited to setting fabric that is sensitive to tension, such as knitted fabrics made of textured polyester.

HT pressure steamer: Pressure steamer in which woven and knitted fabrics are steamed on a continuous basis with saturated steam at temperatures up to 140 °C and can thus be used for pad dyeing/print development.

HT rapid draining device: Controlling the HT rapid draining device on any HT dyeing machine allows the liquor draining temperature to be adjusted as preferred, e.g. to 130–140°C and in a range where oligomers are not attached to the fibre and can be discharged to drain. The hot dye-liquor is mixed with used, cold process water under controlled conditions and eventually led into the effluent discharge system when the temperature is suitable.

HT steam: Steam at a higher temperature (Superheated steam) can also exist under normal pressure and be further used at atmospheric pressure See **High pressure steam, Superheated steam**

HT steaming: This takes place in continuous HT loop steamer. Hot steam above 100 °C is the heat transfer medium. Especially suitable for fixing polyester or triacetate when printing with disperse dyes. It is possible to work tension-free so they are especially suited for processing tension-sensitive knitwear. Operates at temperatures between 160–180 °C for 6–8 minutes. Higher fixation temperature and dwell time require careful selection of dyestuff (above-average fastness to sublimation); as a rule, printing paste contains a fixation accelerator. They are also used, with suitable dyestuff choice and fibre type, with saturated steam (100 °C) so these steamers have a temperature range between 100–180°C. See **HT Loop Steamer.**

Huampo: Bark fibre, yielded by the Cheirostemonplatanoides, in Peru; used for garments by the natives.

Huasi ma: Coarse, reddish-brown bast fibre, yielded by a species of the Guazuma tree in Mexico and Trinidad; used for cordage.

Huchaback: A corruption of huckster-back, meaning originally pedler's ware – Towelling made of all linen, linen and cotton, cotton and wool, either by the yard or as separate towels; the part wool huck always separate towels.

Huck: The nickname for Huckaback; it is more commonly used name. See ***Huck back***.

Huck, towel: A plain weave non-terry foundation product constructed with small warp and weft floats, having hems or selvedge, which is used to dry a person's hands and utensils such as glasses, plates, bowls and flatware.

Huckaback: A linen or cotton cloth with a square weave in which the yarns are twisted to add its absorption qualities as it is used almost exclusively for towels in hotels, schools, etc. It is made in white or plain colours, sometimes with woven contrast stripes or name included. Very hard wearing and fairly stiff and heavy.

Hudson's bay blanket: A well-milled and raised heavyweight blanket made from coarse long-staple wool supplied to the Hudson's bay company since the 17th Century. It is natural with coloured border, or striped all over or solid colour with darker border.

Hue: Shade, tint, colour. (a) The first element in the colour-order system, defined as the attribute by which we distinguish red from green, blue from yellow, etc. Munsell defined five principal hues (red, yellow, green, blue and purple) and five intermediate hues (yellow-red, green-yellow, blue-green, purple-blue and red-purple). These 10 hues (represented by their corresponding initials R, YR, Y, GY, G, BG, B, PB, P and RP) are equally spaced around a circle divided into 100 equal visual steps with the zero point located at the beginning of the red sector. Adjacent colours in this circle may be mixed to obtain continuous variation from one hue to another. Colours defined around the hue circle are known as chromatic colours; (b) The attribute of colour by means of which a colour is perceived to be red, yellow, green, blue, purple, etc. White, black and grey possess no hue.

Hugicion: Fibre yielded by a species of the fig tree in Peru.

Huitoc: Fibrous bast of a species of Genipa in Peru; used for rough clothing by the natives.

Human tissue burn tolerance (in the testing of thermal protective clothing): The amount of thermal energy which causes a second degree burn in human tissue.

Humectants: They are hygroscopic compounds (glycerine, ethylene glycol and urea) which absorb moisture and reduces the brittleness of starch films. They help to reduce shedding caused by fracturing the film. In a few dyeing techniques such as direct application, it is necessary to prevent the dye solution applied to the fabric from drying out before the dye has time to fix to the fibres. A humectant, often urea or occasionally glycerine, may be used for this purpose. Humectants may be unnecessary if it is possible to wrap the fabric in plastic sheeting or put it in a closed plastic container to keep it from drying out.

Humidity: The condition of atmosphere in respect to water vapour. The mass of water vapour present in a unit volume of air. For example, Grams per cubic metre.

Humidity relative: The ratio of the pressure of water vapour present to the pressure of saturated water vapour at the same temperature. The ratio is generally expressed as percentage or as a decimal fraction.

Humum: Old and plain Indian cotton cloth of coarse quality.

Hungry cloth: Loose and transparent knitwear. Not always faulty production. Such cloth is deliberately manufactured according to fashion trend demands in a suitable character (e.g. fine, extremely thin lisle stockings).

Hungry fine: A term used to describe a fine wool caused by poor nourishment as opposed to careful breeding.

Hunnicutt: Early maturing commercial variety of prolific cotton from America, the staple measuring 22–25 millimetres; the yield is 30–32 percent.

Hunter coordinate system: This occurs through converting the CIE tristimulus value to the illuminant D65/2observer according to the following relations:

$$L = 10(Y)1/2$$

$$a = 17.5(1.052.\ X - Y):Y1/2$$

$$b = 7(Y-0.919Z):Y1/2$$

L is the luminosity axis (achromatic axis) analogous to the Y axis in the standard valency system. a and b characterize the chromaticity (and can have positive and negative values). The physical ideal white here has the coordination L = 100 and a = b = 0.

Hunter's green: A dark yellowish-green colour.

Hup dyeing apparatus: Serves to dye fully milled (not well balanced) hat bodies in a floating liquor: relatively low liquor ratio (max. 1:30); liquor is circulated by a propeller which automatically changes its direction of flow; the dye liquor is pushed and drawn through the mass. Attainable dyeing temperatures with closed apparatus around 98 °C. Best results with level dyeing acid dyestuffs.

Hurdle drier: Box drying machine for flock, hank yarn and cross wound spools with perforated floors (cabinet drier).

Husking cloth: Stout and heavy cotton ticking; used for working gloves.

Hussong loose cotton dyeing machine: A machine used in the earlier days to dye loose cotton. It consists of a vat with steam pipes in the bottom and a perforated partition above these pipes and is divided into two compartments, one bigger for holding the cotton and other for liquor. Machine works in the same principle as Schmidt's machine (See **Schmidt's machine**). Liquor is heated after loading the material and guided to the top by working a paddle fitted in the lower compartment and poured over the loose cotton in the next compartment and percolated through the package to the bottom due to the vacuum produced at the bottom. This process continues and the dyeing takes place.

HVM: High wet modulus fibres (HVM fibres).

Hwachow: Dressed silk gros de Naples, made in China; it used to be made at about 24 inches wide. The warp is organzine and the filling a six-ply greige tram.

Hwa mien: Chow cotton poplin, made in China.

Hwasienchow: Very soft silk gros de Naples, made in China; does not crease.

Occasionally made similar to crepe.

Hwayong: A Chinese silk velvet made in green or poppy colours.

Hwayutwan: Wool poplin, made in China.

Hybrid composite: Advanced composite with a combination of different high-strength continuous filaments in the matrix. Also, composite in which continuous and staple fibres are used in the same matrix.

Hybrid fabric: Fabric for composite manufacture in which two or more different yarns are used in the fabric construction. This provides design flexibility to meet performance requirements and controls cost by permitting some lower priced fibres to be used.

Hybrid fibres: Fibres based on polymer blends. The extrusion spinning occurs from a common solution (or melt) of at least two fibre-forming polymers, of which one is a native polymer, e.g. cellulose or casein. Ideally hybrid fibres may be regarded as filament composites. Examples of hybrid fibres (polymer blends and solvents) are

(a) PAI/Cell (HCHO)X/DMSO

(b) PAN/Cell (HCHO)C/DMSO

(c) Cell/PAN N204/DMF

(d) LiCl/DMAC

(e) PAN-Co/Cell $ZnCl_2/H_2O$

(f) PAN-Co/Fib $NaSCN/H_2O$/DMSO

(g) PAN-Co/CA DMF

(PAI = polyamidimide, Cell = cellulose, PAN = polyacrylonitrile,Co = Copolymerisate, Fib = fibroin, CA =acetate).

Hybrid yarn: In aerospace textiles, a yarn having more than one component. Also see **Commingled yarn**.

Hybrid yarn manufacture: Since the development of new high-temperature resistant thermoplasts, interest in using these materials as a matrix with high performance composites has increased particularly in the air and space travel industry. The techniques illustrated above are principally available for the manufacture of suitable prepegs.

Hyacinth blue: A purplish-blue; a colour popular in the 1930's.

Hyacinth red: A reddish-orange colour.

Hyacinthine: The purplish-blue of the hyacinth; blue-red; hyacinth violet or pink.

Hydrangea pink: A yellowish-pink. Also hydrangea blue and hydrangea red.

Hydrate: (Gk.: hydor = water), molecular compound formed through hydration with bound water molecules.

Hydrated cellulose: Regenerated cellulose.

Hydrated lime: Calcium hydroxide, $Ca(OH)_2$.

Hydration no.: A value indicating how many water molecules are associated with each molecule of some other compound in a crystal of that compound. Many chemicals have one or more water molecules associated with each chemical molecule in the commercially-sold form of that chemical. For example, sodium carbonate (soda ash) may be sold as anhydrous (containing no water molecules), monohydrate (one water molecule per sodium carbonate molecule, written as $Na_2CO_3{\bullet}H_2O$) or decahydrate (ten water molecules per sodium carbonate molecule – $Na_2CO_3{\bullet}10H_2O$). This water is not visibly present as water – it does not cause dampness, though it may change the appearance of the dry chemical. The hydration state must be considered when measuring the chemical by weight. For example, 1 gram of sodium carbonate decahydrate contains about 0.37 grams of sodium carbonate, and 0.63 grams of water. If chemicals are exposed to air for long periods, the hydration may either increase or decrease depending on conditions. See **Hygroscopic**.

Hydraulic: A very common driving media used in process house. Hydraulic means 'moved by the pressure of fluid (pumped)'; e.g., high-pressure atomizers for increasing air or fabric humidity (jets) or mechanical drives for step-less speed control (control devices) consisting of driven oil pump and driving fluid motor.

Hydraulic entanglement: See **Hydroentanglement**.

Hydrolytic stability: The ability to withstand the environmental effects of high humidity.

Hydrolysis, sensitivity to: In the USA, it is normal to neutralise alkaline laundering of resin finished goods with the addition of acidic reacting salts (e.g. zinc silicone fluoride) in the final rinsing-bath, in order to prevent yellowing that can easily occur as a result of hot tumble drying. Overdosing of the acidic salt will create acidity on the fabric, which will affect the finished effect. Hence, fabrics destined for the American Market require the best hydrolysis stability.

Hydrolysis, stability to: Resistance of a type of finish, particularly an easy care finish or the resin-finishing agent itself, to acidic or alkaline Hydrolysis. With insufficient stability to acid hydrolysis, splitting of the cross-linked resin-finishing agent may occur, which considerably impairs the finishing effects; there is no risk with resin finishing in the alkaline range.

Hydro (hydros in few countries): Dye house term for sodium hydrosulphite.

Hydro extraction: See **Hydro-extraction**.

Hydro-cellulose: Mineral acid damage to cellulose is shown through water absorption and splitting of oxygen links between glucose building blocks.

Properties– Reduced strength, increased reducing power and modified dye affinity.

Detection – Hydrocellulose alone.

Ammoniacal silver nitrate solution. Also with Prussian blue reaction, Turnbull's blue reaction, Fehling solution, methylene blue and solubility test (which also indicate oxycellulose). No staining with Diamine Blue 2B, Congo red or Benzopurpurine.

Hydrochloric acid: A strong inorganic acid; solution of HCl in water, typically about 37%; often called muriatic acid. Hydrochloric acid is not much used in dyeing, but is used in some processes to strip certain finishes from fabric. It can cause skin burns and the fumes are very noxious. Handle and store it with great care as it is a strong acid.

Hydro-entangling: A Process for forming a fabric by mechanically wrapping and knotting fibres in a web through the use of high-velocity jets or curtains of water. Also see ***Spunlaces Fabric***.

Hydro-extraction: This process removes the water (the water quantity varies according to the type of fibre) dispersed in the fibres by mechanical action; this process aims at reducing energy consumption and is carried out before the final fabric drying or between the various wet processing stages (washing, dyeing). Different methods are available like squeezing, centrifugation, steam pressure (jet force).

Hydro-extraction by centrifuge: Water extraction by centrifuge.

Hydro-extractor: See **Centrifuge**.

Hydro-fixation: Heat setting of polyamide piece goods in HT beam dyeing machines using hot water (up to 140 °C), mostly combined with bleaching, scouring or dyeing.

Hydrogen bond: A hydrogen bond is a type of electrostatic interaction between groups in molecules that have hydrogen atoms bound to electronegative atoms (e.g. so-called Donor, such as oxygen, nitrogen, sulphur, chlorine, fluorine). Hydrogen bonds may therefore be regarded as electrostatic bonding forces between dipole molecules with a high dipole moment, in which the hydrogen atom is at the positive end of one polar group and forms a linkage with the electronegative end of another polar group. A chemical bond in which hydrogen that is already covalently bonded to one atom is electrically attracted to a lone pair of electrons on another atom Some atoms such as oxygen are said to be very electronegative, which means that they strongly draw bonding electrons toward themselves. If hydrogen is bonded to such an atom, the hydrogen "appears" to have some positive charge. Lone pairs (that is, not involved in bonding to other atoms) of electronegative atoms tend to attract hydrogen from other molecules, but a new compound is not formed. Hydrogen bonds are weak to moderate in strength, but there can be very large numbers of hydrogen bonds between molecules of compounds like cellulose. It is hydrogen bonds between cellulose molecules that keep the molecules together as fibres (and it is because they are easily broken and re-formed that cotton wrinkles). Hydrogen bonds play a role in dyeing with direct dyes, and may be present as "temporary" bonds with other dye types.

Hydrogen peroxide: (Hydrogen superoxide, hydroperoxide Peroxide), H_2O_2, normally as a solution in water, structure H–O–O–H. Molecular weight 34.016. Besides water, hydrogen and oxygen form hydrogen peroxide H_2O_2. In its anhydrous state, this is a liquid whose physical properties are very similar to those of water. The boiling point is at 150.2 °C, the freezing point –0.43°C. The molecule is diamagnetic; its structure is reproduced in the metastable original state of H_2O_2 is separated from the product water in the model by an energetic barrier, the "activation peak". Hydrogen peroxide is extensively used in commercial bleaching of textiles, especially cellulose fibres and wool. It is effective and non-polluting. The hydrogen peroxide sold in drug and grocery stores is about a 3% solution. Industrial strength hydrogen peroxide may be 35%, 50% or 70%. At these strengths it is quite dangerous to handle, and spills can cause fire. Hydrogen peroxide can behave as either an oxidizing agent or a reducing agent. It is sometimes used as an antichlor.

Sodium percarbonate and potassium monopersulphate are dry chemicals derived from hydrogen peroxide.

Hydrogen peroxide bleaching: Bleaching of textile fabric using hydrogen peroxide. It can be done in batch or continuous process and in various types of machines. See **Hydrogen peroxide**.

Hydrogen peroxide bleaching regulators: Bleaching systems contain regulators to reduce fibre damage and oxygen release. Buffers are used to maintain the correct pH for activation since alkali is required to neutralize the cellulose breakdown products. In addition, complexing agents are required to chelate any heavy metals present. Bleaching systems with built-in catalyst present do not inevitably cause a corresponding breakdown of cellulose – the catalyst also needs to be activated.

Hydrogenation: The process of passing hydrogen into an unsaturated chemical in the presence of a catalyst to convert the material to a more saturated state (i.e., containing more combined hydrogen).

Hydrolases: The most extensive main group of Enzymes which catalyse the hydrolytic cleavage of C– O and C–N bonds in the breakdown of large molecules.

Hydrolysis: In general, the decomposition of a chemical by reaction with water to form more molecules or ions. A typical example is the hydrolysis of cellulose, i.e., hydrolytic splitting of cellulose, known as hydrocellulose, i.e. hydrolytic splitting of cellulose. Some chemicals, such as reactive dyes, are effectively destroyed by hydrolysis. Others, such as soda ash, are made useful because of hydrolysis. Hydrolysis will occur when susceptible compounds are in aqueous solution, but can also occur because of water absorbed from moist air, without the compound ever appearing wet or damp. This is why many dry chemicals must be kept in sealed containers.

Hydrolysis (reversible): The effect of water on the ions of a dissolved salt lead to a state of equilibrium where ions and molecules of the acid or base forming the salt exist together. If the conditions change, acid or base molecules may change back into ions. This is especially significant with salts of weakly organic acids or weakly organic amines with large hydrophobic tails.

Hydrolysis (sensitivity to):In many countries, it is normal to neutralise alkaline laundering of resin finished goods with the addition of acidic reacting salts (e.g., zinc silicone fluoride) in the final rinsing-bath, in order to prevent yellowing that can easily occur as a result of hot tumble drying. Overdosing of the acidic salt will create acidity on the fabric, which will affect the finished effect.

Hydrometer: A device for measuring the density of liquids consisting of a sealed cylindrical glass tube with a density scale at the upper end and weighted with mercury or lead shot at the lower end. When placed in a liquid, the hydrometer floats upright with the tube partially immersed, the degree of immersion depending on the density of the liquid which can be read off directly from the scale at the liquid/air interface.

Hydrophilic: Having strong affinity for or the ability to absorb water.

Hydrophilic group: Group of molecules whose behaviour towards water is Endophilic.

Hydrophilic property: Tendency of a material to take up water (opposite to ***Hydrophobic***).

Hydrophilic-lyophilic balance (HLB): For a molecule to function as a surfactant, it must have the proper balance of water and oil solubility. The molecule must not be too water soluble otherwise it will not form micelles, yet at the same time it must be sufficiently soluble to do its job. The solubility of a surfactant molecule in water verses oil depends on the water-solubilizing group. Ethoxylated nonyl phenol is a good molecule to describe this concept. Nonyl phenol (considered as the hydrophobe), is slightly soluble in water. It is highly soluble in oils (organic solvents). If one mole of ethylene oxide is added, the water solubility is increased slightly while oil solubility is decreased. As more ethylene oxide is added, water solubility continues to increase at the expense of oil solubility. When 5 to 9 moles of ethylene oxide have been added, the molecule is equally soluble in water and in oil. Above 9 moles of E.O., the water solubility continues to increase at the expense of oil solubility.

Hydrophobic: Property of a material to have a water-repellent effect (opposite Hydrophilic property).

Hydrophobic group: Group of molecules whose behaviour towards water is exophilic.

Hydroplastic: Materials that is deformed in the wet state, e.g. cotton through hydrate formation of the cellulose OH groups, where even slight quantities of water in humid air are enough to break the hydrogen bonds produced by OH groups. See **Thermoplastics**.

Hydroplasticity factor: The wearing properties of textiles made from regenerated cellulose fibres depend, among other things, on the wet and dry crease angles. Both should be in a certain ratio to each other. A good dry crease angle in itself is not sufficient.

Hydros: See **Hydro**.

Hydroscopic: Having the ability to absorb moisture from the atmosphere. All fibres have this property in varying degrees.

Hydro-setting (a setting method): The hydro-setting or aqueous heat-setting of polyester is done with hot water in a high temperature liquor circulating machine at about 130°C. A typical cycle may require 30 minutes. Water (or steam) promote swelling of fibre and may cause some hydrolysis in the ester groups in polyester chain. This causes partial destruction of intermolecular bonds and depolymerisation of fibre with a result in loss in tenacity. However, the hydrosetting temperature can be lowered by adding some selected organic or inorganic substances causing swelling of polyester, but such conditions have not yet found industrial application.

Hydrosulphite: Sodium dithionite.

Hydrosulphite discharges (reductive discharges): Sodium dithionite has not completely replaced tin salt- and zinc-dust stripping (too rapid decomposition in alkaline solution). Against it, formaldehyde sulphoxylates strips are widely used (better stability and protection of cellulose fibres).

Hydrosulphite vat: Reduction vat used in vat dyeing and in indigo-sodium dithionite vats. Sodium dithionite reduction baths produce sulphate as an end-product (concrete-aggressive).

Hydrotrope: Registered trade name for a group of substances which show hydrotropic property and are used in the detergents industry.

Hydroxy ethyl cellulose: Non-ionic and water-soluble cellulose derivatives (cellulose ether) in various states of viscosity.

Hydroxyethyl or hydroxypropyl starches: Important starch ethers as conversion product of starch with ethylene or propylene oxide (in presence of alkali). Degree of etherification of 0.3–0.5 produces good water-solubility and avoids any tendency to reversal of the reaction.

Hydroxyl group (OH group): Typical component of bases, alcohols and acids (as an auxochrome in dyestuffs), caustic soda liquor, phenol. As a dissociated component of water-soluble compounds, the hydroxyl group is the source of the alkaline reaction and degree of alkalinity (hydrogen-ion concentration).

Hydroxyl end group: A polymer chain-terminating (-OH) group.

Hygral expansion: It is a property of fabrics made from fibres that absorbs moisture, in particular fabrics made from wool. It is a reversible change in dimensions which takes place when the moisture regain of a fabric is altered.

Hygrometer: Any instrument for measuring the relative humidity of the atmosphere.

Hygrometer (assmann): This is a more sophisticated instrument than the sling hygrometer in that a fan is used to draw air across the thermometer bulbs at a constant predetermined speed. The temperatures are read when they have reached a steady value.

Hygrometer (hair): Human hair increases or decreases in length as the humidity of the surrounding air increases or decreases. By attaching a bundle of hairs to a suitable lever system, the relative humidity of the atmosphere can be indicated directly and, if required, recorded on a chart. The accuracy of this method is limited to within 3 or 4% of the true value for the range of relative humidity between 30% and 80%. A combined temperature and humidity recording instrument is often used in laboratories and is known as a thermo-hygrograph. The hair hygrometer requires frequent calibration and has a slow response to changes in atmospheric conditions.

Hygrometer (sling or whirling): This instrument works on the same principle as above, but the two thermometers are mounted on a frame with a handle at one end. This allows them to be rotated by hand at a speed of two or three revolutions per second, so giving an air speed of at least 5 m/s past the thermometers. After half a minute of rotation, temperature readings from the two thermometers are taken and the procedure is then repeated until the readings have reached minimum values.

Hygrometer (wet and dry bulb): See **Wet and dry hygrometer.**

Hygroscopic agents: Water-absorbing products as additives for finishes, printing pastes, sizes, to balance and influence the fabric weight and handle.

Hygroscopic: Having a tendency to absorb water (usually meaning absorbing it from air). Some of the commonly used chemicals in dyeing are hygroscopic. If they are left exposed to moist air, they will absorb water from the air. This may lead to caking and increases the ratio of weight to volume of the chemical. For many chemicals there are no visual signs that water has been absorbed. For others (such as copper sulphate), the colour will change. Some, such as sodium hydroxide are deliquescent – they will absorb so much water from air that a liquid solution will form. Hygroscopic chemicals should be stored in tightly sealed and moisture proof containers.

Hypo: Short term for hypochlorite bleach. Don't confuse this use of "hypo" with the use in photographic processing where it means something very different – sodium hyposulphate (more commonly called sodium thiosulfate). See **Bleach (chlorine)**.

Hypochlorite bleach: Bleaching with sodium hypochlorite or calcium hypochlorite (bleaching agent) is known as hypochlorite bleach, even though it is seldom used now. Since peroxide bleaching is associated with a tendency to cause catalytic damage to cellulosic fibres in the presence of heavy metal ions, the inclusion of a prior hypochlorite bleach in the continuous bleaching of woven fabrics has made it possible to employ milder conditions in subsequent peroxide bleaching.

Hypochlorous acid: Hydrolysis of sodium hypochlorite under pH neutral to slightly acid conditions. Hypochlorous acid, though effective at bleaching, is also believed to be very damaging to cellulose. For this reason, chlorine bleach baths are maintained at pH between about 9 and 11.

Hypothetical: As yet unproven scientific assumption, e.g. listing of hypothetical structural formulae.

Hysteresis: (Gk.: remaining behind), elastic aftereffect, i.e. the change in form by which some rubber-like substances do not recover fully or only recover slowly once the elastic stress has ceased to be applied.

I

I and M: Inspection and Maintenance.

IAQ: International Association for Quality (United States based quality control organization).

ICA: International Colour Authority. Colour forecasting publishers, publishing forecasts twice a year and 24 months ahead of the season.

Ice-blue: Greenish-blue; a pale blue.

Ice colours: Produced directly on the fibre by preparing cotton with beta-naphthol followed by development with ice-cooled diazotized aniline. Obsolete process; now substituted by Naphthol AS types.

Ice-green: An extremely pale shade of green.

ICR: Informationsstelle Chemischreinigung, Stuttgart, Germany (German Information Office for Dry-cleaning); Technical and professional organizations.

IC/TC: Intelligent Colour/Trash Co-ordinator.

ICSID: International Council of Societies of Industrial Design. An international, non-profit making, non-governmental organisation, established to advance the discipline of industrial design.

Icteritious: Jaundiced yellow.

Ida canvas: Soft and open face canvas that is made of unbleached linen.

Ideal black (physical): A black colour, which absorbs 100% of the incident light at all wavelengths, i.e. the reflection amounts to zero. See also **Ideal white**.

Ideal white (physical): Also described as normal white; closely approximated with barium sulphate with a reference point approximately equal to 00. In colour space, it corresponds to the Achromatic point to which a colour approximates after bleaching. Ideal white is a borderline case as far as 100% of the incident light is reflected at each wavelength (counter term: Ideal black).

Idle machine time: Useful working life of a machine.

IDT: Intelligent Data Terminal.

IDFC: International Dry-cleaning Research Committee. Contact organization of the dry-cleaning research offices of different countries. Technical and professional organizations.

IEK: Italienische Echtheitskommission (Italian Fastness Commission); technical and professional organizations.

IFAI: Industrial Fabrics Association International.

IFAT: Internationale Fachmesse für Abwasser und Abfalltechnik, Munich, Germany (German international trade fair for waste water and waste technology).

IFATCC: International Federation of Associations of Textile Chemists and Colourists.

IFT: Institut für Textiltechnik der RWTH Aachen, Germany (Institute for Textile Engineering); Technical and professional organizations.

IFVTCC: Internationale Föderation der Vereine der Textilchemiker und Colouristen (International Federation of Associations of Textile Chemists and Colourists); Technical and professional organizations.

Ignition: The initiation of combustion. In the sense of flammability testing, ignition represents the initiation of the burning process. See **Burning behaviour of textiles**.

Ignition loss (in glass textile): The amount of organic material consumed by ignition.

Ignition point: (a) Spontaneous ignition temperature (SIT); (b) The specific point on a textile specimen at which a flammability test is initiated.

Ignition temperature: The lowest temperature to which a given substance can be heated before it ignites (without the application of a spark or flame). It is often called the auto ignition temperature. See **Flash point**.

Ignition time (minimum ignition time): The ignition time is the time required to ignite a textile material under the conditions prescribed in a flammability test. That is, the time of flame contact on the surface of a material just sufficient to ignite the material so that it continues independent combustion after the removal of ignition.

IIC: International Institute for Cotton (GB); technical and professional organizations.

IIRSC: Irish Standards Organization; technical and professional organizations.

IK dyeing method: Dyeing method for vat dyes (IK dyes) which have the highest affinity for cotton at 25–30 °C. Because of inferior fastness properties, the so-called cold-dyeing vat dyes which can only be applied by this method have diminished in importance.

Ikat: The word "ikat" is derived from a Malaysian word *mengikat*, which means to tie, knot, bind or wind. The uniqueness of ikat lies in it being a combination of two methods – 'tie and dye' as well as weaving. In ikat, dyeing takes place in the yarn stage, following which it is woven. This process is either applied on warp or weft when it is called 'single ikat' designs or done on both warp and weft to form an overlapping or 'double ikat' pattern. The beautiful trade mark of ikat – the jagged weave in striking colour sequences, is an outcome of the enormous skill that goes into tying the yarn. Though it is a complex method, the depth of India's textile tradition is visible in the way weavers plot the design with a dexterity matching modern aids. Rarely will any two motifs, in the same running length, be facsimile yet they look strikingly similar and therein lies the beauty of ikat. It's another highlight is that both the sides of the fabric appear similar. The technique is highly developed in Indonesia, India, Madagascar, Central and South America, etc.

Illuminant: (1) Mathematical description of the relative spectral power distribution of a real or imaginary light source — i.e., the relative energy emitted by a source at each wavelength in its emission spectrum. Often used synonymously with "light source" or "lamp," though such usage is not recommended.

(2) An Illuminating material.

Illuminant A (CIE): Incandescent illumination, yellow-orange in colour, with a correlated colour temperature of 2856K. It is defined in the wavelength range of 380 to 770nm.

Illuminant C (CIE): Tungsten illumination that simulates average daylight and bluish in colour with a correlated colour temperature of 6774K.

Illuminants D (CIE): Daylight illuminants, defined from 300 to 830 nm (the UV portion 300–380nm being necessary to correctly describe colours that contain fluorescent dyes or pigments). They are designated as D, with a subscript to describe the correlated colour temperature; D65 is the most commonly used, having a correlated colour temperature of 6504K, close to that of illuminant C. They are based on actual measurements of the spectral distribution of daylight.

Illuminating dye: Dye mixed with discharge paste in printing techniques, also called head dye. In a few discharge printing techniques, dye is mixed with the discharge paste so that the discharged area is given a new colour. Illuminating dyes must be selected carefully, since they must resist the effects of the discharge chemicals.

Illuminating dyes: In textile printing, illuminating dyes are those dyes which are used to substitute the missing shades in a particular dye class, e.g. vat leuco ester dyes as coloured resists under naphthol and aniline black dyeing. In discharge printing, a distinction is made between white discharge printing and coloured discharge printing. In coloured discharge printing, coloured grounds (produced with dischargeable dyes) are printed with discharge printing pastes containing dyes resistant to the discharging agent.

Illuminating colours: An old term for textile colours used in resist and discharge printing. Illuminating colours are produced mainly with brilliant dyes from a different dye class to supplement the shades missing from a particular dye range.

Illumination (in lighting): The density or flux of light on unit area of surface.

Illusion: Fine French silk tulle that is used for trimming; also a net with star mesh, used for making veils and dresses.

Imagdong: A plant in the Philippines. The fibres of this plant are used for making cords and ropes.

Imbabura: Cotton grown in Peru; the staple is fine, white and clean.

Imbe: Fibre yielded by the stem of a species of Philodendron in Brazil; used for ropes.

Imbibition: A measure of the liquid or water-holding capacity of a textile material.

Imidazolines (softeners): Amino esters and amino amides are converted to cyclic imidazolines by heating them under reduced pressure to split out a second mole of water. One nitrogen in the ring reacts as a simple amine, so it can form acid salts to become cationic or it can be quaternized. The imidazolines are less viscous than their parent amino amide and therefore have better softening properties. Imidazolines are less likely to discolour with age than their corresponding parent compound.

Imirat: A bast Indian cotton fabric of plain weave.

Imitation fur (simulated fur, imitation skin): A collective term for numerous imitations of natural furs and skins. Synthetic fibres are predominant as raw materials for the production of simulated fur pile fabrics with increasing use being made of hollow fibres. Imitation fur fabrics are often foam-backed.

Imitation horsehair: Narrow plain woven fabric, made of heavily sized vegetable fibres to imitate horse hair cloth and used for interlining.

Imitation suede (imitation suede leather): These materials may be subdivided into the following categories depending on the manufacturing process –

(a) Imitation suede 1st generation: Artificial leather produced mainly from textile random webs, woven or knitted base fabrics
   (i) either by flock coating or
   (ii) artificial leather coatings (mainly polyurethane and polyvinyl chloride).

(b) Imitation suede 2nd generation: Available commercially since the middle of the 1970's as special fibre laminates in, e.g., polyurethane resin compounds.

Imitation wool: Made by treating Senegal hemp in cold solution of caustic soda, peroxide of sodium and soluble oil or in bath of peroxide of sodium and ammonia in equal parts. The hemp is then treated with an acid solution.

Imitation yarn: Soft-twisted, bulked coarse cotton yarn consisting of short fibres spun in the manner of carded yarn (condenser yarn).

Imi-wax: An abbreviation for imitation wax print. The purpose is to imitate real wax prints by avoiding the laborious wax resist printing process. The crackle or veining effect in imi-wax Africa prints is produced by direct printing and the indigo of real wax prints is imitated by printing blue diazo fast colour salts on naphtholated grounds. In the vast majority of cases, the same typical designs and colours are used as for real wax prints. The production of imi-wax prints is mainly carried out on roller printing machines.

Immature cotton: Cotton picked before it is fully mature. The fibres are not properly formed and the yarn made from them is generally weaker and inferior. Immature cotton tends to form small neps in the spinning process. In contrast to dead cotton, the fibre contains a considerable amount of protoplasm and therefore has different dyeing properties.

Immature fibres: (1) Cotton fibres treated with sodium hydroxide solution. Fibres that either –(a) have swollen and assumed a spiral form or (b) remained flat, thinly outlined and almost transparent. Total wall width is less than the lumen width.

(2) Cotton fibres observed under polarized light. Fibres that appear purple, indigo, or blue, turn orange or yellow-orange upon rotation to the subtractive position, and upon removal of the selenite plate show parallel extinction.

Immediate elastic deformation: Recoverable deformation that is essentially independent of time, i.e., occurring in (a time approaching) zero time and recoverable in (a time approaching) zero time after removal of the applied load.

Immediate elastic recovery: Recoverable deformation which is essentially independent of time, that is, occurring in (a time approaching) zero time and recoverable in (a time approaching) zero time after removal of the applied force.

Immersion bleach process: A process in which the material to be bleached is kept immersed in the bleaching liquor for a predetermined duration, e.g. in a "wet" J-box (or sump) in contrast to other continuous bleaching processes where the fabric is simply impregnated with bleach liquor and plaited down in a storage unit. Even though this method is not very common now as other more efficient methods have been introduced, it has certain advantages as more complete removal of impurities from the substrate, greater uniformity of the achieved effects and the prevention of crease and rope marks, development or preservation of particular fabric characteristics, reduced chemical concentrations with consequent protection of the material against excessive chemical damage and sensitivity.

Immersion centrifuge: Textile material is wetted out in the basket of a particular type of centrifuges and subsequently centrifuged. An immersion centrifuge is used to saturate or impregnate textile materials with liquors by centrifugal force as well as for rinsing, souring and neutralizing. Processing liquors may be collected during centrifuging and re-used.

Immersion jigger: A type of jigger which was often used in the past for dyeing with vat dyes. In this type of jigger, the draw rollers are not mounted above the liquor as in a normal jigger but below the surface of the liquor so that the problem of atmospheric oxidation of the vat dyes during dyeing is prevented.

Immersion squeezing: In order to intensify the liquor exchange in wet-on-wet processing or the wetting out of textile fabrics in wet-on-dry treatments at high production speeds in a padder trough with a low liquor content (i.e. with a short immersion path), squeeze rollers are also employed below the level of the liquor in addition to the usual squeeze rolls mounted above the trough.

Immersion test: A test method for water-repellent impregnations on textiles (especially for yarns and knitted fabrics). Procedure – The textile test specimen is weighed and then submerged in a sieve for 5 min under 10 cm of water (the sieve is jerked a few times to remove collected air). The specimen is then taken out of the water, allowed to drain for approximately 10 min, weighed again and the water uptake is calculated.

Impact resistance: Resistance to fracture under the sudden application of an external force.

Imperfection: A departure of a quality characteristic from its intended level or state.

Imperial blue: A deep blue; a dye.

Imperial cloth: A coating fabric of fine worsted in twill weaves. Firm and durable cloth, usually dyed in navy blue and used for coats, capes and rain coats, if shower proofed.

Imperial green: See **Emerald green**.

Imperial red: A deep bluish-red colour.

Imperial satin: It is a closely woven cotton fabric, made with an eight-leaf weft-face satin weave, two contiguous warp ends raised together. There are at least about twice as many picks (of soft spun filling) than warp ends. They are napped on the back or on the face when they are called lambskins. The reversible imperial is very closely picked, showing the filling at each side.

Imperial serge: Another name for perpetuana.

Imperial yellow: A deep yellow colour originating from the yellow porcelain produced in China. The colour was considered to be the reserve of the imperial court. See also **Yellow**.

Impermeable: Waterproofed.

Impid: A Philippine fibre used for cordage.

Impregnated fabric: A fabric in which the interstices between the yarns are completely filled with the impregnating compound throughout the thickness of the material, as distinguished from sized or coated materials where the interstices are not completely filled.

Impregnating machine: Roller vat or rope device with squeezing unit used for wetting with bleach concentrates in rope bleaching and open-width continuous bleaching, e.g. J-box.

Impregnation: In Latin, the word *impregnare* means "to soak, saturate". In a general sense, impregnation refers to treatment of textiles by impregnation with solutions, dispersions or emulsions. Impregnation is one of the most important operations in finishing, less for yarns than for fabrics and knitwear. It is required for all continuous and semi-continuous processes including boiling off, bleaching and dyeing.

Impressioning machine: A single-colour roller printing machine for the production of test prints as a means of checking the engraving quality of engraved printing rollers, or to produce strike-offs of new designs, as well as small samples.

Impression mark: The relief print-off of defects, e.g. slubs, under excessive rolling tension. Improved multi-volition cross.

Improved prolific: A commercial variety of upland cotton from North Carolina, the staple measuring up to 25 mm; the yield is about 30% lint.

In the grease: Signifies wool in its natural state, as it comes from the sheep's back with all the grease and other impurities attached to it.

In the gum: The natural silk fibre before the gum covering is removed by boiling out.

Inaja: Strong leaf fibre yielded by the Inaja palm in Brazil; used for cordage, etc.

IN dyeing method: A method for dyeing those vat dyes (IN dyes) which have the highest affinity for cotton at 50–60 °C. See **Vat dyeing**.

In between: Name of a curtain fabric which is used because of its construction and optical effect, as a curtain and/or decorative cloth. In between is more transparent and lighter than a normal decoration cloth, but is of high density. In between often have large designs and are used as coloured sails.

Inca brown: A dark brown colour.

Incandescence: The quality of a body which, as a result of its high temperature, emits light.

Incandescent: Shining brightly; emitting light as a consequence of reaching a high temperature. Thus 'incandescence'.

Inch quality: A measure of quality of warp knit fabric, the number of inches of fabric per rack.

Inches per rack (IPR) (in warp knitting): The length of fabric in one rack measured on the machine under operating take-up tension.

Incubation: Holding a sample, microbial culture, etc., for a suitable time at a temperature, appropriate to the investigation, at which microbes are likely to multiply. Typical temperatures are 55 °C for thermophils, 44 °C for ***E. coli***, 37 °C (human body temperature) for the coli form group of bacteria and 20 °C for incubating samples in the BOD test.

Incorrectly woven yarn: A thread which in some parts only of the fabric is not interlaced in the standard way.

Indanthrene IK process: See **Indanthrene IW process**.

Indanthrene IN process: See **Indanthrene IW process**.

Indanthrene IW process: Vat dyes are divided into groups according to their affinity for the fibre and the amount of alkali required for dyeing – IK (I = indanthrene, K = cold), IW (W = warm), IN (N = normal). The IK dyes have a low affinity for the fibre; they are dyed at 20–30 °C and require little alkali. Salt is added to increase the absorption of these dyes. The IW dyes have a higher affinity than IK dyes. Dyeing is performed at 40–45 °C with more alkali and little or no salt. The IN dyes are highly substantive and applied at 60 °C. They require much alkali for vatting, but with no addition of salt.

Indanthren yellow paper (vat yellow paper): Yellow indicator paper (strips) coloured with indanthrene yellow. Indanthrene yellow paper is used to test vats in vat dyeing. If the vat conforms to regulations, then the yellow visibly changes colour to cornflower-blue within 3 s.

Indanthrone: See **Anthraquinone-N-hydroazine**.

Inde blue: A greyish purple; a dye.

Indentation: Permanent pile deformation in upholstery velour fabrics visible as pressure marks resulting from the normal everyday use of upholstered chairs and seats.

Independent variable: An assigned variable associated with a dependent variable so that when the value of the assigned variable is changed, the dependent variable may also change.

Index of refraction: Ratio of the velocity of light in one medium to its velocity in a second medium as the light passes from medium to medium. If a medium is crystalline, the velocity may depend on the direction of the light with respect to the crystalline axes and the substance may have several indexes of refraction, i.e., it may be birefringent. See **Refractive index** and **Birefringence**.

India chintz: A thick, stout chintz with large patterns; used for upholstery.

Indian brown: A dark reddish-brown colour.

Indian ciciclia: Silk brocade with gold and silver flowers over a ribbed ground.

Indian crepe': A cotton crepe of rough unfinished appearance, used for blouses, men's shorts, etc.

Indian dimity: Fine, plain woven, light cotton dress goods, having cross ribs formed by the filling. It is printed in floral warp stripes, and also between the ribs.

Indian mull: A fine, undressed, plain woven mull, dyed in the piece, made originally in India; used for dresses.

Indian orange: A vivid reddish-orange colour.

Indian red: A dark brownish-red originally made from iron oxide. See **Venetian red**.

Indian silk: Hand-loomed silk, not always made in India. This fabric has a slightly wrinkled impressed appearance, but it is a crisp fabric. Made in plain colours, used for sarees, evening wear and sometimes for furnishing, such as cushion covers.

Indian yellow: A vivid golden-yellow pigment known in India for many years, but now synthetically produced. It was originally made from earth on which mango leaf eating cows or camels had urinated over many weeks. Since mango leaves give insufficient nourishment, the export of Indian yellow was banned to avoid cruelty to animals, and by the beginning of the 20th Century it had ceased to exist. Also referred to as purree, puri, peoli and gaugoli.

Indicator: The Latin word *indicare* means "to point out, compare". A reaction-sensitive substance which allows the progress of a chemical change to be followed, but generally a substance which marks a precise stage in certain chemical reactions, e.g. acidic, alkaline, neutral reaction, by a characteristic change in colour. The determination of pH is a typical application for indicators of this kind. Thermopapers, which show when a particular temperature has been exceeded, represent another type of indicator.

Indienne: Indienne was formerly a term used to describe printed cotton fabrics from India. Today it is used as a general term for lightweight cotton prints achieved by specific printing techniques.

Indigo: Originally, a natural blue vat dye extracted from plants, especially of the *Indigofera tinctoria* genus grown principally in eastern India and China. Indigo is probably one of the oldest colourants used for textiles. It represents the oldest blue vat dye. The active component, indigotin, is obtained by fermentation, oxidation and boiling the dried plant leaves; in this process, 100 kg of dried plants yields 1.5–2.0 kg of indigo dye; but now it is usually synthetic. It is of low substantivity, so multiple 'dips' are required to produce strong shades. Unlike other vat dyes, it exhibits quite poor wash-fastness. Natural and synthetic indigo are chemically the same. Natural indigo quickly declined in importance with the development of synthetic indigo in 1880.

Indigo carmine (sodium salt of indigo disulphonic acid): It is a blue water-soluble powder. Very sensitive to oxidizing agents. *Uses:* test for nitrates, chlorates, wool damage.

Indigo dyeing/sizing machines: Continuous plants for dyeing warp yarns with indigo in full open width followed by sizing.

Indigo print (blue print): Cotton fabric with white motifs on a blue ground. The existence of blue prints in Egypt had been reported by Pliny the Elder (c. 23–79 AD). The technique was first introduced into Europe at the end of the 17th Century by Jeremias Neuhofer of Holland via Augsburg in Germany. The technique is still occasionally practised today in Westphalia and Bavaria and more extensively in Slovakia. It involves the production of a white or coloured resist as an illuminating colour under indigo blue. The printing operation is carried out with the aid of wood blocks.

Indigo style in African prints:

(a) *Real wax* – the oldest and most artistic technique involves covering (resisting) parts of the textile material with a wax-colophony mixture. The design is produced by subsequent dyeing with indigo. Since the wax resists are fairly brittle, irregular and extremely fine cracks are formed which results in the typical veining or crackle effects. After dyeing, the wax is removed (e.g. by saponification) and the remaining colours are applied either by repeating the wax resist treatment or by block or screen printing.

(b) *Imi-wax* – the industrial production of real wax prints is extremely laborious and costly. For this reason, a technical imitation under the name imi-wax has been introduced in practice. The imi-wax technique is based on the same designs and colours typically used for real wax prints. However, imi-wax prints are mainly produced on naphtholated grounds in order to imitate indigo with diazo fast colour salts which are also used to obtain all other shades that are technically possible. Other dye classes are used to obtain shades which cannot be achieved with diazo fast colour salts.

(c) *Real indigo* – for this style, the white cotton fabric is printed directly with indigo followed by fixation and washing off. This technique is used to produce high coverage blue/white prints often supplemented by other colours.

(d) *Imi-Indigo* –the real indigo style described above is also imitated industrially. In this process, all the dyes are applied in a single operation and then fixed.

(e) *Plangi* – this is a very old technique for the production of resist effects by gathering, folding, or rolling the fabric, usually held with stitching, to form specific patterns on subsequent dyeing.

Indigo test: It is the most commonly used test for the identification of indigo dyeings. Spotting with concentric acid produces a yellow stain surrounded by green edges (= oxidation of indigo to the yellow coloured satin).

Indigo white: The colourless form to which **indigo** is reduced to enable it to be taken up by wool for dyeing blue.

Indirect costs: Those costs associated with manufacture that cannot be directly attributed to any individual product or job. Such costs are also termed overheads.

Indirect warping: The transference of yarn from package creel on to a swift from which it is subsequently wound onto a beam.

Indirect yarn numbering system: A system that expresses yarn number as the reciprocal of linear density.

INDITECNOR (Sp.): Instituto Nacional de Investigaciones Tecnólogicas y Normalización, Santiago, Chile (Chilean Standards Institute).

Individual Fibrills: Also known as elementary fibrils, basic fibrils, microfibrils, protofibrils, capillaries. This refers to the initial developmental stage of the next highest cell organisation of the fibril bundle or fibrils and is constructed of micelles or crystallites in parallel bundles. Approximate diameter in wool – crystallite 0.005 μm, individual fibril 0.01–0.04 μm, fibril bundle 0.1– 0.3μm.

Indoor furniture: Furniture manufactured for the use in the interior of a building.

Indophenols: All quinoneimine derivatives are commonly called indophenols by dye manufacturers, even though they are often indoanilines or indamines. The following methods can be used to prepare indophenols –

(a) Oxidation of the starting materials with hypochlorite at 0°C, with MnO_2, or with dichromate. (b) Condensation of the starting materials in concentrated H_2SO_4, in some cases at 0.25–0.35°C. (c) Condensation of compounds containing activated chlorine with amines or aminophenols.

(d) Condensation of naphthols with *p*-aminophenol in $NaHSO_3$ solution or by dry heating.

(e) Condensation of amines with hydroquinone.

Industrial waste-water management: The objectives should be to minimize wastewater production and to recover and reuse wastewater here economically viable. This enables the cost of waste-water treatment to be minimised. The options for the industrialist is to

(a) treat or partially treat at source within the site;

(b) treat at a centralised works effluent plant within the industrial complex or

(c) discharge the wastewater to a sewer and pay for off-site treatment. For off-site treatment, the chemical constituents in the industrial effluent must be checked so that problems do not occur with sewer corrosion or toxicity to the existing biological treatment. The extra BOD load from the industrialist can affect the biological treatment and may generate more sludge (many toxic compounds accumulate in the sludge). The discharge of industrial effluents to a sewer must also consider the hydraulic capacity of the sewers (and waste-water treatment works).

Industrial yarn: A yarn composed of continuous filaments, usually of high breaking tenacity, produced with or without twist, and intended for applications in which functional properties are of primary importance; for example, in reinforcing material in elastomeric products (tyres, hose, belting), in protective coverings, and in cordage and webbings, etc.

Industrial fabric: A broad term for fabrics used for non-apparel and non-decorative uses.

They fall into several classes –

(a) A broad group including fabrics employed in industrial processes (e.g., filtering, polishing, and absorption).

(b) Fabrics combined with other materials to produce a different type of product (e.g., rubberized fabric for hose, belting, and tires; fabric combined with synthetic resins to be used for timing gears and electrical machinery parts; coated or enamelled fabrics for automobile tops and book bindings; and fabrics impregnated with adhesive and dielectric compounds for application in the electrical industry).

(c) Fabrics incorporated directly in a finished product (e.g., sails, tarpaulins, tents, awnings, and specialty belts for agricultural machinery, airplanes, and conveyors). Fabrics developed for industrial uses cover a wide variety of widths, weights and constructions and are attained, in many cases, only after painstaking research and experiment. Cotton and manufactured fibres are important fibres in this group, but virtually all textile fibres have industrial uses. The names mechanical fabrics or technical fabrics sometimes have been applied to certain industrial fabrics.

Industrial nylon: A heavy-duty plain weave nylon fabric made in limited range of colours. It is fairly stiff and does not crease. Used for overalls and protective clothing, but is also useful for the bags of men's trouser pockets as it is extremely hard wearing.

Industrial gums: It is also known as industrial printing gum, crystal gum, labiche gum, sheet rubber, drying rubber. Industrially manufactured vegetable gums, preferably from Asia. Gum and Rosaceae rubber which is dissociated under pressure or using oxidation methods, as a result of which the gum is already cold water soluble and suitable for sprinkling in powder form to thicken print dye pastes.

Industrial soaps (technical soaps): Special soaps used for particular technical applications, e.g. textile soaps, etc.

Industrial textiles: (a) A term used in few countries in the sense of technical textiles which are not always used for industrial purposes, but also for consumer goods with a technical function; (b) A term used for rental linen.

Industrial yarn: See **Yarn (industrial)**.

Inert: In Latin, the word *inertia* means "inactive, sluggish". Slow to react, unreactive, indifferent and passive. Polytetrafluoroethylene, for example, is practically inert to all organic solvents and common chemicals.

Infamoda inc.: Fashion and trend forecasting agency.

Inflatable restraint: A vehicular safety device designed to cushion an occupant or equipment during collision; airbag.

Inflatable structure: Structures opened or enlarged by input of air and, once enlarged, able to retain the air to maintain the distended position.

Inflator: A device for generating and directing expansion gases into a cushion.

Inflow quench: Cooling air for extruded polymer filaments that is directed radially inward across the path of the filaments. The thread line is completely enclosed in a quench cabinet in inflow quenching.

Infrared: Electromagnetic radiation with longer wavelengths than visible radiation. The wavelength range is approximately 0.7 μm to 1 mm. Many materials transparent to visible light are opaque to infrared, including glass. Rock salt, quartz, germanium, or polyethene prisms and lenses are suitable for use with infrared. Infrared radiation is produced by movement of charges on the molecular scale, i.e. by vibrational or rotational motion of molecules. Infrared spectroscopy is of particular importance in organic chemistry and absorption spectra are used extensively in identifying compounds. Certain bonds between pairs of atoms (C–C, C=C, C=O, etc.) have characteristic vibrational frequencies, which correspond to bands in the infrared spectrum. Infrared spectra are thus used in finding the structures of new organic compounds by indicating the presence of certain groups. They are also used to 'fingerprint' and thus identify known compounds. At shorter wavelengths, infrared absorption corresponds to transitions between rotational energy levels, and can be used to find the dimensions of molecules (by their moment of inertia). Infrared (IR) radiation is long-wave radiation which is not visible to the human eye which is 400–780 nm. The entire IR-spectrum extends from 780–4,00,000 nm of which only the smaller component of the vibration spectrum between 780–3000 nm is of most interest. (Lat.: *infra* = below, beneath).

Infrared absorbing dyeing: Infrared dyeing are such that can absorb the infrared radiations from the sunlight. Such dyeings are particularly suitable for heat-retaining winter clothing. Since approximately 50% of sunlight is made up of infrared radiation, dyes which have a pronounced absorption capacity for infrared (IR) offer the possibility to convert absorbed radiant energy into heat.

Infrared drying: The drying process where IR radiation is used for drying. In IR drying technology, the invisible IR-radiation from electromagnetic wave radiation, referred to by physicists as infrared radiation, is used by preference. In terms of drying technology, the most significant spectral range is 700–3000 nm. In this range, the particles are displaced in more rapid, random vibrations; thus, warming the product. In order to achieve a high level of profitability when heating materials, IR radiators with emissions essentially within the area around 3000 nm must be used. As the absorption ratios of textile fabrics and water are also very favourable in the order of 3000 nm (virtually 100%), radiation temperatures of 500–800 °C are used.

Infrared fixation: For example, the two-phase infrared process (Hoechst) – printed fabric is padded and rapidly heated up without steam. No condensation occurs, so there is less tendency for the printed fabric to mark-off or bleed. The recommended wavelength is 1300 nm at which a maximum radiation absorption of 80–85% is achieved.

Infrared laboratory dyeing machines: The use of a liquid heat transfer medium (e.g. glycol) in laboratory dyeing machines can be substituted by heating the rotating dye beakers with infrared radiation; a particularly intense absorption of heat is achieved with black beakers by this means.

Infrared pre-driers: Radiation driers are generally only used in the first stage of drying. As they have a high power density and achieve a high spec. evaporative capacity with well-balanced emission and absorption ratios, there is a risk of surface damage to the goods if the wet bulb temperature is not adhered to. Water-tight infrared radiation fields in working widths up to 6 m are employed especially as pre-driers in printing, dyeing and finishing applications. Both electrically-heated and gas-heated radiators are available. See **Infrared drying**.

Infrared radiation: Radiant energy for which the wavelengths of the monochromatic components are greater than those for visible radiation and less than about 1 mm.

NOTE: The limits of the spectral range of infrared radiation are not well-defined and may vary according to the user. Committee E-2.1.2 of the CIE distinguishes in the spectral range between 780 nm and 1 mm –

IR-A 780–1400 nm

IR-B 1.4–3.0 μm

IR-C 3 μm–1 mm

Infrared reflectance: As far as textiles are concerned, infrared reflectance is, in general, the diffuse reflection of light rays with wavelengths between 700–

1200 nm. Infrared reflectance plays an important role in camouflage dyeing and prints.

Infrared-reflecting dyeing: These are dyeing using IR reflecting dyes. These dyeings represent the opposite of infrared-absorbing dyeings. Such dyeings are particularly suitable for cooling summer or tropical wear. Selected dyes with a pronounced reflecting capacity for IR-radiation, which are also described as "cold dyes", are used for this purpose.

Infrared spectroscopy (IR-spectroscopy): An optical technique for measuring the absorption spectra of solid, liquid or gaseous organic compounds in the IR-region. The technique is also used for the determination of isomers and homologues. The IR-absorption curve obtained is compared with the known curves in a catalogue (e.g. Landolt-Börnstein) which permits a rapid identification of constitution.

Infusion technology: An infused polymer construction process that reinforces the fabric of outerwear garments in the places where they take the most abuse – zipper and pocket flaps, and other high-abrasion areas. The technology blends polymers, penetrates deep into the inner fibres, and surrounds them to form a permanent bond. This tough, resilient matrix ensures a highly wear-resistant surface while allowing the fabric to remain lightweight and flexible. The infused polymer process eliminates the need for heavier-weight abrasion overlays, tapes and bindings, and adds increased strength to the most crucial points on the garment, which dramatically extends the life of the garment.

Ingénue: Yellowish-green colour.

Ingrain dyes: An imprecise description, formerly used for all colourants which were produced *in situ* on the fibre by the development or coupling of one or more intermediate compounds.

Inherent flame resistant: Flame resistance that derives from an essential characteristic of the fibre from which the textile is made.

Inherent viscosity: Viscosity of a polymer dissolved in a solvent.

Inherently flame resistant: (1) Having inherent flame resistance.

(2) As applied to textiles, flame resistance that derives from an essential characteristic of the fibre from which the textile is made.

Inhibitor: A substance that retards or prevents a chemical or physical change. In textiles, a chemical agent that is added to prevent fading, degradation, or other undesirable effects.

In house/staff designers: Designers employed by a company, usually on a fulltime basis (although some may be employed part time).

Initial modulus, in stress strain curve: The slope of the initial straight line portion of the stress-strain (or force-elongation) curve. The modulus is the ratio of the change in stress, expressed in Newton's per tex, grams-force per tex, or grams-force per denier, to the change in strain expressed as a fraction of the original length.

Initiator: A chemical added to start a reaction such as polymerization. Unlike catalysts, initiators may be consumed during the reaction.

Injection moulded (button): A method of forming which requires the filling of a cavity under pressure with polymer that will take the form of a cavity under pressure with polymer that will take the form of the mould when cooled.

Injection moulding: The process of forming a material by forcing it, in a fluid state under pressure, through a runner system [sprue, runner, gate(s)] into the cavity of a closed mould.

Injector: A device used in the colour kitchen of a printing plant for the preparation of, e.g. thickeners. The substance to be dissolved is sucked into the injector by low pressure. The low pressure is produced by spraying water through nozzles with an overpressure. By this means, the substance is wetted-out at the same time. The concentrate thus formed then passes from the injector into a mixer turbine.

Ink: (1) A dark blue colour.

(2) Coloured inks are aqueous solutions of coal tar dyes with suitable additives. Both cationic dyes (e.g. rhodamine, fuchsine, and methylene blue) and acid dyes (e.g. eosin, ponceau, nigrosine, and acid blue) are used. Black inks consist mainly of a mixture of Gallic acid, tannic acid and an iron (III) salt (iron gall ink). Uses: dyeing, dry-cleaning and varnishes.

Ink jet: A stream of liquid in the form of a dye solution produced by a spray nozzle.

Inkjet printing: Dye liquors (inks) are squirted through fine microprocessor controlled jet nozzles. This operation is carried out almost silently at high speed on surfaces which can be in any form, e.g. eggs, wadding, metal cans, etc. Special requirements are placed on the inks and their additives both for the ink jet technology itself as well as for the various media to be printed. The inks must be of low viscosity, have adequate fastness to light and water, and a pH as close as possible to neutral. They must also be rapid-drying without, at the same time, drying-out in the printer head, and must also be non-corroding and possess a certain surface tension. The first functionally-reliable apparatus.

Ink-jet printing technology: (a) Continuous ink-jet printing – In this form of ink-jet printing, the continuous stream of dye liquor forced out of a fine jet nozzle is controlled either by deflection or superimposition of a deflector plate (electro-colour process) or deflection by means of compressed air (Millitron process). The excess dye liquor is collected and returned to the storage tank. (b) Drop on demand ink-jet printing – In this case, discrete drops of dye liquor are jetted by switchable electromagnetic valves located in the dye liquor feed tubes in a predetermined sequence determined by the pattern being printed chromotronic process.

(c) Quantitative control of the ink jet as in the Agfa ink-jet process, etc. Ink-jet printing in the non-textile sector.

Inkle: A narrow fabric, a ribbon, tape or webbing.

Inkle loom: A simple form of narrow handloom. It consists of a frame with pegs which hold a narrow continuous warp (a characteristic of an inkle loom).

Inlay: An extra seam allowance inside the garment for letting out if required or for strength at a given position.

Illusion: Made in silk in Gauze or made on bobbinet machine or knotted. A very fine, all-silk tulle which originated in France. It has a cobweb appearance. Hexagonal open mesh. Made in 52 inch and 72 inch widths. Uses: Veiling, particularly for weddings, trimmings.

Inorganic fibres: Also described as mineral fibres. More exacting requirements in the field of environmental protection, the performance of materials, or the productivity of existing systems also calls for the development and use of newer types of fibres. The importance of inorganic fibres of oxide or non-oxide basis is due, above all, to the extremely high temperature resistance of various physical or chemical properties up to temperatures greater than 1100 °C.

Inseam: (1) The inner seam of the leg of a pair of trousers; sometimes also called the “crotch seam”; the forearm of the sleeve.

(2) The distance from the bottom of a trouser leg to the crotch. The measurement is taken along the inside leg seam that joins the front and the back leg panels.

Insect repellent: An odoriferous substance (repellent) for insects which is particularly effective against flying insects (insect protection). Insect repellents are used to produce insect repellent finishes on textiles which are effective for a limited period of time. Not to be confused with insecticide.

Insect repellent finishes: Finishes applied on textiles which can repel mosquitoes. Useable repellents are, as a rule, neutral, viscous oils of low volatility or crystals with low melting points and, almost without exception, a bitter taste. In order to maintain the longest possible activity, such substances must, in addition, not give rise to skin irritations or cause damage to textiles. Suitable products for textile impregnations include, e.g. indalone, undecenoic acid, mandelic acid hexyl ester, N-cyclohexyl-2-(butoxyethoxy) acetamide, etc., and sometimes plant products are also used now a days. Some insect repellent hosiery recommends saturation with quaternary ammonium compounds followed by subsequent drying and heating to approximately 100–150 °C.

Insect screening (in coated glass yarn fabrics): Woven netting having an approximately even spaced mesh of 12 by 12 yarn or more per 25.4 mm (1 inch).

Inserted seam: A seam in which single or number of plies of material is inserted between two others which are turned in. The complete assembly is sewn in one operation.

Insertion lace: This is usually narrow and often with slits for threading ribbon through, but it always has two straight edges. It is inserted as decorations and the backing fabric is usually curt away afterwards.

Inside seam: A seam formed in which the completed seam allowance is located on the interior of the object, usually on the back side of the fabric.

Inside-leg length (in body measurements): The vertical distance from the crotch to the soles of the feet.

Inspection: The process of measuring, examining, testing, gauging, or otherwise comparing a characteristic or property of a material with applicable requirements. In this case, only by visual examination.

Inspection machine: Textile fabrics are examined on inspection machines after completion of the final finishing process. The fabric inspector carries out the following tasks – quality inspection, detection and recording of faults, defect marking. The fabric runs at full width over an inspection table, an inclined table top which, under certain circumstances, is illuminated from below towards the fabric inspector (matt glass plate).

Inspection (in fabric grading): The process of viewing, measuring, examining, or otherwise comparing the visual characteristics of a fabric with applicable requirements.

Institutional fabric: Household textiles used in the contract market.

Instron tensile tester: A high precision electronic test instrument designed for testing a variety of material under a broad range of test conditions. It is used to measure and chart the load-elongation properties of fibres, yarns, fabrics, webbings, plastics, films, rubber, leather, paper, etc. May also be used to measure such properties as tear resistance and resistance to compression.

Instrumental detection level, effluent analysis (IDL): Constituent concentration that produces a signal greater than five times the signal/noise ratio of the instrument. IDL is, in many respects, similar to 'critical level' and 'criterion of detection'. The later level is stated as 1.645 times the standard deviations of analyses.

Insulation: With respect to a fabric, a material that protects from the loss of warmth or the penetration of cold.

Insulation ratio (effective): An indication of the increase in insulation afforded by the fabric in comparison to the uncovered test plate under specified condition of test. See **Effective insulation ratio**.

Intaglio: A method of printing in which the design is engraved beneath the surface of a printing plate or cylinder, e.g. as in roller printing.

Intarsia: It is a special method of producing designs in knitted loops that form self-contained areas of pure colours. Unequalled colour definition is achieved, with a large number of colours and no adverse effect on the physical properties of the structure such as reduction of extensibility.

Integration dyeing: A method of dyeing which involves the maintenance of the concentration of dyestuff in the dye bath at the lowest possible level throughout the dyeing process. This is achieved by gradually introducing a dyestuff to the dye bath in accordance with the known diffusion rate of the dyestuff or dyestuffs in question. Integration dyeing is of particular advantage when dyeing using dyestuffs which have a high affinity to the fibre, but a low diffusion rate which makes possible the use of direct and vat dyes with which level dyeing is difficult.

Integrator (in tensile testing of textiles): A device that calculates the coefficient of variation unevenness or the mean deviation unevenness. A device for obtaining the time integral of the load.

Integrator (linear): See **Linear integrator**.

Integrator (quadraic): See **Quadraic integrator.**

Intensity: (a) The amount of energy per unit (space, charge and time); (b) The brilliance of a colour; (c) The brightness of light.

Interaction: The condition that exists among factors when a test result obtained at one level of a factor is dependent on the level of one or more additional factors.

Intercellular: A term meaning between cells, and refers to the non-crystalline (amorphous) regions, e.g. in fibres.

Interfacial polymerisation: Polymerization in which two reactive monomers, each dissolved in different solvents that are mutually immiscible, react at the interface between the two solutions.

Interfacing: Woven or non-woven fabric used to give body to certain area of the garment. It adds to the outward appearance of the garment if some areas are reinforced, e.g. collars, buttoned openings, cuffs, pocket flaps, etc. And this structure also makes the garment last longer. Variety of interfacings such as fusible or non-fusible depending on the need is available.

Interference (in testing): An effect due to the presence of a constituent or characteristic that influences the measurement of another constituent or characteristic.

Interfibrillar swelling: Swelling which results from the penetration by water (or other polar solvents) of cotton fibres. The swelling agent may break hydrogen bonds between adjacent chain molecules.

Interlace: Interlace is a mechanical yarn cohesion of (carpet) yarns which are manufactured with zero twist and formed by the interlacing of the individual filaments.

Interlacing: The passing of a thread over or around another thread or loop of another thread.

Interlining: (1) Any textile which is intended for incorporation into an article of wearing apparel as layer between an outer shell and inner lining.

(2) More or less heavily finished fabric. Interlinings such as buckram, linen interlining, horsehair cloth, non-woven material, etc., as well as wadding, shoulder padding, etc., which are fused or sewn inside specific parts of a garment (i.e. between the inner and outer layers) to improve shape retention, strength, warmth or bulk.

Interlining (linings): Interlinings are the “skeleton of clothing” which ensure a lasting fit. They are sewn to the outer fabric in the form of fabrics, knitwear or non-woven or combined directly with these (fixed). A layer of material inserted in certain parts of the garment between the inner and outer fabric for shape retention, strength, warmth, or bulk.

Interlock: Interlock materials are fine meshed knit goods produced on the circular knitting machine equipped with two sets of needles. It is double-faced rib-based structure consisting of two 1 × 1 rib fabrics joined by interlocking sinker loops which are used for close-fitting undergarments, etc. An alternative form of this is referred to as cross tubular goods. Only plain stitches are visible on either side of the fabric as the reverse stitches are found inside the knitted fabric.

Interlock knit: To produce an interlock knit, long and short needles are arranged alternately in both the dial and cylinder; the needles in the dial and cylinder are also positioned in direct alignment. When the long and short needles knit in alternate feeds in both needle housings, a fabric with a type of cross 1 × 1 rib effect is produced.

Interlock knitted fabric: Like rib fabrics, there are several types of interlock fabrics such as 1 × 1 interlock, 2 ×2 interlock. 1 × 1 interlock fabrics are thicker, heavier, narrower and more stable and expensive than plain and rib knit fabrics. It has less width-wise recoverable stretch than plain and rib knit fabric. The V letter appearance is dominant on both sides of the fabric. The yarn can be unrobed easily from the last course of the fabric. 1 × 1 interlock fabric does not curl from the sides and the top/bottom. However, it is less productive than rib and plain knitted fabrics.

Inter-looping: The passing of a loop of thread through another loop formed by the same thread.

Intermingled yarn: A multifilament yarn in which cohesion is imparted to the filament bundle by entwining the filaments instead of or in addition to twisting. The effect is usually achieved by passing the yarn under light tension through the turbulent zone of an air jet. *Note*: (a) Some manufacturers describe such product as an interlaced yarn; (b) Intermingling should be distinguished from air-jet texturizing when a high level of entanglement is achieved with the objective of producing texture or bulk.

Intermingling: (a) Use of air jets to create turbulence to entangle the filaments of continuous filaments yarns, without forming loops, after extrusion. Provides dimensional stability and cohesion for further processing but is not of itself a texturing process. It is compatible with high-speed spin-drawing and high-speed take-up. When compared with twisting processes, it also permits increased take-up package size; (b) Combining two or more yarns via an intermingling jet. Can be used to get special effect yarns, i.e., mixing dye variants to get heather effects upon subsequent dyeing.

Intermittent pattern: A pattern occurring in interrupted sequence.

Intermittent roller printing machine: This is a type of roller printing machine which is used in printing of saris, African prints saorngs etc which cannot be printed in normal roller printing because of the speciality of the designs. These machines have two sets of engraved printing rollers. For example taking the case of saris, one set of the rollers will print the body designs for a particular length of the fabric while the second set of rollers are idle and away from contact with the pressure cylinder. When the pallu area comes the second set of the rollers come into action when the previous set of rollers 'jump away' from contact with the pressure cylinder. After the predetermined length of the pallu has been printed again the first set of rollers take offer and the present set of rollers are taken away.

Intermolecular forces: Secondary valence forces among different molecules.

Internal dye variability: The change from point to point in dye uniformity across the diameter and along the length of the individual filaments. Affects appearance of the dyed product and is a function of fibre, dye, dyeing process and dye bath characteristics.

International grey scale: Two series of pairs of chips which show increasing contrast within pairs and are used visually for comparing the differences in colour of textile specimens or the degrees of staining of transfer cloths attached to the test specimens, that occur during colour fastness testing. (a) *Basic scale for assessing change in colour*– a scale that consists of five pairs of non-glossy grey-coloured chips, and in which a fastness rating of 5 (indicating no colour change) at one end of the scale is represented by two identical grey chips, and a fastness rating of 1 (indicating a severe colour change) at the other end of the scale is represented by a pair of grey chips having a colour difference of 13,6 cielab units. (b) *Basic scale for assessing staining* – a scale that consists of one pair of white and four pairs of non-glossy grey- and white-coloured chips, and in which a fastness rating of 5 (indicating severe staining) at one end of the scale is represented by the pair of white chips, and a fastness rating of 1 (indicating severe staining) at the other end of the scale is represented by a pair of chips having a colour difference of 34,1 cie lab units. *NOTE*: Both scales may be augmented to form a 9-step scale by the provision of similar chips that illustrate the perceived colour differences corresponding to the half-step fastness ratings.See **Grey scales for fastness testing**.

International orange: A vivid orange colour.

International system of units: SI System of units.

International Textile Club (ITC): International professional organization of leading specialists from all branches of the textile industry; technical and professional organizations.

International Textile Service (ITS): A Swiss publishing house providing information on textile production. Publishes the quarterly technical journal "International Textile Bulletin" in several languages.

International Wool Secretariat (IWS): Independent organization for publicity and research service in the entire wool economy. Founded in 1937. Supported by the wool producers of Australia, New Zealand and South Africa with headquarters in London and branches in almost all textile countries.

International Wool Textile Organization (IWTO): Head organization of the world wool economy, resident in the United Kingdom.

Intimate blend: A technique of mixing two or more dissimilar fibres in a very uniform mixture usually by mixing the fibres in loose stock form before or at the picker, in contrast to normal mixtures of warp and weft (e.g. polyester warp wool weft, and vice versa).

Intracellular: A term describing processes within the cell (crystallite), e.g. the further penetration of intracellular reactions from the intercellular spaces through the crystalline surface of the cell causing changes in the structure of the cellular building blocks. Of importance in severe fibre damage (acids: cellulose, caustic alkalis: wool).

Intrafibrillar swelling: A more intensive swelling, which can result in a molecule rearrangement of the cellulose chain molecules. Interfibrillar swelling always precedes intrafibrillar swelling.

Intramolecular forces: Secondary valency forces within the same molecule.

Intrinsic clo: See **Clo**.

Intrinsic viscosity: Ratio of the specific viscosity (R.V.-1) of a solution of known concentration to the concentration of solute extrapolated to zero concentration. Also called the limiting viscosity number. It is directly proportional to the polymer-average molecular weight.

Intussusception dyeing: (Lat.: *intus* = within, *susceptio* = deposition). A deposition dyeing in which new dye is progressively inserted into the thickness of an existing fibre wall. The cross-section of an intussusception dyeing is fully penetrated by dye, i.e. homogeneously dyed throughout, e.g. acetate acid wool and cationic dyeing. Compare **Apposition dyeing**.

Inverness: A long, loose overcoat with cape, without sleeves, for full dress wear.

Invisible green: A very dark green colour, so called because it verges on black.

Invisible waste (in cotton testing): Weight loss due to dust, moisture, loose fibres, etc., carried away by the air stream during the test.

Iodine number (iodine value): A measure of the amount of unsaturated acids present in fats and oils (i.e. the number of double bonds). It is obtained by finding the percentage of iodine by weight absorbed by the sample in a given time under standard conditions, i.e. the number of grams of iodine absorbed by 100 g of a fat or oil.

Iodine starch test: The normal concentration of an enzyme in a desizing bath has an activity in the range from 50–300 Kilo-Novo-Units per litre

(KNU/l) depending on the quantity of enzyme preparation used. The activity of desizing liquor can be determined by the following method. The alpha-amylase in the desizing liquor degrades starch substrates. An iodine solution is used as an indicator so that the course of the reaction can be followed, and the time required to complete the degradation process is a measure of the enzyme activity.

Iodine-Potassium Iodide solution (iodine test): 2 g iodine in 10 ml saturated potassium iodide solution (or 3 parts potassium iodide and 2 parts iodine in 45 parts water). Iodine is so loosely bound that it is liberated immediately in reactions.

Iodine yellow: A bright yellow.

Iodo-, iono- (G): Violet.

Ion: An electrically charged particle resulting from adding or removing electrons to an atom or a group of bonded atoms. Many chemicals ionize in solutions in water. For example, sodium chloride in solution ionizes to positively charged sodium ions (Na+) and negatively charged chlorine ions (Cl^-) – one electron that was shared between the sodium and the chlorine in the molecular form of salt leaves the chlorine and goes with the sodium (sodium chloride is actually an ionic solid). Some chemicals used in dyeing, such as *acetic acid* and *sodium carbonate*, partially ionize in solution, with some remaining in intact molecular form. With chemicals of this nature, if some of the ions in solution are used up in reactions, more of the molecules will ionize.

Ion exchange: A process that takes place in certain insoluble materials, which contain ions capable of exchanging with ions having the same charge in the surrounding medium. ZEOLITES, the first ion exchange materials, were used for water softening. These have largely been replaced by synthetic resins made of an inert backbone material, such as polyphenylethene, to which ionic groups are weakly attached. If the ions exchanged are positive, the resin is a CATIONIC RESIN. An ANIONIC RESIN exchanges negative ions. When all available ions on the resin have been exchanged (e.g. sodium ions replaced by calcium ions from the medium) the material can usually be regenerated by passing a concentrated solution containing the original resin ion (e.g. sodium chloride) through it. In the example cited, the calcium ions will be replaced by sodium ions. Ion-exchange techniques are used for a vast range of purification and analytical purposes. For example, solutions with ions that can be exchanged for OH^- and H^+ can be estimated by titrating the resulting solution with an acid or a base.

Ion exchange resins: Synthetic ion-exchange resins consist of various insoluble copolymers having a cross-linked three-dimensional structure to which ionic groups have been attached. An anionic resin has negative ions built into its structure and therefore exchanges positive ions. A cationic resin has positive ions built in and exchanges negative ions. Ion-exchange resins are synthetic organic polymers containing side-groups that can be ionized. Organoliths.

Ionic bond: It is also known as electrostatic bond, polar bond, hetero-polar bond, electrovalent bond. A chemical bonds which is based on the strong electrostatic attraction forces between cations and anions, and which owes its stability to the fact that certain atoms readily give up electrons and other atoms take up such electrons again.

Ionic bond: A chemical bond as a result of electrical attraction between positive and negative ions; also called salt linkage In ionic bonding, typically an electron is transferred from one atom to another, leaving one with a positive charge and one with a negative charge. These electrical charges produce strong attraction between the different charged ions. Ionic bonds are quite strong, but not generally as strong as covalent bonds (where pairs of electrons are shared equally) between nuclei. Such bonds are important with acid dyes and basic (cationic) dyes.

Ionic copolymerisation: In ionic copolymerisation, copolymer is formed by monomer units, which are polymerised by ionic initiators. The fundamental principle of ionic copolymerisation is almost same to free-radical copolymerisation. There is, however, a major difference that in free-radical copolymerisation, the reactivity ratios '1 and '2 are dependent only on the monomer pairs M1 and M2, and temperature of polymerisation, whereas in ionic copolymerisation, the reactivity ratios dependent not only on the monomer pair and temperature, but also on the solvent and the initiators which are used in that system.

Ionic formula: A type of formula which provides information on the composition of ions, their charge and charge sign, e.g. Na^+, Fe^{2+}, S^{2-}, SO_4^{2-}.

Ionic strength: The ionic strength of a solution is measure of the concentration of dissolved chemical constituents.

Ionization: (a) The process of producing ions. There are several ways in which ions may be formed from atoms or molecules. In certain chemical reactions, ionization occurs by transfer of electrons; for example, sodium atoms and chlorine atoms react to form sodium chloride, which consists of sodium ions (Na^+) and chloride ions (Cl^-). Certain molecules can ionize in solution; acids, for example, form hydrogen ions as in the reaction H_2SO_4.$2H^+ + SO_4$.

(b) The 'driving force' for ionization in a solution is solvation of the ions by molecules of the solvent. H^+, for example, is solvated as a hydroxonium (hydronium) ion, $H3O^+$. Ions can also be produced by ionizing radiation; i.e. by the impact of particles or photons with sufficient energy to break up molecules or detach electrons from atoms: $A.A^+ + e^-$. Negative ions can be formed by capture of electrons by atoms or molecules: $A + e^-.A^-$.

Ionization potential: (IP; Symbol: *I*) The energy required to remove an electron from an atom (or molecule or group) in the gas phase, i.e. the energy required for the process: $M.M^+ + e^-$. It gives a measure of the ability of metals to form positive ions. The second ionization potential is the energy required to remove two electrons and form a doubly charged ion: $M.M_2^+ + 2e^-$. Ionization potentials stated in this way are positive; often they are given in electron volts. *Ionization energy* is the energy required to ionize one mole of the substance, and is usually stated in kilojoules per mole (kJ mol^{-1}). In chemistry, the terms 'second', 'third', etc., ionization potentials are usually used for the formation of doubly, triply, etc., charged ions. However, in spectroscopy and physics, they are often used with a different meaning. The second ionization potential is the energy to remove the second least strongly bound electron in forming a single charge ion. For lithium (ls22s1) it would refer to removal of a 1s electron to produce an excited ion with the configuration 1s12s1.

Ionizers: Ionizers are devices which ionize the air by means of high voltage therefore removing the electrical charge on textiles.

Ions: (Gk.: *ienai* = to go, move). Atoms or groups of atoms carrying one or several positive or negative electrical charges. The ions present in an aqueous solution migrate to the oppositely charged electrodes when a direct current voltage is applied.

Ionomer: A polymer having covalent bonds between the constituents of the long-chain molecules and ionic bonds between the chains.

IONTEC: Colombian standards organization; Technical and professional organizations.

IOS: Iraqi Organization for Standardization, Baghdad, Iraq (technical and professional organizations).

IPN: Interpenetrating polymer network. Mutually penetrating networks of polymers at the molecular level.

IPR: Inches per rack.

IR: Infrared.

IRAM: Instituto Argentino de Racionalización de Materiales, Buenos Aires, Argentina (Argentinian Standards Institute, including textile standards).

Iraq red: A deep red colour.

IRCT: Institut de Recherches du Coton et des Textiles Exotiques, Paris, France (French Research Institute for Cotton and Exotic Fibres).

Iron grey: A dark grey colour.

IRS: Irish Standard.,

Irish lawn: A lawn fabric produced from linen yarns. See **Lawn**.

Irish linen: The fine quality linen, full bleached fabric plain woven Ireland using Irish flax. Used for various purposes like handkerchief (fine varieties), towels and glass clothes (coarser varieties), shirts, collars and cuffs; usually 36 inches wide.

Irish point: (a) Lace with needle-point or bobbin-made sprigs sewn to machine made net, this sometimes being cut away under the patterns; (b) Curtain lace, with patterns mounted on a machine made net.

Irish poplin: A poplin originally made from a dyed pure silk organzine warp and a fine worsted weft. Polyester is sometimes used as an alternative to silk.

Irish stitch: In embroidery used for grounding; it is a long stitch, carried over five or more threads and it requires fine foundation canvas.

Irish tweed: See **Tweed**.

Irish work: White embroidery on white ground, mostly on handkerchiefs.

Iron (II) chloride: Ferrous chloride; $FeCl_2$. A compound prepared by passing dry hydrogen chloride gas over heated iron. White feathery anhydrous crystals are produced. Hydrated iron(II) chloride ($FeCl_2.6H_2O$) is prepared by reacting excess iron with dilute or concentrated hydrochloric acid. Green crystals of the hexahydrate are obtained on crystallization from solution. Iron (II) chloride is readily soluble in water, producing an acidic solution due to salt hydrolysis.

Iron (III) chloride: Ferric chloride ($FeCl_3$). A compound prepared in the anhydrous state as dark red crystals by passing dry chlorine over heated iron. The product sublimes and is collected in a cooled receiver as brownish-black iridescent scales. The hydrated salt ($FeCl_3.6H_2O$) is prepared by adding excess iron (III) oxide to concentrated hydrochloric acid. On crystallization yellow-brown crystals of the hexahydrate are formed. Iron (III) chloride is very soluble in water and undergoes salt hydrolysis. At temperatures below 400°C, the anhydrous salt exists as a dimer, Fe_2Cl_6.

Iron (II) sulphate: Ferrous sulphate; green vitriol; $FeSO_4.7H_2O$. A compound that occurs in nature as the mineral *melanterite* (or *copperas*). It is made industrially from iron pyrites. In the laboratory iron(II) sulphate is prepared by dissolving excess iron in dilute sulphuric acid. On crystallization, green crystals of the heptahydrate are obtained. Careful heating of the hydrated salt yields anhydrous iron(II) sulphate; on further heating the sulphate decomposes to give iron(III) oxide, sulphur(IV) oxide, and sulphur(VI) oxide. The hydrated crystals oxidize easily on exposure to air owing to the formation of basic iron (III) sulphate. A freshly prepared solution of iron (II) sulphate absorbs nitrogen monoxide.

Iron sulphate: Iron (II) sulphate, read iron two sulphate, ferrous sulphate, FeSO (archaic terms include copperas, green 4 vitriol and iron vitriol. Iron (II) sulphate is a common *mordant* for natural dyes. Solutions are not very stable, since reaction with air will cause precipitation of insoluble iron oxide. Fabric can be coloured by dipping into a solution of iron sulphate, then hanging in a moist environment to allow conversion to iron oxide (rust). Post-treatments such as *tannic acid* can be used to alter the colour.

Ironing: A method of pressing using a heated hand iron, sometimes together with moisture or steam and a gliding motion. Care labelling for garments in ironing – Hot iron: Highest temperature setting; Warm iron: Medium temperature setting; Cool iron: Lowest temperature setting; Do not iron: Items not to be smoothed or finished with an iron; Iron wrong side only: Article to be turned inside out for ironing or pressing; No steam: Steam in any form not to be used; Steam only: Steaming without contact pressure; Steam press or Steam iron: Use iron at steam setting; Iron damp: Article to be ironed should be made moist before ironing; Use press cloth: Use a dry or damp cloth between iron and fabric.

Ironing fastness: Resistance of the shade to ironing. Test: duration 15 s, temperature 200 °C (cotton), 150 °C (wool, silk, viscose), 120 °C (acetate, polyamide, polyester, acrylic); (a) Dry: The heated upper plate of the heating device is placed on the sample for 15 s; (b) Moist: Place a dry sample on dry cotton fabric, cover with cotton fabric which has been wetted to 100% moisture; 15 s as in I; (c) Wet: The sample and cotton fabric are saturated in water and squeezed off to 100% moisture. The wet sample is placed on a dry cotton fabric and covered with the wet cotton fabric; 15 s as in 1. Where methods 2 or 3 are used, bleeding is also assessed.

Iron-on canvas interfacing: An iron-on woven canvas does not give the shaping advantages of the sew-in varieties. However, it is quick to use and is effective on flat areas such as coat fronts, pocket flaps, hemlines (where added body is needed), etc.

Iron-on cotton: A soft cotton, almost like muslin, with adhesive granules on the wrong side. It would add slight stiffness to sift fabrics.

Irradiance: Radiant power per unit area incident on a receiver, typically reported in watts per square meter, W/ (m^2nm).

Irradiation: The time integral of irradiance expressed in joules per square meter (J/m^2).

Irrorate: Coloured with dots or small speckles of colour.

IRS: Irish Standard.

IRSFC: International Rayon and Synthetic Fibres Committee.

Isabel (Isabella): A dingy yellow-grey colour which chambers suggests is possibly named after either the daughter of Philip II who wore the same underclothes for three years or Isabella of Castile who had a similar predilection for keeping on her apparel.

Isabel yellow: The same as ***Isabel.***

Isabelle: French dress goods made with single warp in eight leaf and six filling repeats.

Isabelline: Greyish-yellow; See **Isabel**.

Isbahani: Medieval silk fabric of Arabic origin.

ISCC-NBS:A US system devised in 1955 for allocating names to colours using the adjectives

'vivid', 'brilliant', 'strong', 'deep', 'light', 'dark' and 'pale'.

Isfahan carpet: Isfahan carpets are woven in the Iranian city of the same name as well as in the neighbouring villages. Isfahan, which is situated in the fertile plain of the Zaiandeh river in the centre of Iran, is the second largest city in the country. Isfahan carpets are woven on vertical looms.

-ish: A very old suffix, first used in the description of origins such as 'Pictish'. Also used to mean 'resembling' as in 'childish'. The suffix is now used with adjectives (particularly colours) to indicate the quality of being 'somewhat', 'nearly' or 'slightly' as in 'reddish' and 'bluish'. Some colours are capable of bearing this suffix including black, white, pink, red, brown, yellow, green, blue, grey, ochre and purple, but not orange, gold, silver, or olive or many other colours. Fielding refers to 'duskish' and Albers to 'violetish'.

ISI: Indian Standard Organization.

'Islands-in-the-sea' fibre: A term used for bi-component fibres similar to the matrix/fibril type and described as a multiple interface or multicomponent fibre. The "islands" are fibrils of one or more polymers embedded in the "sea" (or matrix) consisting of a different polymer. The matrix is often dissolved away to leave ultra-fine filaments. The number and titre of these fine filaments is pre-programmed. Such fibres have been used in ion-exchange products as well as in imitation fur materials and in textile fabrics to achieve a different handle.

Island designs: Prints where the pattern unit making up the designs are surrounded by substantial portions of the ground.

Ismaili: Pale and dark striped cotton fabric with a border of red, white and yellow stripes and a selvage of red, in East Africa.

ISMIU: Acronym for the "Institut por Standardizacia, Merki i Izmeritilni Uredi", i.e. the Bulgarian Institute for Standardization, Metrology and Measuring Instruments; technical and professional organizations.

Isodibenzanthrone or isoviolanthrone: These are Anthraquinonoid dyes with a perylene structure. Example is a dye produced by the chlorination of benzanthrone and treatment with alcoholic potassium hydroxide.

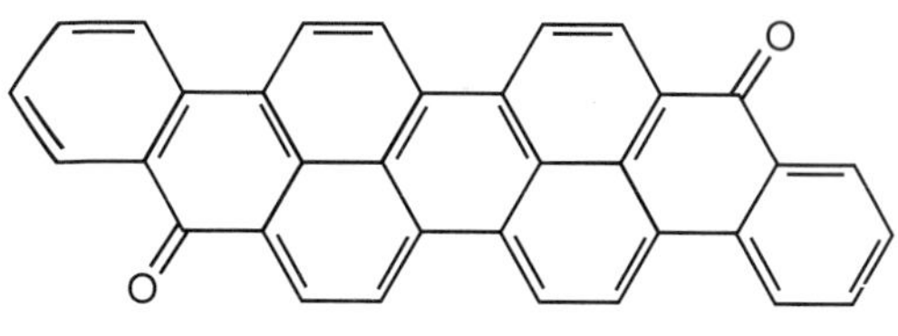

isoviolet anthrone

ISO: International Organization for Standardization Acronym for the International Standards Organization, based in Geneva, Switzerland, founded in 1947. ISO is the international organization responsible for standardization and includes national standards bodies from numerous countries among its members.

ISO 14000 series: A series of international standards on the environment. ISO 14001 gives general guidance on the use of Environmental Management Systems (EMS) and ISO 14004 deals with principles, systems and techniques for EMS. ISO 14010 gives general guidance on auditing environmental management systems. ISO 14011 relates to procedures for auditing environmental management systems. ISO 14012 relates to the qualification of environmental auditors. ISO 14031 details the methodology for environmental performance evaluation.

Isotactic polymer: A polymer structure in which there is a regular spatial or stereo relationship from one repeat unit to the next. A polymer in which all the pendant groups are arranged on the same side of the polymer backbone. Also see **Atactic polymer, Syndiotactic polymer** and **Tactic polymer**.

Isotherm: Constant temperature line used on graphs of climatic conditions or thermodynamic relations, such as pressure-volume relations at constant temperature.

Isotropic: Having the same physical properties in every direction in the plane of a fabric. It is related to the random distribution of fibres in non-woven manufacture.

Isoviolanthrone: See **Isodibenzanthrone**.

Ispahan: Persian woollen rug made with hand tied knots.

Isparta carpet: Most of the carpets known by the name Isparta, or Sparta, have been woven at Smyrna (Izmir) in Turkey. The Persian knot is used. Sparta carpets are among the best from this particular area of Turkey. The decoration and techniques employed in Sparta carpets are very like those of the rough Tabriz carpets, but they can be distinguished from the latter by the different feel of the fabric, i.e. they are limper.

Issits shaker: See **Shoddy shaker**.

Istaberk: An East Indian silk satin, woven of opalescent wild silk.

ITA: Instituto Textil Argentino, Buenos Aires, Argentina (Argentinian Textile Research Institute).

Italian cloth: Smooth, highly lustrous lining fabric (cotton, or cotton warp and worsted weft = half-wool Italian cloth) in satin weave.

Italian ferret: Silk tape or binding for flannels and dressing gowns.

Italian pink: A brilliant yellow lake made from quercitron bark. See **English pink**.

Italian stitch: Same as Holbein stitch.

ITC: International Textile Club.

ITCRA: International Textile Care and Rental Association, United Kingdom.

Item: (1) An object or quantity of material on which a set of observations can be made.

(2) An observed value or test result obtained from an object or quantity of material.

ITF: Institute Textile de France (French Textile Research Institute).

ITMA: Internationale Textilmaschinen-Ausstellung (German) (International Textile Machinery Exhibition).

ITMF: International Textile Manufacturers' Federation, Headquarters – Zürich, Switzerland. Founded in 1978 as the successor to the IF-CATI. Umbrella organization of 32 textile associations in 38 countries on all continents; technical and professional organizations.

ITS: International Textile Service.

IW dyeing method: A method for dyeing those vat dyes (IW dyes) which have the maximum affinity for cotton at 40–50 °C. Vat dyeing.

IWS: International Wool Secretariat.

IWS cold-dwell printing process for wool: A printing process for the production of fast and brilliant shades on wool with reactive dyes of high reactivity which are stable to sodium bisulphite. The print pastes contain approximately 300 g/kg urea and 0–20 g/kg sodium bisulphite depending on the dye concentration. After printing, the fabric is allowed to stand in the wet state for approximately 24 hours. Shorter dwell times are sufficient on chlorinated or Hercosett-treated wool.

IWS finish: A process for the setting of wool fabrics. The fabric is moistened with Si-Ro-Set solution (MEAC/MEAS) by padding and squeezing. The fabric is permanently set by wet steam or gloss decatizing in the still moist state for 3–5 min. Finally, the fabric is dried without tension.

IWS Flame retardent finish of wool: Wool is treated with potassium hexafluorozirconate (K_2ZrF_6) or potassium hexafluorotitanate (K_2TiF_6) from a long liquor at 60 °C or also from short liquors in various combinations with dyeing processes and/or anti-felting crease-resist finishes.

Iws superwash process: See **Superwash**.

IWTO: International Wool Textile Organization.

Ixtle fibre (istle fibre, Tampico fibre): A fibre obtained from various tropical Central American agave and yucca trees (also palm fibres). Ixtle fibre belongs to the group of hard fibres and is similar to Sisal although not a hemp variety. The fibres, of variable length, are stiffer and coarser than sisal. Uses: cords, ropes, brushes, flat carpets and coarse fabrics.

Ivory: The off-white colour of an elephant's tusks.

Ivory black: A black pigment with a brown tinge originally made from carbonised bones. The impressionists stopped using this colour because they considered that black did not occur in nature.

Ivory white: The white of ivory.

Ivory yellow: A light yellow approaching white.

Ivy green: A dark olive-green.

J

J-Acid: 2-amino-5-naphthol-7-sulphonic acid. Organic intermediate used in the manufacture of azo dyes.

Jacinth (e): Reddish-orange; sometimes yellow colour.

Jacinthine: The dark purple colour of the jacinth, and sometimes a brilliant reddish-yellow.

Jacitara: Very strong and elastic bast fibre, yielded by the Jacitara climbing plant in Brazil.

Jack (loom): Part of a dobby head designed to serve as a lever in the operation of the harness of a loom.

Jack (knitting): The jack is a secondary weft knitting element which may be used to provide versatility of latch needle selection and movement. It is placed below and in the same trick as the needle and has its own operating butt and cam system.

Jacket: A textile product, woven or felted into tubular or sleeve form and ready for covering and shrinking on a machine roll.

Jacket: A short coat.

Jackson turbidity unit (JTU), Jackson candle unit (JCU): A measure of turbidity in water. It can be defined in two ways– (a) The turbidity resulting from 1 mg/l of fuller's earth suspended in water; (b) The depth of the column of water that just obscures the image of a burning standard candle viewed vertically through the sample.

Jackson's kier: It is used for cotton scouring. It consists of a horizontal shell and a carriage which supports three rollers. The two outer batch rollers rotate in the fixed bearings at either end of the carriage. The roller or drum is large, hollow and perforated and is carried on oscillating arms pivoted at either end of the base of the carriage. The fabric is saturated with scouring liquor outside in an iron tank. The batch roller is then placed on the carriage and pushed into the kier on guide rails and then door is closed. The fabric is wound from one roller to the other during the scouring process. Hot scouring liquor is circulated by the centrifugal pump and is continuously sprayed over the fabric as it passes from one roller to the other. Scouring is done under a pressure of about 301b/in^2.

Jacobite tartan: It was worn by the Jacobite people in Scotland at the beginning of the 18th Century. The design is composed as follows – Bright yellow stripe; group consisting of white, blue, red, white, red, blue, white stripes, the entire group being of the same width as the yellow stripe and the three white lines being very narrow; yellow stripe as above; group, as above; solid green stripe, of the same width as yellow; group, as above; solid green stripe, of the same width as yellow; group, as above.

Jaconet: A thin plain-weave cotton fabric that is heavier than cambric but similar in type, it is glazed on the right side to produce a high lustre. May be plain or have a stripe or check pattern. May now contain a proportion of polyester fibre. Used for men's shirts, children's clothes and pyjamas.

Jacqueminot: The red colour of the Jacqueminot rose.

Jacquard: This is a weave by which designs are made on the fabric. Jacquard weave fabric can be identified by examining both sides; the design will appear in relief in the underside and long threads will show on both sides where they are carried from one part of the design to the other. Any fibre can be woven in a Jacquard pattern; most designs are elaborate and may be of one colour or multi-coloured. Fabric may fray readily as there are more floats in making the design and more so if silk or other slippery fibres are used. Fabric may be called as cotton jacquard, silk jacquard depending on the material used.

Jacquard drill: Stout drill made with linen face and cotton back.

Jacquard mechanism: (1) Knitting – a term in general use in the knitting industry and applied to mechanisms for selection of knitting elements; (2) Weaving –a shedding (q.v.) mechanism (attached to the loom) that controls up to several hundred warp threads individually and thus enables intricate figured designs to be produced.

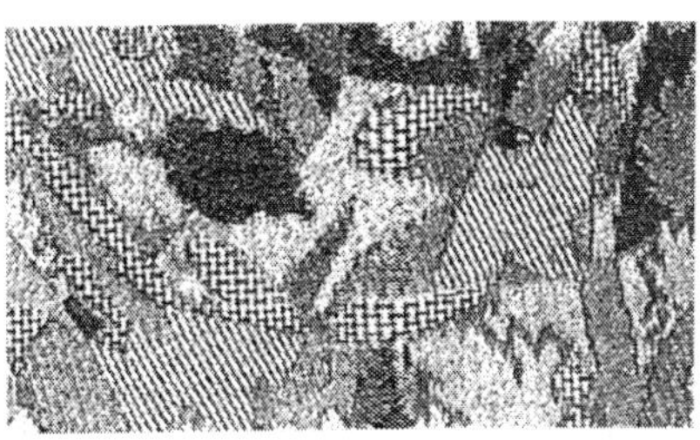

A woven jacquard

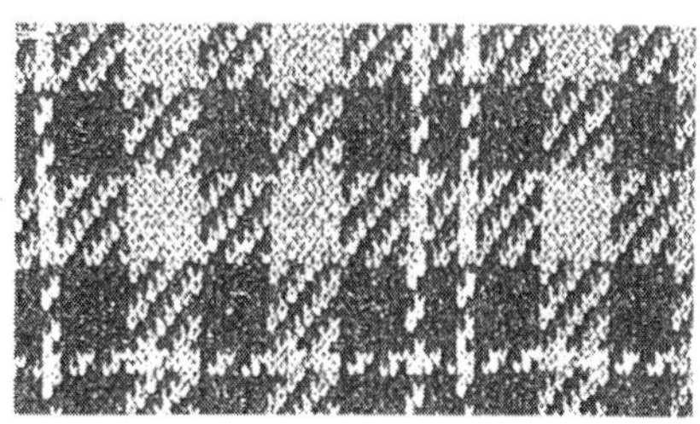

A knitted jacquard

Jacquard fabric: Collective term for a figured fabric woven with a design produced by lifting individual warp threads. Finely drawn and large repeat designs of great variety are obtained by this method which cannot be reproduced by dobby weaving. Typical jacquard fabrics include – damask, brocade, matelassé, and tapestries.

Jacquard harness (in weaving): The series of cords and their attachments.

Jacquard knit: A weft double-knit fabric in which a Jacquard type of mechanism is used. This device individually controls needles or small groups of needles and allows very complex and highly patterned knits to be created.

Jacquard machine: The name Jacquard machines originates from the designer who improved its operation; today the name "Jacquard" is used to identify all machines with a capacity higher than 28–32 threads which are therefore used to produce figured fabrics. The indication of the capacity of the machine, which in the past was used to differentiate the various machine models, has today no significance, as the hook number is no more strictly related to the mechanical structure of the machine.

Jacquard mechanism: A shedding mechanism on a weaving machine which controls the lifting of the warp threads via a harness with healds. Lifting of the ends is controlled by a data carrier, e.g. a chain of punched cards or by a loop of punched paper. The warp threads are lifted individually, corresponding to the required design (analogous to jacquard mechanisms on knitting machines).

Jacquard technology: Knitting and weaving technology for the production of large repeat designs in which each warp thread can be lifted or each loop-forming operation can be individually controlled via a jacquard mechanism. Named after the inventor Joseph Marie Jacquard (1752–1834).

Jade: Green colour in various shades, but mainly yellowish-green.

Jade green: A blue-green or yellowish-green colour.

Jaeger: Knitted or woven fabrics, made of fine, pure, natural wool and camel hair. They are very porous and are considered hygienic.

Jafferson-Walker's kier: Kier of this type is also a vertical kier but utilises a vacuum system for circulation of the kier liquor. In this system, there is a small subsidiary receiver kier between the actual kier and supply pipe. About one-fifth of the liquor in the kier is withdrawn into the receiver kier and this creates a partial vacuum, drawing fabric together. The kier liquor is periodically withdrawn, heated and then returned to the kier. The entire liquor is circulated every 10–12 minutes.

Jagarleinen, Bavarian linen, Strawcloth: In German language, the word *Jager* means "Hunter" and the word *Leinen* means *Linen*. Plain woven, green–brown fabrics from linen, half linen, or cotton. Used for suits and costumes.

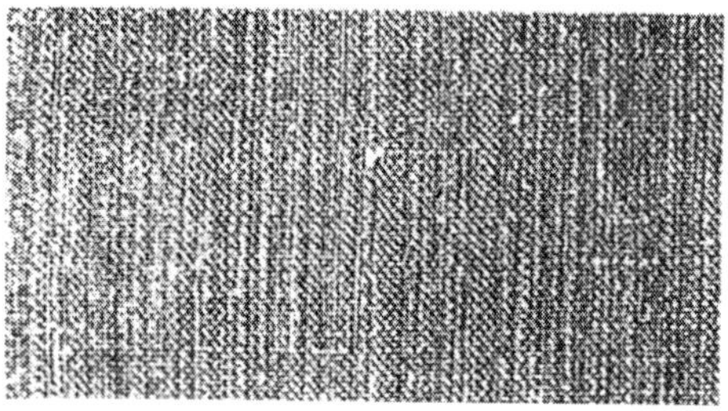

Jagenburg machine: It is one of earliest dyeing machine designed for dyeing loose cotton by Jagenburg. This machine consists of a metal rectangular or conical shaped vat fixed on supports and can be heated from beneath shaft with paddles fixed just above the vat is turned by hand or motor to mix the cotton. The disadvantages of the machines are entanglement of the fibres, un even dyeing, limitation of dyes used, etc.

Jago: A linen cloth.

Jaldai: Indian cotton muslin, figured with net-like designs.

Jamawar: Indian shawl, made of coarse wool with wide stripes.

Jamdani: See **Jamdane**.

Jamadane: Very fine brocaded or embroidered-like muslin made in India by laying short lengths of gold or coloured threads parallel with the warp in the hand looms.

Jamkhana: Indian cotton carpet with generally stripe patterns.

Jangipurl: Inferior quality of Indian jute having a reddish-brown and weak fibre.

Janus cord: Black dyed dress goods with warp ribs, made alike on both sides; made of cotton warp and worsted filling.

Jap silk: Even though the name is Jap silk, it is made in China. It is also called China silk, made in Northern China. It has insufficient weight or durability for outer clothes and is used as lining light weight garments.

Japan blue: Dark blue.

Japanese red: A reddish-brown colour.

Japan silk (Japanese silk): A fine slub-free natural silk fabric with a satin weave produced from absolutely uniform yarns reeled from mulberry silk (warp and weft: raw silk). Also a collective term for fine natural silk fabrics without slubs for apparel use and lampshades. Specific constructions include pongé, Habutai, Helvetia silk, and Ningai. Japanese silk fabrics are sold in specific widths and lengths for export (22.8 × 3.8 m).The quality is indicated by an official quality stamp on the piece ends: red = first quality, blue = second quality and black = reject quality. The fabric weight is given in Japanese momme units per square yard (1 momme = 3.75 g). The lightest Japanese silks weigh 3–4 momme units, which corresponds to 13–18 g/m^2.

Japan wax: Japan wax is a triglyceride. It is obtained from the sumac tree's berries and is high in palmitic acid. Its wax-like properties makes it useful for candles, textile finishes, crayons, cosmetics and the like.

Japanese rugs: Pile or tapestry rugs, made of cotton or jute in Oriental scroll designs.

Japergonsi: Fine East Indian muslin with gold selvage.

Japonette: Printed cotton crepe in Canada.

Japrak: Green, red, orange and blue Smyrna rugs.

Jardiniere: Designs composed of flowers, fruits and leaves.

Jasmine:A pale yellow colour.

Jaspé: (1) A ply-like yarn produced from two differently coloured rovings lightly twisted together; (2) A filament yarn produced from two intermingled or air-jet textured strands of different chemical composition and cross-dyed to achieve two different colours.

(3) A fabric used for suiting, draperies, or upholstery characterized by a series of faint stripes formed by dark, medium, and light yarns of the same colour; (4) A term describing carpets having a faint striped effect.

Jaspé mouliné: A French term for a ply yarn produced from two jaspé threads (cotton and viscose) having a fine flammé appearance.

Jaspee: (1) French term for yarn twisted of several coloured nub or plain yarns; (2) Cloth made of same yarn.

Jasper: Yellowish-green; a blackish-green colour.

Jaspe (jasper): A fabric that has a shaded appearance created by the warp thread colour pattern, may be multi-coloured or different shades of the same colour. Small matching dots are sometimes woven into the fabric. It may be woven plain or figured and is used for bed-spread, chair covers or curtains.

Jasper: Pepper and salt effect by having the warp black and the filling white or vice versa.

Jasper carpet: Carpet having a flame like regular pattern traditionally produced by using pile yarn dip dyed in the hanks to two tones of the same colour or two different colours in fixed proportion of the lengths of the hands in any one lot.

Jato: Jahres-Tonne (tons/annum).

Java: A cotton fabric with red ground and red-yellow and white stripes in East Africa; used for dresses by the native women.

Java canvas: Made of cotton, linen or worsted ply yarns with open face; used for embroidery; the yarns are grouped two or three together each way.

Java jute: See **Kenaf**.

Java print: A type of Africa prints with a particularly wide appeal among the indigenous populations of many African countries. A modern "Java print" may be based on practically any multi-coloured design having no connection with Java. A Java print generally consists of lemon-yellow and gold resists on a green, blue (mostly phthalocyanine), dark-red or brown ground. **Procedure**: Cotton fabric, prepared with a naphthol, is printed with a resist agent (together with reactive dyes for coloured resists). The formation of diazo compounds in those parts of the design printed with the resists is inactivated in subsequent development more quickly than coupling with the naphtholate can take place.

Javanese: Smooth, soft plain weaves viscose fabric with fluid drape. Filament warp and spun weft, giving a lightly ribbed appearance.

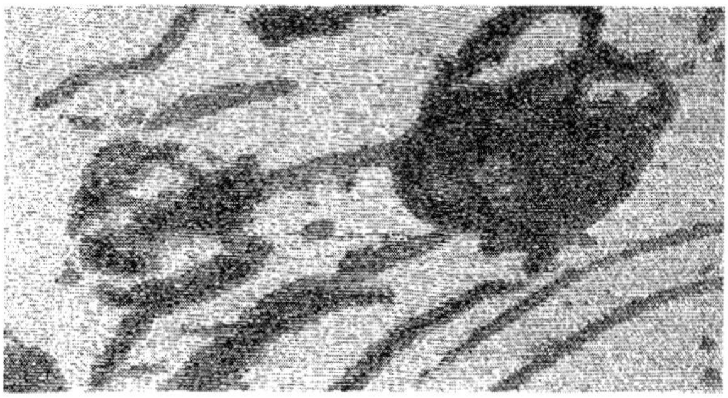

Jaw face (in tensile testing machine): The surface of a jaw which in the absence of a liner contacts the specimen.

Jaw liner (in tensile testing machines): Any material placed between the jaw face and specimen to improve the holding power of the jaws.

Jaws (in tensile testing machine): The elements of a clamp which grip the specimen.

J-box: A J-shaped holding device used in continuous operations to provide varying amounts of intermediate material storage such as in wet-processing of fabrics and in tow production. The material is fed to the top and pleated to fill the long arm before being withdrawn from the short arm.

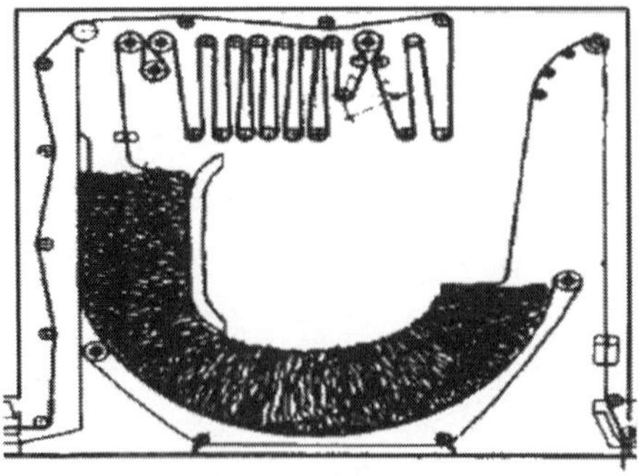

J-box

J-cut: In tufting cut-pile carpet constructions, uneven cutting of the loops caused by poor adjustment of knives and hooks or excessive tension.

Jean: Originally an American term (now international) derived from Gênes (cloth from Genoa). The term was coined in the USA to describe tight-fitting trousers made of cotton twill, usually woven with a dark blue dyed warp and undyed weft (denim), stitched on seams and pockets and having copper rivets here and there. A very well-known fabric/garment all over the world. A very hard wearing cotton fabric, similar to drill but softer to handle. Made in plain colours, especially indigo (blue) colour, sometimes in herringbone weave. Used for overalls, dungarees, protective clothing, children's clothes. It is impossible to imagine today's clothing market without jeans in one form or another, e.g. "stone washed", "moon washed" "used look" or "rodeo washed".

(i) *Stone washed* – (a) the ready-made garment pieces are turned inside out and prewashed or desized; (b) they are then taken out of the washing machine, turned face side out again and washed with chalky sandstone at a ratio of 1:3 (1 kg of garments: 3 kg stones; (c) the garments are taken out of the machine, the stones removed, and bleaching is carried out with sodium hypochlorite (according to the desired shade). To minimise fibre damage, an alternative milder treatment can be given as follows: 1st bath: enzymatic desizing followed by an intermediate rinse. 2nd bath: chemical stone wash treatment with enzymes followed by 2–3 rinses. 3rd bath: application of softener.

(ii) *Snow washed* – (a) stones are soaked for 12 h in a solution of potassium permanganate; (b) the wet stones are then hydro-extracted for 30—50 s in a centrifuge; (c) the hydro extracted stones and garments are loaded into the washing machine for a cold treatment of 15–30 min; d) the stones are removed; (e) the brown stains from the permanganate are washed off the garments with sulphurous acid followed by neutralization; f) a final application of softener is given. To minimise fibre damage, an alternative milder treatment can be given as follows: 1st bath: enzymatic desizing followed by an intermediate rinse. 2nd bath: chemical stone wash followed by 2–3 rinses. 3rd bath: bleaching. 4th bath: application of softener. There are many more other effects which are produced by other methods and innovations.

Jeannet: Stout, coarse warps twilled fabric, made of cotton warp and wool filling; used for working clothes.

Jeannette: (1) In England, a three shaft twilled wool fabric with a warp face; (2) Coarse jean, lighter in texture; (3) name for the one-and-two twill. See also Jean.

Jeans: See Jean

Jeans twill: Three end twills may be woven in two different ways: Warp-faced or weft-faced. The back of warp faced twill is weft faced twill and wise versa. These are the Jeans twill.

Jenappe: See **Genappe**.

Jenfez: A cross-ribbed Turkish silk fabric.

Jenkins: A commercial variety of early maturing American cotton, the staple measuring 22–25 millimetres; the yield of lint is 34–36 percent.

Jennets: Stout, twilled English cottons, come in white, figured or printed.

Jerked in filling: See **Pulled in filling**.

Jerk-in (in woven fabric): An extra filling thread dragged into the shed with the regular pick and extending only part of the way across the cloth.

Jequitiba: Native Brazilian term for the fibrous bast of the Couratari tree; used for blankets, clothing, ropes, etc.

Jerga: Coarse woollen fabric with plaid pattern; used by the natives of Mexico.

Jersey: (1) Originally, the description for a milled wool fabric produced on the British island of Jersey, in the Channel Islands and used for fishermen's clothing. The single weft-knitted fabrics which alone qualify, on technical grounds, for the description "jersey" are nowadays often referred to as "single jersey" and the term is in no way limited to wool fabrics. Today a jersey is warp or weft knitted comfortable fabric made of any fibre including cotton, wool, silk and nylon. Acetate, polyester, triacetate, acrylic, modal, etc. There are many types of jersey which are described under different names.

(2) A tricot fabric made with a simple stitch, characterized by excellent drape and wrinkle recovery properties.

Jersey blue: The blue of the uniform of New Jersey soldiers in Colonial days.

Jersey cloth: Knitted on circular, flat-bed or warp knitted methods (later popular as a tricot knit). Right side has length-wise ribs (wales) and wrong side has crosswise ribs (courses). It is very elastic and with good draping qualities. It also has special crease-resistant qualities due to its construction. It is knitted plain or has many elaborate tweed designs and fancy motifs as well as printed designs, can look very much like woven fabric, wears very well and if washable, it washes very well. First made on the Island on Jersey off the English coast and used for fisherman's clothing. Stretch as you sew. Uses: Dress goods, sportswear, suits, underwear, coats, gloves, sweaters, hats. See **Jersey made of wool, worsted, silk, cotton, rayon,** and **synthetics**.

Jersey (doubleknit): Made of cotton, wool, worsted, silk, rayon, and synthetics in circular or flat-needle bar type. A two-faced cloth, either face may be utilized as the right side. The fabric originated in Milan and Florence. Can be stabilized for shrinkage control and dry cleans satisfactorily.

Jersey gaberdine: The name given to a special effect knitted fabric, is a closely knitted jersey, having a twill effect made of polyester. Usually made in plain colours. Used for blouse tops, jumpsuits, T-shirts, etc.

Jersey stitch: A basic stitch used in weft knitting in which each loop formed in the knit is identical. The jersey stitch is also called the plain, felt, or stockinet stitch.

Jersey velour: It is knit fabric with a smooth ground knit structure and short soft pile on the right side. It has a plush surface and an attractive velvety appearance. Normally made with polyester base with viscose piles. Made in plain, attractive and bright colours.

Jersey (wool): Wool, worsted, silk, cotton, rayon and synthetics. Knitted on circular, flat-bed or warp knitted methods (later popular as a tricot-knit). Right side has length-wise ribs (wales) and wrong side has cross-wise ribs (courses). Very elastic with good draping qualities. Has a special crease-resistant quality due to its construction. It is knitted plain or has many elaborate tweed designs and fancy motifs as well as printed designs. Can look very much like woven fabric. Wears very well and if washable, it washes very well. First made on the Island on Jersey, off the English coast and used for fisherman's clothing. Stretch as you sew. Uses: Dress goods, sportswear, suits, underwear, coats, gloves, sweaters and hats.

Jesuit cloth: Coarse, stout, black, plain woven woollen cloth made of hard twist yarn; used as suiting by religious orders.

Jet: (1) Black; jet-black. From the Greek word *gagates* meaning 'mineral' from Gagas – the Roman name for Lycia where jet was mined. Corrupted from 'jayet' (partridge).

(2) The part of a jet dyeing machine, usually in conical shape, through which the fabric in rope form is passed through and the bath liquor is pumped on to the fabric by a circulation pump and heat exchanger usually through the gap between the inner and outer shell of the jet. During this process, the fabric in the endless form (forming a ring) is passed continuously through the jet with the help of a reel.

A jet system in a jet dyeing machine

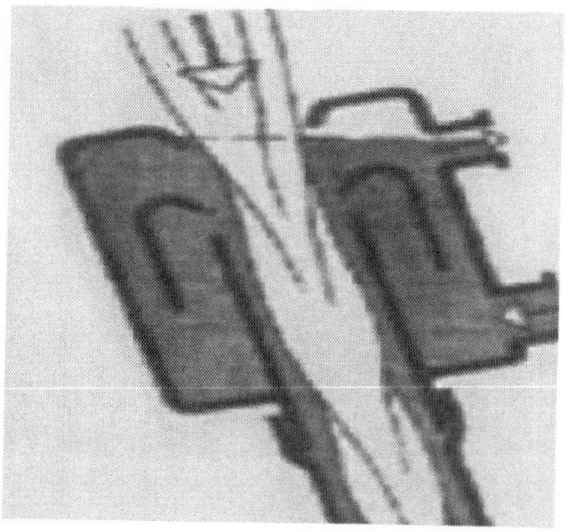

A soft flow jet

Jet black (jet-black): A deep black colour.

Jet driers (steam jet driers): The principle on which these machines are based involves directing jets of a superheated steam/air mixture with a high steam content to both sides of the textile fabric without the introduction of fresh air. The high flow rate of the steam/air mixture generated in the jet nozzles by relaxation of pressure gives a high heat transfer and results in short drying times in the region of 10–20 s for normal fabrics of approximately 150 g/m^2.

Jet dyeing: The process of dyeing in a Jet machine.

Jet dyeing machine: (1) Jet machines are similar to becks in that a continuous loop of fabric circulates through the machine. They differ however in that a stream of liquor is forced through a venturi tube. This provides the force to propel the fabric through the machine. In a jet dyeing machine an endless rope of textile fabric, sewn end-to-end, is transported in a constant direction with the dye liquor in a narrow tube or chamber at temperatures up to approximately 142 °C under pressure. Depending on machine design, the tube system is subdivided into a flow-through section and a dwell compartment (Fig); several jet dyeing machines may be coupled together for high production. Liquor ratios from 5:1 to 10:1 or even lower are usual depending on fabric quality and type of machine.

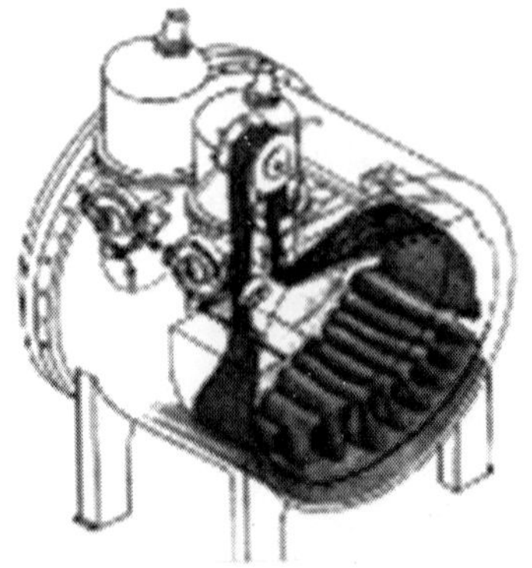

Diagram of jet dyeing machine

(2) (a) A machine for dyeing fabric in rope form and in which the fabric is carried through a narrow throat by dye liquor circulated at high velocity; (b) A machine for dyeing garments and in which the garments are circulated by jets of liquid rather than by mechanical means.

Jet loom: A shuttle-less loom that employs a jet of water or air to carry the filling yarn through the shed. See also **Weft insertion**.

Jet machines: Jet dyeing machines where dyeing is usually done using a jet flow of the dye liquor.

Jet printing: Ink jet printing.

Jet printing techniques: In these methods of textile printing, liquid dye solutions are applied to a moving textile fabric with the aid of controlled jet nozzles (e.g. Chromotronic process, Electrocolour process, Millitron process, Ink jet printing, polychromatic dyeing, TAK dyeing). Jet impingement.

Jet rapid washer: A washing machine for wool piece goods in rope form with an assembly of jet nozzles and perforated tunnel passages before the squeezer rolls. The fabric rope is sprayed with rinsing liquor from the jet nozzles which inflates the rope before the jet nozzle housing and between the jet nozzles and the squeezer rolls. This produces a very efficient washing and rinsing action at considerably higher running speeds than conventional washers.

Jet rope washing machine: The continuous rope wash in jet rope washing machines takes advantage of the excellent liquor exchange of the jet system as well as the dwell time in the storage chambers so that both phases are incorporated into the washing process in an alternating rhythm.

Jet simulator: A laboratory apparatus for pretesting the dispersion stability of disperse dye liquors, especially for use in rapid dyeing machines with large pump capacities and greatly increased rates of liquor circulation, e.g. 60 l/min/kg.

Jet spinning: Also called Air-jet spinning – A system of staple fibre spinning which utilizes air to apply the twisting couple to the yarn during its formation. The air is blown through small holes arranged tangentially to the yarn surface and this causes the yarn to rotate. Majority of systems using this technique produce faciated yarns, but by using two air jets operating in opposite twist directions it is possible to produce yarns with more controlled properties but of more complex structure.

Jet stenter: The drying chamber in a typical jet stenter consists of a number of air ducts each of which as a type of closed air circulation and heating unit so that each drying chamber can be controlled to a specific temperature.

Jet washer: (1) The machine employs slit washing jets with stepless adjustable regulation of the minimum water quantity at speeds of 3–20 m/min (max 100 l/min water for every 1 m working width) before the suction pipe. After deflection, the water flows downwards through the carpet pile into a sump (consisting of circulation becks, filter with automatic lint separator, pump and jet); (2) An element in continuous open-width treatment plants consisting of an immersion trough with liquor circulation pump and a cascade zone from the upper limit of which a coherent surge of water is released through slit jets and flows over the material.

Jet (jetting): A narrow strip of fabric cut warp-wise sewn parallel to the pocket opening and turned over the pocket mouth edge.

Jetee: Lustrous and very strong elastic and durable bast fibre, yielded by the Rajinahal hemp in India; used for bowstrings, fishing nets, etc.

Jetted pocket: A packet having a mouth on which the edges are finished by application of jetting.

Jeypore: Indian cotton rugs having Persian designs with tree and animal forms in red, ivory and blue.

Jhilmeel: Very light East Indian silk fabric, made with an open weave.

Jhuganat: A stout bleached and glazed cotton fabric of inferior grade in East India.

Jig: Jig or Jigger is a batch machine which handles fabric in open width. It is a simple machine consisting of let-off and take-up rolls, a trough containing the processing liquor and guide rolls which allow the fabric to pass from the let-off to the take-up roll. After all the fabric transfers, the driving mechanism is reversed and the fabric makes a second pass through the liquor by returning to the original let-off roll. The cycle can be repeated as many times as it takes to achieve the desired effect. Material is processed/dyed in open width and is transferred repeatedly from one roller to another and passes each time through a dyebath of relatively small volume. *NOTE*: Jigs are also frequently used for scouring, bleaching and finishing.

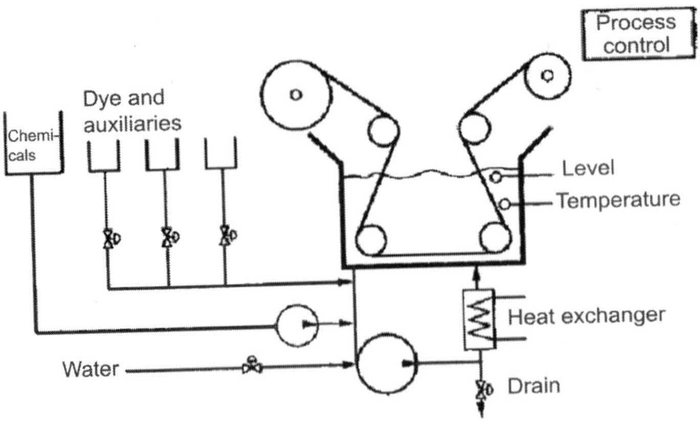

Schematic drawing of an automatic Jig

Jig dyeing: Dyeing fabric in open width using the dyeing machine named Jig or Jigger. In jig dyeing, the fabric is transferred alternately from one batch roller to another through a bath of relatively low liquor capacity during each passage. The process, unlike many others, is economical and universally applicable if the actual running speeds can be individually adapted to the respective requirements of the dyeing process and the particular fabric quality, and the selected fabric speeds and fabric tensions remain constant throughout the entire fabric run from one batch roll to the other.

Jigger dyeing: See **Jig dyeing**.

Jigger: The button inside of a double breasted coat (usually sewn with a long shank).

Jigger (jig): A dyeing machine used for dyeing cloth in full width.

Jigging stenter (jigging tenter): A type of stenter mainly used to "straighten" weft distortion in woven fabrics, e.g. bookbinder's calico. The two side chains are linked, one moving forward whilst the other moves back and vice versa which imparts an oscillating motion to the fabric. This diagonal or "to-and-fro" longitudinal movement helps to straighten the weft threads and reduce any inter yarn bonding between warp and weft, thereby producing a softer fabric.

Jipins rug: Commercial term for draperies and hangings, made of wool and used on doors and windows.

Jircaza: Originally, fine East Indian cotton lawn with woven coloured flowers.

JIS: Japanese Industrial Standard, issued by the Japanese Industrial Standards Committee (JISC), Tokyo, Japan trial activities in Japan. The standardization process is coordinated by the Japanese Industrial Standards Committee and published through the Japanese Standards Association. The JIS in many ways has been Japan's answer to ISO. The JIS is extremely sophisticated and complex and goes beyond the requirements of the ISO 9000 series, but essentially performs the same quality management function. The JIS is more rigorous and comprehensive in standards, making it extremely challenging for an developed by ASTM include test methods, specifications and definitions abrasion resistance, twist determination, fibre maturity, denier and yarn chemicals and therefore is concerned specifically with textile products. This in addition to the development of test methods, AATCC sponsors scientific Japanese Industrial Standards (JIS) specifies the standards used for Indo-organization to successfully implement. This fact has led to its adoptions almost exclusively in Japan and makes its requirement outside Japan very rare. Organizations that have a JIS certification can be considered to be at least as good as if not better than an organization that has an to be at least as good as if not better than an organization that has an ISO 9000 certification, with most JIS systems being closer in scope to ISO 9001.

JIT: Just in time.

Job dyeing: It is also called commissioned dyeing. Fabric which is supplied by the customer in grey or partly processed form is taken for dyeing and finishing on contract basis. Charges are levied on previously agreed rates as per process done on the fabric.

Jockey club: A deep blue colour.

Johnston: Highland tartan, made as follows: Wide dark green bar, split in the centre by three narrow stripes (black at the outside, yellow between) these stripes divided by green stripes of the same width; navy blue bar of the same width as green, split in the centre by three narrow black stripes, spaced their own width from each other.

Joining line: See **Seam**.

Jolocin: Coarse bast fibre, obtained from a species of the Heliocarpus tree in Central America; used for cordage.

Jones Number 1: A commercial variety of cotton from Alabama, the staple measuring 18–22 millimetres; the yield of lint is 33 percent.

Jonote: Coarse bast fibre yielded by a species of the Heliocarpus tree in Central America; used for cordage.

Jonquil: A pale yellow shade after the narcissus flower of the same name; also a vivid yellow.

Joree: Silk worm in Assam that lives on fig trees, yielding a strong and lustrous fibre.

Josephine tricot: A very open crochet work, consisting of closely crocheted rows connected by double yarns at intervals.

Josette: A strong, stout twilled cotton fabric; used for sporting wear.

Joule: Symbol **J**. The SI unit of energy and work, equal to the work done when the point of application of a force of one newton moves one meter in the direction of action of the force. 1 J = 1 N m. The joule is the unit of all forms of energy.

Jours: Open work in embroideries, laces or in fabrics. Designs in open work on fabrics are made by pressing the fabric against a plate having points arranged to form the design and then dress the fabric to hold the holes. See **Modes**.

Jour Deux Place (Jour Trois Place): French term for leno weaves, having two and three rows of holes respectively.

Jouy canvas: Printed cotton or silk fabric, showing small floral designs.

Jowers: Commercial variety of late maturing American cotton, the staple measuring 22–25 millimetres; the yield is 34 percent or more.

JPA: Japanese Standards Association.

JR: Urena fibre.

JS: Rosella fibre (roselle).

J-scray: A J-shaped trough used for short duration intermediate storage of textile fabric (usually in the open-width state) between one process and another. The long arm of the J takes the form of an inclined plane on to which the fabric accumulates before being withdrawn from the short arm after a short dwell period.

JSDC: Journal of the Society of Dyers and Colourists.

JTN: Japan Textile News, a leading Japanese textile journal published in English.

JU: Jute.

Jubbulpore: Hemp Of India, one of the very best grades of hemp, having a long, lustrous and very strong fibre.

Jumbo: Commercial variety of a prolific and early maturing American cotton, the staple measuring 18–22 millimetres; the yield is 32–34 percent.

Jumbo skein (jumbo hank): Hank of yarn of greatly increased weight, e.g. for wool = 2–5 kg per hank.

Jumel: Variety of Egyptian raw cotton. Same as ***Mako***.

Jumper: A knitted or crocheted garment similar to a pullover for covering the upper part of the body, usually made without fastenings at the front.

Jungle green: A deep yellowy-green; a dark or blackish-green colour.

Junker bar (in tufting machine): A yarn guide mounted on the needle bar assembly, for taking up the slack yarn on the needle upstroke of the tufting cycle.

Jupe: Underskirt, made visible by gathering up the overdress (manteau); usually different colour and richly decorated.

Jupon: Plain woven French dress goods made with cotton warp and woollen filling.

Jusi: A fine, sheer and plain woven fabric, made of hemp warp and pineapple fibre filling, or pineapple warp and silk or cotton filling. It comes in coloured stripes and checks and is made by the natives of the Philippines; used for women's dresses.

Just in time (JIT): A production management system of inventory keeping in which stocks are kept to a minimum and products delivered just before they are required. Likewise, finished goods are delivered to customers on demand.

Justaucorps: Elegant, knee length, close fitted men's coat of velvet or brocade and decorated with cords; later with flying tails.

Jute: (1) Soft fibres from the inner bark of the round pod jute (*Corchorus capsularis*, the long pod jute (*Corchorus olitorius*), and from the inner bark of other closely related plants, such as kenaf, sometimes referred to as Meshta (*Hibiscus cannabinus*).

(2) Very long, lustrous and strong bast yielded by the Corchorus in India. It is silver-grey or yellowish-brown; used for bagging, coarse ropes, cheaper carpets, etc. The jute fibre is not durable and rots quickly in water. The chief grades of jute in the native markets are: Uttariya, deswal, deora, desi, naraingunja, chatial, karimganji, bakrabadi, mirganji and jangipuri. In European markets the standard quality is the bale marked M, which is the basis for all quotations; this contains seragjung jute. Another important quality

in bale is marked D, containing Dacca jute. The bales marked CDM contain common jute for sacking. Jute is sorted into three qualities, of which the first one contains from 15 to 20 per cent of the total, the second quality from 25 to 30 per cent and the third from 50 to 60 percent.

Jute bagging (gunny): See **Hessian**.

Jute butts: The flaggy lower ends of the jute fibre 10–25 cm in length cut off in preparing jute for the market used in twines and coarse baggings especially in covering of cotton bales.

Jute count: The weight in pounds of a spindle of 14,400 yards of yarn.

Jute line: Fine yarn produced from long jute fibres after hackling.

JZS: Jugoslavenski Zavod za Standardizacija, Belgrade, Yugoslavia (Yugoslavian Standards Institute).

Encyclopaedic dictionary of textile terms

Encyclopaedic dictionary of textile terms

Volume III

Mathews Kolanjikombil
Senior Textile Processing Technologist

WOODHEAD PUBLISHING INDIA PVT LTD

New Delhi

Published by Woodhead Publishing India Pvt. Ltd.
Woodhead Publishing India Pvt. Ltd.,
303, Vardaan House, 7/28, Ansari Road,
Daryaganj, New Delhi - 110002, India
www.woodheadpublishingindia.com

First published 2018, Woodhead Publishing India Pvt. Ltd.

Reprinted, 2019, 2020

Woodhead Publishing India Pvt. Ltd. ISBN: 978-93-85059-11-7
Woodhead Publishing India Pvt. Ltd. e-ISBN: 978-93-85059-66-7

Typeset by Bhumi Graphics, New Delhi
Printed and bound by Replika Press Pvt. Ltd.

Contents

Volume IV

Preface

It has been told the first thing that a business man should know, is—the little things of his business. This book is meant for textile personae, be it a textile student, textile chemist, a spinner, a weaver, garment maker or a merchandiser who may be an expert in his/her field but he/she may come across many terms in his/her day to day business which he/she is not familiar with but is related to his/her fieldwhich he/she should know, but nine times out of ten does not know. He/She may not have an expert in that field of that term near him/her to clear his doubt about that term. This book comes handy in such circumstances.

There is nothing scholarly in this book, but is a compilation of short easy understandable meanings of the textile terms enough to help the referrer to understand the term. I have come across dictionaries which gives the meaning of the textile terms in one sentence. But it may not be sufficient to give a full idea about it. But this book is a little different. The meaning of the terms is explained in a concise manner even with the help of diagrams or photos, wherever necessary, which is enough to clear his/her doubts. These terms and meanings have been collected right from my college days and throughout my career. I believe it is not complete, but, such as it is, the compilation is reliable. There are further terms which are being collected by the author which will be added in the next edition, probably. The author believes that he has produced a book which may be profitably consulted by all who are either interested or practically engaged in textile trade.

It has to be specially mentioned that, the students can use the present book as a reference guide for his/her immediate needs without going to many text books. For detailed study of any terms he/she can further refer books specialized in that field. I may not suggest this book for a research student. The author also has tried to explain the construction of many fabrics new or old for general knowledge.

Hope this book will be greatly accepted by textile personnel. Any suggestions or corrections are welcome, which will be included in the next editions.

Mathews Kolanjikombil

Senior Textile Processing Technologist

Bangalore

Dedication

This book is dedicated to my Father (Thomas Abraham) and my Mother (Thankamma Abraham) who solely made me what I am today.

K

K index: (1) Index in general is used for characterizing a commercial grade of polyvinylchloride, calculated from its viscosity in solution and indicating its molecular weight. Polyvinylchloride with a K value of approximately 68–80 is usually used for coating. (2) Index is used for the classification of basic (cationic) dyes with respect to their rate of strike (Association index K). (3) Index is used in association with machines for measuring fabric length because of the effect on standardized length measurement due to the stretching of the material.

KA: Casein fibres.

Kabakaraman rugs: Coarse and heavy, small rugs made by the nomad Karamanian in Asia Minor. The warp and weft are of heavy and coarse wool. The long and very loose pile is tied in G-hordes knot. The design is usually that of a prayer rug.

Kabistan rugs: Very fine rugs made in Caucasia with cotton web, occasionally wool warp, and short, wool pile, tied closely in G-hordes knot. The designs are geometrical, stars and diamonds and pearls in rows often occur. The field is often divided into horizontal rows of pears or finely blended stripes. The border often contains conventionalized animal figures. The ends are finished with a narrow selvage and a loose or knotted fringe. The sides are overcast, occasionally having only one warp thread left.

Kabylo: French shawl, made with carded warp and filling.

Kachoji: Mosquito netting in Japan.

Kadu: Sleeping mat in Java, made of palm leaves.

Kaffir hemp: Very strong, white bast fibre, yielded by the South African, the *Grewia occidentalis*; used for rope and cloth toy the Kaffirs.

Kaffir sheet: Very coarse, twilled cotton fabric, with fancy coloured heading; used for garments by the natives of South Africa.

Kaga: Trade term for the medium grade Japanese silk fabrics.

Kagotsuko printing: Typical Japanese resist printing process, comparable with paste resist printing. See **Resist printing**.

Kaikai: A thin, cheap Japanese silk fabric.

Kaimakani: Fine sheer cotton cloth; used in Turkey to bind the turbans with.

Kairens: Turkish wool rug of good quality; used as floor and furniture cover.

Kairuan: Rug from Tunis, made by the natives of wool with hand tied knots.

Kaisarich rugs: Very bright-coloured hand-knotted rugs; the cotton or silk pile is tied in Ghiorde's knot; made in Kaisa, in Asia Minor.

Kakarally: Very fine and thin layers of fibrous bast, obtained from the Monkey-pot tree in South America; used for wrapping, cordage, baskets, etc.

Kakeda fine: Japanese raw silk.

Kalamal: Striped cotton fabric with a white ground; used in Turkestan for dresses.

Kalasiris: Ancient national robes for men and women; narrow shell with straps, richly ornamented, shift style, transparent and finely pleated.

Kalga: Indian applique work; used for curtains and covers.

Kalamkari: Traditional printing with vegetable colours in India.

Kalmuck (fries, swanboy): In general, a 2-sided strongly raised soft cotton fabric, weight and roughened effect are even more marked than in Melton; twill(cross) weave with back pick or as double cloth, striped or with a check pattern. Also long-haired, coarse, thick woollen fabrics. Application for heavy blankets, bed padding, table under-cloths, back-cloth materials, ironing board and mangle covers.

Kaloz process (lime and ozone): It is a combined process for effluent cleaning, as a single stage or sequentially (better results) by the precipitation of organic/inorganic impurities using calcium hydroxide. By the use of the correct techniques, even fused and difficult to oxidise substances be made to oxidise more easily. Particularly suited for the cleaning of textile effluents due to its good decolourising effect.

Kama: Japanese trade term for the cotton cop.

Kambaliya: A *bandhani* design. See **Bandhani**.

Kamdani: Ancient Indian gold embroidery (silk thread wrapped in metal foil) of lighter style on fine fabric.

Kanakin: Generic trade term in Japan for a variety of plain woven cotton goods, shirting and print cloths.

Kanthas (embroidered quilts): Made from old saris and dhotis, for generations, by women of West Bengal, India, using coloured threads from borders for motifs and white and off-white thread from ground fabric for quilting. Rich folklore and landscape of rural Bengal with its green paddy fields, rhythm of seasons, festivals and daily life, rivers with boat and fish, animals, birds and plant life inspire the repertoire of motifs of *kanthas*. A style similar to the *kantha* are made in the neighbouring states of West Bengal called *Ledra* in Jharkhand, and *Sujini* in Bihar, India

Kaolin: China clay, porcelain clay, terra alba, $Al_2O_3 \cdot 2SiO_2 \cdot 2H_2O$, density 2.1–2.6. White, loose powder, practically insoluble in water. Application – with starches and oils for filling and softening of finishes (bleached cotton goods, light coloured lining fabrics); pigment for viscose de-lustring (fixed in particular by sulphonated surfactants); also as spotting powder; as tailor's chalk, etc.

Kapok: Vegetable hair fibres from the fruit pod of tropical cotton of the *Bombaceen* family (kapok, cotton or silk cotton; East and West Africa, India, Java, Sumatra, Mexico, Brazil).The fibre resembling cotton but silkier, it is a seed fibre which is round, smoothand light. It is only used as a bulky soft filling in upholstery, cushions, mattresses, toys, etc.

Karabagh carpets: Knotted carpets from the Southern Caucasus, patterning and colours similar to Kazakh carpets, but shorter pile made of less lustrous wool. Approximately 1,50,000 Turkish knots per m^2.

Karl-Fischer method (KF process): Simple and volumetric process for water determination, based upon the fact that iodine oxidizes sulphur dioxide to sulphuric acid in the presence of water. Particularly suitable in the presence of the smallest quantities of water in organic solvents, dyes, also in textiles.

Kasan: In Germany and Austria, a woollen dress goods that is similar to a stout flannel.

Kasha: (1) Kasha may also be a cotton flannel, with a napped right side, slightly ecru in colour and made as sheeting.

(2) A softly napped fabric with a crosswise streak caused by darker hairs, may be made of fine wool and Tibet goat hair.

Kashan carpets: Knotted carpet from Central Persia. Fine ornamentation on a matt red, violet or dark blue background. Around 3,50,000 Persian knots per m^2. Low, smooth pile made of fine wool.

Kasheda: Indian fabric, made of wild silk, often mixed with cotton, and embroidered.

Kashgar: Coarse cotton rugs with long loose wool pile tied in Senna knot. They are made in Central Asia. The design consists of Chinese fret, dragons, fish, etc., in bright pinks, orange, yellows, etc.

Kashgar cloth: Thick but light dress goods made of Kashgar wool; It has a long, napped face.

Kashkai: Collective name for Kermanshah, Shiraz and Mecca rugs.

Kashmere silk: A fine silk produced in Kashmir, India.

Kashmir rug: Another name for Soumak rugs.

Kashmir shawl: See **Cashmere shawl**.

Kasida: Sort of Dacca muslin (see), embroidered in floss silk; used for scarfs, turbans, etc.

Kassapbatchi: Coarse Turkish skin wool; used for carpets.

Kasuri: See **Kasuri dyeing**.

Kasuri dyeing (Kasuri-Resist): Laborious, old Japanese dyeing technique of a white patterning on coloured ground by the partial resist of warp and weft threads and subsequent over-dyeing.

Kata aya: General trade term in Japan for twilled cotton shirting.

Katox-system: An effluent treatment system that can be used together with other systems to create a complete effluent treatment plant for textile processes. It is based upon an activated carbon system that carries biologically active media in cyclone baths.

Kattun: German for calico.

Kawamatta: Trade term for inferior grades of Japanese silk fabrics.

Kawabata system: System developed in Japan for handle evaluation (Kawabata system of handle evaluation). Describes fabric handle in terms of 0 (slight) to 10 (very marked) with an overall rating scored as 0 (unacceptable) to 5 (excellent). In this system, various characteristics are used for handle evaluation:

(a) Fukurami = fullness and softness
(b) Hari = "non-drape" stiffness
(c) Kishimi = scroopy handle
(d) Koshimi = stiffness and elasticity
(e) Numeri = flexible smoothness
(f) Shinayakasa = suppleness with softness
(g) Sofutosa = soft handle feel

These characteristics can be detected using four devices:

(i) KES-F-1 for measuring fabric behaviour under tensile and shear loading.

(ii) KES-F-2 for measuring bending behaviour.

(iii) KES-F-3 for measuring fabric behaviour under compressive loading.

(iv) KES-F-4 for measuring surface characteristics of the fabric.

A complete evaluation of a fabric can take 6–8 hours depending upon type.

Kawokawo: Silky, yellowish seed hair of the Bombax tree in the Malay States.

Kaya: A mosquito netting made in Japan.

Kazakh carpets: Tightly knotted, long pile carpets from the Caucasus. The large, bold, geometric patterning in bright dark-red, blue, green, yellow and light ivory is typical. Dull silk sheen wool pile. 120000–200000 Turkish knots per m^2. The term Kazakh carpets is often wrongly used as a collective term for Caucasian carpets.

Kazeto: Japanese trade term for a spool containing about 54 feet of hand spun cotton yarn.

Kasuri: A hand woven cotton fabric in Japan, showing irregular white checks on a blue ground. It is made in plain weave of yarn, which is tied up at certain intervals with a string before dyeing. These places remain intact by the dye and form the white spots. It is used mostly for clothes for the native school children.

KE: Kenaf.

Kedis: Stout cotton shirting and lining in Asiatic Turkey.

Kefieh: Printed cotton cloth, measuring about 42 inches square; used as head covering in Asia Minor.

Keith: Commercial variety of early maturing, prolific cotton from Alabama, the staple measuring 24–26 millimetres; the yield is 32–34 percent.

Kekchi: Raw cotton from Guatemala, the staple being of good length and quality.

Kelly: Commercial variety of prolific and late maturing American cotton, the staple measuring 26-30 millimetres; the yield of lint is 30–31 percent. Also called Marston.

Kelly green: A strong mint green colour.

Kelt: Scotch frieze made of natural black faced wool.

Kelvin: The unit of thermodynamic temperature; the SI unit of temperature for which an interval of one Kelvin (K) equals exactly an interval of one degree Celsius (1 °C) and for which a level of 273.15 K equals exactly 0 °C. According to this scale, short wave radiation (light wavelength) has a high colour temperature and, conversely, blue light has a high colour temperature and red light a low colour temperature. In practice direct sunlight is 5000–65000 K, blue diffuse light is 5700–6600 K, white light with a completely overcast sky is 6400–6900 K and with a clear blue sky is 19000–24000 K.

Kemea: Indian all-silk taffeta made with flower patterns.

Kemp fibre: A medullated animal fibre in which the diameter of the medulla is 60%, or more, of the diameter of the fibre.

Kemp: Diseased wool fibres, with unevenly developed medulla which causes streaks in dyeing. It is a coarse, white and undeveloped fibre.

Kemuku: Japanese silk waste, yielded by the outer skin of the cocoon in reeling.

Kenaf: Bimli-jute, Gambo hemp, Ambari hemp, Bombay hemp, Mestha, similar to jute bast fibres (India, Soviet Union, China, Korea, Taiwan, Philippines, South Africa, Brazil, Cuba). The basic fibres have irregular wall thickness, an average length of 2 mm and a width of 20 um. As jute substitute and mixed with jute for the manufacture of packing material, ropegoods and fabric. In addition to the fibre extraction, the plant is also used for the production of oil.

Kendal: A coarse English tweed of green colour in the 14th Century; used for clothing.

Kendyr fibre (Turka fibre): It is of the bast fibres family. A winter resistant plant fibre from Turkestan. The stalk is cut annually; preparation takes place in a similar way to flax. The pure white, 80–120 cm long bast bundle with fibre strength up to 55 cN/tex can be processed by linen and tow spinning; by cottonising the kendyr fibre can also be used for fine spun yarn (mixing with cotton). The most important field of application is in the manufacture of highly rot-resistant fishing nets.

Kennedy: A Highland tartan, composed of a green ground, dark-blue and black checks and red and yellow lines.

Kensington quilt: Has large patterns formed of coarse thread on a fine plain woven ground.

Kenting: Thin, sheer Silesia linen fabric.

Kentucky jean: Very strong, stout, weft face fabric, made with cotton warp and wool filling in satin weave; used for trousers, etc.

Keratin: One of a group of fibrous insoluble sulphur-containing proteins (scleroproteins) found in ectodermal cells of animals, as in hair, horns, and nails. Leather is almost pure keratin. There are two types: α keratins and β keratins. The former have a coiled structure, whereas the latter have a beta pleated sheet structure (Gk.: horn material), High molecular Protein substance (sclera protein) containing cysteine.

Keratose: High-molecular wool oxidation product, consisting of β-keratose (cuticula), α-keratose (microfibrils) and γ-keratose (matrix).

Kerf: In England, the flock resulting from shearing the cloth.

Kermanshah rugs: Usually all-wool Persian rugs. The medium long, close pile is tied in Ghiordes knot. The design usually consists of floral patterns, often small palm leaves in rows.

Kermel: Kermel is a polyamide-imide fibre made from either trimillitic anhydride chloride and a diamine or a diisocyanate. These polymers have reasonable thermal stability, very resilient and have excellent flame-resistance.

Kermer: A shawl made of pure silk, or mixed with cotton, worn by the women in Egypt.

Kermes: Red colouring matter, yielded by the dried bodies of the *coccus ilicis*, a small insect. It is ground up in hot water, producing the carmine. Extensively used in past centuries to dye silk and wool fabrics. Active component – carmine acid, related to Cochineal. Mordant dyes for wool and silk. Light fastness and luminosity of equal value to the expensive purple.

Kerr: A Highland tartan, made as follows – Wide red bar, split in the centre by three narrow black stripes which are spaced their own width from each other; black stripe, about one-sixth in width of the red bar; dark green bar, same in width as red bar, split with a pair of black stripes (as wide as those in the red bar) near each edge, these black stripes being spaced from the edge of the green bar and from each other their own width.

Kersey: (1) A woollen cloth originated in Kersey, Suffolk in 11th Century. Medium to heavy weight, similar to Melton and beaver. It has a lustre caused by the use of cross bred wool and is a very heavy, thick wool coating fabric which has been fulled and also felted. Nap often has direction. Often twill weave, although the face finish that is applied conceals this. It is a conventional cloth in appearance and not much used now except for classic overcoats or in lighter weight for a classic black dress. It may contain some acrylic fibre nowadays. Gives good wear and is dressy looking. Blues, browns and blacks are the most popular colours. Very similar to beaver but it is fulled more, has a shorter nap and a much higher lustre. Uses: Men's overcoating, uniforms, women's coats, and skirts.

(2) A heavy, closely woven cloth with a smooth face and glossy finish.

Kerseymere: A fine, twilled, woollen cloth of peculiar texture, one thread of warp and two of wool being always above.

Kerseynet: Light English fabric, made with cotton warp and woollen filling; used for men's clothes.

KES-F-System: Kawabata's Evaluation System for Fabrics. See **Kawabata system**.

Kettle cloth: Made of cotton and polyester, plain weave stiff fabric, with a dull surface and a slight slub in the weave, which may be coloured. Used for men's light summer jackets, trousers for men and women and other casual dresses.

Kevergik: Turkish skin wool of merino sheep.

Kevlar: Kevlar, modified nylon, poly (p-phynylene terephthalamide) represents a break-through in high modulus aromatic polyamide fibre and mainly used as reinforcement tyres, conveyers belt etc., made by DuPont.

Key-lock principle: In 1894, Emil Fischer developed the theory to explain enzyme specificity, that the surface of an enzyme possesses a pocket (the "lock"), into which the substrate fits sterically like a "key", enabling the process of enzymatic catalysis to function.

Keymo: A finishing process, rendering woollens and worsteds unshrinkable by a bath of sulphuric acid.

Khabbikutah: Collective name for various non-descript short pile, knotted Persian rugs.

Khadi (Khaddar): An Indian handloom fabric usually made with hand spun yarn. It will have a characteristic rough appearance because of the unevenness of the spun yarn due to the hand spinning. Propagated by Mahatma Gandhi during the Indian Independence struggle, to replace the machine made cotton imported from Europe.

Khaddis: Indian name for wooden looms used in earlier days.

Khaiki: A plain woven and washable light Japanese silk fabric.

Khaki: (1) A light, yellow-brown coloured cotton cloth used for army service in hot countries; (2) A khaki-coloured cloth of cotton, wool, or combinations of these fibres with manufactured fibres used primarily in military uniforms and work clothes.

Khali: Natural brown felted Persian fabric, embroidered in coloured silk and silver flowers.

Khandeish: Indian cotton, having a harsh, strong, and very dirty staple of golden colour.

Kharajobi: Name for a great variety of gold and silver embroidery made in India.

Kharjikhan: Indian embroidery work in gold and silver.

Kharwa: Red Indian cotton fabric; used for ticks.

Khasa: Cotton muslin from India.

Kheetee: Indian chintz.

Khemir: Silk shawl from Egypt.

Khersek: Heavy Persian woollen rugs with a shaggy pile.

Khes: Stout Indian cotton fabric made with check patterns or coloured, often gold border.

Khilim rugs: These come in all sizes made in Anatolia, Persia, Turkestan and several of the Balkan States. They are woven by hand and have no pile, the weft being of hard twisted wool. The design, which is alike on both sides, consists of angular geometrical figures in a great variety of colours. They are used for divan covers and portieres.

Khiva: Fine, all-wool rugs of small size, made in Turkestan, the short, close pile being tied in Senna knot. The design consists of octagonal figures or the prayer rug pattern in rich reds, blue, ivory and a little green.

Khokti: Yellowish or brown cotton cloth made in India. It is very durable, smooth, and glossy, the finest grade being similar to the best brown Holland.

Khoktibanga: Variety of East Indian raw cotton, having a yellowish but good, strong staple.

Khonia: Indian cotton shawl with richly embroidered design.

Khorassan:(1) Variety of Persian wool of long and fine staple; (2) Medium and large size Persian rugs with close and medium long fine wool pile tied in Senna knot and clipped unevenly. The pattern consists usually of fish or palm design the latter having smaller ones placed in each large one. The border usually has trailing palm leaves. Magenta and blue are characteristic colours.

Khoseb: Originally a fine muslin from Egypt; used for turbans.

Khudurangi: Native East African name for a coarse cotton fabric, dyed with henna.

Khum: Dyed Terry cloth in Turkey; used for long coats by the natives.

Khuskus, Cuscus: The roots of this grass (*AndropogonMuriaticus*) are used in India for mats and baskets.

Khustka: Shawl from Southern Russia, embroidered with coloured floral designs.

Kibisu: See **Frisons**.

Kick tape: A binding of twill or herringbone twill, usually, 3/1. Originally made with silk warp and polished cotton weft, but now made with mercerized cotton warp or rayon warp and weft and woven in such a way as to give the article a stiff handle.

Kidderminster carpet: Double cloth carpets, in which two layers of cloth, manufactured from two warp and weft systems, lie on top of each other. The patterning is created by the threads of the cloth layers alternating, i.e. the wefts are inserted in tiers, which are sometimes formed by the warp threads of the upper cloth and sometimes by those of the lower cloth. This has the effect on the pattern that a red figure on a blue background on the front appears on the back as a blue figure on a red background (reversible carpet). It is woven without the use of heald shafts exclusively with Jacquard machines, with the depth of the harness being split according to the number of cloth layers. The Scottish or tree-ply carpet consists of three layers of cloth. These flat but thick carpets usually made of wool yarn were first produced in the English town of Kidderminster.

Kidney: Cotton Trade name for Brazilian cotton.

Kienchow: Silk Foulard with pressed in patterns, made in China; is about 16 inches wide.

Kikci or Kikoy: A heavy grey cotton sheeting in the Eastern parts of Africa, having yellow, black and red border stripes; used for garments by African tribes.

Kier: A pressure vessel used for boiling textile material.

Kier boiling: In kier boiling (alkali-boiling) of cotton materials, the cotton is heated with dilute NaOH solution under pressure for several hours, in the absence of oxygen. Small quantities of detergents and solvents are frequently added. The objective is to eliminate the waxy cotton cuticle, the natural oils, waxes and proteins as well as various salts and soluble cellulose and carbohydrate derivatives. It also helps to remove any size residues. The removal of the hydrophobic constituents of the cotton fibres renders them absorbent. In many modern plants, much faster, continuous processes of this type use higher concentrations of NaOH. They are, however, less efficient than the classical kier process because of the short treatment times.

Kier boiling fastness: It is a fastness assessment whether a dyed or printed material is able to withstand weak NaOH under pressure. Test method – treat test piece wound in raw cotton fabric of 10 fold weight for 4 h at 120 °C and 98 100 Pa in the kier boiling liquor with a liquor ratio of 1:50 in the pressure boiler. Solution I: 4 ml/l NaOH 35.5%; Solution II: 4 ml/l NaOH 35.5% + 4 g/l m-sodium m-nitrobenzene sulphonate. Evaluation with grey scale.

Kier decatizing: A finishing process for wool and its blended fabrics for achieving dimensional stability, voluminosity, handle and sheen. These are rolled up with backcloths (satin or molleton fabric) and treated in an evacuated boiler with saturated steam. Remaining shrinkage value is below 1%.

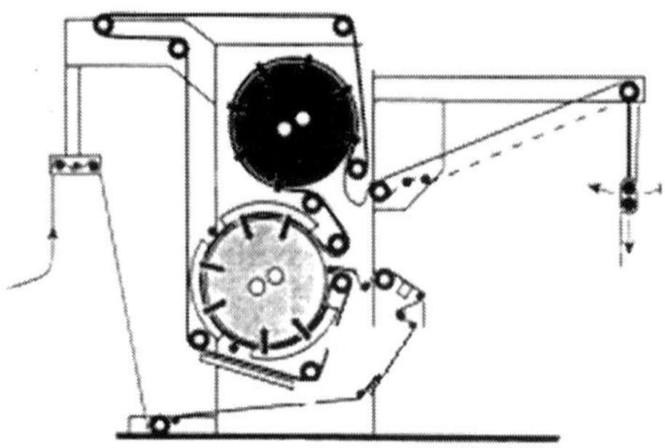

Double drum kier decatizing machine

Kier lining: Treatment of pressure kiers and bleaching boilers with milk of lime, cement and water glass solution to prevent kier and bleach stains when the material touches unprotected boiler walls.

Kier stains: It arises due to unfavourable working conditions, such as hard, turbid water, poor, one-sided liquor circulation, inconsistent packing of the kier boiling material (channelling), wrong rinsing (cold instead of boiling hot), etc. This usually causes brownish deposits and uneven stains, which can only be removed by repeating the kier boiling or hot acid souring and boiling with sodium carbonate.

Kiering: The process of boiling of textile in a large pressure vessel called Kier.

Kilim: See **Khilim**.

Kilim carpets: Woven, bright, originally Turkish flat carpet with coarse wool weft. Sometimes with woven perforations, i.e. small open cracks on the border lines of the multi-coloured pattern. Used as floor carpet on thick felt underlay.

Kilmarnock: (1) A coarse 18th Century Scotch serge; (2) See **Kidderminster**.

Kilo: (Gk.: chílioi = thousand), unit prefix for one thousand times = 1000.

Kilogram: A unit of mass (weight) in the metric system; approximately equal to 2.2 pounds avoirdupois. 1 kg is the mass of the international kilogram prototype of a platinum-iridium body stored in the vicinity of Paris. In processing, since one litre of water weighs one kilogram, it makes calculations such as percentage of a chemical owg, or liquor ratio, etc., easy.

Kinari: Trade term for Persian pile runners of various characters and origin.

Kincob: Fast Indian silk muslin, richly interwoven with gold or silver brocade (zari); used for men's and women's dresses. It was made in Varanasi but until 16th Century, the best of this material was coming from Ahmedabad. Commonly known as kimkhwab (anglicized to kincob, although the root word is Chinese kim-hoo-gold flower). These Ahmedabad brocade were sold four times their weight in gold on African East coast. The beauty of Ahmedabad's kincobs lay in the colourful inter-weaving of silk into designs with gold and silver threads. They were also famous for their strength and durability. First the Portuguese piracy, and then the rise of Surat as a centre of weaving under the English wiped out the craft completely in Ahmedabad.

Kindergarten cloth: Stout, heavy, plain woven cotton fabric, usually made with yarn dyed warp stripes. The warp is single yarn, two ends drawn in. The number of ends is almost three times higher than the number of warps; used for children's clothes.

Kinetic chain-length: It is defined as the average number of monomer molecules contained per radical which initiates a polymer chain.

Kinetic friction: Friction developed between two bodies in motion. Compare ***Static friction***.

King: An improved variety of early maturing commercial variety of prolific upland cotton, the staple measuring 25–28 millimetres; the yield of lint is 32–34 percent.

King bobbin: One-sided conical packages with precision cross-winding.

King spool: A thread cone that has a vertical tube with a horizontal base.

King's blue: A light blue colour, resulting from mixing ultramarine and **Cremnitz white**. See **Cobaltblue**.

King's yellow: A brilliant yellow colour.

Kingfisher blue: A brilliant blue colour. Also referred to simply as 'kingfisher'.

Kinik: Variety of Turkish raw wool.

Kinjishusu: Japanese silk satin with a partly or completely gilded face; used for kimonos.

Kink: A short length of yarn that has spontaneously doubled back on itself to form a loop in the yarn or fabric.

Kink (in fabric): A short length of yarn that has spontaneously doubled back on itself to form a loop.

Kink (in rope): An abrupt bend or loop in the rope, which is the result of an unbalanced twist relationship in the rope structure.

Kinkale: Light bast Indian silk, brocaded with silk or gold threads.

Kinking: The doubling back of yarn on itself to relieve torque imparted by twisting or texturing.

Kinky thread: See **Kink**.

Kiotonan: Chinese satin with damask figures.

Kir-Shehr rugs: Made in Angora, Asia Minor, the warp, and weft are of dyed wool, the long pile is tied in Giordes knot. The Arabic designs are in brilliant greens, reds, and blues. The sides and ends are salvaged.

Kiss dyeing process: Dyeing liquor application onto the fibre tips or on the back, often for carpet material.

Kiss print: Special printing technique used in the rotary screen printing of carpets. Carpet transport on a needled backcloth, which can be adjusted by electric push-button control. By controlling the height between backcloth and template a tunnel is formed, the height and thickness of which correspond with those of the carpet. Because this does not alter the position of the pile, a sharp printing edge is obtained.

Kiss roll (lick roller): See **Kiss roll applicator; Cover roller**.

Kiss roll applicator: A kiss roll applicator consists of a drum rotating in a trough containing the finish. A layer of liquid is picked up by the drum surface. This liquid is transferred onto a fabric "kissing" the exposed section of the drum. Kiss roll applicators are relatively simple devices. Within certain ranges, the wet pick-up can be adjusted to meet whatever is desired by controlling the variables stated above. However, uncontrolled variations in fabric speeds or drum rpm will cause unwanted variations in fabric properties.

Kitay: Fancy coloured Chinese silk and cotton cloth.

Kittool or Kitul: Black, very coarse, straight, smooth, glossy and strong fabric, obtained from the leaves of the kit tool palm found in India and Sri Lanka; used for strong ropes, brushes, mats, etc.

Kleanka: A Russian buckram.

Km: Camel hair; Kilometre.

Knap: Coarse, blue woollen, used for sailors' clothes in England.

Knee (in anatomy): The joint between the lower and upper leg.

Knee break-out test: A method to evaluate the performance of fabrics, especially boys' wear, when subjected to abrasion, stretch, and impact forces under conditions which simulate ordinary wear at the knee.

Knee girth (in body measurements): With the leg straight, the circumference of the knee over the kneecap and parallel to the floor.

Knee lifter: A lever under the sewing machine table which, when operated by the machinists knee, lift the presser foot.

Kneeing: (1) An unstable condition arising in melt spinning wherein the extrudate forms an inflexion on leaving the spinneret instead of drawing down directly from the orifice.The molten filament thus has a knee-like shape, just below the orifice.

(2) Abnormal behaviour of a spinning threadline (especially in melt spinning) in which one or more filaments form an angle (knee).

Kneipp linen: Porous, rough faced, twilled linen fabric, made in Germany and Austria; used for towels and underwear. Originally it was knitted.

Knib: Technical term for knots, or uneven places on the silk fibre.

Knickerbocker: (1) Rough faced wool and cotton mixture dress goods, made with nub yarns. It comes in mixture colors.

(2) In weaving, an effect produced by rough knotted yarns; similar to boucle and bourette.

Knickerbocker yarns: Yarns with coloured nubs made by printing the card sliver.

Knickerbockers: Knee breeches.

Knife (in tufting machine): A flat sharpened blade which operates in conjunction with the looper to cut the formed loops to form cut pile carpet.

Knife-coating principle: The knife application unit (spreader) consists of the doctor blade holder; the doctor blade with lateral limit blocks and possibly a doctor blade liner. The coating substance is applied to the doctor blade which is set at a defined distance between knife and fabric using the doctor blade holder. In continuous coating, the base material moves under the fixed doctor blade which causes the coating of the coating substance on to the material depending on the doctoring speed, doctor blade tilt angle, doctor blade profile, doctor blade distance, base material tension, paste viscosity.

Knife coaters: A knife-coater is a metering device that continuously spreads viscous liquids onto fabric. This type of equipment is commonly used in the coated fabrics industry. It consists of a stationary knife blade positioned over a fabric support. The material applied to the fabric is fed to one side of the knife blade as the fabric moves continuously under the spreader. The amount deposited on the fabric is controlled by the gap formed between the blade and the fabric.

Knife-on-blanket coating: Doctor blade for coating and laminating which is supported on an endless rubber blanket.

Knife pleat: A pleat with folds lying flat and facing in one direction.

Knifing: Carpets with polyacrylonitrile or polyamide cut pile are drawn under a doctor blade to align the pile in warmed state.

Knit fabric: Knit fabric are made from all fibres and blend of fibres. The thickness and weight may be anything from very fine to coat weight.

Knit texturing: See **Texturing**, **Knit**.

Knit-de-knit: See **Texturing**, **Knit-de-knit method**.

Knit-de-knit method (Whitaker process): A knit-pressure-fixation-rewinding process for the texturizing of yarns (similar crinkle).

Knit-miss: A form of tricot knitting in which yarns on each bar of a two-bar machine are knit at alternate courses only. This type of knitting permits the use of heavy-denier yarns without creating undesirable bulkiness in the fabric.

Knits: Knitted material.

Knitted carpet (knitted velvet pile carpet): It is created from knitted fabric bonded to a stabilising fabric. A backing is then attached to this composite material. The pile layer is usually roughened, creating the knitted velvet pile carpet. See **Stitch bonding, Raschel** and **Raschel batt-on-base carpet**.

Knitted composites: Bonded in the interweaving or cross-stitch methods; similar to stitch bonded fabrics. See **Bonded fabrics**.

Knitted curtains: The development of Rachel machines for high production with unlimited possibilities, development of filament yarn which is best suited for the curtains, attractive pricing of knitted materials has caused the manufacture of knitted curtains (warp-knitted fabrics) into an important branch of knitting.

Knitted fabric: A structure produced by interloping one or more ends of yarn or comparable material. Collective term for warp-knitted and knitted fabrics

– Textile materials made of stitches; loop and floating using single or multi-thread systems with a horizontal or vertical thread path. Unlike woven fabric which consists of two thread systems crossing each other at right angles, warp knitted and knitted fabrics are textile materials of which the smallest unit is the stitch.

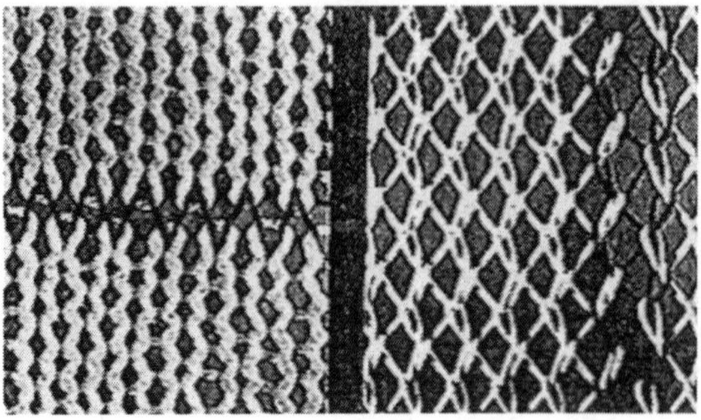

Knitted fabric: See **Fabric (knitted)**.

Knitted loop (in weft knitting): A basic unit of any knitted fabrics consisting of a loop of yarn meshed at its base with a previously formed loop.

Note: The point of mesh with the previously formed loop is usually open but may be crossed. The component part of the knitted loop may be identified as back loop. A knitted loop viewed such that it meshed through the previous loop away from the viewer face or plain (weft knitting); a knitted loop viewed such that it meshes through the previous loop towards the viewer. Needle loop: The upper curved portion of a knitted loop derived from that part of the loop formed by the needle. Sides or legs: The part of the knitted loop that connects the sinker and needle loops. Sinker loops: The lower curved portion of a knitted loop derived from that part of the loop formed by a sinker on machine employing this type of element.

Knitted plush: Plush cloth (e.g. for imitation fur), manufactured knit goods with cut plush loops that are firmly bonded to the knitted base. Manufactured on circular knitting machines and tubular hosing machines.

Knitted welt: A secure edge of a knitted fabric or garment made during or subsequent to knitting usually parallel to the courses and at the starting end of the fabric.

Knitting: A method of constructing fabric by interlocking series of loops of one or more yarns. The two major classes of knitting are warp knitting and weft knitting, as follows:

(a) *Warp knitting* – A type of knitting in which the yarns generally run lengthwise in the fabric. The yarns are prepared as warps on beams with one or more yarns for each needle. Examples of this type of knitting are tricot, milanese, and raschel knitting.

(b) *Milanese knitting* – A type of run-resistant warp knitting with a diagonal rib effect using several set of yarns.

(c) *Raschel knitting* – A versatile type of warp knitting made in plain and Jacquard patterns; the latter can be made with intricate eyelet and lacy patterns and is often used forunderwear fabrics. Raschel fabrics are coarser than other warp-knit fabrics, but a wide range of fabrics can be made. Raschel knitting machines have one or two sets of latch needles and up to thirty sets of guides.

(d) *Tricot knitting* – A run-resistant type of warp knitting in which either single or double sets of yarn are used. See **Tricot**.

(e) *Weft knitting* – A common type of knitting in which one continuous thread runs crosswise in the fabric making, all of the loops in one course. Weft knitting is of two types, i.e. circular and flat knitting.

(f) *Circular knitting* – The fabric is produced on the knitting machine in the form of a tube, the threads running continuously around the fabric.

(g) *Flat Knitting* – The fabric is produced on the knitting machine in flat form, the threads alternating back and forth across the fabric. The fabric can be given shape in the knitting process by increasing or decreasing loops. Full-fashioned garments are made on a flat-knitting machine. See **Flat-knit fabric**.

Knitting machine: A machine for the production of fabric by warp knitting or weft knitting. The different type of warp and weft knitting machines are classified and names primarily, according to (a) the type of fabric or garment they are intended to produce; (b) the type of needles used; (c) the format and arrangement of their needles or their needle beds. Machines are also distinguished by the type of patterning control used and by whether they are hand or power operated.

Knitting machine (double bed): Double-bed circular knitting machines are equipped with two series of needles – one series of needles fits in the circular needle-bed called "cylinder", while the other series is accommodated inside radial grooves positioned at 90° with respect to the cylinder, on a special circular plate called "dial". Double-bed circular knitting machines usually incorporate latch needles, but a few manufacturers also offer machines equipped with compound needles. The cams that command the various needles are fastened to two cam frames, one around the cylinder and the other above the dial.

Knitting machine (large diameter): Large-diameter circular knitting machines are those having diameter from 24 to 40 inches. Large-diameter machines are mostly designed for manufacturing tubular fabrics, while a more restricted number of large-diameter machines are used to manufacture fabrics similar to those manufactured with flat knitting machines.

Knitting machine (medium diameter): Machine diameters from 8 to 22 inches are called medium diameter knitting machines. Medium-diameter machines are designed for the production of body-sized tubular fabrics as well as fabrics with welt and separation thread, ideal for the underwear market.

Knitting machine (single bed): Single-bed circular knitting machines are equipped with only one series of needles sliding in the grooves of a circular needle-bed. The needles are usually latch needles, but a few manufacturers have designed machines equipped with compound needles. The cams, which drive the movement of the needle forming the stitch, are placed outside the needle-bed; each feed system is provided with its own cam group.

Knitting machine (small diameter): These are the machines which having diameters from 3 to 6 inches. Small-diameter machines are designed for manufacturing hosiery. The architecture of these machines differs from standard circular knitting machines above all from a technical point of view, but also for considerable differences in the type of available accessories.

Knitting (double cylinder): A circular knitting machine with two cylinders, one superimposed above the other. It is equipped with one set of double-ended needles that can be caused to knit in either cylinder as required to make plain, rib, and purl knitted structures. This arrangement is commonly used in hosiery knitting machines and garment length knitting machines.

Knitting machine (circular): A knitting machine in which the needles are set radially or in parallel in one or more circular beds. Used without further classification, the term generally refers to a weft knitting machine of this type.

Note: Machines with diameters less than 165 mm are often termed smaller diameter machines.

Knitting machine (hosiery): A knitting machine for the production of hosiery. Most are small diameter latch needle circular knitting machines.

Knitting machines (garment length): A knitting machine adapted for the production of individual garment panels, in series, rather than for continuous production of fabrics. The term is most commonly used to specify this type of circular weft knitting machine.

Knitting notation: It is a simple, easily-understood, and symbolic representation of a knitting repeat sequence and its resultant fabric structure that eliminates the need for time-consuming and possibly confusing sketches and written descriptions.

Knitwear: Garments made out of knitted fabrics.

Knock-over position (knitting): The new yarn is drawn through the old loop, thus new yarn produces a new loop. If knock-over position is shifted vertically, stitch length can be changed.

Knops (knots): (1) Symptom of a fault – caused by knots or knot-like thickenings due to inextricably entwined fibres;(2) Yarn effect – in the form of dot-like or knot-like thickenings, fluffy places, etc., in slub yarns;(3) Weaving effect – as the V- or W-weave loop formation in pile fabrics (pile loops); also similar in pile carpets;(4) Denepping – removal of unevenness in fabrics (knots, spun-in wool flock, curls, trapped pieces of thread, etc.) using tweezers and fabric relaxation. The dyeing of undyed or off-shade dyed weaving points using so-called retouching pencils, this occurs after weaving or before or after wool scour.

Knop yarn: Same as nub yarn in England.

Knopp work: Framework knitting, with two sets of needles and Jacquard attachment, which regulates the accumulation of loops on certain needles and thus forms the design.

Knot: A joining by tying together.

Knot breaking force (in tensile testing): The breaking force of a strand having a specified knot configuration tied in the test specimen portion of a strand mounted between the clamps of a tensile testing machine.

Knot breaking load: It is a deprecated term. Use the preferred term – Knot breaking force.

Knot breaking strength: Strength expressed in terms of knot breaking force.

Knot density: Slub or knot density, number of knots, or slubs in relation to 1 m^2 of an oriental carpet. Top quality pieces contain up to 1 million; normal material 50000–500000; cheap pieces only approximately 15000 knots or slubs.

Knot (long) in raw silk: Knots, which have loose ends from 3 to 25 mm in length. See **Long knot**.

Knot (type 'O'): One in which the bight is below, the bight crosses over the right hand end.

Knot (type 'U'): One in which, when bight is below, the bight crosses under the right hand edge.

Knot (very long) in raw silk: Knots, which have loose ends exceeding 25 mm in length. See very long knot.

Knotted laces: Made in Italy, Dalmatia, etc., by tying lengths of thread into knots by the hands, the knots forming patterns like the macrame.

Knotted work: Laces made by knots; either tatting or macrame.

Knots (1): Used to join ends of yarn or thread together. Most textile companies use "weavers" knots, or air splices. Most knots will cause the thread to break but most air splices will sew through the needle without breaking. There are many types of knots – single weaver's knots, double weaver's knots, tension knots (for the connection of threads coming from opposite directions, e.g. broken ends), cats head knots (connection of parallel threads).

Knots (2): There are two kinds of knots forming the pile in oriental rugs, one is the ***Turkish or Ghiordes***, the other the ***Persian or Senna*** (see). The fineness of the rug is judged by the number of knots to a square inch.

Knotting: A process used for macrame, fishing nets and carpets.

Knotting(in weaving): If a new drawing-in operation is not necessary (this expensive operation is avoided whenever possible) because no style change is needed, the warp is taken from the beam store and brought directly to the weaving room, where it is knotted on board the loom to the warp prepared with the knotting machine.

Knotting machine: See **Knotting (in weaving)**.

Knotted carpets: Pile carpets created by the knotting of brightly coloured threads in a warp system. With the exception of weaving technology, the oldest and most common process for the manual production of carpets. We differentiate between Persian and Turkish knots. See **Oriental carpets**.

Knubbs: In England, the very closely woven, hard and fine inner layer of the cocoon; used for waste silk. See **Frisons**.

Knuckle (in zippers): See **Crimp (in zippers)**.

Ko: Coir fibres.

Ko Hemp: Very durable, soft, fine and silky bast fibre yielded by the *Pueraria thunbergiana* in China, India and Japan; used for summer clothing fabrics.

KOB: Abbreviation for children's outerwear.

Kokura: A hand-woven Japanese cotton fabric made in plain or twill weave and comes in five varieties, all having ribbed effect.

Kokura Hanaoji: A hand-woven Japanese cotton fabric made of ply yarn, with weft cord effect. In weaving, three or four warp ends are drawn through one heddle. This cloth is cut into strips, rolled and sewn together to form cord for the Japanese clogs.

Kokura Kabanji: A cotton duck made on handlooms in Japan, mostly ply yarn being used and weft cord effect produced; used for trunk covers.

Kokura Obiji: A hand-made Japanese fabric, made of silk warp and cotton filling or of all-cotton with weft cords. It is usually plain woven and comes mostly in dark blue with a single plain or ornamented stripes; used for sashes by the natives.

Kolinsky mink: See **Mink**.

Kompon: A plain woven, stout linen in China; used for garments by the natives.

Konieh rugs: Heavy all-wool rugs made in Asia Minor, the medium long pile is tied in Ghiordes knot. Many of the old samples have a hexagonal field and rich colours. Modern K. rugs have a great variety of colours. The sides and ends are finished with a selvage.

Konje: Native African name of the bowstring hemp fibre.

Koomach: Cotton cloth in Russia, usually dyed bright solid red and also indigo or green; used for blouses, women's dresses, etc.

Kora cloth: Colloquial name Greige fabric in India.

Korako: Native name for the New Zealand flax, yielding fibres suitable for fine fabrics.

Kordofan gum: Gum Arabic.

Kotonin: Cottonized flax.

Kotzen: Rugs and laprobes, made in Austria of coarse goat's hair with a very long hair on both sides.

Koujong: Very soft, fine, twilled woollen, made in China.

KP: Kapok.

Kracher: Thick and knotty points in yarn.

Kraft pulps: Pulps prepared in the alkaline liquor consisting of sodium hydroxide, sodium carbonate, and sodium sulphide. Also called sulphate pulp.

Kraft yarn: A yarn made by twisting a strip of paper manufactured from kraft pulp. See **Yarn kraft**.

Kraftcord: This yarn produced by tightly twisted plant fibre is sometimes used in carpet backings.

Krätzen: More or less sharply limited points in hosiery after dyeing, at which hosiery stitches are more dense and irregular, presumably points which, due to bagginess during the fixing of the raw material, did not lie close to the hosiery form and therefore had inadequate tension.

Kron: The best sort of Russian flax. Kuba Oriental all-wool rug, having a long and fine pile in light colours.

Kronos titanium white: Titanium dioxide (from Norwegian ilmenite) in different types with varying TiO_2 content (up to 98%). Chemically barely corrodible, unchanging white with maximum covering effect. Used for printing dyes, delustring, paints.

Kroy shrinkproofing process: Continuous process for shrink proofing wool tops in which there is a direct chlorination step with no intervening chemical reaction followed by anti-chlorination and neutralization. Provides better hand and strength than do conventional shrink proofing.

Kubleka Munk value (K/S): Describes the optical properties of a coloured sample as a double constant made up of K = light absorption (absorption constant of dyes) and S = light dispersal (determined by textile material). If K/S values for the maximum absorption wavelengths are displayed logarithmically as a function of the dyestuff concentration C, then in the ideal case this produces a straight line with a gradient of 45°, with dyeing or printing errors, for example, showing up as measuring points that deviate from the straight line. Calculated using the formula K/S= (1-B) 2/2B, where B= luminosity coefficient.

Kulah rugs: Prayer rugs from the Turkish city of Kula. Velvety smooth pile, approximately 150000 knots per m^2. The large red or blue prayer niche is framed by a border with stylised, often tiny flowers. The *masarlikula* (cemetery kula carpet) is a special type.

Kulkan: Richly brocaded silk shawls, made in Persia, similar to the cashmere shawls.

Kumbi: Soft and silky seed fibre of the *Cochlosper mumgossypium*, a tree in India; used for stuffing.

Kurbelstickerei: German term for machine embroidery, the design being of tape over a net ground; used for curtains.

Kurdistan rugs: (1) Coarse and rough rugs, made in Mesopotamia of all-wool with a heavy, long and loose pile tied in Ghiordes knot. Dark coloured natural wool is often used. The ends are finished with braided fringe; (2) Fine all-wool rugs made in Persia, the close short pile is tied in Ghiordes knot. The pattern consists of small figures covering the field of a conventionalized blossoming tree or a diamond shaped centre medallion in blues and reds. The end selvages contain one strand of coloured wool.

Kuriwata: Japanese trade name for ginned raw cotton.

Kurk: Fine soft wool, yielded by a species of white goat in Persia; used for shawls which are finished with a hairy back.

Kurrijong: Dark, tenacious bast fibre yielded by a species of the *Commersonia* in New South Wales; used for nets.

Kusters dyeing range: Continuous dye range for carpets. The unit wets the carpet, applies dyes and auxiliary chemicals by means of a doctor blade, fixes the dyes in a festoon steamer, and washes and dries the carpet in one pass through the range. An optional auxiliary unit may be installed to randomly drip selected dyes onto the background shade for special styling effects. This process is called TAK dyeing.

Kyrle: Fancy woollen yarn; used for various overcoatings and dress goods.

Kz: Cashmere.

L

'L' Designation: The sequence number given each AATCC Blue Wool Lightfastness standard according to T e number of AATCC Fading Units required to produce a colour change equal to Step 4 on the AATCC Gray Scale for Colour Change.

L*, A*,B*

La: A very strong and light mat, made of the leaves of the pandanus in the Tonga Islands: used for sails.

Label printer: A small flat-film printing machine for textile tapes or fabric patterns to be subsequently cut. Different types are used in a thermal print process, either single-coloured textile labels or multicolour, wash fast, and printed on one or both sides.

Labile complex: An unstable (liable to change) coordination compound.

Labrador blue: The dark blue of the mineral labradorite. The most common colours of labradorite are blue and green although it can also display gold, red, pink, purple and bronze.

The marvelous display of colour is caused by light reflected from the unique spacing of the planes of crystals often producing a three-dimensional image. See **coloured gems**.

Laboratory sample, in wool top: Portions drawn from the lot in accordance with the described procedure.

La uie: Large plant, similar to the pandanus. The leaves are used by the natives of Samoa for clothing mats.

Lac: A fairly fast orange or crimson dye yielded by the dried bodies of the coccus laccae, an insect living on flag trees. It was used formerly as mordant dye, but is very little in use now.

Lace: Open width fabric of various constructions and elaborate designs. Any fibre may be used but the most common are now nylon, viscose and cotton. Made as edging as well as wide-width fabric, it is with few exceptions, machine made. Narrow lace is used flat, pleated or gathered as decoration, wide lace is mainly used for lingerie, night wear, formal dresses, such as wedding outfits. Cheaper varieties are sometimes used as curtains and bed spreads.

Lace bark: Yielded by the Lagetta lintearia tree in Jamaica. It is a fine thin lace-like bast; used for dress trimmings, hats, also for cordage.

Lace, bobbin: Bobbin lace is a kind of plaited structure made from 4 to 400 bobbin yarns. The yarns are worked into a pattern by twisting and crossing the threads. The pattern is outlined by pins stuck in a pillow. The lace may be formed directly on the pins or may be pinned to the pattern on the pillow as it is formed. Also known as pillow lace. Well known examples are laces of Brussels, Chantilly, Antwerp, and Valenciennes.

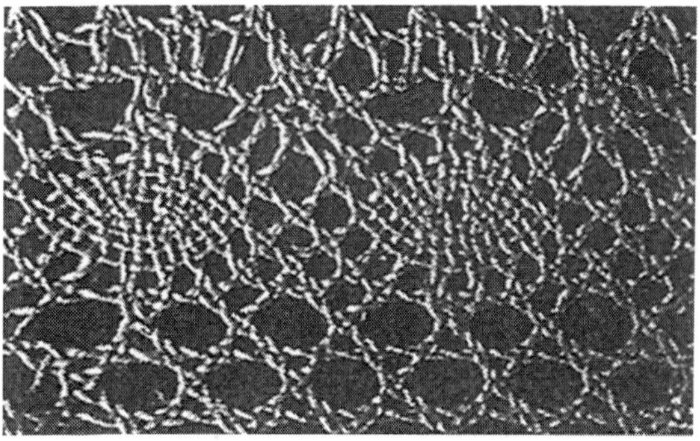

Bobbin lace

Lace, embroidery: Embroidery lace is made by hand or machine application of embroidered patterns to a base fabric. After the pattern has been embroidered, the base fabric may be wholly or partly removed, either by cutting or by using burned out technique.

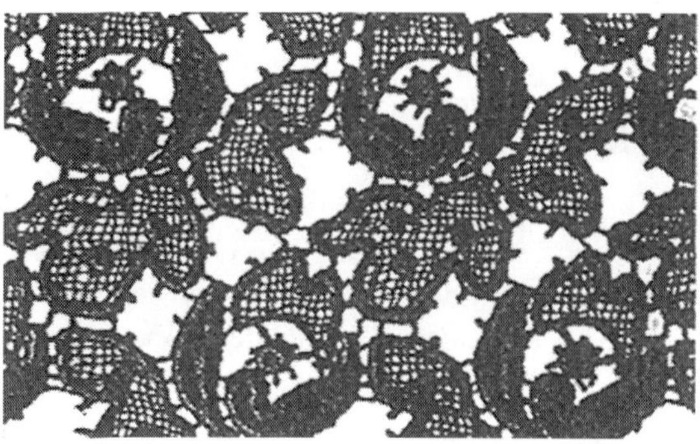

Embroidery lace (burnt out)

Lace, Raschel: Raschel lace is made on a warp knitting machine with latch needles. Most handmade laces can be pretty well imitated on a warp knitting machine. Machine made laces are now much more common than hand made because of the volume and cost advantage over bobbin laces.

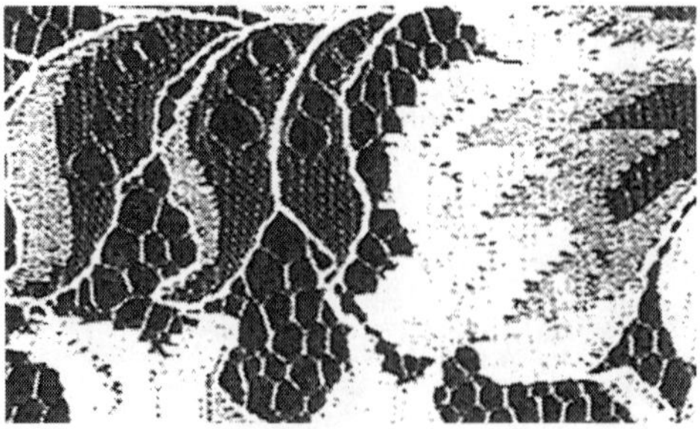

Raschel lace

Lace cloth: A sheer and light fabric, woven of fine yarn in leno or mock leno weave; used for dresses, etc.

Lace galloon: This is a fine, delicate copy of the genuine Valencienne lace on the weaver's lace machine. Knitted ribbons of this type are called galloons. Chiefly made of cotton but also containing a small proportion of polyamide for strength.

Lace machine: The Machines for manufacturing Lace. See **Lace making.**

Lace making: Simultaneous braiding (interlacing) and twisting of threads (yarn, twist) in a diagonal direction for the manufacture of laces, small covers and similar items. These are classified into handmade lace (lace pillow), which in terms of quantity no longer has any significance, and machine-made on lace making machines. In the last case, the bobbins travel around each other in the warp sheet.

Lace stitch: In this knitting stitch structure, loops are transferred from the needles on which they are made to adjacent needles to create a fabric with an open or a raised effect.

Lace Tweed: Very light-weight Raschel-knitted fabrics made of natural silk, crêpe or polyamide warp. The bulky, roving or even wadding-type weft material is predominant and covers the warp (spun thread fine support structure). No qualitative requirements have developed.

Lacklustre, lack-lustre: Lacking in brightness or lustre. See **dull**.

Lacquer-black: A glossy lustrous black colour as in '*Madonna with long lacquer-black tresses*' (*The Evening Standard* 1.2.99).

Lacquer printing: Local gloss print chiefly produced in film printing using film (hole) screens, on thin fabrics and in Rouleaux printing (Chintz prints).

Matt to high gloss effects are achieved with a certain degree of relief effect (depending on the thickness of the metal screen which ranges from approx. 0.3–1 mm).

Lace work: Open work hosiery.

Laced running stitch: Embroidery stitch. Running stitch can be laced with a contrasting colour to form a decorative border. Use a pointed needle for lacing and do not pick up any of the fabric.

Lacet: (1) Silk or cotton braid used to form patterns for laces, and, (2) Lace made of such braid.

Lacets bleu: French coutil (see); used for trousers.

Lacis: Name for darned netting in the fifteenth century.

Lacquers: In most cases substances soluble in volatile solvents which form a thin, adherent, continuous film on the substrate by evaporation or physico-chemical processes. Lacquers are classified according to their composition as follows: (a) Oil- (resin) lacquers: Not used extensively now, these are the oldest type of so-called "long-oil" varnishes with resin (e.g. mix of copal, linseed oil, oil of turpentine) or without resin (e.g. so called enamel lacquers for white internal and external coatings consisting of oil of turpentine, zinc white, boiled linseed oil, and wood oil additive). (b) Cellulose lacquers: First among these are the different high-gloss collodium lacquers corresponding to the oldest type of so-called zapon varnishes (made from dissolved collodium wool or cheap film-formers or celluloid in amyl acetate) and pure nitro lacquers (nitrocellulose dissolved in ethyl or butyl acetate etc. diluted with toluene etc.) used in lacquer printing and in pigment and bronze printing etc. Plasticizers are used as softeners. (c) Synthetic resin lacquers: among these are, for example, phenolic resin lacquers made from phenolic plastic, spirits and another solvent, either as a so-called stove enamel (where infrared drying at 90–150°C follows curing) or "cold cure" lacquers with the later addition of very dilute acid (used as wood lacquers in particular).

Lactic acid: $CH_3CH(OH)COOH$; MW 90. Used as a wool mordant, in the manufacture of numerous dye mordants, as a substitute for tartaric acid and citric acid (leads to the softening of textiles to produce soft, silk handle without weakening the fibre or lustre), as a dye solvent (diphenyl and aniline black pad dyes) and in aniline-black dyeing (aniline lactate), leather dyeing and finishing and for spotting agents, etc.

Ladder: Stitch Used in embroidery and made by running parallel or zigzag bars over an evenly wide space or by working the bars on the material to imitate the rungs of a ladder.

Ladder: See **Run.** Appearance of a flaw in weft knit fabrics, especially women's hosiery. Arises as a result of a wale or wales having been mechanically undone or opened following thread breakage, where in addition to the stitch originally affected, those stitches connected to it also become undone in succession.

Ladder braid: Open work braid, made similar to a ladder; used for laces and for trimmings.

Ladder tape: Stout cotton tape; used for Venetian blinds.

Ladder web: A crochet warp knitted narrow fabric consisting of two wales or columns of stitches spread apart by the width of the slat of the Venetian blind. Weft yarn crosses from one wale to the other from a ladder like structure which separates and positions the slats of the blind. Knitted ladder web is usually made from polyester yarn and headset after manufacture.

Ladder proof finishes: (snag and ladder proof and antisnag finishes). Treatment of knitted fabrics usually made from fine yarns, and stockings against laddering and snags. Antisnag agents are used for this purpose. They are applied during exhaustion or spraying processes.

Ladies cloth: A fine, wide, wool flannel, slightly napped, similar to broadcloth.

Ladik rugs: Small, all-wool rugs, made in Asia Minor; the weft is coloured; the short and close pile is tied in Ghiordes knot. Scarlet red and white are often used. The ends are finished with a red selvage.

Ladines: Eighteenth century woollen fabric in England.

Lagos: Variety of African cotton having a moderately strong and coarse staple of brown colour and very irregular in length.

Lahore: (1) Knotted cotton and wool rugs made in India. The design consists usually of alternate rowe of palm leaves; (2) English dress goods, made of cashmere wool.

Laid fabric: A fabric made without filling yarn, the parallel warp yarns being held together by means of rubber latex or other binding material. See Fabric laid.

Laid-in Fabric: A knit fabric in which an effect yarn is tucked in, not knitted into, the fabric structure. The laid-in yarns are held in position by the knitted yarns.

Laid-in yarn: See **Axial Yarn**.

Laid work Embroidery, consisting of couching.

Laine: French for wool.

Laine brodee A two-fold yarn, made of an open and firm Australia wool, and used as weft in genuine Beauvais and Gobelin tapestries.

Laine de Carmenie: Persian goats' hair.

Laine elastique: All-wool, dull black, light French dress goods made in crepe and corded effect.

Laine de terneaux: Variety of French merino wool.

Laisot: Strong French linen canvas o' good quality, made 48 inches wide.

Laizes: French laces, consisting of a clear net ground and powdered patterns, in dots or small flowers.

Lake, lac: A dark-red crimson resin deposited on certain trees and plants by the insect *coccus laccae* (possibly *laccifer lacca*) originally used as a dye or pigment or glaze. (Compare **cochineal** and **kermes**). The term has for many centuries come to designate any insoluble compound of organic matter produced from plants, vegetables, coal-tar or clay which absorbs colorant and is used to produce pigments of many different colours with a high degree of translucency.

Lake: Fine medieval woollen fabric; used flexible bast fibre, yielded by an Indian shrub, and used for ropes.

Lama: Plain or twilled flannel-like French woollen lining with a slight nap; comes in plain colours or stripes.

Lama-barchent: In Germany and Austria, a very softly finished cotton fabric woven with two sets of filling threads of different colours, one forming the face and the other the back and napped on both sides.

Lambsdown: A plated knit cloth, the face made with very heavy and spongy fleece raised from slack twist woollen yarn, while the back is of hard spun cotton; used for children's coats, carriage robes, etc.

Lambskin: A very closely woven cotton fabric, made in an eight-leaf, weft face satin weave, containing a very large number of picks. The fabric has a woolly nap raised on the face; used for working clothes.

Lamb's wool: Soft resilient wool shorn from lambs upto seven months old. Usually used for knitwear and in small amounts, mixed with other fibres in fabrics.

Lamé: French for "trimmed with leaves of gold or silver" Made with silk or any textile fibre in which metallic threads are used in the warp or the filling. The main fibre of the fabric can be silk, viscose, acetate, triacetate, nylon or polyester also. Lame is also a trade mark for metallic yarns.

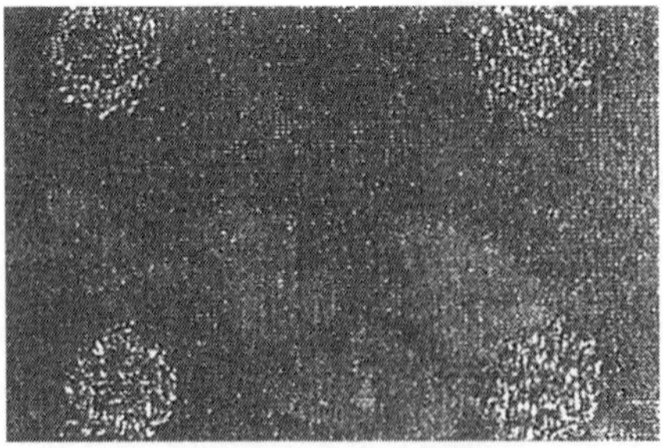

Usually a figured weaves but could be any. Often has pattern all over the surface. The shine and glitter of this fabric makes it suitable for dressy wear. The term comes from the French for "worked with gold and silver wire". The fabric is thin and drape well. Uses: Principally for evening wear.

Lamellar macromolecule: In crystalline areas (in the plane of the paper), the structure of two-dimensional or lamella macromolecules in the sense of so-called chain grids (Pleated sheet structure), for example, in the case of cellulose (lamella), wool (keratin grates), synthetics (crystallite) and silk.

Lamella: (1) A thin leaf or disc. (2) The term membrane lamella is used in fibre histology, for example, in the case of cellulose fibres and synthetic polymers. The two-dimensional system consisting of cellulose chains arranged in parallel and bonded to each other via hydrogen bonds represents the building block for the three-dimensional *Micelle.*

Lamella separator, lamellar plate clarifier, tilted plate separator: A settling device with parallel, steeply inclined plates, about 50 mm apart.

Lament: A Highland tartan, composed of the following colours; *Dark green stripe; white stripe, measuring less than half the width of the green dark green stripe, repeated ; black stripe, as wide as green; stripes of dark blue, black, dark blue, black, each as wide as white one; dark blue stripe as wide as green; stripes of black, dark blue, black, dark blue each as wide as the white; black stripe, as wide as the green; repeat group described above between two *; black stripe, as wide as green one; dark blue stripe, as wide as green one group of three stripes, black, blue black, each as wide as white; blue stripe, as wide as green; black stripe, as wide as green.

Lametta: Thin metal thread used in brocades.

Laminar flow: Streamline flow in a viscous fluid, such as molten polymer, near a solid boundary.

Laminated fabric: A layered fabric structure wherein a face or outer fabric is joined to a continuous sheet material, such as polyurethane foam, in such a way that the identity of the continuous sheet material is retained, either by the flame method or or by an adhesive, and this is in turn normally, but not always, is joined on the back with a backing fabric such as tricot. See Fabric, laminated.

Laminated fabric, in fabric roof systems: A flexible material composed of superimposed layers of textile fabric firmly united by bonding or impregnating with an adherent polymeric material to one or more surfaces.

Laminates: Laminated materials. Made of foam backed fabric, where the fabric is usually inexpensive acrylic jersey, a nylon or acetate jersey and sandwiched between the two, a thin layer of synthetic foam. The layers may be stuck together by adhesive or more usually, all three are passed through a machine which applies flame on both sides of the foam to melt it. The three layers are then pressed together. Used for children's clothes, women's jackets etc.

Laminating: Bonding together of two or more layers of similar or differing flat textile materials such as woven, knit goods, films, felt, paper, nonwovens, fleeces and foam films.

Laminating calendar: This is used to bond textile or paper webs by hot-melt spray coating or with thermoplastic films.

Laminating carpet: Term formerly used for Bonded-pile carpets.

Laminating technology: Bonding textile materials with other non-textile flat materials such as paper and plastics. Techniques:

(a) Adhesive is applied to one laminate component in the form of a paste (coating process), powder (powder point coating process) or melt (roller melt process) and this is pressed with a second or third component.

(b) The surface of a foam is melted using a flame and pressed onto a textile material (Flame laminating).

(c) A (foamable) thermoplastic is first applied to a carrier such as paper (release paper) or continuous metal band and released at the moment of bonding (transfer or reverse process). This process can be used to produce very thin coatings. The flame laminating process, using thermoplastic foams, is widely used, for example, in the manufacture of car upholstery.

Lamot: Native Philippine name for Manila hemp.

Lamparillas: All-wool or linen or silk warp, lightweight fabric in solid colour, stripes, or figures, made in Flanders in the eighteenth century.

Lampas: Originally a fabric from India, (East India) was a printed silk fabric, similar to brocade. But nowadays the name is given to a woven fabric with a rep ground and a satin like pattern formed by the warp yarns. A contrasting effect is achieved, too, with the weft yarns so that the same colours appears in the pattern as in background. Very elaborate designs are be produced. It is a heavy fabric, usually made of cotton, iscose, acrylic, or mixtures now. Used for curtains, and furnishings.

Lampas du Japon: Rich French silk brocade, made with warp ribs; used for dress goods, drapery, etc.

Lana del tambor: A Venezuelan silky seed hair, grown on a species of the Bombax tree; used for stuffing.

Lanameter: An instrument used for measuring fibre fineness including that of chemical fibres. It consists of a desk-shaped housing which has a diagonally arranged measuring matt disc upon which the fibres magnified 500 times are projected. The thickness measurement is accurate to within 0.001 mm.

Lance: Clip-spot (fr: lancer = to throw)A term applied to fabric with an associated group patterning (figuring). This is formed by a second warp and/ or weft system in which threads interlace with only a few warp threads at intervals across the fabric and float over or under the rest of the warp threads

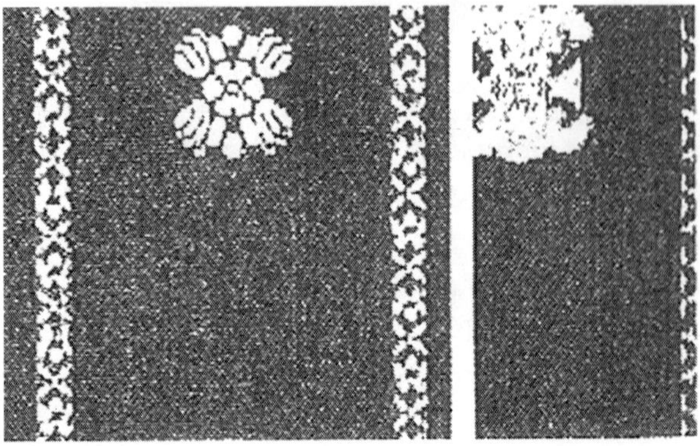

to produce tiny dots or spokes on the ground. The patterning only appears in stripes even if the pattern itself shows no stripe arrangement. The threads not required for patterning on the face float or are woven in at points on the reverse (in contrast to Broché).

Lances: General French term for fabrics in which certain of the filling threads are crossing only a number of warp threads, floating for the rest.

Lanella Fleece: Name for a fibre prepared from a grass; it is carded with shoddy and used for cheap clothing.

Langely: A unit of total solar radiation equivalent to one gram calorie per square centimeter of irradiated surface. NOTE: The internationally recommended units are: Joule (J) for quantity of radiant energy, watt (W) for quantity of radiant power, and meter squared (m2) for area. The following factors are to be used: 1 langley = 1 cal/cm^2; 1 cal/cm^2 = 4.184 J/cm^2 or 41840 J/m^2.

Lanilla: Hard worsted serge of medium quality on the western coast of South America.

Lanolin: Purified wool grease, chiefly a mixture of cholesterol esters. It is used in salves, cosmetic, grease paints, and ointments.

Lanometer: (Rapid Lanometer) – instrument for testing the thickness of fibres, particularly wool. The total cross-section of 100 parallel bundled fibres is measured using a special caliper. Dividing the measurement by 100 gives the average fibre thickness. See **Lanameter**.

Lansdown: Soft dress goods made with silk warp and worsted filling in a three harness twill weave.

Laos: Silk crepe, made with raw silk warp and tussah filling, the latter being alternately two picks of right hand twist and two picks of left hand.

Lap: A continuous, considerably compressed sheet of fibres that is rolled under pressure into a cylindrical package, usually weighing between 40 and 50 pounds. The lap is used to supply the card.

Laps: Loops are termed '*laps*' in warp knitting because the warp guides lap their yarn around the needles in order to form the loop structure. The loops (overlaps) may be open or closed.

Lap splitting: A condition caused by a lap that will not unwind in carding in the same thickness as it was wound in picking. This splitting of the sheet of fibre can result in either a thicker or thinner sheet being fed into the card.

Lap waste: See **Waste, lap.**

Lap, Asbestoes: A flat compressed form of carded or otherwise felted asbestos fibre with or without carrier fibres.

Lap, closed, in warp knitting: *Closed lapis* produced when a subsequent underlap goes in the opposite direction to the preceding overlap, thus lapping the same yarn around the back as well as around the front of the needle.

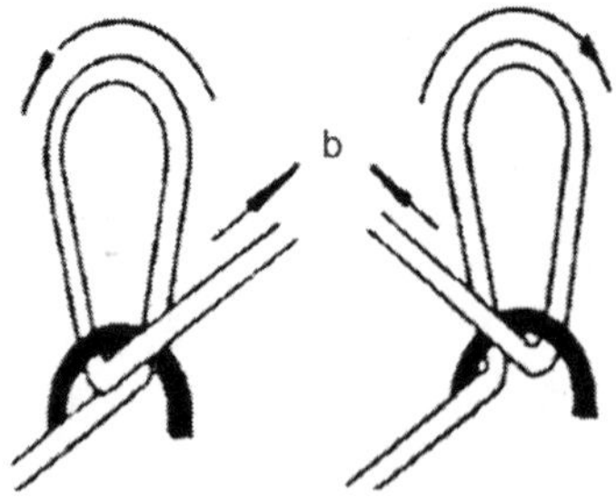

Closed Lap

Lapel: The upper part of the front edge of a Jacket or coat which holds back on to the fore part. The length of the lapel extends from the gorge seam to the break.

Lapel roll: The fall and the curl of the lapel downwards from the break seam of the collar to the first button. The term 'roll' applies chiefly to a soft finish lapel, not one which is pressed almost flat on the crease line.

Lapel seam: The seam caused by sewing the back edge of the lapel strip to the front edge of the fore part becomes the "lapel-seam".

Lap-felled seam: A seam formed with the edge of both plies of material concealed by interlapping. Two or more rows of stitches secure the turned pieces of material.

Lapidary blue: A brilliant blue.

Lapis lazuli: A bright blue; also 'lapis lazuli blue'. A mineral used from the 13th century to

Make ultramarine. For many years the only known deposit was in Badakhstan in Afghanistan. Its vivid blue is the result of its sulphur content. See **Ultramarine**.

Lapped seam: A complex seam formed on the inside of the object with both raw edges enclosed, and having one visible line of topstitching on the face side and and a small fold formed by the topstitching. Used for yolks and applied pieces like gussets, where one section of the garment is lapped over the other for sewing.

Lapped seam: In the lapped seam the edges are folded each within the other or one over the other so that both sides are alike. If made of heavy material, the raw edges are left unturned; in muslin or linen the edges are in turned, lapped, basted and the hem stitched on both edges or hemmed down on both sides by hand.

Lappet: See **Ballooning eye.**

Lappet: A woven fabric with patterns, usually geometric produced by floating threads on the right side of the fabric. These floats are not so long. An example is a zig-zag stripes in white yarn on a coloured plain weave ground. The techniques are also used to form woven spots, produced by cutting the floating yarns between the spots. Any fibres including mixtures are used in this type of fabric.

Lappet weaving: A process of decorating plain woven or gauze fabrics with embroidery-like effect simultaneously with the weaving of the ground. The pattern is formed by an extra warp thread, heavier than the ground, trailing in the direction of the filling.

Lapping: A term describing the movement of yarn guides between needles, at right angles to the needle bar, or laterally in relation to the needle bar, or laterally in relation to the needle bar during warp knitting.

Lappings: Used for iron printing cylinders in roller printing. These can be made from ramie or viscose in the warp and wool in the weft.

Lard: Lard is the soft, white fatty deposit found in hogs. The best grade is called leaf lard and is used for frying foods. The difference between lard and tallow is the presence of the polyunsaturated linoleic acid which is responsible for the softer physical state. The fatty acid composition is 43% oleic, 10% linoleic, 27% palmitic and 14% stearic acid.

Large repeat prints: Large repeat printed articles are on the one hand the Sari and sarong of traditional Far-Eastern ladies' clothing and traditional clothing of the black African population and on the other, table cloths, bed linen, curtains, terry towels, ready-to-wear patterned ladies' outer clothing, flags and printed carpets.

Large Repeat Roatary Screens: These combine the advantage of a high print speed offered by the rotary screen with that of a large format offered by the flat screen. For reasons of stability, the perimeter of the screen can only be increased to 2 m. They are used for printing table cloths, carpets, deco fabric and clothing materials with special patterns.

Larkspur: The pale greenish-blue colour of the plant of the same name; a colour in Winifred Nicholson's '*Chart of Colours*' 1944.

Lase: An acronym for load at specified elongation: the load required to produce a given elongation of a yarn or cord.

Laser: A device for producing an intense beam of coherent light. It is used for cutting, spectroscopy, photography, biomedical investigations, etc.

Abbreviation for "Light Amplification by Stimulated Emission of Radiation", i.e. light amplification by excited emission of radiation. The laser is an instrument for producing very small, concentrated spots of high-intensity light.

Laser screen engraving: An electronically controlled diamond tipped engraving tool is used to engrave/apply the design to the screen directly at high speed and with high precision. The edges of the burnt out holes are sharply contoured and are sealed by the singeing and melting process. The lacquer used for this purpose is not related in any way to the original photosensitive lacquers and can be selected from an existing range of the most durable lacquers (even self-cross-linking).

Lash-In: See **Jerk-in.**

Lashed-In filling: See **Pulled in filling.**

Lashed Pile: Weft pile fabrics, the pile picks interwoven with three warp ends after each float. This secures a fast pile.

Lasting: A narrow and very stout English worsted, woven with double or three-ply warp, and single filling in a five-leaf satin weave; used for shoe tops.

Lastrile fibre: (1) See **Rubber.** (2) A manufactured fibre in which the fibre-forming substance is a copolymer of acrylonitrile and a diene composed of at least 10% by weight, but not more than 50% by weight, of acrylonitrile [-CH_2-CH(CN)-] units (FTC definition). Lastrile fibres are fibres formed from copolymers of acrylonitrile and a diene such as butadiene and contain 10%-50% acrylonitrile units. Lastrile fibres have not been commercially produced. The major proportion of butadiene within the copolymer suggests that the fibre would have extensive elastomeric character.

Latch needle: One of the two types of knitting machine needles. The latch needle has a small terminal hook with a latch that pivots automatically in knitting to close the hook. The fabric loop is cast off. The latch then opens, allowing a new loop to be formed by the hook, and loop-forming and casting-off proceed simultaneously. (Also see **Spring Needle**.)

Latent acids in Dp finishing: Ammonium chloride, amine hydrochloride are used to cure U / F and M/F resins when fast, low temperature curing is wanted. Disadvantages are fabric pH is acidic. Ammonium salts give rise to trimethyl amine, fish odor.

Latent crimp: Crimp in fibres that can be developed by a specific treatment. Fibres are prepared specially to crimp when subjected to specific conditions, e.g., tumbling in a heated chamber or wet processing.

Lateral: A descriptive term for a textile fibre composed of two or more polymers at least two of which have a continuous longitudinal external surface.

Lateral holding strength: The force required to disengage a snap fastener resulting from a pull in the plane parallel to the material to which the snap fastener is attached.

Lateritious, Latericeous: Brick-red.

Latex yarns: Yarns extruded from latex, elastic rubber yarns (Rubber fibres) that are subsequently vulcanized.

Latex: The raw material from which natural rubber is made. It is used to produce stretch yarns, but is of less importance after the introduction of spandex.

Latile: A grass; used for fine braids, etc., by the natives of the Solomon Islands.

Lattice braid: Same as **Ladder braid.**

Lattice stitch: In embroidery, slanting bars are run across a long, narrow, open space, forming latticework.

Lattice-Tulle : The name of a knitted or fine woven curtain with regular lattice-structure. Also named raschel tulle. The dimension stability of Tulle is not very good.

Lauhala: Fine mats, made of pandanus leaves by the natives of Hawaii; used for covers.

Launderability, in buttons: Ability of a button to undergo multiple cycles of launderings without damage such as cracks or loss of finish.

Laundering: A process used to refurbish a textile products or parts thereof by (1) cleaning it in water containing a cleaning agent, and possibly bleach, (2) drying it, and (3) usually ironing or pressing it.

Launder-o-meter: (wash-O-meter) A standard laboratory device for testing a cloth's colorfastness to, and shrinkage in, washing and dry-cleaning. An american standard washing machine consisting of a heated boiler with motorised drive for using 10–20 screw-on bottles (500 ml capacity and 42 rpm) that contain 15 steels balls (6 mm dia) to mechanically manipulate the test samples during testing. This apparatus is especially recommended for laundering and milling durability testing but is also used as a wet fastness testing and HT dyeing machine.

Laundry: Industrial plant (large-scale plants are usually of industrial form) normally engaged in the Washing of Laundry.

Laundry marker: Textile markers (inks, pencils, ballpoint pens, and marker pens) especially designed for the laundry and dry cleaning, with appropriate practical durability but easily removed in case of spotting.

Laurel: A dark yellowy green.

Lauric acid: $C_{11}H_{25}COOH$. Component of coconut oil, laurel oil and sperm oil. Among other things, used for the manufacture of textile auxiliaries such as detergents.

Lava-red: An intense orangey-red.

Laval: Various French linen fabrics.

Lavena: Fine, lightweight woollen dress goods of natural colour.

Lavender: Pale lilac; having the colour of lavender flowers; also lavender grey and lavender blue. Possibly derived from the Latin *lavare,* to wash. Referring, particularly in the US, to homosexuality – see the **Lavender List**.

Laventine: A thin silk used for sleeve linings.

Lawn: (1) The name lawn derives from the name of Laon, France, where it was made first. A very fine smooth fabric. It is lighter in weight than cambric, may be bleached, plain dyed or printed, and has a slightly stiff finished which may be permanent. It is a plain-weave fabric made from cotton or linen and is cool and absorbent and hard-wearing. A typical texture is 72 x 60 with 45s warp and 60s weft, using carded cotton yarns. Fine lawns can be made with combed 70s for warp and 90s weft. Spun synthetics are also used instead of cotton. Lawn is crisper than voile but less crisper than organdy. A very popular fabric for women's summer dresses, baby clothes, blouses, nightwear, underwear etc.

(2) Bleached linen, cotton or half linen fabrics with a linen finish. Mainly used for regularly washed articles.

Lay: See **Lea.**

Lay: (1) Plaiting down a fabric is called laying. It can be a free plaiting as happens in most of the processing machines or can be a precision laying as in the cases of garment production laying for panel cutting.

(2) It can be a set of laid fabric for a purpose.

Lay out, in designing: The arrangement motifs in the framework of the design frame.

Lay out, in sewing: Lay out, in sewing is the cutting chart on the pattern instruction sheet that shows placement of the pattern pieces on the fabric.

Lay, Direction Of: See **Direction of twist.**

Lay-on-air stenter: An air control system for the fabric running in parallel to the fabric pathway and using a combination of slit and jets.

Lay-on-air stenter: An air control system for the fabric running in parallel to the fabric pathway and using a combination of slit and jets. Two fans in a vertical shaft with changeable direction of flow. These ventilators, in each field, allow an over and under-flow of air to be split through regulating flaps to control the air flow. The fabric is carried by the stream of hot air emerging from the jets without the aid of any additional guiding mechanisms.

Laying: (1) The doubling process in manufacturing ropes.

(2) The process of forming a lay. See **Lay.**

Lazouri lake: A blue pigment.

Lazy Kate: The device used to support full bobbins while plying. A fairly traditional design, involves two vertical posts that support the bobbins. Rather like a free-standing ladder. Another form, favored by Schacht, has the bobbins supported horizontally with the addition of a breaking cord to control the backspin. Now, Alden Amos favors a vertical support with the addition of leather washers to help slow down the backspin. At this time, there is no image available.

LC50, LD50: The lethal concentration (LC) or lethal dose (LD) at which 50% of the organisms die when exposed to that concentration or dose of the substance. The exposure time is short at 48 hours or sometimes 96 hours. It is also known as the TLM (median tolerance limit) or MLD (median lethal dose).

LC-50 and LD-50 values: These parameters are used to describe the lethal concentration (Lethal limit) of toxic substances. The acute toxicity is given as the average lethal dose (50% of tests) in mg/kg live body weight for oral toxicity (LD-50) and as the lethal concentration in mg/l for inhalation toxicity (LC-50).

LCP: Short form for Liquid Crystal Polymer.

LDPE: Low-density polyethylene; See **Polyethylene**.

Lé: French carpet-width measurement unit. 1 lé =approx. 70 cm.

Lea, in cotton yarns: The number of 120 yard lengths of yarn over pond; an indirect yarn numbering system.

Lea, in linen yarns: The number of 300 yard lengths of yarn per pound; an indirect yarn numbering system.

Lea, in worsted: A length of 80 yards per pound; an indirect yarn numbering system.

Lea: (1) Measure for wet spun linen yarn; it means the number of 300 yard (274.32 m.) cuts thai weigh one pound. Twelve leas make a hank, 200 leas or 16 & Half hanks make a bundle. See cut and yarn number. (2) A skein used for yarn strength tests. The skein contains 80 turns each 1 & half yards (137.16 cm) in length (A.S.T.M.). (3) Also represents a unit of length in reference to other fibres and varies: 120 yards (109.73 m) of cotton yarn, 120 yards (109.73 m) of spun silk and 80 yards (73.15 m) of worsted.

Lea count constant: See **Break factor.**

Lea product: See **Break factor.**

Lea skein: A skein of length and circumference specified for the specific yarn numbering system involved commonly to determine the linear density and strength of yarns.

Lea strength: Tensile strength giving the no. of lbs tensile load required to break 1 lea of a particular yarn.

Lea Test: Skein strength test (fibre-bundle tensile strength).

Leachate, percolate: When water passes through a ***landfill*** it becomes contaminated with various organic and inorganic pollutants. This wastewater is known as leachate. Leachate generation occurs once the absorbent characteristics of the waste are exceeded. The leachate contains material dissolved from the landfill including Na, Cl and sulphate. Organic material in the landfill biodegrades to simpler organics which may be washed out in the leachate giving a BOD of several thousand mg/l. Some of the organic nitrogen in the fill decomposes to ammonia and the ammonia content of the leachate can be several 100 mg/l. Ferric compounds in the fill are reduced to more soluble ferrous ions and leachate may contain 100 to 1500 mg/l of Fe. The leachate also contains suspended solids and it has a brownish colour. As the organic material in the fill becomes more stabilised, the contaminants in the landfill decrease and the BOD reduces to less than 100 mg/l. The ammonia and iron content of the leachate also reduce substantially as the fill ages.

Leaching: The removal of any substance or dye from textiles by the percolating action of a suitable liquid.

Lead: The colour of the metal lead, particularly in the form used for roofing purposes.

Lead-blue: A greyish blue.

Leader: This is the length of yarn attached to the center core of a bobbin or shank of a spindle to aid in starting your yarn.

Leader cloth: Fabric used to lead the production fabric the machine. Leader cloth is usually a part of the machine. When production cloth has to be started it is stitched on to the leader cloth and run through the machine and taken out. When the machine is to be stopped it is stitched again to the end of the production fabric and left on the machine after the production fabric has been removed at the delivery end of the machine. In jig, it is cheap fabric which is used as the first layer on the jigger and also in drying and steaming and other continuous processes to lead the piece goods. Often made from polypropylene (low dyeability and economic). See **Back grey.**

Leader mark: See **Decatizing mark.**

Leaden: A dull grey colour.

Leaf-green: Having the colour of leaves; sometimes the yellowy-green of young leaves. One of the colours in the **Ostwald circle**.

Leaf fibres: Sub-group of Hard fibres in the form of woody fibres and vascular bundle components from the leaves or leaf sheaths of numerous sub-tropical plant species (such as sisal and manila). Fibre extraction by manual or mechanical methods; fibre preparation by rotting often produces poorer qualities.

Leaf stitch: Embroidery stitch Bring the thread through at the stem point and make a sloping stitch to the one side of the leaf. Bring the thread through at the other point of the stem joining the leaf (next to the first point and make a sloping stitch to the other side of the leaf.

Bring the thread through at point next to the second point on opposite side of the leaf, then continue working alternate stitches on each side in this way until the shape is lightly filled.

When this stitch is used there is usually an outline of Stem Stitch or Chain Stitch worked round the shape.

Leamington Axminster: Small size machine-made Axminster rugs in light colours for bath rooms.

Lean: Wool with a harsh handle; spins unsatisfactorily.

Leas Ties: Also known as lees ties and lease ties. This is such an interesting term that pops in and out of textiles. I asked a longtime weaver which was correct. She said it wasn't so much correct but where you learned to weave. The fact that a term used in measuring linen yarns is "lea" makes me think leas ties came from that direction. So what are they? They are the short threads tied around hanks of yarn to help keep them from tangling while being washed, dyed, and stored. They are also the short threads tied around a warp to allow you carry it from the warping board/mill to the loom. They serve the same process of keeping the threads in order. They are tied by running a thread at right angles to the warp/hank and interweaving through and coming back at opposite angles. Kind of a series of sideways figure 8's.

Lease rods: See **Lease sticks.**

Lease sticks: Lease sticks are two rods or flat sticks longer than the width of the warp. They are placed in the openings of the cross. The cross is maintained in the space between the lease sticks, keeping the threads in order for threading or sleying.

Lease, Weaving: A lease in a warp in a loom at the rear of the healds which is maintained by two traverse or lease rods, often oval in cross section.

Leasing: The renting of capital commodities and long service items or industrial plant by lease financing or lease-purchase agreements. Used increasingly as an alternative means of finance in the textile and finishing industries.

Leasing reed, in sectional warping: A leasing reed forms the shed into which the strings which separate odd from even threads are inserted. To permit this operation, the leasing rod frame is knocked over and two deflection bars bring all the threads to same level. The odd reed dents are interrupted by welding spots, whereas the even dents are completely practicable; the downwards reed movement permits to the welding spots to drag the odd threads downwards. This way the shed for the insertion of the first lease cord is formed. In a second stage the comb moves upwards, dragging the odd threads upwards; thus the second cord can be inserted into the new shed.

Leasing rod frame: A leasing rod frame is one which during warping separates the threads into various layers, so that they can go through the subsequent expanding comb without mutual crushing. It serves also to create the room necessary to insert the leasing strings for the sizing operation.

Least count, in tensile testing: The smallest change in the indicated property that can customarily be determined.

Leather: The hide or skin of animals prepared by Tanning. The thickness and type varies according to the source.

Leather cloth: A coated, light calico cloth, elastic and waterproof for table cloths and the lining of hats, etc.

leather brown: The brown tan colour of shoe leather.

Leather tanning: A high degree of stabilisation of the leather structure is achieved through the bonding of chrome to carboxylic acid groups, thus forming so called binuclear chrome complexes between two carboxylic acid groups of the amino acids in the collagen structure. The stability is so high that the collagen fibres are resistant to boiling. The process is termed collagen tanning and it is this which turns hide into leather.

Leavers lace / Calais lace: A generic term for laces made on Leavers machine. Used for curtain this fabric is a very finely patterned Tulle-lace. Today this material can be perfectly produced on Lace-Raschel machines. Fibres used are : Polyester, Polyester/Cotton blends, Polyester/Acrylic blends. Limited dimension stability, hanging-out is possible. If cotton is used, shrinkage may occur during household washing.

Lechuguilla: Fibre yielded by the plant of same name growing wild in Mexico. The fibre is very strong, coarse and is used for bags, ropes, etc. Also called ixtle.

Ledra: Embroidered quilts made by the natives of Jharkhand, India. See **Kanthas.**

Leek-green: The green of the vegetable of the same name; sometimes an olive green. See **Prasinous**.

Left hand twist: Any single or ply yarn, the final twist of which is from right to left.

Left twill: Any diagonal twill, running to the left.

Left handed twill: See **Twill, left handed, Left twill**.

Leghorn: The yellow colour of straw; from Leghorn in Italy (now Livorno) which produced a particular kind of wheat from which Leghorn straw hats and bonnets were made. The straw when harvested was green in colour, but was bleached before being used.

Legs, in Zippers: The two portions of a separate element that affix the element on the bead.

Lehriya: Striped turban fabric made in Rajasthan, India. It is composed of harmoniously arranged diagonal stripes, they were originally dyed in yellow and red and used for turbans made for Rajput community of Rajasthan. The lehriya was also patronized in the nineteenth and 20th century by Marwari merchant class of Rajasthan, who wore turbans of multicoloured lehriya fabric, mostly featuring pachranga (five coloured) design, considered auspicious because of the special position of number five in the Hindu mythology. The satranga flaunts the seven colours of the rainbow. Unidirectional stripes are known as lehriya, while the diagonal stripes intersecting at right angles to form checks are the Mothra.

Leicester: Good English wool, spinning about 40s to 44s yarns. The staple is very lustrous, light and long.

Leisure: In England the selvage of velvets and silks.

Leisure wear: Style of clothing which, in comparison to the so-called conventional-classical clothing, is designed more to meet the user's requirements and place more emphasis on comfort with light fabrics and a casual style with the trend of fabric compositions towards synthetic mixtures and ease of care; used for shirts, ladies' suits, dresses, men's jackets, trousers, suits and trouser suits.

Lemon: A vivid yellow; the colour of the outer part of the fruit of the same name.

LemonChiffon: One of the colours in the **X11 Colour Set**. It has hex code #FFFACD.

Lemon chrome: A brilliant yellow pigment.

Lemon yellow: A strong yellow composed in a number of different ways one of which includes strontium nitrate

Length analyzer: An instrument which determines the upper-half-mean length and length uniformity index of a test beard of cotton.

Length of lay: The length of one layer of the fabric laid on the cutting table. In the CAD based cutting, the lay length is decided by the software to get the maximum efficiency in cutting taking into consideration of all the sizes if the garment and the packing ratio of different sizes of the garment.

Length, Bending: See **Bending length.**

Length, Breaking: See **Breaking length.**

Length distribution, of staple fibres: A graphic or tabular presentation of the proportion or percentage (by number or by weight) of fibres having different lengths.

Length distribution, of fibres: A graphic or tabular presentation of the proportion or percentage (by number or by weight) of fibres having different lengths.

Length, of a Fabric: The distance from one end of the fabric to the other measured parallel to the side edge of the fabric while it is under zero tension and it is free of folds or wrinkles.

Length, effective fibre: See **Effective fibre length.**

Length group: All fibres, or pulls, whose length fall within a given length interval.

Length interval: A class interval of 3mm (1/8 in.) usually designated by its midpoint length in odd numbered sixteenth of an inch.

Length, Mean: See **Mean length.**

Length of lay, in steel cords and strands: The actual distance required to make one 360' revolution in any element in a strand or cord.

Length of tear, in fabric: The measured distance propogated by a tearing force from the initiation of the force until its termination.

Length, traverse: See **Traverse length.**

Length, skein loop: See **Skein loop length.**

Lengthwise direction, in textile: The direction in a machine made fabric parallel to the direction of movement the fabric followed in the manufacturing machine.

Leno: (1) Leno is the name given to the structure made when the weft is passed between warp ends or a group of ends that have been twisted around each other before the shed is made. The leno weave produces an open-textured fabric that may be sheer or heavy. It is produced by the doup attachment to the basic loom. The doup attachment controls the warp threads horizontally as well as vertically and the unusual warp interlacing prevents slippage of the filling and reduces shrinkage. Examples include mosquito nets, household bags and curtains.. Most often, the designs formed in leno fabrics are stripes, which add textural interest.

(2) Light weight cotton or silk fabrics, having two sets of warp, crossing' each other in the weave; used for dresses, waists, etc. See **Marquisette.**

Leno crepon: A fabric comprising of a cotton crepon background cloth with leno-weave style embroidery may be white or in a contrasting colour and it is usually worked in stripe formation. Used for blouses, shirts and dresses.

Leno fabric: (àjour fabrics, filigree work, gauzes), in most cases, transparent fabrics with Gauze weave, leno weave such as àjour fabrics, etamine, net curtain fabrics, gauzes, marquisette and similar weaves; also used for fashion effects in dress and blouse fabrics.

Leno voile: A plain-weave cotton voile fabric with an open, leno drapy fabric. Used for children's clothes, blouses and nightwear.

Leno weave: A weave in which two adjacent warp yarns cross each other between the picks. This type of weave gives firmness and strength to an open weave fabric and prevents slippage and displacement of warp and filling yarns.

Leontine: French silk dress goods, made with two-ply warp in white, blue and black colours.

Leopard: A buff-coloured, short-haired, flat fur, with black rosette markings. Very hard wearing.

Leslie: Highland tartan, consisting of the following: Wide dark green stripe; narrow black stripe; wide dark green stripe; very fine white line; black stripe, made a little narrower than green one; dark blue of same width as the black; red stripe, as wide as black between the green ones; dark blue stripe repeated; wide black stripe repeated; fine white line.

Let-off: This motion delivers warp to the weaving area at the required rate and at a suitable constant tension by unwinding it from the weavers beam.

Let off mark, in woven fabrics: A corrugated defect pattern distributed across the fabric width. Let off marks are caused by badly adjusted let-off motion in a loom.

Let-off motion: A device for controlling the delivery and tension of the warp during weaving.

Lethal concentration (LC): Constituent concentration estimated to produce death in a specified no. of test organisms in a specified time period (eg., 96-h LC_{50}).

Lethal index: Concentration of a Poison in mg/m3/min. required to lead to the death of an adult fish.

Lethal limit: Lethal limit for poisons: LC-50 and LD-50 values; Death rate for fish in effluent. See **LC-50** and **LD-50 values.**

Lettuce green: The yellowish green colour of lettuce leaves.

Leuco: Leuco refers to vat dyes in the reduced form; in general a prefix meaning white. Vat dyes are converted from the insoluble pigment form to the soluble leuco dye form by means of a reducing agent and an alkali, often sodium hydrosulfite and sodium hydroxide. Sulfur dyes are also converted to leuco form for application. Once on the fibre, the leuco form is converted back to the insoluble pigment by oxidation. In spite of the formal definition of leuco as a prefix, dyes are often still coloured in the leuco form (for example, leuco indigo is yellow-green).

Leuco compounds: Water-soluble colourless reduction compounds, for example, vat and sulphur dyes, in aqueous and alkaline liquors (Vat) that are fibre substantive and are rendered insoluble through oxidation to become fixed in the fibre.

Leuco sulphur dye: Powder-form of (Sulphur dyes) which already contain the reducing chemicals necessary for solution and dyeing. Applied without the addition of sodium sulphide. No longer commercially available.

Leucorhodina: A medieval silk fabric of very pale pink colour.

Levant red: See **Turkey red.**

Levantine: A stout twilled silk, each side finished equally but of different colours.

Levantine pour: Leaf, double-faced, closely woven silk serge, having single or ply warp. Comes mostly in solid colours, but also in stripes.

Level: Uniform in shade over the surface of a piece of dyed fabric or along the length of dyed yarn Level dyeing is usually the objective in commercial processes. Some dyes, such as leveling acid dyes, are easy to accomplish level results with. They do not bind tightly to the fabric in the dye bath, and dye can leave the fibre and re-enter the dye bath. Other dyes, such as reactive dyes, don't level as easily, and greater care is required to achieve level results. In general, level dyeing is promoted by good agitation, careful control of the rate of rise of the temperature of the dye bath, control of pH, and sometimes by use of special leveling agents or *retarders*. It is often the case that the dyes that level most easily are the least *washfast.*

Level loop: A term describing a tufted or woven carpet with uncut, equal length loops composing the pile surface.

Level pile, for pile yarn floor coverings: Pile in which all tuft legs are of substantially same length.

Levelling: (1) The process of correcting an uneven dyeing to an even dyeing by any suitable process. There are auxiliaries/chemicals available which can help in the migration of dye from a higher concentrated area to a lower concentration area to make the dyeing level. There are also other technic of making the dye migrating to the dye bath and redepositing. Sometimes the process in which an unlevel dyeing is completely stripped and new dyeing is done, is also called leveling in a processors point of view even though in technical point of view it is not leveling.

(2) Migration leading to uniform distribution of dye in a dyed material. Leveling may be a property of the dye or it may require chemical assistance.

Levelling acid dye: A class of acid dyes Leveling acid dyes typically offer bright colours and level easily. Unfortunately, the same property that makes it easy to get level results causes wash fastness that is not as good as that of many other acid dye classes. Leveling acid dyes are often used for wool.

Levelling agent: A dye bath additive to promote level dyeing; see **Retarder.**

Levelling capacity of dyes: Levelling properties; also the ability of a dyestuff to cover differences in exhaustion onto a substrate. The property of the dye, on the fibre, to migrate from a site of higher concentration to a site of lower concentration. The levelling capacity is dependent on time, temperature and additions of electrolyte and auxiliaries.

Levelling dyes: General term for acid dyes with outstanding levelling properties for dyeing wool.

Levelling roller, sectional warping: The levelling roller permits to carry out the warping under low thread tension, and to attain at the same time a compact winding. When processing mono-filament or multi-filament yarns in fine counts which do not stand high compression, it is possible to cut out the levelling roller.

Levelness: This term is used to describe the uniformity of distribution of a finishing substance on and within goods after a particular treatment.

Leviathan: (hebr. liwiathan = giant), continuously operating plant for washing loose wool which is sluiced through 5–7 wash vats (2500–4400 l capacity) by means of (fork, drag and shake) rakes in accordance with the pass-on principle.

Leviathan canvas: Coarse, open canvas used for embroidery. At felled seam

Leviathan stitch: Called also Railway stitch; used in embroidery. It is composed of three long stitches next to each other crossed by a fourth in the middle.

Leviathan wool: Thick, soft wool yarn for embroidery.

Levi's Redwire DLX iPod Jeans: The international American brand Levi's has added a music player on their symbolic blue jeans in contrast to tops where you are more likely to find music player items (Comprehensive Merchandising Support 2007).

L/H: Abbrev. denoting an aqueous Emulsion, for example an oil (Gk.: lipos) in water (Gk.: hydro) system.

LI: Flax (linen).

Liage: French term for an extra warp which binds the extra brocade weft thread.

Liberty: (1) Originally an East Indian tussah silk cloth, printed in Europe; (2) Light, pliable and highly finished silk fabric; used for dresses, trimmings, etc.

Liberty print, lawn: High quality, light weight, plain woven cotton fabrics made with fine or very fine yarns. Printed and lightly stiffened for dresses and blouses.

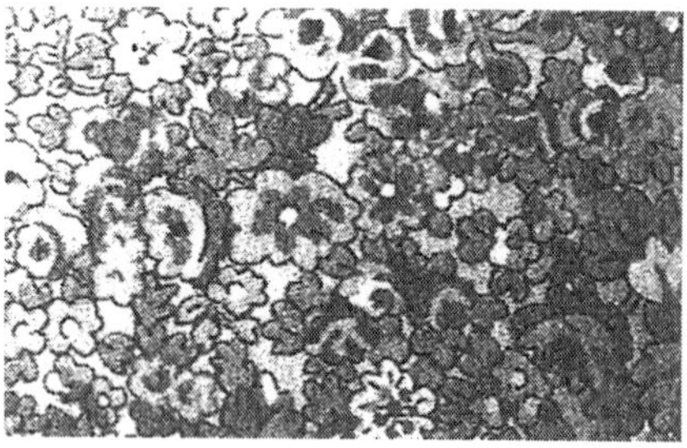

Lichen-green: The whitish-green of the lichen plant.

Lickerin: A part of the feed mechanism of the card. It consists of a hollow, metal roll with a spirally grooved surface containing a special saw-toothed wire. The lickerin opens up the tufts of the picker lap as it is fed to the card and transfers the fibres to the main cylinder.

Lickerin loading: A condition whereby fibres are imbedded in the lickerin wire clothing so as to resist transfer to the cylinder clothing.

Lienoillo: Gray cotton goods in South American countries.

Lienzo: Unbleached cotton sheetings and shirtings in Argentina, Paraguay and Uruguay.

Lif: Name for the fibre yielded by the leaf stalks of the date palm in Arabia and Africa; used for ropes, coarse cloth, etc.

Lifting plan: An indication of the order in which the heald shaft are lifted on each pick in one weave repeat. Also called Peg plan or Chain plan.

Light: (1) Electromagnetic radiation of which a human observer is aware through the visual sensations that arise from the stimulation of the retina of the eye. This portion of the spectrum includes wavelengths from about 380 to 770nm. Thus, to speak of ultraviolet light is incorrect because the human observer cannot see radiant energy in the ultraviolet region. (2) Adjective meaning high reflectance, transmittance or level of illumination as contrasted to dark, or low level of intensity.

Light (colour): **(1)** Lightness. (2) Term used for classification according to "white and pastel shades".

Light cabinet: Colour-matching cabinet

Light damage: See **Photochemical fibre degradation, yellowing of textiles.**

Light edges: See **Edge-to-edge unevenness.**

Light end: See **Fine end.**

Light end: The low boiling fraction in distillation.

Light fading: Exposure to light, through photochemical reactions, causes damage to dyeings; recognizable as changes in the colour tone.

Light fastness: A measure of how resistant a coloring material, such as dye, is to fading due to exposure to light There are a number of textile industry standard methods for valuating how lightfast dyed fabric is. Differences in the light fastness of individual dyes mixed to make a specific colour can result in a colour shift over time as the dyes of lesser fastness fade while others fade much less. Light fastness mostly depends on the molecular structure of the dye itself, but can be influenced by the fibre or contaminants.

Light filling: See **Thin filling.**

Light pick: See **Thin filling.**

Light sensitivity: See **Photochemical reactions**.

Light source: An object that emits light or radiant energy to which the human eye is sensitive. The emission of a light source can be described by the relative amount of energy emitted at each wavelength in the visible spectrum, thus defining the source as an illuminant. The emission also may be described in terms of its correlated colour temperature.

Light weight duck: See **Duck.**

Lightness: Perception by which white objects are distinguished from gray, and light-coloured objects from dark-coloured.

Ligne: A unit of measure for buttons, one ligne equals 0.635 mm (0.025 in.)

Lignin: A mixture of natural aromatic organic compounds found in woody and grassy plant material Lignin occurs in bast fibres such as linen. Lignins are normally removed from fabric to be dyed, since they may dye easily but with poor fastness. Lignins extracted from wood pulp used to make paper are used to make some types of surfactants.

Lilac: The pale purple colour of lilac blossom. See **Anil**. Pertaining to lesbianism; See **Lavender**.

Lille lace: A bobbin lace, the earlier specimens have straight edges, the patterns being outlined with a heavy cordonnet, the hexagonal ground is a very light and fine mesh, each mesh having two sides made of a single thread and four sides of two threads twisted together. A very fine textured bobbin lace, with patterns outlined in thicker threads, characterized by a dotted design. Resembles Mechlin Lace.

Lima: Raw cotton from Peru; the staple is rather coarse and harsh.

Limbacher Artikel: (simplex knit), suede-type finished knitted fabric made from cotton, used for gloves in particular. Strong alkali treatment is used to achieve a more compact structure.

Limbaki: A synthetic fabric for swimwear which has a pile.

Lime: he green or yellowy green of the fruit, lime.

Lime-baryta process: Water softening process used where there is a high level of sulphate hardness and high permanent hardness. Degree of softening up to approx. 1–2°.

Lime blue: A moderate blue.

Lime green: A greenish yellow or olive colour; a pale green.

Lime soap: Lime soap is produced by the displacement principle converting soluble Na stearate soap into insoluble Ca stearate soap (lime soap) according to the following equation:

$$2C_{17}H_{35}COONa + Ca(OH)_2 \rightarrow (C_{17}H_{35}COO)_2Ca + 2\ NaOH$$

109.3 g sodium stearate soap is converted into lime soap by 10 g CaO which corresponds to 1 m^3 water at 1°.

Lime-soda boil: A method of scouring cotton. In this process the fabric is boiled with milk of lime which convert the fatty acid into insoluble lime soaps. The lime soap is then converted into free acids on acidification and calcium is washed away as calcium chloride. The deposited free acid is then converted to soluble soaps on subsequent boiling with sodium carbonate. This method is tedious and not economical as two boils are required to remove the natural impurities.

Lime soda process: Lime water softening process

Lime water: Calcium oxide.

Lime water softening process: Oldest method used for precipitating out hardness salts with calcium hydroxide; also used for de-acidifying water (stoichiometric amounts). Softening is possible up to around 2– 4°d depending on the non-carbonate hardness. Not very suitable for water with a high chloride content as this increases the solubility of calcium carbonate.

Lime wool: Skin wool, (slipe wool) obtained by de-hairing, i.e. by separating the wool from the skin by hydrolysis using calcium carbonate, sodium sulphide. Results in wool which is generally damaged.

Lime yellow: A medium yellow or greenish yellow.

Limestone: A brownish grey colour.

Limerick lace: An Irish hand- or machine-made lace, in which the patterns are embroidered on to the net with a darning stitch. This is a fine lace.

Limiting oxygen Index: A relative measure of flammability that is determined as follows. A sample is ignited in an oxygen/nitrogen atmosphere. The oxygen content is adjusted until the minimum required to sustain steady burning is found. The higher the value, the lower the flammability.

Limiting quality level (LQL), in acceptance sampling: The fraction of non conforming items at which the process average can be considered barely tolerable; the process average at which the risk of acceptance is called the Consumers risk.

Limerick Lace: Irish needlework, executed either by stitching patterns with heavier thread over machine made net ground or by buttonholing the edges of the pattern traced over lawn or muslin, cutting away the ground and applying the whole on machine made net.

Limoges: (1) coarse French bagging made of strong hemp yarn; (2) a cotton and linen cloth, made in coloured stripe for beds in Switzerland; (3) an obsolete French guipure lace.

Lin du pays: Trade term for dew retted flax from Premesques and Beaucamps, France.

Lincere: The finest linen fabrics in ancient Greece, made with double warp and single filling.

Lincoln: Lustrous and thick English wool, used for yarns from 38s to 40s.

Lincoln green: A yellowish green colour being the colour of cloth of the same name and associated with Robin Hood and his band of outlaws.

Linden green: A greenish yellow.

Lindsay: A Highland tartan, made as follows: Wide rose coloured bar, split in the center with a pair of very narrow, dark blue lines; dark blue stripe about one- fifth the width of the rose bar; dark green bar, slightly narrower than rose one, split with a pair of dark blue, narrow stripes near each edge, these stripes being spaced from the edge and from each other their own width; dark blue stripe, about one-fifth the width of the rose bar.

Line: (1) Name for hackled flax, which is numbered either as warp numbers (for fine dressed line), the numbers, ranging from 25 to 100 indicating the lea of a fair warp yarn which can be spun of that line. With the other numbering, used in Scotland, the number indicates the pounds per spindle of 14,400 yards of yarn; dressed line is a thoroughly hackled flax; (2) Standard of measurement for the width of ribbons, being 1-11 inch; (3) The rib on the braid.

Line Flax: Line fax is the long flax fibre that has been drawn off of the hackles. The finest preparation is often used for wet-spun linen, but line can also be dry spun.

Line Fleece: A fleece of wool midway between two grades in quality and length, which can be thrown into either grade.

Line yarn: Linen yarn spun from the longer flax fibres; syun up to 300 leas.

Linear density: Mass per unit length expressed as grams per centimeter, pounds per foot,

or equivalent units. It is the quotient obtained by dividing the mass of a fibre or yarn by its length.

Linear integrator, in textile unevenness testing: An integrator that operates continuously and reports the unevenness for a certain, and unchanging, time past.

Linear lea: An indirect yarn numbering syatem in the linen spinning system equal to the number of 300 yard length per pound.

Linear Polymers: These are the polymers where monomeric units are linked together to form long straight chains. The polymeric chains are stacked over one another to give a well packed structure.

Linear chain

As a result of close packing, such polymers have high densities, high tensile strength and high melting points. Common examples of these type polymers are polyethylene, polyester and nylon etc.

Linear pressure: This refers to the pressure per cm of roller length in the nip of two press-rollers.

Lined work: A twill, made by the symmetrical combinations of the broken diamond twills, like the bird's eye.

Linen: Yarn, thread, or fabric made from flax fibres.

Linen: A very strong yarn of high lustre made from flax fibres. The resulting fabric is also called linen. The yarn can be used for all types of weave and all weights of fabric from fine lawn to suiting and house hold furnishing.

Linen, birds eye: See **Birds eye linen.**

Linen, butcher: See **Butcher Linen.**

Linen canvas: Tailors close, even weave beige coloured canvas of excellent quality. Used in men's coats in conjuction with hair canvas. It is also called Holland or Flax canvas.

Linen canvas

Linen checks: Blue and white striped or checked all linen cloth; used for dresses and aprons.

Linen fabrics: Fabrics made of the fibres of the flax.

Linen-look: Fabrics made to resemble plain-weave linen suiting, which were developed when linen became very expensive for dress wear. They have the typical uneven slub yarn and a crisp even finish. Most have a crease resist or crease recovery finish applied to them. The fibre content may be 50% cotton with nremaining polyester or they may be mainly viscose fibre. Used for shirt-waister dresses, trousers for men and women.

Linen numbering systems: The systems for numbering linen yarn are:

In Scotch, dry spun yarn one spindle (or spangle) contains 2 hesps, or 4 hanks, or 24 heers, or 48 cuts, or 5,760 threads, or 14,400 yards.

In Ireland and England 1 bundle contains 16 2-3 hanks, or 200 cuts, or leas, or 24,000 threads, or 60,000 yards. In Austria one schock contains 12 bundles, or 60 pieces, or 240 hanks, or 4,800 cuts, or 288,000 threads.

The Dorset and Somerset system takes the weight of 21,600 yards (called "dozen" or 12 half hanks).

In France the pacquet contains 360,000 yards.

In Belgium the pacquet contains 180,000 yards.

According to the numbering based on the metric system, the number gives the kilometres (1,000 metres) of yarn contained in one kilogram (2.2 pounds).

Linen twill: A twill fabric made of linen in light weight and usually used for embroidery.

Linen union: Linen union is smooth or figured fabrics (tablewear) made from flax (tow) and cotton yarn, in either warp or weft.

Linen weave: Same as plain weave.

Linen yarn: Spun of flax fibre. Dry spinning, employed in Scotland, gives a very strong yarn.

Linoleum: A floor covering made on burlap base. Oxidized linseed oil is mixed with ground cork and other pigments. This composition is rolled over the burlap base. It comes in plain, printed or inlaid. The plain has a uniform surface in one colour; printed linoleum has patterns printed in colours; inlaid has patterns of different coloured compositions which go through to the burlap base lining the flax fibre is first macerated in hot water, which separates the fibre into its short, ultimate components; this is used for the fine counts.

Linet: A French-make lining fabric of unbleached linen.

Lingerie knit: A very wide tricot fabric of all weights which is easy to sew. 'Power net' is a two-way stretch fabric used for foundation garments.

Lining: Material used to cover part or all of the inside of a garment.

Lining: A lining fabric can be made from silk, viscose, acetate, triacetate. Polyester etc. The fabric suitable for lining are either plain weave, satin or twill and they are usually slippery to enable the garment to to be put on and off with ease. Some fabric like acetate is cheap but not durable, others like silk or polyester satin or pongee is more durable.

Lining fabric: Fabric that is used to cover inner surfaces, especially when the inner surface is of a different material than the outer. May refer to garment lining, lining for boxes, coffins, etc. Generally of smooth, lustrous appearing fabrics, but also of felt and velvet. Both manufactured fibres and natural fibres are used.

Linitest: Wash fastness tester which can also be used as a laboratory dyeing machine; can also be used for HT-dyeing.

Linizing: Permanent finish to make cotton fabrics look like linen e.g. with strong caustic and cuprammonium solutions whereby the fabric surface is partially dissolved and subsequently coagulated.

Link-and-link: See **Purl.**

Linked processes: Refers to the connection of the various steps of fibre-to-yarn processing via pneumatic fibre-transport systems, on-line monitoring, and process control. Process linking results in less labor-intensive processing. A typical linked system might include all stages from bale opening through carding.

Linking: A sewing process used to attach knitted trimmings to body pieces by chain stitching individual stitches to the body.

Links-links fabric, purl fabric: Knitted fabrics which only have purl stitches visible on both fabric faces. The purl stitch courses conceal the adjacent plain stitches, as can be seen in the crosswise extension of the fabric.

Linon: Plain and closely woven fine, very light, glossy, washable cotton or linen fabric; used for dresses, waists, etc. It comes in white or solid colours. It is the French for lawn.

Linon a jour: A gauze-like linen fabric used as dress goods.

Linoxyn: Linseed oil is (air) oxidised and polymerized to form a thin, transparent, skin, e.g. by the oxidation of (linseed-) oil size

Linseed oil: Linseed oil is high in linoleic (60%) and linolenic acid (25%). These oils will polymerize through the double bonds to form hard lustrous finishes on wood. Their major uses are as wood and furniture finishes and in oil based paints.

Linsey: (1) A term used to describe any waste or rags containing wool.

(2) A coarse fabric with a cotton warp and a blended yarn of cotton and waste wool for the weft. The weave is plain twill, usually in stripes across the width. Used for aprons and overalls.

Linsey-Woolsey: A fabric no longer made or used; it was a cheap, coarse cloth of wool and cotton or linen.

Lint: The main seed hair of the cotton plant. A form of waste; Fibre fragments agraded from textile materials; also loose short fibres or fluff.: An inherent fault in silk only apparent after degumming or dyeing. It is charecterised by fine fibrils of fibrillae that become separated from the filament, so giving a speckled, disheveled appearance.

Lint ball: See **Fuzz ball, Balling up.**

Lint content: That portion of a mass of cotton fibre consisting of fibre, including normal, moisture content, but excluding foreign matter.

Lint cotton: Loose cotton fibres in any form, either raw or processed, free of seeds, and not bound together in yarn or faric.

Lint doctor: In roller printing, the Doctor blade removes excess print paste, the lint doctor is located on the opposite side of the printing roller, behind the printing surface formed by the roller and printing cylinder, to remove excess print paste left on the roller after printing. This prevents the print paste from contaminating the following fabric.

Lint, In loose cotton: Fibres mostly of spinnable length.

Lint, ginned: See **Ginned lint**.

Linters: The short fibrous material adhereing to the cotton seed after the spinnable lint has been removed by ginning and which is subsequently recovered from the seed by a process called delinting.

Linthee: A Chinese silk taffeta.

Linton tweed: See **Tweed**.

Lipases: Used in desizing to improve the removal of fats/oils of animal and vegetable origineven at temperatures at which the fatty materials are in solid form. To remove triglyceride-based size lubricants from fabrics.

Lipophilic groups: Lipophilic groups are molecular groups which have an Endophilic reaction in an organic, non-gaseous phase.

Lipophobic groups: Lipophobic groups are molecular groups which have an Exophilic reaction in an organic, non gaseous phase.

Liquid ammonia treatment: A process using anhydrous liquid ammonia (Sanforset) has been commercialized for stabilizing Denim fabrics. Smooth drying ratings and reduced residual shrinkage were improved without losses in strength and abrasion resistance. The improvement was not to the level one would expect of durable press but enough to warrant the expense. A variation of this process has also seen commercial usage. Cotton shirting fabric was passed through the liquid ammonia process and then topped with a low level of DMDHEU. The DP performance was higher than if the resin had been used alone. The losses in strength and abrasion were minimized by the low level resin treatment. The liquid ammonia process involves passing the cotton fabric containing a controlled, low level of moisture into a pad bath of liquid ammonia. The ammoniated fabric is put through a Palmer unit and removing the ammonia while the fabric is pressed flat against the drum.

Liquid crystal: A liquid in which the molecules are oriented parallel to each other resulting in birefringence and interference patterns visible in polarizing light.

Liquid crystal Polymer: Polymers such as aramids or the thermotropic polyesters that form liquid crystals when in the appropriate state, (concentrated solution or melt). Most liquid crystal polymers have in their structure a succession of para-ring structures. The liquid crystal formation is thought to relate to the fact that there is a limiting concentration of rod-like chains that can exist in a random arrangement in a solution or melt. Once this concentration is reached, ordering or alignment of the chains in necessary to accommodate them. Fibres from liquid crystal polymers generally have high modulus and tenacity, good chemical resistance, and high temperature resistance. They are used in a wide range of applications including protective apparel, tire cord, composites, ropes and cables, etc.

Liquid dyes: Liquid paste. Easy to handle, simple to measure, therefore well-suited for automated yarn dyeing and continuous plant. Formulation of Dyestuff.

Liquefied natural gas (LNG): Liquid methane, obtained from natural gas and used as a fuel. See **methane**; **Natural gas**.

Liquefied petroleum gas (LPG): A mixture of liquefied hydrocarbon gases (mainly propane) extracted from petroleum, used as a fuel for internal combustion engines and for heating.

Liquor ratio (or liquor to goods ratio): The ratio of the weight of the dyebath or other processing bath to the weight of the goods being dyed or processed For immersion dyeing in art dyeing processes, a common liquor to goods ratio is 20:1. That is, for each kilogram of fibre to be dyed, 20 kilograms of dyebath are used. In the metric system, this is easy to calculate, since one litre of water weighs one kilogram. High liquor ratios are generally avoided, since they often cause poor exhaustion of the dye, though this is not true for all dye types. Modern commercial dyeing equipment often works with low liquor ratios. Very low ratios may be used for methods where essentially all of the dye solution is to be absorbed by the fibre, such as padding.

Liquor: A solution of dye and/or other chemicals.

Liquor goods ratio: **Liquor-to-goods ratio**: The ratio of goods (kilograms) to liquor (litres), defined more precisely as: the ratio of the dry weight in kg of the goods, including all associated materials, measured immediately prior to the finishing process, to the volume in litres of the entire treatment liquor in the system at a temperature of 20°C at the time of the measurement.

Liquor level: Either the level or the quantity of liquor contained within a dye bath, etc.

Liquor pick up: See **Pick-up**. The amount of liquor picked up by the fabric in any process. The pick-up of liquor in the pad mangle is dependent upon:

– the squeezing pressure of the squeezer,

– the textile material,

– composition of the pad dyeing liquor,

– temperature of the dyeing liquor,

– throughput speed of the goods.

Liquor quantity: Liquor-to-goods ratio.

Liquor ratio: See **Liquor goods ratio.**

Liquor stability: The physical and chemical resistance of dyeing and other treatment liquors.

Liquor wedge: It arises in both horizontal and vertical broadloom washing machines between the guide roller serving as principal washing element (a) and the travelling sheet of fabric (b), i.e. shortly after the installed spray pipe

(c) which constantly applies a film of liquor to the fabric. As a consequence hydrostatic pressure builds within the liquor wedge and forces liquor through the fabric.

Lirelle: Trade name for a polyester fibre developed by Courtalds high degree of wet strength. It is used alone or mixed with other fibres as cotton.

Liro- (G): Pale.

Lisardes: (1) An East Indian and Persian cotton cloth; (2) A coarse Egyptian linen cloth.

Lisere stout: French silk cloth made with weft brocaded flowers and Jacquard figures with the warp.

Lisiere: French for selvage.

Lisle: This fabric, now superseded by finer synthetics, was a knitted material made from mercerized cotton yarns. Originally made in France, it was used in large quantities for womens stockings before the invention of nylon.

Lisle: Hard spun thread, made of long staple, combed cotton, the yarn is gassed and is used for hosiery and underwear.

Lisle yarn: A high-quality cotton yarn made by plying yarns spun from long combed staple. Lisle is singed to give it a smooth finish.

Lisse: (1) French for warp; (2) A silk gauze used for dresses.

List: The selvage.

Listed: Fabrics having damaged selvage.

Listing: An uneven dyeing effect in which there is a variation in colour between that of the selvedge and that of the centre of a piece-dyed fabric.

Listing: Same as **Tailing.**

Listones: Silk and velvet ribbons in Latin-America.

Litmus: A blue pigment obtained from lichen used as an indicator in solution or by colouration of paper. A distinction is made between red litmus paper, which reacts to an alkali by turning blue, and blue litmus paper, which reacts to an acid by turning red. Its neutral colour is violet.

Litre: (typically spelled 'liter' in the U.S.A.) Symbol: l. Fluid measure, i.e. spatial or volumetric measurement equal to 1000 cm^3 of water at 4°C and standard pressure, which equals 1 l or 1000 ml. On this definition, one liter is the same as 1000.028 cubic centimeters. One litre is approximately 1.06 US liquid quarts, or 0.88 imperial quarts. One litre of water weighs, for practical purposes, 1 kilogram.

Litt: Medieval name for dyed fabric.

Liuse: Chienyong Scarlet red cut silk velvet from China.

Livery: The dress or uniform of servants.

Livery Cloth: See **Boxcloth.**

Livery tweed: Very strong and durable whipcord tweed, made of wool in England. Used for uniforms and liveries.

Livid: Bluish; black and blue; lead colour; pale. A word with a history of shifting meanings.

Originally indicating black and blue (the colouring of bruising) from the Latin *lividus*, but used to designate many other colours including lead-colour; grey; ashen; blackish; and purple. Possibly from the Old Slav word '*sliva*', a plum (from which the Russian drink *slivovitz* originates). As from the 20th century the word has taken on the meaning of 'intensely angry' – possibly in reference to the fact that intense anger can give rise to all or any of the above colours. That the colour word has been hijacked in this way is remarkable particularly seeing that anger is usually associated with the colour red (see, for example, **red mist**).

livid- (L): Ashen, black and blue, blueish.

Living ring: See **Revolving spinning ring.**

Lixivitation: The process of separating a soluble substance from an insoluble by the percolation of water.

Lixivium: **(Lye.)** A term often used in old dye books, water impregnated with alkaline salts extracted by lixivitation from wood ashes.

Lizard: A rigid snake skin, not usually available for sewing, except sometimes as a trimming. Mainly used for shoes and handbags.

lizard-green: A shade of green resembling the colour of a lizard.

Llama: The hair of the fleece of various animals found in South America. Types of hair include alpaca, vicuna and guanaco.

Llama fibre: See **Llama, Llama hair.**

Llama hair: South American ruminant, member of the camelidae (camel family); its wool is of various colours, having coarse guard hairs with a diameter up to 150 um and fine under hair with a diameter of 20–35 mm; it has around 100 almost invisible scales per millimetre.

Llanchama: Native Brazilian name for the interior, fibrous bark of the couratari tree; used for clothing, ropes, etc.

LL fabrics: Links-links fabric, purl fabric.

Lm: Llama hair.

Lm: Lumen, SI unit of luminous flux. See **Lux.**

LNG: See **liquefied natural gas**.

Load: Deprecated term, use the preferred term force.

Load: To apply a force.

Load ratio: The ratio of the filling weight of goods in kg to the volume in litres of, e.g., the washing or dry cleaning drum.

Load-at-specified-elongation (LASE): Deprecated term. Yse the preferred term , Force-at-specified-elongation (FASE).

Load, breaking: See **Breaking load.**

Load-deformation curve: A graphical representation of the relationship between the change in dimension (in the direction of the applied force) of the specimen resulting from the application of an external load, and the magnitude of that load. The load may be expressed in units of weight (such as pounds or kilograms) and the deformation in either units of length (such as inches or millimeters) in tension or compression tests, or degrees in shear tests. In a tension test, a load-deformation curve becomes a load-elongation curve. Deprecated term. Use the preferred term, Force-deformation curve.

Load-Elongation curve: Deprecated term. Use the preferred Force-elongation curve.

Load-recovery cycle, In elastic fabric testing: A continuous curve or plot of load versus elongation (with movement stopped momentarily at point of reversal) describing the elongation and recovery of an elastic fabric. Also known as the loading and Unloading cycle.

Load, median: See **Median load.**

Load, peak: See **Peak load.**

Loaded: See **Weighted.**

Loading and unloading cycle: See **Tension recovery cycle.**

Loading of Silk: See **Weighting of silk.**

Loaf Cotton: Raw cotton formerly grown in Montserrat, West Indies.

Localized emboss (colour) printing: Principle of a local synthetic resin application (by means of roller or screen printing), which is embossed in the customary way on the Embossing calendar and then cured. This produces a wash proof design fixed on the synthetic resin sites, similar to *Chintz* printing, and is a coloration process provided that the synthetic-resin precondensate printing paste contains suitable colorants. Printing is followed by pre-dry, emboss, dwell, cure, etc.

Lochrea: Coarse, bleached Irish linen.

Lock: A tuft of wool.

Lock thread canvas: A lattice-effect embroidery canvas, very rigid and kept flat by warp threads locked round the weft threads to avoid movement.

Locknit: See **Locknit fabric.**

Locknit fabric, warp knitted: It is produced by two fully threaded guide bars. Splitting problem is less than that of Tricot fabrics. Locknit fabrics contract widthwise when leaving the knitting zone. It has a curling tendency. Coloured stripes can be obtained by using coloured yarns. The front guide bar generally requires approximately 30% more yarn than the back yarn.

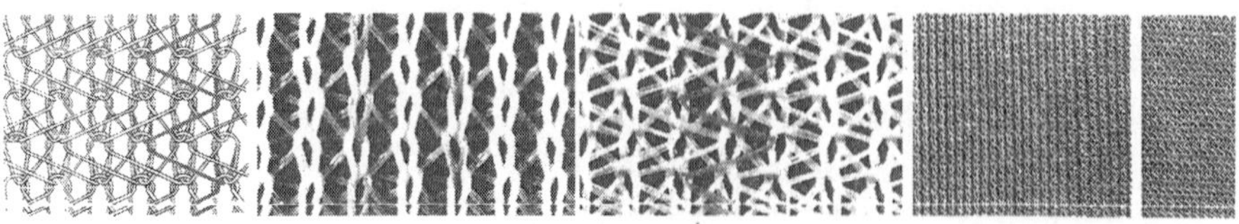

Locknit schematic Locknit Face Locknit back

Lockout: Wool with a stringy formation.

Locks: A term used in wool sorting for short oddments of wool which falls from the skirting tables or rare swept up from the boards. In some countries it can include soiled lefts and pieces from near the rumps of a sheep.

Lockstitch (301 Stitch): A stitch that is formed with a needle thread and a bobbin thread that is inter-locked in the center of the seam being sewn. Even though the same amounts of needle and bobbin thread are consumed, the needle thread requires 5 to 7 times more tension than the bobbin thread. Some of the advantages of using a lockstitch in a seam include: the stitch is reversible, produces the tightest of all seams, and uses the least amount of thread.

Locust bean flour: (fine flour), the ground seeds of the fruit of the locust bean, or carob tree, a leguminous plant native to the Mediterranean, giving a very light, almost dusty whitish or light brown flour.

Loden: (1) Dark green after the cloth of the same name.

(2) A thick, fulled, soft fabric, made in Austria and Germany. Predominantly woollen, coloured blended coarse material made from spun virgin wool and fleece (heavily milled and carded, not cropped), also with admixtures of camelhair, alpaca or mohair for higher quality.

(3) Woven woollen goods in an unfinished state (unsized). Heavily milled to make it dense and durable. Used for suits, costumes and overcoats.

London purple: A shade of purple.

Loden cloth: This is a heavy cloth woven from the fleeces of a mountain sheep. It is sift and thick and characteristically green in colour. The fabric is also waterproof, because the rough wool used is oily and coarse and prevents water penetration. Used for coats, capes, duffle coats especially in mountain countries.

Loft: The properties of firmness, resilience, and bulk of a fibre batting, yarn, fabric, or other textile material.

Lofty: A term applied to an assemblage of fibres to denote a relatively high degree of openness and resilience or large volume of a given mass.

Lofty Wool: Wool that is open, springy, and bulky in comparison to its weight. This type of wool is desirable.

Logan: A Highland tartan, made as follows: Dark green stripe; a group, somewhat narrower, composed of red, black, yellow, black, red stripes of even width; dark green stripe; black stripe, wider than green; dark blue field (over four times wider than black stripe), traversed by fine red stripes, spaced from each other the distance of their width; black stripe.

Logwood: A very deep and permanent black dye obtained from the chipped wood of haematoxylon Campechianum, a tree in Central America and surrounding islands. It is used for dyeing silk, which is mordanted and loaded previously by various salts.

Lohi: Men's shawls in India.

LOI: The original term Limited Oxygen Index has in the meantime been shortened to Oxygen Index, i.e. to OI, and may be described as limiting oxygen level.

London shrinking process: The purposeful restoration of the temporary stretching or tension arising in wool or wool-mix fabrics as a result of the production process by applying the principle of Shrinkage. The fabric is laid in a tension-free state in damp decatizing wrappers for approximately 12 hours with subsequent slow and tension-free drying. In the process a cloth shrinks by approximately 5–10%.

London-shrunk: Shrinking machine for wool and wool-mix fabrics. The fabric travels tension-free from the feed roller in a loose loop to the steam cylinder and then over a sloping, polished heating plate where weft and warp are allowed to relax entirely. The speed of the fabric and the steam intensity may be regulated to achieve any desired shrinkage effect suitable to any fabric.

Londres: Very wide, fulled woollen dress goods of English origin. See also **Londrin**.

Londrin: Light, fulled French and English woollens exported to South America and the Levant.

Long and short stitch: An embroidery stitch. This form of Satin Stitch is so named as all the stitches are of varying lengths. It is often used to fill a shape which is too large or too irregular to be covered by Satin Stitch. It is also used to achieve a shaded effect. In the first row the stitches are alternately long and short and closely follow the outline of the shape. The stitches in the following rows are worked to achieve a smooth appearance.

Long cloth: Plain and closely woven bleached or printed, fine and soft cotton fabric; used for underwear. It has very little sizing and is often gassed.

Long cross stitch: In embroidery a variation of the cross stitch (see) the two stitches crossing each other, forming not a perfect square but an oblong.

Long Draw: This is a woollen-spinning technique.

Long Ell: Twilled English fabric made of hard spun single or two-ply worsted warp and woollen filling.

Long knot, in raw silk: Knots that have loose ends from 3 to 25 mm in length.

Long liquor: Used for a liquor containing a large quantity of fluid, e.g. 1 : 40. Opposite: Short liquor.

Long noil: The best grade of silk noils in England.

Long poll: In England a plush with shaggy pile.

Long slug, in raw silk: A slug which exceeds 10 mm (1/2 in.) in length or which is very much larger in diameter than the yarn.

Long staple: A long fibre. In reference to cotton, long staple indicates a fibre length of not less that 1-1/8 inches. In reference to wool, the term indicates fibre 3 to 4 inches long suitable for combing.

Long staple Fibre: See **Long staple.**

Long Wool: Wool from such breeds as the Lincoln, Leicester, and Cotswold. It is large in diameter and up to 12:15 in. in length.

Longcloth: An old fashioned term referring to the first fabric to be produced in long lengths and rolled. If this term is used now, it simp0ly means a plain weave cotton mainly used for handkerchiefs.

Longotte: Plain woven French cotton cloth, much heavier and stouter than calico.

Longs: Trade term for ready-made clothing cut to fit tall, thin men.

Look: Term used to describe a fashion trend or style, appearance of a garment or fabric. e.g. the used look, safari look, etc.

Loom: A machine for producing fabric by weaving.

(a) *Automatic Loom:* A loom on which the shuttles or pirns are changed automatically.

(b) *Circular Loom:* A loom on which the shuttles travel simultaneously on a circular path through a wave shed (q.v.).

(c) *Shuttle Loom:* A loom that uses a shuttle (q.v.) to insert the weft.

(d) *Shuttleless Loom:* A loom in which the weft is drawn from a stationary supply and is inserted by means other than a shuttle.

There are three main types of shuttleless looms:

(i) Gripper-Projectile (gripper-shuttle) loom in which the weft thread is taken through the shed by a projectile fitted with a jaw that grips the end of the weft thread during insertion of the pick.

(ii) Jet Loom in which the weft thread is taken through the shed by a jet of liquid or air.

Note: Because of the nature of these weft insertion methods, the weft yarn in the fabric is in lengths of one or two picks. Consequently, means are usually provided for forming acceptable edges. (See also **Selvedge**).

(iii) Rapier Loom in which the means for carrying the weft thread through the shed is fixed in the end of a rigid rod or of a flexible ribbon, that (in both cases) is positively driven. Rapier looms may have a single rapier to carry the weft across the full width, or two rapiers that operate from opposite sides of the loom.

Loom barré: A repeated unevenness in the fabric, usually running from selvage to selvage, and caused by uneven let-off or take-up or by a loose crank arm.

Loom beam: See **Weavers beam.**

Loom efficiency: Loom efficiency is defined as

$$\text{Loom efficiency} = \frac{\text{Actual Production}}{\text{Calculated Production}} \times 100$$

Loom Figured: Fabrics having patterns woven in the loom as against printed or embroidered patterns.

Loom finished: A term describing fabric that is sold in the condition in which it comes from the loom.

Loom fly: Waste fibres created during weaving, that are woven into the fabric.

Loom, air jet: See **Weaving machine, jet.**

Loom, Chief motions of: In order to interlace warp and weft thread to produce a fabric on any type of loom three operations are necessary- (a) Shedding: separating the warp threads into two layers to a tunnel known as the shed (b) Picking: Passing the weft thread through the shed, (c) Beating up: Pushing the newly inserted length of weft, known as the pick, into the already woven fabric at a point known as the fell. These three operations are often called the chief (or basic primary) motions of the loom (weaving). These operations must occur in a given sequence, but their precise timing in relation to one another is also of extreme importance on a power loom.

Loom, Shuttleless: A weaving device that carries the filling yarn through the shed by the use of air or water jets and grippers.

Loom, water jet: See **Weaving machine, jet.**

Loom-finished: Fabric marketed in the condition in which it comes from the loom, and which Is given no further finishing.**See Greige Fabric, Loom-state.**

Loom-state fabrics: (loom-state plain weave, loom-state cloth), woven fabric coming from the weaving machine into the trade. The goods as such are not finished but usually merely brushed on a cleaning machine and if necessary steamed. Such loom-state fabrics may be raw fabric (e.g. untreated cotton, homespun etc. = loom-state cloth; hand woven linens = loom-state plain weave) or dyed fabrics (yarn dyed, printed, bleached) or patterned, etc.

Loom-state: Fabric as it comes from the loom, usually unbleached, containing *size*, and maybe a bit dirty Loom state is essentially synonymous with *greige* or *grey*. Sometime it may imply that the fabric has not been inspected for flaws.

Looney: An oblong, flattened pad, used in pressing and giving form to the collar, lapels and shoulders of a coat.

Loop: (1) Any loop or swag of material, e.g. a loop of yarn, a loop of fabric (in a long loop drier or similar device). Loop fabrics are similar to Bouclé fabrics, except that while the latter's decorative threads possess only knot like thickenings or small loops, loop fabrics have larger round loops which, in the case of strong patterns, may have a diameter equivalent to that of a pencil. (2) (pole stitch, plush stitch) a basic element of knitted fabric in combination with Stitches appearing on the reverse side as regular long loops of thread (plush stitches) which are connected with plain stitches.

(3) Loose, plain woven fabric with large loops on the surface, made by weaving with loop yarns in the weft. Used for jackets and costumes.

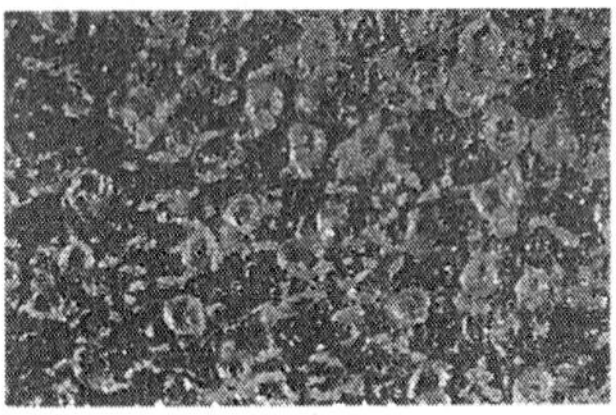

Loop breaking force, in tensile testing: The breaking force of a specimen consisting of two lengths of strands from the same supply looped together in a specified configuration and mounted between the clamps of a tensile testing machine.

Loop breaking load: Deprecated term. Use the preferred term Loop breaking force.

Loop breaking Strength: Strength expressed in terms of loop breaking force (see **Loop breaking force, strength**).

Loop column: See **Tuft.**

Loop drier: Drying machine in which piece goods are moved through the drier in long loops (long loop drier, suspended loop drier) or short loops (Short loop drier) with the reverse side suspended on the rail so that the right side of the goods may dry without contact or tension. Drying systems.

Loop elongation: The maximum extension of a looped yarn at maximum load, expressed as a percentage of the original gauge length.

Loop knot: A snarl or curl produced by a filling yarn due to the excessive length of one or more cocoon filaments.

Loop of yarn, Knitted: A length of yarn bent to shape for incorporation into a knitted structure.

Loop pile, for pile yarn floor covering: Pile in which for each loop a tuft leg of one tuft element is connected to a tuft leg of another tuft element at another binding site so as to form a loop which projects above the backing fabric between the binding sites of the connected tuft elements.

Loop pile Carpets: A pile floor covering in which the pile is of uncut loops.

Loop row: See **Tuft**.

Loop stitch: A type of hand stitch used for over sewing edges so as not to increase their thickness by having an edge turned in.

Loop stitch: Blanket or loop stitch, used to ornament the edge of blankets, etc., and for finishing the edge of stockinet or web material, is worked from left to right, the edge of the material being held towards the worker. Start with three or four running stitches along the edge so the line of stitching will cover them. Insert the needle the desired width from the edge, draw it towards you down over the thread, being careful not to draw the thread too tightly over the edge of the flannel. Fasten the thread by taking running stitches under the last blanket stitch on the wrong side.

Loop strength, in sewing: The strength when one strand of thread is looped with another strand and then broken. This strength test gives an indication of the brittleness of the fibre being tested. Most polyester sewing threads will have loop strength of approximately 1.5 times the single-end breaking strength.

Loop tension, in elastic material testing: The total tension at any specified extension that is exerted on a specimen in a loop formation.

Loop tenacity: The strength of a compound strand formed when one strand of yarn is looped through another strand, then broken. It is the breaking load in grams divided by twice the measured yarn denier or decitex. Loop tenacity, when compared with standard tenacity measurements, is an indication of the brittleness of a fibre.

Loop wheel machine: A bearded needle circular weft knitting machine in which the knitting takes place on a set of vertically mounted needles, the yarn being manipulated with the aid of a bladed wheel.

Loop yarn: A fancy yarn formed by wrapping an effect thread round a twisted core so that loops protrude from the yarn surface.

Loop, Back: **A** knitted loop formed on the dial needles on a knitted machine.

Loop, Face: **A** knitted loop formed on the cylinder needles on a knitted machine.

Loop, Float: A knitted stitch when a needle holds its old loop and does not receive a new yam. It connects two loops on the same course but not in adjacent wales. Also called a miss-loop.

Loop, held: Held loop is an old loop that the needle has retained. It is not released and knocked-over until the next, or a later, yarn feed. A held loop can only be retained by a needle for a limited number of knitting cycles before it is cast-off. A new loop is then drawn through it, otherwise the tension on the yarn in the held loop becomes excessive even though there is a tendency to rob yarn from adjacent loops in the same course.

Loop, Knit: A stitch in a knitted fabric where the yam is formed into a loop shape by the knitting elements. Knitting meshes or interlocks these loops to form a fabric.

Loop, needle: is the basic unit of knitted structure. When tension in the fabric is balanced and there is sufficient take-away tension during knitting, it is an upright noose formed in the needle hook. It consists of a *head and two side limbs* or *legs*. At the base of each leg is a foot (F), which meshes through the head of the loop formed at the previous knitting cycle, usually by that needle. The yarn passes from the foot of one loop into the foot and leg of the next loop formed by it.

Loop, sinker: Sinker loop is the piece of yarn that joins one weft knitted needle loop to the next. On bearded needle weft knitting machines, *loop-forming sinkers* form the sinker loops in succession between the needles – hence the origin of the term sinker loop. On latch needle weft knitting machines, however, the sinker loops are automatically formed as the needles, in succession, draw their new loops.

Loop, Tuck: A knitted stitch when a needle receives a new yam without losing its old loop.

Looped filling: A woven-in loop caused by the filling sloughing off the quill or by the shuttle rebounding in the box.

Looped pile: A pile surface made of uncut looped yarns.

Looped pile floor covering: A pile floor covering in which the pile is composed of uncut loops only.

Looped Yarn: See **Kink.**

Looper, in tufting machine: A hook shaped flat metal plate, which together with the needle, forms the loops during tufting process.

Looper: A stitch forming part which carries the under thread on some types of sewing machines.

Looping: Generally, a method of uniting knit fabrics by joining two courses of loops on a machine called a looper.

Looping bar: A bar inserted in the bottom of an extrusion metier around which the dried filaments pass as they leave the spinning cabinet.

Loop-knot: See **Kink.**

Looper Thread: Refers to the bottom thread used on a serger or over-lock machine, and the threads that cover the edge of an over-edge seam. Looper threads feed directly off the cone into the sewing machine eliminating bobbin changes.

Loopy edge: See **Loopy selvedge.**

Loopy selvage: A weaving defect at the selvage of excessive thickness or irregular filling loops that extend beyond the outside selvages.

Loopy selvedge: An improperly woven selvedge of uneven width or a selvedge containing irregular filling loops extending beyond the outside selvedges. See also **Beaded selvedge, corded selvage, loopy edge, rough selvedge.**

Loopy yarn, in bulk yarns: See **Bulk yarn (textured).**

Loose back: Trade term for quilts, having the stitching warps floated on the back. Also name for welts where the wadding fillings are not interwoven with the warp.

Loose course, in knitted fabrics: A row of loops in the widthwise direction that is larger, looser, or longer than the stitches in the main body of the fabric.

Loose edge: See **Slack selvedge.**

Lost end: An end on a section or tricot beam that has been broken at some stage in warping and has not been repaired by a knot.

Loose filling: A fabric defect that is usually seen as short, loose places in the filling caused by too little tension on the yarn in the shuttle or by the shuttle rebounding in the box. Loose filling can often be felt by an examiner when passing a hand over the surface of the fabric.

Loose pick: See **Slack pick.**

Loose stock: A fine, light and loose structure composed of either smooth or tangled individual fibres (naturally washed, cut or torn). For example, loose viscose stock (or viscose flock) refers to piles of viscose fibres, usually cut to 30–150 mm, delivered for the purpose of yarn manufacture. Similarly, loose cotton is also referred to as loose cotton stock.

Loose stock dyeing: Loose fibres, pressed into a cake in a suitable casing, may be dyed with through flow liquor. The case has both a perforated inner wall and a perforated outer wall. Reversal of the direction of rotation of the axial pump allows the flow direction of the liquor to be switched between inside to outside and outside-to-inside.

Lorna forte: Heavy cotton duck, made in Portugal.

Lot: See **Production lot.** A unit of production.

Lot, acceptance sampling: See **Acceptance sampling lot.**

Lot, in acceptance sampling: That part of a consignment or shipment consisting of material from one production lot.

Lot, in acceptance sampling of cotton: The main stock, supply, source of fibres to be sampled.

Lot, in bonded, fused, or laminated fabric: A single run on the bonding or laminating machine in which the processing is carried out without stopping or changing processing conditions, and consisting of either a single dye lot or a single grey goods lot.

Lot, Production: See **Production lot.** That part of one manufacturer's production made from the same nominal raw material under essentially the same conditions and designed to meet the same specification.

Lot, Sample: One or more shipping units taken at random to represent an acceptance sampling lot and used as source of laboratory samples.

Lot sample, in cotton: A relatively large sample taken in the field to represent a consignment, shipment, or lot, for the use in the preparation of the laboratory samples. See **Sample lot.**

Lot sampling unit: A portion of material taken to represent a lot and used as source of laboratory sampling units or test specimens or both.

Lot size: The number of units in a Lot, batch.

Lot system: Dependent upon the organizational combining of all customer orders falling within a period of time so as to make a single large manufacturing order, which may then proceed as a unit through all manufacturing stages up to dispatching (one-way throughput has time advantages, avoids leaving single items behind and improves production control).

Lot tolerance fraction defective (p2): The process average at which the quality is considered barely tolerable; the process average at which the risk of acceptance is called the consumers risk.

Lot, wool top: The entire quantity, not exceeding 20000 pounds (9100 kg.) of a single combing that comprises a single unit for which a test for neps, vegetable matter, or coloured fibre, or all three combined is desired.

Lotanza: A white linen cloth in Cuba.

Louis Quinze: Lace Imitation tape lace. The patterns are made of braid and connected with bars.

Louisiana: A number of commercial varieties of short staple upland cotton from Louisiana and neighboring States.

Lousine: A plain silk with a glossy texture, and a coarse surface like a very small basket weave. It now often contains synthetic fibres on silk, Used for dresses and coat linings. Hard-wearing.

Lousiness: Flaw in silk or cotton cloth showing speedy spots in the finished goods.

Louver cloth, in coated glass yarn fabrics: A woven netting having an approximately even spaced mesh of fewer than 12 by 12 yarn per 25.4 mm (1 in.).

Lovat: (1) Dull green; from the Scottish town and used in reference to Tweed cloth. Originally 'Lovatt'. Sometimes 'lovat green'.

(2) Although this really describes a colour, it is often used for woollen over coating of good quality in typical lovat colour, which is greeny-grey.

Love: A very sheer, plain silk fabric in England.

Low add-on: Low wet pick-up.

Low-density polyethylene (LDPE): See **Polyethylene**.

Low foaming: The weak foaming characteristic of general purpose washing powders intended for use in a drum washing machine. In resin finishing, it is predominantly compounds based on 4,5-di-hydroxyethene urea which are used, which demonstrably do not split off any formaldehyde in the finishing process, and which in the product itself are set to be low-formaldehyde.

Low/high solids thickeners: Thickening with a small or large amount of Solids content. Low solids thickeners = slow drying, soft elastic film. High solids thickeners = rapid drying, film more or less brittle/hard, contours standing sharply, but danger of splitting off.

Low liquor dyeing: Dyeing process for yarn packages in circulating liquor machines (stacking system) with liquor reduced approx. 50%. Liquor direction exclusively from the inside to the outside.

Low-power stretch: That property of a fabric whereby it exhibits high-fabric stretch and good recovery from low loads.

Low pressure kier: Various designs of kiers are available. Low pressure or open kiers are generally made with a hinged lid, and the scouring liquors are circulated by injector system or by the provision of an external multitubular heater and pump The steam injector located at the centre of the kier permits the heating of the bath and the liquor is circulated from bottom to top before spraying it on the fibre. Open kiers are mainly used for linen, yarn and cloths made of loosely woven goods and knitted materials etc. which will not stand high pressure boiling in alkaline solutions. This type of kiers are also preferred for coloured woven goods which operates at atmospheric pressure below 100°C The main disadvantage of this type of machine is the dilution of the bath by condensed steam on account of direct steam heating system.

Low residual shrimkage: Fabric must shrink less than 2 % when laundered.

Low rows: A carpet defect characterized by rows of unusually low pile height across the width of the goods.

Low water immersion: A dyeing technique where the liquor ratio is very low, with the intent of producing nonuniform mottled appearance LWI is popular for fabrics for quilting. Typically, a piece of fabric is stuffed fairly snugly into a container, and enough dye solution to barely wet the fabric is poured in. The container and contents are set aside to *batch* for some time. Sometimes multiple colours are used. This is an art dyeing technique that is not used commercially.

Low temperature dyeing: Method of dyeing with direct dyes on cellulose fibres by exhaust method, at the lowest possible temperature (40–60°C).

Low wet pick up: Minimum-liquor application; application of finishing liquors to textiles without excess liquor which would subsequently need to be removed. The principal aim is the capillary saturation of the material, e.g. cotton approx. 35% liquor loading, in general < 40%.

Low Wool: Wool of low 1/4 blood or lower in quality. Same as **"coarse wool".**

Lowest-Observed-Effect Concentration (LOEC): Lowest constituent concentration in which the measured values are statically different from the control.

Lowland Wool: These breeds are characterized by producing wool that is generally coarser, and only wavy or quite straight, and of longer staple (over 4 inches). These wools are especially suited for the production of combed yarns, which are worked up into worsted fabrics.

Lowry: Name of an improved cylindrical cotton bale formed from a continuous flat coil, fastened with wire ties and enclosed in 'bagging. Average weight 250 pounds.

LP: Low pressure.

LPG: See **Liquefied petroleum gas.**

LR: Abbrev. for Liquor-to-goods-ratio, in job specifications.

Luana: A fabric characterized by a crosswise rib effect, usually made with a filament yarn warp and a spun yarn filling.

Lubricant in sizing: Lubricants are added to reduce the coefficient of friction and reduce drag as the yarns run over stationary objects. A thorough discussion of these will be found in the section describing oils, fats and waxes.

Lubricant: A material used to reduce friction between surfaces Lubricants are often applied to yarns that are used for knitted fabrics such as jersey. They help protect the yarn from damage and the machinery from wear. Lubricants can interfere with dye uptake and must be removed by thorough scouring.

Lubricity, in sewing: Refers to the frictional characteristics of thread as it passes through the sewing machine and into the seam. Good lubricity characteristics will minimize thread breakage and enhance sew ability.

Lucca: Cloth Medieval fabrics woven of silk and gold or silver in Italy.

Lucet: A layered shaped hand tool of ancient origin, about 70-150 cm. long, made from thinrigid material such as wood, horn ivory etc. It wqas used making square knotted cords with low stretch and good strength characteristics and was widely used until the advent of the industrial revolution when the manufacture of cords and laces became a machine operation.

Lucetted cord: Square knotted cord produced with the aid of a lucet.

luci- (L): Light.

Ludigol: BASF trade name for sodium m-nitrobenzene sulfonic acid

Luftspitze: Is a lace made on shuttle machine in cotton over a wool foundation or in silk or wool on cotton foundation. After the embroidering is done the foundation is destroyed with chemicals which do not affect the work itself thus leaving a lace like product.

Lukchoo: Chinese fabric, about 16 inches wide, made of silk and cotton. It is often blue and is used for clothing.

Lug: See **Bail**.

Lule: Term applied in the Levant markets to very thick Oriental rugs of heterogeneous origin. These rugs are not folded but rolled.

Lumen: Unit of Luminous flux. See **Lux** and **Phot.**

Lumen, in vegetable fibres: The central canal of the fibre.

Luminence factor: Colour measuring spectrophotometers measure the amount of light reflected from an object in relation to the amount of incident light for all wavelengths of interest. It suffices for colorimetric purposes to measure 16 values with an interval of 20 nm in the visible region of the spectrum, from 400-700 nm. These measured values are termed the total reflectance or luminance factor. They can be expressed as a percentage between 0 and 100% or as a ratio between 0 and 1.

Luminescence: Emission of light not caused by incandescence but rather by physiological processes, chemical action, friction or electrical action. (*See* both **Fluorescence** and **Phosphorescence**.)

Luminescent: Most dyes and pigments owe their colour to the selective absorption of incident light. In some compounds, colour can also be observed as a result of the emission of visible light of specific wavelengths. These compounds are referred to as luminescent. The most commonly encountered luminescent effects are fluorescence and phosphorescence.

Lumineux: A lightweight French silk fabric, made in variegated effect and 'finished with a lustre; used for millinery trimmings.

Luminosity function (y) (CIE): A plot of the relative magnitude of the visual response as a function of wavelength from about 380 to 780nm, adopted by CIE in 1924.

Luminous dyes: Fluorescent dyes.

Luminous flux: The derived SI unit is the lumen (unit symbol: lm).

Lump: See **Slub, slug.**

Lump: In the English trade cloths woven 130 yards long and 90 inches wide to toe split and cut into half lengths; also any fabric which is woven double its length it is sold in the market.

Lumps: Plain woven, bleached cotton cloth made in England 32 inches wide, with 64 ends and 64 picks in a square inch; used for calicoes.

Luneburg Flax: A fine variety of German flax.

Luneville Lace: French narrow bobbin lace made of hemp thread with double ground, now obsolete. At the present bobbin-made flower sprigs are sewn to machine-made net.

Lupis: Native name for the fine, white and glossy fibrous layer of the Musa textilis; used for delicate fabrics in the Philippine Islands.

Lurex: It is a metallic yarn produced in various colours by coating thin sheets of aluminium on both sides, using thermoplastic resin. The yarn is slippery and breaks easily but will not tarnish. It can be incorpoarated into a variety of woven and knitted fabric; It is also used for embroidering fabrics and also made into sewing threads.

Lusca: Silk fabric of unknown construction of the Middle Ages.

Lustra fabrics: (from "lustre"). For these goods, highly lustrous wools or hair are used; besides coarse wools, these include mohair and alpaca. In some cases the warp is of viscose and the weft of more lustrous material.

Lustre, Button: The degree of brilliance exhibited in pearlized or pearl buttons.

Lustre: That property of of a textile material by virtue of which the latter exhibits differences in the intensity of light reflected from within a given area of material when reflected from within a given area of the material when the angles of illumination of viewing are changed. The term is frequently associated with the adjectives bright or dull to distinguish between varieties of manufactured fibres.

Lustreine: A fine, wide, wool flannel, slightly napped, similar to broadcloth.

Lustreing: The finishing of yarn or fabric by means of heat, pressure, steam, friction, calendering, etc., to produce lustre.

Lustora: See **Velour.**

Lustre in textile finishing: By chemical addition or by mechanical means the surface reflection or shining can be increased in textiles. Usually mechanical finishing like calendaring supported by heating, friction, shearing, singeing etc. will increase the shining if fabric appreciably.

Lustre lining: In England a lining fabric, made with cotton warp and a mohair or lustrous worsted filling. It is woven in 4 and 1 weft twill.

Lustre orleans: English fabric of the 19th century, made with cotton warp and bright Yorkshire or Lincolnshire wool filling.

Lustre wool: English long wool, having a strong long and glossy staple; used for dress goods. Lincoln and Leicester are included, also the wools grown in Indiana and Kentucky.

Lustre yarn: Glossy hard spun woollen yarn made of long, soft and lustrous wool.

Lustrene: Lightweight, twilled and mercerized cotton lining.

Lute-, luteo- (L): Yellow, yellowish-brown, yellowish-orange

Luteo-virescent: Greenish yellow.

Luteolous: An orangey-yellow.

Lutestring: (1) narrow black silk ribbon, used for eyeglasses; (2) fine, warp ribbed silk dress goods of high finish.

Lux: Unit of Illumination.

Luxor: A soft, ribbed silk satin; used as dress fabric; also obsolete French woollen dress goods.

Lyons thread: Gold filled thread with copper core; used for braids, trimmings, etc.

LWI: Low water immersion.

Lycra: A two way stretch fibre, Lycra is the Trade mark for a spandex fibre made by E.I.du Pont de Nemours & Co. It has remarkable elasticity and instant recovery. It is very strong and withstands repeated wash and wear, It has a high resistance to abrasion but its low absorbency enables articles to dry quickly. Fabrics have a neat, flat appearance. The fibre is resistant to perspiration and is not affected by seawater, hence can be very suitable for sportswear and swimwear. It is mixed with other fibres, only a small percentage is sufficient ot give the necessary stretch, recovery and holding power. Used for foundation garments, exercise wear, swimwear, sportswear etc.

Lye: Sodium hydroxide solution of high strength (rarely, potassium hydroxide)

Lynx: Silky, soft long fur of this wild cat; colours vary from fawn to grey.

Lyocell: A regenerated cellulose fibre made from wood pulp Lyocell is produced with a process that is environmentally friendly in that the chemicals used are nearly completely recoverable and reusable in the process.

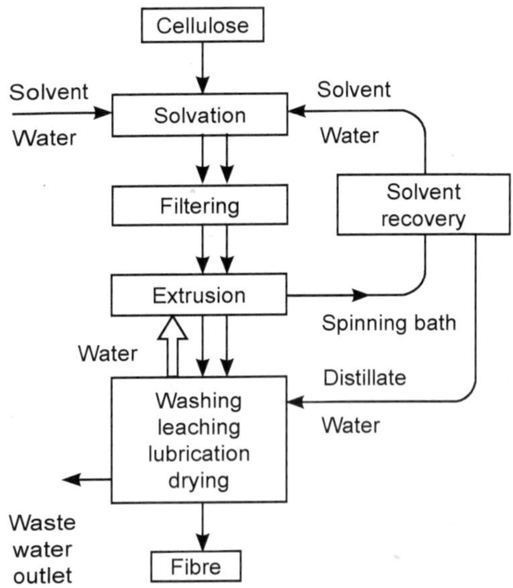

Lyocell manufacturing Process

It is stronger, both wet and dry, than other regenerated cellulose fibres such as viscose rayon, and is also stronger than cotton. In general, lyocell can be dyed much like any other cellulosic fibre. Tencel is one brand-name lyocell fibre.

Characteristics: Lyocell fibre is stronger than other cellulosic fibres. It is inherently absorbent, having a water imbibition of 65%-75%. Lyocell retains 85% of its dry tenacity when wet, making it stronger when wet than cotton. The fibre has a density of 1.15 g/cm^3.

End uses: Lyocell fibre is suitable for blending with cotton or other manufactured fibres.

Because of its molecular structure, lyocell has the tendency to develop surface fibrils that can be beneficial in the manufacture of hydroentabled and other nonwovens, and in specialty papers. For apparel uses, the fibre's unique fibrillation characteristic has enabled the development of fabrics with a soft luxurious hand. The degree of fibrillation is controlled by cellulose enzyme treatment.

Lyogel: Fluid-rich Gel of the jelly type. Opposite: Xerogel.

Lyons velvet: A fine quality thick velvet with a stiff, erect, high pile. The backing may be visible through the pile but it is good quality fabric. The backing is made from silk or sometimes linen, viscose or cotton. The pile is silk.

Lyophilic: Prevailing tendency, in a substance which is distributed in a liquid medium, towards endophilia. Solvent-affinity, solvent-attracting characteristic (similar to *Hydrophilic* property). One also speaks of "lyophilic groups" in connection with textile auxiliaries. Opposite: Lyophobic.

Lyophilic groups: Molecular group, which behaves in an Endophilic manner in relation to a liquid phase.

Lyophobic: Solvent-fleeing, anti-solvent, solvent-repelling characteristic (similar to Hydrophobic). Opposite: Lyophilic. When the solvent is water, the word *hydrophobic* is used. The terms are applied to: (a) Ions or groups on a molecule. In aqueous or other polar solvents, the lyophobic group is nonpolar. For example, the hydrocarbon group on a soap molecule is the lyophobic (hydrophobic) part. (b) The disperse phase in colloids. In lyophobic colloids the dispersed particles are not solvated and the colloid is easily solvated. Gold and sulfur sols are examples. *Compare* lyophilic.

Lyophobic groups: Molecular group which behaves in an Exophilic manner in relation to a liquid phase.

Lyotropic: (Gk.) Phenomenon in which the solubility of a substance which is hard to dissolve in a solvent is increased by the addition of a third substance. This third substance is termed a "lyotropic agent".

Lyotropin Polymer: Polymers that decompose before melting but that form liquid crystals in solution under appropriate condition. They can be extruded from high concentration dopes to give fibres of high modulus and orientation for use in advanced composites, tire cord, ballistic protective devices, etc.

M

MA: (1) Modacrylic fibres, (2) Low wet pick-up.

Ma: Manila fibre.

Ma: Symbol for Myria (ten thousand times).

Mackerel blue: A silvery greenish blue.

mxn lapping fabrics: The most popular mxn lapping fabrics are 1×1, 2×1 and 3×1 lapping fabrics. An increase in the underlaying distance (m) leads to an increase in fabric weight, fabric cover factor, smooth appearance at the technical back of the fabric, extensibility in the wale direction and stability in course direction and leads to a decrease in extensibility in the course direction. The most popular fabric designs (one or two fully threaded) which are also described below are produced by this type of lapping.

MA process: Low wet pick-up process as an impregnation process with only low pick-up of finishing liquor, thereby shorter drying times.

MAC: Modacrylic fibres.

MAC: Maximum Admissible Concentration.

MacAdam unit: Colour difference formulae.

Macalister: A Highland tartan, the composition of which is as follows: The ground is red, traversed by very narrow pale blue and white lines, dark blue stripes and green stripes of two different Widths. These lines and stripes are arranged in groups, in every instance a very narrow stripe of the red ground being visible between the parts of each group. These groups follow in order: *Pale blue line, wide green stripe, pale blue line; stripe of red ground, wide green stripe; white line, wide dark blue stripe, white line, two narrow green stripes, white line, narrow dark blue stripe, white line; strip of the red ground, the width of the wider green stripe; repeat groups described above between the two *, In reversed order; wide strip of the red ground; pale blue line, wide green stripe, pale blue line; strip of the red ground, wide green stripe; white line, narrow green stripe, pale blue line, white line, pale blue line, narrow green stripe; strip of the red ground, the width of the wider green stripe; narrow green stripe, pale blue line, white line, pale blue line, narrow green stripe, white line; strip of red, the width of the wider green stripe; pale blue line, wide green stripe, pale blue line; wide red stripe.

Macalpine: A Highland tartan with a dark green ground, arranged as follows: Wide green strip, split in the middle with a narrow black stripe; a group of stripes (twice as wide as the green strip) composed of two black stripes (about twice the width of the black in the middle of the green) at the edge of the group with two very wide black stripes inside. One of these wide black stripes is split with a white and the other with a yellow narrow stripe in the center, the order of the white and yellow being the same in each repeat.

Macana: A closely and plain-spun fine cotton fabric, made with colored checks.

Macarthur: A Highland tartan, composed of the following on a dark green ground: Yellow stripe with a green bar on each side, six times its width; black stripe, twice as wide as the yellow; green stripe somewhat narrower than this black; black bar, about 6 times as wide as yellow stripe; green stripe, as above; black stripe, twice as wide as the yellow.

Macaulay: A Highland tartan, with a red ground, composed as follows: A very wide red field, split by a black stripe in the middle; dark green stripe, two and-half times as wide as black; red stripe, as wide as black one; dark green bar, the width of the black stripe and half of the red field; this green bar is split in the center by a narrow white line; red stripe, the width of the black one; green stripe, two-and-half times as wide as black one.

Macbean: A Highland tartan, the bright red ground of which is traversed by wide groups of stripes, in which the white, black and green lines are very narrow and of equal width and the dark blue stripes are about half the width of the green and dark red ones. The arrangement of the stripes in a group is: *White line, blue stripe, black line, white line, black line, blue stripe, white line, black line, green stripe, black line, white line, narrow stripe of the ground, dark red stripe (split with green line), narrow stripe of the ground, white line*, green stripe; repeat stripes and lines mentioned between the two *, in reversed order. Red ground, somewhat less than half the width of the entire group.

Macbeth: A Highland tartan, composed as follows: 'Yellow stripe; black stripe, somewhat wider; group of white line, black line, white line, black line, total width same as black stripe; green stripe, twice as wide as black*; red stripe (twice as wide as green), split by two black lines and a white line between, spaced; repeat group described between two *, in reversed order; dark blue stripe, just half the width of the entire complete group of stripes.

Macclesfield silk: A high-textured, hard wearing spun silk. The silk yarn is twisted to give a creepy texture, and it is traditionally striped or small patterned. Used for men's ties and women's classic shirts.

Macdonald: A Highland tartan, one repeat of which is as follows: Very wide green stripe, split by two pairs of red stripes, those on the outside being about twice as wide as the inner ones; black stripe, one-quarter the width of the green; red line; dark blue stripe, the width of the green one and split the same way with two pairs of red stripes.

Macdonald of Clanranald: A Highland tartan one repeat of which is: A wide field of dark green, split in the middle by a white stripe and at each side a narrow red line and a red stripe (wider than the white one) the lines and stripes spaced; black stripe, same in width as the distance between the outer red stripe and the edge of the green .field; narrow red line; dark blue field (about four-fifths the width of the green one), split by a pair of red stripes (as wide as in the green field) the space between these stripes and the edge of the blue field being equal to the width of the black stripe.

Macdonald of Slate: A Highland tartan with a bright red field. Wide, dark green stripes are placed almost four times their own width from each other; one-third the width of these stripes, and on 'both of their sides are very narrow green stripes.

Macdonald of Staffa: A Highland tartan with a bright red field, on which the stripes are arranged as follows: 'black line; red stripe with a narrow green stripe near to each edge; dark blue stripe, half the width of the red; red stripe, as wide as the first red one, split in the center by a narrow white stripe*; green stripe, as wide as the red and split with a fine, narrow line of white; red field, three-and-half times as wide as the green stripe, split with six dark green, two black and four dark blue stripes, arranged as follows: Green near to each edge, in the center are two pairs of blues, with a pair of green stripes on each side, the inner stripes having a black stripe next to them; repeat group described above between two *, in reversed order.

Macdonell of Glengarry: A Highland tartan, made as follows: Black stripe; narrow red line; blue field, four times as wide as black stripe, split by two pairs of narrow red stripes, of which the outer ones are about twice as wide as the inner ones; red line; black stripe, as above; dark green field, as wide as the dark blue, split by 2 pairs of red stripes of same width and position as found in the blue field and with an additional white line between the two pairs.

Macdougal: A Highland tartan composed of the following: 'Fine green line, near to much wider, dark green, red and dark green stripes of even width; gray stripes, narrower than the former, split by a fine red line in the middle; dark blue stripe, wider than green; red stripe (as wide as the first red) split by a fine green line; green field, three times as wide as red stripe*, gray stripe, split with white and edged with red line on each side (this group being as wide as red stripe) ; repeat group described between two *, in reversed order; large field of red (the width of the two green fields and the gray stripe between) split by a fine white line in the center, with a narrow gray stripe on each side of the white line, spaced its own width.

Macduff: A Highland tartan composed as follows on a red ground: 'dark blue, a wider black and a still wider dark green stripe, next to each other*; red tripe (as wide as the above group), split into three even parts by two narrow black stripes; repeat group mentioned above between two *, in reversed order; red field, made somewhat wider than red stripe.

Mace snag test: A test for evaluation of snagging performance. A fabric sample is mounted on a revolving drum in contact with a miniature mace that tracks randomly across the sample. The spikes of the mace effect the snagging. The test predicts results in actual wear.

Mace snag tester: Testing device for checking the formation of snags in chain hosiery.

Maceio: (1) Coarse bobbin lace made of cotton in Brazil; (2) A variety of Brazilian cotton, having a soft, pliable staple.

Macewan: A Highland tartan, composed of dark green bars on blue and black ground and split by red and yellow lines.

Macfarlane: A Highland tartan, composed as follows: 'Dark green stripe, bordered on the outer edge with a black and on the inner edge with a white, line; a narrower red stripe, split with a black line; dark blue stripe, bordered on the outer edge with a fine white line on the inner edge with a heavier 'black stripe; narrow red stripe * ; wide white stripe, split with a dark green stripe; repeat group mentioned between two*, in reversed order; bright red field, being half the width of the entire group of stripes.

Macgillivray: A Highland tartan, composed as follows over a red ground; wide red field, split by a pair of narrow pale blue stripes, each edged with a fine dark blue line on the outside;*pale blue stripe, of same width; very narrow red stripe; dark blue stripe, as wide as pale blue and red stripes combined; very narrow red stripe; dark green stripe, as wide as dark blue and red combined*; red stripe (as wide as dark green, dark blue and two red stripes combined) split by a narrow dark blue stripe in the middle with a narrow pale blue stripe on each side, these stripes spaced their own width; repeat group described between two *, in reversed order.

Macgregor: A Highland tartan, composed as follows, over a bright red field: Group of three dark green stripes, the middle one being a little wider. This one is split by a white stripe, which in turn is outlined by fine black lines.

The green stripes are spaced one-third their own width from each other in the group, the red space between each group being the width of two green stripes and a red between combined.

Machlipatanam Chintz: Printed dress material manufactured at Machalipatanam, Andhra Pradesh, India. The word chintz is apparently rooted in the Sanskrit chitra (meaning variegated or speckled). The Machlipatanam chintz (old name Masulipatanam) were especially prized, and even in the 19th century they could hold their own in the markets of Burma, the Straits and Persian Gulf against the cheap, machine made cloths from Manchester. Mostly block printed, the Machlipatanam Chintz were famous for two traits: firstly, 'the freshness and permanency of dyes, the colour being brighter after each washing" and secondly, a strong Persian influence in the design, probably due to the coastal towns trade with Persia.

Machine Cotton: In England cotton thread used in sewing machines.

Machine direction: The long direction within the place of the fabric, i.e., the direction in which the fabric is being produced by the machine.

Machine gauge: The number of needles on a knitting machine per unit of length (generally, the number of needles in one inch length on the needle bed). For a given machine diameter or width, coarser machine gauges tend to knit narrower fabrics and have fewer feeders (fewer cams) due to fewer needles on the machine and a larger knitting cam, respectively.

Machine Lace: A large variety of cotton, wool or silk laces, made on machines.

Machine made Carpets: Machine produced carpets, in the narrower sense, term for machine spun or knotted carpets.

Machine pitch: This is the distance between two neighbouring needles. Cam system (knitting system): one came together with one feeder forms a cam system or knitting system for weft knitting technology.

Machine stitch, in sewing: The resulting arrangement of sewing threads in a repeating unit formed by a sewing machine. It is also called a seam.

Machine stitches, finishing: See **Finishing Machine stitches**

Machine stitches, Bound: (1) **Self bound seam.** This seam is used to give finishing to the seam margins either on the wrong side or the right side, with help of the extra seam allowance of one piece of the fabric by turning it & placing it over the first seam & giving a edge seam over the fold. (2) **Bounded Seams**: These seams finish is desirable when one is sewing with fabric that fray easily; also behind the seams when sewing with furs. Main bounded seams are (a) **Open Bound Seam**: Binding is an excellent finish for fabrics that tend to fray such as tweed and heavy, coarse weaves, as well as for unlined jackets and coats. Stitch a plain seam. Press the seam, and then press it open. With the Binder on the machine and fine bias seam binding, hind each seam edge, using, either straight stitching or an open zig-zag stitch. To prevent seam edges from stretching or fraying in loosely spun fabrics, place a row of stitching 3 cm (1/8 inch) from the seam edge before applying the binding. (3) **Net Bound Seam**: Delicate fabrics that fray easily, such as chiffon velvet and sheer metallic, may have seam edges bound with nylon net, which prevents fraying without adding bulk. Cut the net into 1 cm strips and insert, attach that bias strip on the seam edges. unfolded, into the Binder with napped or right side up. Stitch, using a medium-width zig-zag stitch. (4) **Plain Bound Seam**: This seam is practical for household items such as cushions, simple loose covers and articles made of plastic. After stitching a plain seam, trim the seam edges to .6 cm (1/4 inch). Press. Insert both edges into the Binder and stitch, keeping the seam edges well into the scroll as you sew.

Machine twist: A hard-twist sewing thread, usually of 3-ply construction spun with S twists and plied with Z twist, especially made for use in sewing machines.

Machine thread, in sewing: Most threads are made of cotton because it is suitable for many different kinds of fabric. Cotton threads are mercerized and available in sizes ranging from No: 20, which is coarse, to No. 80, which is fine. The weight of the fabric determines the size of the thread to be used. The super sheen variety of thread is fine, strong and colorfast and is recommended for most sewing. For terry cotton and other synthetic fabrics, spun, polyester thread may be used. While working with synthetic thread, the upper and lower tension of the sewing machine should be kept slightly loose. When a perfect match of thread is not available, darker shade should be used.

Machine-washable wool: Woollen items, mostly knitted fabric, which are washable in domestic washing machines with a care cycle (ca. 40°C) according to particular Antifelting finish without damaging the appearance of the fabric, handle or wearing characteristics.

Antifelt finishing is generally carried out in the combed top. IWS standards for machine washable items (Superwash).

Machine washable: A label given to garment which can be washed on domestic washing machine, as per care cycle mentioned.

Machine-spun Carpets: Machine spun textile floor coverings like Pile wire; Double plush; Reversible; Axminster carpets.

***Machine zigzagged seam finish**: Used for fabrics that fray. This stitch finish needs a modern machine featuring stitch combinations. Set the stitch regulator to a short stitch length (about 15 stitches per inch) and a medium width stitch. Zigzag stitch as to close the cut edge as possible. It has the stretch which makes it a good seam finish for knits. Used where under arm and crotch seams needs the edges finished together all-in-one after they are double stitched and trimmed.

Mackinac: Using heavy woollen yarn made from recycled wools and wastes, this fabric is of double cloth construction in a pallid design. It is a strong durable cloth, but of medium price and quality. It is a very common fabric with miners, lumber jackets and hunters. It is also used for camping blankets, heavy outdoor shirtings and cold weather jackets.

Mackinaw: Named for MacKinac Island, Michigan. Made of Wool. Ordinary grade of wool and often has shoddy re-used or remanufactured wool mixed in. Sometimes a cotton warp is used. Twill or double cloth. Weave is concealed. A heavy woollen fabric heavily felted and napped on both sides. Construction is like Melton but the structure is not seen due to felting and napping. They are spun in large checks with coarser woollen yarns when compared to Melton or brightly coloured, or different colours on each side.. Used for blankets, cloaks, lumber jackets for cold climates. Also called ski cloth or snow cloth. In addition to cheaper woollen yarns cotton or waste yarns are often mixed. Also called Mackinac. Used by miners, lumbermen, hunters, trappers, fishermen, and cowboys use much of the fabric for jackets, mackinaws and coats. Also used for blankets, shirts, and some heavy sportswear, windbreakers.

Macinnes: A Highland tartan, composed as follows over a bright red ground; red stripe; group of two black and two red stripes of even width, entire width same as former red stripe; black stripe (over twice as wide as first red stripe), split by a narrow pale blue stripe in the middle; group of two red and two black narrow stripes, as above; red stripe, same as first one mentioned; yellow and red stripe of even width; dark blue stripe, as wide as yellow and red together; red and black stripe of even width, as wide as dark blue; dark green, twice as wide as dark blue, next to black, with another black stripe along its other edge*; red stripe, one- third of which taken by a white stripe; repeat group described above between two*, in reversed order.

Macintosh: A Highland tartan with a bright red ground; the design is composed as follows: *dark blue stripe; a narrower red stripe; green stripe, as wide as the first two combined; *red stripe, as wide as the green, split by a narrow blue line in the middle; repeat, in reversed order, group described between two *; wide red bar.

Mackintosh: Obsolete, waterproof coating. It was rubber on cotton with an adhesive to joit the two. The rubber perished in a short time and the fabric could not 'breath' and so caused condensation. Now the name Mackintosh has become generic to a waterproof coat, whatever the fabric.

Macintyre: Highland tartan, composed as follows on a dark green field: Two wide, dark blue stripes, each split by a red stripe, spaced from each other by the width of the red stripes; dark green bar (as wide as two dark blue stripes and space between together), split by a white stripe in the middle.

Mac Iver: Commercial variety of late maturing upland cotton from South Carolina, the staple measuring 22-2'5 millimeters; the yield is 30-32 per cent.

Mackay: A Highland tartan, composed as follows over a dark green ground: A black, two dark blue and another black stripe of equal width, spaced from each other by one-eighth of their own width; green bar (as wide as a black and blue stripe, with a spacing between combined), split by a black stripe, as wide as green spacing.

Mackenzie: A Highland tartan, composed as follows: Dark green bar, split by a narrow white stripe, which is edged on each side by a black line; black stripe and dark blue stripe, each as wide as green on each side of above narrow group; narrow red stripe, edged with black lines, as wide as white and black group; repeat, in reversed order, stripes described between two ; black stripe, as above; dark blue bar, twice as wide as black stripe, split near each edge by a pair of fine black lines; black stripe, as above.

Mackinaw: A heavy double fabric, made in striking colored patterns of all wool or mixed with shoddy; it is more or less felted and finished with a good nap; used for coats.

Mackinaw Blanket: Very heavy, all-wool blanket, dyed red, blue or spun in stripes; used for camping and outdoor life, as it is almost waterproof.

Mackinaw Flannel: Very heavy, napped, woollen fabric, usually red or blue; used for shirts, etc.

Mackinlay: A Highland tartan, composed as follows: dark green stripe, one third of which is occupied, in the center, by a red stripe, edged with black lines; wide black stripe, half of entire green stripe; dark blue stripe, three times as wide as black, split near each edge by a pair of narrow black stripes; wide black stripe, as above; repeat group described between two ; wide black stripe, as above; dark blue stripe (twice as wide as black stripe) split by a pair of narrow black stripes in the middle; wide black stripe.

Mackinnon: Highland tartan, composed as follows, over a red ground; Green stripe; red stripe, split with white and edged with black; next to black dark green stripe, twice as 'wide; red stripe, twice as wide as green; narrow green and wider dark blue, next to each other, as wide together as green stripe;

narrow red stripe; wide green stripe, somewhat wider than wide red stripe; red stripe, less than half the width of former green stripe; group of dark blue and green (of equal width and placed next to each other) as wide as red stripe just mentioned; "group of red, black and red stripes of equal width, each as wide as half of the former group; repeat, in reversed order, groups mentioned between two.

Mackintosh: Cloth treated with rubber; sometimes two layers of cloth are united with pressure, having a rubber layer between; used for raincoats.

Maclachlan: A Highland tartan, composed as follows: Wide rose bar, split by one pair of narrow black stripes placed near one edge and spaced evenly from each other and the edge; wide black stripe, as wide as former two narrow black stripes and two rose spaces; dark blue bar, about three times as wide as wide black stripe, and split by a green stripe in the center; wide black stripe*: rose bar as wide as blue bar, and split in the center by a pair of narrow black stripes; repeat groups mentioned between two, in reversed order.

Maclaine of Loehbuie: A Highland tartan, composed as follows: Two dark green stripes, with a pale green between, of equal width, and next to each other, the pale green being split by a yellow line in the center; red bar of the same width as combined group.

Maclaren: A Highland tartan, composed as follows: Dark green bar, split in the center by a narrow, yellow stripe, which is edged by black lines, each green stripe being further split by a narrow, red stripe, placed nearer to the outer edge; black stripe about one-eighth of green field; dark blue bar. slightly narrower than green field.

Maclean of Duart: A Highland tartan composed as follows: Wide, dark green stripe; a group (narrower than green stripe), containing a fine line of black, white, black, yellow, a stripe of black and line of pale blue *; dark blue stripe, as wide as above group; repeat, in reversed order, groups mentioned between two *; red field (as wide as entire complete group above) split in the center by narrow lines in pale blue with black between, placed very close to each other.

Maclennan: A Highland tartan; see description under **Logan**.

Macleod: A Highland dress tartan, composed as follows: Three wide black stripes, divided by very narrow yellow lines; wide yellow bar (somewhat wider than former group), split in the center by a narrow red stripe.

Macmillan: A Highland tartan, composed as follows: Wide yellow bar, split by a narrow crimson line in the center, each half being split again in the center by a wider crimson line. A crimson bar, as wide as the yellow one, split near to the edge by a yellow stripe and in the middle by a pair of fine lines.

Macnab: A Highland tartan, composed as follows: Wide red stripe, split in the center by a narrow crimson stripe; crimson stripe (more than half the width of the red stripe) split by two narrow dark stripes, and spaced their own width and placed near the edge away from the rep stripe*; dark green stripe, measuring half the width of first red one; repeat group mentioned between two, in reversed order.

Macnaughton: A Highland tartan, composed as follows: Wide green stripe; black stripe, half of the green; dark blue stripe, half of the green; red stripe, about two and a half times as wide as the green, split by a dark ' blue stripe (half of the green), this being split again by a fine black line; repeat group, in reversed order, mentioned between the two. In the filling the wide red stripe is split by a solid dark blue stripe, half the width of the green.

MacNeil: A Highland tartan, composed as follows: Black stripe; dark blue stripe (more than twice as wide as black), split by a white stripe, which leaves on each side a blue stripe equal in width to the black; black stripe; dark green stripe, as wide as the blue, split in the center by a narrow stripe, outlined by blue, the green on each side is as wide as the black stripe.

Macnicol: A Highland tartan, composed of black and green stripes and black, green and pale blue lines on a red ground.

Maco: Variety of raw cotton from Egypt, reddish/brownish/yellow, first-rate Egyptian cotton; very fine, longitudinally stapled fibres, uniform spun yarns.

Maco cotton: See **Maco.**

Maco cotton test: These are tests to identify Maco Cotton. Common tests are: (a) Staple length. (b) Boil sample with thinned nitric acid: real maco turns yellow; reddish tone disappears. If the colour shade becomes considerably, deeper however, simple steaming was carried out. (c) Effect of concentrated sulphuric acid: chromatic shades of colour are produced with direct dyeing. (d) Boiling with tin salt and hydrochloric acid: decolorizing with sulphur dyeing. If you hold lead acetate paper over the reagent glass opening when boiling, dark brown/black dyeing appears with sulphur dyeing. (e) With potassium hexacyanoferrat (II) and hydrochloric acid: blue coloration with iron buff dyeing.

Macphee: A Highland tartan, composed as follows over a bright red ground: Wide dark green bars with narrow green line along each side (spaced its own width); these groups of three are placed from each other farther than their width, each alternate red space being split by a fine white or yellow line.

Macpherson: Several Highland tartans. The dress tartan is composed as follows: Wide red stripe, divided into three even parts by two narrow dark blue lines; green stripe, as wide as one section of the former red and a blue line combined; very fine yellow line; black stripe, as wide as a single section of the red; dark blue about twice as wide as the black, split in the center by a pair of fine black lines; red stripe, as wide as dark blue, split in the center by two fine white and 'between these two fine black lines, all placed very close to each other; repeat, in reversed order, group mentioned between two.

Macquarrie: A Highland tartan with a red ground, composed as follows: Wide stripe of dark green; red field, a little over three times the width of green stripe, split near to each edge by a pair of closely spaced very narrow green stripes.

Macqueen: A Highland tartan, composed as follows: Wide red bar, divided into four even parts by three narrow black stripes; a black bar, considerably wider than the red, split in the center by a narrow yellow stripe.

Macrae: A Highland tartan, composed as follows: Dark blue stripe, split in the center by red line; a narrow white line, separated from the blue stripe by a narrow red line, is along each edge; red stripe, wider than the blue, split in the center by a pair of closely spaced green lines; green stripe, wider than the red, divided into three even parts by two narrow red lines; repeat, in reversed order, groups mentioned between two; red field, as wide as the two wide red and green stripes combined, divided into five equal parts by four groups of closely spaced groups of three dark blue lines each, the middle line in each group being somewhat wider than the lines on the outside and the blue groups measuring the same width as the red, between the groups.

Macrame': A type of knotted construction. Used sometimes in carpets, threads are elaborately knotted by hand from the warp threads which stick out of carpets. Can be used as a needlework technique with typical varied knottings, e.g. for flower garlands wall decoration etc.

Macrame' lace: A revived craft with modern application, this was originally a Spanish knotted lace in geometric patterns, often with fringed ends. Any thread, yarn, wool or string may be used to produce fine or coarse designs. The threads are anchored to a board which may have a paper pattern on it. Motifs may be made or complete items such as mats, bags etc.

Macrofibrils: Macro form of fibres. Normally any basic polymer structure by bonding (for example hydrogen bonding in cellulose) forms microfibres and these micro fibrils are organised to for macro fibrils. These macro fibrils are organized into fibres. In case of wool, designation for cell units of the fibrillar type in Wool structure, e.g. firstly in the sense of spindle cells and secondly where there is no direct connection with the actual fibrils (bundles). Integrated process of sulphur-rich keratin proteins, which show a variable affinity for heavy metals after reduction. Macro fibrils in inter macrofibrillar material are a biological unit in keratin fibrils.

Macrolattice A repeating structure in very small micro fibrils of alternating crystalline and amorphous regions. Yarn properties are thought to be governed by morphology at the macrolattice scale.

Macromolecules: (a) Expression for polymers (coined by Staudinger), high molecular so-called "gigantic molecules" made of at least several hundred atoms with molecular weight > 10 000. Structure made of Monomers, either thickly knotted or cross-linked (starch, aminoplasts, phenolic plastics) or more or less longitudinally stretched Chain molecules (Polymers) as in cellulose, polypeptides, etc. Classification into a) natural macromolecules (e.g. starch; cellulose; pectins; caoutchouc; proteins; enzymes); (b) converted (regenerated) natural macromolecules (e.g. viscose; vulcanized caoutchouc; leather) and (c) synthetic macromolecules (e.g. polymerization-, polycondensation-, polyaddition products). Special natural macromolecules are mostly mixtures of macromolecules similar in construction, but different in size.

Macroscopic: Visible with naked (so-called unarmed) eye; e.g. small individual fibres of a yarn, whose surface structure is however Microscopic.

Madagascar lace: Has the threads twisted into loops and scallops; made by the natives of Madagascar.

Madam: Soft finished white shirting In Turkey.

Madapolan: (1) A bleached or dyed, occasionally printed, plain cotton fabric with a soft finish in a wide range of qualities used for ladies wear. (2) Plain spun, sized cotton fabric, heavier than chiffon; used for embroidery foundation in German, Swiss and Austrian factories, and also for shirts; (3) Various fine bleached cotton muslins in Serve.

Madai: Strong and silky seed hair, grown on the giant Asclepias in India.

Madder: A plant, called Rubia tinctorium, grown in Asia, the root of which yields a rich and fast scarlet dye. Effective component: alizarin (dioxyanthrachinon) Mordant dye for wool and cotton. Total fastness even surpasses indigo. Was already mentioned in the Bible (Egypt, Persia and India). Had its heyday during the 17th century.

Madeira embroidery: Is worked upon fine cambric in eyelet patterns, similar to the modern English embroidery.

Madeira lace: The early specimens are coarse torchon laces; afterward bobbin laces made in imitation of malines and other laces.

Made-to-measure polymers: It is possible to construct polymer molecules chemically using their chemical structure so that they are either extremely flexible and form irregular balls in melting or else are even in liquid crystal form even before solidifying into stiff rods in a strictly parallel packing. The material spectrum of made-to-measure polymers ranges from polyethylene with flexible tangled molecules to polymers with noticeably restricted flexibility (polyester on polyethylene terephthalate base) and also the stiff-chained liquid crystal polyesters (LCP) or even to the extremely rigid structures of carbon polymers, which include, for example, carbon fibres, carbon glass and carbon-fibre-reinforced composites (CFC).

Madras: Originally made from a fine cotton warpand silk weft. Today madras is made from long staple cotton yarns, preferably combed yarns. Sometimes fine spun synthetics are used in weft to resemble the original Madras. 70 x 70 with 40s warp and weft can be considered as a typical construction of an

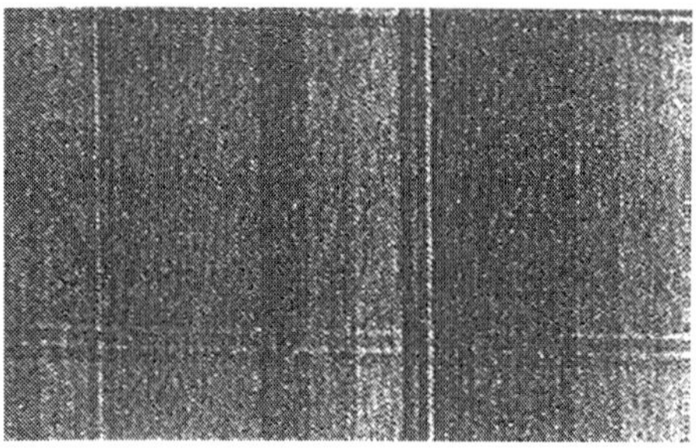

average grade. A fine grade madras made with 2/80s 70 × 70 texture in plain weave with fine stripes, cords or small checks , but are usually accentuated by weaving the decoration ends on a satin basis. Bright shades are favoured in Madras apparel fabrics. Madras curtains are spun with a plain leno weave, with large floral designs inserted using jacquards. Uses: Men's and women's sportswear of all kinds, dresses, separates, shirts.

Madras Cotton: A soft cotton, rather like gingham, but including more colours. It is often rough and unfinished and colours bleed when washed. It is also inclined to fade

Madras gauze: A very light fabric, the foundation of which is in gauze weave, the designs being formed by an extra heavy weft; the floats are cut away; used for dresses, etc.

Madras gingham: See **Gingham.**

Madras Lace: Black and white silk and cotton bobbin lace, made in Madras in Maltese patterns.

Madras muslin: Made in Far East, this has a fine and open gauze foundation of cotton, with figured textures and design in thick, soft spun, weft threads.

IN better quality fabric, two or more different colours are introduced. An inexpensive fabric used for sarees, blouses, nightwear.

Madras shirting: A fine, lightweight, and often colourful, cotton material of better quality than Madras Cotton. The fabric is plain with spun stripes in satin weave, which may being another fibre such as viscose, or in an expensive fabric, like silk. Used for men's shirts, women's dresses.

Magazine screen printing machine: Magazine screen printing machines were the first step towards full mechanization of screen printing. The development started from the fact that a machine with the smallest space required was requested for reasons of space. The printing table is only large enough to enable printing of a report and is equipped with moveable back grey printing in the form of an endless ribbon. The printing screens in question are accommodated in magazines of different design. A distinction is made between staple magazines and magazine screens. So-called "Carousel suspension" and "Serial magazines" are recognized in magazine screen design.

Magenta: A bluish-red colour Magenta is the "official" *subtractive* reddish colour. (Comments paralleling those for *cyan* apply.) In the *MX* reactive dye family, Red MX-8B and Red MX-5B are good magentas. Many dye sellers in the textile arts market call this "fuchsia".

Magnesium chloride: (chlorinated magnesium), $MgCl_2 \cdot 6H_2O$; molecular weight 203. Melted and in crystals, very hygroscopic, water-soluble. Application: finishing additions (hygroscopes, binders and hardening agents), etc.

Magnesium hardness: (MgH), unit per degree of magnesium oxide (MgO) 7.15 mg/l water. See **Water hardness salts.**

Magnesium silicate: Agent for stabilising and controlling the oxygen output in bleaching agents; slightly alkaline to neutral, resistant to temperature and alkalis.

Magnesium sulphate: (epsom salt, sulphuric acid magnesium), $MgSO_4 \cdot 7H_2O$; molecular weight 246; density 1.68. Colourless crystals, often impure, resistant to damp air, weathered in dry air, water-soluble. Application: finishing addition (hardening and sealing agent in particular for cotton goods) in flame-retardant impregnation.

Magnetic rating: An empirical value that reflects the effect of magnetic particles such as magnetic iron compounds, in asbestos material as measured by a magnetic analyzer.

Mahagony or Anacardia gum: Vegetable gums from Brazil, Martinique and Guadeloupe, which are very similar to gum arabic. Reddish colour. Besides mainly arabinose also contain bassorin and dextrin. They are not completely water-soluble.

Mahout: (1) In the Levant trade a fine and light, fulled cloth, made in Europe, of fine Spanish wool; (2) Also a coarse woollen fabric, used in Egypt and Asia Minor for garments.

Maibafi: Very thin Japanese reeled silk.

Maiden: The name for the posts that support the flyer on a spinning wheel. Maiden and the base that supports them is called the mother-of-all.

Maifoss process: Dyeing to achieve "2 shades" on the same wool fibre or on different wool fibres in one dye bath. Particularly selected dyestuffs are used for this; besides these you can use dyeing auxiliaries, which also increase the colour contrast.

Mail cloth: A highly finished silk cloth spun like honeycomb; used for embroidery foundation.

Mailles de bas: Plain French serge dress goods, having eight leaves and four picks in a repeat.

Maize starch: Starch extracted from Maize. Crystals, pieces, powder. White/yellowish/red, a little more gloss than wheat starch. Handle as wheat starch (not so crunchy). Seldom distorted. Water content 12–14%. Heated bloating: swelling at 50°C, start at 55°C, end of glueing at 62.5°C. Paste: Greater adhesive and stiffening strength than wheat and potato starch, less than rice starch; acidic production thinner; alkaline thicker. Reaction against chemicals such as potato starch. Chemical finishing handle: very hard, but very pithy. Application: Sizing of yarn, finishing etc.

Majagua: Very long bast fibre, yielded by the Hibiscus tiliaceus of Central America. Does not deteriorate in water.

Majestic 88 needles: These hand sewing needles are English needles with a special coating allowing easy stitching through cotton batting or difficult - to - needle fabrics. Sizes 9, 10, & 12.

Major defect: A defect other than critical, that judgment and experience indicate is likely to materially reduce the usability of a product for its intended purpose.

Major imperfection, in fabric grading: A deviation in a roll of fabric that judgment and experience indicate is likely to have an adverse effect upon subsequent processing of the fabric.

Makaloa: Fine mats, made with colored geometric patterns of the young leaves of a sedge, Cyperus laevigatus, in Hawaii; used for garments, etc.

Makatlik: Turkish name for Oriental runners; used as divan covers.

Malabar: Indian cotton handkerchiefs printed in brilliant contrasting colors.

Malachra: Long, fine and silky fibre of the Malachra capitata in the West Indies; used for ropes.

Malasap: Coarse fibre used for cordage in the Philippines.

Malborough: See **Malbrouk**.

Malbrouk: Obsolete French wool serge, made very smooth, with small designs. It had a hard twist, single warp. The spacing of the warps and that of the picks was about equal.

Malcolm: A Highland tartan, composed as follows: Dark green stripe; group of the same width as green, composed of black, pale blue, black, yellow and 'black lines of equal width; dark green stripe, as above; black stripe, as wide as green; dark blue stripe more than twice as wide as green, split in the center by a pair of closely spaced fine red lines; 'black stripe, as above.

Malefique: A stout, twilled Belgian worsted fabric; used for bags in pressing oil.

Malida: Indian fabric made of goat's hair; the best grades contain large proportion of the hair of the Thibet goat.

Malifil process: Stitched-bonded composites from bonded threads, e.g. from an untwisted thread with polyamide threads "sewn in".

Malimo: A type of fabric constructed and made in East Germany and produced at a great speed using three sets of yarns, filling yarn placed across the warp and a third system which stitches them together. Technically this fabric is neither spun nor knitted. It is very stable and doesn't move or 'give'. It is often ribbed in appearance. It is mainly acrylic but other fibres may be used. Used for curtains tablecloths for indoors and out, bedspreads and chair covers etc.

Maline: Trade name for hexagonal open mesh, plain net of silk or cotton, usually finished with size; comes in black and white; used for trimming dresses and millinery.

Malines: A stout, plain spun worsted of two or three-ply warp and single filling of a different colour.

Malines Lace: Bobbin lace with sprigs or dots outlined with a heavier cordonnet over a hexagonal or round mesh ground. It is made in one piece of white flax thread.

Malino: Very long, strong leaf fibre, yielded by the aloe in Hawaii; used for cordage.

Malipol fabric: Stitch-bonded fabrics as a combined stitch-bonded yarn composite fabric type. Conventional textile fabrics (e.g. viscose staple fibre) with pile loops (1–7 mm) threaded to one side of the fabric using stitch bonding and the back of the fabric covered with stitches. I.e. chain plush fabric with pile threads bonded using stitch formation in contrast to tufted fabrics, where loops sewn in have to be bonded with adhesive on the back. Application for Terry fabric goods, duvets, furnishing fabrics, carpets, rugs, coat fabrics in types of fleece and velour, car upholstery and technical textiles.

Maliwatt fabric: Stitch-bonded fabrics made out of over sewn nonwovens, with/without binding bonding; for interlinings, felts, upholstering fabrics, coating base for artificial leather, technical purposes; also layered agents for clothing fabrics, raised lining fabrics and artificial leather.

Mallory fatigue test: A test to measure the endurance properties of tire cord.

Malmal: Native East African name for bleached cotton muslin. Also generic term for the finest cotton muslins in India, often embroidered in gold. See also **Mull**.

Malo: A very fine netted fabric in Hawaii, made of olona fibre, and used for loin cloth by the natives. Often feathers are sewed to it.

Malt: Aqueous extract from germinated cereal grains (barley); contains amylase and maltase as diastatic, starch degrading (condensing via dextrines to maltose) enzymes (ferments). Malt diastases.

Maltase: A glucosidase as an enzyme of malt (malt diastases), which the starch decomposition (into maltose) already effected by the amylase continues to form glucose.

Malt diastase: Obtained from germinating barley; consist of α- and β-amylases (Diastases). The effective malt diastatic enzyme decomposes starch into maltose. Effectiveness: PHs of 4.5–6.2, 50–65^{0}C and quantities of 3–20 g/l of sodium chloride addition have an active effect. Alkali, copper, tin and lead salts act as enzyme poisons.

Malt enzymes: See **Malt diastase.**

Maltese lace: Similar in appearance to Cluny Lace, it is heavy and often incorporates regular wheel designs and squares. Used mainly in small areas such as corners of handkerchiefs and table linen.

Mamaki: Fibres yielded by the Pipturus albidus in Hawaii; used for coarse cords and ropes.

Mammoth: Two commercial varieties of late maturing upland cotton from Georgia, the staple measuring 28-30 millimeters; the yield is over 30 per cent.

Mamoudis: Very soft, fine, yellowish linen, originally from Persia.

Manchester Cottons: (1) Originally woollen fabrics made formerly in England, measuring 22 yards in length, three quarter yard in width, and weighing 30 pounds at least; (2) At the present a great variety of cotton fabrics made at Manchester, England.

Manchester velvet: All-cotton velvet made in England with plain weave back.

Manchu Crepe: Cotton or silk crepe, made with very fine warp stripes of colored silk thread.

Mandarine: French fabric, spun with cotton warp and silk filling.

Mandrel: Thread shaft for transmitting a rotary motion.

Mandrel jack: Machine for fitting printing rollers on mandrels onto the steel shaft.

Mandrell: The core around which the impregnated filaments are placed to form a specified shape in composite manufacture.

Mandrenaque: Cloth from the Philippines made of cotton warp and palm fibre filling.

Mandypyta: Raw cotton from Paraguay, yields a reddish brown staple; used for ponchos.

Mandyu: Native name of three kinds of raw cotton in Paraguay, yielding white staple.

Mangle: A system of two rolls which is used to squeeze the fabric in wet processing. It can be same type or a combination of two type rolls. Ebonite/Rubber (Synthetic or natural) or Rubber/Rubber are very common.

Mangling: The process of squeezing the fabric while passing through two rolls. See **Mangle.**

Manila: Fibre obtained from the leaf stalks of the abaca plant. It is generally used for cordage.

Manila fibre: See **Manila.**

Manila hemp: Very light, tenacious and lustrous fibre, yielded by the Musa textilis in the Philippine Islands; used for ropes and for the finest sheer fabrics. The principal classes are current, fair current and brown. The quality grades are marked as: Fair, medium, coarse and coarse brown. The fibre is also called very short (less than 4 feet), short (4 to 5 feet), normal (5 to 8 feet), long (over 8 feet).

Manillese: Embroidered and often knotted drawn work made of agave fibres in the Philippine Islands.

Manirito: A useful fibre, yielded by the bark of a species of the sour-sop in Venezuela.

Man-made fibre: A class name for various genera of filament, tow, or staple produced from fibre forming substances which are chemically synthesized or modified. (a) Made from natural polymers: Regenerated cellulose; Man-made protein fibres; Mineral fibres. (2) Made from synthetic polymers (Synthetic fibres): Polycondensation fibres; Polymerization fibres; Addition polymers.

Man-made protein fibres: (protein man-made fibres, regenerated protein fibres), nozzle-spun manmade fibres made from natural polymers on base of protein solids:

(a) *Vegetable origin:* vegetable protein fibres such as: Zein; Arachin fibre; Glycine.

(b) *Animal origin:* animal protein fibres such as: Casein; Fibroin.

Man-made protein fibres, distinguishing tests: Boiled in 0.1% ninhydrin solution for 1–2 min fibres appear dyed as follows: groundnut protein fibres = greyish brown, zein fibres = grey-blue, milk casein fibres = faint purplish blue, soya fibres = undyed.

Man-made staple fibre: Fibre of spinnable length manufactured directly or by cutting filaments.

Mannose: Belonging to the simplex sugar types. Occurring as a component of similar sugar polysaccharides, e.g. predominantly involved in the structure of Locust bean flour (galactomannan, carob). In its pure state (commodity) white, sweet-tasting crystal powder, easily water soluble.

Manofast: A trade name for thiourea dioxide.

Manometer, pressure gauge: Pressure measuring device for gas, vapour and air pressure (also fluid pressure), which exceeds or falls below normal pressure. There are different types:

(1) Fluid manometer: Simplest type (similar to Barometer) as an open U-tube filled with fluid (mercury, paraffin oil or coloured kerosene), connected on one side or (for differential pressures) on both sides, board mounted, mm pitch, reading off the level stand above and below (rounded distance). Only for low excess pressures (often below 1 bar). (2) Diaphragm pressure gauge: oval housing, airtight clamped diaphragm (corrugated sheet metal), one side with overpressure connection, another side with pushbutton, little gear wheels, pointer and calibrated measuring scale. 3. Tube spring manometer: tube spring

bent in three-quarter circle (Bourdon tube) with oval cross section made of brass or (for maximum pressure) bored steel tube, one side above with excess pressure connection, other side closed with lever-operated transmission, counting segment, gear wheel, pointer and calibrated measuring scale.

Manta: (a) In Central Amèrica, term for gray cotton sheetings; (b) In Colombia, various kinds of cheap cotton fabrics or plain spun goods.

Manta Blanca: Bleached cotton sheeting in Mexico.

Mantel Grijn: Cross ribbed Dutch camlet, made with two-ply goats' hair warp and worsted filling, having warp ribs.

Mantelle: A medieval English worsted.

Manteau: Courtly women's outer robes; bodice and usually open skirt from the same material, used in ancient times.

Mantilla: Shawl, usually triangular covering the shoulders and sometimes the head.

Manto: Plain black shawl, worn by Chilean women; usually made of wool or mixed with cotton.

Manufactured fibre: A class name for various genera of filament, tow or staple produced from fibre forming substance which may be (a) polymers synthesized from chemical compound, (b) modified or transformed natural polymers, (c) glass. The term manufactured usually refers to all chemically produced fibres to distinguish them from the truly natural fibres such as cotton, wool, silk, flax, etc.

Manufacturing based, in quality testing: This definition is concerned with engineering and manufacturing practices based on conformance to requirements or specifications. These specifications are set by design and any deviation from them implies a reduction in quality. Excellence is not necessarily in the eye of the beholder but rather in the standards set by the organisation.

Maolao su Chiyong: Scarlet red Chinese silk velvet.

Marabout: (1) A delicate thin silk fabric made from twisted raw silk. Used for blouses, linings and lampshades. (2) White silk thread used for crepe, made of three strands twisted together very hard, and dyed in the gum; (3) A very light silk dress fabric, or ribbon, similar to the crepe in appearance, spun of marabout yarn in plain weave; 3, five or eight leaf, silk satin; used for millinery. It is made with single warp.

Marabout silk: See **Marabout.**

Maracapas: A Philippine fibre; used for ropes.

Maragnan: Formerly the best grade of raw cotton grown in the West Indies; now less known.

Maranham: Raw cotton from Brazil with glossy, yellowish and strong staple, sometimes quite dirty.

Maratarong: Philippino name for a coarse fibre used for cordage.

Marble cloth: See **Marbled cloth.**

Marble silk: A soft lightweight silk fabric with a mottled appearance. This effect may be achieved by the use of multicoloured yarns or by wrap printing. Used for blouses, as a lining fabric, and for lampshades.

Marbled Cloth: A silk and wool dress goods in England, spun with a. mottled face in various colors, produced by multi-colored weft.

Marbre: A medieval French worsted, spun to imitate the veins of the marble.

Markal I process: Old scouring process by ICI. The dry cloth first passes through trichloroethylene (TCE) in the dewaxing vessel which is divided into number of separate compartments. From the last compartment the cloth passes through a solvent seal and heavy mangle where much of the solvent is removed from the cloth. The cloth then passes through a chamber into which steam is injected. Thus rapid evaporation of solvent in the cloth and the mixture of steam and TCE vapour is condensed and returned to the solvent seal where the solvent and water are separated.

Markal II process: The Markal II process consists of effecting simultaneous scouring and desizing by using suspension of enzyme in TCE and surfactant solution. The treated sample is then passed through steaming chamber for 10-20 sec. where the solvent is flashed off. The fabric is then given normal wash to remove the size. See **Markal I process.**

Markal III process: Markal III process combines bleaching with scouring and desizing by using emulsion of aqueous hydrogen peroxide in the TCE-surfactant solution, again followed by steaming and washing-off. See **Markal I process, Markal II process.**

Marceline: Also called Merceline. A light, thin silk fabric that is almost transparent. Used for lining hats and lightweight clothes.

Marcella: A fine cotton pique; used for Bedding.

Marduff: Native name in East Africa for stout, twilled gray cotton fabrics; used for tents, sails and native dresses.

Marengo: Suit and coat material (similar to fine Loden), not a fabric quality, only a colour description. This concerns very dark-coloured, almost black materials, which are brightened up by a small white component of 1–4% (= radial-line colour). The four-weave symmetrical twill is almost exclusively found as a weave. There are marengos both in worsted yarn warp and weft and in carded spun yarn in both thread systems.

Marengo yarn: Black and white mottled yarn, e.g. made of wool or silk (worsted or carded yarn).

Margherita: Italian embroidered, ma-chine-made net.

Marie antoinette: Curtain, having applique sprays, flowers, and leaves of cord and tape.

Marine fibre: Poseidonia Australia, obtained from the bottom of the gulf in South Australia. The fibre is not very strong and is brittle when dry. It is believed to be New Zealand flax submerged and rotted in salt water. It has good affinity for basic dyestuffs, but acid, salt and sulphur dyes produce little result. This fibre has ‘been discovered only lately and was experimented with as wool substitute for cheap clothing and rugs but no satisfactory result was obtained.

Marine stripes: Good quality English calico shirting of equal stripes in blue and white.

Marker: In the floor coverings industry, a distinctive threadline in the back of a carpet that enables the installer to assemble breadths of carpet so that the pile lays in one direction or so that patterns match.

Marking, in sewing: Marking is the transfer of various pattern symbols from the pattern to the fabric.

Marking chalks: Are used for marking/sighting piece goods in the finishing phase. Consist of dye pigments stored in wash-active substances. Should not be decomposed during operating processes before the piece washing (e.g. steaming, burning at high temperature, etc.); otherwise risk of dye fixation on the fibres.

Marking cotton: Cotton thread, usually dyed blue or red; used to embroider the outlines of a pattern.

Marking inks: Chalk dyes; Laundry markers, but also dyes, which are applied to yarns for labeling without being fixed properly. Must therefore be easily washed out.

Marking off in wet goods: Bleeding or destaining of colorations and printing in fabrics laying on top of each other in a wet or moist state.

Marketing: The business of selling goods, including advertising and packaging.

Marl effect yarn (Filament): Two single continuous filament yarn of different solid colour or dyeing properties (Subsequently dyed) doubled together.

Marl Single or two-ply: Yarn in England, used for filling; It is made in two coloured effect, usually one colour twisted around the other.

Marl yarn: See **Marled Yarn.**

Marled Yarn: Two different colour threads are spun together, not necessarily in equal quantities, to form a marled yarn.

Marmato: A medieval silk and gold brocade; said to be identical with arramas.

Marmot: A cheap fur from a small rodent. It is not used in its original state, but dyed to imitate more expensive furs, such as mink.

Marocain: See **Crêpe Marocain**.

Marocs: French woollen serge, of various qualities, made with a nap on the face.

Marquise: Finish High gloss imparted to fine cotton satins.

Marquisette: An open loose fabric spun in a fancy leno construction, or drop weave sometimes with spots or other designs included. It can be made from cotton, silk or wool, but is now mainly glass fibre nylon or polyester, orlon and made as a curtain fabric. A rayon Marquisette can be made in 20 x 20 to 70 x 50 with 75 to 150 denier yarn in warp and weft. An average nylon Marquisette has 70 denier semi-dull nylon in warp and weft, in 50 x 50 texture. Very lightweight, open, sheer, mesh fabric. Wears very well and launders very well. Comes in white, solid colours and novelty effect. Sometimes with a swivel dot or clip spot (marquisette). Uses: Window curtains, dressy dress wear, such as bridal parties or after 5 wear.

Married fibre clump: A defect that occurs in converter top. It consists of a group of unopened, almost coterminous fibres with the crimp in register.

Marron elastic fabric: An elastic fabric that is less than 150 mm (6 in.) in width.

Marry-muff: An old time coarse fabric worn by the common people in medieval England.

Marseilles: A stiff, double faced, quilted white cotton cloth, similar to pique, made in plain weave with large embossed patterns; used for shirt bosoms, men's vests, women's dresses, bed covers, etc.

Marseilles quilt: Marseilles quilt is a double cotton cloth, composed of two plain spun fabrics, one warp being the stitching warp and one weft a heavy wadding filling. The figures are embossed, formed by Interweaving all threads with each other, but the two fabrics are not united at the ground.

Marseilles soap: Fatty, mild Soap made from olive oil, cotton oil, etc., white, or green, bar or needle shaped.

Marseilles work: Consisted originally of outlining flowers and other ornaments with stitches over a previously padded linen or canvas ground.

Marsella: Heavy, bleached, twilled linen, given a soft finish.

Marselles: A sort of figured pique, used for women's and children's clothes and for men's coats.

Marston: Commercial variety of late maturing cotton from Louisiana, the staple measuring up to 30 millimeters; the yield is 30-31 percent.

Martin: Commercial variety of late maturing, prolific cotton from Louisiana, the staple measuring 26-30 millimeters; the yield is 30 percent.

Martiniques: An 18th century woollen fabric in England.

Masalia: Lightweight cotton fabric, spun in a twill, producing more effect. It is given a smooth, glossy finish.

Mascades: Silk cloth, used in Latin America as head cover.

Mascaret: Loosely spun, high finished worsted satin with spun figures.

Mashru: Fabric made of mixture of silk warp and cotton weft a style of weaving limited to Pathan, Mandvi and Surat in Gujarat, India for the use of orthodox Mohammedans. The fabric was developed as religious customs disallowed certain muslim communities from letting silk touch their body. The "double sided mashru" the term meaning 'permitted' was an intelligent way of wearing garment that had cotton inner and lavish exterior. Mashru is a heavy fabric with a satin glaze appropriate for winters and its characteristic patterns include bold, colourful stripes of various widths as well as pinhead sized dots assembled in floral or geometric pattern.

Masloff: Wide, stout, Russian woollen dress goods.

Mass: The quantity of matter in a body. Mass is determined in two ways: the *inertial mass* of an object determines its tendency to resist change in motion; the *gravitational mass* determines its gravitational attraction for other masses. The SI base unit of mass is the kilogram.

Mass colouration: (spin dyeing). In this process also called jet dyeing, finely distributed dye pigments, which withstand the thermal and chemical demands of the spinning process, are added to the spinning solution/melting of synthetic fibres. Used, for example, in polyamide for upholstery fabrics, carpets, floor carpeting, automobile articles; in polyester for upholstery fabrics, menswear, work clothing.

Mass per unit length: Linear density of fibres and yarns.

Mass per unit volume: Picograms/litre of solution (pg/l), nanograms per litre of solution (ng/l), Micrograms per litre of solution (ug/l), milligrams per litre of solution (mg/l), grams per cubic metre of solution (g/m^3), etc.

Mass ratio: Milligrams/10^9 milligrams (ppb), Milligrams/10^6 milligrams (ppm).

Mass, commercial: See **Commercial mass.**

Massiru: Plain spun, light East Indian silk cloth; used for garments.

Mastic cloth: Embroidery canvas spun in alternate wide stripes of basket weave in cotton and waste silk satin.

Mat: A textile floor covering of predetermined shape and limited dimensions.

Mat Weave: See **Basket weave**, or **Hopsack weave**.

Matabie: General term in East India for all fabrics having spun or printed gold or silver patterns.

Matapalo: Fibrous bark, yielded by a species of fig tree in Peru; used for garments by the Indians.

Matching Process in wool sorting: Consists in grouping the parts of corresponding quality from various fleeces.

Matchings: The different sorts of wool into which the fleece is divided in wool sorting.

Matelasse: French for "cushioned or padded or quilted ". A pooching or quilted effect fabric. Figured made on jacquard or dobby loom, in double cloth weave. In one type, the cloth consists of warp and weft with an extra weft, used as a padding and held in place by an extra stitching warp. The back of the cloth is a fine, loosely spun web of warp and weft interlacing. The fabric is made on a dobby or a jacquard loom. The fibres used may be silk, viscose, acetate, nylon, polyester or combinations. The fabric also may contain metallic threads. The pattern stands out and gives a "pouch" or "quilted" effect to the goods. Crepe yarn in double weave shrinks during finishing causing a blistering effect. in upholstery, coarse yarns cause blistering. Comes in colours, novelty effects, and some with metallic yarns. Gives good wear and drapes well. If washable, it must be laundered with care. It is very attractive and suits quite plain styles. Uses: Some cotton matelassé used for bedspreads, dresses, suits, ensembles.

Material Balance: A mathematical representation of material flow through a reaction system. The input material is accounted for throughout its various transformations.

Material, textile: Fabric, material, e.g. dyeing material, cleaning material, textile material; should always be regarded as fibre loading in the technical finishing sense.

Mathematical model, statistics: An equation or set of equations that describes a system.

Matheson: A Highland tartan, composed as follows: Wide red field, split in the center by a group of five dark green stripes, of which the middle one is much heavier, each pair on the side being fine lines; *dark blue stripe about one-eighth of the red field; dark green stripe (as wide as the blue) split in the center by a pair of narrow red lines; red stripe, somewhat narrower than the blue; two narrow green and red lines alternating; dark blue stripe, as above*; dark green stripe, twice as wide as the blue, divided into three equally wide parts by two narrow red lines; repeat in reversed order, group described between two*.

Matrimonio: Soft finish, bleached cotton bed sheeting in Venezuela; about 52 inches wide.

Matrix: A descriptive term for a textile fibre in which one or more polymeric fibrous material(s) is dispersed in another.

Matrix fibre: (1) A manufactured fibre that is essentially a physical combination or mixture of two or more chemically distinct constituents or components combined at or prior to the time of extrusion (i.e., produced in fibre form), which components if separately extruded would each fall within different definitions of textile fibre. (FTC definition). Matrix fibril fibres have the fibril constituent randomly arranged across the cross section of the matrix. When the fibril component is in high concentration it may actually form a fibrillar network in the matrix. (2) In aerospace textiles, a thermoplastic fibre used with reinforcing fibre to form a composite after consolidation with heat and pressure. (3) In nonwovens manufacture, fibres that are blended with low-melt fibres to form a thermally bonded fabric.

Mats: Pile-free, usually braided or spun underfloor coverings made of bast fibres. Also made of gum, plastic and fine metal filaments as door models.

Matt effect: See **Basket weave**.

Matt jersey: A jersey fabric of almost any weight with a dull surface due to the fact that it is knitted from crepe yarns.

Matt white printing : Printing using pastes, which contain white pigments and binders. Often used on dyed material.

Matta: Short staple cotton grown on the lowlands around Pernambuco, Brazil.

Matter, extractable: See **Extractable matter**.

Matthews: Commercial variety of very prolific, long staple, early maturing American cotton; the staple measuring 35-40 millimeters; the yield is 29 per cent and above.

Mattis: Late maturing commercial variety of American cotton, the staple measuring 25-30 millimeters; the yield of lint is 30-32 per cent.

Mature cotton: Cotton whose fibre wall has thickened to an acceptable level.

Mature fibre, Cotton: Self explained. (1) A mature cotton fibre treated with caustic soda solution – Fibres that have swollen into unconvoluted and almost rod like shapes where total wall width is equal to or greater than the lumen width. (2) Cotton fibres observed under polarized light: Fibres that appear yellow, yellow green or green and are yellow or light yellow upon rotation to to the subtractive position (thro' 90') and show little or no parallel extinction on removal of selenite plate.

Mature fibres: (cotton fibres observed under polarized light)- fibres that appear yellow, yellow green, or green and are yellow or light yellow upon rotation to the subtractive position (through 90') and show little or no parallel extinction on removal of the selenite plate.

Maturity, cotton: An important cotton fibre characteristic which express the relative degree of thickening of the fibre wall. It is sometimes defined as the ratio of the cross sectional area of the circle having the same perimeter as that of the fibre, or the ratio of average wall thickness to the radius of the circle having the same perimeter as that of the fibre. However, in practice, direct measurement of the degree of the wall thickening is seldom carried out and the average maturity of a given of a given sample of cotton is by one or more of several indirect tests which are often used to discover the proportion fibres having a maturity greater than some selected level.

Maturity index: Relative indication of cotton fibre maturity. See **Goldthwait test**.

Maturity, Cotton, test for: See **Goldthwait test**.

Maubois: French droguet made of silk.

Maud: Obsolete term, but it used to describe checked woollen traveling rugs spun in different shades of grey.

Mauritius fibre; Mauritius hemp: A fibre from the leaf of the plant Furcraea giganta.

Mauritius hemp: Trade name for the strong leaf fibre, yielded by the Furcraea gigantea in 'Mauritius (aloe fibre), Brazil (pita fibre), Central America (fique fibre), West Indies and East Africa inter alia; used for cordage, gunny bags; similar to sisal.

Mausari: Thin, open face but coarse cotton fabrics, made in India; used for mosquito netting. Generally made in check pattern.

Mauveine: (aniline purple), first synthetic (cationic) dye, which the then 18 year old Perkin discovered by chance in 1856 (called aniline purple, later mauveine, Perkin's mauve, Perkin violet) and a few years later industrially manufactured on a large scale, whereby he also thoroughly understood how to solve the application problems

Mawata: Silk cocoons that have been simmered and opened onto a wooden frame.

Maxey: Commercial variety of prolific American cotton, the staple measuring 30-35 millimeters; the yield is 30-32 per cent. Also called Meyers Texas.

Maximumum pick up: More than 100% of liquor application (compare: Low wet pick-up; Addition high-wet pick-up). Using special machines, (see Fig.), a fabric is loaded with as much liquor as it can just carry without dripping before it then, for example, goes into a steamer.

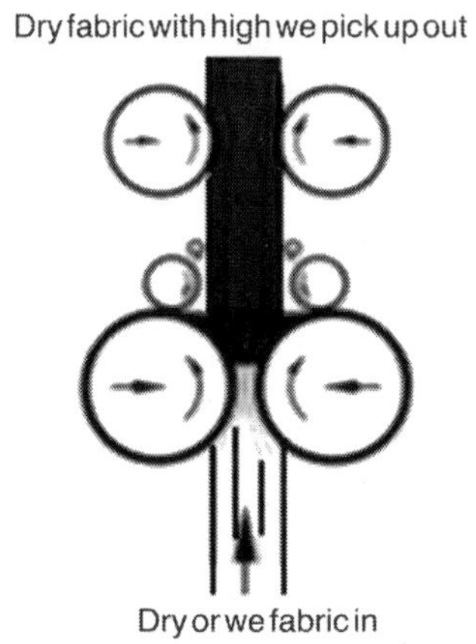

Maxwell: A Highland tartan, composed as follows: A wide green stripe, split by a red stripe in the center; on each side of green stripe, and separated from it by a red stripe is a black stripe, about one-fifth the width of the green stripe;

a red field (as wide as the group measured between the outer edges of the two black stripes), split in the center by a pair of narrow green lines.

Mayenne: A fine, bleached French linen.

Mayo twill: See **Campbell twill**. Name for a twill, producing short, zigzag figures in alternate colours.

Mazamet wool: Sweated wool from Southern French town of Mazamet, where most Australian sheepskins come to be processed.

MBAS, MBS: Methylene blue active substances.

MBF: A moving bed sand filter.

MBBR: A moving bed biological reactor.

MBR: A membrane bioreactor, but can also mean a moving bed biological reactor.

MBT: Mechanical biological treatment.

MCB: Monochlorobenzene.

MCL: Maximum contaminant level.

MCPA: See **Chlorophenoxyacetic acids.**

MCLG: Maximum contaminant level goal. See **Maximum contaminant level.**

MDPE: Medium density polyethylene.

MDR: Multiple drug resistance.

MDTOC: Minimum detectable threshold odour concentration.

MD: Modal fibre.

MDI: Abbreviation for diphenylmethane diisocyanate or diisocyantodiphenymethane.

MDPE: See **Medium Density Polyethylene.**

ME: Metal fibres.

Mean: Average.

Mean deviation unevenness, in textiles: The average of the absolute values of the deviations of the linear densities of the integrated lengths between which unevenness is measured and expressed as percentage of the average linear density for the total length within which unevenness is measured.

Mean Fibre Diameter: The average diameter (thickness) of a group of fibres from an animal.

Mean length, in cotton: The average length of all the fibres in the test specimen based on weight length data.

Mean length, in testing of cotton fibres: The average length of all fibres in the test specimen based on mass length data.

Mean temperature, in thermal transmittance of textiles only: The average of the hot plate temperature and the temperature of the calm, cool air that prevailed during the test.

Measurement process: The act of quantifying a property or dimension.

Measurement value: The numerical result of quantifying a particular property or dimension.

Measure cutting: The cutting of a garment to individual measurements.

Measuring Gauge: Used to mark even widths for hems, pleats, etc. It is available in chart paper and can be made at home in different widths.

Mechanical dewatering of sludge: Dewatering of sludge that uses a mechanical device such as *belt filter press, centrifuge, filter plate press*, tube press, vacuum filter, etc. For wastewater sludges, belt filter presses and centrifuges are common. The filtrate from the dewatering of wastewater sludge can have a high BOD. Dewatering of raw wastewater sludge is very malodorous. Waterworks sludges are difficult to dewater and filter plate presses are often chosen.

Mechanical filtration: This may mean one of several things: ***mechanical dewatering of sludge, microstraining***, use of a ***pressure filter*** (2) ***or screening***.

Mechanical finishing: Changing the appearance or physical properties of a fabric by a mechanical process such as calendaring, embossing, bulking, compacting, or creping.

Mechanical finishing: Mechanical finishing processes can be referred to as those processes generally carried out on open-width dry fabrics, with or without heat application, which give the fabric good dimensional stability (shrink proof and shape retention) and modify the "hand" of the textile product by altering its structure (at least its surface structure) Dry finishing

Calendaring: a lustrous, dense and compact appearance can be obtained by means of friction, pressure and heat.

Ciréing: this calendaring operation is carried out using special calendars and exploiting the combined actions of heat, friction and polishing agents.

Embossing: this particular type of calendaring process allows engraving a simple pattern on the fabric.

Sueding: thanks to this process, the fabric has a much softer hand and an improved insulating effect thanks to the fibre end pulled out of the fabric surface. This process is carried out by means of a roller coated with abrasive material.

Raising: he fibre end pulled out to the fabric surface imparts an insulating effect. This process is carried out by means of hook-needles running in different directions on the fabric.

Shearing: the fibre ends on the fabric surface are cut by using special cutting tools.

Singeing: the fibre ends pulled out to the fabric surface are burnt by means of a flame (see **preliminary treatments**).

Mechanical flocculation, in ETP: A common method of flocculation, capable of improving

the clarity of raw water or wastewater by the gentle movement of a picket fence stirrer or of paddles at speeds of the order of 70 mm/s for 15 minutes. It is usually used after coagulation and then followed by clarification or dissolved air flotation.

Mechlin lace: A soft, filmy lace in which even the denser decoration is filmy. Often used for bridal veils.

Mecklenburgh: (1) 18th century woollen fabric in England, sometimes made with silk flowered patterns; (2) Stout English wool damask. The ground is of colored hard twist warp stripes with colored flower patterns.

Me-Complex dyes: See **Metal complex dyes.**

Mecomba: Native East African name for the cloth-like bast, obtained from the (Brachystegia tree; used by the natives for clothing.

Med fibre: A medullated animal fibre in which the diameter of the medulla is less than 60% of the diameter of the fibre.

Median, for a series of observations: After arranging them in order of magnitude, the value that falls in the middle when the number of observations is odd or the arithmetic mean of the two middle observations when the number of observations is even.

Median load, in tensile testing: That load level that is exceeded by half the recorded peaks and which in turn precedes the other half of the recorded peaks, in a specified distance of cross-head travel.

Medium density polyethylene, m. d. polythene, MDPE: A thermoplastic plastic which has a density of between 0.945 to 0.955 g/cc, which is lower than high density polyethylene (HDPE), but greater than low density polyethylene. It is light and flexible. It may be used for landfill liners but HDPE is more common. MDPE can be used for irrigation pipes, but water pipes and sewers normally use HDPE. MDPE is sometimes used for sliplining as a water pipe rehabilitation technique.

Medallion lace: Motifs are cut or made from a piece of heavy lace and inserted into the cloth. Used for lingerie, table cloth etc.

Medalloion carpets: These carpets have perfectly circular or elliptical motif on self-coloured primary backing in the centre of the carpet. Similar designs can be found along the edge of the carpet or in the corners.

Medical clothing: Medical clothing functions as telemedicine that provides real-time feedback to wearers, and alerts bystanders, if necessary, by applying a sensor or microchip to the clothing or textile product. It can be used to monitor the user's individual heartbeat rate, respiration rate, and temperature, or to record a soldier's wound status in a war, etc.

Medicis: French bobbin lace, similar to the Cluny.

Medium Wools: Usually 1/4, 3/8, and 1/2 blood wools, OR wools grading 50's to 62's, OR wools with a 24 to 31 micron count.

Medley cloth: A mixture cloth, dyed in the wool, originated in the early part of the 17th century in England.

Medulla, in mammalian hair fibres: The more or less continuous cellular marrow inside the cortical layer in most medium and coarse fibres.

Medullated fibre: An animal fibre that in its original state includes a medulla.

Meherjun: Coarse Persian carpet wool.

Meisen: A plain-weave lightweight Japanese Silk, with a blurred pattern achieved by colouring the yarn before weaving them. Used for blouses dresses.

Mercerised yarn: Instead of Mercerising in fabric form, in certain cases the yarn is mercerized and made into fabric, most frequently in knitted form. The method is much costlier than the fabric mercerizing, but the mercerizing effects are usually better as in the yarn form mercerisation more complete due to better absorption of caustic soda.

MEK : Max. permissible concentration of harmful emissions in the exhaust air. Can be determined on the basis of the Maximum emission concentration, in view of the chimney height. So an MEK of ca. 40 kg/h of trichloroethylene or tetrachloroethane and 50 kg/h of benzine is produced from an exhaust gas conduction height of, e.g., 10 m.

Melamine/Formaldehyde: Melamine can react with up to 6 moles of formaldehyde to form a variety of products. Commercially, trimethylol and hexamethylol melamine are the more important condensates. In storage, the hydroxymethyl (N-methylol) groups tend to polymerize and liberate formaldehyde. By converting them to the methoxymethyl derivative, the shelf life is much improved. Most of the commercial products are methylated.

Melange: (1) French for mixture effect; (2) Yarn spun from printed top; (3) Colour effect on fabrics spun from such yarns.

Melange flocking: Flocking or flock printing using blended flock material.

Melange yarn: Mélange yarn made of bleached/unbleached or treated/ untreated fibres, which only appears mottled after dyeing.

Melange' yarn: Yarn made from dyed fibre and undyed fibre to get the melamine effect. The same effect can be brought by mixing two classes of fibre and making into yarn (a blend) and dyeing one component also, but this is possible in case of blended yarn only.

Melded fabric: A nonwoven fabric of a base fibre and a thermoplastic fibre. The web is hot-calendered or embossed at the softening point of the thermoplastic fibre to form the bond.

Meles Rugs: Small, coarse, all-wool rugs made in Asia 'Minor; the loose and short pile is tied in Ghiordes knot. Very bright reds, blues and yellows are usually used in a great variety of designs. The sides and ends are finished with a selvage and there is a fringe at the ends.

Melt: A material in the molten state.

Melt blend: See **Biconstituent fibre.**

Melt blowing: A process in which a polymer is melt extruded through a dye into a high velocity stream of hot air which converts it into finre and relatively short fibres. After quenching by a cold air stream the fibres are collected as a sheet on a moving screen.

Melt flow Flow rate: Melt index measured at 125°C.

Melt index: The weight in grams of a thermoplastic material that can be forced through a standard orifice within a specified time.

Melt spinning: Some polymeric fibres are spun by melting the polymer to a liquid state. The liquid is forced through the spinner opening under pressure and cooled by a jet of air to form the filament. Melt spinning requires no chemical reactions and no solvent recovery system.

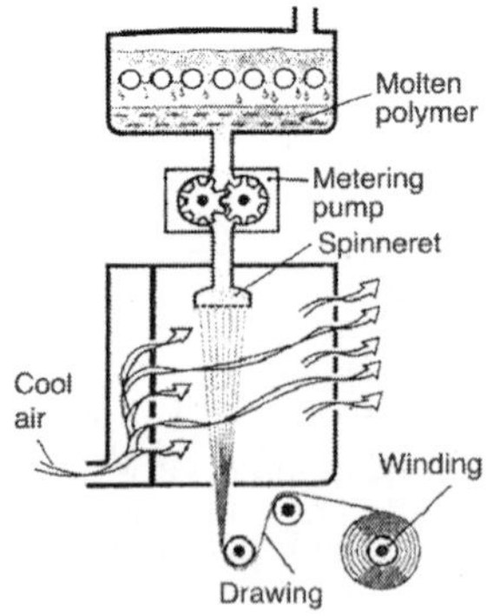

Melt Spinning (schematic)

Melt Strength: Ability of plastic to hold the die shape.

Melt Temperature: Temperature of plastic in an extruder.

Melt viscosity: The resistance of molten polymer to shear deformation. It is primarily a factor of intrinsic viscosity and temperature. It is an apparent polymer viscosity measurement in that it is only true at a specific shear stress and shear rate combination.

Melt-dyed: Synthetic fibres dyed in the melted form during melt spinning.

Melt-fracture: An unstable melt spinning condition in which the surface of the extrudate become rough and irregular.

Melting, in testing thermal protective clothing: A material responds evidenced by softening of the fibre polymer.

Melting point: The temperature at which the solid and liquid states of a substance are in equilibrium; generally, the temperature at which a substance changes from a solid to a liquid.

Melton: A firm medium weight wool cloth in a close plain weave, with a felted nap but not one way nap. It is dull in appearance and comes in white and all men's suiting colours. Rigidly constructed, it is dull, and lusterless and finishing treatments cover up all the interlacings and warp and wefts. It used to be coating fabric, but now it is mainly used as and aid to tailoring, because it does not dray., it is full of backing collars, as the raw edges can be used to reduce bulk. Mostly used for men in overcoating, uniform cloth of all kinds (army, navy, etc., as well as polie and firemen), pea jackets. Also used for heavy outer sports garments and coats for women.

Meltonette: A lightweight fabric, which resembles Melton cloth, and is used for women's wear.

Membrane: **(1)** A skin. A membrane filter is consequently one of skin thickness, but the term usually means a polymer membrane as used in a membrane process. Compare membrane filtration; See **semi-permeable membrane.**

(2) A highly engineered polymer film containing controlled distribution of pores. Membranes act as a barrier permitting the passage of material only upto a certain size , shape and character. Membranes are used as the separation mechanism in reverse osmosis, electrodialysis, ultra filtration, nanofiltration, and microfiltration, as disc filters in laboratories, and as pleated final filter cartridges, particularly in pharmaceutical and electronic applications.

Membrane biological reactor, Membrane bioreactor, MBR : A biological treatment process for wastewater where a membrane process (typically microfiltration or ultrafiltration) is used for separation of the treated effluent from the biomass in the bioreactor. The effluent should contain no suspended solids or micro-organisms, such as bacteria. The membrane separation unit may be separate from, or immersed in the bioreactor. In the separate system, membrane fouling can occur due to the biomass coating the outside of the membrane. Air bubbles on the outside of the membrane or backwashing through the membrane or chemical flushing through the membrane are processes used to minimise membrane fouling. The bioreactor is typically an activated sludge plant, but the bioreactor may be an anaerobic process (membrane separation anaerobic treatment), in which case no air bubbles are used in the bioreactor.

Membrane filtration: (1) A method of direct counting of coliform or other bacteria in water. A measured diluted volume is filtered through a membrane of pore size small enough (0.45m) to hold back bacteria. The membrane is then incubated face upwards in a selective medium at a suitable temperature and after 18 h the colonies that have developed are counted. As in the most probable number method, these counts are only an estimate. Counting is done under a hand lens and suspect colonies can be transferred to liquid or solid medium for confirmation of their properties. Membrane filters are also used for sampling algae or for concentrating bacteria from a water in which they are known to be scarce—e.g. a drinking water, where the expected count is less than about 30 bacteria per ml. Techniques such as bioluminescence and fluorescent staining can be used to adapt membrane filtration to a more rapid detection method. (2) The use of a membrane process to treat water or wastewater.

Membrane fouling: Membranes used in reverse osmosis (RO) or nanofiltration should last 5 years. However, the membrane may become fouled by suspended matter or bacteria growing on its surface. The inlet water may require treatment (e.g. coagulation, filtration and disinfection) prior to RO or nanofiltration. Polyamide membranes used in RO can be attacked by chlorine and so excess chlorine must be removed i.e. dechlorination. Salts may precipitate on the salty side of the membrane due to concentration polarisation. This also causes membrane fouling.

Membrane processes: Various processes use **semi-permeable membranes** to hold back salts or fine solids or pathogens in order to remove these components from water or wastewater. **Reverse osmosis** is a popular method for **Desalination. Microfiltration, Nanofiltration** and ***ultrafiltration*** can be used in water and wastewater treatment for the removal of colloids or colour or pathogens. **Electrodialysis** uses membranes plus a voltage and can be used for desalination. See **Concentration polarisation, Ion exchange membranes, Membrane fouling, membrane biological reactor pervaporation.**

Men chijimi: Cotton crepe in Japan.

Mende: Fine, smooth French serges of various grades; used for lining.

Mending: A process in spun fabric manufacture in which weaving imperfections, tears, broken yarns, and similar defects are repaired after weaving; especially on woollen and worsted fabrics to prepare them for dyeing, finishing, or other processing.

Mending: A finishing process in cloth manufacture in which weaving imperfections, tears, broken yarns, etc. are repaired after the cloth is taken from the loom. Largely done on woollen and worsted fabrics, to prepare them for further finishing.

Menhofu: Cotton duck in Japan.

Menin lace: Bobbin lace, similar to the Valenciennes with the threads of the mesh ground twisted three and a half times. See **Valenciennes**.

Menneru: Trade term in Japan for flannelette.

Menouffieh: Variety of Egyptian cotton, having a good, silky staple.

Menzies: A Highland tartan, composed as follows: Wide red field; "two white stripes with a narrower red between, the group being about one- third the width of the red field; red stripe, as wide as a white and the red stripe together in the previous group*; 'White stripe (as wide as the just mentioned red stripe and the group together) split near each end by a narrow red line; repeat, in reversed order, the stripes mentioned between the two *.

Meraline: Narrow-striped, all-wool dress goods in England.

Mercer, John: Inventor of Mercerisation process for cotton. See **Mercerisation.**

Mercerisation: Treatment of cellulosic textiles in yarn or fabric form with a concentrated solution of sodium hydroxide whereby the fibres are swollen, the strength and dye affinity of the material are increased and their handle is modified; named after its inventor, John Mercer Mercerizing cotton can significantly improve its dye uptake, especially if there are immature fibres present (1844). The additional effect of enhancing the lustre by stretching the swollen material while wet with caustic and then washing off was discovered by Horace Lave (1889). "Slack" mercerizing, which does not require tension, but causes the fabric to shrink, can be done at home, but requires a great deal of care because of the caustic used. Typically it is done using about 20% to 25% sodium hydroxide solution at around 20°C. Fabric becomes very stiff and hard to handle until the solution is all washed out. In industrial practice, in chain mercerizing shrinkage of the fabric is allowed followed by re-stretching and washing on a clip stenter. In chainless mercerizing the fabric is effectively prevented from shrinkage by transporting over rotating drums/rollers.

Mercerised: Fabric or yarn which has undergone the Mercerisation Treatment.

Mercerising: The process of mercerization treatment.

Mercerization: See **Mercerisation.**

Mercerizing: See **Mercerising.**

Mercerising machine; Mercerising plant : The Machine used for mercerizing. Mainly there are two types. (a) Chain mercerizing machine. (b) Chainless mecerising Machine. The main defference is in the satabilisation zone, the other areas like impregnation and washing off zone remains similar, in the former stabilization is done on chain and in the latter stabilization is done on Rollers where the fabric is not allowed shrink by transporting the cloth from Roller to roller giving no chance to shrink. Chain mercerising machine occupies larger space and costlier, while the chainless merceriser occupies less space and cheaper.

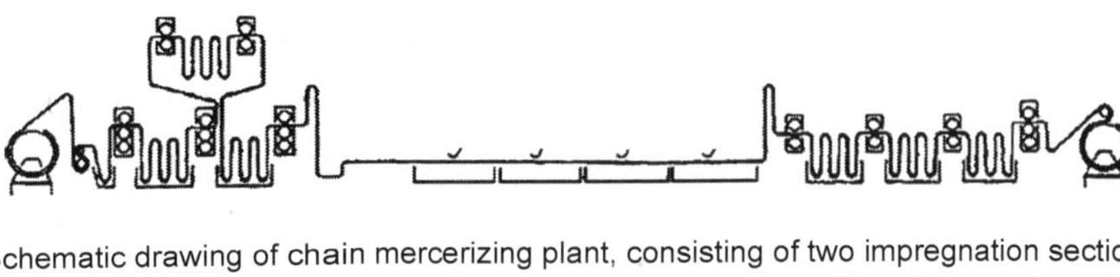

Schematic drawing of chain mercerizing plant, consisting of two impregnation sections with interconnected draw zones. Chain stabilization section with subsequent washing plant.

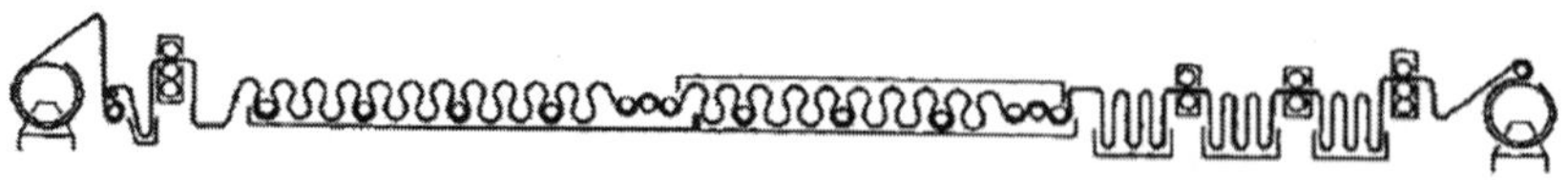

Schematic drawing of chainless Mercerizing plant, consisting of mercerizing section and stabilization section with connected cloth guide and subsequent washing plant.

Mercerized wool: Wool is treated for a brief period at a low temperature in an 80 degree Tw. solution of caustic soda; this gives a high lustre to the wool and strengthens it, but the fibre cannot be felted after.

Merezhiki: All-white hand embroidery over linen, made by the peasants in Ukraine.

Merge: A group to which fibre production is assigned based on properties and dyeability. All fibres within a merge can be expected to behave uniformly, and for this reason, can be mixed or used interchangeably.

Merchandise: Goods for sale.

Merchandise beam: **Same as Cloth roller**, see **Cloth fell.**

Merchandising: Having the right goods of the right quality at the right time, in the right place, in the right quantity, at the right price.

Merchant converter: A supplier of finished fabric who buys loom-state material and commissions its further processing.

Machine gauge: The number of needles on a knitting machine per unit of length (usually this unit length is an inch).

Merino: From pure-bred merino sheep mostly from Australia and South Africa. It gives the finest grade, soft, expensive wool fleece. .Note: Merino wool usually has a fibre diameter of 24μm or less. The wool is used in good quality cloth, blankets etc. Sometimes mixed with other wool to reduce the price.

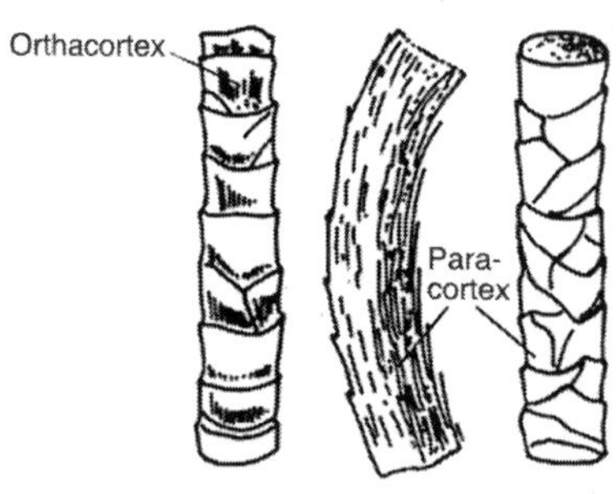

Typical appearance of the merino fibre

Merino: (1) Wool from purebred Merino sheep. Merino wool usually has a mean fibre diameter of 24 microns or less. (2) A yarn of blended wool and cotton fibres.

Merino extrafine wool: A highly refined and superbly soft yarn. Used alone or mixed with other luxurious yarns such as silk, alpaca, cashmere.

Merletto: Italian for lace.

Merveilleux: (1) Diagonal silk lining, given a lustrous finish; (2) A very fine and heavy silk satin, with a twilled back.

Mesh, in coated glass yarn fabrics: The number of warp yarn or ends per linear 25.4mm (1 in.) followed by the number of filling yarns or picks per linear 25.4 mm (1 in.)

Mesh (1): The open spaces in nets, knitting, crocheting and lace.

Mesh (2): Netting fabric, plain or printed, with holes from 3mm (1/8 in.) to 1 cm. (3/5 in.). Some are made from cotton , others from polyester. Used for tops and vests.

Mesh fabrics: A broad term for fabric characterized by open spaces between the yarns. Mesh fabrics may be spun, knit, lace, net, crochet, etc.

Mesh fineness: For the labelling of the fineness of Screen mesh materials, are used for different systems depending on material: silk bolting cloth (screen silk): numbers 0–25 (8–16 for fineness of 50–200 stitches per inch; 8–11 for normal design; 11–14 for sharp contours and 15–16 for the finest half-tone drawings). Phosphor bronze: mesh fineness lie between 90 and 350 threads per inch (110–180 as normal fineness, 200–350 for the finest designs). Synthetic fibre threads: labeling according to threads/cm (12–200).

Mesh number: Parameter for screen mesh, which indicates the number of openings/inch (= 2.55 cm), is however only meaningful using the diameter of the individual openings (see Fig.). As with constant mesh number the screen throughput is proportional to the opening diameter; besides the mesh number the screen manufacturer publishes a whole series of additional information, such as:

- hole diameter,
- number of holes per cm2,
- throughput in %,
- type of hole distribution,
- vertical and horizontal hole geometry.

Mesh sizes: The mesh numbers used in the past to describe sieve openings indicated the number of holes (or wires) per lineal inch; thus, a higher number had smaller openings. The diameter of the wires affects the hole size. Sieves therefore are now defined by the size of opening and the wire diameter in mm.

Meshtester: Device for measuring rotary screens for screen printing with regard to manufacturing-related differences in the open area.

Meshhed rugs: Medium and large sized all-wool Persian rugs with medium long pile tied in Ghiordes knot. The design has usually very large palm leaves placed diagonally and also animal forms. Deep blue and red are the characteristic colours.

Meshi: In the Bible means silk.

Messaline: The name has come from Messalina, wife of Emperor Claudius. She wore a great deal of plain soft silk and this type of fabric was named after her. It is a lustrous, soft, silky fabric in a satin weave, usually in plain colours. It may now be polyester, acetate, triacetate, as well as silk fibre. This is an expensive silk, which drapes well, and is perfect in black for special occasion dresses.

Mestiza: South American name for merino wool yielded by the cross of pure merino and the native creola sheep.

Meta: A chemical prefix, usually abbreviated *m*, that denotes that two substituents on a benzene ring are separated by one carbon atom.

Metachromasy: Property of certain dyes to show a different shade depending on concentration, aggregation, solvents and substrate. If, for example, you place an alcoholic solution of pinacyanol in sufficient concentration in a reagent glass here, this appears dark red. The thin layer, which flows down on the glass walls after the solution is tipped out, is, however, blue. This is a result of different strengths of absorption of different wavelengths of light by the dye. Red is not noticeably absorbed. The strongest absorption of the dye is in the orange parts of the spectrum at about 605 nm. The absorption subsides in the shorter wavelength range, but is still strong enough that with sufficiently large layer density or concentration the eye can no longer distinguish light let through here. So only red light is let through. By reducing the layer density, green and blue rays still have an effect on the eye as well, which results as the colour effect of blue.

Metal cast button: A button produced by the casting of molten metals and metal alloys into single cavity or multiple cavity moulds.

Metal complex dyes 1:1 Type: See **Metal Complex dyes.** Among these dyes primarily the 1:1 chromium complexes containing sulfonic acid groups have achieved commercial importance.

They must be applied from a strongly acid bath, which imposes certain limits on their range of applications. The 1:1 metal complexes are not suitable for polyamide, which is partially decomposed under the dyeing conditions for these products. Their main area of application is in the dyeing of wool, but they are also suitable for leather dyeing.

Metal complex dyes 1:2 Type: See **Metal Complex dyes.** Because of their structure 1:2 metal-complex dyes exhibit anionic character. Those which have gained commercial importance are primarily the ones that are free of sulfonic acid groups and for which adequate water solubility is provided by nonionic, hydrophilic substituents, such as methylsulfone or sulfonamide groups. The introduction of 1:2 metal-complex dyes which are applied from a neutral to weakly acid bath, represented a significant technical advance over the strong-acid-dyeing 1:1 chrome complex dyes. It has led to better protection of the fibre material, simplification of the dyeing process, and improvement of the fastness properties.

1:1:1 metal-complex dyes: mixed complexes made from one coloured, one colourless component and one metal central atom (variation from 1 : 2 complex).

1 : 2 metal-complex dyes, Dispersed: These dyes do not contain any water-soluble groups and are applied as disperse dyes on polyamides. Good light and wet-fast, building up to the deepest tones. Tendency to streaky dyeing in brighter shades. Usually dull tones.

Metal complex dye: (or metallized dye) - dye in which typically one or two dye molecules form a close permanent association or "complex" with a metal atom The metal associated with the dye is held by what is more or less electrical charge attraction of more than one part of the dye molecule for the electrons of the metal atom. The metal is often copper, chromium or cobalt. There six types: 1 : 1 metal-complex dyes:, 1:1:1 metal-complex dyes, 1: 2 metal-complex dyes, Dispersed 1 : 2 metal-complex dyes, Pre-metallized direct dyes, Phthalogen metal-complex dyes.

Metal detecting device: Device for the local monitoring (detecting) of the smallest particles of leading metals to prevent damage to calendar bowls and shearing blades. Used at fabric speeds up to 400 m/min.

Metal embroidered lace: A lace fabric as base, usually of viscose, with embroidery worked in nylon an metal yarns.

Metal fibres: (MTF, ME), textile processed metal filament and staple fibres. Recently spinnable cut forms of metal filaments or Metallized yarns, e.g. up to 5–15% compared with other long staple fibres in fashionable glitter effect yarns. Metal filaments are manufactured from high-grade steel filaments in the nozzle extraction process. Uniformity, high strength, abrasion resistance, electrical conductivity, temperature resistance, low extensibility. Chiefly used as an addition, e.g. in carpet yarns; prevents electrostatic charge. After sizing, 100% yarns can be processed in weaving and knitting. Steel filaments are, e.g., used for tyre cord.

Metal foil printing: Works according to the transfer printing technique. Fabrics are firstly pre-printed with adhesive and afterwards treated together with the metal foil in a transfer calendar

Metal powder printing: Earlier days it was generally called bronze printing because bronze powder was the main metal powder used in printing. Now, metal printing is basically designing technique for the manufacture of gold, silver and graphite printing effects in hand block, screen, spray or roller printing. Precursor of Pigment printing. Fixation of metal powder usually with aqueous plastic dispersions.

Metal salts in DP finishing: Magnesium chloride is a mild catalyst that can be used at high temperatures. It is non-corrosive and presents the fewest side reaction problems, e.g. shade change, fibre damage etc. Zinc nitrate and zinc chloride are more reactive than magnesium chloride. Zinc presents effluent disposal problems. Zinc nitrate cause dye shade changes.

Metal thread fabrics: There are a few fabrics, for example Silk from the east, which have real silver or gold threads spun into them. The problem with these is that the thread tarnishes. Normally advised to keep the garments in tissue paper to reduce the action of the atmosphere.

Metal-complex Dyes: See **Metal Complex Dyes.**

Metallic fibre: A manufactured fibre composed of metal, plastic-coated metal, metalcoated plastic, or a core completely covered by metal (FTC definition). They are available in "yarn" form as well as in staple form for spinning with other fibres. A core yarn with a metal surface is produced by twisting a strip of metal around yarn of natural or manufactured fibres.

The most important characteristic of metallic fibre and the chief reason for its use in textiles is glitter. Metallic fibres are used as a decorative accent in fabrics for apparel, bedspreads, towels, draperies, and upholstery. A relatively new application for metallic fibres is in carpet pile, where they are being used in small percentages for control of static electricity.

Metallic yarn: See **Metallised yarn.**

Metallisable dyes: See **Dyes.**

Metallised yarn: From a textiles point of view, it is more correct to say Metallized yarns.

Metallised yarn; Metallic yarn: A yarn which has fine metal as a component. Note: There are several types, the best known of which are: (a) Yarns in which separate metal fibres metal fibres or filaments are included (b) Metal of narrow strip section usually lustrous. The metal may be coated with film such as viscose, cellulose acetate, butyrate or polyester. The film may be coloured. (c) Yarn on which metal is deposited e.g. by chemically, or by electric arc or by adhesive (cf. tinsel yarn) (d) multiend yarns in which at least one single yarn is metallic.

Metameric colour match: A colour match between two materials in which the colors are identical under some lighting conditions but not under others. Metameric colour matches are common when different pigments or dyestuffs are used to colour the two materials.

Metameric index: Two samples, which, for example, have the same colour in daylight, possibly have a different colour when the illumination is changed, if they have different spectrum reflection curves and are therefore metameric. A metameric index is used for the quantitative labelling of this undesirable property, e.g. in order to discover the sample which shows the smallest metamerism, when comparing several colour recipes for the post-dyeing of a pattern. If both samples do not have a colour difference in daylight D65 ($\Delta E = 0$), the metameric index of this sample pair for an illumination change from D65 to another light source is equal to the colour difference of the samples in this other light source.

Metamerism: A phenomenon exhibited by a pair of colours that match under one or more sets of illuminants (be they real or calculated), but not under all illuminants. Or A marked change in the colour of an object with a change in the spectral composition of the light by which it is viewed. NOTE: Metamerism can be judged only with reference to the changes occurring in other objects in the fields of view as the illumination is changed.

Metameric match: A colour-match that is judged to be satisfactory under a particular illuminant but not under other illuminants of different spectral composition.

Metap weave-knit Process: A technique combining weaving and knitting in one operation with two independent yarn systems wound on warp beams. In the fabrics produced, spun strips are linked together with wales of stitches. Generally, the fabrics have 75-85% spun and 25-15% knitted structure.

Meter: See **Metre.**

Metering: The term "metering" in wet finishing is understood to mean the addition of a given quantity of solid or liquid active material to a liquid treatment medium (dye or treatment liquor) over a period of time (metering time). According to this definition, the term "metering system" covers the metering volume, metering concentration, metering mechanics, metering curve as well as the type and composition of the metered substances.

Metering pump, in spinning: A positive displacement device that pumps a measured amount of polymer solution to the spinnerets.

Metering station: Plant for the metering of particular quantities of dye and chemicals. Supplied as a ring main via mixing station or metering unit directly to the dyeing plant.

Metering Zone, in melt spinning: Section of extruder controlling rate of flow into die.

Methacrylic acid - Ethyl acrylate co-polymers: Monomers containing carboxylic groups can be polymerized with vinyl and acrylic co-monomers to yield a range of co-polymers with varying carboxyl content. Co-polymers of methacrylic or acrylic acid and ethyl acrylate have been found to be particularly useful as soil release agents. An acid content of 70% or less give relatively high molecular weight emulsion polymers whereas higher proportions of acid renders the polymer water soluble and of lower molecular weight. A particularly good combination for soil release is 70% methacrylic acid and 30% ethyl acrylate. The data shows that when a 70/30 MAA/EA co-polymer is added to a typical durable press finish containing DMDHEU, the fabric possesses excellent soil release with fair durability.

Methenamine Pill Test: See **Flammability test.** Flammability test using a methenamine pill.

Methine dyes: Methine dyes and polyene dyes are characterized by a chain of methine groups that forms a system of conjugated double bonds.

Methylhydrogen fluid silicones: Methyl hydrogen dichlorosilane offers a route for making a linear polysiloxane fluid with latent crosslinking potential. Hydrolysis of the dichloro groups will occur rapidly with water to form a linear polymer. Stable emulsions can be prepared, as long as the aqueous pH is maintained between 3-4,. When these emulsions are applied to a fabric with a tin catalyst (e.g. dibutyltin-dilaurate), the Si-H group hydrolyzes to the silanol and condenses to form a crosslink. These offer a way of improving durability.

Methyl cellulose: (cellulose methyl ether), corresponding to the 3 OH groups existing in the cellulose molecule per glucose unit, the etherification using methoxy groups CH_3O can lead to mono-, di- and trimethyl celluloses.

Methyl orange: (golden orange, helianthine, orange III, orange IV D), cationic orange dye; extremely sensitive to acid. Sodium salt of p-dimethyl amino azobenzene-p sulphonic acids. Soluble in water, insoluble in alcohol. Application: methyl orange test (change in colour at pH 4).

Methyl orange test: Methyl orange test is used (with 0.1 g to 100 ml of water) dissolved as an indicator. Reaction to alkali pale yellow; to mineral acid traces rose coloured. Neutralisation point: orange. Change to red/yellow at pH 3.1–4.4.

Methyl red: Cationic red dye, p-dimethyl-aminoazobenzene-o-carboxylic acid. Soluble in alcohol and acetic acid, almost insoluble in water. Used as an indicator (0.5% alcoholic solution) in the titration of weak bases. Yellow in alkali solution, red in acid solution (change in colour at pH 4.2–6.3).

Methyl violet: Cationic violet dye. Mixture of salts containing hydrochloric acid from tetra, penta and hexamethylpararosaniline. Soluble in water; sensitive to acid. Used as cationic dye for dyeing and printing. Dye for inks, ink pads, typewriter ribbons, photocopy paper, etc. As an indicator (Methyl violet test).

Methyl violet test: Indicator for traces of acids, which results in violet/ greenish yellow change of colour occurs in the pH range 3.2–0.1. pH 3 violet; pH 2 blue; pH 1 green; pH 0 yellow.

Methylcellulose test on textiles: Solution of 0.5 g of benzamide indigo blue RRLS to 1 l of distilled water. Load fabric sample, treat for 3 mins, rinse, perhaps dry. Methylcellulose contents are (on every fibre material) immediately dyed dark blue, which is clearly visible under a magnifying glass, but most certainly microscopically. Starch is stained to an extent by this, but not cellulose glycolate.

Methyl dye: Blue and violet cationic dyes.

Methylene blue method: is used for the control and determination of the concentration of cleaning boosters in organic solvents in dry-cleaning. a certain quantity of solution from a cationic substance is added to a mixture of several ml of liquor with anionic dry-cleaning detergents, chloroform and an aqueous acid solution of methylene blue. Two layers are obtained Test method: an upper one containing water with a methylene blue colour and a lower chloroform layer, which should have the same blue colour as the upper layer at the end of the reaction. Pure chloroform, indicator solution (sulphuric acid solution of methylene blue) and titration solution (approx. 1% solution of cetyl pyridinium chloride) are required for titration. Depending on the concentration of dry-cleaning detergent in the liquor, e.g. 5 ml of liquor is added (at 0.5–1% dry-cleaning detergent content) or 1–2 ml (at 2–4% dry-cleaning detergent content).

Methylene blue number: Characterizes the content of the carboxyl groups in damaged cellulose (oxycellulose). Wash out sample in diluted methylene blue solution and determine the dye uptake using colorimetry of the dye bath before and after washing out. More accurate, but more awkward is the Reversible methylene blue method.

Methylene blue test: For the testing of: (a) Oxy- and photocellulose. Wash out using methylene blue (cationic dye), rinsing with hot water. With damaged material, a lasting blue dye occurs (not however to undamaged cellulose). (b) Acid, alkali and mechanically damaged wool. This has improved dye affinity (test by counting the darker and brighter fibres under a microscope). (c) Linen compared with cotton: after washing out and thorough washing, linen remains permanently dyed, cotton does not.

Methylated DMDHEU: Modifying DMDHEU has been the most successful approach for reducing HCHO release. Among the first modified versions were those made from purer starting materials and rigidly controlling the stoichiometry. Later versions led to alkylated DMDHEU and two types have become commercially important products, methylated and glycolated DMDHEU.

Methylene blue testing: See **Methylene blue test.**

Methylolacryamide: This chemical contains an N-methylol group and acrylamide group, toxic. Used, amongst other things, as resin finishing agents in the so-called ambivalent crosslinking, usually together with cyclic N-methylol compounds. Finishings with moderate dry, but outstanding wet creasing angles, soft, smooth fabric handle. After-washing required.

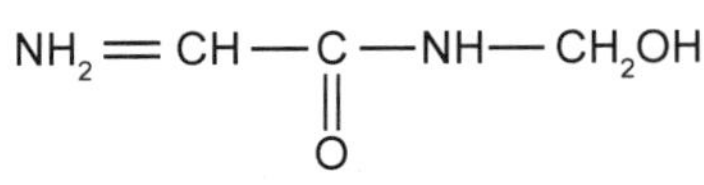

Methylol melamines: Melamine-formaldehyde compounds. Aqueous solutions of methylol melamines, less stable under storage conditions, self-polymerization in storage, therefore usually in powder form; only a few resin finishing products on a methylol melamine base.

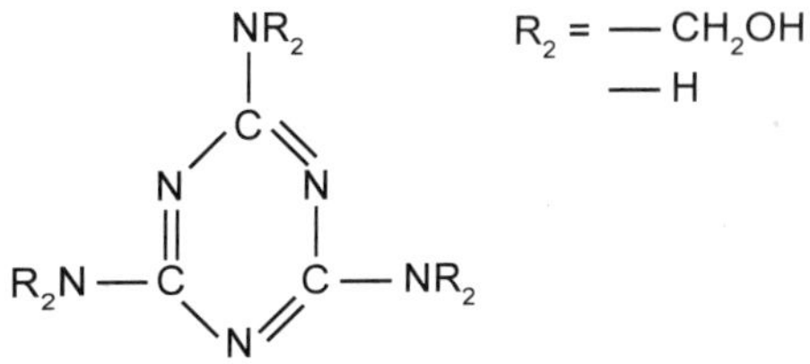

Methylol ureas: See **Urea-formaldehyde compound.**

Methylolation: Introduction of *Methylol* group, e.g. conversion of NH-group containing compounds with formaldehyde to N-methylol compounds.

Metier: A spinning machine for producing manufactured fibres. The bank of cells or compartments and associated equipment used in the dry spinning of fibres, such as cellulose acetate and cellulose triacetate.

Metier twist: The amount of twist present in yarn wound at the metier.

Metre: **(m)** Unit of length; defined as the distance, through which the light in a vacuum passes during an interval of 1/299792458 of a second.

Metric count: An indirect yarn numbering system for silver roving, and yarn, equal to the number of 1000m per kilogram.

Metric number: (metric count, Nm), a mass-related length for the Linear density of textile fibres and yarns. The designation Nm is permitted under the new law of units of measurement, as it derived from SI basic units such as mass and length. A larger Nm corresponds to a finer yarn. As this system is however contrary to the Tex system (a large number indicates a thick yarn) and by using the designation Nm a risk of confusion exists with Nm = Newton metres, the metric number should not be used any more as an indication of yarn fineness.

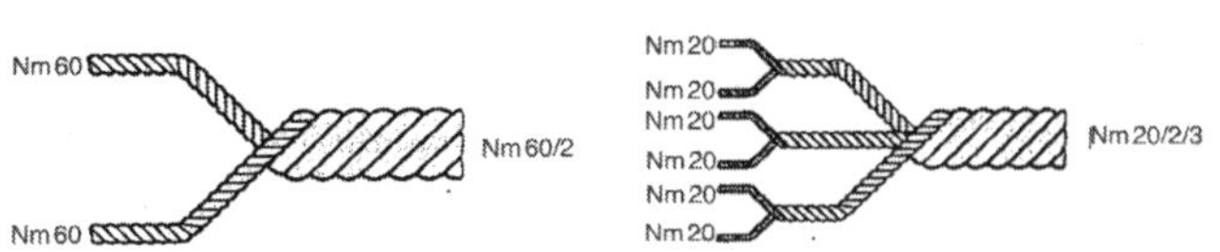

$$Nm = \frac{\text{length (m)}}{\text{mass (g)}}$$

Metric system: In general terms, a system that is internationally consistent, and is based on units that are related to others by powers of ten (for example, a millilitre is 1/1000 of a litre) The international consistency of this system and the general ease of calculations makes it vastly superior to systems with pounds and quarts and feet, etc., which define different quantities in different countries. Also see **SI.**

Mexicaine: French silk dress goods and ribbon made with narrow stripes and small figures on a taffeta foundation.

Mexican Embroidery: Made with ingrain cotton, silk or wool on muslin, cambric or linen; used for dresses, towels, etc. Usually only outlines of the patterns are embroidered.

Mexicans: A variety of gray English cotton goods, made for export; spun with well sized, coarse warp and medium fine filling, containing about 72 threads each way.

Meyers: Texas Commercial variety of prolific upland cotton, the staple measuring 30-95 millimeters; the yield is 31 per cent. Also called Maxey.

Mezzetta: Raw silk from Sicily.

MH: Abbrev. for Mohs' hardness scale, Hardness scale

Micellar bundle: Bundled micelles, which consist of so-called fibrils, for example in Wool structure. One talks of : Micelles; Crystallites; Individual fibrils. of so-called fibrils, for example in Wool structure. One talks of : Micelles; Crystallites; Individual fibrils.

Micelle: When a surfactant is dissolved in water it forms a more or less ordered agglomerates of molecules called micelles. Pure water has a surface tension of about 72×10^{-3} N/m. As the surfactant is dissolved in the water the surface tension of water falls down rapidly, till it reaches a point called critical micelle concentration (CMC) where the surface tension levels off. At this point the surfactant molecules begin to orient themselves into clusters in the solution, the clusters being more or less lamellar or spherical.

Micellar chain: Designation in crystalline cotton structure for: Micelle; Crystallites.

Micelle starches: Starches reduced in their micellar size by steam, pressure, heating or mechanical measures. Particularly well suited to sizing purposes; therefore important for size boiling.

Miroil: French for the gloss given the fabrics in the finish.

Microaerophils: They are group of microorganisms that grow best in the presence of low concentration of oxygen.

Microcrystalline waxes: are higher melting (150^{0}F), have higher melt viscosity and are soft microcrystals obtained by subjecting the wax distillation residue to a solvent crystallization process. The residue is dissolved in a mixture of benzene a methyl ethyl ketone and then allowed to crystallize.

Microbial damage to textile: The damage of fibres and fabric caused by bacteria, mildew and other microbes. It is generally promoted by a high moisture content, relatively high temperature, individual optimum pH (for bacteria pH 7–8.5; for fungi pH 4.5–7.5) and especially in a contaminated state. Cellulose materials are attacked by earth bacilli, fungus, moulds, bacteria especially at higher humidity and temperature of the range 25–35°C (tropical climate). Wool is also damaged by earth bacilli, pathogenic type bacteria and fungus. Since wool are having comparatively higher moisture regain power these type of damage is more frequent. Synthetic materials are more or less unaffected by bacteria.

Microbial growth curve: A graph showing the rate of growth of a particular species of micro-organism for a specific food input. It is not necessarily accurate to use this growth curve to describe population growth in complex microbial systems such as **Activated sludge.** See **Growth phases of microbes.**

Microdenier: Refers to fibres having less than 1 denier per filament or 0.1 tex per filament.

Microdisperse: One commercial form of vat and disperse dyes marketed. These forms will have particle size of 0.2 to a max. 1 mm. and made into fine dispersions for the ease of application.

Microemulsions: Microemulsions are the ones with smaller particle than normal emulsions. Normal emulsion are of particle size of > 300 nm. They are normally white in colour. Microemulsions are the ones with particle size 100 – 10 nm. They are bluish in colour and less than 10 um are as clear as water.

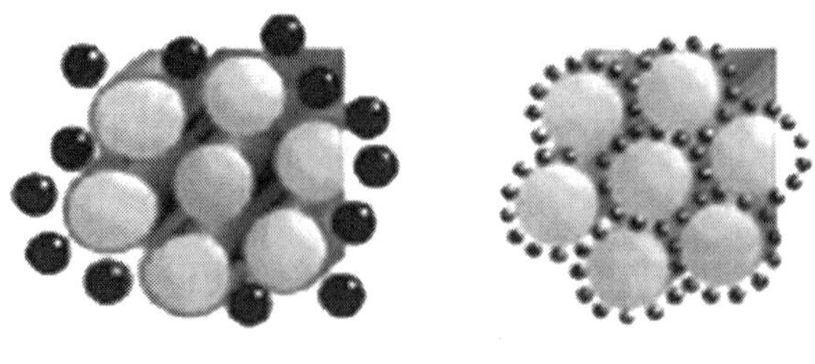

Macro emulsion Microemulsion

Microencapsulation: Enclosing materials in capsules of less than one micron to over 2000 microns in diameter. These can contain polymer additives that can then be released under certain conditions of use or processing.

Micro-encapsulation: It is a designation for the encapsulation of finely dispersed, fluid or solid phases by coating with film-forming polymers.

Microfibres: The usual definition of a microfibre is one that has a count of less that 1.0 denier or decitex. A typical PET yarn might contain 30–50 filaments, each with a denier in the range 2.0–3.0. A microfibre yarn may have well over a hundred filaments of 0.5 denier.

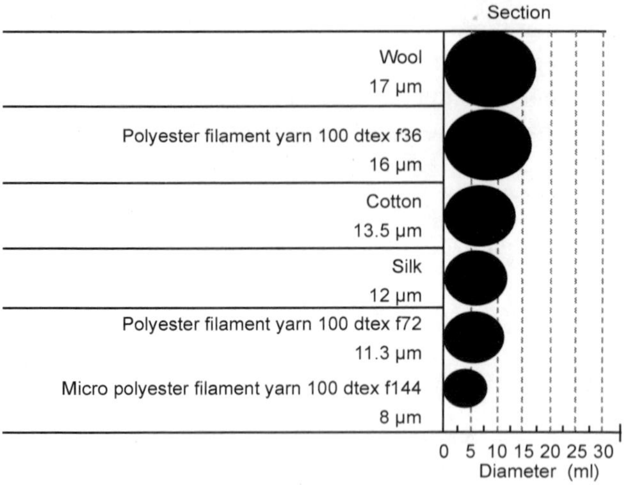

Sizes of different fibres

Microfibre batting: A textile filling material containing fibres, such as polyester or olefin, which have a diameter of less than 10um.

Microfiltration, in ETP: Filtering water through a cartridge of pore size 0.1–0.5 um that can remove particles down to the range of 0.05 to 0.5 um like TSS, turbidity, protozoan oocysts, some bacteria and viruses. The operating pressure is 2–7 bar. The micro-filter membrane requires periodic backwashing. Microfiltration can filter out micro-organisms and it is useful for the removal of ***Giardia*** and ***Cryptosporidium*** from water and the process is used in a number of large scale water treatment plants. It has also been used for the removal of micro-organisms from wastewater effluents.

Micronaire value: A dimensionless number characterizing the fineness of cotton fibre. The value is defined on the Sheffield 60 600 micronaire device (standard device). All different devices must be calibrated to it. The micronaire value scale ranges from 2.3–8.0 corresponding air stream of 9.9 to 60.6 l/min).

Micronaire method: A means of measuring fibre fineness by determining the resistance of a sample to a flow of air forced through it.

Micronire reading: A relative measurement of fibre fineness derived from the porous plug air-flow method.

Micro-organism, microbe: Microscopic organisms that include actinomycetes, algae, bacteria, cyanobacteria, protozoa and some fungi.

Micropore finishing: In this method of finishing coated fabric is made porous for air and sweat to pass through where as heavy rain is kept out. The method is to lay about 0.7mm thick finishing layer on the fabric and the bombarded with electron beam on to the fabric surface to produce 1 million pores per square metre. Polyacrylate, polybutadiene, polyurethane and certain polyvinyl chloride coatings and laminates are particularly suitable. Already manufactured textile materials are not suitable for this process.

Microporous coating: Coatings with breathable properties. The coated fabric will transmit vapors at the same time it will be water proof. Such coatings are produced when coating drying is condensed by the use of calenders and re-waterproofed or given a finishing lacquer . This is absolutely essential, as the pores are somewhat larger than those of the remaining poromerics (15–50 mm) and therefore have relatively poor water impermeability. The number of pores is approx. 500 000 per cm^2. Another process is by products manufactured by "coagulation" (precipitation, clotting) are distinguished by special softness and textile characteristics, unlike any other coating processes. These coatings can be further provided with a closed top film, which increases water impermeability. This top film thereby prevents the contamination of the coating from soiling and pore clogging.

Microtubules, in cotton: Cellulose microfibril orientation angle is an important factor in fibre strength. The orientation of cellulose microfibrils in the cotton fibre secondary cell wall is regulated by the orientation of structures in the cytoplasm collectively called the cytoskeleton. Principal components of the cytoskeleton are the filamentous structures known as microtubules and microfilaments. Microtubules are composed of two proteins, *a*- and *b*-tubulins, and microfilaments are composed of the protein, actin.

Microwave drying: The application of microwaves and high frequencies in fabric drying. The heat transfer in the fabric, from the surface toward the inner part, takes place with a certain difficulty due to the poor thermal conductivity of the fabric with subsequent problems to obtain a temperature uniformity in the whole heated mass, in relatively short times. By means of radio frequency waves, heat develops inside the material in a quantity that is proportional to the water dispersed in it. Radio frequency: 13.56 and 27.12 MHz. Microwaves: 915 and 2,450 MHz.Usually high frequency is applied to very thick fabrics (hanks, packages, bales) or to produce fast binding of latex layers, while microwaves are used for high-speed thermal treatments of yarns and fabrics. A drying unit includes the following elements: a) generator; b) electrodes; c) drying compartment.

Microwave moisture measurement and controls: Microwave technology is used for measuring the moisture content on fabric at various stages like padder (to measure the expression after padding and variation across the width), stenter to control the moisture at the exit of the stenter and there by the speed control of the stenter. The principle is to pass a defined microwave emission beam onto the damp fabric and the proportion of microwaves not absorbed because of its density is measured and related to the humidity by calibration.

Microwax: Used for water proof finishes. They are hydrocarbon wax made from de-paraffined products (paraffin gatch, petroleum gatch, etc.), obtained after de-oiling/cleaning. Colour black/brown/yellow/white; consistency sticky, soft malleable to hard. Has higher setting range/melting point of 55–90°C compared with commercial paraffins and a very fine crystal structure.

Midani Silk: Warp faced fabric with cotton filling, having narrow colored stripes divided by narrow white stripes.

Mid-neck girth, in body measurements: The circumference of the neck approximately 25 mm (1 in.) above the neck.base.

Mid-thigh girth, in body measurements: The circumference of the upper leg between the hip and the knee.

Mi-fils: The finest and thinnest French cambric.

Mi-florence: Light, plain spun silk lining, finished with a high gloss.

Mi-torse: Half twisted French embroidery silk.

Mignardise: Crochet work using narrow braid to form the body of the pattern.

Mignonette': A knitted silk or viscose fabric used for underwear. It is finer than tricot.

Mignonette: (1) Plain cotton netting; used for curtains; (2) French calico with small pattern.

Mignonette Lace: Narrow bobbin lace of lightweight made in the 16th century and 17th century of white flax thread; used for headdress. The mignonette pattern is very small and delicate.

Migration: The movement of an added substance, e.g. a dye or an alkali, from one part of a textile material to another.

Migration: (1) Movement of dye from one area of dyed fabric to another. Includes movement of colour from the dyed area to the undyed area of cloth. This is usually a property of the dye and greatly depends on the molecular size. (2) Movement of fibres which go from the center to the outside surface of yarn and back again periodically. (3) Migration power of a substance in or on the substrate from places of higher to places of lower concentration.

Migration in intermediate drying: This happens due to too little substantivity of the dyes under drying conditions and also to high moisture absorption. Migration tendency is greater in polyester/cellulose blend fabrics than in cotton and is encouraged by uneven drying. The smaller the particles of the dye the stronger the migration during drying. The coarser the yarn or the more densely beaten the fabric, the stronger the migration effect. Recommendations for avoidance: as little liquor pick-up as possible during padding. Resting of the damp fabric after padding or drying. Effective drying evenly across the whole width of the cloth. Avoid shock drying. Use products, which cause a change in the particle size under drying conditions. Only use wetting agents as far as necessary.

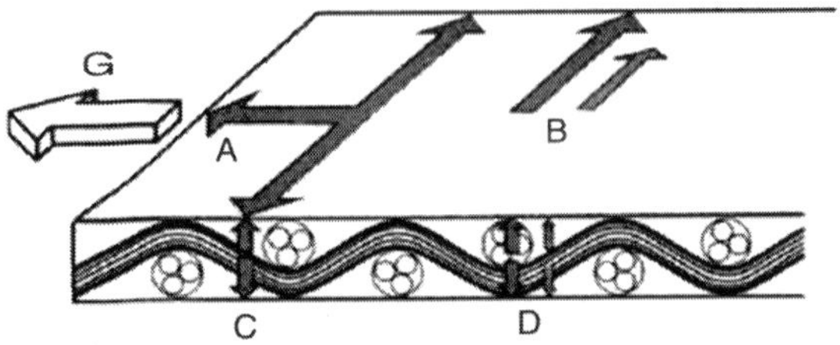

Diagram showing the migration types during pad-dry processes

Migration propensity of dyes: Capacity for Migration, in order to even out concentration differences. The levelling capacity of dyes is not only dependent on dye properties, but also on the textile material and dyeing process. The migration propensity of dyes is jointly responsible for their Levelling behaviour. (Dye migration in drying).

Migration inhibitors Substances: Which are added to the pad liquor in order to prevent the migration of dyes when drying. The alkaline saponification of polyacrylonitrile facilitates the manufacture of water-soluble products, which can be used as migration inhibitors in the thermosol dyeing process.

Migration test: (1) Sandwich test: three fabric sections of 15 x 30 cm are padded and then closely fastened on top of each other on a pin frame. This is then dried in a hot air flow under precisely defined conditions. It is fixed, reductively cleaned or boiled with soap depending on the dye classification. The outside of the fabric (1, top or bottom), the middle fabric layer (2) as well as the inside of the fabric (3, top or bottom) are assessed. (2) Hair-dryer test: a padded fabric is laid onto a wire grid and dried from above using a hot air hair-dryer at a distance of 20 cm. The front and back of the fabric section are assessed with regard to difference in depth of colour or shade of colour after drying. (3) Crease test ("concertina test"): a fabric strip 75 cm long and

6 cm wide is padded and folded in book form. The small fabric staple length is dried between two glass plates in the drying cabinet at a given temperature.

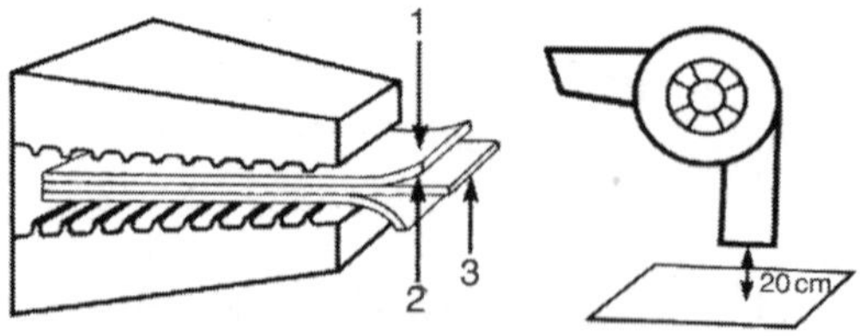

Sandwich test for migration Hair-drier test for migration

Mikado: A fine and light all-silk taffeta.

Mil: A unit of length, 0.001 inch, commonly used for measuring the diameter of wires and textile monofilaments.

Milan braid: A corded flat mohair braid; used for trimming.

Milan point: A needle point tape lace with a picot edge. Once hand made, but the simplicity of the design easy to imitate by the machine. Originally plaited gold and silver lace and reticella. Later fine needle-point laces, made with scroll designs, large flat flowers in cloth stitch. At the present a machine made lace, the design outlined with silk.

Milanaise: Narrow braid or corded fabric in which the cord effect is produced by leno weaving. One end is made to cross a number of ends in an alternate crossed and open shade sequence.

Milas rugs: One of the finest and most beautiful Anatolian Prayer mats. Smoother, rather shorter pile with approx. 150 000 knots per m^2. Colourful, bright colours. Varied patterning varying from small flat stylized flowers to geometric shapes.

Milanese: A low quality cotton fabric of about 17 × 14 tex with approximately 30 ends × 28 picks per cm.

Milanese knitting: See **Knitting.** Knitted fabric with very fine gauge, with almost equal elasticity both ways. It is a warp knitted fabric made with flat bearded needles and thread laying attachment; used for underwear.

Milanese Lace: An embroidered drawn work, made of abaca by the natives of the Philippine Islands.

Milassa: Hand knotted all-wool rugs of Asia Minor; the pile is tied in Ghiordes knot. They come either in striped design or in prayer rug patterns, with old gold as leading colour.

Mildew: A whitish growth caused by spore-forming fungi that grow in a warm, moist, confined atmosphere. The formation of mildew may cause discoloration, tendering, or variation in dyeing properties in cellulosic fibre.

Mildew resistance: The degree to which fabrics are unaffected by certain fungi that cause odor and discoloration.

Mildew-resisting agent: See **Fungicidal finishes; Antimicrobial finishes.**

Military Braid: Flat, coarse ribbed worsted, braid; used for trimming uniforms.

Milium: The trade name for a fabric which is insulkated by applying aluminium flakes to its back. The fibre of the fabric may be cotton, acetate, viscose, nylon, polyester. The insulation keeps out the cold in winter nad heat in summer. It was once a common coat lining, now mainly used for certain lining.

Milk of lime: Calcium oxide.

Mill end: A remnant or short length of finished fabric, like remnants, seconds, short ends of fabrics spun at the mills.

Mill engraving: After producing the master pattern die (Pattern die) the actual mill engraving is carried out on the engraving mill (engraving machine). The hardened relief engraving is thereby pressed against a rotating copper printing roller with a thrust up to 7 tonnes until the required engraving depth has been reached. Mill engraving is precision work, as the repeats in height and width must turn into each other without any transition.

Mill grain, in grain: Grain which is imparted to rubber sheeting while being mixed or conditioned in a rubber mill and which is parallel to the direction the rubber moves in the mill.

Mill run: A yarn, fabric, or other textile product that has not been inspected or that does not come up to the standard quality.

Mill wrinkle: See **Crease**.

Mille fleurs: All over pattern of small flowers. Used in dresses and blouses.

Mille Point: (1) Twilled English woollen, of high finish, with small patterns. (2) Tiny allover dot pattern.

Mille Raye: Originally a percale with many narrow black and white stripes; also a modern cotton and silk dress goods with numerous, very narrow stripes.

Milled composite fabrics: Bonded fabrics made from e.g. nonwoven wool (blends) bonded using felting and milling.

Milli- Symbol: m A prefix denoting 10^{-3}. For example, 1 millimeter (mm) = 10^{-3} meter (m).

Milligram, mg: One thousandth of a gram; 1mg, 1000 micrograms, 10^{-3} gram.

Milligram equivalent per litre, milliequivalent per litre, meq/litre: A way of expressing the concentrations of substances in water instead of using mg/l. The concentration in mg/l is divided by the equivalent weight. A solution with a strength of 1 meq/l has one-thousandth of the strength of a normal solution.

Millimole per litre, mmol/litre: The concentration in mg/l divided by the gram molecular weight of the substance. A solution with a strength of 1 millimole/l has one thousandth of the strength of a molar solution.

Milliners needle: These hand sewing needles are long, narrow needles with a small round eye. They are used in Baltimore Applique, as well as sewing beads and French knots. They range in size from 3 to 9.

Milling: (1) The process of treating fabric in a fulling mill, i.e., fulling. (2) In silk manufacturing the twisting of the filaments into yarn. (3) A grinding process, i.e., ball-milling of dyes and pigments.

Milling auxiliaries: (milling soaps) are used for the support of felt formation (thread end) in the milling of wool. This mostly concerns surfactants or preparations based on them.

Milling dye: a class of acid dye There is no clear distinction between milling acid dyes and super milling acid dyes. They come in a range of colors, including some bright shades and a large choice of "pure" (as opposed to mixtures) colors. Wash fastness is generally quite good. Light fastness varies from poor to very good. Retarders are usually used to help achieve level results. Application temperature is typically at or near the boil. These dyes are used on wool and polyamide (nylon). The term "milling" is said to refer to a process of making wool felt; milling dyes don't wash out badly in the process.

Milling process: Matting and condensing (increased shrinkage in length and width) of wool piece goods (possibly also with a proportion of extraneous fibres) in the hot damp state, preferably of more open carded yarns. Milling is carried out in acid (See **Acid milling**) and/or alkaline liquor.

Milling shrinkage testing: (1) Mark in a metre on grey goods in the sealing department.

After milling, the difference between raw and finished length produces the felting shrinkage (in %, referring to the raw length). (2) Subtract percentage finishing loss (difference between raw and finished fabric weight) from the raw weight; divide the difference by the current finished weight per metre.

Milling Theory: It is also called felting theory. Milling is a process of fibres getting tangled and formation of a homogeneous top layer from loose fibres by use of friction, impact or pressure in the presence of heat and moisture. The main reason of felting is the scaly surface of the wool fibre.

Millimeter of mercury: See **mmHg.**

Millitron® process: Computer-controlled carpet spray printing, whereby dye solution is sprayed into the carpet pile from nozzles. As it is processed without screens, there is no contact with the carpet surface, therefore no deformation of the pile material and, when changing patterns, no machine stoppage.

Millon's reagent: Used for following tests. (1) For the testing of protein substances (finishing analysis); reagent easily decomposable; always prepare freshly before use (consisting of 20 g of mercury and 40 g of nitric acid, heat in water bath until dissolved, leave for 24 hours, pour off precipitate; with protein, is turbid until precipitation, pink hue during heating); optimum pH range for reaction: neutral to weak acid. (2) For the testing of wool damage, i.e. degraded wool (reaction to the amino acid tyrosine). Boil sample with solution of mercury nitrate in nitric acid. This initially results in a white precipitate, which is then dyed reddish-brown (violet). 3. Testing of animal fibres (silk, animal hair fibres, casein) besides man-made fibres. It results in a red to reddish-brown hue when heated (animal fibres).

Milnerised process: Finishing process using application of resin finishing agents, catalysts and additives together with sizing agents and dyes on warp yarns made from polyester/cotton.

Miltons: Thick and well fulled woollen suiting; used for hunting garments in England; came usually in brown, red and blue colours.

mmHg: (millimeter of mercury) A former unit of pressure defined as the pressure that will support a column of mercury one millimeter high under specified conditions. It is equal to 133.322 4 Pa, and is almost identical to the torr.

Mina cloth: Stout twilled cloth of wool and cotton.

Mineral dyes: Inorganic coloured pigments known as so-called paints. Hardly used for textile dyeing and printing any more. The Mineral khaki formed on the fibre is of some interest.

Mineral fibres: A generic term for all non-metallic, inorganic fibres, which may be natural, such as asbestos, or manufactured from such sources as rock, ore, alloys, slag, or glass.

Mineral khaki: An excellent mineral dye produced on the fabric. Method: Using blends of soluble chromium salts (higher quantity used results in greenish tone) and iron salts (higher quantity used results in brownish khaki). Process:

(a) Impregnate on the 3 milling padder with the above blend (density 1.18).

(b) Short steaming or drying (hotflue).

(c) Passage through a) a bath (density 1.06) of 1 part caustic soda liquor 40°Bé and 3 parts sodium carbonate calc. (formation of chromium oxide and iron oxide on the fibre) or b) weak sodium carbonate solution and then diluted hypochlorite solution. Alternatively it is slowly heated to 80°C on the jigger using iron (II) sulphate and chromium salt; then follows rinsing, soaping, cold and hot final rinsing.

Mineral silicate Fibres: Belonging to the group of mineral fibres; industrially manufactured fibres made from the melting of rocks. Only for industrial purposes.

Mineral weighting: Weighting of silk using inorganic salts (tin-phosphate-silicate weighting).

Mineralization: Introduction of metal ion.

Minimum Care: A term describing home laundering methods. A care label used for fabrics which can be washed satisfactorily by a normal home laundering and used or worn after light ironing Minimum care fabrics, garments, and household textile articles can be washed satisfactorily by normal home laundering methods and can be used or worn after light ironing. Light ironing denotes ironing without starching or dampening and with a relatively small expenditure of physical effort.

Mink: A very distinctive long pale brown fur from an animal which can be ranch bred or wild. There are also Chinese Mink or Kolinsk Mink, which is of unusual colourings, e.g. silvery blue and black. Mink is luxurious, but it is also hard wearing. It is often used for hats.

Minor defect: A defect that is not likely to materially reduce the usability of the product from its intended purpose or is departure from established standards having little bearing on the effective use of operation of a product.

Minor imperfections, in fabric grading: A deviation in a roll of fabric that judgement and experience indicate is likely to have no bearing on subsequent processing of the fabric.

Miralene: Polyester fibre, crinkle or boucle type yarn based on Terylene.

Miralon: A process applied on nylon yarn for bulking the yarn. The bulking makes the fibre warmer, as air is introduced, and also softer to the feel.

Mirror velvet: Velvet produced in the normal way, but with a pattern made by pressing the pile flat in different directions and giving a shimmery or mirror effect.

Mirecourt Lace: Originally a French bobbin lace, similar to Lille (see), lately bobbin made sprigs are appliqued on -machine net ground.

Misclip: See **scalloped selvedge.**

Misdraw: See **Wrong draw.**

Mis-Knit, in knitted fabric: A deviation from the designated knitting pattern.

Mispick, in spun fabrics: A pick not properly interlaced which causes a break in the weave pattern. See also **Double pick, Wrong draw.**

Misprint, in printed fabrics: Colours or patterns, or both, either missed, or partially missed, or incorrectly positioned relative to each other.

Misregister, in printed fabrics: Colours or patterns not correctly positioned (compare Misprint).

Missing end: See **End out.**

Missing pick: See **Broken pick.**

Missing thread: A thread or pieces of ground or effect threads which are missing in the fabric weave.

Miss pinning fault: (stenter miss clipping). Appearance of fault on fabrics due to occasional non-gripping of the fabric edges on the part of the clips on the stenter.

Miss-stitch: A knitting construction formed when the needle holds the old loop and does not receive new yarn. It connects two loops of the same course that are not in adjacent wales. Also known as float-stitch.

Mistral: This is a crepe effect worsted cloth. Twisted warp and weft yarns are used.

Mitafi: Variety of Egyptian cotton, having a fine, long, strong staple of dark brown colour; extensively cultivated.

Mitcheline Quilt: A double cloth, spun with two sets of warp and two sets of filling, the figures formed by interchanging the two fabrics. The two fabrics are united together throughout the entire structure.

Miter: is a diagonal seaming at a corner. This is done at the corners of quilt borders to make them look like the corner of a picture frame.

Mitkal: Narrow cheesecloth or cotton sheeting, gray or bleached, made in Russia.

Miwata: Japanese trade name for unginned cotton.

Mixed Checks: English striped or checked fabric, the white stripes being of linen and the colour of cotton.

Mixed end, in spun fabric: A warp yarn differing from that normally being used in the fabric.

Mixed filling: Filling yarn differing from that normally used in the fabric, e.g., yarn with the incorrect twist or number of plies, yarn of the wrong colour, or yarn from the wrong lot.

Mixed Fabrics: Contain more than one kind of fibre.

Mixed filling, in spun fabric: A weft yarn differing from that normally being used in the fabric.

Mixing: The blending of several varieties and grades of cotton or wool to obtain a uniform average as to colour, strength and length of the fibres.

Mixture: (1) Yarn which is spun of fibres in more than one colour but each kind of 'fibre being only of one colour. (2) Fabric spun of such yarn.

Mixture Crepe: Made of silk warp and hard spun cotton filling; used for dresses, waists, etc.

Ml, megalitre One million litres, 1000 m^3, 1000 t of water.

MLC: (1) Minimum lethal concentration. *See* **Minimum lethal dose.** (2) Median lethal dose. *See* **LD50**.

MLE process: Modified Ludzack-Ettinger process.

Mld: Megalitres per day, i.e. 1000 m^3/d.

MLD: Shortened form of minimum lethal dose (lethal limit), i.e. smallest known lethal basis for humans. In inhalation poisons, the criticality limit for gases, vapours and volatile suspended matter is usually given in ppm (cm^3 gas/m^3 air), for non-volatile suspended matter (dust, smoke, aerosol, etc.) in mg matter/m^3 air.

MLR: Material Liquor Ratio. It is most important feature of the dyeing machines. If 100 kg of fabric is dyed in a bath of 1000l dye liquor it its MLR is said to be 1:10. Now a days it is of utmost important because of the limitation of effluent discharges and controls. It is always preferred a machine with lowest possible liquor ratio without affecting the quality of dyeing. Lower liquor ratio means less chemicals, dyes, salt, steam water and energy, less load of waste water, thus overall reducing the dyeing cost. In exhaust dyeing many machines have been developed to reduce MLR, airflow dyeing is one of the most successful of them. In yarn dyeing the air pad method is used to reduce the MLR when dyed with partial loading.

MLSS, mixed liquor suspended solids: The dry suspended solids in mixed liquor (mg/l). Conventional activated sludge plants operate at 2000 to 3500 mg/l but oxygen activated sludge plants may operate at 6000 to 8000 mg/l. See also return activated sludge.

MLSS: Mixed liquor suspended solids The dry suspended solids in *mixed liquor* (mg/l). Conventional *activated sludge* plants operate at 2000 to 3500 mg/l but *oxygen activated sludge* plants may operate at 6000 to 8000 mg/l. *See also* **return activated sludge**.

MLVSS: Mixed liquor volatile suspended solids The dry volatile *suspended solids* in mg/l of *mixed liquor*. MLVSS is often considered to be about 70% of the *MLSS* and may be used in preference to MLSS.

MME: Mechanised metal extraction.

M-Nitrobenzene sulphonic acid, Sodium salt.

Mo: Mohair.

Mobile creel: See **Creel, Mobile.**

Mocado: The mock velvet of the sixteenth and seventeenth centuries; not now made.

Mocha leather: Fine, soft sheepskin leather from Africa and the Middle East. If available, it is easy to sew.

Mock dyeing: A heat stabilization process for yarns. The yarns are wound onto packages and subjected to package dyeing conditions (water, pressure, temperature) but without dye an chemicals in the bath.

Mock flat-felled seam: See **Double welt seam.**

Mock french seam: A complex seam formed on the inside of the object with raw edges enclosed and no stitching rows visible on the face side; similar in appearance to the french seam but constructed differently.

Mock leno: A fabric having weave with a combination of weaves having interlacings that tend to form the warp ends into groups (with empty spaces intervening) in the cloth, thereby giving an imitation of the open structure that is characteristic of leno fabrics. Mock leno fabrics are used for summer shirts, dresses, and other apparel, and as a shading medium in Jacquard designs. See **Mock-leno weave.**

Mock leno weave: A weave in which the warp yarns remain parallel but form open warp stripes by programmed interlacing of warp and filling yarn simulating a leno appearance. A weave that has open spaces between groups of warp yarns and between groups of weft yarns and a similar appearance to that of a leno weave (q.v.).

Mock-space loom: A multitier loom in which all pieces being spun in any one row are so spaced as to lie immediately above or below the landing of those in a vertically adjascent row.

Mock Satin: Strong, stiff weft faced wool satin with flower patterns spun into.

Mock Twist: A fancy single yarn; used for dress goods, similar to the double and twist (see) but the two colors are not outlined as sharply and regularly. It is produced by intermittent feeding of dyed and undyed stock in the spinning frame.

Mock voile: A palin weave cotton fabric spun from hard twisted single yarns, (instead of two fold) spun with one thread per dent.

Mockado or Mokario: Fabrics used for clothing in the 16th and 17th centuries in England. (a) one was a woollen fabric, often mixed with silk, heavily napped and spun with figures; (b) another solid colored napped woollen fabric was also called mock velvet.

Modacrylic fibre: A manufactured fibre in which the fibre forming substance is any long chain synthetic polymer compound of less than 85 % but at least 35 % by weight of acrylonitrile units ($-CH_2-$ CHCN-) Although modacrylics are similar to acrylics in properties and application, certain important differences exist. Modacrylics have superior resistance to chemicals and combustion, but they are more heat sensitive (lower safe ironing temperature) and have a higher specific gravity (less cover). The principal applications of modacrylic fibres are in pile fabrics, flame-retardant garments, draperies, and carpets.

Modal fibres: (abbrev.: CMD, MD), regenerated Cellulosic fibres with a high resistance and high wet modulus, manufactured by the addition of modifiers and by variable spinning speeds (wet elongation 15% at 2.25 cN/dtex load). This includes both Polynosic fibres and High wet modulus fibres. Modal fibres have a higher degree of polymerisation (DP 350–600), higher resistance, wet modulus, dimensional stability and alkaline resistance compared with normal viscose fibres. HWM fibres have lower alkaline resistance and lower brittleness compared with polynosic fibres.

Mode: The value of the variate for which the relative frequency in a series of observations reaches a local maximum.

Modena: Light weight dress goods from Italy, made of mixture of silk waste, cotton and wool.

Moderne: Thin French cloth made of mixture of waste silk with cotton or wool.

Modes: In bobbin and needle-point laces various stitches which fill out the patterns. Also called fillings, jours and lead work.

Modesty piece: Decorated, triangular insert for the neckline of a bodice.

Modica: Raw silk from Sicily.

Modified biochemical oxygen demand test: (BODT, so-called closed bottle test), in which is indicated the measurability of the Biochemical oxygen demand (BOD) utilised in the microbial oxidation of organic substances and the % measuring result of the theoretically necessary calculable oxygen demand for final oxidation, i.e. mineralisation of the test substances (% BODT). With this method, the BOD5 value comes to hand incidentally as it were.

Modified DMDHEU: Modifying DMDHEU has been the most successful approach for reducing HCHO release. Among the first modified versions were those made from purer starting materials and rigidly controlling the stoichiometry. Later versions led to alkylated DMDHEU and two types have become commercially important products, methylated and glycolated DMDHEU.

Modified fibres: Viscose, high tenacity and high wet tenacity fibres with a particular structure caused by spinning in the presence of Modifier (small additions of aliphatic amines, polyamines, quaternary ammonium compounds, polymers of ethylene and propylene oxide amongst others). These altered spinning conditions produce new valuable properties with suitable draw stretching: higher strength, higher working capacity, denser, homogeneous cross-sectional structure (all enveloping structure), lower swelling power and better wet tear strength.

Modified grab test, in fabric testing: A tensile test in which the control part of the width of the specimen is gripped in the clamps and in which lateral slits are made midlength of the specimen serving all yarns bordering that portion of the specimen held between the two clamps.

Modified starch: The working properties of starch solutions (viscosity, retrograding point, penetration into yarns etc.) are influenced by the source of starch. The reader is refered to any of a number of well written books and reviews extolling the virtues of the various natural starches. In addition to naturally occurring variations, there are chemical modifications where some of the natural starch properties are altered to make them more useful.

Modified worsted system: A worsted system for spinning man-made fibres which relies on pin control of fibres during sliver weight reduction, but which bypasses the system of combing required with wool to remove noil.

Modifier: Modification agent, in Modified fibres these assume important structural modifying functions as spinning auxiliaries, which result in the slowing down of the conversion on the spinning bath/ thread surface and thereby encourage the formation of

all cover fibres with improved dry and wet properties.

Module, for inflatable restraints: An assembly composed of an inflator, a cushion, a mounting device, a trigger, and a cover.

Modulus: The property of a material representative to deformation. The ratio of change in stress to change in strain following the removal of crimp from the material being tested; i.e., the ratio of the stress expressed in either force per unit linear density or force per unit area of the original specimen, and the strain expressed as either a fraction of the original length or percentage elongation. Also see **Young's modulus.**

Modulus, Chord: See **Chord Modulus.**

Modulus, Initial: See **Initial modulus.** The slope of the initial straight portion of stress-strain curve (load elongation curve).

Modulus, Secant: See **Secant Modulus.** The ratio of change in stress to change in strain between two points in a stress strain curve.

Modulus, Tangent: See **Tangent modulus.** The ratio of change in stress to change in strain derived from the tangent to any point in a stress strain curve.

Modulus, Textile: See **Textile Modulus.** The ratio of change in strain in the initial straight line portion of the stress-strain curve following the removal of any curve.

Modulus, Young's: See **Young's Modulus.** For perfectly elastic materials: It is the ratio of change in stress to change in strain within the elastic limit of the material.

Mogador: Even though earlier this term was used for a silk fabric used for mens cravats, it is now used for an acetate fabric resembling faille. It is mainly used for mens ties.

Moghan carpets: Knotted carpets from the South-East Caucasus. Moghan Steppe. Medium high wool pile with approximately 150 000 knots/m2. Coloured design with a lot of bright yellow in the bright variation of geometric flourishes, flower and animal motifs.

Mohair: The hair of the Angora goat, Capra species found in Turkey, South Africa and USA.

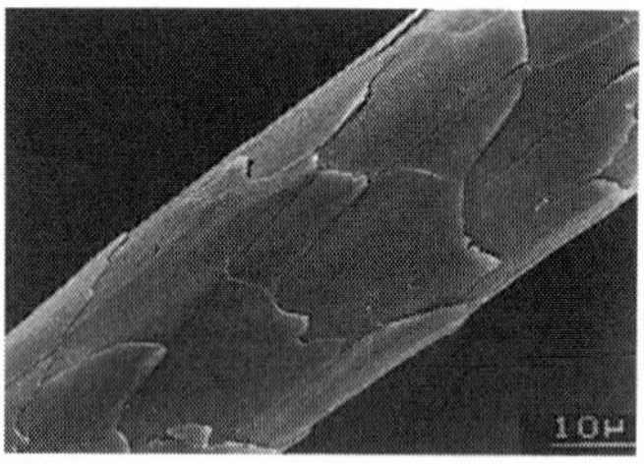

Magnified Mohair fibre

Unlike Sheeps woo this hair is not curly. It is dyeable and having a lustre and stronger than wool. It can be mixed with other yarn to make warm lightweight fabrics.

Mohair suiting: Mohair can be combined with worsted yarn, or spun alone, to produce a very lightweight, shiny, attractive suiting. Used for men's lightweight suits dinner jackets etc.

Mohair wool: Thick, spongy open fabric, in plain weave, with a very hairy texture. Its composition is usually 70 % Mohair and 30 % wool. It is very bulky and can be used for warm, lightweight lined coats, but is best for simple wrap-over jackets, sleeveless waistcoats, etc. which are unlined, and is excellent for shawls and stoles.

Mohr bleach: Combination bleach for cotton fabric in three stages:

(a) Pre-boil grey fabric in (used) peroxide bleach liquor.

(b) Chlorine bleach with sodium hypochlorite.

(c) Peroxide bleach (used liquor is used for each following batch).

Mohwal: Bast fibre of the Bauhinia vahlii in India; used for ropes.

Moiré: A finish with a wavy, watermark effect that can be applied to a variety of fabrics of almost any fibre, but is most used on silk, acetate triacetate. The fabric is passed through engraved cylinders which press the water mark onto the fabric. The finish is not necessarily permanent.

Moire: The "watered" effect given in the finishing process to mostly silk but also cotton and wool fabric. This effect is produced in various ways; (1) see

Moire antique; (2) the fabric is passed between engraved cylinders which press the moire design into the face. This moire is not as lasting as the first one. Ribbed fabrics are better adapted for both these moire effects, although many smooth fabrics, like taffeta, are treated this way. (Besides these moire effect is produced by printing the fabric, the various colors overlapping each other; moire effects are also obtained by certain weaves.

Moire a Pois: Moire silk fabric, spun with small satin dots on the face.

Moire Anglaise: Same as Moire antique.

Moire Antique: A mechanical finish. To produce this finish the fabric is folded lengthwise with face in, the selvages covering each other and stitched together. The fabric is then dampened and passed between hot cylinders. This moire effect is lasting and shows the greatest variety of designs.

Moire' effect: Are produced by interference from two linear systems. Moiré interferometry processes, which are used for the testing of surfaces, offer extensive detection. These processes are based on the fact that the overlay of the transmission functions of two equal, easily bent together, optical gratings produce characteristic intensity patterns in the form of light/ dark stripes In screen printing, moiré effects can in particular be put down to the fact that the thread system of the mesh and the fabric to be printed knit into each other. A certain screen angle of the combining linear systems can prevent the appearance of moiré. Moiré is produced in roller printing if very fine longitudinal stripes are printed on coarser fabric qualities. In order to avoid this, the fabric is calendared before printing. In decatizing, moiré is produced in the interaction of the weft threads of wool fabrics and cotton back grey (worn out).

Moire fabric: (from French: marmoire = marbled), most easily ribbed material (silk or viscose filament) with imprinted (lustrous) pattern effects, in plain, satin or twill weave, predominantly one colour. Antique moiré: with indefinite linear guidance, produced by pressing two material webs lying on top of each other. Renaissance moiré: with stronger "soaking". Application: for garment, furnishing, upholstery fabrics and similar.

Moire francaise: Moire made in stripes. Another variety called moire ronde.

Moire metallique: A frosted watered effect on silks.

Moire' tafetta: Taffeta weave fabric made from silk or synthetic fibres and then embossed with a moiré pattern. The embossing may be or may not be permanent depending on the fibre and processing.

Moirette: Plain spun cotton fabric, made of fine warp and thicker polished filling, finished in a moire effect by pressing.

'Moire' finish: Also called **Watered effect.** A wavy, rippled or watered appearance on a spun rib fabric and that is produced by the action of heat and heavy pressure from rollers.

NOTE: The appearance is caused by differences in the reflection of light by the flattened and the unflattened portions of the ribs, and there is no definite repeat in the pattern.

Moiring: Making moiré effect by any method.

Moist-cure process : Wet crosslinking process.

Moist-dwell process: Fixation of compounds (e.g. resin finishing agents) by dwelling of the fabric in a moist state. See **Wet cross-linking process; Cross-linking**.

Moisture, as used with textiles: Water absorbed, adsorbed or resorbed by a material.

Moisture: (moisture content). Humidity, being moist, light dampness, water (vapour) content.

Moisture as-is: Deprecated term. See **Moisture content**.

Moisture as-received: Deprecated term. See **Moisture content**.

Moisture content: (1) The amount of moisture in a material determined under prescribed conditions and expressed as a percentage of the mass of the moist material, that is, the original mass comprising the oven-dried substance plus any moisture present.

(2) That part of the total mass of a material that is absorbed or adsorbed water, compared to the total mass (compare Moisture pick up, Moisture regain.

Moisture content, at moisture equilibrium: The moisture content of a material in equilibrium with air of known, or specified, temperature and relative humidity.

Moisture equilibrium: The equilibrium reached by a particular material with the surrounding atmosphere with respect to moisture i.e. when it no longer takes up moisture from, or gives up moisture to, the surrounding atmosphere.

Moisture equilibrium, for preconditioning: The moisture condition reached by a material during free exposure to moving air in the standard atmosphere for preconditioning.

Moisture equilibrium, for testing: The condition reached by a sample or specimen during free exposure to moving air controlled at specified conditions.

Moisture equilibrium, for testing- for industrial yarns and tyre cords: The condition reached when, after free exposure to a test atmosphere which is in motion, two successive weighing not less than 4 h apart, show not more than 0.1 % progressive change in mass of the specimen or sample.

Moisture free: The condition of a material that has been exposed in an atmosphere of desiccated air until there is no further significant change in mass.

Moisture free weight: **(1)** The constant weight of a specimen obtained by drying at a temperature of 105°C in a current of desiccated air. **(2)** The weight of a dry substance calculated from an independent determination of moisture content (e.g., by distillation with an immiscible solvent or by titration with Fischer reagent).

Moisture free, in textiles: A descriptive term for a material that (a) has been exposed to a flow of desiccated air at a specified temperature until there is no further significant change in mass, or (b) has been treated by a distillation process using a suitable solvent.

Moisture pick–Up: The mass of absorbed and adsorbed water that is held by the material, compared to the mass of the dried material.

Moisture pick–Up, at atmospheric–equilibrium: The moisture pick up of a material in equilibrium with air of known, or specified, temperature and relative humidity.

Moisture Properties: All fibres when exposed to the atmosphere pick up some moisture; the quantity varies with the fibre type, temperature, and relative humidity.

Measurements are generally made at standard conditions, which are fixed at 65% RH and 70°F. Moisture content of a fibre or yarn is usually expressed in terms of percentage regain after partial drying.

Moisture regain: The amount of moisture in a material determined under prescribed conditions of temperature and humidity, compared to the mass of the dried material.

Moisture regain, of different fibres: The amount of water resorbed by a dried material at specified equilibrium conditions of temperature and humidity, compare to the mass of the dried material.

20.50 Alginate
18.25 Wool combed top, worsted yarn
17.00 Wool carded yarn
15.50 Worsted yarn with 25% wool
14.00 Reclaimed wool
14.00 Caesin fibre
13.75 Jute
13.50 Zein fibre
13.00 Viscose and cupro
12.50 Jute
12.00 Flax, hemp, Ramie
11.00 Natural silk
10.50 Cotton, Mercerized
10.00 Flax, Hemp
9.00 Acetate (de-oiled, desized)
8.50 Cotton
6.25 Polyamide 6 and 6.6 staple fibre
6.00 Acetate (with 3% oil)
5.75 Polyamide 6 and 6.6 filament
3.00 Triacetate
3.00 Polyester filament
2.00 Polyacrylonitrile
1.50 Polyester staple fibre; Polyethylene; Elastane
0.40 Plyvinyl chloride
0.00 Glass Fibe
0.00 Polyethylene

Moisture regain, commercial: See **Commercial moisture regain.**

Moisture, wet-Basis: deprecated term, See **Moisture content**.

Moiting Process: Moiting process consists of picking out all sticks, leaves, etc., from the wool fleece in the sorting.

Moity wool: A term mainly used in UK for wool containing vegetable matter picked up by the sheep while grazing.

Mojo: Very tough, durable bast fibre of good elasticity, yielded by the m. tree in Honduras; used for ropes.

Mokho: Raw cotton grown in Senegambia. The staple is fine, silky and white.

Molaine: In England various fabrics, made of cotton warp and wool filling.

Molar: (1) Denoting a physical quantity divided by the amount of substance. In almost all cases the amount of substance will be in moles. For example, volume (*V*) divided by the number of moles (*n*) is molar volume $Vm = v/n$.

(2) Denoting a solution that contains the mole of solute per cubic decimeter of solvent.

Molarity: A measure of the concentration of solutions based upon the number of molecules or ions present, rather than on the mass of solute, in any particular volume of solution. The molarity (M) is the number of moles of solute in one cubic decimeter (litre). Thus a 0.5M solution of hydrochloric acid contains 0.5 . (1 + 35.5)g HCl per ^{dm3} of solution.

Molded garments: Textiles, which are brought out of the tubular form into the physically suitable form using mechanical/thermal deformation, e.g. for ladies' wear.

Mole: Essentially, an amount of a chemical equal to its molecular mass, but measured in grams, rather than atomic mass units; (more formally, amount of a substance containing approximately 6.02 x 1023 molecules or formula units) For example, sodium chloride has an molecular weight of approximately 28. Twenty eight grams of sodium chloride would be one mole. If this was dissolved enough water to make one litre of solution, the solution would be said to be 1 molar sodium chloride.

Molecular weight: The weight of a molecule; the sum of all atomic weights of all the atoms in the molecule; expressed in atomic mass units (defined as exactly one twelfth the mass of a carbon-12 atom); this term is obsolete an should be replaced with molecular mass or relative molecular mass (r.m.m.) Water, for example, has a molecular weight of about 16 (for the oxygen) plus 2 (for the two hydrogens), or 18. The molecular weight of cellulose is in the millions. The concepts of molecular weight and moles are useful in understanding, among other things, *pH* and variations in effectiveness of different hydration states of a compound.

Moleskin: A strong cotton fabric slightly napped and sheared on the wrong side. It is similar to duvetyne but somewhat longer and closer nap. It is spun with a filling backed weave usually at a ratio of one face pick to two backing picks. A typical construction is 50 x 200 with 2/20s warp and 15s filling. Once only a cloth for protective clothing, particularly dungarees and trousers, because of its durability and additional warmth on the inside, but other fibres such as worsted may be used to produce good quality fabric for men's suits, and it is also made as aligning fabric. The term now refers to any cloth with short mole-like nap.

Molieton: French for Melton.

Molinos: (1) In Austria a plain spun cotton fabric, made of medium fine yarn; it is often printed and is used for shirts, etc. (2) In Mexican a variety of raw cotton, has a yellowish, glossy staple.

Molisch's test: For the testing of carbohydrates and/or their derivatives (e.g. methyl celluloses, cellulose glycolates, etc.). Two drops of a 10% solution of naphthol in chloroform are added to the solution to be tested and carefully lowered using concentrated sulphuric acid; a reddish violet ring is produced on the interface of both liquids. This reaction also takes place in the presence of proteins. Distinction: leave solution with reddish violet ring for 10 mins, then shake it. Deep blue dyeing of the whole solution = carbohydrates and/or their derivatives. Faint purplish brown dyeing = protein.

Mollet: In France, a very narrow fringe of silk or gold.

Molleton: Long-haired, soft wool material (lightly milled) with slightly visible weave under the hair surface; also in cotton, then thicker (e.g.: Beaver cloth; Kalmuck). The cotton molleton materials in practice represent a heavier type of beaver; they are also occasionally described as heavy flannel. They are also found in plain fabric and twill; raw dyed, bleached white and piece-dyed, more rarely designed yarn dyed. Heavier qualities are produced using backing weft. Molleton materials are always raised on both sides.

Mollier diagram: Gives information about steam conditions by means of isobars and isotherms. Describes the correlation between heat content and entropy(s) of the Steam.

Momie: (1) Cloth Black dyed dress goods of cotton or silk warp and wool filling. It is similar to crepe. (2) French for mummy. A cloth of puckered or fluted appearance.

Momme: Japanese weight, equal to 3.675 grams, used to measure and express the weight of silk fabrics. Weight of original Japanese silk for 1 yard length.

Monagum: A trade name for carboxy methyl starch, a starch ether

Moncahiard, Mocayar: Plain or twilled French fabric of silk warp and woollen filling. It is made mostly black.

Mon-Chirimen: A very fine Japanese silk crepe of high lustre, used for embroideries.

Monforizing: Continuous mechanical shrink proof finish on the monforizing system, consisting of padder as a damping system, nozzle dryer as a pre-drying system, palmer with overfeed device and felt calendar for final drying. Machine Manuf.: Monforts.

Monfor-Matic: Computer-controlled guidance system for heat treatment machines (e.g. stenters, hot flue, etc.) with processing sequence of measured values on screen in vertical arrangement (linearly indicating diode instrument); for central analysis and display stages of treatment sections. Mode of operation contact free, maintenance-free. Manuf.: Monforts.

Monk's cloth: See **Abbots cloth.** Made of wool, cotton, linen, silk, rayon, or synthetics in 4 x 4 basket weave. Quite heavy, due to construction. It is difficult to sew or manipulate as the yarns have a tendency to slide, stretch and fray. May sag in time depending on the compactness of the weave. It can also be made in other basket weaves. Quite rough in texture. Uses: (a) Draperies, all types of upholstery and house furnishings. Also used for coats and suits for women and sports coats for men. (b) Medieval English worsted, the piece measuring 12 yards by 45 inches.

Monkey-Jacket: A short jacket cut in the sack form, fitting close to the waist and flaring out at the bottom.

Monochromatic colour scheme: Monochromatic colour schemes comprise two or more colours that are variations of one hue. These colour combinations are considered to be harmonious due to the colour relationship.

Monochromatism: Colour blindness.

Monofil: See **Monofilament.**

Monofilament: A single filament which can function as a yarn in commercial textile operations, that is, it must be strong and flexible enough to be knitted spun or braided.. Any single filament of a manufactured fibre, usually of a denier higher than 14. Instead of a group of filaments being extruded through a spinneret to form a yarn, monofilaments generally are spun individually. Monofilaments can be used for textiles such as hosiery or sewing thread or for nontextile uses such as bristles, papermaker's felts, fishing lines, etc.

Monomer: The simple, unpolymerized form of a compound from which a polymer can be made.

Montagnac cloth: A curly, woollen fabric with an astrakhan like pile, produced by cutting some weft floats and leaving others uncut. It is subsequently brushed to form a very warm and durable fabric.

Montbeliard: Stout French ticking, made with blue and white checks or cross stripes.

Montcayer: Fine French dress goods, mostly in black, made of silk warp and two or three-ply worsted in plain or twilled weave.

Monteiths: English cotton handkerchiefs with white dots over a colored foundation.

Monzome Shusu: Japanese silk satin, spun with stripes.

Mood/theme boards: Visual presentations of ideas to illustrate a mood or a theme.

Moon: Commercial variety of American cotton, maturing in medium time, the lustrous and strong staple measuring 30-35 millimeters; the yield is 31-33 per cent.

Moonga: Species of brown colored wild silk yielded by the Antheraea in Assam and East India. See **Muga.**

Moorish Lace: A drawn work of antique origin, still made in Morocco.

Mops: (type of steaming table for napped fabrics). Designation for the steam table used particularly in the finishing of brushed fabrics, which is built on in many cases.

Moquette: (1) One of the best known and hardest wearing furnishing fabric, used for covering cahirs, upholstery, curtains and drapes, tablecloths, etc., and probably the most popular until the advent of the more luxurious Dralon velvet. Moquette can be made three types, one with cut pile, another incut, and one combining cut and uncut. The pile is worsted, mohair, or nylon, the backing wool or cotton. Its great advantage is that, although apilecloth, it doesnot become flattened by pressure, eg., sitting on it.

(2) (a) Colourfully patterned Plush, manufactured using several pile warps or by printing (print moquette), then with smudged pattern contours; for upholstery covers, divan covers and similar. (b) Single-coloured, densely spun velour wire carpets. Designation also for carpets with corresponding appearance manufactured in a different way, e.g.

flocked carpets.

Moqui: Blanket Plain spun, all-wool blankets made by the Moaui Indians in the U. S. A. The design consists mostly of black, blue and brown stripes.

Morzad: A bandhani design. See **Bandhani.**

Mora hair: Curly fibres yielded by the Southern moss, in the Gulf states and Central and South America; used for stuffing.

Moravian: English sewing cotton of 8 strands.

Mordant: A chemical that aids attachment of a dyestuff to fibres by bonding to both the fibre and the dye A mordant must have high affinity for both the dye and the fibre, acting to attach the dyestuff to the fibre. Mordants are necessary for dyes that have very low or no natural affinity for the fibre. They are often salts of metals such as chromium, copper, tin or iron. Mordants may be applied before, with or after the dye, depending on the nature of the dye, the fibre and the mordant. The mordant (e.g. aluminum hydroxide or chromium salts) is precipitated in the fibres of the cloth, and the dye then absorbs in the particles. Some mordant, especially chromium compounds, are very serious health hazards.

Mordant dye: A dyestuff that requires the use of a mordant There are very few synthetic dyestuffs currently in use that require a separate mordant, except for some dyes for wool, where mordant dyes are still quite popular. Since chromium is almost exclusively used as the mordant on wool, chrome dye has become essentially synonymous with mordant dye. Many natural dyes (plant extracts, etc.) require a mordant. The mordant used can significantly influence the hue produced with a particular dyestuff.

Mordanting: The process of impregnating textiles with some mordant, not dyestuffs themselves but usually salts, which will fasten the dye applied after.

Morea: Variety of raw cotton from Greece.

Moreas: Fancy striped satin of cotton warp and silk filling, finished with high gloss.

Moreen: (1) Originally a Dutch, all-worsted, cross ribbed camlet, with a moire finish; (2) A plain spun stout fabric, made in England, one side ribbed and watered and the other made smooth with a high finish. It is made of hard spun worsted, but also of cotton, in the latter case the filling being polished yarn. Used for skirts (formerly) and for upholstery.

Moresque: A multicolored yarn formed by twisting or plying single strands of different colours.

Morfil: Stout, twilled, Belgian worsted fabric used for bags for pressing oil.

Moriche: Very tough and durable leaf fibre yielded by the Ita palm in Venezuela; used for cordage.

Morocco leather: A term now refers to leather that has been tanned chemically although it once referred to leather only from Morocco.

Morpholine Test: For the rapid identification of certain man-made fibres. The following characteristics result when immersing in morpholine (morpholine): polyvinylidene chloride fibres: slightly darker, after a few minutes morpholine becomes opaque and dyed almost black. Polyvinyl chloride fibres, post-chlorinated: dissolve clearly reddish brown. Polyvinyl chloride fibres, not post-chlorinated and vinyl chloride vinyl acetate copolymer fibres: no effect.

Morphology: The study of the fine structure of a fibre or other material. (Gk.: morphe = form/shape; logos = science), science of structure, shape formation and surface structure, e.g. of textile fibres. The morphology of a fibre loading is of particular importance to the finisher, i.e. for example the order structure and the clearance volume.

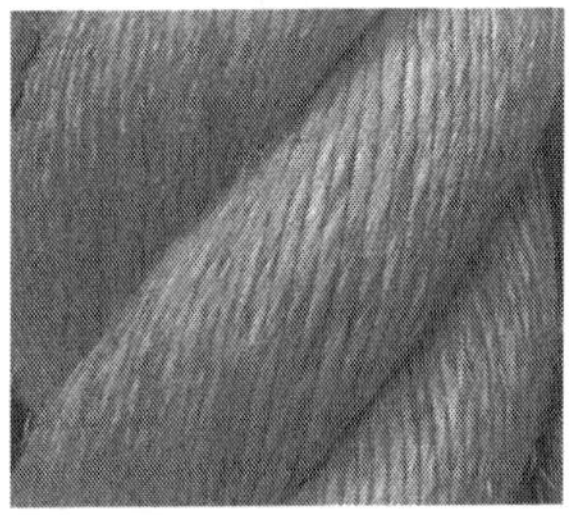

Morphology of a fibre arrangement using the example of a rope.

Morris Rug: Closely spun modern English rugs, dyed with vegetable dyes and having simple floral, usually acanthus designs. Named after William -Morris, its originator.

Morrison: A Highland tartan composed of green, black and blue stripes and white and yellow stripes over a red ground.

Mortling: Name in England for wool taken from dead sheep.

Mosaic Canvas: Very fine embroidery canvas, made of silk or cotton.

Mosaic Lace: Modern Venetian bobbin lace, the patterns being composed of many small sprigs and medallions applied to a net ground.

Mosaic Rug: A cut pile rug, made in England, the pile of which is glued in colored pattern to a canvas foundation, instead of spun to it.

Mosambique: (1) Woollen dress goods with the nap raised in squares, dots or other 'figures; (2) A light, sheer French fabric, made with yarn dyed cotton warp and mohair filling; comes in stripes and checks.

Moscovite: A dress silk, spun with organzine warp and cotton filling, forming ribs; comes mostly in light colors.

Moscow: Heavy, shaggy woollen overcoating.

Moscow canvas: Made in fancy patterns with gold, silver, blue and black threads interwoven, resembling plaited straw. Used for embroidery.

Mosey: An under-jacket, usually made of baize or flannel, and worn instead of an undershirt, usually under the vest, but over the shirt in cold weather, but in moderate or warm weather as an outside garment, and is also called a "wammus."

Mosquito bar: Similar to mosquito netting, having several warp and weft threads placed closer to each other at regular intervals.

Mosquito netting: An open face, very light cotton fabric, nowadays more with synthetic fibres, spun in gauze weave, dyed in solid colours.

Moss: Yarn Coarse woollen yarn of fuzzy or nubbed surface, used for embroideries.

Moss crêpe: (mimosa), small patterned garment crêpe with mossy character made from viscose delustred in spinning.

Mosses: Large hanks of reeled silk, weighing about one pound each, produced by the natives of China in the home industries.

Mossing: In England same as napping.

Mosul embroidery: The Oriental patterns are filled closely with herringbone stitch and are heavily outlined.

Mosul rugs: Made in (Mesopotamia, usually all wool, but warp and weft are, sometimes, of cotton and the soft, silky pile of goat's or camel's hair, tied in Ghiordes knot. Yellow and brown colors are often used. The design consists of various geometrical patterns and several border stripes. The ends are finished with a narrow web or fringe.

Mota: Thick cotton cloth made in India.

Motchenetz: Trade term for Russian, water retted flax.

Mote: A small piece of seed or vegetable matter in cotton. Motes are removed by boiling the fibre or fabric in sodium hydroxide, then bleaching. When not removed, they can leave a dark spot in the fabric.

Mote (2), in raw cotton: Whole developed seed or seed parts of any size or age covered with Fuzz and short fibres and in rare cases, some mature lint fibres.

Mote knife: Doctor blade.

Mote trash: See **Trash.**

Motes: Appearance of fault in cotton in the form of fibre knots felted into each other as growth motes (skin residues, unripe and dead cotton, moisture effects, etc.) or processing motes (torn and spun together, chiefly ripe fibres, which are produced in the ginning and on the carding machine). This also includes the so called small husk motes with fragments of outer husk and with them frequently complete bundles of Cotton fibre tufts.

Mother-of-all: The whole stand that supports the maidens, bobbin, and flyer.

Mothproofing: Mothproofing is a chemical finishing of fabric to prevent the growth of , fur moth (and/or their larvae), the anthremus, attagenus beetle, etc., which live as Textile parasites on keratin-containing substances (wool amongst other protein fibres, fur, duvet feathers, etc.). Also called **Insect repellent finishes**. Mothproofing agents should also protect against carpet beetles amongst other harmful insects. Insecticides are more or less poisonous. The earlier used DDT, Dieldrin etc. is banned in almost all countries. These have been replaced by other chemicals which are not poisonous to human.

Motley: (1) Medieval English mixture worsted, 7 yards long and 45 inches wide: (2) Same as mixture.

Motif: Recurring element in a design.

Motif print: Self-coloured or two-colour, if necessary, also multicoloured print of individual figures, emblems or strokes with the aid of printing screens of different sizes, usually in accordance with the pigment printing process. Predominantly made from tricot on semi-finished or finished products.

Motril: Variety of raw cotton from Spain. It has a white to reddish yellow, lustrous, strong fibre.

Mottle Yarn: Yarn which is mottled.

Mottled: Mixed (bright/dark), speckled, checked. See **Colour blend.**

Moufflin: A double faced thick coating fabric which is soft and has an open, airy surface. The yarns used are mainly wool, or wool mixed with acrylic or mohair. Used for coats and capes.

Moulding: Seamless shaping of textiles made from thermoplastic man-made fibres.

Moule: Soft, thick but light woollen overcoating, made in France.

Moulinage: French for silk reeling.

Mountain flax: Another name for asbestos.

Mountmellick embroidery: Raised Irish embroidery, executed on a heavy, firm fabric with a number of heavy, fancy stitches. The design is in natural or conventionalized flowers, leaves and also scrolls.

Mourning crepe: Is a dull black fabric which may or may not be given a moiré effect.

Mousseline: The French word literally means muslin, but we used it to describe a better quality wool or silk fabric of this nature. General name for a broad range of soft, sheer, light weight, plain spun fabrics made from cotton, silk or finer than muslin.

Mousseline de Laine: Plain spun, soft, light and open fabric, made of fine gassed worsted yarn, often mixed with cotton and somewhat heavier than Mousseline.

Mousseline De Soie: A plain and open weave fine and light fabric, made of silk. It is silk muslin. Sheer, open, and lightweight. It is something like chiffon but with a crisp finish produced by sizing. It does not wear well and it does not launder. *Uses:* Evening wear, and bridal wear. Trimmings. Also used in millinery as a backing.

Mouth width, in Zipper: The measurement between the slider flanges at the point where they bear against the shoulders of the interlocked elements or at the outermost edges of the bead if the bead extends beyond the elements.

Mouth, in Zipper: The opening in a slider that receives the chain.

Movable Retainer, In Zipper: A movable or sliding device performing a similar function to that of the fixed retainer, the pur[pose being to permit seperation of the two stringers from the bottom, without necessity if opening the zipper from the top.

Moving bed biological reactor: M. B. bioreactor, m. b. biofilm reactor, MBR, MBBR, suspended immobilised (immobilized) biomass reactor, SIBR Plastic media is used in an *activated sludge* aeration tank so that biomass can grow on and in the solid media together with the normal suspended bacterial culture in the tank. The plastic media is in suspension in the aeration tank and mixing is normally by diffused air but mechanical mixing may be used. The aeration tank is followed by a sedimentation tank and settled activated sludge from the sedimentation tank may or may not be returned to the start of the aeration tank, depending on the design. The media is often small hollow polyethylene cylinders but other designs have been used including 12 mm cubes of polyurethane pads (LINPOR process). The addition of the media can relieve an overloaded activated sludge plant or allow an existing plant to have extra treatment capacity. When this system is utilised as a *roughing filter* loadings of 4 kg BOD/d per m3 of aeration tank have been achieved, although loadings of 1 kg BOD/d per m3 of aeration tank are more typical for full treatment.

Moving bed scrubber: Floating b. s. A *wet scrubber* containing spheres of plastics, glass, marble, etc., which are fluidised by dirty gas rising up through them. The wash liquor is sprayed down over the spheres. The pressure drop is directly proportional to the flow rate of the liquor. Particles of 0.1um size can be collected, as well as viscous liquids. If the spheres are of hollow plastics, they may need to be held down by an upper constraining plate. There may be more than one bed. These scrubbers are efficient in the absorption of sulphur dioxide when using a caustic scrubbing liquor.

Mouth width: **of zippers**: The measurement between the slider flanges at the point they bear against shoulders of the interlocked elements or at the outermost edges of the bead if the bead extends beyond the elements.

MPI (Multi Product Injection): A dispensing system in which you can programme for dye and chemical addition for dyeing and finishing processes (Thies)

MR Process: (Menschner, Rotta), for wool fabric fixation. An enhancement in the quality of wool fabrics is possible in the MR process through the combination of physical treatment and chemical finishing.

Ms: Mulberry silk.

MSDS: Material Safety Data Sheet. This is a reasonably detailed specification for the possible safety hazards associated with a product. They are prepared by the manufacturer of the product. Unfortunately, they are often not available for products that are regarded as "consumer" products, even if the product poses serious safety hazards. For an excellent Internet resource for MSDS and other safety information, try www.siri.org/msds or www.hazard.com/msds

MSI: Magyar Szabványügyi Intézet, Hungarian Standards Institute.

MSZH: Hungarian Standards Association.

MTF: Metal fibres.

Mucuna: Strong leaf fibre, yielded by the Mucuna urens in Brazil; used for ropes.

Muddai: Very strong, silky fibre, yielded by a species of asclepias in India; used for ropes. It is mixed with cotton when spun; has good affinity for dyes.

Muff: A loose skein of textured yarn prepared for dyeing or bulking. In the bulking operation, the yarn contracts and the resulting skein resembles a muff.

Muff dyeing: See **Dyeing.**

Muga: (1) Species of brown wild silk, yielded by the Antheraea in Assam; (2) Stout, coarse silk fabric, made in India.

Muga cloth: Also called Assam silk. It is obtained from a species of Indian moth, and is wild silk producing a rough surfaced lightweight fabric.

Muga silk: This golden yellow silk is the prerogative of India and is the pride of Assam state. It is obtained from semi-domesticated multivoltine silkworm Antheraea assamensis. These silkworms feed on the aromatic leaves of Som and Soalu plants and are reared on trees similar to that of the tasar. Muga silk is a high-value product and is used in the manufacture of products such as sarees, mekhalas and chaddars.

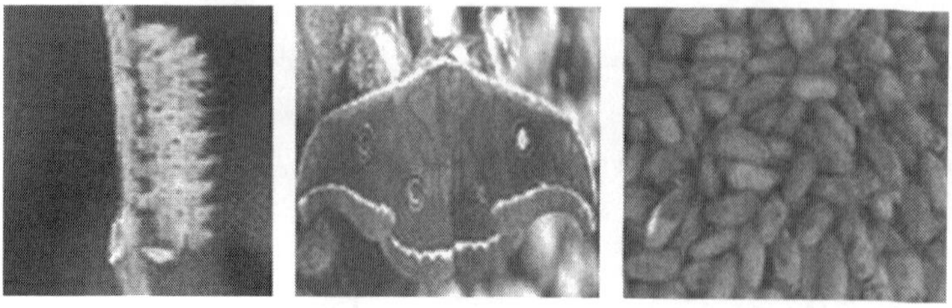

Muka: Native name for the fibre yielded by the New Zealand flax.

Mulberry: A plant whose leaves are eaten by silkworm.

Mule spinning frame: A spinning machine invented by Samuel Crompton in 1782 and termed "mule" because it was a combination of the machines invented by Arkwright and Hargreaves. It was once widely used for spinning wool and to a lesser extent for very fine counts of cotton yarn. Its action was intermittent and slower than that of the more current ring spinning frame. It drew out and twisted a length of yarn and then wound it in the form of a cop, or bobbin, then repeated the cycle. See **Spinning frame.**

Mule yarn: Single, more or less (loose) twisted cotton weft yarn for soft materials, as a filling weft and similar.

Mull: A type of lightweight cotton voile, which was once a dress fabric but is now almost entirely confined to use as an underlining fabric. It is also used for eastern turbans. Lawn, voile and cabric give better service in wear as dresses and shirts, but not mull.

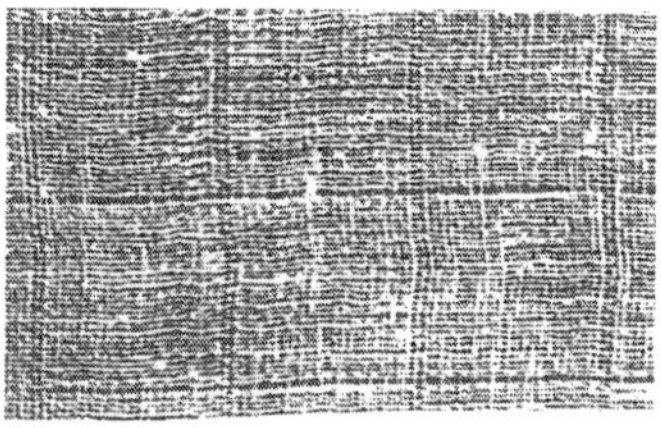

Mull Muslin: A fine, undressed white muslin; used for dresses.

Mulmul: Closely spun East Indian cotton muslin. See **Mull.**

Mullen bursting strength: An instrumental test method that measures the ability of a fabric to resist rupture by pressure exerted by an inflated diaphragm.

Multan: Knotted rugs made in India. They have geometrical designs in deep blues and reds.

Multiaxial fabrics: The structure results from several layers of monoaxial fabrics (fleeces, fabrics; see Fig.) arranged shifted together. The resulting physical properties such as strength and extensibility are clearly above those of monoaxial arrangements.

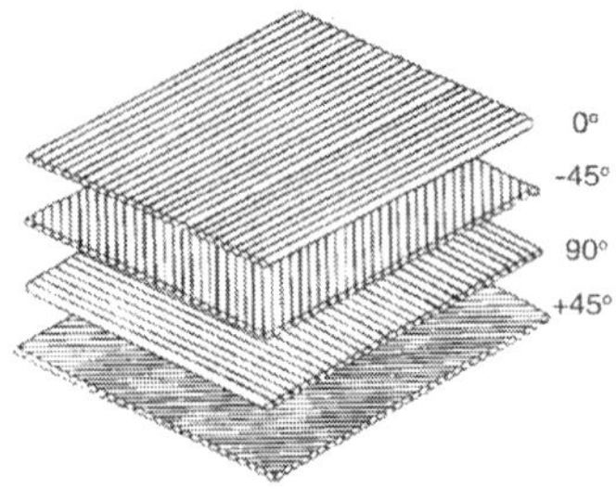

Multiaxial composite material.

Multibath Process: American description for multi-purpose baths, which facilitate various finishing stages in one bath.

Multicellular: Multicell, e.g. viscose fibres with a thin cell wall structure and triangular diameter.

Multicoloured effect: Achievement of heterochromatic marks using discharges, printing, fixation differences, resists, weaving and similar on textile fabrics.

Multicolour effects: Multicolour effects on textile fabrics; e.g. Colour craft process; Differential dyeing or also spray dyeing effects, e.g. on fabrics in special machines

Multicomponent fibre: Bicomponent fibre made from various synthetic fibre components of the matrixfibril type.

Multiend reeling machine

Multi-phase weaving machines: In these looms, during the weaving operation several sheds are opened simultaneously and weft is inserted in them by several carriers. Several phases of the weaving cycle take place at any instant so that several picks are in serted simultaneously.

Multifibre strip: A strip if fabric made using different fibres and is used in washfastness testing to asses the staining rating of a dyeing on different fibres.

Multifilament: A yarn consisting of many continuous filaments or strands, as opposed to monofilament which is one strand. Most textile filament yarns are multifilament.

Multifilament yarn: See **Multifilament.**

Multiflora: Commercial variety of early ripening upland cotton from Alabama, the medium long staple forming clusters of bolls.

Multi-liner, in ETP: A lining system at the base and sides of a ***landfill*** site in which more than two liners are installed. This may be appropriate for landfill sites containing very hazardous waste material. The liners may be three or more separate liners with drainage layers above each liner. Alternatively, the first two layers may be a ***composite liner*** and then a third liner is placed below with a drainage layer between the composite liner and the third liner. Drainage layers are also installed above the composite liner to collect the leachate and below the third layer.

Multi-media filter, multi-layer filter., mixed media filter: A downflow *rapid gravity filter* that uses three or more media, often anthracite at the top (coarsest and least dense), sand in the middle and garnet (finest and most dense) at the bottom. Polystyrene may be added as an extra layer on top and magnetite as an even finer, denser layer at the bottom. Other materials have been used and they should be chosen to suit the water or effluent to be filtered. The grain size varies from 1 to 2 mm at the top to less than 0.5 mm at the bottom to give *coarse to fine filtration*. The filters are backwashed, but because of the differing densities of the various media, the layers remain separated after *backwashing*. Water works may adapt their *rapid gravity sand filters* to become multi-media filters in order to increase the throughput of water.

Multilobal cross section: See **Cross section.**

Multilevel pile: For pile yarn floor covering: Pile in which some tuft legs are substantially longer than others.

Multiple gripper weaving machine: This projectile weaving machine is a versatile unconventional loom a gripper projectile transports a single weft yarn into a shed. It has the advantage of a combination of wide width and high speeds.

Multiple length staple fibres: Manmade staple fibres that are two or more times the nominal cut fibre length.

Multiple stitch zigzag, in sewing: A simple machine stitch pattern of alternating diagonal segments with each segment of two or more stitches having equal length and width.

Multipolymer fibre: See **Copolymer Fibres.**

Multipore fibre: See **Imitation Suede.**

Multi Product Injection: See **MPI.**

Mummy canvas: Similar to Mummy Cloth, but with a coarse mesh and an irregular, rather than creepy surface. Used for embroidery.

Mummy cloth: Also called Grannie cloth and Momie cloth: A fabric with a crinkled surface like crepe, but made with non-crepe yarns and spun on a dobby loom. It can bleached, dyed or printed. The fibres include silk warp and woollen filling or cotton with silk. The fabric lacks lustre, and has been a traditional mourning fabric, but is not used much now. The name is also used for fine linen fabric used in Ancient Egypt, for wrapping Mummies.

Munj: Strong and elastic fibre which stands water well; is yielded by a species of the sugar cane in India; used for ropes, mats, and baskets.

Mungo: (1) Also Called Shoddy. A cheap, poor quality woollen cloth made from mill wastes. Because the fibre staple is too short it does not wear well. Colours are often poor and drab. It is normally not used for clothes but may be used for some types of stage costumes.

(2) Waste yarn from woollen mills which is mixed with other yarn, such as cotton to produce cheap cloth for special use such as backing fabric.

Munroe: A Highland tartan, composed as follows: Wide red stripe, split near each edge by a tine blue and yellow line, the two lines placed next to each other, the blue being on the outside; *green stripe, half the width of the red; red stripe, half the width of the green, split in the center by a narrow blue and yellow line, placed next to each other; dark blue stripe, half the width of the green; *red field, one and one half times wider than the first red stripe mentioned, split near to each edge by a narrow yellow and blue line (placed next to each other with the yellow line on the outside) and split in the center by three green stripes, spaced from each other by their own width; repeat, in reversed order, groups mentioned between the two (*).

Munsell Colour Charts: A three-dimensional colour system developed by Albert Munsell that is based on the attributes Munsell Hue, Munsell Value, and Munsell Chroma.

Munsell colour system: The colour identification of a specimen by its Munsell hue, value and chroma as visually estimated by comparison with the Munsell Book of Colour.

Murata® Spinning: See **Air jet spinning.**

Murga: Native name of the Indian bow string hemp.

Muriatic acid: An antiquated but much-used term for hydrochloric acid Most muriatic acid sold in hardware stores is as strong as commercial hydrochloric acid sold for laboratory purposes. Sometimes it is somewhat diluted, but still is dangerous and must be handled with great care.

Murray of Athole: A Highland tartan, composed as follows: Green stripe, split by a red line in the center; black stripe, half the width of the green; dark blue stripe, a little wider than green one, split in the center by a red stripe, outlined with a fine black line; black stripe, as above; green stripe, split by red, as above; black stripe, as above; dark blue stripe, twice as wide as the black, split near each edge by a pair of narrow black stripes, spaced from the edge and from each other their own width.

Murray of Tullbardine: A Highland tartan, composed as follows: Red stripe; group (twice as wide as red stripe), consisting of fine blue line, fine red line, black stripe, fine red line, fine blue line, red stripe, blue stripe, red stripe, fine blue line, fine red line, blue stripe (narrow), fine red line, fine blue line, red stripe, blue stripe, red stripe, fine blue line, fine red line, folack stripe, fine red line, fine blue line; red stripe, as the first one; *dark blue stripe, little less than half the width of the red; red stripe, as wide as the blue, split by a fine green line near the edge next to the blue; green stripe, half the width of the first wide red*; red stripe, almost twice as wide as the green, split by two blue and a narrower black stripe; repeat, in reversed order, stripes mentioned between the two (*).

Mururuni: Leaf fibre, yielded by a palm in Brazil; used for hats, baskets, etc.

Mushroo: An Bast Indian cotton back silk satin, figured with white or gold flowers, some having wavy stripes of yellow and gold.

Mushroom apparel flammability tester: See **Flammability tests, Mushroom apparel flammability tests.**

Mushroom test: A test developed by American National Bureau of Standards Using Mushroom Apparel Flammability tester

Mushy wool: Dry, fuzzy wool, yielding large percentage of noil in combing.

Muslin: Named after the palce Mosul in mesopotomia. Soft spun cotton, very open weave, with a rough finish. Ideal to use as a pressing cloth. Muslin as a dress fabric is rarely used now mainly due to the heavy shrinkage of this fabric. A characteristic construction is 64 square with 30s warp and 40s filling common widths are 30, 36, and 40 in.

Musimet: Coarse cotton muslin, usually sized.

Muslinette: In England a thick variety of muslin; used for dresses.

Muslin, as applied to bedsheeting: A plain weave fabric with not fewer than 180 yarns/in^2.

Muslin (Silk): Slightly transparent flowing fabric, made of warp of crepe threads with an S or Z twist and weft of crepe threads with 1S/1Z twist.

Musquash: The coat of Musk-rat. It is a long downy, grey fur with darker brown, but is often dyed to various colours including pale honey.

Muss resistance (Wrinkle Resistance): The fabric must not wrinkle when a garment is worn. Wrinkling performance can be expressed in crease recovery terms. Crease recovery angle are used to measure crease resistance or crease recovery.

Mussel silk: (byssus silk, sea silk, etc.), natural silk-like hair of different large ham mussels (Mediterranean, Normandy) belonging to the Natural protein fibres, which are held tightly on the rocky ground with the fibre tuft produced and are torn off there. Reconditioning by kneading by hand using soap, carding, drying, washing using citric acid and hot platelets. Outer smooth, gold shimmering gloss, colour brown to olive green. 2–3 threads of mussel silk are twisted with a barely 200 kg/year. Application: (Italy) for shawls, stockings, gloves, bags, etc. thread of natural silk. Precious, rare material. Production

Mussiness: Surface distortion in a fabric characterized by undesirable unevenness due to many minor deformations.

Mutagenic agents: (Lat.), substances hazardous to health, which damage hereditary factors and trigger mutations, i.e. can cause changes in hereditary makeup.

Mutagenic agents also partly act as Carcinogens.

MUT: (environmental-friendly textile label). Label for textiles kind to the environment, awarded by the Verein für verbraucher- und umweltfreundliche Textilien

MX Dye: A family of "cold" reactive dyes, first developed by Imperial Chemical Industries of Britain, and designated "Procion MX"; chemically MX dyes belong to the dichlorotriazene family A number of companies now manufacture MX dyes. They are by far the most popular dyes for textile artists working with cellulose fibres. They can also be used for wool and nylon in processes where they behave as acid dyes. There are around a dozen "pure" MX dyes in common use, and all can be used quite successful in mixtures with each other. There are magenta, yellow and cyan shades that are excellent "subtractive primary" colors. MX dyes are less popular with industrial dyers than some other reactive dyes, partly because their very high reactivity makes the dye process harder to control, partly because they are fairly expensive, and partly because goods require extensive washing after dyeing to remove hydrolyzed and unfixed dye. Care should be taken to avoid breathing MX dye dust, since it is known to cause respiratory allergies. The use of "Procion" alone to denote MX dyes is incorrect since there are other Procion dye families.

MWS: (metres head of water). Metre water column. Unit for Pressure (-heights, liquid columns) no longer is use. 10 MWS = 9806.65 Pa.

Myrobalans: Tannin-containing, small rock hard fruits (Bengal, India) and/ or fruit parts or their extract. Good myrobalans have a wrinkled exterior and contain 38–40% tannin; worse are smaller, unwrinkled with 25% tannin content. Stoned myrobalans have approx. 50–53% tannin content, 24–34% liquid and 65% solid extracts. Good myrobalans correspond approximately to 1/3 tannin. Used as tannin for tanning.

Mysore Silk: Soft, fine, undressed south Indian silk dress goods, made plain, dyed or printed, mostly in floral patterns.

N

Nabo: Native name of the strong fibre, yielded by the Nauolea in the Philippines; used for cords, ropes.

Naboika: Russian homespun linen, printed with fast colours by means of wooden blocks; used for clothes, religious vestments, covers, etc.

Nacarat: (1) Orange red coloured fine linen in Latin-America; (2) In Portugal a fine crepe or muslin, dyed in flesh colour, which is used by women as rouge.

Nacre': (1) (Fr.), Nacré print, flowing print; like mother-of-pearl, iridescent, changeable effect; (2) Silk fabrics woven in colours producing effects similar to the mother-of- pearl.

Nacre' velvet: A velvet in in which the backing is woven in a different colour from the pile, giving an interesting, changeable appearance in wear. See **Velvet.**

Nae: Hawaiian name for a netting, having a very fine and close mesh; used for garments.

Nail: 2 ¼ inches.

Nainsook: A soft, fine, light cotton fabric made from combed yarns, in plain weave, often mercerized, similar to Batiste in the piece. The construction and the yarns used in nainsook is the same as that of lawn. A lawn fabric which is not given the starch finish to make it crisp is actually nainsook or English nainsook. In case of French nainsook it may be slightly starched and calendared. Once a popular dress fabric but now mainly used for lingerie and baby clothes or for summer dresses and curtains.

Nak: Medieval name for cloths of gold.

NALCC: A leading American specialist textile cleaning organization.

NALI: English textile cleaning specialist organization.

Namad: Felted carpets of Persia and India.

Namazlik: Turkish name for prayer Rugs.

Named: Persian felt rug with patterns of coloured wool pressed into it. See **Namad**.

Nambali: Silk fabric with religious names printed on; used for garments in India.

Namdas: (1) Felted woollen cloth, made in Tibet. Is often embroidered and used for rugs and carpets.

(2) Felt rugs, made in Kashmir, India. Often embroidered by men with chain stich crewel work embroidery.

Namitka: A fine sheer veil, woven in southern Russia from homespun silk by the peasants and used for head ornament.

Nanako: Plain woven Japanese silk fabric.

Nancy: Embroidery French needlework, combining embroidery in coloured silk with drawn work.

Nanduty: Very fine lace made of cotton or pita fibre in South America. It is made in small squares joined together.

Nankeen: (1) A peculiar fabric of a pale dull yellow or orange colour, woven out of the fibrous tissue which lies between the outer and sap-wood of a tree or shrub that grows in the East Indies and especially in China. The name is derived from the city of Nankin. An imitation is made out of cotton, coloured with Annato. The genuine nankeen is never more than eighteen or twenty inches wide and is used for light summer clothing. (2) Cotton cloth in Roumania, having a white warp and pink, red or yellow coloured filling; it is finished with a size. (3) English and French all-cotton, very stout, plain, woven fabric, dyed in the yarn and made in solid colours, stripes, with equal number of threads in the warp and weft in a square inch; used for clothing.

Nankin ticking: See **Ticking.**

Nankinet: (1) Similar to nankin but not woven as close; (2) Fine, fancy coloured percales.

Nano-: Symbol: n A prefix used with SI units, denoting 10^{-9}. For example, 1 nanometer (nm) = 10^{-9} meter (m).

Nanofiltration, in ETP: Filtration of treated water through membrane assembly sieve having pore size lower than 2nm, applying hydrostatic pressure difference to filter small molecules down to a molecular weight of 200, some hardness and viruses. Typical operating range of pore size will be 0.001-0.01µm. The operating pressure is 10–40 bar which is less than ***reverse osmosis*** but higher the ***ultrafiltration***. Nanofiltration can filter out micro-organisms, organic and colour molecules and some larger salt molecules from water.

Nanotechnology: The technology of devices at the nanometer scale. In such small devices the quantum mechanical nature of electrons has to be taken into account fully. Instruments such as the ATOMIC FORCE MICROSCOPE which can identify and manipulate individual atoms are very useful in nanotechnology.

Nanometer: (nm) – Unit of length equal to 10-9 meter (a.k.a. one billionth of a meter, or a milli-micron).

Nansu: Nainsook in Venezuela.

Nap, in raw cotton: A large mass of curled or matted unorganized fibre or fibres that is found after opening and picking, but which disappears during carding and subsequent processing.

Nap: The downy substance, covering either side, entirely or partly, of a woollen or cotton fabric. It is formed by the loose fibres of the warp or weft threads, and is produced by scratching the cloth and thus raising the nap. In this respect it is distinctly different from the pile (see) which is always formed by a cut yarn separate from the ground of the cloth.

Nap cloth: The term used to describe thick, soft double fabrics (900–1000 g/m^2) made from carded yarns usually of high quality merino wool. The typical appearance is achieved by a separately-inserted 3rd weft thread, the nap weft, which is a long float thread, later napped (thus also called the "raised thread") giving rise to fluffy tufts of hair on the surface of the fabric. Depending on weave, comes in ribbed, diagonal or herringbone effect; wool dyed or piece-dyed. Used for high-quality men's coats.

Nap finish machines: Mainly Brushing machines for the production of brushed or napped finishes on grey goods. In the widest sense they also include Raising machines.

Nap velours: A napped fabric, usually wool, in which the raised fibres all lie in a parallel direction. Opposite Upright pile velvet.

Nape: The starting point from which the length and the shoulder measures are taken for a coat. A vertebral protuberance located on the back central part of the neck at its junction with the thorax or trunk.

Napery: Table and household linen.

Naphthalene $C_{10}H_8$: Consisting of two contiguous benzene rings. A product of crude oil (rare) and coal tar. Small white laminations with a distinctive smell. Insoluble in water, soluble in organic solvents.

Naphthalene process: A pad steam process for the continuous dyeing of loose wool, with the addition of emulsified naphthalene.

Naphthol dye: See **Naphthols.**

Naphthol dyeing: Naphthol dyeing is carried out on the fibre by combining (coupling) two soluble components, a naphtholate and a diazotized base. The base is either diazotized in the dye shop itself, a complicated process, or obtained from the dye manufacturer as a stabilized diazotized fast colour base, the dyeing salt. There is a large number of combinations, capable of achieving bright yellow, orange, scarlet and red shades which are not available in the vat-dyeing ranges, and more muted burgundy, brown, navy blue and black shades. Naphthols are characterized by good fastness, in particular fastness to boiling. Some combinations are also weather-resistant. The dyeing technique for naphthol dyeing requires promoting the main reaction to form an azo dye, and avoiding the secondary reaction of hydrolysis to a compound no longer suitable for coupling.

Naphthol dyeing test in UV light: Place a test dye sample in a 50 ml beaker, and fill with sufficient blank vat to cover the sample (36 g/l caustic soda, 35.5% and 24 g/l sodium hydrosulphite); boil until the original dye appears a dirty yellow; remove the sample with a glass rod, squeeze the sample out on the glass side of the beaker, place on a round filter paper (preferably on a glass base) and leave to dry. Examine the test samples under an analytic quartz lamp: strong luminescence indicates naphthol dyeing.

Naphthols; Naphthol dye: (1) Name of hydroxy-derivatives of naphthalene with one or more OH groups. They are used as intermediate products (α-naphthol, β-naphthol) in dye synthesis (Phenols; Cyclic hydrocarbons). (2) Developing dyes (obsolete ice colours), produced on the fibre, usually by means of a two-bath application, they are insoluble azo dyes from naphthols (primers) and fast colour bases or dyeing salts (developers), with around 2000 combinations possible.

Naphthols in direct printing: Naphthols can be directly printed by many mehods. (1) Low affinity naphtols can be padded and dried on a hot flue and printed with diazotized fast colour salt. The ground naphtols are removed in subsequent soaping along with an alkali binder. (2) Low coverage prints can be done by printing naphthols, dried and padded with diazotized fast colour salts. (3) Fabric is printed with naphtolate, antidiazotate, caustic soda and weak oxidizing agent and dried. The non-developed antidiazotate can be converted into a developable diazo compound by steaming and subsequent passage through a hot electrolytic bath containing formic acid, acetic acid or by steaming in acid ager, rinsing and boiling with soap. (4) Fabric is printed with a combination of naphtholate and diazoamino compound, wetting agent and caustic soda (suitably thickened) are printed, dried, with acid bath treatment and finally soaped. (5) Fabric can be printed with a combination of naphtholate and antidiazosulphonate. Also fixed using neutral steam.

Naphthols in discharge printing: Most Naphthol dyeings can be discharge printed because of the ease of splitting of the azo pigments with the help of reducing agents.

Naphthols in resist printing: Mainly of academic interest, naphthols can also be used in resist printing. (a) White resist can be printed with aluminium sulphate along with an organic acid like tartaric acid (suitably thickened). (b) Coloured resists can be printed especially with reactive dyes. The goods are printed, dried, briefly steamed, padded with slightly thickened variamin blue, ventilated, rinsed briefly, immersed in a boiling bath with sodium disulphite, rinsed, boiled with soap with sodium carbonate.

Naphthoquinone: Dyes of the anthraquinone naphthalene derivatives. Dyes can be made by condensation of 5-aminobenzo-α-phenazine-4-sulphonic acid with 2,3-dichloro-1,4-naphthoquinone.

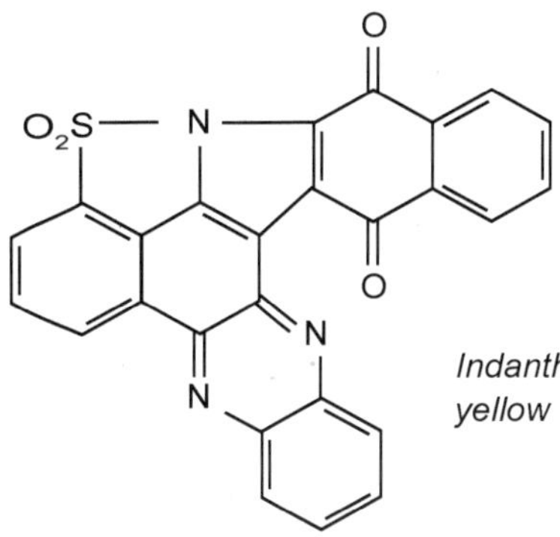

Indanthrene yellow 6 GD

Naphthoylene-benziminazole: Dyes of the anthraquinone naphthalene derivatives. Example is the dye made by heating a mixture of o-phenylenediamine and 1,4,5,8-naphthalene tetracarboxylic acid in glacial acetic acid or pyridine solution.

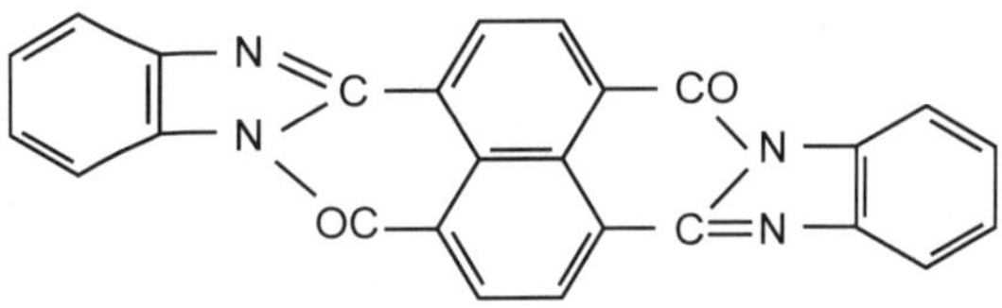

indanthrene brilliant orange GR

Napiei: (1) Double faced coating with wool face and vicuna or goats' hair back. (2) Good grade of floor matting, made of hemp and jute.

Napier Cloth: A once-popular double-face cloth for women's wraps, one side wool, the other of hair.

Napolitaine: A French flannel, originally made with all-woollen warp and filling, loosely woven, not fulled, and dyed in the piece, striped or printed; used for dresses, scarves, etc.

Nappa leather: This is the skin of sheeps or goats that has been tanned by a Californian Process using oil.

Napping: A finishing process that raises the surface fibres of a fabric by means of passage over rapidly revolving cylinders covered with metal points or teasel burrs. Outing, flannel, and wool broadcloth derive their downy appearance from this finishing process. Napping is also used for certain knit goods, blankets, and other fabrics with a raised surface.

Narrow band pass radiometer: A relative term applied to radiometers that have a bandpass width of 20 nm or less at 50% of maximum transmittance and can be used to measure irradiance at wavelengths such as 340 or 420, ± 0.5 nm.

Narrow elastic fabric: An elastic fabric that is less than 150mm or 6 in. in width.

Narrow fabric: A textile fabric not exceeding 300 mm. in width. This includes tapes, ribbons, and webbings. They can be produced from any textile fibres, including elastomers by weaving, braiding, knitting or byany other methods. They can also be made by cutting or slitting wider fabrics into narrow strips.

Narrow fabric dyeing: Narrow fabric lengths are wound into hanks and dyed in a circulating-liquor machine like yarns on sticks or wound directly onto bobbins. Wider narrow goods are dyed by the continuous process.

Narrow Fabric: A textile fabric not exceeding 300 mm. (12 in.) in width

Narrow fabric finishing: Narrow fabrics are woven in working widths of up to 300 warp threads, but usually as narrow textiles substantially less in width, on special looms, and it is in this form that they must be heat-set or crease resistant finished (possibly following pre-treatment and/or dyeing). This is carried out on cylindrical dryers, around which the narrow fabric passes several times.

Narrow-Fall: A small flap closing the opening in the front of trousers, with two or three button holes at the top and corresponding buttons sewed on the waist-bands, to which it was buttoned.

See **Front-Falls**.

Narrow Wale: Narrow diagonal ribs, round or flat, on some woollens and worsteds.

Narrowing, In knitting: The reduction of the number of stitches for the purpose of shaping.

Narumi-shibori: Japanese cotton and silk crepes, dyed as follows: After taken from the loom the fabric has many small knots tied into it by means of wax thread and placed in the dye. After taken out of the dye, the wax thread is removed, leaving behind spots untouched by the dye. These spots form small conical prominences, as the dye also slightly shrinks and crinkles the fabric.

Nate: A French mercerized cotton cloth.

Natte: French term for a basket weave silk fabric made with different coloured warp and filling.

Natural (Lat.): Original, e.g. cotton can be called a natural cellulose fibre in contrast to regenerated cellulose fibres (viscose, acetate). Natural protein fibres are wool and silk, in contrast to man-made protein fibres such as casein fibres.

Natural: Refers to undyed or unbleached cotton or wool.

Natural cellulosic fibres: belong to the group of natural fibre materials of vegetable origin:

(a) Vegetable hair fibres: Cotton; Kapok.

(b) Bast fibres: Flax; Hemp; Jute; Kenaf; Kendyr fibre; Nettle fibres; Ramie; Rosella; Sunn fibre; Typha fibres; Urena fibres.

(c) Hard fibres: Agave fibres; Alfa grass; Bromelia fibres; Coir fibres; Henequen; Manila fibre; Mauritius hemp; Sea jute; New Zealand flax; Palm fibres; Panama; Pine apple fibre; Raffia bast; Sanseviera fibre; Sisal; Yucca fibres.

(d) Others: Paper yarns.

Natural colours: The colour of undyed textile goods prior to dyeing, which needs to be taken into account when formulating the recipe. Sample dyeing is essential.

Natural dyes: Are obtained directly from the various plants and animals, as for instance indigo, cochineal, etc.

Natural dyes: There is a vast number of plants whose extracts can be used for the dyeing of wool, silk cotton and linen; there are also a few examples of dyes of animal origin. Different parts are used depending on the plant, whether the whole plant, the bark, the heartwood, the leaves, the roots or the fruit. For various reasons, e.g. due to the dyeing behaviour, the fastness orthe biological availability, only some of these have gained widespread acceptance as dyes. E.g., (a) Cis-bixin (annato) from the seed pods of the fruit of the rocou tree, for food, oils, margarine, cheese rinds etc. (b) Carminic acid (cochineal) is obtained from the female cochineal insect (coccus cacti) and is used as a wool, silk, food and cosmetics dye. (c) Carotene is obtained mainly from carrots, red palm oil and pumpkin seeds, and is used primarily for food and pharmaceutical purposes. (d) Carthamine is obtained at quantities of 0.2–0.3% from flowers of the safflower (carthamus tinctorius). (e) Curcumin occurs in the root tubers of the curcuma genus, native to east Asia. (f) Euxanthinic acid is obtained from the urine of cattle which have been fed on the leaves of the mango tree. (g) Fustine and fisetine, obtained from young fustic (tanner's sumac). 1 kg wood dust yields approx. 40 g dye. (h). Haematin from logwood (campeachy), the heartwood of haematozylum campechianum. It is obtained from 10–12 year-old trees. (i) Hypericin is found in the ratio 1.2–1.5 g/kg in the dried, ground flowers of St John's wort. (j) Indigo is obtained from East Asian Indigofera genus and European woad (isatis tinctoria) and dyes wool and cotton (principle dye used for jeans). (k) Juglan is found in the ratio approx. 2 g/kg in fresh walnut shells. (l) Kermesic acid is a derivative of anthraquinone, which is obtained from the kermes insect (kermes ilicis or kermes vermilio). (m) Madder dyes are hydroxyanthrachinones, obtained from the root rind of various plants of the Rubia genus, e.g. madder (rubia tinctorum). (n) Laccaic acid, consisting of at least 4 components, occurs at 0.5–0.75% in sticklac, the raw material of shellac, which is obtained from the Coccus laccae insect. (o) Lawson is obtained from henna and is suitable for the dyeing of wool and silk, and as a hair colour. (p) The example of archil, which is obtained from a slow-growing lichen native to southern lands, shows how excessive use and over-exploitation in the Middle Ages has led to an enduring shortage of the dye plant. (q) Phoenician purple (6,6',-dibromic indigo) is obtained from the shell Murex brandaris, or other types, and is used to dye wool and silk. (r) Rutin is contained in the leaves, flowers and fruits of numerous plants, e.g. types of buckwheat and Chinese yellow berries. (s) Saffron is obtained from the stigmas of crocus plants and is used to dye silk as well as a food colouring. Very large quantities of these sources has to be extracted to get even I g. of the dyes which makes it very impractical for industrial uses.

Natural fibre: Textile fibrous material of natural polymers of plant, animal and mineral origin. Fibres; Mineral fibres; Natural cellulosic fibres; Natural protein fibres.

Natural gas: Gas obtained from underground deposits and often associated with sources of petroleum. It contains a high proportion of methane (about 85%) and other volatile hydrocarbons (ethane, propane, and butane).

Natural protein fibres: Fibrous materials of animal origin belonging to the overall category of natural fibres. They are subdivided into the following groups: (a) Wools and hair: sheep's wool (Wool), camel wools (Alpaca; Llama hair; Vicuna; Guanaco), Camel hair; Hare and rabbit fibres; Angora rabbit hair; Goat hairs Mohair; Cashmere; Cashgora; Tibetan cashmere), other hair (Dog hair; Human hair; Cattle hairs; Horse hair, Hog's bristles). (b) Silks: Mulberry silk; Wild silks (Anaphe silk; Eri silk; Fagara silk; Muga silk; Tussah silk; Yamamai silk), other silks (Mussel silk; Spider silk).

Natural textiles: Natural textiles are defined as follows: – Natural textiles: garments made from pure, untreated natural fibres, or those which have been dyed but without chemical treatment. – Natural fibres: fibres of animal or plant origin (wool, silk, natural cellulose fibres, not mineral fibres or regenerated cellulose), preferably cultivated without the use of pesticides. – Untreated natural fibres: neither resin finished nor chemically treated (from antifelt to silk weighting). – Natural fibres without finish: unbleached, plant dyed, unadulterated fibre surfaces for skin contact, without the influence of chemicals, resin coating or toxic substances, on the properties of the natural fibre. Lubricants (paraffin lubricants) and sizes (modified potato starch) must be capable of washing out.

Naturell: In Germany and Austria a very light and soft finished, plain woven cotton fabric; used for underwear.

Naught duck: See **Duck.**

Navel: A component of an open-end spinning machine located on the axis of the rotor through which the yarn is withdrawn from the rotor and which modifies the twist of the yarn inside the rotor.

Navajo Blanket: Heavy, stiff wool blanket, woven with geometrical patterns always in straight or zigzag lines in bright colour combination on primitive, upright frame by the Navajo Indian women. It is very closely woven and waterproof. The first specimens were of native wool or unraveled bayeta, the yarn of which was often twisted harder. Later German town yarn and cotton warp was introduced. Most blankets are alike on both sides.

Navajo Ply: Basically a hand-crocheted loop used to create a thee-ply yarn.

Navy Twill: Heavy weight, navy blue wool twilled flannel; used for working shirts.

NBS: (1) National Bureau of Standards, Washington, an American textile research body; → Technical and professional organizations. (2) New British Standards.

NBS Unit: See **Colour difference formulae**.

NCH: Standard abbrev. for non-carbonate hardness, permanent hardness.

NE: Dutch Patent.

Neale process: The use of sodium hydrosulphite in combination with sodium boron hydride in vat dyeing.

Near-silk: Trade term for several mercerized cotton linings.

Neat: Combing wool taken from the sides of an average lustre fleece; used for yarns from 32s to 36s.

Neatening: Finishing a cut edge to prevent fraying. Various methods include: turning the edge, pinking overedging, or taping.

Neatness in raw silk: An expression of the degree of freedom of raw silk yarn from loops, nibs, and hairiness as defined.

Neck base girth, in body mesurements: The circumference of the neck over the cervicale at the back and at the top of the collar bone at the front.

Necking: (1) The sudden reduction in the diameter of an undrawn manufactured filament when it is stretched. (2) Narrowing in width of a fabric or film when it is stretched.

Neckline: A neckline is an outline or contour of the bodices around the neck or shoulders. It frames the face and directs the eye to or away from the face. A neckline must have an opening or should be large enough to pull through the head. Finishing can be given to the neckline with facing, piping binding and ribbing or with a collar for instance. We have various necklines as jewel, round, 'V' sweet heart etc.

Various Types of Necklines.

Basic/Plain/Jewel: A simple round neckline, which would be faced.

U-Neckline: A deep neckline in the shape of a 'U'.

Scoop: A shallow but wide neckline.

Square: A square neckline, which would be faced with mitred corners.

Horseshoe: A deep horseshoe-shaped neckline, which would be faced.

Built up / Funnel: A high neckline that extends up the neck.

Sweetheart: A feminine, angular, shaped neckline that frames the breasts and neck.

Décolleté: A deep and wide neckline popular in the nineteenth century.

Wide Square: As the name suggests a wide, square shaped neckline that would be faced.

V-Neckline: A faced V-shaped neckline.

Halter: The bodice of a garment is extended round the neck to form straps and a fastening. There is usually only a band of fabric across the back.

Bateau: A wide and shallow neckline that would be faced.

Slashed: A shallow but wide neckline, it is effectively a slit in the garment that is faced.

Sabrina: A wide but shallow neckline that has insets over the shoulder. These may be in a contrast or self-fabric.

One Shoulder: An asymmetrical neckline where the garment is supported by strapping on one shoulder only. The other is bared.

Keyhole: A **jewel neckline** with a slit at the front or the back which is faced and fastened, edge to edge, at the neck with a single button and **rouleau loop**.

Scalloped: A decoratively edged neckline that would be faced.

Inset: A jewel neckline with a slit that has an elasticated insertion.

Racing/ Athletic: The main focus is on the back where the neckline and armholes are cut away to create and 'X' effect. Used in sportswear and casual wear.

Envelope: A wide neckline where the back overlaps over the front and is usually made from a knitted fabric and trimmed with a fine rib.

Crew: A snug-fitting neckline trimmed with a knitted rib as in **tee shirts and sweatshirts.**

Wrap-over Bateau: Similar to the envelope neckline but with a narrower overlap not quite reaching to the sleeve. This example is finished with a knitted rib.

Strapless (Princess Line): A princess line bodice with a band and a notched front but no straps for support.

Off the Shoulder: A feminine neckline where the shoulders are bared revealing the whole neck.

High Cowl: A draped neckline, usually cut on the bias to create more soft folds of draping. This example has a high cowl draping from a funnel or built up type neckline.

Mitred: A square shaped neckline finished with a deep band with mitred corners.

Cowl Inset: See **high cowl.** There is a large inset of fabric, cut on the cross, to allow for soft drapes.

Elasticised: A full neckline that is controlled by elastication.

Drawstring: As the elasticised neckline but here controlled with an adjustable drawstring.

Wide Cowl: See the **high cowl** but here wider and much draped.

Graduated Ruffle: A V- shaped inset of fabric that is trimmed with a pleated and graduated ruffle. The graduation ends at the point of the 'V'.

Sugar Bag: A full neckline that is pulled together with a drawstring or band a few inches down from the edge, this creates a frill effect.

Ruffle Set in Seam: A curved neckline with an inset of fabric outlined with a pleated ruffle

Pleated Ruffle: A V-shaped neckline trimmed by a graduated and pleated frill.

Circular Ruffle: Rather like the Pierrot collar but not as deep.

Necktie: A decorative band of fabric worn around the neck and tied in a knot or a bow.

Needle: (1) A thin, metal device, usually with an eye at one end for inserting the thread, used in sewing to transport the thread. (2) The portion of a knitting machine used for intermeshing the loops. Several types of knitting needles are available. (Also see **Spring needle** and **Latch needle.)** (3) In nonwovens manufacture, a barbed metal device used for punching the web's own fibres vertically through the web.

Needle, in sewing: A thin shaft of drawn steel wire used for sewing pointed at one or both ends with an eye or hook for thread or yarn. The primary stitch-forming device used on all sewing machines to carry thread through a seam. Needles have nine parts including butt, shank, shoulder, blade, groove, scarf, eye, point, and tip. Needles come in a variety of types and sizes depending on the type of sewing machines and the sewing application.

Needle bar, tufting machine: A rigid bar, running the width of the tufting machine in which the tufting needles are mounted at fixed intervals.

Needle Bar: A flat metal plate with slots (tricks) cut into it at regular interval into which needles slide during the knitting process.

Needle bed: Flat metal plate with slots at regular intervals in which the knitting needles slide on the knitting machine.

Needle bonding: Fixing process for Tufted textiles; Needle punched nonwoven fabric, e.g. an alternate insertion and withdrawal of a number of needles in nonwoven fleeces, or similar in the case of napped needle felt fabrics and batt-on-base woven felts.

Needle cooler: A device for blowing a jet of cool air on to the needle while sewing.

Needle cord: A dress fabric, usually of cotton. With very fine cords along the length of the fabric and very short pile. It may be printed or plain, and is hard wearing and usually washable, with one way pile. It is a fine ribbed corduroy.

Needle damage, in sewn fabrics: The partial or complete yarn severance or fibre fusing caused by a needle passing through a fabric during sewing.

Needle feed: The feed mechanism in which the feeding of the material is accomplished or assisted by the needle which moves forward by one stitch while it is through the material.

Needle Felt: See **Punched felt.**

Needle felting, Needle bonding: Fixing process for Tufted textiles; Needle punched nonwoven fabric, e.g. an alternate insertion and withdrawal of a number of needles in nonwoven fleeces, or similar in the case of napped needle felt fabrics and batt-on-base woven felts.

Needle gauge: The distance between needles on a multineedle sewing machine. It is directly connected to the strength of the needle which must bear the stress and strain generated during the various technical cycles of the knitting process. The gauge of the needle is directly proportional to the gauge of the machine.

Needle loom: A machine for bonding a nonwoven web by mechanically orienting fibres through the web. The process is called needling, or needle punching. Barbed needles set into a board punch fibre into the batt and withdraw, leaving the fibres entangled. The needles are spaced in a nonaligned arrangement. By varying the strokes per minute, the advance rate of the batt, the degree of penetration of the needles, and the weight of the batt, a wide range of fabric densities can be made. For additional strength, the fibre web can be needled to a woven, knit, or bonded fabric. Bonding agents may also be used.

Needle loom (Non-Woven): A machine for producing needle felt. A needle beam reciprocates vertically at rates up to 2000 cycles per minute. Felting needles are mounted on a board at a density of 300 to 5000 per metre width and pass through a web or batt which is supported between bed and the stripper plates.

Needle loop: A loop of yarn drawn through a loop made previously. This is the part of the stitch which is composed of loop head and loop leg.

Needle plate: That part of the bed which has openings for needle and feed dog penetrations.

Needle pluck test: A modified burning test for the detection of synthetic fibres spun from the melt (polyamide, polyester) which during the burning test can be drawn out by a needle to form a fine thread following melting, by a heated iron. The needle draw test does not work specifically with polyacrylonitrile, acetate and triacetate.

Needle point: This was originally a type of fine drawn thread work, but it has developed into a craft in which satin stitch and buttonhole stitch are worked over a basic thread.

Needle-point Lace: Laces made with the needle, irrespective of style or design; see bobbin lace.

Needle positioner: A device on a sewing machine which ensures that the needle will stop automatically either in or out of the fabric as required.

Needle punched batting: Textile finishing material that is stabilized but mechanically entangling the fibres.

Needle punched carpets: Textile floor coverings produced by the needling of bonded fibre fleeces (Needle punched felt). May be composed of several different layers (top layer, lining, backing layer). The mechanical needle punching of the nonwoven fabric is known as pre-fixing. In post-fixing, binding agents are used. Nonwoven fabrics with a thermoplastic fibre content are fixed using heat treatment. The fixing needles penetrate the nonwoven fabric either vertically or at a slight angle.

Needle punched nonwoven fabric: Bonded fabrics of the Õ Needle punched felt type and needle punched wadding. Fixed by needling with or without shrinkage or needling with thermoplastic bonding.

Needle punching: The process of converting batts or webs of loose fibres into a coherent nonwoven fabric on a needle loom (q.v.).

Needle set-out: A term that refers to long periods of time when certain needles are removed from the knitting cycle. The process is used to make sweater cuffs.

Needle Size, sewing: Refers to the diameter of the needle measured at the needle eye. Today the most common needle sizing system used around the world is the metric system. The metric number represents the percent of a millimeter. Common needle size application:

Light weight, 60 - 70

Medium weight, 75 - 110

Heavy weight, 100 - 120

Needle slot: A groove that houses a needle in the cylinder or dial of a circular-knitting machine or the needle bed of a flat-bed machine.

Needle specifications, knitting: The typical "European" specifications for a needle includes a word, a number (usually a four digit number) and a final combination of letters and numbers. For example: Vota 78.60 G.02 The capital letter at the beginning of the word ("V"), identifies the origin of the needle (obtained from a wire, pressed or die-cut), the type, the number of butts and the type of tail.

The other capital letters have a very precise meaning, except for the vowels "e" and "a" which are added to make the word pronounceable, and indicate the shape and the height of the butt, the eventual existence of a groove and its size, the length of the tail and some other features of the needle.

The next group of numbers identifies the needle according to the length and the gauge.

The first part (78 in the example) indicates the whole length rounded off to the mm (in our case that makes 78 mm); the second part indicates the gauge of the needle in hundredths of millimetres (in our case the gauge of the needle is equal to 0.60 mm).

The final group of letters and numbers has to be read as follows.

The first capital letter indicates the needle manufacturer (For example Z for Torrington, E for Exeltor, G for Groz-Beckert).

The next number is used to distinguish a specific needle among all the needles produced by the same manufacturer.

The next letter refers to some particular features of the needle: for some needles an "A" indicates that the latch has been fixed with an angular pressed pin while an "R" means that the latch has been fixed with a straight pressed pin.

For other needles, the latch fixing method is indicated by a "0" before the last number.

A "0" indicates that the latch has been fixed with a standard pressed pin; no "0" means that the latch has been fixed with a screw pin.

Needle, knitting: The needle is the basic element of loop formation. There are three most commonly used types of needle: the latch needle, the spring-beard needle and the compound needle. We can divide a needle into three main parts: the hook, which takes and retains the thread to be looped; the hook opening and closing device, which allows the hook to alternatively take a new thread and release the previous one; a system allowing the needle to move and form the loop.

Needle work: All kinds of plain and decorative works (embroidery and lace) executed with the needle by hand.

Needled fabrics: The product of the needle loom (q.v.). Needled fabrics are used for rug pads, papermaker's felts, padding, linings, etc.

Needling; Needle punching; Needle felting; Needle bonding: The use of barbed needle mounted in a needle loom, to entangle a fibre web or batt by mechanical reorientation of some of the fibres within the structure.

Negative feed: When the yam is pulled off the warp beam by the knitting action of the needles during the loop forming step.

Negrepelisse: French woollen cloth, thoroughly fulled with long raised nap; mostly black.

Negretti: A large species of native Spanish sheep, yields fine and soft wool.

Neo-batiks: A technically improved development of Batik. Patterns in molten wax are applied to wellstretched textiles (usually natural silk), and aqueous or alcohol-based dye solutions applied with a cotton wool pad. It is then dried and the wax removed by dissolving in organic solvents, or by ironing between sheets of blotting paper. Then the fabric is steamed to fix the dye. In this way, combinations of wax resistance and different colour schemes can be used to produce unlimited effects.

Nep: A small knot of entangled fibres that usually will not straighten to a parallel position during carding or drafting. Or one or more fibres occurring in a tangled and unorganized mass.

Nephelometer: A device used to measure mainly low turbidity water with results expressed in Nephalometric Turbidity Units (NTU).

Neps: See **Nep.**

Neppy dyeings: Clumps of tangled thin-walled immature cotton fibres (neps) scattered throughout the surface of cotton fabrics, which resist dye uptake and appear as white of pastel-coloured specks on the dyed fabric.

Neri: Grade of waste silk, obtained from the inner smooth skin of the cocoon, left over after the reeling.

Nessler's reagent: Nessler's reagent is used for the detection of ammonia (in water) and for the detection of damaged cellulose (yellow to orange, later grey colouring).

Nessu: Native name in East Africa for nainsook.

Net: A transparent fabric that is neither woven or knitted, but rather the yarns are knotted to form a mesh. Net may be made of cotton, viscose, nylon, polyester or other fibres, according to its ultimate use. Made in varying sizes of mesh and used for curtains and dance dresses, veilings and trimmings. Net can be made by hand or machine in a variety of mesh sizes and weights matched to varying end uses, i.e., veils, curtains, fish nets, and heavy cargo nets.

Net Canvas: Made of cotton or linen with an open texture, in black or white and stiffened with gize; used for embroidery ground.

Net curtain fabrics: Fine, usually transparent fabrics made from suitable yarns. They must be washable and resistant to stretching and ironing; those made from synthetic yarns generally do not require stretching or ironing after washing. There are various different manufacturing methods for net curtain fabrics: looms (fine muslin, voile, etamine, marquisette, Madras), bobbinet machines, a form of bobbin lace machine (bobbinet, tulle and lace curtains), raschel knitting or crocheting machines (raschel or galloon curtains), knotted lacework machines or by hand (lace curtains), and embroidery machines (dotted muslin, burnt-out or air lace).

Net leno: Leno fabrics having the crossing warp (see) floating on the face of the fabric and forming zigzag lines.

Net rate: In a fibre production process the total throughput less waste and inferior or off-grade material.

Net silk: Another name for thrown silk.

Nets: Generally wide-meshed fabrics which are produced by weaving, knitting, crocheting or on raschel looms.

Netting: (1) See **Marquisette.**

(2) In nautics, network made of cord or rope.

(3) The process of knotting threads into meshes that will not ravel.

Nettle cloth: (1) Mentioned in 16th century English manuscript; is of unknown structure.

(2) A light, sheer fabric, woven of the stem fibres of the nettle in Germany.

Nettle fibres: The nettle class of fibres comprise of Common Nettle (*Urtica dioica*), China Grass (*Urtica nivea*), and Ramie (or rhea). Short, fine stem fibre yielded by the nettle; used for twine, cloth, etc., in Austria, Germany, etc.

Network analysis: A generic term used for several project planning methods, of which the best known are PERT (Programme Evaluation and Review Technique) and CPA (Critical Path Analysis).

Neuilly: French machine-made tapestry made in imitation of the real gobelins.

Neutralization: The stoichiometric reaction of an acid and a base in volumetric analysis. The neutralization point or end point is detected with indicators.

Neutrals, in colouring: Neutrals, or neutral colours, are sometimes referred to as earth tones. Each neutral is a combination of all three primary colours of any percentage mix. Neutrals are often muddy in appearance as they are produced by mixing colours that are opposite on the colour wheel – known as complementary colours – which effectively cancels each other out when mixed. Theoretically, this cancellation produces non-colour, i.e. black. In reality, because colours are never 100% pure but contain some impurities, the result is not black but a dirty nondescript colour. For instance, a neutral may be produced by mixing complementary colours orange and blue.

Neutron-absorbing fibre: Polyethylene fibre modified with boron used in the nuclear industry for reducing neutron transmission.

New look: Post warwomen's fashion revival from Paris designed by Christion Dior in the forties.

New Zealand Cotton: Fine, strong bast fibre, yielded by the young branches of the ribbon tree in New Zealand; used for fishing lines.

New Zealand Hemp: Long, soft, white, silky and very strong leaf fibre yielded by the Phormium tenax in Australasia; used for ropes, twine, matting and cloths. The base mark is "fair Wellington."

NF: (1) New Zealand flax or hemp (Phormium), (2) Non-ferrous metals. (3) French standard.

Ngutunui: Native name for a species of the New Zealand flax (see) yielding fibres suitable for the finest fabrics.

NI: Indonesian standard.

Nib: See **Nibs.**

Nibs, in raw silk: Small thichened places less than 3mm. in length.

Nickel screens: Are made for rotary printing using the discharge or rotary screen processes. Nickel screens cannot be recycled, i.e. cannot be re-used for other patterns. → Rotary screens for screen printing.

Niddy-Noddy: A traditional, low-tech way to wind a skein and measure yarn. This image of the elusive niddy-noddy shows a very good quality, homemade one. For the less crafty, or to dry damp skeins, consider making one out of 3/4 or 1" pvc.

Nikki-plush: Circular-knit pile fabric (sheared plush). The grey cloth undergoes shearing during finishing, whereby the loops are cut to give a velvety look. The fabric has the elasticity of jersey. Used for outer clothing; also printed. Circular-knit pile fabric finishing.

Nimes: French piece dyed, wool dress fabric of medium quality, originally having 2,200 warp ends.

Ninon: A fabric sometimes called triple voile, which is plain weave and sheer. It usually made of viscose, acetate or sometimes polyester. A sheer, fairly crisp fabric, heavier than chiffon. Much like voile, but more body. The warp yarns are often grouped in pairs. Washes well, particularly in the synthetics. Uses: Mostly used for curtains, and some for evening or bridal wear.

Nishiki: A very rich gold and silver brocade, made in Japan.

Nip: (1) A defect in yarn consisting of a thin place.

(2) Passage of material (full-width piece goods) between two rollers, in pad dyeing, calendering and similar. The roller nip is the contact line between the two rollers which concentrates the pressure of the two rollers. (3) The gap between two nip rollers (Fig.). There are two types, hard nip rollers, as found in calendars (steel and flexible rollers) and elastic nips between two elastic rollers.

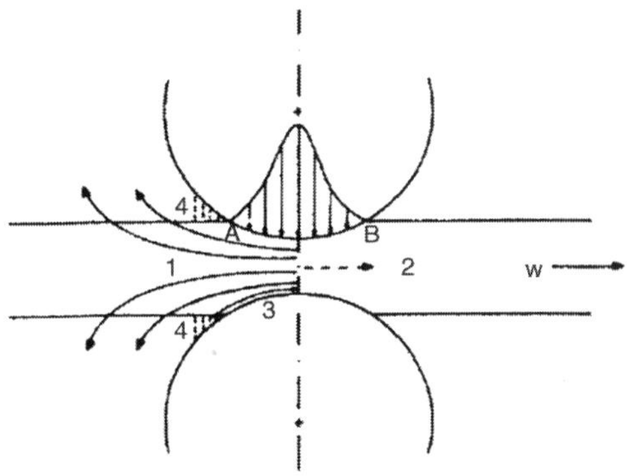

Fig. Diagram showing the squeezing process in the nip. 1 = return flow of liquor; 2 = capillary transportation; 3 = return of liquor to be applied; 4 = liquor wrung out.

Nip creases: Creases occurring at regular intervals along a fabric selvage subsequent to a nipping operation such as calendaring or padding. Such creases are caused by a loosely wound selvage or improper let-off tension, which allows the fabric to fold over or gather at the selvage prior to entering the nip of the rolls.

Nip padding: A particular type of padding. With the two-roller pad the lower roller is wrapped with cloth and dips into the padding liquor; the fabric is passed between both rollers. Used especially in resist printing, as the short contact with the liquor prevents the resist from softening.

Nip pressure profile: With the aid of, e.g. carbon paper, impressions can be obtained of the evenness of the squeezing pressure across the full width of the nip.

Nipco rollers: Deflection-free (elastomer) rollers with hydraulic power transmission Pressure generation with even pressure on outer mantle by means of individual hydraulic supports; load can be applied externally in groups, every 2000 mm, e.g. 5 zones. The Nipco calor roller can be precisely heated from the inside. Used for pads, calendars and other squeezing equipment.

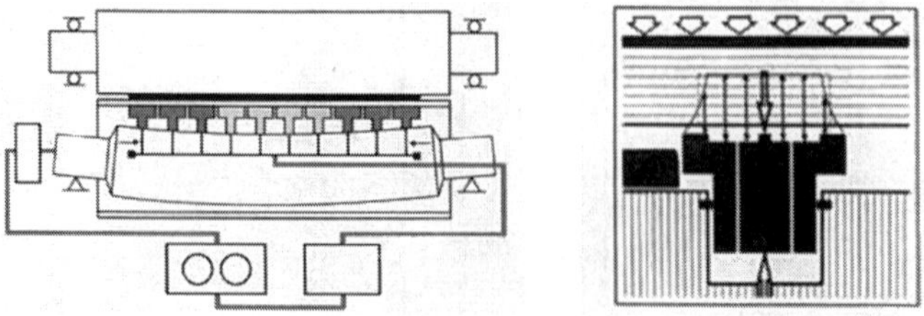

Nipco roller system for the control of an individual preasure stamp (Kleinewefers KTM).

Nitrate discharges: See **Oxidation discharges for indigo ground.**

Nitration: A reaction commonly used in dyes manufacture. Nitration means the introduction of the nitro group, —NO_2, into a molecule. It is accomplished: (a) with dilute or concentrated nitric acid; (b) with mixed acid, i.e., a mixture of nitric and sulfuric acid, sometimes containing some water; (c) by first sulfonating the compound and then nitrating the sulfonic acid, thereby splitting out the sulfo group and replacing it by the nitro group; (d) by oxidizing with dilute nitric acid a previously formed nitroso compound; and (e) by treatment of a diazonium compound with hot, dilute nitric acid, introducing simultaneously a hydroxyl group and a nitro group (e.g., nitrocresol from p-toluidine).

Nitrification, in effluent treatment: The two-step biological process in which ammonia is converted to nitrite and then to nitrate.

Nitrification–denitrification: Nitrification followed by denitrification, resulting in bubbling away of nitrogen gas. The process can be undertaken in a wastewater treatment works by having nitrification in the aerobic biological treatment process followed by a denitrifying filter. Alternatively the process can be incorporated into one system with adaptations of the activated sludge process, as in the AAO process, Bardenpho process or the modified Ludzack–Ettinger process where the nitrified wastewater is recycled back to the anoxic tank at the start of the treatment process. Another alternative is the use of a pre-anoxic filter.

Nitrobacteria: Micro-organisms (e.g. bacterium coli commune, typhus bacillus) which oxidise organic nitrogen compounds (ammonia, nitrites, etc.) in the ground, ground water, etc., to nitrates. Nitrobacteria are of significance in connection with the water supply for textile processes.

Nitro Dyes: These dyes are now of only minor commercial importance, but are of interest for their small molecular structures. The early nitro dyes were acid dyes used for dyeing natural animal fibres such as wool and silk. They are nitro derivatives of phenols, e.g., picric acid or naphthols.

Nitro rayon: (Chardonnet rayon), the first man-made fibre (1884 by the Earl of Chardonnet). Produced from Nitrocellulose, using ethanol and ether

Nitrocellulose: See **Cellulose trinitrate.**

Nitroso dyes: Nitroso dyes are metal-complex derivatives of *o*-nitrosophenols or -naphthols. Tautomerism is possible in the metal-free precursor between the nitrosohydroxy tautomer and the quinoneoxime tautomer.

N-methylol group: See **N-Methylol compounds.**

N-methylolacrylamide: See **Methylolacrylamide.**

N-methylolpolyethyleneureas: A crease-resistant finish; products of the reactions of polyethylene amines or imines, urea and formaldehyde, e.g. from diethylene triamine, urea and formaldehyde:

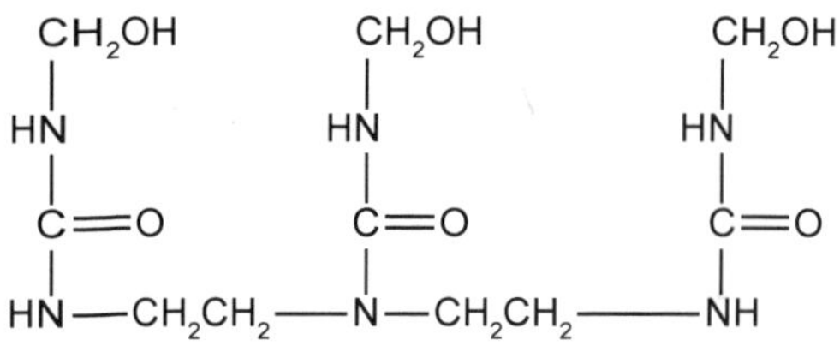

NMMO Process: NMMO stands for N-methyl morpholine oxide, a cyclic amine oxide, in which cellulose is very soluble.The NMMO process is a solvent spinning process in which a Lyocell fibre is produced from cellulose in only a few stages, without the supplementary addition of chemicals. The technique is one of the most ecofriendly process compared with conventional processes which causes of waste water and air emissions and produces much better fibre with comparatively higher dry and wet strength and wet elongation. Moreover, the NMMO solution can be reused.

NOEC (no-effect concentration): When surfactants are biologically degraded, for example, the first stage of the oxidation can be at the hydrophobic end of the chain:

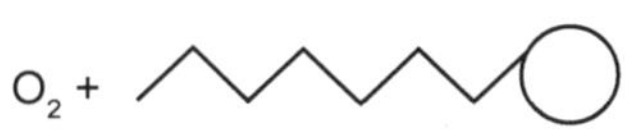

The surfactant loses its effectiveness as a surfaceactive substance, i.e. if during biodegradation the surfactant concentration drops below a certain level, and

it is no longer detectable using the usual tests, and no longer foams or toxic to fish.

No iron: Another term for crease resist finish. The literal expression of the purpose of this finish: no ironing is necessary after the washing process. The most frequently used term is: See **Wash and Wear.**

No-Observed-Effect Concenration (NOEC): Highest constituent concentration at which the measured effect are no different from the control.

Noble Comb: Used commercially in producing worsted yarn.

Node: An English cotton dress fabric haying a boucle face and having twisted knots either in the warp or the filling.

Noil: Short fibres removed while combing, applied particularly for wool, but also to other fibres as cotton, silk, rayon etc.

Nomad carpet: Oriental carpets, plush-like, with pile cropped to a greater or lesser extent, warp, weft and knotted sheep or goat wool pile (100 000–300 000 knots per m_2). Mainly narrow pieces (prayer mats, rugs, mats). Design (work of nomads) differs slightly in pattern, width, colour, often distorted and creased.

Nomex: A modified Nylon, poly(p-phynyleneterephthalamide) which is practically flame proof.

Nominal guage length in tensile testing: (1) The length of a specimen under specified pre-tension measured from nip to nip of the jaws of the holding clamps in their starting position at the beginning of the test, and including any portion of the specimen in contact with bollard or snubbing surfaces.

(2) The length of a specimen under specific pre-tension between frets, in instruments where the specimen is not held by clamps, for example in a vibroscope.

(3) The length of a specimen measured between the points of attachment to the tabs while under specified pre-tension.

Nominal pressure: The maximum permissible Working pressure in bars at 20°C. It is therefore sufficient when denoting the application range of a pipework component (piping, fittings) to give the nominal pressure.

Nominal temperature: The temperature given on measuring instruments at which they should be used. If not given, it is 20'C. If two temperatures are quoted, this means the limits of the temperature range in which the appliance should be used.

Non Battue: Loosely woven French linen canvas of inferior grade.

Noncarbonate hardness: Hardness caused by chlorides, sulphates, and nitrates of calcium and magnesium. Evaporation of waters containing these ions makes the water highly corrosive.

Non-biodegradable: Description of something that is not broken down by microbes and has an ***oxygen demand*** only if it is a chemical reducing agent. It has no ***biochemical oxygen demand***.

Non-conforming: A description of a unit or a group of units that does not meet the unit or group tolerance.

Non-conformity: An occurance of failing to satisfy the requirements of the applicable specification; a condition that results in a nonconforming item.

Nonpareilles: Camlet like French cloth, made either of all wool or mixed with goats' hair or linen yarn. See also **lamparillas**.

Nonpolar compound: A compound that has molecules with no permanent dipole moment. Examples of nonpolar compounds are hydrogen, tetrachloromethane, and carbon dioxide.

Non-chlorine bleach: Bleaching with chemicals other than one containing chlorine.

Nonelastic woven tape: A woven narrow fabric, weighing less than 15 ounces per square yard, made principally of natural and/or manufactured fibres, including monofilaments, but not containing rubber or other similar elastic stands.

Nonformaldehyde DP finishes: Several countries such as Japan and Germany have even more stringent HCHO requirements than the USA. Because of this, there has been and continues to be interest in non-formaldehyde DP finishes. There are products that fit the description "non-formaldehyde" finishes like Dimethyl-4,5 Dihydroxyethylene Urea (DMeDHU), Butanetetracarboxylic acid (BTCA) etc.

Nonionic: Those that develop no ionic charge on the water solubilizing end.

Non-Ionic: Descriptive of a chemical that does not ionize in solution, but remains in intact molecular form Some *surfactants* are non-ionic.

Nonionic surfactants: This is the second most important group of surfactants. They are mainly chemicals whose molecules have the usual hydrophobic 'tail' connected to a short polyethylene oxide chain. This will usually have about five to twenty $-CH_2CH_2O-$ units linked together and forms the polar hydrophilic 'head' to whose oxygen atoms water molecules can hydrogen-bond. A wide range of products is available with different types of hydrophobic groups and varying degrees of polymerization of the polyethylene oxide chain. Some have polypropylene oxide 'heads', usually promoting oil solubility.

Non-lint content: That portion of a mass of cotton fibre which is essentially foreign matter.

Nonmetameric: Shades which are not changing in different lights.

Nonseperable zipper: A zipper having two stringers that are permanently attached to eaxch other at one or both ends.

Non-shrink finish: The finish which prevents shrinking.

Nontorque yarn: See **Textued yarns.**

Non-Wetting agent: The opposite of Wetting agent. A textile-processing agent without specific capillary activity.

Nonwoven blanket: A blanket produced by bonding or interlocking of fibres, or both, accomplished by mechanical, chemical, thermal, or solvent means or a combination thereof.

Nonwoven fabric: An assembly of textile fibres held together by mechanical interlocking in a random web or mat, by fusing of the fibres (in the case of thermoplastic fibres), or by bonding with a cementing medium such as starch, glue, casein, rubber, latex, or one of the cellulose derivatives or synthetic resins. Initially, the fibres may be oriented in one direction or may be deposited in a random manner. This web or sheet of fibres is bonded together by one of the methods described above. Normally, crimped fibres that range in length from 0.75 to 4.5 inches are used. Nonwoven fabrics are used for expendable items such as hospitable sheets, napkins, diapers, wiping cloths, as the base material for coated fabrics, and in a variety of other applications. They can also be used for semi-disposable items and for permanent items such as interlinings.

Nonwoven fabric, Thermally bonded: Textile fabric composed of a web or batt of fibres containing heat sensitive materials, bonded by application of heat with or without pressure.The heat sensitive material may be in the form of fibres, bicomponent fibres or powders.

Nonwoven, wet-laid: Nonwovens of short-staple polymer fibres, produced by spraying from spinning nozzles (polymer solution or melt) onto, e.g. perforated conveyor belt. Used for filter mats, insulation purposes, etc.

Nonwoven pile fabrics: Pile fabric manufactured by the nonwoven fabric manufacturing method.

Nonwovens: See **Non-woven fabric.**

Nonylphenolpolyglycolether: A product of addition of ethylene oxide to nonylphenol (See

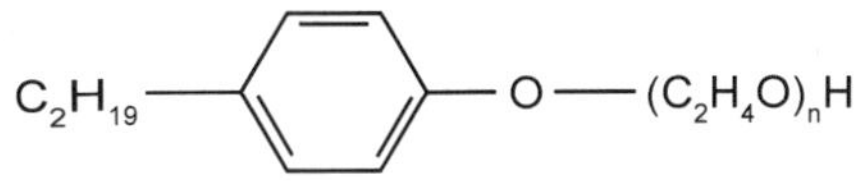

Polyglycol ethers): A washing and wetting agent.

Normal Mixture: Knit goods, made of a mixture of cotton and wool, the colour being black and white mixture.

Normal solutions: Solutions with precisely determined contents for titration. The 1/1 normal solution (1 n solution) contains the Gram equivalent or vol of a chemical substance dissolved in precisely 1 litre distilled water.

Norman Embroidery: Conventional designs fllled with crewel wool, parts of the pattern being covered with open fancy embroidery stitches in floss silk.

Normandy Laces: Bobbin laces, made in imitation of Malines and the Chantilly lace.

Normandy Val: Trade name for machine made lace, similar to the shadow lace.

Norsk Tekstilforskningsinstitut: Norwegian textile research organization.

Northamptonshire Lace: English bobbin laces, made in imitation of Lille, Valenciennes and Brussels laces. The mesh ground is very fine.

Norwegian Yarn: Fine, slack-twist yarn, made of Norwegian lambs' wool in natural white, gray or black; used for hand knitting in England.

Norwich crepe: A 19th century English fabric, made of silk warp and worsted filling in colours different from each other or dyed two shades of the same colour. It is woven both sides alike, without a wale and is finished with a gloss; used for women's dresses. In the 17th century, it was an English worsted crepe dyed in black.

Norwich Shawl: A fine English silk shawl, originally made with checks and stripes and embroidered afterwards; later it was made of printed silk yarn.

None: French for knotted.

Notation systems: Ways of graphically representing woven and knitted structures.

Notches: Small cuts made in the edges of garment parts, which are used as guides during garment assembly.

Notions, in quilting: Are items other than fabric or a pattern that are used to complete sewing projects. Common items include quilting pins, needles, thread and chalk.

Nottingam lace: The term is often used to describe a flat, coarse lace used for curtains, bed spreads and table cloths, but Nottingham is the home of the machine-lace industry, so the word is also used to describe any lace made on a Nottingham-type machine.

Nouveaute: French for novelty; fabrics and trimmings outside of the staple lines.

Novaloid: Novaloid is the designation assigned by the FTC for a class of flame retardant fibres made from cross linked phenol-formaldehyde polymer. The fibres of this class in U.S. production are called Kynol and are manufactured by American Kynol, Inc. The fibre is golden yellow in colour and possesses good physical and chemical properties. The fibre is thought to be formed through spinning phenol-formaldehyde prepolymer, followed by heating the fibre in formaldehyde vapour to crosslink the structure. The fibre has low crystallinity with a tenacity of 1.5-2.5 g/d (14-23 g/tex) and an elongation at break of about 35%. The fibre has a specific gravity of 1.25 and a moisture regain of 4%-8%. It is inert to acids and organic solvents but is more susceptible to attack by bases. The fibre is totally heat resistant up to 150°C. It is inherently flame retardant and chars without melting on exposure to a flame.

Novato: A woollen or silk fabric; used in the 16th and 17th century in England.

Novelty Tweed: A description indicating that the cloth is tweed like, but that is one of the conventional tweeds that are easy to recognize. Usually a cloth with some decoration or distinguishing feature.

Novelty yarn: A yarn produced for a special effect. Novelty yarns are usually uneven in size, varied in colour, or modified in appearance by the presence of irregularities deliberately produced during their formation. In singles yarns, the irregularities may be caused by inclusion of knots, loops, curls, slubs, and the like. In plied yarns, the irregularities may be effected by variable delivery of one or more yarn components or by twisting together dissimilar singles yarns. Nub and slub are examples of novelty yarns.

Novoloid fibres: A manufactured fibre containing at least 85% by weight of a crosslinked phenol formaldehyde polymers or novolac i.e. Novoloid is extremely flame resistant and non melting. Its primary use is flame proof protective garmets.

Noyales: (1) Fine French bleached linen, sometimes mixed with cotton; (2) Unbleached, French hemp sailcloth, the strongest .grade made with a six- ply warp.

Nozzle: (1) The spout through which something is discharged, i.e., oil in finish application or fibres in web laying. (2) A term sometimes used to refer to spinnerets. (3) Part of a jet dyeing machine/jet.

NP: Norma Portuguesa Definitiva; Portuguese standard.

NRC: "National Research Council", Canadian textile research organization

NS: Norsk Standard, Norwegian standard.

NSO: Norwegian standards organization.

NTV process: (low-temperature dyeing process). The dyeing of polyamide carpet piece goods on a wind beck at temperatures below boiling point (e.g. 70– 80°C).

Nub yarn: An irregular thread containing small spots or nubs, very similar to Slub but the irregularities are definitely nubs.

Nucleation: A process by which crystals are formed. Crystals form initially on minute traces of foreign substances that act as the nucleus, then grow by external addition

Nucleophilic addition: A class of reaction involving the addition of a small molecule to the double bond in an unsaturated organic compound. The initial part of the reaction is attack by a nucleophile and the unsaturated bond must contain an electronegative atom, which creates an electron deficient area in the molecule. Nucleophilic addition is a characteristic reaction of aldehydes and ketones where polarization of the C=O carbonyl causes a positive charge on the carbon. This is the site at which the nucleophile attacks. Addition is often followed by the subsequent elimination of a different small molecule, particularly water. See **condensation reaction**.

Nucleophilic substitution: A reaction involving the substitution of an atom or group of atoms in an organic compound with a nucleophile as the attacking substituent. Since nucleophiles are electron rich species, nucleophilic substitution occurs in compounds in which a strongly electronegative atom or group leads to a dipolar bond. The electron-deficient center can then be attacked by the electron-rich nucleophile causing the electronegative atom or group to be displaced. In general terms:

$$R–Le + Nu^- > R–Nu + Le^-$$

where Nu^- represents the incoming nucleophile and Le^- represents the LEAVING GROUP. There are two possible mechanisms for nucleophilic substitution. In the S_N1 (substitution, nucleophilic, monomolecular) reaction the molecule first forms a carbonium ion; for example: $RCH_2Cl > RCH_2^+ + Cl^-$

The nucleophile then attaches itself to this carbonium ion: $RCH_2^+ + OH \rightarrow RCH_2OH$.

In the S_N2 (substitution, nucleophilic, bimolecular) reaction the nucleophile approaches as the other group leaves, forming a transition state in which the carbon has five attached groups. The preferred mechanism depends on several factors: (1) The stability of the intermediate in the S_N1 mechanism. (2) Steric factors affecting the formation of the transition state in the S_N2 mechanism. (3) The solvent in which the reaction occurs: polar solvents will stabilize polar intermediates and so favour the S_N1 mechanism. There is a difference in the two mechanisms in that, for an optically active reactant, the S_N1 mechanism gives a racemic mixture of products, whereas an S_N2 mechanism gives an optically active product (*see* illustration). *See also* **substitution reaction.**

Number duck: See **Duck.**

Nun's Cloth: A very thin, plain woven black woollen fabric, similar to bunting; used for mourning wear, for dresses for nuns, office coats, etc.

Nun's Thread: Very fine, bleached linen thread, made by the nuns in Italy and Flanders since the 16th century; used for laces.

Nun's veiling: A fine, lightweight plain-weave, plain coloured, worsted or silk fabric, which is very soft and thin. It is now used for dresses, but was at one time used only for religious gowns. Originally dye only black, brown and grey, but as a dress fabric it is produced in fashion colours. Used for dresses, blouses and nightwear.

Nutria fur: This comes from a small South American animal and is similar to Beaver.

Nutrients, in effluent treatment: Both Nitrogen and Phosphorous, along with carbon are essential nutrients for growth. When discharged to the aquatic environment, these nutrients can lead to the growth of undesired aquatic life. When discharged in excessive amounts on the land, they can also lead to the pollution of ground water.

Nylon: A manufactured fibre in which the fibre forming substance is any long chain synthetic polyamide having recurring amide groups (-NH-CO-) as an integral part of the polymer chain (FTC definition). The two principal nylons are nylon 66, which is polyhexamethylenedianime adipamide, and nylon 6, which is polycaprolactam. Nylon 66 is so designated because each of the raw materials, hexamethylenediamine and adipic acid, contains six carbon atoms. In the manufacture of nylon 66 fibre, these materials are combined, and the resultant monomer is then polymerized. After polymerization, the material is hardened into a translucent ivory-white solid that is cut or broken into fine chips, flakes, or pellets. This material is melted and extruded through a spinneret while in the molten state to form filaments that solidify quickly as they reach the cooler air. The filaments are then drawn, or stretched, to orient the long molecules from a random arrangement to an orderly one in the direction of the fibre axis. This drawing process gives elasticity and strength to the filaments. Nylon 6 was developed in Germany where the raw material, caprolactam, had been known for some time. It was not until nylon 66 was developed in the United States that work was initiated to convert caprolactam into a fibre. The process for nylon 6 is simpler in some respects than that for nylon 66. Although nylon 6 has a much lower melting point than nylon 66 (a disadvantage for a few applications), it has superior resistance to light degradation and better dye ability, elastic recovery, fatigue resistance, and thermal stability. Two other nylons are: (1) nylon 11, a polyamide made from 11-amino-undecanoic acid; and (2) nylon 610, made from the condensation product of hexamethylenediamine and sebacic acid. Nylon 610 has a lower melting point than nylon 66 and the materials for its manufacture are not as readily available as those for nylon 66. Experimental work has been conducted on other nylons.

CHARACTERISTICS: Although the properties of the nylons described above vary in some respects, they all exhibit excellent strength, flexibility, toughness, elasticity, abrasion resistance, washability, ease of drying, and resistance to attack by insects and microorganisms. END USES: Nylon is used for apparel such as stockings, lingerie, dresses, bathing suits, foundation garments, and wash-and-wear linings; for floor coverings; for tire cord and industrial fabrics; and in-home furnishings such as upholstery fabrics.

Nylon 11: See **Nylon.** $\{-NH-(CH_2)_{10}CO\}_n$ Nylon II is produced by self-condensation of 11-aminoundecanoic acid in Europe and marketed under the name Rilsanite. It is melt spun (mp 189 °C) into fibre and possesses essentially the same properties as nylon 6 and 6,6. It has high strength (5-7.5 g/d), low specific gravity (1.04), and low water regain (1.2%) and melting point 189'C. Nylon II has excellent electrical properties and is used in electrical products as well as in brush bristles, tire cord, lingerie, and hose.

Nylon 3: See **Nylon. $(-CH_2-CH_2-CO-NH-)n$.**

Nylon 4: See **Nylon.** $\{-(CH)_2-CO-NH-)n$. Nylon 4 is produced by polymerization of pyrrolidone using carbon dioxide catalyst. Nylon 4 is melt spun (melting point 273°C) to give a fibre of about 60% crystallinity. The fibre is of moderate dry strength (4.5 g/d) and slightly higher wet strength, with a higher specific gravity (1.25) than nylon 6 and 6,6. The fibre has excellent elongation and recovery properties and a water regain comparable to cotton (8% at 21°C and 65% RH). Nylon 4 is easily laundered and can be dyed easily to give colorfast shades but has limited wrinkle recovery properties and is sensitive to hypochlorite bleaches. Although a major attempt was made to commercialize nylon 4, market conditions limited the penetration of this new fibre type and production was discontinued. Nylon 4 is no longer being produced.

Nylon 7: See **Nylon.** {-NH- $(CH_2)_6$-CO-}n.

Nylon 12: See **Nylon.** {-NH-$(CH_2)_{11}$-CO-}n.

Nylon 6,10: See **Nylon.** {-NH-$(CH_2)_6$-NH-CO-$(CH_2)_8$-CO-}n. Nylon 6,10 is produced by condensation polymerization of he xame thylene diamine and sebacic acid and is melt spun (mp 216°C) into fibres. It resembles nylon 6 and 6,6 in many ways It is primarily used in brush bristles. Moisture absorption is low at 2.14 at 20'C and 65% Relative Humidity.

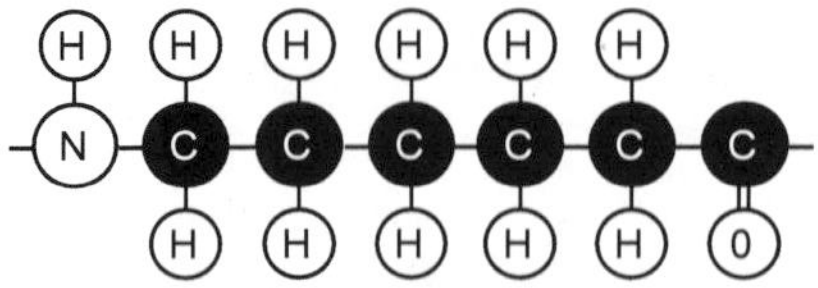

Repeat Unit of Nylon 6

Nylon 66: Polyamides are linear macromolecules containing amide group (-CO-NH-)at regular intervals. Different types of polyamide are made by using starting material (monomers) of different sizes (different numbers of carbonatoms). Nylon 6 has six carbon in repeating unit and Nylon 66 has two sets of six carbons in repeating unit.

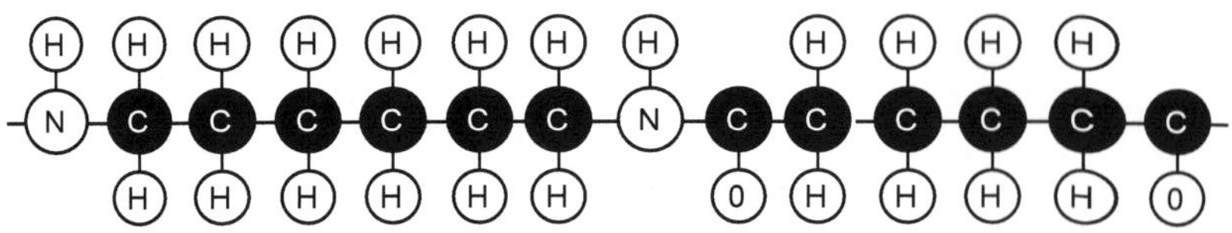

Repeat unit of Nylon 66

Nylon 6T: Modified Nylon. Nylon can be modified in different ways to get different properties including handle,temperature resistance, dyeability, tenacity etc. In Nylon 6T the repeat units are changed i.e., {-NH-$(CH_2)_6$-NH-CO-C_6H_4-CO-}n. It is a very high heat stable fibre, it melts at 370°C.

Nylon cire': Waxing process applied to light weight plain weave nylon, a very thin shining fabric used for rain coats, anoraks , protective clothing for cyclist etc.

Nylon Dyeing assistants: These are auxiliaries used in the dyebath to assist in getting even dyeing of nylon. These are surfactants which have an anionic group and can be ionized in the dyebath. In nylon dyeing the the ionized acid forms positive group in the fibre ($-NH_2 + H^+ \rightarrow NH_3^+$). Acid dye anion is bound on to this positive sites. The binding force is faily strong and hence there will be a rushing of the dye as soon as the positive sites are formed can cause an unlevel dyeing. The Nylon dyeing assistants prevents this rushing up by the formation of electrostic bonds by sulphonate or sulphate groups present in these assistants in competition with the dye anions ($R-SO_3^-$). Since the dye anion bonding is supposted by hydrogen bonds, van de Waals forces etc they are more stable 3while the leveling agents bonds are slowly replaced by dye anions thereby giving a level dyeing. Usually a surfactant like a fatty alcohol sulphate ($R-O-SO_3Na$) like dodecyl benzene sulphonate, Turkey Red Oil or their mixtures can act as a nylon dyeing assistant.

Nylon fibre: See **Nylon.**

Nylon jersey: Similar to tricot but heavier. A very useful lining or mounting fabric. Although in many other linings, polyester fibre has superseded nylon, this is not the case with nylon jersey. It is hard wearing and has a limited amount of 'give' which will prevent the outer fabric from loosing its shape.

Nytril fibre: A manufactured fibre containing at least 85% by weight of a long chain polymer of vinylidene dinitrile [$-CH_2-C(CN)_2-$] and having the vinylidene dinitrile group in no less than every other unit in the polymer chain (FTC definition). Nytril fibres have a low softening point so they are most commonly used in articles that do not require pressing such as sweaters and pile fabrics. They are also blended with wool to improve shrink resistance and shape retention.

NZSR: New Zealand Standards Recommendation; the New Zealand standards mark.

O

O and M, operation and maintenance: O and M is a major problem in some countries for water and waste treatment schemes. A lack of finance for spare parts and for labour may exist. The availability or funding to employ technical staff may be absent. Programme schedules for O and M may not exist or not be implemented. In water supply, poor installation and illegal connections can aggravate the difficulties in maintaining the scheme.

Oak leaf braid: A jacquard woven narrow fabric, traditionally 2.4 to 4 cm. (1 to 1 ¾ in.) in width, having a conventional oak leaf and acorn design contained within a border, woven with a cotton warp and either mohair or continuous filament viscose weft. It is customarily used as uniform cap-band.

Oatmeal: See **Oatmeal cloth.** A sort of armure weave, with a resemblance to oatmeal.

Oatmeal cloth: An old fashioned term which was used to describe any fabric, usually wool. That had a coloured, pebbled effect resembling oatmeal.

Oats: Early maturing commercial variety of American cotton, the staple measuring 20-25 millimeters; the yield is 32-34 per cent.

Obermaier loose cotton dyeing machine: This has got a stainless steel cylindrical vessel and another perforated carrier, also cylindrical vessel less than the dia of the outer vessel. The perforated cylindrical carrier has a solid bottom and a perforated central tube which is connected to a seating which fits into the another seat provided at the bottom of the outside vessel which in turn is connected to a centrifugal pump by suitable pipe connections. Cotton is packed into the carrier and lowered in the pouter vessel carrying the liquor. When the pump is run the dye liquor passes through the central tube and into the package and goes out into the vessel through the perforations. After some time the direction of liquor is changed from inside to outside to outside to inside by changing the direction of the pump and the dyeing takes place.

Observer, in colour chemistry: The human viewer who receives a stimulus and experiences a sensation from it. In vision, the stimulus is a visual one and the sensation is an appearance.

Observer, standard: see **Standard observer.**

Ocelot: This is a fairly coarse textured thin fur, but is very hard wearing. Like Leopard, it has spots, but they are more oval in shape.

Octane number: The percentage of isooctane in a blend of isooctane and *n*-heptane that gives the same anti-knock properties as the fuel being tested. Higher octane normally gave increased engine efficiency but modern petrol (gasoline) engines can be highly efficient with lower octane fuels.

Odjaklik: Turkish name for Oriental hearth rugs, usually having a center field with pointed ends.

OE Spinning: A spinning method. The basis of all OE spinning processes is the separation of a fibre tape feed into individual fibres, which are given a twist through various media, and are spun continuously into an "open end" stripping thread.

OE Yarn: See **Rotor spun yarn.**

Oeil de Perdrix: (1) A fancy ground in old French laces, consisting of brides, ornamented with groups of knots; (2) Plain French serge dress goods, made with eight leaves and four picks in a repeat.

Oeillet: French for eyelet.

Off pressing or pressing off: The final pressing operation on a garment after assembly.

Off square, in woven fabric: The difference between the percentage of warp crimp and the percentage of weft crimp.

Off-clip: See **Scalloped selvedge.**

Offset guide: Guide fitted to the exit of a false twist spindle, which causes a deflection in the yarn path, thus assisting the elimination of tight spots from the yarn.

Offset printing: The patterned coloration takes place indirectly via a rubber blanket, with printing colour paste being applied beforehand to the printing plate or roller, from the roller to the rubber blanket, and only then to the material which is to be printed. During the printing process, the printing mould with the non-reversing image passes through the "moistening and dyeing unit". Through moistening, the non-printing parts become dye-repellent, so that only the printing, water-repellent parts take up the dye paste. This dye paste is "deposited" on a rubber cylinder, which transfers the laterally reversed print image onto the running paper, with slight pressure.

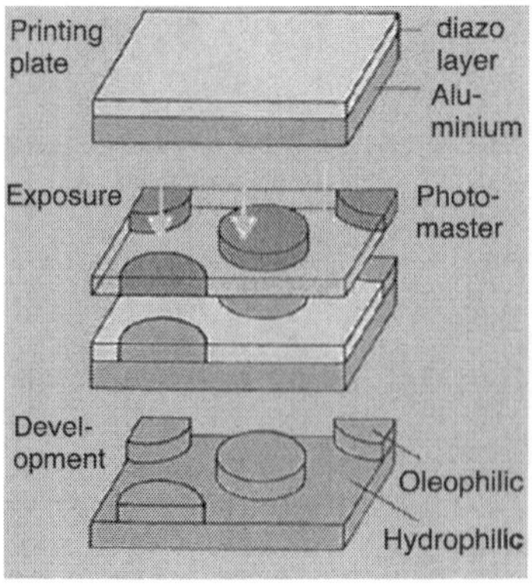

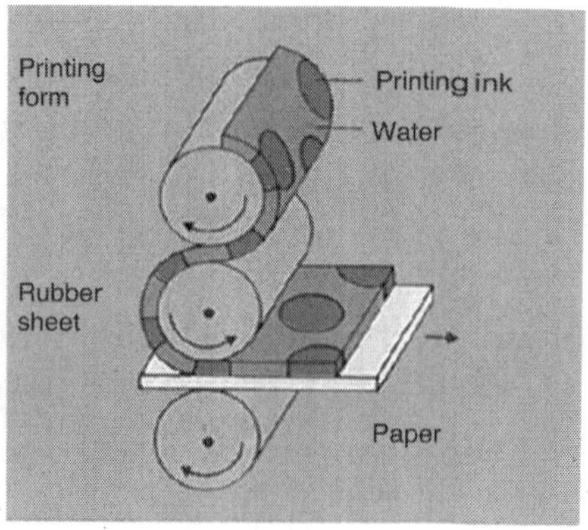

Exposure and development Printing

Off-sorts: The by-products of sorting -- shorts; britch wool, kemp, gray wool, stained wool, etc.

Ogees: Symmetrical onion shapes.

Ogilvie: A Highland tartan of very complicated composition, consisting of wide red and blue stripes, narrower and red stripes and lines of blue, black, red and yellow.

Ohm: Unit of electrical resistance which yields a current strength of 1 A at a voltage of 1 volt; or the resistance of a mercury thread 106.3 cm long and a cross-section of 1 mm≤ at 0°C. The resistance is in direct proportion to the length, and in inverse proportion to the cross-section of a conductor. Resistance in ohms = length (m) multiplied by specific resistance in ohms (resistance of a substance of 1 m in length and 1 mm2 cross-section) divided by the cross-section (mm^2).

Oil circulation heating: (Circulating hot oil unit). In order to obtain a greater evenness of temperature in the stenter frame , one can heat with hot oil, circulated by a pump, which has a flow temperature of around 40 – 50°C higher than the target temperature in the stenter frame, through separate heating , i.e. where the frame temperature is 190°C, the oil temperature is approx. 240°C. Independent of boiler room.

Oil cloth: A cotton fabric that has been treated on one side with a drying oil to make it impervious to water. Usually Linseed oil varnish is used. It is mainly used as a table or shelf covering, before laminated plastic appeared. Also used for bags and raincoats. These type of cloth has been completely superseded by modern coated clothings.

Oil Emulsion: Emulsion system of water-in-oil (W/O).

Oil Finish: Finish with an oily, greasy handle, e.g. for anoraks. Originally, a non-permanent coating with vegetable oils. Permanent effects were only achieved through padding with Silicone elastomers.

Oil of Vitriol: Sulphuric acid.

Oilproof: A term describing fabrics that are impervious to oil.

Oil repellence of textiles: This is the capacity to withstand wetting and penetration by oily liquids. Oil is not absorbed, but is repelled in droplets. The effect is achieved by the application of oil-repellents in the oleophobic or oil-repellent finish. It is brought about by the fact that the forces acting between the oil-repellents and the oil droplet are weak in comparison with the forces acting within a drop of oil.

Oil repellency: Oil Repellency is tested by placing a drop of oil on the fabric and observing whether the drop resides on top the fabric or whether it penetrates. A homologous series of hydrocarbons decreasing in surface tension is used to rate the fabric's oil repellency. The hydrocarbon with the lowest surface tension to remain on top and not penetrate is indicative of the fabric's repellency; the lower the surface tension of the liquid, the better the fabric's resistance to oily stains.

Oil repellent: A term applied to fabrics that have been treated with finishes to make them resistant to oil stains.

Oil repellent finish: The textile auxiliaries used for Water-repellent finishing are not sufficient to protect textiles against grease and oil stains. For this, special products are used, e.g. Fluorocarbon polymers, which are used in the form of emulsion, sometimes in padding, sometimes in the exhaustion method. However, these products in turn do not provide a good water-repellent effect, which is why in practice oil repellents and water-repellents are always used together. Some of the products available on the market provide effects which are resistant to washing and dry cleaning. Used with the Soil release finish.

Oil separator: A tank which separates oil from water or wastewater, such as an API separator, flotation tank, grease trap, inclined plate and tube settlers, oil coalescer, oil grit separator

Oiled silk: (insulating silk). Transparent, impregnated and printed fabric for raincoats, umbrellas, swimsuits, furnishing fabrics, etc. Manufactured according to a special process, such as elasticised urea resins. To avoid surface sticking, it is recommended that less elasticized urea resins are included, and that the mixture is thinned with alcohols or glycol ethers to immersion consistency. Usually dried at 80°C; this can also be accelerated by the addition of weak acids.

Oil emulsion: Emulsion system of water-in-oil (W/O).

Oil finish: Finish with an oily, greasy handle, e.g. for anoraks. Originally, a non-permanent coating with vegetable oils. Permanent effects were only achieved through padding with Silicone elastomers.

Oil-hydraulic batching drive: Flexible drive for all winding operations with hydraulic motors, which guarantee the constant goods tensions for each winding size.

Oil-in-water emulsion: See **Aqueous emulsion.**

Oil repellence: This is the capacity to withstand wetting and penetration by oily liquids. Oil is not absorbed, but is repelled in droplets. The effect is achieved by the application of oil-repellents in the oleophobic or oil-repellent finish. It is brought about by the fact that the forces acting between the oil-repellents and the oil droplet are weak in comparison with the forces acting within a drop of oil.

Oil-repellent: See **Oil-repellent finishing.**

Oil-repellent finishing: The textile auxiliaries used for Water-repellent finishing are not sufficient to protect textiles against grease and oil stains. For this, special products are used, e.g. Fluorocarbon polymers, which are used in the form of emulsion, sometimes in padding, sometimes in the exhaustion method. However, these products in turn do not provide a good water-repellent effect, which is why in practice oil repellents and water-repellents are always used together. Some of the products available on the market provide effects which are resistant to washing and dry cleaning. Used with the Soil release finish.

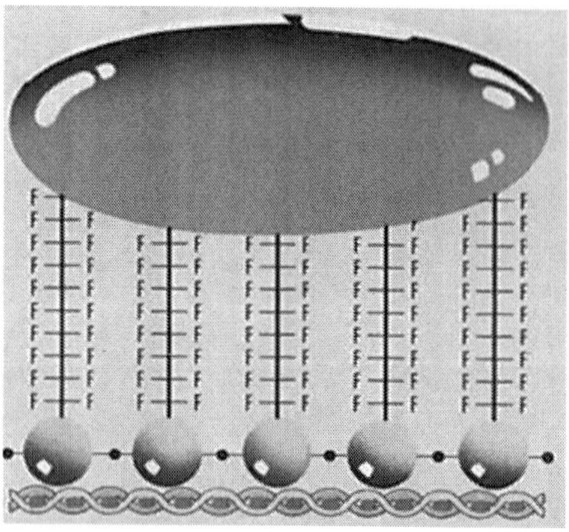

Principlr of oil repellent finish

Oiling: The oiling process aims at minimising resistance during the various operations that transform staple into yarn (e.g. Preparation process for the willowing, carding and spinning of wool fibres or cellulose flocks or blends. Purpose: Increasing sliding properties through Textile lubricants with a film-forming characteristic which ensures an effective protective shell, lowers friction, and encourages stretching ability, flexibility, mouldability, etc. In the case of cellulose, there is in addition the elimination of electrostatic charges) and then yarn into fabric (e.g. for the purpose of better processing in knitting -hosiery, warp knit and weft knit fabrics). Operating conditions: staple is sprayed with oil emulsions in water stabilised with surfactants (favouring the subsequent elimination).

Oilskin: Fabric of any natural or synthetic fibre that has been treated with linseed oil varnish. Used for protective clothing but is obsolete now.

Okinawa Jyofu: Fine and lightweight dress goods made of the leaf fibres of the banana tree in Japan; used for summer dresses.

Okra: (1) White, very light but brittle and not very strong fibre yielded by a species of hibiscus in India, the West Indians, etc.; used for ropes, cordage, etc; (2) Commercial variety of early maturing American cotton, the staple measuring 24-26 millimeters, forming small bolls; the yield is 30-32 percent.

Olala: Native Hawaiian name for the young leaves of a sedge, dried and bleached and used for fine mats.

Olanes: Printed cotton cloth in Cuba with small, usually dark red designs over a white foundation.

Olefin: Generic name for fibres derived from polyethylene or polypropylene. It is resistant to weather, mildew and chemicals, making it useful in rugs, upholstery and furnishings. It is lustrous dyes well and hard wearing.

Olefin Fibre: A manufactured fibre in which the fibre forming substance is any long chain synthetic polymer composed of quantity least 85 % by weight of ethylene, propylene or other olefin units except amorphous polyolefin qualifying under rubber. See **Rubber.**

Olefin sulphonates: a-Olefin Sulphonates are produced by reacting linear a-olefins (usually in the Cn-C range) with SO_3 and air to give mainly 2,3-alkenesulfonic acid and 1,3-sultone. Subsequent alkaline hydrolysis converts the sultones to alkene sulfonates and hydroxyalkane sulphonate. A typical product ratio is 7:3:1 alkene sulphonate/hydroxyalkane sulphonate/ disulphonate. Low levels of vinylidene sulphonates are also present. Compounds with an alkyl chain length in the range below #14 are used in liquid detergents, while those above *C14* are used in spray-dried powders. The hydroxy group gives the product mixture enhanced water solubility, but the double bond makes the mixture unstable to bleach. α-Olefin sulphonates are less sensitive to water hardness than most anionics and are readily biodegradable.

Olefirn (polyolefin/polypropylene): A manufactured fibre characterized by its light weight, high strength, and abrasion resistance. Olefin is also good at transporting moisture, creating a wicking action. End-uses include active wear apparel, rope, indoor-outdoor carpets, lawn furniture, and upholstery.

Oleic acid (Olein, $C_{17}H_{35}$–COOH) : One of the main raw material for the manufacture of soap. It is formed as a by-product of the manufacture of stearic acid. It is a colourless (pure state) to yellowish-brownish oil (unsaturated fatty acids; odourless; solidifies at 4°C, melts at 14°C; water insoluble; forms easily soluble alkali salts to form Soap. Uses: soap manufacture, lubricants for wool (absence of mineral oil, stearin, and similar substances, which are difficult or impossible to saponify).

Olein acid: Oleic acid.

Olein soap: Soap. Olein with special emulsifiers (mostly OE addition compounds), possibly with corrosion- preventing additives. With water, directly yields finely dispersed working emulsions.

Oleophilic: A term describing a substance that has a strong affinity for oils. Opposite of Oil repellent (oleophobic). Some synthetic fibres, as well as cellulose and cotton which have been given a noniron/ crease-proof finish with synthetic resins, are strongly oleophilic, but also Water-repellent (hydrophobic). Such materials are therefore easy to soil with oil, but difficult to wash with an aqueous liquor.

Oleophobic: A term describing a substance that does not have a strong affinity for oils.

Oligofilament: Term used to describe a filament consisting of a few (2–16) continuous chemical fibres.

Oligomer: A polymer molecule consisting of only a few monomer units. (Greek: oligos = few, meros = part), few duplicated parts. Low-molecular attendants in High polymers. Can be thought of as by-products not preventable by reaction kinetics (Pleionomers) in the synthesis of fibre polymers. Can be extracted using organic solvents. The structure is linear and cyclical.

Oligosaccharide: A carbohydrate formed of a small number of monosaccharide units (up to around 20). See **Sugar**.

Olive Oil: Olive oil contains 82% oleic acid and 8% linoleic acid. It is virtually pure triolein. It is a high grade salad oil, remaining liquid when refrigerated. It is also highly prized as a cooking oil.

Olona: Very strong and durable bast fibre yielded by a species of the nettle (Touchardia latifolia) in Hawaii; used for fishing nets, cords and lines.

Omb, Omg, Omf: See **owf, owb, owb**, respectively; the 'm' stands for 'mass', which is preferred to 'weight' in modern scientific language.

Ombré: The French word for 'shadow' used to describe any fabric in which the colour fades gradually from light to dark; sometimes a stripe effect is achieved , sometimes an attractive all-over shading. Fabric printed in this way amy be of any type of fibre.

Ombre printing: Printing with shaded effect. The effect is usually brought by the technique in engraving very effectively done by laser engraving, manual engraving is less effective.

On-all cotton: Raw cotton purchased under a procedure whereby the price, (Points on or off futures) as agreed between buyer as setter, but the actual futures price is left to be fixed within a stipulated period. The buyer has the right to 'call' any time within the stipulated period.

Onde: (1) A French word meaning waved, used in relation to textile fabric to describe a wave effect produced by calendaring or weaving. (2) A light French woollen dress goods; (3) French term for moire effect on silk and wool fabrics.

Ondee: Yarn made of a fine and a heavier strand.

Ondine: A thick, cord bengaline in which every cord is crinkled.

Ondule: This fabric is of similar appearance as in Ombre, i.e. wavy effecting the warp direction, also of French derivation but is produced by groups of warp threads being forced from side to side using an nodule reed that moves up and down during the weaving of a series of picks. The material is often used in curtains and furnishing fabrics.

Ondule' reed: A special reed that has a group of dented or spaced wider at the top and closed at the base and vice versa alternatively used to produce an Ondule' fabric.

O'Neill's H3 Series, Comm. Emt Jacket: Global sportswear company O'Neill is a major brand in the manufacturing of products for wakeboarding, surfing, snowboarding, etc. They have developed the Comm. Emt jacket for snowboarding, using ElekTex's technology. Techniques used by O'Neill are based on ElekTex's techniques, but the Comm. Emt jacket has big, special feathers, which are warm, waterproof, and also maintain the music player and cell phone in many environments like wakeboarding, surfing, snowboarding, etc. These factors show that music player clothes have to develop further to maintain ability not only in everyday life but also under various circumstances.

Onium dye: Cationic dye, in which the central atom effects one atom bond more than corresponds to the number of its valency electrons. The relevant anion is connected to the cationic complex via an ion bond.

Onium compounds: (cation-active compounds), definition as Complex compounds, in which the central atom effects one atom bond more than corresponds to the number of its valency electrons. Since, in onium compounds, the cation represents the active constituent, the onium compounds used in textile finishing are termed cation active (auxiliaries).

On-line: An ERP system in which the data input and output terminals and the central computer which processes the data are directly connected. The output data can be seen online.

On-line controller: A controller which checks and controls all the parameters on continuous runs of material. These include e.g. synchronous controllers, loop and tension controllers and band thickness controllers.

On-line quality control: In a manufacturing process, on-line testing of the goods currently in production is used in order to prevent non-permitted variations in quality being produced at a production location (e.g. colour). On-line testing is thus an early warning system within quality assurance, to prevent "freak values". The online testing can be inter connected to automatic adjustments of the parameters on the machine to get better consecutive results (e.g. moisture content after padding, GSM of fabric in compacting, online pH control etc.

ÖNORM: Austrian standard.

On-stream: The state of having been brought into production. The term is usually used for chemical and metallurgical plants or processes.

Onteora Rug: American rug woven on hand looms. The weft consists of piece ends of coloured denims, forming blocks, stripes and arrows as designs.

Opal batiste: Fine-filament, milky-looking cotton batiste, named after the semi-precious stone opal. Simple cotton yarns, usually maco yarns, are used for this. Finish: pre-mercerising, bleaching, acid treatment and final mercerising without tension. This creates a shrink effect which lends the fabric its milky appearance. Opal comes dyed, printed and embroidered in delicate colours, as well as in white. Used for summer clothing fabric, decorative aprons and in the cleaning industry.

Opaque finish: (1) In the case of thin knits, reduced transparency is often desirable, and this can be achieved through a calendar process. (2) Black-out curtains are intended to prevent any light entering; this is achieved by coating.

Open area (printing screen): See **Printing area.**

Open box washer: Washing machine in which the washing compartments are open.

Open Band: Yarn twisted to the right hand.

Open decatizing: See **Decatizing.**

Open end spinning: The production of spun yarns by a process in which the sliver (q.v.) or roving (q.v.) is opened or separated into its individual fibres or tufts and is subsequently reassembled in the spinning element into a yarn. (See also spinning). This system of spinning is based on the concept of introducing twist into the yarn without package rotation by simply rotating the yarn end at a gap or break in the flow of the fibres between the delivery system and the yarn package. Because the twisting element can be compact and the mass of material to be rotated is small, very high twisting speeds can be attained. The process, in a sense combines the traditional processes of roving and spinning in one operation. Present work is directed toward incorporating the drafting operation into the process by using card sliver as the feedstock. This can facilitate process linking.

Open face: A fabric defect consisting of an open appearance of the fabric which permits the filling to "grin" through the warp ends in the center portion of the fabric.

Open lace: See **Darned lace.**

Open lap: Is produced either when a subsequent underlap is in the same direction as the preceding overlap or an underlap is omitted so that the overlap of the next knitting cycle commences in the needle space where the previous overlap finished.

Open place: See **Crack mark.**

Open place: See **Crack.**

Open-shoulder construction: A method used on better coats that is characterized by hand-sewn lining shoulder seams.

Open width Centrifuge: Centrifuges for piece goods in their open-width state, also for warp- and weft tie beams.

Open width dyeing machines: These systems are used for dyeing open-width and well-flattened fabrics. These systems can be used also for carrying out pre-dying treatments (for example upgrading, bleaching, mercerising), dyeing treatments and wetting operations for both types of treatment. Among the systems used for open-width treatments it is worth pointing out mercerizing machines, jiggers, pad dyeing machines, beam dyeing machines, continuous washing systems, stenters.

Open width processing: The treatment of fabric at its full width in the unfolded state in contrast to rope form processing. It may be carried on rollers through the processing media or be held on a roll as in beam dyeing or cold pad batch.

Open width washers: An open width washer can be a simple box containing a series of vertically stacked rollers where the lower rollers are submerged in the wash water. Fabric enters the box at one end and traverses the box by going over and under each stack. Any number of these boxes can be arranged in series to provide the appropriate amount of rinsing. More sophisticated boxes will be divided into several compartments. Squeeze rolls are placed between them to speed up the removal of impurities. Some are equipped with spray nozzles which also facilitate the flushing action. There are other type designs also.

Open-width dyeing machines: These systems are used for dyeing open-width and well-flattened fabrics. These systems can be used also for carrying out pre-dying treatments (for example upgrading, bleaching, mercerising), dyeing treatments and wetting operations for both types of treatment. Among the systems used for open-width treatments it is worth pointing out mercerizing machines, jiggers, pad dyeing machines, beam dyeing machines, continuous washing systems, stenters.

Open Work Fabrics: Are fabrics like lace and net which can be made by different techniques like leno, bobbinet or warp knitting etc.

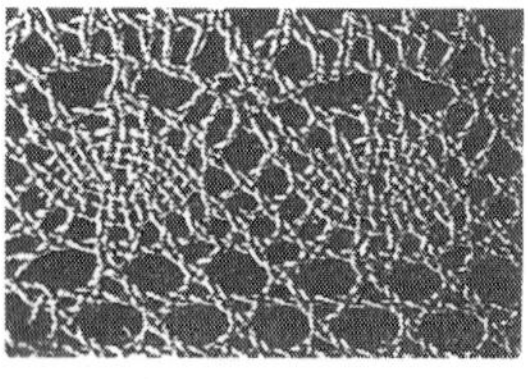

Open Wool: Wool that is not dense on the sheep and shows a distinct part down the ridge or middle of the back. Usually found in the coarser wool breeds.

Opening: The second step in commercial wool processing (after sorting). The purpose is to open up the fleece in order that scouring will be more effective.

Openness value (OV), in fibre mixing: The openness value (OV) is a measure of how 'fluffed up' the fibre mass has become on passing through a beater system, i.e., the effectiveness of the degree of opening. Since we are concerned with changes in tuft density, and to take account of different fibre densities, OV is defined as the product of the specific volume of the fibre mass and the specific gravity (SG) of the constituent fibres (or for a blend of fibres the sum of the product of their relative proportions and their SGs).

Openwork: In embroidery, laces, knitting and crocheting interstices in a ground of closer texture, often to form or enhance a pattern.

Opening: (1) A preliminary operation in the processing of staple fibre. Opening separates the compressed masses of staple into loose tufts and removes the heavier impurities. (2) An operation in the processing of tow that substantially increases the bulk of the tow by separating the filaments and deregistering the crimp.

Open lap: if the overlap and the next underlap are made in the same

direction, an open lap is produced.

Openness, in garments: The provision for leg stance in the angle of the leg seams.

Opera flannel: A narrow and very smoothly finished lightweight wool flannel, usually dyed in light colours; used for women's and children's garments.

Oporto: Coarse Portuguese wool; used for carpets.

Optical brightener: A fluorescent dye used to make white textiles appear brighter; often called just "optical" in dye industry; also called fluorescent brighteners or fluorescent brightening agents (FBA) Optical brighteners are colorless dyes that work by emitting visible light, typically in bluish hue, when exposed to invisible ultraviolet light. This bluish tint helps mask the residual yellowish cast of bleached cellulosic fibres and wool, making the fabric appear a more neutral white. These are "colour substances" which absorb light in the ultra-violet range and emit it again in the visible range. They therefore bring about an optical illusion on textiles, in that they raise the level of whiteness of the goods. This characteristic is associated with their chemical structure, as follows: Cellulose fibre brighteners are derivatives of diamino stilbene disulfonic acid. The speciality of this product is that it has trans-configuration, since only this form has a planar π-electron system enabling mesomerism in the excited state:

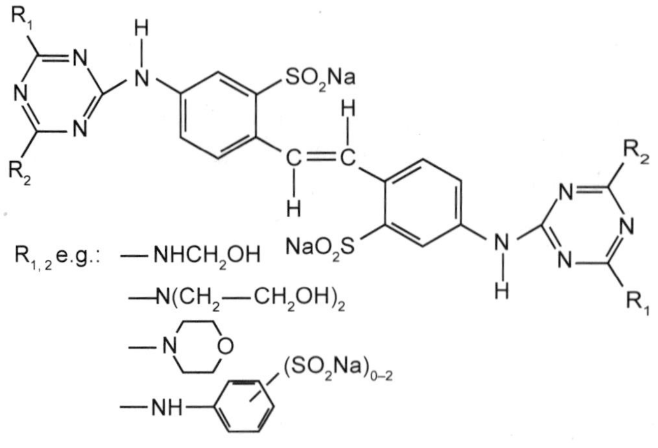

Most fully-bleached commercial white fabrics have been treated with these brighteners. Fabrics and garments that are truly prepared for dyeing should not contain brighteners. Optical brighteners can interfere with some dyes by competing for the "dye sites" on the fibres. It is almost impossible to remove brighteners once they are applied. Most commercial laundry detergents contain optical brighteners, which is primarily the reason that they are not the best choice for scouring fabric prior to dyeing. When detergent containing optical brighteners is used to wash brightener-free fabric that has been dyed to pale shades, noticeable shade change may occur.

Optical brightening: The process of applying optical brightening agent on fibre, fabric or any other material.

Optical fibre: Cladded core fibre made of inorganic or organic glass, whose core (diameter 5–100 mm) has a higher refractive index than the sheath. Through total reflection, light which enters at one end of the fibre and emerges at the other end transferred with as little loss as possible. Used for information transfer, lighting engineering, etc.

Optical finishes: Lustre may be imparted to a fabric by physical means. The techniques basically involve flattening or smoothing of the surface yarns using pressure. Beating of the fabric surface or passing the fabric between hard calendaring rolls under pressure and with some friction will tend to flatten out the yarns and lower light scattering by the fabric surface, thereby improving reflectance and lustre. Lustre may be improved further if the calendaring rolls are scribed with closely spaced lines which will be imprinted on the fabric to reinforce light striking and reflecting from the fibre surface. Similar techniques can be used to impart optical light interference patterns to the fabric (moire'). Thermoplastic fibres which can deform under heat and pressure can most readily be modified to impart lustre.

Optical properties: A general term used to refer to the relations of yarn or fibres with light. It includes such parameters as birefringence, refractive index, reflectance, optical density, etc.

Optimization: The organisation of technical and economic systems or processes in order to achieve a given aim in the best possible way, taking into account all secondary conditions. The optimization of finishing processes must be the aim in every firm. Process control through measuring and control equipment represents the means for this.

Optimum twist: In spun yarns, a term to describe the amount of twist that gives the maximum breaking strength or the maximum bulk at strength levels acceptable for weaving or knitting.

Opus Anglicum: Medieval Latin name for embroidery made in England by chain stitch worked in circular lines.

Opus Araneum: Medieval name for a coarse darned netting.

Opus Consutum: Medieval Latin name for applique.

Opus Filatorum: Medieval name for darning embroidery on a square mesh foundation.

Opus Pectineum: Medieval Latin term for a brocaded silk fabric woven on hand looms with the aid of a comb-like instrument.

Opus Plumarium: Medieval Latin name for embroidery in feather stitch.

Opus Pulvinarium: Medieval Latin name for embroidery made on open canvas ground with silk or wool in cross stitch.

Opus Saracenicum: Medieval Latin term for tapestry.

Opus Scissum: Latin name for the first cut work.

Opus Tiratum: Latin name for drawn work.

Orange peel effect: Local shrinkages in two-layered fabrics such as e.g. synthetic leather; occurs particularly when non-woven fabrics are used as carriers. The cause is likely to be an uneven fibre distribution in the non-woven fabric.

Orbis printing: (Tschekonin method). Patterning by means of a semi-solid coloured (mosaically composed) gelatine mass which has been applied to a metal roller, and which is produced beforehand and cut out in accordance with the pattern. Printing takes place with this, until the printing mass has been used up (i.e. similar to Offset printing). A new roller can then be produced in the same way. Strong prints are produced by this method.

Orchil: Violet dyestuff, used for animal fibres, obtained from a lichen (rocella tinctoria).

Orenburg Shawl: A framework knitted fabric, made by shifting certain loops sideways the distance of several needles and thus forming the design.

Orifice: Hole in end of flyer, directing the yarn to the bobbin, may also be a hook or pig's tail.

Organdie: The sheerest cotton fabric made of very fine 100 % cotton yarn. Its crisp finish is due to the application of a sulphuric acid process. Organdie creases very easily and used in small areas such as collars, ciuffs, bows sashes and cumbernunds, and in India as Sarees. It can be used as an interfacing fabric in lightweight garments and also as a mounting material. It is sometimes used for hats.

Organdy or Organza: (silk) Warp and weft non-degummed Organzine yarn woven in a lattice structure. A light very transparent fabric characterized by a rather stiff handle (Organza is a registered trade mark).

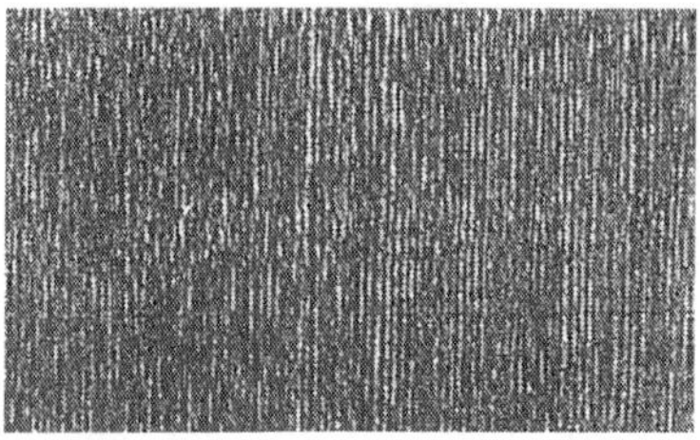

Organdy: It is a light weight, sheer, stiff, transparent plain weave cloth of typical construction 82 x 74 with 80s warp and 100s weft. Sometimes finer yarn counts are also used. Yarns are combed and are even mercerized.

Organdy is always stiff. The stiff crisp finish is obtained either by treatment of the cloth with suphuric acid at certain concentration or by other permanent, semi-permanent or temporary finishes.

Organdy finish: See **Transparent and opal finishes.**

Organic: In this glossary, this term carries the chemist's meaning: a chemical that is derived from carbon and hydrogen. It does not imply that the chemical had a biological origin, or that it is not dangerous.

Organoliths: They are mainly organic ion exchangers and is mainly divided into two groups. (1) Artificial organoliths (carbon exchangers): sulphonated high molecular humus coals (brown coal, peat, wood), as acid-resistant ion exchangers. (2) Synthetic organoliths (exchanger resins, synthetic resin exchangers, resin exchangers): porous condensation and polymerisation synthetic resins with a large inner surface, with numerous reaction-capable, ionforming acid or basic groups. (a) Cation exchangers are, e.g. condensation products of aromatic basic bodies of an acid nature, such as mono-valent or poly-valent phenols, aromatic and aliphatic sulphonic or carboxylic acids, with aldehydes, sulphonated phenol resin, sulphonated styrene, carboxylic acid resin, oxysulphonic acid resin, etc.; e.g. (in 3-dimensional terms):

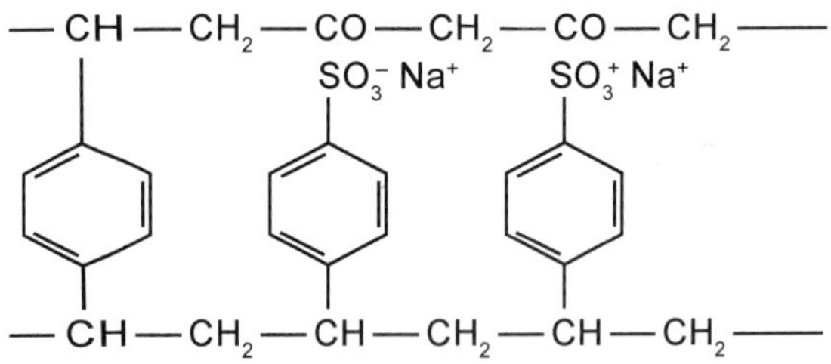

(b) Anionic exchangers are, e.g., condensation products of aromatic and aliphatic amines with aldehydes and related compounds of quaternary polyamines, etc.; e.g. (in 3-dimensional terms):

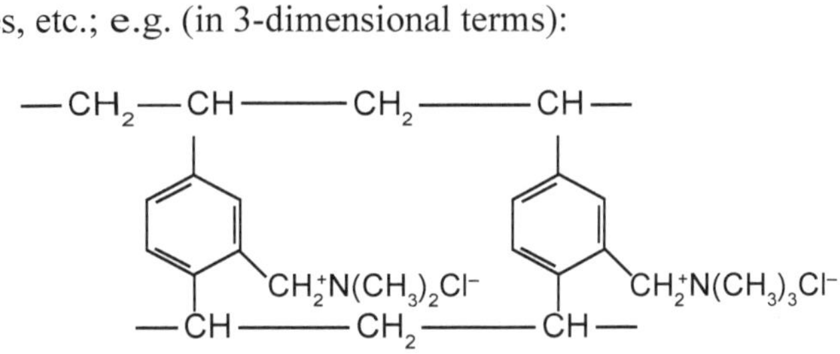

In textile these are mainly used in infeed water treatments, demineralisation (e.g boiler feed water and even in effluent treatment plants.

Organopolysiloxanes: Organic silicon compounds of an oily or resin-like condition. Used for water-impermeable finishes.

Organosiloxanes (organopolysiloxanes): See **Silicones.**

Organza: Similar in appearance to Organdie (Organdy), but made from silk viscose or polyester fibre. It is fine stiff and wiry and creases very easily. Used for evening wear and trimming and may also be used as interfacing in fine fabrics.

Organzine: Silk yarn from two or more twisted greige threads single yarns are given a preliminary twists and two or more are doubled and twisted again in the reverse direction. The preliminary twist consists of 600-700 and the subsequent one 500-600 twists per metre. Organzine yarn has a stiff handle and produces a tightly closed effect. In the boiled off state a soft lustre is produced. It is a preferred yarn for warp for plain and coloured woven fabrics, but is also used for knits, which have recently made increasing headway in to fashion collections.

Organzine yarn: See **Yarn, Silk.**

Oriental carpets: Hand-made carpets of Asia, especially Asiatic Turkey, Persia, Bokhara, Afghanistan and India. They have either hand-knotted pile or are woven similar to tapestry. The material is usually wool, but the finest Persians have silk pile. The design is characteristic of each district and good examples of carpets show beautiful rich and finely blended colours. Genuine, tightly-woven plush carpets, originating from the Orient: usually hand-knotted (typical Knot density), always knotted through (each individual knot and each pattern discernable on the reverse), irregular pattern, often not quite uniform colour (slightly stripey in the background), fringed edge formed of warp threads. Main groups: Smyrna carpets; Persian carpets; Nomad carpet. Dyeing: previously with vegetable dye (usually good fastness to light, frequently not sufficiently fast to water and washing); European imitations of Oriental carpets are more resistant to washing and cleaning.

Orientale: Single faced, ribbed, French silk dress goods with a high finish, made with ply warp. There are 18 warp ends and 18 picks in a repeat.

Orientation: a) The degree of parallelism of fibres, usually as a result of a combing or attenuating action on fibre assemblies that causes the fibres to be substantially parallel to the main axis of the web (q.v.) or strand. b) A preferred direction of linear molecules in the fine structure of fibres and usually caused by so stretching an extruded fibre that the length direction of the molecules tends to lie parallel to the main axis of the fibre. c) In the case of natural fibres, a preferred direction of linear molecules laid down during growth, e.g. a spiral around the fibre axis in cotton.

Orientation, in Button: The degree of order and spatial alignment of pearlescent pigments to the button resin formulation or to the formulation used for coating.

Orifice: Generally, an opening. Used specifically to refer to the small holes in spinnerets through which the polymer flows in the manufacture of fibres.

Original twist: The twist in a single or plied yarn component of a plied or cabled yarn as the component was before incorporation into the more complex structure. See also **Twist, Original.**

Orleantine: (1) French serge dress goods, made with 10 leaves and 10 picks in a repeat; (2) French dress goods made with two-ply warp, having eight leaves and six picks in a repeat.

Orlon: One of the first acrylic fibres made by du Pont. It melts at high heat, but does not catch fire. It is warm and absorbent, blends very well with wool and other fibres, and is used extensively in knitwear and dress fabrics.

Ormuk: A fine, soft fabric, made of the hair of young camels in Turkestan.

Ornamental stitches: The ornamental stitches most frequently used are (a) outline, (b) chain, (c) cat or herringbone, (d) blanket or loop, (e) feather, coral or briar, (f) hemstitching, (g) French knots, (h) button hole, and (i) cross stitch. Excepting the cross stitch, these are all variations of the plain and button hole stitches.

Orraye: Heavy and closely woven silk satin of Persian origin with embroidered flowers; it is made alike on both sides.

Orsey Silk: Same as organzine.

Ortho: A chemical prefix, usually abbreviated *o*, signifying that two substituents appear in adjacent positions on a benzene ring.

Orthopedic and surgical felt: A white, soft, low density, highly resilient felt.

Ortigao: Strong bast fibre, yielded by a species of the nettle in Brazil; used for nets, clothing, etc., by the natives

Ortigues: A coarse French packing canvas.

Osman: A very firm terry cloth made in England, the loops being beaten up on four picks.

Osmotic pressure: Symbol: π The pressure that must be exerted on a solution to prevent the passage of solvent molecules into it when the solvent and solution are separated by a semipermeable membrane. The osmotic pressure is therefore the pressure required to maintain equilibrium between the passage of solvent molecules through the membrane in either direction and thus prevent the process of osmosis proceeding. The osmotic pressure can be measured by placing the solution, contained in a small perforated thimble covered by a semipermeable membrane and fitted with a length of glass tubing, in a beaker of the pure solvent. Solvent molecules pass through the membrane, diluting the solution and thereby increasing the volume on the solution side and forcing the solution to rise up the glass tubing. The process continues until the pressure exerted by the solvent molecules on the membrane is balanced by the hydrostatic pressure of the solution in the tubing. A sample of the solution is then removed and its concentration determined. Osmosis is a colligative property; therefore the method can be applied to the determination of relative molecular masses, particularly for large molecules, such as proteins, but it is restricted by the difficulty of preparing good semipermeable membranes. As the osmotic pressure is a colligative property it is directly proportional to the molar concentration of the solute if the temperature remains constant; thus π is proportional to the concentration n/V, where n is the number of moles of solute, and V the solvent volume. The osmotic pressure is also proportional to the absolute temperature. Combining these two proportionalities gives $\pi V = nCT$, which has the same form as the gas equation, $PV = nRT$, and experimental values of C are similar to those for R, the universal gas constant. This gives considerable support to the kinetic theory of colligative properties.

Osnaburg: A traditional coarse cotton cloth, named after German town Osnaburck where it originated, woven with uneven yarns and often incorporating cotton waste. If the cotton used is fresh (new), it is called Clean Osnaburg. Sometimes it contains partially cotton waste it is called P.W. Osnaburg. A typical construction is 36 x 30 with 7s warp and weft. It is plain weave, resembles crash in appearance, and is the basic fabric from which Cretonne may be made.

Ostads: A twilled and thoroughly fulled woollen cloth with nap raised, shorn and calendered; now obsolete. Originally from Holland.

Ostwald hue number: The number allocated to each of the 24 hues in the Chromatic circle.

Otting process: Multicolour process for colouring carpets, with the work being done without thickening.

Otto witt theory of colour: An early theory of dyes first formulated by O. Witt provided a basis for understanding the reaction between colour and structure of the molecule. According to the O. Witt colour theory a dye is made up of two essential kinds of parts, Chromophores and Auxochromes. He designated a group that produces colour as a chromophore (Greek, Kuroma= colour + Phors = carrier). Chromophores are unsaturated groups. Presence of at least one such group is essential to produce a colour in an organic compound and a molecule containing such a group is called as chromogen.

Ottoman: A heavy fabric with broad, flat crosswise ribs of even size. Some Ottoman have small ribs, and Ottoman cord have alternating wide and narrow ribs.

Ottoman cord: See **Ottoman.** Basic yarn used may be silk or wool, but now more likely to be acetate, viscose, triacetate, or cotton. The weft cords are usually cotton yarn covered by the warp yarns. It is a stiff unyielding fabric. Used for plain dresses, coats and curtains.

Ottoman cloth: Light to heavy, stiff cloth with transverse ribbing, where the ribs are distinct and can be of even or different sizes (approx. 3–5 per cm). Worsted yarns are practically the only suitable type to use as warp yarn, and the weft can either be worsted yarn or carded yarn. Bonding: smooth or patterned ribs. Woollen ottoman cloth is primarily used for ladies' suits and coats, upholstery and soft furnishings. Fabrics used for suits and coats are often raised on the reverse side to provide better heat insulation. Usually they are piece-dyed goods.

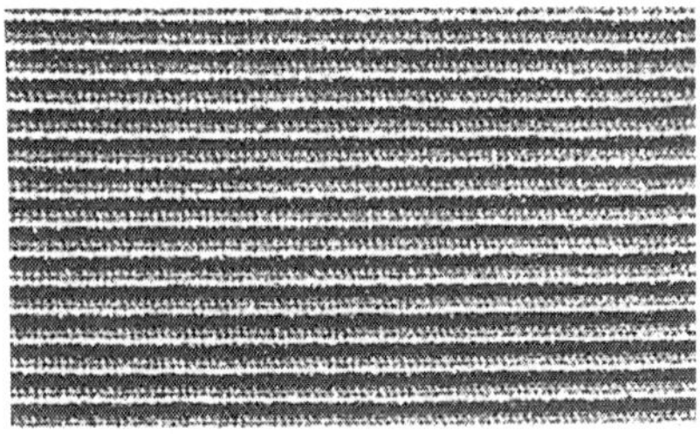

Ottoman rib: Finely ribbed ottoman cloth in particular falls into the épinglé category, and is primarily used for covering upholstered furniture.

Ottoman: A lustrous plain woven silk fabric with heavier cross ribs than faille. The warp completely covers the filling, which is of cotton.

Ottoman Cord: A silk or wool dress fabric, made with very heavy warp, forming ribs and much -finer, hard spun filling, which entirely covers the warp in plain weave.

Ouate Vegetale: French trade term for various tree cotton fibres; used for stuffing.

Ounce: (1) (fluid) - volume equal to approximately 29.57 millilitres in the US, or 28.41 millilitres in Canada and Britain.

(2) (weight, avoirdupois) - approximately 28.35 grams.

Ounce Thread: Fine linen yarn for laces and embroideries. Also called Nun's thread.

Oushak Rugs: Usually large sized, all wool rugs made in Asia Minor. The medium long, soft and loose pile is tied in Ghiordes knot.

Outerwear: Garments such as shirts, blouses, trousers, skirts, jackets, coats, as opposed to underwear. See **DOB; HaKa; KOB.**

Outflow quench: Air for cooling extruded polymer that is directed radially outward from a central dispersion device around which the filaments descend.

Outing cloth: An old fashioned term which was used to describe a variety of flannel-type woollen fabrics. They are plain or striped and often used for blazers, cricket trousers etc.

Outinq Flannel: (1) An all-wool or cotton mixed fabric; used for men's and women's outing garments. It is woven with a four-leaf, even-sided twill, fulled, napped, shorn and pressed; (2) A flannelette, made in imitation of the above.

Outline: Thin, sketchy lines drawn around (print) patterns; contour, bordering. In print designing: Fine line printing. Clarity is mainly dependent on the thickeners used.

Outline stitch: The outline stitch is the simplest of all embroidery stitches. Take a long stitch on the surface, with the needle pointing towards the chest in the line to be covered, and a short back stitch on the under side of the material. The effect of the under or wrong side of the material is exactly that of an ordinary back stitch. The beauty of this stitch depends upon its regularity and in always keeping the thread on the same side of the needle.

Outseam: The distance from the bottom of the trouser leg to the top of the pant at the waist. The measurement is taken along the outside leg seam that joins the front and back leg panels, and includes the width of the waistband.

Outside seam: A seam formed in which in which the completed seam allowance is located on the exterior of the object, usually on the face side of the fabric.

Outside-leg length, inbody measurements: The distance from the side waist to the soles of the feet, following the curve of the body.

Ouvre: French term for fabrics having checks and other small patterns, produced on an ordinary loom. See **Damasse.**

Oven: Enclosed heating equipment used by garment manufacturers to apply heat for the purpose of applying heat to a garment to set, or cure (bake), a durable press finish on the article.

Oven dried wool: Wool dried to moisture equilibrium under specified conditions.

Oven dry: The condition of a material that has been heated under specified conditions of temperature and humidity until there is no significant change in mass.

Oven-dried: A descriptive term for a material that has been heated under prescribed conditions of temperature and humidity until there is no further significant change in the mass of the material.

Oven-dry mass: The constant mass obtained by drying at a temperature of 105-110 °C.

NOTE: A ventilated drying oven (that has a positively induced air current) or other suitable oven must be used for determination of the oven-dry mass.

Oven-dry weight: The constant weight of a specimen obtained by drying in an oven under prescribed conditions of temperature and humidity.

Overhanding: See **Whipping stitch.**

Overall: Work wear usually designed to be worn over everyday clothes to give protection to body and part of the leg of the wearer.

Overcasting: Overcasting is a slanting stitch used to keep raw edges from ravelling. This stitch, like over sewing, may be worked from right to left or from left to right.

Overcheck: Two checks of different colours woven over each other, or a check design introduced over a ground check. Generally, the overcheck is of a different colour.

Overcoat: (top coat) Dark, self-shaded or conservatively patterned (mottled) carded yarn cloths (e.g. Marengos); high-quality fabric in almost all types of finish. The fabric is used for smart, timeless coats of the same name.

Overcoat stitch: Embroidery stitch. Bring the laid threads through at one point and hold with left thumb. The bring through the working thread at the same point and work small Satin Stitches closely over the laid threads, following the line of the design. The laid threads are taken through to the back of the fabric to finish. This stitch resembles a fine cord and is useful for embroidering delicate stems and outlines.

Overcoating: Fabrics woven especially for overcoats—covert, kersey, melton, beaver, frieze, vicuna, whipcord, cheviot, chinchilla, etc., made of both wool and worsted.

Overcut: A staple fibre that is longer than nominal length. Usually, the length is a multiple of 2, 3, or more times the nominal length. An overcut is caused by the failure of filaments to be cut to the desired length during staple manufacture.

Overknit: See **Double pique**

Over drying: Excessive drying of fibre material, so that the capillary water content is removed. Heat and over drying are damaging to a greater or lesser degree for any textile fibres, if allowed to take effect under uncontrolled conditions as opposed to controlled heating.

It can result in thermal splitting of polymer chains. (Drying).

Over dyeing: When dyeing takes place on top of a previous colouring process.

Overedge stitch: A type of chain stitch, in which the needle thread loop is carried from the underside of the fabric over the edge and turned to the line of stitching to be secured by the needle on it in next downward stroke. Varying amount of edge cover can be obtained by using machines employing more threads.

Over edging or over locking: The use of overedge stitching either by hand or machine to bind the edge to acvoid fraying. It is sometimes used as a decoration and often described as overlocking or serging if done by machine.

Over feed devices: Device particularly on pin stenters, to allow lengthways shrinkage during fabric tensioning (see Fig). Over feed devices consist of two pairs of powered rollers located above the clip chains, which feed the fabric to the chains at a higher speed than the chains are capable of. This means that the fabric edges are fed onto pins while they are in a crimped position. Clip stenters have specially constructed clips with slits.

Overfeed systems: See **Over feed devices.** Fabric advancing devices for compression and crimping, e.g. on pin stenters, compacting, sanforizing machines, etc.

Overflow dyeing machines: This rope dyeing machine system for woven cloth and knit goods, is particularly suitable for delicate fabric types, and was designed on the basis of the overflow principle, which is easily manageable.

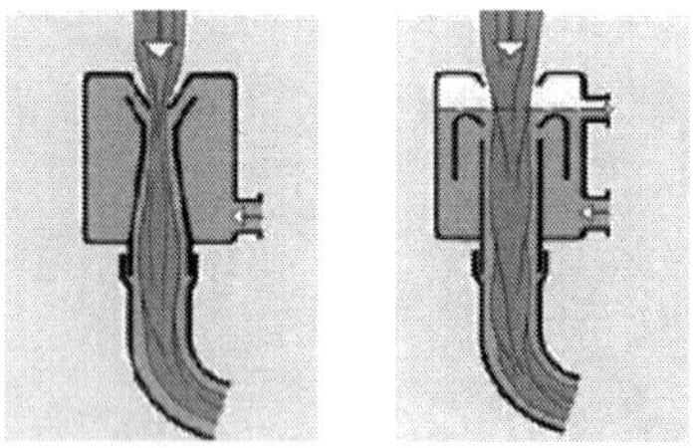

Nozzle System Overflow system

Overlap, in knitting: Overlap refers to the section of stitch in the warp knitted fabric. The yarn that is wrapped around the needle hook is called the overlap. Overlap is rarely taken across two needle hooks.

Overlength: See **Overcut.**

Over locking: A sewing process which cuts and forms stitches round a raw edge, preventing the fabric from unraveling. This is the generic name given to over-edge stitch machines used to trim and cover the rough edges of the fabric in order to present a clean and neat appearance where seam edges are visible. Over lock machines are also used for the assembly of some types of knitted articles such as T-shirts.

Overplaid: Apparently a larger plaid woven on top of a smaller plaid or stripe design.

Overprint: In dyeing, a direct print covering lighter colour shades, not affecting the dye underneath.

Over-reduction Undesirable characteristic of certain vat dyes, caused by factors such as an excessively high dyeing temperature, excessive levels of sodium dithionite in the dye liquor, extended dyeing periods, or a combination of these factors.

Oversack: An overcoat.

Oversewing: See **Whipping stitch.**

Overshirt: An outer or top shirt.

Overshot: See **Float.** Term in England for floats formed by the weft.

Oversleeve: Separate sleeves of almost any material worn by bookkeepers, clerks, etc., to protect shirt or coat sleeves.

Overspraying: A term sometimes used to describe the application, by spraying, of a fibre lubricant to staple fibres during opening and blending.

Overspun: Irregular yarn, showing too much twist at the thin places.

Over-the-counter: A term that usually refers to direct sales to a retail customer in a store, as opposed to wholesale marketing.

Owb(or o.w.b.): On weight of bath; usually expressed as percentage; omb is on mass of bath - preferred modern usage The amount of some constituent of a dyebath or other process bath based on the weight of the bath. For example, something specified as 6% owb would require 6 gram of that item per 100 gram of bath. Since the bath is invariably mostly water, which weighs 1 kilogram per litre, calculations in the metric system are much easier.

Owf (or o.w.f.): On weight of fibre, usually expressed as percentage; omf is on mass of fibre - preferred modern usage Often this is synonymous with owg but distinction may be appropriate when considering a particular fibre in a blend.

Owg (or o.w.g): On weight of goods; usually specified as percentage; omg is on mass of goods - preferred modern usage The amount of dye or auxiliary chemicals used is often based on ratio to the weight of the goods to be dyed. For example, if a formula calls for 3% dye owg, and 400 grams of fabric are to be dyed, the required amount of dye would be 3% of 400 grams, or 12 grams. Owf may be more accurate when blended fibres are considered.

Oxford weave: A modification of plain weave in which two warp yarns weave together as one.

Oxford: Bi/multi-coloured woven cotton shirt fabric with a "sturdy" appearance; usually calico weave; occasionally also available as Panama or cotton twill weaves. Due to its soft, voluminous effect, it is also known as "non-raised flannel". 2-thread warp and coarser weft, causing the Panama effect.

Oxford cloth: A cotton fabric that looks as if it were woven with a basket weave, oxford actually has a fine warp and a coarse filling. A characteristic construction is 76 x 38 with 36s warp and 15s weft. Combed mercerized yarns are used, especially in the warp. Now a days it is made with a cotton viscose or modal mixture cloth, usually in basket weave and often with a white filling yarn. It is an inexpensive fabric more expensive if mercerized cotton is used. Mainly used for shirts, light suits, dresses.

Oxford Gray: Various fabrics, made of yarn containing black and white fibres mixed in various proportions.

Oxidative washing agents: This type of washing agent can consist of soap, alkali (sodium carbonate, phosphates, etc.), stabilizers (magnesium silicate) and a source of oxygen (sodium perborate, percarbonate, perpyrophosphate). Oxygen salts ensure that the alkali level is not too high and will therefore cause no damage. Modern oxidative washing agents contain appropriate synthetic surface active detergents instead of, or in addition to, soap.

Oxidising agent: A chemist's term for any chemical that causes loss of electrons from another chemical with which it reacts; the oxidizing agent is itself reduced (see reducing agent) in the process Oxygen is the oxidizing agent whence the name comes. Chlorine is another strong oxidizing agent. As an example, when sodium reacts with chlorine, sodium chloride (common salt) is formed. Sodium loses an electron in the process (going from Na to Na), and chlorine gains an electron (going from Cl to Cl). Strong oxidizing agents + -must be handled with care, since contact with substances that burn may cause fire.

Oxo Wool: A flax substitute for wool.

Oxycellulose: Cellulose damaged during oxidation (more or less broken down by oxidizing acids, bleaching agents, kier-boiling, etc.). The properties are analogous with those of photocellulose and Hydrocellulose. Test: oxycellulose on its own: Gold purple reaction; Nessler's reagent; Phloroglucinol reaction. In addition, as with hydrocellulose, with: Prussian blue reaction; Fehling's solution; Methylene blue test; Vatting test, Turnbull's blue reaction.

Oxycellulose, test for: See **Gold purple reaction.**

Oxygen demand: (waste water), The usual methods of analyzing waste water pollution are the Chemical oxygen demand (COD), measured by the amount of potassium dichromate ($K_2Cr_2O_7$) used, or the Biochemical oxygen demand (BOD).

Oxygen fading: See **Ozone fading.**

Oxygen number of plumage: A measure of the degree of cleanliness; the amount of oxidisable water soluble and fine suspended matter present in the water extract.

Oxygen depletion: Bacteria are responsible for the depletion of most organic impurities, consuming oxygen at the same time. It is essential that the microorganism/ waste water interface provides water-soluble conditions, and that the molecular weight is low, to ensure that penetration into the bacteria cell is possible. The bacteria cells do not absorb polymers.

$$\text{Organic substance} + O_2 \rightarrow CO_2 + H_2O + \text{new Bacteria (Biomass)}$$

Oyah Lace: Coarse Turkish crochet lace made of coloured silk yarn.

Ozier: Early maturing .commercial variety of upland cotton, the staple measuring 25-28 millimeters, forming medium bolls; the yield is 30-32 percent.

Ozonation, in ETP: The injection of an ***ozone*** air mixture into clean filtered water, normally to disinfect the water but sometimes for other reasons, such as removal of colour or oxidation of manganese. Ozone quickly decomposes to oxygen. It is made on-site in an ***ozoniser***. Ozone is a powerful disinfectant. For water disinfection, 5 to 10 minutes contact time is needed if there is 1% ozone in the ozone-air mixture injected into the water. It is a more efficient disinfectant than chlorine, but ozone decomposes in a few minutes whereas chlorine residuals last many hours. Ozone has a lower tendency to make undesirable ***disinfection by-products*** than chlorine (*see* ***bromate***). The main disadvantage is the cost of making the ozone and the lack of any ***residual disinfectant***. For wastewater disinfection, the lack of any residual ozone can be an advantage, because, in ***chlorination***, the chlorine residual can be a problem for the receiving water

Ozonator: A device which generates ozone by passing a high voltage current through a chamber containing air or oxygen. Used as a didinfection system.

Ozone (O_3), oxygen molecule with three atoms. Individual free oxygen atoms (known as "oxygen in status nascendi", which have an extremely high level of oxidation energy Ozone bleach) accumulate around O_2 molecules and combine with them to form ozone.

Ozone bleach: Industrial development of an old type of bleach known as grass bleach, designed specifically for linen, less suitable for cotton. It has not been implemented on a large scale, both for cost reasons and also because it has no particular advantages over other bleaching processes.

Ozone fading: (oxygen fading), the appearance of oxidative bleaching of blue, red and yellow dispersion dyes caused by the effect of atmospheric ozone, which is similar in appearance to gas fume fading. This effect is most common in acetate, triacetate and polyester dyes. Heat treatment of triacetate and polyester improves fastness to ozone fading. Ozone fading can be prevented, by using anti-oxidation agents such as diphenyl ethylene diamine (also used as a gas fading inhibitor) and para-octylphenol (which has no gas fading inhibition properties), which means that using the methods detailed here, both gas and ozone fading can be prevented at the same time.

Ozone fastness: Common in the USA, meaning the resistance of textile colour to the effects of atmospheric ozone in the following conditions: (a) Air humidity below 65% at normal temperature; (b) High air humidity of at least 80% at a raised temperature (40°C). Samples, and also control samples, are subjected to the effects of ozone in special chambers under the above conditions. The colour change is evaluated by means of a grey scale after a certain number of cycles.

Ozu Aya: In the Japanese trade name cotton jean.

Ozoniser, ozonizer: An electrical discharge unit operating at 5000 to 20 000 V and 50 to 500 Hz, which converts some of the oxygen in air to ozone. It produces air with about 1% ozone. Dust and moisture must be removed from the air before entering the ozoniser, otherwise flashover occurs.

P

P.F.D: Short form for Prepared For Dyeing.

P.F.P: Short form for Prepared For Printing.

PA: Polyamide fibres.

Pack, in spinning: (1) The complete assembly of filters and spinneret through which polymer flows during extrusion. (2) A unit of weight for wool, 240 pounds. (3) Linen yarn measure, equal to 60,000 yards.

Pack duck: In England a coarse, stout linen fabric; used for packing.

Pack life: The time during which a pack assembly can remain in use and produce good quality yarn.

Pack Thread: Very strong twine; used for tying bundles. It is made two or three ply, of hemp or flax, in various thicknesses and fineness.

Package: See **Packages.** Wound package, consisting of a tube and winding, e.g.: Pirn; Crosswound yarn package; Flanged bobbin; Hollow cop.

Package build: A general term that applies to the shape, angles, tension, etc., of a yarn package during winding. Package build affects performance during subsequent processing.

Package density: Density of a package is its g/volume. Uniformity of package density is important.

Package dislocation: Deformation of crosswound yarn packages, yarn slides off the dyeing tube.

Package dyeing: Also called **Pack dyeing**: The forced circulation of dye liquor through packages of fibre, yarn or fabric, without limitation of temperature. Advantages: Short liquor ratio, direct use of the yarn package for warping or warp yarns on the sizing machine. Disadvantages: Absolute evenness cannot be guaranteed in hard packages.

NOTE: The use of the term "pressure dyeing" in this connection is deprecated. (See also **high-temperature dyeing**).

Package dyeing: See **Dyeing.**

Package winding: The winding of a Package (Crosswound yarn package) determines its properties during further processing and thus the quality of the end product and the effectiveness of production. This has a critical influence on characteristics such as for example consistent package density or good run-off characteristics. Traditionally, there are two package winding systems: Random winding and precision winding.

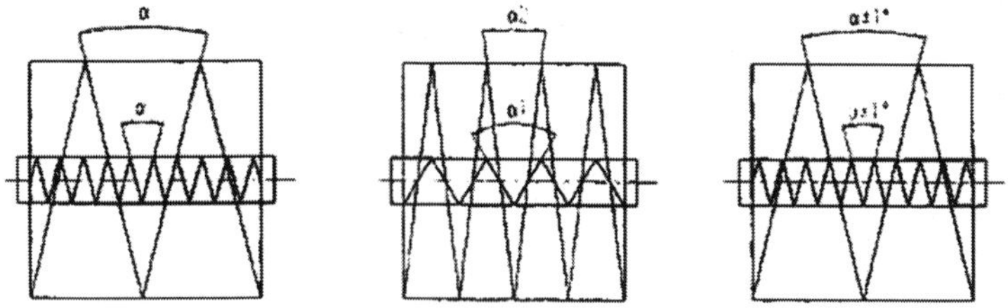

Random winding Precision winding Stepped precision winding

Package dyeing machines: (1) Yarn packages are packed and if necessary, compressed on material carriers, which are transported by conveyors and craneways to the dyeing machine in question. In dyeing machines supplied with a centrifugal pump or an axial pump the dyeing liquor flows through the yarn package columns. Dyeing machines, circulating liquor type. (2) Dyeing machines for crosswound yarn packages, bottle package, combed top bobbins, card sliver and warp beams.

Package hardness testers: The package hardness is an important criterion for faultless full penetration dyeing. A densiometer, which detects the Shore hardness, is used for hardness testing.

Package hydro-extractor: (package centrifuge) Centrifuge hydro-extraction of yarn packages. These are deformed in normal centrifugal baskets, which is why baskets with spool carriers or single spool centrifuges are used.

Packages: A large selection of forms for winding yarn is available to meet the requirements of existing machinery and a variety of package builds is used to ensure suitable unwinding in later stages of manufacturing. Since a package with flanges cannot be unwound easily and quickly by pulling the yarn off overend, most packages are flangeless with self-supporting edges. Some can be unwound at speeds up to 1500 yd/min. The accompanying diagram shows six common types of yarn packages.

Packed bed filter: Any treatment process in which a solid medium is packed into a tank or tower. Examples in water and wastewater treatment include anaerobic filter, pre-anoxic filter, denitrifying filter, nitrifying filter, rapid deep bed filter.

Packed bed scrubber: A type of wet scrubber consisting of a tower packed with media (typically plastic media or metal mesh or wooden slats) down which the reagent (scrubbing or wash water) trickles. The air flow to be cleaned normally flows up through the tower (i.e. countercurrent to the scrubbing liquid). Gases with a high concentration of particulates may clog a packed tower. A packed bed scrubber can also be used as an entrainment separator. *See* **Moving bed scrubber.** (1) A bioscrubber. (2) The term may be used to mean a packed tower aerator.

Pad: (1) Impregnation on a padder. Differentiated according to procedure e.g. Pad-batch process; Pad-dry- process; Pad-jig-process; Pad-Roll-Process; Pad-Steam-Process.

(2) A flat, compact cushion or mattress shaped implement laid upon the lap-board or any other press-board and used in "pressing-off" and giving form and finish to a garment. Its elasticity accommodates it to any inequalities of the garment and therefore the whole surface is reached by the pressure of the iron and made smoother and solider than it could be without its use. See **Looney**.

Pad batch: A sequence of operations involving padding and batching without intermediate drying.

Pad batch process: (Cold pad batch dyeing), performed on the padder; high performance fabric opening and rolling machines guarantee crease-free material flow without edges rolling up from material entry to the nip. On the way from the nip to the rolling point the material is fed over additional rubberized rollers. The fixed material feed prevents the material from stretching lengthways and the edges from rolling up. This requirement to prevent rolling up on the batch roller is achieved by an adjustable hydraulic force on the last transfer roller. The material tension is modified to suit the process using a computer controlled centre roller. The following material qualities can be dyed economically, gently and with reproducible effects using the pad-batch process:

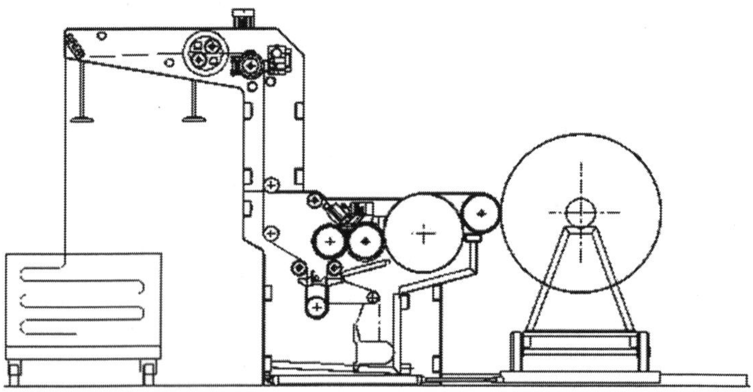

(a) Knitted fabrics: Single jersey, interlock, lining fabrics, nikki plusch,

(b) Woven fabrics: Mercerized and non-mercerized cotton fabric, inlet material, cord, terry fabric.

After the padding and cold dwelling the material must be washed in a suitable manner.

Pad dyeing: See **Dyeing.** Pad dyeing (continuous or semi-continuous systems). This process is carried out using mechanical means (pad-batch wetting). The dyeing liquor is distributed homogeneously onto the fabric (i.e. also the dye is distributed homogeneously). In a second stage the dye penetrates into the fabric and is then fixed. At the end of the process the material is washed.

Pad printing: Complete printing of a material with the Kiss roll; Thousand-point roller or hatching roller.

Pad process: Used for padding (Padding). The material is impregnated with liquor containing the dye or chemical on the padder, whereby the substantivity of the application product is almost eliminated due to shorter submersion time and shorter liquor ratio. Unlike the exhaustion processes, material flow is continuous, as the first work stage, if necessary followed by a second wet treatment (wet-on-wet-padding).

Pad roll: A sequence of operations involving padding and wet batching with a provision for intermediate heating. In this process goods are padded through the scouring solution and then batched open width onto rolls which rotate slowly for several hours ranging from 6 to 24 h depending upon the type of cloth. The goods are then washed and rinsed in open width washers.

Pad-Develop-Thermosol-Process: Pad-dyeing technique for polyester/ cotton using vat leuco ester and disperse dyes.

Pad-dyeing: The Pad process is replacing discontinuous dyeing methods for woven fabrics to an ever greater degree. Function mode of padder with swimming rolls.

Pad-fix-process: Pad-dyeing process for polyester spun tow. The tow, which is padded with the dye dispersion, is either packed into a packing cylinder or wound onto a perforated cylinder and dyed in the liquor in the HT circulation machine.

Pad-steam-roll system: After impregnation with scouring liquor, the fabric is quickly pre-heated in a steam chamber or by an infra-red radiation. When it has reached a high enough temperature, the fabric is introduced into a transportable reaction chamber heated by steam injection, where it is stored in a rotating roller. The reaction time is of the order of 1-1.5 h at 100°C.

Padder: Open-width dyeing machine, padding machine for open-width treatment, padding, impregnating, (padder) delustering etc. of piece goods (loose material on special padders.) The task of the padder is to remove the excess liquor absorbed in the dyeing trough by uniform squeezing across the entire width of the fabric.

Padder-Jig-process: Garment panels are padded with dye solutions and developed or fixed on the jigger. The process is of particular practical interest in the dyeing of cotton and linen woven fabrics using vat dyes and also direct dyes to a lesser extent. Advantage: Better penetration and dyeing of larger batches. Disadvantage: The discontinuous vatting can sometimes cause lack of end-to-end uniformity in the full bath of the jigger, because some of the dye is transferred into the blank vat. These faults can be rectified by the addition of padding liquor to the bath at the start and end of the first passage.

Pad-roll bleach: Semi-continuous process for cotton, cotton/polyester mixes, reclaimed cellulose. Material is impregnated with bleaching liquor and batch rotated and dwelled in a heated chamber.

Pad-Roll-Process: Discontinuous pad-dyeing process for dyeing cellulose fabrics with direct dyes, in batches of up to 5000 m. The woven fabric impregnated with the dye is heated up to approximately 100°C in a pre-chamber by IR emitters and wound in an adjoining travelling and heatable chamber. When winding is complete the chamber is connected to a steam line (direct and indirect), whilst the skein slowly turns ("roll"). Dwell time is dependent upon the colour depth.

Pad-steam continuous dyeing: A continuous dyeing process for many dyes where the dye and chemical is padded in one or two padders and steamed for fixation and the further treatment is done continuously. One of the widely accepted method of dyeing in the open width form.

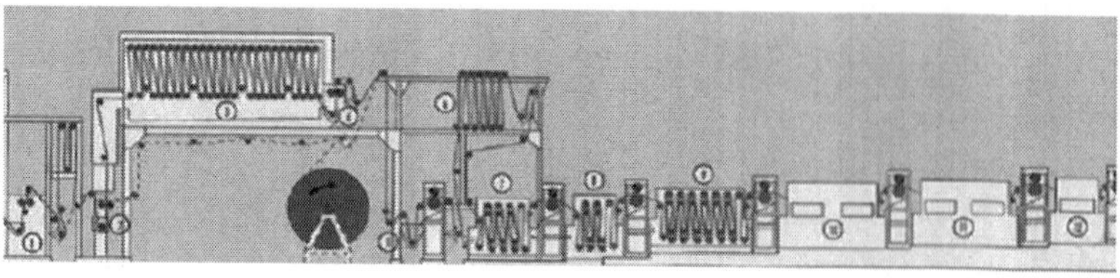

Pad Steam Range

Pad-steam process: A process of continuous dyeing in which the fabric in open width is padded with dyestuff and, if necessary, with a reducing agent, and is then steamed. (See also padding).

Pad-thermofix process: General term for the pad dyeing process using reactive and/or disperse dyes.

Pad transfer technique: Low wet pick-up technique by transfer, e.g. QS-process.

Pad-wet fixation process for reactive dyes: Pad-wet fixation process for reactive dyes (Hoechst), continuous process according to the principle: Padding with neutral dye solution, drying, passage through hot, saline, alkali solution.

Pad-wet process: Dyeing process, in which development or fixation of the dye after padding takes place without an intermediate drying stage.

Pad-winch process: (Pad-Winch beck), process for dyeing difficult to penetrate woven fabrics (above all those made of viscose staple fibres with direct dyes). Material is padded with the dye solution and then treated by boiling with a sodium sulphate or sodium chloride

solution on the winch beck, whereby the dye is fixed.

Padder-jig: Combination Padder with Jig dyeing, whereby the upper padder roller functions as a jigging roller. Used for discontinuous dyeing (Pad jig-process) with vat, vat leuco ester, reactive dyes, etc.).

Padding: (1) The application of a liquor or a paste to textiles, either by passing the material through a bath and subsequently through squeeze rollers, or by passing it through squeeze rollers, the bottom one of which carries the liquor or paste. The entire padding function can be divided into elementary functions: (a) Submersion of the textile material in the liquor for mixing. (b) Squeezing the loaded textile material for proportioning. (c) Transport of the reaction mixture (textile material with defined product application).

(2) A dyeing method with very low liquor to goods ratio, where typically only enough strong dye solution is used to saturate the fabric Padding can have the advantage of high dye yield. Padded goods are usually “batched” - wrapped in plastic and left for some period of time for the dye to attach to the fibre, or steamed to fix the dye quickly. Padding methods are sometimes used for other textile processes such as bleaching.

(3) Felt, cotton or wool used for giving shape to garments.

Padding machine: A machine used for padding. See **Padding, Padder.**

Paddle dyeing machine: A machine used for dyeing garments, hosiery, and other small pieces that are packaged loosely in mesh bags. The unit consists of an open tank and revolving paddles that circulate the bags in the dyebath.

Paddock: (1) It is a worsted cloth which resembles gaberdine in weight and moisture repellency.

(2) A rather long, slightly formfitting overcoat, somewhat similar to the Newmarket.

Pador gum: Lowest quality class of Senegal gum.

Padou: A narrow silk ribbon in France.

Paduasoy: Stout, rich Italian silk fabric; name obsolete, the fabric being known as peau de sole.

Paenula: Ancient oval or diamond shaped cape with head opening, closed all around or slit at the front; often with a hood.

Pailletine: (1) (Yarn) rice straw yarn for furnishing fabrics. (2) (material) Fine clothing material similar to Paillette; 8 Thread, reinforced satin.

Paillette: (1) Sequins, mica; thin perforated metal or glass plates for applying to clothes.

(2) Light, shiny clothing, blouse or lining satin with a soft handle (Satin de Chine); 5 or 7 thread warp satin.

Pailletine: (1) (Yarn) rice straw yarn for furnishing fabrics. (2) (material) Fine clothing material similar to Paillette; 8 thread, reinforced satin.

Paisley: A fine fabric always with scroll designs. Originally only from Scotland, a 'paisely' meant a shawl made in fine soft woollen yarn, with a woven pattern. Original designs are from India. Real paisley cloth is very expensive, but anything of any fibre or mixture printed with the traditional elaborate multicloured scrolls is described as paisley design.

Paisley pattern: Distinctive tear drop/mango shape decorative pattern based on traditional on Indian pine or cone motifs.

Paita: Variety of raw cotton from Peru; see **Payta**.

Pajamas: A two piece ensemble consisting of a top and bottom generally used for, but not limited to, night wear.

Palampores: A kind of chintz bed-cover, sometimes made of beautiful patterns, formerly made at various places in India. The word palampore is originated fron the Hindi/Persian word Palangposh (Bedcover) but palmpores were not just ordinary counterpanes.The printing is done using laborious kalankari (printing with natural vegetable dyes)technique that involved penpainting a part of the overall design with a mordant, then dyeing. The dye would fix to the cloth where the mordant has been applied, and the artist would then sketch another part of the design for dyeing in a different design fir dyeing in a different colour. The process is repeated till the entire design is complete. Since it is time and labour intensive the Palampores is never cheap.

Palempore: See **Palampore.**

Paletot: A long, loose overcoat.

Palla: Ancient rectangular woollen cloth used as a robe by women out of doors; often covered the head.

Pallium: Ancient men's rectangular woollen cloth outer robe, wrapped the body, over the tunic.

Palmer: Open-width tensioning device on the felt calendar for decatizing/ drying open-width materials.

Pallas: French pile fabric; used for coating, made of cotton warp and filling and long goafs' hair warp. It is either dyed black in the piece or printed in fur effect.

Pallav: The end part of Sari worn by Indian women. On wearing the sari the pallav comes out over the shoulder towards behind of the person, or sometimes in the front of the person, in different type of wearing a sari. Pallav will have different pattern compared to the other parts of the sari. The nature of design will be in harmony with the body of the sari. There is always a distinct relationship between the border, body and the pallav. Some saris will have Pallav at both ends (usually the nine yard Saris). The two pallavs are embellished differently. This enable the wearer to drape the garment from either end with different pallav each time.

Palm fibres: Vegetable fibres of numerous palm types belonging to the group of Hard fibres, mainly in the form of so-called vegetable horse hair substitute, for brushes, braiding, upholstery material, etc. We differentiate between the dwarf palm fibre (Algeria, Morocco, Spain), the date palm fibre (similar to coir fibres) and further bassine, crin d'Afrique, kitul fibre, panama fibre, para fibres (the most important), piassava fibre, rafia bast, tecum fibre. They are all of regional importance only.

Palm Oil: Palm oil has a high content of a C16 saturated acid (40% palmitic acid). It also has a high content of oleic acid, 43%. This combination of fatty acids is ideal for making good toilet soaps

Palmer: A Palmer unit is a large heated cylinder with an endless rubber belt pressing against it. Normally used along with Sanforising, Safor-set processing units in the final stage of the finishing process to give an ironing effect with sheen, for the fabric. Open-width tensioning device on the felt calender for decatizing/drying open-width materials. Modern design Pin wheel.

Palmet: Strong black leaf fibre yielded by the Prionium palmita in South Africa; substitute for horsehair.

Palmette: Shawl made in France with a foundation of two-ply wool warp and combed wool filling, and carded wool yarn for the figures.

PAM: Modacrylic fibres.

PAN: Polyacrylonitrile fibres.

Panache: French for high and variegated coloured effect.

Panama: Very lightweight cloth used for men's summer and tropical wear and women's dresses. The yarns used are cotton and worsted in the Traditional Panama cloth, although other fibres such as polyester and viscose are sometimes used. The cloth has recognizable square-weave effect, and is usually in plain colours.

Panama: Plain woven worsted dress goods, dyed in the piece; hopsacking made of coarse yarn in basket weave, made plain or in two colours. There is also a dress goods of cotton warp and double wool filling producing an effect similar to the texture of the Panama hats.

Panama canvas: Thick, cream coloured, basket weave canvas; it is beetled and used for embroidery.

Panama weave: Consists of several warp and weft threads crossing each other at once, producing a mat like effect.

Pandanus: A palm in East Africa, India and Polynesia; the leaves are used for mats, baskets, hats.

Paneled Back: The inside back of a coat made with broad facings of the goods around seams and edges, and filled in with (panels of) silk or other lining.

Pang: A Chinese silk dress goods.

Pangalo: A new variety of Egyptian cotton; the staple is described as brownish, of silky gloss and strong.

Panier: Framework for 18th century hooped skirt; typically very wide, but flattened at front and back.

Panna satin: See **Satin.**

Panné Satin: A satin fabric with an unusually high lustre because of the application of very heavy roll pressure in finishing. Panné satin is made of silk or one of the manufactured fibres.

Panne' velvet: Knitted backing of polyester, silk or acetate lustrous, lightweight velvet fabric, usually made of silk with a short pile on the right side that is pressed flat during manufacture.

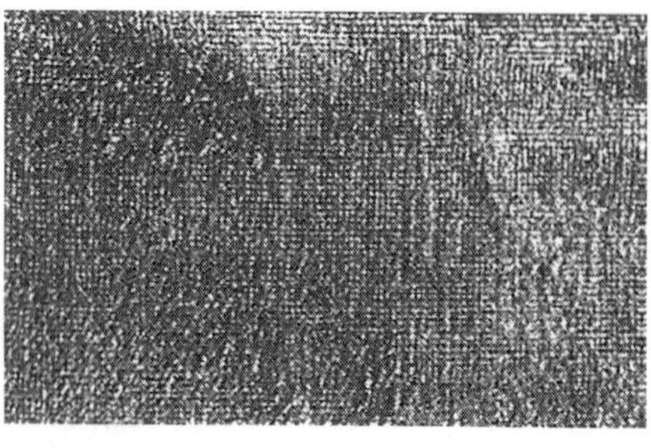

This gives the effect of a shiny shimmery velvet. The fabric is soft and floppy comfortable to wear. It is easiest of all velvets to sew. Used for evening wear, robes, leisure clothes, tops and trousers.

Pannus: A medieval silk fabric, made in Italy.

Pantaloons: Originally calf or ankle length narrow breeches, later full length and held with stirrups.

Panting: Same as trousering.

Pantograph: (micrograph), drawing instrument used in pantograph engraving (Engraving) for the manufacture of calendar rollers. The pattern is transferred from zinc plate onto a calendar roller, which is painted with acid resistant paint, by a heated diamond pantograph, and the engraving is enhanced by etching.

Pantone: A company and its trademark. The Pantone company produces a wide range of colour guides useful in almost any industry dealing with colour. The colours in the guide are widely accepted as standards. Colours are sometime seen described as a Pantone number: Pantone applies a unique number to each colour.

Papeline: Originally a lightweight dress goods, made in Avignon, France, of silk warp and silk waste filling in plain colours, or figured; one side had a selvage of different colour than the warp.

Paper cambric: Lightweight, lustrous cotton lining.

Paper cloth: Cotton, hemp or jute fabric coated with paper on one or both sides, between heavy rollers; used for boxes, etc.

Paper muslin: Light weight cotton muslin, sized and given a lustrous finish; used for lining.

Paper tafetta: A fine-weave, light-weight taffeta which has been treated to make it crisper than usual, hence the name paper. The crispiness is usually a permanent finish.

Paperwork, in designing: Designers initial drawings and design ideas produced on a paper.

Papering: The insertion of cold and heated board elements (paper) prior to pressing, usually in a hydraulic press.

Paper makers felt: A fabric made from wool or manmade fibres or mixtures of both, fabricated as an endless belt for use on a paper making machine to form an endless conveyer belt for pulp or wet paper in its passage through the machine. These products are now also made of various constructions, woven and nonwoven, of manufactured fibres and monofilaments.

Paper yarns: (1) Made of cut narrow strips of spinning paper which (possibly after previous folding) are twisted together on a spinning machine in a moistened state (drilled). Sulphite paper or (tear resistant) sodium or sulphate paper or mixtures of these are used. Finenesses from 33–1000 tex, resistance to tearing from 4–7 cN/tex at 4.5–7% elongation at tear.

(2) Wood pulp yarn is created by twisting together band shaped, wet strips of wood pulp, without previous processing into spinning paper. Finenesses from 33–2000 tex, resistance to tearing 7–9 cN/tex. Application: For carpet underlays, carpet runners, wall coverings, bags, cord, hat braiding, etc.

Papery: Excessive smoothness given to the cloth with the aid of sizes in the finishing process.

Papoon: Plain woven cotton fabric; used in India; made of different coloured warp and filling or in small checks.

Papyrus: The paper reed, Cyperus papyrus, of Egypt: the stem fibres are used for cloth, sails, mats, cords, etc.

Para: (1) A chemical prefix, usually abbreviated ***p***, indicating that two substituents on a benzene ring are separated by two carbon atoms.(2) Variety of raw cotton from Brazil.

Parachute cloth: A close-weave, lightweight, synthetic fibre or silk fabric with high bursting and tearing strengths

Paraffin: Paraffin waxes come from a purified mixture of solid hydrocarbons obtained from the distillation of petroleum. This mixture contains C23 to C33 hydrocarbons. Straight chain hydrocarbons are solids, branched ones are oils. The distillate fraction contains both straight and branch chain isomers since they have identical boiling points. When the distillation fraction is chilled, the straight chain hydrocarbons solidify and can be filtered from the oil.

Paraffin: A wax extracted from crude (petroleum) oil (North American usage; in Britain, paraffin means kerosene) Paraffin wax (actually a number of waxes with somewhat different characteristics) is used, by itself or in mixtures with other waxes such as beeswax, as a dye *resist* in batik. It quite brittle and cracks easily. "Canning wax", sold in North America (and elsewhere, presumably) for sealing home-made jam and jelly and the like, is paraffin. Most candle wax, other than beeswax, is paraffin.

Paraffin waxes: See **Paraffin.**

Paraformaldehyde: Mixture of different polyoxymethylene-dihydrates, DP 3–100, approx. 95% formaldehyde content, develops gaseous formaldehyde from 100°C (space disinfection).

Paraguay lace: Single threads are used to produce spiders web effects/ wheel design, which are then woven together.

Parahyba: Variety of raw cotton from Brazil, having a fairly strong, harsh staple of white colour.

Paring-Off: The process of paring-off the surplus cloth along the margin of a garment, when the edges are made raw.

Paring-Shears: Shears or scissors made expressly for paring the edges of garments.

Paralleling: The process of aligning fibres to produce a more uniform, smoother, stronger yarn.

Parchment cotton: A fine, plain woven colon fabric in England, sized and finished to resemble parchment.

Parellel laying: In the production of a web or batt from single or superimposed card webs laid in the direction of carding for the manufacture of non-woven fabric.

Parallel winding: Even or uneven, slightly helical thread winding, with threads lying side-by-side or one on top of each other. Yarn packages, preparation; Random winding.

Paris Embroidery: Consists of small leaves and berries embroidered with white cord on pique or with floss silk on coloured satin or cloth.

Parisienne: (1) A silk cloth in France made with small patterns; (2) A figured orleans, very fashionable in the middle of the 19th century in France and England; (3) Very soft, lightweight French dress fabric, made black, of merino wool.

Paropa: A fabric made of silk and wool, used in England during the 16th and 17th centuries. Also called peropus.

Part run: A partially filled bobbin that occurs when an end breaks before the completion of a doff cycle. The total weight of yarn normally wound during a cycle is not obtained on the bobbin at the break position. The number of part runs is used as a measure of spinning performance.

Part set: A term that indicates that all guide eyes in a guide bar do not have a **yam** from the warp.

Part wool felt: A felt composed of any one of or a combination of new and recycled wool fibres mixed with one or more man-made fibres, vegetable fibres, or animal fibres other than wool.

Parthenos: French silk velvet dress goods.

Parti-coloured: Hose of different colours, or garments with panels of different colours.

Partial cleavage: In textiles a transverse gouge cut or other crosswise rent in the fibre. Clearly penetrating at least the cuticle of the fibre.

Partially oriented Staple: Staple fibres cut from tow that has been drawn less than normal so that only partial longitudinal orientation of the polymer molecules exists.

Partially oriented yarn: Filament yarn in which polymer molecules are only partially aligned (by reducing the draw ratio).

Particle filteration (PF): Filtration rated in the range of 1 to 75 microns. Typically handled by catridge filter.

Particulate: Minute, separate pieces of matter.

Partlet: Ancient women's broad shoulder collar used to cover a wide neckline.

Partridge Cord: A mottled corduroy.

Pashim, or Pashmina: Very fine, downy wool found under the hair of the cashmere goat in India; it comes in gray, white or drab colours and is used for the finest rugs and shawls.

Pashmina: Woollen shawls having one-sided embroidery (they have distinct obverse and reverse face) made in Kashmir, India. However on some shawls the embroidery appears similar on both sides of the shawl. In the finest shawls, embroidery on each side of the shawl bears different motifs, accomplished by splitting the warp and embroidering a different motif on each side.

Passing: Thread consisting of a narrow flat strip of gold twisted around a silk core. Used for embroidering and tapestry work. The smallest size is called tambour.

Passing off, in designing: Intentionally trying to deceive customers into thinking a design is by a particular individual or company. This happens frequently with designer labels and branded textiles.

Paste resist printing: Synonym of paste resists (Resist printing): Purely mechanical pre-printed resist made of many high solids thickeners, thickener containing a weighting agent (e.g. kaolin, zinc oxide, lead carbonate, etc.), preferably using indigo and vat dyeing.

Paste resists: See **Paste resist printing, Resist printing.**

Pastille: (1) Pattern consisting of dots only. (2) Round dots, usually of velvet or other thick material woven or appliqued to nets, veils, etc.

Pastourelle: Plain French serge, made with eight harnesses and four picks in a repeat.

Patagonian: Long staple, heavy shrinking wool from Southern Chile, washes very white; used for hosiery.

Patch Pockets: Pockets made wholly on the outside of a garment.

Patchwork: Needlework, consisting of joining various coloured and shaped clippings of materials with fancy stitches, to form quilts, covers, etc.

Patent Axminster: Pile carpet woven on power loom, chenille being used for filling.

Patent Beaver: All-wool beaver cloth of very fine quality, filled very thoroughly, making the fabric almost waterproof.

Patent Flannel: A very light and sheer English flannel.

Patent leather: A varnished leather that is very expensive. Used mainly for shoes and hand bags.

Pathogens: Communicable diseases can be transmitted by the pathogenic organisms that may be present in the waste water.

Patole: Bordered silk fabric with printed, embroidered or hand painted patterns, made in India. It is made about a yard wide.

Pattern: (1) An arrangement of form; a design or decoration such as the design of woven or printed fabrics. (2) A model, guide, or plan used in making things, such as a garment pattern.

Pattern attachment: A device which controls the pile yarn feed so as to produce a high and low pile surface.

Pattern co-ordination: Co-ordination usually achieved by using selected shapes and motifs from the main design to produce related patterns.

Pattern direction: The way a pattern lies on a fabric.

Pattern draft, in weaving: **Point paper design**: The diagrammatic representation of the construction is called pattern draft or point paper design. It is marked and read from bottom left to the top right. Vertical columns represent the warp yarns and horizontal rows represent the weft yarn. A mark placed in any cell indicates that the warp is lifted over the weft at that insertion. Absence of a mark indicate the reverse (warp remains below weft). Each cell represent a warp and weft interlacing.

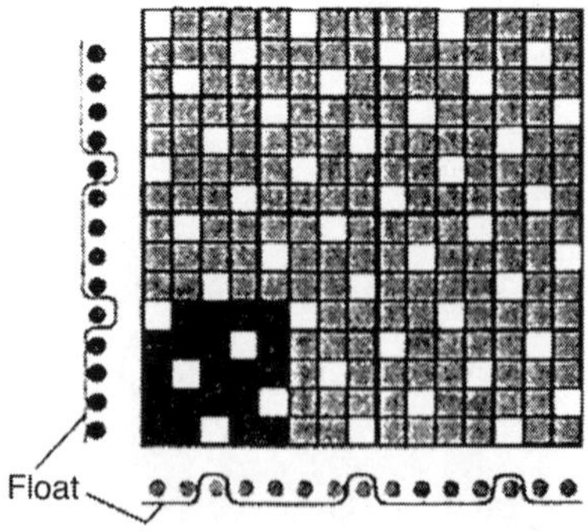

30 – 04 01 – 01 – 03

Pattern Draft of satin weave

Pattern row: Pattern row is a horizontal row of needle loops produced by adjacent needles in one needle bed. In plain weft knitted fabric this is identical to a course but in more complex fabrics a pattern row may be composed of two or more course lengths. In warp knitting, every loop in a course is usually composed of a separate yarn.

Pattern wheel: In a circular-knitting machine, a slotted device for controlling individual needles so that patterns can be knit in the fabric.

Pattu: A pattu is a woollen wrap around woven by women in the hills of Himachal Pradesh, and Kashmir, India. A woven silk material made in Rajasthan, and other parts of India. Pattus are woven on pit – looms , were available only in natural camel or sheep wool shades and had deep contrasting borders. The emergence of synthetic dyes changed that colour coding. Pattu follows a twill weave method but uses a heavy weft- that is wound on small sticks which are shuttled across- for is two ends, lending them an embroidered appearance. The motifs are usually geometric within liear bands. A typical pattu will be always in two segments that are neatly stitched in the centre as the width of the traditional looms are narrow.

Patwa: Very strong bast fibre, yielded by the Bauhenia vahlii, a climbing plant in India; used for cordage; also called mohwal.

Paukpan: Bast fibre yielded by the Aeschyomena aspera, a small bush in Burma; used for cords and hats.

Paunch Mat: A thick mat woven of ropes and flattened; used in ships.

Payta: Variety of raw cotton from Peru, the staple is of grayish colour, little lustre and strong.

PB: Elastodiene fibres.

PBI: Polybenzimidazole fibre. Made by Celanese Corporation. It is obtained by reacting diaminobenzidine and diphenylisophthalate and is used as an alternate to asbestos in high temperature filtration and thermal protection clothing. Its moisture regain is high at 14.4%, which permits high degree of comfort. Polybenzimidazole, or PBI is nonflammable, comfortable, flexible and used in apparel for space and for drogue chutes and under suit for astronauts.

PC: Polyacrylonitrile fibres.

PCP: Penta chlorophenol.

PE: Polyethylene fibres.

Pea Jacket: A short coat of thick closely woven cloth, worn in rough weather.

Peak force, for pile floor covering: The force required to separate two or more layers and registered on a chart as a peak, that is, a value exceeding the value immediately preceding and following it.

Pearl: See **Purl.**

Pearl printing: Performed on the hank (Yarn printing) using a suitable yarn printing machine, in the form of point single colour illuminating dye.

Pearl yarn: (knitted yarn), pearl ply yarn made up of two different (heavily) twisted single threads (usually with opposing twist).

Peasant Combs: Single-pitch and 2-pitch hand combs used to produce a semi-worsted fibre preparation.

Peasecod-belly: Spanish doublet with heavily padded breast, with a ridge down the middle.

Peat fibres: Peat contains residues of grass, swamp plants, mosses etc. that have decomposed in the absence of air. The dark brown or black strands it contains (approx. 15 cm long) consist of cellulose fibres, which provide a spinnable material.

Peau de Cygne: As the name implies, a very soft atin, woven in fine soft yarns with a lustrous finish. Used mainly for expensive lingerie.

Peau de Soie: Originally made from silk, the term is now also used to describe fabrics made from synthetic fibres, such as polyester and acetate, that have the typical look of Peau de Soie. It is an 8-shaft satin with one point added on the right or left to the original spots, giving the fabric a somewhat grainy appearence. With matt finish on both sides., and is usually a firm, heavy fabric in plain colours. Used for formal dresses.

Pebble: A term often used to describe the characteristic appearance of a crepe fabric.

Pebble-weave fabric: A fabric with an irregular or rough surface texture formed by either a special weave or by the use of highly twisted yarns that shrink when they are wet.

PECE: Post-chlorinated vinyl chloride polymer. The post-chlorination process increases chlorine content form 57% to 64%. The resulting polymer is soluble in acetone and can be wet spun.

Pectinases: Specific Enzymes (belonging to the hydrolases and carbohydrases), which gradually break down high molecular pectins (and hemicellulose) to single galacturonic acid. Pectinases play a decisive role in the biological retting of stalk bast fibres. Used in the retting of Flax. Long and controlled is possible with use of enzyme preparation and no bacterial or fungal contamination occurs like that in dew and water retting. Also pretreatment of flax with SO2 brings about sufficient break down of the woody straw to speed up enzyme action and to prevent bacterial and prevent bacterial and fungal contamination.

Pectins: High molecular, carbohydrate-like vegetable substance, composed mainly of methylated polygalacturonic acid in a glucosidic bond

Peel adhesion: The force required to delaminate a structure or to separate the surface layer from a substrate. Peel adhesion is the usual measure of the strength of the bond between fibre reinforcements and rubber in tires and other mechanical rubber goods.

Peeler: In beaming, a defect caused by a portion of an end sticking or remaining on the beam, causing the filament to strip back or peel until it is broken. Although they are often associated with ringers, peelers are not necessarily defects that will circle the beams.

Peg top: Trousers cut wide and full around the hips and narrowing sharply down to the ankle.

Peg-plan: The method of instructing the dobby mechanism on a loom to lift the required shafts.

Pegging: A finishing process for velveteen consisting of applying friction with blocks of wood or soapstone to impart a gloss or sheen to the fabric.

Pegging jets: A technique for freeing a plugged hole in a spinneret by rubbing the face with a piece of wood. Use of the technique has been discouraged because of damage to the spinneret.

Pegging plan: Synonymous with the lifting plan for dobbies where pegs inlays control selection.

Pekin: These are fabrics in which stripes of a different interlacing run in the direction of the warp. In combining these weaves it is advantageous to have them contrast distinctly, for instance, a short weave such as taffeta or Gros de Tours, with a longer and looser one such as satin, sergé or cannele, also changes from warp to filling effects. Care must be taken to arrange the joining of the two weaves so that the last thread of one weave will cross the first thread of the other. This will prevent the threads from either stripe to slide over into the other, and so make a clean cut line.

Pekinse stitch: An Embroidery stitch. Work Back Stitch is usual way, then interlace with a thread to tone or thread of a different colour. But the loops should be pulled slightly when working.

Pelerine: A device for transferring stitches from the cylinder to the dial or vice versa on a circular-knitting machine.

Pelletes: See **Bassini**.

Pelt: A sheep skin with the wool attached.

Pelts: Term for the hairy outer skin of smaller animals in particular, e.g. rabbits, goats, etc.

Pencil Locked: A fleece with narrow staples or lock formation: indicates an open fleece that has less density and probably more vegetable matter. This type of lock formation is genetic and is passed on to offspring.

Pendant groups: Groups attached to the main polymer chain or backbone. An example is the methyl groups in polypropylene.

Penetration: (1) Penetration of a fibrous mass with the applied liquor, which if possible must wet all surfaces if uniformity of product application is to be guaranteed before fixation. (2) Characteristic feature of printed fabric (roughly comparable to full penetration dyeing), which is particularly dependent upon the type of fabric and thickener. A thick, dense fabric or a suitable filter paper is used to evaluate the penetration capacity of a thickener. Penetration percentages are best obtained by a comparatory whiteness measurement. Penetration is = 0% if the printing on a thick fabric remains invisible on the reverse side, i.e. the material still shows its original high, intrinsic degree of whiteness (for thinner fabrics 0% is never achievable, print with very high thickener concentrations is always visible on the reverse). Therefore: Penetration = 100%, the colouring is the same on the reverse and right sides.

Penta-screens: Second generation of Rotary screens for screen printing with mesh values of 125–215. Finer designs can be achieved by using screens with higher mesh values (Mesh number). This is made possible by better mastery of the hole/web ratio by altering the cross-section of the engraving from the sawtooth type to drop-shape. Penta screens require a new special paint.

Improved uniformity and higher printing speed and low doctor blade pressure at the same printing paste viscosity are further advantages.

Pepita: **(Houndstooth)** These are particular types based on 2/2 twill. The former is said to be named after a Spanish actress and doesnot show the prominent jagged edges to the squares, which are a feature of houndtooth.

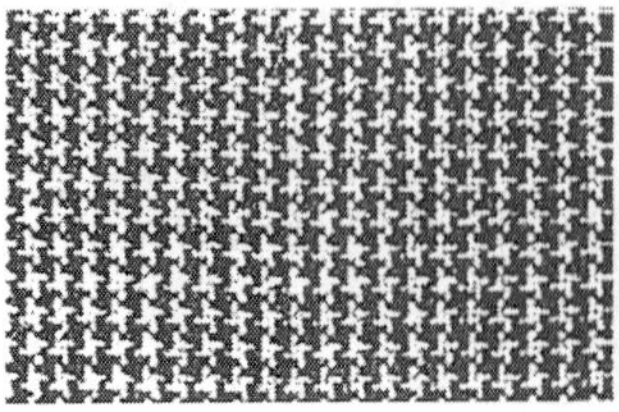

Pepita

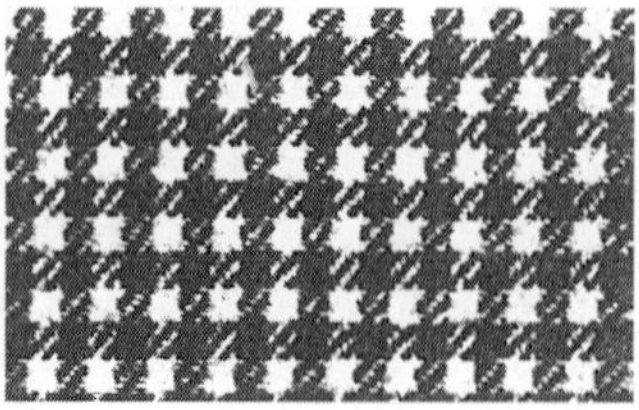

Houndtooth

Peplos: Women's rectangular woollen cloth robe, wrapped around the body and fastened over the shoulders.

Pepper-and-salt: Colour patterning effect, also known as thread by thread, in which a light thread alternates with a dark thread in warp and weft, giving rise to a diagonal stripe. Grey shades are typical, but beige brown and other colour combinations are also possible.

Per cent mean deviation uneveness, in yarns: The average of the absolute values of the deviations of the linear density from the average linear density expressed as a percentage of the average linear density within the tested length of the strand.

Per Cent Mean Range Unevenness, in yarns: The average of the ranges of linear density found within segments of equal length, expressed as a percentage of the average linear density equal to within the tested length of strand.

Peracetic acid: Molecular weight 76. A 40% organic acid with 8.6% oxygen. Is classed as explosive; colourless; slightly ionized at pH 8.2. Commonly used as polyester and polyamide bleach, for the stripping of triacetate dyes and also as an oxidative bleaching and stripping agent for proteins and other alkali sensitive textiles, particularly for bright fabrics, for polyamide/ cotton mixes (optimal 65–80°C; not damaging to fibres) and acetate materials (pH 6–8).

Per acids: Acids created by to the action of hydrogen peroxide (H_2O_2) on normal acids, which contain the –O–O–H group instead of –OH, e.g. caro's acid H_2SO_5; Peroxy disulphuric acid $H_2S_2O_8$; peracetic acid. CH_3COOOH; performic acid. Oxygen releasing principle:

$$H_2SO_5 \rightarrow H_2SO_4 + 1/2\ O_2$$

Percale: A fine print cloth is sometimes called paercale. Originally a trade name for sheeting made from combed yarns, but today the name is used to describe any fine cloth of medium weight that is woven in plain weave, (180 to 250 threads per square inch). A typical percale structure is 80s square with combed 40s warp and combed 50s filling.

Percale: A medium weight, plain weave, low to medium count cotton-like fabric. End-uses include sheets, blouses, and dresses.

Percale, as applied to bed sheeting: A plain weave fabric with no fewer than 180 yarns/in^2.

Percaline: Cloth is similar to percale but woven with still finer yarns and lighter in weight is called a percaline. Light weight cotton print cloth similar to lawn, with a bright soft finish. Usually the yarn is mercerised.

Percent by mass: Mass of solution x 100/ Combined mass of solute+solvent. % (by mass)

Percent by volume: Volume of solutionx100/total volume of solution. % (by vol)

Percent UV blocking: 100 minus the UV Transmission. See ***UV Transmission***

Percentage composition: The fibre content expressed as percentages.

Percentage cover: Cover Factor as a percentage of the maximum possible for a particular weave structure.

Percentage point: A difference of 1 % of the base quantity.

Percentage by weight: (weight per cent), is a percentage mass relationship, i.e. grams in relation to 100 gram units, e.g. dye mass stated as a percentage of material weight. Where percentage figures are given in recipes, solvent concentrations etc., and no further information given, the percentage by weight is usually referred to, in solutions Per cent by volume. Conversion:

Percentage by weight = Per Cent By Volume/ density

Perching: (1) A closely woven, plain-weave, spun fabric used for dress goods and sheeting, generally 80 x 80 threads per inch or better.

(2) Examining and correcting raw woollen cloth for and of knots, burs, holes and other imperfections before filling.

Perchloroethylene: See **Tetrachloroethylene.**

Persian carpets: Oriental carpets of Persian origin. Usually dense knotting of 400 000 to 1 million knots per m^2 with fine patterning. Names such as Bakhtiari carpets; Bijar carpets; Hamadan carpets; Isfahan carpet; Kashan carpets; Shiraz carpets; Senneh carpets; Seraband carpets (Mir) describe the province of origin.

Percieved value: The value of a product seen to have.

Percolating filter: Used in biological Waste water treatment for enriching the waste water with air.

Performance Fabrics: Fabrics made for a variety of end-use applications, which provide functional qualities, such as moisture management, UV protection, anti-microbial, thermo-regulation, and wind/water resistance.

Performance requirements: How a fabric or product has to function.

Perfume treated textiles: Fabrics are treated with perfumes to gice a fresh scent while wearing. In new technology the finishes are capable of refilling with fresh scent when the available scent fades away.

Perkin, Sir William Henry: Founder of the modern dyestuffs industry. (1838–1907), English chemist. In 1856, at the age of 18, discovered the first technically usable coal-tar dye (aniline purple or mauvine) when looking for a quinine synthesis.

Perlon: A branded polyamide fibre.

Permanganate number: (1) Permanganate number. For quantitative determination of fibre damage to cellulose. Oxycellulose and hydrocellulose are dissolved from the fibre by repeated boiling and detected using a potassium permanganate solution. Permanganate number = number of millilitre 0.1 n $KMnO_4$ solution, consumed by 1 g fibre. (2) Waste water permanganate consumption. Method: Boil 1 g material with NaOH and titrate dissolved organic products with 0.1n $KMnO_4$. $KMnO_4$ consumption up to 10 ml normal, 10– 13 ml little damage, above this significant damage. When making the calculation, deduct the basic value for unbleached material. The Permanganate number can also be an important method of detecting cotton materials kier boiled in the presence of air.

Permanent deformation: The change in length of a sample after removal of an applied tensile stress and after the removal of any internal strain (e.g., by boiling off the sample and allowing it to dry without tension). The permanent deformation is expressed as a percentage of the original sample length.

Permanent finish: A term for various finishing treatments, chemical and/or mechanical, applied to fabric so that it will retain certain properties, such as glaze of chintz, crispness of organdy, smoothness of cotton table damask, and crease, crush, and shrinkage resistance of many apparel fabrics during the normal period of wear and laundering.

Permanent finish: Applications or finishes on textiles, which resist the stresses of use and care, e.g. washing, chemical cleaning.

Permanent growth: See ***Secondary creep.***

Permanent hardness: A type of water hardness caused by the presence of dissolved calcium, iron, and magnesium sulphates or chlorides. This form of hardness cannot be removed by boiling. Compare temporary hardness. See also ***Water softening.***

Permanent pleating: There is a wide variety of permanent pleated fabric available. It is usually knife pleated, narrow pleats on fine fabric, wider ones on thicker fabrics. The fabric may be of almost any weave, including knit, and may be closely woven or have a lacy pattern. If the fibre is synthetic the pleating will be permanent and therefore washable, but if natural fibres other than cotton are included the pleating cannot be heat set and is there for probably not permanent. Many fabrics are sold in matching colours, one plain cloth, the other pleated so that garments combining both can be made.

Permanent press: Garment shape is maintained during wearing and washing. Creases must be sharp, flat areas must be smooth. Creases are rated on a 1 to 5 scale. A 5 rating is perfect. See **Durable press.**

Permanent press finishing: Originally used exclusively as a collective term for Resin finishing processes (dry cross-linking), in which the final fixing or stabilizing of the material takes place during garment manufacture. This produces items of clothing with excellent permanent dimensional stability, sharp ironing folds and pleats, nonshrink seams and particularly good smoothness after washing. Fixation is therefore not performed flat, but shaped. Also a common term for conventional resin finishing, however only for items with high-quality finishing effects.

Permanent set: See **Secondary creep.**

Permanent-press finish for wool: Collective term for: Antifelting finish and Non-shrink finish.

Permanent Stitches: Stitches, which are not ripped off after being made, are called permanent stitches. For example-Hemming.

(a) *Back stitch*: It is a firm hand stitch for awkward areas where machining is tricky. It looks like a machine stitch on the right side but stitches overlaps each other on the wrong side. To make this stitch bring the needle out a stitch9 length beyond where the thread emerges. Insert the thread again at the end of the previous stitch.

(b) *Half back stitch:* When one does not have much time and a permanent stitch is needed, one can opt for half back stitch. In this stitch, the needle is carried on half-length of the first stitch

(c) *Prick stitch:* It is a variation of back and half back stitch. It is used to attach Zipper by hand & also to attach sari falls. Back through one or two threads of the fabric, & a bigger distance is taken in front. It is done from the right side of the garment.

(d) *Hemming:* This stitch is used to fold the raw edges of the fabric or to flatten the seam. It also works from the wrong side of the fabric. This stitch works from right to left with the hem draped over the fingers of the left hand. To make this stitch pick up a thread from the fabric just below the folded edge of the hem and then slip this thread from the fold in the hem itself. It is of three types:

(e) *Visible hemming:* This stitch is strong stitch used to fold hemline of a garment. This is called visible hemming because stitch is visible at the right side of the garment as we take three to four strands of fabric to make this stitch.

(f) *Invisible hemming:* This stitch is used to attach facing and piping. This stitch is called invisible hemming because we take only one or two strands of fabric to make this stitch and because of it; the stitch is not visible at the right side of the fabric.

(g) *Slip hemming*: This is an invisible stitch used to draw two edges together invisibly as in a hand-sewn lining. Slip the needle through the fold and immediately below the needle comes out. Take a small stitch of the fabric. Reinsert the needle immediately above.

(h) *Catch stitch:* An invisible stitch worked loosely taking up only a thread of fabric from top and bottom alternately, inside a hem or to hold the inner edge of the seam in interfacing in place.

(i) *Herringbone stitch:* This stitch is used on the hemline or during the hemming. It is used for heavy fabric as thick and heavy fabrics needs to be hold at both folds. To achieve this, sewing proceeds diagonally up and down, as in cross stitch, but the needle is brought back slightly behind the previous stitch each time so that it will cross over it.

Permeable: Allowing some material to pass through.

Permeability: The rate of flow of fluid under differential pressure through the material.

Permeate: That portion of the feed stream which passes through a membrane, leaving behind a more concentrated stream.

Permeator: A hollow fine-fibre membrane element itself consisting of thousands of hollow fibre.

Permutit: (Trademark) A substance used to remove unwanted chemicals that have dissolved in water. It is a zeolite consisting of a complex chemical compound, sodium aluminum silicate. When hard water is passed over this material, calcium and magnesium ions exchange with sodium ions in the Permutit. This is a good example of ION EXCHANGE. Once all the available sodium ions have been used up, the Permutit can be regenerated by washing it with a saturated solution of sodium chloride. The excess sodium ions then exchange with the calcium and magnesium ions in the Permutit.

Peroxide: With reference to textile processes, usually used to mean hydrogen peroxide, though there are many organic and inorganic peroxides.

Peroxide cold bleach: Same process as Peroxide cold-pad-dwell process; Cold pad batch bleach; Cold dwell process. Very suitable as a pre-bleach. Only recommended for clean husk-free raw material.

Peroxide cold-batch dwell process: Bleaching process specifically for cotton circular goods. Method: (a) Impregnate with bleaching and brightening solution on padder using expander for circular material; (b) dwell in heat insulated vehicle; (c) rinse on rope washing machine.

Peroxide pad steam method: The most common process for bleaching cotton fabric. Care has to be taken to avoid heavy metals like iron, which can cause pin hole damage due to the uncontrolled release of nascent oxygen.

Perrotine: Printing machine named after its inventor, Perrot, which can be considered as the first stage in the mechanisation of block printing.

Per salts: Genuine salts of Per acids, such as for example ammonium persulphate or potassium persulphate. Application: Oxidizing agent in dyeing and printing, kier boiling, desizing, bleaching agent; disinfectant.

Persian carpets: Oriental carpets of Persian origin. Usually dense knotting of 400 000 to 1 million knots per m2 with fine patterning. Names such as: Bakhtiari carpets; Bijar carpets; Hamadan carpets; Isfahan carpet; Kashan carpets; Shiraz carpets; Senneh carpets; Seraband carpets (Mir) describe the province of origin.

Persoz reagent: Dissolves only silk from silk/wool mixes at 45°C. Consists of 10 g zinc chloride in 10 g water and 2 g zinc oxide.

Perspiration: Sweating of the human being. There are two types of perspiration by human being: (a) Acidic, (b) Alkaline. See ***Perspiration fastness.***

Perspiration fastness: Resistance of dyeing and printing to human sweat. (a) Determination according to DIN 54 020: one sample each is treated with an undyed accompanying fabric in an alkaline or acidic testing solution for 30 min (liquor ratio 1 : 50) at room temperature (slight mechanical action for better penetration). After the removal of excess testing solution the test pieces are individually exposed to a pressure of 12.5 kPa between two plates and are treated for 4 h in a warming cupboard at 37°C ± 2°C. After opening, the test pieces are hung to dry at max. 60°C. The change in colour and the bleeding are evaluated against the grey scale. (b) Perspiration fastness according to AATCC: a) Alkaline: 10 g/l sodium chloride, 4 g/l ammonium carbonate, 1 g/l disodium phosphate (Na_2HPO_4) and 0.25 g/l histidine monohydrochloride, pH of the solution = 8; b) acid: 10 g/l sodium chloride, 1 g/l lactic acid 85%, 1 g/l disodiumhydrogenphosphate and 0.25 g/l histidine monohydrochloride, pH of the solution = 3.5. Procedure: treat test piece for 15–30 minutes in the solutions (a) and)b) individually, squeeze (squeezing effect 250–300%) and load 10 kg for 6 h (or more) in perspirometer at 38°C.

Perspiration Resistant: A treatment on a fabric which allows a fabric or a dye to resist perspiration.

Perspirometer: A device by AATCC for testing Perspiration fastness. Sample is treated at a pressure of 4.5 kg between glass plates for at least 6 h at 38°C.

Persulphate desizing: Procedures for the removal of starch sizes generally use enzymes because, unlike oxidizing agents, this rules out damage to the cotton. The latter are therefore only found in isolated cases, for example in cases where the size contains enzyme poisons.

Persulphate desizing can be used both after the cold dwell process and after the steaming process. In order to avoid fibre damage, max. 0.5% sodium persulphate can be used in the cold dwell process, and no more than 0.2–0.35% (in relation to the material weight) in the steaming process. Application takes place in an alkaline medium. Without the alkali, complete desizing is impossible.

Persulphates: (peroxysulphates, dioxysulphates) Salts of peroxysulphuric acids ($H_2S_2O_8$ Sulphur oxyacids), such as the highly water soluble ammonium, potassium or sodium persulphates. Application: Bleaching and strong oxidizing agent for vat dyeing and printing; desizing agent; bleaching agent for wool, waxes, fats and oils.

Perylene tetracarboxylic acids: It is a group of anthraquinone dyes comes in anthraquin-onoid dyes with a perylene structure and condensation product of 3,4,9,10-perylene tetracarboxylic acid with p-chloroaniline.

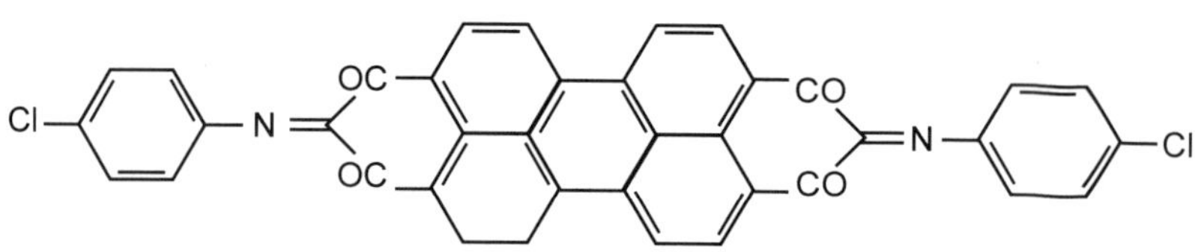

Indanthrene Scarlet R

PES: Polyester.

PET: Often used as an abbreviation for polyester in the Anglo-Saxon language area.

Petersham: (1) Narrow and ribbed belting used for tops or skirts, stiffening waistbands etc. It is in plain weave; the warp is fine and a thick weft gives a warp rib structure. In various widths from 2 to 5 cm. (3/4 to 2 in.)It may be cotton, viscose or polyester, sometimes elastic inserted.

(2) Originally a thick napped overcoating fabric associated with Lord Petersham, but rarely found now.

(3) A heavy woollen, woven or felted kind of cloth for overcoating. It has a rough, knotty surface, and is sometimes called "nigger-head." A very heavy, close-felted kind of this

cloth is also called "cordington" ; the finer, softer kinds "chinchilla."

Petticoat: Stiff, dome shaped underskirt, usually layered and decorated worn by ladies in the fifties.

PETP: Polyethylene terephthalate.

Petroleum sulphonates: Petroleum sulphonates are prepared by sulphonation of hydrocarbons; most commercial products are actually byproducts of petroleum refining. In the case of "natural" petroleum sulphonates, they are produced by sulphonation of crude oil fractions. The purpose of the sulphonation is removal of a portion of the aromatics from the oil and the resulting by product/co-product stream is a complex mixture of sulphonated aliphatic and (mainly) aromatic hydrocarbons. To complicate matters, the term petroleum sulfphnates is also applied to materials produced by sulphonation of synthetic materials, such as alkylbenzenes (detergent alkylate). In this case, a higher boiling alkylbenzene feedstock is used than for the production of surfactants for household detergents. The products therefore have only limited water solubility.

Petticoat finish: In particular in polyamide and polyester items to achieve a permanent stiffening finish using aminoplasts and/or thermoplastic artificial resins.

PFD: (or pfd): Prepared for dyeing.

PFP: (or pfp) Prepared for printing.

pH: A measure of acidity or alkalinity representing the logarithm of the reciprocal (negative logarithm) of the concentrationof hydrogen ions. The pH of the textiles is generally determined on aqueous extract of the textile being tested. A measure of the concentration of hydronium (H O , H attached to a molecule or water) in a solution; log[H 0] (that is, the negative base-10 logarithm of the concentration of hydronium) 3 +Acids have pHs less than 7; bases have pHs greater than 7. A pH of 7 is neutral. The normal pH range is 0 to 14. Note that although the pH scale is logarithmic, it does not mean that 10 times as much of a base/acid in solution will increase/decrease the pH by 1 unit. This is true for strong acids and strong *bases*, but is not true for weak acids and weak bases.

pH of fibre extracts: Determination procedure: Fibre samples are leached for 60 min at room temperature in distilled water in a liquor ratio of 1 : 5, 1 : 10, 1: 20 and 1 : 50 (in polyethylene bottles). The pH of the extracts is measured with a glass electrode and plotted against the associated liquor ratio in a co-ordinate system. The curve obtained is extended to intersect with the pH ordinate. This point is the pH (extrapolated) at a liquor ratio of 1 : 0 and corresponds with the pH of the fibre. It can be determined both graphically and by calculation and is applicable for all fibrous material (for wool only at pH < 3 or > 11).

pH Down: As a spa or swimming pool additive, a chemical to reduce pH

pH Down is almost always sodium bisulfate, which is typically in the form of tiny spherical beads.

pH Up: As a spa or swimming pool additive, a chemical to increase pH pH Up is usually “pure” sodium carbonate (soda ash). Because it can often be purchased at grocery or hardware stores, usually reasonably cheaply, it may a good alternative to buying soda ash from a dye vendor. Sometimes sodium bicarbonate is sold for increasing pool pH. It is not equivalent to soda ash.

Phase Change/Discolour textile materials: Phase change textile material reacts to temperature change. It does not respond to a change in body temperature, but when there is a temperature change outside, phase change materials melt or solidify. Discolor material is also known as Chameleon material. It changes colour when exposed to heat.

Phase change materials: A hydrophilic compound applied to a fibre or fabric which results in superior breathability and a moisture management system within the fabric that helps to maintain a comfortable body temperature when the garment is worn.

Phases beam: A beam on which each of the ends is wound from the same depth of each of the bobbins on the creel. Phased beams are prepared when yarn properties vary from the inside to the outside of the bobbins in order to prevent warp streakiness in the finished fabric.

Phase-separation spinning: See **Spinning.**

Phenolic: (1) Containing or pertaining to phenol. (2) A resin or plastic made by the condensation of a phenol with an aldehyde and used particularly in coatings and adhesives.

Phenyl: A chemical radical, (C6H5-), derived from benzene. It is the basis of many aromatic derivatives.

Phloroglucinol: 1,3,5 trioxybenzene), $C_6H_5(OH)\cdot 2H_2O$. White or yellowish crystals, soluble in water, easily soluble in alcohol and ether; reduced Fehling's solution in heat. Application: Detection of lignin and pentose (Phloroglucinol reaction).

Phosphorescence: Emission of light that persists for a noticeable time after the removal of the excitation source.

Phosphoric acid: An inorganic acid having the formula (H3PO4).

Phosphorylase: Ferment (magnesium adenylic acid protein), which splits starch into glucose (in test tube also artificial starch structure from glucose), whereby inorganic phosphate is inserted at the dissolved link point of the glucose molecule and splits glucose-I-phosphate.

Photo sensitized degradation of Cellulose: Although light of wavelengths greater than 310 nm cannot degrade cellulose directly, some other compounds such as dyes and some metallic oxides can absorb near-ultraviolet radiation or visible light and in their excited states can induce the degradation of cellulose. These reactions are designated photosensitized degradation but do not have a common mechanism. The photochemical tendering of cotton fabrics containing vat dyes has been recognized for many years. Emphasis has been placed on practical problems involving textile performance rather than on mechanistic studies. It is known that this sensitized deterioration depends on the presence of oxygen in contrast to direct photolysis. The cotton cellulose is also subjected to degradation upon exposure to high-energy radiation. The radiation sources have included cobalt-60, fission products in spent fuel rods, and high voltage electrons generated by linear acceleration . High-energy radiation causes oxidative depolymerization of cellulose in the presence or absence of oxygen to yield hydrogen, one-carbon gaseous products, and multicarbon residues. At low doses, fibrous properties are retained.

Photochromism: Photochromism is generally regarded as the reversible interchange of a single compound between two molecular states, which show different absorption spectra when activated by light. Or in simple words in textile context "the change of colour of a dyeing changes on exposure to light, but reverts to its original state after the sample is kept in the dark. The extent of this photochromism is expressed as a grey scale assessment given alongside the rating for fastness to light.

Photodegradation: Degradation caused by the absorption of light (particularly ultraviolet light) and consequent chemical reaction.

Photographic printing: See **Printing.** Photographic prints are made similar to that used in making photograph. A negative is placed on the fabric and light is transmitted to it and the design is developed. The fabric is washed and the design is as permanent as a photo.

Photonic clothing: Photonic clothing is a general term for clothing that has an embedded or detachable illuminator and a digital control device emitting light while preserving the characteristics of the conventional clothing. Due to the chemical properties of clothing materials, the existing clothing undergoes varying degrees of fabric discoloration in accordance with temperature and light of the surrounding environment. By contrast, however, the photonic clothing enables wearers to control the degree of illumination with a light-emitting device.

Photonic clothing, colour-changing: This type of photonic clothing has a function that enables the wearer to choose and control colours and patterns of the garment's luminescent part by regulating the mode of light emission on the optical fibres. The application of this comparatively simple function is suitable for functional sports clothing and safety garments.

Photonic clothing, colour-responding: These are garments that are the kind of photonic clothing with colours that automatically vary with surrounding colours or that allow wearers to adjust them as they please. The photonic clothing with colour recognition capabilities consists of a bare garment made of optical fibre and other fibres, a colour sensor that perceives colours surrounding the clothing, RGB LED that emits light of colour recognized at the optical fibre end point, and a colour control module. Colours that the wearer inputs with the colour sensor change to colours similar to what the optical fibre of the photonic clothing has sensed.

Photonic clothing, sound-responding: It is a kind of smart clothing capable of changing colours or light emission amount by reacting to external vibration or sound. Smart clothing adjustable to external vibration consists of a bare garment made of optical fibre and other fibres, a vibration sensor that perceives vibration external to the clothes, and a colour light control module that adjusts light emission of RGB LED to correspond to the level of vibration sensed on the optical fibre end point. In an environment such as a club where music is present, the wearers are able to experience various effects, with the area and the size of their garment's luminescent parts made to change according to the loudness of external sound.

Photosensitive coatings: Photosensitive coatings are increasingly competing with chrome gelatine in the manufacture of screen printing screens. These coatings mainly use Polyvinyl alcohol which has been sensitized using chromium salts. Such photosensitive coatings have better adherence, contain no water soluble components when illuminated and always have sharp, unchanging contours. The screens are coated cold, in daylight. It should be noted that these coatings require slightly longer illumination times.

Photosensitive coating of screens: Layer of photosensitive lacquer applied to the gauze during the manufacture of screen printing templates, which hardens due to the effect of light.

Photosensitive lacquer: Forms the basis for the creation of printing screens, because once illuminated it takes on lacquer-like solidity and resistance, i.e. after illumination no further reinforcement is necessary (Positive process for screen making).

Photosensitive layers: Photosensitive layers are used in screen making. One of the methods of screen making, water soluble photosensitive layers are coated on the screen or bolting cloth and exposed to light with the design on a transparent sheet covering them. On exposure, the portion where light has fallen water-insoluble compounds (e.g. chrome gelatine) are formed. On washing the unexposed portion, where the photosensitive layer remains water soluble are washed off and the design is exposed on the screen.

Photoropism: (Gk.), colour change due to UV light. Occurs if the dye reflects the UV part of daylight with longer wavelengths than that of visible light. This can be caused by after-treatments, such as anti-crease finishing, chintz finishing, etc. The normal shade generally

only returns after long periods in the dark or when the material is irradiated with red light. See **Photochromism.**

Phthaleins: Dyes derived from xanthene or triphenylmethane-carboxylic Acid. Typical representatives are phenolphthalein, fluorescein, eosin.

Phthalic acid: An organic acid obtained by oxidation of various benzene derivatives and having two adjacent (ortho) acid (COOH) groups on the benzene ring.

Phthalocyanine dyes: The term phthalocyanine was first used by R. P. Linstead in 1933 [1] to describe a class of organic dyes, whose colours range from reddish blue to yellowish green. The name phthalocyanine originates from the Greek terms *naphtha* for mineral oil and *cyanine* for dark blue. In

1930.1940, Linstead et al. elucidated the structure of phthalocyanine (H2Pc) and its metal complexes

Phulkari: (Literally means flower work) Dazzling drapes worked with the darning stitch, were traditionally embroidered by women in the Punjab and Haryana in India, to be gifted to their daughters and daughter-in-law at weddings, ceremonies and festivals. See also **Bagh.**

Physical finishes: Physical finishing methods for textiles include optical finishing, brushing and napping, softening, shearing, and compacting of the textile structure.

Piassava fibre: Palm fibres (Africa, Brazil) for mats, brushes, brooms, ropes, braiding, also stiff fabric. Light to dark coloured, containing silicic acid. Dyeing properties similar to coir.

Pick (n): (a) A single operation of the weft-insertion mechanism in weaving. (b) One or more weft threads inserted between successive beat-ups. See **Beating-up**.

Pick (v): To pass the weft through the warp shed in weaving. See **Picking.**

Pick: An individual weft yarn.

Pick: The number of threads within a given space.

Pick bar: A bar in which the pick spacing is different from that in the normal cloth This is a general term which covers no. of specific faults like starting place, weaving.

Pick count, in woven fabrics: The number of filling/weft yarn per inch of the fabric.

Pick counter: (1) A mechanical device that counts the picks as they are inserted during weaving. (2) A mechanical device equipped with a magnifying glass used for counting picks (and/or ends) in finished fabrics.

Pick glass: See **Pick Counter, Piece Glass.**

Pick-up: Quantity of any liquor taken up by the substrate in any process. It is expressed in percentage of the dry weight of the substrate .

Pick up place: See **Temple mark.**

Pick-out mark: (1) A weftwise band or bar charecterised by a chafed or fuzzy appearance due to the pulling out of the original picks.

(2) A filling wise band or bar characterized by a chafed or fuzzy appearance due to pulled-out picks.

Pick-out place: (1) See **Temple mark.**

(2) A place in the woven Fabric where the pick is missing.

Pick-and-pick: A woven fabric with alternate picks of yarns of two different colours or kinds

Picker: (1) A machine that opens staple fibre and forms a lap for the carding process used in the production of spun yarns. (2) That part of the picking mechanism of the loom that actually strikes the shuttle. (3) A mechanical device used to open fleece before carding. The swing picker looks, for all the world, out of something from an Edgar Allen Poe horror story. Not recommended for the faint-of-heart, children, or those who don't pay attention.

Picker lap: A continuous, considerably compressed sheet of staple that is delivered by the picker and wound into a cylindrical package. It is used to feed the card.

Picker sticks: The two sticks that throw the shuttles from box to box at each end of the brace plate of the loom.

Picking: (1) The second of the three basic motions in weaving, in which the weft is passed through the warp shed. As the harnesses raise the heddles or healds, which raise the warp yarns, the shed is created. The filling yarn is inserted through the shed by a small carrier device called a shuttle. The shuttle is normally pointed at both ends to allow easier passage through the shed. In a traditional shuttle loom the weft or filling yarn is wound onto a quill and mounted inside a shuttle. The filling yarn emerges through a hole in the shuttle as it moves across the loom. A single crossing of the shuttle from one side of the loom to the other side is known as a pick. As the shuttle moves back and forth across the shed it weaves an edge or selvedge on each side of the fabric which prevents the fabric from ravelling.

(2) The rectification of the face and the back of a carpet after manufacture, including insertion of missing tufts, replacement of incorrect ones and repair of broken yarns in the backing (local mending).

(3) A process carried out before the final stage of fabric finishing to remove, by hand, any contamination (such as kemp (see kemp fibres), wrong fibre, coloured hair, etc.) that has not been removed by previous processing. NOTE: This process is carried out in particular during the finishing of suitings, face-finished fabrics and cream or off-white fabrics.

Picking order: The smallest number of picks in colour and/or count that repeats up and down the fabric.

Picklock wool: A term used in wool sorting mainly in UK for the second best sorts from the fleeces.

Picks per inch, finished: Number of picks per inch on the finished woven fabric.

Picks per inch, in loom: Number of picks per in the loom state woven fabric.

Picks per inich, in grey: Number of picks per inch on a grey woven fabric.

Picolay: Compact cotton fabric with a plain weave, embossed with a diamond pattern to resemble diamond pique. Used for childrens clothes, table cloths.

Picot: The term is used to describe a loop edge. This may be found on many types of lace and edgings.

Picot: (1) A small loop woven on the edge of ribbon, or a purl on lace. A picot edge may also be produced by a hemstitching machine. (2) A run-resistant loop usually found at the top of hosiery.

Pictorial and figurative designs: Designs featuring visual representation of objects, people and animals.

PID control: (Proportional-Integral-Differential), generally valid temperature control principle that uses an additional measuring device to rule out variations in temperature control due to the delayed response of the control elements and guarantee an optimal constant temperature.

Piece: A standard length of a fabric, such as 40, 60, 80, or 100 yards.

Piece dyed fabric: Basically means a dyed fabric, which fabric is dyed in the fabric form, i.e., after weaving.

Piece dyeing: (1) See **Dyeing.** The process of dyeing a fabric after weaving.

(2) A section in the processing department where the fabric is dyed in piece form.

Piece dyeing machines: (1) (with fabric movement) The following general principle applies: the piece goods move, whilst the liquor remains stationary; for example Winch beck, Jigger, Padder, Jet dyeing machines, Overflow dyeing machines. However, a clear demarcation of this type cannot be made. In jet dyeing machines, for example, both the material and the liquor are in motion.

(2) (with liquor movement) The following general principle applies: The liquor moves and the item to be dyed remains stationary; for example: HT beam dyeing machines; Star frames. Only the discontinuous working mode is possible.

Piece glass: (pick counter). A magnifying glass with horizontal and vertical measuring scales for determining the yarn count (woven fabrics) or stitch density (knitted fabrics).

Piece goods: See **Piece.**

Piece-dyed goods: A piece of standard length after dyeing.

Piecing: A thick place in a spun yarn caused by poor splicing.

Piecing: See **Slug.**

Piecing: The joining of two or more ends of sliver, roving, yarn, etc.

Pied de poule: See **Check pattern fabric**.

Piecing out: The act of sewing together the pieces, which sometimes occur in the facings of garments, is called "piecing them out." "Dograbbiting."

Pierced Cocoons: Cocoon from which the moth have been allowed to emerge so that they mate and produce eggs.

Pigment: A coloured substance that is insoluble in water, usually in the form of a fine powder Pigments are used to colour many types of paint, including some textile paints, and almost all "inks" used for screen printing ("silkscreen" printing). Pigments need some sort of binder to hold them onto fabric. Azoic dyes and vat dyes actually form pigment inside fibres. These pigments are physically trapped inside the fibre, so no binder is necessary. Inorganic pigments are often oxides of metals, such as iron oxide or titanium dioxide. Many organic pigments are chemically similar to parts of dye molecules, but lack the necessary features to make them soluble and to bond, unaided, to the fibre.

Pigment, leafing (Bronze): So called leaf forming 'fat rich pigments' which float on liquid surface and orientate to give a highly reflecting film.

Pigment, non leafing (Non-Bronze): 'Low fat' pigments which are completely wetted out and therefore do not float but orientate within the binder film.

Pigment dyeing: (1) Colouring fabric with pigments mixed with a binder; this term is considered to be improper, since 'dyeing' is generally restricted to application of colorants that are soluble, and pigments are insoluble Some pigment application processes are much like dipping the fabric in a dilute paint. Newer processes involve pretreating the fabric with substantivity-producing agents that actually make the process more like true dyeing, where the pigment preferentially leaves the bath and attaches to the treated fabric. So-called pigment dyed fabrics often have, by design, poor wash fastness, to produce an aged appearance. Commercially, pigment dyeing is generally used more for finished garments than for fabrics.

(2) In terms of procedure it resembles the classic proofing process with pad-dye impregnating, drying and curing. The pigment dyeing process makes use of water-insoluble pigment dyes of either inorganic or organic nature which

are fixed to the fibres with the aid of a fixing agent. Since no affinity between fibre and dyestuff is involved, dyeing can be carried out on all fibres and fibre blends.

Pigment dyeing Process The dye is padded, the material is dried and heat set. After-treatment is not necessary. Good fastness. Wash fastness depends upon the binder and not the dye. Suitable pretreatment necessary to create good absorbency, as are faultlessly functioning padder and drier (dye migration). Only light shades and possibly some medium shades are possible because the rubbing fastness is insufficient in darker shades. The pigment dyeing process is suitable for all natural and synthetic fibres and is used primarily for cheap qualities.

Pigment printing: Printing technique based upon a system of pigment and binder. Water insoluble white or coloured pigments with no fibre affinity are fixed to the textile material using synthetic resin binders. It differs from normal printing in that the dye is present in the form of insoluble bodies with no affinity to the substrate. See **Printing.**

Pigment tafetta: The only taffeta fabric without shine. Dull pigmented yarns lacking lustre used but the fibres are usually silk or manmade.

Pigmentation process of vat dyeing: In a package dyeing machine, for example, these distribute the non-vatted dye first, in order to then make it soluble at the end of the exhaustion phase in the form of pigment adsorbed on the fibre surface. Normally the following procedure is followed for vat dyeing: Pre-vatting in the liquor.

Pigmented yarn: A dull or coloured yarn spun from a solution or melt containing a pigment. (Also see **DYEING, Mass-Coloured.**)

Pigmenting process, in solvent technology: Breaking down of insoluble product additives into pigments, using mechanical pressure, they are then removed mechanically, for instance by after-print cleaning (polyester, polyamide, triacetate weaves and knits). For fixed prints a high pressure tetrachloroethylene liquor jet is applied, causing insoluble (acrylate) films (print thickeners and synthetic sizes) to be broken down into pigments, which, in an insoluble state, are washed off and flushed away as unfixed print dye residues. Operating speeds in solvent continuous technology are > 50m/min. The print thickeners or sizes accumulate with dye in the form of a thick solvent-free sludge, which is disposed of as hazardous waste

Pigskin: This is easily recognised by the little holes or pores where once the bristles of pigs grew. Very hard wearing. Used for gloves, coats, belts, handbags etc.

Pilate: A deprecated term, see preferred term **Pile lifting.**

Pile, in pile fabrics: The raised loops or tufts (cut loops) that form all or part of the surface.

Pile, in carpet: That part of a carpet consisting of textile yarns or fibres, cut or looped projecting from the substrate and acting as use-surface.

Pile: (1) A fabric effect formed by introducing tufts, loops, or other erect yarns on all or part of the fabric surface. Types are warp, filling, and knotted pile, or loops produced by weaving an extra set of yarns over wires that are then drawn out of the fabric. Plain wires leave uncut loops; wires with a razor-like blade produce a cut-pile surface. Pile fabric can also be made by producing a double-cloth structure woven face to face, with an extra set of yarn interlacing with each cloth alternately. The two fabrics are cut apart by a traversing knife, producing two fabrics with a cut-pile face. Pile should not be confused with nap. Corduroys are another type of pile fabric, where long filling floats on the surface are slit, causing the pile to stand erect. (2) In carpets, pile refers to the face yarn, as opposed to backing or support yarn. Pile carpets are produced by either tufting or weaving. (Also see **Cut pile** and **Loop pile**.)

Pile Bursting: The appearance of a cut pile of a carpet achieved by different process producing a volume of apparently greater than that which was occupied before the application of this process.

Pile Carpets: See **Carpet, pile** Carpets with pile or Pile layer (pile loops), in the form of knotted carpets, tufted carpets, or woven carpets; Axminster carpets; Bouclé carpets; Brussels carpets; Chenille, Plush carpet; Tapestry carpet; Tournay carpet; Velvet carpet.

Pile carpet woven over wires: Pile carpet produced in a single operation on a loom in which steel rods are used to form the pile and to determine its length.

Pile Crush: The bending of upholstery or carpet pile that results from heavy use or the pressure of furniture.

Pile Cutting Machine: Special form within the group of Shearing machine. Whilst classical shearing machines always cut in several passes, new models make only a single shearing passage. Long pile items are cut to the desired pile height in a single stage. This makes it possible to economically reuse the cut off fibres.

Pile Lifting Machine: A machine used for the Shearing of fabrics for the alignment of the pile or fibres to be sheared.

Pile Fabric: Any cloth of any fibre that has a raised surface if individual fibres, making a furry surface. Pile may be uncut in manufacture, e.g., terry toweling, or cut, e.g., Velour. The direction of the pile ion the cloth may be felt or it can sometimes be observed by holding the length of the cloth vertically. Where the pile is running upwards, the cloth will look a different shade from where it is running downwards. The cloth can be made with the pile running in either direction.

Pile firmness: See **Knops resistance**.

Pile floor covering: A pile fabric intended for use as a floor covering. The pile may be in the form of cut loops or loops or both. Either cut loops or loops may vary in height.

Pile formation: Undesired phenomenon in velour materials, particularly carpets. Pile fibres lie to one or other side in patches, giving rise to a blotchy appearance on the surface of the velour.

Pile height: A measurement that uses a small graduated ruler inserted until it touches the backing to measure the pile from the top surface of the primary backing to the top of the tuft.

Pile height, effective: See **Pile length, effective**.

Pile knit: Any fabric with a knitted construction as the base and a lopped or cut pile on the surface. Examples are stretch toweling, Jersey velour.

Pile knitted fabric: Manufactured according to the type of Plush cloth with pile surface, i.e. cut/ shorn, pile quite short up to 14 cm long. Application for lining fabrics, outer cloth, imitation fur, blankets etc.

Pile lay, in floor covering: The direction in which most of the pile fibres lean in the original, uncrushed carpet.

Pile layer: Wear layer in carpet material which contains an additional pile forming material above a special Carpet primary backing.

Pile length, effective, in carpet: The length of the fibre or one leg of a tuft from the place where it emerges from the substrate to its farthest extremity or half the length of a loop measured between the two points where it emerges from the substrate.

Pile lifter: Used in carpet shearing machines, for example, for aligning each pile row for shearing using a pile lifting roller fitted with special needles.

Pile lifting: The process of raising the pile on a textile.

Pile lofting: See preferred term Pile lifting.

Pile lifting machine: Pile lifting machine is used in the Shearing of fabrics for the alignment of the pile or fibres to be sheared.

Pile loops: (1) In pile fabric systems that have closed or cut open yarn loops, typically for Pile carpets. (2) Thread sections of the Pile warp between the cutting/untying or bonding points. Knops.

Pile mass above bacing: The mass of the pile that can be shorn or cut.

Pile mass per unit area: The mass of the pile yarn in the given area including that forming the base of the tufts or held in the backing.

Pile opening: With the aid of a pile opening machine pile threads, which are usually made of twisted yarns, are gently opened in wool plush items, for example. The material is processed wet to minimize the material loss.

Pile patterning: The surface of pile fabrics or pile knitted goods can be patterned using different methods: – stencil brush– whirling brush– partitioned shearing tables, which are rotatable and driven.

Pile rake: A hand tool with smooth circular plastic tines used for pile lifting.

Pile reversal: A persistent change in the direction of pile lay in certain areas, resulting in an apparent visual difference of shade.

Pile root, in carpet: That part of the tuft and or loop, excluding dead yarns which is in the substrate.

Pile thickness, pile yarn floor covering: The difference in the unextended height of the tuft elements above the backing measured as the difference between two parallel plates exerting a specified compression on the pile and the backing and on the backing with the pile removed.

Pile thickness (Effective): The difference in the thickness of carpet before and after the pile above has been shorn away.

Pile thread: Yarns forming the characteristic pile in Velvet or Plush fabrics. This is done by placing the threads placed across thin metal rods (pile wires) during weaving, which are subsequently pulled out, creating loops over the base fabric, which can also be cut open. If the pile yarn is a warp thread this creates warp velvet; if it is a weft thread it creates weft velvet.

Pile volume ratio: Ratio of the actual volume of the pile-forming fibres above the subsrate to the total volume of the effective pile.

Pile warp: Warp ends for the creation of the Pile layer and its bonding into the Carpet primary backing.

Pile weave: Pile fabrics are formed by having the basic plain or twill weave as a backing and a third yarn is woven to yield a surface pile. The pile may be warp pile or weft pile. For making ground fabric, plain or twill weave is used, the extra set of filling yarn floats over three or more warp yarns. The floats are cut and brushed up to form pile. This is called filling pile. Examples are velveteen and corduroy fabrics. If an extra warp yarn floats over the filling yarn, it is called warp pile. Examples are velvet, velour and rug velvet.

Pile weight: (1) Weight in grams of the pile yarn processed per m^2 of carpet. (2) Weight in g/m^2 of the pile material protruding above the carpet base; determined by shearing (shaving) the pile.

Pile wire: A metal rod over which yarn is woven to generate a pile fabric.

Pile yarn floor covering: A textile product in which yarn or yarn segments are attached intermittently to a backing fabric to project above the backing fabric and form a pile, in the form of cut loops, or both, with the yarn entering the backing fabric substantially perpendicular to the plain of the backing fabric.

Pile yarns: Threads in Velvet or Plush, which form the characteristic pile. This occurs by threads being placed across thin metal rods (pile wires) during weaving, which are subsequently pulled out, creating loops over the base fabric, which can also be cut open. If the pile yarn is a warp thread this creates warp velvet; if it is a weft thread it creates weft velvet.

Pile yarn mass, for back coated pile yarn floor covering: The mass per unit area of pile yarn.

Pile, carved: The pile of a carpet in which, after manufacture is subjected to a shearing operation with the object of creating different levels of pile, in general on the periphery of certain elements or designs formed by the pile.

Pile, curled: The pile of a carpet, in which curl has been induced in the pile yarn by over twist or by other means.

Pile, cut: The pile of a carpet consisting of legs or tufts or individual fibres.

Pile, cut loop: The pile of a carpet formed during the manufacture by loops and tufts of different lengths or of the same length loop pile.

Pile, sculptured: A pile in which a pattern is created by having areas of different lengths of pile and/ or by omitting pile in certain areas.

Pile, textured: A pile in which the surface character is varied. E.g. by having areas of different characteristics or by a combination of different yarn or pile types. E.g. Soft and hard twist etc.

Pile, tip-sheared: The pile of a carpet originally consisting of loops of different lengths, which has been subjected after manufacture to a shearing process to cut the tips of the longer pile loops.

Pile, uncut: The pile of a carpet consisting of loops.

Pills: Bunches or balls of tangled fibres which are held to the surface of a fabric by one or more fibres.

Pill: A small accumulation of fibres on the surface of a fabric. Pills, which can develop during wear, are held to the fabric by an entanglement with surface fibres of the material, and are usually composed of the same fibres from which the fabric is made.

Pill build-Up: The pills which are formed on the surface of a fabric due to abrasion, wear etc, are in the form of small balls. This on further wear will cause pulling put more fibres from the body of the fabric and the pills will grow bigger and bigger. This phenomenon is called pile build up. After a certainsize, as per the fibre type, this pills will fall off as the fibre connecting the pill and body of fabric can no longer hold them.

Pill formation: Synthetic fibre fabrics and its blends have got a tendency of forming small balls due to entanglement of the fibres on abrasion and wear of the garment. This process of ball formation is called pill formation.

Pill Test: Method for testing the burning behaviour of carpets. A metheneamine pill is ignited upon the test piece and the combustion effects evaluated.

Pillar stitch, in warp knitting: See warp knitting. Chains of loops in unconnected wales are produced. They must be connected together by yarns supplied for a second guidebar.

Pillar Stitch: So-called single chain stitch seam, in which all stitches consist of one thread system and are easy to undo again, e.g. as decorative stitching and as piece goods sewing.

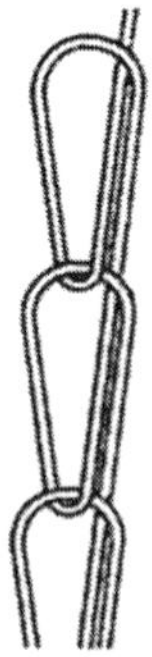

Pilling: (pills, from the Lat. pila = hair balls), faulty, knop-like places that arise at points on clothing that are subject to particular wear due to the undesired rolling together of fibre ends on the surface of the fabric. It gives rise to a rough material surface (pilling effect) which is virtually impossible to dye uniformly. Most common on mixed fabrics and knitted goods made of wool with synthetics (polyamide, polyester or polyacrylonitrile fibres), also in pure synthetics.

Pilling Test: See **Pill Test.**

Pilling resistance: Resistance to the formation of pills on a textile fabric.

Pillow lace: See **Lace, bobbin.**

Pills: Bunches or balls of tangled fibres that are held to the surface of the fabric or yarn by one or more fibres.

Pilot: A woollen cloth generally made in navy blue and used for seamen's coats. It is usually a heavily milled 2/2 twill with a raised, brushed finish.

Pin check, Pinhead, Pick and Pick: Made of worsted, also made in cotton and rayon. A minute check effect caused by a combination of weave and colour. It has the appearance of tiny white dots appearing in rows, vertically, and horizontally. Holds a sharp crease, tailors and wears exceptionally well. In time, it is inclined to shine with wear. A cheaper imitation is made from viscose but this is not hard wearing. Used for men's suitings, womens skirts. In cotton, it usually has a white dot on a blue ground and it is used for work clothes.

Pin clips: See **Clips.**

Pin cord: A very fine needle cord fabric.

Pin drafting: Any system of drafting in which the orientation of the fibres relative to one another in the sliver is controlled by pins.

Pin holes, in fabrics: A very small hole approximately the size of the cross section of a pin, formed due to the breakage of the yarn by mechanical or chemical action.

Pin lock slider, in Zipper: A slider that incorporates a projection on the pull that fits between adjacent interlocking elements of a zipper when a pinlock slider is in the locked position.

Pin locking, in Zippers: Projection on a pull that fits between adjacent scoops of a fastener when a pin lock slider is in the lock position.

Pin mark: A series of holes near the edge parallel with the lengthwise direction of a fabric caused by the holding device on a pin stenter frame. See also **Clip Mark.**

Pin retainer, in Zipper: A tubelike element similar to the separable pin attached over the bead at the bottom end of the stringer opposite to the separable pin. This element is designed to hold the fixed retainer in position.

Pin stripes: The name given to any cloth with very fine pin-width stripes. Usually associated with men's worsted suitings.

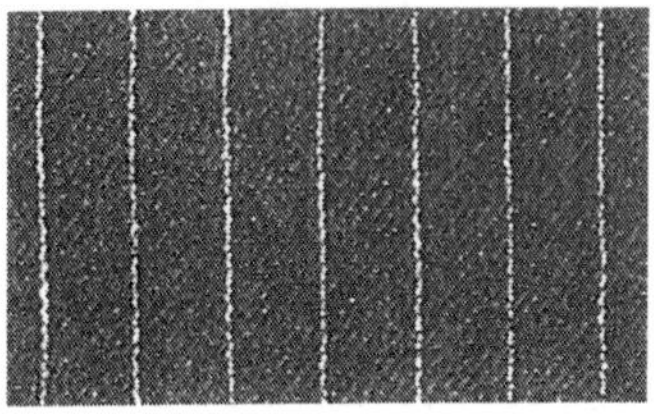

Pin Tucks: Tiny lines of pleats that are stitched down. They can be any distance apart and are purely decorative. They are seen on shirts, blouses, skirts etc., usually on lightweight fabrics.

Pin, Separable, in Zippers: A tube like element attached over the bead at the bottom side of the stringer.

Pin-wale corduroy: See **Corduroy.**

Pin-drafting: A system of drafting in which the fibres are oriented relative to one another in the sliver and are controlled by rolls of pins between the drafting rolls. It is primarily used for long fibres in the semi-worsted and worsted spinning systems.

Pina: The vegetable leaf fibre from the pineapple plant.

Pina cloth: A fine plain woven cloth for which the yarns are produced from the fibres of the leaves of the pineapple plant. It is a stiff wiry fabric and makes a good base for embroidery.

Pineapple cone: Biconical package, conical with spherical face. Crosswound yarn package.

Pineapple fibre: A fibre obtained from the leaves of the plant Ananas comosus, capable of being processed into fine fabrics.

Pineapple fibre: (pineapple leaf fibre) Bromelia fibres.

Pineapple leaf fibre: See **Bromelia fibres.**

Pinhead: A small pinhead-sized opening usually found about 10 to 12 inches from a selvage.

Pinheads usually run in a fairly straight line along the warp and are formed by the shuttle pinching the filling, causing small kinks that show up as small holes in transmitted light.

Pinhole: A very small hole in hosiery or woven fabric.

Pinion barré: A fine, filling wise fabric defect appearing as one or two pick bars in an even repeat. It is caused by a faulty loom pinion.

Pinked seam finish: Pinking a seam edge is only a ravelling stop gap which works best on firmly woven fabrics. You will need a sharp pair of pinking shears for this seam finish. Pink just on the edge of the seam allowance, taking short even strokes.

Pinking: Sewing of a material either to prevent fraying or to decorate at the raw edge.

Pinning; On a Pin Stenter there are two pin chains on to which fabrics are pinned at both selvedges and carried through the stenter chambers for drying, finishing or heat setting of the fabric etc. At the entry of the stenter there is a pinning mechanism to pin the fabric at the selvedges exactly.

Pint: A volume measure equal to half a quart; in the US, 1 pint is equal to 16 fluid ounces, in Canada and Britain, pint equals 20 fluid ounces

Pinwale: A very narrow wale or rib in corduroy. Also written pin wale.

Piping: (1) A strip of material with or without a filling applied folded, normally to the edge of a garment. (2) A narrow fabric having a cord on one edge. (3) The operation of applying piping on a fabric.

Pique: Originally piques were fabrics with cords in the weft direction. As it is undesirable to weave cloth with cords in the weft direction today nobody makes pique of the old name. Today pique is a cloth of fine Bedford cord construction. A weave of real pique is rather complicated. It has two warps: A face warp and a backing warp, usually in a proportion of two ens of face to 1 end back. It has 3 wefts: a face weft which forms the surface of the cords; a cutting filling which forms the grooves between the cords; a coarse stuffer weft which is held between the backing warp and the face warp, under the face warp, to accentuate the cords. An average construction is 100 x 160 with 30s for the face warp and 2/30s for the back warp, 30 for face and cutting weft and 10s for the stuffer. In case of Bedford cord pique, a typical construction can be 150 x 80 with 40s warp and 20s stuffer and 40s filling. Besides Bedford cords there are other pique derivatives. Pique (Fr: stepped), the characteristic stepped property is a result of the weave, sometimes also by an embossing treatment. Genuine piqué (fur piqué) consists of a fine plain weave top cloth

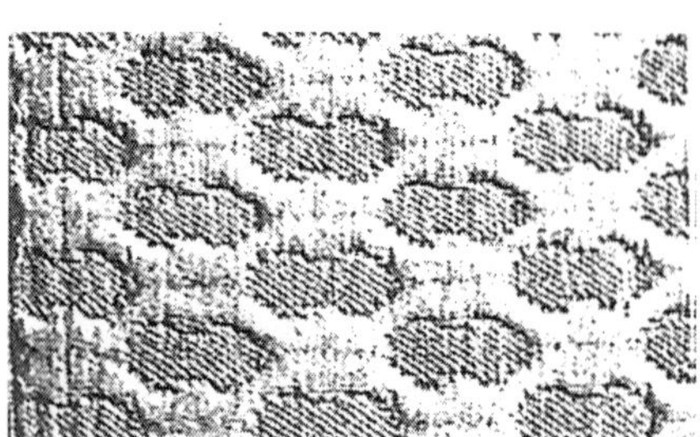

Pique

and a coarse back cloth. The effect is created by filling and figuring wefts made of thick material and bound according to a pattern, which imprint onto the fine threaded surface to form the embossed pattern. In double piqué the material has two right sides. In press piqué, a non-genuine piqué, the figuring pattern is pressed into the smooth fabric by the embossing calender; often made durable by a special synthetic resin finish. Once always made in cotton and in white, but now small amount of other fibres may be used. Piqued fabric an also be made from other fibres also. Cotton pique is stiff and absorbent. Used for tennis clothes, it is hard wearing.

Pique Anglais: Solid coloured French serge, having eight leaves and eight picks in a repeat.

Pique' voile: See **Voile.**

Pique, fast-back: Refer ***Loose-back pique.*** In case of a construction with four shafts one may weave in to the stitching warp, and this is called a fast; back pique.

Pique, loose-back: Simple pique has horizontal ridges that run from selvedge to selvedge. The first two shafts are alternated to start the plain weave, and then a soft squishy wadding weft hat will not show on the face is inserted under the face warps and over the stitching warps. The wadding pick is inserted halfway through the weaving of the face fabric, which positions it nicely in the centre of the cord. The stitching warps float on the back of the fabric. With three shafts this is the only option. This structure is called a loose-back pique.

Piquete: Corded twilled French vesting, made with eight warps and four picks in a repeat.

Piramides: 17th century fine and narrow English worsted, made with two-coloured yarns.

Piripiri: Fine stem fibre of the Pipturus argenteus in Tahiti; used for cords, nets and bags.

Pirl: Gold or silver thread made spiral by winding.

Pirle: An English finishing process, rendering the woollen fabrics waterproof and unshrinkable.

Pirn: (1) A support slightly tapered with or without a conical base on which yarn is wound for use as weft. (2) A single-headed bobbin with cone shaped head and barrel 'which varies in length. The pirn is the yarn-supply for the shuttle used in weaving.. (3) Short or long package, with or without cone, primarily for filling yarn.

Pirn barré: A fabric defect consisting of crosswise bars caused by unequal shrinkage of the filling yarn from different points on the original yarn package.

Pishteh: In the Bible means flax and linen.

Pita: Very long, strong, glossy, white and silky fibre, yielded by the various agaves and pineapples in Central America. Used for threads, fine hammocks, etc.

Pitch: The balanced insertion of sleeves into the garment, controlled by balance marks in the scye line.

Pitch, in woven pile floor covering: The number of pile yarn end in 27" measured in the weft wise direction.

Pitch, rope: The distance parallel to the axis of a rope in which a strand makes one complete spiral.

Pitch: See **Gauge**.

Pitch, for pile floor covering: The number of binding sites in 686 mm (27 in.) of width.

Pittman: Commercial variety of early maturing and prolific cotton from Louisiana, the staple measuring 23-25 millimeters; the yield is about 31 per cent.

Pixel: A tiny picture element that contains red, green, and blue information for colour rendering on a monitor or a scanner. When generating colours, pixels are similar to dots of ink on paper. A monitor resolution description in terms of pixels-per-inch (ppi) is similar to a printer resolution description in terms of dots-per-inch (dpi).

pK value: Is the pH at which a compound splits off a proton by the addition of alkali or attaches a proton by the addition of acid. Plays a role in reactions of amino acid side chains of wool. For example, if the pKS value of per acetic acid lies at pH 8.2; at this pH there are the same concentrations of per acetic acids and per acetate anions.

PKN: Polish standards association Technical and professional organizations.

PKS bleach Process: (peroxide continuous fast process), the continuous immersion bleach process without a padder is suitable for the bleaching of all cotton, cotton mix and linen fabric as well as coloured fabric, particularly where relatively high demands are made of the material quality, such as good absorbency.

PL: Polyester fibres.

Placarder: French term for a resist dyeing process by which the ground will take the dye but the pattern remains white.

Placket: An opening provided in a garment or an extra piece material applied to that opening for reinforcement or as a style feature which may incorporate fastening.

Placket, Tape faced: The tape faced placket is stronger and may be used in children's drawers, etc., in place of a gusset to strengthen the end of the opening. A single piece of tape folded back as for a loop is stitched along all edges, making an opening without a lap. This offers as much resistance as a gusset and is more quickly done.

Placket, Faced: In a third kind of placket, the opening is faced with a continuous piece of tape on both sides and finished with a piece of material on the outside. See illustration. This makes a strong and simple placket. When a tape cannot be used, a hem or facing may be made on the under side of the opening and a facing on the upper side, over which the on-set piece is stitched. The on-set piece and facing may be cut from one piece, but the fitting is more troublesome. In figured goods, the piece set on should match the pattern exactly.

Plagae: Linen sheeting in ancient Rome.

Plaid: Plaids are basically in a checked design, either square or rectangular. The checks are often simple, sometimes produced by only two or three colours. The word plaid was originally used only for the Scottish tartan shawl worn by Highland women of U.K..

Plain curved seam: This is a basic plain seam but curved as is required when stitching the outside edge of rounded collars or cuffs and when applying facings to curved armhole and neckline edges.

Plain dyed fabric: Single coloured, unpatterned.

Plain fabric: Manufactured in plain weave, (usually) smooth material made of flax or cotton of different fineness, density and finish (the finest plain fabric = batiste).

Plain fabric, Weft knitted: A fabric in which all the component loops are of the same sort and meshed in the same manner. The fabric may also be described as a plain single jersey web or stockinet. The appearance may be described with reference to the surface of the structure (a) face; Technical face: The surface of a plain weft knitted fabric that consists wholly of face loops (b) Back; technical back: the surface of a plain weft knitted fabric that consists wholly of back loops

Plain Gauze: Leno weave where the warp threads cross other warps always in the same direction.

Plain net (Bobbinet): On the plain net or bobbinet machine a series of pairs of thread carriers (bobbins) swing backwards and forwards through the warp sheet whist progressing from one side of the machine to the other, and back again. The bobbin yarns are thus looped around the war yarns in a spiral formation leaving a regular series of holes in the fabric, in honey comb effect. The machines can be provided with jacquard mechanisms for figuring and pattern threads can be provided in the warp. Patterns also can also be applied on plain net fabric by subsequent embroidery. Applications: veils, laces, drapes etc.

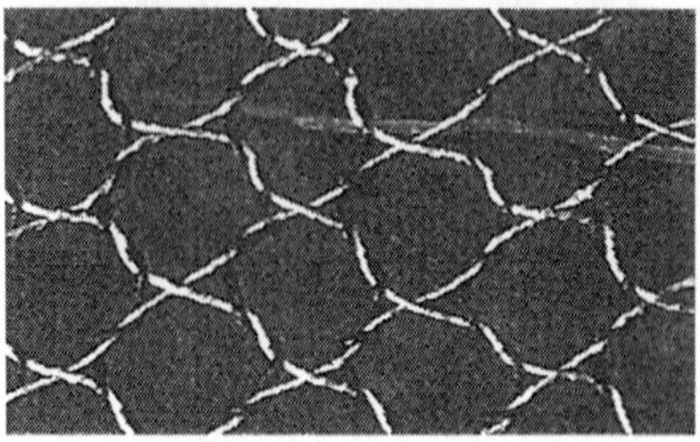

Plain net (Bobbinet)

Palin knit fabric: See **Flat knit fabric.** Single jersey fabric with all face loops on one side and all reverse loops on the other.

Plain loom: Loom used for weaving plain weave, with no attachment for designs, pattern, checks etc.

Plain princess seam: A princess seam is a seam with opposing curves; one seam edge curves out and the other one curves in.

Plain raschel fabric: Raschel-knitted fabrics, in which a guide bar forms the base and a further guide bar inserts a weft in accordance with the pattern, which achieves the side connection of the stitch wales (e.g. weft pattern).

Plain satin: See **Satin.**

Plain seam: A basic seam formed by a single joining line. A seam formed by a row or rows of stitches joining two pieces of materials together face to face. This is the workhorse seam of clothing construction, usually of 10 to 12 stitches per 2.5 cm. (per 1 in.). (a) *Straight:* This is the most common seam used to sew two pieces of fabric together the length of the stitch is between ten to twelve stitches per inch. The width of seam for fashion sewing is usually 5/8 inch and usually ½ inch for industry sewing. (b) *Bias Seam:* Hand-tuck bias seams, leaving the threads loose at the ends. Allow your work to hang overnight before stitching. Always use a shorter stitch length for a bias seam Fabric requires a 2 mm (12) stitch length on a straight seam requires a 1.5

mm (15) stitch length on a bias seam to increase the elasticity of the seam. Stitch with the grain. Bias seams are given on a garment as a style line such as while joining 'v' shaped yokes. In the bodice or in case of 'v' shaped skirt yokes. Bias seams can be given in the form of decorative tucks & seams on a garment. (c) (i) *Curved:* Curved seams are quite commonly used when sewing together a garment. In many cases, the curved seam is the style line of the garment, such as princess seams, bodice yokes or skirt yokes. (d) (ii) *Right-Angle Seam or Corner Turns:* This seam with a 90-degree angle usually found in a style line. The sewing method used to turn a corner can be applied to any angular seam found on many parts of a garment. (e) *Double-Stitched Seam:* Use this seam is sheer fabrics and lace for curved as well as straight seams. Place the first row of stitching on the seamline. Press. Place the second row of stitching within the seam allowance, about .6 cm (1.4 inch) from the first row, using a fine multi-stitch zig-zag. Straight stitching may also be used. Trim the seam allowance. (f) *Lapped Seam:* Use lapped seams when joining sections of interfacing and interlining to eliminate bulk. Lap one edge over the other, with the seam lines meeting in the center. Stitch through the center, using a multi-stitch zigzag or straight stitching. If the seam edges are too wide, trim after stitching. (g) *Lapped seam:* This seam is most often used for yokes & applied parts such as gussets. One section is lapped over the other & topstitched. Fold under the seam allowance on the section to be lapped & press flat. Place the sections together right sides up, matching the fold to the seam line accurately & baste in position. Stitch close to the folded edge through all thickness. (h) *Double seam:* For Double top seam, place a row of seam parallel to the first row of top seam, a pressure foot distance away from it towards the same side. (i) *Abutted Seam:* Use an abutted seam for non-woven interfacings and interfacings of hair canvas. Lawn and similar fabrics. Trim away the seam allowance on both sections. Bring the two edges together and pin over an underlay of a lightweight fabric the underlay should be 2.5 cm (1 inch) wide and slightly longer than the seam. Stitch from the right side, using the multi-stitch, zigzag. Widest stitch width 1.5 – 1 mm (20) stitch length. Backstitch at each end. The abutted line should be aligned win the center of the presser foot. Straight stitching may also be used on each side of the abutment. (j) *Top-Stitched Seam:* Top stitching is used as a styling point along the finished edge of a garment or along the seam within a garment. When preparing a curved. Topstitched seam, stay-stitch near the seam line on both the overlap and the underlap. Fold under the seam allowance on the overlap and pin, tack, trim and press. Lap the folded edge a full seam's width over the underlap. Then pin and tack remove the first tacking. Topstitch from the right side, close to the folded edge. If the stitching line with tacking as instructed for Top Stitching.

Plain stitches: Stitches may be divided into plain and ornamental. The plain stitches are the (a) basting, (b) running, (c) the running and back stitch, (d) half back stitch, (e) back stitch, (f) overhand or whipping stitch, (g) overcast, (h) hemming, and (i) blind or slip stitch.

Plain (Tricot) stitch, in warp knitting: see **warp knitting.** Each yarn works in a zigzag fashion lapping between two neighbouring wales. All laps are closed.

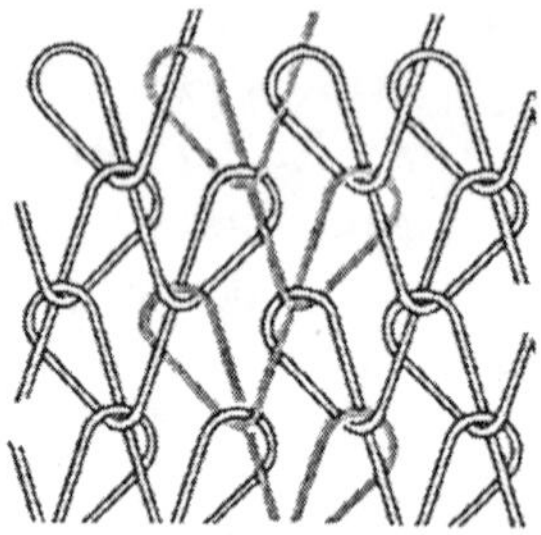

Tricot stitch

Plain straight seam: See **Plain seam**.

Plain surface fabric: A fabric which does not have any intentionally raised fibre or yarn surface such as pile, napped, or tufted surface.

Plain weave: The simplest of all weave interlacings, in which the odd warp threads operate over one and under one weft thread throughout the fabric and the even warp threads reverse this order to under one, over one throughout. A plain weave does not necessarily result in a plain surface effect or plain design in the fabric, e.g. variation in the yarn counts warp to weft or throughout the warp or weft (or both) and variation of the thread spacing warp to weft can produce rib effects (see taffetta, poult, faille and grosgrain), while colour patterning of the warp or weft (or both) results in colour-and-weave effects.

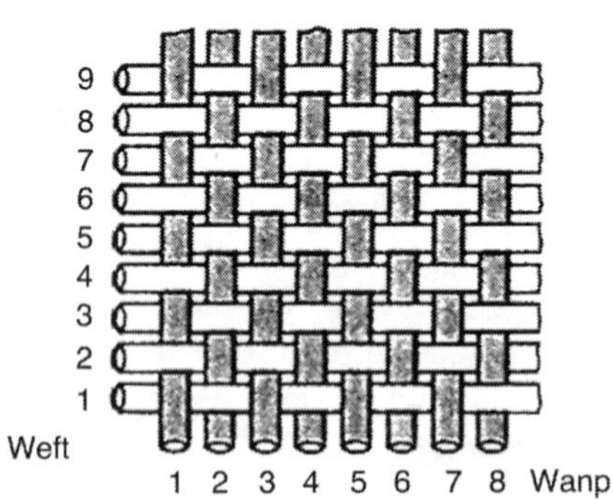

Plain weave cotton fabric: Collective term for plain weave fabrics of different finenesses and thread densities made of single cotton yarn (nettle cloth, coarse cotton cambric), which are sometimes used raw, bleached, dyed or printed. The 3 basic qualities in increasing fineness of the fabric structure are (a) Cretonne; (b) Renforcé; (c) Calico.

Plainback: An English worsted fabric, originated in the 19th century; it is made of single yarn with a twilled face and plain back in imitation of jean.

Plain dyed fabric: A fabric dyed in single coloured, unpatterned.

Plaited goods wagon: Means of internal transport for materials in stacked or rope form.

Plaited stitch: Or herringbone stitch used n embroidery to imitate the herringbone effect.

Plaiter: Device for stacking material in folds.

Plaiting: (1) See **Braiding.** (2) The arrangement of fabric in widthwise folds.

Plaiting down: Stacking of continuous fabric lengths in a zigzag form. Takes up more space than batching.

Plangi: (Malyasian: bright, recessed spot), a resist technique that is still widespread in South East Asia and Africa (like the Japanese Shibori technique) for the patterning of clothing materials. Before dyeing yarn package or binder resistance take place (Ikat), which appears as recessed pattern after dyeing. Stitch plangi is called Tritik. In a wider sense the plangi technique is performed by folding, knotting or weaving, by screens, by resist application (sludge, paste, wax) and by negative resists using dye mordanting. Often several techniques are combined.

Planning: Managing and controlling events to achieve a goal and making the best use of resources.

Planting: A process in weaving by which the various coloured extra warps are interchanged.

Plasma: Plasma is the least organized collection of electrons and atomic nuclei, i.e. it is a mixture of these two components. Material is made of atoms which can be subdivided into electrons and nuclei. In gas phase the matter is in the form of atoms, liquid phase atoms forms unordered bonds and in solid phase they are in ordered bonded state and in most cases symmetrical organization of atoms as in case of crystals. The removal of the atomic bond between electrons and atomic nuclei, typical in plasmas, gives rise to the decisive difference between gases and plasmas.

Plasma Treatments: Plasma treatment is a surface modifying process, where a gas (air, oxygen, nitrogen, argon, carbon dioxide and so on), injected inside a reactor at a pressure of approximately 0.5 mbar, is ionized by the presence of two electrodes between which is a high-frequency electric field. The need to create the vacuum is justified by the necessity to obtain a so-called cold plasma with a temperature no higher than 80 °C. This, with the same energy content that can be reached at atmospheric pressure at a temperature of some thousands of degrees C, permits the treatment of fabrics even with a low melting point such as polypropylene and polyethylene, without causing any form of damage.

Plastic: A high polymer, usually combined with other ingredients such as curatives, plasticizers, and fillers. It can be moulded under heat and pressure and then machined accurately in its hardened state. General term for a wide range of substances.

Plastic-coated textiles: In its widest sense, this group also includes materials of various fibre composition with coatings, rubberizing and coating finishes. More specifically, it is used to describe products in which yarns, knitted goods from cellulose, natural or synthetic fibres are overall coated with synthetic polymers (during manufacture or as a subsequent permanent finish).

Plasticizer: (1) A chemical added to polymers and resins to impart flexibility, workability, or stretchability. (2) A bonding agent that acts by solvent action on fibres.

Plastisol: With reference to textile arts, a type of screen printing ('silkscreen') ink which consists of a coloured pigment, fine particles of polyvinyl chloride, and a plasticizer Plastisol ink does not dry. It must be heated to the point where the PVC particles melt and mix with the plasticizer and pigment to form a flexible coloured plastic film. Plastisol ink stays on fabric simply because is surrounds and encapsulates the fibres mechanically. This is by far the most popular type of ink for commercial screen printing of t-shirts and the like.

Plastisol coating: (paste coating), working method using Plastisols, consisting of synthetic resin, softener, pigment, stabiliser, etc. in premixed form, which is spread upon the fabric or knitted goods, upon which the resin is hardened.

Plate felting machine: For the manufacture of fabric felts, which are usually used as technical felts. In the plate felting machine the upper plate (50°C) describes a circular or elliptical path, whereas the lower plate (40°C) is fixed. The pressure is mechanically applied.

Plate heat exchanger: Reclamation of heat energy by means of compression moulded plates, which are sealed at the edge. The two media involved in the heat exchange are alternately led through gaps. The heat exchange takes place directly via the exchange surfaces (glass, metal, plastic). There is no exchange of moisture nor dirt transfer from the escaping air flow to the external air flow.

Plated fabric, Knitted: A fabric knitted from two yarns of different properties of which are used in the same loop whilst positioned one behind the other. The special feature of the fabric is that each loop exhibits the characteristics of one yarn on the face side and the characteristics of the other yarn on the reverse side. In the plain weft knitted fabrics when all are formed in the same direction characteristic of one yarn is visible on the surface composed of the face loops while the characteristics of the other are only visible on the reverse surface composed of the back loops.

Plated fabric, cross: An arrangement of a plated face stitch (knitted on one set of needle) and of an adjacent plated reverse stitch (knitted in the sane course on the other set of needles) the relative positions of the face yarn and the back yarn in the face stitch being interchanged in the reverse stitch.

Plated fabric, float: The knitting of a plated fabric in which the relationship of the two yarns in is reversed in certain stitches within the same course to give a patterned or fancy effect.

Platille: A fine quality of pure French linen.

Platt: Machine lace made flat without any raised work. The pattern is worked by threads running in zig zag line between straight and parallel threads. Some of these laces have the design outlined by a heavier thread while others are made without this outline.

Pleat: Three layers of fabric involving two folds or reversals of direction; the back fold may be replaced by a seam Pleats are sharp folds of fabric introduced to allow for expansion or to provide style features, Which are pressed flat and stitched to place on the top edge. They are heat set into thermoplastic fabric such as polyester and nylon etc.

Pleating: Creation of very varied shapes and arrangements of folds on textiles using Pleating machines. Permanent, wash-resistant pleating can be produced.

Pleating machine: Used for Pleating: (a) Knife, squeezing and group pleating machines for the generation of a wide range of different patterns (approx. 1 mm to several cm underlay area and fold width). (b) Crystal pleating machines for the generation of vertical folds (approx. 1–5 mm fold height). (c) Accordion pleating machines for upstanding folds (5–50 mm fold height). The pleating temperature is always dependent upon the fibre type (thermofixation):

Cellulose 150–200°C

Fibre mixes 150–210°C

Polyamide 6 180–194°C

Polyamide 6.6 190–208°C

Polyacrylonitrile 120–160°C

Polyester 160–180°C.

Pleats: Pleats are sharp folds of fabric. Which are pressed flat and stitched to place on the top edge. They are heat set into thermoplastic fabric such as polyester and nylon etc.

Pleats, Accordian: Narrow straight pleats usually parallel to one another.

Pleats, Box: The fabric is folded equally on both sides and the edges of the pleat meet at the centre of the back of the pleat. It is the combination of two flat folds in opposite direction.

Pleat, Broomstick: Gathers unevenly spread. As the name suggests they are very uneven pleats.

Pleats, Cartridge: These resemble box pleats, but box pleats are very solid pleats whereas these are soft & free flowing pleats. These pleats generally open up or broaden and have a slight flare unlike box pleats.

Pleats, Crystal: Very fine pleats like accordion pleats but much Smaller.

Pleat, Inverted: A single reversed box pleat where the pleat is not seen from the front, just the two edges meeting together. These have narrow falls towards one side.

Pleat, Kick: A small pleat placed in a garment to allow more movement, usually in a tight skirt.

Pleats, Knife: Pleats of any width, similar to accordion pleats, but all face the same direction when folded.

Pleat, Re-inforced Inverted: As the inverted pleat, but with the addition of re-enforcements to keep the pleat in a certain position.

Pleats, Sunray: Pleats radiating form a single point usually the centre of full circular skirts.

Pleats, Top – stitched: Any kind of pleat, but part of the pleat is topstitched to create a fixed shape that then contrasts with the unstitched pleat.

Pleats, Watteau: Pleats at the back of a gown, known as the Robe a‘ la Française. The pleats are known as Watteau pleats and are stitched down at the back of the neckline and then allowed to fall freely to the ground.

Pleats, Unpressed: Any kind of pleat that has not been pressed. They have a rounded appearance.

Pleures: French term for pulled wool, taken from sheep that died of natural causes.

Plie: Piece material laid in uniform widths; hanging in plie (e.g. when suspending silk goods), i.e. hung folded on the stick. The stick has three notches for suspension loops, onto which the piece ends can be attached.

Plied yarn: A yarn formed by twisting together two or more singles yarns in one operation.

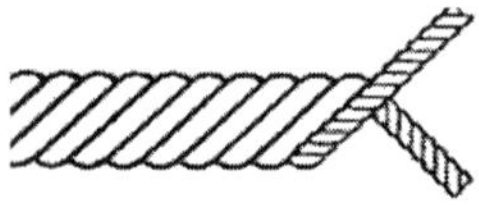

Plied yarn Duck: Duck with plied yarns in both warp and filling (army, numbered and special used ducks).

Plissé: A fabric consisting of alternating wrinkled and plain stripes, often in different colours. The fibre content is usually polyester with cotton. Sometimes crinkled crepe effect can be obtained by special chemical treatments, on ordinary light weight cotton cloth. Caustic soda solution treatment is one type. Caustic soda solution is printed in narrow stripes, whereby those portions are shrunk compared to the unprinted portions. This untreated stripes will pucker causing the crinkling crepe effect. When the stripes in the crinkled crepe are rather wide, the cloth is called plisse. This is a semi-permanent finish which is liable to lessen with constant washing and particularly with ironing. It is a soft cool summer fabric and is used for casual blouses and skirts, beachwear and children's clothing.

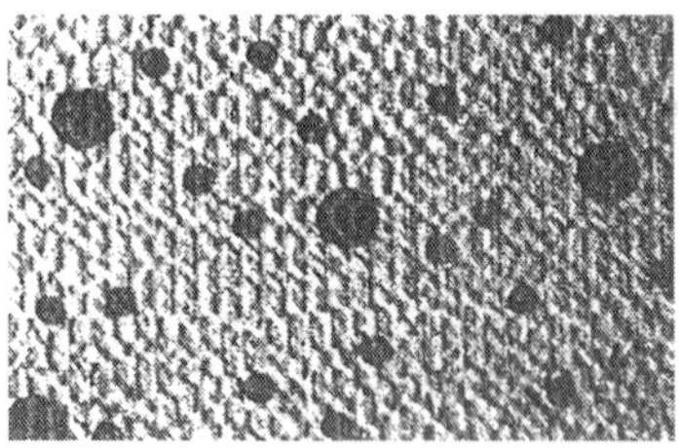

Plisse'; Tisse': Made with very strongly twisted threads which are shrunk in the pretreatment to produce a pleat or relief like fabric appearance.

Plodan: In the 16th century a coarse woollen made in two or three coloured checks worn in Scotland by citizens' wives for cloaks. Believed to be the early form of plaid.

Plommett: A 16th century fabric in England, made of worsted and silk or linen, the piece weighing 4 Ibs. and measuring 14 yards.

Plucked Wool: Shorn from a carcass several days old.

Plucker: (opener, willow). Preparation machines in spinning. Classified as tearing, beating, striking, oil lubrication, mixing, carding plucker, etc. according to application.

Pluckings: Short clean fibre produced at the end of a scotching machine where operatives dress and square pieces of flax ready for selection.

Pluie: Lustrous dress goods and tapestry, made in France of silk or camel hair warp with gold or silver threads interwoven in the filling.

Plugging value: In the manufacture of acetate fibres, a measure of filterability. It is the weight of solids in an acetate dope that can be passed through a fixed area of filter before the filter becomes plugged. It is expressed as weight of solids per square unit of filter area, e.g., g/cm^2.

Plumage: The out growth of fowl, consisting of feathers and down (waterfowl) or feathers only (non-waterfowl).

Plumetis: Embroidery in feather stitch over a clear and light ground.

Plumules: Downy waterfowl plumage with under developed soft and flaccid quill with barbs indistinguishable from those of down.

Plush: (1) Any cut pile fabric that has a pile length of more than 3 mm, a velvet. The yarns can be almost any natural or synthetic usually acrylic, and many of the fabrics are now washable. Made in various weights fir dresses, curtains drapes and upholstery.

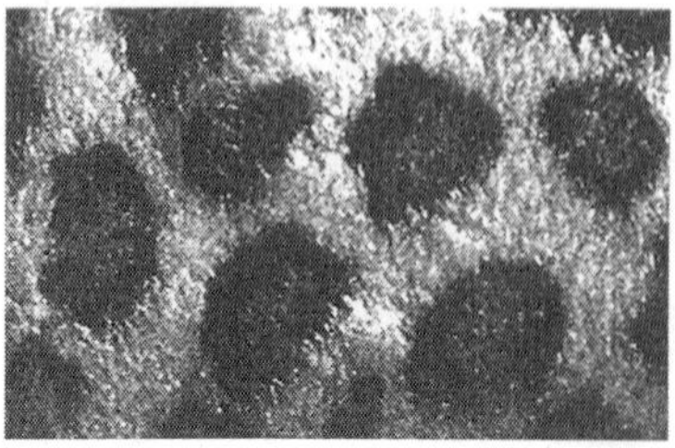

(2) Knitted plush: Plush-like appearance due to pile formation in tricot by a plush loop. Plush cloth. See **Single plush; Double plush**.

Plush Stitch: A stitch applied in Berlin work to form a fringe along the edge of the embroidery and it is left either in loop form or is cut.

Plush carpet: General term for carpets with cut open pile warp (thread pile); either shorter velour-like (velours carpet, Tournay carpet) or longer pile (plush carpet). Named according to material and pattern: Hair yarn, linen, wool, knotted, striped plush carpet, etc. Manufactured in the form of Wire carpets or as Face-to-face carpets (face-to-face tournay carpet).

Plush cloth: (plush fabric, knitted plush), 2–3 thread Knitted fabrics usually made of ground thread and plush thread. The latter are formed into long loops and either closed, curled, frizzy (thermal underwear, jogging suits, lining material, etc.) or cut, shorn for velvet, pelt or fur like pile goods (lining material, outer material, fake fur, blankets). Pile-knitted fabrics.

Plush trimming: Trimming the pile loops of knitted plush fabrics by Shearing.

Ply: (1) The number of single yarn twisted together to form a plied yarn; also the number of plied yarn twisted together to form a cord.

(2) Individual yarn in a plied yarn or in a cord.

(3) One of the several layers of fabric.

Ply adhesion: See **Ply security**.

Ply broad cloth: See **Broad cloth**.

Ply poplin: See **Poplin**.

Ply Security, in sewing thread: Refers to a sewing thread's ability to stay together during the sewing process. The most common type of thread break is caused by a loss of ply security. When the plies open up during the sewing process they are much more susceptible to breaking and causing an unravelled type of thread break.

Ply voile: See **Voile**.

Ply yarn: Thread made of two or more Yarns twisted together. We talk of 2-ply, 3-ply, 4-ply yarn, etc. Soft ply-yarn made of numerous slightly twisted yarns is also called a wick.

Plying: Twisting together two or more singles yarns or ply yarns to form, respectively, ply yarn or cord.

PN: Polska Norma, Polish standard.

Pneumatic guiders: Expander and edge uncurling device. Works with 2 pneumatic blowers and photoelectric sensors; material is scanned and expanded using compressed air. Differs from mechanical guiders due to non-contact, low tension material guidance.

Pneumatic squeeze: In this screen printing squeegee the squeegee rubber is held pneumatically and can be changed quickly and easily.

Pochote: Very fine and lustrous fibre yielded by the Eriodendron anfractuosum in Mexico. It is similar to the kapok. The fibre is used for fine cloth and for hats, in this case mixed with rabbit's fur.

Pocket: A bag for practical or decorative use formed by the inserting or attachment of material to a garment. Pockets may have flaps, jettings or welts. E.g. include- cross pockets, have pocket, jetted pocket, and patch pocket

Pocket, patch: A pocket attached to the outside of the garment and constructed of self-fabric.

Pocket, quarter: The angle from the side seam.

Pocket, rule: A patch pocket attached on the outseam, halfway between the hip and the knee of the garment; usually found on coveralls.

Pocket, serged: A pocket formed by joining two pieces of fabric and joining the edges with safety-stitching.

Pocket, slash: A pocket that must be entered through a slash on the garment. The pocket pouch is suspended from and attached to the slash.

Pocket, stitch and turn: Formed when two pieces of fabric are joined along the edges and turned so that the raw seam margin is inside of the finished pocket.

Pocket, stitched/topstitched: The same as stitch and turn pocket, except with an added row of stitching along the folded edges.

Pocket, swing: The pocket pouch is suspended from and attached.

Pocket, in Zipper: The cavity of an element designed to receive the head.

Pocket, in Zippers: The cavity of a scoop designed to receive the head.

Pocket Drill: Stout, unbleached cotton drill, used for pockets.

Pocket Facing: A piece of shell (outer) material super-imposed on the top of the pocket material at its opening to conceal the lining.

Pocketing: Cotton velvet used for overcoat pockets; also napped cottons used for the same purpose.

pOH: A measure of the concentration of hydroxide ions (OH) in solution - pOH = –log[OH] = (14 – pH); see **pH**.

Poil-grege: (high twist grey pile yarn). Highly twisted Grège threads with approx. 1000–2000 twists/m.

Poile de Chevre: Fine, soft French dress goods made with coloured silk or cotton warp and angora filling.

Point: (1) In hand-made laces denoting fine quality, irrespective of the make; (2) French term for stitch in laces.

Point a I' Aiguille: Lace sprigs made with the needle, irrespective of the design.

Point Anglaise: French for feather stitch.

Point bonding: A method of making thermally bonded nonwoven fabrics in which heat and pressure are applied to specific areas by the use of embossed calendar rollers to cause local bonding.

Point D'Alencon: Same as **Alencon lace**.

Point D' Angleterre: Originated in England as bobbin lace, improved upon by Flemish needle point stitches in the 17th century. The ground shows great variety, the net ground being bobbin-made around the patterns. Often cordonnet or ribs are produced by plaiting the threads and also bobbin made brides or fancy needle jours are employed.

Point d'Esprit: The name describes net fabrics that has embroidered spots or squares on the surface, but the term is often used in relation to any light weight fabric such as voile that is decorated with satin-stitched spots. Unusually polyesters, cotton and sometimes silk First made in France in 1834. Dull surfaced net with various sized holes. Has white or coloured dots individually spaced or in groups. Used for dresses, wedding gowns, curtains, bassinettes, evening gowns, etc.

Point d'Anvers: Same as Antwerp lace.

Point Applique: Applique lace, made of needle-point sprigs attached to a net ground.

Point d'Arabe: Coarse French bobbin curtain lace of Arabian origin, made of ecru cord with large patterns.

Point d'Armes: An embroidery stitch used on transparent materials for leaves and flowers, showing on the face a hemstitch while the threads are crossed in the back in a close lattice fashion.

Point d'IAttache: A variety of stitches in embroidery, by which fancy material is attached to the foundation.

Point de Biais: Embroidery stitch consisting of slanting satin stitches made of different length.

Point de Brabancon: A flat filling used in needle-point laces, consisting of rows of buttonhole stitches linked together.

Point a Brides: Generic term for laces with bars in the ground.

Point Brode: Bobbin lace sprigs, made with raised work.

Point de Bruges: See Bruges lace.

Point de Bruxelles: .See: **Brussels lace**.

Point de Cable: Same as rope stitch.

Point Cam pan: Narrow French bobbin lace edging of the 17th century.

Point a Carreaux: A French bobbin lace made with a simple, trellis-like pattern.

Point de Chainette: Same as **Chain stitch**.

Point de Champ All laces having a mesh ground, irrespective of style.

Point de Chant: See Point de Paris.

Point Chaudieu: In macrame lace a chain bar, formed by looping one thread around the other.

Point Chemin de Fer: Same as **Railway stitch in embroidery**.

Point de Cone: In guipure lace a cone shaped form stretching over four square meshes and filled in with cloth stitch.

Point de Cordova: A filling in needle-point laces, made stretching three threads close to each other and darning dote over.

Point de Cote: Same as **Rope stitch**.

Point Coute: See **Darned lace**.

Point Crochet: A lace made by crocheting; the various sprigs are made separate and joined together with bars, resulting in a guipure lace.

Point Croise: An embroidery stitch which forms in the front rows of continuous short stitches (like the hemstitch) while the thread is crossed in the back.

Point de Croix: Same as **Cross stitch.**

Point de Diable: An embroidery stitch, consisting of a star with eight rays in a square.

Point de Dieppe: Same as **Dieppe point**.

Point Double: See **Point de Paris**.

Point Duchesse: See **Bruges lace**.

Point d'Epine: Same as **Feather stitch**.

Point d'Escaliei: Same as **Ladder stitch**.

Point d'Espagne: Heavy guipure-like needle-point lace of the 17th and 18th centuries somewhat resembling Point de Venise and made of gold or silver thread with thick cordonnet outlines.

Point d'Echelle: A variety of ladder stitch in embroideries in which the bars are stitched across an open space.

Point d'Esprit: (1) Machine net with small dots scattered all over. (2) Light and open stitches in needle guipure laces, consisting of loops forming various patterns in the square meshes.

Point d'Etoile: A stitch over square mesh ground in needle-point laces; it is a more or less ornate star, covering nine or 16 squares.

Point Evantail: In guipure laces fan shaped spots formed by darning stitches.

Point Faisceau: A heavy stitch in needlepoint laces, consisting of herringbone stitches joined with a loop in the centre.

Point de Feston: A filling in needle-point laces, consisting of festoons fastened with a knot at every loop.

Point de Feuillage: In raised macramé lace a bar made with four threads.

Point de Filet: A ground in needle-point laces made by fastening loc-ps to each other in buttonhole stitch, which form a filet mesh.

Point de Flanders: See **Brussels**.

Point de France: Term applied to French needle-point laces similar to the Point de Venise started by Colbert in the 17th century.

Point de Gauze: Very fine needle-point mesh for Brussels lace.

Point de Genes: (1) A species of Aetzstickerei (see), made on a wool ground; (2) A stitch over square mesh in needlepoint lace; two or three threads are stretched diagonally across several squares of the mesh and fastened together by darning.

Point de Gerbe: A stitch in guipure lace, made by looping a thread several times around the opposite sides of a square mesh, forming buttonhole loops on one side; the threads are drawn together at the middle in a buttonhole stitch.

Point de Gibeciere: A bar in macramé lace formed by four threads, divided into two even groups and looped over each alternately.

Point de Gobelin: See **Gobelin stitch**.

Point de Grecque: A ground in needlepoint laces, consisting of darned square spots alternating with octagonal meshes.

Point Guipure: See **Guipure lace**.

Point de Havre: Narrow French needlepoint lace of the 17th and 18th centuries, similar to Valenciennes.

Point d'Hongrie: French rug of hemp warp and silk filling with various figures.

Point d'lrlande: Inferior machine imitation of Venise lace

Point draws: Point draws are a combination of a regular straight draw from back to front and one from front to back, the first and the last shafts only being used once, while the rest receive two ends each in one repeat of the draw.

Point Jesuit: Crochet imitation of Venise laces made in Ireland.

Point de Jours: In embroidery open spaces with buttonholed edge and some filling.

Point Lace: Made by the needle and thread. The term “point” applied also to very fine bobbin laces.

Point Lache: In needle-point laces triangles filling half of the square meshes; they consist of rows of buttonhole stitches linked together.

Point Lance: An embroidery made with short, straight and broken stitches in coloured wool.

Point de Malines: (1) Same as Maline lace. (2) A filling used in needle-point laces, consisting of small circles, buttonholed all around and connected with each other in a zig-zag line.

Point de Marli: A bobbin made net used as ground for bobbin laces in the 18th century.

Point de Medicis: Old name for Italian needle-point lace with raised cordonnet.

Point Mexique: In Mexican embroidery the outlining buttonhole stitch in black or coloured silk.

Point de Milan: Lace with a small mesh ground and large trailing scroll pattern.

Point a la Minute: Cross and star shaped titches; used to fill in small spaces.

Point de Moscow: Early Italian needlepoint laces in Russian designs. See also ***Russian lace***.

Point Natte: (1) Embroidery made with pieces of bright satin appliqued on a dark foundation, the edges fastened with braid; the pieces of satin are embroidered in floss silk or wool yarn. (2) An embroidery stitch arranged in herringbone effect without a center line.

Point Neige: (1) Needle-point lace of very fine design ornamented with various loops and picots. (2) A crochet work, made with regular open places and stitches radiating from each opening; used for quilts and jackets.

Point Net: Net made by the needle and used as ground for applique laces before the invention of the bobbinet.

Point None: Same as **Buttonhole stitch**.

Point Noue: In needle-point laces a knotted buttonhole stitch.

Point Ondule: Double bar in macramé lace.

Point d’Or: See **Point de Pois**.

Point de Paris: Narrow French bobbin lace of the 17th century, similar to Brussels.

Point de Paris: Ground For black bobbin laces, consisting of hexagons and triangles.

Point paper design, in weaving: See **Pattern Draft**.

Point Passe: Same as **Satin stitch in embroidery**.

Point Pecheur: Italian bobbin lace, made in white or black, similar to the Maltese lace.

Point Perle: Same as **Satin stitch**.

Point Plat: In laces such patterns which have no raised parts.

Point Plat: Applique 'Modern Belgium lace of bobbin made sprigs applied to machine-made net.

Point de Plume: A padded satin stitch.

Point Plumetis: Same as **Feather stitch**.

Point de Pois: An embroidery stitch, consisting of small dots.

Point de Poste: See **Point de Pois**.

Point de Pyramide: Same as **Point da Cone**.

Point de Raccroc: Same as **Raccroc stitch**.

Point de Ragusa: Same as **Ragusa lace**.

Point de Repasse: See **Cloth stitch**.

Point de Reprise: A filling in needle-point laces, consisting of darned triangular spots

Point a Reseau: Needle-point lace, the pattern being formed by the meshes of the ground.

Point de Riz: In embroidery short, irregularly scattered stitches with a fancied resemblance to rice.

Point de Rose: (1) one of the most delicate needle laces, made originally in Venice, where the art revived of late. The beautiful and intricate patterns are made in raised effect, connected with bars; (2) a stitch in embroidery, consisting of broad buttonhole stitches over a padded surface.

Point Russe: Short straight stitches in fancy embroidery, forming geometrical patterns, Irke stars, diamonds, crosses, etc.

Point de Sable: See **Point d'Armes.**

Point Saracene: French tapestry made in imitation of Turkish carpets.

Point Serre: A stitch in needle-point laces; the thread is carried diagonally across each square mesh, looped and drawn tight and carried across the following meshes in zigzag line.

Point de Sorrento: Ground in needle-point laces; it consists of a series of long loose loops, each worked around by a number of loose buttonhole stitches.

Point Tiellage: A stitch in needle-point laces; the thread is carried diagonally across a square mesh and twisted half way around the knot in the corner and carried afterwards through the next mesh. A second thread is carried between the same corners of the squares, but as it is twisted the opposite way around the corners, it forms a slight angle with the first thread.

Point de Tigre: Same as **Overcast stitch**.

Point Tire: See **Drawnwork**.

Point de Toile: See **Cloth stitch**.

Point Tresse: 16th century pillow lace of human hair.

Point de Tricot: In crocheting forming large open squares by chain stitches; used for fancy quilts, made of heavy wool yarn.

Point de Tulle: (1) See **Mignonette**. (2) A very fine ground in needle-point laces.

Point Turc: (1) A filling in needle-point laces, consists of rows of interlinked festoons fastened to straight threads. (2) Same as ladder stitch in embroidery.

Point de Valenciennes: (1) Same as Valenciennes lace. (2) A filling in needlepoint laces, consisting of squares separated by open work; the squares are made of rows of buttonhole stitches linked together.

Point de Venise: (1) the finest of the needle-point laces, the characteristics being the padded cordonnets, the design often being raised repeatedly, the great variety and fineness of the stitches and the rich ornamentation of the edges of the patterns and the brides with picots and stars; (2) A filling in needle-point laces, consisting of rows of festoons, the loops in every alternate row being fastened with four buttonhole stitches.

Point paper: A special paper for designing woven fabrics. The standard point paper used is ruled in groups of 8 x 8, separated by thicker bar lines.

Pointed 2/2 straight herringbone: See **Herringbone**.

Pointille: French for small dots.

Poiret twill: Named after poiret, the French designer. This is a firm, twilled worsted material, but its twill is pronounced, and steeply angled. A typical construction is 3/3, 45 degree twill but may be made in a 63 degree twill weave also. The fabric normally has twice as amny ens as it has picks in the construction. Used for women's wear dress goods.

Poise': Out-dated measure for dynamic Viscosity. Replaced in the SI system by Pascal seconds (Pa · s): 1 P = 0.1 Pa·s.

Poke bonnet: Richly decorated bonnet-hat with a broad brim; tied with ribbons under the chin.

Polacrylate sizes: Polyacrylate based sizing agents. Conventional polyacrylate sizes primarily in the form of sodium or ammonium poly salts:

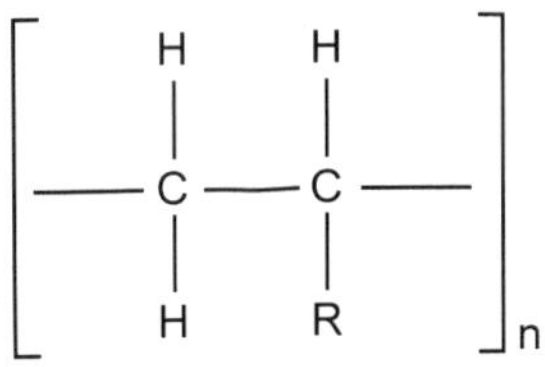

R = $-COO^{-}NH_4^+$ (Na^+),
$-CN$, $-OH$, $-CONH_2$,
$-OOCCH_3$

Polyacryate sizes of very different chemical composition in the form of "universal sizes" are an important class of size products for filament yarns in general. They produce good effects when used in water jet power looms, with a size film that is resistant to the influence of cold, neutral water. After weaving the polyacrylate sizes, present in acidic form, can be made water soluble again at an alkaline pH and washed out. New generation moisture insensitive and thermostable sizes by partial or total substitution of ammonium or sodium in specific copolymers (acrylonotrile/acrylic acid) by earth alkalis (e.g. magnesium, sodium, calcium salts). Such Polyacrylate sizes can be completely washed off even after raw thermofixing because they are soluble in hot and cold water at high concentrations (up to approx. 25%).

Polar: Terminal, working in opposite directions (Dipoles). For example, in textile auxiliary products the polar arrangement of the hydrophilic group in relation to the hydrophobic group. In the wetting process, water molecules are in a polar arrangement to the auxiliary at the boundary layer, if the latter aligns its hydrophilic (negatively charged) groups against the positively charged groups of the surrounding water.

Polar bond: Another name for Ionic bond.

Polar group: Functional group, the electron distribution of which gives the molecule a considerable Dipole moment (Polar). It determines the affinity to markedly polar liquids (particularly to water) and the Hydrophilic molecule character.

Polarin: Curl pile fabrics with a cut mohair pile.

Polemite: Holland and French solid coloured camlet, made of two-ply Angora warp and worsted filling, forming cross ribs.

Polished cotton: A plain-weave cotton fabric, often inexpensive, that has been calendared to give it an attractive shine. The calendaring does not survive laundering so garments have to be starched, unless a resin finish is added, when the finish is permanent. Many weights of cotton may be polished and used for curtains, loose covers, dresses etc. Handle as for curtains, loose covers, dresses etc.

Polished Twine: Two-ply coarse hemp or flax twine with a smooth finish; used In stores to tie packages with.

Polishing: Usually done on commercial sewing thread, the process involves burnishing the plied yarn.

Polishing: The treatment of tanned skins or of fabrics particularly pile fabrics, to increase the lustre by mechanical means, without compressing the material.

Polka Dot: Round printed dots differing in colour from the ground.

Polka Gauze: Swivel embroidered dots scattered over a gauze ground; used for dresses, etc.

Poll: (1) French term for pile; (2) silk thread used as core for gold tinsel, consisting eight or 10 reeled filaments twisted together.

Poll seaux: Coarse, stout French linen canvas of various widths.

Polledavy: Unbleached French hemp sailcloth.

Pollution: Any damaging or unpleasant change in the environment that results from the physical, chemical, or biological side-effects of human industrial or social activities. Pollution can affect the atmosphere, rivers, lakes, seas, and soil. Air pollution is caused by the domestic and industrial burning of carbonaceous fuels, by industrial processes, and by gases in car exhausts. Among recent problems are industrial emissions of sulphur (IV) oxide, a cause of acid rain, and emissions of chlorofluorocarbons (CFCs), previously widely used in refrigeration, aerosols, etc., and linked to the depletion of ozone in the stratosphere. Carbon dioxide, produced by burning fossil fuels, is slowly building up in the atmosphere, which could result in an overall increase in the temperature of the Earth due to its greenhouse effect. Car exhausts also emit carbon monoxide and, if leaded gasoline (petrol) is used, lead. Atmospheric CO has not yet reached dangerous levels, but vegetation near main roads has in the past been found to contain a high proportion of lead, with levels sufficiently high in urban areas to cause concern about the effects on children. Leaded gasoline has thus been banned in many countries. Photochemical smog, caused by the action of sunlight on volatile hydrocarbons and nitrogen oxides from car exhausts, is a problem in several countries. Catalytic converters have helped reduce harmful emissions from vehicle exhausts.

Water pollutants include those that are biodegradable, such as sewage effluent, which cause no permanent harm if adequately treated and dispersed, as well as those that are non-biodegradable, such as certain chlorinated hydrocarbon pesticides (e.g. DDT) and heavy metals, such as lead, copper, and zinc in some industrial effluents, which cause heavy-metal pollution. When these pollutants accumulate in the environment they can become very concentrated in food chains. The pesticides DDT, aldrin, and dieldrin are now banned in many countries. Water supplies can become polluted by leaching of nitrates, pesticides, or animal wastes from agricultural land. The discharge of waste heat can cause thermal pollution of the environment, which can be reduced by the use of cooling towers. In lakes, rivers, and the sea, spillage from tankers and the discharge of inadequately treated sewage effluent are the main problems. Other forms of pollution are noise from aircraft, traffic, and industry and radioactivity from improper disposal of radioactive waste.

Pollution index: (waste water), characteristic figure relating to waste water, derived from measured COD values (mg O_2/l), BOD values and water quantity in l/kg material; See Waste water evaluation.

Polo Cloth: The name given to top quality highly napped fabric that is made into sportswear and polo caps and coats, jackets etc. Mainly produced in expensive looking shades of brown, Made form wool, it may be woven or knitted.

Polonaise: Corded silk dress goods, made in France.

Poly-1,4 cyclohexylene dimethylene terephthalate fibre: Poly-l, 4- resembles polyethylene

terephthalate in most properties: The cyclohexylene group within this fibre provides additional rigidity to the molecular chains, but the packing of adjacent polymer chains may be more difficult due to the complex structure. As a result, the fibre has a lower tenacity than polyethylene terephthalate. It has a tenacity of 2.5-3 g/d (22-27 g/tex) and exhibits lower elongations than polyethylene terephthalate. It has a lower specific gravity (1.22-1.23) than polyethylene terephthalate. The fibre melts at 290°-295°C and is attacked and shrunk by trichloroethylene and methylene chloride. The fibre has good chemical resistance. The major trade name for this polyester is Kodel II. The fibre is somewhat superior to polyethylene terephthalate in certain end-use properties including better recovery from stretch and better resistance to pilling. The fibre has superior resiliency and is particularly suited for use in blend with cellulosics and wool, as a carpet fibre, and as fiber fill.

Poly-m-phenylenedibenzimidazole fibre (PBI): PBI was developed by the U.S. Air Force and Celanese as a flame retardant fibre for use in aerospace applications. The fibre is spun from N,N-dimethylacetamide followed by derivatization with sulphuric acid to form a golden fibre. It has moderate strength (3 g/d or 27 g/tex) and good elongation at break (30%), a moderate density (1.4 g/cm^3), and a high moisture regain (15%). Fabrics from the fibre possess good hand and drape and are stable to attack by ultraviolet light. PBI does suffer from poor dye ability, however. In addition to its low flammability, the fibre has a high degree of chemical and oxidation resistance, does not show appreciable heat shrinkage up to 600°C and generates only small quantities of smoke and toxic gases on ignition. In recent years, it has shown potential as a replacement for asbestos, as a flue gas fi1ter material, and as an apparel fabric in specialized applications.

Poly-p-ethyleneoxybenzoate fibre: This fibre was introduced by a Japanese firm as A-Tell. The properties of the fibre are very similar to the polyesters containing terephthalate units in most respects including tensile properties, specific gravity, melting point, chemical resistance, and sunlight resistance. As a result, the FTC expanded the polyester category to include the polymeric p-ethyleneoxybenzoate esters. The fibre is produced by reaction of ethylene oxide with p-hydroxybenzoic acid, followed by melt spinning. The fibre has a tenacity of 4-5.5 g/d (36-50 g/tex) and an elongation at break of 15%-30% with nearly complete recovery from low elongation. Its specific gravity of 1.34 and moisture regain of 0.4%-5.0% are nearly the same as that of polyethylene terephthalate polyester. The fibre melts at 224°C and softens at about 200°C. The fibre is reported to be even more resistant to attack by acids and bases than terephthalate-based polyesters. The properties of this fibre were not sufficiently different from other polyesters to achieve reasonable market penetration, and the fibre has been discontinued.

Poly tetrfluoroethylene fibres: Fibre made of polytetrafluoroethylene chains. These fibres of high physical and chemical stability are made by melt spin process and cold drawn into extremely fine threads. They are beige-brown in colour, soft slippery, cross-section round. Properties: density 2.3; normal moisture absorbency 0.02% (unwettable); virtually non-flammable (glows in flame); softening point approx. 327°C; at 225°C, 15% fall in tear resistance; at 310°C full strength loss; m.p.400°C with decomposition. ; strength dry/wet 16–27 cN/tex; elongation dry/wet 13–15%; dermatalogically harmless; resistant to mildew, bacteria, insects, weather resistant; almost totally resistant to acids, alkalis, organic solvents; difficult to dye (disperse dyes; spin dyeing). Application: for technical purposes (filter cloth, packing material, protective clothing, dye bags, screen printing gauze, etc.), also for non-wovens

Polyacrylates: Polyacrylates: I. Polyacrylic acid salts; Polyacrylic acid ester of following formula.II.

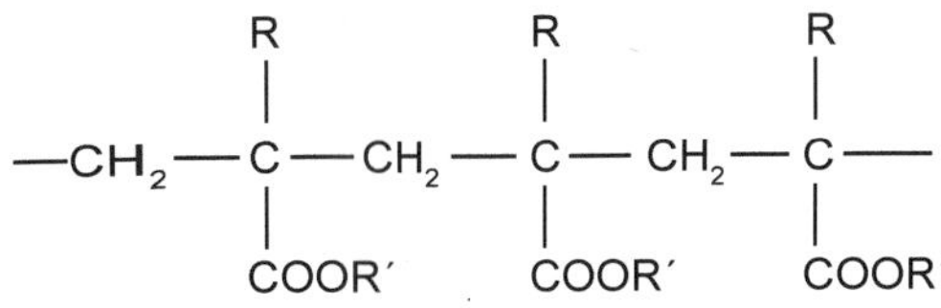

Polyacrylate

Polyacrylate products are for example used as finishing agents, coatings, binders for pigment printing, fibres, laminating agents, adhesives, sizes, thickeners, materials.

Polyacrylic acid. Polyacrylic acid is a water soluble polyelectrolyte that has excellent adhesion to nylon therefore it is used to size filament nylon yarns.

$$[-CH_2-CHCOOH-]_2$$

The affinity is through hydrogen bonding of the –COOH with amide and amine end groups in the nylon polymer. Application for sizes, finishes, adhesives, binders, etc. Polyacrylic acid ester and Polyacrylic acid salts are of greater importance.

Polyacrylic acid salts: Sodium and Ammonium salts of Polyacrylic acid.

$$[-CH_2-CHCOOMe-]_n$$

Water soluble, produce solid, elastic films with antistatic properties. Application: Sizing agents, finishing agents, thickeners for synthetic dispersions in coating finishes, coatings.

Polyacrylonitrile: (polyacrylic acid nitrile, polyvinyl cyanide), polymer of Acrylonitrile.

$$[-CH_2-CHCN-]_n$$

Used in emulsions or dispersions (particularly copolymers with butadiene, styrene, vinyl ether) primarily for wash fast finishes, for coating and laminating,

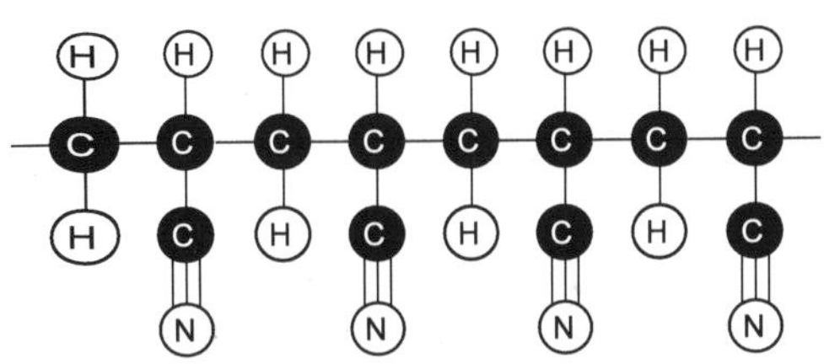

Repeat unit of polyacrylonitrile

as a material and for synthetic fibres (Polyacrylonitrile fibres; Modacrylic fibres). Acrylonitrile polymer emulsions with ammonium chloride as a catalyst can develop, when hardening, a toxic bi-chloromethylether (lung cancer hazard).

Polyacrylonitrile fibres: Group of synthetic polymeride fibres, the chain of which is made up of at least 85% Acrylonitrile. Polyacrylonitrile fibres are produced by dry or wet spinning and drawing (8–12 fold). Further processing from single titre 0.6 dtex upwards. The cross-section varies, usually dumbbell-like (dry spinning process) or kidney-shaped (wet spinning process).

Poly-addition: This operation joins together several molecules and redistributes the valence links existing in the monomers, however without removing secondary products.

Many unsaturated compounds which are characterized by the presence of a double link between two adjacent carbon atoms as ethylene and its derivatives, polymerise according to this reaction; within this category fall e.g. acrylic and polyolefin fibres. Among the most used polymers there are ethylene derived molecules with one or more substitutes of hydrogen atoms.

For example:$CH_2 = CHX$

Where X = H, CH_3, Cl, CN, OH and other groups.

The chains which are thus formed originate from simple openings of double ethylene links and are therefore characterized by links only among carbon atoms.

Polyalcohols: Alcohols with several OH groups, which includes glycols, glycerol $C_3H_5(OH)_3$ and sugar alcohol, but specifically high molecular alcohols such as polyglycol, polyglycerol, polyvinyl alcohol, cellulose (approx. 10 000 -OH groups).

Polyamide: A synthetic polymer family used for fibres and solid plastics; protein fibres are also technically polyamides, but the term is almost always used only for synthetic materials Nylon is a polyamide, and the oldest of the commercial synthetic polymers. Polyamide can be dyed with acid dyes or disperse dyes. Some MX dyes will also work well for nylon if applied as if they were acid dyes (at acid pH and high temperature). Fully imidized, manufactured fibre formed from the condensation polymer of an aromatic dianhydride and an aromatic diisocyanate. The fibre is produced by dry spinning. It is a high-shrinkage fibre used in the formation of mechanically stable nonwoven fabrics. These fabrics are made without binders or resins; bonding apparently results from the local temperature and pressure that develop during shrinkage.

Polyamide dyes: Practical term for dyes for dyeing polyamide fibres. Selected dyes, specifically for dyeing polyamide, are disperse, acid and metal-complex dyes, which are usually classified into specific ranges. Developing, chrome and reactive dyes are also suitable. Pigment dyes can also be used for light shades.

Polyamide fibre: See **Polyamide.**

Polyaramide fibres: See **Aramid fibres**.

Polyarylate: High-temperature-resistant aromatic polyesters from bisphenols.

Polybenzimidazole fibre (PBI): A manufactured fibre in which the fibre-forming substance is a long chain aromatic polymer having recurrent imidazole groups as an integral part of the polymer chain. (FTC definition). The polymer is made from tetraaminobiphenyl and diphenyl isophthalate and is dry spun from a dope with dimethylacetamide as a solvent. Characteristics: A high-performance fibre with high chemical resistance that does not burn in air. It has no melting point and does not drip when exposed to flame. The fibre and fabrics from PBI retain their flexibility, dimensional stability, and significant strength without embrittlement even when exposed to flame or extreme heat. The fibre emits little smoke in extreme conditions. It processes well on conventional textile equipment, having processing characteristics similar to polyester. It can be used in 100% form or blended with other fibres. It has a high moisture regain and low modulus with comfort properties similar to cotton. The natural colour of PBI is a gold-khaki shade, but it can be dyed to almost any medium to dark shade with conventional basic dyes. End uses: With excellent thermal, flame, and chemical resistance, combined with good comfort properties, PBI is a good fibre for many critical uses including: firefighter's protective apparel, aluminized proximity gear, industrial apparel such as pants, shirts and underwear, protective gloves, welder's apparel, aircraft fire-blocking layers, aircraft wall fabrics, rocket motor insulation, race car driver's apparel, and braided packings among others.

Polyblends: See **Biconstituent fibres**.

Polybuteneterephthalate (PBTP): Polytetramethylene terephthalate fibre.

Polycaproamide fibres: Rare term for Polyamide fibres of type polyamide 6 or polyamide 6.6.

Polycaproamides: Group of polyamide fibres of type polyamide 6 (base Caprolactam).

Polycarbonate Fibres: Polycarbonate fibres are linear polyesters of aliphatic or aromatic dihydroxy compounds with carbonic acid and has excellent resistance to heat and weathering.

Polycondensation products: (polycondensate, polycondensation resin, polycondensation plastic), products of Condensation polymerisation, in the form of plastics or synthetic resins, such as : Alkyd resins; Epoxy resins; Urea-formaldehyde compounds; Ketone resins; Melamine-formaldehyde compound; Phenolic plastics; Polyamides; Polycarbonates, Polycondensation fibres; Polysulphides.

Polychlal fibre: A manufactured, bicomponent fibre of polyvinyl alcohol and polyvinyl chloride. Some vinyl chloride is grafted to the polyvinyl alcohol (Japanese Chemical Fibres Association definition). The fibre is emulsion spun into tow and staple. Characterics: Polychlal fibres have a soft, lamb's wool-like hand and moderate moisture regain. The fibres are also characterized by high flame resistance and high abrasion resistance. End uses:: Polychlal fibres are suitable for end uses such as children's sleepwear, blankets, carpets, curtains, bedding, upholstery, nonwovens, and papermaking.

Polychromatic: Multi-coloured, bright. Opposite of Monochromatic.

Polychromatic dyeing: For fabrics and carpets. In principle this consists of aiming separate jets and dye solutions at continuously running lengths of material, before and whilst the material passes through squeeze rollers. The application device consists of a number of colour jets and is attached to two or more rods. The length of the rods is the same as the width of the application machine. The nozzles direct the jets of dye liquor at an inclined plate and from there onto the length of material (Dye Weave) or directly onto one of the two squeezing rollers (Flow Form). The nozzle rods can be moved backwards and forwards and in this manner form patterns and colour mixtures (Space dyeing).

Polychromatic printing: Multi-coloured printing with a printing roller.

Polychrome: Bobbin lace made of fine silk threads of various colours.

Polycondensation: See **Condensation polymerisation, Polycondensation fibres. Polycondensation products**. By polycondensation, two molecules of same type or of different types are joined together to form macromolecules by removing simple secondary products as water, hydrochloric acid, alcohol. The prerequisite for reactions of this type is the presence in the molecule (monomer) of two terminal reactive groups with functional properties. The molecules composed of 2,3,4...n monomers are named dimers, trimers, tetramers (oligomers)...polymers. Some of the mostly used monomers are: (1) Aliphatic di-acids HOOC-R-COOH (used for nylon 6.6). (2) Aliphatic di-amines NH_2-R-NH_2 (used for nylon 6.6) Aliphatic amino acids H_2N-R-COOH (used for nylon 6). (3) Aromatic di-acids HOOC-Ar-COOH (used for polyester). (4) Diols (bi-functional alcohols) HO-R-OH (used for polyester) etc.

Polycondensation, Melt: The production of polyethylene terephthalate from dimethyl terephthalate and ethylene glycol, and nylon 66 are prepared by using melt polycondensation technique. In this method, one of the monomers used is solid, which cannot decompose around its melting point. The reaction has to be carried out in an inert atmosphere of N_2 or CO_2 to avoid side reactions such as oxidation, decarboxylation, etc.

Polycondensation, Solution: In the solution polycondensation technique, all the reactants are dissolved in an inert solvent as a solution, therefore, the whole system is in a homogeneous phase. In this method, the reaction can be carried out at low temperature during which mass and heat transfer processes are comparatively easier. The solvent can also help as an entrapping agent for the by product formed; hence the removal of the by product from the final product and solution becomes easy. This technique is useful in the preparation of liquid polyester resin from the glycols and unsaturated dicarboxylic acid using high boiling aromatic hydrocarbon as solvent.

Polycondensation fibres: Currently the most important main group of Synthetic fibres (natural fibres also belong to the group of polycondensation fibres), which are created by Condensation polymerisation, i.e. by a condensation reaction between the different molecules (intermolecular condensation), whereby smaller molecules, such as water or alcohol, form due to splitting or as a secondary product. Unlike Polymerization fibres, the main chain contains carbon and foreign atoms (usually nitrogen, sulphur, oxygen) or linking polar groups (–COONH–, COO–, etc.).

Polycondensation products: (polycondensate, polycondensation resin, polycondensation plastic), products of Condensation polymerisation, in the form of plastics or synthetic resins, such as : Alkyd resins; Epoxy resins; Urea-formaldehyde compounds; Ketone resins; Melamine-formaldehyde compound; Phenolic plastics; Polyamides; Polycarbonates, Polycondensation fibres; Polysulphides.

Polycotton: A short form for polyester/cotton blended fabric.

Polyelectrolyte: A ***polymer*** that carries ionisable groups. When dissolved in water, they acquire many positive or negative charges (cationic or anionic polyelectrolytes). As a ***coagulant aid***, a polyelectrolyte is added to water in small quantities to help in the ***flocculation*** of fine suspended solids or colour molecules by reducing their surface charges and also by enmeshing fine solids or colour molecules in the long chain molecules of the polyelectrolyte. They are used in the ***conditioning*** of sludges from water or wastewater treatment.

Most of them are ***polyacrylamides*** or compounds of polyacrylic acid. A few are polyamides or quaternary ammonium compounds. Non-ionic polyelectrolytes are a misnomer. They are either ***ampholytic*** or do not ionise at all, and should not be called polyelectrolytes.

Polyene and polymethyne dyes: Polyene and polymethine dyes are two structurally related groups of dyes which contain as their essential structural feature one or more methane (CH=) groups. Polyene dyes contain a series of conjugated double bonds, usually terminating in aliphatic or alicyclic groups. They owe their colour therefore simply to the presence of the conjugated system. In polymethine dyes, electron-donor and electron-acceptor groups terminate either end of the polymethine chain, so that they may be considered as typical donor—acceptor dyes.

Polyester: A man-made fibre of great strength. It is treated in various ways and mixes well with other fibres like cotton. A manufactured fibre in which the fibre-forming substance is any long chain synthetic polymer composed of at least 85% by weight of an ester of dihydric alcohol and terephthalic acid (FTC definition). The polymer is produced by the reaction of ethylene glycol and terephthalic acid or its derivatives. Fibre forms produced are filament, staple, and tow. The process of production resembles that of nylon. Polymerization is accomplished at a high temperature, using a vacuum by one of two methods. (1) The glycol and a terephthalate ester react to form a polymer chain, releasing methanol; or (2) the glycol and terephthalic acid react directly to form the polymer

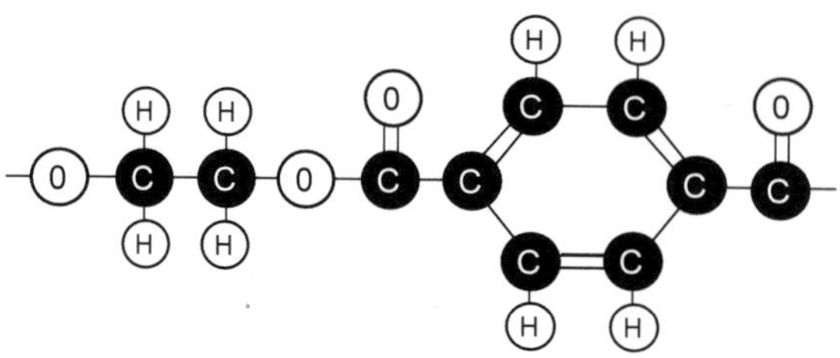

Repeat unit of the Polyester macromolecule

with water as the by-product. As with nylon, the filaments are spun in a melt-spinning process, then stretched several times their original length, which orients the long chain molecules and gives the fibre strength. A somewhat generic term used for a variety of synthetic polymers used both for solid plastics and for fibres; polyethylene terephthalate is probably the most common Polyester is are harder to dye than many other fibre polymers. It is dyed almost exclusively with disperse dyes. Because of its high glass transition temperature, dyeing is usually done at high temperature (around 130°C) in apressure vessel. Carriers can be used for dyeing at the boil. Washfastness of polyester is very high because it is almost impenetrable to water even at the boil. Some polyester is formulated to allow dyeing with basic dyes. Polyester containers (often labelled PET or PETE), very popular for beverages, can be used for dye solutions, at least for short term storage. Polyester is attacked by strong alkalis. A few stitches of brightly-coloured polyester thread can be handy for 'labelling' items being dyed in almost any process other than disperse dyeing, because they will retain their original colour.

Characteristics: Polyester fibres have high strength and are resistant to shrinking and stretching. Fabrics are quick drying and tend to have wrinkle resistance and crease retention, wet and dry. Polyester is used alone and in blends. It has been one of the first fibres to be developed in fabrics with durable-press features. END USES: Polyester is widely used in many types of apparel fabrics such as textured knits and wovens, durable-press blend fabrics, shirtings, dress goods, rainwear, worsted-blend summer suitings, sleepwear, underwear, bloused, and lingerie. It is also used extensively in floor coverings and for tire cord and other industrial uses such as sewing thread. Polyester fiberfill is used in filled items such as quilted jackets, comforters, pillows, furniture cushions, and sleeping bags.

Polyester chiffon: See **Chiffon**.

Polyester cotton: Polyester and cotton are mixed in various percentages according to the weight and type of fabric produced. The advantage over 100 % cotton is that the polyester reduces creasing and provides good draping qualities. It also makes the fabric more hard wearing. It may be plain, or printed or woven in patterns. Used for blouses, shirts, nightwear, children's clothes etc.

Polyester crepe': A soft synthetic crepe in a wide range of plain colours, which has largely replaced other types of blouses, dresses, evening wear, lounging pyjamas, negligees. It is hard wearing.

Polyester dyes: Simplified practical term for dyes for Dyeing of polyester fibres.

Polyester elastomers: (polyester rubber), Elastomers based upon polyester, which require no cross-linking. Some properties: unusual resistance against impact/bending cracks, good fabric adhesion; resistance against diluted acids good, against concentrated acids poor; against solvents excellent/good; against oxidation (ozone) excellent; against sunlight good; against ignition moderate to good.

Polyester fibre: The generic name for fibres made from a synthetic linear polymer that contains, in the chain, at least 85% (m/m) of an ester of a substituted aromatic carboxylic acid , including but not restricted to terephthalate units p(–R–O–CO–C_6H_4–CO–O–)and para substituted hydroxyl benzoate units p(–R–O–C_6H_4–CO–O–).

Polyester fibres in printing: Disperse dyes are used for direct printing procedures. They can be classified on the basis of their fixing properties:

fixation type and time,

– hot air 15–30 s,

– saturated steam 45–60 min,

– high pressure saturated steam 20–30 min,

– superheated steam 2–9 min.

When fixing in superheated steam, hot air or contact heat on hot cylinders fixation accelerators, e.g. based upon alkyl phenols, can be useful.

Polyester gauze: The low moisture absorption, which results in good deformation resistance is particularly beneficial for Screen making. Polyester gauze also possesses good light resistance and low elongation with good elasticity. Polyester gauze has to a large degree supplanted Polyamide gauze, although dye paste penetration and abrasion resistance are somewhat lower.

Polyester georgette: See **Georgette.**

Polyester honan: A fine textured fabric with a silky finish and a slub yarn across it. It resembles a heavy georgette but is softer to touch.

Polyester jersey: Close plain knit fabrics often printed, in various weights from very thin to fairly heavy. The light weight fabrics drape well and are soft, slippery and shiny, although those made from spun polyester are a matt. These fabrics often cause static problems which may be lessened by using appropriate finishes. Used for dresses, robes, evening clothes, light suits.

Polyester polyamide biconstituent fibres: It represent a two-phase system, composed of a polyester matrix containing evenly distributed polyamide fibrils. Acid and metal-complex dyes only dye the polyamide component. As the dyeing effect is only superficial it can be influenced by treating the biconstituent fibres in such a way as to bring about an increase in the polyamide content in the surface layers of the fibres and the formation of micropores, thus promoting the diffusion of the dye solution.

Polyester resins: High molecular cross-linking products made of unsaturated Polyester and unsaturated polymerizable compounds (e.g. styrene). Of interest for lacquers, coatings and materials, because many polyester resins are available as more or less viscose fluids (so-called casting resins), which only harden after the addition of a catalyst (peroxy compounds) and accelerator, e.g. with excellent strength, proven on fibre glass materials as a binder.

Polyester satin: A satin that creases very little, due to its fibre content, and is soft and comfortable to wear and drapes well. It is made in various weights suitable for lingerie, blouses and evening wear. It has less lustre than satins made from shiny fibres like silk, or acetate.

Polyester sheer: Firm vision net used for curtains. Fibres vary but polyester is mainly used, because it does not loose its colour. Fibre combinations include polyester and silk, and polyester, acrylic and nylon.

Polyester voile: Very fine polyester curtain net made in various widths. It is soft, crapes well and keeps its colour.

Polyester-polyamide biconstituent fibres: It Represent a two-phase system, composed of a polyester matrix containing evenly distributed polyamide fibrils. Acid and metal-complex dyes only dye the polyamide component. As the dyeing effect is only superficial it can be influenced by treating the biconstituent fibres in such a way as to bring about an increase in the polyamide content in the surface layers of the fibres and the formation of micropores, thus promoting the diffusion of the dye solution.

Polyetheretherketone fibre (PEEK): A manufactured fibre from polyetheretherketone polymer with high temperature and chemical resistance used in composites as a matrix material and in other industrial applications.

Polyetherimide fibre (PEI): A manufactured fibre spun from polyetherimide polymer having high temperature resistance, excellent processibility, and toughness. Used for matrix materials in composites and in other industrial applications.

Polyethers: By the polymerization of ethylene oxide we obtain the simplest type of aliphatic polyethers, polyethylene oxides (Polyglycol ethers; Polyethylene glycol), which also includes, for example, polyvinyl ethers. Aromatic polyethers usually contain built-in benzene cores. Linear aliphatic and aromatic polyether with a melting point of approx. 130–150°C possess the ability to form synthetic fibres.

Polyethersulfone fibre (PES): High molecular weight fibres from polymers containing sulphone ($-SO_2-$) groups and aromatic nuclei. They demonstrate high thermal stability and chemical inertness.

Polyethylene: ($-CH_2-CH_2-$)n. A synthetic "olefin" polymer (plastic) Polyethylene, particularly high-density polyethylene (HDPE), is a good choice for containers for many dyeing processes. It has good chemical resistance, but generally should not be subjected to boiling water (see polypropylene for an alternative). A great deal of the commercial packaging for household chemicals and food products is made from polyethylene. Some fabric sizing compounds and softening compounds are polyethylene derivatives.

Polyethylene emulsions: Polyethylene emulsions dry down to form hard, waxy films. When the emulsion is applied to fibres, a waxy coating deposits on the surface reducing its coefficient of friction. These coatings offer good protection against needle cutting and thread breakage and improve abrasion resistance and tearing strength.. To be emulsifiable, the polyethylene polymer is first oxidized by passing air through the melt. Oxidation converts some polymer end groups to –COOH and the quantity of carboxyl groups is controlled. Both low and high density polyethylene are processed this way. A number of grades of polyethylene polymers are available .differing in melting point, melt viscosity, molecular weight and carboxyl content. Dispersions with anionic, nonionic and cationic character are made by selecting appropriate auxiliary emulsifier. Selecting an emulsion with the proper ionic character is important otherwise the finishing bath will become unstable and break out. Stable water emulsions with solids up to 20% are commercially available. The alkali salt of the polymer's carboxyl group is an important factor in the stability of the dispersions.

Polyethylene fibre: A manufactured fibre made of polyethylene, often in monofilament form as well as continuous filament yarns and staple. Ethylene is polymerized at high pressures and the resulting polymer is melt spun and cold drawn. It may also be dry-spun from xylene solution. Characteristics: Polyethylene fibres have a low specific gravity, extremely low moisture regain, the same tensile strength wet and dry, and are resistant to attack by mildew and insects. These qualities have made polyethylene fibre suitable for industrial applications, geotextiles, outdoor furniture, and similar applications. Polyethylene fibre does not dye, and in most cases, it is coloured by the addition of pigments and dyes to the material prior to spinning. It has a low melting point, a property that has restricted its use in apparel.

Polymer: A substance composed of long chain molecules, sometimes with more than 1000 atoms, and built up from simple molecules (monomers) linked together. Many are synthetic, but they may be of natural origin, such as ***cellulose***. Different polymers may be formed from different unions of one monomer, so that they have the same formula but different linkages.

Polypropylene, PP: A ***thermoplastic*** material that is commonly used for food packaging. It is also used as a membrane for ***microfiltration***. Its tolerance to chlorine is limited.

Polythene fibre: See ***Polyethylene fibre (Olefin)*** a) high-density polyethylene HDPE), 0.96g/m^2, produced by low-pressure polymerisation; and b) low-density polyethylene (LDPE), 0.93g/m^2, produced by high-pressure polymerisation.

Polyethyleneglycols: Higher molecular weight polyoxyethylene analog are wax- like.

Polyethylene glycol esters: Polyethylene glycol esters of fatty acids have wax like properties.

Polyethylene glycols: Similar to diglycol or triglycol, for example, etherification product of several glycerol molecules in linear structure or in the form of complex polymers. Also mixed glycol-glycerol ethers, which become more similar to higher Polyethylene glycols with increasing oxyethylation. Like Glycerol, proposed for dyeing and printing purposes.

Polyfunctional: Multiple actions, e.g. a chemical compound with several Functional groups, or a multipurpose machine.

Polyglycol ethers: Addition products of ethylene oxide (ethylene oxide addition products) to fatty alcohols, fatty acids, amines, alkyl phenols, alkyl naphthols, etc. Polyglycol ethers form the group of non-ionogenic surfactants. A cationic character is also possible, because the water molecules, in aqueous solution with the residual valency of its atoms, attach to the residual valencies of the ether oxygen atoms. This additive compound then stabilizes to form co-ordination compounds. Polyglycol ethers are textile auxiliaries and due to their chemical resistance are often used as wetting, dispersing, levelling, stripping and, above all, washing agents, as well as emulsifiers and scrooping agents.

Polyhydroxyl fibres: Fibres, which have hydroxyl groups as prosthetic groups, above all cellulose and cellulose regenerate fibres, also appropriately modified synthetic fibres.

Polyimide fibre: See **Aromatic polyimide fibres**. An aromatic polyimide introduced by Upjohn Company for use in flame retardant, high-temperature applications. The fibre is spun from the polymer by wet or dry processing techniques using a polar organic solvent such as **N,N-dimethylformamide** to give a fibre with a round or dog-bone cross section. The highly coloured fibre may be crimped at 325°C using fibre relaxation. The fibre has a tenacity of 2-3 g/d (18-27 g/tex) and a 28%-35% elongation at break. The fibre has a moisture regain of 2.0%-3.0% and melts at about 600°C.

Polymannuronic fibre: Alginate fibre type, so called because Alginic acid represents a polymer dimannuronic acid (with pyranose structure).

Polymer: A macromolecular material formed by the chemical combination of monomers having either the same or different chemical combination. A compound in which there are very large molecules made up of repeating molecular units (monomers). A polymer has a repeated structural unit, known as a mer. Polymers do not usually have a definite relative molecular mass, because there are variations in the lengths of different chains. They may be natural substances (e.g. polysaccharides and proteins) or synthetic materials (e.g. nylon and polyethene). The two major classes of synthetic polymers are thermosetting (e.g. Bakelite) and thermoplastic (e.g. polyethene). The former are infusible, and heat may only make them harder, whereas the latter soften on heating. See **Polymerization.**

Polymer doping: It was found that polyacetylene manufactured according to the so-called Shirakawa process acquires a high electrical conductivity if it is brought into close contact with an oxidizing agent (in this case iodine). This process is now called doping, in an analogy to semiconductor physics. Conductivity rose by a billion fold compared to undoped polyacetylene.

Polymer etching: Low-pressure plasma can be used to chemically change the fibre surface. Oligomers are decomposed from the fibre surface in an inert gas plasma according to the following chemical mechanism (on polyester fibres). In a longer treatment (1 min) and high power input (in watts) the etching is

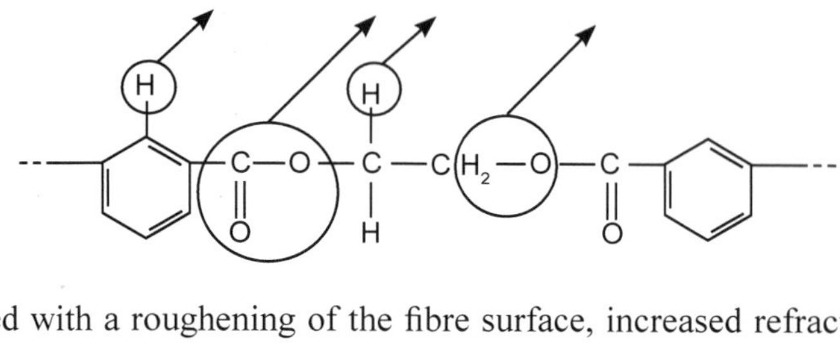

associated with a roughening of the fibre surface, increased refractive index and apparently darker colouring. Etching of wool fibre surfaces creates a sheen due to the partial removal of cuticle irregularities. In a longer treatment (1 min) and high power input (in watts) the etching is associated with a roughening of the fibre surface, increased refractive index and apparently darker colouring. Etching of wool fibre surfaces creates a sheen due to the partial removal of cuticle irregularities.

Polymerization: The process in which one or more compounds react to form a POLYMER. Homopolymers are formed by polymerization of one monomer (e.g. the formation of polyethene). Heteropolymers or copolymers come from two or more monomers (as for nylon). Heteropolymers may be of different types depending on the arrangement of units. An alternating copolymer of two units A and B has an arrangement: –A–B–A–B–A–B– A block copolymer has an arrangement in which blocks of one monomer alternate with blocks of the other; for example: –A–A–A–B–B–B–A–A–A– In a graft copolymer there is a main choice of one monomer (–A–A–A–A–), with short side chains of the other monomer attached at regular intervals (–B–B–). Stereospecific polymers have the subunit repeated along the chain in a regular way. These are tactic polymers. If one particular group is always on the same side of the chain, the polymer is said to be isotactic. If the group alternates in position along the chain the polymer is syndiotactic. If there is no regular pattern, the polymer is atactic. Polymerization reactions are also classified according to the type of reaction. Addition polymerization occurs when the monomers undergo addition reactions, with no other substance formed. Condensation polymerization involves the elimination of small molecules in the formation of the polymer. See also **Cross linkage.**

Polymerisation, Emulsion: Emulsion polymerisation is a most widely used industrial technique. The monomers *e.g.*, butadiene, chloroprene, vinyl acetate, vinyl chloride, acrylates and methacrylates, etc., are used to polymerised by this method. A very high molecular weight product can be achieved by this technique. In this system, the monomer is dispersed in the aqueous phase as a uniform emulsion, and the emulsion is stabilised by surfactants *(i.e.,* surface active agents), protective colloids and by certain buffers.

Polymerisation, Group Transfer (GTP): Group transfer polymerisation (GTP) is a new technique of polymerisation. This process has few advantages over other, as it is found to impart perfect control in producing polymer chains with terminal functional groups. GTP does not require a very low temperature to bring about polymerisation. Many of the examples occur at room temperature. In this technique, initiator, generally organosilicon and a bifluoride catalyst is used. The monomers, *e.g.*, nitriles, α,β-unsaturated esters, carboxamides, ketones can be polymerised by this process. The formed polymer consists of the excellent yields with narrow molecular weight distribution.

Polymerization, Plasma: If gaseous, radical polymerizable gases (with double bonds in monomers) are allowed to flow into an evacuated plasma generating device, which contains synthetic fibres, and a plasma luminous discharge is generated, then plasma polymerization can be observed on the fibre surface by grafting:

More conductive fibres can be generated in this manner, for example by the grafting of organometallic polymers. Fire resistant textiles are created by the use of monomers containing phosphorus. Surface fluoridation is performed by first removing surface oligomers by argon etching, followed by impregnation with fluorocarbon monomers and finally grafting with CF_4 plasma gas.

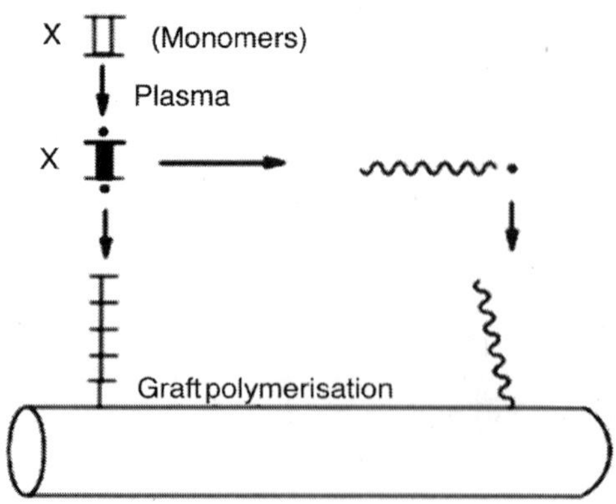

Polymerisation, Suspension: Suspension polymerisation is a heterogeneous system, where polymerisation proceeds to 100% conversion and the product is obtained as spherical beads or pearls. For this reason, this technique is known as beads or pearl polymerisation.

By this technique only water-insoluble monomers can be polymerised. In the form of fine droplets, the monomer molecules are suspended in water and mixed with surface active agents and water-soluble protective colloids. During polymerisation process, stirring remains continue. The initiators are monomer soluble. The continuous aqueous phase separating the monomer droplets acts as an heat transfer medium and hence the exothermicity is quite under control.

Polymerization cross-linker: Reactive compounds, which are capable of polymer formation and cellulose cross-linking, e.g. N-methylol acrylamide, N-methylol methacrylamide, N-methylol polyethylene urea. They yield a favourable ratio of resistance to tearing and abrasion in Resin finishing.

Polymerisation fibres: Extensive main group of synthetic fibres, the macromolecule of which contains the same or different continuous long carbon chains, formed by Polymerization, which is unique to all polymerization fibres (unlike Polycondensation fibres), and which differ from each other due to side chains containing other foreign atoms (chlorine, fluorine, cyanogen, etc.) or groups ($-CH_3$, $-OH$, $-CONH_2$). This group includes e.g. polyacrylonitrile fibres, polybutadiene fibres, polyolefin fibres, polystyrene fibres, polytetrafluorethylene fibres, polyvinyl alcohol fibres, polyvinylacetate fibres, polyvinylchloride fibres, polyvinyl fluoride fibres, polyvinylidene cyanide fibres, polyvinyl copolymerization fibres.

Polymerization products: (polymerizate, polymerized synthetic resins, polymerization plastics), products of Polymerization, either in the form of

Plastics or Synthetic resins, e.g. : Polyacrylic acid; Polyacrylic acid ester; polyacrylates; Polyacrylonitrile; polybutadiene; Polychlorobutadiene; Polyethylene; Polyisobutene; Polymerization fibres; Polymethacrylic acid ester; Polystyrene; Polytetrafluorethylene; polyvinyl alcohol; Polyvinyl ethers; Polyvinyl acetals; Polyvinyl chloride; Polyvinyliden chloride; Polyvinylpyrrolidone.

Polymerise: To undergo polymerization. To react molecules resulting in their combining and forming relatively long-chain, large molecules.

Polymethacrylic acid (PMAA): Poly(methacrylic) acid is completely water soluble and functions as a soil release finish. However the proper amount of cross-linking is necessary before the finish to functions properly. Table 9 shows that the soil release rating are influenced by the inclusion of a diepoxide crosslinking agent. When PMAA is crosslinked with only the diepoxide, marginal SR ratings are obtained. However if a small amount of diepoxide is added with DMDHEU, the soil release ratings are vastly improved. Increasing the amount of diepoxide causes the SR rating to drop again. This data supports the contention that the ultimate properties of the cured film deposited on the surface of the fibre determines soil release.

Polynosic: Derived from the French "polymère non synthéthique" (non-synthetic polymer). Since 1961 the trade name, protected by trademark law, has been "Association Internationale Polynosic" - members for Polynosic fibres. The scope of protection is controversial, because countries with a fibre declaration of "polynosic" do not recognize it as a generic term. The trademark associated with certain modal fabres, it is a manufactured cellulosic fibre with properties similar to cotton, but the fibres are softer. Modified viscose fibres, subgroup of the cotton-like Modal fibres. Differ from High wet modulus fibres due to their good alkali resistance (mercerizability). They possess a characteristic uniform structure of fine fibrils in a rather homogeneous arrangement distributed over the entire round cross-section. Therefore, they have good penetration and reactivity at low swelling (approx. 60%) compared to finishing resins. They have good dimensional stability due to the high wet modulus.The fibre is more difficult to dye than cotton. Polynosic is often mixed with cotton and other fibres.

Polynosic fibre: A regenerated cellulose fibre that is characterised by a high initial wet modulus of elasticity and a relatively low degree of swelling in sodium hydroxide solution.

Polyolefin: Any long chain synthetic polymer composed of at least 85 weight %of ethylene, propylene or other olefin units (monomers), except amorphous (noncrystalline) polyolefin qualifying under Rubber.

Polyolefin fabric: A fabric woven from polyolefin monofilaments.

Polyolefin fibre: See **Polyolefin.**

Polyolefin fibre: A fibre made from a synthetic linear polymer obtained by polymerising an unsaturated hydrocarbon (e.g. ethylene CH^2-CH^2 or propylene $CH_2=CH\text{-}CH_3$) to give a linear saturated hydrocarbon. A poly olefin fibre is generally defined as a manufactured fibre in which tne fibre forming substance is any long chain synthetic polymer compound of at least 85 % by weight of ethylene. Propylene, or other olefin units except 'amorphous' (non-crystalline) polyolefins qualifying under rubber. (See also polyethylene fibre and polypropylene fibre, rubber).

Polyolefin monofilament: A flat single filament of the slit film type, which can function as yarn in commercial textile operations.

Polyoxyamide fibre: Copolymeric fibre with good comfort properties, particularly high moisture absorption and transfer, and intrinsic softness.

Polyoxyethylene thioether: Surfactant of polyoxyethylatedthioether type.

α-polyoxymethylene: $HO–CH_2–O–(CH_2–O)x– CH_2–OH$, a polyoxymethylene-dihydrate with DP = 2 to above 100, reductive action, is split by sodium sulphite.

γ-polyoxymethylene: $H_3C–O–CH_2–O–(CH_2–O)x– CH_2–O–CH_3$, a polyoxymethylene-dimethylether, DP = 2 to above 100, no longer has reductive action, no splitting by sodium sulphite.

Polyphosphates: Phosphates with varying degrees of condensation and a chain-shaped molecular structure. Diphosphate (= pyrophosphate = two chain), triphosphate (= tripolyphosphate = three chain), tetraphosphate (= tetrapolyphosphate = four-chain), etc. up to high molecular polyphosphates = Graham's salt (25 Na3PO4). When dissolved in water, polyphosphates show the typical behaviour of complex formers. They dissociate max. at 30%. In water softening (i.e. bonding e.g. of calcium ions) initially non-dissociated sodium atoms (or H atoms in polyphosphoric acids) are continuously exchanged. The solubility of the complex salts decrease with increasing calcium content. If there are excess calcium ions, dissociated sodium ions are also replaced by calcium and water insoluble calcium polyphosphates. It follows from this that polyphosphates must always be dosed in a certain stoichiometric excess, in order to obtain soluble calcium (or other metal) complexes. Application: for masking water hardness and heavy metal ions, in washing, bleaching, dyeing, printing, resin finishing, as a component of domestic and industrial washing agent. The following general structural formula applies for polyphosphates:

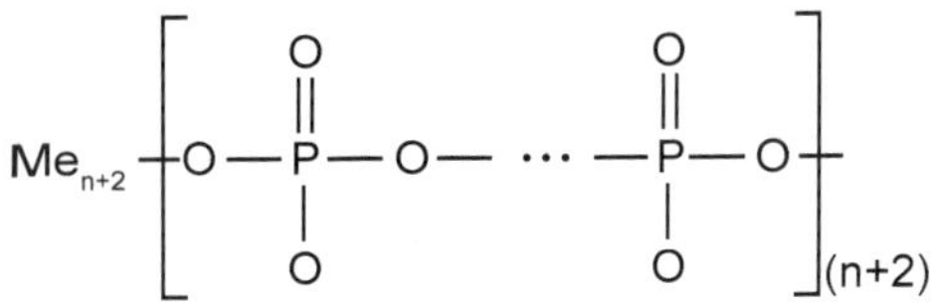

Polypropylene: A synthetic "olefin" polymer (plastic).

Polypropylene fibre: **{-CH_2-$CH(CH_3)$-}$_n$.** Polypropylene containers are a good choice for many dyeing processes. Polypropylene has excellent resistance to most chemicals used in dyeing, with the possible exception of some disperse dye carriers. It can be used at the boiling point of water, though it can't be used on a stove top. A great many household plastic containers are made from this polymer. Polypropylene is also used for fibres, mostly for carpets, but finds some use in garments. It cannot conventionally dyed so it is coloured by "dope dyeing" (pigments are added to the raw material). .A manufactured, olefin fibre made from polymers or copolymers of propylene. Polypropylene fibre is produced by melt spinning the molten polymer, followed by stretching to orient the fibre molecules.Characteristics: Polypropylene fibres have a number of advantages over polyethylene fibres in the field of textile applications. The degree of crystallinity, 72 to 75%, results in a fibre that is strong and resilient, and does not fibrillate like high-density polyethylene.

Polypropylene has a high work of rupture, which indicates a tough fibre, and may be made with tenacities as high as 8.0 to 8.5 grams per denier. The melting point of polypropylene is 165°C, which is low by comparison with nylon or polyester, but is high enough to make it suitable for most textile applications. So light that it actually floats, polypropylene fibre provides greater coverage per pound than any other fibre. It is highly resistant to mechanical abuse and chemical attack. Uses: Polypropylene fibres are widely used in industrial, carpet, and geotextile applications. They have found important uses in fishing gear, in ropes, and for filter cloths, laundry bags and dye bags. The excellent chemical resistance of polypropylene fibre is of advantage in the filtration and protective clothing fields. Fibrillated polypropylene yarns are widely used in indoor-outdoor carpets. Staple fibre finds application in blankets, pile fabrics, underwear, and industrial fabrics; it is being developed for carpets, candlewicks, knitted outerwear, hand-knitting yarns, and upholstery.

Polypropylene glycol: Colourless, non-volatile, does not corrode metals, highly to slightly soluble in water, cold resistant. Application: for lubricants, solvents, non-ionogenic cleaning products, synthetic resins, softeners, etc.

Polyreflets: French plush having patterns produced by the inclination of the pile at different angles.

Polysaccharide: A high-molecular-weight polymer of a monosaccharide. The polysaccharides contain many repeated units in their molecular structures. They can be broken down to smaller polysaccharides, disaccharides, and monoisotactic saccharides by hydrolysis or by the appropriate enzyme. Biopolymers made up of ten or more simple sugar (like glucose) units Starches and cellulose are polysaccharides. (Below 10 sugar units, a specific numerical prefix is usually used, such as in "disaccharide".)Important polysaccharides are heparin, inulin, starch, glycogen (sometimes known as animal starch), and cellulose. See also Carbohydrates; Sugar.

Polysalts: See **Polyelectrolytes.**

Poly-set-process: Two-stage Permanent-press process: Stage I cross-linking in the presence of weak catalysts such as zircon acetate. Stage II after manufacture using strong catalysts such as zinc nitrate.

Polysilicic acids: Higher Silicon oxo acid e.g. of the type $(H_2SIO_3)n$, the most well-known salt of which is water glass or Sodium silicate.

Polysiloxane polymers: See **Silicons.**

Polystyrene: A synthetic polymer made from styrene (phenylethene). Expanded polystyrene is a rigid foam used in packing and insulation.

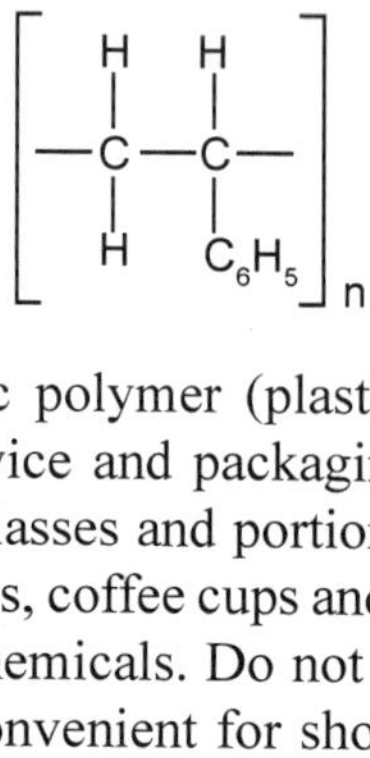

Polystyrene fibres: Synthetic polymer (plastic) Polystyrene is extensively used for disposable food service and packaging items. Solid polystyrene is used for disposable cutlery, glasses and portion cups. Foamed polystyrene is used for plates, packaging trays, coffee cups and the like. Polystyrene has poor resistance to many organic chemicals. Do not use it for storage of dry dyes. Polystyrene items are very convenient for short-term handling dye powders when weighing or mixing dyes.

Polysulphonated dyes: See **Acid dyes.**

Polytetrafluoroethylene fibre (PTFE): Fluorine-containing manufactured fibres characterized by high chemical stability, relative inertness, and high melting point. Made my emulsion spinning, a process that essentially results in fusion of fibrils by passing an emulsion through a capillary, then drawing the resulting fibre. The fibre has a moderate tensile strength and is particularly resistant to the effect of high temperatures and corrosive chemicals. Having very low frictional coefficients, it has a slippery hand. Its principal uses are in packaging and filtration media.

Polytetramethylene terephthalate fibre: Polyester modification, which can be dyed without a carrier, e.g. for domestic textiles. Density 1.31; melting point up to 224°C; glass transition temperature 40°C.

$$[-CO-C_6H_4-CO-(CH_2)_4-O-]_n$$

Polythioethers: Synthetic polymers with Thioethers (alkylpolysulphide), e.g. of type $[-CH_2CH_2S-]_x$. They form the basis for synthetic rubber like products and as synthetic fibre formers (Polymethylene sulphone fibres).

Polythioureas See **Polyureas.**

Polyureas: Organic conversion products of di-isocyanates and diamines, e.g.:

$$2H_2N-(CH_2)_6-NH_2 + OCN-(CH_2)_n-NCO \quad \text{(Monomer)}$$

$$H_2N-(CH_2)_6-\underbrace{NH-CO-NH}_{\text{Urea residue}}-(CH_2)_6-\underbrace{NH-CO-NH}_{\text{Urea residue}}-(CH_2)-NH-$$

$$|-HN-(CH_2)_n-NH-CO-NH-(CH_2)_6-|_m$$

Polyurea

Application: adhesives, cements; animalization; hydrophobing; fibre material (Polyurea fibres); moulded plastics; synthetic leather, etc.

Polyurea fibres: Further development within the group Polyurethane fibres made of spun Polyureas as. The properties of the polyurea fibres largely correspond with those of polyurethanes, having the following advantages compared to the latter: Low density of 1.07 and significantly higher melting point of 237°C.

Polyurethane: A synthetic polymer containing the group –NH–CO–O– linking the monomers. Polyurethanes are made by condensation of isocyanates (–NCO) with alcohols.

Polyurethane fibre: See **Spandex fibre.**

Polyurethane elastomeric fibres: These are, elastomer fibres made of high polymers, which consist primarily of segmented Polyurethanes. The elastomer behaviour owes its elastic stretchability to amorphous soft segments (60–80% by weight), and to crystalline hard segments (40–20% by weight) for chain coherency (elastane fibres). Polyurethane elastomer fibres have rubber-like high elasticity.

Polyurethene rubber: (urethane rubber), synthetic rubber based upon Polyurethane elastomer fibres.

Polyvinyl acetate: Thermoplastic synthetic polymer of vinyl acetate (produced from acetylene and acetic acid). Application: particularly for finishes (weighting, stiffening effects, hat stiffeners, ladder proof finishes, cheap coatings (synthetic leather, nitrate cellulose line end), laminating, pigment print pastes (production shortly before processing due to rapid hardening), lacquer raw material, adhesive, seam free floor covering.

Polyvinyl acetate fibres: Vinyl acetate is represented in copolymers with polyvinyl chloride and polyacrylonitrile fibres as a modification component. Polyvinyl acetate fibres can be dissolved.

Polyvinyl acetate sizes: Polyvinyl acetate, as waterproof sizes for polyester chains for use on water jet weaving machines. Properties: good water solubility and adhesiveness.

Polyvinyl alcohol: Polyvinyl alcohol is manufactured by hydrolysing polyvinyl acetate. The reason is because vinyl alcohol does not exist as a monomer, tautomerization favours the more stable acetaldehyde. Polyvinyl acetate, however, can be hydrolysed into polyvinyl alcohol under acidic or basic conditions. Either method leaves undesirable salts that are difficult to remove. The preferred commercial method of hydrolyzing polyvinyl acetate is to use catalytic amounts of sodium methoxide in methanol. The parent polyvinyl acetate forms branches during polymerization by a chain transfer mechanism at the methyl group of the ester. However, the corresponding polyvinyl alcohol is lower in molecular weight and virtually linear. The branch points are ester linkages which are broken during the hydrolysis step. The branches become linear, lower molecular weight fragments. This accounts for both lower molecular weight and the linearity of the PVA.

Polyvinyl alcohol fibre: A manufactured textile fibre developed in Japan. It is made by dissolving polyvinyl alcohol in hot water and extruding this solution through a spinneret into a sodium sulphate coagulating bath. In Japan, the fibre is used in apparel, household, and industrial fabrics.

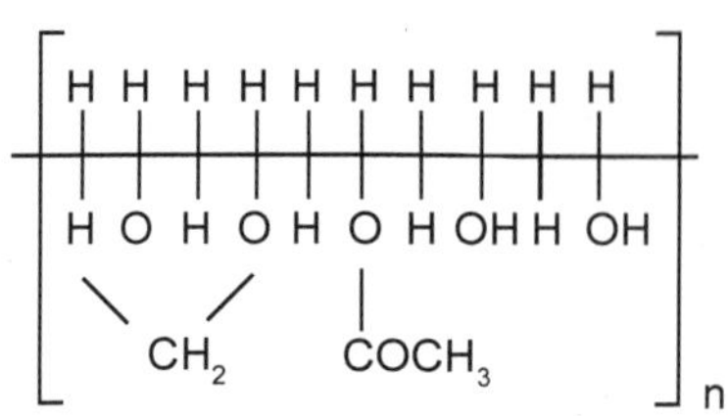

Polyvinyl chloride: A synthetic polymer (plastic) Polyvinyl chloride (PVC) is a very widely used plastic that varies from rigid to very flexible, depending upon additives. It has quite good resistance to most dyeing chemicals, with the exception of carriers for disperse dyes. PVC pipe, intended for plumbing

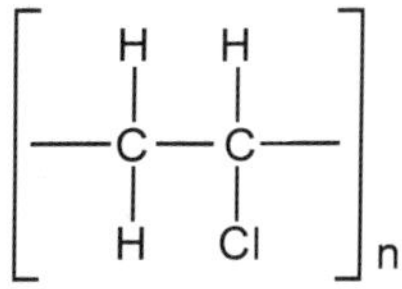

use, is popular for 'forms' around which to wrap fabric for direct dye application techniques such as shibori. Temperature resistance depends on additives and varies considerably. PVC is used extensively as water-proofing coatings on fabrics and for imitation leather.

Polyvinyl chloride fibres: Group of synthetic polymer fibres (the first commercially available synthetic fibres) made of at least 85% Polyvinyl chloride (if lesser proportion: Polyvinyl chloride copolymer fibres), which can also be post-chlorinated, whereby two polyvinyl chloride fibre types can be differentiated:

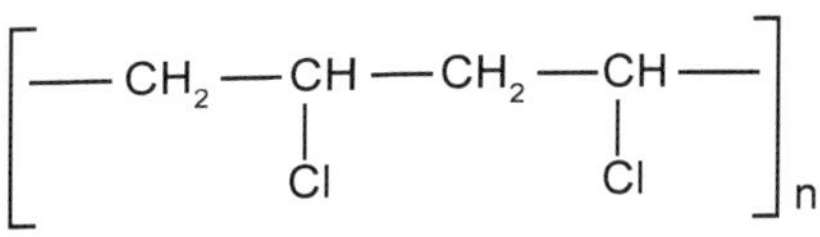

PVC (55–57% Cl)

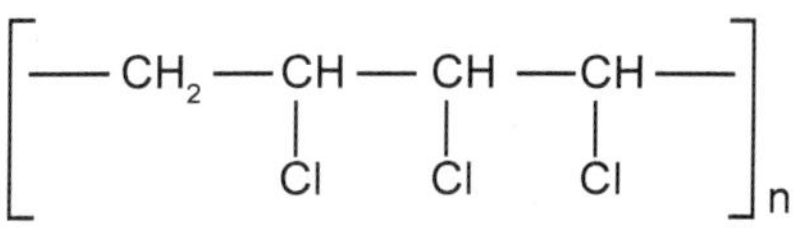

PVC+ (60–64% Cl)

Polyvinyl chloride copolymer fibres: Vinyl chloride copolymers, group of modified polyvinyl chloride types in the form of multipolymer fibres. Application: technical fabrics, protective clothing; also heat-pressure welded with other fibres, however usually mixed due to low strength and high expansion (cotton, wool, viscose staple).

Polyvinyl ethanoate: See **Polyvinyl acetate.**

Polyvinyl ethers: Synthetic polymer of Vinyl ether. Particularly polyvinyl methyl ether. Soluble in water and organic solvents. Resistant to saponification. Reversible coagulation point at 30°C in aqueous solution. Generates full soft handle. Application: sizing agent, finishing agent, printing thickeners (organically soluble); adhesive; lacquer raw material; softener for nitrate cellulose.

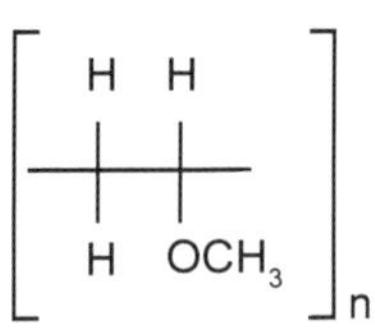

Polyvinyl fluoride: Polymerized vinyl fluoride, produced from acetylene and hydrogen fluoride (polymerization at 810 bar). Polyvinyl fluoride fibres are

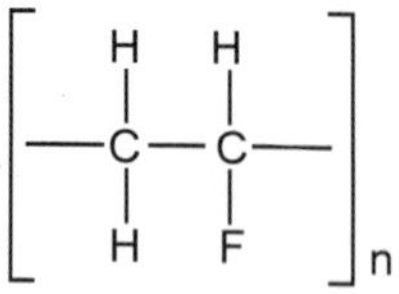

produced in the melt spin process and drawn (100%). Properties: melting point 175–200°C; softening point 130°C; temperature resistance and shrinkage are superior to those of polyvinyl chloride fibres, as is their resistance to light and chemicals; insoluble in xylene; very difficult to dye (thus spin-dyeing). Dyed polyvinyl fluoride is used as a heat resistant fibre glass application.

Polyvinyl formal fibres: Polyvinyl alcohol fibres that are subsequently hardened with formaldehyde (degree of formalization approx. 35–45% of the OH groups) are also termed polyvinyl formal fibres.

Polyvinylidene chloride: Synthetic polymer; of asymmetric dichloro ethylene, produced by, e.g. chlorinated vinyl chloride and HCl splitting. Application: emulsion for lacquer, adhesive, for coating; (acrylonitrile copolymer:) acid resistant pipe and container lining, also tank linings for benzine and petroleum products See **Polyvinylidene chloride fibres**.

Polyvnylidene chloride fibres: See **Polyvinylidene chloride.** Synthetic polymer fibres of min. 80% vinylidene chloride. When spun and drawn, such polyvinylidene chloride fibres mainly form horsehair-like monofilaments (for wear-resistant cushion and grille cloths, filter cloths, sieve fabrics, conveyor belts, ropes, bristles, decorative materials) and also for continuous filament yarn and smooth and curly fibres (for nonwovens, technical felts, upholstery and grille cloth, blankets, carpets, etc. Properties (largely similar to polyvinyl chloride fibres, superior, e.g. against solvents, more recent polyvinylidene chloride fibres also in thermal behaviour): Density 1.65–1.75; normal moisture 0–0.1%; strength dry/wet 13–21 cN/tex; elongation, dry/wet 20–30%; difficult to burn, dirt repellent, resistant against mildew and bacteria, good resistance against cold acids (except H_2SO_4 conc.), alkalis (except concentrated) and most organic solvents including acetone (if they contain oxygen, e.g. dioxane, there is a swelling or softening effect); heat resistance differs depending upon origin: shrinks even at 60–75°C (softens at 90–135°C), sometimes boiling resistant, softens at 115–140°C, shrinking range starts at 145–155°C and melting point 150–175°C. Chlorine resistance poor; dyeable with disperse dyes. Detection: Morpholine test.

Polyvinyl derivatives: They are polymerization products of vinyl derivatives, which have in common the vinyl group CH_2=CH– (polyvinyl group: –CH_2CH–). Normally called plastics or synthetic resins. Imporatnt polyvinyl derivatives are CH_2=CH–CH=CH_2 divinyl (butadiene), CH_2=CH–C≡CH vinyl acetylene, CH_2=CH_2 ethylene, CH_2=CH–HCO acrolein CH_2=CH–Cl vinyl chloride, CH_2=CH–F vinyl fluoride, CH_2=CH–OH vinyl alcohol (unstable) CH_2=CH–OCOCH_3 vinyl acetate, CH_2=CH–OCH_3 vinyl methyl ether, CH_2=C$(CH_3)_2$ isobutylene, CH_2=CH–C_6H_5 styrene (vinyl benzene), CH_2=CH–CN acrylonitrile (vinyl cyanide), CH_2=CHCOOH acrylic acid, CH_2=CH–COOCH_3 methacrylic acid ester CH_2=C–CH_3COOH methacrylic acid, CH_2=C–CH_3COOCH_3 methacrylic acid methyl ester.

Polyvinylidene dinitrile: (polyvinylidene cyanide, ethylidene dicyanide, 1.1 icyanoethene), synthetic polymer of vinylidene cyanide. Used as copolymer (vinylidene cyanide/vinyl acetate 1 : 1) for the production of Polyvinylidene dinitrile fibres.

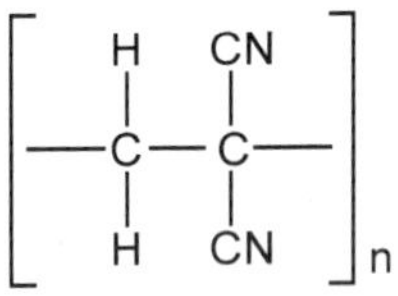

Polyvinylidene dinitrile fibres: (dinitrile fibres), copolymer fibres of type polyvinylidene dinitrile, 50–230% drawn, cashmere-like character. Properties: cross-section kidney shaped, density 1.18; normal moisture take-up 2–3%; softening point 171–176°C; melting point 175°C; breaking strength 20–40 cN/tex (wet 15 cN/tex); elongation at break dry/wet 30–35%; burnable similar to cotton; good weather resistance; good resistance against mildew, rotting, insects; low shrinkage even after repeated boiling wash (max. 1%);Application: pure or mixed (wool, cotton), particularly for knitgoods, high quality pile fabrics (e.g. imitation fur coat materials, downy soft, elegant fall, no pilling), for nonwovens, etc.

Polyvinylidene group: (–CH_2–CR_2–); See **Polyvinylidene chloride** (R = Cl); see **Polyvinylidene Dinitrile** (R = CN)

Polyzymes: Amylases, belonging to the class of bacterial enzymes, however not so heat resistant; effectiveness from 40°C; high liquefaction power, low saccharification power for starches. Bred from cultures of aspergillus orzyae on wheat germ.

Pompadour: A silk term for small floral effects.

Pompadour Serge: English woollen serge with small flower patterns.

Poncho: (1) A narrow woollen blanket, made with cotton warp and woollen .filling napped and well fulled, being made quite waterproof; used in camping; (2) Cotton or woollen fabric, made in plain, twill or rep weave with fringes at the ends, usually with coffee brown stripes; used as overcoat in Argentine, Uruguay and Paraguay, having a slit in the centre for the head.

Pond: No longer permissible unit of force. SI unit is Newton. Conversion: 1 p = 9.80665 mN; 1 kp = 9.80665 N.

Pongee(silk): Greige warp, greige or tram weft. Fabric has discrete lustre. Habota's silks are mostly light fabrics 30-60 g/m^2, often known as pongee.

Pongee (ponge): The word pen-chi, from which pongee has been derived, was originally the name of a Chinese silk. Tussah silk was used for the fabric and was normally woven on hand looms. Light silk fabric of slightly rough feel with excellent draping qualities, but it may now be acetate, triacetate, nylon, or other fibres, all resembling the silk fabric in weight and handling. It is plain weave material notable for its high lustre. Its weight ranges from light to medium. And may be plain dyed or printed. Uneven yarns are used in the cloth and there are usually more picks than ends. Used for blouses, soft dresses, nightwear, and in plain colous as a lining fabric. A typical rayon pongee will have a construction of 64 × 72 with 50s in the warp and 30s in the filling. It is finished usually at 38 in. width and weighs around 3 ½ oz. per sq. yard.

Ponge venezia

Pongee Imperial: A rich, plain woven and highly finished silk dress goods, made of pongee silk.

Pongee Print: A fine and light cotton goods with a soft finish, printed with stripes and dots; used in South Africa.

Ponson Velvet: Heavy velvet, made all silk or only with silk pile; used for women's coats and cloaks.

Ponte di Roma: A common double-knit fabric construction with a four-feed repeat produced with the dial and cylinder needles in interlock gaiting. The fabric is plain and looks the same on both sides.

Pontiac: Dark grey, stoutly knitted and waterproof woollen fabric; used for skirts and coats for outdoor wear.

Poodle cloth: A medium or heavy fabric characterized by loops of yarn on the surface. It may be wool, acrylic or other yarns or mixtures. Mohair is also sometimes included. It is in plain colours or random effects. Used for jackets or coats and warm dresses.

Pooling: See **Pile reversal.**

Poor appearance: Small blemishes that individually do not warrant a string (q.v.) but that, when the fabric is assessed overall, render it unacceptable (in part or in whole).

Poor cover: A faulty fabric in which the warp or weft yarns show through the covering yarns when not so required by the construction.

Popcorn: (1) A special-effect yarn containing short, thick spots. (2) In polymer manufacture a

term used to describe oversize, deformed chip.

Poplin: The name poplin comes from the word papaline, fabric made in the papal town of Avignon, France. It was used for church vestment and hangings. The material is somewhat like broadcloth, but fewer picks per inch and heavier yarn is used for filling. A characteristic construction may be 100 x 40 with 40s warp and 20s filling. All rayon, poplin is made in textures ranging from 80 x 40 to 100 x 50 with warp of 75, 100, 120, or 150 denier rayon and filling 150, 200, 250 or 300 denier rayon. Typical texture may be 92 x 40 with 150 denier warp and 300 denier filling. Even 2/30s or 3/40s cotton or spun synthetic can be used along with 150 denier rayon. Poplins are usually mercerized and then chased for the lustrous appearance. It may be bleached, dyed or printed. It is used for blouses, shirts, gowns, robes, children's suits, rain wear and draperies, and finishes are given according to these end uses. It creases in wear, although less so if some polyester yarn has been included. It is in plain colours or prints, and is very hard wearing, comfortable and absorbent.

Poplin, single, ply, semi: All these are poplins the names derived from the type of yarn used.

Poplinette: Lightweight single poplin is called poplinette. A variety of grenadine, similar to a very light poplin.

Porcelain Lace: Lace steeped in thick solution of kaolin and fired which destroys the fibres but leaves the porcelain behind.

Porcelain printing: In hand printing with zinc plate screens small white, plastic dot patterns applied to dark coloured, usually black, silk.

Pore size of fibres: Pore size can be explained as the the intermicellar gaps (crystallite) of the Pore volume of textiles. The pore pore width of cellulose fibres is approximately 0.6 nm in dry state and in aqueous swollen state it is around 4–6 nm. In swollen cellulose acetate we can only expect pore widths of 5–6 nm. The ratio of pore widths in dry and swollen wool is around 0.6 : 4 nm. The significantly more compact synthetic fibres with their low water take-up or water swellability only achieve pore sizes up to around 0.8–1 nm even in aqueous wetted state.

Pore volume of textiles: The pore volume of textile fabrics is calculated as follows:

$$\frac{\text{density} - \text{volumetric weight}}{\text{density}} \times 100\%$$

Usually, porosity is made up of three elements: porosity within the fibres, so-called intrafibrillary porosity, porosity between the individual fibres, so-called interfibrillary porosity, the gaps between the yarn threads, so-called yarn porosity.

Porosity: The ratio of the volume of air or void contained within the boundaries of a material to the total volume (solid matter plus air or void) expressed as a percentage. Porosity = V1 × 100/V2 where V1 = Volume of voids, V2 = Total volume.

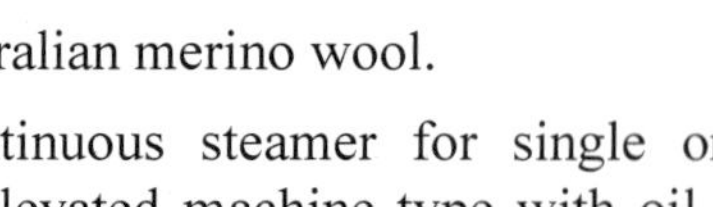

Fabric porosity: the fabric is a porous plate, while the threads are solid.

Port Cabello: Raw cotton from Colombia.

Port Philip: Fine Australian merino wool.

Portal steamer: Continuous steamer for single or two-phase printing, combined underfloor/elevated machine type with oil heated steamer cover. Spiral-shaped roller arrangement in narrow arch or channel shape, without deflecting bar. Guaranteed freedom from drips, air and folds. For pile and stretch fabrics; material guidance usually from inside to outside; water stabilization gates for slow cooling off of polyacrylonitrile pile, also warm air gates. All rollers are driven. The spiral-shaped material flow has proved particularly beneficial when used on a continuous production line, e.g. for printing after-treatment or pile fabric dyeing, thus in continuous dyeing. Material content single track 45–170 m, multitrack correspondingly greater content.

Portuguese Stem Stitch: An embroidery stitch. Commence as for ordinary stem stitch. Pull the thread through and pass the needle under the stitch just made, without entering the fabric. Then pass the needle under the same stitch below the first coll. And make another stem stitch. Pass the needle twice under the stitch just made and under previews stitch.

Positive Feed, in knitting: When the yarn is metered off the warp beam by a metering device.

Post Bed: A bed incorporating a raised post underneath the sewing head which enables sewing to be carried out above the level of the table on which the machine is mounted.

Post-curing: Variously called delayed and deferred curing: Permanent-press process.

Post-mercerising: The abrasion resistance of linen cloth with an anti-crease finish can be significantly improved by mercerization after the hardening of the synthetic resin.

Post-purchase feeling: How a purchaser feels after purchasing something.

Poster Rug: Woven of rags with white warp, having borders in several colours, showing landscapes, flowers, etc.

Pot: A length of usually 10,000 yards, into which pieces of linen are sewed together previous to bleaching.

Pot spinning: A method formerly used for making viscose rayon. The newly spun yarn was delivered into the center of a rapidly rotating, centrifugal pot, where it received twist and centrifugal force caused it to go to the wall of the pot. The yarn package so formed was called a cake.

Potash: A common name for potassium or potassium compounds. Generally used to mean potassium carbonate.

Potash alum: Aluminium potassium sulphate.

Potassium bitartrate: (cream of tartar, potassium tartaric acid, double potassium tartaric acid), $C_4H_5O_6K$, molecular weight 188. Slightly soluble in cold water, better in hot water. Application: No longer used.

Potassium carbonate: (pearl ash; potash) It is used in the laboratory as a drying agent. (carbonate of potassium, potash), $K_2CO_3 \cdot 2H_2O$. Molecular weight 174. A translucent (granular) or white (powder) deliquescent solid known in the anhydrous and hydrated forms. K_2CO_3 (monoclinic; r.d. 2.4; m.p. 891°C) decomposes without boiling. $2K_2CO_3.3H_2O$ (monoclinic; r.d. 2.04) dehydrates to $K_2CO_3.H_2O$ above 100°C and to K_2CO_3 above 130°C.

It is prepared by the Engel Precht process in which potassium chloride and magnesium oxide react with carbon dioxide to give the compound **Engel's salt**, $MgCO_3.KHCO_3.4H_2O$. This is decomposed in solution to give the hydrogencarbonate, which can then be calcined to K_2CO_3. White (anhydrous) granular powder or crystalline mass, both hygroscopic, deliquescent, water soluble, solution strongly alkaline. Application: Milder alkali than sodium carbonate, particularly for wool treatment, in printing, as soap component, etc.

Potassium chlorate: (chlorate of potassium), $KClO_3$, molecular weight 122.5. A colourless crystalline compound, $KClO_3$, which is soluble in water and moderately soluble in ethanol; monoclinic; r.d. 2.32; m.p. 356°C; decomposes above 400°C giving off oxygen. The industrial route to potassium chlorate involves the fractional crystallization of a solution of potassium chloride and sodium chlorate but it may also be prepared by electrolysis of hot concentrated solutions of potassium chloride. It is a powerful oxidizing agent finding applications in weed killers and disinfectants and, because of its ability to produce oxygen, it is used in explosives, pyrotechnics, and matches and particularly for textile printing and aniline black dyeing.

Potassium chromate: (potassium chromic acid, potassium mono chromate), K_2CrO_4, molecular weight 194.5. A bright yellow crystalline solid, soluble in water and insoluble in alcohol; rhombic; r.d. 2.73; m.p. 968.3°C; decomposes decomposes without boiling. It is produced industrially by roasting powdered chromite ore with potassium hydroxide and limestone and leaching the resulting cinder with hot potassium sulphate solution. Potassium chromate is used in leather finishing, as a textile mordant, and in enamels and pigments. In the laboratory it is used as an analytical reagent and as an indicator. Like other chromium(III) compounds it is toxic when ingested or inhaled.

Potassium dichromate: (bi-chromate of potash, Red bi-chromate of potash; double bi-chromate of potassium, potassium bichromate), $K_2Cr_2O_7$, molecular weight 294.5; density 2.7. An orange-red crystalline solid, anhydrous and dry, stable in air, toxic, water soluble, solution weakly acidic, possesses oxidation property, insoluble in alcohol; monoclinic or triclinic; monoclinic changes to triclinic at 241.6°C; m.p. 396°C; decomposes above 500°C. It is prepared by acidification of crude potassium chromate solution (the addition of a base to solutions of potassium dichromate reverses this process). The compound is used industrially as an oxidizing agent in the chemical industry and in dyestuffs manufacture, in electroplating, pyrotechnics, glass manufacture, glues, tanning, photography and lithography, and in ceramic products. Laboratory uses include application as an analytical reagent and as an oxidizing agent. Potassium dichromate is toxic and considered a fire risk on account of its oxidizing properties. Application: after treatment of chrome development dyes to improve fastness; dye mordant for wool; component of metachrome mordant; oxidizing agent for the development of fast dyeing and printing; vigorous decolorizer for wool.

Potassium ferricyanide: See **Potassium hexacyanoferrat III.**

Potassium ferrocyanide: See **Potassium hexacyanoferrat II.**

Potassium hexacyanoferrat II: (yellow prussiate of potash, yellow potassium prussiate), $K_4[Fe(CN)_6]\cdot 3H_2O$. Light yellow crystals, stable in air, water soluble, insoluble in alcohol and ether. Application: Iron indicator.

Potassium hexacyanoferrat III: (red prussiate of potash, red potassium prussiate), $K_3[Fe(CN)_6]\cdot H_2O$. Red crystals; break down in sunlight, gradually decomposing to potassium hexacyanoferrat(II) (yellow prussiate of potash) due to the separation of iron(III) hydroxide; strong oxidizing agent. Application: discharge printing, in analytic chemistry as an iron indicator.

Potassium hexafluorozirconate: (potassium fluorozirconate). K_2ZrF_6. White crystals, soluble in hot water. Application: IWS flame retardant finish for wool.

Potassium hydroxide: (caustic potash, potash lye), KOH. Molecular weight 56, density 2.04. White, highly caustic crystal mass (K_2O) or pieces, fragments, plates, powder; highly hygroscopic (eagerly attracts water and carbon dioxide, therefore store in sealed container); readily deliquescent, readily water soluble and in ethanol and very slightly soluble in ether ; or viscous lye, strongly alkaline, strongest dissociated base ; rhombic; m.p. 360.4°C; b.p. 1320°C. It is prepared industrially by the electrolysis of concentrated potassium chloride solution but it can also be made by heating potassium carbonate or sulphate with slaked lime, $Ca(OH)_2$. It closely resembles sodium hydroxide but is more soluble and is therefore preferred as an absorber for carbon dioxide and sulphur dioxide. It is also used in the manufacture of soft soap, other potassium salts, and in Ni–Fe and alkaline storage cells. Potassium hydroxide is extremely corrosive to body tissues and especially damaging to the eyes.

Potassium iodide starch paper: Filter paper, which is soaked in boiled starch paste and 3–5 g/l potassium iodide before drying. Application: (a) Detection of traces of free chlorine (in bleaches, etc.) = wetted material is dark blue. (b) Detection of acid traces on cellulose goods = dark blue colouring. (c) Testing of diazotization baths (detection of free nitrous acids), if no blue colouring occurs then there is no nitrite or acids. Principle: released iodide determines the formation of blue iodide starch.

Potassium percarbonate $K_2C_2O_6$,: Molecular weight 198.3. Bleaching agent and component of oxidative washing agents.

Potassium permanganate: **potassium manganate(VII)**: $KMnO_4$, a compound, forming purple crystals with a metallic sheen, soluble in water (intense purple solution), acetone, and methanol, but decomposed by ethanol; Density 2.70; decomposition begins slightly above 100°C and is complete at 240°C. The compound is prepared by fusing manganese(IV) oxide with potassium hydroxide to form the manganate and electrolyzing the manganate solution using iron electrodes at about 60°C. An alternative route employs production of sodium manganate by a similar fusion process, oxidation with chlorine and sulphuric acid, then treatment with potassium chloride to crystallize the required product. Potassium manganate(VII) is widely used as an oxidizing agent and as a disinfectant in a variety of applications, and as an analytical reagent. Application in textile: potassium permanganate is a powerful oxidizing agent which has been used in garment washing indigo-dyed denim to produce a worn, faded appearance. It has also found some use in discharge printing of indigo. It is no longer favoured by industry because of environmental concerns. It is quite hazardous, and can be caustic to skin. Care must be used when mixing it with organic materials because it can cause spontaneous combustion.

Potassium persulphate: (potassium sulphate peroxide), $K_2S_2O_8$, molecular weight 270. White, crystalline salt, 98–99%; contains 5.7–5.9% active oxygen, water soluble. Bleaching and oxidizing agent for greases, oils, silk, wool, dyes; starch degradent (desizing); for mordanting, browning, cleaning.

Potassium bi-tartrate: Cream of tartar, potassium acid tartrate, potassium hydrogen tartrate, $KHC_4H_4O_6$ Potassium bi-tartrate is often used with other chemicals as mordant for natural dyes.

Potassium dichromate: Also called potassium bichromate; $K_2Cr_2O_7$ Potassium dichromate is the most commonly used chromium ('chrome') mordant in wool dyeing, both with synthetic and natural dyes. In terms of health risks of repeated exposure, It is one of the most hazardous chemicals used by textile art dyers. It is corrosive to tissues and can cause ulceration. It is known to be a human carcinogen. It poses serious disposal problems.

Potassium mono persulphate: A "peroxygen" compound Potassium monopersulfate can be used as an alternative to hydrogen peroxide for bleaching. It is quite expensive, but a reasonably stable and safe-to-handle dry chemical. Some "non-chlorine shock" chemicals for swimming pools are based on this compound.

Potassium persulphate: (potassium sulphate peroxide), $K_2S_2O_8$, molecular weight 270. White, crystalline salt, 98–99%; contains 5.7–5.9% active oxygen, water soluble. Bleaching and oxidizing agent for greases, oils, silk, wool, dyes; starch degradant (desizing); for mordanting, browning, cleaning.

Potassium thiocyanate: (potassium rhodanide), KCNS, molecular weight 97; density 1.9. Colourless crystals, transparent, deliquescent, hygroscopic, water soluble (in cold solution), also soluble in alcohol. Application: for increasing the dye affinity of wool, analytic (iron indicator = blood red dye).

Potato starch: Powder, pure white; Handle crunchy, floury, cool; usually pure; water content 19–21%. Properties: insoluble in water. If heated from slurry at 46.25°C swelling takes place, at 58.75°C gelatinization begins, at 62.5°C gelatinization ends. Paste: jellylike, slimy, sticky, slightly grey translucent, not very stable, has the lowest stiffening and adhesive power of all starches. Gradual liquefaction if continuously boiled. Boiling with oxidation product (e.g. sodium perborate) leads to (water) soluble starch; degradation by enzymes. Acids lead to dextrose. Alkalis (depending upon concentration) increase the viscosity at normal temperature (great adhesive power), boiling gives rise to very tough, creamy solution (without particular adhesive power). Similar also tannin, salts, paraffin, oils, etc. Finishing handle: mild, soft, supple, full. See ***Starch.***

Potting fastness: Resistance to the effect of boiling water (wet decatizing). Guideline applies for wool and possibly in mixture with other fibres. Roll test piece around a glass rod, treat for 1 h in boiling distilled water with backflow. Dry in air at ≤ 60°C. Evaluation: change of shade, bleeding of white adjacent fabric (wool or cotton) with grey scale.

Potten Kant: (Characteristic pattern of the Antwerp lace (see), consisting of a pot or vase and conventionalized flowers. Originally the pattern pictured the Annunciation.

Poulaine: Shoes with extended (piked) toes.

Poulangy: Coarse and stout French twilled suiting, made in solid colours with linen warp and woollen filling.

Poult: A fabric similar to taffeta but usually heavier and with a more pronounced rib. The fibres used are acetate, triacetate, polyester or mixtures or even silk. It is made in plain colours. It is a stiff fabric liable to crease. Used only for formal dresses, coats, wedding gowns millinery.

Poult de la Reine: Very heavy silk dress goods, woven with a pebbled surface.

Poult de Soie: Solid coloured silk fabric, made in plain weave with a two-ply warp and a heavier filling containing more than two strands and forming cross ribs; used for dresses.

Pounce: To raise the nap on a felt hat with emery paper.

Pouritache: Similar to the soutache braid only much heavier.

Poussin: Very light, narrow French lace, similar to Valenciennes.

Powder cleaner: A cleaning agent in which an absorbent compound is the principal ingredient.

Powder cleaning: See more preferred term, **Dry extraction cleaning.**

Powder-bonded Nonwoven: A manufactured product in which a carded web is produced and treated with a thermoplastic powder that has a melting point less than that of the fibre in the web. The powder is heated to its melting point by through-air and infrared heating or by hot-calendering to effect bonding.

Powder-dot Coating: A method of powder coating whereby the melted poder is made to fall on the heated surface of the material to be coated and pressed so that the powder the powder melts and adheres to the fabric in the form of droplets. Usually, in a dot powder coating machine the indentations of a heated engraved roller are filled with thermoplastic powder which is then

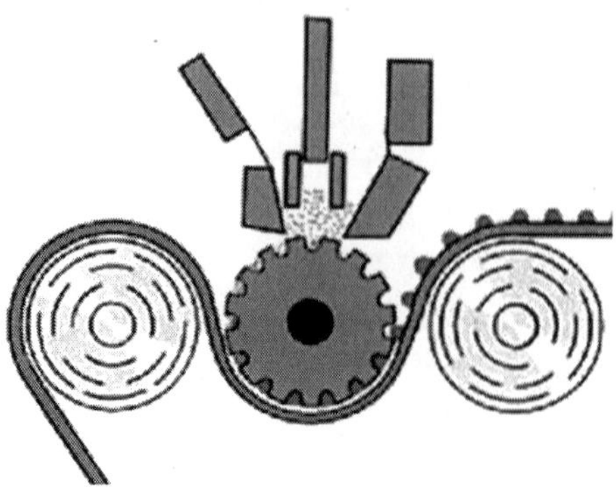

Dot coating sintering

slightly agglomerated. The material is heated to above the melting point of the coating powder on a heated roller. The hot fabric is pressed against the powder filled hollows, so that the melted powder is deposited on the surface of the material in the form of droplets. Subsequently in a further sintering zone the droplets are melted to form hemispherical pearls, which sit cleanly upon the surface of the fabric.

Powdering: Dotted pattern strewn all over the fabric.

Power: As physical variable: Work done per unit time, SI unit is Watt (W) = J/s.

Power bonding: A method of making thermally bonded non-woven fabrics in which a fibre web or batt is bonded by the use of heat sensitive powder dispersed within it.

Power factor: Of an insulating material, the ratio of the power in watts dissipated in a capacitor in which the material is dielectric, to the product of the sinusoidal voltage and current expressed in effective volt-amperes.

Power loom: A loom which is driven by a source of power such as an electric motor. If it is non-automatic, the shuttle change is done manually.

POY: See **Partially Oriented Yarn**. Polyester fibres.

PP: See **Polypropylene fibres.**

Ppb: (parts per billion), billionths = $10^{-9.}$

Ppi, Finished: Picks Per Inch, Finished fabric.

PPI, In grey: Picks Per Inch, grey fabric.

PPI, In loom: Picks Per Inch, loomstate fabric.

Ppm: (parts per million), millionth = $10^{-6.}$

Ppt: Parts per trillion.

PR : Protein fibres.

P.R.: (Pressley-Ratio) Pressley-Index.

Prato wool: Italian cloth export from Prato, previously made primarily of —. Reclaimed wool, which after the Second World War achieved world-wide importance due to its low price (around 50% of production for export was for the Orient, West Germany, England, etc.).

Prayer Rugs/Prayer mats: Oriental ‘ rug of usually small size; used by tie Moslems to kneel on when praying. The chief characteristic is the representation of a niche’ o-r arched doorway at one end, sometimes with the tree of life or a lamp hanging from the middle; occasionally there are only three medallions, two for the knees and one for the head.

Precieuse: Plain woven striped silk dress goods.

Precipitate: In chemistry, a solid formed by a reaction in a solution or as a result of physical changes, such a temperature, of the solution A precipitate has low solubility in the solution from which it was formed, giving a cloudy appearance, or actually settling out of solution.

Precision winding: A method of winding cones and cheeses in which the number of winds per double traverse is constant and angle of wind increases as the diameter increases. It can be used to build a very dense package.

Precision-wound packages: A winding method where throughout the winding process of a cone a regular winding is made at all diameters. The rewinding tube is positively driven and has an adjustable, constant ratio to the

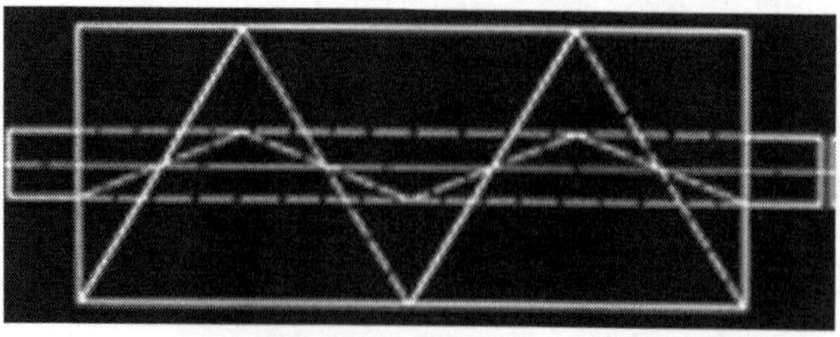

movement of the thread guide at all times. The thread distance can also be selected. The number of windings per winding stroke remains constant from the starting diameter to the full batch diameter. The most advantageous winding can be adapted at will to the yarn material and field of application. See **Crosswound yarn package**.

Pre-coating of screen: In order to facilitate possible painting of screens by hand, the screen is pre-coated e.g. with an auxiliary layer of gelatine or sodium alginate.

Precondition: To bring a sample or specimen of textile material to a relatively low moisture content (Approximate equilibrium in an atmosphere between 5 – 25 5 relative humidity) prior to conditioning in a controlled atmosphere for testing.

Preconditioning: Bringing a sample or specimen of textile material to a relatively low moisture content (approximate equilibrium in an atmosphere between 5 and 25% relative humidity) prior to conditioning in a controlled atmosphere of higher humidity for testing. (While preconditioning is frequently translated as predrying, specimens should not be brought to the overdry state.)

Pre-cure: **A** process where the curing is done at the finishing plant in the flat state. A finishing treatment in which the durable press finish is applied to the fabric and set, or cured, through the use of heat at the mill, prior to shipment of the fabric to the garment manufacturer.

Pre-dip: Pretreatment of yarn or fabric with an auxiliary that promotes the adhesion of the subsequent impregnation.

Predrier: A drying unit for partially drying a wet fabric (usually padded) which should prevent migration in the main dryer due to the mild nature of drying. There are many methods of predrying where the IR drier is considered the best.

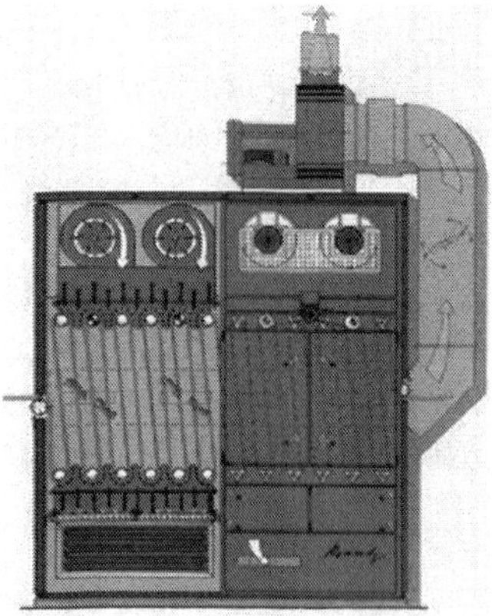

A nozzle pre dier

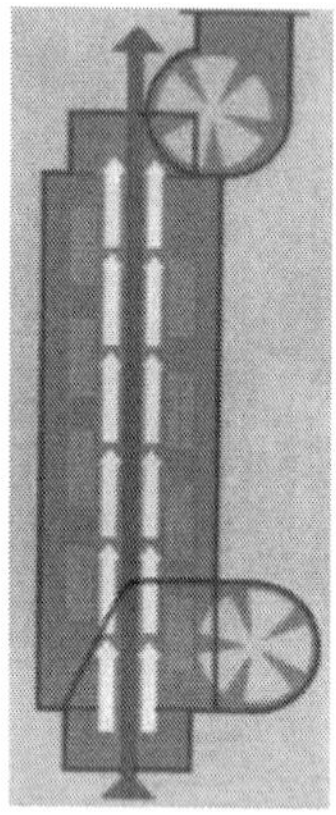

An IR predrier

Prefelting machine: Prefelting of folded fur hats by means of rotating rollers or rollers with strong rope wound around them (ropemill).

Preferential orientation: Preferential orientation of silicones on hydrophilic fibre surface to their water repellent finishing. The hydrophilic part of the silicon polymers align themselves to the fibre surface, whereas the hydrophobic parts point outwards.

Pre-lacquering:Positive process for screen making.

Pre-wound Bobbins: Refers to bobbins that are wound by the thread supplier. From 15 to 20% more thread can be wound by precision winding equipment than by a sewing operator using a sewing machine bobbin winder.

Preform: (1) Fabrics that have been stacked in multiple layers and bonded with a stitching yarn or by other means. The preform may be cut into a shape for subsequent molding into a composite part. (2) A preshaped nonwoven made by distributing fibres over a screen in the approximate contour of the finished part. (3) A preshaped, three-dimensional reinforcement made via braiding, weaving, or knitting or some combination of these. (4) A formed but not fully consolidated stack of prepared layers which have the contour of the finished product. (5) The first stage in the production of blow-molded products such as PET bottles.

Prein: A German process by which the shearing of the goods is eliminated by pressing all irregular fibres into the body of the cloth while it is in the loom.

Prelate: French sailcloth and tent canvas, made of hemp and coated with tar.

Preliminary treatment: Any treatment of wastewater that precedes ***primary sedimentation***.

It is partly aimed at protecting equipment in later treatment. Mainly it is the removal of large solids, rags and grit by ***screens, comminutors, grit channels***, etc. Oil, fat, grease, etc., may be removed in ***pre-aeration***. For wastewater treatment works fed by ***combined sewers***, preliminary treatment includes the proportioning of storm sewage—i.e. ***storm water overflows***.

Pre-metallised dye: A dye that is a complex of dye molecules with a metal atom; metal complex dye The term "pre-metallized" is used primarily to refer to classes of acid dyes, although sometimes considered as apart from acid dyes. The metal is part of the dye structure as it comes from the manufacturer. There are 1:1 metallized dyes, having one dye molecule per metal atom, and 2:1 types, with two dye molecules per metal atom (use is inconsistent here 2:1 and 1:2 mean the same thing). The metal in the complex helps the dye attach to the fibre, typically wool or polyamide. These dyes have excellent wash fastness, but leveling may be difficult. There is a wide range of "pure" colours (as opposed to mixtures) available, though the brightness is the lowest of the acid dye subgroups. Application temperature is typically at or near the boil. 1:1 types are usually applied at very low pH, often with formic or sulphuric acid in quite large quantities, while 2:1, sometimes called "neutral dyeing", are often applied using acid donor chemicals, so the bath is nearly neutral at the start an becomes more acid with continued boiling.

Premium: Commercial variety of early maturing upland cotton, the staple measuring 23-27 millimeters; the yield is 32-33 per cent; also called Peerless.

Pre-mordanting: Chroming.

Pre-sedimentation tanks: In water treatment, a term used for (a) ***grit chambers*** or (b) sedimentation processes used to remove large quantities or suspended material prior to the normal water treatment processes.

Preoxidized fibre: In carbon fibre production, a fibre that results from a relatively low temperature (200-500°C) heat treatment in the presence of oxygen which converts the precursor fibre, PAN or rayon, to an infusible fibre that is stable to further processing.

Preparation% (1) (formulation). Mixtures and solutions made of two or more substances.

(2) The process of making the grey fabric ready for dyeing, printing, finishing or for any other further process. The process generally includes desizing (in case of sized gerige) scouring, bleaching, mercerizing etc. for cotton and suitable process for other materials.

Preparation vessel: In textile finishing machines the treatment liquor is often prepared in a separate container, which is then supplied to the machine by a pump.

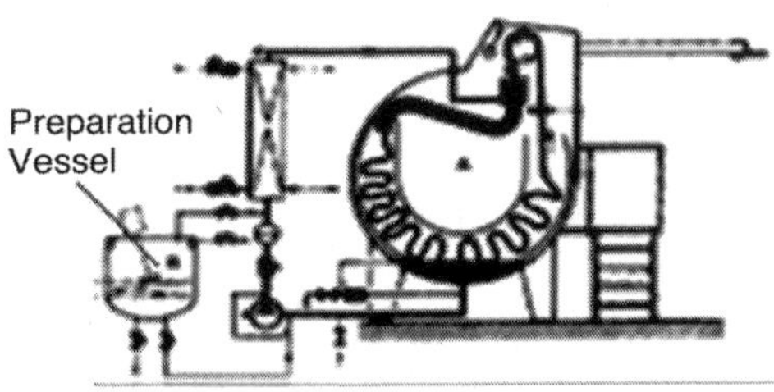

Example of a preparation vessel in jet dyeing

Prepared for dyeing: A fabric or garment that is specially made to be dyed; sometimes "preferred for dyeing"; usually abbreviated PFD, pfd or p.f.d. PFD fabrics have been desized, scoured, and fully bleached, but have been processed without optical brighteners or softeners which can interfere with dye uptake. Often called 'prepared for garment dyeing' or PFGD, such fabrics are usually only carried by specialty suppliers. Most fabric is prepared for dyeing at the same facility that does dyeing. Any fabric can be prepared for dyeing and sold as such, but types other than cotton are extremely rare, and often available only by special order of large quantities. PFD cotton garments are made with PFD fabrics and sewn with cotton thread, so that the thread will dye similarly to the fabric. Sometimes they are made oversize to allow for shrinkage in dyeing processes. A few large manufactures of t-shirts offer PFD shirts. It may be advisable to wash PFD products before dyeing, mostly as a hedge against contaminants picked up in handling or to remove thread lubrication oils that might have been used in garment construction.

Prepared for printing: A fabric that has been processed to be ready for printing, usually meaning printing with dye PFP cotton will be very similar to PFD cotton, but is almost always mercerized to improve the colour uptake. Wool fabrics are usually chlorinated in preparation for printing.

Prepigmentation: Unvatted vat dyes are padded onto fabric as a dispersion and deposit themselves onto the fibres as pigments during drying. Subsequent

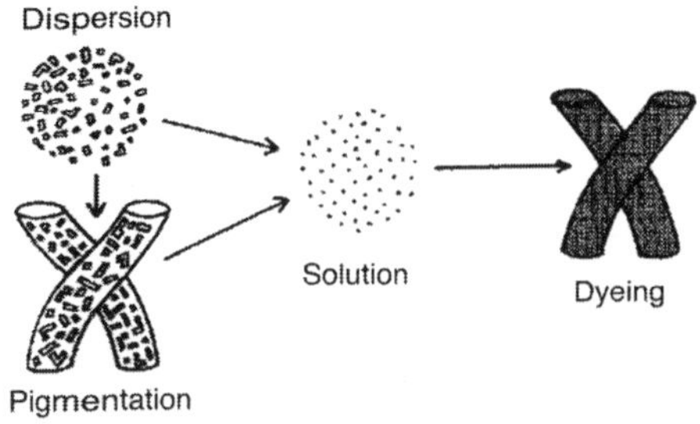

Principle of prepigmentation and dyeing

padding of chemicals (reducing agents and alkali) vats the dye to a solution, which diffuses into the fibre in the subsequent steaming process.

Prepreg: Ready-to-mold, reinforcing material, either fibre, fabric, or mat, that is fully impregnated with resin and in some cases, partially cured. Prepregs are

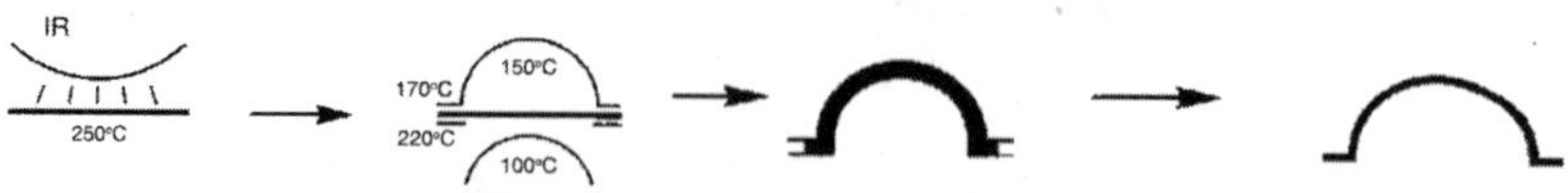

The processing of a prepreg in a deep drawing process.

then used by fabricators in laying-up and molding composites after which curing is completed. Prepregs are not free-flowing, and therefore cannot change shape independently. Prepregs consist of thermoplastic fibres (PA 6.6) or non-meltable fibres (in the field of prepreg forming), such as glass fibres, carbon fibres or para-aramide fibres; this selection is based upon the profile of requirements of the fibre-reinforced plastic component to be produced. The shaping processes that come into consideration are similar to the deep drawing process common in sheet processing.

Pre-printed resist: Half-tone resists; Resist printing.

Preservative: Used to protect textiles and finishing substances (sizes, printing thickeners) against decomposition due to micro-organisms. Common preservatives in textile finishing are fungicides and bacterial poisons, often typical disinfectants, which should be used with care.

Preservatives, in sizing: To prevent the growth of bacteria and fungi in size mixes and on sized yarn. Cresol, phenol, salicylic acid are examples of preservatives.

Presetter: Used for the presetting of unsewn hosiery blanks, for fixing hosiery or socks made of polyamide (Hosiery setting) and for the relaxing of tufted carpets (Carpet pre-steamer). Working method for hosiery in saturated steam process with or without vacuum plant.

Presetting: Heat setting at temperatures that lie below the softening range of synthetic fibres, so that they are protected against creasing and folding in subsequent finishing operations, specifically wet processes.

Preshrink: Steam or hot water pre-treatment for textured yarn for the development of texture.

President: A double woollen fabric, the face made with cotton warp and woollen filling in five-leaf satin weave, the back with an extra mohair filling in two and one weave.

President: Braid A twilled woven braid with diagonal ribs; used for trimming.

Presidents: In Europe napped woollen fabrics, made with a large proportion of shoddy or mungo.

Pre-Shrunk: Quality term for material that has been relaxed to the degree that under the effect of a normal ironing the length and width of the sample decrease by no more than ± 1% (See **Shrinkage; Sanforize**). Such a material may still have a press lustre, for example, (final treatment on the flat press) which is not itself fast to hot pressing or spotting.

Press: A finishing process, consisting in pressing the fabric between heated plates or cylinders.

Press-Off, in knitted fabric: A condition in which the yarn fails to knit and either the fabric falls off the needles or the design is distorted or incomplete.

Pressboard: Special cardboard, which is used as glazed millboard in Flat press and is placed between the layers of material.

Pressed-increase: A sharp crease inserted intentionally in a fabric usually by application of pressure, heat, moisture or a combination of two or three.

Presser foot: An attachment which presses the material being sewn against the lower feed during sewing.

Pressing: See also **Ironing**. Removing or creating of creases or shape in material by heat and/or steam and/or vacuum or a combination of them with pressure.

Pressing marks: The accidental occurrence of "stripy", curved or wavy folds or pleats as a result of unskilled ironing.

Pressing (Finishing): A method of enhancing the appearance of fabric, by application of pressure to increase their lusture and smoothness. *Note:* The equipment used varies with the type of fabric being finished. For e.g. (a) a hydraulic press in which the layers of cloth formed by cutting the material are interleaved with heated or cold press papers and the pile of cloth so formed is subjected to pressure. (b) a rotary press in which fabric is formed by by a smooth rotating heated cylinder and a fixed curved metal plate, pressure between the two elements being applied hydraulically. (c) garments and knitted goods are pressed by being placed on a perforated, fabric covered steam bed and subjected to pressure by lowering a second hinged steam bed on to the material. Steam is blown through the material while under pressure and the fabric can be cooled by drawing air through it. This type of machines is often known as a hoffman press and is the process as Hoffman pressing. Many new machines and methods are introduced from time to time.

Pressing and curing: A variety of industrial pressing and curing treatments used in garment industry.

Pressing off: The final pressing operations on a garment after assembly (or Off pressing).

Pressley Index, in cotton testing: A measure of the strength of fibre bundles determined under prescribed conditions and expressed in an arbitrary units pounds per milligram.

Pressure: Force or load per unit area. Mechanical tension. The derived SI unit is the Pascal (unit symbol: Pa).

1 Pa = 1 N/m^2 = 10^{-4} N/cm^2 ; 1 N/cm^2 = 10 kPa

Special name for the tenth part of the mega pascal

(MPa) is the Bar (unit symbol: bar):

1 bar = 0.1 Mpa = 100 kPa = 102 kPa = 105 Pa

1 bar = 1000 mbar

The previously used Torr, at., and kp/cm^2 units are no longer permitted:

1 mbar = 0.750 062 Torr

1 bar = 1.01971621 kp/cm^2

Pressure drier: In this drier the system is completely closed and pressurised, up to 8 bar. The moisture is extracted from the circulating air via a cooler, heated up and fed back into the boiler; the water is removed by a separator. See **Yarn dryer, air stream.**

Pressure drop: (1) A decrease in pressure that is caused by friction between a flowing liquid and a constricting container. The pressure drop is increased by a reduction in diameter of the container. (2) The change in pressure across a filter.

Pressure dyeing: See **Dyeing.**

Pressure filter: (1) A ***filter plate press***. (2) A rapid ***sand filter*** that is not open to the weather but is contained in a steel shell. It can therefore work at a pressure higher than the gravity head and may be an ***in-line*** treatment. The sand is 0.45 to 1 m deep, with an ***effective size*** between 0.5 and 1 mm and a maximum ***uniformity coefficient*** of 1.7. ***Surface loading rates*** are from 4 to 10 m^3/h per m^2 of filter area, although up to 20m^3/h per m^2 has been used. *Compare* ***rapid gravity filter***.

Pressure kier: Wrought iron autoclaves for the kier boiling of loose stock, yarn, wound packages and piece material.

Pressure mark: See **Bruise.**

Pressure mark: An impression or an area of greater lustre in fabric, caused by irregularities of pressure during the finishing process.

Pressure relief valve: A spring loaded pressure relief valve is designed to open to receive excess pressure and to reclose and prevent further flow of fluid after normal conditions are restored. It is characterized by a rapid rapid-opening 'pop' action or by opening in a manner generally proportional to the increase in pressure over the opening pressure. It may be used for both compressible or non-compressible fluids, depending on the design, adjustment, or application.

Pressure steamer: Continuous steamer for steaming at temperatures above 100°C for developing pad dyeings and prints or HT bleaches. Discontinuous for combed tops (vigoureux) printing, loose material and piece goods. See **HT pressure steamer.**

Pressure steaming process: Fixing, for example, of disperse dyes in the pressure steamer at 130°C with saturated steam. Application on polyester combed tops and spun tow. See **Thermosol dyeing processes.**

Pressure Vessels: All vessels or pipe layouts in which, when in use, there is a positive operating pressure greater than 0.1 bar or a negative operating pressure of under 0.2 bar.

Pretension: The relatively low tension applied to remove the kinks and crimp when mounting a specimen preparatory to making a test and used to establish the nominal gauge length.

Pretreat, in cleaning of textiles: Pre-application of cleaning agent to spots, stains and areas of high soil concentration to maximize activation time and therefore facilitate soil removal during overall cleaning.

Pretreatment: Treatment of a fabric for the next operation like dyeing printing etc.

Pretreatment Jig: Jigger used for the pretreatment of fabric.

Pretreatment, in ETP: (1) For industrial wastewater, partial **on-site treatment**. (2) In wastewater treatment, **preliminary treatment**. (3) For sludge treatment, **conditioning** to improve its **de-waterability**.

Price points: Price at which products will retail.

Primary backing, for tufted pile yarn floor covering: The fabric through which the pile yarn is carried by needles to form tufts; the backing fabric.

Primary cellulose acetate fibre: See **Triacetate fibre.**

Primary colours: Magenta, yellow, and cyan (red, yellow, blue); these are the subtractive primaries used when mixing dyes, paints, etc., to make all other colours. In the CIE (Commission Internationale de l'Eclairage) system of colour measurement, which is the international system most widely used today, the primary colour vectors are red, green, and blue-violet. These are additive primaries based on the perception of colour of reflected light by the human eye.

Primary colours: The visual organ can achieve eight extreme sensory options. In a mixing system (Colour mixtures) for opaque colorants all eight colours are needed: cyan, black, green, violet blue, magenta red, white, orange red and yellow.

Primary creep: The recoverable component of creep. Also see **Delayed deformation**.

Primary salts: Acid salts of multivalent acids, in which only one hydrogen atom is replaced by metal; e.g. primary acid sodium phosphate NaH_2PO_4 derived from phosphoric acid H_3PO_4. Secondary salts; Tertiary salts.

Primary sedimentation: The ***sedimentation*** in wastewater treatment that immediately follows grit removal, releases settled wastewater for biological treatment and removes organic solids as sludge. ***Radial flow sedimentation tanks*** or ***horizontal flow sedimentation tanks*** may be used. The design is often based on the ***surface loading rate***, which should be about 30 to $45m^3m^{-2}d^{-1}$ at maximum flow to ensure 50 to 70% removal of suspended solids and about 30% removal of BOD from domestic wastewater.

Alternatively, the design retention time is 1.5 to 2 h at maximum flow, usually 3 times ***dry weather flow***. ***Weir loadings*** should be less than 250m3m_1 day_1, although this is less critical than the overflow rate. See **primary sludge, sludge hopper**.

Primary treatment: For wastewater treatment, the first major treatment process, which is usually ***primary sedimentation***.

Prime: Merino clothing wool taken from sides of a very fine fleece or shoulder of a good fleece.

Princes lace: An imitation of luxurious Duchesse Lace.

Princesse: Four-leaf silk satin dress goods, made with single warp.

Princesse Cashmere: French cotton dress fabric, made in imitation of cashmere.

Princetta: An English worsted fabric in the 19th century, made with silk warp and worsted filling; originally made of pure worsted.

Print: A fabric with designs applied by means of dyes or pigments used on engraved rollers, blocks, or screens.

Print bonding: A process of making nonwoven fabrics in which there is controlled application of adhesive to specific areas of the fibre web or batt in a discrete pattern by using printing techniques similar to those used for colouration.

Print coverage: In textile printing the ratio of dye applied/printing area.

Print definition: The so-called staying power of a print paste, i.e. the upper limit of possible thickener concentration for printing purposes (Flow properties of thickeners and print pastes) at which the print paste covers the fabric uniformly (not to be confused with full print penetration). At an even higher concentration, the print would already "grin".

Print-Dye process: 2-stage process consisting of printing/drying and dyeing/ rinsing for floor coverings.

Print flushing: Print flushing arises due to excessively deep engraving, inadequate absorbency of the printed material and excessive pressure by the printing roller on the material, either on one side or on both sides. Print flushings show up on the material as blurred, lacerated contours, which usually run opposite to the printing direction.

Print impression method: This was developed for examining the surface of engraved calendar rollers and rotary screens for screen printing, and are based upon the creation of an imprint in polystyrene by means of a silicon rubber negative. The print impression method, originally developed for use with the scanning electron microscope, is now used with the stereomicroscope.

Print paste: The mixture of gum or thickener, dye, and appropriate chemicals used in printing fabrics. Viscosity varies according to the types of printing equipment, the type of cloth, the degree of penetration desired, etc.

Print paste sieving machine: (sieving machine), is used for the: Straining of print pastes; Printing thickeners and chemical finishing products. The method of operation can vary greatly; e.g. brush principle (with 1, 2 or 4 sieving points), pulsation principle, vacuum principle for sucking through the gauze at high sieving speed (vacuum filtering or screening machine), in which blocked screen sections cannot be ruled out however, and the colour grinding mill principle for particularly fine milling. For the purpose of screen printing, when print paste containers with replaceable silk cloth screens are used, the gauze one number finer than the screen gauze is always selected.

Print paste straining centrifuge: The print pastes run into the centrifuge via a funnel. The inner screen mesh basket rotates and sieves the print pastes. In centrifuging, special inserts are used to separate the impurities from the print pastes by centrifugal force.

Print paste wedge: In screen printing, wedge shaped area formed by the angle between blade/rod squeegee and flat/rotary screen. The print paste wedge is a determining factor for print paste application. If the blade squeegee is set at a shallow angle then a long, narrow wedge forms (print paste application is correspondingly lower); alternatively in rod squeegees of different diameters. The wedge is shorter in blade squeegees set at a steep angle. The wedge pressure created is decisive.

Print steamers: After the application of the dye paste and the drying of the printed textile the material passes through the print steamer to transfer the dye from the thickener into the fibre. To achieve this the dried material is collected and passed onto a separate process, e.g. a high performance universal loop steamer with 2 adjacent tracks. If possible the moisture content of the steamer should be constant.

Printcloth: A medium weight, plain-weave fabric made of carded yarns, usually cotton or polyester/cotton blends, with counts from 28's to 42's. Millions of yards of print cloth are printed annually and other millions are finished as white goods. Large amounts of the goods are also used in the greige for bags, containers, and base fabric for coated materials.

Print-dye process: A print technique followed in textile floor coverings. It is 2-stage process consisting of printing/drying and dyeing/rinsing finishing for floor coverings. Chemicals, resist,(half or full tone resist) agents are applied to the material in a colourless print paste using a suitable printing machine. On dyeing the printed portion remains undyed or partially dyed as per the resist agent.

Printed suede: Silk screen printed suede fabric produced as piece goods for high quality women's wear and men's wear.

Printing: The reproduction of a pattern onto a textile material by applying a suitable substance by means of an engraved surface, a stencil or other patterning device.

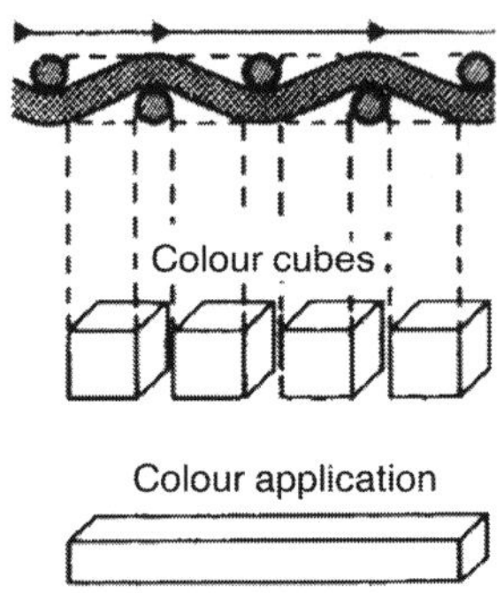

Schematic representation of the printing process.

Printing: A process for producing a pattern on yarns, warp, fabric, or carpet by any of a large number of printing methods. The colour or other treating material, usually in the form of a paste, is deposited onto the fabric which is then usually treated with steam, heat, or chemicals for fixation. Various types of printing are described below: Also see ***Dyeing***.

(1) Methods of Producing Printed Fabrics:

Block Printing: The printing of fabric by hand, using carved wooden or linoleum blocks, as distinguished from printing by screens or roller.

Blotch Printing: A process wherein the background colour of a design is printed rather than dyed.

Burn-Out Printing: A method of printing to obtain a raised design on a sheer ground. The design is applied with a special chemical onto a fabric woven of pairs of threads of different fibres. One of the fibres is then destroyed locally by chemical action. Burn-out printing is often used on velvet. The product of this operation is known as a burnt-out print.

Direct Printing: A process wherein the colours for the desired designs are applied directly to the white or dyed cloth, as distinguished from discharge printing and resist printing.

Discharge Printing: In "white" discharge printing, the fabric is piece dyed, then printed with a paste containing a chemical that reduces the dye and hence removes the colour where the white designs are desired. In "coloured" discharge printing, a colour is added to the discharge paste in order to replace the discharged colour with another shade.

Duplex Printing: A method of printing a pattern on the face and the back of a fabric with equal clarity.

Etching: See **Printing, Burn-Out Printing.**

Extract Printing: See **Discharge Printing.**

Heat Transfer Printing: A method of printing fabric of polyester or other thermoplastic fibres with disperse dyes. The design is transferred from pre-printed paper onto the fabric by contact heat which causes the dye to sublime. Having no affinity for paper, the dyes are taken up by the fabric. The method is capable of producing well-defined, clear prints.

Ink-Jet Printing: Non-contact printing that uses electrostatic acceleration and deflection of ink particles released by small nozzles to form the pattern.

Photographic Printing: A method of printing from photoengraved rollers. The resultant design looks like a photograph. The designs may also be photographed on a silk screen which is used in screen printing.

Pigment Printing: Printing by the use of pigments instead of dyes. The pigments do not penetrate the fibre but are affixed to the surface of the fabric by means of synthetic resins which are cured after application to make them insoluble. The pigments are insoluble, and application is in the form of water-in-oil or oil-in-water emulsions of pigment pastes and resins. The colours produced are bright and generally fat except to crocking.

Resist Printing: A printing method in which the design can be produced: (i) by applying a resist agent in the desired design, then dyeing the fabric, in which case, the design remains white although the rest of the fabric is dyed; or (ii) by including a resist agent and a dye in the paste which is applied for the design, in which case, the colour of the design is not affected by subsequent dyeing of the fabric background.

Roller Printing: The application of designs to fabric, using a machine containing a series of engraved metal rollers positioned around a large padded cylinder. Print paste is fed to the rollers and a doctor blade scrapes the paste from the unengraved portion of the roller. Each roller supplies one colour to the finished design, and as the fabric passes between the roller and the padded cylinder, each colour in the design is applied. Most machines are equipped with eight rollers, although some have sixteen rollers.

Rotary Screen Printing: A combination of roller and screen printing in which a perforated cylindrical screen is used to apply colour. Colour is forced from the interior of the screen onto the cloth.

Screen Printing: A method of printing similar to using a stencil. The areas of the screen through which the colouring matter is not to pass are filled with a waterproof material. The printing paste which contains the dye is then forced through the untreated portions of the screen onto the fabric below.

Warp Printing: The printing of a design on the sheet of warp yarns before weaving. The filling is either white or a neutral colour, and a grayed effect is produced in the areas of the design.

(2) Methods of Producing Printed Carpets:

Millitron Process: A computer-controlled, non-contact spray printing process that allows the production of intricate multi-coloured designs. Although this process was developed for carpets by Milliken & Co., it can also be used for upholstery, pile fabrics, and other textiles.

Mitter Printing Machine: A rotary carpet printing machine with up to eight stainless-steel mesh screens, and with cylindrical squeegees of moderately large diameter in each rotary screen. The unit has a streaming zone for dye fixation.

Stalwart Printing Machine: A carpet printing machine in which colour is applied to the carpet with a neoprene sponge laminated to the pattern. The pattern is cut in a rubber base attached to a wooden roll. It is very similar to relief printing. Used primarily for overprinting random patterns on dyed carpets. Suitable for shags and plush carpets as well as level loop and needle tuft types. ***Zimmer Flatbed Printing Machine (Peter Zimmer)***: A carpet printing machine that uses flat screens and dual, metal-roll squeegees. The squeegees are operated by electromagnets to control the pressure applied. The unit also has a steamer for dye fixation. The Zimmer flatbed machine is normally used for carpets of low to medium pile heights.

Very precise designs are possible, but speeds are slower than with rotary screen printers.

Zimmer Rotary Printing Machine (Johannes Zimmer): A three-step, rotary carpet printing machine consisting of: (i) rotary screens with small diameter steel-roll squeegees inside, with pressure adjusted electromagnetically for initial dyestuff application; (ii) infrared heating units to fix dyes on the tips of the tufts; and (iii) application of low-viscosity print paste, followed by steaming for complete penetration of dyes into tufts. Zimmer Rotary Printing Machine (Peter Zimmer): A rotary carpet printing machine in which each rotary screen has a slotted squeegee inside to feed print pastes through the screens to the carpet. Pressure of the print paste is adjusted by hydrostatic head adjustments.

Printing area: That part of the gauze of a screen printing screen that is permeable to print paste.

Printing blanket: Endless rubber coated fabric, which in roller printing is pulled over the lapping on the main cylinder.

Printing carriage: A movable carriage in which the printing screens are mounted one at a time. The carriage is supported on wheels running along rails at either side of the printing table. The printing screen is raised or lowered by lever action, and the squeegee (used to force print paste through the screen) is driven across the screen by a motor attached to the carriage.

Printing carriages of this type represent a partial mechanization of the hand screen printing process.

Printing gauze: See **Screen mesh, screen gauze.**

Printing rollers With regard to the type of colour paste application a differentiation is made between printing with deep engraved rollers, roller printing and relief printing, in which the rollers have the pattern embossed upon them.

Printing screen locking unit: Printing tables are fitted with a U-shaped rail at the front, upon which repeat stops are mounted. The locking units fastened onto the screens fall into the repeat stops when the screen is lowered and thus hold the screen securely in the repeat.

Printing screens: See **Screens for screen printing.**

Printing style: Fundamental technical possibilities for printing fabrics: Direct printing; Discharge printing; Resist printing; Blotch printing.

Printing table: Support for the fabric to be printed in block screen printing and flat screen printing.

Printing thickener: Finished product of Thickeners for printing (Print pastes). Every composition is dependent upon fibre material, fabric type, print pattern, dye group, printing process, etc.

Printing unit: Finished product of Thickeners for printing (Print pastes). Every composition is dependent upon fibre material, fabric type, print pattern, dye group, printing process, etc.

Printing without screens, Digital printing: See **Digital printing.**

Printing, methods Of: See **Printing.**

Prints on dyed ground: Prints can be done on dyed ground, either as over print or as discharge printing. Over print will give a shade which is a total of print colour and dyed colour. Where printed as a discharge print on dischargeable ground the ground colour is completely discharged and a pure print colour will be seen. By this method the ground can be completely discharged to a white print also. In both cases the ground dyeing is done as per the demands of the print to be undertaken.

Priority pollutants: Organic and inorganic compounds selected on the basis of their known or suspected carcinogenicity, mutagenicity, teratogenicity or high acute toxicity. Many of these compound are found in waste water.

Probability: Statistical frequency of comparable events in Populations. The ambiguous term "probable" expresses the fact that there "appear" to be more arguments in favour of the prediction. The verification value is equal to zero in the individual case. We talk of mathematical probability because statistical probability evaluates the laws of large numbers (See **Stochastic**).

Processing Performance: A general term denoting the processing efficiency of a specific process or covering a group or series of process.

Processing sequence: In the context of a work analysis this consists of individual processes: Operating time, Production resources, Handling time factor. The processing sequence is also divided into a sequence of individual processes.

Processing sequence in textile finishing: Sequence of the individual working steps; these are often determined by the experience of the finisher. If the aims of the finishing process are specified in a logical manner based upon the profile of requirements for the finished material the processing sequence can be determined by factual arguments.

Processing time: (1) Part of the Handling time factor or basic time for all direct processing operations on the material. (2) Operating time, usable life, gel time, Working life; term for the period of time from applying a synthetic resin to when it is still just possible to process it, i.e. shortly before the onset of gelling. Important for condensation resins, because their viscosity falls sharply when the processing time is exceeded, whereupon they gel and then become unusable.

Procion: A trade name, originally belonging to Imperial Chemical Industries of Britain, now to Dystar of Germany, and used for their lines of reactive dyes, including the very popular MX family. Although it is often seen in art and hobby dyeing publications, the term 'Procion dye' is incorrect because it omits the name of the specific family such as MX (usually what is meant) or H or H-E. Dystar apparently no longer manufactures Procion MX, but others, not entitled to use the Procion name, do manufacture "MX" type dyes.

Producer twist: Small amounts of twist, usually ½ turn per inch or less, applied to yarns by the manufacturer to provide cohesion of filaments for further processing.

Producers risk: The probability of rejecting a lot when the process average is at the acceptable quality level (AQL).

Producer-textured yarns: Continuous filament yarns that have been bulked during manufacturing by the fibre producer. (Also see **Texturing**.)

Product based, in testing: In this definition quality is viewed as a quantifiable attribute based on the product's performance in fields such as durability or reliability. Because it is quantifiable then quality can be determined objectively.

Product development: is carried out by the textile designer (known as the designer in the cloth industry). After a creative phase, product development moves into the realisation phase; the pattern collection is developed.

Production: Manufacture, generation of products from raw materials or semi-finished goods, e.g. by textile finishing. In a broader sense, factors of production as commercial service provision: purchasing, transport, stocks, manufacture, administration, checking; without financing and sales.

Production control: (1) The logging of process data and its processing and provision to the sender for decision making and influencing the process as part of manufacturing control.

The automation devices used for production control consist of a central processing unit and a certain number of decentralized recording devices for manual entry or semiautomatic or automatic recording of data directly at work places, whereby the central processing unit incorporates at least a minicomputer. (2) Quality control.

Professionalism: Demonstrated by self-confidence and flair, capability and expertise, rational and systematic thought, creativity and judgement, sensitivity to environment, and to other professions, nationalities and cultures, appreciation of business in global context, and a lifetime commitment to personal education and professional advancement.

Profile reed: A reed the front of which is shaped to form a channel. The reed serves to separate the warp threads, to determine their spacing to guide the weft across the loom in air jet weaving, and also to beat up the weft.

Profiled fibres: This group, which includes hollow profiled fibres, consists of modified cross-section synthetic fibres, produced by spinning the melt from profiled jet orifices usually with a slit-shaped or pointed profile: Star-shaped, trilobal to multilobal, i.e. 3, 5, 6, 10 cornered/pointed, etc. (other cross-sectional variations, such as normal hollow fibre types, the straw type, flat shapes, large letter shapes, etc. are only associated in the broader sense).

Profiled hollow fibres: Hollow fibres based upon polyamide and polyester with additional cross-section profiling. Additional advantages compared to round cross-sections and also compact profile figures: further increased voluminosity (up to approx. 40%), greater stiffness, improved heat retention capacity (up to approx. 20%), reduced transparency of knit goods (up to 50%), double the durability for hosiery.

Profit: The monetary gain from being in business; the excess of revenues over outlays and expenses in a business enterprise.

Prograde mercerizing process: Process patented by Coats under German Patent for the mercerizing of sewing yarn on running threads (150–200 m/min). The yarn runs through liquid ammonia at –33°C for max. 10 s and the ammonia then removed by heat and steam, whilst the yarn is drawn. Ammonia consumption approx. 20–50% of material weight. The structural change of the cotton is similar to Mercerizing with caustic soda liquor. A marked increase in breaking strength and dye take-up is achieved. The lustre, on the other hand, remains the same.

Progressive: (Lat.: progressus = progress), increasing, continuing in stages; e.g. progressive feed of textiles, sequential wet/dry cycles, or progressing onwards from wash to wash, i.e. increasing shrinkage each time (in wool also as felting shrinkage).

Projectile: In projectile weaving machine the weft insertion is carried out by small clamp, called projectile. There are various projectile versions: made of

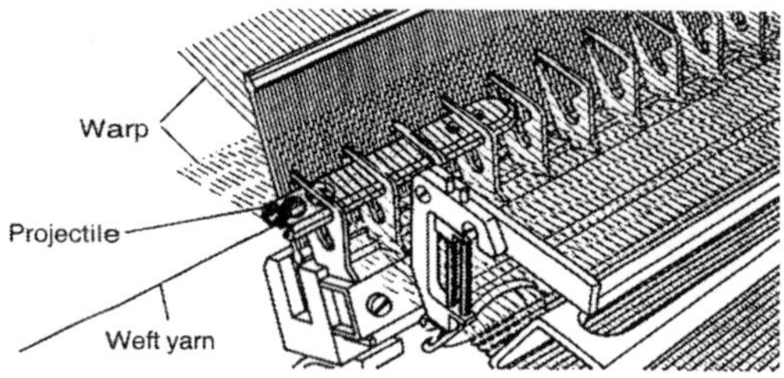

Weft insertion by projectile

steel, 9 cm long and 40 g heavy, with small section, as suitable for yarns of fine to medium count; made of steel, 9 cm long and 60 g heavy, with large cross-section which, thanks to their higher weight and to the larger clamping section of the gripper, are particularly suited for machines with high reed width or when for weft bulky yarns, as e.g. fancy yarns, are used. See **Projectile loom.**

Projectile loom: A shuttle less loom that uses small, bullet-like projectiles to carry the filling yarn through the shed. Fill is inserted from the same side of the loom for each pick. A tucked selvage is formed. (Also see **Weft insertion**.) In this weaving machine the weft insertion is carried out by small clamp projectiles, which number depends on the weaving width and which with their grippers take out the weft yarn from big cross-wound bobbins and insert it into the shed always in the same direction. The projectiles work in sequence, that is they are launched in succession. They run therefore one after the other, describing in the space a continuous, endless route, as if they would be stuck on a conveyor belt.

The first projectile takes and holds in its back the weft in form of a tail; then, pushed by the release of the projectile thrower, it passes through the shed and deposits the weft inside the warp; subsequently the projectile falls and is collected by a device which, by passing under the array of the warp threads, takes it at reduced speed back to the starting point. Here the projectile goes up to take up a new weft; meanwhile the other projectiles have run after each other making the same operation.

Proof: Resistant to a specified agency either by reason of the physical structure or the chemical non-reactivity of the textile, or arising from a treatment designed to impart the desired characteristics.

Note: (a) Proofing treatments should be defined by specified limits ascertained by tests, and the use of the term should be related to the limiting conditions. (b) The indiscriminate use of this term is deprecated, and its substitution by words such as "resistant", "retardant" or "repellent" in the appropriate context is recommended.

Proofed: Descriptive of material that has been treated to render it's resistant to a specified agency. Note: The efficacy of a proofing treatment is normally defined by a limit that is related to a specific test procedure, and the use of the term should be related to limiting conditions. The designation of materials as 'proofed' should indicate that the material conforms to definite standards.

Proofed gabardine: Crisp medium weight fabric, usually cotton or polyester, that is shower proof.

Proofed poplin: This is usually heavier weight than dress poplin but constructed in the same way. It may be polyester or cotton or polycotton. The fabric has been treated to make showerproof. Used for rain coats, anoraks.

Proteases: An enzyme used in the scouring and bleaching of Jute. It degrades the lignin and improves bleachability. Pectins, waxes, colour, residual seed coatings can be removed. These substances, inhibit the natural absorbency of the fibre and prevents dyeing, printing and other finishings. Earlier scouring was done by caustic soda to remove desirable natural lubricants and other materials, which causes heavy effluent disposal problem.

Protective clothing: Protective clothing is the one which gives protection against undesired harmful effects. It can be a particular hazard or a combined requirement of multiple hazards. The multiple hazards can be protected by a single cloth or different layers of cloths. Simplest example is fire protective clothing's.

Protective colloids: Particles in dispersion has a tendency to aggregate. Some Lyophilic Colloid, which has the property of delaying the aggregation of the particles of a hydrophobic dispersion within certain concentration ranges. This tendency permits the protective colloid to attach itself to emulsified or suspended particles (precipitated lime soap, dirt, dye, delustring pigments, paraffin, etc.), envelop them and, because they have the same (negative) charge, to mobilize mutually repellent forces, which act against coagulation and keep them in a finely dispersed suspension. Examples are gelatine, glue, albumin, starch (natural), Sulphite cellulose waste lye, oil sulphonate and fatty acid and protein condensation products and nonionogenic polyether alcohols (synthetic).

Protein: Biological polymers made up of amino acids All hair-based fibres (wool, mohair, etc.) are protein based. Silk is also a protein fibre. Acid dyes are the most common type used for protein fibres. Reactive dyes are used in some cases. Basic dyes can be used, but generally produce very poor fastness on protein fibres.

Protonation: Due to their small ionic radius and their charge, hydrogen ions are mobile if they are attracted to an opposing charge or by free electron pairs. This property of acids is due to the electrolytic dissociation of substances, and thus to the fact that substances in aqueous solution decompose into electrically charged ions. Ions migrate in the electric field and aways transport the same gram-equivalent quantity of electricity, namely 96 494 Coulombs.

Protrusion in fabrics: A yarn or a part of a yarn pulled or plucked from the surface.

Prud'homme black: See **Aniline black.**

Prunella: A fine worsted cloth that has largely gone out of use. It was twill weave and light in weight, made in plain colours only. Often used for children's clothes, but also for clergymen's wear.

Prunelle Batarde: Plain French serge, made with 8 or 10 ends and four or five picks in a repeat.

Prussian Binding: In England, made with silk face and cotton back and is twilled diagonally; is used as binding on waterproofs and coats.

Prussian Shawls: Fringed, twilled cotton shawls printed twilled cotton shawls printed in Oriental designs in England.

Prussian Velvet: A German pile fabric, made with cotton or linen warp and mohair filling, which forms the pile.

Prussiate black: See **Aniline black.**

Prussiate discharges: Oxidation discharges, of the type for para red on indigo background. However, it only has an historical importance.

PS: (1) Pakistan standard.

(2) Horse power. See **Watt.**

(3) Polystyrene.

PSA: Philippine Standards Association,

PSI: Pakistani Standards Institute

PST: Polystyrene fibres.

Pseudo- (Gk.), not genuine, imitation, apparent. Pseudo-hair: A "fake hair", for example the beard of a whale with its elastic-fixed intermeshed, scaly radial structure, which is similar to hair, but differs in its structure and purpose (filtering out plankton as food).

Psychrometer: A variety if hygrometer comprising a dry bulb temperature indicator and a wet bulb temperature indicator which is cooled to the wet bulb temperature by the spontaneous evaporation of moisture.

p/tex: Pond pro tex, obsolete unit of force in relation to fineness. SI unit is cN/ tex, (centinewton).

PTFE: Polytetrafluoroethylene fibres.

PTO fibres (Enkatherm): Enkatherm, a high temperature flame-proof fibre is based on poly (terephthaloyl oxalic-bis-amidrazone) (PTO). This fibre is produced by wet-spinning and then drawn.

PU: (1) Polyurethane, (2) Abbreviation for Liquor pick-up, quoted in %, related to the weight of the dry or wet textile entering the machine.

PUA: Polyurea fibres

Pua Hemp: Very strong bast fibre, resisting water; found in Assam, Burmah, Japan, etc., where it is used for fishing nets, ropes, twine, bags, etc.

Pucker: An undulation in the fabric, caused by wrong conditions during finishing, e.g. during compressive shrinking.

Pucker, in sewing, in garments: Generally unwanted material waviness along the line of stitching.

Puckering: Wavy, three dimensional effect typified by closely spaced wrinkles on either face of the fabric. Puckering may be due to (a) differential shrinkage of the component layers (in laminated and bonded fabric) or (b) difference in tension when the component layers are combined (c) selective lineal delamination.

Pudding Cloth: Plain woven bleached cotton fabric in England; used in cooking.

PUE: Polyurethane elastomer fibres (elastane).

Puke: A woollen fabric worn during the Middle Ages in England.

Pull: A group of fibres grasped by the forceps at one time and drawn from the specimen in the combs.

Pull-in: See **Jerk-in.**

Pull, in Zipper: A part connected to the zipper slider by which the slider is operated.

Pulled in filling: An extra thread dragged in the shed with the regular pick and extending only a part of the way across the cloth. See also **Dragged-in, Jerked-in, Lashed-in, Whipped-in.**

Pulled Wool: Removed from the skin of killed sheep through "painting" the fleshy part with lime.

Pulling Cotton: A test to determine the length, strength and uniformity of length of the fibres, in buying cotton for mills. A small quantity of cotton is pulled apart with two hands, the projecting long fibres separated from the rest and broken between two hands to test the strength.

Pulp: The end product of cooking wood chips, cotton, or some source of cellulose with water and appropriate chemicals. Used in the manufacture of cellulosic fibres, paper, and other cellulose-based products.

Pultrusion: The production of continuous lengths of fibre-reinforced advanced composites.

Fibre are fed as roving, fabric, or mat, or some combination of these, through a resin impregnation process, to a forming guide, then to a heated die to produce a specific shape, and finally to a puller where the structure is pulled through different forming and cooling stages.

Thermoplastic fibres may be substituted for the resin in the pultrusion process

Punched Felt: Wool stock of short staple but good felting property is attached to both sides of a burlap by barbed needles. The wool is then felted. Also called needle felt.

Punjums: A very tough fabric made in Visakhapatanam, Andhrapradesh, India as work wear. Dyed blue at Chennai, India, the punjums found their way to markets in Brazil, the Mediterranean and the West Indies as dress material for the slaves.

Puni: A puni is a tighter-than-normal rolag traditionally used with cotton.

Punpee: Chinese unfinished taffeta. See **Pong'ee.**

Punta Arenas: Very good, spongy and bulky wool, grown in the Terra del Fuego; used for knit goods.

Punto: Spanish and Italian for mesh or lace.

Punto Aquila: A medieval Italian linen lace, made in the Abruzzi mountains. It was a bobbin lace, somewhat similar to the English point, showing designs with raised edges, both the design and the net ground, being made at the same time of fine, bleached, hand spun linen yarn.

Pupa: A Stage in the life cycle of Silk worm.

PUR: Polyurethanes.

Purchase invoices: Invoices of items bought.

Purdah: (1) East Indian, blue and white striped cotton-; used for curtains, often painted by hand; (2) Closely woven, but very fine cotton or linen veil, worn by the women of in India.

Pure Finish: Finish without any weighting or dressing.

Pure linen: Fabric, in which the warp and weft must consist of 100% pure Linen (usually in plain weave) and only in this case may be marked with the "Pure Linen" quality symbol of a hand raised in oath.

Pure new wool: (South African), term for items made of pure Virgin wool.

Pure spectral colour: Optimal colour, which yields the lightest colour for a given Type of colour and the most saturated colour for a given Shade of colour. The colour valency associated with the shade is characterized by maximum brightness. According to Ostwald, every colour corresponds with the sum of pure spectral colour (P) (pure colour), ideal white (W) and ideal black (B) according to the equation P + W + B = 100. The so-called clear bright series is created from pure spectral colour + white and the so-called clear dark series is created from pure spectral colour + black (Chromaticity diagram).

Purl: (1) A picot or small loop that edges needlework, lace, or ribbon. Sometimes spelled pearl. (Also see PICOT.) (2) Coiled gold or silver thread used for embroidery.

Purl fabric: A fabric in which both back and face loops occur in some or all of the wales. The structure may be identified by the particular knitting sequence used, E.g. 1×1 Purl fabric: A fabric which a single course of back loops alternates with a single course of face loops. Other designations like 2x2 purl and 3×3 purl are possible. A variety of patterns can be produced mainly for ladies pullovers and cardigans.

Figured Purl

Purl gaiting: See **Gaiting.**

Purl knit fabric: The semi-circular appearance is dominant on both sides of a fabric. There are several types of purl fabrics such as 1 × 1, 5 × 3, 4 × 4 purl fabrics and the simplest one is the 1 × 1 purl fabric. It is thicker, softer and warmer than a plain knit fabric and it has more elasticity lengthwise than a plain knit fabric. It can be unroved from both ends of the fabric.

Purl Knitting: Framework knitting, the loops formed in reverse direction, both sides of the fabric looking alike.

Purl stitch: Purl stitch (reverse stitch): this is the technical reverse of the stitch where loop legs are below the neighbour stitch and the loop head is above the neighbour stitch. The reverse of a plain stitch is the purl stitch.

Purl, Facy: A general term used to describe patterned purl structures in which a design is formed from back and face loops.

Purple: Reddish violet Natural dyes (vat dye) made of the gland secretion of dead murex snails (Mediterranean and West Indian coasts. 8000 snails provide just 1 g dye). Active component: Dibromoindigo.

Purpurin: (1,2,4-trihydroxyanthraquinone), madder purple. Naturally occuring together with alizarin in the Madder (previous madder dyeing); provides reddish violet chrome lacquer.

Purse Silk: Thick silk thread, made smooth and soft; used for embroidery.

Purshed Velvet: Medieval term for velvet the pile of which was raised in a net pattern.

Purumu: Fine, silky bast fibre, yielded by the Sida carpinifolia in the Canary Islands; used for cordage locally.

Pushmina or Pushum: See **Pashim.**

Pussy willow: A plain weave fabric, characterized by the fine horizontal lines and made from nett silk yarns. The warp sett is much greater than weft sett.

Put: East Indian knotted rugs, made of fine white wool.

Puttee: A legging that looks like a surgical bandage.

Put-Up: Refers to the type, size, and yardage of the cone, king-tube or tube upon which thread is wound.

Putang: Narrow, very coarse homespun cotton cloth made in China.

Puttoo: (1) East Indian inferior wool cloth; used for shawls; (2) East Indian inferior goats' wool; (3) French dress goods and coating of thick but light texture, having a long nap which is of a lighter colour than the body.

Puy Lace: French bobbin lace, made in a great variety, mostly in black and white and also in silk guipures.

Puyuenchow: Another native Chinese term for pongee, made in widths of 30 and 32 inches and boiled out after weaving.

PVA: Polyvinyl alcohol fibres.

PVA+: Polyvinyl acetate fibres.

PVAL: Polyvinyl alcohol fibres.

PVC: Its full name is Polyvinyl Chloride and it is a chemically poroduced thermoplastic material. Most fabrics referred to as PVC are base fabrics, such as plain knitting or plain or printed cotton, that have been sprayed with coloured or colourless polyvinyl chloride. Some are stiff and heavy, some are softer. Used for aprons, protective wear, raincoats, hats, curtains etc.

PVC+: Post-chlorinated Polyvinyl chloride fibres.

PVM: Polyvinyl chloride copolymer fibres.

P.W. Osnaburg: Part Waste Osnaburg. See **Osnaburg**.

Pyjama cloth: A hard wearing cotton (once silk) fabric with a close warp and woven in stripes of a varying widths and colours. The fabric is often brushed on the right side, highly inflammable.

Pyranometer: (also incorrectly called Pyrheliometer), device for the measurement of sunshine in the form of heat generated in Langley units (gcal/ cm^2). Often used in meteorology. Not suitable for use as a substitute for Blue scale in light fastness testing.

Pyranthrone: These are anthraquinonoid dyes with a pyrene structure. Example is a dye which can be made by heating 2,2′-dimethyl-1,1′-bianthraquinone with alcoholic potassium hydroxide – Indanthrene gold orange G.

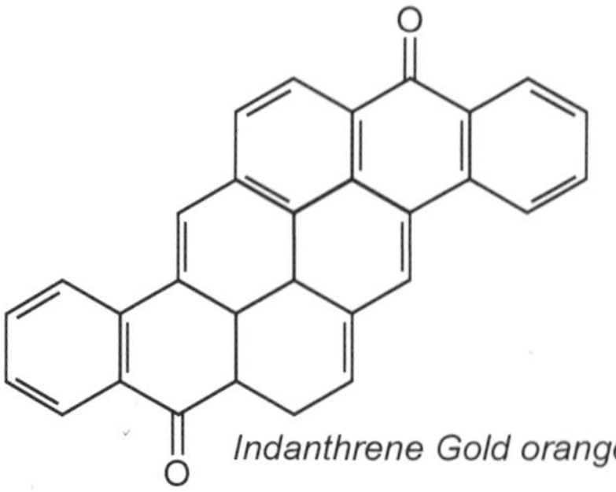

Indanthrene Gold orange G

Pyrheliometer: Light sensitive measuring cell for illumination equipment, a thermoelement photometer consisting of two concentric rings with a black or white surface (hot solder points of a gold-palladium and platinum- rhodium thermoelement battery, attached to the black underside of the ring).

Pyrenean: Wool Coarse Spanish wool; used for carpets.

Pyro-: As a prefix in the designation of inorganic compounds created from water-containing ortho compounds by heating and the emission of water.

Example: Pyrophosphoric acid $H_4P_2O_7$ (from $P_2O_5 + 2H_2O$) by the emission of 1 molecule of water from two molecules of ortho acid. Salts Pyrophosphates.

Pyrocellulose: Cellulose modification that has been damaged due to excessive heating. Always contains aldehyde and carboxyl groups. Fibre damage (cellulosic fibres), tests for.

Pyroset: Original trademark of American Cyanamide Company for a durable fire retardant finish.

Pyrolysis: A chemical change brought about by the action of heat, usually in the absence of a reactive medium. Complex chemical molecules are reduced to simpler chemical units as a result of pyrolysis.

Pyrometer: Measuring instrument for high temperatures

Pyrovatex CP: N-Methyloldimethyl phosphonopropioamide: Pyrovatex CP (Ciba) provides a method of attaching phosphorus to cellulose making use of N-methylol reactivity with cellulose. It is applied with a methylolated melamine resin using a phosphoric acid catalyst by a pad-dry-cure process. The high nitrogen content of melamine provides synergistic activity to the phosphorus of the flame retardant.

Q

QD (in zipper): Quick disassembly zipper.

Qiviut: Down undercoat fibre from the musk ox.

QR(in zipper): Quick release zipper.

QS-process: Squeezing/suction technique in accordance with the principle of Capillary dehydration developed by Pfersee. This is one of the low wet pick-up techniques; liquor reduction in the goods through capillary liquor exchange. Material which has been conventionally padded and squeezed in the first padder nip to the usual liquor absorption is brought together with the same, but dry fabric, and pressed against it in the 2nd nip. This results in removal of the excess liquor (adhering liquor), leading to a reduction in the liquor uptake by a further 15–40%. Liquor treatment squeezing/suction methods on cotton approximately 40%. *Advantages* – Low wet pick-up. Principally used in resin finishing.

Quadrant scales: Yarn scales, for determining the weight of a certain reel length, by which the yarn count is determined.

Quadrille: French for checked (small chequered pattern). Now a general term for an overall design of small squares, woven or printed. The illustration is Satin 'quadrille'.

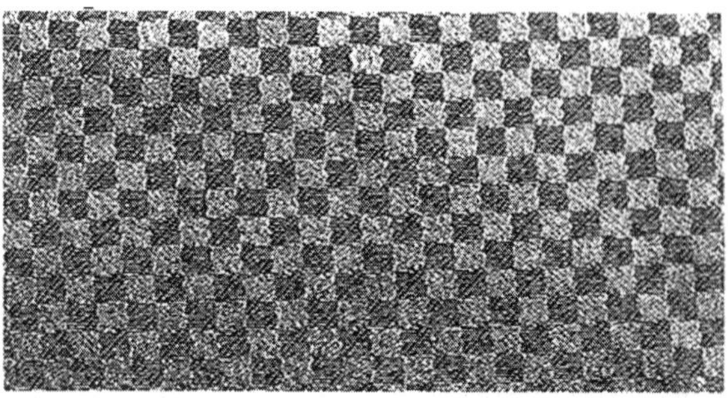

Quadripolymer: A polymer made from four distinct monomers.

Quality: Quality is the sum total of all characteristics and features of a product or an activity which relate to their suitability for fulfilling set requirements.

Quality assurance: This is an essential measure for controlling production, to keep complaints about products within limits, to detect faults and eliminate them. In the textile industry, however, thorough quality assurance comes into conflict with ever shorter batch lengths and project times, and with ever smaller profits from finishing. In order to be able to produce economically in terms of time and costs, in the last few years, many firms have invested in the field of quality assurance, under the motto "quality is profitable".

Quality binding: A strong, twilled, wide tape, made of coarse worsted; used for binding carpets.

Quality assurance: The term quality assurance covers all the processes within a company that contribute to the production of a quality product. It does not just cover the final testing of the product before shipment and it is not solely concerned with testing the product.

Quality control: The term quality control used by itself has a very narrow meaning and it is generally taken to mean the maintenance of product quality by the regular inspection of critical stages in the manufacturing process. The inspection is carried out on a limited number of items selected as being representative of the current production and the results recorded chronologically on control charts.

Quality of conformance: This is a measure of the fidelity with which the product taken at the point of acceptance conforms to the above design. This is the area that is usually thought of as the province of quality assurance. However, the overall quality depends also on the design as performance cannot be introduced at this stage, if it is not present in the original design.

Quality of customer service: This is a measure of factors such as the speed of response to orders, the response to customer returns and complaints, the speed and quality of installation and servicing and the initial availability of the product.

Quality of design: This can be considered as the value inherent in the design. It is a measure of the excellence of the design in relation to the customer's requirements. The production of a quality product starts with its design. The initial meeting of the customer's requirements and the continued functioning of the product throughout its lifetime depend on choice of materials, construction and processes.

Quality of use: This is a measure of the extent to which the user is able to secure continuity of use from the product. Provided material is being produced which conforms to specification, the length of time the product lasts in use depends on the original design.

Quality-related abbreviations:
TQM = Total Quality Management
CAQ = Computer Aided Quality Assurance
QA = Quality Assurance
PSG = Problem-Solving Groups
FMEA = Failure Mode and Effects Analysis
SQC = Statistical Quality Control
SPC = Statistical Production Control
LIMS = Laboratory Information and Management System

Quality sample: Representative sample for the final quality of the goods.

Quality standards: Requirement profiles.

Quarantain: Fine French woollen cloth, originally of 4,000 warp ends.

Quart: Measure of volume; in the United States of America, a liquid quart is 32 fluid ounces or approximately 946 millilitres; in Canada and Britain, a quart is 40 fluid ounces, or approximately 1137 millilitres.

Quarter: (1) English measure for carpet width 1 quarter = 9 inches = approximately 23 cm; (2) German carpet measure, now obsolete, equivalent to 1/4 cubit, or approximately 17 cm.

Quarter-blood wool: One of the grades in the standards for wool.

Quatrefils: Very strong French sail cloth, made with four-ply warp.

Quartz fibre: Pure silica that has been melted and drawn into glass-like fibres. It is used for heat resistance and high dielectric strength. Quartz is a fusion product of silicon oxide which does not show any crystal arrangement when cooled slowly (lattice silicate). The way in which these are obtained corresponds to that for glass fibres. Quartz fibres are stable up to a temperature of 1200–1300 °C, and the melting point lies at around 1700 °C. Resistance to acids is very good, and resistance to alkaline solutions is good.

Quaternary(from the Latin quattuor = 'four'): Quadruple substitution, e.g. by alkyl groups, such as with the most important quaternary ammonium compounds and phosphonium compounds, where the alkyl complex occurs as a cation.

Quaternary ammonium salts(softeners): Quaternary ammonium salts are extremely important fatty acid derivatives. The quat's cationic charge is permanent, being maintained at all pHs. In addition to imparting softness, quats reduce the static charge on synthetic fabrics and inhibit the growth of bacteria. Quats are therefore used as anti-stats and germicides as well as softeners. Cationics containing two C18 fatty tails attached to the nitrogen impart very soft, fluffy hand to textile products. Cationics based on di-tallow amine are used as home laundry rinse-added and dryer-added fabric softeners as well as mill applied softeners.

Quaternization of monochlorotriazine reactive dyes: Normally, monochlorotriazine reactive (MCT) dyes react with cellulose only under alkaline conditions. However if one adds tertiary amines to a neutral printing paste, the reaction can proceed during steaming even below pH7.

Queen: Commercial variety of upland cotton from Arkansas; the staple measuring 22–25 millimetres; the yield is 34–36%.

Queen's cord: The fabric width changes very slightly when leaving the knitting zone. It is a very stable and rigid fabric.

Queen stitch: Used in embroidery, consists of a square inside of another with parallel sides.

Quench: (1) A box filled with water into which fabric is run after singeing to prevent sparks or fires; (2) See **Cabiner** (Also see **Quenching**).

Quench spacer: The "quiet" zone below the spinneret in which thereis no quench airflow. Quench spacer distance is important in controlling fibre orientation and birefringence.

Quenching: The cooling of fibre filaments after extrusion by carefully controlled airflow. See **Crossflow quench**, **Inflow quench** *and* **Outflow quench**.

Quercitron: A bright yellow mordant dyestuff, yielded by the bark of an oak (*Quercustinctoria*) in Southern United States of America.

Quetsch: The nip rollers of a padding machine.

Quick copper detection in effluent or water: 50 ml of the water to be tested is added to 2 ml of a light coloured oleic acid. After the mixture has been thoroughly shaken, the creaming oil layer reveals a weak to strong green discoloration in the presence of copper(copper oleate). The presence of more than 1.2 mg/l of iron upsets this colour reaction.

Quick dyeing: Rapid dyeing process for synthetic fibre material. The reduction in time is enabled by rapid heating of the dyeing liquor and by a rapid cooling phase. It is predominantly polyamide fibres which are dyed by this method.

Quick lime (calcium oxide): White to grey in colour, formed through heating and roasting of calcium carbonate (chalk, limestone, marble) $CaCO_3$. It is fireresistant and melts only at 3000 °C. It is used for furnace lining, etc.

Quill: A light, tapered tube of wood, metal, paper, or plastic on which the filling yarn is wound for use in the shuttle during weaving.

Quill embroidery: See **Canadian embroidery**.

Quill point(in feathers): The section of quill extending beyond the section of barb attachment.

Quill, shaft(in feathers): The section of quill from which the barbs emanate.

Quill(in feathers): The stem or central shaft.

Quilling: The process of winding filling yarns onto filling bobbins, or quills, in preparation for use in the shuttle for weaving.

Quilot: The fine, white and glossy inner fibre, yielded by the abaca plant: used for very fine fabrics by the natives, but only little of it is found on the market.

Quilt: Piqué.

Quilt lining: Lining material from quilted non-woven fleece (padding), wadding or foam material on lining material, duchesse, serge, etc.

Quilt wadding: Goods of the Maliwatt fabrics; for wadding, lining purposes, layer base for tensioning purposes, insulating material, etc.; 200–425 g/m.

Quilting: Plain or printed cotton, viscose or nylon fabric mounted on to a layer of 2 oz wadding (usually polyester) and held with rows of machining in diamond pattern or parallel rows. A backing usually added before quilting-this may be a thin layer of plain coloured knitted nylon or another fabric to contrast with or match the outer material. This is referred to as double-sided quilting. All quilting is warm but light. It is used for making robes, dressing gowns, bed jackets, anoraks, children's clothes, snow suits, jackets, boleros, belts and also small pieces of quilting can be used for collars, pockets, yokes and lining hoods.

Quina: A modified nylon fibre with a trilobal cross-section, is a polycondensate of diamine diphenyl methane and decane dicarboxylic acid. It is distinguished by its silk-like handle and regarded as synthetic substitute for pure silk.

Quintain: Very fine lawn, made in Quintain, France; used for embroideries and laces.

Quintes: The best grade of French linen.

Quomotanetu: Very strong bast fibre, yielded by a species of the Asclepiadaceae in South Africa.

R

RA: Ramie

Ra: Ramie

R.M.M.: Relative molecular mass; proper modern term instead of *molecular weight.* 'Relative' is used because the mass is in proportion to one twelfth the mass as carbon-12.

Rabanna: A coarse fabric made of raffia fibres in Madagascar.

Rabbit hair: (1) The fur from the common or wild rabbit. Used for hats and as substitute of vicuna, for shawls, etc.

(2) Long rabbit hair is used in combination with other fibres such as wool, to add softness and interest to a fabric. The percentage of rabbit hair included is normally specified on the label.

Rabo de Leon: Very strong leaf fibre, similar to Manila hemp, yielded by a species of the Sanseviera in the Philippines; used for cloth by the natives.

Raccroc Stitch: Used in joining the separately executed pieces of lace together into a larger piece.

Race: In pile fabrics the narrow space between two adjoining rows of pile

Race-board: The flat part of a sley in front of the reed.

Raschel Knit: A warp knitted fabric in which the resulting knit fabric resembles hand crocheted fabrics, lace fabrics, and nettings. Raschel warp knits contain inlaid connecting yarns in addition to columns of knit stitches.

Raschel Lace: Raschel Lace fabrics are often made on a base of net fabric with a pattern formrd from inly yarns. They are used for foundation and lingerie, bridal and formal wear, and as trimmings. See **Lace, raschel**.

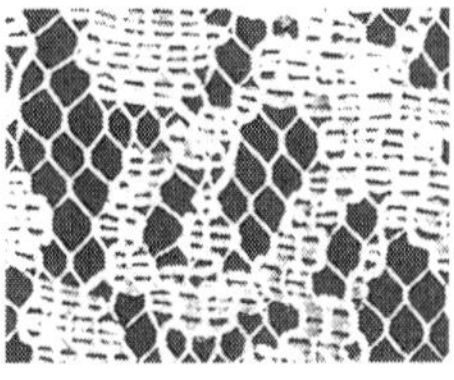

Raschel Lace

Raschel net: Raschel lace fabrics are often made on a base of net fabric with a pattern formed from inlay yarns. They are used for foundation and lingerie, bridal and formal wear, and as trimmings.

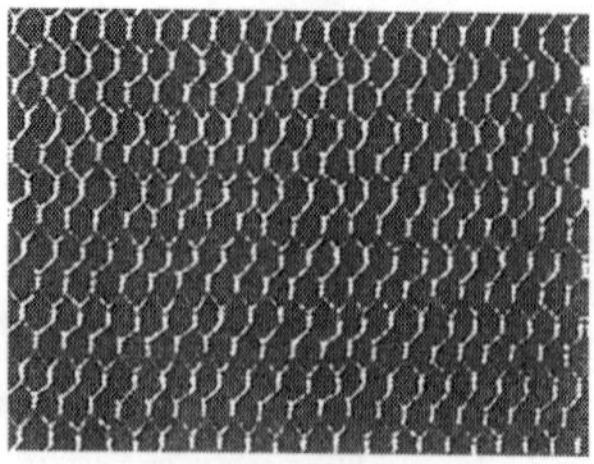

Raschel net

Rack: A warp-knitting measure consisting of 480 courses. Tricot fabric quality is judged by the number of inches per rack.

Rack, in warp knitting: A unit of length measure consisting of 480 courses. This is the working cycle of 480 knitted courses of warp knitting.

Racked Stitch: A knitting stitch that produces a herringbone effect with a ribbed back. It is employed in sweaters for decorative purposes or to form the edge of garments. The racked stitch is a variation of the half-cardigan stitch; it is created when one set of needles is displaced in relation to the other set.

Racking: A term referring to the side-to-side movement of the needles of the needle bed of a knitting machine. Racking results in inclined stitches and reduced elasticity.

Raddle: Raddle is a board at least as long as the width of the warp with evenly spaced pegs or nails (1⁄4"–1" apart). Groups of warp ends are placed in each raddle space and the warp is wound on the warp beam with the raddle secured on the back beam. The width of the warp in the raddle and the number of ends in each space determine the density and width of the warp on the warp beam. (Raddles are used for warping methods in which the warp is wound on the warp beam before being threaded through the heddles.)

Raddle cross: Raddle cross is formed by groups of warp ends on the opposite end of the warping board or reel from the threading cross (which is formed of individual ends). The number of ends in each group is equal to the number of ends to be placed in each dent of the raddle.

Radial balance, in designing: This is when design elements radiate from a central point, as the spokes of a wheel or in the natural form of a daisy.

Radial Dyeing Machine: Packing system dyeing machine with circulating liquor. The material block is lifted after dyeing on the material carrying plate and is spun in a conventional centrifuge, the upper rim of which can be removed. This kind of special centrifuge is no longer in use, since a normal centrifuge can be used after a batch has been dyed in the radial dyeing machine. This includes other batches such as piece goods, hanks, bobbins. Radial dyeing machines are also used for dyeing wool, cotton, viscose, polyamide, acetate, polyacryl nitrile, hair and rags.

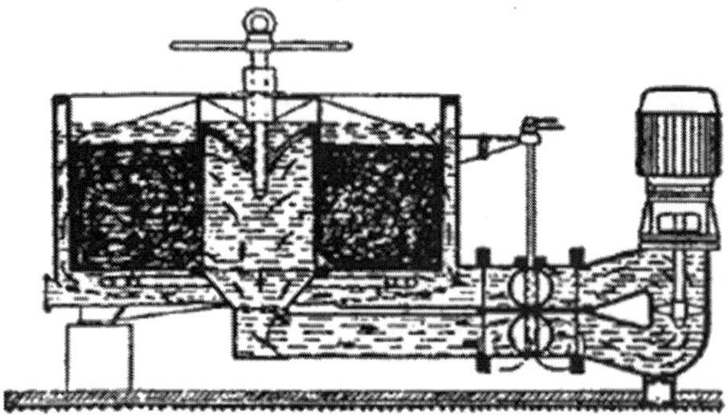

A radial dyeing machine

Radian: The plane angle between two radii of a circle which intersects the circumference of the circle making an arc equal in length to the radius.

Static friction: Friction developed between two touching bodies at the time one body starts to move relative to another.

Radiant Energy: A form of energy consisting of the electromagnetic spectrum, which travels at 299,792 kilometers/second (186,206 miles/second) through a vacuum, and more slowly in denser media (air, water, glass, etc.). The nature of radiant energy is described by its wavelength or frequency, although it also behaves as distinct quanta ("corpuscular theory"). The various types of energy may be transformed into other forms of energy (electrical, chemical, mechanical, atomic, thermal, radiant), but the energy itself cannot be destroyed.

Radiant flux density: Rate of flow of the radiant energy past the material.

Radiant Power: Energy per unit time emitted, transferred, or received as radiation.

Radiation singeing: Indirect Singeing, without contact between material and flames. The effect is achieved by re-radiation from bricks heated to high temperatures.

Radicals: (Lat: radix = root) groups of atoms with free valency that cannot exist under normal conditions (unbound electrons). For example, if triphenyl methyl chloride is treated with finely-distributed silver in benzene in a vacuum, a yellowy solution is produced. If this solution is treated in air, the yellow colour disappears.

Radio: Plain woven lustrous lightweight, sheer French silk dress goods, comes printed or dyed.

Radio-frequency drying: Use of radio-frequency electromagnetic radiation for drying textiles. The application of RF to wet goods results in the selective heating of the water, which has a partial polarity, because the molecule must do work to align in the RF field causing heat generation within the water droplets. Non-polar materials, i.e., fabrics, are unaffected. RF drying in very uniform and energy efficient when airflow patterns through the dryer are properly designed and controlled.

Radio Frequency Technology: This technology is based on the peculiar characteristics of dielectric (insulating) materials such as water. Water molecules are subject to polarization when put in to an electric field, this polarization corresponds to a deformation and reorientation of the molecules in accordance with the field itself. In an alternated (oscillating) electric field water molecules invert their own polarization at each single inversion of the field. Furthermore, water molecules invert their polarization at the same frequency of that of the electric field they are subjected to, dissipating energy due to their inter molecular friction, this energy is called dielectric loss. The intermolecular friction generates heat which increases water temperature, discharging steam in the environment. The more frequent are the oscillations, the higher will be the energy absorbed by the material. This is why the RF driers use upto 27 millions inversions per second.

Radiometer: An instrument used to measure radiant energy.

Radium: A plain weave lustrous fabric that has the draping quality of Crepe, but the crispness of Taffeta. The fibres used include silk, viscose, acetate, triacetate. Used for lingerie, robes, blouses, and as lining fabric.

Radzimir: A very fine, lustrous, stout silk dress fabric in England, made in plain weave but weft ribs; it is usually dyed black and used for mourning.

Raffia: A fibre obtained from the leaves of the raffia palm, Raphiaruffia (South Africa, Madagascar, South America). Flat, brown/yellow bast strands (for gardener bast and as braided bast for arts and crafts), easily divided lengthways to form solid fibre elements (for string and coarse yarns).

Raffia bast: See **Raffia.**

Rag: Rug Made of strong cotton warp and stripes of various coloured rags, forming patterns.

Rags: (1) Are classified: Carpets, skirting (women's dress goods and men's lining), merino (very fine women's dress goods), flannel, linsey (all other wool fabrics containing cotton), serge (braids, tresses, etc), new (new clippings from tailors), rubbish (the lowest grades of rags). Each of this is sorted again according to colours. Ragusa Lace (Dalmatian needle-point lace of early origin, Similar to the Point Venise. At the present patterns formed of picot edge silk or gold tape are united with brides to form laces.

(2) Term for the remains of worn, more or less worn-out items of clothing from households.

Raglan: A loose overcoat, the sleeves of which are continued over the shoulders up to the collar. Named after Lord Raglan, who devised a similar garment for the protection of his soldiers during the Crimean war, the idea being to increase their fighting efficiency by the greater freedom of action afforded by the peculiar shoulder.

Raglan sleeve: A sleeve with armhole line extending from the front and back scye to the neck point so that the shoulder section is joined to the sleeve crown, eliminating the conventional shoulder and sleeve head seams.

Railroad Canvas: Black or white and considerably sized cotton or linen fabric made with open texture; used for embroidery.

Rails: The metal bars on which the spindles of a down twister are mounted.

Railway Stitch: Similar to leviathan stitch.

Rainbow Effect: Produced in calico printing by the different coloured fields shaded into each other at the edges. Raincloth Twilled or satin weave light cotton, wool or silk cloth, dyed in the piece and made waterproof by rubber, oils, etc. used for raincoats. Rainproof same as waterproof.

Rain-Grown-Cotton: Cotton cultivated under natural climatic conditions (Irrigated cotton).

Rainwater harvesting: The collection and use of rainwater by a household for direct use in toilet flushing, cleaning purposes, garden watering, etc., but not normally drinking water.

Rain resistance: Water spotting fastness.

Raincoat fabric: Water proof fabric, prepared for making rain coats.

Raised check: Fabric figured with extra warp on a plain ground.

Raised Colours: In textile printing various dyestuffs which are treated after printing, in a bath which will fix or develop the colour

Raised Embroidery: Made with padded or raised patterns over a flat foundation.

Raised Fabric: Same as napped goods.

Raised finish: (brushed finish, napped finish) Brushed fabric.

Raised jersey: A term used to describe a medium-weight knit fabric of fairly open texture, that has been brushed on right side to add warmth and give an attractive fuzzy effect. Often made in random colour coordinating yarns. The fibre is usually acrylic. Used for sift suits, dresses, coats.

Raised resist print: Printing of pigment white and/or pigment dyes. Subsequent drying is followed by friction texturizing, condensing and raising. Produces a raised pile in non-printed areas, whereas printed areas are not raised.

Raised Stitch: In Berlin work, is called also velvet stitch and is a variety of the plush stitch; it is suitable for raised wool work. The loops are being cut and brushed, imitating the pile of the velvet.

Raised surface fabric, in textile fabrics: A fabric which has an intentionally lifted fibres or yarns such as pile, napped, tufted, flocked, or similar surfaces.

Raised Velvet: Having the pattern formed by a higher pile than that covering the ground. See **Pile over pile.** Raised work in hand-made laces the raised edge of the sprigs.

Raising: (Napping, Sueding) The production of a layer of protruding fibres on the surface of fabrics by brushing, teazling or rubbing. *Note*: The fabric, in open width, is passed between rotating rollers covered with teazles, fine wires, carborundum, etc., whereby the surface fibres are lifted or broken to give the required effect.

Raising teasels: Natural teasels on thistle heads that are used for Raising. The thistle heads can be inserted into rods (rod teasels) or attached to rotating rollers (roller teasels) that themselves rotate on a cylinder.

Rajah: Made of Silk, rayon in plain- warp yarn is 4 thread organzine - filling is heavier. Made from a tussah silk or certain silk wastes. It belongs to the pongee family of silks. Made from irregular yarns, so has slubs and irregularities but thicker than shantung. It is rather compact and strong. Has a pebble-like feel and appearance. Comes in all colours as well as natural ecru shades, but often warp and filling are different colours (irridescent effect). A soft, strong, plain weave silk fabric with rough texture. It was first made, not in the East, but in the US. Tussah silk was normally used, but it is now made from acetate, triacetate, nylon, polyester. It is very similar in appearance and weight to Pongee.

Ram's Wool: Shorn from male sheep; it is stronger than other wool.

Ramage: French for branching patterns.

Ramina: Braiding made from Ramie for hats, bags, arts and crafts, etc.

Rambouillet: French merino wool of lustrous, strong, long staple.

Random Tumble Pilling Tester: Device for testing Pilling behaviour, consisting of six round containers lined with cork or plastic. Three samples of a type of material whose edges have been bonded with glue are placed in each round container, together with 25 mg of cotton lint. The samples are tumbled around for 1 hour by a paddle rotating in the round container (1200 rpm). The cotton lint is used to make the pilling effect visible. Visual evaluation. Graded from 1 (extremely strong pilling) to 5 (no pilling).

Rameses: Commercial variety of early maturing upland cotton, the staple measuring up to 26 millimeters; the yield is 32-33 percent.

Ramie: A strong vegetable fibre, white silky colour, yielded by the Boehmeria plant of the nettle family also known as China Grass, growing in China, America, etc. It is difficult to decorticate. Method of obtaining fibre: manually or automatically (Ramie decorticator) from freshly cut stems and by cottonizing. The Ramie stem structure is comparable to that of the Flax stem structure. Raw fibre: pure white, fine, silky gloss. Individual fibres: 50–260 mm long, ∅ 20–100 mm. Owing to its porousness it is used for underwear. Ramie is expensive to produce and is usually mixed with other fibres, such as wool. Fabrics resemble Lutex.

Rampoor Chuddah: Very fine and soft East Indian twilled woollen dress goods in red, white and gray colours.

Rampur chadars: Woollen hawls made in Ludhiana, India. The name Rampur chadar is after the wool that came from Rampur Bushair in Himachalpradesh, India. Once the city had 2000 shawl weavers and 500 wool manufacturers, even though the place now is famous for the hosiery manufacture. Although shawls were not patterned, they were made in five popular colours ivory white, scarlet, turquoise, blue and grey with slightly decorated ends.

Ran: (1) Rope measure, 20 yards; (2) Silky, long fibre, yielded by the Malachra capitata n tropical Africa and America; used as substitute for jute.

Random tumble pilling tester: Device for testing Pilling behaviour, consisting of six round containers lined with cork or plastic. Three samples of a type of material whose edges have been bonded with glue are placed in each round container, together with 25 mg of cotton lint. The samples are tumbled around for 1 hour by a paddle rotating in the round container (1200 rpm). The cotton lint is used to make the pilling effect visible. Visual evaluation. Graded from 1 (extremely strong pilling) to 5 (no pilling).

Random winding: A method of winding cones and cheeses in which the number of winds per double traverse is constant and the wind decreases as the diameter increases.

Random-laid web: A term sometimes used to describe a web or batt produced by air laying.

Random-sheared carpet: A pile carpet with a textured face produced by shearing some of the loops and leaving others intact.

Random Yarn: Fancy filling yarn, having a strand wound around a different coloured core.

Range: A group of fabrics (or products) designed, developed, and edited to be shown and sold to the market each season.

Range planning: Managing and controlling events to put together a range.

Range Wool: Wool produced under range conditions in the West or the Southwest. With the exception of Texas and California wools, it is usually classified as territory wool.

Rangoon; East Indian cotton, having a short and weak staple of dull and stained brown colour; contains much leaf.

Rankine scale (R): Degrees Rankine is a normal unit of temperature in England and the USA: start of the scale (like Kelvin) at absolute zero is 0°R, freezing point 491.67°R, vapour point 671.67°R. Fundamental difference between freezing point/vapour point is 180°R, which is equivalent to 180°F; 1°R is therefore equal to 1°F:

1 Rankine unit = 1 Fahrenheit unit
= 5/9 Celsius unit
= 5/9 Kelvin unit.

Rap: A skein, containing 120 yards of yarn.

Rapatelle: Open-work horsehair cloth, used for sifting flour.

Rapid ager: Most common continuous steamer especially for brief steaming (1–10 min) and high outputs in roller printing for vat and developing dyes, also for discharge printing. There is a row of driven guide rollers above and below. The fabric runs in a zigzag arrangement, alternately downward is and upwards, each facing the rollers with the reverse and printed side of the fabric, therefore cannot be used as a screen printing steamer. The oven roof is double-walled and steam heated (protection against spotting), with an exhaust duct or fan. Steam temperature usually 101–103°C. The material is preheated and the air is evacuated (particularly necessary for vat printing) in a heated pre-chamber.

Rapid bleach process: All processes, in which the material to be bleached is exposed to the bleaching bath for less than 10 min.

Rapid cleaning of padders: The automatic rapid cleaning of padders makes it is possible to economically dye even small dyeing batches on a continuous dyeing plant or in accordance with the pad batch process. The squeezing and guide rollers are automatically cleaned according to a pre-selected program when the batches are changed. The padding trough is automatically emptied, cleaned and filled with new liquor. The cleaning time is less than 8 min

Rapid cross linking: See ***Shock-cure process***.

Rapid desizing : Continuous Desizing by a large quantity of desizing agent (usually bacterial amylases) and subsequent steaming, whereby the starch is decomposed at an accelerated rate.

Rapid drier: In the rapid drier, in contrast to the Pressure boiler, heated fresh air is blown through cross wound yarn packages and discharged as moist exhaust air and is not reused.

Rapid dyeing machine: They differ from conventional dyeing machines because of the use of high performance pumps (increase in liquor throughput to approx. 150 l/min/kg), heaters (heating-up rate of approx. 15°C/min), heat exchangers, shock dyeing devices, short liquors up to 1 : 5 and less (e.g. due to dividers), change of liquor direction, etc. On the other hand there is an increase in the energy used for heating (HT = although the temperature is slightly higher, time is reduced) and pumps for higher liquor throughput (compensated by shorter pump running time).

Rapid dyeing process Term for shortened dyeing processes. Thanks to reduced treatment times, rapid dyeing processes make it possible to notably reduce finishing costs. However, it is only possible if operating reliability and reproducibility are retained.

Rapid fill: Rapid transfer of dye liquors into a dyeing machine, normally for reducing the dyeing time.

Rapid fixation dyeing units: Small continuous plant for universal application or as washing/rinsing machine for pre-treatment (desizing, bleaching). The steamer can be flooded and used as a washing plant. Counter current principle; extremely low water consumption, Modular structure.

Rapid mixers, rapid stirrers: are high speed mixers. They differ due to the size and shape of the stirrer assembly (mixing heads), upon which the fine distribution depends: stationary, portable or transportable. In textile finishing above all for the application of printing thickeners and dye solutions.

Rapid steam generator: Boiler with up to 1000 kg/h performance.

Rapid wetting Agents: Highly powerful wetting agents like dioctyl sulpho-succinate, sulphonated modified ester. It is anionic stable over pH of 1 – 10, and hardness upto 300ppm. Used as an auxiliary in wetting the fabric and as a dissolving assistant.

Rapier loom: Looms in which either a double or single rapier (thin metallic shaft with a yarn gripping device) carries the filament through the shed. In a single rapier machine, the yarn is carried completely across the fabric by the rapier. In the double machine, the yarn is passed from one rapier to the other in the middle of the shed. The rapier weaving machines are the most flexible machines on the market. Their application range covers a wide variety

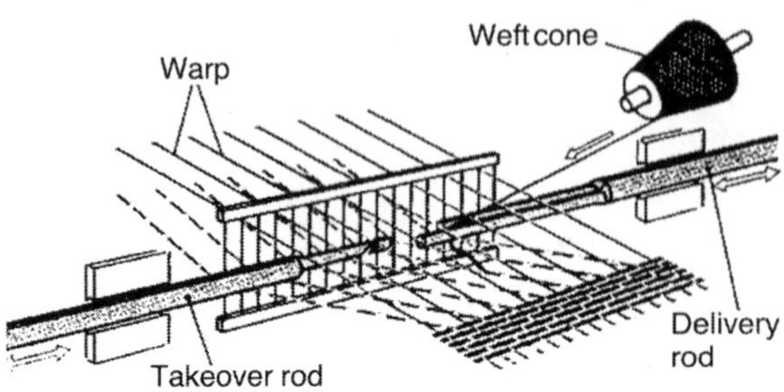

Weft insertion by rapier

of fabric styles. Their present **weaving speed of about 600-700 strokes/min** is the result of the use of a state-of-the-art construction technique, characterized by the use of gear sets without plays and by minimum vibrations of the reed, the slay and the heald frames.

Ras de Cypre: French silk dress goods with cross ribs; comes usually in black.

Ras de Florence: Fine woollen men's wear, woven in fancy coloured twills.

Raso: Italian for satin.

Raschel batt-on-base carpet: Textile floor covering. Instead of pile yarns a fibre nonwoven is used in combination with auxiliary yarn using the raschel technique to form a bouclé carpet, which is then shorn to produce Velvet pile carpets.

Raschel carpet: (knit-woven carpet), a velour carpet manufactured on a raschel machine, where the ground weave is woven at the same time. The pile threads, which are loose in the base material to start with, are fixed in place using reverse-sided coating (latex, artificial resin). Similar appearance to tufted carpets. Usage: e.g. stair carpet material.

Raschel knit: Raschel is the name of a machine invented to produce this type of fabric. The machine uses latch needles set in a vertical plane and produces a wide variety of fabrics, particularly heavier complex structures. Fabrics are not conventional knit construction but appears to be partly woven in texture. These fabrics are often printed, and some are very transparent. Normally made in polyester this fabric is used for blouses, dresses, curtains, bedspreads, depending on the design.

Raschel knitted fabric: Warp-knitted fabrics made on the raschel machine (from the French: Rachel) for curtains (raschel curtains), lace, carpets, tulle, fashion materials (outerwear, underwear), etc.

Raschel knitting: See **Raschel knitted fabric.**

Raschel tulle: Net-like knitted fabric that is manufactured with two retracted guide rails in accordance with a pattern. Each warp thread has a weft thread assigned to it. The effect of the honeycomb-like openings is not shown to its best advantage until finishing has taken place.

Rateen: English all-wool lining serge, similar to frieze.

Rat-tail cord: A small, filled, tubular cord of braided or woven construction, about 3mm in diameter. The woven version is usually of tubular satin weave.

Ratch, spinning: The distance between the nip of the front rollers and that of the back (prior set of) rollers is referred to as the ratch and needs to be carefully selected in relation to the fibre length characteristics of the material being processed. Too low (or short) a ratch leads to many fibres being simultaneously gripped by both sets of rollers and possibly broken, while too great (long) a ratch leads to excessive floating fibres and poor fibre control. Although the nominal, or theoretical, ratch is easy to calculate the actual or effective ratch is more complicated to derive, depending upon a number of factors, such as the sliver thickness (linear density), inter fibre-frictional forces and cohesion.

Ratchet effect: Relates to wool scales and the fact that they hook onto each other as in the so-called directional friction effect (DFE) assumed to play a major role inFelting.

Ratchet lock slider, in Zipper: A slider with a locking mechanism that permits the slider to slip along the chain upon application of a predetermined force so as to prevent damage that would impair either the service or use of either the slider or the chain.

Rate of dyeing: Rate at which the dyeing bath is exhausted. A suitable measure is the inverse of the Time of half-dyeing or the time after which 70% of the dye has been absorbed.

Rate of evaporation: Increases with increasing temperature, because the average molecular speed increases at the same time, and therefore so does the percentage of molecules with above average speed, which can escape from the surface.

Rate of fixation: (1) In reactive dyes the fixation speed is characterized by the period of time in which a stable bond is formed between the dye and fibre molecule. It is dependent upon the reactivity of the reactive group in the dye molecule, alkalinity of the fixation medium and temperature. (2) In disperse dyes the fixation speed depends upon the diffusion speed of the dye in the synthetic fibre, which is determined by the dye concentration and temperature and pressure relationships during fixation.

Ratiné: (1) A French word meaning fuzzy, this a rough pebbly woollen fabric made of novelty yarn with a fancy twist, and similar in finished appearance to Chinchilla. Used for men's overcoats.

(2) A plain-weave, loosely constructed fabric having a rough, spongy texture which is imparted by the use of nubby plied yarns. It is made from worsted, cotton, or other yarns.

(3) A variant of spiral yarns in which the outer yarn is fed more freely to form loops that kink back on themselves and are held in place by a third binder yarn that is added in a second twisting operation.

Ratine' Finish: Woollen fabric character with wave-like appearance on the surface.

Ratine lace: A machine-made lace, the ground work of which consists of heavy loops rather than mesh.

Raumois: Coarse and unbleached French ticking.

Ravel: A type of comb or rail with projecting teeth for separating and guiding warp ends.

Raveled strip test, in fabric testing: A strip test in which the specimen is cut wider than the specified testing width and approximately even number of yarns are removed from each side to obtain the required testing width.

Raveling: The process of undoing or separating the weave or knit of a fabric.

Raven's duck: A very heavy linen fabric in 2 x 1 twill weave, used as sail cloth.

Raw cotton: Ginned lint that has not been subjected any textile manufacturing processes.

Raw cotton composition: 83.71% cellulose, 6.74% water and impurities 0.61% grease, wax (Cotton wax), 5.79% hemicellulose, pectins, 1.50% proteins, 1.65% ash. According to Bowman: 91.35– 90.80% cellulose, 0.40–0.42% wax-oil-grease, 0.53– 0.68% protoplasm, 0.22–0.25% mineral components, 7.50–7.85% water.

Raw fibre: A textile fibre in its natural state, such as silk "in the gum" and cotton as it comes from the bale.

Raw linen: Unbleached linen.

Raw silk: Non-degummed Silk (grège silk), still glued with silk gum and non-twisted individual threads.

Raw wool: Wool or hair of the sheep in the grease, pulled, or scoured state. See Wool, raw. (greasy wool), is shorn Wool, which still contains dirt. It contains 20–50% pure fibres, 6–17% wool fat, 10–30% wool grease, 20–40% impurities and (air dried) up to 18% water.

Rayadillos: Wide cotton goods in the Philippines, made usually in narrow blue and white warp stripes.

Raye: French for striped. Now, a general term for lengthways stripes made by colour weaving, weave interchange, varying warp density, or printing. The illustration shows Satin raye.

Rayleigh: Irregular bars in guipure lace.

Rayon: The earliest man-made fibre, from cellulose such as wood or cotton. Viscose, modal, actetate, cupro all are basically Rayon and hence now specific name are used.

Rayon fibre: A manufactured fibre composed of regenerated cellulose as well as manufactured fibres composed of regenerated cellulose in which substituents have replaced not more than 15 % of the hydrogen of the hydroxyl group.

Reactants: Chemicals taking part in a reaction.

Reaction chamber Chamber, usually mobile, for heat and moisture treatment (hot air and steam heating) of piece ware batch rolls after padding with bleaching agents or dyes on a pad-roll machine.

Reaction rate: A measure of how fast a chemical reaction occurs Many factors affect reaction rates. One very big factor is temperature. Although the range is large, many reaction rates approximately double for each 10°C temperature rise. For example, a reaction may be 4 times as fast at 50°C as it is at 30°C. This is the reason that many dye processes are done at high temperature.

Reactive dye: A dye which attaches to the fibre by forming a *covalent* bond; also called fibre reactive dye Reactive dyes are known for their bright colours and very good to excellent lightfastness and washfastness, though poor resistance to chlorine bleach. There are several broad classes of reactive dyes. Most are intended for cellulose fibres, but some are intended specifically for wool. The most popular reactive dye family for textile artists is the Procion® MX family. Other reactive dye families reasonably available to art dyers include Procion® H, Procion® H-E, Remazol®, and Cibacron® F (equivalents to the MX, H, H-E and some of the Remazol® lines are available from many dye manufacturers; note that Procion®, Remazol® and Cibacron® are trade names of specific companies). Reactive dyes may be more expensive than other dye families suitable for the same fibres, especially when very dark or dull colours are considered. They are the newest class of textile dyes, first introduced commercially, for cellulosic fibres, in 1956 , one hundred years after the development of the first synthetic dye. Within a family, the range of colours available as "pure" dyes (as opposed to mixtures) is typically quite small – a dozen or fewer. Against this, reactive dyes of the same family can generally be mixed to produce a very wide range of colours, while retaining good application characteristics and brightness. The reactivity among families varies widely, so some are easily applied at room temperature, some at boiling temperature and others at intermediate temperature. All types are suitable for *exhaust* dyeing, and many types are suitable for pad-batch (see padding) dyeing and for printing. Apart from cost, the biggest commercial drawbacks to reactive dyes are that they require large amounts of electrolyte (salt) in most processes, and extensive rinsing and hot washing after dyeing to remove unfixed and hydrolyzed dye.

Reactive dyeing: There are two types of reactive dyes and these can be distinguished from each other by their reaction mechanism: those that react by addition and those that react by substitution.

(a) Addition reaction of reactive dyes. On the one hand, these dyes react with the fibre via a nucleophilic addition reaction such as, for example, the vinylsulphone compounds.

(b) Substitution reaction of reactive dyes. Nucleophilic substitution is another reaction route. In this case, the reactants are halogen-substituted, nitrogencontaining heterocyclic groups, the nucleous of which usually consists of a triazine or pyrimidine ring. These active groups are already available to react because the halogen atoms (chlorine and fluorine) are immediately replaced by an active basic group on the fibre.

(c) Combined substitution and addition reaction of reactive dyes. Dyes which react both according to a nucleophilic substitution and a nucleophilic addition at a double bond represent a special case.

Reactive group for dyes: The chromophore component of a reactive dye produces the colour shade and determines its substantivity. The reactive group determines the type of bond between the dye and the fibre and, therefore, some of the dye's performance characteristics.

In addition to the fastnesses, dye/fibre bond and performance characteristics, the reactive groups also determine the fixing value when dyeing with a reactive dye. Basically there are two types of reactive group: (a) The reactive groups which exchange basic nucleofuges for a dissociated hydroxyl group. Among these are the heterocyclic aromatics such as triazine, pyrimidine and benzopyrazine derivatives. (b) The reactive group which can add a functional group and a proton to the fibre. This group is represented by the vinylsulphones.

Reactive resin: (reactive cross linker, reactant). Resin finishing agents which, under certain crosslinking conditions, preferably react with cellulose and to a lesser extent with itself (unlike Self-crosslinking resins) such as certain methylol compounds (e.g, Dimethylolethyleneurea) or epoxides, chlorohydrins and d-ivinyl sulphones.

Reactive silicones: Elastomeric, amino functional and epoxy functional silicones will give a modest boost to DP performance. The elastomeric silicones add resiliency to the fabric improving wrinkle recovery. The amino and epoxy functional improve fibre and yarn slippage, reducing the transmission of the wrinkling force to the fibre. Also wrinkles will fall out easier. These products are often included with conventional resins to boost performance

Reactive systems: The chemical group on which reactive dyes are made. There will be one or more reactive group which can react with the fibre sites. Major reactive groups are given below

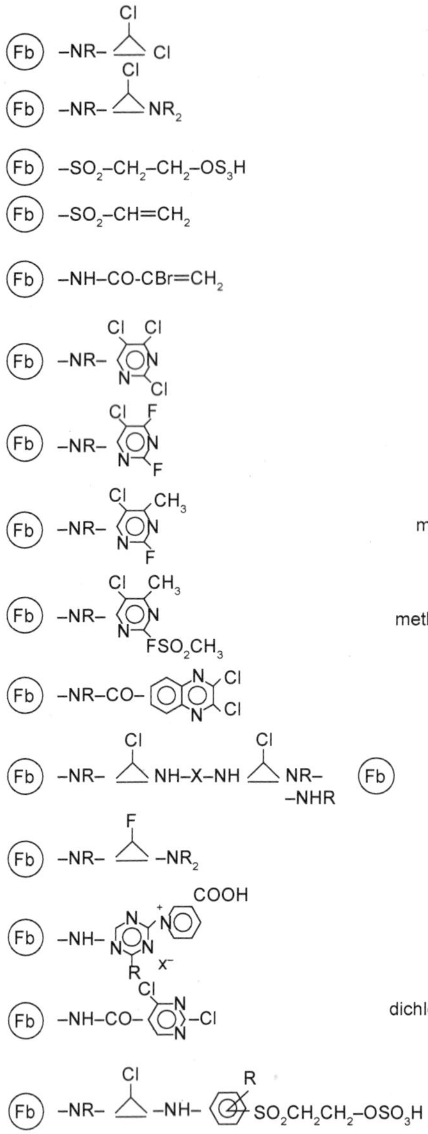

I. dichlorotrlazhe

II. monochlorotriazine

III. sulphatoethylsulphone

vinylsulphone

IV. a-bromoacrylamlde

V. trlchloropyrimidine

VI. fluorochloropyrimidine

VII. methylfluorochloropyrimidine

VIII. methytrychloromethylpyrimdine

IX. dlchloroqulnoxalln

X. dis-monochlorotriazine

XI. monofluorotriazine

XII. triazine-nictinic acid

XIII. dichloropyrimidine carbonamide

XIV. monochlrorotriazine sulphatoetrysulphone

Reaming: Further plying of a two-ply yarn with a singles yarn. Reaming is not the same as plying three singles yarns in one operation.

Rebayn: Medieval English name for fabrics woven with gold patterns over blue silk ground, made at Cologne, Germany.

Rebozo: (1) Shawl, made of knitted silk or wool, worn by Mexican women; (2) Fulled woollen fabric, usually black, navy or dark green, used for skirts by the natives in Colombia.

Reclaimed cotton: (effilochées), cotton reclaimed from cotton rags.

Reclamed wool: Produced from various types of wool products, waste fibres from the spinning mill, weaving mill, knitting mill and hosiery mill as well as (Rags) domestic waste fabrics, waste new fabric from the weaving mill or tailoring shop, etc. Types of woollen spinning yarn recovered by mechanical processing (carbonizing and dyeing, etc.) are classified as follows: Grade I = shoddy; Grade II = Tibet; Grade III = Mungo; Grade IV = alpaca

Recovered wool: (reused wool). Recovered wool is similar to Skin wool but is produced from tanned pelt wastes and wool rags. The quality and extent of damage of recovered wool varies widely (inferior to fleece wool) and is dependent on the amount of wear and tear and degree of tanning. The material is therefore generally unsorted.

Reconstituted fibres: Fibres made from recovered waste polymer or blends of virgin polymer and recovered waste polymer.

Recouvees: Stout French linen of natural reddish colour.

Recoverable elongation of rope: Elongation which may be reclaimed following a period of rope relaxation after the rope was cyclic tensioned.

Recovered wool: Obtained by converting woollen rags into fibres by tearing them up and reworking them again. Shoddy and mungo are recovered wool.

Rectified spirit: A constant-boiling mixture of ethanol and water that contains about 6% water; no more water can be removed by further distillation. It is used as an industrial solvent.

Recupra process: Used for the recovery of polyacrylate finishes. BASF/ Benninger process (Recycling) facilitates recovery of 60–80% of the coated finish. The goods to be desized no longer pass through the desizing bath. Application is by small counter-rotations in three stages and squeezed out and concentrated in a heavy press. The remaining desizing can be combined with the next finishing stage.

Recure: (recuring permanent press process). It is a variant of the pre-cure process (Permanent press process). Fully cured synthetic resin coatings (aminoplasts) are partially hydrolysed by subjecting them to particular cycles of steam ironing ("re-crosslinking" the cotton component) and curing them in the new shape simultaneously.

Red dye, Vegetable: Following plants dye red shade. Birch. '*Betula alba'*. Fresh inner bark. Bed-straw. '*Gallium boreale'*. Roots. Common Sorrel. '*Rumex acetosa'*. Roots, Dyer's Woodruff. '*Asperula tinctoria'*. Roots. Evergreen Alkanet. '*Anchusa sempervirens'*. Gromwell. '*Lithospermum arvense'*. Lady's Bedstraw. '*Gallium verum'*. Roots. Marsh Potentil. '*Potentilla Comarum'*. Roots. Potentil. '*Potentilla Tormentilla'*. Roots. Wild Madder. '*Rubia peregrina'*.

Red mordant: (red liquor) Aluminium acetate.

Red Tape: Cotton tape of red colour used in some government and law offices to tie bundles of papers with.

Red prussiate of potash: Potassium hexacyanoferrate (III).

Redingote: Waisted coat with full skirts or a back vent usually double breasted; riding coat.

Redox: Relating to the process of oxidation and reduction, which are intimately connected in that during oxidation by chemical agents the oxidizing agent itself becomes reduced, and vice versa. Thus an oxidation process is always accompanied by a reduction process. In electrochemical processes this is equally true, oxidation taking place at the anode and reduction at the cathode. These systems are often called *redox systems*, particularly when the interest centers on both compounds. Oxidizing and reducing power is indicated quantitatively by the *redox potential* or standard electrode potential, $E_{\text{ş}}$. Redox potentials are normally expressed as reduction potentials. They are obtained by electrochemical measurements and the values are referred to the $H+/H_2$ couple for which $E_{\text{ş}}$ is set equal to zero. Thus increasingly negative potentials indicate increasing ease of oxidation or difficulty of reduction. Thus in a redox reaction the half reaction with the most positive value of $E_{\text{ş}}$ is the reduction half and the half reaction with the least value of $E_{\text{ş}}$ (or most highly negative) becomes the oxidation half.

Redox potential: (expressed as r_H). "Redox" is an abbreviation of the term reduction-oxidation. The redox potential scale is used for classifying reduction and oxidation processes in solution, the electrical voltage (potential) in question being measured in volts or millivolts against a reference electrode (Standard electrode potential).

Reducing agents: A chemist's term for any chemical that causes gain of electrons by another chemical with which it reacts; the reducing agent is itself oxidized (see oxidizing agent) in the process Many reducing agents are used in dyeing processes. They include thiourea dioxide, sodium bisulfite, sodium formaldehyde sulfoxylate, sodium hydrosulfite and others. They are often used in discharge and stripping processes, and are used for converting insoluble vat or sulfur dyes to the soluble form. Solid-chemical reducing agents are often flammable. See **Reduction**.

Reduction: The gain of electrons by atoms, molecules, ions, etc. It often involves the loss of oxygen from a compound, or addition of hydrogen. Reduction can be effected chemically, i.e. by the use of reducing agents (electron donors), or electrically, in which case the reduction process occurs at the cathode. See also **Redox.**

Reduction clearing: Removal of disperse dye on the surface of fabric by use of a reducing agent Fine particles of disperse dye often remain on the surface of dyed fabric. These particles can cause wash fastness problems, yet are hard to fully remove by washing alone. For polyester, a mixture of about 2 grams per litre each of sodium hydrosulfite and sodium carbonate is used at about 70 degrees Celsius for about 20 minutes. Dye inside the fibre is not effected by this treatment.

Reduction thickener (blend), addition of a thickener with reduced chemical activity for a printing paste (stock colour or full colour) to obtain lighter printing inks, for example, 1 : 9 = 1 part printing paste + 9 parts thickener (1/9).

Reductive bleaching: Bleaching with Reducing agents, predominantly for animal fibres. With other fibres, possibly used in a second bleach bath in combination with oxidising bleaches (Bleaching).

Reductive discharges: Discharges (Discharge printing) based on a reduction effect where the dyes in question, (a) are split up into colourless soluble compounds, i.e. fully destroyed and have to be removed by subsequent washing (for example, direct, acid and mordant dyes or naphthols) or, (b) because of their solubility, particularly in alkaline solution, are more or less removed before the fibre is re-oxidised, i.e. are not destroyed (for example, indigoids and anthraquinoid dyes, such as indigo and vat dyes as well as cationic dyes as derivatives of tri-phenyl methane).

Redyeing: Dye packages may have to be dyed a second time because of faults or stock packages may have to be redyed as they cannot be sold due to their particular shade. In both cases, stripping of dyeings is usually necessary first.

Reed: The warp threads in a loom also go through a reed, a device consisting of several wires closely set between two slates of baulks that may serve any or all of the following purposes separating the warp threads, guiding the shuttle or rapier (if applicable) and beating up the weft. Reeds are sized by length and by the number of spaces—called dents— per inch (i.e., 4, 5, 6, 8, 10, 12, 15, 18 etc.). The reed is placed in the beater and acts to press the weft into the cloth as well as space the warp.

Reed holders: are wooden supports that keep the reed at a vertical position so it can be sleyed at a table rather than on the loom.

Reed hook: A reed hook or sley hook is a flat piece of metal, wood, or plastic with smooth curves at both ends for pulling threads through the dents of the reed.

Reed mark: A warp-way crack or disturbance of the structure in a woven fabric continuous or at intervals, caused by a reed misdraw or a damaged or defective reed. See also **Reedy warp.**

Reed plan: Weaving technology term. The reed plan shows the arrangement of the warp thread in the reed dent.

Reed plate, in tufting machine: A flat comb-like plate with slots on one side spaced in accordance with the machine gauge.

Reedy: Flaw in the cloth, caused by several warp ends running through one dent.

Reedy warp: See **Reed mark.**

Reefer: A short round coat or jacket.

Reel: (1) A revolving frame on which yarn is wound to form hanks or skeins. (2) The frame on which silk is wound from the cocoon. (3) A linen yarn measure of 72,000 yards. (4) The large wheel in a horizontal warper onto which the warp sections are wound in the indirect system of warping. (5) A spool of large capacity used to wind yarn or wire.

Reeled silk: Silk winched from the cocoons of the silk worm as Grèige derived from almost continuous elemental fibres.

Reeling: A preparatory process in the manufacture of silk, consisting in placing the unbroken cocoons in hot water and unwinding the single filaments several of which are joined side by side, without any twist and kept together by the natural gum of the silk.

Re-embroidered lace: A flat lace that has been reworked with another intricate design using a variety of threads. An elaborate and expensive fabric, due to the two operations involved.

REF Fibres: (REF = Roll Embossed Fibre), made from Tape yarns by embossing the yarns with longitudinally-grooved profile rollers followed by width stretching, where the areas on the fabric that have been thinned by the embossing process tear open, forming multi-filaments (elemental filaments with a high degree of uniformity and high strength).

Reference sample: An order sample for the Colour matching of dyed/printed samples.

Reference temperature: A fixed point on a scale of temperatures, for example, the melting point of ice ± 0°C and the boiling point of water + 100°C at ntp (Temperature).

Reference standard: See **Reference sample.**

Refin: French term for the best grades of wools of any certain class.

Reflectance: The ratio of the intensity of reflected radiant flux to that of incident flux. In popular usage, it is considered the ratio of the intensity of reflected radiant energy to that reflected from a defined reference standard.

Reflectance curve: The point associations (from a spectral photometer or automatic recording) of the percentage Reflectance plotted against the wavelength in a coordinate system produces the reflectance curve. In the visible range (400–700 nm), these characterise the ratio of reflected light to incident light, thereby yielding calorimetric information on the physical parameters of each dye or dyeing.

Reflectance factor: The ratio of the light reflected from the specimen to the light reflected from the perfect reflecting diffuser under the same geometric and spectral conditions of measurement.

Reflectance, specular: See **Specular reflectance.**

Reflectance, total: See **Total reflectance.**

Reflection: Component of incident light meeting a surface which is neither subject to Absorption nor Transmission (Reflectance). In the case of reflection, the incident energy is only briefly absorbed by the individual atoms in order to be reflected again as potential energy.

Reflection spectrophotometers: In reflection spectrophotometers, the reflection of light by the sample is measured at each wavelength relative to that of a white standard such as a plate coated with MgO or BaSO4. These standards give almost 100% diffuse reflection (light rays reflected in all directions) between 380 and 750 nm. They are, however, rather fragile and often a working standard such as a ceramic tile of known reflectance is used. Calibration of the instrument with a white standard tile allows calculation of the reflectance of the sample relative to that of a perfect diffuse reflector having 100% reflection of the incident light at all wavelengths.

Refleuret: French term for best grade wool.

Reformdress: Loose, freely draping dress, worn without a croset.

Reform weave: A weave which results in warp-reinforced fabric with main warp.

Refraction: The difference from a straight path undergone by a light ray in passing obliquely from one medium (as air) into another (as glass) in which its velodcity is different.

Refraction Index: The standard scale used for measuring refraction is called the refractive index (refractive exponent): quotient of the sine of the angle of incidence and the sine of the angle of refraction:

$$\sin a/\sin b = n = \text{constant}$$

The refractive index of air (in comparison to vacuum) is practically equal to nD 1; as, for example, the refractive index of solutions varies with concentration, this parameter can be used to determine the concentration. The refractive index is also utilized in fibre microscopy in the so-called immersion system where immersion fluids are used with a refractive index matched to that of the test fibre.

Refraction of light: (refraction), guiding or changing the direction of light as it passes through the boundary between different media such as light/water. While doing so, the long wavelength light (red) is not so strongly refracted as the short-wave light (violet/ blue).

Refractive index: Also called Index of refraction. The ratio of velocity of radiation (as light) in the first of the two media to its velocity in the second as it passes from one into the other.

Refractory fibre: Oxide or non-oxide, amorphous or crystalline, manufactured fibre generally used for applications at temperatures greater the 1063°C in both oxidizing and nonoxidizing atmospheres, i.e., Al_2O_3, ZrO_2, Al_2O_3 SiO_2.

Refractory organics, in waste water treatment: These organics tend to resist conventional methods of waste water treatment. Typical examples include surfactants, phenols and agricultural pecticides.

Refrigeration process: A wool scouring process. In this method raw wool is first exposed to low temperature environment of about-30~ which freezes the wax and vegetable matter and thus becomes hard and brittle. In the second stage these impurities are removed from the fibre by some mechanical means.

Regain: Average normal moisture which the fibre should contain, as for instance: cotton, raw or yarn, 8% per cent; linen, 12 per cent; carded wool and wool waste, 18.4 per cent; wool yarn, 17 per cent; worsted yarn, 18.4 per cent; jute, 13% per cent; silk, 11 percent; noil, 14 per cent.

Regain standard: See **Regain.**

Regatta: (1) A striped or checked cotton fabric, made in England with a two and one, warp face twill and well starched filling; used for aprons, children's dresses, etc; (2) Woollen fabric, made with alternating gray and coloured or blue and white stripes of twill weave.

Regatta Stripes: Good quality, equal blue and white striped calico used in England as dress goods.

Regency Point: A Bedfordshire bobbin lace, made during the 19th century, with a thick edge, originally with tape design on reseau ground, later with plaited ground and raised patterns similar to the Maltese laces.

Regenerated cellulose: A material which begins as cellulose but at some stage in the chemical processing takes the form of another chemical compound, then appears again in its completed state as cellulose. Viscose and cuprammonium rayons are regenerated cellulose.

Regenerated celluolose: "Recovered cellulose" which, unlike native cellulose (cotton) has a lower degree of polymerisation (and correspondingly different characteristic properties). Identical to Cellulosic fibres, regenerated (viscose, acetate and cupro fibres).

Regenerated fibres: These are fibres created by modified natural fibres. The cellulose regenerated fibres include rayon and acetate. The protein-regenerated vegetable fibres include soybean (soylon), peanut (ardil), and corn (vicara). The protein-regenerated animal fibres include casein (aralac), gelatin, and albuminSee Cellulosic fibres, regenerated; Man-made protein fibres

Regenerated protein fibre: See **Man-made protein fibres.**

Register marks (Printing): In roller printing, the pattern being reproduced is broken up into a large number of dots using a reproduction camera with a traverse or shading screen, so that the dots are graded in size according to the shades.

Regrettas: Narrow cotton goods, mostly in narrow blue and white stripes and blue filling; used in the Philippine.

Regular Twill: Trade name for 45 degree twill weave, without any fancy figure.

Regulators for peroxide bleaching: Peroxide decomposition can be influenced by the use of appropriate bleaching agents, known as regulators. Regulators mean substances or substance compounds that are able to preserve the perhydroxy anions (HO^{-2}) responsible for the bleaching effect for as long as possible, and largely suppress decomposition. The essential point of this is to prevent uncontrolled peroxide decomposition during the bleaching process, by means of regulators acting on the reaction sequence.

Reinforced Hosiery: Reinforced Hosiery is knitted at the toe and heel in such a manner as to prevent unraveling of the fabric if the stitches wear out.

Reinforced seam, in sewn fabrics: The seam that includes an additional layer of material on the face or backside of the seam allowance.

Reinforcement fabrics: See **Geotextiles.**

REK: Russian fastness commission.

Related shades: Colours of similar tone in the same or different depths.

Relative humidity of air: The ratio of the pressure of water vapour present to the pressure of saturated water vapour at the same temperature.

Relative humidity: See **Relative humidity of air.**

Relative humidity: The ratio of the actual pressure of the water vapour in the atmosphere to the saturation pressure of water vapour at the same temperature. (The ratio is usually expressed as a percentage).

Relative viscosity: Ratio of the viscosity of the polymer in solution to that of the solvent expressed as time of efflux of the solution divided by the time of efflux of the solvent at constant temperature.

Relative whiteness: Comparison of two whiteness.

Relaxation: The releasing of strains and stresses in textile materials.

Relaxation dimensional change: The dimensional change that occurs when a knitted fabric is immersed in water without agitation and the strains and tensions put to fibres, yarns, or fabrics during previous processing stages such as spinning, knitting and finishing are received.

Relaxation of elastic fibres: Before carrying out any further treatment, it is recommended to relax woven or knitted goods to obtain a uniform shrinkage and avoid stitch distortion or fabric deformation, creases or wrinkles. The fabric relaxation is a crucial step to allow good shrinkage and give excellent elasticity since the fabric width on looms is always bigger than the finished one (tensioned yarns on the loom). Many techniques are used but here are some of the most frequently used ones: table steaming, steaming carried out at the entry of the stenter, scouring carried out with hot solvents, relaxation in hot water with tensionless scouring; these techniques give poorer stabilisation results and do not provide permanent crease resistance to textiles and fabrics.

Relaxation shrinkage: A shrinkage induced by the relaxation of strains present in a textile. **Note**: Strains of a temporary nature can be relaxed to a varied degree, e.g. by steam pressing or by immersion in water. Relaxation shrinkage is the irreversible dimensional change accompanying the release of fibre strains imparted during manufacture which have been set by the combined effects of time, finishing treatments, and physical restraints within the structure.

Relaxed yarn: A yarn treated to reduce tension and produce more uniform shrinkage or torque. Relaxation produces more uniform dyeing characteristics in regular filament yarns of nylon or polyester.

Release point (Rp): Release Point (Rp) is defined as the surface tension of a detergent solution where oily soil just separates from fibre surfaces. This technique for quantifying soil release is based on thermodynamic considerations which state that the work of adhesion (the quantity Yfo-Yfd) must be overcome by the detergent for oily soil to separate from a solid surface. In a detergent solution, when the contact angle becomes 180 degrees and the oil just separates from the fibre, the interfacial tension between the oil/detergent will equal the work of adhesion. Yod=Yfo-Yfd. The quantity Yod is influenced by the adsorbed surfactant. Therefore surface tension of a bath where release just occurs is proportional to the work of adhesion between the oil and fibre surface.

Releasing slider, in Zipper: A slider with a mechanical means for loosening the slider on the chain.

Releasing stop, in Zipper: A device attached at or near the top of the stringer on the separable pin side which limits the travel of the slider at the open end of the chain under normal closing operations.

Relief engraving machine: See **Engraving.**

Relief printed effects: (three-dimensional printing).This does not refer to roller printing with relief rollers, which is widely known, but to the combination of chintz printing or Emboss printing with particles of crêpe prepared in caustic soda on cotton fabrics.

Relief valve: A pressure relief device actuated by inlet static pressure having a gradual lift generally proportional to the increase in pressure over the opening pressure.

Relustering Restoration of gloss finish. After treatment, particularly for modacrylic fibres, partly combined with the drying process.

Remaflam dyeing machine: Continuous dyeing and drying plant for piece goods. Impregnation padder with 36% methanol and required additives, contact-free in short IR heating duct until methanol reaches flash point, ignition in combustion chamber causing water to evaporate, combustion temperature 750°C, fabric temperature 45–70°C (never as high as 100°C if no dye setting), air cooled.

Remaflam process: Drying process developed by Hoechst and Brückner, in which methanol is used as an energy carrier (Methanol/Water flashpoint). Process for padding and drying textiles, in which textiles are soaked in a flammable liquid (e.g. methanol). At the next stage, combustion of the liquid takes place, causing the water to evaporate. Remaflam drying has an initial speed of 10 m/min; after approx. 5 min the speed is 70% of the ultimate speed (which is reached after approx. 20–30 min). The average drying speed is therefore dependent on the length of the cloth run.

Remazol: A family of vinyl sulfone reactive dyes; tradename of Dystar of Germany

Rembrandt Rib: Women's hosiery made with lengthwise rows of five drop stitches alternating with inch-wide strips of plain knitting.

Renaissance lace: Motifs joined by a variety of stitches.

Renforce: Also known as dense plain weave cotton fabric; medium-fine cotton quality, finer than cretonne. Available as raw material, bleached, dyed and printed. See Plain weave cotton fabric. Used for dresses, blouses, bedsheet etc.

Rengue: Fine cloth made of pineapple fibre In the Philippines.

Renter: To restore damaged tapestry through inserting new warp.

Rep (p), Repp: Usually a heavy or medium fabric with very pronounced rib. It differs from poplin in that only every second piock is heavy. For oinstance, a rep for shirting may have a thread count of 100 x 40 with 40s warp and alternating picks of 40s and 10s in the filling. Rep for upho; stery are much coarser. A characteristic construction for these reps is 80 x 20 with 2/30s warp and alternating picks of 15s and 4s. It is made from a variety of fibres, including cotton, viscose and blends of fibres and occasionally from worsted yarn. It is stiff and unyielding but also very hard wearing. Used for curtains, loose covers, bedspreads etc.

Rep Stitch: See **Persian cross-stitch.**

Repeat: The smallest size of a pattern that shows the full pattern.

Repeat artists: Designers who take the designs, put these into a size, and repeat appropriate to the intended end-use.

Repeat for printing: The smallest size of a printing design that shows full design. One or more of this repeat is engraved on to a screen and the printing is done. During screen making each colour from the repeat is engraved on different screens.

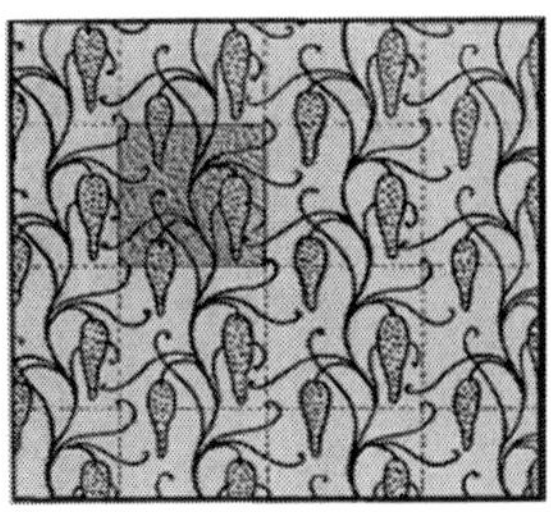

A design with repeats marked

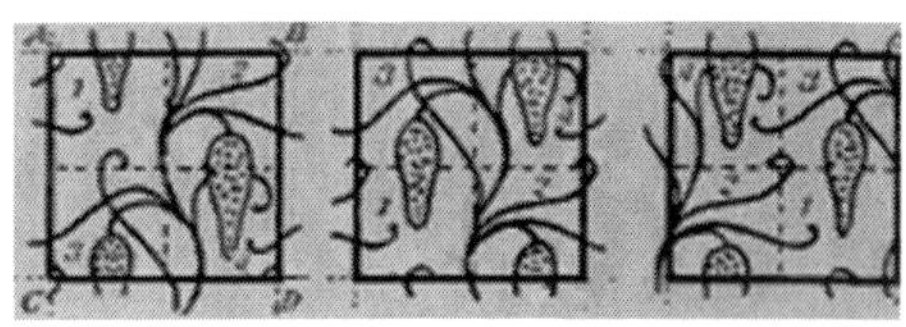

Repeats

Repeat size: The distance covered by a single unit of a pattern that is duplicated over and over, measured along the length of a fabric.

Repellency: The ability to resist wetting and staining by oils, water, soils, and other materials.

Repellent (repellate), substances with specific repellent, but non-damaging, properties, such as Insect repellent finishes.

Repetition, in garment designing: involves the use of patterning to achieve timed movement and visual 'beat'. This repetition may be a clear repetition of elements in a composition, or it may be a more subtle kind of repetition. See **Rhythm.**

Repp or Rep: Wool, worsted, silk, rayon, wool ottoman, cotton or a blend. Crosswise rib. Has a pronounced narrow cylindrical rib in the filling direction - less distinct than bengaline; more distinct than poplin. Sometimes a very distinct rib is alternated with a small rib. It is similar to poplin but heavier in cotton. Can be dyed, printed, or white. Frays badly. Difficult to press (may flatten rib). Uses: Heavy suits, and coats for men's and boy's wear, and also for some women. Also used for upholstery and drapery.

Reprocessed wool: Woollen fibres obtained from existing knitted, or woven garments. Now this fabric is very rare, as it is difficult and very expensive to reprocess blends.

Reproducibility Capacity to repeat the properties defined in the specification, such as colour shade. The evenness of dyeing is assessed on the padder.

Repousse Lace: The design consists mostly of dots, also of flowers and leaves which are raised like blisters and are pushed in in the back.

Reps: (1) French drapery velvet made of cotton and wool; obsolete; (2) A French silk fabric, having organzine warp, the ribs are either warp or cross ribs.

Repp weave: See **Cannele weave.**

Re-reeling: Certain Chinese silks which are given a second reeling after the first one performed by unskilled native workers.

Rere: Fine, white bast fibres of th* Cypholophus macrocephalus in the Pacific Islands; used for clothing mats by the native.

Reseau: See **Net.**

Reserve dyeing: See **Dyeing.**

Reserving agent: Also called restraining agents; a dye bath auxiliary that is typically used to prevent one fibre in a blend from taking up dye intended for the other fibre, or to equalize the uptake When blends are dyed, one fibre may be truly dyed while the other is stained (coloured, but with very poor fastness). Reserving agents can be used to significantly reduced the undesired staining. In blends such as woolnylon, reserving agents can act to reduce the dye uptake by the nylon, so that the nylon and wool ultimately are coloured similarly.

Reservoir fabric: e.g. Made from polyester, functions as re-dyeable dye carrier (typewriter ribbon principle) in the transfer dyeing unit.

Residue, for plumage: Quill pith, quill fragments, trash or foreign matter.

Residual colour difference (Or Residual colour deviation): The residual colour difference is the difference which results when the specimen and the type are both dyed in exactly the same colour strength. Even when the specimen and the type are dyed in one concentration only, the residual colour difference can be precisely computed colorimetrically because any colour strength difference can be computed with the aid of the Kubelka/Munk relation and taken into account. The total residual colour difference can be divided into a hue and a chroma portion. Thus by means of colorimetry an exact coloristic evaluation of the colour strength, hue and chroma is possible based on only one specimen and type dyeing in each case.

Residual elongation of rope: Elongation after cyclic tensioning the rope to a specified force for a specified no. of cycles and allowing the rope to relax for a specified period of time.

Residual moisture content: Moisture content of fabric after the drying process. Pre-driers should be designed so that fabric moisture after drying is maintained at the same level so that, among other reasons, dye migration at the fibre surface is kept constant throughout the dyeing process.

Residual print paste: (print paste returns). Unused print paste left on the print machinery or printing table because a greater quantity than necessary for the meterage was applied for safety reasons.

Residual shrinkage: The potential shrinkage that remains in a fibre, yarn or fabric after treatment designed to reduce or eliminate shrinkage. *Note*: The expression is commonly used with reference to heat-shrinkage properties of synthetic polymer fibre after it has been heat-set.

Residual torsion: Revolutions made by a specified length of cord when one end is held in a flexed position and the other allowed to turn freely.

Resilience: That property of a material by virtue of which it is able to do work against restraining forces during return from a deformed state.

Resiliency: The ability of a fabric to spring back to its original shape after being twisted, crushed, wrinkled, or distorted in any way.

Resin: (1) A general term for solid or semi-solid natural organic substances, usually of vegetable origin and amorphous and yellowish to brown, transparent or translucent, and soluble in alcohol or ether but not in water. (2) Any of a large number of manufactured products made by polymerization or other chemical processes and having the properties of natural resins.

Resin antifelting finish: Antifelting finish of wool on the basis of synthetic resins. See **Hercosett process**.

Resin bonded batting: A textile filling material that is stabilized by spraying it with an acrylic, polyvinyl actetate, or other suitable resin emulsion after which the batting is dried and cured.

Resin dyeing process: Dyeing and finishing combination processes.

Resin finish: See **Resin finishing.**

Resin finishing: General finishing that gives each fibre additional properties depending on requirements (soil-release, moth repellent, antifelting, creasing resistance). Finishing that has a significant practical value, and causes permanent improvement in wear resistance (wash and dry-cleaning resistant), particularly shrinkage stability and crease recovery, of textiles made out of cellulose or cellulose compounds, by means of intercalation and/or modification of the cellulose (Resin finishing processes) with certain finishing products (Resin finishing agents). Also known as: Permanent press process, Wash and wear finishing, Anti-crease finish, Non-shrink finish, Swelling resistant finish, Easy-care finishes, No iron, Non-iron, durable press, minimum-iron, and rapid-iron finishes.

Resin finishing agents: Resin finishing agents are usually water-soluble, low-molecular, polyfunctional products that, in the presence of Catalysts for resin finishing, which work in an acidic or alkaline environment, are converted to either water-soluble, high-molecular compounds that are resistant to washing and dry-cleaning (Self crosslinking resins), and/or crosslink with the cellulose fibres. See ***Reactive resin.***

Resin finishing catalysts: Chemicals which enhances the crosslinking of the resin finishing agents.

Resin finishing of knitted goods: In case of knitted goods, good dimensional stability and good sewability are the crucial factors in resin finishing. N-methylol compounds of ethylene and hydroxyethylene urea (60–90 g/l, 45%) and melamine formaldehyde compounds, usually partially etherized and often as additives with a buffer function, are often used as resin finishing agents for achieving dimensional stability.

Resin finishing process: Process techniques in Resin finishing. Application of Resin finishing agents under specified crosslinking conditions. Largely responsible for defining intended finishing effects. Major processes: Dry cross-linking, moist crosslinking, wet crosslinking. Used alone, they represent multiphase crosslinking processes. When selecting cross linkers and catalysts for a process, there is a range of combination possibilities from a technical point of view. In industrial production there are two basic condensation types – high temperature condensation and cold dwell condensation. In general, condensation in resin finishing is a chemical reaction between cross linkers and cellulose, usually causing water release.

Resin formers: Chemicals which forms resin by any process like condensation etc.

Resin migration: Migration of pre-condensates and catalysts within the yarn and between thread systems during the drying process in resin finishing. The main effect of this is worsening of abrasion resistance in crosslinked cotton fabrics.

Resin resist printing: See **Wax resists.**

Resin-treated: Usually, a term descriptive of a textile material that has received an external resin application for stiffening or an internal fibre treatment (especially of cellulosics) to give wrinkle resistance or permanent press characteristics.

Resist: In dyeing (and some other processes), something applied to fabric to prevent dye from colouring it Resists such as waxes used in batik simply prevent dye from gaining access to the fibre. Tie-dyeing and shibori employ what amount to mechanical resist methods to limit dye access. Chemical resists allow the dye to gain physical access to the fibre, but prevent fixation of the dye to the fibre. These include sulfites such as sodium bisulfite, as a resist for vinyl sulfone dyes, citric acid as a resist for many reactive dyes for cellulosic fibres, and sulphamic acid as a resist for acid dyes on wool.

Resist dyeing: See **Dyeing.** The principle of this process of dyeing follows: The fabric is first printed with some chemical which will resist to the subsequently applied dye on the places where it was applied.

Resist printing: **(resists, reserve print**). There are two different types, pre-printed and over-printed resists, depending on whether the resist print paste was applied before or after preparation of the fabric with dye solution. In both cases, the dyeing of printed areas is prevented. The most suitable context for application is in conjunction with non-dischargeable dyes. There are mechanical resists, chemical resists, and chemical-mechanical resists.

Resist salt: An oxidizing agent added to dye baths or print pastes to prevent *reduction* damage to dyes Under some conditions, especially where air is excluded, such as in steam fixation of dye, some of the dye may be decomposed by reduction (often the fibre itself acting as the reducing agent). Resist salts, such as sodium mnitrobenzene sulfonate, generally used in alkaline conditions, or sodium chlorate, generally used in acid conditions, may be added to prevent this. Resist salts are also sometimes applied to fabric before printing with discharge pastes to protect the ground colour dye and/or improve the sharpness of edges of the discharge. This term is a bit misleading: don't confuse resist salts with chemicals used to keep dye from fixing to the fibre.

Resistance to creasing: Scale for measuring the resistance of textiles to Creasing. See **Crease resistance.**

Resistance to deformation: If an external force acts on a textile, it will be stretched or compressed; the usual measure for resistance to deformation is the modulus of elasticity (Tensile elasticity).

Resistance to slippage, in seam testing: The force required to separate the parts of a standard seam by a specified amount.

Resistance to tear propagation: Required strength when Tensile force is applied to the bearing width for sudden tear propagation of a fabric section from an initial cut. There are various different methods of testing this.

Resistance to yarn slippage: The force required to separate the parts of a standard seam by a specified amount.

Resistance to yarn slippage, at the seam: The force required to displace one or more yarns in a fabric from the original position, causing differences in alignment, or spacing, or both.

Resolution: Optical: interval between two points of the object (specimen), so that they can still be distinguished as separate entities under the microscope, for instance. Calculated by dividing the wavelength of light used for illumination (in the visible spectrum resolution limit approx. 0.4 mm) by the selected numerical aperture (formula for centrally-positioned illumination, almost double the resolution for angled light).

Resorcinol: (metadioxy benzene), $C_6H_4(OH)_2$; molecular weight 110; density 1.283; colourless crystals, slightly sweet tasting, water-soluble; also soluble in alcohol, ether, glycerol; melting point 110/111°C. Gives violet dyeing with iron chloride solution. Use: solvent (phenol considered) for cationic, united chrome dyes and tannic paints; for the fixation of cationic dyes on acetate; saponification of acetate; antiseptic (finishing effectiveness practically the same as phenol).

Resorption: The process by which a material that has given material by desorption takes up some more of the material given up.

Rest position, in knitting: The needle is in rest position with the needle positioned outside the cam. The sinker moves towards the machine centre.

Restoration: Action directed towards returning the condition of an object to its original state.

Restraint systems: An end use for textile fibres; restraint systems are devices such as air bags, seat belts, and shoulder harnesses for passenger protection in automobile, trucks, airplanes, etc.

Resultant linear density: The actual linear density of a plied (folded or cabled) yarn.

Resultant yarn number: The yarn number based on observed mass per unit length of a plied yarn, or a yarn whose number has been changed by processing such as twisting or bulking.

Retainer Pin, in Zipper: A tubelike element, similar to the separable pin, attached over the bead at the bottom end of the stringer opposite to the separable pin and that is designed to hold the fixed retainer position.

Retainer, fixed, in Zipper: See **Fixed retainer, in Zipper.**

Retainer, movable, in Zipper: see Movable retainer, in Zipper

Retainer pin, in zippers: A tube-like element, similar to the separable pin, attached over the bead at the bottom end of the stringer opposite to the separable pin and that is designed to hold the fixed retainer in position.

Retail: The sale of goods in small quantities to consumers.

Retainer, in designing: The fee paid in advance to ensure the services of a designer.

Retarder: A chemical added to a dye bath to reduce the rate at which dye attaches to the fibre; also called leveling agents A retarder may be required to prevent a dye from attaching to fibres so quickly that it would be very difficult to achieve level dyeing. Retarders are often used with acid dyes. They may work by quickly attaching to the fibre thereby temporarily keeping the dye from attaching, or by quickly attaching to the dye, temporarily keeping the fibre from attaching to the dye. Simple chemicals such as sodium sulphate may act as retarders for some dyes in some conditions. There are many retarders on the market that are proprietary mixtures of chemicals, often formulated to be companions for specific dye families. Reserving agents can be thought of as a special class of retarder. The availability to art dyers of retarders is limited.

Retarding: Slowing down, decelerating, braking; e.g. the dye uptake speed, which occurs in all dyes involved in the sense of levelling, but which is also capable of having a selective effect on certain dyes and would at least severely disrupt the levelling of such dye combinations.

Retarding agent: (retention agent), is used to prevent staining of a fibre type when dyeing textile blends. Retarding agents mostly have properties of mordants, which effect the textile fibre to be retarded before or during dyeing and consequently cause a change in the dye affinity (Mordanting).

Reticella lace: A very rarely type of needlepoint lace. It was a combination of drawn-thread lace work and cut work.

Retouching: Removal of blank, yellowed, colourless fabric areas or those altered in colour, caused by intensive Stain removal on non-colorfast parts or by concentrated chemicals, by possible careless stain removal or by the effect of perspiration (armpit areas), etc

Retouching inks: Distinctions are made between liquid, paste and powdery retouching inks. Liquid retouching inks usually contain alcohol or oil soluble dyes. Retouching print pastes, for example, represent substances pulped with oleic acid, which are used directly or possibly diluted with benzine.

Retouching ink powder (Spotting agent) are chiefly so-called coloured Talcum varieties, mostly dust-free, for dusting and rubbing in, for refreshing dry-cleaned gloves, etc.

Retractive force: The tension generated in a textured yarn due to the formation of crimp in the filaments, under specified conditions of crimp development.

Retting (Flax): The subjection of crop or deseeded straw to chemical or Biological treatment to make fibre bundles more easily separable from the woody part of the stem Flax is described as water-retted, dew retted, or chemically retted etc. according to the process employed.

Retting, cold water: Cold water retting in running water is practiced in Belgium. Retting in stagnant water is the method usually employed in Ireland and Russia. The retting in stagnant water is more rapidly done, but there is danger of over-retting on account of the organic matter retained in the water which favours fermentation. In this case the fibre is weakened.

Retting, dew: In dew retting, the flax is spread on the field and exposed to the action of the weather for several weeks[Pg 47] without any previous steeping. This method of retting is practiced in Germany and Russia.

Reused Wool: These are cleaned and shredded into fibres again, and then blended to make utility fabrics. See also **"shoddy"**.

Reutlinger matte: Network made from thin textile fibres for the separation of oil from waste water.

Reutlinger pill standard: (RPG). For the visual assessment of pill formation Pilling behaviour, test methods). Contains 8 stages: RPG 1 = stippled, bobbly fibrous; RPG 8 = maximum pilled.

Reveche: Plain woven fabric, having a long nap, often curled, on one side. Later made in twilled or serge weave with a soft, spongy, fulled body.

Rever: A turned back lapel or cuff.

Reverse coloration method (RCM): The colour is applied onto the pre-padded, wet carpet material from the back side using dye liquor. It is also possible to carry out the dye liquor application repeatedly using different liquors.

Reverse osmosis: (hyperfiltration). Membrane separation process (Permeation), which is used for filtrating waste water. The reverse osmosis results from a pressure (of 40, 60, 80 bar) exerted by the concentrated solution on the semi-permeable membrane (polyamide 6.6- acetate sheets). The solvent (water) migrates from the concentrated solution through the membrane and is described as a permeate. The dissolved substances remain as a concentrate.

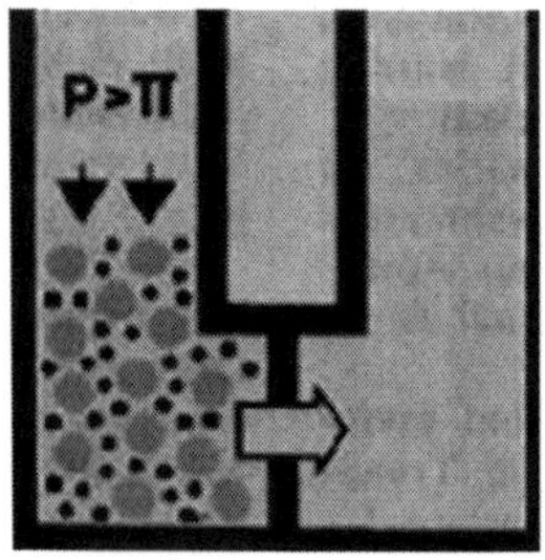

Principle of Reverse Osmosis

Reverse side: The side of a fabric opposite to effect side.

Reversed-tailing effect: Designation for end unevenness in padder dyeings, is caused by the fact that the fabric takes up more dye than corresponds to the absorbed liquid volume; dye concentrate increases in the colour box so that the end of the batch is darker than at the beginning. This process is called "osmosis". After some time, this process reaches a state of equilibrium, at which a hydrostatic pressure is adjusted by the volume shift, which is described as the "osmotic" pressure If a pressure is exerted on the concentrated solution after the state of equilibrium has been achieved, which is higher than the osmotic pressure ($p > \pi$), the process of osmosis can be reversed. Then solvent from the higher concentrated solution migrates into the diluted solution. The membrane retains the dissolved materials, which accumulate into a "concentrate" in the residual solution

Reversed thread: Crossed, exchanged threads or thread pieces;

Reversed Twill: Twill weaves with the warp thread predominating on the face. Used in cotton and linen goods.

Reversible bonded fabric: A bonded structure in which two face fabrics are bonded together so that the two sides may be used interchangeable. There are limitations to the fabrics that may be used because of increased fabric stiffness resulting from bonding.

Reversible carpet: Jacquard patterned, thick Flat carpets made from two or more interwoven warp and weft systems. The design on the top appears colorwise to be a negative of the bottom (Kidderminster carpet, tree ply or double-faced carpet).

Reversible fabric: A fabric that can be made up and worn with the either side as the right side, the two side contrasting in either colour or design or both. The loth is thick, made from wool, acrylic, mohair, or blends. Used for coats, capes, duffle coats etc.

Reversible fabric: Double faced fabrics.

Reversible Imperial or Reversible Satin : A stout cotton fabric woven in 8-leaf satin weave, containing many picks in an inch, the weft forming the face on both sides. It is usually napped on one side.

Reversible twill: See **Reversed Twill.**

Reversing, stenter: Stenter with non-contact, air supported reversing of the material web (see Fig.). Entry and exit of the material length at the same end of the machine. Operating area corresponds to that of the Single-layer stenter.

Revolving spinning ring: A driven ring that rotates in the direction of the traveler on a ring spinning frame. Since both the ring and the yarn package turn when this ring system is used, productivity is increased.

Rewetting: Important part of the washing process, which runs in the opposite direction from previous Wetting, because a fibre, for example initially "wetted" by oil, is wetted by the aqueous detergent solution at the end of the washing process, after removal of the adhesion forces of the oil smuts (ball formation). This phenomenon is described as rewetting.

Rewetting agent: A surfactant which, after application and drying onto textiles, promotes rapid wetting on subsequent exposure to an aqueous solution. Used to improve the wetting of Ducks, tightly woven poplins and twills (0.25 – 0.5 g/l), and in sanforising to improve the moistening of the babric before the entry into the rubber blanket zone See **Rewetting.**

Reworked Wool: Wool that has been previously used. Also called "***shoddy***" and ***"mungo"***.

Reynolds number, *Re*: A dimensionless number that describes ***pipe flow***.

$$Re = vd/v \qquad Re = pvd/n$$

where V = mean velocity; d = diameter of pipe; v = kinematic viscosity; p = dynamic viscosity; *n* = density of the liquid.

RF Drying: See **Radio frequency drier.**

RF heating: (Radio frequency heating) The use of dielectric energy of frequencies 3-300 MHZ for heating.

RF Value: See **Whiteness, degree of.**

RGB: The additive primaries red, green, and blue. See **Additive Primaries.**

RGB Colour measurement system: This so called "international trichromatic RGB system" is based on the law that any four types of radiation can have a known relative light composition, but an unknown light intensity from any two are independent from three types of radiation in the corresponding mixture. The following radiation components were selected for filters in this system:

R = red = 700.0 nm

G = green = 546.1 nm

B = blue = 435.8 nm.

RF value: (degree of wash whiteness) Whiteness, degree of.

rh: Redox potential.

Rhadames: Stout and lustrous silk or cotton mixed dress goods, made with fine diagonal twill and dyed in solid colours.

Rhe: The unit of fluidity; the reciprocal of the unit of viscosity (the poise).

Rheological properties: The properties of viscous substances including polymers that deal with deformation and flow. Includes viscosity and flow rate measurements.

Rheology: The study of the ways in which matter can flow. This topic is of particular interest in the study of polymers(Gk.: rheo = to flow), describes how a body is deformed by externally loaded forces. In the case of print pastes and coatings, for example, it concerns fluids, which are subject to ultimate shearing strains. The paste is made to flow by an applied shear force, i.e. the polymer molecules move past each other. The resistance to this forced irreversible volumetric change is called Viscosity. This is substantially influenced by the molecular weight and density of the particles. See **Rheopexy**; **Thixotropy**; **Viscoelasticity**.

Rheology in printing and coating: Laws of rheology play a crucial role both in printing and in coating as the basis for these textile finishing processes. The flow behaviour of the print pastes influences the print quality. In many cases empirical methods for the assessment of print paste consistency no longer fulfil the requirements in accordance with production safety and reproducibility. It is possible to record the parameters of a print paste which determine quality, viscosity and flow characteristics of printing thickeners and print pastes metrologically using rotary viscometers.

Rheology modifiers: Although viscosity is often treated as a constant, only a few materials such as water and some mineral oils possess a constant "Newtonian" viscosity. In most fluids, viscosity varies with shear stress or shear rate. To completely describe the rheology of a formulation, it is necessary to know how the formulation handles, pours, pumps, coats, penetrates and spreads. Therefore, the correct choice of a thickener or rheology modifier is an important one.

Rheometer: Device for measuring rheological properties, consisting of two coaxially arranged movable cylinders.

Rhinegraves: Wide men's coulette breeches, close fitting at the hips and tied under the knee; richly decorated with loops of ribbon.

Rhodes machine: An earlier loose cotton dyeing machine, it consists of a trough, divided into three compartments by two plates. Lower portions of these compartments and bottom of the middle compartments are perforated. Steam pipes are provided at the bottom of the trough. Loose cotton is loaded in the middle compartment and kept in position by another perforated plate at the top, where it is fixed after loading the cotton. Liquor is added to the side compartment and heated by the steam pipes and heated. The heated liquor is raised through the side compartments and enters the middle compartments and percolates through the cotton to the bottom of the middle compartments and this circulation continues and the dyeing takes place.

Rhythm crepe: The name given to a fabric with regularly spaced puckering. It is made from viscose and resembles Seersucker.

Rib: A usually straight raised cord in textiles, formed by threads which are heavier than the others either in the warp or in weft, formed also by grouping several warps through the same reed or passing more than one llmg through the same shed.

Rib fabrics: Collective name for all materials, which have a ribbed appearance. The ribs may be in a longitudinal or transverse direction (longitudinal or transverse rib), in which the latter predominates; even diagonally in exceptional cases (inclined rib).

Rib, Warp (in woven): A rib running across the width is obtained by high density of warp ends, when two or more weft picks are placed in each shed (fig)warp yarn is usually finer than the weft and cobers the surface of the fabric hence the name warp rib. A rib appearance also can be obtained in plain weave by using coarse weft and fine warp yarns. E,g., Ottoman, rib.

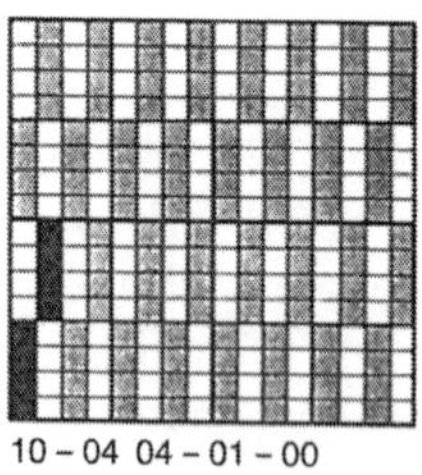

Pattern Draft

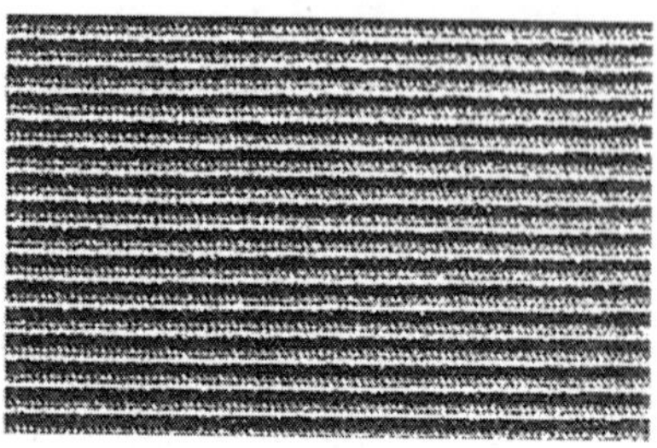

Warp rib

Rib, Weft (in woven): Rib running the length of the cloth are obtained with a high weft density by alternately raising or lowering two or more warp ends over and under the weft picks, with the same number of adjacent end doing the opposite. The high density of weft results in a lower weaving production rate, so these fabrics are not very common.

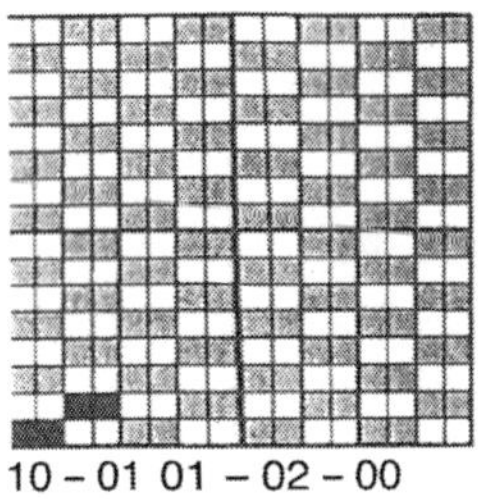

Pattern Draft

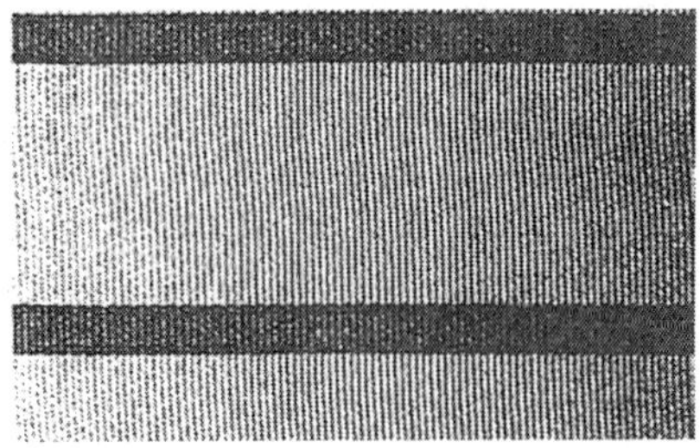

Weft rib

Rib fabric, Weft knitted: A fabric in which both back and face loops occur along the course but is which all the loops contained within any single wale are of the same sort i.e. back or face loops.

Rib Fabrics, 1×1, (English rib): A rib fabric in which single wales of face loops alternate with wales of back loops.

Rib fabric, 2×2 (Swiss rib): A rib fabric in which two adjacent wales of back loops in a series.

Rib gaiting: The needle arrangement on rib machines where the needles are arranged in such a way as to allow them to intermesh when raised.

Rib jacquard: In double jersey jacquard machines, the needles in the two needle beds (cylinder and dial) can be selected either to knit or miss. The pattern is created by selecting cylinder needles or knit or miss each coloured yarn in sequence. When the yarn is not knitting on the cylinder needles it is knitted by the dial needles to form the back of the fabric. The number of feeders required to complete one course depends on the number of colours in the design. These fabrics are used for pullovers, dresses and jackets.

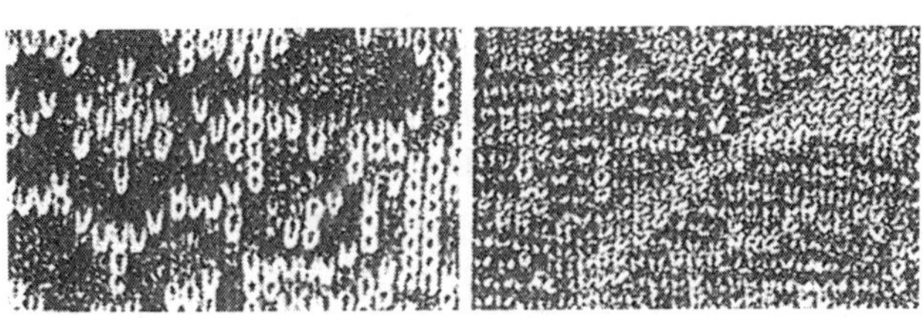

Rib jacquard face and back

Rib knit: A double-knit fabric in which the wales or vertical rows of stitches intermesh alternately on the face and the back. In other words, odd wales intermesh on one side of the cloth and even wales on the other. Ribknit fabrics of this type have good elasticity, especially in the width.

Rib knitted fabric: Rib fabrics are described as 1 × 1, 2 × 2, 6 × 3 rib, etc., and the most simple one is the 1 × 1 rib (a × b; a represents the number of front stitches, b represents the number of back stitches). The 1 × 1 rib fabric is thicker, narrower, heavier, warmer and more elastic and expensive than plain knit fabric. It has higher width-wise recoverable stretch than plain knit fabric. It is also typically knitted using a finer yarn than a plain knit fabric. The yarn can be unravelled easily from the last course of the fabric. Balanced rib fabric such as 1 × 1, 2 × 2, 3 × 3 rib does not curl from the sides and the top/bottom.

Rib velvet: A rarely used name for corduroy.

Rib, Single or simple: These have more than one plain wale but only one rib wale, such as 2/1, 3/1, etc.

Ribs, broad: These have a number of adjacent rib as well as plain wales, for example, 6/3 Derby Rib.

Ribbed spreader rollers: **Spreader rolls used mainly for knitted fabrics on various processing machines for expanding** work. They are provided with externally raised smooth spiral ribs running from the centre to left and to right. The fabric when runs over this rotating roller will be expanded width way due to the movement of the raised ribs outwards. More used to remove the curling of the open width knitted fabrics.

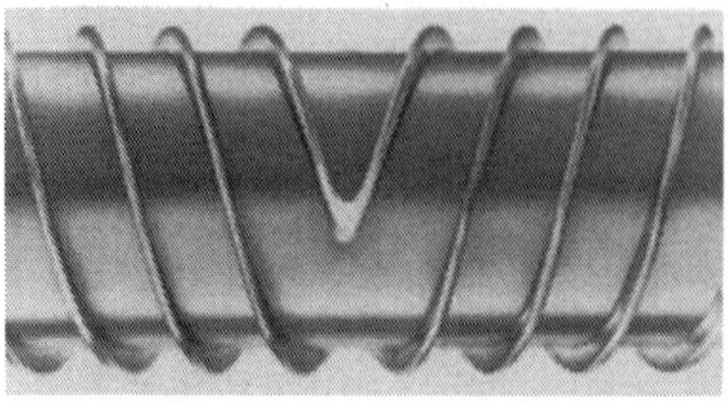

Ribbed spreader rollers

Ribbed weave: Variation of plain weave using fine and coarse weft yarn alternately to get a raised ribbed effect. The rib appearance is produced by using heavy yarns in the warp or filling direction, by grouping yarns in specific areas, or by having more number of yarns in warp than filling. Has a decorative appearance and feel, but not very durable.

Ribbing: A corded effect in a woven fabric that can be either lengthwise, crosswise, or diagonal.

Ribbon: Narrow fabric made in several widths and a variety of weaves and used as a trimming.

Ribbon, in textiles: A fine textured, narrow fabric which weighs less than 2.6 lb/100 yd per inch of width (approximately 15 oz/ sq. yd or 510 g/sq. m) and which is used primarily for trimming or decorative purposes.

Ribbon cutter: Textile or other stiff fabrics may be cut into ribbons using circular knives electrically heated wires or by laser beams and/or ultrasound.

Ribbon embroidered lace: A very elaborate fabric of lace background with matching narrow ribbon ruched and stotched in scroll designs on the right soide of the lace. The ribbon is likely to be nylon or polyester to avoid crushing. The lace could be almost any fibre.

Ribbon weaving machines: The name ribbons identifies fabrics usually with a minimum width between 5 and 20 mm. They are produced today on particular multi-head machines with 2 to 12 heads, which are interchangeable at will and thus permit weaving several ribbons at the same time. The warp threads can be fed by beams or by a bobbin creel. The shed is formed by frames driven by cams (in case of small weft repeats) or by disks bearing linear cams composed of of glieder chains (these are small cams linked together to form closed rings) in case of larger weft repeats. When weaving figured ribbons, the threads are controlled by an electronic Jacquard machine.

Ribbon, Textile: A fine texture narrow fabric which weigh less than 2.6 lb./100 yards per inch width (app. 15 oz/yd^2 or 510 g/m^2) and which is used primarily for trimming or decorative purposes.

Ribknit: See **Rib knit.**

Rice Braid: A braid consisting of a core yarn having alternately thick and thin places, entirely covered with a fine yarn wound around. It comes usually in white, but also in colour and is mostly of cotton; used for trimming and fancy work.

Rice Cloth: A plain woven cotton fabric made with hard spun warp and a fine novelty yarn, the latter having heavier places at close intervals. The cloth is somewhat similar to a fine ratine and comes in solid colour and in printed effect.

Rice net: Woven of coarse cotton thread stiffened by sizing and used as hat foundations.

Rice Stitch: Same as point de riz.

Richelieu: An open – work fabric made in imitation of hand embroidery if the same name. The designs are geometric, the embroidery may be in a yarn which matches or contrasts with the backing fabric. The fibres used may be cotton, viscose, polyester, modal or a mixture of fibres. It is a medium weight crisp fabric. Used for blouses, dresses, children's dresses etc.

Richelieu Embroidery: White embroidery made with padded outlines in imitation of the Venise laces.

Richelieu Guipure: Cutwork with open patterns buttonholed along the edge and joined with bars.

Rickrack: Flat braid in zigzag formation. It is produced by applying different tensions to individual threads during manufacture.

Ricotti: Waste silk obtained from the inner smooth skin of the cocoon after reeling.

Ridgy beam: A beam of yarn on which the ends are not evenly distributed across the barrel, causing a profile of peaks (ridges) and valleys. A ridgy beam can give poor removal characteristics.

Ridgy cloth: See **Baggy Cloth.**

Rigging: The lengthways of folding a fabric.

Right handed twill: **Twill, Right handed.**

Right side: See **Face side.** Right side also called the public side, is the finished side of the fabric.

Rilatti: See **Bassini.**

Rimo: A fine, silky white cotton, grown in the valley of Senegal.

Ring, in hosiery: A narrow band around hosiery, visually different from the rest of the hose. Principal causes: Variations in yarn sizes, dye, absorption, or lustre.

Ring: The device that carries the traveler up and down the package in ring spinning. Also see Ring spinning and Revolving ring spinning.

Ring accumulator: A dwell device for processes which require a set time for continuous implementation. The floor space for a corresponding steamer is considerably larger for the same output.

Ring doubling: See **Ring twisting.**

Ring dyeing: Any dyeing in which the dye has not completely penetrated to the inside (centre) of the fibre. This can be seen in the fibre cross-section as is not completely dyed through. The dye is like a sheath on the outer periphery of the individual fibre; the fibre core is not stained or only stained to a slight degree. Depending on process, fibre type and dye, ring dyeing is a common occurrence, particularly for densely packed, crystalline fibre materials, for large molecular dyes and for short-time dyeing processes. In extreme cases, ring dyeing results in visually higher yields, in poor colour fastness to rubbing and in use in whitish worn areas (cut-pile carpets).

Ring, in buttons: A split ring used to fasten a staple attached button to the substrate.

Ring in hosiery: A narrow, visually different horizontal band.

Ring opening polymerization: Formation of polymers by the opening of rings such as those of ethers or lactams. The formation of Nylon-6 from caprolactam is an example.

Ring spinning: A system of spinning using a ring-and-traveler takeup wherein the drafting of the roving and twisting and winding of the yarn onto the bobbin proceed simultaneously and continuously. Ring frames are suitable for spinning all counts up to 150's, and they usually give a stronger yarn and are more productive than mule spinning frames. The latest innovation in ring spinning involves the use of a revolving ring (Also see Revolving spinning ring) to increase productivity. Ring spinning equipment is also widely used to take-up manufactured filament yarns and insert producer-twist at extrusion.

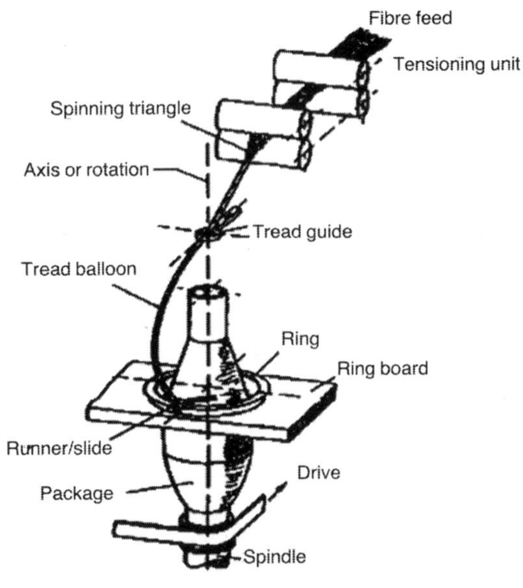

Ring spun: Yarn manufactured by the ring spinning method/machine.

Ring spun yarn: Yarn produced on a ring spinning frame.

Ring twisting: A system of producing folded yarn by twisting together two or more single yarns using ring and traveler as the twisting technique.

Ringer: (1) On a section beam, ringer is a term used for one or more filaments that have left the parent end; as the beam revolves, the filaments continue to unwind, wrapping around the beam (hence the word "ringer"). The severity of a ringer is dependent upon the number of filaments contained therein at the time the filaments break. (2) In slashing, the term ringer is often used when an end breaks on the slasher can, adheres to the can, and continues to wrap around it. This condition should not be confused with ringers on the section beam.

Ring-spinning: See **Ring spinning.**

Rinsing: Thinning, washing out, rinsing away of substances attached to the surface or inside fibres with water or solvent. In practice distinctions are made between the following:

- batch rinsing (15 l/kg water consumption) and rinsing in the overflow (20 l/kg water consumption),
- the counter current principle of rinsing in each individual wash compartment in continuous operation.

Rinzu: Silk satin in Japan.

Rio de Janeiro: Variety of raw cotton from Brazil.

Rip out: Pick out mark.

Rip-stop Nylon: A lightweight, wind resistant, and water resistant plain weave fabric. Large rib yarns stop tears without adding excess weight to active sportswear apparel and outdoor equipment such as sleeping bags and tents.

Ripening: Hydrolysis of cellulose acetate after acetylation to obtain the desired acetyl value.

This is generally accomplished by heat and agitation of the acid cellulose acetate solution under controlled conditions of time, temperature, and acidity. Rapid ripening is accomplished by using increased temperature for the reaction.

Ripout: See **Pick out mark.**

Ripped selvedge: See **Cut selvedge.**

Ripper selvedge: See **Ripped selvedge.**

Ripple cloth: An in expensive plain-weave fabric that has been slightly brushed on both sides and the forcibly shrinked to give it a wavy or rippled surface. Both treatments add warmth. The fabric may be made from wool.

But it is more likely to be acrylicor from cotton and therefore inflammable; in plain colours only. Used for dressing gowns and bed jackets.

Ripple shed: See **Wave shed.**

Rippler: The coarse comb used for removing seeds from flax fibre. The next step would be 'hackling'. The preceeding step would be **"scrutching"**.

Rippling: The process in flax production that removes the seed pods. The tool used is a 'rippler'. After this step of flax fibre preparation, the fibres are 'retted'.

Rise: The length of trouser from the top of the waistband at the fly opening, around the crotch, to the top of the back waistband at the center.

Riser: In textile fabric designing, a coloured or darkened square on the design paper which indicates that the warp end is over the filling pick at that point. The opposite of riser is sinker.

Rising roll batcher: 1. Batching device for open-width fabrics, operating at constant surface speed, with indirect drive. The batching roller lies on two driven rollers located horizontally adjacent to each other its axis is fitted laterally into vertical gliding rails. The roller rises in the rails as the diameter of the batching roller increases. 2. Roll batcher in which the yarn package lies on one or two rollers, of which at least one is driven at constant speed. The winding spindle rises in a guideway as yarn package diameter increases. (Batching drives).

R.M.M.: relative molecular mass; proper modern term instead of *molecular weight* 'Relative' is used because the mass is in proportion to one twelfth the mass as carbon-12.

Roasted starch: See **Dextrines.**

Rob Roy: A Highland tartan adapted and worn by Rob Roy. It is composed of alternate red and black stripes of equal width.

Robbings: Wool of greater length than noil, removed during the combing process.

Robe: Term for women's overdress, also called a manteau. See ***manteau.***

Roberto rolls: Squeeze rollers, with a pressed steel core with porous natural/ man-made fibre. 20–50% improvement in the squeezing effect compared with soft rubber roll. Action: firstly saturated with liquor, then squeezing as other squeeze rollers, at the same time the highly elastic, porous covering is compressed, and a portion of the moisture contained in the covering is pressed out in front of the nip (see Fig.). After the nip, the covering is opened again and thereby additional drawing out of moisture from the fabric takes place using capillary action. – Manuf.: Roberto.

Robertson: A Highland tartan, composed as follows: * dark blue stripe; narrow red line; dark green stripe, as wide as the blue * ; red field, three times wider than the green stripe, split near each edge by a narrow blue stripe (spaced from the edge its own width), and also split in the center by a pair of fine, green lines; repeat, in reversed order, stripes mentioned above between two ‘. Robes 64 square twilled cotton fabric, printed in bright coloured patterns, and left unfinished; used for wrappers,, kimonos, robes, etc., now mostly for drapery purposes.

Robia voile: Robia is the trade name of wide variety of voile fabric made by Tootal. In 100% cotton, the voiles vary from satin stripes in various widths and colours to plain square-effect weave with self colour woven spots.

Rochelles: French linen used for bedding, shirts, etc.

Rocket package: A large version of a super cop used as means of supplying the weft to rapier weaving machine. It may be up to 800 mm. in height and 75 mm. in diameter.

Rod cast button: A button fabricated from a disc sliced or sawed from a cast rod of formulated styrene modified polyester resin.

Rod squeegee: A type of magnetic squeeze used in printing. There are different type of rods are used in practice. See **Doctor blade.**

Rodier: See **Double pique.**

Rolag: The cigar-shapped roll of carded fibre, losely rolled off of the hand cards, used as the fibre source when spinning woollen yarn.

Roll collar: A plain collar without peak or notch; a show collar.

Roll goods: Fabric rolled up on a core after it has been produced. It is described in terms of weight and width of the roll and length of the material on the roll.

Roll lapping: A condition in which groups of fibres attach themselves to the drafting rolls instead of following the normal path through the drafting system. These fibres cause the trailing fibres to wind around the rolls and to bread the end down completely. Cleaning of the rolls is required to remove the accumulated fibre.

Rolled end: See **Sticker.**

Rolled ends: (1) On a section beam, rolled ends are adjacent ends that do not unwind parallel to each other. Rolled ends can be caused by such factors as uneven tension, ridgy beams, and static. (2) The ends can also roll behind the hook reed in slashing and can tangle with each other, resulting in broken ends and ends doubling.

Rolled hem: Rolled hem and whipped gathers are made with the wrong side of the material next the worker. Make a tiny roll of the edge towards the worker, using the left thumb and index finger, rolling an inch at a time (and no more) before hemming. Make fine, even stitches in the roll and goods. Keep the hem perfectly round, firm and not too large. This hem is adapted only to fine material and the edge across the warp is the more easily rolled.

Rolled seam-finish: A finish for the raw edges of the seam allowances of a plain seam, in which both raw edges are enclosed by rolling to one side and hand stitching close to the seam.

Rolled selvage: See **Curled selvedge.**

Rolled selvedge: See **Curled selvedge.**

Roller bed steamer: A steamer in which the fabric is carried in a relaxed state over a bed of individually driven rollers situated in the lower half of the steaming chamber.

Roller card: Generally, any type of card in which rollers do the carding. Usually this refers to a woollen card with a main cylinder and four to seven stripper rolls and worker rolls working in pairs.

Roller coating: Fusible polymers are heated and sealed in a roller until a film from the roller grinding mill capable of application is applied to the fabric to be coated (e.g. in PVC coatings of car covers). Coating pastes can generally be applied with the aid of a spreader, sieve or roller application systems. Roller application technology includes: (a) reverse roll coater; (b) smooth roller coaters; (c) grid roller application devices; (d) nip padding machines; (e) and immersion impregnation.

Roller conveyer steamer: A steamer in which the fabric is supported and carried in a relaxed state on an endless conveyer, made of stainless steel slate; which is driven by rollers at each end.

Roller engraving, electronic: Small bowls are cut with a sharp diamond chisel from the copper surface of the printing form in gravure printing cylinders. The diamond chisel is electronically controlled (electronic signals control position and depth of insertion of the chisel). The small bowls (which are pyramid shaped at the tip) have smooth walls due to the sharp cutting edges of the chisel, which has a positive effect on the transmission of the print paste to the printed goods. Chisel control signals are obtained through optical-electronic scanning of a black and white pattern (extract from a multicoloured design), which also contains developments as grey-scale values. See ***Engraving.***

Roller grinding and polishing machine: This is a lathe-like device, in which the print roller rotates and is moved across a rotating disc. A water tank forms part of the polishing bench, in which the roller body is immersed by approx. 5–10 mm during the polishing process. The following are used as polishing materials: polishing stone, grinding stone, charcoal sized at approx. (5 × 12 × 25) cm^3 .

Roller hardness: The description of the hardness and/or softness for companies anywhere is in accordance with different scales for rubber or plastic rollers for finishing machines amongst other items, whereby confusion continually arises. It should be especially emphasised that the terms DVM softness and Shore hardness conflict with each other and are handled differently to some extent.

Roller mangle: A Mangle, which contains batch rolled piece goods in one and/or two magazines (double roller mangle).

Roller Printing: See **Printing.**

Roller printing: The print paste applicator is composed of the print paste tank (colour box) and the tank in the rotating applicator roller, which transmits the print paste onto the engraved roller A doctor blade lying on top of the latter wipes off the surplus paste so that only the deep engravings still contain print paste and transmit this onto the material. The depth of the engraving is different. Shallow engraving up to 0.09 mm deep is sufficient for surface printing; this increases to 0.2 mm for good imprinting. The depth of the engraving is a criterion for the quantity of absorbed print paste and thereby a cost-determining factor. Shallow engravings give a rather smooth printing. They are pressed onto the cloth running between the printing cylinder and the engraved rollers by highly sprung spindles. During the running of the machine the print paste in the engraved grooves of the print roller is transferred on to the cloth as per the designed engraved.

Roller printing machines: Distinctions are made according to the type of roller printing area: (a) Gravure printing using Roller printing machines: almost exclusive mode of operation using gravure printing rollers, on which print paste is transferred from the print paste container (colour box) by a running roller (furnisher roller). Removal of the surplus print paste by a doctor blade moving to and fro at the side. Such print paste application devices are arranged around the printing cylinder corresponding to the number of print pastes. The fabric runs on a back grey (undercloth) with lappings and a continuous rubber cloth (printing blanket) between the elastic support and the engraved printing rollers. Joint drive of the printing rollers by a repeat wheel. (b) Relief roller printing (high pressure): of less importance. Printing rollers with raised samples (print paste application using continuous felt tapes for example) are used. Still used partly for carpet and wallpaper printing. 3. Mosaic printing, e.g. Conversion print; Orbis printing.

Roller-conveyer steamer: See **Roller conveyer steamer.**

Rollers, Rolls: Rotatable cylindrical bodies. (a) For the guiding or transporting of cloth, either undriven or driven as Guide rollers; Draw rollers. Deflecting rollers may contain the roller pins sealed in a hollow cylinder as a simple equipment practice (b) For the processing of cloth: squeezing rollers, pressure or calender rollers (See **Calendar**), See **Printing rollers; Applicator rollers**.

Roller vat, roller beck: Basic form of all open width washing machines. Principle : Rectangular dyeing vats (approx. 1.1 m wide, 1.2 m deep, 3.5 m long) with guide roller system installed top and bottom, across which the piece good is carried in the zig zag by the liquor after passing an expander (speed of 15–100 m/min). 3–8 such dyeing vats are arranged in combination as required.

Rolling: The winding of finished fabric on tubes.

Rolling machine: For the rolling of textile fabrics during storage, further transport (and also for their protection). Particularly for voluminous textiles or those of such large width, such as carpet lengths for example, special rolling machines are required for the rolling of a polished technique into shape, which form regular yarn packages kind to material on corresponding carriers (tubes).

Roman Numerals In Chemical Names: Many ions can assume different electrical charges in different compounds. For example, iron can have a charge of +2 or +3. For example, in old use, iron chloride (FeCl) where the iron has a charge of +2, would be called 2 “ferrous chloride”, while $FeCl_2$, where the iron has a charge of +3, would be called “ferric chloride”. In modern use, $FeCl_3$ would be called “iron (II) chloride” (read as iron two hloride) and FeCl would be called “iron (III) 2 3 chloride” (read as iron three chloride). The roman numerals directly indicate the charge. In dyeing, this is most likely to be found in aluminum, chromium, copper, iron, tin and zinc compounds.

Rongkhim: Red shawl woven by Yimchunger Naga women of Arunachal Pradesh, India. Normally worn by warriors of great fame of their tribe, as the red is symbolic of the blood. If a man unworthy of the honour wears the shawl, it is believed that he will die of leprosy.

Rongolite C: BASF trade name for sodium formaldehyde sulfoxylate; sometimes seen spelled “Rongalite”

Rooed: The process where fleece is plucked off of Shetland sheep during the spring. This works with Shetland fleece as the sheep produce a weak spot earlier in the season. Sometimes called rooing.

Ropa: Spanish fully tailored overdress, resembling a coat.

Rope: (1) A compact and flexible, generally tortionally balanced continuous structure , greater than 4 mm diameter capable of applying or transmitting tension between two pints. A heavy, strong cord, made from either natural or manufactured fibres or from wire, in a wide range of diameters. Yarns are twisted together to form strands. These strands are then twisted together in the opposite direction to form the rope. The fact that the twist directions alternate at different stages of rope assembly assures that the rope will be twist-stable and will not kink during use. Also called cord. (2) Fabric in process without weft tension, thus having the appearance of a thick rope.

Rope, braided: See **Braided rope.**

Rope creases: Disruptive, longitudinal marks; a problem that occurs in fabric and knitted goods finishing, particularly where dyeing takes place in rope form at higher temperatures. It has been shown in practice that crease inhibitors can minimise this fault. Where impregnation is carried out with plaiting, folds, creases occur either in the padder, when plaited, in the steamer or later during piling.

Rope detwister: A device used in a rope opener to detwist the fabric in rope form to make the rope devoid of twists. The product rope passes between two spring-loaded rollers of the detwister basket, (a) the deflecting roller; (b) and the two sensor rollers; (c) before proceeding to the spreader.

Rope dyeing Machines: These machines process the fabric fed and driven lengthwise to form a rope. The hydrodynamic effect is obtained by means of the motion of the fabric rope, or by means of the simultaneous rope-and-dyebath motion, which ensures a homogeneous contact of the material with the dyeing liquor and a quick exchange of the dyeing liquor dispersed in the material. Machines running according to these operating principle are suitable for treating almost all the fabrics made up of extremely different fibres, woven or knitted fabrics, during preparation and dyeing stages, with only some problems occurring with loose-weave fabrics. During the treatments the fabrics run freely weft-wise and therefore can freely shrink and set thus eliminating almost all tensions. Suitable operating conditions and technical adjustments also reduce to the minimum warp-wise tensions, and continuously move the wrinkles of the rope. An unquestionable benefit obtained with these machines is the extremely soft and fluffy hand, particularly suitable for fabrics to be used for garments. Possible problems are connected to the formation of permanent wrinkles on the fabric, or to uneven dyeing shades, always connected to the problem of the rope wrinkles; for fabrics made up with very delicate or short staple fibres, mechanical stresses can cause losing or extraction of the hair on the surface.

Rope, fibre: See **Fibre rope.**

Rope form: Textile fabric which has been gathered loosely into the form of a rope in the warpwise direction (woven fabrics) or wale direction (knitted fabrics), usually by passage through a ring or pot-eye of suitable diameter for pretreatment and dyeing processes (as opposed to open-width processing).

Rope form processing: The treatment of fabric that has been drawn into the form of a rope, often by passage through a ring (pot-eye) of appropriate diameter.

Rope mark: Also called running mark a long crease mark (q.v.) in a dyed or finished textile and that runs approximately in the length direction. The marks are caused during wet processing in the rope form and may be the result of: (a) for formation of creases along which abrasion or felting may occur; or (b) imperfect penetration or circulation of the processing liquors.

Rope marks, in dye or finished fabrics: A long irregularly shifting longitudinal marking on dye or finished goods or mechanically induced streak.

Rope opener: This spreads all piece goods which are pre-treated in rope form by means of beater rollers, scrollers and spreader rollers.

Rope piler: Device for feeding rope-form piece goods into bleaching vats. It is usually fed in together with a wetting-out liquor. Electric-hydraulic rope pilers have a telescopic tube which can be used for round bleaching vats. Even with varying product speeds from 30–150 m/min (infinitely variable) it ensures an even rope speed without twisting at a variety of rope speeds.

Rope plaiter: Hydrodynamic piece rope plaiter which feeds the rope into the fabric accumulators by means of a rope piling device.

Rope processing: Processing the form of rope, instead of open width. Olden days most of the processing, especially pretreatment, washing etc., was done in the rope form. Nowadays woven fabric is done in the open width form, leave alsone specialized processings. Knitted fabric to some extent is done in the rope dorem in Jet / Sft flow dyeing.

Rope ring: (pot-eye). A ring-shaped component forming part of the Rope form handling device for guiding and supporting the product rope.

Rope soaper: (rope washer). Washing, rinsing and finishing device consisting of e.g. 3–5 adjustable pressure squeeze and guide rollers. As well as the washing, rinsing and blueing of bleached goods, it is also used in particular for the chlorine finishing of e.g. the Ce-Es bleaching of fabrics. Obsolete.

Rope squeezing machine: (rope mangles). Are designed to extract water from piece goods treated in rope form. They can be incorporated into some manufacturing systems as intermediate devices (squeezing units), e.g. washers or autonomous devices (squeezing machines). In the latter case they are usually full-width squeezers.

Rope washing machine: (rope scourers). Are washing machines for washing piece products in rope form. Unlike discontinuous rope washing machines, which consume large amounts of water, continuous rope washing machines operate with smaller amounts of liquid and a specific liquor flow. They almost invariably function on a counter-current basis.

Rope, Plaited: See **Plaited rope.**

Rope, Twisted: See **Twisted rope.**

Roping: A term used for roving in the woollen system of spinning.

Rosella: (1) (pot-eye). A ring-shaped component forming part of the Rope form handling device for guiding and supporting the product rope.

(2) A bast fibre originating in South-East Asia and similar to Kenaf. The length of the basic fibre is about 2–5 mm; its per hectare yield is higher than jute, but it requires more processing. Used as a packaging material.

Rose-point lace: A Venetian needlepoint lace with design connected by a thick thread. It is characterized, as the name suggests, by delicate close flower of leaf designs.

Rot: Rot is chemical decomposition with a mainly alkaline reaction (in contrast to fermentation) and unpleasant odour. It encompasses a variety of indeterminate chemical reactions with the formation of intermediates and soluble end products, which are also released by aerobic and anaerobic bacteria.

Rot resistance: The ability of textile materials to resist physical deterioration resulting from the action of bacteria and other destructive agents such as sunlight or sea water.

Rot steeping: A method of desizing. In this method grey cotton fabric is steeped in water in suitable box at a temperature of about 30-40°C. During the storage micro-organisms develop excreting enzymes which attack the starch. The swollen and hydrolysed starch is thus partially converted into soluble state which are then removed from the fabric by normal washing with water. The main problems in this method are low efficiency due to longer treatment time and degradation of cellulose due to cross-infections of mildew if the fermentation process is not properly controlled.

Rotary cloth press: It should theoretically be possible to achieve a press effect in wool fabrics during continuous processing using calendars that are used in cotton finishing. However, rotary cloth-presses are predominantly employed in the continuous dry finishing of Wool, in order to compress fabrics.

Rotary Cutter, in garmenting: It is a round, very sharp blade attached to a handle and is used to cut out fabric. It should always be used with a rotary cutting mat.

Rotary dryers: Devices commonly used in the chemical industry for the drying, mixing, and sintering of solids. They consist essentially of a rotating inclined cylinder, which is longer in length than in diameter. Gases flow through the cylinder in either a countercurrent or cocurrent direction to regulate the flow of solids, which are fed into the end of the cylinder. Rotary dryers can be applied to both batch and continuous processes.

Rotary drum drier: A sieve-drum drier where the fabric is guided almost all the way around the drum, for the dry treatment and/or thermal treatment of tubular and/or web qualities. Their compact, high-productivity design with little space requirements make them particularly suitable for small and medium-sized shops.

Rotary extraction cleaning: A cleaning procedure which uses a rotating brush machine or a series of rotating jets through which shampoo is fed or sprayed on to the carpet pile and simultaneously removed by suction.

Rotary irons: A further development of Ironing machines and steam ironing dummies used for finishing outerwear in one fully automated process (Steam ironing dummy).

Rotary press: A combination of 4 ironing presses for transfer printing. The transfer printing processes are cyclic, and hence continuous. There are 4 stations: (a) uptake of the material; (b) uptake of the thermal transfer paper; (c) transfer; (d) cooling and removal of the transfer material.

Rotary screen coating machine A device to provide an even coat of screen lacquer to rotary screens. The lacquer is applied in a single process using a variety of different lacquer types on rotary screens from the top downwards with the help of a rubber doctor roll.

Rotary screen coating machine: Rotary screen coating machine A device to provide an even coat of screen lacquer to rotary screens. The lacquer is applied in a single process using a variety of different lacquer types on rotary screens from the top downwards with the help of a rubber doctor roll.

Rotary screen printing The process of screen printing using rotary screens. The screen is not placed flat in a frame, but takes the form of a hollow roller. The advantage over automatic Flat screen printing is a substantially higher level of continuous production (similar to roller printing). The advantage over roller printing is greater ease of changing dye and design, as rotary screens can be inserted and removed by hand, without the need for lifting blocks.

Rotary screen washing: Rotary screen washing is carried out in a screen washing cabinet by means of a spray ring on the outside and a spray head fitted with a nozzle on the inside.

Rotary shampoo: A cleaning procedure which uses a rotating brush machine through which shampoo is fed into the pile of a textile floor covering.

Rotating rod drier: A continuous hank yarn drier with rotating yarn rods which rotate while passing through the machine and which are closely spaced and allow air to flow between them so that the circulated air is fully utilised and the yarn is well aired. Dry treatment in a misty atmosphere at stepped temperatures up to the infeed of the wet product (resulting in the maximum possible protection, easy handling yarn with an even residual moisture content and low vapour consumption).

Rotation cast button: A button fabricated from a disc blanked from a partially polymerized sheet formed in a rotating cylinder.

Rotonde: Long, circular cut cape for women; also called a wrap.

Rotor spinning: See **Open-end spinning.**

Rotor, In open end spinning machine: A device resembling a centrifuge, in which the fibres are assembled in and in which, by virtue of its rotation, real twist is inserted in the forming yarn.

Rotor atomizer: This coating system basically consists of a rotor support, hydraulic control device, liquor container unit and centrifugal pump. The rotors turn at approximately 5000 rpm and distribute the treatment liquid over the product with droplets of approximately 30 mm

Rotor damping: A method for damping fabric and for low wet pick-up. It operates without jet nozzles. The equipment consists of a rotor support, a hydraulic control unit, liquor container and centrifugal pump The liquor is finely dosed. The rotors (approx. 500 rpm) spray a part of the liquor as very flat spray in the form of tiny droplets (diameter approx. 30 mm) at a vertical sheet of fabric. The spray width is approximately 100–400 mm. The staggered sprays and overlapping zones result in a spray width of 180–280 mm for approximately 800 ml/m/min.

Rotor spinning: See ***Open end spinning.*** (OE yarn). Fibre yarn produced by means of rotor or turbine spinning process (OE spinning process). Individual fibres are removed from slubbings and are continuously added to a spinning open yarn end by an air current from a rotating turbine. This method of spinning results in carded -type yarn with typical qualities differing from those of ring-spun yarn Ring-spun/rotor-spun yarns, properties of.

Rotor type open-end spinning machine: An open end spinning machine wherein the assembly of individual fibres and the insertion of real twist are effected by the rotor.

Rotor Washing Machine: Washing machines with a divided inner drum uni-directional. Used in the multi-bath or Flow washing process. Advantages: idle time for reversal of drum rotation is eliminated, thus ensuring a shorter washing process (= levelling for compression treatment reduces washing requirements) and smooth inner drum rotation.

Rotor-Spun Yarn: Open-end spun yarn.

Rouching: Fabric is drawn up with a thread or drawstring to create a full, gathered effect within a garment.

Rough: A descriptive term for a fabric surface which has the feel of sand paper.

Rough: Peruvian Cotton, having very clean, moderately strong but harsh and wiry staple of a light cream colour.

Rough place: See **Temple mark.**

Rough selvage: See **Loopy selvedge.**

Roughness: It is understood as the unevenness of a surface. It can be described geometrically by the size of the roughness elements, or mechanically by the friction coefficient. When touched with a finger, the roughness is not noticeable. However, if a finger is moved across the surface at a constant speed (approx. 3–4 cm/s), the movement receptors in the finger are rhythmically stimulated. These groups of impulses are perceived as roughness, the time between the individual impulses reproducing the roughness. The contact pressure is close to the contact threshold so that the roughness is identified from the friction cause.

Roumanian stitch: An Embroidery stitch. Bring the thread through at the top left of the shape. Carry the thread across and take a stitch on the right side of the shape with the thread below the needle. Take a stitch at left side. Thread above needle These two movements are worked until the shape is filled keep the stitches close together. The size of the center crossing stitch can be varied to make a longer oblique stitch or a small straight stitch.

Round Twill: It is based on the satin weave, forming round diagonal ribs.

Roundabout: A short jacket reaching to the waist.

Rouzet or Roustet: Coarse strong French woollen used for cheap clothing.

Rove, in jute processing: A continuous strand of spun-jute fibre having a spindle number of 50 or greater.

Roving: A loose assemblage of fibres drawn or rubbed in to a single strand, with very little twist. In spun yarn systems, the product of the stage, or stages, just prior to spinning.

Roving, in glass textiles: A multiplicity of filaments or yarns gathered together into an approximately parallel arrangement without twist.

Roving: A loose assemblage of fibres drawn or rubbed into a single strand with very little twist.

Roving: (1) In spun yarn production, an intermediate state between sliver and yarn. Roving is a condensed sliver that has been drafted, twisted, doubled, and redoubled. The product of the first roving operation is sometimes called slubbing.

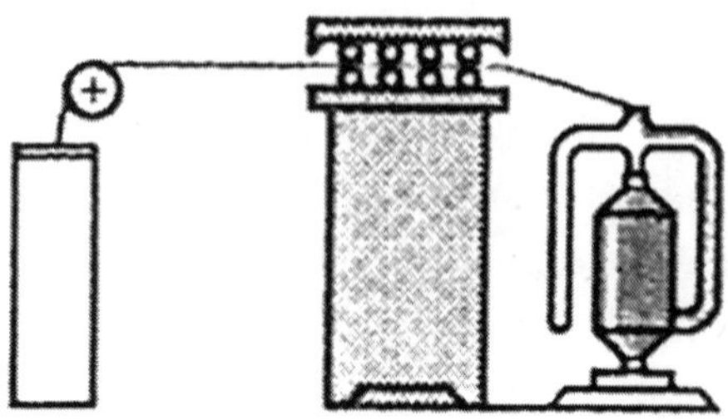

Roving (Schematic)

(2) The operation of producing roving. (3) In the manufacture of composites, continuous strands of parallel filaments.

Roving Construction number, in asbestos: A system designed to show the construction namely, the cut of single roving and whether enforcement is present.

Roving frame: A general name for all of the machines used to produce roving, different types of which are called slubber, intermediate, fine, and jack. Roving frames draft the stock by means of drafting rolls, twist it by means of a flyer, and wind it onto a bobbin.

Roving, reinforced asbestoes: Asbestos roving containing a reinforcement of one or more strands of other fibres or wire.

Rowdy: Flaw in cloth, consisting of streaks.

Rowey: See **Uneven surface.**

Rowing Shirt: A low-necked shirt with very large deep armholes.

Rows: In woven oil floor covering: The average number of pile tufts per inch in warpwise direction.

Royal Armure: A narrow, stout silk dress fabric, woven with a pebbled face.

Royal-Axminster carpet: A carpet having a tufted cut pile, formed by the loom.

Royal cashmere: Fine English summer dress goods of wool warp and worsted filling.

Royal Rib: Same as cardigan.

Royale: Silk dress goods, similar to Gros de Tours, made with two harnesses, four warp ends in each heddle.

Royale weave: Royale is a modification of the regular Gros de Tours, inasmuch as the rib line, which in the latter runs straight across the cloth, is broken off after a given number of warp-threads. These groups, which may comprise 8, 12 or more threads, will interlace each one pick higher than the preceding one.

Royalette: In England a stout fabric made with cotton warp and Botany filling, woven in five-leaf weft satin weave.

Ruanas: Cotton ponchos in Columbia, made in plain weave, usually narrow woven stripes placed three-quarter inch apart.

Rub: See **Abrasion Mark.**

Rubber: A material that is capable of recovering from large deformations quickly and forcibly, and can be, or already is, modified to a state in which it is essentially insoluble (but can swell) in boiling solvent, such as benzene, methyl ethyl ketone, and ethanol-toluene azeotrope.

Rubber: A natural or synthetic polymeric elastic material. Natural rubber is a polymer of methylbuta-1,3-diene (isoprene). Various synthetic rubbers are made by polymerization; for example chloroprene rubber (from 2-chlorobuta-1,3-diene) and silicone rubbers. See *vulcanization*.

Rubber cloth: Cotton fabric, rubberized on one side, used for raincoats in the lighter weights and suit eases and carriage trimmings in the coarser grades.

Rubber coated Textile Fabric: Specialised fabrics made by coating Rubber, examples are water proof shetting etc.

Rubber compound: An intimate mixture of elastomer(s) with all materials necessary for the finished article.

Rubber fibre: A manufactured fibre in which the fibre forming substance is comprised of natural or synthetic rubber including the following categories. (1) A manufacture fibre in which the fibre forming substance is a hydrocarbon such as natural rubber , polyisoprene, polybutadiene, copolymer of diens and hydrocarbons or amorphous (noncrystalline) polyolefins. (2) A manufactured fibre in which the fibre forming substance is a copolymer of acronitrile and a diene (such as butadiene) composed of not more than 50 % but at least10 % weight of acrylic nitrile units –CH_2-CHCN-). TRhe term'lastrile' may be used as ageneric description for the fibre falling in the category. (3) A ,manufactured fibre in which the fibre forming substance is a ploychloroprene or a copolymer of chloroprene in which at least 35 % by weight of the fibre forming substance is composed of chloroprene units (-CH_2-CCl=CH-CH_2-).

Rubber Filament: A filament extruded from natural or synthetic rubber and used as the core of some elastic threads.

Rubber Thread: Thread made of Rubber used mainly before the invention of Elastanes. See **Rubber Filaments.**

Rubber velvet: Consists of a rubberized cloth ground over which coloured flock is blown while the rubber is still soft, giving it a soft, velvety face. Used for raincoats.

Rubberise: See **Rubberising.**

Rubberising: To impregnate or coat with rubber compound or both.

Rubbing fastness testing: (crock fastness testing). A test to check the resistance of textiles and their dyes against soiling and crocking during use; The guidelines apply to all fibre types all standards are almost same. For dry fastness, dry dyed material is rubbed with dry cotton fabric; for wet rub fastness, dry dyed material is rubbed with wet cotton fabric; it can be carried out with a special rub tester (crockmeter). The fabric is rubbed to and fro 10 times over 10 cm for 10 seconds, at a load of approximately 1000 g. The staining of the cotton rubbing cloth is then compared against the grey scale/ stain.

Rubbing fastness, Dry: See **Rubbing fastness testing.**

Rubbing fastness, Wet: See **Rubbing fastness testing.**

Ruby number: A value used for protective colloid effects of textile auxiliaries. It is based on protection against precipitation of a sodium chloride solution, with a solution of Congo ruby A being used as a indicator. The ruby number is the protective colloid in mg, which is just able to prevent the colour change to violet after 10 minutes.

Ruff: Spanish projecting frill of several layers of starched, pleated fabric worn around the neck.

Ruffles: Strips of fabric cut from circles and then opened out and applied inside a band or on a seam. They can be of any width, but may need to be seamed together to create a long ruffle

Rug: (1) Thick and heavy floor covering made of cotton, wool, silk or jute, made with or without any pile, by hand or on the loom, covering of a limited area which is complete in itself and is intended for use as a partial covering of a floor or another floor covering.; also a twilled and figured blanket used for traveling; (2) Coarse woollen fabric worn by the poorer classes in England during the 16th century.

Rug Back: (1) Double shed back, given to carpets and rugs to increase their wearing quality or that part of a rug normally in contact with the floor; (2) the underside of a rug as opposed to the use surface.

Rugby: In England a stout, coloured cotton shirting.

Ruggedness test: An experiment in which environmental or test conditions are deliberately varied to evaluate the effect of such variations.

Rugging: In England a stout, coarse woollen fabric, finished with a thick nap on the face; used for floor covering.

Rugginose: Waste silk, obtained from stained or imperfect cocoons.

Rumal: An old Indian art still practiced today, where fine cotton headscarves and wedding cloths are silk embroidered with figures. Sometimes graced with delicate needle work and motifs, often featuring the Hindu deity Krishna. Infused with the beauty of miniature paintings, theses rumals were traditionally gifted during festivals and ceremonies and also used to cover wedding gifts.

Rumburg Linen: A Fine Austrian linen, the fineness and number of yarns in the warp and weft are alike.

Run: The unit in the American system of woollen yarn count, meaning 1,600 yards of yarn in a hank, the number of hanks giving the yarn count.

Run-in: This is the yarn consumption of each guide bar at warp knitting and it is the length of each yarn knitted into fabric during 480 knitting cycles (one rack).

Run-in ratio: this is the ratio of run-in value of different guide bars at warp knitting.

Run Work: A needlework, consisting in running white or coloured thread through the meshes of a net to form a design.

Run, in experimenting or testing: A single performance or determination using one of a combination using one of a combination of experimental or test conditions.

Run, in knitted fabrics: A series of dropped stitches.

Run, in the wool system: The number of 1600 yd. Lengths of yarn per pound. An indirect yarn numbering system generally used for yarns spun on the woollen system.

Runner: See **End out.**

Runner: A break in the yarn of a knit fabric that causes the stitch to "run" along the needle line (wale) in a vertical direction. See **End out.**

Runner length: In knitting, the number of inches of yarn from a warp to make one rack of fabric.

Runner, carpet: A long narrow length of textile floor covering finished at both ends.

Running and back stich: The running and back stitch is made by taking a few running stitches, drawing out the needle and making a back stitch over the last running stitch to strengthen the seam. Care must be taken not to hold the side next the worker too full and not to miss the under material, but to take the stitches even on both sides.

Running speed: The speed at which a textile product being treated passes through a section, e.g. a finishing machine. It is measured in m/min, while the rotational speed is given in rpm; in weaving machines running speed is measured by weft insertions/min.

Running stitch: A continuous embroidery stitch formed by passing the thread through the material alternately from top to under side and vice versa. Method: Pass the needle over and under the fabric, making the upper stitches of equal length. The under stitches should also be of equal length, but half the size or less of the upper stitches.

Run-of-the-mill: See **Mill run.**

Run-proof: A knitted construction in which the loops are locked to prevent runs.

Run-resistant: A type of knitting stitch that reduces runs.

Rupture: The breaking or tearing apart of a material.

Russel/Russel cord: Cord Union fabric made with cotton warp and worsted filling, two warp ends being run through each dent, forming cords.

Russia: Braid Made of silk or mohair for embroidery purposes.

Russian cord: A colour striped shirt or dress fabric in which the cord stripes are produced by leno weaving. One end is made to cross a number of ends in an alternate crossed and open shed sequence.

Russia Crash: Coarse, strong, plain woven unbleached cloth made of hemp or coarse linen yarn. Used for towels, coats, etc.

Russian Blanket: Very heavy blanket with thick nap raised on both sides, made with wool warp and heavy mohair filling, usually in different coloured stripes. Used for automobile robes.

Russian Duck: Fine bleached linen canvas, used for summer clothing. A heavier grade is used for sail cloth.

Russian hemp Identical to Pita fibre; See **Mauritius hemp.**

Russian Lace: Coarse crochet lace of cotton or linen with simple patterns.

Russian Serge: A very strong, thick all wool serge for women's coats.

Russian Twill: Trade name for fine, clear faced twilled woollen dress goods, woven of right hand twist yarn and dyed in the piece.

Russian Veiling: Cotton veiling made with large meshes and square patterns woven between them.

Rusty Silk: Flaw in white or delicate coloured silk cloth showing fine brownish streaks in the filling.

Rutile: A modification of Titanium dioxide. Used as a white pigment in textile pigment printing.

Rya carpet Scandinavian long-pile carpets in characteristic colours and patterns. They were originally hand-knotted from rough wool (genuine Rya). They are now machine-woven by various processes.

Rhythm, in garment designing: Rhythm is the repeated use of lines or shape to create a pattern. Rhythm also refers to a feeling of movement. Rythmic use of design elements lead the eye through a design giving the garment a continuity e.g.:- use of laced trims and contrasting cuffs and collars lead the eye through the design with a successful rhythm. The eye should travel without effort from one part to the other.

Rhythm, Linear refers to the characteristic flow of individual line. Linear rhythm is not as dependent on pattern, but it is more dependent on timed movement of the viewer's eye. See **Rhythm.**

Encyclopaedic dictionary of textile terms

Encyclopaedic dictionary of textile terms

Volume IV

Mathews Kolanjikombil
Senior Textile Processing Technologist

WOODHEAD PUBLISHING INDIA PVT LTD

New Delhi

Published by Woodhead Publishing India Pvt. Ltd.
Woodhead Publishing India Pvt. Ltd.,
303, Vardaan House, 7/28, Ansari Road,
Daryaganj, New Delhi - 110002, India
www.woodheadpublishingindia.com

First published 2018, Woodhead Publishing India Pvt. Ltd.

Reprinted, 2019, 2020

Woodhead Publishing India Pvt. Ltd. ISBN: 978-93-85059-11-7
Woodhead Publishing India Pvt. Ltd. e-ISBN: 978-93-85059-66-7

Typeset by Bhumi Graphics, New Delhi
Printed and bound by Replika Press Pvt. Ltd.

Contents

Volume IV

Preface

It has been told the first thing that a business man should know, is—the little things of his business. This book is meant for textile personae, be it a textile student, textile chemist, a spinner, a weaver, garment maker or a merchandiser who may be an expert in his/her field but he/she may come across many terms in his/her day to day business which he/she is not familiar with but is related to his/her fieldwhich he/she should know, but nine times out of ten does not know. He/She may not have an expert in that field of that term near him/her to clear his doubt about that term. This book comes handy in such circumstances.

There is nothing scholarly in this book, but is a compilation of short easy understandable meanings of the textile terms enough to help the referrer to understand the term. I have come across dictionaries which gives the meaning of the textile terms in one sentence. But it may not be sufficient to give a full idea about it. But this book is a little different. The meaning of the terms is explained in a concise manner even with the help of diagrams or photos, wherever necessary, which is enough to clear his/her doubts. These terms and meanings have been collected right from my college days and throughout my career. I believe it is not complete, but, such as it is, the compilation is reliable. There are further terms which are being collected by the author which will be added in the next edition, probably. The author believes that he has produced a book which may be profitably consulted by all who are either interested or practically engaged in textile trade.

It has to be specially mentioned that, the students can use the present book as a reference guide for his/her immediate needs without going to many text books. For detailed study of any terms he/she can further refer books specialized in that field. I may not suggest this book for a research student. The author also has tried to explain the construction of many fabrics new or old for general knowledge.

Hope this book will be greatly accepted by textile personnel. Any suggestions or corrections are welcome, which will be included in the next editions.

Mathews Kolanjikombil

Senior Textile Processing Technologist

Bangalore

Dedication

This book is dedicated to my Father (Thomas Abraham) and my Mother (Thankamma Abraham) who solely made me what I am today.

S

S Twist: See **Direction of twist.**

SA 8000: Social standards in textiles and clothing by Social Accountability International.

S-Finish: A term originally applied to the partial surface saponification. A term originally applied to the partial surface saponification of cellulose ester fibres (acetate and triacetate) by alkaline hydrolysis to reduce static charge and improve fabric handle. Alkali treatment of polyester. Nowadays the term has also been adopted for an analogous treatment of polyester fibres in which a controlled partial saponification with alkali (NaOH) is carried out to achieve a silk-like lustre and handle, reduce the buildup of static charges and improve anti-soil properties.

Sable: French for mottled effect.

Sable: A dark brown luxury fur which is very hard wearing.

Sabretache: A leather pocket hung from the left side of the sword hilt; a popinjay ornament.

Sabrina: Applique needlework, the leaves and petals of flowers made of coloured material edged with button hole stitches.

SABS: South African Bureau of Standards

Sacharilla Mull: A very delicate bleached cotton muslin, made of fine yarn with a low construction. It is given a very soft finish. Used for veils and turbans by the Moslems.

Sack: A business coat with pockets, made single and double breasted; a loose Chesterfield overcoat.

Sack Cloth: Very coarse cotton or jute fabric, woven plain, used for bags.

Sacking: See **Bagging and Heavy goods.** (1) Heavy, three or four end twill jute or hemp cloth of double warp and single filling, used for cement and ore bags; (2) Solid coloured flannels for kimonos.

SAD: SAD is a measure for soil pick up which is obtained by reflection measurement.

$$\text{SAD} = \log(\text{R-unsoiled}/\text{R-soiled})$$

Where R = Reflection according to wavelength.

Saddening: Same as after treating.

Saddlecloth: In Arizona and Mexico masses of aloe fibre spread out in regular thickness and tacked to keep them in place. It is not woven.

Sadin: In the Bible means linen cloth.

Sadowa: Woollen dress goods with the nap being raised in circles, dots, squares, etc.

Safety Data Sheets: The ETAD (Ecological and Toxicological Association of the Dyestuffs Manufacturing Industry) has been issuing safety data sheets to affiliated dyestuff manufacturers since 1974 in order to support these efforts to protect environment and to provide information on the regulations to be complied with, details of the physical, toxicological and ecological properties, together with handling, transport, storage and safety aspects. The ETAD (Ecological and Toxicological Association of the Dyestuffs Manufacturing Industry) has been issuing safety data sheets to affiliated dyestuff manufacturers since 1974 in order to support these efforts and to provide information on the regulations to be complied with, details of the physical, toxicological and ecological properties, together with handling, transport, storage and safety aspects. The ecological details are listed under point 8 in the SDS, i.e. whether substances are biodegradable, toxic to fish, harmful to waste water bacteria, and de ails of the water pollutant toxicity classification of the product. More information concerning the exact chemical composition of the products, proper disposal and the by-product level present will be desired in future. If the content of the safety data sheet should change within twelve months of delivery, the supplier must automatically forward the client an updated version of the safety data sheet.

Safety stitch: A stitch formed by an over edge stitch reinforced by a chain stitch (or sometimes a lockstitch) further in from the material edge.

Safety valve: A pressure relief valve actuated by inlet static pressure and characterized by rapid opening or 'pop' up action.

Saffron: A fugitive yellow dyestuff derived from the flowers of the crocus; used formerly to some extent.

Sales invoices: Invoices of the goods sold.

Sail Cloth: A firm fabric in plain colours in plain or basket weave. It is made from cotton, or polyester and cotton, and is stiff and hard wearing. Used for trousers, dresses, children's clothes, men's summer jackets etc.

Sail test: Combined wear and laboratory test for Static charge. A test subject wearing a waist slip made from the fabric to be tested passes a polyester sail fixed in an air conditioning chamber. The slip rubs against the sail and is charged. A second person assesses and records the charging and discharging per unit of time.

Sailcloth: A canvas/duck that is used for the manufacture of sails. Dense and tight, impregnated heavy fabric (linen, cotton, hemp, synthetic fibres) in plain weave. Used for tarpaulins, awnings, sails, window blinds, etc. Laminated fabrics are also finding use in this market.

Sail duck: See **Duck.**

Saint Andrew: In embroidery a stitch forming a St. Andrew cross in a square.

Saint Georges: Unbleached French linen of medium quality.

Saint Jago: Cotton goods in Sierra Leone, Africa.

Saint Jean: Coarse, unbleached French linen made in various widths.

Saint Lucie: Very fine French silk yarn.

Saint Maur: French serge made of pure silk or mixed with wool.

Saint Nicolas: French woollen serge, used by the army.

Saint Omer: Narrow, 17th century English worsted fabric.

Saint Rambert: Unbleached French linen.

Saint Remy: A grade of French organzine or raw silk.

Saint Vincent: Variety of raw cotton from the West Indies.

Sakalleridis: Valuable Egyptian Maco cotton.

Salampore: The tough fabric made in Nellore, India as work wear. Salampores were the preferred dress material for the slaves on the West Indian Plantations. But the West Indian Emancipation Act of 1833 sounded the the death knell of salampores, as "freed negroes refused, very naturally, to wear the grab of their slavery.

Salisbury: White Sort of white English woollen flannel.

Salt: An ionic compound that is formed by a neutralization reaction (reaction of an *acid* with a base) The best known salt, common salt, or sodium chloride (NaCl) is extensively used as an electrolyte in dyeing. Sodium sulphate is also used for this purpose. Many of the chemicals used in dyeing are technically salts, including many dyes.

Salt accumulation: When recycling e.g. rinsing water which is sometimes still cleaned in in-house waste water treatment plants, the salts remain dissolved in the water and are not removed. When the recycled water is reused as process or rinsing water, more salt is added to the salt concentration already present from the first use of the water. If part or all of the water is used several times, the salt water concentration continues to increase imperceptibly and may become a source of interference.

Salt Dyes: See **Direct dyes**.

Saltillo: Woollen blankets made by the natives of Northern Mexico, with a large and richly ornamented medallion, consisting of various coloured concentric bands. This medallion is in the center of a ground covered with small all-over geometrical designs in vertical or diagonal rows. The leading colours of these blankets are blue or red with smaller quantities of green, yellow and black. The weave is usually very fine; the selvages are finished smoothly and without overcasting and the ends are finished with fringes. These blankets are used as ponchos or serapes.

Salting: This refers to incorrect precipitation of dissolved dyestuffs or textile auxiliaries as a result of excessive addition of electrolytes which results in bronzing and insufficient fastness to crocking.

Samardine: Plain French serge, made with eight ends and four picks in a repeat.

Samarkand: Medium size rugs made in Central Asia of wool, silk or cotton web and loose cotton or silk pile, tied in Senna knot. The design shows Chinese influence with five or less medallions, fret-work patterns and odd, stiff flowers. The colours are yellows, blues, reds, browns and white.

Samis or Samit: A medieval fabric made with very heavy silk or hemp warp containing six threads and flat gold filling (tinsel). Originally was made in Italy. It is believed by some to have been similar to velvet.

Sample: (1) A portion of a lot of material which is taken for testing or for record purposes.

(2) A group of specimen used, or observations made, which provide information that can be used for making statistical inferences about the population(s) from which the specimens were drawn.

Sample blankets/section blankets: Woven trial pieces with differing sections across the warp. These sections may vary in colour, yarn type and/or weave.

Sample, bulk: See **Bulk sample and Lot sample.**

Sample, discrete: See **Discrete sample.**

Sample, laboratory: See **Laboratory sample.** A sample taken to represent the lot sample or the original material and used in the laboratory as a souce of test specimen.

Sample, lot: See **Lot sample.**

Sample skein: Skein reeled from the package or beam of the laboratory sample, and used in the laboratory as a source of specimens.

Sampling unit: An identifiable discrete unit or subunit of material that could be taken as part of the sample.

Sampling Vessel: This is a small vessel attached to the dyeing machine, in which quantities of sample are switched to the secondary liquor cycle. These are connected

to the dyeing machine so that the test specimens (samples) can be incorporated into the liquor circulation during dyeing. The interconnecting valves between the dyeing machine and the sampling container should be closed to remove the sample. The liquor should then be drained off from the sampling vessel and this must also be vented if necessary.

Samuhu: (1) Good bast fibre, yielded by a species of the Chorisia tree in Argentine; used for cordage; (2) Silky, flexible but short fibre, yielded by a species of the Bombaceae in Paraguay; used for ponchos, etc. The Indians use the net-like bast for various purposes.

Sancowad Process: Short liquor dyeing process (Sandoz) for fully fashioned articles, stockings, socks, etc. with liquor-to-goods ratios of 1 : 1 to 1 : 2, using special textile auxiliaries which produce microfoam.

Sand: Used as a filter medium in fibre manufacture, particularly used in spinning packs for nylon or polyester production.

Sand Crepe: A type of crepe. A typical sand crepe with apparel weight will have a texture of 90 x 70 and is woven in small broken crepe weave. Short yarn floats in no discernible pattern give an irregular surface texture. Used for dresses and blouses. See **Crepe.**

Sand filter: A layer of sand through which water or wastewater is passed and filtered.

Typically the sand is contained in open concrete tanks. ***Rapid gravitys and filters*** have a downward flow through the sand and are common in water treatment. *See also* **biflow filter, continuous backwash up flow sand filter, moving bed sand filter, pulsed bed sand filter, slow sand filter, upward flow sand filter**. In stormwater management, a 0.5 m deep sand bed, down through which stormwater passes in order to remove much of the suspended matter before the stormwater enters the sewer. The typical sand size is 0.5 mm. The sand filter may be in an underground structure or it may be built just below ground level with a cover of topsoil and grass. Other storm waters and filters include ***linear stormwater sand filter*** or ***peat sand filter***.

Sandarc Gum: Natural Resin from the bark of a North African cypress. Lemon yellow pieces, powdery, fresh surface fracture with glassy lustre. Melting point approx. 135°C. Soluble in alcohol, chloroform, distilled oils. Only used occasionally for silk ribbon finishes and lacquers.

Sanding: Sanding is distinguished from Emerizing in that only one roller is used as a tool per machine, whereas several rollers work on the fabric in emerizing. A guide roller advances the fabric as close as possible to the abrasive roller which operates at high speed. The distance determines the sanding effect and must therefore be accurately adjusted.

Sandwash silk: Firstly, the silk is treated with an enzyme (protease) in the drum washing machine in a weak alkali bath with a low liquor-to-goods ratio. This causes the fabric to peel slightly and at the same time, the mechanical strains on the fabric cause a slight napping of the silk fibres. Together with the appropriate final finish, this achieves the characteristic sandy soft handle. Ready-to-wear items are often finished in this way.

Sandwich: A method of preparing fibre mixtures by layering them horizontally in alternating layers with all elements in the proper proportion. Vertical sections are cut and fed to the next machine in the process, where blending is effected.

Sandwich dyeing: Method of impregnating used predominantly for synthetic fibrous material, mainly to dye light shades whereby the material, rather than being fed through the liquor and squeezed, passes in close contact between two endless absorbent cloths which constantly absorb liquor. Dyestuff from these liquors is thus transferred to the material to be dyed.

Sandwich effects in dyeing (greyness): A dye migration phenomenon in continuous dyeing. The dye migrates from the surface to the interior of the textile material or the dye is washed off the textile surface. As a result, the textile material only appears to be dyed in the middle whilst the surface has a whitish appearance which is often referred to as “greyness”. The problem can be remedied or prevented by using appropriate dyeing systems.

Sanfor Knits: Brand name, which is no longer protected and which is a control standard, relating to cotton knitwear. The standard is achieved by Compressive shrinkage with additional light synthetic resin finish.

Sanfor-set: Brand name which is no longer protected which is a control standard. Limited to effects of textiles which have been finished using the Liquid ammonia process and are characterized by good shrink resistance under drying conditions in a tumbler drier. This process can also be used for lighter cotton qualities.

Sanforise: Mechanical compacting is one method of reducing residual shrinkage. The process forces yarns closer together and the fabric becomes thicker and heavier. As a result of this, the net yardage yield is reduced. A Sanforizer is a fabric compactor developed by Cluett Peabody. The term Sanforized, is their registered trademark and is used to market fabrics that meet certain shrinkage specifications. The term Sanforized is now generally accepted to mean a fabric that has low residual shrinkage and the term Sanforizing is used to describe shrinkproofing processes. While the patents on the machinery have expired, the trademark is actively promoted by Cluett Peabody. The effect of Sanforizing can be seen in figure 68 which shows that open fabric structure has been closed up somewhat. The process, figure 69, consists of a range where the fabric is first moistened with steam, to make it more pliable, run through a short tenter frame (pup tenter) to straighten and smooth out wrinkles, through the compressive shrinkage head and then through a Palmer drying unit to set the fabric. The fabric is wound into large rolls under minimum winding tensions. If the winding tension are excessive, the fabric will be pulled out and the degree of compaction lessened. Usually, a lubricant is added in preceding operations to assist in the realignment of the yarns as the fabric runs through the compactor. Selection of the proper lubricant is critical for some fabrics.

Sanforized: A material which has gone through the Sanforize treatment.

Sanforized Mark: A crimped, rippled, wavy, pebbled, or cockled place in the cloth showing distortion of the texture, produced due to the wrong operation in the Sanforising process.

Sanforset®: A trade mark of Cluett, Peabody & Co., Inc., denoting a controlled standard of shrinkage performance originally developed for denims. Fabrics bearing this trademark will not shrink under home-wash, tumble-dry conditions because they have been subjected to a liquid ammonia treatment and compressive shrinkage.

Sanfor-Set®: See **Ammonia process.**

Sanforising mark: See **Sanforized mark.**

Sangati: One of the finest grades of plain Dacca muslins.

Sangi: Cast Indian satin, made of tussah silk and cotton.

Sanglier: Closely woven French fabric made of hard twist worsted or mohair yarn.

Sanitary wear: In the broadest sense, sanitary wear refers to so-called anti-rheumatic clothing. The alleviating effect is provided in conjunction with electrostatic properties and high thermal retention (heat retention).

Sanitation: Process to control the germ count with the aim of avoiding the risk of undesired germs spreading, achieving an inhibiting effect and/or counteracting damage to material caused by microbes. Antimicrobial finishes; Sanitized finishing.

Sanitised Finishing: Antimicrobial finishes). Protects textiles against the growth of bacteria and mycotoxins, achieving a deodorizing effect at the same time (sanitized finish). Currently used for lining fabrics, sportsgoods, stockings, etc. Sanitized finishing is achieved using e.g. quaternary ammonium compounds, halogenated phenols, salicylanilide derivatives, neomycin sulphate, etc.

Sanseviera Fibre: A fibre obtained from the leaves of various various species of plants of the genius sanseviera, (tropical Africa, Sri Lanka). The sansevieria fibre from Sri Lanka is finer than the African equivalent, similar to Mauritius hemp. Use: for string, matting and coarse cloths.

Santos: Brazilian cotton, with a silky, white staple.

Sanyan: Variety of wild silk from West Africa.

Sanz: New Zealand. Standards organization.

Saponide: Term used abroad, especially in France, to refer to anionic organic detergent base materials, with the exception of soaps.

Saponification: Specifically in relation to manufactured fibres, saponification is the process of removing part or all of the groups from acetate or triacetate fibre, leaving regenerated cellulose.

Saponification number: The saponification number indicates how many mg of potassium hydroxide are required to saponify 1 g fat.

Saponised Yarn: Cellulose acetate filaments are treated to produce very strong yarn.

Sappy: Wool containing a large percentage of natural grease and subject therefore to excessive shrinkage.

Saraband: Small and medium size Persian rugs made of cotton warp and weft and close and short wool pile, tied in Senna knot. The design consists almost without exception of rows of pears and many narrow borders in dark red, blue, etc.

Sarakhs Rugs: Heavy, all-wool Persian rugs, the long and close pile is tied in Ghiordes knot. The design consists of medallion and floral figures chiefly in rich red and blue.

Saran Fibre: A manufactured fibre in which the fibre forming substance is any long chain synthetic polymer compound of at least 80 % by weight of vinylidene chloride units ($-CH_2- CCl_2-$). This fibre, which has an excellent resistance to sunlight and weathering, and is used in lawn furniture, upholstery, and carpets.

Sarcina: Sarcina are earth- and air-borne bacterial microbes (in coccus or cuboidal form, sporadically forming spores). Also occur in perspiration (up to 45% on the upper arm). Sarcina can be harmless but also pathogenic.

Sardasi: Indian velvet, embroidered with gold or silver threads.

Saree: See **Sari.**

Sargia: Medieval Italian twilled fabric, made of wool and silk.

Sargues: French serge made of linen and carded wool.

Sari: A sari is a wrap-around robe used by Indian women which is often printed with many patterns. It is used as plain dyed also.

Sarille: Trade name for Courtalds fibre. Soft cellulosic fibre, developed from viscose and used to produce warm fabrics. Mainly used in dress fabrics, household textiles as blankets, and often mixed with other fibres, such as wool, to reduce the cost.

Sarong: A Sarong is a Malaysian item of clothing for women which often has large characteristic batik patterns.

Sarplar: In England a bag of wool measuring 2,2,40 pounds.

Sarplier: A coarse, strong, plain woven cotton fabric; used in England for baling wool.

Sarpuz: Trade term for Persian and Anatolian knotted wool carpets; used as floor covering.

Sarraux: French linen canvas, made with blue checks; used by sailors for trousers.

Sarsenet: Plain, woven stout piece dyed English cotton cloth finished with high gloss, often calendered to give the appearance of a twill; used for lining, etc.

Sarsonet: Originally a fine silk Arabian fabric, it is now a net or veiling fabric used in millinery. Made from silk, nylon or polyeaster.

Sartex: Sartex is a Swiss association for textile labelling.

Sase: Acronym for stress at specified elongation; the stress experienced by a yarn or cord at a given elongation.

Sash: A waist-scarf.

Sash Ada Tari: Arabic term for bleached tanjibs (see) with a dark blue or lavender striped heading; used for head covering by the natives In Egypt.

Sashing: refers to strips of fabric that are sometimes put between blocks of a quilt to form a pattern or make the quilt larger. They are often made of patchwork.

Sateen: Similar to Satin but made in cotton or spun synthetic. Various satin weaves may be used in making sateen, but the 5 harness warp flush satin weave is by far the most coomon. It is soft and has a sheen and is often of loose construction. It is usually made in plain colours only. In pure Sateen weave the surface of the fabric consists of mostly weft floats only. The most common use of sateen is in lining of coats. It is made in different constructions. One

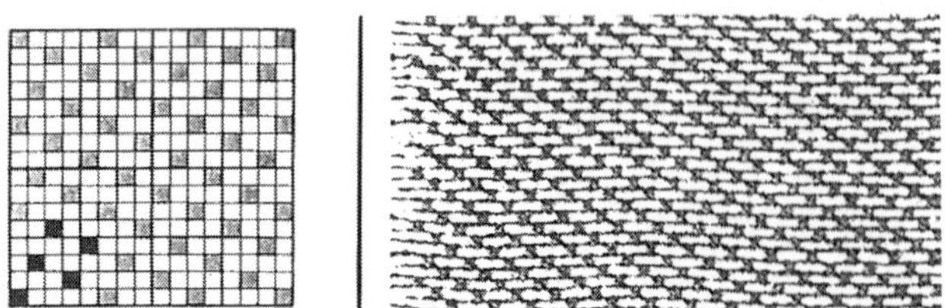

30 – 01 04 – 01 – 02

typical warp flush lining sateen texture is 110 × 60, with 24s carded yarn in the warp and filling. For weft flush lining sateen a construction of 70 × 120 with 30s warp and 40s weft may be an example. The fabric is usually starched and calendered or shreinerised to get the glossy finish. Another use is as curtain linings. It does not wear particularly well; seams tend to pull and crack mark appear where hems are pressed.

Sateen Fabric: A fabric made from yarns with low lustre, such as cotton or other staple length fibres. The fabric has a soft, smooth hand and a gentle, subtle lustre. Sateen fabrics are often used for draperies and upholstery.

Sateen, fine: Fine sateen is used for dress goods and sleeve linings are made from combed mercerized yarns. A thread count of 140 × 90 with 60s warp and 80s filling could be used for such a sateen. Fine heavy sateens, made from 8-harness satin weaves are called Venetian.

Sateen, heavy: Are used for outer wear such as windbreaker jackets and sky suits. In construction, finish, and uses these sateens are quite similar to shee gabardine, but normally they are slightly heavier.

Sateen, tickings: They are used for pillow covers and similar articles. They are heavier than lining sateens, a typical construction being 100 × 50 with 12s warp and 20s filling.

Sateen: (a) A fabric made in sateen weave.

(b) Weave A weft-faced weave in which the binding places are arranged with a view to producing a smooth fabric surface that is free from twill (q.v.).

Note: To prevent confusion with "satin", it is preferable to refer to this as "weft sateen weave".

Sati-drap: French dress goods, made of cotton or silk warp and woollen filling of the same colour, forming a weft satin. It is fulled in the finish; obsolete.

Satin: Examples of satin weave fabrics are cotton damask, sateen, ticking, linen damask, silk, rayon or other synthetic damask, satin.

Satin, envess (silk): Warp of organzine and weft of tram with a 45' diagonal weave structure. Evensided twill with strongly marked slant ribbed effect.

Satin: A fabric that is very smooth on the right side (although sometimes both sides are satin weave). This smoothness is produced by weaving in an interlaced pattern. Satin weave fabrics are produced from all yarns including silk, acetate, nylon, viscose, polyester, cotton.

Satin-back: Any fabric of any fibre where the reverse side is of satin weave.

Satin, duchesse (silk): a thicker, richly lustrous satin.

Satin: (1) Name derived from atlas or satin weave. Generic term for various, mainly lustrous materials, predominantly manufactured using 5-weave satin. Worsted satins are either napless (e.g. satin de Chine) or finished with a fine nap; carded satins almost always have a napped finish. Fabrics with a particularly napped finish have an intense shine which is further emphasised by the weave. (2) Smooth fabric (silk, viscose, wool, flax, cotton, etc.) with high sheen, in atlas or satin weave, e.g. Duchesse satin, cloth; liberty, etc., see 1.

Satin d'Amerique: Satin made of mixture of silk and agave fibre.

Satin Back: Velvets or taffeta and other ribbons made with a reverse side of satin.

Satin Berber: Stout worsted fabric, made in satin weave and finished with a lustre.

Satin Bonjean: French worsted trousering, fulled in the finishing.

Satin de Bruges: Upholstery satin of silk and wool.

Satin Check: Highly finished English cotton fabric, woven in checks.

Satin de Chypre: Obsolete French silk satin.

Satin de Laine: (1) French twilled drapery, made of English worsted, in solid colours, printed or brocaded; (2) French dress goods and cloaking made in various coloured brocaded patterns in satin weave.

Satin de Lyon: Silk satin made with a twilled back, and finely striped face; used for lining.

Satin Cloth: A lightweight woollen dress goods, made in satin weave and lustred.

Satin Damask: (1) the best grade, lustrous linen damask, used for table linen; (2) rich silk satin with fancy Jacquard designs woven either in fancy weaves or in pile.

Satin Delhi: Fine worsted satin.

Satin Duchesse Fine, stout and very lustrous silk satin, woven in eight-end satin **weave.**

Satin Ermine: 40-end French silk satin ribbon, made with two sets of warp, one eet forming a taffeta back, the other the satin surface, similar to the fur or the plush.

Satin Fabric: A traditional fabric utilizing a satin weave construction to achieve a lustrous fabric surface. Satin is a traditional fabric for evening and wedding garments. Typical examples of satin weave fabrics include: slipper satin, crepe-back satin, faille satin, bridal satin, moleskin, and antique satin.

Satin Figaro: Eight-end silk satin dress goods or millinery trimming, the warp being of alternately different coloured threads.

Satin fabric, warp knitted: The technical face of a satin fabric is similar to the face of a locknit fabric. The technical back of the fabric is smoother and shinier due to longer underlap movement of the front guide bar. It contracts in a widthwise direction when leaving the knitting zone due to longer underlap movement of the front guide bar. It has a curling tendency and greater risk of snagging.

Satin finish: High gloss given to cotton, wool or silk fabrics by calendaring.

Satin Foulard: Smooth and highly finished silk foulard.

Satin Grec: Solid coloured silk satin lining or dress goods made with single warp and high finish. Satin Grec is a 12-harness satin, in which a taffeta point is added to each place of interlacing, thus giving the cloth a much firmer hand.

Satin Jean: A stout, heavy cotton jean, made with a highly finished, finely twilled face.

Satin Lisse: Twilled, highly finished cotton dress goods in France, printed with small designs.

Satin Luxor: Rich, stout silk satin dress fabric, made with a corded face.

Satin Marabou: Thin, silk satin made with single marabou yarn (see) for warp.

Satin Merceilleux: A very soft silk dress fabric, woven in a seven-end eatin weave and given a very high finish.

Satin National: Six or eight-end French silk satin dress goods.

Satin Onde: Five or eight-end silk satin, made with single warp; used as dress goods or millinery.

Satin Quilt: Bleached or coloured cotton quilt, made with fine warp and 'filling, woven plain, a second, coarse filling forming raised patterns.

Satin Regence: Stout, rich silk satin dress fabric, made with 'fine runs weftwise.

'Satin a la Reine: Closely woven six-end silk satin.

Satin Rhadames: Fine silk satin dress fabric, made with fine diagonal lines running across the face.

Satin Royal: Double faced, silk satin with twilled stripes crossing the material.

Satin Sheeting: English cotton back thick satin, made of waste silk; used for dresses, etc.

Satin soleil: A version of satin weave, Satin Soleil shows a satin-like surface with a cross line appearance.

Satin stitch, in embroidery: Proceed with Straight Stitches worked closely together across the shape, as shown in the diagram. If desired, Running Stitch or Chain Stitch may be worked first to form a padding, underneath, this gives a raised effect Care must be taken to keep a good edge. Do not make the stitches too long, as they would then be liable to be pulled out of position. This stitch may be worked from right to left or left to right. The number of threads over which the stitches are worked may very depending upon the effect desired.

Satin Stitch: The satin stitch is an over and over stitch and is used on materials of all kinds for marking linen, etc. The *padding* is the first step and should be done in long even stitches placed closely and over one another in the center. The size and proportions of the figure or letters determine the size of the thread. Fine thread gives the best results. The outline should be run twice;

this keeps the edge firm. An even darning or basting stitches, chain stitches or outline stitch may be used if the space is not too small. The padding may be worked in an embroidery hoop to keep it smooth and even. Scallops may be padded in the same way or worked flat.

Satin Stripe: A fabric of almost any fibre, including cotton, that has a satin weave shiny stripe alternating with a contrasting streipe of a different weave of thickness of the yarn , even of different colour.

Satin Sultan: French dress goods and cloaking, also a lustrous East Indian silk fabric.

Satin Surah: Twilled, very soft surah, finished with great lustre.

Satin Tick: Very stout, cotton fabric, made in satin weave; used for upholstery.

Satin Turc: (1) four-end satin in France, made with single warp. It is given a high finish and is used for shoe tops; (2) French fabric made of wool and silk warp and wool filling in seven end satin weave; used for shoe tops, trousers, etc.; similar to lasting. Satin Vigoureux Dress fabric having a satin face or satin stripes, the warp yarns being printed according to the vigoureux process (see), giving a mottled colour effect.

Satin Weave: One of the basic weaves, plain, satin, and twill. The face of the fabric consists almost completely of warp or filling floats produced in the

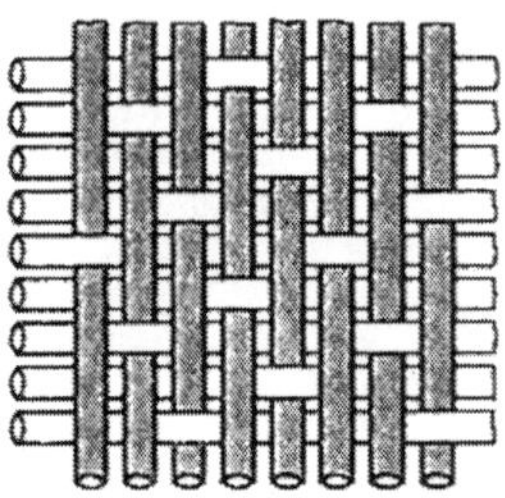

repeat of the weave. The points of intersection are distributed evenly and widely separated as possible. Satin-weave fabric has a characteristic smooth, lustrous surface and has a considerably greater number of yarns in the set of threads, either warp or filling, that forms the face than in the other set.

Satin, all silk: Is made with 300 × 100 construction with 18/20 denier warp and filling.

Satin, crepe-back: 50 to 100 denier acetate warp and 75 to 120 denier dull crepe twist acetate in the filling is used for this fabric. Textures range from 200 × 50 to 400 × 100. A normal construction is 300 × 70 with 75 denier bright yaen in the warp and 100 denier dull crepe twist yarn in the filling.

Satin, plain: (Panne): A typical construction is 110 × 40 to 300 × 80, with 75 to 150 denier lustrous yarn used in the warp and 75-300 denier yarn in the filling. The main feature of plain satin is its lutre and beauty.

Satin Zephyr: French dress goods, made of cotton warp and woollen filling, with a weft plush satin weave, fulled in the finish.

Satine: Twill lining that are given a glossy finish are called satine, sometimes.

Satinet: A cheap clothing material similar to cassimere, made with a cotton warp and a filling of short, inferior, shoddy wool which is mixed with enough long wool to enable it to be spun and woven in a way to bring that filling to the surface of the cloth; afterwards fulled, sheared, and the pattern printed on the face.

Satinette: Satinette is a combination of Satin and finette (cotton twill napped on one side in linen goods). The warp consists of lustrous viscose filament, the weft is cotton; the weave is in the form of 5-weave warp satin. The lustrous warp is generally found on the fabric face, whereas the cotton warp on the reverse side of the fabric is very napped. Used for night-shirts and pyjamas.

Saturated vapour: A vapor that is in equilibrium with its solid or liquid phase. A saturated vapor is at the maximum pressure (the saturated vapor pressure) at a given temperature. If the temperature of a saturated vapor is lowered, the vapor condenses. Under certain circumstances, the substance may stay temporarily in the vapor phase; i.e. the vapor contains more than the equilibrium concentration of the substance. The vapor is then said to be a supersaturated vapor.

Saturation, in colour chemistry: The attribute of colour perception that expresses the amount of departure from a gray of the same lightness. All grays have zero saturation (ASTM). See **chroma/chromaticity.**

Saturation bonding: A process of binding fibres into a nonwoven fabric by soaking the web with an adhesive.

Saturation Bonding: A method of making nonwoven fabrics in which a fibre web or batt is treated by overall application of an adhesive in the liquid form.

Saturation Limit: (1) The maximum intensity or purity of a colour. If the colour is as brilliant as possible, it is at saturation; if the colour is subdued or grayed, it is dull, weak, and low in intensity. (2) The upper limit concentration of a solute in a solvent, i.e., no more solute can be dissolved at a fixed temperature and pressure.

Saturation Value: The maximum amount of dye that can be absorbed by a textile fibre under defined conditions.

Saturator: A machine for thoroughly wetting fabrics with aqueous solutions or allowing interchange of liquor in wet-on wet processing like continuous bleaching, mercerization etc.

Saturator, in ETP: A pressure chamber in which air can be injected into water at (say) 4 atm pressure—e.g. as a part of ***dissolved air flotation***. A saturator can be either unpacked (empty) or packed with stones or plastics medium. Packing helps to break up the bubbles and flow pattern. The retention time is normally from 30 to 60 s.

Sauressig Printing Machine: In this machine, the printing heads are arranged vertically one above the other, unlike conventional roller printing machines where they are positioned around a central cylinder. The colour boxes, doctor blades and furnisher rollers are all mounted on a hinged sub-frame which can be swung to one side of the Saueressig roller printing machine to facilitate the changing of printing rollers.

Saueressig roller printing machine: Roller printing machines in which the central cylinder is replaced by individual cylinders for each printing roller.

Savage: Bleached, stout woven and hard finished cotton shirting in Venezuela, used for collars and cuffs.

Savonnerie: French-make rug made in imitation of Oriental knotted rugs in rococo patterns.

Saw-tooth effect: Defect in screen printing in the form of graduated (stepped) repeated profiles of fine pattern details which run askew; particularly undesirable with fine floral designs or figured patterning, etc. Correction: e.g. by using even finer gauze screen, use of photosensitive screen coatings with a crossed mesh effect and by correctly developing illuminated screens.

SAWTRI: (South African Wool Textile Research Institute).

Saxonienne: French silk armure dress good of small patterns, having the warp in various colours; obsolete. Saxony (a) The finest class of wool, having a short, very fine but strong and elastic staple, with excellent felting properties owing to the large number of serrations; used for the best grades of fulled fabrics; (b) A worsted fabric, originated in England during the 19th century, made with a warp of half-bred, English and Botany wool and the filling of Saxony or South-Down wool; (c) Same as merino in Scotland; 4, means a white flannelette in Canada.

Saxony: Originally made in the province of Saxony, Germany, the name has to be used to describe any soft, plain weave woollen or worsted cloth with a slight nap. High grade yarns are used so the term always implies a good quality cloth. It may be plain or in small checks.

Saxony: (1) A high-grade fabric for coats, made from Saxony Merino wool. (2) A soft woollen with fancy yarn effects, used in sport-coat fabric. (3) A highly twisted worsted knitting yarn. (4) A term describing a cut-pile carpet having highly twisted, evenly sheared, medium-length pile yarns.

Say, Saye: An all-worsted, four-harness serge of black colour, made in England since the middle ages until the 19th century; it was used for linings and shirts by certain religious orders and for aprons by the Quakers. It was usually made of Holland, English or Spanish wool.

Say Cast: Coarse wool taken from the tail part of the fleece.

Sayette: General name in France for various twilled or plain woven goods mixed with little silk. Used for lining and furniture cover.

Sayetterie: French term denoting woollens containing some silk yarn. Schappe See **Shappe**.

SBR: See **Sequencing Batch Reactor.**

Scaffolding effect: A mixed fabric sample (e.g. cotton and polyester) has different burning properties (e.g. in a vertical flammability test) than a corresponding sample made from one of the pure fibre components. Therefore the fibre portion made from polyester fibre for example promotes better burning of the associated cotton as it is worked and spun into the blended yarn almost as a stable scaffold. The scaffolding effect therefore benefits the relative low combustibility of the polyester fibres (at best melting), whereas the relative high flammability of the cotton is increased still further by the "loose" distribution in the scaffolding, which increases the oxygen available.

Scallop: Curves or indentations along the edge of a fabric.

Scalloped Selvedge: An abrupt, narrow indentation in the selvedge of the cloth.

Scarf, in apparel: An oblong or square piece of cloth worn for warmth or as decorative item.

Scanning in photo-engraving: This process refers to the breaking up of coloured areas into colour points of varying sizes, which when observed give the optical impression of different colour intensities due to blending of the colour lying between the points on the printbase.

Scarf Cutting Machine: After the fringes of scarf fabric have been twisted and set, the individual scarves must be cut longitudinally or transversely along the unwoven warp or weft thread which is done on this machine.

Scatter rug: A small rug which is designed to be flexible and is usually cleaned by laundering.

Scattering: Diffusion or redirection of radiant energy encountering particles of different refractive index. Scattering occurs at any such interface, at the surface, or inside a medium containing particles.

Schappe (slivers): These are wool like slivers formed by repeated combing and consist of parallel running filaments of 50-250 mm. in length. An important factor in spinning this material is the regularity of slivers. These are processed on a spreader in a single fleece 3 m. long and then by repeated drawing into a uniform ribbon.

Schappe (Silk): (Chappe, Floret silk). This is taken from the beginnings and ends of the cocoon which can no longer be unwound after the vegetable glue has been decomposed in a putrefaction process. The schappe undergoes further processing in a similar spinning process to that used for worsted yarn. The comber waste which falls away provides Bourette. Schappe is used e.g. for machine sewing silk and for strong silk fabrics. See **Silk.**

Scherli: See **Broche/Scherli.**

Schiffli Embroidery: Shuttle embroidery, the machine being run by a motor instead of by hand. The movement of the carriages is caused by motor power and the pantograph is operated by hand. The cross stitches are visible on both sides of the goods and the work shows the bobbin threads on the back of the embroidery.

Schiffli Lace: The name of a machine (Embroidery) producing various effects on lightweight fabrics.

Schmidt's machine: A machine used to dye loose cotton, consisting of two cylindrical vats one mounted on the other. The outer vat hold the liquor and the inner one with a false bottom holds the material to be dyed. After loading the material its kept in position by a copper wire sieve. A paddle wheel is also provided at the centre of rhe vats to help the liquor to move upwards. The liquor is heated by steam pipes and is raised from the outer vessel and poured over to the inner vessel which percolates to the bottom due to the vacuum produced in the bottom due to the movement of the liquor and the paddle.

Schooner Pants: Spring bottoms, after the sailor style.

Schreiner: A fabric finishing process which smoothens the right side, often impressing a pattern at the same time.

Schreiner Calendar: Often it is desirable to increase fabric lustre without overly thinning the cloth. Schreinering is a method of doing this. Schreinering is actually embossing by the useof a very special pattern. The pattern roll has anywhere from 250 to 350 lines per inch, etched at 26 degrees from the vertical. These lines are lightly embossed into the fabric and being regular, reflect light so as to give the surface a high lustre. This operation gives a silk-like brilliance to cotton fabrics. Schreinering mercerized cotton fabrics gives the nearest resemblance to silk.

Scoop, In Zipper: See **Elements, interlockable.**

Scorch Test: American test method (AATCC) to determine the amount of damage to hardenable synthetic resin finishes due to Chlorine retention on cellulose textiles. Also records the specific buffer capacity for each synthetic resin and each catalyst. Working principle: chlorine treatment (2.5 g/l active chlorine, liquor ratio 1: 50, 25°C, pH 9.5, 15 min), rinse, dry, condition, scorch or press between two metal plates with constant pressure and temperature (185°C, 30 s), condition. Assess by comparing the percentage average decrease in breaking strength and any yellowing which may occur.

Scorching: The tendering of a fibre surface by heat so as to change the colour and texture of the surface.

Scotch Carpets: Pile carpets, similar to the Kidderminster, with design on both sides but in different colours.

Scotch Checks: White muslin with plaid checks in coloured cord. Used for dresses, etc.

Scotch ell: 37.2 inches.

Scotch Fingering: Soft twist woollen yarn for knitting.

Scotch Gingham: Trade name for the finest grades of ginghams.

Scotch Plaid: Coarse, very durable twilled woollen fabric, made of native wool in 'Scotland in various tartan patterns.

Scotch Rug: A rag rug, made with a coarse two-ply cotton warp and long and narrow strips of wool rags.

Scotch Tweed: Woollen tweed made in twill weave using a white warp and highly coloured weft yarns. It has a rough, shaggy appearance, and is made in various weights for suits and overcoats.

Scotchguard: A registered showerproof finish applied to some fabrics.

Scott: A Highland tartan, composed as follows: Wide red field, split in the center by a narrow green stripe with a fine black line near each edge of the green; green stripe (measuring half the width of the distance between the edge of red field and the nearest black line) ; group (as wide as green stripe), composed of three red and two green stripes, the latter being wider and split in the center by a fine white line; green stripe, as above.

Scour: Essentially, thoroughly washing fibres or fabric to remove contaminants Yarns and fabrics may be dirty, contain natural waxes or oils, or have been treated with size or lubricants used in spinning, weaving or knitting. These can all interfere with dyeing, often leading to non-level results. Scouring is a large topic, and the process used depends on the fibre type and its condition. "True" scouring of greige cellulosic fabrics is typically done, after desizing, at the boil or at higher temperature in pressure vessels, with as much as 10 grams sodium hydroxide per litre of water, plus surfactants, and the process may last for several hours. Commercial scouring of wool may use solvents, similar to dry cleaning, as part of the process. White fabrics sold at retail have normally be scoured at the mill; "natural" fabrics usually have not (some "natural" fabrics have been scoured but not bleached).

Art dyeing literature often refers to what amounts to laundering as scouring. This is inadequate for greige fabrics, but often quite acceptable for "white goods". A long machine wash with the hottest water possible, about a gram of *soda ash* per litre of water (about a teaspoon per gallon) and some (preferably optical brightener free) detergent, followed by two rinses is usually acceptable. Sodium hexametaphosphate may be helpful if the water is hard. Woven white cottons often contain starch that will not be removed by such a limited process.

Scoured wool: Wool from the bulk of impurities has been removed by an aqueous or solvent washing process.

Scouring: A process where mill and natural dirt, waxes and grease are removed.

Scratch felt: Cheap quality woollen fabric made to resemble camel cloth. Poor wearing qualities.

Scray: That part of a processing machine where the fabric can be stored. The storage can be for giving time for reaction or synchronize the speed of one part of a machine to another, to avoid tension etc.

Screen: (1) A hollow, cylindrical, coarse-mesh wire device used in pickers and certain openers to form the loose staple stock into a sheet, or lap. The screen is mounted horizontally on a shaft on which it revolves freely. **(2)** A stencil used in screen printing. It is made of fine cloth, usually of silk or nylon, finely perforated in areas to form a design and mounted on a frame. The paste containing the dye is forced through the perforations onto the fabric, leaving the design. A series of screens, one for each colour, is used for multicoloured designs. (Also see **Printing**).

Screen, in ETP: (1) A sieve made of a flat sheet of wire mesh or punched steel plate for separating granular material into sizes ***(classification)***. It can also be made of parallel, wedge shaped wires or bars or round wire either flat or formed into a cylinder, as in the ***trommel***. In a ***mechanical sorting plant***, screens are an inexpensive first step and may also be used intermediately. (2) In the treatment of water or wastewater, a device that removes larger solids out of the flow. At reservoir intakes, screens are usually racks of massive bars. The bars may be vertical, horizontal or at a slope. They may be cleaned mechanically or by hand. At wastewater treatment works, and sometimes water treatment works, screens are part of ***preliminary treatment. Coarse screens, medium screens*** or ***fine screens*** may be used. They are usually designed for a velocity through the bars or mesh of 1.2 to 1.5 m/s at maximum flow. The most common are ***bar screens***. Other types include ***belt screen, cup screen, disc screen, drum screen, mesh screen, vibrating screen***. (3) An embankment or a wire mesh fence, up to 3 m high, built around a ***landfill*** especially on its lee side to catch wind blown refuse.

Screen Coating: The rotary printing screen system is based on seamless rotation screens. If the perforations in the rotary screen are sealed in places using a spec. lacquer (the rotary screen is then engraved) the coating unit can be used as a single-colour printing unit for a pattern. Light squeegee pressure makes it possible to apply a small amount of paste for vertical passage of the cloth. Hence coating will be the basis for the screen engraving to make the design. The coating will be as per the engraving methods adopted. For example if a manual photographic method is adopted the coating will have photographic chemical along with it. Coating is done using a coating machine and a squeeze to apply on the flat or rotary screens. After the coating the screen is cured for the coating to adhere to the screen so that it will withstand the wear and tear during printing operation in unengraved portions.

Screen Engraving: The method of engraving the design (one colou on one screen) on the coated screen. There are different methods available. Like manual photographic methods, Laser engraving, inkjet engraving etc.

Screen Films In Photo Engraving: These are used: (a) In film printing to copy texture or grid effects into the sample pattern. (b) In photo-engraving for printing rollers to copy the Hatching into the negatives of the colour separations by contact printing.

Screen Frames: (film screen frames). The screen gauze is supported by the screen frame. Wood, metal, plastics and combinations of the above, e.g. metal/ plastic, are suitable for the production of screen frames.

Screen Lacquer: This is used for prelacquering (preparation) in Screen making, as a support for photosensitive layers and as a reinforcement coating.

Screen Magazine For Printing: Screen magazines were placed to the side of or in front of the printing table in the first screen printing machines (Hikisch, Teximpex). They are divided into sections which can be raised and lowered in order to bring the desired screen to printing table height. Before printing, the movable screen frame is removed from the screen magazine and is slid back into the same compartment afterwards.

Screen Making: After the printing repeat has been designed and the colour separation has been produced, a flat screen is manufactured using the photochemical method, or alternatively a rotary screen is manufactured using the galvano-plastic method (or less often using the photosensitive lacquer process). The carrier layer is formed by coating with gelatine, which contains ammonia and potassium dichromate. During exposure, the unexposed layers may continue to swell and can be rinsed out at a later stage. Nickel foil cylinders can also be patterned using Laser screen engraving. See **Engraving; Positive process for screen making.**

Screen Mesh: (Screen mesh, screen gauze), (a) Silk: elastic, insensitive to pressure and impact, durable, sensitive to strong alkalis; (b) Phosphor bronze (alloy consisting of 93% copper, 6.75% tin, 0.25% phosphorous): extremely fine wires (0.09–0.03 mm), high resistance to alkalis and acids, low elasticity gives high accuracy of registration, sensitive to pressure and impact (bulges); (c) Polyamide, polyester: fine, even and durable fabric, resistant to chemicals, low moisture absorption, sensitive to highly concentrated caustic soda solutions.

Screen Print: This term is used instead of Screen printing predominantly in the graphics industry. Screen print denotes the printing process in which the dye paste is pressed through the substrate as through a sieve.

Screen Printing: A screen resist is made by covering a frame with bolting cloth of silk, metal or nylon filament yarns. The fabric is covered with a film and the design areas are cut out of the film. Some areas of the mesh are left open to allow the dyestuff to pass through and print the fabric. The frame is laid on the fabric, and the dye is placed at one end of the frame. A rubber knife moves the dye across the screen and forces the dye through the open mesh of the fabric, One screen is prepared for each colour. Screen printing is considered by many textile authorities to be newest method of decorating fabrics.

Screen Printing Carriage: In principle, this is a rigid frame with four runners which are positioned in such a way that the frame can be raised vertically to a height of approx. 100 mm by operating levers. During printing the screen printing carriage is moved along the table with the screen until the repeat device locks into place. After this has engaged, the carriage and screen are lowered so the screen lies against the goods to be printed. After printing, the screen is lifted off again by operating the lever and the carriage is moved to the next repeat or the next repeat but one.

Screen Printing Machine, Automatic: These are machines which carry out the entire printing process automatically when the screens are fixed in a stationary position while the fabric moves.

Screen Printing Squeegees: See **Doctor blade; Squeegee systems in printing.**

Screen Printing Tables: The table, in hand screen printing, on which the item to be printed is affixed, whilst with screen printing machines, the endless printing blanket runs over the screen printing table, on which the piece good, with a backing cloth if necessary) is affixed. Wool felt is often used under the printing blanket as a flexible layer.

Screen Sequence in Screen Printing: Particular care should be taken with the sequence of screens or print pastes as this can have a major influence on the printing result, especially if fine outlines, overlaps or resists are being printed.

Screen Washing: As soon as printing operation is over the screens are to be washed to avoid choke (blocking of the perforations of the engraved portions in the screen due to the drying up of the paste) up of the perforations. This choke up is very difficult to remove. Hence the screens are washed in special screen washing equipments. The outer screen wall is cleaned using a spray ring, and the inside is cleaned with a spray head with jets. Another possibility is the turbo spray washer which is attached to the central wash water feed on the inside of the screen: there are polyamide fibres on the rotating part which are hydro extracted on the screen wall due to the rotating centrifugal force in order to mechanically intensify the washing effect of rinsing with water.

Screw Feeds: Feed systems in which the action of the screw generates pressure that causes flow. The system usually consists of a container with a closely fitting screw unit.

Screw melter: (1) Screw extruder in which frictional forces between the screw and the heated barrel contribute to rapid melting of solid polymer. This configuration is capable of high throughput. (2) System in which a screw feed is used to feed polymer to a melt grid and to maintain a constant pressure at the grid.

Scrim: See **Cheese cloth**: (1) A generic term for A lightweight, open-weave, coarse fabric; the best qualities are made with two-ply yarns. A low-quality plain-weave fabric of the muslin type with traditional cover factors for both warp and weft of about (2) It is very similar to cheese cloth. Normally used for backing other fabric such as fur. Cotton scrim usually comes in white, cream, or ecru and is used for window curtains and as backing for carpets. The mass per unit area of the fabric will vary with the 35-70 g/m^2 when the fabric is made from cotton. (3) Fabric with open construction used as base fabric in the production of coated or laminated fabrics.

Scrimp roller: Also called Scroll roller. Rollers or bars (rails) characterized by grooves or projections inclined at equal and opposite angles to the centerline on each half of the rollers and used for removing folds or creases during finishing operations.

Scroll roller: See **Scrimp roller.**

Scroop: The sound of rustle or crunch that is characteristic of silk. Scroop is a natural property of silk, but may be induced in other fabrics to a degree by various treatments.

Scrooping agents: (1) Common as a finishing agent on fibres to produce specific finishing effects. Different types of scrooping agents are available to take into account water, fibre, type of fabric and intended use. Softening of textiles. (2) These are used to produce a scroopy handle (craquant, See **Silk scroop**). The textile fabric is treated with Marseilles soap, followed by an aftertreatment with lactic or tartaric acid. More recent methods are based on the use of ethoxylated fatty acids or paraffin emulsions.

Scrutching: This is a mechanical operation which, by breaking and beating the retted flax straw, separates the textile fibres in the stem of the plant from the woody matter and the bark. The next step would be 'hackling'.

Sculptured: A term describing a carpet with areas of contrasting depth produced by mixing cut pile and loops.

Scutcher (Finishing): A machine for continuously opening fabric which has previously been in rope form. Automatic and manual machines are available for this.

Scutching: The process of running a fabric in the scutcher. See **Scutcher.**

Scutching (flax): (breaking). This is the first stage in Flax processing after retting and drying, and involves breaking down the brittle woody tissue into shavings. It can be carried out either manually or mechanically using a pair of fluted rollers. It is followed by the scutching process.

Scutching Tow: Is the by-product of the scutching of flax straw, often being rescutched; it is classified in Ireland as coarse, fine and rescutched; used for ropes.

Scye Depth, in body measurements: The distance from the cervicale to a point level with the armpit.

SDC: Society of Dyers and Colourists (British) Their web site is at www.sdc.org.uk.

SE: Silk (mulberry silk).

S.E.B. (Single-End Break), in sewing: Refers to the singleend breaking strength of the thread or tensile strength when stress is applied across a singlestrand of thread until it ruptures. S.E.B. is usually measured in pounds, ounces, or grams.

Sea Island cotton: The best quality cotton fibre in the world. It is strong, sift nad smooth and made into top quality cotton yarn and fabrics in both plain and printed. Since the production is very limited the fabric made out of these fibres are very expensive.

Sea jute: A brown, brittle fibre composed principally of lignocellulose (Lignin) and similar to Peat fibres and being obtained from dead plant material on the Southern Australian coast, where it is obtained in great quantities. Used mainly for insulation and packing.

Sea Silk: Term applied to the strong lustrous fibres yielded by certain algae.

Sea Weed Fibres: yielded by species of algae; used for cordage, fishing lanes etc.

Seal: Black or grey in colour, the hair is shiny and flat with coarse texture; not hard wearing.

Seal Plush: Silk plush cloaking imitating real sealskin, the dyeing material is tipped on the ends of the pile, which has to be straight and slanting in only one direction.

Sealskin: Fur of the Alaskan seal, usually dyed black or brown. It is very hard wearing.

Sealskin Plush: (musquash). An imitation sealskin. Furlike plush consisting of a backing fabric with pile on one side, 3-5 mm thick, extremely lustrous, made from Tussah or Schappe silk, the cheaper grades also of cow hair; dyed in the piece in black to imitate real sealskin; used for coats.

Seam: A line where two or more pieces of fabrics are joined, usually near the edge. A seam is the line of sewing that joins two or more pieces of fabrics, usually near the edge; it may be plain or ornamental. The most important are the overhand, felled, French, slot, lapped, flannel, and beaded.

Seam, book/booking: The raw edge hem done on a blind stitch machine, usually sewn in the side ans back seam outlets, and on the bottom turn-up.

Seam, French: A closure between two pieces of material, made by stitching, turning, and re-stitching so as to conceal all raw edges.

Seam, open gorge: Both the collar and the facing are turned under, basted, and then the seam is felled (edges folded together) from the outside.

Seam, raised: A seam resulting after two pieces of fabric have been joined; one piece is folded back, and a second row of stitching is placed adjacent to the folded edge.

Seam allowance, in sewn fabrics: The width of the fabric used in making the seam assembly bounded by the edge of the fabric and the furthest stitchline. (The distance from the edge of a fabric to the parallel stichline furthest from the edge).

Seam assembly: The composite structure obtained when fabric(s) are joined by means of a seam.

Seam busting: Pressure open seams, often carried out as an intermediate operation during garment assembly.

Seam damage, in sewn fabrics: A reduction in seam efficiency caused by a change in the physical condition of one more of the components in a seam.

Seam detector for jet dyeing machines: This is used for fabric control (monitoring the dyeing process), to stop the passage of the cloth (for sampling), to count the piece cycle and to determine the cycle time. The seam detector consists of a sampling probe, electronic kit and a display unit. As the sampling probe passes the jet pipe (secondary cradle mounting also possible), through which the goods are being transported in rope form in the liquor, the pre-marked seam (marked using a ferromagnetic body Ø 50 mm, 0.1 mm thick) causes the impulse which controls the passage of the goods and stops the piece circulation.

Seam efficiency, in sewn fabrics: The ratio, expressed as a percentage, of the breaking force required to rupture a sewn seam to that required to rupture the fabric.

Seam engineering, in sewn fabrics: The procedures used to select a specific combination of sewing thread, stich type, seam type, and stich density to achieve the maximum sewn seam strength for a particular fabric type.

Seam failure, in sewn fabric: That point at which an external force (a) rupture the sewing thread, (b) rupture the fabric, (c) causes excessive yarn slippage adjacent to the stitches, or (d) causes any combination of these unacceptable conditions.

Seam finish: A treatment of the raw fabric edges of the of the seam allowance in a plain seam.

Seam grin: Gaping of seam under stress usually due to inadequate thread tension at sewing.

Seam interaction, in sewn fabrics: The net effect of the relationship between the combination of fabric, seam type, stitch type and stitch density on seam efficiency.

Seam mark: Also called stitch mark. A particular form of pressure mark (q.v.) in a fabric, and that is produced by the relief print-off of defects such as slubs or seams joining lengths of fabric, under excessive rolling tension or by contraction on the roll during wet processing.

Seam mark, in finished fabric: A pressure mark caused by the thickness of the seam being pressed against the cloth. A defect consisting of separated yarns occurring when sewn fabrics pull apart at the seams. Seam slippage is more prone to occur in smooth-yarn fabrics produced from manufactured filament yarns.

Seam opening: See **Seam busting.**

Seam protection: If a fabric with a seam is passed through the cutting device of a shearing machine, the shearing blade will destroy the fabric along the seam. The cutting device therefore raises up from the shearing table (or preferably vice-versa). As the seam passes through, a seam protection plate automatically slides between the fabric and the cutting device and covers the shearing point until the seam has passed.

Seam pucker: Generally unwanted material waviness along the line of stitching.

Seam ripper: A tool used to undo the seams and removing stitches.

Seam, sewn: See **Sewn seam**.

Seam slippage, in sewn fabric: The partial or complete loss of seam integrity manifested by yarn slippage parallel to, or adjacent to the stitch line.

Seam slippage: The movement of warp and weft threads away from a seamline under traverse stress.

Seam smoothness, in fabrics: The visual impression of planarity of a seamed specimen quantified by comparison with a set of reference standards.

Seam type, in sewn fabrics: An alphanumeric designation relating to the essential characteristics of fabric positioning and rows of stitching in a specified sewn fabric seam.

Seam welding: Any stitchless procedure for joining fabrics based on the use of thermoplastic resins or the direct welding of thermoplastic materials. Seam welding is an alternative to conventional needle-and-thread seaming operations that is extremely popular in the nonwoven field.

Seam, bound: A seam having its material edges bound with a strip of additional material.

Seam, butt: A seam which is sewn with the two material edges abutting.

Seam, flat lock: A butt seam formed using a flatlock stitch.

Seam, french: A seam in which two pieces of material are superimposed, stitched together and then folded over and stitched again to conceal the edges thus producing a flat folded seam with only one row of stitching visible.

Seam, inserted: A seam ion which a single or number of plies of material is inserted between two other which are turned in. The complete assembly is sewn in one operation.

Seam, Lap-Filled: A seam formed with the edges of both plies of material concealed by interlapping. Two or more rows of stitches secure the turned pieces of material.

Seam, Overlock: A seam in which two or more edges of material are joined together, oversewn and edge trimmed in one operation, with the overedge stitches having two or more threads.

Seam, Plain: A seam formed by a row or rows of stitches joining two pieces material together face to face.

Seam, Rolled: A seam where the two edges of the materials to be joined are rolled together and secured by a single row of stitching.

Seam, taped: A seam which includes straight tape. Normally used to prevent or control stretching and for strength.

Seam mark: A particular type of pressure mark in the finished fabric. It is produced during finishing operations by the thickness of the seam used to join pieces for processing. See **Stitch mark.**

Seaming: Joining the overlap of two pieces of fabric, usually near their edges.

Seamless: A term that describes a tubular knit fabric without seams, e.g., seamless hosiery.

Seamless Hose: Made on the circular knitting machine without any seam and in one width throughout the whole piece. It is shaped on drying boards.

Seamless Knitting: A unique process of circular knitting, done on either Santoni or Sangiacomo knitting machines. This circular knitting process essentially produces finished garments with no side seams, which require only minimal sewing to complete the garment. Seamless knitting can transform yarn into complete garments in a fraction of the time it takes for traditional garment manufacturing, by minimizing the traditional labor-intensive steps of sutting and sewing.

Seamless Technology: This term can refer to either "seamless knitting" (See Seamless Knitting), or "welding/bonding technology", which uses a bonding agent to attach two pieces of fabric together, and eliminates the need for sewing threads. (See **welding.**)

Seasonal ranges: Ranges produced for the different seasons Spring/Summer and Autumn/Winter. Mid-season ranges may also be produced.

Seat: Part of trousers or similar garments covering the buttocks.

Seat Angle: The quantity of material (5-8 cm.) allowed at the seat on the underside for ease of movement.

Sebastopol: Fine twilled woollen dress goods with very fine, different coloured narrow runs visible only when the fabric is draped.

Secant Modulus: The ratio of change in stress to change in strain between two points on a stress-strain diagram, particularly the points of zero stress and breaking stress.

Second Cuts: Fribs, or short lengths of wool resulting from cutting wool fibres twice in careless shearing. An excessive number of second cuts decreases the average fibre length, and depreciates spinning quality.

Secondary Backing, in pile yarn floor covering: A material adhered to the backing fabric side of a pile yarn floor covering.

Secondary Colours: Green, orange, and violet, each of which is obtained by mixing two primary colours. Orange is made by mixing equal amounts of red and yellow; green is an equal mix of yellow and blue; and violet is made of blue and red.

Secondary Creep: The nonrecoverable component of creep. (Also see **Delayed deformation**)

Secondary sedimentation tank, s. settling tank, s. clarifier: Sedimentation tanks used after biological treatment of a wastewater to settle out the *biomass* that has formed in the biological reactor from the treated effluent. These tanks are also known as biomass separators, *activated sludge settling tanks* or *humus tanks* (when used after *trickling filters*). *Radial flow* tanks are common but *horizontal flow* tanks are also used. Tanks are typically 2.5 to 4 m deep and the *surface loading rate* is from 20 to 45m3 per day per m2 of tank surface area. In small treatment works, *pyramidal sedimentation tanks*, or *pebble bed clarifiers* may be preferred. *See secondary sludge, solids surface loading rate.*

Secondary sludge: Sludge from secondary sedimentation tanks (return activated sludge or humus sludge). In temperate climates secondary sludge is often returned to the primary sedimentation tanks and is there thickened to give a mixed sludge. In hot climates this may be unsuitable because the secondary sludge becomes anaerobic very quickly and rising sludge is seen in the primary tanks. Flotation may be better than sedimentation for secondary sludges in warm climates. *See* **sludge production**.

Secondary treatment: In wastewater treatment, the treatment process that follows ***primary treatment***. It is used to remove the remaining organic solids that have not been removed in primary treatment together with the 90% or more of the dissolved organics. ***Aerobic biological treatment*** is commonly used. Secondary treatment may also incorporate ***nitrification*** and ***biological phosphorus removal***.

Second-hand filling material: An industry product which contains any filling material which has previously been used should not be offered for sale unless a clear and conspicuous disclosure of that fact is made on the label thereof and in all advertising and invoices relating to such product.

Second-Order Transition Temperature: The temperature at which the noncrystalline (amorphous) portions of polymer melt or become plastic. An inflection point or change is stress-strain properties occurs at this point; however, for most fibres, this change is small.

Seconds: (1) Imperfect fabrics (woven or knitted) containing flaws in the weave, finish, or dyeing, and sold as "seconds." (2) See **Yarn quality.**

Section Beam: (1) A large, flanged roll upon which warp yarn is wound at the beam warper in preparation for slashing. (2) Small flanged or unflanged beams assembled side-by-side on the shaft of a warp beam for further processing.

Section blankets: See **Sample Blankets.**

Section draws: This division of drawing-in drafts is used extensively in all manufacturing; for instance, in all fabrics having a ground warp and a binder warp, also in double-face goods, or where two different weaves are combined in one effect. One or more threads are drawn on the first section, then one or more on a second and third, if the harness is divided in so many sets.

Section Mark, in woven fabrics: Warp bands of different colour, texture or lusture.

Section Mark: Warp stripes/bands of different colour or texture or both that occur at regular intervals across part or all of the fabric width as the result of tension variation in the sections during section warping (q.v.) or because of differential dyeability of the warp yarns.

Section number: Section number = Total no. of threads/Creel loading capacity. If the calculation does not give an exact number, the last section will be produced with a number of threads lower than the other sections, or the number of threads composing each section will be reduced so as to get all sections with one and the same number of threads.

Section warping: See **Sectional warping**.

Sectional warp beams: are warp beams with protruding pegs that separate the beam into sections, usually 1–2" wide each. The warp is wound into each section separately.

Sectional Warping: The parallel winding of several warp threads to produce the warp beam. The yarn is stretched on large frames and guided through grid-like devices in such a way that the threads are wound parallel to one another on the warp beam. For particularly high warp settings and fine material, cone sectional warping machines are used. In sectional warping, it is important to achieve an absolutely uniform tension when winding the individual warp threads. See **Warping Yorkshire Warping, Scotch Warping and Silk-System Warping.**

Section width: Section width = Reed width/number of sections. This way the total number of warp threads will occupy on the dresser a width equal to the width of the weaver's beam on which they will be finally wound.

Sediment oxygen demand: The residual solids discharged with treated waste water will, in time, settle to the bottom of streams and rivers. Because the particles are largely organic, they can be decomposed anaerobically as well as aerobically, depending on conditions. Algae which settle to the bottom will also decomposed, but much more slowly. The oxygen consumed in the aerobic decomposition of material in the sediments represents another Dissolved Oxygen demand in the water body.

Sedimentation: The settling of a suspension, either under gravity or in a centrifuge. The speed of sedimentation can be used to estimate the average size of the particles. This technique is used with an ultracentrifuge to find the relative molecular masses of macromolecules.

Sedimentation tank, settling t.: An important treatment process in which the tank is used to settle out solid material from the water or wastewater. In water treatment, sedimentation tanks are used after ***coagulation*** and ***flocculation*** and are normally known as ***clarifiers. Primary sedimentation*** is common in the treatment of domestic wastewater after ***screening*** and ***grit removal***. Sedimentation tanks are used after biological treatment of a wastewater to settle out the ***biomass*** that has formed in the biological reactor from the treated effluent. These tanks have various names, including ***activated sludge settling tanks, biomass separators, humus tanks*** or ***secondary sedimentation***

tanks. Sedimentation tanks may be ***horizontal flow, radial flow*** or ***upward flow***. They are often designed on ***surface loading rate***. The performance of the tanks may be upgraded by the use of ***inclined plate and tube settlers***. In wastewater treatment, a clarifier is synonymous with ***sedimentation tank***. In sludge treatment, ***gravity thickening*** tanks are a type of sedimentation tank.

Sedjadeh: Turkish name for medium sized Oriental carpets.

Seed Coat Fragment, in cotton: A portion of a cotton seed, usually black or dark brown in colour, broken from a mature or immature seed, and to which fibres and linters may or may not be attached.

Seed Cotton: Cotton, as harvested and before ginning, consisting of seeds with the fibres attatched and usually including measureable amounts of foreign matter.

Seed: Fruit wall hairs.

Seed hair fibres: Vegetable hair fibres, which can be found on the seed husks of various plants. These include Cotton and Akund.

Seeded: Small dots strewn over the face of the fabric; same as powdering.

Seedy Wool: Wool containing numerous seeds or an appreciable amount of vegetable matter.

Seerloop Gingham: Gingham with slack-tension loops of yarn on the surface, often on the white lines only.

Seersucker: Cotton crimp crêpe with vertical stripes, also Craquelé. A fabric with puckerd stripes alternating with the flat ones of various widths only. Stripes may be multicoloured or in a plain coloured fabric with a printed floral pattern on it. A favourite summer fabric is made on twin beam loom, which feeds the yarns at different speeds and the ouckers are therefore woven in and

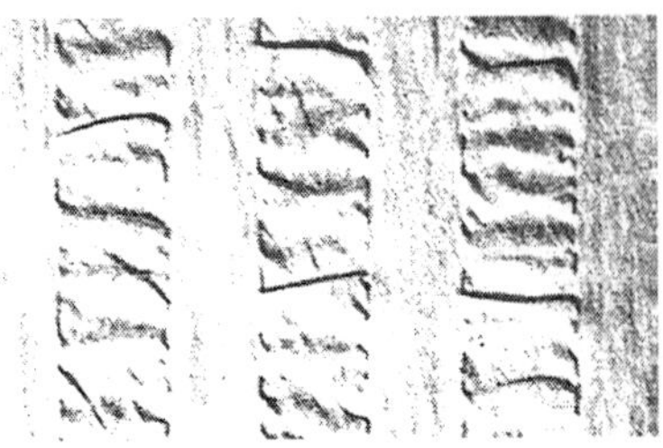

are more or less permanent. Normally made in cotton, polyester/cotton and even Nylon also. A good seersucker construction is 100 x 60 with 30s yarn in the warp and filling. In woven seersucker, finer yarn may be used in the coloured stripes than in the white stripes. Used for lightweight casual cloth such as shirts, skirts, beachwear, aprons and for tablecloths and kitchen curtains. Creases do not show.

(2) From synthetic fibres, e.g. patterned interwoven, shrinkable synthetic spun yarns with different shrinkage properties. At increased temperatures, the high shrinkage material shrinks and creates the embossed design together with the more shrink-resistant adjacent material.

(3) Similar non-fast effects can also be obtained by local application of substances which cause the fibres to swell. When producing a garment made from Crêpe prepared by steeping in caustic soda, a resist print paste which contains a water repellent agent is overprinted in the direction of the warp (as the crêpe effect is only possible in the direction of the warp). The garment is then padded using caustic soda liquor. During this process, a distance of 1–1.5 m should be maintained between the trough and the padding roller so that the caustic soda solution can create a water repellent effect on the areas for resist printing.

Segovie: Very fine French serge, made of Spanish wool with a nap on the face.

SEK, SchEK: Swiss commission for fastness.

Self bonding fibres: Bicomponent fibres consisting of polymers with different melting points. If they are heated to the temperature corresponding to the fibre with the lowest melting point, this acts as an adhesive for the other components.

Self bound seam-finish: A durable seam finish which launders very well is a finish for the raw edges of the seam allowances of a plain seam, in which one seam allowances encloses both raw edges. Stitch a plain seam, right sides of the fabric together, press to lie flat. Trim one seam allowance to 3 mm. (1/8 in.) wide. Turn under the other seam allowance 3 mm. (1/8 in.) and press in a crisp fold. Turn the seam allowance over the trimmed edge. Press in a second fol so that the trimmed edge is encased. Edgestitch the seam allowances together as close to the seam line to the seamline of the garment as possible.

Self-Casing: A piece of fabric where the hem edge is folded over and neatened to create a tunnel. A drawstring or elastic is then threaded through the tunnel to draw up the fabric.

Self cross linker: In contrast to reactive resins, which tend to react more with cellulose and crosslink these, self-cross-linkers tend to react more with themselves and form resin deposits in the pores of the fibre which promote crease resistance. Self crosslinking resins.

Self crosslinking resins: Resin finishing agents which predominantly react with themselves under specific crosslinking conditions (self crosslinkers, self crosslinking agents;), which therefore also only enter into low crosslinking reactions with cellulose (in contrast to Reactive resin) such as Urea and Melamine-formaldehyde compound. See **Cross-linkers.**

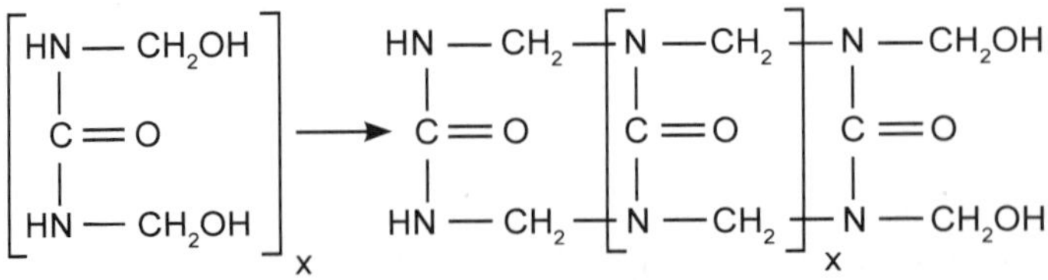

Resin forming self crosslinking mechanism in resin finishing.

Self-extinguishing: Term used in the Flammability test: after the ignition flame has been removed, the test fabric continues to burn for a short time, although the charred area does not increase significantly. The flame extinguishes itself, although there is still non-burned test fabric which could fuel the flame.

Self-goods: When the same material is used as a pocket lining, or in a waistband, collar and fly construction. Also called shell.

Self shade: (solid dyeing, plain dyeing). Monochromatism; occurs in fibre blends as Tone-in-tone dyeing.

Self smoothing fabric: Fabrics with an Anticrease finish.

Self stitching: See **Double stitching.**

Self twist spinning: A method of making a yarn from the roving fed to a drafting unit; the emerging strand of fibre is subjected to a cyclically reversing false twisting action which can be imparted in a number of ways. Two adjacent strands delivered from the false twist system are brought together by guides and the torque in the two strands causes them to wrap about each other. This wrapping action is defined as 'self twist' and produces a self twist pattern of –S-Zero-S-Zero- etc. in the yarn produced. The self twist yarn is then taken up as a cheese. Note: Further twisting may be necessary before this yarn can be used.

Self twist twisted (STT) yarn: A self twist yarn to which unidirectional twist has been added in a subsequent operation.

Self-bonding fibres: Bicomponent fibres consisting of polymers with different melting points. If they are heated to the temperature corresponding to the fibre with the lowest melting point, this acts as an adhesive for the other components.

Self-bound seam finish: A finish for the raw edges of the seam allowances of a plain seam, in which one seam allowance encloses both edges.

Self-cleaning: Rotating filter basket with a rinsing self-cleaning effect to discharge any lint present.

Self-extinguishing: Not defined; ASTM has a ban on this term in their standards. It has no meaning except in association with a specific test method or specific conditions of burning.

Self-fil yarn: This is a single yarn consisting of fibre slubbing (e.g. wool) and two fine synthetic filament threads which loop in alternating directions offset phase around the yarn compound which has been subsequently formed, and thus has the effect of strengthening the yarn.

Self-twist yarn: An inherently twist-stable, two-ply structure having a ply twist that is alternately S- and Z-directed along the yarn.

Selling price: The price the product is sold at.

Selvage: Same as Selvedge. See **Selvedge.**

Selvedge or Selvage: The narrow edge of woven fabric that runs parallel to the warp. It is made with stronger yarns in a tighter construction than the body of the fabric to prevent raveling. A fast selvage encloses all or part of the picks, and a selvage is not fast when the filling threads are cut at the fabric edge after every pick.

Selvedge cutting device: This is a device which is used to cut the selvedges and remove the edge waste.

Selvedge feeler: Non-contact scanning device which operates mechanically or photoelectrically to guide the cloth, particularly on stenters.

Selvedge gassing machine: This is used to remove loose threads on the selvedge by gas singeing.

Selvedge guider: An electronic or mechanical device for presenting fabric straight to stenter pins or clips.

Selvedge mark: A more or less regular mark in the finished cloth along the selvedge produced by folded or doubled selvedge during the finishing operation.

Selvage mark, in finished fabrics: A lengthwise crease mark along the selvage caused by an edge being folded or doubled.

Selvedge monitor: This is a safety device which switches the machine off if one or both of the selvedges is no longer held by clips or the needle bar.

Selvedge printing machines: (selvedge printers). These are generally combined with inspection machines on measuring machines, or less commonly with fabric batching machines. Company logos or advertising slogans are printed on the selvedges using heat transfer printing units.

Selvage or Selvedge: The thin compressed edge of a woven fabric which runs parallel to the warp yarns and prevents raveling. It is usually woven, utilizing tougher yarns and a tighter construction than the rest of the fabric.

Selvage uncurler: A device for straightening selvedges and fabric edges which have rolled during processing or other ways.

Selvedge: The longitudinal edges of a fabric that are formed during weaving with the weft not only turning at the edges but also passing continuously across the width of the fabric from edge.

Note: Selvedges are often up to 20mm wide and may differ from the body of the fabric in construction or weave or both, or they may be of exactly the same construction as the body of the fabric and be separated from it by yarns of a different colour. Although selvedges may contain fancy effects or may have brand names or fabric descriptions woven into or printed on them, their main purposes is to give strength to the edges of the fabric so that it will behave satisfactorily in weaving and subsequent processes. (a) Leno Edge A set of threads that interlace with a leno weave (q.v.) either at the edge or in the body of a fabric. In the latter case, it prevents fraying when the fabric is severed in the direction of the warp.

Note: When in the body of the fabric, a leno edge is often referred to as a "central selvedge".

(b) Sealed Edge. The cut edge of a fabric that has been treated by heat or chemical means to prevent fraying of the edge.

(c) Shuttleless-Loom Edge.

(i) In some cases, either one or both edges are different from the normal woven selvedge in that the weft is held in position at the turn by threads other than the warp threads, e.g. by the use of an independent thread to lock the weft in position at the edge, or by interlocking of the weft threads. In narrow-fabric weaving this type of edge is often called a "needle loom selvedge".

(ii) In other cases, the weft is severed just beyond the edge of the fabric and the cut end is tucked into the shed (q.v.) formed on the next pick.

Selvedge, jacquard: A selvedge that has a jacquard-woven pattern or lettering.

Selvedge cutting device: This is a device which is used to cut the selvedges and remove the edge waste.

Selvedge feeler: Non-contact scanning device which operates mechanically or potoelectrically to guide the cloth, particularly on stenters.

Selvedge gassing machine: This is used to remove loose threads on the selvedge by gas singeing.

Selvedge guider: See **Expander.**

Selvedge mark, in finished cloth: A lengthwise crease mark along the sledge caused by an edge being folded or doubled.

Selvedge monitor: This is a safety device which switches the machine off if one or both of the selvedges is no longer held by clips or the needle bar.

Selvedge trimming Machine: See **Selvedge cutting device.** This is used to remove from the selvedges loose, overhanging nonbound selvedge trailers which occur in weaving. The thread ends are cut off using a special shearing tool and extracted.

Selvedge uncurler: (scrollers). This is a device to uncurl, unfold and stretch fabric and knitwear selvedges which have been rolled or turned up.

Selvedge unroller: These are used to open, unroll and stretch surface sensitive woven and knitwear with folded edges or rolled selvedges.

Selvedge, tight: A continous filament thread that has lost some of its filaments, usually as a result of abrasion or excessive tension during winding or weft insertion and that appears as a thin yarn.

Selvedge printing machine: (selvedge printers). These are generally combined with inspection machines on measuring machines, or less commonly with fabric batching machines. Company logos or advertising slogans are printed on the selvedges using heat transfer printing units.

Selvyt: An unfinished velvet, made of harsh cotton, used for polishing cloth.

Semal: Cotton Silky fibre yielded by the Indian bombax malabaricum. It is straight and flattened and used for stuffing.

Seme: French for powderings or small patterns over the ground of the cloth, lace, etc.

Semi-automatic: Certain machine work stages are initiated manually but run automatically. Only part of the processes are automatic.

Semi-bright Wool: Grease wool that lacks brightness due to the environment under which it is produced, though it is white after scouring.

Semi-broad cloth: See **Broad cloth.**

Semi-collapsed balloon spinning: A system of ring spinning in which the rotating yarn balloon is eliminated at the start of an empty bobbin and is allowed to expand later when the bobbin is becoming filled with yarn. The small-balloon condition is achieved by allowing contact between the rotating balloon and spinning top. As thr bobbin fills contact is broken and the tarn balloon is allowed to reform. Note (a): The system is often used for worsted and semi-worsted spinning. Note (b): Special spindle top extension are used to obtain the required effect.

Semi-continuous bleach: Open-width boiling and bleaching process for larger batches of cotton and cotton/ synthetic blends which are uneconomical for continuous plants. Device: impregnating device, Benteler batching device and open-width scouring machine.

Semi-continuous systems: In these mixed systems several operations are carried out with both continuous and discontinuous machines. For example, a continuous pad-batch machine is used to wet the fabric and a discontinuous system is then used for other treatments. These mixed systems are suitable for processing small and medium lots; they require reasonable start-up costs and grant quite good reproducibility.

Semi-decating: Semi-decating is a batch process requiring three steps: (a) Winding the fabric onto a perforated cylinder between a cotton decating apron, (b) Steaming and followed by cooling the fabric and (c) Unwinding and batching the finished fabric. Proper pressure, heat, moisture, cooling and time are prerequisites for quality results. The procedure requires that the fabric be wound onto a perforated drum between the interleaving cotton decating apron to form a reasonably thick roll. Steam is forced through the roll (inside - out) for several minutes to provide moisture and heat.Compressed air is then blown through the roll in much the same manner as the steam to remove some of the moisture and cool down the fabric. To insure that the effect is uniform from the inside to the outside of the roll, the fabric and blanket are rewound onto another perforated drum so that the outside layers become the inside layers and the cycle is repeated. At the end of the cycle, the fabric and blanket are separated and wound into individual rolls.

Semi-decatizing: See **Semi Decating.**

Semi-felting raising machine: Wire raising machine which guides the napping cylinder in the opposite direction to that of the direction of the cloth by moment reversal.

Semi-permeable membrane: A membrane that, when separating a solution from a pure solvent, permits the solvent molecules to pass through it but does not allow the transfer of solute molecules. Synthetic semipermeable membranes are generally supported on a porous material, such as unglazed porcelain or fine wire screens, and are commonly formed of cellulose or related materials. They are used in osmotic studies, gas separations, reverse osmosis water systems for homes and beverage industries, and in medical applications.

Equilibrium is reached at a semipermeable membrane if the chemical potentials on both sides become identical; migration of solvent molecules towards the solution is an attempt by the system to reach equilibrium. The pressure required to halt this migration is the Osmotic pressure.

Semi-pigmentation process for vat dyes: Principle: the dyebath is prepared cold withfinely dispersed dyestuff, alkali and sodium dithionite and then heated slowly to vatting temperature, if necessary later to temperatures above 100°C (HT process) after reducing agents have been added.

Semi-restraint: Of or relating to a method of mounting that allows an object a limited degree of movement (for example , contaction or expansion of a fabric).

Semi suppressed balloon Spinning: See **Semi collapsed Baloon spinning.**

Semi-combed yarn: A cotton yarn made from slivers that do not have all the nails removed.

Semi-worsted spun: A term applied to yarn spun spun from sliver produced by carding and gilling in which the fibres are substantially parallel, the carded sliver not having been condensed or combed. Alternately, the yarn may be produced from a roving. Note: the above definition is descriptive of processing technique and not fibre content.

Semipermeable: This refers to the characteristic of a membrane which is permeable to specific molecules or ions, and to be impermeable to others. See **Osmotic pressure; Diffusion.**

Semi-poplin: See **Poplin.**

Semi-suppressed balloon spinning: See Semi-collapsed balloon spinning.

Semi-worsted yarn: (a) This refers to soft, full-bodied yarn, loosely or tightly twisted, produced in the semi-worsted spinning process for weaving, knitting, warp-knitting. Intermediate stage between Worsted yarn and Carded yarn. Is fuller bodied, but less strong than worsted yarn, as the short fibres are not separated using combs. Semi worsted yarn is stronger, but less full-bodied than carded yarn, on the other hand; clear finish. II. Fabric with a) a worsted warp and a carded weft or (b) twisted carded and worsted yarn, in the warp and the weft or (c) alternate worsted and carded yarns in the warp and the weft.

Semi-volie: See **Voile.**

Semienyoung: Black coloured Chinese velvet with cotton filling.

Sequestration: The formation of a complex with an ion in solution, so that the ion does not have its normal activity. Sequestering agents are often chelating agents.

Senegal gum: African Vegetable gums with predominant acacia content. Properties similar to those of kordofan gum, described under Gum arabic, but with less adhesive strength. Main varieties: poorest qualities: pador gum, (dark amber to brown, often contaminated with sand and vegetable matter). Galam gums are purer and more water-soluble. The best qualities are found in salabreda gums which can be ground (most similar to kordofan, the purest form, colourless or slightly coloured).

Sengfangtchen: Chinese silk taffeta, white or ecru, with a 'finish; about 20 inches wide. The texture is very 'fine and regular. Used for drapery, painting, etc.

Senna Knot: One of the two kinds of knots found in hand-made Oriental pile carpets. A complete loop is formed by the yarn, thus having a pile extending from every space between the warp threads, thus mak- ing more knots and a denser and evener pile than the Ghiordes knot.

Senneh carpets: (Senna carpets). The finest and most well-known Knotted carpets from Iran. Extraordinarily finely linked patterning in finely graduated, modest colours. 250 000–600 000 Turkish knots/m^2.

Senna Rugs: Very fine Persian rugs made in small sizes, usually with cotton web and very close and short wool pile tied in Senna knots. The design consists usually of small patterns covering the entire field or of a lozenge center medallion. White, red and blue are used the oftenest. Very fine khilims (see) are also made in the same design and colouring.

Separate element Zipper: A zipper consisting of two series of separately formed elements, each attached to one of the opposing edges of two tapes, which are engaged and disengaged by the movement of a slider.

Seperable Pin In Zipper: A tube like element attached over the bead at the bottom end of one striner.

Seperable Zipper: A zipper fitted with special components at the bottom of the chain, so as to complete disengagement and then reengagement of the two stringers.

Separator: A component of some open-end spinning machine located inside the rotor to direct the incoming fibres to the slide surface.

Sequencing batch reactor, SBR: A wastewater treatment that consists of a sequence of different cycles in a reactor, but flow neither enters nor leaves the reactor until the treatment is completed, i.e. it operates on a fill and draw. An example of biological treatment in a SBR for the removal of BOD and nitrification. Other cycles can be added, for example, after filling, the next sequences can be mixed anaerobic conditions, followed by anoxic conditions, followed by aerobic conditions in order to obtain ***biological phosphorus removal***. The next sequence would be sedimentation allowing discharge of the effluent. The final sequence would be the inflow of more wastewater that requires treatment. An anaerobic sequencing batch reactor (ASBR) is when stage 2 is mixed but not aerated and so it is kept anaerobic.

Sequestering Agent(or sequestrant): A chemical compound that tends to bind some 'species' such as metal ions and keep them from being available to participate in other chemical reactions; also called chelating agents Chemicals such as sodium hexametaphosphate, citric acid or EDTA , the choice depending on considerations such as pH and dye type, are used to sequester hardness ions

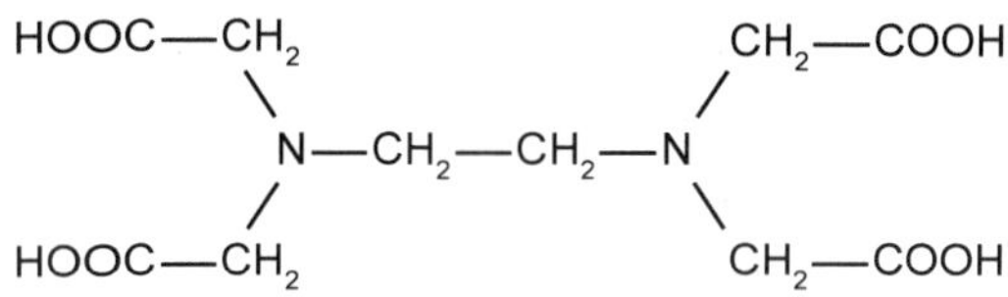

EDTA

such as calcium and magnesium and metal ions such as iron and copper. Care must be taken with pre-metallized dyes, since some powerful sequestering agents can actually remove the metal from the dye complex.

Sequestration: The masking of dissolved metal ions which can form deposits with certain reagents, particularly surfactants. The effect is based on the formation of complexes which remain soluble in the medium in question. Sequestering agents are generally compounds which form chelates with the cation in question.

Sequestrant: See **Sequestering agent**.

Sequin Lace: Crochet lace made of coloured yarn and coloured braid for the design.

Serabend Carpets: (Mir carpets), Persian Knotted carpets from the Sarawan region. Typical flame-effect patterning about the size of an egg on a red, blue or cream background. 120,000–250,000 Turkish knots/m2.

Serapi Rugs: Large, nearly square Persian rugs of cotton web and short, close pile tied in Ghiordes knot. The design consists usually of a cream center medallion, floral patterns and inscriptions.

Serge: (1) A lining of cotton or linen warp and a wool or mohair filling, woven three-leaf twill.

(2) A fine, diagonal, twilled, worsted—both all worsted and with a worsted warp and woollen filling; hard wearing, used for men's and women's suits. A typical good quality serge will have a 70 × 64 construction with 2/ 40s worsted warp and filling. The weight if the fabric will be around 10-20 oz. per yard in

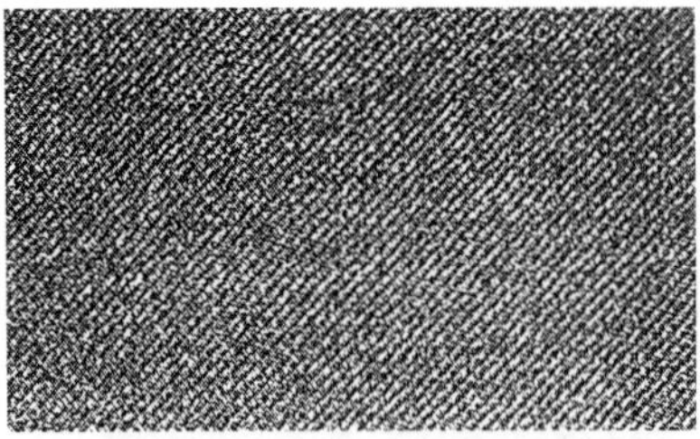

56 in. width. Serge is also made with 2/32s warp and 16s filling. Most serge is made with 2/2 twill weave. Serge usually dyed in Navy and Olive (for army) shades but sometimes made from stock dyed yarns. This fabric is an old favourite for semiformal for farmers.

Serge weave: Same as **Twill weave**.

Serge, double: It is constructed of two warps and two fillings. The weight runs from 12- 20 oz. per yard.

Serge, French: It is a very high type of dress goods with a fine, lofty, springy feel. French serge is superior the average run of serge; the warp may be single or two ply worsted, filling is usually single worsted. The weight run from 6 to 10 oz. per yard and the fabric is usually piece dyed. Usually used for suits for ladies.

Serge, French-back: It is a men's wear of two warps and one filling. It runs from 16-18 oz. per yard in weight and was formerly used in winter suitings.

Serge, storm: This fabric is lower in texture than normal serge, but heavier. Usually made with 2/2 twill weave. Made from poorer grade stock, it has lustre and ids harsh and wiry to the feel. Used as a protective outer wear like storm jackets, hence the name.

Serge d'Aumale: (1) XVIII century French serge, the warp made of slack twist woollen yarn and the filling of harder twisted single wool yarn; (2) Narrow and light French serge, used for lining.

Serge de Berry: In the 19th century a worsted in England, similar to lasting but heavier and woven with seven harnesses.

Serge de Blicourt: 18th century French wool serge made of slack twist warp and harder twisted single filling.

Serge de Boys: 17th century English Worsted.

Serge Cloth: English woollen serge of smooth face and napped back.

Serge Denim: An 18th century pure worsted fabric in England. Believed to be same as serge de Nismes.

Serge de Rome: Piece dyed mostly black fine French serge, made of two-ply warp and very slack twist filling with 8-ends and four picks in a repeat. Made with or without double face. Also made of silk.

Serge De Nimmes: A coarse cotton fabric produced during the last century in Provence. Denim fabric for jeanswear; sail cloth trouser fabric, warp printed in indigo, originally white or brown. Specially sewn and riveted garments which are artificially aged (stonewash; sandblasted, biowash process). First jeans manufacturers to use serge de Nîmes: Lewis, Lee, Wrangler.

Serging: (1) Overcasting the cut edge of a fabric to prevent raveling. (2) Finishing the edge of a carpet by over-sewing rather than binding. Generally, the sides of a carpet are serged and the ends bound.

Sericin: Silk gum. The gelatinous protein that cements the fibroin filaments in a silk fibre which accounts about 19-28% of the raw silk. It is removed in the process called degumming. Composition: 45% carbon, 32% oxygen, 17% nitrogen, 6.5% water. Sericin has a different amino acid structure to that of fibroin (Silk structure, see Tab.), i.e. basic and particularly acid groups in higher concentration (aspartic acid, glutamic acid, serine, threonine, tyrosine). Its isoelectric point is therefore pH 4.1, whereas that of fibroin is pH 4.9.

Sericulture: Rearing of silk worm and prodicing cocoon, including growing mulberry leaves for their food etc. all together is called sericulture.

Serpentine Twill: A twill weave made in wavy ridges.

Serpilliere: The coarsest grade of unbleached, loosely woven French canvas, made of hemp tow; used for bagging.

Served Yarn: In aerospace textiles, a reinforcing yarn such as graphite or glass around which two different yarns are wound, i.e., one in the Z direction and one in the S direction, etc., for protection or compaction of the yarn bundle.

Set: The number of threads found in the fabric within an inch width. In Bradford the number showing how often a beer (40 threads) is found within a width of 36 inches.

Set Checks: Large checks of the same colours are set at certain distance, forming the characteristic feature of the fabric, witht other colours between.

Set Mark: A fabric defect consisting of narrow bars or bands across the full width of the fabric that may appear either as a tight, loose, or corduroy effect caused by loom stops improperly reset by the weaver. Set marks are sometimes caused by the weaver ripping out filling yarn and then not properly adjusting the pick wheel to obtain the proper relation between the fell of the cloth and the reed.

Set Point: An input in process control that defines the desired value or range of values of the variable that is being controlled.

Set Yarns: False-twist yarns stabilized to produce bulk.

Sett, in woven fabrics: The number of warp and weft yarns in fabrics (the number of stitches in knitwear) per unit of measurement, i.e. per 1 or 10 cm, often per cm^2. Magnifying glasses or yarn counters are used to determine the sett. The sett therefore corresponds to the distance from the centre of one yarn to another. Whether the yarns lie closely together or form open stitches is insignificant. Not to be confused with fabric density.

Sett: The warp **sett** is the number of warp ends per inch—usually abbreviated as **epi**. The weft sett is the number of weft picks per inch—usually abbreviated as **ppi.**

Setting: The process of conferring dimensional stability on fibres, yarns or fabrics, generally by means of moist or dry heat. The operation of setting is applied to textile materials of all kinds but assumes special significance in the treatment of synthetic-polymer materials such as nylon, polyester, etc.

Sew: To unite or fasten with stitches made with needle and thread.

Sew together: In order to pass large batches made up of a number of pieces efficiently through the production process of textile finishing, the individual pieces are sewn together. This is carried out on an industrial sewing machine.

Sewability: Ability to sew without skipped stitches or having the thread break. Several factors effect sew ability: improper needle size, wrong thread size, excessive tension, needle heat, worn or defective sewing machine parts, and improper machine settings. Thread likewise plays a key role in sew ability. Factors in thread sew ability include elongation, uniformity, ply security, lubrication, strength, and twist construction.

Sewing force: The force applied to a sewing thread at the needle eye during the penetration of the material by the needle.

Sewing hems: In sewing the hem, the needle should take up only the edge to be hemmed down and just enough to hold on the cloth or lining. In white work the stitches should be fine, showing as little as possible.

Sewing hole, Button: A hole in either the flange or shank used to attach the button to the substrate by means of a needle and thread.

Sewing machine, in sewing: A machine that uses needles and bobbin threads in stitch formation and is primarily for home sewing use rather that industrial use.

Sewing out: Sewing parts of a garment together with the raw edges turned in.

Sewing round: Sewing round the edges of a coat to join the outer material to the facing or the facing and lining.

Sewing Silk: Made of from three to 24 reeled cocoon filaments, twisted together slack in groups of left hand twist and twisted in the reverse direction under tension.

Sewing thread: A flexible small diameter yarn or strand, usually treated with a surface costing lubricant, or both intended to be used to stitch one or more pieces of material or an object to a material.

Sewing thread: See **Thread, sewing.**

Sewing thread lubricant: Due to the increased demand for sewing threads made from synthetic fibres and the use of higher sewing speeds in industrial applications, sewing thread lubricants based on simple organic substances such as waxes or stearates are no longer sufficient to control the development of heat in the sewing needle, i.e. to provide a heat shield. For high sewing speeds, silicone-based products have been found to be particularly suitable.

Sew-through flange button: A button that has two or more holes in its flange for passage of a needle and thread so that the button can be attached to a flexible substrate.

Sew-knit fabric: An alternative turn for stitch-bonded fabric.

Sewn seam, in sewn fabrics: A juncture at which two or more separate plies of material or materials of planar structure such as textile fabrics are joined by sewing, usually near the edge.

Sewn seam strength, in sewn fabrics: The maximum resistance to rupture of the junction formed by stitching together two or more planar structures.

Sew-though flange Button: The button that has two or more holes in its flange for passage of a needle and thread so that the button can be attached to the flexible substrate.

Sew-through shank button: A button attached to one part of a flexible substrate by means of needle and thread passed through a hole or loop in the integral shank and through the substrate.

Seydel converter: Tow-to-top processing equipment. Seydel combines the prestretching and breaking process in one machine.

S-finishing: See **S Finish.**

SFR: Swedish Association of Textile Engineers and Colourists.

SFS: Finnish Standards Organization.

Shade: A common term loosely used to describe broadly a particular colour or depth, e.g. pale shade, 2% shade, mode shade, fashion shade.

Shade (v): To bring about, in dyeing, relatively small modifications in the colour of a substrate by adding further small amounts of dye, especially with the object of obtaining a more accurate match with a required pattern or colour.

Shade Bar: See **Bar; Mixed weft; Mixed filling.**

Shade Buildup: (1) The increase of depth of shade during the dyeing. To achieve a uniform shade build-up in dyeing, combinations of dyes having similar rates of exhaustion should be used so that they will be taken up by the fibre as uniformly as possible in a specified period of time over a specified rise in temperature. (2) Dyeing of polyamide or acrylic fibres with dyes or dye combinations at increasing concentrations. The purpose here is to establish the depth of shade which corresponds to the maximum build up of the particular dye or combination.

Shade Change: Change of colour in any process or a wash fastness test. During a fastness checking there can be a depth difference in the shade and a tonal change. The shade change is graded with the help of a grey scale for shade change and the tonal change is indicated by a letter in brackets along with the grading number. For example a test result is graded as 3 and a tonal change to redder is shown as 3(R).

Shade of colour: (1) (position in colour space). (2) (dyeing) Colour.

Shade number: The number allocated to each of the 24 hues in the Chromatic circle.

Shade reproducibility: The possibility of achieving exact reproducibility of shade from the same dyeing recipe. In practice, two main problems influence shade reproducibility: on the one hand, poor reproducibility may be due to errors in weighing out the dyes specified in the recipe or inaccurate volume metering on the other. Moreover, variations in substrate pretreatment or the quality and/or source of the substrate itself can also influence shade reproducibility.

Shade variation: A coloristic term for a difference in colour (usually between a particular batch and a standard) which can also be expressed numerically by means of computer colour measurement, ΔE.

Shaded Filling: A defect consisting of a bar running across the fabric caused by a difference in appearance of the filling yarn, and occurring at a quill change or knot.

Shading: In cut-pile fabrics, an apparent change in colour when the pile is bent, caused by differences in the way light is reflected off the bent fibres. This phenomenon is a characteristic of pile fabrics, not a defect.

Shading: A side-to-side change in colour across the width of a fabric.

Shading, in carpet: See **Pile reversal.**

Shading, In Carpet: A change in the appearance of a textile floor covering owing to a difference in light reflection because of localized alteration in the orientation of the fibres, tufts or loops. Shading can occur as temperory shading, permanent shading or tracking.

Shading, Permanent: (in carpet) An irreversible localized change in orientation of the pile of textile floor covering (sometimes known as water marking, pooling, or pile reversal). See also Shading in carpet.

Shading, temporary: (in carpet) A reversible localized change in orientation described as a normal characteristic of certain cut pile floor coverings. See also Shading in carpet.

Shadow Embroidery: Shadow embroidery is worked on the wrong side of thin material, using the cat stitch. The outline of the design only shows on the right side, the body of the design being seen dimly through the material.

Shadow Lace: The density of the stitch forms the pattern in this type of machine –made lace. There is no outlining thread to make the design stand out. It is light and flimsy and is used for blouses, evening dresses, etc.

Shaft: (1) A term often used with reference to satins indicating the number of harnesses employed to produce the weave. (2) See **Harness.**

Shaft Mark: A fabric defect characterized by a number of floating ends, usually caused by a broken harness strap on the loom.

Shafty Wool: Strong, dense and well grown wool with good length and spinning characteristics.

Shag Carpet: A loosely tufted carpet construction with cut pile 1 to 5 inches in length and with greater than normal spacing between tufts.

Sham: A fabric covering for a bed pillow, usually of decorative fabric that matches the bed covering.

Shampoo, in cleaning of textiles: A solution of detergent in water formulated for specialized cleaning tasks.

Shank Eye, in Button: The hole or loop in the shank of a sew-through shank button or the hole in the loop of staple of a staple button.

Shank, in Button: That part positioned perpendicular to and at the centre back of the flange, and having a hole or loop for use in attaching the button to one part of a flexible substrate by means of a needle or thread, a ring, or a toggle.

Shantung: Shantung derives its name from the Chinese province where it originated. A rough plain weave silk fabric made of uneven yarns which gives the material an irregular textured appearance due to the use of thicker yarns or long slubs in the weft. Originally made from wild (tussah) silk on hand looms in the Shandong (Shantung) province of China. Nowadays it may be produced from (a) pure silk, in which case the yarns retain all knots, slubs and other imperfections, (b) synthetic fibres or (c) cotton. An imitation silk shantung has also been produced using lower quality bourette silk. A usual texture ranges from 90 × 50 to 160 × 60. It is usually made in 40 – 45 in. wide with 75 to 150 denier rayon or acetate in the filling. In usual practice if acetate is used in the warp, rayon is used in the weft, or vice versa. Average weight will be around 5 oz. per sq. yard. Plain is normally used. A light fabric with a nep structure in the weft; The effect is called 'Shantunn', the name applied to this typical woven fabric, which today is found on the market in practically all fibres like rayon, nylon, acetate etc.

Shape memory: The ability of a garment or other textile article to hold its manufactured shape as a result of durable-press finishing. These materials are able to recover their original set shape again after changes induced by e.g. laundering, since the resin finish is only cured after the fabric has been made up into the final shape of the garment.

Shaped Facing: A garment is finished with a piece of fabric, with seam allowance, that mirrors the garment shape. This example shows the inside of an armhole.

Sharps needles: Fine, medium length hand sewing needles most commonly used for hand sewing and are also used in fine hand embroidery. They range in size from 5 to 12.

Sharkskin: Made from worsted yarn, this is an expensive fabric in fancy or novelty weaves, sometimes mixing colours. It is of fine texture and very hard wearing. Acetate or Viscose Sharkskin is used for sportswear. It is crisp and washes well. Needs to be ironed carefully to avoid shine. Occasionally made in silk also. Worsted sharskin is woven in a 2/2 colour effect twill, with a texture similar to serge. A texture around 70 x 60 with 2/36s worsted yarn in the warp and weft may be considered characteristic. The colour arrangement is usually 1 white, I black or coloured, 1 white and so on in both warp and filling. Usual width 56 in.

Sharkskin, warp knitted: It is more rigid and more stable in a lengthwise direction than satin fabric due to shorter underlap movement of the front guide bar. The technical back of the fabric is rough due to short underlap movement of the front guide bar.

Shaatnez: A fabric mentioned in the Bible, made of wool and linen.

Shabnam: Indian name of a plain, next to the finest grade of Dacca muslin.

Shaded Twills: Twill weaves made in diagonals with increasing or decreasing floats. They are called single or double shaded, according to the diagonals being shaded in one or two directions.

Shadow Check: Patterns produced on various, always solid coloured goods by using right hand twist and left hand twist yarns both for the warp and the filling; stripe patterns are pro-

duced by using these two yarns only in the warp or in the filling.

Shadow Lace: Very light machine-made laces, having a mesh ground and shadow like patterns in closer mesh.

Shafts, in weaving: are the frames that hold the heddles. When a shaft is raised or lowered, all of the warp ends threaded through the heddles on that shaft are raised or lowered. (Harness is sometimes used as a synonym for shaft, although originally it was the name for the mounting that holds all of the shafts).

Shaker Flannel: Soft, well napped white flannel, woollen with cotton warp and woollen filling; used for underwear.

Sham Plush: Made by raising a long nap from a twilled fabric, to imitate pile, or by using chenille filling.

Shamrock Lawn: Lightweight union fabric composed of cotton and linen.

Shanking: Very coarse and short merino clothing wool taken from the legs.

Shappe: Spun silk in Europe, which is partly degummed by fermentation.

Sharak: Arabic term for gray, bleached or dyed doriahs (see) used for outer garments by the natives in Egypt; originally it was made on hand looms.

Shark Skin: (1) A glossy waterproof cloth, used for raincoats.

(2) Rayon (acetate), synthetics, particularly Arnel. Worsted. Made in plain or twill (2 up 2 down). Has a heavy, semi-crisp texture. It is very smooth and slippery. Has a flat look. It is mostly made in white but some also comes coloured. It wears well and launders well particularly in Arnel. Has a tendency to turn yellow with age, but the Arnel remains pure white. Uses: All kinds of summer wear. Dresses, suits, and coats. Used extensively for sportswear, for men, women and children

Shash: (1) In the Bible means cotton; (2) A fine cotton muslin in Arabia; (3) Native East African name for a very thin, bleached cotton muslin.

Shawl: Originated in Cashmere, where it was made of pashmina. Some of the best grades are still being made in India on hand looms, the patterns are being embroidered into the ground.

The best French shawls and the Paisley shawls, made in imitation of these Oriental fabrics, are woven on power looms but the pattern is only on one side of the shawl. Other shawls are crocheted or knitted by hand or by the machine.

Shear rate: See **Rheology.**

Shear strength: The resistance to forces that cause, or tend to cause, two contiguous parts of a body to slide relatively to each other in a direction parallel to their contact.

Shear Strength, in carpet: Ability of adhesive to resist splitting.

Shear, to: Trimming the pile loops of knitted plush fabrics by Shearing.

Shearing: A dry finishing operation in which projecting fibres are mechanically cut or trimmed from the face of the fabric. Woollen and worsted fabrics are almost always sheared. Shearing is also widely employed on other fabrics, especially on napped and pile fabrics where the amount varies according to the desired height of the nap or pile. For flat-finished fabrics such as gabardine, a very close shearing is given.

Shearing blade: A metal cylinder with 12–18 helical knife blades around its circumference is one of the key elements of a shearing machine (Shearing). The milled cut of the cutting blades produces an oblique shearing action so that any lateral yielding of pile fibres can no longer occur.

Shearing brush: A brush roller to prepare the fabric pile for shearing. A brush with a fabric adjusting device in front of each shearing cylinder to achieve optimum alignment of the fibres for shearing, i.e. a pile laying brush.

Shearing Machine: (cropping machine) Shearing is carried out with one or several shearing cylinders accommodated one behind the other in older shearing machines or cylinders mounted one above the other in the modern machines.

Shearing range: A production line consisting of several shearing machines arranged one after the other for continuous Shearing. This is a much more productive alternative than the batchwise shearing of fabrics which would otherwise require several passages on a single machine.

Sheath in feathers: A covering at the quill point end of nestling feathers or nestling down which holds the barbs together.

Sheath Core fibres: Bicomponent fibres of either two polymer types, or two variants of the same polymer. One polymer forms a core and the other surrounds it as a sheath.

Sheath, in feathers: A covering at the quill point end of nestling feathers or nestling down which holds the barbs together.

Sheath-Core: A descriptive term for a multicomponent textile fibre consisting of a continuous envelope which encases a continuous, central, internal region.

Shed: The opening formed when the warp threads are separated in the operation of weaving. See **Warp shed.**

Shed: The opening formed when the warp threads are separated in the operation of weaving. See **Warp shed**.

Shedding: The first of the three basic motions in weaving, in which a shed is formed, by separating the warp into two layers.

Sheen gabardine: Gaberdine where a particularly fine twill line is used in a variety of cloth are called sheen gabardine. This may be made of 24s worsted warp and filing, with a 100 × 56 construction and with a 3/2 63 in. twill weave. Average weight will be 12 oz. Sheen gabardines are quite similar to sateens.

Sheep's wool: The fibrous covering of a sheep classified according to breed and origin (region of cultivation). According to some standards, only the Wool from sheep may be described as wool. Wool that is shorn from sheep bred exclusively for wool production represents a more or less coherent Fleece.

Sheepskin: This is the skin of the sheep or goat with the wool left on. It can be made with the wool on the outside or inside. A suede finish is usually applied to the shin side.

Sheer: (1) Made of any fibre mostly plain but could be various weaves. Any very light-weight fabric (e.g. chiffon, georgette, voile, sheer crepe). Usually has an open weave. They mostly feel cool. Triple Sheers Heavier and flatter than sheers. Almost opaque. Many are made from "Bemberg", which wears, drapes, and washes well. Sheers are used extensively for after 5 wear, as well as afternoon dresses in heavier weights, and some coats, lingerie, curtains, trims, etc. A fabric that is transparently thin or diaphanous.

(2) A warm fabric that is transparently thin or diaphanous such as chiffon, batiste, net, Organdie and voile.

Sheer velvet: See **Velvet.**

Sheers: Transparent, lightweight fabrics of different constructions and yarns, especially those of silk and manufactured fibres. Examples are chiffons, some crepes, georgette, and voile.

Sheet blanket: A thin blanket of cotton or cotton and synthetic blend having a nap on both sides.

Sheet Cast Button: A button fabricated from a cast sheet of formulated styrene modified polyester resin.

Sheet-dye process: (slasher-dyeing process) A conventional continuous dyeing process for the dyeing of cotton warp yarns in open-width form, chiefly with indigo.

Sheet, fitted: See **Fitted sheet.**

Sheet, flat: See **Flat sheet.**

Sheet, in textiles: A large rectangular usually plain woven fabricated product which is used over a mattress on a bed. The product may be carded or combed yarn in a wide range of constructions.

Sheeting: Since the introduction of fitted sheets and duvets, and particularly since coloured and printed sheets became popular, sheeting has been made available to us as piece goods. It can be plain or printed and in all colours. Most of it is 50% Polyester and 50% cotton for easycare. Used for sheets, pillow covers, duvet covers, valances, curtains etc.

Shelf Life: The period of time some product can be stored "on the shelf" before it degrades to some point of reduced effectiveness Some chemicals, including dyes, have limited shelf life. They degrade over time, eventually becoming completely useless for their intended purpose. Useful shelf life may range from hours to hundreds of years. For example, an enzyme used for desizing might be specified as losing about 10% of its strength after storage at 20°C for six months - it is still very useful, just slightly weaker. Shelf life is generally maximized by storage at low temperature (sometimes even freezing, but this can damage some products). Chemicals should also usually be protected from long exposure to bright light, and dry chemicals should be protected from humidity by storage in tightly closed moisture-proof containers

Shell: A fabric from which the garment is made.

Shenti: Loin cloth or hipskirt, highly pleated and decorated.

Sheperd's plaid: See **Sheperd's checks**. Twilled woollen fabric, made with black and white checks, formed by long and cross bars in black over white ground.

Shetland: (1) Originally, the term "Shetland" was a mark of origin for a yarn spun by hand exclusively from the wool of sheep bred and reared in the Shetland Islands. The wool from the Shetland sheep is particularly fine and imparts a soft handle to the textiles produced from it. (2) Currently, the term "Shetland" is applied loosely to all medium to coarse carded yarn fabrics produced from fairly coarse wool qualities (roughly equivalent to C quality), usually with a twill weave construction and a light to medium melton finish.

Shetland Lace: Bobbin lace made of black or white Shetland wool yarn.

Shetland Point: Needle-point lace made of Shetland wool in Italy.

Shetland Shawl: Fine knitted or crocheted light shawl made of Shetland wool.

Shetland Veils or Falls: Shawls, loosely knitted of wool, often containing camel, goat or alpaca hair, and made with scalloped edge.

Shetland Wool: (1) Very fine and lustrous wool, yielded by the Shetland sheep. The real Sh. wool is an undergrowth, found under the longer hairy wool and is not shorn but roo'd (or pulled by hand) in the spring. It comes in white, gray or brown, and is one of the costliest wools known. The wool is scoured and spun by hand, then treated with the fumes of sulphur and made up into hosiery, underwear, crochet work and very fine shawls; (2) English two-strand fine knitting yarn.

Shibori: A Japanese tie-dyeing technique In shibori, fabric, usually silk, is folded or twisted or tied or wrapped around a bamboo pole (often substituted with PVC or ABS pipe), and dye is directly applied. This simple statement does not do justice to the craft and its beautiful results.

Shier, in woven fabric: Fine warpwise cracks randomly distributed across the fabric width.

Shifu: Thread made from paper is an old Japanese tradition and has been used historically in clothing.

Shikhara: A bandhani design. See **Bandhani.**

Shikifugi: Cotton bed sheeting in Japan.

Shima Momen: Striped cotton fabrics in Japan.

Shiner: A relatively short streak caused by a lustrous section of a filament yarn. The principal cause is excessive tension applied to a yarn during processing.

Shipment: Goods or commodities which are transported together as a unit. A quantity of product for which for which a bill of lading has been signed by the carrier.

Shiraz: (Persian, all-wool rugs made in all sizes. The medium long pile is tied in Ghiordes knot. The end selvage is often checked. The design consists of palm patterns, stripes with blue and red as prominent colours. Also called Mecca ruge.

Shiraz carpets: Contrary to expectations, Shiraz carpets are not produced in Shiraz itself but in the whole Fars area of which Shiraz is the capital. The various carpets known by this name are, in fact, woven by different nomadic tribes who have lived for centuries on the Fars tableland. The bazaar in the town of Shiraz is the trading centre for these carpets. Some tribes use the Persian knot whilst others use the Turkish. In both cases, the number of knots per unit area varies considerably, e.g. 100 000–250 000 knots/m^2.

Shiraz Gum: See **Asiatic gums.**

Shirer: See **Crack.**

Shirey Yarn: Flax yarn with a flaw caused by improper setting of the reach.

Shirley Institute: A well-known British textile research institute based in Didsbury, Manchester, formerly the headquarters of the British Cotton Industry Research Association (BCIRA). The Shirley Institute was amalgamated with the Wool Industry Research Association (WIRA) in 1988 to form the British Textile Technology Group (BTTG).

Shirley stiffness test: This test measures the bending stiffness of a fabric by allowing a narrow strip of the fabric to bend to a fixed angle under its own weight. The length of the fabric required to bend to this angle is measured and is known as the bending length.

Shiro-momen: General trade term in Japan for unbleached (similar to nankeen), and bleached plain woven cotton goods. The harrow, plain woven cotton fabrics, made on hand looms and half bleached or often dyed blue are also called by this name; used for socks and clothes for the people.

Shirred Fabrics: A range of fabrics with one edge elasticated to a depth of 6-10 cm. (2 ½ - 4 in.) and sold as ready to sew skirt and sundress fabrics. Often of cotton fabric, but may be of others such as polyester, usually pointed. Used for casual clothes and children's tops and dresses.

Shirring, in garment making: Fabric is drawn up to create fullness using a fine elasticized thread down as shirring elastic. In can be used in a series of lines, depending on the effect required.

Shirting: Any fabric of any fibre that is closely woven and implies absorbent, hard wearing qualities. The most common are made from, cotton, silk, polyester and cotton.

Shirvan Rugs: All-wool rugs made in Caucasia. The warp and weft being of white, gray or dyed wool; the loose pile is tied in Ghiordes knot. The patterns are geometrical in blue, White, yellow and red colours. The ends are finished in long, knotted fringe.

Shirvan carpets: Provenance: the village of Shirvan is situated near the south-west shore of the Caspian Sea and the district of the same name in the southern part of Azerbaijan. Technical details: The warp and weft is in wool in antique specimens, in wool or cotton in those woven between 1850 and 1920, and completely in cotton in modern examples. The pile is always in closely cropped wool. The Turkish knot is used to give a knot density of 150 000–250 000 knots/m^2. Modern examples usually have a higher knot density than the older or antique ones.

Shives: (1) All vegetable matter, except burrs, found entangled in the wool. (2) Short pieces of woody waste in flax which are almost entirely removed during the breaking and scutching operations involved in flax processing.

SHMP: Sodium hexametaphosphate.

Shock catalyst: A strongly acidic catalyst used in the application of resin finishes by the Shock-cure process, e.g. certain organic acids such as citric acid, mixtures of metal salts of strong acids with complex forming hydroxycarboxylic acids.

Shock cooling: Rapid cooling, e.g. of dyed goods, or a more or less exhausted dyebath, by partially running the dyeing machine and adding cold water. Purpose: to control the temperature, e.g. when making further additions of dye for shading. Shock cooling is also carried out after the heat-setting of synthetic fibres by impingement with cold air in order to stabilize the heat-setting effect.

Shock Drying: The shock-like drying of textile fabrics in jet drying (Jet driers).

Shock Dyeing: Rapid transfer of dye liquors into a dyeing machine.

Shock-Cure Process: A variant of the Dry crosslinking process; in this case, crosslinking is achieved by shock curing within a few seconds (20–60 s) either at high temperatures (170–190°C) with conventional resin finishing catalysts, e.g. magnesium chloride or, alternatively, at 140–160°C with so-called Shock catalyst. Both the drying and curing stages are usually

combined in a single pass through a finishing stenter.

Shoddy: (1) The fibrous material made in the wollen trade by pulling down now or old knitted or loosely woven fabric in rag form.

(2) Droppings from woollen cards consisting of very short fibres that may be heavily changed with oil or dust.

Shoddy: Fibres made from ground-up rags and mixed with other fibres to reduce the cost.

Shoddy Shaken: A machine used for shaking shoddy reclaimed from under carding machines. It consists of a revolving cage, covered with perforated polished steel: having a revolving swift with long steel bars attached. The swift and cage revolve in opposite direction

Shoddy Wool: See **Shoddy (2)**

Shoe Cloth: Very strong and durable worsted, woven with corkscrew weave, weighing between 12 and 18 oz. per yard. The warp ends vary from 80 to 150 per inch with picks ranging from 80 to 140 to an inch. The warp is usually double thread and the filling single worsted, sometimes also cotton. The cloth is usually made as single fabric; used for shoe tops.

Shoe Fold: A fabric folded from both ends into twelve or sixteen folds to the piece, the length of the fold depending upon the length of the piece.

Shoe Fold: See Fold, shoe. A manner of folding fabric. The piece is folded from both ends into twelve or sixteen folds. The length of the fold depends upon the length of the piece.

Shog, shogging: See **Warp knitted fabric.**

Shogged Stitch: Rachel stitch. In knit fabrics; used to form the edges of the garments.

Shooda: Commercial name for a lightweight twilled woollen dress fabric.

Shoot: Another name for weft.

Shooting Coat: A sack coat of corduroy, duck, etc., with capacious pockets, used by sportsmen. An old term.

Shooting Jacket: See **shooting coat.**

Shore Hardness: A hardness scale which is in common use internationally for vulcanized materials such as rubber as well as plastic rollers. Measurements are made with a hand-held tester in which the resistance of the material under test to the pressure exerted by a needle is read directly off a scale. It is, however, not altogether accurate. Measurements range between 40–100 Shore which can, however, represent different degrees of roller hardness for various makes of rubber.

Shorn pile: Pile that is removed when a specimen is sheered.

Short Bast Fibres: Cottonized bast fibres; Cottonized flax.

Short-Cure-Process: A 2-stage permanent-press finish. Stage 1: mild curing of finished fabric for a short duration at low temperature. Stage 2: final curing under severe curing conditions after the fabric has been made up into garments.

Short Liquor: In the sense of liquor with a low amount of liquid, e.g. 1 : 5. Opposite: Long liquor.

Short Liquor Jet: Waste water quantities and energy use are substantially decreased in dyeing in an exhaustion process using a short liquor-to-goods-ratio (MLR).

Winch beck	MLR 1: 30
Short liquor jet	MLR 1: 6
Airflow	MLR 1: 3

Short liquor ratio dyeing: Dyeing in accordance with the exhaustion processes in the short liquor-togoodsratio, in general 1 : 10 and below, brings the benefit of less water and energy consumption.

Short Loop Drier: (festoon drier). Universal drying machine for delicate viscose fabric, for tubular jersey fabrics and for the gentle pre-drying of resin impregnated fabrics to avoid pre-curing and migration of the epoxy resin. Tension-free transport of the fabrics in the form of short loops passing in continuous travel over vigorously vibrating rollers, producing constantly changing points of contact.

Short period dyeing technique: Process for increasing productivity with the aim of reducing dyeing time to a minimum without jeopardising the leveling of the dyed goods.

Short period heat setting: (shock setting), takes place on special stenters, on which synthetic fibres are thermofixed in seconds at very high temperature.

Short-Cure Method: A 2-stage permanent-press finish. Stage 1: mild curing of finished fabric for a short duration at low temperature. Stage 2: final curing under severe curing conditions after the fabric has been made up into garments.

Short-Cut Staple: Staple fibre less than 0.75-ingh long. Typically used in wet-laid nonwoven processes to make fabrics, or as reinforcement in plastics, concrete, asphalt, and other materials.

Short-Liquor Dyeing: Dyeing in low MLR.

Shot: (1) Name in England for pick; (2) Another name for changeable or mottled effect.

Shot: The term used to describe the effect created by using one colour of the warp and a totally different colour in the weft. Any fibre may be used, but the most effective are the shiny ones, such as acetate, triacetate, polyester and silk.

Shot, in pile floor covering: The number of filling yarns per row of tufts.

Shot effect: A changeable colour effect on a lustrous or shiny fabric in which the warp yarns and weft yarns are of contrasting colours. The fabric normally has a plain weave or a 2/2 twill weave when this effect is required.

Shot, in pile floor covering: The number of filling yarn per rows of tufts.

Shot Effect: The variable colour seen when certain fabrics are viewed at different angles. The effect is obtained by having warp threads of one colour and weft threads are if a contrasting colour.

Shot Taffetta: One of the most luxurious type of Taffetta, as it is made from two colours and can be seen to change the colour as the wearer moves.

Shotting Yarn: See **Yarn, shotting.**

Shoulder Circumference, in body measurements: With arms down at sides, the maximum distance around the shoulders at the top of the arm.

Shoulder Joint, in anatomy: The junction of collarbone and the shoulder blade.

Shoulder Length, in body measurements: The distance from the side neck base to the arms cye line at shoulder joint.

Shoulder Padding: Shaped inserts used as Interlinings in men's and women's jackets and coats to confer a desired shape. Shoulder padding usually consists of wool or cotton wadding (often lined with muslin) and/or an expanded foam material resistant to dry cleaning.

Shoulder point, in sleeves: The highest point on the sleeve cap; generally located at the center of the sleeve.

Shoulder Slope, in body measurements: The angle formed when the slant of the shoulder line deviates from the horizontal line that originates at the side neck base.

Shoulder, in Zippers: The bearing surface of an interlocking element by which the chain is contained inside the flanges of the slider.

Shower curtain: A hanging fabric used to prevent water spillage from the shower area.

Shower Proof Cotton And Latex: A base fabric of cotton, or polyester/ cotton, resembling cheese cloth in appearance and sprayed on one side with Latex to make it shower proof. The layer of Latex is quite thin so the right side of the fabric is wrinkled like the Cheese cloth. It is soft fairly floppy cloth. Used for raincoats, jackets hats.

Showerproof: Treatment of a textile fabric so as to delay the absorption and penetration of water. In the case of a fabric, a degree of permeability to air is retained.

Showerproofing: A light proofing given to fabric by treating them with metallic salts, insoluble soaps or silicone based preparations. The thermal and ventilating properties and the general appearance are not much affected by these treatments. The term has no precise meaning and heavy rains should be expected to penetrate coats made from such material.

Shredding: The separation of compressed fibres in pulp sheets prior to acetylation in acetate manufacture.

Shrinkage: A decrease in one or more dimensions of an object or material.

Shrinkage: The reduction in a dimension of a fibre, yarn or fabric. Shrinkage may be induced by various treatments, e.g. wetting, steaming, alkali treatment, laundering, dry heat.

Shrinkage A negative change in the dimensions of textile products which involves a reduction in area (length and/or width) of a fibre, yarn, fabric or other textile due to the influence of various agencies, e.g. wetting, laundering, dry cleaning, etc.

Shrinkage in boiling water: Change in the length or width of a specimen immersed in boiling water, distilled or demineralised for a specific time.

Shrinkage of yarn in skein form: The change in loop length of a skein expressed as a percentage of the length prior to exposure to the shrinkage medium fir e.g. boiling water.

Shrinkage in DP finished fabrics: Much of a fabric's residual shrinkage is the result of tensions applied to the fabric during wet processing. Some woven fabrics will shrink both in width and length during preparation and dyeing. These must be pulled out to maintain width and yardage yields. These stresses add to residual shrinkage. Knit goods are inherently wrinkle resistant; however, some are pulled out to a width wider than the fabric's knitted gauge and this too adds to residual shrinkage. Much of the stress induced shrinkage can be eliminated by mechanically compacting the fabric. Compacting will result in reduced yardage yields. Crosslinking also reduces fabric shrinkage. For this reason, chemical stabilization of cellulosic fabric has real economic value. Without resin finishes, the fabrics described here will have excessively high residual shrinkage. Fortunately, a good resin finish will stabilize the fabric and reduce the residual shrinkage to less than 2%. The degree of stabilization required by chemical finishes will depend on the fabric's previous history. In some cases, much more finish is applied than one would reasonably consider simply to keep residual shrinkage within the required tolerances. In these cases, reduced shrinkage is the real reason for DP finish rather than wrinkle resistance or non-iron features.

Shrinkage Force: The force generated by thermoplastic materials when they are subjected to elevated temperatures.

Shrinking Machines: Machines for the shrink-resistant finishing of textiles by mechanical compression in a moist hot state (Compressive shrinkage) or a tension- free treatment in a steam atmosphere. The objective is to achieve residual shrinkage values up to 1% max. in subsequent use.

Shrinkproof finishing: Grammatically speaking, this is a term that should really be restricted only to finishes capable of producing a fabric with a residual shrinkage of zero (an absolutely shrinkproof fabric is not possible in practical finishing) but which is commonly used (erroneously) for finishes which reduce the residual shrinkage of textile fabrics to a level consistent with their fitness for purpose. The correct term for these finishes is Non-shrink finish.

Shrink-Resistant Finish: Non-shrink finish.

Shropshire: A breed of sheep in England and Australia yielding a long, fine, strong and lustrous wool.

Shroud: The last garment we wear.

Shusu: Japanese silk satin.

Shuttle: A boat-shaped device, usually made of wood with a metal tip that carries filling yarns through the shed in the weaving process. It is the most common weft-insertion device. The shuttle holds a quill, or pirn, on which the filling yarn is wound. It is equipped with an eyelet at one end to control rate. The filling yarn is furnished during the weaving operation.

Shuttle chafe mark: A fabric defect that is usually seen as groups of short, fine lines across the fabric, often running for some distance in the piece and usually in the same area. Although these marks run in the direction of the filling, they are actually caused by the shuttle rubbing across and damaging the warp ends, producing a dull, chalky appearance.

Shuttle looms: (shuttle weaving machines). In a shuttle loom, the weft is inserted by means of a Shuttle which is propelled over a race plate through the shed of separated warp yarns. The shuttle is propelled from side to side by a wooden shaft called a "picking stick" so that the weft yarn is inserted alternately from left to right and vice versa thus producing a closed selvedge (in contrast to all of the more modern Weft insertion systems introduced in the meantime).

Shuttle, boat: Boat shuttles (which look a bit like boats) are equipped with a central hinged rod on which a bobbin (a slender spool) wound with the weft thread is placed. The bobbin rotates as the shuttle is thrown, and the weft is pulled snug at the selvedge by the drag of the unwinding thread against the rotating bobbin.

Shuttle, End-feed: An end-feed shuttle has a shaft secured at one end that supports a pirn (like a bobbin but with narrower end). The weft is pulled off the narrow end of the on rotating pirn and tensioned with an adjustable tensioning device at the nose of the shuttle.

Shuttle, Stick: They are flat pieces of wood that are usually notched at each end so that the weft can be wrapped from end to end around the shuttle. The shuttle must be turned over a time or two as it is brought out of the shed to free weft yarn for the next pick. The turn of the thread at the selvedge must be adjusted manually.

Shuttleless Loom: A loom in which some device other than a shuttle is used for weft insertion.

Shyer: See **Crack.**

SI: Le Système International d'Unités" or International System of Units; a modern metric system in which meters, kilograms, seconds, amperes, candelas and moles are the base units internationally adopted metric system of units. It has seven BASE UNITS and two dimensionless units, formerly called supplementary units. DERIVED UNITS are formed by multiplication and/or division of base units. Standard prefixes are used for decimal multiples and submultiples of SI units, along with standard symbols for both units and prefixes.

SI: Israel Standard.

SI: Sisal.

Si: (1) Sisal, 1988; this abbreviation was changed. (2) Symbol for the non-metallic element silicon (atomic number 14).

Siara: Variety of raw cotton from South America.

Sibirienne: Plain woven or twilled, thick woollen fabric with a long nap, finished with a high gloss.

SIBR: Suspended immobilised biomass reactors, i.e. a ***moving bed biological reactor***.

Sibucara: Silky seed hair, grown on a species of the Bombax tree in Venezuela; used for stuffing.

SIC: Israeli Standards Organization.

Sicilan: A plain-woven mohair fabric; see **Brilliantine ; Mohair.**

Sicilian: A lustrous, lightweight fabric, made of fine, hard spun cotton warp and mohair filling of lower count in plain weave.

Sicilienne: This was first made in Sicily as a coarse-weave lining fabric of mohair and cotton. It is now a plain woven fabric with heavy weft ribs, made from silk, cotton and oil mixtures resembling poplin appearance. It is very occasionally used as a dress fabric.

Sida: White, strong and lustrous bast fibre yielded by the sida plants in India, South America and Australia; used for cordage.

Siddo Rags: Rags consisting of interlining from garments the best type of siddo rags are from fabrics made from yarns of hair or blends of hair with wool made on the worsted system.

Sidebands: Fabrics in America, usually printed with a band effect near to one of the selvages. They are used for trimming purposes.

Side-By-Side Fibres: (S/S types) Bicomponent fibres.

Side-to-side shading: Tailing.

Siebenburgen Rugs: Small oriental Knotted carpets, which arrived in Siebenbürgen during the Turkish occupation of South-Eastern Europe and were used to decorate the churches in the manner of the independent principality at that time.

Siemens: Derived SI unit of Electrical conductance, e.g. for electrometric measurements. 1 S is equal to the electrical conductance of a conductor of electrical resistance 1.

Sieuhwakin: Chinese shawls made of embroidered white crepe.

Siglaton: Rich silk dress goods of the Middle Ages, originated from the Orient.

Sign Cloth: Heavily starched, coarse bleached cotton muslin; used for signs.

Sight glass: Built into pipelines, condensate returns, etc. for checking the flow, purity, absence of bubbles.

Sighting dyes: Non-substantive dyes for the particular substrate, which can be added to colorless printing pastes to reveal repeat inaccuracies on blades and rollers and later washed off again (e.g. for white discharges, half-tone reserves, leuco vat ester dyes, stabilized azo dyes, phthalocyanine dyes). Frequently also colourless substances (e.g. fluorescent brightening agents) which fluoresce in UV light.

Silence Cloth: Heavy and thick bleached cotton flannel strongly raised on both sides and napped cotton fabric; used underlay for table cloth ironing boards and beds.

Silesia: A light, close-woven, fine twilled cotton fabric, dye in dark shades and calendared for a glossy finish, used for dress linings, and in tailoring for the bags of pockets etc. Made in a limited range of dark colours.

Silhouette: French, plain woven cloth of cotton warp and a different coloured linen filling, giving a scintillating effect.

Silicate boiler scale: (SiO_2), most dreaded modification of Boiler scale, caused by silicate in boiler feed water (even 5–10 mg/l SiO_2 is a risk to boilers operating at high pressure). Has a particularly low thermal conductivity. 0.2 mm layer thickness causes local overheating, buckling of the tubes, etc.

Silicates: Salts of the meta, ortho and polysilicic acids. Not soluble, non-decomposed in any solvent. The aqueous solution has a limited service life (tendency to gradual flocculation or drying up). Even a small addition of acid causes gelatinous separation (Sodium silicate; Water glass).

Silicone-based dyes: These dyes serve to permanently dye glass fibres in dark shades, as well as to modify the properties of normal silicone oils and resins.

Silicone-based textile auxiliaries: Silicones represent important auxiliaries in the textile industry for processing and finishing. They have many advantages due to their universal application possibilities, problem- free handling, free choice of type of application, gentle cross-linking, great efficiency and slight environmental pollution.

Silicon Carbide: (carborundum), SiC, variously coloured (reddish, greenish blue, black) crystals, very chemical and heat resistant, diamond-like hardness, density 3.2. Is obtained from coal and quartz at over 2000°C. Use: as grinding and polishing material; for fire-resistant bricks and electrical resistance bodies.

Silicone cross-linking: For the silicone crosslinking of reactive-group olydimethylsiloxane at least trifunctional cross-linkers are required so that an elastic network structure is achieved, e.g. methyl hydrogen siloxanes, tri- or tetra- alcoxysilanes and triaminoalkylsilanes. Silicone cross-linking usually takes place in the presence of special catalysts and may drain off as condensation (OH terminal group), addition (vinyl terminal group) and as peroxide cross-linking (methyl terminal group).

Silicone catalysts: Hydrogen methyl polysiloxanes (Silicones) require special catalysts for cross-linking, the choice of which creates the possibility of influencing the overall effects of the fabric.

Silicone elastomers: High-molecular Polydimethylsiloxane with terminal hydroxyl groups, which cross-link with hydrogen methylpolysiloxanes in the presence of catalysts, e.g. tin alcyl compounds or silicic acid esters. Used in solvents or aqueous emulsions (solvent technique enables better effects) as Silicones in finishing, e.g. for elastic finishing of knitwear made of synthetic fibres, antifelting finish and the like.

Silicon Finishes for textiles: Give the textiles, if desired, hydrophobic properties (silicone additive of 1–1.2% of the dry weight of the fabric), protection against wet dirt, a soft handle and improvement to the sewability and abrasion resistance.

Silicone finishes, stripping: May be required with defective silicone finishes. Is difficult, however, particularly if the fault is not detected until after hardening. By means of an acid treatment with 60 g/l hydrofluoric acid (alternatively oxyl acid and suitable polyphosphates and acid-resistant wash-active substance) in 60 min at 80–90°C, if necessary with a subsequent chemical cleaning.

Silicone rubber: Synonym for Silicone elastomers. Poor resistance to abrasion. Resistance to acids is excellent to moderate, to solvents poor (aliphatic/aromatic hydrocarbons), to oxidation, ozone, sunlight ageing, excellent, to flammability moderate to good.

Silicone Softeners: Silicones provide textile substrates with a combination of soft handle, smoothness, gloss, fullness, elasticity and sewability with excellent permanence at the same time. The softening effect of the silicone oils is based on their slide performance both on the fibre surface and in the fibre itself. Amino-functional groups linked to polydimethylsiloxanes enable an improved orientation and substantivity of the silicone on the substrate. This leads to an optimally soft handle and is often described by the term "supersoft".

Silicones: Silicones are Polysiloxane Polymers and fall under the class of materials known as organometallics. The element silicon is considered a metal and is found in abundance in nature as silica, SiO_2. Silicon resembles carbon in that it is tetravalent and forms covalent bond with other elements. Simple tetravalent compounds are called silanes. Silicon forms a stable covalent bond with carbon leading to a class of materials known as organosilanes. For example methyl chloride reacts with silicon to form a mixture of silanes as shown in the box below. The mixture includes silanes containing methyl, chloro and hydrogen groups in varying proportions. Chlorosilanes rapidly react with water to form silanols which further condense to form siloxane linkages. Dimethyldichlorosilane will form linear polysiloxanes which are water clear oils with excellent lubricating properties. The viscosity of the oil will vary with the molecular weight. Utilizing appropriate monomers and reactive groups, polysiloxanes, better known as silicones, are also found as three dimensional resins and high molecular weight elastomers.

Silk: A fine, strong, continuous filament produced by the larva of certain insects, especially the silkworm, when constructing its cocoons. The silkworm secretes the silk as a viscous fluid from two large glands in the lateral part of the body. The fluid is extruded through a common spinneret to form a double filament cemented together. This double silk filament, which is composed of the protein fibroin, ranges in size from 1.75 to 4.0 denier, depending upon the species of worm and the country of origin. The filament of the cocoon is softened and loosened by immersion in warm water and is then reeled off. Although raw silk contains 20 to 30% of sericin, or silk glue, and is harsh and stiff, silk is soft and white when all of the glue has been removed by steeping and boiling in soap baths. Ecru is harsher, as it has only about 5% of the sericin removed. Silk is noted for its strength, resiliency, and elasticity. The major sources of commercial silk are Japan and China.

Silk Bolting Cloth: See **Screen mesh, Screen gauze.**

Silk damage: Testing for damage to natural silk occurs by determining the viscosity of a solution of natural silk in concentrated aqueous lithium bromide solution.

Silk burlap: Silk fabric of linen weight, but feels light and drapes well. Various widths Used for suits and dresses.

Silk Camlet: Silk cloth of two-coloured warp, the filling being of a third colour.

Silk Cotton: A widely used term, applied to the fine and lustrous fibres yielded by the seed pods of a great number of trees and plants.

Silk dusting: Occurs due to the splitting up of silk under the effect of alkali or too great a mechanical stress in the working liquors and gives the fabric surface a whitish appearance, as if sprinkled with flour.

Silk dyeing: The most important dye classes are acid, metal complex (predominantly 1 : 2), direct and reactive dyes. Cationic, after chrome, vat (indigo) and vat leuco ester dyes are only used in special cases. In the high-fashion sector, classic brilliant shades in the red, blue and green range cannot as yet be replaced by more genuine products of equal value.

Silk grass: (Honduras silk grass) Longest type of bast fibres belonging to the Bromelia fibres (Central and South America). Cream-coloured, silky bright, fine and soft (surpassing the Cantala and Mauritius fibre).

Silk Grassy: General term applied to many lustrous fibres of the pineapple or other plants, especially the white, strong and silky fibre yielded by the Furcroea cubensis, in tropical America.

Silk Nankeen: English nankeen having silk satin stripes over a cotton foundation.

Silk Wadding: Waste silk resulting from spinning bourette silk.

Silkeen: A finely ribbed English cotton fabric, printed with coloured pattern over a coloured foundation and highly glazed.

Silk jacquard: Medium weight, luxurious silk patterned fabrics. Depending on the design, the Jacquard weave may give a moire' effect. The fabric drapes well and as silk dyes so well will often be found in jewel-rich colours. Used for evening and bridal wear, luxury suits, blouses and kimonos.

Silk jersey: A fine lightweight knitted fabric, often printed. It drapes beautifully, wears well and always looks superb. Used for special outfits, dresses, blouses and long dresses.

Silk Mousseline: It is firmer than chiffon, and is cool and comfortable, but it does not wear well as laundering spoils it and dry-cleaning is always not satisfactory. Used for evening dresses. See also **Mousseline.**

Silk Noil: An attractive silk fabric of dress or suit weight, its features is that it has small pieces of cocoon woven in it and these appear as dark cream flecks.

Silk ribbon finish: A finish done on yarn dyed pure silk fabric (Taffeta, Liberty structures) to get a full, solid hand feel without stiffness or hardness using resins and ether soluble gums in benzole. This type of finish explained is of historic importance only. Nowadays this finish done using suitable synthetic resins and additives.

Silk Satin: A very soft lustrous fabric for luxury lingerie. It is expensive and drapes beautifully but creases. Used for all items of lingerie, especially in conjunction with lace.

Silk scroop: A special finish applied on the fabric earlier days for a scroopy handle, craquant, crackly finish, crunchy feel, silk finish, silky handle by the treatment of textiles with soap or fat emulsion and organic acids (folic, lactic, acetic, tartaric, citric acid). The silk scroop is not usually stable in storage, leaves an unfavourable flabby handle and a slightly rancid odour due to the eliminated fatty acids due to humidity in the air. Nowadays the finish is imitated by other synthetic chemicals.

Silk Shantung: A medium weight silk fabric woven with irregular yarns. The fabric has a dull appearance and rough texture and is popular for blouses shirts and dresses. It is also referred to as Nankeen or Rajah nd also sometimes carries the name of the province where the yarn is originated.

Silk Twill: A soft twill weave silk fabric without much body, usually printed. Printed squares often made up as scarves. Used for blouses decorative features, possibly pyjamas.

Silk weighting: The weighting process is carried out to increase the silk weight, providing fuller hand, more lustre and bulk, and makig the fibre suitable for the manufacturing of fabrics to be used, for example, for ties. The weight increase is expressed as percentage weighting above or below the parity. Parity weighting means that the fibre regains the original weight it had before the degumming process:

Percentage weighting = (weight after weighting -raw weight) x 100/ raw weight.

There are many types of weighting; till some years ago, a few mills still carried out mineral weighting, but now this process has been abandoned definitively. Today, the most frequently applied type of weighting is synthetic weighting (or chemical linking).

Silk, Boiled Off: Silk with the gum (sericin) removed.

Silk, Raw: see **Raw silk.** Silk as it has been reeled from the cocoon.

Silk, spun: see **Spun silk.** Silk yarn consisting of short filaments obtained from silk wastes spun in a similar manner to worsted yarn.

Silk, Thrown: see **Thrown silk.** Raw silk that has been twisted or doubled and twisted.

Silkaline: A very light, printed, plain woven, glossy cotton fabric, made in the gray and calendered; used for lining, curtains, etc.

Silkworm Gut: Used for fishing lines; the silk worms are immersed in strong vinegar for a couple of hours and then pulled apart, each worm yielding two thick stringe of great strength.

Silk union: A term used for fabrics that contain natural silk or filament yarns similar to silk in one thread system.

Silk waste: See **Bourette; Schappe (silk).**

Siloxanes: Compounds containing Si–O–Si groups with organic groups bound to the silicon atoms. The silicones are polymers of siloxanes. Oxygen compounds of Silicon. Polysiloxanes: See Silicones.

Silver Cloth: French fabric, composed of 4-5 of wool and one-fifth of asclepias cotton

Silver number, silver index: A number to denote the bleach damage on cellulose fibres. The number indicates how much silver is reduced and separated from 100 parts test material numerical expression of bleach damage on cellulose fibres.

Silver test, in textile: Test solution is made as follows: Solution A: 1 g silver nitrate and 10 ml water. Solution B: 4 g sodium thiosulphate and 100 ml water. Mix solutions A and B, and add sufficient thiosulphate until the precipitate is dissolved. Then add solution C (4 g ammoniac and 100 ml water). Bring everything to the boil, filter and preserve the light-sensitive solution in brown bottles. Application: to test for hydrocellulose, especially with subsequent treatment with dilute ammoniac. Yellow-brown colouring according to the degree of damage. Also to test for wool damaged by sunlight (black coloring).

Silver prints: Graphite prints. Carried out with appropriate Metal powders.

SIM: Malaysian standards organization.

Simili Mercerising: A calendering process for increasing the lustre. The effect is similar to that obtained by mercerizing but not permanent.

Simple Machine Stitch Pattern, in sewing: A repeating segment of machine stitches, with each repeat consisting one or more stitches long.

Simplex Fabrics: Velour-type surface processing of knitwear made of cotton or synthetic fibres such as polyester, polyamide, acetate and triacetate. Originally limited primarily to glove qualities, for clothing fabrics including sport and leisure articles.

Simplex machine: One of the earliest machine designed to dye loose cotton. It consists of a rectangular vat fixed on supports. Steam pipes are fixed in the bottom of the trough. Above the steam pipes there is a free perforated plate on which two or more hooks are attached. Ropes are attached to these hooks which is taken above the trough on an eccentric pully. Cotton is loaded above the perforated plate and closed above it with another perforated plate. The liquor is heated and the package of loose cotton is worked up and down by means of the eccentric pully when a more or less even dyeing takes place.

Simulated grass matting: Rot resistant, coarse artificial grass matting made from coloured synthetic fibres, usually polypropylene.

Simulated Spun Yarns: Filament yarns that have been modified to have aesthetics similar to those of spun yarns. Simulated spun yarn have looped or hairy surfaces.

Simulation tests: Degradation test procedure to assess the biological level of waste water treatment plants. The degradation process is simulated in laboratory models under practical operating conditions, as in the aeration tank of a waste water treatment plant.

Sinamay: 'Light, plain woven fabric, made by the natives of the Philippines of abaca fibres. It usually comes in contrasting coloured stripes; used for garments by the natives.

Sinclaii: A Highland tartan, composed as follows: *Green stripe; group (as wide as green stripe) composed of a black stripe, a fine white line and a blue stripe, the latter being wider than the black; "red stripe, being somewhat wider than all the stripes mentioned above; repeat, in reversed order, stripes mentioned between two.*

Singeing: Scorching, gassing of fibre ends protruding from surfaces on yarns, fabrics, etc. if a fluffy fabric is not required. An industrial defibrillation process where rapidly-moving fabric passes over a flame or a very hot plate in order to burn away fibres poking up from the surface. Also see gassing. Singeing is predominantly carried out on textile fabrics manufactured from yarns in the form of knit-goods or wovens. The aim is to achieve a smooth and fibre-free surface, which is essential both for subsequent processing in textile finishing and for desired serviceability properties.

Singeing Machine: Especially used for woven and knitgoods with low-tension controlled throughput involving several singeing positions: tangential, counter roll and counter-fabric. Especially used for the technical pretreatment of knit-goods in tubular form prior to mercerization

Single-bath chrome dyeing process: Chromate dyeing process; Chroming of dyes.

Single-bath chrome mordant: Metal mordants, for mordant and chrome dyes.

Single-bath exhaustion process: Dyeing process in which several components are applied to the fibre in a dye bath, e.g. dispersion dyes and cotton, wool or cationic dyes.

Single-bath process: Treatment, especially for dyeing blended fibres, in a bath, i.e. with dyes for each component. Contrasting process Two-bath process, i.e. treatment in two consecutive baths. Variants: Single bath single-stage process corresponds to singlebath process. Single-bath two-stage process, whereby, at the end of the first treatment stage, the second stageis carried out in the same bath.

Single Breasted: A style of coat or other garment with minimum overlap and a centre front fastening.

Single broadcloth: See **Broad cloth.**

Single Canvas: A popular plain weave cotton, or cotton/polyester, embroidery canvas of fine construction, with easily distinguished holes between the warp and weft threads.

Single Cloth: Is woven with one set of warp and one set of filling, irrespective of the weave.

Single Coarse Raw Silk: Simple, coarse raw silk thread (8–10 cocoon threads) made from inferior cocoons, e.g. as inlay threads for spun metallic yarn.

Single Damask: Both the ground and the pattern, or only the ground is woven in five-leaf satin.

Single end sizing Process involving guiding threads through nozzles which separate several chambers from each other. Residual drying in heated pipes

(< 12 m long), subsequent batching on cross wound yarn packages. Owing to the fast speeds, not suitable for sizing warp ends in industrial standards. Applied to special yarns (e.g. asbestos threads, novelty yarns, sighting thread). Basic disadvantage: nozzle change for each size range.

Single Face Fabric: (plain jersey fabrics) Knitted fabrics with only right-side loops on one side of the fabric and only reverse-side loops on the other side of the fabric.

Single filaments: Monofilament chemical fibres.

Single Jersey: This is often wool, but sometimes may contain acrylic fibre. It is a knitted fabric that is thin and curls at the edges. The right side only has the stocking stitch appearance, the wrong side is the reverse and like hand knitting in effect. Single jersey may be plain or mixed colours, in random stripes. It is soft and drapey. Used for soft gathered dresses and suits.

Single jersey jacquard, weft knitted: A patterned single jersey weft knitted fabric usually made from two or more yarns of differing colour or texture to give a construction that consists essentially of knitted and float loops but may incorporate tuck loops. The surface pattern is derived from the chosen arrangement of the yarns knitted and float loops. The inclusion of tuck loops into the construction is to eliminate long lengths of floating thread from the back of the fabric.

Single jersey tuck jacquard, weft knitted: A patterned single jersey weft knitted fabric usually made from two or more yarns differing in colour or texture to give a construction that consists of knitted and tuck loops. The surface pattern is derived from the chosen arrangement of yarns knitted and tuck loops.

Single lapped seam: A lapped seam used to eliminate bulk when piecing interfacing and interlining edges together. The seam does not produce a finished appearance, and should not be used to seam the garment itself.

Single Level Pile, in floor covering: Having all pile tufts at the same level.

Single Phase Printing Process: The printing paste contains, besides dyestuffs, the chemicals required for the application, e.g. reducing agents with vat dyes or alkalis with reactive dyes.

Single Plush: A plain knitted fabric, made with one face yarn and having the backing yarn almost entirely on the back of the cloth. It is then napped; used for underwear.

Single poplin: See **Poplin.**

Single satin stitch: Embroidery stitch. See **Straight stitch.**

Single squeegee: Screen printing squeegee with one blade.

Single Stitch Zigzag, in sewing: A simple machine stitch pattern made by the needle moving up and down and alternately from one side to the other while the fabric moves through the feed mechanism in either the forward or reverse direction with all segments having equal length and width.

Single Strand Strength: The breaking strength of a single strand of yarn monofilament or cord, not knotted or looped but running straight between the clamp of the testing machine.

Single Twist: The amount of twist in each individual single yarn element in a tyre cord structure based on the length of the element after twist has been removed from the fabric.

Single yarn: See **Yarn, single.**

Single Weft Carpet: Single weft carpet is a Woven carpet, in which each row of knops is fixed in with only one weft.

Single voile: See **Voile.**

Single Yarn: The simplest strand of a textile material suitable for operations such as weaving, knitting etc.

Single Yarn: The simplest strand of textile material suitable for operations such as weaving and knitting. A singles yarn may be formed from fibres with more or less twist; from filaments with or without twist; from narrow strips of material such as paper, cellophane, or metal foil; or from monofilaments. When twist is present, it is all in the same direction. (Also see **Yarn**.)

Single-Knit Fabric: Made on knitting machines with one row of needles producing fabric that smooth and even on the right side, but uneven on the other side. The fabric is lightweight and tend s to curl at the edges. Used for top and trousers.

Single-Phase weaving machines: There is a sequence in the primary motions of weaving and each of them is repeated once in each weaving cycle. The weft insertion which is the principal operation in weaving, takes place only at discrete intervals.

Single-Strand Breaking Force, in tensile testing: The breaking force of one strand that follows a specific path, usually straight line, between the clamps of a tensile testing machine.

Singling: The condition caused by the breaking of one or more strands in a plying operation with resulting unevenness in the finished product.

Singling: A yarn defect caused by the breaking of one or more strands in a plying operation with resulting unevenness in the finished product.

Singonne: Very stout, black, closely woven fulled and coarse woollen with a long nap; it sheds the water and is used in various European countries for winter clothing by the poorer classes.

Sinker: In weave design, a blank square indicating a filling thread over a warp thread at the point of intersection.

Sinker, in knitting: (holding down-knocking over): in textile terminology the sinker commonly used on plain circular knitting machines is called a 'holding down-knocking over' sinker. Its main function is to hold the sinker loop when the needle is moved from the rest position into the clearing position. It is controlled from the sinker butt by a sinker cam. It also takes over the formation of a new loop across the knock over edge (the area of the machine that the sinker loop is formed) at the end of loop formation. They have different shapes and dimensions depending on the machine type and the fabric design.

Sinker loop: this is the part of the stitch which is composed of loop feet belonging to neighbouring stitches.

Sinna Knot: See **Senna knot.**

Sinnerscher Circle: Approximate classification of the significance of four washing parameters (chemistry of the surfactants, washing time, temperature and washing mechanics) to define the cleaning efficiency of a drum washing machine.

Sipacheutoochwongyong: Black, curl pile silk velvet in China, made with serge foundation; used for hats, etc.

Sintering: Forming a bonded mass or fibre by heating the constituents of the mass or fibre without melting.

Siretz: Trade term for Russian uncleaned flax.

SIRO: "Scientific Industrial Research Organisation", Australian textile research organization

Si-Ro-Set colour fastness: Method of dyeing with Si-Ro-Set treatment, i.e. possible effect on the shade and possible staining of white wool and cotton as a result of the treatment. The dye should be selected according to the application of the finished goods, i.e., only reduction-insensitive dyes may be used in dyeing.

Si-Ro-Set process: Australian process for surface fixing and manufacturing of permanent creases and pleats on woollen articles such as trousers, skirts, etc. by treating with a solution of thioglycollic acid derivatives and subsequent shape fixing by steaming.

SiroSpun process: Process developed by CSIRO and IWS, by which a 2-ply yarn is produced at one spinning position, making twining redundant. This process allows a special spun-twisted yarn to be manufactured directly on the ring spinning machine.

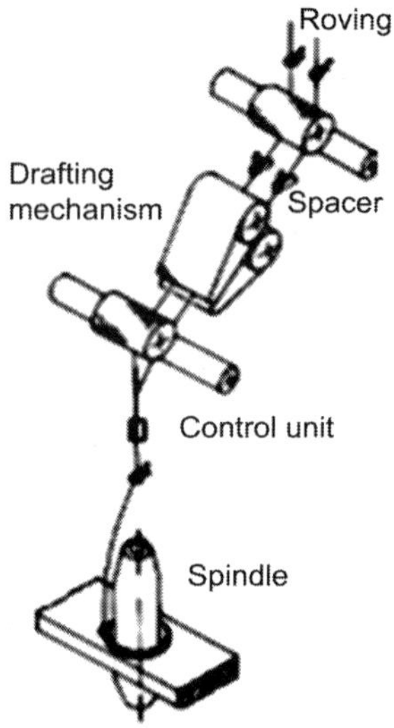

SIS: Swedish standards organization

Sisal: Hard fibres of subtropical agave types (Mexico, Brazil, Africa, Indonesia). Named after the oldest port of exportation of Sisal in Mexico. Snow white to yellowish. Fibre bundle 60–150 cm long. Strength 35–55 cN/dtex. Not very resistant to seawater. Chemical composition: 65.8% cellulose, 10% water, 1.2% lignin, 0.3% fat and wax, 1–4% ash. Dyeable similar to Manila fibre. Application: binding yarn, cord, string, ropes, sacks, sail cloth, carpet industry, decoration materials, also for finer fabric. Sisal tow for upholstered materials.

SISIR: Singapore standards organization

Sister's Thread: Same as **Nun's thread.**

Sistresay: From East India and Turkey, made with two warps, one silk, the other cotton, and a spun silk filling in damask patterns with coloured stripes.

Sivas carpets: Are fine, medium-size Knotted carpets from the north Anatolian city of Sivas. Warp and weft in cotton, pile in low lustre wool; the Turkish knot is used with approx. 280 000 knots/m^2.

Six Quarter: Goods Measuring 54 inches in width.

Sixth Combing: Wool taken from the lower part of the thigh; also called breech.

Size: In textiles, a material applied to yarns or fabrics to make them stiffer or temporarily bind fibres together Sizing is used extensively, especially for cellulose fibres, to make them easier to process or protect them from damage during high-speed weaving or the like. A wide variety of compounds, including starches and other plant derivatives, and synthetic organic compounds, such as polyvinyl alcohol, are used for sizing. Sizing materials can interfere with dyeing, so it is important that they are removed by desizing, usually prior to scouring but sometimes as part of the scouring process.

Size, in button: A unit of measure for button diameter; one ligne is equal to 0.635 mm. (0.025 in).

Size mark: A fabric defect that consists of a rough or frosted spin caused by uneven application or drying of the size.

Size recovery: Size like PVA can be recovered by simply by washing the fabric with hot water and concentrating by suitable methods, say ultrafiltration, evaporation.

Size, Yarn: See **Yarn number.**

Sizing: A gelatinous film-forming substance in solution or dispersion, usually applied to warps but sometimes to wefts, generally before weaving. A generic term for compounds which, when applied to yarn or fabric, from a more or less continuous solid film around the yarn and individual fibre. Varieties applied to yarn (a) sizing applied to warp yarn to bimnd the fibres together and stiffen the yarn. (i) dope: applied to crepe yarn to set the twist and assist creping. (ii) Dressing: applied to sewing thread to bind the strands together and leave a pliable yarn. Varieties applied to fabrics include: (b) Sizing applied to fabric to improve their physical properties such as weight stiffness etc. (i) Dope: Applied to airplane fabrics to make them taut and to balloon fabrics to make them less permeable to gases. (ii) Dressing: applied to fabrics to procedure a glazed lustrous effect.

Note:

– The main types of substance used are carbohydrates and their derivatives, gelatin and animal glues, linseed oil, polyacrylic acid and polyvinyl alcohol.

– The objects of sizing prior to weaving are to protect the yarns from abrasion in the healds and the reed and against each other, to strengthen them and, by the additional of oils and fats, to lubricate them.

– A size may be applied to carpets (e.g. starch) and occasionally to wool fibres (e.g. animal glue).

Sizing system, in garment construction: A method of designating garment sizes.

Sizing, Single End: The application of size to: (a) A thread supplied from a single package; the thread receives a side –traverse as it is wound on to a large drum on which it forms an endless sheet of yarn; this sheet is then cut and wound on to a beam under tension to produce a sample (or short length) warp. (b) A low density sheet of yarn in which the adjacent ends do not touch one another; the yarn is usually supplied either from a cone creel or a single warpers beam; the resulting beam is usually combined with other similar beams in a dry taping process, to produce a weavers beam(s).

Skein: A continuous strand of yarn or cord in the form of a collapsed coil. It may be of any specific length and is usually obtained by winding a definite number of turns on a reel under

prescribed conditions. The circumference of the reel on which yarn is wound is usually 45 to 60 inches. (Also see **Hank.)**

Skein: A continuous strand of yarn in the form of a flexible coil having a large circumference in proportion to its thickness.

Skein Break Factor: The comparative breaking strength of a skein of yarn adjusted for the linear density of the yarn expressed in an indirect system; the product of the breaking strength of the skein and the yarn number expressed in an indirect system.

Skein Breaking Tenacity: The skein breaking strength divided by the product of the yarn number in a direct numbering system and the number of strand placed under tension.

Skein shrinkage: A measure of true or intrinsic yarn shrinkage not including crimp contraction.

Skein strength: The force required to rupture a skein of yarn, expressed in units of force, as breaking strength.

Skeleton: Term applied to a coat made without lining.

Skene: A Highland tartan, composed as follows: Red stripe, split in the center by a green line; dark blue stripe, as wide as the red; red stripe, width and split as above; green stripe, as wide as one red and the blue stripes together; red stripe, width and split as above; green stripe, as above.

Skew straightener: A device for straightening skew distortion in textile fabrics, e.g. with the aid of rollers.

Skewness: A fabric condition resulting when filling yarns or knitted courses are angularly displaced from a linr perpendicular to the selvedge or side of the fabric.

Skewness: A fabric condition resulting when weft yarn or knitted courses are angularly displaced from a line perpendicular to the edge or side of the fabric in which the warp and weft yarns, although straight, are not at right angles to each other.

Skewness: The distance measured parallel to and along a selvedge between the point at which a weft yarn meets this selvedge and perpendicular to the selvedge from the point at which the same weft yarn meets the other selvedge.

Skin back: See **Broken filaments.**

Skin rug: A hand-woven Flat carpet in the form of a weft-figured carpet produced from a single weft consisting of narrow cut strips of skin at a predetermined width (also includes those rugs where the strips of skin are used in the warp with a yarn weft). The edges and ends of the rug are bonded.

Skin wool: See **Pulled wool.** Taken from the skins of slaughtered sheep, either removed by sweating, or by sodium sulphide or by lime (slipe wool). Production: as enzyme wool, as sweated wool (Mazamet wool, lime wool, slipe wool) by (a) shearing, or (b) by extraction with the hair roots, but after a prior treatment with milk of lime to dissolve the hair roots to give so-called lime wool which must consequently always be regarded as chemically damaged wool. See **Recovered wool.**

Skip: See **Float.** Flaw in cloth where a warp thread skips over more filing threads than intended.

Skip draws: The draws coming under this heading are used very extensively in silk weaving, especially for fabrics requiring a heavy warp and a large number of shafts. Enter first the odd and then the even shafts. An 8 harness draw of this kind, runs as follows: 1, 3, 5, 7, 2, 4, 6, 8.

Skipped stitch: A stitch which appears twice as long as the other stitches in a line of stitches. It is caused by failure in the stitch formation allowing the needle thread to return to the surface without being held by the bobbin (lower) thread.

Skirt: That part of a coat , dress or other garment which hangs below the waist.

Skirting: (1) In wool sorting the removal of the stained parts of the fleece, as the legs and the whole edge of the fleece; (2) Rag sorting term, meaning rags of women's dress goods and men's coat linings, containing cotton and wool.

Skyteen: A cotton shirting made in England with a five shaft, warp faced satin weave. It has stripes on a light indigo ground.

Skittery dyeings: An irregular speckled appearance in dyed goods produced by colour differences between adjacent fibres or parts of the same fibre due to a variety of causes (Skittery dyeing wool).

Slack end: (1) A warp yarn woven under insufficient tension. A warp yarn that appears puckered as the result of having been woven under less tension than the adjacent warp yarns.

(2) A warp yarn that appears puckered as the result of having been woven under less tension than the adjacent warp yarns.

Slack filling: See **Slack pick.**

Slack mercerization: A process for producing stretch in cellulosic fabrics. Mercerizing of cotton textiles on chainless mercerizing machines without tension. Fundamental requirements: suitable construction of the textile material (floating weaves), no tension during the mercerization process, the stretch effect is set by subsequent resin finishing. The process is mainly employed to produce weft stretch cotton fabrics and is sometimes referred to as "chemical stretch" or it is applied as part of the process for crease-resistant linen.

Slack pick: A single weft yarn woven under insufficient tension. See also **Loose pick.** A weft thread or part of a weft thread that has been woven into the cloth at a lower tension than the adjacent normal picks. **Slack filling.**

Slack Selvedge: Slack ends in the fabric edge. See also Baggy selvedge (Selvage), Loose edge, Stringy selvedge (selvage) Wavy selvedge (selvage).

Slack Selvage: A self-descriptive fabric defect caused by incorrect balance of cloth structure between the ground and selvage or by the selvage ends being woven with insufficient tension.

Slack Thread: See **Slack end.** A thread or pieces of thread which are slacker than the other pieces/threads.

Slack Twist: Refers to a yarn imperfection where insufficient twist is applied to the thread so it has very poor ply security.

Slack Warp: See **Slack end.**

Slaked lime: See **Calcium hydroxide**

.**Slaking**: See **Calcium hydroxide.**

Slam-Off, in woven fabric: A distortion due to the entrapment of the weft carrier in the shed. Compare **Smash.**

Slanting: Cross Stitch In embroidery a variety of cross stitch and but little used. The first part is same as the cross stitch, the return made like the gobelin stitch (see), can be used only on fine foundation.

Slasher: A machine on which slashing or sizing is done.

Slasher Sizing: A process by which warp yarns are sized during transfer from warpers beams onto loom beams. Two or more size boxes may be used in parallel and/or in tandem if the warp sheet is too close or effective sizing in one box or if it contains yarn with different fugitive tints.

Slashing: It is the process where size is applied to warp yarns for weaving. The purpose of size is to protect the yarn from the abrasive action of the loom.

Slat, in tufting machine: A patterning attachment consisting of two sets of inter meshing metal angular strips, (also called slats) mounted on a continuously moving roller chain. One set has a constant height whilst the other set has a profile machined according to the required pattern.

Slatted expander: Expander combinations consisting of wood laths to maintain a fabric web in the fully open-width state. Depending on requirements, the wood laths are available with or without a rubber covering.

Sleazy: Thin, lacking firmness, open-meshed; usually describes poor-grade fabrics.

Sleazy satin: A cheap, soft satin with high lustre, usually made from acetate or acetate/viscose. It is a thin fabric suitable for lining.

Sleeve: Sleeve is that part of the garment, which covers the arm portion of a body. The appearance of the sleeve can be determined by the position of the armhole, under arm seams, fullness added to any part of the sleeve and to the sleeve hem or cuff. As the arms are mobile, the sleeve should allow enough room for the movement. There are two categories of sleeves: -

(a) Set In Sleeves: Cut separately from the bodice of the garment and fitted into the armhole shape. The standard set-in sleeve should smoothly cover the upper arm and shoulder socket. The armhole seam should fall on the shoulder where the socket joins the arm.

(b) Cut In One Piece Sleeves: Cut in one piece with the bodice of the garment in called cut-in one sleeve, for example, Japanese Kimono sleeve, Raglan sleeve, Batwing, Dolman and Magyar sleeve.

Sleeve Cap: The curved top of the sleeve from front and back.

Sleeve center: Center of the sleeve.

Sleeve Length: The sleeves measured from the center of the neckline in the back to the end of the sleeve or cuff.

Sleeve Tacking: Stitches which attach the sleeve to the lining along the sleeve inseams and elbow seams.

Sleeve Vent: A finished slit or opening in the sleeve. Vents are usually secured by snaps or buttons at the base of the cuff.

Sleeve, Set In: A sleeve which is set into the scye.

Sleeve Lengths: *Long:* A sleeve covering the arm beyond wrist level including some portion of the hand also.

Bracelet: A sleeve covering the arm upto wrist level

Sleeve, 7/8th level: A sleeve covering the arm upto mid of the lower arm portion. Generally found in ladies tops.

Sleeve, 3/4th level: A sleeve covering the arm upto widest portion of the forearm. Generally found in ladies tap.

Sleeve, Elbow: A sleeve covering arm till elbow level. Generally found in ladies blouses.

Short: A short length sleeve covering half portion of upper arm. Generally found in summer wear.

Cap: A very short sleeve covering one-third portion of the upper arm. Generally found in ladies blouses and nightwears.

Sleeveless: It is a garment, which is without sleeve and has a finished armhole. Generally found in kids and ladies summers wear.

Sleeve, Magyar: A sleeve which is cut integrally with the body of the garment, the two sections being joined from neck to wrist. Wedge shaped pieces are inserted under the arm to allow freedom of movement.

Sleeve, Raglan: A sleeve with the armhole line extending from the front and back scye to the neckpoint so that the shoulder section is joined to the sleeve crown, eliminating the conventional shoulder and sleeve head seams.

Sleeve, Set in: The sleeve is shaped around the arm and has a seam on the underarm. The top of the sleeve, the sleeve head, is curved to accommodate the roundness of the shoulder. The sleeve is constructed and then 'set into' the bodice.

Sleeve, Bracelet/Three Quarter: This sleeve reaches half way between the elbow and the wrist.

Sleeve, Padded Shoulder: The shoulder is extended and the sleeve is heightened to accommodate a pad, to achieve a squared or rounded effect in the silhouette.

Sleeve, Raglan with Yoke: See raglan with dart. Here the raglan Is extended to accommodate the top part of the bodice, achieving a 'yoke' effect. There is a seam along the top edge of the sleeve and at the underarm.

Sleeve, Dropped Shoulder: The shoulder is extended and shaped around the shoulder, rather like the capped sleeve. However there is more shaping and the extension is usually longer; also a sleeve is actually set into the armhole. The sleeve construction is modified by removing that part of the head that is now covered by the extension to the shoulder. The sleeve head then looks less curved.

Sleeve, Two Piece Tailored: A shaped sleeve that has a seam down the front and the back to allow for shaping.

Sleeve, Two Piece Tailored: A shaped sleeve that has a seam down the front and the back to allow for shaping. There is no underarm seam; consequently the sleeve is constructed in two pieces.

Sleeve, Capped: The sleeve is an insertion that just covers the shoulder point.

Sleeve, Cape: A full, flared sleeve that is set into the armhole. The sleeve could be cut as a circle to give more flare at the hem.

Sleeve, Capped (Extended Shoulder): This is an extension of the shoulder just covering the shoulder point.

Sleeve, Saddle Raglan: See raglan with dart. Here the seam for applying the sleeve to the bodice is shaped, it narrows along the shoulder line. This sleeve has a seam at the underarm and along the top edge of the arm to the wrist, including the shoulder.

Sleeve, Raglan with dart: The raglan sleeve is derived from the set in sleeve but with the shoulder added to the sleeve head. It was named after Lord Raglan who led the Charge of the Light Brigade in the Crimean War and wore raglan-sleeved jackets. This sleeve is cut in one piece but has a dart at the shoulder to accommodate the roundness of the shoulder.

Sleeve, Raglan with Seam: A normal raglan shaped sleeve with a seam along the top edge of the arm, and at the underarm. See **raglan with dart**.

Sleeve, Man's Shirt: A set in sleeve with two pleats at the buttoned cuff and a placket. Normally seen on men's shirts.

Sleeve, Long Bell: The upper part of the sleeve fits the arm, the lower part flares out into a full, bell shape.

Sleeve, Frill: The upper part of the sleeve fits the arm, the lower part has a deep, gathered frill attached at about elbow length.

Sleeve, Bell: Like the **long bell sleeve,** but much shorter, finishing around elbow length.

Sleeve, Pagoda: An eighteenth century sleeve that is fitted on the upper arm, with tiered frills on the lower part of the arm to the wrist. It can be longer at the back of the wrist. There are usually three tiers giving the appearance of a Chinese pagoda.

Sleeve, Leg of Mutton: A full, gathered sleeve head is set into the bodice and the long sleeve tapers to fit towards the wrist. Popular in the late nineteenth and early twentieth century, fashion frequently re-introduces the leg of mutton.

Sleeve, Juliet: A two-part sleeve similar in effect to the leg of mutton. The top part of the sleeve is full and gathered while the lower part of the sleeve is fitted to the arm and is seamed to the top part above the elbow.

Sleeve, Bishop: The reverse of the leg of mutton - the top part of the sleeve is fitted and then flares out towards the wrist where the sleeve is gathered onto a cuff. A popular 1960's style.

Sleeve, Peasant: A full, short sleeve that is attached to a full bodice with a raglan effect seam. The garment is drawn in with elastic or a drawstring at the neck and at the sleeve hem. There are many variations to this theme.

Sleeve, Lantern: A long set in sleeve constructed in two parts. The top part flares slightly from the sleeve head towards the wrist. The bottom part flares from a fitting wrist to meet the flare of the top. Both are seamed together a few inches above the wrist.

Sleeve, Short Lantern: The same principle as the lantern, but a much shorter version ending just past the elbow.

Sleeve, Drawstring Puffed: A full sleeve, short or long, with a drawstring to draw the sleeve to the desired size. The drawing up forms a small frill at the hem of the sleeve. Here the head is not gathered. (See also **puffed**.)

Sleeve, Draped: A set in sleeve that is slashed open at the top part of the sleeve where fullness is added; the underarm seam remains the original length. The extra fullness created is drawn up with a drawstring or elastic to fix it into place, creating a 'draped' effect.

Sleeve, Square Armhole: The armhole is shaped like a square with a right angle at the corner. The construction is as a set in sleeve.

Sleeve, Dolman/Magyar: This style is named from the Magyars in Hungary and was worn by peasants there. The shoulder seam extends through the top of the sleeve and the underarm Seam follows from the side of the bodice through to the wrist. There are no other seams and any shaping is made from these two.

Sleeve, Petal/Lapped: This sleeve is cut without an underarm seam and is shaped and folded on the upper arm. It was very popular in the 1940s.

Sleeve, Mamaluke/Virago: A long, full sleeve that is partitioned into five, full sections. The five sleeve parts are drawn and seamed together to fit around the arm.

Sleeve, Strapped/Banded: The construction is that of a set in sleeve. From the neck point to the wrist is a narrow band of fabric giving a strapped effect.

Sleeve, Epaulet and Elbow Patch: A strap or tab on the shoulder, normally seen on uniforms, to carry caps. Here there is also a patch on the elbow to protect the garment and prolong its life in heavy-duty use.

Sleeve, Dolman with Gusset: See **dolman/magyar**. This dolman sleeve has a gusset set under the arm to allow for more 'lift'.

Sleeve, Kimono: A long sleeve that is a complete extension of the bodice to the wrist. The seam lines are along the top of the sleeve and at the underarm. A traditionally Japanese sleeve used on the garment with the same name.

Sleeve, Slashed Virago: See mamaluke/virago. This sleeve has the upper part incorporating slashing, a popular sixteenth century decorative effect. The slashing reveals the shirt or tunic underneath, giving a contrast. Here there is extra contrast with the lower part of the sleeve being constructed in a different colour or fabric.

Sleeve, Tippet on Elbow: Popular in the middle Ages. This was a pendant effect hanging from the elbow of a gown or tunic. It was similar in derivation to the liripipe.

Sleeve, Melon/Balloon: A very full, short set in sleeve, padded out to give the effect of a melon or balloon.

Sleeve, Batwing: The same principle as the kimono. This sleeve narrows towards the wrist and has a curved underarm seam. This particular example would be made from a knitted fabric.

Sleeve, Cartwheel/Circle: A short set in sleeve, designed to give an accordion effect.

Sleeve, Wing Ruffle: Like the capped sleeve but here the fabric extension has fullness added, the fabric being gathered and set onto the bodice creating a ruffle effect.

Sleeve, Bag: A long and very full sleeve that is gathered onto the cuff at the wrist. It has the effect of bagging on the lower part of the arm. This sleeve was popular in the fifteenth and sixteenth centuries.

Sleeve, Dalmatian/ Angel: A sleeve that flares towards the wrist extending into a long point.

Sleeve, Buttoned Oversleeve: A medieval sleeve that has a contrasting sleeve layered over the base sleeve. The top sleeve can be unbuttoned to reveal the contrast beneath.

Sleeve, Puffed: A set in sleeve that has fullness at the sleeve head and base. It is set into the bodice and controlled at the base by a cuff or elastication to give a 'puffed' effect.

Sleeve, Double: A puffed sleeve with two layers - the base is opaque and the top sleeve is transparent to reveal the sleeve underneath.

Sleeve, Mahoitres: A fourteenth and fifteenth century sleeve popular in France. The sleeve is padded and bag shaped

Sleeve, Hanging Sleeve: This is a very long sleeve that is open down the front seam and hangs vertically. It is part of a medieval gown or doublet

Sleeve, Trailing Sleeve: This is similar to the hanging sleeve. The sleeve is constructed like a kimono and the opening for the hand is in the same position while the bottom part of the sleeve is stitched up.

Sleeving: Braided, knitted, or woven fabric of cylindrical form having a width less than 100 mm. (4 in.) [circumference less than 200 mm. (8 in.)]

Sley: The number of warp ends per inch of the fabric width exclusive of selvedges.

Sleying the reed: is the placing of the warp threads through the dents of the reed. The number of dents in the reed and the number of warp threads in each dent determine the density—or sett— of the warp.

Slide surface, in rotor of an open end machine: That part of the internal surface of the rotor on which the fibres are deposited and are caused to slide to the collecting surface.

Slide Waste: A yarn defect that is similar in appearance to a slub. It consists of a mass of fibre encircling the yarn end and can be slid freely along the end.

Slider, automatic lock: See **Automatic lock slider**.

Slider, cam lock: See **Cam lock slider.**

Slider, flange lock: See **Flange lock slider.**

Slider, in zippers: The part that opens the zipper when it is moved in one direction and closes the zipper when it is moved in the oppsite direction.

Slider, releasing, in Zipper: See **Releasing slider, in Zipper.**

Slider, automatic lock, in Zipper: See **Automatic lock slider.** A slider that provides automatic positive locking action on the chain when the pull is released.

Slider, cam lock, In Zipper: See **Cam lock slider, in Zipper.**

Slider, flange lock: See **Flange lock slider, in Zipper.**

Slider, flange lock, In Zipper: A slider with notches in the flanges of the slider that block the shoulders of the scoops when the stringers are pulled apart.

Slider, pin lock, in Zipper: See **Pin lock slider, in Zipper.**

Slider, ratchet locker, in Zipper: See **Ratchet locker slider, in Zipper.**

Slider, releasing, in Zippers: A slider with a mechanical means for loosening the slider on the chain.

Slip: Measure for wool, linen and jute yarns in England, equal to 1,800 yards.

Slip cover: A removable, fitted protective textile cover, often decorative and specifically made upholstered furniture.

Slip: (1) An important serviceability property of carpets, rugs, mats, etc. is that they do not slip on floors. (2) The resistance to slippage of warp over weft threads, or vice versa, in a woven fabric.

Slip stitching: Slip-stitching or invisible hemming is done on silk, wool, and thick material. The hem is pressed with an iron, a stitch as fine as possible is taken on the surface of the cloth and the needle slipped under and through the first fold, drawing the thread lightly. The needle and thread used in this stitch must be very fine.

Slipe wool: See **Pulled wool.** Recovered wool obtained by treating pelts with lime and sodium sulphide or some other depilatory. This treatment loosens the wool which can then be pulled away without damaging the hide.

Slippage: Sliding or slipping of the filling threads over the warp ends (or vice versa), which leaves open spaces in the fabric. Slippage results from a loose weave or unevenly matched warp and filling.

Slipper Carpet: Warp-pile fabric made with coloured Jacquard figures, used for bags, slippers, etc.

Slipper satin: A closely woven satin made from good quality yarns, which make it hard wearing. It is less glossy than other satins due to the closeness of the weave, and a more elegant fabric because of this. It is used chiefly for footwear. Textures are high and the material comes coloured, black or white, or richly brocaded efects. Shiniest satin.

Slit tape: A fabric, 12 inches or less in width, made by cutting wider fabric to the desired

width. Slit tapes are made primarily of cotton, linen, jute, glass, or asbestos and are used principally for functional purposes.

Slit-film yarn: Yarn of a flat, tape-like character produced by slitting an extruded film.

Sliver: A continuous strand of loosely assembled fibres that is approximately uniform in cross sectional area and without twist.

Sliver knitted fabric: A single jersey fabric in which untwisted staple fibres are knitted in at each loop to form a pile surface on the technical back of the jersey structure.

Sliver knitting: Circular knitting coupled with the drawing-in of a sliver by the needles to produce a pile-like fabric, usually for high-pile coats or heavy linings.

Slop Padding: A printing process used on chintzes and some calicoes. The fabric is first printed with resist after which the colour is applied to the entire face of the cloth by means of an unengraved roller.

Slope threshold: In this method the slope of the initial linear region is determined, and the point where the slope of the curve decreases to a specified fraction of the initial slope shown

Slops: Baggy, overhanging knee or calf length breeches, often having lengthways slashes with protruding lining.

Slot seam: A complex seam formed on the inside of the object, having a decorative seam underlay slightly visible from the face side held in place by two visible rows of stitching. A slot seam allows you to place a matching or contrasting fabric underlay behind the edges of the seam.

Slot seam: A complex seam formed on the inside of the object, having a decorative seam underlay slightly visible from the face side held in place by two visible rows of stitching. The slot seam, used in cloth dresses and jackets, requires exact basting with silk or very fine thread with small, even stitches. If a coarse thread is used, the material will be badly marked. After basting, press the seam open as if it had been stitched, and baste the strap or under strip of the dress material (which has been cut perfectly straight and even) over the wrong side of the seam, having the center of the seam on the center of the strap. Stitch any width desired beyond the center through the three thicknesses. This will hold the seam in position. Now remove the bastings from the seam and the slot effect is complete. If desired, there may be a double row of stitching, an extra row on the edge of the fold or plait. These seams may be finished at the bottom with arrow heads or stitched designs. The lines of machine stitching should not end without some ornament to *appear* to hold the plait.

Slough off: See **Slug.**

Sloughed filling: See **Looped filling.**

Slough-off, in woven fabrics: A defect caused by several coils of yarn slipping of the weft bobbin simultaneously and being woven into the fabric in a group.

Slow sand filter: One of the earliest filters for water treatment, developed in 1829 by James Simpson. It has a sand layer up to 1.0 m deep over a layer of gravel about up to 0.3 m deep. Perforated concrete or clay underdrains laid in the gravel remove the filtered water. The D10 ***effective*** size of the sand is 0.2 to 0.4 mm, with a ***uniformity coefficient*** of 1.6 to 2.5. The sand is drowned under about 1 m of water. The filter is cleaned every few weeks or months by draining the water and scraping off the top 15 to 25 cm of sand. This is repeated until filtration is no longer efficient, and the sand bed is then renewed. Slow sand filters are capable of removing tastes, odours, some bacteria and some protozoa from the water. The ***surface loading rate*** is 2.5 to $7m^3/d$ per m^2 of filter surface area. In wastewater treatment, slow sand filters can be used for tertiary treatment with a surface loading less than 10m3m_2d_1.

Slub: A abruptly thickened place (in a spun yarn) that has tapering ends and a diameter several times that of the adjacent normal yarn which is irregular and at several places along the length of the yarn, causing the fabric to have a surface interest and broken texture. A yarn defect consisting of a lump or thick place on the yarn caused by lint or small lengths of yarn adhering to it. Generally, in filament yarn, a slub is the result of broken filaments that have stripped back from the end to which they are attached. See **Slug.** Many yarns slub is an imperfection, but slub yarns are deliberately manufactured in other fibres as cotton, polyester, acetate, viscose and their blends. The resulting fabric may be of medium or heavy weight.

Slub Catcher: A mechanical or electronic device designed to aid in the detection and removal of slubs or neps in yarns, usually during coning.

Slub Fabrics: Fabrics made using slub yarn, mainly in knitted fabric form.

Slub silk: Silk yarn with nubs or balls of fibre at intervals. It is woven into fabric with an interesting surface structure. The fabric is usually dress weight and often crisp.

Slub Yarn: Any type of yarn that is irregular in diameter; the irregularity may be purposeful or the result of error. (Also see **Novelty yarn**, **Nub yarn**, and **Slub**.)

Slubber: A machine used in textile processes prior to spinning that reduces the sliver and inserts the first twist.

Slubbing: The name given individually or collectively to relatively thick fibrous strands and also to the strips of web from a condenser card that have been consolidated into a circular cross section by rubbing.

Sludge: Solids settled out from water or wastewater, but still containing high percentage of water. Sludge from wastewater treatment is known as ***biosolids***.

Sludge blanket: The mass of sludge in a ***sedimentation tank*** or ***clarifier***. The bottom of the blanket may rest on the bottom of the tank or it may be a suspension, as in the ***sludge blanket clarifier***. The ***settling regime*** of the sludge blanket is either zone settling or compressive settling.

Sludge blanket clarifier, floc b. c., upward flow c., upward flow floc b. c.: A type of ***solids contact clarifier*** in which the inlet flow passes up through the suspended ***sludge blanket***. The particles from the coagulated water are 'filtered' out by and amalgamate with the blanket. When the sludge level rises too high, some sludge is wasted from the tank. The top of the sludge blanket must be well beneath the top water level. The clarified water flows over weirs at the surface of the clarifier. The ***surface loading rate*** may be 40 to 120m^3/d per m^2 of tank surface area, depending on the quality of the inlet water.

Sludge cake: Sludge that has been thickened to 80% can be picked up with a shovel and is called cake. ***Filter plate presses*** can reduce water works and wastewater sludges to less than 70% water. ***Belt filter presses*** and ***centrifuges*** can reduce wastewater sludges to about 75% water.

Sludge digestion: Usually ***anaerobic sludge digestion***. a treatment that stabilises ***raw sludge***. Fully ***digested sludge*** has little readily biodegradable organic matter. It is not smelly and about 50% of the solids are inorganic. Sludge can also be digested aerobically. *See* ***aerobicdigester***.

Sludge hopper: The lowest part of a ***sedimentation tank***, where the settled sludge collects, either by flowing or being scraped in. This deep part is at the centre of an ***upward flow tank*** or ***radial flow tank*** and usually at the inlet end of a ***horizontal flow tank***. Its capacity, in ***primary sedimentation***, should be 1.4 litres per person, which should provide enough space for 24 h ***sludge production***. With sides sloping at 60 ° to the horizontal, it should be deep enough to ensure some consolidation of the sludge by compression settling.

Sludge production: (1) The mass of ***primary sludge*** accumulated per day in wastewater treatment is estimated by multiplying the wastewater flow by its concentration of suspended solids and by the percentage removal of solids in the ***primary sedimentation*** tank (typically 60%). (2) ***Secondary sludge*** production is often stated as the number of kg of sludge produced per kg BOD5 removed or applied to the ***secondary treatment***, a figure sometimes known as the sludge growth index. For standard rate ***trickling filters*** it is about 0.7. For ***activated sludge*** it is about 0.75 at a ***F:M ratio*** of 0.4 per day and may drop to 0.4 at a F:M of 0.1 per day.

Sludge thickening: The difference between two sludges that are, respectively, 98 and 94% water is that the first, at 2% solids, occupies three times the volume of the second for the same weight of dry solids. Therefore, thickening gives good reductions in volume for small increases in solids content. Sludge may be thickened by a variety of techniques. ***Gravity thickening*** is common. *See* ***belt thickening, centrifuge thickening, deep cone thickening, dissolved air flotation, rotary drum thickening***.

Slug, in glass filament: Unattenuated particles of glass of substantially larger diameter than the average filament diameter.

Slug, in raw silk: A thickened place several times the diameter of the yarn, 3 mm (1/8 in.) or over in length.

Slug general: An abruptly thickened place in the yarn or a bunch of lint entangled in the yarn, cord, or fabric.

Slurry: A watery or solvent suspension; e.g., titanium dioxide mixed with water for addition to polymers.

Small Chain: The binder warp in certain carpets, as Wilton or Brussels.

Smart packaging: A smart packaging is one that is able to sense and respond to a stimulus for a specific functional purpose. The term 'functional purpose' is key here, and responses that are purely aimed to be a visual effect – such as the use of thermochromic inks for visual impact only – are not considered to be within the remit of this section. They are however notable for providing design differentiation,

Smart Textiles/Smart clothing: Smart clothing is a "smart system" capable of sensing and communicating with environmental and the wearer's conditions and stimuli. Stimuli and responses can be in electrical, thermal, mechanical, chemical, magnetic, or other forms Textiles that can sense and react to changes in the environment, such as changes from mechanical, thermal, chemical, magnetic and other sources.

Smash, in woven fabric: A relatively larger hole in the cloth characterized by broken warp ends and floating picks. Compare **Slam-off.**

Smash: A relatively large hole in a fabric and characterised by many broken warp ends and floating picks, or a prominent mark that remains after the repair of such a hole.

Smeared prints: Are formed by inadequately ground doctor blades in roller printing so that the print paste is insufficiently removed locally next to the engraved areas of the roller. The problem can be remedied by regrinding the doctor blade and/or by adjusting the position of the doctor blade against the printing roller.

Smoke Chamber Test, for carpet: Test method to measure smoke generation of carpet.

Smoking Jacket: A fancy coat for house (smoking) wear.

Smoldering: The combustion of a solid material without accompaniment if flame but generally with the production of smoke.

Smooth backing: (smooth coat). A smooth, less structured, back-coating with a lower degree of foaming applied to textile floorcoverings for the purpose of achieving good resistance to slippage and cutting.

Smooth foam backing: (flat foam). A smooth, non-structured, foam carpet back coating.

Smoothen, to: A term used in laundering for Mangling and Pressing. In wool finishing, the term is used for steeping (wet setting, crabbing).

Smoothing agent: is a special type of textile softener in which the smoothing effect is based on reducing friction between individual fibres or particles of finishing agents. Cationic products are more effective as smoothing agents than anionic or non-ionic products.

Smoothing and calibrating calendar: A special calendar for the thickness calibration and surface smoothing of needle felt nonwoven filters for liquid media.

Smoothness appearance: A subjective assessment criterion for the appearance of textile surfaces, especially for durable-press (DP) articles after household washing. The smoothness appearance is assessed by comparison with 3-dimensional standards (Wash and- Wear Standard, DP-Rating and the Monsanto Crease Formation Standards as photographs).

Smoothness effect: Smoothness appearance

Smoothness number: This may be defined as the relative fibre-to-fibre resistance to friction where a low smoothness number corresponds to a smoother effect achieved by the application of Smoothing agent. The untreated fabric, smoothness number = 100.

Smouldering: A slow, flameless, smoking burning of a fabric.

Smyrna: Greek cotton, having a medium v strong, harsh and fairly clean staple of dull white colour.

Smyrna Carpets: (1) A general trade name for Turkish carpets exported to Europe via the Turkish port of Smyrna (today's Izmir). These were usually copies of other well-known types of Turkish carpets produced in designs to satisfy European and American preferences and were often woven with inferior materials with a relatively low knot density. They were made principally in Ushak, Isparta, Sivas and Smyrna itself. (2) Coarse, soft, not very durable, deep-pile Medallion carpets of Greek origin.

Smyrna Rugs: (1) trade name for Turkish rugs made in Asia Minor and marketed through Smyrna; (2) in America, factory-made reversible rugs and carpets, made with chenille filling.

Snag, in fabrics: A yarn or a part of a yarn pulled or plucked from the surface.

Snag: (1) Yarns, fibres or filaments in the form of long loops that have been drawn out from the structure of a fabric by a protruding sharp object.

(2) A textile defect caused by (or due to) the pulling or plucking of yarn(s) or filaments from the fabric surface.

(3) A loop which has been pulled out of fine-knit hosiery due to the smoothness of the fibres (ladder). Drop stitches in smooth weft-knit fabrics are wales which, as a result of a broken thread or needle damage, are present as unformed threads.

(4) (pulled thread), a slub-like structure consisting of individual capillary fibres, 2–8 mm in length, which has been pulled out of the whole thread. It occurs mainly with polyamide yarns and knitted fabrics (hosiery).

Snagging resistance, in textile fabrics: The resistance to the formation of snags.

Snap: See **Snap fasteners.**

Snap action, in: The force required to disengage a snap fastener resulting from a pull in the plane parallel to the material which the snap fastener is attached.

Snap fastener: A device for attaching one material to another consisting of matching male and female parts, each of which is attached to a separate material so that the parts can be joined by a low compressive force and separated by a low perpendicular tensile force. Small snaps (size 4/0 through 1/0) are for light weight fabrics and large sizes (1 through 4) are for heavy ones.

Snarl, See **Kink.**

Snarl: A short length of warp or weft yarn that has twisted on itself owing to lively twist (see twist liveliness) or insufficient tension. The snarling may occur during or prior to the weaving process.

(a) Snarled Yarn (b)Twist Free Yarn

Snicks: Flaws in the yarn, consisting of very thin places.

Snow Ball: See **Fuzz ball, Balling up**.

Snow cloth: The term used to describe any heavy outdoor cloth, particularly those with nap or pile.

Snubber Pin: A stationary pin or guide which induces a localized change in yarn tension. A draw pin is a type of snubber pin.

Soaking: Treatment of rayon yarns in a lubricating and sizing solution preparatory to hard twisting.

Soap: A substance consisting of sodium or potassium compounds of fatty acids used to improve the cleansing properties of water Soap is a surfactant and was the earliest known detergent. Alkali salts of fatty acids with a minimum of 12, and a maximum of 18, carbon atoms. The first signs of surfactant properties are already apparent from 8 carbon atoms upwards. The hydrophilic part of an anionic surfactant is due to the presence of COO– groups, and the hydrophobic part is due to the presence of an alkyl residue, e.g. sodium stearate $C_{17}H_{35}COONa$, . sodium oleate, used as a cleaning agent. The basic method of making soap involved treating animal fat (mainly beef tallow), which is a triglyceride of octadecanoic acid (stearic acid), with caustic soda (sodium hydroxide; NaOH) to produce sodium octadecanoate (sodium stearate). More refined soaps are made from vegetable oils, such as palm oil, which contains hexadecanoic acid (palmitic acid). Liquid soaps (*soft soap*) are made using potassium hydroxide rather than sodium hydroxide.

Soap Flakes: They are actual small pieces of hard soap. Soap Flakes are produced from dried, cut, hard soap.

Soap Handle: The preferred handle for worsted yarn wool fabrics which, in former times, was achieved by using soap in scouring and milling.

Soap powder: Dried, pulverized Soap (produced from soap paste and sodium carbonate) with a total fatty acid and resin acid content of 5–40%.

Soap residues: (fatty residues). Coarse agglomerated residues of soap, fat, lime salts, and dark pigment soils, often appearing as numerous spots which are deposited on garments during washing. They are caused by using a soap with an excessively high fat content or (usually) by carrying out the first rinse with cold water. They can be eliminated by using Syndets or a slightly acidic rinse water

Soaper: See **Soaping Machine.**

Soaping: (or soaping off) - with respect to dyeing, the process of washing dyed fabric with very hot (often boiling) water with surfactants, rarely actually soap, to remove dye that is not fixed to the fibre. Soaping off is important particularly with some reactive dyes, since a good deal of hydrolyzed dye is loosely bonded to the fabric, and must be removed to avoid staining of other garments or fabric in laundering. Soaping is also important in vat dyeing, where there are actual changes, including in hue, in the dye in the fibre. Though surfactants are often used, they may actually contribute almost nothing to the effectiveness of the process.

Soaping Machine: This type of machine consists of a vat with squeeze rolls for wetting out wool piece goods that only require a short milling treatment. When the addition of milling agent is made by pouring it into the machine, this does not result in satisfactory distribution of the product.

Soapstone: A soft type of talc which has a greasy feel and which is easy to carve to make ornaments. It was formerly known as steatite. See **Stearite.**

Soda : See **Sodium Carbonate.**

Soda Ash: Sodium carbonate; an antiquated but much-used term.

Soda-ash boil: The type of alkali used for scouring of cotton depends on the quality of goods.

For example, if coloured yarns present in the fabric, sodium carbonate is ideally suited because of its low pH. Cotton yarns to be dyed in dark shade should be scoured with 1-2% sodium carbonate solution for 30 min in presence of wetting agent.

Soda cellulose: (alkali cellulose). Represents the first stage in viscose rayon manufacture. Cellulose pulp is steeped in warm caustic soda liquor (17–18%)

for 1– 2 h, and then pressed to remove excess solution. The treated cellulose is broken up in a shredder to form powdery crumbs. The crumbs are then aged for several hours during which time the caustic soda reacts with the cellulose to form soda cellulose.

Soda lime: A gray solid produced by adding sodium hydroxide solution to calcium oxide, to give a mixture of $Ca(OH)_2$ and NaOH on evaporation. It is used in the laboratory as a drying agent and as an absorbent for carbon dioxide.

Sodalite, cage zeolite: A blue, grey, yellow or colorless mineral consisting essentially of sodium and aluminium silicates with sodium chloride in a cubic crystalline form of the type Na[Cl(AlSiO)], which occurs in certain basic igneous rocks. The hollow structure of sodalite is important as a lattice unit in Zeolites.

Sodium Acetate: CH_3COONa; a buffer 3 In dyeing, sodium acetate is almost always used together with acetic acid in moderately acid processes (that is, sodium acetate is rarely used without acetic acid, but acetic acid may often be used without sodium acetate). molecular weight 136. Colorless odourless crystals or anhydrous salt. Solutions in water are weakly alkaline. Uses: neutralization of mineral salt esters on cellulosic textiles; reduction of acetic acid acidity (buffering); neutralization of naphthol diazo solutions; aftertreatment of sulphur black (prevents tendering); additive for diazo solutions in dyeing and printing with naphthols, etc.

Sodium alginate: Sodium salt of Alginic acid, $C_{10}H_{18}O_{10}(COONa)_2$; soluble in cold resp. warm water under rapid stirring (it is advisable to allow freshly prepared solutions to stand overnight). Boiling in water reduces the viscosity. Solutions are non-foaming. Stable to alkalis but sensitive to acids (especially below pH 3). The addition of a complexing agent (5–25% of the weight of alginate) is recommended in hard water. Solutions are sensitive to bacterial decomposition (the addition of 0.2% salicylic acid, formaldehyde, etc. can be used as a preservative). Uses: sizing agents, finishing agents, water-repellent impregnations. The main use of sodium alginate is as a thickener for textile printing (discharge pastes, roller and screen printing, easy to wash off, uniform viscosity, excellent penetration in printing, non-foaming, flows readily, good colour yields, sharp outlines). Suitability as a thickener for textile printing: especially important for reactive dyes, azoic diazo dyes, vat leuco ester and vat dyes; not suitable for cationic dyes, chrome dyes and naphthols.

Sodium aluminate: (aluminate of soda). Na_3AlO_3 or $Al(ONa)_3$, molecular weight 144.25. Colorless small lumps or fine powder with 35% Al_2O_3 (stable to air) or 50% Al_2O_3 (somewhat hygroscopic, store in airtight containers);

both forms are readily soluble in water. Uses: waste water treatment, water softening, oil removal from feed water and desilification; mordant for dyeing alizarin red (on cotton), etc.

Sodium Bicrbonate: $NaHCO_3$; more properly called sodium hydrogen carbonate; also called sodium acid carbonate, and most commonly, baking soda Sodium bicarbonate is used as a weak base (alkali) in some dyeing processes, often with reactive dye that is padded onto fabric, then batched for many hours. It has limited stability in solution, decomposing to sodium carbonate, carbon dioxide and water, especially at high temperature, so solutions generally should be made shortly before use. Sometimes this decomposition is used deliberately, usually with highly reactive dyes such as MX. A print paste made with sodium bicarbonate will have a pH around 8, and at this pH the dye will not hydrolyze rapidly and hence become useless. When the printed fabric is steamed, the sodium bicarbonate will decompose and the pH will rise to around 11, facilitating reaction of the dye with the fibre.

Sodium Bisulphate: $NaHSO_4$; also called sodium hydrogen sulphate or sodium acid sulphate. Sodium bisulphate hydrolyzes in water solution, and acts much like sulfuric acid. It can sometimes be used as substitute for sulfuric acid, and because it is a dry chemical, it can be safer to handle, though care is still necessary. Sodium bisulfate is often the basis for etchant pastes for devoré of cotton. "pH Down" for spas and swimming pools is usually sodium bisulfate. Do not confuse this with sodium bisulfite.

Sodium Bisulphite: $NaHSO_3$; a mild reducing agent, most used in dyeing as an antichlor; often actually sodium metabisulfite ($Na_2S_20_3$), which behaves the same way As an antichlor, sodium (meta)bisulfite reacts with sodium hypochlorite and hypochlorous acid, to almost immediately stop bleaching action. The product of the reaction is two acids which need to be thoroughly rinsed from the fabric. Typical use is about a gram (approximately 1/4 tsp) per litre of room-temperature water. Solutions should be prepared shortly before use (preferably within an hour) and with gentle stirring, since oxygen dissolved in the solution will destroy its effectiveness. Discard used solutions. Do not confuse this chemical with sodium bisulphate.

Sodium borate: (borax, sodium tetraborate, sodium pyroborate). $Na_2B_4O_7 \cdot 10H_2O$; molecular weight 382; density 1.7. White crystals or anhydrous powder. Soluble in water; the solution is fairly strongly alkaline. Uses: mild alkali, textile printing, application of alkali blue (wool dyeing); manufacture of hat proofs; addition to finishing liquors (as a preservative); solvent for alizarin dyes and casein; flame-retardant impregnations; component in soaps and gloss starches; flux.

Sodium Carbonate: soda ash; Na_2CO_3; a weak base Sodium carbonate is an xtensively-used alkali in textile preparation and dyeing. It is typically used to adjust the pH of solutions to about 11. It is also used in scouring cellulose fibres. The most common industrial form is the anhydrous type. Sodium carbonate monohydrate ($Na_2CO_3 \cdot H_2O$) is the most common form sold for photographic use. Washing soda, if "pure", is usually sodium carbonate decahydrate ($Na_2CO_3 \cdot 10H_2O$). Soda ash is hygroscopic, so it should be stored in a tightly-closed moisture-proof container. If so protected, it will keep indefinitely.

Sodium Chlorate: $NaClO_3$(chlorate of soda) An oxidizing agent, sometime used as a *resist salt* to protect dyes from reduction in printing processes under acid conditions. Molecular weight 106.5; density 2.5. Colorless crystals, hygroscopic, toxic; water-soluble. Textiles moistened with sodium chlorate are readily ignitable. Uses: powerful oxidizing agent especially in textile printing and aniline black dyeing.

Sodium Chloride: NaCl; common salt Sodium chloride is used in large quantities as an electrolyte in application of direct dyes and reactive dyes

Sodium chlorite: $NaClO_2$; molecular weight 90.5; white crystals or crystalline powder, odourless, very stable product. Acidic solutions decompose very rapidly: Uses: bleaching agent for cellulose (chiefly linen and cotton), cellulose pulp, paper, cupro, acetate and viscose fibres, as well as polyamide, acrylic, polyester and polyvinyl chloride fibres, waxes and straw products, edible and inedible oils, etc.; oxidation of vat and sulphur dyeings.

Sodium Chlorite Bleach: (1) Cellulosic fibres: sodium chlorite produces excellent bleaching effects with a high degree of fibre protection. It can be applied by full bath treatments or by discontinuous and continuous padding methods. Bleaching reaction only proceeds quickly enough in acidic media below pH 5. Damage to the fibre can occur at pH < 3. For full bath bleaching, organic acids (acetic or formic acid) are preferred to adjust the pH. In addition, diammonium or sodium phosphate are added as stabilizers. In pad-steam processes, organic acids are substituted by products which split off acids by the action of heat, i.e. so-called activators (e.g. ethyl acetate, chloral hydrate). (2) Synthetic fibres: with the exception of polyurethane and some individual acrylic fibre types, the sodium chlorite bleach is the suitable bleach for synthetic fibres in which the material is treated exclusively in a long liquor.

Sodium Dithionate: sodium hydrosulfite; dithionite is technically more correct. Uses: the most important reducing agent for stripping dyeings; reducing agent for the dyeing of vat, indigo and sulphur dyes; reducing agent for the printing of vat dyes by the 2-phase process; bleaching agent (especially for wool); antichlor; spotting agent (for dye, fruit, perfume, rust and ink stains, perspiration, etc.). Sodium dithionite-formaldehyde compounds (so-called sulphoxylates) are used for stripping dyes, and as more stable reducing agents for the dyeing and printing of vat dyes, and discharge printing.

Sodium Formaldehyde Sulphoxylate: A reducing agent used in discharge techniques and for stripping dye; Formosul and Rongalit C are trade names; designated in Colour Index as Reducing Agent 2 This chemical is useful for discharge on cellulose and silk, but not usually on wool. It requires alkaline conditions that may damage wool (alkaline conditions can also reduce the sheen of silk). It does not contain free formaldehyde. Also see zinc formaldehyde sulfoxylate.

Sodium Hexameta Phosphate: Na(x+2)PxO(3x+1) , where x = 6 to 21; a *sequestrant* used to treat hard water; SHMP, often referred to in art dyeing literature as "metaphos" Sodium hexametaphosphate softens water by sequestering calcium and magnesium, effectively making them unavailable to participate in other reactions. Typical use would be about 0.5 grams per litre of water. Calgon T is one trade name for sodium exametaphosphate, though retail Calgon products contain other chemicals, and are not the preferred products for dyeing applications.

Sodium Hydrosuphite: $Na_2S_2O_4$; a reducing agent used in discharge techniques, dye stripping, and in vat dyeing; more properly called sodium dithionite; dyehouse term is "hydro" or "hydros" For stripping, it is usually used in an alkaline solution at a temperature near the boiling point of water. This is the active ingredient in some of the "whiteners" or "dye removers" sold in small packages like the "household" dyes in the grocery store. It produces a rather strong sulfurous smell. Sodium hydrosulfite is a flammable powder, and must be handled with care. Do not confuse this with sodium hydrosulphide, which is used with sulfur dyes.

Sodium Hydroxide: NaOH; a strong base; commonly sold as beads or flakes, or as a 50% solution; also called caustic soda or lye, often just called "caustic" in dyeing Sodium hydroxide is used in some dyeing processes requiring very high pH. A solution of 4 grams per litre will have a pH of 13. It is common in vat dyeing, some reactive dye methods, and is used in mercerizing cotton. It is commonly used for industrial *scouring* of cotton. Sodium hydroxide must be handled with great care, since it will cause severe skin burns. It should be dissolved by SLOWLY adding it to stirred VERY COLD water, since a great deal of heat is liberated as it dissolves. Never use aluminum vessels or tools with NaOH: they will be corroded, and explosive hydrogen will be generated. Sodium hydroxide is deliquescent - extremely hygroscopic.

Sodium Hypochlorite: NaOCl Household type chlorine laundry bleach contains about 5% sodium hypochlorite as the active ingredient. Stronger solutions are available for industrial use. See bleach, chlorine. Normally used as the hypochlorite lye: clear, yellow-green, free from calcium salts, stabilised (stable for cool storage during months); density 1.21–1.23 with 140–160 g/l active chlorine, equivalent to 147–168 g/l NaOCl.

Sodium Hypochlorite Bleach: In comparison to Calcium hypochlorite has an advantage in the absence of calcium, hard water salts. Therefore easier rinsing, weaker acidification (cost), softer fabric handle, also retains white effects better and helps to avoid local fibre damage by calcium chlorite particles. Clear disadvantage: AOX loading in effluent.

Sodium Hypophosphite: The most effective catalyst for promoting Durable Press properties. DP rating and shrinkage control equivalent to DMDHEU have been obtained. Catalyst is expensive higher than normal amounts are needed. The hypophosphite is a reducing agent. This is beneficial for maintaining white fabrics, however, it discolors certain dyes especially sulfurs.

Sodium m-nitrobenzene sulphonate: Often referred to by BASF's tradename Ludigol®. This chemical is a mild oxidizing agent that is sometimes used to protect dyes from degradation caused by reducing conditions that exist at high pH and high temperature, especially where air is excluded from the dyeing vessel. Such conditions are quite common in processes for some of the low-reactivity dyes, but very rare in processes for MX dyes, except in steam fixing. Recommended amounts for protecting dyes from reduction range from 1 to 10 grams per litre of dyebath. When used in printing it is sometimes called a resist salt. It is also used as an oxidizing agent for some vat dyes.

Sodium Metabisulphite: See **sodium bisulfite.**

Sodium m-nitrobenzene sulphonate: Used to inhibit the reductive effects of caustic and soda boiling, (bleaching dyed goods), as reserving agent for vat dyestuffs in discharge printing and similar methods.

Sodium N-benzylsulphanilate: Fine white to yellowish powder, easily soluble in water with neutral reaction. Extensively used in dissolving hydrotropic dyestuffs and as a dispersing agent, especially for vat and leuco-vat dyestuffs in printing: effective as a leveling agent, increases the yield and improves fixation.

Sodium nitrite: (nitrite, saltpetre). $NaNO_2$; molecular weight 69; density 2.17, white to pale yellow crystals, weakly hygroscopic, easily soluble in water, slightly soluble in alcohol. Applications: Diazotising azo dyestuffs (1 part sodium nitrite and 3 parts nitric or sulphuric acid; ratio is important to produce the required nitrous acid); dyeing with leuco-vat dyestuffs; in printing, etc.

Sodium number (SN): Expression for the required minimum alkalinity (to avoid hard water deposits) of feedwater in mg/l of NaOH. Calculation:

$$SN = \text{mg NaOH} + 0.222 \text{ mg } Na_2CO_3 + 0.667 \text{ mg}$$
$$Na_3PO_4 \bullet 12H_2O + 0.222 \text{ mg } Na_2SO_3.$$

Sodium Perborate: $NaBO_3$ This is the bleaching compound that is most commonly found in laundry detergents that contain "colour safe" bleach. Such detergents often contain special activators that make the perborate work more effectively at moderate temperatures.

Sodium Percarbonate: A dry chemical made by reacting sodium carbonate with hydrogen peroxide Sodium percarbonate can be used as an alternative to hydrogen peroxide for oxidizing bleaching. It is considerably more expensive, but is safer to handle (relative to industrial-strength hydrogen peroxide). Oxi Clean® is a retail cleaning compound believed to be primarily sodium percarbonate.

Sodium peroxide: (sodium superoxide). Na_2O_2; molecular weight 78; density 2.81. Pale yellow, fine particulate powder (or compact granules of 0.3–1.5 mm ∅) 19–20% active oxygen, very hygroscopic, odourless; 97% purity. Application: Oxidizing agent for vat and sulphur dyestuffs; bleaching agent for silk, wool, cotton, linen, viscose, etc.) In soluble form after the addition of sulphuric acid (to neutralize) and addition of phosphate to make weakly alkaline, stabilizer etc. to produce a non-yellowing full white product (softer fabric handle and less fibre damage).

Sodium Phosphate: Primary sodium phosphate (monosodium phosphate, sodium dihydrogen phosphate) NaH_2PO_4. Not as effective in promoting DP ratings. Requires higher temperature cure. Much less expensive than hypophosphite. Discolors whites at higher temperatures Does not affect dye shade. Application: Water preparation especially for boiler feed as corrective method after base exchange. Dosing ca. 20 g. sodium phosphate crystals/ m3/1° hardness whereby there is a minimum 1° carbonate hardness from sodium hydrogen carbonate. II. Secondary sodium phosphate (disodium hydrogen phosphate). $Na_2HPO_4 \cdot 12H_2O$; density 1.53; molecular weight 358. Application: Water preparation especially for boiler feed (60 g/m^3 and residual hardness) especially after the softening stage; weighting of silk, dyebath additive as a mild alkali (substantive sulphur dyestuffs), fixing agent for aluminium mordant (alizarin red), flame retardant impregnation, builder for laundering detergents. III. Tertiary sodium phosphates Trisodium phosphate.

Sodium sesquiphosphate: $Na_3H_3(PO_4)_2$. Known as a polyphosphate; changes on heating to a neutral mixture of alkaline mono-phosphates at 300–500°C. Application: as detergent and boiler de-scaling agent.

Sodium Silicate: $Na_2Si_3O_7$ and similar compounds, usually as a solution in water Sodium silicate is used as a stabilizer in hydrogen peroxide bleaching. It can also be used in direct application of reactive dyes to cellulose fibres. Here it is applied essentially full-strength after the dye is applied, acting as an alkali to cause the dye to fix to the fibre. Because it is a syrupy liquid, it makes the dye spread less than other postapplied alkalies. Care must be take to avoid making solutions of sodium silicate acidic, since this will result in a water-insoluble gel that is extremely difficult to remove. Sodium silicate solution is also called water glass. It can be very irritating to skin. Application: Degumming of silk; caustic and keir boiling; stabiliser for peroxide bleaching; silk weighting; dyeing of alkali blue (wool); flame retardant impregnation; in printing; matting of viscose; soaping assistant; component in washing powder and liquid soaps; adhesives, etc.

Sodium Sulphate: Glauber's salt; Na_2SO_4, molecular weight 142; density 1.46. Sodium sulphate is used as an alternative to common salt (sodium chloride) in some dyeing processes, such as those involving MX turquoise where it is a simple electrolyte. When used with some acid dyes for wool, it may act as either an electrolyte to enhance exhaustion, or as a retarder to aid leveling, depending on conditions such as pH. It is also used as a diluent for many dyes. Sodium sulphate is often sold in anhydrous form. It will absorb water from the air, so it should be stored in tightly closed containers. In the dye industry, some use "sodium sulphate" only to refer to the anhydrous form, known as "Duisburger sodium sulsulphate", white powder, iron free and 99.5–99.8% pure, water soluble with gentle heating and "Glauber's salt" to refer to the decahydrate form ($Na_2SO_4 \cdot 10H_2O$) is colorless crystals of high purity, easily efflorescent, water soluble with temperature lowering. Application: Salt in the dyehouse (neutral baths: like sodium chloride for salting out dyestuffs, increases dyestuff rate of exhaustion; acid baths: reducing acidity, slows the rate of exhaustion of dyestuffs); in standing baths for increasing the salt content and raising the boiling point; component in many powderform dyestuffs (diluent); additive in weighting formulations, etc..

Sodium stannate: (preserving salt, tin soda). $Na_2SnO_3 \cdot 3H_2O$; molecular weight 267. Easily water soluble crystals, solution (hydrolysed) strongly alkaline. Used as a swelling agent e.g. for viscose filament yarn (by diffusion of salt solution into the fibre) and, by thermal decomposition for the stabilising and storage of stannic acid gels (similar effect and improved by a treatment with barium chloride solution with the formation of barium stannate), as matting pigments. Application: weighting of silk; flame retardant impregnation (textile finish); matting of viscose.

Sodium stearate: (sodium salt of stearic acid). $C_{17}H_{35}COOH$ molecular weight 306.46. Saponified stearic acid. White flakes; soluble in water and alcohol. Application: manufacture of household soaps and tallow soaps, finish component, etc.

Sodium sulphide: (sulphuretted sodium). $Na_2S \cdot 9H_2O$; molecular weight 240. Fused mass or crystals (ratio 1 : 2), colorless (also weakly yellow), smell of hydrogen sulphide, hygroscopic, water soluble with alkaline reaction; gradually decomposes in the presence of air; reducing agent. Application: dissolving and dyeing with sulphur and vat-sulphur dyestuffs; de-hairing agent (tanning), etc.

Sodium sulphite: (sodium salt of sulphurous acid). $Na_2SO_3 \cdot 7H_2O$; molecular weight 252. White crystalline powder, ca 25% SO2, easily efflorescent (with oxidation to sodium sulphate), water soluble, weakly alkaline solution. Application: antichlor; reducing agent (printing); preservative; corrosion inhibitor in process water (binds acids), etc.

Sodium tetraphosphate: $Na_6P_4O_{13}$. Anhydrous sodium tetraphosphate contains ca. 60.4% P_2O_5 and 36% Na_2O, is hygroscopic and reacts weakly alkaline. In contrast tripolyphosphate will bind more calcium ions. Used as builder for paste/highly viscous synthetic detergents.

Sodium tripolyphosphate: (TPP, triphosphate, tripolyphosphate, sodium polyphosphate). Na5P3O10; washing powder additive. Application: chelating agent for water softening; feed-water preparation; hot water supply systems; laundering; milling; after-rinsing, etc.

Sodium Thiosulphate: $Na_2S_2O_3$, a chemical used to neutralize chlorine in water for dye baths The chlorine present in tap water can actually destroy small amounts of dye. This can be a problem, particularly when dyeing very pale shades where loss of a little dye can influence the final shade noticeably. Sodium thiosulfate is preferred for neutralizing chlorine for this application. Typical use would be about 0.15 grams per litre of bath. Sodium thiosulfate is the main ingredient in fixer (types sold as dry powder mixture) for black and white photographic processing. Most antichlor products for aquarium use are sodium thiosulfate.

Sodium tungstate: (sodium salt of wolframic acid), $Na_2WO_4 \bullet 2H_2O$); molecular weight 330; density 3.25. Colorless crystals, water soluble. Application: flame retardant and water repellent finishes.

Sodium Xylenesulphonate: $C_6H_3(CH_3)_2SO_3Na$. Highly effective hydrotropic agent for fat free soaps, scouring agents, washing powders and similar.

Sof: (1) A very fine, plain woven fabric, made by the natives of Kashmir of the finest sort of mohair; obsolete; (2) Very light, changeable or brocaded fine woollen from Asia Minor.

Soft dyeing technique: Term used for gentle handling of fabrics in the dyeing process with overflow or jet dyeing machines having controlled liquor flow.

Soft flow rope dyeing machines: Fabrics with a sensitive surface character that may be subject to creasing or surface roughening during the dyeing process are dyed on machines using the oveflow principle in soft flow dyeing machines. Here the goods are not moved through a jet pipe but fall onto a smooth plate or are twice overflow-treated

Soft Shell: Soft shell fabrics combine the benefits of hard shell fabrics with a breathable, flexible, comfortable fabric. Stretch wovens with a DWR treatment.

Soft silk: A silk from which the natural gum (sericin) has been completely removed by degumming.

Soft soap: See **Soap.**

Soft waste: See **Waste, soft.**

Soft water: Water with little or no ***carbonate hardness*** or ***non-carbonate hardness***. It may originate in the hard rock areas or from peaty moorland, but not from chalk, nor from limestone formations. Soft water may be corrosive, as in ***plumbosolvency***.

Soft window covering: Curtains, draperies, or other accessories on wall or window openings that are either lined or unlined and primarily constructed of textile fabrics.

Softener: A chemical or chemical mixture intended to give fabric a soft *hand* Softeners act primarily as lubricants for fibres, allowing them to slide over each other more easily, giving a softer feel. Some also help reduce static electric charge build up, and some are antimicrobial. Many softeners are surfactants, and they may be cationic, anionic or non-ionic. Household laundry softeners are often based on cationic surfactants. Softeners can interfere with dyeing, so they should be removed by scouring. Some, such as silicones, can be quite difficult to remove.

Softener (Fabric): When the hand is made to drape more or to feel silkier, the fabric is said to have been softened. Chemicals that do this are called Softeners. Many softeners are derived from naturally occurring Fats, Oils and Waxes. Sources and reactions of fats, oils and waxes have been discussed in a Chapter 3. Some softeners are derived from synthetic raw materials. Many of the compounds that work as softeners also function as surfactants or water repellents. These topics are covered in greater detail in other sections. It is hoped that the reader will come to appreciate that certain chemicals can serve many functions as textile finishes and processing auxiliaries. Softeners are divided into three major chemical categories describing the ionic nature of the molecule, namely Anionic, Cationic and Nonionic. Nearly all surfactants are softeners; however, not all softeners are surfactants. Surfactants are two-ended molecule, one end being lyophilic and the other hydrophilic. The lyophile is usually a long hydrocarbon chain, the essence of most lubricants. The ionic portion is responsible for water solubility, (a necessary feature for applying the softeners) and as will be discussed later, in how the molecule aligns itself at the fibre surface. This section will be devoted to describing the chemical structures of important softeners, some of their properties and their fabric uses. It is well to remember that the same chemical structure may describe a surfactant used for other purposes such as detergents, wetting agents, emulsifying agents etc.

Softener (water): Chemical(s) added to water to prevent hardness ions from interfering with other solutes Sodium hexametaphosphate (SHMP) is commonly used as a softener in dyeing. It will hold moderate amounts of hardness ions in solution. Soda ash will soften water, but it does it by formation of insoluble precipitates that can deposit on fabric or equipment, so SHMP is often added before soda ash to prevent this. Water softening systems which use common salt exchange hardness cations for sodium.

Softening point: The temperature at which substances without a sharp melting point change from viscous to plastic flow.

Softening range: The temperature range for synthetic fibres (according to each fibre type) that lies between becoming thermoplastic and the true melting point. It is not identical to the Transformation temperature but lies somewhat above.

Softness, in water: A relative absence of dissolved calcium, magnesium and other salts that react with soluble soaps to form insoluble precipitates.

Softness index: Calculated from the factors of surface smoothness or friction, yarn thickness and compressibility. Fabrics are tested as 1 cm. wide strips. Evaluation is based upon a 10 batch. The softness index indicates e.g. treatment with dist. water 1; with 0.5 gpl soap 1.10; after acidic bleach on cotton 1.36.

Sofu: Plain woven unbleached cotton sheeting in Japan, made usually 36 inches wide and 44/44.

Soie: French for silk.

Soie Mi-serree: French term for loose twist, glossy silk yarn for crochet.

Soie Mitorse: Half twisted silk yarn for embroidery. See **Mitorse silk**.

Soie Ondee: Silk yarn used for gauze; it is made by twisting a fine and coarse thread together.

Soie Ovale: French term for silk embroidery yarn.

Soie Platte: French floss silk yarn used for embroidery, tapestry, etc.

Soie Vegetale: Flax treated to have a high, permanent lustre. It bleaches and dyes well; used for braids, laces, etc. silk.

Soil affinity: The reduction in *Y* value between new and soiled fabrics.

Soil burial Test: A test of resistance of textile material to certain microorganisms present in soil. The samples are buried in soil for an extended period, then removed and measured for strength loss.

Soil Hiding: Ability of carpet to hide soil.

Soil redeposition: Redepositing of removed dirt from another fabric, onto the goods which is clean or relatively clean during washing or dry-cleaning processes. Anti-soil redeposition: Counteracts the re-deposition of dirt during washing.

Soil release: The property of dirt, especially oil based, being more easily released from textiles during cleansing processes. Synthetics and resin finished synthetic/ cellulosic mixtures are poor in this respect in comparison to natural fibres.

Soil release Agent: Soil release products currently in use can be based on the following chemicals, for example: Silicium compounds, carboxymethylcellulose, ethoxylated compounds, polyglycol ester of terephthalic acid, acrylic acid polymers, and fluorochemicals.

Soil release Finish: Application of finishing processes that are specifically designed to ensure more efficient laundering of soil and stains. Good soil release effects are achieved through application of the Dual action principle, using fluorochemicals that are oleophobic and hydrophilic. See **Antisoiling finish**.

Soil repellency: Alternative term for Antisoiling finish Resistance to soiling as a finishing effect, which prevents soil penetration, or makes it difficult. Examples of soiling include dry soil (dust), wet soil (fruit juice, ink), oils and fats (engine oil and skin grease). Should not be confused with Soil release finish. Anti-soiling finishes.

Soil resist agent: Agents which are used in producing the Soil resistant finishes.

Soil resistant: Resistance against water-bound soil, fatty substances and pigment soiling.

Soiled end: Self explanatory.

Soiling: The staining or smudging of textile materials resulting from the deposit of dirt, oil undesirable dye, etc.

Soiling behaviour: Soiling behaviour of textiles is extremely dependent on the type of fibre, Soil release. No standardized tests exist for the soiling behaviour of carpets, because there are so many different types of soiling (dust, mud, lint of all types), and different colours and patterns. However, institutes use internal testing methods that permit absolute conclusions under specified conditions.

Soiling test: Wash or dry clean a test sample of a textile that has been artificially soiled, so that the level of optical brightening (Whiteness, degree of) achieved after washing or dry cleaning can be comparatively assessed. In principle, the soil test incorporates a combination of three soil types, such as carbon pigment, saponifiable fat, mineral oil (standard soil types).

Soil Resistance Merit Rating (SMR): Factor of soil resistance for carpets, taking into account the influence of colour. SMR indicates how high the soil concentration should be in a white test sample, so that the same apparent level of soiling as a specific colour can be achieved. The higher this value is, the greater the soil resistance of that colour. These values are defined on the basis of reflection measurements using a green filter (y value). Evaluation:

1.0–3.9 very bad

4.0–7.9 bad

8.0–15.9 fairly good

16.0–30.9 good

> 31.0 very good

E.g., white 1.0 ; mid-blue 33.0; coffee brown 190.0

Soils: Soils can be defined as unwanted substances at the wrong place. Most common soils fall into one of four categories: (a) water borne stains, (b) oil borne stains, (c) dry particulate soils and (d) composite soils involving oil and grease adsorbed on particulate matter. Water borne stains are not much of a problem, the stains are soluble in the wash water. Food stains and dried blood, although not water soluble, are responsive to proteolytic enzymes found in most commercial detergents. Dry particulate soils such as flour, clay and carbon black are mechanically entrapped in the yarn interstices and reside on the surface of the fibre. Removal of particulate soils depends on overcoming the work of adhesion between the particle and the fibre surface, facilitating the transport of detergent solution to where they reside and transporting the particle into the wash water. Mechanical energy (agitation) is important for latter.

Soisette: The name given to a fine cotton made from mercerized yarns. Its use is confined mainly to good quality nightwear and negligees, due to its softness.

So-Iron: A durable press rating. Fabric must have a smooth appearance after it is laundered dried (line dried or tumble dried). Fabric smoothness is referred to as Durable Press Rating (DP Rating) and is judged on a 1 to 5 scale. A 5 DP rating is the highest (most smooth) and a 1 is the most wrinkled. A 3.5 rating is considered commercially acceptable.

Soleil: The name of a very highly finished wool fabric woven in twill broken-rib effect. The weave and finish give the fabric a shiny appearance. The term is also used to describe any shiny, light reflecting fabrics in any fibres.

Solid braid, in rope: A braided construction in which each strand alternatively passes under and over one or more of the other strand while all strands are rotating around the axis with the same direction of rotation.

Solid Coloured: Fabric dyed in one colour.

Solid content: Dry substance content (in g) per kg of thickener, e.g. locust bean flour 20–25, locust bean flour derivatives 20–250, guar derivatives 40–250, starch ether 100–400, carboxymethyl cellulose 40–200, hydroxymethyl cellulose 15–40, crystal gum 250–330, carragheenates 50–100, polysaccharides 40–200, alginates 30–120, tragacanth 65–80, starch 80–100, polyvinyl products 100–200, polyacrylic products 200–250, British gum 500.

Solid-state polymerization: Reaction of the active end-groups within a solid polymer. It may be intentional as in heating and drying nylon 66 to increase the final degree of polymerization; or it may be undesirable such as that which occurs in fibres under high temperature conditions in tires that leads to increased degree of polymerization, cross-linking, and subsequent brittleness and loss of strength.

Solubilised sulphur Dyes: See **Solubilised vat dyes.** Sulphur dyes are insoluble pigments which are usually converted to soluble form with the addition of sodium sulphide and soda ash, which is the leucoform of the sulphur dye and is not stable. For the ease of application the suphur dye can be converted to solubilised stable form by esterification of the leuco form and these are called solubilised sulphur dyes. These solubilised dyes are applied directly on the fabric and converted to original sulphur dye on the fibre fabric by suitable treatment.

Solubility: Symbol: *S* The amount of one substance that can dissolve in another to form a saturated solution under specified conditions of temperature and pressure. Solubilities are stated as moles of solute per kilogram of solvent (molality), or as kilograms of solute per cubic meter of solvent (density). The solubility of a solid in a liquid generally increases with temperature, whereas that of a gas in a liquid generally decreases. An increase in pressure on a gas above a liquid leads to a proportional increase in the solubility of the gas. See also **concentration**.

Solubilized vat dye: Also called Vat Leuco Ester Dyes. They are basically vat dyes which have been esterified by chemical reaction and thus made soluble for ease of application. The leuco compounds of a Vat or a Sulphur dyes are unstable unless special measures are taken.

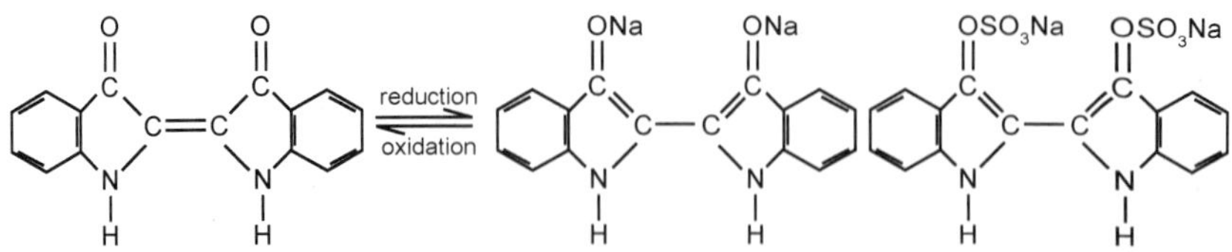

Indigo Dye Leuco form Anthrasol O, Solublised dye by esterification

Dyeing with vat dyes is often accompanied by levelling problems since water-insoluble pigments are involved, which have to be made water-soluble by the addition of appropriate chemicals in order to dye the substrate. Vat leuco ester dyes, on the other hand, are already water-soluble products which produce very level dyeings. In most cases, these dyes have only slight substantivity and are used for dyeing and printing cotton as well as other cellulosic fibres. They are developed to the parent dyes using appropriate treatment after applying these on the fabric / fibre.

Solidonia: Proprietary name for a fibre of gloss and metallic whiteness and harsh feel, used for knit goods as substitute for silk or wool. It is made of a fibrous grass.

Solomon Bar: In macramé lace four threads braided together flat.

Solubility product: The equilibrium constant for a reaction involving a precipitate and its constituent ions is known as the solubility product.

Soluble: Capable of being dissolved, i.e., passing into solution.

Soluble chemical oxygen demand (SCOD): COD can be fractionated practically to particulate and soluble CODs. But there are little standardization on the definition of soluble versus particulate COD. Where filtration is the technique is used to fractionate the sample, the relative distribution between soluble and particulate COD will vary greatly depending on the pore size of the filter. An alternative method used to determine the soluble COD involves precipitation of the suspended solids and a portion of the colloidal material. The COD of the clarified liquid corresponds to the soluble COD.

Solute: A material that is dissolved in solvent to form a solution.

Solution: A liquid system of two or more species that are intimately dispersed within each other at a molecular level. The system is therefore totally homogeneous. The major component is called the solvent (generally liquid in the pure state) and the minor component is called the solute (gas, liquid, or solid). The process occurs because of a direct intermolecular interaction of the solvent with the ions or molecules of the solute. This interaction is called Solvation. Part of the energy of this interaction appears as a change in temperature on dissolution. See also **heat of solution; solid solution; solubility.**

Solution-dyed: A type of fibre dyeing in which coloured pigments are injected into the spinning solution prior to the extrusion of the fibre through the spinneret. Fibres and yarns coloured in this manner are colour-fast to most destructive agents.

Solvent: A liquid capable of dissolving other materials (solids, liquids, or gases) to form a solution. The solvent is generally the major component of the solution. Solvents can be divided into classes, the most important being: *Polar*. A solvent in which the molecules possess a moderate to high dipole moment and in which polar and ionic compounds are easily soluble. Polar solvents are usually poor solvents for nonpolar compounds. For example, water is a good solvent for many ionic species, such as sodium chloride or potassium nitrate, and polar molecules, such as the sugars, but does not dissolve paraffin wax. *Nonpolar*. A solvent in which the molecules do not possess a permanent dipole moment and consequently will solvate nonpolar species in preference to polar species. For example, benzene and tetrachloromethane are good solvents for iodine and paraffin wax, but do not dissolve sodium chloride. *Amphiprotic*. A solvent that undergoes self-ionization and can act both as a proton donator and as an acceptor. Water is a good example and ionizes according to: $2H_2O = H_3O^+ + OH-$ *Aprotic*. A solvent that can neither accept nor yield protons. An aprotic solvent is therefore the opposite to an amphiprotic solvent.

Solvent bleaching: Oxidation bleach with sodium chlorite in organic solvents. Peroxide bleaching of wool using tetrachloroethylene liquors was an example of an exhaustion process in Solvent technology in textile finishing. Aqueous emulsions of H_2O_2 in tetrachloroethylene have 100% attraction to hydrophobic wool, which they bleach during the drying phase, continuing for 24 hours until bleaching is complete.

Solvent bonding: A method of making nonwoven fabrics in which a solvent is used to soften the fibre surfaces in a web or batt and hence cause bonding.

Solvent dye: See **Solvent dyeing.** These water-insoluble but solvent-soluble dyes are devoid of polar solubilizing groups such as sulfonic acid, carboxylic acid, or quaternary ammonium. They are used for colouring plastics, gasoline, oils, and waxes. The dyes are predominantly azo and anthraquinone, but phthalocyanine and triarylmethane dyes are also used.

Solvent dyeing: **(organic solvents)** A process developed in the 1970s, which was designed to replace water as a dyeing medium with suitable solvents, because water was becoming more and more expensive and scarce. Solvent dyeing is only used in specialist dyeing areas as a complementary process for existing aqueous dyeing processes.

Solvent dyeing process: There are two different types of dyeing process: (a) Ionic dyeing process for polyamide and wool: water soluble acidic dyes and special tetrachloroethylene-soluble acidic dyes are used for both fibre types. Since hardly any dyeing occurs without the addition of a "solvent aid", it is essential to add water, because otherwise no ionic bond will be able to form between the dye and the fibre. (b) Non-ionic dyeing process for polyester: in this process, the coloristic advantages and the process advantages stand out particularly in comparison with the aqueous process. Tetrachloroethylene, ethylene glycol, diethylene glycol and glycerine are all examples of a possible solvent medium.

Solvent extraction: (liquid–liquid extraction) A method of removing a substance from solution by shaking it with and dissolving it in a better solvent that is immiscible with the original solvent.

Solvent impregnation: (solvent waterproofing). Specialist technique for waterproofing in organic solvents (benzine, heavy benzine, tetrachloroethylene, etc.) with Solvent impregnation agents. Important in dry-cleaning for re-impregnating garments after cleaning.

Solvent relative humidity: The humidity of air over a drycleaning bath and in equilibrium with the solvent and a small amount of water.

Solvent scouring: Washing off scrooping or solvent preparations (solvent finishing). In the treatment of polyester, it should be remembered that this type of fibre is capable of solvent sorption, which accelerates dye diffusion.

Solvent sizing: Sizing of warp ends (widthwise) using organic solvents. It is carried out in a completely sealed pressure system in a suitably constructed solvent sizing machine. Polystyrene, for instance, is applied cold, can subsequently be removed from the fabric by means of tetrachloroethylene (desizing; extraction possible using ultrafiltration), and long-term experiments in the weaving industry have shown that this method is well suited to sizing.

Solvent spinning: See **Spinning, Dry-spinning.**

Sommiere: All-wool, French serge, made very soft and napped on one or both sides; used for lining of winter garments. Comes in bleached, ecru or dyed in the piece.

Sorbitol: Hexahydric alcohol. Manufactured by reducing d-glucose or dextrose derived from starch or maize. There are different forms: a) white powder or colorless needle-shaped crystals, melting point 100–120°C, easily soluble in water, pyridine, methanol, acetic acid, phenol, not so easily in cold alcohol; b) syrupy solution with 70–80% sorbitol content, density greater than 1.26, neutral, non-reducing, odourless, non-toxic, sweet taste. Application: alternative to glycerine for sizing, finishing, dyeing, printing; for plasticizing rubber and glue; for softening the gelatinous layers on print rollers.

Sorption: The process of taking up or holding a material by adsorption or absorption or both.

Soumak Rugs: All-wool tapestry ruga woven in Transcaucasia. They come in all sizes. The design is geometrical. The hook is often used, the stiches being made in the herringbone fashion. It is also called Kashmir. Modern Soumarks are made in loose weave and with coarse dye.

Souple: Single filament of natural silk with the gum only partly boiled out (also called Mi-cuit).

Sourbassis: White or yellowish Persian raw silk of fine quality.

Source: To find suppliers of the required materials.

Souring: Any treatment of textile materials in dilute acid. Its purpose is the neutralization of any alkali that is present.

Soutache: Narrow rounded braid woven in herringbone effect with odd number of threads made either plain or fancy; used for trimming.

Space dyeing: Sectional printing or dyeing of yarn and cable in varying stages of processing, various colour shades, as well as in different intervals and lengths. In space dyeing it is scarcely possible to differentiate between dyeing and printing. The space dyeing process provides an opportunity to create interesting types of pattern. Multi-coloured, irregular motifs and imaginative effects on the fabric can be achieved by using yarn with partially bordered patches of colour. The washed out looking, asymmetrical patterns on upholstery covers and domestic textiles (floor coverings) are particularly dominant.

Spacer: Spacers are for supporting and sealing the yarn layers in yarn package dyeing.

Spacer Fabric: Spacer fabrics can be described as three-dimensional orthogonal structures with yarns in the *x*, *y* and *z* directions (Fig. below). Numerous definitions of a spacer fabric exist. Roye and Gries (2007) describe such materials as comprising a three dimensional yarn architecture and a three-dimensional textile architecture, produced by the weaving process (woven spacer fabrics), by circular knitting machines (weft-knitted spacer fabrics), or by double needle bar warp knitting processes (warp-knitted spacer fabrics). Two separate fabrics faces knitted independently and then connected by a separate spacer yarn.

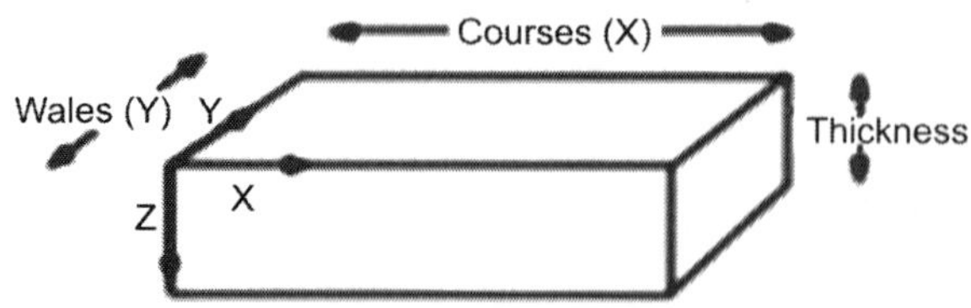

Schematic Diagram of Spacer fabric

These fabrics can be produced on both circular and flat knitting machines. Spacer fabrics have the properties of good breathability, crush resistance, and a 3D appearance.

Span length, in length testing of cotton with Fibrograph: The distance spanned by a specified percentage of the fibres in the test beard, taking the amount reading at the starting point of the scanning as 100%.

Spandex: A synthetic elastic fibre of polyurethane lycra (Invista, formerly Dupont) is a well-known brand of spandex. Spandex is found in fabrics either as bare fine filaments or covered with another fibre. Spandex can be dyed with disperse dyes and some acid dyes.

Usually where bare filaments are used, they are quite inconspicuous even if they are not dyed to match the fabric. Spandex will withstand most preparation and dyeing processes without damage, but is damaged by chlorine bleach.

Spandex fibre: A manufactured fibre in which the fibre forming substance is long chain synthetic polymer comprised of at least 85 % of a segmented polyurethane.

Spanish lace: The most common Spanish lace is made of silk in flat designs, usually floral, and held together with a mesh, but the term Spanish lace also refers to all lace made in Spain.

Spanish Stitch: In embroidery cross stitches arranged in a row to form a line on the face of the fabric and squares on the back.

Spanish Stripes: A lightweight, wide and fulled woollen cloth, originally made of Spanish wool with striped selvage, now made mostly in England. It is light, very soft and well finished with a light nap.

Spatterdash: A sort of overgaiter or legging.

Sparta carpets: Isparta carpets.

Sparterie: Stiff fabric, used in making hat bases as it can easily be shaped.

Spau press: Cold Flat press.

Specialty felt: One of a number of special purpose felt structure available for, but not limited to, specific ude or application.

Specific area, of wool: The ratio of the fibre surface to fibre volume.

Specific gravity: Ratio of the mass of a material to the mass of an equal volume of water at 4°C. The range for modern fibres is not too great and is dependent to some extent on the liquid used as an immersant in measurements, because of fibre swelling and of possible absorption of liquid into fibre voids. (Also see **Density**).

Specific gravity, of felt: The relative mass per unit volume of felt expressed as a percentage of mass per unit volume of water.

Specification: A precise statement of a set of requirements to be satisfied by a material, product, system, or service that indicates the procedures for determining whether each of the requirements is satisfied.

Specimen: A specific portion of a material or a laboratory sample upon which a test is performed or which is selected for that purpose.

Speck: (1) A contaminant in polymer such as gels, metal, or dirt that shows up as a dark spot. (2) A small particle of foreign substance that has not been removed from the stock before spinning.

Speck printing: The Specks that are usually not required in printing are produced artificially by using encapsulated dyes, so that the resulting print profile consists of dots of equal sizes in a mono or multi colour way. This print effect can be carried out on various different types of fibre if suitable dyes are used, for example in transfer printing on polyester.

Speckiness: See **Specky fabric.**

Specking: The removal of burrs, knots, and other objects that impair the finished appearance of woollens and worsteds.

Speckle: Uneven dyeing in yarns or cloths.

Specks, in woollen fabric: Small pieces of undyed vegetable matter which can be removed by carbonizing or can be covered by dyeing or inking.

Specky: A term used to describe dyed woollen fabric with specks of undyed vegetable matter on the face. The specks can be removed by carbonizing or covered by speck dyeing.

Specky cloth: See **Specky fabric.**

Specky fabric: Dyed fabric which shows small specks of undyed vegetable matter in the face.

Spectrophotometer: Photometric device that measures spectral transmittance, spectral reflectance or relative spectral emittance.

Spectral Curve: A colour's "fingerprint"— a visual representation of a colour's spectral data. A spectral curve is plotted on a grid comprised of a vertical axis-the level of reflectance intensity; and a horizontal axis-the visible spectrum of wavelengths. The percentage of reflected light is plotted at each interval, resulting in points that form a curve.

Spectrophotometric curve: A curve measured on a spectrophotometer; a graph with relative reflectance or transmittance (or absorption) as the ordinate, plotted with wavelength or frequency as the abscissa.

Spectroscopy: The study of methods of producing and analyzing spectra using spectroscopes, spectrometers, spectrographs, and spectrophotometers.

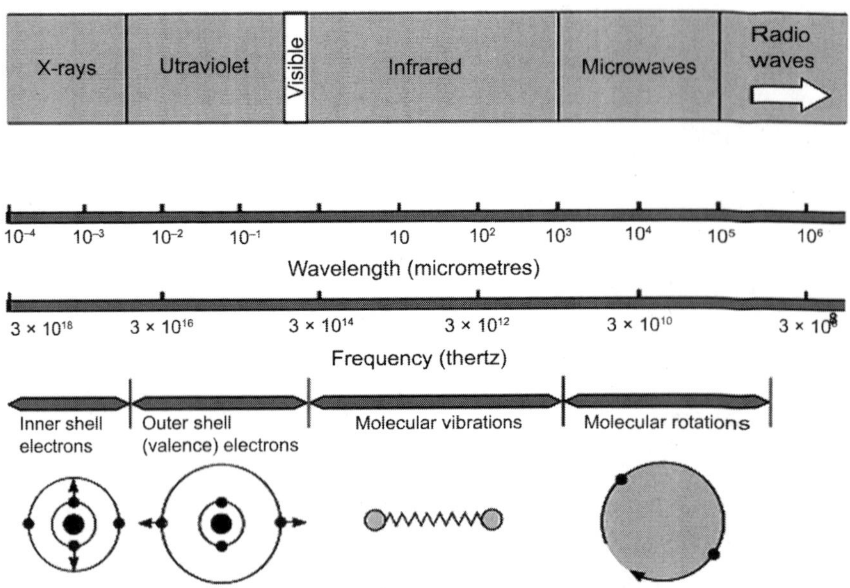

The interpretations of the spectra so produced can be used for chemical analysis, examining atomic and molecular energy levels and molecular structures, and for determining the composition and motions of celestial bodies.

Spectral Data: The most precise description of the colour of an object. An object's colour appearance results from light being changed by an object and reflected to a viewer. Spectral data is a description of how the reflected light was changed. The percentage of reflected light is measured at several intervals across its spectrum of wavelengths. This information can be visually represented as a spectral curve.

Spectral power distribution curve: Intensity of radiant energy as a function of wavelength, generally given in relative power terms.

Spectral energy distribution: The variation of energy due to the source over the wavelength span of the emitted radiation.

Spectral Power Distribution (SPD): The spectral power distribution (SPD) of a light source provides the required numerical description of it. The SPD gives the emitted power (watts per square metre of emitter surface per unit wavelength interval, W m^2 nm–1) as a function of the wavelength.

Spectral transmittance: The percent of incident radiant energy passing through a given material and not absorbed in the process, as a function of wavelength.

Spectral disribution of energy: See **Spectral energy distribution.**

Spectrograph: A spectroscope equipped with a camera or some other device for recording the spectrum. Also see **Spectroscope**.

Spectrometer: (1) An instrument for examining the different wavelengths present in electromagnetic radiation. Typically, spectrometers have a source of radiation, which is collimated by a system of lenses and/or slits. The radiation is dispersed by a prism or grating, and recorded photographically or by a photocell. There are many types for producing and investigating spectra over the whole range of the electromagnetic spectrum. Often spectrometers are called *spectroscopes*. *See also* spectrophotometer.

(2) Any of various other instruments for analyzing the energies, masses, etc., of particles. *See* mass spectrometer.

Spectrophotometer: An instrument used to measure the transmission or reflectance of light as a function of wavelength. A form of spectrometer able to measure the intensity of radiation at different wavelengths in a spectrum, usually in the visible, infrared, or ultraviolet regions.

Spectroscope: An instrument for forming a spectrum for visual examination.

Spectroscopy: The identification of materials by the analysis of their spectra.

Spectrum: Spatial arrangement of components of radiant energy in order of their wavelengths, wave number or frequency.

Specular gloss: The relative luminous fractional reflectance of a specimen in the specular direction. Relative luminous fractional reflectance from a surface in the mirror or specular direction. It is sometimes measured at 60° relative to a perfect mirror.

Reflectance factor: The ratio of the light reflected from the specimen to the light reflected from the perfect reflecting diffuser under the same geometric and spectral conditions of measurement.

Specular reflectance: Reflectance of a beam of radiant energy at an angle equal but opposite to the incident angle; the mirror-like reflectance. The magnitude of the specular reflectance on glossy materials depends on the angle and the difference in refractive indices between two media at a surface. The magnitude may be calculated from Fresnel's Law.

Specular reflectance excluded (SCE): Measurement of reflectance made in such a way that the specular reflectance is excluded from the measurement; diffuse reflectance. The exclusion may be accomplished by using 0° (perpendicular) incidence on the samples. This then reflects the specular component of the reflectance back into the instrument by use of black absorbers or light traps at the specular angle when the incident angle is not perpendicular, or in directional measurements by measuring at an angle different from the specular angle.

Specular reflectance included (SCI): Measurement of the total reflectance from a surface, including the diffuse and specular reflectances.

Specular reflection: The reflection without diffusion, in accordance with the laws of optical reflection, as in a mirror.

SPF (Sun Protection Factor): SPF measures the effectiveness of sunscreen on the body. The test for SPF is done by using a living organism or body to measure the length of time it takes for the skin to redden without coverage or protection.

Spencer: Short, narrow-waisted jacket with lapels; sleeveless or short sleeved.

Spherulite: A common form of polymer crystallization from melts or concentrated solutions. These crystallites show a radial symmetry from a central point and have a distinctive maltese cross pattern of birefringence under the polarizing microscope.

Spider Leno: See **Net leno.**

Spider silk: Product of silk spider (Madagascar).Without silk glue. Extremely light. Strength, glaze and elasticity similar to genuine silk but finer. Spiders silk also known as gossamer is a protein fibre spun by spiders. Spiders use their silk to make webs. Many small spiders use silk threads for ballooning, the scientific term for dynamic kiting spider lings (mostly) used for dispersal. They extrude several threads into the air and let themselves get carried away with upwards wind.

Spider Weave: Name for weaves producing a net-like effect on the face of the cloth by floating and deflecting either the warp or the weft threads.

Spider Wheel: See **Catherine wheel.**

Spiking: Use of spikes fixed to one edge of a hinged table to facilitate the even laying up of fine material. The table top is tilted vertically for hanging the material and back to the horizontal for cutting.

Spin finish: "Oiling agents" consisting of oils or fats and surfactants the behaviour of which, depending on the task in hand and composition, comes somewhere between textile lubricants and textile softening agents. See **Lubricant.**

Spin multiplier: See **Twist Multiplier.**

Spin stretch Ratio: In man-made filament extrusion, the ratio of taken-up or haul off speed to the average speed of the spinning fluid as it leaves the spinneret,

Spin drawing: (1) The reduction of roving during spinning by a roller drafting mechanism similar to that used on the roving frame. (2) Combined spinning and drawing in one operation in melt-spun fibres.

Spindle: (1) Device on spinning and twisting machines for holding packages and the like for imparting a twist to or roll-batching yarn on a spinning frame, roving frame, twister, winder, or similar machine. A bobbin is placed on the spindle to receive the yarn as the spindle is rotated at high speed. (2) Hydrometer. (3) Threaded shaft for the transmission of turning motion.

Spindle tape: A woven narrow fabric of width normally not greater than 50 mm. usually of high warp density and designed for the transmission of power spindle wharves of small diameter.

Spinneret: A metal disc containing numerous minute holes used in manufactured fibre extrusion. The spinning solution or melted polymer is forced through the holes to form the fibre filaments.

Spinnerette: Same as **Spinneret.**

Spinning: The process or processes used in the production of yarns or filaments. (a) This term may apply to the drafting and twisting of natural or man-made fibres (see continuous spinning, intermittent spinning, open-end spinning), to the extrusion of filaments by spiders and silkworms, or to the production of filaments from glass, metals or fibre-forming polymers.

(b) In the spinning of man-made filaments, fibre-forming substances in the plastic or molten state, or in solution, are forced through the holes of a spinning jet (q.v.) or die at a controlled rate (extrusion). There are five general methods of spinning man-made filaments, but a combination of two (or more) of these methods may also be used. They are the following:

(i) *Dispersion Spinning:* The process in which polymers that tend to be infusible, insoluble and generally interactable (e.g. polytetrafluoroethylene) are dispersed as fine particles in a carrier, such a sodium aliginate or sodium xanthate solutions, which permits extrusion into fibres, after which the dispersed polymer is coalesced by a heating process; the carrier is removed either by a heating or by a dissolving process.

(ii) *Dry Spinning:* The process in which a solution of the polymer is extruded into a heated chamber to remove the solvent and leave the solid filament.

(iii) *Melt Spinning:*The process as used in the manufacture of nylon in which the fibre-forming polymer is melted and extruded into air or other gas or a suitable liquid, where it is cooled and solidified.

(iv) *Reaction Spinning:* The process in which polymerisation is achieved during the extrusion through spinning jet (q.v.) system of reactants.

(v) *Wet Spinning:* The process as used in the manufacture of viscose rayon in which the solution of the polymer is extruded into coagulating media where the polymer is regenerated.

(c) In the bast fibre and leaf fibre industries, the terms "dry spinning" and "wet spinning" refer to the spinning of fibres in the drystate and in the wet state, respectively.

Spinning, general: (1) The process or processes used in the production of single yarns:

Yarn from staple fibre: The formation of yarn from sliver or roving by drafting and twisting. *Filament yarn or tow:* The formation of filaments by extrusion of fibre forming substance either in molten form , in solution or in a form suitable for regeneration.

Yarn from filament tow: The formation of yarn from tow by cutting or breaking drafting and twisting in a single operation often called direct spinning.

(2) Spinning: The process or processes used in the production of single yarns or of fabrics generated directly from polymer:

Yarn from Staple Fibre: The formation of a yarn by a combination of drawing or drafting and twisting prepared strands of fibres, such as rovings.

Filament Yarn: In the spinning of manufactured filaments, fibre-forming substances in the plastic or molten state, or in solution, are forced through the fine orifices in a metallic plate called a spinneret, or jet, at a controlled rate. The solidified filaments are drawn-off by rotating rolls, or godets, and wound onto bobbins or pirns. There are several methods of spinning manufactured filaments:

Dry Spinning: The process in which a solution of the fibre-forming substance is extruded in a continuous stream into a heated chamber to remove the solvent, leaving the solid filament, as in the manufacture of acetate.

Gel Spinning: A spinning process in which the primary mechanism of solidification is the gelling of the polymer solution by cooling to form a gel filament consisting of precipitated polymer and solvent. Solvent removal is accomplished following solidification by washing in a liquid bath. The resultant fibres can be drawn to give a product with high tensile strength and modulus.

Melt Spinning: The process in which the fibre-forming substance is melted and extruded into air or other gas, or into a suitable liquid, where it is cooled and solidified, as in the manufacture of polyester or nylon.

Phase-Separation Spinning: Extrusion of polymer and solvent at high temperature into a cooling zone. During the cooling process, a phase separation occurs, usually accompanied by crystallization of the solvent. Solvent can be removed before or after drawing.

Reaction spinning: Process in which an initial prepolymer is formed and then extruded into a reagent bath where polymerization and filament formation occur simultaneously. Spandex fibres can be made by this process.

Wet Spinning: The process in which a solution of the fibre-forming substance is extruded into a liquid coagulating medium where the polymer is regenerated, as in the manufacture of viscose or cuprammonium rayon.

Yarn from Leaf and Bast Fibre: In the manufacture of leaf and bast fibre yarns, the terms "wet spinning" and "dry spinning" refer to the spinning of fibres in the wet state and in the air-dry state, respectively.

Yarn from Filament Tow: The formation of a yarn from filament tow by a combination of cutting or breaking, drafting, and twisting in a single series of operations. Also known as converting.

Nonwoven Fabric: Fabrics can be produced directly from molten or dissolved fibre-forming substances by several continuous processes:

Flash Extrusion: The process in which a fibre-forming substance in a volatile solvent is extruded from a high-temperature, high-pressure environment into lower temperature and pressure conditions, causing the solvent to rapidly evaporate, leaving a lacy, net-like fabric.

Spinning, direct: The manufacture of yarn from sliver on the cotton system, bypassing the roving process.

Spinning, dry: Conversion of a dissolved polymer into filaments by extrusion and evaporation of the solvent from the extrudate. Solvent is removed through flows of warm gas suitably directed to the extruded filaments; gas temperature should be higher than the boiling temperature of the solvent, which will be extracted from the filaments, recovered and recycled.

Spinning, flash: A modification of the accepted dry spinning method in which a solution of a polymer is extruded at a temperature well above the boiling point of the solvent such that on emerging from the spinneret evaporation occurs so rapidly that the individual filaments are disrupted into a highly fibrillar form.

Spinning cake: (spindle crown), a yarn package which has been spun according to the centrifugal or so-called can or pot spinning method (rotating pot resulting in the shape of a ring-shaped cake with the characteristic hole in the centre. These slightly conical spinning cakes occur in different sizes but they are always in the form of compact blocks of yarn tightly wound in almost parallel layers.

Spinning frame: A machine used for spinning staple yarn consisting of a number of spinning positions for converting slivers, slubbings or rovong into yarns.. It drafts the roving to the desired size, inserts twist, and winds the yarn onto a bobbin. The term is generally used to indicate a ring spinning frame, although it does cover flyer spinning and cap spinning on the worsted system.

Spinning limit: The finest yarn number that can be spun satisfactorily form a specified lot of fibre under specified conditions.

Spinning performance: A method for the evaluation of end breakage rate during spinning which is commonly expressed in units of ends down per 1000 spindle –hours, but may be based on other ubnits such as 'spinnable limit' or ends down per specified time other than 1000 spindles-hours, yardage, or condition.

Spinning solution: A solution of a fibre-forming polymer (e.g., cellulose acetate) in a suitable condition to be extruded by either dry spinning or wet spinning.

Spinning test: The processing of a specific lot of stck into yarn for evaluation of processing performance and and product quality at various stages of processing or both.

Spinning twist: The twist added to yarn during spinning to give it strength and other desired characteristics.

Spinning, dry: See **Spinning.**

Spinning, open end: See **Open end Spinning.**

Spinning, break: See **Spinning, open end.**

Spinning, cap: A spinning system in which the lower edge of a stationary cap is supported by the spindle, the lower edge of which guides the yarn on to the revolving spinning package which is traversed.

Spinning, collasped balloon: A system of ring spinning in which the rotating yarn balloon is eliminated by contact with the top of the spindle. Note (a): The system is normally used for economic spinning of semi-worsted and woollen carpet yarns by permitting the use of larger package sizes and/or higher spindle speeds. Note (b): Special spindle top extensions are used to obtain the desired effects.

Spinning, direct: See **Spinning.**

Spinning, double roving: A system in which two rovings are fed to each spindle of a ring frame so producing a pseudo two-fold or two-strand yarn. The rovings are separated in the drafting system by means of special guides and the two drafted strands are combined after the drafting system.

Spinning, flyer: A spinning system in which yarn passes through yarn passes through a revolving flyer leg guide on to the package rotating at slightly different speeds.

Spinning, friction: A method of open-end spinning which uses the external surface of two rotating rollers to collect and twist individual fibres into a yarn. At least one of the rollers is perforated so that air can be drawn through it surface to facilitate the fibre collection. The twisting occurs near the nip of the rollers and because of the relatively large difference between the yarn and roller diameters, high yarn rotational speeds are achieved by the friction between the roller surface and the yarns.

Spinning, hollow spindle: A system of yarn formation in which the feed stock (sliver or roving) is drafted and the drafted twistless strand in wrapped with a yarn as it passes through a rotating hollow spindle. The binder or wrapping yarn is mounted on the hollow spindle and is unwound and wrapped around the core because of the rotation of the spindle. This technique may be used for producing a range of wrap yarns, fancy yarns by using different yarn and fibre feed stock fed to the hollow spindle, at different speeds.

Spinning, jet: (Also called Airjet spinning) A system of staple fibre spinning which utilizes air to apply the the twisting couple to the yarn during its formation, The air is blown through small holes arranged longitudinally to the yarn surface and this causes the yarn to rotate. The majority of systems using this technique produce faciated yarns but by using two air jets operating in opposing twist direction it is possible to produce yarn with more controlled p[roperties but of more complex structure.

Spinning, ring: A spinning system in which the twist is inserted in a yarn by revolving traveller.

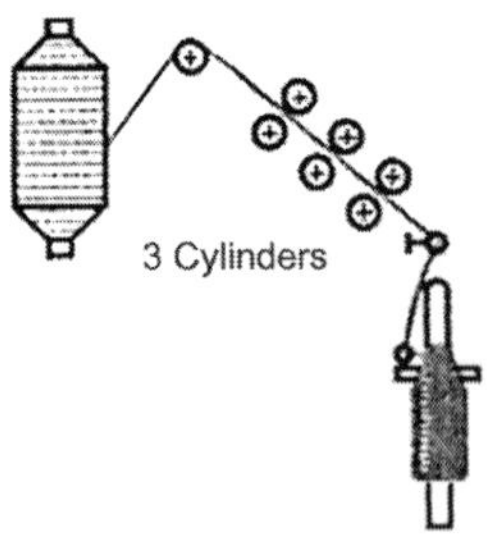

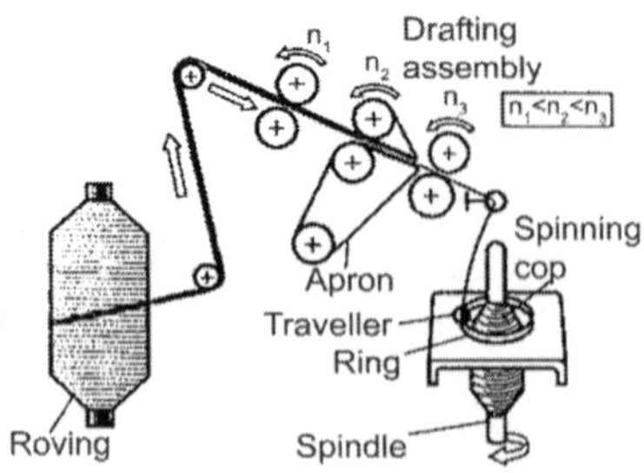

Ring Spinning (Schematic) Principle of ring spinning

The yarn is wound on by the package rotating faster than the traveller.

Spinning, rotor: A method of open-end spinning whichnuses a rotor (high speed centrifuge) to collect and twist individual fibres into a yarn. The fibres on entering the rapidly rotating rotor as distributed around its circumference and temporarily held there due to the centrifuging force. The yarn is withdrawn from the rotor wall and because of the rotation of the rotor, twist is inserted.

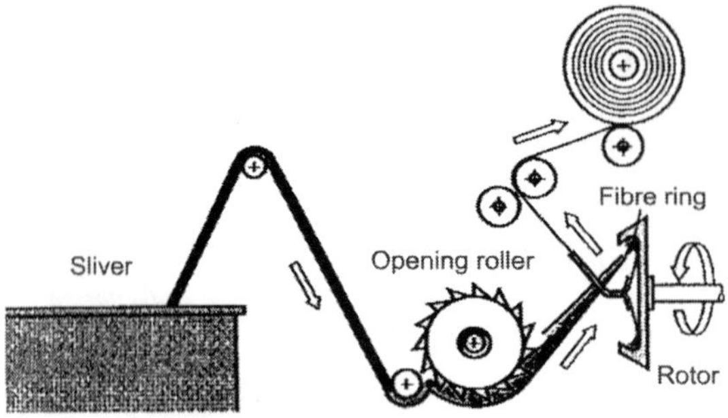

Principle of rotor spinning

Spinning, self twist: A method of making a yarn from rovings fed to a drafting unit; the emerging strand of fibres is subjected to a cyclically reversing false twisting action which can be imparted in a number of ways. Two adjascent strands delivered from from the false twist system are brought together by guides, and the torque in the two strands causes them to wrap about each other. This wrapping action is definesd as 'self twist' and produces a self twist pattern of –S-Zero-S-Zero- etc. in the yarn produced. The self twist yarn is then taken up as a cheese.

Spinning, Semi collapsed ballon: A system of ring spinning in which the rotating yarn balloon is eliminated at the start of an empty bobbin and is allowed to expand later when the bobbin is becoming filled with the yarn. The small ballon condition is achieved by allowing contact between the rotary balloon and the spindle top. As the contact is brocken and the yarn balloon is allowed to reform Notes (a) The system is often used for worsted and semi worsted spinning. (b) Special spindle top extension are used to obtain the required effect.

Spinning, Supressed balloon: See **Collapsed balloon spinning.**

Spinning, Twistless: A system of yarn formation which relies on the use of a permanent or temporary adhesive to bond the fibres together. Note: Where a temporary adhesive is used it is removed during fabric finishing and the yarn (and fabric) strength is obtained through lateral pressure on the yarn produced by the yarn inerlacings in the fabric. A similar fabric construction can be achieved by using warp spun yarns which have been produced with a soluble binder.

Spinning, warp: See **Spinning.**

Spinning, wet: It is the process in which a solution of the polymer is extruded into a liquid when the polymer is regenerated as in the manufacture of viscose or cuprammonium rayon.

Spiral yarn: Specialty yarn made by winding heavier, slackly twisted yarn around a finer yarn with a hard twist to give a slubby appearance.

Splash voile: See **Voile.**

Splice: (1) The joining of two ends of yarn or cordage by intertwining or interweaving strands. (2) The joining of two ends of yarn or cords by means of several knots joining individual yarns or strands the knots being staggered that is, located at some distance from each other along the yarn or the cord. (3) The joining of two ends of a material whereby ends are tapered, lapped and subsequently seized, bound, stitched or stapled.

Splice defective: A slice having irregular final twist, untrimmed ends or excessive thickness or thinness.

Spliced: Reinforced parts of hosiery where the wear is the greatest.

Splicer: Controllable ancillary device on automatic bobbin winders which carry out the splicing operation at the winding head when there is a thread break.

Splicing: (1) The joining of two ends of yarn or cordage. There are several methods used, e.g., by interweaving the strands, by the use of knots, by tapering, lapping, and cementing the ends, etc. Splicing technology is regarded as a genuine alternative to knot-free joining technology. (2) A method of reinforcing knits, e.g., the heels and toes of hosiery, by introducing an additional yarn for strength.

Splinter: Two or more staple fibres adhering together, causing a stiff cluster that resists pulling apart in normal processing, and reacting in the yarn spinning process similarly to higher than nominal denier fibre.

Splinter count: A measure of the number of coalesced fibres, mealy particles, or other such matter in staple fibre.

Split: Two or more lengths of fabric that are woven side by side and subsequently separated from each other by cutting along lines formed by leaving one or more dents. Fraying at the cut edges may be prevented by the use of a leno edge (q.v.) or other suitable means.

Split end: See **Broken filament.** (1) A defect in fabric caused by breakage of some of the singles yarns in a plied warp yarn. (2) A defect in manufactured filament yarn caused by breakage of some of the filaments.

Split stitch: A flat chain stitch used in old church embroidery. Bring the thread through at A and make a small stitch over the design, piercing the working thread with the needle. Split Stitch may be used as a filling where a fine flat surface is required.

Split weft(strained weft): A continous-filament thread that has lost some of its filaments, usually as a result of abrasion or excessive tension during winding or weft insertion and that appears as a thin yarn.

Split Filaments: See **Broken filaments.**

Split Harness: A knot in each double-harness cord below the comber board and above the mail to form a loop long enough to allow a proper depth of shed. A rod is passed through the loops of each long row of harness cords so that each is capable of lifting all the ends in one row of cords independently of the figuring cards. The jacquard lifts the required ends to form the ground weave. The harness is used in weaving jacquards with a finely sett warp and move open weft.

Split-Draft Metier: An extrusion cabinet for dry spinning in which the drying medium (hot air) is introduced between the jet and the yarn outlet and flows in both directions.

Split-Flow Metier: See **Split-draft Metier.**

Split-Stitch, in knitted fabrics: A stitch in which one part of the yarn is knit and the other part is dropped.

Splitting: (1) In the processing of tow, a defect in which the integrity of opened tow is disturbed by separation or division into two or more segments longitudinally. Splitting can be continuous or intermittent, long or short term. (2) In slashing, the separation of sized yarn ends before takeup on the slasher beam.

Splitting resistance, in felt: The force required to ovwercome the interfacial strength of a of a material and specifically to separate a felt into two layers (of approximately equal thickness)

Sponge: (1) Name for a crepe weave made with equal number of warp and weft floats; (2) A honeycomb weave, made with small diamonds on a satin ground, resulting in very small cells.

Sponge Cloth: (1) coarse fabric, made of cotton waste and used for cleaning machinery; (2) fine dress fabric of cotton, wool or silk, made of nub yarn in twill weave.

Sponging: A pre-shrinkage process which involves the dampening with a sponge to woollen and worsted fabrics. The process is accomplished by rolling in moist muslin, or by steaming. This procedure is performed at the fabric mill prior to cutting to insure against a contraction of the material in the garment.

Spool: A flanged wooden or metal cylinder upon which yarn, thread, or wire is wound. The spool has an axial hole for a pin or spindle used in winding. (Also see **BEAM**.)

Spool (Axminster and gripper spool): A double flanged bobbin on which a number of threads of pile yarn are wound in a predetermined order for use in Spool Axminster and Gripper spool loom. The yarns from a spool for all or a part of a row in the carpet design.

Spool rack: Spool rack is a frame with thin horizontal metal rods that can support many spools of thread. Multiple ends can be unwound at the same time from the rotating spools to fill a section of a sectional warp beam or to pass through the slots and holes of a warping paddle.

Spoon bonnet: Small bonnet tied under the chin with a ribbon.

Spot: A small discolored area on, or in, a fabric.

Spot and stain removal: see **Spotting.** A cleaning procedure for localized ares using cleaning agents and mechanical action specific to the removal of the foreign substances present.

Spot bonding: See **Bonding, Point bonding.**

Spot clean: To remove localized spots and stains by treating them with cleaning agents and mechanical actions specific to the fabric, fibre and the product type and the foreign material present.

Spot weave: Spot designs are formed by extra warp or filling yarns. The yarns are inserted the entire length or width of the fabric, spots or dot designs are formed. The long floats on the back side are cut away, leaving the dots. The threads can be pulled easily. Filling threads are easy to cut but warp floats are difficult. Example: **Dotted swiss.**

Sport shirt, for boys: A shirt made in numerical sizes, designed for informal wear and may be worn with or without a jacket.

Sport shirt, for men: A shirt designed for informal wear and made with body sizes such as small, medium, or large, and may be worn with or without a jacket.

Sports jacket: Fashionable men's jacket with sack-shape cut; without shape, used as a summer jacket; shaped, used as a sports jacket (also high-quality, elegant designs, 250–500 g fabric weight) with shape retaining interlining also used as a holiday jacket (See **Topper**).

Sports clothing: Functional clothing for special requirements (Active wear); not to be confused with Sportswear.

Sportswear: Leisure sports and town and weekend clothing in the form of Casual wear in a functional but smart genre, with sports-like attractive details; includes knitted wear and imitation leather.

Spotclean: See **Spotting.**

Spotting: (1) A cleaning procedure for localized areas with cleaning agents and mechanical action specific to the removal of the foreign substances present.

(2) In England same as crabbing.

Spotting agent: Used for removing local contamination (spots, etc.) on textile goods.

Spotting powder: Used to absorb moisture during spotting in order to prevent the appearance of the feared stain borders. Active components: talcum, kaolin, fullers earth and gypsum.

Spray bonding: A process of binding fibres into a nonwoven fabric involving the spray application of a fabric binder to the fibre web or batt. See **Bonding.**

Spray-dampening: Finishing character work; particular usual for washing cloths and wool cloths and the like in order to impart the desired moisture content and heavy, soft handle to the fabric. In most cases with nonsupersaturated moist air after the drying frame (stenter drying) on the spraying machine by spray jets or rotating brush rollers according to the so-called "natural moisture" principle.

Spray dyeing machine: A form of Hank dyeing machine for dyeing hank yarn on perforated yarn-carrier tubes with firmly connected rotating hanks. The machine gently turns suspended hanks. The liquor is sprayed on the hanks through the perforations.

Spray finishing: Finishing with Spray dyeing machine, used particularly for pressure-sensitive silk, viscose filaments and blended fabric articles such as crêpe de chine, marocain crêpe, flamisol crêpe, georgette crêpe, bark crêpe, moss crêpe, blister crêpe, matelassé, hammer blow and even for taffeta and similar smooth articles as well as for dress fabrics and linen fabrics for correcting the final handle.

Spray finishing machine: Special machines derived from older spray dampening machines which apply finishing liquors in the form of powders or low viscosity mists through nozzles either on one or both sides as a form of pray finishing.

Spray printing: The pattern is produced by Stencils for spray printing. The thin and finely matched printing ink solutions are sprayed through the fine stencil voids on to the goods using nozzle atomizers or compressed-air spray guns (air extraction and safety spectacles required).

Spray spinning: Application of a fabric binder.

Spread stitches: See **Pin hole.**

Sprig: Patterns of flowers and leaves in hand -made laces, made separately and appliqued on a net ground.

Spring needle: A knitting machine needle with a long, flexible hook, or beard, that allows the hook to be closed by an action known as pressing so that the loops can be cast off. The hook springs back to its original position when the presser bar is removed. Also see **Latch needle.**

Sprouting: Defect in Brussels and tapestry rugs and carpets, consisting of some of the loops protruding above the surface.

Spunbond fabric: An alternate name for Spun-laid fabric.

Spun-bonded products: Nonwoven fabrics formed by filaments that have been extruded, drawn, then laid on a continuous belt. Bonding is accomplished by several methods such as by hot roll calendering or by passing the web through a saturated-steam chamber at an elevated pressure.

Spun-dyed fibres: Man-made fibres dyed in the spinning paste (spun-dyed).

Spun fabric: A fabric made from staple fibres that may contain one or a blend of two or more fibre types.

Spun Glass: Glass thread of great fineness, dyed in various colours, braided and made into neckwear, as in Venice.

Spun laced fabric: A nonwoven fabric made from a staple fibre web or batt bonded by entanglement using high pressure water jets.

Spun laid fabric: A non woven fabric made by the extrusion of filaments which are laid down in the form of a web and bonded.

Spun polyester: Lightweight woven or knitted fabric that have a soft, wrm feel. Comfortable for sports and casual clothes, dresses and nightwear.

Spun rayon: Yarn spun from rayon staple fibres.

Spun silk: This is a type of silk yarn, although the fabric itself may be labeled this way. The yarn is made by breaking up the short fibres of waste silk and spinning them together, and this yarn is then woven into fabric. Although the fabric should be cheaper than silk as it is made from waste yarn e processing is lengthy and hence expensive. The fabric may be produced of any weight.

Spun yarns: A strand shaped structure made from Staple fibre which is held together by twisting

(spinning) as opposed to Continuous filament yarn.

Spunbond Fabric: Random webs.

Spun-dyed: Man-made fibres dyed in the spinning paste (spun-dyed).

Spun viscose: This is really the name of the yarn, but fabric also carry this title. They are in plain weave and may be plain or printed. These fabric were once considered poor in quality and performance but the fabric finishing techniques have improved so much that they are now good fashion fabrics. They are soft and drape well and are fairly warm, but they crease in wear and highly inflammable. Used for dresses, blouses, shirts nightwear, and childrens clothes.

Spyndle number: See **Yarn numbering system**. Count for dry-spun flax yarn and jute yarn, consisting of 48 cuts (or leas) of 300 yards each, which make up a spyndle of 14,400 yards, the weight in pounds of a spyndle being the count of the yarn.

Spyndle number, in jute: A direct yarn numbering system for jute rove and jute yarns in which the number of pounds per spyndle, or fourteen 400 yd (thirteen 167 m) length, is expressed as pounds per spyndle.

Square construction: See **Balanced fabric.**

Squeegee: The portion of a screen-printing apparatus consisting of a blade that forces the print paste through the screen onto the fabric.

Squeegee pass: The movement of the doctor blade from one side of the screen to the other. This movement can be carried out once or several times depending on the substrate, print style and printing press.

Squeegee pressure: Together with the Squeegee contact angle, this controls the amount of printing paste which is applied during flat and rotary screen printing. In the case of the magnetic-rod squeegee system, it is established by differences in the strength of the magnetic field, but in the case of the blade squeegee, it is determined by appropriately adjusted spring pressure (See Squeegee systems in printing).

Squeegee profile: The squeegee profile is characterized by the cross-section of the squeegee rubber. Depending on requirements, synthetic rubbers of different thicknesses are used which lead out on the side turned towards the screen gauze with the edge rounded off or cut flat or steep. Squeegee systems in printing.

Squeeze rolls: Rolls used to apply pressure for removal of water or chemicals from fabric.

Sress-strain curve: A graphical representation, showing the relationship between the change in dimension (in the direction of the applied stress) of the specimen from the application of an external stress, and the magnitude of that stress. In tension tests of textile materials, the stress can be expressed either in units of force per unit cross-sectional area, or in force per unit linear density of the original specimen, and the strain can be expressed either as a fraction or as a percentage of the original specimen length. See also **Load de-formation curve.**

Srinagar: Knotted rugs made in Kashmir, India, of very fine wool.

s/s type fibres side-by-side, Bicomponent fibres.

ST: (1) Tussah silk. (2) Mineral silicate fibres.

Stabilised Yarn: Yarn which has been subjected to heating and cooling or other setting treatment in order to reduce its tendancy to shrink, contract. Twist Or snarl.

Stability: A term used to describe the tendency of a fibre or fabric to return to its original shape after being subjected to external influence, such as tension, heat, or chemicals.

Stability To Thermal Oxidation, for polyolefin monofilament: The time-to-failure, when polyolefin monofilaments are exposed to circulating air, at 125°C.

Stability To Thermal Oxidation, for polyolefin fabrics: That property of a fabric which resist breaking under a specified tensile strain, when exposed to a current of air at an elevated temperature.

Stable fabric: A textile fabric, the dimension of which does not change significantly during processing, with multiple passes through measuring devices or during use.

Stabilized Fibre: Fibre that is heat or chemically treated to set the fibre properties and

prevent deterioration, shrinkage, etc. Also see **Heat stabilised, Heat setting, and UV Absorber.**

Stabilization, in effluent treatment: The biological process by which the organic matter in the sludge produced in the primary settling and biological treatment of waste water is stabilized, usually by a conversion to gases and cell tissue. Depending on whether the stabilization is carried out under aerobic or anaerobic conditions, the process is known as aerobic or anaerobic digestion.

Stabilizer: In Hydrogen peroxide bleach baths: A chemical compound which when added to an alkaline peroxide bleaching liquor, will control the rate and nature of decomposition of peroxide, thus providing a controlled process of bleaching with minimum tendering of the substrate. These substances are used to regulate the release of oxygen, help prevent oxygen decomposition by catalysts such as copper, manganese and iron, etc. and prevent the loss of oxygen and prevent local fibre damage. Stabilizers are usually not needed in hard water.

Staggered twill: See **Broken twill.**

Stain: An area of discoloration that penetrates the fabric surface.

Stain And Soil Retardancy: Fabric quality of retarding staining and soil.

Stain Blocking: Finishing effect. Active stain blocking prevents penetration of stains and dirt into the substrate (oil-repellent finish). Passive stain blocking eases the removal of dirt which has got on to, or stains which have penetrated into the substrate (Soil-release finishing).

Stain removing table: A table specially made for the removal of stains on the garments or cutpieces. Usually used in garment industries it will have a Steel Table top provided with steam gun for wet and dry steam, spry guns, vacuum extraction and air blowers, various solvents and air blowers for drying the treated area etc. Steel Table top with perforations or even ss steel mesh as table tops are also used. The table is usually operated with the aid of foot pedals.Most of the stain removing table will have a swivelling sleeve board which also has a vacuum connection to an automatic vacuum vibrating valve which switches on when in use and is automatically switched off again when swivelled away from the working area. Spotting brushes are reserved for special purposes only.

Stain Repellency: Stain Repellency is the ability of a treated fabric to withstand penetration of liquid soils under static conditions involving only the weight of the drop and capillary forces.

Stain Repellent Finish: See **Oil repllency, Oil repellent finish.**

Stainblocker: A chemical substance which, when applied to a textile substrate, imparts partial or total resistance to staining.

Stained Cloth: Antiquated term for drapery painted with figures, to imitate tapestry.

Staining: (1) Any adventitious (unwanted) colour, owing to dye, dirt or iron, on textile material. A severe stain is one that will resist processing. (2) The fugitive or permanent coloring of material for identification purposes.

Stainless Steel: A broad class of corrosion-resistant iron alloys Stainless steel is often recommended for vessels for dyeing. It is made of iron alloyed with metals such as chromium, molybdenum, nickel, and others. There are a great many stainless steel alloys. The best alloy for chemical resistance is "316S". "316" is almost as good, followed by "308", "304" and "18-8". You will sometimes see these numbers on the packaging or literature for higher quality stainless ware. High chloride ion concentration, particularly at high temperature, can corrode even the best stainless steel. This is unlikely to cause concern for home dyers. Because there is a possibility that stainless steel pots can become pitted from use in dyeing, and the pits may make it hard to completely remove contaminants, such pots should not be used for food purposes. Beware of aluminum rivets that are sometimes used to fasten handles to inexpensive stainless steel pots.

Stainless-Steel Fibre: (non-rusting metal fibres).Stainless steel metal filaments, e.g. metal filaments produced from stainless steel 18/8 or other stainless variants.

Stalagmometer: An apparatus consisting of a glass capillary with a specified orifice which is used to measure the Surface tension of a liquid by the drop method: suitable for dilute aqueous solutions (1–3 g/l). Measurements are based on a comparison, usually against the drop count of pure water: the results can also be converted into absolute values. Principle: low surface tension = smaller drops = a higher drop count. The apparatus is particularly suitable for determining the efficacy of surface-active textile auxiliaries in terms of the drop count and dependence on concentration (at the same temperature).

Stalk fibres: Bast fibres.

Stamped Velvet: Velvet having patterns stamped into the pile with heated engraved rollers.

Standard: A reference against which instrumental measurements are made. Standards are written agreements setting out technical specifications or other clearly defined criteria that are consistently applied as rules, guidelines or designations of properties to ensure that materials, products, processes and services fulfil the intended purpose.

Standard Atmosphere for preconditioning, in textiles: An atmosphere having a relative humidity of 5-25 %, +/- 2 % tolerance for the selected relative humidity, and a temperature of not over 50^{0}C (122^{0}F), with +/- 1^{0}C (+/-2^{0}F) tolerance for the selected temperature and used to partially dry the material before further treatment or conditioning.

Standard Atmosphere For Testing, in Textiles: An atmosphere for testing in which the air is maintained at a relative humidity of 65 +/– 2 % and a temperature of 21+/– 1°C (70+/–2°F). (a) Standard Temperature Atmosphere An atmosphere having a relative humidity of 65 ± 2% and a temperature of 20 ± 2°C. (b) Standard Tropical Atmosphere. An atmosphere having a relative humidity of 65 ± 2% and a temperature of 27 ± 2°C.

Standard Atmosphere For Testing, in glass textiles: An atmosphere for testing in which the air is maintained at a relative humidity of at least 45% and no greater than 67 % , tolerance of +/– 2 % for the selected relative humidity, and a temperature of at least 20'C (68'F) and not greater than 25°C (77°F) with a tolerance of +/- 1°C (+/– 2°F) at the slected temperature.

Standard Condition: The state of being in moisture equilibrium with the standard atmosphere for testing.

Standard Condition for physical testing: The condition of a textile material that has been dried to approximately constant mass in an atmosphere that has a relative humidity not exceeding 10%, and then kept in the appropriate standard atmosphere for testing (q.v.) until it has reached equilibrium. In cases where a textile material is not likely to lose volatile matter other than water, or to change dimensions, the preliminary drying may be carried out in an oven at 50-60°C situated in the standard atmosphere for testing which is a convenient way of achieving a relative humidity of about 10%. When the oven is supplied with the supplementary standard atmosphere, an oven temperature of 60-70°C is required. Equilibrium with the standard atmosphere for testing may be assumed when successive determinations for mass at intervals of at least 2h show no progressive change exceeding 0,25% in the mass of the textile material.

Standard depth scale, in colour measurement: A series of dyed samples of different hue and chroma that have been accepted to have the same depth.

Standard depth of shade: Since dyeings vary in depth, it is common practice to determine a fastness property at a standard colour depth. Variations in the dye content of commercial dyes do not allow standardisation of colour depth of a dyeing on the basis of the % owf of the dye. The ISO recommends a series of reference colours in 20 different hues ranging from yellow to black. The standard depth is called the standard 1/1 depth. Black and navy are only available in two depths. For the other colours, other depths include the 2/1 standard depths, which are twice as deep as the 1/1 standards, or the 1/3, 1/6, 1/12 or 1/25 standard depths that are successively paler. For fastness testing, a dyeing is produced having a hue and strength matching as closely as possible one of the standard colours.

Standard deviation: Standard deviation is defined as the square root of the average of the squared deviations of the observations from the group average. Squaring the deviations eliminates negative signs and concentrates attention on the magnitude of the deviations and not their sign.

Standard Deviation, of a sample: A measure of the dispersion of variates observed in a sample expressed as the positive square root of the sample variance.

Standard dyeing time: [t_{70}-value, t_{70} (0.5)]. A parameter characterizing the dyeing process which is roughly inversely proportional to the dye diffusion constant. It gives the time in min at 100°C (at a liquor ratio of 40 : 1) required to achieve a bath exhaustion of 70% when dyeing is carried out at half the saturation depth (according to Hoffmann).

Standard Illuminant: Known spectral data established by the CIE for four different types of light sources. When using tristimulus data to describe a colour, the illuminant must also be defined. These standard illuminants are used in place of actual measurements of them light source.

Standard light source: See **Standard illuminant.**

Standard Moisture Regain: The moisture regain of a material at equilibrium with the standard atmosphere for testing textiles.

Standard Moisture Regain: Accepted moisture allowance for textile materials expressed in percentages of their dry weight.

Standard Observer 2,10 (CIE): (1) A hypothetical observer having the tristimulus colour-mixture data recommended in 1931 by the CIE for a 2° viewing angle. A supplementary observer for a larger angle of 10° was adopted in 1964. (2) The spectral response characteristics of the average observer defined by the CIE. Two such sets of data are defined, the 1931 data for the 2° visual field (distance viewing) and the 1964 data for the annular 10° visual field (approximately arm's length viewing). By custom, the assumption is made that if the observer is not specified, the tristimulus data has been calculated for the 1931, or 2° field observer. The use of the 1964 data should be specified.

Standard pressure: An internationally agreed value of 101 325 Pa (approximately 100 kPa), equal in non-SI units to a barometric height of 760 millimeters of mercury at 0°C or one Atmosphere.

Standard solution: A solution that contains a known mass of reagent in a definite volume of solution. A standard flask or volumetric flask is used for this purpose. The solutions may be prepared by direct determination of mass for primary standards. If the reagent is not available in a pure form or is deliquescent the solution must be standardized by titration against another known standard solution. See **primary standard**.

Standard temperature: An internationally agreed value for which many measurements are quoted. It is the melting temperature of water, 0°C (273.15 K). *See also* **STP**.

Standard white: (Ideal white). For comparative reflectance measurements, only those pigments capable of giving a full matt white of 100% white content with as near complete reflection of the incident light rays are used. Relatively few substances are able to satisfy these requirements in practice however. For colour measurement applications, the following two pigments are used as standard whites: (a) barium sulphate (2) freshly prepared magnesium oxide.

Standing bath dyeing: In exhaust dyeing where many batches of the same shades are to be dyed the standing bath dyeing is attractive in both cost reduction and environmental protection. In this method the dyebath is continuously used with the addition of exhausted dye in the previous batch and chemicals if necessary in order to to restore the dyebath to its original state and ensure reproducible results.

Standing dyebaths: Dyeing from a standing bath makes use of exhausted dye liquors to produce further dyeings and therefore represents a worthwhile recycling technique. See **Standing bath dyeing.**

Standing Wire: A broad term describing fixed rods or strips extending through the loom reed, that control the height of the pile in a woven pile fabric.

Stannic chloride: (pink salt, tin tetrachloride, tin perchloride, tin (IV) chloride), $SnCl_4$, molecular weight 260, density 2.229–2.26. Colorless, waterfree, corrosive, fuming liquid, almost chemically pure with a 45.4% tin content. Evolves heat on contact with water. Uses: mordant for alizarin dyeing, weighting of silk, clearing and softening of silk, manufacture of fuchsin, colour lakes.

Stannous acetate: (tin (II) acetate, acetate of tin), $Sn(C_2H_3O_2)_2$, molecular weight 236. Aqueous solution. Used as a reducing agent in discharges on substantive dyes (especially black).

Stannous chloride: (tin crystals, tin salt, tin dichloride, tin (II) chloride), $SnCl_2 \cdot 2\,H_2O$, MW 225, density 2.71. White salt, hygroscopic, soluble in water, powerful reducing agent. Uses: discharge printing, tin mordant (for shading alizarin red and pink on the yellower side to give fiery shades); aftertreatment (with soap) of alizarin red; resists (wool resists); spotting agent.

Stannous oxalate: (tin (II) oxalate), SnC_2O_4, molecular weight 206.5. Uses: dyeing and printing of textiles; mordant for alizarin print pastes and dyeing (more vivid pink and red shades).

Staple: Natural fibres or cut length from filaments. The staple length of natural fibres varies from less than 1 inch as with some cotton fibres to several feet for some hard fibres.

Manufactured staple fibres are cut to a definite length, from 8 inches down to about 1-1/2 inches (occasionally down to 1 inch), so that they can be processed on cotton, woollen, or worsted yarn spinning systems. The term staple (fibre) is used in the textile industry to distinguish natural or cut length manufactured fibres from filament.

Staple crimp, in wool: (1) General- The natural waves in a grease lock. (2) Specific- One complete wave on undulationof crimped lock.

Staple length: Is the average length of Staple fibre which, in the case of man-made fibres, can be any predetermined cut length. The staple lengths (in mm) of some natural and regenerated cellulose fibres are as follows:

1–5 = jute

5–50 = recovered wool

10–25 = hemp

0–50 = cotton

15–50 = bourette silk

25–45 = flax

30–150 = viscose staple fibre

60–150 = schappe silk

70–200 = ramie

50–300 = wool

approx. 1000 = raw silk

Staple, in grease wool: A tuft or lock of fibres which naturally cling together, as found in a fleece.

Staple, in button: A looped metal shank securely positioned perpendicular to and at centre back of the button flange for use in attaching the button to one part of a flexible substrate by means of a needle and thread, a ring or a toggle.

Staple Fabric: See **Spun fabric.**

Staple Fibre, man made: See Man-made staple fibre. Fibre of spinnable length manufactured directly or by cutting filaments.

Staple Fibres, multiple length: See **Multiple length staple fibres.**

Staple Fibre, overlength: Overlength staple fibres. Man-made staple length that are at least 10 % longer than the nominal or average cut length.

Staple glass, yarn: Yarn made from filaments that are nominally 200 to 380 mm (8 to 15 in.)in length.

Staple in grease wool: A tuft or lock of fibres which naturally cling together, as found in a fleece.

Staple length in grease wool: The length of a staple without stretching or disturbing the crimp of the fibres.

Staple: The conversion of staple into spun yarns suitable in evenness, size, twist, and strength for use in the weaving or knitting of fabrics. (Also see **Textile processing.**)

Staple Yarn: See **Spun Yarn.**

Stapled Seam: A seam formed by shaped metal devices such as U-shaped staples.

Star ager: This steaming machine is the oldest steaming system. It consists of a cylindrical heater with a jacket to allow the steam to be blown from the top and the bottom of the cylinder (the steam heats the walls avoiding the formation of drops of condensation). The cover is sealed hermetically to allow the machine to operate under pressure.

Star Stitch: Similar to **Double stitch.**

Star Tape: A coarsely woven tape typically 2/28s cotton count.

Starch: (1) Historically, starches and flours have been the film-formers of choice for textile sizing. The key difference between flours and starches is the gummy substance gluten, starches are flours which have had the gluten removed. Nature produces a wide variety of starches as a white granular substance found in seeds, roots and stem piths of growing plants. Flours or meal is leached with water (to remove the gluten) leaving the white, free-flowing granule which has limited solubility in cold water. Sources of starch are Corn (maize), Tapioca (cassava), Wheat, Sweet potato, Potato (farina), Sago, Rice, Yucca.

(2) A polysaccharide with a much higher molecular weight than that of a sugar, but typically lower than that of cellulose; amylose and amylopectin are the major plant starches. The starches used in the textile industry are derived from plants. In North America, most starch comes from corn (maize), but starch from

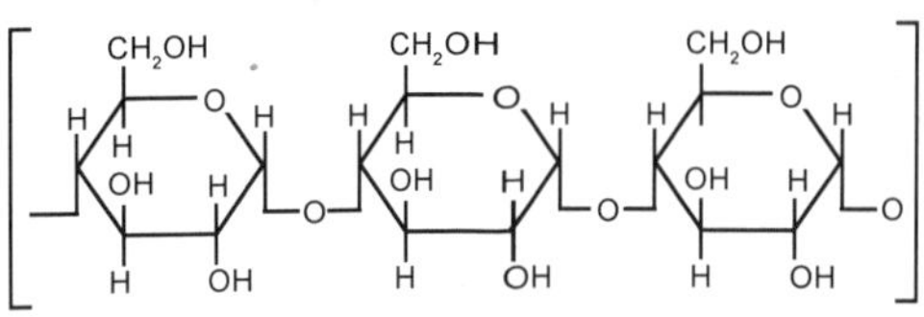

amvlose

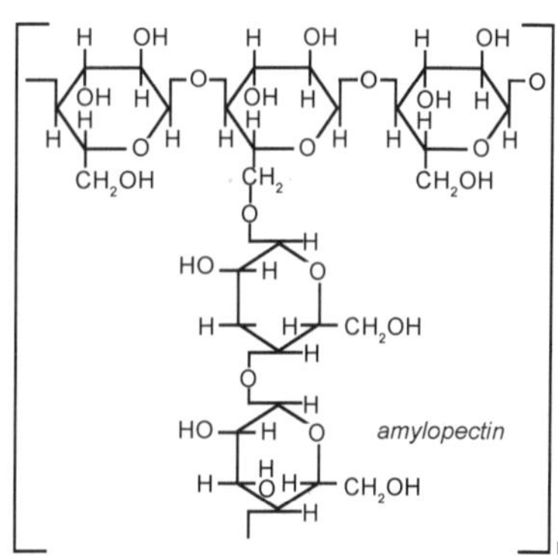

rice or tapioca is also common. Starch is extensively used as a size for warp yarns in commercial weaving. It must be removed before dyeing, because it can interfere with the uptake of dye by the fibre. It isn't appropriate for thickening reactive dye solutions for printing or direct application because much of the dye will react with the starch instead of the fibre. It can be used to as a thickener for some other types of dye. Starches are also chemically processed to produce dextrin, starch ethers (see Monagum) and starch esters. Unmodified starch has very low solubility in water. It is also used in finishing. Its use in all these operations depends on its adhesive or film forming properties and hence different starches (different origin) used for different purposes.

Starch Esters: These are starch products esterified with acetic anhydride, phosphoric acid or other acids which can be used as textile sizing and finishing agents. Starch esters produce more or less free-flowing pastes.

Starch Ethers: Starch ethers are made by reacting the hydroxyl groups in the anhydroglucose ring with appropriate reagents. These reactions add to the hydrophilic nature of the starch and decrease the ability to form hydrogen bonds between polymers, modifying solution and dry film properties. Examples are carboxylated starch, hydroxyethyl and hydroxypropyl starches.

Starch Indicator: Solution that is used to detect starch by a colour reaction Starch indicator is usually a solution of iodine and potassium iodide in water (typically about 2.3% potassium iodide and 0.33% iodine in distilled water). It will react with starch to produce a violet to blue-black colour, depending on the specific type of starch. This can be useful for detecting starch size in fabrics. Mercerized cotton will also give a blue-black colour with this test, though the colour seems to appear slowly, whereas with starch it appears almost instantly.

Starch Lump: See **Hard size.**

Starch paste: External characteristics: opaque, sticky; iodine reaction: dark blue; solubility: insoluble in water; technical qualities: fibre adhering, pore blocking (suitable for use as a printing thickener), good covering properties, subject to chalking, readily forms dust. The starch films are non-transparent, of average strength and low elasticity.

Starch size: Starch is still the most commonly used sizing agent for the weaving of cotton and viscose fabrics and has the advantage of being biodegradable. The question of Size recycling, as in the case of synthetic sizing agents, does not arise with starch sizes.

Starch sugars: (starch syrup). As the end product of starch degradation, starch sugar is identical to Maltose (malt sugar) and Glucose (grape sugar). External characteristics: aqueous, without body; iodine reaction: colorless; solubility: readily soluble in cold water. Technical qualities: starch sugar has neither a stiffening nor a hardening effect and is hygroscopic.

Star dyeing: A process for dyeing (usually delicate) textile fabrics in Star dyeing machines, especially velour upholstery fabrics and other pile fabrics made of cotton, natural silk, acrylic, polyester, acetate, etc.

Star frame: A frame having radial arms resembling the points of a star with hooks placed fairly close together on each arm. It is used to suspend delicate fabrics, like silk in the open width state in Star dyeing machines, either from one edge (vertical star dyeing machine) or both edges (horizontal star dyeing machine). Starting from the centre of the frame, the fabric is uspended from the hooks to form a spiral batch. For fabrics with sensitive edges, a narrow cotton or linen strip is first sewn along the edge/s to be suspended.

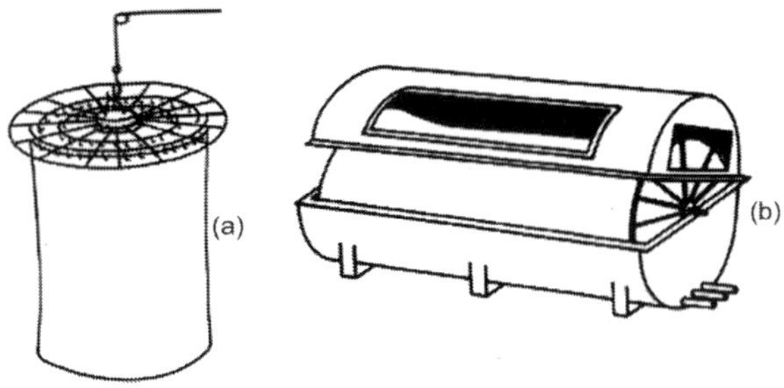

Star frames (a) Vertical (b) Horizontal

Start-Up Mark: See **Set Mark.**

Stash: It is your fabric collection.

Static: An accumulation of negative or positive electricity on the surface of fibres or fabrics because of inadequate electrical dissipation during processing. Static results in an electrical attraction or repulsion of the fibres relative to themselves, to machine parts, or to other materials, preventing the fibre from traveling in a normal path in the process.

Static Adhesion: In tire cord, the measurement of the strength of a cord-to-rubber bond under static conditions or very low strain rate.

Static charge: Whenever two materials are separated a electrostatic charge is developed in between them. The charge can be influenced by various factors like relative humidity, friction, distance between the objects etc. In textile it is important in case of the carpets. When the person walks on a carpet static charges are formed on the shoe and it is distributed all over the person and finally discharged to earth when the person touches any conductive material like door knob or an electronic equipment etc. During the discharge the person can feel a mild shock or it can affect a sensitive electronic equipment. The simplest solution to the problem is to increase the conductivity of the carpet so that when the surfaces are separated the charge is conducted away rapidly and cannot build up further. See **Walk Test.**

Static Crack: See **Shier.**

Static friction: Friction developed between two touching bodies at the time one body starts to move relative to another.

Static load in textile testing: A mass which excerts a force by means a mass alone without motion.

Stature: See **Height.**

Statute Galloons: Narrow cotton or silk braids, used in England for binding flannel underwear.

Stay: A piece of fabric or sometimes a tape used to hold another piece of fabric in place, or to add strength to a seam or tack; corsets.

Staying: Binding Tape to lace women's corsets.

Std Atm. For Preconditioning: See **Standard atmosphere for preconditioning.**

Std. Atm. For Testing: See **Standard atmosphere for testing.**

Steam black: Aniline black produced on the fibre with sodium chlorate as oxidizing agent and yellow prussiate of potash (potassium hexacyanoferrate II) as aniline black catalyst. Steam black is also referred to as prussiate black.

Steam boiler: Various types of steam boiler can be classified a) according to the steam pressure (low-pressure boilers up to 1.5 bar, high-pressure boilers above 1.5 bar); or b) according to the construction (natural circulation; controlled circulation boiler; forced flowthrough boiler).

Steam Calender: A machine used after the drying of finished knitgoods for steaming, compacting and stretching the fabric to the desired tubular width.

Steam Chest: A steam-heated cabinet used in manufactured fibre production. Usually refers to the heated cabinet in which spin-drawing is done or to the cabinet around a stuffer-box crimper.

Steam Cleaning: Often mistakenly used in place of Hot water extraction in the carpet cleaning. See **Hot water extraction.**

Steam Cure Process: A process for the continuous moist crosslinking of cotton with a nitrogen-methylol compound in the presence of a highly active catalyst mixture (magnesium chloride and citric acid) in a steamer supplied with superheated steam for drying and crosslinking (6–10 min at 130—140°C). Reaction takes place in the 15–5% moisture content range. Subsequent neutralization is not necessary.

Steam decatizing: A process in which decatizing is carried out with steam (also known as blowing) in contrast to Wet decatizing.

Steam finisher: An equipment used in garment industry, laundry for batchwise or continuous steaming of outerwear garments It is a steaming cabinet or steaming tunnel through which garments are passed and a notable finish effect is produced especially in the case of blended fabrics.

Steam-setting method: A heatsetting method. Short staple polyester yarns including polyester/cotton blends are normally setby relaxation in saturated steam. The most effective means of stabilising these materials are to steam at 107°C on the ring spinners tube and soft dyeing packages under minimum tension. Steaming is carried out in an autoclave fitted with vacuum pump, e.g. two times 15 rain at 125-135°C with intermediate evacution or alternatively, for 60 rain with saturated steam. Sewing threads receive special settingtreatments, designed to confer stability whilst preserving their high tensile properties.Polyester garments, garment lengths and hosiery are also stabilised by steaming in much the same way as for yams.

Steam trap: Steam traps. A self-acting device which, operating under steam pressure, automatically ejects condensed steam (condensate) from steam pipes, etc. without permitting the escape of steam. Various designs and systems are available, e.g. with float valve, ball valve or slide valve, bimetallic strip control (valve opens with a drop in pressure resp. temperature and closes with a rise in temperature).

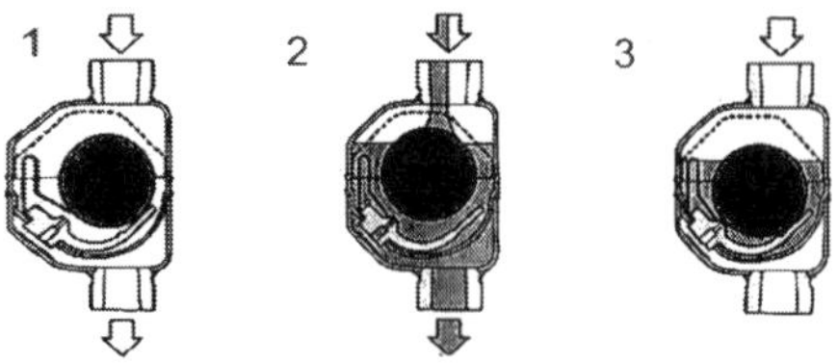

Working of a steam trap. 1. Cold condenstate, 2. Hot condensate, 3. Steam

Steamer: (1) A program-controlled yarn steamer with radial through flow to ensure niformity of the steaming effect. (2) An apparatus for steaming printed or padded piece goods, yarns, tops, hosiery, garments, etc. These steamers are of two types, (a) batchwise or discontinuous steamers and (b) continuous steamers.

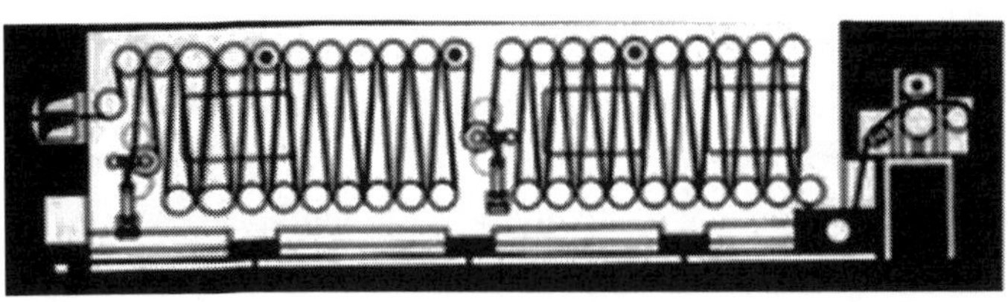

A steamer

Steamer grey: A light woven fabric which is run into a steamer together with printed goods to prevent the dried print pastes on a printed fabric from marking off on to adjacent layers of the same fabric during the steaming process.

Steaming: Finishing process causing general increase in moisture levels, increased swelling, and therefore, further effects, such as improved washing. There are further specific effects on wool-based fibres, including partial alleviation of tension (tension-free steaming), fluffing and raising the pile, removal of creases caused by laying, storage, pressing and decatizing, stripping of shine caused by pressing, etc. The best method to use is to steam for approx. 3 min with saturated steam, but complete saturation of the fibre should be avoided (detrimental effects to the handle).

Steaming at atm. pressure, fastness to: Capacity of the colour shade to resist being affected by steam (e.g. during steaming). The sample is placed between undyed cotton and the adjacent fabric for colour fastness testing. After rolling, the test sample is placed in the neck of a glass cylindrical flask, which is filled with boiling water. It is boiled for 30 min and then dried in warm air (< 60°C) without being touched. Any staining of the undyed adjacent fabric is evaluated using grey scales for fastness testing.

Steaming of prints: Prints are usually set or developed by steaming the various different substrates (details can be found under the relevant dye classification). The steaming process is carried out in open or sealed systems, either at atmospheric pressure or at pressure for temperatures in excess of 100°C.

Stearates: Stearic acid salts, e.g. aluminium stearate, which is used as a water-repellent treatment.

Stearite, soapstone: Mineral mass with a greasy feel, a derivative of Talcum consisting of 60–85% talcum, 5–7% magnesium carbonate, 0–24% barium carbonate, 0–2% lime, 0–2% beryllium oxide. Used to remove grease stains and as tailor's chalk.

Steel cord: A formed structure made by twisting together two or more steel filaments or steel strands.

Steel cord wrap: A filament wound helically around a steel cord.

Steel Fibres: See **Metal fibres.**

Steel strand: A group of steel filaments combined together to form a unit product.

Steeping: In textile processing, steeping is dipping and keeping the material for a stipulated time in water or chemical or a mixed recipe.

Steeple: Tall, conical women's head-dress with a flowing veil.

Steiner Tunnel Test: Method for testing the burning properties of carpets. The specimen is exposed to open flames in a tunnel. See **Flammability test**.

Stem stitch: In embroidery stitches placed next to each other to imitate the twine of a rope. Work from left to right, taking regular slightly slanting stitches along the line of the design. The thread always emerges on the left side of the previous stitch. This stitch is used for flower stems, outlines, etc. It can also be used as a filling, rows of Stem Stitch worked closely together within a shape until it is filled complete.

Stencil printing: Stencil printing was developed by the Japanese. Designs are cut in stencil paper which is coated with wax. The stencil designs are placed on fabric and colour is applied by sponge, air brush or by spray gun. This method is done on minimum fabrics like scarves

and similar products.

Stenter: An open-width fabric-finishing machine in which the selvedges are so held by attachments to a pair of endless travelling chains that the fabric is finished to a specified width.

(a) Attachments may be pins (pin stenter) or clips (clip stenter).

(b) Such machines are used for:

- (i) Drying;
- (ii) Heat-setting of thermoplastic materials;
- (iii) Fixation of chemical finishes.

See **Tenter.** There are different type of stenters: Horizontal stenter, vertical stenter, tier stenter, double return stenter. There are special stenters for knit goods.

Stentering: The process of drying textile fabrics on stenter driers; See **Stenters**.

Step-And-Repeat Machine: The step-and-repeat machine has a print roller pre-prepared with light-sensitive paint, upon which the positive slide is placed and exposed.

Step-Growth Polymers: As the name suggests, the step growth polymerization involves stepwise intermolecular condensation, taking place through a series of independent reactions. Each reaction involves a condensation process involving the loss of a simple molecule like H20, NH3, Hel, ROH etc. This type of polymerisation occurs if the monomer molecules have more than one similar or dissimilar functional groups. The step growth polymerisation starting with two monomers A and B as :

$$\underset{\text{Monomer}}{A + B} \xrightarrow[\text{Step 1}]{\text{Condense}} A - B$$

$$A - B + A \xrightarrow[\text{Step 2}]{\text{Condense}} A - B - A$$

$$A - B - A + B \xrightarrow[\text{Step 3}]{} A - B - A - B$$

The stepwise process of chain growth thus goes on. This process can also occur in another way :

$$A + B \xrightarrow[\text{Step 1}]{\text{Condense}} A - B$$

$$A - B \xrightarrow[\text{Step 2}]{A - B} A - B - A - B.....\underset{\text{Polymer}}{(A - B)_n}$$

Stephanie Lace: Modern handmade lace in imitation of the Point Venise.

Stereochemistry: **A**pplies to the spatial arrangement of atoms within the molecule structure, e.g.Stereoisomerism; Isomeric compounds; Isotactic polymers.

Stereoselective polymerization: In this type of polymerization one type of ordered structure is preferentially formed in contrast to the other.

Stereospecific polymerization: Polymerizations which yield ordered structures (isostatic or syndiotactic).

Stewart: Various Highland tartans, composed as follows:

Stewart, Royal: Wide red stripe; light blue stripe, almost one-eighth of the red; black stripe, wider than pale blue; group of yellow, black, white and black lines; green stripe, about one-quarter of wide red stripe*; red stripe, twice as wide as the green, split by one fine white line (in the center) and two, somewhat heavier black lines, the three lines spaced evenly; repeat, in reversed order groups mentioned between the two*. Old Stewart: Dark brown stripe, edged by red lines and split by a finer red line; dark green stripe, twice as wide as the former, divided into three even parts by two dark blue stripes, each edged by black lines; dark brown stripe, as above; dark blue stripe, split by two groups of narrow stripes, each group consisting of three black and two grey.

Stick-slip: Phenomenon occurring when boundary lubrication is deficient, manifested by alternative periods of sticking and slipping of the surfaces in contact.

Stiffened Fabrics: Stiffening Interlinings between the top cloth and the lining, Front fusing.

Stiffened Linen Extremely strong Buckram, board-like, hard, barely flexible, used for stiffening uniform collars.

Stiffening Agents: Used specifically for finishing and sizing. There are different types (a) Vegetable colloids: starches, mucilages, mosses, algae, gums. (b) Animal colloids: glue, gelatine, casein, protein sizes. (c) Synthetic colloids: polyvinyl products and similar, cellulose derivatives, synthetic resins.

Stiffness: Resistance to bending. (rigidity, flexural rigidity). Flexural rigidity is a measurement of the resistance exhibited by a material when subjected to bending.

Stilb: Unit of Radiant intensity per unit area. Radiant intensity is defined as the quotient of candela and square metres: L = 1 cd/m^2.

Stilbene: (trans 1,2 diphenylethylene) Most of the Optical whitening agents are stilbene derivatives. They are usually derived from diamino stilbene disulphonated acids and as such they don't have any fluorescent property but an amino group is added on to the two benzene structure it shows fluorescent properties.

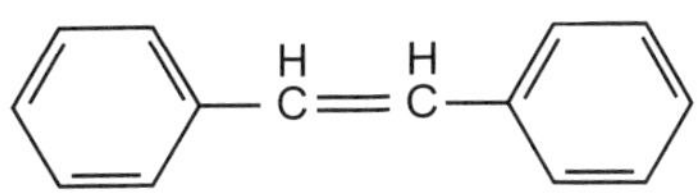

Stilbene

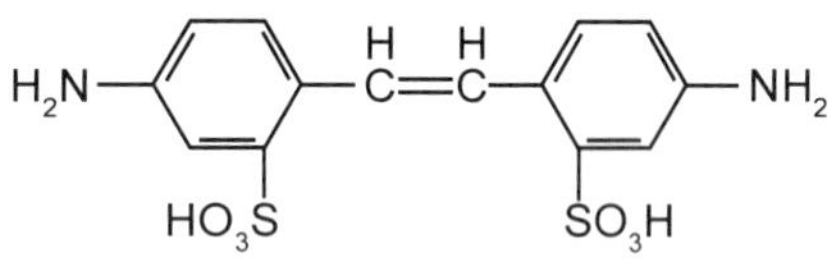

Florescent Dye

Stilbene dyes: Stilbene dyes are in most cases mixtures of dyes of indeterminate constitution that are formed from the condensation of sulfonated nitroaromatic compounds in aqueous caustic alkali, either alone or with other aromatic compounds, typically arylamines. The sulfonated nitrostilbene is the most important nitroaromatic, and the aminoazobenzenes are the most important arylamines.

Stiletto: This is a pointed metal with a wooden handle and is used to make eyelet holes or openings.

Stitch: The repeating unit of sewing thread formation in the production of sewn seams and stitching.

Stitch, in knitting: Stitch: stitch is the smallest unit in knitted fabric. A knitted fabric surface is formed by repeating it, side to side and one on top of the other. It consists of loop head, loop leg and loop feet. Plain stitch: this is the technical face side of stitch where loop legs are above the neighbour stitch and loop head is below the neighbour stitch.

Stitch, in sewn seams: The repeated unit formed by the sewing thread(s) in the production of seams.

Stitch (Backstitch): Used at the beginning and end of stitching to reinforce and prevent raveling. Also called backtack or stay-stitch.

Stitch (aste): A stitching which holds the fabric in place until permanent stitching has been completed.

Stitch, Blind: A stich that is not visible on one side of the fabric.

Stitch, Chain/Class 100: A stitch formed with one or more needle threads, the loops of which are passed through the material and through the loops of the preceding threads.

Stitch, Contrasting: When the stitching thread contrasts the garment colour.

Stitch, Double lock/class 400: A stitch formed with two or more groups of threads that interlace each other. The loops of needle thread are passed through the material where they are secured by looper threads; no bobbins used. This stitching ravels in one direction.

Stitch, Flat seam/class 600: Multi-needle stitches that provide the elasticity necessary for knits.

Stitch, hand/class 200: A stitch formed by hand with one or more needles--- one thread per needle passing in and out of the material.

Stitch, Lock/class 300: A stitch formed with two or more groups of threads that interface each other. The loops of needle threads are passed through the material where they are secured by bobbin threads.

Stitch, overedge/class 500: A stitch formed with one or more groups of threads at least one of which passes around the edge of the material.

Stitch, safety: A combination chain-stitch and overedge stitch made simultaneously on the same sewing machine.

Stitch, Top: A second row of stitching close to the edge of a seam, after two or more pieces of fabric have been sewed together and turned to bury the raw seam margin side.

Stitch, Zig-zag: A stitch made on a sewing machine in which the needle bar comes down alternately on the right and left side of an imaginary center line. Also refers to the type of machine producing this stitch.

Stitch Bonding: A bonding technique for nonwovens in which the fibres are connected by stitches sewn or knitted through the web. Also known as quilting.

Stitch density, in knitted fabrics: Stitch density: this is a product of course density and wale density. It gives a total number of stitches in a square area of fabric. Stitch density tends to give more accurate measurement for fabric dimensions compared to course density and wale density, due to the fact that the adverse effect of tension on the course and wale densities may be eliminated.

Stitch density, in sewn fabrics: The number of stitches per unit length in one row of stitching in the seam.

Stitch gauge, in sewn fabrics: The perpendicular distance between adjascent parallel rows of stitching.

Stich in the ditch: Stich in the ditch is a technique for quilting the quilt top, batting and fabric back together by stitching very close to a seam or appliquéd edge on the top of the quilt.

Sticker: A distortion in the weave charecterised by tight and slack places in the same warp yarns.

Stick-Slip: A phenomenon that occurs when boundary lubrication is deficient and manifests itself by alternate periods of sticking during which time the frictional force slowly rises to a peak value and slipping where the frictional force rapidly decreases to a minimum value.

Stitch type: A numerical designation relating to the essential characteristics of the interlacing of sewing thread(s) in a specified stitch.

Stitch Balance: Refers to the balancing of the sewing machine tension systems so that a proper stitch is formed. Generally, it is desirable to balance the stitch with minimum sewing machine thread tension.

Stitch Guage, in sewn seams: The perpendicular distance between adjascent parallel raws of stitching.

Stitch Bonded Carpets: Loop pile carpets manufactured using the stitch-bonding technique. The stitch-bonded fabrics are then bonded on the reverse side and coated.

Stitch Bonded Fabric: A multicomponent fabric, one component of which is a series of interloped stitches running along the fabric length. The other component may be fibre web or batt, yarn or preformed fabric. Examples of stitch bonded fabrics are: (a) A film web or batt bonded by stitching yarns (arcane, maliwatt); (b) Cross laid yarn with or without machine direction yarn bonded by stitching yarns (malimo); (c) A structure in which either sewing yarns or other threads lying in the machine direction , are taken over sinkers to form a loop pile (araloop, malipol); (d) A structure in which the stitching loops are formed from the fibre s of the web or batt (arabeva).

Stitch-bonded nonwovens: Nonwoven fabric bonded with chain-stitch seams (stitch-bonding technique using long fibres, not sewing thread).

Stitch-bonded pile fabrics: In principle the many different qualities of stitch-bonded pile fabrics can be divided into the following categories: – type used for floor coverings, – type used for rear window shelves, – type used for car boot upholstery.

Stitch-bonded thread-layer fabric: Stitch-bonded fabrics made from criss-crossed thread layers piled on each other as a backing material, bonded with the stitches formed by warp threads that are sewn in.

Stitch-bonded weft pile fabrics: Stitch-bonded fabrics (Stitch-bonded composites) in which threads formed into pile loops are bonded to the backing material by means of sinker loops created by a pillar stitch chain.

Stitch bonding: A method of stitching webs of fibres together, resulting in a non-woven fabric.

Stitch Course: Stitches arranged adjacently to one another. In standard knitting they are formed consecutively, in warp knitting they are formed at the same time. The number of stitches in the stitch course is dependent on the stitch size.

Stitch density, in sewn seams: The number of stitches per unit length in one row of stitching in the seam.

Stitch density: Construction characteristic of all knit fabric types; calculated as the product of the stitch course count and the stitch wale count. This figure represents the number of stitches within an area of 100 cm2.

Stitch gauge: This is the total stitch count (= needle count) per 11/2 inch (Engl.) = 3.81 cm fabric width. This stitch fineness is known as Gauge (g), which can also be abbreviated to gg. The higher the gg number (usually between 36 and 66), the more stitches there are in the knitted fabric, it is finely woven and therefore has better elastic properties. See **Cover factor**.

Stitch length (Loop length): This is the yarn length used in one stitch. It is generally calculated by dividing the course length, which is the yarn length used in one course, into the number of needles used in that course length. Loop length is very important in determining the fabric dimension, since fabric parameters such as course density, wale density and fabric weight are affected by the loop length.

Stitch sewknit fabrics: Variant of Malimo fabric, manufactured for instance from weft layers sewn (using pillar stitch) with woven fabric, Malimo fabric, expanded foam, fleece or knitted fabric as a basis fabric. Synthetic sewing thread is usually used. Use: imitation fur fabrics, friezé for coats, suedette, corduroy, upholstery fabrics.

Stitch type: A numerical designation relating to the essential characteristics of the interlacing of sewing thread(s) in a specified stitch.

Stitch wale: Stitches arranged above each other made by the same needle. The stitch wale count is dependent on the distance between the needles.

Stitch width: The distance between two Stitches, also characterized by the number of stitches per cm or inch. Mesh number. This is important in screen gauzes, for instance.

Stitch, Back-and-fore: A hand stitch employed for sewing linings and pockets. It involves taking a backstitchand the running stitch before the needle is removed.

Stitch, back: A hand stitchused to seam garment parts. So called because the needle, on emerging goes back to be inserted at the end of the previous stitch.

Stitch, blind: A stitch (either by hand or machine) which does not go right through the material.

Stitch, chain: A stitch formed with one or more needle threads and characterized by intralooping. One or more loops of thread are passed through the material and secured by intralooping with a succeeding loop or loops after they are passed through that material.

Stitch, covering chain: A type of chain stitch produced on twin or multineedle machines used for covering seams.

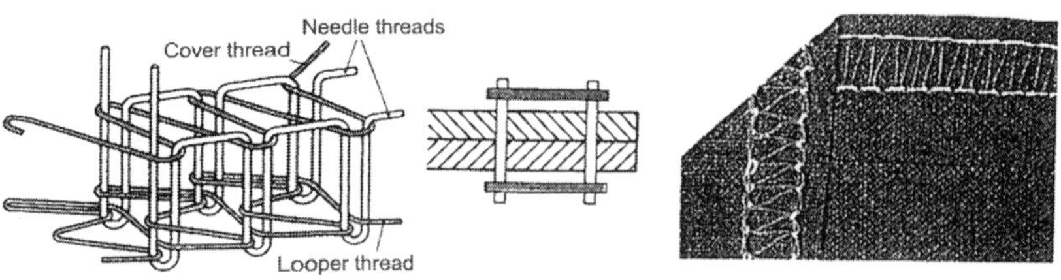

Schematic Symbol Seam appearence

Stitch, cross: A handstitch formed by passing the needle alternately on and off the edge slightly through the material and in advance of the last stitch. Used as a substitute fir falling or ornamentation of raw edges.

Stitch, Fagot: A form of lock stitch using modified zig-zag machine, for joining materials, allowing a narrow gap between two edges.

Stitch, Feather: A decorative stitch, similar to the fagot stitch, but in which the joining of two edges is nort involved.

Stitch, in sewing: The configuration of the interlacing of sewing tread in a specific repeated unit.

Stitch, in knitting: Stitch is the smallest unit in knitted fabric. A knitted fabric surface is formed by repeating it, side to side and one on top of the other. It consists of loop head, loop leg and loop feet.

Stitch, Flat lock: A stitch formed by the 'flat lock' machine with four needles, four loopers and a covering thread.

Stitch, four thread Overedge: See **Stitch, Mock safety.**

Stitch, Lock: The plain stitch in which two separate threads are used. In formation one thread is passed through the material forming a loop on the underside, second is passed through the loop on the underside of the material. This loop is pulled into the material bringing with it the locking thread.

Stitch, Loop: A type of handstitch used for oversewing edges so as not to increase their thickness by having an edge turned in.

Stitch, Mock safety (four thread overedge stitch): A type of overedge stitch which has an appearance similar to that of a safety stitch on the top surface of the material but has a common looper thread joining the two rows of stitching which can be seen underneath the fabric.

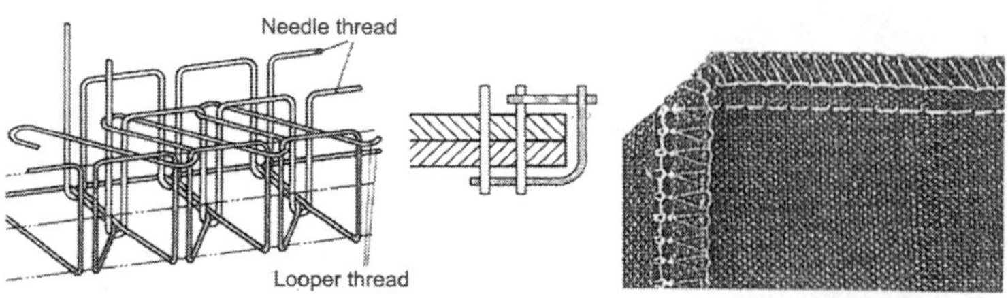

Schematic Symbol Seam appearence

Stitch, Single thread Overedge: A type of chain stitch, in which the needle thread loop is carried from the underside of the fabric over the edge and turned to the line of stitching to be secured by the needle on its next downward stroke. Varying amount amount of edge cover can be obtained by using machines employing more threads.

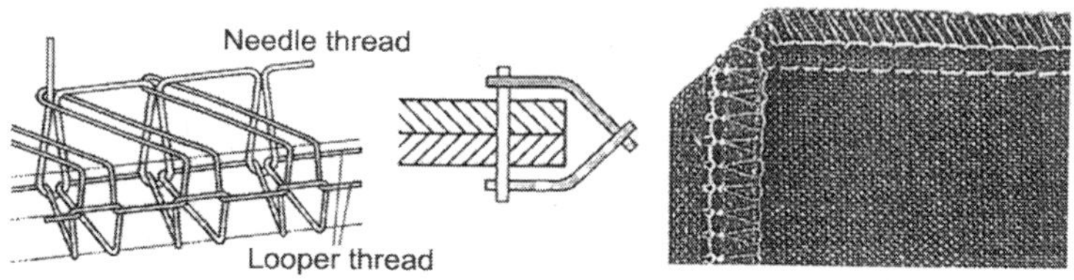

Schematic Symbol Seam appearence

Stitch, three thread Overedge: See **single thread overedge stitching.** Here three threads are used in stitching.

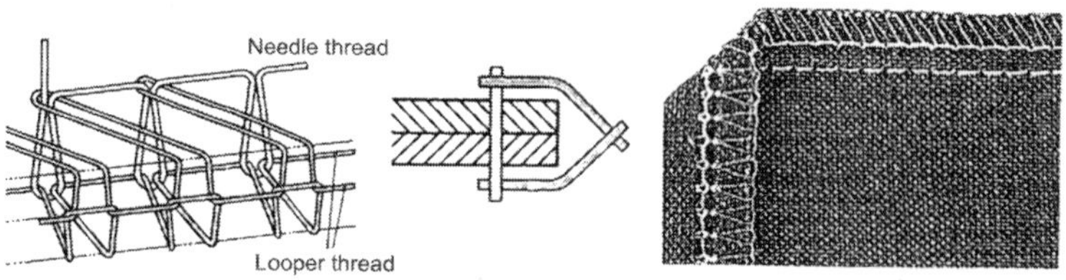

Schematic Symbol Appearence

Stitch, two thread Overedge (edge looping): Here two threads that are used in stitching

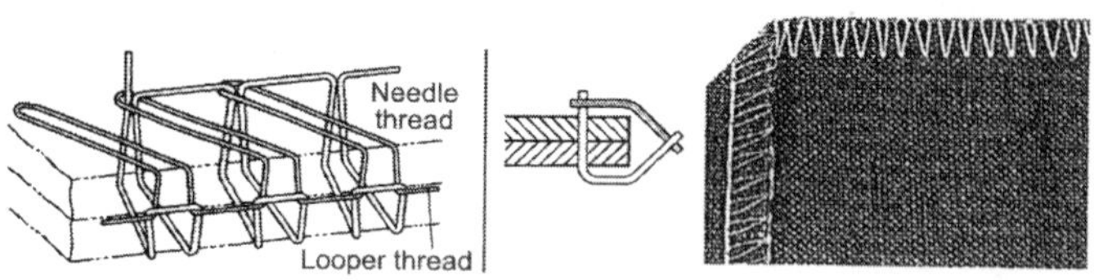

Schematic Symbol Seam appearence

Stitch, two thread Overedge and double chain (Combination safety stitch): Safety seams are a combination of two stitch types. Thread consumption can be upto 16 m thread per 1 m seam.

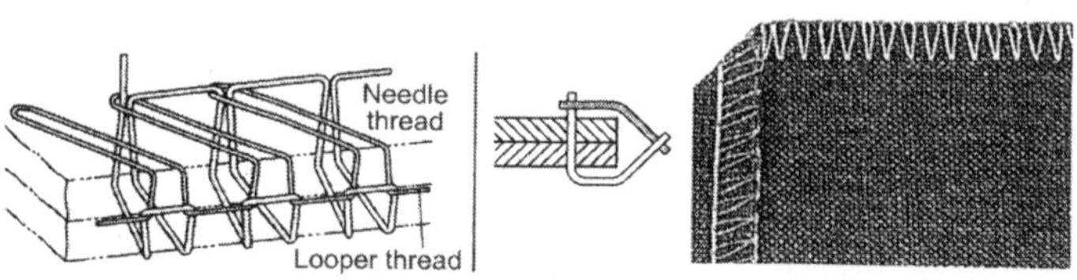

Schematic Symbol Seam appearence

Stitch, two needle multithread chain: Two or more needle are interlaced on the underside by alooper thread on the needle side by a cover thread.

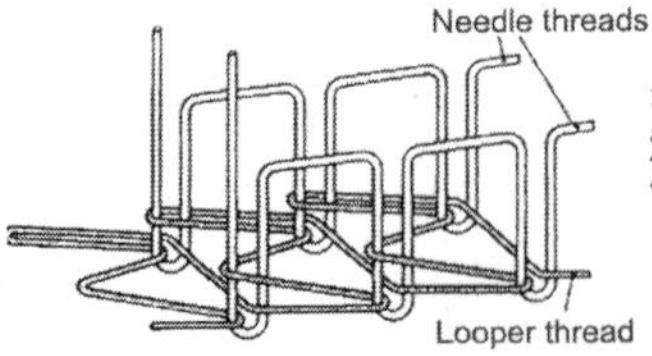

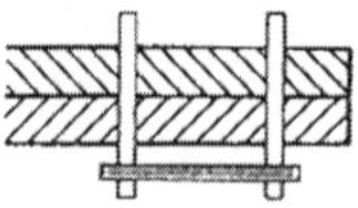

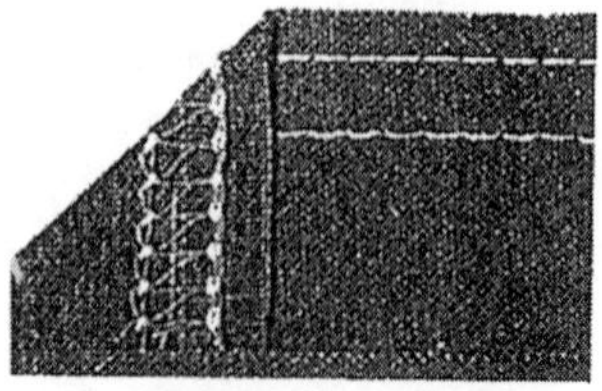

Stitch, plain: This is the technical face side of stitch where loop legs are above the neighbour stitch and loop head is below the neighbour stitch.

Stitch, Prick: A stitch made by passing the needle straight throughthe material at right angles to the surface alternately from one side to the other.

Stitch, Running: A stitch formed by an overedge stitch reinforced by a chain stitch (or sometimes a lock stitch) further in from the material edge.

Stitch, Skipped: A stitch which appears twice as long as the other stitches in a line of stitches. It is caused by failure in the stitch formation allowing the needle thread to return to the surface without being held by the bobbin (lower) thread.

Stitch type, in sewn seams: A numerical designation relating to the essential characteristics of the interlacing of sewing thread(s) in a specified stitch.

Stitched and pinked seam finish: Used for fabric which has a tendency to ravel, it is a finish for the raw edges of the seam allowances of a plain seam, in which each raw seam allowance edge in machine stitched and then cut to a zig-zag raw edge.Using a sewing machine threde with thread to match your fabric and pinking shears stitch aline of straight machine stitches 6 mm. (1/4 in.) in from the cut edge. Use short stitch, 15 stitches per inch. Pink just on the edge of the seam allowance, taking care not to cut into the stitching.

Stitched seam: See **Sewn seam.**

Stitchel: Hair-like wool with little serrations on the surface.

Stitches: Thread loops that are linked together and constitute the smallest units of Knitted fabrics. They consist of 1 head (upper part), 2 shanks (middle part) and 2 feet (lower part); each having 4 bonding points (2 upper and 2 lower, i.e. at the top and bottom).

Stitches Per Inch: Refers to the number of the stitches made in one inch of seam, starting at a needle penetration and measuring the lengths of thread between needle penetrations.

Stitches, in tufted pile floor coverings: The number of pile tufts per inch in the lengthwise direction.

Stitching: A series of stitches embodied in a material or materials of planar structure such as woven textile fabrics, usually for ornamental purposes or finishing an edge, or both.

Stock colour: Print paste with a high concentration of dye (large quantities are usually kept in stock), used in combination with other stock colour pastes or Reduction thickeners as mixes.

Stock dyeing: See **Dyeing.** The process of dyeing fibres in raw state (in the grain) before being spun.

Stock in process, in textiles (spinning): Staple fibres at any stage of manufacture between the opening oft he bale and the completion of the spinning process.

Stock paste Print pastes consisting of all the necessary chemicals except for the dye.

Stock solution: A solution of known strength, made up with the intent of dilution or mixing before final use Stock solutions are a convenient way of avoiding the need to weigh chemicals each time you need to use some. For example, if you need 0.27 grams of Smurf extract in a blue dye formula, and you have a stock solution of 10% extract, you would measure 2.7 milliliters of stock solution to get that amount. Some chemical solutions have limited shelf life.

Stock thickener: (stock thickening). Printing thickeners containing setting chemicals (large quantities are usually kept in stock), to which dye is added at a later point, in powder, paste or solution form.

Stock vat: It is not possible to vat individual vat dyes in a dye bath, they need to be dissolved in concentrated form. This method is known as stock vatting.

Stockinet: A plain, elastic texture made on a knitting frame, used for underwear, etc.

Stockinette: A knit fabric in tubular or flat form made with a plain stitch from yarns of wool, cotton, manufactured fibres, or a combination of these fibres. Stockinette fabrics are used for underwear, industrial applications, and other purposes. In heavier constructions, dyed and napped stockinette finds apparel uses.

Stocks: Solid silicon rubber compounds, which can, for example, be forced through calenders to coat fabrics.

Stola: Womens outer robe in the form of a shift, worn belted over the tunic.

Stoll-quartermaster universal wear tester: A versatile testing apparatus for measuring wear resistance of fabrics, yarns, thread, etc. It can be equipped with either of two testing heads, one for testing abrasion resistance of flat surfaces and the other for testing resistance to flexing and abrasion.

Stolling: Rib trim put on so that the rib is at right angles to the direction of the rib on the body.

Stomacher: Lengthening of the lower bodice front into a pointed or rounded peak.

Stone finish: A finish obtained by vigorous tumbling with pebbles.

Stone polishing: In print roller manufacture, the polishing process on copper print rollers during Copper electroplating. It occurs when agate stones (rolls) are pressed onto rotating rollers, causing the copper crystals that form vertically to be polished.

Stone washed fabric/garment: Denim fabrics (Jeans) dyed with indigo or other dyes are sized with a large quantity of starch and manufactured into garments. Once the textile has been made into a garment it is desized immediately (30 min at 60–70°C) and then treated with sodium hypochlorite. The technique is known as "Stone-Wash" or denim finishes because pumice stones are used as an abrasive agent to achieve special effects (1 kg of stone or sand per kg of fabric). The bleaching effects can be altered by varying the pH and liquor temperature of the sodium hypochlorite bath. The higher the temperature and the lower the pH (pH 8), the stronger the bleaching effect and the attack on the fibre. There are many other methods (mechanical, chemical, enzymes, or a ombination of these) to make stone wash effect made today on any type of dyed materils like pigment dyed, reactive dyed, double dyed etc.

Stop mark: A visible range in the density of the weave across the width of the fabric caused by the tension on the warp not being adjusted properly after the loom has been stopped. See **Set mark.**

Stop motion: Any device that automatically stops a textile machine's operation on the occurrence of a yarn break, a high defect count, etc.

Stop, in Zipper: The device at the top and bottom of the chain or stringer that prevents the slider from leaving the chain.

Stop, Bottom, in Zipper: See **Bottom stop.**

Stop, Bridge top, in Zipper: See **Bridge top stop.**

Stop, Releasing, in Zippers: See **Releasind stop.**

Stop, Top, in Zippers: See **Top stop.**

Stopping motions: Electrical oe mechanical devices employed on many textile machines when a fault develops in raw material feeding, arrangements (openers, scutchers, spinning frames etc.) or a yarn breaks in winding, warping, weaving etc.

Store lighting : (TL 84), type of lighting for colour matching booths, defined as cool, white fluorescent light at a colour temperature of 4400 K. Some customers may ask colour matching of dyed/printed samples in these lights..

Storm Serge: In the United States a very light serge weighing about 7 ounces, made of single warp and filling; used for women's coats.

Storm Shell: Wind proof, wind resistant outerwear.

Stormwater, storm water: Water from any form of ***precipitation***. In wastewater engineering it means the ***overland flow*** that enters the sewers, which can be estimated from the ***rational method*** or from ***unit hydrographs***.

Stormwater drain system: Gutters, ***stormwater sewers, combined sewers***, streams, ditches, etc. that are used to carry stormwater.

Stout: A property of certain fabrics, being the combination of close weave and weight.

Stoving: The process of bleaching raw wool or fabrics with sulphur fumes.

Stoving fastness: Level of resistance against the influence of sulphur dioxide. The guideline applies to wool and silk, even when mixed with other fibres. The fabric sample undergoing testing must be wetted for 5 min at 25°C in soap solution and squeezed at 100% liquor uptake. Fabric sample and control sample is then exposed to an environment containing sulphur dioxide for 16 h. Fabric samples are hung for 2 h in air without rinsing. Textile mixes are rinsed for 10 min after sulphur treatment and then dried at 60°C. The result is analyzed using the grey scale method.

STP: (NTP) Standard temperature and pressure. Conditions used internationally when measuring quantities that vary with both pressure and temperature (such as the density of a gas). The values are 101 325 pascals (Pa) (approximately 100 kPa) and 0°C (273.15 kelvin). See also **Standard pressure; Standard temperature.**

Stradella: A French woollen damask shawl.

Straight draws: These form the simplest and most common method of drawing-in. We begin with the first heddle on the left side of the shaft *nearest to the warp-beam*, then take the first heddle of second shaft and so on until all the shafts the set contains are used in rotation. This completes one "draw," and this operation is repeated until all the warp-threads are taken up.

Straight stitch, in embroidery: This is shown as single spaced stitches worked either in a regular or irregular manner. Sometimes the stitches are of varying size. The stitches should be neither too long nor too loose. This stitch may also be worked on evenweave fabric.

Straight stitch, in sewing: A simple machine stitch pattern of straight, single stitch segments of equal length. The simple machine stitch pattern is by the down and up movement of the needle while the fabric moves through the feed mechanisms in either the forward or reverse direction.

Straightness in steel cord: The property of a cord characreized by a lack of deviation from its central axis over short lengths of a cord.

Strain: Deformation of a material caused by the application of of an external force.

Strain recovery curve: See **Tensile hysterisis curve.**

Strain, Tensile: A relative length deformation exhibited by a specimen subjected to a tensile force strain is expressed as a fraction of normal length; also as a percentage.

Strand: (1) A single fibre, filament, or monofilament; (2) An orderd assemblage of textile fibres having a hifh ration of length to diameter and normally used as a unit, including slivers, rovings, single yarns, plied yarns, cords, braids, ropes, etc.

Strand, Irregularity, in textiles: Variation in a property along the strand.

Strand, in braided rope: One of a number of similar units which are intertwined to produce a braided rope, each unit consisting of one or more yarns which are not twisted together and which follow a prescribed path through the braided rope etc.

Stranfa: Fibre obtained from straw; used in Germany as substitute for jute.

Strapping: Narrow width rib trim, usually 1 x 1 rib or of a half milano construction, applied so that the rib runs along the garment lengthways.

Strasse: Sort of floret silk obtained by converting duppions into waste.

Straw A general term for plant fibres obtained from stems, stalks, leaves, bark, grass, etc.

They are made into hats, bags, shoes, mats, etc., by weaving, plaiting, or braiding.

Strazza: See **Bourre.**

Strazza (Italian): Waste of silk in Italy.

Streak: A stain (rust, oil, dye, grease, soap etc.) extended as an irregular stripe in the cloth.

Streak Stitch: The open veins of leaves in hand-made laces.

Streak: An extended unintentional stripe of narrow width, often a single yarn.

Strength: A generic term for the ability of a fabric to resist strain or rupture induced by external forces.

Strength analyzer: An instrument which determines the tensile strength and elongation at breaking load for a test beard of cotton.

Strength count product: See **Break factor.**

Strength, Breaking: (1) The maximum internal cohesive forces of a material which resist rupture during a tensile test.

(2) The ability or capacity of a specific material to withstand the ultimate tensile load or force required for rupture. Breaking strength is particularly significant as the characteristic of a sample as distinct from a specimen.

Strength, Bursting: (1) General: The ability of a material to resist rupture by pressure. (2) Specific : The force applied at right angles to the plane of the fabric under specified conditions.

Strength, Tearing: See **Tearing strength.**

Strength, Tensile: See **Tensile strength.**

Streptocyanine dyes: Dyes in which both charged terminal atoms are not part of a heterocyclic ring are called streptocyanine dyes.

Stress: The resistance to deformation developed within a material subjected to an external force. Typical examples are tensile stress, shear stress, or compressive stress. Stress usually reaches a maximum at the time of rupture. When a textile material is subjected to a stress below that causing rupture, the stress gradually decreases or decays with time.

Stress-Strain Curve: A graphical representation of the stress and strain relationship of a material under conditions of compression, shear, tension, or torsion. A graphical representation, showing the relationship between the change in dimension (in the direction of the applied stress) of the specimen from the application of an external stress, and the magnitude of that stress. In tension tests of textile materials, the stress can be expressed either in units of force per unit cross-sectional area, or in force per unit linear density of the original specimen, and the strain can be expressed either as a fraction or as a percentage of the original specimen length. (Also see **Load deformation curve**.)

Stretch: General term for elastic textiles made from stretch yarn or stretch fabric, for instance. Slimline clothes are frequently fashionable, which means that stretch garments are popular in the ladies' outerwear fashion sector. Polyurethane elastomer fibres have a better chemical resistance, a higher level of elongation and better recovery properties than rubber thread, which was commonly used in the past.

Stretch Breaking: In conversion of tow-to-top, fibres are hot stretched and broken rather than cut to prevent some of the damage done by cutting.

Stretch cord: Collective term for different qualities of elastic cord. Depending on elasticity they are ideally woollen or cotton and textured polyamide, in addition the warp contains 2–6% elastic thread covered with textured polyamide. Example: Warp – approx. 85% textured polyamide and 15% elastomer yarn (elasticity 15–30 % depending on quality); overall composition: 80 % cotton, 16–17% polyamide, 3–4% polyurethane (elastomer fibre).

Stretch fabric, Woven: See **Stretch woven fabric.**

Stretch fabrics: (elastic fabrics), elastic textiles with comfortable wearing properties (retain shape and are crease-resistant), elasticated either lengthways/warp (ideally) and/or crosswise/weft. The yarns used for this are highly-elasticated synthetic textured yarns, elastomer and rubber thread wrapped with yarn, or corespun yarns, frequently in the form of synthetic mix yarns, with/without natural or synthetic fibres, and also specialist wool and cotton yarns (e.g. with Slack mercerization), etc. See Power stretch, comfort stretch.

Stretch ratio, Draw ratio: (1) Machine draw ratio: In a drawing process, the ratio of the peripheral speed of the draw roller to that of the feed roller (2) True draw ratio: In a drawing process, the ratio pf the linear density of the undrawn yarn ti that of the drawn yarn. (3) Residual draw ratio: The draw ratio required, in draw texturising, to convert a partially oriented yarn into a commercially acceptable product.

Stretch spinning: A term used in the manufacture of rayon. Rayon filaments are stretched while moist and before final coagulation to decrease their diameter and increase their strength.

Stretch stitch, in sewing: A complex macine stitch pattern or of various combinations of straight stiotch, single stitch zigzag, or multiple stitch zig-zag. It is produced by co-ordinated motions of needle and feed mechaniosm in forward or reverse direction.

Stretch towelling: A knit fabric with short loop pile to give it a thin Terry toweling appearance on the right side. It is produced in plain colours and in patterns and stripes. Made from cotton and polyamide, it has a great deal of stretch. Very use ful for baby clothes, children's clothes, sports outfits, jumpsuits, bath robes etc.

Stretch woven fabric: A woven fabric which is capable of at least 20% stretch either in warp or weft direction, or both, under loads and conditions encountered in use, and of almost complete recovery on removal of the load.

Stretch yarn: See **Bulk yarn.** Yarns with exceptional elasticity, e.g. Textured yarns; Core-spun yarns Continuous filament synthetic yarns that have been altered through special treatments or modification to give them elasticity. Techniques include: twisting and untwisting, use of air jets, stuffer boxes, knife blades, crimping, heat setting, curling, steaming, or looping. Use of these yarns gives fabrics a degree of elasticity and comfort.

Stretch yarns, dyeing: Usually in the form of tubeless cross-wound yarn packages (muff) on pack system dyeing machines with a cylindrical yarn carrier in the pack dyeing system (brick-type) or a packing system onto dye tubes.

Stretch, comfort: Stretch fabrics having, lower elasticity, yield on stretching, no body-shaping properties but mould to fit the body, used in ladies' hosiery, underwear, sportswear and outerwear.

Stretch, low power: See **Low-power stretch.**

Stretch, power: Stretch fabrics having high level of elasticity, with bodyshaping properties, for corsetry, swimwear, surgical stockings.

Stretch-spun fibres: Synthetic threads that have been stretched during wet spinning to increase their strength and fineness, e.g. cupro filaments and highstrength viscose fibres.

Stretched filling: See **Tight pick.**

Striations: Streaks or bands or various nature in fibres or fabrics.

Strié: A term describing any cloth having irregular stripes or streaks of practically the same colour as the background.

Strike: Term referring to the speed of dye uptake on the substrate (Half-dyeing, time of). First strike is the ending, which depends on substantivity, in padder dyeing.

Strike off: (1) Sample prints on fabric for pattern and colour approval.

(2) (knocking-over), term originating from old-style hand or block printing. Colour is applied to the print block in the colour trough, and the print is then positioned on the fabric. The print is strengthened by hitting the block with a small hammer, which transfers the dye onto the fabric. Nowadays the term strikeoff is used to mean the printing of a repeat run.

String: (1) Two or three-ply coarse thread of hemp or flax of various thickness and fineness; (2) A unit of ten feet, according to which woollen warps are calculated in Yorkshire, England.

Stringer, In Zippers: The tape, bead and element asmbly that constitutes one side of a chain.

Stringy: (1) Thin, delicate stapled wool; (2) A flaw in the wool, consisting in slight matting, caused by imperfect scouring; (3) Defective raw cotton, the fibres forming strings, caused through the ginning of too wet cotton.

Stringy selvage: See **Slack selvedge.**

Strip Test, in fabric testing: A tensile test in which the full width of the specimen is gripped in the clamps.

Strip test, Cut: See **Cut strip test.**

Strip test, Ravelled: See **Ravelled strip test.**

Stripback: See **Broken filament.**

Stripe Braid: Has stripes, often of different colours or materials interlaced with each other.

Striping Yarn: Any yarn plain or fancy, used singly or as small nimber of threads and made from any textile raw material, to produce stripe effects in a fabric.

Stripper, in textiles: A product usually a reducing agent. That changes the coluring material, dye, or soil stain to reduced colour.

Stripping: Removal of dye from fabric Stripping is usually done with a reducing agent such às thiox, formosul or sodium hydrosulfite and often requires hot to boiling conditions. Some dyes are difficult to strip, and the result often is not white.

Stripping agent: (decolorizing agent). A product for Stripping of dyeings and prints. Reduction bleaching agents on a sodium dithionite basis or hypochlorites are usually used.

Stripping auxiliaries: is used to aid the Stripping of dyeings and printing. Has the task of supporting the brightening effect in the stripping reduction process by restricting the re-uptake of reduced, dissolved dyes. Sometimes also combined with Fibre protective agents, principally in wool. The addition of anthraquinone might be recommended in the stripping of naphthol, vat dyes, etc.

Stroll, in tufting: A pattern attatchment consisting of a number of yarn roller which are capable of being driven at different speeds. The pattern is read from an acetate sheet on a pattern drum.

Strong wool: A term with several meanings including extra long staple wool from any wool clip, an Australian term for crossbred sheep, and description of British lustre wools.

Structured needle Felt: A pile fabric formed by subjecting a previously needled web or batt to a further punching operation with forked, single barb or side-hook needles. Rib, velour and pattern structures may be produced.

Studio: The place where a deigner works. It can be anything from an area set aside on the factory floor to a large smart office, or even a suite of offices.

Strussa: Waste silk, obtained from double cocoons (duppions).

Strusa (Italian): See **Frisons.**

Stubble, in shorn floor covering: The portion of the pile that remains after shearing.

Stubble Height: The distance the stubble extends above the backing fabric.

Stuff: An ols eighteenth century word used to describe any fabric containing worsted yarns.

Stuffer box: A crimping device consisting of a confined space into which a tow, converted tow, a sliver, a yarn or a similar assembly of filaments or fibres is injected by feed rollers or other means such as fluid jets and in which the fibre assembly is packed and compressed into a configuration where the individual filament or fibres became folded or bent.

Stuffer box texturing: See **Texturing, stuffer box.**

Stuffer channel: (compression channel). Is used for compressing the piece goods together in a longitudinal direction when milling on a rotary milling machine for rope treatment or in open-width washing machines to relax the fabric.

Stuffer yarn: An extra backing yarn running in the warp direction through a woven pile floor covering.

Stuffers: Extra yarns running in the warp direction through a woven fabric to increase the fabric's strength and weight.

Stumba: Combing silk obtained from the waste of shappe silk; it is quite coaree and is used for filling yarn, coarse knitted fabrics, etc.

S-Twist: See **Yarn Twisting.**

Stylist: Designer who puts together a range of products.

Stymboline: Felt made of woollen and linen yarn in France.

Styrene: An unsaturated hydrocarbon, (C_6H_5-CH = CH_2), prepared from coal tar. Polystyrene is a colorless, transparent plastic used for molding various articles for insulation, transparent parts, radio parts, etc.

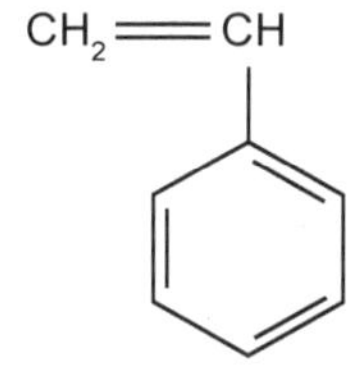

Styryl dyes: The styryl dyes are neutral molecules containing a styryl group $C_6H_5CH{=}C$, usually in conjugation with an *N,N* -dialkylaminoaryl group. Styryl dyes were once a fairly important group of yellow dyes for a variety of substrates.

Sublimation: The conversion of a solid directly to a gas, for instance, at standard pressure iodine, solid carbon dioxide, and ammonium chloride sublime, without passing through a liquid phase. At certain conditions of external pressure and temperature an equilibrium can be established between the solid phase and vapor phase.Some disperse dyes will sublime. This can make dyed fabric subject to fading due to heating, as from ironing at high temperature. Sublimation printing of synthetic fibres, mostly polyester, is used commercially. Typically, the pattern is printed on paper, then the dried pattern is heat transferred to the fabric. The dye gas penetrates the hot fibre, where it becomes physically trapped as it cools back to solid form. Washfastness is very high. There are now a number of special sublimation inks available for computer printers and others for screen printing, both for making transfers. One t-shirt maker also makes shirts that have a polyester outer layer and a cotton inner layer, also specifically for "digital" transfer printing.

Sublimation printing: A form of transfer printing (q.v.) that uses dyes that sublimate readily and have substantivity (q.v.) for the substrate to which they are applied.

Sublimation Fastness: Fastness of a dye or dyeing against sublimation. Usually a property of Disperse dyes which sublimes and stains the adjascent fabric. See **Sublimation**.

Sublistatic® process: A method of applying print designs to fabrics containing manufactured fibres by paper-transfer techniques. Developed by Sublistatic Corp. (Also see Prin Heat Transfer Printing.)

Submerged spin process: Used for dyeing continuous yarn and cross spools with reactive dye substances using the cold residual process. Cross spools are impregnated in a dye bath by a centrifuge. Commercial interest is limited to Space dyeing.

Submerged test: Testing method for water repellent impregnation of textiles (particularly yarn, activity products and knitted fabrics). Technique: weigh textile to be tested, submerge in sieve 10 cm under water for 5 minutes whilst agitating sieve to free trapped air, remove from water, allow to drip for approx. 10 minutes, weigh and ascertain water absorption.

Submerged thermosol process: Follow-on process for (bleaching and) whitening synthetic fibres (also cotton mixtures): stretching process 2–5 min in hot fluorescent brightening liquor in the residence device (immersion accumulator), then squeeze out, dry and thermo fix. When fixing the fibre non-bound or diffused, whitener is thermosol fixed.

Substantivity: The attraction, under the precise conditions of test, between a substrate and a dye (or other substance) where the latter is selectively extracted from the application medium by the substrate. A dye that is substantive will leave the dye bath and be concentrated on the fibre in the bath. Without substantivity, most of the dye would simply remain in solution or dispersion in the bath. Dye substantivity is generally associated with the molecular structure of the dye, and often big molecules have high substantivity, while small molecules have low substantivity. Dye bath conditions, including temperature and additives such as salt influence substantivity. Substantivity is often produced in ways that differ from the final bond of the dye to the fibre. Also see **Affinity.**

Substituent: An atom or group substituted for another in a compound. Often the term is used for groups that have replaced hydrogen in organic compounds. For example, in chlorobenzene (C_6H_5Cl) chlorine can be regarded as a substituent.

Substitute: A cheaper or inferior fibre which takes the place of a more expensive one, as for instance cotton used instead of wool or silk.

Substitution reaction: A reaction in which an atom or group of atoms in an organic molecule is replaced by another atom or group. The substitution of a hydrogen atom in an alkane by a chlorine atom is an example. Substitution reactions fall into three major classes depending upon the nature of the attacking substituent. *Nucleophilic substitution*: the attacking substituent is a nucleophile (i.e. a molecule or ion that can donate electrons). Such reactions are very common with alcohols and halogen compounds, in which the electron-deficient carbon atom attracts the nucleophile and the leaving group readily exists alone. Examples are the hydrolysis of a haloalkane and the chlorination of an alcohol:

$$C_2H_5Cl + OH^- \rightarrow C_2H_5OH + Cl^-$$

$$C_2H_5OH + HCl \rightarrow C2H5Cl + H2O$$

Electrophilic substitution: the attacking substituent is an electrophile (i.e. a molecule or ion that accepts electrons). Such reactions are common in aromatic compounds, in which the electron-rich ring attracts the electrophile. The nitration of benzene in which the electrophile is NO_{2+} is an example:

$$C_6H_6 + NO_2^+ \rightarrow C_6H_5NO_2 + H^+$$

Free-radical substitution: a free radical is the attacking substituent. Such reactions can be used with compounds that are inert to either nucleophiles or electrophiles, for instance the halogenation of an alkane:

$$CH_4 + Cl_2 \rightarrow CH_3Cl + HCl$$

The term 'substitution' is very general and several reactions that can be considered as substitutions are more normally given special names (e.g. esterification, hydrolysis, and nitration). *See also* electrophilic substitution; nucleophilic substitution.

Substrate: Fabric to which coatings or other fabrics are applied. It can be of woven, knit, nonwoven, or weft-insertion construction. Generally, substrate properties are dependent both on fibre type and fabric construction. Usually the fabric is scoured, heat-set, and otherwise finished prior to coating or bonding. Many smooth-surfaced manufactured fibre fabrics require impregnation with a latex prior to coating to ensure adequate adhesion.

Substrate, in effluent treatment: The term used to denote the organic matter or nutrients that are converted during biological treatment or that may be limiting the biological treatment. For example the carbonaceous organic matter in waste water referred to as the substrate that is converted during the bilological treatment.

Substractive, colour chemistry: With reference to colour, removal of colours from light reflected from a surface A surface that is illuminated by white light and that reflects all visible colours will appear white. The surface can be made to appear some other colour than white by altering it with materials that absorb or subtract colours present in the white light. The subtractive primary colours are designated cyan, magenta and yellow. If a white material is coloured with something that absorbs magenta, for example, then cyan and yellow will be reflected, making the material look what is normally called green. If another material is added that absorbs the cyan component "subtracts" cyan, then only the yellow is reflected, and the material looks yellow. If all colours are subtracted or absorbed, then the material appears black. Dyes used on fabric work according to subtractive principals. In theory, if truly pure cyan, yellow and magenta dyes were available, any other colour could be mixed from them. In practice, there are limitations. Colour theory and practice are complex topics.

Subtractive colour mixing: The subtractive primary colours are cyan, yellow and magenta and when mixed together they subtract from the light producing black. When different pairs of subtractive primaries are mixed, the colours red, green and blue are produced.

Subtractive primaries: Cyan, magenta and yellow. Theoretically, when all three subtractive primaries are combined at 100% on white paper, black is produced. When these are combined at varying intensities, a gamut of different colours is produced. Combining two primaries at 100% produces an additive primary, either red, green or blue:

100% cyan + 100% magenta = blue

100% cyan + 100% yellow = green

100% magenta + 100% yellow = red

Suction-drum drier: Sieve-drum drier. Hot-air dryer for loose material, tow (fibre manufacturers), hank yarn, cloth and knitted fabrics. The material is sucked in through the perforated, rotating drum by airflow, soaked heavily and then dried.

Suede: The correct term ios Suede Leather, because it is usually calfskin treated on the inside to give a napped or sueded finish. It is smooth and attractive and can be dyed in a variety of fashin colours.Readily available for sewing and can be made in to a variety of garments.

Suede cloth: See **Suede, Suede Fabric.**

Suede Fabric: Ther is a wide variety of simulated suede or suede fabric available. Most type are fairly expensive, although advantage over real sued is that they donot take on permanent creases or show wear easily, and most of them are washable. The method of manufacture and fibre content vary. It is suitable for dresses, jackets for mn and women, skirts, childrens clothes, sports and leisure outfits.

Suede finish: See **Suede fabric.**

Suède leather: (1) Roughened Leather and is also called genuine suède leather (for example leather made from deer). It is either polished grain (e.g Nubuk) or polished velour (e.g lamb, sheep, calfskin, cow velour). Imitation suede leather Imitation suede. (2) (suède) Leather with a velvety rough or fluffy surface. The flesh side is cut and worn on the outside (for example Danish, suèdc leather). The grainy side is smoothed to obtain a special soft velvety texture (e.g Nubuk Leather).

Suede look: "Swedish Leather" with suède characteristics,for example Duvetine and rough suede jeans material with flocked variants.

Suedette: See **Velour.** This is a five or eight stem reinforced satin material. A velvety effect is obtained by strongly roughing the right side of the fabric. Another name for this is fabric is Duvetine.

Suedine: This is an imitation leather fabric with suède characteristics.

Sueding: (emerizing). This is performed on the surfaces of cloth or knitted fabrics. During emerizing the fabric is covered by a thin pile that does not damage the machine or the structure of the fabric. The effect is dependent on the structure of the yarn and the surface.

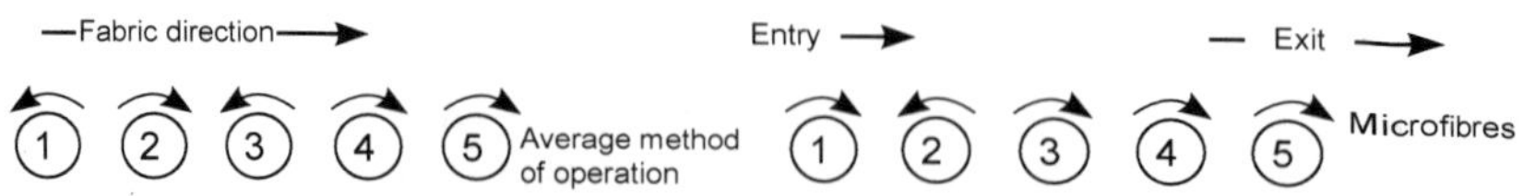

Optimum emerizing effects; direction of emerizing roller rotation relative to fabric direction

The shorter and thinner the fibre in the yarn, the lighter and easier it is to obtain the pile. The initial emerizing refining materials are also influenced by this.

Suèding machines: (emerizing machines). Material with suède leather characteristics is produced on the emerizing machine by the emerizing rollers which split and emerize the fibre. A carrier roller ensures the cloth is guided correctly under constant tension. The emery rollers which are wrapped in a spiral with emery paper are driven by a high performance motor and can be rotated (generally 950. 1200 or 1600 rpm), as desired, with or against the passage of fabric.

Suedoise: French serge, made with 8 harnesses and 4 picks in a repeat.

Suedyne: See **Velour.**

Suessen Heat Setting: Process using dry heat to set twisted yarn.

Suffolk Lace: English bobbin lace of plain patterns, the design usually outlined with a thick thread.

Sugar: (saccharide) One of a class of sweet-tasting simple carbohydrates. Sugars have molecules consisting of a chain of carbon atoms with –OH groups attached, and either an aldehyde or ketone group. They can exist in a chain form or in a ring formed by reaction of the ketone or aldehyde group with an -OH group to form a cyclic hemiacetal. Monosaccharides are simple sugars that cannot be hydrolyzed to sugars with fewer carbon atoms. Two or more monosaccharide units can be linked in disaccharides, trisaccharides, etc., by a glycosidic link. Monosaccharides are also classified according to the number of carbon atoms: a pentose has five carbon atoms and a hex ose six. Monosaccharides with aldehyde groups are aldoses; those with ketone groups are ketoses. Thus, an aldohexose is a hexose with an aldehyde group; a ketopentose is a pentose with a ketone group, etc. The ring forms of monosaccharides are derived by reaction of the aldehyde or ketone group with one of the carbons at the other end of the chain. It is possible to have a six-membered (pyranose) ring or a fivemembered (furanose) ring.

Sugar fibres: Cuban. Raw material fibre from the waste from sugar production.

Suint: Mainly from Wool fat and Wool grease suint (consists of between 8 and 58%).

Suint scouring: In this process the dusted raw wool is steeped in water at 16°C and then the liquor is withdrawn and clarified by sedimentation or centrifuging. The pH of the suint liquor is between 5.5 to 8.8. For removal of wax the suint liquor should be heated at 60°C when wax is emulsified. The material is in then rinsed, washed with soap and finally rinsed in the subsequent bowl.

Suiting lace: The term ised to describe any type of firm lace which would be suitable and durable for formal garments such as jackets and skirts. The yarns are usually thick and matt and firm such as cotton and viscose. The motifs or designs are deliberately close together to provide stability and to prevent fabric from being too see through.

Sujini: Embroidered quilt made by women of Bihar India. See **Kanthas.**

Sulfamic acid and ammonium: Combinations of these compounds also function as non-durable flame retardants.

Sulfur fibre: A manufactured fibre in which the fibre-forming substance is a long chain, synthetic polysulfide in which at least 85% of the sulfide (-S-) linkages are attached to two aromatic rings (FTC definition). The raw material is polyphenylene sulfide which is melt spun and processed into staple fibres. These are high performance fibres with excellent resistance to strong chemicals and high temperature. They show excellent strength retention in harsh environments; are flame retardant; and are non-conducting. They find use in high-temperature filter fabrics, electrical insulation, coal-fired boiler bag houses, papermaker's felt, and highperformance composites.

Sulphate Pulp: Kraft pulp.

Sulphates: A salt or ester of sulphuric(VI) acid. Organic sulphates have the formula R_2SO_4, where R is an organic group. Sulphate salts contain the ion $SO4^{2-}$.

Sulphites: Salts from the sulphuric acids (H2SO4; Sulphur oxyacids). They are either acid salts (e.g. Sodium bisulphite) or normal salts (Sodium sulphite). Only alkali salts are soluble, and as a result have an alykaline hydrolysis reaction. Usage: they are reducers, bleachers and separators.

Sulfonated: A term describing a material that has been reacted with sulfonic acid, usually to impart solubility, dyeability with cationic dyes, or other properties.

Sulfonates: A salt or ester of a sulphonic acid.

Sulphonation: A reaction used in the dyes chemistry. By sulfonation is meant the introduction of an —SO_3H group into a molecule. The operation results in a product which is usually very soluble in water, either in the form of the free sulphonic acid, as is often the case, or in the form of its salts. Of the salts, the inexpensive sodium salt is usually encountered. Sulfonation is effected: (a) With ordinary concentrated sulfuric acid (66°Be); (b) With 100 per cent sulfuric acid; (c) With fuming sulfuric acid (oleum), with the concentration of SO_3 varying from 5 to 70 percent; (d) With chlorosulfonic acid, with or without diluent; or (e) by "baking" (dry heating) the acid sulphate of an amine, often in vacuum.

Sulphonation, degree of: A percentage of the organic bonded sulphuric acid and SO_3 in Fatty alcohol sulphates and similar sulphates. Fats are water-soluble during contact with the Sulphonic (acid) group.

Sulphonic acid: Any acid containing the sulfonic group, (SO3H).

Sulphonyldiethanol: Used as a crosslinking agent for cotton. It can be made by the reaction of di-vinyl sulphone with alkali.

$$CH_2{=}CH_2{-}SO_2{-}CH{=}CH_2 \xrightarrow{OH^-} OH{-}CH_2{-}CH_2{-}SO_2{-}CH_2{-}OH$$

Sulphur dye: a class of dyes made by reacting sulfur with organic compounds; most are of unknown chemical structure Sulfur dyes are insoluble in water, and must be converted to a soluble form for application. The process is a quite similar to that used for *vat dyes*. Sulfur dyes are typically inexpensive, but dull in colour. They generally have good washfastness, but are sensitive to bleaches. Sulfur dyes on fabric, particularly some blacks, may decompose under warm, humid conditions, forming an acid. This can cause *tendering* of cellulose fibres, but can generally be prevented by making the finished fabric slightly alkaline. Sulfur dye is often used commercially to produce a good black at low cost on cellulosic fabrics.

Sulphamic acid Sufamic acid: NH2SO3H Sulfamic acid, applied to wool typically by printing, followed by baking then steaming, fixes to the wool much like an acid dye. It is colorless and prevents the treated wool from taking up other acid dye, so it is an effective

Sulphobetaines: Used for high grade finishing. Examples include Triethylsulphateamine, disodium salt.

Sulphonated fatty acids: These products are used in textile as surfactants. The are made by reaction of fatty acids and sulphuric acids in so doing, finally react (with the efflux of water) with a CH_2-group and which explains, with

Sulphonic acids, the direct bonding of carbon-sulphur (–C–SO_3H–COOH) with the fatty acids- carboxyl group. Sodium salts possess a high level of resistance to acid and alkali, form very soluble lime salts and are (for example with a mixture of Fatty sulphuric acid esters) very active in the processes of saponifying, rinsing, dyeing, softening and modifying finishes. They are also resistant to bitter salts.

Sulphonated fatty acid esters: A good detergent stable to hard water can be produced by the esterification of oleic acid:

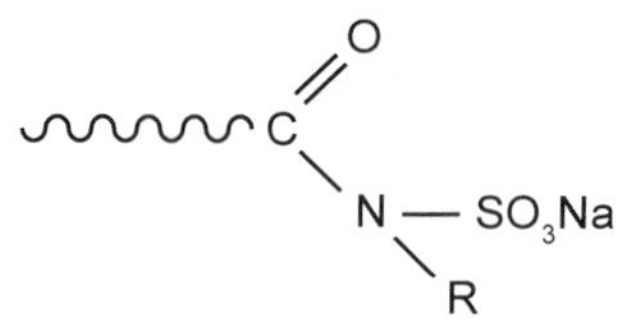

They were originally made by esterification of castor oil and further sulphonation. These products are used as wetting agents and softening agents :

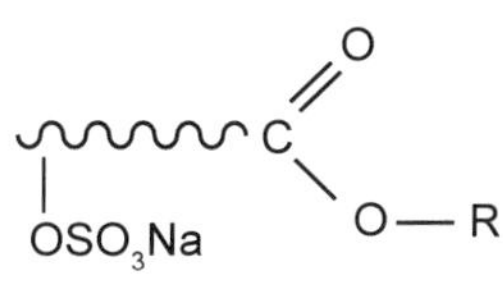

Sulphonated succinic acid esters (sulphosuccinates): One of the highly efficient wetting and detergent of high stability. They are used as mild washing agents and are degradable (by comparison to other products). Sulphuric succinic acid esters from stearic acid monoethanolamides are useful for special after washing in many countries. Washing can be performed without additional electrolytes or alkali to neutralise.

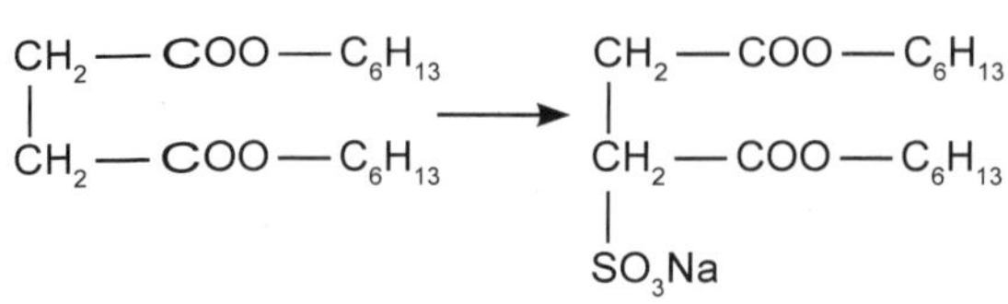

–ester *–sulphosuccinate*

Sulphonic (acid) group: **(SO_3H)**: Monovalent. Characteristic of many textile auxillaries and numerous dyestuffs (auxochrome).

Sulphosuccinamates: They are excellent foaming agents and emulsifiers of the typical formula:

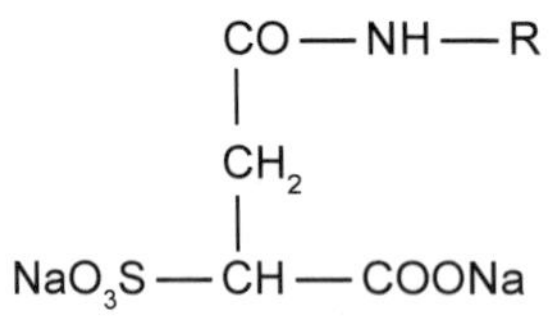

sodium sulphosuccinamate
(R = organic residue)

They are characterized by the presence of two entirely different ionisable groups (sulphonic acid group and carboxyl group) which differentiates them from other surface active agents. The consequent bifunctionality in interaction with the amide group yields specific characteristics which predestine them for use in emulsion systems. Thus certain sulphosuccinamates are used alone or in mixtures with e.g. potassium oleate sulphate or sodium lauryl sulphate as an emulsifier for latex compounds, as well as as an emulsifying and foaming additive in the manufacture of foamed carpet backing coatings.

Sulphoxylate, Sodium formaldehyde: ($NaHSO_2$ $CH_2O \cdot 2H_2O$) mostly 98–99%, stable (if protected against moisture and heat), soluble in cold water; used as: deoxidant in cloth printing (vat dyes), discharging of direct and naphthol dyeings.

Sulphoxylates, Zinc formaldehyde: (a) Primary salt [$Zn(SO_2CH_2OH)_2$], water-soluble; used for: stripping of dyeings, deoxidant in vat-dye printing, discharge printing of acetate fibres, etc. (b) Secondary salt [$Zn(SO_2CH_2O)$], not soluble in water, soluble in ethanoic acid or formic acid bath; used for: stripping of dyeings, discharge agent.

Sulphoxylate, Zinc acetate aldehyde: oxyethane sulphinic acid zinc: [$ZnSO_2(CH_2OH)_2$], difficult to dissolve in water, used for: stripping.

Sulphoxylate, Calcium formaldehyde: deoxidant for printing with vat dyes.

Sulphur dyeing: Dyeing with sulphur dyes. Dyeing is very similar to vat dyes, i.e. dye is applied in the reduced form on the substrate and oxidized to form the original insoluble form on the fibre. Na2S is usually employed as the reducing agent in the presence of sodium carbonate. There are mainly two types of sulphur dyes – Normal and modified types. Normal types should be boiled with 1 to 3 times the amount of sodium sulphide. This solution must be filtered before being added to the dye bath. Modified types, on the other hand, are water-soluble or can be diluted with water and they can be prereduced or otherwise. Method of application as per manufacturers directions. The affinity of these solutions are very less and hence taken up by the substrate slowly. Material can be padded and steamed also. Oxidation is done after rinsing with hydrogen peroxide. It can be oxidized by dichromate, hypochlrites etc which is not well practiced now due to various reasons.

Sulphuric acid: H_2SO_4; A very potent inorganic ("mineral") acid; a strong acid Sulfuric acid is used in some preparation and dyeing process, most often with wool. Concentrated acid will absorb water very rapidly, releasing heat in the process. Skin burns are caused by both this heating and by corrosive action, and can happen within seconds. See comments under acid regarding mixing. Sodium bisulfate (not bisulfite) can sometimes be used as an alternative.

Sultanabad: Medium and large size Persian rugs made with thick pile. The design consists of floral patterns in brilliant blue, red and green colours.

Sumach: Leaves and twigs of several species of Rhus, containing tannic acid. It is sold in the form of crushed leaves or as a powder (15-20% tannin).

Summer Silk: Same as **Louisine.**

Suningchow: Soft silk serge in solid colours, made in China; is about 32 inches wide.

Sunn: A bast fibre obtained from the plant Crotolaria juncea, south Asian, water-retted, coarse and hard. Used for cords and ropes.

Super Light Weight: Term used to describe a fabric used in outerwear, which allows for a minimum pack volume and weight. These lightweight, packable garments offer the most versatile weather protection. Some of these fabrics have a protection layer on the membrane, which provides durability. This means that the garments made from the extra lightweight fabrics need no separate lining.

Super Micro Fibres: The so-called supermicrofibres are finer than Microfibres by a power of ten; supermicrofibres are obtained by dissolving out the polyester "islands" in "islands in the sea" technology. They are generally of less than 0.1detex. In island in the sea technology composite fibres are spun with many super microsize (less than 0.1 dtex) inside another polymer which can be later on dissolved anf super microfibres can be separated. See fig above

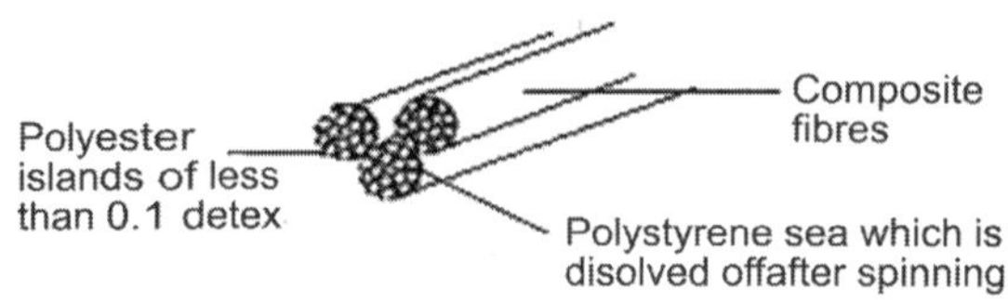

Superabsorbent: A material that can absorb many times the amount of liquid ordinarily absorbed by cellulosic materials such as wood pulp, cotton, and rayon. Absorbent compounds such as highly absorbent polyacrylate powders, can be used in the hygiene sector for the absorption of wetness which occurs suddenly.

Superba Heat Setting: Process using steam and pressure to set twisted yarn.

Supercop: (1) A wooden or plastic conical base on which weft is wound for use in a shuttle loom. (2) A weft package produced by winding a yarn on to the base defined above.

Superdull fibre: Superdull fibres are fibres with 1-3% titanium dioxide, which is used to dullen the fibre.

Superfine wool: The finest, most expensive wools used for mens suitings and luxury knitwear. Mainly from Australian Saxon Merino sheep.

Superimposed yarn layer: Bonded yarn fabrics which are arranged randomly or deliberately.

Supermicrofibres: The so-called supermicrofibres are finer than Microfibres by a power of ten; supermicrofibres are obtained by dissolving out the polyester "islands" in "islands in the sea" technology.

Supermilling: A class of acid dye Supermilling acid dyes offer moderate brightness and good to very good washfastness, but have poor to fair *leveling* tendency. Their leveling characteristics mean that extra care is required in the process to produce level results. There is no clear distinction between milling and supermilling acid dyes. These dyes are used for wool and work well on polyamide (nylon).

Supernatant: Denoting a clear liquid that lies above a sediment or a precipitate.

Superpolyamide fibres: Carothers (the inventor of the polyamide 6.6 fibre), used this name to denote all fibre-forming polyamides with a molecular weight over 10 000.

Superwash: A quality term defined by the International Wool Secretariat (IWS) for wool articles which have been finished to be resistant to machine washing.

Superwash wool: See **Super wash.**

Supewash process: See **Superwash process.**

Suples: Silk yarn dyed with only part of the gum removed.

Supported needle felt: A needled felt that is composed entirely of fibres physically interlocked and reoriented in combination with interlay, scrim, or foundation of knitted, stitched, bonded, or extruded structure.

Supukwenkin: Silk fabric similar to lustring; made in China; used for scarfs.

Surah: (1) Thread dyed warp of organzine and weft of tram with 45' diagonal weave structure. Even sided tewill with srtrongly marked slant-ribbed effect. A twilled silk similar to serge; first made in Surat, India.

(2) A soft twill fabric made from filament yarns, including silk, polyester acetate, triacetate. It is always printed, shiny fabric. It is not hard wearing, and tends to develop slippage at seams amd points if strain, and creases easily. Used for loose dresses, blouses, scarves, ties and as a lining fabric.

Surah de Laine: Fine, twilled, soft dress goods, made of silk and wool.

Surcot: Women's or men's over garment, usually sleeveless and unbelted, often trimmed with fur.

Surface activity: Surface activity of a substance is generally related to the balance between hydrophilic and hydrophobic portions of its molecule.For example, among anionic surfactants C8-C12 alkyl hydrophobes tend to be predominantly wetting agents, while the C12-C15 homologues exhibits better detergency and emulsifying properties. They have different levels of surface activitydue mto the property of the molecules described above.

Surface active agent: See **Surfactant.**

Surface charge: The electrical charge on the surface of a substance.

Surface contour: Divergence of a surface from planeness rough (high) to slippery (low).

Surface energy: (1) The free energy of the surfaces at an interface that arises because of differences in the tendencies of each phase to attract its own molecules. (2) The work that would be required to increase the surface area of a liquid by one unit area.

Surface finishing: Finishing, mechanical or chemical which is made on the surface of a fabric to enhance the quality, appearance etc. In pretreatment, special surface finishing steps such as singeing, emerizing, brushing or roughening can be applied to the dry or wet goods, in order to shape the quality. Raising, shearing peaching, polishing, sanding are other finishes which are applied at various stages of processing which can be considered as the surface finishes

Surface friction: Resistance to slipping offered by surface harsh (high) to slippery (low).

Surface print: Prints with a greater colour intensity resulting from the concentration of dye printed onto the fibre surface. This effect is achieved by using suitable thickeners, which should be in combination with semi-emulsions if reactive dyes are used. For any other type of dye, thickener combinations containing starch should be used.

Surface printing machine: The pattern is printed by means of raised relief-type printing elements. In contrast to Roller printing machines, the surface printing machine can print all wovens using the same rollers. The only difference is that print paste application to the print rollers is varied with a paste application device.

Surface reflectance: A small portion of the incident light (less than 2 % with textile materials and paper) is immediately reflected from the surface due to different optical material/air density. This portion is termed surface reflectance or residual reflectance.

Surface tension: Surface tension is defined as the interfacial tension between a liquid and its vapor. Intermolecular forces acting on the molecules at the free surface of a liquid tend to minimize the surface area of the liquid and give the surface properties similar to those of an elastic skin under tension. When two dissimilar liquids make contact, these intermolecular forces will cause the shape of the interface to change until the potential energy of the entire molecular system is at a minimum. A very simple method of measuring surface tensions is with a du Noy tensiometer. This technique measures the force necessary to pull a platinum ring away from a liquid. For pure water, the force is 72 dynes/cm.

Surface water absorption, in a fabric: The process of removing liquid water from a surface such as human skin, dishes, furniture, car etc.

Surfactant: The word surfactant is coined from the expression "surface active agent". As the phrase implies, a surfactant molecule possesses surface activity, a property associated with the chemical structure of the molecule. The characteristic feature of a surfactant molecule is its two ends attached by a covalent bond The two ends have diametrically opposed polarities. The non-polar end is lyophilic (strongly attracted to organic molecules) while the strongly polar end is lyophobic (having little attraction for organic molecules) yet strongly hydrophilic (water loving). Duality of polarity causes the molecule to align itself with respect to the polar nature of the surfaces it contacts. When used in association with dyeing, this term almost invariably refers to a synthetic detergent. Detergents operate at the surface between a solvent (water) and some material that is to be removed from where it is, and made to enter solution or suspension in the solvent. One end of the surfactant molecule is hydrophilic ("likes" water, and the other is hydrophobic ("fears" or water; sometimes lipophilic - oil loving). Surfactants can be synthesized to have specific properties by varying the structure of the hydrophilic and hydrophobic ends. Surfactants are used to *scour* fibres or fabric, act as wetting agents in dyeing, as retarders in dyeing, and to help remove unfixed dye after dyeing. They may be classified as anionic, non-ionic or cationic. There are even types that can behave as anionic or cationic, depending on conditions.

Some fabric softeners are surfactants. There is a vast array of surfactants on the market.

Suri fibre: Fine animal hair originating from a species of South American camel. From a quality point of view, it is the best type of Alpaca fibre, having an even colour shade and even linear density (approx. 25 mm, about 90–110 scales per mm).

Surikome printing: Ancient Japanese technique, a multi-coloured version of Kasuri dyeing.

Surinam: Variety of raw cotton from Guyana; the fiber is white or yellowish, lustrous and strong.

Surtout: An overcoat cut in the style of a frock coat.

Suspended growth process, in ETP: Biological treatment process in which the microorganisms responsible for the conversion of organic matter or other constituents in the waste water to gases and and cell tissue are maintained in a suspended form within the liquid.

Suspended solids, in waste water: Solids that are held in the waste water, which are not soluble. It can lead to the development of sludge deposits and anaerobic conditions when untreated waste water is discharged in the aquatic environment.

Suspended web: See **Brace web.**

Suspoemulsion: A mixture of anionic and non-ionogenic emulsifiers or anionic dispersants, with specific stabilizers and de-foamers. With a suspoemulsion, it is possible to combine biocides with varying physical and chemical properties, because it has a mixed structure.

Sutherland: A Highland tartan, composed as follows: dark green stripe, split in the center by a very narrow black stripe; black stripe, half as wide as the green; dark blue stripe, as wide as the green, split by a pair of very narrow black stripes, placed near the edges and spaced from each other and from the edge their own width; green stripe, width and split as above; dark blue stripe, as wide as above, split in the center by a single pair of very narrow, black stripes, spaced their own width.

Sutwan: Various Chinese piece dyed silk satins.

Suzeni: Embroidery Persian needlework, consisting of couched silk or gold threads.

SVCC: Schweizerischer Verein der Chemiker-Coloristen (Swiss Association of Chemical Colorists).

Swansdown: A narrow fuzzy decoration originally made from the downy brest feathers of the swan, but now more often made from synthetic fibre. Used to trim nightwear, evening gowns.

Swatch: A piece of fabric used as a representative sample of any fabric.

Sweated wool: Skin wool treated using the sweating process (sweating: controlled rotting process, the oldest biological method of dewooling). See **Mazametwool.**

Sweating: The process of removing wool from the skin, by exposing the skins, which are first soaked in water, to high temperature.

Sweatshirt Fabric: Acrylic or cotton knit fabricwith a bumpy fleecy back in a variety of plain colours and striped designs. The yarns used are often marled. Some fabrics are quite thin; all stretch easily and lose their shape. Not hard wearing. Used for sports and leisure clothes.

Swedish lace: Swedish lace is said to be very very old. Fragments of cloth woven in this structure was found in the marshes in Sweden dating from the time of the Vikings. These structures are effected by the distortion of warps and wefts. The distortion of the warp and weft in this handsome structure results from the interlacing some of the threads from time to time so that the ends/ picks slide to gether. The effect is emphasized by the use of smooth, plain, and particularly hard-twisted yarns and by washing the fabric. The manipulation is involved in the washing process allows the threads to come to the most stable configuration: Threads not held apart come together, It happens every time. Soft woollen yarns may be used to make warm and cuddly shawls, scarves and blankets. The structure of the cloth is somewhat blurred when those yarns are used, but the results are still lovely. Swedish lace needs four shafts. In a typical construction the ends carried on shafts 1 and 4 weaves plain weave everywhere.

Swelling: In textile usage, expanding of a fibre caused by the influence of a solvent or chemical agent. A property often used to facilitate dyeing.

Swelling agent in dyeing: See **Carrier.**

Swelling agent in printing: The absorbency of print pastes is frequently limited in regenerated cellulose fibres, and for this reason swelling agents are added to the print paste.

Swelling shrinkage: Results from the swelling and de-swelling of the constituent fibres of a fabric due to the absorption and desorption of water.

Swift: is an expandable cage that turns on a center rod to hold a skein of yarn so that it can be wound into a ball or directly onto the warping board.

Swim wear: Textile garments intended for wear in fresh, chlorinated or fresh water.

Swimming roller: (S roller), roller construction where the roller floats on a hydraulic cushion. The operation of the hydraulic chamber and the bearing pressure can deflect every S roller positively or negatively, or allow the flexible line of a conventional roller to be followed whilst applying even pressure. For many processes it is sufficient to operate using only one S roller in the padder or drying system. The situation is different in processes with a rapid reaction, such as before a hot flue or steamer, and during drying. Here the most important factor, as before, is that articles are smooth and identical on both sides. These requirements apply to a dye padder with two S rollers arranged horizontally.

Swiss applique: A very light, sheer cotton fabric, having small, separate (not continuous) patterns printed in only one colour. These patterns are raised and consist of finely ground cotton fibres which are stuck to the cloth with glue.

Swiss: Brussels Curtains with patterns outlined in chainstitch by the tambour machine.

Swiss Embroidery: Washable machine and hand embroidery made, mostly white over white, in Switzerland.

Swivel Weave: The Swivel weave differs from lappet in that designs are produced by extra filling yarns. Separate shuttles are placed at each point where the design has to be made. The shed is formed by the pattern, where the shuttle carries the yarn through the shed, the distance of the pattern. The extra filling floats on the back of the fabric, the long floats is cut away after weaving is completed. Example: **Silk sarees.**

Swiss mull: Very thin, bleached and dressed cotton dress goods.

Swiss muslin: Fine, thin cotton muslin, made in Switzerland; it is plain or dotted.

Swissing: Process of calendering bleached muslins between hot rollers.

Swivel Fabrics: Trade term for a variety of silk or cotton fabrics, having relatively heavy Jacquard figures or spots on a very light ground. They are used for dresses, waists, overdreslses, etc. The dots or figures are either woven into the cloth with an extra filling, floating on the back of the cloth between the different patterns, and shorn away in the finishing process, or made as lappet work the extra thread forming a trailing design.

Swivel weaving: Consists of introducing a number of small shuttles besides the fly shuttle, which produce small designs on the foundation. There is one shuttle for each figure, and they do not leave long floats. The result is similar to embroidery.

Swizzing calendars: Swizzing is a British term used to denote that the fabric runs through all of the nips at the same surface speed as the rolls. Swizzing calendars usually consist of seven to ten bowls and are run at ambient temperatures. The fabric effect is closed interstices, a smooth appearance and gloss without the high glaze characteristic of a friction calendar.

Syddo: A fairly stiff but flexible woollen; used for coat fronts in lieu of haircloth.

Symmetrical straight herringbone: See **Herringbone, symmetrical or pointed straight.**

Symmetrical 2/2 vertical herringbone: When the twill is turned 90 degrees to become a symmetrical 2/2 vertical herringbone, three ends floats appear every fourth pick. This twill can be woven on this straight threading; it cannot be woven on a straight threading because a straight threading allows the twill rib to move only in one direction across the weft as a regular 2/2 twill cannot be woven on a point threading.

Synchronisation: (Gk.: synchronos = at the same time), making two processes or machines operate simultaneously.

Syndiotactic polymer: A polymer structure in which the atoms that are not part of the backbone chain are distributed in a symmetrical and recurring manner above and below the backbone chain when the latter is in a single plane. Also see Atactic polymer, Isotactic polymer, Tactic polymer.

Syneresis: (Gk.), contraction; term applying to selfdrying in foaming; Foam performance evaluation.

Synergistic: (Gk.), concurrent influence causing increased interaction, i.e. more than the additive effect of a combination at a specific ratio, e.g. in the case of well-proportioned mixtures of surface-active substances. See Synergistic effect.

Synergistic effect: Effect that causes a combination of two surfactants for example (or other substances) to achieve a specific effect even at a low concentration within certain interactively defined concentration limits in a solution. The concentration is lower than the mix-ratio concentration that corresponds with the linear dependency for this effect. Any substances involved in this effect can also be compounds/mixes.

Syntan: Synthetic tanning agent There are many syntans, and many are proprietary mixtures of chemicals. They are sometimes used as postdyeing treatments for wool or nylon to increase washfastness.

Synthetic dyes: This term can be used to mean man-made dyes (also covers the old-fashioned aniline and tar dyes). They are manufactured using chemical synthesis. As opposed to Natural dyes, which are obtained from plants or animals, synthetic Dyestuff is manufactured on the basis of a chemical reaction involving basic material that was originally obtained from mineral oil (formerly coal).

Synthetic felt: This is a nonwoven fabric, fairly thin but with an interesting texture. It dyes well and is available in a range of bright clear colours. It has no grain but does not stretch. Used for decorations, appliqué etc.

Synthetic fibre: (synthetics) Fibres spun using various different processes from high-molecule, chainstructured substances manufactured synthetically, composed of low-molecule modules of the following main synthetic fibre groups: (a) Polymer fibres, e.g. polyacrylonitrile, polyethylene, polyvinyl chloride. (b) Polycondensation fibres, e.g. polyester, polyamide. (c) Addition polymer fibres, e.g. polyurethane.

Synthetic filaments: Dated term for Elementary filament made from synthetic fibres, Filament; Capillaries; Fibrils, single end threads; but also for Filament yarn.

Synthetic filament yarn: Filament yarn made from synthetic fibres.

Synthetic finishing Agents: Synthetic colloids.

Synthetic leather: (synthetic suede), term for Imitation suede; Artificial leather banned on the grounds that the term is anti-competitive.

Synthetic monofilament: Monofilament made of synthetic fibre with $\varnothing > 0.1$ mm.

Synthetic resins: synthetic resins are substances and preparations that can react primarily with themselves in coating and impregnation, or steeping, such as

– melamine resins,

– urea/formaldehyde resins,

– phenol resins,

– polyester resins.

Monomers that are not able to react with themselves are not categorized as synthetic resins. Use: auxiliaries in all aspects of textile finishing, for resin finishing, wet crease-resistant, hydrophobic, slip-resistant and antisnagging finishes. Permanent handle, stiffening, weighting, backcoating finishes; hat stiffening; permanent embossing, Schreiner, chintz effects; sizing; coating; laminating; permanent impregnation; nonwoven fleece impregnation; flocking adhesive; binding agents for pigment printing, etc.

Synthetic rubber: Typical examples are butadiene polymers, chloroprene polymers or mixed polymers (Polyblends), e.g. made from ethylenepropylene diene, isobutene-isoprene, butadiene-acrylonitrile, butadiene-styrene or sulphochlorinated polyethylene. : Butadiene-acrylonitrile-rubber; Styrene-butadiene rubber; Epichlorohydrine rubber; Isoprene; Polyester elastomers; Polyurethane rubber; Silicone rubber.

Synthetic suede: Synthetic velours.

Synthetic thickener: Weakly crosslinked copolymers of oleofinic monomers containing carboxyl groups with a high molecular weight, which act as polyanionic bodies in the range of pH 7–10 as thickeners with similar rheological properties to an Emulsion thickener.

Synthetic velour: A term covering the following: synthetic leather knit (synthetic leather), imitation suede, imitation leather, microfibre whirled pile, synthetic leather, synthetic suede, suede imitation. "Leather imitations and synthetic leather" is the only generic term considered by certain standards.

Synthetic washing agents: Syndets, i.e. washing agents manufactured from synthetic detergents.

Synthetic waxes: (artificial wax, man-made wax), used as additives in sizing, finishing, waterproofing, etc. Polyethylene glycols (solid), Montan wax or paraffin oxidation products (Fatty acids, synthesis of) in the form of esterized (and natural) fatty acids and alcohols, with or without further additives, also as soft and hard waxes with widely varying melting points (approx. 70–100°C) are used as synthetic waxes.

Synthetics: This applies to (1) Synthetic fibres. (2) Synthetic detergents.

Synthetic detergents: See **Detergents.**

Synthetic weighting: The monomer used for synthetic weighting is often derived from acrylic or methacrylic acid. The silk weighting with acrylonitrile and methymethacrylate has been studied and described thoroughly; in this process, starters are formed by a redox system based on iron salts (Fe++) and hydrogen peroxide, persulphates and other substances.

T

T: Abbreviation for temperature.

TA: Triacetate fibres.

Talc: (French chalk) An extremely soft greenish or white mineral, a hydrous silicate of magnesium ($Mg_3Si_4O_{10}(OH)_2$). Purified it is sold as talcum powder. It is also used as a lubricant, an electrical insulator, a filler (in paint, rubber, and paper) and an ingredient of some ceramics.

T.S.P (trisodium phosphate(V)): (sodium orthophosphate; Na_3PO_4) A white solid prepared by adding sodium hydroxide to disodium hydrogen phosphate. On evaporation white hexagonal crystals of the dodecahydrate ($Na_3PO_4.12H_2O$) may be obtained. These crystals do not effloresce or deliquesce. They dissolve readily in water to produce alkaline solutions owing to salt hydrolysis. Trisodium phosphate is used as a water softener.

Tab, in garment making: A bagged out piece of fabric that is buttoned or studded into place. In can be used on pockets and fastenings or to control fullness.

Tabard: Men's knee or ankle length closed outer robe, draped in many folds, usually unbelted, used ancient times.

Tabaret: Stout, fine silk drapery fabric with alternate stripes of satin and moire in different colours.

Tabbinet: (1) Fine drapery poplin of silk warp and wool filling with moire finish; (2) Also a thin moire taffeta lining.

Tabby: (1) British equivalent of moire; (2) A thick and coarse taffeta or worsted fabric with moire finish; 3. Cotton velvet, made with weft pile and plain ground.

Tabby sample, in tyre cord: The section of the tyre cord fabric between two tabbies that have been woven separately with a distance of 0.5-1.0 m.(18-36 in.) between them.

Tabby, in tyre cord: A short length of plain weave fabric normally woven at each end of a roll of tyre cord fabric for the purpose of supporting and holding the cords in proper position with respect to each other.

Table linen: Fabric used on dining table like Table cloth, napkin etc. It can be white, plain dyed , printed or even with designs,

Table printing: Flat screen printing. High-quality, multi-coloured prints, e.g. in small runs on silk (e.g. scarves), are printed on tables with a moving film printing carriage.

Table raising/polishing/shearing machine: Raises, polishes and shears in a single run. Both sides of the material can also be processed in a single run without the use of a turning device (back-to-back). An electrically-heated polishing cylinder with beaters ensures lustre. The shearing table can be replaced by a roller for heavily-coated or reverse-side-coated articles for example.

Tablet weaving: Thin Persian knotted carpets with short, smooth woollen tufting. Precise, richly decorated, fine patterns in light colour on background colours of sand, red, matt blue and natural white. Also produced as medallion and hunting carpets. 220 000–250 000 Persian knots per square metre.

Tabriz Rugs: Usually large size Persian rugs with cotton warp; the short and very close wool pile is tied in Ghiordes knot. The favorite design consists of a large center medallion with curved outlines and fine floral and animal patterns. Often several small medallions are used with inscriptions.

Tachometer: Revolution (speed) counter for measuring revolutions of machine shafts, rollers, drums etc. Produced as tachometer for measuring number of revolutions per minute and as an automatic counter for any other period of time. Available in various versions for manual use (checks), fixed installation (works checking of pieces or meter count), the latter also provided with electrical contacts, and registration unit etc.(Oscilloscope).

Tack: Condition of adhesive when it wets the surface sufficiently to form a bond.

Tack, for rubber and rubber compounds: A property that causes two layers of these materials, when pressed together, to adhere at the area of contact.

Tack, in testing rubber coated cords: The force required to separate two identical test specimens under controlled pressure and dwell time.

Tack, in glass tyre cord: The force requires to separate two identical test specimens under controlled pressure and dwell time.

Tack tear: The measurement of the resistance of a coated fabric to tearing under conditions simulating an installation that has been tacked in place.

Tackiness: The property of being sticky or adhesive.

Tackiness index: Measure of speed of action of a paper printing paste. This is determined by means of an inkometer or Tackoscope and depends largely on the amount of binder contained. The tackiness index is a relative figure which is influenced by the experimental set-up.

Tacking: The slanting basting stitch or "tacking" is used in dressmaking for holding linings. The needle is pointed towards the worker. Even basting is used for holding several thicknesses of cloth and if the garment is to be fitted, the stitches should be placed rather close. Uneven basting is used for hems and seams to be machine stitched. Several short stitches with one long one are used to baste crape and wiry fabrics, for this method holds them better than stitches of equal length

Tacking cut: Small holes or cuts along the selvedge of the cloth.

Tactel: A family of filament nylon (of type 66) Yarns from ICI Fibres. It is particularly suitable for making into woven and knitted fabric for sportswear.

Tactic polymer: A polymer whose molecular structure exhibits regularity or symmetry of non-backbone side groups rather than random ordering. Also see Atactic polymer, Isotactic polymer, and Syndiotactic polymer.)

Tadpole Eponge: Made of several plain ends alternating with one loop yarn and of plain filling, the knots being scattered irregularly over the surface.

TAED: Short form for the chemical compound N,N,N,N,-tetra-acetylethendiamine $C_{10}H_{16}N_2O_4$) and is used as an activating agent for bleaching agent (usually perborate) in the detergents. It is manufactured by converting ethendiamine with anhydride acetic acid and having a formula

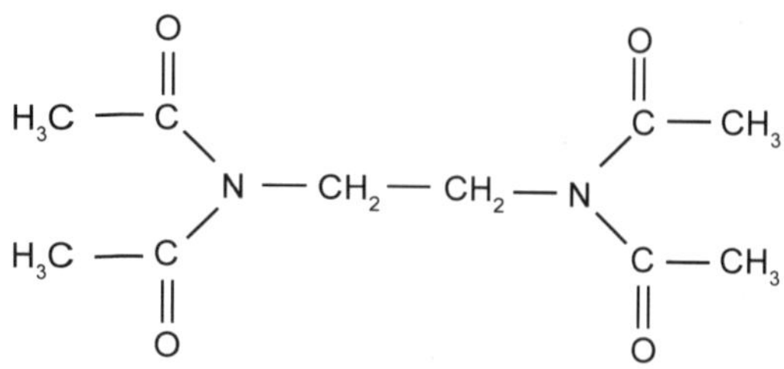

and colourles, odourless crystals having melting point 156–157°C. Easily dissolved in ethanol and a little ice vinegar, less easily dissolved in benzene, difficult to dissolve in petroleum ether. TAED activation of perborate bleaches has proved particularly useful in practice. This process is accelerated in relation to the temperature by the presence of TAED. At moderate temperature perborate and TAED react with each other in the washing bath to form intermediary peracetic acid by residual acyl which has separated from the

TAED combining with H_2O_2. As peri acetic acid is an unstable compound it disintegrates immediately:

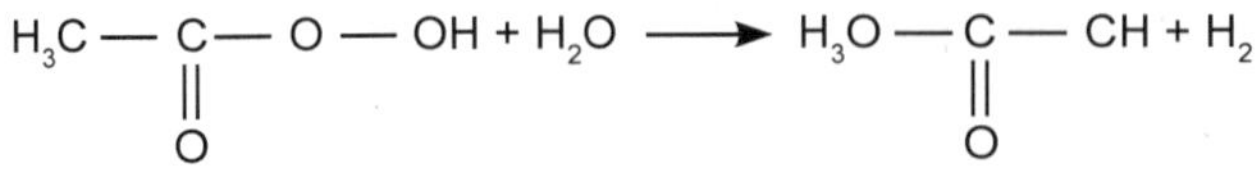

Thus in the lower temperature range TAED produces a significant bleaching effect in the washing bath which at 60°C corresponds virtually in terms of performance to the bleaching effect achieved by perborate at boiling point. The TAED-Perborate bleaching system kills off microbe germs at 40°C and therefore produces hygienic washing results.

Taffet: Taffeta.

Taffeta: A smooth plain weave fabric made from even from yarns of light or medium weight, usually from shiny filament fibres which give the fabric sheen. Taffetta is characteristically crisp and is usually plain coloured but can be printed. Originally a silk fabric but now more likely to be composed of acetate, triacetate, nylon or blends of these. Made in plain or tabby weave, the

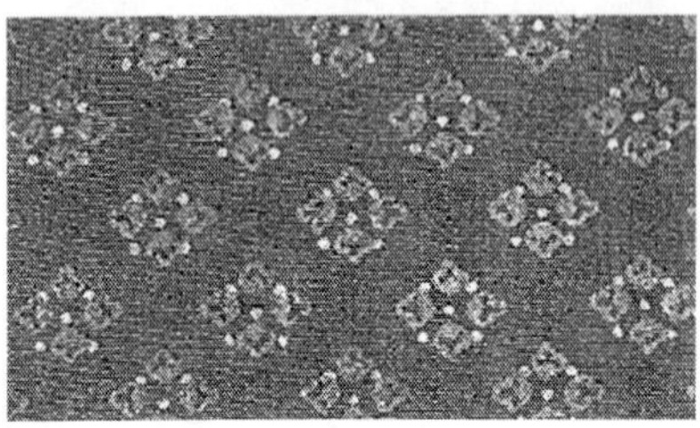

textures range from 60 × 50 uto 200 × 60, will always have more ends than picks per inch. One all rayon taffeta can be made of 150 denier warp and weft with texture of 70 × 46. Another construction may be 90 × 60 with 100 denier rayon in the warp and 150 denier rayon in the weft. It is not har-wearing, hence used mainly for evening wear stiff petticoats, lampshades and drapes, and small items such as cummerbunds, artificial flowers, umbrellas, evening bags, stage costumes and linings.

Taffeta (Silk): Organzine warp (possibly slightly weighted degummed), fabric has a papery handle , with typical rustling. As a rule it is woven with dyed yarns, as further treatments especially at high temperature may cause the fabric to loose its characteristics. In some cases these fabrics are very crease prone.

Taffeta, Faille: Made with a crosswise rib weave. Has a distinct rib effect and is usually quite heavy and firm. Paper Taffeta Plain weave, very light in weight and treated to give a crisp, paper-like finish. Shot Taffeta Usually plain weave, woven with one colour in the warp and another colour in the filling, which gives the fabric an iridescent look. If fabric is moved in the light this colour changes. Silk version of chambray.

Taffeta Flannel: A lightweight, unshrinkable wool fabric, made in plain weave with coloured stripes and checks; used for sporting shirts.

Taffeta Prismatique: Lustrous, French all-silk taffeta. The warp is coloured in the various shades of the rainbow. The filling- is white.

Taffeta, Tissue: Plain weave, very light weight and transparent. Warp-print Taffeta, usually a plain weave, the warp yarns are printed before the filling is inserted. The fabric has a very fuzzy design when design is distorted as fabric is woven.

Taffeta Weave: See **Plain Weave**: Plain weave in silk ware. This is the simplest and oldest method of interlacing. The odd numbers of warp-threads cross the even numbers after every pick; hence of two warp-threads one will always go over the first pick and under the second, and the other end under the first and over the second pick. Taffeta cloth, therefore, has the same appearance on both sides, and in cotton and wool weaving this weave is technically—and properly indeed—called the Plain Weave. It has the smallest repeat, 2 warp-threads and 2 picks, and the exchanging of warp and filling is the most frequent possible. The cloth thus produced is firmer and stronger than that obtained with any other weave.

Taffetine: Plain woven, lightweight lining, made with closely placed organzine warp and coarser cotton, linen or silk filling. It is slightly stiffened.

Tafta: Persian plain woven, rich silk fabric, made of hard spun, ply yarn.

Tag Locks: Large locks of britch wool clotted with dung and dirt.

Tags: Broken or dung-covered wool and other wastes that are swept from the floor of shearing areas.

Tagging: The practice of cutting the dung locks off sheep. Usually this operation is done immediately prior to shearing, and it may be done prior to lambing.

Tahiti: Cotton of the Sea Island type, the staple is good, silky, but irregular in length; contains a large percentage of unripe fibre.

Tahkli: A small, metal-whorl supported spindle.

Tahuari: Native Peruvian name for a thin, fibrous bast of the Couratari tree; used for clothing, blankets, etc.

Tailing: Gradual transition from dark to light shades of the same colour tone in textile printing and dyeing. In the course of this transition one or both edges (listing, side-to-side shading) or the end (ending) of a batch demonstrate lighter or darker colouring as compared with the centre of the fabric.

Tailing and listing: In textile printing and dyeing the gradual transition of a colour shade from dark to light in textile printing and dyeing. See **Tailing**.

Tailors canvas: Buckram; Linen interlining.

Tailors chalk: This is used for marking measurement and pattern details on fabrics. It usually consists of finely pulverised clay or in the form of small plates cut with sharp edges (coloured ones with the addition of mineral dyes), also made of soapstone and talcum.

TAK dyeing: Serves to produce multi-colour effects on carpets. Background colour is first applied to a strip of carpet up to 5 m wide by first dipping/squeezing. Then up to four further batches of dye are floated over the width of the carpet strip using special equipment. To fix the dye, the carpet strip is subjected to steaming and then rinsing processes.

Take-up motion: Loom movement in which the weft is added to the main body of the fabric.

Take up, in fabrics: The difference in distance between two points in a yarn as it lies in the fabric and the same two points after the yarn has been removed from the fabric and straightened under a specified tension, expressed as a percentage of the straightened length.

$$\text{Take up} = [\,(l_y - l_{fab})/\,l_y] \times 100.$$

Take-up, twist: See **Twist take-up.** The change in length of a filament, yarn, or cord caused by twisting, expressed as a percentage of the original (untwisted) length.

Take-up, yarn: See **Yarn take up**. The difference in distance between two points in a yarn as it lies in a fabric and the same two points after the yarn has been removed from the fabric and straightened under specified tension, expressed as a percentage of the straightened length. In this sense, take-up is contrasted to the crimp of a yarn in a fabric, which is expressed as a percentage of the distance between the two points in the yarn as it lies in the fabric. Take-up is generally used in connection with greige fabric.

Take-up Roller, in openend spinning machines: A pair of closely set, continuously rotating cylinders which withdraw the spun yarn from the rotor.

Talanche: Plain or striped coarse cloth of flax and wool in France; used for garments by the poorer classes.

Talcum: (light white, magnesium silicate, talc), $3MgO \cdot 4 SiO_2 \cdot H_2O$. Density 2.7–2.8. Flaky or crystalline substance, shiny appearance like mother of pearl, (non) transparent, colourless or light white in colour, soft and greasy to touch, not soluble in water. It is characterised by its property of rapidly absorbing moisture.

Talitan: Chinese cotton rugs with overcast edges; used as bed covers.

Tall oil: Tall oil is a by-product obtained from converting wood pulp to paper. The black liquor contains a mixture of triglycerides (about 50%) and rosin (about 45%). The triglyceride portion becomes an important source of fatty acids after it is properly refined.

Tallow: Tallow is the hard fatty deposits in beef and mutton. It contains of 30% palmitic, 14% stearic and 48% oleic acid. Its hard, solid, physical state comes from the high saturated acids content. It is an important source of many softeners because of its abundance and low price. Tallows are often hydrogenated to increase the saturated acid content giving rise to a class of materials called hydrogenated tallow.

Tallow soap: Sodium stearate. Derived from congealed fish-oils. Utilized inter alia as a substitute for Marseilles soap, in particular for soap soda rollers. It is suitable for washing temperatures of 40–60°C.

Tamaito: Japanese term for a grade of silk waste obtained from the duppions.

Tambour: (1) The narrowest size of passing embroidery thread; (2) Embroidery, having- the design executed in in chain stitch on a machine made net ground, with the help of a hook. (3) Smooth, hollow roller (printing cylinder; presser roller), sometimes bombarded with rubberised wool or woollen felt fabric in Rouleaux print. Driven by means of pressure rollers. (4) The central, large cylinder on raising machines and in spinning machinery (carding, combing machines).

Tambour muslin: An open and clear muslin; used for embroideries, curtains, etc.

Tambour Work: Tambour work is of Eastern origin, consisting of embroidery in chain and other stitches over a sheer material stretched in a frame.

Tamet: Woven Term in England, denoting fabrics woven both sides alike and without a wale.

Tamis: Plain and open woven and very smoothly -finished worsted; used for sieves.

Tamise: (1), originally an English, all-wool or silk mixed open face, light fabric. (2), French silk dress fabric, made with satin stripes on a sheer, plain woven ground; (3), lightweight, thin, plain woven woollen dress goods with a corded face. See Marquisette.

Tammies: Twilled, highly finished fabrics of worsted and cotton; used for drapery, etc.

Tangent modulus: The ratio of change in stress to change in strain derived from the tangent to any point on a stress-strain curve.

Tangent modulus: The ratio of change in stress to change in strain derived from the tangent to any point on a stress-stain curve.

Tanglelaced fabric: See **Spunlaced fabric**.

Tanner Dirt contaminated scum formed when scouring woollen goods in the scouring or milling machine.

Tanners' Wool: Is removed from the skin of slaughtered sheep through lime.

Tannic acid: A mixture of compounds derived from oak bark, nutgall and other natural sources; no well-defined chemical composition Tannic acid treatment, followed by treatment with tartar emetic, has been used to improve the wash fastness of dyed wool or nylon. Syntans are now used for this purpose.

Tannic acid/tartar emitic after treatment of Polyamide: Dyeing carried out with acid and metal complex dye substances has improved wet-fastness characteristics (water, sweat and washing fastness). Work process: Rinse after dyeing and treat with 2% tannin and 1% acetic acid (80%) for 15 minutes at 70°C. After the tannic acid has been absorbed, add to the same bath 1% cracked wine lees (potassium antimony(III)-oxide of tartrate) solution and continue treatment at same temperature for 15 minutes and rinse.

Tannic acid/tartar emitic mordant: Dyeing stain with sufficient uptake and fixing of cationic dyeing substances on cellulose fibres, particularly for medium to dark colourations. Used before dyeing with cationic dyestuffs in the dyeing of cellulose, wool and silk.

Tannin: (gall tannic acid, tannic acid). A compound used for tanning earlier days because of its capacity to form insoluble substances with metal salts. Now, its uses are limited. A light coloured powder (Density 1.35) which can dissolve in one part water and two parts alcohol, glycerine and related solvents. It has a strong tanning effect; 0.5% solution precipitates glue and gelatine; in combination with metal salts (iron, copper, mercury, lead) it forms insoluble or barely soluble tannins and in combination with cationic dyeing substances it forms insoluble colour varnishes.

High-quality tannin contains 65–76% tannic acid (best quality 75–80%), Used for staining (superseded by Thiophenols) for cationic dyestuffs (particularly water tannin); for colour varnish fabrication (colour fastness increase as an after treatment); decrease in the affinity of wool against certain colour substances.

Tanning: Process for refining animal skins or pelts to make leather. It is carried out using Tanning agents and works on the principle of starting with tannage from weaker solution and finishing with concentrated tannage solutions.

Tanning agents (Tannic acids): Chemicals used for tanning animal skins. They are normally soluble in water or forms colloids; it prevent rotting; precipitate glue and protein solutions and convert animal skins into leather; with lead salts they mostly result in insoluble precipitates.

The presence of numerous phenolic-OH groups in the molecule is characteristic.

Tannin stains, removal of: (1) Warm glycerine; (2) 20% solution of tartaric acid (warm/hot); c) Hot isopropanol.

Tapa cloth: A non-woven cloth, made in the South Seas, of beaten bark fibres of the mulberry tree. Not often used outside the Pacific Islands.

Tape: (1) A narrow, woven fabric not over 8 inches in width. (2) In slide fasteners, a strip of material, along one edge of which the bead and scoops are attached, the bead sometimes being integral with the strip. Also see **Slit tape, Non-elastic woven tape**.

Tape ends, in zippers: The tape extending beyond the stops at either or both end of the stringers or to which continuous monofilament elements are attached.

Tape fibres: Spliced, slit or split fibres. Fabricated of sheet materials, Tape yarns or flat threads, cut lengthwise or through lengthwise split fibrils, and cut into staple fibre lengths.

Tape yarn: Yarn of a flat, tape like character produced by splitting an extruded film.

Tape yarns, stretched: Stretched Tape yarns in degrees of fineness ranging from 300–3000 tex.

Tape width, Exposed in Zipper: See **Exposed tape width.**

Tape, in textiles: A narrow fabric with a mass per unit area of less than 0,5 kg/m^2 (0.1 lb/ft^2 or 15 oz./yd^2 or 510g/m^2) or 26 lb./100 yd per 1 in.(25.4 mm) of width and which is used primarily for utilitarian purposes.

Tape, in Zippers: A strip of material along the edge of which the bead and elements are attached.

Tape, elastic: See **Elastic tape.**

Taped: Two or more separate warp ends run through one heald and woven as one.

Tapestry: A highly ornamental fabric, woven on a jacquard loom, which has an embroidered look. Cotton and worsted yarns are used in many colours, usually showing a picture. A heavy fabric used for upholstery, curtains and bedspreads. At the present tapestry is made either on high (vertically stretched) or low (horizontal) warp, the principle being the same in both cases. The stout warp is stretched, from 8 to 22, within an inch space, and the patterns worked from the wrong side by means of small shuttles. In the design only the filling is visible. Each weft extends continuously only the width of that particular coloured field, the edges of these different coloured fields being properly interlaced with each other to form a continuous fabric. When in use the tapestry is hung with the filling running vertically. Tapestries are ornamental textiles, used mostly for covering walls, curtains and also for upholstery. They are distinguished by the style as verdures, gothic renaissance, etc., and by the origin. Since the 12th century Arras, in France, also Brussels and Lille, were the most famous places for tapestries, succeeded by the Gobelin, Savonnerie, Beauvais and Aubusson tapestries since the 17th century.

Tapestry Back: A single shed back, given to carpets in order in increase their wearing quality.

Tapestry Carpet: It is made with three sets of warps, one forming the loop pile, and only one frame. The pile warp is printed before weaving with the desired pattern in any number of colours, this being the difference between the tapestry carpet and Brussels carpet, although the two are similar in appearance.

Tapestry needles: These hand sewing needles are stout needles with a long eye and a blunt point. They are used for pulled and drawn work and silk ribbon embroidery.

Sizes 13 to 28.

Tapestry Stitch: Similar to gobelin stitch.

Tapestry velvet carpets: A cut pile carpet woven from printed pile warp or single frame of yarn. It was traditionally woven on a tapestry carpet loom with bladed wires.

Tapestry weave: A weft-faced plain weave where the weft threads are packed closely together so that the warp is hardly seen. Tapestry is often associated with the large pictorial wall hangings of medieval and later Europe. However, strictly speaking, tapestry is a distinctive woven structure- a weft-faced plain weave with discontinuous wefts.

Tapeta: Carpets and rugs of ancient Egypt, mentioned by Homer; some made with linen warp and woollen weft.

Tapis: French for 1, carpet; 2, several fancy Oriental fabrics.

Tapisserie: French for tapestry.

Tapisserie d'Auxerre: Consists of net embroidered with soft wool yarn in satin stitches, forming geometrical designs.

Tar stains, removal of: First remove the dark (dark brown or Black) stains with the following chemicals and the again rework the left out yellow stains with the solvents given afterwards. (a) benzene; (b) benzene/carbon tetrachloride (or tri-, tetrachloroethene) 1 : 1, possibly with glycerine (= dye protection) in equal parts; (c) carbon disulphide or acteone; (d) chloroform (ether)/carbon tetrachloride; (e) benzene/carbon tetrachloride 1: 5 with diethyl ether/ethanol. If this does not produce satisfactory results, soak with concentrated grease-solvent soap (possibly overnight). Rework yellow stains: (a) 3% ammoniac; (b) (silk) 20% sodium tetraborate solution, possible reductive post-bleach; (c) particularly for brown residues: Mixture of 75% oleic acid and 25% carbon disulphide, rinse with pure solvent.

Tarare: (1) French, pure or cotton mixed linen of good quality, used for curtains; (2) French hemp canvas, unbleached; made about 27 inches wide and used for furniture covering.

Tare: The mass of all external and internal packing materials (including bobbins, tubes etc.) of a case, bale, or other type of container, varying from less than two to over five per cent of the total weight.

Tarlatan: An open, plain-weave coarse cotton with a starched finish, resembling coarse net. Highly inflammable. Used in millinery and for stiffening belt as well as for extra stiff petticoats or bustle effects. Often used for stage costumes.

Tarlton: Plain woven, very open and light cotton fabric, dyed in the piece. Used for dresses, etc. See **Tarlatan**.

Tarmate: Waste silk, obtained from stained or imperfect cocoons.

Tarpaulin: Water-resistant fabric used to protect loads or materials from the elements. May be a coated fabric, a fabric with waterproof finish, or a fabric that is tightly constructed to prevent water penetration.

Tartan: (1) Authentic Tartan designs belong to individual Scottish clans, although they are now worn by many other people. The cloth is woollen or worsted in twill weave, each tartan is an elaborately coloured check design. The traditional garment is the pleated kilt, but tartan cloth is also used for trousers, shawls and fashion garments. Elaborately checked fabric is available which is not authentic tartan, often made from acrylic and blends of wool and acrylic or cotton. The weight of the cloth varies considerably; handle according to fibre content and thickness. (2) woollen or worsted dress goods woven in twill or basket weave in plain patterns, usually in blue, green, red and yellow colours; (3) In Argentine, Uruguay and Paraguay flannelettes, woven with plaid patterns and napped on both sides. Some of the better grades are also made of wool.

Tartan checks: See **Tartan.**

Tartan fabric: (plaid fabric). Does not denote origin but is a generic term for all boldly-checked fabrics with predominantly green, red and blue colour shades. Woollen tartan fabrics can be fabricated from carded yarn or worsted filaments; important is the strong patterning.

Tartan

There are distinct summer and winter weights. The most frequent weave is the 1:4 twill, in same-sided twill as well as Plain weave. Worsted tartan fabrics are usually sheared; those made of carded yarn have a clear surface to lightly milled, occasionally also finished with long combed finish, causing the colours to flow delicately into one another. Used for dresses, skirts and coats.

Tartanella: Tartan plaids made of wool mixed with cotton or linen.

Tartar emetic: Antimony potassium (or sodium) tartrate This was once commonly used with tannic acid to improve the wash fastness of dyed wool or nylon.

Tartaric acid: (dioxy succinic acid), $C_4O_6H_6$ molecular weight 150; density 1.759. Salts = tartrates. Tartaric acid is a strongly dissociated acid (1% solution pH 2.1; 10% solution pH 1.6). Has a harmful effect on fibres when they are subjected to solution steam; reducing agent. Softening of textiles and silk scroop fabrication (silk, viscose); preservation agents (finishing agents for protein-based substances); spotting agents (lightening, freshening, etc.).

Tasar silk: See **Tussah silk.**

Tassel: Tufts of cotton, wool, silk, metal or chenille yarn, with a more or less ornamental head and a long, open or looped end. The finer qualities are used for dress and coat trimmings, while the heavy grades are used on curtains and upholstered furniture.

Tassel Cloak: Semi-circular shoulder cape closed at the front by a cord or chain on a pair of decorative plates (tassels).

Tassel Stitch: In Berlin wool work for making fringes and is a variety of the plush stitch (see); it is worked with a mesh and the wool doubled.

Tatami mats: Smooth floor covering, frequently used in Japan, made of Tatami straw. This is derived from Tatami grass, a type of rush cultivated in Japan, China and Taiwan.

Tato: Abbreviation of the German Tagestonne dayton(s), tons/day.

Tattersall: Large, loud check woollen cloth. Often in black and white with another colour. Often worn on race course. Used for overcoats, caps, hats, and capes for men.

Tatting: A lacy work of varying coarseness, depending on the thread used (usually crochet thread) It is worked using a shuttle with thread wound on to it, and using the fingers of the other hand in conjunction with it. Usually used only for edgings, motifs or table mats.

Taunton: A medium and coarse grade of English broadcloth, weighing 11 oz. per yard; was made since the 16th century.

Taunton Serge: 18th century worsted serge in England; worn by women.

TCF Process: (textiles continuously formed). Direct technique for fabrication of spun fleeces and spot bonded nonwovens fabricated of viscose. Spun threads contain the components hydrozyl methyl cellulose xanthate (HMCX), cellulose and caustic cellulose in concentrated ringed layers. The thermoplastic component is HMCX, which has a sticking effect during hot calendaring of the spun fleece and is transformed into cellulose during the process. A bonding agent is not required.

Tcharhad: Little square hand knotted rugs in Persia.

Tchechen Rugs: See **Chichi rugs.**

Tchembert: White or fancy cotton muslin in Turkey.

Tcherkess Rugs: See **Circassian rugs.**

Tcheutche: Closely woven, very soft Chinese washable taffeta. It does not crease and is used for garments.

TDI: Abbreviation for toluene diisocyanate.

Tear: Percentage of tops to noils. A 4:1 tear would refer to wool that had 20% waste.

Tear drop, in woven fabrics: A fabric condition characterised by Short elliptical deviations of one or more adjoining picks. Tear drops are most pronounced in taffetas and grosgrain weaves.

Tear resistance, in textiles: The resistance to the tearing force applied on a material.

Tear strength: The force necessary to tear or tear through a strip of fabric in the direction of its length. Measurable with special clips on the tensile strength tester, where values can be read directly from a scale.

Teariness: See **Tear drop.**

Tearing energy: The work done in tearing a material.

Tearing force: The force applied to propagate a tear initiated under specified conditions. It is the work done in tearing the specimen divided by twice the length of tear.

Tearing strength: The force required either (a) to start or (b) to continue or propogate a tear in a fabric under specified conditions.

Tearing strength, in fabrics: The capacity of a material to with stand the ultimate tearing force required to propagate a tear after its initiation.

Tearing tester elemendorf: Used to establish initial tearing and further tearing values (= Elmendorf values) of particularly highly refined cotton fabrics by applying a strain with pendulum stroke testers.

Tease: To open and disentangle fibres prior to carding.

Teasel: (1) Process to raise the nap, especially of woollens, by scratching the cloth.

(2) A plant with a large prickly head (Dipsacus fullonum) used for raising or fulling fabric.

Teasel cards: Dry teasel plant heads for rod teasel machine and roller card raising machines. See **Raising machines.**

Teasel Cloth: Another name for nap faced fabrics.

Teazer: See **Willey.**

Technical grade: A term applied to chemicals sold for general industrial purposes. Technical grade chemicals are usually somewhat less pure than "reagent grade" or "analytical grade" chemicals, and are much cheaper. Technical grade is now rather rare in small packages from laboratory chemical suppliers. Package sizes from industrial chemical vendors often range from about 25 kg or 50 pounds up to rail car lots. "Tech" grade chemicals are very suitable for textile dyeing. Also see grades of chemicals.

Technical textiles: Frequently used term for textiles used for technical purposes, for preventive or auxiliary functions in consumer and utility goods as well as in the production of goods and services (Industrial textiles). These are textiles, whose fibre substance and form, texture or weaving technique, impregnation, coating, rubberising or composite construction render them independent technical products, are used in technical products, or demonstrate technical or special industrial properties. Another definition is textile materials and products manufactured primarily for their technical and performance properties rather than their aesthetic or decorative characteristics'. 2 groups of technical textiles:

- *Agro tech:* Agriculture, Aquaculture, Horticulture and Forestry.
- *Build tech*: Building and Construction.
- *Cloth tech:* Technical components of footwear and clothing.
- *Geo tech*: Geotextiles and Civil engineering.
- *Home tech*: Technical components of furniture, Household textiles and Floorcoverings.
- *Indu tech*: Filtration, conveying, cleaning and other industrial uses.
- *Med tech*: Hygiene and Medical.
- *Mobil tech*: Automobiles, Shipping, Railways and Aerospace.
- *Oeko tech*: Environmental protection.
- *Pack tech*: Packaging.
- *Pro tech*: Personal and property protection.
- *Sport tech:* Sport and Leisure.

Techum fibre: Palm fibres (Brazil), white and solid, similar to flax, sometimes coarser, high moisture absorbency without feeling damp, used particularly for upholstery material.

Tedco mercerising process: Technique developed by the Norwegian Textile Institute and the Institute for Industrial Research in Oslo for treating fabrics in liquid ammonia. The technique is derived from the Prograde process. The fabric is passed through liquid ammonia at –33/35°C causing an immediate reaction to occur. Reaction time is 10 seconds maximum. The fabric is then stretched and fed over heated drums where the ammonia evaporates and is recycled, except for a residue of 3%. The remaining 3% can be removed by steaming. The resulting ammonia steam mixture is incinerated.

TEGEWA: German association of textile and leather aids and staining and washing raw material industry (Verband der Textilhilfsmittel-, Lederhilfsmittel-,Gerbstoff- und Waschrohstoff-Industrie e.V.).

Teklan: Courtalds modacrylic fibre. It is strong and hard wearing but also soft, warm and light and can be bulked. It has good resistance to sunlight, bacteria and chemicals and above all nonflammable. Used mainly in woven and knitted dress materials, and household textiles, such as net curtains, and furnishing materials.

Telomer: See **Telomerization.**

Telomerization: The formation of an addition oligomer. So-called specific Polymerization uses compounds (so-called "telomer") resulting in end groups, which are determined in advance, and renders possible a controllable, stage-by stage increase of polymerization by adduction 1 : 1, 1: 2, 1: 3, 1: x up to several thousand. Telomerization results in a wide range of synthetic waxes, plasticizing agents, lubrication oils, textile aids, plastics, synthetic fibres, etc.

Temperature of zero birefringence: The temperature at which the refractive indexes of a material are equal in two perpendicular directions (longitudinally and transversely for a fibre).

Temple: A part of the loom helping in the draw of the fabric after weaving. See **Temple mark.**

Temple mark, in woven fabrics: Small holes, impressions, marks or distortions adjacent to the selvedge caused by poorly adjusted or improper temples. See also **Bad temple, pick up place, rough place.**

Temporary hardness: The hardness of water due to bicarbonates. This hardness can be removed by boiling of the water, which converts bicarbonates to carbonates, which are not attributing to hardness.

Temporary stitches: Stitches, which are ripped off after a permanent stitch has been made, are called temporary stitches. For example-Basting stitch, Permanent stitch.

Tenacity, in a tensile test: The force exerted in the specimens based on the linear density of the unstrained specimen.

Tenacity: The tensile strength expressed as the force per unit linear density of the unstrained specimen.

Tenacity: The tensile stress when expressed as force per unit of the linear density of the unstrained material. For e.g. gf/tex.

Tenacity The maximum specific strength of a fibre or yarn that is developed in a tensile test taken to rupture point.

Tenacity-at-rupture: The tenacity of the force-at-rupture.

Tenacity-as-specified elongation (TASE): The tenacity of a material at its force-at-specified-elongation.

Tenacity, breaking: See **Breaking tenacity.** The tenacity corresponding to the breaking load for e.g. gf/tex.

Tencel: ®Acordis Fibres (Holdings) Ltd. trademark for their lyocell regenerated cellulose fibres There are two types of TENCEL®. "Conventional" TENCEL®, is subject to vibrillation, which is exploited to produce a "peach skin" finish. TENCEL® A100 is treated to cause cross-linking of cellulose fibres, which prevents fibrillation. TENCEL® has much higher wet strength than other regenerated cellulose fibres such as viscose rayon. It can be dyed much like cotton.

Tender Fleece: A sort of wool separated in wool sorting from others; they have a weak place in the fibre.

Tender Wool: Wool below the normal tensile strength.

Tendering: Weakening of a fibre, normally meaning as a result of chemical degradation Cellulose fibres can be tendered by acids or by excessive action of *oxidative bleaches*.

Teneriffe Work: Consists of making laces similar to the filet with starlike patterns.

Teneriffe Lace: An inexpensive lace typified by a wheel design or spiders web motif. These are often joined to make mats or added to table linen. It is similar to Paraguay Lace.

Tennis Flannel: Same as outing flannel.

Tennis Shirting: Soft, twilled, cotton or woollen shirting, made in fancy stripes, often napped on the back.

Tennis Stripes: Light, twilled woollen dress goods with narrow, coloured stripes.

Tensile: Relating to tension in, or on a material.

Tensile elasticity: The most important type of Elasticity which is measured after pre-tensioning under gradually increasing load (stress-strain behaviour). Evaluation of the figures to attain, for example:

(a) Elastic elongation: Retraction of the test specimen when the load is removed as % of the initial length.

(b) Degree of elasticity: Ratio of elastic elongation to overall elongation, as %.

(c) Modulus of elasticity (E-modulus): Ratio of tensile stress to the elongation unit. Measure of the resistance of a fabric to changes in form.

Tensile force (Load): (tensile load). The force in the direction of traction acting on the test sample at any one time during a simple Tensile strength test: the maximum measured tensile force on rupture of the test sample is known as the maximum tensile force; in relation to fineness = tensile strength; in relation to the initial cross-section = tensile stress.

Tensile hysteresis curve: A complex load elongation or stress strain curve obtained under either (a) when a specimen is successively subjected to the application of a load or force less than that causing rupture, and the removal of the load or force according to a predetermined procedure, (b) When a specimen is stretched less than the breaking elongation and allowed to relax by removal of the strain according to a predetermined procedure.

Tensile recovery curve: See **Tensile hysterisis curve.**

Tensile strain: The strain on a material subjected to tension. The relative length deformation exhibited by a specimen subjected to a tensile force. Strain may be expressed as a fraction of the nominal gauge length or as a percentage. Also see **Elongation.**

Tensile strain recovery: The percent of recoverable extension impressed on a fibre under specified conditions.

Tensile strength: The maximum resistance of a material to deformation in a tensile test carried to rupture; that is, the breaking load, or force, per unit cross sectional area of the unstrained specimen.

Tensile strength: The strength shown by a specimen subjected to tension, as distinct from torsion, compression or shear.

Tensile strength: The strength of a material under tension as distinct from compression, torsion or shear.

Tensile strength: The breaking load (or force) per unit cross sectional area of the unstrained specimen.

Tensile strength tester: Testing machine to find out tensile strength of a material.

Tensile stress: The stress within a material subjected to tension. The resistance to deformation developed within a specimen subjected to tension by external force. The tensile stress is commonly expressed in two ways, either as (a) the tensile strength, i.e., the force per unit cross-sectional area of the unstrained specimen, or as (b) tenacity, i.e., the force per unit linear density of the unstrained specimen. The latter is more frequently used in textile testing.

Tensile stress: In a Tensile strength test the quotient from the Tensile force and the initial cross-section.

Tensile test, in textiles: A test in which a material is extended in one direction to determine one or more of its force-extension, or stress-strain, characteristics; for example, breaking force, elongation at break.

Tensile testing machine: An apparatus designed to impart or transmit, force/extension, or stress/strain, to a material and to measure the effect of the action.

Tension: A uniaxial force tending to cause the extension of a body or balancing force within that body resisting the extension.

Tension-recovery chart, in tension testing: A continuously plotted graph of tension versus extension resulting from a tension recovery cycle.

Tension-recovery cycle, in tension testing: The continuous application of tension on specimen with a momentary hold at the maximum tension, followed by a return to zero tension at a controlled rate.

Tension bar: A bar composed of weft yarn that has been stretched more, or less, than the normal weft prior to, or during, weaving. This abnormal stretch may have been imposed: during winding, by faulty manipulation or by some mechanical fault in the machine; during weaving, by incorrect tensioning in the shuttle; or may have arisen owing to the faulty yarn having been excessively moistened at some stage and consequently stretched more than the normal yarn under the normal applied tensions. Such a bar may appear as a Cockled bar in those cases where stretch has been sufficient to cause cockle on subsequent contraction of this weft.

Tension box, in weaving: When a warp is wound sectionally, it is first threaded in a tension box that tensions each thread evenly and spreads the threads to the width of the section.

Tension, loop: See **Loop tension.**

Tension Control Weave: A type of decorative weave, characterized by a puckered effect which occurs because the tension in the warp yarns is intentionally varied before the filling yarns are placed in the fabric.

Tension test, in textiles: A test designed to measure the tautness in a textile strand or fabric.

Tent fabric: Stout waterproof cotton duck or canvas, used for tents and awnings.

Tent Stitch or Petit Point: In embroidery only half of the cross stitch, repeated.

Tenter frame: A machine that dries fabric to a specified width under tension. The machine consists essentially of a pair of endless chains on horizontal tracks. The fabric is held firmly at the edges by pins or clips on the two chains that diverge as they advance through the heated chamber, adjusting the fabric to the desired width. Also called Stenter frame.

Tenterhook willow: See **Willey.**

Tenter Mark: See **Stenter mark.** A visible deformation on the side edge body of a fabric caused by due to pressure from clips or pins.

Tenter Mark: See **Clip mark.**

Tenuguiji: A soft and plain woven cotton fabric; used as toweling in Japan; it comes in 12 and 15 inch widths and often hand printed. It is also used for table covers, center pieces and extensively for advertising.

Terephthalic Acid: Para-phthalic acid, $[C_6H_4(COOH)_2]$. Used to produce polyester resins, fibres, and films by combination with glycols.

Terinda: This is a special polyester fibre made by ICI.

Terpenes: **Turpentine, pine oil etc. which are handy as stain removers, solvents and emulsifiers comes under the chemical group of Terpenes. They** are cycloaliphatic and hydro-aromatic hydrocarbons with the general formula $(C_5H_8)_n$ Used in components for solvent soaps or emulsions for

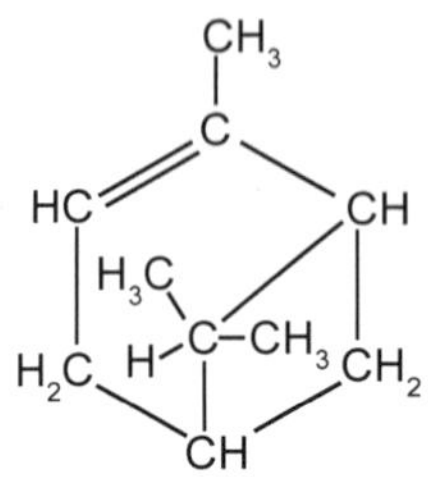

boiling cotton, for linseed oil sizing agents, antifoam agents, for dissolving rubber, polyacryl acid esters and artificial resins (for print thickeners), for stain removing purposes, etc.

Thermal Insulation: The ability of a fabric to retain heat.

Thermoregulation: The ability to maintain a constant temperature independent of dynamic (changing) environmental conditions.

Terpolymer: A product of the polymerization of three different monomers.

Territory Wool: A designation originally given to wools originating in regions west of the Missouri River. Now applies to western range wools, not including Texas and California.

Terry Cloth: First used in terry towels or Turkish towels, terry cloth is used in so many ways as beach wear, sportswear etc. today. It is woven with two sets of warp and one filling, one warp forming rows of loops on the face or back or on both sides, which are not cut. It comes bleached, dyed, in coloured patterns or printed. The loop piles can cover the entire cloth or form patterns. It is made of cotton, linens, wool or silk and used for a great variety of purpose. When made of cotton or linen it is also called Turkish toweling. The warp which forms the loops is dressed on a separate warp beam and is kept very slack. In weaving, a number of .picks are let in, quite far apart from each other, before beaten up hence terry is designated as 3, 4, 5 or 6 pick, according to the number of picks put in in one row of loops. When the picks are beaten up they will slide over the stoutly stretched ground warp but the friction is sufficient to pull the pile warp with it, the section of that warp between each pick, originally far apart, forming a loop on any or both sides when beaten up. A normal terry construction for apparel use in 64 x 24 with 16s cotton warp and filling. Frequently the pile warp yarn is more loosely twisted than ground warp yarn to obtain greater absorbency It comes in stripes, checks, plaids effects in various colours. Used for towels, bathrobes, etc.

Terry cloth identification: (1) *Frotté*: (a) woven on standard weaving machine, (b) looped fabric surface, (c) effect may be produced by using frotté, (d) loops cannot be pulled out. (2) *Terry fabric*: (a) woven on terry weaving machine, (b) loops formed by the pile warp, (c) loop thread can be pulled out in the warp direction.

Terry Fabric: A material with a woven warp pile or a knitted fabric, with uncut loops on both sides, and which is used for such products as toweling, beachwear, and bathrobes.

Terry Towel: A textile product with end hems or fringes and side hems or selvedges which is made with loop piles on one or both sides, generally covering the entire surface or forming stripes, checks or other patterns.

Terry Velour: A pile weave cotton fabric with an uncut pile on one side and a cut pile on the reverse side. Terry velour is valued for its soft, luxurious hand. Typical uses include towels, robes, and apparel.

Terry Velvet: A velvet with a looped pile, produced in the same way as the Terry Toweling.

Tertiary Colours: Shades that are obtained by mixing the three primary colours or by mixing one or more of the secondary colours with gray or black.

Terylene: ICI's polyester fibre, produced from petroleum and its byproducts. It can be successfully bulked. It is extremely hard wearing and mixes well with many other fibres such as cotton, wool, viscose etc. It does not catch light but melts. Used in all types of dresses and furnishing fabrics.

Teshike: Lightweight Japanese silk cloth, treated against perspiration.

Testbeard, in length testing of cotton: A portion of the test specimen that has been combed and brushed into a 'beard' which protrudes from the outside of the comb(s) or clam(s).

Test: (1) A single observation made on one (test) specimen.

(2) A series of observations of which the average, or some other function, ids considered a significant result for a sample.

Test beard, in length testing of cotton: The portion of the test specimen that has been combed and brushed into a 'beard' which protrudes from the outside of the comb(s) or clamp(s).

Test method: A definitive procedure for the identification, measurement, and evaluation of one or more qualities, characteristics, or properties of a material, product, system, or service that produces a rest result.

Test skein: A small skein which has a prescribed length of yarn and is used for determination of linear density or breaking load or both.

Test specimen: The cotton fibres placed randomly on a pair of Fibrograph combs for fibre length measurements.

Test specimen, in cotton length with the length Analyzer: The cotton fibres protruding randomly from the base of a Motion Control inc., specimen clamp for fibre length measurement before being combed or brushed.

Test specimen n cotton maturity tests: The series of slides observed by one technician as one half of the test.

Test specimen, wool top: A length of specified mass taken at random from a length of wool top selected as a lab sample.

Test specimen: See **Specimen.**

Test, tensile: See **Tensile test.**

Test, tension: See **Tension test.**

Tete de Negre: (1) French name for a dark brown colour; (2) French woollen fabric with a knotted face, similar to petersham.

Tetrachloroethylene: (perchloroethylene, per) C_2Cl_4. Colourless fluid, clear as water, neutral, evaporates rapidly, characteristic smell, anaesthetizing, nonflammable. Stabilized, non-decomposed tetrachloro ethylene corrodes metals at temperatures up to 140°C even if water is not present. Advantages of tetrachloro ethylene compared with water for cleaning outer clothing: textile compatibility: no fibre swelling, no shrinkage, causes no deterioration of shape; high density: stable for use in highly-mechanized processes, good dirt removal properties; low surface tension: rapid moistening, rapid dirt penetration; grease dissolving: cleaning without temperature increase; rapid evaporation; rapid, energy-saving drying; low evaporation heat: bath regeneration by means of distillation, waste concentration in residue, well directed disposal. As far as the care of outer clothing is concerned, Dry-cleaning has advantages with regard to maintaining the value of the textiles and to cleaning performance, lower time requirement, lower energy requirement and lower material consumption, since the solvent used is always recovered and the residue is disposed of efficiently. The use of tetrachloro ethylene does, however, involve certain risks.

Tetrafluoroethene: (CF2:CF2) A gaseous organic compound (a fluorocarbon and a haloalkene) used to make the plastic polytetrafluoroethene (PTFE). See **polytetrafluoroethene.**

Tetrafluoroethylene Fibre: See **Polytetrafluoroethylene fibres.**

Tetrakis (Hydroxymethyl) Phosphonium Chloride (THPC): THPC is the most important commercial derivative and is prepared from phosphine, formaldehyde and hydrochloric acid at room temperature. It contains 11.5% phosphorous and is applied by a pad-dry-cure -> oxidize -> scour process. $PH_3+4CH_2O+HCl=(HOCH_2)_4P^+Cl^-$ The compound is highly reducing in character and the methylol groups condense with amines to form insoluble polymers. It is applied with urea, dried and cured. Control of pH and the oxidation state of the phosphorus is important in determining the flame retardant properties and the durability of the finish. The release of HC1 may cause the fabric to tender during curing unless pH is controlled. The final step in finishing requires oxidation of P+3 to P+5 with hydrogen peroxide. This step too must be controlled to prevent excessive tendering of the fabric. An alternative to the THPC is THPS. Sulfuric acid is used instead of HC1 and the corresponding phosphine sulphate is formed in place of the phosphine chloride.

Tetrakis (Hydroxymethyl) Phosphonium derivatives: The bulk of today's durable flame retardant for cellulose centers around the use of derivatives of tetrakis (hydroxymethyl)- phosphonium salts (THP). These derivatives can be applied by padding, drying, curing and oxidizing to yield serviceable flame retardant fabrics. Add-ons are high and the handle of the fabric is stiffer so the finish is normally used for protective clothing applications.

Tetrakis (Hydroxymethyl) Phosphonium Hydroxide (THPOH): THPC is usually partly neutralized with amines, amides and/or alkali. Complete neutralization of THPC with sodium hydroxide yields a compound referred to as THPOH. The distinction between THPC used in a partially neutralized condition and THPOH is difficult to define. If the curing agent is basic as is ammonia, the distinction become meaningless. THPOH-ammonia has received a great deal of commercial attention. The major advantage over THPC is reduced fabric tendering and reduced stiffness. Fabrics padded with THPOH give off formaldehyde during drying.

Tex: A unit for expressing linear density, equal to the mass in grams of yarn, filament, fibre or other textile strand. It is the mass in grams of 1000 metres of fibre, yarn, or other textile strand, that is used in a direct yarn numbering system. The multiple and submultiples recommended for use are the following:

Kilograms per kilometre	kilotex (ktex)
Milligrams per kilometre	millitex (mtex)
Decigrams per kilometre	decitex (dtex)

Example: A yarn of 2.5 km has a mass of 40g. Tt (tex) = 40/2.5 = 16 tex

In case of folded and cabled yarns, individualyarn number is indicated, followed by a multiplication sign and the number of components.

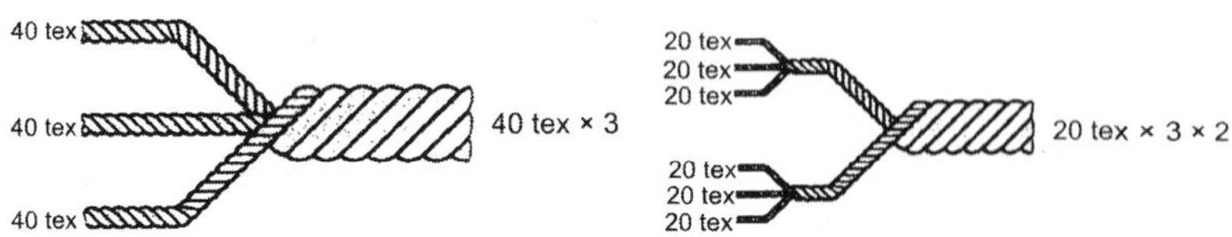

Tex System The direct decimal system based on metric units that has been adopted by ISO as a universal system for designating the linear density of fibres, filaments, slivers and yarns.

Tex ticket number: The designator assigned to a sewing thread to indicate the approximate linear density.

Texas: General trade name of cotton grown in Texas and Oklahoma, the staple measuring from seven-eighths to one inch in length and varying in quality according to the season.

Texas Storm: Proof Commercial variety of late maturing upland cotton, the staple measuring 23-26 millimeters; the yield is 33-35 per cent.

Texas Wood: Commercial variety of medium maturing upland cotton, the staple measuring 22-25 millimeters; the yield is 34-36 per cent; also called Peterkin.

Textile: Originally a woven fabric, now generally applied to; (a) Staple and filament suitable for conversion to or use as yarns, or the preparation of on woven fabrics, (b) Tarns made from natural or man-made fibres, (c) Fabrics and other manufactured products made from fibres as defined earlier , and (d) Garments and other articles fabricated wholly from one or more of the above elements, and articles made principally from the above when the products retain the characteristic flexibility and drape of the original fabrics. Any item manufactured from natural or man-made fibres or filaments, e.g. yarns, threads, cords, ropes, braids, lace, embroidery, nets and fabrics made by weaving, knitting, braiding, felting, bonding and tufting.

Textile Auxiliaries: (assistants, agents processing aids), considered as belonging to Textile chemicals. Textile auxiliaries on the one hand make it possible to or have a positive effect on the execution of a textile process, and on the other hand have a positive effect on quality, efficiency by helping to reduce or eliminate faults and damages. Textile auxiliaries are thus found in all types of textile processes from fibre production, processing, finishing and make-up of finished goods to subsequent textile care in washing, dry cleaning and soil removal.

Textile care: This encompasses the maintenance of efficiency of textiles including amongst other things. See **Textile cleaning.**

Textile chemicals: All chemicals (including Textile auxiliaries), which are used in Textile finishing and should be biodegradable.

Textile fibre, general: A generic term for the various types of matter that form the basic elements of textile fabrics and other textile structures.

Textile fibre, specific: A unit of matter that is characterized by having a length at least 100 times its diameter or width and which can be spun into a yarn or made into a fabric by interlacing in a variety of methods, including weaving, braiding, knitting, felting and twisting.

Textile Fabric: A planar structure consisting yarns or fibres.

Textile finishing: Textile finishing usually includes treatments such as scouring, bleaching, dyeing and/or printing, the final mechanical or chemical finishing operations, that during this stage are carried out on textile products (staple, sliver or top, yarns or filaments, woven or knitted fabrics) to enhance their basic characteristics like dye penetration, printability, wettability, colour, hand, and appearance. By textile finishing, also mean all the processing operations that, though included in the so called finishing stage, are generally applied to the fabrics to improve their appearance, hand and properties, at times in accordance with their field of application.

Textile Floor Covering: A product having a use-surface composed of textile material and generally used for floor covering.

Textile labeling: Details of the raw material composition of textile products. (a) Between processing stages as a rationalization measure. (b) For the consumer as a protective measure when purchasing textiles in accordance with Textile labeling regulations

Textile markers: (1) Textile markers in the form of paste in tubes for marking raw materials with 4 different writing heads (for different types of material), resistant to all finishing stages. (2) (markers, fabric markers, signal markers) are used for Marking of grey fabrics in wet and/or dry processing (fabrics, number strips, garment bags, laundry). Suited to each process type (finishing, dry cleaning, laundering) with appropriate fastness and yet capable of being easily removed (e.g. with soap). Colours usually white, yellow, red and black. Textile marker types: Liquid as inks, semi-liquid in tubes or pens, solid as pencils.

Textile Materials: A general term for fibres, yarn intermediates, yarn, fabrics, and products made from fabrics that retain more or less completely the strength, flexibility, and other typical properties of the original fibre or filaments.

Textile Modulus: Obsolete term, See **Young's Modulus.**

Textile Processing: Any mechanical operation used to translate a textile fibre or yarn to a fabric or other textile material. This includes such operations as opening, carding, spinning, plying, twisting, texturing, coning, quilling, beaming, slashing, weaving, and knitting.

Textile ribbon: See **Ribbon, in textiles.** A fine textured, narrow fabric which weighs less than 2.6 lb. per 100 yd. per inch of width (approximately 15 oz./ yd2 or 510 g/m^2) and which is used primarily for trimming or decorative purposes.

Textile Standards: In many countries it is the responsibility the Specialist Standards Committee of the Textile and Textile Machinery Industry to establish of standards for the entire textile industry, textile machinery constructors and the clothing industry and are are issued by them as standards for that country. In the area of textile testing methods they work closely with the specialist standard committee for "Materials Testing". See **Industrial standards**. These Standards are available everywhere so that one country can refer the requirements of another country when the former is exporting to the latter.

Textile Whiteness Scale: (textile whiteness standard, cotton whiteness standard), derived from the Ciba-Geigy white standard. Division into 18 sections (each comprising 10 units) on a scale from 70 to 240. The lowest section of 70 is applied to max. bleached, non-brightened cotton and the highest level of 240 to intense white effects achieved with optical brighteners.

See **Whiteness scale.**

Textile tape: See **Tape, in textile.**

Textile, webbing: See **Webbing, in textile.**

Textile, non-combustible: See **Noncombustible textile.**

Textile-reinforced concrete. In fibre-reinforced concrete the fibres are short [usually no longer than 2 in. (5 cm)] and discontinuous, textile-reinforced concrete contains continuous woven or knitted mesh or textiles. Conceptually, such reinforcement acts similarly to conventional steel reinforcing bars or welded steel wire fabrics. But these fabric materials are noncorrosive and can have mechanical properties that are superior to those of steel. The fabrics can be pre-manufactured in a wide variety of ways, thereby lending themselves to new applications, especially for repairing or strengthening existing concrete structures.

Textiles: These are understood to include half- and fully-finished fabrics made from spinnable Fibres. They are therefore structures which consist of textile fibres in their smallest components and are further processed by means of textile processes (spinning, weaving, knitting, etc.). Textiles are frequently influenced more by their structure than by the type of fibre.

Textometer: Device for measuring moisture content of textiles which determines the residual moisture using the principle of conductivity. There are various versions of the device: easily transportable battery unit, universal device for monitoring processes as well as a version with electronic regulation for monitoring and automatic control of sizing machines, fabric driers and the like.

Textured Fabric: Fashionable fabrics for underwear, clothing and outerwear with raised or depressed waffle-like pattern effects (lace-like tied warp and weft threads).

Texture: The surface appearance and hand of a textile material. A term describing the surface effect of a fabric, such as dull, lustrous, wooly, stiff, soft, fine, coarse, open, or closely woven; the structural quality of a fabric.

Texture, in carpets or pile floor coverings: The detailed configuration of loops, cut pile ends, and individual fibres in a pile layer.

Textured: An adjective used to describe continuous filament manufactured yarns (and woven and knit fabrics made therefrom) that have been crimped or have had random loops imparted, or that have been otherwise modified to create a different surface texture. Also see **Textured Yarns and Texturing**.

Textured glass yarn: A yarn processed from continuous filament yarn in such a manner to induce bulk to the yarn by disorientation of filaments.

Textured Yarn: (1) A yarn that has been so processed as to introduce durable crimps (q.v.), coils, loops or other fine distortions along the length of the fibres or filaments.

(a) The main texturing processes usually applied to continuous-filament yarns made from or containing thermoplastic fibres, are as follows:

(i) The yarn is highly twisted, heat-set and untwisted, either as a continuous process (false twisting) or as a three-stage process. This continuous method for producing textured yarns utilizes simultaneous twisting, heat-setting, and untwisting. The yarn is taken from the supply package and fed at controlled tension through the heating unit, through a false-twist spindle or over a friction surface that is typically a stack of rotating discs called an aggregate, through a set of take up rolls, and onto a take-up package. The twist is set into the yarn by the action of the heater tube and subsequently is removed above the spindle or aggregate resulting in a group of filaments with the potential to form helical springs. Much higher processing speeds can be achieved with friction false twisting than with conventional spindle false twisting. Both stretch and bulked yarns can be produced by either process. Examples of false-twist textured yarns are Superloft®, Fluflon®, and Helanca®. (Also see **Textured yarns, Coil Yarn**.)

(ii) The yarn is passed through a heated "stuffer box" (stuffer box crimping). The crimping unit consists of two feed rolls and a brass tube stuffer box. By compressing the yarn into the heated stuffer box, the individual filaments are caused to fold or bend at a sharp angle, while being simultaneously set by a heating device. (Also see Textured yarns, Crinkle Yarn.)

(iii) The heated yarn is passed over a knife edge (edge crimping). : In this method of texturing, thermoplastic yarns in a heated and stretched condition are drawn over a crimping edge and cooled. Edge-crimping machines are used to make Agilon yarns. (Also see Textured yarns, Coil Yarn.)

(iv) The heated yarn is passed between a pair of geared wheels or some similar device (gear crimping). In this texturing method, yarn is fed through the meshing teeth of two gears. The yarn takes on the shape of the gear teeth. (Also see Textured yarns, Crinkle Yarn.)

(v) The yarn is knitted into a fabric, heat-set and unravelled (knit-deknit). In this method of texturing, the yarn is knit into a 2-inch diameter hose-leg, heat-set in an autoclave, and then unraveled and wound onto a final package. This texturing method produces a crinkle yarn. (Also see Textured yarns, Crinkle Yarn.)

(vi) Loops are formed in individual filaments by over-feeding into a turbulent airstream (air-textured). Also called Air Jet Method, in this method of texturing, yarn is led through the turbulent region of an air jet at a rate faster than it is drawn off on the far side of the jet. In the jet, the yarn structure is opened, loops are formed, and the structure is closed again. Some loops are locked inside and others are locked on the surface of the yarn. An example of this method is the Taslan process.

(Also see Textured yarns, Core-Bulked Yarn and Entangled Yarn.)

(vii) Bicomponent Fibres (q.v.) are differentially shrunk.

(b) Processes (i) and (iii) above produce yarns of a generally high stretch character. This stretch character is frequently reduced by reheating the yarn in a state where it is only partly relaxed from the fully extended condition, thus producing a yarn with the bulkiness little reduced but with a much reduced retractive power.

(c) Fabrics that contain textured yarns have increased bulk, opacity, and moisture absorbency and improved thermal insulation properties with a warmer handle (q.v.); some textured yarns also confer extensible or "stretch" properties on fabrics made from them.

(2) Yarns that develop stretch and bulk on subsequent processing. When woven or knitted into fabric, the cover, hand, and other aesthetics of the finished fabric better resemble the properties of a fabric constructed from spun yarn. (Also see **Texturing.**)

(a) **Bulked Yarn**: Qualitative term to describe a textured yarn. A bulked yarn develops more bulk than stretch in the finished fabric. (b) **Coil Yarn**: A textured yarn that takes on a coil or spiral configuration when further processed. A coil yarn can be either a torque yarn or a nontorque yarn. A coil yarn can be formed by the false twist or edge crimp methods. Some bilateral fibres become coiled on further processing. (c) **Core-Bulked Yarn**: A bulky or textured yarn composed of two sets of filaments, one of which is straight to give dimensional stability and forms a core around and through which the other set is coiled or looped to give bulk. (d) **Crinkle Yarn**: A torque-free textured yarn that is characterized by periodic wave configurations. Crinkle yarns can be formed by the stuffer box, gear crimping, or knit-de-knit methods. (e) **Entangled Yarn**: A textured yarn of one variant that develops bulk by the air-jet texturing method. (f) **Modified Stretch Yarn**: A stretch yarn that develops more bulk than usual but less bulk than a bulked yarn in the finished fabric. (g) **Non-torque Yarn**: A yarn that does not rotate or kink when permitted to hang freely. A nontorque yarn may be the result of plying two equal but opposite torque yarns. (h) **Set Yarn**: A textured yarn that is heat relaxed to reduce torque. Set yarns are not stretch yarns. (i) **Stretch Yarn**: Qualitative term to describe a textured yarn. A stretch yarn develops more stretch than bulk in the finished fabric. (j) **Torque Yarn**: When a torque yarn is permitted to hang freely, it rotates or kinks to relieve the torque introduced into the yarn during texturing.

Texturing: This term is used to include all process stages downstream of the actual texturing process. In the filament yarn sector these include: Intermingling: normally accomplished by air-jet techniques; Assembling: combination of multiple yarns with zero twist; Coning (winding): transferring the yarn onto a suitable carrier.

Texturing, air jet: The yarn is fed through a turbulent air stream, created by a jet. Usually the air is cold but occasionally hot air or stesam may be used. The air jet causes entangled loops to be formed in the filaments. The resulting textured yarn is very bulky with permanent crimps and loops. Can be used for non-thermoplastic filaments.

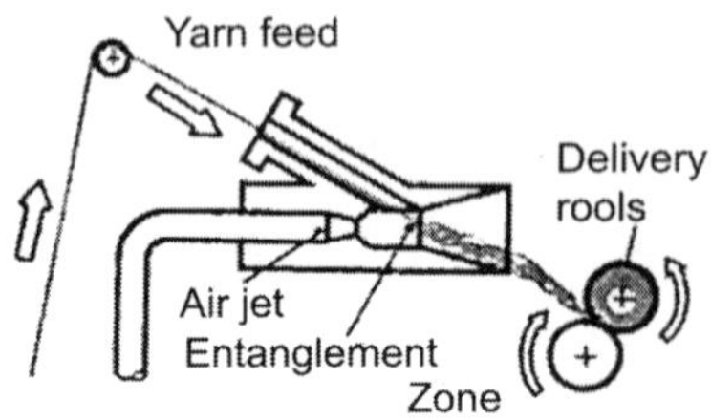

Texturing, False twist: The yarn is drawn through a heated zone whilst a predetermined, high level of false twist is inserted followed by cooling and untwisting. The heat softens the filament and the deformation imparted by the twist is permanently set during cooling. This is the most economical and hence most commonly used.

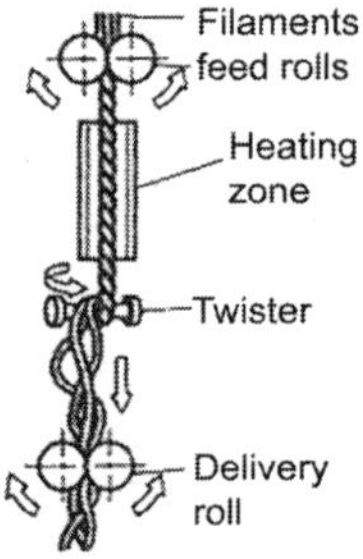

False twist texturing, schematic

Texturing, knit-deknit: The yarn is knitted into a tube on a circular knitting machine. The knitted fabric is heat set and then unraveled. The shape of the knitted loops is set into the yarn which develops a boucle' (crinkled yarn) appearance.

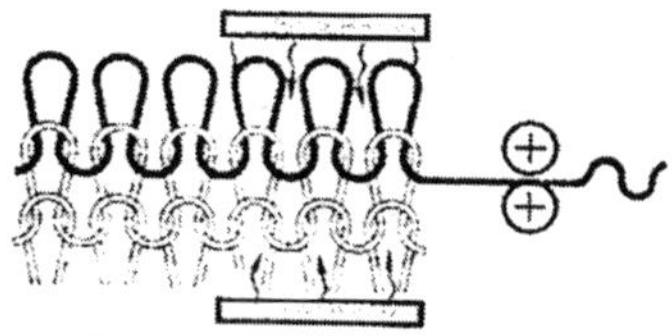

Texturing, stuffer box: The yarn is fed into a heated chamber where it is compressed. The zig-zag deformations are permanently set by subsequent cooling. The filaments can no longer lie close together so the yarn is bulky.

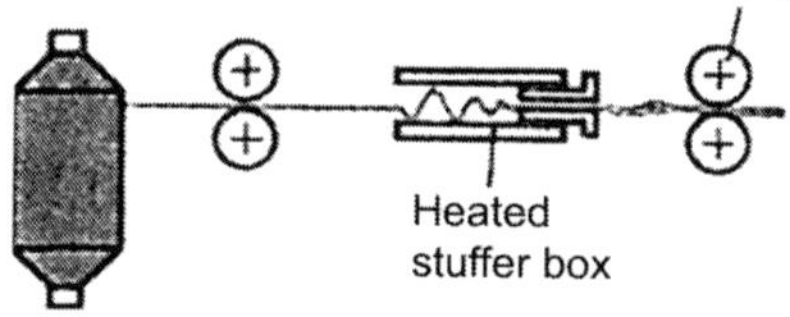

Texturing machines: Machines used for texturing. See **Texturing.**

Texturizing: The treatment of smooth filament yarns by permanently modifying the orientation of the molecular structure giving it more textile character. The changes to texturized yarns are: improved softness, volume and covering power, increased thermal insulation effects and an end-use required increase in elasticity. Thermoplastic filament yarns are particularly suitable for texturizing, i.e. yarns which are deformable under the influence of heat. These are mainly polyamide and polyester yarns.

Theoritical oxygen demand: Organic matter of animal or vegetable origin in waste water is generally a combination of carbon, hydrogen, oxygen and nitrogen. The principal groups of these elements present in the in waste water are, as previously noted, carbohydrates, proteins, oils and grease and products of their decomposition. If you take any of these chemicals, ThOD will be the oxygen required for the complete decomposition (for example we take the case of Glycine (CH_2 (NH_2)COOH- in first step organic carbon and nitrogen are converted to carbon dioxide CO_2) and ammonia (NH_3) respectively. In second and third steps, ammonia is sequentially to nitrite and nitrate. The theoretical Oxygen Demand is the sum of oxygen required for all three steps.

Thermachrome process: (thermacrome process). Transfer print process for transferring paper prints in up to 4 colours onto synthetic fibre textile piece goods. The paper is printed with thermoplastic pigment print pastes which are transferred onto the textile material on passing a heated calender. This process is limited to the printing of labels and emblems.

Thermal blanket: A blanket woven so that cells or openings are created in the fabric so that air warmed by the body is trapped between the yarns, such as textured or leno weaves. The product can be napped or unnapped.

Thermal bonded batting: A textile filling material that contains low melting point fibres or polymers which, when heated fuse the batting material together.

Thermal bonding: **See Thermobonding**.

Thermal character: Apparent difference in temperature of the fabric and the skin of the observer touching it.

Thermal character: A tactile property of a textile material. It is the difference felt in the temperature of the material and the skin of the person touching it.

Thermal conductance: See **Thermal transmittance**.

Thermal conductivity: Time rate of unidirectional heat transfer per unit area, in the steady-state, between parallel planes separated by unit distance, per unit difference of temperature of the planes.

Thermal conductivity coefficient: See **Heat conductivity**.

Thermal conductivity of Textile materials: The technical thermal conductivity (λ) indicates the quantity of heat which flows through two opposite surfaces of a cube of 1m edge length per hour when the temperature difference between both surfaces is 1°C (→ Heat retention).

Thermal conductivity in W/m · °C:

glass	1.000
polyethylene	0.337
polypropylene	0.221–0.302
polyamide	0.209–0.290
m-Aramid	0.128
viscose (filament)	0.071
viscose (spun fibre)	0.063
asbestos	0.058
cotton	0.058
triacetate	0.058
acetate	0.050

polyester	0.046–0.081
wool, silk	0.046
air	0.025

Thermal drying of sludge: Heat can be used to dry wastewater sludge from 25 to 30% solids down to 90% solids. This minimizes the volume of the sludge. Drying is important prior to incineration of wastewater sludge, otherwise an external fuel source has to be used in the incinerator. A common technique is the use of a ***rotary drum dryer.***

Thermal end point, in testing of thermal protective material: The crossing point of the sensor response on the recorder chart and the human tissue burn tolerance criteria overlay.

Thermal fixation See Dyeing. The use of dry heat to achieve a degree of permanence when applying colorants to textile materials.

Thermal resistance: The reciprocal of thermal transmittance.

Thermal resistivity: The reciprocal of thermal conductivity. Time rate of unidirectional heat transfer per unit area, in the steady state between parallel planes, per unit difference of temperature of the planes.

Thermal shock: Heat treatment at 200°C: I. Thermo-stabilization of synthetic fibres within 1 second. II. Rapid fixing of dyes (particularly of reactive dyes) during printing and continuous dyeing within 20–60 seconds.

Thermal shrinkage of textile yarns and cords: Contraction in length caused by heat.

Thermal stability: Stability of a material against temperature variations.

Thermal stability of dyes: Thermal stability when subjected to contact heat (ironing, fixing, drying), hot air (fixing, drying) and steam (steaming, decatizing, pleating).

Thermal stability of fibres: The impact of dry heat can bring about changes to fibre properties, the extent of which depends on the specific thermal stability of the fibre (heat stability) as well as the duration and temperature of treatment. As not only recognizable changes in chem. and technological properties, such as yellowing and greying, can be effected, it is advisable to check for other non-visible signs of temperature impact. This applies particularly to synthetic fibres (Softening range; Melting point).

Thermal transmittance: Time rate of unidirectional heat transfer per unit area, in the steady-state, between parallel planes separated by unit distance, per unit difference of temperature of the planes.

Thermal welding, in garment: Process by which flexible thermoplastic films as well as polyester needle felt and nonwovens are joined by welding in garments. There are two main thermal welding techniques, namely the hot edge and hot air methods. In either case heat is supplied to the inside of the seam, to soften the thermoplastic components, after which the assembly is forwarded between two transport rollers where it is pressed and securely bonded. Starting, stopping and sharp corners are problematic.

Thermally bonded seam: A seam formed by heat and pressure.

Thermally Stabilized: See **Heat stabilized.**

Thermobonding: A technique for bonding fibres of a web with melt able powders or fibres, using infrared heating, hot air, or hot-calendaring. (Also see Bonding, Bonding with Binder Fibres and Powder-bonded non-woven.)

Thermochromism: Thermochromism is the term used to describe a change of colour as a result of a temperature change. Thermochromic dyes and pigments find application where this colour change is used to indicate a temperature change, for example in plastic strip thermometers, medical thermography and non-destructive testing of engineered articles and electronic circuitry. They may also be used in thermal imaging and for a variety of decorative or novelty effects.

Thermocolour markers: (temperature measuring pencil). Usually metal salts of complicated structure whose colour change at certain temperatures is based on oxidation or separation of water, carbonic acid and =ammonia. Temperature measuring dyestuffs are also used as measures of heat radiation.

Thermofixation: Fixation of a dye on the fibre/fabric by application of heat only.

Thermoflush: Procedure for shortening the rinsing process for reactive dyeing by means of intermediate steaming several times, thereby reducing water consumption.

Thermofusion: Procedure whereby dry fleeces are completely bonded just by means of heating and subsequent cooling, e.g Spot-bonded nonwovens. The bonding of nonwovens of bi-components and other fibres or fibres mixtures is carried out using thermos fusion systems, in every case using only hot air without steam or binders.

Thermogravimetric analysis: Analytical technique in which the rate of change in weight of a material undergoing continuous heating versus temperature is plotted. Used in analysis of polymers to provide information on such parameters as degree of crystallinity, glass transition temperature, thermal stability, etc.

Thermolabels: Self-adhesive thermos papers, coated with plastic, printed for 37.8–260°C and for five temperature ranges in 8 different arrangements (reaction speed: 2–5 seconds).

Thermometer papers: Paper strips or rolls with heat-sensitive covering (also self-adhesive) in series of varying heat degree gradations (37.8–254°C). At a certain temperature each strip changes its colour from white to black instantaneously and constantly at a quoted reaction speed of ±1% of the standard value.

Thermomigration: Desorption in the form of migration of dyes, particularly of disperse dyes on polyester under the influence of dry heat and in the presence of surfactants.

Thermoplast print blanket adhesives: Adhesives which are effective from a certain temperature and which are used in film printing where they are heated to the necessary temperature by means of heated rollers. Thermoplast adhesives, vulcanized in rubberized continuous printing blankets, allow the preheated fabric to be printed without glueing and washing using the preheated printing sheet to position the pattern for printing.

Thermoplastic: A textile that is deformable (but not changed chemically) by the application of heat and pressure. The salient feature is that the deformation can be repeated.

Thermoplastic resins: Fibres which demonstrate a softening range and are therefore thermoplastic. This property is also utilized during heat setting, See **Thermoplastics.**

Thermoplastics: Glass-like, hard solids that are thermoplastic are used in textile finishing for improving texture, draping properties, stiffness, durability, scrub-fastness and dimensional stability. See **Thermoplastics.**

Thermoplastics: (plastomers). Substances which are transformed reversibly into a plastic and therefore formable state when heated to a certain temperature (specific softening temperature). This does not involve any chemical change and the substance returns to its original hard or elastic state when it cools down. Thermoplastics can change from one state to the other as frequently as required.

Thermoset: A term used to describe a plastic that, once formed, does not melt when heated.

Thermosetting process: See **Thermo-set Process.**

Thermosol fixation: See **Dyeing, Thermal fixation.**

Thermosol pad steam process: A method which can be followed for dyeing polyester blended fabric. The fixation of disperse dye on synthetic fibre (say, polyester) is done by thermosol method. The cotton dye is fixed by pad steam method. Since reduction clear before cotton dyeing is no possible in many cases a cotton reserve auxiliary is often used to avoid cross staining of disperse dye on cotton.

Thermosol process: See **Dyeing, Thermal Fixation.**

Thermosol stiffening: Fabric stiffness and stiff texture brought about by Thermosol fixation particularly of knitted fabrics in textured yarns which can be removed by subsequent wet treatments or on the mechanical breaking machine.

Thermotropic polymer: Polymer that exhibits liquid crystal formation in melt form.

In thermotropic polymers there must be a balance between having the necessary degree of molecular perfection to preserve the liquid crystal formation and the amount of imperfection to permit melting at workable temperatures. These polymers give high-modulus, highly oriented, extrusion products.

Thermoweave process: Yarn or filament fibre used for producing carpet yarns by the bonding process are rolled and deep-frozen. This roll is sliced and the slices are bonded onto an underlay material.

Therm-set Process: Permanent-press process for cotton and polyester/cotton. The textiles are impregnated and dried with DHDMEU and a weak resin finishing catalyst (glycolic acid) before crosslinking and after-curing with a 1% magnesium chloride hexahydrate catalyst solution to fix pleats and creases in a pressing process.

Theta temperature: A temperature at which a polymer of infinite molecular weight starts to precipitate from a solution.

Threading cross: is the “x” formed by alternate warp ends because they each take a different path over and under a pair of pegs on the warping board: over/under for one end and under/over for the alternate end. If the openings created by the pegs are maintained when the warp is removed from the warping board, the crossing of the threads is also maintained and keeps them from shifting their order.

Terry towel: A textile product with end hems or fringes and side hems or selvedges which is made with loop pile on one or both sides generally covering the entire surface or forming stripes, checks or other patterns.

Thibet: An all-wool stout dress goods or coating, well felted and given a smooth and soft face, showing an indistinct twill. Often made with a cotton warp. It comes usually in solid piece dye or in mixtures.

Thibet Shawl: Made in France of wool and waste silk with various coloured patterns woven into.

Thick and thin places (major defect in fabrics): Places in the fabric where, for 25 cm. (1 in.) or more the count varies more than a specified percentage from the specified count.

Thick and thin places (minor defect in fabrics): Places in the fabric where for less than 25 cm. (1 in.) the count varies more than a specified percentage from the specified count.

Thick filling: See **Coarse pick.**

Thick place, in fabric: Yarn defect characterized by a diameter greater than that of the adjoining segments and extending for 6 mm. (1/4 in.)

Thick place: A textile that is deformable (but not changed chemically) by the application of heat and pressure. The salient feature is that the deformation can be repeated.

Thick weft: See **Coarse pick.**

Thick-and-thin yarn: A novelty yarn of varying thickness.

Thickeners: Usually water-soluble, swell able substances with colloidal character which are used in their thickened state as thickeners for printing with aqueous solutions of dyes, pigments and the like are used to prevent running (flowing) of the printed pattern on textile material (Print paste application).

Thickener, Natural: (1) *Plant-based products*: Starch, particularly wheat starch, swelling starches, roasted maize starch, types of gum (See Vegetable gums) such as gum arabic, industrial and synthetic gum; Locust bean flour, Tragacanth, extracts of lichens, mosses, types of seaweed, such as: Alginates; Agaragar, carragheen. (2) Animal products Albumen; Gelatine; Casein; Glue. (3) Mineral products (earth) China clay; Kaolin; Colloidal solution.

Thickener, artificial: (modified natural products): Cellulose derivatives; Starch ethers; Starch esters; Ethoxylated or monoacetic converted meal.

Thickener, synthetic: (synthetics and synthetic resins): particularly polymerisates on the basis of polyvinyl alcohol, polyacrylate, also chlorinated rubber.

Thickener, emulsion: Used for pigment printing, in a oil -water emulsion.

Thickness of textile fabrics: The distance between one surface and its opposite.

Thickness: The distance between one surface and its opposite; the distance between the upper and lower surfaces of the material, measured under a specified pressure.

Thigh girth, in body measurements: The maximum circumference of the upper leg close to the crotch.

Time-to-break: The time interval during which a specimen is under prescribed condition of tension and is absorbing the energy required to reach the maximum load.

Thin end: See **Fine end.**

Thin filling: A weft yarn smaller in diameter than normally being used in that fabric..

Thin filling bar: See Fine filling bar, Filling bar, fine.

Thin pick: A yarn defect characterized by a segment that is substantially (at least 25 %) smaller in diameter than the average diameter than the average diameter of the yarn.

Thin place: A yarn defect characterized by a segment that is substantially (at least 255) smaller in diameter than the average diameter of the yarn.

Thin place: A prominent band (q.v.) in which there is a decrease in the pick density of a woven fabric or in the stitch density of a knitted fabric, compared with that of the normal fabric.

Thin shade: Term used particularly when sampling dark colours to describe missing Bloom or character.

Thin Spot: See **Crack mark.**

Thio: Greek. "Thios" = Sulphur.

Thiobenzanthrone: Anthraquinone dyes of the Benzanthronoid vat dyes group. Condensation of 3-mercaptobenza-nthrone with chloroacetic acid, fusion with potassium hydroxide and oxidation.

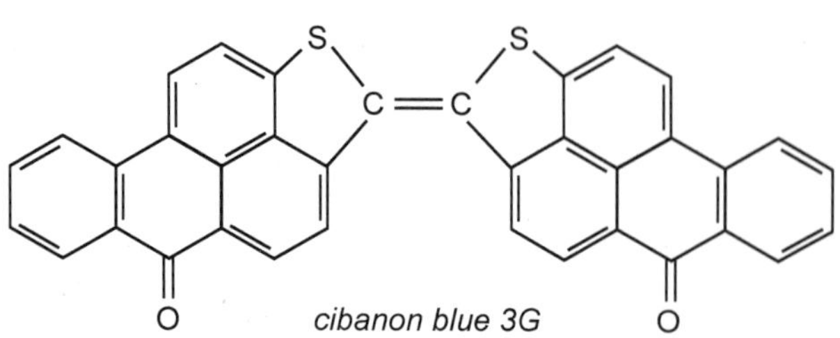

cibanon blue 3G

Thiocarbamide: Thiourea.

Thio compounds: (Gk.: thios = sulphur). They are derived from oxygen compounds whose oxygen was substituted by usually divalent sulphur. Examples are sulphurous acid ($H_2S_2O_3$) and sodium thiosulphate ($Na_2S_2O_3$).

Other thio compounds are, for example, thioethers (alkyl sulphides) of the type H_5C_2–S–C_2H_5 and thiophenols of the type $(C_6H_5)SH$.

Thiocyanates: SCN-type compounds, e.g. thiocyanic acid HSCN and its salts, thiocyanates, such as potassium thiocyanate and KSCN. Thiocyanates are usually colourless and easily soluble; only lead thiocyanate is difficult to dissolve; copper, mercury and silver thiocyanates are not soluble. With iron(III)-chloride solution ($FeCl_3$), soluble thiocyanates result in a typical blood-red colour (iron detection).

Thiocyanate/tin discharges A very aggressive discharge system Tin salt discharges which are carried out under stringent conditions

Thiodiethylene glycol (thiodiglycol): $S(CH_2–CH_2–OH)_2$. Substance with a molecular weight of 122 and b.p. at 282°C which is used as a dye solvent and fixation accelerator for vat printing. Contact with hot hydrochloric acid results in the formation of mustard gas: $S(CH_2–CH_2–Cl)_2$.

Thiodiglycol: Also called thiodiethylene glycol or 2,2'-thiodiethanol Thiodiglycol is sometime used to increase solubility of acid dyes, particularly for making printing pastes. It is very hazardous if mixed with hydrochloric acid. It is difficult to obtain, and there may be restrictions on its sale in some countries.

Thioethers (alkyl sulphides, type: R–S–R1) alkyl-substituted Mercaptans and/or dialkyl derivatives of hydrogen sulphide such as dimethyl sulphide (CH_3–S–CH_3) or methylethyl sulphide (CH_3–S–C_2H_5). They are used as the basis of textile auxiliaries. Thus, for example, the thioether, dodecylmethylethyl sulphide, results in the corresponding cationic sulphonium compound.

Thioglycolate number: Test for evidence of damage to wool (particularly in the case of acid damage) in which the solubility of wool in alkaline thioglycolate solution is determined and quoted as a percentage. Untreated wool has a thioglycolate number of 15.9; after treatment with sodium carbonate: 0.00–6.8; with H_2O_2 : 18.7–25.8, with H_2SO_4 : 27.0–89.5. Determination of the thioglycolate number (based particularly on peptide splitting) is a valuable supplementation of alkali solubility determination (specifically on initiating cystine splitting).

Thionyl chloride: An acid chloride of sulphorous acid: $SOC1_2$; colourless liquid which produces smoke when exposed to air, smells suffocating and is strongly light-refracting. It splits in water to HC1 and SO_2. Thionyl chloride is used among other things for providing evidence of polyvinyl alcohol sizes; pass steam through of an aqueous fibre extract, after cooling add thionyl chloride until yellow coloration indicates total solution has taken place:

change of colour to black-brown, then black precipitation as positive evidence for formation of characteristically coloured polyenes.

Thiourea: Sulphonate urea, sulphonate carbamide, thio carbamide). H_2N–CS–NH_2; molecular weight 76, density 1.40. Thiourea is used as a softening agent to increase the durability of heavy silk; for preventing over-oxidation when developing vat leuco ester dyes; as a vulcanization accelerator; as a padding auxiliary. Thiourea is used less and less frequently for printing polyamide with acid and substantive dyes. As a stable reducing agent in stripping and discharge effects on dyeings formamide sulphinic acid is used from the oxidation of isothiourea.

Thiourea Dioxide: Thiox; a reducing agent used in discharge, stripping and vat dyeing; also called formamidine sulfinic acid or thiox; sometimes abbreviated TUDO; Colour Index Reducing Agent 11 Thiourea dioxide is popular discharge agent. It can also be used for bleaching wool, since will not damage the fibre like chlorine bleach will. Thiourea dioxide is considered to be safe, from a health risk point of view. Thiourea itself, which may exist in very minute proportions in thiox, is known to be a carcinogen. Extra care in handling is warranted. Thiox powder is flammable. Carboxymethyl starch (Monagum) is suitable as a thickener if required.

Thiox: Thiourea Dioxide. A safe replacement for sodium hydrosulfite in vat dyeing, discharge printing, and general colour stripping on cellulose and protein fibres. Good shelf life if stored in a dry location. Five times stronger than sodium hydrosulfite.

Thong hole, in Zipper: The opening at the end of a pull.

Thornproof tweed: A wool type suiting fabric produced from highly twisted yarns containing thorn proof tweed that are closely set to give the cloth a firm and hard handle. Two fold yarns consisting of differently coloured singles are commonly used.

Thousand-point roller: (kiss roller). Has fine point engraving and is used for Kiss roll padding on the printing machine, i.e. for transferring a fine liquid film that may contain additional chemicals and dyestuffs. It is used for pre-moisturizing less absorbent fabrics, for dyeing one side of a fabric and for ludigol treatment, etc.

THPC: Abbreviation for tetrakis-hydroxymethylphosphonium chloride which is used for washing- and dry-cleaning-resistant Flame-retardant finishing of cotton. Originally this finishing process was carried out using a nitrogen methylol compound, such as trimethylolmelamine, under dry crosslinking conditions. Nowadays the finishing process is usually carried out using THPC urea or substituted melamine products and subsequent gas- or aqueous NH3 treatment.

THPOH: Abbreviation for tetrakis-hydroxymethyl phosphonium-hydroxide which is used for flame retardant finishing. THPOH is volatile at higher temperatures and the flame-retardant effect can therefore be partially reduced during machine stops in the finishing process.

THPOH-NH3-Finish: Permanent Flame-retardant finishing for cotton and rayon staple fibre with THPOH or THP salts, such as, for example, THPC.

THPS: Abbrev. for tetrakis ammonium-hydroxymethyl phosphonium sulphate which is used for Flame-retardant finishes of cellulose and cellulose/ polyester mixtures with a minimum proportion of 65% cellulose. The application process involves impregnation, which is carried out first, followed by drying, condensation and after-washing. The finish withstands 50

household washes. When used after the condensation process with the addition of finishing agents it results in a crisp, stiff handle.

Thread: (1) (a) The result of twisting together, in one or more operations, two or more single, folded or cabled yarns (see **under yarn**). (b) A product as defined in (a) above and intended primarily for sewing purposes and known as a sewing thread. (c) A component of silk yarn, and that is the product of winding together, without twist, a number of baves (q.v.), e.g. a 3-thread silk yarn is the result of folding three such products together. 1. The term "thread" is frequently used to describe single yarns.

(2) The term "thread" is also used in such expressions as "threads per unit length", irrespective of their nature.

Thread break, in sewn seams: A mode of failure in sewing evidenced by rupture of the sewing thread.

Thread-by-thread: (Fr.: fil à fil), fabric in plain weave 1/1 or twill 2/2 which is characterized by a special pattern achieved by alternate weaving of light and dark yarns in warp and weft directions. Fil-à-fil fabrics created in this manner are also known as "Pepper and salt articles", black and white and grey-white in point patterns. Thread-by-thread fabrics are used for manufacturing outer clothing.

Thread count: Number of threads in warp or weft direction of a screen gauze per cm (= fabric-no.) or per inch (mesh number).

Thread Fabric: Plain woven, very strong, heavy cotton fabrics, made with ply warp and single filling, the latter placed very far apart from each other. The fabric is used for foundation for tires.

Thread holder: The device on which sewing thread yarn is wound so that it can be transported to and run off smoothly at, the sewing operation. It may be in the form of cone, spool, or tube.

Thread, element attaching, in Zippers: The thread that secures the continuous monofilaments to the tape.

Thread, gut: See **Stuffer yarn, Yarn, Stuffer.**

Thread per sq. cm.: The sum of the number of warps threads per centimetre and the number of weft threads per centimetre in a woven fabric.

Thread per unit length: The number of warp threads (ends) or the number of weft threads (picks) in a specified length of fabric. (a) The unit of length is usually taken as the centimetre, but with fabrics that have less than 10 threads per centimetre, it is advisable to use a unit length of 1 decimetre (10 cm). (b) With fabrics that have more than 10 threads per centimetre, the actual count may be taken over 2 cm, 3 cm or 5cm and the result given by calculation in threads per centimetre. (c) Counting may be done at the following stages of manufacture:

- (i) Finished: The count is taken when no further processing in the piece is prescribed. In all cases, the condition of the fabric at the time the count was taken should be noted.
- (ii) In the Loom: The position of the count should be agreed on. It is usually taken between the fell of the fabric and the take-up roller, with the fabric under weaving tension.
- (iii) Loomstate :The count is taken after the fabric has been removed from the loom and relaxed from weaving tension, but before it is subjected to any further treatment that may modify its dimensions.

Thread out: See **End out.**

Thread Size: Many different thread size systems are used in the world for sewing threads. Generally, the thread size refers to the diameter or thickness of the thread. The Tex Size system can be used universally for all of its thread products. Other thread size systems used include the Cotton Count System; the Metric System, the Denier System and the Silk System used for both Silk and Mercerized Cotton threads.

Thread, Element attaching, In Zipper: Obsolete term.

Thread, silk, for silk threads: A synonym for ply, e.g. 2 thread and two ply, may both be used to describe a silk thread.

Thread sewing: A flexible small diameter yarn or strand usually treated with a surface coating, lubricant or both intended to be used to stitch one or more pieces of material or an object to a material.

Thread-up: Fabric path or thread-up in a Steamer.

Threading hook: A threading hook (or heddle hook) is a long slender piece of flat metal with a handle at one end and a tight curve at the other to catch and pull warp threads through the heddles.

Threading, in weaving: The threading is the order in which each warp end passes through a heddle on a specific shaft.

Treadls: The treadles are pedals attached to shafts to make them go up and/or down in all of the combinations required by the weave structure.

Threeply carpet: (Scottish carpet). Reversible carpet made of three interwoven fabric layers.

Three-Quarter: Goods Measuring 27 inches in width.

Three-quarter cloths: Shorn soft and semi-dull brushed fabric (woollen cloth), belonging to the class of so-called fine cloths, in linen weave, not too heavy, i.e. below 400 g/m^2.

Three-roller yarn: Generally cotton or viscose spun yarns (as opposed to two-roller yarns) from so-called three-roller spinning which works on the principle of the stretching process (3-cylinders as pre-spinning machine, also known as flyer, for stretching, twisting and parallel alignment of tow).

Three-stage drip-in dyeing process: This is used for reactive dyes in exhaust dyeing processes. Especially recommended for jet dyeing machines. Constant temperature of 60°C. Addition of dye in two stages and of trisodium phosphate in three stages.

Threaded-Roll Process: A high-speed method developed by Celanese for converting crimped continuous filament tow into highly bulked, uniformly spread webs of up to 108-inch widths. The webs are useful in a variety of products, such as cigarette filters, sleeping pillows, and battings.

Threadbare: A fabric which lost all its nap and the foundation threads are visible.

Threadlines: The fibre lines of a manufactured fibre in extrusion or subsequent processes.

Thread out: See **End out.**

Threadup: The process of directing or threading fibre or fabric through all machine positions to start or restart a process, or the configuration resulting therefrom.

Thready: Fabric finished to show every thread on the face.

Three-Bar Fabric: A tricot fabric made on a machine equipped with three guide bars.

Three-Dimensional Weaving: To produce three-dimensional textiles, yarns are simultaneously woven in three directions (length, width, and thickness) rather than in the conventional two. The types of structures that can be produced fall into four broad classes: (a) contoured fabrics, (b) expandable fabrics, (c) interwoven fabrics (Also see Double weave), and (d) contoured interwoven fabrics.

Threeply Carpet: (Scottish carpet). Reversible carpet made of three interwoven fabric layers.

Threshold effect: Delayed precipitation of low solubility compounds that is shown by polyphosphates and also by complex phosphates in similar fashion.

Throats, In Zipper: The two opening in a slider that receive the stringers.

Through and Through: Wool fabrics, the face and back of which are made alike.

Throughput: The amount of raw material processed in a specific time. This is the actual amount, not a percentage.

Throw: A removable unfitted protective textile cover, used over upholstered furniture.

Throwing: The operation of doubling or twisting silk or manufactured filament yarns.

Thrown silk: See **Silk, thrown.** Same as reeled silk, and is made into organzine or tram.

Thrown Singles: A single silk filament "thrown" with about 15 turns to the inch.

Throwster: A company that specializes in putting additional twist in yarn. More recently, the term also applies to a company that specializes in texturing yarns.

Thrum: A waste length of warp (yarn) or of fabric, or both, formed during the preparation of loom for weaving. (a) A thrum may be formed as follows:

(i) During the adjustment of a loom at the commencement of the weaving of the warp. When the loom is correctly adjusted, the portion of the warp that contains picks inserted for testing the adjustment of the loom mechanism is cut off.

(ii) During warp replenishment in a loom. The old warp is twisted or knotted to the new warp and, if the new warp is drawn through by weaving, the point in the woven fabric at which the twisted or knitted warp ends occur is called a "through" because the fabric is cut through to remove the thrum containing the imperfect fabric formed by the twisted or knotted warp ends.

(iii) During loom operations away from the loom. In the above cases, a thrum consists of portions of the old and the new warp ends twisted or knotted together.

(b) A thrum may also be: (i) A length of warp ends cut from the warp for the purpose of evaluating the percentage of applied size; and in repairing end-breakages in the warp concerned; (ii) Any loose end(s) of warp; (iii) A bundle of coarse yarns tied together by twine for use in making a mop.

Thunder and Lightning: Same as Oxford gray.

Tiaoyong: Warp printed Chinese silk plush.

Tibet Wool: (Tybet). Term used for unmilled fabrics sorted as Rags and used to produce Reclaimed wool of quality category II.

Tibetan cashmere: is derived from the Tibetan goat which lives in Tibet and is closely related to the cashmere goat. Tibetan cashmere is somewhat rougher and a little more glossy in appearance. Production, properties and utilization are similar to cashmere. Animal hair fibres.

TIC: Abbreviation for "Total Inorganic Carbon"; value used in Biological waste water treatment, quoted in mg/l waste water or in mg carbon/g substance.

Ticket number: The tex ticket number assigned to a sewing thread to designate its approximate linear density of the sewing thread.

Ticking: A very strong, closely woven twill fabric. It is stiff and recognised mainly by the fact that it is woven in narrow stripes of a colour and white, but it can be plain. It is usually made of cotton, although sometimes linen is added. The coarsest ticking are called straw tickings because it was used as covers for straw filled mattresses. A characteristic construction is 64 x 50 in 2/1 twill with 12s warp and 14s filling. Regular tickings usually woven with 2/2 twill with a characteristic construction of 80 x 90 with 12s warp and 16s filling. Finest ticking usually called sa teen ticking are made with the 5 harness warp flush satin weave. The construction will be around 100 x 50 with 14s warp and 20s filling. Not an attractive fabric, normally used loomstate, although it occurs occasionally as a fashion fabric for womens jackets and trousers. Main use of tickings are mattress and furniture covers and as backings for rubberised materials.

Ticking, nankin: Same plain weave fabrics which has the same type uses are sometimes called Nanking tickings. It is made with unbleached yarn in the warp and solid coloured yarn in the weft. A typical texture is 80 x 80 with 22s warp and weft.

Ticking: Densely woven, generally striped patterned linen, cotton or half linen fabrics in twill weave. See **Drill fabrics**, **Ticking, nankin.**

Tie: (tying). This is carried out for patterning of: – warp or weft threads (→: Ikat; Kasuri dyeing), – fabrics (See: Bandhana; Plangi; Shibori technique). Tying is carried out by binding with threads which prevents penetration of the dyes on subsequent dyeing. Machine tying of yarns has also been carried out for some time using wires which give reproducible patterning.

Tie & dye: A technique of dyeing to produce unique patterned effects by tying and or knotting the material befor immersion into a dyeing bath, to rsist pemnetration of dye. Fabrics are made into tiny puffs with some object inside and tied with a waxed thread wherever the dye has to be prevented. The fabric is immersed in the dye solution. If two or more colours are desired the thread is removed and the fabric retied. After drying the object is removed. Other methods of tie and dye include folding the fabric and stitching it and pulling the threads to draw the fabric to resist the dye from penetrating into the fabric, called as tritik. Tie & dye fabrics are quite popular in apparel and home furnishings. See **Bandhani.**

Tie cloth: Any soft or crisp fabric suitable for men's ties or cravats, including tie silk and Faille, Pongee, in twill weave, acetate, polyester or silk fibres.

Tie-back: See **Sticker.**

Tie-up: The tie-up shows the shafts that must be raised and/or lowered by each treadle to make each shed required by the weave structure.

Tieing technique: Fabrics (and in many cases finished items of clothing, particularly T-shirts) are tied in a certain way along the lines of tie-dyed batik or tie dyed technique prior to dyeing resulting in interesting patterns. Dependent on the type of Tie, effects known as the "Mukeka Mutono Special", the "Grill" or the "Tortoise" are given.

Tier drier: Driers through which textiles are moved in several tiers.

Tight end, in woven fabrics: A yarn which was woven under excessive tension or has shrunk more than a normal amount.

Tight filling: See **Tight pick.**

Tight end: A warp thread or part of a warp thread that has less crimp in the fabric than have the adjacent normal ends. This may be owing to weaving under greater tension or to abnormal stretching of the yarn during some process prior to weaving. It may be caused by excess moisture, e.g. during winding, and consequent contraction during finishing.

Tight pick, in woven fabrics: A filing yarn which was woven under excessive tension or has shrunk more than a normal amount, which may cause puckering at the junction with normal picks giving a wavy or ruffled surface appearance.

Tight pick: A weft thread or part of a weft thread that has less crimp than have the adjacent normal picks. This may be owing to a weft yarn having been inserted under greater tension than that imposed on the other weft yarns, or to the relaxation of a weft yarn subsequent to insertion, or to abnormal stretching of a yarn during some process prior to weaving. It may also be caused by the presence of excess moisture, e.g. during winding, and consequent contraction during finishing. (See also **shiner**).

Tight selvedge, in woven fabric: Self descriptive. Selvedge yarn shorter than warp yarn in the body of the fabric.

Tight spot, in textured yarn: Faults in fabric twist textured yarn consisting of small isolated regions of twisted yarn each containing a few turns of twist.

Tight threads: Warp or weft threads caused by too great a tension (usually due to the state of humidity during weaving so that these threads in the fabric shrink during drying) which can result in unevenness in fabrics which has to be eliminated by stitching or at least improved.

Tight/slack thread: A thread or pieces of thread which are tighter or slacker than the other pieces/threads.

Tight twist end: A single end with higher than normal twist.

Tightness factor: See Cover factor.

Tile (or brick) repeat: A simple repeat where the motifs are repeated rather like a simple brick wall pattern. The second row slides halfway across, in a width ways direction.

Tillandsia fibre: (1) (American moss, tree hair, Louisiana moss). This is produced from the jointed and branching stem of the rootless tendril creeper (tree parasite) whose three metre long stem hangs from branches in beard-like fashion. The unpeeled raw fibre is grey-greenish white; the peeled pure fibre is brownish black and is often dyed black artificially. The fibre bundles are separated using boiling sodium carbonate solution. Tillandsia fibre is characterised by great durability and strength and is often used as a substitute for horse hair and as upholstery material. (2) Kenaf.

Time of half-dyeing: Time in minutes which is required for 50% of the amount of a certain dye contained in the liquor to be absorbed by the fibre. Unsuitable as a basis for selection of combination dyes for cationic dyes on polyacrilonitile. Compatibility value is better suited.

Time management: Making the best use time to achieve what is necessary.

Time-to-break: The time interval during which a specimen is under prescribed condition of tension and is absorbing the energy required to reach maximum load.

Timing, Advanced in knitting: is the reverse of delayed timing. The cylinder loops rob from the dial, producing tighter dial loops; advancement can only be about one needle.

This type of timing is sometimes used in the production of figured ripple double jersey fabrics, where selected cylinder needles can rob from the all knitting dial needles

Timing, Delayed, also called rib or interlock timing the dial knock over occurs after about four cylinder needles have drawn loops and are rising slightly to relieve the strain.

Timing, in knitting: The order the needles in the dial and cylinder go through the knitting cycle in relationship to one another. The cylinder needles that correspond to dial needles may go through the knitting cycle before or after the dial needles.

Timing, Needle: is the relationship between the loop-forming positions of the dial and cylinder needles measured as the distance in needles between the two stitch cam knock-over points.

Timing, Synchronized: also known as point, jacquard and 2 × 2 timing, is the term used when the two positions coincide with the yarn being pulled in an alternating manner in two directions by the needles, thus creating a high tension during loop formation.

Tin chloride: (tin (II) chloride, read tin two chloride) SnCl ; also called stannous chloride 2. Formerly, tin (II) chloride was often used as a reducing agent for discharge printing. In most cases sodium formaldehyde sulphoxylate or zinc formaldehyde sulphoxylate has replaced it in modern processes. It still finds some application for discharge on acrylic fabrics. It is also used as a mordant with some natural dyes.

Tin salt discharges: Reductive discharges with tin(II)-chloride. No longer generally used on cellulose fibres. Was used frequently prior to introduction of sulphoxylates. On acetate and triacetate. See **Acetate fibres in discharge printing.**

Tinsalt weighting: Weighting of silk using TinSalts. The disadvantage is the strength reduction of silk at higher weighting.

Tinsel: Very narrow, flat strips of gold, silver or other metal, used as cover for core yarn or used directly for embroidery and brocades.

Tint: (verb) To mix white pigment with absorbing (generally chromatic) colorants.

(noun) The colour produced by mixing white pigment with absorbing (generally chromatic) colorants. The resulting mixture is lighter and less saturated than the colour without the white added.

Tinting Of Bleach Goods: (blueing dyes). Blue, red and violet dyes for application during Tinting of bleached goods. In small quantities tinting dyes achieve a significant increase of the white appearance both with optically-brightened material and with optically brightened ground white. Dyes used for this purpose are as brilliant as possible in order to reduce the absorption range for as narrow a band as possible. On the other hand dull tinting dyes, such as mixtures of a brilliant dye with grey, reduce the brightness unnecessarily.

Tip Definition, in carpet: Visible single ends of cut yarn.

Tip Effects: Tip effects on pile fabrics; vivid patterned woven fabrics or furs are produced by surface coloration of the extreme tips of the pile. See **Emboss printing.**

Tip printing: Only the surface of the embossed sites of the fabric is dyed on the embossing calendar directly after embossing, namely by means of an overdyeing unit situated on the cloth delivery end. An intensified relief effect is relatively easily obtained by this process. Localized tip-dyed effects can also be produced with suitably engraved rollers. See Embossed printing.

Tipping, wool: See **Wool, tipping.**

Tippy Dyeing: The tips of woollen fibres are exposed to light, weather and environmental influences during their growth. For this reason tips and root-ends dye differently (Skittery dyeing wool). The extent of the colour difference between tips and root-ends depends not only on the wool but also on the category of dye and the process used.

Tip-Sheared Carpet: A textured pile carpet similar to a random-sheared carpet, but with a less defined surface effect.

Tiraz: Very rich Arabian silk fabric with names of Sultans and other prominent people interwoven; used as material to write on in medieval Europe, when it was solid coloured.

Tiretaine Serge: Usually made with linen or hemp warp and inferior woollen 'filling, occasionally also of all wool. It is a strong, stout fabric, finished either pressed or with the nap raised; used for working dresses in France.

Tirtey: Reasonably-priced and generally durable fabrics which consist in the warp of fine-threaded strong cotton or rayon staple fibre yarns but of carded yarn (often reclaimed wool with the addition of cellulosics) in the weft. Finished with either Melton or brushed effects.

Tissue: The term used to describe a light transparent cloth in any fibre. It usually means one that has body and is not soft and floppy.

Tissue: (1) A general term used especially in Great Britain for woven fabric, but usually applied to thin, sheer fabrics. The term is also used as an adjective to denote thinness and sheerness. (2) A term applied to 15th century cloths of gold and medieval silk damask with gold and silver threads interwoven. In Jacquard weaving today, tissue is decorative yarn added to fabric in addition to base fabric. Derived from tile French "tissue" which means fabric.

Tissue Taffeta: The finest, lightest taffeta. It is most transparent and normally only used for underwear or for lining fine garments.

Tissute: French serge, having eight ends and four picks in a repeat.

Tissuti: See **Amamee**.

Titan Braid: A coarse, flat military braid made of coarse long wool. Called aleo Hercules braid.

Titanium dioxide: Titanium(IV) oxide, TiO_2; molecular weight 70, density 3.8 . The white powder is not soluble in water or dilute acids. It is used as a white pigment with a high covering capacity in paint coatings and as a matting agent for fibres, in paint and frosted glass. It is usually supplemented by calcium or barium sulphate. A compound (TiO_2) that occurs naturally in three different forms (rutile, anatase, and brookite).

Titer: See **Density, linear.**

Titration: is used for quantitative content determination of a solution to be tested by means of a normal solution of known content which flows into it (using a burette or a pipette), Normal solutions. The end of the reaction (neutralisation, oxidation, reduction, precipitation) is shown by the colour change of an added indicator. The content calculation is made from the relationship between the utilized quantity of solution to be tested and the consumed normal solution.

Titre: French term for the size of the silk thread. It is expressed in deniers in Europe. The international denier being the weight of 10,000 meters of silk yarn. The titre in England and in the United States is expressed in the weight of 1,000 yards in drams.

Titrimetry: See **Titration**.

Tjap print: Japanese stamp batik which is carried out using so-called tjaps, stamps made of copper strips (block fastened). The process is more rapid than the free-hand method using the See **Tjanting.**

TLV: Abbreviation for Threshold Limit Value.

TLV values Maximum threshold limit values in the workplace are the highest permissible concentrations of substances such as gas, steam or other airborne substances which, on the the basis of current knowledge, do not impair the health of employees in the case of repeated long-term exposure and likewise do not place undue strain on them (BAT value).

Tobacco cloth: See **Cheese cloth.** A thin, lightweight, open cloth used to shade and protect tobacco plants.

Tobines: Stout, strong, twilled silk dress goods in France.

TOC: Abbrev. for Total Organic Carbon.

Tochirimen: A cotton crepe made in Japan.

TOD: Abbrev. for Total Oxygen Demand = total content of oxidisable substance (inorganic and organic substance, including nitrogen and sulphur) which is quoted in mg oxygen/l waste water or in mg oxygen/g substance.

Tod: Equal to 28 pounds; used for measuring wool and top in England.

Toe closing: In knitting hosiery, this term refers to closing the toe opening. It may be knit closed, or in tube hosiery, sewn closed.

Toga: Outer robe of the Roman ruling class; an oval woollen cloth folded lengthways and draped artistically around the body.

Toggle, in button: A clip used to fasten a staple button to the flexible substrate.

Toile: This is the French for fabric. (a) A broad term describing many simple plain weave twill fabrics, especially those made from linen. (b) Sheer cotton and linen fabrics.

Toile d'Alsace: Fine thin French linen dress goods, made white or printed.

Toile d'Araigner: Open work French woollen dress goods.

Toile Bleue: Fine linen dyed light blue.

Toile a Bluteau: Sort of bolting cloth in France.

Toile de soile: Warp and weft of schappe twisted thread; a soft rather matt fabric.

Toile a Chapeau: In France highly glazed, and stiffened linen or cotton fabric; used for hat shapes.

Toile de Chinese: Damask table linen in France.

Toile Ciree: Fine French oilcloth

Toile de Coffre: French household linen of good grade.

Toile Colbert: Loosely woven cotton or woollen canvas; used for embroidery.

Toile De Coton: Light dress goods of mixed cotton and linen with woven stripes and printed patterns.

Toile Ecru: Unbleached linen.

Toile d'Emballage: Packing cloth.

Toile d'Embourrure: Linen cloth used as lining.

Toile de Frise: A very fine Holland linen.

Toile de Halles: Stout, unbleached linen.

Toile de Laine: Very soft, light weight, French dress goods, made of merino wool, usually all black, in plain weave.

Toile de Lille: Fine French table linen, often made with coloured stripes.

Toile de Mulquinerie: In France name for the finest lawns, cambrics, linen batiste, etc.

Toile du Nord: French gingham, made in checks with a smooth finish.

Toile d'Orange: Fine stout calico made in France.

Toile d'Ortie: Sort of French batiste made of nettle fibre.

Toile d'Ourville: Unbleached French canvas.

Toile de Religieuse: French for nun's cloth.

Toile Satinee: Very soft and thin 'French cotton cloth, in plain colour or printed.

Toile de Saxe: Plain woven French dress goods, made with cotton warp and worsted filling.

Toile de Sion: Medieval printed linen.

Toile de Sole: Very light, plain woven silk fabric; used for scarfs.

Toile a Tamis: Blue buckram, with prominent stripes.

Toile a Veste: A striped or checked, or solid coloured, plain woven cotton or linen fabric, used for lining in France.

Toile de Vichy: Light French linen dress goods in pink and white or blue and white stripes.

Toile Victoria: All-worsted, light French fabric of plain weave in the 19th century.

Toile a Voile: Sail cloth.

Toile de Vosges: Coarse, stout, plain woven cotton fabric, made in France for the African markets.

Toilet Cloth: Another name for quilt.

Toja: Same as Bun ochra.

Tokhfil: Two-ply silk thread, rereeled from spools by the natives in Central Asia; used for filling.

Tol: A fine and narrow strip of cloth often woven with complicated patterns. Made by the natives of the Caroline Islands and used as girdle, apron, etc.

Tolerance: The amount of acceptable difference between a known correct standard (usually the customer's specifications) and a set of measured samples.

Tolerance, in mathematics: Prescribed limits of variation for specified

Tolerances, when included in a contract: The limits to which the average (or other specified) value observed for a specific characteristic of a lot of material must confirm to constitute a valid delivery. The observed values must have been obtained by specified test methods on sample representatives of the materials to be tested.

Tolotzin: White bast fibre, yielded by the Heliocarpus tree in Mexico.

Tonder Lace: (1) Danish bobbin lace originated in the middle of the 17th century. Early specimens show Flanders influence with trolley latter specimens have fine Malines foundation with various running designs; (2) Drawn work made of fine cambric.

Tone-in-tone dyeing: (1) "Same shade" dyeing in the sense of a single colour shade (Self shade) with mixed-fibre articles and yarn mixtures; both components of the mixture demonstrate the same colour shade. (2) "Colour shift". An example is the Multicolour process in which only one type of fibre is used but of which a section has assumed greater capacity to absorb or to resist dye as a result of appropriate pretreatment. The dyeing process results in different colour shade intensities with the same colour shade.(3) Dyeing of ombré or shaded effects.

Tonga Salempore: A loosely woven, netlike cotton fabric of plain blue colour, made with wide white headings and a stiff finish; used for garments by the natives in South Africa.

Tongue tear strength: The average force required to tear a rectangular sample with a cut in the edge at the center of the shorter side. The two tongues are gripped in a tensile tester and the force required to continue and tear is measured.

Toning: Previously-practiced method of after shading (Shading) on fast dyes, usually vat and sulphur dyes, (due to the lack of suitable products in these groups) with small quantities of brighter dyes from other categories (usually cationic dyes). This has the effect or reducing overall fastness to a greater or lesser extent. Dyes with 1 : 2-metal complex dyes are used predominantly for toning using acid dyes.

Top: After-dyeing of fibres in mixtures, particularly to light dyed fibres in single-bath or two-bath systems. Used predominantly for reserve and for creating two- or multi-colour effects.

Top, in man-made fibres: A sliver obtained by drafting, along with breaking or cutting a multifilament tow. (a) Worsted process: A sliver in which the fibres have been parallelized and usually combed. (b) Man made fibres or tow to top process: A sliver obtained by drafting, along with breaking or cutting a multifilament tow.

Top, in wool: A continuous untwisted strand of wool fibres from which the shorter fibres or noils have been removed by combing.

Top, in worsted processes: A sliver in which fibres have been parellelised, and usually combed.

Top cloth: Generic term for all materials which determine the external appearance of an item of clothing.

Top colours: Colors used on the ground colour to form a design

Top dyeing: (1) The process of covering with an additional dye, not necessarily of the same colour or class, to obtain the desired shade. (2) Fibre in top form is placed in cans and dyed in a batch-dye vessel with reverse cycling capability. An expensive process that is used primarily for fancy yarns.

Top fusing: Term used to describe the direct bonding of lining fabrics for collars and sleeves.

Top stop, in Zipper: A part affixed between or immediately above the interlocking elements on either or both stringers, to prevent the slider from leaving the chain.

Topcoat: A light overcoat.

Topper: (1) Automated machine for trouser waist-band steam pressing machine after dry cleaning. It replaces the time-consuming and technically expensive process of manual steaming or pressing of trouser waist-bands. (2) Garment manufacturing term: a) leisure jacket with the face edge and under seam at right angles; (b) short, elegant gentleman's summer coat, fashionably wide and with lightly padded chest.

Topping dyes: Dyes for topping which as far as possible do not dye the foreign fibre or only slightly. They should be simple to apply.

Topstitching: Often used for decorative purposes, but can add durability. A line of stitching that that shows on the face side in the finished article, usually stitched while having the face side of the fabric up, but when sewn close to the seam edge, it is called edge stitching and can be used instead of understitching. See **Understitching**.

Toque: Small flat hat with a narrow brim or brimless.

Torchon lace: A coarse lace made from cotton or linen thread, often with a shell design. It is inexpensive and not hard wearing, due to the fact that the threads are loosely twisted and it is quickly made.

Torque: A moment (of forces) which produces or tends to produce rotation or torsion.

Torque yarn: See **Bulk yarn.**

Torque yarn: Yarns which are fabricated discontinuously by means of high twist, fixation and un-twisting ("conventional" or "classical" process) or continuously (false twist process). Simple yarns with twisting tendency (Up-twisting of suitable high twist yarns with S- and Z-twist gives compensating possibilities).

Torsion, residual: See **Residual torsion.** Revolutions made by a specified length of cord when one end in a flexed position and the other allowed to turn freely.

Tortin: Low grade French wool carpets.

Tossed pattern: An arrangement of motifs where the motifs in a repeat do not occur at regular intervals. It is as if the designs are dropped in a fairly random way onto the fabric.

Total colour difference DE*: The CIE Lab Colour Space total colour difference between two samples is termed DE* and is equal to the square root of the sum of the squares of DL*, Da* and Db*. Where thendesignation Da* for a difference in red/green value and the designation Db* for a difference in yellow/blue value. CIELAB total colour differences ΔE* can be expressed either as the co-ordinates of colour space or as the correlates of lightness, chroma and hue. Hence

$$\Delta E^* = [(\Delta L^*)^2 + (\Delta a^*)^2 + (\Delta b^*)^2]^{1/2}$$

or

$$\Delta E^* = [(\Delta L^*)^2 + (\Delta C^*)^2 + (\Delta H^*)^2]^{1/2}$$

Total colour management: Special software for recording, monitoring and evaluating dye strength, substrate routine, correlation, transferability from laboratory to production, specification and corrections calculation for continuous and exhaust dyeing, batch production, storage contents and recipe recording, machine programmes, management programmes, colour kitchen software etc.

Total crotch length, in body measurements: The distance from the right shoulder line midway between the neck base and the shoulder joint, down the back through the crotch and over the projection of the right breast to the starting point.

Total denier (Of Tow): The product of the denier per filament times the number of filaments in a tow.

Total elongation of rope: The entire elongation at any given applied force.

Total hardness: The total of Temporary and Permanent Hardness of a water sample.

Total irradiance: Radiant power integrated over all wavelengths at a point in time expressed in watts per square meter (W/m^2).

Total mass, in pile yarn floor covering: The mass of all matter in the mass per unit area expressed in grams per square metre (Ounces per sq. yd.) or in grams per linear metre. (ounces per linear yard).

Total oxygen demand, TOD: TOD can be measured by the depletion of oxygen in A nitrogen oxygen carrier gas in a platinum catalysed combustion chamber.

Total reflectance: Reflectance of radiant flux reflected at all angles from the surface, thus including both diffuse and specular reflectances.

Total reflectance factor: The total reflectance factor is indicates the relation between the reflected and the incident light. It independent of the illuminant used in the measuring instrument . This statement is valid as long as no fluorescence is present. With fluorescent, e.g. optically brightened samples the total reflectance factor heavily depends on the distribution of energy and the intensity of the illuminant used in the photometer. For this reason with optically brightened samples it is essential to use an illuminant similar to daylight.

Total thickness, in floor covering: The distance between upper and lower surfaces of a material including the pile and backing measured under specified conditions (pressure).

Total vertical trunk length, in body measurements: The distance from the right shoulder line midway between the neck base and the shoulder joint, down the back through the crotch and over the projection of the right breast to the starting point.

Total weight: The mass per unit area expressed in ounces per square yard or grams per square meter.

Toughness: The property of a material by virtue of which it can absorb energy.

Toughness, breaking: The actual work per unit volume (or per unit mass) of material that is required to rupture the material. It is proportional to the area under the load elongation curve from origin to breaking point.

Toughness At Rupture: Toughness of a material to breaking or tearing apart.

Tournay carpet: Named after the Belgian town, a wire insertion pile velour carpet with numerous and varied coloured warp pile giving sharply defined jacquard patterning; also seen as Face-to-face Tournay carpet in a double velvet technique.

Tow: (1) The coarse, entangled and shorter flax fibre, separated from the linen in the hackle; used for tow yarn, upholstering, binder twine, lining refrigerator cars, etc. (2) Combed top. 3. 3. spin, fibre tow: manufactured form of chemical fibres which are combined taken off the spinning jets as a bundle of parallel aligned, single endless fibres; virtually no twist Multifilament with counts of > 3 ktex. Further processing is by means of cutting and breaking (converting) or by stretch-break or texturising. (4) Marine rope.

Tow conversion into staple fibre: This is a process, by which fibre is cut into staple yarn employing mechanical cutting devices called "staple cutters" and generally placed in the final stage of the fibre production lines. With the time, various fibre cutting systems have been developed, but only few are now still operative, at least for the production of standard staple fibre for textile use. E.g. "Lummus" staple cutting machine, "Gru-gru" staple cutting machine

Tow dyeing process: Dyeing facilities (perforated-drum or perforated conveyor system) for continuous processing for the dyeing of spinning tow.

Tow Linen: These are the shorter, less desirable flax fibres separated from bast line fibres in hackling. Tow linen is usually carded and spun into a woollen-style yarn. A wonderful use for tow linen is to knit bath mitts -- a situation where you want all of the rough, scratch nature of this yarn.

Tow, in manufactured fibres: A twistless multifilament strand suitable for conversion into staple fibres or sliver, or for direct spinning into yarn. A large strand of continuous manufactured fibre filaments without definite twist, collected in loose, rope-like form, usually held together by crimp. Tow is the form that most manufactured fibre reaches before being cut into staple. It is often processed on tow-conversion machinery into tops, sliver, or yarn, or on tow-opening equipment to make webs for various uses.

Tow yarn: Yarn spun from tow.

Towel: An absorbent textile product used for drying or wiping.

Towelling: Towelling is madder from terry cloth which is given elsewhere. The heavier weight terry cloths are used for towels have atypical construction of 60 × 30 with 2/20s warp and 10s filling. Still heavier construction are used for bath matts. The heavier towels are made in plain weave either from cotton or a mixture of cotton with linen tow. An average construction is about 40 × 36 with 15s warp and 8s filling. Huck back or huck towels have a pebbled surface caused by the weave. A common huck back texture is 60 × 40 with 16s warps and 10s filling.

Towels, glass: These are the costliest of the kitchen towels. They are made of linen or of long staple cotton yarns. After being carefully combed to remove all lint the yarns are spun with high twist and are then singed to remove all protruding fibres. Linen or part linen glass towels are made with plain weave with a typical construction of 60 × 40 with 20s warp and 15s filling and decorated with coloured stripes.

Toxic Unit Acute (TU_a): The TU_a is defined as reciprocal of the waste water concentration that caused the acute effect by the end of the exposure period.

Toxic Unit Chronic (TUC): The TU_C is defined as the reciprocal of the effluent concentration that caused the acute effect by the end of the exposure period, are no different from the control

$$TU_C = 100/\,NOEC$$

where NOEC = No-observed–effect concentration.

Tracery: Name for the raised work in Honiton bobbin lace.

Tracing Braid: Narrow military soutache.

Tracing Cloth: Fine, plain woven cotton or linen cloth finished with heavy dressing, but transparent; used by artists and architects for drawing or tracing.

Tracing wheel: Used to make mark perforations on the fabric to indicate seams, darts, and the position of pockets, necklines, pleat widths, etc. By using a carbon paper, these can be used to transfer markings on both sides of the fabric. The tracing wheel should be made of steel and have sharp edges.

Track suit fabric: A synthetic knit with slightly sueded backing made from polyester or polyester/cotton. It is of medium thickness, and in plain bright colours, black and white. Used for track suits, anoraks, trousers, shorts and it is an excellent fabric for children's clothes.

Tracking (in Carpet): A general change in the appearance of a textile floor covering from edge to middle of a narrow band caused by repeated walking over the same area which may result in a localized change in the orientation and may be irreversible. See also Shading in carpet.

Trademark: Legally protected word and/or symbol denoting goods which have been fabricated in the same way by a certain company and can be differentiated from similar goods, such as, for example, the Indathren trademark. Trademark designation by the letter ® represents the so-called "Registered Trade Mark". (Quality symbols).

Tragacanth: (gum tragacanth), Vegetable gums produced from the southern Astragalus species which has various forms, is up to 20 cm long and yellow-red brown in colour (also bleached white); normally found in powder form. Tragacanth consists of 50–70% Bassorin.

Good quality tragacanth is transparent. The following types are differentiated: Smyrna tragacanth, Morea tragacanth (Greek), Syrian tragacanth

Tram (Silk): Yarn of two or more greige threads which have been doubled without preliminary twisting and twisted with 150-200 twists per metre. A lofty voluminous yarn, soft and supple when boiled off it has an elegent lustrous effect. Dyed as yarn it is a preferred weft material for colour woven material. More strongly twisted it can also be used for knits.

Tramlines: (1) Localized faults in fabrics, in the warp direction (dent stripes). (2) Missed sections in direct printing due to uneven repeat.

Trame: French for tram and for filling in general.

Trammage, in woven crepes: A puckered area in which a filling yarn has twist running in the same direction for several picks instead of alternating S and Z twist.

Tramped Dornoch: Scotch linen cloth.

Transcendent, in testing: This is the meaning that many people connect with the word quality. It implies that the product has an elusive something that makes the product better than all the competing products. Another view is that the product is superior to all competing products in every way possible. The problem with this approach is that quality cannot be defined in a way that can be used for quality management.

Transesterification: In the production of polyester from dimethyl terephthalate and ethylene glycol, the process of exchanging ethylene glycol for the methyl groups to obtain bis-®-hydroxyethyl terephthalate. The methanol generated in the reaction is removed as it is formed to drive the reaction to completion.

Transfer coating: Process for coating or laminating fabrics in which adhesives and coating compounds are not applied directly to the fabric width but are first applied onto a specially prepared paper (release paper) or endless metal band, onto which they are dried and subsequently combined with the fabric. This circumvents the problem of unwanted push-through of coating compounds.

Transfer printing: Indirect Transfer printing processes. Printing does not take place directly onto the fabric but onto paper. During the transfer printing process a paper printed with special disperse dyes is brought into contact with fabrics, knitted fabrics or nonwovens made of synthetic fibres (particularly polyester, triacetate, possibly polyacrylonitrile and polyamide) at a temperature of 190–215°C for 18–30 s. This causes the greater part of the printed dyes to sublime from the paper and be adsorbed onto the fibre material causing it to take on the pattern.

Transfer paper printing/gravure screen size: For printing paper for Transfer printing with water based printing pastes on a vertical roller printing machine a 60 mesh screen is used. Eight colours can be printed onto the paper using the Portalrouleaux printing machine prior to drying in an overhead drier. Disperse dyes are normally used.

Transfer printing calenders: For standard fabric widths with a roller diameter on the electrically heated cylinder of approx. 800 mm, for production speeds of approx. 10 m/min.

Transfer printing Dyes: Non-sublimation-fast Disperse dyes which (when printed on an intermediate carrier) are transferred to the textile substrate during transfer printing at high temperatures.

Transfer printing papers: Are available in two categories: omnibus and exclusive quality. The former quality is supplied at a reduced price to all interested parties without any type of limitation, usually with a minimum quantity of 500 m. The second quality can be reserved for certain companies for the duration of a year or longer at a corresponding surcharge with no minimum quantity being specified. Manufacture is by means of gravure printing, offset printing, screen printing.

Transfer printing Resist Effects: Printing over or under a board with a print paste in which a blocking agent ensures that the dyes on the board are not transferred. There are two possibilities: a) blocking agent in the form of reactant resins, e.g. hexamethoxy methyl melamine and acid catalysts; b) blocking by means of metal salts with complex-forming capability. All salts of multi-valent ions such as iron, chrome, cadmium, aluminium can adopt this function. Copper is the most suitable in the form of acetate in a copper/amine complex.

Transfer tail: A long end of yarn wound at the base of a package that permits increased warping or transfer efficiency by providing an easily accessible connecting point for the succeeding package.

Transition temperature: A temperature at which some radical change, usually a phase change, in the appearance or structure of a substance occurs.

Examples of transition temperatures are melting point, boiling point, and second-order transition temperature.

Transmittance: The ratio of transmitted light to incident light under specified geometric and spectral conditions. Regular transmittance (of transparent materials) is the ratio of undiffused transmitted light to incident light.

Transmittance factor: The ratio of the light transmitted by a specimen and evaluated by a receiver to the light passing through the same optical system and evaluated by the receiver when the specimen is removed from the system.

Transparency: The measure of light transmitted by a transparent medium. It is a function of the wavelength of light and decreases exponentially as the wavelength of the incident light increases. The transmission ratio is the ratio of the transmitted luminous flux to that incident upon a transparent medium. It is the reciprocal of opacity.

Transparent: Describes a material that transmits light without diffusion or scattering.

Transparent prints: Cotton fabrics can be made transparent (parchmented) by printing with suitable chemicals. See Transparent and opal finishes.

Transparent velvet: Lighter in weight than Chiffon Velvet, soft and silky with good draping qualities. Made with a silk viscose or acetate backing and viscose or nylon pile.

Trapezoid tear tester: See **Elemendorf tear tester.**

Trapezoid tearing load: The force required to propagate a tear in a specimen that has been prepared in the shape of a trapezoid and the which has been clamped for testing on the opposing edges of the nonparellel sides.

Trapped end: An end that is unable to unwrap or unwind from the beam. Trapping of an end may be prolonged or intermittent depending upon the cause of trapping (e.g., rolled ends at the selvage, short ends, or mechanical difficulties).

Trasfer paper printing: See **Transfer printing.**

Trash, in cotton: Undeveloped seeds, motes, small bits of seed coat, or particles of leaf appearing as specks.

Trash removal device, in open end spinning machines: A system for removing impurities from the opened feed stock before the fibres conveyed to the rotor.

Trashmeter: An instrument which optically measures the amount of trash on the surface if raw cotton sample as presented to the viewing window.

Traveller: A C-shaped, metal clip that revolves around the ring on a ring spinning frame. It guides the yarn onto the bobbin as twist is inserted into the yarn.

Travers: French term for various weft striped fabrics.

Traverse length: The lateral distance between the points of reversal of the wind on a yarn package.

Traverse ratio: See **Wind ratio.**

Traverse winding: Winding of threads with slight traverse movement resulting in the individual thread layers not being located exactly above one another. This strengthens the spool edges and a better spool form results.

Treadling: The treadling (treadling order, treadling sequence) is the order in which the treadles are depressed, i.e., the order in which each shed is made.

Treated fabric: A fabric to which a finish has been applied in order to change the original level of a specific property or properties for example: water repellency, crease recovery, soil resistant, soft handle, water absorbency etc.

Tree Bark: A term describing the rippled or wavy effect sometimes seen when a bonded fabric is stretched in the horizontal (widthwise) direction. This defect is caused by bias tensions present when two distorted or skewed fabrics are bonded.

Tree beard: Colloquial term for Tillandsia fibre.

Treillis: Leviathan Stitch In embroidery a large, diagonal cross with upright cross stitches in the center and between the arms of the cross 1, French term for the net ground as distinct from the pattern in hand made laces; 2, coarse, stout French canvas, made of unbleached hemp; used for bags, trousers for farmers, etc. Embroidery made with coloured materials, the background being cut away; the pattern is usually a climbing vine. French for fancy braid

Trellis Work: See **Trellis.**

Trentaine: French dress goods made of natural wool, having 3,000 warp ends.

TREX : English textile machinery association.

TRI: Textile Research Institute.

Triacetate fibre: A manufactured fibre produced from cellulose triacetate in the forms of filament yarn, staple, and tow. Cellulose triacetate fibre differs from acetate fibre in that during its manufacture the cellulose is completely acetylated whereas acetate, which is diacetate, is only partially acetylated. The FTC notes that a fibre may be called \ triacetate when not less than 92%

of the hydroxyl groups are acetylated. Fabrics of triacetate have higher heat resistance than acetate fabrics and can be safely ironed at higher temperatures. Triacetate fabrics that have been properly heat-set (usually after dyeing) have improved ease-of-care characteristics because of a change in the crystalline structure of the fibre. (Also see **Acetate fibre**.)

Triacetin: Glycerol triacetate. A type of plasticizer for acetate fibres. It is widely used to add firmness to cigarette filter rods.

Triaxial fabrics: Completely isotropic fabrics made in a weaving process employing three yarns at 60° angles to each other. These fabrics have no stretch or distortion in any direction. With equal sizes and number of yarns in all three directions, the fabric approaches equal strength and stiffness in all directions.

Tricel: A branded acetate fibre.

Tricel suede: Soft, supple knit fabric with pile surface and also may be punched. Used for a variety of fabrics for sports and leisure clothes.

Tricel velour: Acetate yarn (Branded) knitted in combination with nylon or polyester results in a rich pile fabric that is soft and comfortable to wear. Fabrics dye well and therefore carry rich strong colours well and also black. Pale colours tend to look patchy but are very attractive. Used for jumpsuits, skirts, casual clothes, leisure wear, and in heavier weight for drapes and furnishings.

Tricelon: A branded name for a blended yarn of acetate and nylon.

Trichloethane: (1.1.1-trichloroethane, methyl chloroform), CH_3–$CC1_3$; molecular weight 133.42; density 1.31-1.35; boiling point 74°C) (azeotropic mixture with water at boiling point 65°C with 95.7 wt.% trichloroethane). Colourless liquid; volatile; evaporation index 2.6. Not easily flammable. Good solvent for lubricating greases, mineral oils, resins, waxes, tar, vinylidene chloride.

Trichloroethylene: (tri, ethylene trichloride), HC1C=$CC1_2$. Colourless liquid, clear as water, nonflammable, evaporates easily, smell similar to chloroform, strong anaesthetizing effect, prolonged presence on the skin causes "burn blisters". Does not corrode standard construction metals (except light metals) at temperatures under boiling point. Not easily soluble in water; hygroscopic; miscible with almost all organic solvents. Dissolves light and heavy greases, oils, resins, rubber, waxes, tar, oil-based dyes (if not too resinated, otherwise + alcohol + benzene soap), bitumens. Dissolving tendency for certain acetate, naphthol and vat dyes; also for polyvinyl chloride fibres (becoming hard and brittle), certain plastic parts (buttons, zips, buckles and the like) as well as plasticised finishes.

Trichloropyrimidine: The basis of certain commercially available reactive dyes.

Trichlorotrifluoroethane: (R113), $CF_2C1–CFC12$, boiling point 47.6°C; density 1.58. Refrigerant used in chemical cleaning for cleaning leather, furs and delicate items of clothing in special cleaning machinery.

Trichromat: Perception of colour by the normal sighted.

Trichromatic: Three-coloured (Chromatic), the normal perception of colour (Trichromat).

Trichromatic colour coordinates: Colorimetric measures. In this context the following concepts are identical: trichromatic system X,Y,Z = trichromatic components = norm colour values X, Y, Z. Trichromatic coordinates or coefficients = norm colour value proportions x, y, z.

Trichromatic colour values Identical with the so-called trichromatic Colorimetric measures, also known as tristimulus values or tristimuli.

Trichromatic dyeing: Dyeing with triple combination of yellow (or orange), red and blue (basic colours). Widely used in the dyeing of wool and synthetic fibres, not generally used for dyeing cellulose fibre. Colour triangle.

Trichromatic printing: (three colour printing).Use of printing rollers with appropriately graded engraving depth or templates with corresponding grid with which multi-colour pattern is achieved by means of multiple fall-ons. A 4th roller or template is often used for black or grey.

Tricks: In knitting, the slots which space the needles.

Tricomponent fibre: A fibre consisting of three polymers which are chemically different, physically different, or any combination of such differences.

Tricot: Tricot is one of the most popular fabric structures; It is used for lingerie because it is comfortable, long wearing, and resists runs and fraying. It is a light weight, warp knit fabric characterized by vertical ribs or wales on the right side and slight crosswise ribs on the wrong side. Has a thin texture, made from very fine or single yarns. Glove silk is a double bar tricot (very run-resistant). Tricot machines use bearded needles and they can operate extremely high speeds. Almost any fibre can be made into tricot, including silk, viscose, polyester and nylon. Also used for sportswear, bathing suits, gloves.

Tricot Beam: A metal flanged beam, commonly 42 inches in width, on which yarn is wound for use as a supply for the tricot machine.

Tricot de Laine Woollen suiting used for sailors, etc., in France.

Tricot Ecossais: Variety of crochet work, made with stitches taking in five loops at a time.

Tricot Flannel A heavy and thick flannel, which is quite elastic.

Tricot fabric, warp knitted: It is produced by two fully threaded guide bars which have 1 ∞ 1 lapping movement. The fabric has a light weight and it can be split very easily due to short underlaying movement. The loop will not incline towards one direction, the wale therefore will remain vertical. The appearance of tricot fabrics may generally be improved by using a coloured yarn which gives vertical stripes in the fabric.

Tricot Fabric Yield: The number of square yards per pound of greige or finished tricot fabric.

Tricot Knitting: See **Knitting.** A warp knit fabric in which the fabric is formed by interlooping adjacent parallel yarns. The warp beam holds thousands of yards of yarns in a parallel arrangement, and these yarns are fed into the knitting area simultaneously. Sufficient yarns to produce the final fabric width and length are on the beam. Tricot knits are frequently used in women's lingerie items such as slips, bras, panties, and nightgowns.

Tricot Section: See **Tricot beam.**

Tricot weave: (1) Created in fabrics by tying reverse warp threads (warp rib tricot) or reverse weft threads (transverse tricot) as well a special weave (diagonal tricot). (2) A form of ground structure used in warp knitting in which the threads are transferred from one needle to its neighbour and returned.

Tricotine: The name implies a knit fabric, but it is in fact a firm, durable, clean-finished, warp faced twill-weave with a fine steep twill line. Also known as Cavalry Twill. Dress fabric woven as as double twill, giving a pair of diagonal lines on the right side. It may be made from any fibre, including worsted, polyester, acrylic viscose. Used for trousers, uniforms and coats. See **Cavalry Twill.**

Triethanolamine: $N(CH_2–CH_2–OH)_3$; density 1.119–1.124; boiling point 270–279°C. Reaction product of ethane oxide on ammonia. Thick oil similar in consistency to glycerine, slightly fishy odour, sticky, hygroscopic, absorbs carbonic acid from the air, very alkaline.

Used as emulsifiers, pasting agents for vat dyes, solvents for acid dyes. Also used as additives in lubricating oils. Component part of wetting, cleaning, softening, sizing and finishing products.

Triethylsulphateamine, disodium salt: Used as a resin finishing agent for alkaline wet crosslinking and for two-step processes.

Trimellitic anhydride: (TMA). $C_9H_4O_5$, a hemimellitic acid. Colourless crystals. Use: for synthetic resins, adhesives, textile auxiliaries, surfactants, polyvinylchloride softeners.

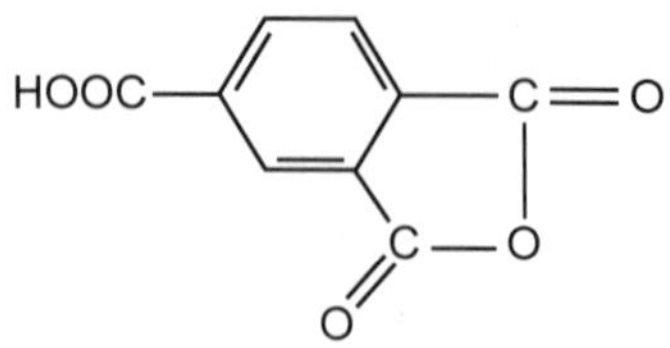

Trim, in textiles: To cut off a portion of a material.

Trimer: A polymer consisting of three monomer units. (Also see Cyclic trimer).

Trimmings: Pieces used to decorate or complete products.

Triphenymethane dyes: See **Diphenymethane dyes.**

Triple-beam balance: A weighing device This term refers a laboratory-type balance scale. "Triple-beam" simply refers to the fact that there are three bars on which weights slide to counterbalance the item on the pan. Typically, one bar has a weight for tenths or hundredths of a gram, one has a weight for grams, and one has a weight for tens of grams. Often additional weights can be hung from pins to increase the total capacity. Triple-beam balances are moderately priced and quite useful in dyeing. A type with a single pan where the mechanism is below the pan (as opposed to a hanging pan type) with a resolution of a tenth of a gram is a good choice for general dyeing use. There are now many electronic scales available that are very easy to use, but typically a good deal more expensive for comparable capacity and resolution.

Triple Cloth: Made with three sets of warps and three sets of fillings, forming three different layers of cloth, tacked together during the course of weaving.

Triskelion cross section: A trilobal cross section in which the radiating arms are curved or bent. (Also see Cross section.)

Tristimulus: Of, or consisting of, three stimuli; generally used to describe components of additive mixture required to evoke a particular colour sensation.

Tristimulus colourimeter: An instrument that measures tristimulus values and converts them to chromaticity components of colour.

Tristimulus Data:The three tristimulus values that combine to define or generate a specific colour, such as R 255/G 255/B 0. Tristimulus data does not completely describe a colour–the illuminant must also be defined. Also, in device-dependent colour models such as RGB, the capabilities of the viewer or colour-rendering device must also be defined.

Tristimulus filters: Optical filters used in conjunction with specific colour lamps to obtain a response function approximating the tristimulus functions of the CIE Standard observer for source C.

Tristimulus Values (CIE): Percentages of the components in a three-colour additive mixture necessary to match a colour; in the CIE system, they are designated as X, Y and Z. The illuminant and standard observer colour-matching functions used must be designated; if they are not, the assumption is made that the values are for the 1931 observer (2° field) and illuminant C. The values obtained depend on the method of integration used, the relationship of the nature of the sample and the instrument design used to measure the reflectance or transmittance. Tristimulus values are not, therefore, absolute values characteristic of a sample, but relative values dependent on the method used to obtain them. Approximations of CIE tristimulus values may be obtained from measurements made on a tristimulus colorimeter that gives measurements generally normalized to 100. These must then be normalized to equivalent CIE values. The filter measurements should be properly designated as R, G and B instead of X, Y and Z.

Tristimulus values X,Y,Z: See **Tristimulus values.**

Tritinum: Medieval silk fabric, made with three-ply warp.

Tritik: (Plangi-Tritik, crimped batik). Plangi technique, in which yarn, threads or bast is drawn into the fabric in rufflets corresponding to the pattern and then gathered before dyeing, in order to produce resists during the dyeing process.

Trolly Lace: Term for bobbin laces having the patterns outlined with a heavy cord.

Tromp: "Tromp" is an old way of saying "step" on a treadle. Tromp as writ means to step on the treadles in the same order as the shafts are threaded (also called "treadle as drawn in). If the shafts are threaded 1-2-3-4; to tromp as writ or treadle as drawn in, you step on the treadles in that same order, 1-2-3-4.

Tropical: Light worsted yarn goods in linen or Panama weave, plain or patterned (yarn or colour-effect patterns); up to 300 g/running m; with napless finish. Used for men and boys' outerwear and women's outerwear summer clothing. Fine hard-twisted worsted fabric in plain weave is wrinkle resistant and breathable and have been termed 'coo-wool'.

Tropical Cloth: Very light men's wear fabric, used for summer clothes, made usually with cotton warp and mohair filling, showing various colours and designs.

Tropical Weight: Suitings weighing from 9 to 11 oz. per yard.

Tropical worsted: A plain or fancy weave worsted cloth, very light in weight and mainly in light colours, beige and white. An excellent cloth for hot weather as the high twist worsted yarn makes it cool. It pleats and creases well. Used for men's suits, women's suits, etc.

Trousering: A large variety of woollen and worsted fabrics, also cotton worsteds, used for trousers.

Trousers: Garments for the lower limbs; widely, but erroneously, known as "pants."

Trubenizing: (Trubenizing process), for the purpose of chemical stiffening of men's laundry (without using finishing products). Method for stiff collars: a lining which has been soaked with cellulose acetate and swollen with solvent (acetone) is inserted between the outer fabrics beforehand and then bonded to them using a hot iron. This type of collar is soft when wet, washes well and always achieves its original stiffness again (without the use of a finish) after ironing.

True: Fibres are called true which have a uniform diameter.

True designs: In weaving, those parts of a section blanket where a weft section is woven on the warp section for which it was intended.

True guage length: A precise length between wee-defined bench marks located on the specimen while under known tension in the unsupported portion between the holding clamps and free from contact with any snubbing surfaces or other surfaces which could result in non-uniform strain.

True rise, in body measurements: The vertical distance (Plumb line) from the waist level at the side to the crotch.

True tensile strength: The maximum tensile stress expressed in force per unit area of the specimen at the time of rupture. (Also see **Tensile strength**.)

Truffette: Fine, narrow, bleached French linen; used for handkerchiefs, shirts, etc.

Trunk: Length Medium long women's hosiery, reaching above the knees, but shorter than opera hose.

Trunk: Double the length of a coverall, from the center of the neck hole at the back to the point of the leg separation on the seat seam.

Trunk hose: Short, heavily padded breeches tied at thigh, often with differently coloured slashes.

Trunnions, In Zipper: The two pivots at the ends of the pull that fit into the bail.

True guage length: A precise length between well-defined bench marks located on the specimen while under known tension in the unsupported portion between the holding clamps and free from contact with any snubbing surfaces or other sources which could result in non-uniform strain.

Truxillo: Inferior Spanish wool.

Tsatlee: Raw silk reeled in northern China, according to the primitive, native methods, producing uneven and irregular yarn.

Tsudzure-no-nishiki: Japanese brocade made of strong silk warp and twisted gold paper filling, made like gobelins.

Tsumugi: A narrow and very expensive Japanese silk fabric made as follows: the silk is spun into a heavy yarn and woven into a coarse cloth with cotton warp. This cloth is run several times through the vat and then buried in the ground. After the cotton warp had rotted, the silk is rewoven into a fabric, the spots which were left un affected in the dye, forming a mottled design; used for kimonos.

Tuanse: A Chinese satin.

Tuareg: Wool rugs made in Morocco.

TS: Turk. Standardlari = Turk. Standard.

T-Shirt Formerly a short-sleeved vest; today, this is a term used for a multitude of different fashionable casual shirts, with or without sleeves (short or long sleeved), coloured, printed, and decorated with motifs, etc.; made from knitwear.

T.S.P: (or TSP) – Na_3PO_4 , trisodium phosphate; also called "sodium phosphate, tribasic" TSP is sometimes used to produce pH in the range of about 12 for dyeing processes. It may be used in scouring. Strong solutions can be very irritating to skin, so handle it with care.

Tsatlee silk: Chinese Grège yarn from live cocoons, products from the cottage industry, non-uniform.

T screens: Rotary screens for printing transfer printing papers. They have a particularly smooth surface and 80 mesh size to compensate for the sensitivity of transfer printing to any unevenness in the screen due to low absorption capacity of paper.

TSE: Turkish Standards Association.

TT process: Thermofixation and thermosol process.

TTZ: (cloud point index) Cloud point index.

Tub: See **Beck, Jet dyeing machine.**

Tub dyeing machine: To shorten the liquor ratio in yarn dyeing, the dyeing volume (otherwise a dyeing boiler) is sectionalised into a number of tubes. A dye column is dyed separately in each tube.

Tube: (1) A cylindrical holder or bobbin used as a core for a cylindrical yarn package. (2) A cylindrical yarn package.

Tubeless cross-wound yarn package: Wound packages without a batch centre (dyeing tubes). Promotes better penetration of dye.

Tubing: Braided, knitted, or woven fabric of cylindrical form having a width of 100mm. (4 in.) or more (Circumference of 200 mm. (8 in.) or more.)

Tubular fabric: A fabric woven or knit in a tube form with no seams, such as seamless pillowcases, some knit underwear fabrics, and seamless hosiery. (Also see Circular knit fabric).

Tubular fabric dye padder: Padder for the continuous dyeing of tubular cotton wear with reactive dyes according to the CPB method. After impregnation of the tube which is width guided through, it is blown up like a balloon by air-injection nozzles and then squeezed out.

A warping device is used to lay the edges of the tube marked by the squeezing process.

Tubular fabric dyeing: (1) Term from piece dyeing which is used for fabric or knitwear with the edges sewn together to form a tube and dyed on the winch as an effective way of countering selvedge curling. (2) Dyeing of tubular knitwear

Tubular fabric mecerising machine: Tubular knitwear which has not been cut open should be mercerized without fixing the fold creases. For this reason, the impregnated hose is blown during stabilisation in the following steps A = roller-vat type impregnation zone using the squeezing unit; B = mercerization cartridge with controllable lateral and longitudinal tension and end squeezing unit; and C = wash, neutralizing and rinsing step.

Tubular fabric slitting machine: (tube openers).They are used to slit open tubular knitted goods to expose the surface of the fabric to the finishing process. The slitting process can be achieved using burners (welding) or cutters. A cut thread or Separating course is provided for this purpose and is picked up by electronically controlled units The tube slitting

machine detects the separating course using a photo electronic device and opens the tubular wear automatically.

Tubular fabric turning machine: Device for turning tubular knitwear using aerodynamics. The tubular wear is pushed on to the turning tube (length 3 000 mm, Ø 180 mm). The end of the piece is drawn into the tube opening aerodynamically and this turns it at the same time. Fabric speed is 600 m/min (loading) and 300 m/min. (unloading).

Tubular felt: (felt sleeves). Industrial wool felt in the form of continuous woven or nonwoven pressed felt for covering rollers (lapping), for sizing and finishing machines, calendars, for transferring printing in roller printing, etc. also used as an underlay for machine foundations and seals, etc.

Tubular gauze: Tubular fabric made from screen gauze used on nickel screens in rotary-screen printing.

Tubular knit goods: See **Tubular fabric.**

Tuck: (1) Lightweight cotton or silk fabric, made with pleats running from selvage to selvage and formed with a separate warp and a filling usually heavier than that used for the ground weave; used for shirts, waists, etc.

(2) A flat fold in a fabric; there is difference between a tuck and a plait.

Tuck, in rope: A free strand of rope placed between rope strands during rope splicing.

Tuck Knitting: Framework knitting, in which a number of loops are accumulated on a needle, thus forming the design.

Tuck stitch: A knitting stitch made when a needle receives a new yarn without losing its old loop. This is the stitch which has the reverse V shape. The loop head of a tuck stitch together with the previous loop head are held by the feet of the following stitch. Normally, a tuck stitch with more than four successive tucks on the same needle should not be used due to high yarn tension and needle damage (not more than six adjacent tucks due to freely floating and snagging). The tuck stitch reduces the lengthwise elasticity of the fabric and the fabric length, while it increases the widthwise elasticity and width of the fabric. It also increases the fabric thickness and needs theoretically less yarn.

Tucked seam: A complex seam formed on the inside of the object with neither raw edge.

Tucking position, in knitting: The needle butt follows the cam track and is pushed upwards by the cam. The old loop moves downwards and opens the latch of the needle. Sinker still moves towards machine centre, thus fabric is held down in the sinker throat. However, the old loop is still on the latch. This position is also used to produce tuck stitch.

Tucks: This is the stitch which has the reverse V shape. The loop head of a tuck stitch together with the previous loop head are held by the feet of the following stitch. Normally, a tuck stitch with more than four successive tucks on the same needle should not be used due to high yarn tension and needle damage (not more than six adjacent tucks due to freely floating and snagging). The tuck stitch reduces the lengthwise elasticity of the fabric and the fabric length, while it increases the widthwise elasticity and width of the fabric. It also increases the fabric thickness and needs theoretically less yarn.

Tucks, in garments: Tucks are folds of fabric stitched down either all or part of the way, such as released tucks. Their purpose is mainly decorative although they may also be used as a shaping device. They are usually folded on the straight and each tuck is formed from two stitching lines that are matched and stitched. The distance from the fold to the matching line determines the tuck's width. Tucks that meet or overlap slightly are called blind tucks; those with predetermined space between them are spaced tucks; and a pin tuck is a very narrow tuck, usually about 3 mm (1/8 in) wide. Light to medium weight fabrics are suitable for tucking, but design and print should be considered before purchasing the fabric. Tucks can be added to plain garments by pre-tucking the fabric before it is cut.

Tucks, corded: Fold the tuck, positioning the cord inside the tuck along the fold. Pin in position and, using a zip foot, stitch close to the cord.

Tucks, cross: First stitch the lengthwise tucks and presses in one direction, then form the cross tucks at right angles to the first set and stitch. Be sure to keep the first set of tucks facing downwards.

Tucks, Released: These dart tucks are usually used to control fullness at the bust or hip. They are mainly formed inside and can be released at one or both ends. Be sure to reinforce the ends. Press carefully to avoid creasing the folds.

Tucks, shell: First stitch the tuck, then hand stitch a few overhand stitches every 1 cm (% in) to form a scallop. Alternatively, blind stitch by machine, with the tuck under the foot and the fold to the left of the needle, allowing the zigzag stitch to scallop the tuck.

Tucking defect, in knitted fabrics: One or more unwanted tuck loops.

TUDO: Thiourea dioxide.

Tuft: (1) A cluster of soft yarns drawn through a fabric and projecting from the surface in the form of cut yarns or loops. (2) The portion of pile-like material that comprises a tufted fabric or carpet. (Also see **Tufted fabric** and **Tufted carpet**.)

Tuft, in pile yarn floor coverings: Those cut or uncut loops which are attached to the backing fabric at one binding site and which form a part of the fabric face.

Tuft bind, in pile fabrics: The force required to pull a tuft from a cut pile floor covering or to pull free one leg of a loop from a looped pile floor covering.

Tuft element, in pile yarn floor covering: A segment of yarn bound to a backing fabric at a binding site so that two portions (legs) of the yarn project above the backing fabric, one portion on each side of the binding site.

Tuft height, for pile yarn floor covering: The length of the tuft leg.

Tuft Leg, for pile yarn floor covering: One of the two portions of a tuft element that project above the backing fabric on the pile side of the floor covering.

Tuft length for pile yarn floor covering: The length of the tuft element measured while extended in a straight line under zero tension.

Tuft bind: Force required to pull a tuft from carpet.

Tufted Carpet: Carpet produced by a tufting machine instead of a loom. It is an outgrowth of hand-tufted bedspreads. Today, broadloom tufting machines produce over 90% of all domestic carpeting. Tufting machines are essentially multi needle sewing machines that push the pile yarns through a primary backing fabric and hold them in place to form loops as the needles are withdrawn. The loops are then either released for loop-pile carpets or cut for cut-pile carpets. The pile yarns may be either pre-dyed or uncolored, in which case, the greige carpet is then piece-dyed or printed. In either case, a latex or other binding agent is applied to the backstitch to lock the tufts in place and to secure the secondary backing fabric. Formerly, all carpets were woven, either by hand or machine. The significantly greater productivity of tufting has revolutionized the carpet industry and made soft floor coverings available to the mass market.

Tufted embroidery: Lace fabric, usually composed mainly of cotton and possibly nylon to create a firm base, with gold, silver or coloured tufts inserted into the right side of the fabric in pattern.

Tufted fabric: A fabric with a pile consisting of tufts or loops formed by inserting yarn into a previously prepared backing fabric. Cotton sheeting, lightweight duck, or other fabric decorated with fluffy tufts of multiple-ply, soft-twist cotton yarns or manufactured fibre yarns closely arranged in continuous lines or spaced at intervals to produce the type of fabric called

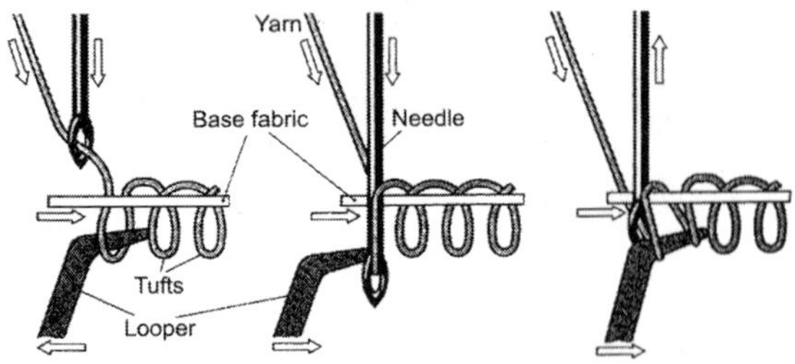

Tufted fabric production schematic

candlewick. The tufts are inserted and cut by machine in previously woven fabric or are woven in by the loom and afterwards cut to form the tufts. They have a chenille-like softness and bulk and are erroneously called chenille. Patterns vary from simple straight lines and elaborate designs to completely covered materials resembling long pile fabrics. The may be white, solid coloured, or multicolored. Tufted fabrics are used for bedspreads, bath mats, and robes, etc.

Tufted fabric: Any fabric made from any yarn or mixture of yarns in which a pile is created by inserting an extra yarn, often thicker than the backing. Some tufts are locked in position and are easily removed (e.g. Candlewick).

Tufted textiles: These are fabrics which are made by inserting yarns, into a usually textile backing, using one or more needles, forming loops (pile loops). The loops may be cut and/or fixed to the backing using an adhesive or by heat sealing.

Tufting: The manufacture of Tufted textiles, in particular Tufted carpets, using the tufting technique.

Tufting, in upholstred furniture: Localised intendation of upholsterd furniture surfaces and cushions, by the use of or appearance of buttons, providing an aesthetic treatment.

Tufting machine: A machine for producing tufted carpets.

Tufting needle: An instrument for inserting tufts into a premade substrate. As in a sewing machine needle the eye and the point are at the same end. The needle is grooved on one side, to take the yarn and chamfered on the other side to lead the looper on to the needle.

Tufts to a carpet backing: Bonding a secondary backing to a primary backing, or increasing fabric body or stiffness.

Tugop: Bast fibre, yielded by a species of the Artocarpus tree in the Philippines; used for twine and cordage.

Tulle: Made in silk, nylon, cotton. Weave is normally Guaze, knotted, leno, made on a lace machine. Name is derived from Tulle, France. First made by Machine in 1768. Has a hexagonal mesh and is stiff. It is difficult to launder. Comes is white and colours, and is very cool, dressy, and delicate. Uses: It is a stately type of fabric when used for formal wear, and weddings. It is also used for ballet costumes and wedding veils.

Tulle Crinoline: Same as pleated tulle, made with gradually increasing meshes, which force the tulle into pleats.

Tulle Grec: French tulle made with meshes larger than in ordinary tulle.

Tulle, warp knitted: Plain net fabrics made on a bobbinet machine or simulated by warp knitting. The honeycomb-type structure may be used as such or as a base for embroidery, for sophisticated blouses and dresses, trimmings, layering or veils.

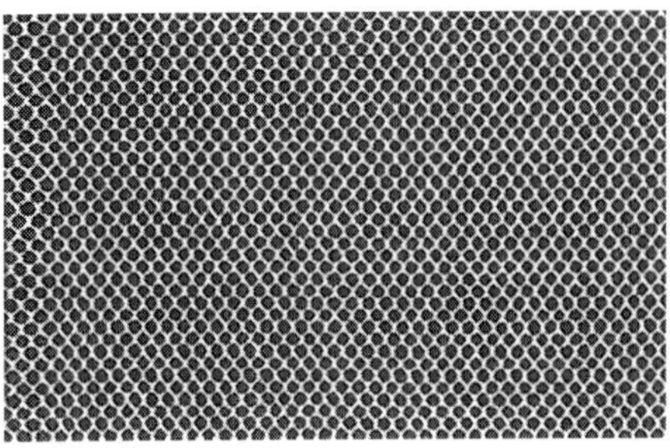

Tumble drier: A drying machine with a rotating, perforated drum, in which the material is spun loosely, without tension, and discontinuously or continuously.

Tumble dry: Drying with hot air in a tumbler, the usual term for the drying method used when testing the fabric smoothness of resin-finished textiles after household washing (Wash-and-Wear Standards).

Tumbler finishes: After treatment agents for washable textiles, for the application of active substances (finishing, anti-static, fragrance) during the drying process in the tumble drier or drum drier. Form of application: as foam spray (carrier: cloth/item being washed), spray (carrier: drum walls), pad impregnated with active substance (fixed by adhesive to tumble drier paddles) and Sheets.

Tumbler shrinkage: Unwanted shrinkage, which can arise when items of clothing made from heavy cotton fabrics are dried excessively in a tumble drier after washing.

Tunic: Women's or men's shift style robe sewn from two pieces of fabric.

Tunis Crochet: A very easily executed crochet, made only in straight work with single ribs.

Tunnel test: See Flammability tests.

Turbehlik: Turkish name for Oriental rugs, intended to be hung over graves; the designs are always tree and flower patterns.

Turbidity: The decrease in optical transparency of a solution because of the presence of particulate matter.

Turbidity, in waste water: Turbidity, a measure of the light transmitting properties of water, is another test used to indicate the quality of waste discharges and natural waters with respect colloidal and residual suspended matter.

Turbidity unit, TU: Various units are used to measure turbidity in water. One method is to use a suspension of silica. Another method is to compare the water to a standard formazin solution as measured in a ***nephelometer***. This gives the ***nephelometric turbidity unit*** (NTU) or ***formazin turbidity unit*** (FTU). The 1998 EU Drinking Water Directive states that the turbidity must be acceptable to the consumer. A value of 1 NTU ex water treatment works is the desirable maximum. The WHO guideline maximum value is an average of 1 NTU for drinking water.

Turbostapler yarns: High-bulk yarns (strongly textured yarns). Manufactured on a turbostapler. A tow made up of high-shrinkage fibres of varying lengths is passed through electrically heated plates then stretched, thus gaining high residual shrinkage.

Turka fibre: Brown or white coloured soft fibre yielded by the bark of the apocyum venetum, grown in Asia and southern Europe; used for twine, baggings, etc.

Turkey red oil: TRO is sulfated castor oil. Ricinoleic acid, the major acid in castor oil has both a hydroxyl group at the $C1_2$ position and a C=C at the C_9 position. Both of these groups are converted to sulphate ester linkages so castor oil can have a degree of substitution up to 6.

Turkey Red: (1) A fast and bright scarlet dye, yielded by the madder and formerly extensively used on wool and cotton preceded by various elaborate processes of mordanting. The modern Turkey red, a fast and brilliant scarlet, is a synthetic dye; (2) plain woven or twilled cotton fabric of various grade, dyed with Turkey red; used for dresses, trimmings, etc.

Turkey red oil: (red oils, T' red oils). Consist of Sulphonated surfactants and similar products. Used as wetting, dye penetration and levelling agents.

Turkish Knot: Used in russ; it is formed by the yarn being twisted about the warp threads, two ends of the pile alternating with every two threads of the warp. See **Ghiordes knot**.

Turkish carpets: Oriental carpets from Anatolia and Asia Minor. Original with traditional patterns of flowers stylised to a greater or lesser extent (tulips, daffodils, hyacinths). Introduced to Europe from 15th century and used in particular as wall hangings and table covers.

Turkish towel: Cotton or linen fabric having loop piles formed by a separate set of warps. Is bleached. Used for towels, bath robes, etc. See under terry cloth.

Turkman Rugs: See **Genghis rugs**.

Turkoman: Made with silk or cotton warp and thick chenille filling; used for drapery.

Turmeric: A fugitive direct yellow dye yielded by the roots of the curcuma tinctoria in India and China.

Turn: The distance parallel to the axis of a yarn or rope in which a strand makes one complete spiral. (Also see TWIST.)

Turned and stitched seam-finish: see **Clean-finished seam-finish.** Fold under a 6 mm. (1/4 in.) edge on the seam allowance and press a crisp edge. Edge stitch next to the folded edge. Use thread matching to the fabric. It is a gooway to finish an unlined Jacket for casual wear. Used on light and medium weight fabrics as it does make the seam allowance bit bulkier close to the seam.

Turn-over: The turned in extremity of trouser legs and sleeves.

Turned-over edge: See **Curled selvedge.**

Turn-up: The bottom of trousers.

Turning: The reversing of two or more pieces of material that are seamed together for pressing or topstitching.

Tuskin: A woollen fabric made in England under Edward VI, believed to have been of blue or gray colour.

Tussah: Made in silk, usually plain but also in twill. Made from wild or uncultivated silkworms. It is coarse, strong, and uneven. Dull lustre and rather stiff. Has a rough texture with many slubs, knots, and bumps. It is ecru or tan in colour and it is difficult to bleach. It usually doesn't take an even dye colour. Wears well and becomes more rough looking with wear. It wrinkles a little, but not as much as some. Various weights. Appears in filament and staple form. Uses: In lighter weights, dresses. In heavier weights, coats and suits and ensembles.

Tussah silk: A coarse silk produced by a wild silkwormThere are three main types (a) Antheraea mytilla (largely indian); (b) Antheraea pernyi (largely Chinese); (c) Antheraea yama-mai (largely japanese). It is brown in colour and usually spun, since most cocoons cannot be reeled. The yarn has an uneven slub which gives a rough texture to the cloth. Used for suits, dresses, skirts, shirts etc.

Tutorette process: A mechanical shrinking process for cotton/polyester blend fabrics, licensed by Bancroft Far East Inc. in Japan and using the Everest Shrinking Machine (Manuf.: Hunt & Moscrop). Tutorette textured seersucker is a finished crêpe. The regular pattern of the textured surface is achieved by a flatbed embossing calendar. An artificial resin finish is used to fix the embossing.

Tuxedo: See **Dinner Jacket.**

Twanse: Stout Chinese silk satin, finished with little gloss. Comes in solid colours or patterns.

Tweed: Much like homespun in appearance, both being either twilled or plain. They are made from rough worsted yarn spun at home. In tweed the yarn is harder twisted, giving a more distinct twill. Tweeds are commonly woven in a 3/2 twill weave other weaves usually used are plain, broken twill and herringbone weaves. Typical plain woven tweed will have a construction 84 × 20 with two run woollen yarn in warp and filling. It is generally more compact, less rough, and better finished than homespun. Commonly used for sportswear, suits, coats, overcoats, and ensembles.

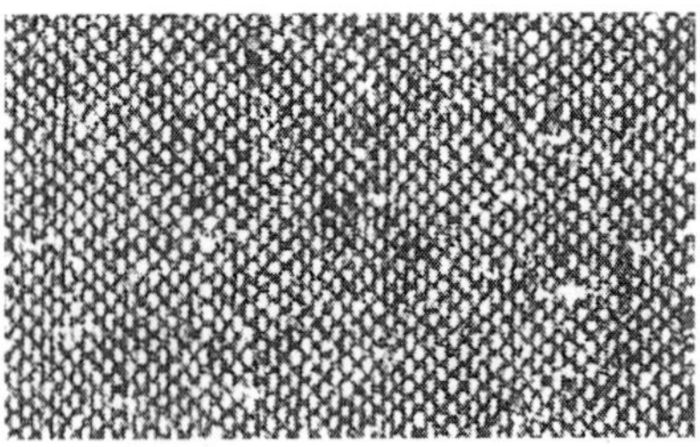

Tweed, Donegal: Plain and herringbone woven similar to Irish tweeds, white nubs are often observed in the warp while filling has nubs of various colours, usually red, blue or green.

Tweed, Harris: Woollen fabric hand woven on Harris and other islands of the outer herbides off western coast of Scotland. The woollen yarn in in this tweeds are made of long, strong wools and the cloth wears exceptionally well.

Tweed, Irish: Usually made with a white warp and a filling of dark shade of blue, brown, black or grey. It is made of 2/2 twill weave often with nub yarns. Irish tweed is normally a little softer and lighter than Harris tweed.

Tweed, Linton: Soft tweed made with Australian Merino wool along with some other animal fibres to make it softer. The fabrics is made only in Carlisle, England.

Tweed, scotch: Similar to Harris tweed in construction white warp and stock dyed fillings are often vivid and highly contrasting.

Tweed Jersey: The old traditional term tweed, which implies hairy, has been applied to a modern fabric. This is a thick knit fabric with fancy knobbles and a hairy surface. It is extremely stretchy and would be suitable for patterns labeled 'knitted fabrics'. The fibre is likely to be acrylic with some wool and nylon. Used for sift suits and dresses.

Twelve-harness satin: A weave similar to eight-harness satin except in warp-faced fabrics warp yarns show on the face of the fabric eleven out of twelve adjacent yarns and in filling-faced fabrics filling yarns show on the face eleven out of twelve adjacent yarns.

Twill: A very rugged fabric woven in 2/2 twill weave with textures ranging from 70 × 40 to 160 × 80 and the width vary from 35 in. to 45 in. The greatest use of twill is in linings and even for infants wear. Usually 150 denier right yarn is used in warp and weft. Sometimes rayon warp and acetate filling or vice versa, is also used. All acetate twill is also uncommon. 110 × 70 may be a typical construction. Cloths in twill weave: Cotton- Jeans, ticking, drill, Canton flannel, denim, gabardine covert cloth, khaki, serge; Linens: Ticking and table and towel twills; Silk: Twill foulard, serge; Wools: serge, worsted cheviot, gabardine, covert, flannel, tweed, unfinished worsted, sharkskin, broadcloth; Rayons, acetate and blends: Gaberdine, foulard, flannel; Polyester: Suitings such as serge and gabardine.

Twill angle: See **Twill Ridge.**

Twill braid, in rope: A braided construction in which one strand of one direction of rotation about the axis passes over two strands of the opposite direction and it in turn passes under the next two strands of opposite direction.

Twill carpet: Flat carpet, woven from coarse yarns in twill weave (Dutch carpets, Parisian, Sayette runners).

Twill damask: See **Damask, twill; False damask.**

Twill ridge: The diagonal ridge in twill weave fabrics.

Twill weave: A weave charcterised by diagonal lines produced by a series of floats staggered in the warp direction. Floats are normally formed by the filling (a weft –faced twill) Examples of twill weave fabrics are cotton denim, gabardine, khaki, ticking, whipcord, wool broadcloth, cashmere, flannel, gabsardine, melton, serge, tweeds, whipcord, silk syrah etc.

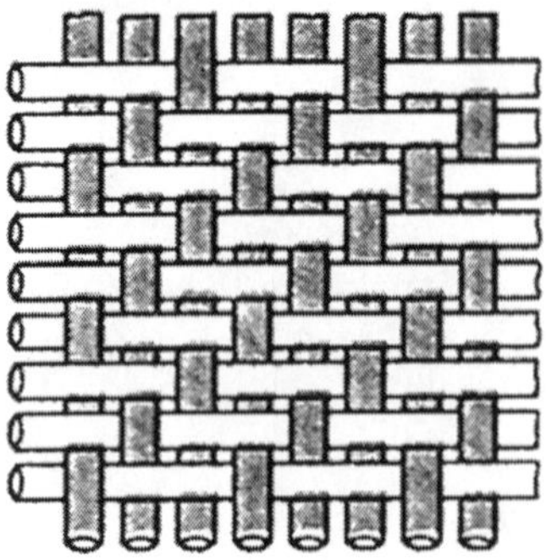

Twill, block: Twills may be woven blocks of warp-faced and weft-faced areas are side by side in both in warp and weft directions. Such a twill is called a block twill.

Twill, broken: Broken twills are formed by reversing the pattern part way through the repeat. Usually the break will be at the centre of the repeat, with only one reversal, but more complicated breaks can be made. (Fig) or in the weft direction, and no twill line will be generated.

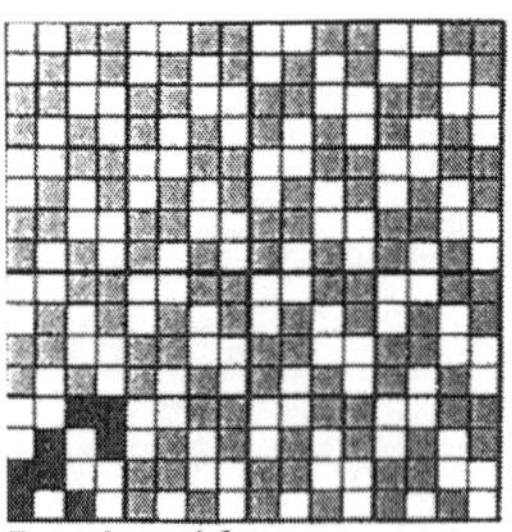

Developed from
20 – 02 02 – 01 – 01

Pattern Draft

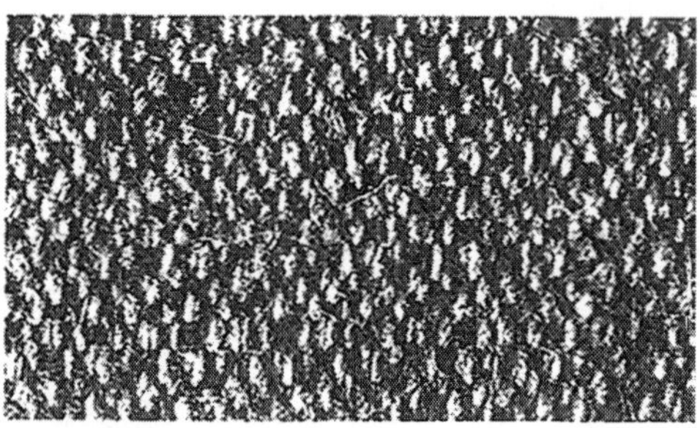

Flattened twill

Twill, even: When the number (or sum of numbers) above the line and the number (or the sum of numbers) below the line are equal the twill is said to be an even twill. In an even twill neither the warp nor the weft predominates.

Twill, Flattened: Flattened twills are usually weft faced, which means that the twill line is formed by the weft yarn and the warp is mainly on the back of thon the back of the fabric.

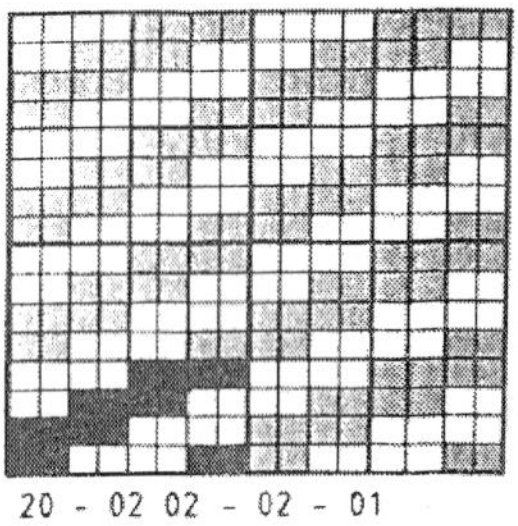

Pattern Draft

Flattened twill

Usually, the weft floats are longer than the move number which, in turn, is greater than one.

Twill, left handed: In a regular twill adjacent picks never float over or under the same set of warp ends. If the first two ends of a 2/1 twill are lifted for the first pick, then the second and third ends may be lifted for the second pick and the first and the third ends lifted for the third pick. Likewise, if the twill is woven as ½ twill, the first end is lifted for the first pick followed by the second and then the third if we maintain in the same direction. Twills may progress to the left or right. The former is called a **left handed twill** and the latter is called a **right-handed twill.**

Twill, steep: With equal densities of warp and weft, the simple twills display a twill line of about 45. A steep twill line can be obtained either by increasing the relative density of warp or by using a move number of two in the pattern, or by special constructions.

Twill, uneven: See **Twill, even.** When the number or sum of numbers above line and the number or the sum of numbers below the line it is called an uneven twill. The warp shows more on one face and the weft on the other face. Because the number above the line is not equal to the number below the line a three-shaft twill is an uneven twill.

Twills, right handed: See **Twills, left handed.**

Twill, waved: By changing the direction of the twill line at regular intervals a wave, or zigzag effect can be created either across the fabric or along its length. Diamond checks can also be made. The twill lines come to a point at the reversals.

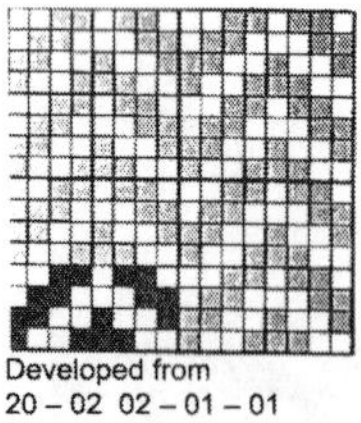

Pattern Draft

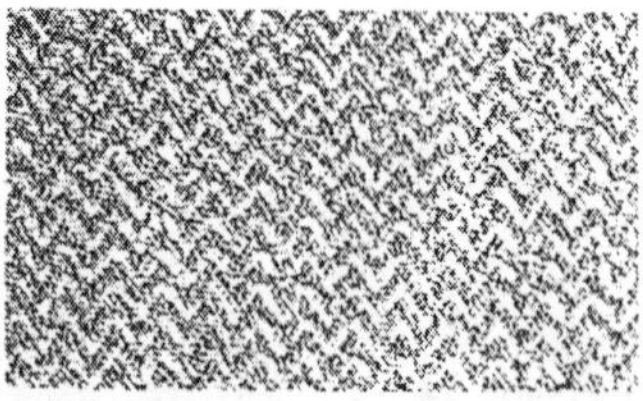

Waved twill

Twills: Materials with the twill structure.

Twilo process: A spinning process in which yarn is made by binding fibres with an adhesive, then removing the adhesive after the yarn is made into fabric.

Twilo yarn: A twist-free yarn, ribbon-like in appearance (viscose staple fibre, HWM-, polynosic fibres) of good uniformity from wet-drawn fibre bundles, bonded with hot-water soluble adhesive or polyvinyl alcohol binding fibres (5–11% by weight). The fabric possesses substantial covering property. Once the binding agent is washed out the yarn strength in the fabric decreases. Twilo yarns possess good strength, covering property, softness, sheen properties.

Twine: (1) A term applied loosely to a variety of textile strands used for tying such articles as parcels, bundles, bales etc. (2) An aggregate of fibres or yarns compacted into a partially or completely balanced twisted structure of indefinite length, generally used for tyting or binding.

Twist, in textile strands: The helical or spiral configurations induced by turning a strand about its longitudinal axis.

Twist: The number of turns about its axis per unit length observed in a yarn or other textile strand.

Twist: Durable plain or twill worsted fabrics, often made with multicoloured folded yarns for a speckled effect. Used for jackets, trousers and suits.

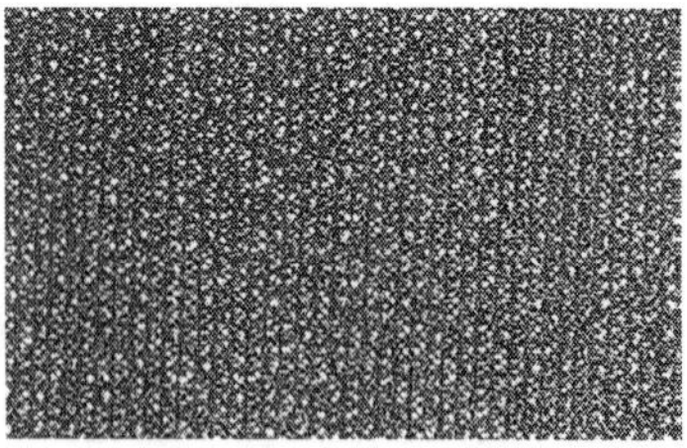

Twist: (a) The spiral disposition of the component(s) of a yarn and that is usually the result of relative rotation of the extremities of the yarn(s).

(b) The number of turns per unit length of yarn, e.g. turns per metre.

NOTE:

Twist Designation:

(i) Twist in Single Yarns
- **S** twist
- **Z** twist

(ii) Twist in Folded Yarns
- **ZS** twist
- **SZ** twist
- **ZZ** twist -on-twist (q.v.)
- **SS** twist -on-twist (q.v.)

(iii) Twist in Cabled Yarns
- **ZSZ** twist (formerly "cabled twist")
- **ZZS** twist (formerly "hawser twist")

The first symbol designates the direction of twist in a single yarn, the second symbol designates the direction of twist in the folding operation, and the third symbol the direction of twist in the cabling operation.

Twist direction: See **Twist, direction of.** The twist of spun yarns denotes the amount of twist applied to a yarn (twist) to fix the bundle. The direction is denoted by S- (right-hand) and Z-twist (left). If the direction of inclination of the fibres appearing at the surface of yarn (or of the single components of a

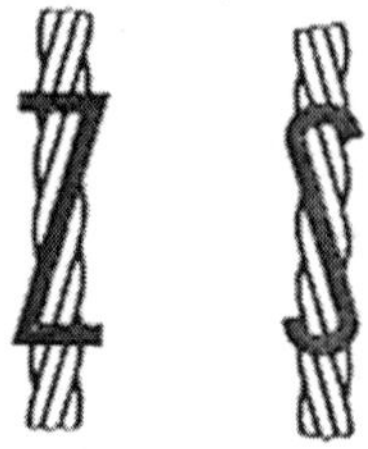

folded yarn) is to the right, when the yarn is held vertically, then it is called Z Twist and when it is to the left it is called S Twist. The number of turns (degree of rotation) relating to 1 m yarn length (T/m) depends on the fineness, raw material and ultimate application of the yarn to be spun and the spinning process. The relationship between twist, raw material and fineness is expressed by the formula $T/m = \alpha_m R(Nm)$.

Twist Fabric: A plain knitted fabric, each loop being twisted as it is formed.

Twist factor: The product obtained when the twist expressed in turns per centimeter is multiplied by the square root of the yarn number expressed in tex.

Twist multiplier: The quotient of the twist expressed in turns per inch and the square root of the yarn number in an indirect system.

Twist of spun yarns: See **Twist, Twist direction.**

Twist factor (Twist multiplier): A measure of the "twist hardness" of a single yarn, determined by the multiplication of the turns per unit length by the square root of the linear density of the tex system .

Typical twist factors:

Fibre Length	**Short**	**Medium**	**Long**
Knitting	-	2400-2875	2010-2500
Weft	3150-3650	2875-3350	2400-2875
Semi-warp	3550-3830	3350-3650	2875-3260
Warp	3830-4790	3650-4300	3260-3750

Twist liveliness: The effect caused by unbalanced torsional forces in any yarn, and of sufficient magnitude to give rise to difficulties in processing or defects in the resulting fabric.

NOTE: Examples of this are snarling (see snarl) in processing and spirality (q.v.) in knitted fabric.

Twist on Twist: English mill parlance for yarns with especially elastic properties, the twist of the two-fold being the same as the twist of the single yarn.

Twist setting: A process for fixing twist in yarns to deaden torque and eliminate kinking during further processing. There are several methods that use steam to condition the packages of yarns.

Twist Stitch: Same as **cord stitch**.

Twist take up: The change in length of a yarn or other textile strand caused by twisting, expressed as a per cent of the original untwisted length.

Twist texturing: A texturing method in accordance with the mechanical/ thermal principles; the two different processes are torsion crimping and friction texturing. The following phases are followed as a continuous process during manufacture:

– high-twisting,

– heating and cooling to fix the twist,

– down twisting (doubling).

The process is essentially based on the twist principle, although it is not only two (or more) yarns being twisted together, but a multifilament being twisted around its own axis. The high-twisted fibres are thermo fixed and the twist again applied. This gives rise to high bulk yarns.

Twist, balanced: See **Balanced twist.**

Twist, cable: See **Cable twist.**

Twist, contraction: See **Twist take-up.**

Twist, cord: See **Cord twist.**

Twist, corkscrew: See **Corkscrew twist.**

Twist, Direction of: See **Direction of twist.** The direction of twist in yarn and other textile strands indicated by the capital letter 'S' abd 'Z'. Yarn has 'S' twist if, when the yarn is held in a vertical position , the visible spirals or helices around its central axis conform in the direction of slope to the to the central portion of the letter 'S' and 'Z' twist if the visible spirals or helices conform in the direction of slope to the central portion of the letter 'Z'.

Twist, final: See **Final twist.** The number of turns per unit length in a single yarn component of the plied yarn or the plied yarn component of plied yarn component of a cabled yarn as the component lies in the more complex structure.

Twist, hawser: see **Hawser twist.** The construction of a cabled yarn, cord or rope in which the single and first ply twist are in the same direction and the second ply twist are in the opposite direction ans S/Z/S or Z/S/Z construction.

Twist, original: The twist in a single or plied yarn component of a plied or cabled yarn as the component was before incorporation in to the more complex structure.

Twist, set: To dissipate or render latent any unbalanced twisting moment of force in a yarn by suitable treatment such as steaming.

Twist, single: The amount of twist in each individual single yarn element in tyre cord structure based on the length of the element after the twist has been removed from the cord.

Twisted or laid rope: Rope made from three or more strands which are laid or twisted together in a twist direction opposite to the twist direction in the strands.

Twisted Multifilament Thread Construction: Thread made from continuous filaments of polyester or nylon that are twisted together into a cohesive bundle and then plied to make the thread.

Twist factor at break: See **Torsional strength**

Twisting: (1) The process of combining filaments into yarn by twisting them together or combining two or more parallel singles yarns (spun or filament) into plied yarns or cords. Cables are made by twisting plied yarns or cords. Twisting is also employed to increase strength, smoothness, and uniformity, or to obtain novelty effects in yarn. (2) A very high level of twist is added to single or plied yarns to make crepe yarns. This operation generally is called creping or throwing. **3.** The process of adding twist to a filament yarn to hold the filaments together for ease in subsequent textile processing, etc.

Twit: A short section of real twist in false-twist yarn that prevents crimp development and hence causes a pinhole effect in fabric. Also called twist bleed or tight spot.

Twitt: A term applied to yarn which is irregular, that is thick and thin, the thin places being below the count required and thick places above. The defect is caused by material being drafted at an irregular rate.

Twitty: Unevenly spun yarn.

Two bath pad-batch process: For reactive dyes. A semi-continuous principle: goods are padded with neutral dye solution, dried, padded with alkali and salt, batched up and given several hours' dwell time

Two bath pad-steam process: Another term for the Pad-steam process. A continuous dyeing process, in which the goods are padded in the 1st bath with dye, in the 2nd with the chemicals necessary for fixing, and then briefly (max. 60 s) fixed in steam. Examples: for vat dyes, caustic soda and sodium hydrosulphite, for reactive dyes a saline alkali or sodium silicate solution.

Two bath process: Working with two baths in series. For example, in the case of polyester/wool blends, the polyester content is dyed first in one bath, with overdyeing of the wool in a fresh bath. Opposite: Single-bath process.

Two Faced: See **Double Faced.**

Two phase flash age process: In this variation of the Two-phase printing process, goods printed with alkali free reactive dyes are fixed in a flash ageing plant which consists of a pad, preferably a Mini-Fluid, and a flash ageing steamer. Fixing involves application from the pad of either an alkaline saline liquor of varying composition or sodium silicate or a combination of sodium silicate and caustic soda.

Two phase mercerising process: A process whereby in the 1st phase cotton is impregnated with 24–25% caustic soda and in the 2nd phase with dilute alkaline solution of approx. 11%.

Two stage twisting: A system of producing a yarn , which consists of stage-1- Inserting a low level of twist into a yarn or yarns by ring twisting, stage 2- taking the product of stage 1 and up and twisting to insert the desired amount of twist.

Two tone effect: E.g. on fabrics (woven, knitted goods) which fashion requires should contain two colours (for more colours Multicolour effects), distributed evenly or randomly; e.g. jaspé, melange, vigoureux, changeant (two different-coloured yarns in warp and weft), thread-by-thread (double twill), mouliné twist.

Two-for-one twister: A twister that inserts twist at a rate of twice the spindle speed.

For example, at a spindle speed of 2,000 rpm, 4,000 turns per minute are inserted in the yarn.

Two-front button: A button in which the face and back shape are identical.

Two-phase flash-age process: In this variation of the Two-phase printing process, goods printed with alkali free reactive dyes are fixed in a flash ageing plant which consists of a pad, preferably a Mini-Fluid, and a flash ageing steamer. Fixing involves application from the pad of either an alkaline saline liquor of varying composition or sodium silicate or a combination of sodium silicate and caustic soda.

Two-phase printing process: The preferred method for vat dyes (Colloresine process; Flash ageing) and reactive dyes. Principle: print on the printing dye paste, dry, pad with fixing chemicals (for vat dyes: alkali + electrolyte), then steam, usually without interim drying (100–140°C, 20 s up to 8 min) and wash. Particular advantages include: unlimited fastness of the dye pastes and prints on printed materials, extremely quick fixing of dyestuffs, low energy consumption for fixing. Disadvantage: high chemical consumption.

Two-phase vat discharge prints The discharge dyeing of direct and reactive dyed backgrounds using vat dyes is only possible using the classic vat printing method with sodium formaldehyde sulphoxylate, which requires long fixing times; with two-phase vat discharge printing the advantages of the Two-phase printing process can also be of benefit to discharge printing. The goods to be printed are padded with reactive dye of the alkali free vinyl sulphone type, dried, and then overprinted with vat-dye paste, which, apart from the dye, only contains formaldehyde sulphoxylate. The alkali necessary for both dye types is padded on and the discharge print completed using the Twophase flash ageing process. The desired rapid decomposition of the formaldehyde sulphoxylate is achieved by a heavy metal complex catalyst, e.g. cobalt dimethylglyoxim, which is added to the printing dye paste.

Two-phase wet fixation process: Hoechst system), for reactive dyes (Wet fixation process, Caustic shock process). Fixing in the two-phase wet fixation process is carried out in the wet fixation trough, a simple, inexpensive device, available in various sizes. The fixation liquor, consisting of a strong alkaline saline solution or, as has proved preferable, 600 ml sodium silicate 38–40°Bé and 200 ml 33% caustic soda, heated indirectly to 95–105°C. Fixation of the goods under these conditions takes approx. 10s.

Two-sided effect: In the case of textiles made from blended fibres, provided two-tone or Multicolour effects are not required, all fibres are dyed by means of tone-in-tone dyeing. Otherwise the goods would have an uneven appearance. In the case of certain fabric types, each side of the fabric has a different shade.

Two-sidedness: See **Two-sided effect.**

Two-tone effects: E.g. on fabrics (woven, knitted goods) which fashion requires should contain two colours (for more colours Multicolour effects), distributed evenly or randomly; e.g. jaspé, melange, vigoureux, changeant (two different-coloured yarns in warp and weft), thread-by-thread (double twill), mouliné twist.

Tying-in: The tail end of the exhausted weavers beam is tyed to the beginning of the new warp. If every end of the new beam is tied to the corresponding end of the old beam, the drawing process can be omitted. Following the tying-in process, all knots are pulled through the drop wires, heddles and the reed. The loom is ready for operation.

Tying: on the warp usually refers to the tying of small groups of warp threads to the front apron rod.

Type I Apparel (AATCC): Apparel defined for general heavy work; it may be subjected to breaking and tearing stresses, indoor or outdoor.

Type II Apparel (AATCC): Apparel designed for light work or leisure activities; it will not be expected to undergo severe physical stresses.

Typha fibres Strong, woolly fibres belonging to the Bast fibres group, obtained from the leaves of the reed mace. Only of localised significance. Spun alone, or (usually) with flax (tow), jute, recovered wool, etc. Used as a replacement for jute and hemp, or for coarse woven fabrics.

Typp: An obsolete yarn numbering system. The number of 1000 yard lengths of yarn per pound.

Tyre bead: That part of a tyre that comes in contact with the rim and that is shaped to secure the tyre to the rim.

Tyre bead wire: A monofilament steel wire with a metallic coating, usually bronze, used in forming a tyre bead.

Tyre builder fabric: Fabric consisting of tire cord in the warp with single yarn filling at extended intervals.

Tyre construction: The geometry of the various layers of tire fabric in the final tyre.

Three constructions are commonly used. (a) **Bias Tire**: In this construction, tire fabric is laid alternately at bias angles of 25 to 40° to the tread direction. An even number of layers (or piles) is used. (b) **Radial Tire**: In a radial tire, tire fabric traverses the body of the tire at 90° to the tread direction. Atop the tire fabric are laid alternating narrow layers of fabric at low angles of 10 to 30° to the tread direction; the belt that is formed around the tire body restricts the movement of the body. (c) **Bias/Belted Tire**: This tire construction combines features of the preceding two. The first layers of fabric are identical to the bias tire. The belt is added in alternating layers at 20° to the tread direction.

Tyre cord: A twisted or formed structure composed of two or more single or plied industrial yarn elements having the same nominal twist, direction of twist, length, and tension.

Tyre cord fabric: A fabric consisting of tyre cord warp woth widely spaced single yarn weft. A loose fabric woven to facilitate large-scale dipping, treating, and calendaring of tyre cords. Usually, 15 to 35 tire cords per inch of warp are woven into a tire fabric by 2 to 5 light filling yarns per inch. In these fabrics, the strength is in the warp and the filling only holds cords in position for processing. The filling yarns are normally broken during tire molding. The warp cords are polyester, rayon, nylon, aramid, glass, or steel and range in strength from 30 pounds to over 100 pounds per cord. A 60-inch fabric would normally have a warp strength of about 7,000 pounds. Such fabrics are used for tyre carcasses and tyre belts. More conventional square woven fabrics are used in certain parts of a tire such as the bead, chafer, and wrapping. (Also see **Tyre cord**.)

Tyre-builder fabric: See **Fabric, tyre builder.**

Tyrian Purple: Red dyestuff, yielded by several species of the murex, a snail in the Mediterranean; used extensively and highly prized by the ancient people. Tyrian purple. A purple colour obtained from certain shell fish, such as Buccinum and Purpura. It is mentioned by Pliny as being discovered in 1400 BC. It was a lost art in the Middle Ages.

Tyrlind Striped: French dress goods, made with silk warp and heavy schappe filling, forming cross ribs.

U

U.L. Down: 'Ultra Light Down' is used in women's and men's jackets. The concept is to make the lightest and warmest insulation layer available. U.L. Down jackets weigh less than a t-shirt, blocks more wind and is warmer than even the heaviest fleece jackets and compress to the size of a water bottle. This outerwear can be used when warmth is critical, minimal weight is paramount, and space is at a premium.

Ulang: In the Chinese markets, stout and strong worsted satins.

Ulster: (1) The name often used to describe a heavy overcoat, but it is in fact the name of a thick cloth napped on the right side. The pile is flattened, but still lies in one direction. Once often used for men's and women's traveling cloaks, but now mainly confined to men's overcoats. A classic ulster is double-breasted; modern sport and short ulster are single-breasted.

(2) A heavy, fulled and rough woollen winter coat material of coarse yarn (often pure virgin wool, but also an admixture of shoddy and rayon) with a Melton finish; usually with a woven-in lining, named after an Irish province. The weave of the outer fabric is usually twill and herringbone; that of the inner fabric is 4-ply evenly spanned twill.

Ultimate oxygen demand(UOD): Either the ***theoretical oxygen demand*** or the ultimate biochemical oxygen demand, the maximum value of the BOD test, ordinarily reached after 25 to 30 days' incubation. The UOD is the sum of the ultimate ***carbonaceous BOD*** and the ultimate ***nitrogenous BOD***.

Ultracentrifuge: Used in colloidal chemistry to determine molecular weight and sedimentation rate of particles suspended in either liquids or gases. It can attain a centrifugal force of 260,000 to 1 million g (motor driven at approximately 30000 rpm).

Ultrafiltration: A membrane extraction process (Permeation) using membranes or cellulose film applying hydrostatic pressure difference to separate colloidally dispersed dissolved substances from molecularly dispersed substances and macromolecules, colloids, most bacteria, some viruses, proteins. Pore diameter from 02–50 nm. Typical operating range will be 0.005–0.200 um. Ultrafiltration is suitable for cleaning waste water from the dyeing and textile industry.

Ultra-light weight: Term used to describe a fabric used in outerwear, which allows for a minimum pack volume and weight. Lightweight packable garments offer the most versatile weather protection. Some of these fabrics have a protective layer on the membrane, which provides durability. This means that the garments made from extra lightweight fabrics need no separate lining.

Ultramarine (-blue, blue powder, washing-blue): Mineral dye obtained by heating together kaolin, sodium carbonate, sulphur and charcoal. A fine brilliant blue powder (of variable composition), also blue-green, violet and red. Insoluble in all solvents.

Usage – component of white conditioner in printing; frequently used in the blueing of bleached goods, underwear, feathers, etc.; dyeing of plastic, rubber, wax cloth, leather, paper; as marking dyes, etc.

Ultra-sonic: Higher than audible sound with a frequency above approximately 16 kHz up to approximately 106 kHz. Infrasound up to 16 Hz; audible sound 16–16,000 kHz; ultrasound above 16,000 Hz. Ultrasonic is a specialist area of more general acoustics which studies the behaviour of mechanical oscillations in solid, liquid and gaseous media (sound). In reference to the characteristics of the human ear, high frequency and no longer audible oscillations are termed ultrasound or hyper-sound.

Ultraviolet: A form of electromagnetic radiation, shorter in wavelength than visible light. Like light, ultraviolet radiation is produced by electronic transitions between the outer energy levels of atoms. However having a higher frequency, ultraviolet photons carry more energy than those of light and can induce photolysis of compounds and photoionization. Light that is just beyond the visible portion of the light spectrum at the blue end. It is primarily ultraviolet light that is responsible for fading of colours, and that makes fluorescent compounds glow. Longwave ultraviolet, "blacklight", with a wavelength of around 365 nano-meters, is the preferred source for exposing many photosensitive materials such as silkscreen emulsions. It is also useful for inspecting fabrics to detect optical brighteners. Longwave UV sources can be found in the form or fluorescent tubes. Tubes that appear white when off also emit considerable visible light and are best as exposure sources. Tubes that appear very dark purple don't produce much visible light and are best for inspection sources. The dark purple type can often be purchased at novelty shops. "Germicidal" lamps are shortwave UV emitters. This wavelength, typically about 254 nano-meters can cause eye damage. In one classification system, the ultraviolet spectrum is referred to by three wavelength classes: UB-A (315–400 nm), UV-B (280–315 nm) and UV-R (280–400 nm). Ordinary glass is fairly transparent to longwave UV, but nearly completely opaque to shortwave UV. Quartz is a much more effective material for making lenses and prisms for use with ultraviolet. See also **Electromagnetic radiation**.

Ultraviolet curing of polymers: Polymer systems capable of being cured by irradiation need not have a complex structure and may comprise of only a single chemical substance. For reactive components such as acrylates, the energy level of accelerated electrons is high enough to begin radical polymerization. However if UV light is used as a source or irradiation, a photo-initiator is usually required in order to generate sufficient radicals to start the polymerization reaction.

Ultraviolet degradation: Weakening or deterioration caused by exposure to ultraviolet rays of sunlight or artificial light.

Ultraviolet fluorescence of textiles: Fibres emit fluorescence in the ultraviolet region, e.g.

Faintly yellowish, whitish (USA)	cotton
Pale yellowish	cotton, mercerised and bleached
Straw yellow	viscose, unbleached
Sulphur yellow with blue shadow	viscose, bleached
Intense yellow	polyester
Yellowish green`polyacrylonitrile	
Blue-green	polyvinylchloride
Bluish violet, brightly luminous	acetate
Reddish white, milky	cuprous
Bluish white	wool, unbleached; silk, scoured; casein
Pale bluish, whitish	wool, bleached.

Ultraviolet lamp (quartz lamp, quartz-mercury lamp, mercury lamp): Artificial source for the generation of ultraviolet radiation. For lamps used in analysis, the visible component of light from such sources is removed using a special filter.

Ultraviolet radiation (UV radiation): Includes light of a wavelength from approximately 140 nm to 360 nm. Ultraviolet radiation is present in sunlight but is nevertheless invisible to the human eye, since it lies beyond the violet portion of the visible spectrum.

Note: The limits of the spectral range of ultraviolet radiation are not well-defined and may vary according to the user. Committee E.2.1.2 of the CIE distinguishes in the spectral range between 400 and 100 nm

UV-A 315–400 nm

UV-B 280–315 nm

UV-C 100–280 nm

Ultraviolet resistance: Ability to retain strength and resist deterioration on exposure to sunlight.

Ultrasuede: Unbleached cotton.

Umbrella fabrics: A thick and light fabric (often printed) made from yarn-dyed polyamide threads and made water repellent.

UMIST: English textile research organisation in Manchester.

UNARI: Burmese Standardization Association.

Unbalanced colourways: When there is no similar colour relationship between colourways, giving the colourways a different appearance to the original design.

Unbleached: Cotton and linen fabrics, left in the natural "grey" state.

Unbleached linen (Écru linen, raw linen): Naturally coloured, unbleached.

Unboiled silk: See **Ecru Silk**. Ecru silk (bast silk) is barely degummed silk (maximum 4% loss) – only grease, wax and resinous substances removed.

Unbonded batting: A textile filling material which is neither needle punched, resin bonded, nor thermal bonded.

UNCO: Una Norma Colombiana, Colombian Standard.

Uncollapsed cake: Rewound cake in crossed layers of yarn, mechanically stronger with improved dyeing characteristics (package and pin up system).

Uncut pile fabrics: Pile fabric before cutting the piles.

Underarm length (in body measurements): With the arm down, the distance from the armpit to the inner wrist bone.

Undercoat hairs (down hairs, wool hairs): Shorter and finer hairs forming the inner layer of an animal's coat, usually unmarked, soft/elastic, curled with good felting quality.

Undercoat: The body-coat.

Undercollar: The underside of a collar; usually, but not necessarily, of softer material.

Underlap: This is the other section of stitch in a warp knitted fabric. The yarn is horizontal and appears at the back of the fabric. It can cover 14 needle spaces or more, depending on the design of machine and fabric.

Underlay: A cushion or padding material used under rugs or carpets to give greater resiliency and longer service.

Under-press: To press the underside of a garment section during manufacturing to open the seams and give it shape.

Under-stitch seam finish: Under stitching is a second row of stitching sewn along the seamline crease on the inside edge of the garment or on the facing edge. It helps to hold the seam allowances in place and keep them from rolling out which would flatten the edge of the garment. Used for neckline areas, on under collars, on cuff facings, on the backside of lapels, on front edge facings or any areas where a sharp seamline crease needs to be maintained.

Undertone: Each hue has an undertone which is its underlying primary colour. It is the undertone of a hue that determines its temperature. A blue undertone gives a hue a cool bias and a red one gives it a warm bias. Each hue can have either a warm or a cool undertone, regardless of the actual hue. Therefore blue, although generally a cool colour, can have a warm bias if the undertone is red. This is also true of the warm hue red, which with a blue undertone becomes cool. Thus each primary hue can have a warm and a cool variation.

Underwear: Clothing worn next to the skin under outer clothes. Underwear is intended to absorb the moisture emanating from the skin through perspiration and release this to the outer air. If the rate of absorption is greater than that of evaporation, the textile remains wet against the skin and this leads to an unpleasant sensation.

Undrawn tow: Tow which is not undergone drawing process. Tow fibre is similar to staple but it is cut up at a later stage in production than staple. The short fibre stock of varying lengths removed in breaking, scutching, hackling, or combing.

Undrawn yarn: Extruded yarn (filaments), the component molecules of which are substantially unoriented. Undrawn yarn exhibits predominantly plastic flow in the initial stages of stretching and represents an intermediate stage in the production of a manufactured yarn.

Undulated twills: Twills with curving diagonal lines are called undulated twills and can be made using four shafts.

UNE: Spanish Standards Association.

Uneven dyeing: Cloth which shows variations in shade due either to incorrect dyeing methods or faulty materials.

Uneven dyeing for fashion effects: Either intentionally dyed unevenly as per requirements of the fashion or make uneven appearance of dyed fabric by mechanical or enzyme treatments or other methods. Example is faded, or vintage look in Jean materials.

Uneven shrinkage: A wavy, warp-wise condition in the fabric that prevents it from lying flat on a horizontal surface.

Uneven surface: An irregular surface characterized by non-uniformity in the physical configuration of the yarns or fibres making up the surface of the fabric.

Uneven yarn: A yarn that varies in diameter to an abnormal degree.

Unevenness (in textiles): Variation in the linear density of a continuous strand or of a portion of a strand.

Unfinished worsted: Despite the name, it is not necessarily wool and is finished. It is characterized by a slight nap. It is a plain weave, heavy weight men's suiting and may be plain or check. Used for men's suits and women's heavy duty trousers.

Unfinished worsted: A worsted fabric with a relatively soft hand and a light nap.

UNI: Italian Standards Association.

Unidirectional fabric: A fabric having reinforcing fibres only in one direction.

Uniform cloth: Cloth suitable for uniforms; usually a stout, fulled, woollen cloth, similar to kersey.

Uniformity: The most important determining factor when assessing goods according to quality, e.g. in terms of pattern, finish, weight, length, thickness, section, strength, stretch, etc. Calculated as: % U (uniformity) = 100 – deviation in % (See **Deviation**).

Uniformity index (in fibre length testing of cotton): The ratio between the mean length and the upper-half-mean length expressed as a percentage of the upper-half-mean length.

Uniformity ratio (in fibre length testing with fibrograph): The ratio between two-span lengths expressed as percentage of the longer length.

Ungummed silk: Which has the natural gum removed by boiling in solution of soap.

Uni: French for single coloured or plain effect.

Union: The term has two meanings – (1) A fabric made from the union of two fibres, e.g. cotton and worsteds, or cotton and linen; (2) A striped fabric from two fibres and then dyed, the fibres taking the dye differently and therefore producing a coloured stripe. This was often produced as a cheap, hard wearing cotton cloth and made into men's working shirts, worn with white cardboard or celluloid collars and cuffs which are removed at home.

Union braid: Worsted or mohair braid, made of more than two cords.

Union carpets: In England, double carpets with pile on both sides; held together by a binder filling.

Union cassimere: Soft and fulled fabric, made with cotton warp and woollen filling; has checks or stripes; used for men's cheap clothing.

Union cloth: A napped shoddy fabric, made with cotton warp in England.

Union cord: Stout, round cord made of cotton and linen; used for lacing.

Union damask: Made with cotton or linen warp and worsted filling, woven with satin figures over satin ground; used for hangings, etc.

Union dye: A dye that is a mixture of two or more different classes of dye; used typically to dye blends of fibres. "Household" dyes, of the sort sold in grocery stores, are usually union dyes containing a direct dye which will work on cellulose fibres, and an acid dye which will work on wool or nylon. Industrially the union dyes may be of the other combinations, such as reactive and disperse dyes for dyeing cotton-polyester blends (often with two distinctly different sub-processes).

Union dyeing: Dyeing of union fabric, i.e., fabric blends containing wool.

Union fabric: The blended fabric containing wool.

Union linen: Made with cotton warp and linen filling.

Union silk: A fabric made usually with cotton warp and silk filling; used as umbrella cloth.

Union yarn: Made of a mixture of cotton and wool.

Unisyn concept: Universal desizing technique for water soluble synthetic size. Water soluble sizes of all types can be washed out cost effectively on a continuous wash production line by using the Unisyn technique (BASF) with the application of

1 g/l wetting agent,

2 g/l dispersing agent,

3 g/l sodium hydroxide.

UNIT: Instituto Uruguayo de Normas Tecnicas, Uruguay Institute for Technical Standards.

Unit sampling (in wool): A portion of material that is taken at one time from one physical location and that is combined with similar portions to make up the laboratory sample.

Universal indicator (multiple-range indicator): A mixture of indicator dyestuffs which shows a gradual change in colour over a wide pH range. A typical formulation contains methyl orange, methyl red, bromothymol blue, and phenolphthalein and changes through a red, orange, yellow, green, blue, and violet sequence between pH 3 and pH 10. Several commercial preparations are available as both solutions and test papers.

Unlevel dyeing: Faults occurring in the dyeing of textiles of all varieties. The dyestuff is not levelled (levelling properties), but instead is spread unevenly over the textile material. Unlevel dyeing may be caused by, among others, any of the following factors: omitted or inadequate pre-treatment, packing too tightly in apparatus, a too weak or a too strong dye bath circulation, inappropriate or faulty dyeing conditions and inappropriate choice of dye.

Unlevelness: The absence of levelness in the dyeing or printing or any other process. See **Unlevel dyeing**.

Unopened staple: Staple fibre in bunches or clusters in the bale in such a condition that it will not process smoothly through carding and subsequent operations in the spun-yarn plant.

Unroved: Unravelled.

Unrelaxed yarn: Yarn in which the tension during the manufacturing process has not been released. The relaxation is done usually by steaming.

Unsaturated steam: Superheated steam. Steam may be either saturated or superheated.

Saturated steam is a vapour and, at a particular pressure, has a definite temperature, the saturation temperature. Superheated steam is a gas and, at a particular pressure, may have any temperature in excess of the saturation temperature.

Unsupported needle felt: A needle felt that is composed entirely of fibres physically interlocked and reoriented with, and of themselves without an interlay, scrim or extruded structure.

Untreated: A descriptive term for glass fibre yarns having no applied chemicals or coatings, other than the minimal lubricant or binder used to control intra-fibre abrasion.

Updraft metier: A dry spinning machine in which the air flows within the drying cabin and is counter current to the yarn path (upward).

UPF (Ultraviolet Protection Factor): The UPF rating indicates how effective a fabric is at blocking out solar ultraviolet radiation from reaching the skin. UPF ratings range from 15 to 50 with higher ratings indicating more effective blocking and therefore better protection for the wearer of a garment. Fabrics that test higher than UPF 50 are rated as UPF50+. UPF testing involves the process of exposing a fabric to ultraviolet radiation (UVR) and measuring how much radiation is transmitted through the sample. Different wave-lengths of radiation in the UVR spectrum have different effects on human skin and this is taken into consideration when calculating the UPF rating. Factors that contribute to the UPF rating of a fabric are as follows:

(1) Composition of the yarns (cotton, polyester, etc.)
(2) Tightness of the weave or the knit (tighter improves the rating)
(3) Colour (darker colours are generally better)
(4) Stretch (more stretch lowers the rating)
(5) Moisture (many fabrics have lower ratings when wet)
(6) Condition (worn and faded garments may have reduced ratings)
(7) Finishing (few fabrics are treated with UV absorbing chemicals)

Upholstered furniture: Furniture covered with such materials as textiles or leather and generally with padding or cushions or both.

Upholstery crepe: A woven crêpe made from cotton, usually printed.

Upholstery fabric: The exterior fabric covering applied to a furniture unit; it can be woven, warp-knitted, raschel-knitted fabrics and stitch-bonded fabrics in wool, polyamide, polyacrylonitrile, cellulose and blends.

UPN: Urad pro Normalizaci a Mereni, Prag; Polish Office of Standardization and Metrology.

Upper-arm girth (in body measurements): The maximum circumference of the arm usually midway between the elbow and the shoulder joint.

Upper-arm length (in body measurements): With the arm bent, the distance from the shoulder joint along the outside of the arm usually midway between the elbow and the shoulder joint.

Upper-half-mean length (in fibre length testing of cotton): The mean length by number, of the longer one half of the fibres by weight.

Upper quartile length, in testing of cotton fibres: That length which is exceeded by 25% of the fibres by weight in the test specimen.

Upright pile finish: Plush fabrics with pile made from cotton fibres are impregnated with synthetic resin, brushed, dried and set. The pile stands up again after being pressed down.

Upright pile velvet: Raised fabric with velvety appearance. The napped fibres stand vertical (opposite Nap velour). See **Pile lifter**.

Uptwister: A machine used for twisting yarns in an upward path from a rotating vertical supply package to a horizontal take-up package. Used for spun yarns and to a small extent for adding twist to a few filament yarns.

Uptwisting: The process of twisting yarn on the uptwister. The yarn to be twisted, which has been wounded on a balanced support package, is placed on a revolving spindle. The yarn from the revolving supply package is fed upward through a gathering eye or guide, over a stop motion and a tension bar or bars, through a traversing guide, and onto the revolving collecting package.

Urban Tool's Groove Rider: Austrian Urban Tool Company has already commercialized iShirts which have a pocket made with elastic fabric for a mobile phone or iPod, so the devices won't slip out. A recently developed product is a shirt with a music player controlled by a smart fabric touchpad. The smart fabric touchpad is one of ElekTex's technology products (Comprehensive Merchandising Support, 2007).

Urea: An organic nitrogen compound (NH_2CONH_2). Urea is used in dyeing for a number of purposes. Urea helps increase the limit of solubility of few dyes, such as MX, in water. This can be useful when strong solutions are to be made for tie-dyeing. From 5% to 20% w/v can be used. Urea is used as a humectant. In tie-dyeing or similar processes, it helps prevent fabric from drying out during the long periods when it is left in the open for the dye to fix. Urea increases the swelling of fibres and can break hydrogen bonds, aiding penetration and mobility of dye. Large amounts of urea, up to 30% w/v, are used in some cold pad-batch dyeing processes for wool. Most urea sold is synthesized from natural gas (Urea is sold as an "organic nitrogen" fertilizer. Pure urea will be designated 46-0-0 when sold as such.).

Urea formaldehyde compound: The reaction of an amide -N-H with HCHO to form a $-NCH_2OH$ is often termed Methylolation because the reaction product is called an N-methylol group. Accordingly urea can be metholylated with up to 4 moles of formaldehyde and the reaction products used as crease resistant finishes. When 2 moles of HCHO are reacted with one mole of urea, dimethylol urea is formed. Being di-functional, it is capable of serving as a cross-linking agent.

Urea formaldehyde resin: See **Urea formaldehyde resin**.

Urea free printing (urea in textile printing): Dye manufacturers try to replace urea as a hydrotropic agent with low molecular surfactants. Since the spray technique, using rotor damping may in the meantime be described as technically dated. The use of this device has proved to be more successful than a print steamer on a semi-technical scale in order to fix reactive prints on viscose without using the waste water problematic urea. In normal setting, yields of 60% effectively fixed dye (referring to the dye used, the K/S value is = 100%) the best colour yields are achieved at approximately 30% rotor damping before entering the steamer.

Urea/Formaldehyde (U/F): See **Urea formaldehyde compounds**.

Urease: Urea-breaking enzyme, belonging to Hydrolases; Amidases. Gradual degradation into ammonia and carbonic acid results with water absorption: in which partially decaying ammonium (hydrogen) carbonate forms again. Urease was of great importance in wool finishing with urine (wool washing, indigo vat). Manufacture in purely crystalline state from the jack and soya bean since 1926. 1 g of urease splits almost 60 g of urea in 1 min at 20°C. Urease is inactivated by low quantities of heavy metal ions (Cu) (immobilization of active groups through complex formation), which can be reversed again by adding cyanides or hydrogen sulphide.

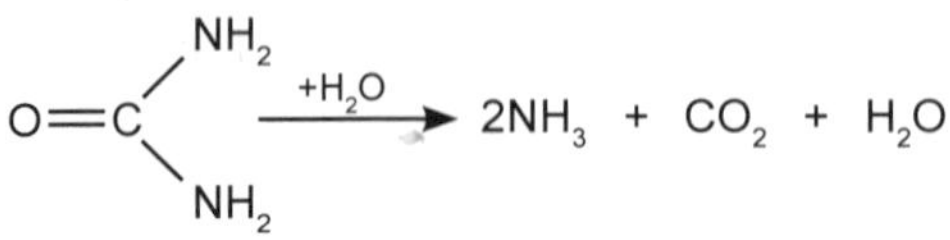

Urena fibres: Bast fibres (e.g., aramina fibre, congo jute) from many tropical countries (Brazil, Madagascar), similar to Rosella as a jute substitute.

Urethane: The name of a group of organic chemical compounds or resins built from isocyanate, a very reactive material that liberates gas during reaction to produce foams of various types. Two types of compounds that react with isocyanate to form foam are polyesters and polyethers. Polyurethanes are used for foams and in other compounds in fibre form. The polyester variety should not be confused with polyester fibres. See **Spandex fibre**.

Urethane elastomer: Polyurethane elastomer fibres.

Urethane group: Basically weaker compared with the Carbamide group (–NHCO), affinity only to disperse and selected acid dyes (below pH 2).

– NH–CO–O–

Urethanes (carbamates, ester amides): CO(OR) (NH_2), the compounds of carbonic acid intended here can be interpreted as esters of Carbamic acid. Due to their chemical relationship to urea they are called urethanes (of Urea), whose structural principle is represented in Polyurethanes.

Urquhart: A Highland tartan, composed as follows: Dark green stripe, split in the centre by a very narrow black stripe; black stripe, less than one third of the green; dark blue stripe, wider than the green, split by a red stripe in the centre (one-sixth the width of the blue) and by a pair of very fine black lines near each edge.

Useable area (carpet): Useable layer in fitted carpets; the surface of fitted carpets needed in use.

Use-surface (for pile yarn floor covering): That part of a textile floor covering directly exposed to the traffic.

User based (in quality testing): In this definition, quality is considered to be an individual matter and the highest quality products are those that best satisfy the customer's preferences. The drawback of this definition is that the consumer preferences vary widely so that it is difficult to aggregate these preferences into products that have sufficiently wide appeal.

USP (A.P.): United States of America Patent (American Patent).

USS: United States Standard.

Uster tester: An instrument that provides a continuous measurement of the variation in weight per unit length of sliver, roving, and yarn.

Usuji Sofu: Light weight and plain woven cotton sheeting made in Japan.

Utility percale: See **Percale**.

Utility stitches: These are the basic stitches found on a sewing machine which are used to sew seams, hem, insert zippers and elastic. Examples of utility stitches are the straight stitch and zig-zag stitch.

Utrecht velvet: A hard wearing upholstery velvet. The Bse cloth is closely covered with the pile which is arranged so that there is no gap.

UV absorber: Polymer additives that absorb light in the UV region or that trap radicals produced in fibre during photo-oxidation. They provide stabilization against actinic degradation. Some critical applications include geotextiles, recreational surface polymers and fibres, tenting tarpaulins, etc.

UV degradation: The breaking down of fibres or fabrics when exposed to ultraviolet rays.

UV fluorescent dye: Fluorescent dyes in combination with UV irradiation. Application – UV marking pen is used for the invisible labelling of material faults and pre-programmable material positions (combination using UV irradiation), also for valuable fabrics (carpets, fleeces, furs, etc.) for invisible sighting using recognition features. Constructed on a wax basis and in the form of Retouching pencils. The fabric spot to be marked must be ironed so that UV dye penetrates into the fabric. Chiefly used for wool and wool blend fabrics, as cotton and synthetics do not absorb any wax.

UV lamp: Ultra Violet Lamp

UV radiation: Ultraviolet radiation.

UV-A type fluorescent UV lamp: A fluorescent UV lamp where radiant emission below 300 nm is less than 2% of its total light output.

UV-B type fluorescent UV lamp: A fluorescent UV lamp where radiant emission below 300 nm is more than 10% of its total light output.**Union linen lawns**: As the name implies, these are made of cotton and linen yarns. The cotton in most cases forms the warp and linen as filling. Union linens are plain woven fabrics made in various textures with regard to ends and picks per inch, and also the quality and counts of yarn. The goods are used for various purposes, principal among which are furniture coverings, summer outing suits and dusters.

Upcycling: Using waste textiles, either pre- or post-consumer, and adding value to them by printing or stitching, and making them into new products.

V

V-bed flat-knitting machine: A latch-needle weft-knitting machine with two needlebeds at a 90° angle to each other in the form of an inverted V. Each needlebed is at a 45° angle to the horizontal. These machines are used primarily to produce collars, sleeves, sweater strips, and rib trims.

Vacuole: (Lat.: vacuus = empty), a cell cavity found especially in plant and algal cells, including vegetable fibres, which takes up a large part of the cellular volume; fibre cavities, intermicellar regions.

Vacuum: (Lat.: vacuus = empty), virtually air-free space, negative pressure. Although, literally, a vacuum is defined as a space totally devoid of any matter, which does not exist on earth, the term is used loosely for an enclosed space within which the pressure of a gas or air lies below that of air at the normal atmospheric air pressure of 1 bar.

Vacuum (v): To clean using an electrically powered machine to create suction in order to remove loose, particulate soil and lint.

Vacuum Breaker: A ventilating valve for steam heated equipment (e.g. feed water tanks) to protect the tank/unit against the build-up of excessive vacuum pressures. A vacuum is formed, e.g. in the feed water de-aerator by the choice of superheated steam to raise the water temperature and introduction of cold water at the same time, due to condensation of uperheated steam which is still present. To prevent the build-up of excessive vacuum pressures, therefore, a flow of air into the de-aerator must be ensured by a suitable device as soon as a negative pressure develops in order to equalize the pressure.

Vacuum Drier: This type of drier is based on the principle of a Suction-drum drier and has been designed particularly for drying yarn on packages or in the form of hanks at temperatures of only 40–70°C. The 3-phase vacuum drier works intermittently by alternating cycles of heating up under pressure and evacuation to allow vaporization at a lower boiling point.

Vacuum Dyeing: Fibres from which all the air has been removed in a vacuum wet out more rapidly and and a draw roller. In the suction cap zone, air is removed from the fabric by a vacuum of 3325 Pa. The cap is sealed against the rotating draw roller. Sealing of the suction cap and the liquid vacuum pump ensure a constant vacuum of 3325 Pa is maintained even during continuous operation. For fabric entry, the vacuum cap allows the fabric to leave the draw roller approx. 170 mm horizontally. The resultant impregnated fabric is squeezed off in the nip of a 2-bowl padder with deformation-resistant bowls. Up to 30% higher liquor pick-up is possible with vacuum impregnation.

Vacuum Extraction Of Piece Goods: Surface water adhering to woven fabrics is easily removed through suction extraction by exploiting the principle of gas kinetics with the aid of a vacuum. In this process, the wet fabric passes over a suction slit

Vacuum Extractor For Yarn Packages: A device for extracting water from yarn packages with a branch circuit to a vacuum station and an interrupter valve. The rapid intermission frequency developed by the latter produces shock-like suction pulses which entrain the bound water. Residual moisture content after extraction is approx. 100%.

Vacuum Extractor in padding: An impregnation technology for applications on the padder which removes the air present in the fibre's intercapillary spaces by means of a vacuum (3325 Pa) immediately before entry into the liquor

Vacuum impregnation: An impregnation technology for applications on the padder which removes the air present in the fibre's intercapillary spaces by means of a vacuum (3325 Pa) immediately before entry into the liquor. After leaving the liquor, the fabric enters the air again. The penetration of liquor is more rapid and complete due to the extraction of air before immersion.

Vacuum impregnator: A machine used for vacuum impregnation. See **Vacuum impregnation.**

Vacuum Mercerisation: Vacuum treatment during mercerization. See also Vacuum treatment during mercerization.

Vacuum Plated Button: A button that is flash metal coated in vacuum chambers and subsequently coloured to simulate other metal finishes.

Vacuum sieving machine: A Print paste sieving machine which operates on the vacuum principle.

Vacuum Strainers: Strainers used for thick solutions which normally does not filter with gravity force. Very commonly used in filtering printing pastes, vacuum applied outside the filter mesh which forces the paste to pass through it.

Vacuum Suction Device: An open-width suction unit designed to reduce the liquor content of a textile fabric before drying, and before or after the application of chemical liquors (wet-on-wet treatment) for the removal of unbound water. Savings in drying energy are achieved. Vacuum suction units can also be retrofitted to existing machines. The use of an under liquor vacuum suction unit allows, e.g., the desizing of tightly woven airbag fabrics.

Vacuum Transfer Printing Machine: The objective of this type of transfer printing machine is to achieve higher production speeds and improved print quality. It is essential that the air-permeable stainless steel conveyor, textile fabric and transfer paper are transported together before entry into the vacuum zone in order that the fabric passage can be checked and corrected if necessary without any intervention in the vacuum zone

Vacuum treatment during mercerization: Immediately before impregnation, entrained air is evacuated from the textile material which allows cold NaOH liquor to be forced into the still unswollen interior of the fibres by external overpressure during immersion.

Vacuum yarn steamer: With automatic dosing of chemical additions. This type of steamer can be preprogrammed from 50–150°C in increments of 1°C with steaming times of 2–120 min (punched card controller).

Vagabundo Effect: In addition to the "stone wash" treatment of denim articles to obtain various effects (e.g. "snow wash", "moon wash", "gold wash", etc.) as well as sand-blasting to produce a fine, velvety handle, another process designed to give denim articles a modish character has established itself.

Val Lace: See **Valenciennes Lace,** Short for of **Valenciennes Lace.**

Valencia vesting: A cloth for waist coats, woven with a cotton binder warp and a silk figuring warp with a worsted weft. It was popular in the early 19th Century.

Valenciennes Lace: A hand made flat type of Bobbin Lace. It is only made of linen thread, but cotton imitations are quite common. The lace is typified by the ground fabric, either roundmesh or diamond, and of the same thread as the decorative design.

Valonia: Acorn cups of certain species of oak from South Europe, containing 25-35% of tannic acid.

Valora: See **Velour.**

Value, in colour chemistry: Indicates the degree of lightness or darkness of a colour in relation to a neutral gray scale. The scale of value (or V, in the Munsell system of colour notation) ranges from 0 for pure black to 10 for pure white. The value scale is neutral or without hue.

van der Waals equation: An equation of state for real gases. For n moles of gas the equation is $(p + n2a/V2)(V - nb) = nRT$ where p is the pressure, V the volume, and T the thermodynamic temperature. a and b are constants for a given substance and R is the gas constant. The equation gives a better description of the behaviour of real gases than the perfect gas equation ($pV = nRT$). The equation contains two corrections: b is a correction for the non-negligible size of the molecules; $a/V2$ corrects for the fact that there are attractive forces between the molecules, thus slightly reducing the pressure from that of an ideal gas. See also **Gas laws**; **Kinetic theory**.

Van Der Vaals Forces: Intermolecular forces as a result of localization of electrical charge within molecules or crystals named after the Dutch physicist and Nobel Prize winner, J. D. van der Waals (1837–1923). They vary inversely as the sixth power of the interatomic distance and are due to momentary dipoles caused by fluctuations in the electronic configuration of the molecules A molecule considered as a whole is electrically neutral. Because of the way in which electrons are held, there may be local areas that appear to have positive or negative electric charge, either permanently or temporarily. These charges lead to attraction between molecules. Van der Waals forces are weak and easily broken, but they can be important in dyeing. They can be important in *affinity*, and hold dye molecules on the fibre near to where a much stronger bond may ultimately be formed.

Vanadium dichloride: (Vanadous chloride). VC12, molecular weight 122. Derivation: from ammonium metavanadate. Used as a catalyst in the dyeing and printing of aniline black (in solution 1: 1000).

Vandyke: A pointed scallop in laces and embroideries.

Vandyke Stitch: In embroidery a raised couching in Vandyke points.

Vane, in feathers: The section that consists of a soild stiff collection of barbs, as distinguished from the section near the quill point that has soft fluffy barbs.

Vanillin(Poly) Ester Fibre: A polyester fibre produced from vanillin (3-methoxy-4- hydroxybenzaldehyde) derived, among other things, from wood lignin (1 kg wood gives 0.5 kg lignin), then converted to protocatechuic acid and subsequently to a condensable oxy acid with epichlorohydrin.

Vankyke stitch: An embroidery stitch in the shape of the fishbone.

Vantean Process A finishing process to produce a high lustre on wool fibres. It involves a preliminary impregnation (mordanting) with a nickel catalyst which, after an intermediate wash, remains in the endocuticle. Subsequent chlorination is accelerated to such an extent that the exocuticle peels off resulting in a smooth fibre surface.

Vapeur: A very fine and loosely woven French cotton muslin.

Vapour Phase Process: See **Gas phase crosslinking.**

Vapour Pressure: The pressure exerted by a vapor. The saturated vapor pressure is the pressure of a vapor in equilibrium with its liquid or solid. It depends on the nature of the liquid or solid and the temperature.

Vaporloc system: The cloth in open width form is first saturated with sodium hydroxide solution (5-9%) along with wetting agent at 70°C padded and then entered into vaporloc reaction chamber which operates at 30 lb/in 2 (2 atm.) at a temperature of 134'C with a reaction time of 90-120 sec. The fabric is deposited on roller-bed system and thus allows the fabric to be in a relaxed state inside the pressure chamber. The fabric passes in between the low friction teflon material and the centre beam special seals enables the generation of necessary pressure inside the steamer. The cloth content inside the steamer is about 200 meters. The production speed is of the order of 120 m/min. The goods after scouring are washed off first with hot water and then with cold water.

Variable: A quantity to which any of the values in a given set may be assigned.

Variable data: Measurements which vary and may take any of a specified set of numerical values.

Variable Temperature Dyeing: Dyeing methods on the basis of drawing out at different dyeing- bath temperatures. Varying degrees of drawing effectiveness are demonstrated at individual temperatures, temperature ranges or stages, i.e. the substantivity, degree of fixation and drawing speed react differently in different temperature ranges. There are temperature stage-dyeing processes using increasing and decreasing temperatures.

Variable-Tension Rolls A system of rolls to adjust fabric tension by changing the wrap angle of the fabric around guide rollers or guide rods. When variable tension rolls are installed at the entry to a continuous processing plant, they are also sometimes described as the fabric feed system (consisting of 2 tubes only).

Variance: In Statistical analysis, variance is a measure of the dispersion of the distribution of a random variable. The square root of the variance is called the Standard deviation(s), and is the most widely used measure of variability.

Variant: A manufactured fibre modified in polymer configuration or by additive during manufacture, resulting in a change in the properties of the fibre. Examples are flame-retardant variants, deep-dyeing variants, high-tenacity variants, low-pilling variants, and cotton-blending or wool-blending variants.

Vase carpets: Oriental carpets with coloured representations of flower vases surrounded by palmettes and stylised floral motifs.

Vasquine: Thick, spongy falbric, made of Che fibres extracted from the pine needles.

Vat: (1) An alkaline sodium dithionite bath either without dye (blank vat) or with Vat dyes (dye vat). In the latter case, the vat dye is present as the soluble Leuco compound of the water-insoluble parent vat dye and is able to exhaust on the fibre in this form (Vat acids).

(2) The term "vat" was originally used to describe a wooden tub used for the fermentation of indigo.

Vat Acids: The leuco compound of a vat or sulphur dye which, in this (water-soluble) form, possesses affinity for the fibre. A vat acid is a very weak acid. Principle: all vat dyes contain the characteristic >C=O group (keto group). In an alkaline sodium dithionite vat, the addition of hydrogen (= reduction, hydrogenation) takes place first of all to form the >C–OH group which corresponds to the leuco form as the vat acid.

Vat Dye: A classification of dyes that are converted from a water-insoluble pigment form to a soluble leuco form (using a reducing agent), applied by immersion to fabric, then converted back to the insoluble form (by oxidation) The name comes from "vatting" which once meant using natural fermentation processes in a vat to produce the reducing conditions to make the dye soluble. Indigo, the blue of blue jeans, is a common vat dye. Vat dyes, with the notable exception of indigo, are generally very lightfast and washfast. Many have very good resistance to chlorine bleach. Multiple applications of dye may be required to build strong shades because of limited substantivity. Sulfur dyes use processes similar to vat dyes, but are distinguished by their sulfur content. Some modern vat dyes are supplied in already-reduced soluble form. Occasionally art dyes will say something is vat dyed when they mean it has been dyed with any dye type in a large volume of solution, as opposed to by direct application of dye or other techniques. This use should be avoided.

Vat Dyeing, Oxidation Of: An important stage in the development of correct shades in vat dyeing. After dyeing, the leuco vat dye which is present in its water soluble reduced form (enol form) must be reconverted into its insoluble keto form again. In exceptional cases this can be achieved by aerial oxidation (e.g. by hanging in air); as a rule, however, the oxidation stage is carried out in baths containing suitable oxidizing agents. The rate of oxidation is dependent above all on the constitution of the dye as well as on the type and concentration of oxidizing agent used and the temperature and pH of the oxidation bath. The most favourable pH for oxidation is from 10–11. The concentration of oxidizing agent and the conditions of oxidation (temperature and time) should be kept to the minimum required in order to prevent damage to the fibre.

Vat Printing: The reduction process necessary for the fixation of vat dyes on the fibre can be carried out in two ways: (a) Potash process ("all-in" process): in this case the print paste contains both a reducing agent (e.g. sodium formaldehyde sulphoxylate) and an alkali (potassium carbonate = potash). The thickeners normally used are starch-ethers, locust bean gums, alginates, etc. After printing during steaming, the shade of a vat dye changes to its so-called vatted colour, e.g. a blue may change to yellow or a green to red (usually the complementary colour is formed). After steaming, the printed fabric is finally oxidized and soaped. (b) Two-phase process: this method is only possible for prints on white (i.e. undyed) fabric and offers optimum process reliability. The water-insoluble vat dyes are printed first as pigments. Thickeners capable of undergoing coagulation are used to prepare the print pastes, e.g. locust bean gums and their modifications. prints are finally developed by impregnation in a bath containing reducing agent/sodium carbonate or sodium dithionite/caustic soda followed immediately by steaming. This bath also contains sodium tetraborate which coagulates the thickener and prevents the vat pigments from bleeding out of the thickener film, thus ensuring sharp printed outlines. Excess liquor is squeezed off through a nip and the printed goods are steamed immediately to reduce the vat dyes to their soluble leuco compounds, in which form they can penetrate the fibre. Finally, the printed goods are oxidized, soaped and washed off as in 1 above.

Vat Yellow Paper: See **Indanthren yellow paper.**

Vatting: The process of converting Vat dyes from the water-insoluble (pigment) into the water-soluble substantive form (Leuco compounds) by reduction, usually with sodium dithionite and caustic soda.

Vatting test: An unreliable test for hydro-, oxy and photocelluloses. Cellulose which has been chemically damaged has reducing properties which are sufficient to convert Indanthren Yellow G (flavanthrene) in alkaline dispersion into its blue leuco compound. Procedure: treat the sample in an alkaline bath with the dye. Formation of a blue colour indicates chemical damage.

V-belts: (Vee belts) Transmission belts, drive belts with a tapered cross-section which run in the corresponding vee-tapered grooves of pulleys. V-belts are usually made of a rubber composition reinforced with V-belt cord. To transfer higher forces, several identical V-belts are used parallel to one another. Advantages: minimum possible distance between shafts, less wear, non-slip characteristics.

VCRS: Abbreviation for the Swiss Federation of Dry Cleaners (Verband Chemischer Reinigungsanstalten der Schweiz).

VDF: Abbreviation for the Society of German Dyers (Verein Deutscher Färber).

Vector dye media for multicolour printing: The usual colour solids conceived as colour ellipsoids can also be represented more simply as colour polyhedrons resp. as polyhedral colour lattices with defined colour edges and corners as colour vectors, e.g. in the form of four, five, six and seven vector colour solids. The seven vector colour solid has, e.g., 448 different coloured edges and 128 coloured corners. It is used for calculating mixing ratios of the yellow, magenta and cyan primary colours employed in typography.

vee notch weir: (1) A vee-shaped notch in a thin plate for measuring the flow of a water. For a 90° notch, the flow, Q in m3/s, is calculated by Q = 1.42H2.5, where H is the measured head (in m) at the weir. (2) A type of ***weir plate***.

VEGAT: Abbrev. for the Federation of Swiss Yarn and Tricot Finishers (Verband Schweizerischer Garnu. Tricotveredler).

Vegetable Dyes: Natural dyes

Vegetable Dyes, Sources: **Yellow**: Following plants dye Yellow. **Agrimony.** '*Agrimonia Eupatoria*'. **Ash.** '*Fraxinus excelsior*'. Fresh inner bark. **Barberry.** '*Berberis vulgaris*'. Stem and root. Birch. Leaves.**Bog Asphodel.** '*Narthecium ossifragum*'. **Bog Myrtle or Sweet Gale.** '*Myrica Gale*'. **Bracken.** '*Pteris aquilina*'. Roots. Also young tops. **Bramble.** '*Rubus fructicosus*'.

Broom. Sarothammus Scoparius. **Buckthorn.** Rhamnus frangula and R. cathartica. Berries and Bark. Common dock. '*Rumex obtusifolius*'. Root. **Crab Apple.** '*Pyrus Malus*'. Fresh inner bark.

Dyer's Greenwood. '*Genista tinctoria*'. Young shoots and leaves. **Gorse.** '*Ulex Europœus*'. Bark, flowers and young shoots.

Vegetable Dyes, Black: Following plants dye Black. Alder. '*Alnus glutinosa*'. Bark, with copperas. Blackberry. '*Rubus fruticosus*'. Young shoots, with salts of iron. Dock. '*Rumex*'. Root.

Elder. Bark, with copperas.

Vegetable Dyes, Blue: Following plants dye blue: Devil's Bit. '*Scabiosa succisa*'. Leaves prepared like woad. Dog's Mercury. '*Mercurialis perennis*'. Elder. '*Sambucus nigra*'. Berries.

Privet. '*Ligustrum vulgare*'. Berries with alum and salt. Red bearberry. '*Arctostaphylos Uva-Ursi*'.Sloe. [A] '*Prunus communis*'. Fruit. Whortleberry or Blaeberry. '*Vaccinium Myrtillus*'. Berries. Woad. Isatis tinctoria. Yellow Iris. Iris Pseudacorus. Roots.

Vegetable Dyes, Brown: The following plants dye Brown: Alder. *Alnus glutinosa.* Bark.

Birch. '*Betula alba*'. Bark. Onion. Skins. Larch. Pine needles, collected in Autumn.

Oak. '*Quercus Robur*'. Bark. Red currants, with alum. Walnut. Root and green husks of nut.

Water Lily. '*Nymphœa alba*'. Root. Whortleberry. '*Vaccinium Myrtillus*'. Young shoots, with nut galls. Dulse. (Seaweed.); Lichens.

Vegetable Dyes, Green: Following plants dye green: Elder. '*Sambucus nigra*'. Leaves with alum. Flowering reed. '*Phragmites communis*'. Flowering tops, with copperas. Larch. Bark, with alum. Lily of the valley. Convalaria majalis. Leaves. Nettle. Urtica dioica and U. Urens. Privet. '*Ligustrum vulgare*'. Berries and leaves, with alum.

Vegetable Dyes, Purple: The following plants dye purple: Byrony. '*Byronia dioica*'. Berries. Damson. Fruit, with alum. Dandelion. '*Taraxacum Dens-leonis*'. Roots.

Danewort. '*Sambucus Ebulus*'. Berries. Deadly nightshade. '*Atropa Belladonna*'.

Elder. '*Sambucus nigra*'. Berries, with alum, a violet; with alum and salt, a lilac colour.

Sundew. '*Drosera*'. Whortleberry or blackberry. '*Vaccinium myrtillus*'. It contains a blue or purple dye which will dye wool and silk without mordant.

Vegetable Fibre: Natural cellulosic fibres; Cellulosic fibres, regenerated; Natural fibres; Fibres.

Vegetable Flannel: A coarse, hygienic cloth, made from the fibres of the pine needles.

Vegetable Gums: (natural gums). Carbohydrate high polymers insoluble in alcohol and other organic solvents which are obtained as dried exudations from various African and Asian trees and shrubs mainly of the acacia type in tropical areas. They are amorphous substances with very complex polysaccharide constitutions and are generally soluble or dispersible in water to give solutions with adhesive characteristics.

Vegetable Hair Fibre: Single cell seed hairs (Cotton; Akund) or fruit wall fibres (Kapok) of various plants the original purpose of which is to disseminate or protect the seeds. Natural cellulosic fibres; See **Fibres.**

Vegetable matter, in wool top: The pieces of burres, seeds, shive, leaves, twigs, grasses etc. which have escaped removal in processing, also foreign vegetable fibres such as hemp, sisal, etc., if present.

Vegetable matter base, in raw wool: Oven-dried scoured burrs, seeds, twigs, leaves, grasses etc., free of mineral matter and alcohol-extractable matter in raw wool.

Vegetable matter present, in raw wool: The weight of vegetable matter base present in the raw wool, adjusted to a moisture content of 12 %, an alcohol extractives content of 1.5 %, and a mineral matter content of 0.5 %.

Vegetable Silk: See **Akund.**

Vegetable tanning: A considerable number of plants and bark woods offer technically economic yields of tanning agents on extraction. Most of these vegetable tanning agents are extracts from horse chestnut trees, wattle, quebracho bark, oak woods, mimosa bark, oak barks, valonia acorns, myrobalan fruits and other plants rich in tanning agents. These vegetable tanning agents are complex compounds of high molecular weight with numerous phenolic hydroxyl groups.

Veiling: An open ended fabric, usually net, made from silk, acetate, viscose, nylon. It is made plain or possibly with flocked spots for millinery, and often embroidered with elaborate patterns for bridal veils.

Vein: The open place in a cloth, caused by a broken warp.

Veining: Long incompletely drawn nodules in spun filaments, e.g. in finished polyamide fabrics caused by soiled polyamide starting materials in spinning. These nodules dye much darker than normal filaments.

Vel line: French woollen winter coating; has a curl pile surface.

Velour: The French for velvet. This is a fabric with short warp pile. There are many types of Synthetic Velour now available for a variety of uses, but originally velour was a thick woven overcoating made from top quality wool with a soft close nap on the right side. A napped surface woven fabric or felt in which the surface fibres are laid in one direction to present a smooth appearance. Generally used as coating material, but in lighter weights and with check patterns, it is also used for dress goods. Usually it is of 10-20 oz. per yard. Various types of yarns are used to make several types of velour. Some velours are made with synthetic and wool blends. There are also many fabrics similar to velour but with a slightly different finish. These are sold under different names as suedyne, suedette, lustora, duvedelaine and valora. An expensive fabric used for men's and women's coats.

Velour, Warp: See **Velour.** A French word derived from the Latin vellosus = "hairy". Velour is a general term for Warp velvet (Velvet). The pile is formed by an additional warp thread system, i.e. two warp thread systems are used, the ground warp and the pile warp, and one weft thread system. Velour is produced as wire and double velvet. Wire velvet can be produced as cut velvet (velour-coupé) with cutting wires, as curled velvet with an uncut pile (velour-frisé) as well as mixed velvet (with alternating cut and uncut pile). Patterned warp velvets are also produced. Numerous modifications exist, e.g. with transverse stripes (velour-chiffon), transparent (velour-transparent), so-called mirror velvet (velour-moiré), or additionally patterned with a swivel embroidery stay.

Velours: French for velvet.

Velours a Deux Polls: French term for velvet with a pile made of two-ply yarn.

Velours a la Reine: Crossribbed silk fabric, made with one set of wrap and two sets of fillings, the ribs alternating with two picks of the liner filling.

Velours Biseautes: Velvet galloon with higher pile on one edge than on the other.

Velours Bombes: French silk velvet, consisting of alternate stripes of cult and uncut velvet.

Velours Broche: Brocaded velvet.

Velours Cameleon: Changeable velvet, made with two sets of pile warps,

each of a different colour.

Velours Chine: Silk velvet, the pile -warp of wh.ch is printed with pictures, etc., before weaving.

Velours Cisele: French for velvet upon velvet; originally -made with two different rods, now the uniform deep pile velvet is subjected to heavy pressure, laying down the pile in certain parts of the fabric. The standing pile is shorn short and the pressed-down pile is steamed and brushed up.

Velours Couche: A velvet woven with a straight silk pile, which in the finishing process is crushed or laid down; used for coats, trimmings and drapery.

Velours d'Angleterre: Silk weft pile velvet.

Velours d'ltalie: Crossribbed fabric made of silk and cotton, the ribs alternating with two fine fillings.

Velours d'Oran: A warp pile French velvet, both sets of warp containing ends ranging from single yarns to five-ply yarns, arranged in natural and reversed order several time the entire width of the fabric. The two sets of warps are placed in such a manner that the five-ply yarn in one set corresponds to the single yarn in the other, and vice versa.

Velours de Gueux: Velvet made of a single set of linen or heavy cotton warp and fine cotton fllling, forming the pile.

Velours de Hollande: Cut velvet originally made of all silk, with good organzine warp and filling.

Velours de Genes: Usually all-silk velvet, the warp, filling and pile made of organzine.

Velours du Nord: Silk velvet having a longer pile than velvet but shorter than plush.

Velours Ecrase: Silk velvet having highly finished, crushed pile, lying in different directions.

Velours Epingle: Lightweight, solid coloured French silk velvet.

Velours Figure: Figured velvet.

Velours Francais: Obsolete French velvet, made with two sets of silk warp of different colours, one being single yarn and the other three-ply yarn. There are also two sets of different coloured boiled off silk filling.

Velours Frappe: Velvet having raised patterns produced by pressing the pile with heated cylinders.

Velours Frise Uni: Lightweight French silk velvet with single, double or triple pile; used for dresses, millinery, etc. comes in solid colours.

Velours Gandin: Silk velvet over satin foundation; used for drapery, etc.

Velours Glace: A velvet having the pile slightly dressed -with size and pressed down in different directions, producing a frosted effect.

Velours Gregoire: French velvet of the first part of the 19th century, showing pictures woven in pile.

Velours Miroii: Glossy silk velvet with a crushed pile.

Velours Ombre: Warp pile velvets, usually ribbons, made in ombrs effects.

Velours Ottoman: French silk fabrics made with two sets of warp, one forming the foundation in plain weave, the other forming floats, half of the warp floating on the face, the other half on the back.

Velours Ras: French for uncut velvet.

Velours Rayes: Striped velvet.

Velours Russe: Velvet dress fabric made -with varicolored diagonal cords.

Velours Sculpte: Velvet having patterns formed with the pile shorn different lengths; see velours cisele.

Velours Simule: Obsolete term for a plain woven fabric made with silk warp and cotton filling. Both the warp and the weft contain threads of various thickness.

Velours Travers: Half wool dress goods velvet with weft stripes.

Veloutine: (1) French dress fabric, made with thick woollen warp, forming cords and soft merino filling; finished with a nap. (2) A lightweight weft-pile fabric with a pronounced twill effect mainly used for linings. (3) A fine cord fabric produced from a silk warp and a wool weft.

Velure: Same as velours.

Velutine: A short pile velveteen, the back sized in the finished.

Velveret: Usually wide cotton velvets made to imitate silk, often having ribs or finished with printed designs. The filling usually crosses two warps at once. Made in England, United States, etc.

Velvet: Velvet, once only silk, may now be cotton, nylon. Acetate, viscose, polyester, modal etc. Velvet is identified as cloth with a pile of no more than 3 mm. length. The fabric is woven with a warp pile and additional yarn. In production wires are inserted to lift this yarn and the when withdrawn, there are loops which are cut or not according to the type of velvet being made. The anchored pile loops can have a V-shaped or a W-shaped form. In the nature of things, the W-form is more firmly anchored. A distinction is also made between warp velvet (Velours) and weft velvet. Velvet is one fabric that dyes truly; There are no bad colours in velvet fabrics, all are rich and attractive. Used for all types of special dresses for men, women, children and there are also special furnishing velvets. An average velvet construction is 150 x 100 with 50 denier bright rayon ground and pile and 75 denier bright rayon filling. It weighs around 4.5 oz. per sq. yard. Some of the varieties of velvet are (a) Bagheera is a fine uncut pile, velvet with a rough surface that makes it crush resist (b) chiffon velvet is a soft light weight velvet with the pile pressed flat (c) Cisele is made by a contrast in cut and uncut loops. (d) Lyons is a stiff thick pile velvet which is used for hats and dresses, suits and evening clothes. (e) Nacre velvet: is one in which back is usually of one colour and the pile another. (f) Transparent or sheer velvet: is lightweight soft velvet with thin silk or rayon back and a heavier rayon pile. For instance 60 denier rayon can be used for ground warp and weft and 150 denier rayon for the pile warp.

Velvet Carpet: A woven carpet in which the pile ends are lifted over wires that are inserted in the same manner as the filling and that cut the pile as they are withdrawn.

Velvet Fabric: A warp-pile woven fabric with short, dense cut pile that produces a rich fabric appearance and soft texture. Two methods are used for weaving velvets. In the double-cloth method, two fabrics are woven face to face with the pile ends interlocking. A reciprocating knife cuts through these pile ends to produce two separate pieces of velvet. In the second method, pile ends are lifted over cutting wires that are inserted with the filling and that are withdrawn to cut the pile. Velvet is produced in a wide range of constructions and types. Originally made of silk, but now also of cotton or manufactured fibres giving fabrics that are sometimes washable. The fabric can be specially finished to make it crush-resistant and water-repellent or it may be embossed or patterned by burn-out printing.

Velvet Finish: Consists in heavily napping woollen fabrics; the nap is shorn or left as it is.

Velvet Pile Carpets: A collective term for carpets produced by various methods of manufacture with a cut plush-like pile. Cut-pile carpets. (a) A cut-pile carpet made in the same manner as tapestry carpets, but substituting cutting wires for looping wires. (b) A non-jacquard carpet with the same construction as a Wilton carpet.

Velvet Rug: Carpets and rugs woven on the same principle as warp pile velvets.

Velvet Satin: A silk fabric made with patterns in pile over a satin foundation.

Velvet Stitch: See **Raised stitch**.

Velvet, cisele: A velvet with a pattern formed by contrast in cut and uncut loops. Faconne Velvet Patterned velvet made by burnt-out print process. The design is of velvet with background plain. Lyons Velvet A stiff, thick pile velvet. Used for hats, coat collars, also for suits, coats and dresses, when thick velvets are fashionable.

Velvet, nacre: The back is of one colour and the pile of another, so that it gives a changeable, pearly appearance. Transparent Velvet (Chiffon Velvet) Lightweight, very soft, draping velvet made with a silk or rayon back and a rayon pile. Panne Velvet Has a longer or higher pile than velvet, but shorter than plush. It is pressed flat and has a high lustre made possible by a tremendous roller-press treatment given the material in finishing. Now often made as knit fabric.

Velveteen: A fabric in twill or plain weave made with a short closely packed weft pile in imitation of velvet. The fabric usually cotton, that is made in the same way as Courduroy, but the surface is completely covered by the short-cut pile. It may be plain or printed and it varies in weight. The colours are good and and it is equally as attractive as velvet. Its great advantage is that it is not difficult to sew.

Velveton: Medium to heavy cotton sateen with an emerised or raised and cropped suede surface, like Duvetine and Moleskin. Firm, durable and washable. Used for trousers, jackets, trimmings and furniture covers.

Venetian: A cloth milled and cropped bare in finish.

Venetian: A well-known fabric term used to describe a highly lustrous twill cloth originally made in Venice from silk. When made in satin weave from wool, it is the cloth worn by Arabs visiting Mecca. Venetian for men's suiting is made from worsted yarn, but women's cloth can be made from synthetic varieties, usually polyester. Used for suits, skirts, trousers- expensive. See **Sateen, fine.**

Venetian Carpet: Originally made with various coloured worsted warp and some cheap filling which was hidden, the pattern being on both sides in warp stripes; of English origin.

Venetian Crepe: It is made with raw silk warp and very coarse tilling, two right hand twist and two left hand twist alternately; used for mourning.

Venetian Embroidery: On batiste foundation, the patterns are outlined with 'buttonhole stitches, the ground is cut away and the parts connected with bars.

Venetian Lace: Floral motifs and designs connected with picots and brides. The effect is irregular. Originally a Needle point Lace.

Venetienne: (1) French dress goods made of fine Italian (yellow) silk with ribbed weave like gros de Tours, and printed or left in one colour; (2) French woollen dress goods, finished with slight fulling.

Venise: Very fine damask table linen made in Holland and France; the pattern consists of large flowers.

Venturi: A particular type of jet in the form of a tube which is narrower in the centre and wider at the ends. This shape produces a higher rate of fluid flow in the central region.

Verandol: In Cuba a pure linen or cotton mixed cloth, white or beige. The white is used as dress goods, the beige as bed cover.

Verdigris: Green basic copper (II) carbonate formed on the surface of copper exposed to moist air (patina).

Verification: The act or process of verifying.

Verify: (1) To determine whether a previously calibrated instrument, standard solution, or other standard is still properly calibrated.

(2) To establish that an operation has been completed correctly.

Vermont: Australian term for wool yielded in Australia by sheep which descended from the merinos.

Verona Serge: Lightweight serge, made of mohair and cotton or worsted and cotton in mixture effect.

Vertical Drier: A sieve-drum drier of vertical design based on the through-flow principle which offers the highest heat transmission coefficient and maximum drying performance in the smallest space. Heat recovery is determined by the machine construction. The specific heat losses in this type of drier are low due to its compact design and effective heat insulation (without thermal bridges).

Vertical Flame Test: See **Flammabiliy test.**

Vertical Flammability Test: See **Flammabiliy test.**

Vertical Migration: (selective migration). The variable rate of migration of a dye combination through the cross-section of a textile fabric. This behaviour is exploited intentionally for the production of Two-sided effects by vertical migration between the face and back sides of a textile material to give either a different colour on each side (two-colour effect) or different depths of the same shade (dark/light effect). See **Migration.**

Vertical Open Width Washing Machine: A machine with a series of open-width compartments for washing off or impregnating fabrics with chemical liquors (e.g. in pretreatment). See **Roller vat.**

Vertical organisation: Mills that take in fibre, spin yarns, dye these and then either weave or knit these into fabric/garments.

Vertical Padder: A Padder with a vertical arrangement of squeeze rolls. See **Padder.**

Vertical roller printing machine: A roller printing machine formerly produced by Some manufacturers with a vertical arrangement and an individual pressure roll for each engraved printing roller or rotary screen. The colour boxes, doctor blades and furnisher rollers are all mounted on a hinged sub-frame which can be swung to one side (like a gate) thereby offering the advantage of easier and quicker servicing compared to conventional roller printing machines.

Vertical space-dyeing printing unit: A development for Space-dyeing (Micro-Space) with vertical fabric transport from bottom to top and up to 6 pairs of printing rollers (print stations) as modular units. Splashing of print pastes at high speeds is prevented with this system and the practical application of lint doctors is possible.

Vertical star dyeing machine: An upright star dyeing machine where vertical movement of the loaded star frame in the dye liquor is actuated by a cam.

Vertical Suedeing Machine: (sanding, emerizing). A surface finishing machine for the Sanding and Emerizing of knitted and woven fabrics. The fabric runs tangentially against the sanding resp. emerizing rollers.

Vervise: A woollen fabric, made in England under Edward VI, believed to have been of blue or gray colour.

Vetterman Drum Test, in carpet: Instrument used to test appearance retention.

VI: (1) Viscose fibres, (2) Abbreviation for Viscosity index.

Vi: Vicuna.

Vibraire process: A continuous felting process for yarns and slivers co-developed by the Fibre Research Institute TNO Delft and the IWS England. The yarn is passed, together with the felting liquor, through a tube which is subjected to intense vibrations (also known as the Periloc process).

Vibroscope: An instrument for determining the mass per unit length of a fibre.

Vicara: See **Cashmere.**

Vichy: (1) Ginghams in Chile, Bolivia and Spain, often in large checks and plaids, and stiffly -finished; (2) Fine ginghams in Servia; (3) General term in Turkey for cheap ginghams and yarn dyed cotton plaids, usually having a stiff, starchy finish. (4) A cotton dress fabric generally made with two different coloured yarns.

Victoria: Sometimes called Victoria Lawn, this is a stiffened cotton 90 cm. (36in.) wide and used as an interfacing.

Victoria Crepe: English, all-cotton crepe of lustrous finish.

Victoria Lawn: Plain and very closely woven heavy English lawn, about 38 inches wide, having as high as 100 by 175 ends in a square inch. Comes white or printed. Used for dresses, the lighter weights for underwear.

Victoria Shawl: French shawl, made on the Jacquard loom with four embroidered borders and warp fringes at both ends; obsolete.

Vicuna: (1) Very long, soft brownish hair yielded by the South American vicuna goat related to the llama and the guanaco (Peru and Bolivia). Only the undercoat hair is processed– deep yellow to reddish brown, soft, silky-glossy, curly and fine (10–20 mm, around 84 scales/mm).; (2) Trade name for yarn composed of coarse wool and cotton or all-cotton finished in imitation of woollen yarn; (3) A soft twilled cloth similar to cheviot, finished with a soft nap, made of the Andes vicuna, an animal threatened with extinction, hence its name. The hair is soft and pale brown and makes the most expensive and luxurious cloth there is. (4) Imitation of the above, made of soft wool, often mixed with cotton, slightly fulled and napped; used for men's suits and overcoats. See also **Vigogne yarn.**

Vienna Cross: Same as Persian cross stitch.

Viewing Booth: A enclosed area with controlled lighting that is used in graphic arts studios, textile manufacturing, service bureaus, and printing companies as a stable environment for evaluating proofs and press sheets. Viewing booths are generally illuminated using graphic arts industry-standard D65 lighting, and are surfaced in neutral gray colours.

Vigogne: French for Vicuna.

Vigogne Yarn: (1) Originally made of a mixture of cotton and wool, of varying proportions (in Europe 20 per cent cotton, 80 per cent wool), used for cheap hosiery, etc.; (2) In Germany, Russia and Italy yarn made of all cotton and cotton waste, finished to look like wool.

Vigoureux Printing: (tops printing). Diagonal or transverse striped printing of combed wool tops (using relief rollers). The twisted and drawn essentially homogenous blends produce mixtures of different coloured fibre. In most cases, black is used for printing which gives grey colour-blend effects depending on the depth of colour in the mixture and cover in non-printed areas. Zebra-striped patterns are produced which blur after drawing and spinning.

Vigoureux yarn: Mottled yarn which is made by printing stripes in a single colour on combed top in Vigoureux printing and then doubled and stretched several times.

Viloft: A versatile viscose fibre, hollow when seen in the cross section, produced by Courtaulds. It is soft, warm, absorbent, and dyes well. Used in blends with polyester, and fabrics often resemble cotton. Made into knitted and woven fabrics and also pile and loop fabrics. Used for sportswear, underwear, toweling, ribbed toweling, and terry toweling.

Vimoutiers: Coarse and loosely woven French unbleached linen, sometimes dyed yellow.

Vinal Fibre: American generic name for polyvinyl alcohol Fibres. A manufactured fibre in which the fibre-forming substance is any long chain synthetic polymer composed of at least 50% by weight of vinyl alcohol units and in which the total of the vinyl alcohol units and any one or more of the various acetal units is at least 85% by weight of the fibre (FTC definition). Vinal fibres show good chemical resistance but soften at comparatively low temperatures. Vinal fibres are used for apparel, industrial goods, and fishnets.

Vinal-Vinyon: Vinal fibres are made from polymers containing at least 50% vinyl alcohol units and in which at least 85% of the units are combined vinyl alcohol and acetal crosslink units. The fibre is inexpensive, resembles cotton in properties, and is produced in Japan. A very soft fibre which is similar to acrylic. It has excellent flame retardency. Used mainly for children's nightwear.

Vincel: Branded cellulosic fibre. It closely resembles cotton, but is slightly weaker. Vincel (modal) is soft and absorbent and on its own is made into fabrics for nightwear, but can be blended with other fibres such as cotton, for additional body.

Vinegar: Dilute acetic acid, typically around 5% Vinegar is a convenient acid for many dyeing processes, although a lot may be required because it is so dilute. Much of the white vinegar sold is made by diluting concentrated acetic acid produced by synthesis from natural gas. Vinegar diluted with 10 parts of water will give a pH of around 4.2.

Vinyl group: The group CH_2:CH–.

Vinyl: Short form for Polyvinyl chloride (PVC).

Vinyl: A univalent radical, (CH_2=CH-), derived from ethylene.

Vinyl Sulphone Dyes: As a Sulphatoethylsulphone dye, the commercial form contains a protected reactive group. In this ester form, they only have moderate substantivity. A type of reactive dyes, generally used for cellulosic fibres but with some use for wool. Vinyl sulphone reactive dyes are intermediate in reactivity, so they are applied above room temperature, but well below the boiling point of water. The actual reactive dye, vinyl sulphone, is not released until the required alkali is added for dyeing to take place so they can be stored as solutions for much longer periods than highly reactive dyes. In this form, the dye is now reactive showing high substantivity. Wash fastness may be somewhat inferior to some other reactive dyes. Vinyl sulfone dyes can be a good choice for dyeing the background colour for discharge, since they are quite easy to discharge with reducing agents. There are also simple chemical resists , that work well with these dyes, preventing fixation to the fibre, so they find application in printing processes. Remazol is a popular brand name.

Vinyl Sulphone : See **Vinyl sulphone dyes.**

Vinyl Sulphone Group: –SO_2–CH=CH_2. Reactive group which reacts with the OH groups of cellulose: –SO_2–CH_2–CH_2–O–Cell in the presence of alkali. See **Vinyl sulphone dyes.**

Vinylal: Generic name for fibres based on Polyvinyl acetate, i.e. acetylated, making the polyvinyl alcohol fibres insoluble in water.

Vinylidene Chloride: A chemical material obtained from ethylene, a petroleum product, and from chlorine. It is used for the manufacture of textile monofilaments and film. It is more commonly identified in the U.S. as Saran. (Also see **Saran fibre**).

Vinylidene fibre: Polyvinylidene chloride fibres.

Vinylon: Since 1948, generic name for polyvinyl alcohol fibres made in Japan.

Vinyon: Generic name for a fibre made from vinyl chloride. Vinyon is defined as a fibre in which at least 85% of the polymerized monomer units are vinyl chloride. Vinyon fibres have high chemical and water resistance, do not burn, but do melt at relatively low temperatures and dissolve readily in many organic solvents, thereby limiting their application. It is resistant to chemicals, softens at low temperature. More commonly referred to as Polyvinyl chloride (PVC).

Vinyon Fibre: A manufactured fibre in which the fibre-forming substance is any long chain synthetic polymer composed of at least 85% by weight of vinyl chloride units.

Violanthrone: See **Dibenzanthrone.**

Virgin Wool: Wool which has never been reclaimed from any spun. Woven, knitted, felted, braided, bonded, or otherwise manufactured or used product.

Virginie: (1) French silk fabric, made In an eight-leaf twill weave; (2) Six or eight-leaf figured serge made with single or ply warp.

Visible matter in cotton testing: Foreign matter deposited in the waste boxes of the machine during the test.

Visible Radiation: Visible light is a narrow band of electromagnetic radiation from 400 to 700 nm (1 nm equals 10^{-9} meters) detected by the human eye. Radiation falling below 400 nm is ultraviolet radiation, and that falling above 700 nm is infrared radiation; both are unseen by the human eye.

Visible Spectrum: The region of the electromagnetic spectrum between 380 and 720 nanometers. Wavelengths inside this span create the sensation of colour when they are viewed by the human eye. The shorter wavelengths create the sensation of violets, purples, and blues; the longer wavelengths create the sensation of oranges and reds.

Viscoelasticity: Property of a substance which, when subjected to shear, compressive and tensile stresses, shows both viscous and elastic properties (Viscosity, Rheology). Plays a decisive role in textile printing. As the polymer molecules are subjected to increased stress, the viscosity of the swollen thickener spontaneously drops and increases again as the stresses are relaxed.

Viscometer: A device designed to measure the viscosity (resistance to flow) of the fluid. Many types exist from simple calibrated glass tubes to extensively instrumented, on-line shear viscometers. (a) Capillary viscometer. (b) Flow cup (c) Falling-ball viscometer (d) Rotary viscometer (e) Vibration viscometer (f) Compression or extension viscometer.

Viscometer, torsion type: (visco printogram). Used to determine the Viscosity of printing thickeners; See **Viscosity of printing pastes.**

Viscose: (1) Hollow fibre usually made from wood pulp. The fibre is soft and absorbent and can be used alone and mixed with other fibres to produce fabrics of all types and weights. Due to its versatility viscose is one of the most commonly used synthetic fibres. (2) Spinning paste, thick, viscous, made from alkali cellulose and carbon disulphide dissolved in caustic soda solution. Starting material for See **Viscose fibres.**

Viscose fibre: The generic name for fibres formed by the regeneration of cellulose from viscose (q.v.) by treatment with a solution of electrolytes (salts and acids). (See also **spinning bath**).

Viscose Embossed Satin: A firm patterned satin for bridal wear but it does not drape well, so designs should be structured; avoid gathers.

Viscose high tenacity fibres: (rayon cord, supercord, all-skin fibres and modal fibres), developed with the aim of increasing strength, dry modulus, fatigue resistance and dimensional stability, etc. This aim was realized to a limited extent using older spinning methods (such as the Lilienfeld and Lanusa principle and stretch spinning similar to cupro). By modifying viscose (Modified fibres) the production of high strength viscose fibres with thickened Covering layer and super-strong viscose fibres in all-skin structure with high-strength and high elongation. Use: as a filament for industrial fabrics, tyre cord, fabric linings, conveyor belts and drive belts. As a staple fibre: alone or in a mixture for rainwear, uniform fabrics, work clothing and protective clothing.

Viscose Modified Fibres: See **Modified Fibres**..

Viscose Process: (1) One of the methods of producing rayon. (Also see **Rayon fibre**).

(2) The chemical process used in the manufacture of cellophane. (Also see **Viscose solution**).

Viscose Rayon: One type of rayon. It is produced in far greater quantity than cuprammonium rayon, the other commercial type. (Also see Rayon fibre.)

Viscose Solution: The solution obtained by dissolving cellulose xanthate in caustic soda, from which viscose filaments and cellophane are produced.

Viscosity (Lat.: viscum): (1) Thickness, internal friction, internal resistance of a fluid against the relative movement of two neighbouring layers, i.e. the resistance a substance offers to flow. (2) Cellulose: A term applied specifically to signify the viscosity (see (1) above) of a standard solution of cellulose in cuprammonium hydroxide solution of specified copper and ammonia content. The flow behaviour of a mixing is best described by a flow curve relating apparent viscosity (in mPa.s) to shearing stress (in Pa). If the shearing stresses operative in the sizing were known, the apparent viscosities of the mixings at these stresses would be related to their sizing behaviour. Without this knowledge, measurements at some arbitrary stress (say 100 Pa) have to be used. These are of value in characterising a particular type of size and can often be related to the take-up of size by the warp.

Viscosity of Print Pastes: The rheological behaviour of print pastes plays a part in textile printing which is not insignificant. Apart from tensile loads, shear forces in particular act on the pastes (Rheology in printing and coating). However, viscosity measurement has been introduced into printing practice indirectly as thickener manufacturers use the flow properties to characterize thickeners and then use this information in formulation recommendations.

Visible radiation: Any radiant energy capable of causing a visual sensation.

NOTE: The limits of the spectral range of visible radiation are not well defined and may vary according to the user. The lower limit is generally taken between 380 and 400 nm and the upper limit between 760 and 780 nm (1 nanometer, 1 nm = 10.9 m).

Visible Region of the Spectrum: See **Visible radiation.**

Visible waste in cotton testing: Foreign matter deposited in the waste boxes of the achine during testing.

Viyella: A trade name describing an old-established, classic fabric. However at times the name Viyells is generically used. The fabric still has the same composition as the original, i.e. 55 per cent lamb's wool and 45 per cent cotton, and it is a fine, soft twill weave material. It was first produced in 1983 in Britain and for a long time was used almost exclusively for children's clothes, babywear and nightwear. Later, it became a popular fabric for blouses and shirts. Now, mostly due to improved designs, it is used for top fashion clothes.

Viyella Pure Wool: This fabric, as well as Viyella Challis and Viyella Pure cotton, has been introduced recently by the manufacturers, William Hollins. Used for skirts and blouses, it is produced in a range of checks, floral and Persian designs, also co-ordinated plain and prints.

Vocational career apparel: Career apparel which is generally subject to abusive wear and for which durability is a more important attribute than appearance.

Voile: Voile is a high twist yarn, but the term Voile refers also to a fine, light, plain weave fabric made in a variety of yarns. A characteristic single voile is 60 x 56 with 50s warp and filling. A ply voile will be similar but will be with 2/100s yarn. Cotton, and also polyester with cotton, produce a soft comfortable fabric used for blouses, dresses, children's clothes and sometime for mens shirts. Voile is also made from silk.

Voile, full: A voile in which warp and weft are ply yarn, See **Voile.**

Voile, imitation: A voile in which warp and weft are simple yarn (Indian mull). See **Voile.**

Voile, semi: A voile in which warp is ply yarn and weft is simple yarn.

Voile, silk: A fine silk voile in which warp and weft natural silk.

Voilette: French machine-made lace, having a very fine mesh ground. Along the edge it is embroidered with light flower design while the body is trimmed 'with dotted powdering or fine trailing lines.

Voile Fabric: A sheer spun cloth that is lightweight and soft. It is usually made with cylindrical, combed yarn. Voile is used for blouses, children's wear, draperies, bedspreads, etc.

Voile Yarn: Fine, hard-twist (gas-singed) cotton or worsted ply yarn.

Volatile: Readily vaporized at a relatively low temperature.

Volatiles: Materials readily vapourisable at relative low temperatures.

Volatility: Property of having a low boiling point or temperature of sublimation at normal pressure. Likewise, having a high vapor pressure at ambient conditions.

Volume ratio: Milliltres/litre (ml/l).

Volume Resistivity: The ration of the potential gradient parallel to the direction of current flow in a compound to the current density after a specified time of voltage application.

Vraio Reseau: In real laces the net ground which was made either by the needle or with the bobbins.

V-Shaped Creel: V-shaped creel: in this creel type, the creel boards are assembled in form of endless chains. While warping is carried out from the outer sides using the already creeled up bobbins, the subsequent yarn lot can be creeled up on the empty spindles positioned inside the creel. This interior room serves at the same time as storage and bobbin exchange station.

The yarn lot can be changed by simply pushing a button, which starts the electrically drive of the chains. The empty bobbins move towards the inside of the creel, the full bobbins towards the outside.

VSM: (1) Swiss standards organization. (2) Swiss textile machine organization.

VST: Association of Swiss Textile Specialists (Vereinigung Schweizer Textilfachleute). United with VeT to form SVT.

VSTV: Association of the Swiss Textile Finishing Industry, Zurich (Verband Schweizerischer Textilveredlungsindustrie).

VTCC : Verein der Textilchemiker und Coloristen. Today VDTF.

Vulcanised Fibre: Cellulose-plastic (plasticized using zinc chloride solution; paper webs "welded" by roller pressure) in the form of slabs, rods and tubes. Horn-like, tough, hard to leather-like soft and pliable. Absorbs little water (also water-proof impregnated). Resistant to oils and hydrocarbons, etc.

Low flammability. Can be machined in any way to make gearwheels, sealing washers and suitcases (and the like).

Vulcanization: An irreversible process, usually accomplished through the application of heat, during which a rubber compound through a change in its chemical structure (for example cross linking) becomes less plastic and more resistant to swelling by organic liquids and elastic properties conferred, improved or extended over a greater range of temperature.

Vulcanization: A process of improving the quality of rubber (hardness and resistance to temperature changes) by heating it with sulfur (about 150°C). Accelerators are used to speed up the reaction. Certain sulfur compounds can also be used for vulcanization.

VY: Vinylal.

W

W: . Abbreviation for Watt.

WA: Angora fibre.

W/O, Water-in-oil emulsions: **(W/O emulsions).** See **Emulsion; Emulsion thickener.**

W/V : Weight by volume.

W/W: Weight by weight.

Wadding: (1) Cotton wadding is thick and lightweight. It is a sandwiched between two papery layers of non-fluffy, shiny covering which hold it together. The wadding can be split to half depth. Use for quilting, shoulder pads etc. Polyester wadding is more open and springy and firmer than cotton. It is available in various thicknesses, i.e. 2 oz. 4 oz. 8 oz. 8 oz. per metre. Use for shoulder pads, quilting, padding embroidery.

(2) Waste silk, obtained from the inner smooth skin of the cocoon, left over after the reeling.

Waffle pique; waffle honeycomb: Woven cotton fabric in which warp and weft threads form ridges and hollows giving a cellular appearance. Fine qualities for blouses and trimmings; heavier types of towels and bathrobes.

Waffle honey comb

Waist, in anatomy: The part of the body at location between the lowest rib and hip identified by bending the body to the side.

Waist girth, in body measurements: The circumference of the waist immediately below the lowest rib.

Waistband (one-piece): A single thickness of fabric that is doubled and stitched to the top of a pant.

Waistband (Two-piece): When two identical pieces of fabric are placed back-to-back at the top of a pant, raw edges turned inside, and joined with two widely spaced rows of titching. The pant body is inserted between and along one edge.

Wale, in knitted fabrics: A column of loops in successive couses that is parallel with the loop axes.

Wale, in woven fabrics: One of a series of raised portions or ribs lying warpwise in the fabrics.

Wale: A diagonal raised line formed by a twill weave on the face of the fabric, usually woollens and worsteds.

Wale count: The number of wales in a knit fabric per unit length. For example: wales per inch.

Wale density: Number of wales per centimetre of fabric width (wale density), this is the number of vertical columns composed of stitches per unit length of fabric through the fabric width (generally, 1 cm of fabric is taken).

Wall-To-Wall Carpeting: A term for carpets that are laid from wall to wall (fitted carpets). Contract carpet.

Walk test: (1) Method of testing wear resistance, such as soiling properties and pile firmness in textile floor coverings. The sample is spread out in an area where people frequently walk, usually together with a comparable carpet that has known properties. Soil pick-up, wear, colour change, etc. are monitored over several months. (2) Test to establish the electrostatic properties of textile floor coverings. (AATCC Test Method 134 – 1979): in this test a person walks backwards and forwards in a specified manner with specified soles and heels over a carpet which has been conditioned for at least 3 days in dry air (23°C and 25% relative humidity). The test person carries a sensor connected to a static voltmeter which registers the voltage absorbed by the body. The test is carried out with three standardized shoe soles: (a) BAM = natural rubber; (b) Neolite = synthetic rubber, and (c) polyvinyl chloride = a typical synthetic shoe sole material. If the voltage exceeds the sensitivity threshold, the carpet has not withstood the Walk Test. The sensitivity threshold is the voltage at which only few people are aware of any static shock or discomfort. In general, 2000–3000 V has been agreed upon and, for this test, the limit has been fixed at 2000 V.

Walking Coat: A compromise between a single breasted sack and a cutaway frock, more nearly resembling the latter, but shorter and with pocket flaps on hips.

Walking foot: Walking foot is a special foot that insures uniform material feed when sewing multiple layers of fabric. This is the best foot to use when quilting and sewing on your binding.

Wammus: A sort of Cardigan jacket.

Wand: A tool used to deliver cleaning solution to carpets and apply vacuum to remove the solution. A *wand* usually consists of an extension handle and a *cleaning head*. Various types are light weight wand, drag wood (heavy duty) and power wand which has a motorized rotating or vibrating part to aid soil release.

Warangal carpets/rugs: Carpets made in Warangal, Andhra Pradesh, India in 17th and 18th centuries or earlier. The Warangal Rugs were prized for two traits. Firstly, their exceedingly fine count of the stitches, about 12000 to the square foot and secondly, they were the only examples in which silk was ever used in carpets with a perfectly satisfactory effect. Birdwood describes a late 17th to an early 17th century Warangal Carpet that took almost 10 years to complete. "It is a marvel of weaving. There are 400 knots to every square inch of it, giving a total of 3.5 million for the entire surface, and so complicated is the pattern that a change of the needles was required for every knot." Cheap 'thug carpets' from the jail factories and the stratospheric price killed the Warangal Carpets.

Wardrobe: Clothing, in particular outerwear, dresses, general outer garments, in French it also applies to underwear.

Warm-dyeing vat dyes: Vat dyes that act in a heated environment on the basis of the IW dyeing method.

Warmth to Weight Ratio: A measurement used to evaluate the effectiveness of an insulated product in relation to weather conditions and the environment. The insulation with the best rating is down. Down provides the best warmth to weight ratio over almost any other insulation material, which is why you will see down garments and sleeping bags as the primary choice for use in almost every high altitude, cold weather expedition.

Warp: A yarn running lengthwise in a woven fabric.

Warp: A group of yarns in long lengths and approximately parallel, put on beams or warp reels for further textile processing including weaving, knitting twisting, dyeing etc.

Warp (n): (1) Threads lengthways in a fabric as woven. (2) A number of threads in long lengths and approximately parallel, which may be in various forms intended for weaving, knitting, doubling, sizing, dyeing or lace-making.

Warp (v): To arrange threads in long lengths parallel to one another preparatory to further processing. In addition to beaming, the following methods of warping are practiced: ball warping, cross-ball warping and chain warping. The primary stage of these methods of warping is the withdrawal of the ends from a warping creel and their assembly in rope form, a form that may conveniently be used for wet processing. For convenience of handling, this rope may be: (a) wound into a ball (ball warping); (b) machine-wound onto a wooden roller into a cross-ball cheese (cross-ball or cheese-ball warping); (c) shortened into a link chain (chain warping). A number of these ropes may be assembled into a complete warp on a beam in a dressing frame, or may be split and dressed and incorporated in warps made by other methods. (See also section warping).

Warp bands: See **Sextion mark.**

Warp beam: A large spool or flanged cylinder around which the warp threads, or ends, are wound in a uniform and parallel arrangement. Also see **Beam.**

Warp beam centrifuge: Centrifuge for beam dyeing machines or card lap beams, usually upright rather than horizontal.

Warp bow: Deviation of the warp yarn from a straight line; alternatively, curvature of the warp yarns.

Warp control: See **Let-off.**

Warp drawing: Warp-drawn fibres may be taken up on packages other than beams.

Warp dyeing process: This process is used for the manufacture of warp patterned carpets. Pile yarns are wound in parallel onto drums with dye liquor application systems, they are dyed in stripes, and finally the dye is set using steam.

Warp elongation and tension: Stretch or tension measured in the warp direction of the fabric.

Warp-faced twill: A twill weave in which the warp yarns produce the diagonal effect.

Warp knit: See **Warp-knitted fabrics.**

Warp knit fabric: A fabric that is knit with the yarns running lengthwise, e.g., tricot, milanese, and Raschel.

Warp knitted Terry: This is a warp knitted fabric and is made with an extra warp sheet of pile yarns which are caused to form loops bound into a ground fabric. Applications include furnishings and bedsheets.

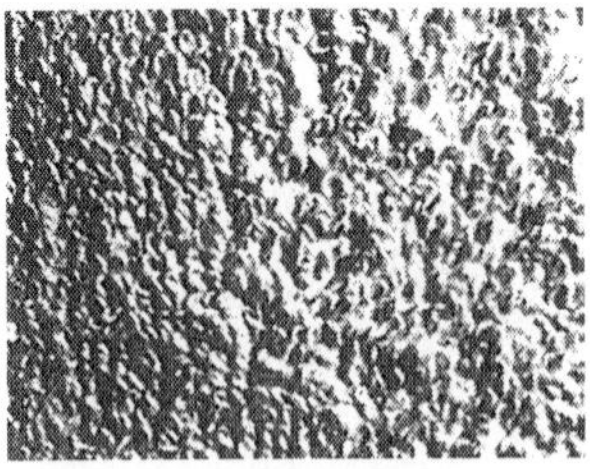

warpknitted terry

Warp knitted plush or velour: In this warp knitted fabric, the pile loops are cut to give a fleecy or velvet like surface. Applications include beach, leisure and sports wear and ladies outer wear.

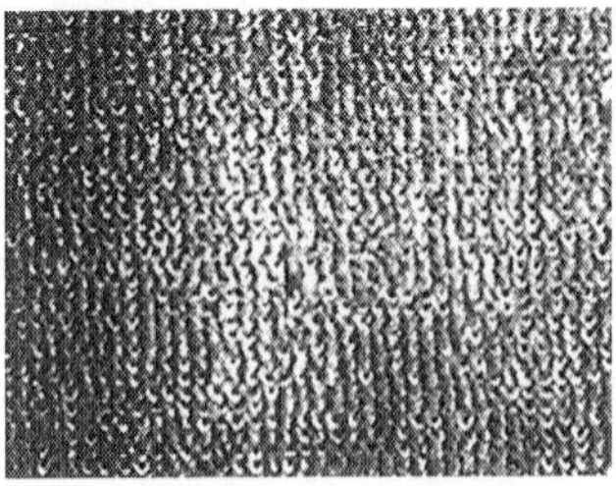

Warpknitted plush or velour

Warp knitted velour: See **Warp knitted plush.**

Warp patterned carpets: Multi-coloured machine-woven carpets in which the pile yarns are dyed using the Warp dyeing process.

Warp pile: The extra set of warp yarns that forms the surface in a double-woven pile fabric, including types such as velvet and velour. Upholstery fabrics such as mohair, plush, and friezé are produced by this method. Also see **Pile** and **Velvet fabric**.

Warp Print: Same as **Chine.** (Chiné printing). Printing the warp thread system of woven fabric with thin print pastes (complete print penetration), drying, and then, in most cases, weaving immediately (to avoid extreme pattern displacement). Dye setting and washing processes are carried out once the fabric has been woven. Small pattern displacements are inevitable, and they produce the characteristic blurred and slightly jagged appearance. This technique is used in particular for articles where complete print penetration is required (e.g. curtains). Two-phase print processes are frequently used for warp printing.

Warp printed carpet: A true warp printed carpet is a wire carpet. The pattern is created before weaving, using the Warp printing technique on the pile warp yarns. This type of warp printed carpet (tapestry or velvet- type carpets) typically exhibits settled patterning.

In contrast to this, multi-coloured printed carpets have a certain degree of pattern bleed after weaving.

Warp printing: See **Warp print.**

Warp printing process: A Variant of Space dyeing.

Warp rib tricot: Tricot with ribs running in the direction of the warp (lengthways). The opposite of this is Transverse tricot.

Warp Ribbed: Fabrics having ribs or cords running in the direction of the warp.

Warp side: In a loom, the back of the machine is where the warp beam is called the warp side.

Warp Stitch: In embroidery, a stitch by which warp threads are drawn together to form a pattern after some of the weft threads have been removed.

Warp sheet: A sheet comprising up to several thousand ends that are combined to make up the warp during preparation for weaving or warp knitting.

Warp Sizing: Sizing of the warp in wovens in preparation for the weaving process. The evenness is particularly useful in practice for reasons relating to technology and economy. For this reason the aim is to reduce unevenness in the sizing along the warp.

Warp Sizing and dyeing: Process for dyeing cotton warp during sizing. Only used for a few types of fabric, such as denim, blue jeans.

Warp streak, in woven fabric: A narrow band running lengthwise and characterized by apparent differences in colour from the adjoining ends. Warp streaks should not be confused with reed marks.

Warp stripes: The one or more faulty threads giving rise to zones of different aspect; it can be due to scraping or rubbing from members of production machines or to inaccurate reeding.

Warp velvet: (wire velvet, velours). Smooth or patterned Velvet, in which the pile layer consists of an additional warp system (pile warp).

Warper-Sizer-System: The warp yarn runs from the warping reel frame straight into the sizing machine with no sectional warping process. This means that each thread in the multi-thread warp is sized individually. This system is particularly suitable for warp preparation of untwisted synthetic fibres (e.g. fine titre of polyamide and polyester threads).

Warpers beam: In a beam warping process as an intermediate stage warpers beam which may contain upto 1000 ends are produced.

Warping: Whereas warp threads with the required sett are wound individually onto the sectional drum or a sectional warp beam in Sectional warping, there is also a type of warping where the warp yarn is wound onto warping beams to the intended weaving width but with a reduced sett. More and more beams of this type are being consolidated into a standard warping beam. This technique is primarily suitable for long warps of a single colour. Multi-coloured yarns are usually warped sectionally.

Warp-knit: Warp-knitted fabrics.

Warp-knit nonwoven pile fabric: Nonwoven knitted fabric consisting of a basic material with additional nonwoven fibre. The nonwoven fibre is bonded into the basic material by forming it into pile loops.

Warp-knit stitch-bonded nonwovens: (stitchbonded nonwovens). Nonwoven knits consisting of fibre fleece, bonded by means of a row of sliding needles with tongues that are pushed into and pulled out of the fabric, causing loops of fibre (stitches) to be formed.

Warp-knitted Fabric: Textile fabric manufactured from one or more thread systems by forming stitches. Warp-knitted fabrics, unlike Weft-knitted fabrics or knit goods (single-thread knits), are formed into stitches from several adjacent threads (warp) at the same time. The process of forming stitches is less time consuming than in weft-knit manufacture, and it is not possible to chafe the fabric.

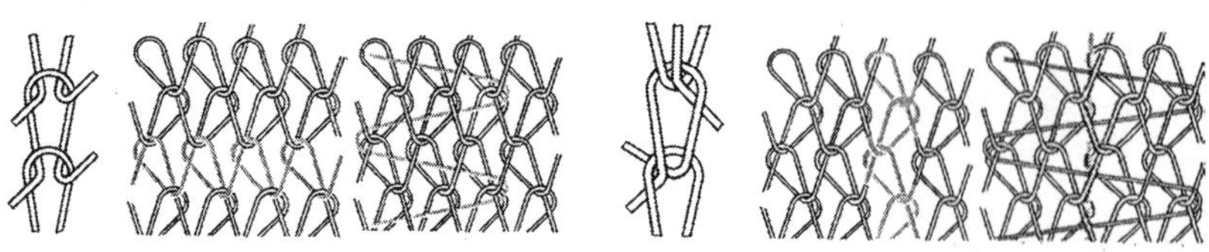

Open Lap Wales Inlay Closed lap Course Vertical Inlay

Warp knitted fabrics are made with at least one sheet of warp yarns. Usually there are two sheets of warp each on its own beam. Each individual yarn is drawn through a guide which is mounted on a guide bar. Movement of the guide bar lapping cause the thread to be lapped around the needle which may be beareded, latch or compound needle. After the yarns are lapped the needle bar is moved so as to cause the loop to be formed simultaneously at all the needles resulting in a whole knitted course. Finally the guide bar is displaced sideways (shogged) by one or more needles before the next cycle produces another course. Shogging of the guide bar determines the structure of the fabric.

Warp-knitted spacer fabrics: Warp-knitted spacer fabrics are normally manufactured on double needle bar Raschel machines (with gauges between 22 and 32 needles per inch) by knitting the face and back layers simultaneously on each needle bed, with guide bars passing between the needles of both beds as they oscillate from the front to the back of the machine. The setup involves guide bars 1 and 2 knitting the front fabric layer on the front needle bar only, and guide bars 5 and 6 knitting the back layer on the back needle bar only. Guide bars 3 and 4 carry the spacer yarns and knit on both needle bars in succession.

Warp-knitting: See **Warp knitted fabric.**

Warp-to-filling seam: A sewn seam in which warp yarns are perpendicular to the sewn seam on one side of the seam and parallel to the seam on the opposite side of that seam.

Warp-to-warp seam: A sewn seam in which yarns in the warp direction on both sides of the seams are perpendicular to the seam.

Warping board: It is a square frame with evenly spaced sturdy pegs along the sides of the frame and pegs that can accommodate a cross along the top and the bottom. Warp threads are wound on the board in order and measured by a designated path around the side pegs.

Warping order: The smallest number of ends in colour and/or count that repeats across the fabric.

Warping paddle: It is a device that allows the winding of multiple warp ends in the same pass around a warping board while keeping them separate and aligned and providing a cross of individual ends. One type of paddle is equipped with two rows of holes that keep the threads separate; the cross is picked by hand at the cross pegs on the warping board. Another type had alternate slots and holes that keep the threads separate and also form a cross of individual ends when the paddle is raised and then lowered at the cross pegs.

Warping reel: Warping reel can also be used to wind the warp. Some reels are vertical; the reel spins on a vertical axis and the warp is wound around it through a path that goes from the top to the bottom of the reel and back again. Some reels are horizontal; the reel spins on a horizontal axis and the warp is wound around it from one end to the other and back again. Reels can accommodate longer warps than boards and can be faster to use (the longer the warp, the greater the savings in time).

Warping, sectional: See **Sectional warping.**

Warping sticks: are thin slats of smooth wood about 1⁄8" thick, 3⁄4" wide, and longer than the width of the warp.

Wash & wear or easy care: A generic term applied to garments which satisfactorily retain their original neat appearance after repeated wear and suitable home laundering with little or no pressing or ironing. See **Wash-and-wear.**

Wash and wear finishing: The finishes applied on cotton, linen, viscose and its blends with synthetic fibres to achieve the effect on the fabric which can be used directly after washing and drying (without ironing). Resin finishing agents, catalysts and additives, applied using low-cost process techniques combined with mechanical finishing phases still ensure the best wash and wear finish (See **Resin finishing**). There are different levels of finishes in this which is ascertained using wash and wear standards. See Wash and wear standards.

Wash and wear standards: Photographic and three-dimensional standards for the assessment of wash and wear properties. AATCC 124-1973: Smoothness of textiles. AATCC 88B-1973: Appearance of garment seams. AATCC 88C-1973: Appearance of ironed creases on garments, in each case after domestic washing. Samples for testing of specified sizes are washed under defined conditions, dried (drip dried, spin dried, tumble dried) and allocated the grades relating to the comparative standard (1–5). 1 is the lowest grade possible, and 5 is the highest (Monsanto crease photographs).

Wash fastness testers: Mechanical instruments for measuring wash fastness. Consist of a water bath, in which sealed stainless steel or glass containers rotate on a horizontal axis. The containers hold the sample with the prescribed test solution, and possibly a set number of steel balls. Usually also suitable for testing other types of fastness (e.g. peroxide washing, hypochlorite washing, milling fastness, dry cleaning fastness, etc.).

Wash fastness testing: Test to establish the resistance to all types of industrial and domestic washing. The specimen is treated together with control fabric samples in a suitable mechanical washing device. After the test is completed, the specimen is rinsed twice in cold distilled water, squeezed, rinsed for 10 min in flowing water, the specimen is opened on 3 sides, dried at max. 60°C and then assessed (using grey scales to check the colour change and level of bleeding).

Wash liquor: An aqueous detergent solution used for the physical removal of extraneous substances from textile materials

Wash/rub test: Test for pigment dyes and prints, by rubbing with the washing solution to check the adhesive properties of the binder.

Wash satin: A soft, fine, inexpensive, white, cream or flesh-coloured satin, made from silk or, more usually, acetate, or triacetate. It washes and irons well.

Wash-and-wear: A generic term applied to garments which satisfactorily retain their neat appearance after repeated wear and suitable home laundering with little or no pressing or ironing. A wash-and-wear garment is essentially free from undesirable wrinkles both during wear and after laundering and retains any original pressed-increases or pleats. The garments should meet

normal consumer demands for durability, colour, stability, and shrinkage. The performance of a wash-and-wear fabric or garment depends on several factors, including the types and amounts (percentages) of fibres used, the fabric construction, the finishing treatment, the presence of a coloured pattern (either woven or printed), and the methods used for washing and drying. These factors determine, in any specific instance, if a fabric or garment's performance will meet customer requirements. Variable conditions result in the varying behaviour of a specific fabric or garment. Garments are labeled to specify the appropriate care for optimal performance.

Wash-and-wear finishing: See **Resin finishing.**

Wash board, in hosiery: A rodgy effect caused by uneven tension between feeds on the knitting machine.

Washable goods: Washable textiles, divided into the following categories: Coloured articles wash; Delicate articles; Laundry for 100°C wash. See: **Wash and Wear; Machine-washable wool.** Care labeling of textiles provides information regarding washability.

Washable suede: Particularly useful if colours are pale. Suitable for most garments.

Washdown: A change in appearance to give a worn or laundered look resulting from washing, scouring, chemical, mechanical, or other treatment, including any combination of such treatments.

Washer wrinkles: See **Crowsfeet marks.**

Washfast acid dye: A vague term for a group of acid dyes that have good washfastness properties This term is sometimes used by dye sellers for a group dyes selected for good washfastness properties. Often the dyes come from the pre-metallized, milling or reactive classes.

Washing: Process in which heavy or slight soiling is removed and transferred to the water in the form of a solution or dispersion. Washing has the effect of cleaning surfaces. The resulting effect is several physical/ chemical processes.

Washing agents: Washing agents or Detergents remove fat and soiling substances from textiles. In the context of textile finishing, detergents are specially adapted to suit each finishing process, e.g. for raw wool scouring, washing after dyeing and printing, etc. Laundry and domestic detergents are specifically included under the category of detergents. Domestic detergents can be classified depending on their use into the following categories: Heavy duty detergent; Mild washing agents; Special washing agents. Depending on the method of manufacture, detergents can be classified as mix and spray (cold and hot sprayed) products. Depending on consistency there is a distinction between detergents in powder form and liquid products.

Washing agent additives: The chemicals and filling agents added in the washing agent formulation mainly for enhancing the versatility of the washing agent. Main additives used are builders (try polyphosphates, Na-Al silicates), bleaching agents (soda perborates), bleach activators (TAED), others CMC, Sodium sulphate, sodium silicates, enzymes, perfumes, pigments, optical brighteners etc.

Washing detergents: Products manufactured from Surfactants; Phosphates; Perborates or alternative products and other complementary components as specified by the Detergent and Cleaning Agent Law. They act in conjunction with water to produce a cleaning effect, or are intended for cleaning (according to an additional suggestion: or products that can be

used as additives and auxiliaries in washing and cleaning processes), which can finish up in environmental waters after use. The definition also includes particular textile and leather auxiliaries and industrial cleaners.

Washing lubricants: See **Textile lubricants.**

Washing machine methods: (1) Machine wash: A process in which products or specimens can be washed, bleached dried and pressed by any customary commercial or home methods, including a sour rinse commonly used in commercial laundering. (a) Hot: Initial water temperature directly from hot water tap from 54-66^{0}C (130-150^{0}F). (b) Warm- initial temperature setting 32-34^{0}F (90-110^{0}F) hand comfortable. (c) Cold: Initial temperature setting same as cold water tap up to 29^{0}C (85^{0}F). (2) Home launder- Same as (1) but exclude commercial laundering. (3) Small load- Smaller than normal washing load. (4) Delicate or gentle cycle: Slow agitation and reduced time. (5) Durable press (Also permanent press) cycle: Cool down rinse or cold rinse before reduced spinning. (6) Separately- Wash alone or with like colours. (7) Wash inside out- turn the product inside out to protect face of the fabric. (8) No bleach- Excludes all bleaches. (9) No chlorine bleach: Permits bleaches other than chlorine. (10) Warm rinse: Initial water temperature setting 32-43

^{0}C (90-110 ^{0}F) (11) Cold rinse: Initial water tempersature setting as cold water tap upto 29 ^{0}C (85 ^{0}F). (12) Rinse thoroughly: Rinse several times to remove detergent and soap. (13) No spin- No spin: Remove material start of the final spin cycle. (14) Donot use roller wring nor wring by hand.

Washing off: The removal of scrooping agents, preparations, sizes, chemicals, textile auxiliaries, non-set dyes and finishing agents. Washing off has a high level of influence on water use, and therefore also on costs and the problem of waste water. With discontinuous washing off, the textile stays in a machine and the washing and rinsing liquors are added or exchanged. Keeping the liquor ratio as short as possible lowers water consumption. With continuous washing off, the textile runs through several individual machines or sections, and the rinsing liquor is directed in counter-flow, i.e. there is a fresh water supply at the point at which the textile leaves the machine. Turbulence is used to ensure that the liquid lamella between the textile and the liquor is constantly broken up (increasing kinetic energy).

Washing powder: Specific term for Washing agents in powder form.

Washing soda: Sodium carbonate. Washing soda, if "pure", is usually sodium carbonate decahydrate ($Na_2CO_3 \cdot 10H_2O$). Retail washing soda may contain additives such as detergents, salt and optical brighteners, and is therefore not a good substitute for soda ash for dyeing.

Wash-n-wear: See **Wash-and-wear.**

Wash-off factor: (C0/Cn), this refers to the fact that the chemical/soil concentration C0 is taken into account at the point of fabric in feed, and the residual concentration Cn at the point of fabric exit. Cn is reduced by the concentration of substances to be washed off the fabric in the machine to the value. C_n=initial concentration/wash-off factor.

Wash-out jeans: Denim garments that change in appearance after a domestic wash cycle. One variant of the pigment dyeing process is the wash-out finish. This consists of pigment dyes to shade depth that are manufactured using reduced quantities of binders. Fabrics are only subjected to a washing process (mill-wash, stonewash) when they have been partially made up into a garment, but usually this is not done until the garment is fully made up. The formulae for these articles, in particular washing methods, vary from firm to firm, and are usually trade secrets.

Washing, Hand methods: (1) Hand wash: Products must be laundered by hand with gentle squeezing action and can be bleached dried and pressed by any customary methods. (a) Cold: Initial water temperature as as cold tap water up to 29°C (85°C). (2) Seperately: Initial water temperature 32-

45°C (90-110°F) (a) Cold: Hand wash alone or with light colours. (3) No Bleach: Excludes all bleaches. (4) No Chlorine Bleach: Permits use of other bleaches. (5) No wring or twist Handle to avoid wrinkles and distortions. (6) Rinse thoroughly: Rinse several times to remove detergent soap and bleach. (7) Damp wipe only: Surface clean with damp cloth or wipe sponge.

Waste: A material removed, rejected, or otherwise lost in various manufacturing processes.

Waste water: Strictly speaking, water that is soiled by domestic, commercial and industrial use, as well as drain precipitation water from built-up land (roofs, roads), contains impurities in solute, colloid and solid forms.

Waste, burr-wool: See **Burr-wool waste.** Waste removed by the burr guard of cards or burr pickers having a very short fibre and full of burr and seeds. The nature of waste varies according to the wool from which the burrs are taken.

Waste, card: See **Card waste.** Short fluffy waste thrown out of the carding machine.

Waste, cotton: See **Cotton waste.** Material removed from seed cotton ginned lint. Or stock in process by any cleaning or processing machinery and usually consisting of undesirable fibres or a mixture of cotton fibres with foreign matter.

Waste, hard: see **Hard waste.** Waste from reeling, warping, winding, weaving and cop bottoms.

Waste, invisible: See **Invisible waste.** Weight loss due to dust, moisture, loose fibre etc. Carried away by the air stream during the cotton testing.

Waste, lap: See **Lap waste.** Waste of long fibres formed in drawing and spinning processes by the breaking of the ends between the front rollers and the bobbins. This may be reprocessed.

Waste, soft: See **Soft waste.** Waste from slubbing, roving and wool tops.

Waste, spinners: See **Spinners waste.** Broken and tangled threads and lengths of yarn left on bobbins and spinning machinery in the manufacture of yarn.

Waste, sweeping: The sweeping from the floors of the various rooms in the mill. This is usually of a short and fluffy character.

Waste, thread: See **Thread waste.** The waste left on bobbins or collected during spinning and weaving (See also waste, thrum; Waste, hard etc.

Waste, thrum: See **Thrum waste.**

Waste, visible, in cotton testing: See **Visible waste.** Foreign matter deposited in the waste boxes of the machines during testing.

Waste water: Strictly speaking, water that is soiled by domestic, commercial and industrial use, as well as drain precipitation water from built-up land (roofs, roads), contains impurities in solute, colloid and solid forms.

Waste water load: Amount of impurities. Pollution in flowing water containing substances that consume oxygen (Waste water; Waste water pollution; Summation parameters). Calculated as a quotient of the limiting value per head of population and low water drainage.

Waste water pollutant load: Calculation of Waste water pollution, waste water quantity/unit time. Conventional value is the Population equivalence.

Waste water pollutants: They are aggressive, toxic and inorganic and organic compounds as well as their ions which impede biodegradation (See **Waste water pollution**). Given as a concentration in mg/l. Permitted waste water threshold limit values for discharging waste water pollutant are aimed at the required protection of the sewerage system, biological and the outfall channel. See **Waste water treatment.**

Waste water pollution: (1) Contaminants accumulating in waste water. Stemming from textile finishing substance burdens in the waste water: dyes (absorbable, non-absorbable), acids, alkalis and neutral electrolytes, oxidants (chrome compounds, bleach liquors, nitrites, nitrates, etc.), reducing agents (sulphur compounds), plastic emulsions/dispersions, finishing agents (e.g. fermentable starch products), metal salts, especially salts from heavy metals (toxic substances) such as copper, chrome(III), chrome(VI), mercury, lead, zinc, zirconium and such ions. Textile auxiliaries/wetting agents, surface-active substances, thickeners, lubricants/oils, sequestering agents and phosphates, tannin-like products amongst others. (2) Results from domestic, technical and agricultural water consumption as well as polluted rainwater.

Waste water pretreatment: Hydrocarbons and colloidally dissolved dyes as well as some auxiliaries can be flocculated by the addition of electrolytes, e.g. iron or aluminium salts. The pH should also be controlled to complete the precipitation and a suitable flocking auxiliary added. This is waste water pretreatment.

Waste water treatment: Waste water treatment methods are classified as follows:

(1) Physical (mechanical): (a) sedimentation, (b) filtration, (c) distillation, (d) sorption, (e) freezing out. (2) Chemical: (a) neutralization, (b) ion exchange, (c) oxidation, (d) reduction, (e) catalysis. (3) Biological: (a) aerobic treatment, (b) anaerobic treatment, (c) activated sludge process, (d) fungus treatment.

(4) Physical/chemical: (a) flocculation and precipitation (coagulation), (b) ventilation, (c) activated carbon filtration, (d) foam fractionation, (e) solvent extraction, (f) combustion, (g) osmosis, h) electrolysis.

Watch Pocket: A small pocket in the garment, typically located just, below the front waistband of men's trousers and used to accomodate change or a pocket watch.

Water: The chemical compound H_2O, widely available anywhere in the world.

Water alkalinity: If the titration of a 100.0 ml water sample with 0.01 M sulphuric acid consumes V ml of acid, the alkalinity is given by:

$$\text{Alkalinity} = 10 \times V \text{ ppm } CaCO_3$$

Water calendar: See **Water mangle.**

Water crimp: See **Crimp, types of.**

Water decarbonisation: Removal of temporary hardness (Water hardness salts) using appropriate. Water softening process, e.g. so-called lime decarbonisation via precipitation process or so-called hydrogen decarbonisation with ion exchangers, which contain carboxyl groups and work according to the principle of desalination.

Water degasification: Release of dissolved gases: air, oxygen, carbonic acid, etc.

1. Mechanical: watering, vacuum. 2. Thermal (especially for atmospheric oxygen):
heating above 100°C (high pressure enclosed vessels or low pressure at 40–60°C). Residual oxygen content approx. 0.5 mg/l, practically no free carbonic acid.

Water demanganizing: Manganese content rarely exceeds 1–2 mg/l, but often becomes unpleasantly noticeable in quantities of 0.1 mg/l in mains water. Manganese makes dark brown stains, which are difficult to remove.

Water desalination process: See **Ion exchanger.**

Water equilibrium: Term in dry cleaning for the final state of water distribution between treatment liquor and textile.

Water fastness: Resistance of dyeings and prints to the effects of water.

(1) Light strain (DIN 54 005-73): test sample with adjacent fabrics well moistened (surplus water poured away), 1 hour weighted with 4.5 kg tested at 20°C, separated and dried at 60°C and assessed using grey scales.
(2) Heavy strain (DIN 54 006-73): as 1., but tested for 4 hours at 37°C and/or 16 hours at 20 0C. AATCC test: 18 hours. Assessment using grey scales.

The perspirometer from AATCC (Atlas Electric) or the hydro test equipment (Schröder) is used as the tester.

Water glass: (Sodium silicate), is produced by melting quartz sand (SiO2) and sodium carbonate, whereby approximately the following ratio exists in the product which is of interest to the textile industry: 8% sodium oxide and 27% silicon dioxide. The exact structure of water glass cannot only be given by 37–40°Bé, but it is well-known that the structures SiO2– 3, a chain structure, and Si4O6– 11, a band structure, represent the main constituents.

Water hammer: **Water hammer** is produced in steam and condensate return piping by rapidly moving/retarded water drops or by combination of steam and hot condensate with colder condensate.

Water hardness: Hardness is defined as the presence of soluble calcium and magnesium salts in the water. If these are present in the form of bicarbonates, the hardness is temporary. Permanent hardness arises when water contains soluble salts of calcium and magnesium such as chlorides and sulphates. It is unaffected by boiling the water. The use of hard water in a textile dyeing or finishing mill can have some serious consequences. These include:

(a) precipitation of soaps;
(b) redeposition of dirt and insoluble soaps on the fabric being washed – this can cause yellowing and lead to unlevel dyeing and a poor handle;
(c) precipitation of some dyes as calcium or magnesium salts;
(d) scale formation on equipment and in boilers and pipelines;
(e) reduction of the activity of the enzymes used in desizing;
(f) decreased solubility of sizing agents;
(g) coagulation of some types of print pastes;
(h) incompatibility with chemicals in finishing recipes.

Water hardness , degree Of: See **Water hardness scale.**

Water hardness scale: The definition of one degree of hardness, however, varied from one country to the next. It is much simpler to express hardness in mg/l or ppm of $CaCO_3$.

Various units used for water hardness

Unit of water hardness	ppm $CaCO_3$
1 British degree	14.3
1 American degree	17.2
1 French degree	10.0
1 German degree	17.9
1 mmol/l $CaCO_3$	100

Water hardness units: See **Water Hardness scale**

Water hardness, Determination of: Complexometry offers a suitable method for total, calcium and magnesium hardness, which works simply, rapidly and accurately (0.05°d). See **Water hardness salts.**

Water hardness salts: Water-dissolved calcium and magnesium salts (total hardness) cause the "hardness" in water. The hydrogen carbonates break down when boiled by releasing the carbonic acid holding them in solution (temporary, carbonic acid or carbonate hardness). The remainder is resistant to boiling (permanent, mineral acid or permanent hardness). Magnesium salts produce magnesium hardness; calcium salts, calcium or lime hardness. The units of hardness are different depending on the country and are usually given in mVal = millival (Water hardness units).

Water jet weaving machines: These machines are produced only by few companies and are used for 'the manufacture of light and medium weight fabrics with standard characteristics and in water repellent fibre materials, primarily multi-filament synthetic yarns. These machines use water jet for weft insertion. Water jet machines are extensively used in East Asia, but have limited importance in other countries. They are characterized in particular by high insertion performance and low energy consumption.

Water lock: The installation of a water seal prior to the delivery of fabric from a steamer ensures an air free passage of the latter through the steamer, e.g. by immersing the lower exit rollers inside the steam chamber in the liquor contained in the 1st after treatment bath

Water mangle, water calendar: A mangle with two squeeze rolls for the dewatering (pre-drying) and simultaneous smoothing of textile fabrics. A water mangle is often used before drying piece goods in rope form after pretreatment.

Water marking: See **Pile reversal.**

Water marks: (1) Differences in light reflection from a textile fabric caused by the action of otherwise harmless water sprays or water drops. The effect is due to a temporary swelling which occurs particularly with regenerated cellulosic fibres sensitive to wetting. (2) Moiré effects caused by the pressure of one layer of fabric on another. (3) Undesired marks due to the presence of wet spots on a textile fabric prior to continuous dyeing (or beam dyeing) which reduce dye uptake in localized areas. (4) Variations in the pile orientation of pile fabrics causing differences in light reflection.

Water of crystallization: (water of hydration). Water which is chemically combined in many crystalline substances. This water may be held in the crystal in various ways. Thus, the water molecules may simply occupy lattice positions in the crystal, or they may form bonds with the anions or the cations present.

Water pockets: If batch rolls containing wet fabric in the open-width state are inadvertently kept standing during, e.g. a prolonged machine stoppage in wet sheets, tenting, tarpaulins, etc. This term must not be confused with the term Water repellent.

Water softening: Removal of ***hardness*** from water. The calcium and magnesium ions can be removed by chemical precipitation. Alternatively, cation ***ion exchange resins*** can be used to exchange the calcium and magnesium for sodium ions. Ion exchange resins that exchange other cations for hydrogen ions, produce an acid water, because the bicarbonates, sulphates and chlorides become carbonic acid, sulphuric acid and hydrochloric acid. However, by mixing this with hard water all the bicarbonates in the raw water can be removed and the acids in the softened water are neutralised—e.g. $Ca(HCO_3)_2 + H_2SO_4 = CaSO_4 + 2H_2O + 2CO_2$

The carbon dioxide can be removed by aeration. This process reduces the ***carbonate hardness*** to zero, although some ***non-carbonate hardness*** remains.

Water treatment: ***Screening***, then ***coagulation, flocculation*** and ***clarification*** or ***dissolved air flotation***, followed by ***filtration*** and ***disinfection***, is a typical treatment which converts raw water to drinking water. It removes colour, suspended and ***colloidal*** matter from reasonably clean water but cannot reduce their ***salinity*** without ***desalination***. A range of other processes may be used in water treatment to remove specific ranges of contaminants in the water. Examples include ***activated carbon filters***, ***ion exchange*** and ***membrane processes***. The ***sedimentation*** that occurs during ***raw water storage*** can also be regarded as part of the water treatment.

Waterproof: A term applied to materials that are impermeable to water; waterproof fabrics have had all their pores closed and are also impermeable to air and very uncomfortable.

Water proof finish: The production of a Water repellent finish (analogous to a waterproof or water pressure resistant finish) on a textile fabric. Waterproof and water repellent finishes.

Waterproofing: The production of a Water repellent finish (analogous to a waterproof or water pressure resistant finish) on a textile fabric. See **Waterproof and water repellent finishes.**

Water quality for dyehouse: The typical water quality for textile processing can be summerised as below:

Typical Dyehouse Water Quality

Component	Permissible Concentrations, mg/l
Hardness	0–25 $CaCO_3$
Iron	0.02–0.1
Manganese	0.02
Silica	0.5–3.0
Alkalinity to pH 4	35–65 $CaCO_3$
Dissolved solids	65–150

Water repellency: The property of a fibre, yarn or fabric characterised by its resistance to wetting by water. Water Repellency is more difficult to define because various static and dynamic tests are used to measure water repellency. Generally speaking water repellent fabrics are those which resist being wetted by water, water drops will roll off the fabric. A fabric's resistance to water will depend on the nature of the fibre surface, the porosity of the fabric and the dynamic force behind the impacting water spray. The conditions of the test must be stated when specifying water repellency. It is important to distinguish between water-repellent and water-proof fabrics.

Water repellency Test: (1) A severe test method for textiles with a hydrophobic (water repellent) finish. Procedure: a specimen of the textile fabric is lightly stretched over a wooden frame inclined at an angle of approx. 15° (attached to a glass plate with a rubber band if necessary) and placed 50 cm below the outlet of a dropping bottle. Water drops (previously adjusted to a rate of 100 drops/s) are allowed to fall on the same area of the textile specimen until a perceptible wetting occurs (as indicated by a darker colour in the wetted area). Good water repellent impregnations should be able to resist at least 20 of these drops without any wetting. Alternative: Spray test. (2) This test method is used for the evaluation of water repellent and waterproof finishes. The test is based on the principle of subjecting a textile test specimen to the action of simulated rain with simultaneous rubbing on the underside of the fabric. The water repellency of the fabric is assessed by measuring the water absorption (%) and the quantity of water which has passed through it (simulated rain values). Pfersee testing instrument; Hydrophobic finishes, evaluation of; Spray test.

Water Repellent: Fabrics that have been treated with a finish which cause them to shed water and resist water penetration, but are still air-permeable. Treatments can include wax coatings, resins, silicones, and fluorine derivatives. Such treatments do not close the pours of the fabric, while waterproof finishes do.

Water repellent fabrics: Water Repellent Fabrics have open pores and are permeable to air and water vapor. Water-repellent fabrics will permit the passage of liquid water once hydro-static pressure is high enough.

Water resistant: A degree by which water is able to penetrate a fabric. Not to be confused with water-repellent. However, the terms are often used interchangeably.

Water resistance: Resistance to wetting and penetration by water.

Water retained, in textiles: The amount of water absorbed by the fibres, adsorbed on the surface of the fibres, and held within the void of the fabric after immersion , measured under specified conditions.

Water retension: The moisture remaining in and on the material after a specified mechanical treatments. In this method water retained by fibre masses includes the water absorbed from the prevailing atmosphere, water imbibed following immersion and moisture adhering to the fibre surfaces after being subjected to 1000 times normal gravitational acceleration for one hour.

Water softening: The removal from water of dissolved calcium, magnesium, and iron compounds, thus reducing the HARDNESS of the water. The compounds are potentially damaging because they can accumulate in pipes and boilers. They also react with, and therefore waste, soap. Temporary hardness can be removed by boiling the water. Permanent hardness can be removed in a number of ways: by distillation; by the addition of sodium carbonate (which causes dissolved calcium, for example, to precipitate out as calcium carbonate); and by the use of ion-exchange products such as Permutit (utilizes zeolites) and Calgon (utilizes polyphosphates).

Water-Proof Fabrics: Water-Proof Fabrics are resistant to the penetration of water under much higher hydrostatic pressure than are water-repellent fabrics. These fabrics have fewer open pores and are less permeable to the passage of air and water vapor. The more waterproof a fabric, the less able it is to permit the passage of air or water vapor. Waterproof is an overstatement, a more descriptive term is impermeable to water. A fabric is made water-repellent by depositing a hydrophobic material on the fibre's surface; however. Waterproofing requires filling the pores as well.

Water softening agents: The function of these products is to eliminate the harmful effects of Water hardness salts and other metal salts by complex formation (use of appropriate sequestering agents). Suitable products are based on salts of polycarboxylic acids, polyphosphates or preparations containing these agents. Such water softening agents are used in all fields of textile dyeing and finishing.

Water softening processes: For the removal of Water hardness salts.

(1) Cultonex process with milk of lime or lime water. In this process, the water hardness is brought down to 2–4°d by stoichiometric dosing. (2) Lime-soda water softening process with high carbonate and non-carbonate hardness. This process brings the water hardness down to 1–2°d. (3) Soda regeneration process, used specifically for non-carbonate hardness (boiler feed water). (4) Trisodium phosphate water softening process: the addition of trisodium phosphate produces a floccular sludge. Water is softened to 0.1°d by this method. V. Baryta process for water with a high non-carbonate hardness (sulphur oxide), residual hardness 1–2°d. (5) Ion-exchange processes for complete water softening

Watt: It is defined as the power when work of 1 J is done in 1 s or an equal heat transfer occurs in 1 s. In electrical applications, 1 W is the product of 1 ampere and 1 volt. Derived units of power, electrical energy and heat are all quotients obtained from a statutory unit of energy, work or quantity of heat, and a statutory unit of time. 1 W = 1 J/s ,1 W = 1 Nm/s.

In the SI system, the watt has now replaced the former unit of mechanical power, i.e. horse power (HP). Conversion: 1 HP = 745.70 W.

Watteline: Watteline is a very loose, soft tricot construction, raised on the back. It is used for waddings and padding between top cloth and lining.

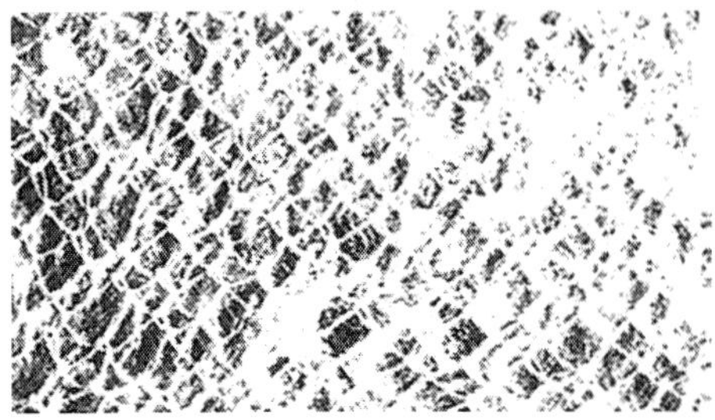

Wave: A physical activity that rises and then falls periodically as it travels through a medium.

Wave shed: A shed in a multiface weaving machine where several sheds are formed in the direction of weft and move laterally. The sheds may be formed by healds driven individually or by a series of narrow heald frames at various stages of heald timing.

Wavy cloth: A cloth that will not lie flat on a cutting table.

Wavy Face: A surface condition characterised by a considerable variation in yarn diameter.

Wavy selvage: See **Slack selvedge.** An undulating selvedge that does not lie flat. Causes: the use of different materials in fabric and selvedge, overturned selvedges due to creased edges, defective beaming, defective temple adjustment on the weaving machine, etc.

Wax: A special group of low melting, thermoplastic organic mixtures generally opaque in appearance. Some are hydrocarbons, others are esters of fatty acids and alcohols. Soluble in most organic solvents, insoluble in water. Saponification of saponifiable waxes yields Fatty acids and Fatty alcohols. Waxes form protective coverings to leaves, stems, fruits, seeds, animal fur, and the cuticles of insects, serving principally as waterproofing For example, waxy deposits on some plant organs add to the efficiency of the cuticle in reducing transpiration, as well as cutting down airflow over the surface and forming a highly reflective surface, thus reducing energy available for evaporation. They may also occur in plant cell walls, e.g. leaf mesophyll. They are used in varnishes, polishes, and candles.

Wax resists: (wax and resin resists) are employed as mechanical resists in resist printing (especially batik articles) for the production of pure untainted white effects that are almost impossible to achieve with reductive discharges. Wax resists consist mainly of natural resins (pine resin, colophony), beeswax, ceresin, stearin, spermaceti, tallow, etc. melted with turpentine oil or rectified petroleum. They are applied to bleached fabric by hand block printing or from heated rollers with deep engravings on a printing machine. To avoid sticking and smearing, the printed material is sprinkled with kaolin, diatomaceous earth or fuller's earth. The goods are then dried by hanging in air for 2–3 days, washed cold and subsequently dyed, usually with cationic or selected acid, direct and mordant dyes as well as indigo. The dyes used must not be soluble in benzine or solvents since, after cold dyeing, rinsing and drying, a passage through benzine must be given to remove the wax (only benzine produced from benzene and toluene is suitable, not petroleum spirit).

Weak acid/Weak base: See **Acid, weak or Base, weak.**

Wear fastness properties: In Colour fastness testing, a distinction is made between those colour fastness properties that are of particular relevance to the end consumer, e.g. colour fastness to light, washing, chlorinated water, sea water, water spotting, perspiration, rubbing, dry cleaning, etc. i.e. properties which relate to the performance of a dyed or printed textile in use, and those which are only important in textile manufacturing processes, e.g. colour fastness to decatizing, stoving, sublimation, vulcanizing, etc.

Wear, wear and tear: When applied to textile materials, the term "wear and tear" refers to the gradual and unintentional wearing a way of a textile surface due to specific mechanical effects such as friction (textile wear resistance. See **Abrasion resistance**).

Wear level: The number of wear refurbishing cycles to which an item has been subjected.

Wear-refurbishing cycle: One complete series of events that may be terminated by laundering or dry-cleaning.

Wear-test: A test in which textiles are subjected to wear service conditions and evaluated for performance.

Weak web: A web of fibre that, when being transferred from the card doffer to the calendar rolls to form sliver, does not have sufficient strength from fibre cohesion or clinging entanglement to hold itself together while forming a continuous bridge in processing.

Weak yarn: A yarn that is found to be either below standard breaking specifications or to be weak enough to cause an abnormally high degree of stops in textile processing.

Wear resistance: See **Textile serviceability.**

Wear service conditions: The specific conditions under which a garment is worn e.g. at school. at work, at leisure etc.

Wear test: A test for fabric wear, abrasion, flexibility, washing, crushing, creasing, etc., in which the fabric is made into a garment, worn for a specific time, then assessed for performance.

Wearable Computing: Wearable computing refers to a computer that is inserted into the personal space of the user, is controlled by the user, and has both operational and interactional constancy, i.e., is always on and always accessible. Most notably, it is a device that is always with the user, and into which the user can always enter commands and execute a set of such entered commands, and the user can do so while walking around or doing other activities (Mann 1998).

Wear, wrinkle resistance: A term applied to textile fabrics which satisfactorily maintain their appearance by recovery from sharp folds, imposed during wear.

Weather: Climatic conditions at a given geographical location, including such factors as sunlight, rain, humidity, and temperature.

Weathering: (1) The action of atmospheric agencies or elements on substances exposed to them. (2) The discoloration, disintergration, etc., that results from this action.

Weathering fastness: (colour fastness to weathering). Determination of the resistance of dyeings and prints on textiles of all kinds to the action of weathering. Accelerated weathering conditions can be achieved in the laboratory by exposing test specimens to conditions

that involve changes in temperature, relative humidity and radiant energy, with or without a direct water spray (simulated rain) in an attempt to reproduce changes in the material similar to those observed after long-term, continuous, outdoor exposure. A suitable test apparatus uses filtered xenon arc radiation and an additional water spray

Weather resistance: Ability of a material to resist degradation of its properties when exposed to climatic conditions.

Weather-o-meter: An instrument used in measuring the weather resistance of textiles. It can simulate various weather conditions such as sunlight, rain, dew, and thermal shock.

Weathering is accelerated to the degree that the effects of years of normal use are attained in only a few days.

Weave: See **Weaving.**

Weather-o-meter: A laboratory testing device for determining the resistance of textile materials to weathering.

Weather resistance: Ability of a material to resist degradation of its properties when exposed to climatic conditions.

Weave pattern: A plan (drawn on graph paper) used as a guide In weaving operations that employ harnesses.

Weave, plain: A fabric pattern in which each yarn of the weft passes alternately over and under a yarn of the warp and each yarn of the warp passes alternately over and under the weft yarn.

Weave, twill: A weave characterized by diagonal lines produced by a series of float staggered in the warp direction, floats are normally formed by the filling.

Weave Structure: The weave structure is the order in which warp and weft threads go over and under each other, i.e., the interlacement. If they interlace alternately, the weave structure is plain weave.

Weavers beam: During the process of warping as an intermediate stage a warpers beam is made having up to 1000 ends are produced. The threads of 6-10 of this warpers beam are combined at the slashing (sizing) stage and wound to a **weavers beam (loom beam).**

Weavers knot: When a yarn breaks or is otherwise discontinuous, it must be repaired and joined back together. In some cases at winding and normally at creeling or at the knitting machine, weaving machine, a "fisherman's" knot or a "weaver's" knot may be used to join the ends of the yarn together.

Weavers side: In a loom, the front of the machine is where the fabric beam is called the weavers side.

Weave repeat: The smallest number of different intersections between warp and weft that, when repeated in either direction, gives the weave of the whole fabric.

Weaving: The conversion of yarn into fabric by interlacing two independent yarn systems, i.e. the warp and the weft, on a weaving machine or loom. In a weaving machine, the warp yarns are passed from a warp beam (wound warp) to the fabric beam (wound fabric).

During this process, each warp yarn is led through an eye on a heald attached to a harness. The harness lifts some of the warp yarns, e.g. the even-numbered warp yarns, and depresses the remainder to form a gap between them known as a "shed" through which the weft

is inserted. This operation is known as shedding and the insertion of weft is called "picking". The sley beats up the weft to the edge of the woven fabric.

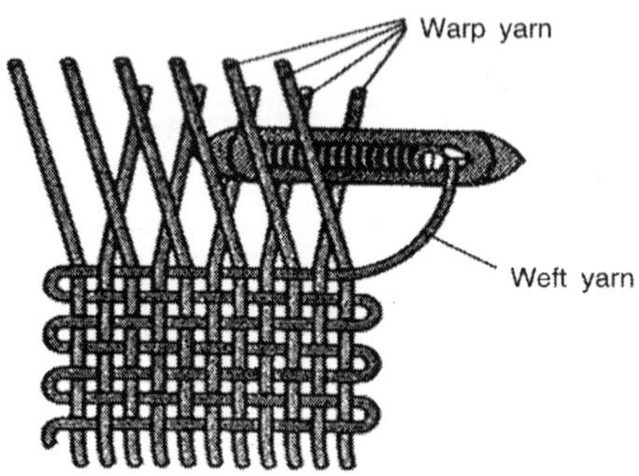

Principle of weaving

Weaving, shaft: warp sheet passes over the back rest roller and lease rods, through heald shafts and reed and over the breast beam to the cloth roller. The heald eyes of each heald shafts are threaded in a set pattern e.g., yarns 1, 3, 5, 7, 9 or yarns 2, 4, 6, 8 etc. By raising and lowering of the healds a shed is formed in the warp yarns through which the weft yarn can be drawn For a plain waeve at least two shafts are needed. After the weft has been introduced the reed beats it up into the fell of the cloth. The range of pattern which can be woven on a shaft loom, is limited by the number of shafts which can be accommodated on the machine.

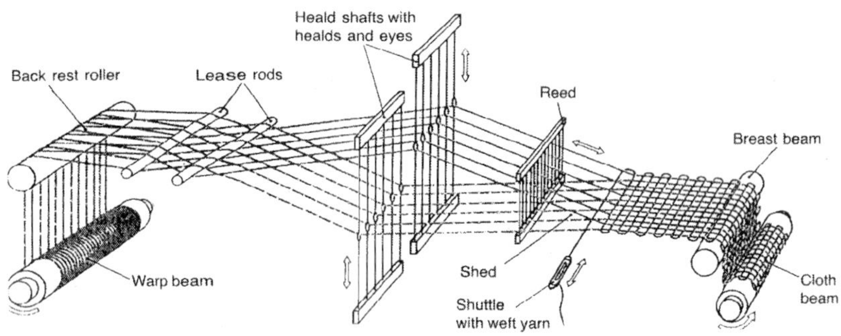

Principle of Shaft weaving

Weaving, Jacquard: The lifting of each warp yarn can be controlled individually. This is effected by either a punched card or an electronic control system according to the required weaving pattern. The technique is named after it's inventor, J.M.Jacquard a Silk weaver of Lyon.

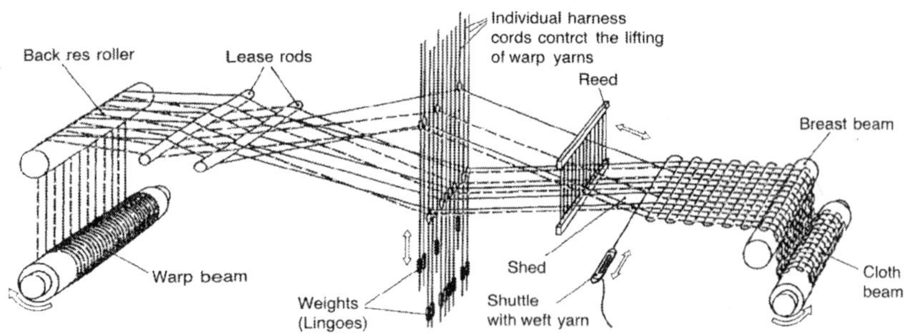

Schematic drawing of Jacquard weaving

Weaving machine, circular: In a circular weaving machine, the warp is circular and there are continuously circulating shuttles running around the periphery in a wave or ripple shed. The shuttles cannot leave the shed and have a continuous motion. Shuttles are driven electromagnetically and each shuttle runs in its own shed.

Weaving machine, double rapier: In double rapier weaving machine, the weft yarn is inserted half way into the shed by one carrier and then taken over in the centre by the other carrier and drawn out to the opposite side of the fabric.

Weaving machine, jet: In a jet loom the weft thread is propelled across the loom by a jet of water or compressed air. In this case the mass of the weft carrier, i. e. water or air jet is redused to a minimum, say 1.5 g.

Weaving machine, multi-shuttle: A multipiece weaving machine in which the weft insertions are by shuttles.

Weaving machine, multipiece: A narrow fabric loom equipped for the simultaneous weaving or two or more pieces.

Weaving machine, single rapier: In a single rapier weaving machine, the weft insertion is done by one raier to the full width of the warp

Weaving, primary motions: Shedding, picking, and beating up together is called the primary motions of weaving.

Web: (1) The wide film of fibres that is delivered from the card. (2) A similar product of other web-forming equipment, such as that formed by air deposition and used to make nonwoven fabrics. (3) A term loosely used for lightweight nonwoven fabrics.

Web slitting machines: (fabric slitting machines), are used to slit running fabrics down the middle or in longitudinal strips of predetermined width. They are designed for installation on production machines.

Webbing: Strong, narrow fabric, closely woven in a variety of weaves and principally used for belts and straps that have to withstand strain (e.g., automobile seat belts, reinforcement of upholstery, suspenders, etc.). Elastic webbing is made with spandex or rubber yarns in part of the warp or filling, or both.

Webbing, elastic: A textile webbing containing rubber or other elastomers to permit rubber like stretch in at least one direction.

Webbing, in textiles: A stout narrow fabric with a mass over unit area of at least 0.5kg/m^2 (0.1 lb/ft2) for each 25.4mm (1.0 in.) in width.

Webby Wool: A thin fleece with poor staple formation and a large number of cross fibres

Weft: **See Filling**.

Weft: (a) Width-way threads as woven in a fabric. (b) Yarn intended for use as in (a) above.

Weft cracks: Cracks present in individual and sometimes neighbouring weft yarns in woven crêpe fabrics produced from viscose or cupro yarns. Causes: embossing conditions too severe, or fibres weakened by oxidized linseed oil sizes.

Weft detector: An electronic device for indicating: (a) The presence of weft during weaving, normally on non-shuttle looms; (b) That the weft on the pirn in the shuttle is becoming exhausted.

Weft distortion: Weft distortion is caused by external mechanical forces acting on a woven fabric (mechanical distortion) as well as internal latent tensions already contributed by the yarns themselves, or which have developed during the processes of manufacture (shrinkage distortion). – Skew distortion: a condition in which one side of the fabric is further advanced in relation to the other thereby causing a diagonal distortion of the weft yarns. Thus, although the warp and weft yarns may still be largely straight, they no longer lie at right angles to each other. – Bow distortion: a curvature in the path of the weft yarns, i.e. although the left and right ends of the weft yarns are at the same height they are displaced either forwards or backwards in the centre of the fabric in relation to the edges. – Combined bow and skew distortion: the weft follows a curved path and the ends of the weft yarns are not at the same height. – S-shaped distortion: the weft is wavy and lies in a series of small curves. It is largely dependent on the weave pattern. – Irregular distortion: the weft yarns may follow any path across the width of a woven fabric. This type of distortion usually involves a ombination of straight skew and various bow-shaped components.

Weft face fabric: Any fabric in which the warp is completely covered with weft.

Weft feeders: These are intermediate feeding devices, which are also called weft storage feeders or weft accumulators, play today an important role in the new age weaving machines where the weft is unwound overhead from the cone and is subjected to abrupt accelerations due to the drawing-off tension exerted by the insertion element (rapier, projectile, fluid). The balloon which is formed at each insertion can cause coil sliding and snarls, owing to the difficulty of braking adequately the yarn and to the high unwinding speed of the yarn from the cone, which results into abrupt stresses, varying with diameter and speed variation. The present weaving speeds made thus absolutely necessary the use of an auxiliary apparatus placed between the cone and the insertion device. This apparatus positions the thread in a way as to favour its unwinding under lower stresses, and at the same time takes off from the package the necessary thread length, also making the most of the dead times between an insertion and the other, therefore with lower unwinding speeds and more continuously.

Weft fork: A mechanical device for monitoring the presence of weft during weaving, normally on shuttle loom.

Weft insertion: (1) Any one of the various methods, shuttle, rapier, water jet, etc., for making a pick during weaving. (2) A marriage of warp knitting and weaving brought about by inserting a length of yarn across the width of the knitting elements and fastening the weft yarn between the needle loop and the underlap.

Weft insertion machine: See **Weft insertion machines.**

Weft insertion systems: A distinction is made between the following weft insertion systems used on weaving machines (See Weaving): (1) Shuttles, as used in Shuttle looms. (2) Projectiles (gripper shuttles) used in projectile weaving machines (3) Rapiers used in Rapier looms (4) Jets: a) Air insertion b)Water: an intermittently directed jet of water is used

Weft knitted fabrics: Knitted fabrics in which one or more thread systems run widthwise across the fabric forming all of the loops in each course. Weft knitted fabrics are produced on flat and circular knitting machines. Typical weft-knit fabrics include double knit, fully fashioned, interlock, jersey, knitted fleece, knitted terry, rib knit, etc.

Weft-knitted spacer fabrics: Weft-knitted spacer fabrics can be manufactured using cylinder and dial machines (circular double jersey) or V-bed machines (electronically controlled flat knitting machines). Fabrics can be produced on both machines where two sets of needles have the ability to create two individual layers of fabric that are held together by tucks. As the spacer stitches are of the tuck variety, the spacer yarn is generally hidden on the technical back of the two plain surface layers. This has the advantage of preventing a rough or harsh feel to the material.

Weft knitting: See **Weft knitted fabrics.**

Weft slubs: Consist mainly of short irregularities in the weft direction (easily visible when viewed against the light). Weft slubs are almost always caused by defective material (impurities), spinning defects (variations in yarn count or titre, differences in twist, irregular yarn, uneven yarn thickness), or defective dyeings (streaky dyeings).

Weft yarn: See **Weft, filling.**

Weft, filling: Threads or yarns running across the width of a Woven fabric (Weaving) from selvedge to selvedge and at right angles to the warp. The weft binds the warp yarns (Warp) by interlacing with them to form a woven fabric structure and also serves as a filling material. Each weft yarn is called a pick.

Weft-knit fabric: See **Weft knitted fabric.**

Weftless fabric: (ref. tyre) A sheet of parallel cords surrounded by uncured rubber compound.

Weftless tape: Advertising tape; Typing tape: A material in strip form that consists of a number of set warp threads held together by an adhesive.

Weight: The force excreted on a body by gravity.

Weight: To determine the mass of a material.

Weight, as used with fabrics: Mass per unit area.

Weight, commercial: See **Commercial mass.**

Weight, in warp knitting: The number of yards per pound of finished fabric.

Weight, fabric, finished: See **Finished fabric weight.**

Weight, fabric: Mass per unit area expressed in grammes per square metre (ounces per square yard). Also expressed in grammes per linear metre (ounces per linear yard) or inversely as metre per kilogram (linear yards per pound)

Weight of cloth: See **Weight, fabric.**

Weight per Metre: Mass per metre of the fabric. In the same construction, weight per metre will vary as per the width of the fabric. Mainly used for finding out the weight of the lot, provided the length of the lot is known. Weight per square metre multiplied by the width in metre gives the weight per metre.

Weight per square metre: Mass per square metre of the fabric. This represents the characteristic, construction, finish etc of the fabric. If we know the weight per metre, weight per metre divide by the metre in width gives the weight per square metre of the fabric.

Weight reduction: Process by which the weight of the fabric is reduced by suitable methods, usually by partly dissolving the fabric (warp or weft or both). Mainly done on synthetic fabric or their blends, where the yarn is partly dissolved, still gives enough strength to make up into garments.

Weighted silk: Silk that has been treated with metallic salts during dyeing and finishing to increase the fabric's weight and improve its drape. Over-weighting can cause deterioration of the fabric.

Weighted silk: See **Weighting of silk.**

Weighting: The addition of metallic salts to silk to increase the weight and impart a firmer handle.

Weighting finishes: See **Weighting, synthetic.**

Weighting of silk: The silk fibre has a remarkable property of absorbing certain metallic salts, still retaining much of its lustre. This process is known as "loading" or "weighting," and gives increased body and weight to the silk. Silk without weighting is known as "pure dye," of which there is little made, as such goods take too much silk.

Weighting, synthetic: Chemical principle for weighting with methacrylamide: The monomer used for synthetic weighting is often derived from acrylic or methacrylic acid. The silk weighting with acrylonitrile and methymethacrylate has been studied and described thoroughly; in this process, starters are formed by a redox system based on iron salts (Fe++) and hydrogen peroxide, persulphates and other substances.

Welded Shell: The outer layer of a bonded or welded garment, such as a jacket.

Welding: There are two basic methods for applying bonding or welded seams. The first method uses an adhesive film, and the application of heat to glue or laminate two substrates together. The second method involves gluing or attaching two fabrics, using ultrasonic technology. The creation and channeling of high frequency vibratory waves cause a rapid buildup of heat in synthetic fabrics to create the bonding.

Well, in buttons: The recess in the centre of a sew through flange button that gives aesthetics and identifiesthe face side.

Welline: (Fr. = waved) A wave-shaped Ratiné effect. It involves carded yarn ratiné material with a wave-like structure. The waves may be oriented longitudinally, transversely or diagonally. The basic material of the cloth is wool velour. In order to achieve a particularly fine effect, cashmere, mohair, camel hair or alpaca is used as upper weft.

Welsh tapestry: Brightly coloured fabrics in geometric designs that resemble tapestry-work. The cloth is double-sided and reversible, plain weave, made from coarse Welsh woollen yarns. The cloth is constructed so that, although the same colours appear on both dides, the arrangement is reversed and gives a different colouring. Used for capes, coats, blankets, shawls, cushion covers.

Welt: (1) A finished edge on knit goods, especially hosiery. In women's stockings, it is a wide

band knitted from heavier yarn than the leg and folded on itself. (2) A small cord covered with fabric and sewn along a seam or border to add strength. (3) A seam made by folding the fabric double, generally over a cord, and sewing it. (4) A term sometimes used for piqué.

Welt Lining: Interlining for pocket welts.

Welt seam: A complex seam formed on the inside of the object with one trimmed raw edge enclosed and one stitching line visible on face side.

Welt stitch: See **Float stitch.**

Welted seam, in upholstered furniture: Seam sewn with a strip of covered cord between two fabric pieces, joined so that the welting shows on the exterior of the furniture unit.

Welting, in upholstered furniture: A cord covered by strips of exterior fabric, used in welted seams of upholstery covering.

West of England: Very high quality woollen cloth, produced in the Cotswold, U.K. area. The type of cloth varies, e.g. it may be flannel, and it often has a characteristic window pane check design. Used for men's suits, trousers, and overcoats, and for women's coats.

Wet and dry bulb hygrometer: If the bulb of a glass thermometer is surrounded by a wet sleeve of muslin in an atmosphere that is not saturated, water vapour will evaporate into the air at a rate proportional to the difference between the actual humidity and 100% humidity. Owing to the latent heat of evaporation, heat is drawn from the thermometer bulb, thus cooling it. This cooling effect has the consequence that the temperature indicated by a wet bulb thermometer is lower than the air temperature. By mounting two identical thermometers together, one with a wet sleeve and one with a normal bulb, the two temperatures can be read directly.

Wet cleaning: The removal of water soluble or emulsifiable soil from garments often carried out on a draining board using water in the presence of a detergent. It is also a process for cleaning garments etc. made from fabrics, e.g. vinyl coated materials liable to be damaged by dry cleaning.

Wet crease resistance: The property of a fabric whereby it does not acquire creases when squeezed in a wet state. Wet crease resistance includes an anti-creasing capacity, in other words the ability of the fabric to lose creases acquired during wear when immersed in water.

Wet crease resistant finishing: A process in resin finishing for improving wet crease resistance. Processes particularly suited for this are those which enable the cross-linking of cellulose when in a swollen state (Wet cross-linking process).

Wet cross linking process: A resin finishing process for cross-linking cellulose in a swollen state by means of hydrolysis resistant resin finishing agents in the presence of strong acid, e.g. hydrochloric acid, or strong alkali, e.g. sodium hydroxide solution, at room temperature.

Wet decatizing fastness: See **Potting fastness.**

Wet development: Fixation of dyeings and prints carried out in an aqueous medium. Prints are with vat leuco ester, naphthol and reactive dyes in particular on cotton, linen and viscose.

Wet dwell (batch) fixation: The fixation of compounds (e.g. resin finishes or reactive dyes or peroxide) by dwelling of the goods in a wet state.

Wet fastness: A collective term generally used for the demand that during use dyeings and prints exhibit colour fastness to Water, Perspiration, Washing and Wet ironing, whereby more specialist demands also include fastness to Alkali fastness of dyeings and prints; fastness to cross dyeing; Fastness to sea water as well as wet light fastness.

Wet finishing: (wet processing). The term encompasses all wet treatment processes such as bleaching, dyeing, printing, impregnation, mercerising, carbonising, milling, etc. in contrast to Dry finishing of fabrics.

Wet fixation process: A special fixation process used for prints in which the dye is applied without or with only a proportion of the fixing chemicals, and in which fixation follows without intermediate steaming and in a bath containing no, or only a proportion of, the fixing chemicals. For example: for reactive dyes, a caustic shock process; for vat leuco ester dyes, a nitrite process.

Wet forming: The production of a nonwoven fabric web from an aqueous suspension of fibres by filtering the short fibres onto a screen belt or perforated drum.

Wet impregnation: (1) See **Impregnation**. (2) See **Water-impermeable finishes.**

Wet lamination: Instead of using foil to stick together the separate layers (dry lamination), an adhesive is used which is applied to both sides of the middle layer (e.g. by spraying) of the lamination package before it enters the two-bowl padder. Once the adhesive has been heat hardened, the laminate is batched.

Wet-laid non-woven: Fabric made by the wet-forming process. The short fibres typically have more random orientation in the web and the web has more isotropic properties than carded webs.

Wet laying: The process of forming a fibre sheet by paper making technique, for nonwoven fabric products.

Wet modulus: Describes the force in cN/dtex which is theoretically required in order to stretch a fibre in a wet state by 100%. The force measured is the force which is required to stretch the fibre in a wet state by 5%, and this figure is then multiplied by a factor of twenty. A high wet modulus serves as a criterion of dimensional stability. *See* **Modal fibres.**

Wet-on-dry finishing: The application of finishing liquor to dry goods.

Wet on wet dyeing process: In a wet on wet continuous dyeing production each of a series of passes through the padder involve the application of dyes or chemical onto the wet goods emerging from the preceding pass. In wet on wet application for example preceding padder can have a for example, 70% pick up (either wet on dry or wet on wet) next padder can have a pick up of 90% (newly acquired pick up of 20%). The advantages of this dyeing process are a saving in drying energy and a high rate of passage of steam, thus providing a significant cost reduction and making the process particularly suitable for dyeing voluminous textiles such as terry and cord fabrics. Example: The dry, absorbent terry material is passed through hot naphthanilide solution on the first padder. The preparation cools during the ventilated stage. Developing with a diazo solution follows wet-on-wet on the second padder. A ventilated stage guarantees complete reaction (diazotisation). Further process can follow as usual.

Wet on wet impregnation: The wet-on-wet technique for coating is greatly favoured in some areas of continuous finishing. The reasons for this are to be found in the undeniable advantages such as saving on drying energy and thus costs, as well as the dispensing with a drying pass, a reduction in textile handling requirements and an increase in drying capacity. The problem of entrapped air (as in dry-on-wet impregnation = wetting) does not arise here, but instead there are other problems.

Wet on wet padding: A padding process in which the material to be padded in wet, e.g. from a previous process stage. Wet on wet impregnation is slightly complicated as there will be a dilution of the bath due to the seepage of the liquor from the fabric to the bath. A proper analysis of the liquor exchanges are carried out and a stronger solution is added to the bath to keep the concentration in the bath same.

Wet on wet printing: The overprinting of two print pastes where the first is not yet dry; a common process.

Wet pick up: This is the pick-up of the bath on to the fabric by the wet on wet padding operation. See wet on wet padding, wet on wet impregnation. Usually the wet fabric will have the 2nd bath solvent on it (usually water). In a usual procedure the fabric is passed through water to even the water on the fabric and give a higher squeeze and passed through actual padding bath and the second mangle where expression is kept high (low squeeze) The pitch up will be the difference between the second mangle and first mangle.

Wet processing: A common term used in various senses, also as wet finishing: 1. Finishing using an aqueous solution (in contrast to Dry finish employing organic solvent liquor).

2. Wet treatment processes in finishing technology such as wet decatizing, kier-boiling, bleaching, mercerising, carbonising, milling, dyeing, etc. 3. As a pre-treatment for all finishing processes which precede dyeing, in other words in addition to those under II. also, for example, singeing, thermofixation, raising, etc.

Wet scrubber: The use of water to assist in the removal of dust particles or polluting gases, such as SO_2, by contacting the dirty gas stream with liquid drops. The liquid flow divided by the gas flow is about 10^{-3}. Wet scrubbers allow particles less than 1μm to be collected on a 50μm droplet of water and then use a simple technique to remove the water droplets. The pumping of water maybe costly and the discharged wash water is polluted. For SO_2 removal, the wash water is alkaline (see **flue gas desulphurisation**). Many types of wet scrubber exist such as ***baffle type scrubber, collision scrubber, cyclonic spray scrubber, ejector Venturi scrubber, low energy scrubber, moving bed scrubber, packed bed scrubber plate tower scrubber, spray chamber***, ***Venturi scrubber***. ***A dry scrubber*** and ***semi-dry scrubber*** are techniques to minimise water usage.

Wet transfer printing: A facility for the wet transfer printing process on acrylic, polyamide and wool and comprising an impregnation padder, transfer calender (of the felt calender type) and a J-wash box. Even light and extremely sensitive qualities are suitable for this wet transfer printing process.

Wet transfer printing process: In contrast to a dry transfer printing process, the transfer of dye in the wet transfer printing process takes place by migration in the aqueous phase or in an organic phase, formed by wetting or other additives, at temperatures of 100–150°C.

Wet sizing: Conventional Sizing in contrast to Solvent sizing.

Wet soiling: The application of soil from a liquid medium.

Wet spinning: A spinning process used in the production of chemical fibres (especially viscose, cupro, polyvinyl alcohol) based on the principal of precipitation fibre formation or coagulation in an aqueous precipitation bath which takes place once the spinning solution has passed out of the spinning nozzle.

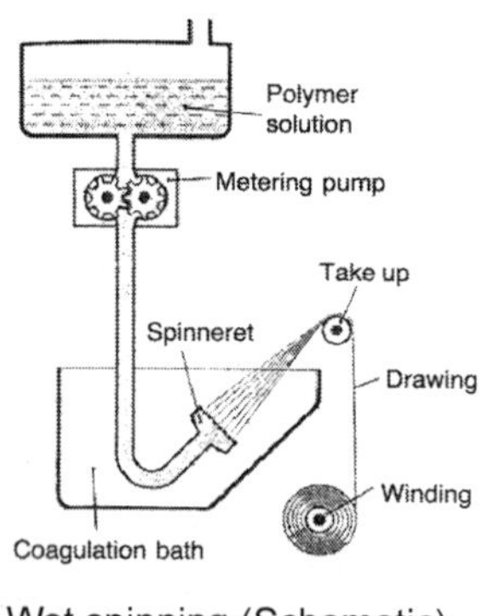

Wet spinning (Schematic)

Wet steam: Steam containing water droplets, in contrast to Dry steam. In the process house the wet steam is made by different method, one being passing the steam through water or by spraying limited fine droplets in to the steam.

Wet steamer: A steamer which can produce and use wet steam for processing. Most of the processing say, bleaching, colur fixation etc needs wet steamer.

Wet strength: The measurement of the strength of a material when it is saturated with water, normally relative to the dry strength.

Weatherometer: A testing machine used to expose thread to accelerated weathering conditions and measure its effect on colour fastness, strength and other physical characteristics.

Wet-laid nonwoven fabric: A fabric made from a fibre sheet formed by paper making techniques, normally followed by adhesive bonding. The material may contain a high percentage of non-textile. E.g. **Wood pulp**.

Wet-on-dry finishing: Most common finishing process, whereby the finishing liquor is applied on dry goods.

Wet-Spun Flax: The process of spinning line flax where the fibres are smoothed by moistening them as they are spun. Traditionally, spinners licked their fingers and drank a lot of beer. More prosaically, you can get little wooden buckets to hold the water and can hang by a leather thong from the wheel. Or even a bowl of water. You should coat the inside part of the bobbin (where the fibres wind around) with paraffin or some other sealant if you don't want to damage your bobbin.

Wether: A male sheep or goat castrated before sexual maturity. Because it's (the operative phrase) isn't caught up breeding, all of its nutrition goes into the fibre thus producing a better quality fibre. Fleeces from weathers cannot be entered into most wool shows but are worth watching for.

Wether Wool: Any fleece clipped after the first shearing is called wether wool. This wool is usually taken from sheep older than fourteen months, and these fleece contains much soil and dust.

Wettability: See **Wetting, Wetting agent.**

Wetting: The ability of a liquid to spread on a smooth solid surface is dependent on the polar nature of the solid and the surface tension of the liquid. (a) Wetting is the tendency of a liquid to spread out across a surface. A Wetting angle of zero corresponds to being fully spread out. The greater the wetting angle between solution and surface, the poorer the wetting. (b) The wetting of a solid body may be described in terms of the angle θ between it and a droplet of liquid, e.g. water, applied to its surface.

Wetting agent: A chemical that helps water penetrate a material or form a film over its surface; usually a *surfactant.* Water containing surfactants will easily spread on paraffin surfaces and have lower contact angles on Teflon. Surfactants used this way are called wetting agents, or penetrating agents when used to wet out repellent fabrics. (A non-polar solid surface such as paraffin wax or Teflon will cause a drop of pure water to bead-up and not spread). Wetting agents are often used in dyeing to help the dye solution penetrate to the individual fibres (they don't usually help the dye penetrate from the surface of the fibre to the interior of the fibre). They can help in some fibre arts dyeing processes, but too much in a dye solution used for direct application may make the dye spread more than is wanted.

Wetting agent test methods: (1) Physical method using measurement of: (a) surface tension (simple to conduct), such as elevation method, bubble pressure method, annular separation method (Surface tension). (b) Wetting angle. Demanding and only of limited validity. (2) measurements of the wetted textile surface: (a) droplet method: a drop is allowed to fall onto a stretched sample of fabric from a micro-burette, then the patch of liquid is measured. (b) elevation method: strips of fabric, weighted at the bottom, are dipped vertically into the wetting fluid and the height of the wet portion of fabric above the fluid surface is measured after a specified time. (3) Sink (immersion) method, simple and quick: (a) (AATCC standard test method 17). (b) Hydrometer method: the increase in apparent specific gravity of the fibre is measured while the wetting liquor penetrates the fibre.

(4) Diverse: the wetting process itself is not measured, but rather the results of it, for example swelling, dimensional change, electrical conductance, etc.

Wetting angle: The angle of contact between a drop of liquid resting in contact with an even surface, the drop of liquid being in a stable state and lens shaped, and the tangent to that surface; or, in a transmission sense, the relationship of a fibre to dirt (wetting angle 90° and often almost 180°). Hydrophilic materials display a wetting angle of 0–90°, and are capable of being wetted by water. Materials which tend to be water repellent display a wetting angle greater than 90° nd are thus hydrophobic (e.g. paraffin, ceresin wax with a wetting angle of 105–110°).

Wetting process: Process in which the fabric is wetted by various methods. It can be with the help of wetting agents or without wetting agents, if fabric is already absorbs water. Wetting process is necessary for many further textile processing.

Wetting promoters: These are textile additives not possessing surface activity but which nevertheless make wetting possible without accompanying generation of foam. They are absorbed by fibres and then impart the wetting process. They are known as special "air removing" wetting agents, in particular phosphoric acid trialkyl ester.

Wetting power: (wetting ability, wetting strength) The degree of capability for Wetting based on the Capillary activity of wetting agents.

Wetting properties: This is the ability of a textile surface to be wetted. A fibre may be well wetted by water if there are chemical groups on its surface which can take up water, for example because of hydrogen bridges, dipolar forces or ionic forces. Groups such as $-COO_{\theta}$ and $-SO_{3\theta}$ are strongly hydrophilic; OH, OR, NH_2 and C=O groups are weakly hydrophilic.

Wevenit: See **Double pique.**

WF: Virgin wool.

WG: Vicuna.

Whale oil: Sperm oil is a liquid wax occurring in the head of a sperm whale. Cetyl alcohol is the major fatty alcohol component of the wax ester while the acids are oleic (14%) palmitoleic (17%), palmitic (11%) and linoleic (9%). The wax oil is very stable to hydrolysis and to oxidation. Its viscosity is low so it makes an excellent lubricant. It was once the main ingredient in synthetic fibre spin finishes.

Wheat starch: A columnar, crystalline, forming a pure white homogeneous "flour" (greyer than potato starch). The handle is similar to maize starch. It is often adulterated using potato starch (detectable under magnifying glass or microscope). Its water content is 15–18%.

Wheel cast button: See **Rotation cast button.**

Wheelchair-resistant: Characteristic of a textile floor covering which enables it to withstand the pressure and abrasion imposed by a wheelchair.

Whip: The extra warp thread which forms the figures in lappet weaving. It is not interwoven with the fabric itself, except at the end of each run.

Whip roll: See **Back rest.**

Whipcord: Atwill weave fabric using bulky yarns to give a raised look to the twill ribs.

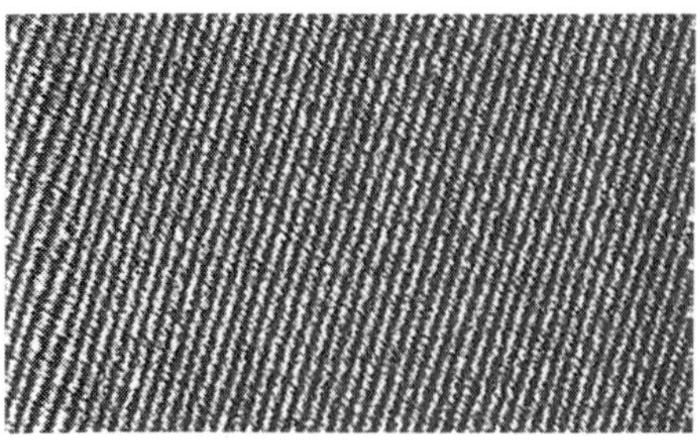

Although originally always wool or worsted, possibly with cotton filling, it is now made in any fibre, including nylon, polyester, acrylic, viscose, cotton and blends. According to fibre content and weight, it is used for making tough riding and sports clothes; light sports clothing such as tennis and Squash shorts; Children's clothes, men's trousers; and car seat covers.

Whipped cream: A type of crepe fabric produced from false-twist textured polyester yarn.

Whipped gathers: To gather, whip the rolled hem without hemming, making overcasting stitches towards you, even and not too fine. Use coarser thread than for hemming. This gathering thread is used to hold down the edge as well as for drawing up the gathers and it not to be taken out, as is the ordinary gathering thread. It should *not* catch in the roll. Have the thread the length of the plain space to which it is to be sewed and regulate the gathers as you do the gathering. After the edge is rolled, whipped and gathered, it is sewed to the garment by the little scallops or raised parts made by the whipping. This is used only for making ruffles or gathering on very fine hand work.

Whipped-in filling: See **Pulled-in filling.**

Whipping stitch: The back stitch is made by placing the needle back to the last stitch, bringing it out once the length of the last stitch, then placing the needle back into the last stitch, and so on, making the stitches follow each other without any space between. This is used in all places that are to bear great strain.

Whirled pile finish: The treatment of raw goods with special, usually round brushes for the manufacture of whirled plush fabrics (woven imitation furs).

Whiskers: Fine fibrils or crystals from polymers, metals, etc.

Whitaker process: Knit-deknit method; Space dyeing.

White: A sensory perception; according to Ostwald, the sum of all Chromatic colours and a body which reflects all visible colours.

White chate' mark: ("white abrasion"), rubbed areas on satin-like goods (especially on silk), and also frequently on poplin clothing in dry cleaning and laundry, caused by faulty mechanical influences.

White goods: A very broad term which implies any goods bleached and finished in the white condition. Some of the cotton white goods are muslin, cambric, dimity, lawn, long cloth, organdy, voile, etc.

White Light: Theoretically, light that emits all wavelengths of the visible spectrum at uniform intensity. In reality, most light sources cannot achieve such perfection.

White-on-White: Some fabrics, such as men's shirting or broadcloth, poplin, madras, etc., are made on a dobby or jacquard loom so the white motifs will appear on a white background.

White spirit: A liquid hydrocarbon resembling kerosene obtained from petroleum, used as a solvent and in the manufacture of paints and varnishes.

White wool: Wool having shade variation from true white to creamy white but free of pigmented, dyed, or otherwise coloured wools.

White, Visual assessment: Visual assessment of white is influenced by the colour of the surroundings and of the background (a light grey - i.e. neutral - plate, coloured ground or bleached and not brightened paper or cotton are best), by the angle of observation (45° to the observer is optimum), by changes of position of the samples and by personal factors. To avoid these the sampling should be done in a cabin kept neutrally grey and under standard lighting. Method a): hold the samples one in each hand at an angle of 45°; exchange the samples several times from hand to hand. Method b): wind the samples around a glass or plastic sheet free of fluorescing or UV absorbing compounds and hold as in a; the samples overlap alternately. Method c): lay the samples next to each other flat on a white or light grey ground free of optical brightener and press flat using a glass sheet which does not absorb UV; change positions several times.

Whitener: Fluorescent brightening agents, Florescent whitening agent.

Whiteness formulae: Used in calculating the Whiteness, degree of Distinction is made between: (a) Whiteness formulae which are based directly on measured physical values, e.g. from Stephansen (not for samples treated with brightening agent): $W_{st} = 2 \cdot R46 - R62$.

(b) Whiteness formulae which are based on colour values, e.g. from Berger: $WB = Y + 3.448 \cdot Z - 3.904 \cdot X$. These correspond better to the characteristics of the human eye.

After years of investigation the CIE has decided to recommend the following whiteness formulae. CIE whiteness formula for D65/10°: $W = Y + 800\,(0.3138 - x) + 1700\,(0.3310 - y)$

Whiteness index: The opposite of Saturation in Chromatic colours and not to be confused with Lightness even though a solid colour may appear brighter the more white it is, for example when it appears under above average illumination which causes an almost total white sensitivity as it approaches the level of being blinding.

Whiteness reduction: When a fluorescent brightening agent is added to a bleached fabric the whitness is increases. Any factor contribute towards a reduction in fluorescence also lead to a reduction in the whiteness effect and are thus undesirable in the brightening process. The phenomenon which appears as a reduction in fluorescence is often referred to in the literature as simply 'Fluorescence quenching' or whiteness reduction. Thus, fluorescence quenching becomes a collective term describing a series of negative final effects which are brought about by, for example, an excessive concentration or the addition of foreign substances (so-called quenchers) to solutions or to absorbing agents.

Whiteness scale: A comparative series of textiles treated with brightening agent and used for the evaluation of the intensity of fluorescent brightening agents.

Whiteness value: A number which characterizes whiteness as a visual perception of white. Whiteness samples of differing appearance can receive the same whiteness value, for which reason the whiteness value can be calculated more precisely from measurements using the Whiteness formulae.

Whiteness, degree of: A dimensionless number giving a measure of White content and used as a scale for evaluating the effect of bleaching, for example, or the brightening of goods involved in dry cleaning or the colourlessness of lubricating oils and the like. The degree of whiteness is determined by various formulae for Whiteness, measurement of.

Whiteness, measurement of: Even though white is also considered as a colour it differs from the chromatic colours in the sense that white products show a high degree of brightness combined with very low colour saturation. Thus the chromaticity of white is very weak. This makes the evaluation

of white samples fundamentally more difficult than that of normal dyeings. This is due to the fact that, that white is a concept of subjective quality subject to a great degree, to the influence of personal taste. A bluish white is preferred to a yellowish white provided that the two differ only in terms of their hue. Absolute evaluation differs from relative evaluation by including an exact definition of the illumination; this means the excitation of the brightening agent corresponding to an average level of daylight, and reference of the degree of whiteness and the hue calculated against an existing scale of whiteness. This method alone offers reproducibility which is adequate to all practical needs and a capacity to compare results obtained using different instruments. An evaluation using the Ganz whiteness formula has the advantage of including a correction for UV.

Whitney: An overcoating cloth made in wool. It is soft and thick with a free-finish on the rightside that produces a wavy line effect. If made from good quality woollen yarn, it is an excellent and hard wearing cloth.

Wholesalers: Essentially, businesses that buy from a manufacturer and, without changing the product, sell it in smaller quantities to retailers or smaller manufacturers.

Wick: A woven or a braided narrow fabric or a yarn or a group of yarns that has/have particularly good capillary properties.

Wickability: The ability of a fibre or a fabric to disperse moisture and allow it to pass through to the surface of the fabric, so that evaporation can take place.

Wicking: (1) Cord, loosely woven or braided tape, or tubing to be cut into wicks. (2) Dispersing or spreading of moisture or liquid through a given area, vertically or horizontally; capillary action in material.

Wicking-test: Test of the absorbency of fabrics following pretreatment for dyeing or finishing. Strips for testing are suspended vertically with their lower ends in the test liquid, and the elevation with time is measured (the wick effect). The test is important in the assessment of fabrics in minimal application techniques.

Wickproof, in tyre fabrics: A term used to describe a fabric that shows no air wicking.

Wide elastic fabric: An elastic fabric that is at least 150mm or 6 in. in width.

Width, of flat knit fabrics: The perpendicular distance between the selvedges when the fabric is under zero tension and free of fold or wrinkles.

Width, of circular/ tubular knit fabrics: Twice the perpendicular distance between the enclosed edges of a flattened tube of fabric that is under zero tension and free from fols and wrinkles.

Width, of raised surface fabrics: The dimension included within the outer limits of the nap or pile, but excluding the selvedges, or otherwise agreed upon by the purchaser and seller.

Width, of fabric: The distance from the outer edge of one selvedge to the outer edge of the other selvedge, measured perpendicular to the selvedges while the fabric is held flat under zero tension and is free of folds and wrinkles.

Width, of fabric woven on a shuttle less loom: The distance between the outer warp on one side to the outer warp on the other side, measured perpendicular to the warp yarns while the fabric is held flat under zero tension and is free of folds or wrinkles.

Widthwise Direction, in textiles: The direction in a machine made fabric perpendicular to the direction of movement of the fabric followed in the manufacturing machine.

Wigan: Named after the English village of the same name the cloth is produced by the converter from light weight sheeting or inexpensive print cloth. A dull-finish, plain weave cotton cloth, usually in dark colours but resembling sheeting. It is highly starched and calendared, to make it stiff. Its main use is as a backing or interfacing in men's jackets, and coats, but it is not easily obtained on the roll.

Wigwan: A converted cotton cloth, dyed black, brown or gray, and given a firm starched, plain calender finish, and used for interlinings in men's and boys' clothing to give body to the garment.

Wild Silks or Tussah Silks: Produced by various silk worms or other insects which are not artificially cultivated. Such silks are often difficult or impossible to reel, are of brownish colour and of uneven diameter and stronger than mulberry silk. Most of the wild silk is torn up for waste silk and is used for pile fabrics or woven into pongee. The most important wild silks are produced in China, Japan, India and Africa.

Willesden Canvas: Plain woven cotton fabric, rendered waterproof by treating it with solution of ammoniacal copper oxide and pressing it between hot calendars.

Williams unit: A wet-processing unit for open-width processing of fabric. The fabric passes up and down over rollers in the liquor. The unit is widely used for dyeing, washing, pretreating, and after treating.

Williams unit: A variant of a type of Roller vat which has an extremely restricted liquor volume. Each assembly comprises four vertical stainless steel containers, each of a U-shaped form. The separation between the casings inside the U is only 0.5 cm, meaning that there is an extremely short liquor ratio. The liquor is heated by a steam jacket to temperatures of 90–95°C (or to temperatures above 100°C if high pressure steam or hot oil is used).

Willow: A stiff fabric woven from esperato grass and cotton; the result is similar to Sparterie. Used for making the foundations of hats.

Willey; willow; teaze: To open and disentangle fibres prior to scouring and/or carding.

Willey; Tenterhook willow; Fearnought, Teaser; Cockspur willey: A machine consisting of bladed or pinned rollers for opening cleaning and mixing staple filment fibre materials as a preliminary to scouring (for greasy wool) or carding (for most animal and synthetic fibres) various machines of different designs exist to meet specific materials requirement. This includes Single cylinder or Double cylinder willey, Dust willey, Wool willey etc.

Wilton carpet: Woven carpet in which the pile yarns are woven in as an integral part of

the carpet, being held in place by the filling, usually made on a loom with a Jacquard head. The pile may be formed by wires and hooks or by weaving between two backings, in which case, the pile ends are cut to form two separate carpets. Wilton carpets are made in two types: (1) cut pile, e.g., tournia, Wilton moquette, plush, and velvet, and (2) loop pile, e.g., Brussels.

Wimple: A band of linen wrapped around a women's head and chin, supplemented with a brow band (fillet) or crown.

Wincey: A strong fabric made of cotton warp and wool filling in plain weave or twilled; used in England for winter underwear and men's shirts.

Winceyette: A plain-weave cotton fabric that has been brushed to give a warm soft feel. It may be plain or printed. It washes well. Used for nightwear, sheets, pillow cases, babywear.

Winch: (reel, windle, winder). A device, mechanically powered or hand driven, of adjustable circumference used to wind on or off (batching, reeling) rope or yarn. According to DIN/ISO a winch is: a rotating guide or transporting structure, the circumference of which is composed of individual carrying elements. The winch may be an open or closed structure.

Winch beck: **Winch with difference.** The principle is as follows: The beck, or vat, has a narrow boiling chamber at the side which can also serve for the addition of dye and the like, and which is separated from the remaining liquor and the goods by means of a perforated dividing wall.

Winch, Wince: (reel, windle, winder). A device, mechanically powered or hand driven, of adjustable circumference used to wind on or off (batching, reeling) rope or yarn. According to a winch is: a rotating guide or transporting structure, the circumference of which is composed of individual carrying elements. The winch may be an open or closed structure.

Wind ratio: The number of wraps that an end or ends make from one side of the wound package back to the same side.

Wind Resistant: The ability of a fabric to act against or oppose the penetration of wind or air, but it is not totally windproof.

Wind way: The number of wraps or turns that an end or ends make from one side of the wound package to the directly opposite side.

Wind, multiple: See **Multiple wind.** The winding of two or more ends to a tube or core in parallel form so that the ends from a compact flat ribbon.

Winding: Winding is the process of transferring yarn or thread from one type of package to another to facilitate subsequent processing. The rehandling of yarn is an integral part of the fibre and textile industries. Not only must the package and the yarn itself be suitable for processing on the next machine in the production process, but also other factors such as packing cases, pressure due to winding tension, etc., must be considered. Basically, there are two types of winding machines: precision winders and drum winders. Precision widers, used primarily for filament yarn, have a traverse driven by a cam that is synchronized with the spindle and produce packages with a diamond-patterned wind. Drum winders are used principally for spun yarns; the package is driven by frictional contact between the surface of the package and the drum.

Winding a warp: Winding a wrap is the process by which all the threads in the warp are measured and aligned in order. Usually this is done by placing each end in a premeasured path on a warping board or warping reel. The order of the ends is maintained by the threading cross.

Winding fallers: Any wool as shorn from the living sheep. The term is in use to distinguish this wool from other forms such as skin wool.

Winding system in open end spinning: A device which forms the yarn package.

Window paning: A fabric defect caused by non-uniform yarn. When thin sections of yarn become grouped together, the resultant increase in the transparency of the fabric is called window paning.

Windows, in Zipper: The opening in the pin lock and cam lock sliders through which the locking pin and cams, respectively, may extend.

Windproof: The ability of a fabric to be non-permeable to wind and air.

Windsor: Plain woven English fabric, made with different coloured tufts, which are introduced into the fabric in the loom with the aid of little hooked needles.

Windsor Brilliant: A washable English cotton dress goods, finished with a high lustre.

Windsor Duck: Printed duck, made In England for summer dresses.

Windsor Louisine Pine: English washable printed cotton dress goods.

Wipe, in upholstery cleaning: To clean the coated upholstery fabric with a spronge or cloth wetted with mild detergent, soap solution, or coated fabric cleaner formulated for the purpose.

WIRA: An English textile research institute

Wire Ground: The separately made raised silk net ground in hand-made Brussels lace.

Wire sleeve: A flexible dyeing sleeve made from stainless steel wire used as a package during the treatment of yarns which may shrink significantly.

Wires per unit length, In piled yarn floor covering: The number of binding sites per unit of the floor covering length. Wires in the widthwise direction being the usual means of forming the pile.

Wiry end: See **Tight twist end**.

Wiry Wool: Wool that is in-elastic and has poor spinning capacity. It is usually straight and may be the result of poor breeding.

Witch Stitch: Same as Herringbone stitch.

Witney Blanket: English woollen blankets, made with dense nap.

Witney Serge: English woollen serge, napped on both sides.

WK: Abbreviation for **Camel hair.**

WL: Abbreviation for **Llama hair.**

WM: Abbreviation for **Mohair.**

WN: Abbreviation for **Rabbit hair.**

WO: Abbreviation for **Wool.**

Wo: Abbreviation for **Wool.**

W/O: Abbreviation for oleaginous. See **Emulsions**, i.e., water in oil system.

Woad: A blue Natural dyes (a vat dye) obtained from the leaves of the golden flowering woad plant (weld, dyer's weed) and squeezed into the form of a ball. Its working ingredient is Indigo. Woad was superseded by indigo, which has 30 times the concentration of dye, at the beginning of the nineteenth century.

Wonder tape: Wonder tape is a double faced transparent tape that eliminates pinning and thread basting. It is especially helpful for positioning your quilt binding to the back, zippers, pockets and trims. Wonder Tape disappears after washing and won't gum up your needle when you sew.

Wood dyes: These are dyes used in the dyeing of various types of wood and are usually cationic dyes, acid dyes and metal complex dyes.

Wood grain: A fabric defect that consists of filling wise streaks resembling the irregular appearance of wood grain in lumber. Wood grain is usually caused by strained filling in quilling, the tension being more pronounced near the butt of the quill.

Wood pulp: The cellulosic raw material for viscose rayon and for acetate.

Wood pulp yarns: Yarns produced from wood pulp.

Woody fibre: These are the fibres in deciduous trees which lend them their strength and also serve to transport air and water. They are 0.5–1.5 mm in length, are narrow and have thick walls. Their ends are pointed. An analogy is found in components of the central lamella in →: Bast fibres; Hard fibres.

Woof: Comes from the Anglo-Saxon "owef". It is another name for the warp or warp yarn. Sometimes in advertising textiles, the word has been used to imply filling yarn, and made to interchange with the other term, weft.

Wool: The fibrous covering of the sheep, Ovis species. The yarn is spun from fibres which are variable in length and randomly oriented to one another. They are intermingled and produce a bulky yarn with a fuzzy surface. Wool is popular not only for its quality of warmth. Because of its breathing properties, wool allows perspiration dry on the skin, which in turn triggers the body's cooling effect. This also prevents perspiration from being absorbed into the fabric as happens with other fibres that have a rapid absorption rate. Very fine fibres weighing only 120-300 g per metre are now used for poplin, gabardine, panama.

Wool top: A continuous untwisted strand of wool fibres from which the shorter fibres or noils have been removed by combing.

Wool, in generic sense: The fibre from the fleece of the sheep or lamb, or Angora or Cahmere goat (and may include the so-called specialty fibres from the hair of the Camel, Alpaca, Ilama, and Vicuna).

Wool base: Oven dried scoured wool free from alcohol extractable matter, mineral matter, vegetable matter, and all impurities.

Wool, carbonized and neutralized: See Carbonised and neurtralised wool. A term descriptive of scoured wool processed to destroy cellulosic impurities by treating with a mineral acid or an acid salt drying and baking, crushing, and dusting out the embrittled cellulosic matter followed by neutralization of the acidified wool.

Wool, carded: See **Carded wool.**

Wool Classer: The person who sorts the "Wool Clip" into the appropriate "Grades".

Wool Clip: The total yield of wool short during one season from the sheep of a particular region.

Wool, combing: Wool that strong and strictly of combing length that is 2" (50 mm.)

Wool Combs: A variety of combs used to produce fibres for worsted spinning. The single-pitch and 2-pitch combs are what I call the peasant combs and are great for producing semi-worsted yarns. English combs are the multi-pitch combs (4- to 5-pitch) used to prepare a true worsted prep.

Wool content: The quantity of new and recycled wool, as defined in the Wool Products Labeling Act, which is determined by chemical analysis.

Wool crepe: Soft, drapy fabric for dresses, usually lightweight but heavier types may be used for ladies suits. A slack weave fabric with typical crepe surface. Sometimes the wrong side appears smooth by comparison.

Wool felt: A felt composed wholly of any one or a combination of new or recycled wool fibres.

Wool, Grease: See **Grease wool.** Wool taken from the living sheep and which has not been commercially scoured.

Wool Mark: The trade mark indicates goods made of pure new wool that meet the quality standards of the International Wool Secretariate. These quality assurance include fastness to light and liquid, abrasion resistance and tensile strength.

Wool, New: See **Virgin wool; Wool, virgin.**

Wool. Pulled: Wool taken from the pelt of a slaughtered sheep and which has not been commercially scoured; also called slipe wool; skin wool.

Wool, Raw: See **Raw wool.** Wool or hair of the sheep, in the grease, pulled or scoured state.

Wool, Recycled: See **Recycled wool.**

Wool, Reprocessed: See **Reprocessed wool.**The resulting fibre when wool has been woven or felted into a wool product which without ever having been utrilized in any way by the ultimate consumer, subsequently have beenmade into fibre state.

Wool, Reused: See **Reused wool.** The resulting fibre when wool or reprocessed wool has been spun, woven, knitted or felted into a wool product which after having been used in any way by the ultimate consumer subsequently has been made into fibrous state.

Wool, scoured: Wool from which the bulk of impurities has been removed by an aqueous or solvent washing process.

Wool, tippy: Wool in which the individual staples has a spear shaped or pyramid shaped tip.

Wool, virgin: The term virgin or new as descriptive of a wool product or any fibre or part thereof shall not be used when the product or part so described is not composed wholly or otherwise manufactured as used product.

Wool Mousseline: It is made from worsted yarn into a plain or printed lightweight woo. Excellent for dresses. See also **Mousseline.**

Wool nonfelting: The nonfelting wool is produced by chlorinating the normal wool.The anti-felting effects achieved by the traditional wool chlorinating process using chlorotriazine-type products have so far not been achieved by any alternative method or product.

Wool or silk poplin: Poplin-type wool or silk material.

Wool Roller: The person in a shearing shed who skirts the fleece, then rolls it. The fleece is then classed.

Wool run: An indirect yarn numbering system, equal to the number of 1600 yd lengths per pound.

Wool stretch cord: Fabric with the appearance of ribbed tweed, but with elastane yarn woven into it. The fabric is hard wearing, comfortable, warm, crease resistant and washable. Used for men's ready-made trousers.

Wool Union: Term used for mixtures of wool and cotton and mixtures of animal hairs and natural or regenerated cellulose fibres in general (wool, mohair and other animal hairs mixed with cotton, viscose and cupro fibres). The mixing ratios of the two types of fibre can vary considerably. The term wool union is used almost exclusively in dyeing technology.

Wool-spun: Of, or pertaining to, material produced by the woollen system of yarn spinning as distinct from materials made by the worsted system of spinning.

Wool/polyester flannel: A thick warm fabric made from wool with polyester and other fibres. Used for coats, jackets, capes, children's clothes.

Woollen: (1) I or pertaining to materials produced by woollen system of yarn spinning as distinct from amterials made by the worsted system of spinning. (2) 'Woollen' applies to wool, reprocessed wool or reused wool.

Woollen blended: Description of yarns spun on the condenser system and having wool as a main component or descriptive of fabrics, garments or yarns made from such yarns.

Woollen card: A type of roller card used in the woollen spinning system, usually consisting of three cards in tandem: the breaker, intermediate, and finisher sections.

Woollen cloth: Fabric made of wool.

Woollen count: The two systems used to determine woollen yarn counts in the U.S. are the run system and the cut system. The run system has a standard of 1600 yards per hand, while the cut system is based on 300 yards per hank.

Woollen Cut: A woollen yarn measure. A 1-cut woollen yarn has 300 yards in one pound of yarn.

Woollen run: See **Wool run.**

Woollen spun: Of or pertaining to, materials produced by the woollen system of yarn spinning as distinct from materials made by the worsted system of spinning.

Woollen system: A spinning system employing a minimum of drafting and producing yarns of low bulk density. Roving is produced by rub aprons in the condenser section at the front of a roller top card. There is roller drafting, or other intermediate process between carding and spinning.

Woollen yarns: Yarns spun from wool fibres which have been carded but not combed or gilled.

Woolmark : International wool mark introduced in 1964 by the International Wool Secretariat (IWS) and lodged in more than 100 countries all over the world for legal protection. Guarantees pure Virgin wool and other quality characteristics. Only valid if the label is firmly sewn in and provided with a code number.

Work: Physics: product of force and distance: SI unit = joule.

Work recovery: The percent of recoverable to the total work required to strain a fibre a specified amount under specified conditions.

Work Recovery curve: See **Tensile hysteresis curve.**

Work-to-break: The total energy expended to tear apart a material.

Work to rupture: The total energy required to rupture the material.

Working elongation of rope: Elongation which is immediately recoverable when tension is removed from the rope.

Working loss: The irrecoverable loss of weight or yardage of a textile material that occurs during a textile process.

Work-to-break: The total energy required to break a measured material.

Workwear: Clothing worn for work, the wearer himself being responsible for its general condition, as opposed to Industrial protective clothing. The rules are different in Germany for personnel working in hospitals and shopping centres, etc. (representation workwear).

In this case, for reasons of safety at the workplace (particularly for personnel working near moving or rotating parts) the following rules apply: workwear must always be in perfect condition including footwear (non-slip soles).

Worn denim look: Fashion trend for casual wear, particularly for Jeans made from denim fabric such as blue warp (grey weft). Gives the impression of previously worn garments which must not only be produced by the "washout" effect (indigo-colour with the typical appearance of frequent washing = washed jeans) but also may be produced by use in wearing ("controlled" abrasion effect). This can also be achieved by gently treating with enzymes.

Worsted: Worsted yarn is spun from combed wool. Combing is a process where the shorter fibres of wool are removed and the longer ones left lying parallel. The result is smooth, even yarn. Manufacture is usually divided amongst several mills to do the combing, spinning, weaving, dyeing and finishing. Worsted yarn blends very well with other fibres.

Worsted blended: Descriptive of yarns in which the fibres are reasonably parallel and having combed wool as the main component or description of fabrics or garments made from such yarns.

Worsted card: A type of roller card user for worsted-system processing. It usually comprises two cards in tandem; the unit has a roller that carries the stock from the first card to the feed-in of the second card.

Worsted count: An indirect yarn numbering system in the worsted system equal to the number of 560 yd lengths per pound.

Worsted fabric: Fabrics made from (single or twist) worsted spun yarns in the warp and the weft. In most cases, relatively fine thread, light to medium weight, smooth and napless cloth qualities. Anything from pure virgin wool, viscose or man-made fibres, including mixtures, can be used. The fabrics can be wool, yarn or piece-dyed.

Worsted spinning: High quality threads prepared from staple fibre. The origin of the textile fibre can be natural or synthetic. In the main, wools and wool mixtures are spun with polyester, polyacrylonitrile or other man-made fibres. The fibres are single or twisted during weaving or knitting to produce the fabrics. The cloth and knitwear is most widely used in women's and men's outer clothing.

Worsted spun: Of or pertaining to, materials produced by the worsted system of yarn spinning as distinct from materials made by woollen system of spinning.

Worsted system: A system of textile processing for manufacturing spun yarns from staple fibres usually over 3 inches in length. The main operations are carding, combing, drafting, and spinning. There are three basic systems of worsted yarn spinning: the Bradford (or English system), the French (Alsatian or Continental system), and the American system.

Worsted yarn: Yarn spun from wool fibres which have been carded, and neither combed or gilled, or both.

Worsted-spun: Of, or pertaining to, materials produced by the worsted system of yarn spinning as distinct from materials made by the woollen system of spinning.

Wound dressings: The application of textiles to wound healing involves products for covering minor wounds, burns, ulcers and other deep skin wounds. Antimicrobial wound dressings manufactured by means of a bi-layer of silver-coated, high-density polyethylene mesh on a rayon adsorptive polyester core delivers nanocrystalline silver from a non-adhering, non-abrasive surface.

Wound (yarn) package: See **Package.**

Woven carpet: Carpets which are woven by machine or on hand looms (Machine-woven carpets).

Woven fabric: A structure produced when at least two sets of strands are interlaced, usually at right angles to each other, according to a predetermined pattern of interlacing, and such that at least one set is parallel to the axis along the lengthwise direction of the fabric.

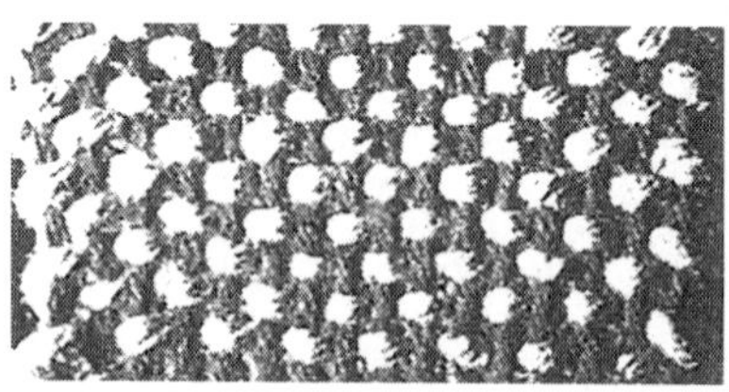

Woven interfacings: These are all grained, even if they are iron-on varieties, so they must be cut to the same grain as the garment section.

Woven-as-Drawn-In: Used often with weaving overshot patterns. This is the order of treadling that will give the finished fabric a squared pattern with a diagonal line running through it.

WP: Alpaca.

Wrap angle, in yarn friction testing: The cumulative angular contact of the test specimen against the friction inducing device, expressed in radians.

Closed faced fabric, I bonded, fused, laminated fabrics: The face or shell fabric of closed construction so that no open face areas appear.

Wrap fibre: The so-called covering fibres of rotor-spun yarns which, among other things, improve the abrasion resistance during practical use. Research has shown that the total number of wrap fibres is smaller for a larger rotor and that coarser and longer fibres form approximately four times as many wrap fibres as finer and shorter fibres.

WRAP/FLA: US based certification programs in Social standards in Textiles and clothing.

Wrap-in, in vinyl coated glass yarn: A method of completing a package after break by wrapping the two ends together on the package without spicing or tying a knot.

Wrapper fibres: Fibres which wrap around the main body of a staple fibre yarn during yarn formation in the production of the open end and fascinated yarns.

Wrinkle: An unwanted crease, generally short and irregular in shape, on the fabric or garments.

Wrinkle: An undesirable crease in a fabric generally short and irregular in shape.

Wrinkle Mark: The marks formed on the garments during wearing.

Wrinkle recovery: That property of a fabric which enables it to resist the formation of wrinkles when subjected to a folding deformation.

Wrinkle resistance: That property of a fabric which enables it to resist the formation of wrinkles when subjected to a folding deformation.

Wrist, in anatomy: The joint which articulates between the end of the lower arm and the hand.

Wrist girth, in body measurements: The circumference over the prominence of the inner and outer forearm bones.

Wrong draft: See **Wrong draw.**

Wrong draw, in woven fabrics: One or more incorrectly drawn warp ends in the harness or reed. See also **Wrong draft, Misdraw.**

Wrong lift: An incorrect interlacing of warp and weft thread caused by a heald shaft or jacquard harness cord being in the incorrect position , as a result of the fault in the pattern chain , (pattern stitching) or faulty mechanical action (machine stitching).

Wrong pick: See **Mispick.**

Wrong side: See **Back side.** It is the side of the fabric that is unfinished and on the inside of the project.

Wronz: Wool Research Organization of New Zealand Inc., New Zealand. Textile research organization. Scope of activities: wool research (See **Wronz wool scouring process**), information service, commercial test projects, annual report and report series.

WRONZ wool scouring process: Process used to achieve slight destabilization of wool-fat emulsions: (a) Small wash trough with 1/3 of the usual bath capacity, vigorous bath circulation and mechanical movement of wool. The system includes the removal of flocks and solid particles, fat recovery and utilization of waste heat. (b) Breaking the emulsion using N-pentanol.

WS : Cashmere fibre.

WU: Guanaco fibre.

WV : Virgin wool.

Wuzzing: A way of removing excess moisture by centrifugal force. Hold onto the end of the skein firmly and spin it around like a helicopter blade.

w/v: Abbreviation for weight/volume Solutions are sometimes specified as being made as some percentage weight/volume. This means that the substance dissolved is measured by weight, and the final solution is measured by volume. For example, 5% sodium chloride w/v would mean that 5 grams of salt would be dissolved in enough water to make a total solution volume of 100 millilitres (or, say, 150g to make 3 litres, etc), and would be labeled "Sodium chloride 5% w/v". Solutions are most commonly made this way, and it can be assumed to be this way unless otherwise specified. Also see **w/v.**

w/w: Abbreviation for weight/weight solutions are sometimes specified as being made as some percentage weight/weight. Both the substance dissolved and the final solution are measured by weight. For example, a 10% solution of urea w/w would be made by dissolving one pound of urea in 9 pounds of water, and would be labeled "Urea 10% w/w". This method is sometimes preferred in industry since automatic mixing equipment is often designed to handle everything by weight rather than volume. Also see **w/v.**

WY: Abbreviation for Yak fibre (hair).

X

X: In the United States wool trade, denotes a full blood or high grade merino wool.

X (in colour chemistry): (a) One of the three CIE tristimulus values, the red primary; (b) Spectral colour-matching functions of the CIE standard observer used for calculating the X tristimulus value; (c) One of the CIE chromaticity coordinates calculated as the fraction ofthe sum of the three tristimulus values attributable to the X value.

Xanthate: A salt of xanthogenic acid. It is produced as a result of the effect of carbon disulphide on sodium alcoholate. See **Cellulose xanthate**.

Xanthating: Introduction of a xanthate group by chemical reaction. See **Xanthate**.

Xanthene dyes: Introduction of an oxygen bridge into the triphenylmethane dye molecule leads to the xanthene dyes. The colour is shifted from blue to red. The restricted rotation of the phenyl groups inhibits non-radiation de-excitation and gives rise to very strong fluorescence. See **Triphenylmethane dyes**.

Xenobiotic: A synthetic chemical foreign to that environment.

Xenon reference fabric: A dyed polyester fabric used for verifying xenon-arc equipment test chamber temperature conditions during a lightfastness test cycle.

Xenon-arc lamp: A type of light source used in fading lamps. It is an electric discharge in an atmosphere on xenon gas at a little below atmospheric pressure, contained in a quartz tube. See **Light fastness testing**; **Xenotest, Light Fastness Tester**.

Xenotest light fastness tester: It is an instrument for light fastness testing. The basis of these instruments, which can vary in size, is 1 or 3 vertically arranged xenon emitters surrounded by variable optic filter systems that allow the spectral distribution to be varied from "sunlight through a window" (light fastness) to "sunlight in the open air" (weather fastness). A high correlation to sunlight can be achieved. This means that the light fastness results achieved correspond closely with those that occur naturally. The advantage is that the high level of radiation means that the test duration is shorter (ratio is approximately 1:9 for the central European climate).

Xerga: Spanish term for serge; twilled woollen blanket in Spain.

Xerogels: Gels that have lost their liquid as a result of vaporization, extraction by suction, etc., and are below their glass temperature state. They are not gels in the true sense (See **Lyogel**). Since they have the spatial structure of a polymer and are filled with air, they can be classified as (hard) foams. The best known xerogel is silica gel.

X-rays: X-rays consist of a bundle of rays (traveling at a speed of 3,00,000 km/s) produced on the same principle as cathode rays (and reflected by the anode opposite the cathode) that can penetrate substances and enable their structure to be detected. They are stopped to a greater or lesser degree by solid substances (e.g., metals), broken down into a clearly defined line spectrum by crystals. The wavelength or oscillation count for each atom is characteristic (in direct ratio to the atomic charge and number).

XX soft: (a) English mill term for very soft and yielding folded thread, the single yarn containing about twice as many twists as the doubled; (b) In the United States, wool trade denotes full blood merino wool.

XXX: In the United States it denotes the wool from a cross between a Saxony and common merino.

Xylodine: Name given to a paper yarn made by twisting and coating with glue and fibre a strip of paper.

Xylolin: (a) Plain woven fabric made of cotton warp and paper yarn filling. Used for cheap working clothes; (b) Paper yarn made in Germany and Austria. It is used for carpets and backings for carpets.

XYZ colour measurement system: See **Colourimetric measures**; **Colourimetry**; **Chromaticity diagram**.

Y

Y (in colour chemistry): (a) One of the three CIE tristimulus values, equal to the luminous reflectance or transmittance, the green primary; (b) Spectral colour matching function of the CIE standard observer used for calculating Y tristimulus value; (c) One of the CIEchromaticity coordinates calculated as the fraction of the sum of the three tristimulus values, attributable to the Y value.

Yachan: Silky seed fibre of the *chorisia insignis*, used for stuffing.

Yacht cloth:Piece dyed fine flannel of a serge weave with a rough finish; used for clothing.

Yak: Hairs that are categorized as fibres of animal origin from the Tibetan yak. These are the long hairs on the belly and tail.

Yama-Mai silk: A species of wild silk yielded by the Antheraea, the Tussah spider in Japan and Chinawhich is extremely similar to real silk in lustre and fineness. It is regular, and is white, yellowish or greenish in colour.

Yard (yd): English unit of length; 1 yd = 0.9144 m; 1 m = 1.0940 yd.

Yard goods: Fabric sold on a retail basis by the running yard.

Yardage: The amount or length of a fabric expressed in yards.

Yarkand: Large and heavy rugs made in Central Asia with cotton web and long, loose, wool pile tied in Senna knot. The Chinese influence is very much marked in the design, consisting of dragons, fret pattern, fish, or other animals, mostly in bright yellows, pinks, reds, blues, greens, etc.

Yarn: A generic term for a continuous strand of textile fibres, filaments, or material in a form suitable for knitting, weaving, or otherwise intertwining to form a textile fabric.Yarn occurs in the following forms:

(a) A number of fibres twisted together;
(b) A number of filaments laid together without twist (zero twist yarn);
(c) A number of filaments laid together with more or less twist;
(d) A single filament with or without twist (a monofilament);
(e) One or more strips made by a lengthwise division of a sheet of material such as a natural or synthetic polymer, a paper or a metal foil used with or without twist in a textile construction.

A textile product of substantial length and relatively small cross-section that consists of fibres or filament(s) (or both) with or without twist:

i. Assemblies of fibres or filaments are usually given other names during the stages that lead to the production of yarn, e.g. tow, slubbing, sliver, roving (q.v.). Except in the case of continuous-filament fibres or tape yarns, any tensile strength possessed by assemblies at these stages would generally be the minimum that would hold them together during processing;
ii. Staple, continuous-filament and monofilament yarns are included;
iii. No distinction is made between single and cabled yarns;
iv. Zero-twist and self-twist staple yarns are included;
v. Zero-twist continuous-filament yarns are included.
vi. By the definition of fibre and filament, paper, metal-film and glass yarns are included.

Yarn (boucle or loop): Boucle or loop yarns are compound yarns made by aspecial folding process which results in wavy or looped projections. Fabrics will have a more or less grainy handle and a textured surface. Examples: Boucle', Frise', Frotte'.

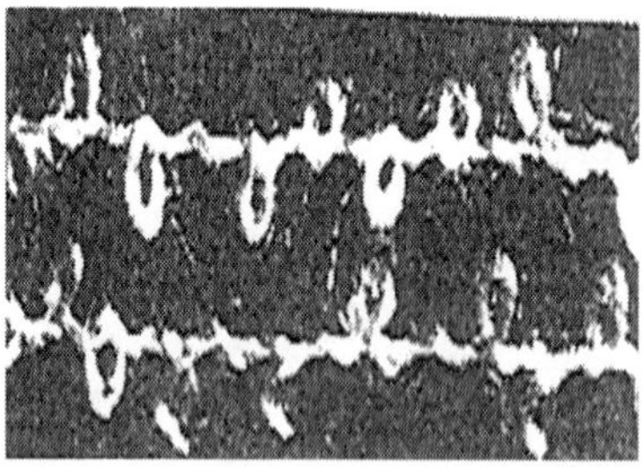

Yarns (bourette or knop): Bourette or knop yarns are folded yarns containing short, often coloured bunches of fibres or yarn at regular or irregular intervals. The knops may be formed during spinning or during folding. Fabrics will have a structured appearance. Fabric example, Donega Tweed.

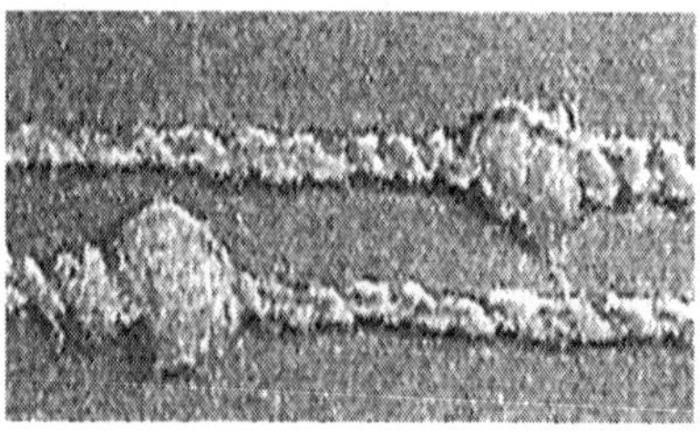

Yarn (chenille): It is a cut pile yarn; it is soft and voluminous. These yarns are made by cutting special fabrics into strips. They are used in furnishing fabrics and knitwear.

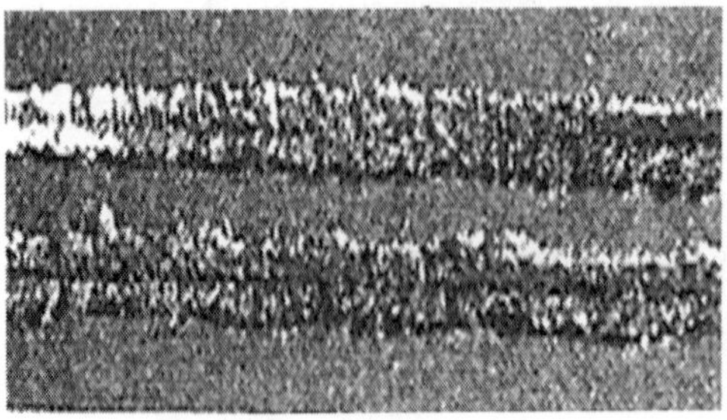

Yarn (crepe'): Crepe yarns are used to make fabrics with a wrinkled surface and a sandy handle. They are made from highly twisted yarns. Fabric examples, crepe' de chine, georgette, crepon, marocain.

Yarn (Jaspe or mouline'): These yarns are made by folding two or more differently coloured yarns made from different fibres with different dyeing behaviour. They give mottled appearance. Fabric example: Fresco.

Yarn (mixture or ingrain): They are made by mixing fibres of different colours during spinning. This results in heather effect. Fabric example: Marengo.

Yarn (mélange or vigourex): Melange or vigourex yarns are spun from combed sliver or top which has been printed with stripes. The appearance is somewhat like mixture. These are also produced by mixing dyed and undyed loose fibres also.

Yarn (Mottle or Marl): These are made by spinning from two-colour roving or from two roving of different colours. The appearance is like mouline, but with less sharp contrast.

Yarn (slub): Slub yarns are single or folded yarns having long thick places, regularly or irregularly disposed. The slub effect is made either in spinning or in folding. Fabric may have the character of linenor wild silk, which is favoured in furnishings.

Yarn appearance: The visual effect obtained by viewing a sample of yarn wound with a designated traverse on a blackboard of designated size. Descriptive terms for yarn appearance are

(a) single yarn —one yarn that contrasts with adjacent yarns;

(b) multiple yarns –two or more yarns that contrast with adjacent yarns;

(c) lean yarn – a yarn having a smaller diameter than the normal population;

(d) bulky yarn –a yarn having a greater diameter than the normal population;

(e) light yarn –a yarn that appears to be lighter than the normal population;

(f) dark yarn –a yarn that appears to be darker than the normal population;

(g) intermittent yarn –a yarn that is non-uniform in colour intensity along its length, appearing as light and dark sequences.

Yarn break(in sewn seams): A mode of failure evidenced by yarns rupturing at the seam or at any other area in the test specimen.

Yarn carriers: This can mean: (a) Perforated inserts on which the yarn is wound; (b) Spindles or perforated tubes onto which the yarn packages are put so that the liquor can flow through them; (c) Moveable material carriers that can be inserted into the dyeing machine, with a hollow base into which the spindles or dyeing tubes can be inserted.

Yarn clearer: A device designed to cut or break when it detects changes in mass or diameter which exceed set limits. Yarn clearers are normally part of, or an addition to, a winding machine.

Yarn composites: Flexible and porous fabrics (bonded fabrics) made from threads or thread layers and bonded by sewing, glueing, plasticizing, fibre bonding, or combinations of these techniques.

Yarn construction: A term used to indicate the number of single yarns and the number of strands combined to form each successive unit of a plied yarn or cord.

Yarn construction number (for asbestoes): A system designed to show the construction namely, the cut of the single yarn, the number of plies and whether reinforcements are present.

Yarn count: Deprecated term, used as specific yarn numbering system. See **Linear density of fibres and yarns**.

Yarn delivery (in knitting machine): Yarn tension presented to a needle hook that has to be uniform, constant and as low as possible, in order to get uniform, good fabric appearance and also to prevent yarn breaks. Therefore, several different types of yarn delivery devices are used. These devices are divided into two main groups, those for constant yarn consumption per unit time on given feeders (positive feeding, yarn metering) and those for variable yarn consumption per unit time on a feeder (negative feeding, yarn storage). The first one is used for simple fabric designs with no large pattern repeat area. The second one is used for jacquard designs that have a large pattern area and most of the feeders have varying yarn consumptions per unit time.

Yarn distortion (in woven fabrics): A condition in which the symmetrical surface appearance of a fabric is altered by the shifting or sliding of warp or filling yarns.

Yarn driers (air stream): This type of dryer is used for drying yarn packages (on perforated inserts) in columns on spindles or perforated tubes as a discontinuous process. If yarn packages with a large quantity of wound yarn are used in the compression column dyeing process, the yarn packages have a high level of flow resistance, and for this reason they are dried using a high frequency dryer after continuous centrifugal drying.

Yarn dyed: Dyed in yarn make-up, e.g. in hank form (dyeing of yarn in hank form), as a yarn package, or spool, and in continuous form.

Yarn dyed fabrics: Fabrics, the yarn of which was dyed before woven.

Yarn dyeing differences: Variations in take-up of dyes by yarns, resulting in streaks in finished fabrics.

Yarn feeding position (in knitting): A new yarn is fed into the needle hook by a yarn feeder, while the needle is moved downwards by following the cam track. Holding down-knocking over sinker begins moving away from the machine centre. The old loop starts to close the latch of needle (castoff) by moving upwards.

Yarn finish: Cotton or linen sewing thread is soaked in starch and wax or paraffin, dried, stretched (lustred) and brushed to give it a smooth surface. See **Glossing**; **Lustring**.

Yarn fleck: A mixture of yarn of spotted and short streaky appearance due to the introduction of a minority of fibres of different colour and/or lustre.

Yarn humidifying agent: These are used for humidifying yarns, to set the required humidity level and maintain it during the course of subsequent textile finishing processes as well as to cause an increase in strength or shape retention. Stabilized wetting agent solutions with added preservatives, bactericides or anti-fungal products are usually used.

Yarn intermediates: A generic term for products obtained during the conversion of fibres to yarns, including card webs, laps, slivers, roving, tops, etc.

Yarn number: The value, actual or nominal, or a yarn in a specific yarn numbering system. A measure of the fineness or size of a yarn expressed either as 'mass per unit length' or 'length per unit mass' depending on the yarn numbering system. Yarn number is an indirect system and are frequently termed yarn counts. For example, No. 20 – cotton count, No. 45 –worsted count. Direct yarn number (equal to linear density) is the mass per unit length of the yarn which is equal to the mass of yarn specimen divided by length of the yarn specimen. Indirect yarn number (equal to the reciprocal of linear density) is the length per unit mass of yarn. Indirect yarn number equals to the length of yarn specimen divided by mass of yarn.

Yarn number (equivalent single): The yarn number based on the observed mass per unit length of plied yarn, a cabled yarn or a yarn whose number has been changed by processing, such as twisting or bulking.

Yarn numbering system: A system for expressing yarn fineness or size as a relationship between length and mass. Two systems are in use –

(a) Direct yarn numbering system: It expresses yarn number in terms of mass per unit length (linear density). Examples of direct yarn numbering system are

(i) *Denier* – Number of grams per 9000 m
(ii) *Grex* – Number of grams per 10000 m (obsolete)
(iii) *Spindle number* – Number of ponds per spindle or 14400 yds. (13167 m) length, expressed as pounds per spindle
(iv) *tex* – The number of grams per 1000 m.

(b) Indirect systems: Expresses yarn number in terms of length per unit mass (reciprocal of linear density). Examples are

(i) *Count, cotton*– The number of 840 yards length per pound; generally used for yarns spun on the cotton system.
(ii) *Count, metric* – The number of metres per 100 gram.
(iii) *Count, worsted* – The number of 560-yard length per pound; Generally used for yarns spun on the worsted system.
(iv) *Cut* – (a) For asbestos, glass: The number of 100 yards lengths per pound; (b) Wool: The number of 300 yards length per pound.

(v) *Hank* – A synonym for the indirect yarn number (count) of slubbing or roving.

(vi) *Lea*(linen) – The number of 300 yds. length per pound.

(vii) *Run* – The number of 1600 yard lengths per pound, generally used for yarn spun on the woollen system.

(viii) The number of 1000 yard lengths per pound (obsolete).

Yarn number (in jute): Mass per unit length of a yarn measured as the number of pounds per 13,167 m (14,400 yd.) and expressed as pound per spindle.

Yarn number(resultant): See **Yarn number (equivalent single)**.

Yarn package: A length or parallel lengths of yarn suitable for handling, storing, or shipping.

Yarn quality:Various grades of yarn designated by the producer with respect to performance characteristics, e.g., first quality, second quality, etc.

Yarn severance: A numerical value expressed on a percentage basis from a test that is used as an index of the degree of cutting of fabric yarns by the sewing machine needle in making sewn seams.

Yarn slippage (in sewn seams in sewn fabrics): The displacement of one or more yarns from the original position, causing differences in alignment, spacing or both.

Yarn take-up (in fabric): The additional length of yarn used to make a given length of fabric.

Yarn (asbestos): A twisted assemblage of (a) asbestos fibre, (b) asbestos and other fibres, (c) asbestos and other fibres with an insert of cotton or other yarn reinforcement.

Yarn, backing (in pile fabrics): The base yarn that holds the pile in place, formed by the stuffer yarn and the warp and weft of the fabric. See **Backing yarn**.

Yarn (blended): A single yarn spun from a blend mixture of different fibre species. See **Blended yarn**.

Yarn (bulk): See **Bulk yarn**.

Yarn (bulked): Same as ***bulk yarn, Yarn, bulk.*** They are texturized yarn which is voluminous with crimps and loops having normal extensibility and elasticity. The term bulked yarn is often used generally to cover all continuous filament textured yarns and bulked staple fibre yarns.

Yarn (cabled): A yarn formed by twisting together two or more plied yarns. The individual plied yarns have the same nominal length and tension.

(a) Combinations of folded yarns and single yarns may be described as cabled yarns, e.g., a single yarn is twisted together with two-folded yarns to give softness to the resulting yarn.

(b) In the tyre-yarn and cord sections of the textile industry, cabled yarns are termed "cable cords" or "cords". These terms include two-fold continuous-filament man-made fibre yarns, a traditional example being 1,830 dtex rayon cord, single twist 480 t/m (Z), and cabled twist 480 t/m (S).

(c) For terms concerning twist designation in cabled yarns, See twist (b). See **Cabled yarn**.

Yarn (carded): Yarns made from fibres that have been carded but not combed in the manufacturing process. See **Carded yarn**.

Yarn (chenille): A type of fancy yarn consisting of a cut pile which may be of one or more of a variety of fibres, helically disposed around axial threads which secure it. Chenille yarns are usually used in the manufacture of furnishing fabrics and other broad and narrow fabrics.

Yarn (cloud): A type of fancy yarn made by using two threads of different colours in such a manner that each thread alternately forms the base and cover to 'cloud' the opposing thread. It is made by alternate fast and sloe deliveries from two pairs of rollers.

Yarn, cockled: See **Cockled yarn**.

Yarn, combed: Yarn made from fibres that have been carded and combed in the manufacturing processed. See **Combed yarn**.

Yarn, combination: A ply yarn twisted from single yarns of different fibres, for e.g. silk and rayon, rayon and acetate, etc. See **Combination yarn**.

Yarn, craft: See **Craft yarn**.

Yarn, doubled: See **Yarn folded**.

Yarn, eccentric: A fancy yarn; an undulating gimp yarn.

Yarn, elastomeric: A non-textured yarn which can be stretched repeatedly at room temperature to at least twice its original length and which after removal of the tensile force will immediately and forcibly return to approximately its original length.See **Elastomeric yarn**.

Yarn, fancy: A compound yarn comprising of a twisted core with an effect yarn wrapped around it so as to produce a wavy projections on its surface. In single yarn, the irregularities may be due to the inclusion of knots, loops, curls, slubs, etc. In plied yarns, the irregularities may be due to variable delivery of one of its components or to twisting together dissimilar single yarns. See **Fancy yarn**, **Boucle' yarn**.

Yarn, filament: A yarn composed of filaments assembled with or without twist. See **Filament yarn**.

Yarn, filler: See **Stiffer yarn**.

Yarn, filling: A yarn intended for the use in the weft (filling). See **Filling yarn**.

Yarn, flat: (a) A multifilament yarn with no twist. The term is still used in respect of these yarn after a small amount of twist has been introduced by subsequent processing, e.g. Add-in over-end winding; (b) A synonym for straw yarn.

Yarn, gimp: A compound fancy yarn comprising a twisted core with an effect yarn wrapped around it so as to produce wavy projections on its surface.

Yarn, high bulk: See **Bulk yarn**.

Yarn, high bulked: Man made staple fibres can also be made into bulky, voluminous yarns. Such high-bulk yarns are usually made by blending acrylic fibres of high and low potential shrinkage. During a subsequent heat treatment, the high-shrinkage fibres contract causing the other fibres to buckle. A similar effect can be obtained by using bicomponent fibres. See **Yarn high bulk**; **Bulk yarn**.

Yarn, industrial: A yarn composed of continuous filament usually of high breakage tenacity, produced without or with twist and indented for application in which functional properties are of primary importance. For e.g., in reinforcing material in elastomeric products (tyres, hose, belting) to protective covering, etc. See **Industrial yarn**.

Yarn, knicker bocker (or knop yarn or nepp yarn or knicker yarn): A fancy yarn. A yarn made on the woollen systemand showing strongly contrasting spots on its surface that are made either by dropping in small balls of at the latter part of the carding process or by incorporating them in the blend and so getting the carding machine that these small lumps are not carded out.

Yarn, kraft: A yarn made by twisting a strip of paper manufactured from the wood pulp by the sulphate process.

Yarn, loop: A compound yarn comprising of twisted core with an effect yarn wrapped around it so as to produce wavy projection on its surface.

Yarn, mercerized: A cotton yarn, which has been treated with a solution of caustic soda under condition of caustic concentration and temperature which effect a permanent or irreversible swelling effect on cellulose. Commercially caustic treatment is supplemented by the application of tension.See **Mercerised yarn**.

Yarn, mock chenille: A double corkscrew yarn.

Yarn, non-torque: See **Bulk yarn**.

Yarn, plied: A yarn formed by twisting together two or more single yarns in one operation.

Yarn, self-blended: A single yarn spun from a blend or mixture of the same fibre species.

Yarn, shotting: A single filling yarn used in weaving gunny sacking.

Yarn, silk: (a) Grenadine – A yarn composed of more than one strand with the twist in the plied yarn in the opposite direction from the twist in the individual components, each twist having a minimum of approximately the following number of turns per inch depending on the number of strands used

Strands	Minimum no. of turns
2 (13 to 15 Denier or equivalent)	32
3	30
4	28

(b) Organzine – Two or raw silk strains twisted in the singles, doubled and twisted in the opposite direction in the ply. Turns per inch in the singles and ply usually within the limits of 10–20 turns generally used as warp; (c) Tram – Raw silk yarns doubled and twisted with a low number of turns per inch and generally used as weft.

Yarn, single: The simplest strand of textile material suitable for operations such as weaving knitting, etc. See **Single yarn**.

Yarn, slub: A yarn in which slubs may be deliberately created to produce a desired effect.

Yarn, snarl: A compound yarn in which snarls or kinks are projecting from the core.

Yarn, spiral: A plied yarn displaying a characteristic smooth spiralling of one component around the other.

Yarn, spun: See **Spun yarn**.

Yarn, stabilized: It is a textured yarn which have been given additional setting treatment to reduce their elasticity and crimp extension.

Yarn, staple glass: See **Staple glass yarn**.

Yarn, stretch: Yarns produced by texturing filament yarns which are highly elastic with crimp extension up to 150–300%. See **Stretch yarn**.

Yarn, stripe: A yarn that contains elongated knops.

Yarn, stuffer: See **Stuffer yarn**.

Yarn, textured: See **Textured yarn**.

Yarn, torque: See **Bulk yarn**.

Yarn, union: A yarn made by twisting together yarns of different fibres.

Yarn, warp: A yarn intended for use or used in the warp of a woven fabric.

Yarn, warp knitting: A yarn intended for use or used in the making of a warp knit fabric.

Yarn, woollen: A yarn spun from wool fibres which has been carded but not combed or gilled. See **Woollen yarn**.

Yarn, worsted: A yarn spun from wool fibres which have been carded and either gilled or combed or both. See **Worsted yarn**.

Yarn, zephyr: A variety of soft worsted yarn characterised by a low twist and spun from wool which is fine or finer in average diameter than United States standard 64s grade tops. See **Zephyr yarn**.

Yarn, zero twist: (a) A continuous filament yarn in which there is no twist; (b) A multi-folded yarn in which there is no folding twist.

Yarn, doubled (folded yarn, plied yarn, doubled yarn): A yarn in which two or more single yarns are twisted together in one operation, e.g. two-fold yarn, three-fold yarn, etc.

Note: In few sections of the textile industry, e.g. the marketing of hand-knitting yarns, these yarns would be referred to as “two-ply”, “three-ply”, etc.

Yarn finishing: Cotton or linen sewing thread is soaked in starch and wax or paraffin, dried, stretched (lustred) and brushed to give it a smooth surface. See **Glossing**; **Lustring**.

Yarn mercerizing: Yarns are mercerized continuously in hank form or as a continuous run of yarn. For yarns in hank form, the following method is still in use in a few cases – yarn is soaked in caustic soda solution 25–50 °Bé (at 10–20 °C), it is squeezed off and tensioned to the original length. Washing takes place under tension.

Yarn number: A measure if the fineness or size of a yarn expressed as mass per unit length or length per unit mass, depending on the yarn numbering system.

Yarn number, jute: Mass per unit length of a yarn measured as the number of pounds per 13 167m (14, 400 yd), and expressed as pounds per spindle.

Yarn numbering system: A system expressing yarn size as a relationship between length and mass. There are two types yarn numbering systems – (a) Direct numbering system in which number of mass per unit length is expressed as the number; (b) Indirect numbering system where number of length units per unit of mass are expressed as the number.

Yarn package: (1) Piece make-up – Piece goods wound onto a yarn package carrier (batching roller); (2) Yarn make-up – Package where the support (spindle or insert) is removed when the process is finished, a yarn spool or winding structure without an insert. Other winding forms for commercial purposes are clews or skeins (for crochet, knitting and embroidery yarns), reels (for sewing silk) and other winding structure for wide-wound yarns.

Yarn package (in dyeing): When dyeing yarn on packages, the type and quality of winding, but also the package and insert system, have a particular influence on the quality of dyeing. Conical yarn inserts continue to be used widely even though they have few disadvantages in comparison with modern cylindrical insert systems – uneven package flow, insert/spacer leaking, low batch weight, unfavourable liquor ratio, high personnel costs for loading and unloading the carrier, unsuitable environment for using robots for loading and unloading.

Yarn package (preparation of): Automatic yarn package winding takes on an important function in the textile production process. A package must be properly wound in order to be suitable for productivity in subsequent stages, whether for twisting, warping, weaving or knitting. "Quality packages" reduce the frequency of out-time and avoid errors.

Yarn package (winding machine): A machine used for winding yarn on to the dyeing tubes.The essential purpose of a yarn package winding machine is to ensure that the yarn package is calibrated from the point of view of density, shape and weight, as well as to the machines and processes for the subsequent finishing phases, so that production is guaranteed. Perforated dye tubes are used as a yarn carrier, usually made of plastic, either disposable or re-usable, and either conical or cylindrical in shape (on axially flexible inserts). There are two different winding methods, random winding and precision winding; it can be done by precision winding or random winding, former being most widely used.

Yarn printing: Yarn printing is done to cause certain fuzzy coloured patterns after weaving, knitting or tufting. For this technique, yarns are printed either in hank form (single shade/pearl printing, or multi-colour/ombré printing), or in warp form (warp printing) or in knitted form (space dyeing).

Yarn printing machine: Parallel winched hanks are printed under tension with relief print rollers (Hank yarn printing machine).

Yarn raising: A yarn raising machine is suitable for raising yarns for hand and machine knitting in the range Nm 20 and coarser. The raising process is carried out by raising rollers on individual threads.

Yarn singeing: In order to remove the hairiness of the yarn and improve the shining of the same, yarn singeing is done. The yarn is passed through a gas flame at high speed where the protruding fibres are burned off.

Yarn singeing machine: See **Gassing machines**.

Yarn slippage(in sewn fabrics): The displacement of one or more yarns from the original position(s) so as to cause differences in alignment, spacing, or both in one or more fabric yarns.

Yarn steamer: See **Batch steamers**.

Yarn steaming: In a free state, the yarn has the tendency to untwist or to form loops and snarls. Cross-wound yarn packages which are processed as weft yarn therefore need to be fixed well. Vapour penetration of wound yarn packages, type of steam, temperature accuracy and reproducibility of programmes have all become more significant in yarn steaming.Yarn fixation is generally carried out in pressure steam chambers with vacuum and excess pressure ranges. The steam is supplied either directly from thesteam network or indirectly via an integrated water bath.

Yarn Tensioner: In winding staple yarns, sufficient yarn tension is used to cause breaks in thin places while the yarn is being wound. This permits the thin places to be cut out, the yarn to be repaired and winding to continue. Variations in running tension alter the level at which the thin places are removed and so affect the yarn regularity of the final product. Here the yarn tensioner is used to take care of the yarn winding regularity. There are various forms of tensioner, the simplest of which works by merely deflecting yarn around fixed posts. This is called capstan.

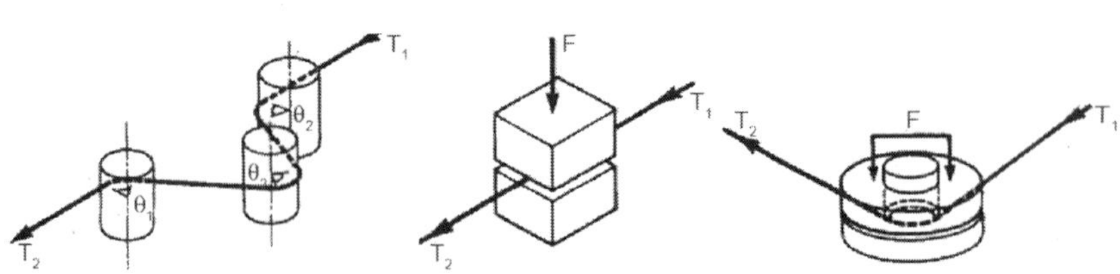

Capstan Tensioner Additive Tensioner Additive Tensioner

Tensioner. Another tensioner is uses a deadweight or spring to give a fixed increment of tension. This type is Tensioner is called additive tensioner. There is also a system in which both these principles are used and it is called a combined tensioner.

Yarn twist: See **Twist of spun yarns**.

Yarn-to-cord conversion efficiency: In tire cord, this is a measurement relating tensile strength of untwisted yarn to tensile strength of cord. Increasing cord twist or increasing yarn diameter lowers conversion efficiency.

Yaws: Flaws in cloth, consisting of thin places.

Yaxci: Very soft, flexible and lustrous sisal hemp from Yucatan.

Yearling: (1)Wool obtained at the second year's shearing.

(2) A sheep or goat that is 12–18 months of age.

Yellow (in colour theory): One of the process ink colours for printing. Pure yellow is the "blueless" colour; it absorbs all wavelengths of blue from light and reflects all red and green wavelengths.

Yellow weed: A mordant dye. See **Natural dyes**.

Yellowing: The propensity of heavily dressed bleached goods to turn yellow as the result of disintegration of the bleaching chemicals.

Yellowing of DP finishes: Yellowing of DP finished fabrics can be caused by a number of conditions. For example, excessive curing temperatures and/or excessive catalyst will scorch cellulosic fabrics. Few reactants have colour bodies which cause yellowing. Usually nitrogenous compounds discolour with heat. Buffers are often added to commercial finishes to combat yellowing problems.

Yellowing of wool: When wool is exposed to radiation from the sun (or similar light sources in the spectrum), two apparently opposite processes can be observed – wavelengths of between approximately 280–400 nm result in yellowing and fibre degradation, whereby the short-wave, higher-energy portion of 280–330 nm (blocked by window glass) has a significantly more damaging effect than the range of approximately 330–400 nm.

Yellowing on storage: In the wide majority of cases of white goods yellowing on storage, three yellow components were found. These yellow components occurred individually, but also in varying blending ratios. These yellow components have been identified as one nitrophenol derivative and two quinone compounds. This yellow body forms easily due to the effect of nitrogen oxides on the sterically hindered phenol 2,6 di-tert. butyl-4-nitrophenol. The formation of compound I is also possible if there is a substituent in opposition, however.

Yield: (1) Number of linear or square yards of fabric per pound of fibre or yarn; (2) The number of finished square yards per pound of greige fabric.

Yield: (1) Wool designated commercial composition obtained by processing a lot of raw wool, expressed as a percentage of the raw wool.

(2) The amount of fabric delivered off a knitting machine in terms of it's weight per unit length or area, or the number of linear units delivered per unit weight. For example, ounces per yard^2 or ounces per square yard or yards per pound.

Yield (of wool): The percentage of a designated commercial composition obtained by processing a lot of raw wool.

Yield strength: The stress at which a material exhibits a specified limiting deviation from the proportionality of stress to strain.

Yield (in raw wool): The combined weight of clean wool fibre present and vegetable matter present as a percentage of the raw wool weight.

Yield, finished (in knitted fabrics): The number of finished square yards per pound (square metres per kilogram) of finished fabric. See **Finished yield(in knitted fabrics); Yield(in knitted fabrics)**.

Yield, greige (in knitted fabrics): The number of finished square yards per pound (sq. m. per kg) of greige fabric. See Greige yield (in knitted fabrics).

Yield point: The point on the load-elongation, or stress strain curve, at which load, or stress, ceases to be linearly proportional to the elongation, or strain; that is, the point beyond which the work is not completely recoverable and permanent deformation takes place.

Yield point (in stress-strain curve): The point beyond which work is not completely recoverable and permanent deformation takes place.

Yerges: Thick, coarse, felted woollen fabric, used for horse blankets.

Yerli: Fine staple wool yielded by the sheep near Smyrna, called kivirdjik, having small horns and a narrow tail. Used for carpets, cloth, blankets, etc.

Yesteklik: Turkish name for small mats, having a soft pile and used for pillows; also called Anatolian mats.

Yolk: The natural grease and suit covering on the wool fibres of the non-scoured fleece, and excreted from glands in the sheep's skin. Usually the finer the wool, the more abundant the yolk. Yolk serves to prevent entanglement of the wool fibres and damage during growth of the fleece.

Yoke (self): The entire back of a garment is one piece and has a single yoke superimposed on the outside.

Yoke (two-pierce): Two identical pieces of fabric are joined to a shortened back piece to produce the total back.

Yokeless shirt: The front and backs of a shirt are joined without a yoke facing.

Yoke (variations of): Three variations of yokes are as follows –

(a) Shoulder yokes

(b) Waist yokes

(c) Hip yokes

Yoke (square): The bodice is split into two, straight across the back or front of a garment using a variety of proportions. It allows contrasting fabrics and other methods of creative construction to be used.

Yoke (cape or storm flap): The top part of this construction echoes the shape underneath, but flare is added to create the cape effect. This part is then laid over the original shape. The storm flap has less flare and simply provides an extra layer of protection during poor weather – found on rainwear and protective garments.

Yoke (tucked): As the square yoke, but a pleat or tuck has been added horizontally and top stitched down.

Yoke with pocket: This pocket is like the epaulette pocket.It is set into the yoke seam.

Yokohama: Crepe that is very fine, close woven crepe, for coats, etc., not transparent.

Yomud: Rugs made of wool and goat hair in Turkestan with medium long and close pile. The design consists of diamonds with angular hooks around the edge in the centre and trailing, angular vine or stripes in the borders. The colours are blues, warm reds, greens, etc.

Youghal lace: The best among the Irish point lace. See **Irish point**.

Young's modulus (in the stress strain curve for an elastic material): The ratio of change in stress to change the strainwithin the elastic regionof the material.

Younge's modulus: This is used to characterise the elasticity of fibres. Gradient of the tangents on the stress-strain curve through its origin (approximately elasticity modulus).

Ypres: (1) Fine medieval worsted from flanders; (2) Bobbin lace similar to the Valenciennes (see), having a square mesh ground, the threads being twisted four times around each other.

Yucca fibres: Similar to Sisal, a leaf bast fibre which is obtained from the 25–60 cm long leaves of the yucca plant (*yucca filamentosa* L.) e.g. from the native family Liliaceae of Southern North America. They are harvested twice a year. Fibres are developed either chemically (lime preparation) or by boiling under pressure, which releases lignin from the yucca fibre to a large extent. Short elementary fibres (approximately 1.3–1.5 mm), breaking length 40–67 cN/tex.

Uses: For twine, agricultural twine, also for carpet fabrics, tarpaulins, etc.

Yuenchingwacheinyong: Chinese figured velvet.

Yuruk: Rugs; small and very durable rugs made by the nomad Yuruks in Asia Minor of goat's hair or dark wool. The long wool pile is tied in Ghiordes knot. The design is composed of very large but simple geometrical devices of brilliant colours over a dark brown field. They are finished with a selvage all around and the ends have a short fringe or braid.

Yutun: Wool camlet made in China in all widths, used by natives for winter clothing. The warp is of silk and the filling of wool.

Yuzen printing (Ju-Sen printing): Artisanal printing process practiced in Japan. A forerunner to modern screen printing. Named after its inventor, the artist Yuzensai Miyasaki (1654–1736). Papery sheets are produced from the bast of the mulberry tree by pasting together several layers, these are stiffened and then made watertight with oil. The screens are manufactured from these by cutting out the motives. Print paste is applied in various ways e.g., hair brush, besom, doctor blade and spray nozzles on paper funnels. This method is used to produce high-quality prints on kimono materials. Depending on the process technique used, direct print or resist effects are obtained.

Z

Z (in colour chemistry): (a) One of the three CIE tristimulus values; the blue primary; (b) Spectral colour-matching function of the CIE standard observer used for calculating the Z tristimulus value; (c) One of the CIE chromaticity coordinates calculated as the fraction ofthe sum of the three tristimulus values attributable to the Z primary.

Z Twist: See **Twist**.

Zacate: Mexican name for various species of fibrous grasses; used for making mats and ropes.

Zackel: Coarse and long carpet wool, yielded by the Hungarian sheep.

Zafiri: Raw cotton of brown staple, grown in Egypt.

Zanella: Smooth and highly lustrous, atlas constructionlining fabric (cotton and also cotton warp +worsted yarn weft = wool/cotton union-Zanella),

Zanella: Serge, made with cotton warp and worsted filling; used for making lining and umbrella covers.

Zanzibar cloth: Grey cotton goods, made in India.

Zappara: In Southern Europe, the agave fibre which is used in braids, horse blankets, etc.

Zapupe: A white, strong, glossy, soft and pliable fibre yielded by a species of the agave in Mexico.It is used for making ropes and coarse fabrics.

Zarasas: Ordinary cotton prints in Latin American countries.

Zardozi: Old Indian gold embroidery (silk yarn wrapped with metal foil) on heavyweight fabrics.

Zarts solution: Dyestuff mixture for differentiating between cupro (violet/blue colouration) and viscose fibres (= pink/red colouration) in fibre blends. Reliable method of differentiation.

Zaza printing method: Yarn sheet printing process (similar to the vigoureux printing process) in which yarn sheets run between a relief and a felt roller which is supplied with printing paste from the printing paste trough via a rubber roller, printing paste beingtransferred to the yarn material in raised places of the relief through felt roller contact pressure. Five shades can be produced with one ground colour and four pairs of rollers.

ZE: Abreviation of Zein fibre.

Zegna sports' iJacket: Ermenegildo Zegna is a worldwide brand. Zegna Sports has produced the iJacket, which has become a prominent Italian product for men. It is compatible with the Apple iPod. During routine chores or exercise such as hiking, the user is able to access the music player through a fabric keypad. This product also uses Elek Tex's interface with Bluetooth so there is no need to reach for one's mobile phone. The fabric keypad on the outside of the jacket can be used to answer or hang up a call. Because this product has characters that are comfortable to wear, and functions of music player and cell phone, we expect positive reactions from consumers.

Zein fibre: Man-made protein fibres from the zein contained in maize meal. Spun by the wet spinning process (precipitating bath: dilute sulphuric acid, acetic acid and zinc sulphate). Dry strength 11 cN/tex at 40% elongation. Used in a blend with cellulosic and other fibres(increases elasticity, improves crease resistance) for warp and weft knitted fabrics and woven pile fabrics.

Zeolite: A member of a group of hydrated aluminosilicate minerals which occur in nature and are also manufactured for their ion-exchange and selective-absorption properties. They are used for water softening and for sugar refining. The zeolites have an open crystal structure and can be used as molecular sieves. See also **Ion exchange; Molecular sieve.**

Zephyr: Originally, a good quality light-weight worsted yarn, but now refers to fine and soft sheer gingham type cottons. Also called ***Zephyr gingham***. See **Gingham**.

Zephyr: Plainweave, very soft, lightweight, sheer fine, woollen flannel, mixed with silk. Often colour woven withstripes on white ground.

Zephyr gingham: Fine soft gingham, finished without dressing.

Zephyr yarn: Soft spun worsted yarn, made of soft merino wool with three or more strands; used for embroidery. A variety of soft worsted yarn characterised by a low twist and spun from wool which is as fine or finer in average diameter than 64's grade tops.

Zerbase: Persian double-faced silk fabric interwoven with gold or silver threads.

Zero discharge layout: Design of an industrial plant to release no effluent, whether cooling water (by closed recirculation system) or industrial effluent.

Zero-moisture: See **Moisture-free (the preferred term).**

Zero-twist: Twist-less, devoid of twist.

Zibelene: A heavy coating fabric with a long, shaggy nap laid in one direction. Usually woollen, but there may be other fibres such as acrylic included. Strong colours are normally used and the cloth is often striped. Made with heavy coarse yarns which can be woollen or blended with mohair. This fabric is often woven as a double cloth so as to obtain added weight, warmth and durability. Used for women's coats, capes, children's coats. Also called Zibeline.

Zig-zag stitch: It is a utility stitch on a sewing machine which allows seams to stretch. It may also be used on the raw edge of fabric to prevent fraying.

Zig-zag twill: A twill weave, producing a zig-zag effect.

Zig-zagged seam-finish: A finish for the raw edges of the seam allowances of a plain seam, in which machine zig-zag stitching is placed 3 to 6 mm (1/8 to ¼ in) from the raw edge.

Zimmer Rotary Printing Machine (Johannes Zimmer): Rotary printing machine manufactured by Zimmer company who has been for many years a worldwide leader among the producers of machines for textile and carpet finishing (digital printing systems, flat screen and rotary screen printing, coating, steaming, washing, drying). Zimmer, Kufstein, Austria. The special features of the machines were Zimmer rotary screen printing machine: this uses a magnet roll squeegee, which guarantees steady rolling of the screens and also makes it possible to use a screen stretching device which is independent of the machine. This makes it possible to fit narrow screens for narrow printing widths on wide rotary printing machines without the need to fit additional units.

Zinc carbonate ($ZnCO_3$): Its molecular weight is 125.4. Use – In textile printing (mordants, white resists, discharge agents in printing, the latter for weighted silks).

Zinc chloride ($ZnCl_2$): Its molecular weight is 136.2. White salt, extremely hygroscopic, easily soluble in water. Easily splits off hydrochloric acid when heated. Use – In textile printing (for white resists on bromine indigo, etc.), (rarely) antiseptic for sizes and chemical finishes, etc.

Zinc chloride/formic acid solution: Used in fibre identification. 100 ml of zinc chloride solution, density 1.566 (produced by dissolving 100 g of moisture-free $ZnCl_2$ in 100 ml of water, and setting the density to 1.566) are mixed with 6 ml of 98–100% formic acid.

Use – (a) Solution cold: Differentiating between 2.5 acetate and triacetate (2.5 acetate = dissolved, triacetate = merely slightly swollen). Polyamide 3, polyamide 6 and polyamide 6.6 produce Koch's contraction reaction at room temperature. Vinylal fibres reveal the same reaction as with zinc chloride-iodine solution, but faster and with no staining. Polyacrylonitrile is insoluble; (b) Solution at 70 °C (10 min) dissolves silk, slowly dissolves regenerated cellulose, does not dissolve wool, regenerated protein fibres, cotton and bast fibres.

Zinc chloride-iodine solution: Dissolve 66 g anhydrous zinc chloride and 6 g potassium iodide in 34 ml water. Next add iodine until the solution is saturated. The solution should be stored in a dark bottle as it is light-sensitive. Reactions with fibres –

(a) Cellulosic fibres: Blue colouring, regenerated cellulose shows a deeper blue colouring than natural cellulose; woody cellulose = yellow; material which has been well broken down = blue. X- and V-transverse splits in bast fibres are heavily dyed. Natural celluloses swell, cottons show a similar reaction to that in cuprammonium.

(b) Animal fibres are not dyed.

(c) Acetate: Dissolution with yellow colouring and no swelling.

(d) Polyacrylonitrile: Dissolution with swelling. Insoluble types also exist. No yellow colouring.

(e) Polyamide, polyurea fibres: Yellow colouring and Koch's contraction reaction without dissolution.

(f) Polyester, polyvinyl chloride: No yellow colouring or dissolution.

(g) Polyvinyl alcohol: Initially greyish-blue, then the fibre core becomes a strong blue and is distorted into a zig-zag formation. Eventually becomes straight again and generally yellow in colour.

Zinc chloride solubility test: This is a solution of zinc chloride (54 °Bé) or 100 g molten zinc chloride + 85 ml water + 4 g zinc oxide at 100 °C dissolved in approximately 15 min. Only dissolves pure silk (not tussah-silk) and acetate (gelatinous type is dissolved) in fibre blends.

Zinc chlor-iodide solution: Compound consisting of iodine, potassium iodide and zinc chloride. Used to test for acetate, cupro and viscose in fibre blends. The following colourings occur – yellow = acetate; brown = cupro; green = viscose; other fibres do not change colour.

Zinc dust: Grey powder, finely divided metallic zinc (some zinc oxide). Strong reducing agent. Used for producing indigo vats and for discharges in textile printing in olden days.

Zinc equivalent: A number related to the contents of ***copper, zinc*** and ***nickel*** insludge, which can be used to estimate the safe amounts of these metals to beapplied to the soil. The zinc equivalent should not exceed 250 mg/l and is calculated by multiplying the copper content of the sludge by 2 and the nickelcontent by 8, and adding them to the zinc content.

Zinc-lime-vat: Used in indigo dyeing, mainly for ordinary cotton yarn, lighter-weight piece goods and paste resist printing (Vat dyes in resist printing). Dye vat: 1000 l contain 400 g of zinc dust (flushed with water) + 2 kg of dissolved lime (mixed to a paste), left to stand overnight, indigo addition as necessary. In the case of yellow vat dye, begin with dyeing.

Zinc oxide (zinc white): ZnO. Molecular weight 81; density 5.78. Loose white powder, practically insoluble in water, soluble in acids (salt formation); temporarily yellowing when heated (slightly luminescent in the cooled state). Use: Much used pigment for viscose delustring (fixated with fatty sulphonates; dyed with cationic or direct dyes for coloured delustring); for sodium dithionite resists; resist for aniline black in printing; for producing damask effects in printing; for neutralizing naphthol-diazo- solution; production of cements, drying agents, etc.

Zinc formaldehyde sulphoxylate: Colour Index Reducing Agent 6, used for discharge.This discharge agent works best in mildly acid conditions, so it is preferred, over sodium formaldehyde sulfoxylate, for work on wool. It is also often used in discharge screen printing of garments.

Zinc sulphate (white vitriol) $ZnSO_4{\cdot}7H_2O$: Molecular weight is 288. Colourless crystals; weakly acid. Use: Alkali binders in naphthol dyeing, for printing mordants, as an antiseptic for sizes and chemical finishes (out of date).

Zins: General trade term in Russia for the best grade of flax, formerly taken by the church as tax.

Zip fastener (tape): A narrow fabric which may be knitted or woven and with high lateral strength, constructed so as to be suitable for the attachment, generally by sewing, or of a nylon or similar spiral along one edge. There are several different methods of incorporating the fastening elements integrally as a part of weaving process. The tape may also be formed with abeaded or thickened edge comprising of one or more suitably sized, cords to which metal or other teeth may be attached.

Zipper: A slide fastener consisting of inter-lockable elements each attached to one of the opposing edges of two tapes and a movable part called a 'slider' that spans the inter-lockable elements, which when moved in one direction causes the elements on one tape to interlock with the elements of the other tape and when moved in the opposite direction, causes the elements to disengage.

Zirkas: In Germany and Austria, a woollen or worsted dress goods, made in four-leaf, even-sided twill in various coloured patterns.

Zirpro: Zripro treatments are a treatment developed by the International wool Secretariat, which improves flame retardance in wool fabrics.

Zirpro process: Flame retardant finishing of wool by applying titanium/ zirconium salt. Application is basically possible in any wool processing stage, i.e. as loose stock, tops, yarn (hank or yarn package), piece goods (woven or knitted fabric) and carpets.

Zone of inhibition: Clear area of no growth of a microorganism, cultured onto the surface of an agar growth medium in proximity to the borders of a specimen placed in direct contact with this agar surface. A zone of inhibition occurs as a result of the diffusion of an antimicrobial agent from the specimen.

Zouave jacket: A short jacket reaching about to the waist and cutaway in front.

Z-polyethylene: Gravure printed polyethylene after its inventor Ziegler.

Z-twist: Spun yarn twist.

Zulu cloth: Twilled and closely woven cloth used for embroidery foundation in England.

Zw: Abbreviation of the no longer admissible concept–Rayon staple fibre.

Zwitterions: Term borrowed from colloquial speech for compounds which have in the same molecule a positively charged group (ammonium and sulphonium ions, more rarely carbenium or phosphonium ions), and also a negatively charged group, and can therefore be understood as intramolecular salts.

Zymosis: (a) Fermentation. (b) Reactions induced by enzymes.